**Lightning
Physics and Effects**

Lightning: Physics and Effects is the first book that covers essentially all aspects of lightning, including lightning physics, lightning protection, and the interaction of lightning with a variety of objects and systems as well as with the environment. It is written in a style that will be accessible to the technical non-expert and is addressed to anyone interested in lightning and its effects. This will include physicists, engineers working in the power industry and in the communications, computer, and aviation industries, meteorologists, atmospheric chemists, foresters, ecologists, physicians working in the area of electrical trauma, and architects. This comprehensive reference volume contains over 300 illustrations, 70 tables containing quantitative information, and over 6000 references and bibliography entries.

VLADIMIR A. RAKOV worked as Assistant Professor of Electrical Engineering at Tomsk Polytechnical University, Russia from 1977 to 1979 and was awarded a Ph.D. in 1983. In 1978 he became involved in lightning research at the High Voltage Research Institute, a division of Tomsk Polytechnic, where, from 1984 to 1994, he held the position of Director of the Lightning Research Laboratory. Professor Rakov received the rank of Senior Scientist in High Voltage Engineering in 1985, was named an Inventor of the USSR in 1986, and received a Silver Medal from the (USSR) National Exhibition of Technological Achievements in 1987. He joined the faculty of the Department of Electrical and Computer Engineering at the University of Florida in 1991 and has also held visiting professorships at the Technical University of Vienna and at the Swiss Federal Institute of Technology in Lausanne. Professor Rakov is the author of over 30 patents and in excess of 200 papers and technical reports on various aspects of lightning.

MARTIN A. UMAN received his Ph.D. degree from Princeton University in 1961 and subsequently worked as an Associate Professor of Electrical Engineering at the University of Arizona in Tucson from 1961 to 1964. Following seven years as a Physics Fellow at Westinghouse Research Laboratories in Pittsburgh, in 1971 he joined the University of Florida Faculty, where he is now Professor and Chair of the Department of Electrical and Computer Engineering. Professor Uman co-founded and served as President of Lightning Location and Protection, Inc. (LLP) from 1975 to 1985. He has been honored many times for his contributions to lightning research including the 1996 IEEE Heinrich Hertz Medal for "outstanding contributions to lightning detection and protection" and the 2001 American Geophysical Union John Adam Fleming Medal for his "outstanding contribution to the description and understanding of electricity and magnetism of the Earth and its atmosphere." Professor Uman has written three other books on the subject of lightning as well as almost 300 journal articles and technical reports.

Lightning

Physics and Effects

VLADIMIR A. RAKOV
AND MARTIN A. UMAN

Department of Electrical and Computer Engineering
University of Florida

CAMBRIDGE UNIVERSITY PRESS
Cambridge, New York, Melbourne, Madrid, Cape Town, Singapore, São Paulo

Cambridge University Press
The Edinburgh Building, Cambridge CB2 2RU, UK

Published in the United States of America by Cambridge University Press, New York

www.cambridge.org
Information on this title: www.cambridge.org/9780521583275

© Vladimir A. Rakov and Martin A. Uman 2003

This publication is in copyright. Subject to statutory exception
and to the provisions of relevant collective licensing agreements,
no reproduction of any part may take place without
the written permission of Cambridge University Press.

First published 2003
Third printing 2005
This digitally printed first paperback version (with corrections) 2006

A catalogue record for this publication is available from the British Library

Library of Congress Cataloguing in Publication data
Rakov, V.A. (Vladimir A.), 1955–
Lightning : physics and effects / V.A. Rakov and M.A. Uman.
    p.   cm.
Includes bibliographical references and index.
ISBN 0 521 58327 6 hardback
1. Lightning.   I. Uman, Martin A.   II. Title.
OC966 .R35   2002
551.56'32–dc21   2002023094

ISBN-13 978-0-521-58327-5 hardback
ISBN-10 0-521-58327-6 hardback

ISBN-13 978-0-521-03541-5 paperback
ISBN-10 0-521-03541-4 paperback

The publisher has used its best endeavors to ensure that the URLs for
external websites referred to in this publication are correct and active at
the time of going to press. However, the publisher has no responsibility
for the websites and can make no guarantee that a site will remain live
or that the content is or will remain appropriate.

The colour figures referred to within this publication have been
removed for this digital reprinting. At the time of going to press
the original images were available in colour for download from
http://www.cambridge.org/9780521035415

Rakov's dedication

To my wife Lucy and our son Sergei

Uman's dedication

To my wife Dorit, our children Mara,
Jon, and Derek, and our grandchildren
Sara, Hunter, Hayden, Summer, and Isabella

Contents

Preface		page x

1. Introduction — 1
1.1. Historical overview — 1
1.2. Types of lightning discharge and lightning terminology — 4
1.3. Summary of salient lightning properties — 6
1.4. The global electric circuit — 6
1.5. Regarding the utilization of lightning energy — 12
1.6. Summary — 12
References and bibliography — 12

2. Incidence of lightning — 24
2.1. Introduction — 24
2.2. Characterization of individual storms and storm systems — 24
2.3. Thunderstorm days — 35
2.4. Thunderstorm hours — 36
2.5. Lightning flash density — 37
2.6. Long-term variations in lightning incidence — 43
2.7. Ratio of cloud flashes to cloud-to-ground flashes — 44
2.8. Characteristics of lightning as a function of season, location, and storm type — 46
2.9. Lightning incidence to various objects — 49
2.10. Summary — 52
References and bibliography — 52

3. Electrical structure of lightning-producing clouds — 67
3.1. Introduction — 67
3.2. Cumulonimbus — 68
3.3. Non-cumulonimbus — 91
3.4. Summary — 93
References and bibliography — 93

4. Downward negative lightning discharges to ground — 108
4.1. Introduction — 108
4.2. General picture — 108
4.3. Initial breakdown — 116
4.4. Stepped leader — 122
4.5. Attachment process — 137
4.6. Return stroke — 143
4.7. Subsequent leader — 164
4.8. Continuing current — 173
4.9. M-component — 176
4.10. J- and K-processes — 182
4.11. Regular pulse bursts — 188
4.12. Summary — 190
References and bibliography — 191

5. Positive and bipolar lightning discharges to ground — 214
5.1. Introduction — 214
5.2. Conditions conducive to the occurrence of positive lightning — 217
5.3. Characterization of positive lightning — 218
5.4. Bipolar lightning discharges to ground — 232
5.5. Summary — 233
References and bibliography — 234

6. Upward lightning initiated by ground-based objects — 241
6.1. Introduction — 241
6.2. General characterization — 244
6.3. Overall electrical characteristics — 247
6.4. Impulsive currents — 250
6.5. Lightning current reflections within tall objects — 252
6.6. Electromagnetic fields due to lightning strikes to tall objects — 259
6.7. Acoustic output — 260
6.8. Summary — 260
References and bibliography — 260

7. Artificial initiation (triggering) of lightning by ground-based activity — 265
7.1. Introduction — 265
7.2. Rocket-triggered lightning — 266
7.3. Other lightning triggering techniques — 296
7.4. Concluding remarks — 299
References and bibliography — 299

8. Winter lightning in Japan — 308
8.1. Introduction — 308
8.2. Formation of winter thunderclouds — 308
8.3. Evolution of winter thunderclouds — 309

8.4.	Characteristics of natural winter lightning	311
8.5.	Rocket-triggered lightning in winter	314
8.6.	Summary	316
References and bibliography		316

9. Cloud discharges — 321
- 9.1. Introduction — 321
- 9.2. General information — 322
- 9.3. Phenomenology inferred from VHF–UHF imaging — 326
- 9.4. Early (active) stage — 329
- 9.5. Late (final) stage — 338
- 9.6. Comparison with ground discharges — 340
- 9.7. Summary — 341
- *References and bibliography* — 341

10. Lightning and airborne vehicles — 346
- 10.1. Introduction — 346
- 10.2. Statistics on lightning strikes to aircraft — 348
- 10.3. Major airborne research programs — 350
- 10.4. Mechanisms of lightning–aircraft interaction — 353
- 10.5. Lightning test standards — 362
- 10.6. Accidents — 364
- 10.7. Summary — 369
- *References and bibliography* — 369

11. Thunder — 374
- 11.1. Introduction — 374
- 11.2. Observations — 374
- 11.3. Generation mechanisms — 377
- 11.4. Propagation — 386
- 11.5. Acoustic imaging of lightning channels — 387
- 11.6. Summary — 389
- *References and bibliography* — 389

12. Modeling of lightning processes — 394
- 12.1. Introduction — 394
- 12.2. Return stroke — 394
- 12.3. Dart leader — 415
- 12.4. Stepped leader — 417
- 12.5. M-component — 419
- 12.6. Other processes — 420
- 12.7. Summary — 420
- *References and bibliography* — 421

13. The distant lightning electromagnetic environment: atmospherics, Schumann resonances, and whistlers — 432
- 13.1. Introduction — 432
- 13.2. Theoretical background — 435
- 13.3. Atmospherics (sferics) — 443
- 13.4. Schumann resonances — 449
- 13.5. Whistlers — 454
- 13.6. Radio noise — 459
- 13.7. Summary — 461
- *References and bibliography* — 461

14. Lightning effects in the middle and upper atmosphere — 480
- 14.1. Introduction — 480
- 14.2. Upward lightning channels from cloud tops — 481
- 14.3. Low-luminosity transient discharges in the mesosphere — 482
- 14.4. Elves: low-luminosity transient phenomena in the lower ionosphere — 492
- 14.5. Runaway electrons, X-rays, and gamma-rays — 493
- 14.6. Interaction of lightning and thundercloud electric fields with the ionosphere and the magnetosphere — 495
- 14.7. Summary — 497
- *References and bibliography* — 497

15. Lightning effects on the chemistry of the atmosphere — 507
- 15.1. Introduction — 507
- 15.2. Mechanism of NO production by return-stroke channels — 511
- 15.3. Laboratory determination of NO yield per unit energy — 514
- 15.4. Ground-based field determination of NO yield per lightning flash — 514
- 15.5. Estimation of global NO production using the flash extrapolation approach (FEA) — 516
- 15.6. Estimation of NO production from airborne measurements — 516
- 15.7. Estimation of NO production from extrapolation of nuclear explosion data — 518
- 15.8. Transport of lightning-produced trace gases — 518
- 15.9. Production of trace gases in the primitive Earth atmosphere and in the atmospheres of other planets — 519
- 15.10. Summary — 520
- *References and bibliography* — 521

16. Extraterrestrial lightning — 528
- 16.1. Introduction — 528
- 16.2. Detection techniques — 530
- 16.3. Venus — 531
- 16.4. Jupiter — 536
- 16.5. Saturn — 543
- 16.6. Uranus — 544
- 16.7. Neptune — 545
- 16.8. Concluding remarks — 546
- *References and bibliography* — 547

17.	**Lightning locating systems**	555
17.1.	Introduction	555
17.2.	Electric and magnetic field amplitude techniques	556
17.3.	Magnetic field direction finding	558
17.4.	Time-of-arrival technique	562
17.5.	The US National Lightning Detection Network	565
17.6.	Interferometry	568
17.7.	Ground-based optical direction finding	570
17.8.	Detection from satellites	570
17.9.	Radar	572
17.10.	Summary	573
References and bibliography		573
18.	**Deleterious effects of lightning and protective techniques**	588
18.1.	Introduction	588
18.2.	Basic mechanisms of lightning damage	588
18.3.	Protection	589
18.4.	Lightning interaction with specific objects and systems	610
18.5.	Lightning test standards	624
18.6.	Summary	626
References and bibliography		626
19.	**Lightning hazards to humans and animals**	642
19.1.	Statistics	642
19.2.	Electrical aspects	644
19.3.	Medical aspects	646
19.4.	Personal safety	648
19.5.	Summary	648
References and bibliography		649
20.	**Ball lightning, bead lightning, and other unusual discharges**	656
20.1.	Introduction	656
20.2.	Witness reports of ball lightning	658
20.3.	Ball lightning statistics	662
20.4.	Ball lightning theories	662
20.5.	Laboratory simulation of ball lightning	663
20.6.	Bead lightning	665
20.7.	Other types of unusual lightning and lightning-like discharges	666
20.8.	Concluding comments	669
References and bibliography		669
Appendix: books on lightning and related subjects		675
Index		679

The colour figures referred to within this publication have been removed for this digital reprinting. At the time of going to press the original images were available in colour for download from http://www.cambridge.org/9780521035415

Preface

In this text, the first monograph to cover essentially all aspects of lightning, every effort has been made to present a balanced review of the present knowledge of lightning physics and lightning effects. The end-of-chapter reference and bibliography lists (a total of over 6000 entries) are essentially complete on most topics up to spring 2002. The Appendix contains a list of other books on lightning and related topics. Each of the authors has contributed to all chapters of the book; however the primary authorship by chapter is as follows. Chapters 1–9 and 12 were written by Rakov; Chapters 10, 13–15, and 18–20 by Uman; and Chapters 11, 16, and 17 were jointly written. General coordination of the work resulting in this book was conducted by Rakov.

Portions of the content of this book have been used for several years as the textbook for a one-semester senior- and graduate-level course on lightning at the University of Florida. The suggested content for a similar one-semester course would include Sections 1.2, 2.5, 2.9, subsections 3.2.1–3.2.7, Chapter 4, Section 7.2, and Chapters 12 and 17. The prerequisite for such a course would be an undergraduate course in electromagnetics or an undergraduate general physics course that covered electromagnetics in moderate detail.

The authors would like to thank colleagues who read various parts of the manuscript and provided useful comments and suggestions, including, in alphabetical order: M. Baker (Chapter 3); C.E. Baum (Chapter 12); M.A. Cooper (Chapter 19); K.L. Cummins (Chapters 2, 17 and sections 6.5, 6.6); R.R. Dickerson (Chapter 15); J.E. Dye (Chapter 15); U. Dyudina (Chapter 16); F. Heidler (Chapter 6); R.L. Holle (Chapter 19); R.H. Holzworth (Chapter 1); V.P. Idone (Chapter 4); U.S. Inan (Chapter 14); E.P. Krider (Chapter 4); N.G. Lehtinen (Chapter 14); T.C. Marshall (Chapter 3); V. Mazur (Chapter 10); D.R. MacGorman (Chapter 5); M. Miki (Chapter 7); J.P. Moreau (Chapter 10); K. Nakamura (Chapter 8); C.A. Nucci (Chapter 18); R.E. Orville (Chapter 2); V.P. Pasko (Chapter 14); D.E. Proctor (Chapter 9); F. Rachidi (Chapter 18); V. Shostak (Sections 6.5, 6.6); R. Solomon (Chapter 3); M. Stolzenburg (Chapter 3); R. Strangeway (Chapter 16); D.B. Walen (Chapter 10); D. Wang (Chapters 7, 8); J.C. Willett (Chapter 7, subsections 4.4.6, 4.4.8, 5.3.2, 5.3.3, and Sections 12.3, 12.4); E.R. Williams (Chapters 2, 13); and W. Zischank (Chapter 18). Of course, the opinions we express in the text should not necessarily be interpreted as the views of these colleagues. Thanks also go to those who provided original figures for the book: J. Autery (cover), P.P. Barker (Fig. 7.31c), Ruth and Ken Bateman (Fig. 19.4), H. Binz and G. Storf (Figs. 6.2–6.4), D.J. Boccippio and H.J. Christian (Fig. 2.12), W. Brooks (Fig. 2.3), D.E. Crawford (Fig. 5.10), S. Cummer (Fig. 14.6), K.L. Cummins (Figs. 2.11 and 17.6), R.J. Fisher (Fig. 7.24), J. Hendry (Fig. 4.1), R. Holle (Fig. 19.3), V.P. Idone (Figs. 7.5–7.7), D.M. Jordan (Figs. 4.37a and 4.53), Z.I. Kawasaki (Fig. 10.1), P.R. Krehbiel and W. Rison (Fig. 3.1), C.T. Mata (Figs. 7.30 and 18.7), K. Nakamura (Fig. 8.8a), T. Nelson (Fig. 14.2), W.D. Rust (Fig. 5.7c), D. Sentman (Fig. 13.11), D. Sentman and E. Wescott (Fig. 14.3), M. Stolzenburg (Figs. 3.11 and 3.15). G.N. Aleksandrov, E.I. Dubovoy, B.N. Gorin, A.V. Orlov, and Ya. M. Shvarts forwarded a number of papers in Russian that were not readily accessible. An exceptional level of secretarial support was provided by R. Crosser. Help with the drafts of some chapters of the book, figures, and references was given by M. Bryant, J.E. Jerauld, A.S. Mata, M. Moore, S.V. Rakov, and K. Thomson. The final figures were prepared by G. Vu.

1 Introduction

> An electrically active thundercloud may be regarded as an electrostatic generator suspended in an atmosphere of low electrical conductivity. It is situated between two concentric conductors, namely, the surface of the earth and the electrosphere, the latter being the highly conducting layers of the atmosphere at altitudes above 50 to 60 km.
>
> D.J. Malan (1967)

In this chapter, we introduce the basic lightning terminology (Section 1.2) and summarize the available quantitative information on various aspects of lightning in tabular form (Section 1.3). Additionally, we give a historical overview of the mythology and science of lightning covering the period from ancient times to the mid twentieth century (Section 1.1) and briefly discuss the global electric circuit that is generally thought to be energized by thunderstorms (Section 1.4). Finally, we consider whether the energy of lightning can be utilized and show that this is impractical (Section 1.5).

1.1. Historical overview

It is likely that lightning was present on Earth long before life evolved on our planet about three billion years ago. Further, it is possible that lightning played a role in producing the organic molecules necessary for the formation of every life form (Oparin 1938; Section 15.9 of this book). Harland and Hacker (1966) reported on a fossil glassy tube, referred to as a fulgurite, created by lightning 250 million years ago. Encounters of early humans with lightning undoubtedly were frightening and fascinating. All ancient civilizations incorporated lightning and thunder in their religious beliefs. Schonland (1964), Prinz (1977), Tomilin (1986), Wahlin (1986), Uman (1987, 2001), and Gary (1994) have reviewed mythological views of lightning in different cultures.

In ancient Egypt, the god Typhon (Seth) hurled the thunderbolts (lightning). A roll seal from Mesopotamia dated about 2200 BC shows a goddess standing on the shoulders of a winged creature and holding a bundle of thunderbolts in each hand (Prinz 1977, Fig. 1). Behind her, in a four-wheeled cart, is a weather god generating thunder with a whip. The similarly equipped weather god Teschup is seen on a Hittite relief found in northern Syria and dated about 900 BC. The thunderbolt is also the emblem of the goddess Tien Mu in Chinese mythology. Tien Mu is among the five dignitaries of the "Ministry of Thunderstorms", which is chaired by Lei Tsu, the God of Thunder, aided by Lei Kung, the drum-beating Count of Thunder. The ancient Vedic books of India describe how Indra, the son of Heaven and Earth, carried thunderbolts on his chariot. Many early statues of Buddha show him carrying in his right hand a thunderbolt with prongs at each end.

According to Prinz (1977), around 700 BC the ancient Greeks began using the lightning symbols of the Middle East in their art, attributing them primarily to Zeus, their supreme god. A lightning flash was one of the chief signs of the displeasure of Zeus in ancient Greece and of Jupiter in ancient Rome. In Rome, the laurel bush was considered, according to Pliny, to be immune from lightning. For this reason, the emperor Tiberius wore a wreath of laurel during thunderstorms. A very powerful political body, the College of Augurs, was formed about 300 BC to determine the views of Jupiter regarding Roman State affairs. This task was accomplished by making observations of three classes of objects in the sky: birds, meteors, and lightning.

In ancient Scandinavia, lightning was believed to be produced by the magic hammer Mjollnir of the god Thor, who hurled it from a chariot rolling thunderously upon the clouds. The Buryats, living in the area of Lake Baikal (Russia), believed that their god produced lightning by throwing stones from the sky. Some Indian tribes of North America, as well as certain tribes in southern Africa, hold the belief that lightning is produced by a magical thunderbird, which dives from the clouds to earth.

There exists a long record of lightning damage to tall structures, particularly churches, covering the period from the Middle Ages to the modern era. For example, the Campanile of St. Mark in Venice, which is about 100 m high, was damaged or destroyed by lightning in 1388, 1417, 1489, 1548, 1565, 1653, 1745, 1761, and 1762. In 1766, a lightning protective system, invented in 1752 by Benjamin Franklin and often referred to as a Franklin rod system, was installed and no further lightning damage has occurred since. In 1718, 24 church towers along the Brittany coast of France were damaged by lightning, apparently during the same storm. In 1769, the Church of St. Nazaire in Brescia, Italy, was struck by lightning. About 100 tons of gunpowder had been placed in the vaults of the church for safekeeping, and when the lightning discharge caused it to explode, about one-sixth of the city was destroyed, more than 3000 people being killed.

There are, however, many historic buildings which have never been seriously damaged by lightning, apparently because they had, in effect, a lightning protective system equivalent to that proposed later by Franklin. For example, the Temple in Jerusalem, originally built by Solomon, experienced no apparent damage from lightning over a period of a thousand years. Other examples are the 62 m high monument erected in 1677 in commemoration of the Great Fire of London and the cathedral of Geneva, the most prominent building in that city. This cathedral was immune from lightning damage for 300 years prior to being equipped with a Franklin rod system. In 1773, Franklin pointed out that "buildings that have their roofs covered with lead or other metal, and spouts of metal continued from the roof into the ground are never hurt by lightning; as whenever it falls on such a building, it passes in the metal and not in the walls" (Schonland 1964, p. 14). More information on the lightning protection of buildings, including historical aspects, is found in Section 18.3.

The practice of ringing church bells during thunderstorms in an attempt to disperse lightning existed for many centuries in Europe. Since bell towers are usually preferred targets for lightning because of their relatively large height (see Section 2.9), this practice caused the deaths of many of those pulling the ropes. In fact, over a period of 33 years, lightning struck 386 church towers and killed 103 bell-ringers, as reported in a book by Fischer published in Munich in 1784 (Schonland 1964, p. 9).

Many ships with wooden masts have been severely damaged or totally destroyed by lightning, as discussed in subsection 18.4.1. Harris (1834, 1838, 1839, 1843) reported that from 1799 to 1815 there were 150 cases of lightning damage to British naval vessels. One ship in eight was set on fire, nearly 100 lower masts were destroyed, about 70 sailors were killed, and more than 130 people were wounded. In 1798, the 44-gun ship *Resistance* exploded as a result of a lightning discharge.

Systematic studies of thunderstorm electricity can be traced back to 10 May 1752 in the village of Marly-la-Ville, near Paris. On that day, in the presence of a nearby storm, a retired French dragoon, acting on instructions from Thomas-François Dalibard, drew sparks from a tall iron rod that was insulated from ground by wine bottles. The results of this experiment, proposed by Benjamin Franklin, provided the first direct proof that thunderclouds contain electricity, although several scientists had previously noted the similarity between laboratory sparks and lightning (Prinz 1977; Tomilin 1986). The Marly experiment was repeated thereafter in several countries including Italy, Germany, Russia, Holland, England, Sweden, and again France. Franklin himself drew sparks from the probably moist hemp string of a kite after the success at Marly, but before he knew about this (Cohen 1990). Not only kites but also balloons, mortars, and rockets were used to extend conducting strings into the electric field of the cloud (Prinz 1977, Fig. 5). In all these experiments, the metallic rod (such as in the experiment at Marly) or the conducting string was polarized by the electric field of the cloud, so that charges of opposite polarities accumulated at the opposite ends of the conductor. As the gap between the bottom end of the conductor and ground was decreased, a spark discharge to ground occurred. The scale and effect of this spark discharge are orders of magnitude smaller than those of lightning. In designing his experiments, Franklin did not consider the possibility of a direct lightning strike to the rod or the kite. Such a strike would almost certainly have killed the experimenter. Thus all those who performed these experiments risked their lives, but there is only one case on record in which a direct strike did occur in such experiments. This happened on 6 August 1753 in St. Petersburg, Russia when Georg Richmann, who had previously done Franklin's experiment, was killed by a direct lightning strike to an ungrounded rod. Interestingly, Richmann was not in contact with the rod, and what caused his death appeared to be a ball lightning that came out of the rod and went to his forehead. This accident is discussed further in Section 20.1.

Franklin also showed that lightning flashes originate in clouds that are "most commonly in a negative state of electricity, but sometimes in a positive state" (Franklin 1774). Even before the experiment at Marly, Franklin had proposed the use of grounded rods for lightning protection. Originally, he thought that the lightning rod would silently discharge a thundercloud and thereby would prevent the initiation of lightning. Later, Franklin stated that the lightning rod had a dual purpose: if it cannot prevent the occurrence of lightning, it offers a preferred attachment point for lightning and then a safe path for the lightning current to ground. It is in the latter manner that lightning rods, often referred to as Franklin rods, actually work (Section 18.3), as suggested by Lomonosov in 1753 (Tomilin 1986); a lightning rod cannot appreciably alter the charge in a cloud. One convincing demonstration of the effectiveness of Franklin rods took place in Siena, Italy, on 18 April 1777. On that day, a large number of people gathered near the 102 m tower of the city hall, which had been repeatedly struck and

1.1. Historical overview

damaged by lightning prior to the installation of lightning rods in 1777. A thunderstorm rumbled into the area, and the crowd saw lightning strike the lightning protective system without damaging the tower. More information on the contributions of Benjamin Franklin to the study of lightning and lightning protection is found, for example, in Dibner (1977) and Krider (1996b).

In 1876, James Clerk Maxwell suggested that Franklin rod systems attracted more lightning strikes than the surrounding area. He proposed that a gunpowder building be completely enclosed with metal of sufficient thickness, forming what is now referred to as a Faraday cage. If lightning were to strike a metal-enclosed building, the current would be constrained to the exterior of the metal enclosure, and it would not even be necessary to ground this enclosure. In the latter case, the lightning would merely produce an arc from the enclosure to earth. The Faraday cage effect is provided by all-metal cars and airplanes. Modern steel-frame buildings with reinforcing metal bars in the concrete foundation connected to the building steel provide a good approximation to a Faraday cage. As the spacing between conductors increases, however, the efficiency of the lightning protection decreases. In practice, a combination of the Franklin rod system concept and the Faraday cage concept is often used. Modern lightning protection schemes for structures containing computers or other sensitive electronics employ a technique known as topological shielding with surge suppression (subsection 18.3.6), which can be viewed as a generalization of the Faraday cage concept.

In the following, we briefly review the history of lightning research from the latter part of the nineteenth century to the middle of the twentieth century. In the late nineteen century, photography and spectroscopy became available as diagnostic tools for lightning research. Among the early investigators of the lightning spectrum were Herschel (1868), Gibbons (1871), Holden (1872), and Clark (1874). It was Herschel who first identified a nitrogen line as being the brightest in the visible spectrum and who noted that the relative intensities of the lines change from spectrum to spectrum. Schuster (1880) made the first systematic identification of the lines in the spectrum of lightning. Dufay (1949) and Israel and Fries (1956) were the first to consider the spectrum of lightning as a source of quantitative information about the physical conditions in and around the lightning channel. Slipher (1917) obtained the first photographic record of the spectrum of lightning and noted that there were both line and continuum emissions. Spectroscopic studies of lightning are reviewed by Uman (1969, 1984, Chapter 5) and Orville (1977).

Among the early investigators who used stationary or moving photographic cameras were Hoffert (1889), Weber (1889), Walter (1902, 1903, 1910, 1912, 1918), and Larsen (1905). Time-resolved photographs showing that lightning flashes often contain two or more strokes, similar to that shown in Fig. 4.1a, were obtained. The invention of the streak camera (Boys 1926) and its further improvement (Boys 1929; Malan 1950, 1957) facilitated the major advances in lightning research made by B.F.J. Schonland, D.J. Malan, and their co-workers in South Africa during the 1930s. For example, it was shown conclusively by the South African researchers that lightning strokes lowering negative charge to ground are composed of a downward leader and an upward return stroke and that the first-stroke leader is stepped (see Fig. 4.2). Schonland (1956) summarized the main results of the South African studies. (Much of the presently used lightning terminology was introduced by the South African researchers.) The results obtained in South Africa have been confirmed and extended by investigators using streak cameras in the United States, Russia, France, Japan, and Switzerland. A review of photographic studies of lightning through the 1960s is found in Uman (1969, 1984, Chapter 2).

The first estimates of lightning peak current, inferred from the residual magnetization of pieces of basalt placed near the strike object, were made by Pockels (1900). Further information on the estimates of lightning currents obtained using magnetizable materials is found in Uman (1969, 1984, Chapter 4). The first oscillographic recordings of lightning current waveforms were obtained using tethered balloons in Russia (Stekolnikov and Valeev 1937) and in England (Davis and Standring 1947), and on the Empire State Building in New York City (McEachron 1939, 1941; Hagenguth and Anderson 1952). The most comprehensive data on lightning current waveforms to date were acquired by K. Berger and his associates on two instrumented towers on Monte San Salvatore in Switzerland, as discussed in Chapters 4, 5, and 6.

C.T.R. Wilson, who received a Nobel Prize for his invention of the cloud chamber to track high-energy particles, was the first to use electrostatic field measurements to infer the charge structure of thunderclouds (subsection 3.2.2) as well as the charges involved in the lightning discharge. Simpson and Scrase (1937) and Simpson and Robinson (1941) made the earliest measurements of electric fields inside thunderclouds and used these measurements to infer cloud charge structure (subsection 3.2.3). The first multiple-station measurements of the electromagnetic fields on ground from relatively close lightning were performed by Workman *et al.* (1942) and Reynolds and Neill (1955). Early measurements of the electric fields of distant lightning in the frequency range from a few to a few tens of kilohertz, called atmospherics, are discussed in Section 13.3. Austin (1926), Appleton *et al.* (1926), Wattson-Watt (1929), Norinder (1936), and Chapman (1939) were among the first to study atmospherics. Norinder and Dahle (1945) made an attempt to relate lightning magnetic field measurements to the current in the lightning channel. The modern era of electric and magnetic field measurements relating to lightning can be traced to the 1970s, when the first field records on microsecond and submicrosecond time scales were reported (see Chapters 4 and 9). More information on the early measurements of the electric and magnetic fields due to lightning is found in Uman (1969, 1984, Chapter 3).

1.2. Types of lightning discharge and lightning terminology

Lightning, or the lightning discharge, in its entirety, whether it strikes ground or not, is usually termed a "lightning flash" or just a "flash". A lightning discharge that involves an object on ground or in the atmosphere is sometimes referred to as a "lightning strike". A commonly used non-technical term for a lightning discharge is "lightning bolt". The terms "stroke" or "component stroke" apply only to components of cloud-to-ground discharges. Each stroke involves a downward leader and an upward return stroke and may involve a relatively low level "continuing current" that immediately follows the return stroke. Transient processes occurring in a lightning channel while it carries continuing current are termed M-components. First strokes are initiated by "stepped" leaders while subsequent strokes following previously formed channels are initiated by "dart" or "dart-stepped" leaders.

From the observed polarity of the charge effectively lowered to ground and the direction of propagation of the initial leader, four different types of lightning discharges between cloud and Earth have been identified. The term "effectively" found here and elsewhere in this book in a similar context (although often it is omitted for simplicity) is used to indicate that individual charges are not transported all the way from the cloud to ground during the lightning processes. Rather, the flow of electrons (the primary charge carriers) in one part of the lightning channel results in the flow of other electrons in other parts of the channel, as discussed by Uman (1987, 2001). For example, individual electrons in the lightning channel move only a few meters during a return stroke that transfers a coulomb or more of charge to ground.

The four types of lightning, illustrated in Fig. 1.1, are (a) downward negative lightning, (b) upward negative lightning, (c) downward positive lightning, and (d) upward positive lightning. Discharges of all four types can be viewed as effectively transporting cloud charge to the ground and therefore are usually termed cloud-to-ground discharges (sometimes referred to as CGs). It is believed that downward negative lightning flashes, type (a), account for about 90 percent or more of global cloud-to-ground lightning, and that 10 percent or less of cloud-to-ground discharges are downward positive lightning flashes (type (c)). Upward lightning discharges, types (b) and (d), are thought to occur only from tall objects (higher than 100 m or so) or from objects of moderate height located on mountain tops (Chapter 6). Rocket-triggered lightning, discussed in Chapter 7, is similar in its phenomenology to the upward lightning initiated from tall objects. The term "initial continuous current," as differentiated from continuing current, is used to denote the relatively low-level current flowing during the initial stage of upward (Chapter 6) and rocket-triggered (Chapter 7) lightning. The downward negative lightning discharge is considered in Chapter 4 and the positive lightning discharge in Chapter 5. Additionally discussed in Chapter 5 are lightning flashes that transfer both negative and positive charges to ground. The majority of lightning discharges, probably three-quarters, do not involve ground. These are termed cloud discharges and sometimes are referred to as ICs. Cloud discharges include intracloud, intercloud, and cloud-to-air discharges and are considered in Chapter 9. Unusual forms of lightning and lightning-like discharges (including ball lightning) are discussed in Chapter 20.

There are three possible modes of charge transfer to ground in lightning discharges. It is convenient to illustrate these for the case of negative subsequent strokes. In negative subsequent strokes these three modes are represented by (a) dart-leader–return-stroke sequences, (b) continuing currents, and (c) M-components. Figure 1.2 schematically shows current profiles corresponding to these three modes, which we now discuss.

(a) In a leader–return-stroke sequence, the descending leader creates a conductive path between the cloud charge source and ground and deposits negative charge along this path. The following return stroke traverses that path, moving from ground toward the cloud charge source, and neutralizes the negative leader charge. Thus, both leader and return-stroke processes serve to transport effectively negative charge from the cloud to ground.

(b) The lightning continuing current can be viewed as a quasi-stationary arc between the cloud charge source and ground. The typical arc current is tens to hundreds of amperes, and the duration is up to some hundreds of milliseconds.

(c) Lightning M-components can be viewed as perturbations (or surges) in the continuing current and in the associated channel luminosity. It appears that an M-component involves the superposition of two waves propagating in opposite directions (see Fig. 1.2). The spatial front length for M-component waves is of the order of a kilometer (shown shorter in relation to the cloud height in Fig. 1.2, for illustrative purposes), while for dart-leader and return-stroke waves the spatial front lengths are of the

1.2. Types of lightning discharge and lightning terminology

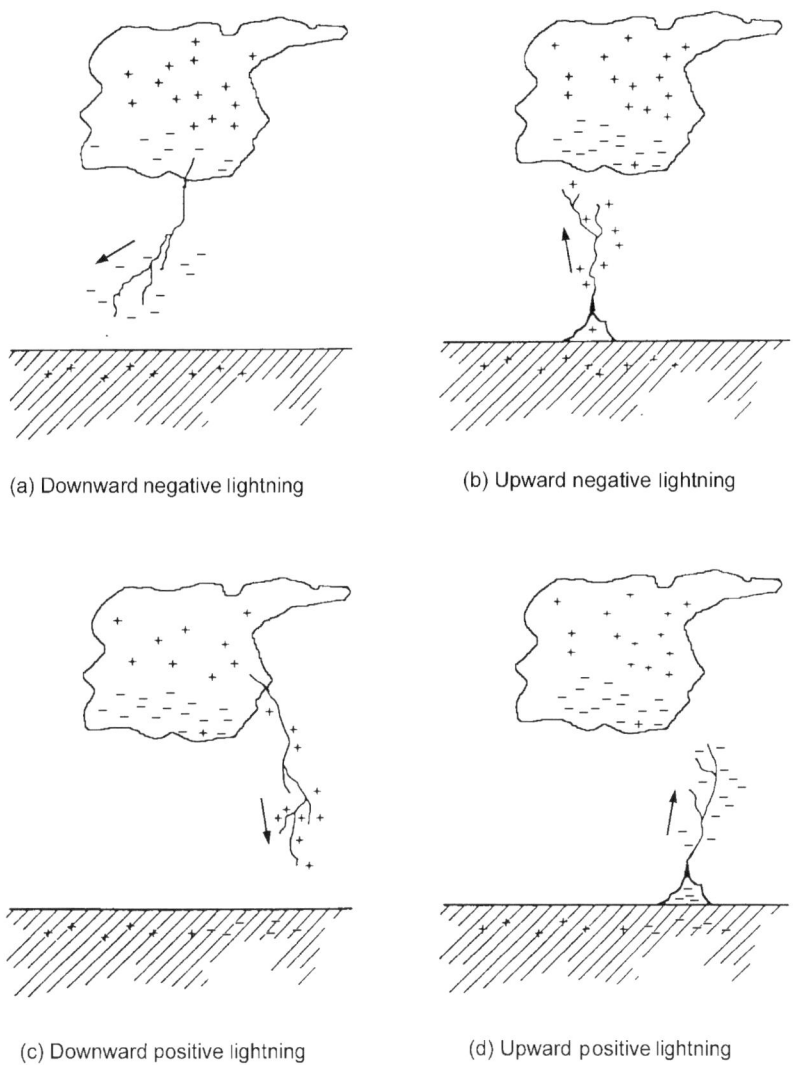

(a) Downward negative lightning (b) Upward negative lightning

(c) Downward positive lightning (d) Upward positive lightning

Fig. 1.1. Four types of lightning effectively lowering cloud charge to ground. Only the initial leader is shown for each type. In each lightning-type name given below the sketch, the direction of propagation of the initial leader and the polarity of the cloud charge effectively lowered to ground are indicated.

order of 10 and 100 m, respectively. The M-component mode of charge transfer to ground requires the existence of a grounded channel carrying a continuing current that acts as a wave-guiding structure. In contrast, the leader–return-stroke mode of charge transfer to ground occurs only in the absence of such a conducting path to ground. In this latter mode, the wave-guiding structure is not available and is created by the leader. For all the processes shown in Fig. 1.2, the channel conductivity is of the order of 10^4 S m^{-1}, except for the channel section between the dart-leader tip and ground shown by a broken line. For this latter channel section, the conductivity is about 0.02 S m^{-1} (Rakov 1998). Thus, the primary distinction between the leader–return-stroke and M-component modes is the availability of a conducting path to ground. It is possible that, as the conductivity of the path to ground decreases, the downward M-component wave can transform to a dart leader.

We now define the terms "leader" and "streamer", as they are used in this book. Any self-propagating electrical discharge creating a channel with electrical conductivity of the order 10^4 S m^{-1} (comparable to that of carbon) is called a leader. Streamers, on the other hand, are characterized by much lower electrical conductivity; the air behind the streamer tip remains essentially an insulator (e.g., Bazelyan et al. 1978). A corona or point discharge consists of numerous individual streamers. Corona discharge is confined to the immediate vicinity of an "electrode" such as a grounded object, a leader tip, the lateral surface of the leader channel,

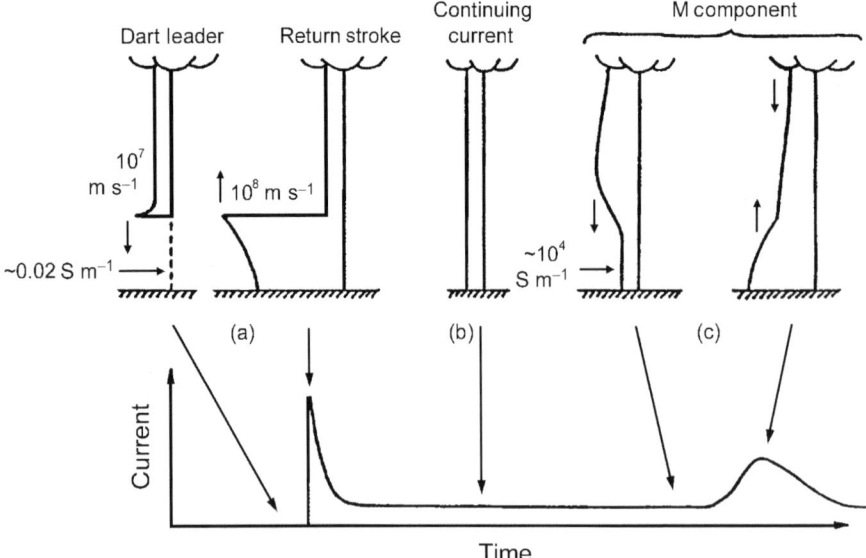

Fig. 1.2. Schematic representation of current versus height profiles for three modes of charge transfer to ground in negative lightning subsequent strokes: (a) dart-leader–return-stroke sequence, (b) continuing current, and (c) M-component. The corresponding current versus time waveform represents the current at the ground.

or a hydrometeor, that is, it is not a self-propagating discharge. It is worth noting that the terms leader and streamer in the lightning literature are sometimes used interchangeably, the term streamer in most cases being used to denote a low-luminosity leader, particularly the upward connecting leader (Section 4.5).

1.3. Summary of salient lightning properties

The salient properties of downward negative lightning discharges, the most common type of cloud-to-ground lightning, are summarized in Table 1.1. Negative lightning is discussed in detail in Chapter 4. Properties of positive and bipolar lightning are discussed in Chapter 5 and those of cloud discharges in Chapter 9. Characteristics of microsecond-scale electric field pulses associated with various lightning processes are summarized in Table 1.2.

Various lightning processes emit electromagnetic signals with a peak in the radio-frequency spectrum at 5 to 10 kHz when observed at distances beyond 50 km or so. At frequencies higher than that of the spectrum peak, the spectral amplitude is approximately inversely proportional to the frequency up to 10 MHz or so and inversely proportional to the square root of frequency from about 10 MHz to 10 GHz (Cianos et al. 1973). The mechanisms of radiation in the high-frequency (HF) region of the spectrum, 3–30 MHz, and above are not fully understood. It is thought that this radiation is caused by numerous small sparks occurring during the formation of new channels, that is, by the electrical breakdown of air rather than by high-current pulses propagating in pre-existing channels.

1.4. The global electric circuit

Shortly after the experiment at Marly that confirmed Franklin's conjecture regarding the electrical nature of thunderstorms (Section 1.1), Lemonnier (1752) discovered atmospheric electrical effects in fair weather. Further research established that the Earth's surface is charged negatively and the air is charged positively, the associated vertical electric field in fair weather being about 100 V m^{-1} near the Earth's surface.

1.4.1. Conductivity of the atmosphere

The atmosphere below about 50 km is conducting, owing to the presence of ions created by both cosmic rays and the natural radioactivity of the Earth. Small ions, those with diameters of 0.1 to 1 nm and lifetimes of about 100 s, are the primary contributors to the conductivity of the lower atmosphere. Free electrons at these heights are attached to neutrals on time scales of the order of microseconds, and their contribution to the conductivity of the atmosphere below about 50 km can be neglected (Gringel et al. 1986; Reid 1986). Above 60 km or so, free electrons become the major contributors to the atmospheric conductivity. The average production rate of ions at sea level is one to 10 million pairs per cubic meter per second. Cosmic rays and natural radioactivity contribute about equally to the production of ions at the land surface. Since large water surfaces have no

1.4. The global electric circuit

Table 1.1. *Characterization of negative cloud-to-ground lightning*

Parameter	Typical value[a]
Stepped leader	
Step length, m	50
Time interval between steps, μs	20–50
Step current, kA	> 1
Step charge, mC	> 1
Average propagation speed, m s^{-1}	2×10^5
Overall duration, ms	35
Average current, A	100–200
Total charge, C	5
Electric potential, MV	∼ 50
Channel temperature, K	∼ 10 000
First return stroke[b]	
Peak current, kA	30
Maximum current rate of rise, kA μs^{-1}	≥10–20
Current risetime (10–90 percent), μs	5
Current duration to half-peak value, μs	70–80
Charge transfer, C	5
Propagation speed, m s^{-1}	$(1-2) \times 10^8$
Channel radius, cm	∼ 1–2
Channel temperature, K	∼ 30 000
Dart leader	
Speed, m s^{-1}	$(1-2) \times 10^7$
Duration, ms	1–2
Charge, C	1
Current, kA	1
Electric potential, MV	∼ 15
Channel temperature, K	∼ 20 000
Dart-stepped leader	
Step length, m	10
Time interval between steps, μs	5–10
Average propagation speed, m s^{-1}	$(1-2) \times 10^6$
Subsequent return stroke[b]	
Peak current, kA	10–15
Maximum current rate of rise, kA μs^{-1}	100
10–90 percent current rate of rise, kA μs^{-1}	30–50
Current risetime (10–90 percent), μs	0.3–0.6
Current duration to half-peak value, μs	30–40
Charge transfer, C	1
Propagation speed, m s^{-1}	$(1-2) \times 10^8$
Channel radius, cm	∼ 1–2
Channel temperature, K	∼ 30 000
Continuing current (longer than ∼ 40 ms)[c]	
Magnitude, A	100–200
Duration, ms	∼ 100
Charge transfer, C	10–20
M-component[b]	
Peak current, A	100–200
Current risetime (10–90 percent), μs	300–500
Charge transfer, C	0.1–0.2
Overall flash	
Duration, ms	200–300
Number of strokes per flash[d]	3–5
Interstroke interval, ms	60
Charge transfer, C	20
Energy, J	10^9-10^{10}

[a] Typical values are based on a comprehensive literature search and unpublished experimental data acquired by the University of Florida Lightning Research Group.
[b] All current characteristics for return strokes and M-components are based on measurements at the lightning channel base.
[c] About 30 to 50 percent of lightning flashes contain continuing currents having durations longer than ∼ 40 ms.
[d] About 15 to 20 percent of lightning flashes are composed of a single stroke.

significant radioactive emanation, the production of ions over oceans is about one-half of that over land. At altitudes of roughly 1 km and greater, cosmic rays are responsible for most of the ions in the fair weather atmosphere, regardless of the presence of land below. The ionization rate depends on magnetic latitude and on the solar activity.

The electrical conductivity of the air at sea level is about 10^{-14} S m^{-1}, and it increases rapidly with altitude. A diagram illustrating conductivity variations up to an altitude of 120 km is shown in Fig. 1.3. The various regions of the atmosphere are named in Fig. 14.1, and typical values of electron density and collision frequency as a function of height are found in Figs. 13.5 and 13.6, respectively.

At a height of 35 km, where the air density is about one percent of that at the Earth's surface, the electrical conductivity is greater than 10^{-11} S m^{-1}, which is more than three orders of magnitude higher than at sea level. For comparison, the average electrical conductivity of the Earth is about 10^{-3} S m^{-1}. As seen in Fig. 1.3, a considerable variation of conductivity exists at the same altitude for different measurements, about six orders of magnitude at 60 km. Above about 80 km, the conductivity becomes anisotropic because of the influence of the geomagnetic field (Section 13.2), and there are diurnal variations due to solar photoionization processes.

Blakeslee *et al.* (1989) reported, from high-altitude U-2 airplane measurements, that the conductivity near 20 km was relatively steady above storms, variations being less than ±15 percent. However, a number of balloon measurements of the electrical conductivity between 26 and 32 km over thunderstorms suggest that some of the time the storm may significantly (up to a factor 2) perturb the

Table 1.2. *Characterization of microsecond-scale electric field pulses associated with various lightning processes. Adapted from Rakov et al. (1996)*

Type of pulses	Dominant polarity[a]		Typical[b] total pulse duration, μs	Typical[b] time interval between pulses, μs	Comments
	Atmospheric electricity sign convention	Physics sign convention			
Return stroke in negative ground flashes	positive	negative	30–90 (zero-crossing time)	60×10^3	3–5 pulses per flash
Stepped leader in negative ground flashes	positive	negative	1–2	15–25	Within 200 μs just prior to a return stroke
Dart-stepped leader in negative ground flashes	positive	negative	1–2	6–8	Within 200 μs just prior to a return stroke
Initial breakdown in negative ground flashes	positive	negative	20–40	70–130	Some milliseconds to some tens of milliseconds before the first return stroke
Initial breakdown in cloud flashes	negative	positive	50–80	600–800	The largest pulses in a flash
Regular pulse burst in both cloud and negative ground flashes	Both polarities are about equally probable		1–2	5–7	Occur later in a flash; 20–40 pulses per burst
Narrow bipolar pulses	negative	positive	10–20	-	Probably associated with the initial breakdown in cloud flashes

[a] The polarity of the initial half cycle in the case of bipolar pulses.
[b] Typical values are based on a comprehensive literature search and unpublished experimental data acquired by the University of Florida Lightning Research Group.

conductivity (Bering *et al.* 1980b; Holzworth *et al.* 1986; Pinto *et al.* 1988; Hu *et al.* 1989). It is usually assumed that the atmosphere above a height of 60 km or so, under quasi-static conditions, becomes conductive enough to consider it an equipotential region. The electrical conductivity increases abruptly above about 60 km because of the presence of free electrons (Roble and Tzur 1986; Reid 1986). This region of atmosphere just above 60 km or so where free electrons are the major contributors to the conductivity is sometimes referred to as the electrosphere (e.g., Chalmers 1967) or "equalizing" layer (Dolezalek 1972). At 100 km altitude (in the lower ionosphere) the conductivity is about 12 ± 2 orders of magnitude (depending on the local time of the day) greater than the conductivity near the Earth's surface, that is, the conductivity at 100 km is comparable to the conductivity of the Earth, whether land or sea (Rycroft 1994).

1.4.2. Fair-weather electric field

As noted above, the electric field near the Earth's surface under fair-weather (also called fine-weather) conditions is about 100 V m^{-1}. The electric field vector is directed downward. This downward-directed field is defined as positive according to the "atmospheric electricity" sign convention. According to the alternative sign convention, sometimes referred to as the "physics" sign convention, a downward directed electric field is negative because it is in the direction opposite to that of the radial coordinate vector of the spherical coordinate system whose origin is at the Earth's center. We will use the physics sign convention in this chapter (both sign conventions are given in Table 1.2) and in Chapters 3 and 8. However, in order to minimize conflict with the existing literature, we will use the atmospheric electricity sign convention in Chapters 4, 5, 6, 7, and 12, and we will employ both the atmospheric electricity and physics

1.4. The global electric circuit

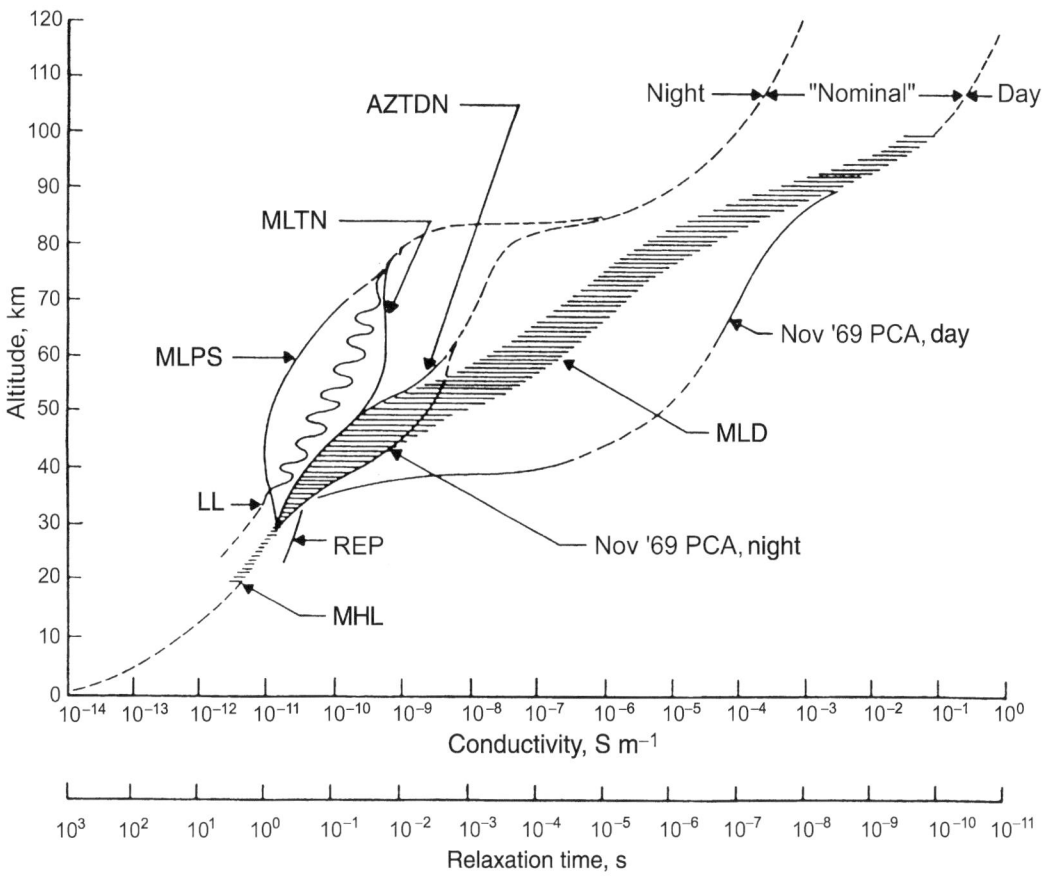

Fig. 1.3. Electrical conductivity σ and corresponding relaxation time $\tau = \varepsilon_0 \sigma^{-1}$, where $\varepsilon_0 = 8.85 \times 10^{-12}$ F m^{-1}, versus altitude under a variety of geophysical conditions. LL, low latitude, "wavy"; MLPS, mid-latitude pre-sunrise (unusual); MLTN, mid-latitude typical night (high-latitude, quiet); AZTDN, auroral zone, typical disturbed night; MLD, mid-latitude day, quiet; MHL, mid-high-latitude, typical of $\sim$ 100 measurements; REP, relativistic electron (energy from a few MeV to 10 MeV) precipitation event (unusual); PCA, polar cap absorption event (an unusually large flux of energetic, $\sim$ 100 MeV solar protons within the polar cap). Adapted from Hale (1984).

sign conventions in Chapter 9. We will indicate explicitly the direction of the electric field vector when appropriate. The magnitude of the fair-weather electric field decreases with increasing altitude. For example, according to Volland (1984),

$$E(z) = -[93.8 \exp(-4.527z) + 44.4 \exp(-0.375z) + 11.8 \exp(-0.121z)] \quad (1.1)$$

where $E(z)$ is the electric field in V m^{-1} (the negative sign indicates that the electric field vector is directed downward) and z is the altitude in kilometers. This equation is valid at mid-latitudes below about 60 km altitude and outside thunderstorms or cloudy areas. According to Eq. 1.1, the electric field at the ground is 150 V m^{-1} and at 10 km altitude decreases to about 3 percent of its value at the ground. The electric field magnitude normally drops to around 300 mV m^{-1} at 30 km at mid-latitudes (Gringel et al. 1986) and to 1 μV m^{-1} or so at about 85 km (Reid 1986).

1.4.3. "Classical" view of atmospheric electricity

The evaluation of the line integral of the electric field intensity from the Earth's surface to the height of the electrosphere yields the negative of the potential of the electrosphere, sometimes termed the ionospheric potential (e.g., Markson 1976), with respect to Earth potential. The potential of the electrosphere is positive with respect to the Earth and its magnitude is about 300 kV, most of the voltage drop taking place below 20 km where the electric field is relatively large. The overall situation is often visualized as a lossy spherical capacitor (e.g., Uman 1974), the outer and inner shells of which are the electrosphere and Earth's surface, respectively. According to this model, the Earth's surface is negatively charged, the total charge magnitude being roughly 5×10^5 C, while an equal positive charge is distributed throughout the atmosphere. Little charge resides on the electrosphere "shell". Further, most of the net positive charge is found within 1 km of the Earth's

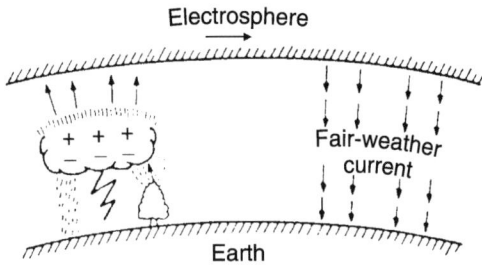

Fig. 1.4. Illustration of the global electric circuit. Shown schematically under the thundercloud are precipitation, lightning, and corona. Adapted from Pierce (1974).

surface, and more than 90 percent of this charge within 5 km (MacGorman and Rust 1998). Because the atmosphere between the capacitor "shells" is weakly conducting, there is a fair-weather leakage current of the order of 1 kA (2 pA m^{-2}; 1 pA = 10^{-12} A) between the shells that would neutralize the charge on the Earth and in the atmosphere on a time scale of roughly 10 minutes (depending on the amount of pollution) if there were no charging mechanism to replenish the neutralized charge. Since the capacitor is observed to remain charged, there must be a mechanism or mechanisms acting to resupply that charge. Wilson (1920) suggested that the negative charge on the Earth is maintained by the action of thunderstorms. Thus all the stormy-weather regions worldwide (on average, at any time a total of about 2000 thunderstorms are occurring, over about 10 percent of the Earth's surface) constitute the global thunderstorm generator, while the fair-weather regions (about 90 percent of the globe) can be viewed as a resistive load. Lateral currents are assumed to flow freely along the highly conducting Earth's surface and in the electrosphere. The fair-weather current, of the order of 1 kA, must be balanced by the total generator current, which is composed of currents associated with corona, precipitation, and lightning discharges. The total current flowing from cloud tops to the electrosphere is, on average, about 0.5 A per thunderstorm (Gish and Wait 1950). The global electric circuit concept is illustrated in Fig. 1.4. Negative charge is brought to Earth mainly by lightning discharges (most of which transport negative charge to ground) and by corona current under thunderclouds. The net precipitation current is thought to transport positive charge to ground, and its magnitude is comparable to the lightning current (Wahlin 1986). Positive charge is presumed to leak from cloud tops to the electrosphere. If we divide the potential of the electrosphere, 300 kV, by the fair-weather current, 1 kA, the effective load resistance is 300 Ω.

The diurnal variation of the fair-weather field as a function of universal time over the oceans, the so-called Carnegie curve, named after the research vessel *Carnegie* on which the measurements were made (Torreson *et al.* 1946), appears to follow the diurnal variation of the total worldwide thunderstorm area (Whipple and Scrase 1936). Both characteristics exhibit maximum values near 1900 UT and minimum values near 0400 UT. However, the annual variation of the fair-weather electric field is not in phase with the annual variation of thunderstorm activity throughout the world (Imyanitov and Chubarina 1967). Füllekrug *et al.* (1999) found that the hourly contribution of global cloud-to-ground lightning activity (as represented by magnetic field measurements in the frequency range 10–135 Hz) to the fair-weather electric field in the Antarctic during December 1992 was about 40 ± 10 percent, and that the contribution to hourly departures from the mean diurnal variation of the electric field was about 25 ± 10 percent. Holzworth *et al.* (1984) showed that significant time variations could occur in the global fair-weather current on time scales of 10 minutes to several hours.

According to the classical picture of atmospheric electricity, the layer of the atmosphere extending from about 15 to 200 km and including the stratosphere, mesosphere, and the lower portion of the thermosphere (Chapter 15 and Fig. 14.1), should be "passive". However, some rocket measurements indicate the existence of strong electric fields (in the volts per meter range, which is orders of magnitude higher than expected in the mesosphere, at altitudes of 50–85 km) of unknown origin (Bragin *et al.* 1974; Tyutin 1976; Hale and Croskey 1979; Hale *et al.* 1981; Maynard *et al.* 1981; Gonzalez *et al.* 1982). Interestingly, these abnormally strong electric fields are observed near the 60 to 65 km height region, where the "equalizing" layer (the electrosphere) is presumed to exist. Their origin remains a subject of controversy. Any plausible explanation of this phenomenon must involve either a local mesospheric field-generation mechanism or a dramatic local decrease in conductivity.

1.4.4. Maxwell current density

The Maxwell current density $\mathbf{J}_M$ associated with a thunderstorm is defined as the sum of four terms (e.g., Krider and Musser 1982):

$$\mathbf{J}_M = \mathbf{J}_E + \mathbf{J}_C + \mathbf{J}_L + \varepsilon_0 \frac{\partial \mathbf{E}}{\partial t} \qquad (1.2)$$

where $\mathbf{J}_E$ is the field-dependent current density, which may include both linear (ohmic, for which $\mathbf{J} = \sigma \mathbf{E}$ where σ is the electrical conductivity) components and nonlinear (corona) components, $\mathbf{J}_C$ is the convection current density, which may include a contribution from precipitation, $\mathbf{J}_L$ is the lightning current density, and the last term is the displacement current density. In planar geometry, the current density $\mathbf{J}_M$ is the same at any height in the atmosphere, as required by the current continuity equation (e.g., Sadiku 1994). Krider and Musser (1982) suggested that the thundercloud, the postulated current source in the global electric circuit, can be viewed as producing a quasi-static Maxwell current density even in the presence of lightning. In this view, the thunderstorm generator in the global electric circuit can, in principle, be monitored through its Maxwell current density.

The total Maxwell current density under thunderstorms has been measured by Krider and Blakeslee (1985), Deaver and Krider (1991), and Blakeslee and Krider (1992). Krider and Blakeslee (1985) found that the amplitude of $\mathbf{J}_M$ is of the order of 10 nA m^{-2} under active storms, while Deaver and Krider (1991) reported amplitudes of the order of 1 nA m^{-2} or less (in the absence of precipitation) under small Florida storms. Above thunderstorms, the convection and lightning terms, $\mathbf{J}_C$ and $\mathbf{J}_L$, in Eq. 1.2 are assumed to be negligible, and the Maxwell current density is expressed as the sum of only two terms:

$$\mathbf{J}_M = \mathbf{J}_E + \varepsilon_0(\partial \mathbf{E}/\partial t) \quad (1.3)$$

The Maxwell current density at altitudes of 16–20 km was measured by Blakeslee *et al.* (1989). They found that $\mathbf{J}_E$ typically accounted for more than half $\mathbf{J}_M$, while at the ground $\mathbf{J}_M$ is generally dominated by the displacement current density during active storm periods, as long as the fields are below the corona threshold of a few kV m^{-1} and there is no precipitation current. It is thought that, under some conditions, the Maxwell current density may be coupled directly to the meteorological structure of the storm and/or the storm dynamics (Krider and Roble 1986, pp. 5–6). However, the lack of simultaneous Maxwell current measurements both on the ground and aloft has prevented the details of this relationship from being determined.

1.4.5. Modeling of the global circuit

Models that can be used to calculate the electric field or the potential distribution around the thundercloud and the current that flows from a thundercloud into the global electric circuit have been developed by Holzer and Saxon (1952), Kasemir (1959), Illingworth (1972a), Dejnakarintra and Park (1974), Hays and Roble (1979), Tzur and Roble (1985b), Nisbet (1983, 1985a, b), Browning *et al.* (1987), Hager *et al.* (1989a, b), Driscoll *et al.* (1992), Stansbery *et al.* (1993), Hager (1998), and Plotkin (1999). Both analytical and numerical models have been developed. Some models (e.g., Holzer and Saxon 1952; Hays and Roble 1979; Tzur and Roble 1985b) have been based on a quasi-static approximation, while others (e.g., Illingworth 1972a; Dejnakarintra and Park 1974; Driscoll *et al.* 1992) have been designed to describe a time-varying problem that includes the effects of lightning. Nisbet (1983) modeled the atmosphere by a network of resistors, capacitors, and switches, where the current through these circuit elements represents conduction, displacement, and lightning currents, respectively. Sometimes corona currents have been taken into account (e.g., Tzur and Roble 1985b). Convection currents, including those due to precipitation, have usually been neglected. These models provide a convenient means of examining, through numerical experiments, the various processes operating in the global circuit. Normally, heights up to 100–150 km are considered, and thunderclouds are represented by two vertically displaced point charges of opposite polarity maintained by a current source. The upward-directed current from thunderclouds is assumed to spread out in the ionosphere of the storm hemisphere and, in some models (e.g., Hays and Roble 1979; Tzur and Roble 1985b; Browning *et al.* 1987; Stansbery *et al.* 1993) to flow along the Earth's magnetic field lines into the conjugate hemisphere. According to Stansbery *et al.* (1993), approximately half the current that reaches the ionosphere flows into the conjugate hemisphere and the other half flows to Earth in the fair-weather regions of the storm hemisphere. Pasko *et al.* (1998d) developed an electrostatic coupling model to examine the effects of the relatively steady electric fields of thunderclouds on the lower ionosphere.

1.4.6. Alternative views of the global circuit

Wilson's suggestion that thunderstorms are the main driving element in the atmospheric global circuit has been questioned by, among others, Dolezalek (1972), Williams and Heckman (1993), Kasemir (1994, 1996), and Kundt and Thuma (1999). Dolezalek (1988) suggested that the net charge on the Earth may be zero or near zero, rather than the 5×10^5 C postulated in the classical model of atmospheric electricity. Kasemir (1994, 1996) proposed an alternative concept of the global circuit in which the only equipotential layer is the Earth. He assumed that the very high negative potential of the Earth with respect to infinity drives the fair-weather current; lightning discharges and corona at the ground are claimed to be "local affairs" that do not contribute to the global circuit. In Kasemir's view, there are two types of generator in the global circuit, convection, acting in both stormy and fair-weather regions, and precipitation, both generators being driven by non-electrical forces. Kundt and Thuma (1999), who estimated a relatively low cloud top voltage, 1 MV relative to the Earth, asserted that thunderstorms are not important in the global electric circuit. However, Marshall and Stolzenburg (2001), from 13 balloon soundings of electric field through both convective regions and stratiform clouds, reported cloud top voltages ranging from −23 to +79 MV, with an average of +25 MV. The average cloud top voltage for the nine cases with positive values was +41 MV. Marshall and Stolzenburg (2001) considered these voltage values as supporting Wilson's hypothesis that thunderstorms drive the global electric circuit. Further, they estimated an average cloud top voltage of +32 MV for electrified stratiform clouds, this finding suggesting that stratiform clouds may make a substantial contribution to the global electric circuit.

1.5. Regarding the utilization of lightning energy

It appears to be impractical to utilize lightning energy. Each cloud-to-ground lightning flash involves an energy of roughly 10^9 to 10^{10} J (Section 4.2). For comparison, the energy required to operate five 100 W light bulbs continuously for one month is

$$5 \times 100 \text{ W} \times 3600 \text{ s} \times 24 \text{ h} \times 30 \text{ days} = 1.3 \times 10^9 \text{ J}$$

or about 360 kilowatt hours (1 kWh = 3.6×10^6 J), which is comparable to the total energy of one lightning flash. Even if it were possible to capture all the energy of a flash (this is not possible since the bulk of this energy is not delivered to the strike point but, rather, is lost to heating the air and producing thunder, light, and radio waves), one would need to attract 12 flashes to the energy storage facility in order to operate the five light bulbs for one year. The probability of a lightning strike to a given point on ground is very low. For example, a 1 m^2 area in Florida is struck by lightning on average, once in 10^5 years. A grounded structure protruding above Earth's surface is more likely to be struck by lightning. A 60 m tower located in Florida is expected to be struck by lightning at a rate roughly between once every other year and once every year (subsection 2.9.2). Thus, one needs 12 to 24 such towers covering a large area of 1 km^2 or so to operate the five 100 W light bulbs, which is obviously impractical. Most of the United States experiences lightning activity that is a factor 2 to 3 lower than in Florida. As a result, the number of lightning-capturing towers needed to operate the five 100 W bulbs in areas of moderate lightning activity would be 24 to 72.

Thus the three main problems with the utilization of lightning energy (leaving aside the issue of energy storage devices) can be formulated as follows.

(i) The power associated with a lightning flash is very high, but it is released in pulses of very short duration (of the order of 10^{-4}–10^{-5} s). As a result, the lightning energy, the integral of the power over the short period of time, is moderate, comparable to the monthly energy consumption, 360 kilowatt hours, of five 100 W light bulbs.

(ii) Not all the lightning energy in a flash is delivered to the strike point. Using a typical value of energy per unit resistance (action integral) of 10^5 A s^2 (Table 4.4) determined from measurements of the current at the negative lightning channel base and an assumed range of resistances at the strike point of 10 to 100 Ω, we estimate the range of the lightning energy delivered to the strike point to be from 10^6 to 10^7 J, which is only 10^{-2} to 10^{-4} of the total energy.

(iii) The capturing of a sufficiently large number of lightning strikes would require the use of a large number of tall towers, which is impractical.

1.6. Summary

In ancient cultures lightning was viewed both as a weapon of the gods, used by them to punish humans, and as a message from heaven used to influence state affairs. Systematic studies of thunderstorm electricity began in 1752 when an experiment proposed by B. Franklin was conducted in France. The results of that experiment and similar following experiments in other countries showed conclusively that thunderclouds were electrified. In the same year that the first thunderstorm electricity experiment was carried out, fair-weather electricity was discovered. According to the "classical" view of atmospheric electricity, the potential of the electrosphere, located at an altitude of 60 km or so, with respect to the Earth's surface is about 300 kV. This potential difference drives a total conduction current of the order of 1 kA in all fair-weather regions on the globe. Thus, the global fair-weather load resistance is about 300 Ω. Thunderstorms are considered as the sources in the global electric circuit. Since there has been no experiment to confirm conclusively this classical picture of atmospheric electricity, it remains a subject of debate. The salient properties of downward negative lightning discharges, the most common type of lightning, are summarized in Table 1.1. It appears to be impractical to utilize lightning energy.

References and bibliography for Chapter 1

Adlerman, E.J., and Williams, E.R. 1996. Seasonal variation of the global electrical circuit. *J. Geophys. Res.* **101**: 29 679–88.

Aikin, A.C., and Maynard, N.C. 1990. A Van de Graaf source mechanism for middle atmospheric vertical electric fields. *J. Atmos. Terr. Phys.* **52**: 695–705.

Allibone, T.E. 1973. Schonland obituary. *The Caian* 57–60.

Anderson, F.J., and Freier, G.D. 1969. Interactions of the thunderstorm with a conducting atmosphere. *J. Geophys. Res.* **74**: 5390–6.

Anderson, R.V. 1967. Measurement of worldwide diurnal atmospheric electricity variation. *Mon. Wea. Rev.* **95**: 899–904.

Anderson, R.V. 1969. Universal diurnal variation in air–Earth current density. *J. Geophys. Res.* **74**: 1697–700.

Anderson, R.V. 1977. Atmospheric electricity in the real world (useful applications of observations which are perturbed by local effects). In *Electrical Processes in Atmospheres*, eds. H. Dolezalek and R. Reiter, pp. 87–99, Darmstadt: Steinkopff.

Anderson, R.V., and Bailey, J.C. 1991. Errors in the Gerdien measurement of atmospheric electric conductivity. *Meteor. Atmos. Phys.* **46**: 101–12.

Anderson, R.V., and Trent, E.M. 1969. Atmospheric electricity measurements at five locations in Eastern North America. *J. Appl. Meteor.* **8**: 707–11.

Appleton, E.V., Watson Watt, R.A., and Herd, J.F. 1926. On the nature of atmospherics, II. *Proc. Roy. Soc. A* **111**: 615–77.

Atkinson, W., Sundquist, S., and Fakleson, U. 1971. The electric field existing at stratospheric elevations as determined by tropospheric and ionospheric boundary conditions. *Pure Appl. Geophys.* **84**: 46–56.

Ault, J.P., and Mauchley, S.J. 1926. Ocean magnetic and electric observations. In *Researches of the Department of Terrestrial Magnetism*, vol. 5, Carnegie Institution of Washington, DC (see the chapter entitled Atmospheric-Electric Results Obtained Aboard Carnegie, 1915–1921.)

Austin, B. 2001. *Schonland, Scientist and Soldier*, 639 pp., Bristol: IoP.

Austin, L.W. 1926. The present status of atmospheric disturbances. *Proc. IRE* **14**: 133–8.

Battan, L.J. 1964. *The Thunderstorm*, 128 pp., New York: Signet.

Baum, C.E. 1992. From the electromagnetic pulse to high-power electromagnetics. *Proc. IEEE* **80**: 789–817.

Bazelyan, E.M., Gorin, B.N., and Levitov, V.I. 1978. *Physical and Engineering Foundations of Lightning Protection*, 223 pp., Leningrad: Gidrometeoizdat.

Bazilevskaya, G.A., Krainev, M.B., and Makhmutov, V.S. 2000. Effects of cosmic rays on the Earth's environment. *J. Atmos. Solar-Terr. Phys.* **62**: 1577–86.

Beasley, W.H. 1995. Lightning research: 1991–1994. In *Rev. Geophys. Suppl.*, pp. 833–43, US National Report to International Union of Geodesy and Geophysics.

Bell, T.H., 1962. *Thunderstorms*, London: Dobson.

Benbrook, J.R., Kern, J.W., and Sheldon, W.R. 1974. Measured electric field in the vicinity of a thunderstorm system at an altitude of 37 km. *J. Geophys. Res.* **79**: 5289–94.

Berger, K. 1955a. Die Messeinrichtungen für die Blitzforschung auf dem Monte San Salvatore. *Bull. Schweiz. Elektrotech. Ver.* **46**: 193–204.

Berger, K. 1955b. Resultate der Blitzmessungen der Jahre 1947–1954 auf dem Monte San Salvatore. *Bull. Schweiz. Elektrotech. Ver.* **46**: 405–24.

Berger, K., 1967a. Novel observations on lightning discharges: results of research on Mount San Salvatore. *J. Franklin Inst.* **283**: 478–525.

Berger, K. 1972. Methoden und Resultate der Blitzforschung auf dem Monte San Salvatore bei Lugano in den Jahren 1963–1971. *Bull. Schweiz. Elektrotech. Ver.* **63**: 1403–22.

Berger, K. 1977. The Earth flash. In *Lightning, vol. 1, Physics of Lightning*, ed. R.H. Golde, pp. 119–90, New York: Academic Press.

Berger, K., and Vogelsanger, E. 1965. Messungen und Resultate der Blitzforschung der Jahre 1955–1963 auf dem Monte San Salvatore. *Bull. Schweiz. Elektrotech. Ver.* **56**: 2–22.

Berger, K., Anderson, R.B., and Kroninger, H. 1975. Parameters of lightning flashes. *Electra* **80**: 223–37.

Bering, E.A. 1995. The global circuit: global thermometer, weather by-product or climatic modulator? *Rev. Geophys. Suppl.* **33**: 845–62.

Bering, E.A., Benbrook, J.R., and Sheldon, W.R. 1977. Investigation of the electric field below 80 km from parachute-deployed payload. *J. Geophys. Res.* **82**: 1925–32.

Bering, E.A., Benbrook, J.R., and Sheldon, W.R. 1980a. Problems with mesospheric electric field measurements. *Nature* **283**: 695–6.

Bering, E.A., Rosenberg, T.J., Benbrook, J.R., Detrick, D., Matthews, D.L., Rycroft, M.J., Saunders, M.A., and Sheldon, W.R. 1980b. Electric fields, electron precipitation, and VLF radiation during a simultaneous magnetospheric substorm and atmospheric thunderstorm. *J. Geophys. Res.* **85**: 55–72.

Bering, E.A., Benbrook, J.R., Liao, B., Theall, J.R., Lanzerotti, L.J., and MacLennan, C.G. 1995. Balloon measurements above the South Pole: study of the ionospheric transmission of ULF waves. *J. Geophys. Res.* **100**: 7807–20.

Bering, E.A., Few, A.A., and Benbrook, J.R. 1998. The global electric circuit. *Physics Today*, October, pp. 24–30.

Blakeslee, R.J. 1984. The electric current densities beneath thunderstorms. Ph.D. dissertation, University of Arizona, Tucson.

Blakeslee, R.J., and Krider, E.P. 1992. Ground level measurements of air conductivities under Florida thunderstorms. *J. Geophys. Res.* **97**: 12 947–51.

Blakeslee, R.J., Christian, H.J., and B. Vonnegut, B. 1989. Electrical measurements over thunderstorms. *J. Geophys. Res.* **94**: 13 135–40.

Boström, R., and Fahleson, U. 1977. Vertical propagation of time-dependent electric fields in the atmosphere and ionosphere. In *Electrical Processes in Atmospheres*, eds. H. Dolezalek and R. Reiter, pp. 529–35, Darmstadt: Steinkopff.

Boys, C.V. 1926. Progressive lightning. *Nature* **118**: 749–50.

Boys, C.V. 1929. Progressive lightning. *Nature* **124**: 54–5.

Bragin, Yu. A. 1973. Nature of the lower D region of the ionosphere. *Nature* **245**: 450–1.

Bragin, Yu. A., Tyutin, A.A., Kocheev, A.A., and Tyutin, A.A. 1974. Direct measurement of the atmospheric vertical electric field intensity up to 80 km. *Cosmic Res. (English transl.)* **12**: 279–82.

Brook, M. 1992. Sferics. In *Encyclopedia of Science and Technology*, pp. 350–2, New York: McGraw Hill.

Brook, M., Holmes, C.R., and Moore, C.B. 1970. Lightning and rockets: some implications of the Apollo 12 lightning event. *Nav. Res. Rev.* **23**: 1–17.

Brooks, C.E.P. 1925. The distribution of thunderstorms over the globe. *Geophys. Mem.* **24**: 147–64.

Browning, G.L., Tzur, I., and Roble, R.G. 1987. A global time-dependent model of thunderstorm electricity, 1, Mathematical properties of the physical and numerical models. *J. Atmos. Sci.* **44**: 2166–77.

Burke, H.K., and Few, A.A. 1978. Direct measurements of the atmospheric conduction current. *J. Geophys. Res.* **83**: 3093–8.

Byers, H.R. 1953. *Thunderstorm Electricity*, 344 pp., Chicago: University of Chicago Press.

Byers, H.R., and Braham, R.R. 1949. *The Thunderstorm*, 287 pp., US Government Printing Office.

Byrne, G.J., Benbrook, J.R., and Bering, E.A. 1991. Balloon observations of stratospheric electricity above the South Pole: vertical electric field, conductivity, and conduction current. *J. Atmos. Solar-Terr. Phys.* **53**(9): 859–68.

Chalmers, J.A. 1967. *Atmospheric Electricity.* 2nd edition, 515 pp., New York: Pergamon Press.

Chapman, F.W. 1939. Atmospheric disturbances due to thundercloud discharges. *Proc. Phys. Soc. London* **51**: 876–94.

Chesworth, E.T., and Hale, L.C. 1974. Ice particulates in the mesosphere. *Geophys. Res. Lett.* **1**: 347–50.

Chew, J. 1987. *Storms Above the Desert – Atmospheric Research in New Mexico 1935–1985*, 153 pp., Albuquerque: University of New Mexico Press.

Cianos, N., Oetzel, G.N., and Pierce, E.T. 1973. Reply. *J. Appl. Meteor.* **12**: 1421–3.

Cianos, N., and Pierce, E.T. 1972. A ground–lightning environment for engineering usage. Stanford Research Institute, Technical Report 1, project 1834, Menlo Park, California 94025.

Clark, J.F. 1958. The fair weather atmospheric electric potential and its gradient. In *Recent Advances in Atmospheric Electricity*, ed. L.G. Smith, pp. 61–74, New York: Pergamon.

Clark, J.W. 1874. Observations on the spectrum of sheet lightning. In *Chem. News, London* p. 28, July 17, 1874.

Clayton, M., and Polk, C. 1977. Diurnal variation and absolute intensity of worldwide lightning activity, September 1970 to May 1971. In *Electrical Processes in Atmospheres*, eds. H. Dolezalek and R. Reiter, pp. 440–9. Darmstadt: Steinkopff.

Cobb, W.E. 1967. Evidence of a solar influence on the atmospheric elements at Mauna Loa Observatory. *Mon. Wea. Rev.* **95**: 905–11.

Cobb, W.E. 1968. Atmospheric electric climate at Mauna Loa Observatory, Hawaii. *J. Atmos. Sci.* **25**: 470–80.

Cobb, W.E. 1977. Atmospheric electric measurements at the South Pole. In *Electrical Processes in Atmospheres*, eds. H. Dolezalek and R. Reiter, pp. 161–7, Darmstadt: Steinkopff.

Cobb, W.E., and Wells, H.J. 1970. The electrical conductivity of oceanic air and its correlation to global-atmospheric pollution. *J. Atmos. Sci.* **27**: 814–19.

Cobb, W.E., Phillips, B.B., and Allee, P.A. 1967. Note on mountaintop measurements of atmospheric electricity in northern United States. *Mon. Wea. Rev.* **95**: 912–16.

Cohen, I.B. 1990. *Benjamin Franklin's Science*, 273 pp., Cambridge, Massachusetts: Harvard University Press.

Cole, R.K. Jr, and Pierce, E.T. 1965. Electrification in the Earth's atmosphere for altitudes between 0 and 100 kilometers. *J. Geophys. Res.* **70**: 2735–49.

Coroniti, S.C. (ed.) 1965. *Problems of Atmospheric and Space Electricity*, 616 pp., New York: American Elsevier.

Coroniti, S.C., and Hughes, J. (eds.) 1969. *Planetary Electrodynamics, I and II*, 503 and 587 pp., New York: Gordon and Breach.

Crichlow, W.Q., Favis, R.C., Disney, R.T., and Clark, M.W. 1971. Hourly probability of world-wide thunderstorm occurrence. Research Report 12, Office of Telecommunications, Boulder, Colorado.

Croskey, C.L., Hale, L.C., Mitchell, J.D., Muha, D., and Maynard, N.C. 1985. A diurnal study of the electrical structure of the equatorial middle atmosphere. *J. Atmos. Terr. Phys.* **47**: 835–44.

Davies, K. 1966. *Ionospheric Radio Propagation*, New York: Dover Publications.

Davis, R., and Standring, W.G. 1947. Discharge currents associated with kite balloons. *Proc. Roy. Soc. A* **191**: 304–22.

Deaver, L.E., and Krider, E.P. 1991. Electric fields and current densities under small Florida thunderstorms. *J. Geophys. Res.* **96**: 22 273–81.

Dejnakarintra, M., and Park, C.G. 1974. Lightning-induced electric fields in the ionosphere. *J. Geophys. Res.* **79**: 1903–10.

Dhanorkar, S., and Kamra, A.K., 1992. Relation between electrical conductivity and small ions in the presence of intermediate and large ions in the lower atmosphere. *J. Geophys. Res.* **97**: 20 345–60.

Dibner B. 1977. Benjamin Franklin. In *Lightning, vol. 1, Physics of Lightning*, ed. R.H. Golde, pp. 23–49, New York: Academic Press.

Dolezalek, H. 1971a. Atmospheric electricity. *IUGG, Trans. Am. Geophys. Union* **52**: 351–68.

Dolezalek, H. 1971b. Introductory remarks on the classical picture of atmospheric electricity. *Pure Appl. Geophys.* **84**: 9–12.

Dolezalek, H. 1972. Discussion of the fundamental problem of atmospheric electricity. *Pure Appl. Geophys.* **100**: 8–43.

Dolezalek, H. 1988. Discussion of Earth's net electric charge. *Meteor. Atmos. Phys.* **38**: 240–5.

Dolezalek, H. 1992. The World Data Centre on atmospheric electricity and global change monitoring. *Eur. Sci. Notes Inform. Bull.* **2**: 1–32.

Dolezalek, H., and Reiter, R., eds. 1977. *Electrical Processes in Atmospheres*. Darmstadt, Germany: Dr Dietrich Steinkopff, Verlag.

Driscoll, K.T., and Blakeslee, R.J. 1996. Comments on "Current budget of the atmospheric electric global circuit: by Heinz W. Kasemir". *J. Geophys. Res.* **101**: 11 037–40.

Driscoll, K.T., Blakeslee, R.J., and Baginski, M.E. 1992. A modeling study of the time-averaged electric currents in the vicinity of isolated thunderstorms. *J. Geophys. Res.* **97**: 11 535–51.

Dufay, M. 1949. Recherches sur les spectres des éclairs, deuxiéme partie: étude du spectre dans les régions violette et ultraviolette. *Ann. Geophys.* **5**: 255–63.

Ette, A.I.I., and Oladiran, E.O. 1980. The characteristics of rain electricity in Nigeria, I, Magnitudes and variations. *Pure Appl. Geophys.* **118**: 753–64.

Exner, F. 1900. Summary of results of recent investigations in atmospheric electricity. *Terr. Magn.* **5**: 167–74.

Few, A.A., and Weinheimer, A.J. 1986. Factor of 2 error in balloon-borne atmospheric conduction current measurements. *J. Geophys. Res.* **91**: 10 937–48.

Few, A.A., and Weinheimer, A.J. 1987a. Reply. *J. Geophys. Res.* **92**(D4): 4338.

Few, A.A., and Weinheimer, A.J. 1987b. Reply. *J. Geophys. Res.* **92**(D9): 11 006–8.

Franklin, B. 1774. *Experiments and Observations of Electricity, Made at Philadelphia in America*, 5th edition, 530 pp., London: F. Newberry.

Freier, G.D. 1962. Conductivity of air in thunderstorms. *J. Geophys. Res.* **67**: 4 683–92.

Freier, G.D. 1979. Time-dependent fields and a new mode of charge generation in severe thunderstorms. *J. Atmos. Sci.* **36**: 1967–75.

Füllekrug, M., Fraser-Smith, A.C., Bering, E.A., and Few, A.A. 1999. On the hourly contribution of global cloud-to-ground lightning activity to the atmospheric electric field in the Antarctic during December 1992. *J. Atmos. Solar-Terr. Phys.* **61**: 745–50.

Gary, C. 1994. *La Foudre. Des Mythologies Antiques à la Recherche Moderne*, 208 pp., Paris: Masson.

Gathman, S.G., and Anderson, R.V. 1977. Aircraft measurements of the geomagnetic latitude effect on air–earth current density. *J. Atmos. Terr. Phys.* **39**: 313–16.

Gelinas, L.J., Lynch, K.A., Kelley, M.C., Collins, S., Baker, S., Zhou, Q., and Friedman, J.S. 1998. First observation of meteoric charged dust in the tropical mesosphere. *Geophys. Res. Lett.* **25**: 4047–50.

Gibbons, J. 1871. Spectrum of lightning. *Chem. News, London* p. 96, August 25, 1871.

Gish, O.H. 1942. Further evidence of latitude effect in potential gradient. *Terr. Magn. Atmos. Electr.* **47**: 323–4.

Gish, O.H. 1944. Evaluation and interpretation of the columnar resistance of the atmosphere. *Terr. Magn. Atmos. Electr.* **49**: 159–68.

Gish, O.H., and Sherman, K.L. 1936. Electrical conductivity of air to an altitude of 23 km. *Nat. Geographic Soc. Stratosphere, Ser. 2*: 94–116.

Gish, O.H., and Wait G.R. 1950. Thunderstorms and the Earth's general electrification. *J. Geophys. Res.* **55**: 473–84.

Goldberg, R.A. 1989. Electrodynamics of the high latitude mesosphere. *J. Geophys. Res.* **94**: 14 661–72.

Goldberg, R.A., and Holzworth, R.H. 1991. Middle atmospheric electrodynamics. *Handbook for MAP* **32**: 63–84.

Golde, R.H. 1973. *Lightning Protection*, 254 pp., London: Edward Arnold.

Golde, R.H. 1977. The lightning conductor. In *Lightning, vol. 2, Lightning Protection*, ed. R.H. Golde, pp. 545–76, New York: Academic Press.

Gonta, I., and Williams E. 1994. A calibrated Franklin chimes. *J. Geophys. Res.* **99**: 10 671–7.

Gonzalez, W.D.A., Pereira, E.C., Gonzalez, A.L.C., Martin, I.M., Dutra, S.L.G., Pinto, O. Jr, Wygant, J., and Mozer, F.S. 1982. Large horizontal electric fields measured at balloon heights of the Brazilian magnetic anomaly and association to local energetic particle precipitation. *Geophys. Res. Lett.* **9**: 567–70.

Grenet, G. 1947. Essai d'éxplication de la charge électrique des nuages d'orages. *Ann. Geophys.* **3**: 306–7.

Grenet, G. 1959. Le nuage d'orage: machine électrostatique. *Météorologie* **I-53**: 45–7.

Griffiths, R.F., Latham, J. and Meyers, V. 1974. The ionic conductivity of electrified clouds. *Q.J.R. Meteor. Soc.* **100**: 181–90.

Gringel, W., Rosen, J.M., and Hoffman, D.J. 1986. Electrical structure from 0 to 30 km. In *The Earth's Electrical Environment*, eds. E.P. Krider and R.G. Roble, pp. 166–82, Washington, DC: National Academy Press.

Gunn, R. 1953. *Thunderstorm Electricity*, University of Chicago Press.

Hagenguth, J.H., and Anderson, J.G. 1952. Lightning to the Empire State Building. *AIEE Trans.* **71**(3): 641–9.

Hager, W.W. 1998. A discrete model for the lightning discharge. *J. Comput. Phys.* **144**: 137–50.

Hager, W.W., Nisbet, J.S., and Kasha, J.R. 1989a. The evolution and discharge of electric fields within a thunderstorm. *J. Comput. Phys.* **82**: 193–217.

Hager, W.W., Nisbet, J.S., Kasha, J.R., and Shann, W.-C. 1989b. Simulation of electric fields within a thunderstorm. *J. Atmos. Sci.* **46**: 3542–58.

Hale, L.C. 1983. Experimentally determined factors influencing electrical coupling mechanisms. In *Weather and Climate Responses to Solar Variations*, ed. B.M. McCormac, Boulder, Colorado: Associated University Press.

Hale, L.C. 1984. Middle atmospheric electrical structure, dynamics, and coupling. *Adv. Space Res.* **4**: 175–86.

Hale, L.C., and Baginski, M.E. 1987. Current to the ionosphere following lightning stroke. *Nature* **329**: 814–16.

Hale, L.C., and Croskey, C.L. 1979. An auroral effect of the fair weather electric field. *Nature* **278**: 239–41.

Hale, L.C., Croskey, C.L., and Mitchell, J.D. 1981. Measurements of middle-atmosphere electric fields and associated electrical conductivities. *Geophys. Res. Lett.* **8**: 927–30.

Halliday, E.C. 1933. Variations in the electric field in the atmosphere measured in Johannesburg, South Africa, during 1929 and 1930. *Terr. Magn.* **40**: 37–53.

Hamelin, J. 1993. Sources of natural noise. In *Electromagnetic Compatibility*, eds. P. Degauque and J. Hamelin, 652 pp., New York: Oxford.

Harland, W.B., and Hacker, J.L.F. 1966. "Fossil" lightning strikes 250 million years ago. *Advancement of Science* **22**: 663–71.

Harris, W.S. 1834. On the protection of ships from lightning. *Nautic. Mag.* **3**: 151–6, 225–33, 353–8, 402–7, 477–84, 739–44, 781–7.

Harris, W.S. 1838. Illustrations of cases of damage by lightning in the British Navy. *Nautic. Mag. (Enlarged Ser.)* **2**: 590–5, 747–8.

Harris, W.S. 1839. Illustrations of cases of damage by lightning in the British Navy. *Nautic. Mag. (Enlarged Ser.)* **3**: 113–22.

Harris, W.S. 1843. *On the Nature of Thunderstorms*, London: Parker.

Havnes, O., Troim, J., Blix, T., Mortensen, W., Naesheim, L.I., Thrane, E., and Tonnesen, T. 1996. First detection of charged dust particles in the Earth's mesosphere. *J. Geophys. Res.* **101**: 10 839–47.

Hays, P.B., and Roble, R.G. 1979. A quasi-static model of global atmospheric electricity, 1. The lower atmosphere. *J. Geophys. Res.* **84**: 3291–305.

Hegai, V.V., and Kim, V.P. 1990. The formation of a cavity in the night-time midlatitude ionospheric E-region above a thundercloud. *Planet. Space Sci.* **38**(6): 703–7.

Helliwell, R.A. 1965. *Whistlers and Related Ionospheric Phenomena*, 349 pp., Stanford, California: Stanford University Press.

Herschel, J. 1868. On the lightning spectrum. *Proc. Roy. Soc.* **15**: 61–2.

Hill, R.D. 1971. Spherical capacitor hypothesis of the Earth's electric field. *Pure Appl. Geophys.* **84**: 67–75.

Hoffert, H.H. 1889. Intermittent lightning flashes. *Phil. Mag.* **28**: 106–9.

Hogg, A.R. 1950. Air–earth current observations in various localities. *Arch. Meteor.* **3**: 40–55.

Holden, E.S. 1872. Spectrum of lightning. *Am. J. Sci. Arts.* **4**: 474–5.

Holzer, R.E., and Saxon, D.S. 1952. Distribution of electrical conduction current in the vicinity of thunderstorms. *J. Geophys. Res.* **57**: 207–16.

Holzworth, R.H. 1977. Large scale dc electric fields in the Earth's environment, Ph.D. dissertation, University of California, Berkeley.

Holzworth, R.H. 1981. High latitude stratospheric electrical measurements in fair and foul weather under various solar conditions. *J. Atmos. Terr. Phys.* **43**: 1115–26.

Holzworth, R.H. 1983. Electrodynamics of the stratosphere using 5000 m^3 superpressure balloons. *Adv. Space. Res.* **3**: 107.

Holzworth, R.H. 1984. Hy-wire measurements of atmospheric potential. *J. Geophys. Res.* **89**: 1395–401.

Holzworth, R.H. 1991a. Conductivity and electric field variations with altitude in the stratosphere. *J. Geophys. Res.* **96**: 12 857–64.

Holzworth, R.H. 1991b. Atmospheric electrodynamics in the US: 1987–1990. *Rev. Geophys., Suppl.*, pp. 115–20, US National Report to Int. Union of Geodesy and Geophysics, 1987–90.

Holzworth, R.H. 1995. Quasistatic electromagnetic phenomena in the atmosphere and ionosphere. In *Handbook on Atmospheric Electrodynamics*, vol. 1, ed. H. Volland, pp. 235–66, Boca Raton, Florida: CRC Press.

Holzworth, R.H., and Chiu, Y.T. 1982. Sferics in the stratosphere. In *Handbook of Atmospherics*, vol. 2, ed. H. Volland, pp. 1–19, Boca Raton: CRC Press.

Holzworth, R.H., and Mozer, F.S. 1979. Direct evidence of solar flare modification of stratospheric electric fields. *J. Geophys. Res.* **84**: 363–7.

Holzworth, R.H., Dazey, M.H., Schnauss, E.R., and Youngbluth, O. 1981. Direct measurement of lower atmospheric vertical potential differences. *Geophys. Res. Lett.* **8**: 783–6.

Holzworth, R.H., Onsager, T., Kintner, P., and Powell, S. 1984. Planetary scale variability of the fair weather vertical electric field. *Phys. Rev. Lett.* **53**(14): 1398–401.

Holzworth, R.H., Kelley, M.C., Siefring, C.L., Hale, L.C., and Mitchell, J.D. 1985. Electrical measurements in the atmosphere and the ionosphere over an active thunderstorm, 2, Direct current electric fields and conductivity. *J. Geophys. Res.* **90**: 9824–30.

Holzworth, R.H., Norville, K.W., Kintner, P.M., and Power, S.P. 1986. Stratospheric conductivity variations over thunderstorms. *J. Geophys. Res.* **91**: 13 257–63.

Holzworth, R.H., Norville, K.W., Kintner, P.M., and Power, S.P. 1988. Reply. *J. Geophys. Res.* **93**: 3915–7.

Holzworth, R.H., Plaff, R.F., Goldberg, R.A., Bounds, S.R., Schmidlin, F.J., Voss, H.D., Tuzzolino, A.J., Croskey, C.L., Mitchell, J.D., von Cossart, G., Singer, W., Hoppe, U.-P., Murtagh, D., Witt, G., Gumbel, J., and Friedrich, M. 2001. Large electric potential perturbations in PMSE during DROPPS. *Geophys. Res. Lett.* **28**(8): 1435–8.

Hoppel, W.A., Anderson, R.V., and Willett, J.C. 1986. Atmospheric electricity in the planetary boundary layer. In *The Earth's Electrical Environment*, eds. E.P. Krider and R.G. Roble, pp. 149–65. Washington, DC: National Academy Press.

Hu, H., Holzworth, R.H., and Li, Y.Q. 1989. Thunderstorm related variations in stratospheric conductivity measurements. *J. Geophys. Res.* **94**: 16 429–35.

Illingworth, A.J. 1971a. The variation of the electric field after lightning and the conductivity within thunderclouds. *Q.J.R. Meteor. Soc.* **97**: 440–56.

Illingworth, A.J. 1971b. Electric field recovery after lightning. *Nature – Physical Science* **229**: 213–14.

Illingworth, A.J. 1972a. Electric field recovery after lightning as the response of the conducting atmosphere to a field change. *Q.J.R. Meteor. Soc.* **98**: 604–16.

Illingworth, A.J. 1978. Charging up a thunderstorm. *New Scientist* **78**: 502–6.

Imyanitov, I.M., and Chubarina, E.V. 1967. *Electricity of the Free Atmosphere*, 210 pp. Available from US Dept of Commerce, Clearing House for Federal Science and Technology Information, Springfield, Virginia 22151.

Imyanitov, I.M., Evteev, B.F., and Kamaldina, I.I. 1969. A thunderstorm cloud. In *Planetary Electrodynamics*, eds. S.C. Coroniti and J. Hughes, pp. 401–25, New York: Gordon and Breach Science.

Imyanitov, I.M., Chubarina, Ye. V., and Shvarts, Ya. M. 1971. *Electricity in Clouds*, 92 pp., Leningrad: Gidrometeoizdat.

Israel, H. 1970. *Atmospheric Electricity, vol. I, Fundamentals, Conductivity, Ions*, 317 pp., Published for the National Science Foundation by the Israel Program for Scientific Translation, Jerusalem.

Israel, H. 1973. *Atmospheric Electricity, vol. II, Fields, Charges, Currents*, 796 pp. published for the National Science Foundation by the Israel Program for Scientific Translation, Jerusalem.

Israel, H., and Fries, G. 1956. Ein Gerät zur spektroskopischen Analyse verschiedener Blitzphasen. *Optik* **13**: 365–8.

Israelsson, S., Knudsen, E., and Tammet, H. 1994. An experiment to examine the covariation of atmospheric electrical vertical currents at two separate stations. *J. Atmos. Electr.* **14**: 63–73.

Israelsson, S., and Tammet, H. 2001. Variation of fair weather atmospheric electricity at Marsta Observatory, Sweden, 1993–1998. *J. Atmos. Terr. Phys.* **63**: 1693–703.

Jura, M. 1977. Relationship between atmospheric electricity and microwave radio propagation. *Nature* **266**: 703–4.

Kamra, A.K., Deshpande, C.G., and Gopalakrishnan, V. 1994. Challenge to the assumption of the unitary diurnal variation of the atmospheric electric field based on observations in the Indian Ocean, Bay of Bengal, and Arabian Sea. *J. Geophys. Res.* **99**: 21 043–50.

Kasemir, H.W. 1955. Measurement of the air–earth current density. In *Proc. Conf. Atmospheric Electricity*, Geophys. Research Paper 42, AFCRC-TR-55-222, Air Force Cambridge Research Center, Bedford, Massachusetts.

Kasemir H.W. 1959. The thunderstorm as a generator in the global electric circuit (in German). *Z. Geophys.* **25**: 33–64.

Kasemir, H.W. 1960. A radiosonde for measuring the air–earth current density. USA SRDL Technical Report 2125, 29 pp., US Army Signal Research and Development Laboratories, Fort Monmouth, New Jersey.

Kasemir H.W. 1963. On the theory of the atmospheric electric current flow, IV. Technical Report 2394, US Army Electronics Research and Development Laboratories, Fort Monmouth, New Jersey.

Kasemir, H.W. 1972. Atmospheric electric measurements in the Arctic and Antarctic. *Pure Appl. Geophys.* **100**: 70–80.

Kasemir, H.W. 1977. Theoretical problems of the global atmospheric electric circuit. In *Electrical Processes in Atmospheres*, eds. H. Dolezalek and R. Reiter, pp. 423–38, Darmstadt: Steinkopff.

Kasemir, H.W. 1979. The atmospheric electric global circuit. In *Proc. Workshop on the Need for Lightning Observations from Space*, NASA CP-2095, pp. 136–47.

Kasemir, H.W. 1994. Current budget of the atmospheric electric global circuit. *J. Geophys. Res.* **99**: 10 701–8.

Kasemir, H.W. 1996. Reply. *J. Geophys. Res.* **101**: 17 033–5 and 17 041–3.

Kasemir, H.W., and Ruhnke, L.H. 1958. Antenna problems of measurement of the air–earth current. In *Recent Advances in Atmospheric Electricity*, pp. 137–47, New York: Pergamon Press.

Kelley, M.C. 1983. Middle atmospheric electrodynamics. *Rev. Geophys. Space Phys.* **21**: 273–5.

Kelley, M.C., Siefring, C.L., and Pfaff, R.F. Jr 1983. Large amplitude middle atmospheric electric fields: fact or fiction? *Geophys. Res. Lett.* **10**: 733–6.

Kelley, M.C., Siefring, C.L., Pfaff, R.F., Kintner, P.M., Larsen, M., Green, R., Holzworth, R.H., Hale, L.C., Mitchell, J.D., and LeVine, D. 1985. Electrical measurements in the atmosphere and ionosphere over an active thunderstorm, 1, Campaign overview and initial ionospheric results. *J. Geophys. Res.* **90**: 9815–23.

Kelley, M.C., Ding, J.G., and Holzworth, R. H. 1990. Intense ionospheric electric and magnetic field pulses generated by lightning. *Geophys. Res. Lett.* **17**: 2221–4.

Kessler, E. (ed.) 1983. The thunderstorm in human affairs. In *Thunderstorms: A Social, Scientific, and Technological Documentary*, vol. 1, 2nd edition, 186 pp., Norman, Oklahoma: University of Oklahoma Press.

Kessler, E. (ed.) 1986. Morphology and dynamics. In *Thunderstorms: A Social, Scientific, and Technological Documentary*, vol. 2, 2nd edition, enlarged, 411 pp., Norman, Oklahoma: University of Oklahoma Press.

Kessler, E. (ed.) 1988. Instruments and techniques for thunderstorm observation and analysis. In *Thunderstorms: A Social, Scientific, and Technological Documentary*, vol. 3, 313 pp., Norman, Oklahoma: University of Oklahoma Press.

Kessler, E. (ed.) 1992. Thunderstorm morphology and dynamics. In *Thunderstorms: A Social, Scientific, and Technological Documentary*, vol. 2, 2nd edition, enlarged, 411 pp., Norman, Oklahoma: University of Oklahoma Press.

Kikuchi, T., Araki, T., Maeda, H., and Maekawa, K. 1978. Transmission of ionospheric electric fields to the equator. *Nature* **273**: 650–1.

Kraakevik, J. 1958. The airborne measurements of atmospheric conductivity. *J. Geophys. Res.* **63**: 161–9.

Kraakevik, J. 1961. Measurements of current density in the fair weather atmosphere. *J. Geophys. Res.* **66**: 3735–48.

Krider, E.P. 1994. Physics of lightning today. *Extraits de la Revue générale de l'Electricité* no. 6, 7 pp.

Krider, E.P. 1996a. 75 years of research on the physics of a lightning discharge. In *Historical Essays on Meteorology, 1919–1995*, ed. J.R. Fleming, pp. 321–50, Boston, Massachusetts: American Meteorological Society.

Krider, E.P. 1996b. Lightning rods in the 18th century. In *Proc. 23rd Int. Conf. on Lightning Protection, Florence, Italy*, pp. 1–8.

Krider, E.P., and Blakeslee, R.J. 1985. The electric currents produced by thunderclouds. *J. Electrostatics* **16**: 369–78.

Krider, E.P., and Musser, J.A. 1982. Maxwell currents under thunderstorms. *J. Geophys. Res.* **87**: 11 171–6.

Krider, E.P., and Roble, R.G. (eds.) 1986. *The Earth's Electrical Environment, Studies in Geophysics*, 263 pp., Washington, DC: National Academy Press.

Krider, E.P. and Uman, M.A. 1995. Cloud-to-ground lightning: mechanisms of damage and methods of protection. *Seminars in Neurology* **15**: 227–32.

Kulkarni, M., and Kamra, A.K. 2001. Vertical profiles of atmospheric electric parameters close to ground. *J. Geophys. Res.* **106**: 28 209–21.

Kundt, W., and Thuma, G. 1999. Geoelectricity: atmospheric charging and thunderstorms. *J. Atmos. Solar-Terr. Phys.* **61**: 955–63.

Larsen, A. 1905. Photographing lightning with a moving camera. *Ann. Rep. Smithsonian Inst.* **60**(1): 119–27.

Latham, J., and Stromberg, I.M. 1977. Point-discharge. In *Lightning, vol. 1, Physics of Lightning*, ed. R.H. Golde, pp. 99–117, New York: Academic Press.

Lemonnier, L.G. 1752. Observations sur l'électricité de l'air. *Mem. Acad. Sci.* **2**: 223.

Lewis, W.W. 1965. *The Protection of Transmission Systems Against Lightning*, 422 pp., New York: Dover.

Lobodin, T.V., and Paramonov, N.A. 1972. Variations of atmospheric-electric field during aurorae. *Pure App. Geophys.* **100**: 167–73.

MacGorman, D.R., and Rust, W.D. 1998. *The Electrical Nature of Thunderstorms*, 422 pp., New York: Oxford University Press.

MacGorman, D.R., Straka, J.M., and Ziegler, C.L. 2001. A lightning parameterization for numerical cloud model. *J. Appl. Meteor.* **40**: 459–78.

Madden, T.R., and Thompson, W. 1965. Low-frequency electromagnetic oscillations of the Earth–ionosphere cavity. *Rev. Geophys.* **3**: 211–54.

Magono, C. 1980. *Thunderstorms*, 261 pp., New York: Elsevier.

Makino, M., and Ogawa, T. 1984. Responses of atmospheric electric field and air-earth current to variations of conductivity profile. *J. Atmos. Terr. Phys.* **46**: 431–45.

Makino, M., and Ogawa, T. 1985. Quantitative estimation of global circuit. *J. Geophys. Res.* **90**: 5961–6.

Makino, M., and Takeda, M. 1984. Three-dimensional ionospheric currents and fields generated by the atmospheric global curcuit current. *J. Atmos. Terr. Phys.* **46**(3): 199–206.

Malan, D.J. 1950. Apparel de grand rendement pour la chronophotographie des eclairs. *Rev. Opt.* **29**: 513–23.

Malan, D.J. 1957. The theory of lightning photography and a camera of new design. *Geofis. Pura Appl.* **38**: 250–60.

Malan, D.J. 1964. *Physics of Lightning*. 176 pp., London: English Universities Press.

Malan, D.J. 1967. Physics of the thunderstorm electric circuit. *J. Franklin Inst.* **283**: 526–39.

Mann, J.E. 1970. Interaction of a thunderstorm with a conducting atmosphere. *J. Geophys. Res.* **75**: 1697–8.

Markson, R. 1971. Considerations regarding solar and lunar modulation of geophysical parameters, atmospheric electricity and thunderstorms. *Pure Appl. Geophys.* **84**: 161–200.

Markson, R. 1976. Ionospheric potential variations obtained from aircraft measurements of potential gradient. *J. Geophys. Res.* **81**: 1980–90.

Markson, R. 1977. Airborne atmospheric electrical measurements of the variation of ionospheric potential and electrical structure in the exchange layer over the ocean. In *Electrical Processes in Atmospheres*, eds. H. Dolzalek and R. Reiter, pp. 450–9, Darmstadt: Steinkopff.

Markson, R. 1978. Solar modulation of atmospheric electrification and possible implications for the sun–weather relationship. *Nature* **273**: 103–9.

Markson, R. 1985. Aircraft measurements of the atmospheric electrical global circuit during the period 1971–1984. *J. Geophys. Res.* **90**: 5967–77.

Markson, R. 1986. Tropical convection, ionospheric potentials and global circuit variation. *Nature* **320**: 588–94.

Markson, R. 1987. Comment on "Factor of 2 error in balloon-borne atmospheric conduction current measurements" by A.A. Few and A.J. Weinheimer. *J. Geophys. Res.* **92**: 11 003–5.

Markson, R., and Muir, M. 1980. Solar wind control of the Earth's electric field. *Science* **208**: 979–90.

Marshall, T.C., and Stolzenburg, M. 2001. Voltages inside and just above thunderstorms. *J. Geophys. Res.* **106**(D5): 4757–68.

Marshall, T.C., Stolzenburg, M., Rust, W.D. 1996. Electric field measurements above mesoscale convective systems. *J. Geophys. Res.* **101**: 6979–96.

Marshall, T.C., Rust, W.D., Stolzenburg, M., Roeder, W.P., and Krehbiel, P.R. 1999. A study of enhanced fair-weather electric fields occurring soon after sunrise. *J. Geophys. Res.* **104**: 24 455–69.

Mason, B.J. 1957. *The Physics of Clouds*, 481 pp., Oxford: Clarendon Press.

Mason, B.J. 1971. *The Physics of Clouds*. 671 pp., Oxford: Clarendon Press.

Mason, B.J. 1972. The Bakerian lecture: the physics of the thunderstorm. *Proc. Roy. Soc. A* **32**: 433–66.

Mateev, L.N., and Velinov, P.I. 1992. Cosmic ray variation effects on the parameters of the global atmospheric electrical circuit. *Adv. Space Res.* **12**(10): 353–6.

Mauchly, S.J. 1923. On the diurnal variation of the potential gradient of atmospheric electricity. *Terr. Magn. Atmos. Electr.* **28**: 61–81.

Maynard, N.C., Croskey, C.L., Mitchell, J.D., and Hale, L.C. 1981. Measurement of volt/meter vertical electric fields in the middle atmosphere. *Geophys. Res. Lett.* **8**: 923–6.

McAdie, A. 1897. Elster and Geitel's resumé of recent papers on atmospheric electricity. *Terr. Magn.* **2**: 128–32.

McEachron, K.B. 1939. Lightning to the Empire State Building. *J. Franklin Inst.* **227**: 149–217.

McEachron, K.B. 1941. Lightning to the Empire State Building. *Trans. AIEE* **60**: 885–9.

Menshutkin, B.N. 1952. *Russia's Lomonosov*. Princeton University Press.

Meyerott, R.E., Reagan, J.B., and Joiner, R.G. 1980. The mobility and concentration of ions and the ionic conductivity in the lower stratosphere. *J. Geophys. Res.* **85**: 1273–8.

Meyerott, R.E., Reagan, J.B., and Evans, J.E. 1983. On the correlation between ionospheric potential and intensity of cosmic rays. In *Weather and Climate Responses to Solar Variations*, ed. B.M. McCormac, pp. 449–60, Boulder: Colorado Associated University Press.

Michnowski, S. 1998. Solar wind influences on atmospheric electricity variables in polar regions. *J. Geophys. Res.* **103**: 13 939–48.

Mitchell, J.D. 1990. Electrical properties of the middle atmosphere. *Adv. Space Res.* **10**: 219–28.

Morita, Y. 1971. The diurnal and latitudinal variation of electric field and electric conductivity in the atmosphere over the Pacific Ocean. *J. Meteor. Soc. Japan, Ser. II*, 49: 56–8.

Morita, Y., Ishikawa, H., and Kanada, M. 1971. The vertical profile of the small ion density and the electric conductivity in the atmosphere in 19 kilometers. *J. Geophys. Res.* **76**: 3431–6.

Mozer, F.S. 1971. Balloon measurement of vertical and horizontal atmospheric electric fields. *Pure Appl. Geophys.* **84**: 32–45.

Mozer, F.S. and Serlin, R. 1969. Magnetospheric electric field measurements with balloons. *J. Geophys. Res.* **74**: 4739–54.

Mühleisen, R. 1977. The global circuit and its parameters. In *Electrical Processes in Atmospheres*, eds. H. Dolezalek and R. Reiter, pp. 467–76, Darmstadt: Steinkopff.

Nickolaenko, A.P., Price, C., and Iudin, D.D. 2000. Hurst exponent derived for natural terrestrial radio noise in Schumann resonance band. *Geophys. Res. Lett.* **27**: 3185–8.

Nisbet, J.S. 1983. A dynamic model of thundercloud electric fields. *J. Atmos. Sci.* **40**: 2855–73.

Nisbet, J.S. 1985a. Thundercloud current determination from measurements at the Earth's surface. *J. Geophys. Res.* **90**: 5840–56.

Nisbet, J.S. 1985b. Currents to the ionosphere from thunderstorm generators: a model study. *J. Geophys. Res.* **90**: 9831–44.

Nisbet, J.S., Barnard, T.A., Forbes, G.S., Krider, E.P., Lhermitte, R., and Lennon, C.L. 1990a. A case study of the Thunderstorm Research International Project storm of July 11, 1978 – 1. Analysis of the data base. *J. Geophys. Res.* **95**: 5417–33.

Nisbet, J.S., Kasha, J.R., and Forbes, G.S. 1990b. A case study of the Thunderstorm Research International Project storm of July 11, 1978 – 2. Interrelations among the observable parameters controlling electrification. *J. Geophys. Res.* **95**: 5435–45.

Norinder, H. 1936. Cathode-ray oscillographic investigations on atmospherics. *Proc. IRE* **24**: 287–304.

Norinder, H., and Dahle, O. 1945. Measurements by frame aerials of current variations in lightning discharges. *Arkiv. Mat. Astron. Fysik* **32A**: 1–70.

Ogawa, T. 1985. Fair-weather electricity. *J. Geophys. Res.* **90**: 5951–60.

Ogawa, T. 1995. Lightning currents. In *Handbook of Atmospheric Electrodynamics*, vol. I, ed. H. Volland, pp. 93–136, Boca Raton, Florida: CRC Press.

Ogawa, T., Tanaka, Y., and Yasuhara, M. 1969. Schumann resonance and world-wide thunderstorm activity. *J. Geomagn. Geoelectr.* **21**: 447–52.

Oparin, A.I. 1938. *The Origin of Life on Earth*. New York: Macmillan (translation from Russian; original published in Moscow in 1936).

Orville, R.E. 1968a. A high-speed time-resolved spectroscopic study of the lightning return stroke: Part I, A qualitative analysis. *J. Atmos. Sci.* **25**: 827–38.

Orville, R.E. 1968b. A high-speed time-resolved spectroscopic study of the lightning return stroke: Part II, A quantitative analysis. *J. Atmos. Sci.* **25**: 839–51.

Orville, R.E. 1968c. A high-speed time-resolved spectroscopic study of the lightning return stroke: Part III, A time-dependent model. *J. Atmos. Sci.* **25**: 852–6.

Orville, R.E. 1977. Lightning spectroscopy. In *Lightning, vol. 1, Physics of Lightning*, ed. R.H. Golde, pp. 281–308, New York: Academic Press.

Orville, R.E. (ed.) 1984. Preprints from *Proc. 7th Int. Conf. on Atmospheric Electricity, June 3–8, 1984, Albany*, New York, American Meteorological Society, 45 Beacon Street, Boston, Massachusetts 02108.

Orville, R.E. (ed.) 1985. Selected papers from *Proc. 7th Int. Conf. on Atmospheric Electricity, June 3–8, 1984, Albany*, New York, *J. Geophys. Res.* **90**(D4).

Paltridge, G.W. 1965. Experimental measurements of the small ion density and electrical conductivity of the stratosphere. *J. Geophys. Res.* **70**: 2751–61.

Paramonov, N.A. 1950a. The unitary variation of the potential gradient of atmospheric electricity, *Dokl. Akad. Nauk* **70**: 37–8.

Paramonov, N.A. 1950b. On the annual variations of the potential gradient of atmospheric electricity, *Dokl. Akad. Nauk* **70**: 39–40.

Park, C.G. 1976. Downward mapping of high-latitude electric fields to ground. *J. Geophys. Res.* **81**: 168–74.

Park, C.G. 1979. Comparison of two-dimensional and three-dimensional mapping of ionospheric electric field, *J. Geophys. Res.* **84**: 960–4.

Park, C.G., and Dejnakarintra, M. 1973. Penetration of thundercloud electric fields into the ionosphere and magnetosphere. 1, Middle and subauroral latitudes. *J. Geophys. Res.* **78**: 6623–33.

Park, C.G., and Dejnakarintra, M. 1977a. Thundercloud electric fields in the ionosphere. In *Electrical Processes in Atmospheres*, eds. H. Dolezalek and R. Reiter, pp. 544–51, Darmstadt: Steinkopff.

Park, C.G., and Dejnakarintra, M. 1977b. The effects of magnetospheric convection on atmospheric electric fields in the Polar Cap. In *Electrical Processes in Atmospheres*, eds. H. Dolezalek and R. Reiter, pp. 536–42, Darmstadt: Steinkopff.

Parkinson, W.C., and Torreson, O.W. 1931. The diurnal variation of the electric potential of the atmosphere over the oceans. *Int. Union. Geomagn. Geophys. Bull.* **8**: 340–5.

Parkinson, W.C., and Torreson, O.W. 1946. The diurnal variation of the electric potential of the atmosphere over the oceans. Ocean atmospheric-electric results, Carnegie Institution of Washington, Publ. 568: 135–6.

Pasko, V.P., Inan, U.S., and Bell, T.F. 1998d. Ionospheric effects due to electrostatic thundercloud fields. *J Atmos. Solar-Terr. Phys.* **60**: 863–70.

Pasko, V.P., Inan, U.S., and Bell, T.F. 2001. Mesosphere–troposphere coupling due to sprites. *Geophys. Res. Lett.* **28**: 3821–4.

Peters, O.S. 1915. Protection of life and property against lightning. Technologic Papers of the Bureau of Standards, no. 56, Washington Government Printing Office, 27 pp.

Pfaff, R., Holzworth, R., Goldberg, R., Freudenreich, H., Voss, H., Croskey, C., Mitchell, J., Gumbel, J., Bounds, S., Singer, W., and Latteck, R. 2001. Rocket probe observations of electric field irregularities in the polar summer mesosphere. *Geophys. Res. Lett.* **28**(8): 1431–4.

Pierce, E.T. 1958. Some topics in atmospheric electricity. In *Recent Advances in Atmospheric Electricity*, ed. L.G. Smith, pp. 5–16, New York: Pergamon.

Pierce, E.T. 1969. The thunderstorm as a source of atmospheric noise at frequencies between 1 and 100 kHz. Stanford Research Institute Technical Report, Project 7045, DASA 2299.

Pierce, E.T. 1974. Atmospheric electricity – some themes. *Bull. Am. Meteor. Soc.* **55**: 1186–94.

Pierce, E.T. 1977. Stratospheric electricity and the global circuit. In *Electrical Processes in Atmospheres*, eds. H. Dolezalek and R. Reiter, pp. 582–6, Darmstadt: Steinkopff.

Pierce, E.T., and Whitson, A.L. 1964. The variation of potential gradient with altitude above ground of high radioactivity. *J. Geophys. Res.* **69**: 2895–8.

Pierce, E.T., and Wormell, T.W. 1953. *Thunderstorm Electricity*, 344 pp., University of Chicago Press.

Pinto, I.R.C.A., Pinto, O. Jr, Gonzalez, W.D., Dutra, L.G., Wygant, J., and Mozer, F.S. 1988. Stratospheric electric field and conductivity measurements over electrified convective clouds in the South American region. *J. Geophys. Res.* **93**: 709–15.

Plotkin, V.V. 1999. Electric fields in the ionosphere due to global lightning activity. *Geomagnetism i Aeronomia* **39**(2): 126–9.

Pockels, F. 1897. Über das magnetische Verhalten einiger basaltischer Gesteine. *Ann. Physik Chem.* **63**: 95–201.

Pockels, F. 1898. Bestimmung maximaler Entladungsstromstärken aus ihrer magnetisierenden Wirkung. *Ann. Physik Chem.* **63**: 458–75.

Pockels, F. 1900. Über die Blitzentladungen erreichte Stromstärke. *Physik. Z.* **2**: 306–7.

Price, C. 2000. Evidence for a link between global lightning activity and upper tropospheric water vapour. *Nature* **406**: 290–3.

Price, C., and Rind, D. 1992. A simple lightning parameterization for calculating global lightning distributions. *J. Geophys. Res.* **97**: 9919–33.

Price, C., and Rind, D. 1994a. Modeling global lightning distributions in a general circulation model. *Mon. Wea. Rev.* **122**: 1930–9.

Price, C., and Rind, D. 1994b. Possible implications of global climate change on global lightning distributions and frequencies. *J. Geophys. Res.* **99**: 10 823–31.

Prinz, H. 1977. Lightning in history. In *Lightning, vol. 1, Physics of Lightning*, ed. R.H. Golde, pp. 1–21, New York: Academic Press.

Prueitt, M.L. 1963. The excitation temperature of lightning. *J. Geophys. Res.* **68**: 803–11.

Rakov, V.A. 1998. Some inferences on the propagation mechanisms of dart leaders and return strokes. *J. Geophys. Res.* **103**: 1879–87.

Rakov, V.A., Uman, M.A., and Thottappillil, R. 1994. Review of lightning properties determined from electric field and TV observations. *J. Geophys. Res.* **99**: 10 745–50.

Rakov, V.A., Thottappillil, R., Uman, M.A., and Barker, P.P. 1995a. Mechanism of the lightning M component. *J. Geophys. Res.* **100**: 25 701–10.

Rakov, V.A., Uman, M.A., Thottappillil, R. 1995b. Review of recent lightning research at the University of Florida. *Elektrotechik and Informationstechinik*, **112**(6): 262–5.

Rakov, V.A., Uman, M.A., Hoffman, G.R., Masters, M.W., and Brook, M. 1996. Bursts of pulses in lightning electromagnetic radiation: observations and implications for lightning test standards. *IEEE Trans. Electromagn. Compat.* **38**: 156–64.

Rakov, V.A., Crawford, D.E., Rambo, K.J., Schnetzer, G.H., Uman, M.A., and Thottappillil, R. 2001. M-component mode of charge transfer to ground in lightning discharges. *J. Geophys. Res.* **106**: 22 817–31.

Randa, J., Gilliland, D., Gjertson, W., Lauber, W., and McInerney, M. 1995. Catalogue of electromagnetic environment measurements, 30–300 Hz. *IEEE Trans. Electromagn. Compat.* **37**: 16–33.

Reagan, J.B., Meyerott, R.E., Evans, J.E., Imhof, W.L. and Joiner, R.G. 1983. The effects of energetic particle precipitation on the atmospheric electric circuit. *J. Geophys. Res.* **88**: 3869–78.

Reid, G.C. 1986. Electrical structure of the middle atmosphere. In *The Earth's Electrical Environment*, eds. E.P. Krider, and R.G. Roble, pp. 183–94, Washington, DC: National Academy Press.

Reiter, R. 1969. Solar flares and their impact on potential gradient and air–earth current characteristics at high mountain station. *Pure Appl. Geophys.* **72**: 259.

Reiter, R. 1971. Further evidence for impact of solar flares on potential gradient and air–earth current characteristics at high mountain stations. *Pure Appl. Geophys.* **86**: 142.

Reiter, R. 1972. Case study concerning the impact of solar activity upon potential gradient and air–earth current in the lower troposphere. *Pure Appl. Geophys.* **94**: 218–25.

Reiter, R. 1973. Increased influx of stratospheric air into the lower troposphere after solar H_α and X-ray flares. *J. Geophys. Res.* **78**: 6167–72.

Reiter, R. 1977a. Atmospheric electricity activities of the Institute for Atmospheric Environmental Research. In *Electrical Processes in Atmospheres*, eds. H. Dolezalek and R. Reiter, pp. 759–96, Darmstadt: Steinkopff.

Reiter R. 1977b. The electric potential of the ionosphere as controlled by the solar magnetic sector structure, result of a study over the period of a solar cycle. *J. Atmos. Terr. Phys.* **39**: 95–9.

Reiter, R. 1992. *Phenomena in Atmospheric and Environmental Electricity*, 541 pp., New York: Elsevier.

Reynolds, S.E., and Neill, H.W. 1955. The distribution and discharge of thunderstorm charge-centers. *J. Meteor.* **12**: 1–12.

Richmond, A.D. 1986. Upper-atmosphere electric-field sources. In *The Earth's Electrical Environment*, eds. E.P. Krider and R.G. Roble, pp. 195–205, Washington, DC: National Academy Press.

Ridley, A.J., Crowley, G., and Freitas, C. 2000. An empirical model of the ionospheric electric potential. *Geophys. Res. Lett.* **27**: 3675–8.

Roble, R.G. 1985. On solar–terrestrial relationships in atmospheric electricity. *J. Geophys. Res.* **90**: 6001–12.

Roble, R.G. 1991. On modeling component processes in the Earth's global electric circuit. *J. Atmos. Terr. Phys.* **53**: 831–47.

Roble, R.G., and Hays, P.B. 1979. A quasi-static model of global atmospheric electricity, 2, Electrical coupling between the upper and lower atmosphere. *J. Geophys. Res.* **84**: 7247–56.

Roble, R.G., and Hays, P.B. 1982. Solar–terrestrial effects on global electrical circuit. In *Solar Variability, Weather, and Climate*, pp. 92–106, NRC Geophysics Study Committee, Washington, DC: National Academy Press.

Roble, R.G., and Tzur, I. 1986. The global atmospheric-electrical circuit. In *The Earth's Electrical Environment*, eds. E.P.

Krider and R.G. Roble, pp. 206–31, Washington, DC: National Academy Press.

Rodger, C.J., Thomson, N.R., and Dowden, R.L. 1998. Are whistler ducts created by thunderstorm electrostatic fields? *J. Geophys. Res.* **103**: 2163–9.

Rosen, J.M., Hofmann, D.J., Gringel, W., Berlinski, J., Michnowski, S., Morita, Y., Ogawa, T., and Olson, D. 1982. Results of an international workshop on atmospheric electric measurements. *J. Geophys. Res.* **87**: 1219–24.

Ruhnke, L.H. 1969. Area averaging of atmospheric electric currents. *J. Geomagn. Geoelectr.* **21**: 453–62.

Ruhnke, L.H., Tammet, H.F., and Arold, M. 1983. Atmospheric electric currents at widely spaced stations. In *Proc. in Atmospheric Electricity*, eds. L. Ruhnke and J. Latham, pp. 76–8, Hampton, Virginia: A. Deepak.

Rust, W.D., Marshall, T.C. 1996. On abandoning the thunderstorm tripole-charge paradigm. *J. Geophys. Res.* **101**: 23 499–504.

Ruttenberg, S., and Holzer, R.E. 1955. Atmospheric electric measurements in Pacific Ocean. *Geophys. Res. Pap.* **42**: 101–8.

Rycroft, M.J. 1994. Some effects in the middle atmosphere due to lightning. *J. Atmos. Terr. Phys.* **56**: 343–8.

Rycroft, M.J., Israelsson, S., and Price, C. 2000. The global atmospheric electric circuit, solar activity and climate change. *J. Atmos. Solar-Terr. Phys.* **62**: 1563–76.

Saba, M.M.F., Pinto, O., and Pinto, I.R.C.A. 1999. Stratospheric conductivity measurements in Brazil. *J. Geophys. Res.* **104**: 27 203–8.

Saba, M.M.F., Pinto, O., Pinto, I.R.C.A., and Mendes, O. 2000. Stratospheric balloon measurements of electric fields associated with thunderstorms and lightning in Brazil. *J. Geophys. Res.* **105**: 18 091–7.

Sadiku, M.N.O. 1994. *Elements of Electromagnetics*, 821 pp., Orlando, Florida: Sounders College.

Salanave, L.E. 1961. The optical spectrum of lightning. *Science* **134**: 1395–9.

Salanave, L.E. 1980. *Lightning and Its Spectrum*, 136 pp., Tucson: University of Arizona Press.

Sapsford, H.B. 1937. Influence of pollution on potential gradient at Apia. *Terr. Magn. Atmos. Electr.* **42**: 153–8.

Schlegel, K., Diendorfer, G., Thern, S., and Schmidt, M. 2001. Thunderstorms, lightning and solar activity – Middle Europe. *J. Atmos. Terr. Phys.* **63**: 1705–13.

Schonland, B.F.J. 1953. *Atmospheric Electricity*, 2nd edition, 95 pp., London: Methuen and Co.

Schonland, B.F.J. 1956. The lightning discharge. In *Handbuch der Physik* **22**: 576–628, Berlin: Springer-Verlag.

Schonland, B.F.J. 1964. *The Flight of Thunderbolts*, 2nd edition., 182 pp., Oxford: Clarendon Press.

Schuster, A. 1880. On spectra of lightning. *Proc. Phys. Soc. (London)* **3**: 46–52.

Scrase, F.J. 1933. The air–earth current at Kew Observatory. *Geophys. Mem.* **58**: Meteorological Office, London.

Shchukin, G.G., Shvarts, Ya. M., and Oguryayeva, L.V. 1992. Recordings of atmospheric electricity in different places. In *Proc. 9th Int. Conf. on Atmospheric Electricity*, St. Petersburg, Russia, pp. 40–5.

Sheftel, V.M., Bandilet, O.I., Yaroshenko, A.N., and Chernyshev, A.K. 1994. Space time structure and reasons of global, regional, and local variations of atmospheric electricity. *J. Geophys. Res.* **99**: 10 797–806.

Sil, J.M. 1938. Some atmospheric-electric observations at Poona. *Terr. Magn.* **43**: 139–42.

Simpson, G.C., and Robinson, G.D. 1941. The distribution of electricity in the thunderclouds. Pt. II. *Proc. Roy. Soc. A* **177**: 281–329.

Simpson, G., and Scrase, F.J. 1937. The distribution of electricity in thunderclouds. *Proc. Roy. Soc. A* **161**: 309–52.

Slipher, V.M. 1917. The spectrum of lightning. *Lowell Obs. Bull. (Flagstaff, Arizona)* **79**: 55–8.

Smith, L.G. (ed.) 1959. *Recent Advances in Atmospheric Electricity*, 631 pp. New York: Pergamon Press.

Sorokin, V.M., Chmyrev, V.M., and Yaschenko, A.K. 2001. Electrodynamic model of the lower atmosphere and the ionosphere coupling. *J. Atmos. Terr. Phys.* **63**: 1681–91.

Stansbery, E.K., Few, A.A., and Geis, P.B. 1993. A global model of thunderstorm electricity. *J. Geophys. Res.* **98**: 16 591–603.

Stekolnikov, I., and Valeev, C. 1937. L'etude de la foudre dans un laboratoire de campagne, CIGRE Report no. 30.

Stergis, C.G., Rein, G.C., and Kangas, T. 1957a. Electric field measurements above thunderstorms. *J. Atmos. Terr. Phys.* **11**: 83–90.

Stergis, C.G., Rein, G.C., and Kangas, T. 1957b. Electric field measurements in the atmosphere. *J. Atmos. Terr. Phys.* **11**: 77–82.

Suszcynsky, D.M., Roussel-Dupré, R., and Shaw, G. 1996. Ground-based search for X rays generated by thunderstorms and lightning. *J. Geophys. Res.* **101**: 23 505–16.

Sverdrup, H.U. 1927. Magnetic, atmospheric-electric, and auroral results, Maud expedition, 1918–1925. In *Researches of the Department of Terrestrial Magnetism, vol. VI*, Carnegie Institution, Washington, DC (see part 4: Observations of the atmospheric electric potential gradient, 1922–5).

Szczerbinski, M. 2000. A discussion of "Faraday cage" lightning protection and application to real building structures. *J. Electrostatics* **48**: 145–54.

Takagi, M. 1977. On the regional effect in the global atmospheric electric field. In *Electrical Processes in Atmosphere*, eds. H. Dolezalek and R. Reiter, pp. 477–81, Darmstadt: Steinkopff.

Thomson, A. 1924. Preliminary report on the atmospheric potential-gradient recorded at the Apia Observatory, Western Samoa, May 1922 to April 1924. *Terr. Magn.* **29**: 97–100.

Thottappillil, R., and Uman, M.A. 1993. Advances in lightning research. *Trends in Geophys. Res.* **2**: 9–26.

Tinsley, B.A. 1996. Correlations of atmospheric dynamics with solar wind induced changes of air–Earth current density into cloud tops. *J. Geophys. Res.* **101**: 29 701–14.

Tomilin, A. 1986. *The Magic of Faunus*, 256 pp., Leningrad: Lenizdat.

Torreson, O.W., Gish, O.H., Parkinson, W.C., and Wait, G.R. 1946. Scientific results of Cruise VII of the Carnegie during 1928–1929 under command of Captain J.P. Ault, Oceanography-III, Ocean atmospheric-electric results. Carnegie Institute of Washington Publication 568, Washington, DC.

Tyutin, A.A. 1976. Mesospheric maximum of the electric field strength. *Cosmic Res. (English transl.)* **14**: 132–3.

Tzur, I., and Roble, R.G. 1985a. Atmospheric electric field and current configurations in the vicinity of mountains. *J. Geophys. Res.* **90**: 5979–88.

Tzur, I., and Roble, R.G. 1985b. The interaction of a dipolar thunderstorm with its global electrical environment. *J. Geophys. Res.* **90**: 5989–99.

Tzur, I., Roble, R.G., Zhuang, H.C., and Reid, R.C. 1983. The response of the Earth's global electrical circuit to a solar proton event. In *Solar–Terrestrial Influences on Weather and Climate*, ed. B. McCormac, pp. 427–35, Boulder: Colorado Associated University Press.

Uchikawa, K. 1977. Annual variations of the ionospheric potential, the air–earth current density and the columnar resistance measured by radiosondes. In *Electrical Processes in Atmospheres*, eds. H. Dolezalek and R. Reiter, pp. 460–3, Darmstadt: Steinkopff.

Uman, M.A. 1969. *Lightning*, 264 pp., New York: McGraw-Hill.

Uman, M.A. 1971. *Understanding Lightning*, 166 pp., Pittsburgh: Bek Technical Publications.

Uman, M.A. 1974. The earth and its atmosphere as a leaky spherical capacitor. *Am. J. Phys.* **42**: 1033–5.

Uman, M.A. 1983. Lightning. US National Report to International Union of Geodesy and Geophysics, 1979–1982. *Rev. Geophys. Space Phys.* **21**: 992–7.

Uman, M.A. 1984. *Lightning*, 298 pp., New York: Dover.

Uman, M.A. 1986. *All About Lightning*, 167 pp., New York: Dover.

Uman, M.A. 1987. *The Lightning Discharge*, 377 pp., San Diego: Academic Press.

Uman, M.A. 1988. Natural and artificially-initiated lightning and lightning test standards. *Proc. IEEE* **76**: 1548–65.

Uman, M.A. 1994. Natural lightning. *IEEE Trans. Ind. Appl.* **30**: 785–90.

Uman, M.A. 2001. *The Lightning Discharge*, 377 pp., Mineola, New York: Dover.

Uman, M.A., Dawson, G.A., and Hoppel, W.A., 1975. Progress in atmospheric electricity. *Rev. Geophys. Space Sci.* **13**: 760–5, 849–53.

Uman, M.A., and Krider, E.P. 1982. A review of natural lightning: experimental data and modeling. *IEEE Trans. Electromagn. Compat.* **24**: 79–112.

Uman, M.A., and Krider, E.P. 1989. Natural and artificially initiated lightning. *Science* **246**: 457–64.

Velinov, P., and Tonev, P. 1993. Penetration of horizontal and vertical components of thundercloud electric fields into the ionosphere – modeling and analysis. *Bulgarian Geophys. J.* **19**(3): 64–72.

Velinov, P.L., and Tonev, P.T. 1995a. Modelling the penetration of thundercloud electric fields into the ionosphere. *J. Atmos. Terr. Phys.* **57**: 687–94.

Velinov, P.L., and Tonev, P.T. 1995b. Thundercloud electric field modelling for the ionosphere–Earth region, 1. Dependence on cloud charge distribution. *J. Geophys. Res.* **100**: 1447–85.

Viemeister, P.E. 1961. *The Lightning Book*, 316 pp., New York: Doubleday and Co.

Volland, H. 1975. Models of global electric fields within the magnetosphere. *Ann. Geophys.* **31**: 154–73.

Volland, H. 1977. Global quasi-static electric fields in the Earth's environment. In *Electrical Processes in Atmospheres*, eds. H. Dolezalek and R. Reiter, pp. 509–27, Darmstadt: Steinkopff.

Volland, H. 1982. Quasi-electrostatic fields within the atmosphere. In *Handbook on Atmospherics*, vol. 1, ed. H. Volland, pp. 65–109, Boca Raton, Florida: CRC Press.

Volland, H. 1984. Atmospheric electrodynamics. In *Physics and Chemistry in Space*, vol. II, eds. L.J. Lanzerotti and J.T. Wasson, 205 pp., New York: Springer-Verlag.

Vonnegut, B. 1953. Possible mechanism for the formation of thunderstorms electricity. *Bull. Am. Meteor. Soc.* **34**: 378–81.

Vonnegut, B. 1965. Thunderstorm theory. In *Problems of Atmospheric and Space Electricity*, ed. S.C. Coroniti, pp. 285–92, New York: Elesevier.

Vonnegut, B., and Moore, C.B. 1988. Comments on "Stratospheric conductivity variations over thunderstorms" by R.H. Holzworth, K.W. Norville, P.M. Kintner, and S.P. Powell. *J. Geophys. Res.* **93**: 3913–14.

Vonnegut, B., Moore, C.B., Espinola, R.P., and Blau, H.H. 1966. Electrical potential gradients above thunderstorms. *J. Atmos. Sci.* **23**: 764–70.

Vonnegut, B., Markson, R., and Moore, C.B. 1973. Direct measurement of vertical potential differences in the lower atmosphere. *J. Geophys. Res.* **78**: 4526–8.

Vonnegut, B., Markson, R., and Moore, C.B. 1987. Comment on "Factor of 2 error in balloon-borne, atmospheric conduction current measurements" by A.A. Few and A.J. Weinheimer. *J. Geophys. Res.* **92**: 4337.

Wahlin, L. 1986. *Atmospheric Electrostatics*. 120 pp. New York: John Wiley.

Wait, G.R. 1927. Preliminary note of the effect of dust, smoke, and relative humidity upon the potential gradient and the positive and negative conductivities of the atmosphere. *Terr. Magn.* **32**: 31–5.

Wait, J.R. 1960. Terrestrial propagation of very-low-frequency radio waves. *J. Res. Nat. Bur. Stand. D* **64**: 152–63.

Walter, B. 1902. Ein photographischer Apparat zur genaueren Analyse des Blitzes. *Physik. Z.* **3**: 168–72.

Walter, B. 1903. Uber die Entstehungsweise des Blitzes. *Ann. Phys.* **10**: 393–407.

Walter, B., 1910. Uber Doppelaufnahmen von Blitzen... *Jahrbuch Hamb. Wiss. Anst.* **27**(5): 81–118.

Walter, B. 1912. Stereoskopische Blitzaufnahmen. *Physik. Z.* **13**: 1082–4.

Walter, B. 1918. Uber die Ermittelung der zeitlichen Aufeinanderfolge zusammengehöriger Blitze sowie über ein bemerkenswertes Beispiel dieser Art von Entladungen. *Physik. Z.* **19**: 273–9.

Watson-Watt, R.A., 1929. Weather and wireless. *Q.J.R. Meteor. Soc.* **55**: 273–301.
Weber, L. 1889. Uber Blitzphotographien. In *Ber. Königliche Akad. (Berlin)*, pp. 781–4.
Whipple, F.J.W. 1929a. On the association of the diurnal variation of electric potential gradient in fine weather with the distribution of thunderstorms over the globe. *Q.J.R. Meteor. Soc.* **55**: 1–17.
Whipple, F.J.W. 1929b. Potential gradient and atmospheric pollution – the influence of summer time. *Q.J.R. Meteor. Soc.* **55**: 351–60.
Whipple, F.J.W., and Scrase, F.J. 1936. Point discharge in the electric field of the Earth. Geophys. Memoir 7, *Meteor. Office, London* **68**: 1–20.
Wiesinger, J. 1972. *Blitzforschung und Blitzschutz*. Munchen: R. Oldenburg.
Willett. J.C. 1979a. Fair weather electric charge transfer by convection in an unstable planetary boundary layer. *J. Geophys. Res.* **84**: 703–18.
Willett, J.C. 1979b. Solar modulation of the supply current for atmospheric electricity? *J. Geophys. Res.* **84**: 4999–5002.
Willett, J.C. 1981a. The influence of corona space charge on a direct measurement of atmospheric potential. *J. Geophys. Res.* **86**: 12 133–8.
Willett, J.C. 1981b. Toward an understanding of the turbulent electrode effect over land. NRL Report 8519, Naval Research Laboratories, Washington, DC.
Willett, J. 1985. Atmospheric-electrical implication of ^{222}Rn daughter deposition on vegetated ground. *J. Geophys. Res.* **90**: 5901–08.
Willett, J.C., and Rust, W.D. 1981. Direct measurements of the atmospheric electrical potential using tethered balloons. *J. Geophys. Res.* **86**: 12 139–42.
Williams, E.R. 1989. The tripole structure of thunderstorms. *J. Geophys. Res.* **94**: 13 151–67.
Williams, E.R. 1992. The Schumann resonance: a global tropical thermometer. *Science* **256**: 1184–7.
Williams, E.R. 1994. Global circuit response to seasonal variations in global surface air temperature. *Mon. Wea. Rev.* **122**: 1917–29.
Williams, E.R. 1995. Schumann resonance measurements as a sensitive diagnostic for global change. In Annual Progress Report, National Institute for Global Environmental Change.
Williams, E.R. 1996. Comment on "Current budget of the atmospheric electric global circuit" by H.W. Kasemir. *J. Geophys. Res.* **101**: 17 029–31.
Williams, E.R. 1999. Global circuit response to temperature on distinct time scales: a status report. In *Atmospheric and Ionospheric Electromagnetic Phenomena Associated with Earthquakes*, ed. M. Hayakawa, pp. 939–49, Tokyo: Terra Scientific (TERRAPUB).
Williams, E.R., and Heckman, S.J. 1993. The local diurnal variation of cloud electrification and the global diurnal variation of negative charge on the Earth. *J. Geophys. Res.* **98**: 5221–34.
Wilson, C.T.R. 1920. Investigations on lightning discharges and on the electric field of thunderstorms. *Phil. Trans. Roy. Soc. A* **221**: 73–115.
Wilson, C.T.R. 1929. Some thundercloud problems, *J. Franklin Inst.* **208**: 1–12.
Woessner, R.H., Cobb, W.E., and Gunn, R. 1958. Simultaneous measurements of the positive and negative light-ion conductivities to 26 km. *J. Geophys. Res.* **63**: 171–80.
Woosley, J.D., and Holzworth, R.H. 1987. Electrical potential measurements in the lower atmosphere. *J. Geophys. Res.* **92**: 3127–34.
Workman, E.J., Holzer, R.E., and Pelsor, G.T. 1942. The electrical structure of thunderstorms. Aero. Technical Note 864, pp. 1–47, Natl. Advan. Comm., Washington, DC.
Wormell, T.W. 1930. Vertical electric currents below thunderstorms and showers. *Proc. Roy Soc. A* **127**: 567–90.
Wormell, T.W. 1953. Atmospheric electricity: some recent trends and problems. *Q.J.R. Meteor. Soc.* **79**: 3–50.
Zadorozhny, A.M., and Tyutin, A.A. 1997. Universal diurnal variation of mesospheric electric fields. *Adv. Space Res.* **20**: 2177–80.
Zadorozhny, A.M., and Tyutin, A.A. 1998. Effects of geomagnetic activity on the mesospheric electric fields. *Ann. Geophys.* **16**: 1544–51.
Zadorozhny, A.M., Tyutin, A.A., Witt, G., Wilhelm, N., Walchli, U., Cho, J.Y., and Swartz, W.E. 1993. Electric field measurements in the vicinity of noctilucent clouds and PMSE. *Geophys. Res. Lett.* **20**: 2299–302.
Zadorozhny, A.M., Tyutin, A.A., Bragin, O.A., and Kikhtenko, V.N. 1994. Recent measurements of middle atmospheric electric fields and related parameters. *J. Atmos. Terr. Phys.* **56**: 321–35.
Zadorozhny, A.M., Vostrikov, A.A., Witt, G., Bragin, O.A., Dubov, D.Yu., Kazakov, V.G., Kikhtenko, V.N., and Tyutin, A.A. 1997. Laboratory and *in situ* evidence for the presence of ice particles in a PMSE region. *Geophys. Res. Lett.* **24**: 841–4.
Zipse, D.W. 1994. Lightning protection systems: advantages and disadvantages. *IEEE Trans. Ind. Appl.* **30**(5): 1351–61.

2　Incidence of lightning

> The worldwide distribution of lightning remains an unknown despite the advancement of ground-based and space-based lightning detection systems... Only part of the land is now covered by lightning detection networks; and the first satellites to measure lightning will be in low orbit, thus measuring only part of the global lightning.
>
> R.E. Orville (1995)

2.1. Introduction

More than 75 years ago Brooks (1925) estimated that the global lightning flash rate, the total of both cloud and ground flashes, was about 100 s^{-1}. He did so by combining the following information: (i) an estimated average of 16 storms per year over the assumed observation area, 512 km^2, of an individual weather station, which was extrapolated to the entire Earth's surface, about 5.1×10^8 km^2 (both land and sea included); (ii) an assumed average storm duration of 1 hr; and (iii) the flash rate of 3.5 min^{-1} observed by Marriott (1908) during a 28 min thunderstorm period in England. More recent estimates of the global flash rate, based on satellite measurements and many assumptions, are within a factor 2 or 3 of this value but are not necessarily more accurate (subsection 2.5.3). The number of flashes per unit area per unit time is called the flash density, and the number of ground flashes per unit area per unit time (usually per square kilometer per year: Section 2.5) is called the ground flash density. A global flash rate of 100 s^{-1} corresponds to a total flash density over the entire Earth's surface of about 6 km^{-2} yr^{-1} and a ground (land or sea) flash density of about 1.5 km^{-2} yr^{-1} if we assume a 3 : 1 global ratio of cloud to cloud-to-ground flashes (Section 2.7). Lightning activity is highly variable on both (i) a temporal scale at a given location, ranging from some tens of years (Section 2.6) to the lifetime of individual thunderstorm cells, which is of the order of one hour (Chapter 3) or shorter, and (ii) a spatial scale ranging from continents and oceans to a few kilometers or less. The ground flash density N_g is the basic information needed for estimation of the lightning strike incidence to a ground-based object or system. Other measures of the lightning incidence to an area include the annual number of thunderstorm days T_D (Section 2.3) and the annual number of thunderstorm hours T_H (Section 2.4). Various characteristics of cloud-to-ground lightning such as the number of strokes per flash and the polarity of the charge lowered to ground apparently depend on season, location, and storm type (Section 2.8). Occurrence statistics and various characteristics of positive lightning are considered further in Chapters 5 and 8.

Clearly, T_D, T_H, and N_g each vary both in space (from one location to another) and in time (e.g., from year to year). Accordingly, there are two different forms in which these quantities are presented, which emphasize either the spatial or the temporal component of their variability. The spatial variability is usually presented in the form of a map for many locations, the temporal variability being suppressed typically by the use of long-term average values (see, for example, Figs. 2.8, 2.9, and 2.11). The temporal variability is usually presented in the form of a time series for a single location (see, for example, Figs. 2.13 and 2.14). When considering correlations among T_D, T_H, and N_g, one should distinguish between these two components of variability.

2.2. Characterization of individual storms and storm systems

2.2.1. Lightning flash rate

Cianos and Pierce (1972) stated that the average flash rate (also called the flashing rate) during the lifetime of any thunderstorm cell (Section 3.1) is about 3 min^{-1}, irrespective of locality. This statement is apparently supported by the finding of Boccippio et al. (2000a) that ocean storms produce lightning at rates that are not dissimilar from those of continental storms. Livingston and Krider (1978) gave statistical data on lightning at the Kennedy Space Center (KSC), Florida for 22 storm days during 1974 and 1975. The data were derived from the electric fields recorded by the KSC field mill network (subsection 17.2.1). The number of cloud and ground discharges per overall storm (individual storm cells could not, in general, be identified) ranged from eight to 1987. About 70 percent of all lightning occurred during what was termed the "active storm period", which represented only about 30 percent of the total storm duration. A strong correlation was found between the logarithm of the total number n of discharges during the active period and the duration in minutes D_A, of that period. The regression equation fitting the 23 data points is

$$\log_{10} n = 0.01 D_A + 1.90 \tag{2.1}$$

The average flash rate per overall storm varied between 0.3 and 9.3 min^{-1}. On five of the storm days, the flash rate averaged over 5 min intervals exceeded 20 min^{-1}, the maximum rate being 26 min^{-1}. Small storms, which perhaps can be regarded as single-cell storms, produced a maximum flash rate of typically 1 to 4 min^{-1}. Piepgrass et al. (1982), also using the KSC field-mill network, presented data on the frequency of lightning occurrence for 79 additional KSC storms during the summers of 1976 through 1980. The number of discharges per storm varied from 1 to 3687. Average flash rates for the storms were between 0.1 and 12.2 min^{-1}. Maximum flash rates, averaged over 5 min intervals, were between 0.2 and 30.6 min^{-1}.

Lhermitte and Krehbiel (1979) reported an abrupt increase in the flash rate from about 10 to 60 min^{-1} in a "normal-size" KSC storm. The discharges were identified using the VHF time-of-arrival source location system called LDAR (subsection 17.4.2). The primary contributors to the increase in flash rate were intracloud discharges that originated at heights ranging from 8 to 10 km (-20 to -35 °C) above the area of maximum radar reflectivity (> 50 dBZ) and were apparently due to the development of a strong updraft and the associated precipitation growth.

Maier and Krider (1982) used a two-station lightning locating system based on the wideband magnetic field direction finding (DF) technique (Section 17.3) to study the occurrence of cloud-to-ground lightning in three severe storms in northern Texas and Oklahoma and 268 air-mass storms (Section 3.1) in south Florida. Severe storms are defined by the US Weather Service as storms that produce at least one of the following three conditions: (i) large hail (diameter ≥ 1.9 cm), (ii) damaging winds (speed ≥ 26 ms^{-1}), or (iii) tornadoes (e.g., Johns and Doswell 1992). Note that lightning is not part of this definition. Sometimes storms are subjectively labeled as severe without reference to the formal definition given above. The three severe storms analyzed by Maier and Krider (1982) produced tornadoes, and therefore they satisfied the U.S. Weather Service criteria. The severe thunderstorms in Maier and Krider's analysis lasted from 2 to 8 hr and produced 200 to 2800 cloud-to-ground flashes over areas of 2000 to 5000 km^2. The Florida air-mass storms rarely lasted longer than 1 to 2 hr, producing about 50 discharges over an area of about 450 km^2. However, there were usually only one or two severe storms per storm day in the Texas–Oklahoma region while in Florida there were 10 to 50 storms per storm day. The severe storms produced cloud-to-ground lightning at a mean rate of about 2 to 6 min^{-1}, while the air-mass storm rate was about 1 min^{-1}. The peak ground flash rate in severe storms approached 20 min^{-1}, averaged over 5 min intervals, while in the 268 air-mass storms it was about 12 min^{-1}. Apparently no correction was applied to the data to account for the lightning locating system's estimated detection efficiency, about 80 percent (Section 17.3).

Peckham et al. (1984) used the same type of ground flash locating system as Maier and Krider (1982) to study cloud-to-ground lightning in the Tampa Bay area of Florida. For 111 storms on eight days, the following parameters were determined: duration, area, number of ground flashes, average ground flash density, average ground flash rate, and maximum ground flash rate averaged over a 5 min period. Note that the flash density definition used in this section to characterize an individual storm or parts of a storm system is different from that used below in Section 2.5 to characterize lightning incidence to an area. The means and standard deviations of the six quantities analyzed by Peckham et al. (1984) are presented in Table 2.1 where the storms are grouped into three categories: (i) single-peak storms, i.e., those with a single peak in their flash rate versus time curves, (ii) multiple-peak storms, i.e., those with multiple-peak flash rates, and (iii) storm systems composed of groups of the first two types of storm, which would be detected as one overall storm by a nondirectional lightning detector such as a lightning flash counter (subsection 2.5.1) or perhaps a network of electric field mills (subsection 17.2.1). Peckham et al. (1984) defined the storm duration as the period of the ground flash rate curve during which there is no more than one 5 min interval without lightning. A map of one day's ground flash activity is shown in Fig. 2.1. The

Table 2.1. *Means and standard deviations (the latter in parentheses) of various storm parameters studied by Peckham et al. (1984)*

Parameter	Single-peak storms	Multiple-peak storms	Storm systems
Duration (min)	41	77	130
	(16)	(26)	(51)
Area (km^2)	103	256	900
	(63)	(154)	(841)
Number of ground flashes	73	270	887
	(69)	(216)	(720)
Ground flash density (km^{-2} min^{-1})	0.018	0.015	0.010
	(0.011)	(0.008)	(0.006)
Average ground flash rate (min^{-1})	1.7	3.4	6.8
	(1.2)	(2.3)	(4.7)
Maximum ground flash rate (min^{-1})	3.7	7.3	14
	(2.6)	(4.4)	(9)

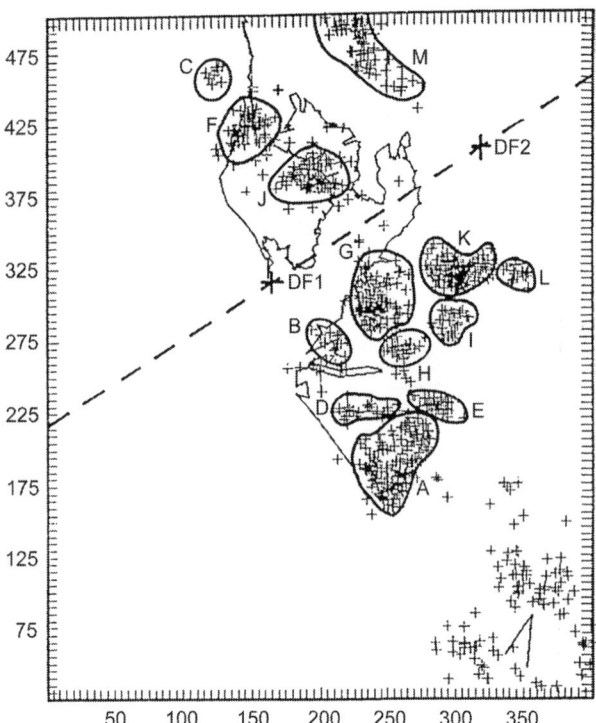

Fig. 2.1. Map of ground flash locations in the Tampa Bay area (Florida) on 8 August 1979 between 13:00 and 16:00 EDT. The two direction finders comprising the lightning locating system are labeled DF1 and DF2. The scale units are thousands of feet (1000 ft = 305 m). Adapted from Peckham et al. (1984).

letters indicate the chronology of the storm starting time, including the single- and the multiple-peak storms. Two storm systems are composed of storms ABDEGHIKL and storms CFJ. As can be seen in Table 2.1, Peckham et al. (1984) found an average ground flash density for storm areas containing lightning of about 0.01 km^{-2} min^{-1} (an underestimate due to the less than 100 percent detection efficiency). They found that the number of ground flashes, n_g, in a storm and the storm duration, D_S, in minutes were related by

$$\log_{10} n_g = 0.014 D_S + 1.2 \qquad (2.2)$$

The corresponding scatter plot is shown in Fig. 2.2.

The maximum flash rates in different storms and storm systems, as reported by various investigators, were summarized by Goodman and MacGorman (1986) and are given in Table 2.2. Most of the maximum flash rate values, some of which are for both cloud and ground lightning discharges while others are for ground discharges only, are in the range from a few to some tens per minute. Recent VHF observations indicate the existence of previously unknown forms of cloud discharge activity in storms in the US Great Plains. Specifically, Krehbiel et al. (2000) observed, in a large supercell storm, what they describe as essentially continuous discharge activity during the 1 min time interval presented. The activity filled a saucer-shaped volume with a diameter over 60 km and with a vertical extent from about 4 to 12 km.

2.2.2. Lightning in the Blizzard of '93

Orville (1993), using data from the US National Lightning Detection Network (NLDN) (Section 17.5), published the characteristics of cloud-to-ground lightning activity in a storm that was termed the Blizzard of '93. The storm occurred on 12–14 March 1993, developed over Texas and the Gulf of Mexico and then moved over Florida and up the US East Coast, breaking records for snowfall and low pressure as it moved toward the Canadian Maritime region. In the 48 hr period from 06 UT 12 March to 06 UT 14 March 1993, this cold-season storm produced over 59 000 cloud-to-ground flashes with a peak lightning rate exceeding 5100 hr^{-1}. For this storm, ground flash densities were in excess of 0.16 km^{-2} in an area just south of Tampa, Florida, while few flashes to ground were recorded north of the Carolinas. The overall percentage of positive flashes averaged about 13 percent, but positive flash percentages exceeded 60 percent in the northern portion of the storm. Estimated median peak currents for negative flashes were 30 kA and for positive flashes 52 kA. A summary of the cloud-to-ground lightning activity for the Blizzard of '93 is plotted in Fig. 2.3 for the period from 16:02 UT 12 March to 10:02 UT 14 March 1993 (52 hours).

2.2.3. Lightning in hurricanes

Molinari et al. (1994), using NLDN data and infrared satellite images, studied the spatial and temporal distribution of cloud-to-ground lightning in the 1992 Hurricane Andrew, one of the most intense hurricanes in United States history. A distinct radial variation was observed in the time-averaged flash density, with a weak maximum in the eye wall, a region of near zero flash density 40 to 100 km from the center, and a steady increase to a large maximum in the outer rainbands (regions of heavier precipitation) 190 km from the center. Similar findings were reported by Lyons and Keen (1994), who additionally noted near-eyewall bursts of cloud-to-ground flashes (detected with a TOA lightning locating system: subsection 17.4.3) preceding periods of intensification in Hurricanes Diana (1984) and Florence (1988). Samsury and Orville (1994), who studied the 1989 Hurricanes Hugo and Jerry, each over an 18 hr period, found that over half the 33 cloud-to-ground flashes in Hurricane Hugo occurred in or near the eyewall, while the outer convective rainbands were the preferred locations of the 691 flashes during Hurricane Jerry. Williams (1995) presented lightning locations for Hurricane Andrew that coincided with

Table 2.2. *Maximum flash rates in various types of storms. Adapted from Goodman and MacGorman (1986)*

Reference	Location	Storm type	Maximum rate, min^{-1}
Shackford (1960)	New England	hailstorms	30 (1800–60 min)
Mackerras (1963)	Subtropics	—	120
Blevins and Maurwitz (1968)	Great Plains	hailstorms	70
Fuquay and Baughman (1969)	Northwest United States	—	23
Israel (1973)	—	—	67 (1000/15 min)
Kinzer (1974)	Oklahoma	squall line	25 (5594/225 min)[a]
Livingston and Krider (1978)	Florida	tornado	26
Piepgrass et al. (1982)	Florida	isolated storms or clusters	7.6 (5 min average)[b]
Peckham et al. (1984)	Florida	isolated storms or clusters	3.7-14 (5 min average)[a]
Johnson and Goodman (1984)	Gulf of Mexico	Hurricane Alicia	53 (3150/60 min)
Goodman and MacGorman (1986)	Oklahoma	MCC	45 (1 h average)[a]
			54 (1 h maximum)[a]
			60 (5 min maximum)[a]

[a] Ground discharges only.
[b] Range from 0.6–30 min^{-1}.

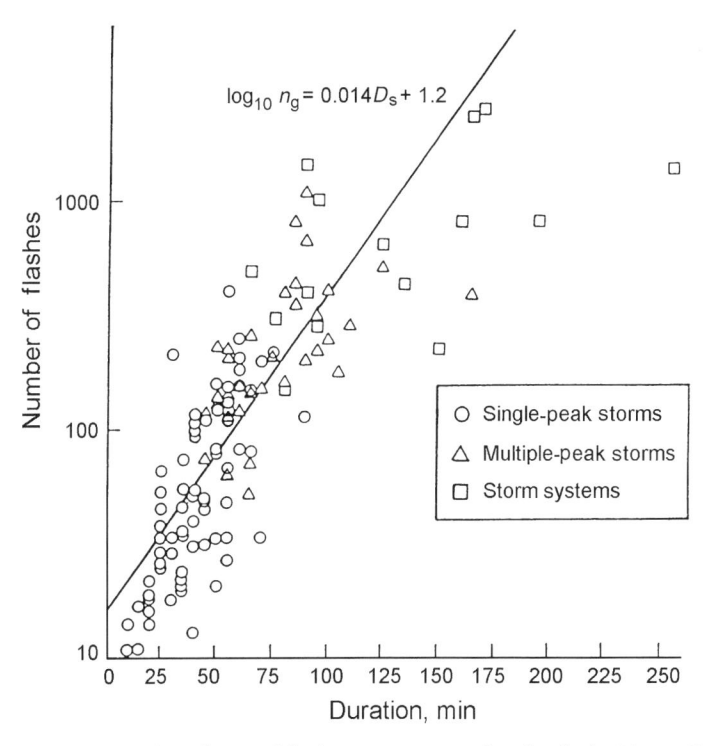

Fig. 2.2. Number of ground flashes versus storm duration in the Tampa Bay area, Florida. In determining the regression line (correlation coefficient $r = 0.78$), the storm system data were excluded; nevertheless these are plotted because the storm systems are a combination of single- and multiple-peak storms and because their definition is subjective. Interestingly, the storm system data lie near the regression line. Adapted from Peckham et al. (1984).

eyewall convection as this storm crossed the warm waters of the Gulf Stream toward the Florida coast. The most extensive study to date of cloud-to-ground lightning in hurricanes was conducted by Molinari et al. (1999). A total of nine Atlantic basin hurricanes were examined, four of which had been studied previously by Samsury and Orville (1994) or Lyons and Keen (1994). A common radial distribution of ground flash density, generally similar to that described earlier in this section, was found: a weak maximum in the eyewall region, a clear minimum 80–100 km

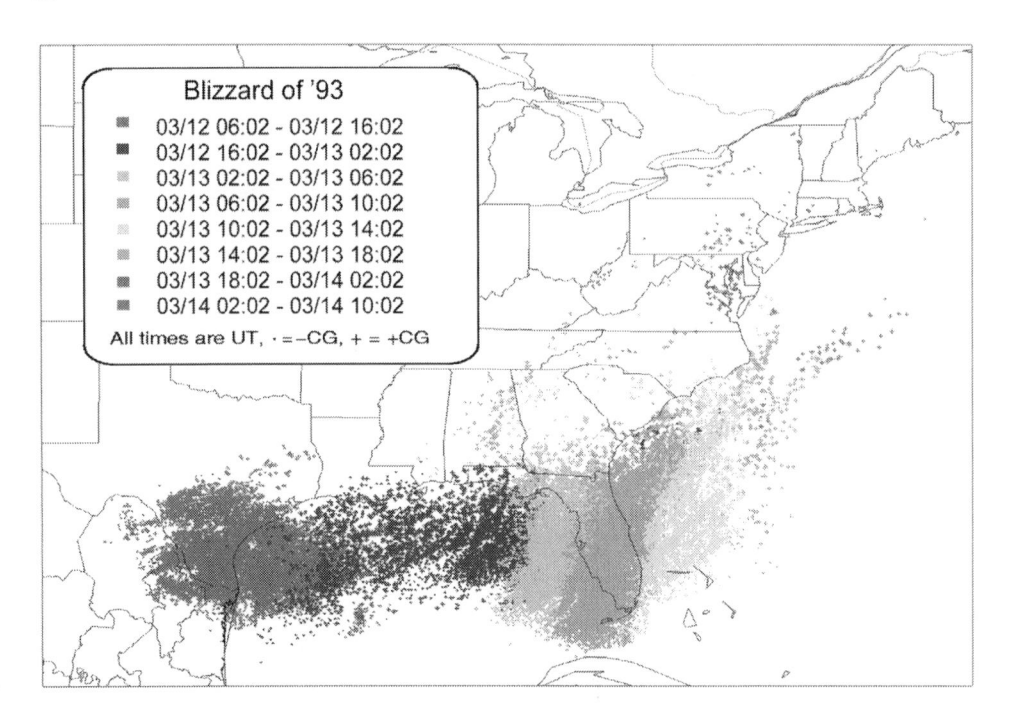

Fig. 2.3. At the time of going to press a colour version of this figure was available for download from http://www.cambridge.org/9780521035415. A summary of the cloud-to-ground lightning activity reported by the NLDN for the Blizzard of '93, from 06:02 UT 12 March to 10:02 UT 14 March 1993. The total number of recorded flashes for this time period was 62 887. The dots indicate negative flashes (55 206) and the plusses indicate positive flashes (7681). Different colors correspond to different times, as shown. Courtesy of Global Atmospherics, Inc.

outside the eyewall, and a strong maximum in the vicinity of the outer rainbands (at a distance of 210–90 km from the center). Molinari *et al.* (1999) considered this flash density distribution as supportive of the division of the convective structure of hurricanes into three distinct regions: the eyewall, the inner bands, and the outer bands. The eyewall is a unique phenomenon but shares some attributes with deep, weakly electrified oceanic monsoonal convection (see, for example, Williams 1995). The inner bands have characteristics similar to the trailing stratiform region of mesoscale convective systems (see subsection 2.2.4), including the production of a relatively high fraction of positive flashes. The outer bands contain the vast majority of the ground flashes in hurricanes.

2.2.4. Lightning in mesoscale convective complexes

Goodman and MacGorman (1986), using the four-station National Severe Storms Laboratory (NSSL) DF lightning locating network (Section 17.3), analyzed lightning discharges to ground in 10 mesoscale convective complexes (MCCs). The MCC is a type of mesoscale convective system (MCS), the latter being described by Goodman and MacGorman (1986) as a weather system having a horizontal extent of 250–2500 km, occurring on a time scale in excess of a few hours, and including significant convection during some part of its life cycle. Examples of MCSs are squall lines, tropical cyclones, and hurricanes. MCSs are discussed in detail in MacGorman and Rust (1998, Section 8.2) and MacGorman and Morgenstern (1998). They are also discussed in relation to the lightning "bipolar pattern" in subsection 2.2.7 below. During the most intense phase of its life cycle, a single MCC can produce one-fourth of the mean annual lightning strikes to ground at any site it passes over. For that reason MCCs, although a relatively infrequent meteorological situation, are considered here in a separate subsection in our characterization of various types of storms and storm systems.

MCCs were identified by Goodman and MacGorman (1986) on the basis of their circular appearance, long duration, and large size of the cloud shield in Geostationary Operational Environmental Satellite (GOES) infrared images. Goodman and MacGorman (1986) found that a typical MCC is characterized by a cloud-to-ground flash rate in excess of 1000 hr^{-1} for nine consecutive hours. Peak ground discharge rates of 60 min^{-1} were not uncommon, the average being 42 min^{-1}. They also compared their results with maximum flash rates published by others (Table 2.2).

2.2.5. Lightning and severe weather phenomena

As defined in subsection 2.2.1, severe weather phenomena include large hail (diameter $\geq$ 1.9 cm), damaging winds (speed $\geq$ 26 m s^{-1}), and tornadoes. Reap and MacGorman (1989), using 1985–86 data from the NSSL network (Section 17.3) for the Great Plains of the United States, found a significant correlation between the occurrence of severe weather and elevated rates of 30 or more positive ground flashes per hour within 48 km × 48 km grid blocks. Also, Rust et al. (1985a), Curran and Rust (1992), Branick and Doswell (1992), Seimon (1993), MacGorman and Burgess (1994), Stolzenburg (1994), and Carey and Rutledge (1998) noted that severe weather occurred in storms in which ground flashes had predominantly positive polarity during significant portions of the storm. Near the beginning of active cloud-to-ground lightning periods of hail-producing thunderstorms in the Great Plains of the United States, Stolzenburg (1994) reported positive lightning flash rates as high as 67 flashes in 5 min and flash densities up to 0.6 km^{-2} hr^{-1}. Nevertheless, many severe storms do not exhibit large numbers of positive ground flashes, and positive flashes are known to occur in non-severe storms (Chapters 5 and 8). Further, there is considerable variability in the evolution of the positive ground flash rate relative to storm evolution and severe weather occurrence (MacGorman and Burgess 1994; Stolzenburg 1994).

Williams et al. (1999) studied lightning activity in severe thunderstorms in Florida. Thunderstorms were identified as severe on the basis of reports by surface observers of dime-size or greater hail, strong winds (trees blown down), or the occurrence of tornadoes. The associated ground flash rates were obtained using NLDN data, and total flash rates were estimated from light detection and ranging (LDAR) records (subsection 17.4.2). In the latter case, any detected VHF source that occurred within 300 ms and 5 km of a previous source was assigned to the same flash as the previous source, the maximum allowed flash duration being 5 s. For more than 10 percent of so-defined flashes only one VHF source was detected. Clearly, in such a grouping of VHF sources into flashes, the decomposition of single large flashes into several smaller ones is possible, the result being an overestimation of the total flash rate. Williams et al. (1999) found that severe storms in Florida were characterized by relatively high maximum total flash rates ranging from about 60 to 500 min^{-1}. The most likely maximum total flash rates associated with small, non-severe thunderstorms in the same study were reported by Williams et al. (1999) to be in the range of 1 to 10 min^{-1}. The maximum ground flash rate was typically more than an order of magnitude lower than the maximum total flash rate. The polarity of ground flashes was not specified by Williams et al. (1999). Various aspects of severe storms, including the production of lightning, are reviewed in Williams (2001).

Changnon (1992) examined ground flash rates and strike locations relative to 48 streaks of crop damage produced by hail (regardless of its size; possibly smaller than required for a storm to be formally classified as severe) in Illinois during six days in June–July 1989. All ground flashes associated with hail streaks were of negative polarity. Ground flashes from hail-producing cells typically began 8–10 min before the first hail damage and exhibited a flash rate and area of coverage that grew rapidly until damage began. Then the ground flash rate decreased, but remained large during hail damage, and decreased more rapidly after hail damage ceased. Ground flash activity typically ended 8 min after the last hail damage. Shafer (1990) observed a storm system in Oklahoma that produced large hail before producing ground flashes.

MacGorman and Rust (1998) summarized the information on lightning activity in tornadic storms. Both total and cloud flash rates in tornadic supercell storms (a supercell can be defined as a long-lived cell with a rotating updraft) were observed to be large at the time of tornado occurrence (MacGorman et al. 1985, 1989; Richard 1992). However, the observed relationship between a tornado and the ground flash rate is rather variable. In three tornadic storms studied by Orville et al. (1982), MacGorman et al. (1989), and Richard (1992), ground flash rates were small before and during tornadoes. In four, studied by Kane (1991), MacGorman and Nielsen (1991), Keighton et al. (1991), and Seimon (1993), ground flash rates peaked during tornadoes, but were usually near a local minimum when the tornadoes began. In two storms studied by MacGorman et al. (1985) and Kane (1991), tornadoes occurred after ground flash rates had begun to decrease from their peaks but before they had decreased to a local minimum. Thirty-four of the 42 tornadoes produced on four days and studied by MacGorman and Burgess (1994) began when ground flash rates were either < 0.5 min^{-1} or near a local minimum value. During seven of the 10 tornadoes that lasted ≥ 20 min, ground flash rates increased while the tornado was occurring. Similar variability was observed by Knapp (1994), who analyzed ground flashes in 264 storms that produced tornadoes east of the Continental Divide of the United States on 35 days from 18 April through 1 July 1991. Perez et al. (1997), using NLDN data, studied cloud-to-ground lightning patterns in 42 tornado-producing supercells that occurred between January 1989 and November 1992. Thirty-one (74 percent) of the 42 storms were characterized by a peak in the ground flash rate preceding tornado formation. Twenty storms (48 percent) exhibited a local minimum ground flash rate coincident with tornado formation. Six of the 42 storms showed a reversal, from positive to negative, in the sign of the charge lowered to ground. Smith et al. (2000) inferred, from studying three tornadic storm events, that the lightning polarity reversals were influenced by the changes in intensity of the storms' updrafts brought about by changes in the buoyancy of ingested surface and/or boundary layer air.

Lightning activity in a tornadic storm in central Oklahoma, as revealed by a VHF TOA lightning imaging system (subsection 17.4.2), is presented by Krehbiel et al. (2000). The lightning activity exhibited a counterclockwise "hook" structure in the vicinity of the tornado, similar to the hook echoes observed by meteorological radars in tornadic storms. Lightning activity within about 3 min of a tornadic thunderstorm observed from space by the optical transient detector (OTD) (Section 17.8) was reported by Buechler et al. (2000).

Most modern studies of lightning in relation to damaging wind in isolated thunderstorms have been concerned with microbursts. These transient strong downdrafts (having a life cycle of 15 to 20 min) produce horizontal winds near the ground that are directed outward from the downdraft and may form one or more vortex rings around the downdraft. The horizontal outflow region is typically 2–4 km across, and the vortex rings may rise about 700 m above the ground. The microburst is particularly hazardous to aircraft during take off or landing. It is worth noting that in most cases the speeds of microburst winds are lower than required for a thunderstorm to be formally classified as severe (see above; Williams et al. 1999).

MacGorman and Rust (1998) summarized the available information on the lightning–microburst relation as follows. In the southeastern United States microbursts are often accompanied by moderate to heavy rain, while in the drier environment of the western United States sometimes they are accompanied only by light rainfall. Goodman et al. (1988a, b) and Williams et al. (1989c) found that microbursts in the southeastern United States tended to occur roughly 10 min after the peak in the total lightning flash rate and shortly after, or roughly coincident, with the peak in the ground flash rate. The storm studied by Goodman et al. (1988b) produced a total of 116 flashes, only six of which were discharges to ground. The peak flash rate, 23 min^{-1}, occurred 7 to 8 min after the initial discharge and 4 min prior to the microburst onset, simultaneously with the peaks in the storm mass, vertically integrated liquid water content, echo volume, and cloud height. The maximum rain flux and rain rate occurred 2 min after the peak flash rate. For the case of microbursts associated with moderate to heavy precipitation, similar lightning–microburst relationships were found in the western United States (Williams et al. 1989b) and in the northeastern United States (Kane 1991). However, Williams et al. (1989b) also observed that storms producing microbursts associated with light precipitation exhibited much smaller lightning flash rates than did the storms with heavier precipitation. While lightning preceded the light-precipitation microbursts, there was no pronounced peak in total flash rate prior to these "dry" microbursts, many of which occurred without any lightning in the parent storms.

2.2.6. Flash rate versus some non-electrical cloud properties

Many investigators have related flash rates to cloud properties determined from radar observations. We review the most important results of those studies; further information can be found in MacGorman and Rust (1998). Some of the results appear to contradict each other, but no attempt is made here to resolve these controversies.

Jacobson and Krider (1976) and Livingston and Krider (1978) showed that total flash rates greater than 10 min^{-1} in Florida are associated only with storms having maximum radar echo tops above 14 km. At the Kennedy Space Center, a total flash rate of 10 min^{-1} is probably indicative of a ground flash rate of roughly 4 min^{-1} (Livingston and Krider 1978). Holle and Maier (1982) reviewed the worldwide literature regarding the cloud top temperature at or after the time of the first lightning. They also reported that the minimum radar echo height (generally assumed to represent the thundercloud height) at which ground flashes were observed in Florida was 7.8 km. Larsen and Stansbury (1974), Marshall and Radhakant (1978), and Cherna and Stansbury (1986) showed that the total flash rate (represented by the occurrence rate of atmospherics at a single station) in Canada was correlated with the area of the storm as measured by radar. The dominant parameter determining the rate of electrical activity, according to Cherna and Stansbury (1986), is the thundercloud height. Goodman et al. (1988b) reported that the peak flash rate in a microburst-producing storm occurred simultaneously with the radar-determined peak in storm mass, vertically integrated liquid water content, echo volume, and cloud height. Solomon and Baker (1994) reported that lightning was correlated with the maximum radar-determined cloud height in New Mexico storms, no lightning being observed in clouds whose top heights did not exceed approximately 9.5 km. Lhermitte and Krehbiel (1979), using a network of three Doppler radars and the LDAR system (subsection 17.4.2), demonstrated that the total lightning flash rate was correlated with in-cloud updraft velocity. Williams (1985) gave a relationship between total flash rate F and cloud height H based on observations in Florida, New Mexico, and New England (see Fig. 2.4), $F = 3.44 \times 10^{-5} H^{4.9}$. Price and Rind (1992) used this relationship for continental thunderstorms and a different one for maritime storms to infer global lightning distributions from satellite-observed cloud heights. Molinié and Pontikis (1995), however, found no correlation between the total flash rate estimated using the SAFIR interferometer (Section 17.6) and the cloud top height inferred from daily meteorological sounding in the coastal region of French Guyana. Michalon et al. (1999) expressed the flash rate as a power function of both the cloud top height and the cloud droplet concentration.

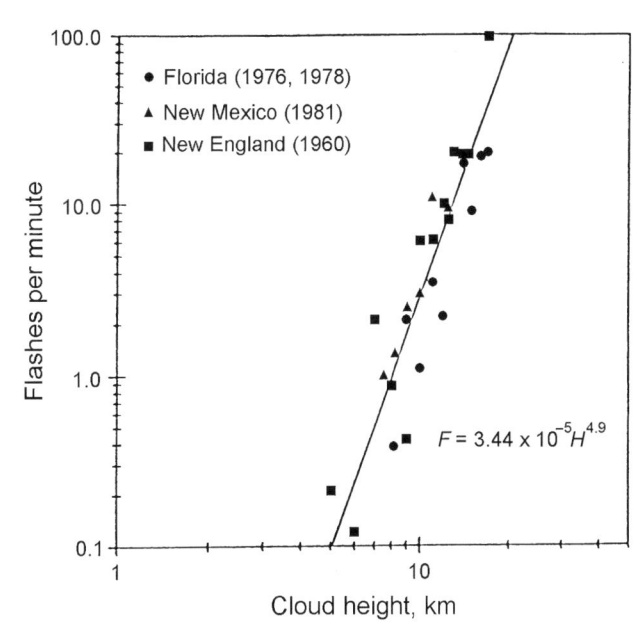

2.2. Characterization of individual storms and storm systems

Fig. 2.4. Relationship between lightning flash rate F and convective cloud top height H. Adapted from Williams (1985). Note that a fifth-power relation between F and H had been predicted previously by Vonnegut (1963).

Reap (1986) compared cloud-to-ground flash rates from the BLM network (Section 17.3) with GOES-determined cloud top temperatures and reported a tendency for cloud-to-ground flashes to be clustered under or near the coldest, that is, the highest, cloud tops. Watson et al. (1995), using data from both DF (Section 17.3) and TOA (subsection 17.4.3) lightning locating systems, studied the relationships between cloud-to-ground lightning and (i) the heights of radar echo tops, (ii) the radar-determined vertically integrated liquid water content. The greater percentage of echoes with lightning occurred when the echo-top heights exceeded about 15 km, and this percentage decreased with decreasing echo-top heights. Holle and Watson (1996), using the NLDN, studied winter storms over the central United States and found that often regions with cloud-to-ground lightning could be better identified by high echo tops than by reflectivity (the radar reflectivity of particles found in clouds is discussed, for example, in subsection 7.1.2 of MacGorman and Rust 1998). However, Petersen et al. (1996), who studied convection and associated cloud-to-ground lightning activity over the tropical western Pacific Ocean, reported that tropical oceanic clouds whose tops were often above 15 km produced little or no lightning (see also Williams et al. 1992; Rutledge et al. 1992; and subsection 2.2.8 below). They attributed this feature to the fact that in such convection the most significant hydrometeor mass, as evidenced by reflectivities greater than 30 dBZ (the reflectivity Z expressed in dBZ is $10 \log_{10} Z$), was often confined to elevations below the height of the freezing level, a condition not conducive to electrification processes (subsection 3.2.6). The oceanic convection that did produce significant lightning activity had reflectivities in excess of 30 dBZ in the mixed-phase region, and the cloud-to-ground flash rate was observed to increase simultaneously with the descent of the precipitation mass bounded by the 30 dBZ reflectivity contour.

Using a polarimetric radar (see, for example, subsection 7.1.2 of MacGorman and Rust 1998), Carey and Rutledge (1996) found that cloud flash rates were proportional to the graupel/hail volume (graupel is millimeter-size soft hail) at temperatures below 0 °C, while for one Colorado storm ground flash rates were proportional to the graupel/hail volume at temperatures higher than 0 °C. Lopez and Aubagnac (1997) found, for an Oklahoma hailstorm with well-developed supercell organization, that the development of graupel above the freezing level was related to the overall increase and decrease in the production of ground flashes, while the descent of small hail (less than 2 cm in diameter) below the height of the graupel region was responsible for the three or four brief increases in ground lightning activity. Carey and Rutledge (2000) reported that, in tropical island convection, the surface electric field, the total lightning flash rate, and the cloud-to-ground lightning flash rate were highly correlated with the radar-inferred ice mass in the mixed-phase region.

MacGorman and Rust (1998: p. 225) noted that a lightning-producing cloud is a very complex system whose electrical state is difficult to determine from measurements of its nonelectrical properties.

2.2.7. Lightning bipolar pattern

Orville et al. (1988) were the first to report on thunderstorm systems that produce positive and negative ground flashes whose ground strike locations tend to be separated in space, forming a so-called bipolar pattern. An example is given in Fig. 2.5. Orville et al. (1988) gave the following characteristics of the bipolar pattern, which occurs in many MCSs (defined in subsection 2.2.4). The pattern has a characteristic length of about 100 km. It is aligned with the upper-level winds and has been observed in all seasons but seems to occur more frequently in the fall and winter. The ratio of the positive flash density (the number of positive ground flashes per unit area accumulated over the life of the storm) to the negative flash density is approximately 0.1 with a range from 0.05 to 0.2. Orville et al. (1988) described the bipolar pattern as a persistent feature that, for the case characterized by them in detail, existed for 11 hours and was oriented from northeast (positive pole) to southwest (negative pole) as the storm moved to the east. The various studies reviewed below have revealed that the positive flashes in the bipole occur predominantly in the latter stage

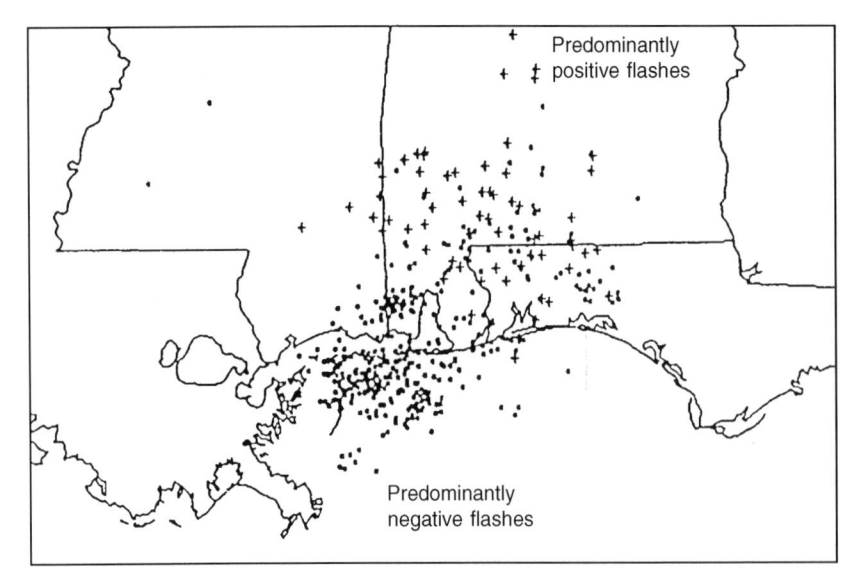

Fig. 2.5. An example of lightning bipolar pattern formed during one hour on 22 February 1987 along the Gulf of Mexico coast of the United States. Positive flashes are shown by plus signs and negative flashes by squares. Adapted from Orville et al. (1988).

of the lifetime of the storm system, so that the change in the predominant flash polarity combined with storm motion can produce the observed spatial bipole. Further, positive flashes are generally associated with regions of relatively shallow convection, such as the trailing stratiform region behind the convective line in summer MCSs and the shallower clouds downwind of the large-scale geostrophic flow (e.g., Houghton 1997) in both winter and summer MCSs. Apparently a bipolar pattern can be also formed by two (or more) isolated storms several tens of kilometers apart, each dominated by a different polarity of ground flashes (e.g., Engholm et al. 1990). Thus it appears that the lightning bipolar pattern is not specific to an MCS or any other type of storm but can occur in a variety of situations.

Examples of winter bipolar patterns over both land and ocean are given by Brook et al. (1989). The over-land storm moved northeast for a period of about 14 hours, the positive strokes occurring in the latter part of the storm life cycle. The resultant spatial bipole, produced via (i) the change of predominant flash polarity during the storm life cycle and (ii) the relatively fast motion of the storm, is oriented in such a way that the positive pole is northeast of the negative pole.

Rutledge and MacGorman (1988) used the NSSL DF lightning locating network (Section 17.3) in conjunction with both conventional and Doppler radar data over a 6 hr period to study the 10–11 June 1985 Kansas MCS. The storm consisted of a 20-km-wide line of convective cells (known as a squall line) trailed by an 80-km-wide stratiform precipitation region and exhibited the bipolar pattern described above, the negative lightning activity being associated mostly with the convective region. The negative

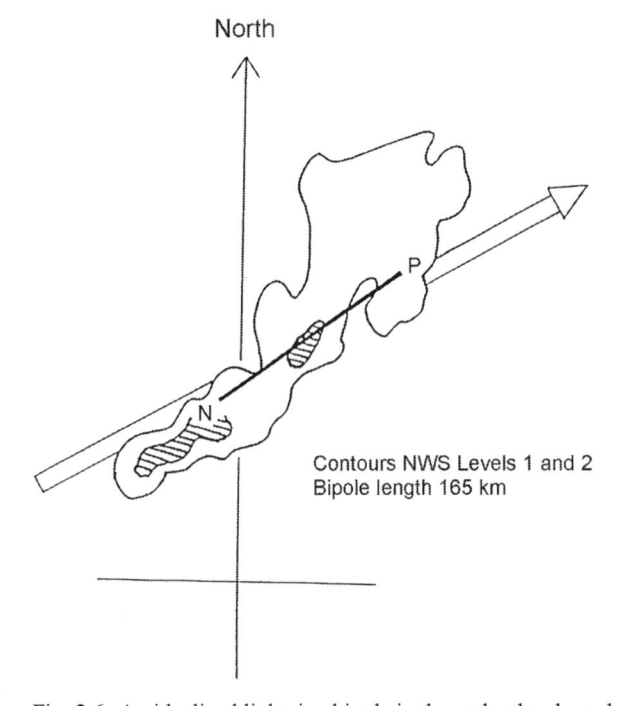

Fig. 2.6. An idealized lightning bipole is shown by the slanted solid line joining the geographic negative and positive centers N and P. The large outlined arrow indicates the geostrophic wind direction at all levels aloft. The contours represent two levels of radar reflectivity in imitation of the National Weather Service radar summary chart. Adapted from Engholm et al. (1990).

lightning flash rate was highest around the period of most intense convective rainfall. The positive lightning activity was confined mainly to the trailing stratiform region, and there was a correlation between the positive lightning

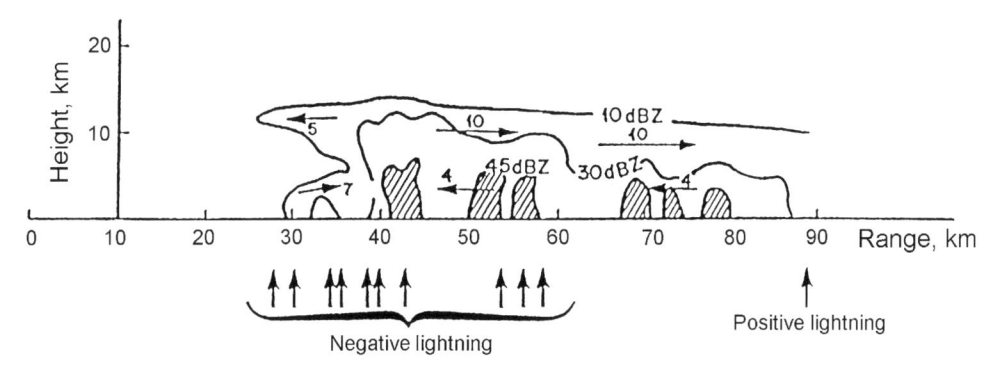

Fig. 2.7. Radar range height indicator (RHI) scan roughly aligned with the upper level geostrophic wind for the thunderstorm cluster on 28–29 June 1986 in Alabama. Doppler radial winds and cloud-to-ground lightning locations are also shown; the wind speeds in m s^{-1} are shown alongside the arrows. Adapted from Engholm et al. (1990).

flash rate and the area-integrated stratiform precipitation. Rutledge and MacGorman concluded from their analysis that the occurrence of positive lightning in the trailing stratiform region was a result of the rearward advection of positively charged ice particles from the upper levels of the convective cells. This charge advection mechanism may be viewed as an extension of the tilted dipole idea, proposed by Brook et al. (1982) as the mechanism for positive lightning production in winter storms in Japan (Chapter 8), from the convective scale (about 10 km) to the mesoscale (a few hundred kilometers). However, the charging of hydrometeors within the stratiform region was also viewed by Rutledge and MacGorman as a possibility. Similar bipolar patterns in both mid-latitude and tropical MCSs were found by Stolzenburg (1990), Engholm et al. (1990), Rutledge et al. (1990), Schuur et al. (1991), Hunter et al. (1992), Petersen and Rutledge (1992), and Rutledge and Petersen (1994). Figure 2.6, taken from Engholm et al. (1990), illustrates, for a generalized bipole in winter (January-March), the relative orientation of the bipole, the wind direction, and the precipitation structure. Figure 2.7, from the same study, presents (i) a vertical radar reflectivity profile, (ii) the Doppler radar winds, and (iii) the locations of positive and negative cloud-to-ground flashes for one thunderstorm cluster in summer. Engholm et al. (1990) concluded that lightning bipoles are aligned with the geostrophic wind (e.g., Houghton 1997) in winter storms and with the vertical wind shear (whose direction may be considerably different from that of the geostrophic wind) in summer storms, with the positive pole downwind (downshear) from the negative pole, the latter being located in proximity to the deepest convection. Hunter et al. (1992, Fig. 11) observed both of these types of bipole occurring simultaneously in summer at a distance of less than 100 km from each other. They and Marshall and Rust (1993) pointed out the lack of understanding of the processes leading to the apparently different types of bipole. Rutledge and Petersen (1994) showed that the number of ground flashes in the stratiform region is strongly correlated with the vertical radar reflectivity profile and view this as an evidence of in situ charging in the stratiform region. However, they stated that the charge advection mechanism cannot be ruled out. Stolzenburg et al. (1994) concluded that no single mechanism seems likely to account for the four or more significant charge layers inferred from the vertical electric field profiles observed in stratiform clouds (Section 3.3). Schuur and Rutledge (2000a, b), from observations and modeling of one MCS, found that noninductive collisional graupel–ice charging (subsection 3.2.6) could account for as much as 70 percent of the total charge in the stratiform region, the remaining 30 percent being contributed by charge advection from the convective region. More information about the electrical charge structure of and electrification processes in stratiform clouds is found in Section 3.3. Hunter et al. (1992) and Marshall and Rust (1993) considered the possibility that positive lightning could originate in the convective region, propagate horizontally into the stratiform region, and produce a channel to ground there.

Holle et al. (1994), using the seven-station NSSL DF lightning locating network (Section 17.3), studied the cloud-to-ground lightning flash characteristics in a series of four MCSs that occurred in Oklahoma and Kansas on 3–4 June 1985. A detailed description of the life cycle of one of the MCSs, lasting about 11 hours, was given, as outlined next. The peak lightning flash rate was reached a few hours after storm initiation. As the storm matured, the fraction of positive flashes increased. During the decaying stage, negative lightning flash rates rapidly decreased. At most, a few percent of the flashes in convective regions lowered positive charge to ground, and positive lightning flash rates in convective regions rapidly decreased during the decaying stage, a trend similar to that observed for negative lightning flash rates. Positive lightning flash rates in the stratiform region, however, tended to increase until early in the decaying stage, positive flashes exceeding 60 percent of all the flashes in the stratiform area. Throughout the lifetime of the storm there were more flashes per convective area than per stratiform area and more negative than positive flashes per area.

Nielsen et al. (1994), using lightning data from the NSSL network along with radar, satellite, and sounding data, examined the life cycle of a distinct mesoscale segment of the 10–11 June 1985 MCS. Positive flashes were dominant when the MCS was first developing. Negative flashes then became dominant during a period of intense leading-line convective activity with high storm tops. Finally, a period of relatively frequent positive flashes within the trailing stratiform region occurred during the demise of the MCS, after the vertical extent of the leading convective line decreased rapidly and markedly (30–40 dBZ echoes occurring mostly below the freezing level).

It is worth noting that not all MCSs show a bipolar pattern. Further, the stratiform region is not the only region of an MCS in which positive flashes can dominate ground flash activity. Several studies in the Great Plains of the United States (Rutledge et al. 1990; Schuur et al. 1991; Holle et al. 1994) have indicated that a substantial fraction, sometimes even a majority, of ground flashes within the convective region of an MCS can have positive polarity. More information on positive lightning occurrence is found in Chapters 5 and 8.

2.2.8. Lightning and rainfall

Thunderstorms typically produce both lightning flashes and rainfall, so the two quantities are expected to be related. Many researchers have attempted to quantify the relationship between lightning and surface rainfall. Battan (1965), using data from a rain gauge network and visual observations of ground flashes for 52 storms in Arizona, estimated an average of roughly 3×10^7 kg (3×10^4 m^3) of rainfall per ground flash with a range of values for individual storms from 3×10^6 to 3×10^8 kg (3×10^3 to 3×10^5 m^3) per flash. Williams et al. (1992), using radar and a lightning locating system in northern Australia, found a range from 2×10^9 to 3×10^{10} kg (2×10^6 to 3×10^7 m^3) per ground flash for monsoon storms and a range from 9×10^7 to 10^9 kg (9×10^4 to 10^6 m^3) per ground flash for continental storms. The average values were about 8×10^9 kg (8×10^6 m^3) and 5×10^8 kg (5×10^5 m^3) per flash for monsoon and continental storms, respectively. These and many other estimates are summarized in Table 7.6 of MacGorman and Rust (1998), the overall range of values extending over four orders of magnitude, from 3×10^6 to 3×10^{10} kg (3×10^3 to 3×10^7 m^3) per ground flash. MacGorman and Rust (1998) also noted that in all the studies the largest observed rainfall yields per flash occurred with the smallest number of ground flashes per storm, typically less than 10–30. Holle et al. (1994) estimated 1.3 ground flashes per 10^6 m^3 (10^9 kg) of rain in a US Great Plains MCS, comparable to the maximum value of one ground flash per 10^6 m^3 (10^9 kg) found by Williams et al. (1992) for continental storms in northern Australia. However, Piepgrass et al. (1982) found only about 10^4 m^3 (10^7 kg) of rain per ground flash in Florida. Soula et al. (1998) reported 3×10^4 m^3 (3×10^7 kg) per flash, and Molinie et al. (1999) reported a range from 3.5×10^3 to 47×10^3 m^3 (3.5×10^6 to 47×10^6 kg) per flash in the Pyrenees. In isolated thunderstorms in Florida, Piepgrass et al. (1982) found that peak rainfall occurred less than 10 min after the peak flash rate, whereas the rainfall of squall lines often peaked 1 to 2 hr after the time of the peak ground flash rate (e.g., Rutledge and MacGorman 1988; Holle et al. 1994; Kane 1993b).

The relation between lightning and rainfall is characterized by large variability and is poorly understood. It apparently depends on location, storm type, stage of storm development, and other meteorological factors. For example, for coastal storms the direction of the prevailing winds interacting with the sea breeze along the coast (Lopez et al. 1991) is likely to influence this relation. Tapia et al. (1998), who analyzed 22 Florida thunderstorms, found that the convective rainfall mass per cloud-to-ground flash varied with the convective regime. There are lightning-producing storms with little or no surface rainfall (e.g., Reap 1986; Watson et al. 1994a), probably because much of the precipitation has evaporated before reaching ground (e.g., Colson 1960), and there are deep convection clouds, in particular marine cumulonimbi, producing rain but little or no lightning (e.g., Williams et al. 1992; Zipser 1994). Zipser (1994), who studied thunderstorm frequency over the oceans using observations from research ships, reported that in regions such as West Africa and South Asia the seasonal rainfall peak is actually accompanied by a thunderstorm minimum. He observed many tropical heavy-rain events that produced no lightning, even when their associated cloud tops reached 17 km, and hypothesized that most storms formed over oceans have updrafts weaker than a presumed threshold value (discussed below) required for the operation of the graupel–ice mechanism of cloud electrification discussed in subsection 3.2.6. As indicated above, Williams et al. (1992), in the Darwin area of Australia (12° S latitude), found that considerably more precipitation fell per ground flash from storms imbedded in monsoonal convection (monsoon storms) than from storms during the period typified by continental convection (continental storms). This latter observation appears to be consistent with Zipser's (1994) finding that clouds formed over the ocean produce less lightning than clouds formed over the land.

According to Zipser and Lutz (1994), lightning is absent or highly unlikely if the vertical updraft speed does not exceed a threshold of roughly 6–7 m s^{-1} (mean) or 10–12 m s^{-1} (peak), regardless of cloud depth. Further, Petersen et al. (1996) attributed the lack of lightning activity

2.3. Thunderstorm days

in oceanic convection to the fact that most of the hydrometeor mass was located below the freezing level, while the cloud tops were often observed to be higher than 15 km (subsection 2.2.6). However, if natural ground radioactivity or corona from ground plays a role in cloud electrification, as postulated by the convection cloud-electrification mechanism (subsection 3.2.6), then the lack of those phenomena over the oceans may explain the observed low electrical activity of the oceanic clouds even when their depth would indicate that they should yield a lightning output similar to that of the same-depth clouds formed over land. According to Boccippio et al. (2000a), who used OTD and lightning imaging sensor (LIS) data (subsection 2.5.3 and section 17.8), the bulk of the observed order-of-magnitude differences between land and ocean regional flash rates are accounted for by differences in storm spacing (density) and/or frequency of occurrence, rather than by differences in the storm instantaneous flash rates. Boccippio et al. (2000a) found that the latter rates only varied by a factor 2 on average.

The relationship between lightning and rainfall is also discussed in subsection 2.2.7 in considering the lightning bipolar pattern and in subsection 2.8.4 in considering the characteristics of lightning as a function of topography. Additional information on the lightning–rainfall relation is found in the works of Shih (1988), Cheze and Sanvageot (1997), Sheridan et al. (1997), Petersen and Rutledge (1998), Tapia et al. (1998), Carey and Rutledge (2000), Grecu et al. (2000) and Nesbitt et al. (2000), and in Section 7.19 of MacGorman and Rust (1998).

Concluding Section 2.2, we note that characteristics of individual storms in relation to season, location, and storm type are discussed in Section 2.8. Additional information on the electrical and other characteristics of thunderstorms is found, for example, in the book by MacGorman and Rust (1998).

In Sections 2.3 through 2.5, three characteristics of lightning activity, the annual number of thunderstorm days T_D, the annual number of thunderstorm hours T_H, and the annual ground flash density N_g, will be discussed. These three quantities are used to characterize the average local lightning activity rather than to characterize individual storms or parts of storm systems.

2.3. Thunderstorm days

The annual number of thunderstorm days T_D, also called the keraunic level, is the only parameter related to lightning incidence for which worldwide data are available extending over many decades. In fact, in some specific locations such data are available for more than a century. A thunderstorm day is defined as a local calendar day during which thunder is heard at least once at a given location. The practical range of audibility of thunder is about 15 km, the maximum range of audibility being typically about 25 km (Section 11.2). Thunderstorm-day data are recorded at most weather stations by human observers, and apparently the first global climatological survey of thunderstorm occurrence based on T_D data was published by Klossovsky (1892) for 439 weather stations. The spatial distribution of long-term average values of T_D is often presented in the form of isokeraunic maps. Examples of such maps are shown in Figs. 2.8 and 2.9, for the entire world and for the United States, respectively.

Considerable effort has been expended to relate the annual number of thunderstorm days T_D to the ground flash density in $km^{-2}\ yr^{-1}$, N_g. Prentice (1977), for example, reviewed 17 proposed relations between these two parameters. Most of the relations are of the form

$$N_g = aT_D^b \quad (2.3)$$

where a and b are empirical constants. Apparently, the most reliable expression is the one proposed by Anderson et al. (1984a),

$$N_g = 0.04 T_D^{1.25} \quad (2.4)$$

This expression is based on the regression equation relating the logarithm of the five-year-average value of N_g measured with CIGRE 10 kHz lightning flash counters (subsection 2.5.1) at 62 locations in South Africa and the logarithm of the value of T_D as reported by the corresponding weather stations. The range for T_D was from 4 to 80, the range for N_g was from about 0.2 to about 13 $km^{-2}\ yr^{-1}$, and the correlation coefficient between the logarithms of N_g and T_D was 0.85. A similar equation, $N_g = 0.036\ T_D^{1.3}$, based on the use of lightning flash counters in the USSR to determine N_g, was proposed earlier by Kolokolov and Pavlova (1972).

Equation 2.4 has been adopted by the Institute of Electrical and Electronics Engineers (IEEE) (Whitehead et al. 1993) and by CIGRE (the French acronym that stands for Conférence Internationale des Grands Réseaux Électriques à Haute Tension, or, in English, the International Conference on Large High Voltage Electric Systems) (CIGRE Document 63 1991) for estimating the lightning performance of power transmission lines. It is important to note that there is a considerable scatter of data points about the regression line corresponding to Eq. 2.4, as illustrated in Fig. 2.10. It is also worth noting that the application of Eq. 2.4, obtained for South Africa, to other areas of the world requires caution since the relation between N_g and T_D may well vary from one region to another (Prentice 1977; Rakov et al. 1990b). As noted above, however, this equation is very similar to the expression obtained in the USSR by Kolokolov and Pavlova (1972).

A characteristic of thunderstorm activity closely related to T_D is the number of thunder events, defined as isolated periods of thunder activity heard at a given weather station (e.g., Changnon 1988b, 1993). The number of thunder

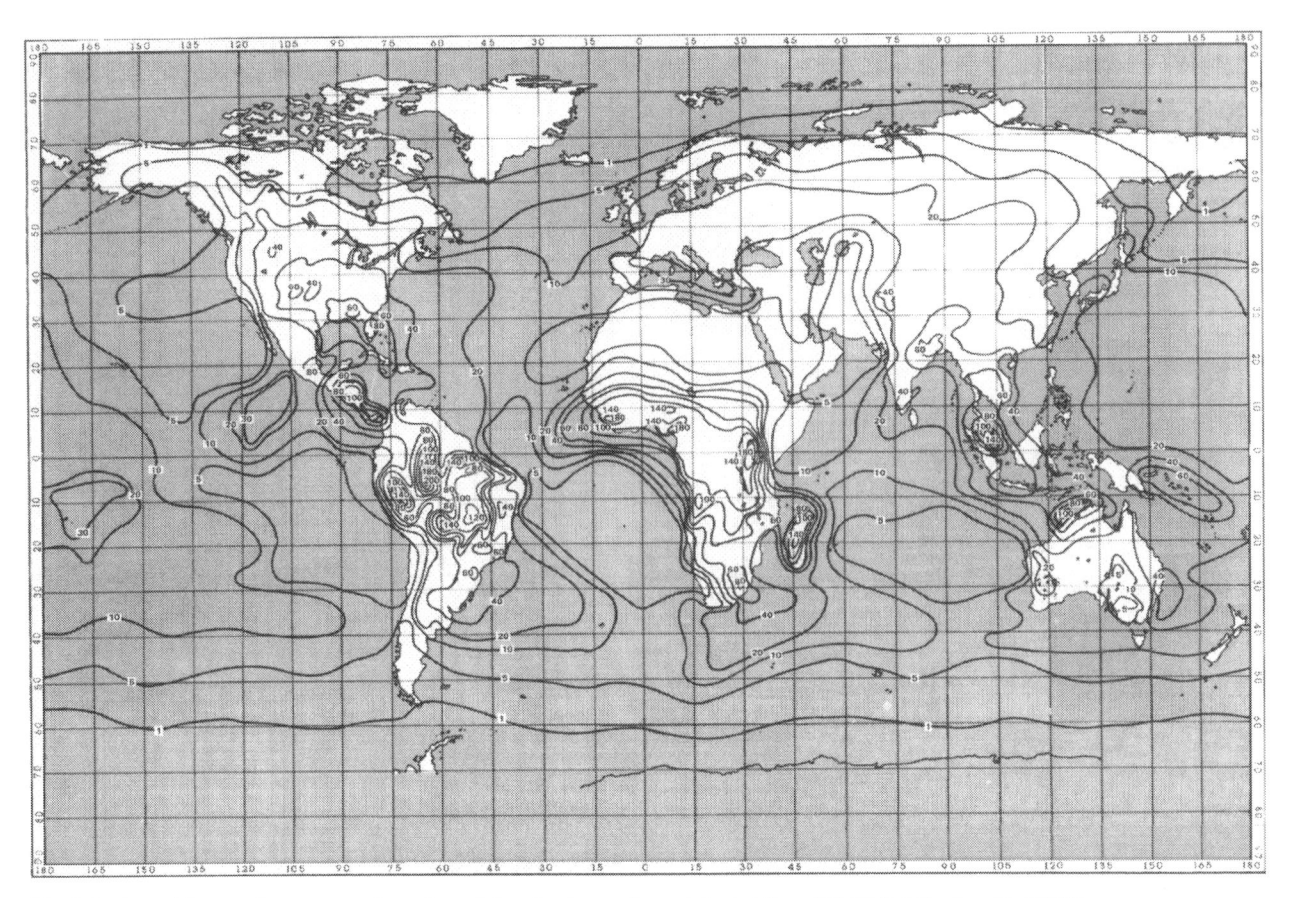

Fig. 2.8. A world map of mean annual number of thunderstorms days. Adapted from WMO Publication 21 (1956).

events can be viewed as the number of thunderstorms that occurred within the observation area of a weather station. A map of the average number of thunder events for the contiguous United States is considered in subsection 2.8.2 in relation to the spatial variation of lightning activity.

2.4. Thunderstorm hours

The annual number of thunderstorm hours T_H is a parameter that is potentially more closely related to the lightning incidence than T_D. Clearly, T_D does not distinguish between a small thunderstorm producing a few lightning flashes in tens of minutes and a large storm lasting for several hours and producing hundreds of flashes, while T_H does. As for T_D, T_H is routinely recorded by human observers at many weather stations, although some stations record only T_D. The frequency distributions of thunderstorm durations for 30 years at 36 stations in the United States were analyzed by Robinson and Easterling (1988). Kolokolov and Pavlova (1972), using weather station reports and lightning flash counter registrations (subsection 2.5.1) in the plain terrain of the European part of the USSR, found a linear relationship between N_g and T_H: $N_g = 0.05 T_H$. MacGorman et al. (1984), using ground flash data from DF systems (Section 17.3) operated in Florida and Oklahoma, inferred values of T_D and T_H that a human observer would report if located at grid points with 25 km spacing and compared the resulting correlations between N_g and T_D and between N_g and T_H. In each location, the correlation with T_H was better than with T_D. MacGorman et al. (1984) constructed an expression for the combined Florida and Oklahoma data,

$$N_g = 0.054 T_H^{1.1} \qquad (2.5)$$

where as before N_g is measured in km^{-2} yr^{-1}. They used this to convert a map for T_H for the contiguous United States into a corresponding map for N_g. Changnon (1993), using 1986–9 data from three DF networks, found the ground flash density in Florida and Oklahoma to be in general agreement with the map of MacGorman et al., while over most of the nation, particularly in the west, the predictions of Eq. 2.5 were a factor 2 higher. Nevertheless, Eq. 2.5 is similar to that previously reported by Kolokolov and Pavlova (1972) and to another relationship, $N_g = 0.04 T_H$, obtained by Eriksson (1987) for the Transvaal Highveld region of South Africa. Note that the exponent on the right-hand side of Eq. 2.5 is close to unity and that in the expressions published by Kolokolov and Pavlova (1972) and by Eriksson (1987) it

2.5. Lightning flash density

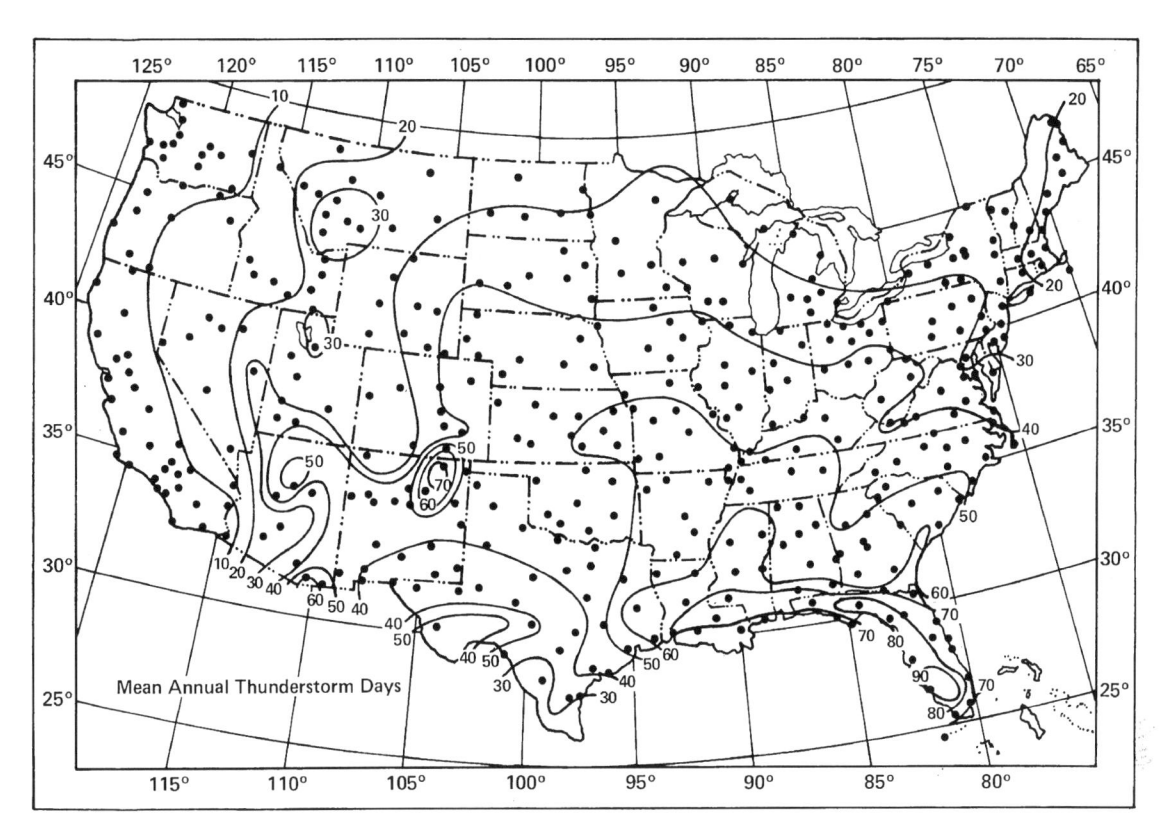

Fig. 2.9. A map of the mean annual number of thunderstorm days for the United States based on data from 450 air weather stations shown as dots (Changery, 1981). Most stations had 30-year records and all had at least 10-year records. Adapted from MacGorman et al. (1984).

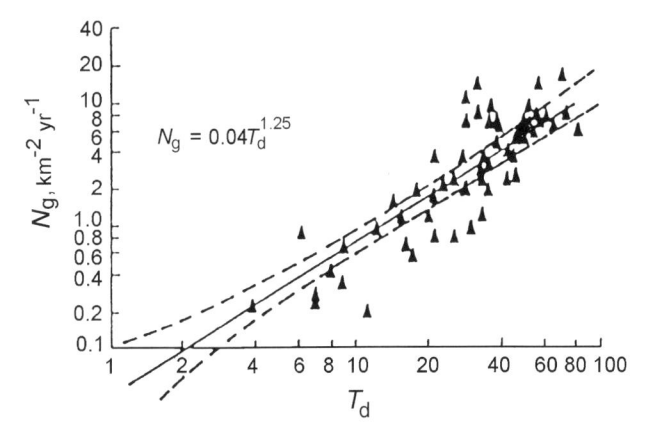

Fig. 2.10. Relationship between annual ground flash density and annual number of thunderstorm days. Adapted from Anderson et al. (1984a).

$N_g = 0.04 T_d^{1.25}$

is unity; that is, N_g is linearly or nearly linearly related to T_H. Dulzon and Rakov (1991) suggested that T_H may not be better than T_D as a predictor of N_g. They found that the long-term annual number of lightning-caused outages of power lines that had similar geometrical and electrical characteristics and that were located in areas with different long-term values of T_D and T_H did not show a better correlation with T_H than with T_D. Perhaps T_H records are more likely to suffer from human error than T_D records. Indeed, in the case of T_D it is just "yes" or "no" for each day, whereas in the case of T_H the observer must record the beginning and the end of storm, the end usually being defined as the time after which thunder is not heard for at least 15 minutes (e.g., Changnon 1988a). If thunder is heard again after, say, 20 minutes, a new beginning is recorded. It is possible that an observer may be hesitant to regard a given thunder as the final one when non-audible (out of thunder-hearing range) but visible lightning activity is present, a situation that is more likely to occur at night (e.g., Reap and Orville 1990).

2.5. Lightning flash density

Lightning flash density is often viewed as the primary descriptor of lightning incidence, at least in lightning protection studies. Flash density has been estimated from records of (i) lightning flash counters and (ii) lightning locating systems (Chapter 17) and can potentially be estimated from records of satellite-based optical or radio-frequency radiation detectors (Section 17.8).

About 50 percent of ground flashes in New Mexico and Florida strike ground at more than one point (Kitagawa et al. 1962; Rakov and Uman 1990b), the spatial separation between individual channel terminations in Florida ranging from 0.3 to 7.3 km as determined using multiple-station TV records (Thottappillil *et al.* 1992). Most measurements of lightning flash density do not account for multiple channel terminations on ground. If all strike points separated by distances of some hundreds of meters or more are of interest, as is the case where lightning damage is concerned, measured values of ground flash density should, in general, be increased. The average number of ground terminations per flash in Florida thunderstorms, based on data reviewed by Rakov *et al.* (1994), is 1.7. Similar values were reported for New Mexico thunderstorms by Kitagawa *et al.* (1962). Thus, at least for these two locations, the correction factor for measured values of N_g to account for multiple channel terminations on ground should be about 1.7.

2.5.1. Lightning flash counters

The lightning flash counter (LFC) is an antenna-based instrument that produces a registration if the electric (or magnetic) field generated by lightning, after being appropriately filtered (the center frequency is typically in the range from hundreds of hertz to tens of kilohertz), exceeds a fixed threshold level. The output of an LFC is the number of lightning events and/or time sequence of lightning events recorded at a given location. The first antenna-based device that was used for the detection of lightning activity was the so-called grozootmetchik (Russian for thunderstorm detector), invented in 1895 by Popov (e.g., Popov 1896). Any LFC will record both ground and cloud lightning discharges, although only ground discharges are needed for the estimation of N_g. In order to "exclude" the undesired (in estimating N_g) cloud discharges from the total number of LFC registrations K, a correction factor Y_g, discussed later, is applied as a multiplier to K. Further, the product $Y_g K$ is converted to N_g by dividing it by the effective operating area πR_g^2 of the counter, where R_g is the effective radius, or range, for ground discharge detection, so that $N_g = Y_g K / (\pi R_g^2)$. Since the electric or magnetic field intensity (in the LFC frequency range) at a fixed distance from the lightning channel varies by two orders of magnitude or so (primarily owing to variations in the return-stroke current), any LFC with a fixed trigger threshold level will fail to register small events nearby while registering large events far away. Accordingly, the effective range for the counter is defined so that the "misses" within the effective range are exactly compensated by the "hits" outside it. An equivalent definition of effective range is the distance within which the number of flashes actually occurring equals the number registered by the counter. For a given LFC, the effective range depends on the field amplitude distribution at the source and may depend on field propagation conditions. In using an LFC, it is assumed that the lightning activity is reasonably homogeneous over its operating range, which is several times greater than the effective range. About 40 percent of ground flashes are registered at a distance equal to the effective range, and lower percentages are registered beyond that range for a typical LFC (Bunn 1968; Prentice and Mackerras 1969; Anderson *et al.* 1979). Many years of measurement are required to obtain a meaningful average flash density from records of lightning flash counters (Section 2.6).

The correction factor Y_g, the fraction of ground flashes in the total number of LFC registrations K, is a function of the ratio z of the actual cloud and ground flash densities (Section 2.7) and of the ratio of the effective ranges, necessarily different, for cloud (R_c) and ground (R_g) discharges. The ratio of the effective ranges depends on the filter and triggering threshold level used. Thus the characteristics needed for proper interpretation of the registrations of any LFC, which are to be determined from independent measurements, include z, R_c, and R_g. Evaluation of these characteristics is not a trivial task since it requires the identification and ranging of both ground and cloud flashes within some tens of kilometers. Additionally, due to their dependence on the source, Y_g, R_c, and R_g can potentially change from one location to another, as discussed by, for example, Rakov (1986) and Rakov and Dulzon (1988).

The two most widely used counters are endorsed by CIGRE (see Section 2.3) and are called the CIGRE 500-Hz counter (Pierce 1956; Barham 1965; Golde 1966; Prentice and Mackerras 1969; Prentice 1972; Barham and Mackerras 1972; Prentice *et al.* 1975; Anderson 1977; Anderson *et al.* 1979) and the CIGRE 10-kHz counter (Anderson *et al.* 1973, 1979; Anderson 1977). The frequency designation refers to the center frequency passed by the filter that precedes the trigger circuit. Another "organization-recommended" counter, called the CCIR (International Radio Consultative Committee) or WMO (World Meteorological Organization) counter (Horner 1960), has not attained widespread use. In contrast with most other LFCs, the CCIR counter was not intended to differentiate between ground and cloud discharges. Its filter is centered at 10 kHz and is similar to that of the CIGRE 10 kHz counter; however, the threshold levels for these two counters differ considerably: 3 V m^{-1} for the CCIR (Horner 1960) and about 20 V m^{-1} for the CIGRE (Anderson *et al.* 1979). The CIGRE 10 kHz counter, employing a 10 kHz filter in conjunction with a 20 V m^{-1} trigger threshold, allows a lower percentage of cloud flashes in the overall counts than does the CIGRE 500 Hz counter, employing a 500 Hz filter in conjunction with a 5 V m^{-1} trigger threshold (Anderson *et al.* 1979). The effective ranges of the CIGRE 500-Hz counter were determined experimentally in Australia to be about 30 km for ground flashes and about 20 km for cloud flashes (Prentice and Mackerras 1969) and in South Africa to be about 37 km for ground flashes and

2.5. Lightning flash density

about 17 km for cloud flashes, about 20 percent of the total registrations being due to cloud flashes (Anderson 1980). The effective ranges of the CIGRE 10 kHz counter in South Africa were found to be about 20 km for ground flashes and about 6 km for cloud flashes, about 5 percent of the total registrations being due to cloud flashes (Anderson *et al.* 1979; Anderson 1980). Apparently, the effective range of the CCIR counter was not been properly determined by measurements, the reported values varying considerably (Horner 1960; Müller-Hillebrand 1963). Cooray (1986) estimated theoretically the effective range of this LFC to be about 140 km. Some comments on the general theory and the inherent weaknesses of LFCs are given by Brook and Kitagawa (1960) and by Bunn (1968), and the operation of the two CIGRE counters and the CCIR counter is discussed in detail by Cooray (1986). If Y_g and R_g are known, LFCs can provide reasonably accurate data on ground flash density. We consider some of these data next.

Prentice (1977) summarized much of the published and unpublished data on average flash density that had been obtained using LFCs, visual observations, and instruments that measure electric field change. The most extensive flash counter studies come from Scandinavia (e.g., Müller-Hillebrand *et al.* 1965), Australia (e.g., Mackerras 1978), and South Africa (e.g., Anderson 1980; Anderson *et al.* 1984a, b). Southeast Queensland, Australia, has a total flash density of 5 km^{-2} yr^{-1}, of which 1.2 km^{-2} yr^{-1} are ground flashes; Norway, Sweden, and Finland have measured ground flash densities between 0.2 and 3 km^{-2} yr^{-1} depending on location, and South Africa has ground flash densities from below 0.1 to about 12 or 13 km^{-2} yr^{-1}, depending on location. Probably the most reliable ground flash density map based of LFC observations is that compiled for South Africa, the observation period covering 11 years (CIGRE Document 63 1991). Ground flash density maps based on more recent LFC observations are presented by Sunoto (1985) for Java, Indonesia, by Romualdo *et al.* (1989) for Mexico, by Diniz *et al.* (1996) for Minas Gerais, Brazil, and by Rakov *et al.* (1990a, b) for the Tomsk region of Russia. Appreciable differences between the spatial distributions (maps) of N_g and T_D are typically noticed. In the study of Rakov *et al.* (1990a, b), the Monte-Carlo technique was used to combine long-term (24-year, on average) statistical distributions of T_D from weather stations and the distributions of N_g per thunderstorm day measured over five years with a network of LFCs.

Crozier *et al.* (1988), using a single-station magnetic direction finding system (essentially an LFC) with a nominal range of 180 km located in Southern Ontario, Canada, estimated ground flash densities of about 1.6 and 2.4 km^{-2} yr^{-1} for 1982 and 1983, respectively. Additional information from the system's output included the lightning flash rates, the percentage of positive flashes, the percentage of multiple-stroke flashes, interstroke intervals, flash duration, and relative stroke amplitudes within the flash.

An attempt to design an LFC capable of recording ground and cloud discharges separately with the same effective range was made by Mackerras (1985). His device, dubbed CGR1, processed electric field changes using various criteria to distinguish the signals produced by ground lightning, by cloud lightning, and by non-lightning sources. First, it was assumed that a 400 V m^{-1} threshold for the overall electric field change assures an effective range of about 14 km for both ground and cloud discharges. Further, the occurrence of lightning must be confirmed either by the detection of "at least one fast positive-going change greater than 40 V m^{-1} or by the detection of a burst of RF noise in a band centered at about 150 kHz, with a detection bandwidth of about 40 kHz". It appears that the detection of cloud flashes largely relies on the RF noise criterion. Since this noise is not always detectable, cloud flashes can be preferentially rejected by the device. Finally, if a signal is recognized as due to lightning, the following three quantities are compared: (i) the overall field change, (ii) the largest fast positive-going field change, and (iii) the sum of fast positive-going field changes. Depending on the results of the comparison (the quantitative criteria as well as the gain of the RF noise channel were set by trial and error), the event is classified as either a ground flash or a cloud flash.

An advanced version of this LFC, dubbed CGR3, has five registers for recording (i) cloud flashes, (ii) positive cloud-to-ground flashes, (iii) negative cloud-to-ground flashes, (iv) so-called distant flashes, and (v) so-called overload conditions (Mackerras and Darveniza 1994). For CGR3, different effective ranges are specified for cloud flashes (12.2 km), for positive ground flashes (16.2 km), and for negative ground flashes (13.6 to 15.4 km, depending on latitude). It is possible that some parameters of the flash identification algorithm, which was apparently adjusted for thunderstorms in Brisbane, Australia are inadequate for other geographical locations. Therefore, the results of the studies based on the use of CGR instruments in various geographical locations, presented in Sections 2.7 and 2.8, should be viewed with caution.

2.5.2. Lightning locating systems

Locating lightning discharges with reasonable accuracy requires the use of multiple-station networks. The principles of operation of multiple-station lightning locating systems (LLS) are described in Chapter 17. In this chapter we will consider only LLSs that employ the radio-frequency electromagnetic fields produced by lightning to determine one location (usually the ground strike point) per stroke or flash, as opposed to systems that are designed to image lightning channels, largely inside the cloud. Ground-strike-point locating systems are presently used in many countries to acquire lightning data that can be used for mapping N_g. Any such LLS fails to detect relatively small cloud-to-ground flashes (particularly near the periphery of the network) and fails to discriminate against some cloud flashes, unwanted in determining N_g. The corresponding system characteristics, the detection efficiency, and the selectivity with respect to ground flashes are influenced by network configuration, position of the lightning relative to the network, the system's sensor gain and trigger threshold, sensor waveform selection criteria, lightning parameters, and field propagation conditions. The interpretation of system output in terms of N_g is subject to a number of uncertainties (e.g., Lopez et al. 1992), but multiple-station lightning locating networks are by far the best available tool for mapping N_g. Difficulties with data interpretation may arise because of the temporal and spatial inhomogeneity of the networks that from time to time undergo upgrading, expansion, adjustment, or reconfiguration. For example, according to Wacker and Orville (1996, 1999a), the 1994 upgrade of the NLDN to employ both magnetic direction finding and time-of-arrival sensors (Section 17.5) resulted in a significant decrease in the measured mean peak current of both positive and negative strokes and a doubling of the annual positive flash count. Huffines and Orville (1999) reported that the number of events identified by the NLDN as positive flashes in 1995–6, after the upgrade, was a factor 2–4 larger than before the upgrade. Wacker and Orville (1999b) attributed the observed decrease in mean peak currents and increase in flash count to the reduction during the 1994 upgrade of the minimum electromagnetic pulse width required in order for an event to be accepted by the system as a stroke to ground. More details on the 1994 upgrade (which was completed in 1995) are found in Section 17.5.

The present NLDN employs 106 sensors (Fig. 17.6), and different versions of the network have provided lightning data covering the continental United States since 1989. From 1989 to 1995, the time and location of flashes have been archived along with estimates of the peak current of the first stroke, the polarity of the first stroke, and the number of strokes in the flash (the so-called "multiplicity"). Since 1 January 1995, the archived data have included the time, location, peak current, and polarity of both first and subsequent strokes detected by the NLDN. The different methods of grouping individually measured strokes into flashes before and after the 1994 upgrade are described in Section 17.5. The ground flash density derived from the locating system's data is computed by dividing the region of interest into small rectangles and accumulating the total number of flashes occurring in each rectangular "grid cell" over the time interval of interest. The gridded data are smoothed before converting them to flash density contours (Byerley et al. 1995).

Orville (1991) published the first N_g map for the contiguous United States based on the data (over 13.4 million ground flashes) from the 1989 NLDN, composed of 114 DF sensors. He assumed a detection efficiency of 70 percent throughout the network (accordingly, the measured numbers were multiplied by 1.4) and used a grid size of 30 by 50 km. A 10-year (1989–98) N_g map for the contiguous United States is shown in Fig. 2.11. As seen in this figure, the variation of N_g from one region to another is more than two orders of magnitude, from 0.1 km^{-2} yr^{-1} or less on the Pacific Coast to 14 km^{-2} yr^{-1} or more in Florida. Maps for 1989–91 for the United States were published by Orville (1994), for 1992–5 by Orville and Silver (1997), for 1989–96 by Huffines and Orville (1999), and for 1995–7 by Orville and Huffines (1999). According to Huffines and Orville (1999), about 25 million cloud-to-ground flashes occur annually over the contiguous United States. Both annual and monthly N_g maps for Florida, based on the 1986–95 data, are found in the work of Hodanish et al. (1997). Values of N_g based on data accumulated over a few years can be affected significantly by the occurrence of just one storm (Orville and Songster 1987; Orville 1991).

The availability of flash characteristics along with locations allows ground flash density studies that were not possible before the implementation of the NLDN and similar networks. For example, Reap and MacGorman (1989) showed the first maps of ground flash density for negative flashes and for positive flashes using data from the NSSL lightning locating network (Section 17.3). Orville (1994) plotted N_g for positive flashes as well as contours of the percentage of positive flashes for the United States for 1989 through 1991. It is even possible to plot N_g for flashes having strokes with inferred peak currents (Section 17.5) greater than, say, 100 kA (e.g., Lyons et al. 1998). Interestingly, lightning locating system data indicate that measured flash densities can vary by an order of magnitude over distances of 20–30 km, partly owing to meteorological effects such as those occurring along the Florida coastline (Maier et al. 1979). Similar observations have been made by Lopez and Holle (1986) for lightning in central Florida and in northeastern Colorado.

Krider (1988), using a three-station DF lightning locating network, studied the distances between successive flashes in three small thunderstorms in Florida. He found that the mean distance between successive flashes ranged from 3.2 to 4.2 km, and the maximum distance ranged from 10 to 12 km. Lopez and Holle (1999), using NLDN data, found that, in storms of small and medium dimensions in Florida and Colorado, the median distance between successive flashes was approximately 4 to 5 km, while for a large mesoscale storm in Oklahoma it was about 9 km.

2.5. Lightning flash density

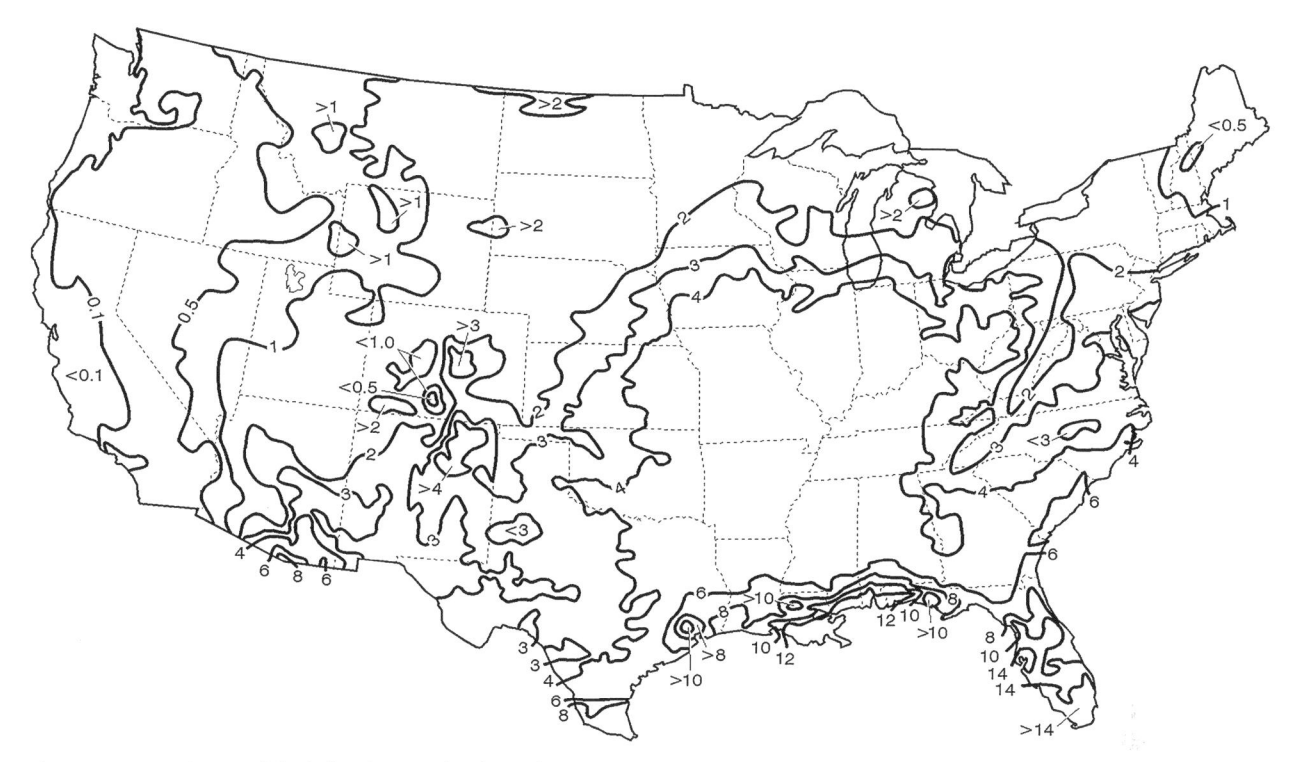

Fig. 2.11. Annual ground flash density map for the United States based on 1989–98 NLDN data. Courtesy of Global Atmospherics, Inc.

2.5.3. Satellite-based detectors

With the advent of Earth-orbiting satellites, it has become possible to map systematically the worldwide lightning activity by detecting the optical or radio-frequency radiation emitted by cloud and ground discharges. It is worth noting that in order to obtain N_g maps from satellite observations, a spatial distribution of the fraction of discharges to ground relative to the total number of lightning discharges, discussed in Section 2.7, is needed. Apparently, the first reports of lightning detected with an instrument flown on a satellite were published by Vorpahl et al. (1970), who used an optical detector on the Orbiting Solar Observatory satellite OSO-2. Satellites in low Earth orbit spend a relatively short time over any given storm and, hence, record only a small fraction of the discharges. In the case of optical detection before 1995, lightning flashes were visible only in photographs taken during night-time passes, and the detection efficiency was often less than two percent (Christian et al. 1989). In the case of radio-frequency detection, which is independent of the time of day, a significant fraction of the data can be contaminated by interference from man-made sources on the ground (e.g., Kotaki et al. 1981a, b; Kotaki and Katoh 1983). More information on satellite-based lightning detectors is found in Section 17.8.

Although the best spatial resolution for optical detectors in orbit before 1995 was of the order of 100 km, it was possible, nevertheless, to estimate total flash densities (both cloud and cloud-to-ground discharges combined) and to determine ratios of activity in different geographical locations and in different seasons. It was not, however, possible to distinguish between cloud and cloud-to-ground discharges. Orville and Henderson (1986), using photographic data from a Defense Meteorological Satellite Program (DMSP) satellite, studied the global distribution of lightning at midnight between 60° S and 60° N for 365 consecutive days (from 1 September 1977 to 31 August 1978). They gave a series of lightning maps for each month and for groups of months in which the locations of the discharges were well correlated with known features of the general atmospheric circulation. These maps show the concentration of lightning over continents and islands as opposed to oceans, the monthly ratio of flash density over land to flash density over oceans ranging from 5.3 to 10 (7.7 on average), which is consistent with the surface hourly observations from research ships in the tropics reported by Zipser (1994). Possible explanations for the lack of lightning over the oceans are discussed in subsection 2.2.8. Orville and Spencer (1979), using dusk and midnight photographic data obtained in 1974 and 1975 with two DMSP satellites, estimated that the global flash rate was 123 s^{-1} at dusk and 96 s^{-1} at midnight, with a potential error of about a factor 2. They reported that there was 1.4 times more lightning during summer in the northern hemisphere than during the southern summer. Turman and Edgar (1982), using DMSP

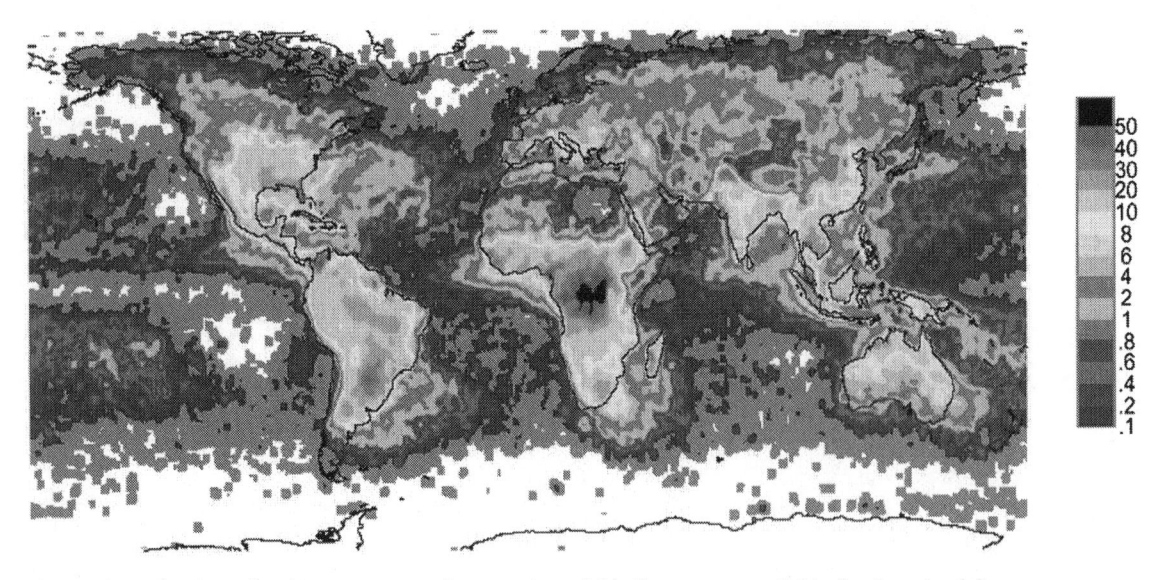

Fig. 2.12. At the time of going to press a colour version of this figure was available for download from http://www.cambridge.org/9780521035415. A global map of total lightning flash density in km^{-2} yr^{-1} based on data from two satellite detectors, OTD (five years) and LIS (three years). Grey areas correspond to a flash density range 0.01–0.1 km^{-2} yr^{-1} and white areas to a flash density <0.01 km^{-2} yr^{-1}. Courtesy of H.J. Christian, NASA/Marshall Space Flight Center.

satellite data, estimated a global flash rate of 40 to 120 s^{-1} with a seasonal variation of about 10 percent. Thirty-seven percent of the global lightning activity originated over the ocean at dawn and 15 percent at dusk. Turman (1978) using an array of 12 photodiodes (each having a field of view of 700 km by 700 km) flown on a DMSP satellite, observed 10 000 flashes from 24 storm complexes during 15 orbits in September 1974 and March 1975. The average flash density within the sensor field of view was 6×10^{-8} km^{-2} s^{-1}, which, if extrapolated to the whole Earth, represents a global flash rate of about 30 s^{-1}. Kotaki *et al.* (1981a, b) and Kotaki and Katoh (1983) detected lightning with the Japanese Ionosphere Sounding Satellite-b (ISS-b), which sensed HF radiation at 2.5, 5, 10, and 25 MHz. They presented worldwide lightning maps for a 2 yr period. They found the global lightning frequency to be 64 s^{-1} for the northern hemisphere spring, 55 s^{-1} for the summer, 80 s^{-1} for the fall, and 54 s^{-1} for the winter. The characteristics of early satellite lightning experiments are summarized by Christian *et al.* (1989, Table 1).

A significant improvement in the optical observation of lightning from space was achieved with the launch, on 3 April 1995, of NASA's Optical Transient Detector (OTD) (Christian *et al.* 1996a). The OTD, also described in Section 17.8, detects lightning by looking for small transient changes in light intensity during both day-time and night-time. An OTD flash is a collection of optical pulse groups (adjacent pixel illuminations within the same 2 ms frame) that are geographically adjacent and that have no "dead time" between groups greater than 333 ms. OTD's spatial errors are, on average, about 20–40 km, and its temporal errors are less than 100 ms (Boccippio *et al.* 2000b). Data averaging over 55 or 110 days is required to remove the diurnal lightning cycle bias. Like previous satellite-based optical detectors, the OTD cannot distinguish between ground and cloud discharges. The global lightning distributions (e.g., Christian and Latham 1998) obtained are similar to those based on earlier satellite studies. Boccippio *et al.* (2000b) compared lightning detected by the OTD and by the NLDN. They found that the OTD detection efficiency for ground flashes was about 46 to 69 percent and is likely to be slightly higher for cloud flashes. Data from the OTD are found on the website http://thunder.msfc.nasa.gov/otd.html. The OTD stopped sending data in April 2000. A follow-on lightning detector, called the Lightning Imaging Sensor (LIS), was launched on 27 November 1997 as part of the Tropical Rainfall Measuring Mission (TRMM). More information on the LIS is found in Section 17.8. Thomas *et al.* (2000) compared lightning detected by the LIS and by a three-dimensional time-of-arrival lightning imaging system, as discussed in Section 17.8. They found that lightning discharges confined below 7 km altitude were less likely to be detected than discharges extending to higher altitudes. Data from the LIS are found on the website http://thunder.msfc.nasa.gov/lis.html. A total lightning flash density map based on data from two satellite detectors, OTD (five years) and LIS (three years), is shown in Fig. 2.12. The data are cross-normalized, and detection efficiency adjustments are applied as a function of detector, local hour, and, for the OTD, sensor threshold setting. An additional correction is applied to the OTD data in the so-called South Atlantic Anomaly. The average values of the detection efficiency estimates are 49 percent for OTD and 82 percent for LIS. The data represent a

2.6. Long-term variations in lightning incidence

0.5 × 0.5 degree composite with a 2.5 degree spatial-moving-average operator applied. The satellite data shown in Fig. 2.12 yield a global flash rate of 45 s^{-1} with an estimated uncertainty of ±5 s^{-1}.

As noted in Section 2.1, Brooks (1925) estimated that the global flash rate was about 100 s^{-1} on the basis of data obtained by human observers. The satellite data are in remarkably good agreement with this value, considering the assumptions that were necessarily made both in analyzing the satellite data and in determining Brook's value.

2.6. Long-term variations in lightning incidence

Since lightning activity varies from year to year, obtaining average values of T_H, T_D, or N_g potentially requires observations over many years, although it is conceivable that stable long-term average values do not exist. Anderson *et al.* (1979) recommended taking data for at least 11 years, one solar cycle, although Freier (1978) found little evidence for coupling between solar activity and thunderstorms. Kitagawa (1989) studied the number of thunderstorm days per month for a period of 100 years in Japan both in winter (November, December, January, and February) over the Japan Sea coast and in summer (July and August) over three different regions, inland plains, inland mountains, and open sea coast. Variations in the number of thunderstorm days over the 100-year period for two locations showing different trends are given in Fig. 2.13.

It has been suggested (e.g., Williams 1992, 1994) that global lightning activity is a sensitive function of the global surface air temperature. Jayaratne (1993), using a CGR3 lightning flash counter (subsection 2.5.1), observed that the monthly lightning activity in Gaborone, Botswana, in summer increased by an order of magnitude for a 2 °C rise in wet bulb temperature. The latter quantity represents the effects of both temperature and humidity and is described, for example, by Williams (1995). Price and Rind (1994a), using cloud top height as a proxy for lightning frequency, estimated an approximate 5–6 percent change in global lightning frequencies for every 1 °C global warming or cooling, while Reeve and Tuomi (1999), who used data from the OTD (subsection 2.5.3), estimated a 40 percent change in lightning activity for a one-degree temperature change. Michalon *et al.* (1999) calculated from a cloud electrification and lightning model that a surface warming of 2 °C would produce a 10 percent increase in the annual global flash frequency. According to Williams (1992), single-station Schumann resonance measurements that represent ELF (3 Hz to 3 kHz) standing waves in the Earth–ionosphere cavity, due to lightning from many storms occurring worldwide, can be used for the estimation of global lightning activity, as discussed in subsection 13.4.3. If the Earth's climate warms owing to an increase in trace gas concentrations in the atmosphere, a temperature increase of

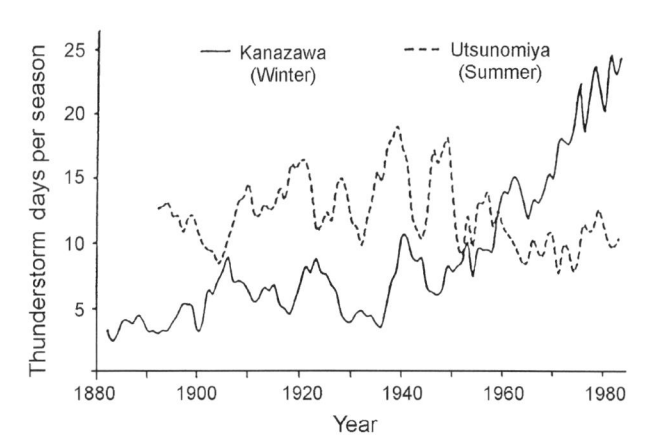

Fig. 2.13. Long-term variation in the number of thunderstorm days per season for winter storms along the Japan Sea coast (Kanazawa) and for summer storms over the inland Kanto Plain (Utsunomiya). Adapted from Kitagawa (1989).

1.5 to 4.5 °C being predicted by the middle of the twenty-first century (Intergovernmental Panel on Climate Change 1990), significant increases in lightning activity should be expected. Price (2000) showed that upper-tropospheric water-vapor variability and global lightning activity are closely linked. Small changes in upper-tropospheric water vapor have a large impact on the greenhouse effect on Earth.

Dulzon (1992) presented T_H data from a weather station in Siberia for a 50-year period between 1936 and 1985. It can be inferred from these data that the mean annual number of thunderstorm hours determined for moving 20-year periods first increased to a maximum value of about 50 percent more than the 50-year mean and then decreased to a minimum value of about 40 percent less than the 50-year mean. According to Changnon (1985), in North America the continental averages of frequency of thunderstorm days exhibited an average increase of 15 percent from 1901 to the 1930s and then a decrease, interrupted by a peak in the 1970s, of about 10 percent from 1940 to 1980. He also reported that northern (≥50° N) stations in Europe, Asia, and North America all exhibit upward trends from 1901 to 1980. Mackerras (1977, Fig. 5) showed a decreasing tendency of annual thunderstorm days in Brisbane, Australia, during the 1911–68 period. Anderson *et al.* (1984a) reported that the annual values of N_g measured using the CIGRE 500 Hz LFC in Pretoria (South Africa) in 1970–1 and in 1972–3 were more than 50 percent in excess of the long-term (15-year) mean value.

Byerley *et al.* (1995) showed N_g maps based on NLDN data for 1992 and 1993 and noted the significant variability in flash density for these two years. In 1992, only 16 million flashes were detected, whereas over 24 million were detected in 1993. The NLDN was stable over this

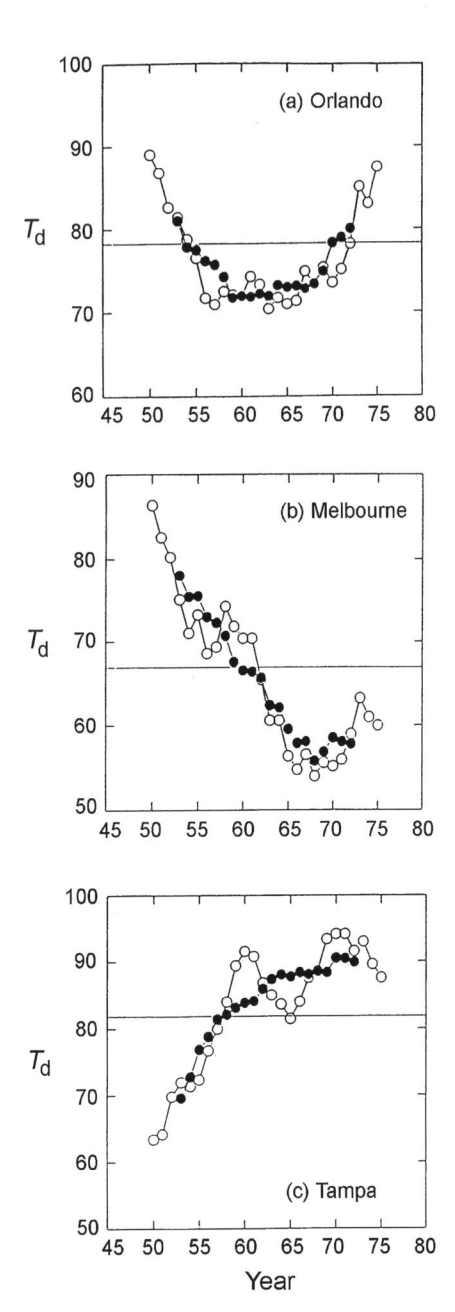

Fig. 2.14. Variation in the annual number of thunderstorm days over a period of approximately 30 years (1948 to 1977) at three Florida weather stations: (a) Orlando, (b) Melbourne, and (c) Tampa. ○, five-year moving average; ●, 11-year moving average; —, 30-year mean. Adapted from Uman *et al.* (1994).

two-year period, in terms of instrumentation and data-processing algorithms.

Figure 2.14 shows the variation in the annual number of thunderstorm days T_D over a period of approximately 30 years (1948 to 1977) at three Florida weather stations: Orlando (Fig. 2.14a), Melbourne (Fig. 2.14b) and Tampa (Fig. 2.14c). To reduce the effect of the year-to-year scatter, running averages with five-year and 11-year windows were performed on the data. The 30-year mean value is indicated by a horizontal line on each of these figures. The corresponding data (not given here) for the annual number of thunderstorm hours T_H show more variation relative to the long-term mean than do the T_D data. Note that, for the same period of time, trends at different stations are different. Some of these variations might be due to changes in the local observation conditions, except for the variations reported by the weather station in Tampa, which has a high-quality thunderstorm record over the period from 1896 to 1995 (Changnon 2001a).

2.7. Ratio of cloud flashes to cloud-to-ground flashes

The literature on the ratio of cloud to cloud-to-ground flashes has been reviewed by Prentice and Mackerras (1977), Livingston and Krider (1978), Rakov and Dulzon (1984), Mackerras and Darveniza (1994), and Mackerras *et al.* (1998). We first present earlier studies that indicated that there is a latitudinal dependence of the ratio, a result that is widely accepted but apparently not confirmed by more recent, although not necessarily more accurate, measurements. Overall, the ratio z of cloud to cloud-to-ground flashes can vary significantly depending on storm type and other factors, the average value being probably about 3.

Prentice and Mackerras (1977) found, from an examination of the literature, that the ratio had mean values of 5.7 for latitudes between 2° and 19°, 3.6 between 27° and 37°, 2.9 between 43° and 50°, and 1.8 between 52° and 69°. They fitted 29 observations with the equation

$$z = N_c/N_g = 4.16 + 2.16 \cos 3\lambda$$

where N_c is the cloud flash density and λ is the latitude. All the experimental data on z fall within a factor 3 of the value given by the above equation, most within a factor 2. Mackerras (1985) employed an automatic device (CGR1; subsection 2.5.1) designed to distinguish between the electric fields of ground and cloud discharges to determine the ratio of cloud flashes to ground flashes for the period September 1982 to May 1984 in Brisbane, Australia ($\lambda = 27.5°$ S). The ratio for about 6100 total flashes was about 3 with a range from 0.9 to 24.7 for individual days having at least 100 total flashes. The ground flash density was 1.2 km^{-2} yr^{-1} and the cloud flash density was 3.7 km^{-2} yr^{-1}, in agreement with the LFC data reported by Prentice (1977) (subsection 2.5.1).

Previously, Pierce (1970) had suggested the use of the following expression to describe the fraction of total flashes that involve ground:

$$p = N_g/(N_c + N_g) = 0.1 + 0.25 \sin \lambda$$

This equation provides the ratios $z = (1 - p)/p = 9$ at the equator and $z = 1.8$ at the poles. Rakov and Dulzon (1984)

2.7. Ratio of cloud flashes to cloud-to-ground flashes

argued that, because of the large scatter in the data, the use of trigonometric functions is not justified and gave a linear regression equation relating z and λ. Price and Rind (1993) found that the value of z in a thunderstorm increases with increasing thickness of the cold cloud region (0 °C level to cloud top). They obtained data on ground flashes from the BLM DF lightning locating network (Section 17.3) and used the radar-determined convective cloud top heights to infer total flash rates. The relation between total flash rate and convective cloud top height is discussed in subsection 2.2.6 and illustrated in Fig. 2.4. Although the relation of z to the cloud thickness above the 0 °C isotherm deduced by Price and Rind (1993) seems to be due in part to their representation of the total flash rate as a function of cloud top height, it apparently explains why the observed values of z are larger in tropical regions than in the mid-latitudes. A similar explanation was previously offered by Rakov and Dulzon (1984) from a comparison of the observed values of z in different parts of the world with typical cloud vertical dimensions and typical isotherm heights as a function of latitude.

Mackerras and Darveniza (1994) reported measurements of z using the CGR3 instrument (subsection 2.5.1) at 14 locations whose latitudes varied from 60° N to 27° S. These measurements, made from 1987 through 1991, the period of observation ranging from two months to five years, in effect refute the hypothesis of the strong latitudinal dependence of z. In contrast to the earlier studies, many of which involved visual observations, with the CGR3 Mackerras and Darveniza found tropical values of z, in the range 0.5 to 3.4 (with a weighted mean of 2.3 for the latitude range 0° to 20°), similar to the values found in subtropical and temperate regions (where the range is from 1.1 to 3.8 and the weighted mean is 2.2 for the latitude range 20° to 40°). It is only at high latitudes (40° to 60°) that their measured values of z (with range 1.0 to 1.5 and weighted mean 1.3) agree with the earlier studies. The stated uncertainty in z-values is about ±50 percent. No dependence on site altitude, which varied from 0 to 2800 m, was observed. Mackerras and Darveniza estimated that the global mean value of z based on measurements at 14 sites is 1.9, the values at 13 of the sites being within a factor 2 of the global mean. Mackerras et al. (1998) reported that the estimates of z presented by Mackerras and Darveniza are underestimates because the assumed value for the effective range for cloud discharges was too high. The adjusted estimates of z given by Mackerras et al. from CGR3 registrations at 11 sites are 4.0 for the latitude range 0° to 20°, 3.2 for the latitude range 20° to 40°, and 1.9 for the latitude range 40° to 60°. As noted in subsection 2.5.1, the performance characteristics of the CGR3 instruments estimated in Brisbane, Australia, may be inapplicable to other regions, even after the introduction of various adjustment factors. Therefore, the estimates of z given by Mackerras et al. (1998) as well as other results obtained using the CGR3 (e.g., Baral and Mackerras 1992, 1993; Mackerras and Darveniza 1994; Jayaratne et al. 1995; Levin et al. 1996; Yair et al. 1998; Jayaratne and Ramachandran 1998) should be viewed with caution. Thus, none of the proposed latitudinal dependencies of z can be accepted with confidence.

In the following, we will consider various factors that influence the z-value at a given location. From field-mill records for summer thunderstorms in Florida ($\lambda = 28°$ N) Livingston and Krider (1978) found that between 42 and 52 percent of all lightning discharges were cloud-to-ground discharges during the "active storm period" (subsection 2.2.1) and that during the final storm period only about 20 percent were to ground. For five storms on three days, 43 percent of 552 total flashes involved ground. Holzer (1953) also noted that the percentage of cloud flashes increases in the latter stage of a storm.

Rutledge et al. (1992) found that the ratio of the cloud flash rate to the cloud-to-ground flash rate was correlated with the total flash rate for MCSs in the Darwin, Australia, area. The ratio varied from 2.7 to 17 as the total flash rate changed from 1 to 40 min^{-1}. Apparently the higher total flash rates were primarily a consequence of increased intracloud discharge activity, consistent with the finding of Lhermitte and Krehbiel (1979) in Florida. However, Jayaratne et al. (1995), using the CGR3 in Gaborone, Botswana, found that the ratio of cloud flashes to ground flashes decreased with increasing total flash rate, implying a larger proportion of ground flashes in the more active thunderstorms. The ratio was generally highest in the early and later stages of individual storms, falling to its lowest values during the more active intermediate stages. The apparent discrepancy between the observations of Rutledge et al. (1992) in Australia and Jayaratne et al. (1995) in Botswana is likely to be due to differences in the methods used for the identification of different types of lightning. Yair et al. (1998), also using the CGR3, observed that the monthly average value of z for the Tel Aviv, Israel, area varied significantly from month to month in a given season and from year to year (1989–6) for a given month. The range of variation was more than an order of magnitude, from less than 0.5 to 5. For MCSs in the United States, Mazur and Rust (1983) and Schuur et al. (1991) found z to be 40 and 8, respectively. Lang et al. (2000) reported on two intense convective storms in Colorado that were characterized by very low cloud-to-ground flash rates (< 1 min^{-1}) while exhibiting relatively high cloud flash rates (> 39 min^{-1}), so that the value of z was greater than 30. They related this high value of z to strong updrafts. Rakov and Dulzon (1986) reported a significantly larger value of z for frontal storms, 4.5, than for air-mass storms, 1.8, in the Tomsk region of Russia (see subsection 2.8.5 below). Boccippio et al. (2001) studied the distribution of z over the continental United States using OTD and NLDN data (Sections 17.8 and 17.5) covering a four-year period to estimate the total number of flashes and the number of cloud-to-ground flashes, respectively. They reported a range of variation in z from 1.0 over the Rocky and Appalachian mountains to 8–9 in the central-upper Great Plains, the average value being between 2.5 and 3. The range of variability at a given location was comparable to the range of values reported from the early studies for regions from the tropics to the mid-latitudes. Thus, the observed large scatter in z at a given latitude is probably due to the dependence of z on various other factors combined with the often limited sample size.

2.8. Characteristics of lightning as a function of season, location, and storm type

Many results in this section are given without attempting to speculate on the physical processes possibly responsible for the various observed features. We refrain from such speculations here because of the complexity and variability of lightning generation mechanisms, some reported results being potentially deficient in the identification of all significant factors involved and in the statistical significance of these results. Additionally, the interpretation of data is often complicated by the inherent limitations of lightning detection techniques, discussed in Chapter 17. Further information on lightning activity in relation to meteorological, geographical, and other factors can be found in Rakov (1992) and Holle and Lopez (1993).

2.8.1. Season

Orville and Songster (1987) and Orville *et al.* (1987), using the US East Coast DF lightning detection network (Section 17.3) in 1984–5, found that the percentage of positive lightning during the summer was less than 5 percent, but in October it began to increase, reaching a maximum of slightly over 80 percent in February and then decreasing to less than 10 percent in April. In the winter, most of the storms were dominated by positive flashes and occurred over the ocean. Orville and Huffines (1999), using 1995–7 NLDN data for the contiguous United States, reported that the minimum percentage of positive flashes occurred in the July–August period and the maximum occurred in January–February, as illustrated in Fig. 5.2. The percentage of positive flashes as a function of month from the studies of both Orville and Silver (1997) and Orville and Huffines (1999) is given in Table 5.2. Further, Orville and Songster (1987) and Orville *et al.* (1987) inferred that the median return-stroke peak current increases by approximately 50 percent in the winter season for both positive and negative flashes. Orville and Huffines (1999) reported, however, from 1995–7 NLDN data, that the median return-stroke peak current for first strokes in negative flashes remained approximately constant from January through November at about 20 kA, increasing to 24 kA in December. The median positive current peak was 25 kA in February, 15 kA in July, and 24 kA in December. Note that the peak current values reported by Orville and Huffines (1999) are averages over the contiguous United States and that the seasonal dependence of peak current may vary from one region to another. Finally, Orville *et al.* (1987) observed that the percentage of single-stroke negative flashes increased from 40 percent in the summer to over 80 percent in the winter (along the US east coast from North Carolina to Maine), although both figures are likely to be overestimates owing to the fact that the locating system fails to detect small subsequent strokes (Rakov *et al.* 1994; Cummins *et al.* 1998).

Moore and Orville (1990), using ground flash data from the State University of New York at Albany (SUNYA) DF lightning detection network (Section 17.3) in conjunction with satellite, surface, upper air, and lake-temperature data, studied the characteristics of Great-Lakes-induced storms. These storms occur in the fall and winter and produce only a few cloud-to-ground flashes, mostly positive (except for one storm). Both the positive and negative flashes were associated with higher peak currents than in the summer. Also in winter, relatively high lightning activity has been found in the rainbands over the warm Gulf Stream off the east coast of the United States near North Carolina during northwesterly offshore flow (e.g., Biswas and Hobbs 1990; Orville 1990b, 1993; Dodge and Burpee 1993).

Brook (1992), using an electric field measuring system with a bandwidth from 0.1 Hz to about 1 MHz, observed that for negative lightning in winter (i) the initial breakdown pulses (Section 4.3) have larger amplitudes and (ii) the stepped leader durations (Section 4.4) are shorter than in summer lightning. No seasonal differences were found for positive lightning. On the basis of the observations for negative lightning, Brook (1992) concluded that winter discharges may be considerably more energetic than summer ones.

Hojo *et al.* (1989), using an LLP (Lightning Locating and Protection) DF lightning locating network, studied the seasonal variations of ground flash density, the distribution of peak magnetic field for first strokes, and the polarity of ground flashes in the coastal area of the Sea of Japan. Ground flash density maps for summer and winter were reported to be markedly different. In winter, ground strikes occurred primarily to the sea surface, and those to the land were no more than 30 km from the coast. For positive lightning, the peak magnetic field was larger in winter. However, for negative lightning no such seasonal variation was observed, the latter finding being in contrast to the results of Orville *et al.* (1987). The seasonal variation of the polarity

2.8. Lightning vs. season, location, and storm type

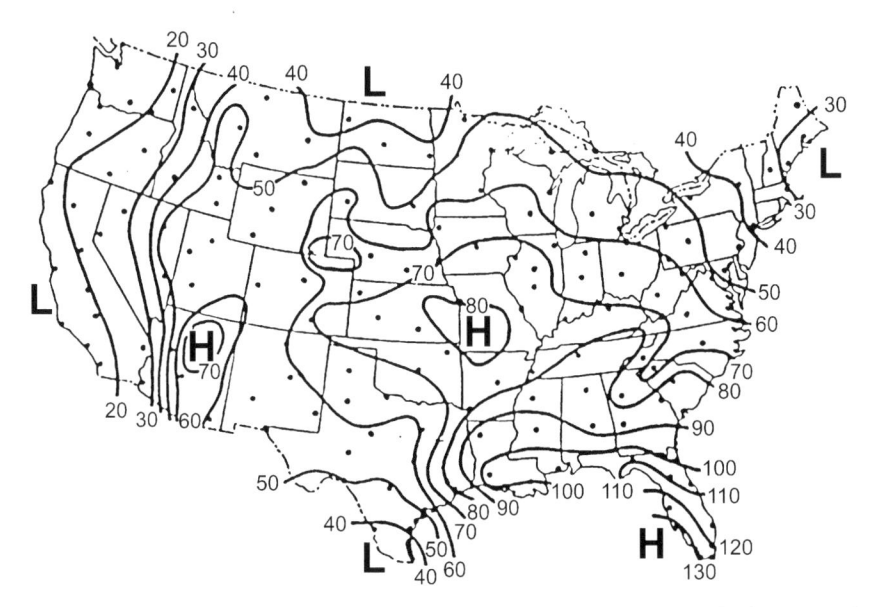

Fig. 2.15. Average annual numbers of thunder events in the contiguous United States. Adapted from Changnon (1988b).

of ground flashes in Japan is similar to that reported by Orville *et al.* (1987) for the northeastern part of the United States. The seasonal variations of the time characteristics of return-stroke field waveforms were analyzed by Ishii and Hojo (1989). They found that at distances of 100–300 km the zero-crossing time (subsection 4.6.3) for both polarities is shorter in winter. More information on winter lightning in Japan is found in Chapter 8.

2.8.2. Region

We discuss here the dependence of lightning characteristics on location in general. The dependences on latitude and topography are discussed separately in subsections 2.8.3 and 2.8.4, respectively.

Cooray and Jayaratne (1994) compared lightning characteristics determined from their electric field measurements in Sri Lanka with similar characteristics observed in Sweden (Cooray and Perez 1994) and in Florida (Rakov and Uman 1990a, b; Thottappillil *et al.* 1992). The geometric-mean interstroke intervals in Sri Lanka, Sweden, and Florida are 57, 48, and 57 ms, respectively. The mean number of strokes per flash and percentage of single-stroke flashes are 4.5 and 21 percent in Sri Lanka, 3.4 and 18 percent in Sweden, and 4.6 and 17 percent in Florida. The similarity of the lightning characteristics given above for the three different geographical regions is rather remarkable.

The literature on the height and magnitude of the negative cloud charge (subsections 3.2.2 and 3.2.3) and the electric dipole moment change per flash was reviewed by Jacobson and Krider (1976), and the literature on the flash duration and the frequency of occurrence of continuing current was reviewed by Livingston and Krider (1978).

Apparently, negative charge centers always occur at about the same cloud-temperature level, typically -10 to $-34\,°C$, although these temperatures occur at different heights above the local terrain in different locations (Fig. 3.7).

Changnon (1988b), using thunderstorm records at 152 first-order weather stations distributed across the contiguous United States for a 30-year period (1948–77), studied "thunder events", which were defined as periods of discrete thunder activity heard at a given weather station. In fact, a thunder event is the thunderstorm, the end of which is usually defined as the time after which thunder is not heard for at least 15 minutes (Section 2.4). The sum of the durations of all thunder events during a year is the annual number of thunderstorm hours, T_H. Since a thunderstorm day T_D (Section 2.3) comprises one or more thunder events (thunderstorms), the two characteristics are closely related, the correlation coefficients ranging, according to Changnon (1988b), from $+0.84$ to $+0.98$.

A map of the average annual number of thunder events is shown in Fig. 2.15. Annual averages of thunder events are high along the Gulf Coast (> 100), in the central United States (Kansas, Missouri, Illinois, with > 75 events), and in the southwest (Arizona, with 60 events). Thunder events are least numerous along the west coast (< 20) and in the northeast (< 30). A peak in thunder events is present in the central United States in all months, and its position is always closely related to the major center of cold frontal activity. The peak in the southwest is related to the summer monsoon intrusion of moist tropical Pacific air and related frontal activity. The peak along the Gulf Coast of Florida is a result of sea-breeze-induced convergence, localized heating, and occasional tropical disturbances in summer and fall. Maps of ground flash density for the contiguous

United States, based on data from the NLDN, are discussed in subsection 2.5.2. These maps show more than two orders of magnitude variability in N_g (see Fig. 2.11).

Westcott (1995), using NLDN data in and around 16 cities for June, July, and August of 1989–92, found an enhancement of 40 to 85 percent in the cloud-to-ground lightning activity over and downwind of many of these cities. Further, Orville et al. (2001), from a similar study for the period from 1989 through 2000, reported elevated ground flash densities, in both summer and winter, centered over and downwind of the metropolitan area of Houston, Texas. They suggested that the observed enhancement of lightning activity could be caused by enhanced convergence associated with the urban heat island and altered microphysical processes associated with anthropogenic pollution.

2.8.3. Latitude

Thomson (1980) reviewed the literature on interstroke time interval and number of strokes per flash and found no statistically significant correlation with latitude. This does not, however, preclude the latitude dependence which some investigators (Pierce 1970; Takeuti et al. 1975a) have inferred should exist, but which others (Harris and Salman 1972) have found not to exist. Rather, Thomson (1980) concluded that "the distributions of both interstroke intervals and the number of strokes per flash are more sensitive to local influences, including measurement and sampling techniques, than they are to latitude effects".

Orville and Huffines (1999) reported that the mean number of strokes per negative flash (also called the "multiplicity" when it is reported by the NLDN) appeared to increase with decreasing latitude in the eastern half of the United States, from 1–1.5 in Maine to over 3 in Florida. As noted earlier, these multiplicity values are probably underestimates due to the NLDN's failure to detect small strokes. Thus, the observed latitudinal trend could be due in part to the fact that both first and subsequent strokes in Florida are, on average, larger than those in higher latitudes, as indicated below for first strokes.

Rakov and Dulzon (1984, 1988) and Pinto et al. (1997), on the basis of their reviews of the literature on lightning current measurements in different countries, and Orville (1990a), on the basis of the analysis of current estimates provided by an LLP DF lightning locating network in the United States, argued that lightning peak currents for first strokes tend to increase as the latitude decreases. Some of the variation in peak current reported by Orville (1990a), however, might be due to instrumental or propagation effects.

A discussion of the possible latitudinal dependence of the ratio of cloud to cloud-to-ground flashes was given in Section 2.7.

2.8.4. Topography

Rakov et al. (1989) used a single-station lightning locating system "Ochag" (Russian for source location) in the North Caucasus region of Russia to estimate the spatial variation in N_g with topography. In this system, azimuth is determined via magnetic direction finding and distance is roughly estimated via the so-called EH ranging algorithm (Kononov et al. 1986; Kononov and Petrenko 1996). This algorithm utilizes the expected difference between the electric and magnetic field waveforms in the distance interval from 15 to 100 km. Rakov et al. (1989) found an average value of N_g that was a factor 1.7 higher for a mountainous area than for a plain-terrain area, the two areas being about equally covered by the lightning locating system. Orville (1994), however, reported lower values of N_g for the Appalachian Mountains than for the neighboring areas to the east and west, from the NLDN data for 1989–91. Kitagawa (1989) reported that the number of thunderstorm days per month (averaged over July and August) in Japan was highest over the inland Kanto Plain, lowest over the open sea coasts, and intermediate over the mountainous regions.

Reap (1986) analyzed cloud-to-ground lightning location data for the summers of 1983 and 1984 from the BLM network (Section 17.3) in the 11 western states of the contiguous United States. The lightning locations and occurrence rates were examined in relation to the topographic features of the generally mountainous western US terrain. Reap (1986) found a high correlation between terrain elevation, which ranged from sea level to near 3 km, and the time of maximum lightning activity: at the higher elevations, maximum lightning activity occurred in the early afternoon to midafternoon whereas at lower elevations it occurred later. The cloud-to-ground lightning activity was also found to increase with increasing terrain elevation, consistent with observations of Rakov et al. (1989) but in contrast with the results of Orville (1994) for the Appalachian Mountains.

King and Balling (1994), using 1989 and 1990 data from the BLM network, analyzed the diurnal variations of lightning flashes in Arizona during the summer monsoon season. In much of the state, the maximum occurred in the mid-to-late afternoon period, but it occurred closer to midnight in the large valley of central Arizona, a finding consistent with that of Reap (1986) discussed above.

Lopez and Holle (1986) used networks of LLP DFs in northeastern Colorado and in central Florida to study the diurnal and spatial variations of the lightning ground flash density (Section 2.5) and to relate those results to the geographic and climatic characteristics of the two regions. The geographic feature of primary importance in Colorado is the mountains and in Florida is the interface between the waters of the Gulf of Mexico or the Atlantic Ocean and the land of the peninsula. For the summer of 1983, detailed maps of the ground flash density as a function of the time of day were given for both locations. Lopez and Holle (1986) showed from these maps that the temporal and spatial distributions of lightning are clearly related to the local topographic features. Reap (1994), using 1987–90 data from the SUNYA DF lightning locating network for Florida, observed organized coastal maxima in lightning activity related to land–sea-breeze convergence zones that form in direct response to the low-level wind flow, with two primary maxima in lightning activity near Tampa and west of Cape Canaveral and a minimum over Lake Okeechobee.

Lucas and Orville (1996), using lightning data from one LLP DF installed at Kavieng, Papua New Guinea, examined the frequency of cloud-to-ground lightning over the ocean as well as its diurnal variation. The data, acquired during 57 days in January and February 1993, indicated that flash counts over the land sector were approximately nine times higher than flash counts over the ocean sector. The highest lightning activity occurred around local midnight for both land and ocean sectors. Hidayat and Ishii (1998), using data from a four-station LLP DF–TOA network on the island of Java, Indonesia, reported the average annual ground flash density over the entire island to be more than an order of magnitude greater than over the Indian Ocean about 100 km south of the island. Further, Hidayat and Ishii (1999) estimated that the median current peak for lightning over the ocean was at least 20 percent higher than for lightning over the land.

Boccippio *et al.* (2000a) analyzed lightning data from the OTD and LIS (subsections 2.5.3 and 17.8) for variability between land and ocean, various geographic regions, and different convective "regimes". They found that ocean storms produce lightning at rates that are not dissimilar from those of continental storms, as noted in subsection 2.2.1. According to Boccippio *et al.* (2000a), the bulk of the observed differences in regional flash rate between land and ocean is accounted for by differences in storm spacing (density) and/or frequency of occurrence.

Takeuti *et al.* (1975a) and Takeuti (1976), using a video camera and an electric field meter, found that cloud-to-sea lightning and cloud-to-ground lightning were similar in the number of strokes per flash and the interstroke interval distribution.

Orville and Huffines (1999) reported, from 1995–97 NLDN data, that median current peaks for first strokes in negative flashes were relatively high, typically exceeding 26 kA, along US continental coastal areas, particularly along the West Coast. Mountainous regions appeared to have lower median negative current peaks of about 15 to 20 kA. Median positive current peaks exceeded 40 kA in the upper midwest but were less than 10 kA in Louisiana and Florida, although the latter value is an underestimate due to misidentification and hence inclusion of some cloud flashes in the ground flash data.

2.8.5. Storm type

For frontal and for air-mass storms, Rakov and Dulzon (1986) compared the statistical distributions of (i) the ratio of cloud to cloud-to-ground discharges, z (Section 2.7), (ii) the return-stroke peak current, I, (iii) the number of strokes per flash, n, averaged over the storm, and (iv) the time interval between strokes, Δt. The distributions of z, I, and n were estimated from electric field measurements in the Tomsk region of Russia, while the distributions of Δt were those previously analyzed for latitudinal dependence by Thomson (1980). Rakov and Dulzon (1986) found that frontal storms are characterized by a larger mean value of z, 4.5 (13 storms) versus 1.8 (22 storms) for air-mass storms, a larger mean value of n, 2.7 (11 storms) versus 1.8 (14 storms), and a smaller geometric mean value of Δt, 51 ms versus 69 ms. The difference between the distributions of I was found to be statistically insignificant, the geometric mean values of I being 15 and 17 kA for frontal and air-mass storms, respectively.

Holzer (1953), working in New Mexico, found that at all stages of the development of an individual thunderstorm cell, in frontal storms ground flashes constituted a larger fraction of the total than in nonfrontal storms. In both classes of storms the cloud flashes became relatively more frequent in the latter stages of the life cycle of the cell. The average number of "repeated elements" (strokes) in cloud-to-ground flashes for air-mass storms was found to be about two, while for the frontal storms the average number was nearly five. The average intervals between strokes were similar, about 50 ms. Schonland (1956) also reported that in South Africa there was a greater number of strokes per flash in frontal storms than in air-mass storms. Kitterman (1980), using a streak camera, observed a greater percentage of "high-order multiple-stroke ground flashes" in frontal storms than in air-mass storms, while flash durations in these two types of storms were similar.

2.9. Lightning incidence to various objects
2.9.1. General information

We first describe briefly how cloud-to-ground lightning "decides" on its ground termination point. More details are given in Section 4.2 and subsection 18.3.2. Ground flashes are normally initiated by stepped leaders that originate in the thundercloud. As the downward-extending leader channel, usually negatively charged, approaches the ground, the enhanced electric field intensity at irregularities of the Earth's surface or at protruding grounded objects increases and eventually exceeds the breakdown value of air. As a result, one or more upward-moving leaders are initiated from those points. When one of the upward-moving leaders from the ground contacts a branch of the downward-moving stepped leader, the point of lightning termination on ground is determined.

Grounded vertical objects produce relatively large electric field enhancement near their upper extremities so that upward-moving connecting leaders from these objects start earlier than from the surrounding ground and, therefore, serve to make the object a preferential lightning termination point. In general, the higher the object, the greater the field enhancement and hence the higher the probability that a stepped leader will terminate on the object. In the limit, when the height (the field enhancement capability, to be more exact) of the object becomes so large that the upward-moving leader from the object tip can be initiated by in-cloud charges or, more likely, by in-cloud discharge processes, as opposed to being initiated by the charge on the descending stepped leader, the object becomes capable of initiating upward lightning, as discussed in Chapter 6. The latter, as opposed to "normal", downward lightning, would not occur if the object were not there. Ground-based objects with heights ranging from about 100 to 500 m experience both downward and upward flashes, the proportion being a function of object height. Eriksson (1987) derived the following equation for the annual lightning incidence N (in yr^{-1}) to ground-based objects, including both downward and upward (if any) flashes:

$$N = 24 \times 10^{-6} H_s^{2.05} N_g \tag{2.6}$$

where H_s is the object height in meters and N_g is the ground flash density in $km^{-2} yr^{-1}$. To do so, he employed (i) observations of the lightning incidence to structures of heights ranging from 20 to 540 m in different countries, (ii) the corresponding local values of the annual number of thunderstorm days T_D, and (iii) Eq. 2.4. It is worth noting that the majority of observed lightning strikes involved in the derivation of Eq. 2.6 correspond to taller structures and that the observed lightning incidence to smaller-height structures might have been affected by the presence of surrounding objects such as buildings and trees. As a result, Eq. 2.6 is probably less reliable for $H_s < 60$ m or so than for greater heights. Eriksson (1978) tabulated the observed percentage of upward flashes as a function of a free-standing structure's height, reproduced in Table 2.3. Eriksson and Meal (1984) fitted the data in Table 2.3 using the following expression:

$$P_u = 52.8 \ln H_s - 230 \tag{2.7}$$

where P_u is the percentage of upward flashes and H_s is the structure height in meters. This equation is valid only for structure heights ranging from 78 to 518 m, since $P_u = 0$ for $H_s = 78$ m and $P_u = 100$ percent for $H_s = 518$ m. Structures with heights less than 78 m are not covered by Eq. 2.7, because they are expected to be struck by downward flashes only, and structures with a height of greater than 518 m are not covered because they are expected to experience upward flashes only.

Table 2.3. *The percentage of upward flashes from tall structures. Adapted from Eriksson (1978)*

Reference	Structure height, m	Percentage of upward flashes
Pierce (1972)	150	23
	200	50
	300	80
	400	91
McCann (1944)	110	8
	180	24
	400	96
Berger (1972)	350[a]	84
Gorin (1972); Gorin et al. (1976)	540	92[b]
Garbagnati et al. (1974)	500[c]	98

[a] An effective height of 350 m was assigned by Eriksson to Berger's 70 m high mountain-top towers to account for the enhancement of the electric field by the mountain, whose top is 640 m above Lake Lugano (914 m above sea level). Pierce (1971) assigned a different effective height, 270 m, to Berger's towers (Section 6.1).
[b] 50 percent of the flashes recorded in this study were classified as "unidentified". The relative incidence of upward flashes is based upon analysis of only the identified data.
[c] The towers of Garbagnati et al. were 40 m high, located on mountain tops 980 and 993 m above sea level (Berger and Garbagnati 1984). See Section 6.1.

In practice, as stated above, it is often assumed that structures having heights less than 100 m or so are struck by downward lightning only and that the upper height limit can be simply taken as 500 m. Accordingly, (i) the total lightning incidence N to a structure is the sum of the downward-flash incidence N_d and the upward-flash incidence N_u if the structure height is in the range from about 100 to 500 m, (ii) $N = N_d$ for structures shorter than 100 m, and (iii) $N = N_u$ for structures taller than 500 m. If both downward and upward flashes are expected, they are often treated separately in estimating the lightning incidence to an object, as described below.

2.9.2. Downward flashes

When the incidence of downward lightning is estimated, it is common to ascribe a so-called equivalent attractive (or exposure) area to the grounded object. The attractive area can be viewed as an area on flat ground surface that would receive the same number of lightning strikes in the absence of the object as does the object placed in the center of that area. In other words, in computing lightning incidence to a structure, the structure is replaced by an equivalent area on ground. For a free-standing structure whose plan-view dimensions are much smaller than its height (such as a mast, tower, or chimney), this area, A, is taken as circular and is generally given by $A = \pi R_a^2$, where R_a is the equivalent attractive radius, discussed later. For straight, horizontally extended structures (such as power lines or their sections), the equivalent attractive area is rectangular and is sometimes termed the "shadow zone" or "attractive swath". For example, if a power line has length l and effective width b (usually taken as the horizontal distance between overhead shield wires or between the outer phase conductors), its equivalent attractive area is generally estimated as $A = l(b + 2R_a)$, where R_a is the equivalent attractive distance, generally thought to be approximately equal to the equivalent attractive radius for a free-standing structure of the same height (Eriksson 1987; Rakov and Lutz 1990). Further, the local ground flash density N_g is assumed to be spatially uniform in the absence of the structure, so that the downward lightning incidence to the structure is found as

$$N_d = A N_g \qquad (2.8)$$

Usually N_g is in km^{-2} yr^{-1}, so that A should be expressed in km^2 to obtain N_d in yr^{-1} (strikes per year).

The equivalent attractive radius (or distance) R_a is usually assumed to be a function of structure height H_s and is generally expressed as

$$R_a = \alpha H_s^\beta \qquad (2.9)$$

where α and β are empirical constants. The procedures used to obtain Eq. 2.9 from data on lightning incidence to structures of different height are given, for example, by Eriksson (1978, 1987). In Eq. 2.9, both H_s and R_a are in meters, and various values of α and β have been proposed. For example, Whitehead et al. (1993) gave $\alpha = 2$ and $\beta = 1.09$ for transmission lines, while the CIGRE Document 63 (1991) recommended $\alpha = 14$ and $\beta = 0.6$ (see also subsection 18.4.3). The attractive radius for individual strikes should depend on the charge carried by the descending leader, this charge being correlated with the associated return-stroke peak current (Section 4.6 and subsection 18.3.2). In this regard, Eq. 2.9 should be understood as representing the entire distribution of peak currents. In the so-called electrogeometric approach (Section 18.3), which is widely used for the estimation of lightning incidence in lightning protection studies (e.g., CIGRE Document 63 1991), the equivalent attractive radius depends explicitly on the statistical distribution of lightning peak currents (e.g., Eriksson 1987; Rakov and Lutz 1988, 1990).

Estimation of N_d from Eq. 2.8 implies a reasonably long-term value of ground flash density and yields a long-term average value of lightning incidence. For example,

Table 2.4. *The number of houses out of a total of 200 (the percentage, if divided by 2) expected to be struck by lightning n times over T years ($N_g = 4$ km^{-2} yr^{-1}; $A = 1200$ m^2)*

Number of years of observation, T	Number of times struck by lightning, n				
	0	1	2	3	>3
10	191	9	0	0	0
20	182	17	1	0	0
30	173	25	2	0	0
40	165	32	3	0	0
50	157	38	5	0	0
60	150	43	6	1	0
70	143	48	8	1	0
80	136	53	10	1	0
90	130	56	12	2	0
100	124	60	14	2	0

if a 60 m tower is located in a part of Florida where $N_g = 10$ km^{-2} yr^{-1} then the long-term average downward lightning incidence will be about 0.5 yr^{-1} (assuming $\alpha = 2$ and $\beta = 1$), that is, on average the tower will be struck every other year. The use of Eq. 2.6 would result in a lightning incidence value of about 1 yr^{-1}. For a house (or other similar structure) located in a region characterized by a moderate ground flash density of 4 km^{-2} yr^{-1}, and having an area of 10×20 m^2 and a height of 5 m so that the equivalent attractive distance is about 10 m (extrapolating Eq. 2.9 with $\alpha = 2$ and $\beta = 1$ to a structure height of 5 m), the approximate equivalent attractive area is $30 \times 40 = 1200$ m^2. Such a house is expected to be struck by lightning 1200 m$^2 \times 10^{-6}$ km^2 m$^{-2} \times 4$ km^{-2} yr$^{-1} = 4.8 \times 10^{-3}$ times a year, or about once every 200 years. Another way to think of this lightning incidence is that, in this region, one in 200 houses will be struck each year, on average.

The probability that a structure, represented by its equivalent attractive area A, will be struck exactly 0, 1, 2, 3, ..., n times in T years can be estimated using the Poisson probability distribution,

$$p(n) = (Z^n/n!) \exp(-Z) \qquad (2.10)$$

where $Z = A N_g T = N_d T$, the average number of strikes expected over T years, provided that N_g remains constant. Continuing the previous example, Table 2.4 gives the number of houses, out of a total of 200, expected to be struck by lightning n times over T years, as predicted by Eq. 2.10. Each number is found as the product of the total number of houses, 200, and $p(n)$ from Eq. 2.10, with subsequent rounding off to the nearest integer. In two cases ($n = 1$, $T = 80$ yr and $T = 100$ yr) the rounded-off number was increased by unity in order to assure that the sum of numbers in each row is 200, the total number of houses considered. It follows from Table 2.4 that, for example, over a 60-year period one should expect that 150 houses will not be struck at all, 50 ($= 43 + 6 + 1$) will receive at least one strike, six houses will be struck twice, and one house will be struck three times. If $N_g = 12$ km^{-2} yr^{-1}, which is characteristic of some areas in Florida, then over a period of 60 years (perhaps the lifetime of a house) only 84 houses out of the 200 will not receive any lightning strikes, and 11 houses will be struck three times or more.

Equation 2.10 was used by Eriksson and Meal (1984) and Anderson (1980) for calculating the number of powerline poles or towers not struck by lightning or struck exactly n times in six or eight years, respectively. The calculated results were found to be in good agreement with experimental data.

2.9.3. Upward flashes

Once the incidence of downward lightning N_d is found from Eq. 2.8 using the concept of an equivalent attractive area, the incidence of upward flashes N_u can be determined by subtracting N_d from N, the latter being found using Eq. 2.6. Recall that if the structure height is less than 100 m or so, it is usually assumed that $N_u = 0$. If only the percentage of upward flashes is sought, Eq. 2.7 can be used.

2.10. Summary

The global lightning flash rate is some tens to a hundred per second or so. The maximum flash rate in different storms varies typically from a few to some tens per minute, and the average flash rate per overall storm is about an order of magnitude lower. The majority of lightning discharges do not involve ground, the ratio z of cloud to cloud-to-ground flashes varying over a wide range, from about 2 to 10 or more. The most common measures of lightning incidence to an area are the annual number of thunderstorm days T_D, the annual number of thunderstorm hours T_H, and the annual ground flash density N_g. For a global flash rate of 100 s^{-1} and a global ratio z of cloud to cloud-to-ground flashes equal to 3, the average flash density over the entire Earth's surface is about 6 km^{-2} yr^{-1}, and the ground (land or sea) flash density is about 1.5 km^{-2} yr^{-1}. The annual ground flash density has been estimated from records of lightning flash counters and lightning locating systems. The observed variation in ground flash density from one region to another in the United States is more than two orders of magnitude. Many flashes strike ground at more than one point. When only one location per flash is recorded, the correction factor for measured values of ground flash density needed in order to take into account multiple channel terminations on ground is about 1.7. Lightning activity varies from year to year, and it is possible that the variation of global lightning activity can be used for studying various aspects of global climate change, such as global temperature change. Ground-based objects with heights ranging from about 100 to 500 m experience both downward and upward flashes, the proportion being a function of the height of the object. Structures having heights less than 100 m or so are often assumed to be struck only by downward lightning, while those with heights greater than 500 m or so can be assumed to experience only upward flashes. A house located in a region characterized by a moderate ground flash density of 4 km^{-2} yr^{-1} and having an area of 10×20 m^2 and a height of 5 m is expected to be struck by lightning roughly once every 200 years.

References and bibliography for Chapter 2

Aina, J.I. 1971. Lightning discharge studies in a tropical area. I. Cloud-to-ground discharges. *J. Geomagn. Geoelectr.* **23**: 347–58.

Aiya, S.V.C. 1981. Some characteristics of tropical lightning. *J. Indian Inst. Sci.* **63**(A): 39–59.

Alexander, G.D., Weinman, J.A., Karyampudi, V.M., Olson, W.S., and Lee, A.C.L. 1999. The effect of assimilating rain rates derived from satellites and lightning on forecasts of the 1993 Superstorm. *Mon. Wea. Rev.* **127**: 1433–57.

Anderson, F.J., and Freier, G.D. 1973. Relation of electric fields to thunderstorm days. *J. Geophys. Res.* **78**: 6359–63.

Anderson, R.B. 1971. The lightning discharge. Ph.D. thesis. University of Cape Town, South Africa (available as Special Report Elek 12, I and II, CRIS, PO Box 395, Pretoria, South Africa).

Anderson, R.B. 1977. Measuring techniques. In *Lightning, vol. 1, Physics of Lightning*, ed. R.H. Golde, pp. 437–63, New York: Academic Press.

Anderson, R.B. 1980. Lightning research in Southern Africa. *Trans. S. Afr. Inst. Electr. Eng.* **71**(4): 75–99.

Anderson, R.B., and Eriksson, A.J. 1980. Lightning parameters for engineering application. *Electra* **69**: 65–102.

Anderson, R.B., Van Niekerk, H.R., and Gertenbach, J.J. 1973. Improved lightning–earth flash counter. *Electron. Lett.* **9**: 394–5.

Anderson, R.B., Van Niekerk, H.R., Prentice, S.A., and Mackerras, D. 1979. Improved lightning flash counters. *Electra* **66**: 85–98.

Anderson, R.B., Eriksson, A.J., Kroninger, H., Meal, D.V., and Smith, M.A. 1984a. Lightning and thunderstorm parameters. In *Lightning and Power Systems*, London: IEE Conf. Publ. no. 236, 5 pp.

Anderson, R.B., Van Niekerk, H.R., Kroninger, H., and Meal, D.V. 1984b. Development and field evaluation of a lightning–earth flash counter. *IEE Proc. A* **131**: 118–24.

Austin, G.I., and Stansbury, E.J. 1971. The location of lightning and its relation to precipitation detected by radar. *J. Atmos. Terr. Phys.* **33**: 841–4.

References and bibliography for Chapter 2

Baker, M.B., Christian, H.J., and Latham, J. 1995. A computational study of the relationships linking lightning frequency and other thundercloud parameters. *Q.J.R. Meteor. Soc.* **121**: 1525–48.

Baral, K.N., and Mackerras, D. 1992. The cloud flash-to-ground flash ratio and other lightning occurrence characteristics in Kathmandu thunderstorms. *J. Geophys. Res.* **97**: 931–8.

Baral, K.N., and Mackerras, D. 1993. Positive cloud-to-ground lightning discharges in Kathmandu thunderstorms. *J. Geophys. Res.* **98**: 10 331–40.

Barham, R.A. 1965. Transistorized lightning-flash counter. *Electron. Lett.* **1**: 373.

Barham, R.A., and Mackerras, D. 1972. Vertical aerial CIGRE-type lightning-flash counter. *Electron. Lett.* **8**: 480–2.

Battan, L.J. 1965. Some factors governing precipitation and lightning from convective clouds. *J. Atmos. Sci.* **22**: 79–84.

Baughman, R.G., and Fuquay, D.M. 1970. Hail and lightning occurrence in mountain thunderstorms. *J. Appl. Meteor.* **9**: 657–60.

Bechini, R., Giaiotti, D., Manzato, A., Stel, F., and Micheletti, S. 2001. The June 4th 1999 severe weather episode in San Quirino, Italy: a tornado event? *Atmos. Res.* **56**: 213–32.

Berger, K. 1972. Mesungen und Resultate der Blitzforschung auf dem Monte San Salvatore bei Lugano, der Jahre 1963–1971. *Bull. SEV* **63**: 1403–22.

Berger, K., and Garbagnati, E. 1984. Lightning current parameters. Results obtained in Switzerland and in Italy. In *Proc. URSI Conf., Florence, Italy*, 13 pp.

Bergh, J.E., and Israelsson, S. 1984. On the occurrence of very low frequency (VLF) electromagnetic radiation and its relation to some important physical conditions of the atmosphere. *Arch. Meteor. Geophys. Bioklim.* **B35**: 113–25.

Bielec, Z. 2001. Long-term variability of thunderstorms and thunderstorm precipitation occurrence in Cracow, Poland, in the period 1896–1995. *Atmos. Res.* **56**: 161–70.

Biswas, K.R., and Hobbs, P.V. 1990. Lightning over the Gulf Stream. *Geophys. Res. Lett.* **17**: 941–3.

Black, R.A., and Hallett, J. 1998. The mystery of cloud electrification. *American Scientist* **86**: 526–34.

Blevins, L.L., and Maurwitz, J.D. 1968. Visual observations of lightning in some Great Plains hailstorms. *Weather* **23**: 192–4.

Bluestein, H.B., and MacGorman, D.R. 1998. Evolution of cloud-to-ground lightning characteristics and storm structure in the Spearman, Texas, tornadic supercells of 31 May 1990. *Mon. Wea. Rev.* **126**: 1451–67.

Boccippio, D.J., Wong, C., Williams, E.R., Boldi, R., Christian, H.J., and Goodman, S.J. 1998. Global validation of single-station Schumann resonance lightning location. *J. Atmos. Solar-Terr. Phys.* **60**: 701–12.

Boccippio, D.J., Goodman, S.J., and Heckman, S. 2000a. Regional differences in tropical lightning distributions. *J. Appl. Meteor.* **39**: 2231–48.

Boccippio, D.J., Koshak, W., Blakeslee, R., Driscoll, K., Mach, D., Buechler, D., Boeck, W., Christian, H.J., and Goodman, S.J. 2000b. The Optical Transient Detector (OTD): instrument characteristics and cross-sensor validation. *J. Atmos. Oceanic Technol.* **17**: 441–58.

Boccippio, D.J., Cummins, K.L., Christian, H.J., and Goodman, S.J. 2001. Combined satellite and surface-based estimation of the intracloud/cloud-to-ground lightning ratio over the continental United States. *Mon. Wea. Rev.* **129**: 108–22.

Branick, M.L., and Doswell, C.A. III 1992. An observation of the relationship between supercell structure and lightning ground strike polarity. *Wea. Forecast.* **7**: 143–9.

Brazier-Smith, P.R., Jennings, S.A., and Latham, J. 1973. Increased rates of rainfall production in electrified clouds. *Q.J.R. Meteor. Soc.* **99**: 776–9.

Brook, M. 1992. Breakdown electric fields in winter storms. *Res. Lett. Atmos. Electr.* **12**: 47–52.

Brook, M., and Kitagawa, N. 1960. Electric-field changes and the design of lightning-flash counters. *J. Geophys. Res.* **65**: 1927–31.

Brook, M., Nakano, M., Krehbiel, P., and Takeuti, T. 1982. The electrical structure of the Hokuriku winter thunderstorms. *J. Geophys. Res.* **87**: 1207–15.

Brook, M., Henderson, R.W., and Pyle, R.B. 1989. Positive lightning strokes to ground. *J. Geophys. Res.* **94**: 13 295–303.

Brooks, C.E.P. 1925. The distribution of thunderstorms over the globe. *Geophys. Mem.* (Air Ministry, Meteorological Office, London) **24**: 147–64.

Buechler, D.E., Wright, P.D., and Goodman, S.J. 1990. Lightning/rainfall relationships during COHMEX. Preprint, *Proc. 16th Conf. Severe Local Storms, Kananaskis Park, AB, Canada*, pp. 710–14, Am. Meteor. Soc., Boston, Massachusetts.

Buechler, D.E., Driscoll, K.T., Goodman, S.J., and Christian, H.J. 2000. Lightning activity within a tornadic thunderstorm observed by the Optical Transient Detector (OTD). *Geophys. Res. Lett.* **27**: 2253–6.

Bunn, C.C. 1968. Application of electric field change measurements to the calibration of a lightning flash counter. *J. Geophys. Res.* **73**: 1907–12.

Burke, W.J., Aggson, T.L., Maynard, N.C., Hoegy, W.R., Hoffman, R.A., Candy, R.M., Liebrecht, C., and Rodgers, E. 1992 Effects of a lightning discharge detected by the DE 2 satellite over Hurricane Debbie. *J. Geophys. Res.* **97**: 6359–67.

Byerley, L.G., Cummins, K.L., Tuel, J., Hagaberg, D.J., and Bush, W. 1995. The measurement and use of lightning ground flash density. In *Proc. Int. Aerospace and Ground Conf. on Lightning and Static Electricity, Williamsburg, Virginia*, pp. 61/1–12.

Camp, J.P., Watson, A.I., and Fuelberg, H.E. 1998. The diurnal distribution of lightning over North Florida and its relation to the prevailing low-level flow. *Wea. Forecast.* **13**: 729–39.

Canosa, E.F., and List, R. 1993. Measurements of inductive charges during drop breakup in horizontal electric fields. *J. Geophys. Res.* **98**: 2619–26.

Canosa, E.F., List, R., and Stewart, R.E. 1993. Modeling of inductive charge separation in rainshafts with variable vertical electric fields. *J. Geophys. Res.* **98**: 2627–33.

Carey, L.D., and Rutledge, S.A. 1996. A multiparameter radar case study of the microphysical and kinematic evolution of a lightning producing storm. *J. Meteor. Atmos. Phys.* **59**: 33–64.

Carey, L.D., and Rutledge, S.A. 1998. Electrical and multiparameter radar observations of a severe hailstorm. *J. Geophys. Res.* **103**: 13 979–14 000.

Carey, L.D., and Rutledge, S.A. 2000. The relationship between precipitation and lightning in tropical island convection: a C-band polarimetric radar study. *Mon. Wea. Rev.* **128**: 2687–710.

Carte, A.E.M. and De Jager, H.C.G. 1979. Multiple-stroke flashes of lightning. *J. Atmos. Terr. Phys.* **41**: 95–101.

Carte, A.E., and Kidder, R.E. 1977. Lightning in relation to precipitation. *J. Atmos. Terr. Phys.* **39**: 139–48.

Cecil, D.J., and Zipser, E.J. 1999. Relationships between tropical cyclone intensity and satellite-based indicators of inner core convection: 85-GHz ice-scattering signature and lightning. *Mon. Wea. Rev.* **127**: 103–23.

Chang, D.-E., Weinman, J.A., Morales, C.A., and Olson, W.S. 2001. The effect of spaceborne microwave and ground-based continuous lightning measurements on forecasts of the 1998 Groundhog Day storm. *Mon. Wea. Rev.* **129**: 1809–33.

Changery, M.J. 1981. National thunderstorm frequencies for the contiguous United States. USNRC Report NUREG/CR-2252, 57 pp.

Changnon, S.A. 1985. Secular variations in thunder-day frequencies in the twentieth century. *J. Geophys. Res.* **90**: 6181–94.

Changnon, S.A. 1988a. Climatography of thunder events in the conterminous United States. Part I: Temporal aspects. *J. Clim.* **1**: 389–98.

Changnon, S.A. 1988b. Climatography of thunder events in the conterminous United States. Part II: Spatial aspects. *J. Clim.* **1**: 399–405.

Changnon, S.A. 1992. Temporal and spatial relations between hail and lightning. *J. Appl. Meteor.* **31**: 587–604.

Changnon, S.A. 1993. Relationships between thunderstorms and cloud-to-ground lightning in the United States. *J. Appl. Meteor.* **32**: 88–105.

Changnon, S.A. 2000. Damaging thunderstorm activity in the United States. *Bull. Am. Meteor. Soc.* **82**(4): 597–608.

Changnon, S.A. 2001a. Assessment of the quality of thunderstorm data at first-order stations. *J. Appl. Meteor.* **40**: 783–94.

Changnon, S.A. 2001b. Thunderstorm rainfall in the conterminous United States. *Bull. Am. Meteor. Soc.* **82**: 1925–40.

Changnon, S.A., Changnon, D., and Pyle, R.B. 1988. Thunder events and cloud-to-ground lightning frequencies. *J. Geophys. Res.* **93**: 9495–502.

Chen, T.-C. 1987. Comments on "Global distribution of midnight lightning: September 1977 to August 1978". *Mon. Wea. Rev.* **115**: 3202.

Cherna, E.V., and Stansbury, E.J. 1986. Sferics rate in relation to thunderstorm dimensions. *J. Geophys. Res.* **91**: 8701–7.

Cheze, J.-L., and Sanvageot, H. 1997. Area-average rainfall and lightning activity. *J. Geophys. Res.* **102**: 1707–15.

Christian, H.J., and Latham, J. 1998. Satellite measurements of global lightning. *Q.J.R. Meteor. Soc.* **124**: 1771–3.

Christian, H.J., Blakeslee, R.J., and Goodman, S.J. 1989. The detection of lightning from geostationary orbit. *J. Geophys. Res.* **94**: 13 329–37.

Christian, H.J., Blakeslee, R.J., and Goodman, S.J. 1992. Lightning imaging sensor (LIS) for the Earth observing system. NASA Technical Memorandum 4350, 36 pp.

Christian, H.J., Driscoll, K.T., Goodman, S.J., Blakeslee, R.J., Mach, D.A., and Buechler, D.E. 1996a. The Optical Transient Detector (OTD). In *Proc. 10th Int. Conf. on Atmospheric Electricity, Osaka, Japan*, pp. 368–71.

Christian, H.J., Driscoll, K.T., Goodman, S.J., Blakeslee, R.J., Mach, D.A., and Buechler, D.E. 1996b. Seasonal variation and distribution of lightning activity. *Eos, Trans. AGU* **77**: F80.

Cianos, N., and Pierce, E.T. 1972. A ground-lightning environment for engineering usage. Stanford Research Institute Project 1834, Technical Report 1, Stanford Research Institute, Menlo Park, California, 136 pp.

CIGRE Document 63 1991. Guide to procedures for estimating the lightning performance of transmission lines, October.

CIGRE Document 172 2000. Characterization of lightning for applications in electric power systems, December.

Clodman, S., and Chisholm, W. 1996. Lightning flash climatology in the southern Great Lakes region. *Atmos. Ocean* **34**(2): 345–77.

Colson, D. 1960. High level thunderstorms of July 31–August 1, 1959. *Mon. Wea. Rev.* **88**: 279–85.

Cooray, V. 1986. Response of CIGRE and CCIR lightning flash counters to the electric field changes from lightning: a theoretical study. *J. Geophys. Res.* **91**: 2835–42.

Cooray, V., and Jayaratne, K.P.S.C. 1994. Characteristics of lightning flashes observed in Sri Lanka in the tropics. *J. Geophys. Res.* **99**: 21 051–6.

Cooray, V., and Perez, H. 1994. Some features of lightning flashes observed in Sweden. *J. Geophys. Res.* **99**: 10 683–8.

Crook, N.A. 2001. Understanding Hector: The dynamics of island thunderstorms. *Mon Wea. Rev.* **129**: 1550–63.

Crozier, C.L., Herscovitch, H.N., and Scott, J.W. 1988. Some observations and characteristics of lightning ground discharges in southern Ontario. *Atmos. Ocean* **26**(3): 399–436.

Cummins, K.L., Bardo, E.A., Hiscox, W.L., Pyle, R.B., and Pifer, A.E. 1995. NLDN '95: a combined TOA/MDF technology upgrade of the US National Lightning Detection Network. In *Proc. Int. Aerospace and Ground Conf. on Lightning and Static Electricity, Williamsburg, Virginia*, pp. 72/1–15.

Cummins, K.L., Murphy, M.J., Bardo, E.A., Hiscox, W.L., Pyle, R.B., Pifer, A.E. 1998. A combined TOA/MDF technology upgrade of the US National Lightning Detection Network. *J. Geophys. Res.* **103**: 9035–44.

Curran, E.B., and Rust, W.D. 1992. Positive ground flashes produced by low-precipitation thunderstorms in Oklahoma on 26 April 1984. *Mon. Wea. Rev.* **120**: 544–53.

Darveniza, M., and Uman, M.A. 1984. Research into lightning protection of distribution systems II – results from Florida field work 1978 and 1979. *IEEE Trans.* **PAS-103**: 673–82.

Davis, M.H., Brook, M., Christian, H., Heikes, B.G., Orville, R.E., Park, C.G., Roble, R.G., and Vonnegut, B. 1983. Some scientific objectives of a satellite-borne lightning mapper. *Bull. Am. Meteor. Soc.* **64**: 114–19.

de la Rosa, F., and Velazquez, R. 1989. Review of ground flash density measuring devices regarding power system applications. *IEEE Trans. Power Del.* **4**: 921–37.

De Pablo, F., and Soriano, L.R. 2002. Relationship between cloud-to-ground lightning flashes over the Iberian Peninsula and sea surface temprature. *Q.J.R. Meteor. Soc.* **128**: 173–83.

Diniz, J.H., Carvalho, A.M., Cherchiglia, L.C.L., de Souza, V.J., Cazetta Filho, A., and Nascimento, C.A.M. 1996. Ground flash densities in Minas Gerais, Brasil. In *Proc. 23rd Int. Conf. on Lightning Protection, Florence, Italy*, pp. 224–9.

Dodge, P.P., and Burpee, R.W. 1993. Characteristics of rainbands, radar echoes, and lightning near the North Carolina coast during GALE. *Mon. Wea. Rev.* **121**: 1936–55.

Doswell, C.A. III, and Brooks, H.E. 1993. Comments on "Anomalous cloud-to-ground lightning in an F5 tornado-producing supercell thunderstorm on 28 August 1990." *Bull. Am. Meteor. Soc.* **75**: 2281–8.

Dotzek, N., Höller, H., Théry, C., and Fehr, T. 2001. Lightning evolution related to radar-derived microphysics in the 21 July 1998 EULINOX supercell storm. *Atmos. Res.* **56**: 335–54.

Driscoll, K.T., Christian, H.J., Goodman, S.J., Bakeslee, R.J., Boccippio, D.J. 1996. Diurnal global lightning distribution as observed by the optical transient detector. *Eos, Trans. AGU* **77**: F92.

Dulzon, A.A. 1992. Lightning performance of power lines – comparison of calculated and observed data. In *Proc. 21st Int. Conf. on Lightning Protection, Berlin, Germany*, pp. 414–18.

Dulzon, A.A. 1996. Lightning as a source of forest fires. *Combustion, Explosion, and Shock Waves* **32**: 587–94.

Dulzon, A.A., and Rakov, V.A. 1990. Spatial inhomogeneity in thunderstorm activity: some possible explanations. In *Proc. 20th Int. Conf. on Lightning Protection, Interlaken, Switzerland*, Paper 1.6P, 3 pp.

Dulzon, A.A., and Rakov, V.A. 1991. A study of power line lightning performance. In *Proc. 7th Int. Symp. on High Voltage Engineering, Dresden, Germany*, pp. 57–60.

Easterling, D.R. 1990. Persistent patterns of thunderstorm activity in the central United States. *J. Clim.* **3**: 1380–9.

Elegbede, A.I., Aina, J.I., and Hudson, R.M. 1991. A logic-controlled lightning flash counter. *Mea. Sci. Tech. UK* **2**: 1138–41.

Elsom, D.M., Meaden, G.T., Reynolds, D.J., Rowe, M.W., and Webb, J.D.C. 2001. Advances in tornado and storm research in the United Kingdom and Europe: the role of the Tornado and Storm Research Organisation. *Atmos. Res.* **56**: 19–29.

Engholm, C.D., Williams, E.R., and Dole, R.M. 1990. Meteorological and electrical conditions associated with positive cloud-to-ground lightning. *Mon. Wea. Rev.* **118**: 470–87.

Eriksson, A.J. 1978. Lightning and tall structures. *Trans. S. Afr. Inst. Electr. Eng.* **69**(8): 238–52.

Eriksson, A.J. 1979. The lightning ground flash – an engineering study. Ph.D. thesis, University of Natal, Pretoria, South Africa. Available as CSIR Special Report ELEK 189 from National Electrical Engineering Research Institute, CSIR, PO Box 39, Pretoria 0001, South Africa.

Eriksson, A.J. 1987. The incidence of lightning strikes to power lines. *IEEE Trans.* **PWRD-2**: 859–70.

Eriksson, A.J., and Meal, D.V. 1984. The incidence of direct lightning strikes to structures and overhead lines. In *Lightning and Power Systems*, London: IEE Conf. Publ. no. 236, pp. 67–71.

Eriksson, A.J., Niekerk, H.R., van Bourn, G.W., and van Zyl, B.B. 1978. Lightning ground flash multiple stroke discriminator. *Proc. IEEE* **125**: 555–7.

Finke, U. 1999. Space–time correlations of lightning distributions. *Mon. Wea. Rev.* **127**: 1850–61.

Forbes, G.S., and Bluestein, H.B. 2001. Tornadoes, tornadic thunderstorms, and photogrammetry: a review of the contribution by T.T. Fujita. *Bull. Am. Meteor. Soc.* **82**: 73–96.

Freier, G.D. 1978. A 10-year study of thunderstorm electric fields. *J. Geophys. Res.* **83**: 1373–6.

Fuquay, D.M., and Baughman, R.G. 1969. Project skyfire-lightning research. Final Report to the National Science Foundation of the United States for 1965–7, US Bureau of Land Management, Boise, Idaho.

Gabriel, K.R., and Changnon, S.A. 1989. Temporal features in thunder days in the United States. *Climatic Change* **15**: 455–77.

Gallagher, F.W. 2000. Distant green thunderstorms – Fraser's theory revisited. *J. Appl. Meteor.* **39**: 1754–61.

Gallagher, F.W. 2001. Ground reflection and green thunderstorms. *J. Appl. Meteor.* **40**: 776–82.

Garbagnati, E., Guidice, E., Lo Piparo, G.B., and Magagnoli, U. 1974. Survey of the characteristics of lightning stroke currents in Italy – results obtained in the years from 1970 to 1973. ENEL Report R5/63-27.

Gay, M.J., Griffiths, R.F., Latham, J., and Saunders C.P.R. 1974. The terminal velocities of charged raindrops and cloud droplets falling in strong electric fields. *Q.J.R. Meteor. Soc.* **100**: 682–7.

Geotis, S.G., Williams, E.R., and Liu, C. 1991. Reply. *J. Atmos. Sci.* **48**: 371–2.

Giaiotti, D., and Stel, F. 2001. A comparison between subjective and objective thunderstorm forecasts. *Atmos. Res.* **56**: 111–26.

Golde, R.H. 1966. A lightning flash counter. *Electron. Eng.* **38**: 164–6.

Golden, J.H., and Bluestein, H.B. 1994. The NOAA–National Geographic Society waterspout expedition (1993). *Bull. Am. Meteor. Soc.* **75**: 2281–8.

Goodman, S.J., and MacGorman, D.R. 1986. Cloud-to-ground lightning activity in mesoscale convective complexes. *Mon. Wea. Rev.* **114**: 2320–8.

Goodman, S.J., Buechler, D.E., and Meyer, P.J. 1988a. Convective tendency images derived from a combination of lightning and satellite data. *Wea. Forecast.* **3**: 173–88.

Goodman, S.J., Buechler, D.E., Wright, P.D., and Rust, W.D. 1988b. Lightning and precipitation history of a microburst-producing storm. *Geophys. Res. Lett.* **15**: 1185–8.

Goodman, S.J., Buechler, D.E., Knupp, K., Driscoll, K., and McCaul, E.W. 2000. The 1997–98 El Niño event and related wintertime lightning variations in southeastern United States. *Geophys. Res. Lett.* **27**: 541–4.

Gorin, B.N. 1972. Lightning discharges to the Ostankino television tower. *Elecktr.* **2**: 24–9.

Gorin, B.N., Levitov, V.I., and Shkilev, A.V. 1976. Distinguishing features of lightning strokes to high constructions. In *Proc. 4th Int. Conf. on Gas Discharges*, IEE Conf. Publ. no. **143**, pp. 271–3.

Grecu, M., Anagnostou, E.N., and Adler, R.F. 2000. Assessment of the use of lightning information in satellite infrared rainfall estimation. *J. Hydrometeor.* **1**: 211–21.

Gremillion, M.S., and Orville, R.E. 1999. Thunderstorm characteristics of cloud-to-ground lightning at the Kennedy Space Center, Florida: a study of lightning initiation signature as indicated by the WSR-88D. *Wea. Forecast.* **14**: 640–9.

Grosh, R.C. 1978. Lightning and precipitation – the life history of isolated thunderstorms. Preprint from *Proc. Conf. on Cloud Physics and Atmospheric Electricity*, pp. 617–24, Am. Meteor. Soc.

Hagemeyer, B.C. 1997. Peninsular Florida tornado outbreaks. *Wea. Forecast.* **12**: 399–427.

Hale, L.C. 1987. Lightning triggering and synchronization. *Nature* **329**: 769.

Hamid, E.Y., Kawasaki, Z.-I., and Mardiana, R. 2001. Impact of the 1997–98 El Niño event on lightning activity over Indonesia. *Geophys. Res. Lett.* **28**: 147–50.

Harris, D.J., and Salman, Y.E. 1972. The measurement of lightning characteristics in northern Nigeria. *J. Atmos. Terr. Phys.* **34**: 775–86.

Heckman, S.J., Williams, E., and Boldi, B. 1998. Total global lightning inferred from Schumann resonance measurements. *J. Geophys. Res.* **103**: 31 775–9.

Hidayat, S., and Ishii, M. 1998. Spatial and temporal distribution of lightning activity around Java. *J. Geophys. Res.* **103**: 14 001–9.

Hidayat, S., and Ishii, M. 1999. Diurnal validation of lightning characteristics around Java Island. *J. Geophys. Res.* **104**: 24 449–54.

Hidayat, S., Sirait, K.T., Pakpahan, P.M., Ishii, M., and Hojo, J. 1999. Lightning characteristics on Java Island, observed by lightning location network. In *Proc. 11th Int. Symp. on High Voltage Engineering*, IEE Publ. no. 467, vol. 2, London, United Kingdom, pp. 192–5.

Hill, R.D. 1988. Interpretation of bipole pattern in a mesoscale storm. *Geophys. Res. Lett.* **15**: 643–4.

Hiser, H.W. 1973. Sferics and radar studies of south Florida thunderstorms. *J. Appl. Meteor.* **12**: 479–83.

Hodanish, S., Sharp, D., Collins, W., Paxton, C., and Orville, R.E. 1997. A 10-yr monthly lightning climatology of Florida: 1986–95. *Wea. Forecast.* **12**: 439–48.

Hohl, R., and Schiesser, H.-H. 2001. Cloud-to-ground lightning activity in relation to the radar-derived hail kinetic energy in Switzerland. *Atmos. Res.* **56**: 375–96.

Hojo, J., Ishii, M., Kawamura, T., Suzuki, F., Komuro, H., and Shiogama, M. 1988. Characteristics and evaluation of lightning field waveforms. *Electrical Engineering in Japan* **108**: 55–65 (translated from *Denki Gakkai Ronbunshi*, **108B**(4), April 1988, 165–72).

Hojo, J., Ishii, M., Kawamura, T., Suzuki, F., Komuro, H., and Shiogama, M. 1989. Seasonal variation of cloud-to-ground lightning flash characteristics in the coastal area of the Sea of Japan. *J. Geophys. Res.* **94**: 13 207–12.

Holle, R.L., and Bennett, S. P. 1997. Lightning ground flashes associated with summer 1990 flash floods and stream flow in Tucson, Arizona: an exploratory study. *Mon Wea. Rev.* **125**: 1526–36.

Holle, R.L., and Lopez, R.E. 1993. Overview of real-time lightning detection systems and their meteorological uses. NOAA Technical Memorandum ERL NSSL-102, National Severe Storms Laboratory, Norman, Oklahoma, 68 pp.

Holle, R.L., and Maier, M.W. 1982. Radar echo height related to cloud–ground lightning in South Florida. Preprint, *Proc. 12th Conf. on Severe Local Storms, San Antonio, Texas*, Am. Meteor. Soc., Boston, Massachusetts, pp. 330–3.

Holle, R.L., and Watson, A.I. 1996. Lightning during two central U.S. winter precipitation events. *Wea. Forecast.* **11**: 599–614.

Holle, R.L., Watson, A.I., Lopez, R.E., MacGorman, D.R., Ortiz, R., and Otto, W.D. 1994. The life cycle of lightning and severe weather in a 3–4 June 1985 PRE-STORM mesoscale convective system. *Mon. Wea. Rev.* **122**: 1798–808.

Holle, R.L., Watson, A.I., and Purcell, D. 1996. A well-organized thunderstorm outflow in Florida. *National Weather Digest* **23**: 18–20.

Holzer, R.E. 1953. Simultaneous measurement of sferics signals and thunderstorm activity. In *Thunderstorm Electricity*, ed. H.R. Byers, pp. 267–5. Chicago, Illinois: University of Chicago Press.

Holzworth, R.H. 1991. Atmospheric electrodynamics in the US: 1987–90. In *Rev. Geophys., Suppl.*, 115–20, US National Report to Int. Union of Geodesy and Geophysics 1987–90.

Hondl, K.D., and Eilts, M.D. 1994. Doppler radar signatures of developing thunderstorms and their potential to indicate the onset of cloud-to-ground lightning. *Mon. Wea. Rev.* **122**: 1818–36.

Hopkins, C.D. 1973. Lightning as background noise for communication among electric fish. *Nature* **242**: 268–70.

Horner, F. 1960. The design and use of instruments for counting local lightning flashes. *Proc. IEE* **107**: 321–30.

Houghton, J.T. 1997. *The Physics of Atmospheres*, 2nd edition, 271 pp., Cambridge, UK: Cambridge University Press.

Houze, R.A., Small, B.F., and Dodge, P. 1990. Mesoscale organization of springtime rainstorms in Oklahoma. *Mon. Wea. Rev.* **118**: 613–54.

Huffines, G.R., and Orville, R.E. 1999. Lightning ground flash density and thunderstorm duration in the continental United States: 1989–96. *J. Appl. Meteor.* **38**: 1013–9.

Hunter, S.M., Schuur, T.J., Marshall, T.C., and Rust, W.D. 1992. Electric and kinematic structure of the Oklahoma mesoscale convective system on 7 June 1989. *Mon. Wea. Rev.* **120**: 2226–39.

Hunter, S.M., Underwood, S.J., Holle, R.L., and Mote, T.L. 2001. Winter lightning and heavy frozen precipitation in the southeast United States. *Wea. Forecast.* **16**: 478–90.

Intergovernmental Panel on Climate Change (IPCC). 1990. *The IPCC Scientific Assessment*, ed. J.T. Houghton, G.J. Jenkins, and J.J. Ephraums, New York: Cambridge University Press.

Ishii, M., and Hojo, J. 1989. Statistics on fine structure of cloud-to-ground lightning field waveforms. *J. Geophys. Res.* **94**: 13 267–74.

Ishii, M., Kawamura, T., Hojo, J., and Iwaizumi, T. 1981. Ground flash density in winter thunderstorm. *Res. Lett. Atmos. Electr.* **1**: 105–8.

Israel, H. 1973. *Atmospheric Electricity, vol. II, Fields, Charges, Currents*. Published for the National Science Foundation by the Israel Program for Scientific Translation, Jerusalem, 796 pp.

Israelsson, S., Schutte, T., Pisler, E., and Lundquist, S. 1987. Increased occurrence of lightning flashes in Sweden during 1986. *J. Geophys. Res.* **92**: 10 996–8.

Jacobson, E.A., and Krider, E.P. 1976. Electrostatic field changes produced by Florida lightning. *J. Atmos. Sci.* **33**: 103–17.

Janischaewskyj, W., and Chisholm, W.A. 1992. Lightning ground flash density measurements in Canada, March 1, 1984 to December 31, 1991. Technical Report 179 T 382, prepared by the University of Toronto, 106 pp., Canadian Electrical Association.

Jayaratne, E.R. 1993. Conditional instability and lightning incidence in Gaborone, Botswana, *Meteor. Atmos. Phys.* **52**: 169–75.

Jayaratne, E.R., and Ramachandran, V. 1998. A five-year study of lightning activity using a CGR3 flash counter in Gaborone, Botswana. *Meteor. Atmos. Phys.* **66**: 235–41.

Jayaratne, E.R., and Saunders, C.P.R. 1984. The rain gush, lightning and the lower positive charge center in thunderstorms. *J. Geophys. Res.* **89**: 11 816–18.

Jayaratne, E.R., Ramachandran, V., and Devan, R.S. 1995. Observations of lightning flash rates and rain-gushes in Gaborone, Botswana. *J. Atmos. Terr. Phys.* **57**: 325–31.

Johns, R.H., and Doswell, C.A. III 1992. Severe local storms forecasting. *Wea. Forecast.* **7**: 588–612.

Johnson, R.L. 1980. Bimodal distribution of atmospherics associated with tornadic events. *J. Geophys. Res.* **85**: 5519–22.

Johnson, R.L., and Goodman, S.J. 1984. Atmospherics electrical activity associated with hurricane Alicia. Preprint, *Proc. 7th Int. Conf. on Atmospheric Electricity, Albany*, New York, pp. 295–8, Am. Meteor. Soc.

Jones, H.L. 1951. A sferic method of tornado identification and tracking. *Bull. Am. Meteor. Soc.* **32**: 380–5.

Jones, H.L., and Hess, P.N. 1952. Identification of tornadoes by observation of waveform atmospherics. *Proc. IRE* **40**: 1049–52.

Jones, P.D., Wigley, M.L., and Kelly, P.M. 1982. Variations in surface air temperature, 1, Northern hemisphere, 1881–1980. *Mon. Wea. Rev.* **110**: 59–70.

Jonsson, H.H., and Vonnegut, B. 1993. Miniature vortices produced by electrical corona. *J. Geophys. Res.* **98**: 5245–8.

Kane, R.J. 1991. Correlating lightning to severe local storms in the northeastern United States. *Wea. Forecast.* **6**: 3–12.

Kane, R.J. 1993a. Lightning–rainfall relationships in an isolated thunderstorm over the mid-Atlantic states. *Natl. Wea. Digest* **18**: 2–14.

Kane, R.J. 1993b. A case study of lightning-radar characteristics in a mesoscale convective complex. Preprint, *Proc. 17th Conf. on Severe Local Storms*, pp. 816–22, Am. Meteor. Soc.

Keenan, T., Rutledge, S., Carbone, R., Wilson, J., Takahashi, T., May, P., Tapper, N., Platt, M., Hacker, J., Sekelsky, S., Moncrieff, M., Saito, K., Holland, G., Crook, A., and Gage, K. 2000. The Maritime Continent Thunderstorm Experiment (MCTEX): overview and some results. *Bull. Am. Meteor. Soc.* **81**: 2433–55.

Keighton, S.J., Bluestein, H.B., and MacGorman, D.R. 1991. The evolution of a severe mesoscale convective system: cloud-to-ground lightning location and storm structure. *Mon. Wea. Rev.* **119**: 1533–56.

Kettler, C.J. 1940. Cameras designed for lightning studies. *Photo Tech.*, May: 38–43.

King, T.S., and Balling, R.C. 1994. Diurnal variations in Arizona monsoon lightning data. *Mon. Wea. Rev.* **122**: 1659–64.

Kinzer, G.D. 1974. Cloud-to-ground lightning versus radar reflectivity in Oklahoma thunderstorms. *J. Atmos. Sci.* **31**: 787–99.

Kitagawa, N. 1989. Long-term variations in thunderday frequencies in Japan. *J. Geophys. Res.* **94**: 13 183–9.

Kitagawa, N., Brook, M., and Workman, E.J. 1962. Continuing currents in cloud-to-ground lightning discharges. *J. Geophys. Res.* **67**: 637–47.

Kitterman, C.G. 1980. Characteristics of lightning from frontal system thunderstorms. *J. Geophys. Res.* **85**: 5503–5.

Klossovsky, A. 1892. Zur Frage über die Verbreitung der Gewitter auf der Erdoberfläche. *Revue Meteor. du Sudouest de la Russie*, **3**: 37.

Klossovsky, A.V. 1918. *Fundamentals of Meteorology*, 3rd edition, Odessa, Russia: Mathesis.

Knapp, D.I., 1994. Using cloud-to-ground lightning data to identify tornadic thunderstorm signatures and nowcast severe weather. *Natl. Wea. Assoc. Digest* **19**: 35–42.

Knupp, K.R., Geerts, B., and Goodman, S.J. 1998. Analysis of a small, vigorous mesoscale convective system in a low-shear environment. Part I: Formation, radar echo structure, and lightning behavior. *Mon Wea. Rev.* **126**: 1812–36.

Kohl, D.A. 1962. Sferics amplitude distribution jump identification of a tornado event. *Mon. Wea. Rev.* **90**: 451–6.

Kohl, D.A., and Miller, J.E. 1963. 500 kc/sec sferics analysis of severe weather events. *Mon. Wea. Rev.* **91**: 207–14.

Kolokolov, V.P., and Pavlova, G.P. 1972. Relations between some thunderstorm parameters. In *Trudy GGO*, no. **277**, *Studies in Atmospheric Electricity*, Gidrometeoizdat, Leningrad, USSR (English translation, Jerusalem, 1974, pp. 33–5).

Kononov, I.I., and Petrenko, I.A. 1996. Experience of lightning location systems elaboration in Russia. In *Proc. 23rd Int. Conf. on Lightning Protection, Florence, Italy*, pp. 236–40.

Kononov, I.I., Petrenko, I.A., and Snegurov, V.S. 1986. *Radiotechnical Methods for Locating Thunderstorms*, 222 pp., Leningrad, Russia: Gidrometeoizdat.

Koshak, W.J., and Solakiewicz, R.J. 1999. Electro-optic lightning detector. *Appl. Optics* **38**: 4623–34.

Koshak, W.J., Stewart, M.F., Christian, H.J., Bergstrom, J.W., Hall, J.M., and Solakiewicz, R.J. 2000a. Laboratory calibration of the Optical Transient Detector and the Lightning Imaging Sensor. *J. Atmos. Oceanic Technol.* **17**: 905–15.

Kotaki, M., and Katoh, C. 1983. The global distribution of thunderstorm activity observed by the Ionosphere Sounding Satellite (ISS-b). *J. Atmos. Terr. Phys.* **45**: 833–47.

Kotaki, M., Kuriki, I., Katoh, C., and Sugiuchi, H. 1981a. Global distribution of thunderstorm activity observed with ISS-b. *J. Radio Res. Lab. Tokyo* **28**: 49–71.

Kotaki, M., Sugiuchi, H., Katoh, C. 1981b. World distribution of thunderstorm activity obtained from Ionosphere Sounding Satellite-b observations June 1978 to May 1980. Radio Research Laboratories, Ministry of Posts and Telecommunications, Japan.

Krehbiel, P.R., Tennis, R., Brook, M., Holmes, E.W., and Comes, R.A. 1984. A comparative study of the initial sequence of lightning in a small Florida thunderstorm. Preprint, *Proc. 7th Int. Conf. on Atmospheric Electricity*, Albany, New York, pp. 279–85.

Krehbiel, P.R., Thomas, R.J., Rison, W., Hamlin, T., Harlin, J., and Davis, M. 2000. GPS-based mapping system reveals lightning inside storms. *Eos, Trans. Am. Geophys. Union* **81**(3): 21–5.

Krider, E.P. 1988. Spatial distribution of lightning strikes to ground during small thunderstorms in Florida. In *Proc. 1988 Int. Aerospace and Ground Conf. on Lightning and Static Electricity*, pp. 318–23, NOAA Special Report.

Kumar, P.P., Manohar, G.K., and Kandalgaonkar, S.S. 1995. Global distribution of nitric oxide produced by lightning and its seasonal variation. *J. Geophys. Res.* **100**: 11 203–8.

Lang, T.J., Rutledge, S.A., Dye, J.E., Venticinque, M., Laroche, P., and Defer, E. 2000. Anomalously low negative cloud-to-ground lightning flash rates in intense convective storms observed during STERAO-A. *Mon. Wea. Rev.* **128**: 160–73.

Larsen, H.R., and Stansbury, E.J. 1974. Association of lightning flashes with precipitation cores extending to height 7 km. *J. Atmos. Terr. Phys.* **36**: 1547–53.

Lethbridge, M.D. 1990. Thunderstorms, cosmic rays, and solar-lunar influences. *J. Geophys. Res.* **95**: 13 645–9.

Levin, Z., Yair, Y., and Ziv, B. 1996. Positive cloud-to-ground flashes and wind shear in Tel-Aviv thunderstorms. *Geophys. Res. Lett.* **23**: 2231–4.

Lhermitte, R., and Krehbiel, P.R. 1979. Doppler radar and radio observations of thunderstorms. *IEEE Trans. Geosci. Electron.* GE-17: 162–71.

Lhermitte, R., and Williams, E.R. 1983. Cloud electrification. *Rev. Geophys.* **21**: 984–92.

Lind, M.A., Hartman, J.S., Takle, E.S., and Stanford, J.L. 1972. Radio noise studies of several severe weather events in Iowa in 1971. *J. Atmos. Sci.* **29**: 1220–3.

Livingston, J.M., and Krider, E.P. 1978. Electric fields produced by Florida thunderstorms. *J. Geophys. Res.* **83**: 385–401.

Livington, E.S., Nielsen-Gammon, J.W., and Orville, R.E. 1996. A climatology, synoptic assessment and thermodynamic evaluation for cloud-to-ground lightning in Georgia: a study for the 1996 Summer Olympics. *Bull. Am. Meteor. Soc.* **77**: 1483–95.

López, L., Marcos, J.L., Sánchez. J.L., Castro. A., and Fraile, R. 2001. CAPE values and hailstorms in northwestern Spain. *Atmos. Res.* **56**: 147–60.

Lopez, R.E., and Aubagnac, J.-P. 1997. The lightning activity of a hailstorm as a function of changes in its microphysical characteristics inferred from polarimetric radar observations. *J. Geophys. Res.* **102**: 16 799–813.

Lopez, R.E., and Holle, R.L. 1986. Diurnal and spatial variability of lightning activity in Northeastern Colorado and Central Florida during the Summer. *Mon. Wea. Rev.* **114**: 1288–312.

Lopez, R.E., and Holle, R.L. 1999. The distance between successive lightning flashes. NOAA Technical Memorandum ERL NSSL-105, National Severe Storms Laboratory, Norman, Oklahoma, 29 pp.

Lopez, R.E., Ortiz, R., Otto, W.D., and Holle, R.L. 1991. The lightning activity and precipitation yield of convective cloud systems in Central Florida. Preprint, *Proc. 25th Int. Conf. on Radar Meteorology, Paris, France*, pp. 907–10, Am. Meteor. Soc.

Lopez, R.E., Holle, R.L., Ortiz, R., and Watson, A.I. 1992. Detection efficiency losses of networks of direction finders due to flash signal attenuation with range. In *Proc. 15th Int. Aerospace and Ground Conf. on Lightning and Static Electricity, Atlantic City, N.J.*, pp. 75/1–18.

Lopez, R.E., Holle, R.L., Watson, A.I., and Skindlov, J. 1997. Spatial and temporal distributions of lightning over Arizona from a power utility perspective. *J. Appl. Meteor.* **36**: 825–31.

Lucas, C., and Orville, R.E. 1996. TOGA COARE: oceanic lightning. *Mon. Wea. Rev.* **124**: 2077–82.

Lyons, W.A., and Keen, C.S. 1994. Observations of lightning in convective supercells within tropical storms and hurricanes. *Mon. Wea. Rev.* **122**: 1897–916.

Lyons, W.A., Uliasz, M., and Nelson, T.E. 1998. Large peak current cloud-to-ground lightning flashes during the summer months in the contiguous United States. *Mon. Wea. Rev.* **126**: 2217–33.

MacGorman, D.R., and Burgess, D.W. 1994. Positive cloud-to-ground lightning in tornadic storms and hailstorms. *Mon. Wea. Rev.* **122**: 1671–97.

MacGorman, D.R., and Morgenstern, C.D. 1998. Some characteristics of cloud-to-ground lightning in mesoscale convective systems. *J. Geophys. Res.* **103**: 14 011–23.

MacGorman, D.R., and Nielsen, K.E. 1991. Cloud-to-ground lightning in a tornadic storm on 8 May 1986. *Mon. Wea. Rev.* **119**: 1557–74.

MacGorman, D.R., and Rust, W.D. 1998. *The Electrical Nature of Storms*, 422 pp., New York: Oxford University Press.

MacGorman, D.R., Maier, M.W., and Rust, W.D. 1984. Lightning strike density for the contiguous United States from thunderstorm duration records, NUREG/CR-3759, Office of Nuclear Regulatory Research, US Nuclear Regulatory Commission, Washington, DC, 44 pp.

MacGorman, D.R., Rust, W.D., and Mazur, V. 1985. Lightning activity and mesocyclone evolution, 17 May 1981. Preprint, *Proc. 14th Conf. Severe Local Storms*, pp. 355–8, Am. Meteor. Soc.

MacGorman, D.R., Burgess, D.W., Mazur, V., Rust, W.D., Taylor, W.L., and Johnson, B.C. 1989. Lightning rates relative to tornadic storm evolution on 22 May 1981. *J. Atmos. Sci.* **46**: 221–50.

Mackerras, D. 1963. Thunderstorm observations related to lightning flash counter performance. Internal Rep. UQ/ERB/4, University of Queensland, Australia.

Mackerras, D. 1977. Lightning occurrence in a subtropical area. In *Electrical Processes in Atmospheres, Proc. 5th Int. Conf. on Atmosphere Electricity, Garmisch-Partenkirchen, Germany*, eds. H. Dolezalek and R. Reiter, pp. 497–502, Darmstadt: Dr. Dietrich Steinkopff.

Mackerras, D. 1978. Prediction of lightning incidence and effects in electrical systems. *Electr. Eng. Trans., Inst. Eng. Aust.* **EE-14**: 73–7.

Mackerras, D. 1985. Automatic short-range measurement of the cloud flash to ground flash ratio in thunderstorms. *J. Geophys. Res.* **90**: 6195–201.

Mackerras, D., and Darveniza, M. 1992. Progress report on worldwide survey of cloud-flash to ground flash ratio using CGR3 instruments. In *Proc. 9th Int. Conf. on Atmospheric Electricity, St. Petersburg, Russia*, pp. 380–3.

Mackerras, D., and Darveniza, M. 1994. Latitudinal variation of lightning occurrence characteristics. *J. Geophys. Res.* **99**: 10 813–21.

Mackerras, D., Darveniza, M., Orville, R.E., Williams, E.R., and Goodman, S.J. 1998. Global lightning: total, cloud and ground flash estimates. *J. Geophys. Res.* **103**: 19 791–809.

Maier, M.W., and Holle, R.L. 1980. Cloud-ground lightning rate dependence on radar echo height. *Eos, Trans. AGU* **61**: 975.

Maier, M.W., and Krider, E.P. 1982. A comparative study of the cloud-to-ground lightning characteristics in Florida and Oklahoma thunderstorms. Preprint, *Proc. 12th Conf. on Severe Local Storms, San Antonio, Texas*, pp. 334–7, Am. Meteor. Soc., Boston, Massachusetts.

Maier, M.W., Boulanger, A.G., and Sarlat, J. 1978. Cloud-to-ground lightning frequency over south Florida. Preprint, *Proc. Conf. on Cloud Physics and Atmospheric Electricity*, Am. Meteor. Soc., pp. 605–10.

Maier, M.W., Boulanger, A.G., and Sax, R.I. 1979. An initial assessment of flash density and peak current characteristics of lightning flashes to ground in South Florida, NUREG/CR-1024, Office of Standards Development, US Nuclear Regulatory Commission, Washington, DC, 43 pp.

Maier, L.M., Krider, E.P., and Maier, M.W. 1984. Average diurnal variation of summer lightning over the Florida Peninsula. *Mon. Wea. Rev.* **112**: 1134–40.

Manohar, G.K., Kandalganonkar, S.S., and Tinmaker, M.I.R. 1999. Thunderstorm activity over India and the Indian southwest monsoon. *J. Geophys. Res.* **104**: 4169–88.

Marriott, W. 1908. Brontometer records at West Norwood, June 4, 1908. *Q.J.R. Meteor. Soc.* **34**: 210–12.

Marshall, J.S. 1953. Frontal precipitation and lightning observed by radar. *Can. J. Phys.* **31**: 194–203.

Marshall, J.S., and Radhakant, S. 1978. Radar precipitation maps as lightning indicators. *J. Appl. Meteor.* **17**: 206–12.

Marshall, T.C., and Rust, W.D. 1993. Two types of vertical electrical structures in stratiform precipitation regions of mesoscale convective systems. *Bull. Am. Meteor. Soc.* **74**: 2159–70.

Marshall, T.C., and Winn, W.P. 1985. Comments on "The rain gush, lightning, and the lower positive center in thunderstorms" by E.R. Jayaratne and C.P.R. Saunders. *J. Geophys. Res.* **90**: 10 753–4.

Mazur, V., and Rust, W.D. 1983. Lightning propagation and flash density in squall lines as determined with radar. *J. Geophys. Res.* **88**: 1495–502.

Mazur, V., Gerlach, J.C., and Rust, W.D. 1984. Lightning flash density versus altitude and storm structure from observations with UHF- and S-band radars. *Geophys. Res. Lett.* **11**: 61–4.

Mazur, V., Gerlach, J.C., and Rust, W.D. 1986. Evolution of lightning flash density and reflectivity structure in a multicell thunderstorm. *J. Geophys. Res.* **91**: 8690–700.

Mazur, V., Ruhnke, L.H., and Rudolph, T. 1987. Effect of E-field mill location on accuracy of electric field measurements with instrumented airplane. *J. Geophys. Res.* **92**: 12 013–9.

McCann, E.D. 1944. The measurement of lightning currents in direct strokes. *AIEE Trans.* **63**: 1157–64.

McCaul, E.W., Buechler, D.E., Hodanish, S., and Goodman, S.J. 2002. The Almena, Kansas, tornadic storm of 3 June 1999: a long-lived supercell with very little cloud-to-ground lightning. *Mon. Wea. Rev.* **130**: 407–15.

Michalon, N., Nassif, A., Saouri, T., Royer, J.F., and Pontikis, C.A. 1999. Contribution to the climatological study of lightning. *Geophys. Res. Lett.* **26**: 3097–100.

Michimoto, K. 1991. A study of radar echoes and their relation to lightning discharge of thunderclouds in the Hokuriku District. Part I: Observation and analysis of thunderclouds in summer and winter. *J. Meteor. Soc. Japan* **69**: 327–35.

Michimoto, K. 1993. A study of radar echoes and their relation to lightning discharge of thunderclouds in the Hokuriku District. Part II: Observation and analysis of "single-flash" thunderclouds in midwinter. *J. Meteor. Soc. Japan* **71**: 195–204.

Michnowski, S., Israelsson, S., Parfiniewicz, J., Enaytollah, M.A., and Pisler, E. 1987. A case of thunderstorm system development inferred from lightning distribution. *Publs. Inst. Geophys. Pol. Acad. Sc. D26*, **198**: 3–54.

Ming, Y., and Cooray, V. 1994. Propagation effects caused by a rough ocean surface on the electromagnetic fields generated by lightning return strokes. *Radio Sci.* **29**: 73–85.

Mohr, K.I., Toracinta, R.E., Zipser, E.J., and Orville, R.E. 1996. A comparison of WSR-88D reflectivities, SSM/I brightness temperatures, and lightning for mesoscale convective systems in Texas. Part II: SSM/I brightness temperatures and lightning. *J. Appl. Meteor.* **35**: 919–31.

Molinari, J., Moore, P.K., Idone, V.P., Henderson, R.W., and Saljoughy, A.B. 1994. Cloud-to-ground lightning in Hurricane Andrew. *J. Geophys. Res.* **99**: 16 655–76.

Molinari, J., Moore, P., and Idone, V. 1999. Convective structure of hurricanes as revealed by lightning locations. *Mon. Wea. Rev.* **127**: 520–34.

Molinié, J., and Pontikis, C.A. 1995. A climatological study of tropical thunderstorm clouds and lightning frequencies on the French Guyana coast. *Geophys. Res. Lett.* **22**: 1085–8.

Molinié, J., and Pontikis, C.A. 1996. Reply. *Geophys. Res. Lett.* **23**: 1703–4.

Molinié, G., Soula, S., and Chauzy, S. 1999. Cloud-to-ground lightning activity and radar observations of storms in the Pyrenees range area. *Q.J.R. Meteor. Soc.* **125**: 3103–22.

Moore, P.K., and Orville, R.E. 1990. Lightning characteristics in lake-effect thunderstorms. *Mon. Wea. Rev.* **118**: 1767–82.

Moore, C.B., and Vonnegut, B. 1991. Comments on "A radar study of the plasma and geometry of lightning". *J. Atmos. Sci.* **48**: 369–70.

Moore, C.B., Vonnegut, B., Stein, B.A., and Survilas, H.J. 1960. Observations of electrification and lightning in warm clouds. *J. Geophys. Res.* **65**: 1907–10.

Moore, C.B., Vonnegut, B., Machado, J.A., and Survilas, H.J. 1962. Radar observations of rain gushes following overhead lightning strokes. *J. Geophys. Res.* **67**: 207–20.

Moore, C.B., Vonnegut, B., Vrabik, E.A., and McCaig, D.A. 1964. Gushes of rain and hail after lightning. *J. Atmos. Sci.* **21**: 646–65.

Müller-Hillebrand, D. 1963. Lightning counters. II – The effect of changes of electric field on counter circuits. *Ark. Geofys.* **4**: 271–92.

Müller-Hillebrand, D., Johansen, O., and Saraoja, E.K. 1965. Lightning-counter measurements in Scandinavia. *Proc. IEE* **112**: 203–10.

Murray, N.D., Orville, R.E., and Huffines, G.R. 2000. Effect of pollution from Central American fires on cloud-to-ground lightning in May 1999. *Geophys. Res. Lett.* **27**: 2249–52.

Nakano, M., Takeuti, T., Funaki, K., Kitigawa, N., and Takahashi, C. 1984. Oceanic tropical lightning at Ponape, Micronesia. *Res. Lett. Atmos. Electr.* **4**: 29–33.

Nesbitt, S.W., Zipser, E.J., and Cecil, D.J. 2000. A census of precipitation features in the tropics using TRMM: radar, ice scattering, and lightning observations. *J. Climate* **13**: 4087–106.

Nickolaenko, A.P., Hayakawa, M., and Hobara, Y. 1999. Long-term periodic variations in global lightning activity deduced from the Schumann resonance monitoring. *J. Geophys. Res.* **104**: 27 585–91.

Nielsen, K.E., Maddox, R.A., and Vasiloff, S.V. 1994. The evolution of cloud-to-ground lightning within a portion of the 10–11 June 1985 squall line. *Mon. Wea. Rev.* **122**: 1809–17.

Oladiran, E.O., Pisler, E., and Israelsson, S. 1988. New lightning flash counter and calibration circuit with improved discrimination of cloud and ground discharges. *Proc. IEE* **135A**: 22–8.

Orville, R.E. 1972. Lightning between clouds and ground. *Weatherwise* **25**: 108–12.

Orville, R.E. 1981. Global distribution of midnight lightning – September to November 1977. *Mon. Wea. Rev.* **109**: 391–5.

Orville, R.E. 1986. Lightning phenomenology. In *The Earth's Electrical Environment*, eds. E.P. Krider and R.G. Robble, pp. 23–9, Washington, DC: National Academy Press.

Orville, R.E. 1990a. Peak-current variations of lightning return stroke as a function of latitude. *Nature* **343**: 149–51.

Orville, R.E. 1990b. Winter lightning along the East Coast. *Geophys. Res. Lett.* **17**: 713–15.

Orville, R.E. 1991. Lightning ground flash density in the contiguous United States – 1989. *Mon. Wea. Rev.* **119**: 573–7.

Orville, R.E. 1993. Cloud-to-ground lightning in the Blizzard of '93. *Geophys. Res. Lett.* **20**: 1367–70.

Orville, R.E. 1994. Cloud-to-ground lightning flash characteristics in the contiguous United States: 1989–91. *J. Geophys. Res.* **99**: 10 833–41.

Orville, R.E. 1995. Lightning detection from ground and space. In *Handbook of Atmospheric Electrodynamics*, vol. 1, ed. H. Volland, pp. 137–49, Boca Raton, Florida: CRC Press.

Orville, R.E. 1999. Comments on "Large peak current cloud-to-ground lightning flashes during the summer months in the contiguous United States". *Mon. Wea. Rev.* **127**: 1937–8.

Orville, R.E., and Henderson, R.W. 1986. Global distribution of midnight lightning: September 1977 to August 1978. *Mon. Wea. Rev.* **114**: 2640–53.

Orville, R.E., and Huffines, G.R. 1999. Lightning ground flash measurements over the contiguous United States: 1995–1997. *Mon. Wea. Rev.* **127**: 2693–703.

Orville, R.E., and Silver, A.C. 1997. Lightning ground flash density in the contiguous Unites States: 1992–95. *Mon. Wea. Rev.* **125**: 631–8.

Orville, R.E., and Songster, H. 1987. The East Coast lightning detection network. *Trans. IEEE* PWRD-2: 899–907.

Orville, R.E., and Spencer, D.W. 1979. Global lightning flash frequency. *Mon. Wea. Rev.* **107**: 934–43.

Orville, R.E., Maier, M.W., Mosher, F.R., Wylie, D.P., and Rust, W.D. 1982. The simultaneous display in a severe storm of lightning ground strike locations onto satellite images and radar reflectivity patterns. Preprint, *Proc. 12th Conf. on Severe Local Storms*, pp. 448–51, Am. Meteor. Soc.

Orville, R.E., Henderson, R.W., and Bosart, L.F. 1983. An east coast lightning detection network. *Bull. Am. Meteor. Soc.* **64**: 1029–37.

Orville, R.E., Weisman, R.A. Pyle, R.B., Henderson, R.W., and R.E. Orville, R.E. Jr 1987. Cloud-to-ground lightning flash characteristics from June 1984 through May 1985. *J. Geophys. Res.* **92**: 5640–4.

Orville, R.E., Henderson, R.W., and Bosart, L.F. 1988. Bipole patterns revealed by lightning locations in mesoscale storm systems, *Geophys. Res. Lett.* **15**: 129–32.

Orville, R.E., Zipser, E.J., Brook, M., Weidman C., Aulich, G., Krider, E.P., Christian, H., Goodman, S., Blakeslee, R., and Cummins, K. 1997. Lightning in the region of the TOGA COARE. *Bull. Am. Meteor. Soc.* **78**: 1055–67.

Orville, R.E., Huffines, G, Nielsen-Gammon, J., Zhang, R., Ely, B., Steiger, S., Phillips, S., Allen, S., and Read, W. 2001. Enhancement of cloud-to-ground lightning over Houston, Texas. *Geophys. Res. Lett.* **28**(24): 2597–600.

Pakiam, J.E., and Maybank, J. 1975. The electrical characteristics of some severe hailstorms in Alberta, Canada. *J. Meteor. Soc. Japan* **53**: 363–83.

Peckham, D.W., Uman, M.A., and Wilcox, C.E. Jr 1984. Lightning phenomenology in the Tampa Bay area. *J. Geophys. Res.* **89**: 11 789–805.

Perez, A.H., Wicker, L.J., and Orville, R.E. 1997. Characteristics of cloud-to-ground lightning associated with violent tornadoes. *Wea. Forecast.* **12**: 428–37.

Petersen, W.A., and Rutledge, S.A. 1992. Some characteristics of cloud-to-ground lightning in tropical northern Australia. *J. Geophys. Res.* **97**: 11 553–60.

Petersen, W.A., and Rutledge, S.A. 1998. On the relationship between cloud-to-ground lightning and convective rainfall. *J. Geophys. Res.* **103**: 14 025–40.

Petersen, W.A., Rutledge, S.A., and Orville, R.E. 1996. Cloud-to-ground lightning observations from TOGA COARE: selected results and lightning location algorithms. *Mon. Wea. Rev.* **124**: 602–20.

Piepgrass, M.V., Krider, E.P., and Moore, C.B. 1982. Lightning and surface rainfall during Florida thunderstorms. *J. Geophys. Res.* **87**: 11 193–201.

Pierce, E.T. 1956. The influence of individual variations in the field changes due to lightning discharges upon the design and performance of lightning flash counters. *Arch. Meteor. Geophys. Bioklim.* **A9**: 78–86.

Pierce, E.T. 1968. The counting of lightning flashes. Special Technical Report 49, Stanford Research Institute, Menlo Park, California, June (available from Defense Documentation Centers as AD 682023).

Pierce, E.T. 1970. Latitudinal variation of lightning parameters. *J. Appl. Meteor.* **9**: 194–5.

Pierce, E.T. 1971. Triggered lightning and some unsuspected lightning hazards. Stanford Research Institute, Menlo Park, California, 20 pp.

Pierce, E.T. 1972. Triggered lightning and some unsuspected lightning hazards. *Naval Res. Rev.* **25**: 14–28.

Pinto, O., Pinto, I.R.C.A., Lacerda, M., Carvalho, A.M., Diniz, J.H., and Cherchiglia, L.C.L. 1997. Are equatorial negative lightning flashes more intense than those at higher latitudes? *J. Atmos. Solar-Terr. Phys.* **59**: 1881–3.

Pinto, O. Jr, Pinto I.R.C.A., Gomes, M.A.S.S., Vitorello, I., Padilha, A.L., Diniz, J.H., Carvalho, A.M., and Cazetta, A. 1999a. Cloud-to-ground lightning in southeastern Brazil in 1993. l. Geographical distribution. *J. Geophys. Res.* **104**: 31 369–79.

Pinto, I.R.C.A., Pinto O. Jr, Rocha, R.M.L., Diniz, J.H., Carvalho, A.M., and Cazetta, A. 1999b. Cloud-to-ground lightning in southeastern Brazil in 1993. 2. Time variations and flash characateristics. *J. Geophys. Res.* **104**: 31 381–7.

Popolansky, F., and Laitinen, L. 1972. Thunderstorm days, thunderstorm duration and the number of lightning flashes in Czechoslovakia and in Finland. *Studia Geoph. et Geod.* **16**: 103–6.

Popov, A.S. 1896. Instrument for detection and registration of electrical fluctuations. *Journal of Russian Physics and Chemistry Society* **XXVIII**, *Physics I*, **1**: 1–14.

Prentice, S.A. 1960. Thunderstorms in the Brisbane area. *J. Inst. Eng. Aust.* **32**: 33–45.

Prentice, S.A. 1972. CIGRE lightning flash counter. *Electra* **22**: 149–71.

Prentice, S.A. 1977. Frequency of lightning discharges. In *Lightning, vol. 1, Physics of Lightning*, ed. R.H. Golde, pp. 465–95, New York: Academic Press.

Prentice, S.A., and Mackerras, D. 1969. Recording range of a lightning-flash counter. *Proc. IEE* **116**: 294–302.

Prentice, S.A., and Mackerras, D. 1977. The ratio of cloud to cloud–ground lightning flashes in thunderstorms. *J. Appl. Meteor.* **16**: 545–50.

Prentice, S.A., and Robson, M.W. 1968. Lightning intensity studies in the Darwin area. *Electr. Eng. Trans., Inst. Eng. Aust.* **4**: 217–26.

Prentice, S.A., Mackerras, D., and Tolmie, R.P. 1975. Development and testing of a vertical aerial lightning flash counter. *Proc. IEE* **122**: 487–91.

Price, C. 2000. Evidence for a link between global lightning activity and upper tropospheric water vapour. *Nature* **406**: 290–3.

Price, C., and Rind, D. 1992. A simple lightning parametrization for calculating global lightning distributions. *J. Geophys. Res.* **97**: 9919–33.

Price, C., and Rind, D. 1993. What determines the fraction of cloud-to-ground lightning in thunderstorms? *Geophys. Res. Lett.* **20**: 463–6.

Price, C., and Rind, D. 1994a. Possible implications of global climate change on global lightning distributions and frequencies. *J. Geophys. Res.* **99**: 10 823–31.

Price, C., and Rind, D. 1994b. Modeling global lightning distributions in a general circulation model. *Mon. Wea. Rev.* **122**: 1930–9.

Proctor, D.E. 1983. Lightning and precipitation in a small multicellular thunderstorm. *J. Geophys. Res.* **88**: 5421–40.

Proctor, D.E. 1984. Correction to "Lightning and precipitation in a small multicellular thunderstorm". *J. Geophys. Res.* **89**: 11 826.

Rakov, V.A. 1986. On the determination of ground flash density. *Elektr.* **3**: 54–6.

Rakov, V.A. 1992. Data acquired with the LLP lightning locating systems. *Meteor. Gidrol.* **7**: 105–14.

Rakov, V.A., and Dulzon, A.A. 1984. On latitudinal features of thunderstorm activity. *Meteor. Gidrol.* **1**: 52–7.

Rakov, V.A., and Dulzon, A.A. 1986. Study of some features of frontal and convective thunderstorms. *Meteor. Gidrol.* **9**: 59–63.

Rakov, V.A., and Dulzon, A.A. 1988. Lightning research in Western Siberia. In *Proc. 8th Int. Conf. on Atmospheric Electricity, Uppsala, Sweden*, pp. 766–9.

Rakov, V.A., and Lutz, A.O. 1988. On estimating the attractive radius for lightning striking a structure. *Elektr.* **9**: 64–7.

Rakov, V.A., and Lutz, A.O. 1990. A new technique for estimating equivalent attractive radius for downward lightning flashes. In *Proc. 20th Int. Conf. on Lightning Protection, Interlaken, Switzerland*, paper 2.2.

Rakov, V.A., and Uman, M.A. 1990a. Long continuing current in negative lightning ground flashes. *J. Geophys. Res.* **95**: 5455–70.

Rakov, V.A., and Uman, M.A. 1990b. Some properties of negative cloud-to-ground lightning flashes versus stroke order. *J. Geophys. Res.* **95**: 5447–53.

Rakov, V.A., Adjiev, A.K., Akchurin, M.M., and Shoivanov, Y.R. 1989. Study of the spatial distribution of ground flash density using the "Ochag" lightning locating system. *Meteor. Gidrol.* **2**: 48–53.

Rakov, V.A., Dulzon, A.A., Shoivanov, Y.R., and Shelukhin, D.V. 1990a. A technique for mapping of ground flash density. *Elektr. Stan.* **3**: 63–6.

Rakov, V.A., Shoivanov, Y.R., Shelukhin, D.V., Lutz, A.O., and Esipenko, R.F. 1990b. Annual ground flash density from lightning flash counter records. In *Proc. 20th Int. Conf. on Lightning Protection, Interlaken, Switzerland*, paper 6.8P.

Rakov, V.A., Uman, M.A., and Thottappillil, R. 1994. Review of lightning properties from electric field and TV observations. *J. Geophys. Res.* **99**: 10 745–50.

Randell, S.C., Rutledge, S.AA., Farley, R.D., and Helsdon, J.H. Jr 1994. A modeling study on the early electrical development of tropical convection: continental and oceanic (monsoon) storms. *Mon. Wea. Rev.* **122**: 1852–77.

Ray, P.S., MacGorman, D.R., Rust, W.D., Taylor, W.L., and Rasmussen, L.W. 1987. Lightning location relative to storm structure in a supercell storm and a multicell storm. *J. Geophys. Res.* **92**: 5713–24.

Reap, R.M. 1986. Evaluation of cloud-to-ground lightning data from the western United States for the 1983–1984 summer seasons. *J. Clim. Appl. Meteor.* **25**: 785–99.

Reap, R.M. 1994. Analysis and prediction of lightning strike distributions associated with synoptic map types over Florida. *Mon. Wea. Rev.* **122**: 1698–715.

Reap, R.M., and MacGorman, D.R. 1989. Cloud-to-ground lightning: climatological characteristics and relationships to model fields, radar observations, and severe local storms. *Mon. Wea. Rev.* **117**: 518–35.

Reap, R.M., and Orville, R.E. 1990. The relationship between network lightning locations and surface hourly observations of thunderstorms. *Mon. Wea. Rev.* **118**: 94–108.

Reeve, N., and Tuomi, R. 1999. Lightning activity as an indicator of climate change. *Q.J.R. Meteor. Soc.* **125**: 893–903.

Renno, N.O., and Ingersoll, A.P. 1996. Natural convection as a heat engine: a theory for CAPE. *J. Atmos. Sci.* **53**: 572–85.

Richard, P. 1992. Application of atmospheric discharges localization to thunderstorm nowcasting. In *Proc. WMO Tech. Conf. Instruments and Methods of Observations*, World Meteor. Organization.

Robinson, P.J., and Easterling, D.R. 1988. The frequency distribution of thunderstorm durations. *J. Appl. Meteor.* **27**: 77–82.

Romualdo, C., Brito, F., Perez, H., de la Rosa, F., and Sarmiento, H.G. 1989. Studying distribution system reliability against lightning. *IEEE Comp. Appl. Power* **2**: 43–7.

Roohr, P.B., and Von der Haar, T.H. 1994. A comparative analysis of the temporal variability of lightning observations and GOES imagery. *J. Appl. Meteor.* **33**: 1271–90.

Rorig, M.L., and Ferguson, S.A. 1999. Characteristics of lightning and wildland fire ignition in the Pacific Northwest. *J. Appl. Meteor.* **38**: 1565–75.

Rust, W.D. 1989. Utilization of a mobile laboratory for storm electricity measurements. *J. Geophys. Res.* **94**: 13 305–11.

Rust, W.D., and Doviak, R.J. 1982. Radar research on thunderstorms and lightning. *Nature* **297**: 461–8.

Rust, W.D., MacGorman, D.R., and Arnold, R.T. 1981a. Positive cloud-to-ground lightning flashes in severe storms. *Geophys. Res. Lett.* **8**: 791–4.

Rust, W.D., Taylor, W.L., MacGorman, D.R., and Arnold, R.T. 1981b. Research on electrical properties of severe thunderstorms in the Great Plains. *Bull. Am. Meteor. Soc.* **62**: 1286–93.

Rust, W.D., MacGorman, D.R., and Goodman, W.J. 1985a. Unusual positive cloud-to-ground lightning in Oklahoma storms on 13 May 1983. Preprint, *Proc. 14th Conf. Severe Local Storms*, pp. 372–5, Am. Meteor. Soc.

Rust, W.D., Taylor, W.L., MacGorman, D.R., Brandes, E., Mazur, V., Arnold, R., Marshall, T., Christian, H., and Goodman, S.J. 1985b. Lightning and related phenomena in isolated thunderstorms and squall line systems. *J. Aircraft* **22**: 449–54.

Rutledge, S.A., and MacGorman, D.R. 1988. Cloud-to-ground lightning activity in the 10–11 June 1985 mesoscale convective system observed during the Oklahoma–Kansas PRE-STORM Project. *Mon. Wea. Rev.* **116**: 1393–408.

Rutledge, S.A., and Petersen, W.A. 1994. Vertical radar reflectivity structure and cloud-to-ground lightning in the stratiform region of MCSs: further evidence for *in-situ* charging in the stratiform region. *Mon. Wea. Rev.* **122**: 1760–76.

Rutledge, S.A., Lu, C., and MacGorman, D.R. 1990. Positive cloud-to-ground lightning in mesoscale convective systems. *J. Atmos. Sci.* **57**: 2085–100.

Rutledge, S.A., Williams, E.R., and Keenan, T.D. 1992. The down under Doppler and electricity experiment (DUNDEE): overview and preliminary results. *Bull. Am. Meteor. Soc.* **73**: 3–16.

Rutledge, S.A., Williams, E.R., and Petersen, W.A. 1993. Lightning and electrical structure of mesoscale convective systems. *Atmos. Res.* **29**: 27–53.

Samsury, C.E., and Orville, R.E. 1994. Cloud-to-ground lightning in tropical cyclones: a study of Hurricanes Hugo (1989) and Jerry (1989). *Mon. Wea. Rev.* **122**: 1887–96.

Sánchez, J.L., Ortega, E.G., and Marcos, J.L. 2001. Construction and assessment of a logistic regression model applied to short-term forecasting of thunderstorms in León (Spain). *Atmos. Res.* **56**: 57–71.

Sartor, D. 1964. Radio observations of the electromagnetic emission from warm clouds. *Science* **143**: 948–50.

Satori, G., and Ziegler, B. 1998. Anomalous behavior of Schumann resonances during the transition between 1995 and 1996. *J. Geophys. Res.* **103**: 14 147–55.

Satori, G., and Ziegler, B. 1999. El Niño related meridional oscillation of global lightning activity. *Geophys. Res. Lett.* **26**: 1365–8.

Saunders, C.P.R., Keith, W.D., and Mitzeva, R.P. 1991. The effect of liquid water on thunderstorm charging. *J. Geophys. Res.* **96**: 11 007–17.

Schonland, B.F.J. 1956. The lightning discharge. In *Handbuch der Physik* **22**: pp. 576–628, Berlin: Springer-Verlag.

Schuur, T.J., Smull, B.F., Rust, W.D., and Marshall, T.C. 1991. Electrical and kinematic structure of the stratiform precipitation region trailing an Oklahoma squall line. *J. Atmos. Sci.* **48**: 825–42.

Schuur, T.J., and Rutledge, S.A. 2000a. Electrification of stratiform regions in mesoscale convective systems. I: Observational comparison of symmetric and asymmetric MCSs. *J. Atmos. Sci.* **57**: 1961–82.

Schuur, T.J., and Rutledge, S.A. 2000b. Electrification of stratiform regions in mesoscale convective systems. II: Two-dimensional numerical model simulations of a symmetric MCS. *J. Atmos. Sci.* **57**: 1983–2006.

Scouten, D.C., and Stephenson, D.T. 1972. A sferic rate azimuth profile of the 1955 Blackwell, Oklahoma, tornado. *J. Atmos. Sci.* **29**: 929–36.

Seimon, A. 1993. Anomalous cloud-to-ground lightning in an F5-tornado-producing supercell thunderstorm on 28 August 1990. *Bull. Am. Meteor. Soc.* **74**: 189–203.

Seity, Y., Soula, S., and Sauvageot, H. 2001 Lightning and precipitation relationship in coastal thunderstorms. *J. Geophys. Res.* **106**: 22 801–16.

Shackford, C.R. 1960. Radar indications of a precipitation–lightning relationship in New England thunderstorms. *J. Meteor.* **17**: 15–19.

Shafer, M.A. 1990. Cloud-to-ground lightning in relation to digitized radar data in severe storms. Master's thesis, University of Oklahoma, 93 pp.

Shafer, M.A., MacGorman, D.R., and Carr, F.H. 2000. Cloud-to-ground lightning throughout the lifetime of a severe storm system in Oklahoma. *Mon. Wea. Rev.* **128**: 1798–816.

Sheridan, S.C., Griffiths, J.F., and Orville, R.E. 1997. Warm season cloud-to-ground lightning–precipitation relationships in the south-central United States. *Wea. Forecast.* **12**: 449–58.

Shih, S.F. 1988. Using lightning for rainfall estimation in Florida. *Trans. Am. Soc. Agric. Eng.* **31**: 750–5.

Shindo, T., and Yokoyama, S. 1998. Lightning occurrence data observed with lightning location systems in Japan: 1992–1995. *IEEE Trans. Power Del.* **13**: 1468–74.

Silberg, P.A. 1965. Passive electrical measurements from three Oklahoma tornadoes. *Proc. IEEE* **53**: 1197–204.

Smith, S.B. 1993. Comments on "Lightning ground flash density in the contiguous United States – 1989." *Mon. Wea. Rev.* **121**: 1572–5.

Smith, S.B., La Due, J.G., and MacGorman, D.R. 2000. The relationship between cloud-to-ground lightning polarity and surface equivalent potential temperature during three tornadic outbreaks. *Mon. Wea. Rev.* **128**: 3320–8.

Snow, J.T., Wyatt, A.L., McCarthy, A.K., and Bishop, E.K. 1995. Fallout of debris from tornadic thunderstorms: a historical perspective and two examples from VORTEX. *Bull. Am. Meteor. Soc.* **76**: 1777–90.

Solomon, R., and Baker, M., 1994. Electrification of New Mexico thunderstorms. *Mon. Wea. Rev.* **122**: 1878–86.

Sommeria, G., and Testud, J. 1984. COPT81: a field experiment designed for the study of dynamics and electrical activity of deep convection in continental tropical regions. *Bull. Am. Meteor. Soc.* **65**: 4–10.

Soriano, L.J.R., de Pablo, F., and Diez, E.L.G. 2001a. Cloud-to-ground lightning activity in the Iberian Peninsula: 1992–1994. *J. Geophys. Res.* **106**: 11 891–901.

Soriano, L.J.R., De Pablo, F., and Diez, E.L.G. 2001b. Meteorological and geo-orographical relationships with lightning activity in Castilla-Leon (Spain). *Meteor. Appl.* **8**: 169–75.

Soriano, L.J.R., De Pablo, F., and Diez, E.L.G. 2001c. Relationship between convenctive precipitation and cloud-to-ground lightning in the Iberian Peninsula. *Mon. Wea. Rev.* **129**: 2998–3003.

Soula, S., and Chauzy, S. 1997. Charge transfer by precipitation between thundercloud and ground. *J. Geophys. Res.* **102**: 11 061–9.

Soula, S., Sauvageot, H., Molinie, G., Mesnard, F., and Chauzy, S. 1998. The CG lightning activity of a storm causing a flash-flood. *Geophys. Res. Lett.* **25**: 1181–4.

Sparrow, J.G., and Ney, F.E. 1971. Lightning observations by satellite. *Nature* **232**: 540–1.

Stansbury, E.F., Cherna, E., and Percy, J. 1979. Lightning flash locations related to the precipitation pattern of the storm. *Atmos. Ocean* **17**: 291–305.

Stolzenburg, M. 1990. Characteristics of the bipolar pattern of lightning locations observed in 1988 thunderstorms. *Bull. Am. Meteor. Soc.* **71**: 1331–8.

Stolzenburg, M. 1994. Observations of high ground flash densities of positive lightning in summertime thunderstorms. *Mon. Wea. Rev.* **122**: 1740–50.

Stolzenburg, M., Marshall, T.C., Rust, W.D., and Smull, B.F. 1994. Horizontal distribution of electrical and meteorological conditions across the stratiform region of a mesoscale convective system. *Mon. Wea. Rev.* **122**: 1777–97.

Stringfellow, M.F. 1974. Lightning incidence in Britain and the solar cycle. *Nature* **249**: 332–3.

Sunoto 1985. Lightning flash counting in Java Island of Indonesia. In *Proc. 18th Int. Conf. on Lightning Protection, Munich, Germany*, pp. 39–41.

Suszcynsky, D.M., Kirkland, M.W., Jacobson, A.R., Franz, R.C., Knox, S.O., Guillen, J.L.L., and Green, J.L. 2000. FORTE observations of simultaneous VHF and optical emissions from lightning: basic phenomenology. *J. Geophys. Res.* **105**: 2191–201.

Szymanski, E.W. Szymanski, S.J., Holmes, C.R., and Moore, C.B. 1980. An observation of a precipitation echo intensification associated with lightning. *J. Geophys. Res.* **85**: 1951–3.

Szpor, S. 1969. Topographie et fréquence de la foudre. I. Le pin carpathique comme enregistreur seculaire de la foudre. *Acta Geophys. Polonica* **17**: 331–58.

Szpor, S. 1971. Topographie et fréquence de la foudre. III. Résultat des Tatras Slovaques, partie occidentale. *Acta Geophys. Polonica* **19**: 127–47.

Szpor, S. 1972. Topographie et fréquence de la foudre. IV. Résultat des Tatras Slovaques, partie orientale. *Acta Geophys. Polonica* **20**: 25–47.

Takagi, N., Watanabe, T., Arima, I., Takeuti, T., Nakano, M., and Kinosita, H. 1986. An unusual summer thunderstorm in Japan. *Res. Lett. Atmos. Electr.* **6**: 43–8.

Takeuti, T. 1976. On cloud-to-sea discharges. *Proc. Res. Inst. Atmos. Nagoya Univ., Japan* **23**: 17–20.

Takeuti, T., and Nagatani, M. 1974. Oceanic thunderstorms in the tropical and subtropical Pacific. *J. Meteor. Soc. Japan* **52**: 509–12.

Takeuti, T., Nakano, M., and Nagatani, M. 1975a. Lightning discharges in Guam and Philippine Islands. *J. Meteor. Soc. Japan* **53**: 360–1.

Takeuti, T., Nagatani, M., and Nakada, H. 1975b. Thunderstorm activities and related meteorological conditions in the northwest subtropical Pacific. *Proc. Res. Inst. Atmos. Nagoya Univ., Japan* **22**: 27–31.

Takeuti, T., Nakano, M., Ishikawa, H., and Israelsson, S. 1977. On the two types of thunderstorms deduced from cloud-to-ground discharges observed in Sweden and Japan. *J. Meteor. Soc. Japan* **55**: 613–16.

Takeuti, T., Nakano, M., Brook, M., Raymond, D.J., and Krehbiel, P. 1978. The anomalous winter thunderstorms of the Hokuriku coast. *J. Geophys. Res.* **83**: 2385–94.

Tapia, A., Smith, J.A., and Dixon, M. 1998. Estimation of convective rainfall from lightning observations. *J. Appl. Meteor.* **37**: 1497–509.

Taylor, W.L. 1973. An electromagnetic technique for tornado detection. *Weatherwise* **26**: 70–1.

Taylor, W.L., Brandes, E.A., Rust, W.D., and MacGorman, D.R. 1984. Lightning activity and severe storm structure. *Geophys. Res. Lett.* **11**: 545–8.

Théry, C. 2001. Evaluation of LPATS data using VHF interferometric observations of lightning flashes during the EULINOX experiment. *Atmos. Res.* **56**: 397–409.

Thomas, R.J., Krehbiel, P.R., Rison, W., Hamlin, T., Boccippio, D.J., Goodman, S.J., and Christian, H.J. 2000. Comparison of ground-based 3-dimensional lightning mapping observations with satellite-based LIS observations in Oklahoma. *Geophys. Res. Lett.* **27**: 1703–6.

Thomson, E.M. 1980. The dependence of lightning return stroke characteristics on latitude. *J. Geophys. Res.* **85**: 1050–6.

Thomson, E.M., Galib, M.A., Uman, M.A., Beasley, W.H., and Master, M.J. 1984. Some features of stroke occurrence in Florida lightning flashes. *J. Geophys. Res.* **89**: 4910–6.

Thottappillil, R., Rakov, V.A., Uman, M.A., Beasley, W.H., Master, M.J., and Shelukhin, D.V. 1992. Lightning subsequent stroke electric field peak greater than the first stroke peak and multiple ground terminations. *J. Geophys. Res.* **97**: 7503–9.

Toracinta, E.R., Mohr, K., Zipser E.J., and Orville, R.E. 1996. A comparison of WSR-88D reflectivities, SSM/I brightness temperatures, and lightning for mesoscale convective systems in Texas. Part I: Radar reflectivity and lightning. *J. Appl. Meteor.* **35**: 902–18.

Toracinta, E.R., and Zipser, E.J. 2001. Lightning and SSM/I-ice-scattering mesoscale convective systems in the global tropics. *J. Appl. Meteor.* **40**: 983–1002.

Trost, T.F., and Nomikos, C.E. 1975. VHF radio emissions associated with tornadoes. *J. Geophys. Res.* **80**: 4117–18.

Turman, B.N. 1978. Analysis of lightning data from the DMSP satellite. *J. Geophys. Res.* **83**: 5019–24.

Turman, B.N., and Edgar, B.C. 1982. Global lightning distributions at dawn and dusk. *J. Geophys. Res.* **87**: 1191–206.

Uman, M.A., Rakov, V.A., and Thottappillil, R. 1994. Is there an appropriate length of time to obtain a meaningful mean lightning ground flash density? Technical Report prepared for EPRI, Palo Alto, California, contract no. 4CH 2891.

Ushio, T., Heckman, S.J., Boccippio, D.J., Christian, H.J., and Kawasaki, Z.-I. 2001. A survey of thunderstrom flash rates compared to cloud top height using TRMM satellite data. *J. Geophys. Res.* **106**: 24 089–95.

van Delden, A. 2001. The synoptic setting of thunderstorms in western Europe. *Atmos. Res.* **56**: 89–110.

Vonnegut, B. 1963. Some facts and speculations concerning the origin and role of thunderstorm electricity. *Meteor. Monogr.* **5**: 224–41.

Vonnegut, B., and Moore, C.B. 1960. A possible effect of lightning discharge on precipitation formation process. *Am. Geophys. Union* **5**: 287–304.

Vonnegut, B., and Moore, C.B. 1965. Nucleation of ice formation in supercooled clouds as the result of lightning. *J. Appl. Meteor.* **4**: 640–2.

Vonnegut, B., and Moore, C.B. 1986. Comments on "'The rain gush', lightning and the lower positive charge center, in thunderstorms", by E.R. Jayaratne and C.P.R. Saunders. *J. Geophys. Res.* **91**: 10 949.

Vonnegut, B., and Moore, C.B. 1987. Comments on "Thunderstorm electrification: the effect of cloud droplets" by E.R. Jayarante and C.P.R. Saunders. *J. Geophys. Res.* **92**: 3139.

Vonnegut, B., Vaughan, O.H., Brook, M., and Krehbiel, P. 1985. Mesoscale observations of lightning from Space Shuttle. *Bull. Am. Meteor. Soc.* **66**: 20–9.

Vorpahl, J.A., Sparrow, J.G., and Ney, E.P. 1970. Satellite observations of lightning. *Science* **169**: 860–2.

Wacker, R.S., and Orville, R.E. 1996. Peak current estimates in the NLDN? *Eos, Trans. AGU*, Fall Meeting Suppl., **77**(47): F89.

Wacker, R.S., and Orville, R.E. 1999a. Changes in measured lightning flash count and return stroke peak current after the 1994 US National Lightning Detection Network upgrade; 1. Observations. *J. Geophys. Res.* **104**: 2151–7.

Wacker, R.S., and Orville, R.E. 1999b. Changes in measured lightning flash count and return stroke peak current after the 1994 US National Lightning Detection Network upgrade; 2. Theory. *J. Geophys. Res.* **104**: 2159–62.

Watkins, D.C., Cobine, J.D., and Vonnegut, B. 1978. Electric discharges inside tornadoes. *Science* **199**: 171–4.

Watkins, N.W., Bharmal, N.A., Clilverd, M.A., and Smith, A.J. 2001. Comparison of VLF sferics intensities at Halley, Antarctica, with tropical lightning and temperature. *Radio Sci.* **36**: 1053–64.

Watkins, N.W., Clilverd, M.A., Smith, A.J., and Yearby, K.H. 1998. A 25-year record of 10 kHz sferics noise in Antarctica: implications for tropical lightning levels. *Geophys. Res. Lett.* **25**: 4353–6.

Watson, A.I., and Holle, R.L. 1996. An eight-year lightning climatology of the southeast United States prepared for the 1996 Summer Olympics. *Bull. Am. Meteor. Soc.* **77**: 883–90.

Watson, A.I., Lopez, R.E., Holle, R.L., and Daugherty, J.R. 1987. The relationship of lightning to surface convergence at Kennedy Space Center: a preliminary study. *Wea. Forecast.* **2**: 140–57.

Watson, A.I., Holle, R.L., Lopez, R.E., Ortiz, R., and Nicholson, J.R. 1991. Surface wind convergence as a short-term predictor of cloud-to-ground lightning at Kennedy Space Center. *Wea. Forecast.* **6**: 49–64.

Watson, A.I., Holle, R.L., and Lopez, R.E. 1994a. Cloud-to-ground lightning and upper-air patterns during bursts and breaks in the southwest monsoon. *Mon. Wea. Rev.* **122**: 1726–39.

Watson, A.I., Lopez, R.E., Holle, R.L. 1994b. Diurnal variations in Arizona monsoon lightning data. *Mon. Wea. Rev.* **122**: 1659–64.

Watson, A.I., Lopez, R.E., and Holle, R.L. 1994c. Diurnal cloud-to-ground lightning patterns in Arizona during the southwest monsoon. *Mon. Wea. Rev.* **122**: 1716–25.

Watson, A.I., Holle, R.L., and Lopez, R.E. 1995. Lightning from two national detection networks related to vertically integrated liquid and echo-top information from WSR-88D radar. *Wea. Forecast.* **10**: 592–605.

Westcott, N.E. 1995. Summertime cloud-to-ground lightning activity around major midwestern urban areas. *J. Appl. Meteor.* **34**: 1633–42.

Whitehead, J.T. *et al.* 1993. Estimating lightning performance of transmission lines II – updates to analytical model. *IEEE Trans.* **PWRD-8**: 1254–66, IEEE Working Group Report.

Willett, J.C., Bailey, J.C., Leteinturier, C., and Krider, E.P. 1990. Lightning electromagnetic radiation field spectra in the interval from 0.2 to 20 MHz. *J. Geophys. Res.* **95**: 20 367–87.

Williams, E.R. 1985. Large-scale charge separation in thunderclouds. *J. Geophys. Res.* **90**: 6013–25.

Williams, E.R. 1992. The Schumann resonance: a global tropical thermometer. *Science* **256**: 1184–7.

Williams, E.R. 1994. Global circuit response to seasonal variations in global surface temperature. *Mon. Wea. Rev.* **122**: 1917–29.

Williams, E.R. 1995. Meteorological aspects of thunderstorms. In *Handbook of Atmospheric Electrodynamics*, vol. 1, ed. H. Volland, pp. 27–60. Boca Raton, Florida: CRC Press.

Williams, E.R. 1996. Comment on "A climatological study of tropical thunderstorm clouds and lightning frequencies on the French Guyana coast" by J. Molinié and C.A. Pontikis. *Geophys. Res. Lett.* **23**: 1701–2 (reply by the authors on pp. 1703–4).

Williams, E.R. 1999. Global circuit response to temperature on distinct time scales: a status report. In *Atmospheric and Ionospheric Electromagnetic Phenomena Associated with Earthquakes*, ed. M. Hayakawa, pp. 939–49, Tokyo: Terra Scientific (TERRAPUB).

Williams, E.R. 2001. The electrification of severe storms. In *Severe Convective Storms, Meteor. Monogr.*, ed. C.A. Doswell, vol. 28 no. 50, pp. 527–61. Am. Meteor. Soc.

Williams, E.R., Geotis, S.G., and Bhattacharya, A.B. 1989a. A radar study of the plasma and geometry of lightning. *J. Atmos. Sci.* **46**: 1173–85.

Williams, E.R., Weber, M.E., and Engholm, C.D. 1989b. Electrical characteristics of microburst-producing storms in Denver. Preprint, *Proc. 24th Conf. Radar Meteorology*, pp. 89–92, Am. Meteor. Soc.

Williams, E.R., Weber, M.E., and Orville, R.E. 1989c. The relationship between lightning type and convective state of thunderclouds. *J. Geophys. Res.* **94**: 13 213–20.

Williams, E.R., Rutledge, S.A., Geotis, S.G., Renno, N., Rasmussen, E., and Rickenbach, T. 1992. A radar and electrical study of tropical "hot towers". *J. Atmos. Sci.* **49**: 1386–95.

Williams, E.R., Boldi, B., Matlin, A., Weber, M., Hodanish, S., Sharp, D., Goodman, S., Raghavan, R., and Buechler, D. 1999. The behavior of total lightning activity in severe Florida storms. *Atmos. Res.* **51**: 245–65.

Williams, E.R., Rothkin, K., Stevenson, D., and Boccippio, D.J. 2000. Global lightning variations caused by changes in thunderstorm flash rate and by changes in the number of thunderstorms. *J. Appl. Meteor.* **39**: 2223–30.

Wilson, J.W., and Megenhardt, D.L. 1997. Thunderstorm initiation, organization, and lifetime associated with Florida boundary layer convergence lines. *Mon. Wea. Rev.* **125**: 1507–25.

Winn, W.P., Hunyady, S.J., and Aulich, G.D. 1999. Pressure at the ground in a large tornado. *J. Geophys. Res.* **104**: 22 067–82.

Winn, W.P., Hunyady, S.J., and Aulich, G.D. 2000. Electric field at the ground in a large tornado. *J. Geophys. Res.* **105**: 20 145–53

WMO Publication 21. 1956. World distribution of thunderstorm days, Part 2, Tables of marine data and world maps. Geneva, Switzerland.

Yair, Y., Levin, Z., and Altaratz, O. 1998. Lightning phenomenology in the Tel Aviv area from 1989 to 1996. *J. Geophys. Res.* **103**: 9015–25.

Ziegler, C.L., and MacGorman, D.R. 1994. Observed lightning morphology relative to modeled space charge and electric field distributions in a tornadic storm. *J. Atmos. Sci.* **51**: 833–51.

Zipser, E.J. 1994. Deep cumulonimbus cloud systems in the tropics with and without lightning. *Mon. Wea. Rev.* **122**: 1837–51.

Zipser, E.J., and Lutz, K.R. 1994. The vertical profile of radar reflectivity of convective cells: a strong indicator of storm intensity and lightning probability? *Mon. Wea. Rev.* **122**: 1751–9.

Zrnic, D.S. 1976. Magnetometer data acquired during nearby tornado occurrences. *J. Geophys. Res.* **81**: 5410–12.

3 Electrical structure of lightning-producing clouds

> The basic difficulty in determining how thunderclouds become electrified lies in the fact that they are large, complex, and short-lived phenomena that need to be examined both as a whole and in detail to understand how they function. The electrical processes are intimately related to the cloud dynamics or motions and to the microphysics of the cloud, namely, to the populations and interactions of the precipitation, cloud droplets, ice crystals, and other particles that make up the cloud.
>
> P. R. Krehbiel (1986)

3.1. Introduction

The primary source of lightning is the cloud type termed cumulonimbus, commonly referred to as the thundercloud. The term "cumulonimbus cloud" is often used in the literature, although it is redundant. Strictly speaking, not every cumulonimbus produces lightning (e.g., Imyanitov et al. 1971), that is, a thundercloud could be more properly defined as a lightning-producing cumulonimbus. Sometimes the term "thunderstorm" is used as a synonym for thundercloud, although a thunderstorm is usually a system of thunderclouds rather than a single thundercloud. Lightning produced by thunderclouds formed over forest fires or contaminated by smoke is considered in Section 5.2. The electrical properties of clouds other than the cumulonimbus (primarily stratiform clouds) are reviewed in Section 3.3, and further information can be found, for example, in the books by Imyanitov et al. (1971) and by MacGorman and Rust (1998: Chapter 2 and Section 8.4). Lightning-like electrical discharges can also be generated in the ejected material above volcanoes, in sandstorms, and in nuclear explosions (Section 20.7). Clouds on planets other than Earth as potential sources of lightning are discussed in Chapter 16.

Before reviewing the electrical structure of thunderclouds it is worth outlining their meteorological characteristics. In effect, thunderclouds are large atmospheric heat engines with input energy from the Sun and with water vapor as the primary heat-transfer agent (Moore and Vonnegut 1977). The principal outputs of such an engine include (but are not limited to) (i) the mechanical work of the vertical and horizontal winds produced by the storm, (ii) an outflow of condensate in the form of rain and hail from the bottom of the cloud and of small ice crystals from the top of the cloud, and (iii) electrical discharges inside, below, and above the cloud, including corona, lightning, sprites, elves, blue starters, and blue jets, the latter four phenomena being discussed in Chapter 14. The processes that operate in a thundercloud to produce these actions are many and complex, most of them being poorly understood.

A thundercloud develops from a small fair-weather cloud called a cumulus, which is formed when parcels of warm, moist air rise and cool by adiabatic expansion, that is, without the transfer of heat or mass across the boundaries of the air parcels. When the relative humidity in a rising and cooling parcel exceeds saturation, moisture condenses on airborne particulate matter within it to form the many small water particles that constitute the visible cloud. The height of the condensation level, which determines the height of the visible cloud base, increases with decreasing relative humidity at ground. This is why cloud bases in Florida are generally lower than in arid locations, such as New Mexico or Arizona. Parcels of warm, moist air can only continue to rise to form a cumulus and eventually a cumulonimbus if the atmospheric temperature lapse rate, that is, the decrease in the temperature with increasing height, is larger than the moist-adiabatic lapse rate, about 0.6 °C per 100 m. The atmosphere is then referred to as unstable, since rising moist parcels remain warmer than the air around them and thus remain buoyant. When a parcel rises above the 0 °C isotherm, some water particles begin to freeze but others (typically smaller particles) remain liquid at temperatures colder than 0 °C. These are called supercooled water particles. At temperatures colder than about −40 °C all water particles will be frozen. In the temperature range 0 °C to −40 °C liquid water and ice particles coexist, forming a mixed-phase region where most electrification is thought to occur (subsection 3.2.6).

The convection of buoyant moist air is usually confined to the troposphere, the layer of the atmosphere that extends from the Earth's surface to the tropopause. The latter is a narrow layer that separates the troposphere from the next layer of the atmosphere, the stratosphere, which extends from the tropopause to a height of approximately 50 km. A plot of atmospheric temperature versus height is found in Fig. 14.1 and a detailed discussion of the regions of the atmosphere in Section 15.1. In the troposphere the temperature decreases with increasing altitude, while in the stratosphere the temperature at first becomes roughly independent of altitude and then increases with altitude. A zero or positive temperature gradient in the stratosphere serves to suppress convection and, therefore, hampers the penetration of cloud tops into the stratosphere. The height of the tropopause varies from approximately 18 km in the tropics in the summer to 8 km or so in high latitudes in the winter. In the case of vigorous updrafts, cloud vertical growth continues into the lower portion of stratosphere, so that cloud tops can reach altitudes up to 20 km.

Although the primary thunderstorm activity occurs in the lower latitudes, thunderclouds are occasionally observed in the polar regions. Thunderstorms commonly occur over warm coastal regions when breezes from the water flow inland after sunrise when the land surface is warmed by solar radiation to a temperature higher than that of the water (Moore and Vonnegut 1977). Similarly, because mountains are heated before valleys, they often aid the onset of convection in unstable air. Further, horizontal wind blowing against a mountain will be directed upward and can aid in the vertical convection of air parcels, a process which is referred to as the "orographic effect". While relatively small-scale convective thunderstorms (also called air-mass thunderstorms) develop in the spring and summer months when the potential for convection is usually the greatest and an adequate water vapor is available, larger-scale storms associated with frontal activity are observed in temperate latitudes at all times throughout the year. These frontal thunderstorms are formed when, for example, a relatively large mass of cold air moves southward over the United States from the high latitudes and slides under warmer moister air.

Lightning is usually associated with convective cloud systems ranging from 3 to 20 km in vertical extent. The horizontal dimensions of active air-mass thunderstorms range from about 3 km to >50 km. Thunderstorms that are seemingly merged may occur in lines along cold fronts extending for hundreds of kilometers. Ordinary thunderstorms are composed of units of convection, typically some kilometers in diameter, characterized by relatively strong updrafts (≥ 10 m s^{-1}). These units of convection are referred to as cells. The lifetime of an individual cell is of the order of one hour. Thunderstorms can include a single isolated cell, several cells, or a long-lived cell with a rotating updraft, called a supercell. For more information on supercells the reader is referred to MacGorman and Rust (1998: Section 8.1). At any given time a typical multicell storm consists of a succession of cells at different stages of evolution. Large frontal systems have been observed to persist for more than 48 hours and to move more than 2000 km (Moore and Vonnegut 1977). Thunderstorms over flat terrain tend to move at an average speed of 20 to 30 km hr^{-1} (Pierce 1977). Further information on thunderstorm morphology and evolution can be found in the books by Ludlam (1980), Houze (1993), and MacGorman and Rust (1998), in the book edited by Kessler (1986a), and in a chapter of the book by Williams (1995a).

3.2. Cumulonimbus
3.2.1. Idealized gross charge distribution
General information The distribution and motion of thunderstorm electric charges, most of which reside on hydrometeors (various liquid or frozen water particles in the atmosphere) but some of which are free ions, is complex and changes continuously as the cloud evolves. Hydrometeors whose motion is predominantly influenced by gravity (with fall speeds ≥ 0.3 m s^{-1}) are called precipitation particles. All other hydrometeors are called cloud particles. The basic features of the cloud charge structure include a net positive charge near the top, a net negative charge below it, and an additional positive charge at the bottom of the cloud. These features appear to be generally accepted and are illustrated in Fig. 3.1, taken from Krehbiel (1986). It is worth noting that charges of both polarities probably always co-exist in any region of the cloud, regardless of the polarity of the net charge in the region (MacGorman and Rust 1998: p. 53). Note also that the lower positive charge is depicted in Fig. 3.1 as carried by descending precipitation. The origin of the lower positive charge center will be discussed in subsection 3.2.7.

Simple model The charge structure in Fig. 3.1 is often approximated by three vertically stacked point charges (or spherically symmetrical charged volumes), positive at the top, negative in the middle, and an additional, smaller positive at the bottom, located above a perfectly conducting ground as illustrated in Fig. 3.2a. The top two charges are usually called the main charges and are often specified to be equal in magnitude. It is thought that the lower positive charge may not always be present. The two main charges form a dipole, said to be positive because the positive charge is above the negative (giving an upward-directed dipole moment). The electric field intensity **E** due to the system of three charges shown in Fig. 3.2a is found by replacing the perfectly conducting ground with three image charges and using the principle of superposition, the total electric field being the vectorial sum of six contributions (three from the actual charges and three from

3.2. Cumulonimbus

Fig. 3.1. An isolated thundercloud in central New Mexico, with a rudimentary indication of how electric charge is thought to be distributed inside and around the thundercloud, as inferred from the remote and *in situ* observations discussed in subsections 3.2.2 and 3.2.3, respectively. Adapted from Krehbiel (1986).

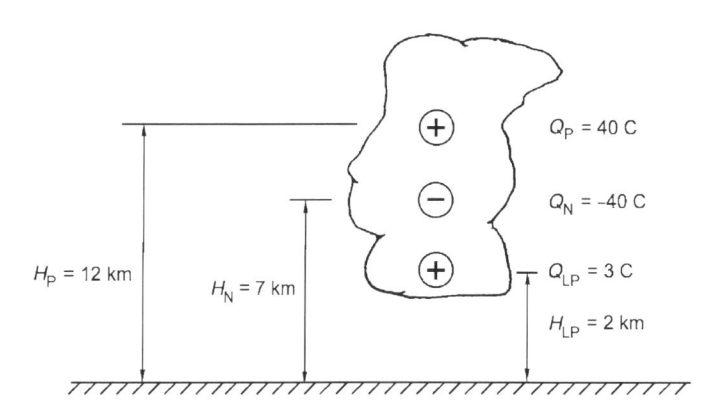

Fig. 3.2a. A vertical tripole representing the idealized gross charge structure of a thundercloud such as that shown in Fig. 3.1; the negative screening layer charges at the cloud top and the positive corona space charge produced at ground are ignored here.

their images). The computation of the electric field on the ground surface due to the main negative charge and its image is shown below, with reference to Fig. 3.2b. Note that in Fig. 3.2b the subscript N is dropped for simplicity, and the charge signs are shown explicitly. Due to the symmetry of the problem with respect to the field point P, the magnitudes of the contributions from the actual charge and its image are equal, and each contribution is found as (e.g., Sadiku 1994)

$$|\mathbf{E}^{(-)}| = |\mathbf{E}^{(+)}| = \frac{|Q|}{4\pi\varepsilon_0(H^2 + r^2)} \quad (3.1)$$

It is clear from Fig. 3.2b that the components of the electric field tangential to the ground plane due to the

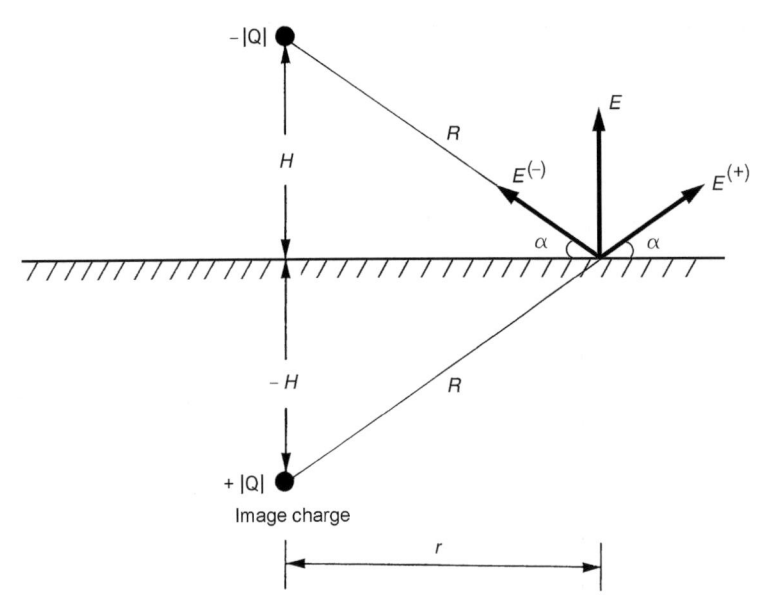

Fig. 3.2b. The method of images for finding the electric field due to a negative point charge above a perfectly conducting ground at a field point located at the ground surface.

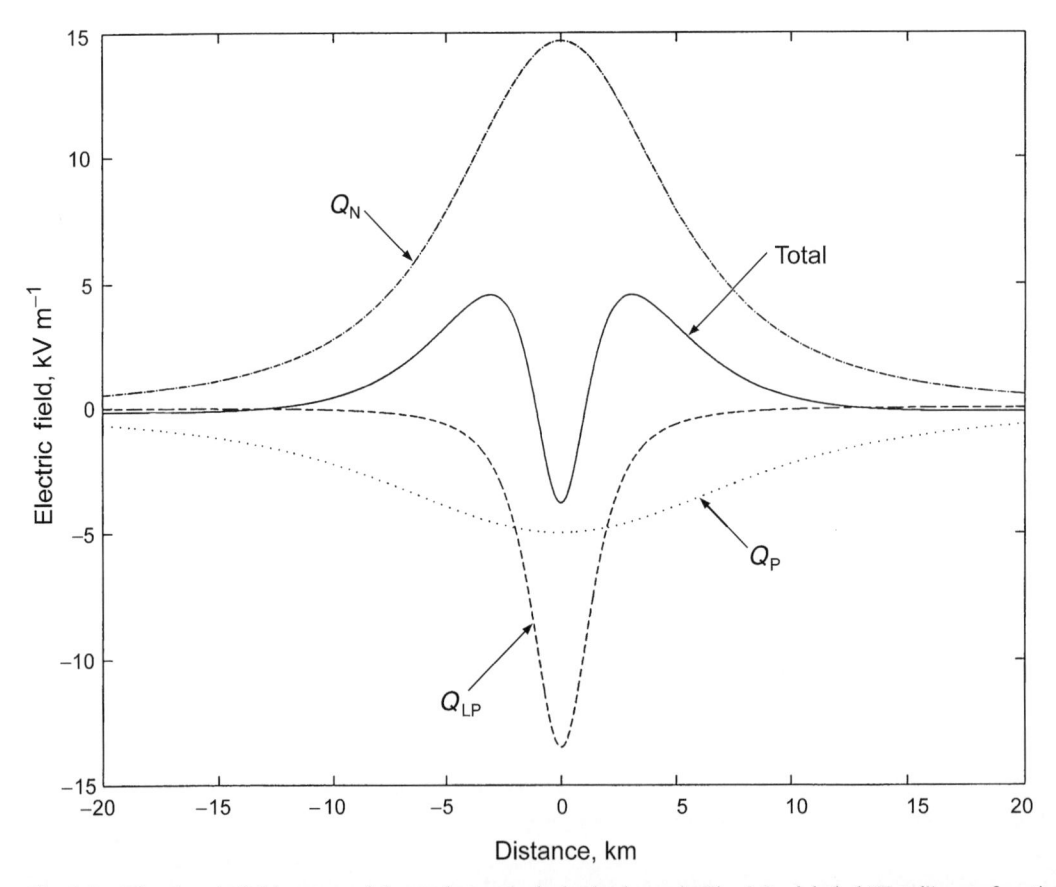

Fig. 3.2c. The electric field at ground due to the vertical tripole shown in Fig. 3.2a, labeled "Total", as a function of the distance from the axis of the tripole. Also shown are the contributions to the total electric field from the three individual charges of the tripole. An upward directed electric field is defined as positive (according to the physics sign convention; subsection 1.4.2).

3.2. Cumulonimbus

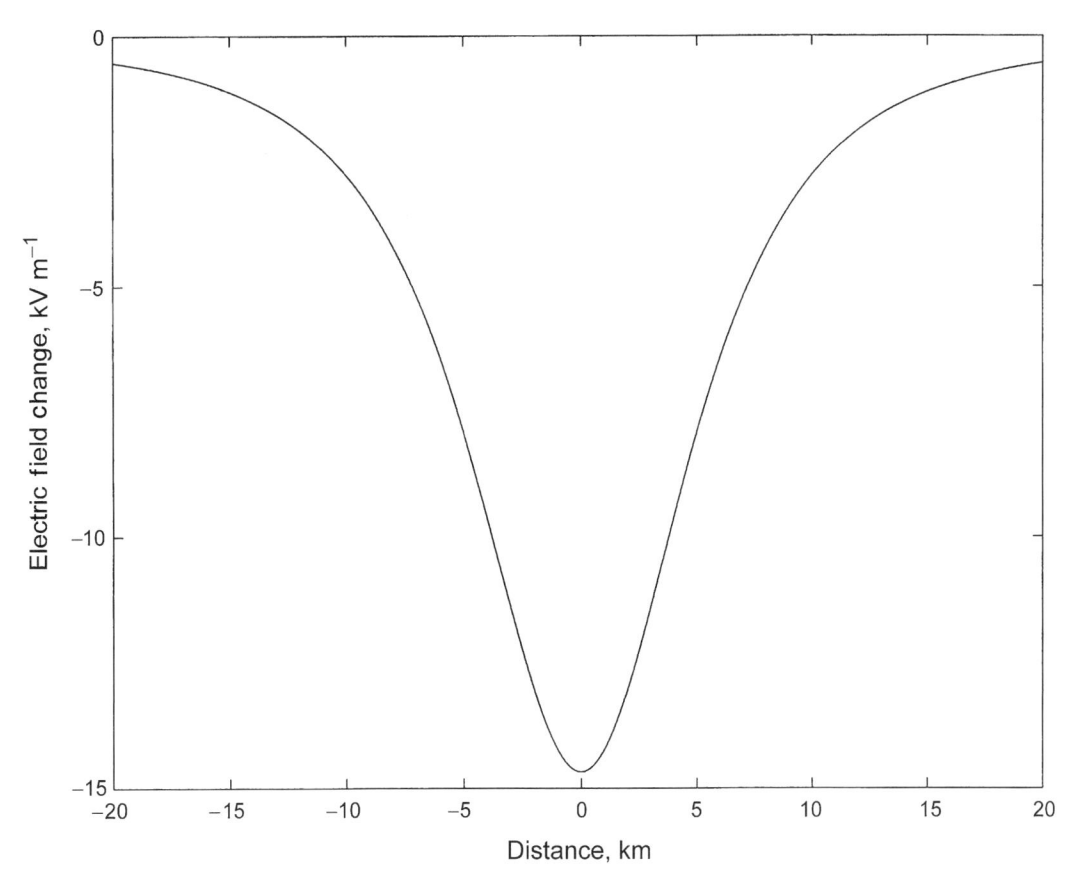

Fig. 3.2d. Electric field change at ground, due to the total removal of the negative charge of the vertical tripole shown in Fig. 3.2a via a cloud-to-ground discharge, as a function of distance from the axis of the tripole. Note that the electric field change at all distances is negative.

actual charge and its image cancel each other, as expected from the boundary condition on the surface of a perfect conductor. The electric field components normal to the ground plane add, the total normal field magnitude being twice the contribution from either the actual charge or its image:

$$|\mathbf{E}| = 2|\mathbf{E}^{(-)}|\cos(90° - \alpha) = 2|\mathbf{E}^{(+)}|\cos(90° - \alpha)$$
$$= 2|\mathbf{E}^{(-)}|\sin\alpha = \frac{|Q|H}{2\pi\varepsilon_0(H^2 + r^2)^{3/2}} \quad (3.2)$$

To facilitate further discussion of the variation in $|\mathbf{E}|$ as a function of H and r ($|Q|$ = constant), we rewrite Eq. 3.2 in the following form:

$$|\mathbf{E}| = k\frac{\sin\alpha}{R^2} \quad (3.3)$$

where $k = |Q|/(2\pi\varepsilon_0)$ and $R^2 = (H^2 + r^2)$. For fixed r, $|\mathbf{E}|$ increases as H increases from zero to $H = D/\sqrt{2}$ (because $\sin\alpha$ increases faster than R^2) and then decreases as H increases further (because now R^2 increases faster than $\sin\alpha$). If we fix H and vary r, $|\mathbf{E}|$ will decrease monotonically with increasing r. The rate of decrease depends on H, being slower for larger H. As a result, for two vertically stacked charges of equal magnitude (say, the main positive and main negative charges in Fig. 3.2a), the relative contribution to $\mathbf{E}$ from each of these two charges will depend on r. The electric field at $r = 0$ is dominated by the lower charge (since it is closer and $\sin\alpha$ is the same for both charges), but as r increases so does the relative contribution from the upper charge. At a certain distance, the contribution from the upper charge becomes dominant, and the total electric field (the sum of the contributions from the two charges) changes its polarity. This distance is called the reversal distance. For the case of two vertically stacked charges of equal magnitude but opposite polarity the reversal distance, D_0, is given by

$$D_0 = \left[(H_P H_N)^{2/3}(H_P^{2/3} + H_N^{2/3})\right]^{1/2} \quad (3.4)$$

where H_P and H_N are the heights of the positive and negative charges, respectively. Thus, for a positive dipole, one might expect that at near ranges the total electric field is negative

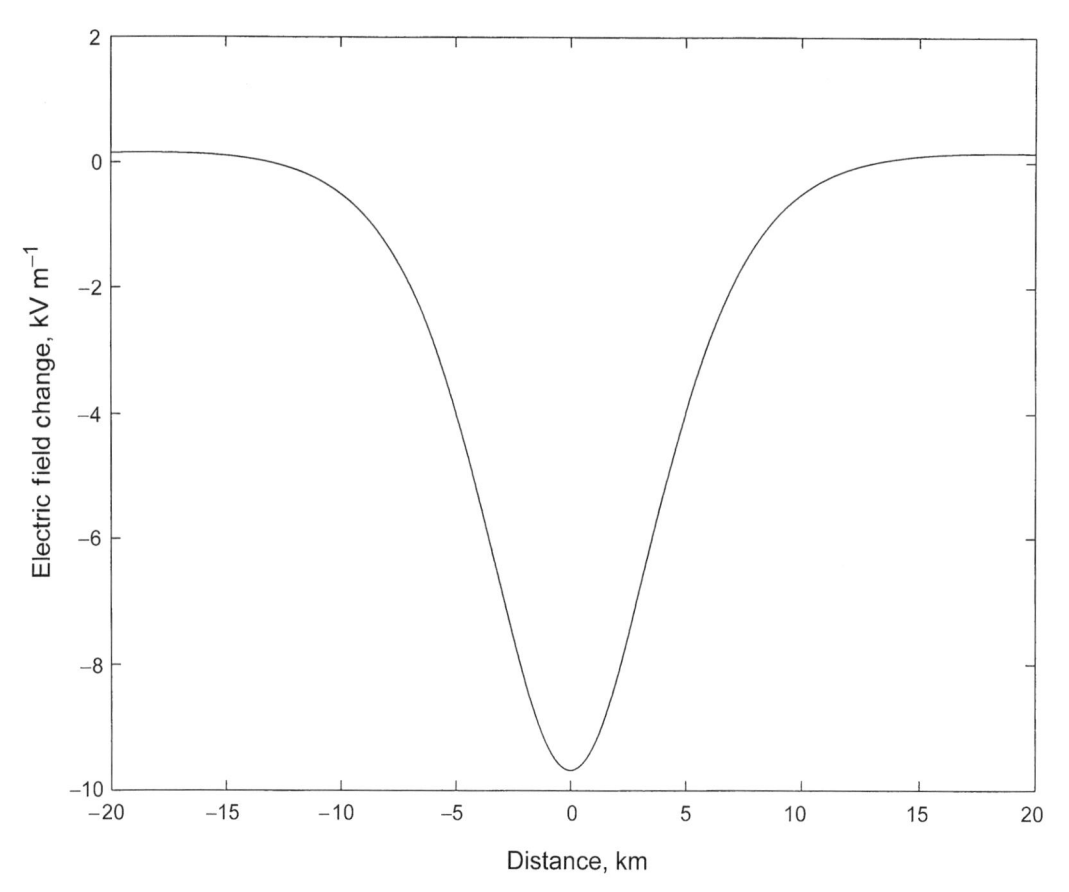

Fig. 3.2e. Same as Fig. 3.2d, but due to the total removal of the negative and upper positive charges via a cloud discharge. Note that the electric field change at close distances is negative, but at far distances it is positive.

(the closer negative charge is more "visible"), while at far ranges it is positive (the positive charge, with larger elevation angle, is more "visible").

The electric field at ground due to the system of three charges (a vertical tripole) shown in Fig. 3.2a, computed assuming that the middle negative and top positive charges are 7 and 12 km above ground, respectively, each having a magnitude of 40 C, and that the bottom positive charge is at 2 km and has a magnitude of 3 C, is shown in Fig. 3.2c. An upward-directed electric field is defined in this chapter as positive (the "physics" sign convention, subsection 1.4.2). As seen in Fig. 3.2c, the total electric field exhibits polarity reversals. Such a variation in electric field versus distance, although model dependent, is qualitatively consistent with the available experimental data, as discussed in subsection 3.2.2. Also shown in Fig. 3.2c are the contributions to the total electric field from each of the three charges.

Now we consider electric field changes due to lightning discharges. In general, an electric field change is found as the difference between the final electric field value (after the charge removal due to lightning) and the initial electric field value due to the original cloud charge distribution. For any charge removed from the cloud, the corresponding electrostatic field change is the negative of the contribution of that charge to the initial electric field. If we assume that the negative cloud charge is completely neutralized as a result of a cloud-to-ground discharge, the resultant net electric field change will be negative at any distance, as shown in Fig. 3.2d, because the upward-directed (positive) electric field due to the negative charge (see Fig. 3.2b) disappears, that is, becomes zero. If both main positive and main negative charges are neutralized via an intracloud discharge, the resultant net field change as a function of distance will exhibit a polarity reversal, as seen in Fig. 3.2e. Note that the positive field values in Fig. 3.2e are considerably smaller than the negative field values. The polarity reversal occurs because the net field change is the negative of the sum of the contributions to the total electric field, shown in Fig. 3.2c, from these two charges, this sum being positive at close ranges (dominated by the negative charge) and negative at far ranges (dominated by the upper positive charge). In other

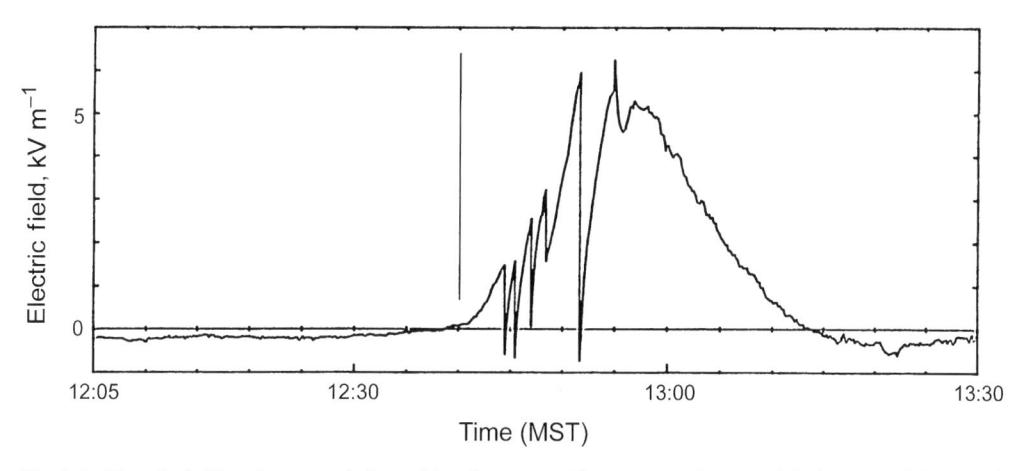

Fig. 3.3. Electric field at the ground about 5 km from a small storm near Langmuir Laboratory, New Mexico, on 3 August 1984. An upward-directed electric field is defined as positive (according to the physics sign convention; see subsection 1.4.2). The large pulses superimposed on the rising portion of the overall electric field waveform are due to lightning. Adapted from Krehbiel (1986).

words, for such an intracloud discharge, the electric field change at close ranges is dominated by the reduction in the positive (upward-directed) electric field and at far ranges by the reduction of the negative (downward-directed) electric field.

In the following two subsections, 3.2.2 and 3.2.3, we will review remote (outside the cloud) and *in situ* (inside the cloud) measurements of electric fields from which the gross charge structure shown in Fig. 3.1 has been inferred. We will see in subsection 3.2.3 that recent *in situ* observations reveal a more detailed structure than presented in Fig. 3.1. Then, we will present the maximum measured in-cloud electric fields (subsection 3.2.4) and the estimated magnitudes of cloud charges and charge densities (subsection 3.2.5). Finally, we will discuss cloud electrification mechanisms (subsection 3.2.6), the possible origin of the lower positive charge center (subsection 3.2.7), and the representation of lightning in numerical models of evolving thunderclouds (subsection 3.2.8).

3.2.2. *Inferences from remote measurements*

Remote electric field measurements fall into two categories: (i) measurements of the slowly varying fields associated with the cloud charges, such as the model-predicted fields shown in Fig. 3.2c, or (ii) measurements of the more rapid field changes associated with the neutralization of a portion of those cloud charges by lightning. Interestingly, Waldteufel *et al.* (1980) presented evidence that charges neutralized by lightning can be located in clear air as opposed to being located inside the thundercloud. Figure 3.3 shows an electric field record that illustrates both the relatively steady cloud electric field and the transient lightning electric field changes for a thunderstorm that grows and dies at a more or less fixed distance of about 5 km from the observation point at ground.

Remote measurements have been made by different researchers both at ground level and over the cloud tops using aircraft or balloons. Here, we will discuss only ground-based measurements. The measurements made above the tops of thunderclouds (e.g., Gish and Wait 1950; Stergis *et al.* 1957; Vonnegut *et al.* 1966; Imyanitov *et al.* 1971; Blakeslee *et al.* 1989; Marshall *et al.* 1995b) generally give results consistent with those of ground-based observations and are reviewed in MacGorman and Rust (1998, pp. 177–8).

Wilson (1916, 1920, 1929) was the first to observe, from ground-based measurements, systematic variations in the polarity of both the relatively steady electric fields of thunderclouds and the transient electric field changes due to lightning. In particular, Wilson found that at close ranges cloud electric fields tended to be positive (upward-directed) and at far ranges negative (downward-directed). Further, he observed that electric field changes due to lightning discharges were more often negative nearby than far away; the reason for this is probably that predominantly vertical intracloud lightning flashes (i) produce negative field change at close ranges and positive field change at far ranges (Fig. 3.2e) and (ii) are more numerous than cloud-to-ground flashes (Section 2.7) most of which, excluding the relatively rare positive discharges, produce negative field changes at all ranges (Fig. 3.2d). In view of these findings, Wilson suggested that thunderclouds typically have positive charge above negative charge, a configuration generally called a positive dipole, as noted in subsection 3.2.1. In the following, we will further discuss the positive dipole and tripole models, referring to Figs. 3.2, 3.3, and 3.4. Note that other

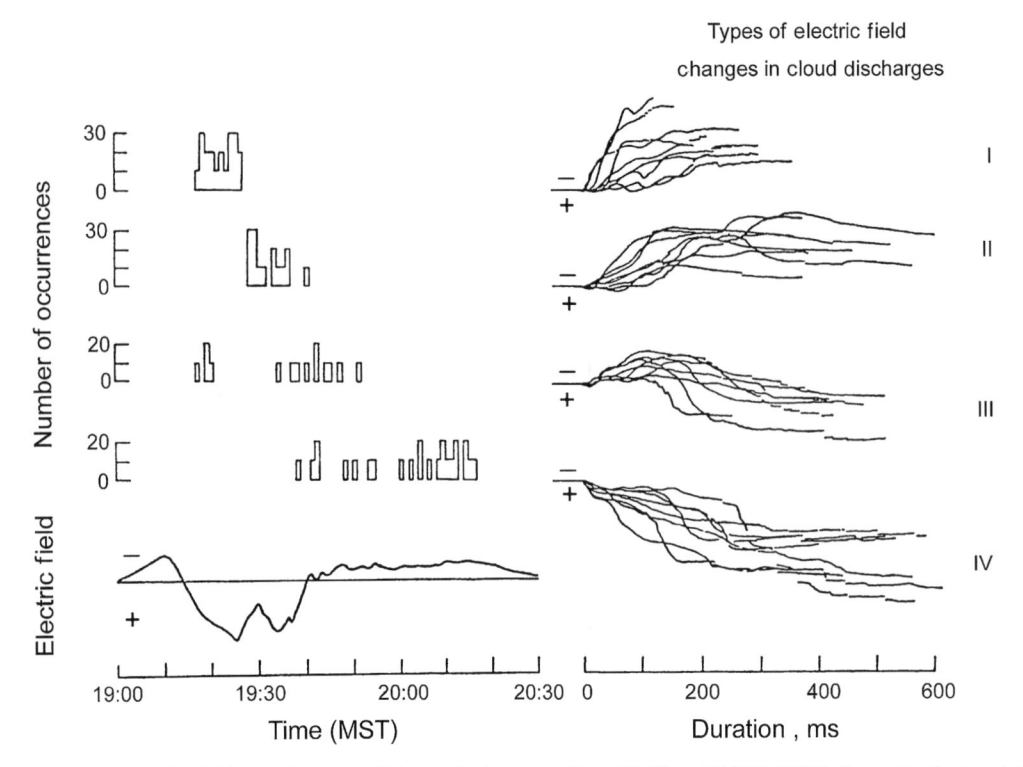

Fig. 3.4. Electric field as a function of time, which ranges from 19:00 to 20:30 MST (left panel at bottom) for the isolated thunderstorm observed on 28 September 1961 in Socorro, New Mexico, and four types of electric field changes (on a 600 ms time scale) due to cloud discharges (right panel) in that thunderstorm. An upward-directed electric field is defined as positive. Positive electric field changes are produced by either an increase of positive electric field magnitude or a decrease of negative electric field magnitude. The occurrence of the four types of electric-field-change waveforms at different times during the thunderstorm is indicated in the upper part of the left panel. The thunderstorm approached the field measuring station to within roughly 4 km between 19:20 and 19:30 MST, when most of the monotonic negative cloud-flash field-change waveforms (type I), were observed; these indicate a reduction in upward-directed (positive) electric field magnitude. Then the thunderstorm receded from the measuring station to roughly 8 km at 19:50 MST, when the cloud-flash electric-field-change waveforms became predominantly monotonic positive (type IV), indicating a reduction in the downward-directed (negative) electric field magnitude. Adapted from Ogawa and Brook (1964).

models (involving more than three charges) may well be consistent with the measurements presented in Figs. 3.3 and 3.4, as discussed later.

As indicated in subsection 1.4.2, in clear-sky (fair-weather) conditions, the electric field vector points downward and is defined in this chapter as negative (according to the "physics" sign convention). The sources of the fair-weather electric field are positive space charge in the atmosphere and negative charge on the Earth's surface. The fair-weather field has a magnitude of about 100 V m^{-1} (subsection 1.4.2). Beneath an active thundercloud the electric field at the ground is usually reversed in sign (directed upward) with respect to the fair-weather field and is considerably larger, 1 to 10 kV m^{-1} on relatively flat terrain. In Fig. 3.3, the fair-weather field is measured from 12:05 to about 12:30 and again after about 13:28, while a large, predominantly upward-directed electric field indicative of a dominant negative charge in the cloud overhead is seen from roughly 12:43 to 13:08. As noted above, the thundercloud whose electric field is shown in Fig. 3.3 was at a distance of about 5 km from the measuring station. It follows from Fig. 3.2c that if the cloud were stationary at $r = 0$, an observer at, say, $r = 20$ km would "see" a negative (downward-directed) electric field, dominated by the larger-elevation-angle upper positive charge center. If the observer moved closer to the vertical axis of the cloud tripole, say to $r = 5$ km, the electric field would now be positive (upward directed), dominated by the closer negative charge center. At $r = 0$, the electric field is strongly influenced by the smaller positive charge at the bottom of the cloud. If the observer moved away from the vertical axis of the cloud tripole, the described electric field features would be "seen" in reverse order. Similar observations would be made if the observer were stationary and the cloud were approaching, passing over, and then receding, as illustrated in Fig. 3.4, adapted from Ogawa and Brook (1964). As a storm approached, the negative (downward-directed) fair-weather electric field increased in magnitude due to the effect of the main positive charge found at the top of the cloud, then decreased and passed through zero to become positive (upward-directed) when the cloud was at the closest position (within a few kilometers) to the observation point. The pronounced field excursion indicating a reduction in the positive (upward-directed) electric field at about 19:30 is likely to have been due to positive charge at the bottom of the cloud. A positive-to-negative (from upward-directed to downward-directed) field polarity reversal occurred at about 7.5 km (about 19:40) as the cloud receded from the observer.

As seen in Fig. 3.3, lightning flashes produce rapid electric field changes associated with the neutralization of charge while the ongoing charging processes (subsection 3.2.6) in the cloud serve to restore and perhaps even to increase further the electric field magnitude. At close ranges, these lightning-related field excursions may cause temporary polarity reversals, as seen in Fig. 3.3, since the magnitude of the excursions may exceed the magnitude of the field beneath a thundercloud. Note that the six pronounced lightning-produced electric field changes in Fig. 3.3 are negative (according to the physics sign convention, subsection 1.4.2). A negative polarity of electric field changes is expected for (i) cloud-to-ground discharges at any range, since they effectively remove negative charge from the cloud, and (ii) for cloud discharges at close ranges (such as 5 km), since the field change due to the removal of negative charge is larger at close ranges than the field change due to the removal of the equal amount of positive charge from a greater altitude. Typically, at 20 km the magnitude of the total lightning electric field change is about 100 V m^{-1}. Figure 3.4 (right-hand panel) illustrates that electric field changes due to cloud discharges tend to indicate a reduction in magnitude of the upward-directed electric field at close ranges and an increase in magnitude of the upward-directed electric field as the distance increases.

We next review inferred cloud charge distributions based on multiple-station measurements of electric field changes (sometimes referred to simply as electric fields, for brevity) on ground. Such measurements were first performed in New Mexico (e.g., Workman *et al.* 1942; Reynolds and Neill 1955) and typically indicated that the negative charge centers neutralized by cloud flashes and individual strokes of ground flashes tended to occur in a layer 1 to 3 km thick, with a lower boundary at a temperature between 0 °C and −10 °C.

Jacobson and Krider (1976), using the KSC electric field-mill network (Section 17.2), found that a total flash charge of −10 to −40 C was lowered to ground from a height of 6 to 9.5 km above sea level, a height where the clear-air temperature was between −10 and −34 °C. In this paper they also summarized much available data from other studies concerning the location and magnitude of the negative charge neutralized by lightning, the bulk of these data being similar to their results from Florida. Maier and Krider (1986), using the same KSC electric field mill network, showed that the altitude from which the negative flash charge is lowered varies very little from flash to flash throughout a given day, but it does vary from day to day.

Krehbiel *et al.* (1979), using an eight-station electric-field-change measuring system in New Mexico, determined the magnitudes of the charges lowered to ground by individual strokes and by continuing currents (Section 4.8) in four multiple-stroke flashes. The charges were displaced primarily horizontally in a relatively narrow range of heights from 4.5 to 6 km (one exception, 3.6 km) above ground (Fig. 3.5) at clear-air temperatures between −9 and −17 °C. Further, Koshak and Krider (1989) presented the results of a lightning field-change analysis (Section 17.2) for a portion of an active storm, based on data from the KSC electric field mill network. These results are reproduced in Fig. 3.6. The circles show the locations of point charge solutions that indicate where cloud-to-ground flashes remove negative charge from or, equivalently, deposit positive charge in the cloud. The numbers in the circles give the magnitudes of those charges in coulombs. The arrows show the locations (the dot in the middle of the arrow), the directions, and the magnitudes of the point-dipole charge moments (Section 17.2); the directions also indicate the positive charge transport by cloud discharges. The relative positions of the circles and the majority of the arrows suggest that a negative charge region is located below a positive charge region. Further, the predominantly upward-directed arrows below the cluster of circles are indicative of a small pocket of positive charge below the negative charge region, a feature first identified from in-cloud measurements (subsection 3.2.3). Additional data similar to those presented in Fig. 3.6 are found in Krider (1989) and Murphy *et al.* (1996).

It has been inferred from a combination of remote and *in situ* (subsection 3.2.3.) measurements that in very different environments negative charge is typically found in the same relatively narrow temperature range, roughly −10 to −25 °C, where the clouds contain both supercooled water and ice. This inference is illustrated in Fig. 3.7. Stolzenburg *et al.* (1998a,b,c), from *in situ* balloon soundings (subsection 3.2.3), indicated that the average temperature of the center of the main negative charge region may depend on storm type: −16 °C in MCS convective region updrafts, −22 °C in supercell updrafts, and −7 °C in New Mexican mountain storm updrafts. These three average temperatures are approximately in the range specified above, the differences apparently being related to

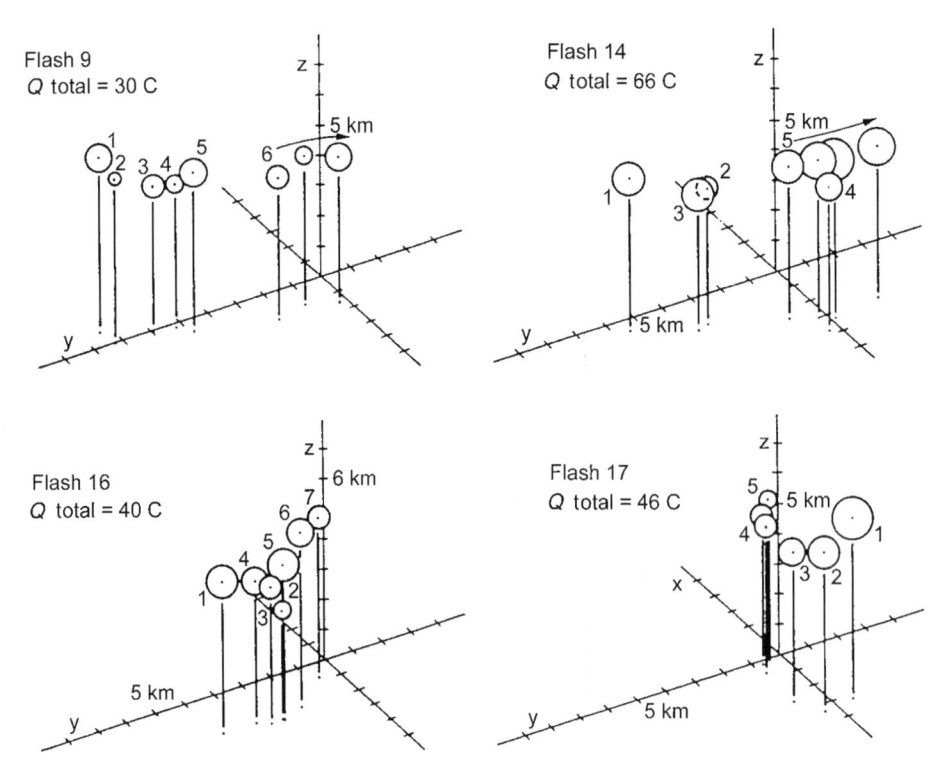

Fig. 3.5. Charge source locations for strokes and continuing currents in four multiple-stroke flashes in New Mexico. All strokes are numbered sequentially. In the case of strokes followed by continuing current, two or more locations appear under one number (this applies to the following: flash 9, event 6; flash 14, event 5; flash 17, event 4). In flashes 9 and 17, the continuing-current charges are incremental while in flash 14 they are cumulative from the beginning of stroke 5, except for the last charge which is incremental. The charge magnitude is represented by the size of the sphere surrounding the charge location point, the radii of the spheres being determined assuming a charge density of 20 nC m^{-3}. The electric fields versus time at eight stations for flash 14 are given in Fig. 4.9. Adapted from Krehbiel et al. (1979).

the updraft speed (the larger the updraft speed, the higher the altitude and the lower the temperature of the main negative charge center). It has been further observed (Krehbiel 1986) that the negative charge center involved in lightning flashes appears to remain at an approximately constant altitude as a storm grows, this feature being illustrated for a small Florida storm in Fig. 3.8. Figure 3.8 shows the heights of the charges neutralized by the first 15 lightning discharges, 13 of which were cloud flashes and two of which were ground flashes. As the storm grew vertically, the positive (upper) charges involved in cloud flashes tended to be found at progressively higher altitudes, increasing in time from 10 km ($-30\,°C$) to 14 km ($-60\,°C$) during the 8 min period of observation. However, the negative (lower) charges involved in cloud flashes and the negative charges neutralized by ground flashes remained at about 7 km altitude ($-15\,°C$). Krehbiel et al. (1979) and Krehbiel (1981) found that the negative charge centers for most cloud and ground flashes were near radar reflectivity cores and usually above the level of maximum reflectivity.

A number of factors are usually neglected when inferring charge locations and magnitudes from electric field measurements on ground, although potentially these factors can compromise the validity of the inferences. First of all, the overall charge of each polarity is not uniformly distributed in a single spherical volume but is localized in pockets of relatively high space-charge concentration. Indeed, Davis and Standring (1947) inferred the existence of oppositely charged regions separated by a relatively small horizontal distance of 300 m by measuring the currents in the cables of kite balloons flying at a height of 600 m under thunderstorm conditions. Bateman et al. (1999) found, from in situ measurements, precipitation particles carrying charge of either polarity at nearly all altitudes. Such a "lumpy" charge distribution, when characterized using remote measurements, is either averaged in some way, as in the case of the electric fields associated with the cloud charges in a simple multipole cloud charge model, or sampled, as in the case of the electric field changes due to individual lightning discharges. Further, besides the charged regions in the cloud interior, appreciable charges can be accumulated at the cloud boundaries, forming "screening charge layers". Such layers must be formed on all surfaces of a cloud because of the conductivity difference across the

3.2. Cumulonimbus

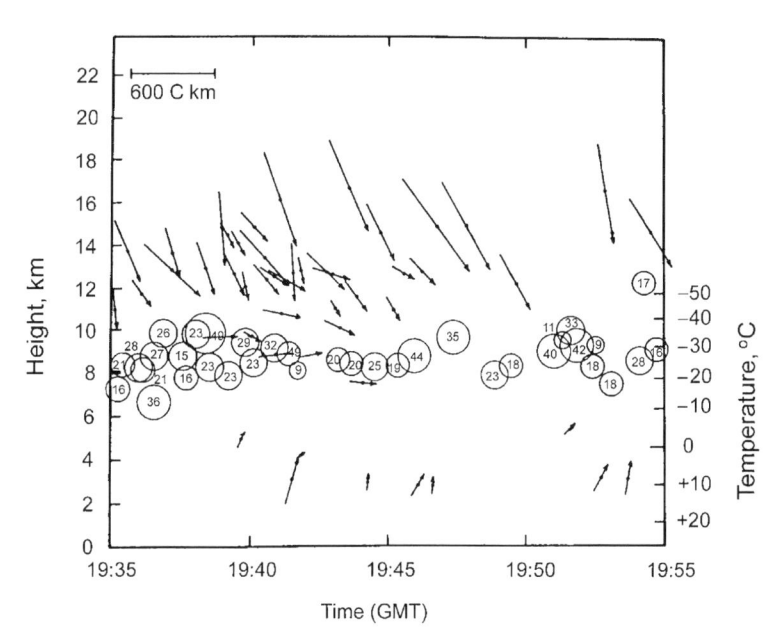

Fig. 3.6. Negative charges (circles) neutralized by ground flashes, and point-dipole charge moments (arrows) describing the effective positive charge transfer by cloud flashes as a function of time for a portion of an active Florida storm on 6 July 1978. The numbers in the circles give the magnitudes of the neutralized charges in coulombs. In cloud flashes, negative charge was effectively transported in the direction opposite to that of the arrow to neutralize a positive charge of equal magnitude. The dot in the middle of each arrow represents the apparent single location of these two charges, the actual locations being indeterminate in the point-dipole solution (subsection 17.2.1). Adapted from Koshak and Krider (1989).

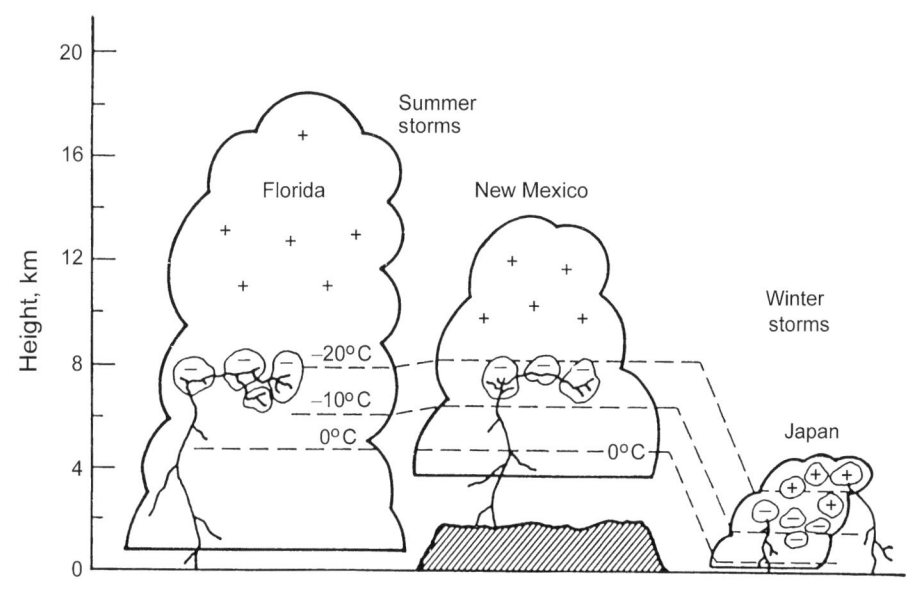

Fig. 3.7. The locations, shown by the small irregular contours inside the cloud boundaries, of ground flash charge sources observed in summer thunderstorms in Florida and New Mexico and in winter thunderstorms in Japan, using simultaneous measurements of electric field at a number of ground stations. More information on the charge structure of winter thunderclouds in Japan is found in Chapter 8. Adapted from Krehbiel (1986).

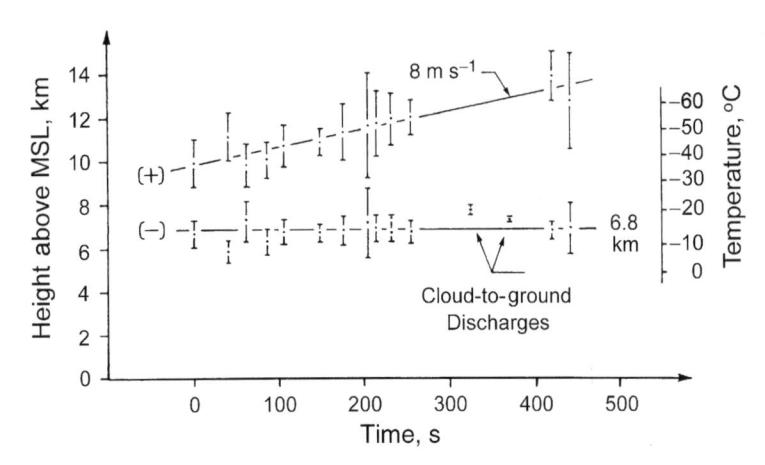

Fig. 3.8. The heights of the charge centers involved in the first 15 lightning discharges versus time (a total of eight minutes) in a small Florida storm. The upper positive charge centers of the 13 cloud flashes increased in altitude as the storm grew, while the negative charge centers remained at constant altitude. The negative charge centers of the two cloud-to-ground discharges that occurred toward the end of the sequence were at about the same altitude as the negative charge centers of the 13 cloud flashes. Adapted from Krehbiel (1986).

cloud boundary, the charge polarity being opposite to that of the charge inside the cloud. Byrne *et al.* (1983), for example, observed both upper negative and lower positive screening layers in New Mexico thunderclouds. The lower positive charge center may in fact be a screening layer at the bottom of the cloud, as discussed along with other possible origins of this charge center in subsection 3.2.7. The experimental evidence for screening layers on the sides of thunderclouds is inconclusive (MacGorman and Rust 1998, p. 71), probably because they are less pronounced and, hence, more difficult to detect. Phillips (1967b) noted that the magnitude of the screening charges should be greatest near the base and top of the cloud, where the outward component of the electric field of the interior dipole charges is a maximum.

The upper (main) positive charge attracts negative ions to the top of the cloud from the electrically conducting clear air (subsection 1.4.1) around the cloud. These ions are produced by the attachment of electrons freed by cosmic rays, high-energy electromagnetic radiation and particles (mostly nuclei of hydrogen and helium) coming from outer space. The ions attach to small cloud particles at the edge of the cloud forming a negative screening layer up to a few hundred meters thick, whose electric field partially cancels the electric field of the interior positive charge and hence partially screens this interior charge from an outside observer. At lower altitudes ionization by cosmic rays diminishes, but ionization caused by the natural radioactivity of Earth can become appreciable (contributing up to half the ions found near ground, see MacGorman and Rust 1998: p. 321).

Finally, the main negative charge causes point discharge, also called corona, from trees and various pointed objects on the ground below the thundercloud; this creates a blanket of positive charge in the air near the Earth's surface, between the cloud and ground-based observer. Because of this corona near the ground, the electric field magnitude at ground level is limited to the typically observed values of 1–10 kV m^{-1} (e.g., Standler and Winn 1979; Chauzy and Raizonville 1982; Soula and Chauzy 1991; Soula 1994), as indicated earlier in this subsection. As evidence of this effect, Soula and Chauzy (1991) measured a maximum electric field of 65 kV m^{-1} at an altitude of about 600 m, while the steady surface electric field did not exceed 5 kV m^{-1}. They attributed the observed altitude dependence of the electric field to ions generated by corona at ground, which modified the electric field profile in the absence of corona. On a lake 500 m from shore Toland and Vonnegut (1977) measured much higher electric fields, ranging from 38 to 130 kV m^{-1} in seven storms, than those typically measured over land. They attributed these relatively high field values to the absence of corona charge over the calm water surface. Chauzy *et al.* (1989) demonstrated that even the magnitude of the lightning electric field change at ground level, viewed on a time scale of some hundreds of milliseconds, could be appreciably affected by corona at the ground. This corona effect can potentially compromise the inferences regarding cloud charge structure made from the measurements of overall lightning electric field changes at ground level.

Figure 3.1, introduced in subsection 3.2.1, shows, besides the previously discussed tripole in the cloud's interior, an upper negative screening layer and a positive corona charge near ground. Nevertheless, the charge distribution depicted in Fig. 3.1 is grossly simplified and, furthermore, is not steady in time (except, perhaps, for the height of the main negative charge region) as is often assumed.

To summarize the results of the remote electric field measurements, the negative charge involved in lightning

3.2. Cumulonimbus

Table 3.1. *Altitude, H, and magnitude, Q, of charge inferred or assumed for the lower positive charge center and the main negative and positive charge centers in thunderclouds. Adapted from MacGorman and Rust (1998)*

Reference	Lower positive charge		Main negative charge		Main positive charge	
	H, km	Q, C	H, km	Q, C	H, km	Q, C
Wilson (1920)	—	—	1	−33	10	33
Simpson and Scrase (1937)	1.7	—	2.7	—	≥4.6	—
Simpson and Robinson (1941)	1.5	4	3	−20	6	24
Gish and Wait (1950)	—	—	3	−39	6	39
Huzita and Ogawa (1976)	3	24	6	−120	8.5	120
Kuettner (1950)	1.5	—	3	—	6	—
Malan (1952)	2	10	5	−40	10	40
Wait (1953)	—	—	3	−39	6.1	39
Tamura (1955)	—	—	5	−120	7	80
Stergis *et al.* (1957)	—	—	5	−20	8	37.5
Kasemir (1965)	—	50	—	−340	—	60
Tzur and Roble (1985)	—	—	5	−50	10	50

Note: Further information on the cloud charge distribution inferred from remote and *in situ* measurements, usually not readily available in the Q–H format, is found in subsections 3.2.2 and 3.2.3, respectively. The magnitudes of cloud charges are discussed additionally in subsection 3.2.5. Various concepts of the origin of the lower positive charge center are reviewed in subsection 3.2.7.

flashes tends to have a relatively small vertical extent that is apparently related to the −10 to −25 °C temperature range, regardless of the stage of storm development, the location, and the season. The main positive charge involved in lightning flashes probably has a larger vertical extent and is located above the negative charge. An additional, smaller, positive charge can be formed below the negative charge. The magnitudes and heights of the two (dipole) or three (tripole) cloud charges from electric field measurements at the ground or aloft (subsection 3.2.3) and from analyses of lightning electric field changes are summarized in Table 3.1. Additional information on the charge structure of winter thunderclouds in Japan is found in Chapter 8.

In the remainder of this subsection, we will consider VHF–UHF, acoustic, and radar observations of lightning activity in the cloud that provide some additional information on the locations of cloud charges. Proctor (1991), using a VHF–UHF TOA lightning locating system (Section 17.4), examined the distribution of origin heights for 773 flashes (including 214 cloud-to-ground flashes) in 13 thunderstorms in South Africa. The distribution was bimodal with peaks at 5.3 and 9.2 km above mean sea level, the lower mode (the low-origin flashes) being associated with a temperature range of about 1 °C to −9 °C and the upper mode (the high-origin flashes) with a temperature range of roughly −25 °C to −35 °C. It is worth noting that the heights of the flash origins are not expected to indicate the heights of the charge centers but, rather, the heights of the boundaries of concentrated-charge regions, where the electric field is a maximum. Based on an analysis of the simultaneous wideband electric field records for 165 flashes, including 17 high-origin flashes, Proctor (1991) inferred that the majority of the 773 flashes, including 342 high-origin flashes, carried negative charge away from the flash origin.

The lower flash origins reported by Proctor (1991) are within a temperature range where the negative charge centers are often found in thunderstorms (Fig. 3.7). Thus it is logical to assume that the lower-origin flashes tap the lower negative charge region. The flashes with high origins probably begin near the interface between the main positive and main negative charge regions. Proctor (1991) suggested that the charge source for some of the high-origin negative flashes was the upper negative screening layer. Lightning activity apparently originating at two different height levels in the cloud has also been reported by MacGorman *et al.* (1983) from acoustic observations, by Mazur *et al.* (1984) from radar studies, and by Ray *et al.* (1987) and Taylor *et al.* (1984) from the VHF imaging of lightning channels. VHF images of so-called bilevel cloud discharges (e.g., Shao and Krehbiel 1993, 1996; Rison *et al.* 1999; Thomas *et al.* 2000) that apparently indicate the positions of the charge regions involved are discussed in subsection 9.3.1.

3.2.3. Inferences from in situ measurements

In situ measurements of electric field inside the cloud have been made using (i) free balloons carrying corona probes (e.g., Simpson and Scrase 1937; Simpson and Robinson 1941; Arabadzhi 1956; Taniguchi *et al.* 1982; Weber *et al.* 1983; Byrne *et al.* 1983, 1987, 1989), (ii) free balloons carrying electric field meters (e.g., Winn and Byerley 1975; Winn *et al.* 1978, 1981; Weber *et al.* 1982;

Marshall and Winn 1982; Marshall et al. 1989, 1995c; Marshall and Rust 1991, 1993), (iii) aircraft (e.g., Gunn 1948; Imyanitov et al. 1971), (iv) rockets (e.g., Winn and Moore 1971; Winn et al. 1974), and (v) parachuted electric field mills (e.g., Evans 1969). Marshall et al. (1995b) used both rocket- and balloon-borne electric field meters. Aircraft measurements are usually made at only a few selected altitudes, while balloons provide a vertical profile without systematic measurements in the horizontal direction. In the rest of this subsection we will primarily discuss measurements made with vertically ascending balloons, since these measurements provide the most revealing data on the electrical properties of thunderclouds.

Simpson and Scrase (1937) and Simpson and Robinson (1941) were the first to infer cloud charge structure from *in situ* measurements. These measurements, made on balloons ascending through thunderstorms in England, included records of the sign of the vertical electric field and of the atmospheric pressure, from which balloon height could be inferred. Their inferred cloud charge structure was composed of three vertically stacked charges (a tripole): a lower positive charge of +4 C at temperatures warmer than 0 °C, a main negative charge of −20 C between 0 °C and −10 °C, and a main positive charge of +24 C at temperatures colder than −10 °C. Note that the algebraic sum of the charges is not zero (Table 3.1). More recent *in situ* measurements, discussed later, are not necessarily in agreement with this widely accepted tripole model (see also Fig. 3.2a).

In situ measurements are superior to remote measurements (subsection 3.2.2) in that a relatively accurate charge height can be determined. However, since the balloon can sense the field only along a more or less straight vertical path and it samples different portions of that path at different times, the charge magnitude can be estimated only if assumptions regarding the size and shape of individual charge regions and the charge variation with time are made. The average volume charge density, ρ_v, in the cloud is generally found by assuming that the charge (i) is horizontally uniform (i.e., that there is a negligible variation with x or y compared to the variation with z), which can be viewed as an assumption of charge layers of infinitely large horizontal extent, and (ii) does not vary in time. Then, according to Gauss's law in point form (e.g., Sadiku 1994), $\rho_v = \varepsilon_0 (dE_z/dz)$, that is, ρ_v is proportional to the rate at which the vertical electric field E_z increases (in a positive charge density region) or decreases (in a negative charge density region) with increasing altitude z as the balloon ascends. Stolzenburg and Marshall (1994) showed that the assumption of charge layers of infinite horizontal extent is a valid approximation for calculating the charge density of finite charge layers. However, the assumption that charge densities are time invariant is questionable. Indeed, the time required for a balloon to traverse a cloud, 30–45 min (Marshall and Rust 1991), is comparable to or exceeds the typical duration of the mature stage of a thunderstorm cell. As a result, the stages preceding and following the mature stage (and associated transitions) are also likely to be involved in the sounding, potentially invalidating the inherent assumption of a steady (time-invariant) charge distribution. From measurements made on aircraft flying repeatedly over a cloud top, Imyanitov et al. (1971) reported many cases of electric field polarity reversal as the cloud proceeded from the initial to the mature stage.

Figure 3.9 shows the results of a vertical sounding of the electric field in a small New Mexico storm that produced no lightning. This electric field profile was obtained up to a height of 10 km above mean sea level using a balloon-borne instrument that measured the corona current from a 1-m-long vertical wire. The corona current and the corresponding vertical electric field reversed sign twice, between 6 and 7 km and above 9 km. The charge structure in Fig. 3.9, a negative charge between −5 and −15 °C with positive charges above and below it, appears to be consistent with the "classical" tripolar charge structure described above.

Marshall and Rust (1991) reported on 12 balloon soundings of the vertical electric field in Alabama, New Mexico, and Oklahoma thunderstorms of different types. Fields were measured with an electric field meter. Using a one-dimensional approximation to Gauss's law, Marshall and Rust (1991) inferred four to 10 charge layers whose vertical extent ranged from 130 m to 2.1 km. These measurements indicated that, in addition to the three charge regions apparently corresponding to the "classical" tripole model, there is a screening layer at the upper cloud boundary and there may be up to six extra charge regions, usually in the lower part of the cloud. An example of the data of Marshall and Rust for a small Alabama thunderstorm, showing four charge layers, is presented in Fig. 3.10. Marshall and Rust (1991) also demonstrated that the interpretation of remote measurements of the electric fields produced by cloud charges is not unique: many different charge distributions can produce similar variations of the remote electric field as a function of distance from the cloud.

Stolzenburg et al. (1998a, b, c) examined and summarized the results from nearly 50 balloon electric field soundings through convective regions of MCSs, isolated supercells, and isolated New Mexico mountain thunderclouds. They noted that these three types of thundercloud may each be characterized by two basic electrical structures, as illustrated in Fig. 3.11. In updrafts of MCS convective regions (average balloon ascent rates of at least 6 m s^{-1}), in strong updrafts (average balloon ascent rates of at least 11 m s^{-1}) of supercells, and in or near the center of convection in New Mexican thunderclouds (average balloon ascent rates from 2.4 to 7.6 m s^{-1}), four charge regions

3.2. Cumulonimbus

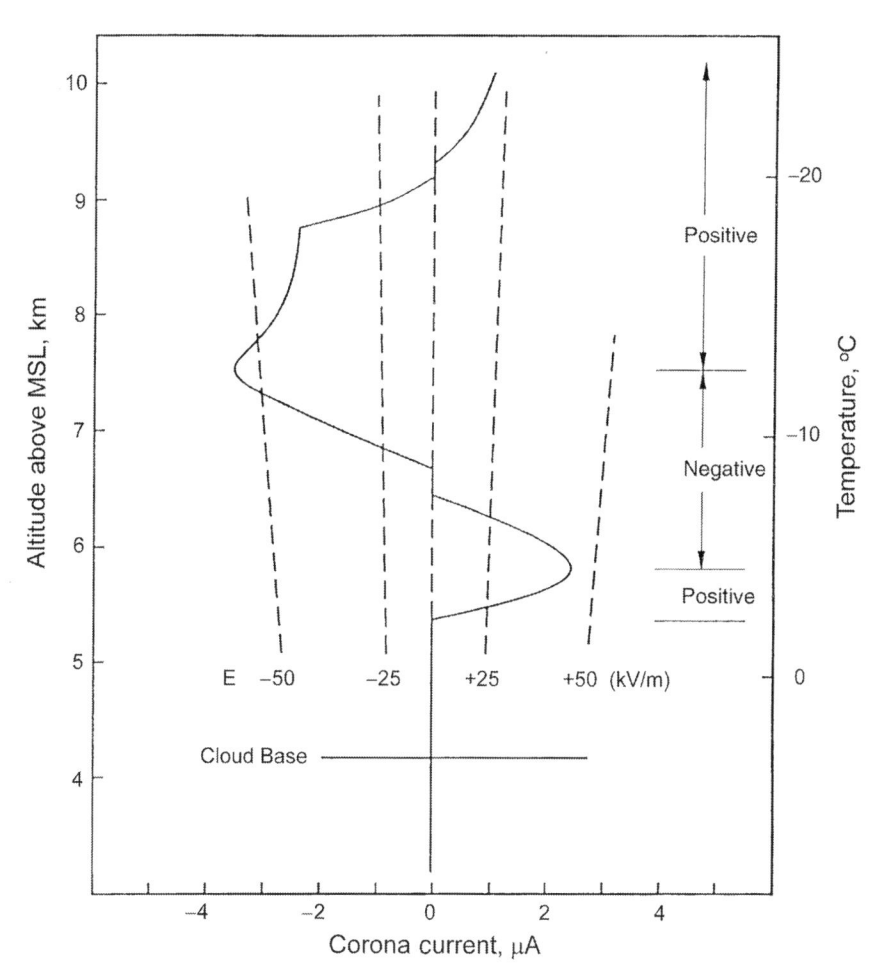

Fig. 3.9. Balloon measurements of corona current and the inferred vertical electric field E versus altitude and air temperature inside a small storm in New Mexico on 16 August 1981, which produced no lightning. The sign convention is that a positive current is induced by an upward-directed electric field, which is defined as positive. The charge regions are labeled positive or negative on the right. The profile is indicative of the "classical" tripole. The total time to acquire the record above cloud base (at roughly 4 km) was about 11 min. Adapted from Byrne et al. (1983).

were identified. These charge regions can be viewed as the tripolar charge structure described above plus an upper negative screening layer. A more complex charge structure was found to exist outside the updrafts of MCS convective regions (average balloon ascent rates of less than 6 m s^{-1}), outside strong updrafts of supercells (average balloon ascent rates from 3.4 to 8 m s^{-1}), and away from the center of convection in New Mexican thunderclouds (average balloon ascent rates from 4 to 5.2 m s^{-1}). In these three situations, Stolzenburg et al. (1998a, b, c) identified six or more charge regions, alternating in polarity, the lowest region being positive.

Byrne et al. (1989) and Marshall et al. (1989) made soundings through the anvils of large storms. Marshall and Lin (1992) reported on the charge structure of two dying thunderstorms. Marshall et al. (1995c) compared the results of soundings in weak and strong updrafts.

Rust and Marshall (1996) re-examined the original *in situ* measurements of Simpson and Scrase (1937) and Simpson and Robinson (1941) and found that the original measurements were not consistent with the "classical" tripolar model in 49 percent of the cases. Further, Rust and Marshall (1996) reported that the soundings made by Marshall and Rust (1991) and others did not fit the tripolar model in at least 86 percent of the storms. Therefore, Rust and Marshall (1996) concluded that the tripolar model is inadequate and that a more detailed model is needed for studying the electrical evolution of thunderstorms. This conclusion is reiterated, based on nearly 50 balloon soundings, by Stolzenburg et al. (1998a, b, c). However, in a gross sense, the "classical" tripolar charge structure of the cloud interior is probably justifiable, particularly in view of the possibility that the "extra" charge regions found by Marshall and Rust (1991) are related to transitions between

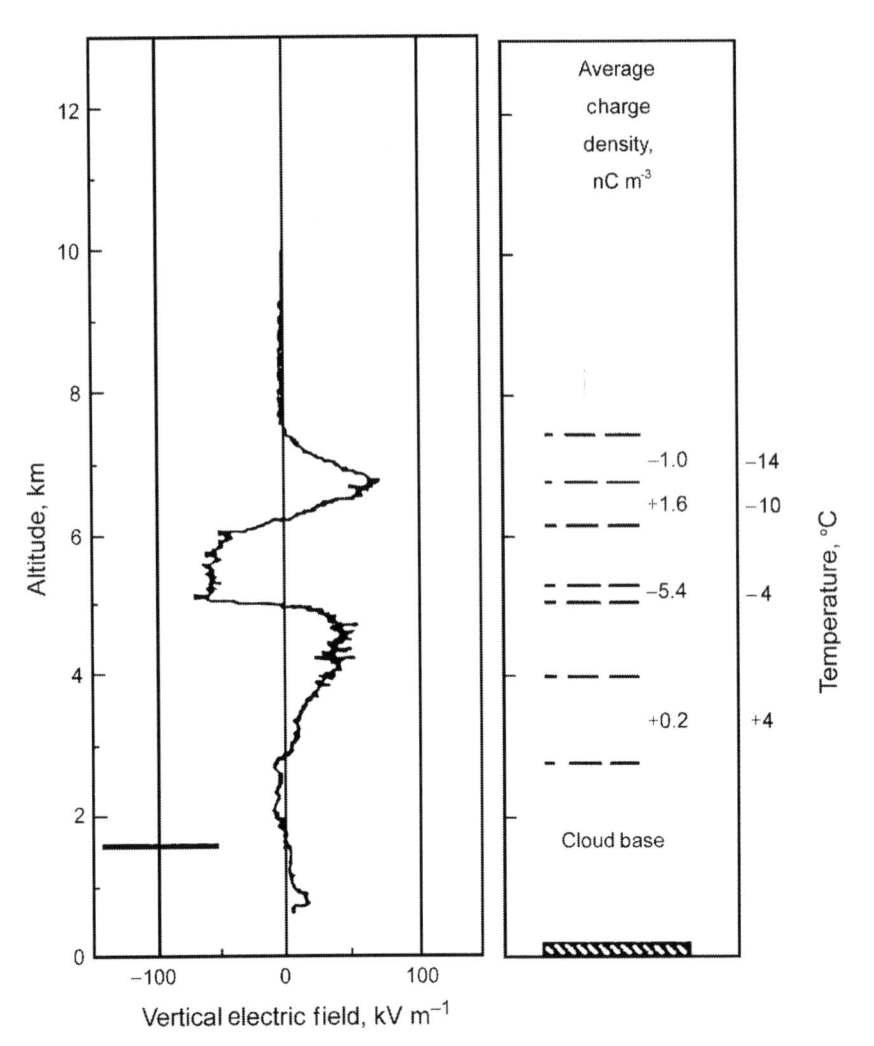

Fig. 3.10. Balloon measurements of the vertical electric field inside a small Alabama thunderstorm. An upward-directed electric field is defined as positive. The values of the inferred average charge density (in nC m^{-3}), assuming that charge regions have large horizontal extent and that the field is steady with time, are shown on the right. The field profile is indicative of a "classical" vertical tripole with an upper negative screening layer. Adapted from Marshall and Rust (1991).

different stages of thunderstorm evolution, as discussed above.

As noted in subsection 1.4.6, Marshall and Stolzenburg (2001), from 13 balloon soundings of electric field through both convective regions and stratiform clouds, estimated cloud top voltages ranging from −23 to +79 MV relative to the Earth. Within clouds, the voltage values ranged from −102 to +94 MV in 15 soundings.

3.2.4. Maximum electric fields: implications for lightning initiation

The value of the electric field intensity for electrical breakdown between two parallel plane electrodes at sea level in dry air is about 3×10^6 V m^{-1} (30 kV cm^{-1}). A smaller ambient field strength is needed for breakdown if the air pressure is reduced, as is the case at cloud altitudes. For example, at an altitude of 6 km the required field strength is 1.6×10^6 V m^{-1} (16 kV cm^{-1}), and this is further reduced by the presence of hydrometeors, which can locally enhance the field strength near their surfaces via polarization and also, in the case of liquid hydrometeors, via elongation in response to the external field (e.g., Malan 1963). For example, at atmospheric pressure the presence of water drops of 1.4 mm diameter or greater reduces the gap breakdown field from the dry-air value, 3×10^6 V m^{-1} (30 kV cm^{-1}), to 1×10^6 V m^{-1} (10 kV cm^{-1}) (Macky 1931). Note that hydrometeors can act both as objects capable of enhancing the electric field at their surface and as net-charge carriers. Under the conditions likely to occur in mature thunderclouds, the minimum fields for the onset of corona streamers, thought to lead to the formation of stepped leaders, from various types of solid and liquid

3.2. Cumulonimbus

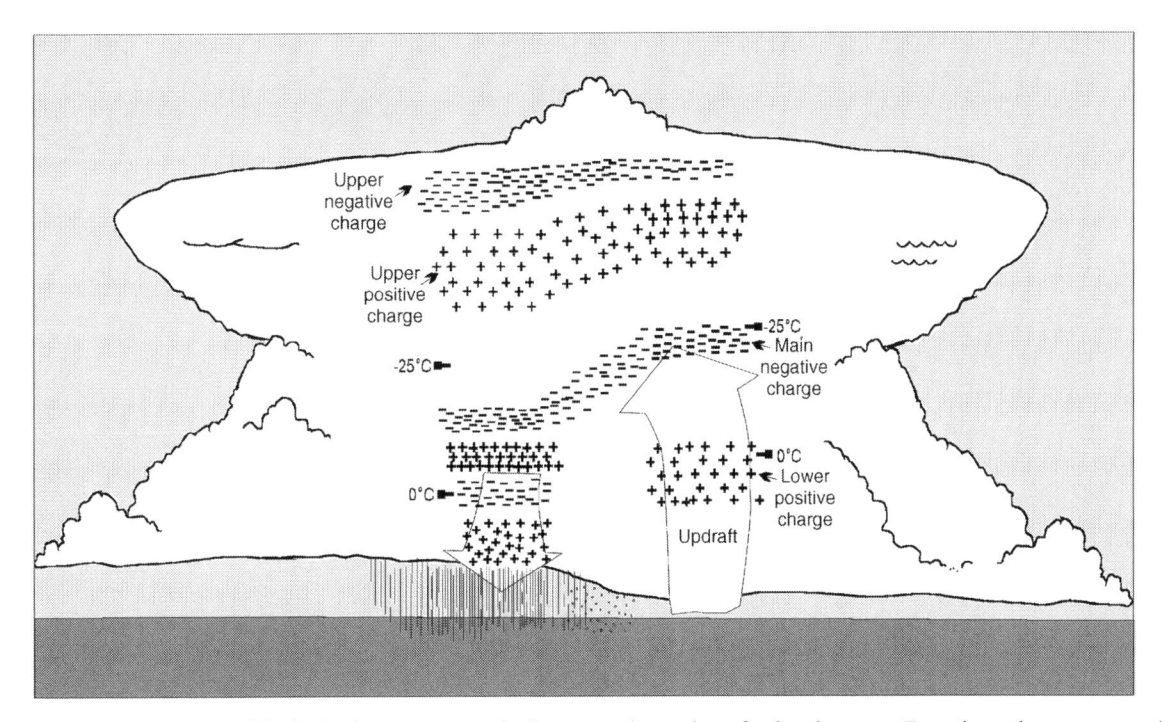

Fig. 3.11. Schematic of the basic charge structure in the convective region of a thunderstorm. Four charge layers are seen in the updraft region, and six charge layers are seen outside the updraft region (to the left of the updraft in the diagram). The charge structure shown applies to the convective elements of mesoscale convective systems (MCS), isolated supercell storms, and New Mexican air-mass storms. Note that there is a variability in this basic structure, especially outside the updraft. Adapted from Stolzenburg *et al.* (1998b).

hydrometeors are in the range from 2.5 to 9.5×10^5 V m^{-1} (2.5 to 9.5 kV cm^{-1}) (Dawson 1969; Richards and Dawson 1971; Crabb and Latham 1974; Griffiths and Latham 1974; Griffiths 1975). Further, Griffiths and Phelps (1976a, b), on the basis of their laboratory experiments and modeling, suggested that corona streamers can propagate and eventually lead to the formation of a stepped leader, as discussed in Section 4.3, if the in-cloud electric field strength is at least 1.5×10^5 V m^{-1} (1.5 kV cm^{-1}) at an altitude of about 6.5 km and 2.5×10^5 V m^{-1} (2.5 kV cm^{-1}) at about 3.5 km.

Maximum magnitudes of thundercloud electric fields measured *in situ* in different experiments are summarized in Table 3.2. Winn *et al.* (1974) reported having measured, with an instrumented rocket, a peak horizontal field of the order of 4×10^5 V m^{-1} (4 kV cm^{-1}) extending over a distance of a few hundred meters at about 6 km altitude above sea level. Gunn (1948) found an electric field of 3.4×10^5 V m^{-1} (3.4 kV cm^{-1}) on the underbelly of an aircraft just before it was struck. He considered this value to be an underestimate of the overall field. Generally, however, measured maximum electric fields are 1 to 2×10^5 V m^{-1} (1 to 2 kV cm^{-1}), as is evident from Table 3.2. Evans (1969), using electric field mills dropped into clouds on a parachute from a typical height of 9 km, reported a maximum measured vertical electric

Table 3.2. *Maximum electric field magnitudes measured in thunderclouds*

Reference	Sounding type	Maximum electric field, V m^{-1}
Gunn (1948)	Aircraft	3.4×10^5
Imyanitov *et al.* (1971)	Aircraft	2.8×10^5
Winn *et al.* (1974)	Rockets	4×10^5
Winn *et al.* (1981)	Balloons	1.4×10^5
Weber *et al.* (1982)	Balloons	1.1×10^5
Byrne *et al.* (1983)	Balloons	1.3×10^5
Fitzgerald (1984)	Aircraft	1.2×10^5
Marshall and Rust (1991)	Balloons	1.5×10^5
Kasemir (as reported by MacGorman and Rust 1998)	Aircraft	3×10^5

Note: The value of 1.2×10^6 V m^{-1} cited by Uman (1987, 2001) as measured by Fitzgerald (1976) is a misprint.

field value 3.9×10^4 V m^{-1} (0.39 kV cm^{-1}), although Vonnegut (1969) viewed the reliability of his measurements as questionable.

Marshall *et al.* (1989) observed a maximum electric field of 7×10^4 V m^{-1} in the anvil of a small thunderstorm in New Mexico. The largest electric field magnitude typically observed at the cloud base is 1 to 5×10^4 V m^{-1} (MacGorman and Rust 1998: p. 53). As indicated in subsection 3.2.2, corona from objects on the ground limits the electric field there to typically 0.1 to 1×10^4 V m^{-1} (e.g., Standler and Winn 1979; Chauzy and Raizonville 1982; Soula and Chauzy 1991; Soula 1994).

Lalande et al. (1999) summarized data for 31 aircraft-initiated lightning events involving the CV-580 research aircraft and 12 involving the C-160 research aircraft (subsections 10.3.3, 10.3.4, and Section 10.4). For the CV-580 the mean ambient electric field just prior to the time of the lightning occurrence was 5.1×10^4 V m^{-1}, with a range from 2.5 to 8.7×10^4 V m^{-1}, and for the C-160 it was 5.9×10^4 V m^{-1} with a range from 4.4 to 7.5×10^4 V m^{-1}. More details will be found in Section 10.4.

Apparently, very strong electric fields are confined to rather small cloud volumes. According to Imyanitov et al. (1971), the average dimensions of cloud regions containing the largest charges in active thunderstorms are of the order of a few hundred meters. Further, the strong-field regions must be short lived, disappearing once they give rise to electrical breakdown. Therefore, the chances of detecting such regions in thunderclouds using balloons, rockets, or aircraft are similar to those of finding the proverbial needle in a haystack. In view of the lack of measured electric fields near the breakdown value in the cloud, Marshall et al. (1995a) suggested that lightning discharges are initiated in a "breakeven field" assumed to be about 10^5 V m^{-1} (1 kV cm^{-1}) at an altitude of 6 km. The hypothetical mechanism that permits lightning initiation in fields lower than the conventional breakdown value involves runaway electrons and is discussed further in Sections 4.3, 14.3, and 14.4.

3.2.5. Charges and charge densities

Winn et al. (1974) inferred, from in situ electric field measurements, a localized charge region with a diameter of order 700 m containing a net charge of about 5 C. The calculated mean charge density in this volume was of order 30 nC m^{-3}. It was at an altitude of about 5.8 km above sea level in the cloud, whose base and top were at altitudes of about 4 and 9 km, respectively. According to Imyanitov et al. (1971), the average charge density in thunderclouds is 0.3–3 nC m^{-3}, while in small regions inside the cloud it can exceed 10 and even 100 nC m^{-3}. Winn and Byerly (1975) estimated the total amount of negative charge in a New Mexico thundercloud to be 120 and 160 C just before two lightning flashes. Proctor (1983), using the total length of the VHF–UHF images of the lightning flashes in two cells and an assumed line charge density on the channels of 0.9 mC m^{-1} (an average he inferred for seven flashes from his measured electric field changes), estimated that one cell discharged 440 C during a 3 min interval and the other 106 C during a 2 min interval. The volumes of the two cells, as enclosed by 30 dBZ radar reflectivity contours, were approximately 260 km^3 and 80 km^3.

Weber et al. (1982), from their balloon sounding in New Mexico, inferred charge densities from 0.6 to 4 nC m^{-3} in the main negative charge center. In a follow-up study, Byrne et al. (1983) estimated average charge densities ranging from 0.7 nC m^{-3} to 1.8 nC m^{-3} for negative charge regions having vertical extent from less than 1 km to 1.6 km in four thunderclouds and from 0.5 nC m^{-3} to 1.7 nC m^{-3} for positive charge regions from 0.8 to 1.5 km thick in three thunderclouds. The negative charge region in these clouds was centered between altitudes where the temperature was 0 °C and -10 °C, and the positive charge region was located above the negative one.

Byrne et al. (1987), for an active cell of a severe storm, reported charge densities of 1.2 nC m^{-3} in the negative charge region, which was centered near the -9 °C ambient temperature level and had a vertical extent of less than 1 km, and 0.15 nC m^{-3} in the upper positive charge region, which had a vertical extent of approximately 6 km. They also found a negative screening layer at the upper cloud boundary, 200–50 m thick and with an average charge density of 1.5 nC m^{-3}, and a concentrated negative charge near the top of the main negative region, at least 40 m thick and with a charge density of approximately 17 nC m^{-3}. Byrne et al. (1989) estimated charge densities of 0.2 nC m^{-3} and 0.1 nC m^{-3} for the lower positive and upper negative charge regions, respectively, in the interior of the anvil. Marshall and Rust (1991), from 12 balloon soundings in Alabama, New Mexico, and Oklahoma, found that the charge density magnitude in the charge regions varied between 0.2 and 13 nC m^{-3}. Krehbiel (1986) gave a range of charge density magnitudes of 1 to 10 nC m^{-3} inside storms, the total charge magnitude ranging from a few coulombs to a few hundred coulombs or more depending on the size and age of the storm.

Bateman et al. (1999) measured charge values on precipitation particles ranging from 2 to 200 pC, the equivalent diameter of the particles ranging from 0.6 to 3.8 mm. These measurements were made using an instrumented free balloon in a convective mountain thunderstorm over Langmuir Laboratory in New Mexico. Bateman et al. (1999) found no correlation between the charge and the size of precipitation particle on which it resides. More details on the measurement of the charges on individual hydrometeors are found in MacGorman and Rust (1998).

3.2.6. Mechanisms of cloud electrification

Any cloud electrification mechanism involves (i) a small-scale process that electrifies individual hydrometeors and (ii) a process that spatially separates these charged hydrometeors by their polarity, the resultant distances between

3.2. Cumulonimbus

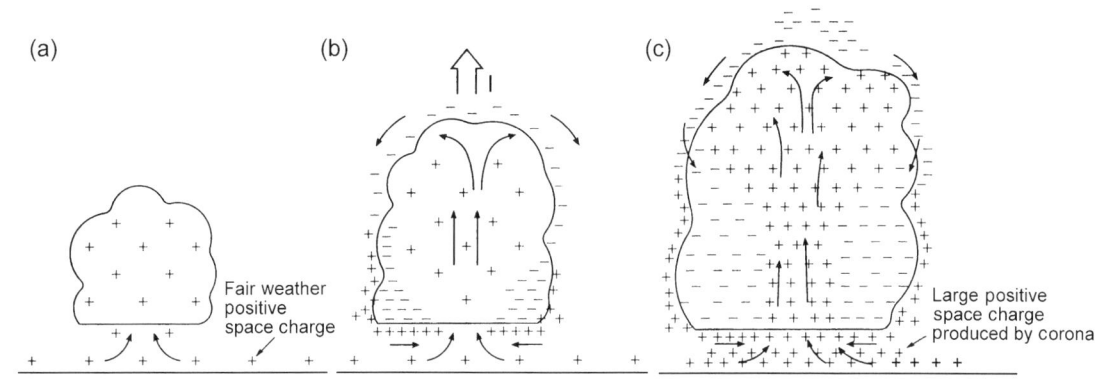

Fig. 3.12. Illustration of the convection mechanism of cloud electrification discussed in subsection 3.2.6. Adapted from MacGorman and Rust (1998).

the charged cloud regions being of the order of kilometers. Since most charges reside on hydrometeors of relatively low mobility, the cloud is a relatively good electrical insulator and leakage currents between the charged regions are thought to have a small effect on the charge separation process.

Reviews of cloud electrification mechanisms are given in Mason (1971), Stow (1969), Moore and Vonnegut (1977), Pruppacher and Klett (1978), Magono (1980), Latham (1981), Illingworth (1985), Williams (1985), Krehbiel (1986), Beard and Ochs (1986), Saunders (1993, 1994, 1995), and MacGorman and Rust (1998). Here, we will consider only the noninductive collisional graupel–ice mechanism and the convection mechanism. The term "noninductive" indicates that hydrometeors are not required to be polarized by the ambient electric field. There is a growing consensus that the graupel–ice mechanism is the dominant electrification mechanism.

Convection mechanism In this mechanism the electric charges are supplied by external sources: fair-weather space charge and corona near the ground and cosmic rays near the cloud top. Organized convection provides large-scale separation. The convection mechanism has been proposed or advocated by, for example, Lomonosov (1753), Grenet (1947, 1959), Vonnegut (1953, 1994), Wilson (1956), Phillips (1967b), Moore and Vonnegut (1977), Wagner and Telford (1981), and Williams and Lhermitte (1983). According to this mechanism, illustrated in Fig. 3.12, warm air currents (updrafts) carry positive fair-weather space charge (Section 1.4) to the top of the growing cumulus. Negative charge, produced by cosmic rays above the cloud, is attracted to the cloud's boundary by the positive charges within it. The negative charge attaches, within a second or so, to cloud particles to form a negative screening layer. These charged cloud particles carry much more charge per unit volume of cloudy air than is carried by precipitation particles (Krehbiel 1986). Downdrafts, caused by cooling and convective circulation, are assumed to carry the negative charge down the sides of the cloud toward the cloud base, this negative charge serving to produce positive corona at the Earth's surface. Corona generates additional positive charge under the cloud and, hence, provides a positive feedback to the process. The convective mechanism results in a positive cloud-charge dipole, although it seems unlikely that the negative charge region formed by this mechanism would lie in a similar temperature range for different types of thunderstorms, as suggested by the observations illustrated in Fig. 3.7. Note that in the convection model there is no role for precipitation in forming the dipole charge structure.

Moore *et al.* (1989) attempted to demonstrate the feasibility of the convection mechanism by releasing large amounts of negative charge beneath clouds, in contrast with the positive space charge normally present and postulated to lead to the formation of a positive dipole. If a cloud were to ingest negative space charge and eventually become electrified in the form of a negative dipole (main negative charge above main positive charge), the convection mechanism would be validated. The results of this experiment were inconclusive.

Chauzy and Soula (1999), using a numerical model and measured surface electric fields in Florida and in France, estimated the amount of corona charge that could be transferred from ground to the lower part of cloud by conduction and convection during the lifetime of a thunderstorm. They argued that their estimates of some tens to a few hundred coulombs over an area of 10×10 km^2 for the entire thunderstorm lifetime were comparable to the charge involved in a single lightning flash and, hence, did not give support to the convection mechanism of overall cloud electrification. Additionally, Chauzy and Soula suggested that the corona charge transported from the ground surface to the cloud may be responsible for the formation of the lower positive charge center (see subsection 3.2.7).

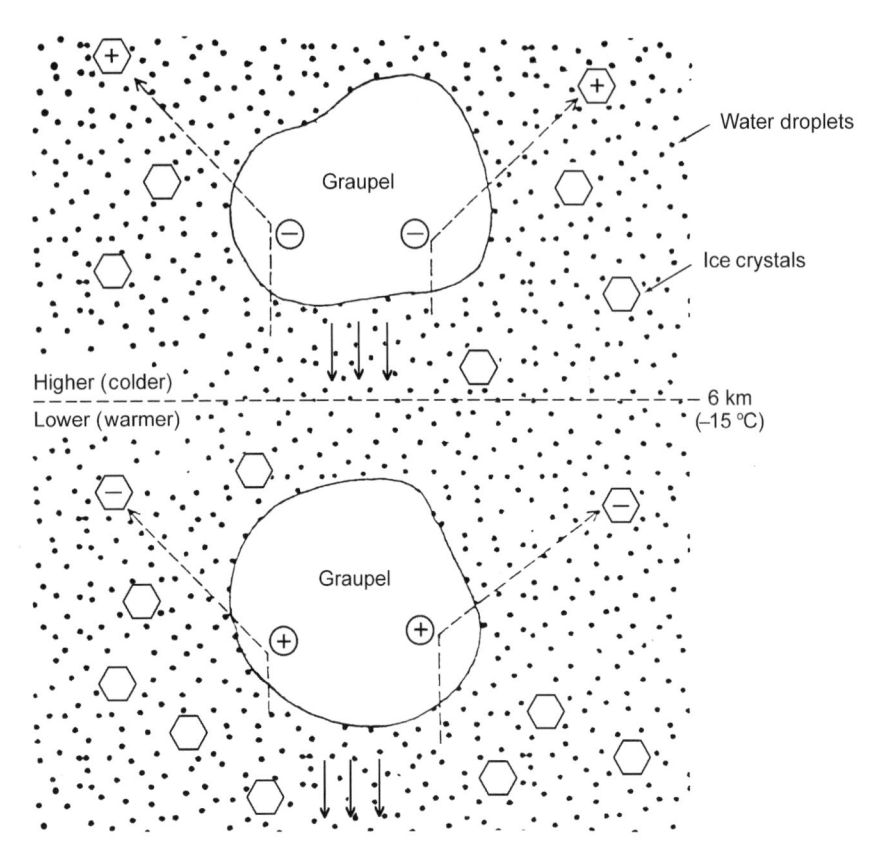

Fig. 3.13. Charge transfer by collision in the graupel–ice mechanism of cloud electrification discussed in subsection 3.2.6. It is assumed that the reversal temperature T_R is $-15\,°C$ and that it occurs at a height of 6 km.

Graupel–ice mechanism In this mechanism the electric charges are produced by collisions between precipitation particles (graupel) and cloud particles (small ice crystals). Recall that precipitation particles are defined as hydrometeors that have an appreciable fall speed (≥ 0.3 m s^{-1}), those that have a lower fall speed being termed cloud particles (subsection 3.2.1). Precipitation particles are generally larger than cloud particles, although there is no absolute demarcation in size to distinguish precipitation particles, which are falling out of the cloud, from cloud particles, which remain essentially suspended or move upward in updrafts. The large-scale separation of charged particles is provided by the action of gravity.

In the graupel–ice mechanism, which appears to be capable of explaining the "classical" tripolar cloud charge structure, the electrification of individual particles involves collisions between graupel particles and ice crystals in the presence of water droplets. The presence of water droplets is necessary for significant charge transfer, as shown by the laboratory experiments of Reynolds *et al.* (1957), Takahashi (1978a), Gaskell and Illingworth (1980), and Jayaratne *et al.* (1983). A simplified illustration of this mechanism is given in Fig. 3.13; additional factors influencing the sign and magnitude of transferred charge will be discussed later.

The heavy graupel particles (two of which are shown in Fig. 3.13) fall through a suspension of smaller ice crystals (hexagons) and supercooled water droplets (dots). The droplets remain in a supercooled liquid state until they contact an ice surface, whereupon they freeze and stick to the surface in a process called riming. Laboratory experiments (e.g., Jayaratne *et al.* 1983) show that when the temperature is below a critical value called the reversal temperature, T_R, the falling graupel particles acquire a negative charge in collisions with the ice crystals. At temperatures above T_R they acquire a positive charge. The charge sign reversal temperature T_R is generally thought to be between $-10\,°C$ and $-20\,°C$, the temperature range characteristic of the main negative charge region found in thunderclouds. Jayaratne and Saunders (1984) suggested that graupel that picks up positive charge when it falls below the altitude of T_R could explain the existence of the lower positive charge center in the cloud, as discussed later. Figure 3.14 shows the charge acquired by a simulated riming hail (graupel) particle during collisions with ice crystals, as a function of temperature, from the laboratory experiments of Jayaratne *et al.* (1983). In general, the sign and magnitude of the electric charge separated during collisions between vapor-grown ice crystals and graupel depends on, besides the

3.2. Cumulonimbus

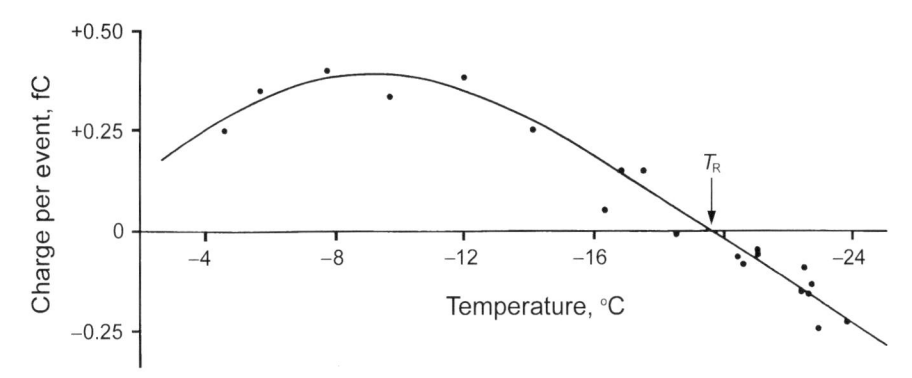

Fig. 3.14. The charge acquired by a riming hail particle (simulated by an ice-covered cylindrical metal rod with diameter 0.5 cm) during collisions with 50 μm ice crystals, as a function of the temperature of the rime in the laboratory. The velocity of impact was 2.9 m s^{-1}. The cloud liquid water content was approximately 1 g m^{-1}, and the mean diameter of the water droplets was 10 μm. Adapted from Jayaratne et al. (1983).

temperature, a number of other factors including cloud water content, ice crystal size, relative velocity of the collisions, chemical contaminants in the water, and the supercooled droplet size spectrum. Jayaratne (1998a) showed that if the droplets are restricted to sizes smaller than 10 μm there may be more than one charge sign reversal temperature (there would be four if the droplets were smaller than 4 μm in diameter). He argued that this result could possibly explain the multi-layer cloud charge distributions reported by Marshall and Rust (1991) and Marshall et al. (1995c) and discussed in subsection 3.2.3. Avila and Pereyra (2000) found from laboratory experiments that both large values of cloud water content and small sizes of droplet were conducive to predominantly positive charging.

It is believed that the polarity of the charge that is separated in ice–graupel collisions is determined by the rates at which the ice and graupel surfaces are growing. The surface that is growing faster acquires a positive charge (Baker et al. 1987). There is no consensus on the detailed physics involved. One reasonable model has been proposed by Baker and Dash (1989, 1994). They suggested that there might be a liquid-like layer (LLL) on ice surfaces and, if so, there should be an excess negative charge in the outer portion of this layer in the form of OH$^-$ ions. If two ice surfaces collide then the thicker LLL will transfer some of its mass, together with some of the negative ions, to the thinner LLL and hence leave positive charge behind. Since the surface that is growing faster would tend to have the thickest LLL, this mechanism at least qualitatively describes most of the laboratory data, although results of recent laboratory experiments of Mason and Dash (1999a, b, 2000) show that the theory of Baker and Dash (1989, 1994) is only partly correct.

It is worth pointing out that some laboratory results may be specific to the experimental techniques employed and the range of simulated cloud conditions covered.

Williams (1995b) noted an apparent discrepancy between the laboratory results of Takahashi (1978a), on the one hand, and those of the Manchester group (Jayaratne et al. 1983 and Saunders et al. 1991) on the other. Pereyra et al. (2000) attributed this discrepancy to the differences in the experimental techniques used. Takahashi (1978a), as well as Pereyra et al. (2000), mixed droplets and ice crystals grown in separate chambers before impact on a riming target, while in the Manchester experiments the ice crystals grew in the same chamber as the supercooled droplets used to rime the target.

More detailed information about the microphysical, electrical, and dynamical properties of thunderclouds is needed before the ice–graupel electrification mechanism can be tested further. There are still major uncertainties about how many collisions actually occur in different regions of the cloud, what the crystal sizes and collision velocities are in these regions, what the temperatures and liquid water contents are in these regions, and what charges are actually present on the various hydrometeors throughout the cloud. Studies aimed at obtaining such information using *in situ* measurements often in conjunction with radar and other observations have been conducted in various types of clouds (e.g., Gaskell et al. 1978; Takahashi 1978b, 1990; Christian et al. 1980; Marshall and Winn 1982; Taniguchi et al. 1982; Dye et al. 1986, 1989; Weinheimer et al. 1991; Marsh and Marshall 1993; Marshall and Marsh 1993, 1995; Willis et al. 1994; Bateman et al. 1995; French et al. 1996; Ramachandran et al. 1996; Marshall and Stolzenburg 1998; Stolzenburg and Marshall 1998; Bateman et al. 1999; Takahashi et al. 1999).

Perhaps the most significant outcome of these studies is the finding that, at least in some regions of the cloud, the charge density derived from the measured precipitation particle charges can be comparable to the charge density inferred from the gradient of the measured electric field. In other words, the charge carried by precipitation in thunderclouds can contribute significantly to the observed *in situ* electric fields. Nevertheless, in other regions of the cloud, typically at relatively high altitudes, the electric field has been inferred to be produced largely by charges on the smaller cloud particles. The latter inference is based on the assumption that the net charge density associated with cloud particles is the difference between the net charge density associated with all hydrometeors (cloud particles and precipitation particles) and the net charge density associated with precipitation particles. The charge density associated with all hydrometeors was estimated from *in situ* measured electric fields, and the charge density associated with precipitation particles was deduced from *in situ* measurements of charges on individual particles.

Using this approach, Marshall and Stolzenburg (1998) and Bateman *et al.* (1999), who studied convective thunderstorms in New Mexico, inferred that the charge of the main positive charge region is carried by cloud particles, the charge of the main negative charge region is carried by a mixture of negative cloud and negative precipitation particles, and the charge of the lower positive charge region is carried primarily by precipitation. Interestingly, Marshall and Stolzenburg (1998) found that the upper part of the main negative charge region was dominated by precipitation charge, while the lower part was dominated by cloud-particle charge. They also inferred that the charge in the negative screening layer near the cloud top was typically on cloud particles. Clearly the approach used by Marshall and Stolzenburg (1998) and Bateman *et al.* (1999) for estimation of the charges carried by cloud particles and by precipitation can be subject to considerable errors because, for instance, many precipitation particles may carry a charge less than the minimum detectable charge. In the study of Bateman *et al.* (1999), this minimum charge was 2 pC, and 65 percent of the precipitation particles were either uncharged or carried a charge below the lower measurement limit. Direct measurements of the charge carried by cloud particles are needed, although this is a much more difficult task than measuring the charge on the precipitation particles, which are larger. The velocity of particle motions has been studied using Doppler radars (e.g., Lhermitte and Williams 1985), and the dominant type of particle (e.g., rain, hail, graupel) in a given region of a thundercloud has been found using polarimetric (also called multiparameter) radars (e.g., Goodman *et al.* 1988; Carey and Rutledge 1996, 2000; Lopez and Aubagnac 1997).

It is possible that the primary electrification mechanism changes once a storm becomes strongly electrified (Krehbiel 1986). For example, collisions between ice crystals and graupel could initiate the electrification, and then the larger convective energies of the storm could continue it.

It is also possible that important electrification mechanisms are still unrecognized.

3.2.7. Origin of the lower positive charge center

A number of hypotheses have been proposed regarding the origin of the lower positive charge. As noted earlier, Jayaratne and Saunders (1984) suggested that graupel, which charges positively at temperatures warmer than the reversal temperature, is responsible for the formation of the lower positive charge center. However, Marshall and Winn (1985) and Marsh and Marshall (1993) argued that the mechanism of Jayaratne and Saunders (1984) would require an unlikely combination of extreme values of the various charging model parameters to explain the observed charges in the lower positive charge center. Marshall and Winn (1982) presented evidence that one lower positive charge center may have been formed when positive charge was deposited by lightning in the lower part of the cloud. However, Marsh and Marshall (1993) observed such a charge center before the first lightning flash. Malan (1952) suggested that the lower positive charge center contains the charge that is produced by corona at ground and is subsequently carried into the cloud by conduction or convection, and Chauzy and Soula (1999) presented calculations in support of this hypothesis. Phillips (1967a) speculated that on theoretical grounds the lower positive charge center could be explained completely by a positive screening layer at the lower cloud boundary. As noted in subsection 3.2.6, Marshall and Stolzenburg (1998) and Bateman *et al.* (1999) found, for New Mexico thunderstorms, that the charge of the lower positive charge center was carried primarily by precipitation, although Marshall and Stolzenburg (1998) reported that in two soundings positively charged cloud particles contributed 20 to 50 percent of the total charge density.

3.2.8. Lightning representation in numerical cloud models

The cloud models discussed below include cloud dynamics, microphysics, and electrical processes. They have been developed primarily to gain better insight into the interaction of the processes involved in the evolution of a thundercloud. MacGorman and Rust (1998) listed the following specific issues that have been addressed by cloud models: (i) the validity of various electrification mechanisms, (ii) the dependence of electrification on updraft characteristics and storm microphysics, (iii) the factors influencing lightning location in thundercloud and flash rates, and (iv) the effects of lightning on the thundercloud charge distribution. The first physical models of thunderclouds did not include lightning discharges and, as a result, were useful only for modeling the early stages of electrification, before the first lightning flash. Subsequent versions of the models have included lightning discharges, initially with the sole purpose of limiting the maximum electric field to the observed values, which are of order 10^5 V m^{-1} (Table 3.2). Thundercloud models without lightning have been developed by, for example, Takahashi (1984), Helsdon and Farley (1987a, b), Ziegler et al. (1991), Norville et al. (1991), Solomon and Baker (1994), and Randell et al. (1994) and models including lightning by Rawlins (1982), Takahashi (1987), Helsdon et al. (1992), Ziegler and MacGorman (1994), Baker et al. (1995), Solomon and Baker (1996, 1998), Mazur and Ruhnke (1998), and MacGorman et al. (2001). The lightning components of these latter models will be briefly discussed in this subsection. Further information on the subject, in review form, is found in Levin and Tzur (1986), MacGorman and Rust (1998: Chapter 9), and MacGorman et al. (2001).

The thundercloud is a very complex system, which necessarily requires simplified mathematical descriptions of the various processes involved. The behavior of a parcel of moist air in a cloud is described by an equation of motion, thermodynamic equations, and a continuity equation. The water substance in all three phases in the cloud is divided into several categories, and the behavior of each category is tracked, usually at grid points fixed with respect to Earth, as it moves and interacts with other categories throughout the cloud's evolution. Electrical processes in thundercloud models include electrification and lightning and are integrated with other cloud processes. The graupel–ice mechanism of electrification described in subsection 3.2.6 is usually employed. Once charged hydrometeors are produced, the resultant electric field is calculated at the grid points. Detailed treatment of the modeling of the electrical processes is found in MacGorman and Rust (1998: Section 9.4). In the following, we will outline the representation of lightning in the most recent thundercloud models. Note that the lightning component of a cloud model is often called the lightning parameterization.

Takahashi (1987), who used an axisymmetric, two-dimensional model, assumed that a flash is initiated when the electric field exceeds 3.4×10^5 V m^{-1} (the maximum value observed by Gunn 1948; see Table 3.2). The model then (i) locates the maximum positive and negative charge densities, (ii) determines the smallest volume about each charge density maximum that contains the assumed lightning flash charge, a charge of just over 20 C, and (iii) deposits -20 C in the positive charge volume and $+20$ C in the negative charge volume, apportioning charge equally over all charged particles in the respective volume. Takahashi (1987) used this model to examine factors influencing the location of lightning in thunderclouds.

Helsdon et al. (1992) developed a model of an intracloud lightning discharge, as part of their two-dimensional thundercloud model, in which the initiation, propagation direction (geometry), and termination of the discharge were controlled by the ambient electrostatic field. In this model, a discharge is initiated when the field exceeds 4×10^5 V m^{-1}. The discharge channel is assumed to be a conductor polarized in the electric field (with net charge equal to zero) and this is allowed to extend in either direction with respect to the origin. The electric field due to the polarization charges induced on the conductor by the ambient field is omitted, which is a physically unrealistic approach. The model computes the angle of the ambient electric field vector with respect to the vertical and determines whether the channel should extend to an adjacent grid point along the diagonal or along the side of the grid box. The discharge extension is terminated when the ambient electric field at the simulated channel ends falls below 1.5×10^5 V m^{-1}. Once the discharge path is determined, the line charge densities along the positively and negatively charged segments of the simulated channel are estimated and then converted to volume charge densities in some region about the channel. Further, the volume charge density at each involved grid point created by lightning is converted to an ion number density (assuming that all ions are singly ionized), which is added to the existing ion number density at that grid point. Finally, the ambient electric field is recalculated, and free ions produced by the lightning discharge are allowed to interact with hydrometeors. In spite of the many arbitrary and sometimes physically unrealistic assumptions, including omission of the electric field due to charges induced on the channel and also the use of slab symmetry in the two-dimensional model, which forces the simulated lightning channel to extend to infinity in either direction along the axis perpendicular to the plane of the model, the lightning model of Helsdon et al. (1992) produces reasonable values of charge per unit lightning channel length, total charge transfer, and dipole moment change.

Ziegler and MacGorman (1994) developed a simple approach to treating the cumulative effect of several cloud flashes in their three-dimensional thundercloud model. This lightning model was used in their cloud model to investigate hypothetical explanations of the observed high intracloud lightning flash rates in an Oklahoma tornadic storm. The simulated lightning events (each comprising one or more flashes) occurred at a rate of 12 min^{-1}, similar to the maximum observed intracloud flash rate of 13 min^{-1}. When the electric field exceeded an assumed threshold of 2×10^5 V m^{-1}, the algorithm developed by Ziegler and MacGorman (1994) neutralized some charge at all grid points where the magnitude of the charge density exceeded some preset threshold value. The threshold charge density and the fraction of charge in excess of the threshold value, which was neutralized, which are apparently adjustable parameters in the model, were set at 0.5 nC m^{-3} and 0.33, respectively. An adjustment procedure was included to ensure that equal amounts of positive and negative charge were neutralized. The neutralization of charge at each grid point was achieved by adding charge of opposite sign to all hydrometeors, the charge being apportioned to each hydrometeor category according to its relative surface area. Most of the lightning-related charge was captured by the smaller cloud particles, which greatly outnumber the precipitation particles.

MacGorman et al. (1996) developed a lightning model in which they used a height-dependent "breakeven" electric field (Section 14.5) for lightning initiation. Their modeling of channel development was similar to that used by Helsdon et al. (1992). However, they traced the electric field lines to a smaller field magnitude than that used by Helsdon et al. (1992), and if the grid point at which this threshold was reached was within the region of sufficient charge density then they assumed that the flash branched throughout the region of charge. Procedures for determining how much charge was neutralized by the flash and for adding the opposite-polarity charge to hydrometeors at the grid points involved in the flash, to carry out the neutralization, were similar to those used by Ziegler and MacGorman (1994).

Baker et al. (1995) presented a rather crude one-dimensional thundercloud model in which they prescribed the end points of both intracloud and cloud-to-ground lightning channels as well as the amount of charge transferred per flash. They assumed that lightning is initiated when the electric field reaches a critical value of 3×10^5 V m^{-1} and that the direction of propagation, and thereby the type of lightning discharge, is controlled by the gradient of the vertical electric field immediately above and immediately below the height at which the electric field reaches the critical value. If the gradient is smaller in the upward direction, an intracloud flash is initiated, while a cloud-to-ground flash occurs when the gradient is smaller in the downward direction. The initiation point is always very near the top of the negative charge center. The charge transferred by a flash is deposited on small cloud particles (ice crystals and droplets) that are ascending in updrafts. This model was used to study the sensitivity of the lightning flash rate to a number of cloud characteristics including updraft speed, cloud top temperature, precipitation rate, radar reflectivity, cloud width, ice-crystal concentration, and the prevailing glaciation mechanism.

Solomon and Baker (1996) developed a lightning model that can be used to study both intracloud and cloud-to-ground flashes in the framework of an axially symmetric cylindrical thundercloud model. When the vertical electric field at some point on the z-axis exceeds a predetermined critical value, 2.5×10^5 V m^{-1}, a vertical ellipsoidal conductor that simulates an embryonic lightning channel is assumed to be placed along the z-axis, centered at the point where the field exceeded the critical value. This simulated channel has a total length 600 m and a maximum radius 2 m. The channel is allowed to extend, both upward and downward, provided that the total electric field (the sum of the ambient field and that due to the charges induced on its surface) exceeds 3×10^5 V m^{-1} at a distance arbitrarily set at 25 m from the end of the channel.

It is worth noting that a similar bidirectional extension of the lightning channel was also considered by Mazur and Ruhnke (1998). In this latter study, the channel is initially represented by a 1-km-long vertical conductor at a height of 5 km above ground, placed along the axis of a cylindrical volume containing a specified distribution of stationary charges.

The channel extension criterion in Solomon and Baker's (1996) model is applied to the upper end and lower end of the channel individually, so that the extension could be unidirectional. In the case of an intracloud flash, the extension stops when the total field 25 m from either end falls below 3×10^5 V m^{-1}, the net charge on the channel being equal to zero. If the lower end contacts the ground, a net charge is then placed on the channel to bring the channel potential to zero. In this case (a cloud-to-ground flash), the downward extension is terminated, while the upward extension might continue. In this model, all discharges initiated above the lower charge center are of the intracloud type, and all those initiated below that center are of the cloud-to-ground type.

MacGorman et al. (2001) developed a new lightning model that allows one to simulate more realistically the location and structure of individual lightning flashes. To do this, three aspects of previous models were modified. (i) To account for sub-grid-scale variations, the initiation point is chosen randomly from among grid points at which the electric field magnitude is above a threshold value, instead of being assigned always to the grid point having the maximum electric field magnitude. (ii) The threshold value for initiation can either be constant, as in previous models, or can vary with height to allow different flash initiation hypotheses to be tested. (iii) Instead of stopping at specified, relatively large ambient electric field magnitudes, extensive flash development can continue in regions having a weak ambient electric field but a substantial charge density. Initial simulations show that, by permitting the development of flashes in regions of substantial charge density and weak ambient electric field, the new model produces a flash structure much like that of observed flashes.

All lightning models considered in this section necessarily contain many largely arbitrary assumptions, owing to the lack of reliable information on even the basic characteristics of the processes associated with the initiation and in-cloud development of lightning discharges. For example, the lightning initiation criterion remains an open

Table 3.3. *Median and maximum measured values of $|E_z|$ in various types of stratiform clouds. Adapted from Imyanitov et al. (1971)*

	Median value of $|E_z|$, V m^{-1}			
Type of cloud	St Petersburg, Russia (60° N)	Kiev, Ukraine (50° N)	Tashkent, Uzbekistan (41° N)	Maximum measured value of $|E_z|$, V m^{-1}
Stratus (St)	100	120	130	$(2–3) \times 10^3$
Stratocumulus (Sc)	80	100	100	$(2–3) \times 10^3$
Altocumulus (Ac)	40	80	70	5×10^3
Altostratus (As)	100	150	350	2×10^4
Nimbostratus (Ns)	150	250	500	4×10^4

question: *in situ* measurements are thus far incapable of detecting localized and short-lived "igniting cells", while the use of the laboratory results of Griffiths and Phelps (1976a, b) is a gross extrapolation from a scale of tens of centimeters to a scale of kilometers. Further, the extension criteria for the positive and negative ends of the lightning channel are likely to be different, as indicated by, for example, the dramatic difference in the lengths of positive and negative upward connecting leaders (subsection 5.3.1) reported by Berger (1967) and Berger and Vogelsanger (1969). Because of the lack of pertinent information on exactly what is happening electrically in the cloud, all the lightning models discussed above are very difficult to validate.

Thundercloud models used in global-circuit studies are discussed in subsection 1.4.5.

3.3. Non-cumulonimbus

The presence of a cloud in the clear atmosphere locally reduces the conductivity of the atmosphere, owing to the capture and, therefore, immobilization of ions by more massive hydrometeors. Since the current flow is continuous across boundaries in the steady state, there will be an accumulation of charge at the cloud boundaries associated with the fair-weather conduction current (Section 1.4), as discussed by MacGorman and Rust (1998). As a result, all clouds are electrified to some extent. However, observations indicate that internal charging mechanisms also exist inside many types of clouds other than the cumulonimbus discussed above in Section 3.2. There has been considerable interest in the electrical structure of stratiform clouds in relation to the characteristics of mesoscale convective complexes (subsection 2.2.4), positive lightning (Section 5.2), airborne vehicle safety (Sections 10.2 and 10.6), and lightning interaction with the middle atmosphere (subsection 14.3.3). The vertical electric fields in stratiform clouds were extensively measured in the USSR by Imyanitov et al. (1971) using penetrating aircraft. More recent soundings have predominantly employed balloons. The results of the pioneering studies of Imyanitov et al. (1971) and of more recent observations are briefly outlined below. Additional information on the electrical and other characteristics of non-cumulonimbi can be found in MacGorman and Rust (1998: Chapter 2 and Section 8.4).

Median and maximum measured absolute values of vertical electric field E_z in various types of stratiform clouds (see WMO 1969 and MacGorman and Rust 1998 for descriptions), as reported by Imyanitov et al. (1971), are summarized in Table 3.3. The relation between cloud thickness (in meters) and the number of charge layers, as observed in 1958–9 in St. Petersburg by Imyanitov et al. (1971), is illustrated in Table 3.4. On average, the charge density in stratus (St) and stratocumulus (Sc) is of

Table 3.4. *Average thickness (m) of various types of stratiform cloud. Adapted from Imyanitov et al. (1971)*

Type of cloud	Single charge layer		Two charge layers		Multiple charge layers
	+	−	±	∓	
St	200	200	450	450	700
Sc	260	250	400	450	700
As	650	700	800	900	1500
Ns	650	700	950	1600	2000

order 10^{-2} nC m^{-3}. In nimbostratus (Ns), charge densities of order 0.1 and even 1 nC m^{-3}, comparable to values in thunderclouds (subsection 3.2.5), have been observed. Davis and Standring (1947), from their measurements of currents in the cables of kite balloons, reported charge transfers of the order of coulombs associated with corona current (in the range of milliamperes) when the balloon was under nimbostratus with continuous rain. Imyanitov et al. (1971) stated that aircraft have been damaged by lightning in non-thunderstorm clouds such as nimbostratus and altostratus.

Chauzy et al. (1980) and Chauzy et al. (1985), using balloons, obtained electric field profiles for the stratiform region of mesoscale systems over southwestern France and West Africa, respectively. These profiles, which extended to an altitude of approximately 5.5 km above mean sea level, revealed negative charge layers at the cloud base.

Schuur et al. (1991) presented the results of an electric field sounding through the transition zone between an Oklahoma squall line and its associated trailing stratiform region. Electric field magnitudes as high as 1.13×10^5 V m^{-1} were observed, and charge densities with magnitudes 0.2–4.1 nC m^{-3} and vertical extents 130–1160 m were inferred using a one-dimensional approximation of Gauss' law, these values being comparable to those characteristic of thunderclouds (subsection 3.2.5).

Marshall and Rust (1993) reported on electric field soundings in the stratiform regions and transition zones of MCSs. They identified only two types (labeled A and B) of vertical electric field profile, in contrast with the much greater variability in the profiles for isolated storms (subsection 3.2.3). Both types of profile exhibited at least four major charge layers, each characterized by a thickness of some hundreds of meters and charge densities of up to 5.3 nC m^{-3}. The maximum electric field in the soundings was typically 1×10^5 V m^{-1}. The lowest charge region was negative for both types; however, the charge polarity at 0 °C for type B was positive while for type A both polarities were observed (see also Shepherd et al. 1996). MacGorman and Rust (1998: p. 264) indicated that other types of electric field profile (different from A and B) may occasionally occur in the stratiform regions of MCSs. Marshall and Rust (1993) estimated that the total charge in a single layer of the trailing stratiform region (an area of 80 km × 200 km) of one MCS was of order 20 000 C. This value is two orders of magnitude larger than the charge thought to exist in the main charge regions of typical isolated thunderstorms (Table 3.1).

Stolzenburg et al. (1994), from their analysis of the kinematic, radar-reflectivity, and electric field sounding data for the 2–3 June 1991 MCS stratiform region, found that the basic electrical charge structure of this region consisted of four horizontally extensive layers. These include, from top to bottom, (i) a region of positive charge, 1–3 km deep and of density about 0.2 nC m^{-3}, between 6 and 10 km, (ii) a region of negative charge, of density about 1.0–2.5 nC m^{-3}, between 5 and 6 km, (iii) positive charge of density about 1.0–3.0 nC m^{-3} near 0 °C, and (iv) negative charge, of density about 0.5 nC m^{-3}, near cloud base. A fifth, negatively charged, layer existed near the echo top in the part of the stratiform region closest to the convective region. In addition, there was a low-density positive charge below the cloud. The basic charge structure reported by Stolzenburg et al. (1994) conforms to the type-B structure defined by Marshall and Rust (1993). The conceptual model of MCS electrical structure, including both convective and stratiform regions, proposed by Stolzenburg et al. (1998c) is reproduced in Fig. 3.15. Stolzenburg et al. (1994, 1998c) infer that charge advection from the MCS convective region is important, especially at upper levels, and that one or more in situ charging mechanisms are also important, especially at low levels. Stolzenburg et al. (1994) consider three in situ charging mechanisms, drop-breakup charging (e.g., Canosa and List 1993), melting charging (e.g., Drake 1968), and noninductive collisional graupel–ice charging (subsection 3.2.6), which could be responsible for the positive charge layer at or near 0 °C. They state that they are uncertain which, if any, of these three mechanisms actually operates. Shepherd et al. (1996) suggested that the positive charge at or near 0 °C is caused by a melting charging mechanism. Schuur and Rutledge (2000a, b), from observation and modeling of one MCS, found that noninductive collisional graupel–ice charging can account for as much as 70 percent of the total charge in the stratiform region, the remaining 30 percent being contributed by charge advection from the convective region. They also studied the evaporation–condensation charging mechanism (e.g., Dong and Hallett 1992) and the melting charging mechanism and found these to be insignificant.

Marshall and Stolzenburg (2001), from balloon soundings of the electric field through two MCS stratiform regions and one thunderstorm anvil, estimated that the average cloud top voltage was +32 MV relative to the Earth. This voltage value was slightly larger than the average cloud top voltage for thunderstorm convective regions, suggesting that stratiform clouds may be important contributors to the global electric circuit (Section 1.4).

The electric charge distribution in winter clouds in Japan (Sections 8.2 and 8.3) has been studied by Taniguchi et al. (1982), Kitagawa (1992), Michimoto (1993), and Kitagawa and Michimoto (1994). Both H_2O and non-H_2O clouds as possible sources of lightning on planets other than Earth are discussed in Chapter 16.

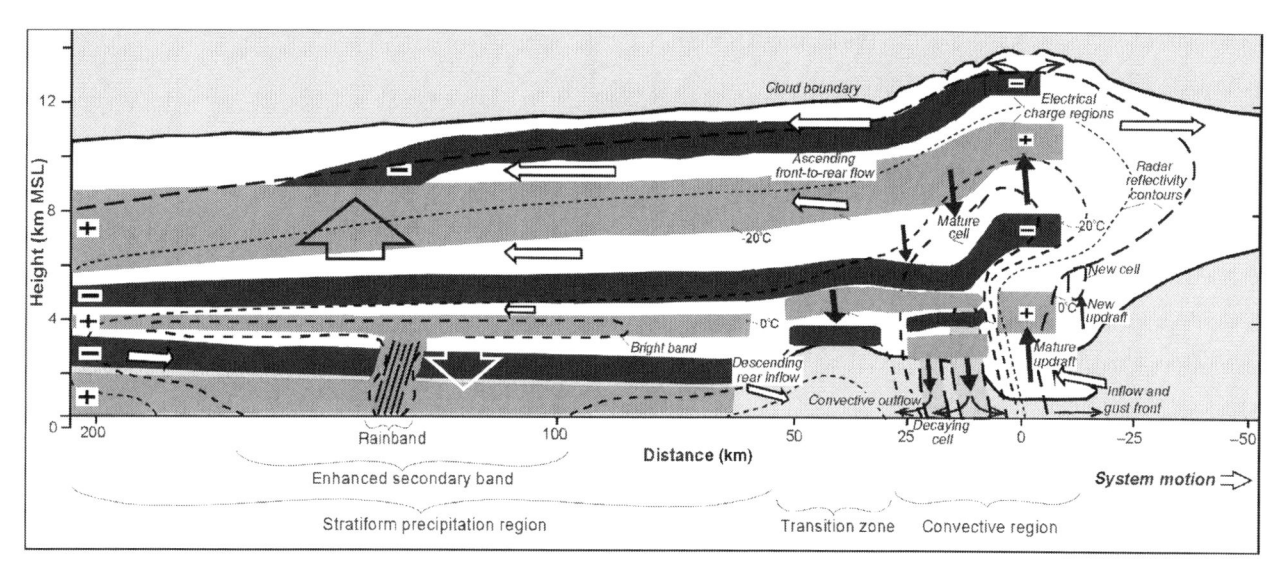

Fig. 3.15. Conceptual model of the charge structure of an MCS. Positive charge layers are indicated by the light grey shading and negative layers are indicated by the dark shading. The broken lines are radar reflectivity contours. In the convective region and the transition zone, the thick solid arrows depict convective updrafts and downdrafts, and the thin solid arrows show divergent outflows. The smaller open arrows represent system-relative flows, which are mainly horizontal. The mesoscale updraft and downdraft in the stratiform region are depicted by large open arrows (black and white outlines, respectively). There are four horizontally extensive cloud charge layers in the part of the stratiform precipitation region farthest behind the convective region, the fifth (lowermost) charge layer being seen in the stratiform region entirely below the cloud. An additional (negative) charge layer extends from the convective region through the nearest part of the stratiform region above all the other layers. Adapted from Stolzenburg et al. (1998c).

3.4. Summary

The typical thundercloud charge structure is often approximated by three vertically stacked point charges (or spherically symmetrical charged volumes): main positive at the top, main negative in the middle, and lower positive at the bottom. The magnitudes of the main positive and negative charges are typically some tens of coulombs, while the lower positive charge is probably about 10 C or less. The typical average charge density in any of the charge regions is probably a few nC m^{-3} or so. The negative charge region tends to have a relatively small vertical extent that is apparently related to the -10 to -25 °C temperature range, regardless of the stage of storm development, location, and season. Recent *in situ* measurements suggest that at least one more charge region, the negative screening layer near the upper cloud boundary, should be added to the "classical" tripolar charge structure. The maximum electric fields typically measured in thunderclouds are 1 to 2×10^5 V m^{-1}, which is lower than the expected breakdown field, of the order of 10^6 V m^{-1}.

Many cloud electrification theories have been proposed. There is growing consensus that the graupel–ice mechanism is the dominant mechanism, at least at the initial stages of cloud electrification. In this mechanism, the electric charges are produced by collisions between graupel and small ice crystals in the presence of water droplets, and the large-scale separation of the charged particles is provided by the action of gravity. It is possible that other cloud electrification mechanisms, including the convection mechanism, become important at later stages of the thunderstorm development. There is no consensus regarding the origin of the lower positive charge center. There exist a number of numerical models designed to study the evolution of thunderclouds, including the production of lightning discharges. Stratiform clouds, such as nimbostratus and altostratus, can be strongly electrified and can produce lightning discharges.

References and bibliography for Chapter 3

Allee, P.A., and Phillips, B.B. 1959. Measurements of cloud-droplet charge, electric field, and polar conductivities in supercooled clouds. *J. Appl. Meteor.* **16**: 405–10.

Allen, N.L., and Ghaffar, A. 1995. The conditions required for the propagation of a cathode-directed positive streamer in air. *J. Phys. D: Appl. Phys.* **28**: 331–7.

Al-Saed, S.M., and Saunders, C.P.R. 1976. Electric charge transfer between colliding water drops. *J. Geophys. Res.* **81**: 2650–4.

Anderson, F.J., and Frier, G.D. 1969. Interaction of the thunderstorm with a conducting atmosphere. *J. Geophys. Res.* **74**: 5390–6.

Andreeva, S.I., and Evteev, B.F. 1974. The potential gradient of the electric field in nimbostratus clouds. In *Studies in Atmospheric Electricity*, eds. V.P. Kolokolov and T.V. Lobodin, translated from Russian, Israel Progr. for Sci. Transl., Jerusalem, pp. 1–5.

Appleton, E.V., Watson-Watt, R.A., and Herd, J.F. 1920. Investigations on lightning discharges and on the electric fields of thunderstorms. *Proc. Roy Soc. A* **221**: 73–115.

Arabadzhi, V.I. 1956. The measurement of electric field intensity in thunderclouds by means of radiosonde. *Dokl. Akad. Nauk. SSSR* **111**: 85–8.

Aufdermaur, A.N., and Johnson, D.A. 1972. Charge separation due to riming in an electric field. *Q.J.R. Meteor. Soc.* **98**: 369–82.

Avila, E.E., and Caranti, G.M. 1994. A laboratory study of static charging by fracture in ice growing by riming. *J. Geophys. Res.* **99**: 10 611–20.

Avila, E.E., and Pereyra, R.G. 2000. Charge transfer during crystal–graupel collisions for two different cloud droplet size distributions. *Geophys. Res. Lett.* **27**: 3837–40.

Avila, E.E., Caranti, G.M., and Lamfri, M.A. 1988. Charge reversal in individual ice–ice collisions. In *Proc. 8th Int. Conf. Atmospheric Electricity, Uppsala, Sweden*, pp. 245–250.

Avila, E.E., Aguirre Varela, G.G., and Caranti, G.M. 1995. Temperature dependence of static charging in ice growing by riming. *J. Atmos. Sci.* **52**: 4515–22.

Avila, E.E., Aguirre Varela, G.G., and Caranti, G.M 1996a. Reply. *J. Geophys. Res.* **101**: 9537–8.

Avila, E.E., Aguirre Varela, G.G., and Caranti, G.M. 1996b. Charging in ice–ice collisions as a function of the ambient temperature and the larger particle average temperature. *J. Geophys. Res.* **101**: 29 609–14.

Avila, E., Caranti, G., Castellano, N., and Saunders, C. 1998. Laboratory studies of the influence of cloud droplet size on charge transfer during crystal–graupel collisions. *J. Geophys. Res.* **103**: 8985–96.

Avila, E.E., Pereyra, R.G., Aguirre Varela, G.G., and Caranti, G.M. 1999. The effect of the cloud-droplet spectrum on electrical-charge transfer during individual ice–ice collisions. *Q.J.R. Meteor. Soc.* **125**: 1669–79.

Avila, E.E., Pereyra, R.G., Castellano, N.E., Saunders, C.P.R. 2001. Ventilation coefficients for cylindrical collector growing by riming as a function of the cloud droplet spectra. *Atmos. Res.* **57**: 139–50.

Baginski, M.E., Hodel, A.S., and Lankford, M. 1996. An investigation of the reconfiguration of the electric field in the stratosphere following a lightning event. *J. Electrostat.* **36**: 331–47.

Baker, M.B., and Dash, J.G. 1989. Charge transfer in thunderstorms and the surface melting of ice. *J. Cryst. Growth* **97**: 770–6.

Baker, M.B., and Dash, J.G. 1994. Mechanism of charge transfer between colliding ice particles in thunderstorms. *J. Geophys. Res.* **99**: 10 621–6.

Baker, B., Baker, M.B., Jayaratne, E.R., Latham, J., and Saunders, C.P.R. 1987. The influence of diffusional growth rates on the charge transfer accompanying rebounding collisions between ice crystals and soft hailstones, *Q.J.R. Meteor. Soc.* **113**: 1193–215.

Baker, M.B., Christian, H.J., and Latham, J. 1995. A computational study of the relationships linking lightning frequency and other thundercloud parameters. *Q.J.R. Meteor. Soc.* **121**: 1525–48.

Baranski, P. and Michnowski, S. 1987. Variations of the electric field and precipitation measured under thunderclouds in Warsaw. *Publs. Inst. Geophys. Pol. Acad. Sci.* **198**: 59–74.

Barnard, V. 1951. The approximate mean height of the thundercloud charge taking part in a flash to ground. *J. Geophys. Res.* **56**: 33–5.

Bateman, M.G., Rust, W.D., and Marshall, T.C. 1994. A balloon-borne instrument for measuring the charge and size of precipitation particles inside thunderstorms. *J. Atmos. Oceanic Technol.* **11**: 161–9.

Bateman, M.G., Rust, W.D., Smull, B.F., and Marshall, T.C. 1995. Precipitation charge and size measurements in the stratiform region of two mesoscale convective systems. *J. Geophys. Res.* **100**: 16 341–56.

Bateman, M.G., Marshall, T.C., Stolzenburg, M., and Rust, W.D. 1999. Precipitation charge and size measurements inside a New Mexico mountain thunderstorm *J. Geophys. Res.* **104**: 9643–53.

Baughman, B.G., and Fuquay, D.M. 1970. Hail and lightning occurrence in mountain thunderstorms. *J. Appl. Meteor.* **9**: 657–60.

Beard, K.V.K., and Ochs, H.T. 1986. Charging mechanisms in clouds and thunderstorms. In *The Earth's Electrical Environment*, eds. E.P. Krider and R.G. Robble, pp. 114–30, Washington, DC: National Academic Press.

Beard, K.V., Durkee, R.I., and Ochs, H.T. 2002. Coalescence efficiency measurements for minimally charged cloud drops. *J. Atmos. Sci.* **59**: 233–43.

Berger, K. 1967. Novel observations on lightning discharges: results of research on Mount San Salvatore. *J. Franklin Inst.* **283**: 478–525.

Berger, K., and Vogelsanger, E. 1969. New results of lightning observations. In *Planetary Electrodynamics*, eds. S.C. Coroniti and J. Hughes, pp. 489–510, New York: Gordon and Breach.

Black, R.A., and Hallett, J. 1998. The mystery of cloud electrification. *American Scientist* **86**: 526–34.

Black, R.A., and Hallett, J. 1999. Electrification of the hurricane. *J. Atmos. Sci.* **56**(12): 2004–28.

Blakeslee, R.J., Christian, H.J., and Vonnegut, B. 1989. Electrical measurements over thunderstorms. *J. Geophys. Res.* **94**: 13 135–40.

Blakeslee, R.J., and Krider, E.P. 1992. Ground level measurements of air conductivities under Florida thunderstorms. *J. Geophys. Res.* **97**: 12 947–51.

Blyth, A.M., Christian, H.J., and Latham, J. 1998. Corona emission thresholds for three types of hydrometeor interaction in thunderclouds. *J. Geophys. Res.* **103**: 13 975–7.

Blyth, A.M., Christian, H.J., Driscoll, K., Gadian, A.M., and Latham, J. 2001. Determination of ice precipitation rates and thunderstorms anvil ice contents from satellite observations of lightning. *Atmos. Res.* **59–60**: 217–29.

Bourdeau, C., and Chauzy, S. 1989. Maximum electric charge of a hydrometeor in the electric field of a thunderstorm. *J. Geophys. Res.* **94**: 13 121–26.

Bringi, U.N., Knupp, K., Detweiler, A., Lu, L., Caylor, I.J., and Black, R.A. 1997. Evolution of a Florida thunderstorm during the Convection and Precipitation/Electrification Experiment: the case of 9 August 1991. *Mon. Wea. Rev.* **125**: 2131–60.

Brooks, I.M., and Saunders, C.P.R. 1994. An experimental investigation of the inductive mechanism of thunderstorm electrification. *J. Geophys. Res.* **99**: 10 627–32.

Brooks, I.M., and Saunders, C.P.R. 1995. Thunderstorm charging: laboratory experiments clarified. *Atmos. Res.* **39**: 263–73.

Brooks, I.M., Saunders, C.P.R., Mitzeva, R.P., and Peck, S.L. 1997. The effect on thunderstorm charging of the rate of rime accretion by graupel. *Atmos. Res.* **43**: 277–95.

Brown, K.A., Krehbiel, P.R., Moore, C.B., and Sargent, G.N. 1971. Electrical screening layers around charged clouds. *J. Geophys. Res.* **76**: 2825–36.

Browning, G.L., Tzur, I., and Roble, R.G. 1987. A global time-dependent model of thunderstorm electricity. Part I: Mathematical properties of the physical and numerical models. *J. Atmos. Sci.* **44**: 2166–77.

Brylev, G.B., Gashina, S.B., Yevteyev, B.F., and Kamaldyna, I.I. 1989. *Characteristics of Electrically Active Regions in Stratiform Clouds*, 303 pp. USAF translation, FTD-ID(RS)T-0698-89, of Kharakteristiki Elektricheski Aktivnykh Zon v Sloistoobraznykh Oblakakh, Gidrometeoizdat, Leningrad, 160 pp.

Buser, O., and Aufdermaur, A.N. 1977. Electrification by collisions of ice particles on ice or metal targets. In *Electrical Processes in Atmospheres*, eds. H. Dolezalek and R. Reiter, pp. 294–300, Darmstadt, Germany: Dr. Dietrich Steinkopff.

Byrne, C.J., Few, A.A., and Weber, M.E. 1983. Altitude, thickness and charge concentration of charged regions of four thunderstorms during TRIP 1981 based upon *in situ* balloon electric field measurements. *Geophys. Res. Lett.* **10**: 39–42.

Byrne, C.J., Few, A.A., and Stewart, M.F. 1986. The effects of atmospheric parameters on a corona probe used in measuring thunderstorm electric fields. *J. Geophys. Res.* **91**: 9911–20.

Byrne, C.J., Few, A.A., Stewart, M.F., Conrad, A.C., and Torczon, R.L. 1987. *In situ* measurements and radar observations of a severe storm: electricity, kinematics, and precipitation. *J. Geophys. Res.* **92**: 1017–31.

Byrne, C.J., Few, A.A., and Stewart, M.F. 1989. Electric field measurements within a severe thunderstorm anvil. *J. Geophys. Res.* **94**: 6297–307.

Canosa, E.F., and List, R. 1993. Measurements of inductive charges during drop breakup in horizontal electric fields. *J. Geophys. Res.* **98**: 2619–26.

Canosa, E.F., List, R., and Stewart, R.E. 1993. Modeling of inductive charge separation in rainshafts with variable vertical electric fields. *J. Geophys. Res.* **98**: 2627–33.

Caranti, J.M., and Illingworth, A.J. 1980. Surface potentials of ice and thunderstorm charge separation. *Nature* **284**: 44–6.

Caranti, J.M., and Illingworth, A.J. 1983. The contact potential of rimed ice. *J. Phys. Chem.* **87**: 4125–30.

Caranti, J.M., Illingworth, A.J., and Marsh, S.J. 1985. The charging of ice by differences in contact potential. *J. Geophys. Res.* **90**: 6041–46.

Caranti, J.M., Avila, E., and Re, M. 1991. The charge transfer during individual collisions in vapor growing ice. *J. Geophys. Res.* **96**: 15 365–75.

Carey, L.D., and Rutledge, S.A. 1996. A multiparameter radar case study of the microphysical and kinematic evolution of a lightning producing storm. *J. Meteor. Atmos. Phys.* **59**: 33–64.

Carey, L.D., and Rutledge, S.A. 1998. Electrical and multiparameter radar observations of a severe hailstorm. *J. Geophys. Res.* **103**: 13 979–4 000.

Carey, L.D., and Rutledge, S.A. 2000. The relationship between precipitation and lightning in tropical island convection: a C-band polarimetric radar study. *Mon. Wea. Rev.* **128**: 2687–710.

Censor, D., and Levin, Z. 1973. Electrostatic interaction of axisymmetric liquid and solid aerosols. Publ. ES 73–015, Dept of Environ. Sci., Tel Aviv University, Israel.

Chalmers, J.A. 1967. *Atmospheric Electricity*, 2nd edition, 515 pp., Oxford: Pergamon.

Changnon, S.A. 1992. Temporal and spatial relations between hail and lightning. *J. Appl. Meteor.* **31**: 587–604.

Chauzy, S., and Kably, K. 1989. Electric discharges between hydrometeors. *J. Geophys. Res.* **94**: 13 107–14.

Chauzy, S., and Raizonville, P. 1982. Space charge layers created by coronae at ground level below thunderclouds: measurements and modelling. *J. Geophys. Res.* **87**: 3143–8.

Chauzy, S., and Raizonville, P. 1983. Electrostatical screening below thunderstorms due to coronae at ground level. In *Proc. Conf. on Atmospheric Electricity*, ed. L.H. Ruhnke and J. Latham, pp. 184–7, Hampton, Virginia: A. Deepak.

Chauzy, S., and Soula, S. 1987. General interpretation of surface electric field variations between lightning flashes. *J. Geophys. Res.* **92**: 5676–84.

Chauzy, S., and Soula, S. 1999. Contribution of the ground corona ions to the convective charging mechanism. *Atmos. Res.* **51**: 279–300.

Chauzy, S., Raizonville, P., Hauser, D., and Roux, F. 1980. Electrical and dynamical description of a frontal storm deduced from LANDES 79 experiment. *J. Rech. Atmos.* **14**: 457–67.

Chauzy, S., Chong, M., Delannoy, A., and Despiau, S. 1985. The June 22 tropical squall line observed during COPT 81 experiment: electrical signature associated with dynamical structure and precipitation. *J. Geophys. Res.* **90**: 6091–8.

Chauzy, S., Soula, S. and Despiau, S. 1989. Ground coronae and lightning. *J. Geophys. Res.* **94**: 13 115–19.

Chauzy, S., Medale, J.-C., Prieur, S., and Soula, S. 1991. Multilevel measurement of the electric field underneath a thundercloud, 1. A new system and the associated data processing. *J. Geophys. Res.* **96**: 22 319–26.

Chiu, C.S. 1978. Numerical study of cloud electrification in an axisymmetric, time-dependent cloud model. *J. Geophys. Res.* **83**: 5025–49.

Chiu, C.S., and Klett, J.D. 1976. Convective electrification of clouds. *J. Geophys. Res.* **81**: 1111–24.

Christian, H.J., Holmes, C.R., Bullock, J.W., Gaskell, W., Illingworth, A.J., and Latham, J. 1980. Airborne and ground-based studies of thunderstorms in the vicinity of Langmuir Laboratory. *Q.J.R. Meteor. Soc.* **106**: 159–74.

Chubarina, Ye. V. 1977. Large electric fields in the clouds of laminar form. *Trudy GGO* **350**: 80–6.

Cole, R.K. Jr, Hill, R.D., and Pierce, E.T. 1966. Ionized columns between thunderstorms and the ionosphere. *J. Geophys. Res.* **71**: 959–64.

Colgate, S.A., and Romero, J.M. 1970. Charge versus drop size in an electrified cloud. *J. Geophys. Res.* **75**: 5873–81.

Connor, J.W., and Hastie, R.J. 1975. Relativistic limitations on runaway electrons. *Nucl. Fusion* **15**: 415–24.

Coquillat, S., Chauzy, S., and Medale, J.-C., 1995. Microdischarges between ice particles. *J. Geophys. Res.* **100**: 14 327–34.

Cotton, W.R., and Anthes, R.A. 1989. *Storm and Cloud Dynamics*, 883 pp., London: Academic Press.

Crabb, J.A., and Latham, J. 1974. Corona from colliding drops as a possible mechanism for the triggering of lightning. *Q.J.R. Meteor. Soc.* **100**: 191–202.

Dash, J.G. 1989. Surface melting. *Contemp. Phys.* **30**: 89–100.

Davies, A.J., Evans, C.J., and Llewellyn-Jones, F. 1964. Electrical breakdown of gases: the spatio-temporal growth of ionization in fields distorted by space charge. *Proc. Roy. Soc. A* **281**: 164–83.

Davis, R., and Standring, W.G. 1947. Discharge currents associated with kite balloons. *Proc. Roy. Soc. (London) A* **191**: 304–22.

Dawson, G.A. 1969. Pressure dependence of water-drop corona onset and its atmospheric importance. *J. Geophys. Res.* **74**: 6859–68.

Dawson, G.A. 1973. Charge loss mechanism of highly charged water droplets in the atmosphere. *J. Geophys. Res.* **78**: 6364–69.

Dawson, G.A. and Duff, D.G. 1970. Initiation of cloud-to-ground lightning strokes. *J. Geophys. Res.* **75**: 5858–67.

Deaver, L.E., and Krider, E.P. 1991. Electric fields and current densities under small Florida thunderstorms. *J. Geophys. Res.* **96**: 22 273–81.

Dejnakarintra, M., and Park, C.G. 1974. Lightning-induced electric fields in the ionosphere. *J. Geophys. Res.* **79**: 1903–10.

Despiau, S., and Houngninou, E. 1996. Raindrop charge, precipitation, and Maxwell currents under tropical storms and showers. *J. Geophys. Res.* **101**: 14 991–7.

Dinger, J.E., and Gunn, R. 1946. Electrical effects associated with a change of state of water. *Terr. Magn. Atmos. Electr.* **51**: 477–94.

Dolezalek, H. 1988. Discussion on the Earth's net electric charge. *Meteor. Atmos. Phys.* **38**: 240–5.

Dong, Y., and Hallett, J.H. 1992. Charge separation by ice and water drops during growth and evaporation. *J. Geophys. Res.* **97**: 20 361–71.

Dotzek, N., Höller, H., Théry, C., and Fehr, T. 2001. Lightning evolution related to radar-derived microphysics in the 21 July 1998 EULINOX supercell storm. *J. Atmos. Res.* **56**: 335–54.

Doyle, A., Moffet, D.R., and Vonnegut, B. 1964. Behavior of evaporating electrically charged droplets. *J. Coll. Sci.* **19**: 136–43.

Drake, J.C. 1968. Electrification accompanying the melting of ice particles. *Q.J.R. Meteor. Soc.* **94**: 176–91.

Dreicer, H. 1959. Electron and ion runaway in a fully ionized gas: I. *Phys. Rev.* **115**: 238–49.

Dreicer, H. 1960. Electron and ion runaway in a fully ionized gas: II. *Phys. Rev.* **117**: 329–42.

Dress, J. and Trinks, H.W. 1967. Runaway-Ströme hoher Intensität in einer toroidalen Entladung. *Zeitschrift für Physik* **100**: 410–18.

Driscoll, K.T., Blakeslee, R.J., and Baginski, M.E. 1992. A modeling study of the time-averaged electric currents in the vicinity of isolated thunderstorms. *J. Geophys. Res.* **97**: 11 535–51.

Dutton, J., Haydon, S.C. and Llewellyn-Jones, F. 1953. Photo-ionization and the electrical breakdown of gases. *Proc. Roy. Soc., A* **218**: 206–23.

Dye, J.E., Jones, J.J., Winn, W.P., Carnti, T.A., Gardiner, B., Lamb, D., Pitter, R.L., Hallett, J., and Saunders, C.P.R. 1986. Early electrification and precipitation development in a small, isolated Montana cumulonimbus. *J. Geophys. Res.* **91**: 1231–327, 6747–50.

Dye, J.E., Jones, J.J., Weinheimer, A.J., and Winn, W.P. 1988. Observations within two regions of charge during initial thunderstorm electrification. *Q.J.R. Meteor. Soc.* **114**: 1271–90.

Dye, J.E., Winn, W.P., Jones, J.J., and Breed, D.W. 1989. The electrification of New Mexico thunderstorms, 1. Relationship between precipitation development and the onset of electrification. *J. Geophys. Res.* **94**: 8643–56.

Eack, K.B., Beasley, W.H., Rust, W.D., Marshall, T.C., and Stolzenburg, M. 1996. Initial results from simultaneous observation of X rays and electric fields in a thunderstorm. *J. Geophys. Res.* **101**: 29 637–40.

Ecker, von G., and Müller, K.G. 1961. Runaways in neutralgas. *Sonderabdruck aus der Zeitschrift für Naturforschung* **16A**(3): 246–52.

Elbaum, M., Kipson, S., and Dash, J.G. 1992. Optical study of surface melting of ice. *J. Cryst. Growth* **129**: 491.

Elster, J., and Geitel, H., 1888. Über eine Methode, die elektrische Natur der atmosphärischen Niederschläge zu bestimmen (About a method for determining the electric nature of atmospheric precipitation). *Meteor. Z.* **5**: 95–100.

Engholm, C.D., Williams, E.R., and Dole, R.M. 1990. Meteorological and electrical conditions associated with positive

cloud-to-ground lightning. *Mon. Wea. Rev.* **118**: 470–87.
Ette, A.I.I., and Olaofe, G.O. 1982. Theoretical field configurations for thundercloud models with volume charge distributions. *Pure Appl. Geophys.* **120**: 117–22.
Evans, W.H. 1965. The measurement of electric fields in clouds. *Rev. Pure Appl. Geophys.* **62**: 191–7.
Evans, W.H. 1969. Electric fields and conductivity in thunderclouds. *J. Geophys. Res.* **74**: 939–48.
Fitzgerald, D.R. 1967. Probable aircraft "triggering" of lightning in certain thunderstorms. *Mon. Wea. Rev.* **95**: 835–42.
Fitzgerald, D.R. 1976. Experimental studies of thunderstorm electrification. Air Force Geophysics Laboratory: Report AFGL-TR-76-0128, AD-A322374, Environmental Research Papers no. 567, 40 pp.
Fitzgerald, D.R. 1984. Electric field structure of large thunderstorm complexes in the vicinity of Cape Canaveral. Preprint, *Proc. 7th Int. Conf. on Atmospheric. Electricity, Albany, New York*, pp. 260–2, American Meteor. Soc., Boston, Massachusetts.
Fleischer, R.L. 1975. Search for neutron generation by lightning. *J. Geophys. Res.* **36**: 5005–9.
Foster, H. 1950. An unusual observation of lightning. *Bull. Am. Meteor. Soc.* **31**: 140–1.
Frankel, S., Highland, V., Sloan, T., Van Dyck, O., and Wales, W. 1966. Observation of X-rays from spark discharges in a spark chamber. *Nuclear Instruments and Methods* **44**: 345–8.
Freier, G.D. 1972. Comments on "Electric field measurements in thunderclouds using instrumented rockets". *J. Geophys. Res.* **77**: 505, and Reply by authors, **77**: 506–8.
French, J.R., Helsdon, J.H., Detwiler, A.G., and Smith, P.L. 1996. Microphysical and electrical evolution of a Florida thunderstorm – 1. Observations. *J. Geophys. Res.* **101**: 18 961–77.
Gardiner, B., Lamb, D., Pitter, R.L., Hallett, J., and Saunders, C.P.R. 1985. Measurements of initial potential gradient and particle charges in a Montana summer thunderstorm. *J. Geophys. Res.* **90**: 6079–86.
Gaskell, W. 1981. A laboratory study of the inductive theory of thunderstorm electrification. *Q.J.R. Meteor. Soc.* **107**: 955–66.
Gaskell, W., and Illingworth, A.J. 1980. Charge transfer accompanying individual collisions between ice particles and its role in thunderstorm electrification. *Q.J.R. Meteor. Soc.* **106**: 841–54.
Gaskell, W., Illingworth, A. J., Latham, J., and Moore, C.B. 1978. Airborne studies of electric fields and the charge and size of precipitation elements in thunderstorms. *Q.J.R. Meteor. Soc.* **104**: 447–60.
Geotis, S.G., Williams, E.R., and Liu, C. 1991. Reply. *J. Atmos. Sci.* **48**: 371–2.
Gish, O.H., and Wait, G.R. 1950. Thunderstorms and the Earth's general electrification. *J. Geophys. Res.* **55**: 473–84.
Goodman, S.J., Buechler, D.E., Wright, P.D., and Rust, W.D. 1988. Lightning and precipitation history of a microburst-producing storm. *Geophys. Res. Lett.* **15**: 1185–8.

Graciaa, A., Creux, P., and Lachaise, J. 2001. Charge transfer between colliding hydrometeors: role of surface tension gradients. *J. Geophys. Res.* **106**(D8): 7967–72.
Grard, R. 1998. Electrostatic charging processes of balloon and gondola surfaces in the Earth atmosphere. *J. Geophys. Res.* **103**: 23 315–20.
Grenet, G. 1947. Essai d'éxplication de la charge électrique des nuages d'orages. *Ann. Geophys.* **3**: 306–7.
Grenet, G. 1959. Le nuage d'orage: machine éléctrostatique. *Météorologie* **I-53**: 45–7.
Griffiths, R.F. 1975. The initiation of corona discharges from charged ice particles in a strong electric field. *J. Electrostat.* **1**: 3–13.
Griffiths, R.F., and Latham, J. 1974. Electrical corona from ice hydrometers. *Q.J.R. Meteor. Soc.* **100**: 163–80.
Griffiths, R.F., Latham, J. and Meyers, V. 1974. The ionic conductivity of electrified clouds. *Q.J.R. Meteor. Soc.* **100**: 181–90.
Griffiths, R.F., and Phelps, C.T. 1976a. The effects of air pressure and water vapor content on the propagation of positive corona streamers, and their implications to lightning initiation. *Q.J.R. Meteor. Soc.* **102**: 419–26.
Griffiths, R.F., and Phelps, C.T. 1976b. A model of lightning initiation arising from positive corona streamer development. *J. Geophys. Res.* **31**: 3671–6.
Gunn, R. 1948. Electric field intensity inside of natural clouds. *J. Appl. Phys.* **19**: 481–4.
Gunn, R. 1954. Diffusion charging of atmospheric droplets by ions and the resulting combination coefficients. *J. Meteor.* **11**: 339–47.
Gunn, R. 1956. Electric field intensity at the ground under active thunderstorms and tornadoes. *J. Meteor.* **13**: 269–73.
Gunn, R. 1957. The electrification of precipitation and thunderstorms. *Proc. IRE* **45**: 1331–58.
Gunn, R. 1965. The electric field intensity and its systematic changes under an active thunderstorm. *J. Atmos. Sci.* **22**: 498–500.
Gunn, R., and Kinzer, G.D. 1949. The terminal velocity of fall for water droplets in stagnant air. *J. Meteor.* **6**: 243–8.
Gurevich, A.V. 1961. On the theory of runaway electrons. *Soviet Phys. JETP* **12**: 904–12.
Gurevich, A.V., Milikh, G.M., and Roussel-Dupré, R. 1992. Runaway electron mechanism of air breakdown and preconditioning during a thunderstorm. *Phys. Lett. A* **165**: 463–7.
Hacking, C.A. 1954. Observations on the negatively-charged column in thunderclouds. *J. Geophys. Res.* **59**: 449–53.
Hager, W.W. 1998. A discrete model for the lightning discharge. *J. Comput. Phys.* **144**: 137–50.
Hager, W.W., Nisbet, J.S., and Kasha, J.R. 1989a. The evolution and discharge of electric fields within a thunderstorm. *J. Comput. Phys.* **82**: 193–217.
Hager, W.W., Nisbet, J.S., Kasha, J.R., and Shann, W.-C. 1989b. Simulation of electric fields within a thunderstorm. *J. Atmos. Sci.* **46**: 3542–58.
Hale, L.C., and Baginski, M.E. 1987. Current to the ionosphere following lightning stroke. *Nature* **329**: 814–16.

Handel, P.H. 1985. Polarization catastrophe theory of cloud electricity – Speculation on a new mechanism for thunderstorm electrification. *J. Geophys. Res.* **90**: 5857–63.

Hays, P.B., and Roble, R.G. 1979. A quasi-static model of global atmospheric electricity, 1, The lower atmosphere. *J. Geophys. Res.* **84**: 3291–305.

Helsdon, J.H. Jr 1980. Chaff seeding effects in a dynamical electrical cloud model. *J. Appl. Meteor.* **19**: 1101–25.

Helsdon, J.H. Jr, and Farley, R.D. 1987a. A numerical modeling study of a Montana thunderstorm: 1. Model results versus observations involving nonelectrical aspects. *J. Geophys. Res.* **92**: 5645–59.

Helsdon, J.H. Jr, and Farley, R.D. 1987b. A numerical modeling study of a Montana thunderstorm: 2. Model results versus observations involving electrical aspects. *J. Geophys. Res.* **92**: 5661–75.

Helsdon, J.H. Jr, Wu, G., and Farley, R.D. 1992. An intracloud lightning parameterization scheme for a storm electrification model. *J. Geophys. Res.* **97**: 5865–84.

Helsdon, J.H. Jr, Wojcik, W., and Farley, R.D. 2001. An examination of thunderstorm-charging mechanisms using a two-dimensional storm electrification model. *J. Geophys. Res.* **106**: 1165–92.

Hendry, A., and McCormick, G.C. 1976. Radar observations of the alignment of precipitation particles by electrostatic fields in thunderstorms. *J. Geophys. Res.* **81**: 5353–7.

Hill, R.D. 1963. Investigation of electron runaway in lightning. *J. Geophys. Res.* **68**: 6261–6.

Hill, R.D. 1988. Interpretation of bipole pattern in a mesoscale storm. *Geophys. Res. Lett.* **15**: 643–4.

Holzer, R.E., and Saxon, D.S. 1952. Distribution of electrical conduction current in the vicinity of thunderstorms. *J. Geophys. Res.* **57**: 207–16.

Holzworth, R.H., Norville, K.W., Kintner, P.M., and Power, S.P. 1986. Stratospheric conductivity variations over thunderstorms. *J. Geophys. Res.* **91**: 13 257–63.

Hoppel, W.A., and Phillips, B.B. 1971. The electrical shielding layer around charged clouds and its role in thunderstorm electricity. *J. Atmos. Sci.* **28**: 1258–71.

Houze, R.A. Jr 1993. *Cloud Dynamics*, 573 pp., San Diego, California: Academic Press.

Hunter, S.M., Schuur, T.J., Marshall, T.C., and Rust, W.D. 1992. Electric and kinematic structure of the Oklahoma mesoscale convective system of 7 June 1989. *Mon. Wea. Rev.* **12**: 2226–39.

Huzita, A., and Ogawa, T. 1976. Charge distribution in the average thunderstorm cloud. *J. Meteor. Soc. Japan* **54**: 285–8.

Illingworth, A.J. 1971a. The variation of the electric field after lightning and the conductivity within thunderclouds. *Q.J.R. Meteor. Soc.* **97**: 440–56.

Illingworth, A.J. 1971b. Electric field recovery after lightning. *Nature – Physical Science* **229**: 213–14.

Illingworth, A.J. 1972a. Electric field recovery after lightning as the response of the conducting atmosphere to a field change. *Q.J.R. Meteor. Soc.* **98**: 604–16.

Illingworth, A.J. 1972b. Comments on "the electrical shielding layer around charged clouds and its role in thunderstorm electricity" by Hoppel and Phillips. *J. Atmos. Sci.* **29**: 1217.

Illingworth, A.J. 1985. Charge separation in thunderstorms: small scale processes. *J. Geophys. Res.* **90**: 6026–32.

Illingworth, A.J., and Caranti, J.M. 1985. Ice conductivity restraints on the inductive theory of thunderstorm electrification. *J. Geophys. Res.* **95**: 6033–9.

Illingworth, A.J., and Latham, J. 1977. Calculations of electric field growth, field structure and charge distributions in thunderstorms. *Q.J.R. Meteor. Soc.* **103**: 277–98.

Imyanitov, I.M., and Chubarina, Y.V. 1967. Electricity of the free atmosphere. NASA Technical Translation NASA TTF-425, TT 67-51374, of Elektrichestvo Svobodnoy Atmosfery, Gidrometeosizdat, Leningrad, 1965, NTIS Accession no. N68-10079, 212 pp.

Imyanitov, I.M., Chubarina, Y.V., and Shvarts, Y.M. 1971. Electricity of clouds. 92 pp., Leningrad: Gidrometeoizdat (NASA Technical Translation from Russian NASA, TT-F-718, 1972).

Israel, H. 1973. *Atmospheric Electricity, vol. II, Fields, Charges, Currents*, 796 pp., Published for the National Science Foundation by the Israel Program for Scientific Translation, Jerusalem.

Jacobson, E.A., and Krider, E.P. 1976. Electrostatic field changes produced by Florida lightning. *J. Atmos. Sci.* **33**: 113–7.

Jayaratne, E.R. 1993. Temperature gradients in ice as a charge generation process in thunderstorm. *Atmos. Res.* **29**: 247–60.

Jayaratne, E.R. 1993. The heat balance of a riming graupel pellet and the charge separation during ice–ice collisions. *J. Atmos. Sci.* **50**: 3185–93.

Jayaratne, E.R. 1998a. Possible laboratory evidence for multipole electric charge structures in thunderstorms. *J. Geophys. Res.* **103**: 1871–8.

Jayaratne, E.R. 1998b. Density and surface temperature of graupel and the charge separation during ice crystal interactions. *J. Geophys. Res.* **103**: 13 957–61.

Jayaratne, E.R., and Griggs, D.J. 1991. Electric charge separation during the fragmentation of rime in an airflow. *J. Atmos. Sci.* **48**: 2492–5.

Jayaratne, E.R., and Saunders, C.P.R. 1984. The rain gush, lightning and the lower positive charge center in thunderstorms. *J. Geophys. Res.* **89**: 11 816–18.

Jayaratne, E.R., and Saunders, C.P.R. 1985a. Reply. *J. Geophys. Res.* **90**: 10 755.

Jayaratne, E.R., and Saunders, C.P.R. 1985b. Thunderstorm electrification: the effect of cloud droplets. *J. Geophys. Res.* **90**: 13 063–6.

Jayaratne, E.R., and Saunders, C.P.R. 1986. Reply. *J. Geophys. Res.* **91**: 10 950.

Jayaratne, E.R., Saunders, C.P.R., and Hallett, J. 1983. Laboratory studies of the charging of soft-hail during ice crystal interactions. *Q.J.R. Meteor. Soc.* **109**: 609–30.

Jayaratne, R., Peck, S.L., and Saunders. C. 1996. Comment on "A laboratory study of static charging by fracture in ice growing by riming" by Eldo E. Avila and Giorgio M. Caranti. *J. Geophys. Res.* **101**: 9533–5.

References and bibliography for Chapter 3

Jennings, S.G. 1975. Charge separation due to water droplet and cloud droplet interactions in an electric field. *Q.J.R. Meteor. Soc.* **101**: 227–34.

Jensen, J.C. 1993. The branching of lightning and the polarity of thunderclouds. *J. Franklin Inst.* **216**: 707–47.

Jones, J.J. 1990. Electric charge acquired by airplanes penetrating thunderstorms. *J. Geophys. Res.* **95**: 16 589–600.

Jones, J.J., Winn, W.P., and Han, F. 1993. Electric field measurements with an airplane: problems caused by emitted charge. *J. Geophys. Res.* **98**: 5235–44.

Jonsson, H.H. 1990. Possible sources of errors in electrical measurements made in thunderclouds with balloon–borne instrumentation. *J. Geophys. Res.* **95**: 22 539–45.

Jonsson, H.H., and Vonnegut, B. 1995. Comment on "Negatively charged precipitation in a New Mexico thunderstorm" by Thomas C. Marshall and Stephen J. Marsh and "Charged precipitation measurements before the first lightning flash in a thunderstorm" by Stephen J. Marsh and Thomas C. Marshall. *J. Geophys. Res.* **100**: 16 867–8.

Kachurin, L.G., Karmov, M.I., and Medaliyev, K.K. 1974. The principal characteristics of the radio emission of convective clouds. *Izv. Acad. Sci. USSR Atmos. Oceanic Phys. (English translation)* **10**: 1163–9.

Kasemir H.W. 1959. The thunderstorm as a generator in the global electric circuit (in German). *Z. Geophys.* **25**: 33–64.

Kasemir, H.W. 1960. A contribution to the electrostatic theory of a lightning discharge. *J. Geophys. Res.* **65**: 1873–8.

Kasemir, H.W. 1965. The thundercloud. In *Problems of Atmospheric and Space Electricity*, ed. S.C. Coroniti, pp. 215–35, New York: American Elsevier.

Kasemir, H.W. 1984. Theoretical and experimental determination of field, charge and current on an aircraft hit by natural and triggered lightning. Preprint, *Proc. Int. Aerospace and Ground Conf. on Lightning and Static Electricity, Orlando*, Florida: National Interagency Coordinating Group.

Kasemir, H.W., and Perkins, F. 1978 Lightning trigger field of the orbiter. Final Report, Kennedy Space Center Contract CC 69694A.

Keenan, T., Rutledge, S., Carbone, R., Wilson, J., Takahashi, T., May, P., Tapper, N., Platt, M., Hacker, J., Sekelsky, S., Moncrieff, M., Saito, K., Holland, G., Crook, A., and Gage, K. 2000. The Maritime Continent Thunderstorm Experiment (MCTEX): overview and some results. *Bull. Am. Meteor. Soc.* **81**: 2433–55.

Keeney, J. 1970. Observations on the microwave emission from colliding charged water drops. *J. Geophys. Res.* **75**: 1123–6.

Keith, W.D., and Saunders, C.P.R. 1988. Light emission from colliding ice particles. *Nature* **336**: 362–4.

Keith, W.D., and Saunders, C.P.R. 1989. Charge transfer during multiple large ice-crystal interactions with a riming target. *J. Geophys. Res.* **94**: 13 103–6.

Keith, W.D., and Saunders, C.P.R. 1990. Further laboratory studies of the charging of graupel during ice crystal interactions. *J. Atmos. Sci.* **25**, 445–64.

Kessler, E. (ed.) 1986a. *Thunderstorm Morphology and Dynamics*, 2nd edition, 411 pp., Norman, Oklahoma: University of Oklahoma Press.

Kessler, E. 1986b. Thunderstorm origins, morphology, and dynamics. In *The Earth's Electrical Environment*, eds. E.P. Krider and R.G. Robble, pp. 81–9, Washington, DC: National Academy Press.

Kieffer, L.J., and Dunn, G.H. 1966. Electron impact ionization cross-section data for atoms, atomic ions, and diatomic molecules: I. Experimental data. *Rev. Mod. Phys.* **38**: 1–35.

Kikuchi, K. 1965. On the positive electrification of snow crystals in the process of their melting, III, IV. *J. Meteor. Soc. Japan* **43**: 343–50, 351–8.

Kitagawa, N. 1992. Charge distribution of winter thunderclouds. *Res. Lett. Atmos. Electr.* **12**: 143–53.

Kitagawa, N., and Michimoto, K. 1994. Meteorological and electrical aspects of winter thunderclouds. *J. Geophys. Res.* **99**: 10 713–21.

Klett, J.D 1972. Charge screening layers around electrified clouds. *J. Geophys. Res.* **77**: 3187–95.

Koshak, W.J., and Krider, E.P. 1989. Analysis of lightning field changes during active Florida thunderstorms. *J. Geophys. Res.* **94**: 1165–86.

Koshak, W.J., and Krider, E.P. 1994. A linear method for analyzing lightning field changes. *J. Atmos. Sci.* **51**: 473–88.

Krehbiel, P.R. 1981. An analysis of the electric field change produced by lightning. Ph.D. dissertation, University of Manchester Institute of Science and Technology, Manchester, England. (Available as Report T-11, Geophys. Res. Ctr. New Mexico Inst. Mining and Tech., Socorro, New Mexico)

Krehbiel, P.R. 1986. The electrical structure of thunderstorms. In *The Earth's Electrical Environment*, eds. E.P. Krider and R.G. Roble, pp. 90–113, Washington, DC: National Academy Press.

Krehbiel, P.R., Brook, M., and McCrory, R.A. 1979. An analysis of the charge structure of lightning discharges to the ground. *J. Geophys. Res.* **84**: 2432–56.

Krehbiel, P.R., Brook, M., Lhermitte, R.L., and Lennon, C.L. 1983. Lightning charge structure in thunderstorms. In *Proc. Conf. on Atmospheric Electricity*, eds. L.H. Ruhnke and J. Latham, pp. 408–11, Hampton, Virginia: Deepak.

Krehbiel, P.R., Chen, T., McCrary, S., Rison, W., Gray, G, and Brook, M. 1996. The use of dual channel circular-polarization radar observations for remotely sensing storm electrification. *Meteor. Atmos. Phys.* **59**: 65–82.

Krider, E.P. 1989. Electric field changes and cloud electrical structure. *J. Geophys. Res.* **94**: 13 145–9.

Krider, E.P., and Blakeslee, R.J. 1985. The electric currents produced by thunderclouds. *J. Electrostat.* **16**: 369–78.

Krider, E.P., and Musser, J.A. 1982. Maxwell currents under thunderstorms. *J. Geophys. Res.* **87**: 11 171–6.

Kruskal, M.D., and Bernstein, I.B. 1964. Runaway electrons in an ideal Lorentz plasma. *Phys. Fluids* **7**: 407–18.

Kuettner, J. 1950. The electrical and meteorological conditions inside thunderclouds. *J. Meteor.* **7**: 322–32.

Kuettner, J. 1956. The development and masking of charge in thunderstorms. *J. Meteor.* **13**: 456–70.

Kuettner, J.P., Levin, Z., and Sartor, J.D. 1981. Thunderstorm electrification – Inductive or noninductive? *J. Atmos. Sci.* **38**: 2470–84.

Lalande, P., Bondiou-Clergerie, A., and Laroche, P. 1999. Analysis of available in-flight measurements of lightning strikes to aircraft. In *Proc. 1999 Int. Conf. on Lightning and Static Electricity, Toulouse, France*, pp. 401–8.

Latham, D. 1991. Lightning flashes from a prescribed fire-induced cloud. *J. Geophys. Res.* **96**: 17 151–7.

Latham, J. 1981. The electrification of thunderstorms. *Q.J.R. Meteor. Soc.* **107**: 277–98.

Latham, J., and Dye, J.E. 1989. Calculations on the electrical development of a small thunderstorm. *J. Geophys. Res.* **94**: 13 141–4.

Latham, J., and Mason, B.J. 1961. Electric charge transfer associated with temperature gradients in ice. *Proc. Roy. Soc.* A**260**: 523–36.

Latham, J., and Mason, B.J. 1962. Electric charging of hail pellets in a polarizing electric field. *Proc. Roy. Soc.* A**26**: 387–401.

Latham, J., and Stromberg, I.M. 1977. Point-discharge. In *Lightning, vol. 1, Physics of Lightning*, ed. R.H. Golde, pp. 99–117, New York: Academic Press.

Latham, J., and Warwicker, R. 1980. Charge transfer accompanying the splashing of supercooled raindrops on hailstones. *Q.J.R. Meteor. Soc.* **106**: 559–68.

Levin, Z. 1976. A refined charge distribution in a stochastic electrical model of an infinite cloud. *J. Atmos. Sci.* **33**: 1756–62.

Levin, Z., and Tzur, I. 1986. Models of the development of the electrical structure of clouds. In *Earth's Electrical Environment*, eds. E.P. Krider and R.G. Roble, pp. 131–45, Washington, DC: National Academy Press.

Lhermitte, R., and Williams, E. 1983. Cloud electrification. *Rev. Geophys. Space Sci.* **21**: 984–92.

Lhermitte, R., and Williams, E. 1985. Thunderstorm electrification: a case study. *J. Geophys. Res.* **90**: 6071–8.

Livingston, J.M., and Krider, E.P. 1978. Electric fields produced by Florida thunderstorms. *J. Geophys. Res.* **83**: 385–401.

Liu, X.-S., and Krehbiel, P.R. 1985. The initial streamer of intracloud lightning flashes. *J. Geophys. Res.* **90**: 6211–18.

Lomonosov, M.V. 1753. Atmospheric phenomena stemming from electricity. Am. Meteor. Soc. (translated from Russian by D. Kraus), 1963, 45 pp.

Lopez R.E., and Aubagnac, J.-P. 1997. The lightning activity of a hailstorm as a function of changes in its microphysical characteristics inferred from polarimetric radar observations. *J. Geophys. Res.* **102**: 16 799–813.

Ludlam, F.H. 1980. *Clouds and Storms*, 405 pp., University Park: Pennsylvania State University Press.

MacGorman, D.R., Few, A.A., and Teer, T.L. 1981. Layered lightning activity. *J. Geophys. Res.* **86**: 9900–10.

MacGorman, D.R., and Rust, W.D. 1998. *The Electrical Nature of Thunderstorms*, 422 pp., New York: Oxford University Press.

MacGorman, D.R., Taylor, W.L., and Few, A.A. 1983. Lightning location from acoustic and VHF techniques relative to storm structure from 10-cm radar. In *Proc. Conf. on Atmospheric Electricity*, eds. L.H. Ruhnke and J. Latham, pp. 377–80. Hampton, Virginia: A. Deepak.

MacGorman, D.R., Ziegler, C.L., and Straka, J.M. 1996. Considering the complexity of thunderstorm electrification. In *Proc. 10th Int. Conf. Atmospheric Electricity, Osaka, Japan*, pp. 128–31.

MacGorman, D.R., Straka, J.M., Ziegler, C.L. 2001. A lightning parameterization for numerical cloud models. *J. Appl. Meteor.* **40**: 459–78.

Macky, W.A. 1931. Some investigations on the deformation and breaking of water drops in strong electric fields. *Proc. Roy. Soc.* A **133**: 565–87.

Maekawa, Y., Fukao, S., Sonoi, Y., and Yoshino, F. 1993. Distribution of ice particles in wintertime thunderclouds detected by a C band dual polarization radar: a case study. *J. Geophys. Res.* **98**: 16 613–22.

Magono, C. 1980. *Thunderstorms*, 261 pp., New York: Elsevier.

Magono, C. and Kikuchi, K. 1965. On the positive electrification of snow crystals in the process of their melting. *J. Meteor. Soc. Japan* **41**: 331–42.

Magono, C. and Koenuma, S. 1958. On the electrification of water drops by breaking due to the electrostatic induction under a moderate electric field. *J. Meteor. Soc. Japan* **36**: 108–11.

Maier, L.M., and Krider, E.P. 1986. The charges that are deposited by cloud-to-ground lightning in Florida. *J. Geophys. Res.* **91**: 13 279–89.

Malan, D.J. 1952. Les décharges dans l'air et la charge inférieure positive d'un nuage orageuse. *Ann. Geophys.* **8**: 385–401.

Malan, D.J. 1963. *Physics of Lightning*, 176 pp., London: English University Press.

Malan, D.J., and Schonland, B.F.J. 1951. The distribution of electricity in thunderclouds. *Proc. Roy Soc.* A **209**: 158–77.

Marsh, S.J., and Marshall, T.C. 1993. Charged precipitation measurements before the first lightning flash in a thunderstorm. *J. Geophys. Res.* **98**: 16 605–11.

Marshall, B.J.P., Latham, J., and Saunders, C.P.R. 1978. A laboratory study of charge transfer accompanying the collision of ice crystals with a simulated hailstone. *Q.J.R. Meteor. Soc.* **104**: 163–78.

Marshall, J.S., and Palmer, W.M. 1948. The distribution of raindrops with size. *J. Meteor.* **5**: 165–6.

Marshall, T.C., and Lin, B. 1992. Electricity in dying thunderstorms. *J. Geophys. Res.* **97**: 9913–18.

Marshall, T.C., and Marsh, S.J. 1993. Negatively charged precipitation in a New Mexico thunderstorm. *J. Geophys. Res.* **98**: 14 909–16.

Marshall, T.C., and Marsh, S.J. 1995. Reply. *J. Geophys. Res.* **100**: 16 869–71.

Marshall, T.C., and Rust, W.D. 1991. Electric field soundings through thunderstorms. *J. Geophys. Res.* **96**: 22 297–306.

Marshall, T.C., and Rust, W.D. 1993. Two types of vertical electrical structures in stratiform precipitation regions of mesoscale convective systems. *Bull. Am. Meteor. Soc.* **74**: 2159–70.

Marshall, T.C., and Stolzenburg, M. 1998. Estimates of cloud charge densities in thunderstorms. *J. Geophys. Res.* **103**: 19 769–75.

Marshall, T.C., and Stolzenburg, M. 2001. Voltages inside and just above thunderstorms. *J. Geophys. Res.* **106**(D5): 4757–68.

Marshall, T.C., and Winn, W.P. 1982. Measurements of charged precipitation in a New Mexico thunderstorm: lower positive charge centers, *J. Geophys. Res.* **87**: 7141–57.

Marshall, T.C., and Winn, W.P. 1985. Comments on "The 'rain gush', lightning, and the lower positive center in thunderstorm" by E.R. Jayaratne and C.P.R. Saunders. *J. Geophys. Res.* **90**: 10 753–4.

Marshall, T.C., Rust, W.D., Winn, W.P., and Gilbert, K.E. 1989. Electrical structure in two thunderstorm anvil clouds. *J. Geophys. Res.* **94**: 2171–81.

Marshall, T.C., McCarthy, M.P., and Rust, W.D. 1995a. Electric field magnitudes and lightning initiation in thunderstorms. *J. Geophys. Res.* **100**: 7079–103.

Marshall, T.C., Rison, W., Rust, W.D., Stolzenburg, M., Willett, J.C., and Winn, W.P. 1995b. Rocket and balloon observations of electric field in two thunderstorms. *J. Geophys. Res.* **100**: 20 815–28.

Marshall, T.C., Rust, W.D., and Stolzenburg, M. 1995c. Electrical structure and updraft speeds in thunderstorms over the southern Great Plains. *J. Geophys. Res.* **100**: 1001–15.

Marshall, T.C., Stolzenburg, M., and Rust, W.D. 1996. Electric field measurements above mesoscale convective systems. *J. Geophys. Res.* **101**: 6979–96.

Marshall, T.C., Rust, W.D., Stolzenburg, M., Roeder, W.P., and Krehbiel, P.R. 1999. A study of enhanced fair-weather electric fields occurring soon after sunrise. *J. Geophys. Res.* **104**: 24 455–69.

Marshall, T.C., Stolzenburg, M., Rust, W.D., Williams, E.R., and Boldi, R. 2001. Positive charge in the stratiform cloud of a mesoscale convective system. *J. Geophys. Res.* **106**: 1157–63.

Mason, B.J. 1953. A critical examination of theories of charge generation in thunderstorms. *Tellus* **5**: 446–98.

Mason, B.J. 1957. *The Physics of Clouds*. Oxford: Clarendon Press.

Mason, B.J. 1971. *The Physics of Clouds*, 2nd edition, 671 pp., London: Oxford University Press.

Mason, B.J. 1972. The physics of thunderstorms, *Proc. Roy. Soc. A* **327**: 433–66.

Mason, B.J. 1988. The generation of electric charges and fields in thunderstorms. *Proc. Roy. Soc. A* **415**: 303–15.

Mason, B.L., and Dash, J.G. 1999a. An experimental study of charge and mass transfer during ice contact interactions. In *Proc. 11th Int. Conf. on Atmospheric Electricity, Guntersville, Alabama*, pp. 264–7.

Mason, B.L., and Dash, J.G. 1999b. Surface melting of ice and thunderstorm electrification. In *Ice Physics in the Natural Environment*, eds. J.S. Wettlaufer, J.G. Dash, and N. Untersteiner, Nato ASI Series I, vol. 56, pp. 321–4, New York: Springer Verlag.

Mason, B.L., and Dash, J.G. 2000. Charge and mass transfer in ice–ice collisions: experimental observations of a mechanism in thunderstorm electrification. *J. Geophys. Res.* **105**: 20 185–92.

Masuelli, S., Scavuzzo, C.M., and Caranti, G.M. 1997. Convective electrification of clouds: a numerical study. *J. Geophys. Res.* **102**: 11 049–59.

Masuelli, S., Caranti, G.M., and Scavuzzo, C.M. 1998. Axisymmetric numerical study of convective cloud electrification. *J. Atmos. Solar-Terr. Phys.* **60**: 573–83.

Mathpal, K.C., and Varshneya, N.C. 1982. Riming electrification mechanism for charge generation within a thundercloud of finite dimensions. *Ann. Geophys.* **38**: 167–75.

Mazur, V. 1986. Rapidly occurring short duration discharges in thunderstorms as indicators of a lightning-triggering mechanism. *Geophys. Res. Lett.* **13**: 355–8.

Mazur, V., and Ruhnke, L.H. 1993. Common physical processes in natural and artificially triggered lightning. *J. Geophys. Res.* **98**: 12 913–30.

Mazur, V., and Ruhnke, L.H. 1998. Model of electric charges in thunderstorms and associated lightning. *J. Geophys. Res.* **103**: 23 299–308.

Mazur, V., Gerlach, J.C., and Rust, W.D. 1984. Lightning flash density versus altitude and storm structure from observations with UHF- and S-band radars. *Geophys. Res. Lett.* **11**: 61–4.

Melnik, O., and Parrot, M. 1998. Electrostatic discharge in Martian dust storms. *J. Geophys. Res. – Space* **103**(A12): 29 107–17.

Mendez, D.J. 1969. Optical polarization induced by electric fields of thunderstorms, *J. Geophys. Res.* **74**: 7032–7.

Michimoto, K. 1993. A study of the charge distribution in winter thunderclouds by means of network recording of surface electric fields and radar observation of clouds structure in the Hokuriku district. *J. Atmos. Electr.* **13**: 33–46.

Michnowski, S. 1974. Transient variation of electric field after sudden discharge of an electrical pole above a conductive plane in a nonhomogeneous medium. *Arch. Met. Geophys. Biokl., Ser. 1* **23**: 333–47.

Miller, K., Gadian, A., Saunders, C., Latham, J., and Christian, H. 2001. Modeling and observations of thundercloud electrification and lightning. *Atmos. Res.* **58**: 89–115.

Mitzeva, R., and Saunders, C.P.R. 1990. Thunderstorm charging: calculations of the effect of ice crystal size and graupel velocity. *J. Atmos. Terr. Phys.* **52**: 241–5.

Mo, Q., Ebneter, A.E., Fleischhacker, P., and Winn, W.P. 1998. Electric field measurements with an airplane: a solution to problems caused by emitted charge. *J. Geophys. Res.* **103**: 17 163–73.

Mo, Q., Feind, R.E., Kopp, F.J., and Detwiler, A.G. 1999. Improved electric field measurements with the T-28 armored research airplane. *J. Geophys. Res.* **104**: 24 485–97.

Moore, C.B. 1975. Rebound limits on charge separation by falling precipitation. *J. Geophys. Res.* **80**: 2658–62.

Moore, C.B. 1976. Reply to "Further comments on Moore's criticisms of precipitation theories of thunderstorm electrification", by B.J. Mason. *Q.J.R. Meteor. Soc.* **102**: 935–9.

Moore, C.B., and Vonnegut, B. 1977. The thundercloud. In *Lightning, vol. 1, Physics of Lightning*, ed. R.H. Golde, pp. 51–98. New York: Academic Press.

Moore, C.B., and Vonnegut, B. 1991. Comments on "A radar study of the plasma and geometry of lightning". *J. Atmos. Sci.* **48**: 369–70.

Moore, C.B., Vonnegut, B., Rolan, T.D., Cobb, J.W., Holden, D.N., Hignight, R.T., McWilliams, S.M., and Cadwell, G.W. 1986. Abnormal polarity of thunderclouds grown from negatively charged air. *Science* **233**: 1413–16.

Moore, C.B., Vonnegut, B., and Holden, D.N. 1989. Anomalous electric fields associated with clouds growing over a source of negative space charge. *J. Geophys. Res.* **94**: 13 127–34.

Murphy, M.J., Krider, E.P., and Maier, M.W. 1996. Lightning charge analyses in small convection and precipitation electrification (CAPE) experiment storms. *J. Geophys. Res.* **101**: 29 615–26.

Nguyen, M.D., and Michnowski, S. 1996a. On the initiation of lightning discharge in a cloud, 1. The high field regions in a thundercloud. *J. Geophys. Res.* **101**: 26 669–73.

Nguyen, M.D., and Michnowski, S. 1996b. On the initiation of lightning discharge in a cloud, 2. The lightning initiation on precipitation particles. *J. Geophys. Res.* **101**: 26 675–80.

Nisbet, J.S. 1983. A dynamic model of thundercloud electric fields. *J. Atmos. Sci.* **40**: 2855–73.

Nisbet, J.S. 1985a. Thundercloud current determination from measurements at the Earth's surface. *J. Geophys. Res.* **90**: 5840–56.

Nisbet, J.S. 1985b. Currents to the ionosphere from thunderstorm generators: a model study. *J. Geophys. Res.* **90**: 9831–44.

Nisbet, J.S., Barnard, T.A., Forbes, G.S., Krider, E.P., Lhermitte, R., and Lennon, C.L. 1990a. A case study of the Thunderstorm Research International Project storm of July 11, 1978–1. Analysis of the data base. *J. Geophys. Res.* **95**: 5417–33.

Nisbet, J.S., Kasha, J.R. and Forbes, G.S. 1990b. A case study of the Thunderstorm Research International Project storm of July 11, 1978–2. Interrelations among the observable parameters controlling electrification. *J. Geophys. Res.* **95**: 5435–45.

Norville, K., Baker, M., and Latham, J. 1991. A numerical study of thunderstorm electrification: model development and case study. *J. Geophys. Res.* **96**: 7463–81.

Ogawa, T. 1985. Fair-weather electricity. *J. Geophys. Res.* **90**: 5951–60.

Ogawa, T. 1993. Initiation of lightning in clouds. *J. Atmos. Electr.* **13**: 121–32.

Ogawa, T., and Brook, M. 1964. The mechanism of the intracloud lightning discharge. *J. Geophys. Res.* **69**: 514–9.

Ogawa, T., and Brook, M. 1969. Charge distribution in thunderstorm clouds. *Q.J.R. Meteor. Soc.* **95**: 513–25.

Orville, H.D. 2000. Millennium perspectives. *Bull. Am. Meteor. Soc.* **81**: 588.

Paluch, J.R., and Sartor, J.D. 1973a. Thunderstorm electrification by the inductive charging mechanism: I, Particle charges and electric fields. *J. Atmos. Sci.* **30**: 1166–73.

Paluch, J.R., and Sartor, J.D. 1973b. Thunderstorm electrification by the inductive charging mechanism: II, Possible effects of updraft on the charge separation process. *J. Atmos. Sci.* **30**: 1174–7.

Papadopoulos, K., Milikh, G., and Valdivia, J. 1996. Comment on "Can gamma radiation be produced in the electrical environment above thunderstorms". *Geophys. Res. Lett.* **23**: 2283–4.

Park, C.G., and Dejnakarintra, M. 1973. Penetration of thundercloud electric fields into the ionosphere and magnetosphere, 1, Middle and subauroral latitudes. *J. Geophys. Res.* **78**: 6623–33.

Pereyra, R.G., Avila, E.E., Castellano, N.E., and Saunders, C.P.R. 2000. A laboratory study of graupel charging. *J. Geophys. Res.* **105**: 20 803–12.

Petersen, W.A., Cifelli, R.C., Rutledge, S.A., Ferrier, B.S., and Smull, B.F. 1999. Shipborne dual-Doppler operations during TOGA COARE: integrated observations of storm kinematics and electrification. *Bull, Am. Meteor. Soc.* **80**: 81–97.

Phillips, B.B. 1967a. Charge distribution in a quasi-static thundercloud model. *Mon. Wea. Rev.*, **95**: 847–53.

Phillips, B.B. 1967b. Convected cloud charge in thunderstorms. *Mon. Wea. Rev.* **95**: 863–70.

Phillips, B.B., and Kinzer, G.D. 1958. Measurements of the size and electrification of droplets in cumuliform clouds. *J. Meteor.* **15**: 369–74.

Pierce, E.T. 1977. Lightning warning and avoidance. In *Lightning, vol. 2, Lightning Protection*, ed. R.H. Golde, pp. 497–519, New York: Academic Press.

Pringle, J.E., Orville, H.D., and Stechmann, T.D. 1973. Numerical study of thunderstorm electrification: model development and case study. *J. Geophys. Res.* **78**: 4508–14.

Proctor, D.E. 1983. Lightning and precipitation in a small multicellular thunderstorm, *J. Geophys. Res.* **88**: 5421–40.

Proctor, D.E. 1984. Correction to "Lightning and precipitation in a small multicellular thunderstorm". *J. Geophys. Res.* **89**: 11 826.

Proctor, D.E. 1991. Regions where lightning flashes began. *J. Geophys. Res.* **96**: 5099–112.

Proctor, D.E. 1997. Lightning flashes with high origins. *J. Geophys. Res.* **102**: 1693–706.

Pruppacher, H.R., and Klett, J.D. 1978. *Microphysics of Clouds and Precipitation*, 2nd edition, 954 pp., Kluwer.

Qie, X., Soula, S., and Chauzy, S. 1994. Influence of ion attachment on the vertical distribution of the electric field and charge density below a thunderstorm. *Ann. Geophys.* **12**: 1218–28.

Rai, J., Kumar, K., Hazarika, S., Parashar, J., and Kumar R. 1993. High speed photographic analysis of intracloud lightning radiation fields. *Ann. Geophys.* **11**: 518–24.

Ramachandran, R., Detwiler, A, Helsdon J., and Smith, P.L. 1996. Precipitation development and electrification in Florida thunderstorm cells during Convection And Precipitation/Electrification Project. *J. Geophys. Res.* **101**: 1599–619.

Randell, S.C., Rutledge, S.A., Farley, R.D., and Helsdon J.H. 1994. A modeling study on the early electrical development of tropical convection: continental and oceanic (monsoon) storms. *Mon. Wea. Rev.* **122**: 1852–77.

Rawlins, F. 1982. A numerical study of thunderstorm electrification using a three dimensional model incorporating the ice phase. *Q.J.R. Meteor. Soc.* **108**: 779–800.

Ray, P.S., MacGorman, D.R., Rust, W.D., Taylor, W.L., and Rasmussen, L.W. 1987. Lightning location relative to storm structure in a supercell storm and a multicell storm. *J. Geophys. Res.* **92**: 5713–24.

Reiter, R. 1958. Observations on the electricity of nimbostratus clouds. In *Recent Advances in Atmospheric Electricity*, ed. L.G. Smith, pp. 435–7, New York: Pergamon.

Reiter, R. 1965. Precipitation and cloud electricity. *Q.J.R. Meteor. Soc.* **91**: 60–72.

Reynolds, S.E. 1953. Thunderstorm-precipitation growth and electrical-charge generation. *Bull. Am. Meteor. Soc.* **34**: 117–23.

Reynolds, S.E., Brook, M., and Gourley, M.F. 1957. Thunderstorm charge separation. *J. Meteor.* **14**: 426–36.

Reynolds, S.E., and Neill, H.W. 1955. The distribution and discharge of thunderstorm charge centers. *J. Meteor.* **12**: 1–12.

Richards, C.N., and Dawson, G.A. 1971. The hydrodynamic instability of water drops falling at terminal velocity in vertical electric fields. *J. Geophys. Res.* **76**: 3445–55.

Rison, W., Thomas, R.J., Krehbiel, P.R., Hamlin, T., and Harlin, J. 1999. A GPS-based three-dimensional lightning mapping system: initial observations in central New Mexico. *Geophys. Res. Lett.* **26**: 3573–6.

Roulleau, M., and Desbois, M. 1972. Study of evaporation and instability of charged water droplets. *J. Atmos. Sci.* **29**: 565–9.

Roussel-Dupré, R.A., Gurevich, V., Tunnel, T., and Milikh, G.M. 1994. Kinetic theory of runaway air breakdown. *Phys. Rev. E* **49**(3): 2257–71.

Ruhnke, L.H. 1972. Atmospheric electron cloud modeling. *Meteor. Res.* **25**: 38–41.

Rust, W.D. 1973. Electrical conditions near the bases of thunderclouds. Ph.D. dissertation, New Mexico Institute of Min. and Tech., Socorro, New Mexico.

Rust, W.D., and Marshall, T.C. 1996. On abandoning the thunderstorm tripole-charge paradigm. *J. Geophys. Res.* **101**: 23 499–504.

Rust, W.D., and Moore, C.B. 1974. Electrical conditions near the bases of thunderclouds over New Mexico. *Q.J.R. Meteor. Soc.* **100**: 450–68.

Rustan, P.L., Uman, M.A., Childers, D.G., Beasley, W.H., and Lennon, C.L. 1980. Lightning source locations from VHF radiation data for a flash at Kennedy Space Center. *J. Geophys. Res.* **85**: 4893–903.

Rutledge, S.A., and Hobbs, P.V. 1984. The mesoscale and microscale structure and organization of clouds and precipitation in mid-latitude cyclones. XII: A diagnostic modeling study of precipitation development in narrow cold-frontal rainband. *J. Atmos. Sci.* **41**: 2949–72.

Rutledge, S.A., and MacGorman, D.R. 1988. Cloud-to-ground lightning activity in the 10–11 June 1985 mesoscale convective system observed during Oklahoma-Kansas prestorm project. *Mon. Wea. Rev.* **116**: 1393–408.

Rutledge, S.A., Williams, E.R., and Keenan, T.D. 1992. The Down Under Doppler and Electricity Experiment (DUNDEE): overview and preliminary results. *Bull. Am. Meteor. Soc.* **73**: 3–14.

Rycroft, M.J. 1994. Some effects in the middle atmosphere due to lightning. *J. Atmos. Terr. Phys.* **56**: 343–8.

Rydock, J., and Williams, E.R. 1991. Charge separation associated with frost growth. *Q.J.R. Meteor. Soc.* **117**: 409–20.

Sadiku, M.N.O. 1994. *Elements of Electromagnetics*, 821 pp., Orlando, Florida: Saunders College.

Sapkota, B.K., and Varshneya, N.C. 1988. Electrification of thundercloud by an entrainment mechanism. *Meteor. Atmos. Phys.* **39**: 213–22.

Sartor, J.D. 1954. A laboratory investigation of collision efficiencies, coalescence and electrical charging of simulated cloud droplets. *J. Meteor.* **11**: 91–103.

Sartor, J.D. 1967. The role of particle interactions in the distribution of electricity in thunderstorms. *J. Atmos Sci.* **24**: 601–15.

Sartor, J.D. 1981. Induction charging of clouds. *J. Atmos. Sci.* **38**: 218–20.

Saunders, C.P.R. 1993. A review of thunderstorm electrification processes. *J. Appl. Meteor.* **32**: 642–55.

Saunders, C.P.R. 1994. Thunderstorm electrification laboratory experiments and charging mechanisms. *J. Geophys. Res.* **99**: 10 773–9.

Saunders, C.P.R. 1995. Thunderstorm electrification. In *Handbook of Atmospheric Electrodynamics*, vol. I, ed. H. Volland, pp. 61–92. Boca Raton, Florida: CRC Press.

Saunders, C.P.R., and Brooks, I.M. 1992. The effects of high liquid water content on thunderstorm charging. *J. Geophys. Res.* **97**: 14 671–6.

Saunders, C.P.R. and Hosseini, A.S. 2001. A laboratory study of the effect of velocity on Hallett-Mossop ice crystal multiplication. *Atmos. Res.* **59–60**: 3–14.

Saunders, C.P.R., and Peck, S.L. 1998. Laboratory studies of the influence of the rime accretion rate on charge transfer during crystal/graupel collisions. *J. Geophys. Res.* **103**: 13 949–56.

Saunders, C.P.R., Keith, W.D., and Mitzeva, R.P. 1991. The effect of liquid water on thunderstorm charging. *J. Geophys. Res.* **96**: 11 007–17.

Saunders, C.P.R., Hickson, H., Malone, M.D., and von Richtofen, J. 1993. Charge separation during the fragmentation of rime and frost. *Atmos. Res.* **29**: 261–70.

Saunders, C.P.R., Avila, E.E., Peck, S.L., Castellano, N.E., and Aguirre Varela, G.G. 1999. A laboratory study of the effects of rime ice accretion and heating on charge transfer during ice crystal/graupel collisions. *Atmos. Res.* **51**: 99–117.

Savchenko, B.I. 1970. Possible mechanism for discharges in thunderstorms. *Soviet Physics – Tech. Phys.* **14**: 1079–82.

Scavuzzo, C.M., Masuelli, S., Caranti, G.M., and Williams, E.R. 1998. A numerical study of thundercloud electrification by graupel–crystal collisions. *J. Geophys. Res.* **103**: 13 963–73.

Schmidt, D.S., Schmidt, R.A., and Dent, J.D. 1998. Electrostatic force on saltating sand. *J. Geophys. Res.* **103**: 8997–9001.

Schonland, B.F.J. 1928. The polarity of thunderclouds. *Proc. Roy. Soc. London A* **118**: 233–51.

Schonland, B.F.J., and Collens, H. 1933. Development of the lightning discharge. *Nature* **132**: 407–8.

Schonland, B.F.J., and Craib, J. 1927. The electric fields of South African thunderstorm. *Proc. Roy. Soc. A* **114**: 229–43.

Schultz, D.M., Steenburgh, W.J., Trapp, R.J., Horel, J., Kingsmill, D.E., Dunn, L.B., Rust, W.D., Cheng, L., Bansemer, A., Cox, J., Daugherty, J., Jorgensen, D.P., Meitín, J., Showell, L., Small, B.F., Tarp, K., and Trainor, M. 2002. Understanding Utah winter storms: The Intermountain Precipitation Experiment. *Bull. Am. Meteor. Soc.* February, 189–210.

Schuur, T.J., and Rutledge, S.A. 2000a. Electrification of stratiform regions in mesoscale convective systems. I: Observational comparison of symmetric and asymmetric MCSs. *J. Atmos. Sci.* **57**: 1961–82.

Schuur, T.J., and Rutledge, S.A. 2000b. Electrification of stratiform regions in mesoscale convective systems. II: Two-dimensional numerical model simulations of a symmetric MCS. *J. Atmos. Sci.* **57**: 1983–2006.

Schuur, T.J., Smull, B.F., Rust, W.D., and Marshall, T.C. 1991. Electrical and kinematic structure of the stratiform precipitation region trailing an Oklahoma squall line. *J. Atmos. Sci.* **48**: 825–42.

Scott, W.D., and Levin, Z. 1975. A stochastic electrical model of an infinite cloud: charge generation and precipitation development *J. Atmos. Sci.* **32**: 1814–28.

Shao, X.M., and Krehbiel, P.R. 1993. Radio interferometric observations of intracloud lightning (Abstract). *Eos, Trans. AGU* **74**, Fall meeting suppl., p. 165.

Shao, X.M., and Krehbiel, P.R. 1996. The spatial and temporal development of intracloud lightning. *J. Geophys. Res.* **101**: 26 641–68.

Shchukin, G.G., Stepanenko, V.D., Yegorov, A.D., Galperin, S.M., and Karavayev, D.M. 1999. Radiophysical studies of atmosphere and underlying surface. In *Contemporary Investigation at Main Geophysical Observatory*, vol. 1, eds. M.E. Berlyand and V.P. Meleshko, pp. 172–90, St. Petersburg, Russia: Gidrometeoizdat.

Shepherd, T.R., Rust, W.D., and Marshall, T.C. 1996. Electric fields and charges near 0 °C in stratiform clouds. *Mon Wea. Rev.* **124**: 919–38.

Sheve, E.L. 1970. Theoretical derivation of atmospheric ion concentrations, conductivity, space charge density, electric field, and generation rate from 0 to 60 km. *J. Atmos. Sci.* **27**: 1186–94.

Simpson, G.C. 1909. On the electricity of rain and snow. *Proc. Roy. Soc. London A* **83**: 392–404.

Simpson, G.C. 1949. Atmospheric electricity during disturbed weather. *Geophys. Mem. London* **84**: 1–51.

Simpson, G.C., and Robinson, G.D. 1941. The distribution of electricity in the thunderclouds. Pt. II. *Proc. Roy. Soc. London A* **177**: 281–329.

Simpson, G., and Scrase, F.J. 1937. The distribution of electricity in thunderclouds. *Proc. Roy. Soc. London A* **161**: 309–52.

Singh, P., Verma, T.S., and Varshneya, N.C. 1986. Some theoretical aspects of electric field and precipitation growth in a finite thundercloud. *Proc. Indian Acad. Sci.* **95**: 293–8.

Solomon, R., and Baker, M. 1994. Electrification of New Mexico thunderstorms. *Mon. Wea. Rev.* **122**: 1878-86.

Solomon, R., and Baker, M. 1996. A one-dimensional lightning parameterization. *J. Geophys. Res.* **101**: 14 983–90.

Solomon, R., and Baker, M. 1998. Lightning flash rate and type in convective storms. *J. Geophys. Res.* **103**: 14 041–57.

Soula, S. 1994. Transfer of electrical space charge from corona between ground and thundercloud: measurements and modeling. *J. Geophys. Res.* **99**: 10 759–65.

Soula, S., and Chauzy, S. 1986. The effects of ground coronae during lightning flashes. *Ann. Geophys.* **4**(B6): 613–24.

Soula, S., and Chauzy, S. 1991. Multilevel measurement of the electric field underneath a thundercloud 2. Dynamical evolution of a ground space charge layer. *J. Geophys. Res.* **96**: 22 327–36.

Soula, S., and Chauzy, S. 1997. Charge transfer by precipitation between thundercloud and ground. *J. Geophys. Res.* **102**: 11 061–9.

Soula, S., Despiau, S., and Chauzy, S. 1987. Comment on "Electric field recovery and charge regeneration after lightning discharge" by P. Pradeep Kumar and J. Rai. *Ann. Geophys.* **5B**: 541–4.

Soula, S., Sauvageot, H., Saissac, M.P. and Chauzy, S. 1995. Observation of thunderstorm by multilevel electric field measurement system and radar. *J. Geophys. Res.* **100**: 5025–35.

Standler, R.B. 1980. Estimation of corona current beneath thunderclouds. *J. Geophys. Res.* **85**: 4541–4.

Standler, R.B., and Winn, W.P. 1979. Effects of corona on electric fields beneath thunderclouds. *Q.J.R. Meteor. Soc.* **105**: 285–302.

Stansbery, E.K., Few, A.A., and Geis, P.B. 1993. A global model of thunderstorm electricity. *J. Geophys. Res.* **98**: 16 591–603.

Stergis, C.G., Rein, G.C., and Kangas, T. 1957. Electric field measurements above thunderstorms. *J. Atmos. Terr. Phys.* **11**: 83–91.

Stith, J.L. 1992. Observations of cloud-top entrainment in cumuli. *J. Atmos. Sci.* **49**: 1334–47.

Stolzenburg, M. 1996. An observational study of electrical structure in convective regions of mesoscale convective systems. Ph.D. dissertation, University of Oklahoma, Norman, 137 pp.

Stolzenburg, M., and Marshall, T.C. 1994. Testing models of thunderstorm charge distributions with Coulomb's law. *J. Geophys. Res.* **99**: 25 921–32.

Stolzenburg, M., and Marshall, T.C. 1998. Charged precipitation and electric field in two thunderstorms. *J. Geophys. Res.* **103**: 19 777–90.

Stolzenburg, M., Marshall, T.C., Rust, W.D., and Smull, B.F. 1994. Horizontal distribution of electrical and meteorological conditions across the stratiform region of a mesoscale convective system. *Mon. Wea. Rev.* **122**: 1777–97.

Stolzenburg, M., Rust, W.D., and Marshall, T.C. 1998a. Electrical structure in thunderstorm convective regions. 2. Isolated storms. *J. Geophys. Res.* **103**: 14 079–96.

Stolzenburg, M., Rust, W.D., and Marshall, T.C. 1998b. Electrical structure in thunderstorm convective regions. 3. Synthesis. *J. Geophys. Res.* **103**: 14 097–108.

Stolzenburg, M., Rust, W.D., Smull, B.F., and Marshall, T.C. 1998c. Electrical structure in thunderstorm convective regions. 1. Mesoscale convective systems. *J. Geophys. Res.* **103**: 14 059–78.

Stolzenburg, M., Marshall, T.C., and Rust, W.D. 2001. Serial soundings of electric field through a mesoscale convective system. *J. Geophys. Res.* **106**(D12): 12 371–80.

Stow, C.D. 1969. Atmospheric electricity. *Rep. Prog. Phys.* **32**: 1–67.

Straka, J.M., Zrnic, D.S., and Ryzhkov, A.V. 2000. Bulk hydrometeor classification and quantification using polarimetric radar data: synthesis of relations. *J. Appl. Meteor.* **39**: 1341–72.

Suszcynsky, D.M., and Roussel-Dupré, R. 1996. Ground-based search for X rays generated by thunderstorms and lightning. *J. Geophys. Res.* **101**: 23 505–16.

Suzuki, T. 1992. Long term observation of winter lightning on Japan sea coast. *Res. Lett. Atmos. Electr.* **12**: 53–6.

Takahashi, T. 1974. Numerical simulation of warm cloud electricity. *J. Atmos. Sci.* **31**: 2160–81.

Takahashi, T. 1966. Thermoelectric effect in ice. *J. Atmos. Sci.* **23**: 74–7.

Takahashi, T. 1969. Electric potential of liquid water on an ice surface. *J. Atmos. Sci.* **26**: 1253–8.

Takahashi, T. 1978a. Riming electrification as a charge generation mechanism in thunderstorms. *J. Atmos. Sci.* **35**: 1536–48.

Takahashi, T. 1978b. Electrical properties of oceanic tropical clouds at Ponape, Micronesia. *Mon. Wea. Rev.* **106**: 1598–612.

Takahashi, T. 1979. Warm cloud electricity in a shallow axisymmetric cloud model. *J. Atmos Sci.* **36**: 2236–58.

Takahashi, T. 1983. A numerical simulation of winter cumulus electrification. Part 1: Shallow cloud. *J. Atmos. Sci.* **40**: 1257–80.

Takahashi, T. 1984. Thunderstorm electrification – a numerical study *J. Atmos. Sci.* **41**: 2541–58.

Takahashi, T. 1986. Electric charge separation and accumulation processes in thunderstorms. *J. Korean Meteor. Soc.* **22**: 26–42.

Takahashi, T. 1987. Determination of lightning origins in a thunderstorm model. *J. Meteor. Soc. Japan* **65**: 777–94.

Takahashi, T. 1990. Near absence of lightning in torrential rainfall producing Micronesian thunderstorms. *Geophys. Res. Lett.* **17**: 2381–4.

Takahashi, T., Tajiri, T., and Sonoi, Y. 1999. Charges on graupel and snow crystals and the electrical structure of winter thunderstorms. *J. Atmos. Sci.* **56**: 1561–78.

Takeuchi, N., Narita, K.-I., and Goto, Y. 1994. Wavelet analysis of meteorological variables under winter thunderclouds over the Japan Sea. *J. Geophys. Res.* **99**: 10 751–7.

Tamura, Y. 1955. An analysis of electric field after lightning discharges. *J. Geomagn. Geoelectr.* **6**: 34–46.

Taniguchi, T., Magono, C., and Endoh, T. 1982. Charge distribution in active winter clouds. *Res. Lett. Atmos. Electr.* **2**: 35–8.

Taylor, W.L., Brandes, E.A., Rust, W.D., and MacGorman, D.R. 1984. Lightning activity and severe storm structure. *Geophys. Res. Lett.* **11**: 545–8.

Thomas, R.J., Krehbiel, P.R., Rison, W., Hamlin, T., Boccippio, D.J., Goodman, S.J., and Christian, H.J. 2000. Comparison of ground-based 3-dimensional lightning mapping observations with satellite-based LIS observations in Oklahoma. *Geophys. Res. Lett.* **27**: 1703–6.

Thomas, R.J., Krehbiel, P.R., Rison, W., Hamlin, T., Harlin, J., and Shown, D. 2001. Observations of VHF source powers radiated by lightning. *Geophys. Res. Lett.* **28**: 143–6.

Tinsley, B.A. Rohrbaugh, Hei, M., and Beard, K.V. 2000. Effects of image charges on the scavenging of aerosol particles by cloud droplets and on droplet charging and possible ice nucleation processes. *J. Atmos. Sci.* **57**: 2118–34.

Toland, R.B., and Vonnegut, B. 1977. Measurement of maximum electric field intensities over water during thunderstorms. *J. Geophys. Res.* **82**: 438–40.

Twomey, S. 1956. The electrification of individual cloud droplets. *Tellus* **7**: 445–51.

Tzur, I., and Levin, Z. 1981. Ions and precipitation charging in warm and cold clouds as simulated in a one-dimensional, time-dependent model. *J. Atmos. Sci.* **38**: 2444–6.

Tzur, I., and Roble, R.G. 1985. The interaction of a dipolar thunderstorm with its global electrical environment. *J. Geophys. Res.* **90**: 5989–9.

Uman, M.A. 1969a. *Lightning*, 264 pp., New York: McGraw-Hill.

Uman, M.A. 1984. *Lightning*, 298 pp., New York: Dover.

Uman, M.A. 1987. *The Lightning Discharge*, 377 pp., Orlando, Florida: Academic Press.

Uman, M.A. 2001. *The Lightning Discharge*, 377 pp., Mineola, New York: Dover.

Vonnegut, B. 1953. Possible mechanism for the formation of thunderstorm electricity. *Bull. Am. Meteor. Soc.* **34**: 378–81.

Vonnegut, B. 1965. Thunderstorm theory. In *Problems of Atmospheric and Space Electricity, Proc. 3rd Int. Conf. Atmospheric and Space Electricity, Montreaux, Switzerland*, ed. S.C. Coroniti, pp. 285–95, New York: Elsevier.

Vonnegut, B. 1969. Discussion of paper by W.H. Evans "Electric fields and conductivity in thunderclouds", *J. Geophys. Res.* **74**: 7053.

Vonnegut, B. 1982. The physics of thunderclouds. In *Handbook of Atmospherics*, vol. I, ed. H. Volland, pp. 1–22, Boca Raton, Florida: CRC Press.

Vonnegut, B. 1983. Deductions concerning accumulations of electrified particles in thunderclouds based on electric field

changes associated with lightning. *J. Geophys. Res.* **88**: 3911–2.
Vonnegut, B. 1984. Reduction of thunderstorm electric field intensity produced by corona from a nearby object. *J. Geophys. Res.* **89**(D1): 1468–70.
Vonnegut, B. 1994. The atmospheric electricity paradigm. *Bull. Am. Meteor. Soc.* **75**: 53–61.
Vonnegut, B., and Moore, C.B. 1965. Nucleation of ice formation in supercooled clouds as the result of lightning. *J. Appl. Meteor.* **4**: 640–2.
Vonnegut, B., and Moore, C.B. 1986. Comments on "The 'rain gush' lightning, and the lower positive charge center in thunderstorms" by E.R. Jayaratne and C.P.R. Saunders. *J. Geophys. Res.* **91**: 10 949.
Vonnegut, B., and Moore, C.B. 1987. Comments on "Thunderstorm electrification: The effect of cloud droplets" by E.R. Jayarante and C.P.R. Saunders. *J. Geophys. Res.* **92**: 3139.
Vonnegut, B., and Rechnitzer, B.W. 1974. Instrument for measuring maximum thunderstorm electric field intensity. *Rev. Sci. Instr.* **45**: 1172–4.
Vonnegut, B., Moore, C.B., Semonin, R.G., Bullock, J.W., Staggs, D.W., and Bradley, W.E. 1962. Effect of atmospheric space charge on initial electrification of cumulus clouds. *J. Geophys. Res.* **67**: 3909–21.
Vonnegut, B., Moore, C.B., Espinola, R.P., and Blan, H.H. Jr 1966. Electric potential gradients above thunderstorms. *J. Atmos. Sci.* **23**: 764–70.
Vonnegut, B., Vaughan, O.H. Jr, and Brook, M. 1989. Nocturnal photographs taken from a U-2 airplane looking down on tops of clouds illuminated by lightning. *Bull. Am. Meteor. Soc.* **70**: 1263–71.
Vonnegut, B., Latham, D.J., Moore, C.B., and Hunyady, S.J. 1995. An explanation for anomalous lightning from forest fire clouds. *J. Geophys. Res.* **100**: 5037–50.
Wagner, P.B., and Telford, J.W. 1981. Charge dynamics and an electric charge separation mechanism in convective clouds. *J. Rech. Atmos.* **15**: 97–120.
Wait, G.R. 1953. Aircraft measurements of electric charge carried to ground through thunderstorm. In *Thunderstorm Electricity* (Chapter 10), ed. H.R. Byers, pp. 231–7, Chicago: University Chicago Press.
Waldteufel, P., Metzger, P., Boulay, J.-L., Laroche, P., and Hubert, P. 1980. Triggered lightning strokes originating in clear air. *J. Geophys. Res.* **85**: 2861–8.
Weber, M.E., Christian, H.J., Few, A.A., and Stewart, M.F. 1982. A thundercloud electric field sounding: charge distribution and lightning. *J. Geophys. Res.* **87**: 7158–69.
Weber, M.E., Stewart, M.F., and Few, A.A. 1983. Corona point measurements in a thundercloud at Langmuir Laboratory. *J. Geophys. Res.* **88**: 3907–10.
Weinheimer, A.J. 1987. The electrostatic energy of a thunderstorm and its rate of change. *J. Geophys. Res.* **92**: 9715–22.
Weinheimer, A.J., and Few, A.A. Jr 1981. Comments on "Contributions of cloud and precipitation particles to the electrical conductivity and the relaxation time of the air in thunderstorms" by A.K. Kamra. *J. Geophys. Res.* **86**: 4302–4.
Weinheimer, A.J., and Few, A.A. 1987. The electric field alignment of ice particles in thunderstorms. *J. Geophys. Res.* **92**: 14 833–44.
Weinheimer, A.J., Dye, J.E., Breed, D.W., Spowart, M.P., Parrish, J.L., Hoglin, T.L., and Marshall, T.C. 1991. Simultaneous measurements of the charge, size, and shape of hydrometeors in an electrified cloud. *J. Geophys. Res.* **96**: 20, 809–29.
Whipple, F.J.W., and Chalmers, J.A. 1944. On Wilson's theory of the collection of charge by falling drops. *Q.J.R. Meteor. Soc.* **70**: 103–19.
Whipple, F.J.W., and Scrase, F.J. 1936. Point-discharge in the electric field of the Earth. *Geophys. Mem.* **7**(68): 1–20.
Willett, J.C., Davis, D.A., and Laroche, P. 1999. An experimental study of positive leaders initiating rocket-triggered lightning. *Atmos. Res.* **51**: 189–219.
Williams, E.R. 1985. Large scale charge separation in thunderclouds. *J. Geophys. Res.* **90**: 6013–25.
Williams, E.R. 1989. The tripole structure of thunderstorms. *J. Geophys. Res.* **94**: 13 151–67.
Williams, E.R. 1995a. Meteorological aspects of thunderstorms. In *Handbook of Atmospheric Electrodynamics*, vol. I., ed. H. Volland, pp. 27–60, Boca Raton, Florida: CRC Press.
Williams, E.R. 1995b. Comments on "Thunderstorm electrification laboratory experiments and charging mechanisms". *J. Geophys. Res.* **100**: 1503–5.
Williams, E.R. 1998. The positive charge reservoir for sprite-producing lightning. *J. Atmos. Solar-Terr. Phys.* **60**: 689–92.
Williams, E.R., and Lhermitte, R.M. 1983. Radar tests of the precipitation hypothesis for thunderstorm electrification. *J. Geophys. Res.* **88**: 10 984–92.
Williams, E.R., Cooke, C.M., and Wright, K.A. 1985. Electrical discharge propagation in and around space charge clouds. *J. Geophys. Res.* **90**: 6059–70.
Williams, E.R., Cooke, C.M., and Wright, K.A. 1988. The role of electric space charge in nuclear lightning. *J. Geophys. Res.* **93**: 1679–88.
Williams, E.R., Geotis, S.G., and Bhattacharya, A.B. 1989a. A radar study of the plasma and geometry of lightning. *J. Atmos. Sci.* **46**: 1173–85.
Williams, E.R., Weber, M.E., and Orville, R.E. 1989b. The relationship between lightning type and convective state of thunderclouds. *J. Geophys. Res.* **94**: 13 213–20.
Williams, E.R., Zhang, R., and Rydock, R. 1991. Mixed-phase microphysics and cloud electrification. *J. Atmos. Sci.* **48**: 2195–203.
Williams, E.R., Rutledge, S.A., Geotis, S.G., Renno, N., Rasmussen, E., and Rickenbash, T. 1992. A radar and electrical study of tropical hot towers. *J. Atmos. Sci.* **49**: 1386–95.
Williams, E.R., Zhang, R., and Boccippio, D. 1994. Microphysical growth state of ice particles and large-scale electrical structure of clouds. *J. Geophys. Res.* **99**: 10 787–92.
Williams, J.C. 1958. Some properties of the lower positive charge in thunderclouds. In *Recent Advances in Atmospheric Electricity*, ed., L.G. Smith, pp. 425–9, New York: Pergamon Press.

Willis, P.T., Hallet, J., Black, R.A., and Hendricks, W. 1994. An aircraft study of rapid precipitation development and electrification in a growing convective cloud. *Atmos. Res.* **33**: 1–24.

Wilson, C.T.R. 1916. On some determinations of the sign and magnitude of electric discharges in lightning flashes. *Proc. Roy. Soc. A* **92**: 555–74.

Wilson, C.T.R. 1920. Investigations on lightning discharges and on the electric field of thunderstorms. *Phil. Trans. Roy. Soc. A* **221**: 73–115.

Wilson, C.T.R. 1929. Some thundercloud problems. *J. Franklin Inst.* **208**: 1–12.

Wilson, C.T.R. 1956. A theory of thundercloud electricity. *Proc. Roy. Soc. A* **236**: 297–317.

Winn, W.P. 1992. Electrification of thunderclouds by the transport of bare charge: implications of Grenet's and Vonnegut's hypothesis. *Eos, Trans. AGU*, Fall meeting supplement, **73**: 110.

Winn, W.P., and Byerley, L.G. 1975. Electric field growth in thunderclouds. *Q.J.R. Meteor. Soc.* **101**: 979–94.

Winn, W.P., and Moore, C.B. 1971. Electric field measurements in thunderclouds using instrumented rockets. *J. Geophys. Res.* **76**: 5003–18.

Winn, W.P., Schwede, G.W., and Moore, C.B. 1974. Measurements of electric fields in thunderclouds. *J. Geophys. Res.* **79**: 1761–7.

Winn, W.P., Moore, C.B., Holmes, C.R., and Byerly III, L.G. 1978. Thunderstorms on July 16, 1975, over Longmuir Laboratory: a case study. *J. Geophys. Res.* **83**: 3079–91.

Winn, W.P., Moore, C.B., and Holmes, C.R. 1981. Electric field structure in an active part of a small, isolated thundercloud. *J. Geophys. Res.* **86**: 1187–93.

Winn, W.P., Han, F., Jones, J.J., Raymond, D.J., Marshall, T.C., and Marsh, S.J. 1988. Thunderstorm with anomalous charge. In *Proc. 8th Int. Conf. on Atmospheric Electricity, Uppsala, Sweden*, pp. 590–5.

Wisner, C., Orville, H.D., and Meyers, C. 1972. A numerical model of a hail-bearing cloud. *J. Atmos. Sci.* **29**: 1160–81.

[WMO] World Meteorological Organization 1969. *International Cloud Atlas (Abridged Atlas)*. WMO, Geneva, 62 pp., 72 plates.

Workman, E.J. 1967. The production of thunderstorm electricity. *J. Franklin Inst.* **283**: 540–57.

Workman, E.J., and Reynolds, S.E. 1948. A suggested mechanism for the generation of thunderstorm electricity. *Phys. Rev.* **74**: 709–9.

Workman, E.J., and Reynolds, S.E. 1950. Electrical phenomena occurring during the freezing of dilute aqueous solutions and their possible relationship to thunderstorm electricity. *Phys. Rev.* **78**: 254–9.

Workman, E.J., Holzer, R.E., and Pelsor, G.T. 1942. The electrical structure of thunderstorms. NASA Technical Note 864, 1–47.

Workman, E.J., Brook, M., and Kitagawa, N. 1960. Lightning and charge storage. *J. Geophys. Res.* **65**: 1513–17.

Wormel, T.W. 1930. Vertical electric currents below thunderstorms and showers. *Proc. Roy. Soc. A* **127**: 567–90.

Wormel, T.W. 1939. The effect of thunderstorms and lightning discharges on the Earth's electric field. *Phil. Trans. Roy. Soc. A* **328**: 249–303.

Ziegler, C.L. 1985. Retrieval of thermal and microphysical variables in observed convective storms: Part I. Model development and preliminary testing. *J. Atmos. Sci.* **42**: 1487–509.

Ziegler, C.L. 1988. Retrieval of thermal and microphysical variables in observed convective storms: Part II. Sensitivity of cloud processes to variation of the microphysical parameterization. *J. Atmos. Sci.* **45**: 1072–90.

Ziegler, C.L., and MacGorman, D.R. 1994. Observed lightning morphology relative to modeled space charge and electric field distributions in a tornadic storm. *J. Atmos. Sci.* **51**: 833–51.

Ziegler, C.L., Ray, P.S., and MacGorman, D.R. 1986. Relations of kinematics, microphysics and electrification in an isolated mountain thunderstorm. *J. Atmos. Sci.* **43**: 2098–114.

Ziegler, C.L., MacGormam, D.R., Dye, J.E., and Ray, P.S. 1991. A model evaluation of noninductive graupel–ice charging in the early electrification of a mountain thunderstorm. *J. Geophys. Res.* **96**: 12 833–55.

Zipser, E.J., and Lutz, K.R. 1994. The vertical profile of radar reflectivity of convective cells: a strong indicator of storm intensity and lightning probability? *Mon. Wea. Rev.* **122**: 1751–9.

Ziv, A., and Levin, Z. 1974. Thundercloud electrification: cloud growth and electrical development. *J. Atmos Sci.* **31**: 1652–61.

4 Downward negative lightning discharges to ground

> The frequency distribution of lightning current amplitudes in earth flashes is reasonably well established. Over most of its range it can be said to follow a log-normal distribution but the frequency of occurrence of exceptionally high and low crest values remains to be established by greatly increased numbers of observation. Further study is also required of current amplitudes in tropical storms and in flashes to very tall structures. The most vital gap in present knowledge concerns the spatial and temporal variations of the current along the entire length of the discharge channel from ground to cloud.
>
> R. H. Golde (1977)

4.1. Introduction

Downward negative lightning discharges, that is, discharges that are initiated in the cloud, initially develop in an overall downward direction, and transport negative charge to ground, probably account for about 90 percent of all cloud-to-ground discharges (Section 1.2). The overall cloud-to-ground lightning discharge, termed a flash, is composed of a number of processes, some of which involve channels that emerge from the cloud while others involve channels that are confined to the cloud volume. Only processes occurring in channels outside the cloud render themselves to optical observations that can be used to determine channel geometry, extension speed and other pertinent features of those channels. All lightning processes are associated with the motion of charge and, therefore, can be studied via measurement of the electric and magnetic fields associated with that charge motion. VHF–UHF radiation associated with breakdown processes, particularly those involved in the formation of negatively charged leader channels, can be used to track channel extension, as discussed in subsections 17.4.2 and 17.6. Major insights into downward negative lightning processes that have been gained from triggered lightning experiments are included in this chapter; a detailed discussion of triggered lightning will be found in Chapter 7. Emphasis here is placed on the current understanding of the lightning processes rather than on presenting a complete evolution of views on each of these processes. Readers interested in a more complete history of lightning research are referred to Uman (1969a, 1984, 1987, 2001), Krider (1996), and Section 1.1 of this book.

4.2. General picture

In this section we present a general picture of downward negative lightning. In order to make the presentation more readable, information will often be given without specific citation. We first introduce, referring to Figs. 4.1a, b, the basic elements of the lightning discharge, termed component strokes or just strokes; each flash typically contains three to five strokes, the observed range being one to 26. Then we will introduce, referring to Figs. 4.2a, b, the two major lightning processes comprising a stroke, the leader and the return stroke. These occur as a sequence, the leader preceding the return stroke. Finally, we review, referring to Fig. 4.3, the entire spectrum of lightning processes in their chronological order, including those occurring both prior to and after the leader–return-stroke sequences. Each of the processes occurring in a negative cloud-to-ground flash will be described individually in Sections 4.3 through 4.11.

Two photographs of a negative cloud-to-ground discharge are shown in Figs. 4.1a, b. The image in Fig. 4.1a was obtained using a stationary camera while the image in Fig. 4.1b was captured with a separate camera that was moved horizontally during the time of the flash, while its shutter was open (Hendry 1993). As a result, the latter image is time resolved, showing seven separate luminous channels between the cloud and ground. The dark intervals between these channels are typically of the order of tens of milliseconds and explain why lightning often appears to the human eye to "flicker". Each luminous channel in Fig. 4.1b corresponds to an individual stroke, the first stroke being on the far right (time advances from right to left). Each stroke is composed of a downward-moving leader and an upward-moving return stroke; these are unresolved in Fig. 4.1b and

4.2. General picture

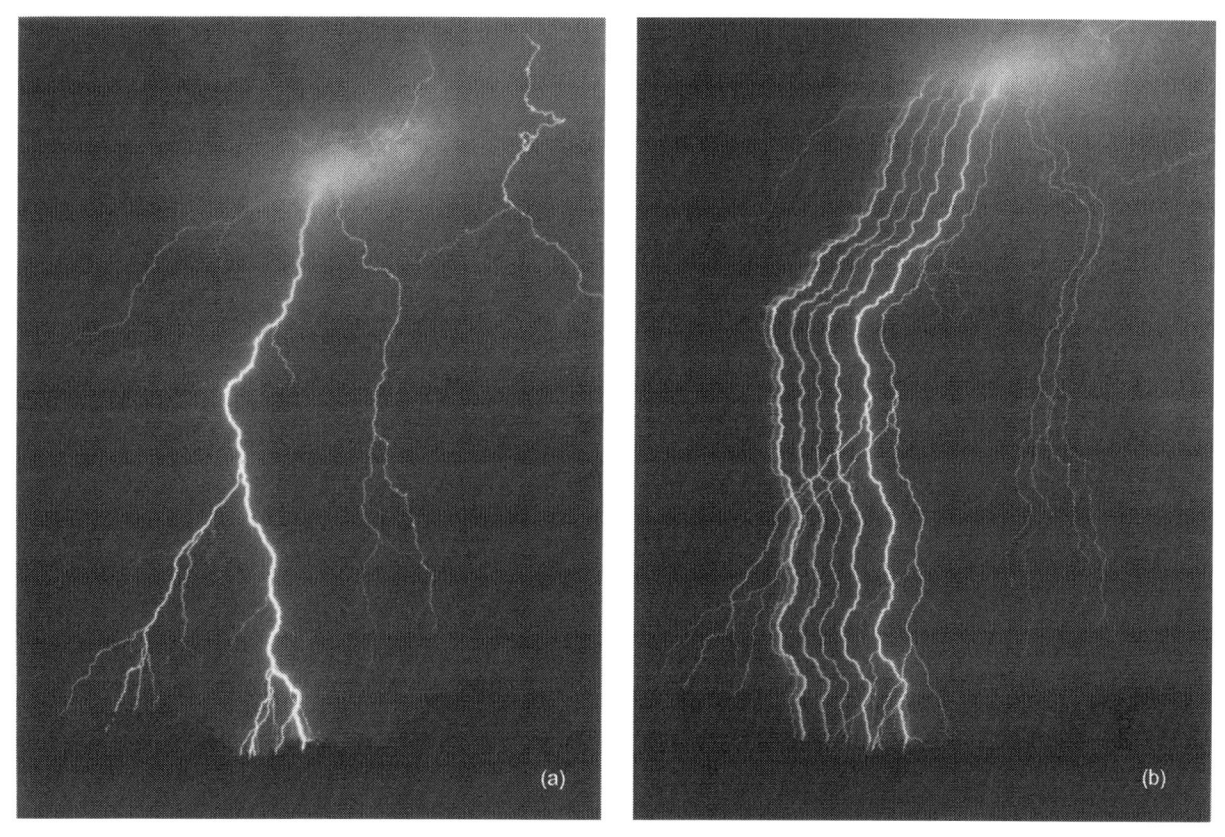

Fig. 4.1. A lightning flash that appears to have at least seven (perhaps as many as 10) separate ground strike points: (a) still-camera photograph, (b) moving-camera photograph. Some of the strike points are associated with separate branches of the same stroke while others are associated with the fact that different strokes may take different paths to ground. The first and the second strokes exhibit unconnected branches. The second and the third strokes, second and third from the right on the streaked photograph, are brighter than the first stroke, on the far right. Adapted from Hendry (1993).

are discussed below. The first two strokes are branched, and the downward direction of the branches indicates that this is a downward lightning flash. The polarity of this multiple-stroke flash is inferred to be negative since positive flashes are typically composed of only one stroke (subsection 5.3.1).

Now we consider the diagrams of still and time-resolved images of the three-stroke lightning flash shown in Figs. 4.2a, b, respectively. The corresponding current at the channel base is shown in Fig. 4.2c. In Fig. 4.2b, time advances from left to right, and the time scale is not continuous. Each of the three strokes in Fig. 4.2b, represented by its luminosity as a function of height above ground and time, is composed of a downward-moving process, termed a leader (SL or DL), and an upward-moving process, termed a return stroke (RS). The leader creates a conducting path between the cloud charge source and ground and effectively distributes negative charge from the cloud source along this path, and the return stroke traverses that path moving from ground toward the cloud charge source and neutralizes the negative leader charge deposited along the conducting path. Thus, both leader and return-stroke processes serve effectively to transport negative charge from the cloud to ground, as further discussed in the next paragraph. As seen in Fig. 4.2b, the leader (SL) initiating the first return stroke differs from the leaders (DL) initiating the two subsequent strokes (all strokes other than the first are termed subsequent strokes). In particular, the first-stroke leader SL appears to be an optically intermittent process, hence the term "stepped leader", while the tip of a subsequent-stroke leader DL appears to move continuously. A continuously moving subsequent-stroke leader tip appears on streak photographs as a downward-moving "dart", hence the term dart leader. The apparent difference between the two types of leader is related to the fact that the stepped leader develops in virgin air while the dart leader follows the pre-conditioned path of the preceding stroke or strokes.

In the following, referring to Fig. 4.3, we present a more complete sequence of the processes involved in a typical negative downward lightning flash. The source of lightning is usually a cumulonimbus (Section 3.2), whose idealized charge structure is shown in Fig. 4.3 as three vertically stacked regions labeled P and LP for the main positive

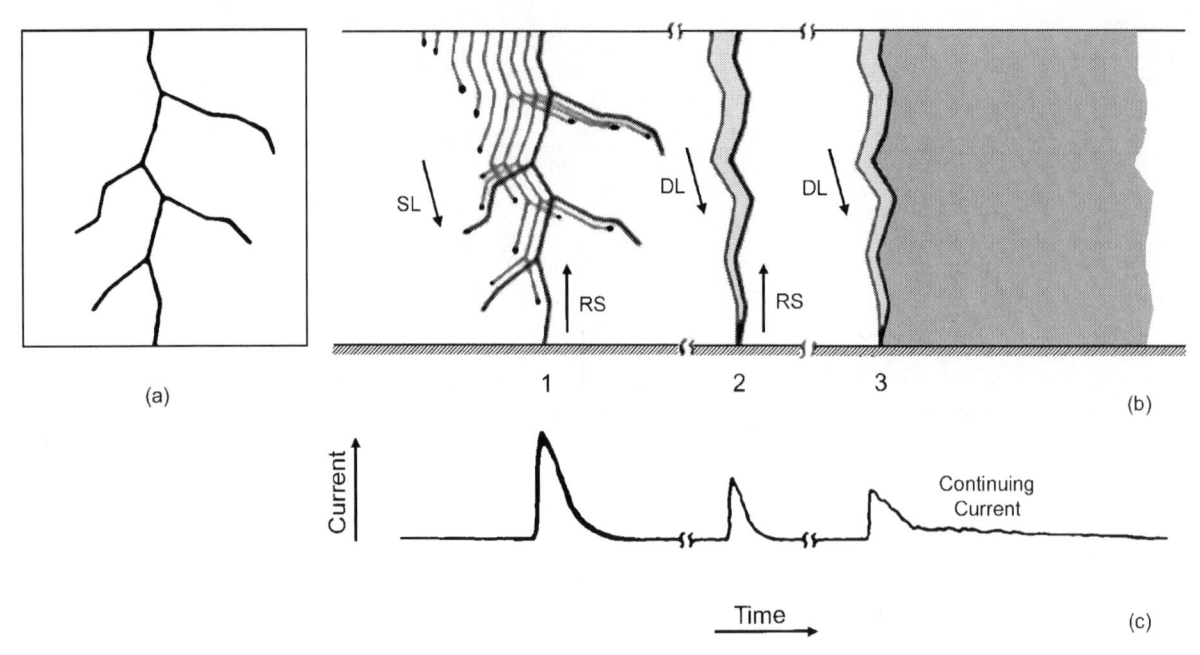

110   4. Downward negative lightning discharges to ground

Fig. 4.2. Diagram showing the luminosity of a three-stroke ground flash and the corresponding current at the channel base: (a) still-camera image, (b) streak-camera image, and (c) channel-base current.

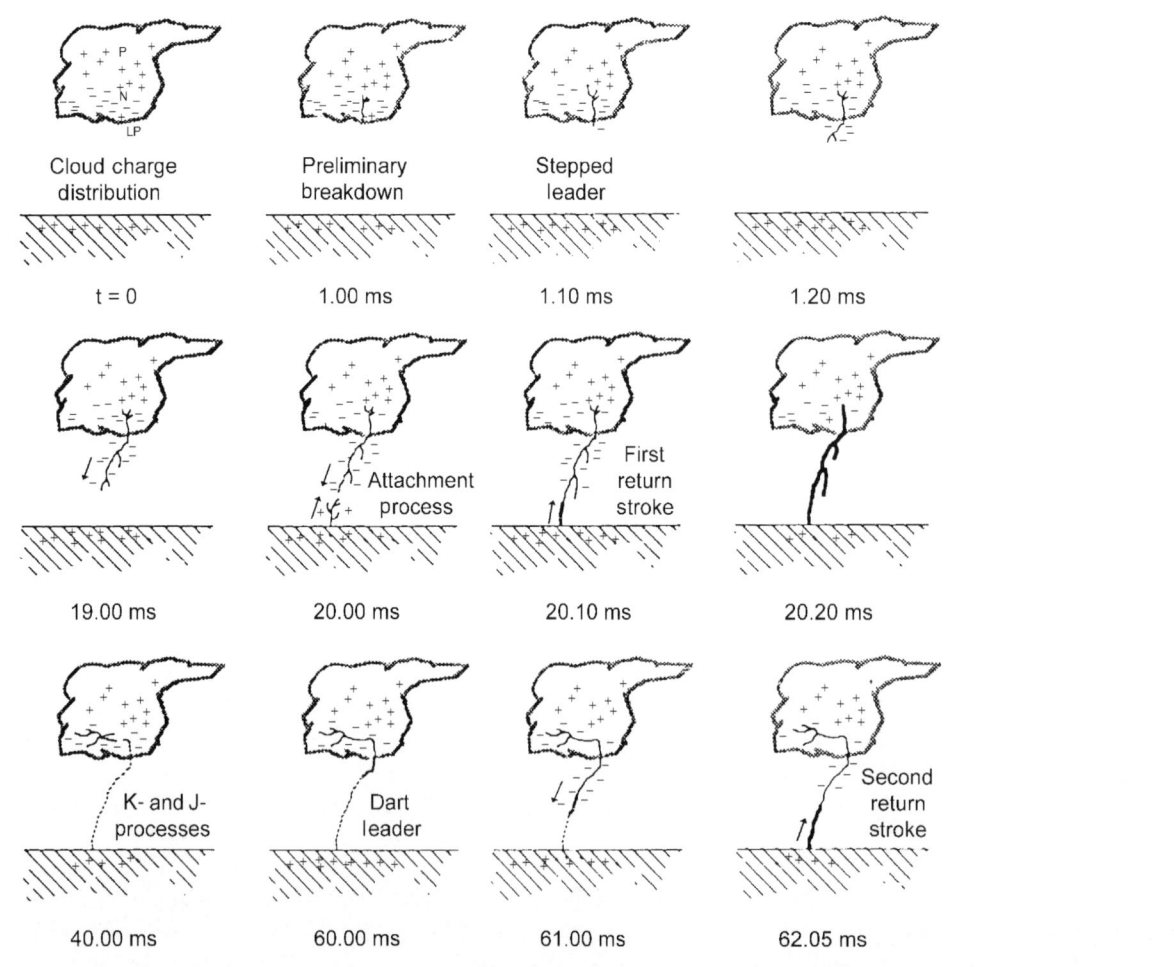

Fig. 4.3. Various processes comprising a negative cloud-to-ground lightning flash. Adapted from Uman (1987, 2001).

and the lower positive charge regions and N for the main negative charge region. The stepped leader is preceded by an in-cloud process called the preliminary or initial breakdown. There is no consensus on the mechanism of this process. It may be a discharge bridging the main negative and the lower positive charge regions, as shown in Fig. 4.3. The initial breakdown may last from a few milliseconds to some tens of milliseconds and serves to provide conditions for the formation of the stepped leader. The latter is a negatively charged plasma channel extending toward the ground at an average speed of 2×10^5 m s^{-1} in a series of discrete steps. From high-speed time-resolved photographs, each step is typically 1 μs in duration and tens of meters in length, the time interval between steps being 20 to 50 μs. The peak value of the current pulse associated with an individual step has been inferred to be 1 kA or greater. The stepped leader serves to form a conducting path or channel between the cloud charge source and ground. Several coulombs of negative charge are distributed along this path, including downward branches (Fig. 4.1). Thus the leader may be viewed as a process removing negative charge from the source and depositing this charge onto the downward extending channel. The stepped-leader duration is typically some tens of milliseconds, and the average leader current is some hundreds of amperes.

The electric potential difference between a downward-moving stepped-leader tip and ground is probably some tens of megavolts (e.g., Bazelyan *et al.* 1978), which is comparable to or a considerable fraction of that between the cloud charge source and ground. The magnitude of the potential difference between two points, one at the cloud charge source and the other on ground, is the line integral of the electric field intensity between those points.

The upper and lower limits for the potential difference between the lower boundary of the main negative charge region and ground can be estimated by multiplying, respectively, the typical observed electric field in the cloud, 10^5 V m^{-1}, and the expected electric field at ground under a thundercloud immediately prior to the initiation of lightning, 10^4 V m^{-1}, by the height of the lower boundary of the negative charge center above ground, 5 km or so. The resultant range is 50 to 500 MV. Marshall and Stolzenburg (2001), from 15 balloon soundings of the electric field through thunderstorms, estimated a maximum value of the potential within a cloud relative to the Earth in the range from -102 to $+94$ MV.

As the leader approaches ground, the electric field at the ground surface, particularly at objects or relief features protruding above the surrounding terrain, increases until it exceeds the critical value for the initiation of one or more upward-connecting leaders. The initiation of an upward connecting leader from ground in response to the descending stepped leader marks the beginning of the attachment process. This process ends when contact is made between the downward and upward moving leaders, probably some tens of meters above ground (more above a tall structure), whereafter the first return stroke begins. The return stroke serves to neutralize the leader charge, in other words, to transport the negative charges stored on the leader channel to the ground. It is worth noting that the return-stroke process may not neutralize all the leader charge or may deposit some excess positive charge onto the leader channel and into the cloud charge source region.

The final stage of the attachment process and the initial stage of the return-stroke process are complex and will be described in Sections 4.5 and 4.6, respectively. The net result of those stages is a fully formed return stroke, which is somewhat similar to the potential discontinuity that would travel upward along a vertical, negatively charged transmission line if the lower end of the line were connected to the ground.

The speed of the return stroke, averaged over the visible channel, is typically between one-third and one-half the speed of light. There is no consensus about whether or how the first return-stroke speed changes over the lower 100 m or so but, over the entire channel, the speed decreases with increasing height, dropping abruptly after passing each major branch. At the same time, a transient enhancement of the channel luminosity below the branch point, termed a branch component, is often observed.

The first return-stroke current measured at ground rises to an initial peak of about 30 kA in some microseconds and decays to half-peak value in some tens of microseconds while exhibiting a number of subsidiary peaks, probably associated with the branches. This impulsive component of current may be followed by a current of some hundreds of amperes lasting for some milliseconds. The return stroke effectively lowers to ground the several coulombs of charge originally deposited on the stepped-leader channel, including that on all the branches.

The high-current return-stroke wave rapidly heats the channel to a peak temperature near or above 30 000 K and creates a channel pressure of 10 atm or more, resulting in channel expansion, intense optical radiation, and an outward propagating shock wave that eventually becomes the thunder (sound wave) we hear at a distance.

When the first return stroke, including any associated in-cloud discharge activity (discussed later), ceases, the flash may end. In this case, the lightning is called a single-stroke flash. However, more often the residual first-stroke channel is traversed downwards by a leader that appears to move continuously, a dart leader. During the time interval between the end of the first return stroke and the initiation of a dart leader, J (for junction) and K-processes occur in the cloud. K-processes can be viewed as transients occurring during the slower J-process. The J-processes amount to a redistribution of cloud charge on a time scale of tens of milliseconds, in response to the preceding return stroke. There is controversy as to whether these processes, which apparently act to extend the return-stroke channel further into the cloud, are necessarily related to the initiation of a following dart leader. The J-process is often viewed as a relatively slow positive leader extending from the flash origin into the negative charge region, the K-process then being a relatively fast "recoil streamer" that begins at the tip of the positive leader and propagates toward the flash origin. Both the J-processes and the K-processes in cloud-to-ground discharges serve to transport additional negative charge into and along the existing channel (or its remnants), although not all the way to the ground. In this respect, K-processes may be viewed as attempted dart leaders. The processes that occur after the only stroke in single-stroke flashes and after the last stroke in multiple-stroke flashes are sometimes termed F (final) processes. These are similar, if not identical, to J-processes.

The dart leader progresses downward at a typical speed of 10^7 m s^{-1}, typically ignores the first stroke branches (in this regard, the second stroke in Fig. 4.1b is somewhat unusual), and deposits along the channel a total charge of the order of 1 C. The dart-leader current peak is about 1 kA. Some leaders exhibit stepping near ground while propagating along the path traversed by the preceding return stroke, these leaders being termed dart-stepped leaders. Additionally, some dart or dart-stepped leaders deflect from the previous return-stroke path, become stepped leaders, and form a new termination on the ground. Leaders that deflect from the previously formed channel are not termed dart-stepped leaders in this book, to avoid the ambiguities found in the previous literature (e.g., Uman 1969a, 1984).

When a dart leader or dart-stepped leader approaches the ground, an attachment process similar to that described for the first stroke takes place, although it probably occurs over a shorter distance and consequently takes less time, the upward connecting-leader length being of the order of some meters. Once the bottom of the dart or dart-stepped leader channel is connected to the ground, the second (or any subsequent) return-stroke wave is launched upward and again serves to neutralize the leader charge. The subsequent return-stroke current at ground typically rises to a peak value of 10 to 15 kA in less than a microsecond and decays to half-peak value in a few tens of microseconds. The upward propagation speed of such a subsequent return stroke is similar to that of the first return stroke, although due to the absence of branches the speed variation along the channel does not exhibit abrupt drops.

The impulsive component of the current in a subsequent return stroke is often followed by a continuing

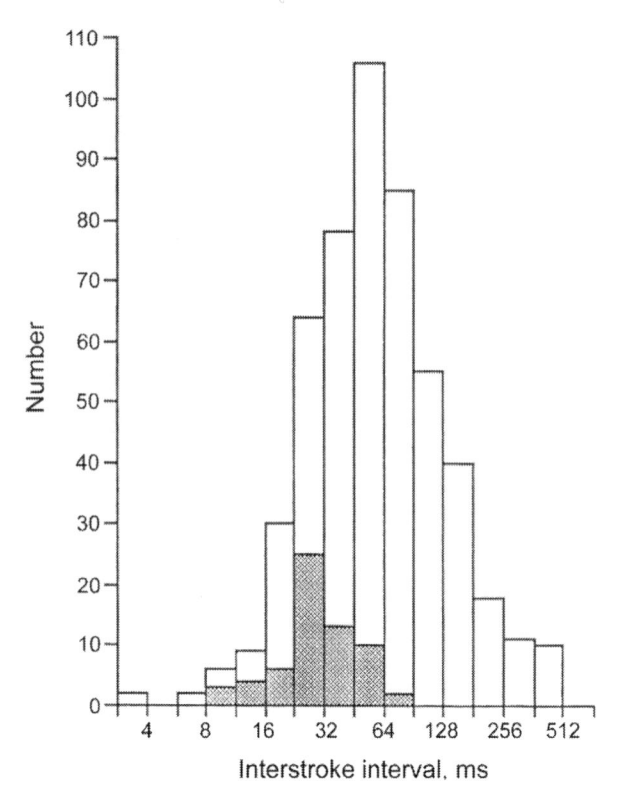

Fig. 4.4. Histogram of 516 interstroke intervals in 132 flashes in Florida and New Mexico. Intervals preceding strokes that initiated long continuing currents (Section 4.8) are shown shaded. Adapted from Rakov and Uman (1990a).

current that has a magnitude of tens to hundreds of amperes and a duration up to hundreds of milliseconds. Continuing currents with a duration in excess of 40 ms are traditionally termed long continuing currents. Between 30 and 50 percent of all negative cloud-to-ground flashes contain long continuing currents. The source for continuing current is the cloud charge, as opposed to the charge distributed along the leader channel, the latter charge contributing to at least the initial few hundred microseconds of the return-stroke current observed at ground. Continuing current typically exhibits a number of superimposed surges that rise to a peak and fall off to the background current level in some hundreds of microseconds, the peak being generally in the hundreds of amperes range but occasionally in the kiloamperes range. These current surges are associated with enhancements in the relatively faint luminosity of the continuing-current channel and are called M-components.

The time interval between successive return strokes in a flash is usually several tens of milliseconds (see Fig. 4.4), although it can be as large as many hundreds of milliseconds if a long continuing current is involved and as small as one millisecond or less. The total duration of

4.2. General picture

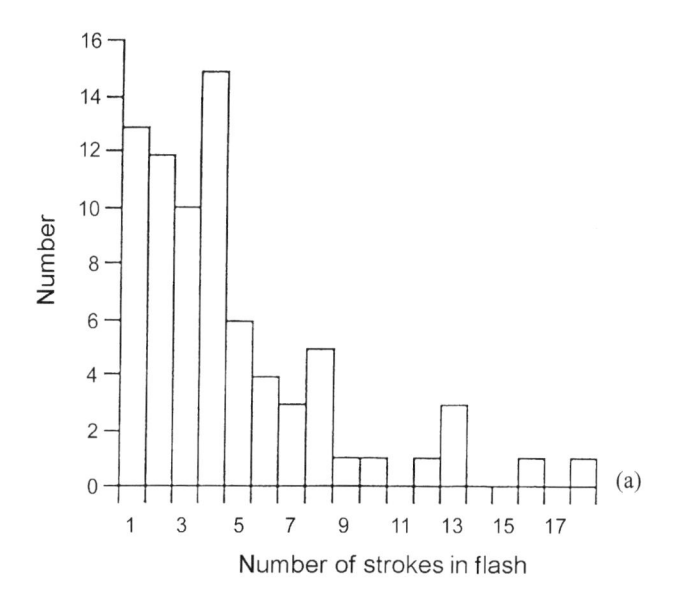

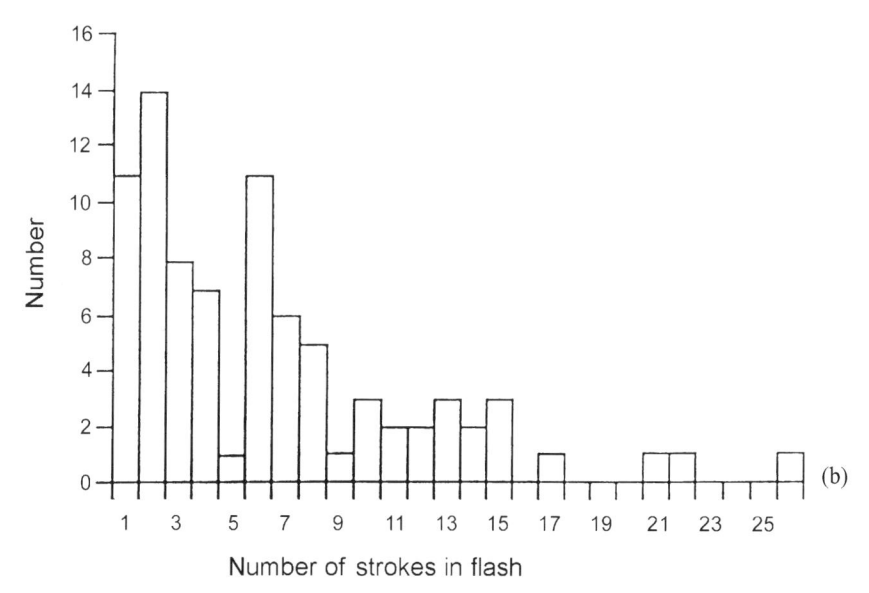

Fig. 4.5. Histograms of number of strokes per flash: (a) 76 flashes in Florida (Rakov and Uman 1990a) and (b) 83 flashes in New Mexico (Kitagawa et al. 1962). The mean number of strokes per flash is 4.6 in (a) and 6.4 in (b). Adapted from Rakov and Uman (1990d).

a flash is typically some hundreds of milliseconds, and the total charge lowered to ground is some tens of coulombs. Ogawa (1995) gives a range for the total ground flash duration from 10 ms to 2 s with a typical value of 300 ms, and for the total ground flash charge he gives a range from 1 to 400 C with a typical value of 20 C. Histograms of the number of strokes per flash from observations made in Florida and New Mexico are given in Fig. 4.5. The percentage of single-stroke flashes in Florida (Rakov and Uman 1990d) and New Mexico (Kitagawa et al. 1962) was found to be 17 and 14 percent, respectively, similar to the 21 percent observed in Sri Lanka (Cooray and Jayaratne 1994) and the 18 percent in Sweden (Cooray and Perez 1994b). Thus, the overwhelming majority (about 80 percent or more) of negative cloud-to-ground flashes contain more than one stroke. Examples of simultaneous photographic and electric field records of two negative multiple-stroke flashes that additionally illustrate the various lightning processes described above are given in Fig. 4.6.

In the following, we give a brief overview of detailed studies of the structure of individual flashes as determined by Rakov and Uman (1990a, b, c, d, 1991, 1994), Rakov

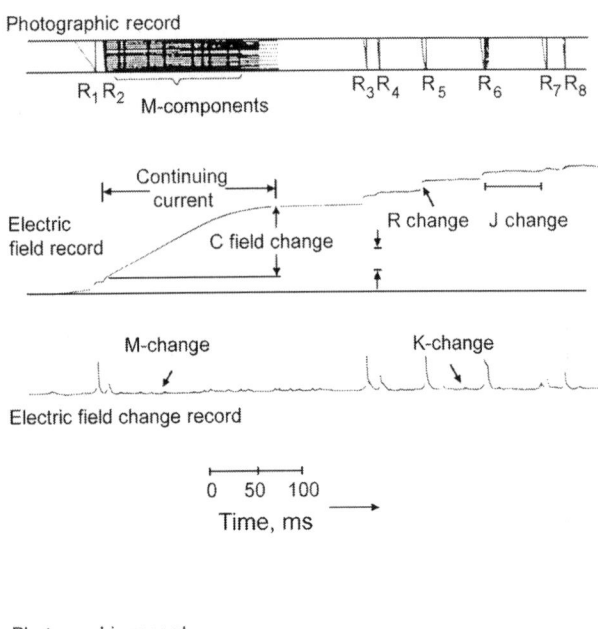

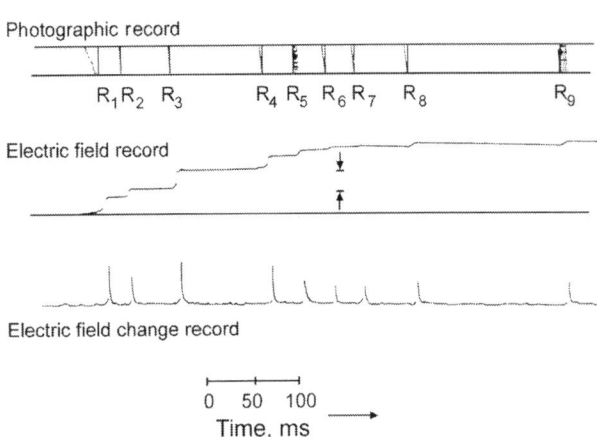

Fig. 4.6. Simultaneous photographic and electric field measurements for two multiple-stroke ground flashes in New Mexico. The upper three diagrams relate to a flash with continuing current (flash no. 106, 20 km distant) and the lower three diagrams to a flash without continuing current (flash no. 109, 19 km distant). Two electric field records are shown for each flash. One, labeled "Electric field record", was obtained with a measuring system having a decay time constant of 4 s and a relatively low gain. This, and other systems with decay time constants of the order of seconds, are sometimes referred to as "slow antenna" systems. The other electric field record, "Electric field change record", was obtained with a measuring system having a decay time constant of 70 μs and a relatively high gain. The unlabeled vertical interval indicated by two arrows in each electric field record corresponds to 1 V cm^{-1}. This, and other systems with submillisecond decay time constants, are sometimes referred to as "fast antenna" systems. Adapted from Kitagawa et al. (1962).

et al. (1990, 1991, 1994), and Thottappillil et al. (1992). It has been found in these Florida studies that (i) about one-half the negative cloud-to-ground flashes create more than one termination on ground (see also Fig. 4.1), the spatial separation between the channel terminations being up to many kilometers (Fig. 4.7), (ii) about one-third of multiple-stroke flashes have at least one subsequent stroke that is larger (in terms of its initial field peak and, by inference, current peak) than the first stroke in the flash, (iii) about 20 percent of flashes contain return strokes separated by relatively short time intervals, of some tens to hundreds of microseconds, and (iv) early subsequent strokes (second to fourth) differ from later (higher-order) strokes, as illustrated in Table 4.1. In particular, the leaders of early subsequent strokes are more likely to deflect from the previously formed channel and create a new termination on the ground and also are more likely to exhibit stepping while developing in a previously formed channel, even if that channel is not particularly "aged". These and other observed differences suggest that the status of the channel depends on the number of strokes that have participated in its cumulative conditioning. Some confirmation of this hypothesis, proposed by Rakov and Uman (1990b), comes from the findings of Davis (1999), as discussed in Section 4.7 in regard to dart-stepped leaders. Additional information on the structure of individual flashes is found in Kitagawa et al. (1962), Brook et al. (1962), Cooray and Jayaratne (1994), and Cooray and Perez (1994b).

In the remainder of this section, we discuss briefly the energy and the power (time rate of change of energy) associated with lightning discharges. Further information on lightning energy is found in Section 1.5 and in subsection 12.2.6.

An approximate range for the electrostatic energy available for a lightning flash lowering a charge Q to ground can be evaluated by multiplying Q by the upper and lower limits for V, the magnitude of the potential difference between the lower boundary of the cloud charge source and ground. Assuming that $Q = 20$ C, thought to be typical for a cloud-to-ground flash, and using the range of V from 50 to 500 MV estimated earlier in this section, we find that each flash dissipates an energy of roughly 1 to 10 GJ (gigajoules). Note that a flash is typically composed of three to five strokes, and that the first stroke is usually a factor 2 to 3 larger (in terms of peak current and peak field) than a subsequent stroke, that is, any stroke other than the first. The above energy range inferred from electrostatic considerations is for all processes involved in a lightning discharge. Specifically, this energy estimate may well be dominated by the energy dissipated in the formation of numerous filamentary channels in the cloud that serve, in effect, to funnel cloud charges into the narrow channel to ground. Marshall and Stolzenburg (2001), from their

4.2. General picture

Table 4.1. *Properties of subsequent leaders and return strokes versus stroke order. Adapted from Rakov et al. (1994). The numbers in parentheses are the sample sizes*

Feature	Stroke order		
	2	2–4	5–18
Probability of creating new termination, %	37 (63)	25 (155)	0 (115)
Occurrence of leader stepping in previously formed channel, %	36 (36)	21 (86)	4.5 (88)
Occurrence of apparently inactive final portion in leader field waveform, %	29 (63)	24 (155)	6.1 (115)
Occurrence of leader–return-stroke sequence with negative net electric field change, %	0 (63)	1.3 (155)	14 (115)
Geometric mean return-stroke field peak normalized to 100 km, V m^{-1}	3.4 (63)	3.3 (155)	2.3 (115)
Geometric mean duration of leader in previously formed channel, ms	1.2 (24)	1.5 (71)	2.2 (83)
Geometric mean preceding interstroke interval, ms	56 (63)	66 (155)	54 (115)

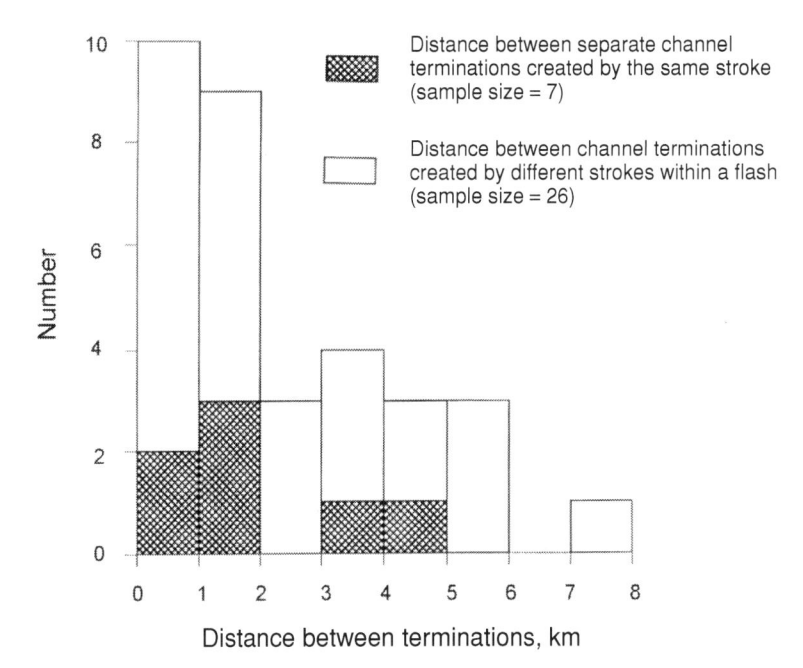

Fig. 4.7. Histogram of the distances between the multiple terminations of 22 individual ground flashes in Florida. Adapted from Thottappillil *et al.* (1992).

balloon soundings of the electric field through thunderstorms and assumed minimum and maximum values of charge transfer, estimated the energy available for lightning to be in the range from 10 MJ to 10 GJ, the energy available for intracloud flashes (Chapter 9) being usually larger than that available for ground flashes. There is no consensus regarding the proportion in which the total return stroke energy is converted to thunder, hot air, light, and radio waves. According to Paxton *et al.* (1986), who used a gas dynamic model of the lightning return stroke (subsection 12.2.2), almost 70 percent of the total energy input to the channel is optically radiated from the channel. However, Few (1995), in his theory of thunder (subsection 11.3.2), assumes that essentially all the input energy is delivered to a shock wave that subsequently is heard as thunder. As discussed in the first part of subsection 12.2.6, the total lightning energy input estimates of Paxton *et al.* (1986) and others, who employed gas dynamic models, differ from that of Few (1995) by two orders of magnitude or so.

Krider and Guo (1983) and Krider (1992) estimated that the radio-frequency power radiated by a subsequent return stroke at the time of the field peak, 3 to 5 GW, is about two orders of magnitude greater than the optical power radiated in the 0.4 to 1.1 μm range at the time of the field peak. The average zero-to-peak risetime of the subsequent-stroke field waveforms was 2.8 μs. The total optical power, however, was found to dominate at later times, the peak optical power occurring about 60 μs after the electric field peak (because the risetime of the optical signal was determined by the geometrical growth of the return-stroke channel).

4.3. Initial breakdown
4.3.1. General information

The initial breakdown, often referred to as the preliminary breakdown, of a cloud-to-ground flash is the in-cloud process that initiates or leads to the initiation of the downward-moving stepped leader. In the early lightning studies in South Africa, the existence of the initial breakdown as a unique lightning process was generally inferred from (i) observations of the luminosity produced by thunderclouds for a hundred or more milliseconds before emergence of the stepped leader from the cloud base, (ii) observations of relatively long electric field changes, exceeding 100 ms in duration, prior to the first return stroke, and (iii) the assumption that the stepped-leader duration was unlikely to exceed a few tens of milliseconds, based on the measured leader speed below the cloud and the cloud charge height (subsection 3.2.2).

Clarence and Malan (1957) suggested, on the basis of single-station electric field measurements of the type illustrated in Fig. 4.8, that the initial breakdown (labeled B by them) is a vertical discharge between the main negative charge center and the lower positive charge center, which has a duration of 2 to 10 ms (subsection 3.2.1). This inference was made from the observed polarity reversal of the B portion of the electric field waveforms, of the type shown in Fig. 4.8, in the distance range from 2 to 5 km.

According to Clarence and Malan (1957), the initial breakdown is followed by the stepped leader (labeled L), either immediately or after a so-called intermediate stage (labeled I), which may last up to 400 ms. The intermediate stage was interpreted by Clarence and Malan (1957) as being due to negative charging of the vertical channel of the initial breakdown until the field at the bottom of the channel was high enough to launch a stepped leader that initiated, on its arrival at ground, a return stroke, labeled

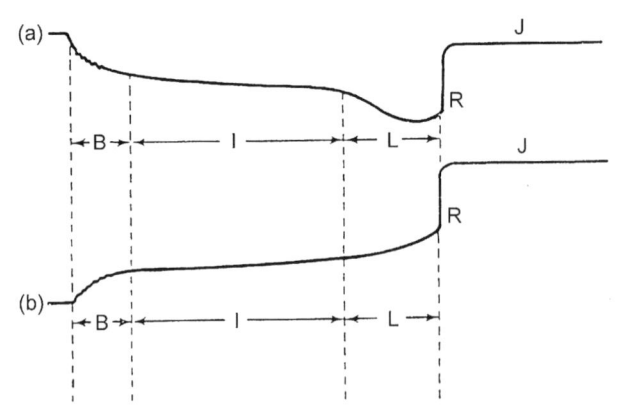

Fig. 4.8. Examples of electric field waveforms of the type used by Clarence and Malan (1957) to introduce the so-called BIL structure of the electric field prior to the first return stroke and to interpret the B stage as a vertical discharge between the main negative charge center and the lower positive charge center in the cloud. (a) The electric field at 2 km, (b) the electric field at 5 km. Adapted from Clarence and Malan (1957).

R in Fig. 4.8. Clarence and Malan's (1957) scenario of the initial breakdown, inferred from single-station electric field measurements, is not confirmed by more recent studies based on multiple-station electric field measurements or on VHF–UHF channel imaging in conjunction with electric field records, as explained below. These recent studies suggest that the initial breakdown can be viewed as a sequence of channels extending in seemingly random directions from the cloud charge source. One of these events evolves into the stepped leader which bridges the cloud charge source and the ground.

Krehbiel *et al.* (1979), from eight-station electric field measurements (Section 17.2), interpreted the initial breakdown processes in two New Mexico flashes, one of which is presented in Fig. 4.9, as a succession of "breakdown events" with considerable horizontal extent, which occurred prior to the development of a leader to ground. These events were associated in part with the negative charge volume that appeared to be the charge source for the following first stroke, and they effectively transported negative charge away from that volume. For the two breakdown events A and B illustrated in Fig. 4.9, negative charges of 2.0 and 3.6 C were effectively moved over 1.5 and 3.8 km, respectively. The corresponding average currents were 44 and 70 A.

Rhodes and Krehbiel (1989) used a 274 MHz interferometer to image the channels of a single-stroke flash. The flash began with a sequence of three predominantly horizontal breakdown events developing in different directions, the final one of which culminated in the development of a stepped leader. The initial breakdown lasted for about 100 ms, and the beginning of the stepped leader was

4.3. Initial breakdown

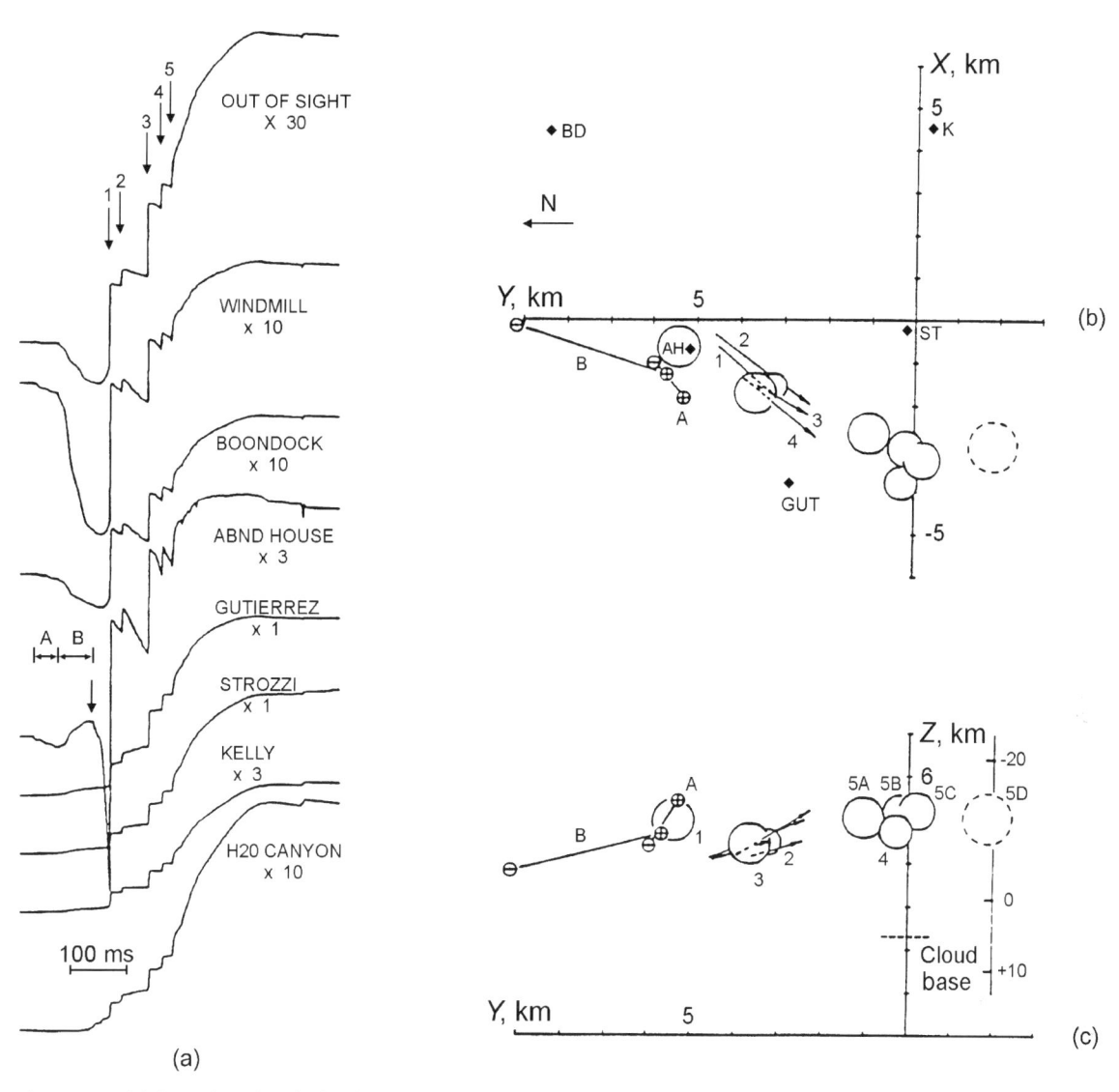

Fig. 4.9. Multiple-station electric field observations of a cloud-to-ground flash in New Mexico: (a) the field changes observed at different locations produced by a five-stroke cloud-to-ground discharge, (b) plane and (c) vertical projection views showing spherical representations of the charges neutralized by ground strokes and by continuing current following stroke 5, and also the dipolar charge rearrangement during the initial breakdown events A and B (the positive and negative pairs) and during the interstroke intervals (arrows). In (b) and (c), the signs of the dipolar charges are those effectively added to the cloud by the process, and the arrows point in the direction of effective positive-charge motion (subsection 17.2.1). The arrow above the Gutierrez waveform in (a) indicates the apparent end of the initial breakdown and beginning of the stepped leader. The vertical scale on the right in (c) indicates the environmental temperature in °C. Adapted from Krehbiel et al. (1979).

marked by relatively large pulses similar to the "characteristic" pulses identified by Beasley et al. (1982) and discussed in the next subsection.

Proctor et al. (1988), who used a 355 MHz TOA lightning channel imaging system (subsection 17.4.2) to image the channels of 47 ground flashes in South Africa, interpreted the BIL-type electric field changes (see Fig. 4.8) prior to first return strokes as being associated with the development of stepped leaders, because the 355 MHz radiation pattern remained the same during the entire waveform. Some of these leaders were preceded or accompanied by cloud discharges starting from approximately the same origin as the leaders. The breakdown (B) portion of the BIL waveform was interpreted by Proctor et al. (1988) as "the start of the leader" while the intermediate (I) portion was found to occur sometimes owing to "inadequacies in the rate of supply of charge at the source", sometimes due to the geometry of the stepped leader, and sometimes due to

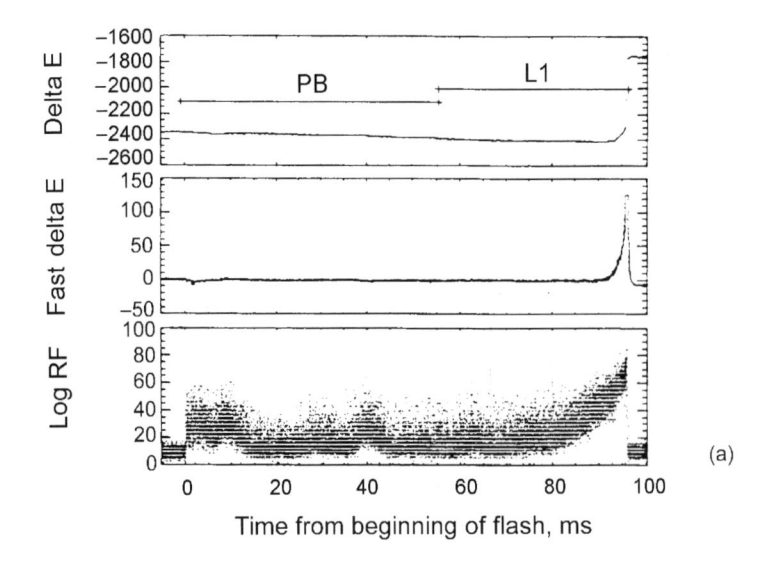

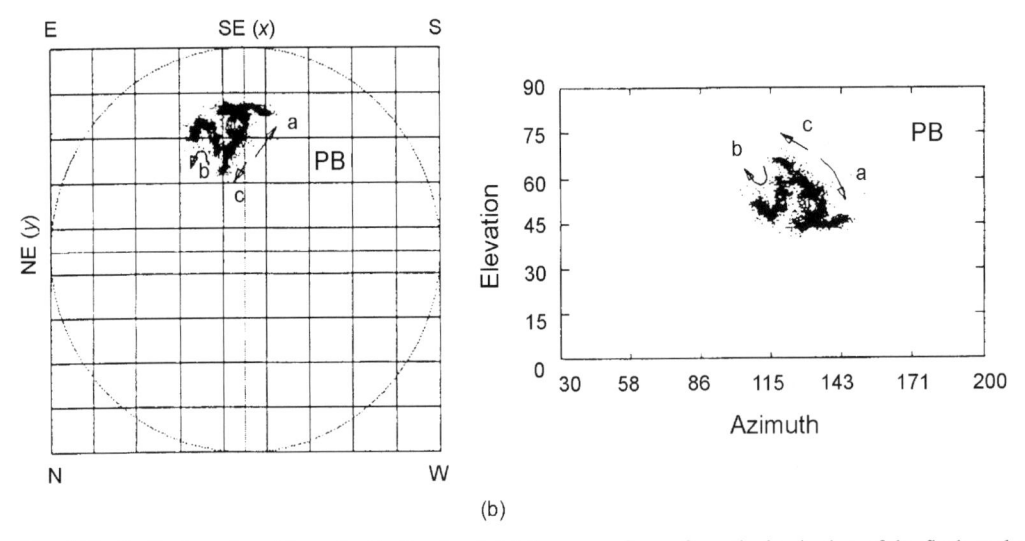

Fig. 4.10. Preliminary breakdown inside the cloud: (a) time waveforms from the beginning of the flash to the first return stroke with the initial (preliminary) breakdown indicated by PB and leader indicated by L1, (b) the radiation sources of PB. Adapted from Shao (1993).

overlapping cloud discharges. Additionally, if, as Proctor et al. (1988) claimed, the initial breakdown is not a unique process but merely the start of the stepped leader, the intermediate portion of the BIL-type electric field change may be due to the slowing of the negative leader as it encounters a positively charged region, for instance, the positive charge center near the cloud base (subsections 3.2.1 and 3.2.7), as suggested by Isikawa et al. (1958). In contrast with the findings of Proctor et al. (1988), Rustan et al. (1980) reported that (i) the stepped leader is characterized by VHF (30–50 MHz) pulses of much lower amplitude and shorter duration than those of the preceding initial breakdown and (ii) no significant electric field change occurs during the initial breakdown. Further, Rhodes and Krehbiel (1989) reported a clear difference between the VHF (274 MHz) interferometric channel images due to initial breakdown events and those due to the stepped leader, the leader being more dispersed and characterized by "less well-defined motion". However, the latter apparent discrepancy might be related to the differences in the instrumentation used by Proctor et al. (a 355 MHz TOA lightning channel imaging system) and that used by Rhodes and Krehbiel (a 274 MHz interferometer). Indeed, Mazur et al. (1997) found a significant difference in the representation of lightning VHF radiation sources by TOA (LDAR) and interferometric (SAFIR) systems (Section 17.6).

Figure 4.10 shows 274 MHz interferometric images of the initial breakdown in a six-stroke Florida flash (Shao 1993). The process develops in three successive branches originating from the volume that appears to also be the

4.3. Initial breakdown

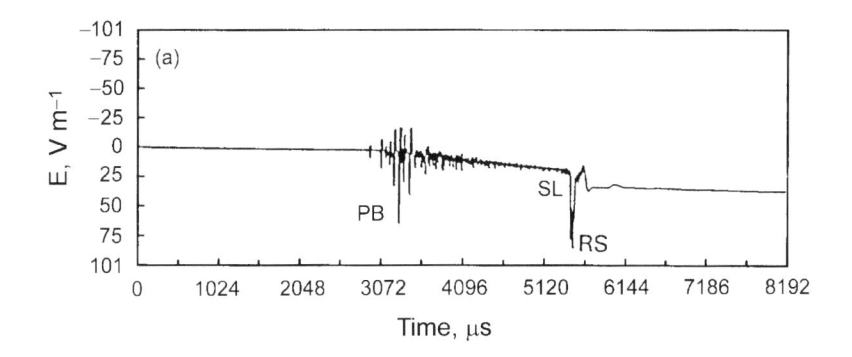

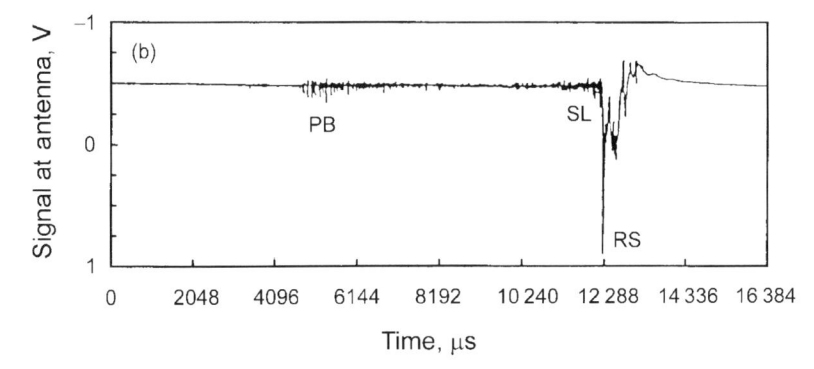

Fig. 4.11. Examples of electric fields due to negative first strokes in cloud-to-ground lightning: (a) winter lightning at about 25 km, (b) summer lightning at an unknown distance. PB stands for preliminary breakdown, SL for stepped leader (just prior to its attachment to ground), and RS for return strokes. The overall time scales in Figs. 4.11a, b are about 8.2 and 16.4 ms, respectively. Note that the separation between PB and RS, the duration of the stepped leader, is smaller than usual, particularly in Fig. 4.11a. The atmospheric electricity sign convention (subsection 1.4.2) is used here. Adapted from Brook (1992).

source for the following stepped leader. The latter is not shown in Fig. 4.10, but its VHF image does not appear more dispersed than the initial breakdown image, in contrast with the observation of Rhodes and Krehbiel (1989). The duration of the entire process is 50 to 60 ms.

It is worth noting that considerable intracloud discharge activity that is apparently not related directly to the initiation of stepped leaders can precede ground flashes, as manifested by a so-called cloud-type portion of the "first-leader field change" (Kitagawa and Brook 1960), a "pre-preliminary discharge" (Isikawa 1961), a long-duration preliminary field change (Krehbiel et al. 1979), and large negative "prestroke field changes" outside the reversal distance (Thomson 1980). Furthermore, sometimes strokes to ground appear to be a minor branch of an extensive cloud flash (e.g., Richard et al. 1986, Fig. 6; Proctor et al. 1988, Fig. 17).

4.3.2. Initial breakdown pulses

It has been observed by many investigators (e.g., Kitagawa 1957a; Clarence and Malan 1957; Kitagawa and Kobayashi 1959; Kitagawa and Brook 1960; Krider and Radda 1975; Beasley et al. 1982; Gomes et al. 1998) that the first-return-stroke electric field waveform (Section 4.6) may be preceded by a train of relatively large microsecond-scale bipolar pulses, as illustrated in Figs. 4.11a, b. The duration of the train is of the order of 1 ms. Since (i) the time interval between the pulse train and the return-stroke waveform is usually several milliseconds or more and (ii) the pulse train (termed the characteristic pulses by Beasley et al. 1982) has been observed to occur at the beginning of the millisecond-scale electric field waveform indicative of the stepped leader (Section 4.4), these pulses are generally thought to mark a transition from the initial breakdown to the stepped leader. The pulses are usually bipolar, the initial polarity being the same as that of the following return-stroke pulse. The amplitude of the initial breakdown pulses can be comparable to that of the first return stroke, as seen in Fig. 4.11a. However, in some records the initial breakdown pulses are either undetectable or have negligible amplitude compared to the following return stroke pulses.

Some investigators (e.g., Hodges 1954; Kitagawa 1957a; Isikawa et al. 1958) considered the occurrence of these pulses followed by a period of many milliseconds of relatively small, if any, pulses as indicative of a β-type leader (subsection 4.4.1). Brook (1992) found that the initial breakdown pulses have larger amplitudes in winter negative lightning than in summer negative lightning (see Figs. 4.11a, b and Section 2.8) and attributed the disparity to the difference in the precipitation mixes in summer and winter clouds. Interestingly, Clarence and Malan (1957) found that, out of their total 407 first-stroke electric field waveforms at distances beyond 200 km, 16 percent were of the so-called large amplitude, fast β-type. The latter, recorded at Johannesburg, South Africa, probably in summer, are very similar to Brook's (1992) "winter"-type waveforms recorded in the United States. Further, similar waveforms were recorded in summer storms by Weidman and Krider (1979, Fig. 1) at 50–100 km in Florida, by Ogawa (1993, Figs. 8 and 9) at some hundreds of kilometers in Japan, and Cooray and Perez (1994b, Fig. 5) within about 50 km in Sweden. Some evidence that the occurrence of initial breakdown pulses may depend on the type of storm or on the stage of the storm's life cycle was found by Clarence and Malan (1957) and by Kitagawa (1957a). Other factors that can potentially influence the amplitude and, hence, the detectability of initial breakdown pulses include the geometry of their associated channels with respect to the observer and propagation effects.

Three examples of initial breakdown pulses as they appear in electric field records taken 50–100 km from the lightning discharge are given in Fig. 4.12. The individual pulses are characterized by a total duration of 20–40 μs (Rakov et al. 1996). The average number of pulses per train is about 10 and the pulse rise time is typically 10 μs. Usually there are two or three smaller pulses superimposed on the rising portion of the initial breakdown pulse, while the falling portion and the opposite polarity overshoot are smooth (Weidman and Krider 1979). The frequency spectra for such pulses are given in Weidman et al. (1981, Fig. 4) and Willett et al. (1990, Fig. 3); the latter is reproduced in Fig. 4.42. It is worth noting that the initial breakdown pulses are distinctly different from the stepped-leader pulses produced near ground (Section 4.4), as discussed in the next section. The differences are seen in terms of (i) the overall waveshape, the former being bipolar with fine structure and the latter essentially unipolar and smooth, (ii) the total duration, 20–40 μs as against 1–2 μs, and (iii) the interpulse interval, 70–130 μs as against 15–25 μs. The characteristics, including the polarity of the initial half-cycle, of the initial breakdown pulses in negative ground flashes differ from those in cloud flashes (subsection 9.4.2).

Optical radiation apparently associated with initial breakdown pulses occurring 25 ms or more before the return stroke and lasting for about 3 ms was observed by Brook and Kitagawa (1960, Fig. 5). Similar observations were reported by Isikawa (1961). Brook and Kitagawa (1964,

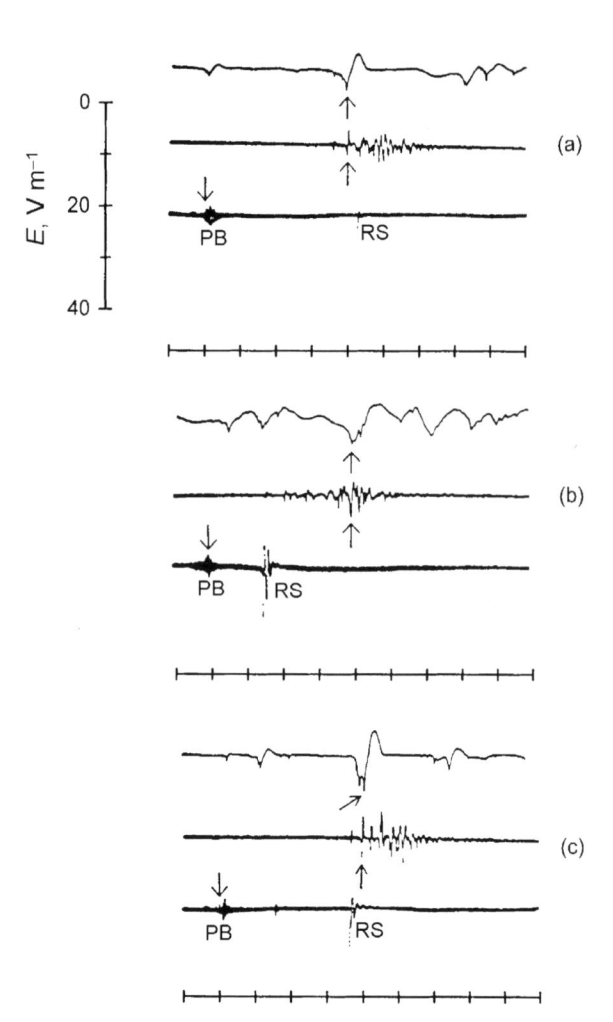

Fig. 4.12. Electric fields radiated by three cloud-to-ground lightning discharges at distances of 50–100 km. The arrows indicate the same event on time scales of 2 ms (bottom traces), 0.4 ms (middle traces), and 40 μs (top traces) per division. The atmospheric electricity sign convention (subsection 1.4.2) is used here. PB stands for preliminary breakdown and RS for return stroke. Adapted from Weidman and Krider (1979).

Fig. 1) recorded appreciable radiation at 420 and 850 MHz associated with initial breakdown pulses. Rust et al. (1979) observed bursts of 2200 MHz radiation some tens of milliseconds prior to the first return stroke. The bursts lasted for 5 to 10 ms and were interpreted as being due to initial breakdown passing through or near the main beam of the antenna. Beasley et al. (1982) reported that the envelope of VHF (30–50 MHz) radiation showed clearly the same pulses as the characteristic pulses observed in the wideband electric field records. However, Proctor (1983) stated that he had never observed similar pulses at VHF. Cooray and Perez (1994a) reported that the 3 MHz radiation during the initial breakdown is comparable to that during the

4.3. Initial breakdown   121

return stroke for both negative and positive lightning in Sweden.

4.3.3. Lightning initiation in thunderclouds

As discussed in subsection 3.2.4, it appears that the maximum values of the large-scale electric fields generated in thunderclouds are of insufficient magnitude to explain the initiation of lightning. This apparent discrepancy has led some researchers to speculate that lightning is initiated via the emission of positive corona from the surface of precipitation particles, highly deformed by strong electric fields in the case of raindrops, coupled with some mechanism whereby the electric field is locally enhanced to support the propagation of corona streamers. Positive streamers are much more likely to initiate lightning than negative ones because they can propagate in substantially lower fields. The most detailed hypothetical scenario of lightning initiation was described by Griffiths and Phelps (1976b), who considered a system of positive streamers developing from a point on a hydrometeor (defined at the start of subsection 3.2.1) where the electric field exceeds the corona onset value of 2.5 to 9.5×10^5 V m^{-1} (2.5 to 9.5 kV cm^{-1}) (subsection 3.2.4). The developing streamers are assumed to form a conical volume that grows longitudinally. The ambient electric field in the thundercloud required to support the propagation of corona streamers, E_0, was found by Griffiths and Phelps (1976a) from laboratory experiments to be 1.5×10^5 V m^{-1} (1.5 kV cm^{-1}) at about 6.5 km and 2.5×10^5 V m^{-1} (2.5 kV cm^{-1}) at about 3.5 km. If the ambient electric field is higher than E_0 then the streamer system will intensify, carrying an increasing amount of positive charge on the propagating base of the cone, which simulates the positive streamer tips, and depositing an equally increasing amount of negative charge in the conical volume that represents the trails of the positive streamers. As a result, an asymmetric conical dipole is formed, which presumably serves to enhance the existing electric field at the cone apex representing the origin of the positive streamers on the surface of the hydrometeor.

Further, Griffiths and Phelps (1976b) suggested that several conical streamer systems may develop sequentially, each one passing into the debris of its predecessors, to achieve the field enhancement required for breakdown. For representative values of ambient electric field and E_0 at 6.5-km altitude, they report that a series of three to seven such systems can give rise to local enhancement of the ambient electric field up to a value of 1.5×10^6 V m^{-1} (15 kV cm^{-1}) over a distance of a few meters, which is sufficient to ensure dielectric breakdown and perhaps, eventually, to lead to formation of the stepped leader. The number of passages required to achieve a certain field value depends on the assumed value of the potential of the streamer tip. Griffiths and Phelps (1976b) used a value of 10 kV, based on the results of laboratory experiments reported by Phelps (1974a), for most of their calculations. These calculations involved an extrapolation from the relatively small (up to 1 m) gaps used in laboratory experiments (Griffiths and Phelps 1976a) to the relatively large distances (of the order of 100 m) over which streamers might travel in a thundercloud.

It is worth noting that the hypothetical mechanism proposed by Griffiths and Phelps (1976b) might describe the preparation stage for the initial breakdown rather than the initial breakdown itself. Griffiths and Phelps (1976a) noted that the majority of the hypothesized streamer systems must fail to provide the degree of field intensification required for the formation of thermally ionized discharge channels (lightning). These abortive events serve to increase the conductivity and leakage currents, thereby dissipating the existing electric field. Griffiths and Phelps (1976a) hypothesized that such events might be the source of radio noise emissions from developing thunderclouds observed by Zonge and Evans (1966) and Harvey and Lewis (1973). Latham and Stromberg (1977) suggested that if the region of intense electric field were highly localized then corona currents could greatly increase the conductivity in that region and possibly inhibit the onset of lightning. If, however, the intense field existed over a large volume of the cloud then corona discharge would tend to lead to a lightning discharge.

Nguyen and Michnowski (1996b) considered the effects of many closely spaced hydrometeors in lightning initiation. Their hypothetical mechanism involves a bidirectional streamer development assisted by a chain of precipitation particles, as opposed to the generally accepted scenario that invokes the propagation of positive streamers alone.

Gurevich *et al.* (1997) suggested that runaway electrons may play an important role in lightning initiation. In order to "run away", an electron must gain more energy from the electric field between collisions with air particles than it loses in a collision. The so-called breakeven electric field, which must be exceeded for runaway to occur, depends on altitude (air particle density) and initial electron energy (Section 14.5). According to Suszcynsky *et al.* (1996), for 200 keV electrons the breakeven field at an altitude of 10 km is 10^5 V m^{-1} (1 kV cm^{-1}), which is about an order of magnitude lower than the breakdown field at this altitude (subsection 3.2.4). It has been further suggested that the runaway breakdown mechanism might operate during the stepped leader process, each step representing a runaway-dominated discharge that produces bursts of X-rays. Moore *et al.* (2001) reported bursts of radiation with energies in excess of 1 MeV that began between 1 and 2 ms before, and

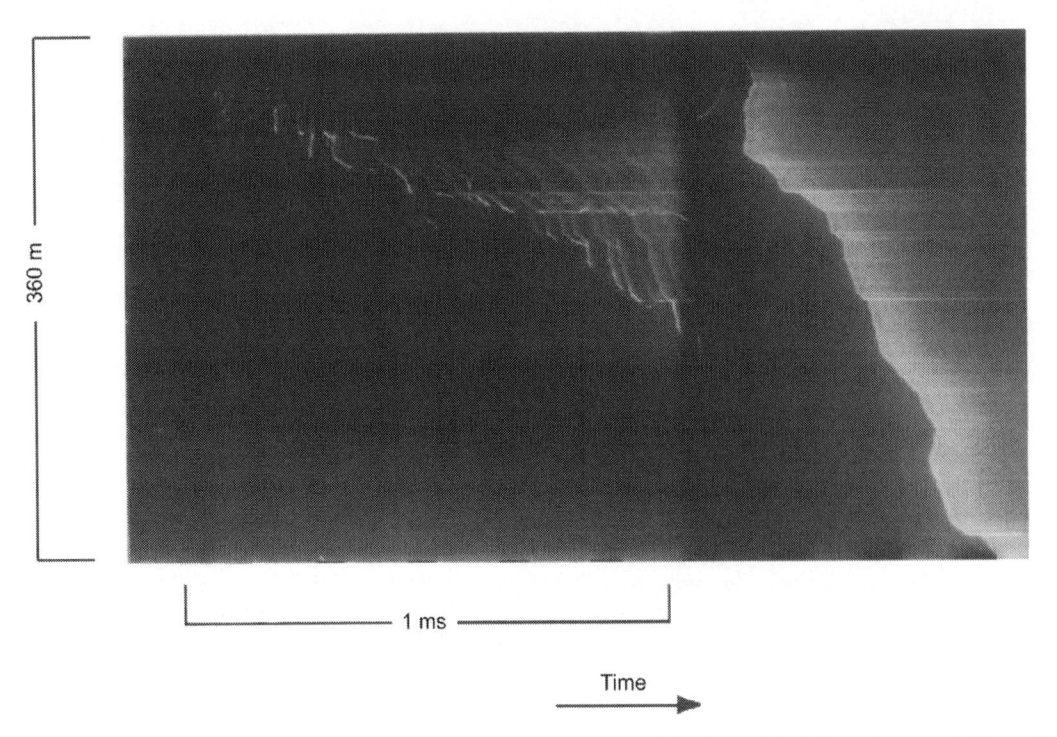

Fig. 4.13. Streak-camera photograph of a downward negative stepped leader within 360 m of ground. Time advances from left to right. The left part of the photograph was overexposed in the reproduction process in order to enhance the intensity of the early portion of the leader image. Adapted from Berger and Vogelsanger (1966).

continued until, the onset of the first return strokes of three negative ground flashes in New Mexico.

4.4. Stepped leader
4.4.1. General information

A streak photograph of a negative downward stepped leader is shown in Fig. 4.13. It is worth noting that in the studies resulting in the bulk of the available information on stepped leaders, leader steps were difficult to photograph and could be reproduced well only by intensifying the negative (e.g., Schonland 1956). In early lightning studies (e.g., Schonland 1938; Schonland et al. 1938a, b), the variety of observed stepped-leader forms was organized in two large categories: α-type leaders and β-type leaders. The majority, 55 percent (Schonland 1956) to 70 percent (Schonland 1938), of photographed leaders were of the α-type. These are characterized by a uniform downward speed of the order of 10^5 m s^{-1} and steps that do not vary appreciably in length or brightness and are shorter and much less luminous than for β-type leaders. Further, β-leaders appear to have two stages in their development outside the cloud. They begin beneath the cloud base, or occasionally emerge from the side of the cloud, with relatively long, bright steps and high average speed, of the order of 10^6 m s^{-1}, and exhibit extensive branching near the cloud base. As a β-leader approaches the ground, it assumes the characteristics of α-leaders, that is, it exhibits a lower propagation speed and shorter and less luminous steps. Both α- and β-leaders exhibit an increased average speed and increased step brightness as they approach the ground.

Schonland et al. (1938b) further considered two subgroups, labeled β_1 and β_2, within the β-type leaders. β_1-leaders are defined as having an abrupt discontinuity in downward speed at some point in their trip from the cloud to ground. The second stage occurs within the bottom kilometer or so of the channel and is similar to a normal α-leader. It appears that the β_1-leader is the same as the previously defined β-leader. The β_2-leader is similar to the β_1-leader except that during the second, α-leader, stage the channel is traversed by one or more luminosity waves continuously propagating from the cloud to the stepped-leader tip. These waves are separated by time intervals of order 10 ms, and each resembles the dart leader to be discussed in Section 4.7. As a much faster moving and much more luminous "dart" catches up with the downward-extending bottom end (tip) of the stepped leader, profuse outward branching occurs momentarily at the tip.

β_2-leaders are probably rare: only four cases were analyzed by Schonland et al. (1938b), who also analyzed

4.4. Stepped leader

10 β_1-type leaders, and one case was documented by Workman *et al.* (1936); apparently none was observed in the high-speed photographic studies of Berger and Vogelsanger (1966), Orville and Idone (1982), and Jordan (1990). It is quite possible that stepped leaders classified as α-leaders are actually similar to β-leaders whose more branched, faster, and brighter upper part is hidden inside the cloud (e.g., Beasley *et al.* 1982).

4.4.2. Speed and duration

Schonland (1956) reported that average two-dimensional speeds for 60 of the most common α-type stepped leaders in South Africa were between 0.8×10^5 and 8×10^5 m s^{-1} in the bottom 2 to 3 km of the channel. About 82 percent of the measured speed values were between 0.8×10^5 and 4×10^5 m s^{-1} and 45 percent were between 0.8×10^5 and 2×10^5 m s^{-1}. Schonland (1956) also estimated the minimum three-dimensional speed value to be about 1×10^5 m s^{-1}. Berger and Vogelsanger (1966) found that average vertical speeds for 14 stepped leaders (apparently of type α) terminating on ground in Switzerland ranged from 0.9×10^5 to 4.4×10^5 m s^{-1} in the final 1.3 km above ground. Additionally, Berger and Vogelsanger (1966) reported average vertical speeds for four stepped leaders terminating on their towers (Section 6.3) ranging from 1.9×10^5 to 2.2×10^5 m s^{-1} over the bottom 100 m of the channel. For two stepped leaders in Florida, Orville and Idone (1982) found average two-dimensional speeds over the last few steps to be 15×10^5, 5.9×10^5, and 11×10^5 m s^{-1} over 161, 188, and 45 m, respectively. Chen *et al.* (1999), who used the ALPS optical imaging system (Yokoyama *et al.* 1990), reported on two downward negative stepped leaders, one of which exhibited a propagation speed ranging from 4.5×10^5 to 11×10^5 m s^{-1} (mean value 7.3×10^5 m s^{-1}) between the heights of 367 and 1620 m above ground, and the other from 4.9×10^5 to 5.8×10^5 m s^{-1} between 33 and 102 m. The former stepped leader was observed in Australia and the latter in China. Thomson *et al.* (1985), from 10 stepped-leader electric field waveforms recorded in Florida, inferred a range of two-dimensional speeds between 1.3 and 19×10^5 m s^{-1} with a mean of 5.7×10^5 m s^{-1} for the bottom 0.6 to 2.0 km of the channel. They assumed that the two-dimensional speed was 30 percent higher than the one-dimensional speed determined from their leader model.

Beasley *et al.* (1983) estimated that stepped-leader speeds within 100 m of ground (three cases in Florida) were between 0.8×10^6 and 3.9×10^6 m s^{-1}. Isikawa (1961, Table 15) found, from photographic measurements in Japan, a median value 3.1×10^5 m s^{-1} for 13 leader trunks (main channels) and 3.2×10^5 m s^{-1} for 14 leader branches. Proctor *et al.* (1988, Table 1), using a VHF TOA lightning-channel imaging system (subsection 17.4.2) in South Africa, measured channel extension speeds for 66 stepped leaders ranging from 3×10^4 to 4.2×10^5 m s^{-1} with median value 1.3×10^5 m s^{-1}. Shao *et al.* (1995), using a VHF interferometer (Section 17.6) to image lightning channels in Florida, found that initial leaders to ground progressed at a typical speed 2×10^5 m s^{-1}. Nagai *et al.* (1982), using streak photography, observed that a stepped leader in a summer single-stroke flash in Japan gradually accelerated during its trip from the cloud base to ground, the speed change being from 1.7×10^5 to 4.7×10^5 m s^{-1} over about 1000 m.

The typical duration of the stepped leader estimated from measured stepped-leader electric field waveforms is some tens of milliseconds, as seen in Table 4.2. Rakov and Uman (1990c) reported distributions of leader duration for strokes of various orders in Florida, reproduced in Fig. 4.14. They found a geometric-mean leader duration of 35 ms for stepped leaders. Using this latter value of leader duration and assuming an average stepped-leader speed of 2×10^5 m s^{-1} (see above), one can estimate a channel length of 7 km, a reasonable estimate for Florida lightning.

4.4.3. Electrical characteristics

We will now discuss the overall electrical characteristics of the stepped leader such as total charge, charge per unit channel length (line charge density), and average current. According to Berger *et al.* (1975), the median value of the so-called impulse charge lowered to ground by negative first return strokes (arbitrarily defined so as to exclude charge associated with continuing current) is 4.5 C. The corresponding 95 percent and five percent values, that is, values which are exceeded with probabilities of 0.95 and 0.05, respectively, are 1.1 and 20 C. Although the charge neutralized by the return stroke is not necessarily equal to the charge stored on the leader channel, the above values also represent, certainly within an order of magnitude, the stepped-leader charge. If the total leader charge is 5 C and the channel length is 7 km then the average charge per unit length is about 7×10^{-4} C m^{-1} (0.7 mC m^{-1}). Additionally, stepped-leader charge can be estimated from leader electric field measurements if a reasonable distribution of charge on the leader channel is assumed. We will show later that this distribution within some hundreds of meters of the ground is likely to be more or less uniform. Brook *et al.* (1962) found from single-station electric field measurements in New Mexico, using a point-charge approximation, that the total charge neutralized by 24 first strokes ranged from 3 to 20 C with a geometric mean value of 6 C, consistent with the charges neutralized by first return strokes reported by Berger *et al.* (1975). First strokes

Table 4.2. *Median duration of stepped leaders from electric field measurements. In some cases the geometric mean was used as an estimate of the median value. Adapted from Rakov and Uman (1990c)*

Reference	Geographical location	Sample size	Median leader duration, ms	Comments
Schonland et al. (1938a)	South Africa	72	9–12	As derived from the histogram in Fig. 9A of Schonland et al. (1938a)
Malan and Schonland (1951b)	South Africa	20	8	Geometric mean
Pierce (1955)	England	332	43	Median values of the durations for $L(\alpha)$- and $L(\beta)$-type variations were combined, assuming those durations to be log-normally distributed
Clarence and Malan (1957)	South Africa	234	30	As derived from the cumulative distribution A in Fig. 5 of Clarence and Malan (1957)
Kitagawa (1957a)	Japan	41	26	Geometric mean
Kitagawa and Brook (1960)	New Mexico	282	~40	As derived from the histogram in Fig. 11 of Kitagawa and Brook (1960)
Isikawa (1961)	Japan	185	~35	As derived from the histogram in Fig. 4b of Isikawa (1961)
Krehbiel et al. (1979)	New Mexico	4	36	Geometric mean
Thomson (1980)	Papua New Guinea	53	~21	As derived from the cumulative distribution in Fig. 10 of Thomson (1980)
Beasley et al. (1982)	Florida	79	18	
Richard et al. (1986)	Ivory Coast	3	23	Geometric mean
Proctor et al. (1988)	South Africa	About 60	~46	Estimated from the median values of stepped-leader speed and stepped-leader length (Table 1 of Proctor et al. 1988)

in multiple-stroke flashes are apparently characterized by larger charge transfer than the solitary strokes of single-stroke flashes: 7.2 C versus 4.6 C in the data of Brook et al. (1962). Krehbiel et al. (1979), from multiple-station electric field measurements, estimated charges for four first strokes ranging from approximately 5 to 20 C.

Proctor et al. (1988) estimated the total charge deposited on the VHF-imaged channels of 15 stepped leaders in South Africa. It was assumed that each located VHF source was associated with the deposition of an incremental charge which was instantaneously transported to the source from a fixed point near the flash origin. The magnitude of the incremental charge depended on the number of VHF sources located and was chosen such that it equalled the final values of the calculated and measured leader electric field changes. The total stepped-leader charges, found in each case as the sum of all the incremental charges, ranged from 3.6 to 57 C with a median value of 11 C (for 15 events). Eight of the 15 stepped-leader charges were between 5 and 15 C. The leader charge was proportional to the channel length, which ranged from 3 to 13 km with a median value of 6 km (for 60 events). Proctor (1997) reported that the line charge density for stepped leaders in so-called lower-origin flashes, including both cloud and cloud-to-ground discharges that

4.4. Stepped leader

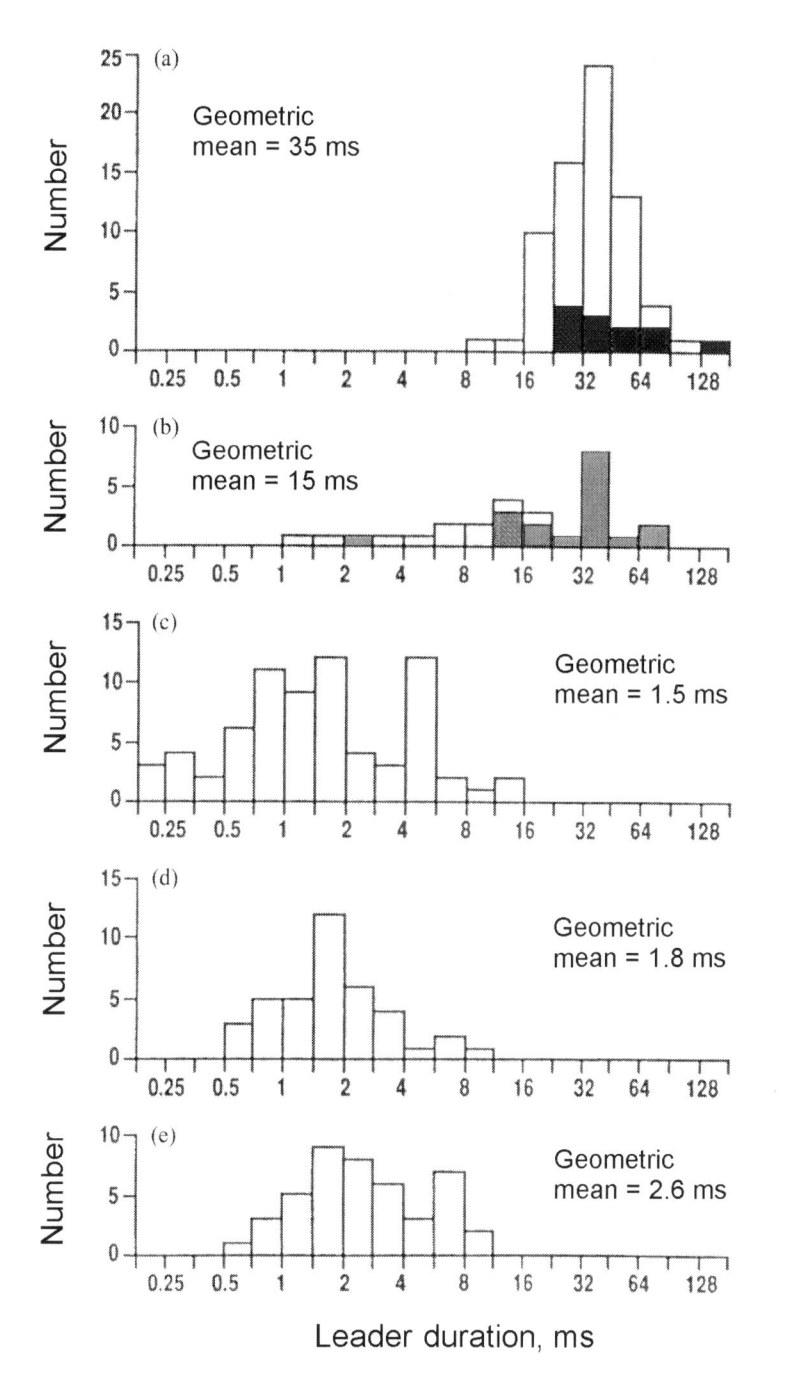

Fig. 4.14. Histograms of leader duration for five sets of strokes: (a) 71 first strokes; (b) 28 second, third, and fourth strokes creating a new termination on ground; (c) 71 second, third, and fourth strokes following the same channel as the preceding stroke; (d) 39 fifth, sixth, and seventh strokes, and (e) 44 strokes of order 8 through 18. Shown shaded in Fig. 4.14a are the data for single-stroke flashes, while in Fig. 4.14b the shading indicates a new termination on ground involving a path between cloud base and ground that is completely separate from that of the preceding stroke as opposed to having a common portion of the channel seen beneath the cloud. Adapted from Rakov and Uman (1990c).

began at 1 to 7.4 km above mean sea level, was about 10^{-3} C m^{-1} (1 mC m^{-1}) (Table 9.2).

An average current flow of 143 A would be required for a charge of 5 C to be lowered in 35 ms (Fig. 4.14a). Williams and Brook (1963), using remote magnetic field measurements, roughly estimated the average stepped-leader currents in two flashes to be 50 A and 63 A. Thomson et al. (1985), using 62 stepped-leader electric field

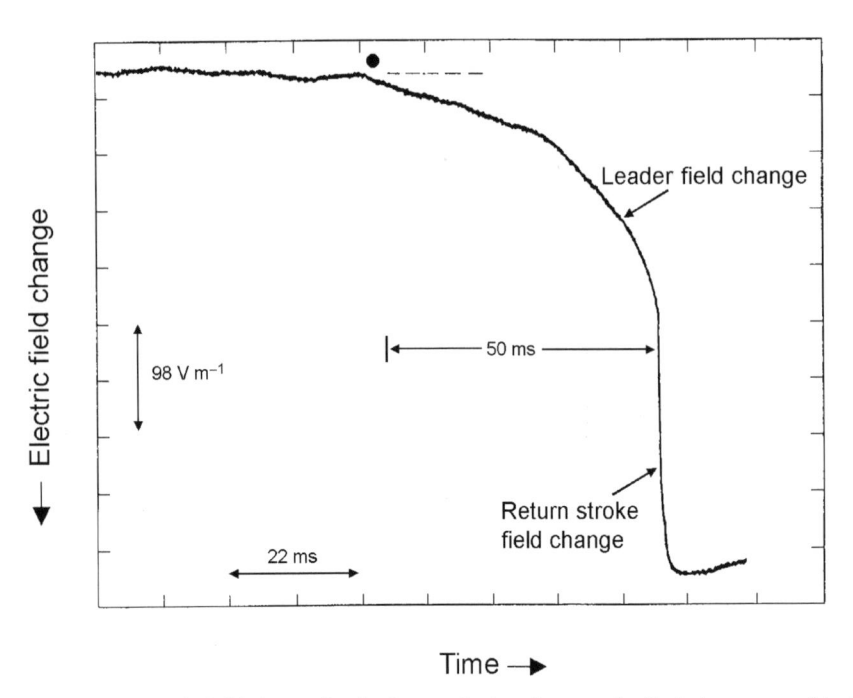

Fig. 4.15. Electric field change for the first stroke in a three-stroke flash that occurred in 1979 in Tampa, FL at a distance of about 12 km. A small solid circle marks the starting point of the leader field change. The time interval labeled 50 ms shows the leader field change duration. Electric field changes due to the leader and due to the return stoke are labeled. The atmospheric electricity sign convention (subsection 1.4.2) is used here. The leader field change is monotonic positive. Adapted from Rakov and Uman (1990c).

waveforms recorded in Florida, inferred stepped-leader currents near ground ranging from 100 A to 5 kA with a mean of 1.3 kA. They also computed the line charge density for 10 leaders near ground as the ratio of the leader current to the final one-dimensional leader speed (see above). The charge density varied from 0.7 to 32×10^{-3} C m^{-1} with geometric mean 3.4×10^{-3} C m^{-1}. Krehbiel (1981) found, from multiple-station electric field measurements in Florida, that final stepped-leader currents averaged over milliseconds were in the range 200 A to 3.8 kA with mean 1.3 kA.

4.4.4. Overall electric and magnetic fields

Examples of the electric field changes produced by stepped leaders together with the corresponding return-stroke field changes as a function of time are given in Figs. 4.15 and 4.16 (left panel, the Delta E waveforms), and the net leader field change as a function of distance is shown in Fig. 4.17. Fig. 4.16 also shows a VHF image of the stepped-leader channel (right panel, top image labeled L1).

We will discuss next, with reference to Fig. 4.18, equations that can be used to describe the stepped-leader electric and magnetic fields. These field equations are also applicable to the subsequent leaders discussed in Section 4.7.

The leader is assumed to create a channel extending vertically downward with constant speed v from a stationary and spherically symmetrical charge source at height H_m (see Fig. 4.18 and the description of this simple leader model in Section 12.3). Ground is assumed to be a perfect conductor.

Here, we will consider only the electrostatic approximation for the leader electric field change and the magnetostatic approximation for the leader magnetic field, which are valid when the significant wavelengths of the electric and magnetic fields are much larger than the dimensions of the overall system of the lightning and the observer. These approximations are expected to be applicable to the overall fields produced by close lightning leaders. The total leader electric field comprises electrostatic, induction, and radiation field components and the total magnetic field comprises magnetostatic and radiation field components.

Fig. 4.16 (opposite). Electric field (Delta E and Fast delta E), logarithmic RF radiation amplitude (log RF), and radiation sources mapped in elevation–azimuth coordinates for a stepped leader (L1), three "attempted" leaders (AL1, AL2, and AL3), and a subsequent leader creating a new termination on ground (L2), which occurred during the first 120 ms of a multiple-stroke cloud-to-ground Florida flash. All leaders started at the same location. Adapted from Shao et al. (1995).

4.4. Stepped leader

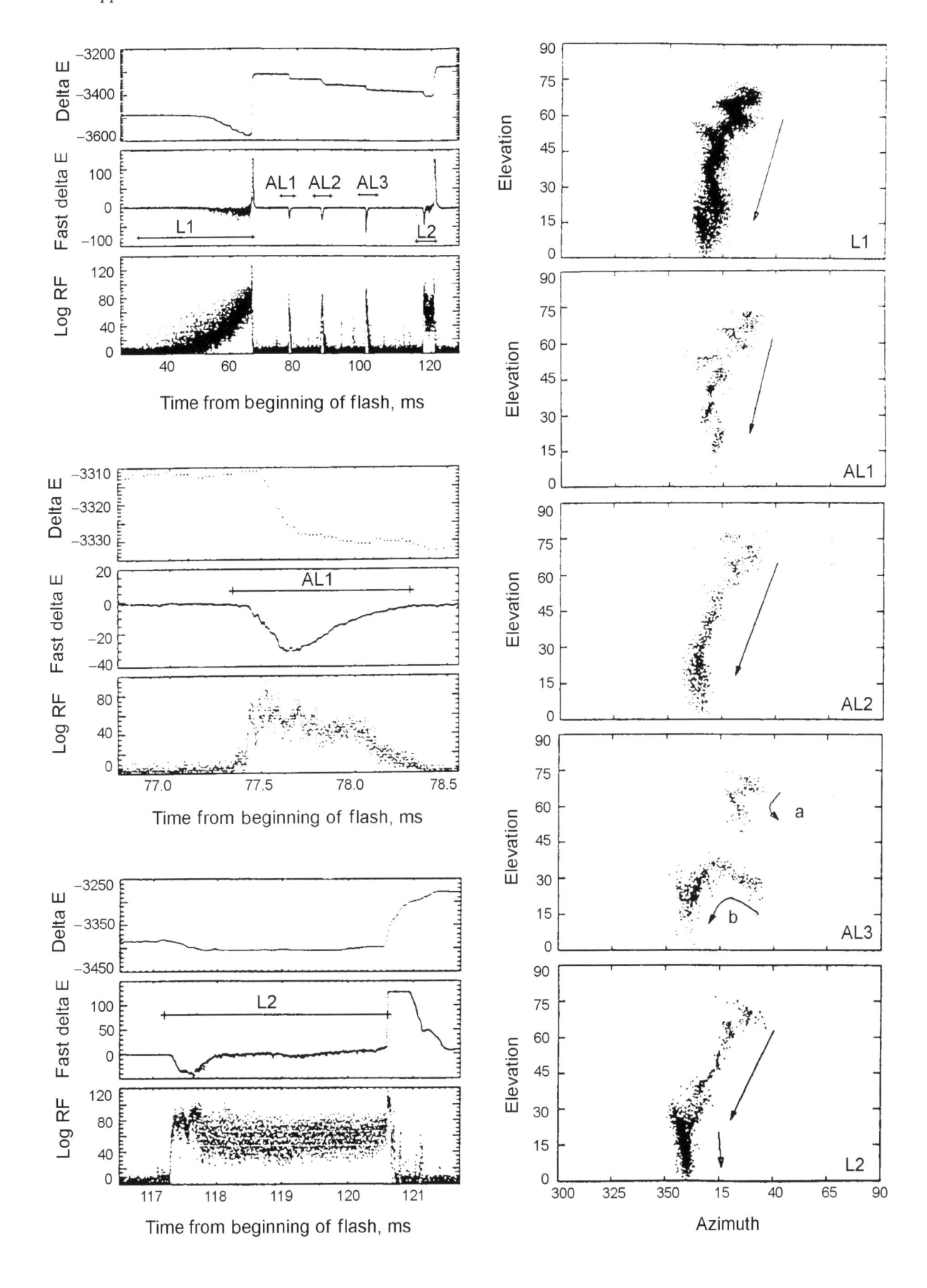

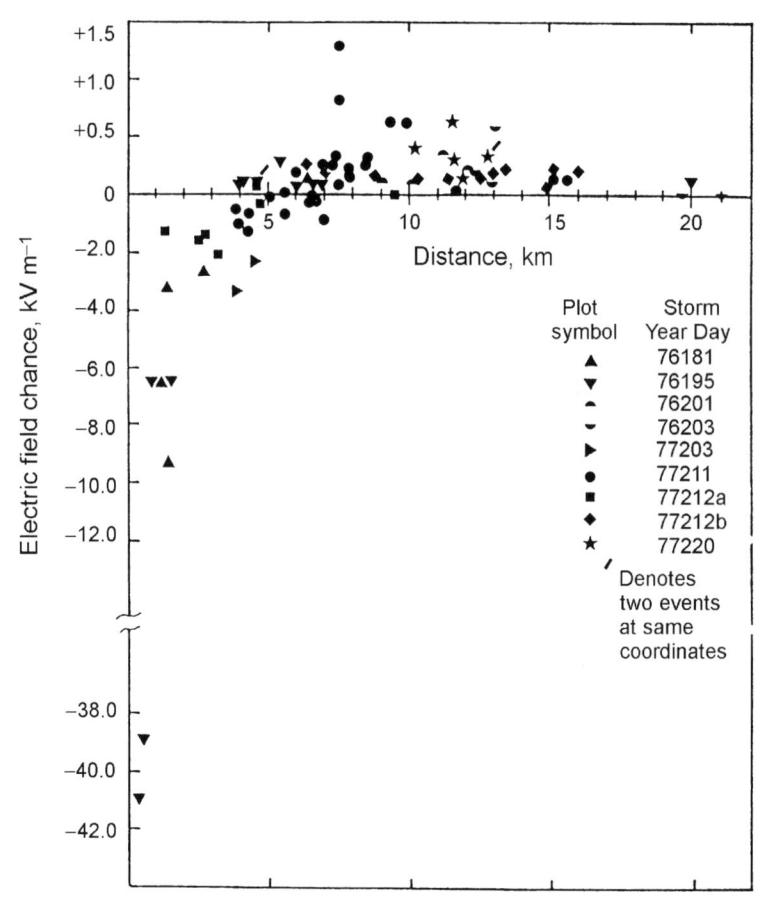

Fig. 4.17. Stepped-leader net electric field change as a function of distance for 80 leaders in nine Florida storms. Adapted from Beasley et al. (1982).

Expressions for the electric and magnetic fields are found in Thottappillil et al. (1997, Eqs. B38 and B39, respectively), where detailed derivations of these equations are presented.

We will start with the general expression for the electrostatic field given by Thottappillil et al. (1997),

$$E_z(r,t) = \frac{1}{2\pi\varepsilon_0} \int_{h(t)}^{H_m} \frac{z'}{R^3(z')} \rho_L\left(z', t - \frac{R(z')}{c}\right) dz' \\ - \frac{1}{2\pi\varepsilon_0} \frac{H_m}{R^3(H_m)} \int_{h(t)}^{H_m} \rho_L\left(z', t - \frac{R(z')}{c}\right) dz' \quad (4.1)$$

where $R(H_m) = (H_m^2 + r^2)^{1/2}$ (see Fig. 4.18) and $h(t)$ is the height at which the observer "sees" the lower end of the leader channel; $h(t)$ is given by the solution of

$$t = \frac{H_m - h(t)}{v} + \frac{\sqrt{h^2(t) + r^2}}{c} \quad (4.2)$$

The first term of Eq. 4.1 represents the field change due to the charge on the downward-extending leader channel, and the second term represents the field change due to the depletion of the charge at the cloud charge source as it is "drained" by the extending leader channel. The total charge on the leader channel at any time is equal to the total charge removed from the cloud charge source, so that the net charge on the overall channel–source system is zero at all times.

We now assume that the maximum difference in propagation time from any source on the channel to the observer is much less than the time required for significant variation in the sources (i.e., retardation effects (e.g., Sadiku 1994) are negligible), and rewrite Eq. 4.1 as

$$E_z(r,t) = \frac{-1}{2\pi\varepsilon_0} \int_{H_m}^{z_t} \left[\frac{z'}{R^3(z')} - \frac{H_m}{R^3(H_m)}\right] \rho_L(z', t) \, dz' \quad (4.3)$$

where $z_t = H_m - vt$ is the height of the leader tip at time t and v is the leader speed, assumed to be constant. Thomson derived a similar equation (Eq. 5 of Thomson 1985) based on Coulomb's law, although he assumed that the charge density distribution on the leader channel behind the leader tip does not vary with time. Thomson's assumption is equivalent to the assumption of a uniform

4.4. Stepped leader

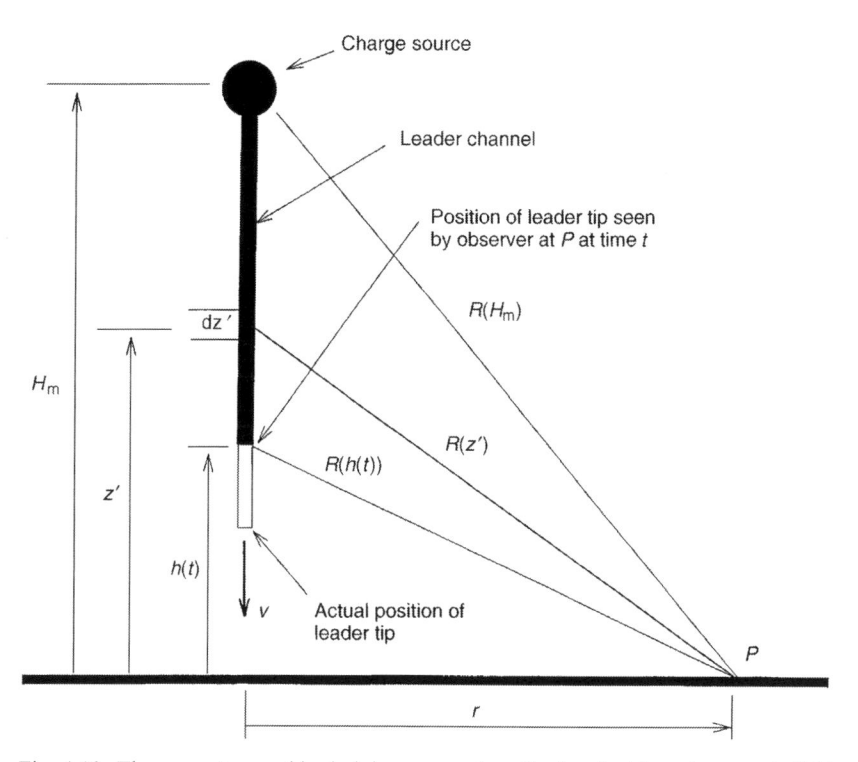

Fig. 4.18. The geometry used in deriving expressions for the electric and magnetic fields at a point P on Earth a horizontal distance r from the vertical lightning leader channel extending downward with speed v. Adapted from Thottappillil et al. (1997).

current in the channel between the cloud charge source and leader tip, as follows from the current continuity equation. In this respect, Thomson's leader model is similar to the BG model for return strokes (subsection 12.2.5), while Eq. 4.3 permits any other leader model (e.g., a TCS type; subsection 12.2.5).

If we assume that $\rho_L(z', t) = \rho_L = $ constant, which corresponds to a uniformly charged leader channel, we can rewrite Eq. 4.3 in the following form:

$$E_z(r, t) = \frac{\rho_L}{2\pi\varepsilon_0 r} \left[\frac{1}{\left(1 + z_t^2/r^2\right)^{1/2}} - \frac{1}{\left(1 + H_m^2/r^2\right)^{1/2}} \right.$$

$$\left. - \frac{(H_m - z_t)H_m}{r^2 \left(1 + H_m^2/r^2\right)^{3/2}} \right] \quad (4.4)$$

The leader electric field changes computed from Eq. 4.4 are shown in Fig. 4.19 as a function of normalized time t/T, where $T = H_m/v$, for different normalized distances r/H_m from the channel. If, for example, $H_m = 5$ km then the distance range represented in Fig. 4.19 is from 3.5 to 7 km. Leader field changes calculated by Rubinstein et al. (1995) at $r = 30$ m and $r = 500$ m are shown in Figs. 4.20a, b. For a very close field point, where $H_m \gg 2r$, and for either $z_t = 0$ (leader touching the ground) or $z_t^2 \ll r^2$ and $z_t \ll H_m$ (leader close to the ground), Eq. 4.4 becomes approximately (Rubinstein et al. 1995)

$$E_z(z_t = 0) \approx \frac{\rho_L}{2\pi\varepsilon_0 r} \quad (4.5)$$

That is, very close to the channel the vertical electrostatic field change at ground due to the fully developed leader channel falls off with distance as r^{-1}, as opposed to the r^{-3} variation far from the channel ($H_m^2 \ll r^2$). Interestingly, Eq. 4.5 is exactly the same expression as that for the radial field produced by an infinitely long, uniform line charge in free space.

The magnetostatic approximation for the leader magnetic fields is usually written in terms of the current $I(t)$, which is assumed to be slowly varying and the same at all heights along the vertical lightning channel (e.g., Uman 1987, 2001). A less familiar formulation, in terms of the charge density ρ_L, for the leader magnetic field was given by Thottappillil et al. (1997). The magnetostatic field of a vertical leader channel, the top end of which is at height H_m and the bottom end at height $z_t = h(t)$ (Fig. 4.18), is given by

$$B_\phi(r, t) = \frac{\mu_0}{2\pi r} \left[\frac{H_m}{R(H_m)} - \frac{z_t}{R(z_t)} \right] I(t) \quad (4.6)$$

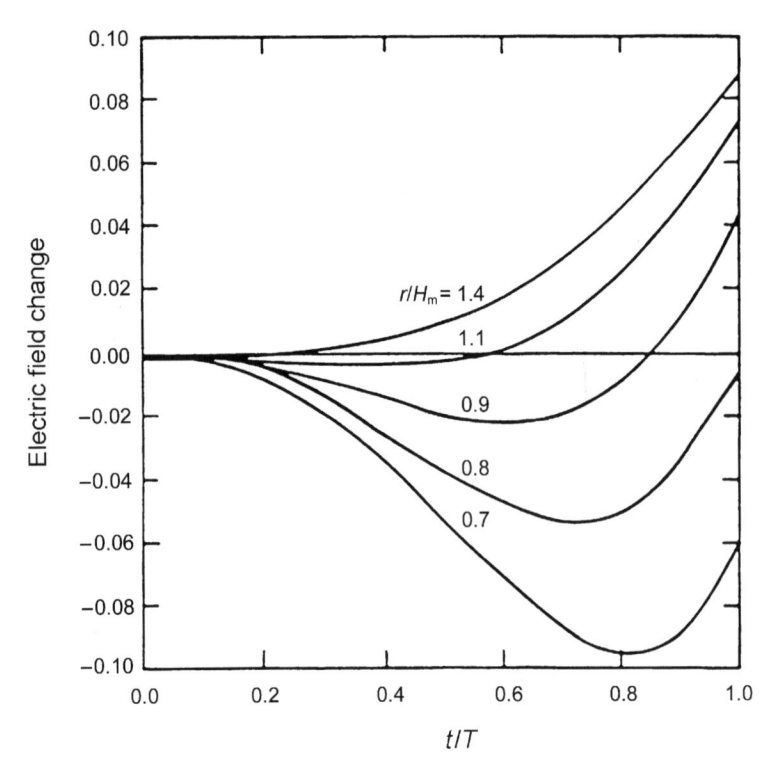

Fig. 4.19. Waveshapes of leader electric field change at different distances r in the kilometer range. H_m is the height of the cloud charge source, t is the time variable, and $T = H_m/v$ is the time required for a leader propagating at a constant speed v to reach ground. In order to obtain V m^{-1}, multiply the ordinate by $\rho_L(2\pi\varepsilon_0 r)^{-1}$, where line charge density ρ_L is in C m^{-1}, $\varepsilon_0 = 8.85 \times 10^{-12}$ F m^{-1}, and r is in meters.

For a fully developed leader channel, that is, for $z_t = 0$,

$$B_\phi(r,t) = \frac{\mu_0}{2\pi r}\left[\frac{H_m}{R(H_m)}\right]I(t) \quad (4.7)$$

Equation 4.7 is the expression for the magnetostatic field of a vertical current-carrying line, the bottom end of which is at ground (a perfect conductor) and the top end at height H_m. If the observation point is very close to the channel base, so that $r \ll H_m$, Eq. 4.7 can be further simplified to give

$$B_\phi(r,t) = \frac{\mu_0 I(t)}{2\pi r} \quad (4.8)$$

which is the same equation as that for an infinitely long current-carrying line in free space (e.g., Sadiku 1994).

4.4.5. Ratio of leader to return-stroke electric field changes

The ratio of the net leader to return-stroke electric field changes has been studied by Schonland et al. (1938a), Beasley et al. (1982), and Rakov et al. (1990). The results of the latter two studies, including both the experimental data and some model-based calculations, are shown in Figs. 4.21 and 4.22. The field ratio is usually computed (i) assuming that the return stroke removes all the leader charge and does not deposit any additional charge either on the channel or in the cloud and (ii) using a simple, uniformly charged leader model, which is illustrated in Fig. 4.18 and discussed in Section 12.3. With these assumptions, the leader electric field change is given by Eq. 4.4, and the return-stroke electric field change is given by the negative of the first two terms of Eq. 4.4. These two terms represent the contribution to the electric field from sources on the channel, while the third term accounts for the removal of that charge from the source (or, what is equivalent, the deposition of a charge of equal magnitude and opposite polarity at the source). The field ratio computed in this manner is shown as a function of H_m/r in Fig. 4.23. At relatively large ranges ($H_m/r < 1.27$), the leader and return-stroke field changes have the same polarity, so that the ratio is positive and approaches unity at distances $r \gg H_m$. The latter result can be visualized as being due to the following two-stage process. In the first (leader) stage, a point charge equivalent to the total leader charge moves from its original position in the cloud to a lower position, which, for the assumed uniform leader charge distribution, is exactly half-way between this original position and ground. In the second (return-stroke) stage, the equivalent charge moves from the intermediate

4.4. Stepped leader

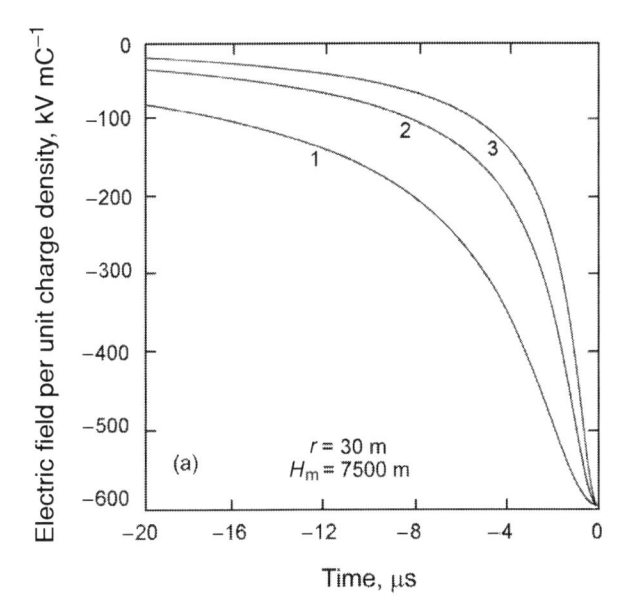

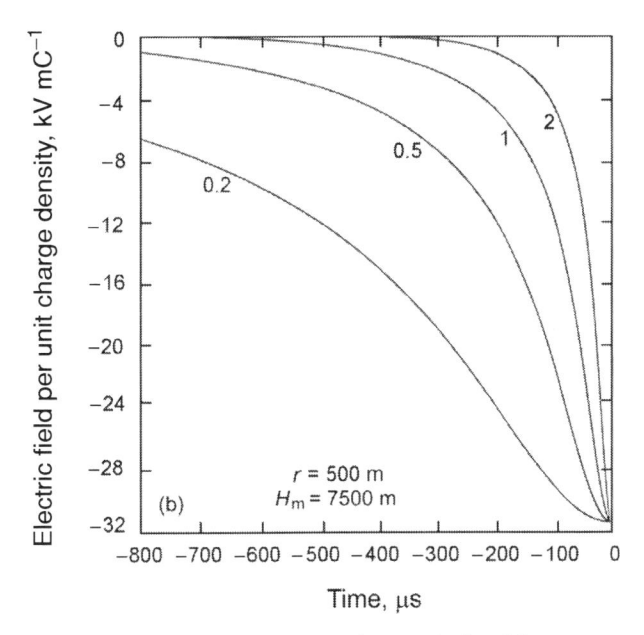

Fig. 4.20. Leader electric field waveforms calculated for (a) $r = 30$ m and (b) $r = 500$ m using a uniformly charged vertical-line leader model with constant speed. The height of the charge source is assumed to be $H_m = 7.5$ km, although the close fields are relatively insensitive to the value of H_m. The numbers on the curves give, in units of 10^7 m s^{-1}, the values of the speed v used in the calculations. At $t = 0$ the leader attaches to the ground. Adapted from Rubinstein et al. (1995).

position to the ground. At far ranges ($r \gg H_m$), these two stages of point-charge lowering produce equal net electric field changes. At relatively close ranges ($H_m/r > 1.27$), the leader and return-stroke field changes have opposite polarities. The ratio is negative and approaches -1 at distances $r \ll H_m$. The latter result is due to the fact that the third,

cloud-source term in Eq. 4.4 makes a negligible contribution to the leader electric field change at close ranges compared to that of the first two terms, representing the channel sources.

4.4.6. Leader steps

The lengths of the individual steps in stepped leaders were reported by Schonland (1956) to vary from 10 to 200 m and the interstep intervals to range from 40 to 100 μs; both the step lengths and their brightness increase as the leader speed increases. According to Schonland (1956), the tortuous path of the negative lightning channel is caused by quasi-random changes in the direction of successive steps. Branching occurs by the division of a step into a fork. Berger and Vogelsanger (1966) reported step lengths between 3 and 50 m and interstep intervals between 29 and 52 μs. According to Schonland (1956), the lowest steps showed increased lengths and reduced interstep intervals as the ground was approached. As noted in subsection 4.4.2, Chen et al. (1999) observed two downward negative stepped leaders using the ALPS optical imaging system (Yokoyama et al. 1990). For one of these leaders, imaged in Australia, the step length ranged from 7.9 to 20 m between the heights of 367 and 1620 m, and the interstep interval ranged from 5 to 50 μs. For the other leader, imaged in China, the step length was about 8.5 m between 33 and 102 m, and the interstep interval ranged from 18 to 21 μs. The measurements by Chen et al. (1999) were characterized by a spatial and time resolution of 108–35 m and 0.5 μs in Australia and 17 m and 0.1 μs in China.

Beasley et al. (1983) were apparently the first to demonstrate the correspondence between the light pulses and the electric field pulses produced by individual leader steps. They examined leader light pulses from approximately 12-m-long channel sections about 80 m above ground and electric field pulses occurring 50 to 100 μs immediately prior to the first return-stroke pulse. Both light and electric field pulses (in three Florida flashes) were separated by time intervals of 5 to 20 μs, although not every pulse in the electric field record after the first light pulse had a corresponding pulse in the light record. Thus, electric field pulses commonly observed (e.g., Kitagawa 1957a; Krider and Radda 1975; Krider et al. 1977; Cooray and Lundquist 1985) within some hundreds of microseconds of the first return stroke are likely to be manifestations of the luminous-step-formation process near ground. Examples of these electric field pulses are shown in Fig. 4.24.

Before characterizing stepped-leader pulses, it is worth reviewing the general pulse structure typically observed prior to the return stroke, illustrated in Fig. 4.11. The initial, larger, pulse train, generally attributed to the initial breakdown, has a duration of a few milliseconds and is followed, after several milliseconds to tens of milliseconds

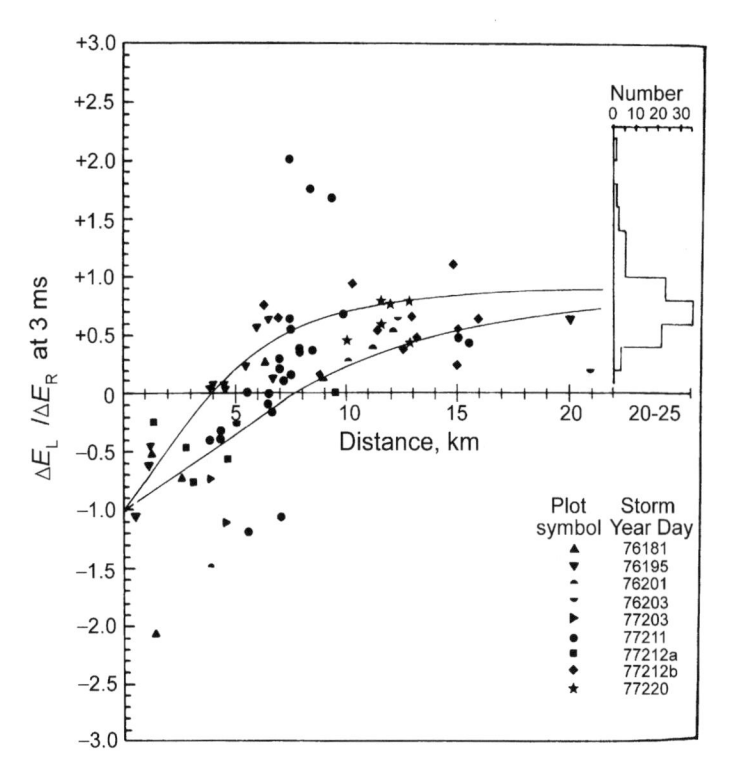

Fig. 4.21. Ratio of the stepped-leader to return-stroke electric field changes for the events whose net leader electric field changes are shown in Fig. 4.17. The histogram at the upper right-hand side shows additional data for the distance range from 20 to 50 km. The upper and lower curves represent the model-predicted ratio for charge source heights $H_m = 5$ km and $H_m = 10$ km, respectively. Adapted from Beasley et al. (1982).

of relatively low and irregular pulse activity, by another pulse train. The latter pulse train has a duration of some tens to some hundreds of microseconds. The characteristics of the two pulse trains appear to be distinctly different (Rakov et al. 1996, Table 3), although some researchers (e.g., Kitagawa 1957a; Kitagawa and Brook 1960) presented combined statistics for all pulses, including the initial breakdown pulses, preceding the first return stroke. In the following we will consider the characteristics of the pulses occurring within a millisecond or so prior to the return-stroke pulse, that is, probably within 200 m or so above the ground (assuming a leader speed of 2×10^5 m s^{-1}). Note that at such heights there is no distinction between α- and β-leaders (subsection 4.4.1).

Krider (1974a), using a photoelectric detector in Arizona, recorded five light pulses produced by a stepped leader within about 70 μs prior to the return stroke pulse. The time intervals between the pulses ranged from 17 to 32 μs, and the time interval between the last step pulse and the return-stroke pulse was 9 μs. According to Krider et al. (1977), the mean time interval between the last stepped-leader pulse and the return-stroke pulse in the electric field records in Florida was 11 μs. Measured interpulse intervals within a millisecond or so of the return-stroke pulse are summarized in Table 4.3. As seen in Fig. 4.24 and noted in early studies (e.g., Kitagawa 1957a), the pulse peaks increase as the leader approaches ground. The last stepped-leader pulse is generally the largest, and its peak, on average, is about 0.1 of the return-stroke pulse peak (Krider et al. 1977).

The 10–90 percent risetimes of the individual step waveforms are on average 0.2–0.3 μs, and the half-peak width is typically 0.4–0.5 μs under conditions such that the distortion of pulses due to propagation effects is minimal

Fig. 4.22 (opposite). Ratio of the leader and return-stroke electric field changes $\Delta E_L/\Delta E_R$ as a function of distance for five different sets of strokes: (a) 60 first strokes; (b) 23 second, third, and fourth strokes creating a new termination on ground; (c) 57 second, third, and fourth strokes following the same channel as the preceding stroke; (d) 35 fifth, sixth, and seventh strokes, and (e) 43 strokes of order 8 through 18. Both the solid and the broken curves (in (a) and (b)–(e), respectively) show the ratio predicted by a simple model having a vertical, uniformly charged channel and a cloud charge source centered at 7.5 km. The data were obtained near Tampa, Florida during three thunderstorms in 1979: •, 79196; ○, 79199; □, 79208. Adapted from Rakov et al. (1990).

4.4. Stepped leader

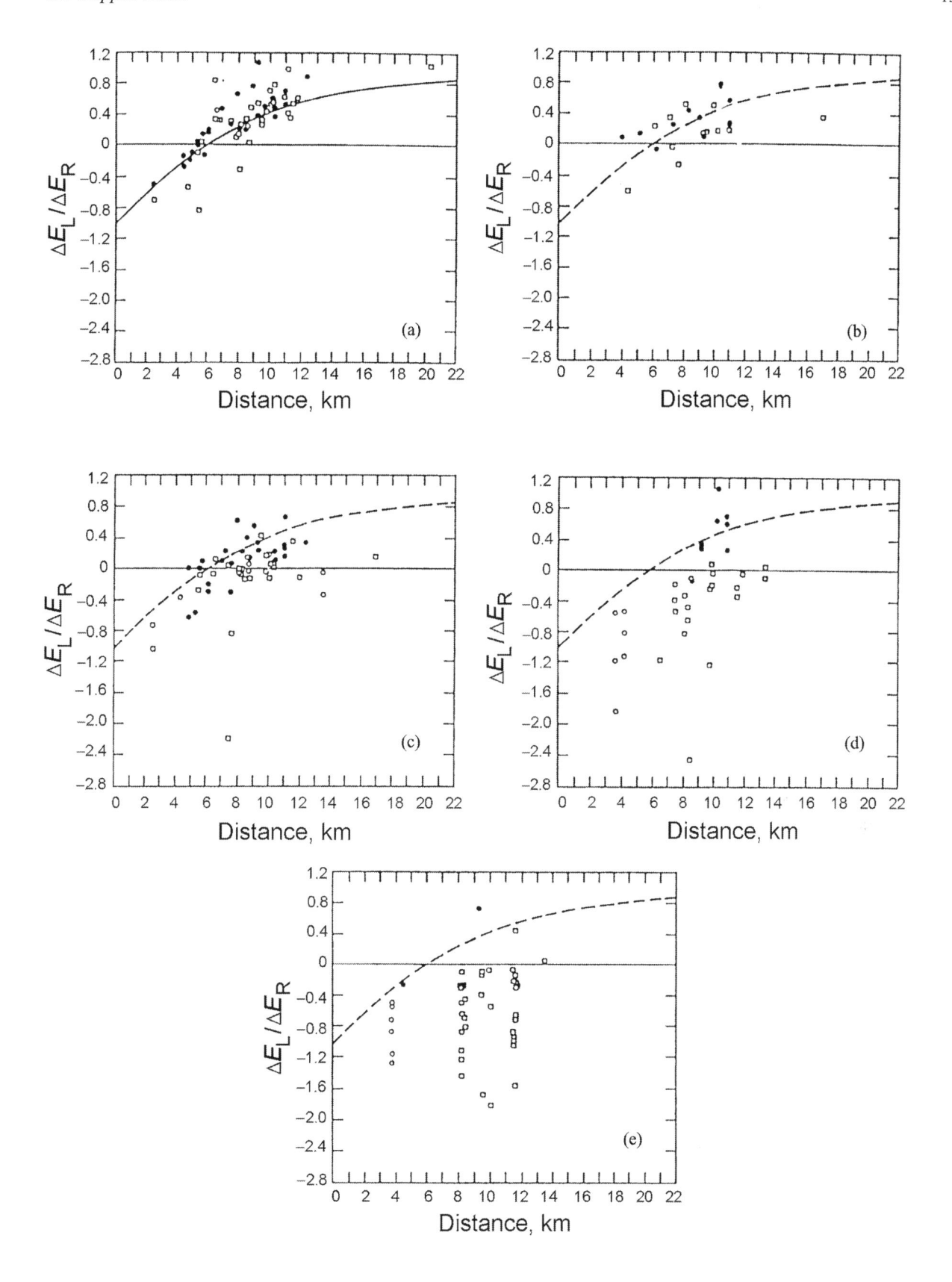

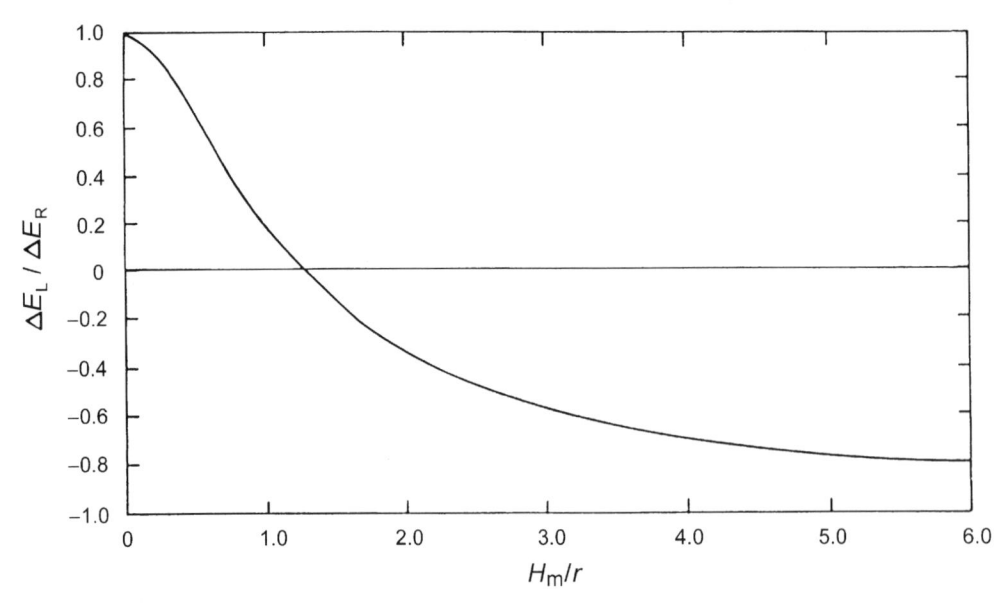

Fig. 4.23. The model-predicted ratio of the leader and return-stroke electric field changes as a function of the ratio of the height H_m of the charge center and the horizontal distance r from the charge center. Adapted from Uman (1987, 2001).

(Krider et al. 1977). Weidman and Krider (1980b) reported that the 10–90 percent risetimes of individual step pulses were similar to those of the fast transitions of return-stroke pulses (Section 4.6), 40 to 200 ns with a mean of 90 ns, when the propagation paths were over salt water. Krider et al. (1992) reported the mean peak of dE/dt (the time derivative of the electric field) for a step waveform, range-normalized to 100 km, to be 13 V m^{-1} μs^{-1} (17 values) and the mean half-peak width to be 69 ns (eight values). Willett and Krider (2000) estimated the mean half-peak width of dE/dt to be 54 ns for 114 stepped-leader pulses, 64 ns for 24 dart-stepped leader pulses, and 65 ns for 25 initial-breakdown pulses. Baum et al. (1980, 1982) observed electromagnetic field pulses, which they attributed to stepped leaders. The pulses were produced by lightning within 5 km, and the propagation was over New Mexico mountainous terrain; the interpulse intervals were as short as a few microseconds, which is unusual for pulses associated with a stepped leader.

Frequency spectra in the interval from about 100 kHz to 20 MHz for individual-step electric field pulses propagating over salt water were given by Weidman et al. (1981, Fig. 3) and Willett et al. (1990, Fig. 4). According to Willett et al., the frequency spectrum for stepped-leader steps (65 events) is nearly identical to that for dart-stepped-leader steps (15 events) and very similar to the spectrum for initial-breakdown pulses (18 events); further, the shapes of the spectra for leader steps and for return strokes are remarkably similar above 2–3 MHz, although the magnitudes of the leader spectra are about five times smaller.

From measurements of the electric field pulses radiated by leader steps, Krider et al. (1977) inferred that the peak step current is at least 2–8 kA close to the ground, the maximum rate of change of step current is 6–24 kA μs^{-1}, and the minimum charge involved in the formation of a step is $(1-4) \times 10^{-3}$ C. These current and charge values are similar to the estimates given by Rakov et al. (1998) for the steps of a dart-stepped leader (subsection 4.7.5) in triggered lightning.

From a photographic spectrum of a stepped leader obtained in Switzerland, Orville (1968d) estimated the peak temperature of the leader step to be about 30 000 K, assuming that the leader is an optically thin source in local thermodynamic equilibrium. He further suggested that the channel temperature behind the leader tip between illuminations does not fall below 15 000 K. Orville also estimated the diameter of the leader channel defined by NII spectral emissions to be less than 0.5 m. Schonland (1953), from photographic measurements in South Africa, reported luminous stepped-leader diameters between 1 and 10 m. Note that the stepped-leader channel is likely to consist of a thin core (probably less than 1 cm in diameter) that carries the longitudinal channel current, surrounded by a radially formed corona sheath whose radius is typically several meters.

Chen et al. (1999), who used the ALPS imaging system (Yokoyama et al. 1990), reported that zero-to-peak risetimes of luminosity pulses associated with steps for one leader in Australia ranged from 0.5 to 3.5 μs, with mean value 1.7 μs, and for one leader in China ranged from 0.6 to 1.2 μs. From earlier photographic and photoelectric

4.4. Stepped leader

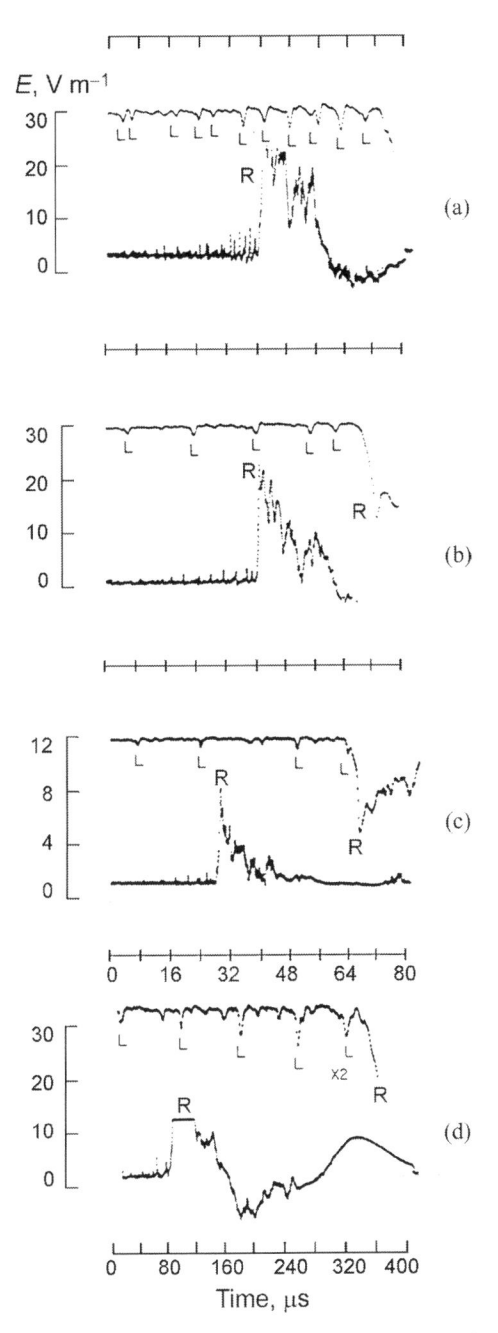

Fig. 4.24. Electric field waveforms produced by four negative first strokes in Florida at distances of some tens of kilometers. Each record is displayed on two time scales, 8 μs per division (upper trace) and 40 μs per division (lower trace), the two traces being inverted with respect to each other. Note that the abrupt return-stroke transitions (clipped in the lower trace of Fig. 4.24d), labeled R, are preceded by small pulses characteristic of leader steps, each labeled L. The vertical scale is shown on the left (this is to be reduced by a factor 2 for the 8 μs per division (upper) trace in Fig. 4.24d). Adapted from Krider *et al.* (1977).

measurements, the luminosity of the step was found to rise to peak in about 1 μs and fall to half this peak in roughly the same time (Schonland *et al.* 1935; Schonland 1956; Orville 1968d; Beasley *et al.* 1983). These values might be affected by the scattering of the step light by aerosol or cloud particles. Since a step becomes luminous in 1 μs or less and has a typical length of 50 m, the rate with which this luminosity fills the step (the process is unresolved by ordinary streak cameras) is in excess of 5×10^7 m s^{-1} (Schonland *et al.* 1935; Orville and Idone 1982).

Baum, C.E. (1999), assuming that the stepped-leader channel extends essentially only during the step formation process, suggested that the ratio of leader extension speed during the step formation process to the average stepped-leader speed can be estimated as the ratio of the average time interval between leader step pulses in electric or magnetic field records and the average duration of the leader step pulse. If we assume that the average interpulse interval is 20 μs (see Table 4.3) and that the average duration of a leader pulse is equal to its typical half peak width of 0.4 μs (Krider *et al.* 1977), we obtain a ratio equal to 50. Using this ratio and the typical measured overall stepped leader speed, 2×10^5 m s^{-1}, we estimate the extension speed during the step formation process to be 1×10^7 m s^{-1}.

An increase in channel brightness toward the leader tip is typically observed (Schonland 1956). Wang *et al.* (1999b), using a digital optical imaging system, ALPS, having 100 ns time resolution and 30 m spatial resolution, found that the luminosity pulses formed at the dart-stepped leader tip in the step-formation processes in triggered lightning propagated toward the cloud at a mean speed 6.7×10^7 m s^{-1} and exhibited a decrease in peak to about 10 percent of the original value within the first 50 m. Note that the spatial resolution of 30 m in the experiment of Wang *et al.* (1999b) was insufficient to resolve the step-formation process.

4.4.7. Streamer zone

Berger (1967a), from streak photography, reported two cases of a brush-like corona occurring ahead of the upward-moving negative leader tip in upward positive lightning (subsection 6.2.2). In one case, the extent of the visible corona was estimated to be at least 3 m. It appeared that corona developed in less than the 5 μs resolution of the streak camera, essentially simultaneously with the formation of each step; that is, the corona did not appear to develop continuously between steps (Berger and Vogelsanger 1969). No corona at the tip of downward negative leaders or at the tip of either upward or downward positive leaders (the positive leaders showed less distinct, if any, steps) was reported by Berger (1967a), probably because of the much lower luminosity of those leaders compared to negative upward

Table 4.3. *Measured intervals between stepped-leader electric field pulses within a millisecond or so of the return-stroke pulse*

Reference	Location	Sample size	Interpulse interval, μs	Remarks
Clarence and Malan (1957)	South Africa		5–10	
Kitagawa (1957a)	Japan		13–14	Mean value
Isikawa et al. (1958)	Japan		20	
Krider and Radda (1975)	Florida, Arizona		10–20	
Krider et al. (1977)	Florida	997	16	Mean values; within 200 μs[a]
	Arizona	130	25	
Beasley et al. (1983)	Florida		5–20	Within 50–100 μs
Cooray and Lundquist (1985)	Sri Lanka		12	Mean value

[a] This remark refers to both Florida and Arizona.

leaders. However, Schonland et al. (1935) reported that for one downward negative stepped leader there was faint luminosity extending downward about 30 m below the bottom of a bright step, the last one seen in the streak photograph before the occurrence of the return stroke. They also noted "a similar effect, but to a much smaller extent" in a few other steps of this leader. Idone (1992) presented a near-ultraviolet streak photograph showing a "diffuse, hemispherical corona brush" extending over a distance of about 5–10 m above the bright steps of one upward-moving positive leader in altitude-triggered lightning. This photograph is reproduced in Fig. 7.7. He observed the "corona brush" in almost all his near-ultraviolet recordings of positive leaders but not in the two recordings of negative leaders in altitude-triggered lightning imaged with the same system. One of these negative leaders is shown in Fig. 7.6. Bazelyan et al. (1978) suggested that the length of the streamer zone is about 100–200 m in front of a negative-stepped-leader tip carrying an electric potential of 50–100 MV.

4.4.8. Step-formation mechanism

In the following, we will briefly discuss a possible step formation mechanism for negative downward leaders. The step-formation process in lightning is not resolved in ordinary high-speed photographic records. However, there appears to be a qualitative similarity between a negative stepped leader in lightning and in a long laboratory spark. The latter type of leader is much better studied via the use of electronic image-converter cameras in conjunction with measurement of the current through the air gap. The negative long spark leader exhibits distinct steps when the gap length is several meters or more. Gorin et al. (1976), for example, reported that a 6 m rod–plane gap was bridged by a negative leader in three to five steps. It is worth noting that stepping is observed in lightning negative leaders regardless of whether they are initiated in the cloud (downward leaders) or at the grounded object (upward leaders). This fact

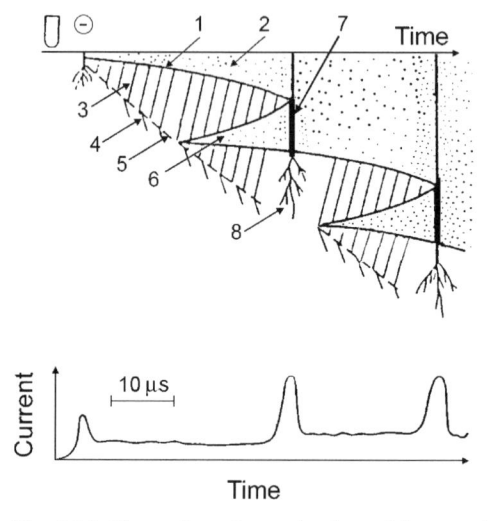

Fig. 4.25. Illustration of a mechanism of the negative stepped leader in a long laboratory spark. The negative electrode is seen at the left of the figure. 1, leader tip; 2, primary leader channel; 3, positive streamers of the streamer zone; 4, negative streamers of the streamer zone; 5, space stem; 6, secondary channel; 7, leader step; 8, burst of negative streamers. Adapted from Gorin et al. (1976).

suggests that the mechanism of formation of a step is determined primarily by the processes at the leader tip and in the leader channel rather than by the source (cloud charge for lightning and impulse generator circuitry for long laboratory sparks). Since the electric potential of a lightning leader, some tens of megavolts (e.g., Bazelyan et al. 1978), is much higher than that of a long-spark leader, a lightning leader should have a larger charge per unit channel length, a larger leader current, and a higher propagation speed than a long-spark leader.

The development of the negative stepped leader in a long laboratory spark, based on a description given by Gorin et al. (1976), is illustrated in Fig. 4.25, which shows schematically a time-resolved picture including an initial impulsive corona from the negative high-voltage electrode and the first two steps, along with the corresponding current through the gap between the electrodes. The initial impulsive corona, a system of branched filamentary channels (streamers), serves to heat the air near the high-voltage electrode (labeled with a minus inside a small circle) and to form the initial section of the leader plasma channel. This process produces the first pulse in the current record shown in Fig. 4.25, the other two current pulses being associated with the two leader steps. The initial section of the leader channel extends from the high-voltage electrode into the gap. The leader tip is brighter than the channel behind it and its path is shown in Fig. 4.25 as a slightly curved solid line labeled 1. The channel behind the moving leader tip is labeled 2. In front of the leader tip is the so-called streamer zone, which, in the case of a negative leader, is composed of both positive and negative streamers. The positive streamers develop toward the leader tip (they are shown in Fig. 4.25 by longer, positively sloped solid lines labeled 3), and the negative streamers develop into the gap (these are shown in Fig. 4.25 by the shorter, negatively sloped lines labeled 4). Both positive and negative streamers appear to start from a plasma formation, termed a space stem, which moves into the gap in front of the leader tip. The downward-moving space stem is shown in Fig. 4.25 as a negatively sloped broken line labeled 5. The air in the region through which the space stem passes apparently remains an insulator. It is presently unknown how a space stem is formed ahead of the leader tip (Bazelyan and Raizer 2000a). When the space stem is sufficiently heated, it gives rise to, in effect, a section of the leader channel, labeled 6 in Fig. 4.25, that is not connected to either of the electrodes and extends in both upward and downward directions. A bidirectional channel extension is shown in Fig. 4.25 as a pair of slightly curved and diverging solid lines drawn from a common origin on the slanted broken line labeled 5. Each such pair of solid lines is an unconnected (space) channel, a sort of bidirectional leader, termed a secondary channel by Bazelyan et al. (1978) and a space leader by Larigaldie et al. (1992). The upward-extending part of this bidirectional leader is charged positively, and the downward-extending part is charged negatively.

The first leader step, labeled 7 in Fig. 4.25, is formed at the instant when the upward-moving positive end of the unconnected leader channel labeled 6 makes contact with the downward-moving negative tip of the primary (since it is connected to the high-voltage electrode) leader channel labeled 2. At that moment, the very high potential (which is close to the potential of the high-voltage electrode) of the primary leader channel is rapidly transferred to the lower end of the secondary channel. As a result, a burst of negative streamers, labeled 8 in Fig. 4.25, is produced at the bottom of the newly added channel section. Such a breakdown generates a current pulse that propagates toward the high-voltage electrode and briefly illuminates the entire channel. The formation of the next step begins with the formation of a new space stem ahead of the newly formed leader step. Thus, in the negative leader, the occurrence of each luminous step is caused by the connection of a secondary (space) channel to the primary leader channel.

It is worth noting that the formation of positive-leader steps, discussed in subsection 5.3.3, does not involve the space channel discussed above, and therefore the steps in positive and in negative leaders are of different natures. However, according to Bazelyan and Raizer (2000a), the positive part of the "bidirectional leader" moves faster than either its negative part or the downward-moving tip of the primary (negative) leader channel. As a result, the overall extension of the negative-leader channel is largely facilitated by the extension of a series of positive leaders that originate from the space stem and propagate in the direction opposite to that of the overall negative leader extension.

Descriptions similar to the above of the processes involved in the formation of negative-leader steps were given by Larigaldie *et al.* (1992), Bondiou-Clergerie *et al.* (1996), Raizer (1997), and Bazelyan and Raizer (1998, 2000a). Various lightning stepped-leader models are discussed in Section 12.4.

The final stages of the step formation process described above for a long laboratory spark, including the burst of streamers (impulsive corona) from the leader tip and the illumination of both the step itself and the channel behind it, have been observed in lightning also. As noted in subsection 4.4.6, Wang *et al.* (1999b), for a downward dart-stepped leader in negative triggered lightning, observed and characterized luminosity waves associated with individual steps that propagated in the direction opposite to that of the advancement of the leader (see also subsection 4.7.5). Similar observations for two downward negative stepped leaders in natural lightning were reported by Chen *et al.* (1999). The initial stages of the negative-leader step-formation process, including the bidirectional development of an unconnected (space) secondary channel, have never been documented for lightning, possibly due to the much lower luminosity of those stages.

4.5. Attachment process

The process of lightning leader attachment to ground or to a grounded object is one of the least understood and most poorly documented processes of the cloud-to-ground lightning discharge. It is generally assumed that the attachment process begins when, in response to an approaching downward-moving leader, an upward-moving leader is initiated at the ground or, more likely, at the tip of an object protruding above ground. It is possible that two or more upward leaders are launched from the ground toward the descending leader, perhaps in response to different branches of the descending leader. Multiple channel terminations on ground created by the same leader process, such as in Fig. 4.1, must involve multiple upward leaders from ground. An upward leader that makes contact with a branch of a downward leader is called an upward connecting leader (sometimes called an upward connecting streamer, although the latter term, apparently arising from the expected relatively low luminosity of its channel, is not consistent with our definition of "streamer", found in Section 1.2). An upward leader that fails to make such a contact is called an unconnected upward leader (or discharge).

The process by which the extending plasma channels of the upward and downward leaders make contact is called the break-through phase or final jump. This process begins when the relatively low-conductivity streamer zones ahead of the two propagating leader tips meet to form a common streamer zone. The subsequent accelerated extension of the two relatively high-conductivity plasma channels toward each other takes place inside the common streamer zone. The break-through phase can be viewed as a switch-closing operation that serves to launch two return-stroke waves from the point of junction between the two plasma channels. One wave moves downward, toward the ground, and the other upward, toward the cloud. The downward-moving return-stroke wave quickly reaches the ground, and the resulting upward-reflecting wave from ground follows and probably catches up with the upward-moving return-stroke wave from the junction point. The reflected wave from the ground propagates in the return-stroke-conditioned channel above the junction point and, hence, is likely to move faster than the upward wave from the junction point that propagates along the leader-conditioned channel (Rakov 1998). When the waves bouncing between the ends of the growing return-stroke channel decay, a single upward-moving wave is formed. Thus, the lightning attachment process involves two plasma channels growing toward each other, initially in air (the upward-connecting leader phase) and then inside the streamer zone (the break-through phase). It is a matter of definition whether the very short-lived bidirectional return-stroke wave should be considered part of the attachment process or part of the return stroke, the latter process being discussed in Section 4.6.

The attachment process occurs in both first and subsequent lightning strokes. In the former case, it occurs in virgin air and in the latter case usually in the remnants of a previous channel (in the warm air left by the previous lightning processes). The attachment process in first strokes is more pronounced than in subsequent strokes, owing to the larger electric potential of the stepped leader. It will be shown below that the length of an upward connecting leader involved in a first stroke is some tens of meters if that leader is launched from the ground, and it can be several hundred meters long or even longer if it is initiated from a tall object. The length of the upward-connecting leader initiated in response to a descending dart leader in a subsequent stroke is of order 10 m or less. The break-through phase has never been observed in lightning and has been inferred to exist by analogy with observations of this phase in long laboratory sparks. As noted in subsection 4.4.7, Bazelyan *et al.* (1978) suggested that the length of the streamer zone is about 100–200 m in front of the negative-stepped-leader tip, which carries an electric potential of 50–100 MV.

In the following, we will present the available experimental evidence for the existence of upward connecting leaders in both first and subsequent lightning strokes. This evidence is organized in two categories: (i) time-resolved optical images and (ii) still photographs. Since it is generally thought that the beginning of the attachment process signifies the instant when the strike point is determined, this process is of great importance for the design of lightning protection for various objects and systems. The practical aspects of the attachment process, including the so-called striking distance, are considered in Section 18.3. Gorin (1985) attempted to model the attachment process, including the break-through phase, based on observations of the break-through phase in long sparks and as an integral part of his nonlinear distributed-circuit model of the return stroke (subsection 12.2.4).

4.5.1. Time-resolved optical images

Time-resolved optical images actually showing upward connecting leaders were reported by Yokoyama *et al.* (1990) for six natural lightning strokes in winter storms in Japan, and by Wang *et al.* (1999a) for one dart leader–return stroke sequence in classical triggered lightning in Florida. All other available images for either natural or triggered lightning give only indirect evidence of such upward leaders. Laroche *et al.* (1991) and Lalande *et al.* (1998) measured the currents associated with upward connecting positive leaders in altitude-triggered lightning (subsection 7.2.3). These currents exhibit pulses indicative of stepping, with pulse amplitudes of order 10 A and inter-pulse intervals of a few tens of microseconds.

First strokes Yokoyama *et al.* (1990), using the ALPS optical imaging system, observed upward-connecting leaders in six lightning discharges to a 80 m microwave telecommunication tower in Japan. All six lightning events occurred during winter storms. For five events the length of the upward connecting leader varied from 25 to 125 m, and for the sixth event the length was greater than 150 m. Yokoyama

4.5. Attachment process

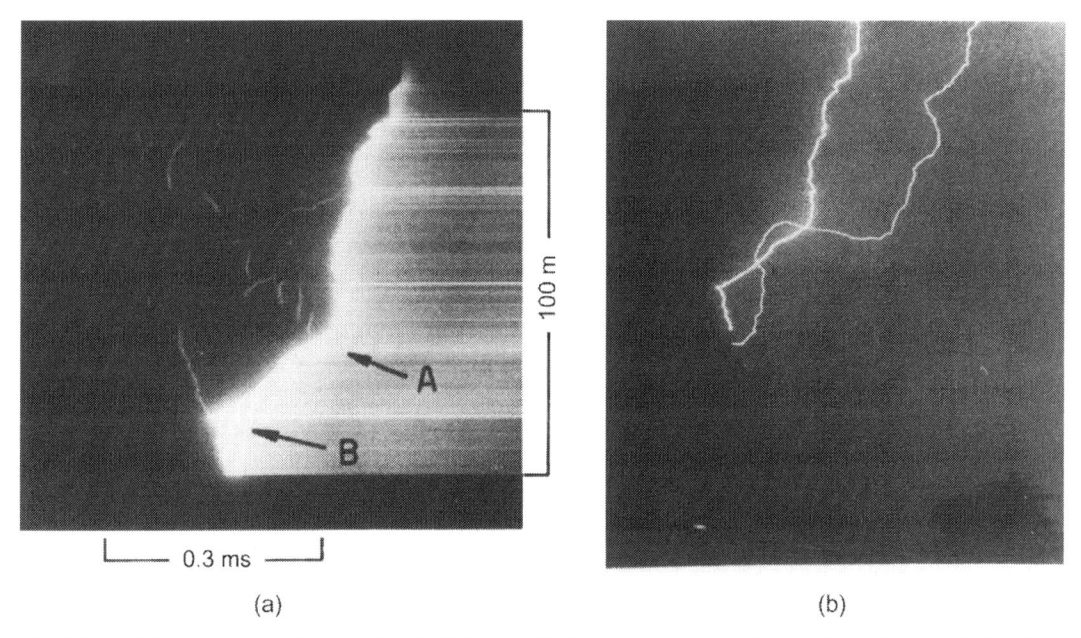

Fig. 4.26. (a) Streak-camera photograph of a lightning discharge to a tower on Monte San Salvatore, Switzerland, showing evidence of an upward connecting leader. (b) Still photograph of the same flash and another flash that attached to the tower below its top. Adapted from Berger and Vogelsanger (1966).

et al. estimated the average progression speeds for three upward connecting leaders as 0.8×10^5, 1.0×10^5, and 2.7×10^5 m s^{-1}. These speeds are similar to those reported for upward positive leaders in natural upward lightning flashes (Table 6.2). They are also similar to the speeds of both upward positive leaders in rocket-triggered lightning (subsection 7.2.2) and downward negative stepped leaders (subsection 4.4.2). Additionally, Yokoyama *et al.* (1990) estimated an average progression speed of 2.9×10^5 m s^{-1} for the downward leader in an event whose upward-connecting leader extended at a speed of 0.8×10^5 m s^{-1}.

Golde (1947) and Wagner (1967) both published a sketch of a streak-camera photograph attributed to D.J. Malan, in which the stepped leader appears to end about 50 m above flat terrain. The gap between the ground and the leader end-point must have been bridged by an upward-moving connecting leader that was probably too faint to be imaged. Berger and Vogelsanger (1966, Fig. 13) and Berger (1977, Fig. 15) published a streak photograph, reproduced in Fig. 4.26a, in which the downward stepped leader ends about 40 m above and 40 m horizontally away from the lightning-channel terminus on the tower top. The position of the leader tip at this time is labeled A in Fig. 4.26a while the tower top is represented in this time-resolved photograph by the lower (horizontal) boundary of the luminous image. The connection between the tower top and point A must have been made via an upward connecting leader that is not imaged in Fig. 4.26a because of its insufficient luminosity. This upward connecting leader apparently forked at point B, the right branch making contact with the descending stepped leader and the left branch remaining unconnected. A still photograph of the event is shown in Fig. 4.26b. A streak photograph of another first stroke terminating on the same tower, similar to that shown in Fig. 4.26a, is given by Berger and Vogelsanger (1966, Fig. 12) and Berger (1967a, Fig. 42). Current oscillograms corresponding to first strokes, including the one whose time-resolved photograph is shown in Fig. 4.26a, exhibit a characteristic concave front. The initial, slowly rising portion of that current front is often interpreted as being due to the upward connecting leader.

Orville and Idone (1982), from their Florida streak-camera photographs, infer upward connecting-leader lengths of approximately 20 and 30 m for two of three first strokes in natural lightning, although those leaders were not seen in the photographs. No upward connecting-leader length could be estimated for the third event.

Upward negative connecting leaders occurring at tower tops in response to descending positive leaders are discussed in subsection 6.2.2. These upward connecting leaders are considerably longer than their positively charged counterparts discussed here. It is possible that some of the upward connecting leaders are so long that they enter the cloud and, as a result, the point of connection with the causative downward leader (the junction point) is not observable with optical techniques. As a result, there is some difficulty in classifying such lightning discharges as either downward or upward flashes (Section 5.1).

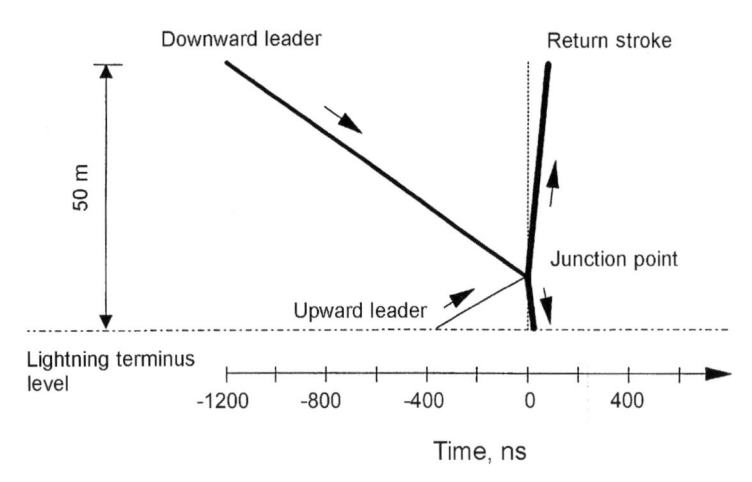

Fig. 4.27. A streaked-image diagram of the attachment process of a dart leader–return-stroke sequence in a rocket-triggered lightning flash (Section 7.2). Adapted from Wang et al. (1999a).

Subsequent strokes Orville and Idone (1982), from streak photography of natural lightning in Florida and New Mexico, found that one dart-stepped leader in Florida did not propagate completely to the ground before the initiation of a return stroke, while three New Mexico dart-stepped leaders apparently did. They inferred the existence of an upward connecting leader of roughly 20 m in length in the Florida event. Further, Orville and Idone (1982) reported that they did not observe any evidence of upward connecting leaders in association with 21 dart leaders (14 in Florida and 7 in New Mexico). From a streak-photographic study of three triggered lightning flashes in New Mexico, Idone et al. (1984) inferred upward connecting-leader lengths of 20 and 30 m in two out of the three flashes. No images of the upward connecting leaders were obtained. Note that strokes in triggered lightning are similar, if not identical, to subsequent strokes in natural lightning.

Idone (1990), using photographic recordings having 0.3 m spatial and 0.5 μs temporal resolution, estimated the upper bound for lengths of inferred (none was imaged) upward connecting leaders to be of order 10 to 20 m for nine strokes in four triggered lightning flashes in Florida.

Wang et al. (1999a), using the digital optical imaging system ALPS (Yokoyama et al. 1990) with 3.6 m spatial and 100 ns time resolution, observed an upward connecting leader in one triggered-lightning stroke and inferred the existence of such a leader in another one. A sketch of the time-resolved image for the former event is shown in Fig. 4.27. In both events, the return stroke was initially a bidirectional process that involved both upward- and downward-moving waves that originated at 7–11 m (in the event with the imaged upward connecting leader) and 4–7 m (in the event with no imaged upward connecting leader). Note that an initially bidirectional extension of the return-stroke channel had been previously hypothesized by Wagner and Hileman (1958), Uman et al. (1973b), Weidman and Krider (1978), Willett et al. (1988, 1989a) and Leteinturier et al. (1990). In the case illustrated in Fig. 4.27, the measured return-stroke peak current was about 21 kA, and in the other case studied by Wang et al. (1999a) it was 12 kA. The imaged upward connecting leader had a light intensity one order of magnitude lower than its associated downward dart leader, a length of 7–11 m, and a duration of several hundred nanoseconds. The propagation speed of the upward connecting leader was estimated to be about 2×10^7 m s^{-1}, similar to the typical speed of downward dart leaders (subsection 4.7.2) and about two orders of magnitude higher than upward connecting-leader speeds in first strokes (see above).

4.5.2. Still photographs

In still photographs it is generally impossible to distinguish between positive and negative flashes, and first and subsequent strokes are usually unresolved. Since the overwhelming majority of flashes are negative, and first strokes are expected to involve more pronounced upward connecting leaders than subsequent strokes, the evidence discussed in this section is usually attributed to first strokes in negative flashes. The following features have been used to infer the presence of upward connecting leaders: (i) a split or loop in the channel image, (ii) upward branches in the lower part of the channel and downward branches in the upper part of the channel, (iii) unconnected upward discharges, and (iv) an abrupt change (kink) in the shape of the lower part of the channel that may be interpreted as being due to the attractive effect of the upward connecting leader.

A split or loop in the channel A still photograph showing this feature was published by Golde (1967) and is

4.5. Attachment process

from a distance of 60 m in Switzerland. Downward-directed branches in the upper part of the lightning channel and upward-directed branches in its lower part indicate a junction point about 12 m above the tree top. The tree was unharmed by this direct lightning strike, as is the case for most trees struck by lightning (subsection 18.4.2). The photograph of a lightning strike to a beach taken at 30 m in New Jersey, shown in Fig. 4.30, exhibits both upward and downward branches originating from the main channel and unconnected upward discharges originating from the ground near the strike point, the latter feature being discussed next.

Unconnected upward discharges Krider and Ladd (1975) published a photograph of two lightning discharges striking mountainous terrain in southern Arizona, that photograph being reproduced in Fig. 4.31. The flash on the left was 810 m away and was accompanied by two unconnected upward discharges of 8 and 10 m length originating from the ground about 10 and 15 m from the main channel, respectively. The flash on the right was 1560 m away and exhibited a single unconnected upward leader (not resolved in Fig. 4.31) of 10 m length about 15 m away from the main channel. The presence of unconnected upward discharges in the immediate vicinity of the strike point suggests that an upward connecting leader of greater length

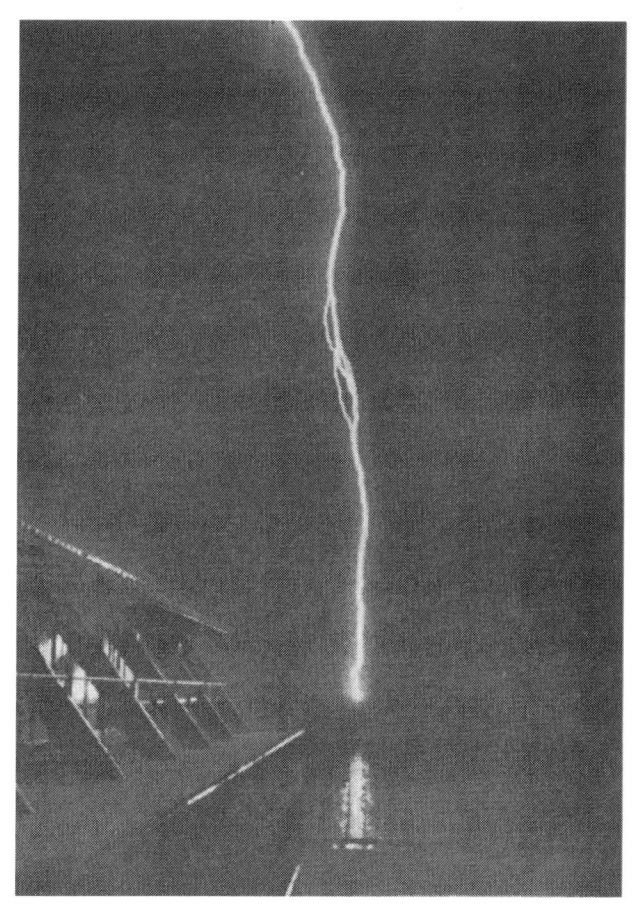

Fig. 4.28. A photograph of a lightning strike to a chimney pot showing a split in the channel, interpreted as evidence of an upward connecting leader. Adapted from Golde (1967).

reproduced in Fig. 4.28. Seen in the photograph is the lower end of a lightning channel attached to the chimney of a building in Helsinki, Finland. The distance from the camera to the channel was about 70 m. The imaged channel is clearly split into at least three separate segments at a height of about 9 m above the chimney. In analogy with long-spark still photographs where time-resolved images were also available, the split was interpreted as being formed by the downward-branched descending leader and the upward-branched ascending connecting leader. Hagenguth (1947) describes a photograph of a lightning strike to a patch of weeds in a lake. The lightning channel was split between about 3 to 9 m above the water, with a maximum separation between the individual segments of over 1 m. A photograph of an identified positive lightning flash with a loop in the lower portion of the channel is given in Fig. 5.9c.

Both upward and downward branching Orville (1968e) published the remarkable photograph shown in Fig. 4.29 of a lightning strike to a 7-m-tall European ash tree taken

Fig. 4.29. Photograph of a lightning strike to a European ash tree in Lugano, Switzerland showing both downward and upward branching. From Orville (1968e).

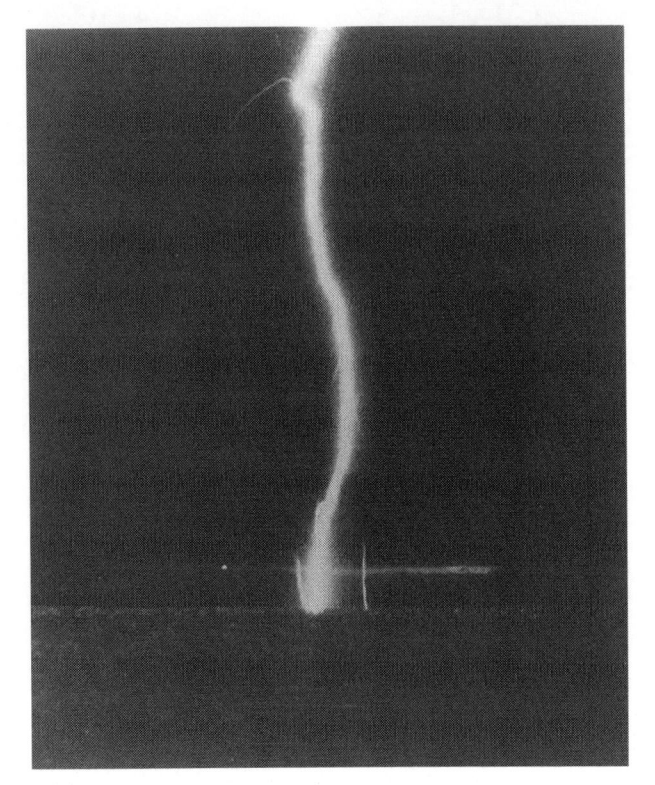

Fig. 4.30. Photograph by Robert Edwards of a lightning strike to the sand at Manasquan Beach, New Jersey taken in July 1934 from a distance of about 30 m. Note both downward and upward branching, and an unconnected discharge from the ground to the right of the main lightning channel. Courtesy of Galloway, New York.

from the tip of the strike rod installed on the rocket launcher. The discharges apparently occurred in response to a leader–return-stroke sequence whose channel was attached to the rocket launcher tube, about 2.5 m below the tip of the strike rod. The second image, recorded about 170 ms after the first one, shows upward branches extending from the main channel attached to the tip of the strike rod. Note that the latter event was associated with the formation of a new strike point, the rod tip as opposed to the rocket launcher tube.

Fisher and Schnetzer (1991) presented two video images, recorded two frames apart (that is, separated by 34 to 68 ms), which were somewhat similar to the pair of images published by Idone (1990) and described above. The second image shows the only return-stroke channel in the flash, attached to the strike object tip with two bright unconnected discharges apparently originating from the same point. The first image contains two relatively faint unconnected discharges from the tip of the strike object. These apparently occurred in response to an impulsive process, during the initial continuous current (Section 7.2), whose channel, attached to the launcher tube, was outside the field of view of the video camera but was imaged by another camera (R.J. Fisher, personal communication, 1998).

We now discuss unconnected upward discharges that develop from various objects in response to relatively than the unconnected discharges must have been involved in the attachment of each of the two flashes to ground. Two upward unconnected discharges are seen in the remarkable photograph of a lightning strike to a 20 m sycamore tree taken in 1984 by Johnny Autery of Dixons Mills, Alabama. This photograph is reproduced on the cover of this book (see also Uman 1991). One of the upward unconnected discharges originated from the tree to the left of the strike point and the other from the television antenna of the farmhouse at left. Interestingly, the latter unconnected discharge of about 14 m length reportedly damaged one of the TV sets in the farmhouse (Newcott 1993). The antenna has a height of 21 m and is about 45 m from the tree. The tree suffered no permanent damage (Uman 1991).

In triggered lightning, unconnected upward discharges have been observed by Idone (1990), Fisher and Schnetzer (1991), and by the authors of this book (unpublished photographs taken at Camp Blanding, Florida). Idone (1990) presented two telephoto images obtained with a millisecond-scale time resolution at the Kennedy Space Center, Florida. One of the images shows multiple unconnected upward discharges up to 1.6 m in length emanating

Fig. 4.31. Photograph showing unconnected discharges from the ground near the strike point of a lightning flash in southern Arizona. Taken from Krider and Ladd (1975).

4.6. Return stroke

distant or nearby lightning discharges that do not terminate on the object in question. One example of such discharges is the unconnected discharge from the television antenna located about 45 m from the lightning strike point in the photograph of Johnny Autery (see the cover of this book) discussed above. Photographs of lightning strikes to the Empire State Building in New York (Section 6.2) show simultaneous upward discharges from nearby buildings (McEachron 1941). Unconnected upward discharges were commonly observed at or near the top of the Ostankino TV tower (Section 6.4), both accompanied and not accompanied by lightning strikes to the tower. Berger (1967a) reported unconnected discharges from tall structures apparently occurring at the instant of a relatively distant ground or cloud discharge. Berger termed these discharges "secondary strokes" since they appeared to occur in response to "primary strokes" elsewhere. While often being too faint to be imaged by a streak camera, the "secondary strokes" were readily seen in still photographs. "Secondary strokes" can be viewed as uncompleted upward flashes (Chapter 6).

An abrupt change in channel shape near the ground
Walter (1937) noticed that most of the photographs available to him showed a pronounced kink in the shape of the lightning channel at a relatively low height above the object that was struck. Between this kink and the terminus point, the channel was observed to be relatively straight and occasionally almost horizontal. An example is given in Fig. 4.32, which shows a lightning strike to a guy wire 14 m below the top of an 80 m TV tower. The channel abruptly becomes almost horizontal at about 70 m from the guy wire. The presence of an unconnected upward discharge 96 m in length from the tower top suggests that initiation of an upward discharge from a tall structure does not necessarily determine the strike point.

Photographs of lightning flashes striking the 540 m Ostankino TV tower in Moscow (Section 6.4), taken simultaneously from two angles, show that the majority of downward flashes approach the tower almost horizontally from a distance of about 200–300 m and terminate up to 215 m below the tower top (Bazelyan *et al.* 1978).

4.6. Return stroke

The return stroke has been the most studied lightning process, owing to both practical considerations (it is the return-stroke current that is thought to produce most of the damage attributable to lightning) and the fact that of all the processes comprising a lightning flash, the return stroke lends itself most easily to measurement. Indeed, the return stroke is the optically brightest lightning process visible outside the cloud, and it produces the most readily identifiable electromagnetic signature.

Fig. 4.32. Photograph of a lightning strike to a TV tower guy wire showing an abrupt change in channel shape near the attachment point. Taken from Krider and Alejandro (1983).

In this section, we present experimental data on negative return strokes. Those data are grouped in four categories: (i) parameters derived from channel-base-current measurements (subsection 4.6.1), (ii) luminosity variation along the channel and propagation speed (subsection 4.6.2), (iii) electric and magnetic fields (subsection 4.6.3), and (iv) properties of the return-stroke channel including radius, temperature, pressure, and electron density (subsection 4.6.5). The calculation of electric and magnetic fields is considered in subsection 4.6.4. Models of the return stroke that are used in field calculations are discussed in detail in Section 12.2.

4.6.1. Parameters derived from channel base current measurements

The most complete characterization of the return stroke in the negative downward flash, the type that normally strikes flat terrain and structures of moderate height, shorter than 100 m or so (Section 2.9), is due to Karl Berger and co-workers (e.g., Berger 1955a, b, 1962, 1967a, b, 1972, 1980; Berger and Vogelsanger 1965, 1969; Berger and Garbagnati 1984; Berger *et al.* 1975). The data of Berger were derived from oscillograms of current measured using resistive shunts installed at the tops of two

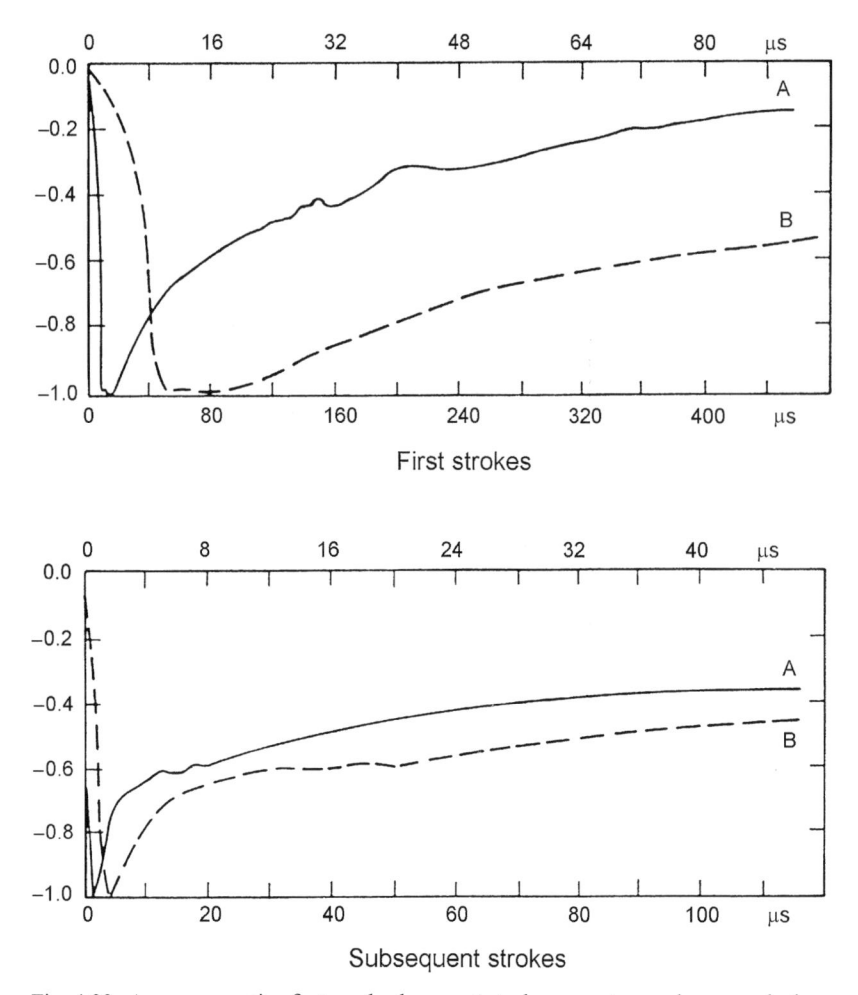

Fig. 4.33. Average negative first- and subsequent-stroke current waveshapes each shown on two time scales, A and B. The lower time scales (A) correspond to the solid curves, while the upper time scales (B) correspond to the broken curves. The vertical (amplitude) scale is in relative units, the peak values being equal to negative unity. Adapted from Berger et al. (1975).

70-m-high towers on the summit of Monte San Salvatore in Lugano, Switzerland. The summit of the mountain is 915 m above sea level and 640 m above the level of Lake Lugano, located at the base of the mountain. The towers are of moderate height, but because the mountain contributed to the electric field enhancement near the tower tops, the effective height of each tower was estimated by Eriksson (1978a) to be 350 m (Table 2.3). As a result, the majority of lightning strikes to the towers were of the upward type to be discussed in Chapter 6. Here we only consider return strokes in negative downward flashes. A total of 101 are included in the summary by Berger et al. (1975). Berger's data were additionally analyzed by Anderson and Eriksson (1980). Fisher et al. (1993) compared the return-stroke parameters published by both Berger et al. (1975) and Anderson and Eriksson (1980) with their counterparts in triggered lightning, the results of the comparison being found in Section 7.2.

The results of Berger et al. (1975) are still used to a large extent as the primary reference source for both lightning protection and lightning research. These results are presented in Figs. 4.33 and 4.34 and in Table 4.4. Fig. 4.33 shows, on two time scales, A and B, the average current waveshapes for negative first and subsequent strokes. The averaging procedure involved the normalization of waveforms from many strokes to their respective peak currents (so that all have peaks equal to unity) and subsequent alignment using the 0.5 peak point on the initial rising portion of the waveforms. The overall duration of the current waveforms was some hundreds of microseconds. The rising portion of the first-stroke waveform has a characteristic concave shape, the initial slower part likely being due to the upward connecting leader (Section 4.5). The averaging procedure masked secondary maxima typically observed in first-stroke waveforms and generally attributed to major branches. Figure 4.34 shows the cumulative statistical

4.6. Return stroke

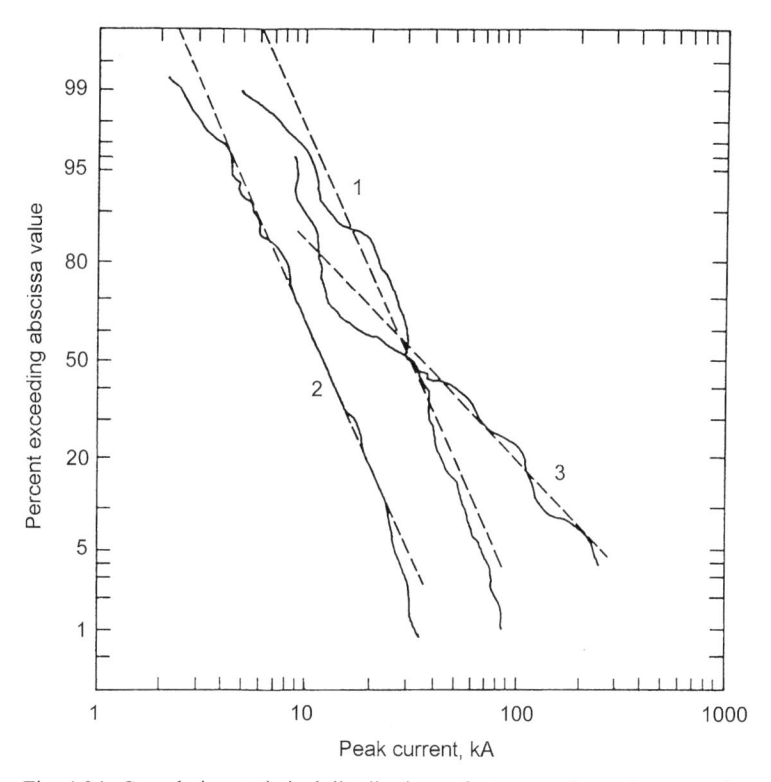

Fig. 4.34. Cumulative statistical distributions of return-stroke peak current from measurements at a tower top (solid curves) and their log-normal approximations (broken lines) for (1) negative first strokes, (2) negative subsequent strokes, and (3) positive first (and only) strokes, as reported by Berger *et al.* (1975).

distributions (the solid curves) of return-stroke peak currents for (i) negative first strokes, (ii) negative subsequent strokes, and (iii) positive strokes (each of which was the only stroke in a flash), the latter being considered further in subsection 5.3.5. These empirical results are approximated by log-normal distributions (the broken lines) and are given as they appear on cumulative-probability-distribution graph paper (see, for example, Fig. B.2 of Uman 1987, 2001), on which a Gaussian (normal) cumulative distribution appears as a slanted straight line, the horizontal (peak current) scale being logarithmic (to base 10). The ordinate gives the percentage of peak currents exceeding the corresponding value on the horizontal axis. The vertical scale is symmetrical with respect to the 50 percent value and does not include the zero and 100 percent values; it only asymptotically approaches those. For a log-normal distribution the 50 percent (median) value is equal to the geometric mean value. The lightning peak current distributions for negative first and subsequent strokes shown in Fig. 4.34 are also characterized by their 95, 50, and five percent values based on the log-normal approximations in Table 4.4, which contains a number of other parameters derived from the current oscillograms. The minimum peak current value included in the distributions is 2 kA. Clearly, the parameters of statistical distributions can be affected by the lower and upper measurement limits. Rakov (1985) showed that, for a log-normal distribution, the parameters of a measured, "truncated" distribution and a knowledge of the lower measurement limit can be used to recover the parameters of the actual, "untruncated" distribution. He applied the recovery procedure to the various lightning peak current distributions found in the literature and concluded that the peak current distributions published by Berger *et al.* (1975) can be viewed as practically unaffected by the lower measurement limit of 2 kA. Note from Fig. 4.34 and Table 4.4 that the median return-stroke current peak for first strokes is two to three times higher than that for subsequent strokes. Also, negative first strokes transfer about a factor of four larger total charge than do negative subsequent strokes. On the other hand, subsequent return strokes are characterized by three to four times higher current maximum steepness (current maximum rate of rise). Note that the maximum rate of rise in current reported by Berger *et al.* (1975) and given in Table 4.4 is an underestimate of the actual value, owing to the limited time resolution of oscillographic data, as discussed later in this subsection. Only a few percent of negative first strokes are expected to exceed 100 kA, while about 20 percent of positive strokes have been observed to do so. However, the 50 percent (median) values of the current distributions for negative and positive strokes are similar. It is worth noting

Table 4.4. *Parameters of downward negative lightning derived from channel-base current measurements. Adapted from Berger et al. (1975)*

Parameters	Units	Sample size	Percentage exceeding tabulated value		
			95%	50%	5%
Peak current (minimum 2 kA)	kA				
First strokes		101	14	30	80
Subsequent strokes		135	4.6	12	30
Charge (total charge)	C				
First strokes		93	1.1	5.2	24
Subsequent strokes		122	0.2	1.4	11
Complete flash		94	1.3	7.5	40
Impulse charge (excluding continuing current)	C				
First strokes		90	1.1	4.5	20
Subsequent strokes		117	0.22	0.95	4
Front duration (2 kA to peak)	μs				
First strokes		89	1.8	5.5	18
Subsequent strokes		118	0.22	1.1	4.5
Maximum dI/dt	kA μs^{-1}				
First strokes		92	5.5	12	32
Subsequent strokes		122	12	40	120
Stroke duration (2 kA to half peak value on the tail)	μs				
First strokes		90	30	75	200
Subsequent strokes		115	6.5	32	140
Action integral ($\int I^2 dt$)	A^2 s				
First strokes		91	6.0×10^3	5.5×10^4	5.5×10^5
Subsequent strokes		88	5.5×10^2	6.0×10^3	5.2×10^4
Time interval between strokes	ms	133	7	33	150
Flash duration	ms				
All flashes		94	0.15	13	1100
Excluding single-stroke flashes		39	31	180	900

that the directly measured current waveforms of either polarity found in the literature do not exhibit peaks exceeding 300 kA or so, although less reliable peak current estimates from the residual magnetization of ferromagnetic detectors (magnetic links) and inferences from remotely measured electric and magnetic fields (considered later in this section) suggest the existence of currents up to 500 kA and even higher (e.g., Le Boulch and Plantier 1990). Lyons *et al.* (1998), using NLDN data for 14 selected summer months from 1991–5, reported that the largest current peaks were 957 and 580 kA for negative and positive flashes, respectively. It is important to note that the peak currents reported by the NLDN and by other similar systems are estimated from measured magnetic radiation field peaks using an empirical formula, the validity of which has been tested only for negative subsequent strokes with peak currents not exceeding 60 kA, as discussed in Section 17.5. The action integral (also referred to as the specific energy) in Table 4.4 represents the energy that would be dissipated in a 1 Ω resistor if the lightning current were to flow through it. It is thought that the heating of electrically conducting materials and the explosion of nonconducting materials is, to a first approximation, determined by the value of the action integral (Section 18.2). Note that the interstroke interval in Table 4.4 was probably mislabeled in Berger *et al.* (1975) and is actually the no-current interval, that is, the interstroke interval excluding any continuing current, as discussed by Fisher *et al.* (1993). On the one hand, essentially no correlation was found between the current peak and the maximum current rate of rise (the correlation coefficients reported by Berger *et al.* (1975)) for negative first and subsequent strokes are 0.36 and 0.11, respectively). On the other hand, similar data for triggered-lightning strokes suggest that a moderate correlation does exist between the current peak and either the 10–90 or the 30–90 percent rate of rise (Fisher *et al.* 1993), and that a relatively strong correlation exists between the

4.6. Return stroke

current peak and the maximum rate of rise (Leteinturier et al. 1990, 1991). The triggered-lightning results are discussed in more detail in subsection 7.2.4. Anderson and Eriksson (1980), using Berger's data, found a weak correlation between the logarithms of peak current and either the 10–90 percent or the 30–90 percent rate of rise.

Direct measurements of natural lightning return-stroke currents on instrumented towers have been made also by researchers in the United States (McCann 1944), in Italy (Garbagnati et al. 1978; Garbagnati and Lo Piparo 1982), in Russia (Gorin et al. 1977; Gorin and Shkilev 1984), in South Africa (Eriksson 1978a), in Canada (Hussein et al. 1995; Janischewskyj et al. 1997), in Germany (Beierl 1992; Fuchs et al. 1998), in Japan (Miyake et al. 1992; Goto and Narita 1992, 1995), in Switzerland (Montandon 1992), in Austria (Diendorfer et al. 2000), and in Brazil (Lacerda et al. 1999). In most studies, the towers experienced predominantly upward discharges (see Table 2.3), the type discussed in Chapter 6. The measurements for negative downward flashes in Italy yielded results similar to those of Berger et al. (1975), although the downward and upward flashes were separated solely on the basis of the early part of the first-stroke current waveshapes, as discussed in more detail in Section 6.3.

The measurements in South Africa were made on a 60-m-high tower (the CSIR research mast; Table 6.1) located on a hill some 80 m above the surrounding relatively flat terrain. The tower was insulated from ground via a short spark gap that operated during the lightning discharge, and the current was measured at the bottom of the tower using a current transformer (a Rogowski coil). Most of the other published tower measurements have been made at the top of the tower. All flashes recorded at the South African tower lowered negative charge to ground. The 50-percent peak current for 11 first return strokes was 41 kA, about 30 percent higher than in Switzerland. However, Melander (1984) estimated that the current peaks measured at the bottom of the South African tower were a factor of about 1.6 (60 percent) greater than the actual peaks, owing to current reflections within the tower (Section 6.5). In one downward flash composed of three strokes, whose currents are shown in Fig. 4.35, a maximum current steepness of 180 kA µs^{-1} was measured for the second stroke, appreciably higher than the five percent value of 120 kA µs^{-1} (Table 4.4) reported for subsequent strokes by Berger et al. (1975). The corresponding peak current was also unusually large, over 70 kA.

As reported by Berger and Garbagnati (1984), the smallest measurable time in the oscillograms of Berger et al. (1975) was 0.5µs (0.5 mm on the 30 µs oscilloscopic sweep). Note that in Table 4.4 the 95 percent value of 0.22 µs for the front duration for subsequent strokes is a prediction of the log-normal approximation, not an experimental value. It follows from the above that the risetimes reported

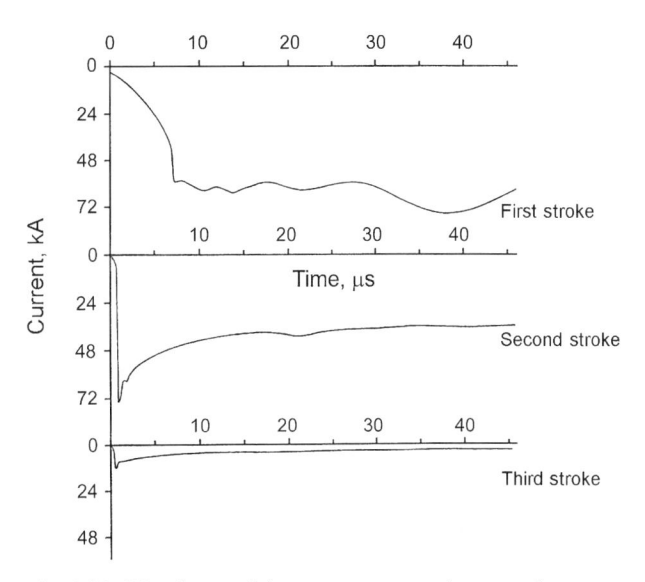

Fig. 4.35. Waveforms of the current measured at tower bottom for the first stroke and two subsequent strokes in a flash in South Africa. Adapted from Eriksson (1978a).

by Berger et al. (1975) are likely to be biased toward larger values and the rates of rise toward lower values. The median 10–90 percent risetime estimated for subsequent strokes by Anderson and Eriksson (1980) from the oscillograms of Berger et al. (1975) is 0.6 µs, comparable to the median values, ranging from 0.3 to 0.6 µs, for triggered lightning strokes (Leteinturier et al. 1991; Fisher et al. 1993). The median 10–90 percent current rate of rise reported for natural subsequent strokes by Anderson and Eriksson (1980) is 15 kA µs^{-1}, almost three times lower than the corresponding value of 44 kA µs^{-1} in the data of Leteinturier et al. (1991) and more than twice lower than the value of 34 kA µs^{-1} found by Fisher et al. (1993). The largest value of the maximum rate of rise, 411 kA µs^{-1}, was measured by Leteinturier et al. (1991, Fig. 4) for a triggered-lightning stroke terminating on a launcher grounded to salt water. The corresponding directly measured current was greater than 60 kA, the largest reported for summer triggered lightning. The mean value of the current derivative reported by Leteinturier et al. (1991) is 110 kA µs^{-1}. That the observed values of current rate of rise are higher for triggered-lightning return strokes than for natural return strokes is likely to be due to better instrumentation (digital oscilloscopes with better upper frequency response), although the influence of lightning triggering conditions has not yet been ruled out.

A variety of indirect *in situ* methods have been used to obtain crude estimates of lightning currents, typically peak value and maximum rate of rise, particularly in the 1930s to 1950s in the USSR (e.g., Stekolnikov and Lamdon 1942), United States (e.g., Lewis and Foust 1945), and Germany (e.g., Baatz 1951). These methods, including those using magnetizable materials placed near the current carrying conductor, are summarized by Uman (1969a, 1984). Additionally, there have been many attempts to infer the values of lightning currents from remotely measured electric and magnetic fields (e.g., Norinder and Dahle 1945; Uman and McLain 1970b; Uman et al. 1973a, b; Dulzon and Rakov 1980; Krider et al. 1996; Cummins et al. 1998). For example, Krider et al. (1996), using measured electric field derivatives and the simple transmission line model (subsection 12.2.5), inferred that average values of the peak of the dI/dt waveform and its duration at half-peak value for first return strokes were 115 kA μs^{-1} and 75 ns, respectively. Such "remote" measurements are model dependent and, therefore, inferior to direct measurements. However, recent triggered-lightning studies indicate that a reasonable estimate of the subsequent return-stroke peak current I can be obtained from the electric field E measured at a distance D using the following regression equation (Rakov et al. 1992b):

$$I = 1.5 - 0.037ED \quad (4.9)$$

where I is in kA and taken as negative, E is positive and in V m^{-1}, and D is in km. The triggered-lightning data (28 events) used to derive Eq. 4.9 were obtained by Willett et al. (1989a), and the range of peak field values was from 1.9 to 11 V m^{-1} (normalized to 100 km). For 18 out of 28 strokes used to derive Eq. 4.9 the two-dimensional return-stroke speed, averaged over 457–625 m, was measured and found to vary from 1.2×10^8 to 1.9×10^8 m s^{-1}, with mean value about 1.5×10^8 m s^{-1}. The field-to-current conversion equation used in the NLDN is discussed in Section 17.5.

4.6.2. Luminosity variation along the channel and propagation speed

The variation in return-stroke luminosity along the channel is thought to reflect the variation in current. Since the latter variation is impossible to measure directly, the luminosity profile along the channel, typically obtained using streak photography, is generally viewed as representative of the current variation.

Another characteristic often determined from streak photographs is the propagation speed. The method of finding the return-stroke propagation speed is straightforward for an idealized straight and vertical channel, but in practice requires a second channel image, either still or streaked in the opposite direction (see Idone and Orville 1982). Figure 4.36 illustrates the time development of the luminosity in the lower portion of a typical first stroke, determined using streak photography in South Africa. In this event, the upward two-dimensional speed along the main trunk was 1.6×10^8 m s^{-1} from the ground to point A,

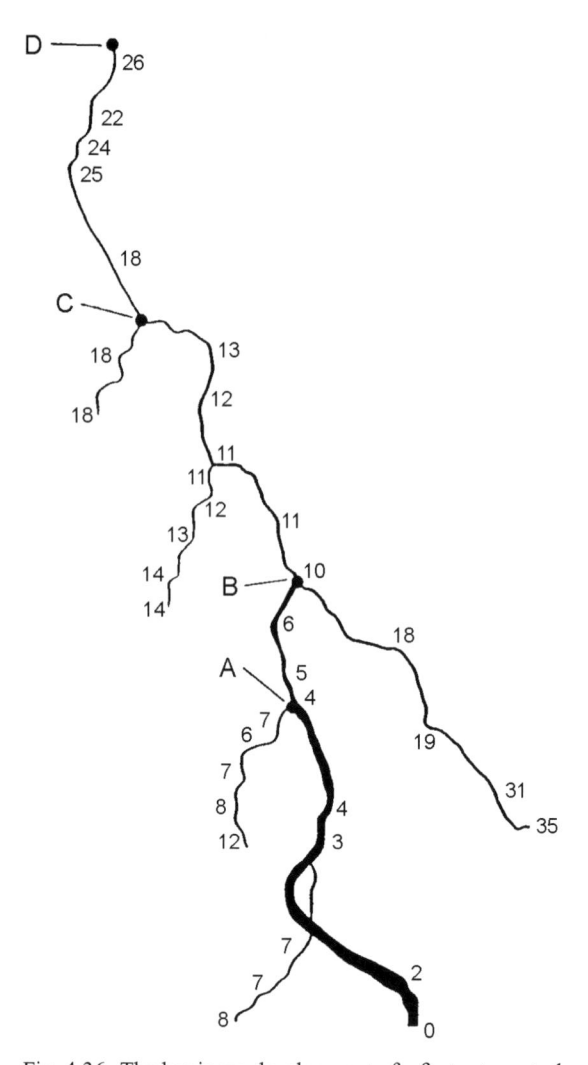

Fig. 4.36. The luminous development of a first return stroke. The numbers indicate the time of arrival in microseconds of the upward-propagating return-stroke front at various points on both the main channel and branches. Adapted from Schonland (1956).

2.1×10^8 m s^{-1} from A to B, 9.7×10^7 m s^{-1} from B to C, and 5.5×10^7 m s^{-1} from C to D (Schonland 1956). When the luminosity reached the large branch at B, it appeared to pause for about 5 μs, and then it proceeded both down the branch and upward with less intensity.

When a return stroke reaches a branch, there is usually a brightening of the channel below that point, this brightening being termed a branch component. It is generally thought that branch components are due to a rapid discharge of a branch, previously charged by the stepped leader, through the channel section below the branch point to ground. Branch components are thought to be responsible for the secondary maxima usually observed in the channel-base current waveforms (subsection 4.6.1). Channel brightening can also occur after the first return stroke has entered the cloud and in typically branchless (at least outside the

4.6. Return stroke

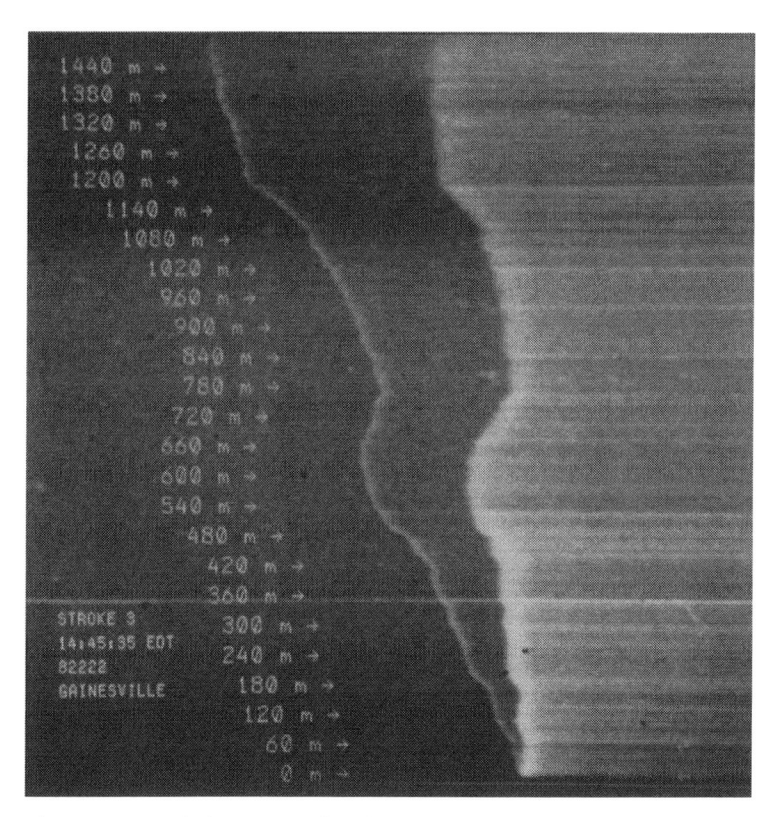

Fig. 4.37a. Streak photograph of stroke 3 in a flash that occurred in Gainesville, Florida on 10 August 1982. The image on the left is due to the dart leader, and the brighter image on the right is due to the return stroke. Adapted from Jordan *et al.* (1997).

cloud) subsequent strokes. In the latter case, brightening often occurs many milliseconds to tens of milliseconds after the return stroke onset, during the continuing current stage. When no branches are seen, a channel brightening is called an M-component (Section 4.9). The channel brightening during continuing current is unlikely to be due to in-cloud branches created by the leader but rather due to concentrated fresh charge made available to the continuing current channel. Because of the lack of branches, the light profile along a subsequent return-stroke channel is relatively simple, usually showing a gradual intensity decay with height as illustrated in Figs. 4.37a, b, c. Jordan and Uman (1983) found for seven subsequent return strokes an exponential decrease in the luminosity peak with height with a decay constant of 0.6–0.8 km, resulting in a luminosity-peak decrease at height 480 m to 45–55 percent of its value at ground and to 9.7–17 percent at height 1400 m. Further, Jordan *et al.* (1995) reported, from a different experiment, a decrease to 33 percent at 600 m and to 19 percent at 1100 m for one subsequent return stroke which they examined in relation to their luminosity-versus-height analysis of an M-component (Section 4.9). Finally, from their analysis of the light profiles of three dart leader–return-stroke sequences, Jordan *et al.* (1997) found that in two out of three events (one of those two is shown in Figs. 4.37a, b, c) the return-stroke luminosity peak at 480 m and 1400 m had decayed to, respectively, 70–75 percent and 25–30 percent of its value at the bottom of the channel and in the third event to, respectively, 90–95 percent and about 70 percent of the channel-bottom value.

As seen in Figs. 4.37b, c, besides a decrease in peak value, the return-stroke light pulses also exhibit an appreciable increase in risetime (that is, a degradation of the front) with increasing height. Jordan and Uman (1983) reported that the 20–80 percent risetime of the return-stroke light signal was between 1 and 4 μs near ground and increased by an additional 1 to 2 μs by the time the return stroke reached the cloud base, a height between 1 and 2 km. Jordan *et al.* (1997) reported that the 20–80 percent risetime increased from 1.5 to 4.0 μs (mean values) as the return stroke propagated from ground to the cloud base at about 1400 m. The return-stroke relative light intensity 30 μs after the initial peak tends to be relatively constant with height, being 15–30 percent of the initial peak near the ground and 50–100 percent of the initial peak near the cloud base (Jordan and Uman 1983). Wang *et al.* (1999a), using the ALPS optical imaging system, observed a

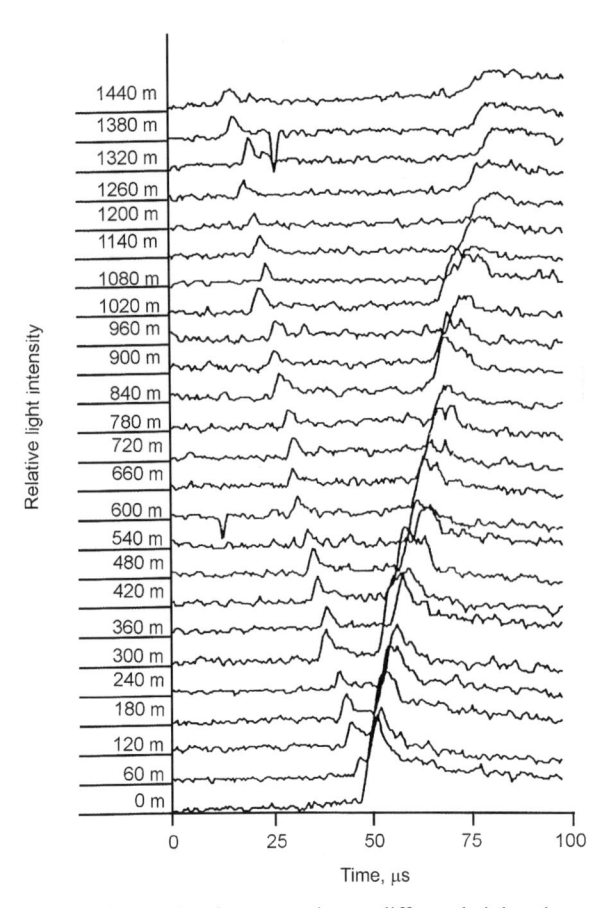

Fig. 4.37b. Luminosity versus time at different heights above ground corresponding to the streak photograph in Fig. 4.37a.

pulse peak was strongly correlated with the channel-base current peak, the latter ranging from 1.6 to 21 kA. The linear correlation coefficients for the two flashes were 0.97 and 0.92.

The optically measured return-stroke speed probably represents the speed of the region of the upward-moving return-stroke tip where power losses are greatest, the power per unit length being the product of the current and the longitudinal electric field in the channel. It is likely that the peak of the power loss wave occurs earlier in time than the peak of the current wave (e.g., Gorin 1985). Since the shape of the return-stroke light pulse changes significantly with height (see Figs. 4.37b, c), there is always some uncertainty when tracking the propagation of such pulses for a speed measurement. For example, if the light-pulse peak is tracked then an increase in pulse risetime, seen in Figs. 4.37b, c, translates into a lower speed value than that obtained if an earlier part of the light pulse is tracked. It is thought that the error involved in identifying the time of the initial exposure on streak photographs, as a considerable decrease in the light-pulse peak within the first tens of meters of propagation for return strokes in triggered lightning.

There have been several attempts to relate the return-stroke light-pulse characteristics to the corresponding characteristics of current and remote field pulses. Jordan and Uman (1983), using streak photography, found that the logarithm of the light-pulse initial peak near ground was linearly related to the initial electric field peak. Ganesh et al. (1984) used a photoelectric system to isolate the light from a section of a return-stroke channel in an attempt to verify this latter result. They presented correlated light and electric field waveforms for both close and distant lightning but were not able to determine the best type of regression equation relating the peak values. Ganesh et al. (1984) also measured light-pulse risetimes ranging from 1 to 60 µs, the larger values probably being due to a combination of multiple scattering of the optical signal off aerosols and the viewing of higher channel portions in the case of more distant lightning. Idone and Orville (1985) found, for 39 strokes in two triggered-lightning flashes, that the return-stroke light-

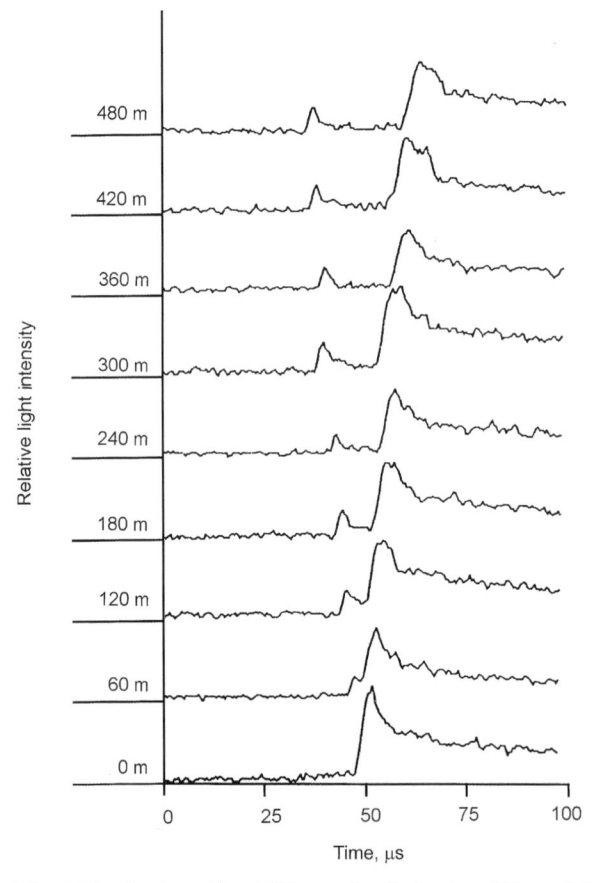

Fig. 4.37c. Same as Fig. 4.37b, but for the bottom 480 m of the channel only.

4.6. Return stroke

Table 4.5. *Summary of measured return-stroke speeds in natural and triggered lightning. Adapted from Rakov et al. (1992b)*

Reference	Min. speed, m s^{-1}	Max. speed, m s^{-1}	Mean speed, m s^{-1}	Standard deviation, m s^{-1}	Sample size	Comments
Natural lightning						
Boyle and Orville (1976)	2.0×10^7	1.2×10^8	0.71×10^8	2.6×10^7	12	Streak camera, 2-D speed
Idone and Orville (1982)	2.9×10^7	2.4×10^8	1.1×10^8	4.7×10^7	63	Streak camera, 2-D speed
Mach and Rust (1989a, Fig. 7)	2.0×10^7	2.6×10^8	$(1.3 \pm 0.3) \times 10^8$	5×10^7	54	Long channel
	8.0×10^7	$>2.8 \times 10^8$	$(1.9 \pm 0.7) \times 10^8$	7×10^7	43	Short channel (Photoelectric, 2-D)
Triggered lightning						
Hubert and Mouget (1981)	4.5×10^7	1.7×10^8	9.9×10^7	4.1×10^7	13	Photoelectric, 3-D speed
Idone et al. (1984)	6.7×10^7	1.7×10^8	1.2×10^8	2.7×10^7	56	Streak camera, 3-D speed
Willett et al. (1988)	1.0×10^8	1.5×10^8	1.2×10^8	1.6×10^7	9	Streak camera, 2-D speed
Willett et al. (1989a)	1.2×10^8	1.9×10^8	1.5×10^8	1.7×10^7	18	Streak camera, 2-D speed
Mach and Rust (1989a, Fig. 8)	6.0×10^7	1.6×10^8	$(1.2 \pm 0.3) \times 10^8$	2×10^7	40	Long channel
	6.0×10^7	2.0×10^8	$(1.4 \pm 0.4) \times 10^8$	4×10^7	39	Short channel (Photoelectric, 2-D)

basis for the speed measurements, is not large, especially near ground. Techniques for measuring the return-stroke speed are discussed by, for example, Idone and Orville (1982).

Schonland et al. (1935) found that the first-return-stroke speed at the channel base was typically near 1×10^8 m s^{-1}, and at the top of the main channel it was typically near 5×10^7 m s^{-1}. A summary of more recently measured return-stroke speeds, averaged over the lowest few hundreds of meters of the channel, for both natural and rocket-triggered lightning is given in Table 4.5. In natural lightning, the two-dimensional return-stroke speed (for both first and subsequent strokes combined) was reported by Idone and Orville (1982) from streak-camera measurements to vary from 2.9×10^7 to 2.4×10^8 m s^{-1}, almost an order of magnitude. The sample of Idone and Orville (1982) includes speeds for 17 first and 46 subsequent return strokes, the mean values within about 1.3 km being 9.6×10^7 m s^{-1} and 1.2×10^8 m s^{-1}, respectively. Idone and Orville (1982) found that the return-stroke speed usually decreased with height, for both first and subsequent strokes, by 25 percent or more over the visible part of channel relative to the speed near ground. Boyle and Orville (1976) reported return-stroke speeds from 12 strokes varying from 2.0×10^7 to 1.2×10^8 m s^{-1}. A similar wide speed range for natural lightning was found from photoelectric measurements by Mach and Rust (1989a, Fig. 13). The more recently measured return-stroke speeds presented in Table 4.5 are generally higher than the earlier results of Schonland et al. (1935), probably owing in part to the fact that the recent measurements were made closer to the ground, where the return-stroke speed tends to be higher.

In triggered lightning, the return-stroke speed range was found to be 6.7×10^7 to 1.7×10^8 m s^{-1} from streak-camera measurements (three-dimensional speed) (Idone et al. 1984) and 6×10^7 to 1.6×10^8 m s^{-1} from photoelectric measurements in the lowest channel section longer than 500 m (the "long-channel" two-dimensional speed) (Mach and Rust, 1989a, Figs. 8 and 14). Accompanying photoelectric measurements in channel sections less than 500 m in length (the "short-channel" two-dimensional speed) resulted in a somewhat wider range of speeds, from 6×10^7 to 2×10^8 m s^{-1} (see Fig. 8 of Mach and Rust 1989a). From earlier photoelectric measurements, Hubert and Mouget (1981) reported a three-dimensional return-stroke speed range of 4.5×10^7 to 1.7×10^8 m s^{-1}.

Wang et al. (1999b) reported on two-dimensional speed profiles within 400 m of ground for two return strokes in triggered lightning. The speed profiles were obtained using the digital optical imaging system ALPS, which had a time resolution of 100 ns and a spatial resolution of 30 m. The return-stroke speeds within the bottom 60 m of the channel were found to be 1.3×10^8 and 1.5×10^8 m s^{-1}. Weidman (1998) reported mean return-stroke speeds in the lowest 100 m of the lightning channel of 7.8×10^7 and 8.8×10^7 m s^{-1} for nine natural-lightning and 14 triggered-lightning strokes, respectively.

Some researchers (e.g., Lundholm 1957; Wagner 1963) have suggested that the return-stroke speed should increase with increasing current peak. This suggestion, implying that the return-stroke wave is highly nonlinear, so that the wave speed is a function of wave amplitude, is not supported by the experimental data. In particular, Willett et al. (1989a) and Mach and Rust (1989a) found a lack of correlation between the return-stroke propagation speed and the return-stroke peak current in triggered lightning. Idone et al. (1984) did observe "a nonlinear relationship" between these two parameters in triggered lightning, but it disappears if one excludes the relatively small events that are characterized by return-stroke peak currents less than 6–7 kA, in order to make the sample of Idone et al. (1984) similar to those of Willett et al. (1989a) and Mach and Rust (1989a).

If there were a relationship between the return-stroke speed and return-stroke current, as might be expected on physical grounds, then it would be influenced by many factors and, as a result, characterized by a large scatter. Rakov (1998) inferred, from a comparison of the behavior of traveling waves on a lossy transmission line and the observed characteristics of the lightning return-stroke process, that the return stroke is similar to a "classical" (linear) traveling wave. Ionization does occur during the return-stroke process but has a relatively small effect on the wave propagation characteristics, which, according to Rakov (1998), are primarily determined by the transmission-line parameters ahead of the front as opposed to being determined by the wave magnitude. As a result, the return-stroke wave suffers appreciable attenuation and dispersion.

4.6.3. Measured electric and magnetic fields

In this subsection, we will discuss measurements of the vertical and horizontal electric fields and the horizontal magnetic fields produced by negative return strokes on microsecond and submicrosecond time scales. Return-stroke vertical electric fields on a millisecond time scale are found in Figs. 4.6, 4.9, 4.11, 4.12, 4.15, and 4.16. Typical vertical electric and horizontal magnetic field waveforms at distances ranging from 1 to 200 km for both first and subsequent strokes were published by Lin et al. (1979). These waveforms, which are drawings based on many measurements acquired in Florida, are reproduced in Fig. 4.38.

The electric fields of strokes observed within a few kilometers of the flash, shown in Fig. 4.38, are dominated, after the first few tens of microseconds, by the electrostatic component of the total electric field, the only field component that is nonzero after the stroke current has ceased to flow. The individual field components that comprise the electric and magnetic fields are discussed at the start of subsection 4.6.4. The close magnetic fields at similar times are dominated by the magnetostatic component of the total magnetic field, the component that produces the magnetic field humps seen in Fig. 4.38. The distant electric and magnetic fields have essentially identical waveshapes and are usually bipolar, as illustrated in Fig. 4.38. The data of Lin et al. (1979) suggest that at a distance of 50 km and beyond, both electric and magnetic field waveshapes are dominated by their respective radiation components.

The initial field peak evident in the waveforms of Fig. 4.38 is the dominant feature of the electric and magnetic field waveforms beyond about 10 km; this initial peak also is a significant feature of waveforms from strokes between a few km and about 10 km and can be identified, with some effort, in waveforms for strokes as close as a kilometer. The initial field peak is due to the radiation component of the total field and, hence, as discussed at the start of subsection 4.6.4, decreases inversely with distance in the absence of significant propagation effects (Lin et al. 1979, 1980). The field peaks produced by different return strokes at known distances can be range-normalized for comparison, for example, to 100 km by multiplying the measured field peaks by $r/10^5$, where r is the stroke distance in meters. Statistics on the normalized initial electric field peak, derived from selected studies, are presented in Table 4.6, and histograms of normalized initial electric field peak for strokes of different order recorded within about 20 km in Florida are given in Fig. 4.39. The mean of the electric field initial peak value, normalized to 100 km, was generally found to be in the range 6–8 V m^{-1} for first strokes and 3–6 V m^{-1} for subsequent strokes. The higher observed mean values are likely to be an indication of the fact that small strokes were missed because the equipment trigger threshold was set too high (e.g., Krider and Guo 1983). Since the initial electric field peak appears to obey a log-normal distribution, the geometric mean value (equal to the median value for a log-normal distribution) may be a better characteristic of the statistical distribution of this parameter than the arithmetic mean value. Note that the geometric mean value for a log-normal distribution is lower than the corresponding (arithmetic) mean value and higher than the modal (most probable) value.

A statistical summary of the return-stroke initial electric field peaks for different types of stroke in Florida lightning, along with the preceding interstroke intervals and leader durations, is given in Table 4.7. As stated in Section 4.2, about one-third of multiple-stroke flashes have at least one subsequent stroke that is larger than the first stroke in the flash. The characteristics of such larger subsequent strokes are summarized in Table 4.8. Interestingly, no subsequent stroke with peak field exceeding the first had a preceding interstroke interval less than 35 ms (Thottappillil et al. 1992). Note from Table 4.7, that first strokes in multiple-stroke flashes exhibit a significantly higher geometric-mean electric field peak than that for the only stroke in single-stroke flashes.

From their measured initial field peaks, Krider and Guo (1983) calculated that the electromagnetic power

4.6. Return stroke

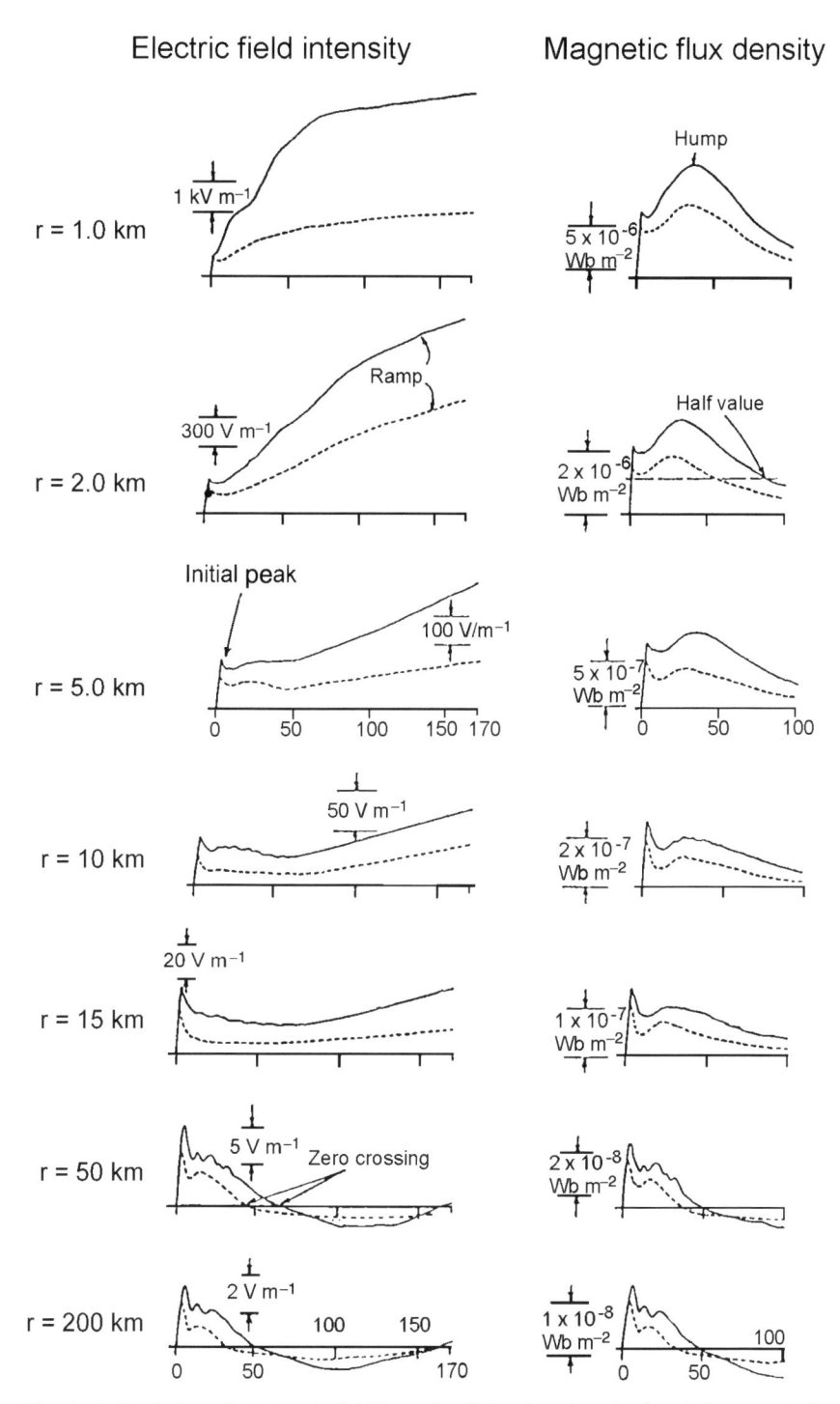

Fig. 4.38. Typical vertical electric field intensity (left column) and azimuthal magnetic flux density (right column) waveforms for first (solid line) and subsequent (broken line) return strokes at distances of 1, 2, 5, 10, 15, 50, and 200 km. The scales in the third diagrams from the top and in the bottom diagrams are in μs. Adapted from Lin et al. (1979).

radiated at the time of those peaks had a mean of 2×10^{10} W for first strokes and 3×10^9 W for subsequent strokes. Krider (1992), using the transmission line model (subsection 12.2.5), calculated that for peak current of 30 kA and a return-stroke speed of 1.5×10^8 m s^{-1} the peak power radiated into the upper hemisphere is 10^{10} W.

Details of the shape of the return-stroke field's rise to peak and the fine structure after the initial peak are shown

Table 4.6. *Parameters of microsecond-scale electric field waveforms produced by negative return strokes. SD is the standard deviation*

Parameter	Location	First strokes			Subsequent strokes		
		Sample size	Mean	SD	Sample size	Mean	SD
Initial peak (V m^{-1}) (normalized to 100 km)							
Rakov and Uman (1990b)	Florida	76	5.9 (GM)		232[a]	2.7(GM)	
					38[b]	4.1(GM)	
Cooray and Lundquist (1982)	Sweden	553	5.3	2.7			
Lin et al. (1979)	KSC	51	6.7	3.8	83	5.0	2.2
	Ocala	29	5.8	2.5	59	4.3	1.5
Zero-crossing time (μs)							
Cooray and Lundquist (1985)	Sweden	102	49	12	94	39	8
	Sri Lanka	91	89	30	143	42	14
Lin et al. (1979)	Florida	46[c]	54	18	77[c]	36	17
Zero-to-peak rise time (μs)							
Master et al. (1984)	Florida	105	4.4	1.8	220	2.8	1.5
Cooray and Lundquist (1982)	Sweden	140	7.0	2.0			
Lin et al. (1979)	KSC	51	2.4	1.2	83	1.5	0.8
	Ocala	29	2.7	1.3	59	1.9	0.7
10–90 percent rise time (μs)							
Master et al. (1984)	Florida	105	2.6	1.2	220	1.5	0.9
Slow front duration (μs)							
Master et al. (1984)	Florida	105	2.9	1.3			
Cooray and Lundquist (1982)	Sweden	82	5.0	2.0			
Weidman and Krider (1978)	Florida	62	4.0	1.7	44	0.6	0.2
		90	4.1	1.6	120	0.9	0.5
					34[d]	2.1	0.9
Slow front, amplitude as percentage of peak							
Master et al. (1984)	Florida	105	28	15			
Cooray and Lundquist (1982)	Sweden	83	41	11			
Weidman and Krider (1978)	Florida	62	50	20	44	20	10
		90	40	20	120	25	10
					34[d]	40	20
Fast transition, 10–90 percent risetime (ns)							
Master et al. (1984)	Florida	102	970	680	217	610	270
Weidman and Krider (1978)	Florida	38	200	100	80	200	40
		15	200	100	34	150	100
Weidman and Krider (1980a, 1984); Weidman (1982)	Florida	125	90	40			
Peak time derivative (normalized to 100 km) (V m^{-1}μs^{-1})							
Krider et al. (1996)	Florida	63	39	11			
Width of time derivative pulse at half-peak value (ns)							
Krider et al. (1996)	Florida	61	100	20			

If not specified otherwise, multiple lines for a given source at the same location correspond to different thunderstorms. GM, geometric mean value, a better characteristic of the distribution of initial field peaks since this distribution is approximately log-normal.

[a] Strokes following previously formed channel.

[b] Strokes creating new termination on ground.

[c] Both electric and magnetic fields.

[d] Subsequent strokes initiated by dart-stepped leaders. Other subsequent strokes studied by Weidman and Krider (1978) were initiated by dart leaders.

4.6. Return stroke

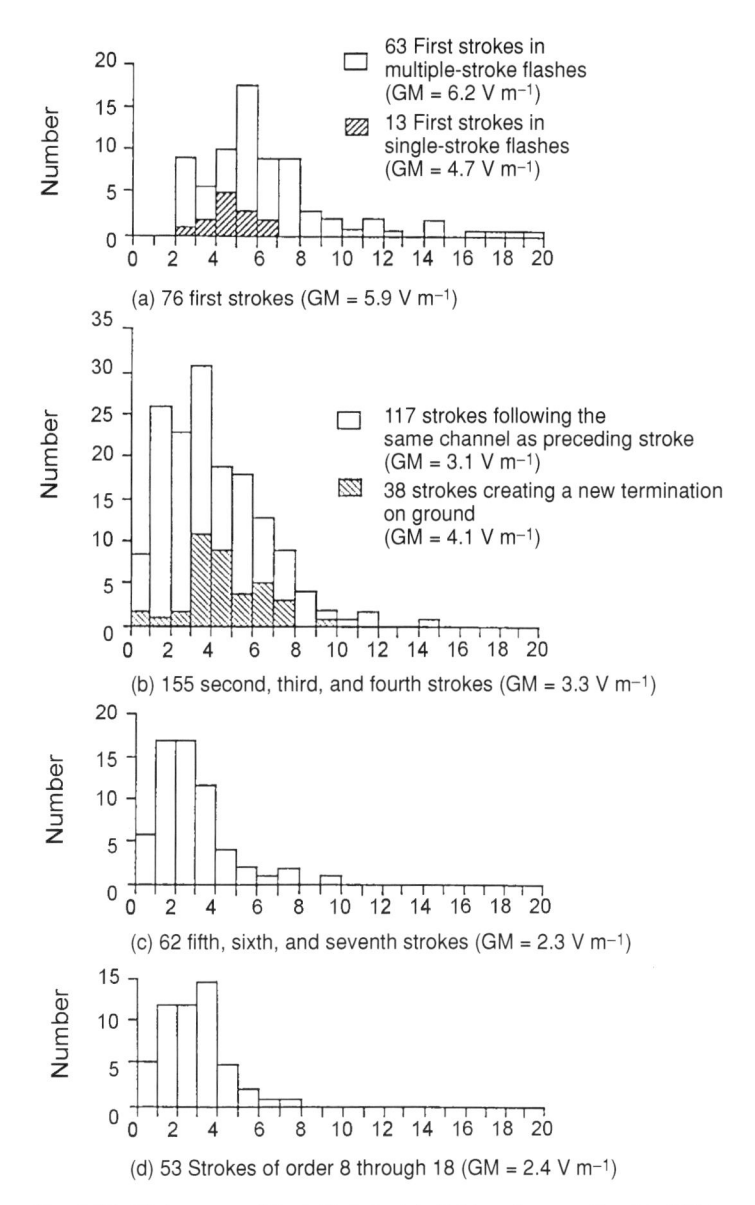

Fig. 4.39. Histograms of the initial electric field peak normalized to 100 km, along the horizontal axis, for strokes of different order in 76 flashes in Florida. GM stands for geometric mean. Adapted from Rakov and Uman (1990b).

in Fig. 4.40, and some measured characteristics are summarized in Table 4.6. Note that also shown in Fig. 4.40 are stepped-leader and dart-stepped-leader pulses occurring prior to the return stroke pulse. As illustrated in Fig. 4.40, first-return-stroke field waveforms have a "slow front" (below the broken line in Fig. 4.40 and labeled F) that rises in a few microseconds to an appreciable fraction of the field peak. Weidman and Krider (1978) found a mean slow-front duration of about 4.0 μs with 40 to 50 percent of the initial field peak attributable to the slow front. Master *et al.* (1984) found a mean slow-front duration of 2.9 μs, about 30 percent of the initial field peak being due to the slow front. Cooray and Lundquist (1982) reported corresponding values of 5 μs and 41 percent.

The slow front is followed by a "fast transition" (labeled R in Fig. 4.40) to peak with a 10–90 percent risetime of about 0.1 μs when the field propagation path is over salt water. Weidman and Krider (1980a, 1984) and Weidman (1982) reported a mean risetime of 90 ns with a standard deviation of 40 ns for 125 first strokes. As illustrated in Fig. 4.40, fields from subsequent strokes have fast transitions similar to those of first strokes except that these transitions account for most of the rise to peak, while the slow fronts are of shorter duration than for first strokes, typically

Table 4.7. *Characterization of downward negative lightning in Florida. N is the sample size, GM is the geometric mean, and SD is the standard deviation of the logarithm (base 10) of the parameter. Adapted from Rakov et al. (1994)*

Characterization of strokes	Electric field peak at 100 km			Interstroke interval			Leader duration		
	N	GM, V m^{-1}	SD log	N	GM, ms	SD log	N	GM, ms	SD log
All first strokes	76	5.9	0.22	—	—	—	70	35	0.20
First strokes in multiple-stroke flashes	63	6.2	0.23	—	—	—	58	33	0.19
First strokes in single-stroke flashes	13	4.7	0.12	—	—	—	12	46	0.23
All subsequent strokes	270	2.9	0.30	270	60	0.35	182	2.5	0.52
Subsequent strokes creating a new termination	38	4.1	0.23	38	92	0.30	28	15	0.47
Subsequent strokes in previously formed channel	232	2.7	0.30	232	56	0.35	154	1.8	0.38
Subsequent leader–return-stroke sequence with negative net electric field charge*	18	1.1	0.33	18	47	0.52	18	2.8	0.29

*Usually, there is a net positive electric field change (atmospheric electricity sign convention, subsection 1.4.2), owing to a leader–return-stroke sequence lowering negative charge to ground. A net negative field change means that not all the leader charge is neutralized by the return stroke. Note the very small return-stroke electric field peak associated with the sequence exhibiting a negative net field change.

Table 4.8. *Geometric mean values for various parameters of larger-than-first-stroke subsequent strokes and of all subsequent strokes, both types of stroke occurring in the same channel as the first stroke. The numbers in the parentheses are the sample sizes. Adapted from Rakov et al. (1994)*

Parameter	Larger strokes	All strokes
Return-stroke field peak (at 100 km), V m^{-1}	7.7 (13)	2.6 (176)
Return-stroke current peak[a], kA	−27	−8.1
Preceding interstroke interval, ms	98 (13)	53 (176)
Leader duration, ms	0.55 (8)	1.8 (117)
Ratio of subsequent-stroke to first-stroke field peak	1.2 (13)	0.39 (176)

[a] Inferred from return-stroke field peak at 100 km using Eq. 4.9.

0.5–1 µs, and only comprise about 20 percent of the total rise to peak (Weidman and Krider 1978).

The higher-frequency content of the return-stroke fields is preferentially degraded in propagating over a finitely conducting earth (Uman et al. 1976; Lin et al. 1979; Weidman and Krider 1980a, 1984; Cooray and Lundquist 1983), so that the fast transition and other rapidly changing fields can only be adequately observed if the propagation path from the lightning to the antenna is over salt water, a relatively good conductor. It is for that reason that the fast transition time observed by Master et al. (1984) for lightning in the 1 to 20 km range over land is an order of magnitude greater than that observed over salt water by Weidman and Krider (1980a, 1984) and Weidman (1982) (Table 4.6). Lin et al. (1979) reported from two-station measurements that normalized field peaks are typically attenuated by 10 percent in propagating over 50 km of Florida soil and 20 percent in propagating over 200 km. Uman et al. (1976) reported on field risetimes observed both near a given stroke and 200 km from it. For typical strokes, zero-to-peak risetimes (Table 4.6) are increased by an amount of order 1 µs in propagating 200 km across Florida soil. Willett et al. (1990), who studied lightning electromagnetic radiation field spectra between 0.2 and 20 MHz, reported that propagation over less than 35 km or so of seawater had little effect on the observed spectral amplitudes, while about 45 km

4.6. Return stroke

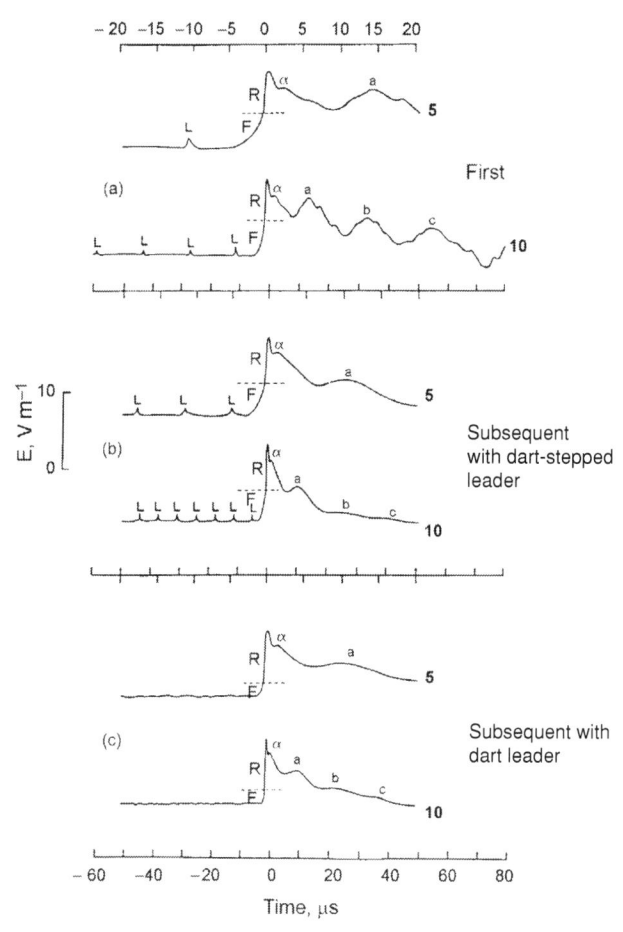

Fig. 4.40. Electric field waveforms of (a) a first return stroke, (b) a subsequent return stroke initiated by a dart-stepped leader, and (c) a subsequent return stroke initiated by a dart leader, showing the fine structure both before and after the initial field peak. Each waveform is shown on two time scales, 5 μs per division (labeled 5) and 10 μs per division (labeled 10). The fields are normalized to a distance of 100 km. L, individual leader pulses; F, slow front; R, fast transition. Also marked are the small secondary peak or shoulder, α, and the larger subsidiary peaks a, b, and c. Adapted from Weidman and Krider (1978).

of such propagation did appear to produce significant attenuation above 10 MHz.

We now discuss measurements of the time derivative of the vertical electric field for natural-lightning return strokes. Krider et al. (1996) reported on such measurements made in 1984 with a system "designed to provide a time resolution of about 10 ns". They studied 63 negative first-return strokes that occurred in one thunderstorm over the Atlantic Ocean about 35 km east of the measuring site, which was located on the easternmost tip of Cape Canaveral, Florida. An example of the measured time derivative waveform is shown, along with the corresponding electric field waveform, in Fig. 4.41. Krider et al. (1996) found a mean peak electric field derivative of 39 V m^{-1} μs^{-1}, normalized

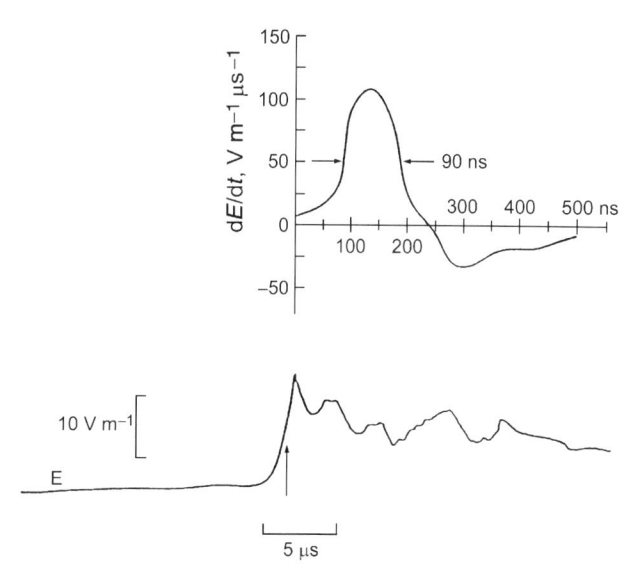

Fig. 4.41. An example from 5 September 1984 at 20:56:49.78 UT of (top) the time derivative of the electric field intensity dE/dt and (bottom) the electric field intensity E itself, produced by a first return stroke at a distance of about 36 km over the Atlantic Ocean. The propagation path was almost entirely over salt water. The vertical arrow under the E record shows the time of the dE/dt trigger. Adapted from Krider et al. (1996).

to 100 km (using the inverse distance dependence valid for radiation fields propagating over a perfectly conducting ground), with a derivative pulse width of about 100 ns at half maximum (see Table 4.6) and about 200 ns at zero level. Employing the methodology of Zeddam et al. (1990) for evaluating propagation effects, Krider et al. (1996) estimated that the peak field derivative would have been reduced by about 20 percent and the derivative pulse width increased by about 30 percent in propagating over a calm sea surface, and they calculated mean peak and mean width values, corrected for propagation, equal to 46 V m^{-1} μs^{-1} and 75 ns, respectively. They did not correct for potential propagation effects caused by a rough ocean surface (Ming and Cooray 1994) since any waves that might have been present were of unknown height; nor did they correct for propagation effects over the few tens of meters of beach from the shoreline to the antenna location (Cooray and Ming 1994), which they argue are small. These additional corrections would potentially increase further the estimated peak derivative value and decrease the estimated derivative pulse half-peak width.

Willett et al. (1990, 1998) reported on 1985 measurements of the electric field derivative for negative lightning strokes in thunderstorms over the Atlantic Ocean near Cape Canaveral that were similar to the 1984 measurements reported by Krider et al. (1996). The antenna was 45 m from the Atlantic Ocean on Playalinda Beach, a narrow barrier island 5 km or so east of the Florida mainland. For 131 first strokes in the range 6 to 40 km, Willett et al. (1998) found a mean peak derivative of 42 V m^{-1} μs^{-1} normalized to 100 km and a mean derivative pulse width at half-peak value equal to 64 ns, both values corrected for propagation effects by using the same approach as Krider et al. (1996), described above. That the mean derivative pulse width was 15 percent smaller than that found by Krider et al. (1996) was attributed by Willett et al. (1998) to the use of a "more modern digitizer" and the fact that the waveforms "were analyzed with more precision". The system of Willett et al. (1998) had a stated bandwidth of 30 MHz, a sampling interval of 10 ns, and eight bit amplitude resolution.

Le Vine et al. (1989) reported that the mean peak field derivative normalized to 100 km for 16 subsequent strokes in natural negative lightning flashes at about 35 km over the Atlantic Ocean in 1987 was 48 V m^{-1} μs^{-1}, the mean width at half-peak value being 78 ns; both values are uncorrected for propagation effects. Le Vine et al. (1989) also showed that natural subsequent-stroke field derivative waveforms were very similar to the field derivative waveforms of triggered-lightning strokes at 5.16 km.

A detailed review of various studies of electric field derivative waveforms for both natural and rocket-triggered lightning was given by Uman et al. (2000). Measurements of electric field derivative waveforms radiated by various natural negative lightning processes, including first and subsequent return strokes, stepped and dart-stepped leader steps just prior to return strokes, and initial breakdown pulses, were presented by Willett and Krider (2000). According to Willett and Krider (2000), the electric field derivative waveforms are characterized by half-peak widths that are similar for the various processes they considered, ranging from 54 ns for stepped-leader steps to 79 ns for subsequent return strokes.

We now discuss the fine structure after the initial peak in the electric field waveform. Weidman and Krider (1978) attributed most of this structure to the effects of branches, since it was more pronounced in the waveforms produced by first strokes than in those produced by subsequent strokes (Fig. 4.40). Small secondary peaks or shoulders within 2 μs of the initial peak, labeled α in Fig. 4.40, were interpreted by Weidman and Krider (1978) as being due to reflections from the ground (Section 4.5). Willett et al. (1995) observed that return strokes following channels created by stepped leaders produced electric field derivative (dE/dt) waveforms that appeared noisy after the initial peak, while return strokes in channels traversed by earlier strokes produced waveforms that were "quiet" after the initial peak. Interestingly, similar features are seen in the horizontal electric field waveforms published by Thomson et al. (1988). Davis (1999), using multiple-station dE/dt measurements for 16 negative flashes in Florida, found that the duration of the fine structure after the initial peak was related to the length of the channel traversed by the stepped leader. Specifically, he reported that for first strokes, whose entire channels are created by stepped leaders, the median duration was greater than 141 μs, while for those subsequent strokes in which only the lowest portion of the channel (the median length was estimated to be 1.1 km) was created by the stepped leader the median duration was 37 μs. In the latter case, the sources of the fine-structure pulses were located between the height at which the leader deflected from the previously formed channel and ground. Apparently in support of Davis' (1999) results, Cooray and Perez (1994a) reported that the median duration of 3 MHz radiation associated with negative first strokes was 137 μs.

The average frequency spectra of 74 first return strokes and 55 subsequent strokes from 0.2 to 20 MHz, as reported by Willett et al. (1990), are given in Fig. 4.42. Return strokes are the strongest sources of radio-frequency radiation from cloud-to-ground lightning in the frequency interval 0.2 to 20 MHz. Within that interval, return-stroke radiation decreases somewhat faster than f^{-1} in spectral amplitude (f^{-2} in energy spectral density) up to about 5 MHz, and then there is a broad "knee" centered around 9 MHz. Above about 12 MHz the spectral amplitude appears to decrease as f^{-5} (Willett et al. 1990). For close lightning, the spectra below 10 kHz are dominated by induction and electrostatic field components, while the distant spectra, associated primarily with radiation field components, exhibit a peak between 1 and 10 kHz (Preta et al. 1985; Uman 1987, 2001).

Le Vine and Krider (1977) made narrowband measurements of electromagnetic radiation at 3, 30, 139, and 295 MHz correlated with wideband electric field measurements. They found that first strokes have strong radiation at all these frequencies but that the radiation does not peak until 10–30 μs after the start of the wideband return-stroke electric field waveform, and they suggested that the HF and VHF radiation may be due to the effects of branches in first strokes and cloud processes. Supporting this suggestion is their observation that subsequent strokes, which usually do not have branches, generate little HF or VHF radiation. The observations of Le Vine and Krider (1977) are in reasonable agreement with those of Takagi, M. (1969a, b), who found a spectrum of delays between wideband return-stroke fields and narrowband radiation at 60, 150, and 420 MHz, with a peak in the first-stroke delay distribution at about 10 μs and peaks in the subsequent-stroke delay distribution at about 10 and 50 μs. However, Brook and Kitagawa (1964) reported that radiation at 420 and 850 MHz was always delayed by 60–100 μs in the 50 percent of return strokes for which narrowband radiation was detected, and they suggested that the electromagnetic radiation at those frequencies was due to breakdown at the top of the return stroke channel. Cooray (1986) and Le Vine et al. (1986) argued

4.6. Return stroke

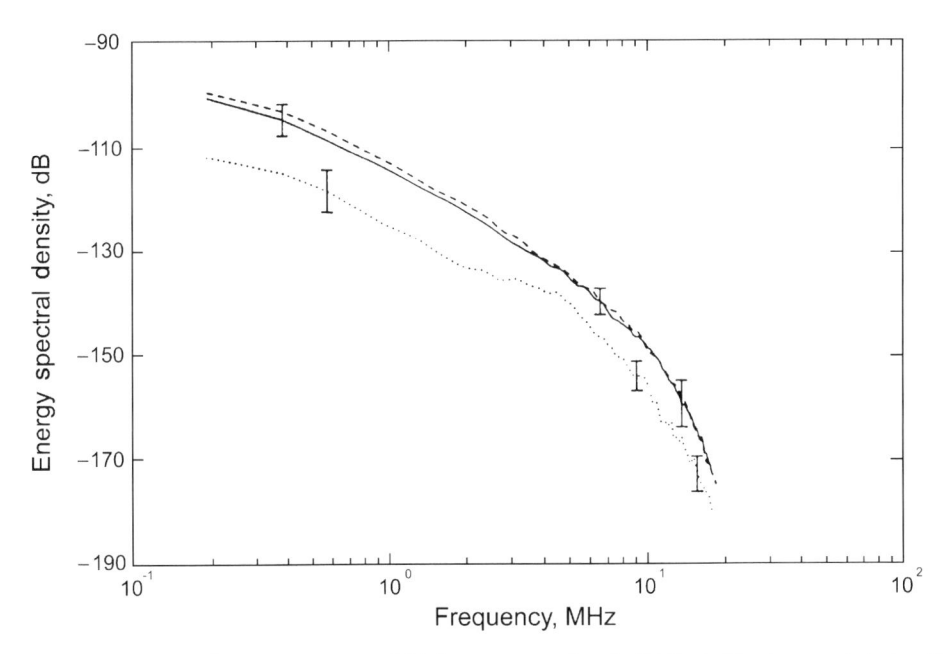

Fig. 4.42. Average frequency spectra of 74 first return strokes (solid line), 55 subsequent strokes (broken line), and 18 initial-breakdown pulses (dotted line). The data were taken from the storms over the Atlantic Ocean on August 8, August 10, and August 14, 1985, at ranges less than 35 km. The error bars show a standard deviation ±1 about the average energy spectral densities for first strokes and initial-breakdown pulses at representative frequencies. Adapted from Willett *et al.* (1990).

that the observed time delay between the narrowband radiation and the wideband electric field was due to propagation effects. According to them, the higher-frequency radiation from the lower part of the channel is more strongly attenuated in propagating over a finitely conducting earth than is the higher-frequency radiation from more elevated channel sections, which, due to the finite upward speed of the return stroke, is radiated later. In support of this suggested propagation effect, Cooray (1986) presented measurements of the 3 MHz radiation and the wideband electric field for lightning over both salt water and earth. In the former case, there is no time delay in the arrival of the HF signals; in the latter case, the HF signal increases slowly to peak following the start of the wideband return-stroke field.

Narrowband radiation often exhibits a quiet period following the return-stroke propagation phase. Malan (1958) first reported these periods of decreased higher-frequency activity, which lasted 5–20 ms. Clegg and Thomson (1979), from their review of several previous studies of the quiet period and from their own work, reported that the distribution of the duration of the quiet period at 10 MHz followed closely the K-change interval distribution (Section 4.10), having a most probable interval in the range from 6 to 9 ms.

We now briefly discuss measurements of the horizontal component of the return-stroke electric field at ground level. Thomson *et al.* (1988) presented simultaneously measured horizontal and vertical electric fields for 42 return strokes in 27 Florida flashes at distances ranging from 7 to 43 km. The horizontal field waveforms exhibited relatively narrow initial peaks, having a mean width at half-peak level of 0.52 μs. The mean ratio of the horizontal field peak to the corresponding vertical field peak was found to be 0.030, and the standard deviation was 0.007. The horizontal electric field waveshape appears to be similar to the derivative of the vertical electric field. Note that the horizontal electric field at ground level would be zero if the earth were a perfect conductor, as discussed in the last part of subsection 4.6.4.

4.6.4. Calculation of electric and magnetic fields

Field equations The most general equations for computing the vertical electric field E_z and azimuthal magnetic field B_ϕ due to an upward-moving return stroke for the case of a field point P on perfectly conducting ground (Fig. 4.43) are given by Thottappillil *et al.* (1997):

$$E_z(r,t) = \frac{1}{2\pi\varepsilon_0}\int_0^{H(t)}\left[\frac{2z'^2-r^2}{R^5(z')}\int_{\frac{z'}{v_f}+\frac{R(z')}{c}}^{t} I\left(z',\tau-\frac{R(z')}{c}\right)d\tau \right.$$
$$+ \frac{2z'^2-r^2}{cR^4(z')} I\left(z', t-\frac{R(z')}{c}\right)$$
$$\left. - \frac{r^2}{c^2 R^3(z')}\frac{\partial I(z', t-R(z')/c)}{\partial t}\right]dz'$$
$$- \frac{1}{2\pi\varepsilon_0}\frac{r^2}{c^2 R^3(H(t))} I\left(H(t), \frac{H(t)}{v_f}\right)\frac{dH(t)}{dt} \quad (4.10)$$

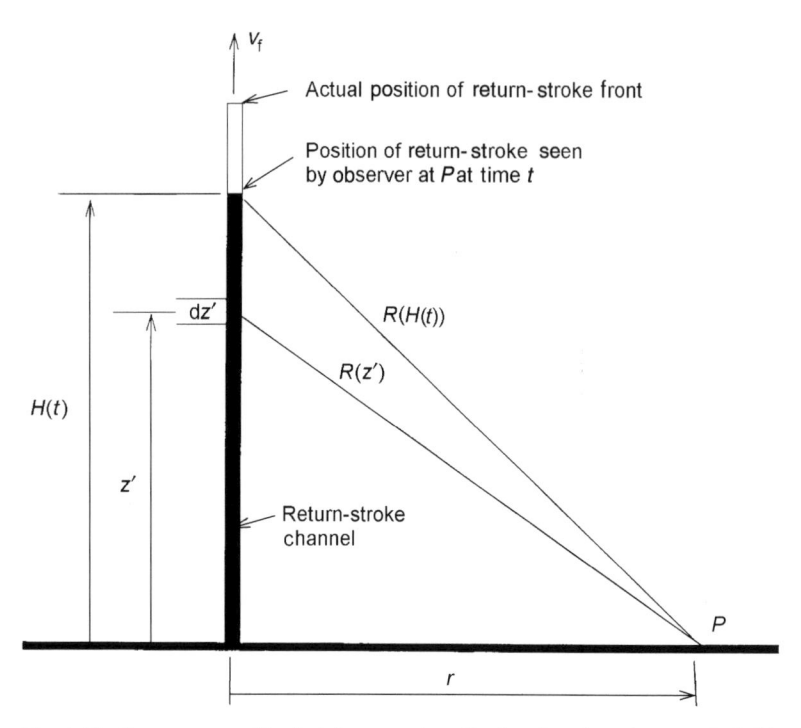

Fig. 4.43. Geometry used in deriving equations for the electric and magnetic fields at a point P on Earth (assumed to be perfectly conducting) a horizontal distance r from the vertical lightning return-stroke channel extending upward with speed v_f. Adapted from Thottappillil *et al.* (1997).

$$B_\phi(r,t) = \frac{\mu_0}{2\pi} \int_0^{H(t)} \left[\frac{r}{R^3(z')} I\left(z', t - \frac{R(z')}{c}\right) \right.$$
$$\left. + \frac{r}{cR^2(z')} \frac{\partial I(z', t - R(z')/c)}{\partial t} \right] dz'$$
$$+ \frac{\mu_0}{2\pi} \frac{r}{cR^2(H(t))} I\left(H(t), \frac{H(t)}{v_f}\right) \frac{dH(t)}{dt}$$
(4.11)

where $H(t)$ is the height of the front as "seen" by the observer at time t (Fig. 4.43). This height can be found from the following equation:

$$t = \frac{H(t)}{v_f} + \frac{R(H(t))}{c} \quad (4.12)$$

Thottappillil *et al.* (1997) also give the equivalent electric and magnetic field equations in terms of the line charge density ρ instead of the current I. A lightning return-stroke model is needed to specify $I(z', t)$.

Equations 4.10 and 4.11 are suitable for computing fields at ground using the electromagnetic, distributed-circuit, or "engineering" return-stroke models to be discussed in Section 12.2. Some of the engineering models include a current discontinuity at the moving front. Such a discontinuity is an inherent feature of the BG and TCS models, even when the current at the channel base starts from zero. The transmission-line-type models may include a discontinuity at the front if the channel-base current starts from a non-zero value. The DU model does not allow a current discontinuity either at the upward-moving front or at the channel base.

The first three terms in Eq. 4.10, referred to as the electrostatic, induction, and electric radiation field components, respectively, and the first two terms in Eq. 4.11, referred to as the magnetostatic (or induction) and magnetic radiation field components, respectively, describe the field due to sources below the upward-moving front. The last term in each of these two equations accounts for a possible current discontinuity at the moving front. The front discontinuity produces only a radiation field component, no electrostatic or induction field components. Note that Eqs. 4.10 and 4.11, when used for numerical computations, take proper account of retardation effects, as shown by Thottappillil *et al.* (1998) and, therefore, do not require any correction such as that via the so-called F-factor considered by Rubinstein and Uman (1990, 1991), Le Vine and Willett (1992), and Krider (1992).

At far ranges, the radiation field components decrease inversely with increasing distance in the absence of significant propagation effects. The electric radiation

4.6. Return stroke

field far from the channel, E_z^{rad}, with current I specified using the transmission-line return-stroke model (subsection 12.2.5) can be expressed as

$$E_z^{\text{rad}}(r, t) = -\frac{v}{2\pi\varepsilon_0 c^2 r} I(0, t - r/c) \qquad (4.13)$$

The corresponding magnetic radiation field can be found from $|B_\phi^{\text{rad}}| = |E_z^{\text{rad}}|/c$.

Krider (1994), using the transmission-line return-stroke model (subsection 12.2.5), computed the peak electric fields radiated into the upper half-space for different values of the return-stroke speed v relative to the speed of light c. He found that the largest fields are radiated in the upward direction when the return-stroke speed is very close to the speed of light. For example, if the peak current is 30 kA, $v/c = 0.9$, and there is no attenuation in the atmosphere or on ground, the peak electric field at 100 km is 21 V m^{-1}, the peak power density is 1.1 W m^{-2}, and the corresponding polar angle, θ, measured from the vertical, at which the peak field is radiated, is 29°. For small v/c ratios (less than 0.7 or so) the peak field is radiated along the ground ($\theta = 90°$).

Thottappillil et al. (1997) derived an electrostatic field equation in terms of charge density, ρ_L, for a very close observation point, such that $r \ll H(t)$, assuming that (i) retardation effects are negligible and (ii) the return-stroke line charge density does not vary significantly with height within the channel section contributing to the field at r:

$$E_z(r, t) \approx -\frac{\rho_L(t)}{2\pi\varepsilon_0 r} \qquad (4.14)$$

Equation 4.14 indicates that the electrostatic field produced by a very close return stroke is approximately proportional to the charge density on the bottom part of the channel. Note that the right-hand side of Eq. 4.14 is the negative of the right-hand side of Eq. 4.5, which was derived for a very close uniformly charged and fully developed leader.

Channel-base current equation For the "engineering" models (subsection 12.2.5), in which a vertical lightning channel and a perfectly conducting ground are assumed, the information on the source required for computing the fields usually includes (i) the channel-base current (either measured or assumed from typical measurements) and (ii) the upward return-stroke front speed, typically assumed to be constant and in the range 1×10^8 to 2×10^8 m s^{-1} (Table 4.5). The typical subsequent-stroke current waveform at the channel base is often approximated by the Heidler function (Heidler 1985):

$$I(0, t) = \frac{I_0}{\eta} \frac{(t/\tau_1)^n}{(t/\tau_1)^n + 1} e^{-t/\tau_2} \qquad (4.15)$$

where I_0, η, τ_1, n, and τ_2 are constants. This function allows one to change conveniently the current peak, maximum current derivative, and associated electrical charge transfer nearly independently by changing I_0, τ_1, and τ_2, respectively. Equation 4.15 reproduces the observed concave rising portion of a typical current waveform, as opposed to the once more commonly used double-exponential function, introduced independently by Bruce and Golde (1941) and Stekolnikov (1941), which is characterized by an unrealistic convex wavefront with a maximum current derivative at $t = 0$. A current equation capable of reproducing a concave, convex, or linear wavefront was used by Rakov and Dulzon (1987). Sometimes the sum of two Heidler functions with different parameters is used to approximate the desired current waveshape. Diendorfer and Uman (1990), for example, described the subsequent-stroke current waveform at the channel base as the sum of two functions given by Eq. 4.15. The first function was characterized by $I_0 = 13$ kA, $\eta = 0.73$, $\tau_1 = 0.15$ μs, $n = 2$, and $\tau_2 = 3.0$ μs and the second by $I_0 = 7$ kA, $\eta = 0.64$, $\tau_1 = 5$ μs, and $\tau_2 = 50$ μs. The resultant current peak is 14 kA, and the maximum current rate of rise is 75 kA μs^{-1}. The channel-base current waveform used by Nucci et al. (1990), Rakov and Dulzon (1991), Thottappillil et al. (1997), and Moini et al. (2000) for the calculation of lightning return-stroke electric and magnetic fields was the sum of a Heidler function and a double-exponential function. For this latter waveform, reproduced in Fig. 12.10, the current peak is 11 kA, and the maximum current rate of rise is 105 kA μs^{-1}.

Channel tortuosity and branches In most computations of the fields due to the return stroke, the return-stroke channel is assumed to be straight, while in fact it is known to be tortuous on scales ranging from less than 1 m to over 1 km (e.g., Evans and Walker 1963; Salanave 1980; Hill 1968, 1969). This "microscale" tortuosity of triggered-lightning channels, which includes geometric features with lengths of the order of 10 cm or less, was examined in detail by Idone (1995). Le Vine and Willett (1995) presented experimental evidence suggesting that the channel geometry is a factor in determining the fine structure observed during the first 10 μs of the electric fields produced by return strokes in both natural and triggered lightning. In the case of natural lightning, only subsequent strokes were considered. The effects of channel tortuosity on return-stroke radiation fields have been studied theoretically, using a piecewise linear representation of the lightning channel and simple return-stroke models, by Hill (1969), Le Vine and Meneghini (1978a, b), Le Vine et al. (1986), Le Vine and Kao (1988), Gardner (1990), Cooray and Orville (1990), Vecchi et al. (1994), and Lupò et al. (2000a, b). The length of an individual linear channel segment in the studies listed above is typically some tens to hundreds of meters.

In general, the effect of tortuosity is to introduce fine structure into the time-domain radiation field waveform and consequently to increase the higher-frequency content, above 100 kHz or so according to Le Vine and Meneghini (1978a), of the waveform. At each kink, that is, each point at which the linear segments join, there is a change in direction of the propagation of the current wave, and such changes introduce rapid variations in the radiation field. The amount of fine structure due to channel tortuosity depends on the current waveshape. Significant variations in the radiation field are produced when the risetime of the current waveform is smaller than the time required for the current wave to propagate between kinks (Le Vine and Kao 1988; Cooray and Orville 1990). When the risetime is significantly larger than the propagation time between kinks, then more than one kink contributes to the radiation field during the rising portion of the current wave and, as a result, an averaging (smoothing) effect of the overall field occurs. Cooray and Orville (1990) demonstrated theoretically that the amount of fine structure due to channel tortuosity is significantly reduced if lengthening of the current wave front during its propagation along the channel is allowed. Further smoothing should occur due to propagation effects (e.g., Le Vine et al. 1986).

Measured first-stroke electric and magnetic fields exhibit a more pronounced fine structure than subsequent strokes, a fact generally attributed to the presence of branches in the first strokes. The effects of channel branches on return-stroke radiated fields have been theoretically studied by Le Vine and Meneghini (1978a), Vecchi et al. (1997), Lupò et al. (2000b), and Zich and Vecchi (2001).

Propagation effects If the observation point of the lightning fields is located on the ground surface, and the ground is assumed to be perfectly conducting, only two field components exist: the vertical electric field and the azimuthal magnetic field. The horizontal electric field component is zero as required by the boundary condition on the surface of a perfect conductor. At an observation point above a perfectly conducting ground, a non-zero horizontal electric field component exists. A horizontal electric field exists both above ground and on (and below) its surface in the case of finite ground conductivity. The horizontal (radial) electric field at and below the ground surface is associated with a radial current flow and resultant ohmic losses in the earth. Propagation effects include preferential attenuation of the higher-frequency components in the vertical electric field and the azimuthal magnetic field waveforms. A good review of the literature on the effects of finite ground conductivity on lightning electric and magnetic fields is given by Rachidi et al. (1996).

Two approximate equations, namely, the wave-tilt formula (Zenneck 1915) and the Cooray–Rubinstein formula (Cooray 1992; Rubinstein 1996), both in the frequency domain, are commonly used for computation of the horizontal electric field in air within 10 m or so above a finitely conducting earth. The term "wave tilt" originates from the fact that when a plane electromagnetic wave propagates over a finitely conducting ground, the total electric field vector at the surface is tilted from the vertical because of the presence of a non-zero horizontal (radial) electric field component. The tilt is in the direction of propagation if the vertical electric field component is directed upward and in the direction opposite to the propagation direction if the vertical electric field component is directed downward, the vertical component of the Poynting vector being directed into the ground in both cases.

The wave-tilt formula states that, for a plane wave, the ratio of the Fourier transform of the horizontal electric field $E_r(j\omega)$, where $j = \sqrt{-1}$, to that of the vertical electric field $E_z(j\omega)$ is equal to the ratio of the propagation constants in the air and in the ground (Zenneck 1915). Therefore,

$$E_r(j\omega) = E_z(j\omega) \frac{1}{\sqrt{\varepsilon_{rg} + \sigma_g/(j\omega\varepsilon_0)}} \quad (4.16)$$

where σ_g and ε_{rg} are the conductivity and relative permittivity of the ground, respectively, and ω is the angular frequency. Equation 4.16 is a special case, valid for grazing incidence, of the formula giving the reflection of electromagnetic waves off a conducting surface and, hence, is a reasonable approximation only for relatively distant lightning or for the early microseconds of close lightning when the return-stroke wavefront is near ground. Typically, $E_z(j\omega)$ is computed assuming a perfectly conducting ground or is measured.

The Cooray–Rubinstein equation can be expressed as follows (Cooray 1992; Rubinstein 1996):

$$E_r(r, z, j\omega) = E_{rp}(r, z, j\omega) - H_{\phi p}(r, 0, j\omega)$$
$$\times \frac{c\mu_0}{\sqrt{\varepsilon_{rg} + \sigma_g/(j\omega\varepsilon_0)}} \quad (4.17)$$

where μ_0 is the permeability of free space, $E_{rp}(r, z, j\omega)$ and $H_{\phi p}(r, 0, j\omega)$ are the Fourier transforms of the horizontal electric field at height z above ground and the azimuthal magnetic field at ground level, respectively, both computed for the case of a perfectly (subscript "p") conducting ground. The second term is equal to zero for $\sigma_g \to \infty$ and becomes increasingly important as σ_g decreases. A generalization of the Cooray–Rubinstein formula has been offered by Wait (1997a). Measurements of the horizontal electric field are discussed in subsection 4.6.3.

Cooray and Lundquist (1983) and Cooray (1987), using an analytical time-domain attenuation function proposed by Wait (1956), have calculated the effects of a finitely conducting earth in modifying the initial portion of the vertical electric field waveforms from the values expected over

4.6. Return stroke

an infinitely conducting earth. The results are in good agreement with the measurements of Uman *et al.* (1976) and Lin *et al.* (1979), discussed in subsection 4.6.3. As noted earlier, Uman *et al.* (1976) observed that zero-to-peak risetimes for typical strokes increase of order 1 μs in propagating 200 km across Florida soil, and Lin *et al.* (1979) reported that normalized peak fields were typically attenuated 10 percent in propagating over 50 km of Florida soil and 20 percent in propagating 200 km. It is thought that minimal distortion of the fast transition in the field wavefront and other rapidly changing portions of the measured field waveforms can be assured when the propagation path is almost entirely over salt water, a relatively good conductor. Nevertheless, Ming and Cooray (1994) found from theory that for frequencies higher than about 10 MHz, the attenuation caused by the rough ocean surface can be significant. For the worst cases considered, they reported that the peak of the radiation field derivative was attenuated by about 35 percent in propagating 50–100 km. Cooray and Ming (1994) considered theoretically the case of propagation partly over sea and partly over land and found that propagation effects on the electric radiation field derivative are significant unless the length of the land portion of the propagation path is less than a few tens of meters. They found that propagation effects on the peak of the radiation field can be neglected if the length of the over-land propagation path is less than about 100 m.

4.6.5. Properties of the return-stroke channel

In this section, we will present physical properties of the lightning channel, such as temperature, electron density, and pressure, that can be estimated from the time-resolved optical spectra of return strokes (Orville 1968a, b, c). For reviews of lightning spectroscopy see Orville and Salanave (1970), Krider (1973), Orville (1977), Uman (1969a, 1984), and Salanave (1980). Additionally, we will discuss estimates of the return-stroke channel radius and measurements of the optical power radiated by return-stroke channels.

From an analysis of 10 time-resolved spectra of return strokes (the time resolution was either 2 or 5 μs), Orville (1968b) obtained the channel temperature and the electron density, each as a function of time. Typical peak temperatures, determined from the ratios of the intensities of spectral lines, were of the order of 28 000–31 000 K. No temperatures exceeded 36 000 K. In two of the 10 strokes, the temperature appeared to rise to a peak value during the first 10 μs (the time resolution was 5 μs) and to decay thereafter. In the remaining eight strokes (including two with 2 μs time resolution), the temperature decreased monotonically from the initial value. These peak temperatures were obtained assuming that the radiating gas was optically thin and in local thermodynamic equilibrium, although Hill (1972) argued that for some spectral lines the assumption of an optically thin channel may be invalid.

Since the time resolution was a few microseconds, if the initial temperatures varied more rapidly than the time resolution then the peak temperature values represent some sort of average over that time (Uman 1969b). The return-stroke temperatures for all strokes studied by Orville (1968b) were below 30 000 K after 10 μs and were near or below 20 000 K after 20 μs. The temperature decay of a lightning channel during the interstroke interval was examined theoretically by Uman and Voshall (1968) and Picone *et al.* (1981). The results of these studies are discussed in relation to the dart-leader mechanism, in subsection 4.7.8. Aleksandrov *et al.* (2000) examined theoretically the variation in channel characteristics during both the return stroke and the following continuing current (Section 4.8).

Electron densities in lightning return strokes were determined by Orville (1968b) from a comparison of the measured Stark width of the H_α line radiated by hydrogen atoms with theory. The Stark width of the H_α line is primarily dependent on the electron density, is independent of the population of the bound atomic energy levels, and is only a weak function of the electron temperature (Uman and Orville 1964). According to Orville (1968b), the electron density was 8×10^{17} cm^{-3} in the first 5 μs, decreasing to $(1-1.5) \times 10^{17}$ cm^{-3} at 25 μs, and remaining approximately constant at 50 μs. Additionally, using Gilmore's tables for the composition of dry air in thermodynamic equilibrium, Orville (1968c) found that the channel is characterized by an average pressure of 8 atm in the first 5 μs and attains atmospheric pressure at approximately 20 μs.

Channel radii were found photographically by Schonland (1937) to be in the 5.5 to 11.5 cm range, by Evans and Walker (1963) to be in the 1.5 to 6 cm range, and by Orville *et al.* (1974) to be in the 2 to 4 cm range. Orville *et al.* (1974) reviewed some of the theory (Braginskii 1958; Oetzel 1968; Plooster 1971; Hill 1971, 1972) and records of physical damage caused by lightning (Schonland 1950; Hill 1963; Uman 1964a; Taylor 1965; Jones 1968) that generally support the values they obtained for the return-stroke channel radius after the initial 10 μs or so of the channel expansion. Idone (1992) reported that a direct measurement of the optical size of the channel, which was definitely not overexposed, yielded radii of 1–1.5 cm for six return strokes in two triggered-lightning flashes. A number of characteristics of a lightning channel associated with various subsequent-stroke processes have been estimated by Rakov (1998) and are given in Table 4.9.

Guo and Krider (1982, 1983) gave absolute values for the light intensity of first and subsequent return strokes in the 0.4 to 1.1 μm wavelength range and presented correlated electric field data. The peak optical power had a mean of 2.3×10^9 W for first strokes, 4.8×10^8 W for subsequent strokes preceded by dart leaders, and 5.4×10^8 W for subsequent strokes preceded by dart-stepped leaders

Table 4.9. *Estimated characteristics of the lightning channels associated with various processes of the lightning discharge. Adapted from Rakov (1998)*

Channel characteristics[a]	Pre-dart-leader channel (ahead of dart-leader front)	Pre-return-stroke channel (behind dart-leader front and ahead of return-stroke front)	Return-stroke channel (behind return-stroke front)
Temperature, K	~ 3000	$\geq 20\,000$	$\geq 30\,000$
Conductivity, S m^{-1}	~ 0.02	$\sim 10^4$	$\sim 10^4$
Radius, cm	~ 3	~ 0.3	~ 3
R, Ω m^{-1}	$\sim 18\,000$	~ 3.5	~ 0.035

[a] For comparison, the electrical conductivity of carbon is 3×10^4 S m^{-1}, of seawater is 4 S m^{-1}, and of copper is 5.8×10^7 S m^{-1} (Sadiku 1994); the temperature of the solar interior is 10^7 K and of the solar surface is 6000 K, and the temperatures at which tungsten and lead melt are 3600 K and 600 K, respectively (Halliday and Resnick 1974).

(Guo and Krider 1982). The average time at which the peak optical power occurs is about 60 μs after the initial electric field peak, because the risetime of the optical signal is determined by the geometrical growth of the return-stroke channel (Guo and Krider 1982). Guo and Krider (1982) also compared their data with similar measurements of Krider (1966a), Mackerras (1973), and Turman (1978). The time- and space-averaged mean radiance near the channel bottom was found by Guo and Krider (1983) to be 1.3×10^6 W m^{-1} for first strokes and 3.9×10^5 W m^{-1} for subsequent strokes initiated by dart leaders, and, by Guo and Krider (1982), to be 1×10^6 W m^{-1} for first strokes and 2.5×10^5 and 4.5×10^5 W m^{-1} for subsequent strokes initiated by dart and dart-stepped leaders, respectively. By making use of the streak-photography data of Jordan and Uman (1983), Guo and Krider (1983) estimated that the average peak radiance for subsequent strokes is 2 to 4 times the time- and space-averaged radiance given above. Krider and Guo (1983) showed that the radio-frequency power radiated by a return stroke, at the time of the field peak, has a mean of 2×10^{10} W for first strokes and 3×10^9 W for subsequent strokes, similar to the values that were estimated from the theory by Krider (1992). As discussed in Section 4.2, the radio-frequency power is about two orders of magnitude greater than the optical power radiated in the 0.4 to 1.1 μm band at the time of peak field. The optical power, however, dominates at later times.

Turman (1977, 1978) measured the peak optical return-stroke power observed from a satellite in Earth orbit and found that the median value was 10^9 W, 2 percent of the strokes being above 10^{11} W. Turman (1977) used the term "superbolt" to describe those powerful light emitters with peak powers in the 10^{11} to 10^{13} W range. Five flashes out of a total of 10^7 were found to exceed an optical power of 3×10^{12} W. Some or many of the superbolts may be positive strokes (Turman 1977; Hill 1978a, b; Uman 1978).

Positive lightning discharges are discussed in Chapter 5. Various theoretical estimates of lightning input energy are considered in the first part of subsection 12.2.6 (see also Section 4.2).

4.7. Subsequent leader
4.7.1. General information

Return strokes subsequent to the first are usually initiated by dart leaders. An example of a dart leader–return-stroke sequence imaged by a streak camera is found in Fig. 4.37a. The image on the left is due to the dart leader, and the brighter image on the right is due to the return stroke. As seen in Fig. 4.37a, the dart leader appears, unlike a stepped leader (see Fig. 4.13), to move continuously, that is, the lowest portion of the leader channel, called the dart, remains luminous during the channel extension from the cloud to ground. Luminosity versus time at different heights above ground, corresponding to the streak photograph in Fig. 4.37a, is shown for the entire visible channel and for the bottom 480 m in Figs. 4.37b and 4.37c, respectively. Many subsequent return strokes (more than one-third of the second strokes in channels that are not too old; see Table 4.1 and Fig. 4.48 below) are initiated by leaders that exhibit pronounced stepping in the bottom portion of the channel. Such leaders produce regular pulse sequences that are observed just prior to the return-stroke pulse in distant electric or magnetic field records and are called dart-stepped leaders. Some subsequent strokes are initiated by "chaotic leaders", a term introduced by Weidman (1982); these are identified, in distant electric or magnetic field records, by irregular pulse sequences just prior to the return stroke. Dart and dart-stepped leaders are often not distinguished in analyses of their propagation speeds (e.g., Jordan *et al.* 1992). When a subsequent leader deflects from the previously formed channel, it continues as a stepped

4.7. Subsequent leader

Table 4.10. *Summary of optically measured downward dart-leader speeds in natural and triggered lightning*

Investigators	Location	Sample size	Mean speed 10^6 m s^{-1}
Natural lightning			
Schonland et al. (1935)	South Africa	55	5.5
McEachron (1939)	USA	17	13
Brook and Kitagawa (Winn 1965)	New Mexico	96	9.9
Orville and Idone (1982)	Florida and New Mexico	21	11
Jordan et al. (1992)	Florida	11	14
Mach and Rust (1997)	Oklahoma, Alabama, Florida	17	19
Triggered lightning			
Hubert and Mouget (1981)	France	10	11
Idone et al. (1984)	New Mexico	32	20
Jordan et al. (1992)	Florida (KSC, 1987, 1989)	36	16
Mach and Rust (1997)	Florida (KSC, 1986)	20	13

leader; more than one-third of second-stroke leaders exhibit this behavior (see Table 4.1). Interestingly, Davis, S.M. (1999) found, from multiple-station dE/dt measurements in Florida, that leaders creating a new termination on ground were of the dart-stepped type when they propagated in the previously formed channel, that is, before they transformed into stepped leaders. The heights at which subsequent leaders deflected from the previously formed channel in Davis' study ranged from 0.7 to 3.4 km. Subsequent strokes creating a new termination on ground are intermediate in many of their characteristics between first strokes initiated by stepped leaders, discussed in Section 4.4, and subsequent strokes following a previously formed channel (see Table 4.7). It is common practice to term all the leaders preceding subsequent return strokes in previously formed channels "dart leaders" (e.g., Schonland 1956; Uman 1987, 2001).

4.7.2. Speed and duration

A survey of measurements of dart-leader propagation speed is given in Table 4.10. A histogram showing the data on dart-leader speed measured by Orville and Idone (1982) is given in Fig. 4.44. Most mean values in Table 4.10 are between 1×10^7 and 2×10^7 m s^{-1} with the exception of that of Schonland et al. (1935), whose mean of 5.5×10^6 m s^{-1} is about half to a quarter of the dart-leader speed reported by the other investigators. About one-third of the 55 measurements of Schonland et al. (1935) fall into the 1 to 3×10^6 m s^{-1} range.

Apparently, the dart-leader speed can either increase or decrease as the leader approaches ground. In natural lightning, Schonland et al. (1935), working in South Africa, never observed the dart-leader speed to increase as the leader approached ground, while Orville and Idone (1982)

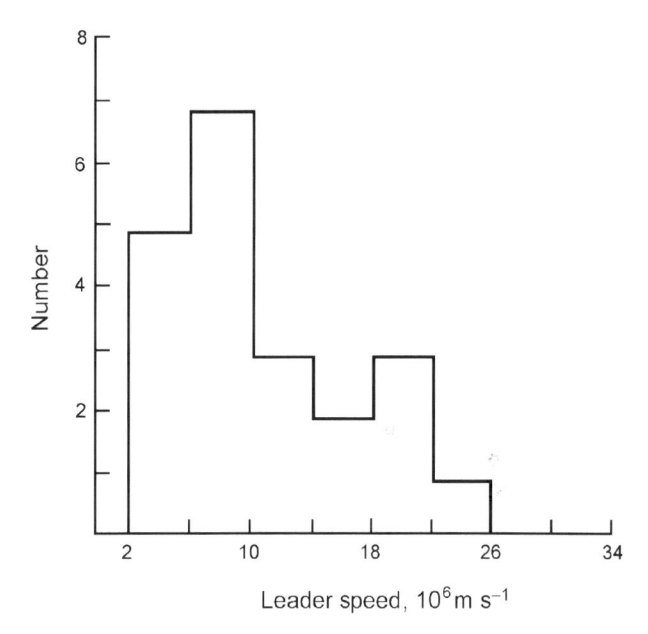

Fig. 4.44. Histogram of the two-dimensional speed of 21 dart leaders in 12 flashes in Florida (14 leaders in 10 flashes) and New Mexico (seven leaders in two flashes). Speeds were measured in the lowest 800 m. The mean value is 1.1×10^7 m s^{-1}. Adapted from Orville and Idone (1982) and Idone et al. (1984).

reported four cases (one from Florida and three from New Mexico) out of a sample of 16, for which dart-leader speeds increased toward ground. However, both Schonland et al. (1935) and Orville and Idone (1982) observed dart leaders whose speed decreased near ground. Jordan et al. (1992) reported an average tendency for dart-leader speed to decrease near ground in 12 triggered-lightning strokes in Florida. It is worth noting that since the photographic measurements of Schonland et al. (1935), Orville and Idone (1982), and

Plot symbol (flash)	Geographical location	Lightning type
■ (220651)		
◆ (224645)		
● (171714)	Florida	natural
▲ (184535)		
▼ (184656)		
□	Florida	
○	New Mexico	triggered
△	France	

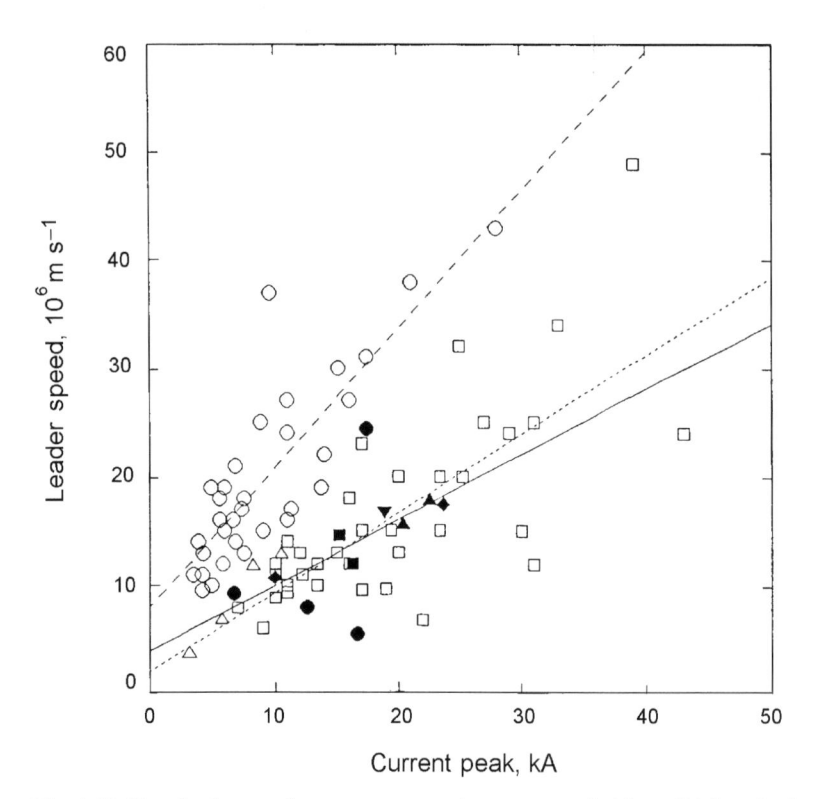

Fig. 4.45. Dart-leader speed versus return-stroke current peak. The solid, long-broken, and short-broken regression lines are for Florida natural lightning, New Mexico triggered lightning, and Florida triggered lighting, respectively. Adapted from Jordan et al. (1992).

Jordan et al. (1992) were two dimensional, it is possible that some of the apparent speed changes were due to the channel geometry. The three-dimensional speed measurements for 29 leaders in New Mexico triggered lightning suggest that dart leaders slow down as they near ground (Jordan et al. 1992). Mach and Rust (1997), using a photoelectric detector, observed no significant change in the dart-leader two-dimensional speed with height for either natural or rocket-triggered lightning. Wang et al. (1999b), using the ALPS optical imaging system, found that one dart leader and one dart-stepped leader in two triggered-lightning flashes accelerated in the lowest channel section, which was some hundreds of meters in length.

Jordan et al. (1992), from a data base consisting of (i) correlated optical and electric field measurements for 11 natural-lightning strokes in Florida, (ii) correlated optical and current measurements for 32 triggered-lightning strokes in New Mexico, and (iii) correlated optical and current measurements for 36 triggered lightning strokes in Florida, examined dart-leader speed as a function of the following return-stroke current peak and of the duration of the previous interstroke interval (excluding the duration of continuing current, if present). Return-stroke electric field peaks in Florida natural lightning were converted to current peaks using Eq. 4.9. Scatter plots illustrating the relation between dart-leader speed and the following return-stroke current and between dart-leader speed and the preceding interstroke interval for the three data sets listed above are shown in Figs. 4.45 and 4.46, respectively. These figures also contain data taken by Jordan et al. (1992) from

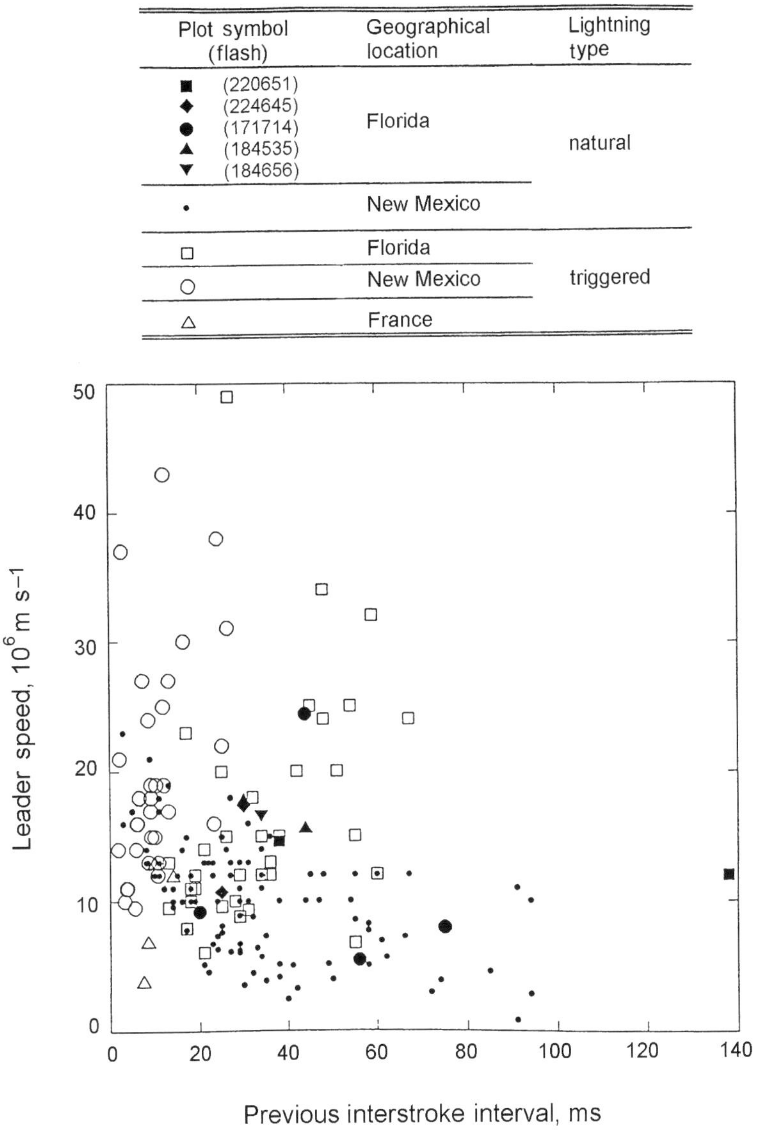

4.7. Subsequent leader

Plot symbol (flash)	Geographical location	Lightning type
■ (220651)		
♦ (224645)		
● (171714)	Florida	natural
▲ (184535)		
▼ (184656)		
·	New Mexico	
□	Florida	
○	New Mexico	triggered
△	France	

Fig. 4.46. Subsequent-leader speed versus previous interstroke interval. Adapted from Jordan et al. (1992).

the literature. Additionally shown in Figs. 4.45 and 4.46 are data for four triggered lightning strokes published by Hubert and Mouget (1981). Also shown in Fig. 4.46 are the data for 96 natural lightning strokes in New Mexico recorded by Brook and Kitagawa and published by Winn (1965).

For each of the three data sets analyzed in detail by Jordan et al. (1992), dart-leader speed and the following return-stroke current peak are positively correlated. The relations between leader speed and current peak for Florida triggered lightning and for Florida natural lightning are similar, indicating a similarity between the dart leaders and return strokes in natural and in triggered lightning at the same geographic location. However, the leaders in New Mexico triggered lightning are, for the same value of return-stroke current peak, about twice as fast as those in both triggered and natural lightning in Florida; the difference is likely to be associated with the relatively short previous interstroke intervals in New Mexico triggered lightning. For all the triggered and natural lightning data taken together there is a weak but statistically significant tendency for lower leader speeds to be associated with longer previous interstroke intervals. However, neither the New Mexico triggered lightning data nor the Florida triggered lightning data, when taken separately, show this tendency, and each of the data sets presented in Fig. 4.46 exhibits a large scatter. McEachron (1939) noted a lack of correlation between the downward dart-leader speed and the previous interstroke interval in upward flashes (Chapter 6) initiated from the Empire State Building.

Table 4.11. *Median duration[a] of subsequent leaders from electric field measurements. Adapted from Rakov and and Uman (1990c).*

Reference	Geographic location	Sample size	Median leader duration, ms	Maximum leader duration, ms	Comments
Schonland et al. (1938a)	South Africa	114	<1.5	15	As derived from histogram in Fig. 10 of Schonland et al. (1938a)
Kitagawa (1957a)	Japan	35	0.6	1.7	Geometric mean
Malan and Schonland (1951b)	South Africa	10	1	3	Geometric mean
Workman et al. (1960)	New Mexico	18	1.8	13	Geometric mean within one flash
Isikawa (1961)	Japan	273	0.9	1.9	Median values for strokes of different order were combined assuming the leader duration to be log-normally distributed
Krehbiel et al. (1979)	New Mexico	10	~1.0	10	Geometric mean

[a] In some cases the geometric mean was used as an estimate of the median value.

Rakov (1998) interpreted the experimental results of Jordan et al. (1992) as confirming his inference (further discussed in subsection 4.7.8) that the downward propagation characteristics of the dart-leader wave are determined primarily by the wave magnitude as opposed to the parameters of the channel ahead of the wavefront. These channel parameters are probably determined by the age of the channel, which is represented in Fig. 4.46 by the previous interstroke interval. As seen in Fig. 4.46, only the upper speed limit appears to be dependent on the channel age.

The typical duration of the subsequent leader as estimated from measured electric field waveforms is of order 1 to 2 ms, as seen in Table 4.11. Histograms of leader duration for subsequent strokes of various orders in Florida reported by Rakov and Uman (1990c) are given in Figs. 4.14b, c, d, e. For subsequent strokes of any order following a previously formed channel, they found a geometric mean of 1.8 ms (see also Table 4.7), while for subsequent strokes creating a new termination on ground, the geometric-mean leader duration was much longer, 15 ms. As seen in Figs. 4.14c, d, e, the leader duration for strokes following a previously formed channel tends to increase with stroke order. If leader speed is not influenced by stroke order, this tendency implies a progressive increase in the channel length with stroke order. Further, the ratio of the leader to the return stroke electric field change in the same data set (see Fig. 4.22) tended to be more negative as the stroke order increased, suggesting that the charge sources for successive strokes were, on average, progressively farther from both the observer and the ground strike point.

4.7.3. Electrical characteristics

Idone and Orville (1985) estimated dart-leader peak currents for 22 leaders in two rocket-triggered flashes using two different optical techniques. In method (i), the ratio of the dart-leader and return-stroke currents was taken as equal to the ratio of the dart-leader and return-stroke speeds; this assumes a simple model in which an equal charge per unit length is involved in each process. The speed ratio and the return-stroke current were measured, allowing a calculation of the dart-leader current.

In method (ii) the relation between return-stroke peak current I_R and return-stroke peak relative light intensity L_R in each of two flashes ($L_R = 1.5 I_R^{1.6}$ and $L_R = 6.4 I_R^{1.1}$) was applied to the dart-leader relative light intensities in that flash to determine the dart-leader current. The two techniques produced very similar results, a mean current of 1.8 kA for method (ii) and 1.6 kA for method (i). Individual values ranged from 100 A to 6 kA. The ratio of dart-leader to return-stroke current ranged from 0.03 to 0.3 with a mean of 0.17 from method (ii) and 0.16 from method (i). The largest dart-leader to return-stroke current ratios were associated with the largest return-stroke currents and relative light intensities. Idone and Orville (1985) discussed the validity of the techniques used to find dart-leader currents, which, as they stated, are certainly open to question.

Guo and Krider (1985) found that 39 of 726 multiple-stroke flashes had one or more dart leaders whose light output per unit length was comparable to that of the following return stroke. Dart leaders with such anomalous light outputs were not observed by Idone and Orville (1985), who reported, in the study discussed above, that the ratio of the maximum light output from a dart leader to that of the following return stroke varied from 0.02 to 0.23, with a mean of 0.1. Orville (1975), studying the optical spectra of five dart leaders, found that the dart-leader peak radiation was a factor of 10 less intense than the return-stroke peak radiation. Orville and Idone (1982) reported that, for strokes

4.7. Subsequent leader

within a given flash, there is a positive correlation between the dart-leader luminosity and the following return-stroke luminosity.

The total charge lowered by the dart leader can be determined from a measurement of either the dart-leader field change or the combined dart-leader and return-stroke field change, if a model for the leader charge distribution is assumed. Brook et al. (1962) reported that the minimum charge brought down by strokes subsequent to the first is 0.21 C, the most frequent value being between 0.5 and 1 C. Thus a dart leader would appear to carry less charge than does a stepped leader (Section 4.4). If the median value, 0.95 C, of the charge transferred to ground in negative subsequent strokes excluding continuing current (see Table 4.4) were deposited on the dart-leader channel in 1.8 ms, the geometric mean duration of subsequent leaders that follow the previously formed channel (see Table 4.7), then the resulting average dart-leader current would be approximately 0.5 kA. As discussed above, peak currents between 0.1 and 6 kA were estimated for 22 dart leaders in two triggered-lightning flashes by Idone and Orville (1985) using optical techniques.

Schonland et al. (1938a) inferred from the ratios of 46 dart-leader and 26 stepped-leader field changes and their associated return-stroke field changes that dart-leader channels tend to be uniformly charged. A uniform distribution of dart-leader charge in the lower portion of the channel has also been inferred from multiple-station electric field measurements at distances from 10 to 500 m from a triggered-lightning channel, although in some cases the leader charge density appeared to decrease toward ground (subsection 7.2.6).

4.7.4. Overall electric fields

Examples of the overall electric field changes of subsequent leaders together with the corresponding return-stroke field changes are found in Fig. 4.16 (left panel, the bottom set of waveforms) and Fig. 4.47. The stroke presented in Fig. 4.47 was the fifth stroke in a flash. That stroke followed a previously formed channel. The subsequent stroke presented in Fig. 4.16 was the second stroke in a flash, which created a new termination on ground. Figure 4.16 also presents a VHF image of the subsequent leader channel (right panel, the bottom image, labeled L2). Electric field waveforms due to dart leaders in triggered lightning recorded at distances ranging from 10 to 500 m are found in subsection 7.2.6. Leader electric field waveforms as a function of distance and stroke order in natural lightning were studied in detail by Rakov and Uman (1990c). They found that for strokes of order 8 through 18, 90 percent of 49 leader waveforms were hook-shaped with a net negative field change (see, for example, Fig. 4.47, although for a fifth stroke) regardless of distance (ranging from 3 to 14 km).

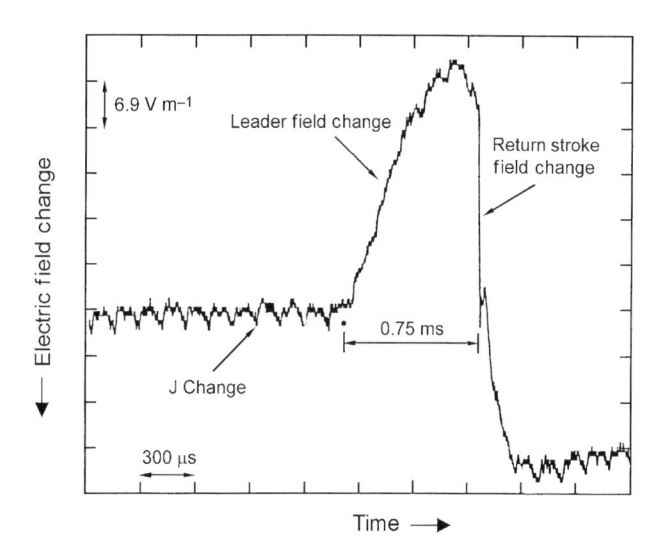

Fig. 4.47. Electric field change of the fifth stroke in a five-stroke flash that occurred in Florida at a distance of 7.6 km. A positive field change (atmospheric electricity sign convention, subsection 1.4.2) deflects downward. The small solid circle near the middle of the figure marks the starting point of the leader field change. The time interval labeled 0.75 ms indicates the leader field change duration. Note that the leader electric field change is hook-shaped. Adapted from Rakov and Uman (1990c).

At distances closer than 5 km, 92 percent of 38 leader waveforms were hook-shaped with a net negative field change (see, for example, Fig. 4.60) regardless of stroke order. Krehbiel et al. (1979) and Rakov and Uman (1990c) found that leader electric field waveshapes are strongly influenced by the leader-channel geometry. The length of the horizontal in-cloud portion of the subsequent leader channel apparently increases, on average, with stroke order. This tendency has been inferred from charge source locations for successive strokes (Fig. 3.5), from imaging lightning as a source of VHF radiation (Fig. 4.61), and from the dependence on stroke order of the ratio of the leader and return-stroke electric field changes as a function of distance (Fig. 4.22).

4.7.5. Dart-stepped leader

A streak photograph of a dart-stepped leader in a triggered lightning flash is presented in Fig. 7.5d. The percentage of dart-stepped leaders occurring after an interstroke interval of 100 ms or less as a function of stroke order is shown in Fig. 4.48. Second strokes are initiated by dart-stepped leaders over five times more often than all higher-order strokes taken together. Rakov and Uman (1990b) reported that the geometric mean interstroke interval preceding 13 second strokes initiated by dart-stepped leaders was 54 ms, shorter than the typical mean interstroke interval of 60 ms for all subsequent strokes without selection (Table 4.7). Davis, S.M. (1999), using multiple-station

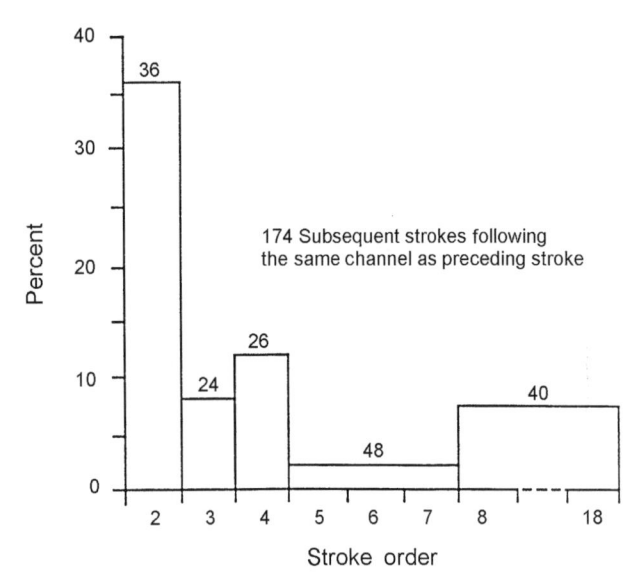

Fig. 4.48. Occurrence of a dart-stepped leader after an interstroke interval of 100 ms or less versus stroke order: the vertical axis gives the percentage of strokes showing step pulses just prior to a return-stroke pulse in the electric field records. The horizontal axis gives the order of the return stroke. Adapted from Rakov and Uman (1990b).

dE/dt measurements in Florida, found that dart-stepped leaders that initiated second strokes occurred after a typical interstroke interval of 61 ms, while those that occurred after two or more strokes in a channel were preceded by a typical interstroke interval more than twice as long, 140 ms. This observation supports the hypothesis of Rakov et al. (1994) regarding the cumulative conditioning of the lightning channel by consecutive strokes. The geometric mean initial electric field peak normalized to 100 km for all 22 strokes initiated by dart-stepped leaders in the study of Rakov and Uman (1990b) was 4.8 V m^{-1}, which is comparable with the geometric mean for subsequent strokes creating a new termination on ground (see Table 4.7). Consistent with this finding is the report of Guo and Krider (1982) that higher light intensity was associated with strokes initiated by dart-stepped leaders than with "normal" subsequent strokes (those without microsecond-scale pulses prior to the return-stroke initial field peak).

Consistent with Fig. 4.48, all six dart-stepped leaders analyzed by Schonland (1956, Table 3) were associated with second strokes. The average speeds for these six leaders varied between about 0.5 and 1.7×10^6 m s^{-1}, step lengths were typically 10 m, and the time interval between steps was about 10 μs. Orville and Idone (1982) analyzed four dart-stepped leaders in detail, including the variation in the step length and the interstep time interval with height and the luminous structure of the steps. One of the dart-stepped leaders showed a decrease in propagation speed and an increase in interstep time interval as ground was approached. Another exhibited the opposite effect. Davis, S.M. (1999), from multiple-station dE/dt measurements in Florida, reported on seven dart-stepped leaders that exhibited a decrease in speed (by a factor of 4.5) during propagation from their origins at heights of 3 to 6 km to ground, as discussed later in this subsection. He also reported on two dart-stepped leaders that showed an increase in speed by a factor of 5 to 10 during their descent. In one of these two cases, there was evidence that the cause of the speed increase was that the leader had joined a newer (less aged) section of channel. Wang et al. (1999b) observed, for a dart-stepped leader in triggered lightning, an increase in downward propagation speed from 2×10^6 to 8×10^6 m s^{-1} as the leader descended from 200 to 40 m.

The luminous structure of several of the individual steps studied by Orville and Idone (1982) exhibit a distinct bright tip in the lower portion of the step that "fans out into a symmetrically diffuse image" in the upper portion of the step. Overall, the step lengths and interstep time intervals are in good agreement with those reported by Schonland (1956). The interstep time intervals of both Schonland (1956) and Orville and Idone (1982) are consistent with the time intervals between electric field pulses identified by Krider et al. (1977) as being associated with dart-stepped-leader steps. For a period of 200 μs prior to the return stroke, Krider et al. (1977) found mean time intervals of 6.5 and 7.8 μs for dart-stepped-leader electric field pulses in Florida and Arizona, respectively. Examples of these electric field pulses are shown in Fig. 4.40b. The mean interpulse intervals for stepped leaders initiating first strokes in the same study were 16 and 25 μs in Florida and Arizona, respectively (Table 4.3). Electric field pulses produced by stepped leaders initiating first strokes are shown in Figs. 4.24 and 4.40a. Further, Krider et al. (1977) reported that the electric field pulses due to individual steps in dart-stepped leaders were essentially the same as individual stepped-leader pulses (subsection 4.4.6) and were very similar to the small pulses in the regular pulse bursts discussed below in Section 4.11. Interestingly, Davis, S.M. (1999) found that the electric field pulses (obtained by integration of measured dE/dt pulses) due to dart-stepped leaders and pulses in regular pulse bursts were more or less similar to those reported by Krider et al. (1977), while the dE/dt pulses due to stepped leaders were undetectable with his wideband measuring system. The mean ratio of the largest dart-stepped-leader pulse peak and the following return-stroke pulse peak was about 0.1 (Krider et al. 1977), similar to that ratio for stepped leaders (subsection 4.4.6).

Davis, S.M. (1999) studied the evolution of various characteristics of seven dart-stepped leaders during their entire lifetime, that is, from their apparent initiation in the cloud to the beginning of the return stroke at ground. He

4.7. Subsequent leader

found a clear tendency for the downward leader speed to decrease, from a mean of 1.6×10^7 m s^{-1} at heights of 3 to 6 km to a mean of 3.5×10^6 m s^{-1} within the lowest kilometer or so. Davis also reported a tendency for the time interval between dE/dt pulses to increase with decreasing height, from 1.6 μs near the leader origin to 4.1 μs near ground (these are mean values of median interpulse intervals determined for recording windows of 205 μs at an amplitude threshold level of 20 percent of the largest pulse peak). Davis observed an inverse relationship between the dart-stepped leader speed and the interstep interval within individual leaders as they accelerated or slowed down during their descent. The near-ground interstep interval value, 4.1 μs, is comparable to the corresponding values, 6.5 and 7.8 μs, reported by Krider *et al.* (1977) from electric field measurements in Florida and Arizona, respectively. Davis also noted the existence of many smaller dE/dt pulses, below the 20 percent amplitude threshold arbitrarily set by him for identifying step pulses, that were produced by sources on the lightning channel, as determined by the locations of the sources of the larger dE/dt pulses. When an amplitude threshold of eight percent was used, the interpulse interval was less than 1 μs near the origins of the leaders in the cloud and 2.2 μs near ground.

Idone and Orville (1984), from photographic observations in New Mexico, reported on two leaders in a rocket-triggered lightning flash that each exhibited an abrupt transition from a continuous to a stepped mode of propagation upon entering the lowest 400 m long section of the channel, the section formed along the triggering-wire trace. The interstep intervals ranged from 2 to 8 μs, and the step lengths from 5 to 10 m, although the last clearly visible step in one of the events had a length of approximately 25 m. Seventeen following leaders in this flash appeared to propagate continuously along the same path all the way to the ground. Note that the flash studied by Idone and Orville (1984) is one of three unusual, 24-stroke New Mexico flashes that were characterized by a geometric mean return-stroke current peak of 5.6 kA and a geometric mean interstroke interval of 8.5 ms (Idone *et al.* 1984), each considerably smaller than its counterpart in either natural lightning or other triggered lightning.

As noted in subsections 4.4.6 and 4.4.8, Wang *et al.* (1999b) found that the optical pulses (luminosity waves) associated with individual steps of one particular dart-stepped leader in rocket-triggered lightning appeared to originate at the downward-moving leader tip and to propagate in the upward direction. The pulses were observed to attenuate to about 10 percent of the original luminosity value after traveling 50 m or so.

Rakov *et al.* (1998), from two-station measurements of the electric and magnetic fields of a dart-stepped leader in triggered lightning, estimated that the formation of each step is associated with a charge of a few millicoulombs and a current of a few kiloamperes.

4.7.6. "Chaotic leader"

As noted earlier, the term "chaotic leader" was introduced by Weidman (1982) to describe the irregular electric field pulse sequences observed prior to the return-stroke field peak. Rakov and Uman (1990b) reported on 15 subsequent strokes that exhibited such chaotic leaders. They found chaotic leaders to be associated with relatively large subsequent strokes. For these strokes, the geometric mean initial electric field peak normalized to 100 km was 4.3 V m^{-1} (compare with the values in Table 4.7). The average power spectral density of 15 chaotic leaders was presented by Willett *et al.* (1990, Fig. 6). Davis, S.M. (1999) observed that chaotic leaders tended to occur after relatively short interstroke intervals and to be associated with relatively high average speeds. He reported a geometric-mean previous interstroke interval of 23 ms for 12 second strokes initiated by chaotic leaders, and for two of them he gave average downward propagation speeds 3.2×10^7 and 5.0×10^7 m s^{-1}. Note that these speeds are close to the highest speeds observed for dart leaders using optical techniques (Jordan *et al.* 1992).

4.7.7. Narrowband radiation

The dart leader produces considerable radiation at VHF (Takagi 1969a, b; Le Vine and Krider 1977; Rustan 1979; Rust *et al.* 1979; Hayenga 1979; Rustan *et al.* 1980) and in the microwave region from 400 to over 2000 MHz (Brook and Kitagawa 1964; Rust *et al.* 1979). This radiation emanates primarily from the cloud rather than from the channel to ground (Proctor 1971, 1976; Rustan 1979; Hayenga 1979, 1984; Rustan *et al.* 1980). The dart-leader narrowband radiation observed by Le Vine and Krider (1977) began an average of 265 μs before the return stroke. Takagi (1969a, b) reported that the dart-leader radiation began 100–1000 μs prior to the return stroke. At frequencies above about 100 MHz, the dart-leader radiation ceased about 100 μs prior to the return stroke, according to Le Vine and Krider (1977) and Brook and Kitagawa (1964), while at 3 MHz Le Vine and Krider (1977) found that it often continued up to and during the return stroke. Takagi (1969a, b) found that about one-third of the dart leaders ceased radiating in the frequency range 60–420 MHz by 50 μs prior to the return stroke, one-third ceased at about the start of the return stroke, and the remaining one-third radiated during the return stroke.

4.7.8. Inferences on the dart-leader mechanism

Fisher *et al.* (1993), studying triggered lightning in Florida and Alabama, found that all subsequent return strokes were invariably preceded by a time interval without measurable current flowing to ground (the minimum detectable current level was less than 2 A), implying that a complete cutoff in channel current is a prerequisite for the formation of a subsequent leader–return-stroke sequence. This finding is consistent with the observations of McCann (1944) and Berger (1967a), who reported that the current between strokes fell below their systems' minimum detectable levels of 0.1 A and 1 A, respectively, and it is contrary to the hypothesis of Brook et al. (1962) that a small continuing current, of order 10 A, is needed for the development of a dart leader. Further, Uman and Voshall (1968) presented calculations to show that, in the absence of additional energy input, the lightning-channel temperature decay due to thermal conduction is slow enough to maintain the channel temperature at a value an order of magnitude higher than the ambient temperature at the end of a typical interstroke interval of 50 ms. Uman and Voshall (1968) therefore argued that no small dark currents are required to keep the channel in a state such that a dart leader can be initiated along the same channel after an interstroke interval of some tens of milliseconds. Channels of larger radius take longer to cool than channels of smaller radius. The calculations of Uman and Voshall (1968) were extended by Picone et al. (1981) to include convective energy loss, with essentially similar results.

Fisher et al. (1993) observed from their movie-camera records having a time resolution of 3 or 5 ms that channel luminosity ceases first at the channel bottom while the upper channel sections remain luminous. Further channel decay is characterized by fragmentation throughout the entire visible length of the channel and more or less simultaneous extinction of the individual fragments separated by dark gaps. Additionally, Krehbiel et al. (1979) inferred from multiple-station electric field measurements that the return-stroke current is first cut off near the ground. Rakov and Uman (1994) argued that sometimes the resultant gap can be bridged by a secondary leader facilitated by the charge accumulated on the upper part of the channel, this part being effectively disconnected from ground. Some confirmation of this scenario comes from the observed occurrence, in the same channel, of two leader–return-stroke sequences within a time interval of the order of one millisecond or less (Rakov and Uman 1994; Idone and Davis 1999). The absence of current flow to ground prior to the development of a dart leader appears to be a crucial element of the dart-leader propagation mechanism discussed below. If a grounded, current-carrying channel does exist and it becomes energized at the top, the leader–return-stroke sequence is not needed to accomplish the transfer of charge to ground and a different mode of charge transfer, called the M-component (Section 4.9), takes place instead. It is worth noting that the above statement implies the existence of a single channel to ground. In the case of multiple channels having a common section near ground, as often observed in upward lightning (Chapter 6), the cloud sources supplying current to these channels or branches appear to be essentially independent (e.g., Gorin et al. 1975; Berger 1967a). As a result, a dart leader can occur in one branch while another branch is still carrying an appreciable current.

Jordan et al. (1997), using high-speed photography, examined in detail the relative light intensity as a function of time and height (within the bottom 1.4 km) for three dart-leader–return-stroke sequences in two flashes, one of these sequences being presented in Figs. 4.37a, b, c. Dart-leader light waveforms appear as sharp pulses with 20–80 percent risetimes of about 0.5–1 µs and widths of 2–6 µs followed by a more or less constant light level (plateau). The plateau continues (for up to some tens of microseconds) until it is overridden by the return-stroke light waveform, suggesting that a steady leader current flows through any channel section behind the downward-moving leader tip before, and perhaps for some time after, the return-stroke front has passed that channel section. Similar waveforms of dart-leader luminosity versus time were also reported, from photoelectric measurements, by Mach and Rust (1997) for both natural and triggered lightning.

Return-stroke light waveforms near ground in the data of Jordan et al. (1997) exhibit 20–80 percent risetimes of about 1–2 µs, somewhat longer than the 20–80 percent risetimes for dart-leader waveforms. Further, there is a significant difference between return strokes and dart leaders in terms of the variation in luminosity along the channel. As indicated in subsection 4.6.2, the 20–80 percent risetime of the return-stroke light pulses increases from 1.5 to 4.0 µs (mean values), and the pulse peak decays, as the return-stroke front propagates from ground to the cloud base at about 1.4 km, whereas the risetime of the dart-leader light pulses is essentially constant with height and the pulse peak is either more or less constant or increases as the leader front approaches ground. Mach and Rust (1997) also found that there was no significant difference between the dart-leader optical characteristics at the bottom and at the top of the visible channel. On the basis of the observed risetimes of the luminosity pulses produced by dart leaders and return strokes and on their propagation speeds, Jordan et al. (1997) estimated that the average electric field intensity across a dart-leader front, with length of order 10 m, was at least an order of magnitude higher, about 1–2 MV m^{-1} (10–20 kV cm^{-1}), than the average electric field intensity, about 50–100 kV m^{-1} (0.5–1 kV cm^{-1}), across the return-stroke front, with length of order 100 m. The average electric field intensity across the dart-leader front should be sufficient for electron-impact ionization.

Rakov (1998) compared the behavior of traveling waves on linear *RLC* transmission lines representing dart leaders and return strokes with the observed properties of the luminosity waves associated with those two processes (Jordan and Uman 1983; Jordan et al. 1997) and with other observed characteristics of dart leaders (Jordan et al. 1992) and return strokes (Willett et al. 1989a; Mach and Rust 1989a). Rakov (1998) inferred that, owing to the poor conductivity of the channel typically available for dart leaders (Table 4.9), the "classical" traveling wave mechanism is not possible for the dart leader. Since dart leaders in fact do propagate on such channels and there is no degradation of the light wave front as the leader approaches ground, there must be significant ionization at the leader front. This ionization counteracts (more successfully than in the case of the return stroke) the front degradation (the losses), a nonlinear effect perhaps equivalent to generation of the dart-leader current at the leader front. Current generation at the leader front has been considered also by Bazelyan (1995), Cooray (1996b), and Thottappillil et al. (1997). The downward progression of the dart leader is apparently maintained by breakdown at the leader front and is therefore controlled by the front electric field, which in turn is largely determined by the charge density near the bottom of the leader channel. Since the leader charge density appears to be positively correlated with the return-stroke current peak (Rubinstein et al. 1995; Rakov et al. 1998) and the channel parameters ahead of the front depend on the previous interstroke interval, the dart-leader propagation mechanism discussed above is consistent with the observation of Jordan et al. (1992) relative to the dependence of the dart-leader speed on the following return-stroke current peak and the apparent lack of dependence on the previous interstroke interval, as discussed in subsection 4.7.2.

If the dart-leader current is generated at the downward-moving leader front, this current should propagate upward, toward the cloud. In this case, the upward progression of the dart-leader current will be controlled by the properties of the pre-return-stroke channel formed behind the dart-leader front. For the upward progression, the phase velocity for frequency components between 100 kHz and 1 MHz (characteristic of the dart-leader front) is about 2×10^8 m s^{-1} (similar to measured return-stroke speeds), and the attenuation distance is of the order of some hundreds of meters, while frequency components between 100 Hz and 1 kHz move at characteristic dart-leader speeds, apparently all the way to the cloud charge source (Rakov 1998). It is possible that the microsecond-scale dart-leader light pulse (see Figs. 4.37b, c) and the corresponding downward-moving luminous channel section, or dart, some tens of meters in length (see below), are associated with the relatively short-lived upward-propagating higher-frequency components, and that the light plateau following the dart-leader light pulse is associated with the relatively long-lived lower-frequency components. Orville and Idone (1982) estimated the dart length by multiplying the total width of the dart-leader track on their streak photographs by the measured dart-leader speed. The average dart length was 34 m, with a range of 7 to 75 m. Similar measurements by Schonland and Collens (1934) yielded an average of 54 m with a range of 25 to 112 m (as given by Orville and Idone 1982), and Schonland (1956) reported a typical value of 40 m. Idone et al. (1984), for 19 dart leaders in triggered lightning, found a mean dart length of 50 m with a range from 15 to 90 m. Orville and Idone (1982) showed that dart length and dart-leader speed are positively correlated.

4.8. Continuing current

The continuing current is usually defined as the relatively low-level current of typically tens to hundreds of amperes that immediately follows a return stroke, in the same channel to ground, and typically lasts for tens to hundreds of milliseconds. This current can be viewed as a quasi-stationary arc between the cloud charge source and ground along the path created by the preceding leader–return-stroke sequence or sequences. Relatively short perturbations in the continuing current, typically lasting for a few milliseconds or less, are called M-components (Section 4.9). Owing to their relatively large charge transfers, continuing currents are responsible for most serious lightning damage associated with thermal effects, such as burned holes in the metal skins of aircraft (Sections 10.1 and 10.2), burned-through ground wires of overhead power lines (Nakahori et al. 1982; subsection 18.4.3), and forest fires (subsection 18.4.2). According to McEachron (1939, 1941), most continuing currents in grounded-neutral electrical systems can flow readily to ground through the transformer windings without causing any damage but may have a considerable effect in blowing the fuses used to protect distribution transformers.

In this section, we will consider only the continuing currents in strokes that are initiated by downward negative leaders. Continuing currents in positive lightning are discussed in subsection 5.3.1. Low-level and long-lasting currents initiated by upward-going leaders from grounded objects are usually referred to as continuous (as opposed to continuing) currents and are discussed in Section 6.3 (natural upward lightning) and subsection 7.2.3 (rocket-triggered lightning). Most of the published information on continuing currents in negative lightning, particularly information regarding the occurrence of this lightning process, concerns only "long-continuing" currents. The latter were arbitrarily defined by Kitagawa et al. (1962) as currents to ground lasting in excess of 40 ms. This criterion was chosen apparently because the most probable interstroke interval reported by Schonland (1956) was about 40 ms (see also Fig. 4.4). In this section, if not specified otherwise, the term "continuing current" means long-continuing current (continuing longer

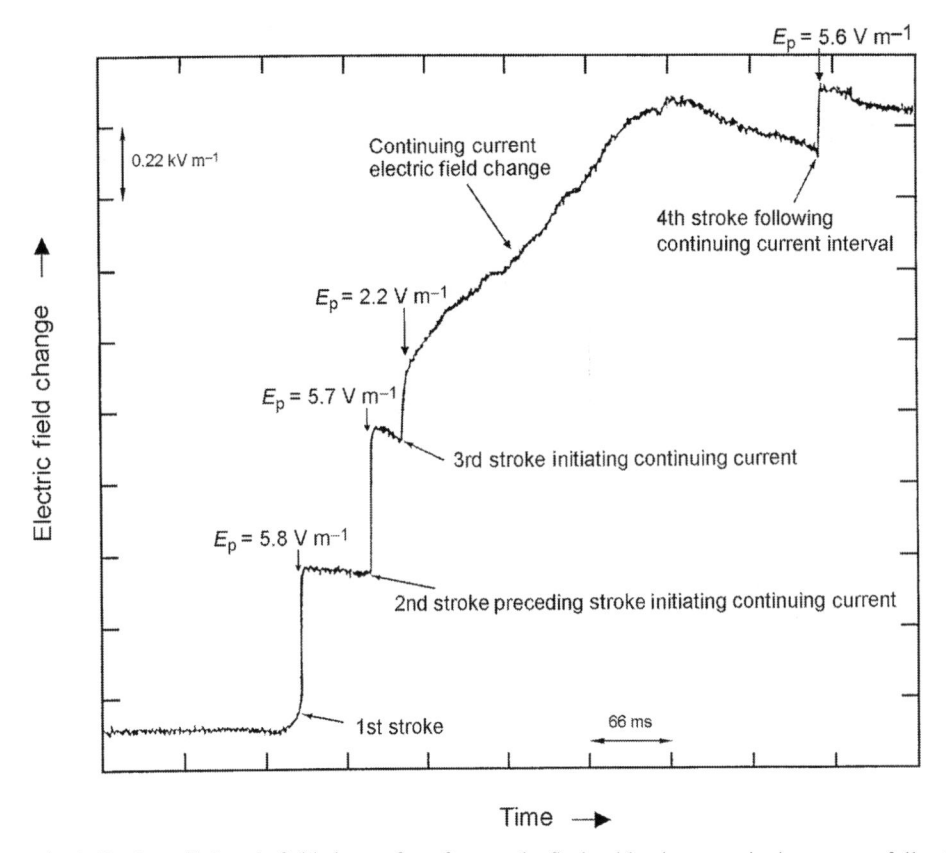

Fig. 4.49. Overall electric field change for a four-stroke flash with a long continuing current following the third stroke. The flash occurred in Florida on 27 July 1979 at 2240:14 UT and at a distance of 6.5 km. Microsecond-scale initial electric field peaks are not resolved in this figure, but their values E_p (normalized to 100 km) are given. A positive electric field change (atmospheric electricity sign convention, subsection 1.4.2) deflects upward. Adapted from Rakov and Uman (1990a).

than 40 ms). Continuing currents with a duration of 40 ms or less were studied by Shindo and Uman (1989).

Kitagawa *et al.* (1962) and Brook *et al.* (1962), from simultaneous photographic and electric field observations in New Mexico, identified the characteristic electric field signature of continuing current (labeled C field change in Fig. 4.6, the upper set of records): a large slow electric field change of the same polarity as that of the preceding return stroke. This finding has enabled other investigators to infer the existence of continuing current from single-station electric field records (e.g., Livingston and Krider 1978). An example of the electric field change of a four-stroke flash with a long (about 200 ms) continuing current following the third stroke is given in Fig. 4.49. A detailed study of long-continuing current, including most of the previously published data, was presented by Rakov and Uman (1990a, 1991).

The overwhelming majority of long-continuing currents are initiated by the subsequent strokes of multiple-stroke flashes as opposed to either the first stroke in a multiple-stroke flash or the only stroke in a single-stroke flash, as illustrated in Fig. 4.50. There appears to be a pattern

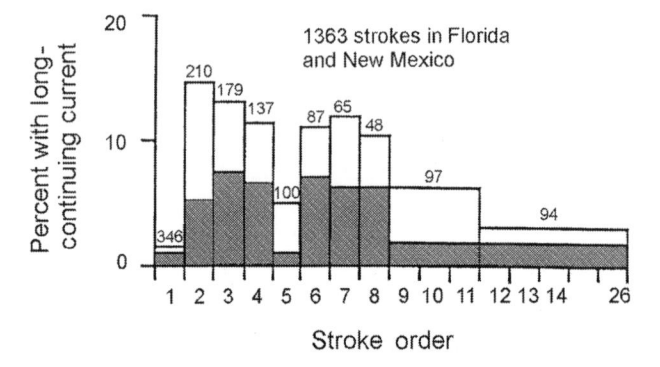

Fig. 4.50. Occurrence of long-continuing current in strokes of different order. The data for long-continuing current occurring at the end of a flash are shown shaded. The numbers above the histogram denote the total number of strokes in each histogram bin. Adapted from Rakov and Uman (1990a).

in the initiation of long-continuing currents (Shindo and Uman 1989; Rakov and Uman 1990a, 1991). First, strokes initiating long-continuing current tend to have a smaller initial electric field peak than regular strokes (Fig. 4.51), the latter defined as neither initiating long-continuing

4.8. Continuing current

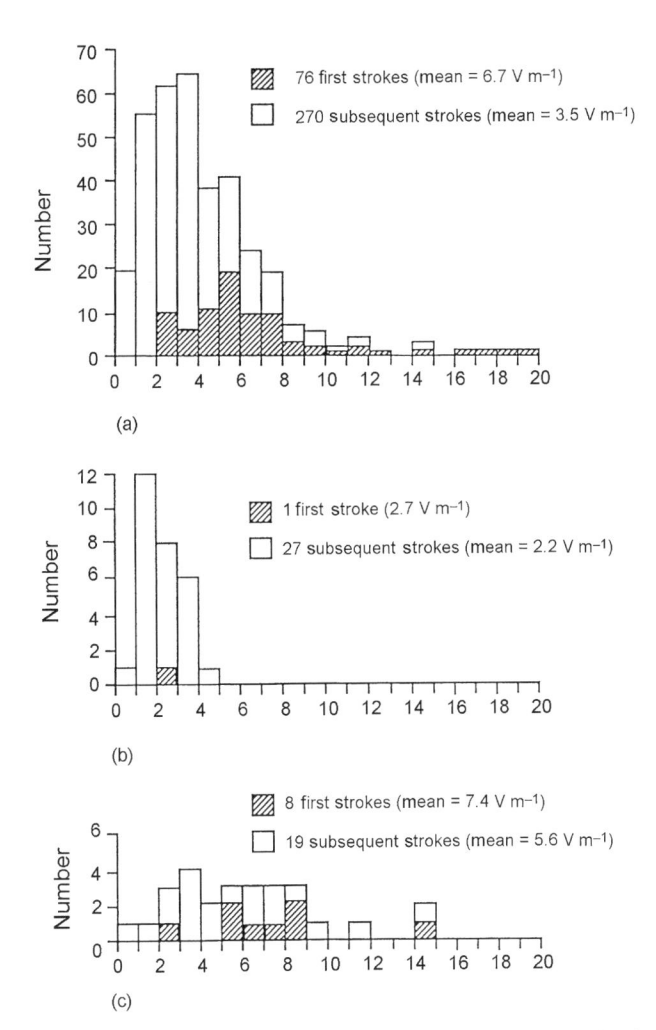

Fig. 4.51. Histograms of the initial electric field peak normalized to 100 km, in V m^{-1}, for 346 strokes in 76 flashes recorded in Florida in 1979. (a) All 346 strokes (mean = 4.2 V m^{-1}); (b) 28 strokes initiating long-continuing current (mean = 2.2 V m^{-1}); (c) 27 strokes preceding those that initiate long-continuing current (mean = 6.1 V m^{-1}). Adapted from Rakov and Uman (1990a).

current nor preceding those doing so nor following a long-continuing current interval. Second, the strokes that precede those initiating a long-continuing current are more likely to have a relatively large field peak than regular strokes (Fig. 4.51). The data of Brook et al. (1962) on the charge lowered by individual strokes show the same tendency, that is, a long-continuing current is usually initiated by a stroke with relatively small charge transfer that follows a stroke with relatively large charge transfer. Hypotheses as to why larger strokes often do not initiate long-continuing currents were given by Rakov and Uman (1990a, 1991). Finally, the strokes that initiate long-continuing current are usually preceded by relatively short interstroke intervals (Fig. 4.4). All these features, besides possibly providing some insights into the physics of the continuing-current process, can potentially be used for studying the relative occurrence of hazardous long-continuing currents in different geographical locations, seasons, types of storms, or stages of storm life cycle with commercially available lightning locating systems (Chapter 17).

Krehbiel et al. (1979), from multiple-station electric field measurements, reported that the negatively charged regions discharged by both individual strokes and by successive portions of long-continuing current are displaced primarily horizontally from one another (Fig. 3.5). The charge centers associated with continuing current tend to be located in the older, more mature regions of a thunderstorm while first strokes in three of four flashes analyzed were initiated in the newer, developing, cells. Krehbiel et al. also found that the J-process (Section 4.10) did not initiate the continuing current but, rather, that the continuing current was initiated directly by the return stroke.

Proctor et al. (1988) reported that in only three of nine flashes in which continuing current could be inferred from electric field records was the continuing current accompanied by "many trains of Q-noise" (VHF radiation) whose sources were scattered over large volumes. In one case, these sources occupied an irregular volume equal to $15 \times 5 \times 10$ km^3 and occurred in groups without an obvious spatial relationship between the groups. In another flash, the sources pervaded a volume equal to $8 \times 8 \times 9$ km^3 and were mostly located progressively further from both the channel to ground and the regions active during the earliest processes in the flash. The observations of Proctor et al. (1988) suggest that profuse branching of the in-cloud portion of the lightning channel is conducive to establishing and maintaining a continuing current.

Rakov and Uman (1990a, 1991) analyzed electric field records for 154 subsequent strokes that followed a previously formed channel, including 14 strokes initiating long-continuing current. They found no difference between these 14 and all other strokes in terms of leader duration.

Fisher et al. (1993), studying triggered lightning in Florida and Alabama, found that continuing currents having durations exceeding 10 ms exhibit a variety of waveshapes that can be grouped into four broad categories, as illustrated in Fig. 4.52. Waveshapes of types I and II were most common (25 out of a total of 30 cases). Fisher et al. (1993) also report that continuing currents longer than 10 ms appeared to start with a current pulse characteristic of the M-component, on average 1.4 ms after the return stroke initial current peak. Similarly, Rakov et al. (1990) reported that in about 95 percent of electric field changes observed within 20 km of natural-lightning return strokes followed by continuing currents (not necessarily long), there was a well-marked change in slope within 3 ms after the return-stroke initial field peak. At relatively close ranges, this change in

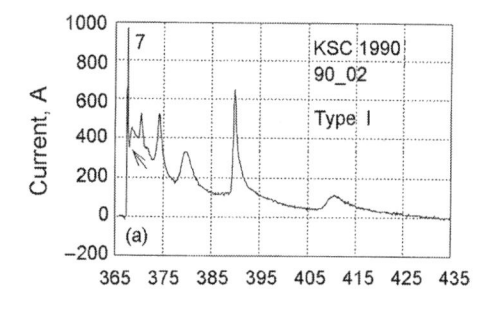

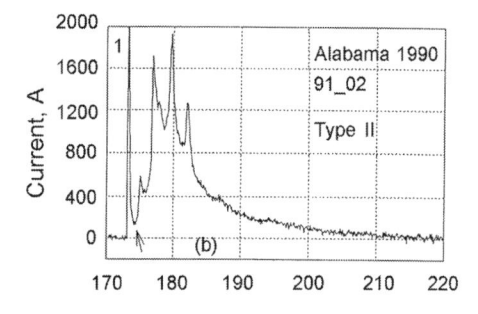

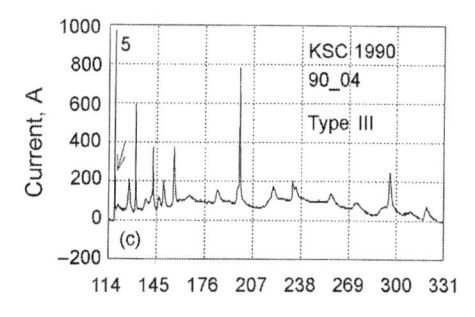

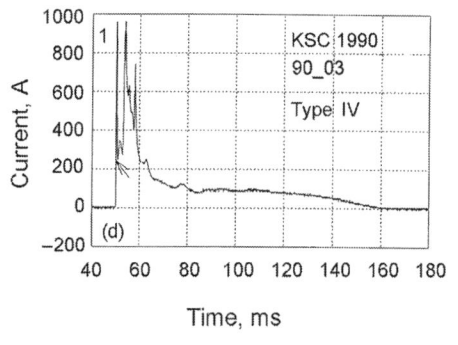

Fig. 4.52. Examples of continuing-current waveshapes in triggered lightning. In each figure, the arrow indicates the assumed beginning of the continuing current. The number in the upper left corner indicates the order of the return stroke in the flash. (a) Type I, more or less exponential decay with superimposed M-current pulses; (b) type II, a hump with superimposed M-current pulses followed by a relatively smooth decay; (c) type III, a slow increase and decrease in current, with superimposed M-current pulses throughout; (d) type IV, a hump with superimposed M-current pulses followed by a steady plateau without pronounced pulse activity. Adapted from Fisher et al. (1993).

slope was commonly accompanied by a hook-shaped electric field waveform indicative of an M-component (such as those seen in Figs. 4.57 and 4.60); M-components are discussed further in Section 4.9.

The distinction between return-stroke current and continuing current is apparently related to the source of the charge transported to ground by these two lightning processes. The return stroke removes charge deposited on the channel by a preceding leader, whereas continuing current is likely to be associated with the tapping of fresh charge regions in the cloud. It is generally thought that the maximum duration of the return-stroke stage is about 3 ms (Malan and Schonland 1951a; Beasley et al. 1982; Rakov et al. 1990). The bulk of the leader charge is stored in the radial corona sheath surrounding the relatively narrow channel core that carries longitudinal current. Pierce (1958) and Rao and Bhattacharya (1966) estimated that the time of the collapse of the radial-corona charge during the return-stroke process is of the order of a millisecond, consistent with the assumed maximum return-stroke duration.

Estimates of continuing currents in sprite-producing lightning discharges from their distant electromagnetic fields are considered in subsections 13.4.3 and 14.3.3.

4.9. M-component

4.9.1. General information

M-components are perturbations (or surges) in the relatively steady continuing current and in the associated channel luminosity. The "M" stands for D.J. Malan, who was the first to study this lightning process. As for leader–return sequences and for continuing current, M-components serve to transport negative electric charge from the cloud to ground. The M-component mode of charge transfer is likely to occur also (i) in the formation of upward-leader steps (subsection 7.2.2) and (ii) during the initial continuous-current stage of object-initiated and rocket-triggered lightning (subsection 7.2.3). As discussed in Section 1.2, the M-component mode differs from the dart-leader–return-stroke mode in that the former requires the presence of a current-carrying channel to ground, while the latter apparently occurs along the remnants of the previously formed channel when there is essentially no current flowing to ground (subsection 4.7.8).

M-components were originally identified by Malan and Collens (1937) as temporary enhancements in luminosity of the faintly luminous continuing current channel. Fisher et al. (1993) were apparently the first to report on correlated current and optical records of M-components, although in triggered rather than natural lightning. Correlated current, electric field, and magnetic field records of M-components, also in triggered lightning, were reported by Rakov et al. (1998). VHF images of M-component channels, primarily in the cloud, have been studied by Shao

4.9. M-component

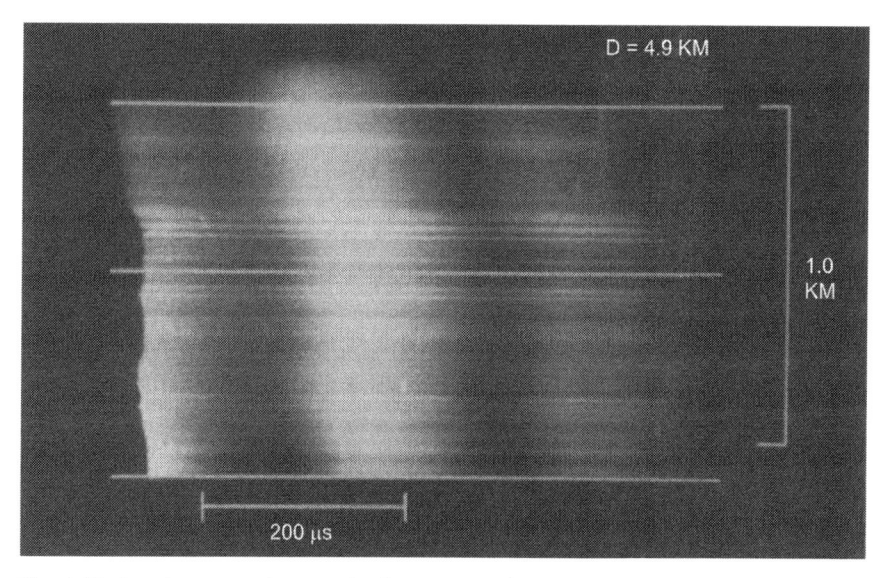

Fig. 4.53. Streak-camera photograph of a return stroke and two associated M-components for a flash that occurred in Florida on 29 July 1978 at distance 4.9 km. The heights indicated by the three horizontal lines correspond to the relative light intensity profiles shown in Figs. 4.54a, b, c. Adapted from Jordan et al. (1995).

(1993), Shao et al. (1995), and Mazur et al. (1995). Rakov et al. (2001) observed that M-component-type processes in triggered lightning produced acoustic signals with peak pressures of the same order of magnitude as those from leader–return-stroke sequences. A detailed characterization of M-components, including a critical review of the previous literature, is found in Thottappillil et al. (1990, 1995), Rakov et al. (1992a, 1995, 2001), and Jordan et al. (1995).

4.9.2. Luminosity

A streak photograph showing two M-components following a return stroke in a natural lightning flash is presented in Fig. 4.53, and the corresponding relative light intensity versus time at three different heights, ground level, 600 m, and 1100 m, are given in Figs. 4.54a, b, c. The M-component light intensity has a character very different from that of the return-stroke light intensity. As seen in Figs. 4.54a, b, c, the return-stroke light pulse (RS) exhibits a relatively fast rise to peak at ground level, appreciably degrading with height (the typical behavior for return strokes; subsection 4.6.2). The first M-component light pulse (M1) is almost invariant with height, at least below the cloud base, at about 1 km. The second M-component pulse (M2) does exhibit some decrease in amplitude with increasing height. The return-stroke pulse has maximum amplitude at ground level (Fig. 4.54c), whereas in the upper half of the visible channel section (Figs. 4.54a, b) the pulse M1 is dominant. The M-component light pulses are more or less symmetrical, with a risetime and falltime of the order of many tens of microseconds. Similar symmetrical waveshapes were reported by Fisher et al. (1993, Fig. 4b) for M-component current pulses measured at the bottom of triggered-lightning channels, as discussed next. A streak photograph showing an M-component in a triggered-lightning flash is presented in Fig. 7.5c.

4.9.3. Current

An example of a channel-base current record showing one return stroke followed by several M-components in a triggered-lightning flash is presented in Fig. 4.55. A summary of M-component current-pulse parameters in triggered lightning reported by Thottappillil et al. (1995) is given in Table 4.12, the definitions of the parameters being illustrated in Figs. 4.55 and 4.56.

A typical M-component is characterized by a more or less symmetrical current pulse having (i) an amplitude of 100–200 A, two orders of magnitude lower than that of a return-stroke current pulse, (ii) a 10–90 percent risetime of 300–500 μs, three orders of magnitude longer than that for a return stroke, and (iii) a charge transfer to ground of 0.1–0.2 C, one order of magnitude smaller than that for a subsequent return-stroke pulse. Some M-components have current peaks in the kiloamperes range (see, for example, Fig. 4.58a), comparable to current peaks of smaller return strokes. About one-third of M-components, which outnumbered return strokes by 4 to 1 in the data of Thottappillil et al. (1995), transferred a charge greater than the minimum charge associated with negative subsequent leader–return-stroke sequences (the so-called "impulse charge"

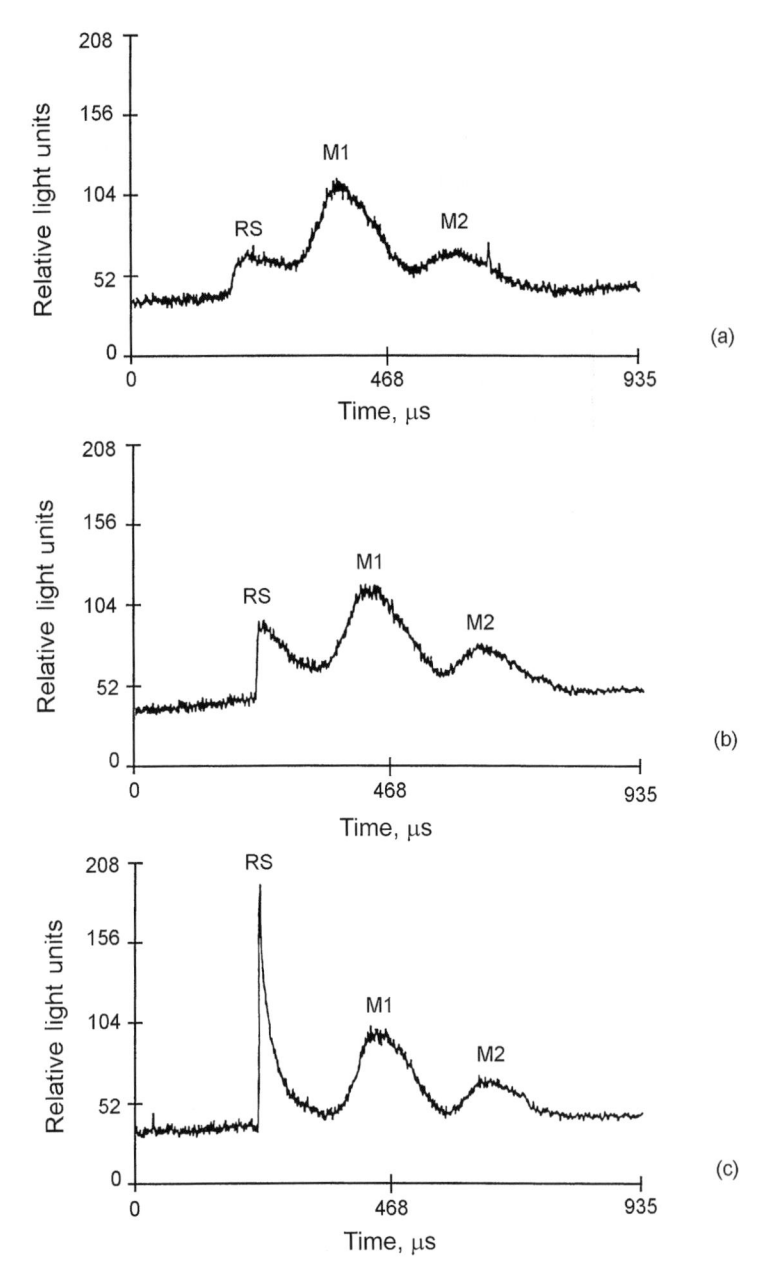

Fig. 4.54. The relative light intensity profiles at (a) 1100 m above ground, (b) 600 m above ground, and (c) ground level corresponding to the streak-camera photograph shown in Fig. 4.53. The profiles show a return-stroke light pulse (RS) followed by two M-component light pulses (M1 and M2). Adapted from Jordan *et al.* (1995).

in Table 4.4). M-current pulses were observed when the continuing current at the channel bottom was greater than 30 A or so, with one exception, in 140 cases, for which the current was about 20 A. The first M-component virtually always (with one exception in 34 cases) occurred no later than 4 ms after the return stroke. Thottappillil *et al.* (1995) found that the M-current magnitude and the M-charge transferred to ground (these are moderately correlated with each other) are each independent of the elapsed time from the return stroke or from the preceding M pulse, the continuing current level, and the shape of the M-current pulse. M-current pulses occurring with larger delays after the return stroke are separated by longer time intervals and tend to be wider.

4.9.4. Electric fields

From correlated electric field and photographic observations in South Africa, Malan and Schonland (1947)

4.9. M-component

Table 4.12. *Parameters of M-components in triggered lightning derived from channel-base current measurements. GM and SD are the geometric mean value and standard deviation, respectively. The parameters are defined in Figs. 4.55 and 4.56. Adapted from Thottappillil et al. (1995)*

Parameter	Sample size	GM	SD $\log_{10}(x)$	Cases exceeding tabulated value		
				95%	50%	5%
Magnitude, A	124	117	0.50	20	121	757
Risetime, µs	124	422	0.42	102	415	1785
Duration, ms	114	2.1	0.37	0.6	2.0	7.6
Half-peak width, µs	113	816	0.41	192	800	3580
Charge, mC	104	129	0.32	33	131	377
CC level, A	140	177	0.45	34	183	991
M interval, ms	107	4.9	0.47	0.8	4.9	23
Elapsed time, ms	158	9.1	0.73	0.7	7.7	156

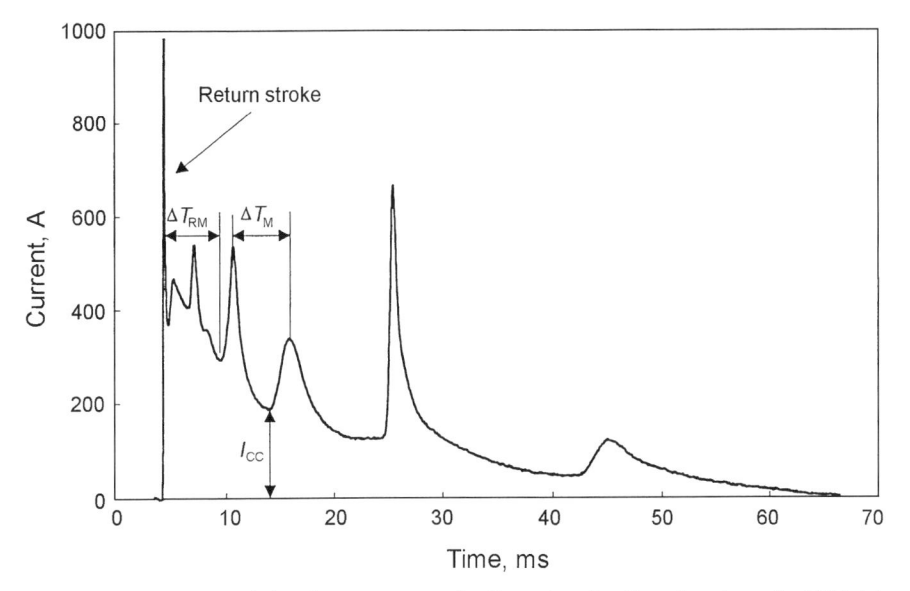

Fig. 4.55. Current record showing a return stroke (intentionally clipped at about the 1000 A level) followed by several M-components in Florida triggered lightning. Shown are the measurements of the continuing current level I_{CC}, M-interval ΔT_M, and elapsed time ΔT_{RM}. Adapted from Thottappillil *et al.* (1995).

reported that field changes associated with M-components, recorded at close range (mostly 6 km or less) with a measurement system having decay time constants of 1 to 21 ms, exhibited a characteristic hook-like shape. They estimated that the duration of the hook-shaped M-field changes was typically from 0.2 to 0.8 ms, although values up to 1.6 ms were observed. The most recent review of the millisecond-scale characteristics of M-field changes was given by Thottappillil *et al.* (1990), who found, in Florida, natural ground flashes occurring within 20 km or so, a geometric-mean M-change duration of 0.9 ms and a geometric mean time interval between M changes of 2.1 ms. An electric field record showing four pronounced M-components is presented in Fig. 4.57. Thottappillil *et al.* (1990) reported that 84 percent of the 38 first M-components occurred within 3 ms of the return stroke. They also found that most of the hooklike field changes occurred in continuing-current field changes of less than 40 ms duration. Rakov *et al.* (1992a) observed that the beginning of the millisecond-scale M-field hook was often marked by microsecond-scale pulses whose shape was dissimilar from and whose polarity was predominantly opposite to the corresponding features of the preceding return-stroke field pulse. They also suggested that the pulses

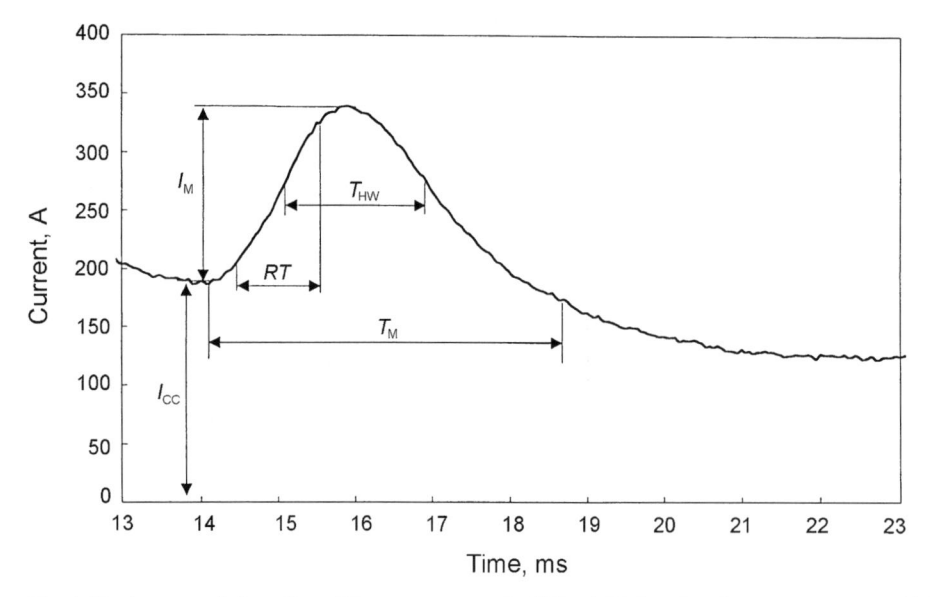

Fig. 4.56. An expanded portion of the current record of Fig. 4.55 showing the measurements of M-current magnitude I_M, 10–90 percent risetime RT, duration T_M, and half-peak width T_{HW}. I_{CC} is the continuing-current level. Adapted from Thottappillil *et al.* (1995).

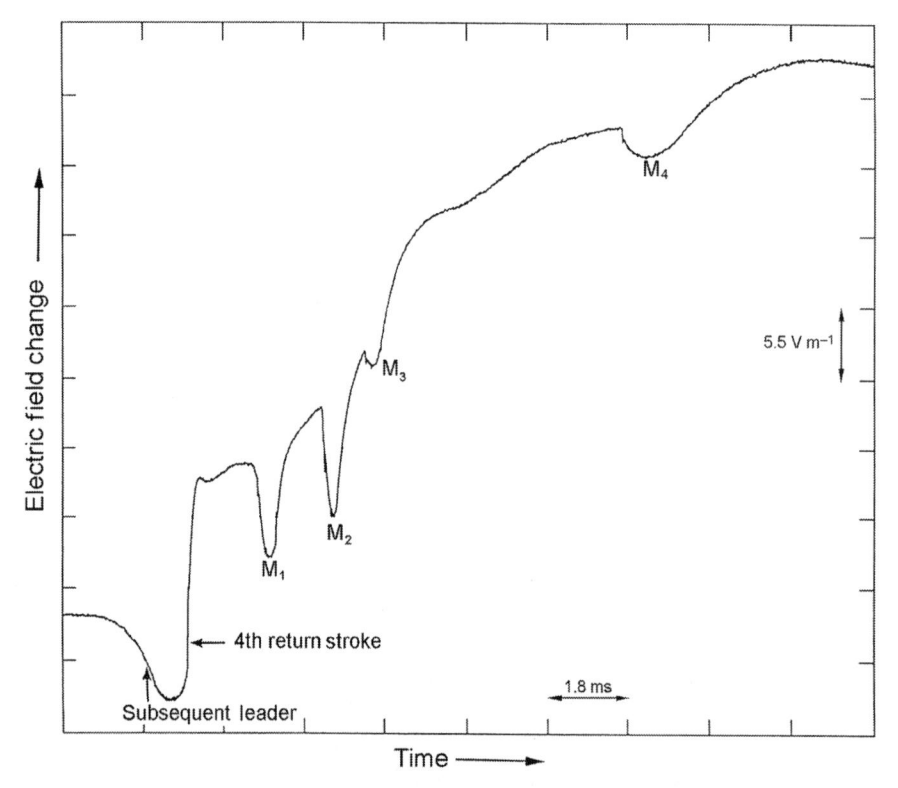

Fig. 4.57. Hook-shaped electric field changes due to M-components during the continuing current initiated by the fourth return stroke in a flash that occurred in Florida on 11 August 1984 at 1857:05 UT at a range of probably about 5 km. A positive field change (atmospheric electricity sign convention, subsection 1.4.2) deflects upward. Adapted from Rakov *et al.* (1992a).

4.9. M-component

were associated with the initiation, at the top of the channel, of charge transfer along the channel to ground, an inference later confirmed by Shao *et al.* (1995), who obtained electric field measurements in conjunction with VHF images of lightning channels. M changes occurring during continuing-current field changes of relatively short duration (less than 20 ms or so) are more likely to have pulses than M changes occurring during continuing-current field changes of longer duration. Electric and magnetic fields due to M-components in triggered lightning at distances ranging from tens to hundreds of meters have been presented by Rakov *et al.* (1995, 1998, 2001). Rakov *et al.* (2001) reported on one M-component in triggered lightning for which the electric field 45 km from the lightning channel was measured together with the current and close fields. One example, showing field waveforms at a distance of 280 m along with the channel-base current, is given in Fig. 4.58. These will be discussed later in this section in relation to the mechanism of the lightning M-component proposed by Rakov *et al.* (1995). A comparison of the electric field changes due to M-components with those due to K-processes is found in Section 4.10.

4.9.5. VHF–UHF imaging

Shao *et al.* (1995), using interferometry (Section 17.6), reported that M-components were typically initiated by what they called fast (10^6–10^7 m s^{-1}) negative in-cloud streamers hitting the upper extremity of the conducting channel to ground, attachment being associated with microsecond-scale field pulses similar to those studied by Rakov *et al.* (1992a). Note that in a later interferometric study by Mazur *et al.* (1998) these so-called streamers were referred to as leaders, the more appropriate term for a self-propagating discharge (Section 1.2). Shao *et al.* (1995) also observed M-components that were apparently initiated by fast (10^7 m s^{-1}) positive streamers developing outward from the upper extremity of the conducting channel to ground and followed by even faster ($> 10^7$ m s^{-1}) negative recoil processes back into the channel. It is possible that the positive streamer is launched from a pocket of excessive positive charge introduced by the preceding return stroke into the leader source region (see, for instance, Krehbiel *et al.* 1979) and does not involve the lower sections of the channel to ground. If this is true, the two types of M-component reported by Shao *et al.* (1995) may differ only in the way in which the negative charge source is connected to the conducting channel to ground, not in the processes occurring in that channel.

4.9.6. Mechanism of the lightning M-component

Although M-components have been observed for more than half a century, until recently there was no con-

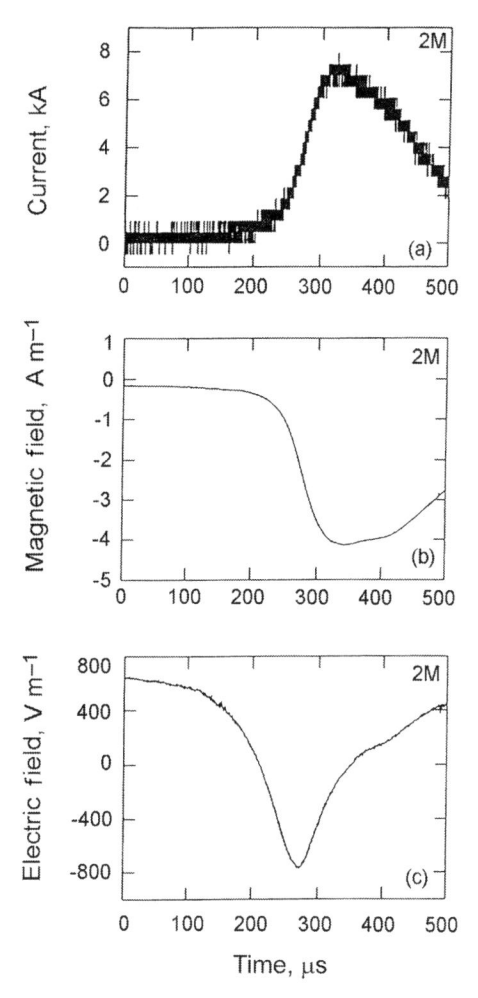

Fig. 4.58. (a) Current, (b) magnetic field, and (c) electric field for a large M-component that followed the second stroke of a rocket-triggered lightning flash (Section 7.2) at Camp Blanding, Florida. The fields were recorded at a distance of 280 m from the lighting channel. Note that the electric field peak occurs appreciably earlier than the current and magnetic field peaks. Adapted from Rakov *et al.* (1998).

sensus regarding the mechanism for this lightning process. Malan and Schonland (1947, p. 498) claimed that the M process is "a minor form of subsequent stroke," which lowers negative charge to ground but that "the charge is usually not large enough, and the other conditions are not always such as, to give rise to a return stroke from the ground". Thus, according to Malan and Schonland (1947), the M process is similar to a downward leader that is unable to produce a discernible upward return stroke. Later, however, Schonland (1956, p. 596), in a review paper, described M-components as manifestations of "processes of unknown nature, probably [involving] branching inside the cloud". Kitagawa *et al.* (1962) suggested that M-components are due to in-cloud

K-processes that occur at a time when there is a relatively low-level current flowing in the channel to ground and described the M-component as "a momentary current increase without involving the leader process". Also, Uman (1987, 2001, pp. 172–3) stated that "downward-moving leaders have not been observed to precede M-components". The latter view is clearly different from the interpretation of the M-component as a leader without a discernible return stroke given by Malan and Schonland (1947). Recent VHF observations of lightning (e.g., Shao et al. 1995) show the development of in-cloud channels that make contact with conducting channels to ground and sometimes also the downward progression of the incident M-component wave. Note that little if any VHF radiation is produced by waves propagating along conducting channels (Sections 1.3 and 17.4), making the VHF imaging of M-components difficult.

From a comparison of the electric fields 30 m from the triggered-lightning channel and the corresponding channel-base currents of lightning M-components in conjunction with modeling (Section 12.5; Figs. 12.19a, b), Rakov et al. (1995) proposed a mechanism for this lightning process. It allows reconciliation of the previous, seemingly contradictory, views noted above. According to Rakov et al. (1995), an M-component is a guided-wave process that involves a downward-progressing incident wave (the analog of a leader) and an upward-progressing wave that is a reflection of the incident wave from the ground (the analog of a return stroke), as illustrated in Figs. 1.2, 12.18a and 12.18b. The amplitudes of these two waves are approximately equal, their expected propagation speeds are between 10^7 and 10^8 m s^{-1}, and the spatial front lengths are comparable to the distance between the lower cloud boundary and the ground. Ground is sensed by the incident M wave as a short circuit, so the reflection coefficient for current at ground is close to $+1$, and the reflection coefficient for the associated charge density is close to -1. At each channel section the two waves are shifted in time, the time shift being small near ground and increasing toward the cloud (Fig. 12.18b). An example of a streak photograph of an M-component in a triggered-lightning flash that apparently exhibited these two M waves is found in Fig. 7.5c. The two M waves are often optically indistinguishable below the cloud base (Fig. 4.53), but the presence of the two waves manifests itself in the observed features of the M-component electric and magnetic fields at ranges from tens to hundreds of meters. Because the reflection coefficients for the traveling waves of current and charge density are different, the incident and reflected current waves (which determine the close magnetic fields) add for each channel section, while the incident and reflected charge-density waves (which determine the close electric fields) subtract. As a result, at close ranges the M-component magnetic field has an overall waveshape similar to that of the channel-base current whereas the M-component electric field has a waveform that appears to be the time derivative of the channel base current, as illustrated in Fig. 4.58. These characteristic features, in addition to the presence of background continuing current immediately prior to the current pulse, allow M-components to be distinguished from leader–return-stroke sequences.

The guided-wave mechanism of the lightning M-component predicts that at close ranges (i) M-component magnetic field magnitudes vary as the inverse distance from the lightning channel, and (ii) M-component electric field magnitudes are relatively insensitive (compared to leader electric field magnitudes) to the distance from the lightning channel. These two predictions were tested using multiple-station electric and magnetic field measurements at distances ranging from 5 to 500 m from the triggered-lightning channel (Rakov et al. 2001). The shapes and magnitudes of the measured close electric and magnetic fields are generally consistent with the guided-wave mechanism of the lightning M-component. Specifically, the M-component electric field peak exhibits a logarithmic distance dependence, $\ln(kr^{-1})$, indicative of a line charge density that is zero at ground and increases with height. Such a distribution of charge is distinctly different from the more or less uniform charge density characteristic of the dart leaders in triggered lightning, as inferred from close electric field measurements (Crawford et al. 2001). The M-component magnetic field peak decreases as inverse distance (i.e., r^{-1}), which is generally consistent with the existence of a uniform current within the lowest kilometer or so of channel. Figure 4.59 shows currents and electric fields at 50, 110, and 500 m for a triggered-lightning stroke, labeled RS1, followed by four M-components, labeled M3a, M3b, M4, and M4a. It is clear from this figure that the magnitude of the M-component electric field pulses relative to the magnitude of the leader–return-stroke electric field pulse increases with distance, as predicted by the guided-wave M-component mechanism.

4.10. J- and K-processes
4.10.1. General information

The J- or "junction" process takes place in the cloud during the time interval between return strokes. It is identified by a relatively steady electric field change on a time scale of tens of milliseconds. The J change has either the same or opposite polarity as that of the return-stroke electric field change. In the former case, it is generally smaller than the field change due to continuing current (Section 4.8), and, as differentiated from the continuing-current field change, is not associated with a luminous channel between cloud and ground. Relatively rapid electric field variations termed

4.10. J- and K-processes

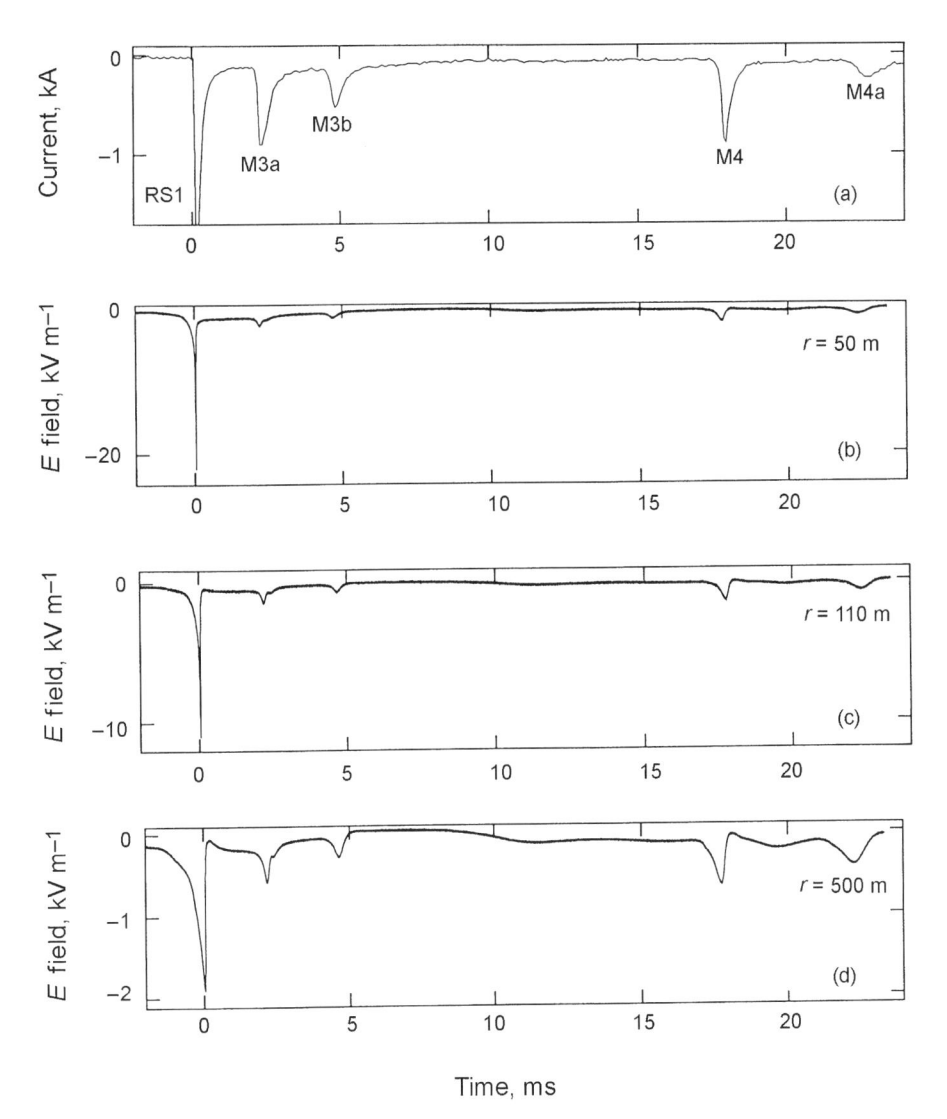

Fig. 4.59. A leader–return-stroke sequence (RS1) followed by four pronounced M-components (M3a, M3b, M4, and M4a) in a rocket-triggered lightning flash (Section 7.2) at Camp Blanding, Florida: (a) the current, (b),(c),(d) the electric field at 50 m, 110 m, and 500 m, respectively. Note that the magnitude of the M-component electric field pulses relative to the magnitude of the leader–return-stroke electric field pulse increases with distance. Adapted from Crawford et al. (2001).

K-changes also occur between strokes, generally at intervals ranging from some milliseconds to some tens of milliseconds and, hence, appear to be superimposed on the overall electric field change associated with the J-process. The K in the term K-change stands either for kleine (the German for small) or for N. Kitagawa and M. Kobayashi, who were the first to study this lightning process in detail (Brook and Ogawa 1977). The electric field change following the last stroke (after the end of continuing current, if any) is probably of the same nature as the J change discussed above.

The J-field change occurring between the strokes of a ground discharge is almost always negative (atmospheric electricity sign convention; subsection 1.4.2) for flashes within a few kilometers and can be either positive or negative for discharges beyond about 5 km (Malan and Schonland 1951a; Malan 1965). Examples of J changes exhibiting different polarity at different ranges are seen in Fig 4.9a. Not only may the J-field change from primarily negative at close range to mixed polarity with increasing distance from the discharge; it may reverse polarity within a single flash. When this happens, early interstroke intervals usually exhibit positive (atmospheric electricity sign convention) J-changes and later ones exhibit negative J-changes. This type of reversal occurs typically when observations are made at intermediate ranges (Malan and Schonland 1951b).

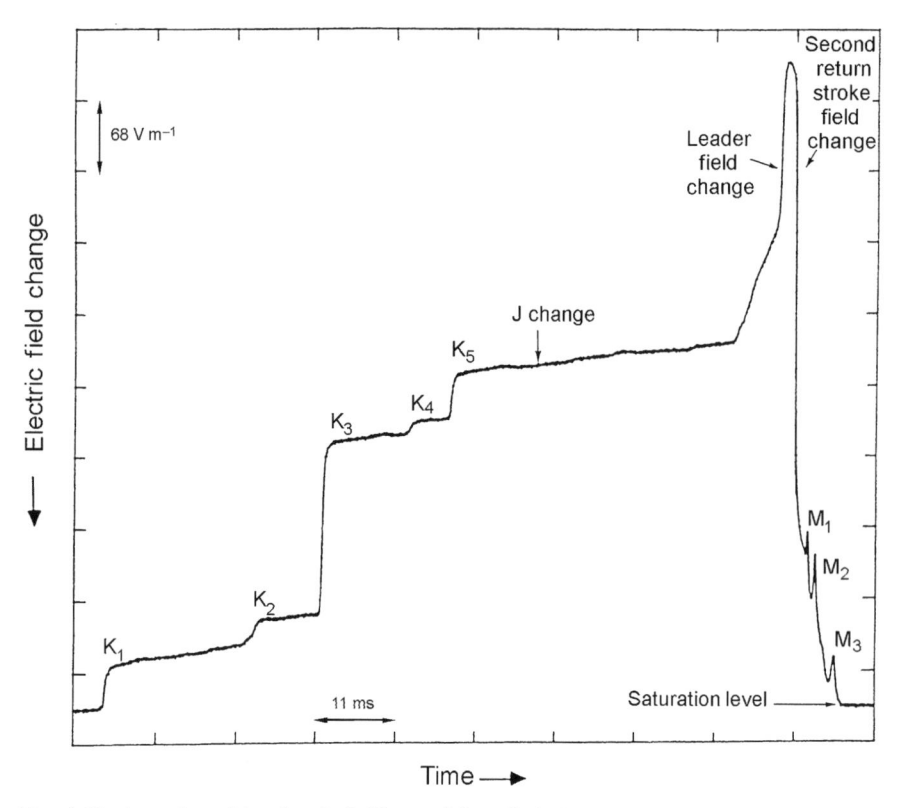

Fig. 4.60. A portion of the electric field record for a flash that occurred in Florida in 1979 at 2228:43 UT at a distance of 2.5 km. Labeled are five pronounced K-changes (K_1 through K_5), a J-change, leader and return-stroke field changes, and three field changes due to M-components (M_1 through M_3). A positive field change (atmospheric electricity sign convention, subsection 1.4.2) deflects downward. Adapted from Thottappillil et al. (1990).

Krehbiel et al. (1979) deduced from their multiple-station electric field measurements (see the example in Fig. 4.9a) that the J-process usually moves negative charge horizontally toward the top of a previous stroke in a manner that probably should not be referred to as a "junction process", since this negative charge is not necessarily the same charge that is involved in the next stroke to ground. Thus the interstroke field changes observed by Krehbiel et al. (1979) are indicative of charge motion that does not always link successive stroke volumes. In fact, charge transport between strokes was observed to persist in the direction of an earlier stroke while subsequent strokes discharged more distant regions of the cloud. Krehbiel et al. (1979) found that the leader and return-stroke field changes appeared to be superimposed upon the interstroke field change, as if independent of it, implying that the coupling between the field change associated with the J-process and the following leader–return-stroke sequence was not strong.

In the following, we will discuss the K-processes that occur during interstroke intervals or after the final stroke, when the channel to ground is not luminous. Similar processes in cloud discharges are considered in Section 9.5. There is controversy and considerable confusion in the literature regarding the definition of a K-change. We present below three definitions used by different researchers, starting with the definition we consider as the least ambiguous.

(i) Thottappillil et al. (1990) and Rakov et al. (1992a, 1996) identified K-changes in ground flashes as the step-like (or ramp-like) electric field changes that occur during interstroke intervals and after the last stroke, have the same polarity as the J-change, and have a 10–90 percent risetime of 3 ms or less. Examples of such K-changes are seen in Fig. 4.60. A similar definition has been used by Ogawa and Brook (1964), Bils et al. (1988), and Villanueva et al. (1994) for K-changes in the cloud lightning discharges discussed in Section 9.5. This definition probably causes a bias toward larger, more pronounced events.

(ii) Kitagawa and Brook (1960) and Kitagawa et al. (1962) identified K-changes as "pulses" occurring during the interstroke interval and after the last stroke in their electric field records, which were obtained with a high-gain

4.10. J- and K-processes

measurement system having a decay time constant of 70 μs, much shorter than the duration of the K-changes, of order 1 ms. Examples of K-changes defined in this way are found in Fig. 4.6 (the bottom traces for each of the two events presented in this figure). Kitagawa and Brook (1960) and Kitagawa et al. (1962) also measured electric fields with a system having a decay time constant of 4 s, adequate to record K-changes without distortion, but it had much lower gain than the system mentioned above, so that only the largest K-changes could be identified (using definition (i)) in those records. The latter records are represented by the middle traces for each of the two events shown in Fig. 4.6. It is important to note that the "K-pulses" seen in Fig. 4.6 are actually instrumentation-distorted step-like field changes, not the microsecond-scale pulses sometimes associated with K-changes and discussed later in this section. The definition of Kitagawa and Brook (1960) and Kitagawa et al. (1962) causes a bias toward smaller events and has led to an erroneous representation of the microsecond-scale signature of K-changes (e.g., Arnold and Pierce 1964), as discussed in Rakov et al. (1992a).

(iii) Rhodes and Krehbiel (1989), Mazur et al. (1995), and Shao et al. (1995) identified K-changes in their time-resolved VHF-images as in-cloud processes that occurred during the interstroke interval and could not be attributed to any other known lightning process. When a K-process was observed to move down the channel to ground it was called an "attempted" leader (Fig. 4.16, left panel, the middle set of waveforms and right panel, the three images labeled AL1, AL2, and AL3). Note that in definitions (i) and (ii) no distinction is made between K-processes and attempted leaders. The five pronounced K-changes (particularly K_3) shown in Fig. 4.60, and perhaps most of the K-changes based on definition (i), would probably have been classified by Shao et al. (1995) as "attempted" subsequent leaders. Using definition (iii), Shao et al. (1995) and Mazur et al. (1995) concluded that K-processes are indistinguishable from M-processes and dart leaders, except that the latter two propagate all the way to ground; it is important to note, however, that this similarity between the three processes applies only to their observed behavior in the source thundercloud whereas the propagation mechanisms in the channel to ground for dart leaders and M-components, discussed in subsections 4.7.8 and 4.9.6, respectively, are apparently quite different.

4.10.2. Properties of K-processes

Krehbiel et al. (1979) reported that ground flash K-changes recorded simultaneously at eight stations had, at each station, the same polarity as the slower J-field change on which the K-changes were superimposed, in spite of the fact that the J-field changes could be of different polarity at different stations. A similar tendency has been found in single-station studies (e.g., Kitagawa 1957b; Kitagawa and Brook 1960; Rakov et al. 1992a), although electric field changes similar to K-changes but having polarity opposite to that of the J-change (approximately 10 percent of all K-type field changes studied by Rakov et al. 1992a) have also been observed.

Thotappillil et al. (1990) found, in Florida ground flashes and using definition (i), a geometric mean K-change duration of 0.7 ms and a geometric mean time interval between K-changes of about 13 ms. Rakov et al. (1992a) found that there was no characteristic microsecond-scale electric field signature associated with K-changes. The majority of K-changes do not contain pronounced pulses. If they do, in most cases the pulses do not occur at the beginning of the step (see also Shao et al. 1995); this finding suggests that the pulses are not associated with the initiation of K-process charge transfer. Sometimes, microsecond-scale pulses associated with K-changes appear in the form of the regular pulse bursts discussed in Section 4.11.

K-changes have been found from all-sky photomultiplier records to be accompanied by pulses of luminosity (Kitagawa and Kobayashi 1959). From photographic records they have been observed occasionally to be associated with a failed leader or an air discharge below the cloud base (Kitagawa et al. 1962).

We now review the characteristics of the VHF–UHF radiation associated with K-processes. Note that different names are given by different researchers to this radiation. Brook and Kitagawa (1964) found strong microwave radiation (from 400 to 1000 MHz) associated with K-changes in both intracloud and cloud-to-ground discharges. A number of other investigators, using VHF–UHF interferometric or time-of-arrival systems for imaging lightning sources of radio wave emission, have observed radiation apparently associated with the K-process. Richard et al. (1986), using an interferometric system, found that their "burst" phenomenon, which accompanied a typical K-change (they labeled the latter as an "intracloud discharge") after the last stroke to ground (see their Fig. 8), propagated toward the ground but did not reach it. The "burst" phenomenon followed the channel defined by the sources associated with the previous stroke. Hayenga (1984), also using an interferometer, reported his fast bursts to "travel through or on the edge of regions previously radiating at VHF". The burst following the first return stroke shown in Fig. 4 of Hayenga (1984) was reported to progress downward along the path to ground defined by the sources before the return stroke (see his Fig. 5). D. E. Proctor (personal communication, 1989) sometimes observed his "Q-noise" sources going down a channel toward ground, though not all the way to ground.

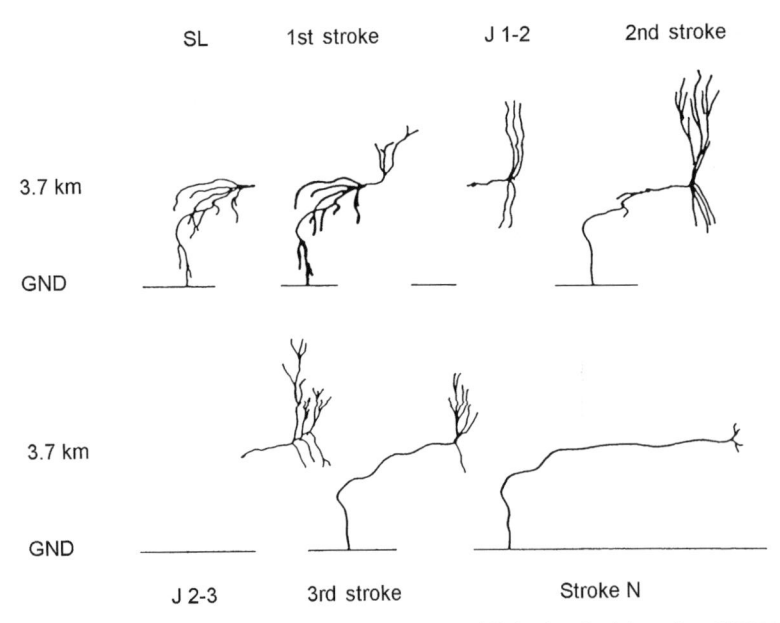

Fig. 4.61. Conceptual model of a cloud-to-ground lightning flash based on VHF imaging of lightning channels. SL denotes the stepped leader and J denotes interstroke processes. Ground level (GND) is 1.4 km above mean sea level. Adapted from Proctor *et al.* (1988).

Rhodes and Krehbiel (1989), as noted earlier, observed what they called a "K-type event" and described it as a sequence of two VHF-tracked intracloud streamers, each of them being associated, at the time that it reached its destination point, with a small microsecond-scale electric field pulse and VHF radiation back at the starting point of the event. No information was presented on whether that event was accompanied by a millisecond-scale field step in their "slow" field record.

VHF–UHF Q-noise, or the fast-burst phenomenon, has usually been reported to precede K-changes in wideband electric field records (Proctor 1981; Hayenga 1984), whereas Rakov *et al.* (1992a) observed that microsecond-scale field variations were commonly delayed with respect to the starting point of the millisecond-scale step-like field changes. It can be inferred from the model of a ground flash proposed by Proctor *et al.* (1988) and illustrated in Fig. 4.61 that "Q-streamers", considered to be the sources of Q-noise, are typically intermittent vertical discharges lasting less than 100 μs with an interdischarge time of a few milliseconds. They are heavily branched in both upward and downward directions and presumably funnel cloud negative charges into the extremity of the intracloud channel. This channel apparently extends horizontally away from the region of the positive charge introduced into the cloud by the previous return stroke (Krehbiel *et al.* 1979). The "Q-streamers" occur in progressively more distant cloud regions in front of the outward-propagating positive channel, probably facilitating its further extension at a speed of the order of 10^4 m s^{-1} (Proctor *et al.* 1988). There could also be channel progression during quiet intervals between the "Q-streamer" bursts. Proctor *et al.* (1988) sometimes observed smooth J-changes not accompanied by any VHF radio noise; perhaps these were "classical" positive J-leaders slowly propagating outward from the extremity of the previous return-stroke channel (e.g., Malan and Schonland 1951a, pp. 145–63). Each burst of "Q-streamers" apparently leads to the effective movement of negative charge along the channel, the latter giving rise to a K-change. In the above view, the "Q-streamers" could be considered as the initial part of a recoil process, in that negative charge is supplied to a developing positive in-cloud channel, in keeping with the recoil-streamer picture of the K-process suggested by Kitagawa (1957b) and Ogawa and Brook (1964) and discussed below in subsection 4.10.3. One common feature of the in-cloud VHF sources associated with the K-process (and also with the M-process and with the subsequent leader) is a propagation speed that tends to be of order 10^7 m s^{-1}, similar to the typical optically measured dart-leader speed in the channel to ground (subsection 4.7.2).

4.10.3. Inferences on the K-process mechanism

The physical process giving rise to what later (Kitagawa 1957b) would be termed a K-change was apparently first inferred by Malan and Schonland (1951a, pp. 161–2) from their electric field and photographic observations of ground flashes in South Africa. They interpreted the step-like field changes (see their Fig. 9d), which occurred between strokes and after the last stroke and which were not accompanied by any detectable luminosity in the

4.10. J- and K-processes

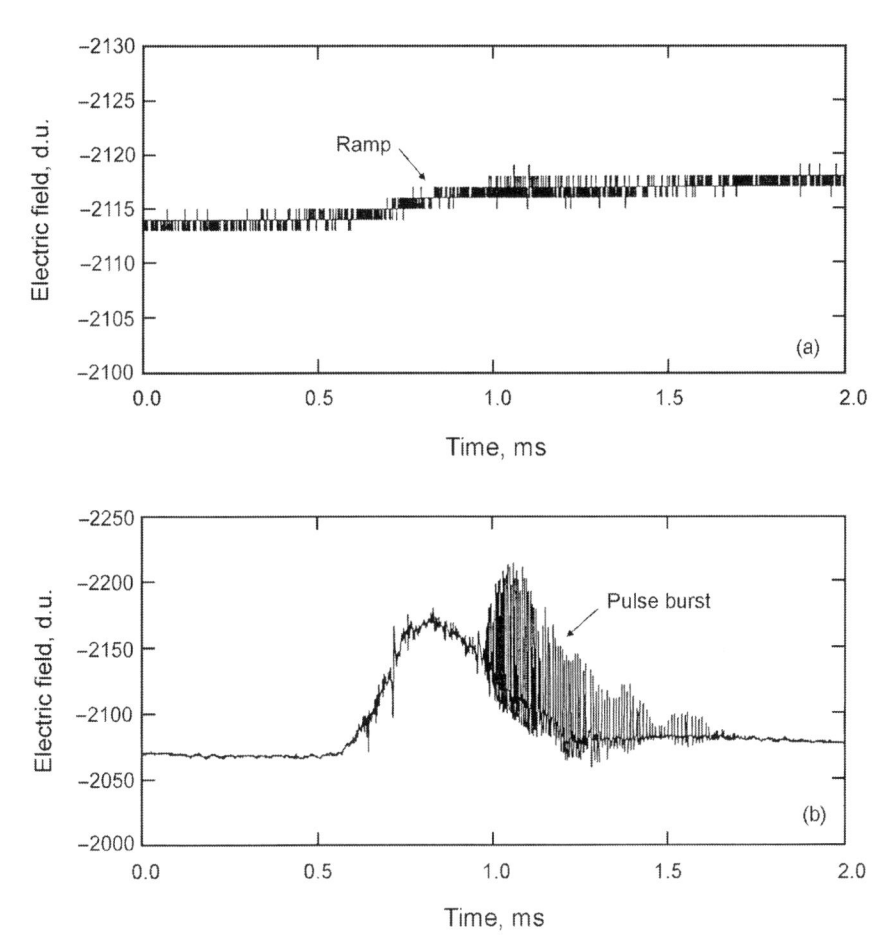

Fig. 4.62. A regular burst of microsecond-scale pulses associated with a ramp-like millisecond-scale field change (K-change) in a ground flash (KSC9122246): (a) low-gain system, decay time constant 10 s; (b) high-gain system, decay time constant 150 μs. (The pulse burst is shown on an expanded time scale in Fig. 4.63). Here and in Figs. 4.63–4.65 the atmospheric electricity sign convention (subsection 1.4.2) is used, and "d.u." denotes digitizer units. Adapted from Rakov et al. (1996).

channel between cloud base and ground, as being due to a leader that was not able to reach ground and instead produced "a small readjustment of charge within the cloud". Subsequently, the K-process in both ground and cloud flashes was hypothesized (e.g., Kitagawa 1957b; Ogawa and Brook 1964) to be a negative "recoil streamer" that occurs when a positive J-type leader propagating within the cloud encounters a region of concentrated negative charge. Clegg and Thomson (1979) found that gaps of typically 6 to 9 ms in the HF (10 MHz) radiation following return strokes were invariably terminated by a K-change. They suggested that these quiet periods were associated with nonradiating positive J-leaders propagating from the top of the preceding return-stroke channel and producing a recoil K-process upon encountering a negative charge region, a view in support of that of Kitagawa (1957b) and Ogawa and Brook (1964), described above.

It appears that the K-process serves to charge the lightning channel that has been previously disconnected electrically from ground and to extend it via breakdown at its extremities. It appears further that K-process waves travel on this channel system and may experience reflections at various discontinuities, particularly at the channel ends, as observed in VHF imaging of lightning channels by Rhodes and Krehbiel (1989) and Shao et al. (1995). If this view of the K-process is correct, then one should expect a great variety of signatures of this lightning process, as is apparently confirmed by the variability of the millisecond-scale electric field waveforms (Thottappillil et al. 1990), by the lack of a characteristic microsecond-scale electric field signature (Rakov et al. 1992a), and by the different propagation patterns of VHF–UHF sources, discussed in subsection 4.10.2.

As noted in subsection 4.10.2, the K-process has in some cases been observed from VHF imaging, as well as from photographic observations, to propagate down a previous return-stroke channel (Kitagawa et al. 1962), and apparently the process can reach ground and produce a return

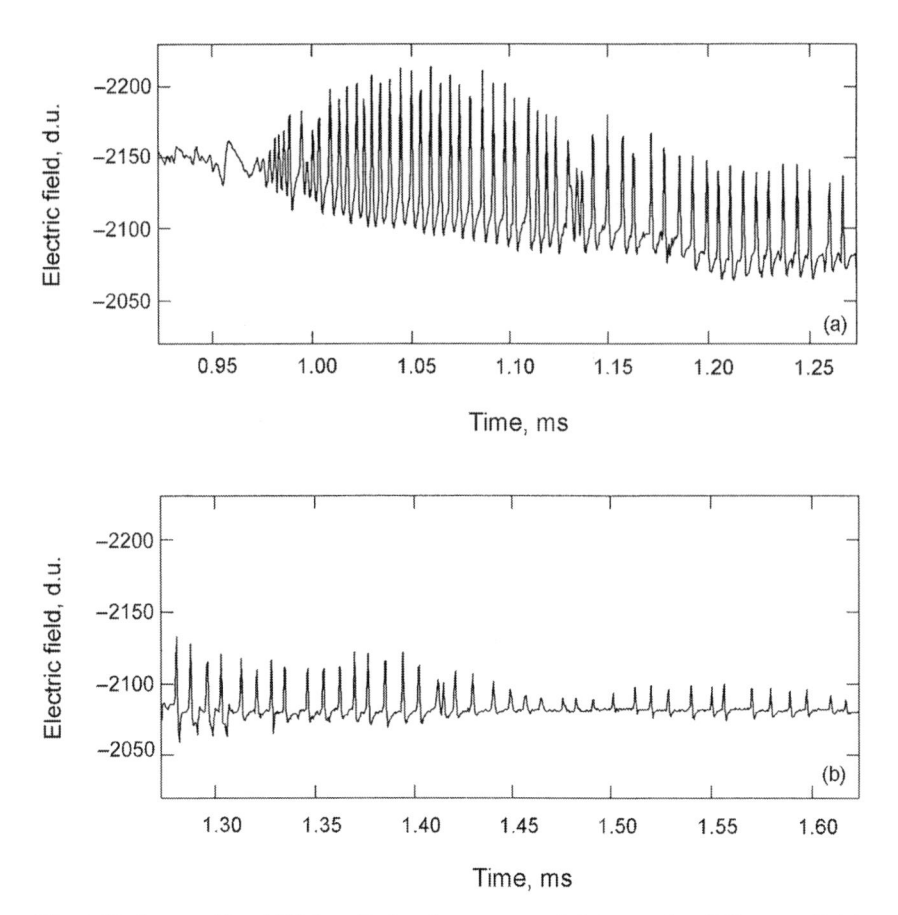

Fig. 4.63. Same as Fig. 4.62b but displayed on an expanded time scale of 50 μs per division. The end of the time scale in Fig. 4.63a is the beginning of the time scale in Fig. 4.63b. Adapted from Rakov et al. (1996).

stroke. Further, Hayenga (1984), Rhodes and Krehbiel (1989), Shao et al. (1995), and Mazur et al. (1995) found that VHF radiation sources associated with K-events during interstroke intervals of cloud-to-ground flashes behave just like dart leaders, except that the K-events do not develop all the way to ground. Thus, probably most of the K-processes in a ground flash can be viewed as unsuccessful or attempted subsequent leaders, as was first suggested by Malan and Schonland (1951a, pp. 145–63).

4.11. Regular pulse bursts

Regular pulse bursts occurring in cloud flashes were first studied by Krider et al. (1975). Similar bursts in both cloud-to-ground and cloud flashes, in relation to corresponding millisecond-scale electric field variations, were analyzed by Rakov et al. (1996). Only regular pulse bursts in cloud-to-ground flashes are presented in this section. A description of the bursts in cloud flashes is found in subsection 9.5.2.

Examples of the regular pulse bursts in a ground flash are given in Figs. 4.62 through 4.65. The microsecond-scale pulse burst shown in Figs. 4.62b and 4.63a,b occurs in the latter portion of a ramp-like millisecond-scale field change characteristic of a K-change, while the bursts in Figs. 4.64 and 4.65 appear to be associated with a hook-shaped field change characteristic of an M-component (Section 4.9). In the trace shown in Fig. 4.62a (low gain, decay time constant of 10 s), no regular pulse burst is seen because the gain is insufficient. What is seen instead is a small ramp characteristic of the K-process in both ground and cloud flashes. In the trace shown in Fig. 4.62b (high gain, decay time constant of 150 μs), the ramp is distorted owing to the fact that the system decay time constant is comparable to the duration of the ramp; nevertheless a pulse burst is readily observable in the later portion of the ramp. Figures 4.63a, b show two portions of the burst on an expanded time scale (350 μs as against 2 ms in Figs. 4.62a, b).

In about 40 percent of all the cases studies by Rakov et al. (1996) the bursts were associated with ramplike field changes and typically occurred in the second half of the ramp, as seen in Fig. 4.62. In 55 percent of the cases there was no detectable relatively slow field change associated with the burst, although an existing ramp might have been too small to be detected with the instrumentation gain used.

4.11. Regular pulse bursts

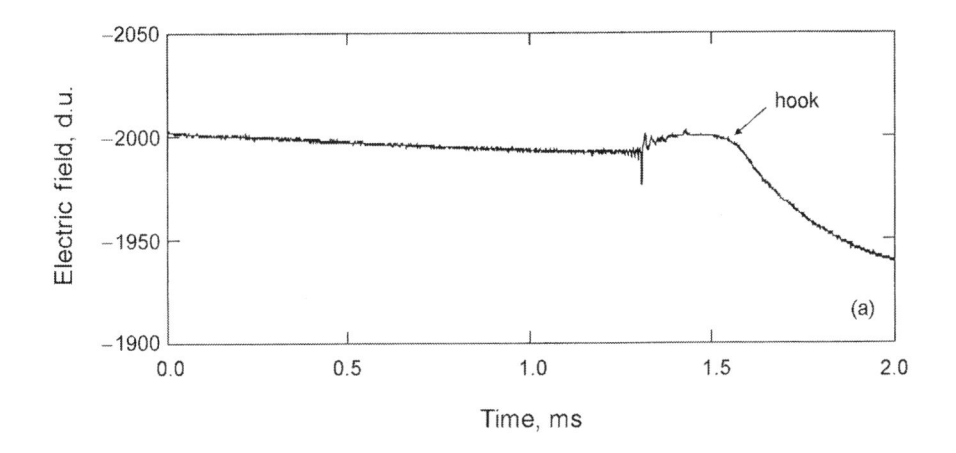

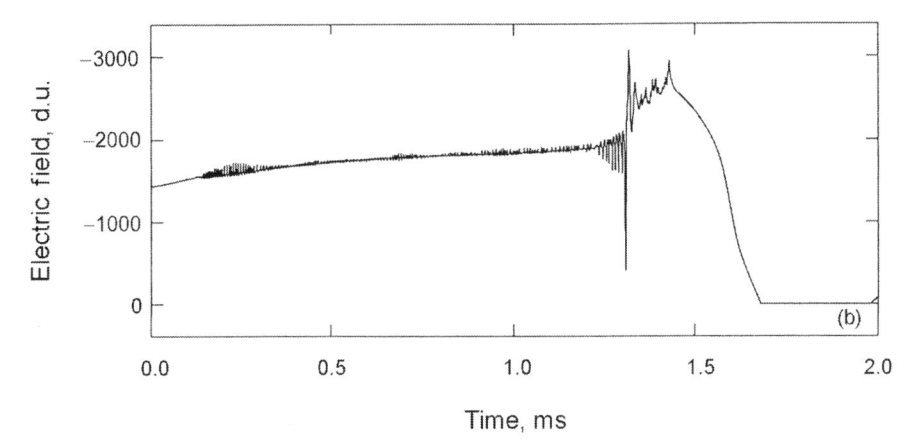

Fig. 4.64. Regular bursts of microsecond-scale pulses associated with a hook-shaped millisecond-scale electric field change (M change) for the same flash as in Figs. 4.62, 4.63. (a) Low-gain system, decay time constant 10 s; (b) high-gain system, decay time constant 150 μs. The pulses are shown on an expanded time scale in Fig. 4.65. Adapted from Rakov et al. (1996).

One should not attach too much significance to the percentages given above since the detectability of relativity slow, predominantly electrostatic field changes depends on the location (orientation and distance) of the lightning channel with respect to the observer in addition to the amount of charge transferred. Note that many ramps in the study of Rakov et al. (1996) did not contain detectable regular pulse bursts, although some contained irregular pulses activity, also observed by Rakov et al. (1992a).

In Figs. 4.64 and 4.65, one of the bursts culminated in the hook-shaped field change characteristic of an M-component (Section 4.9). Note from Figs. 4.64a, b that the pulses in that burst (also shown in Fig. 4.65b) are large enough to appear in both the low-gain (Fig. 4.64a) and the high-gain (Fig. 4.64b) records (the trailing edge of the field "hook" is clipped in the high-gain trace). The large pulse in Fig. 4.65b, which marks the end of the regular pulse activity and is immediately followed by the field hook (see Figs. 4.64a, b), is a typical feature of the hook-shaped M-component field waveforms (Rakov et al. 1992a). The amplitude of this pulse is larger than that of one of the smaller return-stroke pulses in the same flash (Rakov et al. 1996).

Rakov et al. (1996) reported that there were, on average, 28–39 pulses per burst in the three ground flashes analyzed. The average interpulse interval varied from 6.1 to 7.3 μs and the average burst duration from 173 to 235 μs. Usually all pulses within a burst had the same polarity. Positive and negative pulse polarities were about equally probable. The pulse peaks were approximately two orders of magnitude smaller than the return-stroke initial field peaks in the same flash.

The following are some observations reported by Rakov et al. (1996) regarding the position of the regular pulse bursts with respect to the return-stroke pulses in ground flashes. Except for the bursts apparently associated with one pronounced M-component hook (see Figs. 4.64 and 4.65), there was usually a delay in excess

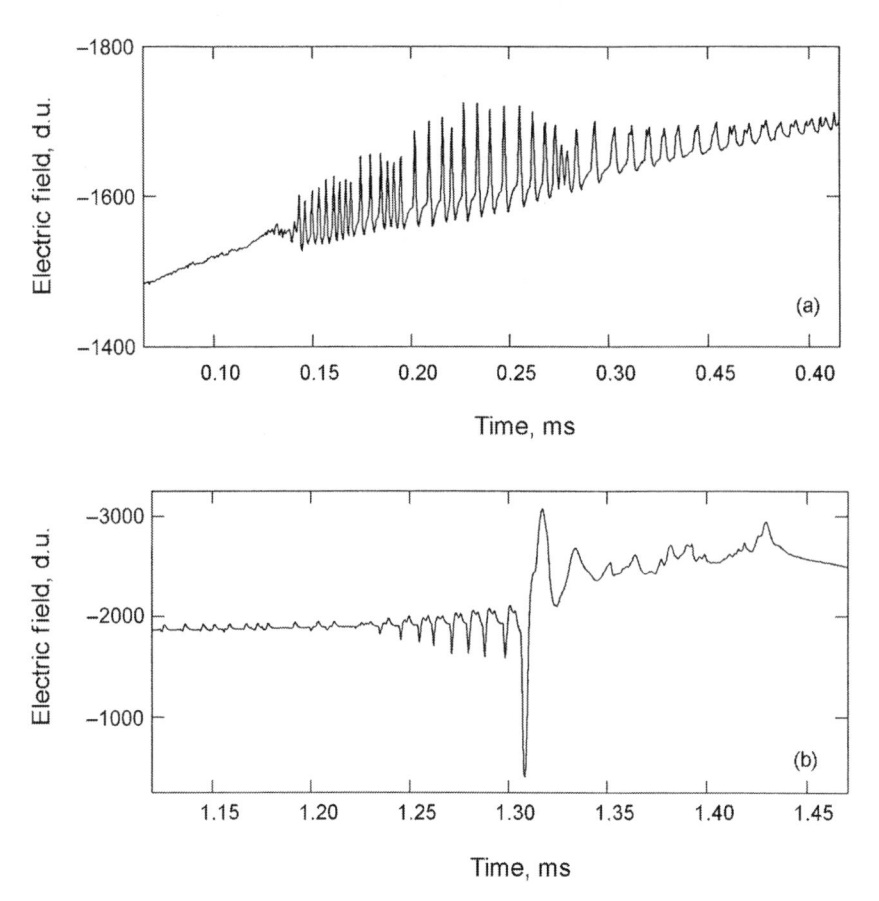

Fig. 4.65. Two fragments of the electric field record shown in Fig. 4.64b but on an expanded time scale of 50 μs per division. Adapted from Rakov *et al.* (1996).

of 10 ms between the preceding return stroke and the pulse burst. Further, the bursts showed a clear tendency not to occur before the first stroke or between the first and the second strokes. In the two ground flashes containing seven and nine strokes, the occurrence of bursts after the fourth stroke gradually decreased. Note that the pulse bursts shown in Figs. 4.64 and 4.65 occurred during the time that continuing current flowed in the channel to ground.

4.12. Summary

The downward negative lightning discharge to ground is the most common and most studied type of cloud-to-ground lightning. Each ground flash typically contains three to five strokes. The maximum number of strokes per flash observed to date is 26. The overwhelming majority, about 80 percent or more, of flashes contain more than one stroke. The time interval between successive return strokes in a flash is usually several tens of milliseconds, although it can be as large as many hundreds of milliseconds if a long continuing current is involved or as small as one millisecond or less. The total duration of a flash is typically some hundreds of milliseconds. The total charge lowered to ground is some tens of coulombs. About half of all negative ground flashes create more than one termination on ground, the spatial separation between the channel terminations being up to many kilometers. Each stroke is composed of a downward-moving leader and an upward-moving return stroke. The leader creates a conductive path between the cloud charge source and ground and deposits negative charge along this path, while the return stroke traverses that path, moving from ground toward the cloud charge source, and neutralizes the negative leader charge. The stepped leader initiating the first return stroke moves intermittently while subsequent-stroke leaders usually appear to move continuously.

Following the initial breakdown, possibly between the main negative and lower positive charge regions in the cloud, the stepped leader propagates toward the ground at an average speed of 2×10^5 m s^{-1}. Below the lower cloud boundary each leader step is typically 1 μs in duration and tens of meters in length, with a time interval between steps of 20 to 50 μs. The average stepped-leader current is between 100 and 1000 A, and the peak value of the current pulse associated with an individual step is at least 1 kA. The transition from the leader stage to the return-stroke stage is referred to as the attachment process. The upward-propagation speed of a return stroke below the lower cloud boundary is typically between one-third and one-half of the speed of light, that is, about three orders of magnitude higher than the stepped-leader speed. The first return-stroke current measured at ground rises to an initial peak of typically about 30 kA (the median value) in some microseconds and decays to a half-peak value in some tens of microseconds. The return stroke lowers to ground the several coulombs of charge originally deposited on the stepped-leader channel. The return-stroke current wave rapidly heats the channel to a peak temperature near 30 000 K and creates a channel pressure of order 10 atm or more, the results being channel expansion, intense optical radiation, and an acoustic shock wave that eventually becomes the thunder we hear at a distance. Subsequent strokes occur after the cessation of current flow to ground. In-cloud processes termed J-processes involve a redistribution of cloud charges on a tens of milliseconds time scale in response to the preceding return stroke. Transients occurring during the slower J-process are referred to as K-processes. Both J-processes and K-processes in cloud-to-ground discharges effectively serve to transport fresh negative charges into and along the existing channel (or its remnants), although not all the way to ground. The dart leader progresses downward at a typical speed of 10^7 m s^{-1} and deposits a total charge along the channel of order 1 C. The dart-leader peak current is about 1 kA. Subsequent return-stroke current measured at ground rises to a peak value of 10 to 15 kA in less than a microsecond and decays to half-peak value in a few tens of microseconds. The maximum current derivative is typically of order 100 kA μs^{-1}. The average upward propagation speed of subsequent return strokes is similar to that of first return strokes. The return-stroke component of a subsequent stroke current is often followed by a continuing current that has a magnitude of tens to hundreds of amperes and a duration up to hundreds of milliseconds. Transient processes that occur during the continuing-current stage and serve to transport negative charge to ground are referred to as M-components. The M-component mode of charge transfer to ground, as opposed to the leader–return-stroke mode, requires the existence of a grounded current-carrying channel.

References and bibliography for Chapter 4

Abbas, I., and Bayle, P. 1981. Non-equilibrium between electrons and field in a gas breakdown ionizing wave, I, Macroscopic model. *J. Phys.* D**14**: 549–60.

Aina, J.I. 1971. Lightning discharge studies in a tropical area. I. Cloud-to-ground discharges. *J. Geomagn. Geoelectr.* **23**: 347–58.

Aiya, S.V.C. 1954. Measurements of atmospheric noise interference to broadcasting. *J. Atmos. Terr. Phys.* **5**: 230.

Aiya, S.V.C. 1955. Noise power radiated by tropical thunderstorms. *Proc. IRE* **43**: 966–74.

Aiya, S.V.C. 1962. Structure of atmospheric radio noise. *J. Science Ind. Res.* **21D**: 203–20.

Aiya, S.V.C. and Sonde, B.E. 1963. Spring thunderstorms over Bangalore. *Proc. IEEE* **51**: 1493–501.

Aiya, S.V.C., Sastri, A.R.K., and Shiraprasad, A.P. 1972. Atmospheric radio noise measured in India. *Ind. J. Rad. Space Phys.* **1**: 1–8.

Albright, N.W., and Tidman, D.A. 1972. Ionizing potential waves and high voltage breakdown streamers. *Phys. Fluids* **15**: 86–90.

Aleksandrov, N.L., Bazelyan, E.M., and Shneider, M.N. 2000. Effect of continuous current during pauses between successive strokes on the decay of the lightning channel. *Plasma Phys. Rep.* **26**: 952–60.

Alizade, A.A., Muslimov, M.M., and Khydyrov, F.L. 1976. Study of electric field strength due to lightning stroke currents. *Elektrichestvo* **11**: 67–8.

Allen, N.L., and Ghaffar, A. 1995. The conditions required for the propagation of a cathode-directed positive streamer in air. *J. Phys. D: Appl. Phys.* **28**: 331–7.

Allibone, T.E. 1979. Velocities of leader-strokes to lightning and spark discharges. *J. Franklin Inst.* **36**: 35–7.

Allibone, T.E., and Dring, D. 1977. Lightning and the long spark: significance of leader-stroke velocity. *Proc. Roy. Soc.* A**357**: 15–35.

Anderson, R.B., and Eriksson, A.J. 1980. Lightning parameters for engineering application. *Electra* **69**: 65–102.

Appleton, E.V., Watson-Watt, R.A., and Herd, J.R. 1920. Investigations on lightning discharges and on the electric fields of thunderstorms. *Proc. Roy. Soc.* A**221**: 73–115.

Armstrong, H.R., and Whitehead, E.R. 1968. Field and analytical studies of transmission line shielding. *IEEE Trans. Pt. III* **PAS-87**: 270–81.

Arnold, H.R., and Pierce, E.T. 1964. Leader and junction process in the lightning discharge as a source of VLF atmospherics. *Radio Sci.* **68D**: 771–6.

Atlas D. 1958. Radar lightning echoes and atmospherics in vertical cross-section. In *Recent Advances in Atmospheric Electricity*, ed. L.G. Smith, pp. 441–59, Oxford: Pergamon.

Baatz, H. 1951. Blitzeinschlag-Messungen in Freileitungen. *Elektrotech. Z. Ausg.* A**72**: 191–8.

Baginski, M.E., Hodel, A.S., and Lankford, M. 1996. An investigation of the reconfiguration of the electric field in the stratosphere following a lightning event. *J. Electrostat.* **36**: 331–47.

Baker, L., Gardner, R.L., and Paxton, A.H., Baum, C.E., and Rison, W. 1987. Simultaneous measurement of current, electromagnetic fields, and optical emission from a lightning stroke. *Electromagnetics* **7**: 441–50.

Barasch, G.E. 1968. The 1965 ARPA–AEC joint lightning study at Los Alamos, vol. 2, The lightning spectrum as measured by collimated detectors, atmospheric transmission, spectral intensity radiated. Los Alamos Sci. Lab. Report LA-3755.

Barasch, G.E. 1969. The 1965 ARPA–AEC joint lightning study at Los Alamos, vol. 4, Discrimination against false triggering of air fluorescence detection systems by lightning. Los Alamos Science Laboratory Report LA-3757.

Barasch, G.E. 1970. Spectral intensities emitted by lightning discharges. *J. Geophys. Res.* **75**: 1049–57.

Barker, P.P., and Mancao, R.T. 1992. Lightning research advances with digital surge recordings. *IEEE Comput. Appl. Pow.* April, 11–6.

Barker, P.P., Mancao, R.T., Kvaltine, D.J., Parrish, D.E. 1993. Characteristics of lightning surges measured at metal oxide distribution arresters. *IEEE Trans. Pow. Del.* **8**: 301–10.

Barlow, J.S., Frey, G.W. Jr, and Newman, J.B. 1954. Very low frequency noise power from the lightning discharge. *J. Franklin Inst.* **27**: 145–63.

Baum, C.E. 1999. Leader-pulse step-formation process. Lightning Phenomenology Notes. Airforce Research Laboratory, Note 20, 8 June 1999, 20 pp.

Baum, C.E., and Gardner, R.L. 1986. An introduction to leader tip modeling. *Electromagnetics* **6**: 111–15.

Baum, C.E., Breen, E.L., O'Neill, J.P., Moore, C.B., and Hall, D.L. 1980. Measurements of electromagnetic properties of lightning with 10 nanosecond resolution. In *Lightning Technology*, NASA Conf. Publ. 2128, FAA-RD-80-30, pp. 39–84.

Baum, C.E., Breen, E.L., O'Neill, J.P., Moore, C.B., and Hall, D.L. 1982. Measurements of electromagnetic properties of lightning with 10 nanosecond resolution (revised). Lightning Phenomenology Notes, Air Force Weapons Laboratory, Note 3, 5 February 1982, 263 pp.

Baum, R.K. 1980. Airborne lightning characterization. In *Lightning Technology*, NASA Conf. Publ. 2128, FAA-RD-80-30, pp. 153–72.

Bazelyan, E.M. 1995. Waves of ionization in lightning discharge. *Plasma Physics Reports* **21**: 470–8.

Bazelyan, E.M., and Raizer, Yu. P. 1998. *Spark discharge*, 294 pp., Boca Raton, Florida: CRC Press.

Bazelyan, E.M., and Raizer, Yu. P. 2000a. *Lightning Physics and Lightning Protection*, 325 pp., Bristol: IOP Publishing.

Bazelyan, E.M., and Raizer, Yu. P. 2000b. Lightning attraction mechanism and the problem of lightning initiation by lasers. *UFN* **170** (7): 753–69.

Bazelyan, E.M., Gorin, B.N., and Levitov, V.I. 1978. *Physical and Engineering Foundations of Lightning Protection*, 223 pp., Leningrad: Gidrometeoizdat.

Beasley, W.H., Uman, M.A., and Rustan, P.L. 1982. Electric fields preceding cloud-to-ground lightning flashes. *J. Geophys. Res.* **87**: 4883–902.

Beasley, W.H., Uman, M.A., Jordan, D.M., and Ganesh, C. 1983. Simultaneous pulses in light and electric field from stepped leaders near ground level. *J. Geophys. Res.* **88**: 8617–9.

Beasley, W.H., Eack, K.B., Morris, H.E., Rust, W.D., and MacGorman, D.R. 2000. Electric-field changes of lightning observed in thunderstorms. *Geophys. Res. Lett.* **27**: 189–92.

Beierl, O. 1992. Front shape parameters of negative subsequent strokes measured at the Peissenberg tower. In *Proc. 21st Int. Conf. on Lightning Protection, Berlin, Germany*, pp. 19–24.

Bell, E., and Price, A.L. 1931. Lightning investigation on the 220-kV system of the Pennsylvania Power and Light Company (1930). *Trans. AIEE* **50**: 1101–10.

Bensema, W.D. 1969. Pulse spacing in high-frequency atmospheric noise bursts. *J. Geophys. Res.* **74**: 2780–2.

Berger, K. 1947. Lightning research in Switzerland. *Weather* **2**: 231–8.

Berger, K. 1955a. Die Messeinrichtungen für die Blitzforschung auf dem Monte San Salvatore. *Bull. Schweiz. Elektrotech. Ver.* **46**: 193–204.

Berger, K. 1955b. Resultate der Blitzmessungen der Jahre 1947–1954 auf dem Monte San Salvatore. *Bull. Schweiz. Elektrotech. Ver.* **46**: 405–24.

Berger, K. 1962. Front duration and current steepness of lightning strokes to Earth. In *Gas Discharges and the Electricity Supply Industry*, eds. J.S. Forrest, P.R. Howard, and D.J. Littler, pp. 63–73, London: Butterworths.

Berger, K. 1967a. Novel observations on lightning discharges: results of research on Mount San Salvatore. *J. Franklin Inst.* **283**: 478–525.

Berger, K. 1967b. Gewitterforschung auf dem Monte San Salvatore. *Elektrotechnik (Z-A)* **82**: 249–60.

Berger, K. 1972. Methoden und Resultate der Blitzforschung auf dem Monte San Salvatore bei Lugano in den Jahren 1963–1971. *Bull. Schweiz. Elektrotech. Ver.* **63**: 1403–22.

Berger, K. 1977. The Earth flash. In *Lightning, vol. 1. Physics of Lightning*, ed. R.H. Golde, pp. 119–90, New York: Academic Press.

Berger, K. 1980. Extreme Blitzströme und Blitzschutz. *Bull. Schweiz. Elektrotech. Ver.* **71**: 460–4.

Berger, K., and Garbagnati, E. 1984. Lightning current parameters. Results obtained in Switzerland and in Italy. URSI Commission E, Florence, Italy, 13 pp.

Berger, K., and Vogelsanger, E. 1965. Messungen und Resultate der Blitzforschung der Jahre 1955–1963 auf dem Monte San Salvatore. *Bull. Schweiz. Elektrotech. Ver.* **56**: 2–22.

Berger, K., and Vogelsanger, E. 1966. Photographische Blitzuntersuchungen der Jahre 1955–1965 auf dem Monte San Salvatore. *Bull. Schweiz. Elektrotech. Ver.* **57**: 599–620.

Berger, K., and Vogelsanger, E. 1969. New results of lightning observations. In *Planetary Electrodynamics, 1*, eds. S.C. Coroniti, and J. Hughes, pp. 489–510, New York: Gordon and Breach.

Berger, K., Anderson, R.B., and Kroninger, H. 1975. Parameters of lightning flashes. *Electra* **80**: 223–37.

Bhattacharya, J. 1963. Amplitude and phase spectra of radio atmospherics. *J. Atmos. Terr. Phys.* **25**: 445.

Bils, J.R., Thomson, E.M., Uman, M.A., and Mackerras, D. 1988. Electric field pulses in close lightning cloud flashes. *J. Geophys. Res.* **93**: 15 933–40.

Bondiou-Clergerie, A., Bacchiega, G.L., Castellani, A., Lalande, P., Laroche, P., and Gallimberti, I. 1996. Experimental and theoretical study of the bi-leader process. Part I:

Experimental investigation. In *Proc. 10th Int. Conf. on Atmospheric Electricity, Osaka, Japan*, pp. 244–7.

Borovsky, J.E. 1995. An electrodynamic description of lightning return strokes and dart leaders: guided wave propagation along conducting cylindrical channels. *J. Geophys. Res.* **100**: 2697–726.

Boyle, J.S., and Orville, R.E. 1976. Return stroke velocity measurements in multistroke lightning flashes. *J. Geophys. Res.* **81**: 4461–6.

Boys, C.V. 1926. Progressive lightning. *Nature* **118**: 749–50.

Boys, C.V. 1928. Progressive lightning. *Nature* **122**: 310–11.

Boys, C.V. 1929. Progressive lightning. *Nature* **124**: 54–5.

Bradley, P.A. 1964. The spectra of lightning discharges at very low frequencies. *J. Atmos. Terr. Phys.* **26**: 1069–73.

Bradley, P.A. 1965. The VLF energy spectra of first and subsequent return strokes of multiple lightning discharges to ground. *J. Atmos. Terr. Phys.* **27**: 1045–53.

Bradley, P.A., and Clarke, C. 1964. Atmospheric radio noise and signals received on directional aerials at high frequencies. *Proc. IEE, London* **111**: 1534–40.

Braginskii, S.I. 1958. Theory of the development of a spark channel. *Sov. Phys. JETP (English transl.).* **34**: 1068–74.

Brantley, R.D., Tiller, J.A., and Uman, M.A. 1975. Lightning properties in Florida thunderstorms from video tape records. *J. Geophys. Res.* **80**: 3402–6.

Brook, M. 1992. Breakdown electric fields in winter storms. *Res. Lett. Atmos. Electr.* **12**: 47–52.

Brook, M., and Kitagawa, N. 1960. Electric-field changes and the design of lightning-flash counters. *J. Geophys. Res.* **65**: 1927–31.

Brook, M., and Kitagawa, N. 1964. Radiation from lightning discharges in the frequency range 400 to 1000 Mc/s. *J. Geophys. Res.* **69**: 2431–4.

Brook, M., and Ogawa, T. 1977. The cloud discharge. In *Lightning, vol. 1, Physics of Lightning*, ed. R. Golde, pp. 191–230, London: Academic Press.

Brook, M., and Vonnegut, B. 1960. Visual confirmation of the junction process in lightning discharges. *J. Geophys. Res.* **65**: 1302–3.

Brook, M., Kitagawa, N., and Workman, E.J. 1962. Quantitative study of strokes and continuing currents in lightning discharges to ground. *J. Geophys. Res.* **67**: 649–59.

Brook M., Rhodes, C., Vaughan, O.H. Jr, Orville, R.E., and Vonnegut, B. 1985. Nighttime observations of thunderstorm electrical activity from a high altitude airplane. *J. Geophys. Res.* **90**: 6111–20.

Brook, M., Henderson, R.W., and Pyle, R.B. 1989. Positive lightning strokes to ground. *J. Geophys. Res.* **94**: 13 295–303.

Brown, G.W., and Whitehead, 1969. Field and analytical studies of transmission line shielding, Pt. II. *IEEE Trans. Pow. Appar. Syst.* **88**: 617–26.

Bruce, C.E.R. 1941. The lightning and spark discharges. *Nature* **147**: 805–6.

Bruce, C.E.R. 1944. The initiation of long electrical discharges. *Proc. Roy. Soc. A* **183**: 228–42.

Bruce, C.E.R., and Golde, R.H. 1941. The lightning discharge. *J. Inst. Elec. Eng.* **88**: 487–520.

Burke, C.P., and Jones, D.K. 1996. On the polarity and continuing currents in unusually large lightning flashes deduced from ELF events. *J. Atmos. Terr. Phys.* **58**: 531–40.

Cannon, J.B. 1918. A note on the sprectrum of lightning. *J. Roy. Astron. Soc. Can.* **12**: 95–7.

Cerri, G., Chiarandini, S., Costantini, S., De Leo, R., Primiani, V.M., and Russo, P. 2002. Theoretical and experimental characterization of transient electromagnetic fields radiated by electrostatic discharge (ESD) currents. *IEEE Trans. Electromagn. Compat.* **44**: 139–147.

Chapman, F.W. 1939. Atmospheric disturbances due to thundercloud discharges, pt. 1. *Proc. Phys. Soc. London* **51**: 876–94.

Chauzy, S., and Kably, K. 1989. Electric discharges between hydrometeors. *J. Geophys. Res. Res.* **94**: 13 107–24.

Chen, M., Takagi, N., Watanabe, T., Wang, D., Kawasaki, Z.-I., and Liu, X. 1999. Spatial and temporal properties of optical radiation produced by stepped leaders. *J. Geophys. Res.* **104**: 27 573–84.

Chernov, E.N., Lupeiko, A.V., and Petrov, N.I. 1992. Repulsion effect in orientation of "lightning" discharge. *J. Phys. III (France)* **2**: 1359–65.

Chowdhuri, P., and Kotapalli, A.K. 1989. Significant parameters in estimating the striking distance of lightning strokes to overhead lines. *IEEE Trans. Pow. Del.* **4**: 1970–81.

Choy, L.A., and Darveniza, M. 1971. A sensitivity analysis of lightning performance calculations for transmission lines. *IEEE Trans. Pow. Appar. Syst.* **90**: 1443–51.

Christian, H.J., Frost, R.L, Gillaspy, P.H., Goodman, S.J., Vaughan, O.H. Jr, Brook, M., Vonnegut, B., and Orville, R.E. 1983. Observations of optical lightning emissions from above thunderstorms using U-2 aircraft. *Bull. Am. Meteor. Soc.* **64**: 120–3.

Christian, H.J., Blakeslee, R.J., and Goodman, S.J. 1989. The detection of lightning from geostationary orbit. *J. Geophys. Res.* **94**: 13 329–37.

Cianos, N., Oetzel, G.N., and Pierce, E.T. 1972. Structure of lightning noise – especially above HF. In *Proc. Conf. on Lightning and Static Electricity*, pp. 50–56, Wright Patterson AFB.

Cianos, N., Oetzel, G.N., and Pierce, E.T. 1973. Reply. *J. Appl. Meteor.* **12**: 1421–3.

Clarence, N.D., and Malan, D.J. 1957. Preliminary discharge processes in lightning flashes to ground. *Q.J.R. Meteor. Soc.* **83**: 161–72.

Clarke, C. and Bradley, P.A. 1966. Discussion on (1) atmospheric radio noise and signals received on directional aerials at high frequencies and (2) characteristics of atmospheric radio noise observed at Singapore. *Proc. IEE, London* **113**: 752–4.

Clarke, C., Bradley, P.A., and Mortimer, D.E. 1965. Characteristics of atmospheric radio noise observed at Singapore. *Proc. IEE, London* **112**: 849–60.

Clegg, R.J. 1971. A photoelectric detector of lightning. *J. Atmos. Terr. Phys.* **33**: 1431–9.

Clegg, R.J., and Thomson, E.M. 1979. Some properties of EM radiation from lightning. *J. Geophys. Res.* **84**: 719–24.

Connor, T.R. 1967. The 1965 ARPA–AEC joint lightning study at Los Alamos, vol. 1, The lightning spectrum, charge transfer in lightning, efficiency of conversion of electrical energy into visible radiation. Los Alamos Sci. Lab. Report LA-3754.

Connor, T.R. 1968. Stroke- and space-resolved slit spectra of lightning. Los Alamos Sci. Lab. Report LA-3754 Addendum.

Connor, T.R., and Barasch, G.E. 1968. The 1965 ARPA–AEC joint lightning study at Los Alamos, vol. 3, Comparison of the lightning spectrum as measured by all-sky and narrow-field detectors, propagation of light from lightning to all-sky detectors. Los Alamos Sci. Lab. Report LA-3756.

Cooray, V. 1984. Further characteristics of positive radiation fields from lightning in Sweden. *J. Geophys. Res.* **89**: 11 807–15.

Cooray, V. 1986. Temporal behaviour of lightning HF radiation at 3 MHz near the time of first return strokes. *J. Atmos. Terr. Phys.* **48**: 73–8.

Cooray, V. 1987. Effects of propagation on the return stroke radiation fields. *Radio Sci.* **22**: 757–68.

Cooray, V. 1992. Horizontal fields generated by return strokes. *Radio Sci.* **27**: 529–37.

Cooray, V. 1996a. Possible influence of the mechanism of return stroke initiation on the remote sensing of lightning current parameters through first return stroke radiation fields. *J. Atmos. Electr.* **16**: 133–44.

Cooray, V. 1996b. A model for dart leaders in lightning *J. Atmos. Electr.* **16**: 145–59.

Cooray, V. 2001. Underground electromagnetic fields generated by the return strokes of lightning flashes. *IEEE Trans. Electromagn. Compat.* **43**: 75–84.

Cooray, V., and Jayaratne, K.P.S.C. 1994. Characteristics of lightning flashes observed in Sri Lanka in the tropics. *J. Geophys. Res.* **99**: 21 051–6.

Cooray, V., and Lundquist, S. 1982. On the characteristics of some radiation fields from lightning and their possible origin in positive ground flashes. *J. Geophys. Res.* **87**: 11 203–14.

Cooray, V., and Lundquist, S. 1983. Effects of propagation on the rise times and the initial peaks of radiation fields from return strokes. *Radio Sci.* **18**: 409–15.

Cooray, V., and Lundquist. S. 1985. Characteristics of the radiation fields from lightning in Sri Lanka in the tropics. *J. Geophys. Res.* **90**: 6099–109.

Cooray, V., and Ming, Y. 1994. Propagation effects on the lightning-generated electromagnetic fields for homogeneous and mixed sea-land paths. *J. Geophys. Res.* **99**: 10 641–52. (Correction: *J. Geophys. Res.* **104**: 12 227, 1999).

Cooray, V., and Orville, R.E. 1990. The effects of variation of current amplitude, current risetime, and return stroke velocity along the return stroke channel on the electromagnetic fields generated by return strokes. *J. Geophys. Res.* **95**: 18 617–30.

Cooray, V., and Perez, H. 1994a. HF radiation at 3 MHz associated with positive and negative return strokes. *J. Geophys. Res.* **99**: 10 633–40.

Cooray, V., and Perez, H. 1994b. Some features of lightning flashes observed in Sweden. *J. Geophys. Res.* **99**: 10 683–8.

Cooray, V., Fernando, M., Sörensen, T., Götschl, T., and Pedersen, Aa. 2000. Propagation of lightning generated transient electromagnetic fields over finitely conducting ground. *J. Atmos. Solar-Terr. Phys.* **62**: 583–600.

Coquillat, S., Chauzy, S., and Medale, J.-C., 1995. Microdischarges between ice particles. *J. Geophys. Res.* **100**: 14 327–34.

Crabb, J.A., and Latham, J. 1974. Corona from colliding drops as a possible mechanism for the triggering of lightning. *Q.J.R. Meteor. Soc.* **100**: 191–202.

Crawford, D.E., Rakov, V.A., Uman, M.A., Schnetzer, G.H., Rambo, K.J., and Stapleton, M.V. 1999. Multiple-station measurements of triggered-lightning electric and magnetic fields. In *Proc. 11th Int. Conf. on Atmospheric Electricity, Guntersville, Alabama*, pp. 154–7.

Crawford, D.E., Rakov, V.A., Uman, M.A., Schnetzer, G.H., Rambo, K.J., Stapleton, M.V., and Fisher, R.J. 2001. The close lightning electromagnetic environment: dart-leader electric field change versus distance. *J. Geophys. Res.* **106**: 14 909–17.

Croom, D.L. 1961. The spectra of atmospherics and propagation of very low frequency radio waves, Ph.D. dissertation, University of Cambridge, England.

Croom, D.L. 1964. The frequency spectra and attenuation of atmospherics in the range 1–15 kc/s. *J. Atmos. Terr. Phys.* **26**: 1015–46.

Cummins, K.L., Krider, E.P., and Malone, M.D. 1998. The US National Lightning Detection Network™ and applications of cloud-to-ground lightning data by electric power utilities. *IEEE Trans. Electromagn. Compat.* **40** (II): 465–80.

Currie, J.R., Choy, L.A., and Darveniza, M. 1971. Monte Carlo determination of the frequency of lightning strokes and shielding failures on transmission lines. *IEEE Trans. Pow. Appar. Syst.* **90**: 2305–12.

Cvetic, J., Heidler, F., and Schwab, A. 1998. Light intensity emitted from the lightning channel: comparison of different return stroke models. *J. Phys. D. Appl. Phys.* **31**: 1–10.

Darveniza, M. 1986. Measurement of impulse currents using premagnetised tape. *J. Electr. Electron. Eng., Australia*, **6**(2), 5 pp.

Darveniza, M., and Mercer, D.R. 1993. Laboratory studies of the effects of multipulse lightning currents on distribution surge arresters. *IEEE Trans. Pow. Del.* **8**: 1035–44.

Darveniza, M., Popolansky, F., and Whitehead, E.R. 1975. Lightning protection of UHV lines. *Electra* **41**: 39–69.

Davies, A.J., Evans, C.J., and Llewellyn-Jones, F. 1964. Electrical breakdown of gases: the spatio-temporal growth of ionization in fields distorted by space charge. *Proc. Roy. Soc. A* **281**: 164–83.

Davis, M.H., Brook, M., Christian, H., Heikes, B.G., Orville, R.E., Park, C.G., Roble, R.G., and Vonnegut, B. 1983. Some scientific objectives of a satellite-borne lightning mapper. *Bull. Am. Meteor. Soc.* **64**: 114–9.

Davis R. 1962. Frequency of lightning flashover on overhead lines. In *Gas Discharges and the Electricity Supply Industry*, eds. J.S. Forrest, P.R. Howard, and D.J. Littler, pp. 125–38, London: Butterworths.

Davis, R., and Standring, W.G. 1947. Discharge currents associated with kite balloons. *Proc. Roy. Soc. A* **191**: 304–22.

Davis, S.M. 1999. Properties of lightning discharges from multiple-station wideband electric field measurements. Ph.D. dissertation, University of Florida, Gainesville, 228 pp.

Dawson, G.A. 1969. Pressure dependence of water-drop corona onset and its atmospheric importance. *J. Geophys. Res.* **74**: 6859–68.

Dawson, G.A., and Duff, D.G. 1970. Initiation of cloud-to-ground lightning strokes. *J. Geophys. Res.* **75**: 5858–67.

Dawson, G.A., and Winn, W.P. 1965. A model for streamer propagation. *Z. Phys.* **83**: 159–71.

Dellera, L., and Garbagnati, E. 1990. Lightning stroke simulation by means of the leader progression model. *IEEE Trans. Pow. Del.* **5**: 2023–9.

Dennis, A.S., and Pierce, E.T. 1964. The return stroke of the lightning flash to Earth as a source of VLF atmospherics. *Radio Sci.* **68D**: 779–94.

Depasse, P. 1994a. Statistics on artificially triggered lightning. *J. Geophys. Res.* **99**: 18 515–22.

Devan, K.R.S. 1986. Electric field spectra of lightning return-strokes in the interval from 2 to 500 kHz. *Inst. Elec. Telecom. Eng.* **32**: 382–8.

Diachuk, V.A., and Muchnik, V.M. 1979. Corona discharge of watered hailstones as a basic mechanism of lightning initiation. *Doklady Akad. Nauk SSSR* **248**: 60–3

Diendorfer, G. 1990. Induced voltage on an overhead line due to nearby lightning. *IEEE Trans. Electromagn. Compat.* **32**: 292–9.

Diendorfer, G., and Uman M.A. 1990. An improved return stroke model with specified channel-base current. *J. Geophys. Res.* **95**: 13 621–44.

Diendorfer, G., Mair, M., Schulz, W., and Hadrian, W. 2000. Lightning current measurements in Austria – experimental setup and first results. In *Proc. 25th Int. Conf. on Lightning Protection, Rhodes, Greece*, pp. 44–7.

Diviya, and Rai, J. 1987. A reply. *Geoexploration* **24**: 3.

Djebari, B., Hamelin, J., Leteinturier, C., and Fontaine, J. 1981. Comparison between experimental measurements of the electromagnetic field emitted by lightning and different theoretical models – influence of the upward velocity of the return stroke. In *Proc. 4th Int. Symp. on Electromagn. Compat., Zurich, Switzerland*, pp. 511–16.

Drellishak, K.S. 1964. Partition functions and thermodynamic properties of high temperature gases. Defense Documentation Center AD, 428 210.

Dubovoy, E.I. 1974. Simultaneous measurements of electric-field pulses and radio reflections from lightning and experimental test of the results of numerical simulation. *Izvestiya AN SSSR-Fizika Atmosfery i Okeana* **33**(1): 122–31.

Dubovoy, E.I., Mikhailov, M.S., Pryazhinsky, V.I., Ogon'kov, A.L., Adjiev, A.Kh., Derkach, V.M., and Sigachev, S.M. 1993. The simultaneous measurements of electric field impulses and radar wave reflection from lightning discharge and comparison with results of numerical modeling. *Izvestiya AN-Fizika Atmosfery i Okeana* **29**(3): 364–8.

Dubovoy, E.I., Mikhailov, M.S., Ogonkov, A.L., and Pryazhinsky, V.I. 1995. Measurement and numerical modeling of radio sounding reflection from a lightning channel. *J. Geophys. Res.* **100**: 1497–502.

Dufay, M. 1926. Spectres des éclairs. *Compt. Rend.* **182**: 1331–3.

Dufay, M. 1947. Sur le spectre des éclairs dans les règions violette et ultraviolette. *Compt. Rend.* **225**: 1079–80.

Dufay, M. 1949. Recherches sur les spectres des éclairs, deuxième partie: étude du spectre dans les régions violette et ultra-violette. *Ann. Geophys.* **5**: 255–63.

Dufay M., and Dufay, J. 1949. Spectres des éclairs photographiés au prisme objectif. *Compt. Rend.* **229**: 838–41.

Dufay, M., and Tcheng, M. 1949a. Spectres des éclairs, de 3830 a 6570 Å. *Compt. Rend.* **228**: 330–2.

Dufay, J., and Tcheng, M. 1949b. Recherches sur les spectres des éclairs, première partie: etude des spectres, de 3830 a 6570 Å au moyen de spectrographes à fente. *Ann. Geophys.* **5**: 137–49.

Dulzon, A.A., and Rakov, V.A. 1980. Estimation of errors in lightning peak current measurements by frame aerials. *Izv. Vuzov SSSR, ser. Energetika* **11**: 101–4.

Dutton, J., Haydon, S.C., and Llewellyn-Jones, F. 1953. Photoionization and the electrical breakdown of gases. *Proc. Roy. Soc. A* **218**: 206–23.

Edano, Y., Kawamata, S., and Nagai, Y. 1982. Observation of simultaneous-bi-stroke flashes. *Res. Lett. Atmos. Elec.* **2**: 49–52.

Eriksson, A.J. 1978a. Lightning and tall structures. *Trans. South African IEE* **69**: 238–52.

Eriksson, A.J. 1978b. A discussion on lightning and tall structures. CSIR Special Report ELEK 152, National Electrical Engineering Research Institute, Pretoria, South Africa.

Eriksson, A.J. 1980. The lightning ground flash – an engineering study. Ph.D. dissertation, University of Natal, South Africa (available as CSIR Special Report ELEK 189, National Electrical Engineering Research Institute, Pretoria, South Africa).

Eriksson, A.J. 1982. The CSIR lightning research mast – data for 1972–1982. NEERI Internal Report no. EK/9/82, National Electrical Engineering Research Institute, Pretoria, South Africa.

Eriksson, K.B.S. 1958. The spectrum of the singly-ionized nitrogen atom. *Arkiv Fysik* **13**: 303–29.

Evans, W.H., and Walker, R.L. 1963. High speed photographs of lightning at close range. *J. Geophys. Res.* **68**: 4455–61.

Faidley, W.E., and Krider, E.P. 1989. A lucky strike. *Weatherwise* **42**: 136–42.

Fernsler, R.F. 1984. General model of streamer propagation. *Phys. Fluids* **27**: 1005–12.

Few, A.A. 1995. Acoustic radiations from lightning. In *Handbook of Atmospheric Electrodynamics*, vol. II, pp. 1–31, Boca Raton, Florida: CRC Press.

Fieux, R.P., Gary, C.H., Hutzler, B.P., Eybert-Berard, A.R., Hubert, P.L., Meesters, A.C., Perroud, P.H., Hamelin, J.H., and Person, J.M. 1978. Research on artificially triggered

lightning in France. *IEEE Trans. Pow. Appar. Syst.* **97**: 725–33.
Fisher, F.A., and Plumer, J.A. 1977. Lightning protection of aircraft. NASA References Publication 1008, NASA Lewis Research Center.
Fisher, R.J. 1998. Personal communication.
Fisher, R.J., and Schnetzer, G.H. 1991. 1990 Sandia rocket-triggered lightning field tests at Kennedy Space Center, Florida. Technical Report, SAN90-2926, 106 pp.
Fisher, R.J., and Uman, M.A. 1972. Measured electric field risetimes for first and subsequent lightning return strokes. *J. Geophys. Res.* **77**: 399–406.
Fisher, R.J., Schnetzer, G.H., Thottappillil, R., Rakov, V.A., Uman, M.A., and Goldberg, J.D. 1993. Parameters of triggered-lightning flashes in Florida and Alabama. *J. Geophys. Res.* **98**: 22 887–902.
Fitzgerald, D.R. 1967. Probable aircraft "triggering" of lightning in certain thunderstorms. *Mon. Wea. Rev.* **95**: 835–42.
Foust, C.M., and Kuehni, H.P. 1932. The surge-crest ammeter. *Gen. Elec. Rev.* **35**: 644–8.
Fowler, R.G. 1974. Non-linear electron acoustic waves, I. In *Adv. Electronics Electron Physics*, vol. 35, pp. 1–86.
Fowler, R.G. 1976. Non-linear electron acoustics waves, II. In *Adv. Electronics Electron Physics*, vol. 41, pp. 1–72.
Fox, P. 1903. The spectrum of lightning. *Astrophys. J.* **14**: 294–6.
Fuchs, F., Landers, E.U., Schmid, R., and Wiesinger, J. 1998. Lightning current and magnetic field parameters caused by lightning strikes to tall structures relating to interference of electronic systems. *IEEE Trans. Electromagn. Compat.* **40**: 444–51.
Funaki, K., Sakamoto, K., Tanaka, R., and Kitagawa, N. 1981. A comparison of cloud and ground lightning discharges observed in South-Kanto summer thunderstorms, 1980. *Res. Lett. Atmos. Electr.* **1**: 99–103.
Fuquay, D.M., Baughman, R.G., Taylor, A.R., and Hawe, R.G. 1967. Characteristics of seven lightning discharges that caused forest fires. *J. Geophys. Res.* **72**: 6371–3.
Fuquay, D.M., Taylor, A.R., Hawe, R.G., and Schmidt, C.W. Jr 1972. Lightning discharges that have caused forest fires. *J. Geophys. Res.* **77**: 2156–8.
Galejs, J. 1967. Amplitude statistics of lightning discharge currents and ELF and VLF radio noise. *J. Geophys. Res.* **72**: 2943–53.
Ganesh, C., Uman, M.A., Beasley, W.H., and Jordan, D.M. 1984. Correlated optical and electric field signals produced by lightning return strokes. *J. Geophys. Res.* **89**: 4905–9.
Garbagnati, E., and Lo Piparo, G.B. 1970. Stazione sperimentale per il rilievo delle caratteristiche dei fulmini. *L'Elettrotecnica* **57**: 288–97.
Garbagnati, E., and Lo Piparo, G.B. 1973. Nuova stazione automatica per il rilievo delle caratteristiche dei fulmini. *L'Energia Elettrica* **6**: 375–83.
Garbagnati, E., and Lo Piparo, G.B. 1982. Parameter von Blitzstroemen. *ETZ-A* **103**: 61–5.
Garbagnati, E., Giudice, E., Lo Piparo, G.B., and Magagnoli, U. 1975. Rilievi delle caratteristiche dei fulmini in Italia. Risultati ottenuti negli anni 1970–1973. *L'Elettrotecnica* **62**: 237–49.

Garbagnati, E., Giudice, E., and Lo Piparo, G.B. 1978. Measurement of lightning currents in Italy – results of a statistical evaluation. *ETZ-A* **99**: 664–8.
Garbagnati, E., Marinoni, F., and Lo Piparo, G.B. 1981. Parameters of lightning currents. Interpretation of the results obtained in Italy. In *Proc. 16th Int. Conf. on Lightning Protection, Szeged, Hungary*.
Gardner, R.L. 1981. Effect of the propagation path on lightning-induced transient fields. *Radio Sci.* **16**: 377–84.
Gardner, R.L. 1990. Effect of the propagation path of lightning – induced transient fields. In *Lightning Electromagnetics*, pp. 139–53, New York: Hemisphere.
Gardner, R.L., Baker, L., Gilbert, J.L., Baum, C.E., and Andersh, D.J. 1985. Comparison of published HEMP and natural lightning on the surface of an aircraft. In *Proc. 10th Int. Aerospace and Ground Conf. on Lightning and Static Electricity, Paris, France*, pp. 108.
Garg, M.B., Mathpal, K.C., Rai, J., and Varshneya, N.C. 1982. Frequency spectra of electric and magnetic fields of different forms of lightning. *Ann. Geophys. T.* **38**: 177–88.
Geldenhys, H.T., Eriksson, A.J., and Bourn, G.W. 1989. Fifteen years of data of lightning current measurements on 60 m mast. *Trans. South African IEEE* **80**: 130–58.
Ghosh, D., and Khastgir, S.R. 1972. Time-variation of the return stroke current in the cloud-to-ground lightning discharges. *Ind. J. Pure Appl. Phys.* **10**: 556–7.
Gilchrist, J.H., and Thomas, J.B. 1975. A model for current pulses of cloud-to-ground lightning discharges. *J. Franklin Inst.* **229**: 199–210.
Gladshteyn, N.D. 1967. Statistical properties of noise from the predischarge part of a lightning discharge. *Geomagn. Aeronomy* **5**: 741–2.
Godlonton, R. 1896. A remarkable discharge of lightning. *Nature* **53**: 272.
Golde, R.H. 1945. The frequency of occurrence and the distribution of lightning flashes to transmission lines. *AIEE Trans.* **64**: 902–10.
Golde, R.H. 1947. Occurrence of upward streamers in lightning discharges. *Nature* **160**: 395–6.
Golde, R.H. 1963. The attractive effects of a lightning conductor. *J. Inst. Electr. Eng. (London)* **9**: 212–13.
Golde, R.H. 1967. The lightning conductor. *J. Franklin Inst.* **283**: 451–77.
Golde, R.H. 1973. *Lightning Protection*, 254 pp., London, Edward Arnold.
Golde. R.H. 1977. The lightning conductor. In *Lightning, vol. 2, Lightning Protection*, ed. R.H. Golde, pp. 545–74, New York: Academic Press.
Golde, R.H. 1978. Lightning and tall structures. *Proc. IEE (London)* **125**: 347–51.
Gomes, C., and Cooray, V. 1998. Correlation between the optical signatures and current waveforms of long sparks: application in lightning research. *J. Electrostat.* **43**: 267–74.
Gomes, C., Cooray, V., and Jayaratne, C. 1998. Comparison of preliminary breakdown pulses observed in Sweden and in Sri Lanka. *J. Atmos. Solar-Terr. Phys.* **60**: 975–9.
Goodman, S.J., Christian, H.J., and Rust. W.D. 1988. A comparison of the optical pulse characteristics of intracloud and

cloud-to-ground lightning as observed above clouds. *J. Appl. Meteor.* **27**: 1369–81.

Gorin, B.N. 1985. Mathematical modeling of the lightning return stroke. *Elektrichestvo* **4**: 10–16.

Gorin, B.N., and Markin, V.I. 1975. Lightning return stroke as a transient process in a distributed system. *Trudy ENIN* **43**: 114–30.

Gorin, B.N., and Shkilev, A.V. 1984. Measurements of lightning currents at the Ostankino tower. *Elektrichestvo* **8**: 64–5.

Gorin, B.N., Sakharova, G.S., Tikhomirov, V.V., and Shkilev, A.V. 1975. Results of studies of lightning strikes to the Ostankino TV tower. *Trudy ENIN* **43**: 63–77.

Gorin, B.N., Levitov, V.I., and Shkilev, A.V. 1976. Some principles of leader discharge of air gaps with a strong non-uniform field. In *Gas Discharges*, IEE Conf. Publ. 143, pp. 274–8.

Gorin, B.N., Levitov, V.I., and Shkilev, A.V. 1977. Lightning strikes to the Ostankino tower. *Elektrichestvo* **8**: 19–23.

Goto, Y., and Narita, K. 1992. Observations of winter lightning to an isolate tower. *Res. Lett. Atmos. Electr.* **12**: 57–60.

Goto, Y. and Narita, K. 1995. Electrical characteristics of winter lightning. *J. Atmos. Terr. Phys.* **57**: 449–59.

Griem, H.R. 1964. *Plasma Spectroscopy*. New York: McGraw-Hill.

Griffiths, R.F., and Phelps, C.T. 1976a. The effects of air pressure and water vapour content on the propagation of positive corona streamers, and their implications to lightning initiation. *Q.J.R. Meteor. Soc.* **102**: 419–26.

Griffiths, R.F., and Phelps, C.T. 1976b. A model of lightning initiation arising from positive corona streamer development. *J. Geophys. Res.* **31**: 3671–6.

Griffiths, R.F., and Vonnegut, B. 1975. Tape recorder photocell instrument for detecting and recording lightning strokes. *Weather* **30**: 254–7.

Grigor'ev, A.I., and Shiryaeva, S.O. 1996. The possible physical mechanism of initiation and growth of lightning. *Physica Scripta* **54**: 660–6.

Griscom, S.B. 1958. The prestrike theory and other effects in the lightning stroke. *Trans. AIEE (pt. 3)* **77**: 919–33.

Griscom, S.B., Caswell, R.W., Graham, R.E., McNutt, H.R., Schlomann, R.H., and Thornton, J.K. 1965. Five-year field investigation of lightning effects on transmission lines. *Trans. IEEE* **PAS-84**: 257–80.

Gross, I.W., and Cox, J.H. 1931. Lightning investigation on the Applachian Electric Power Company's transmission system. *Trans. AIEE* **50**: 1118–31.

Guerrieri, S., Nucci, C.A., Rachidi, F., and Rubinstein, M. 1998. On the influence of elevated strike objects on directly measured and indirectly estimated lightning currents. *IEEE Trans. Pow. Del.* **13**: 1543–55.

Guo, C., and Krider, E.P. 1982. The optical and radiation field signatures produced by lightning return strokes. *J. Geophys. Res.* **87**: 8913–22.

Guo, C., and Krider, E.P. 1983. The optical power radiated by lightning return strokes. *J. Geophys. Res.* **88**: 8621–2.

Guo, C., and Krider, E.P. 1985. Anomalous light output from lightning dart leaders. *J. Geophys. Res.* **90**: 13 073–5.

Gupta, S.N. 1973. Distribution of peaks in atmospheric radio noise. *IEEE Trans. Electromagn. Compat.* **15**: 100–3.

Gupta, S.P. 1987. VLF spectral characteristics of leader pulses. *IEE Proc. A, Phys. Sci. Meas. Instrum. Manage. Educ. Rev. (UK)* **134**: 789–92.

Gupta, S.P., Rao, M., and Tantry, B.A.P. 1972. VLF spectra radiated by stepped leaders. *J. Geophys. Res.* **77**: 3924–37.

Gupta, S.P., Rao, M., and Tantry, B.A.P. 1974. Radiation resistance characteristics of the multiple return stroke lightning. *Ann. Geophys.* **30**: 435–40.

Gurevich, A.V., Milikh, G.M., and Roussel-Dupré, R.A. 1992. Runaway electron mechanisms of air breakdown and preconditioning during a thunderstorm. *Phys. Lett. A* **165**: 463–8.

Gurevich, A.V., Milikh, G.M., and Roussel-Dupre, R.A. 1994. Nonuniform runaway air breakdown. *Phys. Lett. A* **187**: 197–203.

Gurevich, A.V., Milikh, G.M., and Valdivia, J.A. 1997. Model of X-ray emission and fast preconditioning during a thunderstorm. *Phys. Lett. A* **231**: 402–8.

Hagenguth, J.H. 1947. Photographic study of lightning. *Trans. Am. Inst. Electr. Eng.* **66**: 577–85.

Hagenguth, J.H., and Anderson, J.G. 1952. Lightning to the Empire State Building. *AIEE Trans.* **71**(3): 641–9.

Hale, L.C., and Baginski, M.E. 1987. Current to the ionosphere following lightning stroke. *Nature* **329**: 814–16.

Hallgren, R.E., and McDonald, R.B. 1963. Atmospherics from lightning from 100 to 600 MHz. IBM Federal Systems Division, Report no. 63-538-89.

Halliday, D., and Resnick, R. 1974. *Fundamentals of Physics*, New York: John Wiley.

Hamelin, J. 1993. Sources of natural noise. In *Electromagnetic Compatibility*, eds. P. Degauque and J. Hamelin, 652 pp., New York: Oxford.

Harris, D.J., and Salman, Y.E. 1972. The measurement of lightning characteristics in northern Nigeria. *J. Atmos. Terr. Phys.* **34**: 775–86.

Hart, J.E. 1967. VLF radiation from multiple stroke lightning. *J. Atmos. Terr. Phys.* **29**: 1011–14.

Harvey, R.B., and Lewis, E.A, 1973. Radio mapping of 250 and 925 megahertz noise sources in clouds. *J. Geophys. Res.* **78**: 1944–7.

Harwood, J., and Harden, B.N. 1960. The measurement of atmospheric radio noise by an aural comparison method in the range 15–500 kc/s. *Proc. IEE (London) B***107**: 53–9.

Hasbrouck, R.T. 1989. Lightning – understanding it and protecting systems from its effects. Technical Report UCRL-53925, Lawrence Livermore National Laboratory, University of California, 41 pp.

Hatakeyama, H. 1936. An investigation of lightning discharge with the magnetograph. *Geophys. Mag.* **10**: 309–19.

Hatakeyama, H. 1958. The distribution of the sudden change of electric field on the Earth's surface due to lightning discharge. In *Recent Advances in Atmospheric Electricity*, ed. L.G. Smith, pp. 289–98, New York: Pergamon Press.

Hawe, R.G. 1968. Electrostatic trigger used for daylight lightning photography. *Photogr. Sci. Eng.* **12**: 219–21.

Hayenga, C.O. 1979. Positions and movement of VHF lightning sources determined with microsecond resolution by interferometry. Ph.D. dissertation, University of Colorado, Boulder.

Hayenga, C.O. 1984. Characteristics of lightning VHF radiation near the time of return strokes. *J. Geophys. Res.* **89**: 1403–10.

Hayenga, C.O., and Warwick, J.W. 1981. Two-dimensional interferometric positions of VHF lightning sources. *J. Geophys. Res.* **86**: 7451–62.

Heckman, S.J., and Williams, E.R. 1989. Corona envelopes and lightning currents. *J. Geophys. Res.* **94**: 13 287–94.

Heidler, F. 1985. Traveling current source model for LEMP calculation. In *Proc. 6th Int. Symp. on Electromagnetic Compatibility, Zurich, Switzerland*, pp. 157–62.

Heidler, F. 1996. Coordination of surge protection devices using the current data from lightning field measurements. *ETEP* **6**: 441–4.

Heidler, F., Cvetic, J.M., and Stanic, B.V. 1999. Calculation of lightning current parameters. *IEEE Trans. Pow. Del.* **14**(2): 399–404.

Hendry, J. 1993. Panning for lightning (including comments on the photos by M.A. Uman). *Weatherwise* **45**(6): 19.

Herodotou, N., Chisholm, W.A., and Janischewskyj, W. 1993. Distribution of lightning peak stroke currents in Ontario using an LLP system. *IEEE Trans. Pow. Del.* **8**: 1331–9.

Herschel, J. 1868. On the lightning spectrum. *Proc. Roy. Soc.* **15**: 61–2.

Hewitt, F.J. 1957. Radar echoes from inter-stroke processes in lightning. *Proc. Phys. Soc.* **70**: 961–79.

Hill, E.L., and Robb, J.D. 1969. Spectroscopic and thermal temperatures in the return stroke of lightning. *J. Geophys. Res.* **74**: 3426–30.

Hill, R.D. 1963. Determination of charges conducted in lightning strokes. *J. Geophys. Res.* **68**: 1365–75.

Hill, R.D. 1968. Analysis of irregular paths of lightning channels. *J. Geophys. Res.* **73**: 1897–905.

Hill, R.D. 1969. Electromagnetic radiation from erratic paths of lightning strokes. *J. Geophys. Res.* **74**: 1922–9.

Hill, R.D. 1971. Channels heating in return stroke lightning. *J. Geophys. Res.* **76**: 637–45.

Hill, R.D. 1972. Optical absorption in the lightning channel. *J. Geophys. Res.* **77**: 2642–7.

Hill, R.D. 1975. Comments on 'Quantitative analysis of a lightning return stroke for diameter and luminosity changes as a function of space and time' by R.E. Orville, J.H. Helsdon Jr, and W.H. Evans. *J. Geophys. Res.* **80**: 1188.

Hill, R.D. 1977. Energy dissipation in lightning. *J. Geophys. Res.* **82**: 4967–8.

Hill, R.D. 1978a. Comment on 'Detection of lightning superbolts' by B.N. Turman. *J. Geophys. Res.* **83**: 1381–2.

Hill, R.D. 1978b. Reply. *J. Geophys. Res.* **83**: 5524.

Hill, R.D. 1990. Lightning channel decay. *Phys. Fluids B* **2**: 3209–11.

Himley, R.O. 1969. VLF radiation from subsequent return strokes in multiple stroke lightning. *J. Atmos. Terr. Phys.* **31**: 749–53.

Hodges, D.B. 1954. A comparison of the rates of change of current in the step and return processes of lightning flashes. *Proc. Phys. Soc. B* **67**: 582–7.

Hojo, J., Ishii, M., Kawamura, T., Suzuki, F., Komuro, H., and Shiogawa, M. 1988. Characteristics and evaluation of lightning field waveforms. *Electr. Eng. Japan* **108**: 55–65.

Hoole, P.R.P., and Hoole, S.R.H. 1993. Simulation of lightning attachment to open ground, tall towers and aircraft. *IEEE Trans. Pow. Del.* **8**: 732–40.

Horii, K., and Sakurano, H. 1985. Observation on final jump of the discharge in the experiment of artificially triggered lightning. *IEEE Trans. Pow. Appar. Syst.* **104**: 2910–17.

Horner, F. 1961. Narrow band atmospherics from two local thunderstorms. *J. Atmos. Terr. Phys.* **21**: 13–25.

Horner, F., 1962. Atmospherics of near lightning discharges. In *Radio Noise of Terrestrial Origin*, ed. F. Horner, pp. 16–17, New York: American Elsevier.

Horner, F. 1964. Radio noise from thunderstorms. In *Advances in Radio Research 2*, ed. J.A. Saxton, pp. 122–215, New York: Academic Press.

Horner, F. and Bradley, P.A. 1964. The spectra of atmospherics from near lightning. *J. Atmos. Terr. Phys.* **26**: 1155–66.

Horner, F. and Clarke, C. 1958. Radio noise from lightning discharges. *Nature* **181**: 688–90.

Howe, R.G. 1968. Electrostatic trigger used for daylight lightning photography. *Photogr. Sci. Eng.* **12**: 219–21.

Hu, R. 1960. The lightning spectra in the visible and ultra-violet regions with grating spectrograph. *Sci. Rec. (Peking)* **4**: 380–3.

Hubert, P., and Mouget, G. 1981. Return stroke velocity measurements in two triggered lightning flashes. *J. Geophys. Res.* **86**: 5253–61.

Hussein, A.M., Janischewskyj, W., Chang, J.-S., Shostak, V., Chisholm, W.A., Dzurevych, P., and Kawasaki, Z.-I. 1995. Simultaneous measurement of lightning parameters for strokes to the Toronto Canadian National Tower. *J. Geophys. Res.* **100**: 8853–61.

Hylten-Cavallius, N., and Stromberg, A. 1956. The amplitude, time to half-value, and the steepness of the lightning currents. *ASEA J.* **29**: 129–34.

Hylten-Cavallius, N., and Stromberg, A. 1959. Field measurement of lightning currents. *Elteknik* **2**: 109–13.

Idone, V.P. 1990. Length bounds for connecting discharges in triggered lightning. *J. Geophys. Res.* **95**: 20 409–16.

Idone, V.P. 1992. The luminous development of Florida triggered lightning. *Res. Lett. Atmos. Electr.* **12**: 23–8.

Idone, V.P. 1995. Microscale tortuosity and its variation as observed in triggered lightning channels. *J. Geophys. Res.* **100**: 22 943–56.

Idone, V.P., and Davis, D.A. 1999. Photographic documentation of two return strokes separated by about a millisecond. In *Proc. 11th Int. Conf. on Atmospheric Electricity, Guntersville, Alabama*, pp. 50–3.

Idone, V.P., and Henderson, R.W. 1982. An unusual lightning ground strike. *Weatherwise* **35**: 223–4.

Idone, V.P., and Orville, R.E. 1982. Lightning return stroke velocities in the Thunderstorm Research International Program (TRIP). *J. Geophys. Res.* **87**: 4903–15.

Idone, V.P., and Orville, R.E. 1984. Three unusual strokes in a triggered lightning flash. *J. Geophys. Res.* **89**: 7311–16.

Idone, V.P., and Orville, R.E. 1985. Correlated peak relative light intensity and peak current in triggered lightning subsequent return strokes. *J. Geophys. Res.* **90**: 6159–64.

Idone, V.P., and Orville, R.E. 1987. The propagation speed of a positive lightning return stroke. *Geophys. Res. Lett.* **14**: 1150–3.

Idone, V.P., Orville, R.E., Hubert, P., Barret, L., and Eybert-Berard, A. 1984. Correlated observations of three triggered lightning flashes. *J. Geophys. Res.* **89**: 1385–94.

Illingworth, A.J. 1971. The variation of the electric field after lightning and the conductivity within thunderclouds. *Q.J.R. Meteor. Soc.* **97**: 440–56.

Illingworth, A.J. 1971. Electric field recovery after lightning. *Nature, Phys. Sci.* **229**: 213.

Ishii, M. 1992. Lightning-induced voltages on overhead wire. *Res. Lett. Atmos. Electr.* **12**: 77–81.

Ishii, M., and Hojo, J.-I. 1989. Statistics on fine structure of cloud-to-ground lightning field waveforms. *J. Geophys. Res.* **94**: 13 267–74.

Ishii, M., Hojo, J., and Michishita, K. 1992. Recent observation of field waveforms associated with return strokes of natural lightning. *Res. Lett. Atmos. Electr.* **12**: 11–15.

Ishikura, K., Negishi, M., and Kitagawa, N. 1981. On the distribution and variation characteristics of lightning currents in the ground around the strike points (4th report). *Res. Lett. Atmos. Electr.* **1**: 113–18.

Isikawa, H. 1961. Nature of lightning discharges as origins of atmospherics. *Proc. Res. Inst. Atmos. (Nagoya Univ.)* **8A**: 1–273.

Isikawa, H., Takagai, M., and Takeuti, T. 1958. On the leader waveforms of atmospherics near the origin. *Proc. Res. Inst. Atmos. (Nagoya Univ.)* **5**: 1–11.

Israel, H., and Fries, G. 1956. Ein Gerät zur spektroskopischen Analyse verschiedener Blitzphasen. *Optik* **13**: 365–8.

Israel, H., and Wurm, K. 1941. Das Blitzspektrum. *Naturwissenschaften* **52**: 778–9.

Israel, H., and Wurm, K. 1947. Das Spektrum der Blitze. *Wiss. Arb. DMD-ZFO* **1**: 48–57.

Iwata, A. and Kanada, M. 1967. On the nature of the frequency spectrum of atmospheric source signal. *Proc. Res. Inst. Atmos.* **14**: 1–6.

Jacobson, E.A., and Krider, E.P. 1976. Electrostatic field changes produced by Florida lightning. *J. Atmos. Sci.* **33**: 103–17.

Janischewskyj, W., Hussein, A.M., Shostak, V., Rusan, I., Li, J.-X., and Chang, J.-S. 1997. Statistics of lightning strikes to the Toronto Canadian National Tower (1978–1995). *IEEE Trans. Pow. Del.* **12**: 1210–21.

Jayaratne, K.P.S.C., and Cooray, V. 1994. The lightning HF radiation at 3 MHz during leader and return stroke processes. *J. Atmos. Terr. Phys.* **56**: 493–501.

Jones, R.C. 1968. Return stroke core diameter. *J. Geophys. Res.* **73**: 809–14.

Jones, H.L., Calkins, R.L., and Hughes, W.L. 1967. A review of the frequency spectrum of cloud-to-ground and cloud-to-cloud lightning. *IEEE Trans. Geosci. Electr.* (GE-5) **1**: 26–30.

Jordan, D.M. 1990. Relative light intensity and electric field intensity of cloud to ground lightning. Ph.D. dissertation, Univ. Florida, Gainesville, 231 pp.

Jordan, D.M., and Uman, M.A. 1983. Variation in light intensity with height and time from subsequent lightning return strokes. *J. Geophys. Res.* **88**: 6555–62.

Jordan, D.M., Idone, V.P., Rakov, V.A., Uman, M.A., Beasley, W.H., and Jurenka, H. 1992. Observed dart leader speed in natural and triggered lightning. *J. Geophys. Res.* **97**: 9951–7.

Jordan, D.M., Idone, V.P., Orville, R.E., Rakov, V.A., and Uman, M.A. 1995. Luminosity characteristics of lightning M components. *J. Geophys. Res.* **100**: 25 695–700.

Jordan, D.M., Rakov, V.A., Beasley, W.H., and Uman, M.A. 1997. Luminosity characteristics of dart leaders and return strokes in natural lightning. *J. Geophys. Res.* **102**: 22 025–32.

Jose, P.D. 1950. The infrared spectrum of lightning. *J. Geophys. Res.* **55**: 39–41.

Jurenka, H., and Barreto, E. 1982. Study of electron waves in electrical discharge channels. *J. Appl. Phys.* **53**: 3581–90.

Jurenka, H., and Barreto, E. 1985. Electron waves in the electrical breakdown of gases, with application to the dart leader in lightning. *J. Geophys. Res.* **90**: 6219–24.

Kachurin, L.G., Karmov, M.I., and Medaliyev, K.K. 1974. The principal characteristics of the radio emission of convective clouds. *Izv. Acad. Sci. USSR Atmos. Oceanic Phys.* (English Transl.) **10**: 1163–9.

Kawasaki, Z.-I., Nakano, M., Takeuti, T., Nagatani, M., Nakada, H., Mizuno, Y., and Nagai, T. 1987a. Fourier spectra of positive lightning fields during winter thunderstorms. *Res. Lett. Atmos. Electr.* **7**: 29–34.

Kawasaki, Z.-I., Takeuti, T., and Nakano, M. 1987b. Group velocity of subsequent return strokes in triggered lightning. *IEE Japan* **107**: 47–53.

Kawasaki, Z.-I., Matsuura, K., and Israelsson, S. 1990. Calculations of lightning return stroke electric fields above ground with finite conductivity. *Res. Lett. Atmos. Electr.* **10**: 69–79.

Keeney, J. 1970. Observations on the microwave emission from colliding charged water drops. *J. Geophys. Res.* **75**: 1123–6.

Kekez, M.M., and Savic, P. 1983. Contribution to continuous leader channel development. In *Electrical Breakdown and Discharges in Gases*, part A, eds. E.E. Kunhardt and L.H. Luessen, pp. 419–344, New York: Plenum.

Kekez, M.M., and Savic, P. 1976. Laboratory simulation of the stepped leader in lightning. *Can. J. Phys.* **54**: 2216–24.

Kettler, C.J. 1940. Cameras designed for lightning studies. *Photo Technique*, May, 38–43.

Khastgir, S.R. 1957. Leader stroke current in a lightning discharge according to the streamer theory. *Phys. Rev.* **106**: 616–17.

Khastgir, S.R., and Ghosh, D. 1972. Theory of stepped-leader in cloud-to-ground electrical discharges. *J. Atmos. Terr. Phys.* **34**: 109–13.

Kirkland, M.W., Suszcynsky, D.M., Guillen, J.L.L., and Green, J.L. 2001. Optical observations of terrestrial lightning by the FORTE satellite photodiode detector. *J. Geophys. Res.* **106**: 33 499–509.

Kimpara, A. 1965. Electromagnetic Energy Radiated from Lightning. In *Problems of Atmospheric and Space Electricity*, ed. S.C. Coroniti, pp. 352–65, New York: American Elsevier.

Kitagawa, N. 1957a. On the electric field change due to the leader processes and some of their discharge mechanism. *Pap. Meteor. Geophys. (Tokyo)* **7**: 400–14.

Kitagawa, N. 1957b. On the mechanism of cloud flash and junction or final process in flash to ground. *Pap. Meteor. Geophys. (Tokyo)* **7**: 415–24.

Kitagawa, N. 1965. Types of lightning. In *Problems of Atmospheric and Space Electricity*, ed. S.C. Coroniti, pp. 337–47, New York: American Elsevier.

Kitagawa, N., and Brook, M. 1960. A comparison of intracloud and cloud-to-ground lightning discharges. *J. Geophys. Res.* **65**: 1189–201.

Kitagawa, N., and Kobayashi, M. 1958. Distribution of negative charge in the cloud taking part in a flash to ground. *Pap. Meteor. Geophys. (Tokyo)* **9**: 99–105.

Kitagawa, N., and Kobayashi, M. 1959. Field changes and variations of luminosity due to lightning flashes. In *Recent Advances in Atmospheric Electricity*, ed. L.G. Smith, pp. 485–501, Oxford: Pergamon.

Kitagawa, N., Brook, M., and Workman, E.J. 1960. The role of continuous discharges in cloud-to-ground lightning. *J. Geophys. Res.* **65**: 1965.

Kitagawa, N., Brook, M., and Workman, E.J. 1962. Continuing currents in cloud-to-ground lightning discharges. *J. Geophys. Res.* **67**: 637–47.

Klingbell, R., and Tidman, D.A. 1974. Theory and computer model of the lightning stepped leader. *J. Geophys. Res.* **79**: 865–9.

Klingbell, R., and Tidman, D.A. 1974. Reply to comment on the brief report "Theory and computer model of the lightning stepped leader" by C.T. Phelps. *J. Geophys. Res.* **79**: 5669–70.

Klingbell, R., Tidman, D.A., and Fernsler, R.F. 1972. Ionizing gas breakdown waves in strong electric fields. *Phys. Fluids* **15**: 1969–73.

Kobayashi, M., Kitagawa, N., Ikeda, T., and Sato, Y. 1958. Preliminary studies of variation of luminosity and field change due to lightning flashes. *Pap. Meteor. Geophys. (Tokyo)* **8**: 29–34.

Komelkov, V.S. 1941. Investigation of the maximum rate of rise of lightning currents. *Elektrichestvo* **5**: 34–9.

Komelkov, V.S. 1947. Structure and parameters of the leader discharge. *Bull. Acad. Sci. USSR, Tech. Sci. Sect.* **8**: 955–66.

Komelkov, V.S. 1950. The development of electric discharges in long gaps. *Bull. Acad. Sci. USSR, Tech. Sci. Sect.* **6**: 851–65.

Kosarev, E.L., Zatsepin, V.G., and Mitrofanov, A.V. 1970. Ultra-high frequency radiation from lightnings. *J. Geophys. Res.* **75**: 7524–30.

Kosche, H., Fischer, H.J. and Mühleisen, R. 1974. Untersuchungen der Feinstruktur des Luftelektrischen Feldes bei Blitzentladungen. *Meteor. Res.* **4**: 116–18.

Koshak, W.J., Solakiewicz, R.J., Phanord, D.D., and Blakeslee, R.J. 1994. Diffusion model for lightning radiative transfer. *J. Geophys. Res.* **99**: 14 361–71.

Krehbiel, P.R. 1981. An analysis of the electric field change produced by lightning. Ph.D. dissertation, University of Manchester Institute of Science and Technology, Manchester, England. (Available as Report T-11, Geophys. Res. Ctr. New Mexico Inst. Mining and Tech., Socorro)

Krehbiel, P.R., Brook, M., and McCrory, R. 1979. An analysis of the charge structure of lightning discharges to the ground. *J. Geophys. Res.* **84**: 2432–56.

Krehbiel, P.R., Brook, M., Lhermitte, R.L., and Lennon, C.L. 1983. Lightning charge structure in thunderstorms. In *Proceedings in Atmospheric Electricity*, eds. L.H. Ruhnke and J. Latham, pp. 408–11, Hampton, Virginia: A. Deepak.

Krehbiel, P.R., Brook, M., Khanna-Gupta, S., Lennon, C.L., and Lhermitte, R. 1984. Some results concerning VHF lightning radiation from the real-time LDAR system at KSC, Florida. In *Proc 7th Int. Conf. on Atmospheric Electricity*, Albany, New York, pp. 388–93.

Krider, E.P. 1965a. Time-resolved spectral emissions from individual return strokes in lightning discharges. *J. Geophys. Res.* **70**: 2459–60.

Krider, E.P. 1965b. The design and testing of a photoelectric photometer for selected lines in the spectrum of lightning. M.S. thesis, Department of Physics, University of Arizona, Tucson.

Krider, E.P. 1966a. Some photoelectric observations of lightning. *J. Geophys. Res.* **71**: 3095–8.

Krider, E.P. 1966b. Comment on paper by Leon E. Salanave and Marx Brook, "Lightning photography and counting in daylight, using H_α emission". *J. Geophys. Res.* **71**: 675.

Krider, E.P. 1973. Lightning spectroscopy. *Nuclear Instruments and Methods* **110**: 11–19.

Krider, E.P. 1974a. The relative light intensity produced by a lightning stepped leader. *J. Geophys. Res.* **79**: 4542–4.

Krider, E.P. 1974b. An unusual photograph of an air lightning discharge. *Weather* **29**: 24–8.

Krider, E.P. 1975. An all-sky camera for time-resolved lightning photography. *J. Appl. Meteor.* **14**: 2249–52.

Krider, E.P. 1992. On the electromagnetic fields, Poynting vector, and peak power radiated by lightning return strokes. *J. Geophys. Res.* **97**: 15 913–17.

Krider, E.P. 1994. On the peak electromagnetic fields radiated by lightning return strokes toward the middle-atmosphere. *J. Atmos. Electr.* **14**: 17–24.

Krider, E.P. 1996. 75 years of research on the physics of a lightning discharge. In *Historical Essays on Meteorology, 1919–1995*, ed. J.R. Fleming, pp. 321–50, Boston, Massachusetts: American Meteorological Society.

Krider, E.P., and Alejandro, S.P. 1983. Lightning, an unusual case study. *Weatherwise* **36**: 71–5.

Krider, E.P., and Guo, C. 1983. The peak electromagnetic power radiated by lightning return strokes. *J. Geophys. Res.* **88**: 8471–4.

Krider, E.P., and Ladd, C.G. 1975. Upward streamers in lightning discharges to mountainous terrain. *Weather* **30**: 77–81.

Krider, E.P., and Marcek, G.S. 1972. A simplified technique for the photography of lightning in daylight. *J. Geophys. Res.* **77**: 6017–20.

Krider, E.P., and Noggle, R.C. 1975. Broadband antenna systems for lightning magnetic fields. *J. Appl. Meteor.* **14**: 151–6.

Krider, E.P., and Radda, G.J. 1975. Radiation field waveforms produced by lightning stepped leaders. *J. Geophys. Res.* **80**: 2653–7.

Krider, E.P., and Wetmore, R.H. 1987. Upward streamers produced by a lightning strike to radio transmission towers. *J. Geophys. Res.* **92**: 9859–62.

Krider, E.P., Radda, G.J., and Noggle, R.C. 1975. Regular radiation field pulses produced by intracloud lightning discharges. *J. Geophys. Res.* **80**: 3801–4.

Krider, E.P., Weidman, C.D., and Noggle, R.C. 1977. The electric field produced by lightning stepped leaders. *J. Geophys. Res.* **82**: 951–60.

Krider, E.P., Weidman, C.D., and Le Vine, D.M. 1979. The temporal structure of the HF and VHF radiation produced by intracloud lightning discharges. *J. Geophys. Res.* **84**: 5760–2.

Krider, E.P., Leteinturier, C., and Willett, J.C. 1992. Submicrosecond field variations in natural lightning processes. *Res. Lett. Atmos. Electr.* **12**: 3–9.

Krider, E.P., Leteinturier, C., Willett, J.C. 1996. Submicrosecond fields radiated during the onset of first return strokes in cloud-to-ground lightning. *J. Geophys. Res.* **101**: 1589–97.

Kumar, P.P. 1992. Power spectrum analysis of sferics from lightning. *Ind. J. Radio Space Phys.* **21**: 149–52.

Labaune, G., Richard, P., and Bondiou, A. 1987. Electromagnetic properties of lightning channels formation and propagation. *Electromagnetics* **7**: 361–93.

Lacerda, M., Pinto, O., Pinto, I.R.C.A., Diniz, J.H., and Carvalho, A.M. 1999. Analysis of negative downward lightning current curves from 1985 to 1994 at Morro do Cachimbo research station (Brazil). In *Proc. 11th Int. Conf. on Atmospheric Electricity, Guntersville, Alabama*, pp. 42–5.

Lalande, P., Bondiou-Clergerie, A., Laroche, P., Eybert-Berard, A., Berlandis, J.-P., Bador, B., Bonamy, A., Uman, M.A., and Rakov, V.A. 1998. Leader properties determined with triggered lightning techniques. *J. Geophys. Res.* **103**: 14 109–15.

Lanzerotti, L.J., Thomson, D.J., Maclennan, C.G., Rinnert, K., Krider, E.P., and Uman, M.A. 1989. Power spectra at radio-frequency of lightning return stroke waveforms. *J. Geophys. Res.* **94**: 13 221–7.

Larigaldie, S. 1979. Linear gliding discharge over dielectric surfaces. *J. Physique* **40**(C7): 429–30.

Larigaldie, S., Labaune, G., and Moreau, J.P. 1981. Lightning leader laboratory simulation by means of rectilinear surface discharges. *J. Appl. Phys.* **52**: 7114–20.

Larigaldie, S., Roussaud, A., and Jecko, B. 1992. Mechanisms of high-current pulses in lightning and long-spark. *J. Appl. Phys.* **72**(5): 1729–39.

Laroche, P., Idone, V., Eybert-Berard, A., and Barret, L. 1991. Observations of bidirectional leader development in triggered lightning flash. In *Proc. Int. Conf. on Lightning and Static Electricity, Cocoa Beach, Florida*, pp. 57/1–10.

Larsen, A. 1905. Photographing lightning with a moving camera. *Ann. Rep. Smithsonian Inst.* **60**(1): 119–27.

Latham, D.J. 1980. A channel model for long arcs in air. *Phys. Fluids* **23**: 1710–15.

Latham, J., and Stromberg, I.M. 1977. Point-discharge. In *Lightning, vol. 1, Physics of Lightning*, ed. R.H. Golde, pp. 99–117, New York: Academic Press.

Le Boulch, M., and Hamelin, J. 1985. Rayonnement en ondes métriques et décimétriques des orages. *Ann. Telecom.* **40**: 277–313.

Le Boulch, M., and Plantier, T. 1990. The Meteorage thunderstorm monitoring system: a tool for new EMC protection strategies. In *Proc. 20th Int. Conf. on Lightning Protection, Interlaken, Switzerland*, paper 6.13P, 8 pp.

Le Boulch, M., Hamelin, J., and Weidman, C. 1987. UHF–VHF radiation from lightning, *Electromagnetics* **7**: 287–331.

Lee, L.D., Finelli, C.B., Thomas, M.E., and Pitts, F.L. 1984. 1980 to 1982 statistical analysis of direct-strike lightning data. NASA Technical Memorandum 2252, Langley Research Center, Hampton, Virginia.

Lee, S.C., Lim, K.K., Meiappa, M., and Liew, A.C. 1979. Determination of lightning current using frame aerials. *IEEE Trans. Pow. Appar. Syst.* **98**: 1669–75.

Les Renardières Group 1981. Negative discharges in long air gaps at Les Renardières, 1978 results. *Electra* **74**: 67–216.

Leteinturier C. and Hamelin, J. 1987. Experimental study of the electromagentic characteristics of lightning discharge in the 200 Hz 20 MHz band. *Electromagnetics* **7**: 423–39.

Leteinturier, C., and Hamelin, J. 1990. Rayonnement électromagnétique des décharges orageuses. Analyse submicroseconde. *Rev. Phys. Appl.* **25**: 139–46.

Leteinturier, C., Weidman, C., and Hamelin, J. 1990. Current and electric field derivatives in triggered lightning return strokes. *J. Geophys. Res.* **95**: 811–28.

Leteinturier, C., Hamelin, J.H., and Eybert-Berard, A. 1991. Submicrosecond characteristics of lightning return-stroke currents. *IEEE Trans. Electromagn. Compat.* **33**: 351–7.

Le Vine, D.M. 1977. The effect of pulse interval statistics on the spectrum of radiation from lightning. *J. Geophys. Res.* **82**: 1773–7.

Le Vine, D.M. 1980a. Sources of the strongest RF radiation from lightning. *J. Geophys. Res.* **85**: 4091–5.

Le Vine, D.M. 1980b. The spectrum of radiation from lightning. In *Proc. IEE Int. Symp. on Electromagnetic Compatability*, pp. 249–53.

Le Vine, D.M. 1987. Review of measurements of the RF spectrum of radiation from lightning. *Meteor. Atmos. Phys.* **37**: 195–204.

Le Vine, D.M., and Kao, M. 1988. The effects of current risetime on radiation from tortuous lightning channels. In *Proc. 8th Int. Conf. on Atmospheric Electricity, Uppsala, Sweden*, pp. 509–14.

Le Vine, D.M., and Krider, E.P. 1977. The temporal structure of HF and VHF radiations during Florida lightning return strokes. *Geophys. Res. Lett.* **4**: 13–16.

Le Vine, D.M., and Meneghini, R. 1978a. Electromagnetic fields radiated from a lightning return stroke: application of an exact solution to Maxwell's equations. *J. Geophys. Res.* **83**: 2377–84.

Le Vine, D.M., and Meneghini, R. 1978b. Simulation of radiation from lightning return strokes: the effects of tortuosity. *Radio Sci.* **13**: 801–9.

Le Vine, D.M., and Willett, J.C. 1992. Comment on the transmission-line model for computing radiation from lightning. *J. Geophys. Res.* **97**: 2601–10.

Le Vine, D.M., and Willett, J.C. 1995. The influence of channel geometry on the fine scale structure of radiation from lightning return strokes. *J. Geophys. Res.* **100**: 18 629–38.

Le Vine, D.M., Gesell, L., and Kao, M. 1986. Radiation from lightning return strokes over a finitely conducting Earth. *J. Geophys. Res.* **91**: 11 897–908.

Le Vine, D.M., Willett, J.C., and Bailey, J.C. 1989. Comparison of fast electric field changes from subsequent return strokes of natural and triggered lightning. *J. Geophys. Res.* **94**: 13 259–65.

Lewis, W.W., and Foust, C.M. 1945. Lightning investigation on transmission lines, Part 7. *Trans. AIEE* **64**: 107–15.

Liaw, Y.P., Cook, D.R., and Sisterson, D.L. 1996. Estimation of lightning stroke peak current as a function of peak electric field and the normalized amplitude of signal strength: corrections and improvements. *J. Atmos. Oceanic Tech.* **13**: 769–73.

Liew, A.C., and Darveniza, M. 1982. Calculation of the lightning performance of unshielded transmission lines. *IEEE Trans. Pow. Appar. Syst.* **101**: 1471–7.

Liew, A.C., and Darveniza, M. 1982. Lightning performance of unshielded transmission lines. *IEEE Trans. Pow. Appar. Syst.* **101**: 1478–82.

Lin, Y.T., and Uman, M.A. 1973. Electric radiation fields of lightning return strokes in three isolated Florida thunderstorms. *J. Geophys. Res.* **78**: 7911–15.

Lin, Y.T., Uman, M.A., Tiller, J.A., Brantley, R.D., Beasley, W.H., Krider, E.P., and Weidman, C.D. 1979. Characterization of lightning return stroke electric and magnetic fields from simultaneous two-station measurements. *J. Geophys. Res.* **84**: 6307–14.

Lin, Y.T., Uman, M.A., and Standler, R.B. 1980. Lightning return stroke models. *J. Geophys. Res.* **85**: 1571–83.

Lind, M.A., Hartman, J.S., Takle, E.S., and Stanford, J.L. 1972. Radio noise studies of several severe weather events in Iowa in 1971. *J. Atmos. Sci.* **29**: 1220–3.

Light, T.E., Suszcynsky, D.M., and Jacobson, A.R. 2001. Coincident radio frequency and optical emissions from lightning observed with the FORTE satellite. *J. Geophys. Res.* **106**: 28 223–31.

Little, P.F. 1978. Transmission line representation of a lightning return stroke. *J. Phys. D: Appl. Phys.* **11**: 1893–910.

Liu, X.-S., and Krehbiel, P.R. 1985. The initial streamer of intracloud lightning flashes. *J. Geophys. Res.* **90**: 6211–8.

Livingston, J.M., and Krider, E.P. 1978. Electric fields produced by Florida thunderstorms. *J. Geophys. Res.* **83**: 385–401.

Loeb, L. 1965. Ionizing waves of potential gradient. *Science* **148**: 1417–26.

Loeb, L. 1966. The mechanism of stepped and dart leaders in cloud-to-ground lightning strokes. *J. Geophys. Res.* **71**: 4711–21.

Loeb, L. 1968. Confirmation and extension of a proposed mechanism of the stepped leader lightning stroke. *J. Geophys. Res.* **73**: 5813–17.

Loeb, L. 1970. Mechanism of charge drainage from thunderstorm clouds. *J. Geophys. Res.* **75**: 5882–9.

Lopez, R.E., Maier, M.W., and Holle, R.L. 1991. Comparison of the signal strength of positive and negative cloud-to-ground lightning flashes in northeastern Colorado. *J. Geophys. Res.* **96**: 22 307–18.

Love, E.R. 1973. Improvements on lightning stroke modelling and applications to the design of EHV and UHV transmission lines, M.S. thesis, University of Colorado.

Lundholm, R. 1957. Induced overvoltage-surges on transmission lines and their bearing on the lightning performance at medium voltage networks. *Trans. Chalmers Univ. Technol.* **120**, 117 pp.

Lundquist, S., and Scuka, V. 1970. Some time correlated measurements of optical and electromagnetic radiation from lightning flashes. *Arkiv Geophysik* **5**: 585–93.

Lupò, G., Petrarca, C., Tucci, V., and Vitelli, M. 2000a. EM fields generated by lightning channels with arbitrary location and slope. *IEEE Trans. Electromagn. Compat.* **42**: 39–53.

Lupò, G., Petrarca, C., Tucci, V., and Vitelli, M. 2000b. EM fields associated with lightning channels: on the effect of tortuosity and branching. *IEEE Trans. Electromagn. Compat.* **42**: 394–404.

Lyons, W.A., Uliasz, M., and Nelson, T.E. 1998. Large peak current cloud-to-ground lightning flashes during the summer months in the contiguous United States. *Mon. Wea. Rev.* **126**: 2217–23.

Mach, D.M., and Rust, W.D. 1989a. Photoelectric return-stroke velocity and peak current estimates in natural and triggered lightning. *J. Geophys. Res.* **94**: 13 237–47.

Mach, D.M., and Rust, W.D. 1989b. A photoelectric technique for measuring lightning-channel propagation velocities from a mobile laboratory. *J. Atmos. Oceanic Tech.* **6**: 439–45.

Mach, D.M., and Rust. W.D. 1997. Two-dimensional speed and optical risetime estimates for natural and triggered dart leaders. *J. Geophys. Res.* **102**: 13 673–84.

Mackerras, D. 1968. A comparison of discharge processes in cloud and ground lightning flashes. *J. Geophys. Res.* **73**: 1175–83.

Mackerras, D. 1973. Photoelectric observations of the light emitted by lightning flashes. *J. Atmos. Terr. Phys.* **35**: 521–35.

Mackerras, D., Darveniza, M., and Liew, A.C. 1997. Review of claimed enhanced lightning protection of buildings by early streamer emission air terminals. *IEE Proc. Sci. Meas. Tech.* **144**: 1–10.

Maier, L.M., and Krider, E.P. 1986. The charges that are deposited by cloud-to-ground lightning in Florida. *J. Geophys. Res.* **91**: 13 279–89.

Malan, D.J. 1950. Appareil de grand rendement pour la chronophotographie des éclairs. *Rev. Opt.* **29**: 513–23.

Malan, D.J. 1952. Les décharges dans l'air et la charge inférieure positive d'un nuage orageux. *Ann. Geophys.* **8**: 385–401.

Malan, D.J. 1954. Les décharges orageuses intermittentes et continues de la colonne de charge négative. *Ann. Geophys.* **10**: 271–81.

Malan, D.J. 1955a. La distribution verticale de la charge négative orageuse. *Ann. Geophys.* **11**: 420–6.

Malan, D.J. 1955b. Les décharges lumineuses dans les nuages orageux. *Ann. Geophys.* **11**: 427–34.

Malan, D.J. 1956a. The relation between the number of strokes, stroke interval, and the total durations of lightning discharges. *Pure Appl. Geophys.* **34**: 224–30.

Malan, D.J. 1956b. Visible electrical discharges inside thunderclouds. *Geofis. Pura Appl.* **34**: 221–3.

Malan, D.J. 1957. The theory of lightning photography and a camera of new design. *Geofis. Pura Appl.* **38**: 250–60.

Malan, D.J. 1958. Radiation from lightning discharges and its relation to discharge processes. In *Recent Advances in Atmospheric Electricity*, ed. L.G. Smith, pp. 557–63, London: Pergamon Press.

Malan, D.J. 1963. *Physics of Lightning.* 176 pp., London: The English Universities Press.

Malan, D.J. 1965. The theory of lightning. In *Problems of Atmospheric and Space Electricity*, ed. S.C. Coroniti, pp. 323–31, New York: American Elsevier.

Malan, D.J., and Collens, H. 1937. Progressive lightning III – the fine structure of return lightning strokes. *Proc. Roy. Soc.* A**162**: 175–203.

Malan, D.J., and Schonland, B.F.J. 1947. Progressive lightning, Part 7, Directly correlated photographic and electrical studies of lightning from near thunderstorms. *Proc. Roy. Soc.* A**191**: 485–503.

Malan, D.J., and Schonland, B.F.J. 1951a. The electrical processes in the intervals between the strokes of a lightning discharge. *Proc. Roy. Soc.* A**206**: 145–63.

Malan, D.J., and Schonland, B.F.J. 1951b. The distribution of electricity in thunderclouds. *Proc. Roy. Soc.* A**209**: 158–77.

Marney, G.O., and Shanmugam, K. 1971. Effect of channel orientation on the frequency spectrum of lightning discharges. *J. Geophys. Res.* **76**: 4198–202.

Marshall, S.V. 1967. An analytical model for the fluxgate magnetometer. *IEEE Trans. Magnetics* **3**: 459–63.

Marshall, S.V. 1973. Impulse response of a fluxgate sensor-application to lightning discharge location and measurement. *IEEE Trans. Magnetics* **9**: 235–8.

Marshall, T.C., and Stolzenburg, M. 2001. Voltages inside and just above thunderstorms. *J. Geophys. Res.* **106**(D5): 4757–68.

Marshall, T.C., McCarthy, M.P., and Rust, W.D. 1995. Electric field magnitudes and lightning initiation in thunderstorms. *J. Geophys. Res.* **100**: 7097–103.

Master, M.J., Uman, M.A., Beasley, W.H., and Darveniza, M. 1984. Lightning induced voltages on power lines: experiment. *IEEE Trans. PAS* **103**: 2519–29.

Mastrup, F., and Wiese, W. 1958. Experimentelle Bestimmung des Oszillatorenstarken einiger nii und oii Linien. *Z. Astrophys.* **44**: 259–79.

Maxwell, E.L. 1967. Atmospheric noise from 20 Hz to 30 kHz. *Radio Sci.* **2**: 637–44.

Maxwell, E.L., and Stone, D.L. 1963. Natural noise fields from 1 cps to 100 kc. *IEEE Trans. Antenna Propag.* **AP-11**: 339–43.

Mazur, V. 1986. Rapidly occurring short duration discharges in thunderstorms, as indicators of a lightning-triggering mechanism. *Geophys. Res. Lett.* **13**: 355–8.

Mazur, V., and Ruhnke, L.H. 1993. Common physical processes in natural and artificially triggered lightning. *J. Geophys. Res.* **98**: 12 913–30.

Mazur, V., Krehbiel, P.R., and Shao, X.-M. 1995. Correlated high-speed video and radio interferometric observations of a cloud-to-ground lightning flash. *J. Geophys. Res.* **100**: 25 731–53.

Mazur, V., Ruhnke, L.H., and Laroche, P. 1995. The relationship of leader and return stroke processes in cloud-to-ground lightning. *Geophys. Res. Lett.* **22**: 2613–16.

Mazur, V., Williams, E., Boldi, R., Maier, L., and Proctor, D.E. 1997. Initial comparison of lightning mapping with operational time-of-arrival and interferometric systems. *J. Geophys. Res.* **102**: 11 071–85.

Mazur, V., Shao, X.M., and Krehbiel, P. R. 1998. "Spider" lightning in intracloud and positive cloud-to-ground flashes. *J. Geophys. Res.* **103**: 19 811–22.

McCann, D.G. 1944. The measurement of lightning currents in direct strokes. *Trans. AIEE* **63**: 1157–64.

McDonald, T.B., Uman, M.A., Tiller, J.A., and Beasley, W.H. 1979. Lightning location and lower ionospheric height determination from two station magnetic field measurements. *J. Geophys. Res.* **84**: 1727–34.

McEachron, K.B. 1939. Lightning to the Empire State Building. *J. Franklin Inst.* **227**: 149–217.

McEachron, K.B. 1940. Wave shapes of successive lightning current peaks. *Electrical World* **56**: 428–31.

McEachron, K.B. 1941. Lightning to the Empire State Building. *Trans. AIEE* **60**: 885–9.

McEachron, K.B. 1947. Photographic study of lightning. *AIEE Trans.* **66**: 577–85.

McEachron, K.B., and Morris, W.A. 1936. The lightning stroke: mechanism of discharge. *Gen. Electr. Rev.* **39**: 487–96.

Meinel, A.B., and Salanave, L.E. 1964. N_2^+ emission in lightning. *J. Atmos. Sci.* **21**: 157–60.

Melander, B.G. 1984. Effects of tower characteristics on lightning arc measurements. In *Proc. 1984 Int. Conf. on Lightning and Static Electricity, Orlando, Florida*, pp. 34/1–12.

Meyer, G. 1894. Ein Versuch, das Spektrum des Blitzes zu photographieren. *Ann. Physik Chem.* **51**: 415–16.

Michishita, K., Ishii, M., and Hojo, J.-I. 1996. Measurement of horizontal electric fields associated with distant cloud-to-ground strokes. *J. Geophys. Res.* **101**: 3861–7.

Ming, Y., and Cooray, V. 1994. Propagation effects caused by a rough ocean surface on the electromagnetic fields generated by lightning return strokes. *Radio Sci.* **29**: 73–85. (Correction, *Radio Sci.* 1998, **33**: 635)

Misra, R.P., Rai, J., and Banerjee, P.C. 1977. Radiation fields of the multiple return stroke lightning. *J. Inst. Electron. Telecom. Eng.* **23**: 81–3.

Miyake, K., Suzuki, T., and Shinjou, K. 1992. Characteristics of winter lightning current on Japan Sea coast. *IEEE Trans. Pow. Del.* **7**: 1450–6.

Moini, R., Kordi, B., Rafi, G.Z., and Rakov, V.A. 2000. A new lightning return stroke model based on antenna theory. *J. Geophys. Res.* **105**: 29 693–702.

Montandon, E. 1992. Lightning positioning and lightning parameter determination experiences and results of the Swiss PTT research project. In *Proc. 21st Int. Conf. on Lightning Protection, Berlin, Germany*, pp. 307–12.

Montandon, E. 1995. Messung und Ortung von Blitzeinschlaegen und ihren Auswirkungen am Fernmeldeturm "St. Chrischona" bei Basel der Schweizerischen Telecom PTT. *Elektrotechnik und Informationstechnik* **112**: 283–9.

Moore, C.B., and Vonnegut, B. 1977. The thundercloud. In *Lightning, vol. 1, Physics of Lightning*, ed. R.H. Golde, pp. 51–98, New York: Academic Press.

Moore, C.B., Eack, K.B., Aulich, G.D., and Rison, W. 2001. Energetic radiation associated with lightning stepped-leaders. *Geophys. Res. Lett.* **28**: 2141–4.

Moreau, J.P., and Rustan, P.L. 1987. A study of lightning initiation based on VHF radiation. *Electromagnetics* **7**: 333–52.

Moreau, J.P., and Rustan, P.L. 1990. A study of lightning initiation based on VHF radiation. *Lightning Electromagnetics*, ed. R.L. Gardner, pp. 257–76, New York: Hemisphere.

Motoyama, H., Janischewskyj, W., Hussein, A.M., Rusan, R., Chisholm. W.A., and Chang, J.-S. 1996. Electromagnetic field radiation model for lightning strokes to tall structures. *IEEE Trans. Pow. Del.* **11**: 1624–32.

Mousa, A.M., and Srivastava, K.D. 1989. The implications of the electrogeometric model regarding effect of height of structure on the median amplitude of collected lightning strokes. *IEEE Trans. Pow. Del.* **4**: 1450–60.

Müller-Hillebrand, D. 1962. The magnetic field of the lightning discharge. In *Gas Discharge and the Electricity Supply Industry*, eds. J.S. Forrest, P.R. Howard, and D.J. Littler, pp. 89–111, London: Butterworths.

Nagai, Y., Kawamata, S., and Edano, Y. 1982. Observation of preceding leader and its downward traveling velocity in Utsunomiya district. *Res. Lett. Atmos. Electr.* **2**: 53–6.

Nakahori, K., Egawa, T., and Mitani, H. 1982. Characteristics of winter lightning currents in Hokuriku district. *IEEE Trans. Pow. Appar. Syst.* **101**: 4407–12.

Nakai, T., 1977. On the time and amplitude properties of electric fields near sources of lightning in the VLF, HF, and LF bands. *Radio Sci.* **12**: 389–96.

Nakai, T., 1978. The random scatter of spectral parameters of VLF atmospherics. *J. Atmos. Terr. Phys.* **40**: 61–71.

Nakano, M., Nagatani, M., Nakada, H., Takeuti, T., and Kawasaki, Z. 1987. Measurements of the velocity change of a lightning return stroke with height. *Res. Lett. Atmos. Electr.* **7**: 25–8.

Nanevicz, J.E., Vance, E.F., and Hamm, J.M. 1987. Observation of lightning in the frequency and time domains. *Electromagnetics* **7**: 267–8.

Nanevicz, J.E., Vance, E.F., Radasky, W., Uman, M.A., Soper, G.K., and Pierre, J.M. 1988. EMP susceptibility insights from aircraft exposure to lightning. *IEEE Trans. Electromagn. Compat.* **30**: 463–72.

Narita, K., Goto, Y., Komuro, H., and Sawada, S. 1989. Bipolar lightning in winter at Maki, Japan. *J. Geophys. Res.* **94**: 13 191–5.

Nemzek, R.J., and Winckler, J.R. 1989. Observation and interpretation of fast sub-visual light pulses from the night sky. *Geophys. Res. Lett.* **16**: 1015–18.

Newcott, W.R. 1993. Lightning. Nature's high-voltage spectacle. *National Geographic* **184**(1): 83–103.

Nicolet, M. 1943. Le spectre des éclairs. *Ciel Terre* **59**: 91–8.

Nguyen, M.D., and Michnowski, S. 1996a. On the initiation of lightning discharge in a cloud 1. The high field regions in a thundercloud. *J. Geophys. Res.* **101**: 26 669–73.

Nguyen, M.D., and Michnowski, S. 1996b. On the initiation of lightning discharge in a cloud 2. The lightning initiation on precipitation particles. *J. Geophys. Res.* **101**: 26 675–80.

Norinder, H. 1954. The wave-forms of the electric field in atmospherics recorded simultaneously by two distant stations. *Arkiv Geofysik* **2**(9): 161–95.

Norinder, H., and Dahle, O. 1945. Measurements by frame aerials of current variations in lightning discharges. *Arkiv. Mat. Astron. Fysik* **32A**: 1–70.

Nucci, C.A., Diendorfer, G., Uman, M.A., Rachidi, F., Ianoz, M., and Mazzetti, C. 1990. Lightning return stroke current models with specified channel-base current: a review and comparison. *J. Geophys. Res.* **95**: 20 395–408.

Obayashi, T. 1960. Measured frequency spectra of VLF atmospherics. *J. Res. National Bureau of Standards* **64D**: 41–8.

Oetzel, G.N. 1968. Computation of the diameter of a lightning return stroke. *J. Geophys. Res.* **73**: 1889–96.

Oetzel, G.N., and Pierce, E.T. 1969. Radio emissions from close lightning. In *Planetary Electrodynamics*, vol. 1, eds. S.C. Coroniti and J. Hughes, pp. 543–71, New York: Gordon and Breach.

Ogawa, T. 1982. The lightning current. In *Handbook of Atmospherics*, vol. 1, ed. H. Voland, pp. 23–63, Boca Raton, FL: CRC Press.

Ogawa, T. 1993. Initiation of lightning in clouds. *J. Atmos. Electr.* **13**: 121–32.

Ogawa, T. 1995. Lightning currents. In *Handbook of Atmospheric Electrodynamics*, vol. I, ed. H. Volland, pp. 93–136, Boca Raton, FL: CRC Press.

Ogawa, T., and Brook, M. 1964. The mechanism of the intracloud lightning discharge. *J. Geophys. Res.* **69**: 5141–50.

Ogawa, T., and Brook, M. 1969. Charge distribution in thunderstorm clouds. *Q.J.R. Meteor. Soc.* **95**: 513–25.

Oh, L.L. 1969. Measured and calculated spectral amplitude distribution of lightning sferics. *IEEE Trans. Electromagn. Compat.* **11**: 125–30.

Ollendorff, F. 1968. Einige Eigenschaften des Blitzfeldes. *Archiv Elektrotechnik* **52**: 137–47.

Orville, R.E. 1966a. High-speed, time-resolved spectrum of a lightning stroke. *Science* **151**: 451–2.

Orville, R.E. 1966b. A spectral study of lightning strokes, Ph.D. dissertation, University of Arizona, Tucson.

Orville, R.E. 1967. Ozone production during thunderstorms, measured by the absorption of ultraviolet radiation from lightning. *J. Geophys. Res.* **72**: 3557–62.

Orville, R.E. 1968a. A high-speed time-resolved spectroscopic study of the lightning return stroke: Part I, A qualitative analysis. *J. Atmos. Sci.* **25**: 827–38.

Orville, R.E. 1968b. A high-speed time-resolved spectroscopic study of the lightning return stroke: Part II, A quantitative analysis. *J. Atmos. Sci.* **25**: 839–51.

Orville, R.E. 1968c. A high-speed time-resolved spectroscopic study of the lightning return stroke: Part III, A time-dependent model. *J. Atmos. Sci.* **25**: 852–6.

Orville, R.E. 1968d. Spectrum of the lightning stepped leader. *J. Geophys. Res.* **73**: 6999–7008.

Orville, R.E. 1968e. Photograph of a close lightning flash. *Science* **162**: 666–7.

Orville, R.E. 1975. Spectrum of the lightning dart leader. *J. Atmos. Sci.* **32**: 1829–37.

Orville, R.E. 1977. Lightning spectroscopy. In *Lightning, vol. 1. Physics of Lightning*, ed. R.H. Golde, pp. 281–308, New York: Academic Press.

Orville, R.E. 1980. Daylight spectra of individual lightning flashes in the 370–690 nm region. *J. Appl. Meteor.* **19**: 470–3.

Orville, R.E. 1990. Peak-current variations of lightning return strokes as a function of latitude. *Nature* **343**: 149–51.

Orville, R.E. 1991. Calibration of a magnetic direction finding network using measured triggered lightning return stroke peak currents. *J. Geophys. Res.* **96**: 17 135–42.

Orville, R.E. 1999. Comments on "Large peak current cloud-to-ground lightning flashes during the summer months in the contiguous United States". *Mon. Wea. Rev.* **127**: 1937–8.

Orville, R.E., and Henderson, R.W. 1984. Absolute spectral irradiance measurements of lightning from 375 to 880 nm. *J. Atmos. Sci.* **41**: 3180–7.

Orville, R.E., and Idone, V.P. 1982. Lightning leader characteristics in the Thunderstorm Research International Program (TRIP). *J. Geophys. Res.* **87**: 11 177–92.

Orville, R.E., and Salanave, L.E. 1970. Lightning spectroscopy – photographic techniques. *Appl. Optics* **9**: 1775–81.

Orville, R.E., and Uman, M.A. 1965. The optical continuum of lightning. *J. Geophys. Res.* **70**: 279–82.

Orville, R.E., Uman, M.A., and Sletten, A.M. 1967. Temperature and electron density in long air sparks. *J. Appl. Phys.* **38**: 895–6.

Orville, R.E., Helsdon, J.H. Jr, and Evans, W.H. 1974. Quantitative analysis of a lightning return stroke for diameter and luminosity changes as a function of space and time. *J. Geophys. Res.* **79**: 4059–67.

Orville, R.E., Lala, G.G., and Idone, V.P. 1978. Daylight time resolved photographs of lightning. *Science* **201**: 59–61.

Pathak, P.P. 1987. Parallel and perpendicular electric field components of a lightning discharge – a discussion. *Geoexploration* **24**: 3.

Pathak, P.P., Rai, J., and Varshneya, N.C. 1982. VLF radiation from lightning. *Geophys. J.R. Astr. Soc.* **69**: 197–207.

Paxton, A.H., Gardner, R.L., and Baker, L. 1986. Lightning return stroke: a numerical calculation of the optical radiation. *Phys. Fluids* **29**: 2736–41.

Pawsey, J.L. 1957. Radar observation of lightning on 1.5 meters. *J. Atmos. Terr. Phys.* **11**: 289–90.

Peckham, D.W., Uman, M.A., and Wilcox, C.E. Jr 1984. Lightning phenomenology in the Tampa Bay area. *J. Geophys. Res.* **89**: 11 789–805.

Petersen, W.A., and Rutledge, S.A. 1992. Some characteristics of cloud-to-ground lightning in tropical northern Australia. *J. Geophys. Res.* **97**: 11 553–60.

Peterson, B.J., and Wood, W.R. 1968. Measurements of lightning strikes to aircraft. Report SC-M-67-549, to Dept of Transportation, Federal Aviation Admin., Sandia National Labs., Albuquerque, New Mexico.

Petrie, W., and Small, R. 1951. The near infrared spectrum of lightning. *Phys. Rev.* **84**: 1263–4.

Phelps, C.T. 1974a. Positive streamers system intensification and its possible role in lightning initiation. *J. Atmos. Terr. Phys.* **36**: 103–11.

Phelps, C.T. 1974b. Comments on brief report by R. Klingbell and D.A. Tidman: theory and computer model of the lightning stepped leader. *J. Geophys. Res.* **79**: 5669.

Pickering, E.C. 1901. Spectrum of lightning. *Astrophys. J.* **14**: 367–9.

Picone, J.M., Boris, J.P., Greig, J.R., Raleigh, M., and Fernsler, R.F. 1981. Convective cooling of lightning channels. *J. Atmos. Sci.* **38**: 2056–62.

Pierce, E.T. 1955. Electrostatic field changes due to lightning discharges. *Q.J.R. Meteor. Soc.* **81**: 211–28.

Pierce, E.T. 1957. Recent advances in meteorology: lightning. *Sci. Progr. (London)* **45**: 62–75.

Pierce, E.T. 1958. Some topics in atmospheric electricity. In *Recent Advances in Atmospheric Electricity*, ed. L.G. Smith, pp. 5–16, New York: Pergamon.

Pierce, E.T. 1967a. Spherics (sferics). In *Encyclopedia of Atmospheric Sciences and Astrogeology*, ed. R.W. Fairbridge, pp. 935–9, Reinhold.

Pierce, E.T. 1967b. Atmospherics: their characteristics at the source and propagation. In *Progress in Radio Science 1963–1966*, Part 1, pp. 987–1039, Berkeley, California: Int. Sci. Radio Union.

Pierce, E.T. 1969. The thunderstorm as a source of atmospheric noise at frequencies between 1 and 100 kHz. Stanford Research Institute Technical Report, Project 7045, DASA 2299.

Pierce, E.T. 1972. Triggered lightning and its application to rockets and aircraft. In *Proc. 1972 Lightning and Static Electricity Conf.*, paper AFAL-TR-72-325, Wright-Patterson Air Force Base, Ohio.

Pierce, E.T. 1977. Atmospherics and radio noise. In *Lightning, vol. 1. Physics of Lightning*, ed. R.H. Golde, pp. 351–84, New York: Academic Press.

Pierce, E.T., and Wormell, T.W. 1953. Field changes due to lightning discharges. In *Thunderstorm Electricity*, ed. H.R. Byers, pp. 251–66, Chicago, Illinois: University of Chicago Press.

Pinto, O. Jr, Pinto, I.R.C.A., Lacerda, M., Carvalho, A.M., Diniz, J.H., and Cherchiglia, L.C.L. 1997. Are equatorial negative lightning flashes more intense than those at higher latitudes? *J. Atmos. Solar-Terr. Phys.* **59**: 1881–3.

Pitts, F.L. 1982. Electromagnetic measurements of lightning strikes to aircraft. *J. Aircraft* **L9**: 246–50.

Pitts, F.L., and Thomas M.E. 1981. 1980 direct strike lightning data. NASA Technical Memorandum 81946, Langley Research Center, Hampton, Virginia.

Pitts, F.L., and Thomas, M.E. 1982. 1981 direct strike lightning data. NASA Technical Memorandum 83273, Langley Research Center, Hampton, Virginia.

Plooster, M.N. 1971. Numerical model of the return stroke of the lightning discharge. *Phys. Fluids* **14**: 2124–33.

Pockels, F. 1897. Über das magnetische Verhalten einiger basaltischer Gesteine. *Ann. Physik Chem.* **63**: 195–201.

Pockels, F. 1898. Bestimmung maximaler Entladungsstromstärken aus ihrer magnetisierenden Wirkung. *Ann. Physik Chem.* **63**: 458–75.

Pockels, F. 1900. Über die Blitzentladungen erreichte Stromstärke. *Physik. Z.* **2**: 306–7.

Podgorski, A.S., and Landt, J.A. 1987. Three dimensional time domain modelling of lightning. *IEEE Trans. Pow. Del.* **2**: 931–8.

Podosenov, S.A. 2001. Comments on "Transient radiation of traveling-wave wire antennas". *IEEE Trans. Electromagn. Compat.* **43**: 246–8.

Popolansky, F. 1972. Frequency distribution of amplitudes of lightning currents. *Electra* **22**: 139–47.

Preta, J., Uman, M.A., and Childers, D.G. 1985. Comment on "The electric field spectra of first and subsequent lightning return strokes in the 1- to 200-km range" by Serhan et al. *Radio Sci.* **20**: 143–5.

Proctor, D.E. 1971. A hyperbolic system for obtaining VHF radio pictures of lightning. *J. Geophys. Res.* **76**: 1478–89.

Proctor, D.E. 1976. A radio study of lightning. Ph.D. dissertation, Univ. Witwatersrand, Johannesburg, South Africa.

Proctor, D.E. 1981. VHF radio pictures of cloud flashes. *J. Geophys. Res.* **86**: 4041–71.

Proctor, D.E. 1983. Lightning and precipitation in a small multicellular thunderstorm. *J. Geophys. Res.* **88**: 5421–40 (Correction, 1984, **89**: 11 826).

Proctor, D.E. 1989. Personal Communication.

Proctor, D.E. 1997. Lightning flashes with high origins. *J. Geophys. Res.* **102**: 1693–706.

Proctor, D.E., Uytenbogaardt, R., and Meredith, B.M. 1988. VHF radio pictures of lightning flashes to ground. *J. Geophys. Res.* **93**: 12 683–727.

Prueitt, M.L. 1963. The excitation temperature of lightning. *J. Geophys. Res.* **68**: 803–11.

Rachidi, F., Nucci, C.A., Ianoz, M., and Mazzetti, C. 1996. Influence of a lossy ground on lightning-induced voltages on overhead lines. *IEEE Trans. Electromagn. Compat.* **38**: 250–64.

Radda, G.J., and Krider, E.P. 1974. Photoelectric measurements of lightning return stroke propagation speeds. *Trans. Am. Geophys. Union* **56**: 1131.

Rai, J., and Bhattacharya, P.K. 1971. Impulse magnetic flux density close to the multiple return strokes of a lightning discharge. *J. Phys. D, Appl. Phys.* **4**: 1252–6.

Rai, J., Rao, M., and Tantry, B.A.P. 1972. Bremsstrahlung as a possible source of UHF emissions from lightning. *Nature Phys. Sci.* **238**: 59–60.

Rai, J., Bhattacharya, P.K., and Razdan, H. 1973a. UHF power from expanding return stroke channel of a lightning discharge. *Int. J. Electron.* **35**: 679–84.

Rai, J., Singh, A.K., and Saha, S.K. 1973b. Magnetic field within the return stroke channel of lightning. *Ind. J. Rad. Space Phys.* **2**: 240–2.

Rai, J., Gupta, S.P., and Bhattacharya, P.K. 1974. Electromagnetic field of multiple return stroke lightning. *Int. J. Electron.* **36**: 649–54.

Rai, J., Kumar, K., Hazarika, S., Parashar, J., and Kumar, R. 1993. High speed photographic analysis of intracloud lightning radiation fields. *Ann. Geophys.* **11**: 518–24.

Raizer, Yu. P. 1997. *Gas Discharge Physics*, 2nd edition, 449 pp., Berlin: Springer.

Rakov, V.A. 1985. On estimating the lightning peak current distribution parameters taking into account the lower measurement limit. *Elektrichestvo* **2**: 57–9.

Rakov, V.A. 1998. Some inferences on the propagation mechanisms of dart leaders and return strokes. *J. Geophys. Res.* **103**: 1879–87.

Rakov, V.A. 1999c. Lightning discharges triggered using rocket- and-wire techniques. In *Recent Res. Devel. Geophysics*, vol. 2, pp. 141–71, Research Signpost, India.

Rakov, V.A. 1999d. Lightning electric and magnetic fields. In *Proc. 13th Int. Symp. on Electromagnetic Compatibility, Zurich, Switzerland*, pp. 561–6.

Rakov, V.A. 2001. Characterization of lightning electromagnetic fields and their modeling. In *Proc. 14th Int. Symp. on Electromagnetic Compatibility, Zurich, Switzerland, Supplement*, pp. 3–16.

Rakov, V.A., and Dulzon, A.A. 1984. On latitudinal features of thunderstorm activity. *Meteorologiya i Gidrologiya* **1**: 52–7.

Rakov, V.A., and Dulzon, A.A. 1987. Calculated electromagnetic fields of lightning return stroke. *Tekh. Elektrodinam.* **1**: 87–9.

Rakov, V.A., and Dulzon, A.A. 1991. A modified transmission line model for lightning return stroke field calculations. In *Proc. 9th Int. Zurich. Symp. on Electromagnetic Compatibility, Zurich, Switzerland*, pp. 229–35.

Rakov, V.A., and Uman, M.A. 1990a. Long continuing current in negative lightning ground flashes. *J. Geophys. Res.* **95**: 5455–70.

Rakov, V.A., and Uman, M.A. 1990b. Some properties of negative cloud-to-ground lightning flashes versus stroke order. *J. Geophys. Res.* **95**: 5447–53.

Rakov, V.A., and Uman, M.A. 1990c. Waveforms of first and subsequent leaders in negative lightning flashes. *J. Geophys. Res.* **95**: 16 561–77.

Rakov, V.A., and Uman, M.A. 1990d. Some properties of negative cloud-to-ground lightning. In *Proc. 20th Int. Conf. on Lightning Protection, Interlaken, Switzerland*, paper 6.4, 4 pp.

Rakov, V.A., and Uman, M.A. 1991. Long continuing currents in negative cloud-to-ground lightning flashes: occurrence statistics and hypothetical mechanism. *Proc. USSR Academy of Science (Izvestiya AN SSSR, ser. Fizika Atmosfery i Okeana)* **27**: 376–90.

Rakov, V.A., and Uman, M.A. 1994. Origin of lightning electric field signatures showing two return-stroke waveforms separated in time by a millisecond or less. *J. Geophys. Res.* **99**: 8157–65.

Rakov, V.A., Uman, M.A., Jordan, D.M., and Priore, C.A. III, 1990. Ratio of leader to return stroke field change for first and subsequent lightning strokes. *J. Geophys. Res.* **95**: 16 579–87.

Rakov, V.A., Uman, M.A., Thottappillil, R., and Shindo, T. 1991. Statistical characteristics of negative ground flashes as derived from electric field and TV records. *Proc. USSR Academy of Sciences (Izvestiya AN SSSR, Ser. Energetika i Transport)* **37**: 61–1.

Rakov, V.A., Thottappillil, R., and Uman, M.A. 1992a. Electric field pulses in K and M changes of lightning ground flashes. *J. Geophys. Res.* **97**: 9935–50.

Rakov, V.A., Thottappillil, R., and Uman, M.A. 1992b. On the empirical formula of Willett *et al.* relating lightning return-stroke peak current and peak electric field. *J. Geophys. Res.* **97**: 11 527–33.

Rakov, V.A., Uman, M.A., and Shelukhin, D.V. 1992c. On the possibility to improve an accuracy of the field amplitude lightning-ranging technique. *Proc. USSR Academy of Sciences (Izvestiya AN SSSR, Ser. Radiotekhnika i Elektronika)* **37**: 237–9.

Rakov, V.A., Uman, M.A., and Thottappillil, R. 1994. Review of lightning properties determined from electric field and TV observations. *J. Geophys. Res.* **99**: 10 745–50.

Rakov, V.A., Thottappillil, R., Uman, M.A., and Barker, P.P. 1995. Mechanism of the lightning M component. *J. Geophys. Res.* **100**: 25 701–10.

Rakov, V.A., Uman, M.A., Hoffman, G.R., Masters, M.W., and Brook, M. 1996. Bursts of pulses in lightning electromagnetic radiation: observations and implications for lightning test standards. *IEEE Trans. Electromagn. Compat.* **38**: 156–64.

Rakov, V.A., Uman, M.A., Rambo, K.J., Fernandez, M.I., Fisher, R.J., Schnetzer, G.H., Thottappillil, R., Eybert-Berard, A., Berlandis, J.P., Lalande, P., Bonamy, A., Laroche, P., and Bondiou-Clergerie, A. 1998. New insights into lightning processes gained from triggered-lightning experiments in Florida and Alabama. *J. Geophys. Res.* **103**: 14 117–30.

Rakov, V.A., Crawford, D.E., Rambo, K.J., Schnetzer, G.H., Uman, M.A., and Thottappillil, R. 2001. M-component mode of charge transfer to ground in lightning discharges, *J. Geophys. Res.* **106**: 22 817–31.

Randa, J., Gilliland, D., Gjertson, W., Lauber, W., and McInerney, M. 1995. Catalogue of electromagnetic environment measurements, 30–300 Hz. *IEEE Trans. Electromagn. Compat.* **37**: 16–33.

Rao, M. 1967. Corona currents after the return stroke and the emission of ELF waves in a lightning flash to earth. *Radio Sci.* **2**: 241–4 (Correction, **2**: 1394, 1967).

Rao, M. 1970. The dependence of dart leader velocity on the interstroke time interval in a lightning flash. *J. Geophys. Res.* **75**: 5868–72.

Rao, M., and Bhattacharya, H. 1966. Lateral corona currents from the return stroke channel and slow field change after the return stroke in a lightning discharge. *J. Geophys. Res.* **71**: 2811–14.

Ratnamahilan, P., Hoole, P., Ratnajeevan, S., and Hoole, H. 1993. Simulation of lightning attachment to open ground, tall towers and aircraft. *IEEE Trans. Pow. Del.* **8**: 732–40.

Reising, S.C., Inan, U.S., and Bell, T.F. 1996. Evidence for continuing current in sprite-producing cloud-to-ground lightning. *Geophys. Res. Lett.* **23**: 3639–42.

Rhodes, C., and Krehbiel, P.R. 1989. Interferometric observations of a single stroke cloud-to-ground flash. *Geophys. Res. Lett.* **16**: 1169–72.

Rhouma, A.B., and Auriol, P. 1997. Modelling of the whole electric field changes during a close lightning discharge. *J. Phys. D. Appl. Phys.* **30**: 598–602.

Richard, P., Delannoy, A., Labaune, G., and Laroche, P. 1986. Results of spatial and temporal characterization of the VHF–UHF radiation of lightning. *J. Geophys. Res.* **91**: 1248–60.

Rinnert, K., Lauderdale, R. II, Lanzerotti, L.J., Krider, E.P., and Uman, M.A. 1989. Characteristics of magnetic field pulses in Earth lightning measured by the Galileo probe instrument. *J. Geophys. Res.* **94**: 13 229–35.

Roussel-Dupré, R.A., Gurevich, V., Tunnel, T., and Milikh, G.M. 1994. Kinetic theory of runaway air breakdown. *Phys. Rev. E* **49**(3): 2257–71.

Rubinstein, M. 1996. An approximate formula for the calculation of the horizontal electric field from lightning at close, intermediate, and long range. *IEEE Trans. Electromagn. Compat.* **38**: 531–5.

Rubinstein, M., and Uman, M.A. 1990. On the radiation field turn-on term associated with traveling current discontinuities in lightning. *J. Geophys. Res.* **95**: 3711–13.

Rubinstein, M., and Uman, M.A. 1991. Transient electric and magnetic fields associated with establishing a finite electrostatic dipole, revisited. *IEEE Trans. Electromagn. Compat.* **33**: 312–20.

Rubinstein, M., Rachidi, F., Uman, M.A., Thottappillil, R., Rakov, V.A., and Nucci, C.A. 1995. Characterization of vertical electric fields 500 m and 30 m from triggered lightning. *J. Geophys. Res.* **100**: 8863–72.

Rühling, F. 1972. Modelluntersuchungen über den Schutzraum und ihre Bedeutung für Gebäudeblitzableiter. 1972. *Bull. Schweiz. Elektrotech. Ver.* **63**: 522–8.

Rust, W.D., Krehbiel, P.R., and Shlanta, A. 1979. Measurements of radiation from lightning at 2200 MHz. *Geophys. Res. Lett.* **6**: 85–8.

Rust, W.D., MacGorman, D.R., and Taylor, W.L. 1985. Photographic verification of continuing current in positive cloud-to-ground flashes. *J. Geophys. Res.* **90**: 6144–6.

Rustan, P.L. Jr 1979. Properties of lightning derived from time series analysis of VHF radiation data. Ph.D. dissertation, University of Florida, Gainesville.

Rustan, P.L. Jr 1986. The lightning threat to aerospace vehicles. *AIAA J. Aircraft* **23**: 62–7.

Rustan, P.L. 1987. Description of an aircraft lightning and simulated nuclear electromagnetic pulse (HEMP) threat based on experimental data. *IEEE Trans. Electromagn. Compat.* **29**: 49–63.

Rustan, P.L., Uman, M.A., Childers, D.G., Beasley, W.H., and Lennon, C.L. 1980. Lightning source locations from VHF radiation data for a flash at Kennedy Space Center. *J. Geophys. Res.* **85**: 4893–903.

Sadiku, M.N.O. 1994. *Elements of Electromagnetics*, 821 pp. Orlando, Florida: Sounders College.

Saha, S.K., Rai, J., and Singh, A.K. 1974. Luminosity of the return stroke lightning. *Ind. J. Meteor. Geophys.* **25**: 485–8.

Salanave, L.E. 1961. The optical spectrum of lightning. *Science* **134**: 1395–9.

Salanave, L.E. 1962. The ultraviolet spectrum of lightning: first slitless spectra down to 3000 angstroms. *Trans. Am. Geophys. Union* **43**: 421–32.

Salanave, L.E. 1964. The optical spectrum of lightning. *Adv. Geophys.* **10**: 83–98.

Salanave, L.E. 1971. Astronomers look at lightning. *Astronom. Soc. Pacific* **10**: 1–8.

Salanave, L.E. 1980. *Lightning and Its Spectrum*, 136 pp., Tucson: University of Arizona Press.

Salanave, L.E., and Brook, M. 1965. Lightning photography and counting in daylight, using H_α emission. *J. Geophys. Res.* **70**: 1285–9.

Salanave, L.E., Orville, R.E., and Richards, C.N. 1962. Slitless spectra of lightning in the region from 3850 to 6900 angstroms. *J. Geophys. Res.* **67**: 1877–84.

Sargent, M.A. 1972. The frequency distribution of current magnitudes of lightning strokes to tall structures. *IEEE Trans. Pow. Appar. Syst.* **91**: 2224–9.

Sastry, A.R.K. 1970. Duration of a lightning flash. *J. Atmos. Terr. Phys.* **32**: 1841–3.

Savchenko, B.I. 1970. Possible mechanism for discharges in thunderstorms. *Soviet Physics – Technical Physics* **14**: 1079–82.

Schafer, J.P., and Goodall, W.M. 1939. Peak field strengths of atmospherics due to local thunderstorms at 150 megacycles. *Proc. IRE* **27**: 202–7.

Schonland, B.F.J. 1933. Development of the lightning discharge. *Nature* **132**: 407–8.

Schonland, B.F.J. 1937. The diameter of the lightning channel. *Phil. Mag.* **23**: 503–8.

Schonland, B.F.J. 1938. Progressive lightning, part 4, The discharge mechanisms. *Proc. Roy. Soc.* A**164**: 132–50.

Schonland, B.F.J. 1950. *The Flight of Thunderbolts*, 63 pp., New York: Oxford University Press.

Schonland, B.F.J. 1953. The pilot streamer in lightning and the long spark. *Proc. Roy. Soc.* A**220**: 25–38.

Schonland, B.F.J. 1956. The lightning discharge. In *Handbuch der Physik*, vol. 22, pp. 576–628, Berlin: Springer-Verlag.

Schonland, B.F.J. 1962. Lightning and the long electric spark. *Adv. Science* **19**: 306–13.

Schonland, B.F.J., and Collens, H. 1934. Progressive lightning. *Proc. Roy. Soc.* A**143**: 654–74.

Schonland, B.F.J., and Craib, J. 1927. The electric fields of South African thunderstorms. *Proc. Roy. Soc.* A**114**: 229–43.

Schonland, B.F.J., Malan, D.J., and Collens, H. 1935. Progressive lightning II. *Proc. Roy. Soc.* A**152**: 595–625.

Schonland, B.F.J., Hodges, D.B., and Collens, H. 1938a. Progressive lightning, part 5, A comparison of photographic and electrical studies of the discharge process. *Proc. Roy. Soc.* A**166**: 56–75.

Schonland, B.F.J., Malan, D.J., and Collens, H. 1938b. Progressive lightning, part 6. *Proc. Roy. Soc.* A**168**: 455–69.

Schuster, A. 1880. On spectra of lightning. *Proc. Phys. Soc. (London)*. **3**: 46–52.

Scuka, V. 1969. Electronic optical system for lightning research. *Arkiv Geophysik* **38**: 569–84.

Sen, A.K., and Das Gupta, M.K. 1987. Atmospherics in relation to source phenomena and radio wave propagation in the VHF, UHF, microwave and millimetre wave bands. *Ind. J. Radio Space Phys.* **16**: 127–35.

Serhan, G.I., Uman, M.A., Childers, D.G., and Lin, Y.T. 1980. The RF spectra of first and subsequent lightning return strokes in the 1–200 km range. *Radio Sci.* **15**: 1089–94.

Shao, X.M. 1993. The development and structure of lightning discharges observed by VHF radio interferometer. Ph.D. dissertation, New Mexico Inst. of Min. and Technol., Socorro.

Shao, X.M., and Krehbiel, P.R. 1996. The spatial and temporal development of intracloud lightning. *J. Geophys. Res.* **101**: 26 641–68.

Shao, X.M., Krehbiel, P.R., Thomas, R.J., and Rison, W. 1995. Radio interferometric observations of cloud-to-ground lightning phenomena in Florida. *J. Geophys. Res.* **100**: 2749–83.

Shao, X.M., Holden, D.N., and Rhodes, C.T. 1996. Broad band radio interferometry for lightning observations. *Geophys. Res. Lett.* **23**: 1917–20.

Shindo, T., and Uman, M.A. 1989. Continuing current in negative cloud-to-ground lightning. *J. Geophys. Res.* **94**: 5189–98.

Shumpert, T.H., Honnell, M.A., and Lott, G.K. Jr 1982. Measured spectral amplitude of lightning sferics in the HF, VHF, and UHF bands. *IEEE Trans. Elecromagn. Compat.* **24**: 368–72.

Slipher, V.M. 1917. The spectrum of lightning. *Lowell Obs. Bull. (Flagstaff, Ariz.)* **79**: 55–8.

Smeloff, N.N., and Price, A.L. 1930. Lightning investigation on 220-kV system of the Pennsylvania Power and Light Company (1928 and 1929). *J. AIEE* **49**: 771–5.

Smith, A.J., and Jenkins, P.J. 1998. A survey of natural electromagnetic noise in the frequency range $f = 1$–10 kHz at Halley station, Antarctica: 1. Radio atmospherics from lightning. *J. Atmos. Solar-Terr. Phys.* **60**: 263–77.

Smith, G.S. 2001. Teaching antenna radiation from a time-domain perspective. *Am. J. Phys.* **69**(3): 288–300.

Solomon, R., and Baker, M. 1996. A one-dimensional lightning parameterization. *J. Geophys. Res.* **101**: 14 983–90.

Sporn, P., and Lloyd, W.L. Jr 1930. Lightning investigation on 132-kV system of the Ohio Power Company. *J. AIEE* **49**: 259–62.

Sporn, P., and Lloyd, W.L. Jr 1931. Lightning investigations on the transmission system of the American Gas and Electric Company. *Trans. AIEE* **49**: 1111–17.

Srivastava, C.M., and Khastgir, S.R. 1955. On the maintenance of current in the stepped leader stroke lightning discharge. *J. Sci. Ind. Res.* **14B**: 34–5.

Srivastava, K.M.L. 1966. Return stroke velocity of a lightning discharge. *J. Geophys. Res.* **71**: 1283–6.

Srivastava, K.M.L., and Tantry, B.A.P. 1966. VLF characteristic of electromagnetic radiation from the return stroke of lightning discharge. *Ind. J. Pure Appl. Phys.* **4**: 272–5.

Stanford, J.L. 1971. Polarization of 500 kHz electromagnetic noise from thunderstorms: a new interpretation of existing data. *J. Atmos. Sci.* **28**: 116–19.

Stanford, J.L., Lind, M.A., and Takle, G.S. 1971. Electromagnetic noise studies of severe convective storms in Iowa: the 1970 storm season. *J. Atmos. Sci.* **28**: 436–48.

Steadworthy, A. 1914. Spectrum of lightning. *J. Roy. Astron. Soc. Can.* **8**: 345–8.

Stekolnikov, I.S. 1941. The parameters of the lightning discharge and the calculation of the current waveform. *Elektrichestvo* **3**: 63–8.

Stekolnikov, I.S., and Lamdon, A.A. 1942. Lightning currents in USSR supply systems during the period 1937–1940. *J. Tech. Phys.* **12**: 204–10.

Steptoe, B.J. 1958. Some observations on the spectrum and propagation of atmospherics. Ph.D. dissertation, University of London, England.

Suszcynsky, D.M., Roussel-Dupré, R., and Shaw, G. 1996. Ground-based search for X rays generated by thunderstorms and lightning. *J. Geophys. Res.* **101**: 23 505–16.

Szpor, S. 1969. Comparison of Polish versus American lightning records. *IEEE Trans. Pow. Appar. Syst.* **88**: 646–52.

Szpor, S. 1970. Review of the relaxation theory of the lightning stepped leader. *Acta Geophys. Polonica* **18**: 73–7.

Szpor, S. 1971a. Electrodynamic considerations of lightning problems, II. *Acta Geophysica Polonica* **19**: 35–48.

Szpor, S. 1971b. Courbes internationales des courants de foudre. *Acta Geophysica Polonica* **19**: 365–9.

Szpor, S. 1972a. Photographic studies on lightning by means of a stationary camera. *Archiwum Elektrotechniki* **21**: 13–15.

Szpor, S. 1972b. Steps in the cloud and air discharges. *Archiwum Elektrotechniki* **21**: 19–20.

Szpor, S. 1977. Critical comparison of theories of stepped leaders. *Archiwum Elektrotechniki* **26**: 291–9.

Szpor, S., and Kotlowski, J. 1972. Photographic studies on lightning and air discharge by means of a rotating camera – III. South America. *Archiwum Electrotechniki* **21**: 3–7.

Szpor, S., and Turkowski, W. 1968. Laboratory corroboration of the relaxation theory of the lightning stepped leader. *Archiwum Elektrotechniki* **17**: 405–7.

Szpor, S., Zaborowski, B., Bylick, E., and Kardacz, K. 1972. Photographic studies on lightning by means of a rotating camera – IV. Northern Poland. *Archiwum Elektrotechniki* **21**: 9–11.

Takagi, M. 1961. The mechanism of discharges in a thundercloud. *Proc. Res. Inst. Atmos., Nagoya Univ., Japan* **88B**: 1–105.

Takagi, M. 1969a. VHF radiation from ground discharges. In *Planetary Electrodynamics*, eds. S.C. Coroniti and J. Hughes, pp. 535–8, New York: Gordon and Breach.

Takagi, M. 1969b. VHF radiation from ground discharges. *Proc. Res. Inst. Atmos., Nagoya Univ., Japan* **16**: 163–8.

Takagi, M. 1975. Polarization of VHF radiation from lightning discharge. *J. Geophys. Res.* **80**: 5011–4.

Takagi, M., and Takeuti, T. 1963. Atmospherics radiation from lightning discharge. *Proc. Res. Inst. Atmos., Nagoya Univ., Japan* **10**: 1–11.

Takagi, N., and Takeuti, T. 1983. Oscillating bipolar electric field changes due to close lightning return strokes. *Radio Sci.* **18**: 391–8.

Takagi, N., Watanabe, T., Arima, I., Takeuti, T., Nakano, M., Kawasaki, Z.-I., and Kinosita H. 1988. The risetime of lightning return stroke fields in summer and winter thunderstorms. *Res. Lett. Atmos. Electr.* **8**: 15–20.

Takagi, N., Sugiura, T., Watanabe, T., Arima, I., Takeuti, T., Nakano, M., Shimizu, M., Katuragi, Y., and Nonaka, F. 1993. The magnetic field change associated with the very close lightning ground discharge struck to an electric power transmission line. *J. Atmos. Electr.* **13**: 145–50.

Takagi, N., Wang, D., Watanabe, T., Arima, I., Takeuchi, T., Simizu, M., Katuragi, Y., Yokoya, M., and Kawashima, Y. 1998. Expansion of the luminous region of the lightning return stroke channel. *J. Geophys. Res.* **103**: 14 131–4.

Takahashi, T. 1987. Determination of lightning origins in a thunderstorm model. *J. Meteor. Soc. Japan* **65**: 777–94.

Takeuti, T. 1966. Studies on thunderstorm electricity, II: Ground discharge. *J. Geomagn. Geoelectr.* **18**: 13–22.

Takeuti, T. 1992. The preliminary discussion on the distribution of lightning striking points on the ground. *Res. Lett. Atmos. Electr.* **12**: 144–9.

Takeuti, T., and Nakano, M. 1970. Preliminary observation of the lightning channel in the thundercloud. *Proc. Res. Inst. Atmos., Nagoya Univ., Japan* **17**: 107–9.

Takeuti, T., Ishikawa, H., and Takagi, M. 1960. On the cloud discharge preceding the first ground stroke. *Proc. Res. Inst. Atmos., Nagoya Univ., Japan* **7**: 1–6.

Takeuti, T., Muehleisen, R., and Fischer, H.J. 1969. Bemerkenswerte Gewitter in Sueddeutschland. *Proc. Res. Inst. Atmos., Nagoya Univ., Japan* **16**: 155–61.

Takeuti, T., Hashimoto, and Takagi, N. 1993. Two dimensional computer simulation of the natural stepped leader in summer. *J. Atmos. Electr.* **13**: 9–14.

Taylor, A.R. 1965. Diameter of lightning as indicated by tree scars. *J. Geophys. Res.* **70**: 5693–5.

Taylor, W.L. 1963. Radiation field characteristics of lightning discharges in the band 1 kc/s to 100 kc/s. *J. Res. Nat. Bureau Standards* **67D**: 539–50.

Taylor, W.L. 1965. Lightning characteristics as derived from spherics. In *Problems of Atmospheric and Space Electricity*, ed. S.C. Coroniti, pp. 388–404, New York: American Elsevier.

Taylor, W.L. 1973. Electromagnetic radiation from severe storms in Oklahoma during April 29–30, 1970. *J. Geophys. Res.* **78**: 8761–77.

Taylor, W.L., and Jean, A.G. 1959. Very low frequency radiation spectra of lightning discharges. *J. Res. Nat. Bureau Standards* **63D**: 199–204.

Teer, T.L., and Few, A.A. 1974. Horizontal lightning. *J. Geophys. Res.* **79**: 3436–41.

Théry, C. 2001. Evaluation of LPATS data using VHF interferometric observations of lightning flashes during the EULINOX experiment. *Atmos. Res.* **56**: 397–409.

Thomas, H.A., and Burgess, R.E. 1947. Survey of existing information and data on radio noise over frequency range 1–30 Mc/s. Radio Res. Special Report no. 15.

Thomas, M.E. 1985. 1983 direct strike lightning data. NASA Technical Memorandum 86426, Langley Research Center, Hampton, Virginia.

Thomas, M.E., and Carney, H.E. 1986. 1984 direct strike lightning data. NASA Technical Memorandum 87690, Langley Research Center, Hampton, Virginia.

Thomas, M.E., and Pitts, F.L. 1983. 1982 direct strike lightning data. NASA Technical Memorandum 84626, Langley Research Center, Hampton, Virginia.

Thomas, R.J., Krehbiel, P.R., Rison, W., Hamlin, T., Harlin, J., and Shown, D. 2001. Observations of VHF source powers radiated by lightning. *Geophys. Res. Lett.* **28**: 143–6.

Thomason, L.W., and Krider, E.P. 1982. The effects of clouds on the light produced by lightning. *J. Atmos. Sci.* **39**: 2051–65.

Thomson, E.M. 1978. Photoelectric detector for day-time lightning. *Electron. Lett.* **14**: 337–9.

Thomson, E.M. 1980. Characteristics of Port Moresby ground flashes. *J. Geophys. Res.* **85**: 1027–36.

Thomson, E.M. 1985. A theoretical study of electrostatic field wave shapes from lightning leaders. *J. Geophys. Res.* **90**: 8125–35.

Thomson, E.M. 1999. Exact expressions for electric and magnetic fields from a propagating lightning channel with arbitrary orientation. *J. Geophys. Res.* **104**: 22 293–300.

Thomson, E.M., Uman, M.A., and Beasley, W.H. 1985. Speed and current for lightning stepped leaders near ground as determined from electric field records. *J. Geophys. Res.* **90**: 8136–42.

Thomson, E.M., Medelius, P., Rubinstein, M., Uman, M.A., Johnson, J., and Stone, J.W. 1988a. Horizontal electric fields from lightning return strokes. *J. Geophys. Res.* **93**: 2429–41.

Thomson, E.M., Medelius, P., and Uman, M.A. 1988b. A remote sensor for three components of transient electric fields. *IEEE Trans. Ind. Electr.* **35**: 426–33.

Thottappillil, R. 2002. Electromagnetic pulse environment of cloud-to-ground lightning for EMC studies. *IEEE Trans. Electromagn. Compat.* **44**: 203–13.

Thottappillil, R., and Rakov, V.A. 2001. On different approaches to calculating lightning electric fields. *J. Geophys. Res.* **106**: 14 191–205.

Thottappillil, R., Rakov, V.A., and Uman, M.A. 1990. K and M changes in close lightning ground flashes in Florida. *J. Geophys. Res.* **95**: 18 631–40.

Thottappillil, R., Rakov, V.A., Uman, M.A., Beasley, W.H., Master, M.J., and Shelukhin, D.V. 1992. Lightning subsequent-stroke electric field peak greater than the first stroke peak and multiple ground terminations. *J. Geophys. Res.* **97**: 7503–9.

Thottappillil, R., Goldberg, J.D., Rakov, V.A., Uman, M.A., Fisher, R.J., and Schnetzer, G.H. 1995. Properties of M components from currents measured at triggered lightning channel base. *J. Geophys. Res.* **100**: 25 711–20.

Thottappillil, R., Rakov, V.A., and Uman, M.A. 1997. Distribution of charge along the lightning channel: relation to remote electric and magnetic fields and to return-stroke models. *J. Geophys. Res.* **102**: 6987–7006.

Thottappillil, R., Uman, M.A., and Rakov, V.A. 1998. Treatment of retardation effects in calculating the radiated electromagnetic fields from the lightning discharge. *J. Geophys. Res.* **103**: 9003–13.

Thum, P.C., Liew, A.C., and Wong, C.M. 1982. Computer simulation of the initial stages of the lightning protection mechanism. *IEEE Trans. Pow. Appar. Syst.* **101**: 4370–7.

Tiller, J.A., Uman, M.A., Lin, Y.T., Brantley, R.D., and Krider, E.P. 1976. Electric field statistics for close lightning return strokes near Gainesville, Florida. *J. Geophys. Res.* **81**: 4430–4.

Trouvelot, E.T. 1888. Sur la forme des décharges électriques sur les plaques photographique. *Lumière Eléc.* **30**: 269–73.

Turman, B.N. 1977. Detection of lightning superbolts. *J. Geophys. Res.* **82**: 2566–8.

Turman, B.N. 1978. Analysis of lightning data from the DMSP satellite. *J. Geophys. Res.* **83**: 5019–24.

Turman, B.N., Cummins, T.B., and Deabenderfer, P.R. 1976. Measurements of optical power radiated by Florida lightning. Headquarters, US Air Force. Air Force Technical Applications Center, Report no. 76-6.

Tyahla, L.J., and Lopez, R.E. 1994. Effect of surface conductivity on the peak magnetic field radiated by first return strokes in cloud-to-ground lightning. *J. Geophys. Res.* **99**: 10 517–25.

Udo, T. 1993. Estimation of lightning current wave front duration by the lightning performance of Japanese EHV transmission line. *IEEE Trans. Pow. Del.* **8**: 660–71.

Udo, T. 1998. Contents of large current flashes among all the lightnings measured on transmission lines. *IEEE Trans. Pow Del.* **13**: 1432–6.

Uman, M.A. 1963. The continuum spectrum of lightning. *J. Atmos. Terr. Phys.* **25**: 287–95.

Uman, M.A. 1964a. The diameter of lightning. *J. Geophys. Res.* **69**: 583–5.

Uman, M.A. 1964b. The peak temperature of lightning. *J. Atmos. Terr. Phys.* **26**: 123–8.

Uman, M.A. 1966. Quantitative lightning spectroscopy. *IEEE Spectrum* **3**: 102–10.

Uman, M.A. 1969a. *Lightning*, 264 pp., New York: McGraw-Hill.

Uman, M.A. 1969b. On the determination of lightning temperature. *J. Geophys. Res.* **74**: 949–57.

Uman, M.A. 1978. Criticism of "Comment on 'Detection of lightning superbolts' by B.N. Turman" by R.D. Hill. *J. Geophys. Res.* **83**: 5523.

Uman, M.A. 1979. Discussion of "Determination of lightning currents using frame aerials" by S.C. Lee, K.K. Lim, M. Meiappa, and A.C. Liew. *IEEE Trans. Pow. Appar. Syst.* **98**: 1675.

Uman, M.A. 1984. *Lightning*, 298 pp., New York: Dover.

Uman, M.A. 1985. Lightning return stroke electric and magnetic fields. *J. Geophys. Res.* **90**: 6121–30.

Uman, M.A. 1987. *The Lightning Discharge*, 377 pp., Orlando, Florida: Academic Press.

Uman, M.A. 1988. Natural and artificially-initiated lightning and lightning test standards. *Proc. IEEE* **76**: 1548–65.

Uman, M.A. 1991. The best lightning photo I've ever seen. *Weatherwise* **44**(3): 8–9.

Uman, M.A. 2001. *The Lightning Discharge*, 377 pp., Mineola, New York: Dover.

Uman, M.A., and Krider, E.P. 1982. A review of natural lightning: experimental data and modeling. *IEEE Trans. Electromagn. Compat.* **24**: 79–112.

Uman, M.A., and Krider, E.P. 1984. The electromagnetic characteristics of lightning. *J. Defense Res.* **84-1**: 343–61.

Uman, M.A., and McLain, D.K. 1969. Magnetic field of the lightning return stroke. *J. Geophys. Res.* **74**: 6899–910.

Uman, M.A., and McLain, D.K. 1970a. Radiation field and current of the lightning stepped leader. *J. Geophys. Res.* **75**: 1058–66.

Uman, M.A., and McLain, D.K. 1970b. Lightning return stroke current from magnetic and radiation field measurements. *J. Geophys. Res.* **75**: 5143–7.

Uman, M.A., and Orville, R.E. 1964. Electron density measurement in lightning from Stark-broadening of H_α. *J. Geophys. Res.* **69**: 5151–4.

Uman, M.A., and Orville, R.E. 1965a. The optical continuum of lightning. *J. Geophys. Res.* **70**: 279–82.

Uman, M.A., and Orville, R.E. 1965b. The opacity of lightning. *J. Geophys. Res.* **70**: 5491–7.

Uman, M.A., and Voshall, R.E. 1968. The time-interval between lightning strokes and the initiation of dart leaders. *J. Geophys. Res.* **73**: 497–506.

Uman, M.A., McLain, D.K., Fisher, R.J., and Krider, E.P. 1973a. Electric field intensity of lightning return stroke. *J. Geophys. Res.* **78**: 3523–9.

Uman, M.A., McLain, D.K., Fisher, R.J., and Krider, E.P. 1973b. Currents in Florida lightning return strokes. *J. Geophys. Res.* **78**: 3530–7.

Uman, M.A., Brantley, R.D., Lin, Y.T., Tiller, J.A., Krider, E.P., and McLain, D.K. 1975. Correlated electric and magnetic fields from lightning return strokes. *J. Geophys. Res.* **80**: 373–6.

Uman, M.A., Swanberg, C.E., Tiller, J.A., Lin, Y.T., and Krider, E.P., 1976. Effects of 200 km propagation in Florida lightning return stroke electric fields. *Radio Sci.* **11**: 985–90.

Uman, M.A., Beasley, W.H., Tiller, J.A., Lin, Y.T., Krider, E.P., Weidman, C.D., Krehbiel, P.R., Brook, M., Few, A.A., Bohannon, J.L., Lennon, C.L., Poehler, H.A., Jafferis, W., Gulick, J.R., and Nicholson, J.R. 1978. An unusual lightning flash at the Kennedy Space Center. *Science* **102**: 9–16.

Uman, M.A., Master, M.J., and Krider, E.P. 1982. A comparison of lightning electromagnetic fields with the nuclear electromagnetic pulse in the frequency range 10^4 to 10^7 Hz. *IEEE Trans. Electromagn. Compat.* **24**: 410–16.

Uman, M.A., Rakov, V.A., Schnetzer, G.H., Rambo, K.J., Crawford, D.E., and Fisher, R.J. 2000. Time derivative of the electric field 10, 14, and 30 m from triggered lightning strokes. *J. Geophys. Res.* **105**: 15 577–95.

Ushio, T.-O., Kawasaki, Z.-I., Matsu-Ura, K., and Wang, D. 1998. Electric fields of initial breakdown in positive ground flash. *J. Geophys. Res.* **103**: 14 135–9.

Vance, E.F., and Uman, M.A. 1988. Differences between lightning and nuclear electromagnetic pulse interactions. *IEEE Trans. Electromagn. Compat.* **30**: 54–62.

Vassy, A. 1954. Comparaison des spectres d'étincelles de grande longueur dans l'air et du spectre de l'éclair. *Compt. Rend.* **238**: 1831–3.

Vecchi, G., Labate, D., and Canavero, F. 1994. Fractal approach to lightning radiation on a tortuous channel. *Radio Sci.* **29**: 691–704.

Vecchi, G., Zich, R.E., and Canavero, F.C. 1997. A study of the effect of channel branching on lightning radiation. In *Proc. 12th Int. Symp. on Electromagn. Compat., Zurich, Switzerland*, pp. 65–70.

Vereshchagin, I.P., Orlov, A.V., Temnikov, A.G., Antsupov, K.V., Koshelev, M.A., and Makaisky, L.M. 1996. Discharge development inside the charged aerosol formations. In *Proc. 23rd Int. Conf. on Lightning Protection, Florence, Italy*, pp. 42–6.

Villanueva, Y., Rakov, V.A., Uman, M.A., and Brook, M. 1994. Microsecond-scale electric field pulses in cloud lightning discharges. *J. Geophys. Res.* **99**: 14 353–60.

Vonnegut, B., and Passarelli, R.E. 1978. Modified cine sound camera for photographing thunderstorms and recording lightning. *J. Appl. Meteor.* **17**: 1079–81.

Vonnegut, B., Vaughan, O.H. Jr, and Brook, M. 1983. Photographs of lightning from the Space Shuttle. *Bull. Am. Meteor. Soc.* **64**: 150–1.

Vonnegut, B., Vaughan, O.H. Jr, and Brook, M. 1989. Nocturnal photographs taken from a U-2 airplane looking down on tops of clouds illuminated by lightning. *Bull. Am. Meteor. Soc.* **70**: 1263–71.

Wadhera, N.S., and Tantry, B.A.P. 1967a. VLF characteristics of K-changes in lightning discharges. *Ind. J. Pure Appl. Phys.* **5**: 447–9.

Wadhera, N.S., and Tantry, B.A.P. 1967b. Audio frequency spectra of K-changes in a lightning discharge. *J. Geomagn. Geoelectr.* **19**: 257–60.

Wagner, C.F. 1963. Relation between stroke current and velocity of the return stroke. *AIEE Trans. Power Appar. Syst.* **82**: 609–17.

Wagner, C.F. 1967. Lightning and transmission lines. *J. Franklin Inst.* **283**: 558–94.

Wagner, C.F., and Hileman, A.R. 1958. The lightning stroke (1). *AIEE Trans.* **77**(3): 229–42.

Wagner, C.F., and Hileman, A.R. 1961. The lightning stroke (2). *AIEE Trans.* **80**(3): 622–42.

Wait, J.R. 1956. Transient fields of a vertical dipole over a homogeneous curved ground. *Can. J. Phys.* **34**: 27–35.

Wait, J.R. 1997a. Concerning the horizontal electric field of lightning. *IEEE Trans. Electromagn. Compat.* **39**: 186.

Wait, J.R. 1997b. On the wave tilt at high frequencies – a personal view. *IEEE Trans. Electromagn. Compat.* **39**: 65.

Wait, J.R. 2000. Ground wave of an idealized lightning return stroke. *IEEE Trans. Antennas Propag.* **48**: 1349–53.

Wallace, L. 1960. Note on the spectrum of lightning in the region 3670 to 4280 Å. *J. Geophys. Res.* **65**: 1211–14.

Wallace, L. 1964. The spectrum of lightning. *Astrophys. J.* **139**: 944–98.

Walter, B. 1902. Ein photographischer Apparat zur genaueren Analyse des Blitzes. *Physik. Z.* **3**: 168–72.

Walter, B. 1903. Über die Entstehungsweise des Blitzes. *Ann. Physik* **10**: 393–407.

Walter, B., 1910. Über Doppelaufnahmen von Blitzen... *Jahrbuch Hamb. Wiss. Anst.* **27**(5): 81–118.

Walter, B. 1912. Stereoskopische Blitzaufnahmen. *Physik. Z.* **13**: 1082–4.

Walter, B. 1918. Über die Ermittelung der zeitlichen Aufeinanderfolge zusammengehöriger Blitze sowie über ein bemerkenswertes Beispiel dieser Art von Entladungen. *Physik. Z.* **19**: 273–9.

Walter, B. 1937. Von wo ab steuert der Blitz auf seine Einschlagstelle zu? *Z. für Techn. Physik* **18**: 105.

Wang, C.P. 1963. Lightning discharges in the tropics – 2. Component ground strokes and cloud dart streamer discharges. *J. Geophys. Res.* **68**: 1951–8.

Wang, D., Kawasaki, Z.I., Yamamoto, K., and Matsuura, K. 1995. Luminous propagation of lightning attachment to CN tower. *J. Geophys. Res.* **100**: 11 661–7.

Wang, D., Rakov, V.A., Uman, M.A., Takagi, N., Watanabe, T., Crawford, D.E., Rambo, K.J., Schnetzer, G.H., Fisher, R.J., and Kawasaki, Z.-I. 1999a. Attachment process in rocket-triggered lightning strokes. *J. Geophys. Res.* **104**: 2143–50.

Wang, D., Takagi, N., Watanabe, T., Rakov, V.A., and Uman, M.A. 1999b. Observed leader and return-stroke propagation characteristics in the bottom 400 m of the rocket triggered lightning channel. *J. Geophys. Res.* **104**: 14 369–76.

Wang, D., Takagi, N., Watanabe, T., Rakov, V.A., and Uman, M.A. 2000. Luminosity waves in branched channels of two negative lightning flashes. *J. Atmos. Electr.* **20**: 91–7.

Watt, A.D., and Maxwell, E.L. 1957. Characteristics of atmospherics noise from 1 to 100 kc. *Proc. Inst. Radio Eng.* **45**: 55–62.

Weber, L. 1889. Über Blitzphotographien. *Ber. Königliche Akad. (Berlin)*, 781–4.

Weidman, C.D. 1982. The submicrosecond structure of lightning radiation fields, Ph.D. dissertation, University of Arizona, Tucson.

Weidman, C.D. 1998. Lightning return stroke velocities near channel base. In *Proc. 1998 Int. Lightning Detection Conf.*, 25 pp., GAI, 2705 East Medina Road, Tucson, Arizona 85706-7155.

Weidman, C.D., and Krider, E.P. 1978. The fine structure of lightning return stroke wave forms. *J. Geophys. Res.* **83**: 6239–47. (Correction, 1982, **87**: 7351).

Weidman, C.D., and Krider, E.P. 1979. The radiation field wave forms produced by intracloud lightning discharge processes. *J. Geophys. Res.* **84**: 3159–64.

Weidman, C.D., and Krider, E.P. 1980a. Submicrosecond risetimes in lightning return-stroke fields. *Geophys. Res. Lett.* **7**: 955–8.

Weidman, C.D., and Krider, E.P. 1980b. Submicrosecond risetimes in lightning radiation fields. In *Lightning Technology*, pp. 29–38, NASA Conf. Publ. 2128, FAA-RD-80-30.

Weidman, C.D., and Krider, E.P. 1984. Variations à l'échelle submicroseconde des champs électromagnétiques rayonnés par la foudre. *Ann. Telecomm.* **39**: 165–74.

Weidman, C.D., and Krider, E.P. 1986. The amplitude spectra of lightning radiation fields in the interval from 1 to 20 MHz. *Radio Sci.* **21**: 964–70.

Weidman, C.D., Krider, E.P., and Uman, M.A. 1981. Lightning amplitude spectra in the interval 100 kHz to 20 MHz. *Geophys. Res. Lett.* **8**: 931–4.

Weidman, C., Hamelin, J., Leteinturier, C., and Nicot, L. 1986. Correlated current derivative (dI/dt) and electric field-derivative (dE/dt) emitted by triggered lightning. In *Proc. 11th Int. Aerospace and Ground Conf. on Lightning and Static Electricity, Dayton, Ohio*, 10 pp.

Weidman, C., Boye, A., and Crowell, L. 1989. Lightning spectra in the 850- to 1400-nm near-infrared region. *J. Geophys. Res.* **94**: 13 249–57.

Whitehead, E.R. 1977. Protection of transmission lines. In *Lightning, vol. 2., Lightning Protection*, ed. R.H. Golde, pp. 697–745, New York: Academic Press.

Wiese, W.L. 1965. Line broadening. In *Plasma Diagnostic Techniques*, eds., R.H. Huddlestone and S.L. Leonard, pp. 265–317, New York: Academic Press.

Wiesinger, J. 1995. Blitzmessung und Blitzsimulation. *Elektrotechnik und Informationstechnik* **112**: 273–9.

Willett, J.C., and Krider, E.P. 2000. Rise times of impulsive high-current processes in cloud-to-ground lightning. *IEEE Trans. Ant. Propag.* **48**: 1442–51.

Willett, J.C., Idone, V.P., Orville, R.E., Leteinturier, C., Eybert-Berard, A., Barret, L., and Krider, E.P. 1988. An experimental test of the "transmission-line model" of electromagnetic radiation from triggered lightning return strokes. *J. Geophys. Res.* **93**: 3867–78.

Willett, J.C., Bailey, J.C., Idone, V.P., Eybert-Berard, A., and Barret, L. 1989a. Submicrosecond intercomparison of radiation fields and currents in triggered lightning return strokes based on the transmission-line model. *J. Geophys. Res.* **94**: 13 275–86.

Willett, J.C., Bailey, J.C., and Krider, E.P. 1989b. A class of unusual lightning electric field waveforms with very strong HF radiation. *J. Geophys. Res.* **94**: 16 255–67.

Willett, J.C., Bailey, J.C., Leteinturier, C., and Krider, E.P. 1990. Lightning electromagnetic radiation field spectra in the interval from 0.2 To 20 MHz. *J. Geophys. Res.* **95**: 20 367–87.

Willett, J.C., Le Vine, D.M., and Idone, V.P. 1995. Lightning-channel morphology revealed by return-stroke radiation field waveforms. *J. Geophys. Res.* **100**: 2727–38.

Willett, J.C., Krider, E.P., and Leteinturier, C. 1998. Submicrosecond field variations during the onset of first return strokes in cloud-to-ground lightning. *J. Geophys. Res.* **103**: 9027–34.

Williams, D.P., and Brook, M. 1962. Fluxgate magnetometer measurements of continuous currents in lightning. *J. Geophys. Res.* **67**: 1662.

Williams, D.P., and Brook, M. 1963. Magnetic measurement of thunderstorm currents, 1: Continuing currents in lightning. *J. Geophys. Res.* **68**: 3243–7.

Williams, J.C. 1959. Thunderstorms and VLF radio noise. Ph.D. dissertation, Harvard University, Cambridge, Massachusetts.

Wilson, C.T.R. 1916. On some determinations of the sign and magnitude of electric discharges in lightning flashes. *Proc. Roy. Soc. A***92**: 555–74.

Wilson, C.T.R. 1920. Investigations on lightning discharges and on the electric field of thunderstorms. *Phil. Trans. Roy. Soc. A***221**: 73–115.

Winn, W.P. 1965. A laboratory analog to the dart leader and return stroke of lightning. *J. Geophys. Res.* **70**: 3265–70.

Wolfe, W.L. 1983. Aircraft-borne lightning sensor. *Optical Eng.* **22**: 456–9.

Workman, E.J., Beams, J.W., and Snoddy, L.B. 1936. Photographic study of lightning. *Physics* **7**: 345–79.

Workman, E.J., Brook, M., and Kitagawa, N. 1960. Lightning and charge storage. *J. Geophys. Res.* **65**: 1513–17.

Wu, D., and Ruan, C. 1999. Transient radiation of traveling-wave antennas. *IEEE Trans. Electromagn. Compat.* **41**: 120–3.

Wurden, G.A., and Whiteson, D.O. 1996. High-speed plasma imaging: a lightning bolt. *IEEE Trans. Plasma Sci.* **24**: 83–4.

Yokoyama, S., Miyake, K., Suzuki, T., and Kanao, S. 1990. Winter lightning on Japan Sea coast – development of measuring system on progressing feature of lightning discharge. *IEEE Trans. Pow. Del.* **5**: 1418–25.

Yokoyama, S., Miyake, K., and Gouzu, H. 1992. Development of measuring system on progressing feature of lightning discharge. *Res. Lett. Atmos. Electr.* **12**: 17–21.

Zaepfel, K.P., and Carney, H.K., 1988. 1985 and 1986 direct strike lightning data, NASA Technical Memorandum 100533, parts 1 and 2, NASA, Langley Research Center, Hampton, Virginia.

Zaepfel, K.P., Fisher, B.D., and Ott, M.S. 1985. Direct strike lightning photographs, swept-flash attachment patterns, and flight conditions for storm hazards '82. NASA Technical Memorandum 86347, Langley Research Center, Hampton, Virginia.

Zeddam, A., and Degauque, P. 1987. Current and voltage induced on telecommunication cables by a lightning stroke. *Electromagnetics* **7**: 541–64.

Zeddam, A., Leteinturier, C., and Degauque, P. 1990. Influence of the ground conductivity on the electric field derivative measurements. In *Proc. 20th Int. Conf. on Lightning Protection, Interlaken, Switzerland*, paper 6.2, 4 pp.

Zenneck, J. 1915. *Wireless Telegraphy*, New York: McGraw-Hill. (English transl., A.E. Seelig.)

Zich, R.E., and Vecchi, G. 2001. Lightning discharge on a branched channel. In *Proc. 14th Int. Symp. on Electromagnetic Compatibility, Zurich, Switzerland*, pp. 299–304.

Zonge, K.L., and Evans, W.H. 1966. Prestroke radiation from thunderclouds. *J. Geophys. Res.* **71**: 1519–23.

5 Positive and bipolar lightning discharges to ground

> On account of the heavy charges involved, the positive strokes are of primary importance in relation to protection against lightning.
>
> K. Berger and E. Vogelsanger (1969)

5.1. Introduction

Positive flashes are defined as those transporting positive charge from cloud to Earth. It is thought that less than 10 percent of global cloud-to-ground lightning is positive. The first well-documented study of positive lightning was due to K. Berger and co-workers, who presented the detailed statistical characteristics of 26 positive lightning discharges observed on Monte San Salvatore in Switzerland (Berger et al. 1975). It is interesting to follow the evolution of the views of Berger and co-workers regarding these discharges. Initially, Berger and Vogelsanger (1969) and Berger et al. (1975) considered the discharges as downward positive flashes that involved long, upward-connecting, negatively charged leaders. Later, Berger (1977) expressed his uncertainty in identifying those discharges as either downward or upward positive flashes. Finally, Berger and Garbagnati (1984) assigned all 67 positive flashes observed on Monte San Salvatore to the upward discharge category, perhaps as an indication that such flashes were not expected to occur in the case of objects of moderate height (less than 100 m or so) located on flat terrain.

Interestingly, Berger and Garbagnati reported on five upward positive and seven downward positive discharges from tower measurements in Italy, the lightning type (upward or downward) being identified solely on the basis of the recorded current waveforms. Overall current waveforms characteristic of downward negative and upward negative lightning discharges, which apparently also illustrate the difference between downward positive and upward positive discharges, are shown in Figs. 4.2c and 6.1c, respectively. The characteristics of the 26 positive discharges reported by Berger et al. (1975) are summarized in Table 5.1. This table characterizes positive flashes with impulsive current components having magnitudes in the return-stroke range. Characteristics of positive flashes both containing and not containing such large impulsive components, as reported by Berger (1978), are given in Table 6.5 (in the chapter concerned with upward lightning discharges). Previously, Hagenguth and Anderson (1952) had reported on five positive current pulses observed over a 10-year period in the Empire State Building studies. These occurred in the bipolar discharges that are discussed in Section 5.4. One of the positive current pulses at the Empire State Building had a peak of 58 kA, the largest ever measured at the building, with an apparently "normal" return-stroke risetime. Another had a peak of 12 kA and an unusually long risetime, in excess of 80 µs. Gorin and Shkilev (1984) reported that 9 of 90 flashes (10 percent) initiated from the Ostankino tower transported positive charge to the ground, and Podgorski and Landt (1987) reported that only one of 15 CN-tower flashes recorded in 1985 was positive. No positive discharges were observed in the 1972–82 lightning studies in South Africa, although for 27 percent of 44 flashes the polarity could not be identified (Eriksson 1982). Three upward positive discharges (without leader–return-stroke sequences) were reported from the Peissenberg tower studies in Germany (Fuchs et al. 1998). Only one positive flash (versus 22 negative flashes; Lacerda et al. 1999) was reported from the 1985–94 tower current measurements in Brazil (M. Lacerda, personal communication, 1999).

Positive discharges have attracted considerable attention for the following four reasons.

(i) The highest directly measured lightning currents (near 300 kA) and the largest charge transfers to ground (hundreds of coulombs or more) are thought to be associated with positive lightning; examples of current waveforms produced by such extreme positive lightning events are shown in Fig. 5.1. As a result, positive lightning often causes more severe damage to various objects and systems than negative lightning (e.g., Nakahori et al. 1982; Idone et al. 1984; Le Boulch and Plantier 1990) and is apparently more likely to start forest fires (subsection 18.4.2) (e.g., Fuquay et al. 1967). Current-viewing resistors (shunts) designed to measure lightning current have been destroyed by positive discharges in lightning studies on Monte San Salvatore (Berger 1977) and in triggered-lightning experiments (Section 7.2) at the

5.1. Introduction

Table 5.1. *Lightning current parameters for positive flashes. Adapted from Berger et al. (1975)*

Parameters	Units	Sample size	Percent exceeding tabulated value		
			95%	50%	5%
Peak current (minimum 2 kA)	kA	26	4.6	35	250
Charge (total charge)	C	26	20	80	350
Impulse charge (excluding continuing current)	C	25	2.0	16	150
Front duration (2 kA to peak)	µs	19	3.5	22	200
Maximum dI/dt	kA µs^{-1}	21	0.20	2.4	32
Stroke duration (2 kA to half-peak value on the tail)	µs	16	25	230	2000
Action integral ($\int I^2 dt$)	A^{2}s	26	2.5×10^4	6.5×10^5	1.5×10^7
Flash duration	ms	24	14	85	500

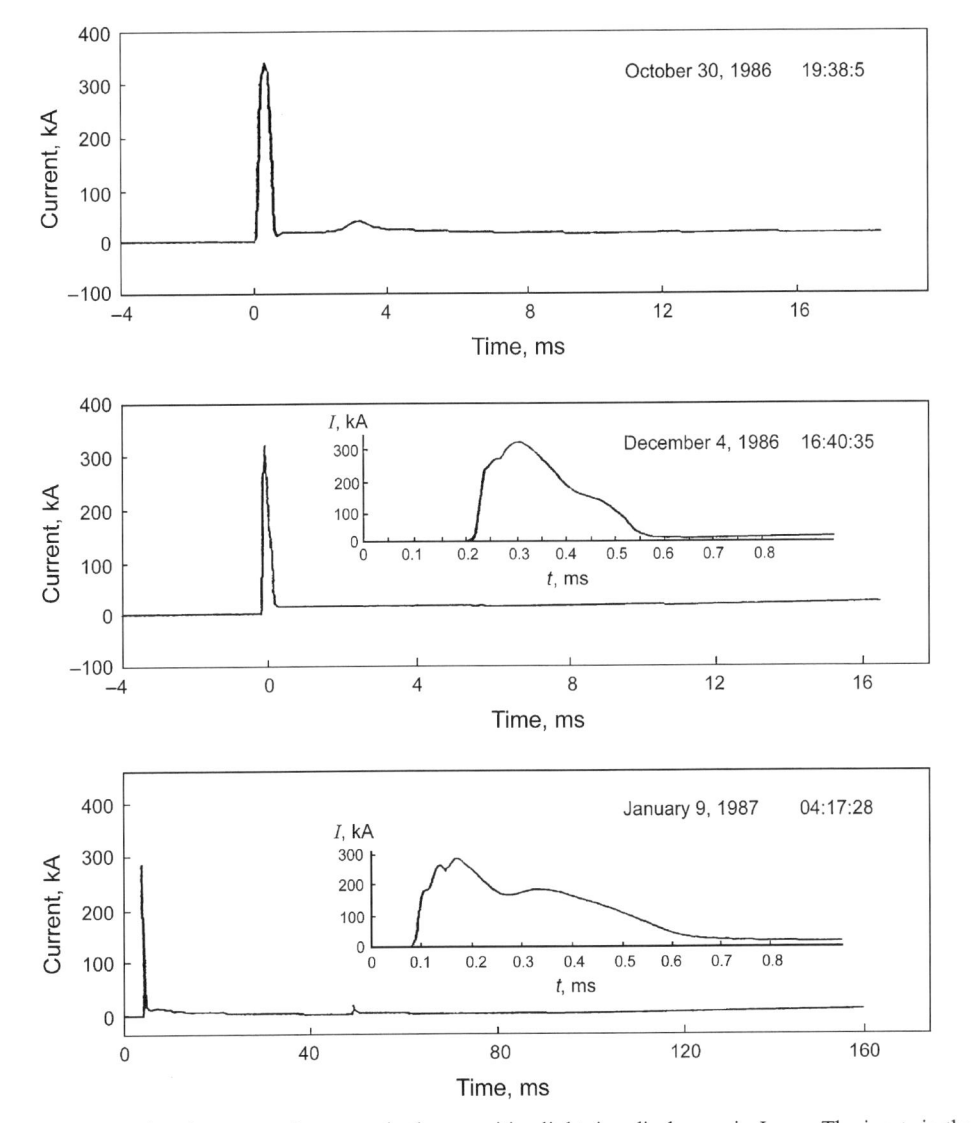

Fig. 5.1. Directly measured currents in three positive lightning discharges in Japan. The insets in the middle and bottom diagrams show the current on an expanded scale. Note the very large peaks, from top to bottom 340, 320, and 280 kA, of the initial pulses, which are followed by continuing currents. The transferred charges are 330, 180, and 400 C, respectively. Adapted from Goto and Narita (1995).

Table 5.2. *Monthly and annual percentages of positive flashes over the contiguous United States for the years 1992 through 1997. The percentages are based on data from the NLDN. See also Fig. 5.2. Adapted from Orville and Silver (1997) and Orville and Huffines (1999)*

	Percentage of positive flashes					
Month	1992	1993	1994	1995	1996	1997
January	13.4	10.3	15.6	20.2	24.5	19.8
February	11.9	10.3	12.6	14.3	19.5	22.7
March	9.1	8.7	11.3	14.9	15.8	19.2
April	6.6	9.4	7.8	17.2	15.9	15.4
May	4.6	6.0	6.3	14.7	14.3	12.6
June	4.1	5.3	4.0	8.2	8.6	11.1
July	3.5	3.7	4.0	6.5	8.0	8.1
August	3.0	3.2	3.7	6.6	7.3	6.9
September	3.9	3.6	4.6	7.0	10.6	7.0
October	5.4	5.9	8.7	12.1	15.1	14.1
November	9.4	12.6	15.2	18.4	17.6	15.9
December	21.5	24.9	16.5	19.3	17.5	19.3
Annual average	4.2	4.6	4.9	9.3	10.2	10.1

Kennedy Space Center, Florida (Eybert-Berard *et al.* 1988) and at Camp Blanding, Florida (Fernandez 1997).

(ii) Positive lightning can be the dominant type of cloud-to-ground lightning during the cold season (subsection 2.8.1, Table 5.2, and Fig. 5.2), during the dissipating stage of any thunderstorm, and in other situations discussed in Section 5.2.

(iii) Positive lightning has been found recently to be related preferentially to luminous phenomena in the middle atmosphere known as sprites (e.g., Boccippio *et al.* 1995; Lyons *et al.* 1998b), as discussed in subsection 14.3.3.

(iv) Reliable identification of positive discharges by lightning locating systems such as the US National Lightning Detection Network (NLDN) (Section 17.5) has important implications for various meteorological and other studies, considered in Sections 2.2, 2.5, and 2.8 (see also Fig. 5.3), that depend on data from these systems (e.g., Petersen and Rutledge 1992; Seimon 1993; Stolzenburg 1994).

Five situations that appear to be conducive to the occurrence of positive lightning are reviewed in Section 5.2. A characterization of positive lightning is given in Section 5.3,

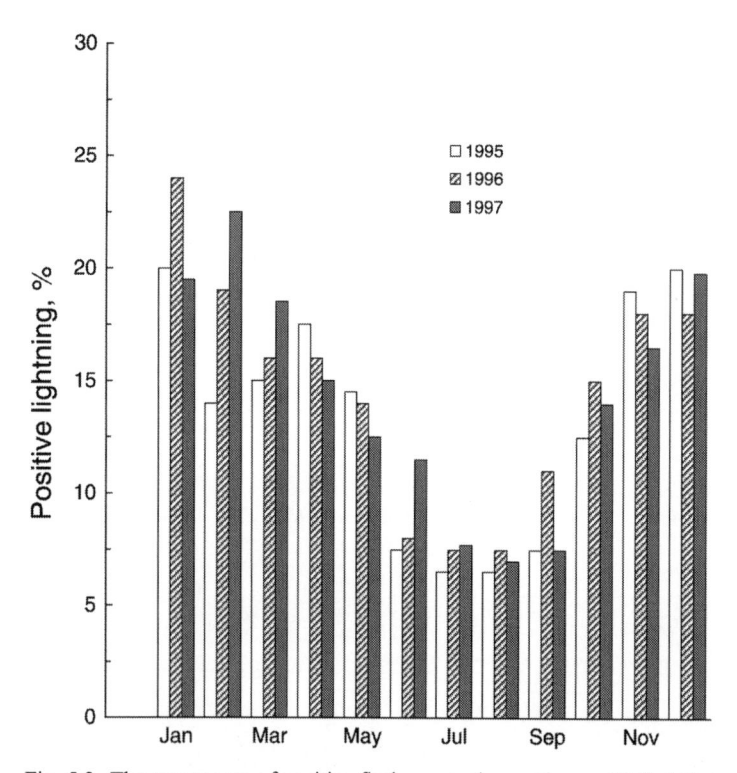

Fig. 5.2. The percentage of positive flashes over the contiguous United States as a function of month for each of the years 1995 through 1997. The minimum occurs in late summer and the maximum is in January–February. See also Table. 5.2. Adapted from Orville and Huffines (1999).

including a comparison of positive and negative leaders, regardless of the polarity of charge transferred to ground by the overall lightning discharge, and a description of the mechanism of the positive leader inferred from laboratory long-spark experiments. More information on positive lightning in winter storms in Japan is found in Chapter 8. Positive lightning is also considered in subsections 2.2.2, 2.2.3, 2.2.5, 2.2.7, 2.8.1, 2.8.4, 6.2.2, 13.4.3, 14.3.3, and 18.4.2. Lightning discharges that transfer to ground both positive and negative charge, that is, bipolar discharges, are discussed in Section 5.4 and subsection 6.2.3.

5.2. Conditions conducive to the occurrence of positive lightning

As noted earlier, positive lightning discharges are thought to account for 10 percent or less of global cloud-to-ground lightning activity. Berger and Garbagnati (1984), from direct measurements of lightning currents on towers in Switzerland and Italy, reported that only 5 to 10 percent of flashes lowered positive charge to ground. Fuquay (1982), from electric field measurements made in conjunction with optical observations in Montana, found that about 3 percent of cloud-to-ground lightning was positive. Orville and Silver (1997), using NLDN data for 1992–5, found that the annual percentage of positive flashes over the contiguous United States ranged from 4.2 to 4.9 percent in 1992–4 and was 9.3 percent in 1995 (Table 5.2). It is likely that the factor 2 or so increase in 1995 was caused by the false identification of some cloud flashes as positive cloud-to-ground flashes, as a result of modifications to the NLDN made in 1994–5 (Section 17.5). Mackerras and Darveniza (1994) reported on the percentage of positive lightning discharges for 14 sites in 11 countries from measurements made with the CGR3 counter (subsection 2.5.1), an instrument that is claimed to be capable of identifying negative and positive cloud-to-ground lightning discharges on the basis of their electric field signatures. The percentage varied from two to 28 percent (but note that 12 out of the 14 values were <15 percent), all four tropical sites being characterized by low values. No clear trend toward either high or low values could be discerned for the non-tropical sites.

Spatial distributions of the larger positive flashes detected by the NLDN, those with inferred first-stroke peak currents ≥ 75 kA and ≥ 200 kA, were studied for 14 selected summer months from 1991–5 by Lyons et al. (1998b). They found that positive flashes with peak currents ≥ 75 kA were "tightly concentrated in the High-Plains–upper-Midwest corridor in which the majority of optical observations of sprites and elves has been obtained". However, positive flashes with peak currents ≥ 200 kA (a total of 2121) did not exhibit such a regular spatial pattern. It is important to note that peak current estimates reported by the NLDN for either positive or negative lightning are based on an empirical formula relating measured fields to currents that has only been validated for negative subsequent strokes (triggered-lightning strokes) with peak currents not exceeding 60 kA (Orville 1999).

Although the overall percentage of positive lightning discharges is relatively low, there are five situations, listed below, that appear to be conducive to the more frequent occurrence of such discharges.

(i) *The dissipating stage of an individual thunderstorm.* The tendency for positive lightning to occur toward the end of a thunderstorm was reported, for example, by Fuquay (1982) and by Orville et al. (1983). Pierce (1955b) suggested that positive flashes are initiated from the upper (main) positive charge region of thunderclouds (subsection 3.2.1) after much of the main negative charge has been removed by negative ground flashes. However, Krehbiel (1981) reported on three positive flashes in Florida which apparently involved, or were a byproduct of, long (longer than 40 km) horizontal lightning discharges that effectively removed positive charge from a layer near the 0° C isotherm (that is, from a region considerably lower than the main positive charge region) where frozen precipitation was melting.

(ii) *Winter thunderstorms.* The production of large positive lightning discharges by winter storms in Japan was reported by Takeuti et al. (1978) and by Brook et al. (1982) from multiple-station electric field measurements in conjunction with optical observations. Brook et al. (1982) observed that positive flashes constituted about 40 percent of the total number of cloud-to-ground flashes and suggested that positive flashes originated from the upper positive charge that was displaced horizontally by vertical wind shear from the lower negative charge and thereby exposed to the ground, as illustrated in Fig. 3.7. Later studies, however, indicated that some winter thunderclouds in Japan may contain predominantly positive charge during most of their life cycle (Section 8.3). About one-third of winter lightning currents recorded by Miyake et al. (1992) on two towers in Japan with heights 88 m and 200 m were of positive polarity.

Orville et al. (1987), using one year's data from the US East Coast lightning locating network (Sections 17.3 and 17.5) found that positive lightning accounted for about 80 percent of all ground discharges in the northeastern United States in February but for less than 5 percent during the summer. From the NLDN data for 1992–5, Orville and Silver (1997) reported that the monthly percentage of positive flashes over the contiguous United States ranged from 3 percent (August 1992) to about 25 percent (December 1993). Orville and Huffines (1999), using NLDN data for the contiguous United States for 1995–7, reported on the percentage of positive flashes as a function of month; their results are reproduced in Fig. 5.2. The smallest percentage (6.5 percent) was observed in July 1995, and the largest (about 25 percent) in January 1996. The percentage of positive flashes as a function of month from the studies of both Orville and Silver (1997) and Orville and Huffines (1999) is given in Table 5.2. Figure 5.2 and Table 5.2 show a clear tendency for the largest percentage of positive flashes to occur during the cold season. As noted above, some of the increase in the percentage of positive flashes in 1995–7 seen in Table 5.2 is likely to be related to the modification of the NLDN in 1994–5. Additional information on the percentage of positive flashes in winter storms is found in subsection 2.8.1.

Moore and Orville (1990) observed that storms induced by the Great Lakes in fall and winter tend to produce positive lightning, although usually only a few flashes were observed per storm. Holle and Watson (1996), studying NLDN-detected lightning during two central US winter precipitation events, found the percentage of positive flashes to be 59 percent for the first event (a total of 27 flashes) and 29 percent for the second one (a total of 2417 flashes). The latter event lasted for 16 hours, and the percentage of positive flashes during the first four hours was 52 percent.

In the coastal area of the Sea of Japan, the percentage of positive flashes, as determined using a direction finding (DF) lightning locating system, reached a maximum of 60 percent in December (Hojo *et al.* 1989). In France, also from DF lightning locating system data, a maximum of 44 percent was observed in February (Le Boulch and Plantier 1990).

(iii) *Shallow clouds such as the trailing stratiform regions of MCSs.* The production of predominantly positive flashes by shallow clouds in mesoscale convective systems has been observed in both winter and summer seasons (e.g., Engholm *et al.* 1990; see also Fig. 2.7). This observation might be due largely to the tendency for the occurrence of negative flashes to decrease dramatically with decreasing cloud depth. As discussed in subsection 2.2.7, some thunderstorm systems produce positive and negative flashes whose ground strike locations tend to be separated in space by polarity, thereby forming the "bipolar" pattern illustrated in Fig. 2.5. It has been found that most positive flashes are associated with the shallow-convection region of the system, while negative flashes tend to occur in the deepest-convection region. Perhaps winter thunderstorms can be included in the shallow convection category since their depth is usually small. Interestingly, the inner-band region of hurricanes (subsection 2.2.3) has been found to have characteristics similar to those of the trailing stratiform region of MCSs, including a relatively large fraction of positive flashes (Molinari *et al.* 1999).

(iv) *Severe storms.* The occurrence of infrequent, widely scattered positive flashes in the mature and later stages of severe springtime storms over the Great Plains in the United States was first observed by Rust *et al.* (1981), who used electric field measurements in conjunction with optical observations. More recently, some severe storms have been observed (e.g., Seimon 1993; MacGorman and Burgess 1994; Perez *et al.* 1997) in which positive flashes detected by lightning locating systems such as the NLDN occurred relatively frequently and outnumbered negative ground flashes for more than 30 minutes. The period of a storm's lifetime in which positive flashes dominated varied, although often it occurred during the earlier stages of the storm, as illustrated in Fig. 5.3. The cause of this behavior and its relationship to severe weather production is unclear.

(v) *Thunderclouds formed over forest fires or contaminated by smoke.* Vonnegut and Orville (1988), using data from the BLM DF lightning locating system, found that 25 percent of about 50 cloud-to-ground flashes apparently associated with the forest fires in Yellowstone National Park lowered positive charge to Earth. Latham (1991), studying lightning discharges from a cloud generated by a prescribed forest fire, reported that, in a sample of 28 lightning events, 19 (two-thirds) were identified by the Ontario Province (Canada) DF lightning-locating system as positive flashes. A relatively high percentage of positive flashes in the central United States detected by the NLDN in Spring 1998 (see Table 5.3 and Fig. 5.4) has been associated with cloud contamination by smoke from massive forest fires in Mexico that were up to thousands of kilometers away (Lyons *et al.* 1998a; Murray *et al.* 2000).

5.3. Characterization of positive lightning

In subsection 5.3.1 we present a general discussion of the various properties of positive lightning and compare them with the properties of negative lightning. A comparison of the properties of positive and negative leaders, regardless of the polarity of charge transferred to ground by the overall lightning discharge, is given in subsection 5.3.2, and a mechanism for the positive leader is considered in subsection 5.3.3. A discussion of microsecond-scale electric and magnetic field waveforms is found in subsection 5.3.4 and of peak currents in subsection 5.3.5. Subsection 5.3.6 contains information on measurements of the speed of positive return strokes.

5.3.1. General information

A comparison of the median (50 percent) values of several parameters reported from direct current measurements by Berger *et al.* (1975) for positive and negative discharges is given in Fig. 5.5. These parameters include peak current, charge transfer, maximum rate of rise of current, stroke duration, and flash duration. As seen in Fig. 5.5, the median charge transfer by positive flashes

5.3. Characterization of positive lightning

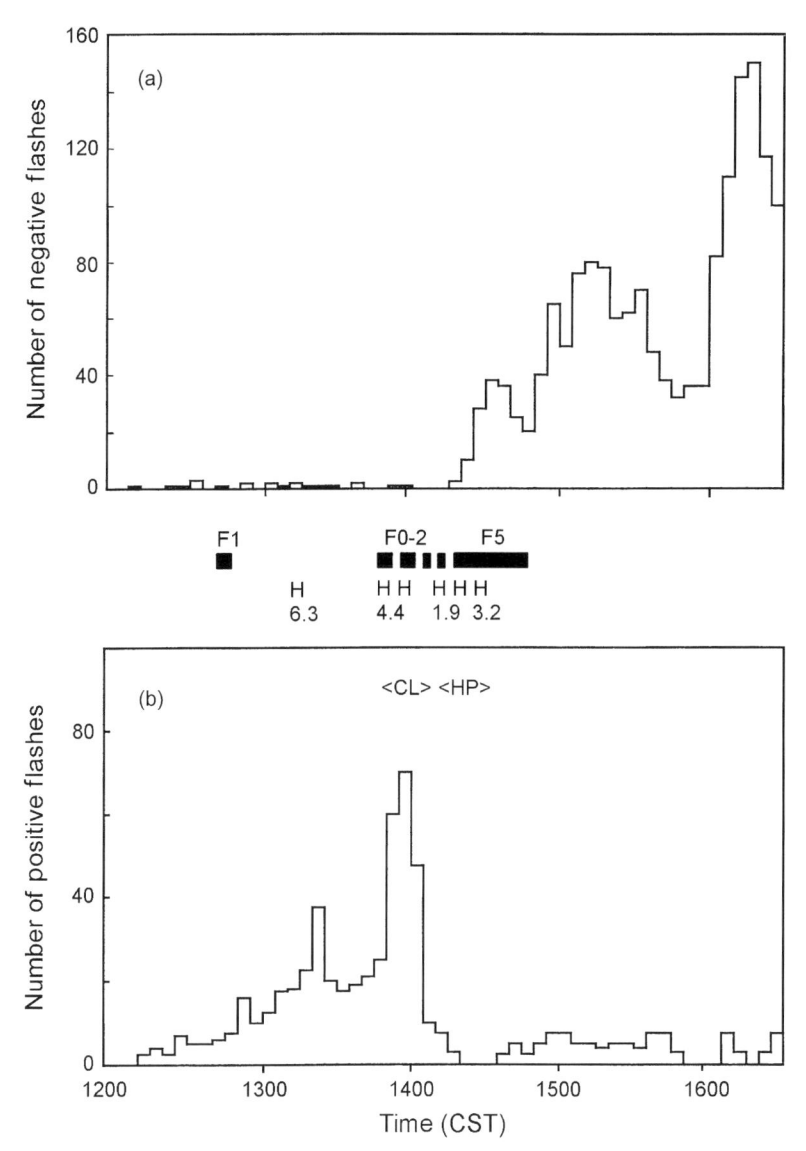

Fig. 5.3. Five-minute counts of (a) negative and (b) positive cloud-to-ground lightning flashes over a 4.5-hour period of the storm that produced the Plainfield, Illinois, F5 tornado on 28 August 1990. The bars between the two plots indicate the approximate time of the tornadoes, labeled with the F-scale rating of damage given by Storm Data. The F5 tornado began between 1415 and 1420 CST and traveled approximately 26 km over a period of 25–30 minutes. The left-hand edges of the H's mark the time of large hail reports; an H that overlaps a symbol indicating an earlier report is omitted. The hail reports are labeled with the hail diameter in centimeters. When the diameters of two or more reports overlap, the diameter of only the larger or largest report is given. The bracketed periods labeled CL and HP in (b) indicate the times at which the storm was classified as a classic supercell or a heavy-precipitation supercell, respectively. Note that positive discharges are clearly dominant at the beginning of the storm. Adapted from MacGorman and Burgess (1994).

is about an order of magnitude greater than that by negative flashes. The difference between the median peak currents for positive discharges and negative first strokes is small (35 kA as against 30 kA); however, the five percent peak current for positive discharges is 250 kA (Table 5.1), more than a factor of three greater than the 80 kA five percent peak current for negative first strokes (Table 4.4). All current waveforms observed by Berger et al. (1975) for positive lightning can be divided into two categories. The first category includes microsecond-scale waveforms similar to those for negative lightning, and the second category includes millisecond-scale waveforms with risetimes up to hundreds of microseconds. Examples of the two types of current waveform along with illustrations of the processes presumably leading to the formation of these waveforms are given in Figs. 5.6a, b. While microsecond-scale

Table 5.3. *NLDN-reported lightning data for different periods of time, the storms in Spring 1998 apparently being influenced by smoke from forest fires in Mexico. SP, southern plains (25°–40° N, 94°–107° W); CG, cloud-to-ground lightning discharges; +CG, positive discharges, as identified by the NLDN (Section 17.5). Adapted from Lyons et al. (1998a)*

	1996–7 Entire NLDN 8 April–7 June	1998 Entire NLDN 8 April–7 June	1998 Entire NLDN 14–18 May	1996–7 SP region 8 April–7 June	1998 SP region 8 April–7 June
No. CG flashes	4 809 921	4 401 008	190 990	1 401 229	424 343
No. days ≥ 25% +CG	0.5	14	5	1	15
% +CGs	13.7	20.1	39.5	12.9	37.3
% +CGs ≤ 10 kA	21.3	24.7	5.6	15.1	5.1
% +CGs ≥ 76 kA	4.8	6.9	11.0	4.0	13.1

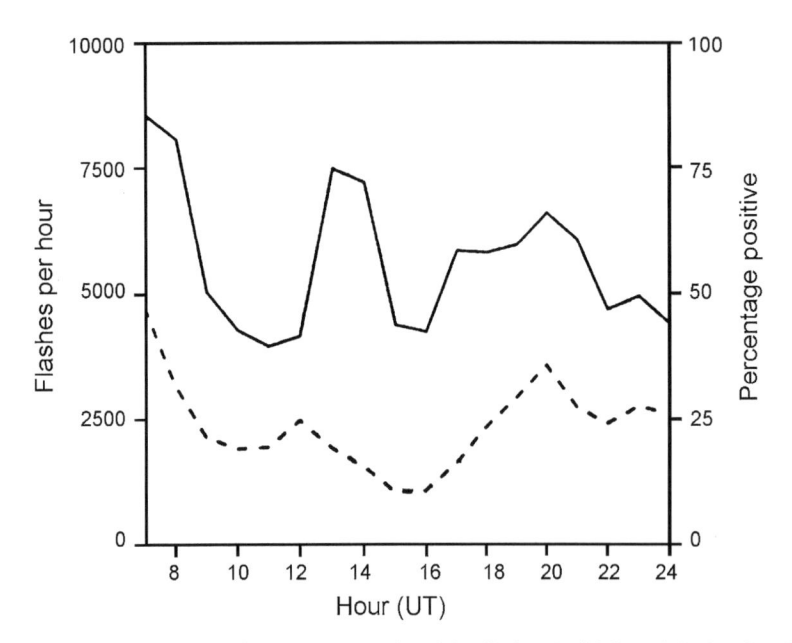

Fig. 5.4. Percentage of NLDN-reported positive flashes (solid line; right-hand scale) as a function of time for storms that occurred over the United States from 0700 to 2400 UT 15 May 1998 and that were apparently affected by smoke from forest fires in Mexico. Also shown is the total cloud-to-ground flash rate (broken line; left-hand scale) as a function of time. Adapted from Lyons et al. (1998a).

waveforms are probably formed in a manner similar to that in downward negative lightning, millisecond-scale waveforms are likely to be a result of the M-component mode of charge transfer to ground (Section 4.9), as discussed further below.

Berger and Vogelsanger (1969) could record channels only within 1 km of the tops of their instrumented towers. However, they inferred the lengths of upward negative connecting leaders to be up to 2 km or so using (i) the duration of the upward connecting leader current (a few hundred amperes) prior to the large current pulse (tens of kiloamperes or more) and (ii) the leader speed measured within the bottom 1 km of the channel, assuming that this speed remained the same for higher channel sections. For downward negative lightning they observed upward connecting leaders of 20 to 70 m in length. Berger and Vogelsanger suggested that very long upward connecting leaders in positive lightning are responsible for the slower current wavefronts measured at the towers. Indeed, if a downward current wave originates at a height of 1 to 2 km as a result of connection of

5.3. Characterization of positive lightning

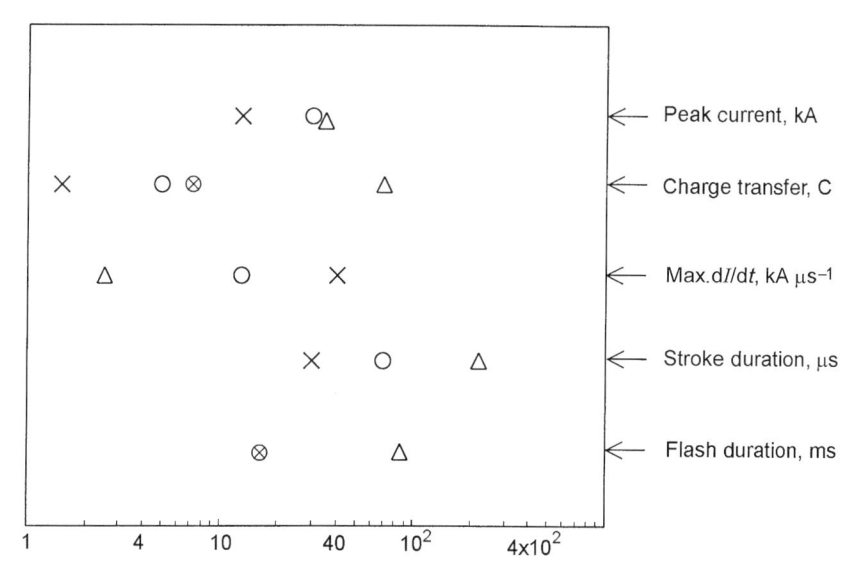

Fig. 5.5. Comparison of the median (50 percent) values of some parameters of positive and negative lightning reported by Berger *et al.* (1975): ○, negative first strokes; ×, negative subsequent strokes; ⊗, negative single-stroke flashes; △, positive flashes. Adapted from Ogawa (1995).

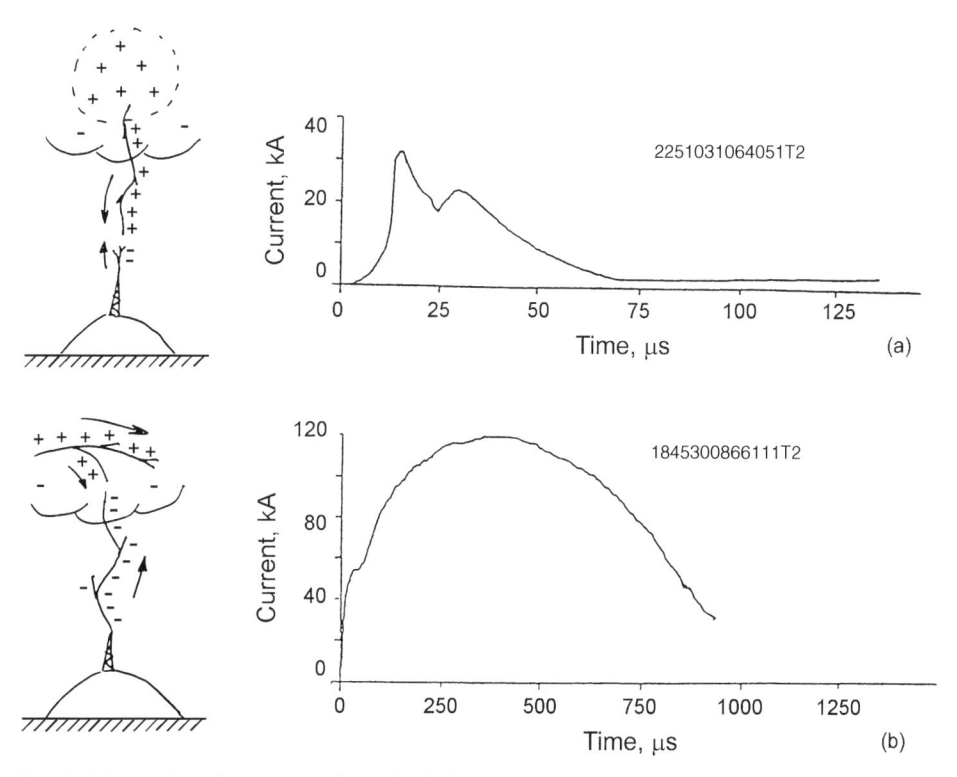

Fig. 5.6. Examples of two types of positive lightning current waveforms observed by Berger: (a) a microsecond-scale waveform (right-hand panel) and a diagram (left-hand panel) illustrating the type of lightning that might have led to its production; (b) a millisecond-scale waveform (right-hand panel) and a sketch (left-hand panel) illustrating the type of lightning that might have led to its production.

the upward connecting leader to a charged in-cloud channel (see the diagram on the left in Fig. 5.6b), the charge transfer to ground associated with this wave is likely to be a process of M-component type, that is, the process characterized by a relatively slow current front at ground (subsection 4.9.3). Berger and Vogelsanger (1969) noted a tendency for slower current wavefronts to be associated with longer upward connecting leaders. In one case, they observed what they called a downward positive discharge that attached to the tower apparently without any upward connecting leader. The corresponding current waveform exhibited a risetime of approximately 10 µs, similar to that for first strokes in downward negative lightning. In the absence of tall man-made structures, most positive lightning is likely to be initiated by a downward positively charged leader (Berger and Vogelsanger 1966; Takeuti et al. 1978; Rust et al. 1981; Fuquay 1982; Brook et al. 1982), rather than by an upward negatively-charged leader. As of this writing, there exists apparently only one documented case, reported by Takeuti et al. (1976), of a positive discharge inferred (from upward branching) to be initiated by an upward negatively charged leader from a relatively small object, one less than 15 m in height.

The triggering of a positive return stroke using the classical rocket-and-wire technique (subsection 7.2.1) has so far been unsuccessful, probably owing to the tendency for positive lightning to have a single component (stroke) per channel instead of multiple components, that is, to exhibit an uninterrupted charge transfer to ground. This tendency, in the case of rocket-triggered lightning, translates to a positive discharge composed of the initial stage only; that is, it contains no following return strokes (e.g., Gary et al. 1975). Sometimes an initial positive continuous current in rocket-triggered lightning is followed by leader–return-stroke sequences transferring negative charge to ground. As noted in subsection 7.2.2, Wang et al. (1999) reported on a positive flash initiated using the altitude-triggering technique (subsection 7.2.1) from a summer thunderstorm in China. This is the only documented triggering of a positive lightning using the altitude-triggering technique. For this flash, the length of the grounded intercepting wire was 35 m and the length of the insulating cable was 86 m. The flash was apparently initiated when the rocket was at an altitude of 550 m.

The following is a list of five observed properties that are thought to be characteristic of positive lightning discharges.

(i) Positive flashes are usually composed of a single stroke, whereas about 80 percent of negative flashes contain two or more strokes, as discussed in Section 4.2. Multiple-stroke positive flashes do occur, but they are relatively rare. Heidler et al. (1998), from electric field measurements in 1995–7 in Germany, found that, of 36 positive flashes, 32 contained one stroke and four contained two strokes. The mean interstroke interval in these four cases was relatively long, 102 ms. Further, Lyons et al. (1998b), using NLDN data for 14 selected summer months from 1991–5, reported that 1002 (about 0.04 percent) of 2.7 million positive flashes were composed of more than 10 strokes. However, it is likely that at least some of these multiple-stroke events were actually misidentified cloud discharges. According to Pierce (1955b), the occurrence of a subsequent negative stroke after the first positive stroke is more likely than the occurrence of a subsequent positive stroke. Such bipolar lightning discharges are discussed in Section 5.4 below. Ishii et al. (1998) observed that subsequent strokes in multiple-stroke positive flashes in winter storms in Japan always create a new termination on ground.

(ii) Positive return strokes tend to be followed by continuing currents that typically last for tens to hundreds of milliseconds (e.g., Rust et al. 1981, 1985; Fuquay 1982; Beasley et al. 1983; Beasley 1985). Brook et al. (1982), from multiple-station electric field measurements, inferred continuing currents in positive flashes in excess of 10 kA, an order of magnitude or more larger than for negative flashes, for periods up to 10 ms. In Fig. 5.1, directly measured positive continuing currents in the kiloamperes to tens of kiloamperes range in winter lightning in Japan are seen following the initial current pulses. Such large continuing currents are probably responsible for the unusually large charge transfers by positive flashes. Brook et al. (1982), for one positive lightning flash in a winter storm in Japan, inferred a charge transfer in excess of 300 C during the first 4 ms. Charge transfers during the first 2 ms estimated by Berger (1967) for summer positive lightning in Switzerland were of the order of tens of coulombs. Charge transfers of order 1000 C were reported from direct current measurements by Miyake et al. (1992) for both positive and negative winter lightning in Japan. However, these events may well be unusual forms of lightning discharges, since the top of the grounded strike object was very close to or inside the cloud, as discussed in Section 8.4.

(iii) From electric field records, positive return strokes often appear to be preceded by significant in-cloud discharge activity lasting, on average, in excess of 100 ms (Fuquay 1982) or 200 ms (Rust et al. 1981), as illustrated in Fig. 5.7a. This observation suggests that a positive discharge to ground may be initiated by a branch of an extensive cloud discharge.

(iv) Several researchers (e.g., Fuquay 1982; Rust 1986) have reported that positive lightning discharges often involve long horizontal channels, up to tens of kilometers in extent. A photograph of one such event is published by Takeuti et al. (1977).

(v) It appears that positive leaders can move either continuously or in a stepped fashion, as determined from

5.3. Characterization of positive lightning

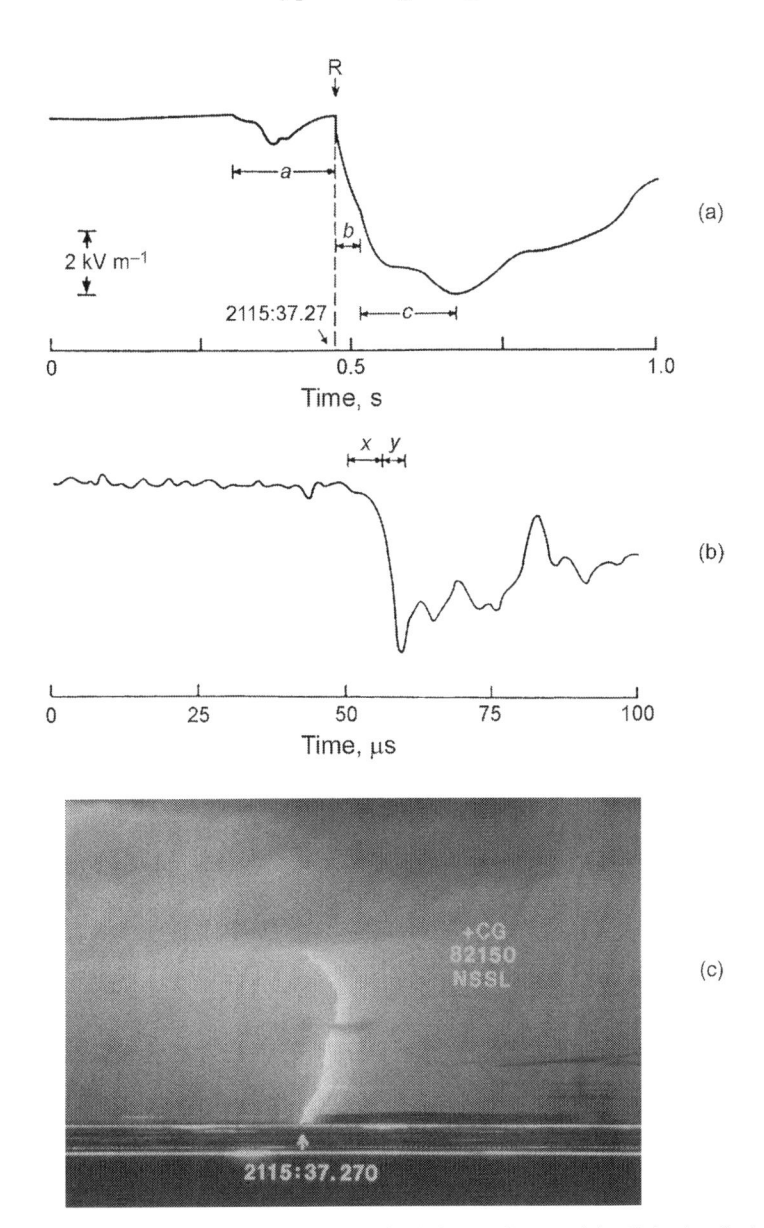

Fig. 5.7. (a) Typical overall electric-field change for a positive lightning flash recorded at 2115:37.270 CST on 30 May 1982. The distance to the flash was about 4 km. The time at the center of the scale (0.5 s) is 2115:37.3 CST. The interval a is the preliminary breakdown and leader; the interval b is the time from the return stroke through the end of the continuing current seen on the streak photograph shown in (c); the interval c is a period of possible continuing current not seen on the streak photograph. The return stroke is labeled R. (b) A microsecond-scale electric field waveform for the return stroke of the positive flash in (a). The slow ramp, interval x, is followed by a faster transition to peak, interval y; these intervals are typical of both positive and negative return strokes (subsection 4.6.3). (c) Streak photograph for the return stroke and the following continuing current of the positive flash whose electric field waveforms are shown in (a) and (b). Time advances from left to right. The channel luminosity persists for about 60 ms and corresponds to interval b in (a). The leader luminosity was apparently too faint to be imaged by the streak camera. Adapted from Rust et al. (1985).

time-resolved optical images. This is in contrast to negative leaders, which are always optically stepped when they propagate in virgin air. Further, distant (radiation) electric and magnetic field waveforms due to positive discharges are less likely to exhibit step pulses immediately prior to the return stroke waveform than are first strokes in negative lightning. Finally, positive leaders usually do not radiate at VHF as strongly as negative leaders. A more detailed discussion of positive and negative leaders (both downward- and upward-going) is given in subsection 5.3.2. A further discussion of upward positive and upward negative leaders in object-initiated lightning is found in subsections 6.2.1 and 6.2.2, respectively, and of upward positive leaders in rocket-triggered lightning in subsection 7.2.2.

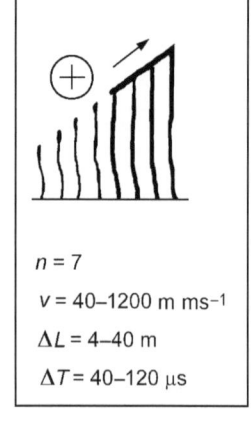

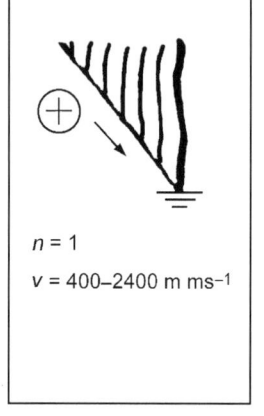

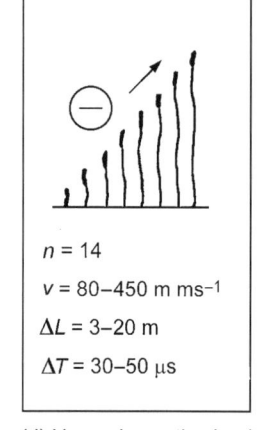

5.3.2. Comparison of positive and negative leaders

In the case of a positive leader, electrons present or produced ahead of the leader tip move toward the tip since they are attracted to the positive charge on it, and the resultant ionization occurs in the strong field near the tip. In the case of a negative leader, electrons tend to "run ahead" of the moving leader tip to where the field is relatively low since they are repelled by the negative charge on the leader tip. Thus ionization occurs under less favorable conditions for the negative leader than for the positive leader. As a result, streamer zone formation in the negative leader requires a higher tip potential than for the positive leader. However, the characteristics of positive and negative leader channels, such as temperature and conductivity, are similar (Bazelyan and Raizer 2000a). In this section, we compare experimental data for positive and negative leaders in lightning discharges. A mechanism for the formation of steps in negative leaders, based on long laboratory spark experiments, is described in subsection 4.4.8, and a mechanism for positive leaders is discussed in subsection 5.3.3.

Figure 5.8, adapted from Berger and Vogelsanger (1969), shows schematic streak-camera images for four types of lightning leaders, (a) downward negative, (b) upward positive, (c) downward positive, and (d) upward negative. These leaders correspond to the four types of lightning illustrated in Fig. 1.1. Note that the upward positive leader (Fig. 5.8b) initiates upward *negative* lightning (Fig. 1.1b), and the upward negative leader (Fig. 5.8d) initiates upward *positive* lightning (Fig. 1.1d). Also given in Fig. 5.8 are the leader propagation speed v, the step length ΔL, and the interstep interval ΔT (only the speed v is given for the downward positive leader, which shows fluctuations in luminosity, but no distinct steps; Figs. 5.8c and 5.9a).

The characteristics of the three types of stepped leaders illustrated in Figs. 5.8a, b, d are similar within a factor of 2 to 3, while the downward positive leader shown in Fig. 5.8c, which is the only optically observed leader of this type, appears to be considerably faster than those illustrated in Figs. 5.8a, b, d. Corresponding characteristics for the positive and negative leaders in rocket-triggered lightning are given in subsection 7.2.2. As noted in subsection 5.3.1, positive leaders appear in time-resolved optical images as either stepped or continuously moving, while negative leaders in virgin air are always optically stepped. In other words, negative leaders appear in streak photographs as a series of leader channel "snap shots" separated by clear no-luminosity gaps (see Figs. 5.8a, d), while positive leaders usually exhibit a continuously luminous channel image

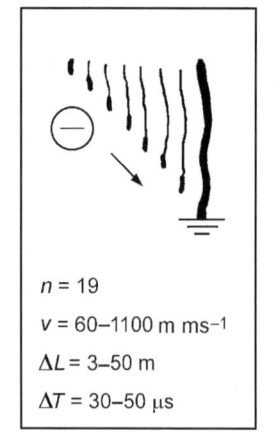

(a) Downward negative leader
$n = 19$
$v = 60–1100$ m ms^{-1}
$\Delta L = 3–50$ m
$\Delta T = 30–50$ μs

(b) Upward positive leader
$n = 7$
$v = 40–1200$ m ms^{-1}
$\Delta L = 4–40$ m
$\Delta T = 40–120$ μs

(c) Downward positive leader
$n = 1$
$v = 400–2400$ m ms^{-1}

(d) Upward negative leader
$n = 14$
$v = 80–450$ m ms^{-1}
$\Delta L = 3–20$ m
$\Delta T = 30–50$ μs

Fig. 5.8. Four different types of lightning leader: (a) downward negative leader, (b) upward positive leader, (c) downward positive leader, and (d) upward negative leader. For each leader type, the polarity refers to that of the leader charge, as opposed to that of the cloud charge effectively lowered to ground by lightning, as in Fig. 1.1. Given for each type of leader are the sample size n and the range of propagation speeds v. Additionally given, for clearly stepped leaders, are the range for average step lengths ΔL and the mean time interval between steps ΔT. Adapted from Berger and Vogelsanger (1969).

either with superimposed steps in the form of luminosity enhancements (Figs. 5.8b, c) or without such steps. Streak photographs of upward positive and upward negative leaders in object-initiated lightning are shown in Figs. 6.2a and 6.3a, respectively. Interestingly, Avansky *et al.* (1992) reported that whether a positive leader in a long laboratory spark exhibits stepping depends on humidity, higher humidity being conducive to stepping. A similar effect was previously observed by the Les Renardières Group (1977). Note that, depending on such factors as the leader light intensity, the spectral sensitivity of the film, and the film processing, streak photographs with varying degrees of detail

5.3. Characterization of positive lightning

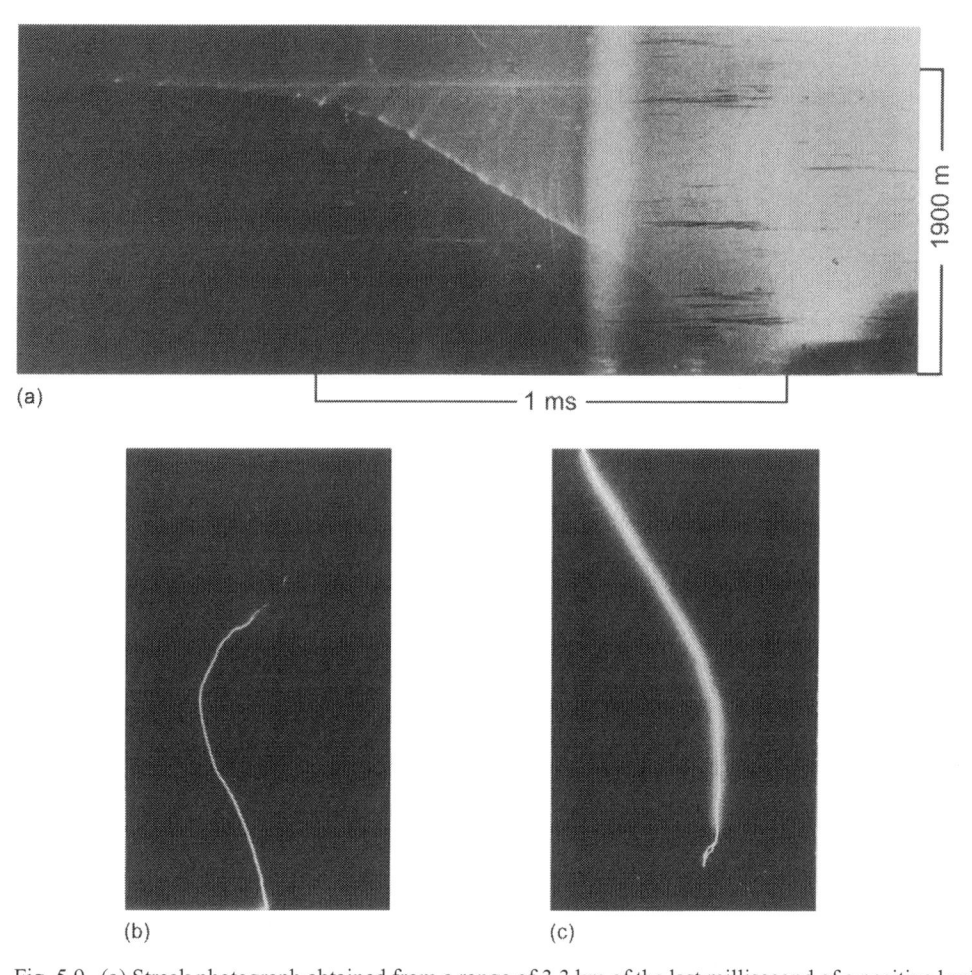

Fig. 5.9. (a) Streak photograph obtained from a range of 3.3 km of the last millisecond of a positive leader followed by a return stroke, (b) a still camera view from the same location, (c) a different still camera view showing the strike point on Lake Lugano and the "loop" indicative of an upward connecting leader (subsection 4.5.2). Adapted from Berger and Vogelsanger (1966).

can be obtained. In many streak-camera images, only the brightest portion of the leader channel, near its tip at the time of step formation, is discernible (see, for example, Figs. 4.13, 4.26, and 7.6 for negative leaders and Figs. 6.2, 7.5a, and 7.7 for positive leaders). Positive leaders are considerably less luminous than negative leaders (e.g., Berger 1977); no leader image is seen in the streak photograph of the positive flash shown in Fig. 5.7c. Additional discussion of upward positive leaders in object-initiated lightning is found in subsection 6.2.1.

The only available streak photograph of a downward-moving positive leader, obtained by Berger and Vogelsanger (1966), is reproduced in Fig. 5.9a. The schematic representation of a downward positive leader, shown in Fig. 5.8c, is based on this photograph alone. The corresponding still pictures taken from two different angles are shown in Figs. 5.9b, c. No distinct steps are seen in Fig. 5.9a, only some fluctuations in intensity. The propagation speed of this positive leader increased from 4×10^5 to 2.4×10^6 m s^{-1} as it approached ground, while for 19 downward negative leaders the range of speeds was 6×10^4 to 1.1×10^6 m s^{-1} (Berger and Vogelsanger 1969). It is worth noting that since the lightning shown in Figs. 5.9a, b, c did not strike an instrumented tower, it was identified by Berger and Vogelsanger (1966) as positive solely on the basis of the similarity of the luminous characteristics of its leader to those of identified positive leaders developing upward from the towers. Interestingly, while downward negative lightning channels typically exhibit multiple downward branches (see, for example, Fig. 4.1), no branches appear in the still photographs shown in Figs. 5.9b, c (except for those forming the loop near ground in Fig. 5.9c). A similar lack of branches is seen in the still photograph of a positive flash shown in Fig. 5.7c and in video images of three positive flashes studied by Beasley et al. (1983, Fig. 7). However, the still photograph of a positive flash published by Fuquay (1982) does show branches.

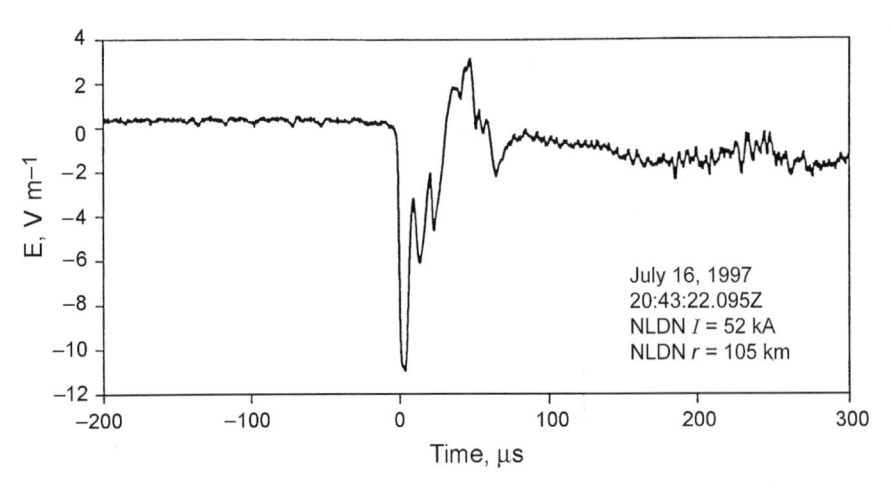

Fig. 5.10. Example of a typical electric field waveform (atmospheric electricity sign convention; subsection 1.4.2) due to the only stroke of a positive flash from a distant thunderstorm as recorded at Camp Blanding, Florida. The return stroke starts at $t = 0$. The event was detected by the NDLN (Section 17.5) and its NLDN-reported characteristics are given on the plot.

Next we discuss evidence of the stepped mode of propagation for positive leaders from measurements of electric and magnetic fields. Further discussion of microsecond-scale electric and magnetic field waveforms due to positive lightning is found in subsection 5.3.4. A significant fraction (26–30 percent; Hojo et al. 1985) of such waveforms exhibit pulses indicative of a leader stepping process prior to the return-stroke waveform, an example being given in Fig. 5.10. Note that these pulses could be due to a downward positive leader, an upward connecting negative leader, or both. Cooray and Lundquist (1982) reported that the mean time interval between the electric field pulses just preceding return strokes in Sweden was 26 µs for positive lightning as against 14 µs for negative lightning.

The initial leader in a natural positive cloud-to-ground flash apparently does not radiate in the frequency range from 25 to 500 MHz, at least during the last few milliseconds before the return stroke (Shao et al. 1999). This could be one of the reasons why downward positive leaders were not detected in Florida positive flashes with a VHF interferometer (Mazur et al. 1998). However, Rhodes et al. (1994) and Shao et al. (1996) observed in-cloud positive leaders that radiated at VHF at least as strongly as negative leaders. Further, Proctor (1997) reported that positive stepped leaders emitted VHF–UHF pulses "rather more strongly than negative stepped leaders". He observed positive stepped leaders in two out of 175 flashes in South Africa.

High-frequency radiation at 3 MHz associated with both positive and negative return strokes was studied by Cooray and Perez (1994a). They found that the duration of the 3 MHz radiation associated with negative return strokes ranged from 60 to 600 µs with a mean value of 137 µs. The duration of the 3 MHz radiation associated with positive return strokes was much longer, ranging from 5 to 25 ms.

In the following, we present some characteristics of positive and negative leaders obtained from triggered-lightning experiments (see also subsection 7.2.2). Streak photographs of upward positive and upward negative leaders in rocket-triggered lightning are found in Fig. 6.3.1 of Horii and Nakano (1995). Fieux et al. (1975) noted that the triggering of positive lightning in summer required an electric field at the ground higher than that required for triggering negative lightning. They also reported that an upward negative leader that initiates a positive discharge tends to propagate faster (at about 10^5 m s^{-1}) than the upward positive leader (initially at about 2×10^4 m s^{-1}). In contrast with this observation, Berger and Vogelsanger (1969) reported similar speeds for upward positive and upward negative leaders in object-initiated lightning (see Figs. 5.8b, d, respectively). Fieux et al. (1975) further reported from triggered-lightning experiments in France that positive lightning discharges initiated by upward negative leaders were associated with fewer upward branches than were negative discharges initiated by upward positive leaders. Interestingly, Laroche et al. (1985) noted (presumably for negative rocket-triggered lightning discharges initiated by upward positive leaders) that while flashes in New Mexico exhibited low-altitude upward branching, flashes in Florida did not do so. Horii and Ikeda (1985) reported, for winter lightning, that upward positive leaders were characterized by a lower current than upward negative leaders, this observation being apparently consistent with the reported lower luminosity of positive leaders (Berger 1977).

5.3.3. Mechanism of the positive leader

In this section we briefly describe the processes involved in both a continuously moving and a stepped positive leader, as inferred from laboratory long-spark

5.3. Characterization of positive lightning

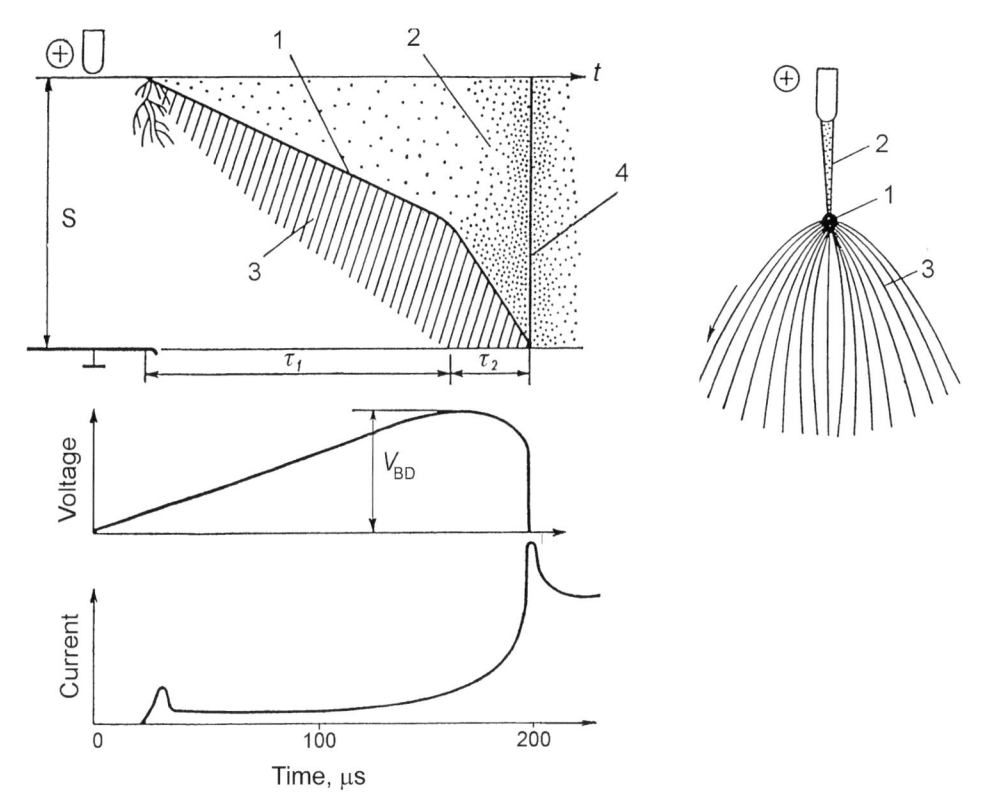

Fig. 5.11. Mechanism and structure of a continuously moving positive leader: 1, leader tip; 2, leader channel; 3, streamer zone; 4, return-stroke channel. V_{BD} is the breakdown voltage. See the main text for details. Adapted from Gorin *et al.* (1976).

experiments. This description, similar to that given in subsection 4.4.8 for a negative stepped leader, is based on the work of Gorin *et al.* (1976). The left-hand panel of Fig. 5.11 shows schematically a time-resolved picture of a continuously moving positive leader, followed by a return stroke, in a long rod–plane gap, labeled S. Also shown in the left-hand panel is the corresponding voltage across the gap and the current through the gap. Shown in the right-hand panel of Fig. 5.11 is a "snap-shot" showing the structure of a positive leader. Bazelyan and Raizer (1998; pp. 206–7) argued that a continuously moving positive leader would appear stepped if imaged with sufficient spatial resolution. In fact, McEachron (1939, p. 191) had suggested that the time intervals between steps could become short enough to give the positive leader a continuous appearance.

As seen in the right-hand panel of Fig. 5.11, the positive leader consists of the leader tip, labeled 1 (also called the leader head), which occupies a volume of about 1 cm^3, a leader channel, labeled 2, about 1 mm in diameter, and a streamer zone, labeled 3. The latter is a bundle of extending and arrested streamers that originate from the leader tip. The leader tip is considerably brighter than either the channel or the streamer zone. The leader channel is a plasma formation characterized by a temperature of order 10 000 K and a relatively high conductivity, of order 10^4 S m^{-1} (Gorin and Inkov 1962; Bazelyan *et al.* 1978), close to the conductivity of carbon (e.g., Sadiku 1994). As a result, the electric potential of the leader tip is close to that of the high-voltage electrode, a condition that is necessary for the formation of a self-propagating discharge. In contrast, according to Bazelyan *et al.* (1978), the streamer zone is characterized by a gas temperature of 1000 K or less, and the air in the bulk of the streamer zone is essentially an insulator. The latter inference is confirmed by the fact that contact of the streamer zone with the grounded electrode at the end of time interval τ_1 in Fig. 5.11 does not complete the breakdown of the gap. The leader channel radiates primarily in the visible and near-infrared region, while the streamer zone radiates in the UV (e.g., Bazelyan and Raizer 1998). The streamer zone can be viewed as the vehicle by which the energy of the electric field in the gap is made available to the extending leader channel. More information on streamers in spark discharges is found in the books of Bazelyan and Raizer (1998, 2000a).

As seen in the left-hand panel of Fig. 5.11, the positive leader is preceded by an initial impulsive corona (also called the first corona; Les Renardières Group 1977), a burst of streamers that serves to heat the air near the high-voltage electrode (labeled by a plus inside a small circle) and to form the initial section of the leader plasma channel. In the left-hand panel, the leader tip 1 is shown by a curved (two-section) solid line. During the time interval τ_1 (the initial stage), the lengths of both the channel 2 and the streamer zone 3 increase, and the charge associated with the leader system remains in the gap. During the time interval τ_2 (the so-called break-through stage), the streamer zone is in contact with the grounded electrode, making possible the removal of charge from the gap. As a result, the rate of extension of the leader channel and the current through the gap increase. The break-through stage is also called the final jump (e.g., Les Renardières Group 1977). The characteristics of this latter stage in laboratory spark discharges are largely determined by the external circuit. When the leader tip makes contact with the grounded electrode, a return stroke, labeled 4, is initiated. This completes the breakdown of the gap, and the voltage across the gap drops to nearly zero.

It has been observed that, depending on the rate of rise of voltage across the gap, the positive leader in a long spark can move either continuously, as shown in Fig. 5.11 (left panel), or intermittently (Gorin et al. 1976). The intermittent form of positive leader development, which occurs at relatively low rates of rise of voltage across the gap, can be viewed as a kind of stepping, although not of the same nature as the stepping in lightning positive leaders. As noted in subsection 5.3.2, the occurrence of "steps" (also called reilluminations or restrikes, Les Renardières Group 1977) in a positive leader may be influenced by humidity. Each "step" involves an impulsive corona burst, similar to the initial corona burst from the high-voltage electrode seen in Fig. 5.11, left-hand panel, but originating from the leader tip. Then the leader channel extends continuously for a time period of the order of some microseconds, after which the extension stops for some tens to hundreds of microseconds until a new corona-burst from the leader tip takes place.

Similar descriptions of the processes involved in the positive leader of a laboratory long spark are found in Allibone (1977), Les Renardières Group (1977), Gallimberti (1979), and Bazelyan and Raizer (1998, 2000a). Note that the step-formation mechanism in the positive leader outlined in the previous paragraph is distinctly different from that of the negative leader, discussed in subsection 4.4.8.

A model for the positive part of a bidirectional leader has been proposed by Bondiou-Clergerie et al. (1996) and a model for the downward positive lightning leader by Fofana and Beroual (1998) and Fofana et al. (1998).

5.3.4. Microsecond-scale electric and magnetic field waveforms

In this section, we discuss electric and magnetic field waveforms produced by the preliminary breakdown and return strokes of positive lightning. Field pulses presumed to be associated with leader steps in positive lightning discharges have been previously considered in subsection 5.3.2. We first discuss return-stroke fields. A number of researchers have reported microsecond-scale electric field waveforms that are very similar to those characteristic of negative first return strokes, except for their initial polarity (see Figs. 5.7b and 5.10). For discharges to ground, the (initial) polarity of the electric field waveform is indicative of the sign of charge transferred to ground.

In the studies of Rust et al. (1981, 1985), optical images of channels to ground associated with microsecond-scale electric field waveforms indicative of positive cloud-to-ground discharges were recorded (see Fig. 5.7), and those waveforms were interpreted as due to positive return strokes. Cooray and Lundquist (1982) and Cooray (1984), working in Sweden, characterized electric field waveforms that they attributed to distant positive return strokes on the basis of the overall similarity of the fields to those of negative return strokes (except for polarity) and the dissimilarity of the fields to those due to any known cloud discharge processes. Cooray (1986a, b) showed that correlation of these waveforms with the onset and temporal characteristics of the associated 3-MHz radiation indicated that the waveforms were radiated from near the ground and hence were due to positive return strokes. Hojo et al. (1985) provided a characterization of positive return strokes similar to that of Cooray and Lundquist (1982) and Cooray (1984), but for both winter and summer storms in Japan. Beasley et al. (1983), working in Florida, showed that microsecond-scale electric fields similar to those usually attributed to positive lightning (e.g., Cooray and Lundquist 1982; Cooray 1984; Fig. 5.10) are accompanied by microsecond-scale optical signals radiated from near ground. In this regard, it is worth noting that some pulses produced by cloud discharges have shapes not much different from those characteristic of positive return strokes. Some of these cloud-flash pulses may be misidentified by lightning locating systems such as the NLDN as being due to positive cloud-to-ground discharges. Published parameters of the radiation field waveforms produced, or thought to be produced, by positive return strokes are summarized in Table 5.4.

Note from Tables 5.4 and 4.6 that the initial electric field peak (mean value) normalized to 100 km for positive first strokes is about a factor 2 larger than that for negative first strokes (11.5 V m^{-1} as against 5.3 V m^{-1}; Cooray and Lundquist 1982). Cooray et al. (1998) reported that the mean peak electric field derivative (normalized to 100 km) for positive return strokes, 22 V m^{-1} µs^{-1}, was a factor 1.5 to 2 lower than its counterpart for negative strokes, 39 V m^{-1} µs^{-1}, found by Krider et al. (1996). In both these studies the field propagation path was almost entirely over salt water. However, Heidler and Hopf (1998) found (for lightning over land in Germany) that the mean electric

5.3. Characterization of positive lightning

Table 5.4. *Parameters of microsecond-scale electric and magnetic field waveforms produced by positive lightning*

Parameter	Location	Sample size	Mean	SD	Range of values
Initial electric field peak normalized to 100 km, V m^{-1}					
Cooray and Lundquist (1982)	Sweden	58	11.5	4.5	4.5–24.3
Cooray et al. (1998)	Denmark				
summer		22	13.9	5.8	6.5–26
Zero-to-peak risetime, μs					
Rust et al. (1981)	USA	15	6.9	1.9	4–10
Cooray and Lundquist (1982)	Sweden[a]	64	13	4	5–25
		52	12	3	5–25
Cooray (1986a, b)	Sweden	20	8.9	1.7	4–12
Hojo et al. (1985)	Japan				
winter		—	22.3	—	—
Ishii and Hojo (1989)	Japan				
summer		32	13.2	3.6	6–22
winter		123	21.2	5.8	8–44
Ushio et al. (1998)	Japan				
winter		19	18	—	—
10–90 percent risetime, μs					
Beasley et al. (1983)	Florida	6	—	—	1.6, 2.0, 4.5 1.2, 2.8, 4.0
Hojo et al. (1985)	Japan				
summer		44	6.7	—	—
winter		32	8.7	—	—
Cooray (1986a, b)	Sweden	15	6.2	1.4	3–9
Slow front duration, μs					
Cooray and Lundquist (1982)	Sweden	63	10	4	3–23
		33	9	3	3–19
Cooray (1986a, b)	Sweden	20	8.2	1.7	3–11
Hojo et al. (1985)	Japan				
winter		—	19.3	—	—
Cooray et al. (1998)	Denmark				
summer		23	8.4	2.8	4–16
Slow front amplitude relative to the peak, percent					
Cooray and Lundquist (1982)	Sweden[a]	67	38	11	10–70
		31	44	14	10–80
Cooray (1986a, b)	Sweden	20	45	7	30–60
Cooray et al. (1998)	Denmark				
summer		23	64	5	50–80
Fast-transition 10–90 percent risetime for propagation over salt water, ns					
Cooray (1986a, b)	Sweden	20	560	70	400–800
Ishii and Hojo (1989)	Japan				
winter (20) and summer (4)		24	110	62	30–400
Cooray et al. (1998)	Denmark				
summer		23	250	120	<100–600
Zero-crossing time, μs					
Ishii and Hojo (1989)	Japan				
summer		34	151	50	80–280
winter		89	93	32	30–160
Opposite-polarity overshoot relative to the peak, percent					
Ishii and Hojo (1989)	Japan				

cont.

Table 5.4. (cont.)

Parameter	Location	Sample size	Mean	SD	Range of values
summer		9	24	11	10–60
winter		57	40	17	10–90
Width of dE/dt pulse at half-peak value, ns					
Cooray et al. (1998)	Denmark				
summer		21	150	100	<100–600
Peak electric field derivative normalized to 100 km, $V\,m^{-1}\,\mu s^{-1}$					
Cooray et al. (1998)	Denmark				
summer		22	22	8.7	10–42

[a] The two lines of data refer to two different thunderstorms.

field derivative peak for positive strokes was appreciably higher than that for negative strokes. Cooray et al. (1998) reported that the mean duration of the "slow front" and the 10–90 percent risetime of the "fast transition", 8.4 µs and 250 ns, respectively, for positive return strokes were each about a factor 2 greater than for negative return strokes.

The characteristics of initial breakdown pulse bursts in electric field records for positive lightning have been studied in Sweden in summer by Gomes et al. (1997) and by Gomes and Cooray (1998), and in Japan in winter by Ushio et al. (1998). The individual pulses are usually bipolar, the initial polarity being in most cases the same as the polarity of the following return-stroke pulse (negative according to the atmospheric electricity sign convention; subsection 1.4.2). The duration of a pulse burst is of the order of a few milliseconds, similar to its counterpart in negative lightning. Ushio et al. (1998), who analyzed 19 positive flashes, reported the range of individual pulse widths to be from 5 to 50 µs with a mean value of 18 µs and the range of interpulse intervals to be from 10 to 180 µs with a mean value of 54 µs. Both these mean values are slightly below the range of their counterparts in negative lightning (Table 1.2). Gomes and Cooray (1998), who analyzed 70 positive flashes, noted a greater variability in pulse characteristics than for negative lightning. Gomes et al. (1997) examined only three positive flashes.

5.3.5. Peak current

A reliable distribution of positive-lightning peak currents applicable to objects of moderate height on the flat ground is presently unavailable. As indicated in subsection 5.3.1, the sample of 26 directly measured positive-lightning currents analyzed by Berger et al. (1975) and presented in Table 5.1 is based apparently on a mix of (i) discharges initiated as a result of a junction occurring between a descending positive leader and an upward connecting negative leader within some tens of meters of the tower top and (ii) discharges initiated when a very long (1 to 2 km) upward negative leader from the tower makes contact with an oppositely charged channel inside the cloud (a descending-positive-leader or positively charged in-cloud channel). Because the upward negative leaders of these two types of positive discharge differ so much from each other in length, the two types are expected to produce very different current waveforms at the tower, as illustrated in Figs. 5.6a, b. Perhaps the two types of positive discharge should be viewed as the limiting cases, all intermediate forms being possible.

As noted in subsection 5.3.1, the "microsecond-scale" current waveform shown in Fig. 5.6a is probably a result of processes similar to those in downward negative lightning, whereas the "millisecond-scale" current waveform shown in Fig. 5.6b is probably a result of the M-component mode of charge transfer to ground (Section 4.9). It is possible that such millisecond-scale waveforms are characteristic of tall objects capable of generating very long upward connecting leaders. Therefore, the best-known distribution of positive lightning peak currents, that published by Berger et al. (1975), may not be applicable to objects of moderate height on flat ground. Further, the distributions of positive lightning peak currents inferred from electric or magnetic fields recorded by multiple-station lightning locating systems, such as the NLDN, are influenced by uncertainties in conversion of the measured field to current (Section 17.5). Additionally, the lower end of the positive-lightning peak current distributions inferred from the NLDN data after the 1994 system upgrade is contaminated by misidentified cloud-flash pulses.

Matsumoto et al. (1996) reported on three positive rocket-triggered flashes in winter in Japan that had peak currents of about 32, 27, and 54 kA. These three had durations in excess of 60 ms and apparently contained no leader–return-stroke sequences. Matsumoto et al. (1996) also observed a flash having a positive peak current that occurred as long as 13 s after launching a triggering rocket.

5.3. Characterization of positive lightning

Table 5.5. *Seasonal variation of the median peak current for first strokes in positive and negative lightning*

Reference	Location	Season	Polarity	
			positive	negative
Orville *et al.* (1987)	USA, northeast	All year	45 (17 694)	30[a] (720 284)
		February	90	65
Orville and Huffines (1999)	Contiguous USA	July	15	20
		February	25	20
Hojo *et al.*(1989)	Japan	Summer	44 (485)	33 (6408)
		Winter	66 (397)	29 (747)
Le Boulch and Plantier (1990)	France	All year	81	34
		May–October	74–82	34–37
		November–April	98–108	36–48

[a] Value arbitrarily set by Orville *et al.* (1987) to be equal to the median peak current for negative lightning (first strokes) observed by Berger *et al.* (1975).

The initial part of the current waveform appeared as two cycles of damped oscillation about the zero current level and had a duration of 110 μs. The risetime to the initial positive peak of 132 kA was 20 μs or so. The oscillatory part of the waveform was followed by a positive continuing current having a magnitude of about 10 kA with a duration of 35 ms (see also Section 5.4). Bejleri (1999) reported on a positive rocket-triggered flash in summer in Florida (Camp Blanding) that produced a current waveform with an overall risetime of about 25 μs and a peak value of about 6 kA. The waveform had a duration of several tens of milliseconds. Interestingly, the initial rise in current to a value of about 2 kA, less than the peak of 6 kA, occurred in a microsecond or so. An example of an initial-stage current waveform in an upward positive flash initiated from the Peissenberg tower in Germany is found in Fig. 4 of Fuchs (1998). This waveform has an amplitude of about 6.5 kA and a duration of about 100 ms.

The median value of the positive-lightning peak current reported by DF lightning locating systems is found to be greater in winter than in summer, as seen in Table 5.5. There is controversy, however, regarding whether a similar seasonal variation exists for negative lightning (Table 5.5). Interestingly, Orville and Huffines (1999) reported, from 1995–7 NLDN data, that median positive peak currents exceeded 40 kA in regions of the north central United States, but were less than 10 kA in Louisiana and Florida.

Petersen and Rutledge (1992), using data from a four-station DF lightning locating system in Australia, observed a tendency for positive peak current maxima (determined over 30 min time intervals) to occur in the trailing stratiform regions of MCSs (subsection 2.2.7). Conversely, the positive peak current minima tended to occur in the convective regions of the MCSs. Further, the positive peak current maximum in their study appears to vary during the storm life cycle and reaches its largest value when the stratiform region is most intense in terms of its radar reflectivity.

MacGorman and Morgenstern (1998) examined 25 MCSs that occurred within the range of the NSSL DF network in 1985. They found that the distribution of positive lightning peak currents varied widely from one MCS to another, unlike those for negative flashes, which varied little. MacGorman and Morgenstern divided the 25 MCSs into three categories. The median positive lightning peak current was 14 kA for the first category (which included six MCSs), 36 kA for the second (nine MCSs), and 55 kA for the third (10 MCSs), and the overall distributions for the three categories were quite different. The first category had only a small fraction of flashes with peak currents above 20 kA, and the third category had only a small fraction of flashes with peak currents below 20 kA. The second category was between these two. Furthermore, they determined that these differences could not be explained by propagation effects, in part because the different distributions were maintained when data were restricted to MCSs in regions close to the DF stations. In contrast, median negative lightning peak currents for the same three categories of MCSs were similar, 27, 28, and 31 kA, respectively.

5.3.6. Return-stroke speed

Mach and Rust (1993), from photoelectric measurements, reported on two-dimensional propagation speeds for seven positive and 26 negative natural-lightning return strokes. They presented their speed measurements in two groups: one included values averaged over channel segments less than 500 m (four positive flashes were analyzed over segments 332 to 433 m long) and the other included values averaged over channel segments greater than 500 m (seven positive flashes were analyzed over segments 569 m to 2300 m long). For the short-segment group, Mach and Rust (1993) found average speeds of 0.8×10^8 m s^{-1} for positive return strokes and 1.7×10^8 m s^{-1} for negative return strokes. Two-dimensional measurements of positive return-stroke speed were also reported by Idone et al. (1987) for one positive return stroke that was part of an eight-stroke rocket-triggered lightning flash (Section 7.2) in Florida (KSC), the other seven strokes being negative, and by Nakano et al. (1987, 1988) for one natural positive lightning stroke in winter in Japan. The measurements of Idone et al. (1987) yielded a value about 10^8 m s^{-1} for the positive stroke and values ranging from 0.9×10^8 to 1.6×10^8 m s^{-1} for the seven negative strokes, all averaged over a channel segment 850 m in length near ground. Nakano et al. (1987, 1988) reported a significant decrease in two-dimensional speed with increasing height over a 180 m section of the channel, from 2×10^8 m s^{-1} at 310 m to 0.3×10^8 m s^{-1} at 490 m, while Mach and Rust (1993) found little speed variation along the channel. Clearly, more data on positive return-stroke speed are needed.

5.4. Bipolar lightning discharges to ground

Lightning current waveforms exhibiting polarity reversals were first reported by McEachron (1939, 1941) from his studies at the Empire State Building. According to Hagenguth and Anderson (1952), the number of bipolar flashes observed over a 10-year period was 11 (14 percent) from a total of 80 flashes for which polarity could be determined. The charge transfer was reported to be greater for negative polarity, probably owing to the fact that the initial-stage current was mostly negative. Interestingly, no flashes transferring only positive charge to ground were observed in these studies. Berger (1978) reported that 72 (six percent) of 1196 discharges observed in the years 1963–73 at Monte San Salvatore were bipolar, 68 of them being of the upward type. For 30 bipolar events (Table 6.6), he found median current peaks for the negative and positive parts of the waveform to be 350 A and 1.5 kA, respectively. The corresponding median charge transfers were 12 and 25 C. Gorin and Shkilev (1984) reported that six (6.7 percent) of 90 upward discharges initiated from the Ostankino tower (Table 6.1) were bipolar; all these initially transported negative charge to ground. The total number of bipolar lightning discharges observed on the Peissenberg tower was two (Heidler et al. 2000), both of which initially transported negative charge to ground. Many bipolar current waveforms have been observed in winter lightning studies in Japan, the reported frequency of occurrence ranging from five to 33 percent (Table 8.1). At least one bipolar lightning discharge was reported from each of the triggered-lightning experiments in France, Japan, New Mexico, Florida, and China.

There are basically three types of bipolar lightning discharge, although some events may belong to more than one category.

(i) *The first type of bipolar discharge* is associated with a polarity reversal during a slowly varying (millisecond-scale) current component, such as the initial continuous current in object-initiated lightning (Fig. 6.1) or in rocket-triggered lightning (Fig. 7.8a). Examples of such waveforms observed in winter storms in Japan are found in Figs. 8.4, 8.5, and 8.9b. The polarity reversal may occur one or more times and may involve an appreciably long no-current interval between opposite-polarity portions of the waveform (see, for example, Fig. 8.9b). An example of a bipolar current waveform lacking such a no-current interval is found in Fig. 8.4 (middle trace). A bipolar, millisecond-scale current waveform associated with a summer rocket-triggered lightning discharge in southeastern China is given by Liu and Zhang (1998; Fig. 4a). A similar waveform is found in Fig. 19 of McEachron (1939). Also, Laroche et al. (1985) noted a polarity reversal from negative to positive during the initial stage in one rocket-triggered flash in Florida. Hubert et al. (1984, p. 2515) reported a bipolar, millisecond-scale current waveform produced by a rocket-triggered lightning discharge in New Mexico. Superimposed on this waveform were both a large positive pulse with a peak of 17 kA and a risetime of 40 µs and many microsecond-scale negative current pulses with amplitudes up to several kiloamperes. The magnitudes of positive and negative charge transfers were similar, 235 and 240 C, respectively. The positive current component appeared to be associated with a branch of large horizontal extent below the cloud.

A reversal of polarity of the continuing current from negative to positive at the end of an otherwise negative rocket-triggered lightning discharge in France is seen in Fig. 2 of Waldteufel et al. (1980) and in Fig. 4 of Hubert and Mouget (1981). The occurrence of positive continuing current at the end of an otherwise negative flash initiated from the Peissenberg tower in Germany was reported by Fuchs (1998).

Davis and Standring (1947), who measured currents in the cables of kite balloons flying at a height of 600 m under thunderstorm conditions, reported a polarity change from negative to positive in a current record apparently associated with the initial stage of an upward discharge.

(ii) *The second type of bipolar discharge* is characterized by different polarities of the initial-stage current and of the following return stroke or strokes. An example of such a current waveform is shown in Fig. 23b of Berger and Vogelsanger (1966) and Fig. 6a of Berger (1978). The initial-stage current in this waveform, reproduced in Fig. 6.4b, is negative, with an amplitude of some hundreds of amperes and a total charge transfer of 40 C, and the return-stroke current is positive with a peak value of 27 kA. The positive return stroke was followed by continuing current, the total charge transfer being 90 C. The positive return-stroke current was separated from the negative initial-stage current by a zero-current time interval of about 100 ms. Berger (1978) gave one more example of a bipolar current waveform in his Fig. 6b. In this latter case, the positive initial-stage current, approaching 4 kA, was followed by a negative current pulse having a peak of 6.5 kA, possibly due to a return stroke. The negative pulse exhibited fluctuations on its tail, including a brief polarity reversal, and was followed by a slow, positive waveform having a peak approaching 3 kA. Nakahori *et al.* (1982) observed, in a lightning discharge to a 200 m smoke stack during a winter storm in Japan, a negative initial-stage current with superimposed pulses up to 20 kA or so in amplitude followed by a positive return-stroke current pulse having peak 31 kA. Fernandez (1997) reported on a positive initial-stage current in triggered lightning at Camp Blanding, Florida that was followed by leader–return-stroke sequences transferring negative charge to ground. Apparently a similar event was reported in triggered-lightning experiments during the winter in Japan by Akiyama *et al.* (1985, Fig. 8a).

(iii) *The third type of bipolar discharge* involves return strokes of opposite polarity. All the documented bipolar discharges in this category are of the upward type. Examples of current waveforms produced by such discharges are found in Fig. 17 of McEachron (1939) and in Fig. 18 of Berger and Vogelsanger (1965). Janischewskyi *et al.* (1999) observed three return strokes in an upward discharge initiated from the CN tower in Toronto (Table 6.1) with currents of −10.6, +6.5, and −8.9 kA. The time interval between the first and second strokes was 300 ms, and that between the second and the third strokes was 335 ms. All three strokes followed the same channel, as observed within about 535 m of the tower top. The waveshape characteristics (except for polarity) of all three strokes were similar. As discussed in subsection 5.3.6, Idone *et al.* (1987) studied a Florida rocket-triggered flash that contained eight return strokes, one of which was positive and the other seven negative. The positive stroke was the third in the flash, the preceding and following interstroke intervals being 374 and 369 ms, respectively. The time intervals between the negative strokes in this flash ranged from 39 to 101 ms. Gary *et al.* (1975), from summer triggered-lightning experiments in France, reported on a negative discharge with peak current of 19 kA and a risetime of 1.5 μs followed by a positive discharge with a peak current of 4 kA and an unspecified risetime. It is possible that these two events were two return strokes of opposite polarity. The positive event occurred 280 ms after the beginning of the negative event and 140 ms after cessation of the luminosity of the negative event.

For winter lightning in Japan, Narita *et al.* (1989) suggested that, in a bipolar discharge, currents of both polarities follow the same channel to ground but from different, oppositely charged, regions in the cloud, as illustrated in Fig. 8.3. The discharge channels and the associated current record, shown in Figs. 8.9a, b, respectively, for a winter bipolar lightning flash triggered by the rocket-and-wire technique appear to be consistent with this suggestion. It is likely that the explanation of bipolar current waveshapes suggested by Narita *et al.* (1989) for winter lightning also applies to summer bipolar lightning.

It is not clear whether a polarity change from positive to negative occurs less often than from negative to positive, although some data in Table 8.1 appear to suggest so. For all 30 bipolar flashes characterized by Berger (1978) (Table 6.6) the initial current polarity was negative. Horii (1986) noted differences between the negative and positive portions of the bipolar waveforms in winter triggered lightning in Japan, as illustrated in Fig. 8.9b. In this figure, the negative portion is characterized by multiple pulses with risetimes of the order of microseconds, while the positive portion is dominated by a single, wider pulse with a risetime of the order of tens of microseconds.

As noted in subsection 5.3.5, Matsumoto *et al.* (1996) observed a current waveform whose initial part appeared as two cycles of damped oscillation about the zero current level. The oscillatory part of the waveform had a duration of 100 μs and was followed by a positive continuing current with a duration of 35 ms. The first cycle of the oscillation had an initial, positive peak of 132 kA and a following negative peak of about 45 kA. The magnitude of the positive continuing current was about 10 kA. The current waveform was measured at the top of a test transmission line tower in Japan during winter.

5.5. Summary

Our knowledge of the physics of positive lightning remains much poorer than that of negative lightning. Although positive lightning discharges account for 10 percent or less of global cloud-to-ground lightning activity, there are five situations that appear to be conducive to a more frequent occurrence of positive lightning. These situations include (i) the dissipating stage of an individual thunderstorm, (ii) winter thunderstorms, (iii) shallow clouds such as the trailing stratiform regions of mesoscale convective systems, (iv) some severe storms, and (v) thunderclouds formed over forest fires or contaminated by smoke. The highest directly measured lightning currents (near 300 kA) and the largest charge transfers (hundreds of coulombs or more) are thought to be associated with positive lightning. Impulsive positive current waveforms with risetimes of order 10 μs, comparable to those for first strokes in negative lightning, as well as waveforms with longer risetimes, up to hundreds of microseconds, were observed by K. Berger and co-workers; the latter waveforms were apparently associated with very long, 1 to 2 km, upward negative connecting leaders. Positive flashes are usually composed of a single stroke. Positive return strokes often appear to be preceded by significant in-cloud discharge activity, tend to be followed by continuing currents, and involve long horizontal channels. In contrast to negative leaders, which are always optically stepped when they propagate in virgin air, positive leaders seem to be able to move either continuously or in a stepped fashion. The positive return-stroke propagation speed averaged over the lowest hundreds of meters of the lightning channel is of order 10^8 m s^{-1}. Bipolar lightning discharges are usually initiated by upward leaders from tall objects. It appears that positive and negative charge sources in the cloud are tapped by different upward branches of the lightning channel.

References and bibliography for Chapter 5

Akiyama, H., Ichino, K., and Horii, K. 1985. Channel reconstruction of triggered lightning flashes with bipolar currents from thunder measurements. *J. Geophys. Res.* **90**: 10 674–80.

Allibone, T.E. 1977. The long spark. In *Lightning, vol. 1, Physics of Lightning*, ed. R.H. Golde, pp. 231–80, London: Academic Press.

Anderson, F.J., and Freier, G.D. 1974. Relative amounts of positive and negative charge in lightning flashes. *J. Geophys. Res.* **79**: 5057–8.

Avansky, V.R., Chernov, E.N., Lupeiko, A.V., and Petrov, N.I. 1992. Experimental investigations of lightning phenomena in laboratory. In *Proc. 9th Int. Conf. on Atmospheric Electricity, St. Petersburg, Russia*, pp. 783–7.

Avila, E.E., and Pereyra, R.G. 2000. Charge transfer during crystal–graupel collisions for two different cloud droplet size distributions. *Geophys. Res. Lett.* **27**: 3837–40.

Baral, K.N., and Mackerras, D. 1992. The cloud flash-to-ground flash ratio and other lightning occurrence characteristics in Kathmandu thunderstorms. *J. Geophys. Res.* **97**: 931–8.

Baral, K.N., and Mackerras, D. 1993. Positive cloud-to-ground lightning discharges in Kathmandu thunderstorms. *J. Geophys. Res.* **98**: 10 331–40.

Bazelyan, E.M., Gorin, B.N., and Levitov, V.I. 1978. *Physical and Engineering Foundations of Lightning Protection*, 223 pp., Leningrad: Gidrometeoizdat.

Bazelyan, E.M., and Raizer, Yu. P. 1998. *Spark discharge*, 294 pp. Boca Raton, FL: CRC Press.

Bazelyan, E.M., and Raizer, Yu. P. 2000a. *Lightning Physics and Lightning Protection*, 325 pp., Bristol: IOP Publishing.

Bazelyan, E.M., and Raizer, Yu. P. 2000b. Lightning attraction mechanism and the problem of lightning initiation by lasers. *UFN* **170**(7): 753–69.

Beasley, W.H. 1985. Positive cloud-to-ground lightning observations. *J. Geophys. Res.* **90**: 6131–8.

Beasley, W.H., Uman, M.A., Jordan, D.M., and Ganesh, C. 1983. Positive cloud to ground lightning return strokes. *J. Geophys. Res.* **88**: 8475–82.

Bejleri, M. 1999. Triggered-lightning testing of an airport runway lighting system. Master's thesis, University of Florida, Gainesville, 233 pp.

Berger, K. 1967. Novel observations on lightning discharges: results of research on Mount San Salvatore. *J. Franklin Inst.* **283**: 478–525.

Berger, K. 1972. Methoden und Resultate der Blitzforschung auf dem Monte San Salvatore bei Lugano in den Jahren 1963–1971. *Bull. Schweiz. Elektrotech, Ver.* **63**: 1403–22.

Berger, K. 1973a. Kugelblitz und Blitzforschung. *Naturwissenschaften* **60**: 485–92.

Berger, K. 1973b. Oszillographische Messungen des Feldverlaufs in der Nähe der Blitzeinschläge auf dem Monte San Salvatore. *Bull. Schweiz. Elektrotech. Ver.* **64**: 120–36.

Berger, K. 1977. The Earth Flash. In *Lightning*, vol. 1, *Physics of Lightning*, ed. R.H. Golde, pp. 119–90, New York: Academic Press.

Berger, K. 1978. Blitzstrom-Parameter von Aufwärtsblitzen. *Bull. Schweiz. Elektrotech. Ver.* **69**: 353–60.

Berger, K. 1980. Extreme Blitzströme und Blitzschutz. *Bull. Schweiz. Elektrotech. Ver.* **71**: 460–4.

Berger, K., and Garbagnati, E. 1984. Lightning current parameters. Results obtained in Switzerland and in Italy. In *Proc. URSI Conf., Florence, Italy*, 13 pp.

Berger, K., and Vogelsanger, E. 1965. Messungen und Resultate der Blitzforschung der Jahre 1955–1963 auf dem Monte San Salvatore. *Bull. Schweiz. Elektrotech. Ver.* **56**: 2–22.

Berger, K., and Vogelsanger, E. 1966. Photographische Blitzuntersuchungen der Jahre 1955–1965 auf dem Monte San Salvatore. *Bull. Schweiz. Elektrotech. Ver.* **57**: 599–620.

Berger, K., and Vogelsanger, E. 1969. New results of lightning observations. In *Planetary Electrodynamics*, eds. S.C. Coroniti and J. Hughes, pp. 489–510, New York: Gordon and Breach.

Berger, K., Anderson, R.B., and Kroninger, H. 1975. Parameters of lightning flashes. *Electra* **80**: 23–37.

Biswas, K.R., and Hobbs, P.V. 1990. Lightning over the Gulf Stream. *Geophys. Res. Lett.* **17**: 941–3.

Bluestein, H.B., and MacGorman, D.R. 1998. Evolution of cloud-to-ground lightning characteristics and storm structure in the Spearman, Texas, tornadic supercells of 31 May 1990. *Mon. Wea. Rev.* **126**: 1451–67.

Boccippio, D.J., Williams, E.R., Heckman, S.J., Lyons, W.A., Baker, I.T., Boldi, R. 1995. Sprites, ELF transients, and positive ground strokes. *Science* **269**: 1088–91.

Bondiou-Clergerie, A., Bacchiega, G.L., Castellani, A., Lalande, P., Laroche, P., and Gallimberti, I. 1996. Experimental and theoretical study of the bi-leader process. Part II: Theoretical investigation. In *Proc. 10th Int. Conf. on Atmospheric Electricity, Osaka, Japan*, pp. 676–9.

Branick, M.L., and Doswell, C.A., III, 1992. An observation of the relationship between supercell structure and lightning ground-strike polarity. *Wea. Forecast.* **7**: 143–9.

Brook, M. 1992. Breakdown electric fields in winter storms. *Res. Lett. Atmos. Electr.* **12**: 47–52.

Brook, M., Nakano, M., Krehbiel, P., and Takeuti, T. 1982. The electrical structure of the Hokuriku winter thunderstorms. *J. Geophys. Res.* **87**: 1207–15.

Brook, M., Krehbiel, P., MacLaughlin, D., Takeuti, T., and Nakano, M. 1983. Positive ground stroke observations in Japanese and Florida storms. In *Proc. Conf. on Atmospheric Electricity*, eds. J. Latham and L. Ruhnke, pp. 365–9, Hampton, Virginia: A. Deepak.

Brook, M., Henderson, R.W., and Pyle, R.B. 1989. Positive lightning strokes to ground. *J. Geophys. Res.* **94**: 13 295–303.

Bruce, C.E.R., and Golde, R.H. 1941. The lightning discharge. *J. Inst. Electr. Eng. (London)* **88**: 487–520.

Burke, C.P. and Jones, D.K. 1996. On the polarity and continuing currents in unusually large lightning flashes deduced from ELF events. *J. Atmos. Terr. Phys.* **58**: 531–40.

Cooray, V. 1984. Further characteristics of positive radiation fields from lightning in Sweden. *J. Geophys. Res.* **89**: 11 807–15.

Cooray, V. 1986a. A novel method to identify the radiation fields produced by positive return strokes and their submicrosecond structure. *J. Geophys. Res.* **91**: 7907–11.

Cooray, V. 1986b. Correction to "A novel method to identify the radiation fields produced by positive return strokes and their submicrosecond structure". *J. Geophys. Res.* **91**: 13 318.

Cooray, V. 2000. The modeling of positive return strokes in lightning flashes. *J. Atmos. Solar-Terr. Phys.* **62**: 169–87.

Cooray, V., and Lundquist, S. 1982. On the characteristics of some radiation fields from lightning and their possible origin in positive ground flashes. *J. Geophys. Res.* **87**: 11 203–14.

Cooray, V., and Perez, H. 1994a. HF radiation at 3 MHz associated with positive and negative return strokes. *J. Geophys. Res.* **99**: 10 633–40.

Cooray, V., Fernando, M., Gomes, C., Sorensen, T., Scuka, V., and Pedersen, A. 1998. The fine structure of positive return stroke radiation fields: a collaborative study between researchers from Sweden and Denmark. In *Proc. 24th Int. Conf. on Lightning Protection, Birmingham, UK*, pp. 78–82.

Cooray, V., Fernando, M., Sörensen, T., Götschl, T., and Pedersen, Aa. 2000. Propagation of lightning-generated transient electromagnetic fields over finitely conducting ground. *J. Atmos. Solar-Terr. Phys.* **62**: 583–600.

Curran, E.B., and Rust, W.D. 1992. Positive ground flashes produced by low-precipitation thunderstorms in Oklahoma on 26 April, 1984. *Mon. Wea. Rev.* **120**: 544–53.

Cummer, S.A., and Füllekrug, M. 2001. Unusually intense continuing current in lightning produced delayed mesospheric breakdown. *Geophys. Res. Lett.* **28**: 495–8.

Davis, R., and Standring, W.G. 1947. Discharge currents associated with kite balloons. *Proc. Roy. Soc. A***191**: 304–22.

Engholm, C.D., Williams, E.R., and Dole, R.M. 1990. Meteorological and electrical conditions associated with positive cloud-to-ground lightning. *Mon. Wea. Rev.* **118**: 470–87.

Eriksson, A.J. 1982. The CSIR lightning research mast-data for 1972–1982. National Electric Engineering Research Institute, South Africa, Report no. Ek/9/82, 26 pp.

Eybert-Berard, A., Barret, L., and Berlandis, J.P. 1988. Campagne d'experimentations foudre RTLP 87, NASA Kennedy Space Center, Floride, USA (in French). STT/LASP 88-21/AEB/LB/JPB-pD, Cent. D'Etud. Nucl. de Grenoble, Grenoble, France.

Fernandez, M.I. 1997. Responses of an unenergized test power distribution system to direct and nearby lightning strikes. M.S. thesis, University of Florida, Gainesville, 249 pp.

Fieux, R., Gary, C., and Hubert, P. 1975. Artificially triggered lightning above land. *Nature* **257**: 212–14.

Fieux, R.P., Gary, C.H., Hutzler, B.P., Eybert-Berard, A.R., Hubert, P.L., Meesters, A.C., Perroud, P.H., Hamelin, J.H., and Person, J.M. 1978. Research on artificially triggered lightning in France. *IEEE Trans. Pow. Appar. Syst.* **PAS-97**: 725–33.

Fofana, I., and Beroual, A. 1998. Induced effects on an overhead line due to nearby positive lightning downward leader. *Electric Power Systems Research* **48**: 105–19.

Fofana, I., Ben Rhouma, A., Beroual, A., and Auriol, P. 1998. Modeling of positive lightning downward leader to study its effects on engineering systems. *IEEE Proc.: Generation, Transmission and Distribution* **145**: 395–403.

Fuchs, F. 1998. Lightning current and LEMP parameters of upward discharges measured at the Peissenberg tower. In *Proc. 24th Int. Conf. on Lightning Protection, Birmingham, UK*, pp. 17–22.

Fuchs, F., Landers, E.U., Schmid, R., and Wiesinger, J. 1998. Lightning current and magnetic field parameters caused by lightning strikes to tall structures relating to interference of electronic systems. *IEEE Trans. Electromagn. Compat.* **40**(4): 444–51.

Funaki, K., Kitagawa, N., Nakano, M., and Takeuti, T. 1981. Multistation measurements of lightning discharges to ground in Hokuriku winter thunderstorms. *Res. Lett. Atmos. Electr.* **1**: 19–26.

Fuquay, D.M. 1982. Positive cloud-to-ground lightning in summer thunderstorms. *J. Geophys. Res.* **87**: 7131–40.

Fuquay, D.M., Baughman, R.G., Taylor, A.R., and Hawe, R.G. 1967. Characteristics of seven lightning discharges that caused forest fires. *J. Geophys. Res.* **72**: 6371–3.

Gallimberti, I. 1979. The mechanism of the long spark formation. *J. de Phys. Coll.* **C7**(40): 193–250.

Gao, L., Akyuz, M., Larsson, A., Cooray, V., and Scuka, V. 2000. Measurement of the positive streamer charge. *J. Phys. D: Appl. Phys.* **33**: 1861–5.

Garbagnati, E., Giudice, E., Lo Piparo, G.B. 1978. Measurement of lightning currents in Italy – results of a statistical evaluation. *ETZ-A* **99**: 664–8.

Gary, C., Cimador, A., and Fieux, R. 1975. La foudre: étude du phénoméne. Applications à la protection des lignes de transport. *Revue Générale de l'Electricité* **84**: 24–62.

Gomes, C., and Cooray V. 1998. Radiation field pulses associated with the initiation of positive cloud to ground lightning flashes. In *Proc. 24th Int. Conf. on Lightning Protection, Birmingham, UK*, pp. 365–70.

Gomes, C., Thottappillil, R., and Scuka, V. 1997. Bipolar electric field pulses in lightning flashes over Sweden. In *Proc.*

12th Int. Zurich Symp. on Electromagn. Compat., Zurich, Switzerland, pp. 163–6.

Gorin, B.N., and Inkov, A.Y. 1962. Investigation of a spark channel. Sov. Phys. Tech. Phys. **7**: 235–41.

Gorin, B.N., and Shkilev, A.V. 1984. Measurements of lightning currents at the Ostankino tower. Elektrichestvo **8**: 64–5.

Gorin, B.N., Levitov, V.I., and Shkilev, A.V. 1976. Some principles of leader discharge of air gaps with a strong non-uniform field. In Gas Discharges, IEE Conf. Publ. 143, pp. 277–8.

Goto, Y., and Narita, K. 1992. Observations of winter lightning to an isolate tower. Res. Lett. Atmos. Electr. **12**: 57–60.

Goto, Y., and Narita, K. 1995. Electrical characteristics of winter lightning. J. Atmos. Terr. Phys. **57**: 449–59.

Hagenguth, J.H., and Anderson, J.G. 1952. Lightning to the Empire State Building – Part III. AIEE Trans. **71**: 641–9.

Halliday, E.C. 1932. The polarity of thunderclouds. Proc. Roy. Soc. **A138**: 205–29.

Heidler, F., and Hopf, C. 1998. Measurement results of the electric fields in cloud-to-ground lightning in nearby Munich, Germany. IEEE Trans. Electromagn. Compat. **40**(4): 436–43.

Heidler, F., Drumm, F., and Hopf, Ch. 1998. Electric fields of positive earth flashes in near thunderstorms. In Proc. 24th Int. Conf. on Lightning Protection, Birmingham, UK, pp. 42–7.

Heidler, F., Zischank, W., and Wiesinger, J. 2000. Statistics of lightning current parameters and related nearby magnetic fields measured at the Peissenberg tower. In Proc. 25th Int. Conf. on Lightning Protection, Rhodes, Greece, pp. 78–83.

Hill, R.D. 1988. Interpretation of bipole pattern in a mesoscale storm. Geophys. Res. Lett. **15**: 643–4.

Hojo, J., Ishii, M., Kawamura, T., Suzuki, F., and Funayama, R. 1985. The fine structure in the field change produced by positive ground strokes. J. Geophys. Res. **90**: 6139–43.

Hojo, J., Ishii, M., Kawamura, T., Suzuki, F., Komuro, H., and Shiogama, M. 1988. Characteristics and evaluation of lightning field waveforms. Electr Eng. Japan **108**: 55–65.

Hojo, J., Ishii, M., Kawamura, T., Suzuki, F., Komuro, H., and Shiogama, M. 1989. Seasonal variation of cloud-to-ground lightning flash characteristics in the coastal area of the Sea of Japan. J. Geophys. Res. **94**: 13 207–12.

Holle, R.L., Lopez, R.E., Howard, K.W., Cummins, K.L., Malone, M.D., and Krider, E.P. 1997. An isolated winter cloud-to-ground lightning flash causing damage and injury in Connecticut. Bull. Am. Meteor. Soc. **78**: 437–41.

Holle, R.L., and Watson, A.I. 1996. Lightning during two central U.S. winter precipitation events. Wea. Forecast. **11**: 599–614.

Horii, K. 1986. Experiment of triggered lightning discharge by rocket. In Proc. Korea–Japan Joint Symp. on Electrical Material and Discharge, Cheju Island, Korea, paper 8-3, 6 pp.

Horii, K., and Ikeda, G. 1985. A consideration on success conditions of triggered lightning. In Proc. 18th Int. Conf. on Lightning Protection, Munich, Germany, paper 1-3, 6 pp.

Horii, K., and Nakano M. 1995. Artificially triggered lightning. In Handbook of Atmospheric Electrodynamics, vol. 1., ed. H. Volland, pp. 151–66. Boca Raton, Florida: CRC Press.

Huang, E., Williams, E., Boldi, R., Heckman, S., Lyons, W., Taylor, M., Nelson, T., and Wong, C. 1999. Criteria for sprites and elves based on Schumann resonance observations. J. Geophys. Res. **104**: 16 943–64.

Hubert, P., and Mouget, G. 1981. Return stroke velocity measurements in two triggered lightning flashes. J. Geophys. Res. **86**: 5253–61.

Hubert, P., Laroche, P., Eybert-Berard, A., and Barret, L. 1984. Triggered lightning in New Mexico. J. Geophys. Res. **89**: 2511–21.

Huffines, G.R., and Orville, R.E. 1999. Lightning ground flash density and thunderstorm duration in the continental United States: 1989–96. J. Appl. Meteor. **38**: 1013–19.

Hunter, S.M., Schuur, T.J., Marshall, T.C., and Rust, W.D. 1992. Electric and kinematic structure of the Oklahoma mesoscale convective system of 7 June 1989. Mon. Wea. Rev. **120**: 2226–39.

Hunter, S.M., Underwood, S.J., Holle, R.L., and Mote, T.L. 2001. Winter lightning and heavy frozen precipitation in the southeast United States. Wea. Forecast. **16**: 478–90.

Huse, J., and Olsen, K. 1984. Some characteristics of lightning ground flashes observed in Norway. In Lightning and Power Systems, IEE Conf. Publ. 236, pp. 72–6, London and New York.

Idone, V.P., and Orville, R.E. 1982. Lightning return stroke velocities in the Thunderstorm Research International Program (TRIP). J. Geophys. Res. **87**: 4903–15.

Idone, V.P., Orville, R.E., and Henderson, R.W. 1984. Ground truth: a positive cloud-to-ground lightning flash. J. Clim. Appl. Meteor. **23**: 1148–51.

Idone, V.P., Orville, R.E., Mach, D.M., and Rust, W.D. 1987. The propagation speed of a positive lightning return stroke. Geophys. Res. Lett. **14**: 1150–3.

Ishii, M., and Hojo, J. 1989. Statistics on fine structure of cloud-to-ground lightning field waveforms. J. Geophys. Res. **94**: 13 267–74.

Ishii, M., Shimizu, K., Hojo, J., and Shinjo, K. 1998. Termination of multiple-stroke flashes observed by electromagnetic field. In Proc. 24th Int. Conf. on Lightning Protection, Birmingham, UK, pp. 11–16.

Janischewskyi, W., Shostak, V., Hussein, A.M., and Kordi, B. 1999. A lightning flash containing strokes of opposite polarities. CIGRE SC 33.99 (COLL), 2 pp.

Jensen, J.C. 1932. The relation of branching of lightning discharges to changes in the electrical field of thunderstorms. Phys. Rev. **40**: 1012–14.

Jensen, J.C. 1933. The branching of lightning and the polarity of thunderclouds. J. Franklin Inst. **216**: 707–47.

Kawasaki, Z-I., and Mazur, V. 1992a. Common physical processes in natural and triggered lightning in winter storms in Japan. J. Geophys. Res. **97**: 12 935–45.

Kawasaki, Z-I., and Mazur, V. 1992b. Common physical processes in natural and triggered lightning in winter storms in Japan. Res. Lett. Atmos. Electr. **12**: 61–70.

Kawasaki, Z.-I., Nakano, M., Takeuti, T., Nagatani, M., Nakada, H., Mizuno, Y., and Nagai, T. 1987. Fourier spectra of positive lightning fields during winter thunderstorms. *Res. Lett. Atmos. Electr.* **7**: 29–34.

Kirkland, M.W., Suszcynsky, D.M., Guillen, J.L.L., and Green, J.L. 2001. Optical observations of terrestrial lightning by the FORTE satellite photodiode detector. *J. Geophys. Res.* **106**: 33 499–509.

Knupp, K.R., Geerts, B., and Goodman, S.J. 1998. Analysis of a small, vigorous mesoscale convective system in a low-shear environment. Part I: Formation, radar echo structure, and lightning behavior. *Mon Wea. Rev.* **126**: 1812–36.

Krehbiel, P.R. 1981. An analysis of the electric field change produced by lightning. Ph.D. thesis, University of Manchester Institute of Science and Technology, Manchester, England (available as Report T-11, Geophys. Res. Ctr., New Mexico Inst. Min. and Tech., Socorro).

Krehbiel, P.R., Brook, M., and McCrory, R. 1979. An analysis of the charge structure of lightning discharges to the ground. *J. Geophys. Res.* **84**: 2432–56.

Krider, E.P., Leteinturier, C., and Willett, J.C. 1996. Submicrosecond fields radiated during the onset of first return strokes in cloud-to-ground lightning, *J. Geophys. Res.* **101**: 1589–97.

Lacerda, M. 1999. Personal communication.

Lacerda, M., Pinto, O., Pinto, I.R.C.A., Diniz, J.H., and Carvalho, A.M. 1999. Analysis of negative downward lightning current curves from 1985 to 1994 at Morro do Cachimbo research station (Brazil). In *Proc. 11th Int. Conf. on Atmospheric Electricity, Guntersville, Alabama*, pp. 42–5.

Laroche, P., Eybert-Berard, A., and Barret, L. 1985. Triggered lightning flash characteristics. In *Proc. 10th Int. Aerospace and Ground Conf. on Lightning and Static Electricity, Paris, France*, pp. 231–9.

Latham, D. 1991. Lightning flashes from a prescribed fire-induced cloud. *J. Geophys. Res.* **96**: 17 151–7.

Le Boulch, M., and Plantier, T. 1990. The Meteorage thunderstorm monitoring system: a tool for new EMC protection strategies. In *Proc. 20th Int. Conf. on Lightning Protection, Interlaken, Switzerland*, paper 6.13P, 8 pp.

Les Renardières Group 1977. Positive discharges in long air gaps at Les Renardières, 1975 results and conclusions. *Electra* **53**: 31–153.

Levin, Z., Yair, Y., and Ziv, B. 1996. Positive cloud-to-ground flashes and wind shear in Tel-Aviv thunderstorms. *Geophys. Res. Lett.* **23**: 2231–4.

Lewis, W.W., and Foust, C.M. 1945. Lightning investigation on transmission lines, Part 7. *Trans. AIEE* **64**: 107–15.

Liu, X.-S., Wang, C., Zhang, Y., Xiao, Q., Wang, D., Zhou, Z., and Guo, C. 1994. Experiment of artificially triggering lightning in China. *J. Geophys. Res.* **99**: 10 727–31.

Liu, X., and Zhang, Y. 1998. Review of artificially triggered lightning study in China. *Trans. IEE Japan* **118-B**(2): 170–5.

Lopez, R.E., Holle, R.L., Watson, A.I., and Skindlov, J. 1997. Spatial and temporal distributions of lightning over Arizona from a power utility perspective. *J. Appl. Meteor.* **36**: 825–31.

Lopez, R.E., Maier, M.W., and Holle, R.L. 1991. Comparison of the signal strength of positive and negative cloud-to-ground lightning flashes in northeastern Colorado. *J. Geophys. Res.* **96**: 22 307–18.

Lyons, W.A. 1996. Sprite observations above the U.S. high plains in relation to their parent thunderstorm systems. *J. Geophys. Res.* **101**: 29 641–52.

Lyons, W.A., Nelson, T.E., Williams, E.R., Cramer, J.A., and Turner, T.R. 1998a. Enhanced positive cloud-to-ground lightning in thunderstorms ingesting smoke from fires. *Science* **282**: 77–80.

Lyons, W.A., Uliasz, M., and Nelson, T.E. 1998b. Large peak current cloud-to-ground lightning flashes during the summer months in the contiguous United States. *Mon. Wea. Rev.* **126**: 2217–33.

MacGorman, D.R., and Burgess, D.W. 1994. Positive cloud-to-ground lightning in tornadic storms and hailstorms. *Mon. Wea. Rev.* **122**: 1671–97.

MacGorman, D.R., and Morgenstern, C.D. 1998. Some characteristics of cloud-to-ground lightning in mesoscale convective systems. *J. Geophys. Res.* **103**: 14 011–23.

MacGorman, D.R., and Nielsen, K.E. 1991. Cloud-to-ground lightning in a tornadic storm on 8 May 1986. *Mon. Wea. Rev.* **119**: 1557–74.

MacGorman D.R., and Taylor, W.L. 1989. Positive cloud-to-ground lightning detection by a direction-finder network. *J. Geophys. Res.* **94**: 13 313–18.

Mach, D.M, and Rust, W.D. 1993. Two-dimensional velocity, optical risetime, and peak current estimates for natural positive lightning return strokes. *J. Geophys. Res.* **98**: 2635–8.

Mackerras, D. 1968. A comparison of discharge processes in cloud and ground lightning. *J. Geophys. Res.* **73**: 1175–83.

Mackerras, D. 1973. Photoelectric observations of the light emitted by lightning flashes. *J. Atmos. Terr. Phys.* **35**: 521–35.

Mackerras, D., and Darveniza, M. 1994. Latitudinal variation of lightning occurrence characteristics. *J. Geophys. Res.* **99**: 10 813–21.

Marshall, T.C. 2000. Comment on "'Spider' lightning in intracloud and positive cloud-to-ground flashes" by Vladislav Mazur, Xuan-Min Shao, Paul R. Krehbiel. *J. Geophys. Res.* **105**: 7397–9; Reply, *ibid.*, 7401–2.

Marshall, T.C., Stolzenburg, M., and Rust, W.D. 1996. Electric field measurements above mesoscale convective systems. *J. Geophys. Res.* **101**: 6979–96.

Marshall, T.C., Stolzenburg, M., Rust, W.D., Williams, E.R., and Boldi, R. 2001. Positive charge in the stratiform cloud of a mesoscale convective system. *J. Geophys. Res.* **106**: 1157–63.

Matsumoto, Y., Sakuma, O., Shinjo, K., Saiki, M., Wakai, T., Sakai, T., Nagasaka, H., Motoyama, H., Ishii, M. 1996. Measurement of lightning surges on test transmission line equipped with arresters struck by natural and triggered lightning. *IEEE Trans. Pow. Del.* **11**: 996–1002.

Mazur, V., Shao, X.M., and Krehbiel, P. 1998. "Spider" lightning in intracloud and positive cloud-to-ground flashes. *J. Geophys. Res.* **103**: 19 811–22.

McEachron, K.B. 1939. Lightning to the Empire State Building. *J. Franklin Inst.* **227**: 149–217.

McEachron, K.B. 1941. Lightning to the Empire State Building. *AIEE Trans.* **60**: 885–90.

Miyake, K., Suzuki, T., and Shinjou, K. 1992. Characteristics of winter lightning current on Japan Sea coast. *IEEE Trans. Pow. Del.* **7**: 1450–6.

Mohr, K.I., Toracinta, E.R., Zipser, E.J., and Orville, R.E. 1996. A comparison of WSR-88D reflectivities, SSM/I brightness temperatures, and lightning for mesoscale convective systems in Texas. Part II: SSM/I brightness temperatures and lightning. *J. App. Meteor.* **35**: 919–31.

Molinari, J., Moore, P., and Idone, V. 1999. Convective structure of hurricanes as revealed by lightning locations. *Mon. Wea. Rev.* **127**: 520–34.

Molinie, G, Soula, S., and Chauzy, S. 1999. Cloud-to-ground lightning activity and radar observations of storms in the Pyrenees range area. *Q.J.R. Meteorol. Soc.* **125**: 3103–22.

Moore, P.K., and Orville, R.E. 1990. Lightning characteristics in lake-effect thunderstorms. *Mon. Wea. Rev.* **118**: 1767–82.

Murray, N.D., Orville, R.E., and Huffines, G.R. 2000. Effect of pollution from Central American fires on cloud-to-ground lightning in May 1998. *Geophys. Res. Lett.* **27**: 2249–52.

Nakahori, K., Egawa, T., and Mitani, H. 1982. Characteristics of winter lightning currents in Hokuriku district. *IEEE Trans. Pow. Appar. Syst.* **PAS-101**: 4407–12.

Nakano, M. 1979. Initial streamer of the cloud discharge in winter thunderstorms of the Hokuriku coast. *J. Meteor. Soc. Japan* **57**: 452–8.

Nakano, M., Nagatani, M., Nakada, H., Takeuti, T., and Kawasaki, Z. 1987. Measurements of the velocity change of a lightning return stroke with height. *Res. Lett. Atmos. Electr.* **7**: 25–8.

Nakano, M., Nagatani, M., Nakada, H., Takeuti, T., and Kawasaki, Z. 1988. Measurements of the velocity change of a lightning return stroke with height. In *Proc. 1988 Int. Aerospace and Ground Conf. on Lightning and Static Electricity, Oklahoma City, Oklahoma*, pp. 84–6.

Nakano, M., Kawasaki, Z.-I., Takeuti, T., Sirai, T., and Kawamata, Y. 1989. VHF/UHF radiation from positive and negative lightning. In *Proc. URSI Symp. on Environmental and Space Electromagnetics, Tokyo, Japan*, pp. 459–65.

Narita, K., Goto, Y., Komuro, H., and Sawada, S. 1989. Bipolar lightning in winter at Maki, Japan. *J. Geophys. Res.* **94**: 13 191–5.

Norinder, H. 1956. Magnetic field variations from lightning strokes in vicinity of thunderstorms. *Ark. Geofys.* **2**: 423–51.

Ogawa, T. 1982. The lightning current. In *Handbook of Atmospherics*, vol. 1, ed., H. Volland, pp. 23–63, Boca Raton, Florida: CRC Press.

Ogawa, T. 1995. Lightning currents. In *Handbook of Atmospheric Electrodynamics*, vol. I, ed. H. Volland, pp. 93–136, Boca Raton, Florida: CRC Press.

Orville, R.E. 1993. Cloud-to-ground lightning in the Blizzard of '93. *Geophys. Res. Lett.* **20**: 1367–70.

Orville, R.E. 1994. Cloud-to-ground lightning flash characteristics in the contiguous United States: 1989–1991. *J. Geophys. Res.* **99**: 10 833–41.

Orville, R.E. 1999. Comments on "Large peak current cloud-to-ground lightning flashes during the summer months in the contiguous United States". *Mon. Wea. Rev.* **127**: 1937–8.

Orville, R.E., and Huffines, G.R. 1999. Lightning ground flash measurements over the contiguous United States: 1995–1997. *Mon. Wea. Rev.* **127**: 2693–703.

Orville, R.E., and Silver, A.C. 1997. Lightning ground flash density in the contiguous United States: 1992–95. *Mon. Wea. Rev.* **125**: 631–8.

Orville, R.E., Henderson, R.W., and Bosart, L.F. 1983. An east coast lightning detection network. *Bull. Am. Meteor. Soc.* **64**: 1029–37.

Orville, R.E., Weisman, R.A., Pyle, R.B., Henderson, R.W., and Orville, Jr, R.E. 1987. Cloud-to-ground lightning flash characteristics from June 1984 through May 1985. *J. Geophys. Res.* **92**: 5640–4.

Orville, R.E., Henderson, R.W., and Bosart, L.F. 1988. Bipole patterns revealed by lightning locations in mesoscale storm systems. *Geophys. Res. Lett.* **15**: 129–32.

Orville, R.E., Zipser, E.J., Brook, M., Weidman, C., Aulich, G., Krider, E.P., Christian, H., Goodman, S., Blakeslee, R., and Cummins, K. 1997. Lightning in the region of the TOGA COARE. *Bull. Am. Meteor. Soc.* **78**: 1055–67.

Perez, A.H., Wicker, L.J., and Orville, R.E. 1997. Characteristics of cloud-to-ground lightning associated with violent tornadoes. *Wea. Forecast.* **12**: 425–37.

Petersen, W.A., and Rutledge, S.A. 1992. Some characteristics of cloud-to-ground lightning in tropical northern Australia. *J. Geophys. Res.* **97**: 11 553–60.

Phelps, C.T. 1971. Field-enhanced propagation of corona streamers. *J. Geophys. Res.* **76**: 5799–806.

Phelps, C.T., and Griffiths, R.F. 1976. Dependence of positive corona streamer propagation on air pressure and water vapor content. *J. Appl. Phys.* **47**: 2929–34.

Pierce, E.T. 1955a. Electrostatic field changes due to lightning discharges. *Q. J. R. Meteor. Soc.* **81**: 211–28.

Pierce, E.T. 1955b. The development of lightning discharges. *Q. J. R. Meteor. Soc.* **81**: 229–40.

Pierce, E.T. 1976. Winter thunderstorms in Japan – a hazard to aviation. *Naval Res. Rev.* **29**(6): 12–16.

Pinto, O. Jr, Pinto I.R.C.A., Gomes, M.A.S.S., Vitorello, I., Padilha, A.L., Diniz, J.H., Carvalho, A.M., and Cazetta, A. 1999a. Cloud-to-ground lightning in southeastern Brazil in 1993: 1. Geographical distribution. *J. Geophys. Res.* **104**: 31 369–79.

Pinto, I.R.C.A., Pinto O. Jr, Rocha, R.M.L., Diniz, J.H., Carvalho, A.M., and Cazetta, A. 1999b. Cloud-to-ground lightning in southeastern Brazil in 1993: 2. Time variations and flash characateristics. *J. Geophys. Res.* **104**: 31 381–7.

Podgorski, A.S., and Landt, J.A. 1987. Three dimensional time domain modelling of lightning. *IEEE Trans. Pow. Del.* **2**: 931–8.

Proctor, D.E. 1981. VHF radio pictures of cloud flashes. *J. Geophys. Res.* **86**: 4041–71.

Proctor, D.E. 1997. Lightning flashes with high origins. *J. Geophys. Res.* **102**: 1693–706.

Rakov, V.A. 1993. Data acquired with the LLP lightning locating systems. *Meteorologiya i Gidrologiya* **7**: 105–14.

Rakov, V.A. 1998. Comparison of positive and negative lightning. In *Proc. 1998 Int. Lightning Detection Conf., GAI, 2705 East Medina Road, Tucson*, Arizona 85706-7155, 19 pp.

Rakov, V.A. 1999. Lightning electric and magnetic fields. In *Proc. 13th Int. Symp. on Electromagnetic Compatibility, Zurich, Switzerland*, pp. 561–6.

Rakov., V.A. 2000a. Positive and bipolar lightning discharges: a review. In *Proc. 25th Int. Conf. on Lightning Protection, Rhodes, Greece*, pp. 103–8.

Rakov, V.A. 2001b. Positive Blitzentladungen. *ETZ Elektrotech. Autom.* **122**(5): 26–9.

Rakov, V.A., Thottappillil, R., and Uman, M.A. 1992. On the empirical formula of Willett *et al.* relating lightning return stroke peak current and peak electric field. *J. Geophys. Res.* **97**: 11 527–33.

Rakov, V.A., Uman, M.A., and Thottappillil, R. 1994. Review of lightning properties from electric field and TV observations. *J. Geophys. Res.* **99**: 10 745–50.

Rakov, V.A., Thottappillil, R., Uman, M.A., and Barker, P.P. 1995. Mechanism of the lightning M component. *J. Geophys. Res.* **100**: 25 701–10.

Rakov, V.A., Uman, M.A., Hoffman, G.R., Masters, M.W., and Brook, M. 1996. Bursts of pulses in lightning electromagnetic radiation: observations and implications for lightning test standards. *IEEE Trans. Electromagn. Compat.* **38**: 156–64.

Reap, R.M., and MacGorman, D.R. 1989. Cloud-to-ground lightning – climatological characteristics and relationships to model fields, radar observations, and severe local storms. *Mon. Wea. Rev.* **117**: 518–35.

Reising, S.C., Inan, U.S., and Bell, T.F. 1996. Evidence for continuing current in sprite-producing cloud-to-ground lightning. *Geophys. Res. Lett.* **23**: 3639–42.

Rhodes, C.T., Shao X.M., Krehbiel, P.R., Thomas, R.J., and Hayenga, C.O. 1994. Observations of lightning phenomena using radio interferometry. *J. Geophys. Res.* **99**: 13 059–82.

Rust, W.D. 1986. Positive cloud-to-ground lightning. In *The Earth's Electrical Environment*, eds. E.P. Krider and R.G. Roble, pp. 41–5. Washington, DC: National Academy Press.

Rust, W.D., MacGorman, D.R., and Arnold, R.T. 1981. Positive cloud to ground lightning flashes in severe storms. *Geophys. Res. Lett.* **8**: 791–4.

Rust, W.D., MacGorman, D.R., and Taylor, W.L. 1985. Photographic verification of continuing current in positive cloud-to-ground flashes. *J. Geophys. Res.* **90**: 6144–6.

Rutledge, S.A., and MacGorman, D.R. 1988. Cloud-to-ground lightning activity in the 10–11 June 1985 mesoscale convective system observed during Oklahoma–Kansas pre-storm project. *Mon. Wea. Rev.* **116**: 1393–408.

Rutledge, S.A., and Petersen, W.A. 1994. Vertical radar reflectivity structure and cloud-to-ground lightning in the stratiform region of MCSs: further evidence for *in situ* charging in the stratiform region. *Mon. Wea. Rev.* **122**: 1760–76.

Rutledge, S.A., Lu, C., and MacGorman, D.R. 1990. Positive cloud-to-ground lightning in mesoscale convective systems. *J. Atmos. Sci.* **47**: 2086–100.

Sadiku, M.N.O. 1994. *Elements of Electromagnetics*, 821 p. Orlando, Florida: Sounders College.

Schonland, B.F.J., and Allibone, T.E. 1931. Branching of lightning. *Nature* **128**: 794–5.

Schuur, T.J., Smull, B.F., Rust, W.D., and Marshall, T.C. 1991. Electrical and kinematic structure of the stratiform precipitation region trailing an Oklahoma squall line. *J. Atmos. Sci.* **48**: 825–42.

Seimon, A. 1993. Anomalous cloud-to-ground lightning in an F5-tornado-producing supercell thunderstorm on 28 August 1990. *Bull. Am. Meteor. Soc.* **74**: 189–203.

Shao, X.M., and Krehbiel, P.R. 1996. The spatial and temporal development of intracloud lightning. *J. Geophys. Res.* **101**: 26 641–68.

Shao, X.M., Krehbiel, P.R., Thomas, R.J., and Rison, W. 1995. Radio interferometric observations of cloud-to-ground lightning phenomena in Florida, *J. Geophys. Res.* **100**: 2749–83.

Shao, X.M., Holden, D.N., and Rhodes, C.T. 1996. Broad band radio interferometry for lightning observations. *Geophys. Res. Lett.* **23**: 1917–20.

Shao, X.M., Rhodes, C.T., and Holden, D.N. 1999. RF radiation observations of positive cloud-to-ground flashes. *J. Geophys. Res.* **104**: 9601–8.

Silfverskiöld, S., Thottappillil, R., Ye, M., Cooray, V., and Scuka, V. 1999. Induced voltages in a low-voltage power installation network due to lightning electromagnetic fields: an experimental study. *IEEE Trans. Electromagn. Compat.* **41**(3): 265–71.

Smith, S.B., La Due, J.G., and MacGorman, D.R. 2000. The relationship between cloud-to-ground lightning polarity and surface equivalent potential temperature during three tornadic outbreaks. *Mon. Wea. Rev.* **128**: 3320–8.

Stolzenburg, M. 1990. Characteristics of the bipolar pattern of lightning locations observed in 1988 thunderstorms. *Bull. Am. Meteor. Soc.* **71**: 1331–8.

Stolzenburg, M. 1994. Observations of high ground flash densities of positive lightning in summertime thunderstorm. *Mon. Wea. Rev.* **122**: 1740–50.

Takagi, N., Takeuti, T., and Nakai, T. 1986a. On the occurrence of positive ground flashes. *J. Geophys. Res.* **91**: 9905–9.

Takagi, N., Watanabe, T., Arima, I., Takeuti, T., Nakano, M., and Kinosita, H. 1986b. An unusual summer thunderstorm in Japan. *Res. Lett. Atmos. Electr.* **6**: 43–8.

Takeuti, T., Nakano, M., Nagatani, M., and Nakada, H. 1973. On lightning discharges in winter thunderstorms. *J. Meteor. Soc. Japan* **51**: 494–6.

Takeuti, T., Nakano, M., and Yamamoto, Y. 1976. Remarkable characteristics of cloud-to-ground discharges observed in winter thunderstorms in Hokuriku area, Japan. *J. Meteor. Soc. Japan* **54**: 436–9.

Takeuti, T., Nakano, M., Ishikawa, H., and Israelsson, S. 1977. On the two types of thunderstorms deduced from cloud-to-ground discharges observed in Sweden and Japan. *J. Meteor. Soc. Japan* **55**: 613–16.

Takeuti, T., Nakano, M., Brook, M., Raymond, D.J., and Krehbiel, P. 1978. The anomalous winter thunderstorms of the Hokuriku coast. *J. Geophys. Res.* **83**: 2385–94.

Takeuti, T., Israelsson, S., Nakano, M., Ishikawa, H., Lundquist, S., and Astrom, E. 1980. On thunderstorms producing positive ground flashes. *Proc. Res. Inst. Atmos. Nagoya Univ., Japan* **27-A**: 1–17.

Takeuti, T., Funaki, K., and Kitagawa, N. 1983. A preliminary report on the Norwegian winter thunderstorm observation. *Res. Lett. Atmos. Electr.* **3**: 69–72.

Takeuti, T., Kawasaki, Z., Funaki, K., Kitagawa, N., and Huse, J. 1985. On the thundercloud producing the positive ground flashes. *J. Meteor. Soc. Japan* **63**: 354–58.

Taylor, W.L. 1963. Radiation field characteristics of lightning discharges in the band 1 kc/s to 100 kc/s. *J. Res. Nat. Bur. Stand.* **67D**: 539–50.

Théry, C. 2001. Evaluation of LPATS data using VHF interferometric observations of lightning flashes during the EULINOX experiment. *Atmos. Res.* **56**: 397–409.

Thomas, R.J., Krehbiel, P.R., Rison, W., Hamlin, T., Harlin, J., and Shown, D. 2001. Observations of VHF source powers radiated by lightning. *Geophys. Res. Lett.* **28**: 143–6.

Toracinta, E.R., Mohr, K.I., Zipser, E.J., and Orville, R.E. 1996. A comparison of WSR-88D reflectivities, SSM/I brightness temperatures, and lightning for mesoscale convective systems in Texas. Part I: Radar reflectivity and lightning. *J. Appl. Meteor.* **35**: 902–18.

Uman, M.A. 1969a. *Lightning*, 264 pp., New York: McGraw-Hill.

Uman, M.A. 1984. *Lightning*, 298 pp., New York: Dover.

Ushio, T., Kawasaki, Z.-I., Matsuura, K., and Wang, D. 1998. Electric fields of initial breakdown in positive ground flash. *J. Geophys. Res.* **103**: 14 135–9.

Vereshichagin, I.P., Orlov, A.V., Temnikov, A.G., Antsupov, K.V., Koshelev, M.A., and Makaisky, L.M. 1996. Discharge development inside the charged aerosol formations. In *Proc. 23rd Int. Conf. on Lightning Protection, Florence, Italy*, pp. 42–6.

Vonnegut, B., and Orville, R.E. 1988. Evidence of lightning associated with the Yellowstone Park forest fire (Abstract). *Eos, Trans. AGU* **69**: 1071.

Vonnegut, B., Latham, D.J., Moore, C.B., and Hunyady, S.J. 1995. An explanation for anomalous lightning from forest fire clouds. *J. Geophys. Res.* **100**: 5037–50.

Waldteufel, P., Metzger, P., Boulay, J.L., Laroche, P., and Hubert, P. 1980. Triggered lightning strokes originating in clear air. *J. Geophys. Res.* **85**: 2861–8.

Wang, C., Yan, M., Liu, X., Zhang, Y., Dong, W., and Zhang, C. 1999. Bidirectional propagation of lightning leader. *Chinese Sci. Bull.* **44**(2): 163–6.

Williams, E.R. 1998. The positive charge reservoir for sprite-producing lightning. *J. Atmos. Solar-Terr. Phys.* **60**: 689–92.

Winckler, J.R., Lyons, W.A., Nelson, T.E., Nemzek, R.J. 1996. New high-resolution ground-based studies of sprites. *J. Geophys. Res.* **101**: 6997–7004.

Yoda, M., Miyachi, I., Kawashima, T., and Katsuragi, Y. 1992. Lightning current protection of equipments in winter. *Res. Lett. Atmos. Electr.* **12**: 117–21.

Yos, J.M. 1963. Transport properties of nitrogen, oxygen, and air to 30 000 K. Tech. Mem. RAD-TM-63-71, Avco Corp., Wilmington, Delaware.

6 Upward lightning initiated by ground-based objects

> It is realized that when making use of a tall structure, like the Empire State Building, the building acts as a great needle point, and may have a marked influence upon the character of the discharge.
>
> K.B. McEachron (1939)

6.1. Introduction

In this chapter, we consider lightning discharges initiated by leaders that originate from stationary grounded objects, usually tall towers, and propagate upward toward charged clouds overhead. Upward lightning, as opposed to "normal" downward lightning, would not occur if the object were not present and, hence, can be considered to be initiated by the object. As noted in subsection 2.9.1, objects with heights ranging from approximately 100 to 500 m experience both downward and upward flashes, the fraction of upward flashes increasing with the height of the object. The observed percentage of upward flashes for structures of different heights is given in Table 2.3. Structures having heights less than 100 m or so are usually assumed to be struck only by downward lightning, and structures with heights greater than 500 m or so are usually assumed to experience only upward flashes. In other words, upward flashes are usually neglected for structures lower than 100 m, and downward flashes are neglected for structures taller than 500 m. Apparently, all lightning discharges recorded by Davis and Standring (1947), who measured currents in the cables of kite balloons flying at a height of 600 m in England, were of the upward type. If a structure is located on the top of a mountain, then an effective height that is greater than the structure's physical height is often assigned to the structure in order to account for the additional field distortion due to the presence of the mountain on which the structure is located. For example, the two towers used by Berger in his lightning studies on Monte San Salvatore in Switzerland each had a physical height of about 70 m while their effective height was independently estimated to be 270 m by Pierce (1971) and 350 m by Eriksson (1978a). Eriksson's estimate is based on the observed percentage of upward flashes initiated from the towers, while Pierce's estimate is based on the observed overall increase in lightning incidence to these mountain-top towers over that to similar towers on the flat ground. For a 60 m tower located on a hill some 80 m above the surrounding terrain near Pretoria, South Africa, Eriksson (1978a) estimated the effective height to be 148 m. This latter estimate of effective height was based on the observation that 20 percent of discharges to the tower during the years 1972–7 were of the upward type. (Eriksson (1982) reported that out of a total of 44 discharges to the South African tower observed in 1978–82, 14 percent were upward, 32 percent were downward, and about 54 percent were indeterminate.) For two 40 m TV towers used for lightning studies on mountain tops in Italy (most of the published data having been obtained on the tower on Monte Orsa, located only 10 km from Monte San Salvatore), Eriksson (1978a) gave an effective height of 500 m (see Table 6.1). This latter estimate was apparently based on his assumption, with reference to Garbagnati et al. (1974), that 98 percent of the flashes observed on the Italian towers were of the upward type. However, Berger and Garbagnati (1984) reported, for the same towers in Italy, that only about 60 percent of the flashes were of the upward type. In the absence of optical images, which would allow the unambiguous identification of upward and downward flashes, the fraction of upward discharges in the total number of flashes strongly depends on what features of the measured current waveform are assumed to be characteristic of an upward flash. The concept of effective height was apparently introduced for the estimation of the number of downward discharges to a ground-based structure (subsection 2.9.2). Probably, for this concept to be meaningful there must exist a large gap between the upper extremity of the structure and the cloud charge center. If this gap is small or non-existent, as, for example, in winter storms in Japan (Section 8.4), no such effective height can be determined.

It appears that upward flashes more often transport negative than positive charge to ground (Berger 1978). The phenomenology of upward negative lightning, shown in Fig. 6.1 by schematic diagrams illustrating the

Table 6.1. *An overview of lightning studies conducted at instrumented tall objects*

Object	Location	Height, m	Terrain	Effective height, m	Selected references
Empire State Building	New York City, USA	410	Flat	410	McEachron (1939, 1941), Hageguth and Anderson (1952)
Two towers 400 m apart[a]	Monte San Salvatore, Lugano, Switzerland	70	Mountain 640 m above Lake Lugano, 912 m above sea level	270 (Pierce 1971), 350 (Eriksson 1978a)	Berger and Vogelsanger (1965, 1966, 1969), Berger (1967, 1972, 1977, 1978), Berger et al. (1975)
Ostankino TV tower	Moscow, Russia	540	Flat	540	Gorin et al. (1975; 1977), Gorin and Shkilev (1984)
Two TV towers[b]	Sasso di Pale, near Foligno, central Italy and Monte Orsa, near Varese, northern Italy	40	Mountains 980 and 993 m above sea level	500 (Eriksson 1978a)	Garbagnati and Lo Piparo (1970, 1973, 1982a, b), Garbagnati et al. (1974; 1975; 1978; 1981)
CSIR research mast	Pretoria, South Africa	60	Hill 80 m above surrounding terrain, 1400 m above sea level	148 (Eriksson 1978a)	Eriksson (1978a, 1982)
CN Tower	Toronto, Canada	553	Flat	553	Hussein et al. (1995), Janischewskyj et al. (1997)
Peissenberg tower	Hoher Peissenberg, Munich, Germany	160	Mountain about 288 m above surrounding terrain, 988 m above sea level[c]	Unknown	Beierl (1992), Fuchs et al. (1998)
St. Chrischona tower	Basel, Switzerland	248	Mountain 493 m above sea level	Unknown	Montandon (1992, 1995)
Cachimbo tower	Brazil	60	Mountain 200 m above surrounding terrain, 1600 m above sea level	Unknown	Lacerda et al. (1999)
Gaisberg tower	Salzburg, Austria	100	Mountain 1287 m above sea level	Unknown	Diendorfer et al. (2000)

[a] The first tower, made of wood and equipped with a grounded lightning rod, was erected in 1943. The second tower, made of steel, was erected in 1950. In 1958, the wooden tower was replaced by a tower made of steel (F. Heidler, personal communication, 2000).
[b] Most data have been obtained on Monte Orsa which is only 10 km from Monte San Salvatore.
[c] The tower is located below the mountain top, at about 937 m above sea level (F. Heidler, personal communication, 1999).

still and time-resolved photographic records and by the corresponding current record, is similar to that of negative lightning triggered using the classical rocket-and-wire technique (subsection 7.2.1). In the latter case, the thin triggering wire plays the role of the grounded object, one that is rapidly erected and then replaced by the plasma channel of the upward leader. Thus, much of the information on the properties of classical negative rocket-triggered lightning obtained from optical and current observations and presented in subsections 7.2.2 and 7.2.3 is applicable to the upward negative lightning discussed here. Since rocket-triggered lightning has been studied more comprehensibly than object-initiated lightning, it might be advantageous to read subsections 7.2.2 and 7.2.3 before reading subsections 6.2.1, 6.3, and 6.4. Downward-leader–upward-return-stroke sequences in upward lightning (Fig. 6.1), and their counterparts in rocket-triggered lightning (Fig. 7.1), are similar to subsequent strokes in natural downward lightning (Fig. 4.2). For this reason, leader–return-stroke sequences in upward (object-initiated) lightning and in

6.1. Introduction

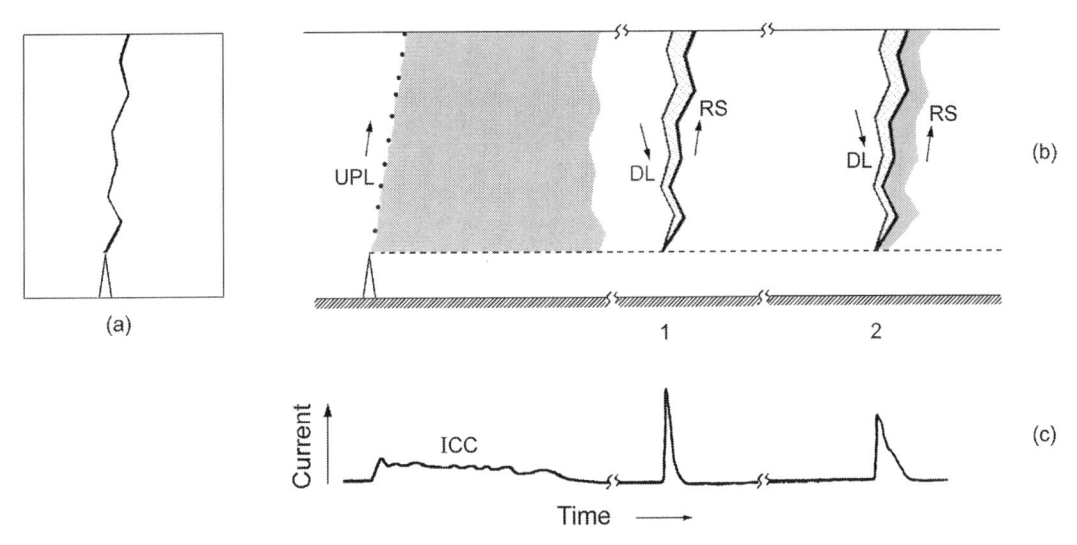

Fig. 6.1. Schematic diagram showing the luminosity of an upward negative flash and the corresponding current at the channel base. (a) Still-camera image, (b) streak-camera image, (c) current record. The flash is composed of an upward positive leader, UPL, followed by an initial continuous current, ICC, and two downward-dart-leader–upward-return-stroke (DL–RS) sequences. UPL and ICC constitute the initial stage of an upward negative flash.

rocket-triggered lightning are sometimes referred to as subsequent strokes.

Before the advent of rocket-triggered lightning, instrumented tall objects were the primary vehicle for the detailed study of lightning currents, usually in conjunction with optical observations. Upward lightning was first characterized by McEachron (1939), who recorded lightning currents and associated time-resolved photographic images at the Empire State Building in New York. This and other projects that have yielded information on upward lightning in summer, and, in most cases, also on downward lightning, are summarized in Table 6.1. The characteristics of downward negative lightning obtained on instrumented towers are presented in subsections 4.4.2, 4.4.6, 4.5.1, and 4.6.1, and the characteristics of positive and bipolar lightning are presented in Sections 5.3 and 5.4, respectively. Upward lightning discharges initiated from grounded objects during winter thunderstorms in Japan are discussed in Section 8.4.

An object-initiated discharge begins when the electric field intensity over some critical distance from the tip of the object exceeds the breakdown value. According to Bazelyan *et al.* (1978), this distance for objects with heights ranging from 50 to 300 m is about 15 m. The existence of an electric field intensity higher than the breakdown value at only the object tip or over a shorter distance than the critical distance is insufficient for the formation of a self-propagating discharge. If this were not the case, such a minor and commonly occurring form of electrical discharge as corona would necessarily result in upward lightning.

Interestingly, Berger and Vogelsanger (1969), who reported on their observations of lightning during the years 1957–66 in Switzerland, which included optical observations in which eight still cameras recorded all lightning flashes in the area at night, stated that most of the observed upward discharges originated from man-made structures rather than from mountain peaks. These man-made structures apparently always had grounded metallic conductors. Only once was an upward lightning observed to occur from a rock outcrop. Further, Berger and Vogelsanger (1969) suggested that the high electric field needed for the initiation of upward lightning is rapidly created by an in-cloud discharge, rather than being a consequence of the slower charge build-up in the cloud produced by the cloud electrification processes discussed in subsection 3.2.6. Apparently a slower charge buildup allows time for corona to inhibit the development of an upward-going leader from the object, as suggested by Brook *et al.* (1961) and discussed in Section 7.1.

We first present, in Section 6.2, a general characterization of upward lightning discharges including optical images and associated current records. All these were obtained in the lightning studies at Monte San Salvatore. The optical images for upward negative and upward positive lightning include both still and time-resolved photographs, while only a still picture is shown for upward bipolar lightning, since no time-resolved optical images of bipolar flashes are available. As defined in Section 1.2, upward *negative* lightning discharges are initiated by upward *positive* leaders and transport *negative* charge to ground. Further, upward *positive* lightning discharges are initiated by upward *negative* leaders and transport *positive* charge to ground. Finally, *bipolar* lightning discharges are initiated by upward leaders of *either* polarity and transport *both* negative and positive charge to ground during the course of the discharge.

Apparently the majority of upward bipolar lightning flashes are initiated by upward positive leaders, so that negative charge transfer to ground is followed by positive charge transfer to ground. Section 6.2 also contains a discussion of the optically determined properties of upward leaders, including propagation speed, step length, and interstep interval. In Section 6.3, we describe the structure of the overall current waveform of the most common upward lightning event, the upward negative flash, and consider a number of electrical characteristics of upward lightning such as the magnitude of the slowly varying (millisecond-scale) current components, the charge transfer to ground, and the duration of the event. Impulsive current components in upward discharges, including return-stroke current waveforms, are characterized in Section 6.4. In Section 6.5, we discuss the transient behavior of tall objects struck by lightning, and in Section 6.6 the electromagnetic fields due to strikes to tall objects. The material in Sections 6.5 and 6.6 concerns only the impulsive current components, primarily the return-stroke current pulses. Since, as noted earlier, downward-leader–upward-return-stroke sequences in upward lightning are similar to the subsequent leader–return-stroke sequences in downward lightning, the material in Sections 6.5 and 6.6 is also applicable to the downward lightning discharges considered in Chapter 4. The available information on the acoustic output of upward lightning is presented in Section 6.7.

6.2. General characterization

In this section, we present examples of streak and still photographs of upward lightning discharges and the associated current waveforms measured at the strike object. We also discuss a number of properties of upward leaders such as propagation speed, step length, and interstep interval. Further discussion of upward leaders is found in subsections 5.3.2 and 7.2.2.

6.2.1. Upward negative lightning

As defined in Sections 1.2 and 6.1, upward negative discharges are initiated by upward positive leaders from the tops of grounded objects. As in rocket-triggered negative flashes (Section 7.2), the upward positive leader bridges the gap between the object and the negative charge source in the cloud and serves to establish an initial continuous current, typically lasting for some hundreds of milliseconds. The upward positive leader and initial continuous current constitute the initial stage of an upward flash. The initial stage can be followed, after a no-current interval, by one or more downward-leader–upward-return-stroke sequences, as illustrated in Fig. 6.1. Figure 6.2 gives data on two concurrent upward flashes from the two towers on Monte San Salvatore. The two towers are separated horizontally by about 400 m. The time-resolved (streak) photograph in Fig. 6.2a shows the luminous development of the upward positive leaders from both towers. The corresponding still photograph is shown in Fig. 6.2b where tower 1 is on the right and tower 2 is on the left. The leaders in Fig. 6.2a exhibit stepping and are similar to the upward positive leader in rocket-triggered lightning shown in Fig. 7.5a. Figure 6.2c shows the overall current for the flash initiated from tower 1. This flash was composed of an initial stage only, that is, no downward-leader–upward-return-stroke sequences occurred. The initial stage lasted for 150 ms and was characterized by an initial current peak of 1.5 to 2 kA and an average current of some hundreds of amperes. The initial current pulse in Fig. 6.2c has some similarity, perhaps, to the initial current variation discussed in subsection 7.2.3.

The propagation speed, step length, and interstep interval of upward positive leaders, as reported by McEachron (1939) and Berger and Vogelsanger (1966, 1969), are summarized in Table 6.2 (see also Fig. 5.8). The speeds given in Table 6.2 for leaders observed at the Empire State Building are based on a sample of 20 events, whereas the step lengths and interstep intervals correspond to a single leader that exhibited 23 steps. No relation was found between step length and interstep interval in the Empire State Building studies. McEachron (1939) also reported average interstep intervals of 23, 18.6, and 24 μs for three other upward positive leaders. The leader characteristics in Table 6.2 reported by Berger and Vogelsanger (1966, 1969) are based on a sample of seven events. Steps were observed in the height range 40 m to approximately 1 km above the tower top, although at heights greater than 150 m above the tower top the channel remained faintly luminous between steps. Step length, interstep interval, and overall propagation speed were evaluated for three consecutive height intervals. Step lengths and interstep intervals were averaged over 5–30 steps. There was a tendency for step length to decrease with height. There was also a tendency for upward propagation speed to increase with height, which is similar to the trend observed in rocket-triggered lightning (see Fig. 7.7).

It is clear from Table 6.2 that the optically determined characteristics, speed, step length, and interstep interval, of upward positive leaders are similar for lightning at the Empire State Building in New York and at the two towers on Monte San Salvatore in Switzerland. They are also similar to the corresponding values for both upward positive leaders in rocket-triggered lightning (subsection 7.2.2) and downward negative stepped leaders (subsections 4.4.2 and 4.4.6). Note, however, that not all positive leaders exhibit pronounced stepping (see subsection 5.3.2).

6.2. General characterization

Table 6.2. *Characteristics of upward positive leaders determined from optical observations. Mean values are given in parentheses*

Study	Reference	Speed, m s^{-1}	Step length, m	Interstep interval, μs
Empire State Building	McEachron (1939)	$5.2 \times 10^4 - 6.4 \times 10^5$ (2.5×10^5)	6.2–23 (8.2)	20–100 (30)
Monte San Salvatore	Berger and Vogelsanger (1966, 1969)	$4 \times 10^4 - 1 \times 10^6$	4–40	40–120

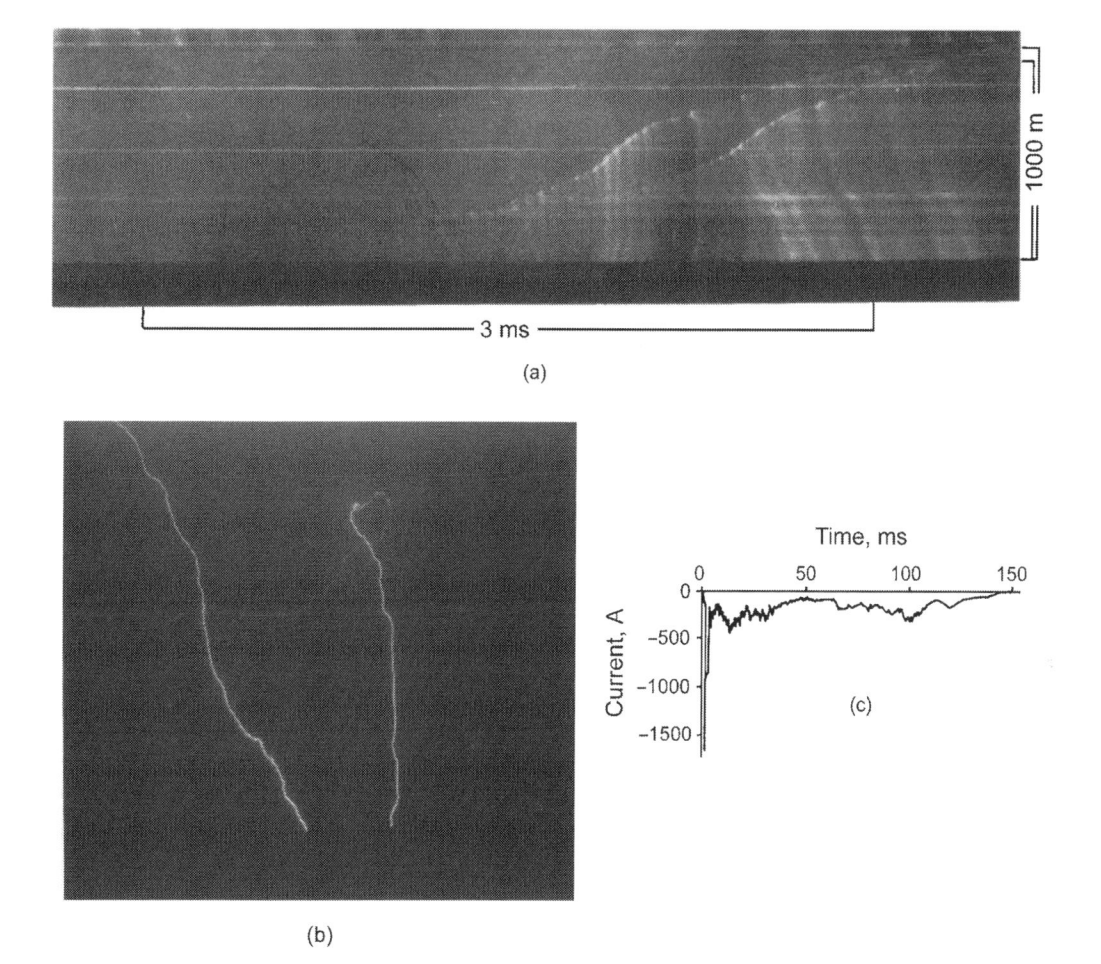

Fig. 6.2. Concurrent upward negative flashes from towers 1 and 2 on Monte San Salvatore (event 6558). (a) Streak photograph showing upward positive leaders from towers 1 and 2, (b) still photograph of the flashes (tower 1 is on the right), and (c) the current of the flash from tower 1, this flash being composed of the initial stage only (no return strokes). The streak photograph corresponds to the first 3–4 ms of the current record. Adapted from Berger and Vogelsanger (1966).

6.2.2. Upward positive lightning

Upward positive lightning discharges are initiated by upward negative leaders from the top of a grounded object. If such leaders are initiated in response to in-cloud discharges, as discussed in Section 6.1, (see also subsection 5.3.1), the upward negative leader from the object can be viewed as an extremely long upward connecting discharge that bridges the gap between the object and the in-cloud discharge channel. In this scenario, a current wave is expected to originate at a height of order 1 km or more above the top of the object, where the junction point between the upward negative leader and the cloud-discharge channel is to be found, and to propagate down the upward-negative-leader channel toward the object (there will be also

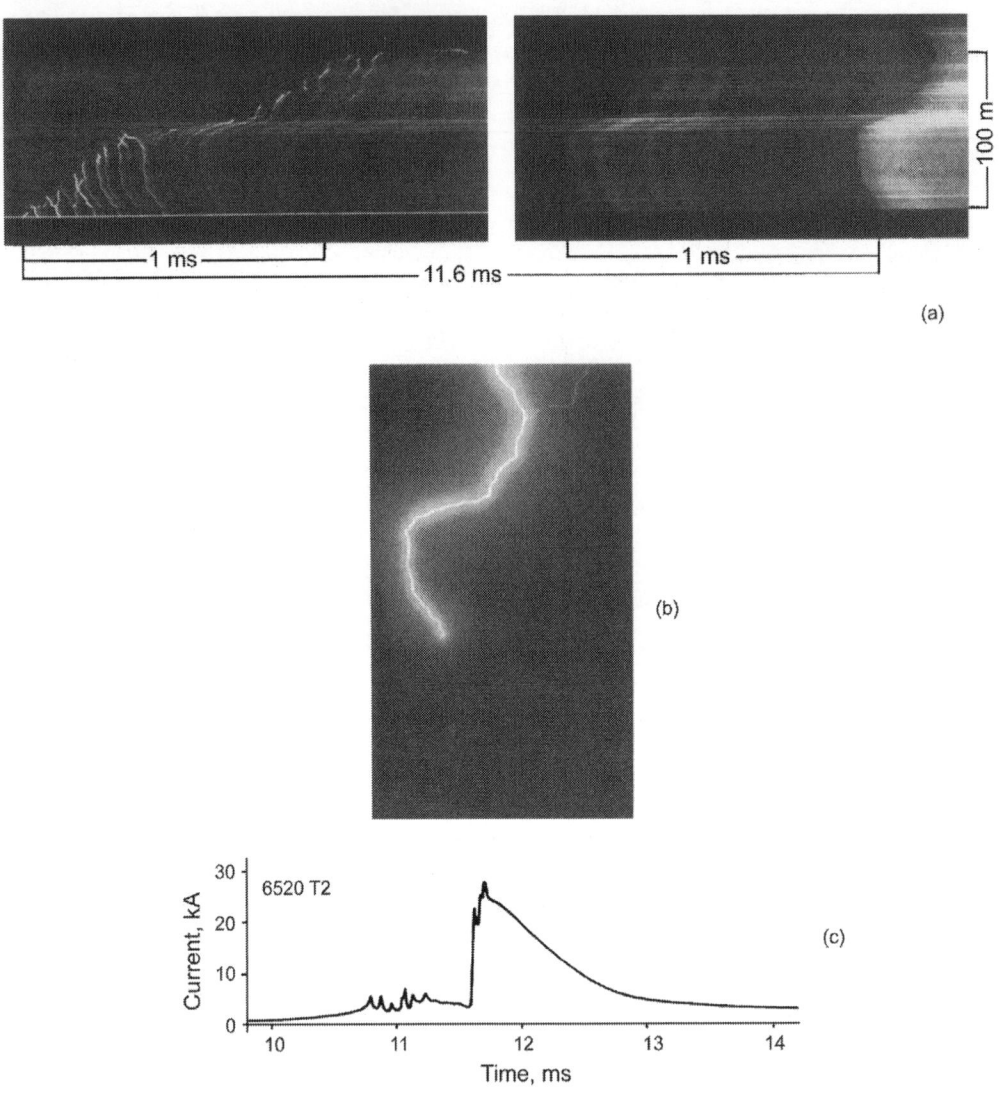

Fig. 6.3. Upward positive flash from tower 2 on Monte San Salvatore (event 6520). (a) Two segments of a streak photograph showing an upward negative leader and a bright channel illumination 11.6 ms after the beginning of the leader, (b) still photograph of the flash, and (c) the current measured at the tower. The large current pulse at 11.6 ms in (c) corresponds to the bright channel illumination in (a), right-hand panel. Berger and Vogelsanger (1969) estimated that the length of the upward negative leader before its contact with the charged in-cloud channel, which gave rise to the large current pulse, was about 1.2 km. In the original of (a), left-hand panel, a corona-like discharge is seen at the tip of the upward extending leader. Adapted from Berger (1977).

a wave propagating along the cloud-discharge channel). As discussed in subsection 5.3.1, such a current wave, called by Berger (1977) a return stroke because of its large magnitude, is more likely to result in the M-component mode of charge transfer to ground (Section 4.9) than an actual return stroke. The largest currents and charge transfers observed in the Monte San Salvatore studies are apparently associated with very long upward negative leaders making contact with in-cloud channel systems. As noted in Section 5.1, Berger and Garbagnati (1984) classified all 67 positive flashes in Switzerland as upward discharges, although some of those were probably downward discharges, as discussed in subsection 5.3.1. They also reported on five upward positive and seven downward positive discharges from tower measurements in Italy, the lightning type being identified solely on the basis of the recorded current waveforms.

Figure 6.3 gives data on an upward positive discharge from tower 2 on Monte San Salvatore. This flash was apparently composed of an upward negative leader that lasted for 11.6 ms (see the streak photographs in Figs. 6.3a) followed by a process that produced a current pulse having a peak value 28 kA, a maximum rate of rise 1 kA μs^{-1}, and a time to half-peak value 700 μs (see Fig. 6.3c). A still photograph of the flash is shown in Fig. 6.3b. We estimate from Fig. 6.3c that the risetime of the current pulse to its peak is of order 100 μs, although the rise to peak is not monotonic. A total charge of 35 C was transferred during the first 2 ms of the pulse. Berger and Vogelsanger (1969) inferred that the downward current wave that gave rise to the measured pulse originated at a height of about 1.2 km. The luminosity associated with the large current pulse shown in Fig. 6.3c appears in Fig. 6.3a, right-hand panel.

Optically determined characteristics of upward negative leaders from the Monte San Salvatore studies are given in Fig. 5.8d. The upward propagation speed typically ranged from 8×10^4 to 4.5×10^5 m s^{-1}, the step length from 3 to 20 m, and the interstep interval from 30 to 50 μs.

In the original photograph reproduced in Fig. 6.3a, left-hand panel, and in some other streak photographs of upward negative leaders, a corona-like discharge is observed at the tip of the leader. Further discussion of this feature is found in subsections 4.4.7 and 7.2.2. Berger and Vogelsanger (1969) noted that the corona discharge did not develop continuously between consecutive steps. It seems that corona discharge from the tip of a newly formed step was an integral part of the stepping process that was not resolved at their photographic-system time resolution, 5 μs. It appears from the description of one of the upward negative leaders given by Berger and Vogelsanger (1969) that the visible corona discharge extended from the leader tip over a distance of about 3 m, comparable to a typical step length of 3 to 20 m.

6.2.3. Upward bipolar lightning

All documented bipolar lightning discharges, those discharges during which charges of both polarities are transferred to ground, were initiated by upward-going leaders. Judging from channel-base current records, the leaders initiating these discharges can be either positively or negatively charged. No streak-camera photographs of an upward leader in a bipolar flash are presently available. Figure 6.4a shows a still photograph of two upward flashes from the two instrumented towers on Monte San Salvatore. Figure 6.4b shows the corresponding current records, both having a duration of about 550 ms. The flash seen on the left in Fig. 6.4a is an upward negative discharge from tower 2 (the type discussed in subsection 6.2.1) as evidenced by the bottom current trace in Fig. 6.4b. This flash apparently contained three downward-leader–upward-return-stroke sequences. In the current record (the bottom trace in Fig. 6.4b), the return-stroke pulses, having peaks of −7, −10, and −7 kA, are preceded by an initial-stage current lasting in excess of 300 ms and a no-current interval lasting some tens of milliseconds. The flash seen on the right in Fig. 6.4a is an upward bipolar flash initiated by an upward positive leader from tower 1. This flash first transferred negative charge and then positive charge to ground, as seen in the top current trace in Fig. 6.4b.

Bipolar lightning discharges are discussed in detail in Section 5.4.

6.3. Overall electrical characteristics

Like rocket-triggered flashes (Section 7.2), object-initiated discharges can be composed of only an initial stage or of an initial stage followed by downward-leader–upward-return-stroke sequences. Table 6.3 gives, in the last column, the percentage of object-initiated lightning discharges that contain return strokes, as observed in three different studies. The initial stage is often associated with multiple upward branches that may pervade extensive charged regions in the cloud, the processes occurring in each individual branch being largely independent of processes in the other branches. This latter feature might explain why object-initiated lightning discharges are capable of transferring larger charges to ground than downward discharges, and it also explains the observation that apparently it is only object-initiated discharges that can tap both negatively charged and positively charged cloud regions. The branched channel system created during the initial stage apparently can support both the M-component mode and the leader–return-stroke mode of charge transfer to ground (Section 1.2), depending on whether the branch in question carries continuous current.

Parameters of upward negative, upward positive, and upward bipolar discharges, as reported by Berger (1978), are summarized in Tables 6.4, 6.5, and 6.6, respectively. The median values of maximum initial-stage current, total flash charge, and total flash duration observed for upward negative discharges from three different studies are compared in Table 6.3. Impulsive current components in upward discharges, including return-stroke current waveforms, are characterized in Section 6.4 below.

The median values of total flash duration in Table 6.3 range from 114 to 338 ms and are comparable, within a factor of 2 to 4, to the median flash durations of 350 and 470 ms reported by Hubert (1984) for rocket-triggered lightning in France and New Mexico, respectively. Natural downward cloud-to-ground and cloud flashes also typically last some hundreds of milliseconds (Sections 4.2 and 9.2). Wang et al. (1999a) found the geometric mean duration of the initial-stage current to be 279 ms for rocket-triggered lightning in Florida (Camp Blanding) and Alabama. They found little difference in this parameter for flashes containing and not containing return strokes.

The median values of total flash charge in Table 6.3 range from 5.7 to 23 C as against the values 50 and 32 C reported by Hubert (1984) for rocket-triggered lightning in

Table 6.3. *Comparison of 50 percent values of maximum initial-stage current, total flash charge, and total flash duration for upward negative flashes from different studies*

Reference	Object	Maximum initial-stage current, A	Total charge, C	Total flash duration, ms	Percentage of flashes with return strokes
Hagenguth and Anderson (1952)	Empire State Building, New York	250	19	270	50
Berger (1978)	Two towers on Monte San Salvatore, Switzerland	203^a 248^b	12^a 23^b	163^a 338^b	20–5
Gorin and Shkilev (1984)	Ostankino TV tower, Moscow	146^a 300^b	5.7^a 10^b	114^a 220^b	27

^{a} Flashes not containing return strokes.
^{b} Flashes containing return strokes.

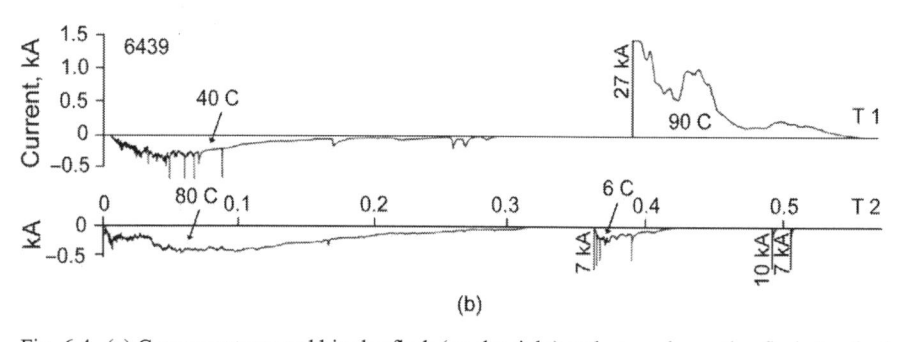

Fig. 6.4. (a) Concurrent upward bipolar flash (on the right) and upward negative flash (on the left) from towers 1 and 2, respectively, on Monte San Salvatore (event 6439). (b) The currents measured at tower 1 (top trace) and at tower 2 (bottom trace). The time scale (which is in seconds) on the bottom trace also applies to the top trace. Adapted from Berger and Vogelsanger (1966).

6.3. Overall electrical characteristics

Table 6.4. *Lightning current parameters for upward (object-initiated) negative flashes. Adapted from Berger (1978)*

Parameter	Unit	Sample size	Percentage exceeding tabulated value 90%	50%	10%
Maximum initial-stage current in flashes without return strokes	A	639	40	203	1030
Maximum initial-stage current in flashes with return strokes	A	195	47	248	1310
Maximum return-stroke current	kA	176	4.2	10	25
Initial-stage charge in flashes without return strokes	C	638	1.9	12	69
Total charge in flashes with return strokes	C	172	5.4	23	100
Return-stroke charge	C	579	0.14	0.77	4.1
Front duration for return strokes	μs	696	0.3	1	4
Maximum dI/dt for return strokes	kA μs^{-1}	710	5.6	26	123
Action integral for return strokes	A^2 s	398	5×10^2	2.3×10^3	10^4
Duration of flashes without return strokes	ms	639	65	163	407
Duration of flashes with return strokes	ms	212	144	338	791
Return-stroke duration	ms	888	0.57	3.6	22

Table 6.5. *Lightning current parameters for upward (object-initiated) positive flashes. Adapted from Berger (1978). Note that some of the impulsive current components represented in this table were considered by Berger et al. (1975) as occurring in downward positive as opposed to upward positive flashes, as discussed in Section 5.1*

Parameter	Unit	Sample size	Percentage exceeding tabulated value 90%	50%	10%
Maximum current in flashes without large impulsive components	kA	132	0.21	1.5	11
Maximum current in flashes with large impulsive components	kA	35	10	36	127
Charge in flashes without large impulsive components	C	137	3.7	26	187
Charge in flashes with large impulsive components	C	35	20	84	348
Front duration for impulsive components	μs	23	4.5	39	340
Maximum dI/dt for impulsive components	kA μs^{-1}	24	0.28	1.9	12
Action integral for impulsive components	A^2 s	35	5×10^4	6.6×10^5	9×10^6
Duration of flashes without large impulsive components	ms	138	24	72	215
Duration of flashes with large impulsive components	ms	34	19	68	240

Table 6.6. *Lightning current parameters for upward (object-initiated) bipolar flashes. The initial current polarity was negative in all cases. Adapted from Berger (1978)*

Parameter	Unit	Sample size	Percentage exceeding tabulated value 90%	50%	10%
Maximum negative current	A	30	50	350	2400
Maximum positive current	kA	30	0.18	1.5	13
Negative charge	C	30	0.8	12	181
Positive charge	C	30	1.5	25	345

France and New Mexico, respectively. Natural downward flashes also typically transfer charges of the order of tens of coulombs (Section 4.2). For Florida rocket-triggered lightning, the geometric mean values of the charge transferred to ground during the initial stage are 30 and 21 C for flashes containing and not containing return strokes, respectively (Wang *et al.* 1999a). The median values of the maximum initial-stage current in Table 6.3 range from 146 to 300 A. Wang *et al.* reported that the geometric mean values of the average initial-stage current in Florida rocket-triggered lightning were 110 and 69 A for flashes containing and not containing return strokes, respectively.

Note that the median value of the maximum initial-stage current in negative flashes without return strokes, 203 A (Table 6.4), is considerably smaller than the median value of the maximum initial-stage current in positive flashes without large impulsive components, 1.5 kA (Table 6.5). In agreement with this observation, Horii and Ikeda (1985) reported, for winter lightning in Japan, that the upward positive leaders that initiate upward negative flashes are characterized by lower currents than the upward negative leaders that initiate upward positive flashes. Also, Berger (1977) observed that upward positive leaders are less luminous than upward negative leaders.

Berger and Vogelsanger (1969) pointed out that the charge transfer in upward flashes of either polarity tends to be larger than in downward flashes. They attributed the greater charge transfer to the large number of upward branches that can facilitate access to the cloud charge, distributed over an area of several square kilometers. However, McEachron (1939) reported that most of the upward discharges from the Empire State Building, all initiated by upward positive leaders, showed no branching. In Berger's studies, upward negative flashes with return strokes lasted two to three times longer and lowered to ground charge amounts two to three times larger than did downward negative flashes.

6.4. Impulsive currents

Current pulses measured at the strike object during an upward lightning discharge can be divided into three main categories. The first category includes pulses superimposed on the initial-stage (IS) current. Most of these pulses are probably the same as the initial continuous current (ICC) pulses studied by Wang *et al.* (1999a) in rocket-triggered lightning (subsection 7.2.3). The second category includes return-stroke pulses that are necessarily preceded by an essentially zero-current interval, as discussed in subsection 4.7.8. The third category includes the M-component pulses that are superimposed on the continuing currents following some return strokes. In rocket-triggered lightning, ICC pulses were found to be similar to M-component pulses (Wang *et al.* 1999a). In the following,

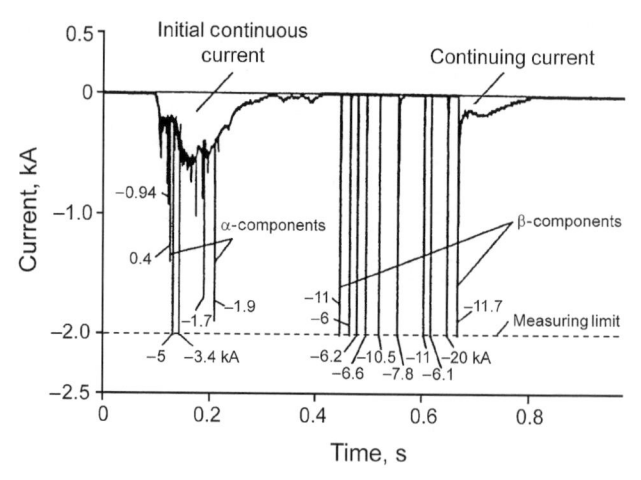

Fig. 6.5. Overall current record for an upward negative flash initiated from the Peissenberg tower in Germany. The record was intentionally clipped at -2 kA level in order to accentuate the lower-current components. The record begins with the slowly varying initial-stage current followed, after a no-current interval, by 10 return-stroke pulses, labeled β-components. Superimposed on the initial-stage current are pulses, labeled α-components. The last return-stroke pulse is followed by long-continuing current. Adapted from Fuchs *et al.* (1998).

we will refer to both M-component pulses and ICC pulses as M-component-type pulses. An example of a current record for an upward flash showing the first two types of pulse is given in Fig. 6.5. This record is taken from Fuchs *et al.* (1998) who labeled ICC pulses as α-components and return-stroke pulses (a total of 10) as β-components. The last return-stroke pulse in Fig. 6.5 is followed by continuing current, but no pronounced pulses are superimposed on this current. The characteristics of return-stroke pulses and M-component-type pulses usually differ considerably, as discussed in Section 4.9. Nevertheless, in some studies of lightning at tall objects (e.g., Miyake *et al.* 1992) and in some triggered-lightning studies (e.g., Hubert *et al.* 1984), no distinction is made between these two types of pulse; statistics are given for all current pulses exceeding a specified amplitude level.

In the following, we present statistics for various parameters of ICC pulses (α-components) and return-stroke pulses (β-components), based on current-derivative measurements made at the top of the Peissenberg tower in Germany (Fuchs *et al.* 1998). In doing so, we use the original symbols introduced by Fuchs *et al.* (1998), some of which are different from those used elsewhere in the book. The current waveforms were obtained by numerical integration of the measured current-derivative waveforms. Statistics are given in Table 6.7 for the peak current i_{max}, the maximum current derivative, the 10–90 percent steepness of the initial part of double-peak current waveforms, $S_{i/1}$, and the 10–90 percent steepness of the single-peak

6.4. Impulsive currents

Table 6.7. *Statistical characterization of impulsive currents in negative lightning initiated from the Peissenberg tower, Germany. Impulsive currents with amplitudes lower than 1 kA or with risetimes of about 10 μs or larger were generally not included in this statistical characterization (F. Heidler, personal communication, 2000). Adapted from Fuchs et al. (1998)*

Parameter	Unit	Sample size	Max. value	Min. value	Mean value	Percentage exceeding tabulated value			
						99%	50%	5%	1%
Peak current, i_{max}									
α and β	kA	125	21	1.1	6.1	0.95	4.8	15	25
α	kA	90	16	1.1	4.8	0.87	3.9	11	17
β	kA	35	21	2.0	9.5	2.7	8.5	20	28
Maximum current derivative									
α and β	kA μs^{-1}	125	155	0.97	23	0.58	11	87	207
α	kA μs^{-1}	89	74	0.97	11	0.69	6.8	34	67
β	kA μs^{-1}	36	155	1.2	51	2.7	36	219	465
10–90 percent steepness of the initial part of double-peak current waveforms, $S_{i/1}$									
α and β	kA μs^{-1}	93	52	0.69	8.2	0.64	5.4	24	45
α	kA μs^{-1}	62	20	0.69	5	0.74	3.9	13	21
β	kA μs^{-1}	31	52	1.2	15	1.1	10	48	92
10–90 percent steepness of single-peak current waveforms, $S_{i/max}$									
α and β	kA μs^{-1}	33	7.7	0.29	2	0.22	1.5	5.6	9.6

current waveforms, $S_{i/max}$. The definitions of the latter two parameters are indicated on the two typical current waveforms shown in Fig. 6.6. Impulsive lightning currents were found in 43 out of 80 upward flashes initiated from the Peissenberg tower. The total number of current pulses was 323, the maximum number of pulses per flash being 43. A majority of the pulses, 249, occurred during the initial stage; these were classified as α-components. The remaining 74 pulses, probably associated with downward-leader–upward-return-stroke sequences, were classified as β-components. The statistics in Table 6.7 are given for the α- and β-components separately and combined for all parameters except $S_{i/max}$, for which the statistics are given only for α- and β-components combined because the number of β-components suitable for measuring $S_{i/max}$ was only five. The statistics include maximum and minimum values, mean values, and 99 percent, 50 percent, five percent, and one percent values based on approximation of the data by a log-normal distribution. The percentages indicate the probability that the corresponding value of the parameter will be exceeded. Interestingly, both maximum current derivatives and current peaks in the Peissenberg tower data were somewhat lower in winter (when the top of the tower was often inside the cloud) than in summer (F. Heidler, personal communication, 2000).

As seen in Table 6.7, the return-stroke pulses (the β-components) are characterized by i_{max} values about a factor 2 higher, maximum current derivative values a factor 5 higher, and $S_{i/1}$ values a factor 3 higher than corresponding values for the ICC pulses (the α-components). The observation that ICC pulses have smaller peaks and smaller rates of rise to peak is consistent with other measurements for object-initiated and rocket-triggered lightning, as discussed next. Berger (1977) reported that the pulses superimposed on the ICC were relatively small, less than 10 kA, while the return-stroke current pulses had peaks mostly in the range from 10 to 30 kA. Further, Hubert et al. (1984) reported that ICC pulses had a relatively slow risetime; for 50 percent it was greater than about 20 μs. Wang et al. (1999a) reported on the detailed characteristics of ICC pulses in rocket-triggered lightning flashes in Florida (Camp Blanding) and Alabama. They found that the geometric mean ICC pulse peak was 144 A, the geometric-mean 10–90 percent risetime was 528 μs, and the geometric mean charge transfer was 143 mC; these values were similar to their counterparts for M-component pulses (subsection 4.9.3) and distinctly different from those for return-stroke pulses. The current peak and 10–90 percent steepness for return-stroke pulses in different types of flash are compared in Tables 6.8 and 6.9, respectively. It appears that the parameters of the β-components studied by Fuchs et al. (1998) are of the same order of magnitude, although somewhat lower (particularly the 10–90 percent rate of rise), as the corresponding parameters of (i) return-stroke pulses in object-initiated lightning from other studies, (ii) subsequent return-stroke pulses in natural downward lightning, and (iii) return-stroke pulses in rocket-triggered lightning. However, the current peaks for α-components studied by Fuchs et al. (1998) are typically

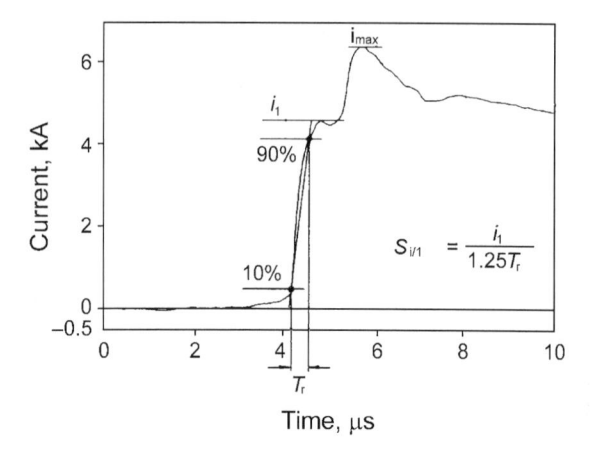

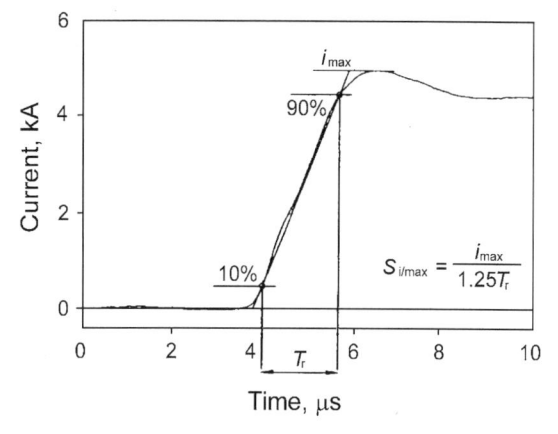

Fig. 6.6. Typical impulsive current waveforms recorded at the top of the Peissenberg tower. Definitions of the various parameters, presented in Table 6.7, are shown. The waveforms are due to negative return strokes labeled β-components in Table 6.7. Adapted from Fuchs et al. (1998).

about an order of magnitude higher than those for the ICC pulses studied by Wang et al. (1999a). Further, the risetime of the order of microseconds reported by Fuchs et al. (1998) for α-components is about two orders of magnitude smaller than the risetime of the order of hundreds of microseconds found by Wang et al. (1999a) for ICC pulses. There are three possible reasons for the apparent dissimilarity between the α-components of Fuchs et al. and the ICC pulses studied by Wang et al. Firstly, Wang et al. considered all measurable pulses, while Fuchs et al. analyzed only the larger pulses. Indeed, the smallest peak reported by Fuchs et al. for α-components was about 1.1 kA, whereas many pulses with peaks less than 1 kA (absolute value) are seen in the example given in Fig. 6.5. Secondly, multiple upward branches could have facilitated the simultaneous occurrence of a continuous current in one branch and a downward leader in another branch, in the Peissenberg tower flashes, as noted in subsection 4.7.8 for Monte San Salvatore and Ostankino tower flashes. This hypothesis implies that triggered-lightning flashes in Florida and Alabama are less likely than lightning at the Peissenberg tower to have upward branches originating near the strike point. Perhaps in support of this hypothesis is the observation of Laroche et al. (1985) regarding a lack of low-altitude branching in Florida (KSC) triggered lightning, as opposed to that in triggered lightning in New Mexico. Thirdly, the charge sources for α-current pulses in thunderclouds over the Peissenberg tower might have been located closer to the lightning attachment point than the sources of ICC pulses in Florida and Alabama. In this case, because of the longer propagation path between the in-cloud source and the lightning attachment point, the fronts of downward-propagating current waves in the Peissenberg tower flashes might have suffered less degradation due to dispersion and attenuation than their counterparts in Florida and Alabama rocket-triggered flashes. The peak rate of rise of the return-stroke current in object-initiated lightning is compared with its counterparts in natural downward lightning and in rocket-triggered lightning in Table 6.10. A similar comparison for the return-stroke charge transfer is given in Table 6.11.

6.5. Lightning current reflections within tall objects

A lightning current waveform, or any waveform for that matter, can be viewed as the superposition of a large (strictly speaking infinite) number of frequency components, each having a wavelength $\lambda = cf^{-1}$ where c is the speed of light in free space and f is the frequency in hertz. If the height of a strike object is comparable to or greater than some of these wavelengths, usually the shortest wavelengths that are associated with the initial rising portion of the lightning return-stroke current waveform, then the object behaves (at the corresponding frequencies) as a distributed, as opposed to lumped, circuit. As a result, different current waveforms will be observed at different points along such an "electrically long" object. Indeed, as shown in Fig. 6.7, very different typical lightning current waveforms were reported by Gorin and Shkilev (1984) from their measurements at heights of 47, 272, and 533 m on the 540-m-high Ostankino tower in Moscow. The median peak currents from their measurements at 47 and 533 m were 18 and 9 kA, respectively, as seen in Table 6.8. Gorin and Shkilev used current oscillograms recorded near the tower top (at 533 m) for the cases in which the current risetime was smaller than the time (about 3.5 µs) required for a current wave to travel (at the speed of light) from the tower top to its base and back, in order to estimate the equivalent impedance of the lightning channel, assumed to be a real number. Their estimates varied from 600 Ω to 2.5 kΩ when the characteristic impedance of the tower was assumed to be 300 Ω and the grounding resistance was assumed to be zero (the low-current, low-frequency value was about 0.2 Ω ; Gorin et al. 1977). Thus, Gorin and

Table 6.8. *Comparison of return-stroke current peaks (in kA) in natural upward (object-initiated), natural downward, and rocket-triggered lightning that effectively transported negative charge to ground*

Reference	Location	Sample size	Percentage exceeding tabulated value					
			99%	95%	90%	50%	10%	5%
Return strokes in object-initiated flashes								
Fuchs et al. (1998)	Germany	35 (β components)	2.7	—	—	8.5	—	20
Gorin and Shkilev (1984)	Russia	58 (measurements at 533 m above ground)	—	—	4	9	19	—
		76 (measurements at 47 m above ground)	—	—	4[a]	18[a]	40[a]	—
Berger (1978)	Switzerland	176	—	—	4.2	10	25	—
Hagenguth and Anderson (1952)	New York	84[b]	—	4	5	10	27	35
Subsequent return strokes in natural downward flashes								
Anderson and Eriksson (1980)	Switzerland	114	—	4.9	—	12	—	29
Return strokes in rocket-triggered flashes[c]								
Fisher et al. (1993)	Florida (KSC) and Alabama	45	—	4.7	—	13	—	29
Depasse (1994)	Florida (KSC)	305	—	4.7	—	12	—	31
	France	54	—	4.5	—	9.8	—	22

[a] Overestimates due to transient processes in the tower (see Section 6.5).
[b] Two events out of 84 were of positive polarity.
[c] Additional information on current peaks in rocket-triggered lightning is found in subsections 7.2.4 and 7.2.5.

Table 6.9. *Comparison of 10–90 percent rates of rise (in kA μs^{-1}) of return-stroke current in natural upward (object-initiated), natural downward, and rocket-triggered lightning that effectively transported negative charge to ground*

Reference	Location	Sample size	Percentage exceeding tabulated value					
			99%	95%	90%	50%	10%	5%
Return strokes in object-initiated flashes								
Fuchs et al. (1998)	Germany	31[a]	1.1	—	—	10	—	48
Hagenguth and Anderson (1952)	New York	71	—	—	2	13	34	—
Subsequent return strokes in natural downward flashes								
Anderson and Eriksson (1980)	Switzerland	114	—	3.3	—	15	—	72
Return strokes in rocket-triggered flashes								
Fisher et al. (1993)	Florida (KSC) and Alabama	43	—	5.4	—	34	—	83
Leteinturier (as reported by Fisher et al. 1993)	Florida (KSC)	36	—	16	—	44	—	72

[a] β components.

Shkilev (1984) inferred that the characteristic impedance of the tower was appreciably lower than the equivalent impedance of the lightning channel (Kostenko 1995).

We now discuss the transient behavior of tall towers deduced from measurements of the time rate of change of the current, that is, the derivative of the current with respect to time, in the tower. It is generally easier to identify the current-derivative peak than the current peak during a transient process in the tower because the derivative pulse is sharper. When the current derivative with respect

Table 6.10. *Comparison of maximum rates of rise (in kA μs^{-1}) of return-stroke current in natural upward (object-initiated), natural downward, and rocket-triggered lightning that effectively transported negative charge to ground*

			Percentage exceeding tabulated value				
Reference	Location	Sample size	95%	90%	50%	10%	5%
Return strokes in object-initiated flashes							
Berger (1978)	Switzerland	710	—	5.6	26	123	—
Subsequent return strokes in natural downward flashes							
Berger et al. (1975)	Switzerland	122	12	—	40	—	120
Return strokes in rocket-triggered flashes[a]							
Depasse (1994)	Florida (KSC)	134	28	—	91	—	298
Depasse (1994)	France	47	15	—	37	—	91

[a] Additional information on maximum rates of rise in the return-stroke current in rocket-triggered lightning is found in subsection 7.2.4.

Table 6.11. *Comparison of the charge (in C) of return strokes in natural upward (object-initiated), natural downward, and rocket-triggered lightning that effectively transports negative charge to ground*

			Percentage exceeding tabulated value				
Reference	Location	Sample size	95%	90%	50%	10%	5%
Return strokes in object-initiated flashes							
Hagenguth and Anderson (1952)	New York	83[a]	—	—	0.15	1.3	—
Berger (1978)	Switzerland	579	—	0.14	0.77	4.1	—
Subsequent return strokes in natural downward flashes							
Berger et al. (1975)	Switzerland	117	0.22	—	0.95	—	4.0
Return strokes in rocket-triggered flashes							
Depasse (1994)	France	24	0.27	—	0.59	—	1.3

[a] This sample includes one or two events of positive polarity. Charge was determined only up to half-peak value on the tail of the current waveform.

to time is measured near the tower top, the arrival of the current wave reflected from the bottom of the tower to the tower top is manifested by a second pronounced current-derivative pulse, separated from the first current-derivative pulse associated with the incident current wave by the time required for the current wave to make a round trip down and up the tower at nearly the speed of light. Of course, this feature is seen only if the incident current risetime is smaller than the current round-trip time. Two examples of such current-derivative records and the associated current records, obtained on the 160-m-high Peissenberg tower near Munich in Germany, are given in Fig. 6.8. Additionally given in Fig. 6.8 are corresponding magnetic- and electric-field-derivative waveforms measured about 200 m from the tower and magnetic and electric field waveforms obtained by integration of the respective measured field-derivative waveforms. The separation between the negative current derivative peaks in Fig. 6.8 is 1.15 μs, which corresponds to twice the time needed for a wave to traverse the 160-m-high tower at a speed of 278 m μs^{-1}, slightly less than the speed of light, 300 m μs^{-1}. If the foundations of the Peissenberg tower, which extend about 10 m below the ground surface, are taken into account, the propagation speed becomes nearly equal to the speed of light (F. Heidler, personal communication, 2000).

Montandon and Beyeler (1994) presented, for two lightning events, current-derivative waveforms measured at heights of 248 m and 175 m above ground on the 248-m-high St Chrischona tower near Basel, Switzerland. For one of the events the separation between the current-derivative pulses was 1.66 μs at 248 m and 1.16 μs at 175 m above ground, consistent with a wave propagation speed nearly equal to the speed of light. For the second event presented by Montandon and Beyeler (1994), the current-derivative waveforms at both heights appeared as singly peaked pulses.

6.5. Lightning current reflections within tall objects

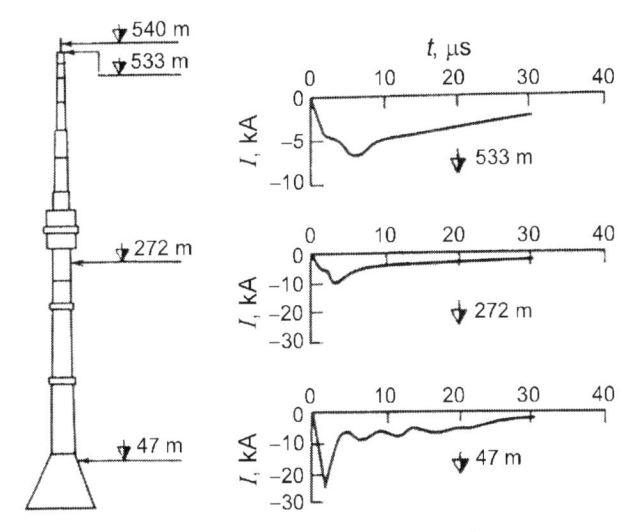

Fig. 6.7. Typical impulsive (probably return-stroke) current waveforms for upward negative lightning recorded near the top (at 533 m), in the middle (at 272 m), and near the bottom (at 47 m) of the 540-m-high Ostankino tower in Moscow. Adapted from Gorin and Shkilev (1984).

Shostak et al. (1999b) presented current-derivative waveforms showing evidence of reflections, measured 474 m above ground on the 553-m-high CN tower in Toronto, Canada, and the associated electric and magnetic field waveforms measured a distance of 2 km from the tower. They also showed the current waveforms obtained by integration of the measured current-derivative waveforms.

Janischewskyj et al. (1996), from their analysis of five current waveforms measured 474 m above ground on the CN tower, inferred average current reflection coefficients at the bottom and at the top of the tower of about 0.40 and −0.37, respectively. The reflection coefficient at the tower bottom varied from 0.34 to 0.43, while that at the top varied from −0.27 to −0.49. The corresponding ratios of the equivalent impedance of the lightning channel and the characteristic impedance of the CN tower ranged from 1.7 to 2.9. The reflection coefficients reported by Janischewskyj et al. (1996) were estimated assuming that the CN tower could be represented by a uniform lossless transmission line. Rusan et al. (1996) considered CN-tower models composed of one to three sections of lossless transmission lines with different parameters. It appears from their analysis that the inclusion of reflections from discontinuities in the tower can alter the inferred reflection coefficients at the tower extremities. Information on the modeling of lightning strikes to the CN tower is found in subsection 12.2.7 (second part).

Beierl (1992), from his analysis of current waveforms measured near the top of the 160-m-high Peissenberg tower, estimated that the current reflection coefficient was around unity at the bottom of the tower and about −0.5 at the top of the tower. The latter implies that the equivalent impedance of a lightning channel (Kostenko 1995) was a factor 3 or so larger than the characteristic impedance of the tower, consistent with the findings of Gorin and Shkilev (1984) for the Ostankino tower. Fuchs (1998b), from 13 simultaneous current measurements at the top and bottom of the Peissenberg tower, found that the average current reflection coefficients at the bottom and top of the tower were 0.70 and −0.53, respectively. The range of variation was from 0.64 to 0.81 at the bottom and from −0.39 to −0.68 at the top.

All the current reflection coefficients given above were estimated assuming that the tower could be represented by a lossless transmission line (or lines), that the reflection coefficients are each a constant real number, and that reflections within the lightning channel could be neglected. In reality, the magnitude of the reflection coefficient at the tower top is expected to decrease as the equivalent impedance of the lightning channel decreases with increasing lightning current (e.g., Kostenko 1995). The reflection coefficient at the tower bottom can be influenced by nonlinear effects in the soil that depend on current magnitude (subsection 18.3.4). Each of the two reflection coefficients is in general a function of frequency. Interestingly, Fuchs (1998b, Figs. 10 and 11) found that the current reflection coefficients at the bottom and at the top of the Peissenberg tower estimated for 11 lightning strokes were apparently independent of either lightning current peak (ranging from about 1 to 8 kA) or maximum current rate of rise (ranging from about 5 to 60 kA μs^{-1}).

Wang et al. (1995), using an eight-channel photodiode system, recorded luminosity waveforms as a function of time within 180 m of the top of the CN tower. They obtained, in 1991–2, a total of 45 luminosity profiles (apparently one per flash) that were suitable for analysis. No simultaneous current or electromagnetic field measurements were available for these events. Out of the 45 events, seven exhibited light profiles characteristic of dart-leader–return-stroke sequences (subsection 4.7.1). The remaining 38 events exhibited relatively slowly rising light waveforms similar to those observed for M-components (subsection 4.9.2). For 35 of these 38 slow waveforms, the risetimes were estimated to vary from 9 to 102 μs, with a geometric mean value of about 35 μs. Out of the 38 events with relatively slow light waveforms, six appeared to propagate upward from the tower top while 32 exhibited a bidirectional progression originating at a height several tens of meters above the tower top. No luminosity waves moving either up or down were detected prior to the bidirectional processes. The upward propagation speed (one-dimensional) was estimated for 38 events and ranged from 8×10^6 to 2.9×10^8 m s^{-1}, with a geometric mean value of about 2.8×10^7 m s^{-1}. The downward propagation speed, estimated for only 22 events,

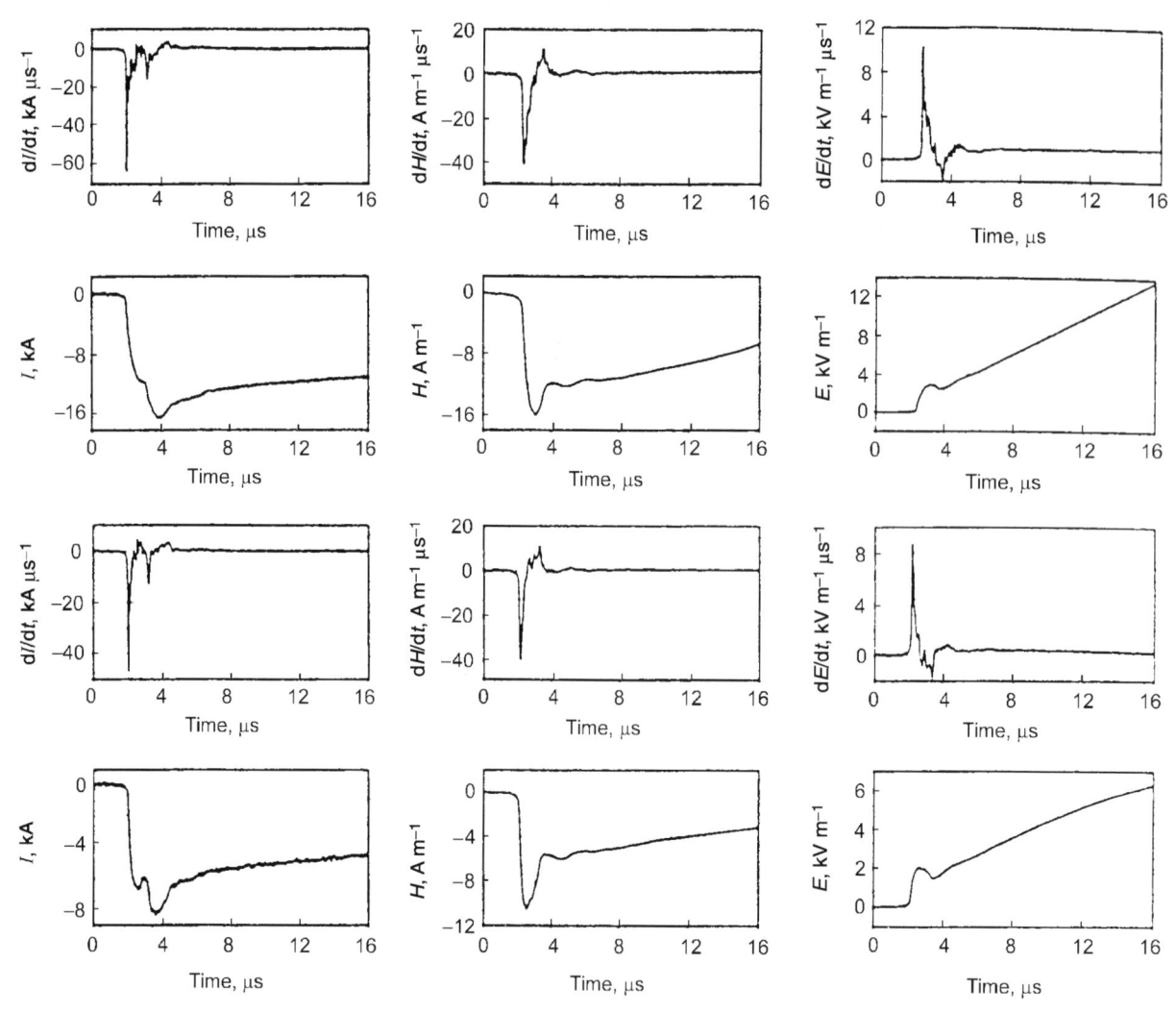

Fig. 6.8. Waveforms for current derivative dI/dt, current I, magnetic field derivative dH/dt, magnetic field H, electric field derivative dE/dt, and electric field E, for the two impulsive events (probably return strokes) in an upward negative flash initiated from the Peissenberg tower on 19 July 1993. The dI/dt and I waveforms were measured at the tower top, and the dH/dt and dE/dt waveforms were measured about 200 m from the tower. The H and E waveforms were obtained by numerical integration of the dH/dt and dE/dt waveforms and, therefore, could have become distorted after some microseconds. Adapted from Zundl (1994b).

varied from 10^7 to 2.5×10^8 m s^{-1}, with a geometric mean value of about 4.9×10^7 m s^{-1}. The ratio of the upward and downward propagation speeds, for 22 events, varied from 0.17 to 1.5, with a geometric mean value of about 0.60. The physics of the bidirectional events is unclear. It is possible that they occurred during the initial stage of upward flashes.

As follows from the above discussion, impulsive lightning currents measured on tall objects can be influenced ("contaminated") by the transient processes that occur in the tower in response to the higher-frequency components of the current waveform. These higher frequencies are associated with the current rise to peak and may be associated with other parts of the current waveform. As a result, the statistics on lightning-current parameters based on tall-object measurements may be not representative of lightning discharges directly to ground. The estimation of this potential bias is important since current measurements for natural lightning are usually performed on relatively tall towers (Table 6.1). In the following, we will estimate the magnitude of such a bias. Most importantly, we will show that the tower-top measurements of peak currents on Monte San Salvatore, which are usually viewed as the primary reference for both lightning research and lightning protection design, are not significantly affected by the transient response of the 70-m-high towers.

6.5. Lightning current reflections within tall objects

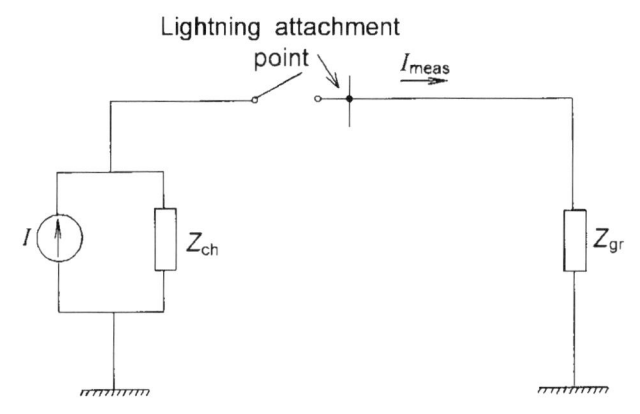

Fig. 6.9. Equivalent circuit for the case of a lightning strike to ground or to an object of negligible height. The lightning is represented by a Norton equivalent circuit composed of an ideal current source (the short-circuit current I) in parallel with a lightning-channel equivalent impedance Z_{ch}. Z_{gr} is the effective grounding impedance at the lightning attachment point, and I_{meas} is the current that would be measured at the attachment point.

Let us first consider the case of lightning attachment to ground either directly or via an object of negligible height. If we represent the lightning by a Norton equivalent circuit (e.g., Carlson 1996), as also discussed in subsection 7.2.5 and illustrated in Fig. 6.9, the lightning current measured at ground is given by $I_{meas} = I Z_{ch}/(Z_{ch} + Z_{gr})$, where Z_{ch} is the equivalent impedance of the lightning channel, Z_{gr} is the grounding impedance at the lightning attachment point, and I is the lightning current that would be measured at the attachment point if Z_{gr} were equal to zero (the short-circuit current). In the following, we will assume that Z_{gr} is purely resistive, that is, $Z_{gr} = R_{gr}$, and that Z_{ch} is a real number. Further, we will neglect nonlinear processes that could lead to the dependence of Z_{ch} and Z_{gr} on lightning current. If the lightning were an ideal current source ($Z_{ch} = \infty$) then the lightning current measured at ground I_{meas} would always be equal to the short-circuit current I and, hence, independent of Z_{gr} (as long as it was finite), that is, independent of the conditions at the strike point. As discussed later in this section, for the case $Z_{ch} = \infty$ even the presence of a tall ("electrically long") strike object has no effect on the current measured at the lightning attachment point (at the top of the object). In the following, we will refer to the short-circuit current as the "undisturbed" current. It is this current that is used to build a Norton equivalent source for modeling direct lightning strikes to various objects and systems. In reality, Z_{ch} ranges from hundreds to thousands of ohms, and the measured current is nearly equal to the short-circuit current ($I_{meas} \approx I$) when $Z_{gr} \ll Z_{ch}$, the two idealized special cases being $Z_{ch} = \infty$ (an ideal current source) and $Z_{gr} = 0$ (ideal grounding). It follows that, in practice, the "undisturbed" current would be that measured at a well-grounded object of negligible height. As will be discussed in subsection 7.2.5, triggered-lightning return-stroke currents measured under very different grounding conditions are similar, suggesting that lightning is capable of lowering the grounding impedance encountered at the attachment point to a value much lower than Z_{ch}. Such lowering of the grounding impedance is apparently facilitated by plasma channels that develop in the ground and along the ground surface.

We now extend the Norton equivalent circuit shown in Fig. 6.9 to include a tall strike object represented by a lossless transmission line with characteristic impedance Z_0. In doing so, we neglect the presence of any upward connecting leader from the strike object. The extended equivalent circuit is shown in Fig. 6.10a. Also given in Fig. 6.10a are expressions for the reflection coefficients for the current at the top, ρ_T, and at the bottom, ρ_B, of the object. Both ρ_T and ρ_B are assumed to be constants, that is, any nonlinear processes resulting in the dependence of Z_{ch} or Z_{gr} on current are neglected. If the switch closes at $t = 0$, then the first current reflection occurs at the bottom of the object at $t = \tau$, where τ is the time required for a wave to traverse the length of the object. For $t < \tau$, the downward moving wave "sees" the characteristic impedance of the object Z_0, and the magnitude of the initial current wave, I_0, is found from the equivalent circuit shown in Fig. 6.10b. Note that the "undisturbed" current is still I, as in the case of an object of negligible height, considered above, and that usually $I_0 < I$.

We will now assume that (i) the short-circuit current is a step function of magnitude I, (ii) $Z_{gr} = 0$, so that $\rho_B = 1$, and (iii) $Z_{ch} = 3Z_0$, as found from measurements on both the Ostankino and Peissenberg towers, so that $\rho_T = -0.5$. Under these assumptions, the currents at the top, I_T, and at the bottom, I_B, of the strike object versus time are as shown in Figs. 6.11a, b, respectively. In both cases the current magnitude asymptotically approaches the "undisturbed" value I indicated by the horizontal broken lines; it is related to the magnitude I_0 of the initial-current wave propagating down the tower by the equation given in Fig. 6.10b. Note that the current at the top of the object I_T varies from $0.75I$ to $1.13I$ and that the current at the bottom of the object varies from $0.75I$ to $1.5I$ (from I_0 to $2I_0$). As seen in Fig. 6.11a, the "undisturbed" lightning current I is greater than the injected current I_0 but smaller than the maximum value of the current I_T. If the lightning could be represented by an ideal current source ($Z_{ch} = \infty$ or $Z_{ch} \gg Z_0$, $\rho_T = -1$) then the current at the top of the object I_T would be equal to I at all times regardless of object height and its grounding. Indeed, in this case, (i) the current injected into the object would be equal to the source current ($I_0 = I$) and (ii) upward-moving current reflections from the bottom of the object would be necessarily canceled at the top of the object by downward-moving current reflections from the lightning channel.

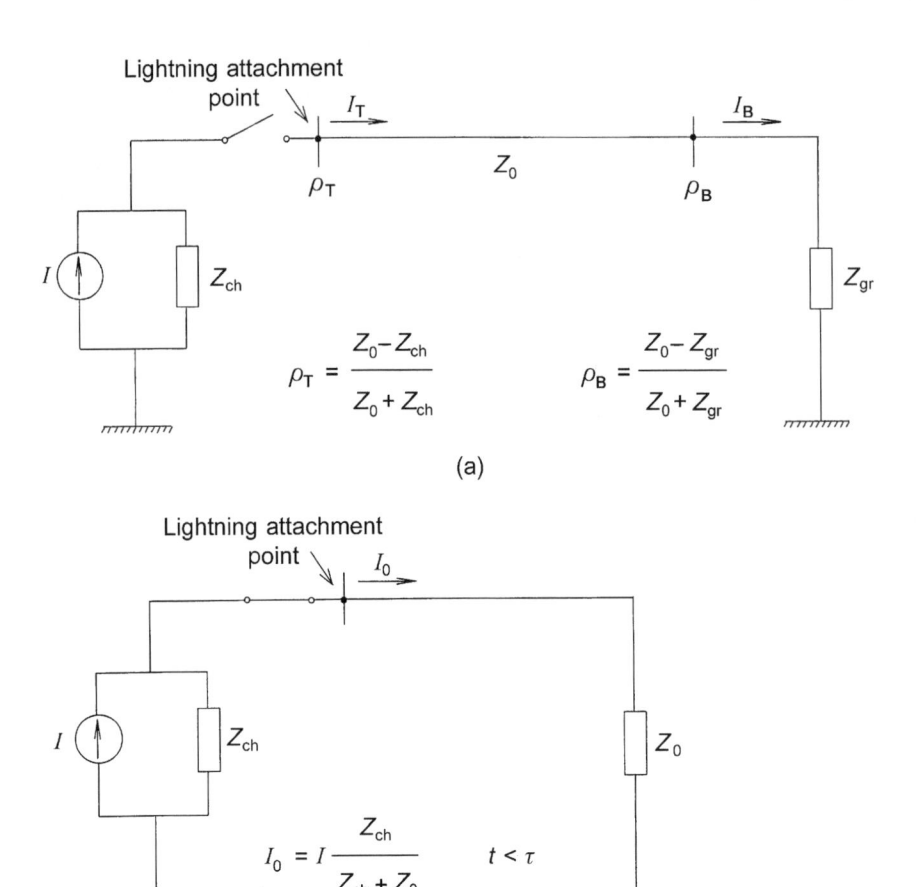

Fig. 6.10. (a) Equivalent circuit for the case of a lightning strike to a tall grounded object whose characteristic impedance is Z_0 and (b) the same as (a) but for $t < \tau$, where τ is the time required for a wave to traverse the length of the object. Z_{gr} in (a) is the effective grounding impedance of the object, ρ_T and ρ_B are the current reflection coefficients at the top and at the bottom of the object and I_T and I_B are the currents that would be measured at the top and at the bottom of the object. See also the caption of Fig. 6.9. I_0 in (b) is the magnitude of the initial downward-propagating current wave.

We now compare the statistical distributions of return-stroke peak currents measured on objects of different heights and show that these distributions are similar, suggesting that the measured current peaks are not significantly influenced by the presence of the object. The distributions, represented by their 99, 95, 90, 50, 10, and 5 percent values are summarized in Table 6.8 for the following three categories of event: (i) return-strokes in natural upward (object-initiated) lightning with strike object heights ranging from 70 to 540 m, (ii) subsequent return strokes in natural downward lightning with a strike-object height equal to 70 m, and (iii) return strokes in rocket-triggered lightning with strike-object heights ranging from about 5 to 20 m. These three types of return stroke are initiated by downward dart leaders and are expected to have similar characteristics. As noted in Section 6.1, the return-strokes in categories (i) and (iii) are sometimes referred to as subsequent return strokes, by analogy with the return strokes in category (ii). All data in Table 6.8 for object-initiated and natural downward lightning were obtained at the top of the strike object, except in the case of the Ostankino tower for which measurements both at the top and in the lower part of the tower are included. As seen in this table, the distributions of peak currents measured at objects with heights ranging from about 5 to 540 m are not much different from each other.

Further, Rakov et al. (1998) summarized the geometric mean peak currents from six different triggered-lightning experiments in Florida and Alabama, some of which are given in Table 7.6. They found a relatively narrow range of currents, 9.9 to 15 kA, and no systematic variation with triggering structure height ranging from 4.5 to 20 m or with lower measurement limit ranging from 1 to 5 kA. Thus the experimental data obtained at objects with heights ranging from 4.5 to 540 m suggest that current peaks are not significantly affected by the presence of a tall object (as long as measurements are taken at the top of

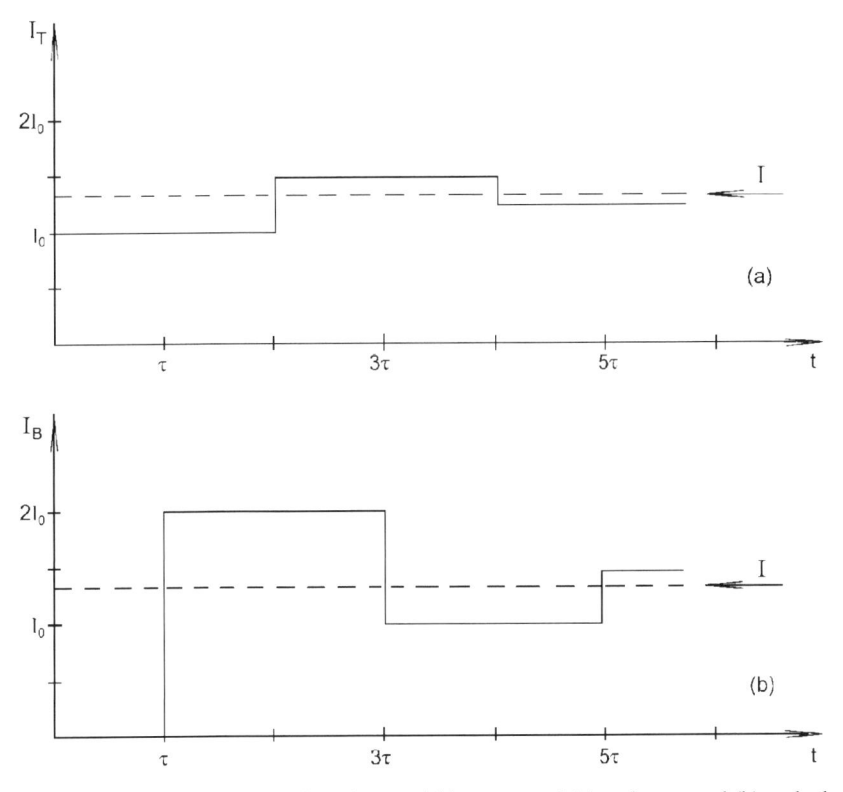

Fig. 6.11. The current versus time that would be measured (a) at the top and (b) at the bottom of a tall object if the short-circuit current I were a step function, for the cases $\rho_B = 1$ and $\rho_T = -0.5$. The "undisturbed" current magnitude, I, is indicated by the horizontal broken dashed lines. The magnitude of the initial downward-propagating current wave, I_0, is found from the equation given in Fig. 6.10b.

the tall object). This inference is consistent with the modeling results of Melander (1984) who estimated the influence of the presence of towers on current measurements by Berger *et al.* (1975), Garbagnati and Lo Piparo (1982a), and Eriksson (1978a). In doing so, she represented the towers by linear distributed circuits and used a distributed-circuit model of the lightning return stroke, the resistance per unit length being determined by a gas-dynamic model, as proposed by Strawe (1979). Melander (1984) found from her modeling that the current measurements of Berger *et al.* (1975) and Garbagnati and Lo Piparo (1982a) obtained at the top of, respectively, 70-m and 40-m towers are essentially unaffected by the presence of the towers. However, the current peaks measured by Eriksson (1978a) at the bottom of a 60-m tower were found to be overestimated by a factor of about 1.6.

6.6. Electromagnetic fields due to lightning strikes to tall objects

Diendorfer and Schulz (1998) reported that current peaks estimated by the Austrian lightning locating system (Section 17.5) tended to be somewhat higher for lightning within 1 km of towers located on mountains than for lightning within 1 to 10 km from the towers. It is important to note that current peaks reported by lightning locating systems are inferred from electromagnetic field peaks, usually assuming a direct proportionality between the two (Section 17.5). Diendorfer and Schulz (1998) found that, for the first detected strokes in 81 negative flashes located within 1 km of towers, the median current peak was about 12 kA while for 686 negative flashes at distances ranging from 1 to 10 km from the towers the median current peak was 9.8 kA. To explain this difference in median current peaks, Diendorfer and Schulz (1998) assumed that, because of lightning-locating system errors of about 1 km, most flashes reported by the system as striking ground within 1 km of towers actually terminated on the towers. A similar effect was noted by Byerley *et al.* (1999), who used NLDN (Section 17.5) data to examine lightning strikes to and in the vicinity of tall towers in Oklahoma. Interestingly, Diendorfer and Schulz (1998) also observed a considerably larger number of strokes per flash in strikes to towers than in strikes to ground.

The transient behavior of the strike object depends on the current waveshape, and, as a result, different electric or magnetic field signatures can be produced by lightning discharges characterized by the same current peak but different current waveshapes (Rachidi *et al.* 1998, 2001;

Janischewskyj et al. 1998, 1999a). Abdel-Rahman et al. (1998) compared two groups of measured current pulses (presumably associated with return strokes) in upward flashes initiated from the CN tower, the first group including the first pulse in a sequence recorded for each flash, and the second group including all pulses other than the first. Abdel-Rahman et al. found that the current pulses in the second group were characterized by smaller amplitudes, but they tended to produce larger observed magnetic field peaks at a distance of 2 km from the tower. A lack of proportionality between the CN-tower current peaks and either the electric or magnetic field peaks at a distance of 2 km from the tower was also reported, from measurements, by Janischewskyj et al. (1999a) and Shostak et al. (1999b).

If the lightning short-circuit current (Section 6.5) is not influenced by the presence of towers, the higher field peaks in the case of lightning strikes to towers might be due to (i) the existence of two wavefronts propagating simultaneously in opposite directions from the junction point between the descending leader and upward connecting leader from the tower top, the downward-moving wavefront traversing both the upward-connecting-leader channel and the tower (neglecting current reflection at the tower top), (ii) the near doubling of the downward-moving current wave on reflection at the bottom of the tower (Section 6.5), and (iii) the propagation speed of current waves on the tower being nearly equal to the speed of light and higher than that of the return-stroke current wave in a natural lightning channel (subsection 4.6.2).

Diendorfer and Uman (1990) were apparently the first to compute lightning fields using a return-stroke model extended to include a strike object. Further efforts to model the interaction of lightning with tall objects have been made by Zundl (1994a), Guerrieri (1996, 1998, 2000), Rusan et al. (1996), Motoyama et al. (1996), Rachidi et al. (1998, 2001), Janischewskyj et al. (1998, 1999a), Shostak et al. (1999a, 2000), Goshima et al. (2000), and Baba and Ishii (2001). Rachidi et al. (1998, 2001) noted, from their modeling of lightning strikes to the CN tower using the MTLE model (subsection 12.2.5), that the radiation field component of the total field is the component most affected (enhanced) by the presence of the tower. According to Janischewskyj et al. (1999a), the contribution from the 553-m-high CN tower to the total electric field at a distance of 2 km from the tower is almost a factor 2 greater than the contribution from the lightning channel.

6.7. Acoustic output

It appears that some upward flashes produce a sound in the vicinity of the strike object that is not the usual thunder. McEachron (1939) suggested that thunder is produced by return strokes and that lightning without the usual thunder is composed of initial continuous current only. Note that cloud discharges do not contain return strokes but do produce thunder. Interestingly, Davis and Standring (1947), who measured lightning currents in the cables of kite balloons, observed a few discharges with currents less than 1 kA that were heard and seen and several discharges with currents of about 100 A that were not detected by human observers. A discussion of thunder-generation mechanisms is found in Section 11.3.

6.8. Summary

Very tall objects (higher than 200 m or so) located on flat terrain and objects of moderate height (some tens of meters) located on mountain tops experience primarily upward lightning discharges that are initiated by upward-propagating leaders. Upward, or object-initiated, lightning discharges always involve an initial stage that may or may not be followed by downward-leader–upward-return-stroke sequences. The latter are similar to subsequent-leader–return-stroke sequences in downward lightning and to downward-leader–upward-return-stroke sequences in rocket-triggered lightning. The initial stage in object-initiated lightning is similar to the initial stage in rocket-triggered lightning and can be viewed as being composed of an upward leader and an initial continuous current. The initial stage, in a sense, replaces the downward-stepped-leader–upward-return-stroke sequence characteristic of downward lightning.

Lightning-current waveforms measured on tall objects may be affected by transient processes in the object that involve wave reflections from object extremities and from impedance discontinuities within the object. The similarity of the statistical distributions of subsequent-return-stroke peak currents in (i) natural upward (object-initiated) lightning, (ii) natural downward lightning, and (iii) rocket-triggered lightning, measured at objects with heights ranging from 4.5 to 540 m, suggests that current peaks are not significantly influenced by the presence of a tall object, provided that measurements are taken at the top of the object. The peak current measured at the bottom of a tall object is usually more strongly influenced by the transient process in the object than is the peak current at the top. A tall metallic strike-object replacing the lower part of the lightning channel apparently serves to enhance the lightning radiated electromagnetic fields relative to the fields due to similar strikes to ground, this effect probably being more pronounced for sharper lightning-current pulses.

References and bibliography for Chapter 6

Abdel-Rahman, M., Janischewskyj, W., Hussein, A.M., Rachidi, F., Chang, J.S. 1998. Statistical analysis of magnetic fields due to CN tower multistroke flashes. In *Proc. 24th Int. Conf. on Lightning Protection, Birmingham, UK*, pp. 107–12.

Anderson, R.B., and Eriksson, A.J. 1980. Lightning parameters for engineering application. *Electra* **69**: 65–102.

Asakawa, A., Miyake, K., Yokoyama, S., Shindo, T., Yokota, T., and Sakai, T. 1997. Two types of lightning discharges to a high stack on the coast of the Sea of Japan in winter. *IEEE Trans. Pow. Del.* **12**: 1222–31.

Axup, P.R. 1983. The effects of towers on lightning current measurements. Master's thesis, Air Force Institute of Technology.

Baba, Y., and Ishii, M. 2001. Numerical electromagnetic field analysis of lightning current in tall structures. *IEEE Trans. Pow. Del.* **16**(2): 324–8.

Bazelyan, E.M., Gorin, B.N., and Levitov, V.I. 1978. *Physical and Engineering Foundations of Lightning Protection*, 223 pp., Leningrad: Gidrometeoizdat.

Beierl, O. 1991. Lightning current measurements at the Peissenberg tower. In *Proc. 7th Int. Symp. on High Voltage Engineering, Dresden, Germany*, paper 81.02.

Beierl, O. 1992. Front shape parameters of negative subsequent strokes measured at the Peissenberg tower. In *Proc. 21st Int. Conf. on Lightning Protection, Berlin, Germany*, pp. 19–24.

Berger, K. 1961. Gewitterforschung auf dem Monte San Salvatore. *ETZ-A* **82**: 249–60.

Berger, K., 1967. Novel observations on lightning discharges: results of research on Mount San Salvatore. *J. Franklin Inst.* **283**: 478–525.

Berger, K. 1972. Methoden und Resultate der Blitzforschung auf dem Monte San Salvatore bei Lugano in den Jahren 1963–1971. *Bull. Schweiz. Elektrotech. Ver.* **63**: 1403–22.

Berger, K. 1977. The Earth flash. In *Lightning, vol. 1, Physics of Lightning*, ed. R.H. Golde, pp. 119–90, New York: Academic Press.

Berger, K. 1978. Blitzstrom-Parameter von Aufwartsblitzen. *Bull. Schweiz. Elektrotech. Ver.* **69**: 353–60.

Berger, K., and Garbagnati, E. 1984. Lightning current parameters. Results obtained in Switzerland and in Italy. In *Proc. URSI Conf., Florence, Italy*, 13 pp.

Berger, K., and Vogelsanger, E. 1965. Messungen und Resultate der Blitzforschung der Jahre 1955–1963 auf dem Monte San Salvatore. *Bull. Schweiz. Elektrotech. Ver.* **56**: 2–22.

Berger, K., and Vogelsanger, E. 1966. Photographische Blitzuntersuchungen der Jahre 1955–1965 auf dem Monte San Salvatore. *Bull. Schweiz. Elektrotech. Ver.* **57**: 599–620.

Berger, K., and Vogelsanger, E. 1969. New results of lightning observations. In *Planetary Electrodynamics*, eds. S.C. Coroniti and J. Hughes, pp. 489–510, New York: Gordon and Breach.

Berger, K., Anderson, R.B., and Kroninger, H. 1975. Parameters of lightning flashes. *Electra* **80**: 223–37.

Bosart, L.F., Chen, T.J., Orville, R.E., and Roesli, H.B. 1974. A preliminary study of the synoptic conditions associated with upward and downward lightning flashes over Mt San Salvatore, Lugano, Switzerland. *Tellus* **26**: 495–505.

Brook, M., Armstrong, G., Winder, R.P.H., Vonnegut, B., and Moore, C.B. 1961. Artificial initiation of lightning discharges. *J. Geophys. Res.* **66**: 3967–9.

Byerley, L.G., Brooks, W.A., Noggle, R.C., and Cummins, K.L. 1999. Towers, lightning and human affairs. In *Proc. 11th Int. Conf. on Atmospheric Electricity, Guntersville, Alabama*, pp. 180–3.

Carlson, A.B. 1996. *Circuits*, 838 pp., New York: John Wiley and Sons.

Chang, J.S., Beuthe, T.G., Hu, G.G., Stamoulis, G., and Janischewskyj, W. 1985. Thundercloud electric field measurements in the 553-m CN tower during 1978–1983. *J. Geophys. Res.* **90**: 6087–90.

Chang, J.S., Beuthe, T.G., Seto, L., Duft, A., Hayashi, N., Chisholm, W., and Janischewskyj, W. 1989. An investigation of the possible relationships between thundercloud electric fields and the lightning parameters for tall structures. *J. Geophys. Res.* **94**: 13 197–205.

Chen, M., Takagi, N., Watanabe, T., Wang, D., Kawasaki, Z.-I., and Liu, X. 1999. Spatial and temporal properties of optical radiation produced by stepped leaders. *J. Geophys. Res.* **104**: 27 573–84.

Crawford, D.E. 1998. Multiple-station measurements of triggered lightning electric and magnetic fields. M.S. thesis, Univ. Florida, Gainesville, 282 pp.

Davis, R., and Standring, W.G. 1947. Discharge currents associated with kite balloons. *Proc. Roy. Soc. A***191**: 304–22.

Depasse, P. 1994. Statistics on artificially triggered lightning. *J. Geophys. Res.* **99**: 18 515–22.

Diendorfer, G., and Schulz, W. 1998. Lightning incidence to elevated objects on mountains. In *Proc. 24th Int. Conf. on Lightning Protection, Birmingham, UK*, pp. 173–5.

Diendorfer, G., and Uman M.A. 1990. An improved return stroke model with specified channel-base current. *J. Geophys. Res.* **95**: 13 621–44.

Diendorfer, G., Schulz, W., and Fuchs, F. 1998a. Comparison of correlated data from the Austrian lightning location system and measured lightning currents at the Peissenberg tower. In *Proc. 24th Int. Conf. on Lightning Protection, Birmingham, UK*, pp. 168–72.

Diendorfer, G., Schulz, W., and Rakov, V.A. 1998b. Lightning characteristics based on data from the Austrian lightning locating system. *IEEE Trans. Electromagn. Compat.* **40**(II): 452–64.

Diendorfer, G., Mair, M., Schulz, W., and Hadrian, W. 2000. Lightning current measurements in Austria – experimental setup and first results. In *Proc. 25th Int. Conf. on Lightning Protection, Rhodes, Greece*, pp. 44–7.

Eriksson, A.J. 1978a. Lightning and tall structures. *Trans. S. Afr. IEE* **69**: 238–52.

Eriksson, A.J. 1978b. A discussion on lightning and tall structures. CSIR Special Report Elek 152, National Electrical Engineering Research Institute, Pretoria, South Africa.

Eriksson, A.J. 1982. The CSIR lightning research mast-data for 1972–1982, NEERI Internal Report no. EK/9/82, National Electrical Engineering Research Institute, Pretoria, South Africa.

Fisher, R.J., Schnetzer, G.H., Thottappillil, R., Rakov, V.A., Uman, M.A., and Goldberg, J.D. 1993. Parameters of triggered-lightning flashes in Florida and Alabama. *J. Geophys. Res.* **98**: 22 887–902.

Fuchs, F. 1998a. Lightning current and LEMP parameters of upward discharges measured at the Peissenberg tower. In *Proc. 24th Int. Conf. on Lightning Protection, Birmingham, UK*, vol. 1, pp. 17–22.

Fuchs, F. 1998b. On the transient behaviour of the telecommunication tower at the mountain Hoher Peissenberg. In *Proc. 24th Int. Conf. on Lightning Protection Birmingham, UK*, vol. 1, pp. 36–41.

Fuchs, F., Landers, E.U., Schmid, R., and Wiesinger, J. 1998. Lightning current and magnetic field parameters caused by lightning strikes to tall structures relating to interference of electronic systems. *IEEE Trans. Electromagn. Compat.* **40**: 444–51.

Garbagnati, E., and Lo Piparo, G.B. 1970. Stazione sperimentale per il rilievo delle caratteristiche dei fulmini. *L'Elettrotecnica* **57**: 288–97.

Garbagnati, E., and Lo Piparo, G.B. 1973. Nuova stazione automatica per il rilievo delle caratteristiche dei fulmini. *L'Energia Elettrica* **6**: 375–83.

Garbagnati, E., and Lo Piparo, G.B. 1982a. Results of 10 years investigation in Italy. In *Proc. Int. Aerospace Conf. on Lightning and Static Electricity, Oxford, UK*, paper A1, 12 pp.

Garbagnati, E., and Lo Piparo, G.B. 1982b. Parameter von Blitzstroemen. *ETZ-A* **103**: 61–5.

Garbagnati, E., Giudice, E., Lo Piparo, G.B., and Magagnoli, U. 1974. Survey of the characteristics of lightning stroke currents in Italy – results obtained in the years from 1970 to 1973. ENEL Report R5/63-27.

Garbagnati, E., Giudice, E., Lo Piparo, G.B., and Magagnoli, U. 1975. Rilievi delle caratteristiche dei fulmini in Italia. Risultati ottenuti negli anni 1970–1973. *L'Elettrotecnica* **62**: 237–49.

Garbagnati, E., Giudice, E., and Lo Piparo, G.B. 1978. Measurement of lightning currents in Italy – results of a statistical evaluation. *ETZ-A* **99**: 664–8.

Garbagnati, E., Marinoni, F., and Lo Piparo, G.B. 1981. Parameters of lightning currents. Interpretation of the results obtained in Italy. In *Proc. 16th Int. Conf. on Lightning Protection, Szeged, Hungary*, paper R-1.03, 16 pp.

Geldenhys, H.T., Eriksson, A.J., and Bourn, G.W. 1989. Fifteen years of data of lightning current measurements on 60 m mast. *Trans. South African IEE* **80**: 130–58.

Golde, R.H. 1978. Lightning and tall structures. *IEE Proc.* **125**: 347–51.

Gorin, B.N., and Shkilev, A.V. 1984. Measurements of lightning currents at the Ostankino tower. *Elektrichestvo* **8**: 64–5.

Gorin, B.N., Sakharova, G.S., Tikhomirov, V.V., and Shkilev, A.V. 1975. Results of studies of lightning strikes to the Ostankino TV tower. *Trudy ENIN* **43**: 63–77.

Gorin, B.N., Levitov, V.I., and Shkilev, A.V. 1977. Lightning strikes to the Ostankino tower. *Elektrichestvo* **8**: 19–23.

Goshima, H., Motoyama, H., Asakawa, A., Wada, A., Shindo, T., and Yokoyama, S. 2000. Characteristics of electromagnetic fields due to winter lightning stroke current to a high stack. *Trans. IEE Japan* **120-B**(1): 44–8.

Guerrieri, S., Heidler, F., Nucci, C.A., Rachidi, F., and Rubinstein, M. 1996. Extension of two return stroke models to consider the influence of elevated strike objects on the lightning return stroke current and the radiated electromagnetic field: comparison with experimental results. In *Proc. Int. Symp. on Electromagnetic Compatibility (EMC '96 ROMA), Rome, Italy*, pp. 701–6.

Guerrieri, S., Nucci, C.A., Rachidi, F., and Rubinstein, M. 1998. On the influence of elevated strike objects on directly measured and indirectly estimated lightning currents. *IEEE Trans. Pow. Del.* **13**: 1543–55.

Guerrieri, S. Krider, E.P., and Nucci, C.A. 2000. Effects of traveling-waves of current on the initial response of a tall Franklin rod. In *Proc. 25th Int. Conf. on Lightning Protection, Rhodes, Greece*, pp. 94–9.

Hagenguth, J.H., and Anderson, J.G. 1952. Lightning to the Empire State Building. *AIEE Trans.* **71** (3): 641–9.

Heidler, F. 2000. Personal communication.

Heidler, F., and Zundl, T. 1995. Influence of tall towers on the return stroke current. In *Proc. 1995 Int. Conf. on Lightning and Static Electricity, Williamsburg, Virginia*, pp. 67/1–10.

Horii, K., and Ikeda, G. 1985. A consideration on success conditions of triggered lightning. In *Proc. 18th Int. Conf. on Lightning Protection, Munich, Germany*, paper 1–3, 6 pp.

Hubert, P. 1984. Triggered lightning in France and New Mexico. *Endeavour* **8**: 85–9.

Hubert, P., Laroche, P., Eybert-Berard, A., and Barret, L. 1984. Triggered lightning in New Mexico. *J. Geophys. Res.* **89**: 2511–21.

Hussein, A.M., Janischewskyj, W., Chang, J.-S., Shostak, V., Chisholm, W.A., Dzurevych, P., and Kawasaki, Z.-I. 1995. Simultaneous measurement of lightning parameters for strokes to the Toronto Canadian National tower. *J. Geophys. Res.* **100**: 8853–61.

Janischewskyj, W., Shostak, V., Barratt, J., Hussein, A.M., Rusan, R., and Chang, J.-S. 1996. Collection and use of lightning return stroke parameters taking into account characteristics of the struck object. In *Proc. 23rd Int. Conf. on Lightning Protection, Florence, Italy*, pp. 16–23.

Janischewskyj, W., Hussein, A.M., Shostak, V., Rusan, I., Li, J.-X., and Chang, J.-S. 1997. Statistics of lightning strikes to the Toronto Canadian National tower (1978–1995). *IEEE Trans. Pow. Del.* **12**: 1210–21.

Janischewskyj, W., Shostak, V., and Hussein, A.M. 1998. Comparison of lightning electromagnetic field characteristics of first and subsequent return strokes to a tall tower: 1. Magnetic field. In *Proc. 24th Int. Conf. on Lightning Protection, Birmingham, UK*, pp. 245–51.

Janischewskyj, W., Shostak, V., and Hussein, A.M. 1999a. Lightning electric field characteristics of first and subsequent return strokes to a tall tower. In *Proc. 11th Int. Symp. on High Voltage Engineering, London, UK*, IEE Publ. no. 467, vol. 1, pp. 270–4.

Janischewskyj, W., Shostak, V., Hussein, A.M., and Kordi, B. 1999b. A lightning flash containing strokes of opposite polarities. CIGRE SC 33.99 (COLL), 2 pp.

Kitterman, C.G. 1981. Concurrent lightning flashes on two television transmission towers. *J. Geophys. Res.* **86**: 5378–80.

Kostenko, M.V. 1995. Electrodynamic characteristics of lightning and their influence on disturbances of high-voltage lines. *J. Geophys. Res.* **100**: 2739–47.

Krider, E.P., and Wetmore, R.H. 1987. Upward streamers produced by a lightning strike to radio transmission towers. *J. Geophys. Res.* **92**: 9859–62.

Lacerda, M., Pinto, O., Pinto, I.R.C.A., Diniz, J.H., and Carvalho, A.M. 1999. Analysis of negative downward lightning current curves from 1985 to 1994 at Morro do Cachimbo research station (Brazil). In *Proc. 11th Int. Conf. on Atmospheric Electricity, Guntersville, Alabama*, pp. 42–5.

Laroche, P., Eybert-Berard, A., and Barret, L. 1985. Triggered lightning flash characteristics. In *Proc. 10th Int. Aerospace and Ground Conf. on Lightning and Static Electricity, Paris, France*, pp. 231–9.

McCann, D.G. 1944. The measurement of lightning currents in direct strokes. *Trans. AIEE* **63**: 1157–64.

McComb, T.R., Cherney, E.A., Linck, H., and Janischewskyj, W. 1980. Preliminary measurements of lightning flashes to the C.N. tower in Toronto. *Can. Elec. Eng. J.* **5**: 3–9.

McEachron, K.B. 1939. Lightning to the Empire State Building. *J. Franklin Inst.* **227**: 149–217.

McEachron, K.B. 1941. Lightning to the Empire State Building. *Trans. AIEE* **60**: 885–9.

Melander, B.G. 1984. Effects of tower characteristics on lightning arc measurements. In *Proc. 1984 Int. Conf. on Lightning and Static Electricity, Orlando, Florida*, pp. 34/1–34/12.

Miyake, K., Suzuki, T., and Shinjou, K. 1992. Characteristics of winter lightning current on Japan Sea coast. *IEEE Trans. Pow. Del.* **7**: 1450–6.

Montandon, E. 1992. Lightning positioning and lightning parameter determination experiences and results of the Swiss PTT research project. In *Proc. 21st Int. Conf. on Lightning Protection, Berlin, Germany*, pp. 307–12.

Montandon, E. 1995. Messung und Ortung von Blitzeinschlaegen und ihren Auswirkungen am Fernmeldeturm "St. Chrischona" bei Basel der Schweizerischen Telecom PTT. *Elektrotechnik und Informationstechnik* **112**: 283–9.

Montandon, E., and Beyeler, B. 1994. The lightning measuring equipment on the Swiss PTT telecommunications tower at St. Chrischona, Switzerland. In *Proc. 22nd Int. Conf. on Lightning Protection, Budapest, Hungary*, paper R1c-06.

Motoyama, H., Janischewskyj, W., Hussein, A.M., Rusan, R., Chisholm, W.A., and Chang, J.-S. 1996. Electromagnetic field radiation model for lightning strokes to tall structures. *IEEE Trans. Pow. Del.* **11**: 1624–32.

Mousa, A. 1986. A study of the engineering model of lightning strokes and its application to unshielded transmission lines. Ph.D. dissertation, University of British Columbia, Vancouver, Canada.

Müller-Hillebrand, D. 1960. On the frequency of lightning flashes to high objects. A study on the Gulf of Bothnia. *Tellus* **12**: 444–9.

Orville, R.E., and Berger, K. 1973. An unusual lightning flash initiated by an upward propagating leader. *J. Geophys. Res.* **78**: 4520–5.

Pierce, E.T. 1971. Triggered lightning and some unsuspected lightning hazards. Stanford Research Institute, Menlo Park, California, 20 pp.

Pierce, E.T. 1972. Triggered lightning and some unsuspected lightning hazards. In *Naval Research Reviews*, pp. 14–18.

Pinto, O., Pinto, I.R.C.A., Lacerda, M., Carvalho, A.M., Diniz, J.H., and Cherchiglia, L.C.L. 1997. Are equatorial negative lightning flashes more intense than those at higher latitudes? *J. Atmos. Solar-Terr. Phys.* **59**: 1881–3.

Podgorski, A.S., and Landt, J.A. 1987. Three dimensional time domain modeling of lightning. *IEEE Trans. Pow. Del.* **2**: 931–8.

Popolansky, F. 1965. Study of lightning strokes to high objects in Czechoslovakia. English transl., ERA Trans./IB 2291.

Rachidi, F., Janischewskyj, W., Hussein, A.M., Nucci, C.A., Guerrieri, S., and Chang, J.S. 1998. Electromagnetic fields radiated by lightning return strokes to high towers. In *Proc. 24th Int. Conf. on Lightning Protection, Birmingham, UK*, pp. 23–8.

Rachidi, F., Janischewskyj, W., Hussein, A.M., Nucci, C.A., Guerrieri, S., Kordi, B., and Chang, J.S. 2001. Current and electromagnetic field associated with lightning-return strokes to tall towers. *IEEE Trans. Elecromagn. Compat.* **43**(3): 356–67.

Rakov, V.A. 2001. Transient response of a tall object to lightning. *IEEE Trans. Electromagn. Compat.* **43**: 654–61.

Rakov, V.A., and Dulzon, A.A. 1987. Calculated electromagnetic fields of lightning return stroke. *Tekh. Elektrodinam.* **1**: 87–9.

Rakov, V.A., and Dulzon, A.A. 1991. A modified transmission line model for lightning return stroke field calculations. In *Proc. 9th Int. Zurich Symp. on Electromagnetic Compatibility, Zurich, Switzerland*, pp. 229–35.

Rakov, V.A., Uman, M.A., Rambo, K.J., Fernandez, M.I., Fisher, R.J., Schnetzer, G.H., Thottappillil, R., Eybert-Berard, A., Berlandis, J.P., Lalande, P., Bonamy, A., Laroche, P., and Bondiou-Clergerie, A. 1998. New insights into lightning processes gained from triggered-lightning experiments in Florida and Alabama. *J. Geophys. Res.* **103**: 14 117–30.

Rusan, R., Janischewskyj, W., Hussein, A.M., and Chang, J.-S. 1996. Comparison of measured and computed electromagnetic fields radiated from lightning strikes to the Toronto CN tower. In *Proc. 23rd Int. Conf. on Lightning Protection, Florence, Italy*, pp. 297–303.

Rustan, P.L., and Axup, P.R. 1984. Analysis of lightning current measurements. In *Proc. 1984 Int. Conf. on Lightning and Static Electricity, Orlando, Florida*, pp. 24/1–7.

Shindo, T., and Aihara, Y., 1993. A shielding theory for upward lightning. *IEEE Trans. Pow. Del.* **8**: 318–24.

Shindo, T., Aihara, Y., and Suzuki, T. 1990. Model experiment of upward leaders – shielding effects of tall objects. *IEEE Trans. Pow. Del.* **5**: 716–23.

Shostak, V., Janischewskyj, W., Hussein, A.M., Chang, J.-S., and Kordi, B. 1999a. Return-stroke current modeling of lightning striking a tall tower accounting for reflections within the growing channel and for upward-connecting discharges. In *Proc. 11th Int. Conf. on Atmospheric Electricity, Guntersville, Alabama*, pp. 123–6.

Shostak, V., Janischewskyj, W., Hussein, A.M., and Kordi, B. 1999b. Characteristics of return stroke current and electromagnetic field waveforms observed in multistroke lightning flashes to a tall tower. In *Proc. 11th Int. Symp. on High Voltage Engineering, London, UK*, IEEE Publ. no. 467, vol. 2, pp. 389–392.

Shostak, V., Janischewskyj, W., Hussein, A.M., and Kordi, B. 2000. Electromagnetic fields of lightning strikes to a tall tower: a model that accounts for upward-connecting discharges. In *Proc. 25th Int. Conf. on Lightning Protection, Rhodes, Greece*, pp. 60–5.

Standler, R.B. 1975. The response of elevated conductors to lightning. Masters thesis, New Mexico Institute of Mining and Technology.

Strawe, D.F. 1979. Non-linear modeling of lightning return strokes. In *Proc. Federal Aviation Administration/Florida Institute of Technology Workshop on Grounding and Lightning Technology, Melbourne, Florida*, Report FAA-RD-79-6, pp. 9–15.

Szpor, S., Wasilenko, E., Samula, J., Dytkowski, E., Suchocki, J., and Zaborowski, B. 1964. Results of lightning stroke registrations in Poland. CIGRE Report 319.

Torres, H., Trujillo, O., Amortegui, F., Herrera, F., Pinzon, G., Quintana, C., Gonzalez, D., Rondon, D., Salgado, M., and Avila, D. 1999. Experimental station to measure directly lightning parameters in tropical zone. In *Proc. 11th Int. Symp. on High Voltage Engineering, London, UK*, IEE Publ. no. 467, vol. 2, pp. 177–80.

Uman, M.A., 1971. *Understanding Lightning*, 166 pp. Pittsburgh: Bek Technical Publications.

Uman, M.A. 1987. *The Lightning Discharge*, 377 pp., San Diego: Academic Press.

Uman, M.A. 2001. *The Lightning Discharge*, 377 pp., Mineola, New York: Dover.

Wang, D., Kawasaki, Z.I., Yamamoto, K., Matsuura, K., Chang, J.-S., and Janischewskyj, W. 1995. Luminous propagation of lightning attachment to CN tower. *J. Geophys. Res.* **100**: 11 661–7.

Wang, D., Rakov, V.A., Uman, M.A., Fernandez, M.I., Rambo, K.J., Schnetzer, G.H., and Fisher, R.J. 1999a. Characterization of the initial stage of negative rocket-triggered lightning. *J. Geophys. Res.* **104**: 4213–22.

Wang, D., Takagi, N., Watanabe, T., Rakov, V.A., and Uman, M.A. 1999b. Observed leader and return-stroke propagation characteristics in the bottom 400 m of the rocket triggered lightning channel. *J. Geophys. Res.* **104**: 14 369–76.

Zundl, 1994a. Lightning current and LEMP calculations compared to measurements gained at the Peissenberg tower. In *Proc. 22nd Int. Conf. on Lightning Protection, Budapest, Hungary*, paper R1c-08.

Zundl, T. 1994b. First results of the coordinated lightning current and LEMP measurements at the Peissenberg tower. In *Proc. 22nd Int. Conf. on Lightning Protection, Budapest, Hungary*, paper R1c-09.

7 Artificial initiation (triggering) of lightning by ground-based activity

> Triggering lightning at will, at a predetermined place and time, is the old Promethean dream which seems related more to legend than to science.
>
> P. Hubert (1984)

7.1. Introduction

This chapter is concerned primarily with the artificial initiation (triggering) of lightning discharges from natural thunderclouds by means of the "rocket-and-wire" technique. This technique involves the launching of a small rocket that extends a thin wire (either grounded or ungrounded) into the gap between the ground and a charged cloud overhead. Additionally, other types of ground-based activity used to trigger lightning will be briefly reviewed. We limit the use of the term "triggered lightning" to those discharges intentionally stimulated to occur by ground-based activity, the energy source being a naturally electrified cloud. Laboratory discharges between metallic electrodes, which are often used to simulate lightning effects and whose length can reach some tens and even hundreds of meters (e.g., Anisimov *et al.* 1988; Mrazek 1998) and discharges of up to 3 m length produced using artificially charged aerosol clouds (e.g., Vereshchagin *et al.* 1989, 1996; Antsupov *et al.* 1990, 1991) draw their energy from man-made sources and therefore are not considered here. Upward lightning discharges between a stationary, ground-based object (usually of large height) and a naturally electrified cloud, which are sometimes (e.g., Pierce 1974) referred to as triggered lightning, are considered in Chapter 6. We choose to call these discharges object-initiated lightning or upward lightning. As noted in Section 6.1, it is likely that the occurrence of most upward flashes is stimulated by in-cloud discharge activity, the grounded object being a passive element of the system. Further, lightning discharges initiated by airborne vehicles, such as aircraft and spacecraft, are discussed in Chapter 10.

Finally, we do not consider here lightning that is initiated unintentionally by ground-based human activity, an example being the discharge between a water plume created by an underwater explosion in Chesapeake Bay and an electrified cloud overhead, shown in Fig. 20.5. Such rare events are reviewed in Chapter 20, which also includes a description of upward, lightning-like discharges from grounded structures caused by the detonation of the first thermonuclear device in the Marshall Islands in 1952. In the latter case the charge producing the electrical environment was primarily generated by gamma rays emanating from the detonation and interacting with the atmosphere to create outward-moving Compton electrons, as discussed in subsection 20.7.3.

The possibility of artificially initiating lightning by ground-based activity was apparently first discussed by Newman (1958) and by Brook *et al.* (1961). Brook *et al.* showed that, in the laboratory, a spark discharge could be triggered by the rapid introduction of a thin wire into an electric field, whereas the steady presence of the wire did not result in a spark. They suggested that the corona discharge from a stationary conductor acts to shield the conductor so that the high fields necessary to initiate electrical breakdown are not obtained, whereas the field enhancement due to the rapid introduction of a conductor is not significantly reduced by corona since there is insufficient time for its development.

The first triggered-lightning discharges were produced in 1960 by launching small rockets trailing thin grounded wires from a research vessel off the west coast of Florida (Newman 1965; Newman *et al.* 1967; Newman and Robb 1977). The first triggering over land was accomplished in 1973, at Saint-Privat d'Allier in France (Fieux *et al.* 1975, 1978). In the following decades, a number of triggered-lightning programs have been developed in different countries, as summarized in Table 7.1. Rocket-triggered lightning experiments in France have been reviewed by Fieux *et al.* (1978), in Japan by Horii (1982), Kito *et al.* (1985), and Nakamura *et al.* (1991, 1992), in New Mexico by Hubert *et al.* (1984), at the Kennedy Space Center, Florida, by Willett (1992), at Camp Blanding, Florida, by Uman *et al.* (1997) and Rakov *et al.* (1998, 2000b), and in China by Liu *et al.* (1994) and Liu and Zhang (1998). Triggered-lightning experiments conducted in different countries have been reviewed by Uman (1987), Horii and Nakano (1995), and Rakov (1999c).

Table 7.1. *An overview of major triggered-lightning programs*

Experimental site[a]	Height above sea level, m	Years of operation[b]	Wire material	Location of wire spool	Selected references
Saint Privat d'Allier, France	1100	1973–96	steel or copper	ground or rocket	Fieux et al. (1978), SPARG (1982)
Kahokugata, Hokuriku coast, Japan	0	1977–85	steel	ground	Horii (1982), Kito et al. (1985)
Langmuir Laboratory, New Mexico	3230	1979–present	steel	ground	Hubert et al. (1984), Idone et al. (1984)
KSC, Florida (south of Melbourne, Florida in 1983)	0	1983–91	copper	rocket	Eybert-Berard et al. (1986, 1988), Willett (1992)
Okushishiku, Japan	930	1986–98	steel	ground or rocket	Nakamura et al. (1991, 1992)
Four sites in northern and southeastern China	Various	1989–present	steel or copper	ground or rocket	Liu et al. (1994), Liu and Zhang (1998)
Fort McClellan, Alabama	190	1991–5	copper	rocket	Fisher et al. (1993), Morris et al. (1994)
Camp Blanding, Florida	20–5	1993–present	copper	rocket	Uman et al. (1997), Rakov et al. (1998)
Cachoeira Paulista, Brazil	570	1999–present	copper	rocket	Saba et al. (2000)

[a] Additionally, triggered-lightning experiments have been conducted in Germany (Hierl 1981), in Indonesia (Horii et al. 1990), and in Russia (Beituganov and Zashakuev 1992).
[b] As of the time of writing and not necessarily continuous.

In all published experiments, the triggering wires were made of either steel or copper with a diameter of typically about 0.2 mm wound on a spool located either on the ground or on the rocket. Various rockets made of plastic and of steel have been used, the rocket length being typically about 1 m. Most of the experiments in Japan were conducted in the winter (Chapter 8), the several attempts made to trigger in the summer months being unsuccessful. At Camp Blanding, Florida, lightning has been triggered in both summer and winter storms. All other triggering sites have apparently been operated only during the summer.

The results from these programs have made possible a number of new insights into the various lightning processes and effects. Some of these findings are discussed later in this chapter while others have been presented in Chapters 4, 5, and will be presented in Chapter 18. Since the properties of triggered lightning processes, after the initial stage, are thought to be similar to the properties of natural downward lightning processes following the first stroke, somewhat arbitrary decisions were made by the authors regarding which findings to present here and which to present in the chapters concerning natural lightning. We chose to discuss elsewhere in this book (i) the observation of an upward connecting leader in a dart-leader–return-stroke sequence (subsection 4.5.1) and (ii) the mechanism of the M-component mode of charge transfer to ground along current-carrying lightning channels (Section 4.9). Additionally, we chose to present along with their natural lightning counterparts both optically measured return-stroke speeds (subsections 4.6.2 and 5.3.6) and also the various properties of dart and dart-stepped leaders inferred from optical measurements (Section 4.7). Further, some characteristics of upward positive and upward negative leaders in triggered lightning were discussed in subsection 5.3.2, and measurements of thunder from triggered lightning will be discussed in Section 11.2. Finally, the use of triggered-lightning data in testing the validity of various lightning models is discussed in subsection 12.2.6 and Section 12.5, in providing ground truth for the NLDN in Section 17.5, and in testing the susceptibility to lightning of overhead power lines and underground cables in subsections 18.4.3 and 18.4.4, respectively.

Descriptions of the classical and altitude rocket-and-wire triggering techniques are given in subsection 7.2.1. Probably close to a thousand lightning discharges have been triggered to date using these techniques. Properties of rocket-triggered lightning are reviewed in subsections 7.2.2 through 7.2.6. The use of rocket-triggered lightning for testing various objects and systems is described in subsection 7.2.7. An overview of "non-conventional" lightning-triggering techniques is given in Section 7.3.

7.2. Rocket-triggered lightning
7.2.1. Triggering techniques

Two techniques for triggering lightning with a small rocket that extends a thin wire in the gap between a thundercloud and the ground are discussed here, "classical"

7.2. Rocket-triggered lightning

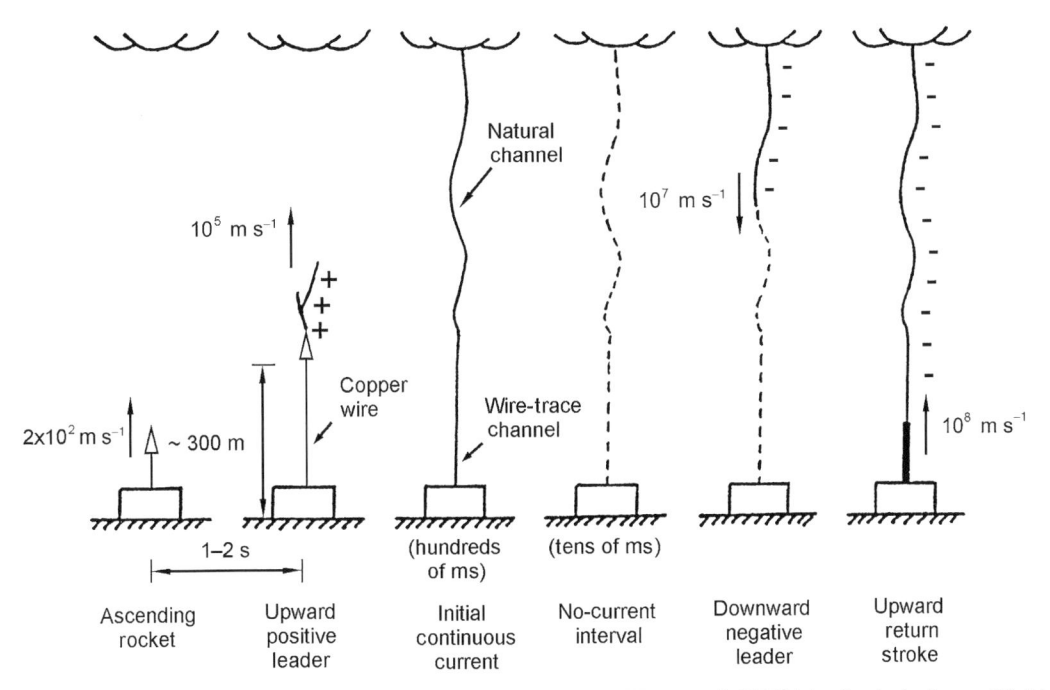

Fig. 7.1. Sequence of events (except for the attachment process; Wang et al. 1999b) in classical triggered lightning. The upward positive leader and initial continuous current constitute the initial stage. Adapted from Rakov et al. (1998).

triggering and "altitude" triggering. These descriptions primarily apply to triggering negative lightning.

Classical triggering The most effective technique for triggering lightning involves the launching of a small rocket trailing a thin grounded wire toward a charged cloud overhead. This triggering method is usually called classical triggering and is illustrated in Fig. 7.1. Still photographs of classical triggered lightning flashes are shown in Fig. 7.2. To decide when to launch a triggering rocket, the cloud charge is indirectly sensed by measuring the electric field at ground; field values of 4 to 10 kV m^{-1} are generally good indicators of favorable conditions for negative-lightning initiation in Florida, as seen in Fig. 7.3. However, other factors, such as the general trend of the electric field and the frequency of occurrence of natural lightning discharges, are usually taken into account in making the decision to launch a rocket. The triggering success rate is generally relatively low during very active periods of thunderstorms, one reason being that during such periods the electric field is more likely to be reduced by a natural lightning discharge before the rocket rises to a height sufficient for triggering.

When the rocket, ascending at about 200 m s^{-1}, is about 200 to 300 m high, the enhanced field near the rocket tip results in a positively charged leader that propagates upward toward the cloud. This upward positive leader (UPL) vaporizes the trailing wire, bridges the gap between the cloud and ground, and establishes an initial continuous current (ICC) with a duration of some hundreds of milliseconds that effectively transports negative charge from the cloud charge source to the triggering facility. The ICC can be viewed as a continuation of the UPL when the latter has reached the main negative charge region in the cloud. At that time the upper extremity of the UPL is likely to become heavily branched. The UPL and ICC constitute the initial stage (IS) of a classical triggered-lightning discharge. After cessation of the initial continuous current, one or more downward dart-leader–upward-return-stroke sequences may traverse the same path to the triggering facility. The dart leaders and the following return strokes in triggered lightning are similar to dart-leader–return-stroke sequences in natural lightning, although the initial processes in natural downward and in classical triggered lightning are distinctly different.

In summer, the triggering success rate for positive lightning is apparently lower than for negative lightning (e.g., Fieux et al. 1978), one known exception being the triggered lightning experiment in northern China (Liu et al. 1994; Liu and Zhang 1998), although all discharges triggered there were composed of an initial stage only, that is, no leader–return-stroke sequences occurred.

There is contradictory information regarding whether the height H of the rocket at the time of lightning triggering depends on the electric field intensity E at ground at the time of launching the rocket. Hubert et al. (1984) found a strong correlation (with correlation coefficient -0.82) between H and E for triggered lightning in

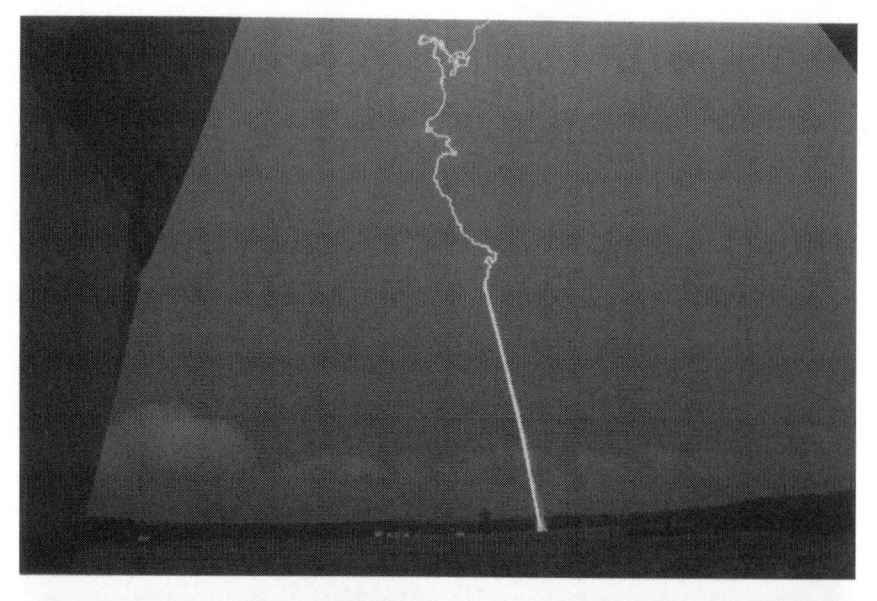

Fig. 7.2. Photographs of lightning flashes triggered in 1997 at Camp Blanding, Florida. Top, a distant view of a strike to the test runway; bottom, a close-up view of a strike to the test power system.

New Mexico. They gave the following equation relating H (in meters) and E (in kV m^{-1})

$$H = 3900 E^{-1.33} \qquad (7.1)$$

In the study by Hubert *et al.* (1984), E varied from about 5 to 13 kV m^{-1} and H from about 100 to 600 m, with a mean value of 216 m. However, in winter triggered-lightning studies at the Kahokugata site in Japan (Table 7.1), no clear relation was observed between H and E for either sign of E (Horii and Nakano 1995, Fig. 6.2.3).

Willett *et al.* (1999b), who used electric-field-sounding rockets in Florida, studied the ambient-field conditions required to initiate and sustain the propagation of upward positive leaders in triggered lightning. It was found that lightning can be initiated with grounded triggering wires approximately 400 m long when the ambient fields aloft are as small as 13 kV m^{-1}. When lightning occurred, ambient potentials with respect to earth at the triggering-rocket altitude were 3.6 MV (negative with respect to earth). These potentials were referred to as triggering potentials by Willett *et al.* (1999b). The first measurable current pulses at the bottom of the triggering wires ("precursors", subsection 7.2.3) were observed at similar fields aloft but at wire heights only about half as large, the corresponding potential being 1.3 MV.

7.2. Rocket-triggered lightning

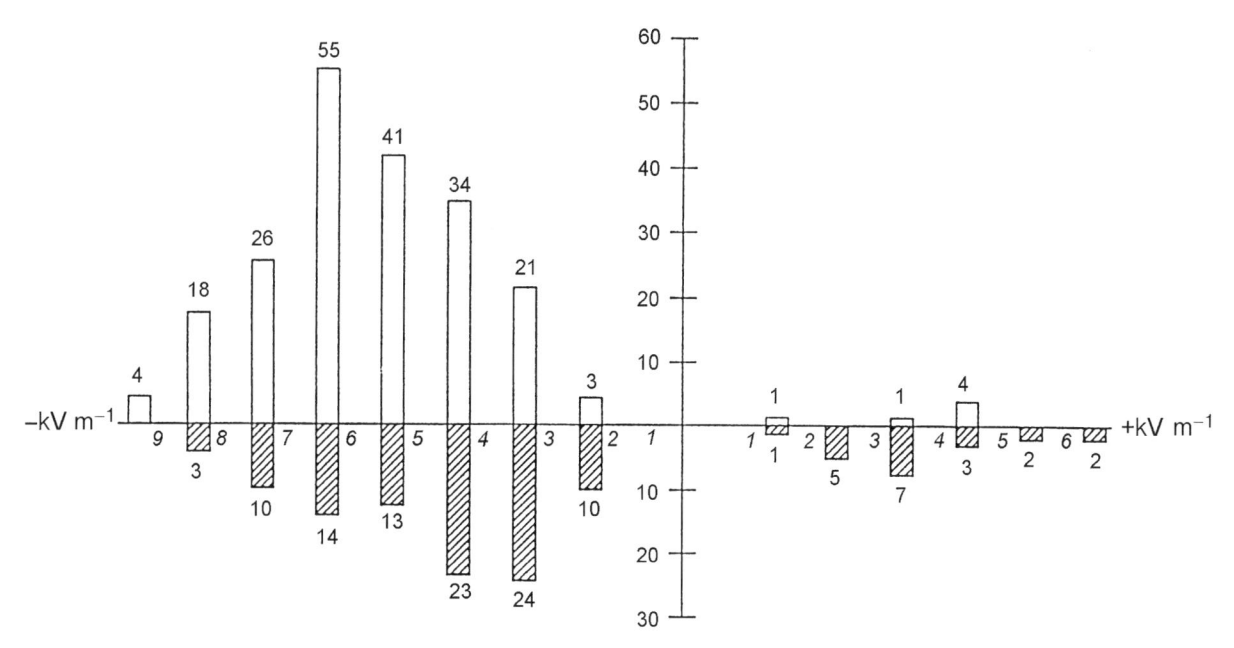

Fig. 7.3. Histograms of successful (above the horizontal axis) and unsuccessful (below the horizontal axis) classical triggering attempts in 1983 to 1991 at the NASA Kennedy Space Center. 242 rockets triggered lightning (readings not available for 34) and 172 rockets did not trigger lightning (readings not available for 50). The individual histogram bins, indicated on the horizontal axis, correspond to different positive and negative values of the surface electric field at the time of rocket launch. An upward-directed field is considered negative (atmospheric electricity sign convention, subsection 1.4.2). Adapted from Jafferis (1995).

Altitude triggering A stepped leader followed by a first return stroke in natural downward lightning can be reproduced to some degree by triggering lightning via a metallic wire not attached to the ground. This ungrounded-wire technique is usually called altitude triggering; as shown in Fig. 7.4, a bidirectional (positive charge up and negative charge down) leader process is involved in initiating the first return stroke from ground. Note that the "gap", in this case, the length of the insulating kevlar cable, between the bottom end of the upper (triggering) wire and the top end of the grounded (intercepting) wire is some hundreds of meters. Altitude triggering can also be accomplished without using an intercepting wire, whose only function is to increase the probability of lightning attachment to the instrumented rocket-launching facility.

In some triggered-lightning experiments, the bottom end of the triggering wire has been attached to an air gap of up to 10 m in length (e.g., Nakamura *et al.* 1992). Such triggering is not considered as being of the altitude type, since it is not intended to produce the downward stepped leader (discussed below) from the bottom of the triggering wire. However, altitude triggering may also occur as a result of accidental breakage of the wire during classical triggering, so that the wire connection to ground is unintentionally lost. Additionally, altitude triggering has been accomplished using a two-stage rocket system in which the two rockets separated in the air, the triggering wire extending between them (e.g., Nakamura *et al.* 1992). The properties of altitude-triggered lightning are discussed by Laroche *et al.* (1991), Lalande *et al.* (1996, 1998), Uman *et al.* (1996), and Rakov *et al.* (1996a, 1998).

In the following, we briefly discuss the sequence of processes involved in altitude-triggered lightning, as illustrated in Fig. 7.4. A downward negative leader is usually launched from the lower end of the elevated triggering wire some milliseconds after the initiation of the upward positive leader from the upper end of the wire (Lalande *et al.* 1998; Fig. 6); the downward negative leader shown in Fig. 6 of Lalande *et al.* (1998) was apparently initiated after two unsuccessful attempts. As a negative downward leader approaches the triggering facility, an upward connecting leader (not shown in Fig. 7.4) is initiated from the grounded intercepting wire. Once the attachment between the two leaders is made, the return stroke is initiated. Since (i) the length of the channel available for the propagation of the first return stroke in altitude triggered lightning is relatively small (of the order of 1 km) and (ii) the return-stroke speed is two to three orders of magnitude higher than that of the leader, the return stroke catches up with the tip of the upward leader within 10 μs or so. As a result, the upward leader becomes strongly intensified, an example of such an intensified leader being given in Fig. 7.7.

The processes that follow, the initial continuous current and downward-leader–upward-return-stroke

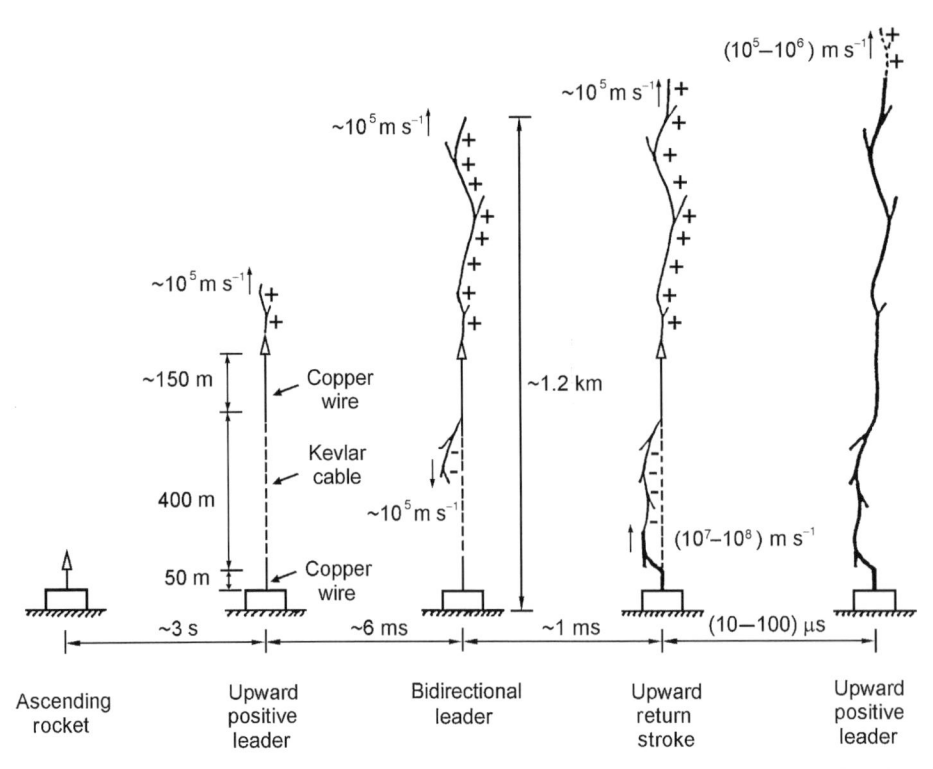

Fig. 7.4. Sequence of events in altitude-triggered lightning leading to the establishment of a relatively low-resistance connection between the upward-moving positive leader tip and the ground (except for the attachment process; Lalande et al. 1998). The diagrams are based on the event described by Laroche et al. (1991). The processes that follow the sequence shown, an initial continuous-current and possibly one or more downward-leader–upward-return-stroke sequences, are similar to their counterparts in classical triggered lightning (see Fig. 7.1). The rocket speed is of order 10^2 m s^{-1}. Adapted from Rakov et al. (1998).

sequences, are probably similar to those in classical triggered lightning (Fig. 7.1). Thus the downward-moving negative leader of the bidirectional leader system and the resulting return stroke in altitude-triggered lightning, serve to provide a relatively low-resistance connection between the upward-moving positive leader tip and the ground. The initial stage of altitude-triggered lightning can be viewed as being composed of an initial upward leader, a bidirectional leader, part of which is a continuation of the initial upward leader, an attachment process, an initial-stage return stroke, an intensified upward leader, and an initial continuous current.

7.2.2. Optically observed characteristics

Classical triggering Streak photographs of an upward positive leader, a downward negative dart-leader–upward-return-stroke sequence and a downward negative dart-stepped-leader–upward-return-stroke sequence are shown in Figs. 7.5a, b, d, respectively. Shown in Fig. 7.5c is a streak photograph of an M-component, a lightning process discussed in Section 4.9. Dart and dart-stepped leaders are considered in Section 4.7. The upward positive leader in Fig. 7.5a shows pronounced stepping and increasing propagation speed and luminosity with increasing height. The speed increased from 1.2×10^5 m s^{-1} when first imaged to 6.5×10^5 m s^{-1} as the leader exited the field of view. The mean upward propagation speed over a channel length of 530 m was 3.6×10^5 m s^{-1}. The time intervals between the luminous leader steps seen in Fig. 7.5a are about 20 µs. Laroche et al. (1988), for a different flash, reported a similar mean interstep interval and a mean step length of 14 m at a height of about 1 km above ground. The corresponding leader propagation speed was about 5×10^5 m s^{-1}. For further comparison, we note that Berger and Vogelsanger (1969; Fig. 6) reported, for seven upward positive leaders in tower-initiated lightning, propagation speeds ranging from 4×10^4 to 1×10^6 m s^{-1}, mean interstep intervals ranging from 40 to 120 µs, and average step lengths ranging from 4 to 40 m (see Table 6.2). Luminosity waves associated with the individual leader steps in Fig. 7.5a propagate back down the channel, estimated lower bounds of the speed for two steps being 5×10^7 m s^{-1} and 6×10^7 m s^{-1} over 600 m and 400 m, respectively. McEachron (1939, p. 191) reported, for one upward positive leader from the Empire State Building, that the luminosity

7.2. Rocket-triggered lightning

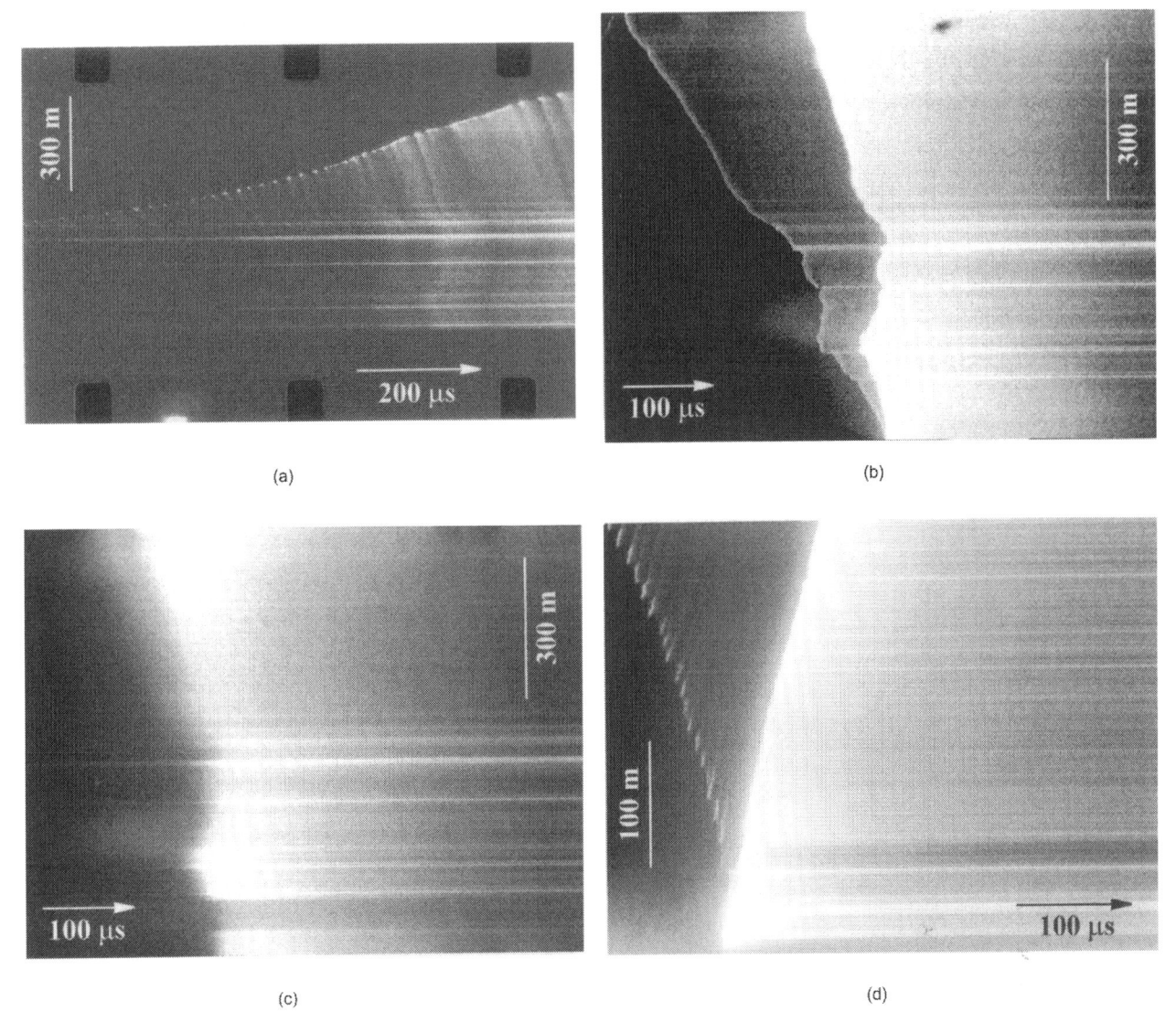

Fig. 7.5. Streak-camera records of (a) the upward positive leader, (b) and (d), two subsequent-leader–return-stroke sequences, and (c) an M-component, in negative rocket-triggered lightning. Time advances from left to right. Records (a), (b), and (c) are from Kennedy Space Center (KSC) flash 8827; record (d) is from Camp Blanding flash 9734. Note the pronounced stepping evident in the upward positive and downward negative leaders of (a) and (d). All images courtesy of V.P. Idone, University at Albany, SUNY.

enhancements associated with individual steps propagated upward from the top of the tower at a speed of approximately 6×10^7 m s^{-1}, and that usually there was a slight decrease in speed toward the upward-leader tip.

It appears that under certain conditions the upward positive leader can propagate continuously, without pronounced stepping, an example being given by Horii and Nakano (1995, Fig. 6.3.1). The upward positive leader can also switch from a stepped to a continuous propagation mode and back, as seen in Fig. 7.5a and also in Fig. 14 of Hagenguth and Anderson (1952); the latter figure illustrates upward lightning initiated from the Empire State Building. A more detailed comparison of the properties of positive and negative leaders is presented in subsection 5.3.2.

The characteristics, including the attachment process, of leader–return-stroke sequences in classical triggered lightning, obtained using optical techniques are presented in subsections 4.5.1, 4.5.2, 4.6.2, and 4.7.2.

Altitude triggering Altitude-triggered lightning is considerably less well studied than classical triggered lightning, in part because it is more difficult to produce. The upward positive and downward negative leaders of the bidirectional leader system have never been documented for the same lightning flash. A streak photograph of a negative downward leader that developed from the lower end of a triggering wire is shown in Fig. 7.6. From streak photography of the negative downward leader in a different altitude-triggered lightning flash, Lalande *et al.* (1998) estimated that the step length was 3–5 m and the mean interstep interval was 21 μs. For comparison, Berger and Vogelsanger (1969) reported, for 19 downward negative leaders, propagation speeds ranging from 6×10^4 to 1.1×10^6 m s^{-1}, mean interstep intervals ranging from 30 to 50 μs, and average step lengths ranging from 3 to 50 m. Upward positive leaders in altitude-triggered lightning are very faint and hence are difficult to image with streak cameras (in general, positive leaders produce much less light than negative leaders, only 15 percent of them being detectable in streak photographs; Berger and Vogelsanger 1969). Idone (1992) presented a near-ultraviolet streak photograph of an upward positive leader that was strongly intensified by the initial-stage return stroke. This photograph is shown in Fig. 7.7. The individual leader steps in Fig. 7.7 are described by Idone (1992) as each consisting of a bright, thin "stem", typically of length 3–5 m, and a diffuse, hemispherical "corona brush" that appears to extend from the stem in the direction of propagation over a distance of about 5–10 m. This corona brush (streamer zone) is most apparent in the steps directly to the right of the scale marker in the upper left corner in Fig. 7.7. A similar streamer zone was observed for upward negative leaders by Berger (1967), as discussed in subsections 4.4.7 and 6.2.2, and a faint luminosity at the lower end of its bright steps was reported by Schonland *et al.* (1935) for one downward stepped leader. The structure of positive and negative leaders, including the streamer zone, is described in subsections 5.3.3 and 4.4.6–4.4.8, respectively.

Wang *et al.* (1999d) reported on a positive flash that was initiated using the altitude-triggering technique from a summer thunderstorm in China. This is the first documented triggering of positive lightning using the altitude-triggering technique. For this flash, the length of grounded intercepting wire was 35 m and the length of insulating cable was 86 m. The flash was apparently initiated when the rocket was at an altitude of 550 m, so that the length of the ungrounded triggering wire was 429 m.

7.2.3. *Overall current waveforms*

In this subsection, we discuss the currents measured at the rocket launcher. For both classically triggered and altitude-triggered lightning, the emphasis will be placed on the initial stage, characterization of the current waveforms due to return strokes (primarily from classically triggered lightning) being presented in subsection 7.2.4. Initial-stage return strokes in altitude-triggered lightning are discussed

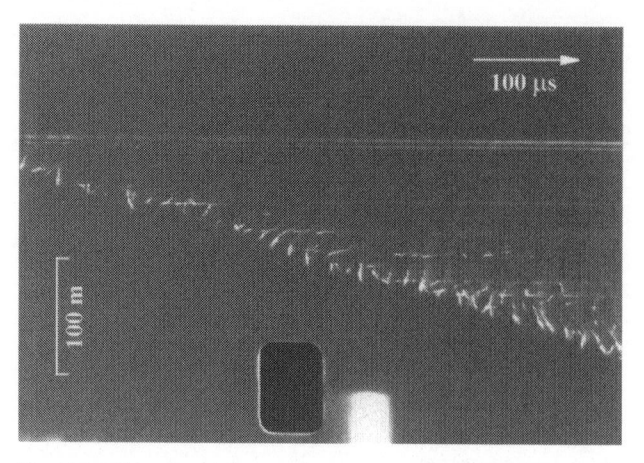

Fig. 7.6. Near-ultraviolet streak-camera record of the downward negative leader in KSC altitude-triggered lightning flash 9119. Time advances from left to right. Adapted from Idone (1992).

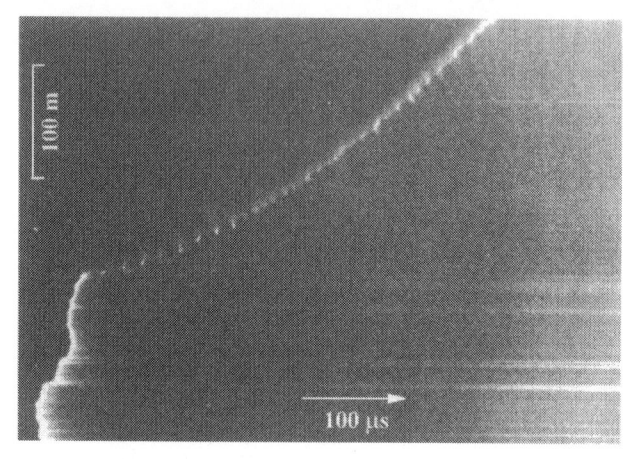

Fig. 7.7. Near-ultraviolet streak-camera record of the intensified upward positive leader in KSC altitude-triggered lightning flash 8911. Time advances from left to right. Adapted from Idone (1992).

in this subsection. For classically triggered lightning, the initial-stage current flows through the triggering wire until the wire is destroyed and replaced by the upward positive leader channel. For altitude-triggered lightning, a current value exceeding some amperes is first measured when an upward connecting leader (not shown in Fig. 7.4) emanates from the launcher (or from a grounded intercepting wire) in response to the approaching downward-extending, negative part of the bidirectional leader system.

Classical triggering The overall current record for a typical negative classically triggered lightning flash is presented in Fig. 7.8a, and portions of this record are shown on expanded time scales in Figs. 7.8b, c (Wang *et al.* 1999a). The record in Fig. 7.8a is intentionally clipped at the 2 kA level in order to accentuate the current components in the hundreds of amperes range. Other researchers

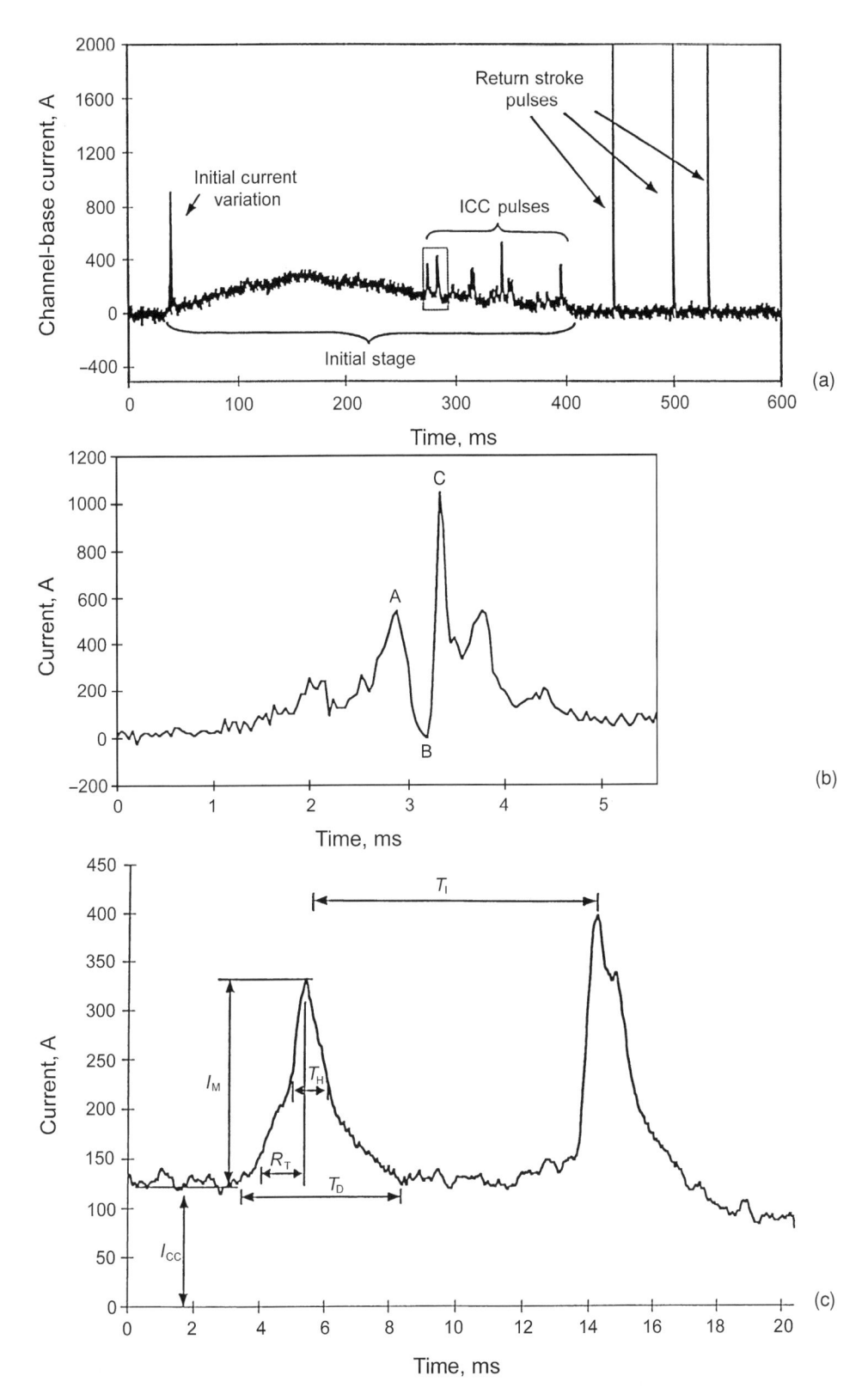

Fig. 7.8. (a) Example of the overall current record of a triggered-lightning flash at Camp Blanding, Florida on 24 June 1996 at 17:04:56 EDT, containing an initial stage and three return strokes. The initial tens of milliseconds of the initial stage are due to the upward positive leader (UPL), while the rest of the initial stage is due to the initial continuous current (ICC). The record is intentionally clipped at about 2 kA. (b) Initial current variation of (a) on an expanded time scale. (c) First two ICC pulses of (a) on an expanded time scale. This illustrates the definitions of the ICC pulse magnitude I_M, 10–90 percent risetime R_T, duration T_D, half-peak width T_H, interpulse interval T_I, and preceding continuous-current level I_{CC}. All these parameters have been found to be similar to the corresponding parameters of M-component current pulses analyzed by Thottappillil et al. (1995). Adapted from Wang et al. (1999a).

(e.g., Eybert-Berard et al. 1986, 1988) have used recorders with a logarithmic scale in order to be able to view both small currents and large currents on the same record. Median values of the overall flash duration from triggered lightning experiments in France and New Mexico are 350 and 470 ms (Hubert 1984), respectively. The median flash charges from the same studies are 50 and 35 C, respectively. Both the flash duration and charge transferred are comparable, within a factor 2 to 4, to their counterparts in object-initiated lightning (Section 6.3) and in natural downward lightning (Section 4.2).

We first consider the overall characteristics, that is, the duration, the charge transfer, and the average current, of the initial stage (IS), and then discuss (i) the current variation at the beginning of the IS, termed the initial current variation (ICV), and (ii) the current pulses superimposed on the later part of the IS current, referred to as initial continuous-current (ICC) pulses. Finally, we consider the microsecond-scale pulses that occur prior to the ICV and at the onset of the ICV. These pulses, termed precursor current pulses, are unresolved in Figs. 7.8a, b, primarily because of an insufficiently small sampling interval (40 μs; also, the noise level was approximately 20 A). The parameters of the return-stroke current pulses (three pulses are shown in Fig. 7.8a) that often follow the initial-stage current are discussed in subsection 7.2.4.

Wang et al. (1999a), based on data from Fort McClellan, Alabama and Camp Blanding, Florida (see Table 1 of that paper), reported that the initial stage had a geometric-mean (GM) duration of 279 ms and lowered to ground a GM charge equal to 27 C. The average initial-stage current in an individual lightning discharge varied from a minimum of 27 A to a maximum of 316 A with GM value 96 A. In many cases, the initial-current variation includes a current drop, as illustrated in Fig. 7.8b. This current drop is probably associated with the disintegration of the copper triggering wire (the abrupt current decrease from A to B) and is followed by a current re-establishment (the abrupt current increase from B to C). Fieux et al. (1975, 1978), from triggered-lightning experiments in which the steel-wire spool was located on ground, reported that the wire first disintegrated near the spool, where the mechanical stress was a maximum. The resultant gap between the spool and ascending lower end of the wire was bridged by an arc whose length at the time of vaporization of the wire typically reached 5 m. Note that, in the experiments in Florida and Alabama, the triggering wire was made of copper and the spool was attached to the rocket.

The GM time interval between onset of the initial-current variation and the abrupt decrease in current was found by Wang et al. (1999a) to be 8.6 ms, and the GM current level just prior to the current decrease was 312 A. The abrupt current decrease took typically several hundred microseconds and was followed, immediately or after a time interval of up to several hundred microseconds, by a pulse with a typical peak of about 1 kA and a typical risetime of less than 100 μs.

Crawford (1998), using a different data set from Camp Blanding, Florida, identified the signature of the initial current variation in the electric field waveforms recorded within some hundreds of meters of the lightning channel. This signature appears as a sharp V-shaped waveform superimposed on a millisecond-scale ramp apparently produced by the upward positive leader. Crawford (1998) also observed that, at the time of the onset of the V-shaped pulse, the magnetic field 500 m from the lightning channel decreased rapidly to nearly zero and then increased to its peak value, a signature similar to that of the channel-base current discussed above.

The initial continuous current, usually includes impulsive processes, Fig. 7.8c, that resemble the M processes observed during the continuing currents that often follow return strokes in both natural and triggered lightning (Sections 4.8 and 4.9). Wang et al. (1999a), from a comparison of various characteristics of the initial continuous-current pulses with the characteristics of the M-component current pulses analyzed by Thottappillil et al. (1995), concluded that these two types of pulses are similar and hence likely to be due to similar lightning processes. Like M-component pulses, the initial continuous-current pulses sometimes have amplitudes in the kiloamperes range.

Next we consider the current pulses, unresolved in Figs. 7.8a, b, that occur prior to the initial current variation (the precursor current pulses) and at the onset of this current variation. These current pulses, recorded at the bottom of the triggering wire before it is melted by the current of the upward positive leader, have been studied by Laroche et al. (1985, 1988), Lalande et al. (1998), Davis and Laroche (1999), and Willett et al. (1999b). According to Lalande et al. (1998), the precursor current pulses appear (see Fig. 7.9) as damped oscillations lasting for about 10 μs and have initial amplitudes of the order of tens of amperes. The average time interval between the pulses in the example shown in Fig. 7.9 is 5 ms. Laroche et al. (1985) reported that the time interval between these precursor current pulses may range from 10 to 100 ms. The "precursors" may first occur some hundreds of milliseconds before the onset of the upward leader, and, as the rocket continues ascending, they may come in groups of two, three, or more (Laroche et al. 1988), although not in the example shown in Fig. 7.9. The pulses within a group are separated by time intervals of 20 to 30 μs (Willett et al. 1999b). Lalande et al. (1998) estimated that each isolated pulse is associated with a charge of several tens of microcoulombs. They interpreted these current pulses as due to attempted inceptions of an upward positive leader from the wire.

7.2. Rocket-triggered lightning

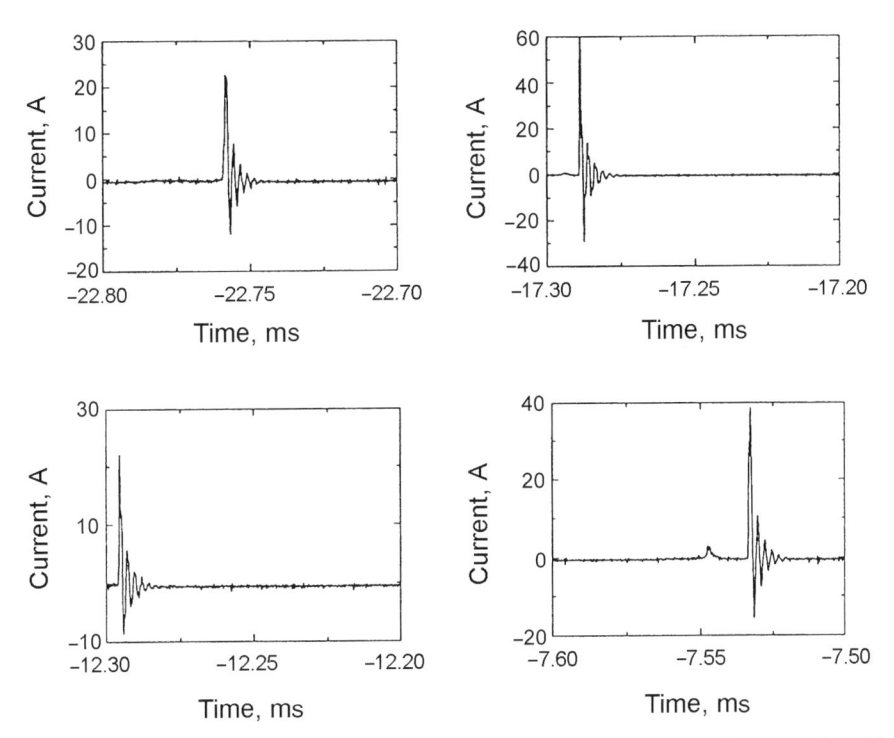

Fig. 7.9. Oscillating current pulses associated with attempted inceptions of an upward positive leader in classical triggered lightning flash 9519 at Camp Blanding, Florida. Time $t = 0$ corresponds to the beginning of the stable propagation phase of the upward positive leader (Fig. 7.10). Adapted from Lalande et al. (1998).

The onset of an upward positive leader, identified by a gradually increasing steady current at the bottom of the triggering wire at $t = 0$, is shown in Fig. 7.10a. As seen in this figure, the steady-current increase is marked by a series of current pulses, shown on expanded time scales in Figs. 7.10b, c, that are similar to precursor pulses (see Fig. 7.9) in their waveshape and amplitude. The former pulses, however, are separated by time intervals of a few tens of microseconds, as opposed to the time intervals of 10 to 100 ms between the precursor pulses or precursor pulse groups. Note that Fig. 7.10c represents the first 50 μs of Fig. 7.10b, the latter being the first 300 μs of the current record in Fig. 7.10a, starting from $t = 0$. Since the interpulse intervals, seen in Figs. 7.10b, c, are similar to those observed between optical upward positive-leader steps (see the first part of subsection 7.2.2), the current pulses occurring at the onset of the upward positive leader have been attributed to leader steps. It is worth noting, however, that no synchronized observations of optical leader steps and current pulses at the bottom of the wire have been reported. Current pulses of the type illustrated in Fig. 7.10 have been associated with steps (or ramps) in the electric field records obtained at close range (e.g., Laroche et al. 1988). The oscillatory shape of the current pulses is apparently due to the transient process excited in the wire by the step-formation process (subsection 5.3.3) at the upper end of the wire or at the upward-moving leader tip. As seen in

Fig. 7.10b, after about 200 μs there is a change in the shape of the current pulses from oscillatory (bipolar) to unipolar (although two unipolar pulses occurred before the last oscillatory pulse), presumably owing to the increased damping effect of the leader-channel resistance as this channel extends farther above the upper end of the wire (Lalande et al. 1998).

Altitude triggering As noted in subsection 7.2.1, the initial stage of altitude-triggered lightning includes an initial upward leader, a bidirectional leader (which includes a continuation of the initial upward leader), an attachment process, an initial-stage return stroke, an intensified upward leader, and an initial continuous current. Since the triggering wire is ungrounded, no current can be directly measured at ground during the initial upward leader and bidirectional leader stages. Shown in Fig. 7.11b is the current associated with an upward positive connecting leader initiated in response to the approaching downward negative leader of the bidirectional leader system (Fig. 7.4), the corresponding electric field 50 m from the lightning attachment point being shown in Fig. 7.11a. This current record, reported by Lalande et al. (1998), suggests that the upward positive connecting leader is stepped, the interstep interval being 20 μs or so. When contact is established between the downward leader and the upward connecting leader, the initial-stage return stroke begins. The current

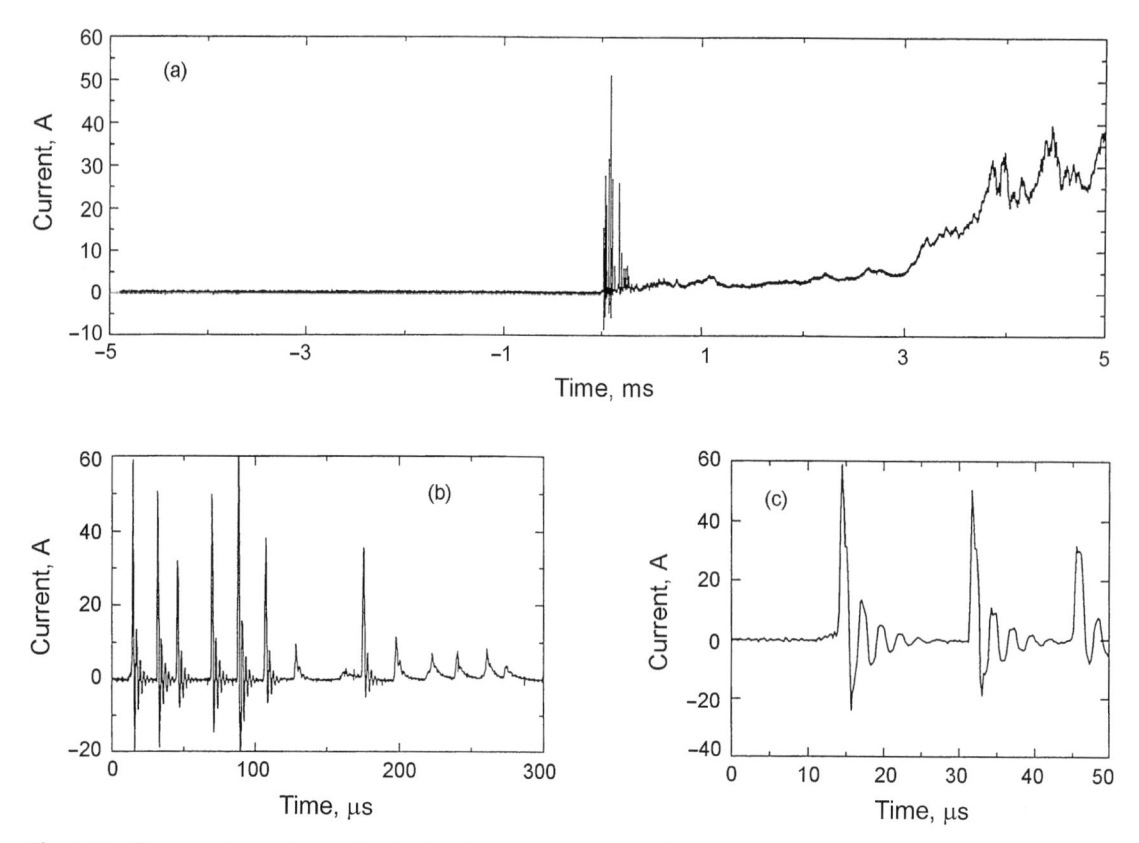

Fig. 7.10. The current associated with the development of the upward positive leader in classical triggered lightning flash 9519 at Camp Blanding, Florida. Time $t = 0$ corresponds to the beginning of the stable propagation phase of the upward positive leader; (b) and (c) are expansions of (a). Adapted from Lalande et al. (1998).

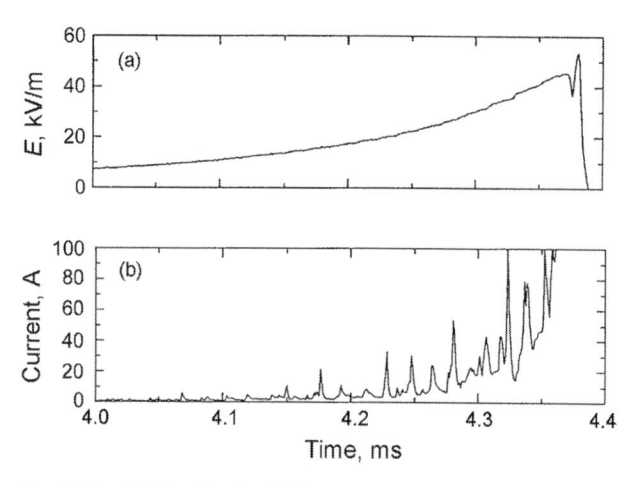

Fig. 7.11. (a) The electric field measured 50 m from the lightning attachment point and (b) the current produced by the upward connecting positive leader from the grounded 50 m wire in altitude-triggered lightning flash 9516 at Camp Blanding, Florida. Adapted from Lalande et al. (1998).

waveform of this return stroke differs appreciably from a typical return-stroke current waveform in that the former appears to be chopped soon after reaching its peak value (see, for example, Fig. 7c of Hubert 1984). As a result, the width of the current waveform produced by the initial-stage return stroke is appreciably smaller than that of the following return strokes in the same flash. As discussed in the second part of subsection 7.2.1, the initial-stage return-stroke front catches up with the upward-moving leader tip after 10 μs or so. This is likely to produce an opposite-polarity downward-moving reflected current wave that is presumably responsible for the chopped shape of both the channel-base current and the close-magnetic-field waveforms. Examples of the latter are shown, along with waveforms produced by "normal" return strokes, in Fig. 7.12. The initial-stage characteristics of altitude-triggered lightning, after the return stroke has established a relatively low-resistance connection between the upward-moving positive leader tip and ground (see Fig. 7.4), are apparently similar to their counterparts in classically triggered lightning (Rakov et al. 1996b). Further, the downward-leader–upward-return-stroke sequences that follow the initial stage in altitude-triggered lightning are thought to be similar to those in classically triggered lightning.

7.2.4. Parameters of return-stroke current waveforms

In this section, we discuss return-stroke current peak and current waveform parameters such as risetime, rate of

7.2. Rocket-triggered lightning

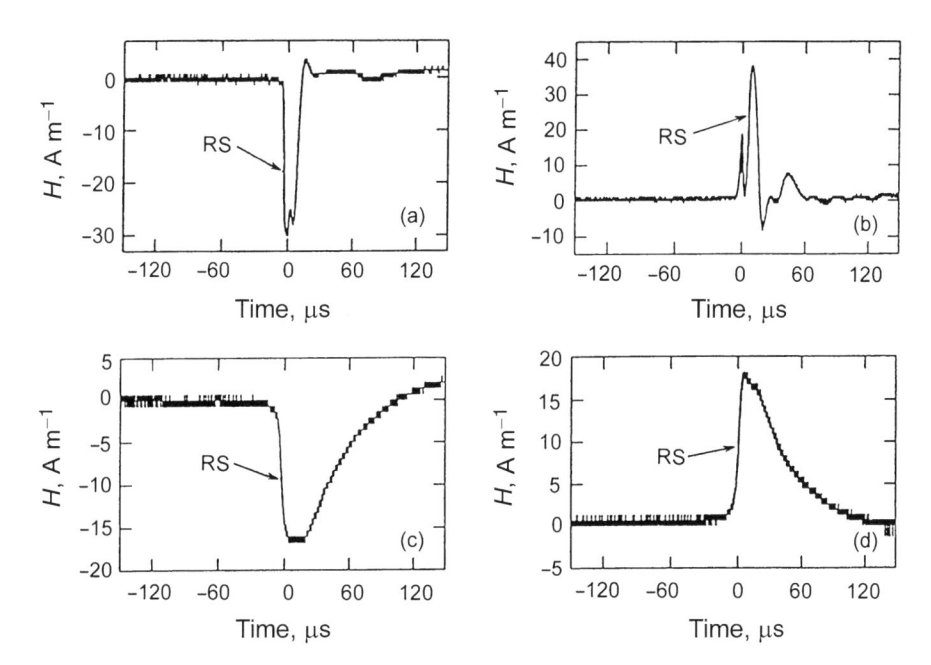

Fig. 7.12. The magnetic fields produced by the first two strokes of each of the Camp Blanding altitude-triggered lightning flashes 9514 ((a), first stroke, (c), second stroke, four strokes total) and 9516 ((b), first stroke, (d), second stroke, four strokes total). The first flash was at distance $r = 90$ m and the second was at $r = 50$ m. RS stands for return stroke. In each case, the waveshapes of all the higher-order strokes were similar to the second-stroke waveshape. The measuring system's decay time constant was about 120 μs. The difference in polarity of the waveforms is due to different positions of the lightning channel with respect to the magnetic field antenna; all strokes lowered negative charge to ground. Note that the first-stroke magnetic field pulses in (a) and (b) are appreciably shorter than the corresponding second-stroke magnetic field pulses in (c) and (d), respectively. Adapted from Rakov *et al.* (1998).

rise (steepness), and half-peak width. Additionally, we will consider interstroke intervals and characteristics that may involve both the return-stroke current component and the following continuing current component, such as the total stroke duration, the total stroke charge, $\int I(t)\,dt$, and the total stroke action integral, $\int I^2(t)\,dt$. The action integral is measured in A^2 s, which is the same as $J\,\Omega^{-1}$, and represents the joule or ohmic heating energy dissipated per unit resistance at the lightning attachment point. The action integral is also called the specific energy. We will also discuss correlations among the various parameters listed above. The characterization of the return-stroke current waveforms presented in this section is based primarily on data for classical triggered lightning. It is possible that some of the samples on which the statistics presented here are based contain a small number of initial-stage return strokes from altitude-triggered lightning, but their exclusion would have essentially no effect on the statistics.

Some researchers (e.g., Hubert *et al.* 1984; Horii and Ikeda 1985), in presenting statistics on triggered-lightning currents, do not distinguish between current pulses associated with return strokes and those produced by other lightning processes, such as M-components and the processes giving rise to the initial current variation and initial continuous current pulses described in subsection 7.2.3. In this section, we consider only return-stroke current pulses. These can usually be distinguished from other types of pulse by the absence of a steady current immediately prior to a pulse (Fisher *et al.* 1993). Further, we do not consider here three unusual New Mexico triggered-lightning flashes, each of which contained 24 return strokes (Idone *et al.* 1984). For these three flashes, the geometric means of the return-stroke current peak and interstroke interval were 5.6 kA and 8.5 ms, respectively, each considerably smaller than its counterpart in either natural lightning or in triggered lightning flashes, discussed below.

We first review measurements of the peak values of current and current derivative. Summaries of the statistical characteristics, from triggered-lightning experiments in Florida and France, of measured return-stroke current peaks, I, and time derivative peaks, dI/dt, are given in Tables 7.2 and 7.3, respectively. Additionally, relations between dI/dt and I, and the correlation coefficients for the logarithms of dI/dt and I are included in Table 7.3. According to Depasse (1994a), the statistical distributions of I and dI/dt for individual years represented in Tables 7.2 and 7.3 obey log-normal distributions, except for the dI/dt distributions obtained in Florida in 1987. As seen in Table 7.2, the

Table 7.2. *Characterization of return-stroke current peak from triggered-lightning experiments in Florida and France. Adapted from Depasse (1994a)*

Location	Years	Sample size	Min., kA	Max., kA	Mode, kA	Median, kA	Mean, kA	SD, kA	Percentage of cases exceeding tabulated value		
									95%	50%	5%
Kennedy Space Center, Florida	1985, 1987–91	305	2.5	60.0	8.7	12.1	14.3	9.0	4.7	12.1	31.3
Saint-Privat d'Allier, France	1986, 1990–1	54	4.5	49.9	7.8	9.8	11.0	5.6	4.5	9.8	21.5

median values of I are 12.1 and 9.8 kA in Florida (1985, 1987–91) and France (1986, 1990–1), respectively. These two values are similar to the median value of 12 kA reported by Anderson and Eriksson (1980) for subsequent strokes in natural lightning. The year-to-year variation in the median currents in Florida was from 10 to 14.5 kA, and in France this variation was from 9.0 kA (1990–1) to 13.6 kA (1986) (Depasse 1994a, Table 7). The median values of dI/dt are 91.4 and 36.8 kA μs^{-1} in Florida and France, respectively, different by a factor 2.5. However, the power relations between dI/dt and I, derived from linear regression equations relating the logarithms of dI/dt and I, for Florida and France are very similar. The year-to-year variation in the median dI/dt was from 58.3 to 109.2 kA μs^{-1} in Florida, and in France it was from 33.1 kA μs^{-1} (1990–1) to 57.5 kA μs^{-1} (1986) (Depasse 1994a, Table 7). The largest measured value of dI/dt was 411 kA μs^{-1}, as reported from Florida (KSC) studies by Leteinturier *et al.* (1991). The corresponding measured peak current was greater than 60 kA, the largest value of this parameter reported for summer triggered lightning to date.

Scatter plots of dI/dt peak versus I peak from the triggered lightning experiments in Florida (1985, 1987, and 1988) and in France (1986) are shown in Fig. 7.13. The correlation coefficients are 0.87, 0.80, and 0.70 for the 1985, 1987, and 1988 Florida data, respectively, and 0.78 for the 1986 data from France. Also shown in Fig. 7.13 are the linear regression line and the regression equation for each of the four data subsets. Note that the correlation coefficients for the logarithms of dI/dt and I for the same data were found to be lower: 0.79, 0.56, and 0.60 for the 1985, 1987, and 1988 Florida data, respectively, and 0.71 for the 1986 data from France (Depasse 1994a, Table 10).

A number of parameters of the triggered-lightning current waveforms recorded in France in 1990–1 are summarized in Table 7.4, I and dI/dt here indicating the peak values of the current and the current derivative, respectively. This table additionally includes parameters of lightning electric field waveforms that will be discussed in subsection 7.2.6. Relations between the parameters are summarized in Table 7.5. Note that there is a relatively weak positive correlation between the logarithms of dI/dt and I (correlation coefficient 0.46) and relatively strong positive correlations between the logarithms of the impulse charge, $\int I(t)\,dt$, and I (correlation coefficient 0.77) and between the logarithms of specific energy (action integral), $\int I^2(t)\,dt$, and I (correlation coefficient 0.92).

The correlation coefficient, 0.46, between the logarithms of dI/dt and I for the 1990–1 data from France is appreciably lower than the value 0.71 determined for the 1986 French data, and lower than the values 0.79, 0.56, and 0.60 determined for the 1985, 1987, and 1988 Florida data, respectively (see above). However, a similar relatively low correlation coefficient, 0.47, was found between the logarithms of dI/dt and I for the 1989 Florida data (Depasse 1994a, Table 10). The considerable observed year-to-year variation in this correlation coefficient at each location is apparently associated with a greater year-to-year variability in dI/dt than in I.

Fisher *et al.* (1993) compared a number of return-stroke current parameters for classically triggered-lightning strokes from Florida and Alabama with the corresponding parameters for natural lightning in Switzerland reported by Berger *et al.* (1975) and Anderson and Eriksson (1980). This comparison is given in Figs. 7.14 through 7.22. Recall that triggered-lightning strokes are considered to be similar to subsequent strokes in natural lightning. Therefore, the comparison in Figs. 7.14 through 7.22 applies only to subsequent strokes. Both Berger *et al.* (1975) and Anderson and Eriksson (1980) fitted a straight line representing a log-normal approximation to the experimental statistical distribution in order to determine the percentages (95, 50, and 5 percent) of cases exceeding the tabulated values, while Fisher *et al.* (1993) used the nearest experimental point instead. There appear to be appreciable differences between the triggered-lightning data of Fisher *et al.* (1993) and the natural-lightning data of Berger *et al.* (1975) and Anderson and Eriksson (1980) in terms of current wavefront parameters, half-peak width, and stroke charge. The shorter risetime and higher average slope

Table 7.3. *Characterization of the peak derivative of return-stroke current with respect to time from triggered-lightning experiments in Florida and France. Adapted from Depasse (1994a)*

Location	Years	Sample size	Minimum, kA μs^{-1}	Maximum, kA μs^{-1}	Mode, kA μs^{-1}	Median, kA μs^{-1}	Mean kA μs^{-1}	Standard deviation, kA μs^{-1}	Percentage of cases exceeding tabulated value			Relation between dI/dt and I	Correlation coefficient
									95%	50%	5%		
Kennedy Space Center, Florida	1985, 1987–91	134	5.0	411.0	54.6	91.4	118.3	97.1	28.1	91.4	297.7	dI/dt = 2.6 $I^{1.34}$	0.45
Saint-Privat d'Allier, France	1986, 1990–1	47	12.8	139.0	27.2	36.8	42.8	25.4	14.9	36.8	90.9	dI/dt = 2.0 $I^{1.28}$	0.62

Table 7.4. *Summary of return-stroke parameters from triggered-lightning experiments in France in 1990–1. The values of I and dI/dt are peak values, measured by the Centre d'Etudes Nucléaires de Grenoble and the Centre National d'Etudes des Télécommunications. E_{50m} and E_{77m} are the maximum electric field changes due to the return stroke, whose waveshapes are similar to those seen in Fig. 7.26, recorded by Faculte Polytechnique de Mons and Ecole Centrale de Lyon, respectively. Here and in Table 7.5, the subscripts 50 m and 77 m indicate the distances at which E or dE/dt were measured. $Q_{impulse}$ is the charge transferred by the stroke excluding continuing-current components, if any. Δt_{X-Y} is the time between X percent and Y percent of the maximum value; for the impulsive waveforms (I, dI/dt, and dE/dt), if $X = Y$ then the percentages are on both sides of the peak. $\Delta t_{10-90} I$ and $\Delta t_{50-50} I$ are the same as T-10 and the half-peak width in Figs. 7.18 and 7.22, respectively. Adapted from Depasse (1994a)*

Parameter	Sample size	Min.	Max.	Mode	Median	Mean	SD	Percentage of cases exceeding tabulated value		
								95%	50%	5%
I, kA	43	4.5	49.9	7.5	9.0	9.9	4.6	4.4	9.0	18.5
dI/dt, kA μs^{-1}	38	12.8	102.9	26.4	33.1	37.1	18.6	15.2	33.1	72.2
E_{50m}, kV m^{-1}	14	14.5	41.9	22.0	24.6	26.0	9.0	14.2	24.6	42.8
E_{77m}, kV m^{-1}	13	11.4	25.2	15.2	16.7	17.5	5.5	10.1	16.7	27.7
dE/dt_{50m}, kV m^{-1}μs^{-1}	14	4.6	143.6	15.5	47.1	82.2	17.6	8.3	47.1	267.2
$Q_{impulse}$, C	24	0.31	1.6	0.47	0.59	0.66	0.33	0.27	0.59	1.29
$\int I^2(t)dt$, kJ Ω^{-1}	24	0.65	41.8	1.0	2.8	4.7	6.4	0.5	2.8	15.1
$\Delta t_{10-90} I$, μs	37	0.25	4.9	0.42	0.82	1.14	1.10	0.22	0.82	3.1
$\Delta t_{50-50} I$, μs	24	14.7	103.2	37.8	45.4	49.8	22.4	22.4	45.4	92.0
Total duration, μs	24	63.8	305.2	180.8	202.1	213.7	73.3	116.7	202.1	349.9
$\Delta t_{10-10}\, dI/dt$, μs	17	0.07	2.01	0.17	0.30	0.40	0.21	0.08	0.30	1.07
$\Delta t_{10-90} E_{50m}$, μs	12	1.12	2.06	1.36	1.43	1.47	0.34	0.98	1.43	2.07
$\Delta t_{10-90} E_{77m}$, μs	9	1.73	2.53	2.08	2.11	2.13	0.26	1.73	2.11	2.58
$\Delta t_{10-10}\, dE/dt_{50m}$, μs	12	0.14	2.31	0.31	0.61	0.87	0.86	0.16	0.61	2.40

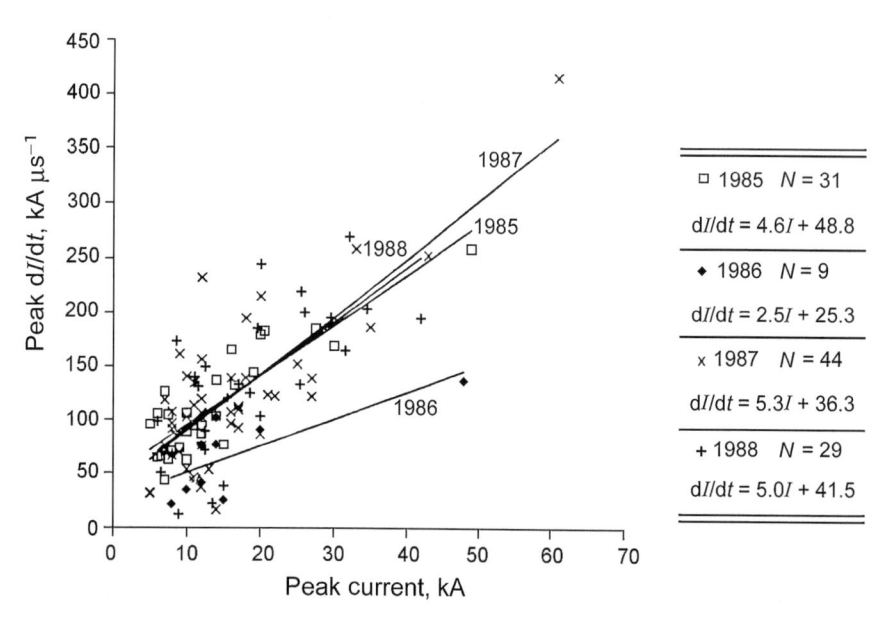

Fig. 7.13. Relation between the peak rate of current rise, dI/dt, and the peak current I, from triggered-lightning experiments conducted at the NASA Kennedy Space Center, Florida, in 1985, 1987, and 1988 and in France in 1986. The regression line for each year is shown; the sample size N and the regression equation are given in the table. Adapted from Leteinturier et al. (1991).

7.2. Rocket-triggered lightning

Table 7.5. *Relations between parameters of triggered-lightning strokes in France (1990–1). The current peak I is expressed in kA, the current derivative peak dI/dt in $kA\ \mu s^{-1}$, the maximum electric field change E due to the return stroke in $kV\ m^{-1}$, the electric field derivative peak dE/dt in $kV\ m^{-1}\ \mu s^{-1}$, the impulse charge Q_{impulse} in C, and the specific energy $\int I^2(t)\,dt$ in $kJ\ \Omega^{-1}$. Correlation coefficients are determined for the logarithms of Y and X. See also the caption for Table 7.4. Adapted from Depasse (1994a)*

Sample size	Y	X	Correlation coefficient	Relation between Y and X
38	dI/dt	I	0.46	$dI/dt = 1.65 I^{1.37}$
14	dE/dt_{50m}	E_{50m}	0.55	$dE/dt_{50m} = 0.002 E_{50m}^{3.14}$
11	E_{50m}	I	0.82	$E_{50m} = 6.3 I^{0.69}$
12	E_{77m}	I	0.88	$E_{77m} = 4.7 I^{0.59}$
7	E_{77m}	E_{50m}	0.88	$E_{77m} = 0.68 E_{50m}^{0.97}$
8	dE/dt_{50m}	dI/dt	0.57[a]	—
22	Q_{impulse}	I	0.77	$Q_{\text{impulse}} = 0.09 I^{0.89}$
22	$\int I^2(t)\,dt$	I	0.92	$\int I^2(t)\,dt = 0.05 I^{1.89}$

[a] Not significant.

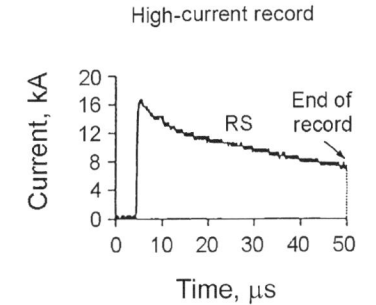

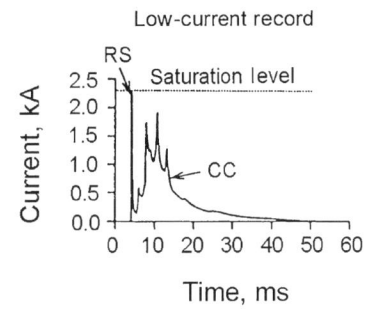

Parameter	Reference	Sample size	Unit	GM	SD	Percentage of cases exceeding tabulated value		
						95%	50%	5%
$\int I(t)\,dt$	Fisher *et al.* (1993)	65	C	2.5	0.55	0.38	2.1	15
	Berger *et al.* (1975)	122		—	—	0.2	1.4	11

Fig. 7.14. The curves show the current records for classically triggered lightning strokes; the table gives values deduced for the total stroke charge. RS, return stroke; CC, continuing current; GM, geometric mean; SD, standard deviation of the logarithm (to base 10) of a parameter. Adapted from Fisher *et al.* (1993).

(greater steepness) in the triggered-lightning data may be explained by better time resolution of the measuring systems used in the triggered-lightning studies. The Swiss data were recorded as oscilloscopic traces, the smallest measurable time being 0.5 μs (Berger and Garbagnati 1984).

Fisher *et al.* (1993) also studied relations among some return-stroke parameters; the results are shown in Fig. 7.23. They found a relatively strong positive correlation between 10–90 percent average steepness S-10 and current peak I (correlation coefficient $r = 0.71$) and between 30–90 percent average steepness S-30 and current peak I (correlation coefficient $r = 0.74$). Note that Anderson and Eriksson (1980) found a weak correlation between the logarithms of S-10 and I and between the logarithms of S-30 and I, for natural lightning in Switzerland. As seen in Fig. 7.23a, there is essentially no linear correlation between current peak and 10–90 percent risetime. Crawford (1998), from the 1997 triggered-lightning experiment at Camp Blanding,

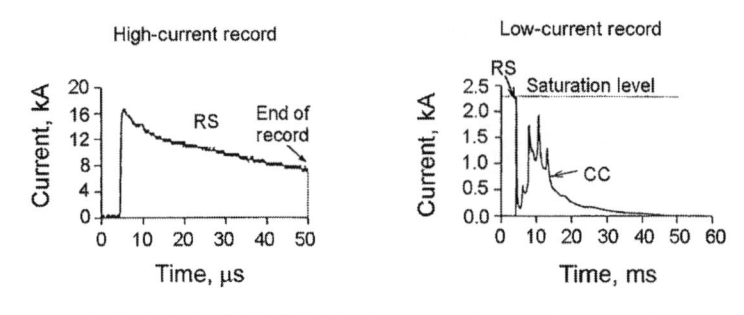

High-current record / Low-current record

Fig. 7.15. As in Fig. 7.14, but for the total-stroke action integral. Adapted from Fisher et al. (1993).

Parameter	Reference	Sample size	Unit	GM	SD	Percentage of cases exceeding tabulated value		
						95%	50%	5%
$\int I^2(t)\,dt$	Fisher et al. (1993)	65	$A^2 s$	3.5×10^3	0.54	0.4×10^3	3.8×10^3	20×10^3
	Berger et al. (1975)	88		—	—	0.55×10^3	6.0×10^3	52×10^3

Fig. 7.16. As in Fig. 7.14, but for the interstroke interval. Adapted from Fisher et al. (1993).

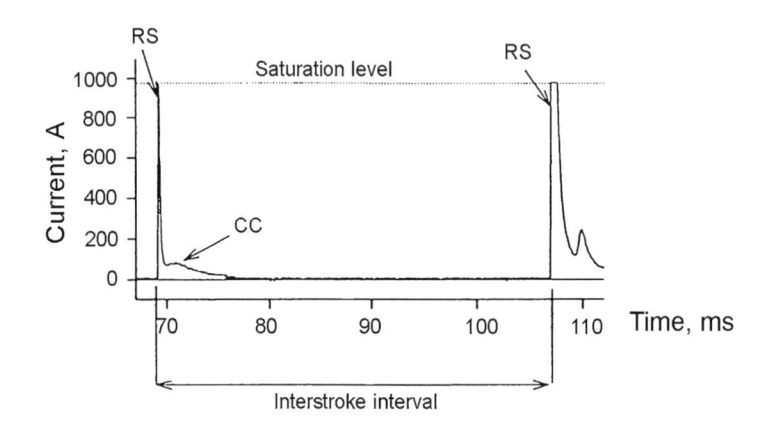

Symbol	Reference	Sample size	Unit	GM	SD	Percentage of cases exceeding tabulated value		
						95%	50%	5%
Δt	Fisher et al. (1993)	52	ms	47	0.48	5.0	48	215
	Berger et al. (1975)	133		—	—	7.0	33	150

Florida, reported a correlation coefficient $r = 0.8$ (which is statistically significant) between the logarithms of return-stroke current peak and 10–90 percent risetime.

The 50-percent (median) values of current peak (Fig. 7.17), 10–90 percent risetime (Fig. 7.18), and half-peak width (Fig. 7.22) for triggered lightning in Florida (KSC) and Alabama can be compared with their counterparts for triggered lightning in France given in Table 7.4. While the median current peaks differ by 40 percent or so, the French peak being smaller, there is more than a factor 2 difference in terms of the median 10–90 percent risetime and the median half-peak width, the French current waveforms rising to peak slower and being wider.

7.2.5. Return-stroke current peak versus grounding conditions

In examining the lightning current flowing from the bottom of the channel into the ground, it is convenient

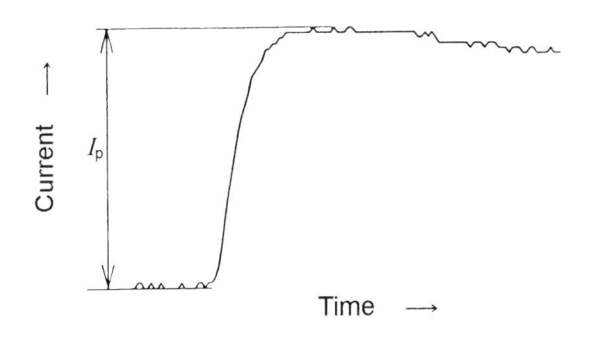

7.2. Rocket-triggered lightning

Fig. 7.17. As in Fig. 7.14, but for the current peak. Adapted from Fisher et al. (1993).

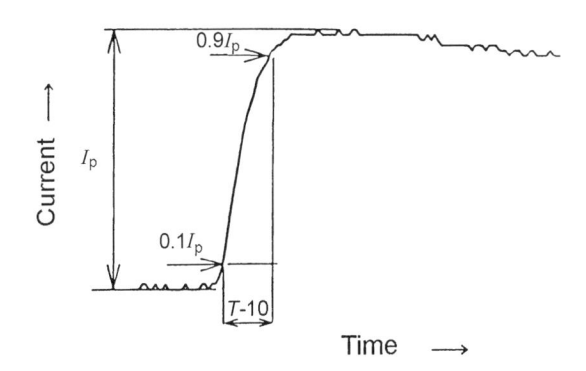

Symbol	Reference	Sample size	Unit	GM	SD	Percentage of cases exceeding tabulated value		
						95%	50%	5%
I_p	Fisher et al. (1993)	45	kA	12	0.28	4.7	13	29
	Anderson and Eriksson (1980)	114		—	—	4.9	12	29

Fig. 7.18. As in Fig. 7.14, but for the 10–90 percent rise time. Adapted from Fisher et al. (1993).

Symbol	Reference	Sample size	Unit	GM	SD	Percentage of cases exceeding tabulated value		
						95%	50%	5%
T-10	Fisher et al. (1993)	43	μs	0.37	0.29	0.20	0.32	1.1
	Anderson and Eriksson (1980)	114		—	—	0.1	0.6	2.8

to approximate lightning by a Norton equivalent circuit (Carlson 1996), i.e., by a current source consisting of the lightning current that would be injected into the ground if that ground were perfectly conducting (a short-circuit current) in parallel with a lightning-channel equivalent impedance Z_{ch}, assumed to be constant (see Fig. 6.9). The lightning grounding impedance Z_{gr} is a load connected in parallel with the lightning Norton equivalent. Thus the "short-circuit" lightning current I effectively splits between Z_{gr} and Z_{ch} so that the current measured at the lightning-channel base is $I_{meas} = I Z_{ch}/(Z_{ch} + Z_{gr})$. The source characteristics I and Z_{ch} both vary from stroke to stroke, and Z_{ch} is a function of channel current, the latter nonlinearity being in violation of the linearity requirement

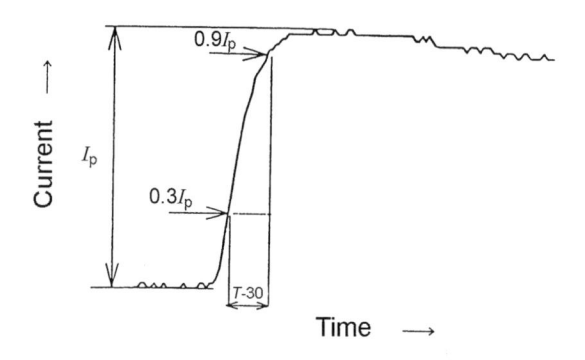

Fig. 7.19. As in Fig. 7.14, but for the 30–90 percent rise time. Adapted from Fisher et al. (1993).

Symbol	Reference	Sample size	Unit	GM	SD	Percentage of cases exceeding tabulated value		
						95%	50%	5%
T-30	Fisher et al. (1993)	43	μs	0.28	0.28	0.14	0.24	0.96
	Anderson and Eriksson (1980)	114		—	—	0.1	0.4	1.8

Fig. 7.20. As in Fig. 7.14, but for the 10–90 percent average slope (steepness). $S\text{-}10 = 0.8 I_p / T\text{-}10$. Adapted from Fisher et al. (1993).

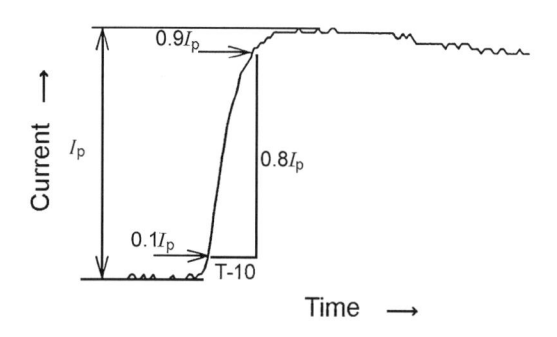

Symbol	Reference	Sample size	Unit	GM	SD	Percentage of cases exceeding tabulated value		
						95%	50%	5%
S-10	Fisher et al. (1993)	43	kA μs^{-1}	28	0.37	5.4	34	83
	Anderson and Eriksson (1980)	114		—	—	3.3	15	72

necessary for obtaining the Norton equivalent circuit. Nevertheless, if we are concerned only with the current peak, and assume that for a large number of strokes the average peak value of I and the average value of Z_{ch} at current peak are each more or less constant, then the Norton equivalent becomes a useful tool for studying the relation between lightning current peak and the corresponding values of Z_{ch} and Z_{gr}. For instance, if the measured channel-base current peak statistics are similar under a variety of grounding conditions, then Z_{ch} must always be much larger than Z_{gr} at the time of the current peak.

The representation of lightning by a Norton equivalent circuit was also used in analyzing lightning discharges to tall objects in Section 6.5. In the following, we will

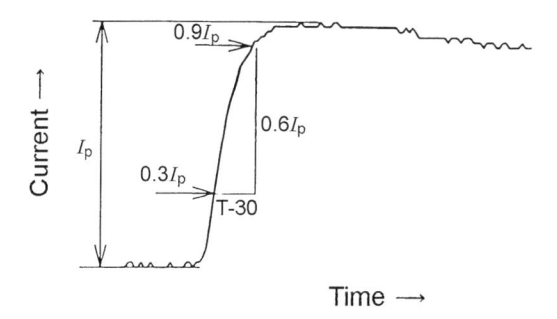

7.2. Rocket-triggered lightning

Symbol	Reference	Sample size	Unit	GM	SD	Percentage of cases exceeding tabulated value		
						95%	50%	5%
S-30	Fisher et al. (1993)	43	kA μs^{-1}	28	0.36	4.6	31	91
	Anderson and Eriksson (1980)	114		—	—	4.1	20	99

Fig. 7.21. As in Fig. 7.14, but for the 30–90 percent average slope (steepness). $S\text{-}30 = 0.6I_p/T\text{-}30$. Adapted from Fisher et al. (1993).

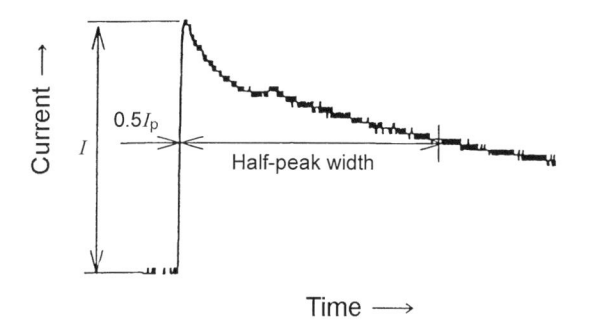

Parameter	Reference	Sample size	Unit	GM	SD	Percentage of cases exceeding tabulated value		
						95%	50%	5%
$\Delta t_{50\text{-}50}$	Fisher et al. (1993)	41	μs	18	0.30	5.0	20	40
	Anderson and Eriksson (1980)	115		—	—	6.5	32	140

Fig. 7.22. As in Fig. 7.14, but for the half-peak width. Adapted from Fisher et al. (1993).

compare the geometric mean current peaks from four triggered lightning experiments in which similar rocket launchers having a relatively small height of 4–5 m were used but grounding conditions differed considerably. All the information needed for this comparison is given in Table 7.6.

As seen in Table 7.6, Camp Blanding measurements of lightning currents that entered sandy soil with a relatively poor conductivity of 2.5×10^{-4} S m^{-1} and without any grounding electrode resulted in a value 13 kA for the geometric mean return-stroke current peak. This is similar to the geometric mean value, 14 kA, estimated from measurements at KSC made in 1987 using a launcher of the same geometry that was much better grounded into salt water with a conductivity of 3–6 S m^{-1} via underwater braided metallic cables. Additionally, fairly similar

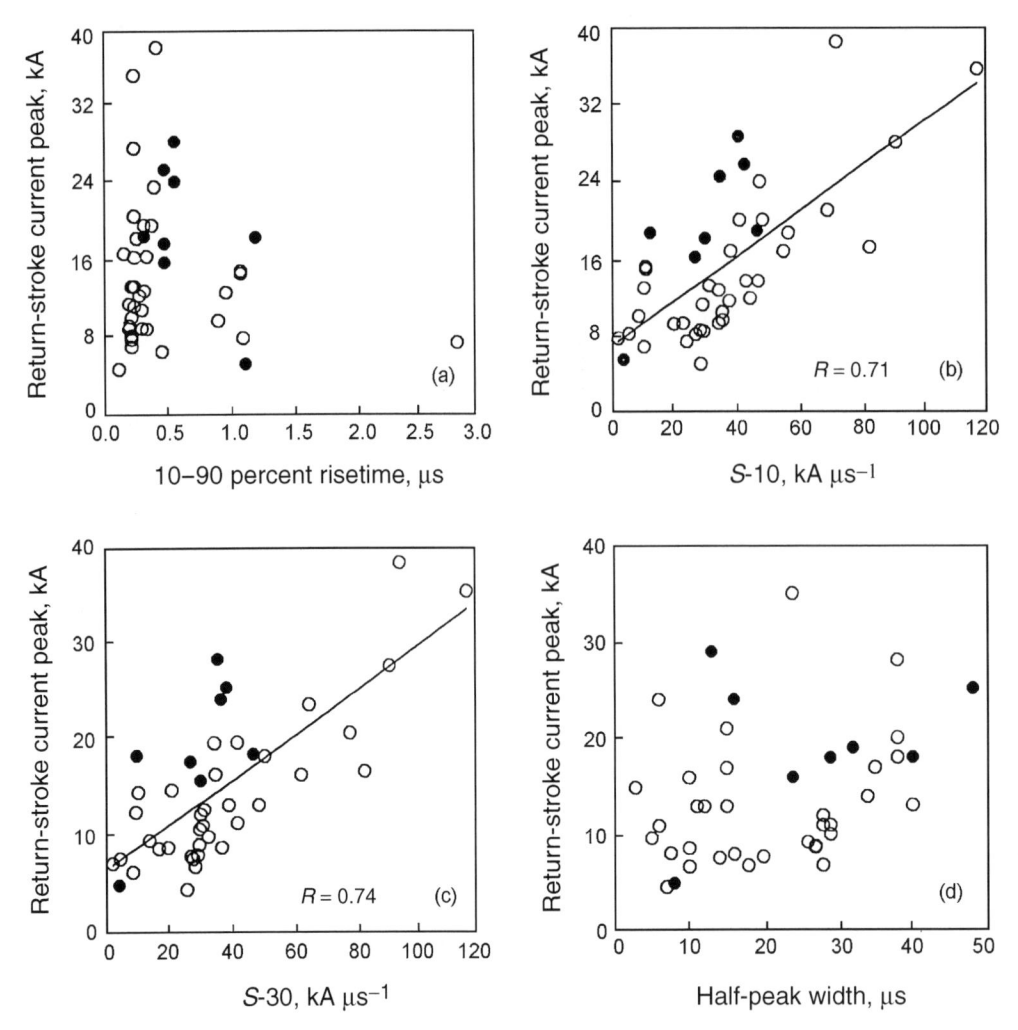

Fig. 7.23. Scatterplots relating various return-stroke parameters. The solid circles represent 1990 data from KSC, Florida, and the open circles represent 1991 data from Fort McClellan, Alabama. (a) Current peak versus 10–90 percent risetime; (b) current peak versus S-10; (c) current peak versus S-30; (d) current peak versus half-peak width. Regression lines, and correlation coefficients R are given in (b) and (c). Adapted from Fisher et al. (1993).

geometric mean values were found from the Fort McClellan, Alabama, measurements using a poorly grounded launcher (10 kA) and the same launcher well grounded (11 kA) in 1993 and 1991, respectively. The median current peaks in the triggered-lightning experiments in Florida (KSC) and France differed by only 20 percent or so, although there was a larger year-to-year variation at each location (subsection 7.2.4). Also, Ben Rhouma et al. (1995) gave arithmetic mean values of return-stroke current peaks in the range from 15 to 16 kA for the Florida triggered-lightning experiments at Camp Blanding in 1993 and at KSC in 1987, 1989, and 1991.

The values of grounding resistance (probably the dominant component of Z_{gr}) given in Table 7.6 should be understood as the initial values encountered by a lightning downward leader before the onset of any breakdown processes in the soil or along the ground surface associated with the return stroke. Note from Table 7.6 that the grounding resistance varies from 0.1 Ω to 64 kΩ, while Z_{ch}, assumed to be a real number, was estimated from analysis of the current waves traveling along the 540-m-high tower to be in the range from hundreds of ohms to some kilohms (Gorin et al. 1977; Gorin and Shkilev 1984).

The observation that the average return-stroke current is not much influenced by the level of man-made grounding, ranging from excellent to none, implies that lightning is capable of lowering the grounding impedance it initially encounters (Table 7.6) to a value that is always much lower than the equivalent impedance of the main channel. On the basis of (i) the evidence of the formation of plasma channels (fulgurites) in the sandy soil at Camp Blanding (Uman et al. 1994a, 1997; Rakov 1999b)

Table 7.6. *Geometric mean peak current versus grounding conditions from different triggered-lightning experiments. The values of grounding resistance are determined by the geometry of the grounding electrode (or the geometry of the contact surface between the channel and the ground in the absence of grounding electrode) and soil conductivity. They were measured under low-frequency, low-current conditions and should be understood as the initial values of resistance encountered by lightning before the onset of any breakdown processes in the soil or along the ground surface. Adapted from Rakov et al. (1998)*

Experiment	Reference	Trigger threshold, kA	Sample size	GM peak current, kA	Soil	Artificial grounding	Grounding resistance, Ω
KSC[a], Florida, 1987	Eybert-Berard et al. (1988), Leteinturier et al. (1991), as reported by Fisher et al. (1993)	5	36	14	0.5-m deep salt water ($3-6$ S m^{-1})	1.2×1.2 m^2 metal plane connected, by three 0.5-m-long wires at the four corners, to salt water	0.1
Fort McClellan, Alabama, 1991	Fisher et al. (1993)	2 (two strokes below 2 kA from continuous tape record included)	37	11	clay (3×10^{-3} S m^{-1})	Rebar framework of the munition storage bunker interconnected with lightning protection system including air terminals, down conductors and buried counterpoise	Presumably low
Camp Blanding, Florida, 1993	Uman et al. (1994a, 1997)	3.3 and 4.2	37	13	sand (2.5×10^{-4} S m^{-1})	None. The launcher was based on two parallel concrete slabs, 15 m long, 2 m apart, above three unenergized power cables buried 1 m deep and 5 m apart	64×10^3 (assuming that the contact surface between the channel and ground was a hemisphere with 1 cm radius)
Fort McClellan, Alabama, 1993	Fisher et al. (1994)	~4	31	10	heavy red clay (1.8×10^{-3} S m^{-1})	Single vertical grounding rod 0.3-m or 1.3-m long	260

[a] Kennedy Space Center.

Fig. 7.24. At the time of going to press a colour version of this figure was available for download from http://www.cambridge.org/9780521035415. Photograph of surface arcing associated with the second stroke (current peak of 30 kA) of flash 9312 triggered at Fort McClellan, Alabama. The lightning channel is outside the field of view. One of the surface arcs approached the right edge of the photograph, a distance of 10 m from the rocket launcher. Adapted from Fisher et al. (1994).

and (ii) optical records showing arcing along the ground surface at both Camp Blanding and Fort McClellan (Fisher et al. 1994; Rakov et al. 1998), it can be inferred that surface and underground plasma channels are important means of lowering the lightning grounding impedance, at least for the types of soil at the lightning-triggering sites in Florida and Alabama (sand and clay, respectively). A photograph of surface arcing during a triggered-lightning flash from Fort McClellan, Alabama, is shown in Fig. 7.24, and evidence of surface arcing in natural lightning is presented in Fig. 18.17. The injection of laboratory currents up to 20 kA into loamy sand in the presence of water sprays simulating rain resulted in surface arcing that significantly reduced the grounding resistance at the current peak (M. Darveniza, personal communication, 1995).

The fulgurites (glassy tubes produced by lightning in sand, Figs. 7.31 and 18.18) found at Camp Blanding usually show that the in-soil plasma channels tend to develop toward the better conducting layers of soil or toward buried metallic objects that, when contacted, serve to lower further the grounding resistance. The percentages of return strokes producing optically detectable surface arcing versus return-stroke current peak, from the 1993 and 1995 Fort McClellan experiments, are shown in Fig. 7.25. The surface arcing appears to be random in direction and often leaves little if any evidence on the ground. Even within the same flash, individual strokes can produce arcs developing in different directions. In one case, it was possible to estimate the current carried by one arc branch that contacted the instrumentation. That current was approximately 1 kA, or five percent of the total current peak in that stroke.

The observed horizontal extent of surface arcs was up to 20 m, which was the limit of the photographic coverage during the 1993 Fort McClellan experiment. No fulgurites were found in the soil (red clay) at Fort McClellan, only concentrated current exit points at several spots along the 0.3- or 1.3-m steel earthing rod (Table 7.6). It is likely that the uniform ionization of the soil usually postulated in studies of the behavior of grounding electrodes subjected to lightning surges is not an adequate assumption, at least not in the southeastern United States, where distinct plasma channels in the soil and on the ground surface appear to contribute considerably to lowering the grounding resistance.

7.2.6. Close electric fields

In this section, we discuss the vertical electric fields due to leader–return-stroke sequences at distances ranging from several meters to several hundred meters from the classically triggered lightning channel. Rubinstein et al. (1992, 1995) measured and analyzed electric field waveforms at 500 m for 31 leader–return-stroke sequences and at 30 m for two leader–return-stroke sequences in lightning flashes triggered at the Kennedy Space Center, Florida, in 1986 and 1991, respectively. They found that, at tens to hundreds of meters from the lightning channel, the leader–return-stroke vertical electric field waveforms appear as asymmetrical V-shaped pulses, the negative slope of the leading edge being lower than the positive slope of the trailing edge. The bottom of the V is associated with the transition from leader (the leading edge of the pulse) to return stroke (the trailing edge of the pulse).

7.2. Rocket-triggered lightning

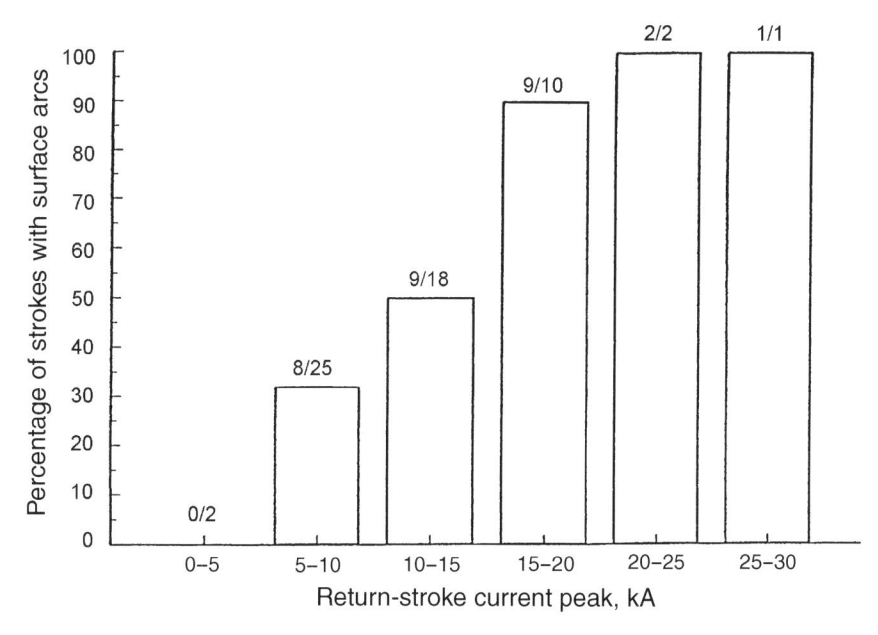

Fig. 7.25. Percentages of return strokes producing optically detectable surface arcing as a function of return-stroke current peak (Fort McClellan, Alabama, 1993 and 1995). The numbers above each histogram column indicate the number of strokes producing optically detectable arcing (numerator) and the total number of strokes in that current peak range (denominator). Adapted from Rakov et al. (1998).

Table 7.7. *Characterization of electric field waveforms due to triggered-lightning leader–return-stroke sequences at distances r equal to 30, 50, and 110 m from the lightning channel, recorded in 1993 at Camp Blanding, Florida. Adapted from Crawford et al. (1999). Here, and in Table 7.8, ΔE_L is the leader electric field change, T_{HPW} is the half-peak width of the electric field waveform. The numbers 30, 50, and 110 in the subscripts indicate the distances (in meters) at which the waveforms were measured. In the rightmost column, r is in meters and ΔE_L is in kV m^{-1}.*

Flash	Stroke	I, kA	ΔE_{L30}, kV m^{-1}	T_{HPW30}, μs	ΔE_{L50}, kV m^{-1}	T_{HPW50}, μs	ΔE_{L110}, kV m^{-1}	T_{HPW110}, μs	Expression for ΔE_L
9313	RS2	9.7	23	2.6	20	4.3	16	9.1	$61r^{-0.28}$
9313	RS3	11	25	2.2	21	4	17	7.5	$69r^{-0.30}$
9313	RS4	13	27	2.2	23	3.5	18	7.3	$76r^{-0.30}$
9313	RS5	11	23	3.5	21	5.1	17	9.1	$56r^{-0.25}$
9320	RS1	9.6	30	8.5	24	14	16	29	$172r^{-0.51}$
9320	RS2	8.4	24	8.4	20	15	14	30	$102r^{-0.42}$

The first multiple-station electric field measurements within a few hundred meters of the triggered-lightning channel were performed in 1993 at Camp Blanding, Florida (Uman et al. 1994b) and at Fort McClellan, Alabama (Fisher et al. 1994). Detailed analyses of these data were presented by Rakov et al. (1998). The results of Rakov et al. (1998) for distances r equal to 30, 50, and 110 m are shown in Table 7.7, where I, ΔE_L, and T_{HPW} are the channel-base current peak, the net electric field change due to the leader, and the half-peak width of the V-shaped pulse, respectively. Electric field waveforms for the fourth leader–return-stroke sequence (RS4 in Table 7.7) of the five-stroke flash 9313 are shown on two different time scales in Fig. 7.26. As seen in Fig. 7.26, at 30 m the rate of change of the leader electric field near the bottom of the V can be comparable to that of the return-stroke field, while at 110 m the leader rate of change is considerably lower than that of the return stroke. Uman et al. (2000), who measured the electric field derivative (dE/dt) waveforms at 10, 14, and 30 m from the triggered-lightning channel, reported that for one event the dE/dt peak for the dart leader at 14 m was more than 70 percent of the dE/dt peak for the corresponding return stroke. From the 1993 experiment, the geometric mean width of the V at half-peak value (T_{HPW}) was 3.2 μs at 30 m, 7.3 μs at 50 m, and 13 μs at 110 m, a distance dependence close to linear.

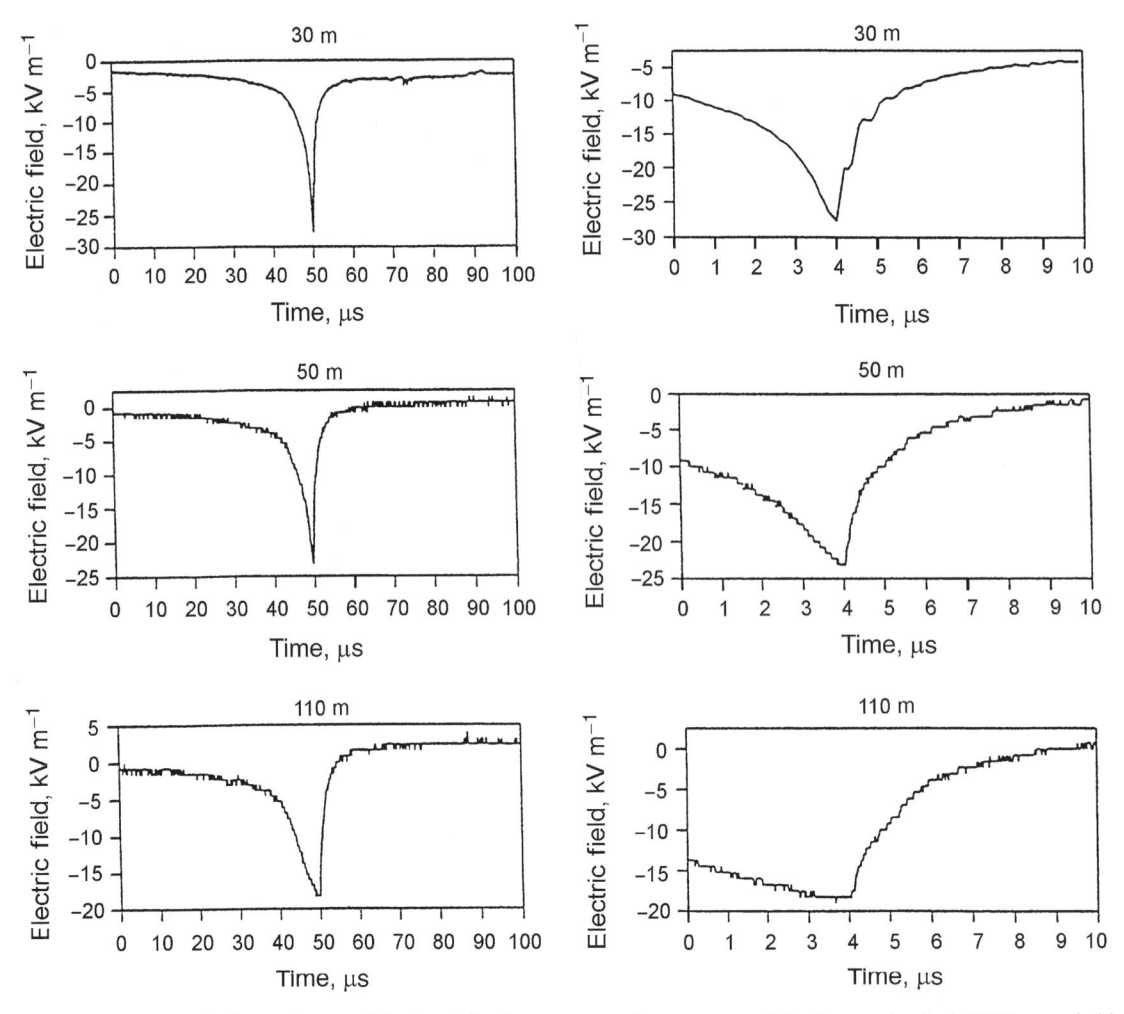

Fig. 7.26. Electric field waveforms of the fourth leader–return-stroke sequence of the five-stroke flash 9313 recorded in 1993 at 30, 50, and 110 m at Camp Blanding, Florida, shown on 100 μs and 10 μs time scales (left-hand and right-hand columns, respectively). The initial downward-going portion of the waveform is due to the dart leader, and the upward-going portion is due to the return stroke. Adapted from Rakov et al. (1998).

In 1997, the multiple-station field-measuring experiment at Camp Blanding, Florida, was extended to include seven stations at distances of 5, 10, 20, 30, 50, 110, and 500 m from the triggered-lightning channel (Crawford et al. 1999). Most of the data obtained at 5 m appeared to be corrupted, possibly owing to ground surface arcs (see subsection 7.2.5) and are not considered here. The leader–return-stroke electric field waveforms in one flash (S9721) simultaneously measured at 10, 20, 30, 50, 110, and 500 m are shown in Fig. 7.27. The evolution of the leader–return-stroke electric field waveform as distance increases is consistent with previous measurements (Rubinstein et al. 1995; Rakov et al. 1998) and reflects an increasing contribution to the field from progressively higher channel sections.

As seen in Table 7.7, the 1993 multiple-station electric field measurements suggest a variation in ΔE_L with distance that is slower than an inverse proportionality. These data were obtained at distances 30, 50, and 110 m. To facilitate direct comparison between the 1993 and 1997 data, we first present the 1997 data for the same three distances (three leader–return-stroke events out of a total of five) in Table 7.8. As seen in the last column of Table 7.8, the leader electric field change varies in a way that is approximately inversely proportional to the distance from the lightning channel. The data on leader electric field change as a function of distance for all five 1997 events and for all distances ranging from 10 to 500 m are presented in Table 7.9. For four out of five 1997 events the leader field change varied approximately as r^{-1} in the distance ranges 50 to 500 m (S9711), 20 to 500 m (S9718), 30 to 500 m (S9720), and 10 to 500 m (S9721). For one 1997 event (S9712) the field change varied approximately as $r^{-0.56}$ between 10 and 30 m from the channel. Thus, one of five events recorded in 1997

7.2. Rocket-triggered lightning

Table 7.8. *Characterization of electric field waveforms due to triggered-lightning leader–return-stroke sequences, 30, 50, and 110 m from the lightning channel, recorded in 1997 at Camp Blanding, Florida. Adapted from Crawford* et al. *(1999). For an explanation of the symbols, see the caption to Table 7.7*

Flash	Stroke	I, kA	ΔE_{L30}, kV m^{-1}	T_{HPW30}, μs	ΔE_{L50}, kV m^{-1}	T_{HPW50}, μs	ΔE_{L110}, kV m^{-1}	T_{HPW110}, μs	Expression for ΔE_L
S9718	RS1	12	43	1.9	27	3.2	12	7.5	$1.4 \times 10^3 r^{-1.02}$
S9720	RS1	21	56	1.7	36	2.7	16	6.4	$1.7 \times 10^3 r^{-0.99}$
S9721	RS1	11	40	4.1	27	7.2	12	23	$0.94 \times 10^3 r^{-0.93}$

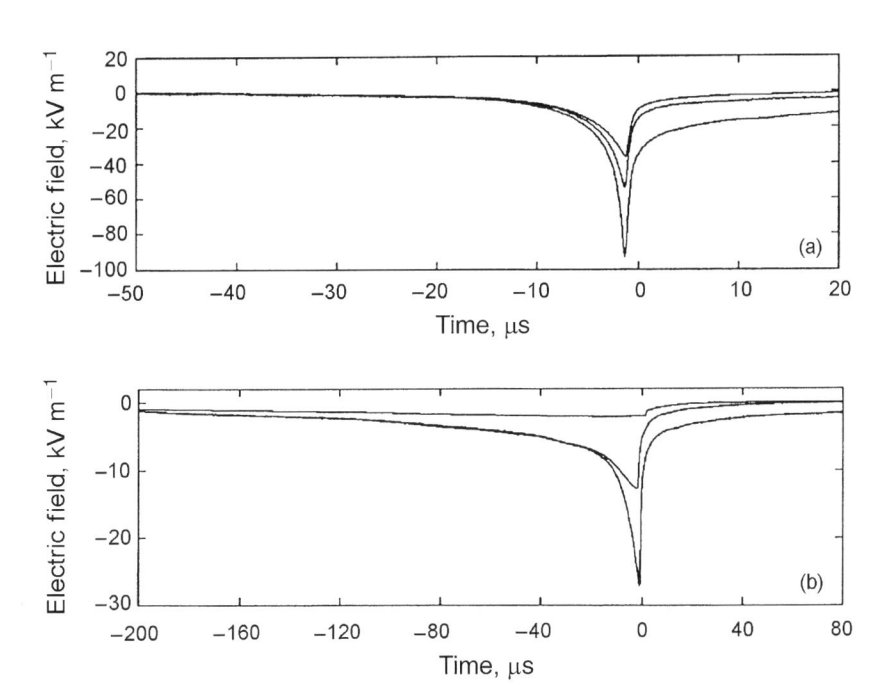

Fig. 7.27. Electric field waveforms of the first leader–return-stroke sequence of flash S9721, as recorded in 1997 at distances (a) 10, 20, and 30 m (bottom to top) and (b) 50, 110, and 500 m (bottom to top) at Camp Blanding, Florida; note the change in scale between (a) and (b). The initial downward-going portion of the waveform is due to the dart leader, and the upward-going portion is due to the return stroke. Adapted from Crawford *et al.* (1999).

exhibited a behavior somewhat similar to that observed for the six events recorded in 1993.

A variation in ΔE_L as r^{-1} is consistent with a more or less uniform distribution of charge along the bottom portion of the fully formed leader channel (see Eq. 4.5). A variation in ΔE_L slower than r^{-1} implies a decrease of leader charge density with decreasing height. There appears to be no evidence that the observed discrepancy between the 1993 and 1997 data given in Tables 7.7 and 7.8, respectively, is due to the use of different instrumentation in those two years. A variation of ΔE_L slower than r^{-1} was also inferred from the Fort McClellan measurements (a total of five flashes) at about 10 and 20 m (Fisher *et al.* 1994), while the 1998 and 1999 electric field measurements at Camp Blanding indicate an r^{-1} distance dependence (Crawford *et al.* 2001). Perhaps the 1993 Camp Blanding and Fort McClellan flashes, for which distance dependences slower than r^{-1} were observed, were not typical triggered-lightning events. Measurements for event S9712 from the 1997 experiment confirmed that the variation of ΔE_L with distance may occasionally be appreciably slower than r^{-1} (Table 7.9).

Next we consider the leader electric field change as a function of the return-stroke current peak. There is a positive linear correlation between the leader electric field change and the succeeding return-stroke current peak (absolute values). For the 1993 Camp Blanding experiment, the correlation coefficients were 0.98, 0.98, and 0.87 at 30, 50, and 110 m, respectively, and for the 1993 Fort McClellan experiment, 0.70 and 0.95 at about 10 (9.3) and about 20 (19.3) m, respectively. A similar relatively strong correlation between the leader field change and the return-stroke

Table 7.9. *Leader electric field change ΔE_L as a function of distance r for all 1997 data obtained at Camp Blanding, Florida. In the rightmost column, r is in meters and ΔE_L is in kV m^{-1}. Adapted from Crawford (1998)*

Flash	Stroke	I, kA	ΔE_L, kV m^{-1} at various distances						Expression for ΔE_L
			10 m	20 m	30 m	50 m	110 m	500 m	
S9711	RS1	6.5	—	—	—	21	11	1.8	$1.6 \times 10^3 r^{-1.1}$
S9712	RS1	5.3	36	24	19	—	—	—	$1.4 \times 10^2 r^{-0.59}$
S9718	RS1	12	—	62	43	27	12	1.7	$2.1 \times 10^3 r^{-1.1}$
S9720	RS1	21	—	—	56	36	16	2.6	$2.6 \times 10^3 r^{-1.1}$
S9721	RS1	11	98	58	40	27	12	1.8	$1.3 \times 10^3 r^{-1.0}$

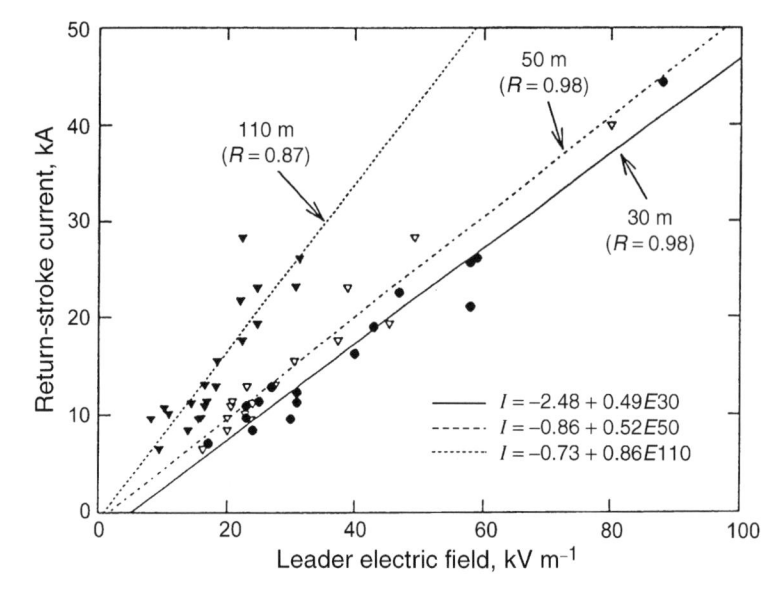

Fig. 7.28. Return-stroke current peak I versus leader electric field change at 30 m (E_{30}), 50 m (E_{50}), and 110 m (E_{110}) from the 1993 experiment at Camp Blanding, Florida. The data for 30 m are represented by solid circles, for 50 m by open triangles, and for 110 m by solid triangles. Also given are the correlation coefficients (R) and the regression equations. Adapted from Rakov *et al.* (1998).

current peak (correlation coefficient 0.8) was also found by Rubinstein *et al.* (1995) at 500 m at KSC. Scatterplots for 30, 50, and 110 m from the 1993 Camp Blanding experiment are given in Fig. 7.28 and for 10 (9.3) and 20 (19.3) m from the 1993 Fort McClellan experiment in Fig. 7.29. The intercepts in the current versus field regression equations for the measurements at Camp Blanding (see Fig. 7.28) are all negative, while those for the measurements at Fort McClellan (see Fig. 7.29) are all positive. If we neglect the intercept and take the vertical electric field E_z at ground as that due to a vertical uniformly charged line whose bottom end is at ground, $E_z = \rho_L/(2\pi\varepsilon_0 r)$ (Eq. 4.5), where r is the distance and ρ_L is the line charge density, we find from each current versus field regression equation (see Figs. 7.28 and 7.29) a proportionality coefficient v between ρ_L and peak current I, the corresponding relation being $I = v\rho_L$. This proportionality coefficient has the dimension of speed, and, interestingly, its value is of the order of the typically measured speed for return strokes (subsection 4.6.2).

7.2.7. *Studies of the interaction of lightning with various objects and systems*

In the first two parts of this subsection we will consider the triggered-lightning testing of power distribution lines and power transmission lines, respectively. In the third part we will briefly review the use of triggered lightning for testing the components of power systems, different types of lightning rod, and other objects and also for measuring step voltages and for making fulgurites. Additional information is found in subsections 18.4.3 and 18.4.4.

Power distribution lines Most of the published studies concerned with the responses of power distribution lines

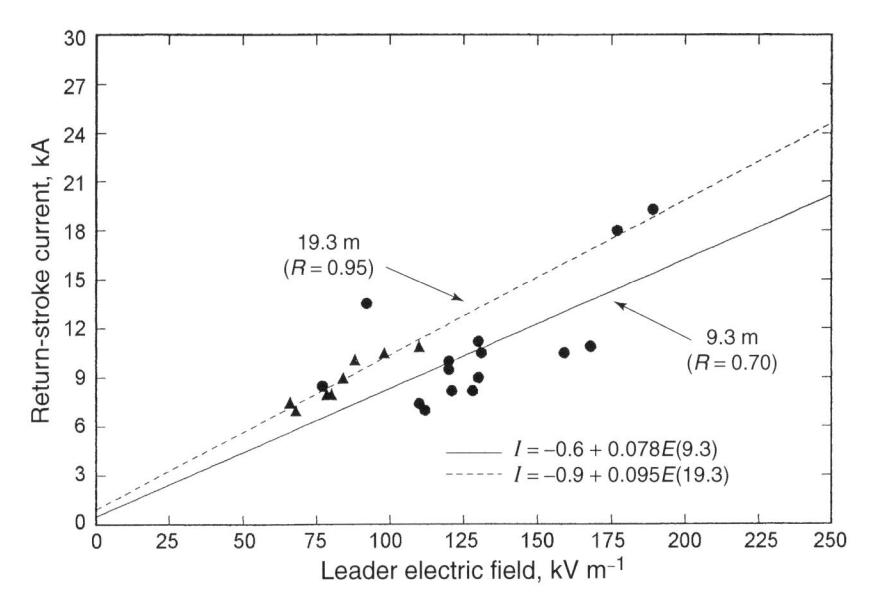

Fig. 7.29. Return-stroke current peak I versus leader electric field change E at about 10 (9.3) m and about 20 (19.3) m from the 1993 experiment at Fort McClellan, Alabama. The data for 10 (9.3) m are represented by circles and for 20 (19.3) m by triangles. Also given are the correlation coefficients (R) and the regression equations. Adapted from Rakov *et al.* (1998).

to direct and nearby triggered-lightning strikes have been conducted in Japan and in Florida.

From 1977 to 1985, a test power distribution line at the Kahokugata site in Japan (see Table 7.1) was used for studying the induced effects of close triggered-lightning strikes to ground (Horii 1982). Both negative- and positive-polarity flashes were triggered. The wire simulating the phase conductor was 9 m above ground, and the minimum distance between the test line and the rocket launcher was 77 m. The peak value of the induced voltage was found to be linearly related to the peak value of the lightning current; 25–30 kV corresponded to a 10 kA stroke. Installation of a grounded wire 1 m above the phase conductor resulted in a reduction in the induced voltage peak of about 40 percent. Horii and Nakano (1995) showed a photograph (their Fig. 6.4.2) of the test distribution line being struck directly during the induced-effect experiments. All triggered-lightning experiments in Japan were performed in winter.

In 1986, the University of Florida lightning research group studied the interaction of triggered lightning with an unenergized, three-phase 448 m overhead test line at the NASA Kennedy Space Center. Lightning was triggered 20 m from one end of the line, and the acquired data included the induced voltages on the top phase (10 m above ground) and the fields at a distance of 500 m from the lightning channel (Rubinstein *et al.* 1994). Two types of induced-voltage waveforms were recorded: oscillatory and impulsive. The former exhibited peak values that ranged from tens of kilovolts to about 100 kV, while the latter showed peak voltages nearly an order of magnitude larger. The oscillatory nature of the waveforms was due to multiple reflections at the ends of the line. Both types of voltage waveform were observed to occur for different strokes within a single flash. The time-domain technique of Agrawal *et al.* (1980), as adopted by Master and Uman (1984), Rubinstein *et al.* (1989), and Georgiadis *et al.* (1992), was used to model the observed voltages. Some success was achieved in the modeling of the oscillatory voltage waveforms, whereas all attempts to model the impulsive waveforms failed, probably because these measurements had been affected by a flashover in the measuring system. Rubinstein *et al.* (1994) used only the return-stroke electric field as the source in their modeling, assuming that the contribution from the leader was negligible. In a later analysis of the same data, Rachidi *et al.* (1997) found that the overall agreement between calculated and measured voltages of the oscillatory type was appreciably improved by taking into account the electric field of the dart leader.

Since 1993, studies of the interaction of triggered and natural lightning with power distribution systems have been conducted at the research facility named in 1995 the International Center for Lightning Research and Testing (ICLRT) at Camp Blanding, Florida. An overview of the facility in 1997 is given in Fig. 7.30.

In 1993, an experiment was conducted at Camp Blanding to study the effects of lightning on underground power distribution systems. All three cables shown in Fig. 7.30 were used in this experiment. The cables were 15 kV coaxial cables with polyethylene insulation between the center conductor and the outer concentric shield (neutral). One cable (cable A) had an insulating jacket and was placed

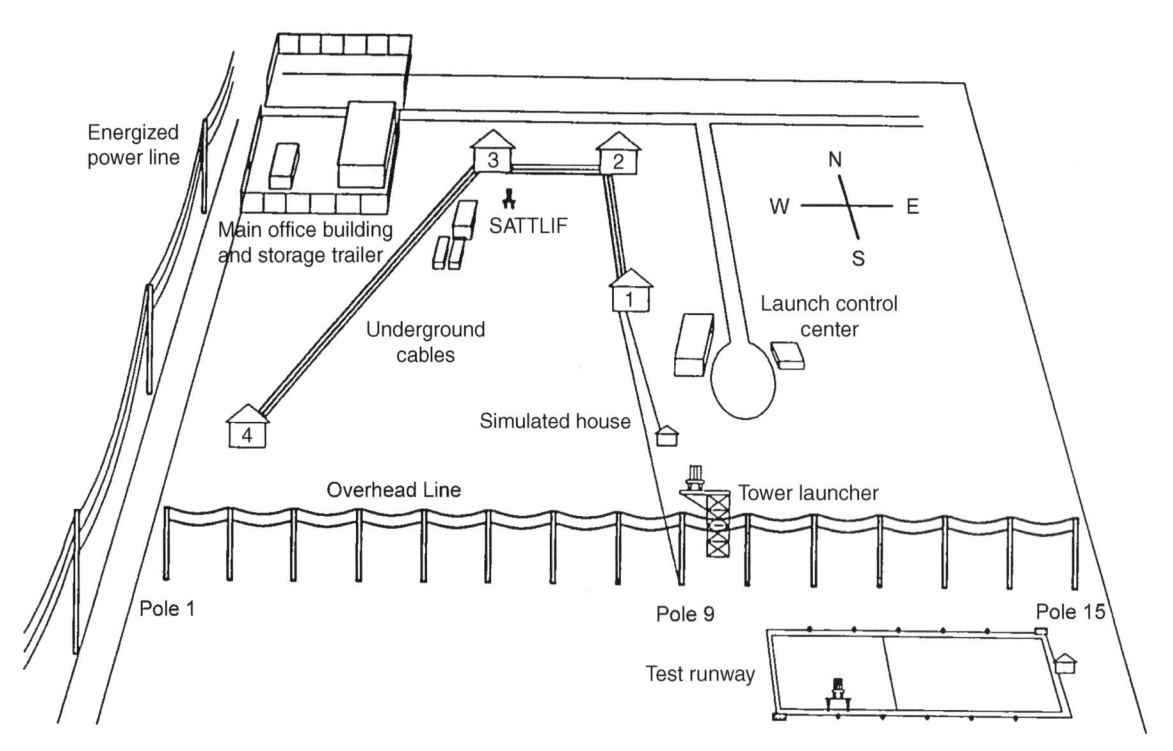

Fig. 7.30. Overview of the International Center for Lightning Research and Testing (ICLRT) at Camp Blanding, Florida, 1997. 1, 2, 3, and 4 are instrument stations. Artwork by C.T. Mata.

in PVC conduit, another (cable B) had an insulating jacket and was directly buried, and the third (cable C) had no jacket and was also directly buried. The three cables thus specified were buried 5 m apart at a depth of 1 m.

Thirty lightning flashes were triggered, and lightning current was injected into the ground directly above the cables, the current injection point being approximately equidistant from instrument stations 1 and 2 (see Fig. 7.30) but at different positions with respect to the cables. The cables were unenergized. Transformers at the instrument stations 1, 2, 3, and 4 were connected to cable A. More details on this test system configuration can be found in Fernandez et al. (1998a).

Barker and Short (1996a, b, c) reported the following results from the underground power distribution system project (see also subsection 18.4.4). After lightning attachment to ground, a substantial fraction of the lightning current flowed into the neutral conductor of the cable, approximately 15 to 25 percent of the total lightning current (measured at the rocket launcher) being detected 70 m in either direction from the strike point at instrument stations 1 and 2. The largest voltage measured between the center conductor and the concentric neutral of the cable was 17 kV, which is below the cable's basic insulation level (BIL) rating. Voltages measured at the transformer secondary were up to 4 kV. These could pose a threat to residential appliances. The underground power cables were excavated by the University of Florida research team in 1994. Lightning damage to each of these three cables is illustrated in Fig. 7.31.

Additionally, during the 1993 experiment at Camp Blanding, the voltages induced on the overhead distribution line shown in Fig. 7.30 were measured at poles 1, 9, and 15. The line had a length of about 730 m and was not connected to the underground distribution system. The distance between the line and the triggered-lightning strikes over the cables was 145 m. The line was terminated at both ends with a resistance of 500 Ω, and its neutral (the bottom conductor, see Fig. 7.30) was grounded at poles 1, 9, and 15. The results of this experiment were reported by Barker et al. (1996) and we now review them briefly (see also the second part of subsection 18.4.3). Waveforms for the induced voltages and for the total lightning current were obtained for 63 return strokes from the 30 triggered flashes. A strong correlation was observed between the peak values of the return-stroke current, ranging from 4 to 44 kA, and the voltage induced at pole 9, ranging from 8 to 100 kV, with correlation coefficient 0.97. The voltages induced at the terminal poles were typically half the value of the voltage induced at pole 9.

In 1994–7, the test distribution system at Camp Blanding shown in Fig. 7.30 was subjected to both direct and nearby triggered-lightning strikes. A large number of system configurations were tested, and several important

Cable A

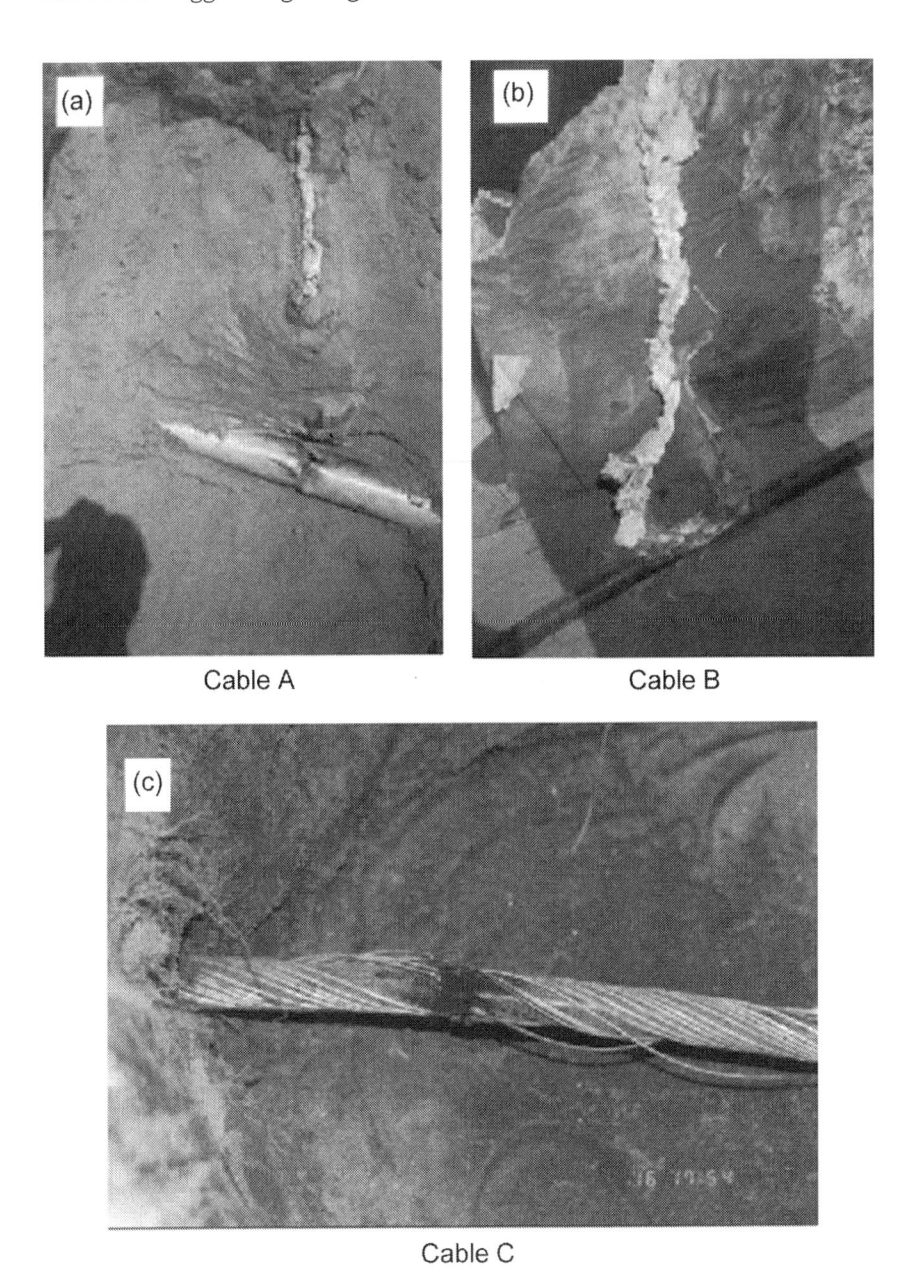

Cable B

Cable C

Fig. 7.31. Lightning damage to underground power cables. (a) Coaxial cable in an insulating jacket inside a PVC conduit; note the section of vertical fulgurite in the upper part of the picture (the lower portion of this fulgurite was destroyed during excavation) and the hole melted through the PVC conduit. (b) Coaxial cable in an insulating jacket, directly buried; note the fulgurite attached to the cable. (c) Coaxial cable whose neutral was in contact with earth; note that many strands of the neutral are melted through. The cables were tested at Camp Blanding, Florida in 1993. Photographs (a) and (b) were taken by V.A. Rakov, and photograph (c) was provided by P.P. Barker.

results were obtained. It was observed, for example, that when lightning strikes Earth at tens of meters from the system's ground connections, an appreciable fraction of the total lightning current enters the system from Earth (Fernandez 1997; Fernandez et al. 1998b, c). The observed peak values of current entering the system from Earth, as a percentage of the total lightning current peak, were (for three different events) 10 percent at 60 m, 5 percent at 40 m, and 18 percent at 19 m from the ground strike point. Further, vertical ground rods in sandy soil subjected to lightning currents appeared to exhibit a capacitive behavior rather than the often expected resistive behavior (Rakov et al. 2001). More details on findings from the 1994–7 experiments at Camp Blanding can be found in Uman et al. (1997), Fernandez (1997), Fernandez et al. (1998a, 1999), and Mata et al. (2000).

Power transmission lines Extensive studies of the interaction of triggered lightning with an unenergized power transmission line, the Okushishiku test line, were performed in Japan (Section 8.5). The line was designed to operate at a voltage of 275 kV and had six phase conductors and one ground wire suspended on seven steel 60-m towers. The total length of the test line was 2 km. All experiments were conducted in winter, primarily using the altitude-triggering technique. A photograph of a triggered-lightning strike to the Okushishiku test line is shown in Fig. 8.8a.

In Fig. 8.8a, the channel of the altitude-triggered lightning discharge terminates on the ground wire. The wire was damaged by the strike and broke in a few days (Horii and Nakano 1995). The distribution of classically triggered lightning current injected into the tower top among the four tower legs and the overhead ground wire was studied. The currents through the four legs were not equal, presumably because of the differences among the grounding impedances of the individual legs. It was observed that the higher-frequency components of current tended to flow to ground through the struck tower while the lower-frequency components appeared to travel to other towers along the ground wire. The currents in the phase conductors and the voltages between each phase conductor and the tower were also measured.

Miscellaneous experiments Besides the power line tests described above, triggered-lightning experiments have been performed in order to study the interaction of lightning with a number of miscellaneous objects and systems and for a variety of other reasons. We briefly review some of those studies now. Triggered lightning has been used to test power transformers (e.g., Horii 1982), lightning arresters (e.g., Horii and Nakano 1995; Kobayashi *et al.* 1997; Barker *et al.* 1998), overhead ground wires (e.g., Horii and Nakano 1995), lightning rods (Section 18.3) including so-called early-streamer-emission rods (e.g., Eybert-Berard *et al.* 1998; subsection 18.3.7) and high-resistance (tens to hundreds of kilo-ohms), current-limiting rods (e.g., Teramoto *et al.* 1996), explosive materials (e.g., Fieux *et al.* 1978), explosives storage facilities (Morris *et al.* 1994), and an airport runway lighting system (Rakov 1999a; Bejleri *et al.* 2000).

Various aspects of lightning safety have been studied using a mannequin with a hairpin on the top of its head and a metal-roof car with a live rabbit inside (e.g., Horii 1982). The car was confirmed to be a lightning-safe enclosure. Step voltages (Section 19.2) have been measured within a few tens of meters of the triggered-lightning strike point (Horii 1982; Fisher *et al.* 1994; Schnetzer and Fisher 1998). Voltages have been measured across a single overhead power-line tower and between the tower footing and remote ground (over a distance of 60 m), along with the lightning current injected into the tower (Gary *et al.* 1975). Additionally, triggered lightning has been used to make fulgurites (Kumazaki *et al.* 1993; Davis *et al.* 1993; Uman *et al.* 1994a; Rakov 1999b). Photographs of fulgurites are found in Figs. 7.31 and 18.18.

7.3. Other lightning triggering techniques

The rapid extension of a thin vertical wire, grounded or ungrounded, under a thundercloud is now a well-established approach to triggering lightning discharges. There have also been attempts or suggestions of the possibility of triggering lightning using (i) laser beams, (ii) microwave beams, (iii) water jets, and (iv) transient flames. In the following, we will review these "non-conventional" approaches to lightning triggering, none of which has produced, as of this writing, a single lightning discharge. Laser plasma channels, water jets, and flames are all inferior to a metallic wire in terms of their capability to distort the ambient electric field and of their susceptibility to adverse meteorological factors (wind, rain, snow). The conductivity of a laser-produced plasma channel is probably of order 10^{-3} S m^{-1} (Borisov *et al.* 1999b), similar to the conductivity of fresh water, while for copper it is 5.8×10^7 S m^{-1} (e.g., Sadiku 1994). A motivation for studying the feasibility of the "non-conventional" lightning triggering approaches (particularly those involving lasers) is the desire to develop active lightning protection systems that would prevent lightning from striking critical systems or structures by discharging thunderclouds to a designated point on ground (e.g., Wang *et al.* 1995c; Zhao *et al.* 1995). The use of rocket-and-wire techniques, which have proved to be very effective in lightning research, would be impractical for this purpose.

7.3.1. Laser beams

Two types of laser have been used in attempting to develop lightning-triggering systems: (i) high-energy (typically tens to hundreds of joules per pulse) infrared lasers and (ii) relatively low-energy (of the order of millijoules or less per pulse) ultraviolet lasers. The capability of infrared lasers (usually carbon dioxide, CO_2, or neodymium, Nd, lasers with an operating wavelength of 10.6 μm or 1.06 μm, respectively) to initiate and guide electrical discharges has been tested extensively in a laboratory in the United States (e.g., Koopman and Wilkerson 1971; Koopman and Saum 1973; Greig *et al.* 1978; Caressa *et al.* 1979), in Russia (e.g., Aleksandrov *et al.* 1980; Asinovsky *et al.* 1987, 1988; Vasilyak *et al.* 1991), and in Japan (e.g., Shindo *et al.* 1993a, b; Miki *et al.* 1993, 1999; Wang *et al.* 1994a, b, 1995a, b, c). The results of these laboratory studies, such as the lowering of the breakdown voltage in a gap and the observed guiding of electrical discharges, are not directly related to the feasibility of triggering lightning. Wang *et al.*

(1995c), Yasuda *et al.* (1997), and Uchida *et al.* (1999) reported on the field testing of an infrared laser system designed to assist in triggering upward lightning from a tall tower. Ultraviolet laser systems are presently at the early stages of laboratory experimentation (e.g., Yamabe *et al.* 1992; Nakamura *et al.* 1993; Zhao and Diels 1993; Zhao *et al.* 1995; Miki and Wada 1996; Rambo *et al.* 1999). There is a growing consensus among researchers concerned with laser-triggered lightning that multiple laser beams will be required in order to obtain a laser plasma channel whose properties could conceivably be suitable for triggering lightning (Rambo *et al.* 1999; Uchida *et al.* 1999; Borisov *et al.* 1999b; Savel'eva 1999).

Infrared lasers When an infrared laser beam is focused in air, a high-level of ionization, called optical breakdown, may occur at a number of points along the beam. Aerosol particles that are present in the atmosphere (primarily those with sizes greater than 1 μm) act as nuclei for such optical breakdown. Optical breakdown at many points along the laser beam can produce a "beaded" plasma channel up to 60 m or more in length (Parfenov *et al.* 1976). The conductivity in the center of an individual bead is thought to be about 1 S m^{-1} (of the order of the conductivity of seawater, 4 S m^{-1}; Sadiku 1994) and to decrease exponentially with distance (Gromovenko and Danilov 1999). The number of beads per unit channel length usually does not exceed some tens per meter or so because the laser radiation is blocked by these beads (Vasilyak *et al.* 1991; Savel'eva 1999), although Danilov and Tul'skii (1999) reported a value 300–400 m^{-1} for the dense part of a 1.5-m-long channel produced by a Nd laser. The radiation-blocking effect of dense plasma formations for the so-called light-detonation regime of optical air breakdown becomes significant in tens to hundreds of nanoseconds (Borisov *et al.* 1999b; Savel'eva 1999), this time interval being apparently related to the rate of expansion of plasma formations in this regime. As a result, the energy delivered during the initial tens to hundreds of nanoseconds strongly influences the line density of plasma formations and the length of the laser-produced plasma channel. For example, Savel'eva (1999) reported, for a 70 J CO_2-laser pulse having a submicrosecond initial spike and an overall width of 6 μs, that the plasma channel lengths were 4 and 17 m when, respectively, 10 and 30 percent of the energy were in the initial spike. However, if the radiation intensity exceeds the threshold for the so-called superdetonation regime of optical air breakdown, the rate of expansion of plasma formations sharply increases, reducing the time for the radiation-blocking effect to become significant (Savel'eva 1999).

The beaded plasma channel produced by a laser beam is probably incapable of distorting the ambient electric field to a degree sufficient for triggering lightning, as discussed below. An attempt to use a powerful infrared CO_2 laser for triggering lightning during the 1978–9 Thunderstorm Research International Program (TRIP) failed (e.g., Barnes and Berthel 1991; Barnes and Kozma 1995).

Borisov *et al.* (1999b) distinguished between two modes of laser control of lightning: (i) the lightning triggering in which a discharge that would not have occurred otherwise is initiated via the formation of a laser plasma channel in the thundercloud electric field and (ii) the interception of a descending leader independently initiated in the cloud by an upward connecting leader from ground assisted by a laser beam. They consider primarily the interception mode, (ii) above, and conclude from their theoretical analysis that an initial conductivity of 1 to 10 S m^{-1} is needed to assure both the necessary polarization of the long laser-plasma channel, such that the channel's ends become oppositely charged causing a distortion of the ambient electric field, and also transformation of the laser-plasma channel into a thermally ionized leader channel in a time of about 1 μs.

Borisov *et al.* (1999a) and Gromovenko and Danilov (1999) considered the dynamics of polarization processes in a laser-plasma channel. Presently available lasers apparently produce plasma channels with an initial conductivity of 10^{-3} S m^{-1} or so, which is three to four orders of magnitude below the required value, according to Borisov *et al.* (1999b). In order to increase the conductivity of the laser-plasma channel, Borisov *et al.* (1999b) and Savel'eva (1999) proposed the use of the radiation of a Nd laser to ionize further the channel created by the CO_2 laser, as described below. They argued that, since the breakdown threshold at a wavelength of 10.6 μm is substantially lower than at 1.06 μm, the initial optical breakdown of the air will be produced primarily by the CO_2 laser. However, the resultant plasma formations will primarily block the 10.6 μm radiation of the CO_2 laser, while remaining transparent for the 1.06 μm radiation of the Nd laser, which is expected to facilitate further ionization.

Wang *et al.* (1994a, b, 1995a, b, c) argued that a laser beam alone is incapable of producing the required distortion of the ambient electric field and therefore suggested the use of a combination of a tall metallic tower and a laser beam for triggering lightning, as illustrated in Fig. 7.32. In their view, the tall tower would provide a distortion of the thundercloud electric field that was significant but not necessarily sufficient for a self-propagating upward leader to be initiated from the tip of the tower. The function of the laser beam is to aid in the transformation of the corona-like discharge processes at the tower top, which are otherwise incapable of leaving the region of highly enhanced electric field in the immediate vicinity of the tower tip, into a self-propagating leader (Section 6.1).

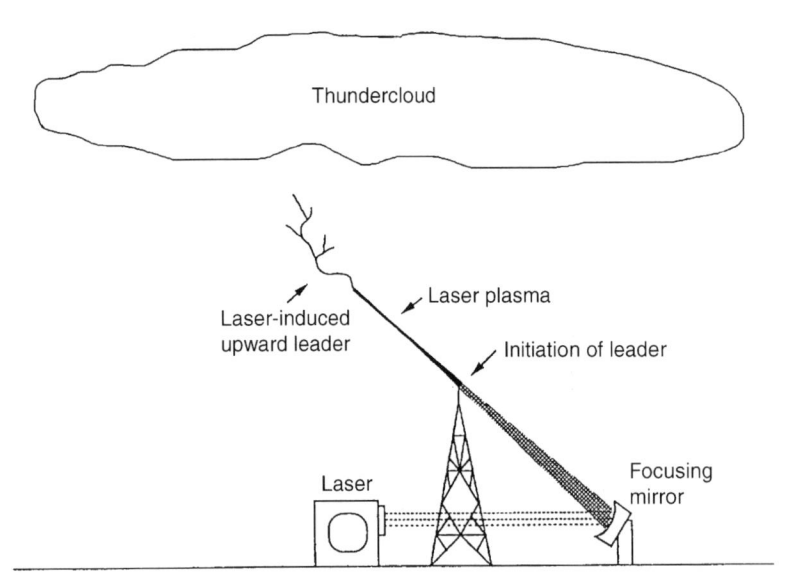

Fig. 7.32. Illustration of a proposed approach to the triggering of an upward lightning flash from a tall tower using a laser. Adapted from Wang et al. (1995c).

Wang et al. (1995c) and Yasuda et al. (1997) reported on the results of full-scale testing of this approach to lightning triggering. They used a 50 m tower on top of a 200-m-high hill, a CO_2 laser system that delivered two 1 kJ–50 ns pulses, and large-aperture focusing telescopes. In fine (clear sky) weather conditions, the system produced a 13-m-long plasma channel at the top of the tower, the channel being composed of two sequential sections created by the two lasers beams, 8 and 5 m long. In snowy conditions, the number of beads per unit volume of the laser plasma channel was found to decrease to 20 percent or less of the number in fine weather (Wang et al. 1995c). This decrease was attributed both to the reduction in the density of aerosol particles that served as nuclei for optical breakdown and also to attenuation of the laser radiation. Uchida et al. (1999) described a version of the same experimental installation, in which a total of four laser beams were used.

Yasuda et al. (1997) claimed that they succeeded in triggering lightning with their system in winter using a laser triggering circuit activated by initial breakdown pulses (subsection 4.3.2). They reported that an upward lightning discharge (no return strokes) occurred at the time when a relatively short laser plasma channel, 1 m in length, was created at the top of the tower. It is likely that a cloud discharge triggered both an upward lightning from the tower and also the laser system and that the 1 m plasma channel near the tower top had little to do with the development of the upward lightning discharge.

Ultraviolet lasers As opposed to infrared lasers, which produce, via avalanche ionization, a high-density plasma that can block further propagation of the laser radiation, ultraviolet lasers create, via multiphoton ionization, a low-density plasma transparent to the laser radiation. The New Mexico group (Zhao and Diels 1993; Zhao et al. 1995) was apparently the first to propose the use of ultrashort pulses (subpicosecond-scale; 1 picosecond (ps) = 10^3 femtoseconds (fs) = 10^{-12} s) of ultraviolet radiation to achieve very high peak intensities with relatively low pulse energies, of the order of millijoules or less. They argued that such a pulse propagating through the atmosphere is capable of creating an extended conducting channel. However, the conductivity of this path needs to be maintained using an external energy source, since the electrons freed by the laser beam will attach to oxygen to form low-mobility O_2^- and O^- ions within some tens of nanoseconds (e.g., Koopman and Saum 1973). The use of a second, visible-light, laser beam with a pulse width of several microseconds directed along the ultraviolet laser beam was proposed for the purpose of detaching electrons from the O_2^- and O^- ions (Zhao and Diels 1993; Zhao et al. 1995; Rambo et al. 1999). It is not clear whether such a system could create a sufficiently conducting laser-plasma channel for a sufficiently long time, particularly in the presence of precipitation and wind, the usual conditions during thunderstorms. In a laboratory experiment, 200-fs-long bursts of ultraviolet (248 nm) radiation with an energy of about 0.2 mJ were used to initiate and guide electrical discharges in a 25-cm parallel-plane gap. The New Mexico group plans to use a UV laser beam in conjunction with a "photodetaching", visible-light laser beam for guiding electrical discharges in 5 to 15 m gaps (Rambo et al. 1998). Laboratory experiments with ultraviolet lasers have been also conducted by Yamabe et al. (1992), Nakamura et al. (1993), and Miki and Wada (1996).

7.3.2. Microwave beam

Shiho *et al.* (1996) proposed using a high-power microwave beam produced by a free-electron laser (FEL) to create a weakly ionized plasma channel between ground and a thundercloud. They consider four frequencies, 35, 90, 140, and 270 GHz, since there are atmospheric transmission windows at these frequencies. Shiho *et al.* stated that the advantage of microwave radiation is that it is less attenuated in the presence of rain or snow than the radiation of the lasers discussed in subsection 7.3.1. At the time of writing there are no further publications on the use of microwave beams for lightning triggering.

7.3.3. Water jet

As noted in Section 7.1, the lightning discharge shown in Fig. 20.5 and discussed in Section 20.6 was unintentially initiated by a plume of water resulting from an underwater explosion. This observation has motivated some researchers to investigate in the laboratory the effects of water jets on electrical discharges. Gaps up to 10 m and water of different resistivity (65 Ω m and 0.8 Ω m) have been used (Wada 1994). For the case of water with resistivity 0.8 Ω m (salt water), the 50 percent flashover voltage was found to be half that without the water jet. As in the case of the laser experiments, the results obtained in laboratory gaps are not directly related to the feasibility of triggering lightning. Horii and Nakano (1995) speculated that a water jet of 50 m length may be sufficient to trigger lightning.

7.3.4. Transient flame

Watanabe *et al.* (1996) proposed using a transient flame for triggering lightning. A composite solid propellant was used in a field experiment in which a transient flame of 10 m length was produced with a rocket engine at the top of a 50-m-high tower under winter thunderstorm conditions in the Hokuriku coast, Japan (Chapter 8). The flame extended at a speed of 1000 m s^{-1}. No lightning was triggered.

7.4. Concluding remarks

The rocket-and-wire technique has been routinely used since the 1970s to initiate (trigger) lightning artificially from natural thunderclouds for the purposes of research and testing. Leader–return-stroke sequences in triggered lightning are similar in most (if not all) respects to subsequent-leader–return-stroke sequences in natural downward lightning and to all such sequences in object-initiated lightning. The initial processes in triggered lightning are similar to those in object-initiated (upward) lightning and are distinctly different from the first-leader–return-stroke sequence in natural downward lightning. The results of triggered-lightning experiments have provided considerable insight into natural lightning processes that would not have been possible from studies of natural lightning owing to its random occurrence in space and time. Among such findings are observation of an upward connecting leader in a dart-leader–return-stroke sequence, evidence of the guided-wave mechanism of the lightning M-component (these two findings being discussed in subsections 4.5.1, second part, and 4.9.6, respectively), the observation of a lack of dependence of the return-stroke current peak on grounding conditions, and characterization of the electromagnetic environment within tens to hundreds of meters of the lightning channel. Triggered-lightning experiments have contributed significantly in testing the validity of various lightning models (subsection 12.2.6) and in providing ground-truth data for the NLDN (Section 17.5). Triggered lightning has proved to be a very useful tool for studying the interaction of lightning with various objects and systems. A number of new approaches to triggering lightning, including those using lasers, microwave beams, water jets, and transient flames, have been proposed, but it appears unlikely that any of these "non-conventional" approaches will lead to an operational technique in the near future.

References and bibliography for Chapter 7

Agrawal, A.K., Price, H.J., and Gurbaxani, S.H. 1980. Transient response of multiconductor transmission lines excited by a nonuniform electromagnetic field. *IEEE Trans. Electromagn. Compat.* **22**: 119–29.

Aihara, Y., Shindo, T., Miki, M., and Suzuki, T. 1992. Laser-guided discharge characteristics of long air gaps and observation of the discharge progressing feature. *Trans. IEE Japan* **112-B**: 668–76.

Aleksandrov, G.N., and Kadzov, G.D. 1999. Possibility of using a laser spark to enhance the reliability of lightning protection. *J. Opt. Technol.* **66**(3): 183–6.

Aleksandrov, G.N., Ivanov, V.L., Kadzov, G.D., Parfenov, V.A., Pakhomov, L.N., Petrunkin, V. Yu., Podlevskii, V.A., and Selesnev, Yu. G. 1977. Study of the effect of the highly ionized channel created by a powerful laser on the development of a discharge in a long air gap. *Zh. Tekh. Fiz.* **47**: 2122–4 (*Sov. Phys. Tech. Phys.* **22**: 1233–4).

Aleksandrov, G.N., Ivanov, V.L., Kadzov, G.D., Parfenov, V.A., Pakhomov, L.N., Petrunkin, V. Yu., Podlevskii, V.A., and Selesnev, Yu. G. 1980. The possibility of enhancing the efficiency of the protective action of lightning rods by means of a laser spark. *Elektrichestvo* **2**: 47–8.

Aleksandrov, G.N., Ivanov, O.G., Ivanov, O.P., Kadzov, G.D., Okunev, R.I., Pakhomov, L.N., and Petrunkin, V. Yu. 1989. Orientation of electrical discharge along a long laser spark. *Pis'ma Zh. Tekh. Fiz.* **15**: 19–23.

Anderson, R.B., and Eriksson, A.J. 1980. Lightning parameters for engineering application. *Electra* **69**: 65–102.

Anisimov, E.I., Bogdanov, O.V., Gaivoronsky, A.S., Gol'tsov, V.A., and Ovsyannikov, A.G. 1988. Superlong negative spark as the analog of natural lightning. *Elektrichestvo* **11**: 55–7.

Antsupov, K.V., Vereshchagin, I.P., Koshelev, M.A., Lupeiko, A.V., Makalsky, L.M., Sysoev, V.S., and Chernov, E.N. 1990. A study of spark discharges from a cloud of charged aerosol particles. *Proc. USSR Academy Sciences (Izvestiya AN SSSR, Ser. Energetika i Transport)* **4**: 158–62.

Antsupov, K.V., Vereshchagin, I.P., Koshelev, M.A., Makalsky, L.M., and Sysoev, V.S. 1991. Discharges from cloud of charged aerosol. In *Proc. 7th Int. Symp. on High Voltage Engineering, Dresden, Germany*, pp. 15–17.

Asinovsky, E.I., Vasilyak, L.M., and Nesterkin, O.P. 1987. Impulse electrical discharge in the air at atmospheric pressure, that is guided by a long laser spark. *Pis'ma Zh. Tekh. Fiz.* **13**(4): 249–54.

Asinovsky, E.I., Vasilyak, L.M., and Nesterkin, O.P. 1988. On the ability of laser spark to guide electrical discharge. *Pis'ma Zh. Tekh. Fiz.* **14**(1): 41–4.

Avansky, V.R., Chernov, E.N., Lupeiko, A.V., and Petrov, N.I. 1992. Experimental investigations of lightning phenomena in laboratory. In *Proc. 9th Int. Conf. on Atmospheric Electricity, St. Petersburg, Russia*, pp. 783–7.

Ball, L.M. 1974. The laser lightning rod system: thunderstorm domestication. *Appl. Opt.* **13**: 2292–6.

Barker, P.P., and Short, T.A. 1996a. Lightning effects studied: the underground cable program. In *Transmission and Distribution World*, May, pp. 24–33.

Barker, P.P., and Short, T.A. 1996b. Lightning measurements lead to an improved understanding of lightning problems on utility power systems. In *Proc. 11th CEPSI, Kuala Lumpur, Malaysia*, vol. 2, pp. 74–83.

Barker, P., and Short, T. 1996c. Findings of recent experiments involving natural and triggered lightning. Panel Session Paper presented at 1996 Transmission and Distribution Conference, Los Angeles, California, September 16–20.

Barker, P.P., Short, T.A., Eybert-Berard, A.R., and Berlandis, J.P. 1996. Induced voltage measurements on an experimental distribution line during nearby rocket triggered lightning flashes. *IEEE Trans. Pow. Del.* **11**: 980–95.

Barker, P., Short, T., Mercure, H., Cyr, S., and O'Brien, J. 1998. Surge arrester energy duty considerations following from triggered lightning experiments. In *Proc. IEEE PES, 1998 Winter Meeting*, Panel Session on Transmission Line Surge Arrester Application Experience, 19 pp.

Barnes, A.A., and Berthel, R.O. 1991. A survey of laser lightning rod techniques. In *Proc. Int. Aerospace and Ground Conf. on Lightning and Static Electricity*, vol. 1, pp. 53/1–6, NASA Conf. Publ. 3106, Cocoa Beach, Florida.

Barnes, A.A., and Kozma, M.A. 1995. Using lasers to trigger/guide lightning. In *Proc. Int. Aerospace and Ground Conf. on Lightning and Static Electricity, Williamsburg, Virginia*, pp. 64/1–4.

Basov, N.G., Boiko, V.A., Krokhin, O.N., and Sklizkov, G.V. 1967. Formation of a long spark in air under the action of weakly focused laser radiation. *Dokl. Akad. Nauk SSSR* **173**: 538 (*Sov. Phys. Dokl.* **12**: 248, 1967).

Baum, C.E., O'Neill, J.P., Breen, E.L., Hall, D.L., and Moore, C.B. 1987. Electromagnetic measurement of and location of lightning. *Electromagnetics* **7**: 395–422.

Bazelyan, E.M., and Raizer, Yu. P. 2000a. *Lightning Physics and Lightning Protection*, 325 pp., Bristol: IOP Publishing.

Bazelyan, E.M., and Raizer, Yu. P. 2000b. Lightning attraction mechanism and the problem of lightning initiation by lasers. *UFN* **170**(7): 753–69.

Bazelyan, E.M., Gorin, B.N., and Levitov, V.I. 1978. *Physical and Engineering Foundations of Lightning Protection*, 223 pp., Leningrad: Gidrometeoizdat.

Beituganov, M.N., and Zashakuev, T.Z. 1992. The experience of initiation of lightning and spark discharges. In *Proc. 9th Int. Conf. on Atmospheric Electricity, St. Petersburg, Russia*, pp. 283–6.

Bejleri, M. 1999. Triggered-lightning testing of an airport runway lighting system. M.S. thesis, University of Florida, Gainesville, 307 pp.

Bejleri, M., Rakov, V.A., Uman, M.A., Rambo, K.J., Mata, C.T., and Fernandez, M.I. 2000. Triggered lightning testing of an airport runway lighting system. In *Proc. 25th Int. Conf. on Lightning Protection, Rhodes, Greece*, pp. 825–30.

Ben Rhouma, A., Auriol, A.P., Eybert-Berard, A., Berlandis, J.-P., and Bador, B. 1995. Nearby lightning electromagnetic fields. In *Proc. 11th Int. Symp. on Electromagn. Compatbility, Zurich, Switzerland*, pp. 423–8.

Berger, K., 1967. Novel observations on lightning discharges: results of research on Mount San Salvatore. *J. Franklin Inst.* **283**: 478–525.

Berger, K. 1977. The Earth Flash. In *Lightning, vol. 1, Physics of Lightning*, ed. R.H. Golde, pp. 119–190. New York: Academic Press.

Berger, K., and Garbagnati, E. 1984. Lightning current parameters. Results obtained in Switzerland and in Italy. In *Proc. URSI Conf., Florence, Italy*, 13 pp.

Berger, K., and Vogelsanger, E. 1969. New results of lightning observations. In *Planetary Electrodynamics*, eds. S.C. Coroniti and J. Hughes, pp. 489–510, New York: Gordon and Breach.

Berger, K., Anderson, R.B., and Knoninger, H. 1975. Parameters of lightning flashes. *Electra* **41**: 23–37.

Bondiou-Clergerie, A., Lalande, P., Laroche, P., Willett, J.C., Davis, D., and Gallimberti, I. 1999. The inception phase of positive leaders in triggered lightning: comparison of modeling with experimental data. In *Proc. 11th Int. Conf. on Atmospheric Electricity, Guntersville, Alabama*, pp. 22–5.

Borisov, M.F., Gromovenko, V.M., Danilov, M.F., and Lapshin, V.A. 1999a. Interaction of a long laser spark with an atmospheric electric field. *J. Opt. Technol.* **66**(3): 210–14.

Borisov, M.F., Gromovenko, V.M., Lapshin, V.A., Rezunkov, Yu.A., Saveleva, V.P., and Stepanov, V.V. 1999b. Long laser spark for controlling the trajectory of an atmospheric electric discharge. *J. Opt. Technol.* **66**(3): 203–9.

Brook, M., Armstrong, G., Winder, R.P.H., Vonnegut, B., and Moore, C.B. 1961. Artificial initiation of lightning discharges. *J. Geophys. Res.* **66**: 3967–9.

Caressa, J.-P., Autric, M., Dufresne, D., and Bournot, Ph. 1979. Experimental study of CO_2-laser-induced air breakdown over long distances. *J. Appl. Phys.* **50**(11): 6822–5.

Carlson, A.B. 1996. *Circuits*, 838 pp., New York: John Wiley & Sons.

Chauzy, S., Medale, J.-C., Prieur, S., and Soula, S. 1991. Multilevel measurement of the electric field underneath a thundercloud, 1. A new system and the associated data processing. *J. Geophys. Res.* **96**: 22 319–26.

Chen, H., and Liu, X. 1992. Triggered lightning flashes at Beijing Lightning Trigger Laboratory. In *Proc. Int. Aerospace and Ground Conf. on Lightning and Static Electricity, Atlantic City, New Jersey*, pp. 54/1–7.

Cooray, V., Fernando, M., Sörensen, T., Götschl, T., and Pedersen, Aa. 2000. Propagation of lightning generated transient electromagnetic fields over finitely conducting ground. *J. Atmos. Solar-Terr. Phys.* **62**: 583–600.

Crawford, D.E. 1998. Multiple-station measurements of triggered lightning electric and magnetic fields. M.S. thesis, Univ. Florida, Gainesville, 282 pp.

Crawford, D.E., Rakov, V.A., Uman, M.A., Schnetzer, G.H., Rambo, K.J., and Stapleton, M.V. 1999. Multiple-station measurements of triggered-lightning electric and magnetic fields. In *Proc. 11th Int. Conf. on Atmospheric Electricity, Guntersville, Alabama*, pp. 154–7.

Crawford, D.E., Rakov, V.A., Uman, M.A., Schnetzer, G.H., Rambo, K.J., Stapleton, M.V., and Fisher, R.J. 2001. The close lightning electromagnetic environment: dart-leader electric field change versus distance. *J. Geophys. Res.* **106**: 14 909–17.

Danilov, O.B., and Tul'skii, S.A. 1999. An electrical discharge initiated by a long laser spark. *J. Opt. Technol.* **66**(3): 187–9.

Darveniza, M. 1995. Personal communication.

Davis, D.A., and Laroche, P. 1999. Positive leaders in rocket-triggered lightning: propagation velocity from measured current and electric field derivative at ground. In *Proc. 11th Int. Conf. on Atmospheric Electricity, Guntersville, Alabama*, pp. 158–61.

Davis, D.A., Murray, W.C., Winn, W.P., Mo, Q., Buseck, P.R., and Hibbs, B.D. 1993. Fulgurites from triggered lightning. *Eos Trans. AGU* **74**(43), Fall Meeting Suppl.: 165.

Depasse, P. 1994a. Statistics on artificially triggered lightning. *J. Geophys. Res.* **99**: 18 515–22.

Depasse, P. 1994b. Lightning acoustic signature. *J. Geophys. Res.* **99**: 25 933–40.

Diels, J.-C., Bernstein, R., Stahlkopf, K.E., and Zhao, X.M. 1997. Lightning control with lasers. *Scientific American*, August: 50–5.

Djebari, B., Hamelin, J., Leteinturier, C., and Fontaine, J. 1981. Comparison between experimental measurements of the electromagnetic field emitted by lightning and different theoretical models – influence of the upward velocity of the return stroke. In *Proc. 4th Int. Symp. on Electromagn. Compat., Zurich, Switzerland*, pp. 511–16.

Eybert-Berard, A. 1996. Laser-triggered lightning. *Science et Avenir*, November.

Eybert-Berard, A., Barret, L., and Berlandis, J.P. 1986. Campagne foudre aux ETATS-UNIS Kennedy Space Center (Florida), Programme RTLP 85* (in French). STT/ASP 86-01, Centre d'Etudes Nuclaire de Grenoble, Grenoble, France.

Eybert-Berard, A., Barret, L., and Berlandis, J.P. 1988. Campagne d'experimentations foudre RTLP 87, NASA Kennedy Space Center, Florida, USA (in French). STT/LASP 88-21/AEB/JPB-pD, Centre d'Etudes Nuclaire de Grenoble, Grenoble, France.

Eybert-Berard, A., Lefort, A., and Thirion, B. 1998. On-site tests. In *Proc. 24th Int. Conf. on Lightning Protection, Birmingham, UK*, pp. 425–35.

Fernandez, M.I. 1997. Responses of an unenergized test power distribution system to direct and nearby lightning strikes. M.S. thesis, University of Florida, Gainesville, 249 pp.

Fernandez, M.I., Mata, C.T., Rakov, V.A., Uman, M.A., Rambo, K.J., Stapleton, M.V., and Bejleri, M. 1998a. Improved lightning arrester protection results, final results. Technical Report, TR-109670-R1 (Addendum AD-109670-R1), EPRI, 3412 Hillview Avenue, Palo Alto, California 94304.

Fernandez, M.I., Rambo, K.J., Stapleton, M.V., Rakov, V.A., and Uman, M.A. 1998b. Review of triggered lightning experiments performed on a power distribution system at Camp Blanding, Florida, during 1996 and 1997. In *Proc. 24th Int. Conf. on Lightning Protection, Birmingham, UK*, pp. 29–35.

Fernandez, M.I., Rakov, V.A., and Uman, M.A. 1998c. Transient currents and voltages in a power distribution system due to natural lightning. In *Proc. 24th Int. Conf. on Lightning Protection, Birmingham, UK*, pp. 622–9.

Fernandez, M.I., Rambo, K.J., Rakov, V.A., and Uman, M.A. 1999. Performance of MOV arresters during very close, direct lightning strikes to a power distribution system. *IEEE Trans. Pow. Del.* **14**(2): 411–18.

Fieux, R., and Hubert, P. 1976. Triggered lightning hazards. *Nature* **260**: 188.

Fieux, R., Gary, C., and Hubert, P. 1975. Artificially triggered lightning above land. *Nature* **257**: 212–14.

Fieux, R.P., Gary, C.H., Hutzler, B.P., Eybert-Berard, A.R., Hubert, P.L., Meesters, A.C., Perroud, P.H., Hamelin, J.H., and Person, J.M. 1978. Research on artificially triggered lightning in France. *IEEE Trans. Pow. Appar. Syst.* **PAS-97**: 725–33.

Fisher, R.J., Schnetzer, G.H., Thottappillil, R., Rakov, V.A., Uman, M.A., and Goldberg, J.D. 1993. Parameters of triggered-lightning flashes in Florida and Alabama. *J. Geophys. Res.* **98**: 22 887–902.

Fisher, R.J., Schnetzer, G.H., and Morris, M.E. 1994. Measured fields and earth potentials at 10 and 20 meters from the base of triggered-lightning channels. In *Proc. 22nd Int. Conf. on Lightning Protection, Budapest, Hungary*, paper R1c-10, 6 pp.

Fujiwara, E., Izawa, Y., Kawasaki, Z.-I., Matsuura, K., and Yamanaka, C. 1991. Laser triggered lightning. *Laser Res.* **19**: 528–37.

Gardner, R.L., Baum, C.E., and Rison, W. 1987. Resistive leaders for triggered lightning. *Kiva Memos* **7**, November, 10 pp.

Gary, C., Cimador, A., and Fieux, R. 1975. La foudre: étude du phénomène. Applications à la protection des lignes de transport. *Revue Générale de l'Électricité* **84**: 24–62.

Georgiadis, N., Rubinstein, M., Uman, M.A., Medelius, P.J., and Thomson, E.M. 1992. Lightning-induced voltages at both ends of a 450-m distribution line. *IEEE Trans. Electromagn. Compat.* **34**: 451–60.

Gorin, B.N., and Shkilev, A.V. 1984. Measurements of lightning currents at the Ostankino tower. *Elektrichestvo* **8**: 64–5.

Gorin, B.N., Levitov, V.I., and Shkilev, A.V. 1977. Lightning strikes to the Ostankino tower. *Elektrichestvo* **8**: 19–23.

Greig, J.R., Koopman, D.W., Fernsler, R.F., Pechacek, R.E., Vitovitsky, I.M., and Ali, A.W. 1978. Electrical discharges guided by pulsed CO_2-laser radiation. *Phys. Rev. Lett.* **41**: 174–7.

Gromovenko, V.M., and Danilov, M.F. 1999. Local enhancement of atmospheric electrical field by long laser spark. In *Proc. 11th Int. Symp. on High-Voltage Engineering, London, UK*, vol. 2, pp. 419–22, IEE Conf. Publ. no. 467.

Gromovenko, V.M., Rezunkov, Yu. A., Saveljeva, V.P., and Kadzov, G.D. 1998. Study of long laser spark generation for protection systems against a lightning. In *Proc. 24th Int. Conf. on Lightning Protection, Birmingham, UK*, pp. 463–7.

Hagenguth, J.H., and Anderson, J.G. 1952. Lightning to the Empire State Building – part III. *AIEE Trans.* **71** (III): 641–9.

Hamelin, J. 1993. Sources of natural noise. In *Electromagnetic Compatibility*, eds. P. Deguauque and J. Hamelin, 652 pp., New York, Oxford.

Hamelin, J., Karczewsky, J.F., and Sene, F.X. 1978. Sonde de mesure du champ magnétique dû à une décharge orageuse. *Ann. des Télécommun.* **33**: 198–205.

Hamelin, J., Leteinturier, C., Weidman, C., Eybert-Berard, A., and Barret, L. 1986. Current and current-derivative in triggered lightning flashes – Florida 1985. In *Proc. Int. Conf. on Lightning and Static Electricity, NASA, Dayton, Ohio*, 10 pp.

Hierl, A. 1981. Strommessungen der Blitztriggerstation Steingaden. In *Proc. 16th Int. Conf. on Lightning Protection, Szeged, Hungary*, paper R-1.04, 10 pp.

Honda, C., Takuma, T., Muraoka, K., Akasaki, M., Kinoskita, F., and Katakira, O. 1993. Characteristic of discharge induced by laser-generated plasmas in atmospheric air. *Trans. IEE Japan* **113-B**: 994–1002.

Horii, K. 1982. Experiment of artificial lightning triggered with rocket. *Mem. Faculty Eng., Nagoya Univ. Japan* **34**: 77–112.

Horii, K., and Ikeda, G. 1985. A consideration on success conditions of triggered lightning. In *Proc. 18th Int. Conf. on Lightning Protection, Munich, Germany*, paper 1–3, 6 pp.

Horii, K., and Nakano, M. 1995. Artificially triggered lightning. In *Handbook of Atmospheric Electrodynamics*, vol. 1. ed. H. Volland, pp. 151–66, Boca Raton, Florida: CRC Press.

Horii, K., and Sakurano, H. 1985. Observation on final jump of the discharge in the experiment of artificially triggered lightning. *IEEE Trans. Pow. Appar. Syst.* **PAS-104**: 2910–17.

Horii, K., Wada, A., Nakamura, K., Yoda, M., Kawasaki, Z., Sirait, K.T., Soekarto, J., Sunoto, A.M. 1990. Experiment of rocket-triggered lightning in Indonesia. *Trans. IEE Japan* **110-B**: 1068–9.

Hubert, P. 1981. Triggered lightning at Langmuir Laboratory during TRIP-81. Service d'Electronique Physique, DPH/EP/81/66.

Hubert, P. 1984. Triggered lightning in France and New Mexico. *Endeavour* **8**: 85–9.

Hubert, P. 1985. A new model of lightning subsequent stroke – confrontation with triggered lightning observations. In *Proc. 10th Int. Conf. on Lightning and Static Electricity, Paris, France*, paper 4B4.

Hubert, P., Laroche, P., Eybert-Berard, A., and Barret, L. 1984. Triggered lightning in New Mexico. *J. Geophys. Res.* **89**: 2511–21.

Hubert, P., and Mouget, G. 1981. Return stroke velocity measurements in two triggered lightning flashes. *J. Geophys. Res.* **86**: 5253–61.

Idone, V.P. 1992. The luminous development of Florida triggered lightning. *Res. Lett. Atmos. Electr.* **12**: 23–8.

Idone, V.P. 1995. Microscale tortuosity and its variation as observed in triggered lightning channels. *J. Geophys. Res.* **100**: 22 943–56.

Idone, V.P., and Orville, R.E. 1984. Three unusual strokes in a triggered lightning flash. *J. Geophys. Res.* **89**: 7311–16.

Idone, V.P., and Orville, R.E. 1985. Correlated peak relative light intensity and peak current in triggered lightning subsequent return strokes. *J. Geophys. Res.* **90**: 6159–64.

Idone, V.P., and Orville, R.E. 1988. Channel tortuosity variation in Florida triggered lightning. *Geophys. Res. Lett.* **15**: 645–8.

Idone, V.P., Orville, R.E. Hubert, P. Barret, L., and Eybert-Berard, A. 1984. Correlated observations of three triggered lightning flashes. *J. Geophys. Res.* **89**: 1385–94.

Jafferis, W. 1995. Rocket triggered lightning – Kennedy Space Center and beyond. In *Proc. 1995 Int. Conf. on Lightning and Static Electricity, Williamsburg, Virginia*, pp. 57/1–57/20.

Jordan, D.M., Idone, V.P., Rakov, V.A., Uman, M.A., Beasley, W.H., and Jurenka, H. 1992. Observed dart leader speed in natural and triggered lightning. *J. Geophys. Res.* **97**: 9951–7.

Kadzov, G.D., and Rezunkov, Yu. A. 1998. Promising techniques of increased reliability in a lightning protection system. In *Proc. 24th Int. Conf. on Lightning Protection, Birmingham, UK*, pp. 468–72.

Kawasaki, Z. 1999. UHF radiation and laser triggered lightning experiment in the field (in Japanese). *Trans. IEE Japan* **119-B**(10): 1115–20.

Kawasaki, Z.-I., and Mazur, V. 1992. Common physical processes in natural and triggered lightning in winter storms in Japan. *Res. Lett. Atmos. Electr.* **12**: 61–70, *J. Geophys. Res.* **97**: 12 935–45.

Kawasaki, Z.-I., Kanao, T., Matsuura, K., Nakano, M., Horii, K., and Nakamura, K.-I. 1991. The electric field changes and UHF radiations caused by the triggered lightning in Japan. *Geophys. Res. Lett.* **18**: 1711–14.

References and bibliography for Chapter 7

Kawasaki, Z.-I., Matsuura, K., Fujiwara, E., Izawa, Y., Nakamura, K., and Yamanaka, C. 1992. Long laser-induced discharge in atmospheric air. *Res. Lett. Atmos. Electr.* **12**: 139–42.

Kikuchi, H. 1985. Overview of triggered lightning EMP. In *Nonlinear and Environmental Electromagnetics*, ed. H. Kikuchi, pp. 347–50, Amsterdam: Elsevier.

Kito, Y., Horii, K., Higasbiyama, Y., and Nakamura, K. 1985. Optical aspects of winter lightning discharges triggered by the rocket-wire technique in Hokuriku district of Japan. *J. Geophys. Res.* **90**: 6147–57.

Kobayashi, M., Sasaki, H., and Nakamura, K. 1997. Rocket-triggered lightning experiment and consideration for metal oxide surge arresters. In *Proc. 10th Int. Symp. on High Voltage Engineering, Montreal, Quebec, Canada*, 4 pp.

Koopman, D.W., and Saum, K.A. 1973. Formation and guiding of high-velocity electrical streamers by laser-induced ionization. *J. Appl. Phys.* **44**: 5328–36.

Koopman, D.W., and Wilkerson, T.D. 1971. Channeling of an ionizing electrical streamer by a laser beam. *J. Appl. Phys.* **42**: 1883–6.

Krider, E.P., Noogle, R.C., Uman, M.A., and Orville, R.E. 1974. Lightning and the Apollo/Saturn V exhaust plume. *J. Spacecraft Rockets* **11**: 72–5.

Kumazaki, K., Nakamura, K., Naito, K., and Horii, K. 1993. Production of artificial fulgurite by utilizing rocket triggered lightning. In *Proc. 8th Int. Symp. on High Voltage Engineering, Yokohama, Japan*, pp. 269–72.

La Fontaine, B., Vidal, F., Comtois, D., Chien, C.-Y., Desparois, A., Jonston, T., Kieffer, J.-C., Mercure, H.P., Pepin, H., and Rizk, F.A.M. 1999a. The influence of electron density on the formation of streamers in electrical discharges triggered with ultrashort laser pulses. *IEEE Trans. Plasma Science* **27**(3): 688–700.

La Fontaine, B., Vidal, F., Jiang, Z., Chien, C.-Y., Comtois, D., Desparois, A., Johnston, T.W., Kieffer, J.-C., Pepin, H., and Mercure, H.P. 1999b. Filamentation of ultrashort pulse laser beams resulting from their propagation over long distances in air. *Phys. Plasma* **6**(5): 1615–21.

Lalande, P., Bondiou-Clergerie, A., Laroche, P., Eybert-Berard, A., Berlandis, J.P., Bador, B., Bonamy, A., Uman, M.A., and Rakov, V.A. 1996. Connection to ground of an artificially triggered negative downward stepped leader. In *Proc. 10th Int. Conf. on Atmospheric Electricity, Osaka, Japan*, pp. 668–71.

Lalande, P., Bondiou-Clergerie, A., Laroche, P., Eybert-Berard, A., Berlandis, J.-P., Bador, B., Bonamy, A., Uman, M.A., and Rakov, V.A. 1998. Leader properties determined with triggered lightning techniques. *J. Geophys. Res.* **103**: 14 109–15.

Lapshin, V.A., Rublev, A.V., Sakyan, A.S., Sidorovskii, and Starchenko, A.N. 1999. Complex for diagnosing the photometric characteristics of a lightning leader. *J. Opt. Technol.* **66**(3): 219–22.

Laroche, P., Eybert-Berard, A., and Barret, L. 1985. Triggered lightning flash characteristics. In *Proc. 10th Int. Aerospace and Ground Conf. on Lightning and Static Electricity, Paris, France*, pp. 231–9.

Laroche, P., Eybert-Berard, A., Barret, L., and Berlandis, J.P. 1988. Observations of preliminary discharges initiating flashes triggered by the rocket and wire technique. In *Proc. 8th Int. Conf. on Atmospheric Electricity, Uppsala, Sweden*, pp. 327–33.

Laroche, P., Idone, V., Eybert-Berard, A., and Barret, L. 1991. Observations of bi-directional leader development in a triggered lightning flash. In *Proc. Int. Conf. on Lightning and Static Electricity, Cocoa Beach, Florida*, pp. 57/1–10.

Leteinturier, C., and Hamelin, J. 1990. Rayonnement électromagnétique des décharges orageuses. Analyse submicroseconde. *Revue Phys. Appl.* **25**: 139–46.

Leteinturier, C., Weidman, C., Hamelin, J. 1990. Current and electric field derivatives in triggered lightning return strokes. *J. Geophys. Res.* **95**: 811–28.

Leteinturier, C., Hamelin, J.H., and Eybert-Berard, A. 1991. Submicrosecond characteristics of lightning return-stroke currents. *IEEE Trans. Electromagn. Compat.* **33**: 351–7.

Le Vine, D.M., Willett, J.C., and Bailey, J.C. 1989. Comparison of fast electric field changes from subsequent return strokes of natural and triggered lightning. *J. Geophys. Res.* **94**: 13 259–65.

Lhermitte, R. 1982. Doppler radar observations of triggered lightning. *Geophys. Res. Lett.* **9**: 712–15.

Liaw, Y.P., Cook, D.R., and Sisterson, D.L. 1996. Estimation of lightning stroke peak current as a function of peak electric field and the normalized amplitude of signal strength: corrections and improvements. *J. Atmos. Ocean. Tech.* **13**: 769–73.

Lippert, J.R. 1978. Laser-induced lightning concept experiment. Air Force Flight Dynamics Lab., Technical Report AFFDL-TR-78-191, AD A065 897.

Liu, X.-S., and Zhang, Y. 1998. Review of artificially triggered lightning study in China. *Trans. IEE Japan* **118-B**(2): 170–5.

Liu, X.-S., Wang, C., Zhang, Y., Xiao, Q., Wang, D., Zhou, Z., and Guo, C. 1994. Experiment of artificially triggering lightning in China. *J. Geophys. Res.* **99**: 10 727–31.

Mach, D.M., and Rust, W.D. 1989. Photoelectric return-stroke velocity and peak current estimates in natural and triggered lightning. *J. Geophys. Res.* **94**: 13 237–47.

Mach, D.M., and Rust, W.D. 1997. Two-dimensional speed and optical risetime estimates for natural and triggered dart leaders. *J. Geophys. Res.* **102**: 13 673–84.

Master, M.J., and Uman, M.A. 1984. Lightning induced voltages on power lines: theory. *IEEE Trans. Pow. App. Syst.* **103**: 2505–17.

Mata, C.T. 2000. Interaction of lightning with power distribution lines. Ph.D. dissertation, University of Florida, Gainesville, 388 pp.

Mata, C.T., Fernandez, M.I., Rakov, V.A., Uman, M.A., Bejleri, M., Rambo, K.J., and Stapleton, M.V. 1998. Overvoltages in underground systems, phase 2 results. Technical Report TR-109669-R1 (Addendum AD-109669-R1), EPRI, 3412 Hillview Avenue, Palo Alto, California 94304.

Mata, C.T., Fernandez, M.I., Rakov, V.A., and Uman, M.A. 2000. EMTP modeling of a triggered-lightning strike to the phase conductor of an overhead distribution line. *IEEE Trans. Pow. Del.* **15**(4): 1175–81.

Matsumoto, Y., Sakuma, O., Shinjo, K., Saiki, M., Wakai, T., Sakai, T., Nagasaka, H., Motoyama, H., and Ishii, M. 1996. Measurement of lightning surges on test transmission line equipped with arresters struck by natural and triggered lightning. *IEEE Trans. Pow. Del.* **11**: 996–1002.

Mazur, V., and Ruhnke, L. 1993. Common physical processes in natural and artificially triggered lightning. *J. Geophys. Res.* **98**: 12 913–30.

McEachron, K.B. 1939. Lightning to the Empire State Building. *J. Franklin Inst.* **227**: 149–217.

Miki, M., and Wada, A. 1996. Guiding of electrical discharges under atmospheric air by ultraviolet laser-produced plasma channel. *J. Appl. Phys.* **80**(6): 3208–14.

Miki, M., Aihara, Y., and Shindo, T. 1993. Development of long gap discharges guided by a pulsed CO_2 laser. *J. Phys. D: Appl. Phys.* **26**: 1244–52.

Miki, M., Shindo, T., and Aihara, Y. 1996. Mechanisms of guiding ability of CO_2 laser-produced plasmas on pulsed discharges. *J. Phys. D: Appl. Phys.* **29**: 1984–96.

Miki, M., Wada, A., and Shindo, T. 1999. Characteristics of laser-guided discharges in long air gaps. *J. Opt. Technol.* **66**(3): 190–3.

Miki, M., Rakov, V.A., Uman, M.A., Rambo, K.J., and Schnetzer, G.H. 2000a. Measuring electric fields near the lightning channel using Pockels sensors. (abstract), *Eos, Trans. Suppl., AGU* **81**(48): F49.

Miki, M., Wada, A., and Shindo, T. 2000b. Model experiments of the laser-triggered lightning using an intense CO_2 laser pulse. In *Proc. Soc. Photo-Optical Instrumentation Engineers*, vol. 3886, pp. 680–90.

Miyachi, I., Horii, K., Muto, S. Ikeda, G., and Aiba, S. 1979. Experiment of long gap discharge by artificially triggered lightning with rocket. In *Proc. 3rd Int. Symp. on High Voltage Engineering, Milan, Italy*, paper 51.10, 4 pp.

Miyake, K., Suzuki, T., Takashima, M., Takuma, M., and Tada, T. 1990. Winter lightning on Japan Sea coast – lightning striking frequency to tall structures. *IEEE Trans. Pow. Del.* **5**: 1370–6.

Morris, M.E., Fisher, R.J., Schnetzer, G.H., Merewether, K.O., and Jorgenson, R.E. 1994. Rocket-triggered lightning studies for the protection of critical assets. *IEEE Trans. Ind. Appl.* **30**: 791–804.

Mortensen, P. 1993. Laser plasma channel safely handcuffs lightning bolts. In *Laser Focus World*, p. 50.

Motoyama, H., Shinjo, K., Matsumoto, Y., and Itamoto, N. 1998. Observation and analysis of multiphase back flashover on the Okushishiku test transmission line caused by winter lightning. *IEEE Trans. Pow. Del.* **13**: 1391–8.

Mrazek, J. 1998. The thermalization of the positive and negative lightning channel. In *Proc. 24th Int. Conf. on Lightning Protection, Birmingham, UK*, pp. 5–10.

Nakamura, K., Horii, K., Kito, Y., Wada, A., Ikeda, G., Sumi, S., Yoda, M., Aiba, S., Sakurano, H., and Wakamatsu, K. 1991. Artificially triggered lightning experiments to an EHV transmission line. *IEEE Trans. Pow. Del.* **6**: 1311–18.

Nakamura, K., Horii, K., Nakano, M., and Sumi, S. 1992. Experiments on rocket triggered lightning. *Res. Lett. Atmos. Electr.* **12**: 29–35.

Nakamura, K., Suzuki, T., Yamabe, C., and Horii, K. 1993. Fundamental research for lightning trigger experiment by using UV lasers. *Trans. IEE Japan* **113-B**: 1265–73.

Nakano, M., Takagi, N., Kawasaki, Z., and Takeuti, T. 1983. Return strokes of triggered lightning flashes. *Res. Lett. Atmos. Electr.* **3**: 73–8.

Newman, M.M. 1958. Lightning discharge channel characteristics and related atmospherics. In *Recent Advances in Atmospheric Electricity*, ed. L.G. Smith, pp. 475–84, New York: Pergamon Press.

Newman, M.M. 1965. Use of triggered lightning to study the discharge channel. In *Problems of Atmospheric and Space Electricity*, pp. 482–90, New York: Elsevier.

Newman, M.M., and Robb, J.D. 1977. Protection of aircraft. In *Lightning, vol. 2: Lightning Protection*, ed. R. H. Golde, pp. 659–96, London: Academic Press.

Newman, M.M., Stahmann, J.R., Robb, J.D., Lewis, E.A., Martin, S.G., and Zinn, S.V. 1967. Triggered lightning strokes at very close range. *J. Geophys. Res.* **72**: 4761–4.

Okano, D. 1999. Guided discharge path by weakly ionized region between two plasmas produced by YAG laser in an atmospheric air gap with nonuniform dc electric field. *IEEE Trans. Plasma Science* **27**(1): 124–5.

Orville, R.E. 1991. Calibration of a magnetic direction finding network using measured triggered lightning return stroke peak currents. *J. Geophys. Res.* **96**: 17 135–42.

Orville, R.E., and Idone, V.P. 1982. Lightning leader characteristics in the Thunderstorm Research International Program (TRIP). *J. Geophys. Res.* **87**: 11 177–92.

Parfenov, V.A., Pakhomov, L.N., Petrunkin, V. Yu., and Podlevskii, V.A. 1976. A study of the possibility to produce a very long optical breakdown in atmosphere. *Pis'ma Zh. Tekh. Fiz.* **2**: 731–4 (*Sov. Tech. Phys. Lett.* **2**: 286, 1976).

Pierce, E.T. 1971. Triggered lightning and some unsuspected lightning hazards. Stanford Research Institute, Menlo Park, California, 20 pp.

Pierce, E.T. 1972. Triggered lightning and some unsuspected lightning hazards. In *Naval Research Reviews*, pp. 14–28.

Pierce, E.T. 1974. Atmospheric electricity – some themes. *Bull. Amer. Meteor. Soc.* **55**: 1186–94.

Rachidi, F., Rubinstein, M., Guerrieri, S., and Nucci, C.A. 1997. Voltages induced on overhead lines by dart leaders and subsequent return strokes in natural and rocket-triggered lightning. *IEEE Trans. Electromagn. Compat.* **39**(2): 160–6.

Rakov, V.A. 1999a. Rocket-triggered lightning experiments at Camp Blanding, Florida. In *Proc. Int. Conf. on Lightning and Static Electricity, Toulouse, France*, pp. 469–81.

Rakov, V.A. 1999b. Lightning makes glass. In *1999 Journal of the Glass Art Society*, pp. 45–50.

Rakov, V.A. 1999c. Lightning discharges triggered using rocket-and-wire techniques. In *Recent Res. Devel. Geophysics*, vol. 2, pp. 141–71, Research Signpost, India.

Rakov, V.A. 2001a. Characterization of lightning electromagnetic fields and their modeling. In *Proc. 14th Int. Symp. on Electromagn. Compatbility, Zurich, Switzerland*, Suppl., pp. 3–16.

Rakov, V.A. 2001b. Positive Blitzentladungen. *ETZ Elektrotech. Autom.* **122**(5): 26–9.

Rakov, V.A., Thottappillil, R., Uman, M.A., and Barker, P.P. 1995a. Mechanism of the lightning M component. *J. Geophys. Res.* **100**: 25 701–10.

Rakov, V.A., Uman, M.A., and R. Thottappillil 1995b. Review of recent lightning research at the University of Florida. *Electrotechnik und Informationstechnik* **112**: 262–5.

Rakov, V.A., Uman, M.A., Fernandez, M.I., Thottappillil, R., Eybert-Berard, A., Berlandis, J.P., Rachidi, F., Rubinstein, M., Guerrieri, S., and Nucci, C.A. 1996a. Observed electromagnetic environment close to the lightning channel. In *Proc. 23rd Int. Conf. on Lightning Protection, Florence, Italy*, pp. 30–5.

Rakov, V.A., Uman, M.A., Rambo, K.J., Fernandez, M.I., Eybert-Berard, A., Berlandis, J.P., Barker, P.P., Fisher, R.J., and Schnetzer, G.H. 1996b. Initial processes in triggered lightning. *Eos, Trans. AGU*, **77**, Fall Meeting Suppl.: F86.

Rakov, V.A., Uman, M.A., Rambo, K.J., Fernandez, M.I., Fisher, R.J., Schnetzer, G.H., Thottappillil, R., Eybert-Berard, A., Berlandis, J.P., Lalande, P., Bonamy, A., Laroche, P., and Bondiou-Clergerie, A. 1998. New insights into lightning processes gained from triggered-lightning experiments in Florida and Alabama. *J. Geophys. Res.* **103**: 14 117–30.

Rakov, V.A., Uman, M.A., Wang, D., Rambo, K.J., Crawford, D.E., Schnetzer, G.H., and Fisher R.J. 1999. Some results from recent experiments at the International Center for Lightning Research and Testing at Camp Blanding, Florida (abstract). *Eos, Trans. Suppl., AGU* **80**(46): F203.

Rakov, V.A., Uman, M.A., Rambo, K.J., Schnetzer, G.H., and Miki, M. 2000a. Triggered-lightning experiments conducted in 2000 at Camp Blanding, Florida (abstract). *Eos, Trans. Suppl., AGU* **81**(48): F90.

Rakov, V.A., Uman, M.A., Wang, D., Rambo, K.J., Crawford, D.E., and Schnetzer, G.H. 2000b. Lightning properties from triggered-lightning experiments at Camp Blanding, Florida (1997–1999). In *Proc. 25th Int. Conf. on Lightning Protection, Rhodes, Greece*, pp. 54–9.

Rakov, V.A., Uman, M.A., Fernandez, M.I., Mata, C.T., Rambo, K.T., Stapleton, M.V., and Sutil, R.R. 2002. Direct lightning strikes to the lightning protective system of a residential building: triggered-lightning experiments. *IEEE Trans. Pow. Del.*, **17** (2): 575–86.

Rambo, P., Biegert, J., Kubecek, V., Schwartz, J., Bernstein, A., and Diels, J.-C. 1998. Laboratory tests of laser induced lightning discharge, laboratory and field experiments. In *Proc. 1st Int. Workshop on Lightning Protection by Lasers, Sosnovy Bor, Leningrad region, Russia*, pp. 23–4.

Rambo, P., Biegert, J., Kubecek, V., Schwartz, J., Bernstein, A., and Diels, J.-C. 1999. Laboratory tests of laser-induced lightning discharge. *J. Opt. Technol.* **66**(3): 194–8.

Rubinstein, M., Tzeng, A.Y., Uman, M.A., Medelius, P.J., and Thomson, E.M. 1989. An experimental test of a theory of lightning-induced voltages on an overhead wire. *IEEE Trans. Electromagn. Compat.* **31**: 376–83.

Rubinstein, M., Uman, M.A., Thomson, E.M., Medelius, P., and Rachidi, F. 1992. Measurements and characterization of ground level vertical electric fields 500 m and 30 m from triggered-lightning. In *Proc. 9th Int. Conf. on Atmospheric Electricity, St. Petersburg, Russia*, pp. 276–8.

Rubinstein, M., Uman, M.A., Medelius, P.J., and Thomson, E.M. 1994. Measurements of the voltage induced on an overhead power line 20 m from triggered lightning. *IEEE Trans. Electromagn. Compat.* **36**(2): 134–40.

Rubinstein, M., Rachidi, F., Uman, M.A., Thottappillil, R., Rakov, V.A., and Nucci, C.A. 1995. Characterization of vertical electric fields 500 m and 30 m from triggered lightning. *J. Geophys. Res.* **100**: 8863–72.

Rühling, F. 1974. Gezielte Blitzentladung mittels Raketen. *Umschau in Wissenschaft und Technik* **74**: 520–1.

Rühling, F. 1974. Raketengetriggerte Blitze im Dienste des Freileitungsschutzes vor Gewitterüberspannungen. *Bull. SEV* **65**: 1893–8.

Saba, M.M.F., Pinto, O., Pinto, I.R.C.A., Pissolato, F.J., Eybert-Berard, A., Lefort, A., Potvin, C., Heine, L.F., and Chauzy, S. 2000. An international center for triggered and natural lightning research in Brazil. In *Proc. 2000 Int. Lightning Detection Conf.*, paper 40, 7 pp., GAI, 2705 East Medina Road, Tucson, Arizona 85706-7155.

Sadiku, M.N.O. 1994. *Elements of Electromagnetics*, 821 pp., Orlando, Florida: Sounders College.

Savel'eva, V.P. 1999. Optical shaping systems for generating a long laser spark. *J. Opt. Technol.* **66**(3): 215–18.

Schnetzer, G.H., and Fisher, R.J. 1998. Earth potential distributions within 20 m of triggered lightning strike points. In *Proc. 24th Int. Conf. on Lightning Protection, Birmingham, UK*, pp. 501–5.

Schnetzer, G.H., Fisher, R.J., Rakov, V.A., and Uman, M.A. 1998. The magnetic field environment of nearby lightning. In *Proc. 24th Int. Conf. on Lightning Protection, Birmingham, UK*, pp. 346–9.

Schonland, B.F.J., Malan, D.J., and Collens, H. 1935. Progressive Lightning II. *Proc. Roy. Soc. A* **152**: 595–625.

Schubert, C.N. 1997. The laser lightning rod system: a feasibility study. Air Force Flight Dynamics Lab., Technical Report AFFDL-TR-78-60, AD A063 847.

Schubert, C.N., and Lippert, J.R. 1979. Investigation into triggering lightning with a pulsed laser. In *Proc. 2nd IEEE Int. Pulse Power Conf., Lubbock, Texas*, pp. 132–5.

Shiho, M., Watanabe, A., Kawasaki, S., Ishizuka, H., Takayama, K., Kishiro, J., Shindo, T., and Fujioka. 1996. Lightning control system using high power microwave FEL. *Nuclear Instruments and Methods in Phys. Res. A* **375**: 396–400.

Shimada, Y., Uchida, S., Yasuda, H., Motokoshi, S., Ishikubo, Y., Kawasaki, Z.-I., Yamanaka, T., Adachi, M., and Yamanaka, C. 1999. Research on field experiments of laser triggered lightning (in Japanese). *Trans. IEE Japan* **119-A**(7): 990–5.

Shindo, T., and Ishii, M. 1995. Japanese research on triggering lightning. *Elektrotechnik und Informationstechnik* **112**(6): 265–8.

Shindo, T., Aihara, Y., and Miki, M. 1992. Model experiments of laser-triggered lightning. *Res. Lett. Atmos. Electr.* **12**: 135–8.

Shindo, T., Aihara, Y., Miki, M., and Suzuki, T. 1993a. Model experiments of laser-triggered lightning. *IEEE Trans. Pow. Del.* **8**: 311–7.

Shindo, T., Miki, M., Aihara, Y., and Wada, A. 1993b. Laser-guided discharges in long gaps. *IEEE Trans. Pow. Del.* **8**: 2016–22.

Soula, S., and Chauzy, S. 1991. Multilevel measurement of the electric field underneath a thundercloud. 2. Dynamical evolution of a ground space charge layer. *J. Geophys. Res.* **96**: 22 327–36.

SPARG (1982): Saint-Privat-d'Allier Research Group (SPARG) 1982. Eight years of lightning experiments at Saint-Privat-d'Allier. *Extrait de la Revue Générale de l'Electricité*, September, Paris.

Teramoto, M., Yamada, T., Nakamura, K., Matsuoka, R., Sumi, S., and Horii, K. 1996. Triggered lightning to a new type lightning rod with high resistance. In *Proc. 10th Int. Conf. on Atmospheric Electricity, Osaka, Japan*, pp. 341–4.

Thottappillil, R., Goldberg, J.D., Rakov, V.A., and Uman, M.A. 1995. Properties of M components from currents measured at triggered lightning channel base. *J. Geophys. Res.* **100**: 25 711–20.

Uchida, S., Shimada, Y., Yasuda, H., Tsubakimoto, K., Motokoshi, S., Yamanaka, C., Kawasaki, Z.-I., Yamanaka, T., Matsuura, K., Ushio, T., Adachi, M., and Ishikubo, Y. 1996. Laser triggered lightning experiments in the field. In *Proc. 10th Int. Conf. on Atmospheric Electricity, Osaka, Japan*, pp. 680–63.

Uchida, S., Shimada, Y., Yasuda, H., Motokoshi, S., Yamanaka, C., Yamanaka, T., Kawasaki, Z,-I., and Tsubakimoto, K. 1999. Laser-triggered lightning in field experiments. *J. Opt. Technol.* **66**(3): 199–202.

Uchiyama, T., Hirohashi, M., Miyata, H., and Sakai, H. 1988. Study of triggering lightning by using TEA CO_2 laser. *Rev. Laser Eng.* **16**: 267.

Uman, M.A. 1987. *The Lightning Discharge*, 377 pp., San Diego, California: Academic Press.

Uman, M.A. 2001. *The Lightning Discharge*, 377 pp., Mineola, New York: Dover.

Uman, M.A., Seacord, D.F., Price, G.H., and Pierce, E.T. 1972. Lightning induced by thermonuclear detonations. *J. Geophys. Res.* **77**: 1591–6.

Uman, M.A., Cordier, D.J., Chandler, R.M., Rakov, V.A., Bernstein, R., and Barker, P.P. 1994a. Fulgurites produced by triggered lightning. *Eos Trans. AGU* **75**(44), Fall Meeting Suppl.: 99.

Uman, M.A., Rakov, V.A., Versaggi, J.A., Thottappillil, R., Eybert-Berard, A., Barret, L., Berlandis, J.-P., Bador, B., Barker, P.P., Hnat, S.P., Oravsky, J.P., Short, T.A., Warren, C.A., and Bernstein, R. 1994b. Electric fields close to triggered lightning. In *Proc. Int. Symp. on Electromagn. Compat. (EMC '94 ROMA), Rome, Italy*, pp. 33–7.

Uman, M.A., Rakov, V.A., Rambo, K.J., Vaught, T.W., Fernandez, M.I., Bach, J.A., Su, Y., Eybert-Berard, A., Berlandis, J.-P., Bador, B., Lalande, P., Bonamy, A., Audran, F., Morillon, F., Laroche, P., Bondiou-Clergerie, A., Chauzy, S., Soula, S., Weidman, C.D., Rachidi, F., Rubinstein, M., Nucci, C.A., Guerrieri, S., Hoidalen, H.K., and Cooray, V. 1996. 1995 triggered lightning experiment in Florida. In *Proc. 10th Int. Conf. on Atmospheric Electricity, Osaka, Japan*, pp. 644–7.

Uman, M.A., Rakov, V.A., Rambo, K.J., Vaught, T.W., Fernandez, M.I., Cordier, D.J., Chandler, R.M., Bernstein, R., and Golden, C. 1997. Triggered-lightning experiments at Camp Blanding, Florida (1993–1995). *Trans. IEE Japan* **117-B**: 446–52.

Uman, M.A., Rakov, V.A., Schnetzer, G.H., Rambo, K.J., Crawford, D.E., and Fisher, R.J. 2000. Time derivative of the electric field 10, 14, and 30 m from triggered lightning strokes. *J. Geophys. Res.* **105**: 15 577–95.

Vasilyak, L.M., Vetchinin, S.P., and Polyakov, D.N. 1991. Laser initiation of electrical breakdown in a long gap. *Elektrichestvo* **1**: 59–61.

Vereshchagin, I.P., Koshelev, M.A., Makalsky, L.M., and Sysoev, V.S. 1989. Electrical discharge in charged aerosol. *Proc. USSR Academy Sciences (Izvestiya AN SSSR, Ser. Energetika i Transport)* **4**: 100–6.

Vereshchagin, I.P., Orlov, A.V., Temnikov, A.G., Antsupov, K.V., Koshelev, M.A., and Makalsky, L.M. 1996. Discharge development inside the charged aerosol formations. In *Proc. 23rd Int. Conf. on Lightning Protection, Florence, Italy*, pp. 42–6.

Wada, A. 1994. Discussion on lightning protection by artificial triggered lightning. In *Proc. 22nd Int. Conf. on Lightning Protection, Budapest, Hungary*, paper R2-06, 5 pp.

Waldteufel, P., Metzger, P., Boulay, J.L., Laroche, P., and Hubert, P. 1980. Triggered lightning strokes originating in clear air. *J. Geophys. Res.* **85**: 2861–8.

Wang, D., Kawasaki, Z.-I., Matsuura, K., Shimada, Y., Uchida, S., Yamanaka, C., Fujiwara, E., Izawa, Y., Simokura, N., and Sonoi, Y. 1994a. A preliminary study on laser-triggered lightning. *J. Geophys. Res.* **99**: 16 907–12.

Wang, D., Takagi, N., Watanabe, T., Ushio, T., Kawasaki, Z.-I., Matsuura, K., Shimada, Y., Uchida, S., and Yamanaka, C. 1994b. The study of the possibility of lightning triggered by means of a laser. *J. Atmos. Electr.* **14**: 49–55.

Wang, D., Takagi, N., Watanabe, T., Ushio, T., Kawasaki, Z.-I., Matsuura, K., Shimada, Y., Uchida, S., and Yamanaka, C. 1995a. Progression of leader system in long gap laser triggered discharges. *J. Atmos. Electr.* **15**: 67–74.

Wang, D., Ushio, T., Kawasaki, Z.-I., Matsuura, K., Shimada, Y., Uchida, S., Yamanaka, C., Izawa, Y., Sonoi, Y., and Simokura, N. 1995b. A possible way to trigger lightning using a laser. *J. Atmos. Terr. Phys.* **57**: 459–66.

Wang, D., Ushio, T., Kawasaki, Z.-I., Uchida, S., Shimada, Y., Yasuda, H., Yamanaka, T., Yamanaka, C., Matuura, K., Ishikubo, Y., and Adachi, M. 1995c. Field experiments on laser triggered lightning. In *Proc. Int. Aerospace and Ground Conf. on Lightning and Static Electricity, Williamsburg, Virginia*, pp. 78/1–10.

Wang, D., Rakov, V.A., Uman, M.A., Fernandez, M.I., Rambo, K.J., Schnetzer, G.H., and Fisher, R.J. 1999a. Characterization of the initial stage of negative rocket-triggered lightning. *J. Geophys. Res.* **104**: 4213–22.

Wang, D., Rakov, V.A., Uman, M.A., Takagi, N., Watanabe, T., Crawford, D.E., Rambo, K.J., Schnetzer, G.H., Fisher, R.J., and Kawasaki, Z.-I. 1999b. Attachment process in rocket-triggered lightning strokes. *J. Geophys. Res.* **104**: 2141–50.

Wang, D., Takagi, N., Watanabe, T., Rakov, V.A., and Uman, M.A. 1999c. Observed leader and return-stroke propagation characteristics in the bottom 400 m of the rocket triggered lightning channel. *J. Geophys. Res.* **104**: 14 369–76.

Wang, C., Yan, M., Liu, X., Zhang, Y., Dong, W., and Zhang, C. 1999d. Bidirectional propagation of lightning leader. *Chinese Sci. Bull.* **44**(2): 163–6.

Watanabe, T., Takagi, T., Wang, D., Takeda, A., Yasue, N., Kawasaki, Z.-I., Serizawa, C., Kato, M., Suda, M., and Katuragi, Y. 1996. Feasibility study on triggering lightning with a transient flame. In *Proc. 10th Int. Conf. on Atmospheric Electricity, Osaka, Japan*, pp. 248–51.

Willett, J.C. 1992. Rocket-triggered-lightning experiments in Florida. *Res. Lett. Atmos. Electr.* **12**: 37–45.

Willett, J.C., Davis, D.A., and Laroche, P. 1999a. Positive leaders in rocket-triggered lightning: propagation velocity from measured current and ambient-field profile. In *Proc. 11th Int. Conf. on Atmospheric Electricity, Guntersville, Alabama*, pp. 58–61.

Willett, J.C., Davis, D.A., and Laroche, P. 1999b. An experimental study of positive leaders initiating rocket-triggered lightning. *Atmos. Res.* **51**: 189–219.

Yamabe, C., Nakamura, K., Susuki, T., Horii, K., Takagi, S., Sato, S., Goto, T., and Matsuoka, R. 1992. Laboratory experiments on laser-triggered lightning by using ultraviolet lasers. *Res. Lett. Atmos. Electr.* **12**: 127–33.

Yasuda, H., Uchida, S., Shimada, Y., Motokoshi, S., Yamanaka, C., Onuki, J., Ushio, T., Kawasaki, Z.I., Tsubakimoto, K., Yamanaka, T., Ishikubo, Y., and Adachi, M. 1997. First observation of laser triggered lightning in field experiment. In *Proc. Pacific Rim Conf. on Lasers and Electro-optics (CLEO/Pacific Rim'97), Chiba, Japan*, paper PD1.14.

Yoda, M., Miyachi, I., Kawashima, T., and Katsuragi, Y. 1992. Lightning current protection of equipments in winter. *Res. Lett. Atmos. Electr.* **12**: 117–21.

Young, G.A. 1962. A lightning strike of an underwater explosion plume. US Naval Ordnance Laboratory, NOLTR 61–43.

Zhao, X.M. 1993. Femtosecond ultraviolet pulses triggering of ground to cloud lightning. Ph.D. dissertation, University of New Mexico, Albuquerque.

Zhao, X.M., and Diels, J.-C. 1993. How lasers might control lightning strikes. *Laser Focus World*, November, pp. 113–23.

Zhao, X.M., Diels, J.-C., Wang, C.Y., and Elizondo, J.M. 1995. Femtosecond ultraviolet laser pulse induced lightning discharges in gases. *IEEE J. Quantum Electron.* **31**: 599–612.

8 Winter lightning in Japan

> ...there remains the dilemma of how these relatively small clouds make and store such large amounts of electricity.
> M. Brook, M. Nakano, P. Kriehbiel, and T. Takeuti (1982)

8.1. Introduction

There are two primary reasons for considering winter lightning in Japan in a separate chapter of this book. First, lightning discharges produced by winter storms in the coastal area of the Sea of Japan exhibit a number of features, discussed in Section 8.4, that have not been observed in the summer season in Japan or in any season in other geographical locations. It is these discharges in the Japan Sea coastal area (see Figs. 8.1a and b) that are usually referred to as winter lightning in Japan, although winter lightning in other regions of Japan may not have similar characteristics. Second, a considerable amount of data on winter lightning in Japan has been published over the last two to three decades. These data are scattered over many papers in different journals and, therefore, are in need of a consolidating review.

It is possible and even likely that the conditions conducive to the production of lightning discharges similar to those observed in winter storms in Japan exist in other seasons and/or locations. If so, further lightning research may lead to the replacement of the term "winter lightning in Japan" by another term that provides more insight into the nature of this phenomenon. As a step in this direction, information about individual lightning characteristics that are observed elsewhere but resemble the corresponding characteristics of winter lightning in Japan is included, where appropriate, in this chapter.

The first observations of winter lightning that indicated the unusual electrical behavior of winter thunderstorms in Japan were due to Takeuti et al. (1973, 1976) and Takeuti and Nakano (1977). In these studies, most lightning discharges to ground were found to be positive, whereas in summer the majority of cloud-to-ground discharges are negative (Table 5.2 and Fig. 5.2). Further studies, based on multiple-station electric field measurements in conjunction with video recordings in the Hokuriku area of Japan, confirmed that ground strokes in these winter storms generally lower positive charge to earth (Takeuti et al. 1978; Brook et al. 1982). Positive lightning discharges in both winter and summer storms in different locations are discussed in detail in Chapter 5. Other unusual features of winter lightning in Japan include, but are not limited to, a relatively high frequency of occurrence of (i) bipolar flashes, also discussed in Section 5.4, (ii) upward flashes, even in the case where the strike object is of moderate height, also discussed in Chapter 6, and (iii) very large, up to tens of kiloamperes, slowly varying lightning currents of either polarity. Many studies of cold-season storms in Japan and in other countries have been performed using multiple-station lightning locating systems (Sections 17.3 and 17.5); some of the results can be found in subsections 2.2.2, 2.2.7, 2.8.1, and 5.2. The dependence of lightning characteristics on season is considered in subsection 2.8.1.

8.2. Formation of winter thunderclouds

The meteorological conditions leading to the formation of winter thunderclouds in Japan and the characteristics of these clouds, as summarized by Kitagawa and Michimoto (1994), are listed below.

(i) In winter, trains of convective clouds are formed over the Japan Sea by the advection of dry polar air masses from Siberia. In the case that the tropospheric instability caused by the advection is intense enough, some of the clouds develop into thunderclouds over the warm Tsushima current (see Fig. 8.1a) and over the Japan Sea coast (Goto and Narita 1991; Kitagawa 1992).

(ii) When winter thunderstorms develop, the vertical wind shear through the whole troposphere is as strong as in the case of severe thunderstorms developing in the summer season (Kitagawa 1992).

(iii) While the tropopause (the boundary between the troposphere and the stratosphere, below which the temperature decreases with increasing altitude; Sections 3.1, 15.1, and Fig. 14.1) is at about 16 km altitude in summer, it is only at about 10 km altitude in winter, limiting the vertical extent of convective activity in winter to lower heights than in summer.

(iv) The $-10\ °C$ temperature level, where the charge-separation process is thought to be most efficient

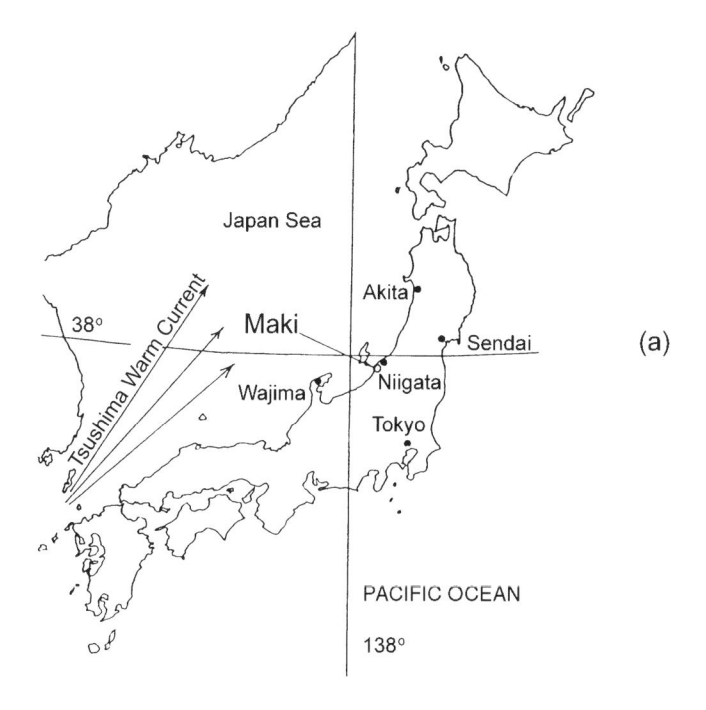

8.3. Evolution of winter thunderclouds

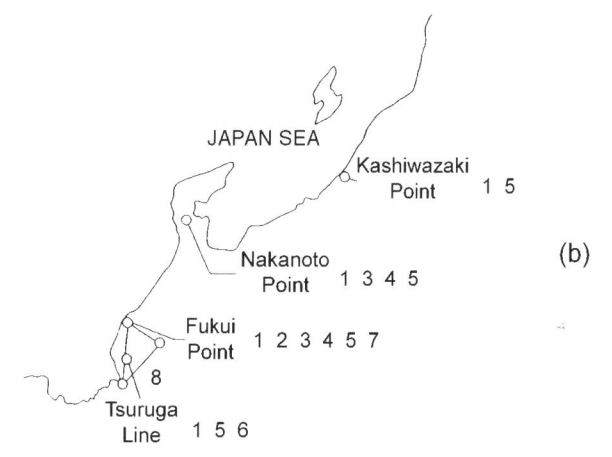

Fig. 8.1. Locations of the lightning observation stations in the coastal area of the Sea of Japan. (a) Map of Japan showing the position of the 150-m-high meteorological tower at Maki used for lightning research by Narita et al. (1989) and Goto and Narita (1992, 1995), (b) part of the northwestern coast of Honshu Island facing the Sea of Japan showing positions of the various observation systems used for lightning research by Nakahori et al. (1982), Miyake et al. (1990, 1992), Yokoyama et al. (1990), and Suzuki (1992). The two figures are adapted from Narita et al. (1989) and Miyake et al. (1992), respectively.

(subsection 3.2.6, second part), is closer to the Earth's surface in winter than in summer, and the velocity of updrafts at this level is very low (Kitagawa 1992). According to Michimoto (1993a), the rate of winter lightning activity is related to the altitude of the $-10\,°C$ temperature level in the following way. When the altitude of the $-10\,°C$ temperature level is higher than 1.8 km, the clouds exhibit relatively strong lightning activity. When it lies between 1.8 and 1.4 km, the clouds exhibit either weak or no lightning activity. When it is lower than 1.4 km, the clouds produce no natural lightning discharges.

Hojo et al. (1989), using data from a lightning locating system employing magnetic direction finding (Section 17.3) in a coastal area of the Sea of Japan, observed that in winter the lightning activity does not move inland farther than 20–30 km and that most of the flashes have their ground terminations on the sea surface.

8.3. Evolution of winter thunderclouds

Kitagawa and Michimoto (1994), using data from a network of electric field mills and from radar observations, inferred the electrical structure of winter clouds at

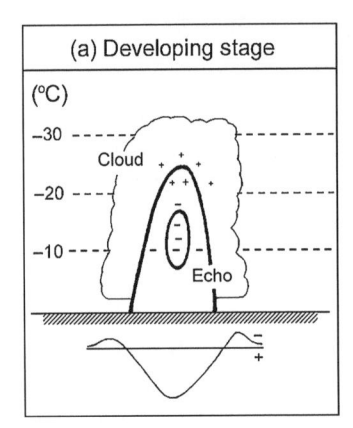

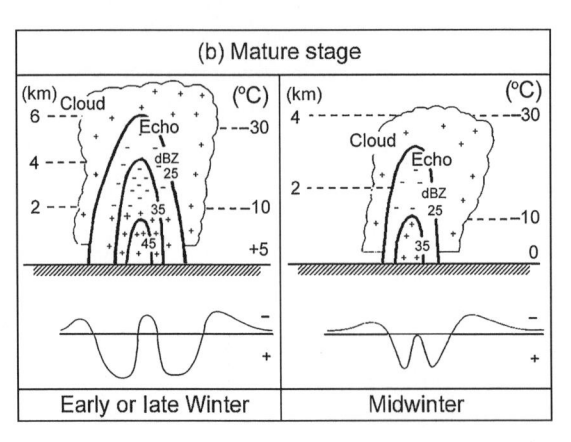

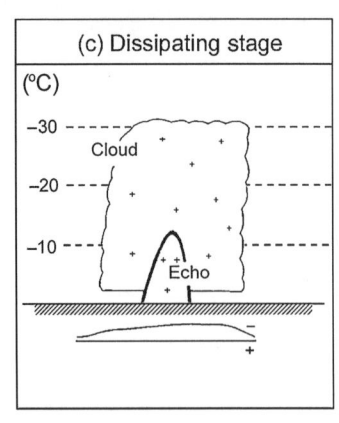

Fig. 8.2. Schematic representation of the stages of the evolution of winter thunderclouds in Japan for the case when the −10 °C temperature level is higher than 1.4 km. (a) Developing, with dipolar charge structure, (b) mature, with tripolar charge structure, and (c) dissipating, with monopolar charge structure. The radar echo levels are indicated by the heavy solid lines. Ground level is indicated by the line with hatching. Below each figure is shown (qualitatively) the corresponding temporal variation in the electric field (physics sign convention; subsection 1.4.2) at ground. Adapted from Kitagawa and Michimoto (1994).

three stages of their development for the case where the altitude of the −10 °C temperature level is higher than 1.4 km. These three stages, developing, mature, and dissipating, are illustrated in Figs. 8.2a, b, c, respectively, and are discussed below. Note that the description of charge-structure evolution offered by Kitagawa and Michimoto (1994) is inferred from remote observations (subsection 3.2.2) and is in need of confirmation by *in situ* measurements (subsection 3.2.3).

(i) During the developing stage (Fig. 8.2a), the electric charge distribution is a positive dipole (subsection 3.2.1, second part), the net positive charge being found in the upper portion of the cloud and the net negative charge (carried by graupel) being distributed between temperature levels −20 °C and −10 °C. This dipolar charge structure was inferred by Kitagawa and Michimoto (1994) from their observation that when a cloud at this stage passed over an electric field mill the temporal variation of electric field shown in Fig. 8.2a was observed. The production of such an electric field variation by a positive dipole was discussed in subsection 3.2.1.

(ii) During the mature stage (Fig. 8.2b), winter thunderclouds exhibit a tripolar electrical charge structure, but only for a short period of time. The overall period characterized by dipolar or tripolar charge structure is less than 10 minutes in early or late winter and less than several minutes in midwinter. The lower positive charge appears at heights below the −10 °C temperature level, this charge being carried close to the Earth's surface by falling graupel. Sometimes this lower positive space charge is apparently capable of initiating ground flashes. Figure 8.2b represents clouds in their mature stage in early or late winter (left-hand panel) and in midwinter (right-hand panel). The tripolar charge structure was inferred by Kitagawa and Michimoto (1994) from their observation that when a cloud system at this stage passed over an electric field mill, the temporal variations of electric field shown in Fig. 8.2b were observed. The production of such electric field variations by three vertically stacked charges (two positive, at the top and the bottom, and one negative, in the middle) is discussed in subsection 3.2.2. Takahashi *et al.* (1999), from their *in situ* measurements of charges on graupel and snow crystals in Hokuriku winter cumuli, reported that the reversal in sign of the charge on graupel occurred at the −11 °C level, promoting formation of the tripolar charge structure.

(iii) During the dissipating stage (see Fig. 8.2c), positive charge is dominant in the whole cloud, this charge being carried by ice crystals and snowflakes. This stage, characterized by a positive monopolar charge structure, has a much longer duration than the preceding two stages and thus accounts for most of the lifetime of winter thunderclouds. The monopolar positive charge structure was inferred by Kitagawa and Michimoto (1994) from their observation that the electric field under clouds at this stage had a relatively constant and low negative (physics sign convention; subsection 1.4.2) value, as indicated in Fig. 8.2c.

Perhaps the most important point in the above description of the evolution of winter thunderclouds in Japan is that graupel particles, which are thought to be both the main carriers of negative charge and also the carriers of the lower positive charge, are present in the cloud for only a relatively short period of time. As a result, the lifetime of the dipolar (Fig. 8.2a) or tripolar (Fig. 8.2b) charge structure is very short, as noted above, usually less than 10 minutes in early or late winter and less than several minutes in midwinter. Note that the description of charge structure evolution offered by Kitagawa and Michimoto

8.4. Characteristics of natural winter lightning

Table 8.1. *Observed polarity of currents in winter lightning in Japan. N_+, N_-, and $N_\pm$ are the numbers of positive, negative, and bipolar currents, corresponding percentages being given in parentheses*

Source	Strike object	Observation period	N_+ (percent)	N_- (percent)	$N_\pm$ (percent)	Total (percent)
Miyake et al. (1992)	88-m weather observation tower at Kashiwazaki and 200-m stack at Fukui	1978–86	41 (33)	78 (62)	6[a] (5)	125 (100)
Goto and Narita (1995)	150-m meteorological tower at Maki	1982–93?	25 (17)	91 (63)	29 (20)	145 (100)
Wada et al. (1996a, b)	200-m stack at Fukui	1989–94	4 (9)	36 (80)	5[b] (11)	45 (100)
Nagai et al. (1996)	500-kV Genden-Tsuruga transmission line tower	1986–92	1 (4)	15 (63)	8 (33)	24 (100)

[a] Five changes from negative to positive and one change from positive to negative.
[b] Changes from negative to positive (Wada et al. 1996a).

(1994) is apparently not consistent with the charge structure of winter thunderclouds in Japan depicted in Fig. 3.7, as discussed later in this section.

For the case where the altitude of the $-10\,°C$ temperature level is between 1.8 and 1.4 km, Kitagawa and Michimoto (1994) observed that occasionally winter thunderclouds temporarily exhibited a tripolar electrical charge structure and yet produced no lightning discharges. When the $-10\,°C$ level was lower than 1.4 km, the charge separation in winter convective clouds resulted in a positive-dipole charge distribution, the negative charge being localized in a very small portion of cloud volume, even in the mature stage. The temporal variations in the electric field shown in Fig. 8.2b were not observed and, as noted in Section 8.2, no lightning was produced.

Takeuti et al. (1978) and Brook et al. (1982) inferred, from multiple-station measurements of electric field changes due to lightning discharges (subsection 3.2.2), that the charge structure in winter thunderclouds in Japan is a positive dipole. They also suggested that the frequent occurrence of positive flashes in winter storms in Japan can be explained by the effect of vertical wind shear (see Fig. 3.7 and Section 5.2). Kitagawa (1992), however, argued that the dipolar or tripolar charge distribution exists for too short a period of time for this to be the explanation and that the tilted-dipole cloud charge structure model is not confirmed by the radar observations of Michimoto (1991). Interestingly, Hojo et al. (1989) found a negative correlation between the monthly percentage of positive flashes and the average radar echo height in Japan, that is, the percentage of positive flashes tends to increase as the cloud depth decreases. This tendency could be due to an increased occurrence of positive flashes, and/or a decreased occurrence of negative flashes.

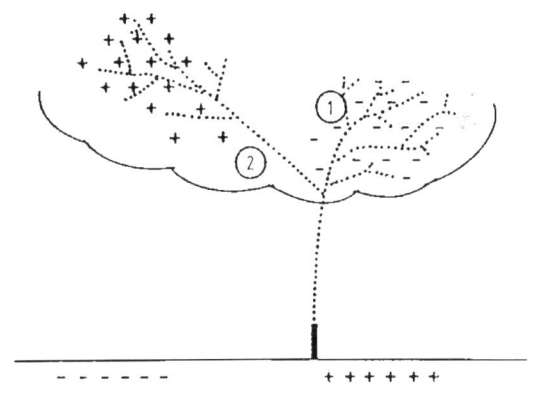

Fig. 8.3. Diagram illustrating a possible explanation of observed bipolar lightning currents. The negative charge source is labeled 1, and the positive charge source is labeled 2. See also Fig. 8.9, Section 5.4. Adapted from Narita et al. (1989).

8.4. Characteristics of natural winter lightning

The characteristics of winter lightning in Japan, some of which are also observed in other geographic locations while others appear to be unique, include the following.

(i) The occurrence of lightning flashes is much lower than in summer thunderstorms. Storms that exhibit only a few lightning flashes, sometimes referred to as "single-flash storms", are commonly observed (Michimoto 1993a). A relatively low occurrence of lightning in cold-season thunderstorms has been observed in other geographical locations also (subsection 2.8.1 and Section 5.2).

(ii) The incidence of lightning strikes to tall structures relative to that for summer storms is greater (e.g., Suzuki 1992; Miyake et al. 1990).

(iii) A very large percentage, 98 percent according to Miyake et al. (1990), of upward lightning discharges, even

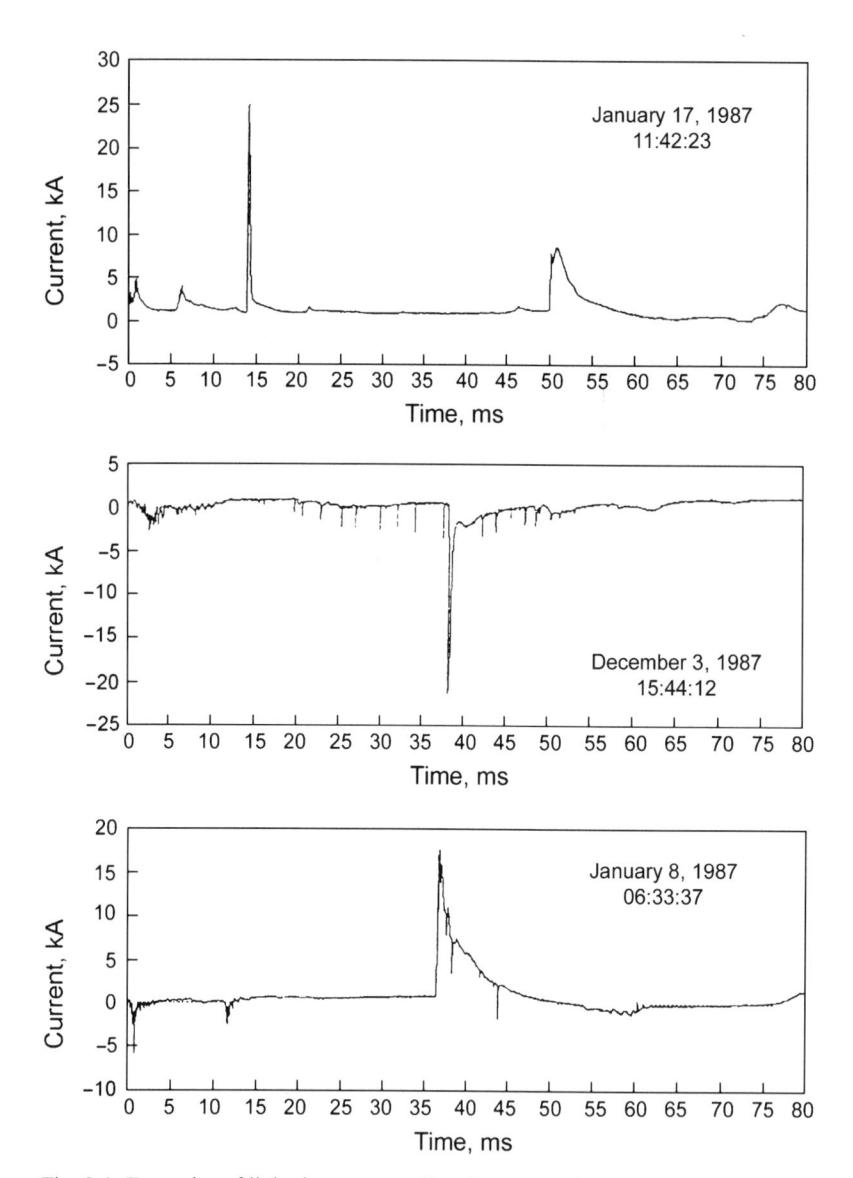

Fig. 8.4. Examples of lightning currents directly measured at the 150-m-high meteorological tower at Maki (see Fig. 8.1a). Note that the middle and bottom current waveforms are bipolar. Adapted from Goto and Narita (1995).

when the strike object is of moderate height (some tens of meters) and is located on relatively flat terrain (e.g., Suzuki 1992; Ishii and Shindo 1995). In other locations, upward lightning is characteristic of strike objects with actual or effective heights (Section 6.1) in excess of 100 m or so (subsection 2.9.1). Fuchs et al. (1998) reported that all flashes (a total of 118) observed in both cold and warm seasons on the 160-m-high Peissenberg tower in Germany were of the upward type.

(iv) The occurrence of lightning discharges apparently initiated nearly simultaneously by upward leaders from two or more grounded structures is relatively frequent (e.g., Miyake et al. 1990; Suzuki 1992). According to Miyake et al. (1990), up to 20 percent of strikes to transmission line towers involve more than one tower, these towers being separated by distances from some hundreds of meters to more than 2 km. Such multiple discharges, sometimes called concurrent flashes, have also been observed to occur from tall objects in summer (e.g., McEachron 1939; Berger and Vogelsanger 1968; Kitterman 1981). Multiple upward discharges are likely to be initiated by the same in-cloud discharge, whose horizontal extent largely determines the number of, and the distances between, the objects involved. Predominantly horizontal lightning discharges have been observed to extend over more than 40 km in summer storms (Section 5.2).

(v) The percentage of ground flashes that lower positive charge to ground is relatively high, about 33 percent,

8.4. Characteristics of natural winter lightning

while the percentage of positive lightning in summer storms in Japan is only 10 percent (Suzuki 1992). A higher percentage of positive lightning flashes in winter than in summer has been also observed in other geographical locations (Section 5.2). There is, however, a considerable variation in the percentage of positive lightning striking tall objects in winter in Japan, as illustrated in Table 8.1, where this percentage ranges from 4 to 33 percent. Interestingly, Ishii *et al.* (1998) observed that subsequent strokes in multiple-stroke positive flashes in winter storms in Japan always create a new termination on ground.

(vi) Reversal of the polarity of charge transferred to ground during one flash is relatively frequent (e.g., Nakahori *et al.* 1982; Narita *et al.* 1989; Miyake *et al.* 1992; Goto and Narita 1995). The frequency of occurrence of bipolar flashes in winter in Japan is summarized in Table 8.1. Narita *et al.* (1989) suggested that in bipolar lightning events, currents of both polarities follow the same channel to ground, but from different, oppositely charged regions in the cloud, as illustrated in Fig. 8.3 (see also Fig. 8.9). Further discussion of bipolar flashes in both winter and summer storms is found in Section 5.4.

(vii) There is a relatively large variety of current waveshapes (Miyake *et al.* 1992; Goto and Narita 1995). Examples of directly measured current waveforms in winter lightning in Japan are shown in Figs. 8.4 and 8.5. Many of these waveforms, bipolar or unipolar, appear to lack any regularity. Somewhat similar waveforms were reported for summer lightning discharges to the Empire State Building (see Figs. 14, 17, and 19 of McEachron 1939 and Fig. 3 of McEachron 1941) and to the towers on Mount San Salvatore (see Figs. 17 and 18 of Berger and Vogelsanger 1965 and Fig. 6b of Berger 1978). However, the fraction of "unusual" current waveforms from studies of winter lightning in Japan is considerably larger than that from any study of summer or winter lightning in other geographical locations. Beierl (1991, 1992), from measurements on the Peissenberg tower in Germany, reported that the microsecond-scale current waveforms for winter and summer lightning were different. Specifically, the current waveforms in winter tended to have longer and more complex initial rising portions than their summer counterparts. Ishii and Hojo (1989) observed that return-stroke radiation field waveforms in winter storms in Japan had longer slow fronts (subsection 4.6.3) and smaller widths than did their summer counterparts.

(viii) The occurrence of very large, slowly varying (millisecond-scale) lightning currents, up to tens of kiloamperes, is relatively frequent. These currents, of either polarity, examples of which are found in Figs. 8.4 and 8.5, are two to three orders of magnitude higher than the steady currents (initial continuous currents if they occur prior to the first-leader–return-stroke sequence or

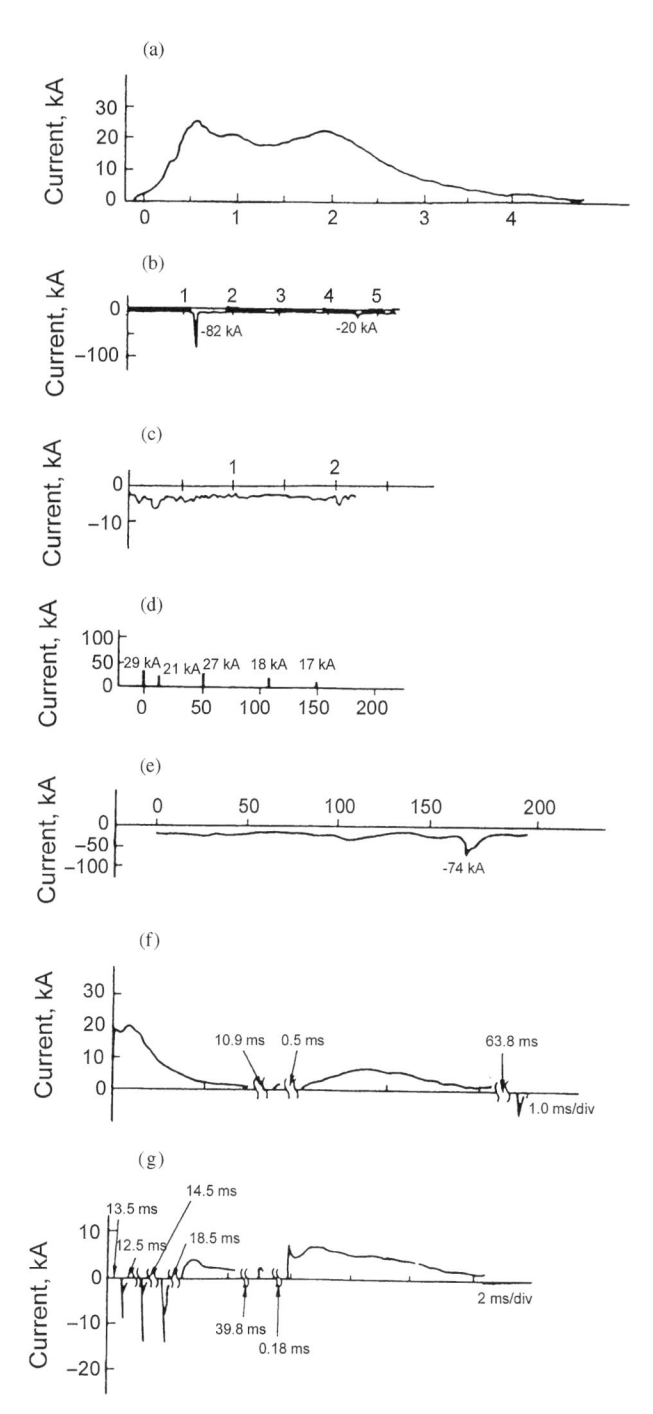

Fig. 8.5. Examples of lightning currents directly measured at the 88-m-high weather observation tower at Kashiwazaki and the 200-m-high stack at Fukui (see Fig. 8.1b). (a) Positive long-duration wave; (b) negative pulse; (c) negative long-duration wave; (d) positive pulse; (e) negative long-duration wave with pulse; (f) positive wave followed by a negative pulse; (g) negative pulse followed by positive wave. The time scales for (a)–(c) are in ms and for (d), (e) are in μs. Note that traces (f) and (g) exhibit polarity changes. Adapted from Miyake *et al.* (1992).

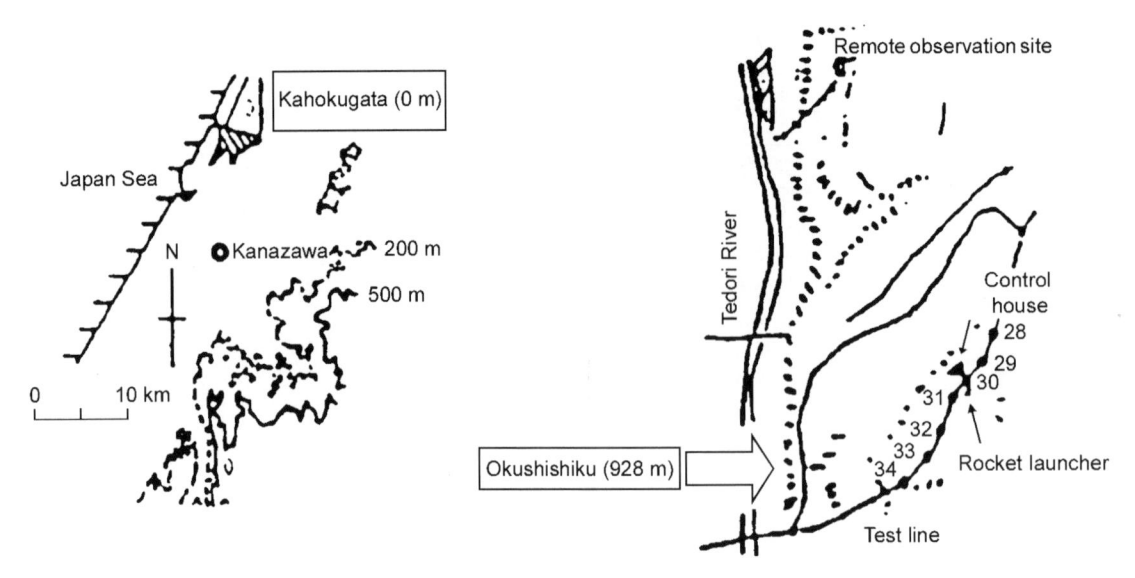

Fig. 8.6. Locations and heights above sea level of two lightning triggering sites in Japan. The Kahokugata site (left) was used from 1977 to 1985 and the Okushishiku site (right) was used from 1986 to 1998. Adapted from Yoda et al. (1997).

continuing currents if they are initiated by return strokes) found in negative lightning discharges observed elsewhere (Figs. 4.52, 6.2c, 6.4b, 6.5, and 7.8a). Charge transfers in excess of 1000 C have been reported (Miyake et al. 1992).

One possible explanation for the above features of winter lightning in Japan is the proximity of the cloud charge to the ground and the relatively large horizontal extent of that charge. The latter conjecture is invoked to explain the observed large charge transfers from relatively shallow clouds. Davis and Standring (1947), from their measurements of currents in the cables of kite balloons flying at a height of 600 m under thunderstorm conditions, inferred the presence of local charged regions at heights less than 1 km above ground in summer thunderstorms in England. In winter in Japan, the upper extremities of taller objects are sometimes located inside the cloud, since the cloud base (its lower visible boundary) is usually at 200 to 800 m above sea level (e.g., Goto and Narita 1992). Under such conditions, discharges may occur inside a cloud penetrated from the bottom by a grounded metallic object. One can speculate that this mode of discharge differs from the "classical" lightning discharge to ground, which typically involves a few kilometers of clear air between the cloud base and the ground strike point, although this gap can contain heavy precipitation.

In the remainder of this section we consider some differences between winter and summer lightning inferred from electric field measurements in the United States, these differences possibly being relevant to lightning activity in Japan. Brook (1992) observed a tendency for the initial breakdown pulses in the negative flashes (subsection 4.3.2)

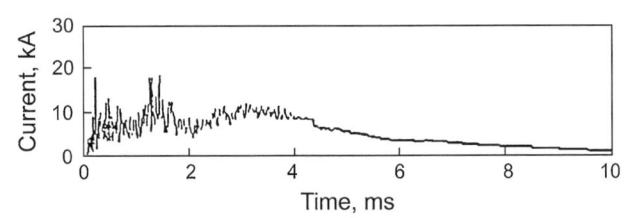

Fig. 8.7. Current of a positive lightning discharge (number 87-05) triggered in winter at the Okushishiku site, Japan. The discharge terminated at the lightning rod at the top of the transmission line tower. Adapted from Nakamura et al. (1991).

to be larger in winter than in summer. Also, the negative stepped leaders tended to have shorter duration (estimated from electric field records as the time interval between the initial-breakdown pulses and the return-stroke pulse) in winter than in summer. No similar seasonal variations were observed for positive lightning. Brook (1992) inferred from his observations that negative stepped leaders in winter were faster than in summer and that the electric field in winter thunderclouds was higher than in summer thunderclouds.

8.5. Rocket-triggered lightning in winter

A variety of experiments involving the artificial initiation of lightning from natural thunderclouds using the rocket-and-wire technique (Section 7.2) has been performed in Japan since 1977 (e.g., Miyachi et al. 1979; Miyachi and Horii 1982; Horii 1982). These experiments have been conducted only in winter. From 1977 to 1985, the lightning-triggering site was located at Kahokugata on

8.5. Rocket-triggered lightning in winter

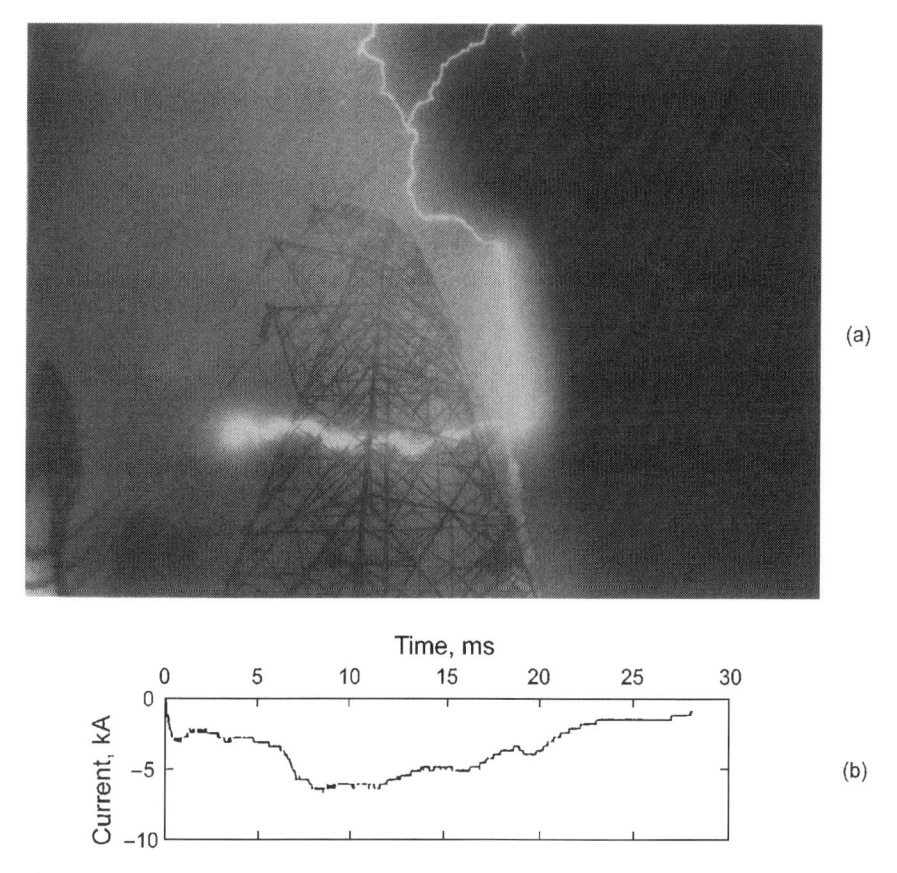

Fig. 8.8. An example of a negative lightning discharge (number 87-08) triggered in winter at the Okushishiku site, Japan. (a) A still photograph of the discharge, which terminated on the ground wire 25 m from the transmission line tower top, (b) the current through the ground wire (on one side of the strike point). Adapted from Nakamura et al. (1991).

reclaimed land bordering the Sea of Japan (see Fig. 8.6). A total of 71 flashes were triggered at that site (e.g., Nakamura et al. 1991). In 1986, the lightning-triggering site was moved to the top of Okushishiku mountain, 930 m above sea level, and was operated through 1998. Since the heights of the visible cloud bases in winter are low, the triggering site was often within the clouds (e.g., Nakano 1996). The Okushishiku site was equipped with an unenergized test transmission line whose six phase conductors and one ground wire were supported by seven towers, numbered 28 through 34, as illustrated in Fig. 8.6. The distance between the terminal towers, 28 and 34, was 2103 m (Nakamura et al. 1991). Several rocket launchers were set 10 to 15 m from the base of tower 30, which was 59.4 m in height and had a 2.6-m high lightning rod on its top (Nakamura et al. 1991, 1997). Yoda et al. (1997) reported that a total of 90 flashes were triggered at the Okushishiku site, and Nakamura et al. (1997) stated that 31 flashes were triggered using the altitude-triggering technique (subsection 7.2.1).

As for natural lightning in winter, a large variety of current waveforms, including negative, positive, and bipolar, was observed. In the following, we summarize the observations and try to identify some common features.

(i) It appears that classical rocket-triggered lightning (subsection 7.2.1) in winter in Japan is usually composed of an initial stage only, that is, it does not contain following downward-leader–upward-return-stroke sequences. Altitude-triggered lightning (Nakamura et al. 1997) does contain the form of the leader–return-stroke sequence that occurs during its initial stage, as observed in other triggered-lightning experiments (subsections 7.2.1 and 7.2.3).

(ii) The percentage of triggered flashes that are identified as positive ranges from about 36 to 74 percent. For 96 flashes triggered from 1977 to 1988, Nakamura et al. (1989) assigned negative polarity to 54 events and positive to 30 events. Of 31 altitude-triggered flashes presented by Nakamura et al. (1997), five were negative and 14 positive (the current for the remaining 12 was not measured). Nakano (1996) reported that of 81 flashes triggered in 1986–95, 18 were identified as negative and 38 as positive. He also gave values of the maximum measured currents, -77 kA for negative lightning and $+132$ kA for positive lightning, and the maximum values of charge transfer, -243 C and

+640 C, respectively. The statistical distribution for positive peak currents is characterized by a larger dispersion than that for negative peak currents.

(iii) Yoda *et al.* (1997) reported that positive lightning current waveforms exhibit oscillations or noise superimposed on the front and a relatively smooth tail, as illustrated in Fig. 8.7. The overall duration of the waveforms is more than 10 ms, and the average current magnitude is of the order of 10 kA, in the peak current range of return strokes. The resultant charge transfer is often greater than 100 C.

(iv) There are fewer examples found in the literature of currents from negative rocket-triggered lightning flashes than from positive ones. An example of such a current waveform from negative lightning is given by Nakamura *et al.* (1991, 1992) and is reproduced in Fig. 8.8a. The corresponding charge transfer was more than 100 C. This flash terminated on the ground wire of the Okushishiku transmission line and melted several strands of this wire. The current shown in Fig. 8.8b was measured on one side of the strike point. Another example (accompanied by a streak photograph and a low-level current record) is given in Nakamura *et al.* (1989, Fig. 8) and in Horii (1982, Fig. 32). This latter current had a peak of about 3 kA and a duration of about 400 ms.

(v) Horii (1982, Fig. 42; 1986, Fig. 4), Akiyama *et al.* (1985, Fig. 8b), and Nakamura *et al.* (1989, Fig. 14) gave examples of bipolar current waveforms; the waveform from Horii (1982, 1986) and Akiyama *et al.* (1985) is reproduced in Fig. 8.9b. Bipolar waveforms, from both triggered- and natural-lightning experiments, are discussed in Section 5.4.

8.6. Summary

Lightning discharges produced by winter storms in the coastal area of the Sea of Japan exhibit a number of features that have not been observed in the summer season in Japan or in any season in other geographical locations. These features include, but are not limited to, a relatively high frequency of occurrence of (i) flashes that lower both positive and negative charges to ground, (ii) upward flashes, even where the strike object is of only moderate height, and (iii) very large, slowly varying lightning currents of either polarity. Charge transfers in excess of 1000 C have been observed. Many of the features of winter lightning in Japan are apparently related to the proximity of the cloud charge to the ground and the large horizontal extent of that charge. At the time of writing, there is no consensus regarding the evolution of the cloud charge distribution in winter storms in Japan; *in situ* measurements are needed to resolve the controversy. Classical rocket-triggered lightning in Japan is usually composed of the initial stage only, that is, it does not contain downward-leader–upward-return-stroke sequences. As for natural winter lightning in Japan, the percentage of triggered events lowering positive charge to ground is relatively high. It is possible that conditions conducive to the production of lightning discharges similar to those observed in winter storms in Japan also exist in other seasons and/or locations, so that further lightning research may lead to replacement of the term "winter lightning in Japan" by a more descriptive term.

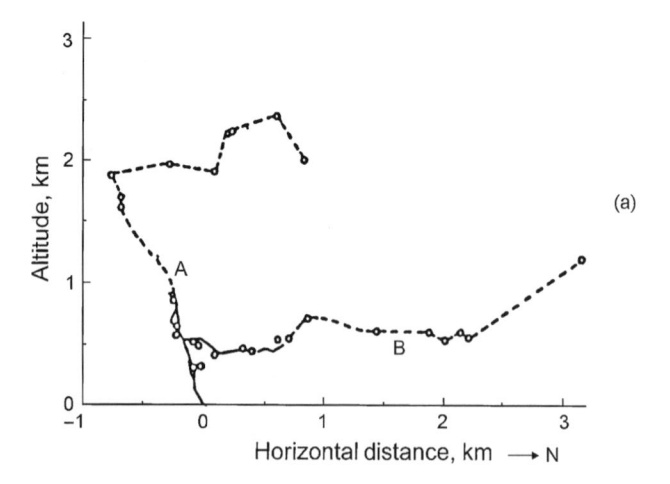

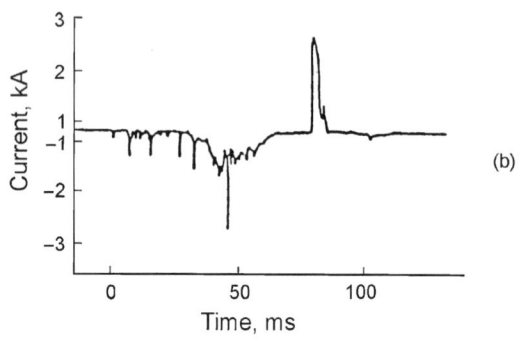

Fig. 8.9. An example of bipolar lightning (number 81-11) triggered in winter at the Kahokugata site, Japan. (a) The discharge channels, photographically observed (solid line) or reconstructed from three-station acoustic measurements (broken line); (b) the current to ground. The negative current was supplied by branch a and the positive current, which follows, by branch b (see also Fig. 8.3). Adapted from Horii (1982, 1986) and Akiyama *et al.* (1985).

References and bibliography for Chapter 8

Akiyama, H., Ichino, K., and Horii, K. 1985. Channel reconstruction of triggered lightning flashes with bipolar currents from thunder measurements. *J. Geophys. Res.* **90**: 10 674–80.

Asakawa, A., Miyake, K., Yokoyama, S., Shindo, T., Yokota, T., Sakai, T. 1997. Two types of lightning discharges to a high stack on the coast of the Sea of Japan in winter. *IEEE Trans. Pow. Del.* **12**: 1222–31.

Beierl, O. 1991. Lightning current measurements at the Peissenberg tower. In *Proc. 7th Int. Symp. on High Voltage Engineering, Dresden, Germany*, paper 81.02.

Beierl, O. 1992. Front shape parameters of negative subsequent strokes measured at the Peissenberg tower. In *Proc. 21st Int. Conf. on Lightning Protection, Berlin, Germany*, pp. 19–24.

Berger, K. 1967. Novel observations on lightning discharges: results of research on Mount San Salvatore. *J. Franklin Inst.* **283**: 478–525.

Berger, K. 1978. Blitzstrom-Parameter von Aufwartsblitzen. *Bull. Schweiz. Elektrotech. Ver.* **69**: 353–60.

Berger, K., and Vogelsanger, E. 1965. Messungen und Resultate der Blitzforschung der Jahre 1955–1963 auf dem Monte San Salvatore. *Bull. Schweiz. Elektrotech. Ver.* **56**: 2–22.

Berger, K., and Vogelsanger, E. 1968. New results of lightning observations. In *Proc. Int. Conf. on Large High Tension Electric Systems (CIGRE) Paris, France*, paper 33-03, 11 pp.

Brook, M. 1992. Breakdown electric fields in winter storms. *Res. Lett. Atmos. Electr.* **12**: 47–52.

Brook, M., Nakano, M., Krehbiel, P., and Takeuti, T. 1982. The electrical structure of the Hokuriku winter thunderstorms. *J. Geophys. Res.* **87**: 1207–15.

Brook, M., Krehbiel, P., MacLaughlin, D., Takeuti, T., and Nakano, M. 1983. Positive ground stroke observations in Japanese and Florida storms. In *Proc. Int. Conf. on Atmospheric Electricity*, eds. J. Latham and L. Ruhnke, pp. 365–9, Hampton, Virginia: Deepak.

Brook, M., Henderson, R.W., and Pyle, R.B. 1989. Positive lightning strokes to ground. *J. Geophys. Res.* **94**: 13 295–303.

Cherrington, M., Breed, D.W., Yarnell, P.R., and Smith, W.E. 1998. Lightning injuries during snowy conditions. *Br. J. Sports Med.* **32**: 333–5.

Davis, R., and Standring, W.G. 1947. Discharge currents associated with kite balloons. *Proc. Roy. Soc. A***191**: 304–22.

Engholm, C.D., Williams, E.R., and Dole, R.M. 1990. Meteorological and electrical conditions associated with positive cloud-to-ground lightning. *Mon. Wea. Rev.* **118**: 470–87.

Fuchs, F., Landers, E.U., Schmid, R., and Wiesinger, J. 1998. Lightning current and magnetic field parameters caused by lightning strikes to tall structures relating to interference of electronic systems. *IEEE Trans. Electromagn. Compat.* **40**: 444–51.

Funaki, K., Kitagawa, N., Nakano, M., and Takeuti, T. 1981. Multistation measurements of lightning discharges to ground in Hokuriku winter thunderstorms. *Res. Lett. Atmos. Electr.* **1**: 19–26.

Goodman, S.J., Buechler, D.E., Knupp, K., Driscoll, K., and McCaul, E.W. 2000. The 1997–98 El Nino event and related wintertime lightning variations in southeastern United States. *Geophys. Res. Lett.* **27**: 541–4.

Goshima, H., Motoyama, H., Asakawa, A., Wada, A., Shindo, T., and Yokoyama, S. 2000. Characteristics of electromagnetic fields due to winter lightning stroke current to a high stack. *Trans. IEE Japan* **120-B**(1): 44–8.

Goto, Y., and Narita, K. 1991. The meteorological conditions for the occurrence of winter thunderstorms. *Res. Lett. Atmos. Electr.* **11**: 61–9.

Goto, Y., and Narita, K. 1992. Observations of winter lightning to an isolate tower. *Res. Lett. Atmos. Electr.* **12**: 57–60.

Goto, Y. and Narita, K. 1995. Electrical characteristics of winter lightning. *J. Atmos. Terr. Phys.* **57**: 449–59.

Hobara, Y., Iwasaki, N., Hayashida, T., Hayakawa, M., Ohta, K., and Fukunishi, H. 2001. Interrelation between ELF transients and ionospheric disturbances in association with sprites and elves. *Geophys. Res. Lett.* **28**(5): 935–8.

Hojo, J., Ishii, M., Kawamura, T., Suzuki, F., and Funayama, R. 1985. The fine structure in the field change produced by positive ground strokes. *J. Geophys. Res.* **90**: 6139–43.

Hojo, J., Ishii, M., Kawamura, T., Suzuki, F., Komuro, H., and Shiogama, M. 1988. Characteristics and evaluation of lightning field waveforms. *Electr. Eng. Japan* **108**: 55–65.

Hojo, J., Ishii, M., Kawamura, T., Suzuki, F., Komuro, H., and Shiogama, M. 1989. Seasonal variation of cloud-to-ground lightning flash characteristics in the coastal area of the Sea of Japan. *J. Geophys. Res.* **94**: 13 207–12.

Holle, R.L., and Watson, A.I. 1996. Lightning during two central U.S. winter precipitation events. *Wea. Forecast.* **11**: 600–14.

Holle, R.L., Lopez, R.E., Howard, K.W., Cummins, K.L., Malone, M.D., and Krider, E.P. 1997. An isolated winter cloud-to-ground lightning flash causing damage and injury in Connecticut. *Bull. Am. Meteor. Soc.* **78**: 437–41.

Holle, R.L., Cortinas, J.V., and Robbins, C.C. 1998. Winter thunderstorms in the United States. In *Proc. 16th Conf. on Weather Analysis and Forecasting, Phoenix, Arizona*, pp. 298–300.

Horii, K. 1982. Experiment of artificial lightning triggered with rocket. *Mem. Faculty Eng., Nagoya Univ., Nagoya, Japan* **34**: 77–112.

Horii, K. 1986. Experiment of triggered lightning discharge by rocket. In *Proc. Korea–Japan Joint Symp. on Electrical Material and Discharge, Cheju Island, Korea*, paper 8-3, 6 pp.

Horii, K., and Ikeda, G. 1985. A consideration on success conditions of triggered lightning. In *Proc. 18th Int. Conf. on Lightning Protection, Munich, Germany*, paper 1-3, 6 pp.

Horii, K., and Sakurano, H. 1985. Observation on final jump of the discharge in the experiment of artificially triggered lightning. *IEEE Trans. Power Appar. Syst.* **PAS-104**: 2910–17.

Hunter, S.M., Underwood, S.J., Holle, R.L., and Mote, T.L. 2001. Winter lightning and heavy frozen precipitation in the southeast United States. *Wea. Forecast.* **16**: 478–90.

Inoue, S., and Kanao, S-I. 1996. Observation and analysis of multiple-phase grounding faults caused by lightning. *IEEE Trans. Pow. Del.* **11**: 353–60.

Ishii, M., and Hojo, J.I. 1989. Statistics on fine structure of cloud-to-ground lightning field waveforms. *J. Geophys. Res.* **94**: 13 267–74.

Ishii, M., and Shindo, T. 1995. Recent topics on research of lightning and lightning protection in Japan. *Elektrotechnik und Informationstechnik* **112**(6): 269–72.

Ishii, M., Shimizu, K., Hojo, J., and Shinjo, K. 1998. Termination of multiple-stroke flashes observed by electromagnetic field. In *Proc. 24th Int. Conf. on Lightning Protection, Birmingham, UK*, pp. 11–16.

Kawasaki, Z.-I., and Mazur, V. 1992a. Common physical processes in natural and triggered lightning in winter storms in Japan. *J. Geophys. Res.* **97**: 12 935–45.

Kawasaki, Z.-I., and Mazur, V. 1992b. Common physical processes in natural and triggered lightning in winter storms in Japan. *Res. Lett. Atmos. Electr.* **12**: 61–70.

Kawasaki, Z.-I., Nakano, M., Takeuti, T., Nagatani, M., Nakada, H., Mizuno, Y., and Nagai, T. 1987. Fourier spectra of positive lightning fields during winter thunderstorms. *Res. Lett. Atmos. Electr.* **7**: 29–34.

Kawasaki, Z.-I., Kanao, T., Matsuura, K., Nakano, M., Horii, K., and Nakamura, K-I. 1991. The electric field changes and UHF radiations caused by the triggered lightning in Japan. *Geophys. Res. Lett.* **18**: 1711–14.

Kitagawa, N. 1992. Charge distribution of winter thunderclouds. *Res. Lett. Atmos. Electr.* **12**: 143–53.

Kitagawa, N., and Michimoto, K. 1994. Meteorological and electrical aspects of winter thunderclouds. *J. Geophys. Res.* **99**: 10 713–21.

Kito, Y., Horii, K., Higashiyama, Y., and Nakamura, K. 1985. Optical aspects of winter lightning discharges triggered by the rocket-wire technique in Hokuriku district of Japan. *J. Geophys. Res.* **90**: 6147–57.

Kitterman, C.G. 1981. Concurrent lightning flashes on two television transmission towers. *J. Geophys. Res.* **86**: 5378–80.

Krehbiel, P.R. 1981. An analysis of the electric field change produced by lightning. Ph.D. thesis, University of Manchester Institute of Science and Technology, Manchester, UK (available as Report T-11, Geophys. Res. Ctr., New Mexico Inst. Mining and Tech., Socorro, NM 87801, 1981).

Krider, E.P., and Wetmore, R.H. 1987. Upward streamers produced by a lightning strike to radio transmission towers. *J. Geophys. Res.* **92**: 9859–62.

Lopez, R.E., Holle, R.L., Watson, A.I., and Skindlov, J. 1997. Spatial and temporal distributions of lightning over Arizona from a power utility perspective. *J. Appl. Meteor.* **36**: 825–31.

Maekawa, Y., Fukao, S., Sonoi, Y., and Yoshino, F. 1993. Distribution of ice particles in wintertime thunderclouds detected by a C band dual polarization radar: a case study. *J. Geophys. Res.* **98**: 16 613–22.

Magono, C. 1980. *Thunderstorms*, 261 pp., Elsevier: Amsterdam.

Magono, C., Endoh, T., and Shigeno, T. 1983. The electrical structure of snow clouds, I: Relationship with its precipitation physics. *J. Meteor. Soc. Japan* **61**: 325–38.

Magono, C., Sakamoto, H., Endoh, T., and Taniguchi, T. 1984. The electrical structure of snow clouds, II: Vertical distribution of precipitation charge. *J. Meteor. Soc. Japan* **62**: 323–34.

Matsumoto, Y., Sakuma, O., Shinjo, K., Saiki, M., Wakai, T., Sakai, T., Nagasaka, H., Motoyama, H., and Ishii, M. 1996. Measurement of lightning surges on test transmission line equipped with arresters struck by natural and triggered lightning. *IEEE Trans. Pow. Del.* **11**: 996–1002.

McEachron, K.B. 1939. Lightning to the Empire State Building, N.Y. *J. Franklin Inst.* **227**: 147–217.

McEachron, K.B. 1941. Lightning to the Empire State Building. *Trans. AIEE* **60**: 885–9.

Michimoto, K. 1990. Relation between surface electric field and radar echo due to winter thunderclouds over the Hokuriku district. *Res. Lett. Atmos. Electr.* **10**: 17–23.

Michimoto, K. 1991. A study of radar echoes and their relation to lightning discharge of thunderclouds in the Hokuriku District, I: Observation and analysis of thunderclouds in summer and winter. *J. Meteor. Soc. Japan* **69**: 327–36.

Michimoto, K., 1993a. A study of radar echoes and their relation to lightning discharges of thunderclouds in the Hokuriku District, II: Observation and analysis of "single-flash" thunderclouds in midwinter. *J. Meteor. Soc. Japan* **71**: 195–204.

Michimoto, K. 1993b. A study of the charge distribution in winter thunderclouds by means of network recording of surface electric fields and radar observation of clouds structure in the Hokuriku District. *J. Atmos. Electr.* **13**: 33–46.

Michimoto, K. 1993c. Statistics of lightning strikes to aircraft in winter around Komatsu airbase, Japan. *J. Atmos. Electr.* **13**: 47–58.

Miyachi, I., and Horii, K. 1982. Five years' experiences on artificially triggered lightning in Japan. In *Proc. 7th Int. Conf. on Gas Discharges and Their Applications, London, UK*, pp. 468–71.

Miyachi, I., Horii, K., Muto, S. Ikeda, G., and Aiba, S. 1979. Experiment of long gap discharge by artificially triggered lightning with rocket. In *Proc. 3rd Int. Symp. on High Voltage Engineering, Milan, Italy*, paper 51.10, 4 pp.

Miyake, K., Suzuki, T., Takashima, M., Takuma, M., and Tada, T. 1990. Winter lightning on Japan Sea coast – lightning striking frequency to tall structures. *IEEE Trans. Pow. Del.* **5**: 1370–6.

Miyake, K., Suzuki, T., and Shinjou, K. 1992. Characteristics of winter lightning current on Japan Sea coast. *IEEE Trans. Pow. Del.* **7**: 1450–6.

Montandon, E., and Beyeler, B. 1994. The lightning measuring equipment on the Swiss PTT telecommunications tower at St. Chrischona, Switzerland. In *Proc. 22nd Int. Conf. on Lightning Protection, Budapest, Hungary*, paper R1c-06, 6 pp.

Motoyama, H., Shinjo, K., Matsumoto, Y., and Itamoto, N. 1998. Observation and analysis of multiphase back flashover on the Okushishiku test transmission line caused by winter lightning. *IEEE Trans. Pow. Del.* **13**: 1391–8.

Nagai, T., Matsui, T., Sonoi, Y., Adachi, M., Sekioka, S., and Sugimoto, O. 1996. Observation on winter lightning at a tall transmission line tower in Hokuriku district. In *Proc. 10th Int. Conf. on Atmospheric Electricity, Osaka, Japan*, pp. 476–9.

Nakahori, K., Egawa, T., and Mitani, H. 1982. Characteristics of winter lightning current in Hokuriku District. *IEEE Trans. Pow. Appar. Syst.* **PAS-101**: 4407–12.

Nakamura, K., Horii, K., and Aiba, S. 1989. Discharge currents in the experiment of artificially triggered lightning for winter thunderclouds. In *Proc. 1989 Int. Symp. on Electromagnetic Compatibility, Nagoya, Japan*, paper 9p2–C5, pp. 664–9.

Nakamura, K., Horii, K., Kito, V., Wada, A., Ikeda, G., Sumi, S., Yoda, M., Aiba, S., Sakurano, H., and Wakamatsu, K. 1991. Artificially triggered lightning experiments to an EHV transmission line. *IEEE Trans. Pow. Del.* **6**: 1311–18.

Nakamura, K., Horii, K., Nakano, M., and Sumi, S. 1992. Experiments on rocket triggered lightning. *Res. Lett. Atmos. Electr.* **12**: 29–35.

Nakamura, K., Horii, K., and Sakurano, H. 1997. Lightning discharge parameters by rocket triggered lightning to transmission line in winter. In *Proc. 10th Int. Symp. on High Voltage Engineering, Montreal, Quebec, Canada*, 4 pp.

Nakano, M. 1979. Initial streamer of the cloud discharge in winter thunderstorms of the Hokuriku coast. *J. Meteor. Soc. Japan* **57**: 452–8.

Nakano, M. 1996. Rocket-triggered lightning experiment in Okushishiku, Japan. In *Proc. Int. Symp. on Winter Lightning in Hokuriku, Kanazawa, Japan*, p. 56.

Nakano, M., Takagi, N., Kawasaki, Z., and Takeuti, T. 1983. Return strokes of triggered lightning flashes. *Res. Lett. Atmos. Electr.* **3**: 73–8.

Nakano, M., Nagatani, M., Nakada, H., Takeuti, T., and Kawasaki, Z. 1987. Measurements of the velocity change of a lightning return stroke with height. *Res. Lett. Atmos. Electr.* **7**: 25–8.

Nakano, M., Nagatani, M., Nakada, H., Takeuti, T., and Kawasaki, Z. 1988. Measurements of the velocity change of a lightning return stroke with height. In *Proc. 1988 Int. Aerospace and Ground Conf. on Lightning and Static Electricity, Oklahoma City, Oklahoma*, pp. 84–6.

Nakano, M., Kawasaki, Z.-I., Takeuti, T., Sirai, T., and Kawamata, Y. 1989. VHF/UHF radiation from positive and negative lightning. In *Proc. URSI Symp. on Environmental and Space Electromagnetics, Tokyo, Japan*, pp. 459–65.

Narita, K., Goto, Y., Komuro, H., and Sawada, S. 1989. Bipolar lightning in winter at Maki, Japan. *J. Geophys. Res.* **94**: 13 191–5.

Orville, R.E. 1990. Winter lightning along the east coast. *Geophys. Res. Lett.* **17**: 713–15.

Orville, R.E. 1991. Lightning ground flash density in the contiguous United States – 1989. *Mon. Wea. Rev.* **119**: 573–7.

Orville, R.E. 1993. Cloud-to-ground lightning in the Blizzard of '93. *Geophys. Res. Lett.* **20**: 1367–70.

Pierce, E.T. 1976. Winter thunderstorms in Japan – a hazard to aviation. *Naval Res. Rev.* **29**(6): 12–16.

Rakov, V.A. 1993. Data acquired with the LLP lightning locating systems. *Meteorologiya i Gidrologiya* **7**: 105–14.

Rakov, V.A. 1998. Comparison of positive and negative lightning. In *Proc. 1998 Int. Lightning Detection Conf.*, 19 pp., GAI, 2705 East Medina Road, Tucson, Arizona 85706-7155.

Rakov, V.A. 2000a. Positive and bipolar lightning discharges: a Review. In *Proc. 25th Int. Conf. on Lightning Protection, Rhodes, Greece*, pp. 103–8.

Rakov, V.A. 2001b. Positive Blitzentladungen. *ETZ Elektrotech. Autom.* **122**(5): 26–9.

Rakov, V.A., Uman, M.A., Wang, D., Rambo, K.J. Crawford, D.E. and Schnetzer, G.H. 2000a. Lightning properties from triggered-lightning experiments at Camp Blanding, Florida (1997–1999). In *Proc. 25th Int. Conf. on Lightning Protection, Rhodes, Greece*, pp. 54–9.

Sakurano, H., Matsubara, R., Murashita, N., and Kito, Y. 1989. Presumption of regional distribution of the electric field polarity on the ground induced by winter thundercloud in Hokuriku district. *Trans. IEE Japan* **109-B**: 355–60.

Schultz, D.M., Steenburgh, W.J., Trapp, R.J., Horel, J., Kingsmill, D.E., Dunn, L.B., Rust, W.D., Cheng, L., Bansemer, A., Cox, J., Daugherty, J., Jorgensen, D.P., Meitín, J., Showell, L., Smull, B.F., Tarp, K., and Trainor, M. 2002. Understanding Utah winter storms: the Intermountain precipitation experiment. *Bull. Am. Meteor. Soc.*, February, 189–210.

Shindo, T., and Ishii, M. 1995. Japanese research of triggering lightning. *Elektrotechnik und Informationstechnik* **112**(6): 265–8.

Suzukawa, M., Tomine, K., Abe, S., and Michimoto, K. 1987. Case studies on the distribution of electric field under thunderstorms in winter in the area surrounding Komatsu (in Japanese). *Tenki* **34**: 443–52.

Suzuki, T. 1992. Long term observation of winter lightning on Japan Sea coast. *Res. Lett. Atmos. Electr.* **12**: 53–6.

Takagi, N., and Takeuti, T. 1983. Oscillating bipolar electric field changes due to close lightning return strokes. *Radio Sci.* **18**: 391–8.

Takagi, N., Takeuti, T., and Nakai, T. 1986a. On the occurrence of positive ground flashes. *J. Geophys. Res.* **91**: 9905–9.

Takagi, N., Watanabe, T., Arima, I., Takeuti, T., Nakano, M., and Kinosita, H. 1986b. An unusual summer thunderstorm in Japan. *Res. Lett. Atmos. Electr.* **6**: 43–8.

Takahashi, T., Tajiri, T., and Sonoi, Y. 1999. Charges on graupel and snow crystals and the electrical structure of winter thunderstorms. *J. Atmos. Sci.* **56**: 1561–78.

Takeuchi, N., Narita, K.I., and Goto, Y. 1994. Wavelet analysis of meteorological variables under winter thunderclouds over the Japan Sea. *J. Geophys. Res.* **99**: 10 751–7.

Takeuti, T., and Nakano, M. 1977. On lightning discharges in winter thunderstorms. In *Electrical Processes in Atmospheres*, eds. H. Dolezalek and R. Reiter, pp. 614–17, Darmstadt, West Germany: Steinkopff.

Takeuti, T., Nakano, M., Nagatani, M., and Nakada, H. 1973. On lightning discharges in winter thunderstorms. *J. Meteor. Soc. Japan* **51**: 494–6.

Takeuti, T., Nakano, M., and Yamamoto, Y. 1976. Remarkable characteristics of cloud-to-ground discharges observed in winter thunderstorms in Hokuriku area, Japan. *J. Meteor. Soc. Japan* **54**: 436–9.

Takeuti, T., Nakano, M., Ishikawa, H., and Israelsson, S. 1977. On the two types of thunderstorms deduced from cloud-to-ground discharges observed in Sweden and Japan. *J. Meteor. Soc. Japan* **55**: 613–6.

Takeuti, T., Nakano, M., Brook, M., Raymond, D.J., and Krehbiel, P. 1978. The anomalous winter thunderstorms of the Hokuriku coast. *J. Geophys. Res.* **83**: 2385–94.

Takeuti, T., Israelsson, S., Nakano, M., Ishikawa, H., Lundquist, S., and Astrom, E. 1980. On thunderstorms producing positive ground flashes. *Proc. Res. Inst. Atmos. Nagoya Univ., Japan* **27-A**: 1–17.

Takeuti, T., Funaki, K., and Kitagawa, N. 1983. A preliminary report on the Norwegian winter thunderstorm observation. *Res. Lett. Atmos. Electr.* **3**: 69–72.

Takeuti, T., Kawasaki, Z., Funaki, K., Kitagawa, N., and Huse, J. 1985. On the thundercloud producing the positive ground flashes. *J. Meteor. Soc. Japan* **63**: 354–8.

Ushio, T., Kawasaki, Z.-I., Matsuura, K., and Wang, D. 1998. Electric fields of initial breakdown in positive ground flash. *J. Geophys. Res.* **103**: 14 135–9.

Wada, A., Asakawa, A., and Shindo, T. 1996a. Characteristics of lightning strokes to a 200 m high stack in winter. In *Proc. 10th Int. Conf. on Atmospheric Electricity, Osaka, Japan*, pp. 464–7.

Wada, A., Asakawa, A., and Shindo, T. 1996b. Characteristics of lightning flash initiated by an upward leader in winter. In *Proc. 23rd Int. Conf. on Lightning Protection, Florence, Italy*, pp. 85–90.

Wang, D., Rakov, V.A., Uman, M.A., Fernandez, M.I., Rambo, K.J., Schnetzer, G.H., and Fisher, R.J. 1999b. Characterization of the initial stage of negative rocket-triggered lightning. *J. Geophys. Res.* **104**: 4213–22.

Yoda, M., Miyachi, I., Kawashima, T., and Katsuragi, Y. 1992. Lightning current protection of equipments in winter. *Res. Lett. Atmos. Electr.* **12**: 117–21.

Yoda, M., Nakajima, T., Miyachi, I., Ieda, M., and Nakamura, K. 1997. Measurement of lightning current and emission in winter. In *Proc. 10th Int. Symp. on High Voltage Engineering, Montreal, Quebec, Canada*, 4 pp.

Yokoyama, S., Miyake, K., Suzuki, T., and Kanao, S. 1990. Winter lightning on Japan Sea coast – development of measuring system on progressing feature of lightning discharge. *IEEE Trans. Pow. Del.* **5**: 1418–25.

Zundl, T. 1994b. First results of the coordinated lightning current and LEMP measurements at the Peissenberg tower. In *Proc. 22nd Int. Conf. on Lightning Protection, Budapest, Hungary*, paper R1c-09, 6 pp.

9 Cloud discharges

> One flash produced five short radial streamers; in another flash, pulse sources were distributed almost randomly, like raisins in fruit cake.... Two flashes produced streamers that appeared to encircle their origins several times.... Only two flashes produced only one stepped leader channel each; 19 extended along multiple paths.
>
> D.E. Proctor (1997)

9.1. Introduction

The term cloud discharges is used to denote three types of lightning: (i) intracloud discharges, those occurring within the confines of a thundercloud, (ii) intercloud discharges, those occurring between thunderclouds, and (iii) air discharges, those occurring between a thundercloud and clear air. It is thought that the majority of cloud discharges are of the intracloud type, although no reliable statistical data are found in the literature to confirm that this is the case. Often the abbreviation IC (for intracloud flash) is used to refer to all cloud flashes. Ogawa and Brook (1964) reported that intracloud and cloud-to-air lightning discharges produced similar overall electric field changes. As discussed in Section 2.7, approximately three-quarters of lightning discharges do not contact ground, although this fraction depends on storm type, the stage of storm development, and possibly other factors. The early stages of thunderstorm development tend to be dominated by cloud discharges. According to Williams et al. (1989), 10 or more cloud flashes may occur before the first cloud-to-ground flash.

On the one hand, from the point of view of the lightning hazard to ground-based objects and systems, cloud discharges are generally considered to be of relatively little or no consequence. Obviously, cloud lightning does not pose a threat of death or injury to human beings and animals on the ground, nor does it ignite forest fires. On the other hand, cloud discharges are of great practical interest to those concerned with the safety of aircraft, spacecraft, and other vehicles that may pass through or approach clouds. A discussion of the interaction of aircraft and launch vehicles with both cloud and ground discharges is found in Chapter 10.

Cloud discharges have been less well studied than cloud-to-ground discharges because of (i) the difficulty of securing photographic records of in-cloud channels, although optical observations have certainly provided useful information about cloud flashes (e.g., Sourdillon 1952; Isikawa 1961; Takagi 1961; Ogawa and Brook 1964; Mackerras 1968), and (ii) the inability to measure directly the currents and charge transfers associated with cloud discharges, as has been done on towers for ground discharges. Historically, ground-based single-station and multiple-station electric field measurements have been the primary means for studying cloud discharges. Some studies have been concerned with both electric field pulses and light emissions (e.g., Kobayashi et al. 1958; Kitagawa and Kobayashi 1959; Takagi and Takeuti 1963). More recently, VHF–UHF lightning locating systems (subsections 17.4.2 and 17.6) have made it possible to obtain time-resolved images of in-cloud channels (e.g., Proctor, 1981, 1997; Mazur 1989; Shao and Krehbiel 1996; Rison et al. 1999; Thomas et al. 2000). Additional information on cloud discharges has been obtained using *in situ* electric field and acoustic measurements (e.g., Weber et al. 1982) and radar observations (e.g., Mazur et al. 1984).

In Section 9.2, we give general information on the cloud discharge. The phenomenology of the cloud discharge inferred from VHF–UHF channel imaging is presented in Section 9.3. The development of intracloud discharges that apparently bridge the vertically stacked negative and positive cloud charge centers is discussed in subsection 9.3.1, and a similar discussion of predominantly horizontal cloud discharges, which exhibit a large variety of forms, is found in subsection 9.3.2. A characterization of the early (active) and the late (final) stages comprising a cloud discharge is found in Sections 9.4 and 9.5, respectively. The general characteristics of the electric and magnetic field pulses that occur during the early stage and the late stage are discussed in subsections 9.4.2 and 9.5.2, respectively. A comparison of cloud and ground discharges is given in Section 9.6. In order to minimize conflict with the existing literature, we use both the atmospheric electricity and the physics sign conventions (subsection 1.4.2) in this chapter. Specifically, the atmospheric electricity sign convention is used in Table 9.3 and in Figs. 9.1, 9.2, 9.8, 9.9, and 9.13, while the physics sign convention is used in Table 9.4 and in Figs. 9.10, 9.11, and 9.12.

9.2. General information

In this section we first give a general picture of the cloud discharge based on a variety of studies. Then we present the various overall cloud flash characteristics inferred from measured electric fields and briefly discuss optical observations that illustrate the complexities of such "electrodeless" discharges. Finally, we consider the distribution of the heights of the flash origin in a cloud.

Cloud discharges can be viewed as being composed of an early (or active) stage and a late (or final) stage. The early stage has a duration of some tens to a few hundreds of milliseconds, and the remainder of the flash constitutes the late stage. The early stage typically involves a negatively charged channel extending in an intermittent manner with an average speed of order 10^5 m s^{-1}. Overall, the early-stage processes are probably similar to the initial breakdown and stepped-leader processes in negative cloud-to-ground lightning, the differences being discussed in Section 9.6. In the following, we will apply the terms initial breakdown and negative stepped leader to the corresponding processes in cloud lightning.

In general, the upper and lower boundaries of a negative charge region, where the electric fields are highest, are the most likely places for a cloud flash to begin. It is thought that cloud flashes often bridge the main negative and upper positive charge regions. Other scenarios may involve the lower positive charge region or may be indicative of a cloud charge distribution that is different from a simple tripolar model (subsection 3.2.1). The beginning of a cloud discharge is typically marked by the largest microsecond-scale pulses in its wideband electric field record (see Figs. 9.1a, b). These early-stage pulses, in analogy to the initial pulses in ground flashes, are usually referred to as initial-breakdown pulses and include the narrow bipolar pulses that are sometimes attributed to "compact intracloud discharges" (subsection 9.4.2). Some characteristics of the initial-breakdown and stepped-leader processes depend on the height of the origin of the cloud discharge. It is likely that a cloud discharge begins as a bidirectional leader, the positive section of this leader pervading the negative charge region and effectively supplying negative charge through the discharge origin to the negative section that extends into the positive charge region. The transition from the early stage to the late stage is thought to be associated with the loss of connection between the positive and negative sections (or between the negative and positive charge regions) and with the cessation of the extension of the negative stepped-leader channel.

The late-stage processes apparently serve to transport negative charge to the region of the flash origin from progressively more remote sources in the negative charge region. The late (or final) stage of cloud discharges is also called the J-type stage, this term implying that the associated physical process is similar to the J-process (junction process) in ground discharges (Section 4.10). The J-process is often visualized as a positively charged channel (or channels) extending from the region of the flash origin at a relative low speed of order 10^4 m s^{-1}. The various transient processes occurring during the late stage are referred to as K-processes and sometimes "recoil streamers". Their form depends on the origination point and on conditions along the propagation path and at the termination point. Successive K-processes can retrace the same path several times. Rapid electric field variations associated with K-processes (see, for example, the pronounced step-like field change between 300 and 320 ms in Fig. 9.1a(i)) are referred to as K-changes. The observation that many K-changes are accompanied by regular pulse bursts (subsection 9.5.2.) suggests that a process similar to the dart-stepped leader in cloud-to-ground flashes is involved. Note that some researchers (e.g., Kitagawa and Kobayashi 1959; Krehbiel 1981) have apparently labeled any impulsive (transient) process occurring during a cloud discharge as a K-process. K-processes in cloud discharges are thought to be similar to those in ground discharges. Ambiguities in the definition of K-processes in ground discharges, which apply also to K-processes in cloud discharges, are discussed in subsection 4.10.1.

As shown by Villanueva et al. (1994), the three-stage structure of electric fields postulated for cloud discharges by Kitagawa and Brook (1960, Fig. 6), and until recently generally accepted as valid, is in fact not supported either by more recent measurements or by the original data of Kitagawa and Brook (1960) from which the three-stage structure was erroneously inferred. The incorrect feature of the three-stage structure is the alleged occurrence of small pulses at the beginning of the flash and well before the largest pulses in the flash. Kitagawa and Brook (1960) termed the first stage of the cloud flash, which, they claimed, was characterized by smaller pulses, the "initial stage" and the following stage, which was characterized by larger pulses, the "active stage." As noted above, the largest pulses tend to occur at the beginning of the flash. The third, "final" stage of the three-stage structure of Kitagawa and Brook (1960) is similar to the late (or final) stage of the two-stage structure described above.

Examples of the overall electric fields due to cloud flashes are found in Figs. 3.4, 9.1a and 9.2. The polarity of the overall electric field waveform for a vertical cloud discharge depends on distance, as discussed in

9.2. General information

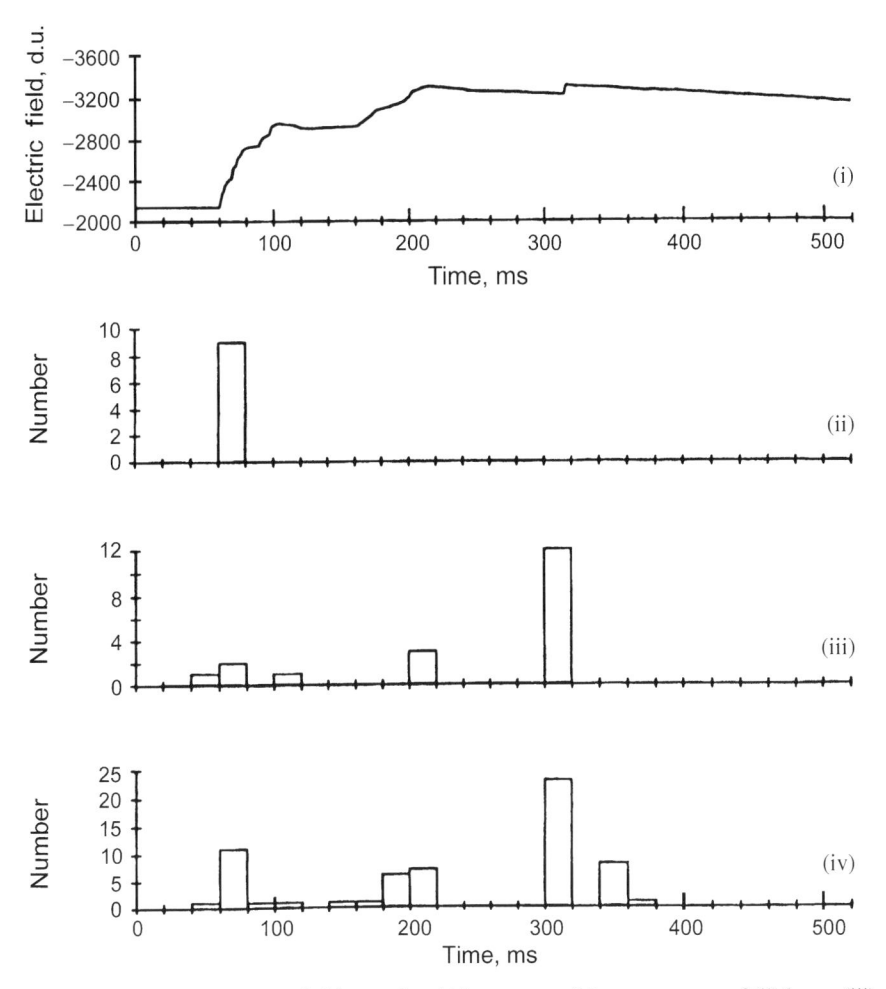

Fig. 9.1a. (i) Overall electric field record and histograms of the occurrence of (ii) large, (iii) medium, and (iv) small electric field pulses in different parts of this record for cloud flash 64 on day 231 in 1991 at the Kennedy Space Center, Florida. The field and histograms are displayed on the same time scale; d.u. indicates "digitizer units". The atmospheric electricity sign convention (subsection 1.4.2) is used here. Adapted from Villanueva et al. (1994).

subsection 3.2.1 and illustrated in Figs. 3.2e and 3.4. At close ranges the cloud-flash electric field change tends to have the same polarity as the electric field change due to negative cloud-to-ground flashes. As distance increases, a polarity reversal is observed, as illustrated in Fig. 3.4. The overall characteristics of cloud discharges, the flash duration and the charge transfer, estimated from single-station and multiple-station measurements of electric field changes are summarized in Table 9.1. As seen in Table 9.1, cloud discharges have a typical duration of some hundreds of milliseconds and a typical charge transfer of some tens of coulombs, similar to cloud-to-ground discharges.

Cloud discharges inherently exhibit more variability than ground discharges, since the latter involve a relatively well-conducting ground "electrode", while cloud discharges are "electrodeless". Sourdillon (1952) and Isikawa (1961) observed luminosity surges that traveled in either direction along the lightning channel; that is, some propagated toward the channel extremity while others apparently began there and propagated back toward the origin. Ogawa and Brook (1964) presented a time-resolved photographic record of an air discharge that exhibited a continuously extending, branched leader channel that was repeatedly retraced by luminosity surges originating at the channel extremities. Sourdillon (1952) reported that some initial leaders were stepped, that some appeared to begin as continuously extending, similarly to the two types of air discharge reported by Schonland et al. (1935, p. 621), and that most initial leaders emitted very faint luminosity, which was difficult to detect.

The height of origin of a cloud flash influences the characteristics of the initial leader of the flash (subsection 9.3.2). Proctor (1991), from studies in South Africa, gave the distribution of the heights at which

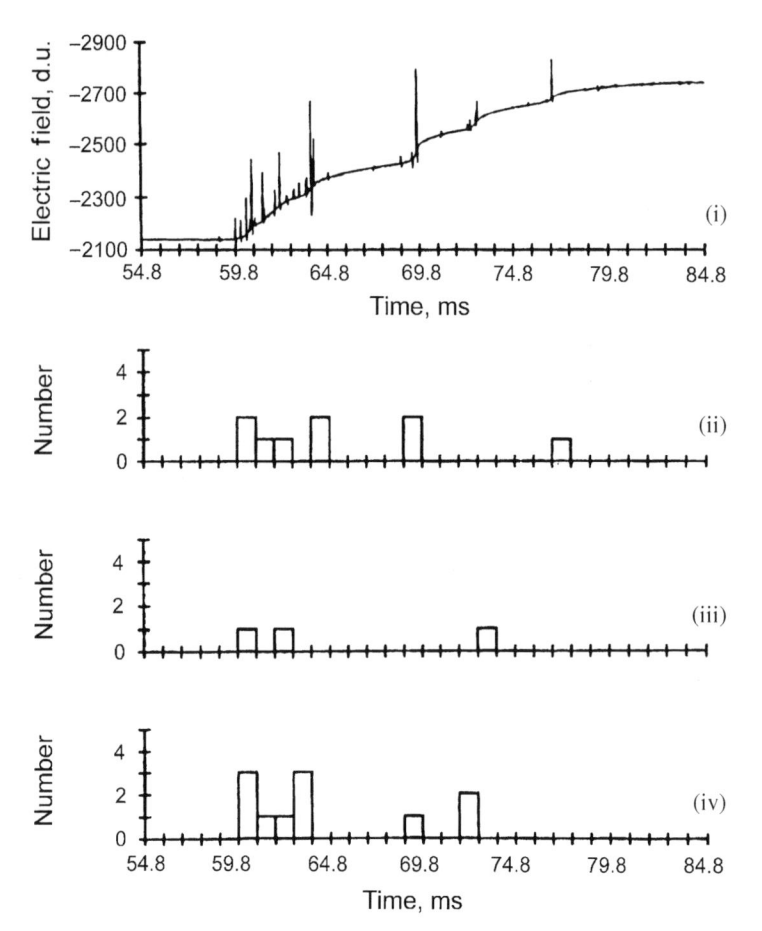

Fig. 9.1b. Same as Figure 9.1a but for the first 25 ms of the flash. Adapted from Villanueva et al. (1994).

773 cloud and cloud-to-ground lightning flashes originated. These heights ranged from 1 to 12 km. The distribution is bimodal with peaks at 5.3 km and 9.2 km above mean sea level (amsl) where the average environmental temperatures were $-3.3\,°C$ and about $-28\,°C$, respectively. Ground level was 1.43 km above sea level.

The distribution appears to be a superposition of two normal distributions that intersect at 7.4 km amsl. Proctor (1991) attributed 342 flashes (44 percent) of a total 773 flashes to the higher-origin group and 431 flashes (56 percent) to the lower-origin group. Bimodal distributions of the heights at which lightning flashes occur have also been reported from acoustic measurements by Weber (1980) and MacGorman et al. (1981), from radar observations by Mazur et al. (1984), and from the location of VHF sources by Taylor et al. (1984) and Ray et al. (1987). Additionally, Mackerras (1968), using photography and time-to-thunder measurements, reported on two height ranges in which cloud flashes occurred: 4 to 12 km and 1 to 8 km, flashes in the latter range being more common. Proctor (1991) stated that lightning-flash origins exhibited a propensity to occur at or near boundaries enclosing regions where the radar reflectivities were 20 dBZ and suggested that these regions contained excess negative charge. Of the 773 flashes studied by Proctor (1991), 214 (about 28 percent) were cloud-to-ground flashes, essentially all of which (210 out of 214) belonged to the lower-origin group. If we remove all 214 cloud-to-ground flashes from Proctor's (1991) data set, so that only 559 cloud flashes remain, the bimodal shape of the distribution of flash origins is preserved. Thus the bimodal shape of the distribution of lightning-origin heights is not solely due to the different heights of origin of ground and cloud discharges. For cloud discharges only, 338 flashes (60 percent) belong to the higher-origin group and 221 flashes (40 percent) to the lower-origin group.

Considering both cloud and ground discharges, Proctor (1991) reported that in only three storms from a total of 13 studied were higher-origin flashes in the majority. In the 10 remaining storms the percentages of higher-origin flashes ranged from 0 to 36 percent. The percentage of higher-origin flashes changed during the lifetimes of the storms, and it usually decreased as the storm aged, although, according to Proctor (1991), other factors, not specified by him, influenced the occurrence of higher-origin flashes.

9.2. General information

Table 9.1. *Overall characteristics of cloud flashes inferred from measurements of electric fields. The numbers in parentheses are sample sizes*

Reference	Location	Average or typical value	Remarks
Flash duration, ms			
Pierce (1955a)	England	245 (685)	Mean value; negative field changes[a]
Takagi (1961)	Japan	300	
Isikawa (1961)	Japan	420	
Ogawa and Brook (1964)	New Mexico	500	Excluding short-duration field changes at very close ranges
Mackerras (1968)	Australia	480	
Bils *et al.* (1988)	Florida	660 (89)	Mean value
Charge transfer, C			
Workman and Holzer (1942)	New Mexico	32 (16)	Eight stations
Reynolds and Neill (1955)	New Mexico	21 (35)	Mean value; 11 stations
Isikawa (1961)	Japan	32	
Wang (1963a)	Singapore	15	
Ogawa and Brook (1964)	New Mexico	30	

[a] Atmospheric electricity sign convention (subsection 1.4.2).

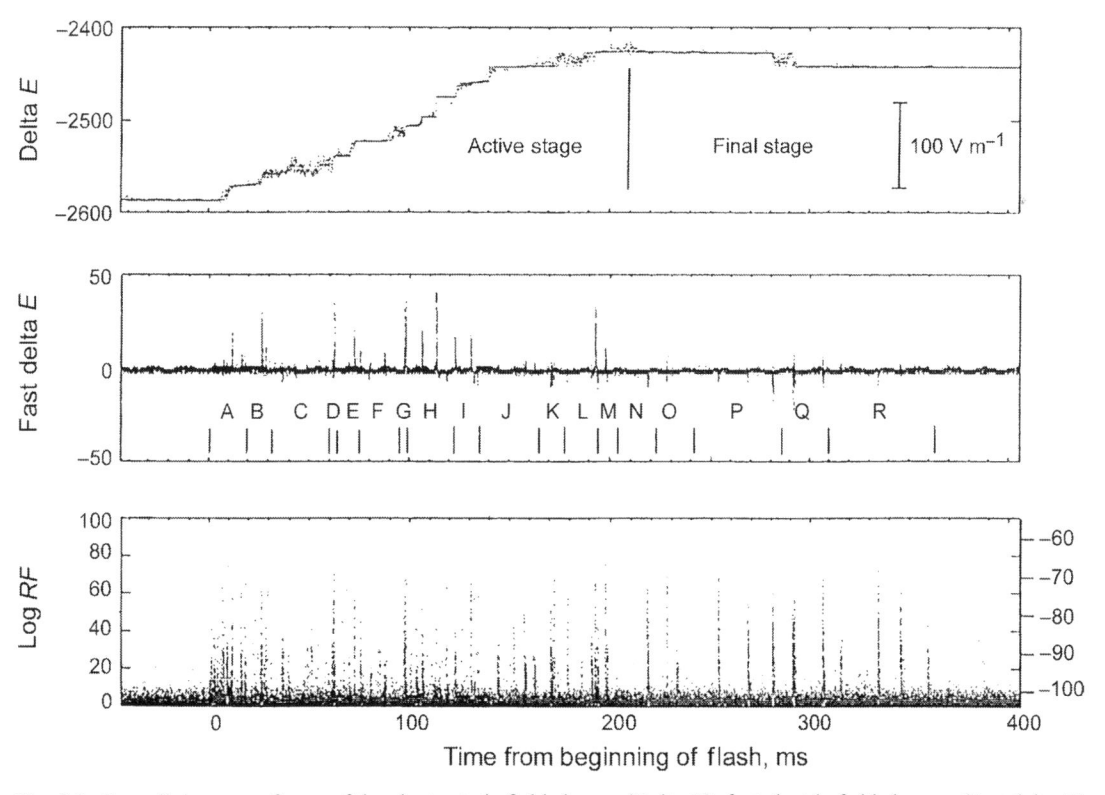

Fig. 9.2. Overall time waveforms of the electrostatic field change (Delta E), fast electric field changes (Fast delta E), and RF radiation amplitude (Log RF) for intracloud flash that occurred at 1941:54 UT on 1 September 1992 at a distance 5–10 km from the observation site near Orlando, Florida. The right-hand vertical scale in the bottom figure is in dB m^{-2} MHz^{-1}. The atmospheric electricity sign convention (subsection 1.4.2) is used here. Adapted from Shao and Krehbiel (1996).

9.3. Phenomenology inferred from VHF–UHF imaging

Cloud discharges are often viewed as occurring between the main negative and main positive charge regions (subsection 3.2.1). "Bilevel" flashes that conform to this view are described in subsection 9.3.1, in a discussion based on the work of the New Mexico Institute of Mining and Technology (NMIMT) group (Shao and Krehbiel 1993, 1996; Rison et al. 1999). However, many cloud discharges do not exhibit a pronounced bilevel structure, and many appear to be predominantly horizontal. Discussion of this latter type of cloud discharge, based on the work of Proctor (1981, 1997) and on the observations of the Office National d'Etudes et de Recherches Aérospatiales (ONERA) group, as reported by Mazur (1989), is found in subsection 9.3.2. It is worth noting that there exist a number of forms of cloud discharges that are presently not well understood; they are not specifically discussed in this book. One such form involves extensive horizontal channels near the cloud base, sometimes referred to as "spider" lightning (Mazur et al. 1998; Marshall 2000). At the time of writing, there are no reliable statistics on the occurrence of "spider" lightning, but many flashes photographed by Mackerras (1968) in Brisbane, Australia, occurred near the lower cloud boundary, which was at 2 to 4 km above ground level. Further, recent VHF observations indicate the existence of new forms of cloud discharge activity in storms in the US Great Plains. Specifically, Krehbiel et al. (2000) observed, in a large supercell storm, what they describe as essentially continuous discharge activity during the one-minute time interval presented. The activity filled a saucer-shaped volume with a diameter over 60 km and with a vertical extent from about 4 to 12 km. The storm also exhibited lightning discharge activity at an altitude of 15–16 km. Krehbiel et al. (2000) associated these high-altitude discharges with "overshooting convective tops that briefly penetrate the base of the stratosphere".

In reading subsections 9.3.1 and 9.3.2, it might be helpful to refer to the general picture of the cloud discharge given in Section 9.2. It is also worth noting that VHF–UHF radiation is associated with air-breakdown processes, as opposed to processes occurring in already created lightning channels.

9.3.1. Bilevel flashes

In this section we present a description of the bilevel intracloud discharge, using detailed case studies of six events in Florida by Shao and Krehbiel (1993, 1996), who obtained two-dimensional 274-MHz interferometric images of lightning channels in conjunction with single-station electric field measurements. Electric field waveforms and records of the amplitude of VHF radiation as a function of time for one of the two cloud flashes

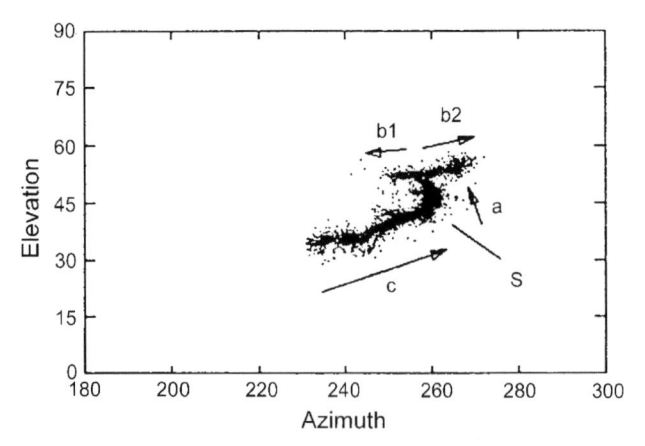

Fig. 9.3. All the VHF radiation sources in azimuth-elevation format for the flash whose electric field changes and RF radiation amplitude are shown as a function of time in Fig. 9.2. The sources exhibited a two-level structure, the upper-level channels, b1, and b2, and the lower-level channel, c, being connected by a vertical channel, a. S denotes the region where the flash started and the arrows indicate the direction of motion of negative charge. Channel a is fully formed during time interval A, channels b1 and b2 during time intervals A to H, and channel c during time intervals I to R (see Fig. 9.2). Adapted from Shao and Krehbiel (1996).

presented by Shao and Krehbiel are given in Fig. 9.2, and the corresponding two-dimensional VHF image in azimuth-elevation coordinates is given in Fig. 9.3. According to Shao and Krehbiel (1996), cloud discharges often exhibit a single vertical channel that bridges the lower-level (presumably main negative) and upper-level (presumably main positive) charge regions in the cloud. The length of this channel was estimated by Shao and Krehbiel (1996) to be 2 to 3 km. The discharge appeared to originate at the lower level (presumably at the upper boundary of the main negative charge region) and to transport negative charge to the upper level. VHF sources above and below the vertical channel tended to form predominantly horizontal layers at heights presumably corresponding to the positive and negative charge regions, respectively.

The initial, upward-developing negative leader reached the upper level in 10 to 20 ms, so that its propagation speed is about $(1.5–3) \times 10^5$ m s^{-1}, which is similar to the stepped-leader speed in downward negative cloud-to-ground flashes. During the following few hundreds of milliseconds the vertical channel serves to transport negative charge from the lower level to the upper level via both steady-current process and upward-moving transient processes. Initially (during time intervals A to H, that is, during the first 100 ms or so for the flash shown in Fig. 9.2), negative charge is tapped near the flash origin, labeled S in Fig. 9.3, and transported to the upper level, as indicated by the arrows a, b1, and b2 in Fig. 9.3. Later, during time intervals I to N, that is, during the next 100 ms or so for the flash shown in Fig. 9.2, further negative charge transfer to the upper level is facilitated by progressively more remote sources at the lower level, as illustrated by arrow c in Fig. 9.3. The time period during which negative charge is transported from the lower level to the upper level (from A to N in Fig. 9.2) was referred to by Shao and Krehbiel (1996) as the active stage. During the remainder of a cloud flash (after time N in Fig. 9.2), the continued negative charge transfer to the flash origin (arrow c in Fig. 9.3) was usually not followed by its further transport to the upper level (time interval Q in Fig. 9.2 was an exception), which is indicative of a lack of connection between the lower and upper levels. This latter part of the cloud flash, marked by the cessation of increase in the electric field, seen in Fig. 9.2, top trace, was referred to by Shao and Krehbiel (1996) as the final stage. According to Proctor (1981), the final stage is associated with no further extension of the initial leader channel, although different branches of the same flash can be at different stages of development (subsection 9.3.2). Shao and Krehbiel (1996) noted that "not all flashes exhibit a final stage". The active and final stages defined by Shao and Krehbiel roughly correspond to the early and late stages, respectively, defined by Villanueva et al. (1994). The bilevel cloud discharge described by Shao and Krehbiel appears to be somewhat similar to an inverted negative cloud-to-ground discharge, except for the lack of a return stroke from the upper positive charge region.

Rison et al. (1999) and Thomas et al. (2000), who used a three-dimensional 63 MHz TOA lightning locating system (subsection 17.4.2) in central New Mexico and Oklahoma, respectively, reported on bilevel cloud discharges similar to those described by Shao and Krehbiel (1996). Most of the detected radiation sources were located at the upper level, presumably because the breakdown produced when the negative leader penetrated the upper positive charge region was "noisier" at RF than the breakdown produced when the positive leader penetrated the negative charge region. Rison et al. (1999) stated that radiation sources detected at the lower level were probably associated with a negative polarity breakdown along the path of an undetected positive leader in the negative charge region. A similar scenario had been suggested previously by Mazur (1989), as discussed in subsection 9.3.2.

The bilevel intracloud discharge discussed in this subsection represents one type of cloud flash. In subsection 9.3.2, we discuss another type of cloud flash, the predominantly horizontal cloud discharge reported by Proctor (1981, 1997) and by Mazur (1989).

9.3.2. Predominantly horizontal flashes

Proctor (1981, 1997) studied cloud discharges in South Africa using, in different years, a 253 MHz (VHF) or a 355 MHz (UHF) time-of-arrival lightning imaging system. In Proctor's view, a cloud flash involves (i) a predominantly horizontal stepped leader that propagates in the direction defined as the "forward" direction and (ii) so-called Q-streamers that appear to extend the channel in the opposite or "reverse" direction. Q-streamers emit light and are usually attributed, following Proctor, to K-processes, although often they do not produce measurable electric field changes (Proctor 1981, 1997). Additionally, Proctor noted that sometimes the stepped-leader extension is preceded by some preliminary activity that "ionizes" the origin.

Proctor (1997) analyzed in detail 21 flashes whose origin heights ranged from 7.6 to 9.9 km amsl, that is, those of the higher-origin type (Section 9.2). Further, detailed descriptions of one higher-origin and four lower-origin cloud flashes were given by Proctor (1981), one of the lower-origin flashes being initiated by a predominantly horizontal positive leader. The characteristics of higher-origin flashes and lower-origin flashes are summarized in Table 9.2. As seen in this table, the higher-origin flashes emit VHF–UHF pulses at a much lower rate than the lower-origin flashes. Proctor (1997) apparently interpreted these pulses as being due to a stepping process and stated that the principal difference between flashes with higher and lower origins is related to the step length and interstep interval, the steps for higher-origin flashes being longer and less frequent. According to Proctor (1997), the stepped leaders in lower-origin cloud flashes are indistinguishable from the stepped leaders in cloud-to-ground flashes. In fact, the characteristics of lower-origin flashes in Table 9.2 are probably based on data for stepped leaders in both cloud and ground flashes, since, in studying the distribution of flash origin heights, Proctor (1991) did not separate cloud and ground discharges.

Table 9.2. *Comparison of the stepped leaders in cloud flashes with low and high origins. Adapted from Proctor (1997)*

Attribute	Flashes with low origins	Flashes with high origins	Units
Origin height	1 to 7.4	7.4 to 12	km amsl[a]
Pulse rate	2×10^5 to 2×10^6	2×10^3 to 4×10^4	s^{-1}
Typical source length	60	300	m
Pulse source NND[b]	30 to 60	60	m
Speed of channel extension	6×10^4 to 5×10^5	1.3×10^4 to 1.3×10^5	m s^{-1}
Line charge density	10^{-3}	7×10^{-4} to 8.7×10^{-3}	C m^{-1}

[a] Above mean sea level.
[b] Nearest neighbor distance.

In the following, we will give a description of the higher-origin cloud flashes, based on the work of Proctor (1997). The overall phenomenology of the lower-origin cloud flashes is apparently similar to the former (Proctor 1981), although some of their characteristics are different, as seen in Table 9.2. Possibly the lower positive charge region plays the same role in lower-origin flashes as the upper positive charge region in higher-origin cloud flashes. In contrast with the attempt of the NMIMT group to present a unified picture of the cloud flash, as discussed in subsection 9.3.1, Proctor (1997) emphasized the considerable variation in cloud-flash morphology. He reported that only two of 21 cloud flashes exhibited a single stepped-leader channel, while the other 19 extended along multiple paths. Relatively short branches often extended simultaneously, while relatively long branches usually formed in succession. One flash produced five short radial branches. In another flash, pulsed radiation sources were distributed almost randomly in the cloud. Two flashes produced leaders that appeared to encircle their origins several times. Of 93 stepped-leader channels in 21 cloud flashes studied by Proctor (1997), 53 (57 percent) were predominantly horizontal, 26 (28 percent) extended mainly upward, and 14 (15 percent) extended mainly downward. Similarly, Mackerras (1968) reported that 55 percent of 94 cloud flashes photographed in Brisbane, Australia showed mainly horizontal channels and 33 percent showed extensive branching or multiple channels. Predominantly horizontal lightning discharges have also been reported from acoustic channel imaging (Section 11.5) and from radar observations (Ligda 1956; Section 17.9). The horizontal extent of the flashes studied by Proctor (1997) ranged from 0.5 to about 23 km, and the vertical extent ranged from 0.5 km to about 13 km. Note that two of four flashes described in detail by Proctor clearly occurred in growing (as opposed to dissipating) storms, within squall lines, and one occurred in a supercell storm (Section 3.1). In all 21 flashes, the stepped leaders initially transported negative charge away from the origin, but one leader changed its polarity during its progression through the cloud.

Plan and elevation views of the 355 MHz image of a cloud discharge described by Proctor (1997) as "the nearest to being typical" are presented in Figs. 9.4a, b, respectively. The image is based on the "pulsed radiation" attributed by Proctor (1997) to a stepped-leader process, while the sources of radiation called Q-noise by Proctor are not plotted. The Q-noise is attributed by Proctor (1997) to Q-streamers, which are discussed later in this section. The flash transported negative charge along two main branches that extended almost horizontally from a common origin marked by the large cross. The formation of these branches was preceded by some preliminary activity that involved "vertical scanning" of the region of origin of the flash and served to expand that region to a volume equal to

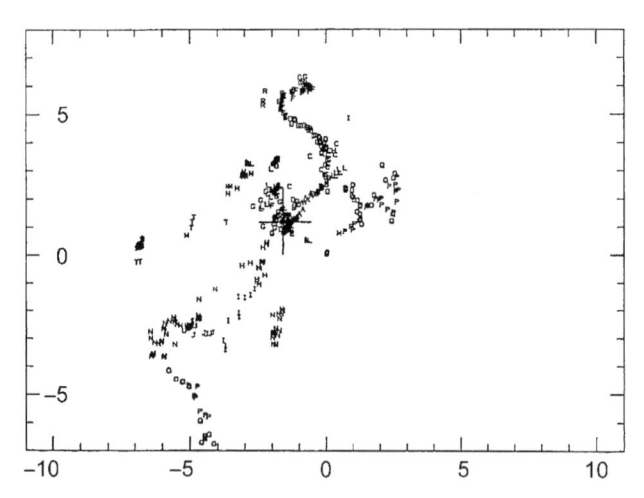

Fig. 9.4a. Plan view of UHF (355 MHz) pulsed radiation sources for flash 151, which occurred at 1502:12.3 on 26 March 1985 in South Africa; the scales are in km. The time scale is indicated by alphabetical symbols that increment every 20 ms. Q-noise sources have been omitted. The large cross indicates the position of the origin of the flash. Adapted from Proctor (1997).

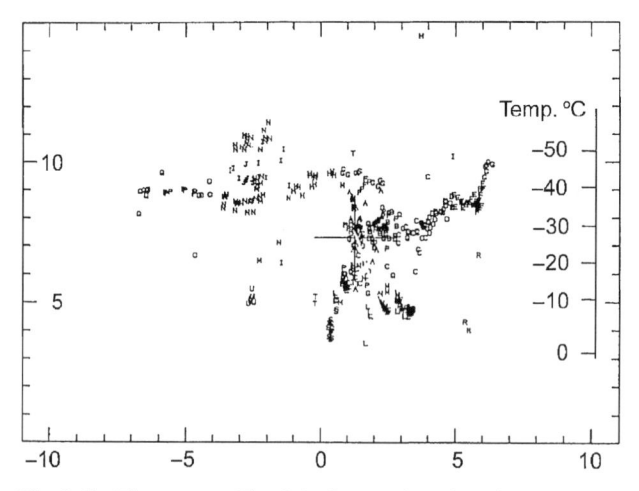

Fig. 9.4b. The same as Fig. 9.4a, but an elevation view. The rms height errors were 175 m at the origin and 233 m at the left-hand extremity, Q. Adapted from Proctor (1997).

$1.5 \times 0.5 \times 2.5$ km^3. Proctor (1997) called these preliminary processes the initial stage and stated that it occurred in roughly 50 percent of the higher-origin flashes and never in the lower-origin flashes. Villanueva et al. (1994) speculated that this difference between flashes originating at relatively high and relatively low altitudes might be related to different precipitation mixes (concentrations of large water drops) at those altitudes. The first main branch in Figs. 9.4a, b developed from A to G at a speed of about 6×10^4 m s^{-1}. The second main branch developed toward Q. Additional branches formed below the flash origin.

As noted above, the overall channel system shown in Figs. 9.4a, b was thought to be formed by a stepped leader, and the radiation sources associated with Q-streamers are

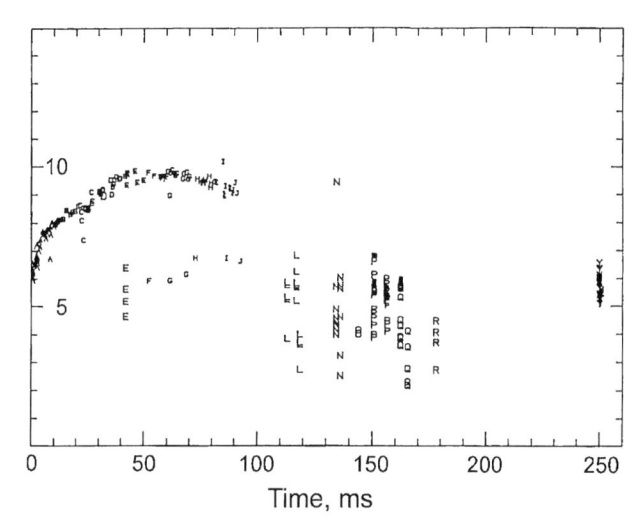

Fig. 9.5. Heights above ground in km of UHF sources for flash 165, which occurred at 1654:40.8 on 23 December 1987 in South Africa, plotted as a function of time. The smaller letters denote pulsed radiation sources; the larger letters represent Q-noise sources. Adapted from Proctor (1997).

not plotted. The extension of the cloud-flash channel system by both the stepped leader and by Q-streamers is illustrated for a different, predominantly vertical flash, in Fig. 9.5; plotted is radiation source height versus time, the smaller letters denoting sources associated with the stepped leader and the larger letters denoting sources associated with Q-streamers. The origin of this flash was about 6.2 km above ground. Proctor (1997) specifically stated that the flash represented in Fig. 9.5 exhibited no initial stage, as he defined it (see above), that is, in this flash the stepped-leader extension was the first activity detected at 355 MHz. The stepped leader extended upward along a curved path of 6.8 km length and reached the points mapped as J at about 92 ms. Four bursts of Q-noise occurred during that period of time (at the times labeled A, D, and E). The stepped leader extended at an average speed 7.3×10^4 m s^{-1}. Its speed during the first 10 ms was about 3×10^5 m s^{-1}. No 355 MHz radiation was detected from 92 to 115 ms, and 17 Q-streamers, labeled L to R, were recorded during the next 65 ms (apparently the same letter was used to label more than one Q-streamer). These Q-streamers appeared to extend the lightning-channel system in the downward direction at an overall speed of 7×10^4 m s^{-1}. Individual Q-streamers developed much more rapidly. For example, Q-streamer N propagated at 1.5×10^7 m s^{-1}. Some of the Q-streamers appeared to propagate toward the flash origin while others moved away from it (D.E. Proctor, personal communication, 2000). Additionally, some radiation occurred at around 250 ms. As seen in Fig. 9.5, most stepped-leader sources are located above the flash origin, while the majority of Q-streamer sources are below the origin. The upward ("forward") extension was facilitated by the stepped leader, and the following downward ("reverse") extension by the Q-streamers. Proctor (1997) noted a similarity of this cloud flash, which occurred in a growing storm, to an inverted cloud-to-ground flash (see Fig. 4.61).

The "forward" extension stage of a cloud flash was called by Proctor (1981, 1997) the very active stage and the following part of the flash, characterized by no "forward" extension, the final or J-type stage. The very active and final stages defined by Proctor (1981, 1997) roughly correspond to the early and late stages, respectively, defined by Villanueva et al. (1994).

We now discuss UHF-interferometer observations of extensive horizontal cloud discharges by the ONERA group in France, as presented by Mazur (1989). These observations showed that during the first half of a flash the negative leader progressed away from the flash origin in a "forward" direction at a speed of 10^4 to 10^5 m s^{-1}. Subsequently, a number of "recoil streamers" propagated at a speed of about 10^7 m s^{-1} toward and into the flash-origin region from the opposite side, as illustrated in Fig. 9.6. This sequence of events was interpreted by Mazur (1989) as indicating that the cloud flashes were initiated by a bidirectional leader whose negative and positive sections extended in opposite directions from the origin. Further, the positive part was postulated to be undetectable at VHF–UHF, and the recoil streamers were assumed to originate at the positive leader tips and propagate toward the flash origin (Mazur 1999, Fig. 12.2). The detectability of positive and negative leaders at VHF–UHF is considered in subsection 5.3.2. Note that Shao and Krehbiel (1996) argued that recoil streamers propagating toward the flash origin in their VHF-imaged bilevel flashes, described in subsection 9.3.1, were "not necessarily evidence of undetected positive breakdown". However, Rison et al. (1999) later interpreted similar "recoil streamers" as, probably, processes retracing undetected paths of positive leaders.

9.4. Early (active) stage

In this section, we characterize the early (active) stage of a cloud discharge. The overall characteristics of this stage, including the stepped-leader propagation speed, line charge density, and average current, are presented below in subsection 9.4.1. The characteristics of early-stage electric and magnetic field pulses in cloud flashes, including narrow bipolar pulses, are discussed in subsection 9.4.2. A comparison of these early-stage pulses with the initial-breakdown pulses in ground flashes is found in Section 9.6.

9.4.1. Overall characteristics

Ogawa and Brook (1964) measured about 600 waveforms of electrostatic field changes due to cloud discharges in a small thunderstorm in New Mexico. The approximate

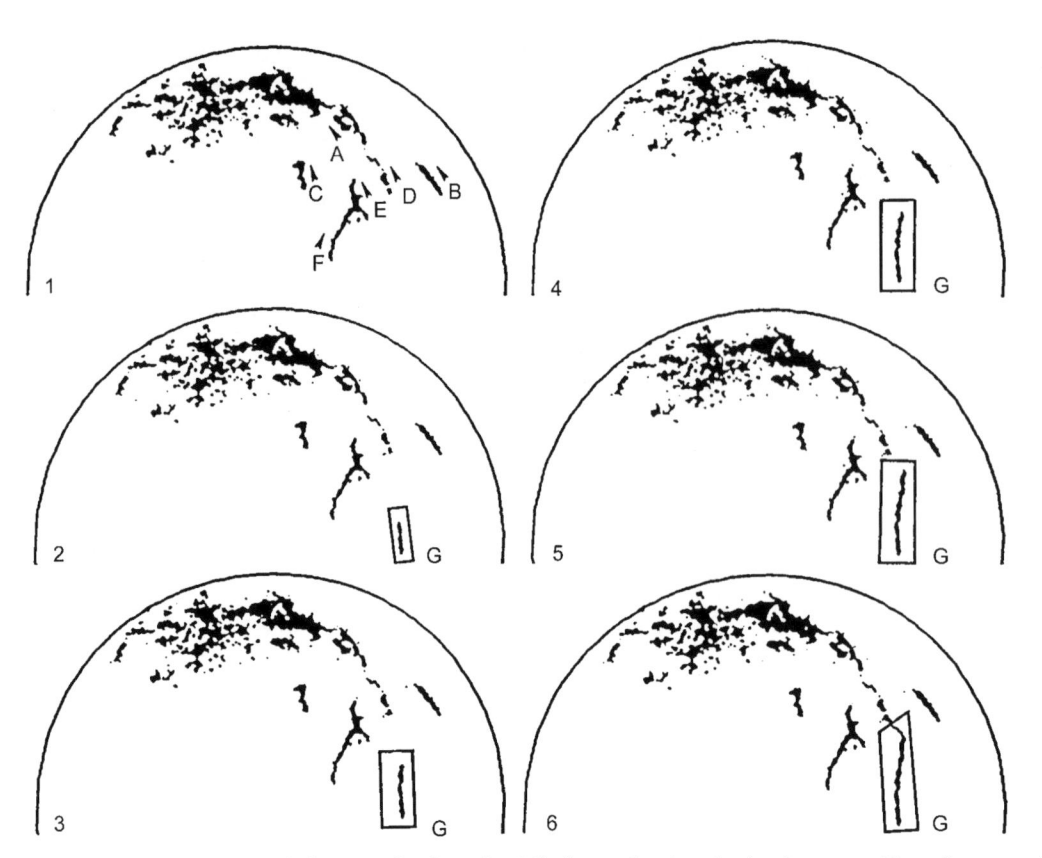

Fig. 9.6. UHF interferometric images of an intracloud discharge showing the development of "recoil streamer" G toward the flash origin. Time increases from image 1 toward image 6. Adapted from Mazur (1989).

distance to the storm as it passed by the observation station was estimated from the time between flash and thunder. The observed dependence of the polarity of the field changes on distance (see Fig. 3.4) suggests that the discharges had large vertical components. Comparing the measured electric field waveforms recorded at different distances with the fields calculated for vertical in-cloud channels, Ogawa and Brook (1964) inferred, erroneously as is now thought, that the initial half of the overall cloud-flash electric field change was due to a descending positive leader. Brook and Ogawa (1977) estimated that the positive leader propagated downward at a speed of about 10^4 m s^{-1}, assuming a total leader-channel length of 2 km and a leader duration of 250 ms, and carried an average current of about 120 A, from an inferred average neutralized charge of 30 C and an assumed leader duration of 250 ms. As will be seen in the following discussion (see also Section 9.3), the descending-positive-leader hypothesis proposed by Ogawa and Brook (1964) and Brook and Ogawa (1977) is not supported by other observations, some of which are more recent. The electric field measurements of Smith L.G. (1957) at two stations in Florida and of Nakano (1979a, b) at three to seven stations in Japan suggested that cloud flashes tend to be initiated by an upward negative leader. Takeuti (1965), from three-station electric field measurements in Japan, inferred both vertical and horizontal discharges, 24 vertical discharges being initiated by a descending positive leader and 14 by an ascending negative leader. Weber *et al.* (1982), from in-cloud electric field measurements and thunder-source reconstruction in New Mexico, reported on seven predominantly horizontal lightning discharges. They estimated a mean propagation speed of about 5×10^4 m s^{-1} for a mean distance of about 5 km, a mean current of 390 A, and a mean line charge density of about 7 C km^{-1}.

Krehbiel (1981) analyzed electric field change measurements at nine to 11 stations for 21 cloud discharges and three hybrid discharges (cloud discharges with a cloud-to-ground component) in Florida using time-varying point charge and point dipole models (subsection 17.2.1). All but four discharges were found to be predominantly vertical. The flashes appeared to reach their full vertical extent via an upward-propagating negative leader in a few tens of milliseconds or less, as opposed to the 250 ms assumed by Ogawa and Brook (1964) for a descending positive leader. Accordingly, the speed of the leader propagation was estimated by Krehbiel (1981) to be of order 10^5 m s^{-1}, an order of magnitude higher than that inferred by Ogawa and Brook (1964) and most the other earlier investigators. Four of the

9.4. Early (active) stage

cloud flashes studied by Krehbiel (1981) were reanalyzed by Liu and Krehbiel (1985) using a piecewise-linear model for the initial leader. They found that the initial 15–30 ms of the observed fields were best modeled by a negative leader that developed upward from the negative charge region at a speed of 1×10^5 to 3×10^5 m s^{-1}. The negative charge density on the leader channel was estimated to be 1–4 C km^{-1}. Liu and Krehbiel stated that they could not determine the development of the negative leaders after the initial 15–30 ms with their model, but, as noted above, vertical cloud flashes reached their full extent in a time short compared with the duration of the flash. Krehbiel (1981), and Liu and Krehbiel (1985) argued that small growing thunderclouds produce primarily vertical cloud discharges while large mature or dissipating storms are more likely to produce extensive horizontal cloud discharges. Three of the four predominantly horizontal cloud discharges observed by Krehbiel (1981) occurred during the dissipating stage of a large storm system; the fourth was the tenth flash in a sequence of 15 flashes produced by a small storm. However, as discussed above in subsection 9.3.2, Proctor (1997), using UHF–VHF imaging, observed predominantly horizontal cloud discharges in relatively small thunderstorms in South Africa, some of these storms being in their initial growing stage.

We now present estimates of the propagation speeds, line charge densities, and average currents obtained by Proctor (1997) for flashes that began at heights more than 7.4 km (7.4 to 12 km) above mean sea level. A histogram of the speeds of 34 stepped leaders in cloud flashes is shown in Fig. 9.7a, and a histogram of 13 average stepped-leader currents in Fig. 9.7b. Each current was calculated by taking the products of channel extension speed and line charge density. The line charge densities, estimated using measured electric field changes and corresponding 355-MHz channel images, ranged from 0.7 to 8.7 C km^{-1}, the median value being 3.3 C km^{-1}. Stepped-leader speeds ranged from 1.3×10^4 to 1.3×10^5 m s^{-1} and currents from 37 to 630 A. The median speed was 7.5×10^4 m s^{-1}, and the median current was 130 A. Note from Table 9.2 that leaders in the lower-origin flashes are characterized by higher speeds and lower charge densities than those in the higher-origin flashes. However, the currents, 60 to 500 A, computed for the lower-origin flashes using the charge density and speed values given in Table 9.2, are similar to the current values (37 to 630 A) reported by Proctor (1997) for the higher-origin flashes.

9.4.2. Electric and magnetic field pulses

As noted in Section 9.2, the largest wideband electric and magnetic field pulses in the cloud flash are associated with the early stage and tend to mark the beginning of the discharge. Pierce (1955a) reported that radiation pulses as-

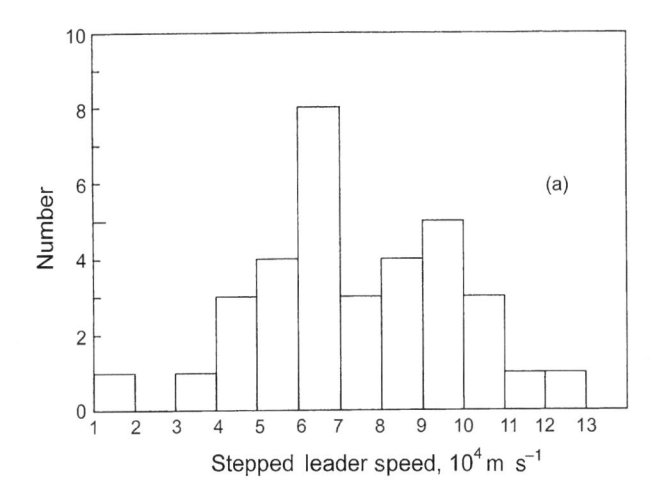

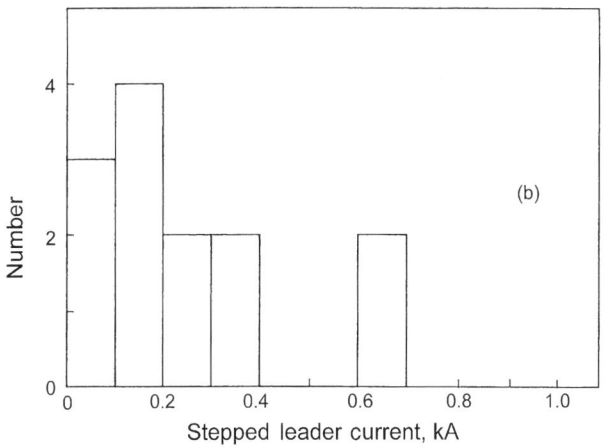

Fig. 9.7. Histograms of (a) channel-extension speeds (34 observations) and (b) average currents (13 observations) for stepped leaders in high-origin cloud flashes observed in South Africa. Adapted from Proctor (1997).

sociated with cloud flashes in Great Britain at ranges of 40 to 100 km were usually confined to the first 10 to 20 ms of the flash. The characteristics of large bipolar electric and magnetic field pulses produced by cloud discharges are summarized in Table 9.3. In the following, we will assume that all these pulses occur during the early stage and refer to them as initial-breakdown pulses. We now give a general characterization of initial-breakdown pulses. The narrow bipolar pulses observed in the early stage are then discussed further.

General characterization Table 9.3 includes two main types of initial breakdown pulse: (a) relatively slow-rising, wide bipolar waveforms with several small pulses superimposed on the initial half-cycle, as illustrated in Fig. 9.8, and (b) relatively narrow and smooth singly peaked or multiply peaked bipolar waveforms, illustrated in Fig. 9.9. The singly peaked pulses are sometimes termed narrow bipolar pulses. When these pulses are accompanied by large HF emissions,

Table 9.3. *Characteristics of large bipolar electric and magnetic field pulses (from wideband recordings) associated with cloud lightning discharges. Atmospheric electricity sign convention (subsection 1.4.2) is used throughout the table. The numerical data are, if not specified otherwise, the mean value, or the mean value ± standard deviation (when the latter is available), or the range of typical values. The numbers in parentheses are the sample sizes*

References	Characterization of pulses	Total pulse duration including overshoot, μs	Total width of initial half-cycle, μs	Half-peak width of initial half-cycle, μs	Ratio of initial peak to overshoot	Remarks
Isikawa (1961)	Mostly negative-initial-polarity bipolar pulses	40[a] (616)	—	—	—	Electric field system with an upper bandwidth of 100 kHz. Storms in Japan, 1956–9.
Weidman and Krider (1979)	Bipolar waveforms usually with two or three narrow pulses superimposed on the initial half cycle. The initial polarity is negative. The statistics are given for negative pulses only, some of these pulses being followed by a ground return stroke.	63 ± 39 (137)	About one-half the total duration (excluding singly peaked pulses, for which it is one-quarter the total duration)	—	3.6 ± 1.8 (65) 5.8 ± 2.0 for "fast-rising" pulses	Triggered system with useful bandwidth of 1 kHz to 2 MHz. Summer air-mass stroms in Florida, 1976–8.
Le Vine (1980)	Negative-initial-polarity bipolar pulses. Both singly peaked and multiply peaked initial half-cycles were observed.	10 to 20	10 or less	—	—	Electric field system triggered on the output of an RF detector. Summer storms in Florida in 1978, and in Virginia in 1979.
Cooray and Lundquist (1985)	Negative-initial-polarity bipolar pulses with a smooth rise to peak	75 ± 26 (23)	13 ± 4.9 (26)	—	3.3[b] (26)	Triggered system with bandwidth of some tens of Hz to 600 kHz. Ten night-time convective storms in Sri Lanka.
Bils et al. (1988)	Negative-initial-polarity bipolar pulses superimposed on millisecond-scale field changes of the same polarity. Observed in the early stage of the cloud flash. The initial half-cycle could be either singly peaked or multiply peaked.	74 (13)	—	9.3 (13) (mean) 2.7 (13) (median)	3.6[b] (13)	Continuous tape recording with upper 3-dB bandwidth of 500 kHz, 15-min period of 8–16 km-distant summer storm in Florida in 1984.

Reference	Waveform description	Col 3	Col 4	Col 5	Comments
Willett et al. (1989)	Negative- and positive-initial-polarity bipolar pulses with smooth initial rise to peak	Some[c] tens	10 to 15[c]	2.4 ± 1.4 (18) 1.9 ± 1.0 (6) (for two different storms)	8.8 ± 5.2 (18) 9.1 ± 2.0 (6) (for two different storms)	Electric field and field derivative systems triggered on the output of an HF receiver. The dE/dt system had >30 MHz bandwidth. Summer storms in Florida in 1985 and 1987.
Medelius et al. (1991)	Negative and positive (initial polarity) bipolar pulses with a smooth initial rise to peak	>13[d] (156, negative) >22[d] (10, positive)	>4.7[e] (156, negative) >7.7[e] (10, positive)	1.8 ± 0.63 (156, negative) 1.6 ± 0.87 (10, positive)	4.6 ± 2.1 (156, negative) 4.2 ± 1.5 (10, positive)	Triggered electric field and field derivative systems, and continuous tape recording. The dE/dt system had a bandwidth of about 100 MHz. Summer and fall storms in Florida in 1989.
Villanueva et al. (1994); Rakov et al. (1996)	Negative-initial-polarity bipolar pulses, both singly peaked and multiply peaked. Tended to occur in the early stage of the cloud flash.	53 ± 35 (17, singly peaked) 61 ± 36 (6, multiply peaked)	25 ± 13 (17, singly peaked) 27 ± 8.1 (6, multiply peaked)	2.8 ± 1.0 (17, singly peaked) 3.3 ± 2.5 (5, multiply peaked)	5.7 ± 2.1 (14, singly peaked) 4.3 ± 1.6 (5, multiply peaked)	Triggered electric field system with a useful upper frequency response of about 1 MHz. Three flashes from summer storms in Florida in 1991.

[a] Median value. For another ("minor") group of cloud-flash pulses the median value was 90 μs.
[b] Inverse of the mean ratio of the overshoot and the initial peak.
[c] Estimated from the composite pulse waveform.
[d] Estimated as the sum of the average values of the 10 percent to peak risetime, the peak to zero time, and the total overshoot width.
[e] Estimated as the sum of the average values of the 10 percent to peak and peak to zero times.

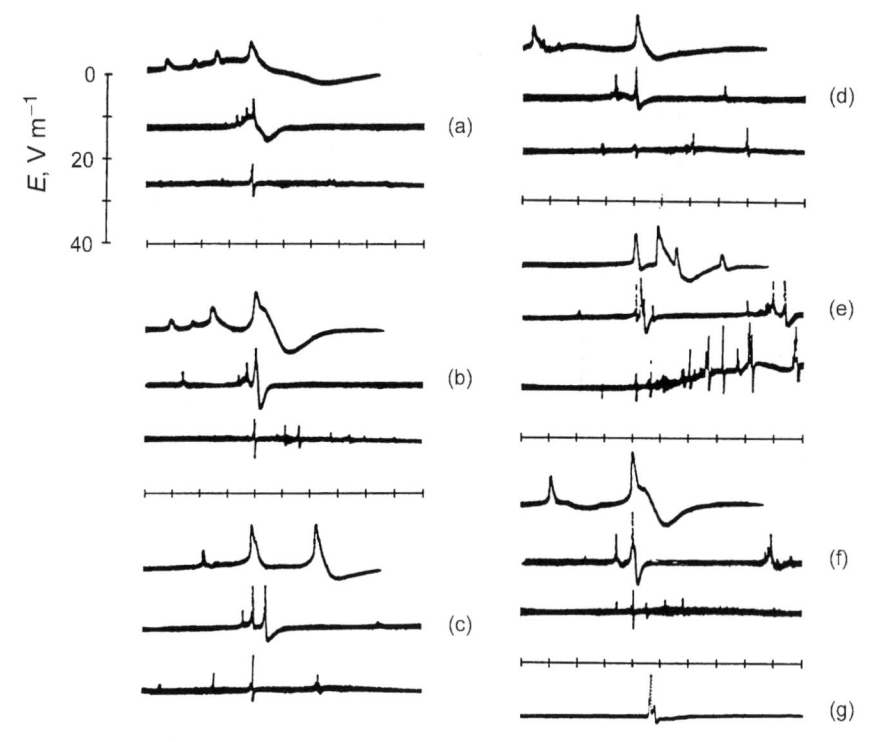

Fig. 9.8. Electric field waveforms apparently associated with the initial breakdown in cloud discharges at distances of 15–30 km. Each waveform is shown on time scales with 2 ms, 40 µs, and 8 µs per division (bottom, middle, and top traces in (a)–(f)) except for (g), which is shown just on a time scale with 40 µs per division. The atmospheric electricity sign convention (subsection 1.4.2) is used here. Adapted from Weidman and Krider (1979).

they are sometimes interpreted as being due to "compact intracloud discharges" (see below). The initial polarity of the large bipolar electric field pulses in cloud discharges is usually opposite to that of the initial-breakdown pulses in negative cloud-to-ground discharges, the latter polarity being usually the same as that of the following return-stroke pulse.

Individual initial-breakdown pulses in cloud discharges are characterized by a typical total duration of 50–80 µs (Table 1.2), about a factor 2 greater than the typical total duration of initial-breakdown pulses in negative cloud-to-ground flashes (subsection 4.3.2). Kitagawa and Brook (1960) reported that cloud-flash pulses appeared in groups separated by intervals ranging from 0.3 to 10 ms. The typical time interval between pulses is 600–800 µs (Table 1.2), considerably greater than the 70 to 130 µs intervals for initial-breakdown pulses in cloud-to-ground discharges (subsection 4.3.2).

Krider et al. (1979) studied the HF and VHF radiation associated with the initial-breakdown pulses in cloud flashes. They found that radiation at 3, 69, 139, and 295 MHz tended to peak during the initial half-cycle of the pulse.

Narrow bipolar pulses As noted above, narrow bipolar pulses are sometimes attributed to "compact intracloud discharges", a term coined by Smith D.A. et al. (1999b). The term compact intracloud discharge is based on the inference of Smith et al. (1999a, b) from a simple model that the spatial extent of this in-cloud process must be relatively small, 300 to 1000 m. Since this inference is model dependent, it should be viewed as tentative. Compact intracloud discharges are thought to produce, besides narrow bipolar pulses lasting for 10–15 µs, HF emissions whose amplitudes are 10 times larger than the HF emissions from typical cloud-to-ground and other cloud discharges. These two features are illustrated in Fig. 9.10. Additionally, compact intracloud discharges are apparently responsible for the "trans-ionospheric pulse pairs" (TIPPs) observed from space (Holden et al. 1995; Massey and Holden 1995; Massey et al. 1998; Zuelsdorf et al. 1997, 1998a, b, 2000; Jacobson A.K. et al. 1999, 2000; Suszcynsky et al. 2000) and "subionospheric pulse pairs" (SIPPs) observed on the ground (Smith D.A. and Holden 1996). TIPPs are also considered in subsections 14.6.1 and 17.8. The characteristics of narrow bipolar pulses are summarized, along with those of other initial-breakdown pulses, in Table 9.3.

9.4. Early (active) stage

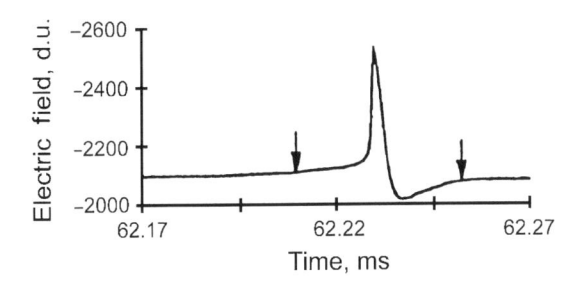

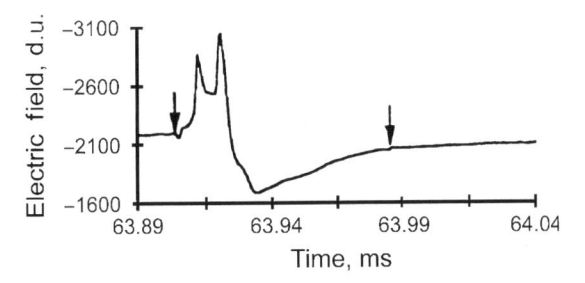

Fig. 9.9. Examples of (a) singly peaked and (b) multiply peaked electric field pulses in cloud flashes. Both waveforms are from the flash whose electric field is shown, on longer time scales, in Figs. 9.1a, b. The waveforms belong to the large-pulse category and occurred within the first 5 ms of the discharge; d.u. indicates "digitizer units". The atmospheric electricity sign convention (subsection 1.4.2) is used here. The arrows show the starting point and ending point chosen for waveform duration measurements presented in Table 9.3. Adapted from Villanueva et al. (1994).

We begin with a discussion of the narrow bipolar electric field pulses that have been recorded, in conjunction with their associated HF–VHF radiation. Such observations were first reported by Le Vine (1980) who used an electric field measuring system triggered when the associated HF–VHF signal, in the range from 3 to 300 MHz, exceeded a relatively high threshold level. The experiments were conducted at the NASA Kennedy Space Center, Florida during the Thunderstorm Research International Project (TRIP) in summer 1978 and at Wallops Island, Virginia in August 1979. The initial polarity of the bipolar pulses was negative (atmospheric electricity sign convention, subsection 1.4.2), that is, opposite to that of electric field pulses due to return strokes in negative cloud-to-ground lightning. The pulse amplitudes were of the order of one-third those of the return-stroke peaks recorded at about the same time. One pulse (event C in Fig. 7 of Le Vine 1980) was preceded and followed by pulses indicative of in-cloud lightning processes, while others were observed to occur after the first return stroke in cloud-to-ground flashes. Le Vine (1980) attributed the observed narrow bipolar pulses accompanied by strong HF–VHF radiation to K-processes, a hypothesis not confirmed by subsequent studies.

Willett et al. (1989), working at KSC in 1985 and 1987, obtained recordings of both electric field (E) and electric field derivative (dE/dt) for narrow bipolar pulses. Their measuring system was triggered by the output of an

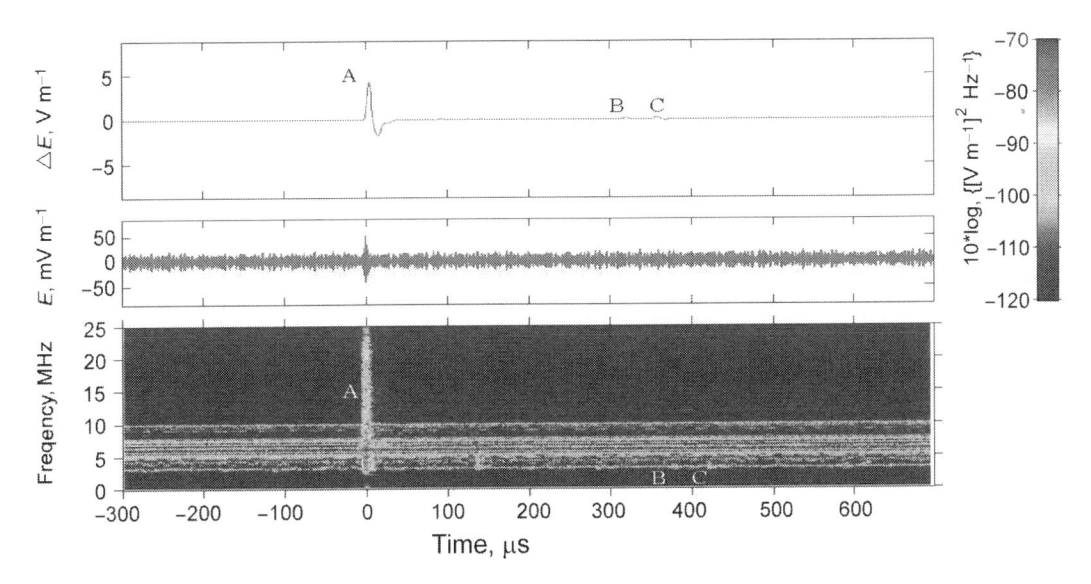

Fig. 9.10. At the time of going to press a colour version of this figure was available for download from http://www.cambridge.org/9780521035415. Electric field change waveform (top panel), broadband HF waveform (middle panel), and spectrogram (bottom panel) for a narrow bipolar pulse recorded at distance 103 km. The physics sign convention, which is opposite to the atmospheric electricity sign convention (subsection 1.4.2), is used here. Thus the electric field pulse has the same initial polarity as pulses shown in Figs. 9.8 and 9.9. The source of this event was located at a height of 7.1 km above ground level (9.2 km amsl). The features labeled A in the top and bottom panels are the ground-wave and the line-of-sight signals from the source. The pulses labeled B and C in the top panel are reflections from the ionosphere and Earth. The color-scale units are $10 \log (V^2 m^{-2} Hz^{-1})$. Adapted from Smith et al. (1999a).

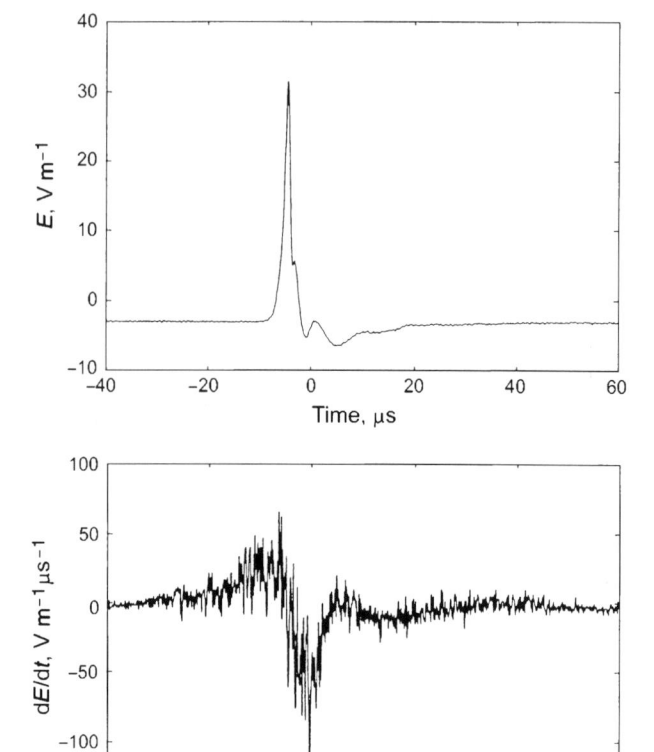

Fig. 9.11. Electric field (E) and electric field derivative (dE/dt) waveforms for a typical narrow bipolar pulse with positive initial polarity (physics sign convention, subsection 1.4.2), labeled NPBP by Willett et al. (1989), from an isolated thunderstorm (203928.628) centered 45 km away from the measuring site. This pulse has the same initial polarity as the pulses shown in Figs. 9.8, 9.9, and 9.10. Note the noisy appearance of dE/dt (shown on an expanded time scale) in comparison with E. Adapted from Willett et al. (1989).

HF receiver, which could be tuned to any center frequency between 3 and 18 MHz. Both polarities of the initial half-cycle of the narrow bipolar pulses were observed, negative polarity (atmospheric electricity sign convention, subsection 1.4.2) being more frequent. Note that Willett et al. (1989) (and Smith D.A. et al. 1999a, whose work is discussed later in this subsection) used the physics sign convention, which is opposite to the atmospheric electricity sign convention employed by other researchers who have reported on narrow bipolar pulses, including Weidman and Krider (1979), Le Vine (1980), Bils et al. (1988), Medelius et al. (1991), Villanueva et al. (1994), and Rakov et al. (1996). Examples of waveforms of each polarity recorded by Willett et al. (1989) are shown in Figs. 9.11 and 9.12. For 18 waveforms, Willett et al. (1989) estimated the mean values of the E and dE/dt peaks normalized to 100 km as 8 V m^{-1} and 20 V m^{-1}μs^{-1}, respectively. Both values are comparable to their counterparts for first return strokes observed in the same experiment. The overall pulse width was 20 to 30 μs. Spectral analysis indicated that the sources of the narrow bipolar pulses radiated much more strongly than first return strokes at frequencies from 10 to at least 50 MHz. At 18 MHz the energy spectral density, measured in (V m^{-1} Hz^{-1})^2, for these pulses was nearly 16 dB higher than that for first return strokes at the same distance.

Smith D.A. et al. (1999a), who used a multiple-station electric field change measuring system in concert with an HF (3 to 30 MHz) TOA lightning locating system, presented a detailed characterization of the narrow bipolar pulses in three thunderstorms at distances greater than 80 km in New Mexico and west Texas, including locations of their sources in the cloud. The relative times of arrival of signal reflections from the ionosphere and from the Earth were used to estimate source heights. The sources were located in the most active regions of the thunderstorms, in

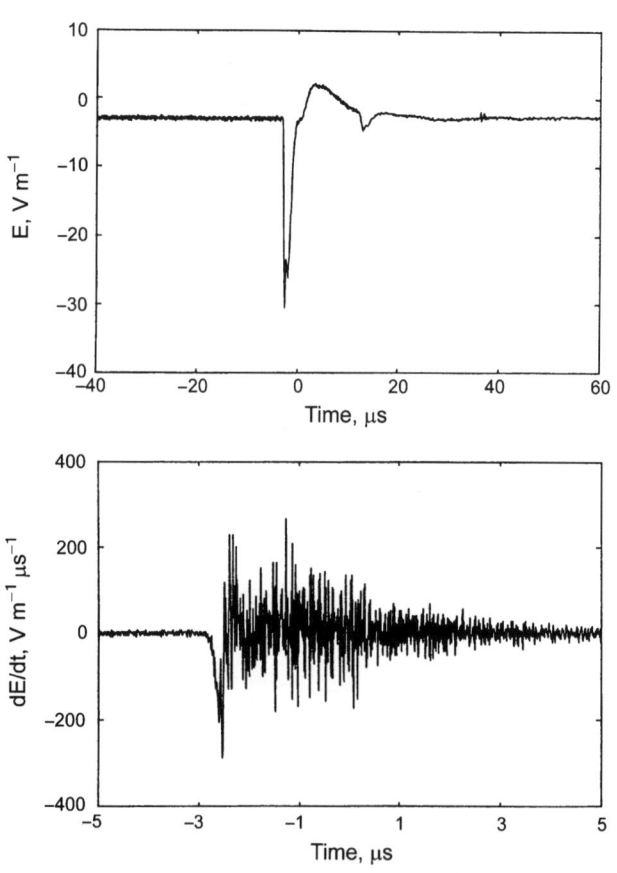

Fig. 9.12. Same as Fig. 9.11 but for a narrow bipolar pulse with negative initial polarity (physics sign convention, subsection 1.4.2). This event, labeled NNBP by Willett et al. (1989), occurred at unknown range. Adapted from Willett et al. (1989).

9.5. Early (active) stage

Table 9.4. *Characteristics of narrow bipolar electric field pulses and associated HF (3–30 MHz) radiation reported by Smith et al. (1999a)*

	Mean ± Std. Dev.
Electric field pulse characteristics	
Risetime (10–90 percent)	2.3 ± 0.8 μs
Half-peak width	4.7 ± 1.3 μs
Pulse duration	25.8 ± 4.9 μs
Initial peak[a]	9.5 ± 3.6 V m^{-1}
Opposite-polarity overshoot[a]	−3.9 ± 1.6 V m^{-1}
Ratio of initial peak to opposite-polarity overshoot	2.7
Ratio of peaks for narrow bipolar pulses and return-stroke pulses	0.71
Ratio of peaks for narrow bipolar pulses and cloud-flash pulses	2.6
HF radiation characteristics	
Duration	2.8 ± 0.8 μs
Peak[b]	2.4 ± 1.1 mV m^{-1}
Ratio of peaks for narrow bipolar pulses and return-stroke pulses	9.9
Ratio of peaks for narrow bipolar pulses and cloud-flash pulses	29

[a] Normalized to 100 km; physics sign convention (subsection 1.4.2).
[b] Normalized to 10 km and 1 kHz bandwidth.

close proximity to high-reflectivity (in excess of 40 dBZ) cores, and at altitudes between 8 and 11 km above mean sea level.

The characteristics of the 24 narrow bipolar electric field pulses and associated HF (3 to 30 MHz) radiation studied by Smith D.A. *et al.* (1999a) are summarized in Table 9.4. Smith *et al.* reported that the pulse peaks were comparable to those of return-stroke waveforms. The pulse polarity was always opposite to that of negative return-stroke pulses in cloud-to-ground flashes. The lack of opposite-polarity pulses was probably due to the use of unipolar triggering circuits. Nearly all the pulses studied by Smith *et al.* were the only events within the field record having a length of typically 4 to 10 ms. According to Smith *et al.*, the HF (3 to 30 MHz) emissions associated with narrow bipolar pulses had a duration of only a few microseconds and were typically 10 times more powerful than the HF emissions from "normal lightning discharges".

Rison *et al.* (1999), who used a three-dimensional VHF (63 MHz) TOA lightning locating system, found that the discharge process giving rise to narrow bipolar pulses occurred between the main negative and main positive charge regions of the cloud and was the initial event of an "otherwise normal intracloud discharge". This observation appears to be consistent with the previous findings of Bils *et al.* (1988) and Villanueva *et al.* (1994), discussed later in this subsection, that narrow bipolar pulses tend to occur early in the cloud discharge. Thus the compact intracloud discharge is likely to be an element of the initial breakdown process of a cloud flash, although the role of this element is presently unclear. Rison *et al.* (1999) reported that the peak VHF radiation from compact intracloud discharges was typically 30 dB stronger than that from other lightning discharge processes and corresponded to a source power in excess of 100 kW over a 6 MHz bandwidth centered at 63 MHz.

Observations of narrow bipolar electric field pulses without simultaneous recordings of associated HF–VHF radiation have been reported by Weidman and Krider (1979, Fig. 4g), Cooray and Lundquist (1985, Fig. 10), Bils *et al.* (1988, Figs. 4 and 12), Medelius *et al.* (1991), Villanueva *et al.* (1994, Fig. 4a), Rakov *et al.* (1996), and Rakov (1999, Fig. 2c). The characteristics of pulses from most of these studies are found in Table 9.3. Note that a necessary feature of the narrow bipolar pulses discussed earlier in this subsection is strong HF–VHF radiation. It is presently unknown if there exist narrow bipolar pulses that are not accompanied by such radiation. Therefore, it is conceivable that the pulses considered in the remainder of this section are due to discharge processes that do not necessarily produce strong HF–VHF radiation and hence may be different from those considered above.

The pulses reported by Cooray and Lundquist (1985) were recorded in Sri Lanka and had amplitudes of about 0.2 to 0.5 of the return-stroke pulse amplitudes observed at approximately the same time. The total duration of these pulses was about 75 μs, a factor 3 to 4 larger than that reported by Le Vine (1980), the initial half-cycle duration was 13 μs, and the ratio of the initial peak to opposite-polarity overshoot was 3.3. The corresponding values reported by Villanueva *et al.* (1994) from measurements in Florida were 53 μs, 25 μs, and 5.7. Both Bils *et al.* (1988) and Villanueva *et al.* (1994) noted that the waveforms occurred in the early stage of cloud flashes. Weidman and Krider (1979) noted that the first signals detected from developing storms usually were "fast negative-polarity wave forms with relatively short pulse widths". Medelius *et al.* (1991) analyzed 156 negative and 10 positive (atmospheric electricity sign convention, subsection 1.4.2) narrow bipolar pulses that occurred during overhead thunderstorms at the Kennedy Space Center, Florida. They reported that about two-thirds of 30 pulses identified in continuous magnetic tape records occurred "more than one second apart from any burst activity typical of lightning", apparently in contrast with other studies discussed above, in which narrow bipolar pulses were usually found to occur in the early stage of cloud discharges.

9.5. Late (final) stage

As noted in Section 9.2, the late-stage processes apparently serve to transport negative charge to the region of the flash origin from progressively more remote sources in the negative charge region. The overall characteristics of the late (final) stage are presented in subsection 9.5.1 and the late-stage microsecond-scale pulses, including regular pulse bursts, in subsection 9.5.2.

9.5.1. Overall characteristics

The electric field changes of the late (or final) stage of the cloud flash are similar to the field changes between strokes and after the last stroke of the cloud-to-ground flash (Section 4.10). Indeed, in each case, on relatively slow electric field changes (J-changes) are superimposed step-like changes (K-changes), typically lasting for 1 ms or so and separated by time intervals of order 10 ms (Ogawa and Brook 1964; Thottappillil et al. 1990). It is thought that the corresponding processes (J- and K-processes) in cloud and ground flashes are similar. Thus, the discussion of J- and K-processes in ground discharges found in Section 4.10 is likely to be applicable to cloud discharges. As noted in Section 9.2, the J-process in ground flashes is often visualized as a positively charged channel (or channels) extending from the region of the flash origin (following the arrival there of the return stroke) at a speed of order 10^4 m s^{-1}, which is equivalent to supplying negative charge to the origin. K-processes can be viewed as transients occurring during the slower J-process. The late stage of a cloud flash has a duration of 50–200 ms (Ogawa and Brook 1964).

In a view that is now generally considered incorrect, Ogawa and Brook (1964) argued that the K-process is a negative "recoil streamer" that occurs when the initial downward-propagating positive leader that they postulated as initiating a cloud flash encounters a region of concentrated negative charge. They deduced from photographic studies of air discharges that the "recoil streamer" originated at the tip of the advancing leader and propagated back along the previously formed channel, the luminosity decreasing as it did so. Successive "recoil streamers" were observed to retrace the same channel several times. Thus, according to Ogawa and Brook (1964), the "recoil streamer" originates at the lower extremity of the downward-moving positive-leader channel and propagates upward, the overall leader–"recoil-streamer" process being much like the leader–return-stroke sequence in ground flashes. In fact, Ogawa and Brook (1964) even stated that the K-process "must be viewed as an intracloud return stroke". An interpretation of K-changes similar to that of Ogawa and Brook (1964), but extended to include either polarity, was offered by Khastgir and Saha (1972). Further, Ogawa and Brook (1964) inferred that the J-change is "effectively the sum of the individual K-changes". These hypothetical views of K-processes are not supported by more recent studies. As stated in Section 9.2, the initial leader in cloud discharges is usually negative, and Thottappillil et al. (1990) demonstrated, although for ground discharges, that appreciable J changes can occur without step-like K-changes. On the one hand, Ogawa and Brook (1964) found that on average there were six K-changes per cloud flash and noted that K-changes did not occur during the first 100–300 ms of the discharge. On the other hand, Krehbiel (1981) concluded that K-changes occur throughout the duration of a cloud flash, thus extending the term "K-change" to transient processes occurring during the early stage of a cloud discharge, when the initial negative leader is extending in the "forward" direction. Krehbiel (1981) reported that the early K-processes are associated with vertical charge transfer and the later ones with horizontal charge transfer at the negative charge level. The interpretation of K-processes by Mazur (1989) as "recoil streamers" propagating toward the flash origin along channels created by positive leaders is discussed in subsection 9.3.2. As noted in Section 9.2, K-processes in cloud discharges can apparently occur in a variety of forms, depending on the origination point, conditions along the propagation path, and conditions at the termination point.

From single-station electric field change measurements, Brook and Ogawa (1977) estimated that the propagation speed of the "recoil streamers", assumed to be vertical, was of order 10^6 m s^{-1} (from an inferred "recoil streamer" channel length of 1.3 km and an observed typical K-change duration of 1 ms). They also inferred that the charge transfer and the average current of "recoil streamers" were about 1.4 C and 1.4 kA, respectively.

9.5.2. Wideband electric and magnetic field pulses

Microsecond-scale pulse activity during the late stage is usually attributed to K-processes. Bils et al. (1988), who studied the electric field changes of six cloud flashes at distances of 8 to 16 km in Florida, reported that the K-change pulses were irregular in shape, some having opposite polarity to that of the step-like K-changes within which they occurred. Microsecond-scale pulses occurring during the final stage of cloud flashes, most of which are probably associated with K-processes, are characterized by smaller amplitudes than the initial breakdown pulses (Villanueva et al. 1994). Many K-changes are associated with microsecond-scale pulse activity in the form of trains of relatively small pulses separated by time intervals of 5 μs or so. These trains are usually referred to as regular pulse bursts.

Regular pulse bursts occurring in cloud flashes were first studied in detail by Krider et al. (1975). Similar bursts in both cloud-to-ground and cloud flashes have been analyzed in relation to the corresponding millisecond-scale electric field changes by Rakov et al. (1996). Only regular

9.5. Late (final) stage

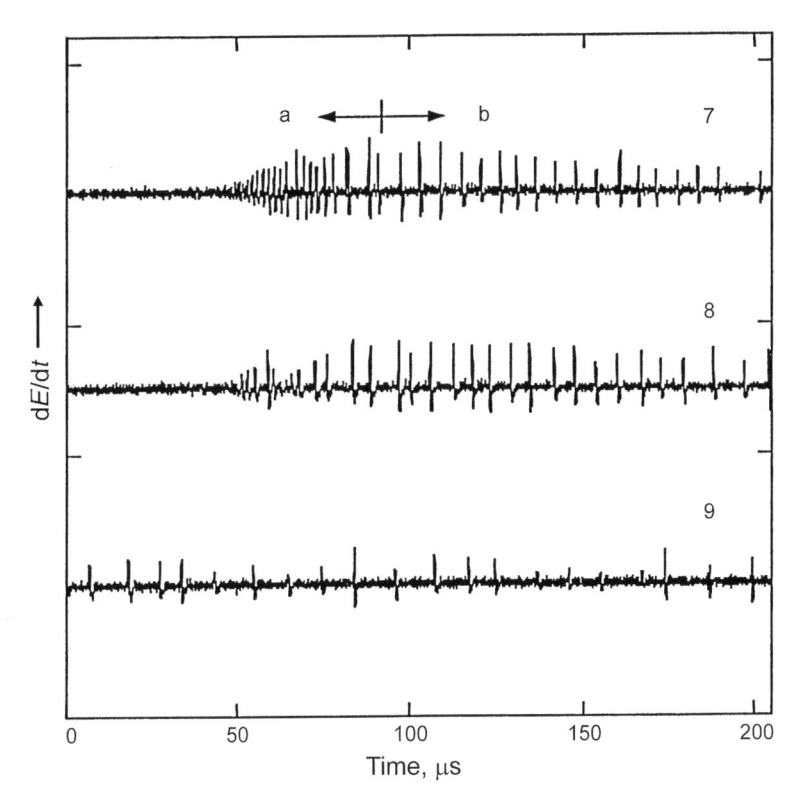

Fig. 9.13. Three dE/dt records (labeled 7, 8, and 9) of two pulse bursts in a cloud flash. Records 8 and 9 are separated by a time interval of 30 μs and are considered to be due to the same pulse burst. Adapted from Davis (1999).

pulse bursts in cloud flashes are presented in this section, although the characteristics of regular pulse bursts in ground discharges are similar to those in cloud discharges. A description of the regular pulse bursts in cloud-to-ground flashes is found in Section 4.11.

Rakov et al. (1996) reported that there were on average 18 to 24 electric field pulses per burst in the three cloud flashes they analyzed. The average interpulse interval varied from 6.4 to 7.2 μs, and the average burst duration from 117 to 161 μs. As in ground flashes, there was a tendency for the bursts to occur in the latter stages of the discharge, and positive and negative pulse polarities were about equally probable; likewise, many bursts in cloud flashes occurred in the latter part of K electric field changes, as illustrated for ground flashes in Fig. 4.62, although the majority of bursts studied by Rakov et al. (1996) occurred when there was no detectable slow field change.

Davis (1999), using multiple-station dE/dt measurements in Florida and a time-of-arrival (TOA) source locating technique, studied 35 regular pulse bursts in 11 cloud discharges. The sources of pulses were located at heights ranging from 5.5 to 8.2 km. In one case, the sources appeared to retrace the path defined by the previous regular pulse burst. Davis (1999) observed different pulse polarities for bursts recorded at different stations and also polarity reversals at a single station. He showed that pulse polarities were related to channel orientation. The average propagation speed was found to range from 1.2×10^6 to 5.6×10^6 m s^{-1}, with a mean value of 2.7×10^6 m s^{-1}, which is comparable to the average speed of dart-stepped leaders near ground (subsection 4.7.5). The average interpulse interval was about 5 μs, and there was a tendency for the speed to decrease with increasing interpulse interval. An example of two bursts is shown in Fig. 9.13. The source locations for these bursts are shown in Fig. 9.14. Record 7 represents the first burst, and records 8 and 9, which are separated by a 30 μs measuring system dead-time gap, represent the second one. The average speed was computed for each of the two segments of record 7, designated a and b in Fig. 9.13. The average speed over segment a, lasting 33 μs, was 5.4×10^6 m s^{-1} while that over segment b, lasting 92 μs, was 2.2×10^6 m s^{-1}. Note that the decrease in speed in record 7 corresponds to an increase in average interpulse interval from 2.1 to 5.7 μs. The average speeds for records 8 and 9 were 2.9×10^6 and 1.8×10^6 m s^{-1}, respectively. The corresponding average interpulse intervals are about 6 and 10 μs. Multiplying the average speed for a burst of dE/dt pulses by its duration and dividing the result by the number of pulses, Davis (1999) estimated the average distance between the sources of consecutive

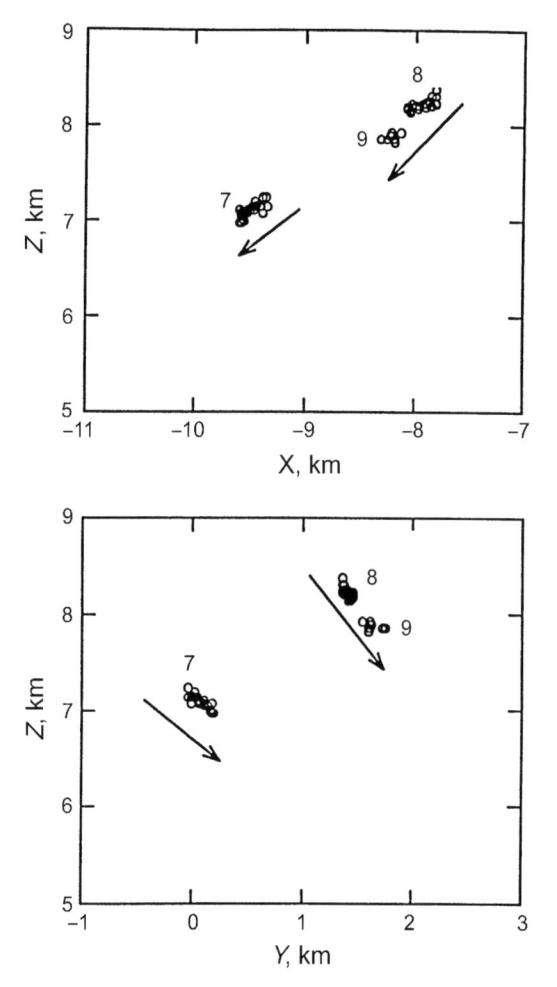

Fig. 9.14. Locations of sources for the pulse bursts shown in Fig. 9.13. The arrows indicate the directions of movement of the sources. Adapted from Davis (1999).

pulses, which he interpreted as the average step length. For an amplitude threshold level of 30 percent of the maximum pulse peak, the step lengths varied from 7 to 30 m with a mean of 14 m, similar to the step lengths, of the order of 10 m, observed optically for dart-stepped leaders near ground (subsection 4.7.5). For the two sections of record 7 shown in Fig. 9.13 the step lengths were about 13 and 12 m, and for records 8 and 9 they were about 16 and 26 m, respectively. Thus the in-cloud process giving rise to the regular pulse-burst signature appears to be generally similar to the dart-stepped leader process, as first suggested by Krider et al. (1975).

9.6. Comparison with ground discharges

The stepped-leader currents in cloud flashes, tens to hundreds of amperes as estimated by Proctor (1997) (subsection 9.4.1), are comparable to the stepped-leader currents in ground flashes (subsection 4.4.3).

We now compare the initial breakdown processes in cloud and ground flashes. As indicated in Table 1.2, the dominant polarity of the initial half-cycle of wideband electric field pulses associated with the initial breakdown in cloud flashes is negative, while in cloud-to-ground flashes the dominant polarity of the initial-breakdown pulses is positive (atmospheric electricity sign convention, subsection 1.4.2). Kitagawa and Brook (1960) observed that the intervals between electric field pulses occurring during the first 10 ms or so of cloud discharges were typically 680 μs versus 40 to 60 μs for cloud-to-ground discharges. Further, Rakov et al. (1996), who summarized, for cloud discharges and for negative ground discharges, the characteristics of initial breakdown pulses found in the literature together with those deduced from their own electric field records, reported that the typical interpulse intervals were 600–800 and 70–130 μs, respectively (Table 1.2). Apparently in support of these findings, Shao and Krehbiel (1996) reported that the VHF radiation associated with the initial breakdown in cloud flashes is "much more intermittent" than that in cloud-to-ground flashes. As noted in subsection 9.4.2, individual initial-breakdown pulses in cloud discharges are characterized by a typical total duration that is a factor 2 greater than the typical total duration of initial-breakdown pulses in negative ground discharges. However, Proctor (1981, 1991, 1997) found that (i) cloud flashes originated at both higher and lower altitudes while ground flashes only originated at lower altitudes (Section 9.2) and that (ii) the characteristics of VHF–UHF radiation due to the initial breakdown and the stepped-leader processes (he does not distinguish between the two) are determined by the flash-origin height, so that ground flashes and lower-origin cloud flashes are indistinguishable at VHF–UHF. Further, Proctor (1997) stated that the characteristics of the initial breakdown in ground and in lower-origin cloud flashes "differ in no way that can be detected at MF, at HF, or at VHF". Indeed, a low-origin initial-breakdown process has no way of "knowing" whether the following stages of the flash will involve ground or not. Shao and Krehbiel (1996) reported that the propagation speeds of initial leaders in cloud and ground flashes were similar, 1×10^5 to 3×10^5 m s^{-1}. Further, Proctor (1997) found progression speeds of the stepped leaders in lower-origin cloud flashes and in ground flashes to be similar, the speed range being from 6×10^4 to 5×10^5 m s^{-1} (Table 9.2). Note from Table 9.2 that the initial-leader progression speeds for the higher-origin cloud flashes appear to be a factor 4 to 5 lower than for either lower-origin cloud flashes or ground flashes.

As noted in subsection 9.5.2, the regular pulse bursts that are often associated with K-processes are similar for cloud and ground flashes.

Malan (1958), who studied electromagnetic radiation from cloud and ground discharges at frequencies ranging from 3 kHz to 12 MHz, found that the overall radiation patterns of cloud lightning resemble those of the interstroke and final stages (after the last stroke) of ground lightning. He also estimated the ratios of the amplitudes of the return-stroke radiation in ground discharges and the most intense radiation pulses in cloud discharges at different frequencies. There is a tendency for this ratio to decrease with increasing frequency, from 20 to 40 at 3 kHz to 1 at 1.5–20 MHz. Brook and Kitagawa (1964) reported that cloud discharges appeared to be radiators at 420 and 850 MHz, as strong or stronger than ground discharges.

9.7. Summary

The typical cloud discharge exhibits two stages. The first stage has a duration of some tens to a few hundreds of milliseconds and is referred to as the early or active stage. The second stage, also lasting for some tens to a few hundreds of milliseconds, is termed the late or final stage. The early (active) stage typically involves a negatively charged channel extending in a manner presumably similar to that of the stepped leader in cloud-to-ground discharges, the extension speed being of order 10^5 m s^{-1}. Cloud flashes are most likely to begin near the upper and lower boundaries of the main negative charge region and in the former case often bridge the main negative and main positive charge regions in the cloud. Other scenarios may involve the lower positive charge region or may be indicative of a cloud charge distribution different from a simple tripolar model. The beginning of a cloud discharge is marked by the largest pulses in the electric field record. These pulses are usually attributed to the initial breakdown processes that lead to the formation of the stepped leader, although the distinction between the initial breakdown and the stepped leader is not clear. Some characteristics of the initial breakdown and stepped-leader processes apparently depend on the height of the origin of the cloud discharge. It is likely that a cloud discharge begins as a bidirectional leader, the positive section of this leader pervading the negative charge region and effectively supplying negative charge through the flash origin to the negative section extending into the positive charge region. The transition from the early stage to the late stage is associated with a loss of connection between the positive and negative sections (or between the negative and positive charge regions in the cloud) and with cessation of the extension of the negative stepped leader channel. The late-stage processes apparently serve to transport negative charge to the flash origin from progressively more remote sources in the negative charge region.

The initial polarity of the initial-breakdown pulses in cloud flashes is usually opposite to that of the initial-breakdown pulses in cloud-to-ground flashes. Individual initial-breakdown pulses in cloud discharges are characterized by a total duration of 50 to 80 μs, about a factor 2 greater than the typical total duration of initial-breakdown pulses in cloud-to-ground discharges. The typical time interval between initial-breakdown pulses in cloud discharges is 600 to 800 μs, considerably greater than the 70 to 130 μs intervals for initial-breakdown pulses in cloud-to-ground discharges. "Compact intracloud discharges" produce relatively smooth, singly peaked bipolar electric field pulses lasting for 10 to 15 μs, accompanied by HF emissions whose amplitudes are 10 times larger than the HF emissions from typical cloud-to-ground and other cloud discharges. Compact intracloud discharges can be, and perhaps most often are, elements of the initial-breakdown process of cloud flashes.

The processes that occur in the late (final) stage of a cloud discharge (J- and K-processes) are similar to those between strokes and after the last stroke in a cloud-to-ground discharge. Both J- and K-processes serve to transport negative charge to the region of the flash origin from progressively more remote sources in the negative charge region. K-processes in cloud flashes, which can be viewed as transients that occur during the relatively slow J-process, appear to take a variety of forms depending on the origination point, conditions along the propagation path, and conditions at the termination point. The characteristics of regular pulse bursts in cloud discharges are similar to those in ground discharges.

References and bibliography for Chapter 9

Aina, J.I. 1971. Lightning discharge studies in a tropical area. II. Discharges which do not reach the ground. *J. Geomagn. Geoelectr.* **23**: 359–68.

Aina, J.I. 1972. Lightning discharge studies in a tropical area. III. The profile of the electrostatic field changes due to non-ground discharges. *J. Geomagn. Geoelectr.* **24**: 369–80.

Arnold, H.R., and Pierce, E.T. 1964. Leader and junction processes in the lightning discharge as a source of VLF atmospherics. *Radio Sci.* **68D**: 771–6.

Baral, K.N., and Mackerras, D. 1992. The cloud flash-to-ground flash ratio and other lightning occurrence characteristics in Kathmandu thunderstorms. *J. Geophys. Res.* **97**: 931–8.

Bils, J.R., Thomson, E.M., Uman, M.A., and Mackerras, D. 1988. Electric field pulses in close lightning cloud flashes. *J. Geophys. Res.* **93**: 15 933–40.

Bondiou, A., Richard, P., Taudiere, T., and Helloco, F. 1986. Preliminary correlation between 3-dimensional lightning discharge mapping and radar measurements. In *Proc. 1986 Int. Conf. on Lightning and Static Electricity, Dayton, Ohio.*

Brantley, R.D., Tiller, J.A., and Uman, M.A. 1975. Lightning properties in Florida thunderstorms from video tape records. *J. Geophys. Res.* **80**: 3402–6.

Brook, M., and Kitagawa, N. 1960. Electric-field changes and the design of lightning-flash counters. *J. Geophys. Res.* **65**: 1927–31.

Brook, M., and Kitagawa, N. 1964. Radiation from lightning discharges in the frequency range 400 to 1000 Mc/s. *J. Geophys. Res.* **69**: 2431–4.

Brook, M., and Ogawa, T. 1977. The cloud discharge. In *Lightning, vol. 1, Physics of Lightning*, ed. R. Golde, pp. 191–230, London: Academic Press.

Carey, L.D., and Rutledge, S.A. 1998. Electrical and multiparameter radar observations of a severe hailstorm. *J. Geophys. Res.* **103**: 13 979–14 000.

Clegg, R.J., and Thomson, E.M. 1979. Some properties of EM radiation from lightning. *J. Geophys. Res.* **84**: 719–24.

Cooray, V., and Lundquist, S. 1985. Characteristics of the radiation fields from lightning in Sri Lanka in the tropics. *J. Geophys. Res.* **90**: 6099–109.

Davis, S.M. 1999. Properties of lightning discharges from multiple-station wideband electric field measurements. Ph.D. dissertation, University of Florida, Gainesville, 228 pp.

Dotzek, N., Höller, H., Théry, C., and Fehr, T. 2001. Lightning evolution related to radar-derived microphysics in the 21 July 1998 EULINOX supercell storm. *Atmos. Res.* **56**: 335–54.

Funaki, K., Sakamoto, K., Tanaka, R., and Kitagawa, N. 1981. A comparison of cloud and ground lightning discharges observed in South-Kanto summer thunderstorms, 1980. *Res. Lett. Atmos. Electr.* **1**: 99–103.

Galvan, A., Cooray, V., and Scuka, V. 1999. Interaction of electromagnetic fields from cloud and ground lightning flashes with an artificial low-voltage power installation. *IEEE Trans. Electromagn. Compat.* **41**: 250–7.

Goshima, H., Motoyama, H., Asakawa, A., Wada, A., Shindo, T., and Yokoyama, S. 2000. Characteristics of electromagnetic fields due to winter lightning stroke current to a high stack. *Trans. IEE Japan* **120-B**(1): 44–8.

Hager, W.W., and Wang, D. 1995. An analysis of errors in the location, current, and velocity of lightning. *J. Geophys. Res.* **100**: 25 721–9.

Hamelin, J. 1993. Sources of natural noise. In *Electromagnetic Compatibility*, eds. P. Degauque and J. Hamelin, 652 pp., New York: Oxford.

Hayenga, C.O. 1984. Characteristics of lightning VHF radiation near the time of return strokes. *J. Geophys. Res.* **89**: 1403–10.

Heckman, S. 1992. Why does a lightning flash have multiple strokes? Ph.D. dissertation, Massachusetts Institute of Technology, Cambridge, Massachusetts.

Helsdon, J.H. Jr, Wu, G., and Farley, R.D. 1992. An intracloud lightning parameterization scheme for a storm electrification model. *J. Geophys. Res.* **97**: 5865–84.

Holden, D.N., Munson, C.P., and Devenport, J.C. 1995. Satellite observations of transionospheric pulse pairs. *Geophys. Res. Lett.* **22**: 889–92.

Huzita, A., and Ogawa, T. 1976a. Charge distribution in the average thunderstorm cloud. *J. Meteor. Soc. Japan* **54**: 285–8.

Huzita, A., and Ogawa, T. 1976b. Electric field changes due to tilted streamers in the cloud discharge. *J. Meteor. Soc. Japan* **54**: 289–93.

Isikawa, H. 1961. Nature of lightning discharges as origins of atmospherics. *Proc. Res. Inst. Atmospherics, Nagoya Univ., Japan* **8A**: 1–274.

Isikawa, H., and Takeuchi, T. 1966. Field changes due to lightning discharge. *Proc. Res. Inst. Atmospherics, Nagoya Univ., Japan* **13**: 59–61.

Jacobson, A.R., Knox, S.O., Franz, R., and Enemark, C.D. 1999. FORTE observations of lightning radio-frequency signatures: capabilities and basic results. *Radio Sci.* **34**: 337–54.

Jacobson, A.R., Cummins, K.L., Carter, M., Klingner, P., Roussel-Dupré, D., and Knox, S.O. 2000. FORTE radio-frequency observations of lightning strokes detected by the National Lightning Detection Network. *J. Geophys. Res.* **105**: 15 653–62.

Jacobson, E.A., and Krider, E.P. 1976. Electrostatic field changes produced by Florida lightning. *J. Atmos. Sci.* **33**: 113–17.

Jayaratne, E.R., Ramachandran, V., and Devan, K.R.S. 1995. Observations of lightning flash rates and rain-gushes in Gaborone, Botswana. *J. Atmos. Terr. Phys.* **57**: 325–31.

Kelley, N.C., Baker, S.D., Holzworth, R.H., Argo, P., and Cummer, S.A. 1997. RF and MF observations of the lightning electromagnetic pulse at ionospheric altitudes. *Geophys. Res. Lett.* **24**: 1111–14.

Khastgir, S.R., and Saha, S.K. 1972. On intra-cloud discharges and their accompanying electric field-changes. *J. Atmos. Terr. Phys.* **34**: 115–26.

Kitagawa, N. 1957. On the mechanism of cloud flash and junction or final process in flash to ground. *Pap. Meteor. Geophys. (Tokyo)* **7**: 415–24.

Kitagawa, N., and Brook, M. 1960. A comparison of intracloud and cloud-to-ground lightning discharges. *J. Geophys. Res.* **65**: 1189–201.

Kitagawa, N., and Kobayashi, M. 1959. Field changes and variations of luminosity due to lightning flashes. In *Recent Advances in Atmospheric Electricity*, ed. L.G. Smith, pp. 485–501, Oxford: Pergamon.

Kobayashi, M., Kitagawa, N., Ikeda, T., and Sato, Y. 1958. Preliminary studies of variation of luminosity and field change due to lightning flashes. *Pap. Meteor. Geophys. (Tokyo)* **9**: 29–34.

Krehbiel, P.R. 1981. An analysis of the electric field change produced by lightning. Ph.D. thesis, University of Manchester Institute of Science and Technology, Manchester, England (available as Report T-11, Geophys. Res. Ctr., New Mexico Inst. Mining and Tech., Socorro, NM 87801, 1981).

Krehbiel, P.R., Brook, M., and McCrory, R, 1979. An analysis of the charge structure of lightning discharges to the ground. *J. Geophys. Res.* **84**: 2432–56.

Krehbiel, P.R., Thomas, R.J., Rison, W., Hamlin, T., Harlin, J., and Davis, M. 2000. GPS-based mapping system reveals lightning inside storms. *Eos, Trans, Am. Geophys. Union* **81**(3): 21–5.

Krider, E.P. 1974. An unusual photograph of an air lightning discharge. *Weather* **29**: 24–7.

Krider, E.P., Radda, G.J., and Noggle, R.C. 1975. Regular radiation field pulses produced by intracloud lightning discharges. *J. Geophys. Res.* **80**: 3801–4.

Krider, E.P., Weidman, C.D., and Le Vine, D.M. 1979. The temporal structure of the HF and VHF radiation produced by intracloud lightning discharges. *J. Geophys. Res.* **84**: 5760–2.

Leteinturier, C., and Hamelin, J. 1990. Experimental study of the electromagnetic characteristics of lightning discharge in the 200 Hz–20 MHz band. In *Lightning Electromagnetics*, ed. R.L. Gardner, pp. 423–39, New York: Hemisphere.

Le Vine, D.M. 1980. Sources of the strongest RF radiation from lightning. *J. Geophys. Res.* **85**: 4091–5.

Ligda, M.G.H. 1956. The radar observation of lightning. *J. Atmos. Terr. Phys.* **9**: 329–46.

Light, T.E., Suszcynsky, D.M., and Jacobson, A.R. 2001. Coincident radio frequency and optical emissions from lightning observed with the FORTE satellite. *J. Geophys. Res.* **106**: 28 223–31

Liu, X.-S., and Krehbiel, P.R. 1985. The initial streamer of intracloud lightning flashes. *J. Geophys. Res.* **90**: 6211–18.

Lupo, G., Petrarca, C., Tucci, V., and Vitelli, M. 2000. EM fields generated by lightning channels with arbitrary location and slope. *IEEE Trans. Electromagn. Compat.* **42**: 39–53.

MacGorman, D.R., and Rust, W.D. 1998. *The Electrical Nature of Storms*. 422 pp., New York: Oxford Univ. Press.

MacGorman, D.R., Few, A.A., and Teer, T.L. 1981. Layered lightning activity, *J. Geophys. Res.* **86**: 9900–10.

Mackerras, D. 1968. A comparison of discharge processes in cloud and ground lightning flashes. *J. Geophys. Res.* **73**: 1175–83.

Malan, D.J. 1955a. La distribution verticale de la charge négative orageuse. *Ann. Geophys.* **11**: 420–6.

Malan, D.J. 1955b. Les décharges lumineuses dans les nuages oraguex. *Ann. Geophys.* **11**: 427–34.

Malan, D.J. 1958. Radiation from lightning discharges and its relation to the discharge processes. In *Recent Advances in Atmospheric Electricity*, ed. L.G. Smith, pp. 557–63, London: Pergamon.

Marshall, T.C. 2000. Comment on "'Spider' lightning in intracloud and positive cloud-to-ground flashes" by Vladislav Mazur, Xuan-Min Shao, and Paul R. Krehbiel. *J. Geophys. Res.* **105**: 7397–9; Reply, *ibid*, 7401–2.

Massey, R.S., and Holden, D.N. 1995. Phenomenology of transionospheric pulse pairs. *Radio Sci.* **30**: 1645–59.

Massey, R.S., Holden, D.N., and Shao, X.-M. 1998. Phenomenology of transionospheric pulse pairs. Further observations. *Radio Sci.* **33**: 1755–61.

Mazur, V. 1989. Triggered lightning strikes to aircraft and natural intracloud discharges. *J. Geophys. Res.* **94**: 3311–25.

Mazur, V. 1999. Lightning and aviation. In *Aviation Weather Serveillance Systems*, ed. P. Mahapatra, pp. 407–27, copublished by the Institution of Electrical Engineers (IEE) and the American Institute of Aeronautics and Astronautics (AIAA).

Mazur, V., Gerlach, J.C., and Rust, W.D. 1984. Lightning flash density versus altitude and storm structure from observations with UHF- and S-band radars. *Geophys. Res. Lett.* **11**: 61–4.

Mazur, V., Shao, X.-M., and Krehbiel, P.R. 1998. "Spider" lightning in intracloud and positive cloud-to-ground flashes. *J. Geophys. Res.* **103**: 19 811–22.

Medelius, P.J., Thomson, E.M., and Pierce, J.S. 1991. E and dE/dt waveshapes for narrow bipolar pulses in intracloud lightning. In *Proc. 1991 Int. Conf. on Lightning and Static Electricity, Cocoa Beach, Florida*, vol. I, pp. 12-1–10, NASA Conf. Publ. 3106.

Moreau, J.P., and Rustan, P.L. 1990. A study of lightning initiation based on VHF radiation. In *Lightning Electromagnetics*, ed. R.L. Gardner, pp. 257–76, New York: Hemisphere.

Müller-Hillebrand, D. 1962. The magnetic field of the lightning discharge. In *Gas Discharge and the Electricity Supply Industry*, eds. J.S. Forrest, P.R. Howard, and D.J. Littler, pp. 89–111, London: Butterworths.

Murphy, M.J., Krider, E.P., and Maier, M.W. 1996. Lightning charge analyses in small convection and precipitation electrification (CAPE) experiment storms. *J. Geophys. Res.* **101**: 29 615–26.

Nakano, M. 1979a. The cloud discharge in winter thunderstorms of the Hokuriku coast. *J. Meteor. Soc. Japan* **57**: 444–51.

Nakano, M. 1979b. Initial streamer of the cloud discharge in winter thunderstorms of the Hokuriku coast. *J. Meteor. Soc. Japan* **57**: 452–7.

Norinder, H. 1954. The wave-forms of the electric field in atmospherics recorded simultaneously by two distant stations. *Arkiv för Geofysik* **2**(9): 161–95.

Ogawa, T. 1982. The lightning current. In *Handbook of Atmospherics*, vol. 1, ed., H. Voland, pp. 23–63, Boca Raton, Florida: CRC Press.

Ogawa, T. 1993. Initiation of lightning in clouds. *J. Atmos. Electr.* **13**: 121–32.

Ogawa, T. 1995. Lightning currents. In *Handbook of Atmospheric Electrodynamics*, vol. I, ed. H. Volland, pp. 93–136, Boca Raton, Florida: CRC Press.

Ogawa, T., and Brook, M. 1964. The mechanism of the intracloud lightning discharge. *J. Geophys. Res.* **69**: 5141–50.

Petterson, B.J., and Wood, W.R. 1968. Measurements of lightning strikes to aircraft. Report no. SC-M-67-549 to Dept of Transportation, Federal Aviation Administration, Sandia Labs., Albuquerque, New Mexico.

Picone, J.M., Boris, J.P., Grieg, J.R., Rayleigh, M., and Fernsler, R.F. 1981. Convective cooling of lightning channels. *J. Atmos. Sci.* **38**: 2056–62.

Pierce, E.T. 1955a. Electrostatic field changes due to lightning discharges. *Q.J.R. Meteor. Soc.* **81**: 211–28.

Pierce, E.T. 1955b. The development of lightning discharges. *Q.J.R. Meteor. Soc.* **81**: 229–40.

Prentice, S.A., and Mackerras, D. 1977. The ratio of cloud to cloud-ground lightning flashes in thunderstorms. *J. Appl. Meteor.* **16**: 545–50.

Price, C., and Rind, D. 1993. What determines the cloud-to-ground lightning fraction in thunderstorms? *Geophys. Res. Lett.* **20**: 463–6.

Proctor, D.E. 1971. A hyperbolic system for obtaining VHF radio pictures of lightning. *J. Geophys. Res.* **76**: 1478–89.

Proctor, D.E. 1974a. Sources of cloud-flash sferics. CSIR Special Report no. TEL 118, Pretoria, South Africa.

Proctor, D.E. 1974b. VHF radio pictures of lightning. CSIR Special Report no. TEL 120, Pretoria, South Africa.

Proctor, D.E. 1976. A radio study of lightning. Ph.D. thesis, University of Witwatersrand, Johannesburg, South Africa.

Proctor, D.E. 1981. VHF radio pictures of cloud flashes. *J. Geophys. Res.* **86**: 4041–71.

Proctor, D.E. 1983. Lightning and precipitation in a small multicellular thunderstorm. *J. Geophys. Res.* **88**: 5421–40.

Proctor, D.E. 1984. Correction to "Lightning and precipitation in a small multicellular thunderstorm". *J. Geophys. Res.* **89**: 11 826.

Proctor, D.E. 1991. Regions where lightning flashes began. *J. Geophys. Res.* **96**: 5099–112.

Proctor, D.E. 1995. Radio noise above 300 kHz due to natural causes. In *Handbook of Atmospheric Electrodynamics*, ed. H. Volland, vol. I, pp. 311–58, Boca Raton, Florida: CRC Press.

Proctor, D.E. 1997. Lightning flashes with high origins. *J. Geophys. Res.* **102**: 1693–706.

Proctor, D.E. 2000. Personal communication.

Rai, J., Kumar, K., Hazarika, S., Parashar, J., Kumar, R. 1993. High speed photographic analysis of intracloud lightning radiation fields. *Ann. Geophys.* **11**: 518–24.

Rakov, V.A. 1999. Lightning electric and magnetic fields. In *Proc. 13th Int. Symp. on Electromagnetic Compatibility, Zurich, Switzerland*, pp. 561–6.

Rakov, V.A., Thottappillil, R., and Uman, M.A. 1992. Electric field pulses in K and M changes of lightning ground flashes. *J. Geophys. Res.* **97**: 9935–50.

Rakov, V.A., Uman, M.A., Hoffman, G.R., Masters, M.W., and Brook, M. 1996. Bursts of pulses in lightning electromagnetic radiation: observations and implications for lightning test standards. *IEEE Trans. Electromagn. Compat.* **38**: 156–64.

Rao, M., Khastgir, S.R., and Bhattacharya, H. 1962. Electric field changes. *J. Atmos. Terr. Phys.* **24**: 989–90.

Ray, P.S., MacGorman, D.R., Rust, W.D., Taylor, W.L., and Rasmussen, L.W. 1987. Lightning location relative to storm structure in supercell storm and a multicell storm. *J. Geophys. Res.* **92**: 5713–24.

Reynolds, S.E., and Neill, H.W. 1955. The distribution and discharge of thunderstorm charge-centers. *J. Meteor.* **12**: 1–12.

Rhodes, C. 1989. Interferometric observations of VHF radiation from lightning. Ph.D. dissertation, New Mexico Inst. of Min. and Technol., Socorro.

Richard, P., and Auffray, G. 1985. VHF–UHF interferometric measurements, applications to lightning discharge mapping. *Radio Sci.* **20**: 171–92.

Richard, P., Delannoy, A., Labaune, G., and Laroche, P. 1986. Results of spatial and temporal characterization of the VHF–UHF radiation of lightning. *J. Geophys. Res.* **91**: 1248–60.

Rinnert, K., Lauderdale, R., II, Lanzerotti, L.J., Krider, E.P., and Uman, M.A. 1989. Characteristics of magnetic field pulses in Earth lightning measured by the Galileo probe instrument. *J. Geophys. Res.* **94**: 13 229–35.

Rison, W., Thomas, R.J., Krehbiel, P.R., Hamlin, T., and Harlin, J. 1999. A GPS-based three-dimensional lightning mapping system: initial observations in central New Mexico. *Geophys. Res. Lett.* **26**: 3573–6.

Russell, C.T., Zuelsdorf, R.S., Strangeway, R.J., and Franz, R. 1998. Identification of the cloud pulse responsible for a trans-ionospheric pulse pair. *Geophys. Res. Lett.* **25**: 2645–8.

Rust, W.D. 1989. Utilization of a mobile laboratory for storm electricity measurements. *J. Geophys. Res.* **94**: 13 305–11.

Schonland, B.F.J. 1956. The lightning discharge. *Handbuch der Physik*, vol. 22, pp. 576–628, Berlin: Springer-Verlag.

Schonland, B.F.J., Malan, D.J., and Collens, H. 1935. Progressive Lightning II. *Proc. Roy. Soc. London* A**152**: 595–625.

Schonland, B.F.J., Hodges, D.B., and Collens, H. 1938. Progressive lightning, Pt. 5, A comparison of photographic and electrical studies of the discharge process. *Proc. Roy. Soc. London* A**166**: 56–75.

Shao, X.M., and Krehbiel, P.R. 1993. Radio interferometric observations of intracloud lightning (Abstract). *Eos Trans. AGU* **74**, Fall meeting suppl., p. 165.

Shao, X.M., and Krehbiel, P.R. 1996. The spatial and temporal development of intracloud lightning. *J. Geophys. Res.* **101**: 26 641–68.

Smith D.A. 1998. Compact intracloud discharges. Ph.D. dissertation, University of Colorado, 272 pp.

Smith D.A., and Holden, D.N. 1996. Ground-based observations of subionospheric pulse pairs. *Radio Sci.* **31**: 553–71.

Smith, D.A., Shao, X.M., Holden, D.N., Rhodes, C.T., Brook, M., Krehbiel, P.R., Stanley, M., Rison, W., and Thomas, R.J. 1999a. A distinct class of isolated intracloud lightning discharges and their associated radio emissions. *J. Geophys. Res.* **104**: 4189–212.

Smith, D.A., Massey, R.S., Wiens, K.C., Eack, K.B., Shao, X.M., Holden, D.N., Argo, P.E. 1999b. Observations and inferred physical characteristics of compact intracloud discharges. In *Proc. 11th Int. Conf. on Atmospheric Electricity, Guntersville, Alabama*, pp. 6–9.

Smith, L.G. 1957. Intracloud lightning discharges. *Q.J.R. Meteor. Soc.* **83**: 103–11.

Sourdillon, M. 1952. Etude à la chambre de Boys de "l'éclair dans l'air" et du "coup de foudre à cime horizontale". *Ann. Geophys.* **8**: 349–64.

Steptoe, B.J., 1958. Some observations on the spectrum and propagation of atmospherics. Ph.D. thesis, University of London, UK.

Suszcynsky, D.M., Kirkland, M.W., Jacobson, A.R., Franz, R.C., Knox, S.O., Guillen, J.L.L., and Green, J.L. 2000. FORTE observations of simultaneous VHF and optical emissions from lightning: basic phenomenology. *J. Geophys. Res.* **105**: 2191–201.

Takagi, M. 1961. The mechanism of discharges in a thundercloud, *Proc. Res. Inst. Atmospherics, Nagoya Univ., Japan* **8B**: 1–106.

Takagi, M., and Takeuti, T. 1963. Atmospherics radiation from lightning discharges. *Proc. Res. Inst. Atmospherics, Nagoya Univ., Japan* **10**: 1–11.

Takeuti, T. 1965. Studies on thunderstorms electricity, 1, Cloud discharges. *J. Geomagn. Geoelectr.* **17**: 59–68.

Tamura, Y., Ogawa, T., and Okawati, A. 1958. The electrical structure of thunderstorms. *J. Geomagn. Geoelectr.* **10**: 20–7.

Taylor, W.L., Brandes, E.A., Rust, W.D., and MacGorman, D.R. 1984. Lightning activity and severe storm structure. *Geophys. Res. Lett.* **11**: 545–8.

Teer, T.L., and Few, A.A. 1974. Horizontal lightning. *J. Geophys. Res.* **79**: 3436–41.

Tepley, L.R. 1961. Sferics from intracloud lightning strokes. *J. Geophys. Res.* **66**: 111–23.

Thomas, R.J., Krehbiel, P.R., Rison, W., Hamlin, T., Boccippio, D.J., Goodman, S.J., and Christian, H.J. 2000. Comparison of ground-based 3-dimensional lightning mapping observations with satellite-based LIS observations in Oklahoma. *Geophys. Res. Lett.* **27**: 1703–6.

Thomas, R.J., Krehbiel, P.R., Rison, W., Hamlin, T., Harlin, J., and Shown, D. 2001. Observations of VHF source powers radiated by lightning. *Geophys. Res. Lett.* **28**: 143–6.

Thomson, E.M., Medelius, P.J., and Davis, S. 1994. System for locating the sources of wideband dE/dt from lightning. *J. Geophys. Res.* **99**: 22 793–802.

Thottappillil, R., Rakov, V.A., and Uman, M.A. 1990. K and M changes in close lightning ground flashes in Florida. *J. Geophys. Res.* **95**: 18 631–40.

Uman, M.A. 1987. *The Lightning Discharge*, 377 pp., Orlando, Florida: Academic Press.

Uman, M.A. 2001. *The Lightning Discharge*, 377 pp., Mineola, New York: Dover.

Villanueva, Y., Rakov, V.A., Uman, M.A., and Brook, M. 1994. Microsecond-scale electric field pulses in cloud lightning discharges. *J. Geophys. Res.* **99**: 14 353–60.

Vonnegut, B., Vaughan, O.H. Jr, and Brook, M. 1989. Nocturnal photographs taken from a U-2 airplane looking down on tops of clouds illuminated by lightning. *Bull. Am. Meteor. Soc.* **70**: 1263–71.

Wadehra, N.S., and Tantry, B.A.P. 1967a. VLF characteristics of K changes in lightning discharges. *Ind. J. Pure and Appl. Phys.* **5**: 447–9.

Wang, C.P. 1963a. Lightning discharges in the tropics – 1. Whole discharges. *J. Geophys. Res.* **68**: 1943–9.

Wang, C.P. 1963b. Lightning discharges in the tropics – 2. Component ground strokes and cloud dart streamer discharges. *J. Geophys. Res.* **68**: 1951–8.

Weber M.E. 1980. Thundercloud electric field soundings with instrumented free balloons. Ph.D. dissertation, Rice University, Houston, Texas.

Weber, M.E., Christian, H.J., Few, A.A., and Stewart, M.F. 1982. A thundercloud electric field sounding: charge distribution and lightning. *J. Geophys. Res.* **87**: 7158–69.

Weidman, C.D., and Krider, E.P. 1979. The radiation field wave forms produced by intracloud lightning discharge processes. *J. Geophys. Res.* **84**: 3159–64.

Weidman, C.D., Krider, E.P., and Uman, M.A. 1981. Lightning amplitude spectra in the interval 100 kHz to 20 MHz. *Geophys. Res. Lett.* **8**: 931–4.

Willett, J.C., Bailey, J.C., and Krider, E.P. 1989. A class of unusual lightning electric field waveforms with very strong HF radiation. *J. Geophys. Res.* **94**: 16 255–67.

Williams, D.P., and Brook, M. 1963. Magnetic measurements of thunderstorm currents, l. Continuing currents in lightning. *J. Geophys. Res.* **68**: 3243–7.

Williams, E.R., Weber, M.E., and Orville, R.E. 1989. The relationship between lightning type and convective state of thunderclouds. *J. Geophys. Res.* **94**: 13 213–20.

Wong, C.M., and Lin, K.K. 1978. The inclination of intracloud lightning discharges. *J. Geophys. Res.* **83**: 1905–12.

Workman, E.J., and Holzer, R.E. 1942. A preliminary investigation of the electrical structure of thunderstorms. Technical Notes of the National Advisory Committee on Aeronautics no. 850.

Workman, E.J., Holzer, R.E., and Pelsor, G.T. 1942. The electrical structure of thunderstorms. Technical Notes of the National Advisory Committee on Aeronautics no. 864.

Zaffo, P.A. 1990. Lightning above a large thunderstorm. *Weatherwise* **43**: 363.

Ziegler, C.L., and MacGorman, D.R. 1994. Observed lightning morphology relative to modeled space charge and electric field distributions in a tornadic storm. *J. Atmos. Sci.* **51**: 833–51.

Zuelsdorf, R.S. Strangeway, R.J., Russell, C.T., Casler, C., and Christian, H.J. 1997. Trans-ionospheric pulse pairs (TIPPs): their geographic distributions and seasonal variations. *Geophys. Res. Lett.* **24**: 3165–8.

Zuelsdorf, R.S., Casler, C., Strangeway, R.J., Russell, C.T., and Franz, R. 1998a. Ground detection of trans-ionospheric pulse pairs by stations in the National Lightning Detection Network. *Geophys. Res. Lett.* **25**: 481–4.

Zuelsdorf, R.S., Strangeway, R.J., Russell, C.T., and Franz, R. 1998b. Trans-ionospheric pulse pairs (TIPPs): their occurrence rates and diurnal variation. *Geophys. Res. Lett.* **25**: 3709–12.

Zuelsdorf, R.S., Franz, R.C., Strangeway, R.J., and Russell, C.T. 2000. Determining the source of strong LF/VLF TIPP events: implications for association with NPBPs and NNBPs. *J. Geophys. Res.* **105**: 20 725–36.

10 Lightning and airborne vehicles

> ... when a large conductor, such as an aircraft, intrudes into the cloud, the field intensification it creates is sufficient to artificially initiate or 'trigger' a lightning flash.
>
> E.T. Pierce (1976)

10.1. Introduction

Most lightning strikes to aircraft in flight are initiated by the aircraft, as opposed to the aircraft's having intercepted a discharge already in progress, although this was not convincingly demonstrated to be the case until the 1980s. Even in the case of the interception of a natural lightning flash, there is likely to be a significant discharge initiated from the aircraft in response to the naturally occurring channel. A video frame showing evidence of the initiation of lightning by an aircraft at relatively low altitude is found in Fig. 10.1. That the lightning shown in Fig. 10.1 was initiated at the aircraft is indicated by the different directions of channel branching above and below the aircraft. The apparently similar artificial initiation of lightning by an ungrounded wire trailing behind a small rocket, so-called "altitude triggering", is discussed in subsection 7.2.1. Early arguments that aircraft could initiate lightning were based primarily on the many observed cases of lightning strikes to aircraft inside or near clouds that previously had not produced natural lightning (e.g., Harrison L.P. 1946; Harrison H.T. 1965; Fitzgerald 1967; Clifford and Kasemir 1982). The first direct evidence of the initiation of a lightning strike by an aircraft was provided by UHF radar echoes of lightning channel formation during strikes to the NASA F-106B research aircraft (subsection 10.3.2) showing that the initial leader channels originated at or very near (the radar resolution was 150 m) the F-106B and propagated away from it (Mazur et al. 1984a). In about 80 percent of 49 observed F-106B strikes, the lightning echo initially moved in two opposite directions away from the aircraft echo, and in about 5 percent of the strikes the lightning echo moved in one direction away from the aircraft. A less common event, the interception of a lightning flash by the F-106B, as inferred from radar, is described in Mazur et al. (1986).

The fact that aircraft serve to initiate lightning has also been inferred from the analysis of measured electric field waveforms on the surface of aircraft. For example, Reazer et al. (1987) showed that in 35 of 39 strikes to the CV-580 research aircraft (subsection 10.3.3) the characteristics of the electric field waveforms were consistent with an aircraft-initiation hypothesis, although their suggested physical interpretation of the typical waveform does not represent the current consensus view (e.g., Mazur 1989b), while the other four waveforms were clearly different and could be interpreted as due to an aircraft's intercepting a naturally initiated flash. Associated current measurements on the CV-580 and the C-160 research aircraft (subsection 10.3.4) provided further confirmation of aircraft initiation, as did high-speed video records of channel formation (e.g., Moreau et al. 1992; Mazur 1989b). The electric and magnetic field, the current, and the photographic instrumentation on research aircraft during aircraft-initiated and aircraft-intercepted lightning strikes are considered in subsections 10.3.1–10.3.4, and the resulting data and their interpretation are considered in Section 10.4.

The mechanism for lightning initiation by a conducting object not attached to the Earth is often referred to as the "bidirectional leader" theory (Kasemir 1950), and its application to lightning initiation by aircraft and other airborne vehicles has been considered, for example, by Clifford and Kasemir (1982), Mazur (1989a, b), Mazur (1992), and Mazur and Moreau (1992). In an ambient electric field typically near 50 kV m^{-1}, a common value in thunderclouds (subsection 3.2.4), the CV-580 and C-160 research aircraft flying near 5 km altitude are inferred, from interpretation of the measurements noted in the previous paragraph, to launch a positive leader in the direction of the electric field from one aircraft extremity and, a few milliseconds later, a negative leader in the opposite direction from a different extremity (Section 10.4). Similar bidirectional leader development is inferred by Mazur (1989b) in the case of the F-106B, except that the initial positive leader was apparently preceded by corona or other processes that caused a millisecond-duration field change opposite in polarity to that caused by the positive leader (Section 10.4). It is reasonable to expect that a positive leader would occur first in the bidirectional leader development since, in general, positive leaders are initiated and can propagate in lower electric

10.1. Introduction

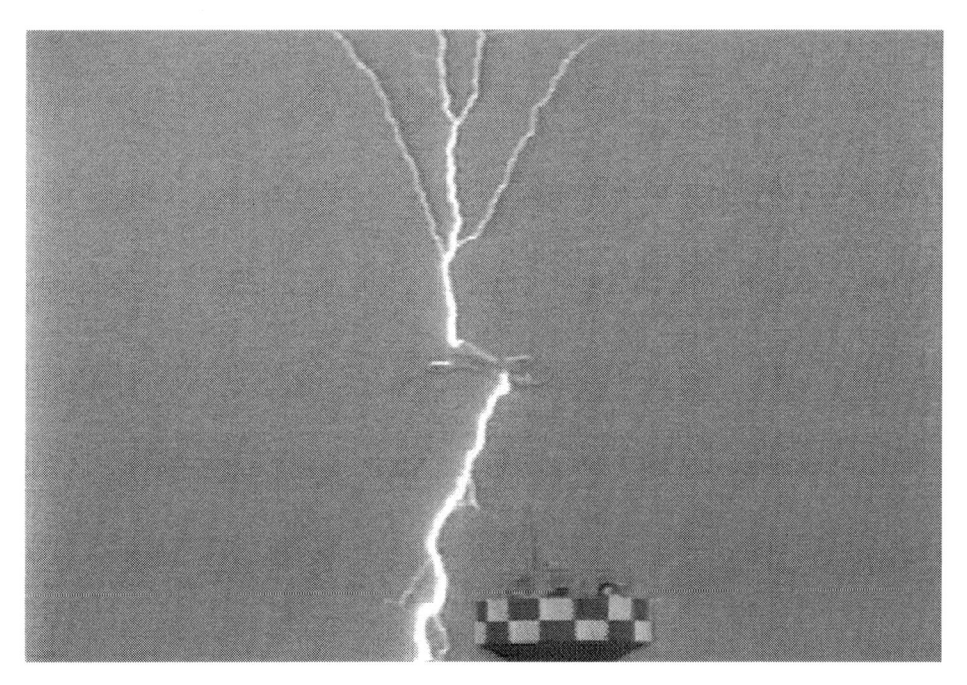

Fig. 10.1. Video frame of a lightning strike to an aircraft on takeoff from the Kamatsu Air Force Base on the coast of the Sea of Japan during winter. Courtesy of Z.I. Kawasaki.

fields than negative leaders. After positive-leader initiation, the electric field at the aircraft is apparently enhanced by the positive leader's removal of positive charge from the aircraft and also by elongation of the overall conducting system, thereby resulting in negative-leader initiation. Although no instance appears to have been recorded, according to the available literature, there is no obvious reason why a negative leader could not be emitted from an aircraft prior to a positive leader if the field enhancement at the extremity launching the negative leader were much greater than at the extremity launching the positive leader, the ratio of the field enhancements being a function both of the detailed shape and orientation of the aircraft and also of the effects of corona processes which could potentially reduce field enhancement. The aircraft extremities provide the region of high electric field needed to initiate a lightning discharge by enhancing the ambient electric field to breakdown values, 3×10^3 kV m^{-1} near sea level and about half that value at 6 km altitude (subsection 3.2.4). Thus, at flight altitudes, a reasonable aircraft enhancement factor, of order 10, is required to initiate lightning in the observed ambient fields. The shape of the aircraft is the most important factor in determining the increase in the local electric field at, for example, the wing tips or the vertical stabilizer to magnitudes that make the initiation of lightning possible. After the initial stage of the discharge, which is characterized by impulsive currents near 1 kA apparently associated with the steps of the negative stepped leader, the observed current through the research aircraft is generally composed of a steady component and a variety of impulses, probably not unlike a natural intracloud flash (e.g., Mazur 1989a). Occasionally, aircraft initiate or otherwise become involved in cloud-to-ground lightning, this being more likely when they are closer to the Earth (e.g., Mazur et al. 1986, 1990; Mazur and Moreau 1992). Clearly, if an aircraft initiates lightning at low enough altitudes, say, soon after takeoff, as for the case shown in Fig. 10.1, that aircraft will necessarily be involved in a ground flash.

According to Harrison (1965), who studied 99 lightning strikes to United Air Lines aircraft, electrical discharges to aircraft in flight exhibit three common features:

(i) a bright flash, sometimes blinding;
(ii) a loud explosive "boom", sometimes muffled;
(iii) minor damage to aircraft in one-third to one-half all cases.

Pilots often distinguish between two types of lightning–aircraft interaction, which they call, in layman's terms, "static discharge" and "lightning". The former, "static discharge", is characterized by radio static on the pilot's earphone of some seconds duration and a corresponding corona discharge on the aircraft prior to the major observed electrical discharge (when it is dark, the luminous corona is visible and is commonly referred to as St Elmo's fire). The latter, "lightning", is an electrical discharge that occurs without much prior warning. Static discharges are a much more common occurrence and apparently correspond to aircraft-initiated lightning. Interestingly, many

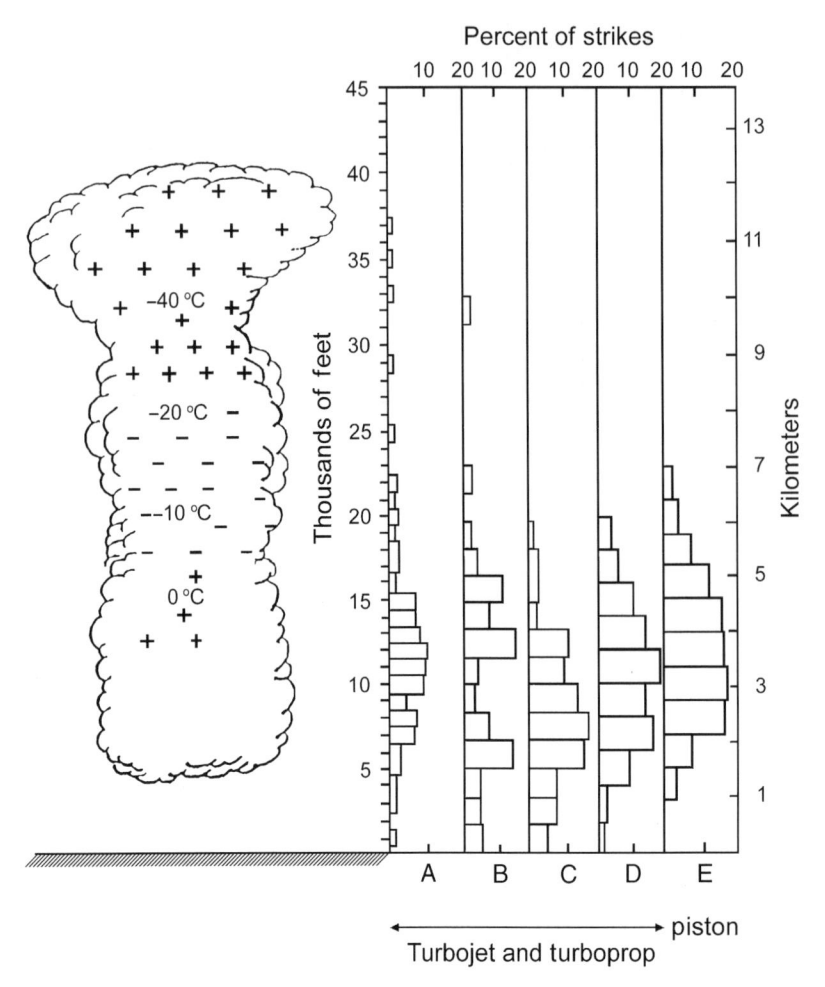

Fig. 10.2. Aircraft lightning incidents versus altitude. Adapted from Fisher et al. (1999) with a correction to their typical summer thunderstorm charge distribution (subsection 3.2.1). A, USA (Plumer 1971–5); B, Europe/S.A. (Anderson 1966–74); C, USSR (Trunov 1969–74); D, UK/Europe (Perry 1959–75); E, USA (Newman 1950–61).

pilots view events in the static-discharge category as being caused by discharge of the charge stored on the aircraft, which is typically of the order of 1 mC (Section 10.4) and thus much too small to produce the damage often observed to the aircraft skin after these discharges have occurred. Typically, this skin damage involves a sequence of burn marks or burn holes due to the changing lightning-channel attachment point as the aircraft moves relative to the lightning, the so-called swept-stroke phenomenon (Larsson et al. 2000a, b; Testé et al. 2000). Such aircraft skin damage coupled with laboratory testing provides evidence that so-called static discharges can transfer charges similar to those of natural lightning and hence are likely to be aircraft-initiated lightning. The "lightning" category apparently includes primarily flashes initiated independently of the aircraft which the aircraft then intercepts.

This chapter is organized as follows. In the next section we examine the statistics on lightning strikes to aircraft. In Section 10.3, we discuss the major airborne research programs, and in Section 10.4 we summarize their salient results. In Section 10.5, we review the lightning test standards developed for aircraft. Finally, in Section 10.6, we look at some high-profile accidents involving lightning strikes to aircraft and launch vehicles.

10.2. Statistics on lightning strikes to aircraft

Figure 10.2 summarizes the results of five studies of the altitude at which aircraft–lightning incidents have occurred. These studies took place between the early 1950s and the mid-1970s. The statistics are similar for all types of aircraft. Older piston aircraft, which cruise at 10 000 to 15 000 feet (about 3 to 4.5 km), show a pattern of occurrence of strikes as a function of altitude similar to that of the newer jet aircraft, which cruise at much higher altitudes. For modern commercial jets, most strikes occur either in climbing to a cruising altitude, generally near 30 000 feet (about

10.2. Statistics on lightning strikes to aircraft

9 km), or in landing, when the aircraft passes through the region of the cloud where the temperature is near 0 °C. According to Fisher et al. (1999) and Plumer et al. (1985), the overwhelming majority of strikes occur when the aircraft is within a cloud, only a few percent of strikes taking place when the aircraft is below or beside the cloud; the majority of strikes are associated with local airmass instability (27 percent) and organized fronts including squall-line activity (53 percent); and the vast majority of strikes are associated with turbulence and precipitation, 70 percent with rain and another 12 percent with a mixture of rain and snow, sleet, or hail.

A typical thundercloud charge distribution is shown in Fig. 10.2, but we do not intend to imply that all lightning strikes are associated with such clouds. For example, strikes have been recorded in clouds described as composed of ice crystals, and according to Harrison H.T. (1965) roughly 40 percent of all discharges occur in areas where no thunderstorms are reported, thunder or lightning being reported in the general area in the remaining 60 percent. Evidence for a thunderstorm at the spot of discharge is present in only 33 percent of all reported strikes. Harrison further states that any weather situation producing precipitation appears to be capable of causing electrical discharges to aircraft in flight. As we shall see in subsection 10.4.3, most strikes to the three research aircraft, the F-106 B, the CV-580, and the C-160, have occurred in conditions of light turbulence, light precipitation, and relatively infrequent lightning, although the research aircraft intentionally penetrated storms whereas commercial aircraft generally do not do so.

Data on the frequency of strikes to US commercial aircraft from 1950 to 1974 are found in Table 10.1. During the period studied, a typical commercial plane was struck once for each 3000 flight hours, or about once a year.

Murooka (1992) provided statistics on lightning strikes to commercial jets in Japan for the period 1980 to 1991. Data on over 1000 strikes are shown separately for summer and winter in Figs. 10.3a, b. Murooka found that the bulk of the strikes in both summer and winter occurs in the same temperature range, -5 to 0 °C. The data from summer and winter are combined in Fig. 10.4. In Japanese winter storms (Chapter 8) the freezing level is near the ground and the cloud top is near 5 km. Hence, in winter the strikes occur at considerably lower altitudes than those in summer, as indicated in Fig. 10.3. Michimoto (1993) provided similar data for lightning strikes to military aircraft in Japanese winter storms for the period from 1961 to 1990. Goto and Narita (1986) compared the altitude and temperature of winter strikes in Japan with similar data from all seasons in South Africa, the United States, and the USSR. They found the greatest similarity between events in Japan and in the USSR.

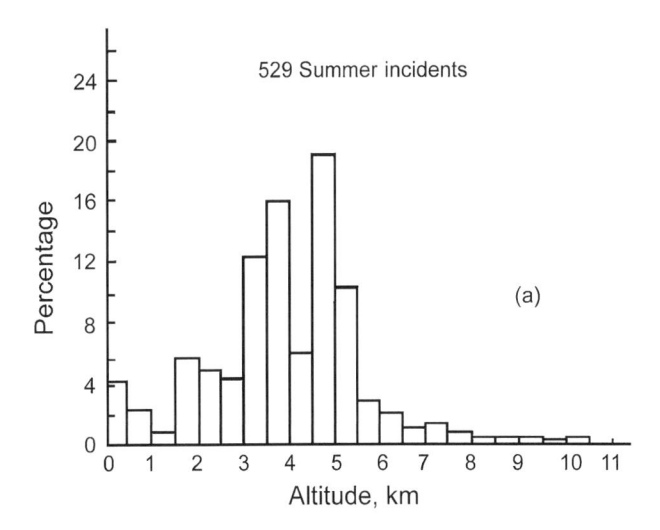

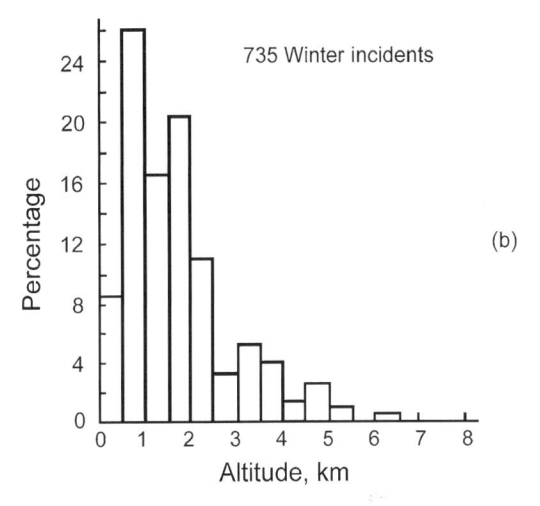

Fig. 10.3. (a) Aircraft lightning incident rate versus altitude in summer for commercial aircraft in Japan. (b) Aircraft lightning incident rate versus altitude in winter for commercial aircraft in Japan. Adapted from Murooka (1992).

Anderson and Kroninger (1975) examined South African Airways lightning-strike records from 1948 to 1974. Most strikes occurred 3 to 5 km above sea level. The number of strikes reported per 10 000 hours of flying time for different years varied between about 1 and 4, which is consistent with the data in Table 10.1.

The effects of lightning on aircraft are generally minimal, although occasionally the consequences of the interaction can be catastrophic, as we shall see in Section 10.6. Lightning damage is usually divided into "direct" and "indirect" (or "induced") effects. Direct effects occur at the points of the lightning contact and include holes in metal skins, puncturing or splintering of nonmetallic structures such as the plastic radomes that cover the radars located at the front of aircraft, welding or roughening of moveable hinges and bearings, damage

Table 10.1. *Incidence of reported lightning strikes to commercial aircraft. Adapted from Fisher et al. (1999)*

	Newman (1950–61)		Perry (1959–74)		Total		
	Strikes	Hours	Strikes	Hours	Strikes	Hours	Hours per strike
Piston	808	2 000 000	—	—	808	2 000 000	2 475
Turboprop	109	415 000	280	876 000	389	1 291 000	3 320
Pure jet	41	427 000	480	1 314 000	521	1 741 000	3 340
All	958	2 842 000	760	2 190 000	1 718	5 032 000	2 930

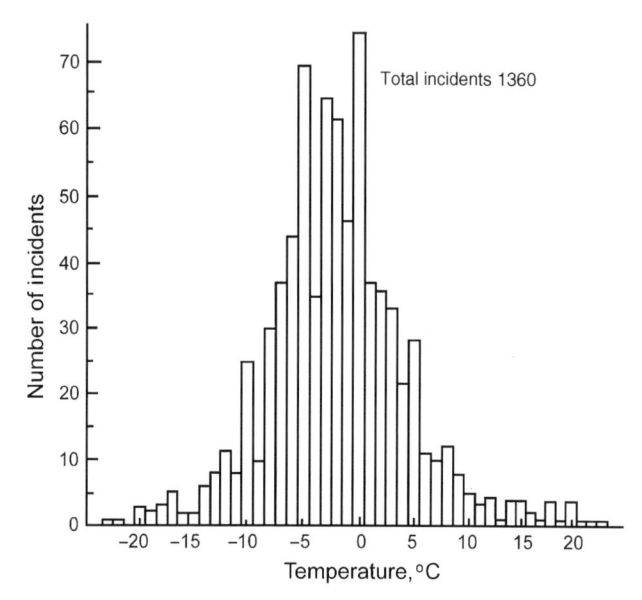

Fig. 10.4. Number of aircraft lightning incidents during all seasons versus ambient temperature for commercial aircraft in Japan. Adapted from Murooka (1992).

Table 10.2. *Incidence of indirect effects in commercial aircraft during 214 lightning strikes. Adapted from Fisher et al. (1999)*

	Interference	Outage
HF communication set	—	5
VHF communication set	27	3
VOR receiver	5	2
Compass (all types)	22	9
Marker beacon	—	2
Weather radar	3	2
Instrument landing system	6	—
Automatic direction finder	6	7
Radar altimeter	6	—
Fuel flow gauge	2	—
Fuel quantity gauge	—	1
Engine rpm gauges	—	4
Engine exhaust gas temperature	—	2
Static air temperature gauge	1	—
Windshield heater	—	2
Flight director computer	1	—
Navigation light	—	1
AC generator tripoff	(six instances of tripoff)	
Autopilot	1	—

to antennas and lights located at aircraft extremities, and fuel ignition. Indirect effects are those produced by the deleterious voltages and currents induced within the aircraft by the lightning electric and magnetic fields; they include upset or damage to any of the aircraft electronic systems. Table 10.2 gives some statistics on indirect effects that occurred from 1971 to 1984, involving 20 percent of 851 reported strikes (Fisher et al. 1999). Fisher et al. presented additional details on various forms of damage to aircraft from lightning, including photographs of damage due to direct effects. Anderson and Kroninger (1975) reported that aircraft-frame or instrument damage occurred in 40 percent of the 245 recorded strikes to aircraft in South Africa between 1948 and 1974.

10.3. Major airborne research programs

Four major airborne lightning-characterization studies involving four different instrumented aircraft, an F-100F, an F-106B, a CV-580, and a C-160 are discussed in this section. Additional minor airborne lightning studies using other aircraft are reviewed in Reazer et al. (1987).

10.3.1. Air Force Cambridge Research Laboratories, Rough Rider project, 1964–6

The Rough Rider project, involving the F-100F, was described by Fitzgerald (1967) and by Petterson and Wood (1968). Airborne studies of Florida thunderstorms were conducted from 1964 to 1966 using three instrumented aircraft, the F-100F, a C-130, and a U-2. The C-130, flying outside the Florida storms, measured ambient electric fields (not lightning fields) and conducted radar observations at medium altitudes. The U-2 obtained photographic, infrared,

10.3. Major airborne research programs

and ambient electric field data from above the storms. The F-100F, a single-engine jet, penetrated the storms to measure turbulence and to obtain lightning photographic, shock wave, and electrical current records. Current measurements on the F-100F were made on the nose boom, wing tips, and vertical stabilizer, using 5 mΩ current-viewing resistors with a 20 MHz upper frequency response. The current data were displayed on oscilloscopes and recorded on photographic film. Data were recorded for 49 lightning discharges.

Petterson and Wood (1968) indicated that there were difficulties in recording the current risetime and rate of rise since often the oscilloscopes would trigger too soon and miss the potential event of interest, or the event of interest would occur during film advance. The F-100F shock wave overpressure system did not trigger on any of the 49 strikes. The trigger level for the measurement was near 50 psi (about 3.3 atm). Petterson and Wood (1968) noted that the threshold level probably should have been set near or below 1 atm (15 psi) (subsections 11.3.1 and 12.2.2).

10.3.2. NASA storm hazards program, 1980–6.

As part of the NASA storm hazards program, a NASA F-106B, a delta wing, single-engine jet aircraft of 21.5 m length including a sharp 3 m nose boom (a slender metal extension projecting from the plane's nose), flew about 1500 thunderstorm traversals at altitudes ranging from 5000 to 40 000 feet (1.5 to 12 km) and was struck by lightning 714 times. Almost 10 times as many strikes were obtained for the high altitudes (>6 km) as for the low, although the number of high and low-cloud penetrations, a total of near 1500, was not much different (Fisher et al. 1986). Statistics were compiled for aircraft-surface electric and magnetic field derivatives and for the lightning current and current derivative flowing through the aircraft. Detailed information on the instrumentation and data obtained are found in Trost and Zaepfel (1980), Pitts and Thomas (1981,1982), Pitts (1982), Thomas and Pitts (1983), Lee et al. (1984), Mazur et al. (1984a, 1986, 1990), Thomas (1985), Thomas and Carney (1986), Fisher et al. (1986), Pitts et al. (1987, 1988), Yang et al. (1987), Zaepfel and Carney (1988), and Mazur (1989a, b).

Sensors for measuring the current, the rate of change of current, and the rates of change of the electric and magnetic fields on the F-106B were mounted on the nose, on the vertical tail, under the forward and aft portions of the fuselage, at the base of the vertical tail, and under each wing. Induced voltages on a few internal wires were also measured. Three types of recording instruments were used: (i) continuous analog recorders, (ii) digital transient recorders, and (iii) peak recording instruments.

(i) The *continuous analog recorders* had bandwidths of 400 Hz to 100 kHz. On the one hand they could not record fast current pulses and electromagnetic fields with acceptable fidelity but on the other hand they were capable of recording data during the entire storm penetration. Although their nominal lower frequency response was 400 Hz, the analog recorders could record steady (dc) currents using frequency-division multiplex recording techniques.

(ii) Each of the *digital transient recorder's* 12 channels could store a maximum of 65 356 8-bit samples. Thus, the data window at the fastest digitization rate, yielding a 5-ns sampling interval, was 327 μs. At the maximum sampling rate, the digital transient recorders had an upper frequency response of 100 MHz.

(iii) The *peak recorders* supplemented the analog and digital recorders by capturing the maximum values on selected sensors during a flight. Usually, they monitored the peak rate of change of current in the nose boom and the peak rate of change of electric flux density (displacement current density) under the forward fuselage. In general, it was not possible to tell where these peak values occurred in the overall sequence of events in the flash.

Approximately 2500 individual time-domain waveforms were logged from these various sensors during the strikes, and 130 peak recorder readings were obtained during about 400 of the strikes.

The F-106B was equipped with a total of eight cameras. Two black and white video cameras were mounted in the cockpit, which was covered by a clear glass canopy. One camera faced forward and the other backward. Two other black and white video cameras were installed in a pod on the upper surface of the left wing tip and together gave overlapping views of the whole airplane, from just ahead of the nose boom to the trailing edge of the vertical tail. A high-speed 16 mm movie camera, normally loaded with color film, was mounted under a fairing on the left side of the fuselage, looking to the rear with a view including the left wing tip and vertical tail. Three 70 mm still cameras loaded with color film shared the cockpit video cameras' platform, two facing forward for stereo views and the third providing much the same view as the rear-facing cockpit video camera. The video cameras were operated continuously during flights, while the movie camera and the three still cameras were triggered automatically by inputs from light-sensitive diodes mounted throughout the cockpit.

Three radars at the NASA Goddard Space Flight Center's Wallops Flight Facility and one radar in the F-106B were used for guiding the flight crew, documenting storm structure, and confirming lightning strikes to the aircraft. One of the ground-based radars, an S-band radar (10 cm wavelength), was used to determine the reflectivity profile of the clouds. Another, a UHF radar (70.5 cm wavelength), was used to locate lightning flashes by detecting the signals reflected from ionized lightning channels. These radars and their operation are discussed in Mazur et al. (1984b). Both radars followed the F-106B by being slaved to the third ground-based radar which tracked a special transponder mounted on the F-106B. The intention was to guide the F-106B through the storms so that it encountered lightning while avoiding potentially damaging hail or heavy rain. So, when approaching a storm, the F-106B pilots used their onboard 3-cm-wavelength weather radar to observe and avoid regions of strong precipitation. Ideally, the F-106B would fly through a storm of interest along a pre-selected heading directly toward or away from the 10-cm-wavelength radar, which conducted a series of vertical scans along the heading to define the storm conditions surrounding airplane lightning strikes. The data from these scans were used to reconstruct the evolution and structure of the storms for comparison with the electromagnetic and turbulence data recorded on the airplane. Generally, the F-106B was flown within 150 nautical miles (275 kilometers) of the NASA Langley Research Center in order to maintain line-of-sight communications.

The F-106B study provided the first statistical data on the lightning currents and fields encountered by an aircraft at various flight altitudes. It was able to do so partly because of the development of more advanced recording instrumentation than was available for the earlier (1964–6) F-100F study. In addition, the F-106B study under discussion provided information on the probability of lightning strikes as a function of ambient turbulence, ambient precipitation, and frequency of natural lightning. The CV-580 and C-160 studies to be discussed next benefited even more from the continuing advances in modern digital data acquisition and processing techniques.

10.3.3. USAF/FAA lightning characterization program, 1984–5, 1987

The USAF/FAA CV-580, a two-engine turboprop transport aircraft of 24.7 m length, was instrumented as described by Rustan (1986), Reazer et al. (1987), Mazur (1989b), Mazur and Moreau (1992), and Lalande et al. (1999). Five sensors that measured the time derivative of the surface electric field intensity (or the electric flux density) and five sensors that responded to the rate of change of the surface current density were mounted at various positions on the CV-580. An electric field derivative sensor on the forward upper fuselage was combined with an active integrator to provide measurement of electric field intensity with a lower-frequency response of 1 Hz and with relatively high sensitivity. Electric field sensors with lower sensitivity were mounted on the wing tips and left side of the vertical stabilizer. Five field mills were mounted at various locations on the fuselage, making possible measurement of the ambient field and the charge on the aircraft. A magnetic field sensor was mounted at the end of a 10-foot-long (about 3 m) horizontal boom attached to the tail of the aircraft. Current sensors were located at the base of the tail boom and at the bases of horizontal booms installed on the two wing tips. The current sensor outputs were split into two recording channels having different gains in order to increase the dynamic range. A VHF antenna acted as a narrowband electric field sensor. Outputs of all sensors were recorded continuously on a wideband analog recorder. Digital records were obtained of individual pulses or sequence of pulses during the lightning attachments.

Eight-bit waveform digitizers were triggered simultaneously by a common pulse from the trigger system, so that all the digital records were synchronized. The digitizers were operated at their highest sampling rate to yield a 10 μs window with 2048 data points at 5 ns intervals. The analog recorder used had 28 channels, the FM channels having a bandwidth of dc to 500 kHz and the direct-record channels having a bandwidth of 400 Hz to 2 MHz.

The trigger system was activated by either a positive or a negative signal from any surface current sensor. It would then send a common pulse to the digitizers and to the analog recorder. Synchronization between the analog and digital recordings was via the trigger pulse recorded on one analog channel. An IRIG B time code signal was recorded on the analog recorder for time correlation with the data obtained by instrumentation on the ground.

The digital system required 500 ms to store each 10 μs data set and to rearm. Since only one such data set could generally be obtained per flash, the digital system was set to trigger only at the high current levels expected after the lightning had attached to the aircraft. The analog recorder, in contrast, provided 15 minutes of continuous data at a tape speed of 120 ips (about 3 m s^{-1}).

10.3.4. French Transall program, 1984, 1988

The C-160 research aircraft used during the Transall field programs in France was a two-engine aircraft similar to the CV-580 but somewhat larger, 32.4 m in length versus 24.7 m for the CV-580. According to Lalande et al. (1999) the capacitance of the C-160 is 1 nF and of the CV-580 is 0.74 nF. The C-160 program is described in Moreau et al. (1992), Mazur (1992), Mazur and Moreau (1992), Lalande and Bondiou-Clergerie (1997) and Lalande et al. (1999). There are apparently no published data from the 1984 program. The 1988 program was conducted in the south of France by the Office National d'Etudes et de Récherches Aérospatiales (ONERA) and the Centre d'Essais en Vol (CEV). For the 1988 study the C-160 was instrumented specifically for investigation of the initial processes of lightning attachment. The instruments used were a network of five electric field mills, a network of seven capacitive antennas with active integrators, current shunts, and a high-speed (200 frames per second) video system. Moreau et al. (1992)

10.4. Mechanisms of lightning–aircraft interaction

gave the amplitude and frequency ranges for all sensors. The bandwidth of the electric field mill system on the C-160 was too narrow (0 to 40 Hz) to characterize the field variation within the first several milliseconds of strike initiation but was used to determine the ambient field value and the charge on the aircraft prior to the strike. The capacitive antenna network had a bandwidth from 1.5 Hz to 5 MHz, was equipped with 100 MHz 10 bit digitizers, and was used to characterize the electric field on the aircraft surface on submicrosecond to tens-of-millisecond time scales.

A video camera having a recording speed of 200 frames per second and a fish-eye lens with a 197° viewing angle was located in a pylon under the right wing, 10 m from the fuselage. The camera, with a vertical resolution of 262 lines and a horizontal resolution of 200 pixels, simulated a still camera with a 5 ms time exposure and a 400 μs interval between frames. The video signal was recorded with a special video cassette recorder having a 12 MHz bandwidth. The light sensitivity of the camera at the F/16 aperture was about 16 000 lux, which corresponds to a brightness of 8 W m^{-2} for a light source 15 m from the camera. The video recording was synchronized to the electrical measurements to within 1 s.

The Transall program provided the low-altitude fields, the currents, and the high-speed video records that, in concert with the data from the F-106B and CV-580, make possible a description of the physical processes common to most aircraft lightning interactions.

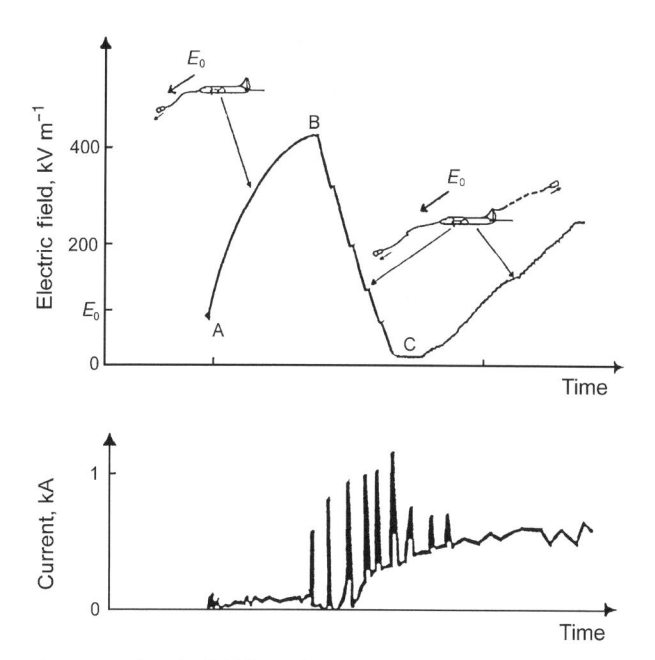

Fig. 10.5. Electric field intensity, current, and schematic representation of leader development during the initial phase of a typical aircraft-initiated lightning. E_0 is the ambient electric field at lightning initiation. The interval between the two scale markings on the upper time axis is 5 ms, and applies to both time axes. Adapted from Lalande *et al.* (1999).

10.4. Mechanisms of lightning–aircraft interaction
10.4.1. Aircraft initiation

Mazur (1989b), from an analysis of electric field and current records obtained on the F-106B and the CV-580, proposed a physical model of lightning initiation by aircraft that has been accepted by most researchers. Moreau *et al.* (1992) and Lalande *et al.* (1999) have reviewed and summarized the data from the CV-580 and C-160 studies, including the high-speed video observations on the C-160 of Moreau *et al.* (1992), and the two groups have described the lightning initiation process in essentially the same way. Figure 10.5 shows the typical electric field waveform observed on either the CV-580 or the C-160 during events interpreted as aircraft-initiated lightning, along with the typical time-correlated current through the aircraft. Also indicated in Fig. 10.5 is the bidirectional leader development inferred from the field, current, and photographic measurements. Measured electric field and current waveforms from the C-160 experiment are shown in Fig. 10.6. From examinations of the correlated electric field and current records, both Moreau *et al.* (1992) and Lalande *et al.* (1999) inferred that about 90 percent of the lightning strikes to the CV-580 and the C-160 were triggered by the aircraft.

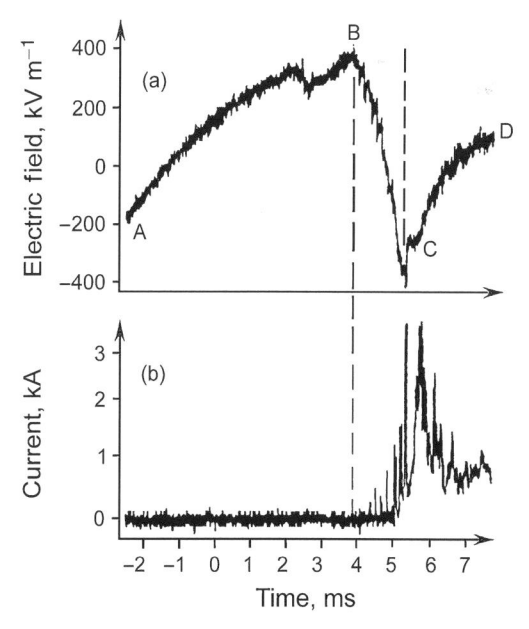

Fig. 10.6. Correlated electric field and current records on the C-160 aircraft: (a) electric field at the forward upper fuselage sensor and (b) current through the nose boom shunt. Adapted from Moreau *et al.* (1992).

The following description of the physical processes occurring in aircraft-initiated discharges to the CV-580 and C-160 is taken from Mazur (1989b), Moreau et al. (1992), and Lalande et al. (1999). Aircraft-initiated lightning events can be divided into two phases. The first phase involves the initiation and development of a bidirectional leader; this begins when the aircraft flies into a region of the cloud where the ambient electric field is typically near 50 kV m^{-1}. Interestingly, the orientation of the ambient field during strikes to the C-160 was mostly vertical while for the CV-580 it was mostly horizontal, which according to Lalande and Bondiou-Clergierie (1997) may be due to the fact that the C-160 flew in France at an altitude of 4.6 km (ambient temperature -5 °C), which was inside the main negative charge region of the French thunderclouds, while the CV-580 flew in Florida at 4.5 km (ambient temperature 0 °C), the same altitude as the bottom of the main negative charge region in Florida thunderclouds. In a sufficiently high field, the first discharge-related event on the CV-580 or C-160 was inferred to be the initiation of a positively-charged leader from the aircraft, as shown in Fig. 10.5. The leader propagates in the direction of the ambient field. During the development of this positive leader, from A to B in Figs. 10.5 and 10.6, a net negative charge increases on the aircraft owing to the removal of positive charge by the propagating positive leader, and the field enhancement on the aircraft increases owing to the increase in length of the overall conducting system of aircraft plus positive leader. The increase in negative charge on the aircraft produces an increase in the electric field pointing toward the aircraft surface at all points on the aircraft surface. In Figs. 10.5 and 10.6 this increase (A to B) is plotted as a positive field change although the vector direction depends on where on the aircraft the field is measured. Further, the relation between the directions of the ambient field and the field change AB is also aircraft-position dependent.

A few milliseconds after the initiation of the positive leader, the electric field value on the aircraft necessary for launching a negative leader, the field near point B in Figs. 10.5 and 10.6, is reached. The negative leader develops from an extremity of the aircraft opposite to that where the positive leader develops, and it propagates in a direction opposite to both the ambient electric field and the direction of extension of the positive discharge. The negative-leader development serves to reduce the negative charge on the aircraft, leading to a reduction in the electric field pointing toward the aircraft surface, from B to C in Figs. 10.5 and 10.6, although the negative leader may be initiated prior to B. According to Lalande et al. (1999), as the negative leader propagates, the positive leader accelerates and branches, producing a positive increase in the aircraft electric field after C in Figs. 10.5 and 10.6.

In the view of Lalande et al., from B to C the negative leader is more efficient in removing charge from the aircraft than the positive leader, whereas after C the positive leader is the more efficient because of its branching and higher speed, but the details of the physics of the bidirectional leader development are certainly unclear. According to Lalande et al., currents of only a few amperes are associated with the initial positive leader, the current level being deduced from electric field change measurements. Moreau et al. (1992) presented, as noted earlier, a 200-frame-per-second video that shows images of a positive leader (phase AB) prior to negative leader initiation, although the video and electrical measurements were only synchronized to 1 s. Moreau et al. estimated from the magnetic field variations during the AB phase observed on the CV-580 that the steady current in the positive leader is about 1 A, and they noted that the current-measuring shunts used on the research aircraft had insufficient sensitivity to observe this small value. The total evidence for the existence of the positive leader is apparently (i) the electric and magnetic field variations during the AB phase, which could be subject to other interpretations, (ii) the high-speed video imaging that was inferred to be associated with the AB phase, and (iii) the fact that laboratory studies of positive leaders in long gaps indicate currents increasing in a few milliseconds to a value of order 1 A with a current rate of rise 6.6×10^2 A s^{-1}. According to Moreau et al. (1992), the latter value is close to the current rate of rise for positive leaders in rocket-triggered lightning. During the first few milliseconds of negative leader formation, identified by Moreau et al. as phase BC in Figs. 10.5 and 10.6, there are typically 10 or so impulses of current of nearly 1 kA amplitude separated by a mean time interval of 250 μs and superimposed on a relatively steady current, which increases to about 300 A.

Lalande et al. (1999) summarized data for 31 aircraft-initiated events involving the CV-580 and 12 involving the C-160. Additional details from this analysis are found in the report by Lalande and Bondiou-Clergerie (1997). The average duration of all the aircraft-initiated flashes was 400 ms, with a minimum of 140 ms and a maximum of 1 s. For the CV-580 the mean ambient electric field just prior to the time of the lightning occurrence was 51 kV m^{-1} with a range from 25 to 87 kV m^{-1} and for the C-160, 59 kV m^{-1} with a range from 44 to 75 kV m^{-1}. This ambient field value (E_0) is the field at point A in Figs. 10.5 and 10.6. For the CV-580 the electric field change attributed to the positive leader, from A to B in Figs. 10.5 and 10.6, had a mean value 342 kV m^{-1} and occurred in a mean time 3.9 ms; for the C-160 the corresponding values are 551 kV m^{-1} in 4.3 ms. The field change from A to B for the combined data varied from about 200 to 800 kV m^{-1} and the time interval from about 1 ms to about 9 ms. During the period B to C, in which the negative stepped leader is assumed to be initiated and propagating away from the aircraft, the electric field on the surface of the aircraft is reduced to near zero in a mean time of 1 ms for the CV-580 and 2 ms for the C-160. From the combined data from the two aircraft, the mean duration of the steady current was 188 ms, with mean amplitude 330 A, and mean maximum value 910 A. For the combined data, the mean charge, the integral of the current that flowed through the aircraft during the total duration of its interaction with the lightning, was 60 C.

Moreau et al. (1992) presented seven examples of electric field waveforms from the C-160 and one from the CV-580 during aircraft-initiated strikes, one of these being shown in Fig. 10.6 Additionally, Moreau et al. provided statistical data, similar to those presented by Lalande et al. (1999), on 33 lightning events inferred to be initiated by the CV-580 and 16 by the C-160, that is, two more events than provided by Lalande et al. (1999) for the CV-580 and four more than for the C-160. Moreau et al. stated that the data are from storm penetrations in central Florida (CV-580) and in southern France (C-160), at altitudes of 6 km and lower, while Lalande and Bondiou-Clergerie (1997) stated that generally the CV-580 flew at an altitude of 4.5 km and the C-160 at 4.6 km. Why the later data analysis of Lalande et al. (1999) contains fewer events than the earlier analysis of Moreau et al. (1992) is not explicitly stated, but one might assume that Lalande et al. (1999) excluded some data that they felt were not of sufficient quality for analysis; Lalande and Bondiou-Clergerie (1997) stated that the analysis involved waveforms previously printed on paper since the original tape-recorded data were degraded or otherwise not available for analysis.

Lalande et al. (1999) presented their statistical data in tables, Moreau et al. (1992) presented theirs in the form of histograms. The values presented by Moreau et al. and Lalande et al. for aircraft-initiated lightning are generally similar, but Moreau et al. apparently presented additional data with shorter time duration and smaller field change for the AB and BC phases (Figs. 10.5 and 10.6) than did Lalande et al.

The first correlated electric field and current waveforms of the type illustrated in Fig. 10.5 and 10.6 and interpreted as indicating aircraft-initiated lightning were apparently recorded on the CV-580 in 1984 (Rustan 1986; Reazer et al. 1987). Reazer et al. gave three examples of 35 such correlated pairs of field and current waveforms obtained on the CV-580 in 1984 and 1985, one pair being reproduced on two time scales in Figs. 10.7a, b; these waveforms are similar to those in Figs. 10.5 and 10.6. Reazer et al., however, did not interpret the AB phase of the waveforms in Figs. 10.5 and 10.6 as being due to a positive leader, as have most subsequent investigators. Rather, they considered the lightning initiation to begin with a negative stepped leader occurring at about the time B. They stated that at time C the negative leader connects with a region "of positive charge, producing larger current pulses of the same polarity and continuing current flow". Reazer et al. in their Fig. 9, show a correlated record of initial current and electric field in which the current pulses increase in magnitude and continue for a millisecond or so after C, in contrast with the situation illustrated in Figs. 10.5 and 10.6.

Although Reazer et al. stated that there is more than one mechanism that can account for the electric field variation AB, they favor rapid triboelectric negative charging of the aircraft in the presence of a positive charge center that produces "streamers" approaching the aircraft, a mechanism which the authors of this book have difficulty understanding. Mazur (1989b) interpreted Reazer et al. as attributing the AB phase to a variation in charge near the aircraft caused by an approaching positive leader; he stated that in the view of Reazer et al. the aircraft initiated an intracloud discharge "that was about to happen" by flying within a certain distance of a cloud charge region. This distance is calculated as the product of an assumed leader velocity, 1.5×10^5 m s^{-1}, and the duration of the electric field variation, a few milliseconds, and is therefore equal to a few hundred meters. Mazur argued against such an interpretation from two points of view: (i) there is no reason why a discharge in the proximity of an aircraft should begin more readily on a hydrometeor than on an aircraft extremity; (ii) according to the polarization mechanism which we will discuss later when intercepted lightning is considered, opposite electric-field-polarity changes should be observed at positions on the airplane located near to and far from the approaching leader. Such an effect is not seen, however, in the records of the four field mills on the CV-580 (Anderson and Bailey 1987) for the lightning strikes analyzed by Mazur. During the initial period of each strike, the electric field changes in all field mill records were of the same polarity.

Pitts et al. (1987), whose research on the NASA F-106B (subsection 10.3.2) was primarily aimed at providing statistics and maximum values for the derivatives of the current, electric flux density, and magnetic flux density, also discussed lightning initiation by the F-106B for both the case where the aircraft approaches a positive charge center and also the case where it approaches a negative charge center, and they discuss their field derivative data as they relate to lightning initiation by the F-106B. They suggested that corona processes will occur first on the nose boom followed by breakdown from other extremities of the aircraft, but primarily they considered only the first few microseconds of such processes. Research results from the

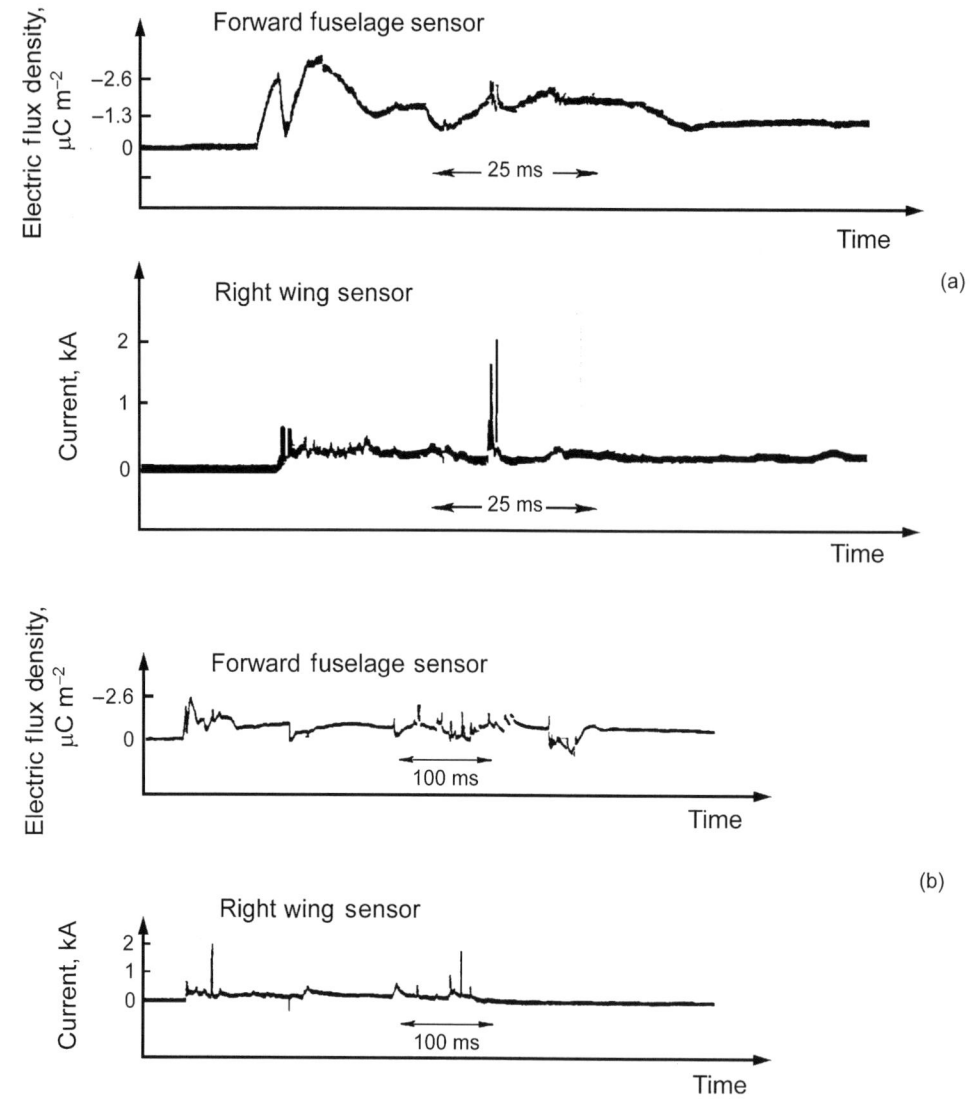

Fig. 10.7. (a) Electric field and current waveforms during the first 100 ms of lightning initiation by the CV-580. Adapted from Reazer et al. (1987). (b) The same data as in (a) but on a longer time scale.

F-106B study found in Pitts et al. (1987) include measured values as follows: maximum current rate of change (current derivative) of 3.8×10^{11} A s^{-1}, a maximum current of 54 kA, and a maximum rate of change of electric flux density (displacement current density) of 97 A m^{-2}, which was the upper limit of the measurement. All these measurements were obtained with peak recorders (subsection 10.3.2). The maximum measured current rate of change was about four times greater than that in the existing aircraft test standard for that parameter and prompted an increase in that value in the standard (Section 10.5). The current rate of change is an important parameter because the indirect effects of lightning strikes (Section 18.2 and Table 10.2) are thought to be related to the magnitude of the current rate of change.

Mazur (1986), from the F-106B study, stated that the current pulses observed during the initial discharge period of F-106B-initiated lightning have the following characteristics:

(i) a pulse repetition rate ranging from one pulse every 100 μs to one pulse every 20 ms;
(ii) a current-pulse duration ranging from a fraction of a microsecond to several microseconds;
(iii) a current-pulse amplitude in the range 2–20 kA.
(iv) a duration of the pulse series in the range 2–35 ms.

The pulse waveforms are unipolar and asymmetric and sometimes contain superimposed fine structure. The steady-current amplitude ranges from hundreds of amperes to 3 kA. The duration of the steady current is from tens to hundreds

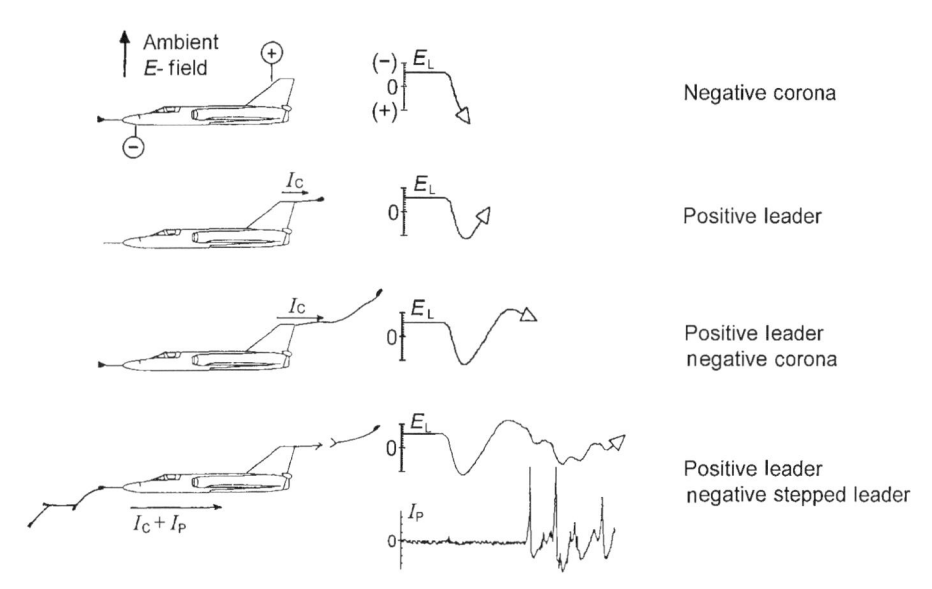

10.4. Mechanisms of lightning–aircraft interaction

Fig. 10.8. The processes involved in the initiation of lightning by the F-106B (the first 3 ms or so are shown). The strike attachment points are the nose boom and the tail fin cap. The circled plus and minus signs near the aircraft identify the polarities of the induced charges on the airplane in the vertical ambient electric field; E_L is the local field on the left forward part of the fuselage; I_C and I_P are the continuous (steady) current and the pulse current, respectively. Adapted from Mazur (1989b).

of milliseconds as estimated from video observations of lightning channels attached to the tail and wing tips of the F-106B.

Mazur (1989b) compared the lightning fields and currents measured on the F-106B with those measured on the CV-580 and C-160 and found them to be generally similar except that, prior to the AB phase of lightning initiation on the CV-580 or C-160 shown in Figs. 10.5 and 10.6, there is, on the F-106B, a negative field change, illustrated in Fig. 10.8. Mazur attributed this negative field change to positive charging of the F-106B by negative corona from one of the aircraft's extremities (see also Pitts et al. 1987). In his view, no negative leader, only negative corona, occurs at this time because the pulses attributed by him to negatively charged leader steps in the current waveform did not occur until phase 4 in Fig. 10.8, as for the BC phase in Figs. 10.5 and 10.6. Mazur attributed the differences between the F-106B initiation processes and those on the CV-580 and C-160 to the sharp nose boom on the F-106B, which supports early corona. He did not discuss whether the field and current signatures during aircraft initiation were a function of altitude. Thus it is not clear whether the initiation process high in the cloud, where the temperature is near $-40\,°C$ and the air density is relatively low, is similar to that in and beneath the primary negative charge region, where the temperature is near $0\,°C$. He did state, however, that the particular F-106B data presented in Mazur (1989b) were for aircraft altitudes below 6 km, as is apparently the case for all the CV-580 and C-160 data (Moreau et al. 1992).

According to Lalande et al. (1999), the second phase of the aircraft-initiated discharge begins roughly 50 ms after positive-leader initiation, apparently tens of milliseconds after C in Figs. 10.5 and 10.6, and is characterized by groups of current impulses, called bursts, separated by a few tens of milliseconds, as illustrated as Fig 10.9 and evident in Figs. 10.7a, b. Mazur (1989a) argued that the second phase of aircraft-initiated lightning is similar to the so-called junction (Section 4.10) or late (final) stage of natural intracloud flashes (Section 9.5). Some of the current bursts shown in Fig. 10.9 and Figs. 7a, b are apparently superimposed on steady current. Lalande et al. characterized the current pulses in the second phase as due to "recoil streamers", traditionally thought to be in-cloud miniature return strokes generated when a leader encounters a pocket of opposite charge, a view described in the discussion of the interception process in subsection 10.4.2. However, Lalande et al. attributed these current pulses to "conductivity instabilities" within the positive-leader channel, although they state that the prediction, following from this hypothesis, of more recoil streamers with lower steady current is not supported by the data. The peak current in the current bursts in the second phase of the discharge is indicated in Fig. 10.9 as being up to a factor 3 or so larger than the 1 kA or so peak current shown for the negative-leader pulses occurring between B and C (Figs. 10.5 and 10.6). On the C-160, the highest measured current derivative after C was 2×10^{10} A s^{-1} (with mean 6.5×10^9 A s^1) and the highest current was 20 kA (with mean 4.8 kA), but Lalande et al. cautioned that these values might not be representative because of the relatively

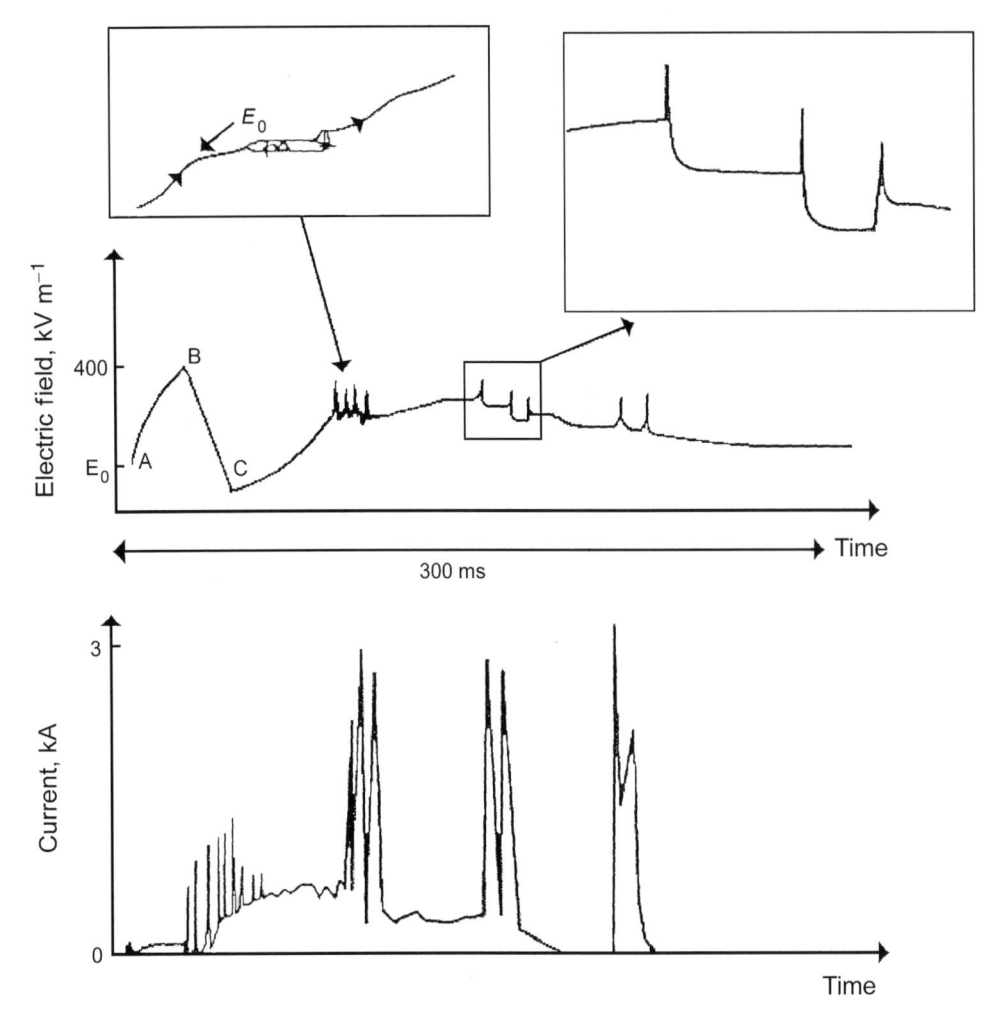

Fig. 10.9. Electric field intensity and current during the total duration of a typical aircraft-initiated lightning flash. E_0 is the ambient electric field at lightning initiation. Adapted from Lalande *et al.* (1999).

few events recorded and the difficulty in measuring the peak current. In recent aircraft–lightning test standards, the pulse-burst specification, the so-called component H, derived from interpretation of the F-106B, CV-580, and C-160 airborne measurements, describes pulses with an amplitude of 10 kA (Section 10.5). Bursts of pulses observed in the electromagnetic fields of both cloud and ground flashes, which must be associated with similar channel currents, and their relation to these test standards are discussed in Rakov *et al.* (1996). Additional information on the electromagnetic pulse structure of cloud discharges is found in Chapter 9.

Processes occurring in the second phase of the aircraft initiation process that are similar, according to Mazur and Moreau (1992), to processes occurring in the latter part of aircraft-intercepted lightning are considered in the discussion of aircraft-intercepted lightning in subsection 10.4.2 below.

Larigaldie *et al.* (1992) presented a model, based on laboratory studies, of the formation of individual negative-leader steps; they showed an example of the computed current associated with a step and compared the model pulse to a step current pulse observed on the C-160, the latter having a 500 A peak and width about 300 ns at half-peak value.

Numerical modeling has provided quantitative information on the threshold ambient electric field for lightning initiation as a function of the aircraft size and shape (Pitts *et al.* 1987; Lalande *et al.* 1999). In general, the larger the aircraft the lower the ambient field threshold for leader inception, owing to the generally higher field enhancement factors for larger aircraft, although if a larger plane has exactly the same shape as a smaller plane then the field enhancement in a uniform field is the same. However, larger aircraft have a larger capacitance and therefore acquire larger surface charges. The sign of this charge is apparently always observed to be negative for the CV-580 and

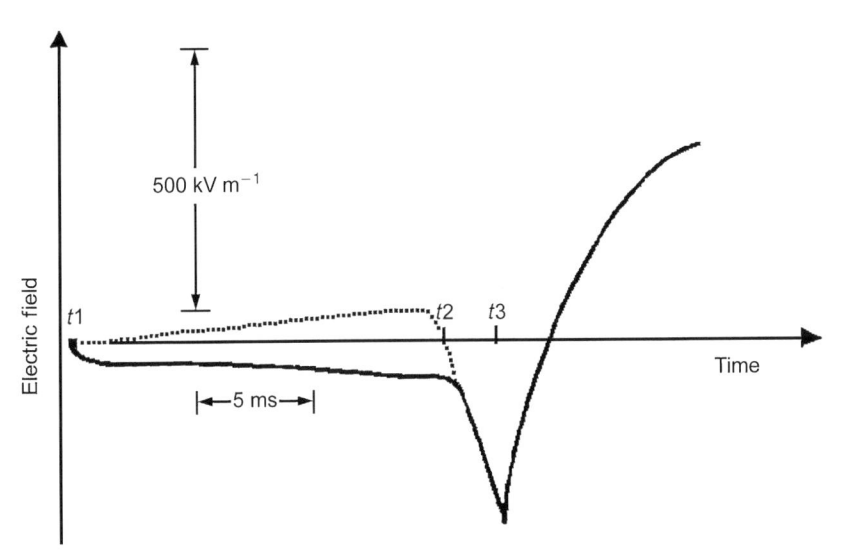

10.4. Mechanisms of lightning–aircraft interaction

Fig. 10.10. Electric field waveforms observed by two sensors (dotted and solid lines) during the first part of a lightning strike intercepted by an aircraft. From $t1$ to $t2$ the electric field changes have different polarities depending on sensor location. Adapted from Lalande *et al.* (1999).

the C-160; aircraft charging is discussed in Jones (1990). Probable aircraft charging mechanisms are aircraft-surface interaction with precipitation and charge removal by engine exhaust. The electric field associated with the negative charge on the aircraft serves to inhibit the development of an initial positive leader. The ambient field threshold to be reached for positive-leader initiation is therefore higher than it would be in the absence of a net charge on the aircraft. According to Lalande *et al.* (1999), considering the capacitance of the CV-580, 0.74 nF, and of the C-160, 1 nF, which is larger (32.4 m in length versus 24.7 m), it appears that the net-charge effect dominates the field enhancement. This leads to a higher ambient field threshold in the case of the larger C-160, an average of 59 kV m^{-1} versus 51 kV m^{-1} for the CV-580, although it is certainly not clear that there are enough data on the ambient field threshold for triggering or that the threshold field is determined accurately enough to draw such a conclusion.

10.4.2. The interception process

We consider now the case of aircraft-intercepted lightning. As illustrated in Fig. 10.10, which is a schematic diagram of the typical field changes of intercepted flashes, the millisecond-scale electric field variation from $t1$ to $t2$ observed on the CV-580 and C-160 research aircraft had a different sign for the sensors at different locations on the aircraft, as determined from the data for three strikes to the CV-580 and three to the C-160. An example of actual data from the C-160 experiment is given in Fig. 10.11. The different field change polarity observed by different sensors is generally interpreted as indicating that an externally applied electric field due to an approaching lightning channel produced a polarizing effect on the aircraft (Moreau *et al.* 1992): negative charge was induced on one part of the aircraft and positive charge on the opposite part, rather than a change in charge of the same sign everywhere on the aircraft, as in the AB phase of aircraft-initiated strikes (Figs. 10.5 and 10.6). According to Lalande *et al.* (1999), the fact that the electric field change from $t2$ to $t3$ in Fig. 10.10 from the different electric field sensors is similar in magnitude indicates that the aircraft has acquired a net positive charge. This result, according to Lalande *et al.* (1999) could be due to the nearly simultaneous inception of a positive and a negative leader, the later injecting a higher charge in the aircraft, or even to the inception of a single negative stepped leader. After $t3$, the electric field observed by all sensors increases toward a positive value as observed for the time following C in the aircraft-initiated case (Fig. 10.9). The remainder of the intercepted discharge is generally similar in its characteristics to the aircraft-initiated case.

Moreau *et al.* (1992) gave a different interpretation of the interception process from an analysis of waveforms such as those shown in Fig. 10.11. They considered that charge separation (polarization) on the aircraft due to an approaching leader occurs until point B in Fig. 10.11 ($t3$ in Fig. 10.10), which they postulate is the time of attachment. According to Moreau *et al.*, the rapid positive field change on both sensors at B indicates attachment and charging by a negative leader, and the rapid negative field change at C in Fig. 10.11 indicates that the negative leader has exited the aircraft.

Reazer *et al.* (1987) showed examples of correlated current and electric field change for electric field changes of the type shown in Figs. 10.10 and 10.11, one case being

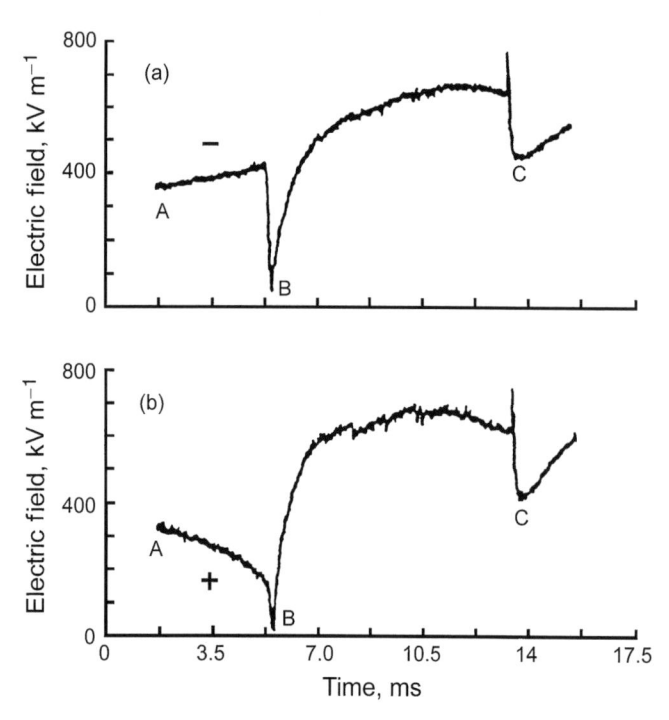

Fig. 10.11. Electric field variation on the C-160 for an intercepted lightning flash at (a) the rear fuselage sensor and (b) the front fuselage sensor. Note the positive field change at the rear sensor corresponding to induced negative charge (indicated by the minus sign) and the negative field change at the front sensor corresponding to induced positive charge (indicated by the plus sign) during the AB phase ($t1$–$t2$ in Fig. 10.10). Adapted from Moreau *et al.* (1992).

presented in Fig. 10.12. The current waveform is noisy, but it appears to indicate that a current pulse or pulses of a few hundred amperes probably occurred near $t3$ of Fig. 10.10 and B of Fig. 10.11, with no large pulses in the millisecond or so before or after $t3$ (Fig. 10.10) or B (Fig. 10.11). Reazer *et al.* interpreted the "hooked shape of the electric field" as having been "produced by the leader as it approaches the aircraft", an interpretation of the electric field waveform different from those of Lalande *et al.* (1999) and Moreau *et al.* (1992) discussed above.

Mazur and Moreau (1992) examined seven strikes to the CV-580 and 11 to the C-160 in order to try to understand "processes taking place during the intracloud propagation of lightning strikes initiated on or intercepted by the airplane". They identified "recoil streamers," dart-leader–return-stroke sequences, and "secondary initiations of new discharges". Recoil streamers were defined by Mazur and Moreau as current pulses that originate near the tip of the positive leader and propagate back toward the aircraft. The fields and currents attributed to recoil streamers occur generally in the latter part of the flash and are associated with the deposition of a negative charge on the aircraft. The current pulses attributed to recoil streamers generally occur in bursts with a typical time between the pulses in a burst equal to a few milliseconds, although sometimes the pulses are single. These current pulse bursts are illustrated

in Fig. 10.9 and have also been attributed to recoil streamers by Lalande *et al.* (1999) and others, as discussed earlier. Mazur and Moreau hypothesized that bursts of recoil streamers are associated with rapidly branching channels encountering opposite-sign charges. Another view of recoil streamers is given by Lalande *et al.* (1999), as discussed in subsection 10.4.1.

Mazur and Moreau (1992) gave data for two strikes to the CV-580 which include field and current waveforms that can be interpreted as dart-leader–return-stroke sequences, which is similar to the interpretation of Mazur *et al.* (1990) considered in the last paragraph of subsection 10.4.3. An example of such a sequence from Reazer *et al.* (1987) is shown in Fig. 10.13. The so-called "secondary initiations of new discharges" have currents and fields resembling the primary initiation processes but occur during the overall discharge development and have positive and negative leaders inferred to be of shorter duration than the initial ones. These inferred secondary initiations can occur more than once during the overall discharge, each time producing bursts of pulses presumably due to negative stepped leaders launched from the aircraft.

10.4.3. Other inferences and results

Petterson and Wood (1968) presented many photographs of lightning channels attached at two extremities of

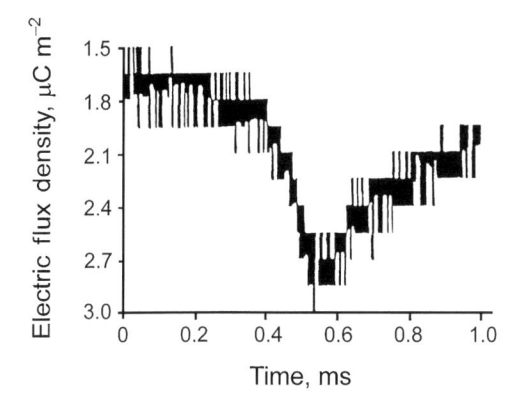

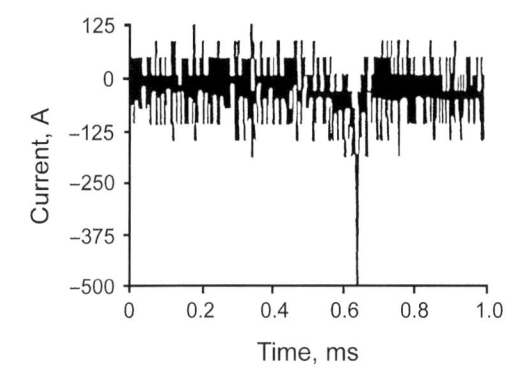

10.4. Mechanisms of lightning–aircraft interaction

Fig. 10.12. Initial electric field and current records on the CV-580 for an intercepted flash at 15 000 feet (about 5 km). Upper figure, front fuselage sensor; lower figure, left-wing sensor. Adapted from Reazer *et al.* (1987).

the F-100F (subsection 10.3.1). The peak current derivative measured was 2×10^9 A s^{-1} with most measurements being between 0.5×10^9 and 2×10^9 A s^{-1}. The maximum current measured was 22 kA, values in the range 1 to 5 kA being most common. Altogether, data were recorded for 49 strikes at altitudes mostly near 30 000 feet (about 9.1 km), where the temperature was near -40 °C, with a few strikes near 15 000 to 21 000 feet (about 4.5 to 6.4 km). Twenty-six of 29 strikes to the nose boom exhibited positive currents (e.g., negative leaders leaving the nose boom), which Petterson and Wood (1968) suggest is due to the aircraft "leaving a negative cell and approaching a positive cell which was centered at some higher level". The F-100F made most cloud traversals at higher altitudes than the CV-580 and C-160 could fly, where commercial aircraft would not be likely to enter a thunderstorm, and may well have encountered a different environment relative to the mechanisms of aircraft-initiated lightning. As noted in Section 10.2, most strikes take place to commercial aircraft in the 3 to 5 km altitude range on ascent or descent, or to the older propeller planes at cruise altitudes, a region probably just below the primary negative cloud charge location.

Mazur *et al.* (1984a) described several aspects of the 1982 NASA F-106B program and gave data for 36 cloud penetrations by the F-106B. They presented evidence from radar echos that the F-106B initiates lightning, and they argued that aircraft-initiated strikes closely resemble intracloud flashes in their radar characteristics (a view expanded upon by Mazur (1989a) by also considering the current and electric field waveforms observed on the F-106B and CV-580 aircraft). Mazur *et al.* (1984a) showed that the greatest probability of initiating lightning with an aircraft of the F-106B type exists in the upper portions of a thunderstorm where the ambient temperature is -40 °C or colder, when turbulence and precipitation are light to negligible and when the lightning flash rate is less than 10 min^{-1}. Fisher *et al.* (1986) gave detailed statistics on the number of strikes versus ambient temperature and pressure for the 1980–5 F-106B program including the number of cloud penetrations by year at high and low altitude and the resultant strike statistics. The dividing altitude between high and low is apparently 6 km. For 175 missions in total there were 839 high penetrations of thunderclouds and 539 low ones. The high penetrations resulted in 615 strikes, whereas the low ones resulted in 75.

Mazur *et al.* (1986) described aspects of the 1984 F-106B program, during which storm penetrations were made at 6 to 8 km altitude. There were 34 strikes of which 10 were apparently related to cloud-to-ground flashes as inferred from the East Coast Lightning Detection Network, a predecessor of the NLDN (Section 17.5). In four of these cases, strikes to the F-106B preceded ground return strokes by 10 to 110 ms and in the remaining six cases they followed return strokes after 20 to 100 ms. The majority of strikes to the F-106B occurred during the decaying stage of the storm when turbulence and rain intensity were low. The probability of a strike increased with decreasing flash rate. The conditions for aircraft-initiated events at the lower altitudes are apparently similar to the high-altitude strike case discussed in the previous paragraph.

Mazur *et al.* (1990) described a multiple-stroke cloud-to-ground lightning discharge triggered by the F-106B when it was flying at an altitude of 5 km where the ambient temperature was -1 °C, there was light turbulence, and no precipitation was observed. The East Coast Lightning Detection Network registered six return strokes while eight events were interpreted as dart-leader–return-stroke sequences in the airborne data, at least three passing through the F-106B. Previously, Reazer *et al.* (1987) had provided evidence that the CV-580 was involved with two cloud-to-ground events, one in the main channel-to-ground of a subsequent return stroke; and, as noted in the previous paragraph, Mazur *et al.* (1986) had reported a correspondence within about 100 ms between ground return strokes and strikes to the F-106B. In the aircraft-initiated flash investigated by Mazur *et al.* (1990), the F-106B apparently initiated the lightning about 70 ms before strokes

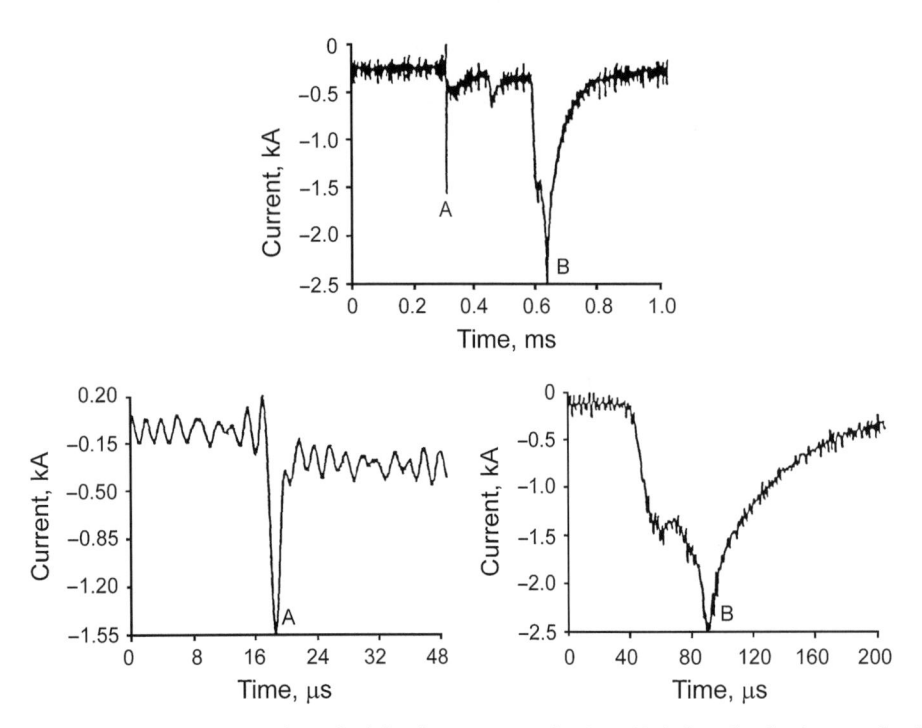

Fig. 10.13. Current pulses from the left-wing sensor on the CV-580, inferred to be due to a dart leader (A) and subsequent return stroke (B) passing through the aircraft as part of a cloud-to-ground flash. Adapted from Reazer et al. (1987).

to ground were observed. Mazur et al. (1990) admitted that "some interpretation of lightning processes made in this paper may seem questionable in view of using the limited resolution airborne data that characterizes processes only in the time domain". The dart-leader–return-stroke sequence was identified by two sequential current pulses of the same polarity within a time sufficient for the dart leader to reach the ground and the return stroke to travel to the F-106B, as can be seen in the data shown in Fig. 10.13 for the CV-580.

10.5. Lightning test standards

Three major standards for the lightning protection of aircraft were published in 1999 by the Society of Automotive Engineers (SAE) in coordination with the European Organization for Civil Aviation Equipment (EUROCAE) and are listed below:

(i) ARP5412 (ED 84), Aircraft lightning environment and related Test waveforms, http://www.sae.org
(ii) ARP5413 (ED 81), Certification of aircraft electrical/electronic systems for the indirect effects of lightning, http://www.sae.org
(iii) ARP5414 (ED 91), Aircraft lightning zoning, http://www.sae.org

The acronym ARP stands for aerospace recommended practices and ED for EUROCAE document. Some of the ARP and the equivalent ED documents listed above differ in the material in their appendices. There are numerous earlier versions of these standards published by the SAE, FAA, EUROCAE, and various military and other organizations, many of which are referenced in the documents listed above. A review and discussion of the test standards existing in the early 1990s was given by Plumer (1992).

The first document listed above, ARP5412, specifies a series of idealized voltage and current waveforms with which aircraft are to be tested for the effects of lightning. The idealized test current waveforms are labeled A,B,C,D,D/2 and H and are illustrated in Figs. 10.14a, b, c. From the known characteristics of lightning, components A through D/2 represent severe currents in cloud-to-ground lightning (although component C, the continuing current, is more likely to follow component D in natural flashes than to follow A and B as in Fig. 10.14a, e.g., Rakov and Uman, 1990), which an aircraft would only encounter at very low altitude. ARP5412 also specifies the allowed approximations to the idealized waveforms of Figs. 10.14a, b, c that can be used in the laboratory. As noted in subsection 10.4.1, component H is derived from airborne F-106B, CV-580, and C-160 measurements and is intended to describe the multiple current bursts observed on an aircraft in flight.

Specification of the numbers of pulses, the numbers of bursts, and the time intervals between the pulses and bursts for component H has evolved with time. A

10.5. Lightning test standards

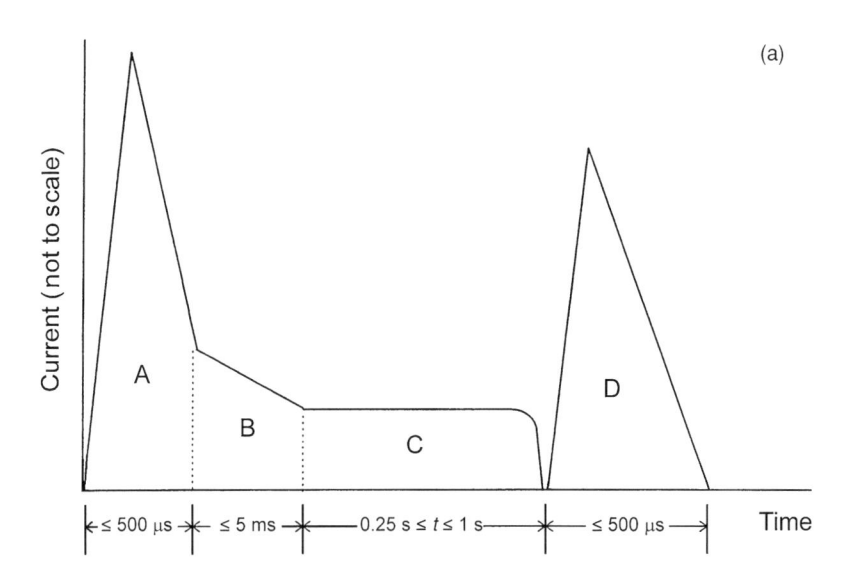

Component A (first return stroke)
- peak amplitude: 200 kA (+10%)
- action integral: $2 \times 10^6 \, A^2 \, s$ (±20%) (in 500 μs)
- duration: ≤ 500 μs

Component B (intermediate current)
- max. charge transfer: 10 C (±20%)
- average amplitude: 2 kA (±20%)
- duration: ≤ 5 ms

Component C (continuing current)
- amplitude: 200–800 A
- charge transfer: 200 C (±20%)
- duration: 0.25 to 1 s

Component D (subsequent return stroke)
- peak amplitude: 100 kA (±10%)
- action integral: $0.25 \times 10^6 \, A^2 \, s$ (±20%) (in 500 μs)
- duration: ≤ 500 μs

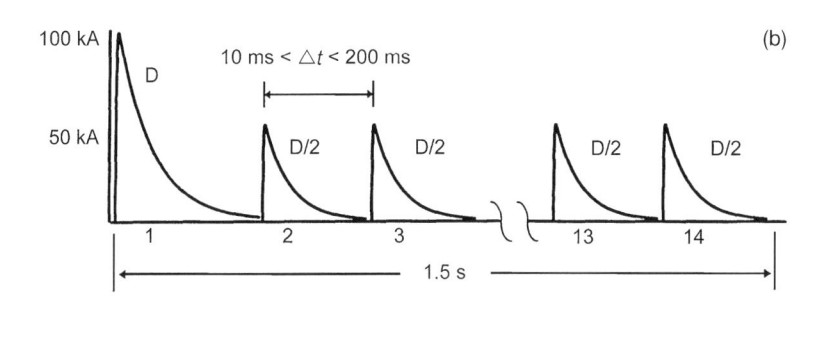

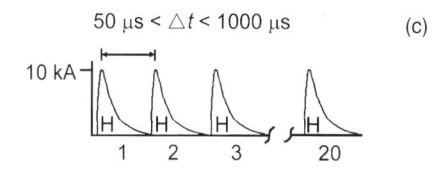

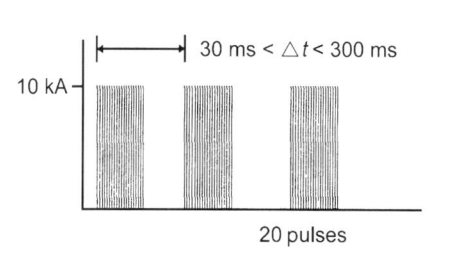

Fig. 10.14. (a) Current components A through D simulating the first two strokes in a cloud-to-ground flash, (b) multiple-stroke waveform simulating the second and additional strokes, and (c) multiple-burst (H-component) waveform showing the pulse waveshape (top) and pulse burst structure (bottom). Adapted from SAE ARP5412.

discussion is found in Rakov et al. (1996). The first specification was derived from F-106B measurements at altitudes near 10 km. The data recording system could provide only one burst per flash (from its triggered transient recorder, see subsection 10.3.2) with a sufficient high frequency response to characterize the burst properly, while similar bursts were inferred to occur from the tape recorder data in which the pulse waveforms were likely to be significantly distorted. Fisher et al. (1999) showed, in their Fig. 3.27, a F-106B pulse-burst waveform, apparently occurring during the initiation process, of the type that provided the original impetus for specifying component H. The larger pulses in the burst had amplitudes near 10 kA. The CV-580 and C-160 experiments provided additional data on pulse bursts. The burst currents associated with the initiation process are typically of about 1 kA amplitude and those occurring later in the flash are of indeterminate amplitude (Lalande et al. 1999). Mazur and Moreau (1992) suggested that the initiation process shown in Figs. 10.5 and 10.6 can be repeated several times per flash (subsection 10.4.2), and Moreau (personal communication, 2000) views these multiple initiations as the source of the multiple pulse bursts specified in component H; he regards an amplitude in the test waveform that is an order of magnitude greater than the 1 kA typical of the measured values as being conservative, this being the case for the other components in the test waveform too. Mazur and Moreau (1992) reported on current pulses occurring later in the flash and attributed them to recoil streamers that exhibit greater duration and interpulse interval than the H-component specification. The H-component can probably best be viewed as a conservative, compromise standard test waveform that accounts for both the current-pulse bursts associated with negative stepped leaders from the multiple initiation processes at the aircraft and also the current pulses flowing through the aircraft from so-called recoil streamers originating at a distance from the aircraft.

Four voltage test waveforms are found in ARP5412 that are intended to identify lightning-attachment points and dielectric-breakdown paths through non-conducting surfaces or structures. The voltages are to be imposed between an external electrode and the grounded airframe. The four different waveforms have been chosen for their ability to produce various forms of damage observed on aircraft involved with lightning. The voltage waveform specified in ARP5412 as waveform A increases linearly at 1000 kV m s^{-1}, waveform B has a 1.2 μs (± 20 percent) risetime to peak and a 50 μs (± 20 percent) time to half-peak value, waveform C is a linear rising voltage chopped to zero at 2 μs (± 50 percent), and waveform D has a risetime to peak of 50 to 250 μs and a time to half-peak value of about 2 ms.

Different zones of the aircraft are expected to experience different levels of lightning severity, and these zones are defined in ARP5414. Zone 1A is that portion of the aircraft that can be expected to encounter a direct first-return stroke but not the remainder of the flash, owing to subsequent motion of the channel across the aircraft surface. Zone 1B is expected to encounter a first stroke and the rest of the flash. Zone 1C is expected to encounter a reduced-amplitude first stroke only and not the remainder of the flash. A subsequent stroke is likely to be swept into zone 2A by channel motion but with low probability that the remainder of the flash will occur there, while zone 2B is likely to encounter a swept subsequent stroke and the remainder of the flash. Zone 3 is expected to receive conducted current only, not direct channel attachment. Different test current and test voltage waveform components are to be applied to the different zones, as specified in Table 3 of ARP5412, for testing against both direct and induced effects. Procedures to test for indirect effects are considered in ARP5413.

10.6. Accidents

Lightning damage to aircraft varies from minor pitting of the aluminum skin to complete destruction of the aircraft. Most lightning–aircraft interactions are isolated occurrences. However, sometimes weather conditions are apparently such as to make lightning triggering by aircraft more likely, and then several aircraft may be involved. This was apparently the case on 24 February 1987 when, in a period of a few hours, at least six aircraft were struck by lightning arriving or departing airports in the Los Angeles area. The winter storm system present that day exhibited rain showers and occasional lightning. Four Boeing 727s, flying between 3800 and 8000 feet (between about 1.1 and 2.4 km), suffered lightning-caused holes in their radomes, and a Boeing 737 suffered unspecified damage at 3200 feet (about 1 km) (V. Mazur, personal communication, 2000: a copy of a letter received from the Air Line Pilots Association). A NASA T-38A jet flown by two astronauts suffered a lightning-induced in-flight explosion at 2500 feet (about 0.75 km) followed by a fire that extensively damaged the center fuselage. The T-38A, still on fire, landed at a military base near Los Angeles. The crew escaped injury. The official report describing the T-38A incident is found in McMurtry (1987).

In the remainder of this section we will examine a number of crashes or near-crashes of commercial aircraft where lightning did play or may have played a role. We will also discuss an extraordinary lightning strike to a small commercial aircraft on takeoff, as well as two related incidents, and the initiation of lightning, and its effects, by two space vehicles, Apollo 12 and Atlas-Centaur 67, during their launches from the Kennedy Space Center and the adjacent Cape Canaveral Air Force Station, respectively.

10.6. Accidents

10.6.1. Boeing 707 in 1963

On 8 December 1963, a Pan American World Airways Boeing 707-121 was in a holding pattern at 5000 feet (about 1.5 km) near Elkton, Maryland. The aircraft was built in 1958 and had flown less than 15 000 hours. There was thunderstorm activity in the area. Ninety-nine witnesses reported a cloud-to-ground lightning flash near or on the aircraft at about the time it burst into flames. All aboard, 73 passengers and eight crew members, were killed. An investigation determined that three fuel tanks had exploded and that there were lightning strike marks and holes on the left wing tip. Photographs of this lightning damage are found in Uman (1986). Evidence indicated that the left reserve fuel tank, the outermost fuel tank in the left wing, exploded first, followed by the center and right reserve fuel tanks. There was lightning damage about 30 cm from the edge of the left-reserve-fuel-tank vent outlet. The largest single indication of lightning was an irregular-shaped hole about 4 cm in diameter burned through the top of the wing. Exactly how the fuel tanks were ignited could not be determined. Possibly an attached lightning channel burned through the wing surface into a fuel tank (the fuel tank container is the wing skin in some areas) or sufficiently heated the inside surface to cause the explosion, or possibly lightning ignited combustible fumes at the left-reserve-fuel-tank vent outlet. Further, laboratory tests showed that lightning-like currents injected over fuel filler caps and access plates on the 707 wing could produce sparks inside the fuel tanks.

After the accident and as a result of further research (Kofoid 1970), the thickness of the aluminum skin enclosing the fuel on 707s and on other aircraft was increased and fuel filler caps and access plates were better bonded to the airframe. The official report, Aircraft Accident Report, Boeing 707-121 N709PA Pan American World Airways, Inc., near Elkton, Maryland, 8 December 1963, Civil Aeronautics Board File no. 1-0015, 25 February 1965, attributes the disaster to "lightning-induced ignition of the fuel/air mixture in the no. 1 reserve fuel tank with resultant explosive disintegration of the left outer wing and loss of control". Subsection 10.6.2 contains some further comments on this accident.

10.6.2. Boeing 747 in 1976

On 9 May 1976, an Imperial Iranian Air Force B-747, Flight ULF48, was struck by lightning near Madrid, Spain with catastrophic results. The aircraft was on a military logistics flight to McGuire Air Force Base, United States, from Teheran, Iran, with an intermediate stop in Madrid, Spain. The plane crashed at approximately 4:30 pm local time or 2:30 GMT. The last radio contact was made as the aircraft was descending to 5000 feet (about 1.5 km) in clouds, probably near an altitude of 6000 feet (about 1.8 km). Since the type of aircraft involved, the Boeing 747, was used extensively in commercial operations worldwide at that time (and subsequently), in view of the nature of the accident the US National Transportation Safety Board requested and was granted permission to assist in the investigation. The resultant report is labeled NTSB-AAR-78-12, October 1978: Special Investigation Report – Wing Failure of Boeing 747-131, Near Madrid, Spain, May 9, 1976, from which the discussion in this section is taken.

At the time of the accident, the weather near Madrid was cloudy with rain and lightning; visibility was good. There were severe thunderstorms in the area. Two witnesses reported seeing lightning strike the aircraft. Some witnesses stated that they saw an in-flight fire confined to the no. 1 engine. Other witnesses reported seeing an in-flight explosion and fire followed by the separation of aircraft parts. Pitting and localized burn areas typical of lightning-attachment damage were found on the left wing tip and on the vertical fin. No holes were burned into any of the fuel tanks. The left wing had separated into 15 major pieces before ground impact and parts of it were found at a number of locations.

The first significant event on the cockpit voice recorder was the exclamation "We're in the soup!" Approximately 3 s later a signal characteristic of an electrical transient occurred on the tape, which has been interpreted by the investigating team as indicating that the aircraft was struck by lightning. An explosion occurred 0.2 s after the electrical transient. A sound interpreted as thunder was heard before the explosion.

Several motor-operated valves were present in the fuel tanks, and the electric motors that operated these valves were mounted on the outside surfaces of the front or rear spar. The motors were connected to the valves by mechanical couplings or drive shafts that penetrated the spars. The motor for the valve in the no. 1 fuel tank was never recovered. The drive shaft was found and was determined to be electrically insulated at the spar penetration. The mechanical coupling and drive-shaft arrangement could have provided a path for an electric current to enter the tank and cause a spark. The level of residual magnetization in this area of the valve was indicative of high currents.

The evidence (i) that the explosion in the no. 1 tank occurred in the immediate area of a motor-driven fuel valve, (ii) that the motor was never recovered, (iii) that a high level of residual magnetization existed in the ferrous material in this area, (iv) that certification tests showed this area to be a probable lightning-attachment point, (v) that lightning strikes are known to have disabled the motors on other aircraft, and (vi) that no other possible ignition source could be determined, provided the foundation for the hypothesis that the tank explosion was likely to have been ignited by a spark at this motor-driven valve.

The official report (NTSB-AAR-78-12), from which we quote directly, states that "assuming that a lightning strike can generate a source of ignition to fuel vapors, aircraft fuel explosions could occur more frequently. However, events must combine simultaneously to create the explosion, and this combination would occur rarely. In this case, the events were (1) an intermittently conductive path which closed and opened an electrical loop, (2) a lightning-induced current of sufficient intensity flowed in this path and formed a spark, and (3) a flammable vapor surrounded this spark. Possibly this combination of events has occurred a number of times before, in the following accidents: Milan, Italy (Constellation); Elkton, Maryland (B-707); Madrid, Spain (USAF KC-135); KSC, Florida (USAF F-4); Pacallpa, Peru (L-188)". The second accident was discussed in subsection 10.6.1.

10.6.3. Fairchild Metro III in 1988 and Fokker F28 MK 0100 in 1998

In the previous two subsections we have discussed two cases in which aircraft exploded when fuel vapor was ignited by lightning. In the present subsection we examine two accidents in which the lightning damage was less direct but potentially as fatal: the lightning-caused failure of the electrical system in a Fairchild Metro III, which led to the loss of the aircraft and the deaths of its occupants; and the lightning-caused failure of the hydraulic system of a Fokker F28, which nearly caused similar results.

On 8 February 1988 a Fairchild Metro III commuter airliner powered by two turboprop engines and carrying 19 passengers and two crew members on a flight from Hannover to Düsseldorf, Germany, was struck by lightning and subsequently crashed, killing all on board. The airliner was approaching Düsseldorf at an altitude of about 3000 feet (about 0.9 km). There were thunderstorms in the area. The pilot had lowered the landing gear although the copilot had argued that he should not do so. When the gear was lowered the plane fell and rose in altitude between 2500 and 3000 feet (between about 0.75 and 0.9 km) as the pilots tried to trim the aircraft for proper descent. The cockpit voice recorder provided a record of the pilot and copilot's conversation. As they were stabilizing the aircraft, lightning struck it and apparently disconnected all batteries and generators from the aircraft's electrical system, also terminating the cockpit voice recorder record. Without electrical power, the pilots evidently had no control of the landing gear and limited control of the flaps. The aircraft was inside a cloud and had no cockpit lights so the pilots would probably not have been able to read their instruments. Emergency flashlights apparently were not present in the aircraft as they were supposed to be, or at least none was found at the crash scene. Observers on the ground saw the aircraft dive out of the cloud base and then climb again into the cloud, this pattern being repeated two or three times. On one of these oscillations in altitude, the right landing gear was torn from the aircraft, further destabilizing it. The subsequent aircraft motion resulted in a wing being separated from the aircraft. The Fairchild went into a spiral dive and crashed. A reconstruction of the electrical system failure pointed to the failure of a critical relay.

Overall, the accident was probably due to a combination of insufficient pilot judgement or skill and the lightning-caused electrical failure. Whether the electrical system was properly designed, that is, whether lightning should have been able to cause it to fail completely, is also an issue. The official report of the accident is found in "Bericht über die Untersuchung des Flugunfalles mit dem Flugseug SA Z27–AC, Metro III, D-CABB, am 8. Februar 1988 bei Kettwig AZ.: 1X001/88, Flugunfalluntersuchungsstelle beim Luftfahrt-Bundesamt, Bundesrepublik Deutschland."

On 26 February 1998, a US Airways Fokker F28 MK 0100 flying from Charlotte, NC to Birmingham, Alabama carrying 87 passengers and five crew members was struck by lightning with no immediate effect. However, within a few minutes the aircraft suffered a failure of both of its hydraulic systems. In order to make an emergency landing, the landing gear and flaps were extended via an alternate method but without control of the nose landing-gear steering. A number of brake applications were also possible in an alternate mode to the hydraulic. On landing, the aircraft traveled about 1100 feet (about 330 m) in the grass off the left side of the runway. The nose landing-gear separated from the aircraft and the nose section came to rest on a taxiway about 540 feet (about 160 m) from the aircraft. Airport personnel reported finding pieces of the main landing-gear tires on the runway, and the left main landing-gear shimmy damper reservoir was found on the left side of the runway. Examination of the two hydraulic system reservoirs of the airplane revealed that both were empty and hydraulic fluid was noted on the vertical stabilizer. When the hydraulic systems were pressurized, leakage occurred from a hole in the no. 1 elevator pressure line approximately three-quarters of the way up the vertical stabilizer and from a second hole in the no. 2 elevator return line, this hole being located behind the rudder flutter damper approximately half way up the vertical stabilizer. Examination of the airframe revealed that the right exterior fuselage skin exhibited approximately 103 lightning burn marks which ranged in size from 1/16 inch to 5/8 inch (0.16 cm to 1.6 cm) in diameter. Additionally, the right stabilizer showed evidence of scorching at the outboard corner of the upper surface at the trailing edge. The outboard static wick on the right stabilizer was missing, with evidence of heat at its base. Additionally, a bonding strap that provided an electrical connection between the horizontal and vertical stabilizers failed and the strap was discolored. The tail of the airplane had been designated by

10.6. Accidents

the aircraft manufacturer as a "swept stroke area", lightning zone 2B, and the interface between the horizontal and vertical stabilizers was designated as zone 3 (see Section 10.5). Apparently, the trailing edges should have been designated zone 1B since a hinge bonding strap used on the tail assembly of the Fokker could fail when subjected to lightning currents at or below the zone-3 current specifications, according to the accident report referenced below. The bonding strap was located near the hydraulic tubes, on the left side of the vertical stabilizer. It appears that lightning current flowing in the bonding strap between the vertical and horizontal stabilizers side-flashed to the hydraulic lines, burning through them and releasing the hydraulic fluid.

A report on this accident by the US National Transportation Safety Board is found at www.ntsb.gov/aviation/MIA/98A089.htm.

10.6.4. Aircraft struck by lightning at very low altitude

Figure 10.1 shows a commercial aircraft initiating lightning at low altitude after take off from an airport in Japan during winter. At the time of writing, a video of the event is found at http://lightning.pwr.eng.osaka-u.ac.jp/lrg/temp/plane.html. Frame 1 shows the aircraft, apparently a few hundred meters above ground, without any lightning evident. Frame 2 is given in Fig. 10.1. This frame shows evidence of a downward branched leader below the aircraft and an upward branched leader above the aircraft, both of which are likely to have been illuminated by a return stroke that has propagated from ground (from the bottom of the downward leader) upward through the aircraft, catching up with the top of the upward-extending leader channel, as has been observed to be the scenario in altitude rocket-triggered lightning (subsection 7.2.1). The next four frames (3 to 6) in the video show the decay in luminosity of the initial continuous current (ICC) channel; no channel branches are apparent by frame 6. Frame 7 is overexposed, probably because of either a dart-leader–return-stroke sequence or a large ICC pulse. Frame 8 shows a single channel of high and uniform brightness between the top and bottom of the frame, through the aircraft to Earth, as would be expected from continuing current. Because of the limited time resolution of the video system, other interpretations than those given above are possible.

Vonnegut (1966) reproduces a pilot's report of a lightning strike to a small commercial aircraft during take off, while the aircraft was still above the runway. The event involved a Convair aircraft, Flight 517, taking off from the Salt Lake City Airport on 15 October 1965. At the time of the event, there was some light rain in the area but apparently no lightning other than the event to be described. During take off, an extremely loud noise occurred. The first officer stated to the pilot that he believed they had sustained a lightning strike, subsequently confirmed by observers in the control tower, based on his observation of a blue-white glow around the nose of the aircraft at the time of the explosion. The aircraft returned to the airport. Three large holes were found in the runway which matched the exact dimensions of the two main landing-gear systems and the nose wheel. The largest hole, under the right main gear, was nearly 2 m in diameter and 15 to 20 cm deep. Pieces of asphalt as large as 0.3 m had been hurled 30 to 50 m down the runway. The aircraft suffered numerous burns to the wheel rims and fuselage just aft of the nose wheel-well. The rotating beacon, the grounding wire on the right main gear, and the fixed vertical-stabilizer cap were burned off. The fact that there was little if any lightning in the area at the time of the strike to the Convair would imply that the aircraft had initiated the lightning.

The excessive damage to the runway just described would imply a relatively large current and a relatively large action integral, such as apparently occurs, for example, in Japanese winter storms (Chapter 8) and has been observed in strikes to airborne vehicles in Europe in winter. One example is a lightning strike to a glider at 2500 feet (about 760 m) in England, April 1999: AAIB Bulletin no.: 12/99 Ref: EW/C99/04/02 Category: 3.0 at www.open.gov.uk/aaib/dec1999htm/bga3705.htm. Another example is a lightning strike to a helicopter at 3000 feet (about 910 m) over the North Sea, January 1995: AAIB Accident Report no.: 2/97 (EW/C95/1/1) at www.open.gov.uk/aaib/gtigk/gtigk.htm. In the case of the glider, a hollow tube that was part of the wing structure was crushed by the lightning. Laboratory tests involving currents over 300 kA and action integrals in excess of 2.5×10^7 A^2s have not been able to reproduce this damage. In the case of the helicopter, damage to portions of the main rotor blade assembly indicated extremely large action integrals.

A Boeing 727, Eastern Air Lines Flight 66, with 124 passengers and crew struck the approach light towers near the end of the runway at about 4 pm on 24 June 1975 while making its final approach to New York's John F. Kennedy International Airport. A violent thunderstorm was in progress. The pilot had been warned of severe windshear near an altitude of 500 feet (about 150 m) by the pilot of an aircraft that had previously taken the same approach path and landed successfully. According to an Associated Press report (e.g., *Gainesville Sun*, 25 June 1975), a Nassau County policeman saw a lightning bolt hit the plane: "It tilted to the right and went about 20 more yards, then hit the ground." Another witness who said he was about 150 m from the crash said "It was almost like lightning hit it and blew it up in a ball of fire." A number of eyewitnesses, most of them motorists on nearby Rockaway Boulevard, said they saw a bolt of lightning which appeared to hit the plane just before it burst into flames. Nevertheless, the official report (see below) found no evidence of "in-flight fire, explosion, bird strike, or lightning strike." One hundred and thirteen individuals were killed in the crash, but 11 others survived. A discussion of the possible indirect effects of lightning on the 727 control electronics is found in "Postmortem for Flight 66", *IEEE Spectrum* **12**: 35, July 1975. The official report of the crash (NTSB-AAR-76-8, dated March 1976, Eastern Airlines, Inc., Boeing 727-225, John F. Kennedy International Airport, Jamaica, New York, June 24, 1975) attributes the crash to "adverse winds associated with a very strong thunderstorm".

In an incident similar to Eastern Flight 66, Delta Flight 191, a Lockheed L-1011 jumbo jet was descending through stormy weather toward the Dallas Fort Worth Airport on 2 August 1985 and was about a mile (1.6 km) away at 1000 feet (about 300 m) altitude when, according to a witness (*Newsweek*, August 12, 1985, p. 30), it was struck by lightning, turned incandescent orange, and almost simultaneously plunged abruptly downward where it hit cars on Highway 114, skimmed along the ground hitting ground-based structures, and broke up in a ball of fire. The official report (see below) contains no such eyewitness accounts of a lightning strike but does indicate that an examination was made of the limited wreckage remaining for evidence of lightning and that none was found. One-hundred and thirty-four individuals died. Twenty-nine others survived. The official report of the crash, NTSB-AAR-86-05, dated August 15, 1986 Delta Airlines, Inc., Lockheed L-1011-3 85-1, N726DA, Dallas Fort Worth International, June 24, 1975, attributes it to "microburst-induced, severe wind shear from a rapidly developing thunderstorm" (see subsection 2.2.5).

10.6.5. Apollo 12 in 1969

The Apollo 12 space vehicle was launched from the NASA Kennedy Space Center (KSC), Florida on 14 November 1969. Within a minute of lift off, major electrical disturbances occurred that were later determined to be due to two separate vehicle-initiated lightning events. Nine non-essential instrumentation sensors were permanently damaged. Temporary upsets of equipment included momentary loss of communications, disturbances on instruments, illumination of various warning lights and alarms in the crew compartment, disconnection of three fuel cells from their busses, loss of attitude reference by the inertial platform, and disturbances to various clocks. All critical system problems were subsequently corrected, and the mission successfully delivered two astronauts to the surface of the Moon and returned them to Earth.

At the time of launch (11:22 am EST) a cold front was passing through the launch area. The tops of isolated cumulus congestus within 50 km reached a maximum height of 23 000 feet (about 7 km). In the vicinity of the launch complex, broken clouds were reported at 800 feet (about 0.25 km) with a solid overcast from about 10 000 to 21 000 feet (about 3 km to 6 km). The freezing level was near 12 400 feet (about 3.8 km). No lightning was reported in the KSC area six hours prior to or after the launch, although the instrumentation available for detecting lightning was primitive.

The vehicle apparently initiated a lightning discharge to ground 36.5 s after launch when it was at an altitude of about 6400 feet (about 1.9 km) and then initiated a cloud discharge at 52 s when it was at about 14 400 feet (about 4.4 km). In the 20 minutes prior to launch the vertical electric field at ground near the launch site was rapidly varying, but the crude electric field measuring devices used at the time were not calibrated. The possibility that the Apollo vehicle could initiate lightning had not been previously considered, according to Godfrey *et al.* (1970), the official report which presented the findings of the team that investigated the incident. The realization that Apollo 12 had initiated lightning led to a very significant round of funding for research into triggered and natural lightning and for the development of a variety of modern instruments to monitor the electrical characteristics of clouds and to determine lightning locations and characteristics.

According to the calculations found in Godfrey *et al.* (1970), if a 300-m-long (including the total exhaust plume) Saturn V vehicle with 5 m radius and 10 cm radius-of-curvature top cap were placed in an electric field of 7.5 kV m^{-1} then the field at the top cap would be enhanced 320 times to produce a breakdown field of 2.4 MV m^{-1} at an altitude of 6000 feet (about 1.8 km). The Saturn vehicle was 110 m long, its opaque exhaust was about 40 m, and its total visible exhaust about 200 m (Krider *et al.* 1974). The effective electrical length was probably between 150 and 300 m, since it is not clear how much of the exhaust plume contributed to the overall electrical length. Breakdown fields at the vehicle tip could easily be achieved for an enhancement factor between 100 and 300 in rather moderate cloud fields (commonly observed field values in cumulonimbus are 50 to 100 kV m^{-1}, subsection 3.2.4). Once the field at the pointed upper extremity of the vehicle exceeds the breakdown field, a positive discharge emanates from that location of the vehicle toward the cloud charge, assuming the cloud charge to be negative, similarly to the case of altitude-triggered lightning using the rocket and wire technique (subsection 7.2.1). This positive leader would then further enhance the field both at the tip of the upward-propagating discharge and at the opposite end (or exhaust) of the vehicle, resulting in a downward-propagating negative-stepped leader from the exhaust. In the cases of Apollo 12 and of Atlas Centaur 67, to be discussed in the next subsection, the vehicle-initiated discharges were apparently very similar in their characteristics to natural downward lightning (Chapter 4).

10.6.6. Atlas Centaur 67

The Atlas Centaur 67 vehicle was launched on 26 March 1987 at 4:22 pm local time from the Cape Canaveral Air Force Station, Florida, adjacent to the Kennedy Space Center. Weather conditions were similar to those at the time of the Apollo 12 launch. There was a broad cloud mass covering most of Florida and the Gulf of Mexico, and a nearly stationary cold front, oriented southwest–northeast, extended across northern Florida well north of Cape Canaveral. A weak squall line, also oriented southwest–northeast, was centered over the eastern Gulf of Mexico and was moving eastward over the Florida peninsula. This squall line produced substantial amounts of cloud-to-ground lightning activity of both negative and positive polarity throughout the day, but almost without exception this activity was well west of the Cape. At the launch site there was heavy rain, and layer clouds were reported at altitudes between 8000 and 20 000 feet (about 2.4 and 6.1 km). No cloud-to-ground lightning had been observed within 5 nautical miles (9.3 km) of the launch site in the 42 minutes prior to launch, and only one discharge was within 10 nautical miles (18.5 km) during this time. A cloud discharge apparently occurred about 2 minutes prior to launch, undetected by KSC lightning detection instrumentation, but reported to one of the authors (M.A.U.) after the launch by members of the press corps. At the time of the launch, the electric field at the launch site was -7.8 kV m^{-1}. There were no electric field constraints on the unmanned launch although, as discussed in Section 7.2, small rockets with trailing wires would have been capable of triggering lightning in the field that was present as well as in lower fields. Forty-nine seconds after launch, when the vehicle was at an altitude of about 12 000 feet (about 3.6 km), a lightning flash was observed below cloud base. That flash produced at least four strokes to ground, which were recorded by television cameras. The first two strokes followed one channel to ground and the latter two followed separate and different channels to ground. At the time of lightning initiation, the vehicle was at a height near 12 000 feet (about 3.6 km) where the temperature was $+4$ °C, while the freezing level was at 14 400 feet (about 4.4 km) and inside a cloud having a radar echo level of 10 dBZ, far below the value of 40 dBZ generally observed in thunderstorms in that area. From the magnetic field signal recorded by the KSC lightning-locating system (Section 17.3), the first stroke current was determined to be of negative polarity and was estimated to have a peak value of 20 kA.

At the time of the lightning strike there was a memory upset in the part of the vehicle guidance system called the digital computation unit, leading to an unplanned vehicle rotation. The stresses associated with this motion caused the vehicle to begin breaking apart. About 70 s after lift-off, the range safety officer ordered the Atlas-Centaur to be destroyed. Substantial portions of the fiberglass-honeycomb structure that covered the front 6 to 7 m of the vehicle were subsequently recovered from the Atlantic Ocean. These showed physical evidence of lightning attachment. Approximately 40 percent of the telemetry outputs showed anomalous electrical behavior at the time of the event.

The Atlas Centaur vehicle, which was about 40 m in length, served to enhance any uniform electrical field in which it was immersed by a factor of about 30 to 50 (Bussey 1987). Thus a breakdown field of nearly 2 MV m^{-1} would exist at the nose of the vehicle in an ambient field of 50 to 80 kV m^{-1}.

All the information given in this subsection and further details of the Atlas-Centaur 67 event, including radar echoes from the vehicle-initiated lightning and reference to lightning events on two earlier Atlas-Centaur vehicles, are found in Bussey (1987), the official report on the incident, and in Christian *et al.* (1989).

10.7. Summary

About 90 percent of the lightning discharges to aircraft are thought to be initiated by the aircraft itself. The initiation apparently involves a bidirectional leader whose positive and negative parts develop from opposite sides of the aircraft. The fields and currents associated with flashes initiated by aircraft at 5 to 6 km altitude, beneath or in the lower part of the main negative charge region, are fairly well established, at least for the initiation stage, from the F-106B, CV-580, and C-160 research programs. The characteristics of strikes to aircraft relatively high in the cloud, near a temperature of -40 °C and an altitude of 10 km, apparently above the main negative charge region, are available from the F-100F and F-106B studies, but these events are much less well documented. About 10 percent of strikes to aircraft involve interaction with an already occurring natural flash, an inference from the CV-580 and C-160 programs. Aircraft can become part of the lightning path of either cloud or ground discharges, the probability of the latter increasing with lower flight altitude, although there are no reliable statistics on this issue, only a few observations. Lightning damage to aircraft is generally minimal but can be occasionally catastrophic. There are two well-documented cases where lightning was initiated by large rockets launched from Earth, the Saturn V vehicle of NASA's Apollo 12 and US Air Force's Atlas-Centaur 67. The latter suffered damage that led to the loss of the vehicle and its payload.

References and bibliography for Chapter 10

Aircraft Accident Report, Boeing 707-12, N709PA Pan American World Airways, near Elkton, Maryland, 8 December 1963, published 1965. Civil Aeronautic Board File no. 1-0015.

Alexander, L.L. 1956. Three aircraft simultaneously struck by lightning. *Meteor. Mag., London* **85**: 246–8.

Anderson, R.B., and Kroninger H. 1975. Lightning phenomena in the aerospace environment. Part II: Lightning strikes to aircraft. *Trans. South African Inst. Electr. Eng.* **66**: 166–75.

Anderson, R.V., and Bailey, J.C. 1987. Vector electric fields measured in a lightning environment. NRL Memo. Rep. 5899.

Baskin, D. 1952. Lightning without clouds. *Bull. Am. Meteor. Soc.* **33**: 348.

Bellaschi, P.L. 1941. Lightning strokes in field and laboratory – III. *Trans. AIEE* **60**: 1248–56.

Bils, J.R., Thomson, E.M., Uman, M.A., and Mackerras, D. 1988. Electric field pulses in close lightning cloud flashes. *J. Geophys. Res.* **93**: 15 933–40.

Bosart, L.F. 1971. Weather at the launch of Apollo 12. *Weather* **26**: 19–23.

Brook, M., Holmes, C.R., and Moore, C.B. 1970. Lightning and rockets: some implications of the Apollo 12 lightning event. *Nav. Res. Rev.* **23**: 1–17.

Bussey, J. 1987. Report of Atlas/Centaur-67/FLTSATCOM F-6 investigation board, vol. II, NASA.

Castellani, A., Bondiou-Clergerie, A., Lalande, P., Bonamy, A., and Gallimberti, I. 1998a. Laboratory study of the bi-leader process from an electrically floating conductor. Part I: General results. *IEE Proc. Sci. Meas. Technol.* **145**: 185–192.

Castellanti, A., Bondiou-Clergerie, A., Lalande, P., Bonamy, A., and Gallimberti, I. 1998b. Laboratory study of the bi-leader process from an electrically floating conductor. Part II: Bi-leader. *IEE Proc. Sci. Meas. Technol.* **145**: 193–9.

Christian, H.J., Frost, R.L, Gillaspy, P.H., Goodman, S.J., Vaughan, O.H. Jr, Brook, M., Vonnegut, B., and Orville, R.E. 1983. Observations of optical lightning emissions from above thunderstorms using U-2 aircraft. *Bull. Am. Meteor. Soc.* **64**: 120–3.

Christian, H.J., Mazur, V., Fisher, B.D., Ruhnke, L.H., Crouch, K., and Perala, R.P. 1989. The Atlas/Centaur lightning strike incident. *J. Geophys. Res.* **94**(D11): 13 169–77.

Clifford, D.W. 1980. Aircraft mishap experience from atmospheric electricity hazards, atmospheric electricity–aircraft interaction. NATO AGARD-LS-110, National Technical Information Service, Springfield, Virginia, 22161.

Clifford, D.W., and Kasemir, H.W. 1982. Triggered lightning. *IEEE Trans. Electromagn. Compat.* **24**: 112–22.

Clifford, D.W., Crouch, K.E., and Schulte, E.H. 1982. Lightning simulation and testing. *IEEE Trans. Electromagn. Compat.* **24**: 209–24.

Cobb, W.E., and Holitza, F.J. 1968. A note on lightning strikes to aircraft. *Mon. Wea. Rev.* **96**: 807–8.

Corbin, J.C., ed. 1982. Special issue on lightning and its interaction with aircraft. *IEEE Trans. Electromagn. Compat.* **24**.

FAA 1985. Protection of aircraft fuel systems against fuel vapor ignition due to lightning. Federal Aviation Agency Advisory Circular AC 20-53A, Federal Aviation Administration, Department of Transportation, Washington, DC, 12 April 1985.

Fisher, B.D., Brown, P.W., and Plumer, J.A. 1986. Summary of NASA storm hazards lightning research, 1980–1985. In *Proc. Int. Conf. on Lightning and Static Electricity, June 24–26, Dayton, Ohio.*

Fisher, F.A., Plumer, J.A., and R.A. Perala 1999. Lightning protection of aircraft. Report, Lightning Technologies Inc., 10 Downing Parkway, Pittsfield, Massachusetts 01201, second printing.

Fitzgerald, D.R. 1965. Measurement techniques in clouds. In *Problems of Atmospheric and Space Electricity*, pp. 199–214, ed. S.C. Coroniti, New York, American Elsevier.

Fitzgerald, D.R. 1967. Probable aircraft "triggering" of lightning in certain thunderstorms. *Mon. Wea. Rev.* **95**: 835–42.

Fitzgerald, D.R. 1976. Experimental studies of thunderstorm electrification. Air Force Geophy. Lab Report AFGL-TR-76-0128, AD-A0322374.

Fowler, R.T. 1970. Ion collection by electrostatic probe in a jet exhaust. Graduate thesis, Air Force Institute of Technology, Wright-Patterson Air Force Base, Ohio.

Gifford, T. 1950. Aircraft struck by lightning, *Meteor. Mag.* **79**: 121–2.

Godfrey, R., Mathews, E.R., and McDivitt, J.A. 1970. Analysis of Apollo 12 lightning incident. NASA Report MSC-01540.

Goto, Y., and Narita, K. 1986. Lightning interaction with aircraft and winter lightning in Japan. *Res. Lett. Atmos. Electr.* **6**: 27–34.

Hagenguth, J.H. 1949. Lightning stroke damage to aircraft. Pt II. *Trans. AIEE* **68**(II): 1036–46.

Harrison, H.T. 1965. United Air Line Turbojet experience with electrical discharges. UAL Meteorological Circular, no. 57.

Harrison, L.P. 1946. Lightning discharges to aircraft and associated meteorological conditions. Technical Note no. 1001, National Advisory Committee for Aeronautics, Washington, DC, 149 pp.

Hourihan, B.I. 1975. Data from the airlines lightning strike reporting project. June 1971 to November 1974, Summary Report GPR-75-004, High Voltage Laboratory, Electromagnetic Unit, Corporate Research and Development General Electric Company, Pittsfield, Massachusetts.

Hubert, P. 1981. Triggered lightning at Langmuir Laboratory during TRIP-81. Centre d'Etude Nucléaires de Saclay Report no. Dph/EP/81-66, 23 November, France.

Hurst, G.W. 1956. Aircraft struck by lightning. *Meteor. Mag. London* **85**: 248–9.

Imyanitov, I.M. 1971. Aircraft electrification in clouds and precipitation. USAF Foreign Tech. Div. Report FTD-HC-23-544-70. (Published originally in *Elektrizatsiya Samoletov v Oblakakh i Osadkakh*, pp. 1–211, 1970.)

Jones, J.J. 1990. Electric charge acquired by airplanes penetrating thunderstorms. *J. Geophys. Res.* **95**: 16 589–600.

Jones, J.J., Winn, W.P. and Han, F. 1993. Electric field measurements with an airplane: problems caused by emitted charge. *J. Geophys. Res.* **98**: 5235–44.

Kasemir, H.W. 1950. Qualitative Uebersicht ueber Potential-Feldund Ladungsverhaltnisse bei einer Blitzentladung in der Gewitterwolke. In *Das Gewitter*, ed. Hans Israel, Akad. Verlags. Ges. Geest and Portig K.-G, Leipzig, Germany.

Kofoid, M.J. 1970. Lightning discharge heating of aircraft skins. *J. Aircraft* **7**: 21–6.

Krider, E.P., Noggle, R.C., Uman, M.A., and Orville, R.E., 1974. Lightning and the Apollo/17 Saturn V exhaust plume. *J. Spacecraft Rockets* **11**: 72–5.

Lalande, P., and Bondiou-Clergerie, A. 1997. Collection and analysis of available in-flight measurement of lightning strikes to aircraft. Report AI-95-SC.204-RE/210-D2.1, ONERA (France) Transport Research and Technological Development Program DG VII, 24 February 1997.

Lalande, P., Bondiou-Clergerie, A., and Laroche, P. 1999. Studying aircraft lightning strokes. *Aerospace Engineering (publisher: SAE Aerospace)*: 39-42. See also Analysis of available in-flight measurements of lightning strikes to aircraft, in *Proc. 1999 Int. Conf. on Lightning and Static Electricity, Toulouse, France*, pp. 401–8.

Lamb, N. 1975. Electrical phenomena observed at night in the tropics from a Hercules aircraft. *Meteor. Mag.* **104**: 56–7.

Larigaldie, S., Roussaud, A. and Jecko, B. 1992. Mechanisms of high-current pulses in lightning and long-spark stepped leaders. *J. Appl. Phys.* **72**: 1729–39.

Larsson, A., Lalande, P., Bondiou-Clergerie, A., and Delannoy, A. 2000a. The lightning swept stroke along an aircraft in flight. Part I: thermodynamic and electric properties of lightning arc channels. *J. Phys. D: Appl. Phys.* **33**: 1866–75.

Larsson, A., Lalande, P., and Bondiou-Clergerie, A. 2000b. The lightning swept stroke along an aircraft in flight. Part II: numerical simulations of the complete process. *J. Phys. D: Appl. Phys.* **33**: 1876–83.

Lee, L.D., Finelli, G.B., Thomas, M.E., and Pitts, F.L. 1984. Statistical analysis of direct-strike lightning data. NASA Report TP-2252, 27 pp., NASA Langley Research Center, Hampton, Virginia.

FAA 1967. Lightning strike survey report for the period of January 1965 through December of 1966. Federal Aviation Agency Report of the Conference on Fire Safety Measures for Aircraft Fuel Systems, Appendix II, Department of Transportation, Washington, DC.

Little, P.L. 1980. Lightning hazards in the air. *Spectrum* **167**: 7–10.

Mason, D. 1964. Lightning strikes on aircraft – II. *Weather* **19**(8): 248–55.

Mazur, V. 1986. Rapidly occurring short duration discharges in thunderstorms, as indicators of a lightning-triggering mechanism. *Geophys. Res. Lett.* **4**: 333–358.

Mazur, V. 1989a. Triggered lightning strikes to aircraft and natural intracloud discharges. *J. Geophys. Res.* **94**: 3311–25.

Mazur, V. 1989b. A physical model of lightning initiation on aircraft in thunderstorms. *J. Geophys. Res.* **94**: 3326–40.

Mazur, V. 1992. Physics of lightning-aircraft interaction. *Res. Lett. Atmos. Electr.* **12**: 107–15.

Mazur, V. 1993. Lightning threat to aircraft: do we know all we need to know? *J. Aircraft* **30**: 156–9.

Mazur, V. 1999. Lightning and aviation, Chapter 12 in *Aviation Weather Surveillance Systems*, by P. Mahapatra with contributions from R.J. Doviak, V. Mazur, and D.S. Zrnic, published jointly by The Institution of Electrical Engineers and The American Institute of Aeronautics and Astronautics (ISBN 0 85296 9376).

Mazur, V. 2000. Personal communication: a copy of a letter received from the Air line Pilots' Association.

Mazur, V., and Moreau, J.P. 1992. Aircraft-triggered lightning: processes following strike initiation that affect aircraft. *J. Aircraft* **29**: 575–80.

Mazur, V., Ruhnke, L.H. 1993. Common physical processes in natural and artificially triggered loghtning. *J. Geophys. Res.* **98**: 12 913–30.

Mazur, V., Fisher, B.D., and Gerlach, J.C. 1984a. Lightning strikes to an airplane in a thunderstorm. *J. Aircraft* **21**: 607–11.

Mazur, V., Fisher, B.D., and Gerlach, J.C. 1984b. Lightning flash density versus altitude and storm structure from observations with UHF- and S-band radars. *Geophys. Res. Lett.* **11**: 61–4.

Mazur, V., Fisher, B.D. and Gerlach, J.C. 1986. Lightning strikes to a NASA airplane penetrating thunderstorms at low altitudes. *J. Aircraft* **23**: 499–505.

Mazur, V., Ruhnke, L.H., and Rudolph, T. 1987. Effect of E-field mill location on accuracy of electric field measurements with instrumented airplane. *J. Geophys. Res.* **92**(D10): 12 013–19.

Mazur, V., Fisher, B.D., and Brown, P.W. 1990. Multistroke cloud-to-ground strike to the NASA F-106B airplane. *J. Geophys. Res.* **95**: 5471–84.

McEachron, K.B., and Hagenguth, J.H. 1942. Effect of lightning on thin metal surfaces. *Trans. AIEE* **61**: 559–64.

McKague, L. 1977. Lightning-hazard assessment: a first pass probalistic model. *J. Aircraft* **14**: 1022–4.

McMurtry, T.C. 1987. NASA 914 T-38A jet trainer lightning strike investigation report. Date of mishap Feb. 24, 1987. NASA Johnson Space Center document, 6 July 1987.

Michimoto, K. 1993. Statistics of lightning strikes to aircraft in winter around Komatsu airbase, Japan. *J. Atmos. Electr.* **13**: 47–58.

Mo, Q., Ebneter, A.E., Fleischhacker, P., and Winn, W.P. 1998. Electric field measurements with an airplane: a solution to problems caused by emitted charge. *J. Geophys. Res.* **103**: 17 163–73.

Moreau, J.-P. 2000. Personal communication.

Moreau, J.-P., Alliot, J.-C., Mazur, V. 1992. Aircraft lightning initiation and interception from in situ electric measurements and fast video observations. *J. Geophys. Res.* **97**: 15 903–12.

Murooka, Y. 1992. A survey of lightning interaction with aircraft in Japan. *Res. Lett. Atmos. Electr.* **12**: 101–6.

Nanevicz, J.E., Pierce, E.T., and Whitson, A.L. 1972. Atmospheric electricity and the Apollo series. Note 18, Stanford Research Institute, Menlo Park, California.

Nanevicz, J.E., Vance, E.F., Radsky, W., Uman, M.A., Soper, G.K., and Pierre, J.M. 1988. EMP susceptibility insights from aircraft exposure to lightning. *IEEE Trans. Electromagn. Compat.* **30**: 463–72.

Petterson, B.J. and Wood, W.R. 1968. Measurements of lightning strokes to aircraft. Sandia Laboratory Report SC-M-67-549, Sandia Laboratories, Albuquerque, New Mexico (also Report DS-68-1 of the Department of Transportation, Federal Aviation Administration, Washington, DC 10590).

Pierce, E.T. 1970. Atmospheric electric and meteorological environment of aircraft incidents involving lightning strikes. Special Interim Report I, Stanford Research Institute, Menlo Park, California.

Pierce, E.T. 1976. Winter thunderstorms in Japan – a hazard to aviation. *Naval Res. Rev.* **29**(6): 12–16.

Pifer, A.E., and Krider, E.P. 1972. The optical temperature of the Apollo 15 exhaust plume. *J. Spacecraft and Rockets* **9**: 847–8.

Pitts, F.L. 1982. Electromagnetic measurements of lightning strikes to aircraft. *J. Aircraft* **19**: 246–50.

Pitts, F.L., and Thomas, M.E. 1981. 1980 direct strike lightning data. NASA Technical Memorandum 81946, Langley Research Center, Hampton, Virginia.

Pitts, F.L., and Thomas, M.E. 1982. 1981 direct strike lightning data. NASA TM-83273, Langley Research Center, Hampton, Virginia.

Pitts, F.L., Perala, R.A., Rudolph, T.H., and Lee, L.D. 1987. New results for quantification of lightning/aircraft electrodynamics. *Electromagnetics* **7**: 451–85.

Pitts, F.L., Fisher, B.D., Vladislav, V., and Perala, R.A. 1988. Aircraft jolts from lightning bolts. *IEEE Spectrum* **25**: 34–8, July.

Plumer, J.A. 1981. Investigation of severe lightning strike incidents to two USAF F-106A aircraft. NASA Report CR-165794, September 1981.

Plumer, J.A. 1992. Aircraft lightning protection design and certification standards. *Res. Lett. Atmos. Electr.* **12**: 83–96.

Plumer, J.A., Rasch, N.O., and Glynn, M.S. 1985. Recent data from the airlines lightning strike reporting project. *J. Aircraft* **22**: 429–33.

Rakov, V.A., and Uman, M.A., 1990. Long continuing current in negative lightning ground flashes. *J. Geophys. Res.* **95**: 5455–70.

Rakov, V.A., Uman, M.A., Hoffman, G.R., Masters, M.W., and Brook, M. 1996. Bursts of pulses in lightning electromagnetic radiation: observations and implications for lightning test standards. *IEEE Trans. Electromagn. Compat.* **38**: 156–64.

Reazer, J.S., Serrano, A.V., Walko, L.C., and Burket, H.D. 1987. Analysis of correlated electromagnetic fields and current pulses during airborne lightning attachments. *Electromagnetics* **7**: 509–39.

Rudolph, R., and Perala, R.A. 1983. Linear and nonlinear interpretation of the direct strike lightning response of the NASA F-106 thunderstorm research aircraft. NASA Report CR-3746.

Rudolph, R., Perala, R.A., McKenna, P.M., and Parker, S.L. 1985. Investigations into the triggered lightning response of the F-106 thunderstorm research aircraft. NASA Report CR-3902.

Rudolph, R., Perala, R.A., Easterbrook, C.C., and Parker, S.L. 1986. Development and application of linear and nonlinear methods for interpretation of lightning strikes to inflight aircraft. NASA Report CR-3974.

Rustan, P.L. 1986. The lightning threat to aerospace vehicles. *AIAA J. Aircraft* **23**: 62–7.

Rustan, P.L. 1987. Description of an aircraft lightning and simulated nuclear electromagnetic pulse (HEMP) threat based on experimental data. *IEEE Trans. Electromagn. Compat.* **29**: 49–63.

Rustan, P.L., Kuhlman, B.P., Burket, H.D., Reazer, J., and Serrano, A. 1987. Low altitude lightning attachment to an aircraft. Flight Dynamics Laboratory, Wright Patterson AFB, Report AFWAL-TR-86-3009.

Schowalter, J.S. 1983. Direct lightning strikes to aircraft. M.S. thesis, Air Force Institute of Technology.

Shuttle Launch Commit Criteria and Background Document 1988. NASA JSC-16007, Revision D.

Smith, F.T., and Gatz, C.R. 1963. Chemistry of ionization in rocket exhausts. In *Ionization in High Temperature Gases – Progress in Astronautics and Aeronautics*, ed. K.E. Shuler, vol. 12, pp. 302–16, New York: Academic Press.

Stern, A.D., Brady, R.H. III, Moore, P.D., and Carter, G.M. 1994. Identification of aviation weather hazards based on the integration of radar and lightning data. *Bull. Am. Meteor. Soc.* **75**: 2269–80.

STS Operational Flight Rules – All Flights. Flight Rule 4-64D, JSC-12820.

Suzuki, S., Koyama, K., Hayami, T., and Murooka, Y. 1992. On the lightning protection of ground crews. *Res. Lett. Atmos. Electr.* **12**: 123–6.

Testé, Ph., Leblanc, T., Uhlig, F., and Chabrerie, J.-P. 2000. 3D modeling of the heating of a metal sheet by a moving arc: application to aircraft lightning protection. *Eur. Phys. J. AP* **11**: 197–204.

Thomas, M.E. 1985. 1983 direct strike lightning data. NASA Report TM-86426.

Thomas, M.E., and Carney, H.K. 1986. 1984 direct stroke lightning data. NASA Report TM-87690, parts 1, 2, and 3.

Thomas, M.E., and Pitts, F.L. 1983. 1982 direct stroke lightning data. NASA Report TM-84626.

Trost, T.F., and Zaepfel, K.P. 1980. Broadband electromagnetic sensors for lightning research. NASA Report CP-2128, FAA-RD-8-30, *Lightning Technology*, p. 131, NASA Langley Research Center.

Uman, M.A. 1986. *All About Lightning*, 167 pp., New York: Dover Press.

Villanueva, Y., Rakov, V.A., Uman, M.A., and Brook, M. 1994. Microsecond-scale electric field pulses in cloud lightning discharges. *J. Geophys. Res.* **99**: 14 353–60.

Vonnegut, B. 1965. Electrical behavior of an airplane in a thunderstorm. A.L. Little, Cambridge, Massachusetts,

References and bibliography for Chapter 10

Defense Documentation Center Report AD-614-914.

Vonnegut, B. 1966. Effects of a lightning discharge on an aeroplane. *Weather* **21**: 277–9.

Winn, W.P. 1993. Aircraft measurement of electric field: self-calibration. *J. Geophys. Res.* **98**: 7351–65.

Yang, F.C., Lee, K.S.H., Andersh, D.J., and Steil, J. 1987. Lightning response of aircraft. *Electromagnetics* **7**: 487–507.

Zaepfel, K.P., and Carney, H.K. 1988. 1985 and 1986 direct strike lightning data. NASA Report TM-100533, parts 1 and 2, NASA, Langley Research Center, Hampton, Virginia 23665.

Zaepfel, K.P., Fisher, B.D., and Ott, M.S. 1985. Direct stroke lightning photographs, swept-flash attachment patterns, and flight conditions for storm hazards '82. NASA Report TM-86347.

11 Thunder

> First let me talk with this philosopher – what is the cause of thunder?
>
> W. Shakespeare, *King Lear*

11.1. Introduction

Thunder can be defined as the acoustic emission associated with a lightning discharge. It appears that all impulsive processes in both cloud-to-ground and cloud flashes, including M-component-type processes (Section 4.9), produce thunder. The significant part of the thunder spectrum extends from a few hertz or less to a few kilohertz. It is the general view that audible thunder (above 20 Hz or so) is a series of degenerated shock waves produced by the gas dynamic expansion of various portions of the rapidly heated lightning channel, while infrasonic thunder (approximately 20 Hz and below) is associated with the sudden contraction of a relatively large volume of the thundercloud when lightning rapidly removes the charge from that volume.

11.2. Observations

There have been remarkably few measurements of the properties of thunder in the two decades or more since the detailed experiments of the groups from Rice University, Texas (e.g., Few *et al.* 1967; Few 1968, 1969a; Teer 1972, 1973; Few and Teer 1974) and from the New Mexico Institute of Mining and Technology (e.g., Holmes *et al.* 1971a; McCrory 1971). The most significant observations, including recent measurements of the acoustic signatures of rocket-triggered lightning (Depasse 1994), are reviewed below in this section.

11.2.1. Time to and duration of thunder

The electromagnetic (including optical) radiation from the lightning channel propagates at about $300 \text{ m } \mu\text{s}^{-1}$, the speed of light, and hence arrives at an observer, say, 3 to 4 km away in about 10 μs. The corresponding thunder, an acoustic or sound wave, travels at about 340 m s^{-1} for an air temperature of 20 °C and atmospheric pressure and hence arrives in about 10 s. Therefore, the time interval between the arrival of the electromagnetic signal and the thunder is essentially determined by the distance to the channel divided by the speed of sound. This "flash-to-bang" time is approximately 3 seconds per kilometer of distance to the closest audible point on the lightning channel. "Thunder ranging" is the name given to the technique, widely used by both the layman and the researcher, of determining the lightning distance from the time interval between the electromagnetic signal, at either optical or radio-frequency, and the first sound of thunder.

Thunder can seldom be heard from lightning flashes more than about 25 km distant, an observation apparently first due to De L'Isle (1738). Veenema (1917, 1918, 1920) studied nearly every thunderstorm occurring near him during the years 1895 to 1916 with the aim of determining how far thunder could be heard. He confirmed De L'Isle's conclusion but documented occasional examples of thunder heard from lightning up to and over 100 km distant. Isolated reports of thunder heard at distances greater than about 25 km or so have been given by Cave (1919), Brooks (1920), Page (1944), Taljaard (1952), and Thomson (1980). However, Ault (1916), captain of the research ship Carnegie, reported that a thunderstorm at sea became inaudible beyond a distance of about 8 km. As we shall discuss in Section 11.4, the distance at which any sound produced in the atmosphere can be heard at ground level depends primarily on the height of the source and the variations with height of both the atmospheric temperature and the horizontal wind, the latter variation being called the wind shear.

The duration of the thunder is a measure of the difference in the distances between the closest and farthest audible points on the lightning channel, thus representing the minimum possible length for the channel. An approximation to the actual channel length would be obtained if the observer were located at one end of the channel, assuming that the channel were more or less straight and that the atmosphere were more or less homogeneous and isotropic. In fact, even for ground flashes, the thunder duration usually has little to do with channel length, because of the large horizontal extent of the in-cloud part of the lightning channel (Teer and Few 1974; Proctor *et al.* 1988; Krehbiel *et al.* 1979; Rakov *et al.* 1990; Shao *et al.* 1995; Section 11.5). Teer and Few (1974) gave statistics on thunder duration, without regard to range or type of flash, at a number of geographic locations. The median thunder

11.2. Observations

duration they reported was about 15 s in Socorro, New Mexico, 29 s in Roswell, New Mexico, 18 s in Tucson, Arizona, and 41 s in Houston, Texas.

11.2.2. The sounds of thunder

The terms used to describe the various sounds of thunder are based on subjective descriptions and therefore are usually poorly defined. Nevertheless, it is worth giving a brief overview of the major types of sounds produced by lightning. The terms "clap", "peal", "roll", and "rumble" are most commonly used to describe the sounds of thunder. Claps are sudden loud sounds lasting from about 0.2 to 2 s (Latham 1964). They are apparently produced when individual thunder signals from portions of sufficiently long sections of lightning channel arrive more or less simultaneously at the observer's location (Few 1970, 1982, 1995). Thus, a clap apparently comes from a section of channel having a relatively large overall length that is oriented more or less perpendicularly to the observer's line of sight. The time interval between claps has been reported to be typically 1 to 3 s (Latham 1964), and there are generally two to four claps per flash (Latham 1964; Uman 1987, 2001). Histograms of relative clap amplitudes were given by Latham (1964) and by Uman and Evans as reported by Uman (1987, 2001). Peals are loud sounds that change in frequency and/or amplitude. The terms peal and clap are often used interchangeably. The term roll is used to describe irregular sound variations of moderate amplitude. Rumble is a relatively weak sound of long duration and relatively low frequency. As with clap and peal, the terms roll and rumble are often used synonymously. Of course, none of these terms (clap, peal, roll, or rumble) is quantitative, and there could be a range of definitions for each. Therefore researchers prefer to avoid them whenever possible.

We now discuss the sounds of thunder produced by very close lightning. Many observers of close lightning have reported certain sounds prior to the first loud clap. When lightning strikes several hundreds of meters away, the first sounds may be like those of tearing of cloth (e.g., Malan 1963). This tearing sound can last an appreciable fraction of a second and merges into the louder sound, not dissimilar from a cannon shot, presumably produced by the return stroke in the vertical channel section near ground. The origin of the tearing sound has been attributed (i) to a discharge in a single, very straight channel section whose length is of the same order of magnitude as the distance to the observer (M. Brook, as quoted in Hill 1977b) and (ii) to a number of simultaneous upward-going connecting leaders from Earth (Malan 1963). When the lightning is within about 100 m, according to Malan (1963) one first hears a click, then a whiplike crack, and finally a continuous rumbling thunder. Malan (1963) viewed the click as due to the major upward connecting leader, the crack as due to the return stroke in the closest part of the lightning channel, and the rumble as the sound from the higher sections of the tortuous channel.

According to MacEachron (1939), upward lightning discharges not involving return strokes do not make the sounds we regard as thunder (Section 6.7). However, Hubert (personal communication, 1985) argued that, regardless of the presence of return strokes, rocket-triggered lightning initiated by an upward-going leader produces relatively little noise at about 100 m while beyond 1 km the noise is similar to normal thunder. Davis and Standring (1947) observed a few discharges to the cables of kite balloons with associated currents less than 1 kA that were both heard and seen and several discharges with currents of about 100 A that were not detected by human observers (Section 6.7). Note that cloud discharges do not contain return strokes (they do contain other impulsive processes, Chapter 9) but do produce thunder. M-components and initial continuous current pulses in rocket-triggered lightning (Section 7.2) have been reported to produce acoustic signals whose magnitudes are comparable with those generated by leader–return-stroke sequences in the same flash (Rakov et al. 2001). Thunder from the lightning-like electrical discharges of less than a kilometer length that originate in the material ejected from some volcanoes (subsection 20.7.1) is reported to take the form of "a sharp noise like the firing of artillery" (Anderson et al. 1965).

11.2.3. Frequency spectrum

The most comprehensive work on the measured frequency spectrum of thunder is due to Holmes et al. (1971a). A variety of prior published data, most containing significant errors, has been reviewed by Uman (1984, 1987, 2001). Holmes et al., who worked on a New Mexico mountain top about 3 km above sea level, found, for 40 thunder records, that the thunder power spectrum peaked at frequencies from less than 4 to 125 Hz, with mean peak values 28 Hz and 50 Hz for five cloud and 30 ground flashes, respectively. The ratio of the peak frequency and the width of the power spectrum at half amplitude varied between 0.5 and 2. Figure 11.1 gives a histogram of the peak frequency in the power spectrum of the thunder from 24 ground flashes. Two major peaks, at 0–20 Hz and 40–60 Hz, can be seen. The infrasonic (0–20 Hz) peak is caused primarily by wind noise; nevertheless, some essentially wind-free ground flash spectra and most cloud-flash spectra have been found to peak in the infrasonic range.

A typical power spectrum with peak frequency in the audible range near 100 Hz is shown in Fig. 11.2, along with the spectrum of the ambient wind noise measured during a 2-s interval prior to the thunder. A spectrum that peaks in the infrasonic range (below 20 Hz) is shown in Fig. 11.3a. The peak power flux in the recorded spectra, examples of which are shown in Figs. 11.2 and 11.3a, ranged from 4×10^{-4}

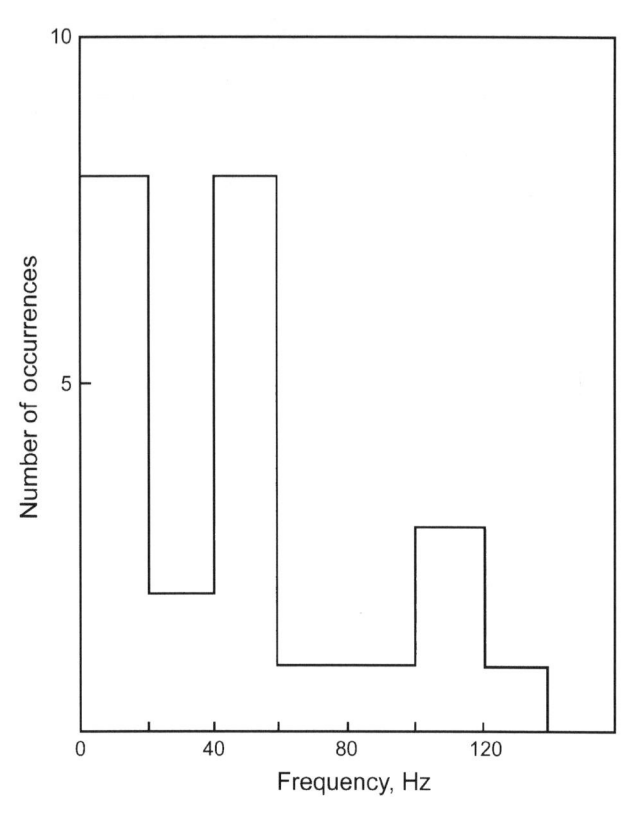

Fig. 11.1. Histogram showing the peak frequency of the acoustic power spectrum for 24 ground flashes. Adapted from Holmes et al. (1971a).

three less than that for ground flashes, 6.3×10^6 J. The total acoustic energy W in joules was determined from the expression

$$W = \int P(t) 4\pi R^2(t) \, dt \qquad (11.1)$$

where $P(t)$ is the recorded total power flux as a function of time t in J m^{-2} s^{-1}, $R(t)$ is the distance to the acoustic source, found as $v(t - t_0)$, where v is the speed of the acoustic wave, and t_0 is the time at which the lightning occurs as identified from electric field records. For all the thunder data, the average total power flux ranged from 0.17×10^{-3} to 19.3×10^{-3} J m^{-2} s^{-1}. Among the assumptions made in using Eq. 11.1 is that atmospheric attenuation and refraction are small and that a spherical (isotropic) acoustic wave is radiated from each point on the channel. For ground flashes, Holmes et al. (1971a) calculated the efficiency for the conversion of electrical energy to acoustic energy as follows. They assumed that the total energy dissipated per unit length of channel by a first stroke was 2.3×10^5 J m^{-1} (Krider et al. 1968), a value, discussed in the first

to 3×10^{-6} J m^{-2} s^{-1} Hz^{-1}. The thunder power spectrum is time varying, as illustrated in Fig. 11.3b, where the spectrum was calculated for successive 1 s time windows. As seen in Fig. 11.3b, at 13 s, as the first thunder arrives, the dominant frequencies are 50, 110 and 135 Hz whereas at 21 s the peak frequency is about 5 Hz (infrasonic). Holmes et al. (1971a) viewed this shift in spectral peak frequency with time as indicative of the presence of more than one mechanism of thunder production.

Depasse (1994) measured the acoustic signals 70 m from rocket-triggered lightning in France and found that the dominant frequency ranged from 205 to 1775 Hz for entire flashes (12 cases) and from about 300 to 900 Hz for individual strokes (18 cases). An example of the spectrum obtained by Depasse for a 50 kA triggered-lightning stroke is shown in Fig. 11.4.

11.2.4. Energy

Holmes et al. (1971a), from their analysis of 40 cloud and ground flashes, most of which apparently occurred at a range of some kilometers, found a significant difference in total acoustic energy between the thunder from cloud and from ground flashes. The mean total acoustic energy radiated by cloud flashes was 1.9×10^6 J, a factor of

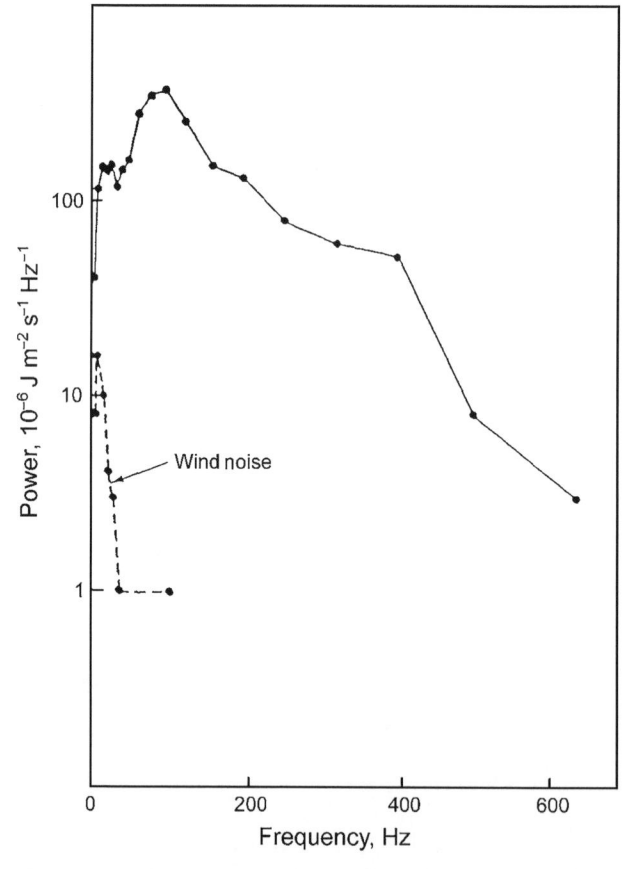

Fig. 11.2. A typical thunder spectrum with peak frequency near 100 Hz for a lightning flash. Also shown (broken line) is the spectrum of the wind noise. Adapted from Holmes et al. (1971a).

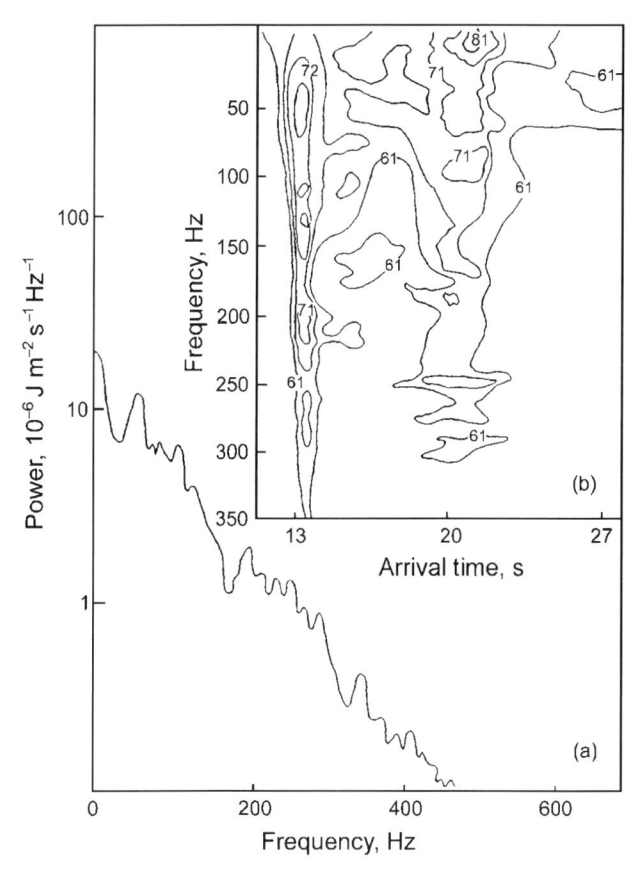

Fig. 11.3. (a) Typical thunder spectrum with peak frequency in the infrasonic range. The wind noise was insignificant. (b) The power spectrum as a function of thunder arrival time, calculated for time windows of 1 s duration and displayed as power contours in decibels (dB) above a standard level 10^{-12} J m^{-2} s^{-1} Hz^{-1}. The unlabeled contours are 5 dB above or below the nearest labeled contour. Adapted from Holmes et al. (1971a).

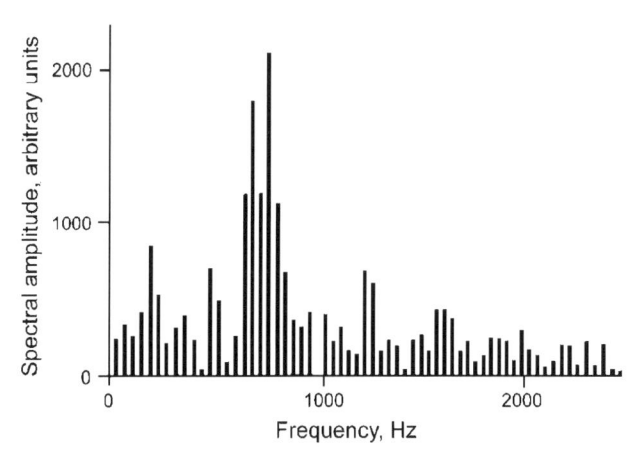

Fig. 11.4. Acoustic spectrum with peak frequency near 700 Hz for a 50 kA stroke in rocket-triggered lightning at 70 m. Adapted from Depasse (1994).

part of subsection 12.2.6, that is a subject of controversy. The length of an average lightning channel was assumed to be 4 km, yielding a first-stroke energy of 9.2×10^8 J. All subsequent strokes together were assumed to have the energy of the first stroke, the total flash energy being about 1.8×10^9 J. This total electrical energy was divided into the average measured acoustic energy of 11 selected ground flashes, 3.26×10^6 J, resulting in an acoustic efficiency of 0.18 percent. Since the actual total energy input per unit length might conceivably be as low as 10^3–10^4 J m^{-1} (subsection 12.2.6) rather than the assumed value, 2.3×10^5 J m^{-1}, the actual acoustic efficiency could be as high as 2 to 20 percent.

11.2.5. Pressure

For both cloud and ground flashes at a probable range of some kilometers, Holmes et al. (1971a) found that the average rms pressure was from 0.22 to 2.4 N m^{-2}. (Note that a pressure of one standard atmosphere, 1 atm = 1 bar, is about 10^5 N m^{-2}; one N m^{-2} is also called a pascal.) According to Few et al. (1970), observed overpressures at 1 km are generally less than 10^{-4} atm (10 N m^{-2}). Hill E.L. and Robb (1968) and Newman et al. (1967a, b) reported on a measurement of the overpressure 35 cm from a spark channel that bridged a 10 cm gap placed in series with a rocket-triggered lightning channel. The maximum overpressure was 2 atm (2×10^5 N m^{-2}), with a range for four discharges from 0.3 to 2 atm. Uman (1969) noted that current waveforms corresponding to the measured pressure pulses had millisecond-scale risetimes and therefore that the results of Hill and Robb (1968) and Newman et al. (1967a, b) might not be representative of normal return strokes, whose currents have microsecond- or submicrosecond-scale risetimes (Section 4.6). Some comments on the interpretation of these pressure measurements offered by Dawson et al. (1968b) are given in subsection 11.3.2. Depasse (1994), from acoustic signals measured 70 m from rocket-triggered lightning in France, found an oscillating pressure signal having a maximum of about 5 N m^{-2} with an initial underpressure (rarefaction), in contrast with natural lightning and laboratory spark measurements, which invariably show an initial overpressure (compression). Measured overpressures near long laboratory sparks are discussed in subsection 11.3.2 in relation to testing the validity of proposed thunder-generation mechanisms.

11.3. Generation mechanisms

As indicated in Section 11.1, thunder is usually divided into two categories: (i) audible, acoustic energy that we can hear; (ii) infrasonic, acoustic energy that is below the frequency that the human ear can detect, generally 20 Hz or so. This division, based on the properties of the human ear, does not necessarily translate into different thunder-generation mechanisms. The origin of most audible thunder is thought to be expansion of the rapidly heated lightning channel, although some infrasonic thunder must also be generated by this mechanism. The origin of most infrasonic thunder is postulated to be conversion to sound of the energy stored in the electrostatic field of the thundercloud when lightning rapidly reduces that cloud field, although this mechanism can also produce audible thunder. The expanding-hot-channel mechanism predicts an initial positive pressure change (an overpressure) while the electrostatic-pressure-relief mechanism predicts an initial negative pressure change (an underpressure). Holmes *et al.* (1971b) pointed out that the relation of the initial pressure variation (compression or rarefaction) to the mechanism of thunder production was likely to remain ambiguous until measurements were made with several microphones placed close to the source under examination, an experiment that apparently has still not been performed.

11.3.1. *The acoustic emission from rapidly heated channels*

As discussed in subsections 4.6.5 and 12.2.2, the return stroke heats the channel created by the preceding stepped or dart leader from near 10 000 K to near 30 000 K or more in several microseconds or less. The return-stroke channel pressure must increase in response to this rapid temperature rise since there is insufficient time for the channel particle density to decrease appreciably. The spectroscopic data (Orville 1968c) indicate an average channel pressure of about 10 atm (10^6 N m^{-2}) during the first 5 μs. Such a channel overpressure will result in an expansion of the luminous channel and the formation of a shock wave that propagates outward and eventually beyond the luminous channel, which attains pressure equilibrium with the surrounding atmosphere within tens of microseconds (Orville 1968c). The shock wave differs from an acoustic wave (thunder) in that it compresses and heats the air and, as a result, propagates at supersonic speeds. The initial propagation speed of the shock wave is probably about 10 times the speed of sound (e.g., Few 1995), that is, of order 3 mm μs^{-1}, but it decreases rapidly. After the bulk (probably about 99 percent) of the energy delivered to the shock wave has been expended in performing thermodynamic work on the surrounding atmosphere, the shock wave is transformed, within a few meters or less from the lightning channel, into an acoustic wave that propagates at the velocity of sound (e.g., Few 1975). Thus, the heated-channel thunder-generation mechanism involves the production and evolution of the shock wave, which is typically characterized by its pressure as a function of radial coordinate at different instants of time. Such pressure profiles, resulting from the assumed ohmic heating of the leader channel as predicted by the gas dynamic return-stroke models, are shown in Fig. 12.3.

The physics of the gas dynamic channel expansion outlined above applies to short sections of lightning channel. However, in modeling thunder one must also take account of the overall channel geometry and, in particular, the tortuosity of the lightning channel. Hill R.D. (1968), for measured cloud-to-ground channel "straight" segments between 5 and 70 m in length, found that the direction changes for successive segments were randomly distributed, essentially independently of segment length, with a mean absolute value of channel direction change from segment to segment of about 16 degrees. Tortuosity on a much smaller scale is evident in close photographs such as those of Evans and Walker (1963) and Idone (1995).

To take account of the observed characteristics of the tortuosity, Few *et al.* (1967, 1970) and Few (1969a, b, 1982, 1995) proposed that, for the purpose of thunder generation, the lightning channel could best be modeled as a connected series of short cylindrical segments, each one treated as an independent acoustic source. At a radius smaller than the length of a given segment, cylindrical shock-wave theory was postulated to apply; at a radius larger than the approximate length of the channel segment, a more-or-less spherical divergence of the shock wave was assumed. We will consider further the details, predictions, and tests of the validity of this approach in subsection 11.3.2 below. Some researchers (Jones *et al.* 1968; Troutman 1969, 1970; Remillard 1969; Plooster 1968, 1970a, b) considered the lightning channel as a straight line producing a cylindrical shock wave. In subsection 11.3.2 we will compare the predictions of the spherical model (e.g., Brode 1956) and the two cylindrical models of Plooster (1968, 1970a, b) and of Plooster (1971a) with pressure measurements made near a long laboratory spark (Uman *et al.* 1970). In the spherical model and the first cylindrical model it is assumed that the total input energy is infinitely concentrated and instantaneously released; the second cylindrical model is a detailed gas dynamic model for the return stroke (subsection 12.2.2). In addition, we will compare the predictions of these models with the limited data on natural and rocket-triggered lightning given above in subsections 11.2.3–11.2.5.

Lightning flashes typically contain three to five return strokes, separated by time intervals ranging from milliseconds to hundreds of milliseconds, with a geometric mean of about 60 ms (Section 4.2). During an interstroke interval of 60 ms, the acoustic pulse from a preceding stroke will have propagated about 20 m from the channel, so that the newly generated shock wave can be treated as independent of the previous one (Few 1974). However, Rakov and Uman (1990) found that subsequent strokes produce cumulative effects in heating the lightning channel (and the surrounding atmosphere), a result that could potentially influence the thunder-generation conditions for later strokes.

11.3.2. Few's thunder theory

The thunder theory of Few occupies a special place in the literature on thunder because it was the first detailed and generally accepted model of thunder. Before we present Few's theory, some preliminary comments are in order.

There appears to be a contradiction between Few's thunder theory and the predictions of the gas dynamic models of the lightning return stroke discussed in subsection 12.2.2. The gas dynamic models are based on a more rigorous description of the physics of the return-stroke process, including (i) a gradual energy input to the channel, in accordance with the measured return-stroke current waveforms (e.g., Berger *et al.* 1975) and (ii) the incorporation of energy losses from the channel via electromagnetic radiation and, as a result, temperature and pressure profiles more or less consistent with the spectroscopic measurements for return strokes (subsection 12.2.6), while Few's model assumes that (i) the energy input is instantaneous and (ii) all the input energy is delivered to the shock wave.

However, the gas dynamic models consider only one energy deposition mechanism, ohmic heating for a gas in thermodynamic equilibrium, while in Few's model no specific mechanism of energy deposition is assumed. Furthermore, since both the leader and return stroke participate in heating the channel, the input energy in Few's model effectively accounts for both these processes, while the gas dynamic models only account for the stroke input energy that is expended during the return-stroke stage. In other words, Few's model is more general than the gas dynamic models in terms of energy input, but involves some gross simplifications of the processes involved. Because of the difference in the assumed energy balance as indicated above, the gas dynamic models predict a total return-stroke input energy of 10^3–10^4 J m^{-1}, of which only a few percent, perhaps 10–100 J m^{-1}, goes to the shock wave, while the theory of Few appears to work for lightning only when the input energy (all delivered to the shock wave) is 10^5–10^6 J m^{-1}.

The basic ideas of Few's theory are outlined below with a view toward predicting the power spectrum of thunder.

(i) *Overall channel geometry.* The thunder signal is a linear superposition of basically identical pressure pulses produced simultaneously by the deposition of a specified amount of energy into a large number of cylindrical segments connected in series and randomly oriented with respect to the average channel direction.

(ii) *Energy balance.* Essentially all the input electrical energy per unit length W is delivered to the shock wave and expended in performing thermodynamic PV work (expansion) on the surrounding atmosphere within a radial distance R_0 (the so-called relaxation radius) found from the condition $P_0 V(R_0) = WL$, where P_0 is the ambient pressure, V is the cylindrical volume, and L is the channel segment length. Thus $R_0 = [W/(\pi P_0)]^{1/2}$. For an input energy of 10^3 to 10^4 J m^{-1}, the range predicted by the gas dynamic models, the relaxation radius is 5.6 to 18 cm and for an input energy of 10^5 to 10^6 J m^{-1}, the range predicted by the theory of Few, it is 56 cm to 1.8 m.

(iii) *Sound-radiating-channel length.* The length of the elemental cylindrical channel section (the acoustic radiator) is approximately equal to R_0 and therefore is not an independent parameter but is a function of the input energy per unit length W and the ambient pressure P_0. Shorter channel sections will be engulfed by the expanding shock wave during the energy deposition stage, while longer sections will be composed of more than one radiator, the latter condition being potentially inconsistent with the random orientation of elemental radiators; see (i) above.

(iv) *Transformations of the shock wave* (see Fig. 11.5a). For each elemental acoustic radiator, the shock wave diverges cylindrically during the thermodynamic work stage (within R_0), and after that stage (beyond R_0) it makes the transition to spherical divergence. This transition is claimed to be needed in order to account properly for the effects of channel tortuosity (Few 1969b). Thus at the end of the thermodynamic work stage each elemental channel section can be replaced by a volume of hot air behind a spherically diverging shock wave, a "string-of-pearls" model. Pressure profiles for the spherical case were computed by Brode (1956) up to a distance of $10.5 R_0$ from the source. For an energy input of 10^3 to 10^4 J m^{-1} (R_0 ranging from 5.6 to 18 cm) this distance is from about 59 cm to 1.9 m, and for 10^5 to 10^6 J m^{-1} (R_0 ranging from 56 cm to 1.8 m) it is from about 5.9 to 19 m. The distance at which the transition from the shock wave to acoustic wave occurs is not well defined but is probably between $2R_0$ and $5R_0$ (Few *et al.* 1967), 11 to 89 cm for an input energy of 10^3 to 10^4 J m^{-1} and 1.1 to 8.9 m for an input energy of 10^5 to 10^6 J m^{-1}.

(v) *Formation of N-shaped pressure pulse.* As the thermodynamically driven shock-wave front passes the relaxation radius R_0, the pressure at the channel axis is reduced to the ambient value. The momentum gained by the gas during the thermodynamic-work stage carries it beyond R_0 and forces the pressure at the channel axis momentarily to decrease below ambient. As a result, a characteristic N-wave is formed. The N-wave is so named for its N-like shape, an initial overpressure half-cycle with a relatively fast risetime and a slower pressure decay followed by a half-cycle of rarefaction (Fig. 11.5b). This characteristic pressure signature is preserved by the resulting acoustic wave, although it diminishes in amplitude and lengthens with distance. The most distant N-shaped acoustic wave from a hot sphere, computed by Brode (1956) for a distance of $10.5 R_0$, had a length $\Delta r = 2.6 R_0$, which in Few's theory is assumed not to change up to distances of the

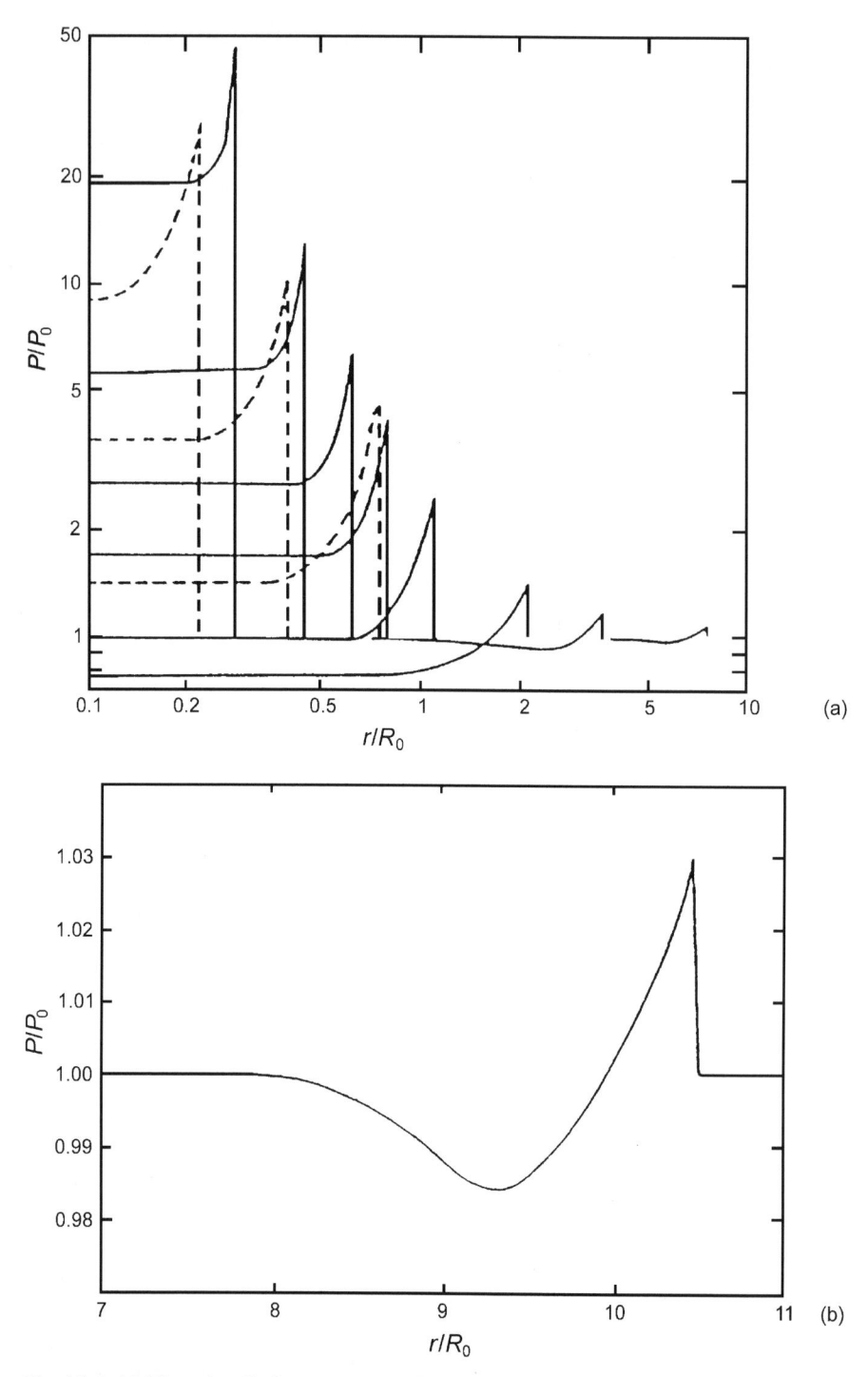

Fig. 11.5. (a) "Snapshots" of pressure normalized to atmospheric, P/P_0, as a function of radial distance normalized to the relaxation radius, r/R_0, from spherical (solid lines) and cylindrical (broken lines) sources. For cylindrical sources $R_0 = [W/(\pi P_0)]^{1/2}$ and for spherical sources $R_0 = [3W_t/(4\pi P_0)]^{1/3}$, where W is the energy per unit length and W_t is the total energy. Note that out to $r/R_0 = 1$ the initially formed strong shock wave ($P/P_0 \gg 1$) is thermodynamically driven, but then it "relaxes" and for $r/R_0 = 2$–5 degrades to an acoustic wave. Adapted from Few (1995). (b) Computed N-shaped pressure wave from a spherical source whose front is at $r/R_0 = 10.5$. The length of the wave is about $2.6R_0$. Adapted from Few (1995).

11.3. Generation mechanisms

order of kilometers, where thunder is usually measured. For an input energy of 10^3 to 10^4 J m^{-1}, Δr is 15 to 46 cm, and for an input energy of 10^5 to 10^6 J m^{-1}, it is 1.5 to 4.6 m. Uman et al. (1970) clearly demonstrated that wave stretching occurs beyond 10.5 R_0 (Fig. 11.7).

(vi) *Power spectrum of thunder.* If the channel geometry is as described in (i) and (iii) above and if each elemental acoustic radiator can be replaced by a spherically diverging wave then the normalized power spectrum of thunder from the entire tortuous channel should be identical to that from a single elemental radiator (Few 1969b). Thus the Fourier transform of the appropriately scaled most distant N-shaped acoustic wave (see Fig. 11.5b) computed by Brode (1956) should yield the power spectrum of the thunder signal produced by a uniform, randomly tortuous lightning channel. From such an analysis Few (1969b) derived an equation for the frequency (in hertz) of the power spectrum maximum:

$$f_m = 0.63 C_0 (P_0/W)^{1/2} \quad (11.2)$$

where C_0 is the speed of sound in m s^{-1}, P_0 is the ambient pressure in N m^{-2}, and W is the input energy in J m^{-1}. Note that Eq. 11.2 is essentially equivalent to the ratio of C_0 and $\Delta r = 2.6 R_0$, used by Few et al. (1967) for estimating the "lower limit for the dominant frequency of thunder". In general, the constant 0.63 in Eq. 11.2 should be a function of the length Δr of the acoustic wave relative to the relaxation radius R_0 (this ratio was 2.6 in the analysis of Few), and the observed peak frequencies should decrease with distance from the channel owing to lengthening of the sound wave (Otterman 1959; Few 1969a, 1982, 1995; Wright and Medendorp 1967; Section 11.4). Assuming $C_0 = 343$ m s^{-1}, we find from Eq. 11.2 that for an input energy of 10^3 to 10^4 J m^{-1} the range in f_m predicted by gas dynamic models is 2.2 kHz to 683 Hz, and for an input energy of 10^5 to 10^6 J m^{-1} the range in f_m predicted by the theory of Few is 216 to 68 Hz.

The predictions of the thunder theory of Few have been compared with experimental data for natural cloud-to-ground lightning (Holmes et al. 1971a), for rocket-triggered lightning (Hill E.L. and Robb 1968; Dawson et al. 1968b; Depasse 1994), and for long laboratory sparks (Dawson et al. 1968a; Uman et al. 1970). Holmes et al. (1971a) used their measured average acoustic efficiency, 0.18 percent (based on an assumed average total flash energy 1.8×10^9 J and the measured average acoustic energy per flash 3.26×10^6 J, as discussed in subsection 11.2.4), together with the measured acoustic energy for each flash and an assumed 4 km channel length, to determine the energy per unit length to use in Eq. 11.2. For one set of data, with measured peaks in the acoustic power spectrum between 40 and 100 Hz, the agreement between theory and measurement was reasonably good, although the calculated peak frequencies were always larger (up to about a factor 2) than the measured values. However, this result does not represent an independent theory test since it is scaled, through the acoustic efficiency, to a value of input energy of order 10^5 J m^{-1}. To explain the existence of measured peaks at 5 Hz and lower in some ground flashes, the input energy would have to be at least 100 times greater than the assumed input energy, of order 10^5 J m^{-1}. As noted in Section 11.1, the infrasonic peaks are postulated to be due to a source other than the hot channel; we shall discuss this further in subsection 11.3.4. Holmes et al. (1971a) also pointed out that while the acoustic energy received from cloud flashes was a factor 3 less than that from ground flashes, the average peak acoustic frequency for cloud flashes was about a factor 2 lower than for ground flashes.

Dawson et al. (1968a) related the measured input energy to a 4 m laboratory spark, 5×10^3 J m^{-1}, to the observed dominant frequency in the acoustic signal, between 1350 and 1650 Hz, and found that the data were reasonably fitted by an equation of the form of Eq. 11.2 but with the constant 0.63 replaced by unity. Uman et al. (1970) Fourier-transformed a typical shock waveform from the spark and found a peak frequency of 1400 Hz. Some additional support to the theory of Few is supplied by the detailed pressure-waveshape measurements of Uman et al. at distances from 0.34 to 16.5 m from a tortuous 4 m laboratory spark. Assuming that the shock wave degenerates to an acoustic wave at a distance of $5 R_0$ (Few et al. 1967), where $R_0 = 0.13$ m for the spark, we see that the distance range studied covers both the shock-wave and sound-wave modes of propagation. The measured overpressure data are reproduced in Figs. 11.6 and 11.7, as are the theoretical curves for cylindrically and spherically diverging waves. In the cylindrical case, for distances less than 2 m both the magnitude of the overpressure and the duration of the overpressure are a factor between 1.5 and 5 less than the values predicted by shock-wave theory. Close to the spark a single wave was observed, as illustrated in Fig. 11.8, whereas farther away several waves, generally three or four, overlapped (see Fig. 11.9), as might be expected if individual channel segments were generating separate shock waves. At 16.5 m (Fig. 11.10), the number of individual waves was less than at intermediate distances, apparently because various points on the spark channel become more nearly equidistant to the point of measurement and hence previously separated shock waves merge together.

As can be seen in Fig. 11.6, the data on the overpressure magnitude are fairly well fitted by assuming that they are due to spherical shock waves with total input energy $WL = 2.5 \times 10^3$ J derived from 0.5 m segments of spark channel (the middle solid line). Further, the duration of the initial half-cycle (compression) of the observed pressure

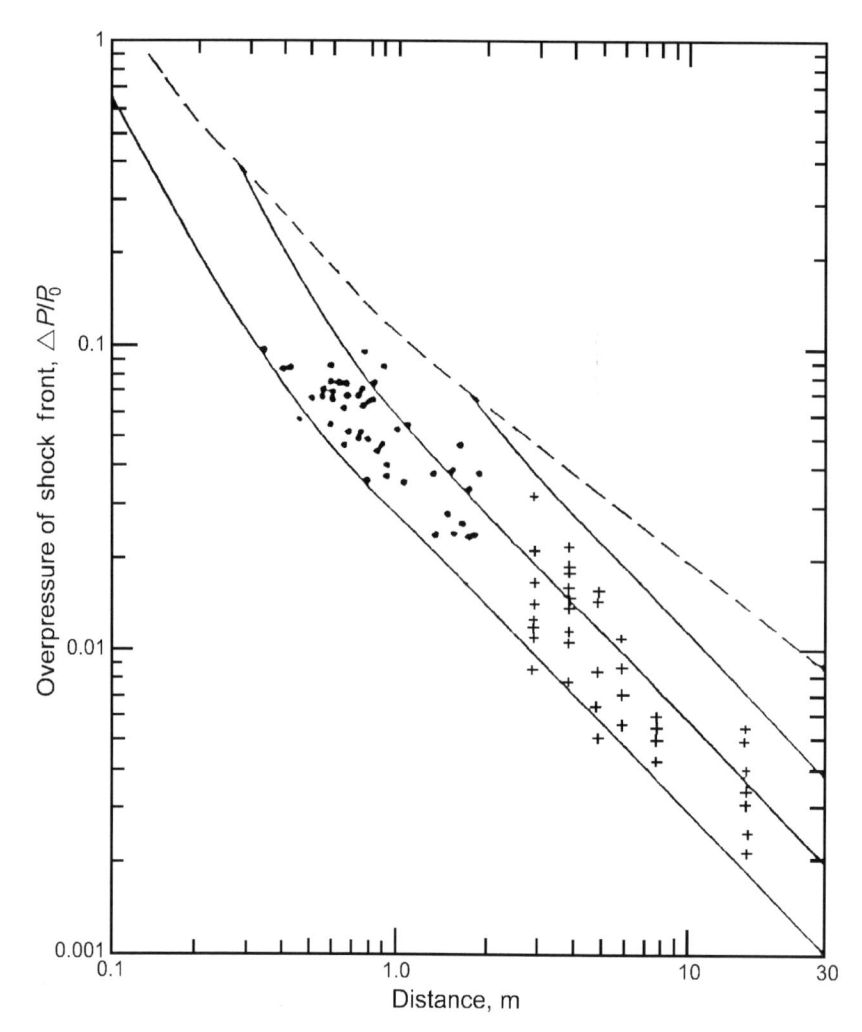

Fig. 11.6. Initial overpressure normalized to atmospheric, $\Delta P/P_0$, as a function of the distance from a 4 m laboratory spark. Dots, data obtained with a piezoelectric microphone; crosses, data obtained with a capacitor microphone; broken line, cylindrical source with $W = 5.0 \times 10^3$ J m^{-1} (Plooster, 1968); top, middle, and bottom solid lines, spherical sources with respective values $WL = 2 \times 10^4$ J, $L = 4.0$ m, $WL = 2.5 \times 10^3$ J, $L = 0.5$ m and $WL = 3.1 \times 10^2$ J, $L = 6.25$ cm (Brode 1956). The total electrical energy per unit length computed from measurements of the spark voltage and current is 5×10^3 J m^{-1}. Adapted from Uman et al. (1970).

waves is somewhat better fitted by the spherical shock-wave theory than by cylindrical shock-wave theory (Fig. 11.7). However, Uman et al. (1970) noted that the cylindrical shock-wave theory would provide a reasonable fit to the data if it were assumed that not all the input energy is delivered to the shock wave but only 10 to 20 percent.

The duration of the wave's negative half-cycle (rarefaction) differs from the predictions of either the cylindrical or the spherical theory. The latter discrepancy, as well as the somewhat shorter model-predicted initial overpressure duration (Fig. 11.7), was attributed by Few (1995) to the fact that the energy input to the spark was gradual, rather than instantaneous as assumed in the models considered. Uman et al. (1970) reported that, from measurements of spark voltage and current, 90 percent of the total input energy was absorbed by the 4 m spark in a period of 3 μs, while for lightning return strokes the energy deposition rate due to ohmic heating in gas dynamic models was lower, about 50 percent in 10 μs according to Plooster (1971b). Few et al. (1970) criticized the cylindrical-shock-wave approach on the grounds that it predicts overpressures between 10^{-3} and 10^{-2} atm (10^2 and 10^3 N m^{-2}) at a distance of 1 km if the input energy is between 10^5 and 10^6 J m^{-1}, while the observed overpressures are generally less than 10^{-4} atm (10 N m^{-2}) at this distance. However, it appears that the use of an input energy of 10^4 J m^{-1} would result in overpressures from a cylindrical source more consistent with the observations.

11.3. Generation mechanisms

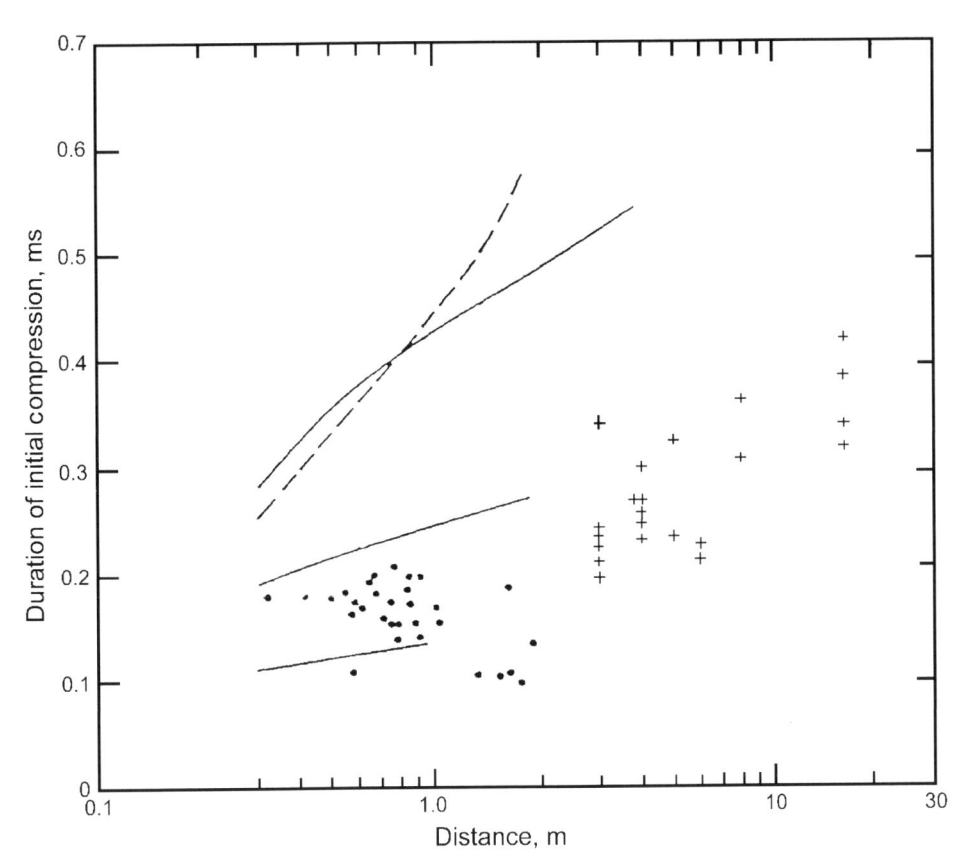

Fig. 11.7. Same as Fig. 11.6, but the ordinate is now the duration of the initial overpressure. Adapted from Uman et al. (1970).

Plooster (1971a) attempted to model the laboratory spark data of Uman et al. (1970) using a detailed gas dynamic model (subsections 12.2.2 and 12.2.6); in earlier papers (Plooster 1968, 1970a, b) he had considered an instantaneous energy input to a cylindrical shock wave and had found that, when the current measured by Uman et al. (1970) was employed as an input in this model, only about 10 percent of the energy input measured by Uman et al. was dissipated in raising the channel temperature to values consistent with those measured by Orville et al. (1968b). Plooster (1971a) argued that a cylindrical energy input to the spark of 4.2×10^2 J m^{-1}, an order of magnitude below that measured but consistent with his calculations and with the measured input current, would result in a cylindrical-theory curve that would pass through the data points shown in Fig. 11.6. He therefore concluded that Uman et al. must have made an order-of-magnitude error in determining the spark input energy. Hill R.D. (1977a, b), however, suggested that Uman et al. had measured the input energy correctly but that only a small fraction of that energy subsequently appeared in the hot spark channel producing a cylindrical shock wave, the remainder being diffusely deposited around the spark channel by processes preceding the spark's return stroke and producing no appreciable acoustic emission. It is possible, as noted above, that both the leader and return stroke participate in the production of thunder, and, if so, the calculations of Plooster (1971a) made specifically for the return stroke, as opposed to those of Plooster (1968, 1970a, b), the latter made without specifying an energy deposition process, cannot be directly compared with the experimental data in Fig. 11.6.

Depasse (1994) recorded 25 acoustic signals at 70 m from triggered-lightning strokes (Section 7.2) using a microphone with frequency bandwidth from 50 Hz to 15 kHz. The measured channel-base currents ranged from 4.5 to 50 kA. The microphone was placed about 1 m above ground, the axis of its presumably narrow directional pattern making an angle of 15 degrees with the horizontal plane. Depasse argued that the microphone received acoustic signals only from the bottom few tens of meters of the channel, the contribution from the higher channel sections, as well as from other sources (including reflections), being negligible. The lower part of the channel seen by the microphone was not as tortuous as in natural lightning so that it was possible to identify acoustic signatures produced by individual lightning strokes. It is important to note that strokes in triggered lightning typically involve branchless dart leaders and are similar to subsequent strokes in natural lightning,

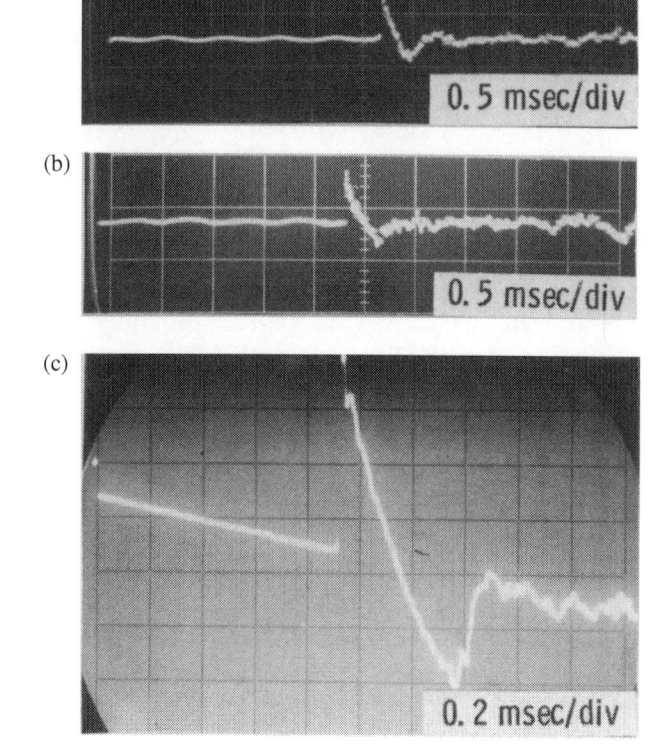

Fig. 11.8. Three typical pressure waves observed close to a 4 m laboratory spark: (a) piezoelectric microphone record 88 cm from the spark at midgap height, $\Delta P/P_0 = 0.085$. (b) Piezoelectric microphone record 84 cm from the spark at midgap height, $\Delta P/P_0 = 0.075$. (c) Capacitor microphone record 3 m from the spark at a height 1 m above the bottom electrode, the time delay between spark initiation and oscilloscope triggering being 8 ms; $\Delta P/P_0 = 0.020$. The acoustic signal is superimposed on the electrical response of the microphone to ambient electric fields. Adapted from Uman et al. (1970).

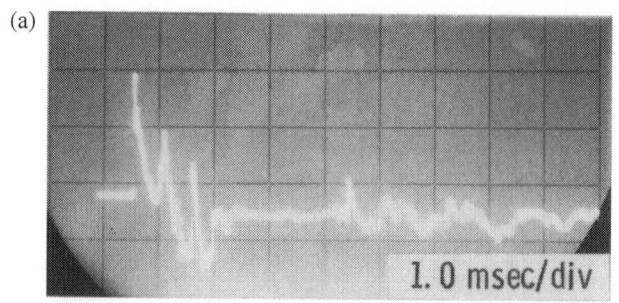

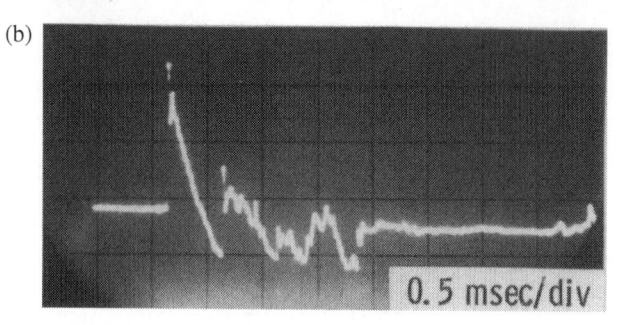

Fig. 11.9. Two typical pressure waves recorded 8 m from the laboratory spark at midgap height by a capacitor microphone. Reflections from the ground plane arrive at the microphone about 4 ms after the direct signal. There is a 23 ms delay between spark initiation and oscilloscope triggering. (a) $\Delta P/P_0 = 0.0049$, from the same spark whose pressure wave at 84 cm is shown in Fig. 11.8b, the capacitor microphone being on the opposite side of the spark from the piezoelectric microphone; (b) as in (a) but with $\Delta P/P_0 = 0.0057$. Adapted from Uman et al. (1970).

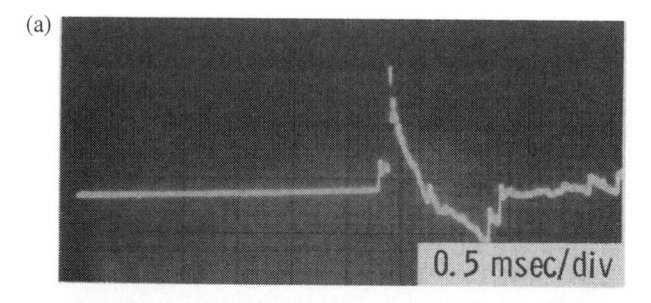

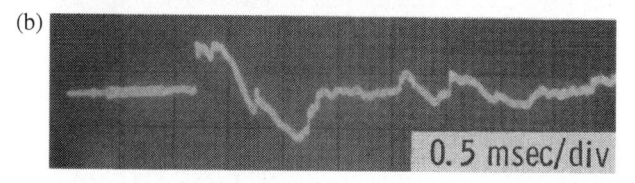

Fig. 11.10. Two typical pressure waves recorded 16.5 m from the laboratory spark at midgap height by a capacitor microphone. The reflections from the ground plane arrive about 2 ms after the direct signal; (a) $\Delta P/P_0 = 0.0050$, time delay to oscilloscope triggering 48 ms; (b) $\Delta P/P_0 = 0.0021$, time delay to oscilloscope triggering 49 ms. Adapted from Uman et al. (1970).

and that the channel and air in the vicinity of the channel might contain particles of vaporized triggering wire. The recorded acoustic waveforms exhibited an initial rarefaction followed by a compression, the maximum pressure being a compression in 17 cases and a rarefaction in eight cases. The median maximum pressure and median acoustic wave duration were 4.7 N m^{-2} and 3.0 ms, respectively. The initial rarefaction duration in two examples given by Depasse (1994) was about 0.6 ms. Interestingly, this initial rarefaction duration measured 70 m from a triggered-lightning channel is more or less consistent with the trend observed by Uman et al. (1970) for the initial overpressure duration versus distance from a 4 m laboratory spark (Fig. 11.7). However, the median maximum pressure of 4.7 N m^{-2} at 70 m is one to two orders of magnitude lower than indicated by extrapolation of the 4 m spark data of Uman et al.

(1970) shown in Fig. 11.6. It is also much smaller than the 100 N m^{-2} (0.001 atm) or so predicted by the cylindrical and spherical shock-wave theories at 70 m for an input energy, respectively, as low as 10^2 J m^{-1} (Jones et al. 1968) and 2.5×10^3 J (Uman et al. 1970).

Depasse et al. (1994) computed acoustic frequency spectra by applying Fourier transform technique to (i) the acoustic records of triggered flashes lasting for about 100 ms and containing one to six strokes, and (ii) the acoustic records of individual strokes (an example of the stroke spectrum is shown in Fig. 11.4). The maximum of the frequency spectrum for 12 flashes was found at frequencies ranging from 205 to 1775 Hz, which translates, via Eq. 11.2, to an input energy per unit length ranging from 1.5×10^3 to 1.1×10^5 J m^{-1} (less than 10^4 J m^{-1} in six out of 12 cases). The total energy per unit length inferred from Eq. 11.2 was found to be correlated with the specific electrical energy (the action integral) determined by adding up specific energies for all strokes within the 100 ms interval. The latter finding allowed Depasse to conclude that Few's theory of thunder was confirmed, although Eq. 11.2 was derived for a single return stroke while the action integral was computed by Depasse for all strokes comprising the flash. However, the dominant frequency for 18 individual strokes ranging from approximately 300 to 900 Hz was found not to be correlated with the stroke action integral, a finding apparently in contradiction with Eq. 11.2. The length of the elemental radiating channel segment corresponding to the median inferred energy per unit length, about 10^4 J m^{-1}, is a few tens of centimeters, implying a random tortuosity of the triggered-lightning channel on this scale.

The results of Depasse (1994) are very interesting, but they are probably not directly comparable with the natural-lightning data, at least in part because of the lack of branched stepped leaders in classical triggered lightning (Section 7.2).

Dawson et al. (1968b) analyzed the close overpressure measurements made 35 cm from a spark channel bridging a 10 cm gap placed in series with a triggered-lightning channel by Hill E.L. and Robb (1968) and Newman et al. (1967a, b). These measurements were discussed in subsection 11.2.5. The measured maximum overpressure, about 2 atm (2×10^5 N m^{-2}) at a distance of 35 cm, is more consistent with the spherical shock wave theory than with the cylindrical shock wave theory; these theories predict 1.7 atm and 3.8 atm, respectively, for the same assumed energy input of 10^5 J m^{-1}.

11.3.3. Effects of tortuosity and branches

Wright (1964) and Wright and Medendorp (1967) characterized the acoustic waveforms produced by a 1 cm laboratory spark as a function of angle from the spark. Observations made perpendicularly to the spark yielded a single acoustic N-shaped wave, similar to those shown in Fig. 11.8, since all the spark-channel sections were essentially equidistant from the observer. The waveform changed its shape when measured out of the perpendicular plane. Few (1974) computed, using the experimental data of Wright and Medendorp (1967), that 80 percent of the acoustic energy from the 1 cm spark was confined to within ± 30 degrees of the plane perpendicular to the spark. This directed acoustic radiation pattern was interpreted by Few (1982, 1995) as one of the important factors for the production of loud thunder claps, further discussed below.

Ribner and Roy (1982) used the acoustic waveshapes observed by Wright and Mendendorp (1967) as the starting point for a computer generation of acoustic signals from model tortuous channels. Few (1974, 1982, 1995) described a similar study. The calculated acoustic signals have much in common with observed thunder and illustrate the crucial role played by tortuosity and channel orientation in producing various features of thunder.

Few (1982, 1995) discussed in detail the generation of thunder claps that are to be associated with the sound emitted by relatively long sections of the main channel and channel branches approximately perpendicular to the line of sight of the observer, a fact that had been verified experimentally by, for example, Few himself (1970). It is reasonable to expect the branches of first strokes to be powerful sources of sound since, according to the data of Malan and Collens (1937), branches may be instantaneously brighter than the channels above those branches. Further, it is interesting to note that there are roughly as many claps per thunder as there are branches in a first stroke (Schonland et al. 1935), so that the branches may well account for a significant fraction of the claps.

11.3.4. The acoustic emission due to relief of electrostatic pressure

The origin of the bulk of the infrasonic thunder is thought not to be a hot expanding channel, because of the unreasonably large input energies, about 10^7 J m^{-1} and greater (Holmes et al. 1971a) required to produce waves of such low frequency. Holmes et al. established that the peak acoustic power could be in the audible at one time during a thunder record and in the infrasonic at another time, as illustrated in Fig. 11.3b. Measurements have indicated that infrasonic thunder is preferentially observed beneath thunderstorms, arrives in discrete pulses, and is characterized by an initial compression followed by a larger rarefaction, the latter observation being contrary to the model predictions described below (Bohannon et al. 1977; Balachandron 1983). The conversion of stored electrostatic energy to acoustic energy, as an explanation for the infrasonic component of thunder, has been examined from a theoretical point of view by Wilson (1920), McGehee (1964), Dessler

(1973), and Few (1985). We consider some aspects of that theory now.

The mutual repulsion of the charged hydrometeors causes atmospheric pressure to be reduced within the charged regions of the cloud. If a lightning flash discharges a cloud region rapidly, then the atmospheric pressure equilibrium tends to be restored in that region, and, as a result, an acoustic rarefaction pulse is produced. The amplitude of this pulse is related to the electric field at the edge of the charged volume, and the dominant frequency to the dimensions of the region.

Wilson (1920) first suggested that the electrical stress in a cloud, when relieved by lightning, would provide a "by no means negligible contribution to thunder". Analyses adopting this physical model have been provided for a variety of geometries in the papers referenced above. McGehee (1964) considered spherical charge volumes in the cloud. Dessler (1973) considered spherical, cylindrical, and disk geometries of the cloud charge. Few (1982, 1995) gave a generalized expression for the electrostatic pressure experienced by a volume of air characterized by a constant charge density,

$$P_\mathrm{E} = -(n+1)\frac{\epsilon_0 E^2}{2} \quad (11.3)$$

where $n = 0$, 1, or 2, respectively, for plane, cylindrical, and spherical geometries, and E is the electric field intensity. The analysis of Dessler (1973) is of particular interest in that he found that the infrasound emission from the collapse of the cloud electric field was highly directional, primarily propagating upward and downward. He suggested that this fact might explain the variability in the observations of infrasound, in that one needs to be beneath a thundercloud for efficient detection, a prediction apparently experimentally verified by both Bohannon *et al.* (1977) and Balachandron (1983). However, Dessler (1973), as well as all others who have modeled the cloud charge collapse as an origin for infrasound, predicted an initial infrasonic rarefaction. In fact, measurements indicate a compression. Few (1985) attempted to remedy this inconsistency by proposing a model similar to Dessler's but with the addition of an extensive network of small electrical discharges in the cloud, the heating of whose channels is responsible for the initial compression.

Colgate and McKee (1969) examined the acoustic effect of the ions moving in the radial electric field of the stepped leader (Section 4.4). The electrostatic sound predicted by their analysis, for a stepped-leader radius of 2.5 m and a stepped-leader charge density of about 4×10^{-4} C m^{-1}, corresponding to a stored electrostatic energy of 10^3 to 10^4 J m^{-1}, and an initial ion speed of 420 m s^{-1}, had a dominant frequency of 130 Hz, clearly not infrasonic. The initial overpressure was more than two orders of magnitude smaller than that of the acoustic wave from the hot channel of the ensuing return stroke, with an assumed input energy 5×10^5 J m^{-1}. A larger-radius stepped leader would yield a sound pulse having a larger magnitude and a lower dominant frequency.

11.4. Propagation

Acoustic signals generated by lightning propagate for large distances through an atmosphere that is nonhomogeneous, anisotropic, and turbulent. Few (1982, 1995) reviewed propagation effects on sound waves in air as they pertain to thunder. We now consider briefly the more salient of these effects.

The three primary propagation effects of interest are the waveshape change associated with finite-amplitude sound waves, the sound wave attenuation, and the thermal refraction. In principle all these effects can be considered satisfactorily in a general theory of thunder propagation. Refraction due to wind shear, the change in wind velocity with height, which does not vary appreciably with time, can also be modeled if the horizontal wind velocity is known as a function of height. The effects of other factors influencing propagation paths, for example, transient winds, aerosols, turbulence, and reflections from irregular terrain such as mountains, are difficult to handle analytically.

We first consider thermal refraction. Fleagle (1949) showed that the inaudibility of thunder beyond 25 km can be ascribed to the upward curving of sound rays resulting from the usual atmospheric temperature decrease with height. Since the speed of sound is proportional to the square root of the temperature, Snell's law indicates that sound waves will normally be refracted upward. Fleagle calculated that, in the presence of a linear lapse rate (temperature decreasing linearly with height), those sound rays which leave the channel and at some point become tangent to the ground plane exhibit a trajectory that is very nearly parabolic. For a typical lapse rate, 7.5 K km^{-1}, sound that originates at a height of 4 km has maximum range of audibility 25 km if wind shear is ignored. That is, the sound rays from a 4 km height are tangent to the ground 25 km distant from the discharge channel. All sounds originating from heights below 4 km will not be heard at 25 km; sounds originating from above 4 km will be heard. It follows from these calculations that only the very close observer can hear sound from the base of the lightning channel. Fleagle also showed that wind shear can provide a refraction of the thunder of the same order of magnitude as that due to the temperature gradient. Sound rays may be refracted upward or downward, depending on the relation between the wind shear and the sound-ray direction. A wind shear of 4 m s^{-1} km^{-1} can yield a sound-ray trajectory almost equivalent to a lapse rate of 7.5 K km^{-1}. Similar analyses are found in Fleagle and Businger (1963) and in Few (1982, 1995).

Fleagle (1949) cautioned that factors other than the lapse rate and wind shear may affect the audibility of thunder. For example, a region of temperature inversion will tend to increase the range of audibility, and features of the terrain that hinder the essentially horizontal propagation of the critical sound ray in its final several kilometers will decrease the range of audibility.

We now consider the lengthening of sound waves due to nonlinear or finite-amplitude propagation. As a large-amplitude spherically diverging acoustic wave propagates through air, the shape of the wave must change. Otterman (1959) published a theoretical treatment of the propagation of large-amplitude acoustic signals that can be applied to thunder. An initial shock-wave pulse evolves into an N-wave of the type observed in the laboratory by Wright and Medendorp (1967) and Uman et al. (1970) and illustrated in Fig. 11.8. The length of the N-wave increases with propagation distance. Few (1982, 1995) reproduced an expression due to Otterman (1959) for the pulse length and used it to evaluate how the Brode waveshape at a distance of $10.5R_0$, the most distant waveshape calculated by Brode (1956), lengthens for larger distances from the source: between Brode's most distant calculation and about 1 km, the N-wave lengthens by about a factor 2, owing to finite-amplitude propagation. Beyond that range its length remains roughly constant, as expected in small-amplitude linear propagation. Few (1982) viewed the factor 2 increase in pulse length as the maximum lengthening, since the attenuation caused by energy losses, not included in Otterman's (1959) theory, would be expected to counteract the pulse lengthening. If the pulse length did increase by a factor 2 then the constant in Eq. 11.2, derived using the assumption of linear propagation beyond $10.5R_0$, would be halved and the resultant energy per unit length, 2×10^6 J m^{-1}, found by Few (1969a) for a 40 Hz maximum frequency in the thunder power spectrum, would be reduced by a factor 4 to the value 5×10^5 J m^{-1}, in better agreement with the value of order 10^5 J m^{-1} for that parameter inferred by Krider et al. (1968) but still about two orders of magnitude higher than the values predicted by gas dynamic models (subsections 12.2.2 and 12.2.6).

Finally, we consider the attenuation caused by the dissipation of wave energy. The attenuation of acoustic signals in air is primarily due to the interaction of the sound waves with air molecules and is a function of the water vapor content of the air. Harris (1967) produced tables of attenuation constants for the decrease in acoustic signal amplitude with propagation distance as a function of signal frequency, atmospheric temperature, and humidity. Few (1982, 1995) used this and other literature to calculate the attenuation for typical atmospheric conditions and found that there is little attenuation for frequencies below about 100 Hz for distances as large as 10 km. For a frequency of 1 kHz and a range of 10 km, the attenuation is a factor 2. However, Bass and Losely (1975) calculated that, for 50 percent humidity at 20 °C and a range of 5 km, the attenuation at 400 Hz is about a factor 3 but at 50–100 Hz is negligible. They also showed that changing the relative humidity from 20 to 100 percent for a discharge at 2 km increases the attenuation at 400 Hz by about a factor 3. It follows that observed thunder frequency spectra will be affected by attenuation (and also by the lengthening of acoustic waves discussed above) at the high-frequency end, and hence the peak frequency is likely to be distance dependent. According to Few (1995), at distances beyond a few kilometers from a lightning channel one never hears the high-frequency components that are heard close to the channel. Further, differences in atmospheric conditions at different times or locations could result in different thunder spectra for similar sources and similar ranges.

Attenuation due to the scattering of acoustic waves from cloud particles has been discussed by Few (1982, 1995). This type of scattering also preferentially attenuates the higher frequencies. A variety of other processes, such as turbulence, that can result in attenuation were also discussed by Few (1982, 1995). It appears that propagation effects can modify the original acoustic signature of lightning to such an extent that the derivation of the source characteristics from remote thunder measurements becomes virtually impossible, a situation somewhat similar to atmospherics (Sections 13.1 and 13.3), which contain more information on their propagation paths than on their sources.

11.5. Acoustic imaging of lightning channels

If the same significant feature of a given thunder record can be recognized at three or more measuring stations, then a knowledge of the arrival time of that thunder feature at each microphone relative to the lightning electromagnetic signal allows a determination of the three-dimensional location of its source. Two different techniques have been used to find thunder sources. The technique that is more accurate and capable of giving many locations per thunder event is called ray tracing. The time difference between the arrival of significant features at different microphones in a network, typically tens of meters apart, is used to determine the direction of the incoming sound wave at the network, and that directional ray is mathematically traced back to the source given the atmospheric conditions and the time between the arrival at the network of the lightning electromagnetic signal and the particular acoustic feature. A discussion of the accuracy of ray tracing was given by Few and Teer (1974). Lightning channels reconstructed using ray tracing have been published, for example, by Few (1970), Nakano (1973, 1976), Few and Teer (1974), Teer and Few (1974), Winn et al. (1978), Christian et al. (1980), MacGorman et al. (1981), and Weber et al. (1982). In ray tracing, the signals on separate microphones are very similar because the microphones are located relatively close together. A second, less accurate source-location technique is called thunder ranging, defined in subsection 11.2.1. In thunder ranging, at least three non-collinear microphones are separated by a relatively large distances, of the order of a kilometer. To thunder-range on a feature of the acoustic waveform such as a clap, the difference between the electromagnetic field and the clap arrival times at each station is used to define a spherical surface of possible source locations. From three microphones, the three spherical surfaces intersect at a single point, the clap location. The implementation of this technique was discussed by Bohannon (1978). The technique of thunder ranging was used to reconstruct the channel of lightning that struck the weather tower at KSC, which was studied in detail by Uman et al. (1978). According to Few (1982, 1995), thunder signals become spatially incoherent at microphone separations greater than about 100 m due to differences in perspective and propagation path. However, gross features such as claps remain coherent for microphone separations of the order of kilometers.

Perhaps the most interesting feature of the lightning channels that have been reconstructed from thunder records is the fact that the sources in the cloud are generally oriented more horizontally than vertically (e.g., MacGorman et al. 1981), although vertical distributions of thunder sources do occur (e.g., Christian et al. 1980). That the horizontal orientation is more common is probably associated with the horizontal extent of the negative charge generally found between -10 and -25 °C (subsection 3.2.2). Few (1970) observed a horizontal channel about 20 km in extent at about 5 km altitude. Nakano (1973, 1976) found that horizontal channels at 7 or 8 km height were oriented mainly along the wind direction. Few and Teer (1974) reported on the general agreement of thunder-channel reconstructions with lightning photographs. Teer and Few (1974), for 17 cloud-to-ground and 20 cloud discharges occurring during 30 minutes at the end of a storm found that the typical ratio of the longer horizontal extent, the shorter horizontal extent, and vertical extent for cloud flashes and the in-cloud parts of ground flashes was 3: 2: 1. The cloud flashes and the in-cloud portions of the cloud-to-ground flashes were generally aligned in the same direction. Winn et al. (1978) showed thunder-source locations as part of an overall study of one thunderstorm. MacGorman et al. (1981) reconstructed all the lightning channels in three storms, one each in Arizona, Colorado, and Florida (examples for Colorado are given in Figs. 11.11 and 11.12). They found that the in-cloud lightning activity was in layers 2 to 3 km thick and that in the Arizona and Florida storms there appeared to be two separate layers of activity. They interpreted the two layers as being associated with the upper positive and lower negative charges of the cloud dipole structure (see Section 3.2). In

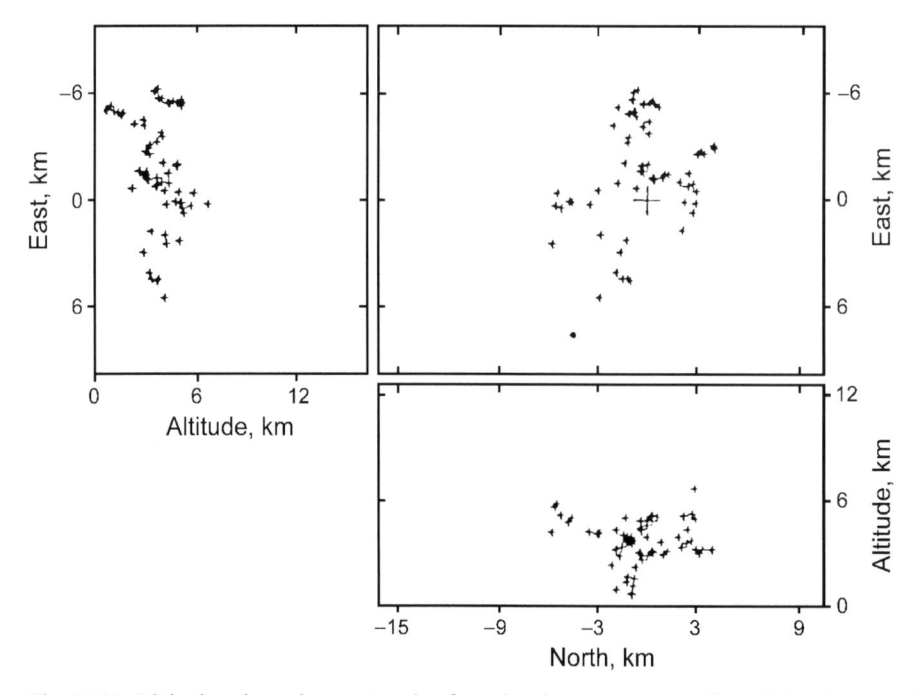

Fig. 11.11. Lightning channel reconstruction from thunder measurements for a Colorado ground flash (25 July 1972, 18:09:29 MDT). The lightning acoustic sources are plotted in three orthogonal projections. Altitude is measured relative to ground level, the latter being approximately 1450 m above mean sea level. The origin is marked by a large cross in the east–north projection. Adapted from MacGorman et al. (1981).

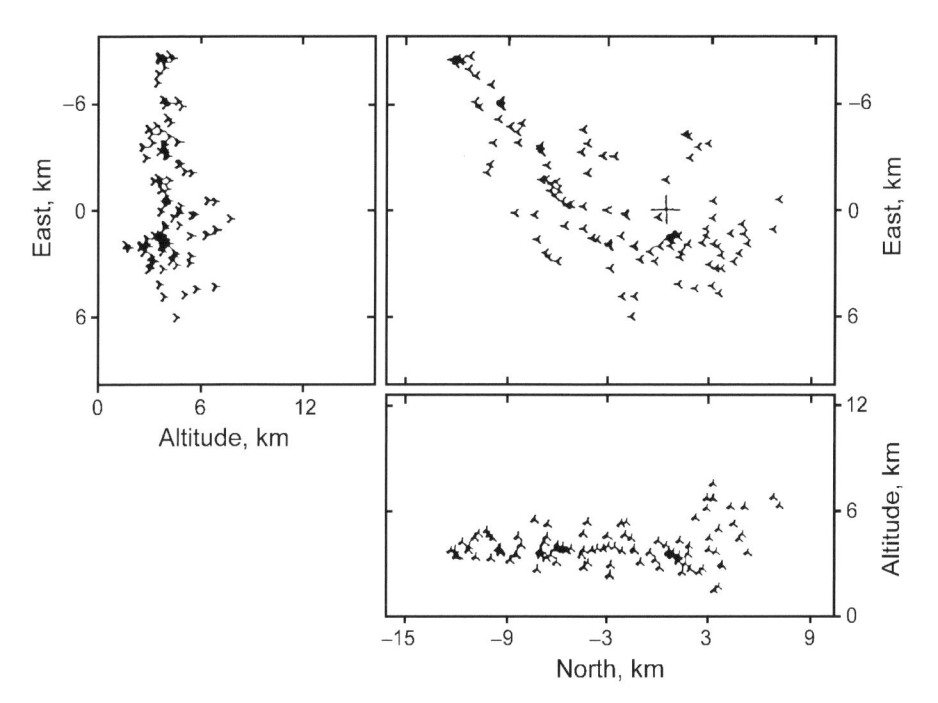

Fig. 11.12. Same as Fig. 11.11, but for a Colorado cloud flash (25 July 1972, 18:26:11 MDT). The acoustic sources are plotted as Y's, rather than crosses as in Figure 11.11, to indicate that they were reconstructed from a different microphone array. Note that this lightning flash has a greater horizontal extent than the ground flash in Fig. 11.11. Adapted from MacGorman et al. (1981).

the Colorado storm, there was a single layer having a lower boundary near the 0 °C isotherm. Channels of a bipolar lightning, triggered in winter in Japan, reconstructed from three-station acoustic measurements are shown in Fig. 8.9a.

11.6. Summary

The peak of the power spectrum of thunder produced by natural cloud-to-ground lightning discharges typically occurs at frequencies of some tens of hertz. Some natural cloud-to-ground flashes and most cloud flashes have thunder power spectra that peak in the infrasonic range, below about 20 Hz. Dominant frequencies in the acoustic signals generated by rocket-triggered lightning are typically of the order of hundreds of hertz. The mean total acoustic energy radiated by cloud-to-ground flashes is about 6 MJ, about a factor 3 greater than that radiated by cloud flashes. For both cloud-to-ground and cloud flashes at distances of the order of kilometers, the average rms pressure is 10^{-6}–10^{-5} atm. It appears that all impulsive processes in the lightning discharge, including M-component-type processes, produce thunder. The origin of most audible thunder is thought to be the expansion of the rapidly heated lightning channel, although some infrasonic thunder must also be generated by this mechanism. The origin of most infrasonic thunder is apparently associated with the conversion to sound of the energy stored in the electrostatic field of the thundercloud when lightning rapidly reduces that cloud field, although this mechanism can also produce audible thunder. There still exists a lack of detailed understanding of the dominant mechanism or mechanisms of the generation of thunder. Thunder recordings at three or more stations can be used for the imaging of lightning channels.

References and bibliography for Chapter 11

Ajayi, N.O. 1972. Acoustic observation of thunder from cloud-ground flashes, *J. Geophys. Res.* **77**: 4586–7.

Akiyama, H., Ichino, K., and Horii, K. 1985. Channel reconstruction of triggered lightning flashes with bipolar current from thunder measurements. *J. Geophys. Res.* **90**: 10 674–80.

Anderson, R., Bjornsson, S., Blanchard, D.C., Gathman, S., Hughes, J., Jonasson, S., Moore, C.B., Survilas, H.J., and Vonnegut, B. 1965. Electricity in volcanic clouds. *Science* **148**: 1179–89.

Arabadzhi, V. 1952. Certain characteristics of thunder. *Dokl. Akad. Nauk SSSR* **82**: 377–8. (English translation RJ-1058 is available from Associated Technical Services, Glen Ridge, New Jersey.)

Arabadzhi, V. 1957. Some characteristics of the electrical state of thunderclouds and thunderstorm activity. *Uch. Zap. Minsk. Gos. Ped. Inst. im. A.M. Gorkogo*, Yubil. Vypusk,

Ser. Fiz-Mat., no. 7. (English translation RJ-1315 is available from Associated Technical Services, Glen Ridge, New Jersey.)
Arabadzhi, V. 1965. The spectrum of thunder. *Priroda* **54**: 74–5.
Arabadzhi, V. 1968. Acoustical spectra of electrical discharges. *Soviet Physics – Acoustics* **14**: 92–3.
Arnold, R.T., Bass, H.E., and Atchley, A.A. 1984. Underwater sound from lightning strikes to water in the Gulf of Mexico. *J. Acoust. Soc. Am.* **76**: 320–2.
Ault, C. 1916. Thunder at sea. *Sci. Am.* **114**: 525.
Balachandran, N.K. 1979. Infrasonic signals from thunder. *J. Geophys. Res.* **84**: 1735–45.
Balachandran, N.K. 1983. Acoustic and electric signals from lightning. *J. Geophys. Res.* **88**: 3879–84.
Bass, H.E. 1980. The propagation of thunder through the atmosphere. *J. Acoust. Soc. Am.* **67**: 1959–66.
Bass, H.E., and Losely, R.E. 1975. Effect of atmospheric absorption on the acoustic power spectrum of thunder. *J. Acoust. Soc. Am.* **57**: 822–3.
Bates, E.L. 1903. The cause of thunder again. *Sci. Am.* **88**: 115.
Beasley, W.H., Georges, T.M., and Evans, M.W. 1976. Infrasound from convective storms: an experimental test of electrical source mechanism. *J. Geophys. Res.* **81**: 3133–40.
Berger, K., Anderson, R.B., and Kroninger, H. 1975. Parameters of lightning flashes. *Electra* **80**: 23–37.
Bhartendu, H. 1964. Acoustics of thunder, Ph.D. dissertation, Physics Department, University of Saskatchewan, Saskatoon, Canada.
Bhartendu, H. 1968. A study of atmospheric pressure variations from lightning discharges. *Can. J. Phys.* **46**: 269–81.
Bhartendu, H. 1969a. Thunder – a survey. *Naturaliste Can.* **96**: 671–81.
Bhartendu, H. 1969b. Audio frequency pressure variations from lightning discharges. *J. Atmos. Terr. Phys.* **31**: 343–7.
Bhartendu, H. 1971a. Comments on paper "On the power spectrum and mechanisms of thunder" by C.R. Holmes, M. Brook, P. Krehbiel, and R. McCrory. *J. Geophys. Res.* **76**: 7441–2.
Bhartendu, H. 1971b. Sound pressure of thunder. *J. Geophys. Res.* **76**: 3515–16.
Bhartendu, H., and Currie, B.W. 1963. Atmospheric pressure variations from lightning discharges. *Can. J. Phys.* **41**: 1929–33.
Bohannon, J.L. 1978. Infrasonic pulses from thunderstorms, M.S. thesis, Rice University, Houston, Texas.
Bohannon, J.L. 1980. Infrasonic thunder explained. Ph.D. dissertation, Department of Space Physics and Astronomy, Rice University, Houston, Texas.
Bohannon, J.L., Few, A.A., and Dessler, A.J. 1977. Detection of infrasonic pulses from thunderclouds. *Geophys. Res. Lett.* **4**: 49–52.
Braginskii, S.I. 1958. Theory of the development of a spark channel. *Sov. Phys. JETP* **34**: 1068–74.
Brode, H.L. 1955. Numerical solutions of spherical blast waves. *J. Appl. Phys.* **26**: 766–75.
Brode, H.L. 1956. The blast wave in air resulting from a high temperature, high pressure sphere of air. Rand Corp. Res. Mem. RM-1825–AEC.

Brook, M. 1969. Discussion on the Few–Dessler paper. In *Planetary Electrodynamics*, eds. S.C. Coroniti and J. Hughes, vol 1., p. 579, New York: Gordon and Breach.
Brooks, C.F. 1920. Another case. *Mon. Wea. Rev.* **48**: 162.
Brown, E.H., and Clifford, S.F. 1976. On the attenuation of sound by turbulence. *J. Acoust. Soc. Am.* **60**: 788–94.
Cave, C.J.P. 1919. The audibility of thunder. *Nature (London)* **104**: 132.
Christian, H., Holmes, C.R., Bullock, J.W., Gaskell, W., Illingworth, A.J., and Latham, J. 1980. Airborne and ground based studies of thunderstorms in the vicinity of Langmuir Laboratory. *Q. J. Roy. Meteor. Soc.* **106**: 159–74.
Colgate, S.A., and McKee, C. 1969. Electrostatic sound in clouds and lightning. *J. Geophys. Res.* **74**: 5379–89.
Davis, R., and Standring, W.G. 1947. Discharge currents associated with kite balloons. *Proc. Roy. Soc. A* **191**: 304–22.
Dawson, G.A., Richards, C.N., Krider, E.P., and Uman, M.A. 1968a. The acoustic output of a long spark. *J. Geophys. Res.* **73**: 815–16.
Dawson, G.A., Uman, M.A., and Orville, R.E. 1968b. Discussion of paper by E.L. Hill, and J.D. Robb "Pressure pulse from a lightning stroke." *J. Geophys. Res.* **73**: 6595–7.
De L'Isle, J.N. 1738. Memoires pour Servir a l'Histoire et au Progres de l'Astronomie de la Geographie et de la Physique. *L'Imprimerie de l'Academie des Sciences*, St Petersburg.
Depasse, P. 1994. Lightning acoustic signature. *J. Geophys. Res.* **99**: 25 933–40.
Dessler, J. 1973. Infrasonic thunder. *J. Geophys. Res.* **78**: 1889–96.
Drabkina, S.I. 1951. The theory of the development of the spark channel. *J. Exper. Theoret. Phys.* **21**: 473–83. (English translation, AERE LIB/Trans. 621, Harwell, Berkshire, UK.)
Evans, W.H., and Walker, R.L. 1963. High speed photographs of lightning at close range. *J. Geophys. Res.* **68**: 4455–61.
Few, A.A. 1968. Thunder, Ph.D. dissertation, Rise University, Houston, Texas.
Few, A.A. 1969a. Reply to letter by W.J. Remillard. *J. Geophys. Res.* **74**: 5556.
Few, A.A. 1969b. Power spectrum of thunder. *J. Geophys. Res.* **74**: 6926–34.
Few, A.A. 1970. Lightning channel reconstruction from thunder measurements. *J. Geophys. Res.* **75**: 7517–23.
Few, A.A. 1974. Thunder signatures. *Trans. AGU* **55**: 508–14.
Few, A.A. 1975. Thunder. *Scientific American* **233**: 80–90.
Few, A.A. 1982. Acoustic radiations from lightning. In *Handbook of Atmospherics*, ed. H. Volland, vol. II., pp. 257–290, Boca Raton, Florida: CRC Press.
Few, A.A. 1985. The production of lightning-associated infrasonic acoustic sources in thunderclouds. *J. Geophys. Res.* **90**: 6175–80.
Few, A.A. 1995. Acoustic radiations from lightning. In *Handbook of Atmospheric Electrodynamics*, ed. H. Volland, vol. II, pp. 1–31, Boca Raton, Florida: CRC Press.
Few, A.A., and Teer, T.L. 1974. The accuracy of acoustic reconstructions of lightning channels. *J. Geophys. Res.* **79**: 5007–11.

Few, A.A., Dessler, A.J., Latham, D.J., and Brook, M. 1967. A dominant 200–hertz peak in the acoustic spectrum of thunder. *J. Geophys. Res.* **72**: 6149–54.

Few, A.A., Garrett, H.B., Uman, M.A., and Salanave, L.E. 1970. Comments on letter by W.W. Troutman "Numerical calculation of the pressure pulse from a lightning stroke". *J. Geophys. Res.* **75**: 4192–5.

Fleagle, R.G. 1949. The audibility of thunder. *J. Acoust. Soc. Am.* **21**: 411–12.

Fleagle, R.G., and Businger, J.A. 1963. *An Introduction to Atmospheric Physics*, New York: Academic Press.

Georges, T.M. 1973. Infrasound from convective storms: examining the evidence. *Rev. Geophys. Space Phys.* **11**: 571–94.

Georges, T.M., and Beasley, W.H. 1977. Refraction of infrasound by upper-atmospheric winds. *J. Acoust. Soc. Am.* **61**: 28–34.

Goyer, G.G., and Plooster, M.N. 1968. On the role of shock waves and adiabatic cooling in the nucleation of ice crystals by the lightning discharge. *J. Atmos. Sci.* **25**: 857–62.

Graneau, P. 1989. The cause of thunder. *J. Phys. D: Appl. Phys.* **22**: 1083–94.

Harris, C.M. 1967. Absorption of sound in air versus humidity and temperature. NASA Report CR-647, Columbia University, New York.

Higham, J.B., and Meek, J.M. 1950. The expansion of gaseous spark channels. *Proc. Phys. Soc. (London) B* **63**: 649–61.

Hill, E.L., and Robb, J.D. 1968. Pressure pulse from a lightning stroke. *J. Geophys. Res.* **73**: 1883–8.

Hill, R.D. 1968. Analysis of irregular paths of lightning channels. *J. Geophys. Res.* **73**: 1897–905.

Hill, R.D. 1971. Channel heating in return stroke lightning. *J. Geophys. Res.* **76**: 637–45.

Hill, R.D. 1975. Comments on "Quantitative analysis of a lightning return stroke for diameter and luminosity changes as a function of space and time" by R.E. Orville, J.H. Helsdon Jr, and W.H. Evans. *J. Geophys. Res.* **80**: 1188.

Hill, R.D. 1977a. Energy dissipation in lightning. *J. Geophys. Res.* **82**: 4967–8.

Hill, R.D. 1977b. Comments on "Numerical simulation of spark discharges in air" by M.N. Plooster. *Phys. Fluids* **20**: 1584–6.

Hill, R.D. 1977c. Thunder. In *Lightning, vol. I, Physics of Lightning*, ed. R.H. Golde, pp. 385–408, New York: Academic Press.

Hill, R.D. 1979. A survey of lightning energy estimates. *Rev. Geophys. Space Phys.* **17**: 155–64.

Hill, R.D. 1985. Investigation of lightning strikes to water surface. *J. Acoust. Soc. Am.* **78**: 2096–9.

Hirn, M. 1888. The sound of thunder. *Sci. Am.* **59**: 201.

Holmes, C.R., Brook, M., Krehbiel, P., and McCrory, R. 1971a. On the power spectrum and mechanism of thunder. *J. Geophys. Res.* **76**: 2106–15.

Holmes, C.R., Brook, M., Krehbiel, P., and McCrory, R. 1971b. Reply to comments by Bhartendu "On the power spectrum and mechanisms of thunder". *J. Geophys. Res.* **76**: 7443.

Holmes, C.R., Szymanski, E.W., Szymanski, S.J., and Moore, C.B. 1980. Radar and acoustic study of lightning. *J. Geophys. Res.* **85**: 7517–32.

Hubert, P. 1985. Personal communication.

Idone, V.P. 1995. Microscale tortuosity and its variation as observed in triggered lightning channels. *J. Geophys. Res.* **100**: 22 943–56.

Jones, D.L. 1968a. Intermediate strength blast wave. *Phys. Fluids* **11**: 1664–7.

Jones, D.L. 1968b. Comments on paper by A.A. Few, A.J. Dessler, D.J. Latham, and M. Brook, "A dominant 200-hertz peak in the acoustic spectrum of thunder". *J. Geophys. Res.* **73**: 4776–7.

Jones, D.L., Goyer, G.G., and Plooster, M.N. 1968. Shock wave from a lightning discharge. *J. Geophys. Res.* **73**: 3121–7.

Kappus, M.E., and Vernon, F.L. 1991. Acoustic signature of thunder from seismic records. *J. Geophys. Res.* **96**: 10 989–11 006.

Kinney, G.F. 1962. *Explosive Shocks in Air*. 198 pp., New York: Macmillan.

Kitagawa, N. 1965. Discussion of thunder. In *Problems of Atmospheric and Space Electricity*, ed. S.C. Coroniti, pp. 350–1, New York: Elsevier.

Krehbiel, P.R., Brook, M., and McCrory, R. 1979. An analysis of the charge structure of lightning discharges to the ground. *J. Geophys. Res.* **84**: 2432–56.

Krider, E.P., Dawson, G.A., and Uman, M.A. 1968. The peak power and energy dissipation in a single-stroke lightning flash. *J. Geophys. Res.* **73**: 3335–9.

Latham, D.J. 1964. A study of thunder from close lightning discharges. M.S. thesis, Physics Department, New Mexico Institute of Min. and Technol., Socorro.

Lin, S.C. 1954. Cylindrical shock waves produced by an instantaneous energy release. *J. Appl. Phys.* **25**: 54–7.

Lucretius, T. (98–55 B.C.). *On the Nature of Things*, Book VI, H.A.J. Munro, Trans., *Great Books of the Western World*, p. 81, Chicago: William Benton.

Lyon, J.A. 1903. The cause of thunder again. *Sci. Am.* **88**: 191.

MacGorman, D.R. 1977. Lightning location in a Colorado thunderstorm. M.S. thesis, Rice University, Houston, Texas.

MacGorman, D.R. 1978. Lightning location in a storm with strong wind shear. Ph.D. dissertation, Department of Space Physics and Astronomy, Rice University, Houston, Texas.

MacGorman, D.R., Few, A.A., and Teer, T.L. 1981. Layered lightning activity. *J. Geophys. Res.* **86**: 9900–10.

Malan, D.J. 1963. *Physics of Lightning*. 176 pp., London: The English Universities Press.

Malan, D.J., and Collens, H. 1937. Progressive lightning III – the fine structure of return lightning stokes. *Proc. Roy. Soc.* **A162**: 175–203.

McCrory, R.A. 1971. Thunder and its relationship to the structure of lightning. Ph.D. dissertation, New Mexico Institute of Mining and Technology, Socorro.

McCrory, R.A., and Holmes, C.R. 1968. Comment on paper by Bhartendu "A study of atmospheric pressure variations from lightning discharges". *Can. J. Phys.* **46**: 2333–4.

McEachron, K.B., 1939. Lightning to the Empire State Building, *J. Franklin Inst.* **227**: 149–217.

McGehee, R.M. 1964. The influence of thunderstorm space charges on pressure. *J. Geophys. Res.* **69**: 1033–5.

Mershon, R.S. 1870. A theory of thunder. *Sci. Am.* **23**: 68–9.

Nakano, M. 1973. Lightning channel determined by thunder. *Proc. Res. Inst. Atmospherics, Nagoya Univ., Japan* **20**: 1–9.

Nakano, M. 1976. Characteristics of lightning channel in thunderclouds determined by thunder. *J. Meteor. Soc. Japan* **54**: 441–7.

Nakano, M., and Takeuti, T. 1970. On the spectrum of thunder. *Proc. Res. Inst. Atmospherics, Nagoya Univ., Japan* **17**: 111–13.

Newman, M.M., Stahmann, J.R., and Robb, J.D. 1967a. Experimental study of triggered natural lightning discharges. Report DS-67-3, Project 520–002–03X, Federal Aviation Agency, Washington, DC.

Newman, M.M., Stahmann, J.R., Robb, J.D., Lewis, E.A., Martin, S.G., and Zinn, S.V. 1967b. Triggered lightning strokes at very close range. *J. Geophys. Res.* **72**: 4761–4.

Orville, R.E. 1968a. A high-speed time-resolved spectroscopic study of the lightning return stroke: Part II, A quantitative analysis. *J. Atmos. Sci.* **25**: 839–51.

Orville, R.E. 1968b. A high speed time-resolved spectroscopic study of the lightning return stroke: Part III, A time-dependent model. *J. Atmos. Sci.* **25**: 852–6.

Orville, R.E., Uman, M.A., and Sletten, A.M. 1967. Temperature and electron density in long air sparks. *J. Appl. Phys.* **38**: 895–6.

Otterman, J. 1959. Finite-amplitude propagation effect on shock-wave travel times from explosions at high altitudes. *J. Acoust. Soc. Am.* **31**: 470–4.

Page, D.E. 1944. Distance to which thunder can be heard. *Bull. Am. Meteor. Soc.* **25**: 366.

Pain, H.J., and Rogers, W.E. 1962. Shock waves in gases. *Rep. Progr. Phys.* **25**: 287–336.

Plooster, M.N. 1968. Shock waves from line sources. Technical Report NCAR-TN-37, National Center for Atmospheric Research, Boulder, Colorado.

Plooster, M.N. 1970a. Erratum: shock waves from line sources. Numerical solutions and experimental measurements. *Phys. Fluids* **13**: 2248.

Plooster, M.N. 1970b. Shock waves from line sources. Numerical solutions and experimental measurements. *Phys. Fluids* **13**: 2665–75.

Plooster, M.N. 1971a. Numerical simulation of spark discharges in air. *Phys. Fluids* **14**: 2111–23.

Plooster, M.N. 1971b. Numerical model of the return stroke of the lightning discharge. *Phys. Fluids* **14**: 2124–33.

Plooster, M.N. 1972. On freezing of supercooled droplets shattered by shock waves. *J. Appl. Meteor.* **11**: 161–5.

Proctor, D.E., Uytenbogaardt, R. and Meredith, B.M. 1988. VHF radio pictures of lightning flashes to ground. *J. Geophys. Res.* **93**: 12 683–727.

Rakov, V.A., and Uman, M.A. 1990. Some properties of negative cloud-to-ground lightning flashes versus stroke order. *J. Geophys. Res.* **95**: 5447–53.

Rakov, V.A., Uman, M.A., Jordan, D.M., and Priore, C.A. III 1990. Ratio of leader to return stroke electric field change for first and subsequent lightning strokes. *J. Geophys. Res.* **95**: 16 579–87.

Rakov, V.A., Uman, M.A., and Thottappillil, R. 1994. Review of lightning properties determined from electric field and TV observations. *J. Geophys. Res.* **99**: 10 745–50.

Rakov, V.A., Crawford, D.E., Rambo, K.J., Schnetzer, G.H., Uman, M.A., and Thottappillil, R. 2001. M-component mode of charge transfer to ground in lightning discharges, *J. Geophys. Res.* **106**, 22 817–31.

Reed, J.W. 1972. Airblast overpressure decay at long ranges. *J. Geophys. Res.* **77**: 1623–9.

Remillard, W.J. 1960. The acoustics of thunder. Technical Memorandum 44, Acoustics Research Laboratory, Division of Engineering and Applied Physics, Harvard University, Cambridge, Massachusetts.

Remillard, W.J. 1969. Comments on paper by A.A. Few, A.J. Dessler, D. J. Latham, and M. Brook "A dominant 200–hertz peak in the acoustic spectrum of thunder". *J. Geophys. Res.* **74**: 5555.

Remillard, W.J. 1976. Pressure disturbances from a finite cylindrical source. *J. Acoust. Soc. Am.* **59**: 744–8.

Reynolds, R.V. 1903. The cause of thunder. *Sci. Am.* **88**: 44.

Ribner, H.S., and Roy, D. 1982. Acoustics of thunder: a quasilinear model for tortuous lightning. *J. Acoust. Soc. Am.* **72(B)**: 1911–25.

Ribner, H.S., Lam, F., Leung, K.A., Kurtz, D., and Ellis, N.D. 1976. Computer model of the lightning–thunder process with audible demonstration. *Progress in Astronautics and Aeronautics* **46**: 77–87.

Rouse, C.A. 1959. Theoretical analysis of the hydrodynamic flow in exploding wire phenomena. In *Exploding Wires*, eds. W.G. Chace and H.K. Moore, New York: Plenum.

Sakurai, A. 1953. On the propagation and structure of the blast wave (1). *J. Phys. Soc. Japan* **8**: 662–9.

Sakurai, A. 1954. On the propagation and structure of the blast wave (2). *J. Phys. Soc. Japan* **9**: 256–66.

Sakurai, A. 1955a. On exact solution of the blast wave problem. *J. Phys. Soc., Japan* **10**: 827–8.

Sakurai, A. 1955b. Decrement of blast wave. *J. Phys. Soc., Japan* **10**: 1018.

Sakurai, A. 1959. On the propagation of cylindrical shock waves. In *Exploding Wires*, eds. W.G. Chace and H.K. Moore, New York: Plenum.

Schmidt, W. 1914. Über den Donner. *Meteorol. Z.* **31**: 487–98.

Schonland, B.F.J., Malan, D.J., and Collens, H. 1935. Progressive lightning II. *Proc. Roy. Soc. A* **152**: 595–625.

Shao, X.M., Krehbiel, P.R., Thomas, R.J., and Rison, W. 1995. Radio interferometric observations of cloud-to-ground lightning phenomena in Florida. *J. Geophys. Res.* **100**: 2749–83.

Swigart, R.J. 1960. Third order blast wave theory and its application to hypersonic flow past blunt-nosed cylinders. *J. Fluid Mech.* **9**: 613–20.

Taljaard, J.J. 1952. How far can thunder be heard? *Weather* **7**: 245–6.

Taylor, G.I. 1950. The formation of a blast wave by a very intense explosion, 2, the atomic explosion of 1945. *Proc. Roy. Soc. A* **201**: 175–86.

Teer, T.L. 1972. Acoustic profiling: a technique for lightning channel reconstruction, M.S. thesis, Rice University, Houston, Texas.

Teer, T.L. 1973. Lightning channel structure inside an Arizona thunderstorm, Ph.D. dissertation, Rice University, Department of Space Physics and Astronomy, Houston, Texas.

Teer, T.L., and Few, A.A. 1974. Horizontal lightning. *J. Geophys. Res.* **79**: 3436–41.

Temkin, S. 1977. A model for thunder based on heat addition. *J. Sound and Vibration*, **52**: 401–14.

Thomson, E.M. 1980. Characteristics of Port Moresby ground flashes. *J. Geophys. Res.* **85**: 1027–36.

Troutman, W.S. 1969. Numerical calculation of the pressure pulse from a lightning stroke. *J. Geophys. Res.* **74**: 4595–6.

Troutman, W.S. 1970. Reply to comments by A.A. Few, H.B. Garrett, M.A., Uman, and L.E. Salanave on "Numerical calculations of the pressure pulse from a lightning stroke". *J. Geophys. Res.* **75**: 4196.

Uman, M.A. 1969. *Lightning*, 264 pp., New York: McGraw-Hill.

Uman, M.A. 1984. *Lightning*, 298 pp., New York: Dover.

Uman, M.A. 1987. *The Lightning Discharge*, 377 pp., San Diego: Academic Press.

Uman, M.A. 2001. *The Lightning Discharge*, 377 pp., Mineola, New York: Dover.

Uman, M.A., McLain, D.K., and Myers, F. 1968. Sound from line sources with application to thunder. Westinghouse Res. Lab. Report 68-9E4-HIVOL-R1.

Uman, M.A., Cookson, A.H., and Moreland, J.B. 1970. Shock wave from a four-meter spark. *J. Appl. Phys.* **41**: 3148–55.

Uman, M.A., McLain, D.K., and Krider, E.P. 1975. The electromagnetic radiation from a finite antenna. *Am. J. Phys.* **43**: 33–8.

Uman, M.A., Beasley, W.H., Tiller, J.A., Lin, Y.T., Krider, E.P., Weidman, C.D., Krehbiel, P.R., Brook, M., Few, A.A., Bohannon, J.L., Lennon, C.L., Poehler, H.A., Jafferis, W., Gulick, J.R., and Nicholson, J.R. 1978. An unusual lightning flash at the Kennedy Space Center. *Science* **201**: 9–16.

Veenema, L.C. 1917. Die Hörweite des Gewitterdonners. *Z. Angew. Meteorol., Wetter* 127–30, 187–92, 258–62.

Veenema, L.C. 1918. Die Hörweite des Gewitterdonner. *Z. Angew. Meteorol., Wetter* 56–68.

Veenema, L.C. 1920. The audibility of thunder. *Mon. Wea. Rev.* **48**: 162.

Weber, M.E., Christian, H.J., Few, A.A., and Stewart, M.F. 1982. A thundercloud electric field sounding: charge distribution and lightning. *J. Geophys. Res.* **87**: 7158–69.

Whitham, G.B. 1956. On the propagation of weak shock waves. *J. Fluid Mech.* **1**: 290–318.

Wilson, C.T.R. 1920. Investigations on lightning discharges and on the electric field of thunderstorms. *Phil. Trans. Roy. Soc. A* **221**: 73–115.

Winn, W.P., Moore, C.B., Holmes, C.R., and Byerley, L.G. 1978. Thunderstorm on July 16, 1975, over Langmuir Laboratory: a case study. *J. Geophys. Res.* **83**: 3079–92.

Wright, W.M. 1964. Experimental study of acoustical N waves. *J. Acoust. Soc. Am.* **36**: 1032.

Wright, W.M., and Medendorp, N.W. 1967. Acoustic radiation from a finite line source with N-wave excitation. *J. Acoust. Soc. Amer.* **43**: 966–71.

Zhivlyuk, Yu., and Mandel'shtam, S.L. 1961. On the temperature of lightning and force of thunder. *Soviet Phys. JEPT* (English transl.) **13**: 338–40.

12 Modeling of lightning processes

> Models are conceptual constructs that can be used to make predictions about the outcomes of measurements.
> D.A. Randall and B.A. Wielicki (1997)

12.1. Introduction

Any lightning model is a mathematical construct designed to reproduce certain aspects of the physical processes involved in the lightning discharge. No modeling is complete until the model predictions are compared with experimental data; that is, model testing, often called validation, is a necessary component of any modeling. In the following, we will present various models of the different lightning processes, some models being applicable to more than one lightning process.

12.2. Return stroke

12.2.1. General overview

We define four classes of lightning return-stroke models. Most published models can be assigned to one, or sometimes two, of these four classes. The classes are primarily distinguished by the type of governing equations.

(i) The first class of models comprises the gas dynamic or "physical" models, which are primarily concerned with the radial evolution of a short segment of the lightning channel and its associated shock wave. These models typically involve the solution of three gas dynamic equations (sometimes called hydrodynamic equations), representing the conservation of mass, of momentum, and of energy, coupled to two equations of state. Principal model outputs include temperature, pressure, and mass density as a function of the radial coordinate and time.

(ii) The second class of models comprises the electromagnetic models, which are usually based on a lossy, thin-wire antenna approximation to the lightning channel. These models involve a numerical solution of Maxwell's equations to find the current distribution along the channel from which the remote electric and magnetic fields can be computed.

(iii) The third class of models comprises the distributed-circuit models, which can be viewed as an approximation to the electromagnetic models and which represent the lightning discharge as a transient process on a vertical transmission line characterized by resistance (R), inductance (L), and capacitance (C), all per unit length. The distributed-circuit models (also called RLC transmission-line models) are used to determine the channel current versus time and height and can therefore be used also for the computation of remote electric and magnetic fields. Some distributed-circuit models incorporate a gas dynamic model, the latter being used to find R as a function of time.

(iv) The fourth class of models comprises the engineering models, in which a specification of the spatial and temporal distribution of the channel current (or the channel line charge density) is made based on such observed lightning return-stroke characteristics as the current at the channel base, the speed of the upward-propagating front, and the channel luminosity profile. In these models, the physics of the lightning return stroke is deliberately downplayed, and the emphasis is placed on achieving agreement between the model-predicted electromagnetic fields and those observed at distances from tens of meters to hundreds of kilometers. A characteristic feature of the engineering models is their small number of adjustable parameters, usually only one or two besides the specified channel-base current.

Outputs of the electromagnetic, distributed-circuit, and engineering models can be used directly for the computation of electromagnetic fields, while the gas dynamic models can be used for finding R as a function of time, R being one of the parameters of the electromagnetic and distributed-circuit models. Since the distributed-circuit and engineering models generally do not consider lightning channel branches, they best describe subsequent strokes or first strokes before the first major branch has been reached by the upward-propagating return stroke, a time that is usually longer than the time required for the formation of the initial current peak at ground. If not otherwise specified, we will assume that the lightning channel is straight and vertical and has no branches. Channel tortuosity and branches are discussed in relation to the calculation of electric and magnetic fields in the third part of subsection 4.6.4 and propagation effects in the fourth part of subsection 4.6.4.

12.2. Return stroke

We will attempt to maintain a balance between completeness and an emphasis on the primary features of modern lightning return-stroke modeling. Thus, we will not consider several engineering models found in the older literature that are trivial special cases of the major models reviewed here (e.g., the Lundholm 1957 or Norinder and Dahle 1945 models). Further, we will not consider the Master–Uman–Lin–Standler (MULS) model (Master *et al.* 1981), since Rachidi and Nucci (1990) demonstrated that the modified transmission line model with exponential current decay with height (the MTLE model), which we do discuss here, is essentially a more conveniently formulated equivalent of the MULS model. At the same time, we shall consider brief descriptions of several recently published models, most of them rather cumbersome to use, for the purpose of reflecting the scope of current efforts in lightning return-stroke modeling. These include generalizations of the DU and TCS models (subsection 12.2.5), Cooray's model (subsection 12.2.5), and the electromagnetic model of Borovsky (subsection 12.2.3). Further information on lightning return-stroke models is found in Nucci *et al.* (1990), Thottappillil and Uman (1993), Thottappillil *et al.* (1997), Rakov and Uman (1998), Gomes and Cooray (2000), and in references given in these papers.

12.2.2. Gas dynamic models

Gas dynamic models describe the behavior of a short segment of a cylindrical plasma column driven by the resistive heating caused by a specified time-varying current. Some models of this type were developed to describe laboratory spark discharges in air but have been used for, or thought to be applicable to, the lightning return stroke (e.g., Drabkina 1951; Braginskii 1958; Plooster 1970, 1971a, b).

Drabkina (1951), assuming the spark-channel pressure to be much greater than the ambient pressure, the so-called strong-shock approximation, described the radial evolution of a spark channel and its associated shock wave as a function of the time-dependent energy injected into the channel. Braginskii (1958) also used the strong-shock approximation to develop a spark channel model giving the time variation of such parameters as radius, temperature, and pressure as a function of the input current. For a current $I(t)$ linearly increasing with time t, he obtained the following expression for channel radius $r(t)$ as presented by Plooster (1971b): $r(t) \approx 9.35 I(t)]^{1/3} t^{1/2}$, where $r(t)$ is in centimeters, $I(t)$ in amperes, and t in seconds. In the derivation of this expression, presumably applicable to the early stages of the discharge, Braginskii (1958) set the electrical conductivity σ of the channel at 2.22×10^4 S m^{-1} and assumed the ambient air density to be 1.29×10^{-3} g cm^{-3}. For a known $r(t)$, the resistance per unit channel length can be found as

$$R(t) = [\sigma \pi r^2(t)]^{-1}$$

and the energy input per unit length as

$$W(t) = \int_0^t I^2(\tau) R(\tau) \, d\tau$$

More recent "physical" modeling algorithms, published by Hill R.D. (1971, 1977a), Plooster (1970, 1971a, b), Strawe (1979), Paxton *et al.* (1986, 1990), Bizjaev *et al.* (1990), and Dubovoy *et al.* (1991a, b, 1995), can be briefly outlined as follows. It is assumed that (i) the plasma column is straight and cylindrically symmetrical, (ii) the algebraic sum of the positive and negative charges in any volume element is zero, and (iii) local thermodynamic equilibrium exists at all times. Initial conditions that are meant to characterize the channel created by the lightning leader include its temperature (of order 10 000 K), channel radius (of order 1 mm), and either pressure equal to ambient (1 atm) or mass density equal to ambient (of order 10^{-3} g cm^{-3}), the latter two conditions representing, respectively, the older and the newly created channel sections. The initial condition assuming pressure equal to ambient probably best represents the upper part of the leader channel, since that part has had sufficient time to expand and attain equilibrium with the surrounding atmosphere; the initial condition assuming mass density equal to ambient is more suitable for the recently created, bottom part of the leader channel. In the latter case, variations in the initial channel radius and the initial temperature are claimed to have little influence on the model predictions (e.g., Plooster 1971b; Dubovoy *et al.* 1995). The input current is assumed to rise to about 20 kA in some microseconds and then decay in some tens of microseconds.

At each time step, (i) the electrical energy sources and (ii) the radiation energy sources, and sometimes (iii) the Lorentz force (Dubovoy *et al.* 1991a, b, 1995) are computed (these three quantities are discussed in more detail below), and the gas dynamic equations are solved numerically for the thermodynamic and flow parameters of the plasma. The exact form of the gas dynamic equations and the set of variables for which the equations are solved differ from one study to another. Plooster (1970, 1971a, b), for example, used five equations, comprising equations for the conservation of mass, momentum, and energy, a definition for the radial gas velocity, and an equation of state for the gas. These equations were solved for the following five variables: radial coordinate, radial velocity, pressure, mass density, and internal energy per unit mass.

Electrical energy sources. The electrical energy deposited into the channel is determined in the following manner. The plasma column is divided into a set of concentric annular zones, in each of which the gas properties are assumed to be constant. For a known temperature and mass density, tables of the computed properties of air in thermodynamic equilibrium (Hill 1971; Dubovoy *et al.*

1991a, b, 1995) or the Saha equation directly (Plooster 1970, 1971a, b; Paxton *et al.* 1986, 1990) provide the plasma composition. Given the plasma composition, temperature, and mass density, one can compute the plasma conductivity for each of the annular zones. The total input current is apportioned to all the annular zones as if they were an array of resistors connected in parallel. Using this cross-sectional distribution of current and plasma conductivity, one finds the electrical energy input (the joule heating) in each of the annular zones. Most of this energy is thought to be spent in radiation, ionization, and expansion of the channel (Paxton *et al.* 1986, 1990).

Radiation energy sources. The electrical energy deposited in the channel in the form of heat is then transported from the hot conducting gas in the inner part of the channel to the cooler gas in the outer part of the channel. Radiation is the dominant mechanism of energy redistribution at temperatures above 10 000 K or so, thermal conduction is usually neglected (Paxton *et al.* 1986, 1990). The radiative properties of air are complex functions of frequency and temperature. Radiation of a given frequency can be absorbed and reradiated a number of times in traversing the channel in the outward direction. Photons with wavelengths of 1000–2000 Å or shorter (e.g., in the UV) are absorbed at the edge of the hot channel and contribute to enlarging the plasma column, while longer-wavelength photons (mostly having optical wavelengths) eventually escape from the system. Paxton *et al.* (1986, 1990) and Dubovoy *et al.* (1991a, b, 1995) used tables of the radiative properties of hot air to determine absorption coefficients (opacities) as a function of temperature for a number of selected frequency (wavelength) intervals (for example, 10 intervals were used by Dubovoy *et al.* 1991a, b, 1995) and so solved the equation of radiative energy transfer in the diffusion approximation. Less detailed radiative transport algorithms have also been employed (Hill R.D. 1971; Plooster 1971a, b; Strawe 1979).

Lorentz force. Dubovoy *et al.* (1991a, b, 1995) included in their model the pinch effect due to the interaction of a current with its own magnetic field. They computed the Lorentz force, which counteracts the channel's gas dynamic expansion, and included this force in the momentum and energy conservation equations. Inclusion of the interaction of the channel current with its own magnetic field is claimed to result in a 10 to 20 percent increase in the input energy for the same input current.

Perhaps the most advanced and complete gas dynamic model to date is that of Paxton *et al.* (1986, 1990). Their results for the temperature, mass density, pressure, and electrical conductivity versus radial coordinate at different instants of time are shown in Figs. 12.1 through 12.4. Return-stroke input-energy estimates predicted by various gas dynamic models as well as an estimate of Krider *et al.* (1968) (derived from a comparison of the

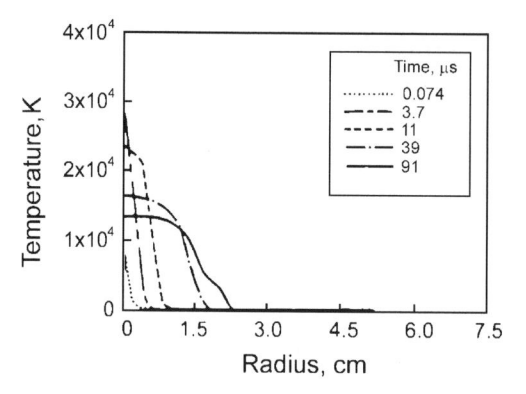

Fig. 12.1. Temperature versus radius (radial distance from the channel axis) at five instants of time ranging from 0.074 to 91 μs, as predicted by the gas dynamic model of Paxton *et al.* (1986, 1990) for an input current linearly rising to 20 kA in 5 μs and thereafter exponentially decaying with a time constant of 50 μs. Adapted from Paxton *et al.* (1986, 1990). According to Paxton *et al.* (1987), the profile at 3.7 μs should be interpreted as having a constant value equal to that at the channel axis out to a radius of 0.36 cm.

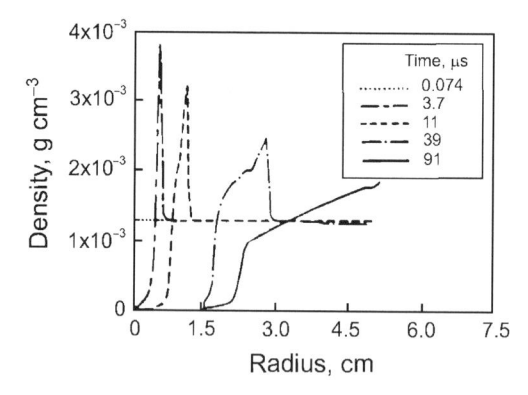

Fig. 12.2. Same as Fig. 12.1 but for the mass density profiles.

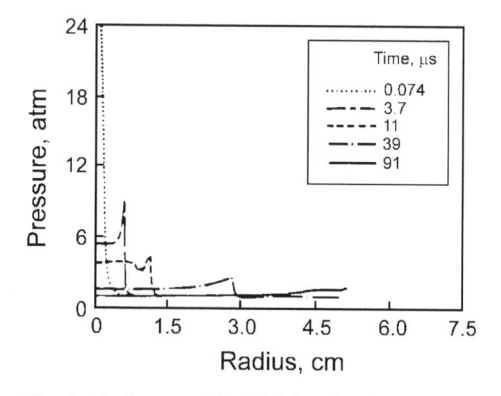

Fig. 12.3. Same as Fig. 12.1 but for the pressure profiles.

optical radiation produced by lightning with that of a laboratory spark of known input energy) and estimates based on the electrostatic considerations of Uman (1987, 2001) and Borovsky (1998) are summarized in Table 12.1.

12.2. Return stroke

Table 12.1. *Lightning energy estimates*

Source	Current peak, kA	Input energy, J m^{-1}	Percentage converted to kinetic energy	Percentage of energy radiated	Comments
Hill (1971, 1977a)	21	1.5×10^4 ($\sim 3 \times 10^3$)	9 (at 25 μs)	$\sim 2^a$ (at 25 μs)	Underestimation of electrical conductivity resulted in overestimation by a factor of 5 or so of input energy. Corrected value is given in parentheses
Plooster (1971b)	20	2.4×10^3	4 (at 35 μs)	~ 50 (at 35 μs)	Crude radiative transport mechanism adjusted to the expected temperature profile
Paxton *et al.* (1986, 1990)	20	4×10^3	2 (at 64 μs)	69 (at 64 μs)	Individual temperature-dependent opacities for several wavelength intervals
Dubovoy *et al.* (1991a, b, 1995)	20	3×10^3	—	25 (at about 55 μs)	Individual temperature-dependent opacities for 10 wavelength intervals. Magnetic pinch effect is taken into account
Borovsky (1998)	—	$2 \times 10^2 - 1 \times 10^4$	—	—	Electrostatic energy stored on a vertical channel assuming a line charge density of 100–500 μC m^{-1}
Krider *et al.* (1968)	Single-stroke flash	2.3×10^5	—	0.38^b	Measured optical energy is converted to the total energy using energy ratios observed in laboratory spark experiments
Uman (1987, 2001)	—	$(1-10) \times 10^5$	—	—	From electrostatic considerations (lowering 5 C of charge from a height of 5 km to ground, assuming a potential difference of 10^8-10^9 V between the Earth and the charge center)

^{a} Estimated by subtraction of the internal and kinetic energies from the input energy shown in Fig. 1 of Hill (1977a), probably incorrect as well as the percent of energy radiated from the channel due to a factor of 20–30 error in electrical conductivity.
^{b} Only radiation in the wavelength region from 0.4 to 1.1 μm.

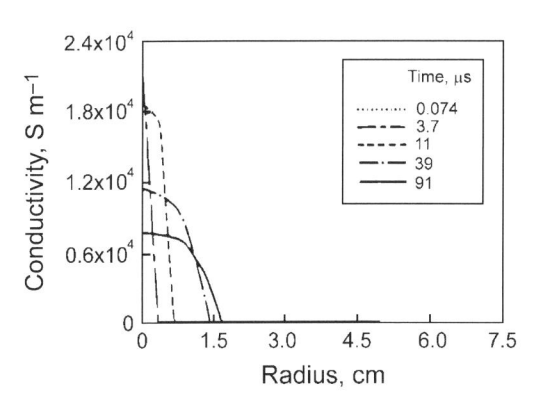

Fig. 12.4. Same as Fig. 12.1 but for the electrical conductivity profiles.

Additionally given for some models are the percentage of the input energy converted to kinetic energy of gas motion (shock-wave and conducting-channel expansion) and the percentage of the energy radiated from the channel.

As noted earlier, the gas dynamic models do not consider the longitudinal evolution of the lightning channel. Usually, they also ignore the electromagnetic skin effect, which Plooster (1971a) found to be negligible, the corona sheath, which presumably contains the bulk of the leader charge, and any heating of the air surrounding the current-carrying channel by the preceding lightning processes. An attempt to include the previous heating of the surrounding air in a gas dynamic model was made by Bizjaev *et al.* (1990). Aleksandrov *et al.* (2000) used a gas dynamic model to study both the return stroke and the following continuing current (Section 4.8).

12.2.3. Electromagnetic models

Electromagnetic return-stroke models based on representation of the lightning channel as a lossy antenna have been proposed by Podgorski and Landt (1987), Moini et al. (1997, 2000), and Baba and Ishii (2001). These models involve a numerical solution of Maxwell's equations using the method of moments (MOM) (e.g., Sadiku 1994), which yields a complete solution for the channel current, including both the antenna-mode current and the transmission-line-mode current (e.g., Paul 1994). The resistive loading used by Podgorski and Landt was $0.7\ \Omega\ m^{-1}$ and that used by Moini et al. was 0.065 or $0.07\ \Omega\ m^{-1}$. Baba and Ishii used both resistive ($1\ \Omega\ m^{-1}$) and inductive ($3\ \mu H\ m^{-1}$) loading. In order to simulate the effect on the return-stroke speed of the radially formed corona surrounding the current-carrying channel core and presumably containing the bulk of the channel charge, Moini et al. set the permittivity ε of the air surrounding the equivalent antenna to a value greater than ε_0 for computation of the current distribution along the antenna. As a result, even without resistive loading the phase velocity $v_p = (\mu_0 \varepsilon)^{-1/2}$ of the electromagnetic wave guided by the antenna was reduced with respect to the velocity of light $c = (\mu_0 \varepsilon_0)^{-1/2}$. The resistive loading should further reduce v_p. For $\varepsilon = 5.3\varepsilon_0$ and $R = 0.07\ \Omega\ m^{-1}$ Moini et al. (2000) obtained $v_p = 1.3 \times 10^8\ m\ s^{-1}$. The current distribution computed assuming that the surrounding air had permittivity ε and that the antenna was resistively loaded was then allowed to radiate electromagnetic fields into free space characterized by $\varepsilon = \varepsilon_0$, $\mu = \mu_0$. In the model of Baba and Ishii (2001), the combined resistive and inductive loading resulted in $v_p = 1.5 \times 10^8\ m\ s^{-1}$. The models of Moini et al. and Baba and Ishii consider a straight vertical channel and ignore any nonlinear effects, while the model of Podgorski and Landt (1987) deals with a three-dimensional channel of arbitrary shape and, reportedly, can include branches, the strike object, the upward connecting discharge, and nonlinear effects during the attachment process.

Borovsky (1995) used Maxwell's equations to describe both dart-leader and return-stroke processes as guided waves propagating along conducting cylindrical channels. The resistance per unit length of the channel guiding the return-stroke wave was assumed to be $16\ \Omega\ m^{-1}$. No current distribution along the channel was calculated since the dart leader and return stroke were each represented by a single, dominant sinusoid and only a middle section of the lightning channel, undisturbed by the conditions at the channel ends, was considered.

12.2.4. Distributed-circuit models

Distributed-circuit models describe the lightning channel as an RLC transmission line for which the voltage V and current I are solutions of the telegrapher's equations:

$$-\frac{\partial V(z', t)}{\partial z'} = L\frac{\partial I(z', t)}{\partial t} + RI(z', t) \quad (12.1)$$

$$-\frac{\partial I(z', t)}{\partial z'} = C\frac{\partial V(z', t)}{\partial t} \quad (12.2)$$

where R, L, and C are, respectively, the series resistance, series inductance, and shunt capacitance, all per unit length, z' is the vertical coordinate specifying the position on the lightning channel, and t is the time. For a vertical lightning channel for which the return path for current is the vertical channel image (assuming a perfectly conducting ground), L and C are each a function of z'. However, the dependence is weak (logarithmic) and often neglected. Baum and Baker (1990) represented the lightning-channel "return path" by a cylinder coaxial with and enclosing the lightning channel. Clearly, the radius of the artificial cylindrical return path affects the L and C values of such a coaxial RLC transmission line model, although the dependence is weak. Note that the telegrapher's equations 12.1 and 12.2 are the same for any two-conductor transmission line (including a coaxial one), all information on the actual line geometry being contained in L and C. The equivalent transmission line is usually assumed to be charged by the preceding leader to a specified potential and then closed at the ground end with a specified grounding resistance to initiate the return stroke. The second of the telegrapher's equations is equivalent to the continuity equation. Equations 12.1 and 12.2 can be derived from Maxwell's equations assuming that the electromagnetic waves propagating on and guided by the line exhibit a quasi-transverse electromagnetic (quasi-TEM) field structure and that R, L, and C are constant (e.g., Agrawal et al. 1980). Note that the term "quasi-transverse electromagnetic field structure" implies that the transverse component of the total electric field is much greater than the z-directed component associated with a non-zero value of R (Paul 1994). The telegrapher's equations can be also derived from the equivalent circuit shown in Fig. 12.5 using Kirchhoff's laws (e.g., Sadiku 1994).

In general, each transmission line parameter representing the return-stroke channel is a function of time

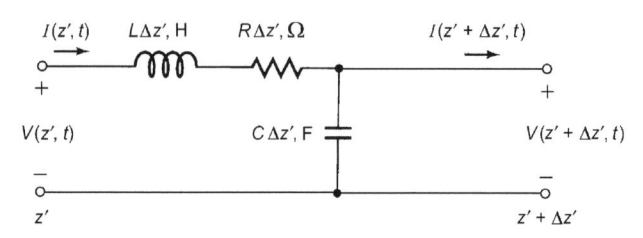

Fig. 12.5. The equivalent circuit for an elemental section of an RLC transmission line from which the telegrapher's equations (Eqs. 12.1 and 12.2) can be derived using Kirchhoff's laws in the limit as $\Delta z' \to 0$.

12.2. Return stroke

and space; that is, the transmission line is nonlinear and non-uniform (e.g., Rakov 1998). The channel inductance changes with time owing to variations in the radius of the channel core that carries the z-directed channel current. The channel resistance changes with time owing to variations in the electron density, the heavier-particle density, and the radius of the channel core. The channel capacitance changes with time, mostly owing to neutralization of the radially formed corona sheath that surrounds the channel core and presumably contains the bulk of the channel charge deposited by the preceding leader. For the case of a nonlinear transmission line, Eqs. 12.1 and 12.2 are still valid if L and C are understood to be the dynamic (as opposed to the static) inductance and capacitance, respectively (e.g., Gorin 1985): $L = \partial\phi/\partial I$, $C = \partial\rho/\partial V$, where ϕ is the magnetic flux linking the channel and ρ is the channel charge, both per unit length.

An exact closed form solution of the telegrapher's equations can generally be obtained only for the case in which R, L, and C are constant. There is at least one exception: the nonlinear distributed-circuit model described by Baum and Baker (1990) and Baum (1990b), in which C is specified as a function of charge density in order to simulate the effects of the radial-corona sheath. However, the telegrapher's equations representing this model admit exact solutions only if $R = 0$. Linear distributed-circuit models have been used, for instance, by Oetzel (1968), Price and Pierce (1977) ($R \approx 0.06\ \Omega\ m^{-1}$), Little (1978) ($R = 1\ \Omega\ m^{-1}$), and Takagi and Takeuti (1983) ($R = 0.08\ \Omega\ m^{-1}$). Rakov (1998) found that the behavior of electromagnetic waves guided by a linear RLC transmission line representing the pre-return-stroke channel formed by a dart leader and having $R = 3.5\ \Omega\ m^{-1}$ is consistent with the observed luminosity profiles for the return stroke (Jordan and Uman 1983). If the line nonlinearities are taken into account, solution of the telegrapher's equations requires the use of a numerical technique, for instance, a finite-difference method (Quinn 1987). Attempts to take into account the lightning channel nonlinearities using various simplifying assumptions have been made by Gorin and Markin (1975), Gorin (1985), Baum and Baker (1990), Baum (1990b), Mattos and Christopoulos (1988, 1990), and Kostenko (1995). The results presented by Gorin and Markin (1975) and Bazelyan et al. (1978) are shown, as an example, in Figs. 12.6 and 12.7.

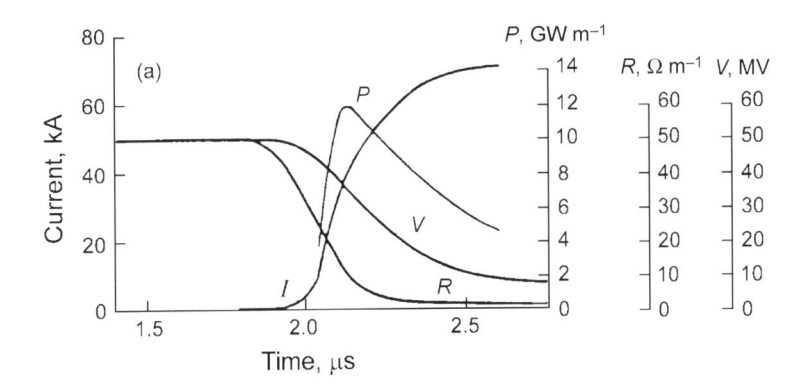

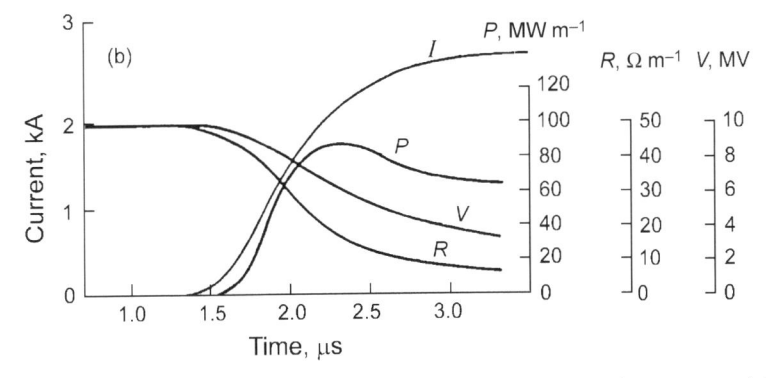

Fig. 12.6. Current I, voltage V, power per unit length P, and resistance per unit length R as a function of time t at a height of 300 m above ground as predicted by the distributed-circuit model of Gorin and Markin (1975). Profiles are given for (a) $V_0 = 50$ MV and an instantaneously discharged corona sheath, and (b) $V_0 = 10$ MV, and no corona sheath; V_0 is the initial uniform voltage on the channel due to charges deposited by the preceding leader. Adapted from Gorin and Markin (1975) and Bazelyan et al. (1978).

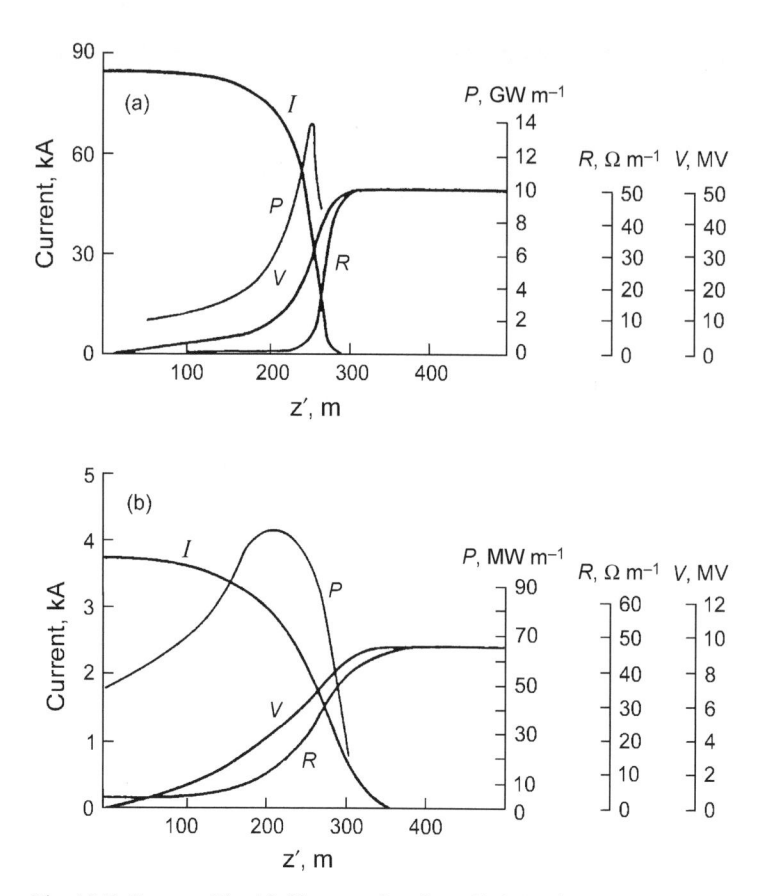

Fig. 12.7. Same as Fig. 12.6 but as a function of height z' along the channel at $t = 1.8$ μs.

Strawe (1979) proposed two versions of a distributed-circuit model that differ in the way that the value of R as a function of channel current and channel electrical conductivity is computed. In the first version, the conductivity is assumed to be constant, so that R varies only because of channel expansion. In the second version, the conductivity is a function of channel temperature and pressure, which were found using a model of the gas dynamic type. In both versions, L and C were assumed constant. An upward-moving connecting discharge from earth of 100 m length was simulated as an RLC transmission line as well. The second version of Strawe's (1979) model is actually a combination of a gas dynamic model and a distributed-circuit model. Such a combination model was also proposed by Baker (1990), although this model was not described in detail.

12.2.5. Engineering models

An engineering return-stroke model, as defined here, is simply an equation relating the longitudinal channel current $I(z', t)$ at any height z' and any time t to the current $I(0, t)$ at the channel origin, $z' = 0$. An equivalent expression in terms of the line charge density $\rho_L(z', t)$ on the channel can be obtained using the continuity equation (Thottappillil et al. 1997). Thottappillil et al. defined two components of the charge density at a given channel section, one component being associated with the return-stroke charge transferred through the channel section and the other with the charge deposited at the channel section. As a result, their charge density formulation reveals new aspects of the physical mechanisms behind the models that are not apparent in the longitudinal-current formulation. We first consider the mathematical and graphical representations of some simple models and then categorize and discuss the most used engineering models in terms of their implications regarding the principal mechanism of the return-stroke process. Rakov (1997) expressed several engineering models by the following generalized current equation:

$$I(z', t) = u(t - z'/v_f) P(z') I(0, t - z'/v) \qquad (12.3)$$

where u is the Heaviside function, equal to unity for $t \geq z'/v_f$ and zero otherwise, $P(z')$ is the height-dependent current attenuation factor introduced by Rakov and Dulzon (1991), v_f is the upward-propagating return-stroke-front speed, and v is the current-wave propagation speed. Table 12.2 summarizes $P(z')$ and v for five engineering models, namely: the transmission-line model, TL (Uman

12.2. Return stroke

Table 12.2. *P(z') and v in Eq. 12.3 for five engineering models*

Model	P(z')	v
TL (Uman and McLain 1969)	1	v_f
MTLL (Rakov and Dulzon 1987)	$1 - z'/H$	v_f
MTLE (Nucci *et al*. 1988a)	$\exp(-z'/\lambda)$	v_f
BG (Bruce and Golde 1941)	1	∞
TCS (Heidler 1985)	1	$-c$

and McLain 1969), not to be confused with the *RLC* transmission-line models discussed above; the modified transmission line model with linear current decay with height, MTLL (Rakov and Dulzon 1987); the modified transmission line model with exponential current decay with height, MTLE (Nucci *et al*. 1988a); the Bruce–Golde model, BG (Bruce and Golde 1941); and the traveling-current-source model, TCS (Heidler 1985). In Table 12.2, H is the total channel height, λ is the current decay constant (assumed by Nucci *et al*. 1988a to be 2000 m), and c is the speed of light. If not specified otherwise, v_f is assumed to be constant. Front speeds that decay exponentially with time, which is equivalent to decaying linearly with height, as shown by Leise and Taylor (1977), have also been used in attempts to model the first stroke in a flash (e.g., Bruce and Golde 1941; Uman and McLain 1969; Dulzon and Rakov 1980). The three simplest models, TCS, BG, and TL, are illustrated in Fig. 12.8 and the TCS and TL models additionally in Fig. 12.9. We consider first Fig. 12.8. For all three models we assume the same current waveform at the channel base ($z' = 0$) and the same front speed represented in z', t coordinates by the slanting line labeled v_f. The current-wave speed is represented by the line labeled v, which coincides with the vertical axis for the BG model and with the v_f line for the TL model. Shown for each model are current versus time waveforms at the channel base ($z' = 0$) and at heights z'_1 and z'_2. Because of the finite front-propagation speed v_f, the current at height, say, z'_2 experiences a delay z'_2/v_f with respect to the current at the channel base. The dark portion of the waveform indicates the current that actually flows through a given channel section, the blank portion being shown for illustrative purpose only. As can be seen in Fig. 12.8, the TCS, BG, and TL models are characterized by different current profiles along the channel, the difference being, from a mathematical point of view, due to the use of different values of v, listed in Table 12.2, in the generalized Eq. 12.3 with $P(z') = 1$. It also follows from Fig. 12.8 that if the channel-base current were a step function then the TCS, BG, and TL models would be characterized by the same current profile along the channel, although established in an apparently different way in each of the three models.

The relation between the TL and TCS models is further illustrated in Fig. 12.9, which shows that the spatial

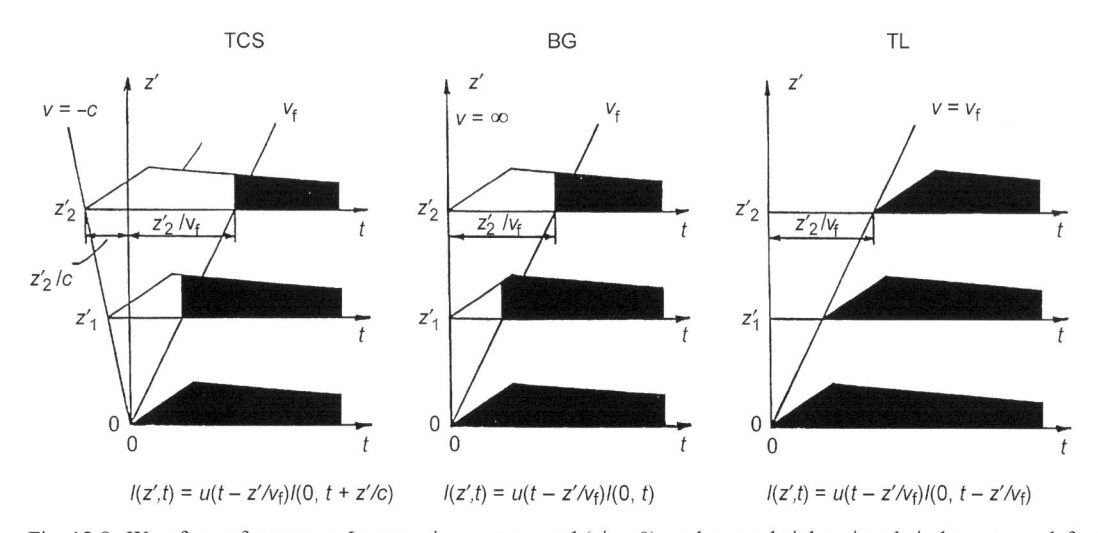

Fig. 12.8. Waveforms for current I versus time t at ground ($z' = 0$), and at two heights z'_1 and z'_2 above ground, for the TCS, BG, and TL return-stroke models. The slanting lines labeled v_f represent the upward speed of the return-stroke front, and the lines labeled v represent the speed of the return-stroke current wave. The dark portions of the waveforms indicate where the current actually flows through a given channel section. Note that the current waveform at $z' = 0$ and the front speed v_f are the same for all three models. The Heaviside function $u(t - z'/v_f)$ equals zero for $t < z'/v_f$ and unity for $t \geq z'/v_f$. Adapted from Rakov (1997).

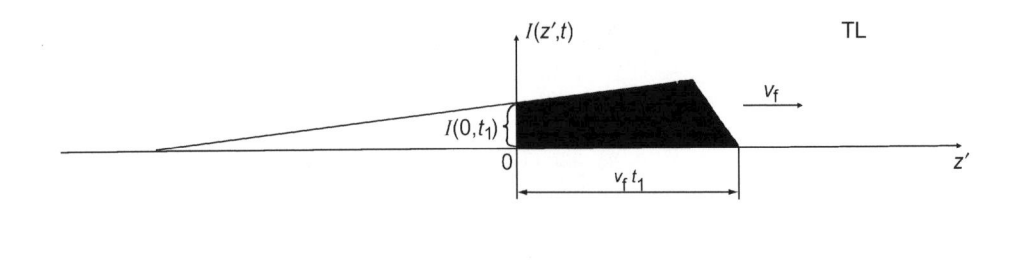

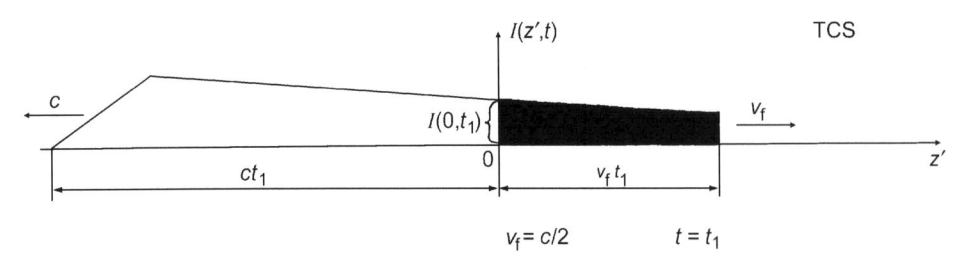

Fig. 12.9. Current versus height z' above ground at an arbitrary instant of time $t = t_1$ for the TL and TCS models. Note that the current at $z' = 0$ and v_f are the same for both models. Adapted from Rakov (1997).

current wave moves in the positive z' direction for the TL model and in the negative z' direction for the TCS model. Note that in Fig. 12.9, the current at ground ($z' = 0$) and the upward-moving front speed v_f are the same for both the TL and the TCS models. As in Fig. 12.8, the dark portion of the waveform indicates the current that actually flows in the channel, the blank portion being shown for illustrative purpose only.

The most used engineering models can be grouped in two categories: transmission-line-type models and traveling-current-source-type models, summarized in Tables 12.3 and 12.4, respectively. Each model is represented in Tables 12.3 and 12.4 by both current and charge density equations. Table 12.3 includes the TL model and its two modifications, the MTLL and MTLE models. Rakov and Dulzon (1991) additionally considered modified transmission-line models with current attenuation factors other than the linear and exponential functions used in the MTLL and MTLE models, respectively. The transmission-line-type models can be viewed as incorporating a current source at the channel base which injects a specified current wave into the channel. This wave propagates upward in the TL model without either distortion or attenuation and in the MTLL and MTLE models without distortion but with specified attenuation, as seen from the corresponding current equations given in Table 12.3.

Table 12.4 includes the BG model (Bruce and Golde 1941), the TCS model (Heidler 1985), and the DU model (Diendorfer and Uman 1990). In the traveling-current-source-type models, the return-stroke current may be viewed as being generated at the upward-moving return-stroke front and then propagating downward. In the TCS model, the current at a given channel section turns on instantaneously as the front passes this section, while in the DU model the current turns on gradually (exponentially with a time constant τ_D if $I(0, t + z'/c)$ is a step function). The channel current in the TCS model may be viewed as a single downward-propagating wave, as illustrated in Fig. 12.9. The DU model involves two terms (Table 12.4), one being the same as the downward-propagating current in the TCS model, which exhibits an inherent discontinuity at the upward-moving front (see Fig. 12.8), and the other being an opposite-polarity current that rises instantaneously to a value equal in magnitude to the current at the front and then decays exponentially with time constant τ_D. The second current component in the DU model may be viewed as a "front modifier". It propagates upward with the front and eliminates any current discontinuity at the front.

The time constant τ_D is the time during which the charge per unit length deposited at a given channel section by the preceding leader reduces to $1/e$ (about 37 percent) of its original value. Thottappillil and Uman (1993) and Thottappillil et al. (1997) assumed that $\tau_D = 0.1\,\mu s$. Diendorfer and Uman (1990) considered two components of charge density, each released with its own time constant, in order to match the model-predicted fields with the measured fields. If $\tau_D = 0$ then the DU model reduces to the TCS model. In both the TCS and DU models, the downward-propagating current wave speed is equal to the speed of light. The TCS model reduces to the BG model if the downward current-wave speed is set equal to infinity instead of the speed of light. Although the BG model could

Table 12.3. *Transmission-line-type models for* $t \geq z'/v_f$. $Q(z', t) = \int_{z'/v}^{t} I(0, \tau - z'/v) \, d\tau$, $v = v_f$ = constant, H = constant, λ = constant

TL	$I(z', t) = I(0, t - z'/v)$
(Uman and McLain 1969)	$\rho_L(z', t) = \dfrac{I(0, t - z'/v)}{v}$
MTLL	$I(z', t) = \left(1 - \dfrac{z'}{H}\right) I(0, t - z'/v)$
(Rakov and Dulzon 1987)	$\rho_L(z', t) = \left(1 - \dfrac{z'}{H}\right) \dfrac{I(0, t - z'/v)}{v} + \dfrac{Q(z', t)}{H}$
MTLE (Nucci et al. 1988a)	$I(z', t) = e^{-z'/\lambda} I(0, t - z'/v)$
	$\rho_L(z', t) = e^{-z'/\lambda} \dfrac{I(0, t - z'/v)}{v} + \dfrac{e^{-z'/\lambda}}{\lambda} Q(z', t)$

Table 12.4. *Traveling-current-source-type models for* $t \geq z'/v_f$. $v^* = v_f / (1 + v_f/c)$, v_f = constant, τ_D = constant

BG	$I(z', t) = I(0, t)$
(Bruce and Golde 1941)	$\rho_L(z', t) = \dfrac{I(0, z'/v_f)}{v_f}$
TSC	$I(z', t) = I(0, t + z'/c)$
(Heidler 1985)	$\rho_L(z', t) = -\dfrac{I(0, t + z'/c)}{c} + \dfrac{I(0, z'/v^*)}{v^*}$
DU	$I(z', t) = I(0, t + z'/c) - e^{-(t - z'/v_f) \tau_D^{-1}} I(0, z'/v^*)$
(Diendorfer and Uman 1990)	$\rho_L(z', t) = -\dfrac{I(0, t + z'/c)}{c} - e^{-(t - z'/v_f) \tau_D^{-1}} \left[\dfrac{I(0, z'/v^*)}{v_f} + \dfrac{\tau_D}{v^*} \dfrac{dI(0, z'/v^*)}{dt}\right]$
	$+ \dfrac{I(0, z'/v^*)}{v^*} + \dfrac{\tau_D}{v^*} \dfrac{dI(0, z'/v^*)}{dt}$

also be viewed mathematically as a special case of the TL model with v replaced by infinity, we choose to include the BG model in the traveling-current-source-type model category. Thottappillil et al. (1991a) mathematically generalized the DU model to include a variable upward front speed and a variable downward current-wave speed, both separate arbitrary functions of height (this model was dubbed MDU, where M stands for "modified"). A further generalization of the DU model (Thottappillil and Uman 1994) involves a single, height-variable, time constant τ_D. Generalizations of the TCS model are discussed later in this subsection.

The principal distinction between the two types of engineering model, formulated in terms of current, is the direction of propagation of the current wave: upward for the transmission-line-type models ($v = v_f$) and downward for the traveling-current-source-type models ($v = -c$ except for the BG model), as can be seen for the TL and TCS models, respectively, in Fig. 12.9. As noted above, the BG model can be viewed mathematically as a special case of either the TCS or the TL model. The BG model includes a current wave propagating at an infinitely large speed and, as a result, the wave's direction of propagation is indeterminate. As in all the other models, the BG model includes a front moving at a finite speed v_f. Note that even though the direction of propagation of the current wave in a model can be either up or down, the direction of the current is the same: charge of the same sign is transported to ground in both types of engineering model.

The TL model predicts that, provided (i) the height above ground of the upward-moving return-stroke front is much smaller than the distance r between the observation point on ground and the channel base, so that all contributing channel points are essentially equidistant from the observer, (ii) the return-stroke front propagates at a constant speed, (iii) the return-stroke front has not reached the top of the channel, and (iv) the ground conductivity is high enough that propagation effects (see the last part of subsection 4.6.4) are negligible, the vertical component E_z^{rad} of the electric radiation field and the azimuthal component of the magnetic radiation field are proportional to the channel-base current I (e.g., Uman et al. 1975). The equation for the electric radiation field E_z^{rad} is as follows,

$$E_z^{\text{rad}}(r, t) = [-v/(2\pi\varepsilon_0 c^2 r)] I(0, t - r/c) \qquad (12.4)$$

where ε_0 is the permittivity of free space, v is the upward propagation speed of the current wave, which is the same as the front speed v_f in the TL model as well as in the MTLL and MTLE models, and c is the speed of light. For the most common return stroke, which lowers negative charge to ground, the sense of the positive charge flow is upward so that the current I, assumed in deriving Eq. 12.4 (same as

Eq. 4.13) to be upward directed, is by convention positive and E_z^{rad} is, by Eq. 12.4, negative; that is, the electric field vector points in the negative z direction. Taking the derivative of this equation with respect to time, one obtains

$$\frac{\partial E_z^{\text{rad}}(r,t)}{\partial t} = -\frac{v}{2\pi\varepsilon_0 c^2 r}\frac{\partial I(0, t-r/c)}{\partial t} \quad (12.5)$$

Equations 12.4 and 12.5 are commonly used, particularly the first one and its magnetic radiation field counterpart, found from $|B_\phi^{\text{rad}}| = |E_z^{\text{rad}}|/c$, for estimation of the peak values of the return-stroke current and its time derivative, subject to the assumptions listed prior to Eq. 12.4. Equations 12.4 and 12.5 have been used, as discussed further in the third part of subsection 12.2.7, for the estimation of v from measured values of E_p/I_p and $(dE/dt)_p/(dI/dt)_p$, respectively, where the subscript z and superscript "rad" have been dropped, and the subscript "p" refers to peak values. Expressions relating the channel-base current and the electric radiation field far from the channel for the BG, TCS, and MTLE models were given by Nucci et al. (1990). General equations for calculating electric and magnetic fields at ground are considered in subsection 4.6.4.

As stated in subsection 12.2.1, a characteristic feature of the engineering models is the small number of adjustable parameters, usually one or two besides the channel-base current. In these models, the physics of the lightning return stroke is deliberately downplayed, and emphasis is placed on achieving agreement between the model-predicted electromagnetic fields and those observed at distances from tens of meters to hundred of kilometers.

In the rest of this section we will briefly describe Cooray's (1993) model and generalizations of the TCS model by Heidler and Hopf (1994, 1995, 1996) and by Cvetic and Stanic (1997).

Cooray (1993) proposed a model with a large number of adjustable parameters, the values for many of which are at present unknown. In this model, a charge density distribution along the channel at $t = 0$ is specified separately for the inner part of the channel (the channel core and the so-called hot-corona sheath) and for the outer part of the channel (the so-called cold-corona sheath). Four adjustable parameters are used. Further, the dynamics of charge release by the return-stroke front is assumed and involves four more adjustable parameters. The return-stroke speed profile is predicted by the model (as opposed to the engineering models, in which it is specified on the basis of optical measurements) but requires one more adjustable parameter: the longitudinal electric field intensity in the pre-return-stroke channel. It is not clear whether Cooray's model, which therefore includes a total of nine adjustable parameters, is an improvement on the engineering models from the standpoint of the model-predicted electromagnetic fields. The model proposed by Cooray was intended to describe subsequent return strokes. An extension of this model to negative first-return strokes was presented by Cooray (1997a).

Heidler and Hopf (1994) modified the TCS model to take account of wave reflections at ground and at the upward-moving front, using the traveling-current-source current as an input to the model. Both upward and downward waves behind the upward-moving front propagate at the speed of light, and the resultant reflection coefficient at the front is a function of v_f and $v = c$. The channel-base current in this model depends on the reflection coefficient at the strike point and on the initial charge density distribution along the channel. Heidler and Hopf (1995) further modified the TCS model by expressing the source current, and therefore the initial charge density distribution along the channel, in terms of the channel-base current and the current reflection coefficient at ground. Cvetic and Stanic (1997) proposed a model from which the TCS and DU models can be derived as special cases. Within the concept of the TCS model, they specified independently the channel-base current and the initial charge density distribution along the channel. The resultant current distribution along the channel was determined using the equation of current continuity.

12.2.6. Testing model validity

Gas dynamic models Attempts to test the gas dynamic models of lightning return strokes have been made by comparing their predictions with (i) the temperature, electron density, and pressure versus time curves published by Orville (1968a, b, c), (ii) the radiated optical power observed by Guo and Krider (1982, 1983), and (iii) the input electrical energy estimated by Krider et al. (1968).

As discussed in subsection 4.6.5, Orville (1968b), from an analysis of 10 time-resolved spectra of return strokes (the time resolution was either 2 or 5 μs), obtained the channel temperature and the electron density, each as a function of time. Typical peak temperatures, determined from the ratios of the intensities of spectral lines, were of order 28 000–31 000 K. No temperatures exceeded 36 000 K. In two of the 10 strokes, the temperature appeared to rise to a peak value during the first 10 μs (the time resolution was 5 μs) and to decay thereafter. In the remaining eight strokes (including two with 2 μs time resolution), the temperature decreased monotonically following its initial maximum value. The electron density, determined from the Stark broadening of the H_α line, was 8×10^{17} cm^{-3} in the first 5 μs, decreasing to $(1-1.5) \times 10^{17}$ cm^{-3} at 25 μs, and remaining approximately constant to 50 μs. Using Gilmore's tables for the composition of dry air in thermodynamic equilibrium, Orville (1968c) found that the channel was characterized by an average pressure of 8 atm in the first 5 μs and attained atmospheric pressure at approximately 20 μs. Hill R.D. (1971, Fig. 1), Plooster (1971b, Fig. 1), and

12.2. Return stroke

Paxton et al. (1990, p. 56 and Fig. 8, the latter reproduced in Fig. 12.1 here) showed that their model-predicted temperature versus time curves were generally consistent with those of Orville (1968b). Plooster (1971b, Figs. 2 and 3) did so also for electron density and pressure.

Guo and Krider (1982, 1983), using a photoelectric detector, found that the time- and space-averaged mean radiance in the 0.4 to 1.1 µm wavelength range (involving essentially optical power) for first strokes was of order 10^6 W m^{-1} (subsection 4.6.5). Paxton et al. (1986, 1990), using their model and current waveform with a peak of 20 kA, calculated that the average radiated optical power (over the first 10 µs) in the 0.4 µm to 1.2 µm range was essentially equal to the value found by Guo and Krider (1982, 1983).

Using the measured radiated energy in the optical wavelength range from 0.4 to 1.1 µm from a single-stroke lightning of 870 J m^{-1} and that from a long laboratory spark of known input energy, Krider et al. (1968) deduced that the input energy to the lightning return stroke was 2.3×10^5 J m^{-1}. The percentage of the total energy that was optically radiated from the channel was 0.38 percent, according to Krider et al. (1968). The gas dynamic models predict (Table 12.1) lightning input-energy values about two orders of magnitude lower than the value deduced by Krider et al. and percentages of radiated energy correspondingly higher. For example, Dubovoy et al. (1991a, b, 1995) computed for a 20-kA return stroke that the energy lost (after 55 µs or so) by radiative processes was 700 J m^{-1}, which is roughly 25 percent of the input-energy value, 3×10^3 J m^{-1}, computed by them. Dubovoy et al. (1993) and Dubovoy (1997), using the model developed by Dubovoy et al. (1991a, b, 1995) in conjunction with measured electric field waveforms (the TL model was used to obtain the corresponding parameters of the current waveforms) and 35 cm radar cross-sections of lightning discharges, estimated input-energy values ranging from 250 to 1000 J m^{-1}. These values were attributed to return strokes, although some of the events could have been misidentified cloud discharges.

It is important to note that the relatively high energy value, 1.5×10^4 J m^{-1} (for a gas dynamic model) reported by Hill R.D. (1971) is an overestimate, because Hill used an equation for the electron–neutral collision frequency that is invalid for the most important temperature range, 8000 to 30 000 K, as pointed out by both Paxton et al. (1987) and Dubovoy et al. (1991a, b, 1995). As a result, Hill's values of the channel electrical conductivity are 20–30 times lower than those observed experimentally in this temperature range, leading to the erroneous value of input energy mentioned above. Since the input energy varies roughly as the inverse square root of the conductivity (Plooster 1971a), the corrected value of energy for Hill's model is about 3×10^3 J m^{-1}, which is consistent with values predicted by other gas dynamic models (see Table 12.1).

Note that other results of Hill (1971) (e.g., the commonly quoted values for pressure versus radius and time (Uman 1987, 2001, Fig. 15.11) must be affected by Hill's error in the conductivity computations and, therefore, should be viewed accordingly. It follows from the above discussion that there exists a discrepancy of about two orders of magnitude between the lightning input energy predicted by the gas dynamic models and that deduced by Krider et al. (1968) from a comparison of the optical output of one lightning stroke with long-laboratory-spark measurements. This disparity remains the subject of controversy. Plooster (1971b), in particular, argued that with a channel electrical conductivity near 2×10^4 S m^{-1}, the radius of the conducting channel would have to be less than 0.15 cm for the entire duration of the current waveform, instead of rapidly increasing by an order of magnitude or so, in order to give a total energy input of 10^5 J m^{-1}. Paxton et al. (1986, 1990) viewed Plooster's (1971b) argument as strong evidence that the value of 10^5 J m^{-1} was a significant overestimate of the lightning return-stroke energy. Further, the input energy predicted by the gas dynamic models appears to be consistent with the estimate of Borovsky (1998), which was based on a computation of the electrostatic energy stored on a lightning channel with a line charge density of 100–500 µC m^{-1} (Table 12.1). Finally, Hill R.D. (1977a) suggested that only about one-thirtieth of the input electrical energy supplied to the laboratory spark whose characteristics were used by Krider et al. (1968) for the calibration of their measurement of lightning energy was dissipated in the hot return-stroke channel. According to Hill, the bulk of the input energy was dissipated in the "plasma ahead of the advancing secondary streamer", that is, during the preceding leader processes. Cooray (1997b), from electrostatic considerations, estimated that two-thirds of the subsequent-stroke input energy is dissipated in the dart-leader stage and one-third in the return-stroke stage, whereas for first strokes roughly one-third of the input energy is dissipated in the stepped-leader stage and two-thirds in the return-stroke stage. If an appreciable portion of the input energy is indeed dissipated during the leader process, the energy estimate of Krider et al. (1968) cannot be compared directly with the gas dynamic model predictions, which consider only the return-stroke process. However, a lightning input energy of order 10^5 J m^{-1} appears to be consistent with the thunder theory of Few (1969, 1995), discussed in subsection 11.3.2, although this theory itself remains a subject of debate. Further, the value 10^5 J m^{-1} is comparable with that inferred for the total electrostatic energy available to a lightning first stroke (Section 4.2, Table 12.1), although it is likely that a significant fraction of this energy is dissipated by processes other than the return stroke. These include the in-cloud discharge processes that serve to collect charges from isolated hydrometeors in volumes measured in cubic kilometers and to transport those charges into the developing leader channel. Mackerras (1973) estimated that about 1.7 percent of the total electrostatic energy available to a lightning stroke is converted to optical radiation. Additional experimental data and modeling are needed to resolve the two-orders-of-magnitude uncertainty in the value of the lightning input energy; this has important implications, for example, for the estimation of the worldwide production of nitric oxide (NO) by lightning (Chapter 15) and in testing the validity of the proposed thunder-generation mechanisms (Section 11.3).

Electromagnetic models For the electromagnetic models, as well as for the distributed-circuit and engineering models, the most appropriate test of model validity would appear to be a comparison of the model-predicted electromagnetic fields with the measured fields. The measured electric and magnetic fields due to natural lightning at 1 to 200 km presented by Lin *et al.* (1979) and the electric fields due to triggered lightning at 10 to 500 m published by Uman *et al.* (1994, 1997), Rakov *et al.* (1998), and Crawford *et al.* (2001) are presently the most useful data for such an evaluation. These are reproduced in Figs. 4.38, 7.26, and 7.27.

Podgorsky and Landt (1987) do not give any model-predicted fields. Moini *et al.* (1997, 2000) have demonstrated fairly good agreement between the model-predicted and typical measured electric fields at distances ranging from tens of meters to tens of kilometers. At 100 km their model does not predict a field zero-crossing within 200 μs or so, and hence it is inconsistent with the published measured fields at this distance (Fig. 4.38). The significance of the zero-crossing time as a criterion of model validity is discussed in the last part of this subsection. Baba and Ishii (2001, Fig. 2) presented electric and magnetic field waveforms (for the initial 7 μs) at a distance of 2 km that are consistent with observation.

Distributed-circuit models The electromagnetic fields calculated by Takagi and Takeuti (1983, Figs. 12 and 13) and Price and Pierce (1977, Fig. 4), who used linear distributed-circuit models, and by Mattos and Christopoulos (1990, Figs. 7–9) and Baker (1990, Figs. 3 and 6), who used nonlinear distributed-circuit models, are largely inconsistent with typical measured fields (Fig. 4.38). Other authors do not present model-predicted electromagnetic fields.

Engineering models Two primary approaches to model testing have been used. The first approach involves using a *typical* channel-base current waveform and a *typical* return-stroke propagation speed as model inputs and then comparing the model-predicted electromagnetic fields with *typical* observed fields.

The second approach involves using the channel-base current waveform and the propagation speed measured for the same *individual* event and comparing the computed fields with the measured fields for that same *specific* event. This approach is able to provide a more definitive answer regarding model validity, but it is feasible only in the case of triggered-lightning return strokes or natural lightning strikes to tall towers where the channel-base current can be measured. In the field calculations, the channel is generally assumed to be straight and vertical with its origin at ground ($z' = 0$), conditions that are expected to be valid for subsequent strokes, but potentially not for first strokes. The channel length is usually not specified unless it is an inherent feature of the model, as is the case for the MTLL model (e.g., Rakov and Dulzon 1987). As a result, the model-predicted fields and associated model validation may not be meaningful after 25–75 μs, the expected time it takes for the return-stroke front to traverse the distance from ground to the cloud charge source.

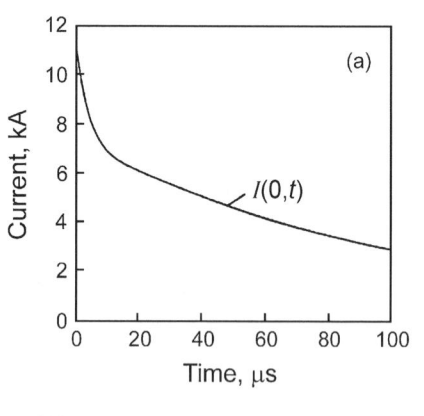

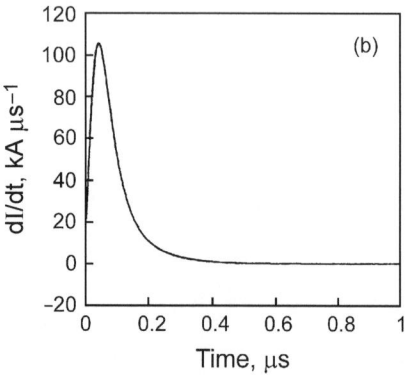

Fig. 12.10. (a) The current at ground level and (b) the corresponding current derivative used by Nucci *et al.* (1990), Rakov and Dulzon (1991), Thottappillil *et al.* (1997), and Moini *et al.* (2000) for testing the validity of return-stroke models by the "typical-return-stroke" approach. Adapted from Nucci *et al.* (1990).

12.2. Return stroke

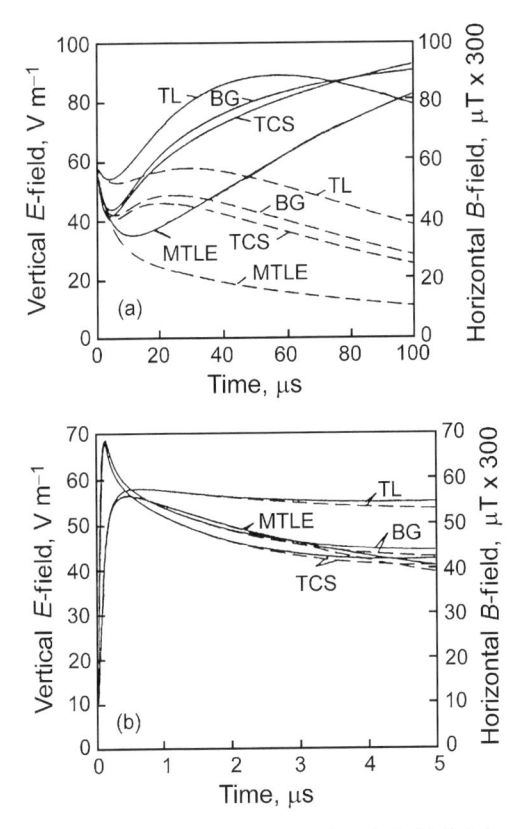

Fig. 12.11. Calculated vertical electric field (left-hand scale, solid lines) and horizontal (azimuthal) magnetic field (right-hand scale, broken lines) for four return-stroke models at a distance $r = 5$ km displayed on (a) 100 μs and (b) 5 μs time scales. Adapted from Nucci et al. (1990).

We now discuss these two approaches in turn. The *typical-return-stroke* approach has been adopted by Nucci et al. (1990), Rakov and Dulzon (1991), and Thottappillil et al. (1997). Nucci et al. identified four characteristic features in the fields at 1 to 200 km measured by Lin et al. (1979) (Fig. 4.38) and used those features as a benchmark for their validation of the TL, MTLE, BG, and TCS models (and also the MULS model, not considered here). The characteristic features include (i) a sharp initial peak, which varies approximately as the inverse distance beyond a kilometer or so in both electric and magnetic fields, (ii) a slow ramp following the initial peak and lasting in excess of 100 μs for electric fields measured within a few tens of kilometers, (iii) a hump following the initial peak in the magnetic fields within a few tens of kilometers, the maximum of which occurs between 10 and 40 μs, and (iv) a zero-crossing within tens of microseconds of the initial peak in both electric and magnetic fields at 50 to 200 km. For the current (Fig. 12.10) and for other model characteristics used by Nucci et al. (1990), feature (i) is reproduced by all the models examined, feature (ii) by all the models except for the TL model, feature (iii) by the BG, TL and

TCS models but not by the MTLE model, and feature (iv) only by the MTLE model, not by the BG, TL, and TCS models, as illustrated in Figs. 12.11 and 12.12. Diendorfer and Uman (1990) showed that the DU model reproduces features (i), (ii), and (iii), and Thottappillil et al. (1991b) demonstrated that a relatively insignificant change in the channel-base current waveform (well within the range of typical waveforms) allows the reproduction of feature (iv), the zero-crossing, by the TCS and DU models. Rakov and Dulzon (1991) showed that the MTLL model reproduces features (i), (ii), and (iii). The observed sensitivity of the distant-field waveforms predicted by the TCS and DU models to the variations in the channel-base current waveform has important implications for tests of the validity of these models. Indeed, since appreciable variation in the current waveform is a well-documented fact (Table 4.4), the relatively narrow range of observed zero-crossing times (Table 4.6) appears inconsistent with the TCS and DU models. On the other hand, the experimental field data of Lin et al. (1979) might be biased toward earlier zero-crossing times and more pronounced opposite-polarity overshoots, for the following two reasons. First, the oscilloscope sweep

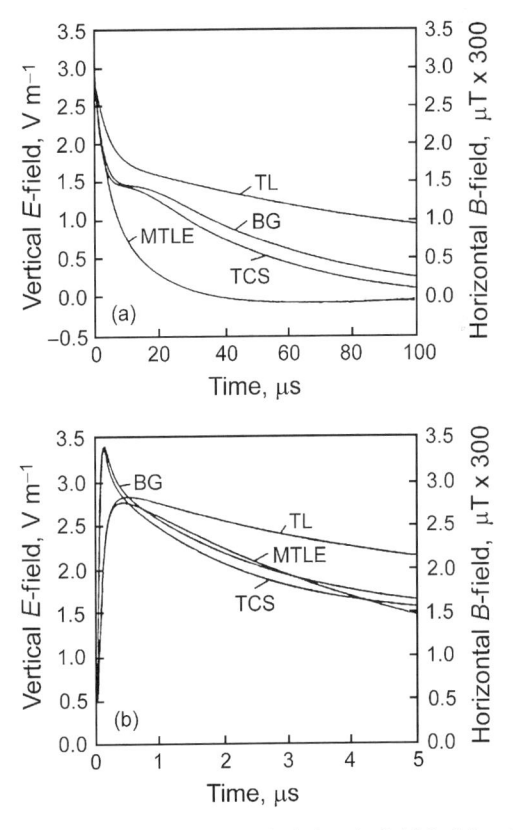

Fig. 12.12. Calculated vertical electric field (left-hand scale) and horizontal (azimuthal) magnetic field (right-hand scale) for four return-stroke models at a distance $r = 100$ km displayed on (a) 100 μs and (b) 5 μs time scales. Adapted from Nucci et al. (1990).

of 200 μs that they used was insufficient to measure zero-crossing times greater than about 200 μs. Second, the initial rising portion of the waveform was not always completely recorded since the delay line they used allowed recording only 2.5 μs prior to the time of oscilloscope trigger. As a result, the zero-field level, which was set to the first recorded point, was potentially higher than the actual zero field level. Nucci *et al.* (1990) concluded from their study that all the models evaluated by them, using measured fields at distances ranging from 1 to 200 km, predict reasonable fields for the first 5–10 μs and that all the models except the TL model do so for the first 100 μs.

Thottappillil *et al.* (1997) noted that the measured electric fields at tens to hundreds of meters from triggered lightning (e.g., Uman *et al.* 1994, 1997; Rakov *et al.* 1998; Crawford *et al.* 2001) exhibited a characteristic flattening within 15 μs or so, as seen in Figs. 7.26 and 7.27. The electric fields predicted at 50 m by the BG, TL, MTLL, TCS, MTLE, and DU models are shown in Fig. 12.13. As follows from this figure, the BG, MTLL, TCS, and DU models, but not the TL and MTLE models, are consistent with the measured fields presented in Figs. 7.26 and 7.27. Additionally, the MTLE model is inconsistent with the observed ratio of leader-to-return-stroke electric field change at far ranges. Beasley *et al.* (1982), for 97 first-stroke leaders at distances of approximately 20 to 50 km, reported for this ratio a mean value 0.8 (see also Fig. 4.21), in support of the BG, MTLL, TCS, and DU models, whereas the MTLE model predicts a ratio value near 3.

The *specific-return-stroke* approach was adopted by Thottappillil and Uman (1993), who compared the TL, TCS, MTLE, DU, and MDU models. They used 18 sets of three simultaneously measured features of triggered-lightning return strokes: the channel-base current, the return-stroke propagation speed, and the electric field at about 5 km from the channel base, the data previously used by Willett *et al.* (1989) for their analysis of the TL model. Comparisons made for the three strokes shown in Fig. 12.14, which are characterized by somewhat different rising portions of the channel-base current, are given in Figs. 12.15 through 12.17 for the TL, MTLE, TCS, and DU models. It was found that the TL, MTLE, and DU models each predict the measured initial electric field peaks within an error whose mean absolute value is about 20 percent, while the TCS model has a mean absolute error about 40 percent.

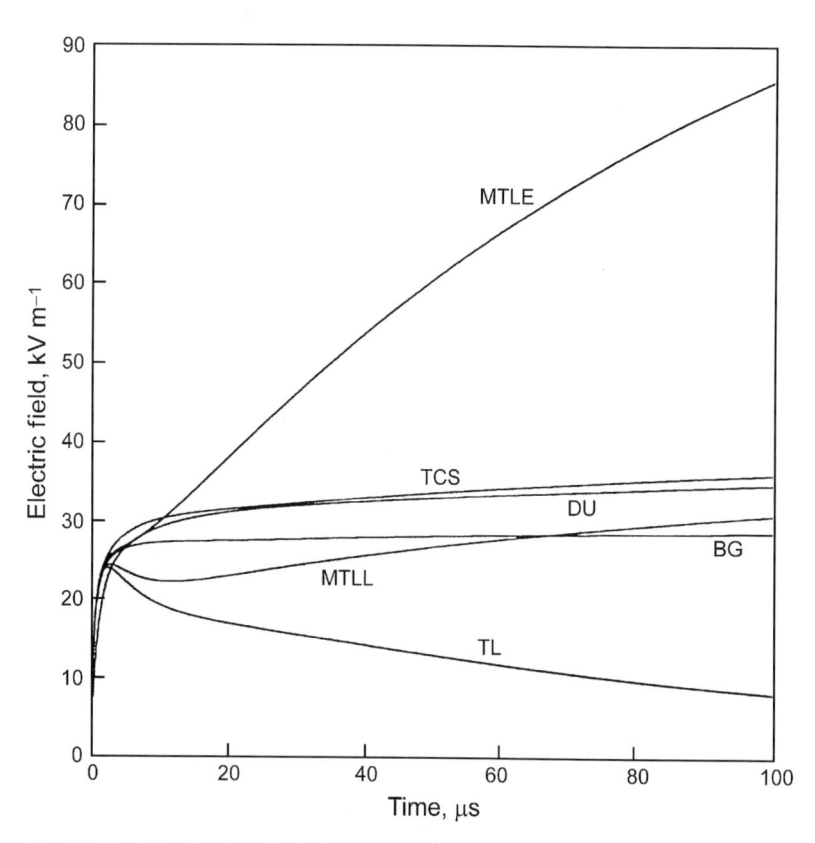

Fig. 12.13. Calculated vertical electric fields for six return-stroke models at a distance $r = 50$ m, to be compared with the typical measured return-stroke fields at 50 m presented in Figs. 7.26 and 7.27. Note that only the upward-going portion of the waveforms shown in Figs. 7.26 and 7.27 is due to the return stroke, the downward-going portion being due to the preceding dart leader. Adapted from Thottappillil *et al.* (1997).

12.2. Return stroke

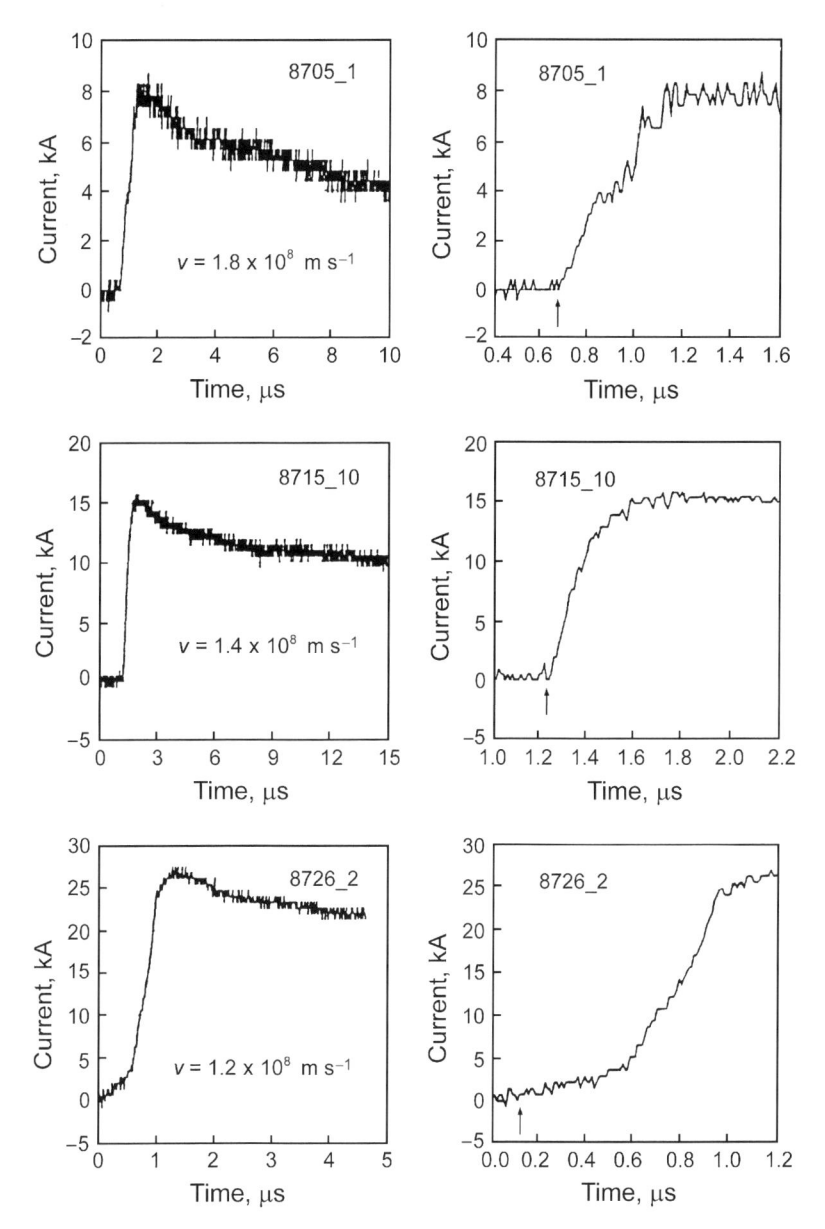

Fig. 12.14. The current waveform at the base of the channel (left-hand column) and a close-up of the current wave front on an expanded time scale (right-hand column) for three different triggered-lightning return strokes, 8705_1, 8715_10, and 8726_2, used by Thottappillil and Uman (1993) for testing the validity of return-stroke models by means of the "specific-return-stroke" approach. Also given for each stroke is the measured return-stroke speed. Adapted from Thottappillil and Uman (1993).

The overall results for the validity of the engineering models can be summarized as follows.

- The relation between the initial field peak and the initial current peak is reasonably well predicted by the TL, MTLL, MTLE, and DU models.
- The electric fields at tens of meters from the channel after the first 10–15 μs are reasonably reproduced by the MTLL, BG, TCS and DU models, but not by the TL and MTLE models.
- From the standpoint of the overall field waveforms at 5 km (the only distance at which the specific-return-stroke model-testing approach has been used) all the models should be considered less than adequate.

Based on the entirety of the testing results and mathematical simplicity, we rank the engineering models in the following descending order: MTLL, DU, MTLE, TCS, BG, and TL. However, the TL model is recommended for estimation of the initial field peak from the current peak or, conversely, the

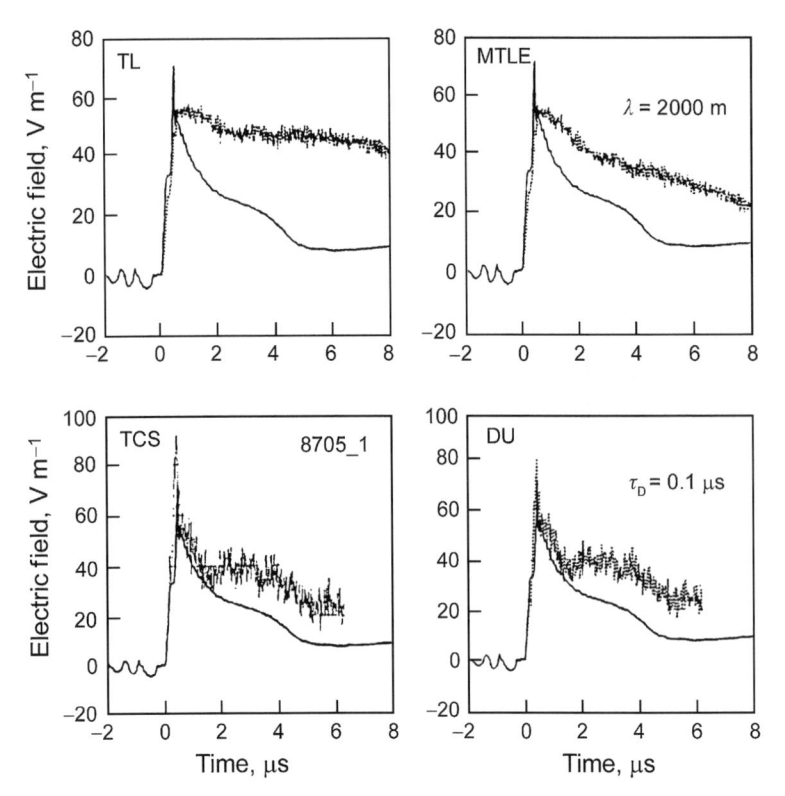

Fig. 12.15. The calculated vertical electric fields (dotted lines) from the TL, MTLE, TCS, and DU models shown together with the measured field (solid lines) at 5.16 km, for return stroke 8705_1. $\lambda = 2000$ m and $\tau_D = 0.1$ μs. The measured current at the channel base and the measured return stroke speed are given in Fig 12.14, top two panels. Adapted from Thottappillil and Uman (1993).

current peak from the field peak, since it is the mathematically simplest model, with a predicted relationship between the peak field and the peak current that has an accuracy equal to or greater than that of the more mathematically complex models.

12.2.7. Further topics in return-stroke modeling

Treatment of the upper, in-cloud portion of the channel It is the common view (e.g., Lin et al. 1980) that subsequent return strokes are easier to model than first return strokes. First return strokes are usually branched, may involve an upward connecting leader from ground of appreciable length, perhaps many tens of meters, and typically exhibit a significant variation in propagation speed along the channel. This view is correct provided that the lightning channel is predominantly vertical, a condition that is less likely to be satisfied for subsequent than for first return strokes after the return stroke reaches cloud charge height, typically after 25–75 μs, assuming that the return-stroke front propagation speed in the cloud is approximately the same as that below the cloud base. Subsequent strokes are expected to follow predominantly horizontal paths in the cloud charge region (Krehbiel et al. 1979; Rakov et al. 1990).

None of the engineering models except the MTLL model specifies the boundary conditions at the channel top. In general, a reflection should be produced when the return-stroke front encounters an impedance discontinuity at the channel top. Some indirect evidence of channel-top reflection apparently comes from the VHF interferometric studies of Shao et al. (1995, p. 2759), who observed VHF bursts, indicative of breakdown, at the preceding-leader starting point when the return stroke arrived there. Further, the channel-base current waveshape for the first strokes in altitude-triggered lightning appears to be significantly modified by a reflection from the upper end of the channel at 1 km or so (Rakov et al. 1998).

Various boundary conditions at the channel top have been considered in the distributed-circuit models, including an open circuit (Strawe 1979; Baker 1990), a capacitor or an LC transmission line (Takagi and Takeuti 1983), and an RC network (Mattos and Christopoulos 1988, 1990). When only the first few microseconds of the field waveforms are of interest, the treatment of the channel top is unimportant.

Boundary conditions at ground In the transmission-line-type engineering models, the boundary conditions at ground

12.2. Return stroke

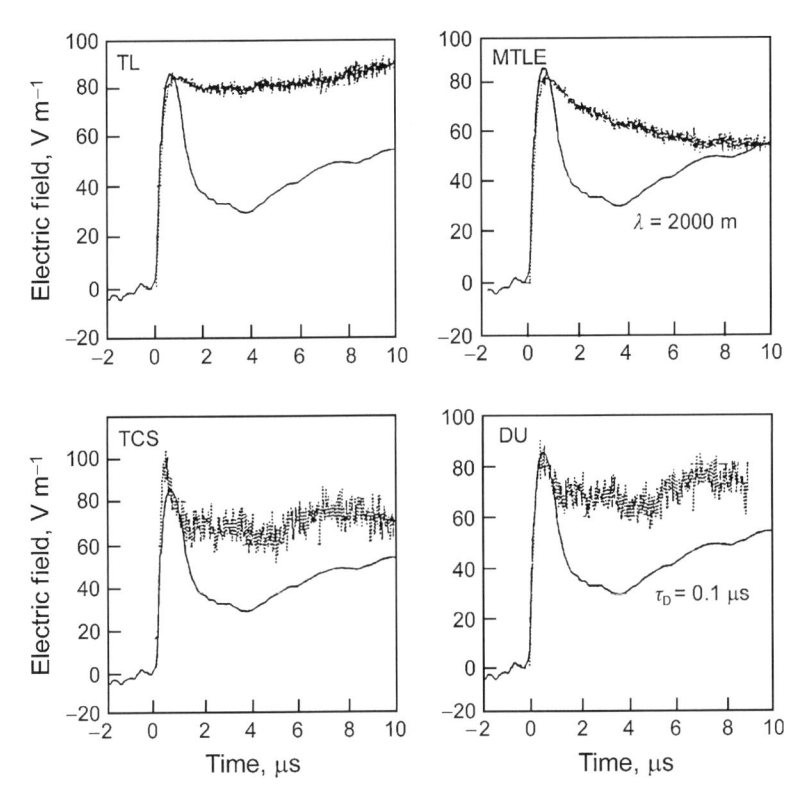

Fig. 12.16. The calculated vertical electric fields (dotted lines) from the TL, MTLE, TCS, and DU models shown together with the measured field (solid lines) at 5.16 km for return stroke 8715_10. The measured current at the channel base and the measured return stroke speed are given in Fig. 12.14, middle two panels. Adapted from Thottappillil and Uman (1993).

are determined by the specified channel-base current, that is, by the current source at the channel bottom. In the TCS and DU models, which assume that the return-stroke current is generated at the upward-moving front and propagates toward ground, it is usually implied that the channel is terminated at ground in its characteristic impedance so that the current reflection coefficient at ground is equal to zero. This implication is invalid for the case of a lightning strike to a well-grounded object, where an appreciable reflection from ground is expected. Extensions of the TCS model to include reflection at ground and at the upward-moving front were considered above, in subsection 12.2.5. In the distributed-circuit return-stroke models, the boundary conditions at ground are specified explicitly, a terminating resistor (typically tens to hundreds of ohms) being used to simulate the connection to ground.

Some engineering models have been extended to include a grounded strike object, modeled as an ideal transmission line that supports the propagation of waves at the speed of light without attenuation or distortion. Such an extension results in a second current wave front which propagates from the top of the object toward ground at the speed of light. This current wave front either produces no reflection on its arrival there, implying that the grounding impedance is equal to the characteristic impedance of the object (e.g., Diendorfer and Uman 1990), or is allowed to bounce between the top and bottom ends of the object and, in general, to produce transmitted waves at either end (e.g., Zundl 1994; Guerrieri et al. 1996, 1998; Motoyama et al. 1996; Rachidi et al. 1998, 2001). Shostak et al. (1999) additionally included in their model an upward connecting leader from the object and reflections from the upward-moving return-stroke front. The electromagnetic model developed by Podgorski and Landt (1987) included simulation of the strike object, the 553 m high CN (Canadian National) tower in Toronto, by a three-dimensional wire structure. Baba and Ishii (2001) used their electromagnetic model for studying lightning strikes to the CN tower and to the 200 m stack at Fukui, Japan (see Fig. 8.1).

The transient behavior of tall objects under direct lightning-strike conditions is discussed in Section 6.5. For the simple example of a *non-ideal* current source attached to the top of the object and generating a step-function current wave, the magnitude of the wave injected into the object depends on the characteristic impedance of the object (Section 6.5). Specifically, the total source current divides between the source impedance and the object inversely to the values of the source impedance and the characteristic impedance of the object. However, after a sufficiently long period of time, the current magnitude at any point on the

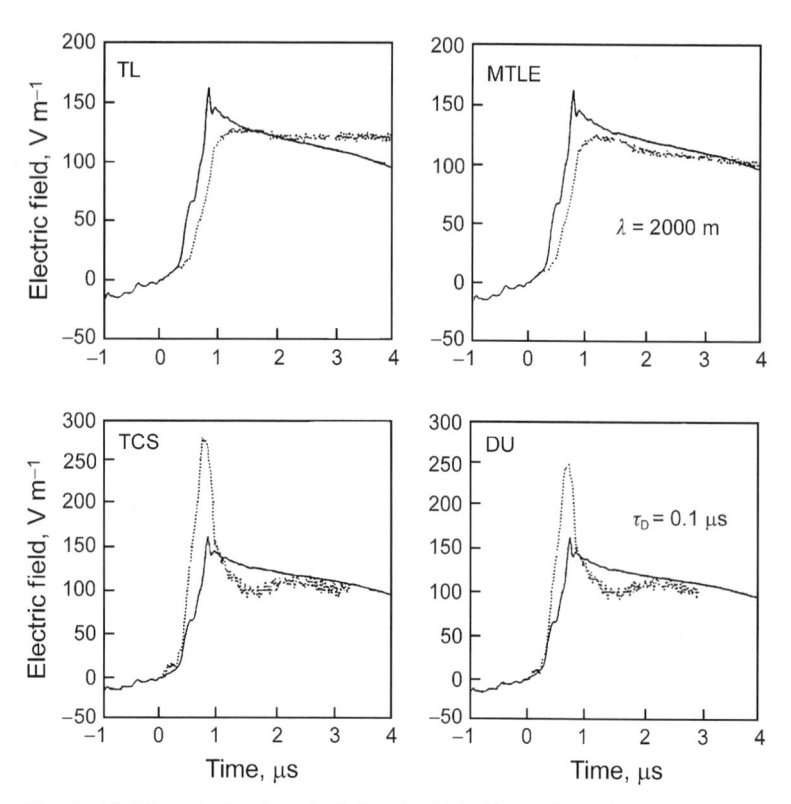

Fig. 12.17. The calculated vertical electric fields (dotted lines) from the TL, MTLE, TCS, and DU models together with the measured field (solid lines) at 5.16 km for return stroke 8726_2. $\lambda = 2000$ m and $\tau_D = 0.1\,\mu$s. The measured current at the channel base and the measured return stroke speed are given in Fig. 12.14, bottom two panels. Adapted from Thottappillil and Uman (1993).

object will be equal to the current magnitude that would be injected directly into the grounding impedance of the object from the same current source if the object were absent. Note that the above example applies only to a step-function current wave; the current distribution along the object is more complex for the impulsive current waveform characteristic of a lightning return stroke. If the lightning current-wave round-trip time on the strike object is appreciably longer than the risetime of the current measured at the top of the object, the current peak reflected from the ground is separated from the incident-current peak in the overall current waveform, at least in the upper part of the object, as discussed in Section 6.5.

The presence of a vertically extended strike object may substantially increase the initial peaks of both the electric and magnetic fields and also the electric and magnetic field derivatives relative to the case where the return stroke is initiated at ground level. Janischewskyj et al. (1999) used the modified transmission line model with exponential current decay with height (MTLE) to simulate lightning strikes to the CN tower and to compute the associated vertical electric fields at a distance of 2 km from the tower. The MTLE model was extended to include multiple reflections both within the tower and within the upward-extending return-stroke channel. The propagation speed along the tower was assumed to be equal to the speed of light. Similar calculations were previously reported for magnetic fields at 2 km from the tower by Janischewskyj et al. (1998) and for both electric and magnetic fields by Rachidi et al. (1998, 2001). It was found from these calculations that the electric or magnetic field peak 2 km from the 553 m tower was dominated by its radiation component and that the contribution to the total field from the current in the tower was considerably larger than the contribution from the current in the lightning channel. Janischewskyj et al. (1999a) reported that the presence of the tower caused an increase in the electric field peak by a factor 2–3. Current reflections within the return-stroke channel (from the upward-moving return-stroke front and from the tower top) were found to have little effect on the electric field peak. Further discussion of the electromagnetic fields radiated by a source composed of a tall metallic object and a lightning channel attached to the object top is found in Section 6.6.

Note that when the shortest significant wavelength in the lightning current is much longer than the height of the strike object, there is no need to consider the distributed-circuit behavior of such an object. For example, if the

12.2. Return stroke

minimum significant wavelength is 300 m (corresponding to a frequency of 1 MHz), objects whose heights are about 30 m or less may be considered as lumped, in most cases as a short-circuit between the lightning channel base and the grounding impedance of the object; the corresponding equivalent circuit is shown in Fig. 6.9.

Return-stroke front speed at early times Baum (1990b) argued that at the instant of return-stroke initiation the geometry of the bottom few tens of meters of the leader channel is an inverted circular cone because the corona closer to ground has not had enough time for its full development. Propagation speeds of radial-corona streamers from conductors subjected to negative high voltage in the laboratory have been reported to be about 10^5 m s^{-1} (0.1 m μs^{-1}) (Cooray 1993), so that some microseconds are required for the development of a corona sheath with a radius of the order of meters. For stepped leaders, the average downward propagation speed is also of order 10^5 m s^{-1}, so that there is a relatively short delay in the corona-sheath formation as a stepped leader moves toward ground, although it is not clear what is occurring during the attachment process. For dart leaders, the downward propagation speeds (typically 10^7 m s^{-1}) are about two orders of magnitude higher than the radial-streamer speeds, so that the delay in the corona-sheath formation may be appreciable. The charge density in Baum's (1990b) model is zero at ground and increases linearly with height. This conical model of the leader charge distribution in the bottom part of the channel predicts an initial return-stroke speed of nearly c, the speed of light, because both the longitudinal channel current and the channel charge near the ground are confined in a volume of approximately the same radial dimension.

The speed was predicted by Baum to decrease in some hundreds of nanoseconds to approximately one-third of the speed of light when the return-stroke front has reached a height of the order of tens of meters where the corona sheath is fully developed (and the channel geometry is cylindrical), that is, where the radii of the current-carrying channel core and of the charge-containing corona sheath differ appreciably from each other. However, the optically measured return-stroke speed versus height profiles within 400 m of ground reported for two triggered-lightning strokes by Wang *et al.* (1999b) indicate an initial upward speed of the order of one-third to one-half the speed of light with apparently no systematic variation in the bottom 100 m or so of the channel. Wang *et al.* used the digital optical imaging system ALPS (Yokoyama *et al.* 1990) with 100 ns time resolution, and the spatial resolution of his measurements was about 30 m. Further, Weidman (1998) reported mean return-stroke speeds in the lowest 100 m of the lightning channel equal to 7.8×10^7 and 8.8×10^7 m s^{-1} for nine natural- and 14 triggered-lightning strokes, respectively.

Some researchers (e.g., Willett *et al.* 1989; Leteinturier *et al.* 1990; Uman *et al.* 2000, 2002) have attempted to estimate the return-stroke speed using Eqs. 12.4 and/or 12.5. Such estimates are necessarily model dependent and are often difficult to interpret, as discussed below. Leteinturier *et al.* (1990) estimated the return-stroke speed using (i) the measured peak values of the time derivative of the channel-base current, (ii) the measured peak values of the time derivative of the electric field at 50 m, and (iii) Eq. 12.5. They reported the speed values to be on average near c, 14 out of 40 values apparently being greater than the speed of light. Uman *et al.* (2000), using measured time derivatives of the electric fields at 10, 14, and 30 m, current-derivatives, obtained by numerical differentiation of the current records, and Eq. 12.5 estimated mean speed values as follows: 1.7×10^8 m s^{-1} (10 m, seven events); 3.1×10^8 m s^{-1} (14 m, three events); and 2.9×10^8 m s^{-1} (30 m, seven events).

Baum (1990b) invoked the model-dependent speed estimates of Leteinturier *et al.* (1990) in support of his theoretical prediction that the initial return-stroke speed is nearly equal to the speed of light. Similar estimates of speed using peak electric field derivatives measured at about 5 km gave a mean value of approximately two-thirds the speed of light (Willett *et al.* 1989). Further, the use of (i) the measured channel-base current peak, (ii) the measured electric field peak at about 5 km, and (iii) Eq. 12.4 led to a mean return-stroke speed of about one-half the speed of light, consistent with the corresponding optical speed measurements over the bottom 400–600 m of the channel (Willett *et al.* 1989). It is possible that, since the peak derivative precedes the peak of electric field or current, the speed estimates using Eq. 12.5, that is, using the peak values of the time derivatives of electric field and current, are representative of a somewhat lower channel section than those based on Eq. 12.4, that is, on the peak electric field and the peak current. However, this conjecture implies a very rapid speed decay within the bottom 100 m or so while, as noted above, the measurements of Wang *et al.* (1999b) do not appear to indicate a systematic speed variation near the bottom of the channel. Additionally, such a rapid speed decay would probably render invalid Eqs. 12.4 and 12.5, derived assuming a constant v.

One possible explanation for the discrepancy between the speeds inferred from Eq. 12.5 using the 14 to 50 m data (nearly c) and using the 5 km data ($c/3$) is the contribution of the induction and electrostatic field components at 14 to 50 m to the total electric-field-derivative peak (Cooray 1989a; Leteinturier *et al.* 1990; Uman *et al.* 2000, 2002). This contribution is not accounted for in Eq. 12.5, which was derived for the radiation field component only.

Uman et al. (2002), using the measured current-derivative waveforms for two triggered-lightning strokes and the transmission-line model, computed the total electric and magnetic field derivative waveforms 15 m from the lightning channel for three assumed values of return-stroke speed, 1×10^8, 2×10^8 and 3×10^8 m s^{-1}. They found that for the magnetic field derivative a reasonable model fit to the data in waveshape and amplitude is achieved for a return-stroke speed near 2×10^8 m s^{-1}. The electric-field-derivative waveshape matches were found to be reasonable for a similar or higher speed, but the amplitudes of the calculated derivatives were somewhat less than the observed values. Uman et al. (2001) also estimated the return-stroke speeds using their measured current and 15-m field derivative peaks together with Eq. 12.5 and its counterpart for the magnetic field derivative, for the same two triggered-lightning strokes. The results were 2.9×10^8 and 2.7×10^8 m s^{-1} for the two electric-field-derivative peaks and 1.9×10^8 and 1.8×10^8 m s^{-1} for the two magnetic-field-derivative peaks.

Additional discussion of speed estimates from Eqs. 12.4 and 12.5, related to the initial bidirectional extension of the return-stroke channel, is found below.

Return-stroke speed estimates from Eqs. 12.4 and 12.5 involve the assumption that this speed is constant over the radiating channel section. A different return-stroke speed profile was suggested by Gorin (1985). According to his distributed-circuit model for a first stroke, the speed initially increases to its maximum over a channel length of the order of some hundreds of meters and thereafter decreases. The initial speed increase in Gorin's model is associated with the so-called break-through phase (also called the final jump or switch-closing phase), thought to be responsible for the formation of the initial rising portion of the return-stroke current pulse (see also Rakov and Dulzon 1991 and Rakov et al. 1992).

Using the experimental data published by Schonland (1956), Srivastava (1966) proposed a bi-exponential expression for the first-return-stroke speed as a function of time, according to which the speed rises from zero to its peak and falls off afterwards. More experimental data on the attachment process and on the early stages of the return-stroke process are needed to deduce typical first- and subsequent-return-stroke speed profiles near ground.

Initial bidirectional extension of the return-stroke channel
The initial bidirectional extension of the return-stroke channel from the junction point between the descending leader and an upward connecting leader from the ground or a grounded object has been considered by, for example, Wagner and Hileman (1958), Uman et al. (1973b), Weidman and Krider (1978), Willett et al. (1988, 1989), and Leteinturier et al. (1990). General descriptions of this process are found in Section 4.5 and subsection 18.3.2. The first direct experimental evidence of such an extension was presented by Wang et al. (1999a). Using the ALPS with 3.6 m spatial resolution and 100 ns time resolution, they observed an upward connecting leader in one triggered-lightning stroke and inferred the existence of such a leader in another stroke. In both events, the return stroke was initially a bidirectional process, the upward- and downward-moving waves originating at height 7–11 m for the event that had an imaged upward connecting leader and at height 4–7 m for the event that did not have an imaged upward connecting leader. In the former case, the duration of the initial bidirectional extension of the return-stroke channel was estimated by Wang et al. (1999a) to be of the order of tens of nanoseconds, which is comparable to geometric mean risetimes reported by Uman et al. (2000) for electric field derivative waveforms at distances 10 to 30 m from the triggered-lightning channel.

Both upward- and downward-moving wavefronts necessarily contribute to the measured electric and magnetic fields, while the current measured at the channel base is thought to be associated only with the downward wave and its reflection at ground. Equations 12.4 and 12.5 are derived for a single wave and hence, in general, are invalid during the time of the initial bidirectional extension of the return-stroke channel. Indeed, two wavefronts moving in opposite directions from the junction point at, say, $v = 0.5c$ may produce a radiation field that appears as being due to a single wavefront moving upward at $v = c$. This effect of two wavefronts may explain at least some of the physically unrealistic values of the return-stroke speed, values greater than the speed of light, inferred using Eq. 12.5.

The electromagnetic model of Podgorski and Landt (1987) and the distributed-circuit model of Strawe (1979) include an upward connecting-discharge channel, which facilitates the initial bidirectional development of the return-stroke process.

Relation between leader and return-stroke models Usually the lightning leader and the lightning return stroke are modeled independently; the dart-leader and stepped-leader models are considered in Sections 12.3 and 12.4, respectively. Such an approach is based on the implicit assumption that the leader process ends or its effects become negligible when the return stroke begins. However, the return stroke operates on the charge deposited onto the channel by the preceding leader, and therefore these two lightning processes should be strongly coupled. Indeed, Rubinstein et al. (1995) and Rakov et al. (1998) inferred from triggered-lightning experiments that the return-stroke current peak at the channel-base is largely determined by the dart-leader charge density within the bottom tens to hundreds of meters of the channel, a height consistent with the typical return-stroke front speed and the typical channel-base current risetime. The product of these two quantities gives the height of the return-stroke front at the time when the channel-base current peak is formed. The formulation of a return-stroke model in terms of charge density (Thottappillil et al. 1997; see Tables 12.3 and 12.4) provides a direct link to the dart-leader model, assuming that all leader charge is neutralized by the return stroke and that the latter does not deposit any additional charge on the channel. Further, Rakov (1998) suggested that a subsequent return stroke could be viewed as a ground "reflection" of the dart-leader. Interestingly, Idone and Orville (1984) observed, in a New Mexico rocket-triggered lightning flash, the partial "reflection" (an upward-propagating luminosity wave) of the downward-propagating dart-leader luminosity wave from the junction between the upper, natural section of the channel and its lowest 400-m-long section formed along the triggering wire trace.

12.3. Dart leader

We first consider the "atomic physics" models (Schonland 1938, 1956; Loeb 1965, 1966; Jurenka and Barreto 1982, 1985). These models describe the ionization processes within the dart-leader front that facilitate the advancement of the leader. The principal output of these models that can be compared to observations is the front propagation speed.

Jurenka and Barreto (1982, 1985) assumed that, prior to the dart leader, the lightning channel is composed of a weakly ionized gas, and they considered this gas to be a mixture of three "fluids": neutral particles, ions, and electrons. They described the electron fluid using one-dimensional hydrodynamic equations (the equations of continuity, momentum, and energy conservation) in conjunction with Poisson's equation, in order to obtain the propagation characteristics of electron density gradients that simulated dart leaders moving along defunct lightning channels. The neutral and ion fluids were assumed to be stationary in relation to the electron fluid. Jurenka and Barreto claimed that the electron pressure wave, basically a sharp increase in the electron density at the wavefront, can propagate into a weakly ionized gas without appreciable attenuation and with a velocity exceeding the electron acoustic velocity of about 10^6 m s^{-1} in the gas and hence similar to observed dart-leader velocities (subsection 4.7.2). Borovsky (1995) criticized this electron-pressure driven-wave model on the grounds that (i) it actually cannot explain the observed dart-leader speeds (often in excess of 10^7 m s^{-1}; Section 4.7) because an electron-pressure shock wave cannot propagate faster than the electron thermal velocity behind the wavefront, which is about 7×10^5 m s^{-1} for a temperature as high as 30 000 K, (ii) electron pressure waves would only propagate for distances much less than 1 cm in lightning channels, and (iii) the electron pressure cannot transport energy fast enough to account for the air ionization and heating thought to be produced by dart leaders.

Schonland (1938, 1956) proposed a simple formula for the speed of an ionizing wave as a function of the initial electron density in the front, the electron drift velocity (which is considerably lower than the front velocity), and the wave-front length. Another simple speed equation was proposed by Loeb (1965). Loeb's formula uses the same input parameters as that of Schonland plus the final electron density (after the wave has traversed a distance equal to the wave-front length) and the number of new electrons created per unit length by a drifting electron (the first Townsend ionization coefficient). Both Schonland's and Loeb's formulas were discussed by Uman (1969, 1984). Schonland's formula is based on an arbitrary assumption that the time required for an ionizing wave to traverse a distance equal to the wave-front length is the same as the time necessary for each electron to travel the average distance (much smaller than the front length) between electrons. In Loeb's formula, the former time is equal to the time required for many electron avalanches to produce a specified increase in electron density within the front. Both formulas require a knowledge of quantities that are generally not known, and therefore they are of little practical value.

We now consider the electromagnetic model of the dart leader proposed by Borovsky (1995). He used Maxwell's equations to simulate both dart-leader and return-stroke processes as guided electromagnetic waves propagating along conducting cylindrical channels. The resistance per unit length of the channel guiding the dart-leader wave was assumed to be constant and equal to 600 Ω m^{-1}. The dart-leader wave was represented by a single, dominant sinusoid (about 160 kHz) for which various propagation characteristics were found from the model. Only a middle section of the lightning channel, undisturbed by the conditions at the channel ends, was considered. Rakov (1998) argued that Borovsky's model is not adequate because it predicts an attenuation of more than an order of magnitude within 100 m or so, contrary to the experimental data on dart-leader luminosity profiles (Jordan et al. 1997). Note that the attenuation distance given by Borovsky is based on an amplitude decay to less than 0.2 percent of the original value, instead of the generally assumed 37 percent. In summary, it appears that the electromagnetic model of Borovsky is not suitable for describing the dart leader.

Bazelyan (1995) modeled the dart leader as a transient process initiated by impressing a prescribed voltage at one end of an *RLC* transmission line short-circuited at the other end, with *L* and *C* assumed constant and *R* assumed to vary as a function of current. Thus, Bazelyan's *RLC* (distributed circuit) model is nonlinear in that it takes account of the reduction in R due to the joule heating of the channel by the dart-leader wave. The range of initial values of R used by Bazelyan was from 10 to 100 Ω m^{-1}, these values being two to three orders of magnitude lower than that estimated by Rakov (1998) for a pre-dart-leader channel. Since the current pulses at ground predicted by Bazelyan's model were characteristic of M-components (Section 4.9) rather than return strokes, Bazelyan concluded that electron impact ionization should be taken into account when modeling dart leaders. Thus, it appears that one cannot adequately describe the dart leader by RLC transmission line theory, even considering the reduction in R due to joule heating.

Bazelyan (1995), Cooray (1996b), and Thottappillil et al. (1997) suggested that the dart-leader current can be viewed as being generated at the downward-moving leader front and propagating in the upward direction. Cooray (1996b) presented a model based on this idea. In this model, the total current behind the dart-leader front, the front being of zero spatial extent, is the sum of the currents supplied by elemental current sources distributed along the channel and turned on by the downward-moving front. The upward propagation speed of current pulses behind the dart-leader front was assumed to be equal to 2×10^8 m s^{-1}. The downward propagation speed of the dart-leader front as a function of height was found by equating an expression for the electric field at the front to the critical breakdown field in the warm air ahead of the front, computed as a function of temperature and pressure. The model involves assumptions regarding the final charge density distribution along the dart-leader channel and regarding the dynamics of the charge deposition onto the channel. Cooray (1996b) performed a numerical experiment to determine the dependences of the dart-leader speed on various factors, but he did not present any model-predicted electric field waveforms that could be compared with the available experimental data.

We now discuss dart-leader models in which the emphasis is put on achieving agreement between the model predictions and the measured electric fields, as opposed to the previously discussed models, which were primarily concerned with dart-leader propagation characteristics. The most used model of the present type portrays the dart leader as a uniformly charged line emerging from the center of a volume of charge in the cloud and extending at a constant speed vertically toward the ground (e.g., Schonland et al. 1938; Uman 1987, Appendix A; Rubinstein et al. 1995). This charge transfer model also applies to stepped leaders (Section 12.4 below) and is illustrated in Fig. 4.18. The model can be described mathematically by the following charge density equation:

$$\rho_L(z', t) = \rho_{L0} u\left(t - \frac{H_m - z'}{v}\right) - \rho_{L0} v t \delta(H_m - z') u(t) \quad (12.6)$$

Here $\rho_L(z', t)$ is the line charge density as a function of the height z' along the channel and the time t, H_m is the height of the charge source above ground, ρ_{L0} is the constant line charge density, v is the propagation speed of the dart-leader front, which is assumed to have zero (negligible) spatial extent. The Heaviside function $u[t - (H_m - z')/v]$ ensures that $\rho_L(z', t) = \rho_{L0}$ at the front and at all points on the channel above the front and that $\rho_L(z', t) = 0$ below the downward-moving front; the Heaviside function $u(t)$ ensures that the second term of the equation exists only when $t \geq 0$. The spatial Dirac delta function $\delta(H_m - z')$ is equal to zero for all values of z' except for $z' = H_m$; its integral over z', including $z' = H_m$, is equal to unity. A less formal description of the model is found in subsection 4.4.4.

The first term of Eq. 12.6 represents the uniform charge density distribution along the channel from $z' = H_m - vt$ to $z' = H_m$, while the second term represents (after integration over z') a point charge of opposite polarity, whose magnitude is equal to the total charge on the channel, placed at the origin ($z' = H_m$) in order to satisfy the conservation of charge. Thus, the net charge of the system obtained as the integral of $\rho_L(z', t)$ over the entire channel length, including the source at $z' = H_m$, is zero at all times. The placement of a point charge at the origin simulates the more or less spherically symmetrical development of a large number of branched channels whose function is to "funnel" negative charges (in the case of a negative flash, the more common type) from the hydrometeors (defined in the first part of subsection 3.2.1) to the downward-propagating channel. One can view the entire leader system as being composed of negative and positive sections, the negative section extending more or less vertically toward the ground and the positive section being heavily branched inside the cloud.

Mazur and Ruhnke (1993) simulated the upper section of the leader system by a single-channel positive leader extending vertically upward at the same speed as the negative section of the leader system. They found that the charge density on such a vertically symmetrical bidirectional leader, regarded as a vertical conductor polarized in a uniform electric field, varied linearly with height, with zero charge density at the origin. The model of Mazur and Ruhnke predicts a ratio of the leader and the return-stroke electric field changes at 20 to 50 km to be between approximately 0.2 and 0.3, a factor 2 to 3 lower than the average value observed for first strokes at these distances by Beasley et al. (1982), while the uniformly-charged-leader model described by Eq. 12.6 is consistent with the experimental data (Fig. 4.21). Further testing of the bidirectional leader model is complicated by the well-documented absence of appreciable HF and VHF radiation from positive leaders (e.g., Shao et al. 1999). Apparently as a result of this absence, the positive section of a bidirectional leader has never been detected by either VHF–UHF-interferometer or time-of-arrival systems, although other types of positive in-cloud leaders have been observed to radiate at VHF as strongly as negative leaders (Rhodes et al. 1994; Shao et al. 1995; Shao and Krehbiel 1996). Further information on the characteristics of positive and negative leaders is found in subsection 5.3.2. Results of laboratory studies of the bidirectional leader have been reported by Castellani et al. (1998a, b). Bidirectional leaders initiated from aircraft in flight are discussed in Sections 10.1 and 10.4.

Rubinstein et al. (1995) used the leader model described by Eq. 12.6 to compute vertical electric fields and to infer ρ_{L0} and v from a comparison of the model-predicted fields with measured electric fields produced by triggered-lightning leaders at distances of 30 and 500 m. The results of their calculations are shown in Figs. 4.20a, b. Leader electric field changes predicted by this model at larger distances are shown in Fig. 4.19. Rakov et al. (1990) considered an inverted-L channel geometry and found that the horizontal channel section is important in explaining the observed evolution of leader electric field waveforms with increasing stroke order. Thomson (1985) generalized the simple model discussed above by considering inclined leader channels and three-dimensional charge distributions (the latter were intended to account for the effects of branching, typically not observed in dart leaders) but assuming, as for the simple model, that the charge density distribution behind the leader tip does not vary with time. Thottappillil et al. (1997) relaxed the latter assumption while considering only a vertical channel.

A uniform distribution of charge along the leader channel is consistent with the modified-transmission-line return-stroke model with linear current decay with height (MTLL model), discussed in subsection 12.2.5. Other engineering return-stroke models require a non-uniform distribution of leader charge along the channel (Thottappillil et al. 1997). Crawford et al. (2001) reported that the distance dependence of the dart-leader electric field change typically observed at distances ranging from tens to hundreds of meters from the lightning channel is close to an inverse proportionality. This finding is consistent with a uniform distribution of leader charge along the bottom kilometer or so of the channel (see Eq. 4.5).

12.4. Stepped leader

It is generally thought that the stepped leader involves both a stepping process and a more or less continuous charge transfer along the channel, the latter process probably being the cumulative result of many steps (e.g., Bazelyan et al. 1978; Bazelyan and Raizer 2000). The dart-leader model described by Eq. 12.6 can also serve as a charge transfer model for stepped leaders. In the following, we first discuss step models and then various stepped-leader models that do not describe the stepping process. The discussion in this section is primarily concerned with negative stepped leaders. Positive stepped leaders were considered in subsection 5.3.3, and a comparison of positive and negative leaders was given in subsection 5.3.2.

Attempts to develop a qualitative description of the stepping mechanism have been made by Schonland (1938, 1953, 1956), Bruce (1941, 1944), Komelkov (1947, 1950), Wagner and Hileman (1958, 1961), and Loeb (1966), based either directly on laboratory spark experiments or on those experiments in conjunction with Schonland's photographic observations of lightning stepped leaders. A step-formation mechanism based on more recent laboratory spark experiments (e.g., Gorin et al. 1976) is described in subsection 4.4.8.

A quantitative model of the stepping process based on a fluid dynamic approach (e.g., Fowler 1982) in conjunction with the phenomenological characteristics of various processes in laboratory sparks was proposed by Bondiou-Clergerie et al. (1996), who modeled a bidirectional leader, including a downward negative stepped leader. The results of the modeling were reported to be consistent with the observations of the bidirectional leader in altitude-triggered lightning (Laroche et al. 1991). The model of Bondiou-Clergerie et al. (1996) requires the initial electric field distribution in the gap as an input and relies on the assumed similarity between a lightning discharge and a laboratory spark in terms of the various processes comprising the leader and the initiation criteria for each of these processes.

An approach somewhat similar to that of Bondiou-Clergerie et al. was used by Larigaldie et al. (1992), who derived most of the basic physical concepts used in their lightning stepped-leader model from laboratory experiments with surface discharges and long sparks. They considered the time variation in the resistance per unit channel length (due to the radial channel expansion) and employed the method of moments (MOM) to find the currents in individual leader steps. The model was applied to the negative section of the bidirectional leader initiated from an in-flight instrumented C-160 aircraft, and the calculated results were reported to be in overall agreement with the lightning current pulses observed at the aircraft (subsection 10.3.4). Uman and McLain (1970a) applied the BG and TL models discussed above in subsection 12.2.5 to a short vertical channel segment simulating an individual leader step. A current source was applied at the top end of the segment, and the reflections from the segment-ends in the TL model were neglected.

We now discuss stepped-leader models that do not include a description of the stepping process. Klingbeil and Tidman (1974a) proposed a fluid dynamic model (see also Phelps 1974b; Klingbeil and Tidman 1974b), which is similar to that of Jurenka and Baretto (1982, 1985) for dart leaders but in which the wave moves into a neutral gas as opposed to moving into a weakly ionized gas. Photoionization plays an important role in the wave propagation in this model.

Baum (1981, 1990a) considered the stepped leader as waves of current I and charge density ρ_L propagating on a lossless coaxial RLC transmission line, although, as discussed in Section 12.3, the presence of a conducting path ahead of the leader tip is not a good assumption even for the dart leader. The transmission line's inductance per unit length L was assumed to be constant and its capacitance per unit length C was assumed to be a function of line charge density ρ_L. The dependence of C on ρ_L was introduced in order to simulate a radially formed corona surrounding the channel core. Baum gave an illustrative example in which both the current and charge waves were represented as step functions. Further, assuming (i) that the effective breakdown field was 2 MV m^{-1}, (ii) the radius of the channel core was 1 mm, and (iii) the radius of the fictitious outer return conductor of the coaxial system was 10 m, and varying the overall channel radius (including the radial corona sheath) from 0.1 to 2 m, Baum obtained leader propagation speeds varying from $0.62c$ to $0.26c$, charge density values varying from 11 to 220 μC m^{-1}, and a longitudinal channel current varying from 2.1 to 17 kA. These

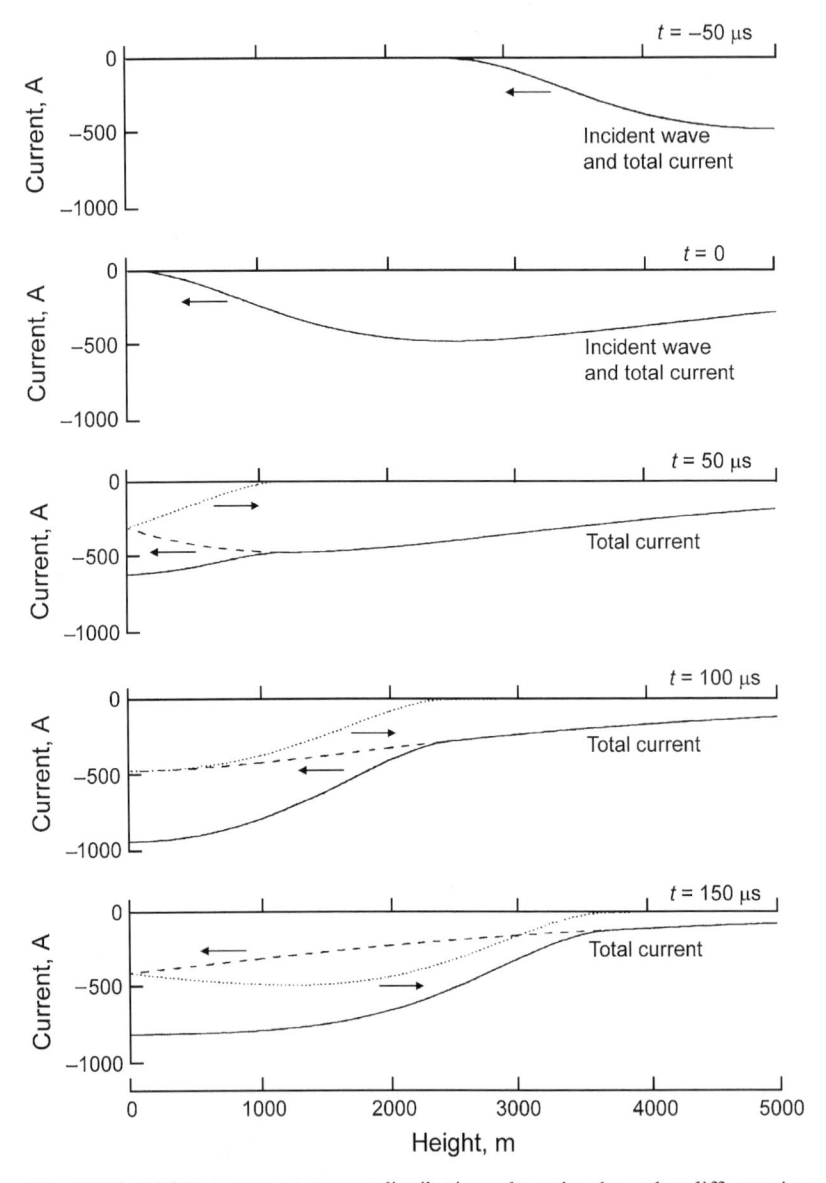

Fig. 12.18. (a) M-component current distributions along the channel at different times as predicted by the "two-wave" model (Eqs. 12.7 and 12.8). At $t = 0$, the incident wave front arrives at ground. The solid, broken, and dotted lines indicate respectively the total current, the incident wave and the wave. $v = 2.5 \times 10^7$ m s^{-1}. Adapted from Rakov et al. (1995).

12.5. M-component

et al. (1989), Dellera and Garbagnati (1990a, b), Takeuti (1992), Takeuti *et al.* (1993), Petrov and Petrova (1993), Dulzon *et al.* (1996, 1999), Kawasaki and Matsuura (2000), and Charalambakos *et al.* (2000).

12.5. M-component

As discussed in Sections 1.2 and 4.9, an M-component is thought to involve a downward-propagating incident wave (the analog of a leader) followed by an upward-propagating reflected wave (the analog of a return stroke). Before the downward incident wave makes contact with ground it is the only wave traveling along the channel, while after the contact is made both the incident and reflected waves exist simultaneously in the channel. Based on this mechanism, Rakov *et al.* (1995) proposed a "two-wave" (later termed "guided-wave") model for the M-component, in which the M-component current $I(z', t)$ at any height z' along a straight vertical channel of height H at time t is expressed as follows:

$$I(z', t) = I(H, t - (H - z')/v) \quad \text{if} \quad t < H/v \tag{12.7}$$

$$I(z', t) = I(H, t - (H - z')/v) + I(H, t - (H + z')/v) \quad \text{if} \quad t \geq H/v \tag{12.8}$$

where v is the M-wave propagation speed and $I(H, t)$ is the current injected at the top of the channel (at height H), equal to one-half of the total current measured at the channel base. As follows from Eqs. 12.7 and 12.8, each of the two M-component waves is described by the TL model (subsection 12.2.5), and the current reflection coefficient at ground is assumed to be equal to unity. The M-component model is illustrated in Figs. 12.18a, b, showing, respectively, current distributions (incident, reflected, and total) along the channel at different times and current waveforms (incident, reflected, and total) at different heights along the channel.

Rakov *et al.* (1995) tested the validity of the model using (i) the measured channel-base M-component current, (ii) the measured electric field 30 m from the channel, and (iii) an adjustable speed v, which essentially controls the M-component electric field magnitude but has relatively little effect on the field waveshape. The model-predicted electric fields 30 m from the channel for two M-components, along with measured fields and currents and with the value of v that provides the best field-magnitude match with measurements, are shown in Figs. 12.19 and 12.20. There is fairly good agreement between the calculated and observed electric fields. Additional information on the method of testing the validity of the M-component model proposed by Rakov *et al.* (1995), including the characteristic dependence

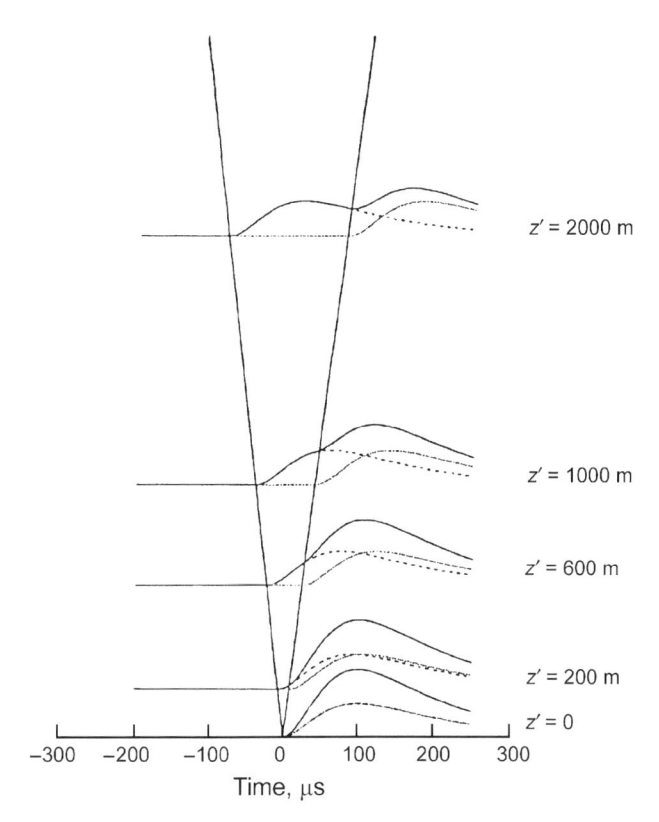

Fig. 12.18. (b) M-component current waveforms at different heights along the channel, as predicted by the "two-wave" model (Eqs. 12.7 and 12.8). At $t = 0$, the incident wave front arrives at ground. The solid, broken, and dotted lines indicate respectively the total current, the incident wave and the reflected wave. At $z' = 0$, the incident and reflected waves coincide with each other; note that the current waveforms have been inverted, for illustrative purposes, relative to (a). The slopes of the slanting lines in (b) give the speeds of the incident and reflected waves, 2.5×10^7 and $-2.5 \times 10^7 \text{m s}^{-1}$, respectively. Adapted from Rakov *et al.* (1995).

characteristics appear to be consistent with those expected for individual leader steps (subsection 4.4.6). Baum (1999) assumed that the leader channel extends essentially only during the step-formation process and used the experimental data of Baum *et al.* (1982, 1987, 1990) for a stepping process attributed by these researchers to an upward positive leader in rocket-triggered lightning. Baum estimated that the leader extension speed during the step formation process was about 10^7 to 10^8 m s^{-1}, two orders of magnitude greater than the average stepped-leader extension speed.

There exist a number of stepped-leader models that can be viewed as path-formation models. These models are primarily concerned with the geometry of the stepped-leader channel, including its tortuosity and branching. Such models have been proposed by Niemeyer (1987), Kawasaki

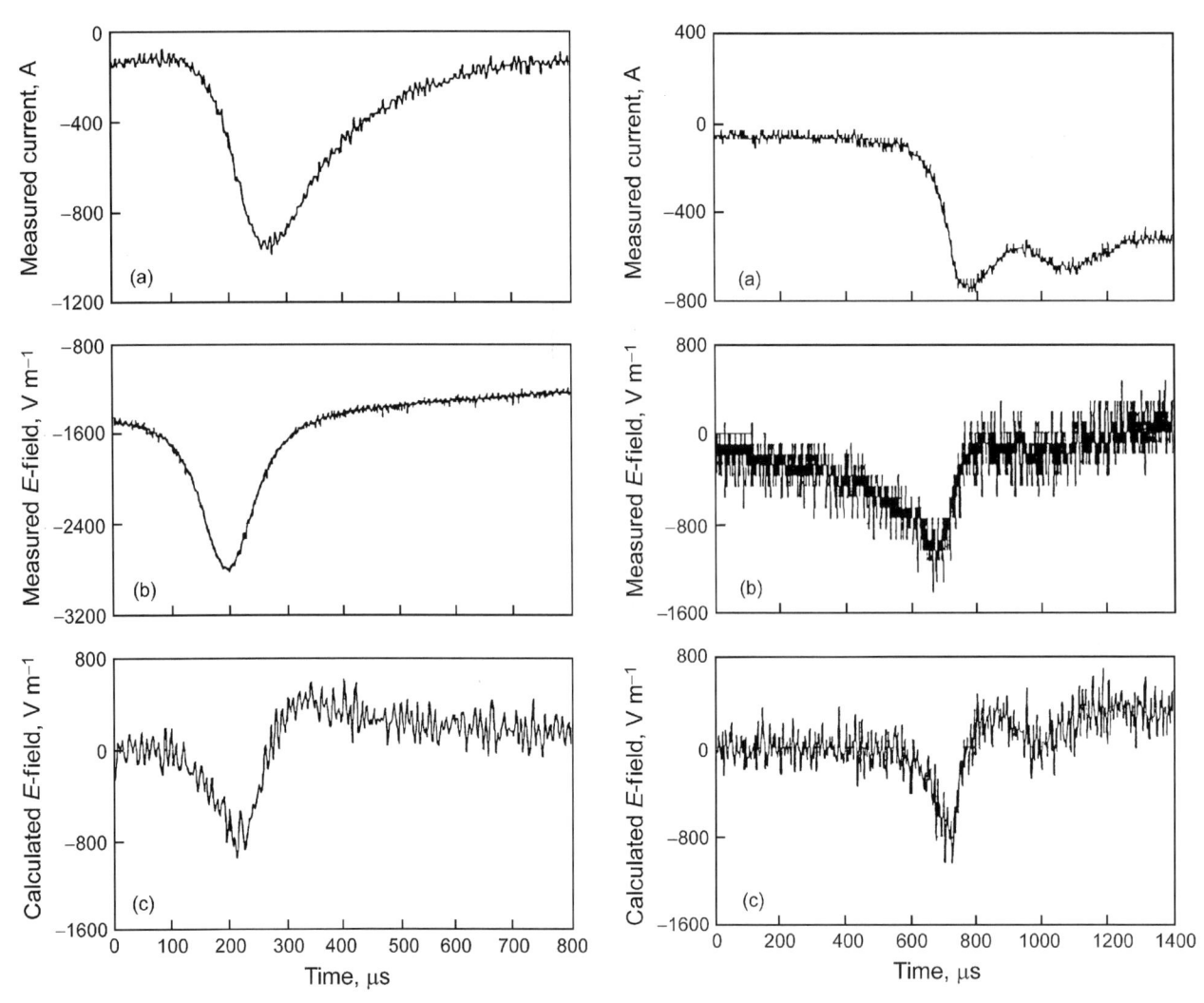

Fig. 12.19. (a) Measured channel-base current, (b) measured 30 m electric field, and (c) calculated 30 m electric field for an M-component that occurred about 0.6 ms after the first stroke of the seven-stroke flash 9320 triggered at Camp Blanding, Florida. The electric field in (c) was computed using the measured channel-base current in (a) and an equation similar to Eq. 4.10, the current distribution along the channel being specified by Eqs. 12.7 and 12.8. H was set at 5 km, and v was an adjustable parameter. The value of v providing the best field-magnitude match with the measurement in (b) was 3.0×10^7 m s^{-1}. Adapted from Rakov et al. (1995).

Fig. 12.20. Same as Fig. 12.19 but for an M-component that occurred about 2 ms after the last stroke of the five-stroke flash 9313 triggered at Camp Blanding, Florida. Here $v = 2.5 \times 10^7$ m s^{-1}. Adapted from Rakov et al. (1995).

of the M-component electric field on distance, is found in subsection 4.9.6.

12.6. Other processes

So far, the modeling of lightning processes other than return strokes, leaders, and M-components has been done only at the conceptual level, some aspects of such "models" being described in Sections 4.3, 4.5, 4.8, 4.10, and 4.11. Models of transient phenomena occurring between thundercloud tops and the ionosphere (blue starters, blue jets, and red sprites) and in the lower ionosphere (elves) are discussed in Chapter 14. Elves (Section 14.4) and sprites (subsection 14.3.3) are associated with individual lightning discharges (predominantly positive ground flashes in the case of sprites), while blue starters and blue jets (subsection 14.3.2) are apparently not accompanied by ordinary lightning discharges.

12.7. Summary

Most lightning return-stroke models can be assigned to one, sometimes two, of the following four classes: (i) gas dynamic models, (ii) electromagnetic models, (iii) distributed-circuit models, and (iv) engineering models. The testing of model validity is a necessary component of modeling. For the gas dynamic models, testing is based on the observed optical power and spectral output from lightning. The electromagnetic, distributed-circuit, and engineering models are most conveniently tested using measured electric and magnetic fields from natural and triggered lightning. Based on the entirety of the testing results and on mathematical simplicity, we rank the engineering models in the following descending order: MTLL, DU, MTLE, TCS, BG, and TL. When only the relation between the initial peak values of the channel-base current and the remote electric or magnetic fields is concerned, the TL model is preferred. The most used model of the dart or stepped leader portrays the leader process as a uniformly charged line emerging from the center of a volume of charge in the cloud and extending at a constant speed vertically toward the ground. This model includes the placement of a point charge of opposite sign at the origin whose magnitude is equal to the charge on the leader channel. The point charge simulates a large number of branched channels whose function is to "funnel" cloud charges to the downward-extending channel. The lightning M-component is modeled initially as an incident, downward-propagating wave and, after the arrival of this wave at ground, as the superposition of the incident wave and a reflected, upward-propagating wave, each of these two M-component waves being described by the TL model.

References and bibliography for Chapter 12

Abbas, I., and Bayle, P. 1981. Non-equilibrium between electrons and field in a gas breakdown ionizing wave, I, Macroscopic model. *J. Phys. D* **14**: 549–60.

Agrawal, A.K., Price, H.J., and Gurbaxani, S.H. 1980. Transient response of multiconductor transmission lines excited by a nonuniform electromagnetic field. *IEEE Trans. Electromagn. Compat.* **22**: 119–29.

Aleksandrov, N.L., Bazelyan, E.M., and Shneider, M.N. 2000. Effect of continuous current during pauses between successive strokes on the decay of the lightning channel. *Plasma Phys. Rep.* **26**: 952–60.

Amoruso, V., and Lattarulo, F. 1993. The electromagnetic field of an improved lightning return-stroke representation. *IEEE Trans. Electromagn. Compat.* **35**: 317–28.

Amoruso, V., and Lattarulo, F. 1994. Reply to comments on "The EM field of an improved lightning return-stroke representation" by J.R. Wait. *IEEE Trans. Electromagn. Compat.* **36**: 258–59.

Andreotti, A., Delfino, F., Girdinio, P., and Verolino, L. 2001a. An identification procedure for lighting return strokes. *J. Electrostat.* **51–2**: 326–32.

Andreotti, A., De Martinis, U., and Verolino, L. 2001b. An inverse procedure for the return stroke current identification. *IEEE Trans. Electromagn. Compat.* **43**(2): 155–60.

Andreotti, A., De Martinis, U., and Verolino, L. 2001c. Comparison of electromagnetic field for two different lightning pulse current models. *European Trans. Electr. Pow.* **11**: 221–6.

Andreotti, A., Delfino, F., Girdinio, P., and Verolino, L. 2001d. A field-based inverse algorithm for the identification of different height lightning return strokes. *Int. J. Comp. Math. Electr. Electron. Eng.* **20**: 724–31.

Baba, Y., and Ishii, M. 2001. Numerical electromagnetic field analysis of lightning current in tall structures. *IEEE Trans. Pow. Del.* **16**(2): 324–8.

Baker, L. 1987. Return-stroke transmission line model. *Electromagnetics* **7**: 229–40.

Baker, L. 1990. Return-stroke transmission line model. In *Lightning Electromagnetics*, ed. R.L. Gardner, pp. 63–74, New York: Hemisphere.

Baker, L., Gardner, R.L., Paxton, A.H., Baum, C.E., and Rison, W. 1987. Simultaneous measurement of current, electromagnetic fields, and optical emission from a lightning stroke. *Electromagnetics* **7**: 441–50.

Barreto, E., Jurenka, H., and Reynolds, S.I. 1977. The formation of small sparks. *J. Appl. Phys.* **48**: 4510–20.

Baum, C.E. 1981. Properties of lightning-leader pulses. Lightning phenomenology notes, Air Force Weapons Laboratory, Note 2, 22 December 1981, 23 pp.

Baum, C.E. 1990a. Properties of lightning-leader pulses. *Lightning Electromagnetics*, ed. R.L. Gardner, pp. 3–16, New York: Hemisphere.

Baum, C.E. 1990b. Return-stroke initiation. In *Lightning Electromagnetics*, ed. R.L. Gardner, pp. 101–14, New York: Hemisphere.

Baum, C.E. 1999. Leader-pulse step-formation process. Lightning phenomenology notes, Air Force Research Laboratory, Note 20, 8 June 1999, 20 pp.

Baum, C.E., and Baker, L. 1987. Analytic return-stroke transmission-line model. *Electromagnetics* **7**: 205–28.

Baum, C.E., and Baker, L. 1990. Analytic return-stroke transmission-line model. In *Lightning Electromagnetics*, ed. R.L. Gardner, pp. 17–40, New York: Hemisphere.

Baum, C.E., and Gardner, R.L. 1986. An introduction to leader tip modeling. *Electromagnetics* **6**: 111–15.

Baum, C.E., Breen, E.L., O'Neill, J.P., Moore, C.B., and Hall, D.L. 1980. Measurements of electromagnetic properties of lightning with 10 nanosecond resolution. In *Lightning technology*, NASA Conf. Publ. 2128, FAA-RD-80-30, pp. 39–84.

Baum, C.E., Breen, E.L., O'Neill, J.P., Moore, C.B., and Hall, D.L. 1982. Measurements of electromagnetic properties of lightning with 10 nanosecond resolution (revised). Lightning phenomenology notes, Air Force Weapons Laboratory, Note 3, 5 February 1982, 263 pp.

Baum, C.E., O'Neill, J.P., Breen, E.L., Hall, D.L., and Moore, C.B. 1987. Electromagnetic measurement of and location of lightning. *Electromagnetics* **7**: 395–422.

Baum, C.E., O'Neill, J.P., Breen, E.L., Hall, D.L., and Moore, C.B. 1990. Electromagnetic measurement of and location of lightning. In *Lightning Electromagnetics*, ed. R.L. Gardner, pp. 319–46, New York: Hemisphere.

Bazelyan, E.M. 1995. Waves of ionization in lightning discharge. *Plasma Phys. Rep.* **21**: 470–8.

Bazelyan, E.M., and Raizer, Yu. P. 2000. *Lightning Physics and Lightning Protection*, 325 pp., Bristol: IOP Publishing.

Bazelyan, E.M., Gorin, B.N., and Levitov, V.I. 1978. *Physical and Engineering Foundations of Lightning Protection*, Gidrometeoizdat, Leningrad, 223 pp.

Beasley, W.H., Uman, M.A., and Rustan, P.L. 1982. Electric fields preceding cloud to ground lightning flashes. *J. Geophys. Res.* **87**: 4884–902.

Bizjaev, A.S., Larionov, V.P., and Prokhorov, E.H. 1990. Energetic characteristics of lightning channel. In *Proc. 20th Int. Conf. on Lightning Protection, Interlaken, Switzerland*, pp. 1.1/1–13.

Bondiou, A., and Gallimberti, I. 1994. Theoretical modelling of the development of the positive spark in long gaps. *J. Phys. D: Appl. Phys.* **27**: 1252–66.

Bondiou-Clergerie, A., Bacchiega, G.L., Castellani, A., Lalande, P., Laroche, P., and Gallimberti, I. 1996. Experimental and theoretical study of the bi-leader process. Part II: Theoretical investigation. In *Proc. 10th Int. Conf. on Atmospheric Electricity, Osaka, Japan*, pp. 676–9.

Bondiou-Clergerie, A., Lalande, P., Laroche, P., Willett, J.C., Davis, D., and Gallimberti, I. 1999. The inception phase of positive leaders in triggered lightning: comparison of modeling with experimental data. In *Proc. 11th Int. Conf. on Atmospheric Electricity, Guntersville, Alabama*, pp. 22–5.

Borovsky, J.E. 1995. An electrodynamic description of lightning return strokes and dart leaders: guided wave propagation along conducting cylindrical channels. *J. Geophys. Res.* **100**: 2697–726.

Borovsky, J.E. 1998. Lightning energetics: estimates of energy dissipation in channels, channel radii, and channel-heating risetimes. *J. Geophys. Res.* **103**: 11 537–53.

Braginskii, S.I., 1958. Theory of the development of a spark channel. *Sov. Phys. JETP (English transl.)* **34**: 1068–74.

Bruce, C.E.R. 1941. The lightning and spark discharges. *Nature* **147**: 805–6.

Bruce, C.E.R. 1944. The initiation of long electrical discharges *Proc. Roy. Soc. A* **183**: 228–42.

Bruce, C.E.R., and Golde, R.H. 1941. The lightning discharge. *J. Inst. Electr. Eng.* **88**: 487–520.

Castellani, A., Bondiou-Clergerie, A., Lalande, P., Bonamy, A., and Gallimberti, I. 1998a. Laboratory study of the bi-leader process from an electrically floating conductor. Part 1: General results. *IEE Proc. Sci. Meas. Technol.* **145**: 185–92.

Castellani, A., Bondiou-Clergerie, A., Lalande, P., Bonamy, A., and Gallimberti, I. 1998b. Laboratory study of the bi-leader process from an electrically floating conductor. Part 2: Bi-leader properties. *IEE Proc. Sci. Meas. Technol.* **145**: 193–9.

Cerri, G., Chiarandini, S., Costantini, S., De Leo, R., Primiani, V.M., and Russo, P. 2002. Theoretical and experimental characterization of transient electromagnetic fields radiated by electrostatic discharge (ESD) currents. *IEEE Trans. Electromagn. Compat.* **44**: 139–147.

Chang, D.C. 1973. Electromagnetic pulse propagation over a conducting earth. Technical Report no. 8, Dept Electr. Eng., University of Colorado, Boulder, Colorado (sponsored by the National Oceanic and Atmospheric Administration).

Chang, D.C., and Fisher, R.J. 1974. A unified theory on radiation of a vertical electric dipole above a dissipative Earth. *Radio Sci.* **9**: 1129–38.

Chang, D.C., and Wait, J.R. 1970. Appraisal of near-field solutions for a Hertzian dipole over a conducting half-space. *Can. J. Phys.* **48**: 738–43.

Charalambakos, V., Kupershtokh, A.L., Agoris, D., Karpov, D.I., and Danikas, M. 2000. An approach in modeling of lightning processes using cellular automata. In *Proc. Int. Conf. on Lightning Protection, Rhodes, Greece*, pp. 72–7.

Chowdhuri, P., and Kotapallil, A.K. 1989. Significant parameters in estimating the striking distance of lightning strokes to overhead lines. *IEEE Trans. Pow. Del.* **4**: 1970–81.

Cooray, V. 1987. Effects of propagation on the return stroke radiation fields. *Radio Sci.* **22**: 757–68.

Cooray, V. 1989a. Derivation of return stroke parameters from the electric and magnetic field derivatives. *Geophys. Res. Lett.* **16**: 61–4.

Cooray, V. 1989b. A return stroke model. In *Proc. 1989 Int. Conf. on Lightning and Static Electricity, University of Bath, UK*, paper 6B.4, 6 pp.

Cooray, V. 1992. Horizontal fields generated by return strokes. *Radio Sci.* **27**: 529–37.

Cooray, V. 1993. A model for subsequent return strokes. *J. Electrostat.* **30**: 343–54.

Cooray, V. 1994. Calculating lightning-induced overvoltages in power lines: a comparison of two coupling models. *IEEE Trans. Elecromagn. Compat.* **36**: 179–82.

Cooray, V. 1996a. Possible influence of the mechanism of return stroke initiation on the remote sensing of lightning current parameters through first return stroke radiation fields. *J. Atmos. Electr.* **16**: 133–44.

Cooray, V. 1996b. A model for dart leaders in lightning flashes. *J. Atmos. Electr.* **16**: 145–59.

Cooray, V. 1997a. A model for negative first return strokes in lightning flashes. *Physica Scripta* **55**: 119–28.

Cooray, V. 1997b. Energy dissipation in lightning flashes. *J. Geophys. Res.* **102**: 21 401–10.

Cooray, V. 1998. Predicting the spatial and temporal variation of the electromagnetic fields, currents, and speeds of subsequent return strokes. *IEEE Trans. Electromagn. Compat.* **40**: 427–35.

Cooray, V. 2000. The modeling of positive return strokes in lightning flashes. *J. Atmos. Solar-Terr. Phys.* **62**: 169–87.

Cooray, V. 2001. Underground electromagnetic fields generated by the return strokes of lightning flashes. *IEEE Trans. Electromagn. Compat.* **42**: 75–84.

Cooray, V., and Gomes, C. 1998. Estimation of peak return stroke currents, current time derivatives and return stroke

velocities from measured fields. *J. Electrostat.* **43**: 163–72.
Cooray, V., and Lundquist, S. 1983. Effects of propagation on the rise times and the initial peaks of radiation fields from return strokes. *Radio Sci.* **18**: 409–15.
Cooray, V., and Ming, Y. 1994. Propagation effects on the lightning-generated electromagnetic fields for homogeneous and mixed sea–land paths. *J. Geophys. Res.* **99**: 10 641–52. (Correction, *J. Geophys. Res.* **104**: 12 227, 1999.)
Cooray, V., and Orville, R.E. 1988. Modeling of the return strokes. University of Uppsala Report UURIE: 208-88, 44 pp., Uppsala, Sweden.
Cooray, V., and Orville, R.E. 1990. The effects of variation of current amplitude, current risetime, and return stroke velocity along the return stroke channel on the electromagnetic fields generated by return strokes. *J. Geophys. Res.* **95**: 18 617–30.
Cravath, A.M., and Loeb, L.B. 1935. The mechanism of the high velocity of propagation of lightning discharges. *Physics (now J. Appl. Phys.)* **6**: 125–7.
Crawford, D.E., Rakov, V.A., Uman, M.A., Schnetzer, G.H., Rambo, K.J., Stapleton, M.V., and Fisher, R.J. 2001. The close lightning electromagnetic environment: dart-leader electric field change versus distance. *J. Geophys. Res.* **106**: 14 909–17.
Cvetic, J.M., and Stanic, B.V. 1997. LEMP calculation using an improved return stroke model. In *Proc. 12th Int. Symp. on Electromagnetic Compatibility, Zurich, Switzerland*, pp. 77–82.
Cvetic, J., Heidler, F., and Schwab, A. 1999. Light intensity emitted from the lightning channel: comparison of different return stroke models. *J. Phys. D, Appl. Phys.* **31**: 273–82.
Dellera, L., and Garbagnati, E. 1990a. Lightning stroke simulation by means of the leader progression model. Part I: Description of the model and evaluation of exposure of free-standing structures. *IEEE Trans. Pow. Del.* **5**: 2009–20.
Dellera, L., and Garbagnati, E. 1990b. Lightning stroke simulation by means of the leader progression model. Part II: Exposure and shielding failure evaluation of overhead lines with assessment of application graphs. *IEEE Trans. Pow. Del.* **5**: 2023–9.
Dennis, A.S., and Pierce, E.T. 1964. The return stroke of the lightning flash to earth as a source of VLF atmospherics. *Radio Sci.* **68D**: 779–94.
Diendorfer, G. 1990. Induced voltage on an overhead line due to nearby lightning. *IEEE Trans. Electromagn. Compat.* **32**: 292–9.
Diendorfer, G., and Uman M.A. 1990. An improved return stroke model with specified channel-base current. *J. Geophys. Res.* **95**: 13 621–44.
Drabkina, S.I. 1951. The theory of the development of the spark channel. *J. Exper. Theoret. Phys.* **21**: 473–83. (English translation, AERE LIB/Trans. 621, Harwell, Berkshire, UK.)
Dubovoy, E.I. 1997. Simultaneous measurements of electric-field pulses and radio reflections from lightning and experimental test of the results of numerical simulation. *Izvestiya AN-Fizika Atmosfery i Okeana* **33**(1): 122–31.
Dubovoy, E.I., Pryazhinsky, V.I., and Chitanava, G.I. 1991a. Calculation of energy dissipation in lightning channel. *Meteorologiya i Gidrologiya* **2**: 40–5.
Dubovoy, E.I., Pryazhinsky, V.I., and Bondarenko, V.E. 1991b. Numerical modeling of the gasodynamical parameters of a lightning channel and radio-sounding reflection. *Izvestiya AN-Fizika Atmosfery i Okeana* **27**: 194–203.
Dubovoy, E.I., Mikhailov, M.S., Pryazhinsky, V.I., Ogon'kov, A.L., Adjiev, A.Kh., Derkach, V.M., and Sigachev, S.M. 1993. The simultaneous measurements of electric field impulses and radar wave reflection from lightning discharge and comparison with results of numerical modeling. *Izvestiya AN-Fizika Atmosfery i Okeana* **29**(3): 364–8.
Dubovoy, E.I., Mikhailov, M.S., Ogonkov, A.L., and Pryazhinsky, V.I. 1995. Measurement and numerical modeling of radio sounding reflection from a lightning channel. *J. Geophys. Res.* **100**: 1497–502.
Dulzon, A.A., and Rakov, V. A. 1980. Estimation of errors in lightning peak current measurements by frame aerials. *Izvestiya VUZ-Energetika* **11**: 101–4.
Dulzon, A.A., Noskov, M.D., Lopatin, V.V., and Shelukhin, D.V. 1996. The strike points distribution from fractal model of the stepped leader. In *Proc. 10th Int. Conf. on Atmospheric Electricity, Osaka, Japan*, pp. 260–3.
Dulzon, A.A., Lopatin, V.V., Noskov, M.D., and Pleshkov, O.I. 1999. Modeling the development of the stepped leader of a lightning discharge. *Tech. Phys.* **44**(4): 394–8.
Fernsler, R.F. 1984. General model of streamer propagation. *Phys. Fluids* **27**(4): 1005–12.
Few, A.A. 1969. Power spectrum of thunder. *J. Geophys. Res.* **74**: 6926–34.
Few, A.A. 1995. Acoustic radiations from lightning. In *Handbook of Atmospheric Electrodynamics*, vol. II, pp. 1–31, Boca Raton, Florida: CRC Press.
Fofana, I., and Beroual, A. 1998. Induced effects on an overhead line due to nearby positive lightning downward leader. *Electric Power Systems Res.* **48**: 105–19.
Fofana, I., Ben Rhouma, A. Beroual, A., and Auriol, P. 1998. Modelling a positive lightning downward leader to study its effects on engineering systems. *IEE Proc. Gen. Transm. Distr.* **145**(4): 395–403.
Fowler, R.G. 1974. Nonlinear electron acoustic waves, part I. *Adv. Electronics Electron Phys.* **35**: 1–86.
Fowler, R.G. 1976. Non-linear electron acoustics waves, part II. *Adv. Electronics Electron Phys.* **41**: 1–72.
Fowler, R.G. 1982. Lightning. *Appl. Atmos. Collision Phys.* **5**: 31–67.
Gallimberti, I. 1979. The mechanism of the long spark formation. *J. de Phys. Coll. C7*, **40**: 193–250.
Gardner, R.L. 1980. A model of the lightning return stroke. Ph.D. dissertation, University of Colorado.
Gardner, R.L. 1981. Effect of the propagation path on lightning-induced transient fields. *Radio Sci.* **16**: 377–84.

Gardner, R.L. 1990. Effect of the propagation path of lightning – induced transient fields. In *Lightning Electromagnetics*, pp. 139–53, New York: Hemisphere.

Gardner, R.L., Frese, M.H., Gilbert, J.L., and Longmire, C.L. 1984. A physical model of nuclear lightning. *Phys. Fluids* **27**(11): 2694–8.

Gomes, C., and Cooray, V. 2000. Concepts of lightning return stroke models. *IEEE Trans. Electromagn. Compat.* **42**: 82–96.

Gorbachev, L.P., and Fedorov, V.F. 1977. Electromagnetic radiation from a return streamer of lightning. *Geomagn. Aeron.* **17**: 641–2.

Gorin, B.N. 1985. Mathematical modeling of the lightning return stroke. *Elektrichestvo* **4**: 10–16.

Gorin, B.N. and Markin, V.I. 1975. Lightning return stroke as a transient process in a distributed system. *Trudy ENIN* **43**: 114–30.

Gorin, B.N., and Shkilev, A.V. 1984. Measurements of lightning currents at the Ostankino tower. *Elektrichestvo* **8**: 64–5.

Gorin, B.N., Levitov, V.I., and Shkilev, A.V. 1976. Some principles of leader discharge of air gaps with a strong non-uniform field. In *Gas Discharges*, IEE Conf. Publ. 143, pp. 274–8.

Gorin, B.N., Levitov, V.I., and Shkilev, A.V. 1977. Lightning strikes to the Ostankino tower. *Elektrichestvo* **8**: 19–23.

Goshima, H., Motoyama, H., Asakawa, A., Wada, A., Shindo, T., and Yokoyama, S. 2000. Characteristics of electromagnetic fields due to winter lightning stroke current to a high stack. *Trans. IEE Japan* **120-B**(1): 44–8.

Grover, M.K. 1981. Some analytic models for quasi-static source region EMP: application to nuclear lightning. *IEEE Trans. Nuclear Sci.* **28**: 990–4.

Guerrieri, S., Heidler, F., Nucci, C.A., Rachidi, F., and Rubinstein, M. 1996. Extension of two return stroke models to consider the influence of elevated strike objects on the lightning return stroke current and the radiated electromagnetic field: comparison with experimental results. In *Proc. Int. Symp. on Electromagnetic Compatibility (EMC '96 ROMA), Rome, Italy*, pp. 701–6.

Guerrieri, S., Nucci, C.A., Rachidi, F., and Rubinstein, M. 1998. On the influence of elevated strike objects on directly measured and indirectly estimated lightning currents. *IEEE Trans. Pow. Del.* **13**: 1543–55.

Guerrieri, S. Krider, E.P., and Nucci, C.A. 2000. Effects of traveling-waves of current on the initial response of a tall Franklin rod. In *Proc. 25th Int. Conf. on Lightning Protection, Rhodes, Greece*, pp. 94–9.

Guo C., and Krider, E.P. 1982. The optical and radiation field signatures produced by lightning return strokes. *J. Geophys. Res.* **87**: 8913–22.

Guo C., and Krider, E.P. 1983. The optical power radiated by lightning return strokes. *J. Geophys. Res.* **88**: 8621–2.

Gupta, S.P., Rai, J., and Tantry, B.A.P. 1974. Radiation resistance characteristics of the multiple return stroke lightning. *Ann. Geophys.* **30**: 435–40.

Hager, W.W., and Wang, D. 1995. An analysis of errors in the location, current, and velocity of lightning. *J. Geophys. Res.* **100**: 25 721–9.

Haldar, M.K., and Liew, A.C. 1987. Validation of Rusck's scalar and vector potential expressions due to a return stroke in a lightning channel. *IEE Proc.* **134C**: 366–7.

He, S., Popov, M., and Romanov, V. 2000. Explicit full identification of a transient dipole source in the atmosphere from measurement of the electromagnetic fields at several points at ground level. *Radio Sci.* **35**(1): 107–17.

Heckman, S.J., and Williams, E.R. 1989. Corona envelopes and lightning currents. *J. Geophys. Res.* **94**: 13 287–94.

Heidler, F. 1985. Traveling current source model for LEMP calculation. In *Proc. 6th Int. Symp. on Electromagnetic Compatibility, Zurich, Switzerland*, pp. 157–62.

Heidler, F., and Hopf, Ch. 1994. Lightning current and lightning electromagnetic impulse considering current reflection at the Earth's surface. In *Proc. 22nd Int. Conf. on Lightning Protection, Budapest, Hungary*, paper R 4-05.

Heidler, F., and Hopf, Ch. 1995. Influence of channel-base current and current reflections on the initial and subsidiary lightning electromagnetic field peak. In *Proc. 1995 Int. Aerospace and Ground Conf. on Lightning and Static Electricity, Williamsburg, Virginia*, pp. 18/1–10.

Heidler, F., and Hopf, Ch. 1996. On the influence of the ground conductivity, the current reflections and the current generation on the electric field in a general TCS-model. In *Proc. 23rd Int. Conf. on Lightning Protection, Florence, Italy*, pp. 316–21.

Heidler, F., Cvetic, J.M., and Stanic, B.V. 1999. Calculation of lightning current parameters. *IEEE Trans. Pow. Del.* **14**: 399–404.

Hill, E.L. 1957. Electromagnetic radiation from lightning strokes. *J. Franklin Inst.* **263**: 107–9.

Hill, R.D. 1966. Electromagnetic radiation from the return stroke of a lightning discharge. *J. Geophys. Res.* **71**: 1963–7.

Hill, R.D. 1969. Electromagnetic radiation from erratic paths of lightning strokes. *J. Geophys. Res.* **74**: 1922–9.

Hill, R.D. 1971. Channel heating in return stroke lightning. *J. Geophys. Res.* **76**: 637–45.

Hill, R.D. 1972. Optical absorption in the lightning channel. *J. Geophys. Res.* **77**: 2642–7.

Hill, R.D. 1973. Lightning induced by nuclear bursts. *J. Geophys. Res.* **78**: 6355–8.

Hill, R.D. 1975. Comments on "Quantitative analysis of a lightning return stroke for diameter and luminosity changes as a function of space and time" by R.E. Orville, J.H. Helsdon Jr, and W.H. Evans. *J. Geophys. Res.* **80**: 1188.

Hill, R.D. 1977a. Energy dissipation in lightning. *J. Geophys. Res.* **82**: 4967–8.

Hill, R.D. 1977b. Comments on "Numerical simulation of spark discharges in air" by M.N. Plooster. *Phys. Fluids* **20**: 1584–6.

Hill, R.D. 1987. Comments on "Lightning return stroke. A numerical calculation of the optical radiation". *Phys. Fluids* **30**: 2585–6.

Himley, R.O. 1969. VLF radiation from subsequent return strokes in multiple stroke lightning. *J. Atmos. Terr. Phys.* **31**: 749–53.

Hoole, P.R.P. 1993. Modeling the lightning earth flash return stroke for studying its effects on engineering systems. *IEEE Trans. Magn.* **29**: 1839–44.

Hoole, P.R.P., and Balasuriya, B.A.A.P. 1993. Lightning radiated electromagnetic fields and high voltage test specifications. *IEEE Trans. Magn.* **29**: 1845–8.

Hoole, P.R.P., and Hoole, S.R.H. 1988. Guided waves along an unmagnetized lightning plasma channel. *IEEE Trans. Magn.* **24**: 3165–7.

Hoole, P.R.P., and Hoole, S.R.H. 1993. Simulation of lightning attachment to open ground, tall towers and aircraft. *IEEE Trans. Pow. Del.* **8**: 732–40.

Hubert, P. 1985. A new model of lightning subsequent stroke – confrontation with triggered lightning observations. In *Proc. 10th Int. Conf. on Lightning and Static Electricity, Paris, France*, paper 4B4.

Idone, V.P., and Orville, R.E. 1984. Three unusual strokes in a triggered lightning flash. *J. Geophys. Res.* **89**: 7311–16.

Iwata, A. 1970. Calculations of waveforms radiating from return strokes. *Proc. Res. Inst. Atmospherics, Nagoya Univ., Japan* **17**: 115–23.

Janischewskyj, W., Shostak, V., and Hussein, A.M. 1998. Comparison of lightning electromagnetic field characteristics of first and subsequent return strokes to a tall tower: 1. Magnetic field. In *Proc. 24th Int. Conf. on Lightning Protection, Birmingham, UK*, pp. 245–51.

Janischewskyj, W., Shostak, V., and Hussein, A.M. 1999a. Lightning electric field characteristics of first and subsequent return strokes to a tall tower. In *Proc. 11th Int. Symp. on High Voltage Engineering, London, UK*, IEE Publ. no. 467, vol. 1, pp. 270–4.

Jones, D.L. 1970. Electromagnetic radiation from multiple return strokes of lightning. *J. Atmos. Terr. Phys.* **32**: 1077–93.

Jones, R.D., and Watts, H.A. 1975. Close-in magnetic fields of a lightning return stroke. Sandia Laboratories Report, SAND75-0114, 34 pp.

Jordan, D.M., and Uman, M.A. 1983. Variation in light intensity with height and time from subsequent lightning return strokes. *J. Geophys. Res.* **88**: 6555–62.

Jordan, D.M., Rakov, V.A., Beasley, W.H., and Uman, M.A. 1997. Luminosity characteristics of dart leaders and return strokes in natural lightning. *J. Geophys. Res.* **102**: 22 025–32.

Jurenka, H., and Barreto, E. 1982. Study of electron waves in electrical discharge channels. *J. Appl. Phys.* **53**: 3581–90.

Jurenka, H., and Barreto, E. 1985. Electron waves in the electrical breakdown of gases, with application to the dart leader in lightning. *J. Geophys. Res.* **90**: 6219–24.

Kasemir, H.W. 1960. A contribution to the electrostatic theory of a lightning discharge. *J. Geophys. Res.* **65**: 1873–8.

Kawasaki, Z., and Matsuura, K. 2000. Does a lightning channel show a fractal? *Applied Energy* **67**: 147–58.

Kawasaki, Z., Matsuura, K., Hasegawa, T., Takeuti, T., and Nakano, M. 1989. Fractal model for the leader of lightning. *Res. Lett. Atmos. Electr.* **9**: 63–71.

Kekez, M.M., and Savic, P. 1976. Laboratory simulation of the stepped leader in lightning. *Can. J. Phys.* **54**: 2216–24.

Kekez, M.M., and Savic, P. 1983. Contributions to continuous leader channel development. In *Electrical Breakdown and Discharges in Gases*, Part A, eds. E.E. Kunhardt and L.H. Luessen, pp. 419–55, New York: Plenum.

Kerroum, K., Amri, A., Chandezon, J., and Fontaine, J. 1988. Propagation d'impulsions électromagnétiques au-dessus d'un terrain irrégulier ou non homogène. *Ann. Télécommun.* **43**: 665–74.

Khastgir, S.R. 1957. Leader stroke current in a lightning discharge according to the streamer theory. *Phys. Rev.* **106**: 616–17.

Khastgir, S.R., and Ghosh, D. 1972. Theory of stepped-leader in cloud-to-ground electrical discharges. *J. Atmos. Terr. Phys.* **34**: 109–13.

Kline, L.E., and Siambis, J.G. 1972. Computer simulation of electrical breakdown in gases. *Phys. Rev. A* **5**: 794–805.

Klingbeil, R., and Tidman, D.A. 1974a. Theory and computer model of the lightning stepped leader. *J. Geophys. Res.* **79**: 865–9.

Klingbeil, R., and Tidman, D.A. 1974b. Reply to comment on the brief report "Theory and computer model of the lightning stepped leader" by C.T. Phelps. *J. Geophys. Res.* **79**: 5669–70.

Klingbeil, R., Tidman, D.A., and Fernsler, R.F. 1972. Ionizing gas breakdown waves in strong electric fields. *Phys. Fluids* **15**: 1969–73.

Komelkov, V.S. 1947. Structure and parameters of the leader discharge. *Bull. Acad. Sci. USSR, Tech. Sci. Sect.* **8**: 955–66.

Komelkov, V.S. 1950. The development of electric discharges in long gaps. *Bull. Acad. Sci. USSR, Tech. Sci. Sect.* **6**: 851–65.

Kostenko, M.V. 1995. Electrodynamic characteristics of lightning and their influence on disturbances of high-voltage lines. *J. Geophys. Res.* **100**: 2739–47.

Krasnitsky, Y.A. 1994. Evaluation of lightning current pulse parameters from spherics waveforms. *J. Geophys. Res.* **99**: 10 723–5.

Krehbiel, P.R., Brook, M., and McCrory, R. 1979. An analysis of the charge structure of lightning discharges to the ground. *J. Geophys. Res.* **84**: 2432–56.

Krider, E.P. 1992. On the electromagnetic fields, Poynting vector, and peak power radiated by lightning return strokes. *J. Geophys. Res.* **97**: 15 913–17.

Krider, E.P., Dawson, G.A., and Uman, M.A. 1968. The peak power and energy dissipation in a single-stroke lightning flash. *J. Geophys. Res.* **73**: 3335–9.

Krider, E.P., Leteinturier, C., and Willett, J.C. 1992. Submicrosecond field variations in natural lightning processes. *Res. Lett. Atmos. Electr.* **12**: 3–9.

Krider, E.P., Leteinturier, C., and Willett, J.C. 1996. Submicrosecond fields radiated during the onset of first return strokes in cloud-to-ground lightning. *J. Geophys. Res.* **101**: 1589–97.

Kuester, E.F., and Chang, D.C. 1979. Evaluation of Sommerfeld integrals associated with dipole sources above Earth. Scientific Report no. 43, Electromagnetics Laboratory, Dept Electr. Eng., University of Colorado at Boulder.

Kumar, U., and Nagabhushana, G.R. 2000. Novel model for the simulation of lightning stepped leader. *IEE Proc., Sci. Meas. Technol.* **147**(2): 56–64.

Larigaldie, S. 1979. Linear gliding discharge over dielectric surfaces. *J. de Phys. C 7*, 40: 429–30.

Larigaldie, S., Labaune, G., and Moreau, J.P. 1981. Lightning leader laboratory simulation by means of rectilinear surface discharges. *J. Appl. Phys.* **52**: 7114–20.

Larigaldie, S., Roussaud, A., and Jecko, B. 1992. Mechanisms of high-current pulses in lightning and long-spark stepped leaders. *J. Appl. Phys.* **72**: 1729–39.

Laroche, P., Idone, V., Eybert-Berard, A., and Barret, L. 1991. Observations of bidirectional leader development in triggered lightning flash. In *Proc. 1991 Int. Conf. on Lightning and Static Electricity, Cocoa Beach, Florida*, pp. 57/1–10.

Larsson, A., Lalande, P., Bondiou-Clergerie, A., and Delannoy, A. 2000a. The lightning swept stroke along an aircraft in flight. Part I: Thermodynamic and electric properties of lightning arc channels. *J. Phys. D, Appl. Phys.* **33**: 1866–75.

Larsson, A., Lalande, P., and Bondiou-Clergerie, A. 2000b. The lightning swept stroke along an aircraft in flight. Part II: Numerical simulations of the complete process. *J. Phys. D, Appl. Phys.* **33**: 1876–83.

Labaune, G., Richard, P., and Bondiou A. 1987. Electromagnetic properties of lightning channels formation and propagation. *Electromagnetics* **7**: 361–93.

Latham, D.J. 1980. A channel model for long arcs in air. *Phys. Fluids* **23**(8): 1710–15.

Latham, D.J. 1986. Anode column behavior of long vertical air arcs at atmospheric pressure. *IEEE Trans. Plasma Sci.* **PS-14**: 220–7.

Le Vine, D.M., and Kao, M. 1988. The effects of current risetime on radiation from tortuous lightning channels. In *Proc. 8th Int. Conf. on Atmospheric Electricity, Uppsala, Sweden*, pp. 509–14.

Le Vine, D.M., and Meneghini, R. 1978a. Electromagnetic fields radiated from a lightning return stroke: application of an exact solution to Maxwell's equations. *J. Geophys. Res.* **83**: 2377–84.

Le Vine, D.M., and Meneghini, R. 1978b. Simulation of radiation from lightning return strokes: the effects of tortuosity. *Radio Sci.* **13**: 801–9.

Le Vine, D.M., and Meneghini, R. 1983. A solution for the electromagnetic fields close to a lightning discharge. In *Proc. 1983 Int. Aerospace and Ground Conf. on Lightning and Static Electricity, Fort Worth, Texas*, pp. 70/1–10.

Le Vine, D.M., and Willett, J.C. 1992. Comment on the transmission-line model for computing radiation from lightning. *J. Geophys. Res.* **97**: 2601–10.

Le Vine, D.M., Gesell, L., and Kao, M. 1986. Radiation from lightning return strokes over a finitely conducting earth. *J. Geophys. Res.* **91**: 11 897–908.

Lefferts, R.E. 1978. A statistical simulation of ground-wave atmospherics generated by lightning return strokes. *Radio Sci.* **13**: 121–30.

Lefferts, R.E. 1979. Probabilistic model for the initial peaks of ground wave atmospherics generated by lightning return strokes. *Radio Sci.* **14**: 1017–26.

Leise, J.A., and Taylor, W.L. 1977. A transmission line model with general velocities for lightning. *J. Geophys. Res.* **82**: 391–6.

Leteinturier, C., Weidman, C., and Hamelin, J. 1990. Current and electric field derivatives in triggered lightning return strokes. *J. Geophys. Res.* **95**: 811–28.

Lin Y.T. 1978. Lightning return-stroke models. Ph.D. thesis, University of Florida, Gainesville.

Lin, Y.T., Uman, M.A., and Standler, R.B. 1980. Lightning return stroke models. *J. Geophys. Res.* **85**: 1571–83.

Lin, Y.T., Uman, M.A. Tiller, J.A., Brantley, R.D., Beasley, W.H., Krider, E.P., and Weidman, C.D. 1979. Characterization of lightning return stroke electric and magnetic fields from simultaneous two-station measurements. *J. Geophys. Res.* **84**: 6307–14.

Little, P.F. 1978. Transmission line representation of a lightning return stroke. *J. Phys. D: Appl. Phys.* **11**: 1893–910.

Little, P.F. 1979. The effect of altitude on lightning hazards to aircraft. In *Proc. 15th European Conf. on Lightning Protection*, vol. 2, Institute of High Voltage Research, Uppsala University, Sweden.

Loeb, L.B. 1965. Ionizing waves of potential gradient. *Science* **148**: 1417–26.

Loeb, L.B. 1966. The mechanism of stepped and dart leaders in cloud-to-ground lightning strokes. *J. Geophys. Res.* **71**: 4711–21.

Loeb, L.B. 1968. Confirmation and extension of a proposed mechanism of the stepped leader lightning stroke. *J. Geophys. Res.* **73**: 5813–17.

Lundholm, R. 1957. Induced overvoltage-surges on transmission lines and their bearing on the lightning performance at medium voltage networks. Trans. of Chalmers University of Technology, Gotheburg, Sweden, 117 pp.

Lupò, G., Petrarca, C., Tucci, V., and Vitelli, M. 2000a. EM fields generated by lightning channels with arbitrary location and slope. *IEEE Trans. Electromagn. Compat.* **42**: 39–53.

Lupò, G., Petrarca, C., Tucci, V., and Vitelli, M. 2000b. EM fields associated with lightning channels: on the effect of tortuosity and branching. *IEEE Trans. Electromagn. Compat.* **42**: 394–404.

Mackerras, D. 1973. Photoelectric observations of the light emitted by lightning flashes. *J. Atmos. Terr. Phys.* **35**: 521–35.

Maier, W.B. II, Kadish, A., Sutherland, C.D., and Robiscoe, R.T. 1990. A distributed parameter wire model for transient electrical discharges. *J. Appl. Phys.* **67**(12): 7228–39.

Master, M.J., and Uman, M.A. 1983. Transient electric and magnetic fields associated with establishing a finite electrostatic dipole: an exercise in the solution of Maxwell's equations. *Am. J. Phys.* **51**: 118–26.

Master, M.J., Uman, M.A., Lin, Y.T., and Standler, R.B. 1981. Calculations of lightning return stroke electric and magnetic fields above ground. *J. Geophys. Res.* **86**: 12 127–32.

Mattos, M.A. da F., and Christopoulos, C. 1988. A nonlinear transmission line model of the lightning return stroke. *IEEE Trans. Electromagn. Compat.* **30**: 401–6.

Mattos, M.A. da F., and Christopoulos, C. 1990. A model of the lightning channel, including corona, and prediction of the generated electromagnetic fields. *J. Phys. D, Appl. Phys.* **23**: 40–6.

Mazur, V., and Ruhnke, L. 1993. Common physical processes in natural and artificially triggered lightning. *J. Geophys. Res.* **98**: 12 913–30.

Mazur, V., Ruhnke, L.H., and Laroche, P. 1995. The relationship of leader and return stroke processes in cloud-to-ground lightning. *Geophys. Res. Lett.* **22**: 2613–16.

Mazur, V., Ruhnke, L.H., Bondiou-Clergerie, A., and Lalande, P. 2000. Computer simulation of a downward negative stepped leader and its interaction with a ground structure. *J. Geophys. Res.* **105**: 22 361–9.

Melander, B.G. 1984. Effects of tower characteristics on lightning arc measurements. In *Proc. 1984 Int. Conf. on Lightning and Static Electricity, Orlando, Florida*, pp. 34/1–12.

Meneghini, R. 1984. Application of the Lienard–Wiechert solution to a lightning return stroke model. *Radio Sci.* **19**: 1485–98.

Ming, Y., and Cooray, V. 1994. Propagation effects caused by a rough ocean surface on the electromagnetic fields generated by lightning return strokes. *Radio Sci.* **29**: 73–85. (Correction, *Radio Sci.* **33**: 635, 1998.)

Moini, R., Rakov, V.A., Uman, M.A., and Kordi, B. 1997. An antenna theory model for the lightning return stroke. In *Proc. 12th Int. Symp. on Electromagnetic Compatibility, Zurich, Switzerland*, pp. 149–52.

Moini, R., Kordi, B., Rafi, G.Z., and Rakov, V.A. 2000. A new lightning return stroke model based on antenna theory. *J. Geophys. Res.* **105**: 29 693–702.

Motoyama, H., Janischewskyj, W., Hussein, A.M., Rusan, R., Chisholm, W.A., Chang, J.-S. 1996. Electromagnetic field radiation model for lightning strokes to tall structures. *IEEE Trans. Pow. Del.* **11**: 1624–32.

Müller-Hillebrand, D. 1965. The theory of the stepped and dart leader. In *Problems of Atmospheric and Space Electricity*, ed. S.C. Coroniti, pp. 332–6, Amsterdam: Elsevier.

Niemeyer, L. 1987. A stepped leader random walk model. *J. Phys. D* **20**: 897–906.

Norinder, H. 1935. Lightning currents and their variations. *J. Franklin Inst.* **220**: 69–92.

Norinder, H., and Dahle, O. 1945. Measurements by frame aerials of current variations in lightning discharges. *Arkiv Mat. Astron. Fysik* **32A**: 1–70.

Nucci, C.A., Mazzetti, C., Rachidi, F., and Ianoz, M. 1988a. On lightning return stroke models for LEMP calculations. In *Proc. 19th Int. Conf. on Lightning Protection, Graz, Austria*, pp. 463–9.

Nucci, C.A., Mazzetti, C., Rachidi, F., and Ianoz, M. 1988b. Analyse du champ électromagnétique dû à une décharge de foudre dans les domaines temporel et fréquentiel. *Ann. Télécommun.* **43**: 625–37.

Nucci, C.A., Diendorfer, G., Uman, M.A., Rachidi, F., Ianoz, M., and Mazzetti, C. 1990. Lightning return stroke current models with specified channel-base current: a review and comparison. *J. Geophys. Res.* **95**: 20 395–408.

Nucci, C.A., Rachidi, F., Ianoz, M.V., and Mazzetti, C. 1993a. Lightning-induced voltages on overhead lines. *IEEE Trans. Electromagn. Compat.* **35**: 75–86.

Nucci, C.A., Rachidi, F., Ianoz, M., and Mazzetti, C. 1993b. Corrections to "Lightning-induced voltages on overhead lines". *IEEE Trans. Electromagn. Compat.* **35**: 488.

Nucci, C.A., Ianoz, M., Rachidi, F., Rubinstein, M., Tesche, F.M., Uman, M.A., and Mazzetti, C. 1995. Modelling of lightning-induced voltages on overhead lines: recent developments. *Electrotechnik und Informationstechnik* **112**(6): 290–6.

Oetzel, G.N. 1968. Computation of the diameter of a lightning return stroke. *J. Geophys. Res.* **73**: 1889–96.

Orville, R.E. 1968a. A high-speed time-resolved spectroscopic study of the lightning return stroke: Part I, A qualitative analysis. *J. Atmos. Sci.* **25**: 827–38.

Orville, R.E., 1968b. A high-speed time-resolved spectroscopic study of the lightning return stroke: Part II, A quantitative analysis. *J. Atmos. Sci.* **25**: 839–51.

Orville, R.E., 1968c. A high-speed time-resolved spectroscopic study of the lightning return stroke: Part III, A time-dependent model. *J. Atmos. Sci.* **25**: 852–6.

Pan, E., and Liew, A.C. 2000. Analysis of a novel method of current sharing in a resistive lightning protection terminal. *IEEE Trans. Pow. Del.* **15**: 948–52.

Papet-Lepine, J. 1961. Electromagnetic radiation and physical structure of lightning discharges. *Ark. Geophys.* **3**: 391–400.

Paxton, A.H., Gardner, R.L., and Baker, L. 1986. Lightning return stroke: a numerical calculation of the optical radiation. *Phys. Fluids* **29**: 2736–41.

Paxton, A.H., Baker, L., and Gardner, R.L. 1987. Reply to comments of Hill. *Phys. Fluids* **30**: 2586–7.

Paxton, A.H., Gardner, R.L., and Baker, L. 1990. Lightning return stroke: a numerical calculation of the optical radiation. In *Lightning Electromagnetics*, ed. R.L. Gardner, pp. 47–61, New York: Hemisphere.

Paul, C.R. 1994. *Analaysis of milticonductor transmission lines*, 559 pp., New York: Wiley Interscience.

Petrov, N.I., and Petrova, G.N. 1993. Modelling of the trajectory of leader discharge development. In *Proc. 8th Int. Symp. on High Voltage Engineering, Yokohama, Japan*, pp. 101–4.

Phelps, C.T. 1974a. Positive streamers system intensification and its possible role in lightning initiation. *J. Atmos. Terr. Phys.* **36**: 103–11.

Phelps, C.T. 1974b. Comments on brief report by R. Klingbeil and D.A. Tidman: theory and computer model of the lightning stepped leader. *J. Geophys. Res.* **79**: 5669.

Picone, J.M., Boris, J.P., Grieg, J.R., Rayleigh, M., and Fernsler, R.F. 1981. Convective cooling of lightning channels. *J. Atmos. Sci.* **38**: 2056–62.

Pierce, E.T. 1960. Atmospherics from lightning flashes with multiple strokes. *J. Geophys. Res.* **65**: 1867–71.

Plooster, M.N. 1970. Shock waves from line sources: numerical solutions and experimental measurements. *Phys. Fluids* **13**: 2665–75.

Plooster, M.N. 1971a. Numerical simulation of spark discharges in air. *Phys. Fluids* **14**: 2111–23.

Plooster, M.N. 1971b. Numerical model of the return stroke of the lightning discharge. *Phys. Fluids* **14**: 2124–33.

Podgorski, A.S., and Landt, J.A. 1987. Three dimensional time domain modelling of lightning. *IEEE Trans. Pow. Del.* **2**: 931–8.

Popov, M., and He, S. 2000. Identification of a transient electric dipole over a conducting half space using a simulated annealing algorithm. *J. Geophys. Res.* **105**: 20 821–31.

Popov, M., He, S., and Thottappillil, R. 2000. Reconstruction of lightning currents and return stroke model parameters using remote electromagnetic fields. *J. Geophys. Res.* **105**: 24 469–81.

Price, G.H., and Pierce, E.T. 1977. The modeling of channel current in the lightning return stroke. *Radio Sci.* **12**: 381–88.

Quinn, D.W. 1987. Modeling of lightning. *Mathematics and Computer Simulation* **29**: 107–18.

Rachidi, F., and Nucci, C.A. 1990. On the Master, Uman, Lin, Standler, and the modified transmission line lightning return stroke current models. *J. Geophys. Res.* **95**: 20 389–94.

Rachidi, F., and Thottappillil, R. 1993. Determination of lightning currents from far electromagnetic fields. *J. Geophys. Res.* **98**: 18 315–21.

Rachidi, F., Nucci, C.A., Ianoz, M., and Mazzetti, C. 1996. Influence of a lossy ground on lightning-induced voltages on overhead lines. *IEEE Trans. Electromagn. Compat.* **38**: 250–64.

Rachidi, F., Janischewskyj, W., Hussein, A.M., Nucci, C.A., Guerrieri, S., and Chang, J.S. 1998. Electromagnetic fields radiated by lightning return strokes to high towers. In *Proc. 24th Int. Conf. on Lightning Protection, Birmingham, UK*, pp. 23–8.

Rachidi, F., Janischewskyj, W., Hussein, A.M., Nucci, C.A., Guerrieri, S., Kordi, B., and Chang, J.S. 2001. Current and electromagnetic field associated with lightning-return strokes to tall towers. *IEEE Trans. Elecromagn. Compat.* **43**(3): 356–67

Rai, J. 1978. Current and velocity of the return lightning stroke. *J. Atmos. Terr. Phys.* **40**: 1275–85.

Rai, J., and Bhattacharya, P.K. 1971. Impulse magnetic flux density close to the multiple return strokes of a lightning discharge. *J. Phys. D: Appl. Phys.* **4**: 1252–6.

Rakov, V.A. 1997. Lightning electromagnetic fields: modeling and measurements. In *Proc. 12th Int. Symp. on Electromagnetic Compatibility, Zurich, Switzerland*, pp. 59–64.

Rakov, V.A. 1998. Some inferences on the propagation mechanisms of dart leaders and return strokes. *J. Geophys. Res.* **103**: 1879–87.

Rakov, V.A. 2001. Characterization of lightning electromagnetic fields and their modeling. In *Proc. 14th Int. Symp. on Electromagnetic Compatibility, Zurich, Switzerland, Supplement*, pp. 3–16.

Rakov, V.A., and Dulzon, A.A. 1987. Calculated electromagnetic fields of lightning return stroke. *Tekh. Elektrodinam.* **1**: 87–9.

Rakov, V.A., and Dulzon, A.A. 1991. A modified transmission line model for lightning return stroke field calculations. In *Proc. 9th Int. Symp. on Electromagnetic Compatibility, Zurich, Switzerland*, pp. 229–35.

Rakov, V.A., and Uman, M.A. 1998. Review and evaluation of lightning return stroke models including some aspects of their application. *IEEE Trans. Electromagn. Compat.* **40**: 403–26.

Rakov, R.A., Uman, M.A., Jordan, D.M., and Priori III, C.A. 1990. Ratio of leader to return-stroke electric field change for first and subsequent lightning strokes. *J. Geophys. Res.* **95**: 16 579–87.

Rakov, V.A., Thottappillil, R., and Uman, M.A. 1992. On the empirical formula of Willett et al. relating lightning return-stroke peak current and peak electric field. *J. Geophys. Res.* **97**: 11 527–33.

Rakov, V.A., Thottappillil, R., Uman, M.A., and Barker, P.P. 1995. Mechanism of the lightning M-component. *J. Geophys. Res.* **100**: 25 701–10.

Rakov, V.A., Uman, M.A., Rambo, K.J., Fernandez, M.I., Fisher, R.J., Schnetzer, G.H., Thottappillil, R., Eybert-Berard, A., Berlandis, J.P., Lalande, P., Bonamy, A., Laroche, P., and Bondiou-Clergerie, A. 1998. New insights into lightning processes gained from triggered-lightning experiments in Florida and Alabama. *J. Geophys. Res.* **102**: 14 117–30.

Rakov, V.A., Crawford, D.E., Rambo, K.J., Schnetzer, G.H., Uman, M.A., and Thottappillil, 2001. M-component mode of charge transfer to ground in lightning discharges, *J. Geophys. Res*, **106**: 22 817–31.

Randall, D.A., and Wielicki, B.A. 1997. Measurements, models, and hypotheses in the atmospheric sciences. *Bull. Amer. Meteor. Soc.* **78**: 399–406.

Rao, M. 1967. Notes on the corona currents in a lightning discharge and the emission of ELF waves. *Radio Sci.* **2**: 1394.

Rao, M., 1970. The dependence of dart leader velocity on the interstroke time interval in a lightning flash. *J. Geophys. Res.* **75**: 5868–72.

Rao, M., and Bhattacharya, H. 1966. Lateral corona currents from the return stroke channel and slow field change after the return stroke in a lightning discharge. *J. Geophys. Res.* **71**: 2811–14.

Rao, M., and Khastgir, S.R. 1966. The physics of the return stroke and the time-variation of its current in a lightning discharge. *Trans. Bose Res. Inst.* **29**: 19–24.

Rhodes, C.T., Shao, X.M., Krehbiel, P.R., Thomas, R.J., and Hayenga, C.O. 1994. Observations of lightning phenomena using radio interferometry. *J. Geophys. Res.* **99**: 13 059–82.

Rizk, F.A.M. 1990. Modeling of transmission line exposure to direct lightning strokes. *IEEE Trans. Pow. Del.* **5**: 1983–97.

Rubinstein, M. 1996. An approximate formula for the calculation of the horizontal electric field from lightning at close, intermediate, and long range. *IEEE Trans. Electromagn. Compat.* **38**: 531–5.

Rubinstein, M., and Uman, M.A. 1989. Methods for calculating the electromagnetic fields from a known source distribution: application to lightning. *IEEE Trans. Electromagn. Compat.* **31**: 183–9.

Rubinstein, M., and Uman, M.A. 1990. On the radiation field turn-on term associated with traveling current discontinuities in lightning. *J. Geophys. Res.* **95**: 3711–13.

Rubinstein, M., and Uman, M.A. 1991. Transient electric and magnetic fields associated with establishing a finite electrostatic dipole, revisited. *IEEE Trans. Electromagn. Compat.* **33**: 312–20.

Rubinstein, M, Rachidi, F., Uman, M.A., Thottappillil, R., Rakov,V.A., and Nucci, C.A. 1995. Characterization of vertical electric fields 500 m and 30 m from triggered lightning. *J. Geophys. Res.* **100**: 8863–72.

Sadiku, M.N.O. 1994. *Elements of Electromagnetics*, 821 pp. Orlando, Florida: Sounders College.

Safaeinili, A., and Mina, M. 1991. On the analytical equivalence of electromagnetic fields solutions from a known source distribution. *IEEE Trans. Electromagn. Compat.* **33**: 69–71.

Schonland, B.F.J. 1938. Progressive lightning, Part 4, The discharge mechanisms. *Proc. Roy. Soc. A* **164**: 132–50.

Schonland, B.F.J. 1953. The pilot streamer in lightning and the long spark. *Proc. Roy. Soc. A* **220**: 25–38.

Schonland, B.F.J. 1956. The lightning discharge. In *Handbuch der Physik* **22**: 576–628, Berlin: Springer-Verlag.

Schonland, B.F.J. 1962. Lightning and the long electric spark. *Adv. Sci.* **19**: 306–13.

Shao, X.M., and Krehbiel, P.R. 1996. The spatial and temporal development of intracloud lightning. *J. Geophys. Res.* **101**: 26 641–68.

Shao, X.M., Krehbiel, P.R., Thomas, R.J., and Rison, W. 1995. Radio interferometric observations of cloud-to-ground lightning phenomena in Florida. *J. Geophys. Res.* **100**: 2749–83.

Shao, X.M., Rhodes, C.T., and Holden, D.N. 1999. RF radiation observations of positive cloud-to-ground flashes. *J. Geophys. Res.* **104**: 9601–8.

Shostak, V., Janischewskyj, W., Hussein, A.M., Chang, J.-S., and Kordi, B. 1999. Return-stroke current modeling of lightning striking a tall tower accounting for reflections within the growing channel and for upward-connecting discharges. In *Proc. 11th Int. Conf. on Atmospheric Electricity, Guntersville, Alabama*, pp. 123–6.

Shostak, V., Janischewskyj, W., Hussein, A.M., and Kordi, B. 2000. Electromagnetic fields of lightning strikes to a tall tower: a model that accounts for upward-connecting discharges. In *Proc. 25th Int. Conf. on Lightning Protection, Rhodes, Greece*, pp. 60–5.

Smirnova, E.I., Mareev, E.A., and Chugunov, Yu. V. 2000. Modeling of lightning generated electric field transitional processes. *Geophys. Res. Lett.* **27**: 3833–6.

Smyth, J.B., and Smyth, D.C. 1976. Lightning and its radio emission. *Radio Sci.* **11**: 977–84.

Smyth, J.B., and Smyth, D.C. 1977. Critique of the paper. *Am. J. Phys.* **45**: 581–2.

Srivastava, C.M., and Khasgir, S.R. 1955. On the maintenance of current in the stepped leader stroke of lightning discharge. *J. Sci. Ind. Res.* **14B**: 34–5.

Srivastava, K.M.L. 1966. Return stroke velocity of a lightning discharge. *J. Geophys. Res.* **71**: 1283–6.

Srivastava, K.M.L., and Tantry, B.A.P. 1966. VLF characteristics of electromagnetic radiation from the return stroke of lightning discharge. *Indian J. Pure Appl. Phys.* **4**: 272–5.

Strawe, D.F. 1979. Non-linear modeling of lightning return strokes. In *Proc. Federal Aviation Administration/Florida Institute of Technology Workshop on Grounding and Lightning Technology*, pp. 9–15, Melbourne, Florida, Report FAA-RD-79-6.

Suzuki, T. 1977. Propagation of ionizing waves in glow discharge. *J. Appl. Phys.* **48**: 5001–7.

Szpor, S. 1970. Review of the relaxation theory of the lightning stepped leader. *Acta Geophys. Polonica* **18**: 73–7.

Szpor, S. 1972. Steps in the cloud and air discharges. *Archiwum Elektrotechniki* **21**: 19–20.

Szpor, S. 1977. Critical comparison of theories of stepped leaders. *Archiwum Elektrotechniki* **26**: 291–9.

Szpor, S., and Turkowski, W. 1968. Laboratory corroboration of the relaxation theory of the lightning stepped leader. *Archiwum Elektrotechniki* **17**: 405–7.

Takagi, N., and Takeuti, T. 1983. Oscillating bipolar electric field changes due to close lightning return strokes. *Radio Sci.* **18**: 391–8.

Takeuti, T. 1992. The preliminary discussion on the distribution of lightning striking points on the ground. *Res. Lett. Atmos. Electr.* **12**: 155–9.

Takeuti, T., Hashimoto, T., and Takagi, N. 1993. Two dimensional computer simulation on the natural stepped leader in summer. *J. Atmos. Electr.* **13**: 9–14.

Thomson, E.M. 1985. A theoretical study of electrostatic field wave shapes from lightning leaders. *J. Geophys. Res.* **90**: 8125–35.

Thottappillil, R., and Rakov, V.A. 2001. On different approaches to calculating lightning electric fields. *J. Geophys. Res.* **106**: 14 191–205.

Thottappillil, R., and Uman, M.A. 1993. Comparison of lightning return-stroke models. *J. Geophys. Res.* **98**: 22 903–14.

Thottappillil, R., and Uman, M.A. 1994. Lightning return-stroke model with height-variable discharge time constant. *J. Geophys. Res.* **99**: 22 773–80.

Thottappillil, R., McLain, D.K., Uman, M.A., and Diendorfer, G. 1991a. Extension of the Diendorfer–Uman lightning return stroke model to the case of a variable upward return stroke speed and a variable downward discharge current speed. *J. Geophys. Res.* **96**: 17 143–50.

Thottappillil, R., Uman, M.A., and Diendorfer, G. 1991b. Influence of channel base current and varying return stroke speed on the calculated fields of three important return stroke models. In *Proc. 1991 Int. Conf. on Lightning and Static Electricity, Cocoa Beach, Florida*, pp. 118.1–9.

Thottappillil, R., Rakov, V.A., and Uman, M.A. 1997. Distribution of charge along the lightning channel: relation to remote electric and magnetic fields and to return stroke models. *J. Geophys. Res.* **102**: 6887–7006.

Thottappillil, R., Uman, M.A., and Rakov, V.A. 1998. Treatment of retardation effects in calculating the radiated electromagnetic fields from the lightning discharge. *J. Geophys. Res.* **103**: 9003–13.

Thottappillil, R., Schoene, J., and Uman, M.A. 2001. Return stroke transmission line model for stroke speed near and equal that of light. *Geophys. Res. Lett.* **28**: 3593–6.

Tidman, D.A., and Fernsler, R.F. 1972. Ionizing gas breakdown waves in strong electric fields. *Phys. Fluids* **15**: 1969–73.

Turcotte, D.L., and Ong, R.S.B. 1968. The structure and propagation of ionizing wave fronts. *J. Plasma Phys.* **2**: 145–55.

Uman, M.A. 1969. *Lightning*. 264 pp., New York: McGraw-Hill.

Uman, M.A. 1977. Reply to Smyth and Smyth. *Am. J. Phys.* **45**: 582.

Uman, M.A. 1984. *Lightning*, 298 pp., New York: Dover.

Uman, M.A. 1985. Lightning return stroke electric and magnetic fields. *J. Geophys. Res.* **90**: 6121–30.

Uman, M.A. 1987. *The Lightning Discharge*, 377 pp., San Diego: Academic Press.

Uman, M.A. 2001. *The Lightning Discharge*, 377 pp., Mineola, New York: Dover.

Uman, M.A., and McLain, D.K. 1969. Magnetic field of the lightning return stroke. *J. Geophys. Res.* **74**: 6899–910.

Uman, M.A., and McLain, D.K. 1970a. Radiation field and current of the lightning stepped leader. *J. Geophys. Res.* **75**: 1058–66.

Uman, M.A., and McLain, D.K. 1970b. Lightning return stroke current from magnetic and radiation field measurements. *J. Geophys. Res.* **75**: 5143–7.

Uman, M.A., and Voshall, R.E. 1968. The time-interval between lightning strokes and the initiation of dart leaders. *J. Geophys. Res.* **73**: 497–506.

Uman, M.A., Seacord, D.F., Price, G.H., and Pierce, E.T. 1972. Lightning induced by thermonuclear detonations. *J. Geophys. Res.* **77**: 1591–6.

Uman, M.A., McLain, D.K., Fisher, R.J., and Krider, E.P. 1973a. Electric field intensity of lightning return stroke. *J. Geophys. Res.* **78**: 3523–9.

Uman, M.A., McLain, D.K., Fisher, R.J., and Krider, E.P. 1973b. Currents in Florida lightning return strokes. *J. Geophys. Res.* **78**: 3530–7.

Uman, M.A., McLain, D.K., and Krider, E.P. 1975. The electromagnetic radiation from a finite antenna. *Am. J. Phys.* **43**: 33–8.

Uman, M.A., Master, M.J., and Krider, E.P. 1982. A comparison of lightning electromagnetic fields with the nuclear electromagnetic pulse in the frequency range 10^4 to 10^7 Hz. *IEEE Trans. Electromagn. Compat.* **24**: 410–16.

Uman, M.A., Rakov, V.A., Versaggi, V., Thottappillil, R., Eybert-Berard, A., Barret, L., Berlandis, J.-P., Bador, B., Barker, P.P., Hnat, S.P., Oravsky, J.P., Short, T.A., Warren, C.A., and Bernstein, R. 1994. Electric fields close to triggered lightning. In *Proc. Int. Symp. on Electromagnetic Compatibility (EMC '94 ROMA), Rome, Italy*, pp. 33–7.

Uman, M.A., Rakov, V.A., Rambo, K.J., Vaught, T.W., Fernandez, M.I., Cordier, D.J., Chandler, R.M., Bernstein, R., and Golden, C. 1997. Triggered-lightning experiments at Camp Blanding, Florida (1993–1995). *Trans. IEE Japan* **117-B**: 446–52.

Uman, M.A., Rakov, V.A., Schnetzer, G.H., Rambo, K.J., Crawford, D.E., and Fisher, R.J. 2000. Time derivative of the electric field 10, 14, and 30 m from triggered lightning strokes. *J. Geophys. Res.* **105**: 15 577–95.

Uman, M.A., Schoene, J., Rakov, V.A., Rambo, K.J., and Schnetzer, G.H. 2001. Correlated time derivatives of current, electric field intensity, and magnetic flux density for triggered lightning at 15 m. *J. Geophys. Res.* **107** (D13), 10. 1029/2000JD000249.

Vecchi, G., Labate, D., and Canavero, F. 1994. Fractal approach to lightning radiation on a tortuous channel. *Radio Sci.* **29**: 691–704.

Vecchi, G., Zich, R.E., and Canavero, F.C. 1997. A study of the effect of channel branching on lightning radiation. In *Proc. 12th Int. Symp. on Electromagnetic Compatibility, Zurich, Switzerland*, pp. 65–70.

Volland, H. 1981a. A waveguide model of lightning currents. *J. Atmos. Terr. Phys.* **43**: 191–204.

Volland, H. 1981b. Waveform and spectral distribution of the electromagnetic field of lightning currents. *J. Atmos. Terr. Phys.* **43**: 1027–42.

Volland, H. 1982. Simulation of a lightning channel by a prolate spheroid. *Radio Sci.* **17**: 445–52.

Wagner, C.F. 1960. Determination of the wave front of lightning stroke currents from field measurements. *AIEE Trans.* **79**(3): 581–9.

Wagner, C.F. 1963. Relation between stroke current and velocity of the return stroke. *AIEE Trans. Power Appar. Syst.* **82**: 609–17.

Wagner, C.F., and Hileman, A.R. 1958. The lightning stroke (1). *AIEE Trans.* **77**(3): 229–42.

Wagner, C.F., and Hileman, A.R. 1961. The lightning stroke (2). *AIEE Trans.* **80**(3): 622–42.

Wagner, C.F., and Hileman, A.R. 1962. Surge impedance and its application to the lightning stroke. *AIEE Trans.* **80**(3): 1011–22.

Wait, J.R. 1956. Transient fields of a vertical dipole over a homogeneous curved ground. *Can. J. Phys.* **34**: 27–35.

Wait, J.R. 1988a. Determining the strength and orientation of an elevated dipole. *Electronics Lett.* **24**: 32–4.

Wait, J.R. 1988b. Determining transient dipole source from observed field waveforms. *Electronics Lett.* **24**: 282–3.

Wait, J.R. 1988c. A multipole expansion for the EM fields of a linear radiator. *IEEE Trans. Electromagn. Compat.* **30**: 413–15.

Wait, J.R. 1988d. Letter to the editor. *Geoexploration* **25**: 173.

Wait, J.R. 1994. Comments on "The EM field of an improved lightning return stroke representation". *IEEE Trans. Electromagn. Compat.* **36**: 82.

Wait, J.R. 1997. Concerning the horizontal electric field of lightning. *IEEE Trans. Electromagn. Compat.* **39**: 186.

Wait, J.R. 1999a. Upward traveling current wave excitation of overhead cable. *IEEE Trans. Electromagn. Compat.* **41**(1): 75–7.

Wait, J.R. 1999b. Influence of finite ground conductivity on the fields of a vertical traveling wave of current. *IEEE Trans. Electromagn. Compat.* **41**(1): 78.

Wait, J.R. 2000. Ground wave of an idealized lightning return stroke. *IEEE Trans. Ant. Propag.* **48**: 1349–53.

Wait, J.R., and Teschan, P. 1990. A basic limitation in interpreting field waveforms of lightning return strokes at a distant point. *IEEE Trans. Electromagn. Compat.* **32**: 249–50.

Wang, D., Rakov, V.A., Uman, M.A., Takagi, N., Watanabe, T., Crawford, D., Rambo, K.J., Schnetzer, G.H., Fisher, R.J., and Kawasaki, Z.I. 1999a. Attachment process in rocket-triggered lightning strokes. *J. Geophys. Res.* **104**: 2141–50.

Wang, D., Takagi, N., Watanabe, T., Rakov, V.A., and Uman, M.A. 1999b. Observed leader and return-stroke propagation characteristics in the bottom 400 m of the rocket-triggered lightning channel. *J. Geophys. Res.* **104**: 14 369–76.

Wang, D., Takagi, N., Watanabe, T., Rakov, V.A., and Uman, M.A. 2000. Luminosity waves in branched channels of two negative lightning flashes. *J. Atmos. Electr.* **20**: 91–7.

Weidman, C.D. 1998. Lightning return stroke velocities near channel base. In *Proc. 1998 Int. Lightning Detection Conf.*, 25 pp., GAI, 2705 East Medina Road, Tucson, Arizona 85706-7155.

Weidman, C.D., and Krider, E.P. 1978. The fine structure of lightning return stroke wave forms. *J. Geophys. Res.* **83**: 6239–47.

Weidman, C.D., and Krider, E.P. 1980. Submicrosecond risetimes in lightning return-stroke fields. *Geophys. Res. Lett.* **7**: 955–8.

Weidman, C.D., and Krider, E.P. 1982. Correction. *J. Geophys. Res.* **87**: 7351.

Weidman, C.D., and Krider, E.P. 1984. Variations à l'échelle submicroseconde des champs électromagnetiques rayonnés par la foudre. *Ann. Telecomm.* **39**: 165–74.

Weidman, C., Hamelin, J., Leteinturier, C., and Nicot, L. 1986. Correlated current derivative (dI/dt) and electric field-derivative (dE/dt) emitted by triggered lightning. In *Proc. 11th Int. Aerospace and Ground Conf. on Lightning and Static Electricity, Dayton, Ohio*, 10 pp.

Willett, J.C., Idone, V.P., Orville, R.E., Leteinturier, C., Eybert-Berard, A., Barret, L., and Krider, E.P. 1988. An experimental test of the "transmission-line model" of electromagnetic radiation from triggered lightning return strokes. *J. Geophys. Res.* **93**: 3867–78.

Willett, J.C., Bailey, J.C., Idone, V.P., Eybert-Berard, A., and Barret, L. 1989. Submicrosecond intercomparison of radiation fields and currents in triggered lightning return strokes based on the transmission-line model. *J. Geophys. Res.* **94**: 13 275–86.

Winn, W.P. 1965. A laboratory analog to the dart leader and return stroke of lightning. *J. Geophys. Res.* **70**: 3265–70.

Winn, W.P. 1967. Ionizing space-charge waves in gases. *J. Appl. Phys.* **38**: 783–90.

Wu, D., and Ruan, C. 1999. Transient radiation of traveling-wave antennas. *IEEE Trans. Electromagn. Compat.* **41**: 120–3.

Yokoyama, S., Miyake, K., Suzuki, T., and Kanao, S. 1990. Winter lightning on Japan Sea Coast – development of measuring system on progressing feature of lightning discharge. *IEEE Trans. Pow. Del.* **5**: 1418–25.

Zeddam, A., and Degauque, P. 1987. Current and voltage induced on telecommunication cables by a lightning stroke. *Electromagnetics* **7**: 541–64.

Zeddam, A., Degauque, P., and Laray, R. 1988. Etude des perturbations induites par une décharge orageuse sur un câble de télécommunication. *Ann. Telecommun.* **43**: 638–48.

Zundl, 1994. Lightning current and LEMP calculations compared to measurements gained at the Peissenberg tower. In *Proc. 22nd Int. Conf. on Lightning Protection, Budapest, Hungary*, paper R1c-08.

13 The distant lightning electromagnetic environment: atmospherics, Schumann resonances, and whistlers

> ...taking into consideration the large number of lightning strokes simultaneously in existence around the globe and the fact that over much of their frequency spectrum [their] fields will be trapped within the Earth–ionosphere waveguide, there will be a continuous mean noise or "atmospheric noise".
>
> J. Hamelin (1993)

13.1. Introduction

In Chapters 4, 5, 7, and 9, we examined the electric and magnetic fields associated with lightning at distances less than a few hundred kilometers. At these ranges, the fields propagate from the lightning source to the observation point either (i) in the atmosphere along the finitely conducting Earth boundary by way of a ground wave or (ii) directly through the atmosphere by line-of-sight propagation. The features of the field waveforms from relatively close lightning are determined primarily by the characteristics of the lightning. A discussion of the effects of the conductivity and permittivity of the Earth in changing the characteristics of the ground wave was given in the last part of subsection 4.6.4. In the present chapter we examine the characteristics of the electric and magnetic fields of distant lightning, observed at ranges from a few hundred kilometers to tens of thousands of kilometers. At such distances the field waveforms are significantly influenced by the properties of the propagation path.

We will subdivide our discussion of lightning's distant electromagnetic environment into three parts, depending on the main types of observed signals: (i) atmospherics, time-domain signals typically in the VLF (3 kHz to 30 kHz) frequency band, which propagate in waveguide modes in the spherical Earth–ionosphere cavity (Section 13.3); (ii) Schumann resonances, signals in the spherical Earth–ionosphere waveguide at discrete frequencies in the ELF (3 Hz to 3 kHz) frequency range, the lowest of which occurs at about 8 Hz (Section 13.4); and (iii) whistlers, signals in the ELF and VLF bands that travel out through the ionosphere along the Earth's magnetic field lines and back through the ionosphere into the atmosphere (Section 13.5). (Note that in some literature sources atmospherics are defined, in a much broader way, as any lightning electromagnetic signal observed in the cavity between the Earth's surface and the ionosphere.) Two types of time-domain signal in the ELF band, slow tails and Q-bursts, will be examined as part of our discussion of atmospherics and Schumann resonances.

Atmospheric radio noise, discussed in Section 13.6, is a term generally used to denote all sources of ELF and VHF signals in the Earth–ionosphere waveguide, including both those produced by lightning and also the chorus and hiss produced, particularly at high latitudes, by charged particles in the ionosphere and magnetosphere. Lightning RF emissions observed at satellite altitudes in the VHF band (30–300 MHz) are discussed in Sections 9.4, 14.6, and 17.8.

It should be noted that for this chapter there are three separate reference lists, for Sections 13.1–13.3 and 13.6, for Section 13.4, and for Section 13.5.

(i) *Atmospherics*, also called sferics, are defined here as lightning-produced electric and magnetic fields whose spectrum spans frequencies from a few kilohertz to a few hundred kilohertz, although as noted above other definitions are found in the literature. Such sferics are easily observed at distances up to several thousand kilometers from the causative lightning. (The Earth's circumference is about 40 000 km.) Typical time-domain sferics from distant sources have frequency content in the VLF range, 3 to 30 kHz, as mentioned earlier, but the sferics frequency spectrum can extend upward into the LF range, 30 to 300 kHz, and above. The propagation of sferics occurs in the Earth–ionosphere cavity primarily by way of multiple reflections from the ionosphere and the Earth's surface, similarly to electromagnetic wave propagation in a metal-wall waveguide. At sferics frequencies the atmosphere is a reasonably good conductor above an altitude of 70 to 90 km. There is a frequency-dependent dissipation of the energy of the sferic in the processes of reflection and refraction (transmission) at the ionosphere boundary, which causes distortion of the time-domain sferic waveshape. An observed atmospheric at a given range is composed, in general, of a ground wave followed by a sequence of "skywaves". The ground wave arrives first (via the shortest path) but,

13.1. Introduction

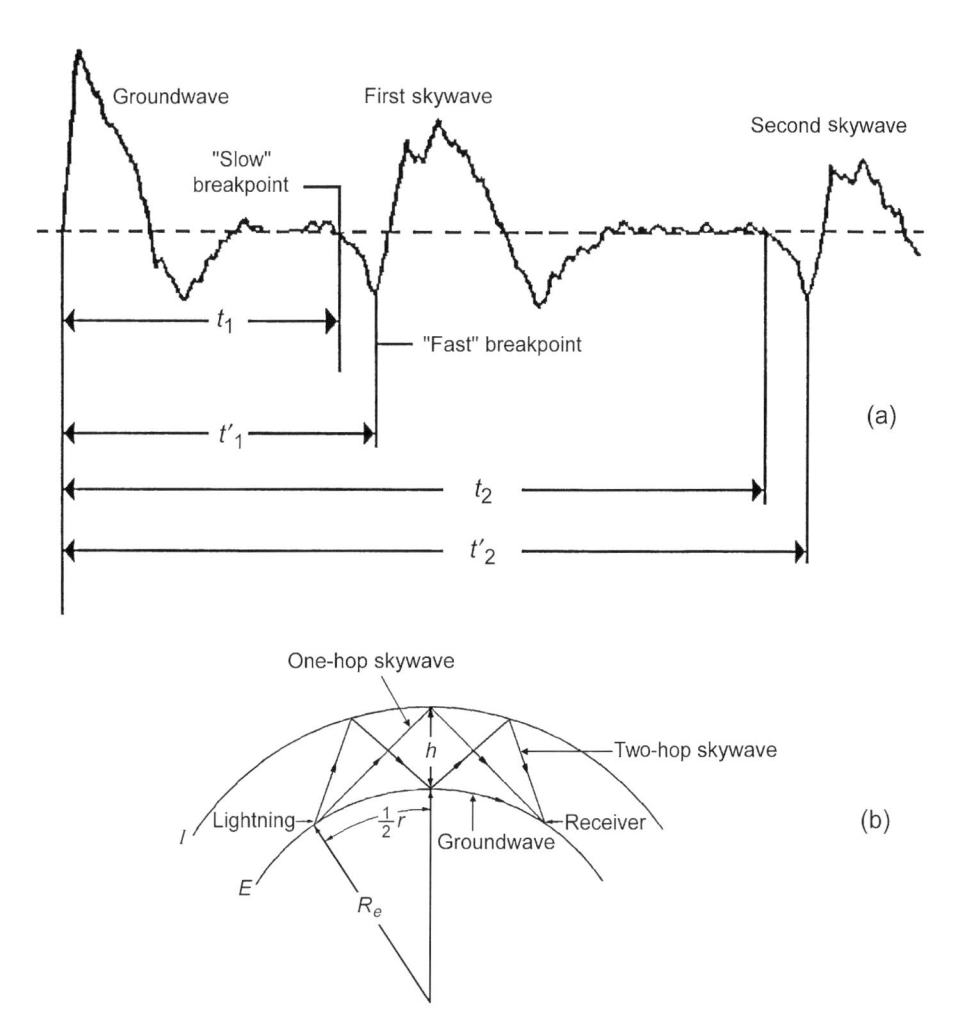

Fig. 13.1. (a) Typical groundwave and skywaves recorded at both Gainesville and KSC, Florida for storms over the Gulf of Mexico and the Atlantic Ocean about 300 km distant. The broken line is the baseline for the electric field. For illustrative purposes, the one- and two-hop skywaves have been compressed in time by a factor 2 relative to the ground wave, whose typical duration is about 100 μs. (b) Geometry and parameters for the groundwave and for the one- and two-hop skywaves. E, Earth's surface; I, ionosphere. Adapted from McDonald *et al.* (1979).

at large distances, may have an amplitude below the local noise level owing to propagation losses in the finitely conducting Earth. The term "skywave" can represent, with increasing time, one reflection off the ionosphere (a one-hop skywave), or a reflection off the ionosphere, ground, and the ionosphere again (a two-hop skywave), or a higher-order-hop skywave.

The electric field waveform of the groundwave and the first two skywaves of a typical sferic observed in Florida near midnight from lightning about 300 km distant is shown in Fig. 13.1a, and the propagation paths between Earth and the ionosphere of the groundwave and the first two skywaves are illustrated in Fig. 13.1b. Since sferics, as we have defined them, propagate in waveguide modes, they cannot contain frequency components below about 1 to 3 kHz, the cutoff frequency for the Earth–ionosphere waveguide.

Below about 1 kHz, however, there is another mode of propagation, the quasi-TEM (quasi-transverse-electromagnetic) mode, which allows lightning-produced signals in the ELF band (3 Hz to 3 kHz) to be present in the Earth–ionosphere cavity. ELF slow-tail waveforms and Q-bursts follow some sferics waveforms. We choose not to refer to these slow-tail time-domain signals as sferics, although some authors do. For example, in Brook (1992) the slow tail is considered to be the latter portion of the overall waveform of some sferics. Slow tails generally resemble several cycles of a damped sine wave having a primary frequency below about 500 Hz. Q-bursts are electromagnetic oscillations at even lower frequencies and are discussed in the next paragraph.

(ii) *Schumann resonance fields* are ELF standing waves in the spherical waveguide cavity formed by the Earth and the layer of the atmosphere at about 45 to 60 km, at and above which the atmosphere appears to be a reasonable conductor at ELF frequencies. Lightning ELF signals with wavelengths of the order of the Earth's circumference and fractions of that circumference exhibit a resonant effect, that is, have spatial coherence and an enhanced amplitude in the aforementioned spherical cavity. Schumann resonances are observed at nominal frequencies of 7.8, 14, 20, 26, 33, 39, and 45 Hz, there being a variation in these frequencies of ±0.5 Hz. This variation occurs primarily because of the changing properties of the upper part of the spherical waveguide. The amplitude of the Schumann resonances at any time and location is determined by the integrated effect of the worldwide lightning activity at that time and, therefore, the amplitude varies diurnally, from day to day, from season to season, and from year to year. Every few minutes, the Schumann time-domain field from an individual "large" lightning, that is, a large individual time-domain signal spanning the frequency range of the Schumann resonances, can be detected above the usual background of the resonance fields. Such individual Schumann events are called Q-bursts. Examples of Q-bursts are shown in Fig. 13.2. Q-bursts have been generally studied with instrumentation sensitive primarily to ELF frequencies, so that the preceding sferic may not have been observed. Q-bursts and the slow tails that follow some sferics are essentially the same phenomena except that the Q-burst spectrum is in the lower part of the ELF band while the slow-tail spectrum is in its upper part.

(iii) Whistlers (more properly, "ducted" whistlers) are lightning-generated electromagnetic signals with a frequency spectrum in the range from about 100 Hz to over 10 kHz, which traverse the ionosphere into the magnetosphere and then propagate along the Earth's magnetic field lines to the opposite hemisphere. Here the waves again traverse the ionosphere, this time in the downward direction, and are observable on the Earth's surface. Propagation along the geomagnetic field lines is possible when the frequency of the propagating wave is less than both the electron plasma frequency (defined in the next section, Eq. 13.34) along those field lines, determined by the local electron density, and the electron gyrofrequency (also defined in the next section, Eq. 13.6), determined by the local magnetic field. The gyrofrequency is near 1 MHz in the equatorial plane of the ionosphere and the magnetosphere, corresponding to a magnetic flux density of about 30 μT. The magnetic field increases by a factor 2 or so toward the poles. The plasma frequency ranges from about 10 MHz in the ionosphere (corresponding to an electron density of 10^{12} m^{-3} or 10^{6} cm^{-3}) to 1 kHz in the upper magnetosphere (an electron density of 10^{8} m^{-3} or 10^{2} cm^{-3}). The physics of the propagation of the typical ducted whistler in the magnetospheric plasma is such that the higher frequencies in the signal

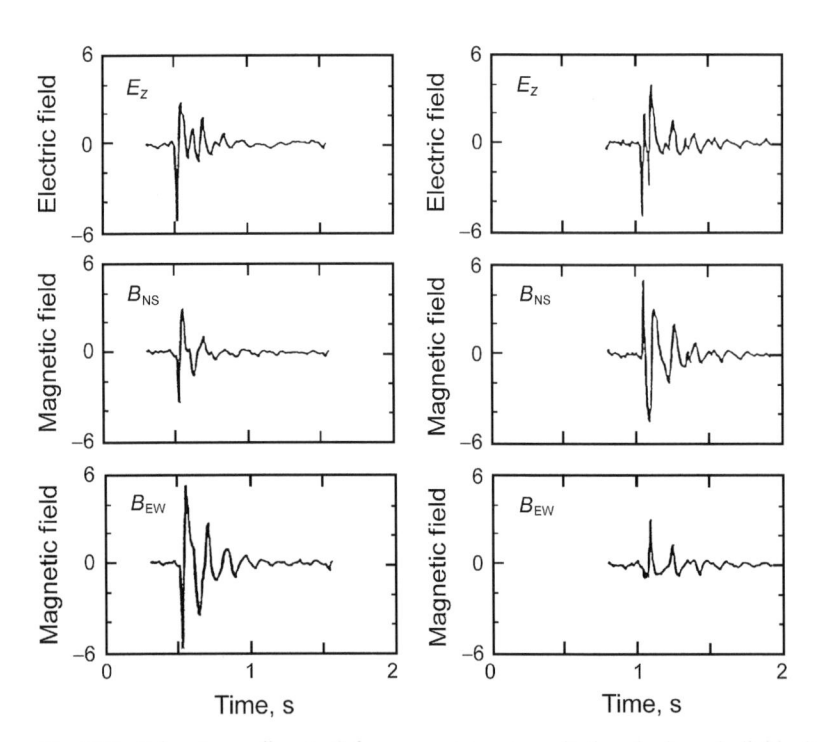

Fig. 13.2. Q-bursts are discrete Schumann resonance excitations by large individual lightning events. Plotted versus time are the vertical component E_z of the electric field and the orthogonal horizontal components B_{NS} and B_{EW} of the magnetic field of two events from 10 October, 022207 UT (left) and 11 October, 010416 UT (right) 1987. The amplitudes are in arbitrary units. The Q-burst duration is typically about 0.5 s. Adapted from Sentman (1989).

13.2. Theoretical background

travel faster than lower ones, and hence the higher frequencies arrive at an observation point on the Earth's surface earlier than the lower frequencies. If the whistler electromagnetic signal is converted into an audio signal, the audio signal will sound like a whistle. A less commonly observed type of whistler is the so-called "nose" whistler, for which there is a minimum propagation time, corresponding to a maximum propagation speed, at one "nose" frequency, the frequencies below and above that value traveling slower and arriving increasingly later. The more commonly observed whistler represents the lower-frequency portion of the nose whistler. Apparently, only a small fraction of lightning flashes produce ducted whistlers, but probably a majority of flashes produce electromagnetic signals that propagate through the ionosphere and then propagate in the magnetosphere both across and along the magnetic field lines as "non-ducted" whistlers, traveling in the magnetosphere until their energy is dissipated. It follows that non-ducted whistlers can only be observed in the magnetosphere and not on the Earth's surface. Diagrams showing idealized paths taken by ducted and non-ducted whistlers are presented in Figs. 13.3a, c, d, respectively. An idealized time-domain whistler waveform is shown in Fig. 13.4a and idealized plots of frequency versus time in Figs. 13.3b and 13.4b. The square root of inverse frequency versus time is plotted in Fig. 13.4c. Figures 13.3 and 13.4 will be discussed further in Section 13.5.

In the next section, Section 13.2, we first give some salient properties of the ionosphere and magnetosphere and then present the theoretical background needed to read the pertinent literature on atmospherics, Schumann resonances, and whistlers.

13.2. Theoretical background

13.2.1. Characterization of the ionosphere and magnetosphere

Heaviside and Kennelly (Kennelly 1902) postulated the existence of an atmospheric "conducting" layer in order to explain the observed reflection of radio waves. Such a layer was called, at that time, the Kennelly–Heaviside layer. The first verification of the existence of such a conducting layer, the ionosphere, is attributed to Appleton and Barnett (1925). We now know that the ionosphere has a complex

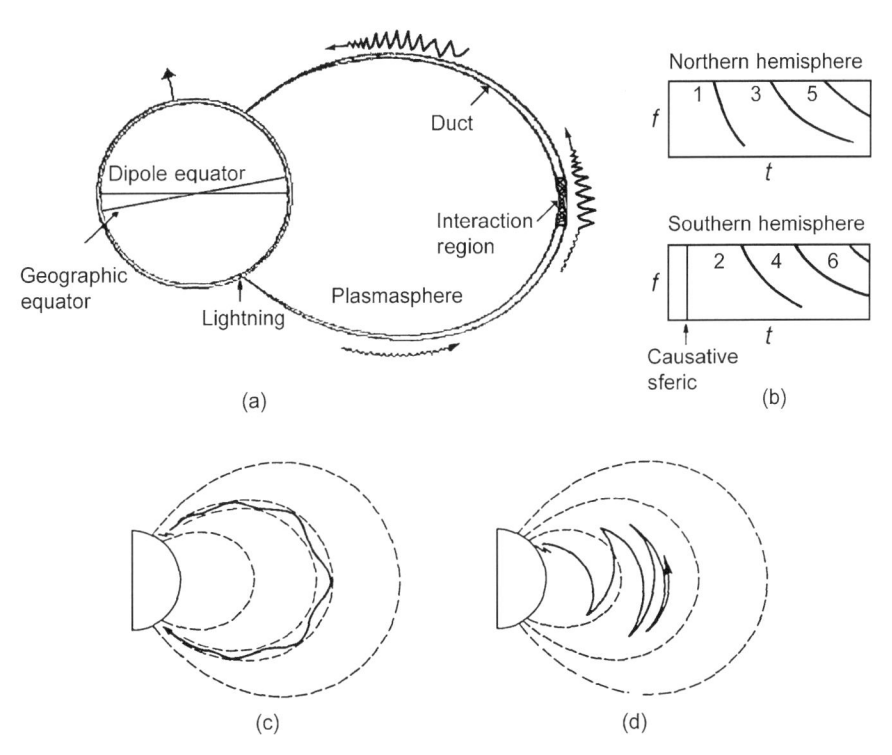

Fig. 13.3. (a) Typical path of a ducted whistler within the plasmasphere, shown together with the waveforms of the ducted signal before, during, and after its nonlinear interaction with cyclotron-resonant electrons in the region at the dipole equator. Adapted from Helliwell (1997). (b) Frequency versus time curves for a whistler traversing a path that echoes from hemisphere to hemisphere along a geomagnetic field line as observed at conjugate points in the northern and southern hemispheres. The causative sferic is indicated; it was produced by lightning in the southern hemisphere. Adapted from Hayakawa et al. (1995). Potential ray paths (solid lines) for ducted (c) and un-ducted (d) whistler-mode propagation in the magnetosphere. Adapted from Park and Carpenter (1978) and Park (1982).

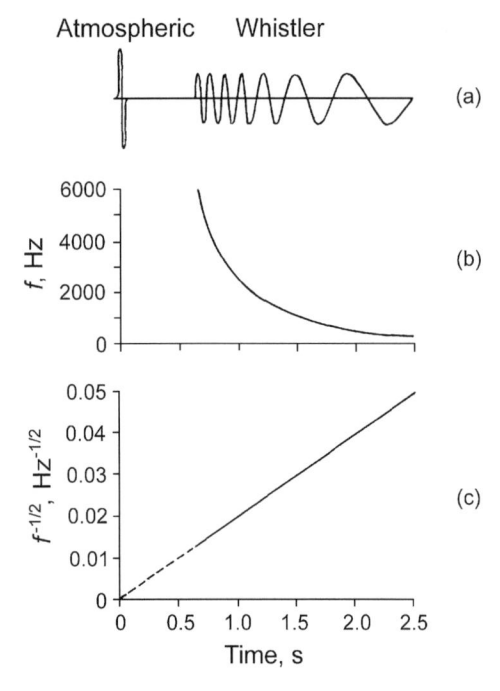

Fig. 13.4. Idealized time-domain waveform and frequency content of a ducted whistler as a function of time. (a) The waveform; each cycle represents 400 cycles on the original. (b) The actual frequency f versus time; (c) $f^{-1/2}$ versus time. Adapted from Helliwell (1965).

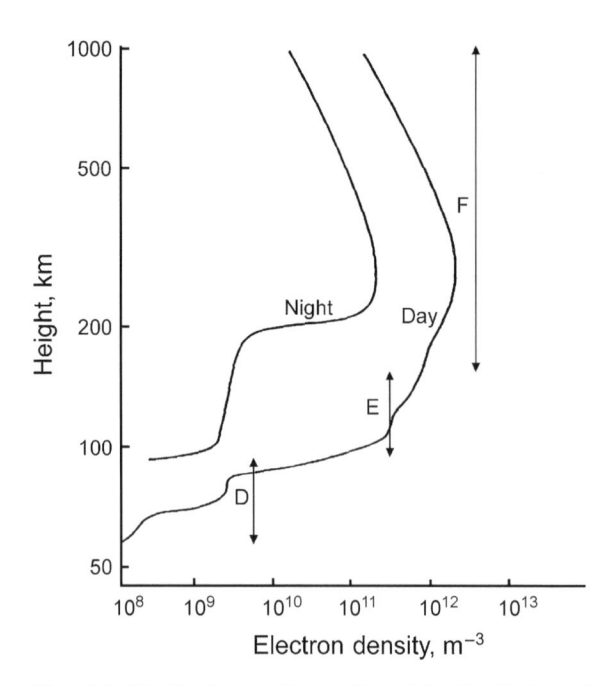

Fig. 13.5. Idealized ionization profiles of the Earth's ionosphere at temperate latitudes near sunspot maximum. Adapted from Yeh and Liu (1972).

structure composed of three major regions or layers, the D, E, and F regions, as illustrated in Fig. 13.5 for temperate latitudes near sunspot maximum. The lowest layer, the D-region, extends in height from about 40 to 90 km. Its typical electron density is of order 10^9 m^{-3} in the daytime and diminishes to a negligible value after sunset. The E region of the ionosphere extends between about 90 and 160 km. The electron density in this region typically has a value above 10^{11} m^{-3} in the day-time, but at night it is about two orders of magnitude lower. Above the E region is the F region, which extends to a height of 1000 km or so. The peak F-region electron density has an average value of about 2×10^{12} m^{-3} during the day and 2×10^{11} m^{-3} at night. As shown in Fig. 13.5, the ionospheric electron density varies over four orders of magnitude. Further, because of the decrease in the density of atoms and molecules with height, from the D region to the upper F region, the medium changes from one in which collisions between electrons and the heavier particles significantly influence the character of the wave propagation to one in which collisions have a negligible effect. The effective collision frequency for momentum transfer between electrons and ions and neutral atoms is plotted as a function of height in Fig. 13.6. Additionally, where such collisions are not important, the conductivity of the ionosphere is highly anisotropic as a result of the presence of the geomagnetic field, as we shall see later in this section.

Above the F-region of the ionosphere, at heights over about 1000 km (see Fig. 13.5), is the magnetosphere, where the motions of the electrons and ions, and hence the characteristics of the electromagnetic fields that propagate through this region, are strongly influenced by the geomagnetic field. The mixture of essentially equal densities of electrons and ions in the ionosphere and in the magnetosphere is called a plasma. Plasma interactions with electromagnetic waves are considered in subsection 13.2.2 below. The inner part

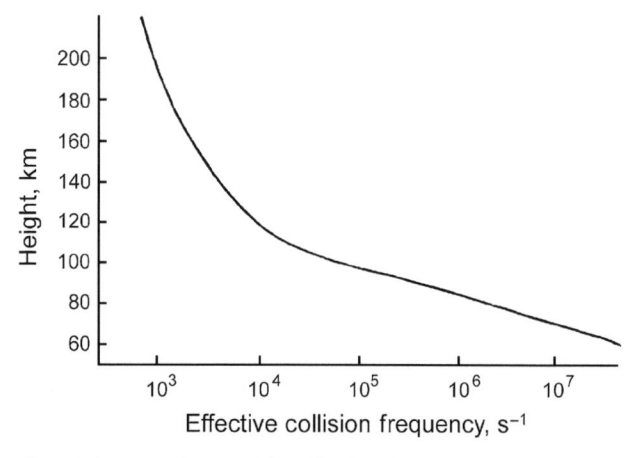

Fig. 13.6. Dependence of the effective electron collision frequency for momentum transfer v on height z. Adapted from Budden (1988).

13.2. Theoretical background

of the magnetosphere is called the plasmasphere. It extends outward by four or five Earth radii (the Earth's radius is 6370 km at the equator) to the plasmapause, where the plasma density abruptly decreases, as discussed further in subsection 13.5.1.

13.2.2. General equations

We give now a general theoretical overview of the interaction of electromagnetic waves with the charged particles of the ionosphere and magnetosphere. We consider a particle having charge q and mass m which moves with velocity $\mathbf{v}$ and (i) collides with other nearby particles with a constant effective collision frequency ν for momentum transfer, (ii) interacts with a uniform static magnetic field $\mathbf{B}_{0z}$ pointing in the z direction (the model geomagnetic field), and (iii) interacts with a uniform time-varying electric field $\mathbf{E}$ (the model electric field of an electromagnetic wave propagating in the ionospheric or magnetospheric plasma). We assume a time-harmonic electric field of the form $e^{-i\omega t}$, where ω is the angular frequency. Note that, in the interests of simplicity, our formulation does not include density or pressure gradients, although the effects of these could be important in some of the phenomena that we discuss later. We are also justified in neglecting the force on the charged particle from the time-varying magnetic field that accompanies the time-varying electric field. The force due to the time-varying magnetic field is generally negligible in comparison with the force due to the electric field, since the ratio of the forces is roughly v/c and $v/c \ll 1$, where c is the speed of light and v is the magnitude of $\mathbf{v}$. The equation for the particle interactions described above is

$$-i\omega m \mathbf{v} = q\mathbf{E} + q\mathbf{v} \times \mathbf{B}_{0z} - m\nu \mathbf{v} \tag{13.1}$$

In Eq. 13.1 and elsewhere in this chapter $i = \sqrt{-1}$, while in other chapters this quantity is denoted as j; i is used here rather than j in order to avoid confusion with the symbol for the current density, to be introduced in Eq. 13.3. Equation 13.1 has been used to describe many types of charged-particle behavior in plasmas (e.g., Ratcliffe 1959; Spitzer 1962; Stix 1962; Ginzburg 1970b; Yeh and Liu 1972; Budden 1988); it can be solved for the three components of particle velocity, v_x, v_y, and v_z, and expressed in matrix notation as

$$\begin{bmatrix} v_x \\ v_y \\ v_z \end{bmatrix} = \frac{q}{m} \begin{bmatrix} \frac{\nu - i\omega}{(\nu - i\omega)^2 + \omega_b^2} & \frac{\omega_b}{(\nu - i\omega)^2 + \omega_b^2} & 0 \\ \frac{-\omega_b}{(\nu - i\omega)^2 + \omega_b^2} & \frac{\nu - i\omega}{(\nu - i\omega)^2 + \omega_b^2} & 0 \\ 0 & 0 & \frac{1}{\nu - i\omega} \end{bmatrix} \begin{bmatrix} E_x \\ E_y \\ E_z \end{bmatrix} \tag{13.2}$$

where $\omega_b = qB_{0z}/m$ is called the cyclotron frequency or the gyrofrequency and is the angular frequency at which the charged particle makes circular orbits in a plane perpendicular to the static magnetic field, in the absence of collisions. The current density in the plasma due to both ions and electrons is

$$\mathbf{j} = (Z_i|e|n_i\mathbf{v}_i - |e|n_e\mathbf{v}_e) \tag{13.3}$$

where n_i and n_e are the ion and electron densities, $\mathbf{v}_i$ and $\mathbf{v}_e$ are the ion and electron velocities, Z_i is the degree of ionization of the ions, and $|e|$ is the magnitude of the electron charge. We can therefore write the current density as the product of a conductivity tensor $\langle \sigma \rangle$ and an electric field vector $\mathbf{E}$:

$$\mathbf{j} = \langle \sigma \rangle \cdot \mathbf{E} \tag{13.4}$$

where the conductivity tensor $\langle \sigma \rangle$ is given by

$$\langle \sigma \rangle = \frac{Z_i^2|e|^2 n_i}{m_i} \begin{bmatrix} \frac{\nu_i - i\omega}{(\nu_i - i\omega)^2 + \omega_{bi}^2} & \frac{\omega_{bi}}{(\nu_i - i\omega)^2 + \omega_{bi}^2} & 0 \\ \frac{-\omega_{bi}}{(\nu_i - i\omega)^2 + \omega_{bi}^2} & \frac{\nu_i - i\omega}{(\nu_i - i\omega)^2 + \omega_{bi}^2} & 0 \\ 0 & 0 & \frac{1}{\nu_i - i\omega} \end{bmatrix}$$
$$+ \frac{|e|^2 n_e}{m_e} \begin{bmatrix} \frac{\nu_e - i\omega}{(\nu_e - i\omega)^2 + \omega_{be}^2} & \frac{\omega_{be}}{(\nu_e - i\omega)^2 + \omega_{be}^2} & 0 \\ \frac{-\omega_{be}}{(\nu_e - i\omega)^2 + \omega_{be}^2} & \frac{\nu_e - i\omega}{(\nu_e - i\omega)^2 + \omega_{be}^2} & 0 \\ 0 & 0 & \frac{1}{\nu_e - i\omega} \end{bmatrix} \tag{13.5}$$

The ion and electron cyclotron frequencies are, respectively,

$$\omega_{bi} = \frac{Z_i|e|B_{0z}}{m_i} \quad \text{and} \quad \omega_{be} = -\frac{|e|B_{0z}}{m_e} \tag{13.6}$$

and m_i and m_e are the ion and electron masses, respectively.

Consider now Maxwell's equations for the case of time-harmonic fields that vary in space as

$$\exp[i(\mathbf{k} \cdot \mathbf{r} - \omega t)] \tag{13.7}$$

where we have used the time variation previously assumed and now include also a spatial variation, specified by the term $i\mathbf{k} \cdot \mathbf{r}$, where $\mathbf{k}$ is the vector propagation constant or wave number whose direction indicates the direction of propagation and whose magnitude, k, equals 2π divided by the wavelength of the wave. The corresponding phase velocity υ_p and group velocity υ_g in the direction of propagation of the wave can be determined as

$$\upsilon_p = \frac{\omega}{k} \tag{13.8}$$

$$\upsilon_g = \frac{\partial \omega}{\partial k} \tag{13.9}$$

The determination of $\mathbf{k}$ as a function of ω is one of the primary objectives in most wave analyses, since $\mathbf{k}$ carries information on whether a wave will propagate, on its phase and group velocities, and on its dispersion characteristics. For example, if $\mathbf{k}$ is imaginary then the spatial exponent in

Eq. 13.7 is real and negative, indicating a non-propagating wave that decays exponentially with distance along the direction of propagation.

For an electromagnetic wave with temporal and spatial variation given by Eq. 13.7, we can write Maxwell's equations as

$$i\varepsilon_0 \mathbf{k} \cdot \mathbf{E} = \rho \qquad (\nabla \cdot \mathbf{D} = \rho) \qquad (13.10)$$

$$\mathbf{k} \times \mathbf{E} = \mu_0 \omega \mathbf{H} \qquad \left(\nabla \times \mathbf{E} = -\frac{\partial \mathbf{B}}{\partial t}\right) \qquad (13.11)$$

$$i\mathbf{k} \times \mathbf{H} = \mathbf{j} - i\varepsilon_0 \omega \mathbf{E} \qquad \left(\nabla \times \mathbf{H} = \mathbf{j} + \frac{\partial \mathbf{D}}{\partial t}\right) \qquad (13.12)$$

$$\mu_0 \mathbf{k} \cdot \mathbf{H} = 0 \qquad (\nabla \cdot \mathbf{B} = 0) \qquad (13.13)$$

where we assume that $\mathbf{D} = \varepsilon_0 \mathbf{E}$ and $\mathbf{B} = \mu_0 \mathbf{H}$. With $\mathbf{j} = \langle \boldsymbol{\sigma} \rangle \cdot \mathbf{E}$, the right-hand side of Eq. 13.12 may be written

$$\langle \boldsymbol{\sigma} \rangle \cdot \mathbf{E} - i\varepsilon_0 \omega \mathbf{E} = -i\omega \langle \boldsymbol{\varepsilon} \rangle \cdot \mathbf{E} \qquad (13.14)$$

where Eq. 13.14 defines the permittivity tensor $\langle \boldsymbol{\varepsilon} \rangle$, so that

$$\langle \boldsymbol{\varepsilon} \rangle = \varepsilon_0 \left(\langle 1 \rangle - \frac{\langle \boldsymbol{\sigma} \rangle}{i\varepsilon_0 \omega}\right) \qquad (13.15)$$

Here $\langle 1 \rangle$ is a diagonal tensor whose elements are equal to unity. It follows that Eq. 13.12 can be rewritten as

$$\mathbf{k} \times \mathbf{H} = -\omega \langle \boldsymbol{\varepsilon} \rangle \cdot \mathbf{E} \qquad (13.16)$$

which we will use later in this subsection as one of Maxwell's equations describing electromagnetic waves in a plasma. Since $\mathbf{k} \cdot \mathbf{k} \times \mathbf{H}$ is zero, Eq. 13.12 can also be written as

$$0 = \mathbf{k} \cdot \mathbf{j} - i\varepsilon_0 \omega \mathbf{k} \cdot \mathbf{E} \qquad (13.17)$$

After substituting $\mathbf{k} \cdot \mathbf{E}$ from Eq. 13.10 into Eq. 13.17, we obtain

$$\rho = \frac{\mathbf{k} \cdot \mathbf{j}}{\omega} \qquad (13.18)$$

Equation 13.18 gives the relation between ρ and $\mathbf{j}$; it is the current continuity equation. It can also be written as

$$\rho = \frac{\mathbf{k} \cdot \langle \boldsymbol{\sigma} \rangle \cdot \mathbf{E}}{\omega} \qquad (13.19)$$

If we substitute ρ from Eq. 13.19 into Eq. 13.10, we find that

$$\mathbf{k} \cdot \langle \boldsymbol{\sigma} \rangle \cdot \mathbf{E} - i\varepsilon_0 \omega \mathbf{k} \cdot \mathbf{E} = 0 \qquad (13.20)$$

From a comparison of Eq. 13.20 with Eq. 13.14, it is apparent that

$$\mathbf{k} \cdot \langle \boldsymbol{\varepsilon} \rangle \cdot \mathbf{E} = 0 \qquad (13.21)$$

We can therefore rewrite Maxwell's equations, Eqs. 13.10 to 13.13, in the following form:

$$\mathbf{k} \cdot \langle \boldsymbol{\varepsilon} \rangle \cdot \mathbf{E} = 0 \qquad (\nabla \cdot \mathbf{D} = \rho) \qquad (13.22)$$

$$\mathbf{k} \times \mathbf{E} = \mu_0 \omega \mathbf{H} \qquad \left(\nabla \times \mathbf{E} = -\frac{\partial \mathbf{B}}{\partial t}\right) \qquad (13.23)$$

$$\mathbf{k} \times \mathbf{H} = -\omega \langle \boldsymbol{\varepsilon} \rangle \cdot \mathbf{E} \qquad \left(\nabla \times \mathbf{H} = \mathbf{J} + \frac{\partial \mathbf{D}}{\partial t}\right) \qquad (13.24)$$

$$\mathbf{k} \cdot \mathbf{H} = 0 \qquad (\nabla \cdot \mathbf{B} = 0) \qquad (13.25)$$

where the permittivity tensor is defined by Eqs. 13.15 and 13.5. Equations 13.22 to 13.25 specify the properties of an electromagnetic wave as it interacts with a uniform plasma in the presence of a uniform static magnetic field. It is convenient to combine Eqs. 13.23 and 13.24 into a wave equation, which can then be solved to yield these properties.

If we take the cross product of $\mathbf{k}$ with both sides of Eq. 13.23, this equation becomes

$$\mathbf{k} \times \mathbf{k} \times \mathbf{E} = \omega \mu_0 \mathbf{k} \times \mathbf{H} \qquad (13.26)$$

and can then be combined with Eq. 13.24 to yield

$$\mathbf{k} \times \mathbf{k} \times \mathbf{E} + \omega^2 \mu_0 \langle \boldsymbol{\varepsilon} \rangle \cdot \mathbf{E} = 0 \qquad (13.27)$$

Equation 13.27 is the wave equation for the anisotropic medium. By expanding $\mathbf{k} \times \mathbf{k} \times \mathbf{E}$ and then collecting the terms, it is possible to write

$$\mathbf{k} \times \mathbf{k} \times \mathbf{E} = \langle \boldsymbol{\Lambda} \rangle \cdot \mathbf{E} \qquad (13.28)$$

where the tensor $\langle \boldsymbol{\Lambda} \rangle$ is given by

$$\langle \boldsymbol{\Lambda} \rangle = \begin{bmatrix} -k_y^2 - k_z^2 & k_x k_y & k_x k_y \\ k_y k_x & -k_z^2 - k_x^2 & k_y k_z \\ k_z k_x & k_z k_y & -k_z^2 - k_y^2 \end{bmatrix} \qquad (13.29)$$

and k_x, k_y, and k_z are the components of $\mathbf{k}$. Equation 13.27 can therefore be written as

$$(\langle \boldsymbol{\Lambda} \rangle + \omega^2 \mu_0 \langle \boldsymbol{\varepsilon} \rangle) \cdot \mathbf{E} = 0 \qquad (13.30)$$

The solution to any equation of this form is found by setting to zero the determinant of the second-order tensor that multiplies $\mathbf{E}$:

$$\|(\langle \boldsymbol{\Lambda} \rangle + \omega^2 \mu_0 \langle \boldsymbol{\varepsilon} \rangle)\| = 0 \qquad (13.31)$$

The vector propagation constant can be found as a function of frequency from Eq. 13.31. This result is inserted into Eq. 13.30, which can then be solved for the electric field. Once both the electric field and the propagation constant are known, the magnetic field can be calculated from Eq. 13.23.

The solution of Eq. 13.31 has been used to predict the occurrence and behavior of a variety of electromagnetic and electrostatic waves in plasmas (e.g., Ratcliffe 1959; Stix 1962; Spitzer 1962; Ginzburg 1970b; Yeh and Liu 1972; Budden 1988).

13.2. Theoretical background

13.2.3. Four special cases

If the medium in which the wave travels is isotropic, that is, a plasma without a static magnetic field or one in which the electron effective collision frequency far exceeds the cyclotron frequency, then we can determine the propagation constant as a function of frequency from the right-hand side of Eq. 13.24. In an isotropic dielectric medium other than free space, the magnitude of the phase velocity of an electromagnetic wave is given by $v_p = (\mu_0 \varepsilon)^{-1/2}$. It follows, then, that $v_p/c = (\varepsilon_0/\varepsilon)^{1/2}$ or $v_p/c = \varepsilon_r^{-1/2}$, where $\varepsilon_r = \varepsilon/\varepsilon_0$ is the dielectric constant of the medium. Since, in the direction of propagation, $k = \omega/v_p$, the magnitude of the propagation constant k can be written in terms of the dielectric constant:

$$k = \frac{\omega}{c}\sqrt{\varepsilon_r} \qquad (13.32)$$

Next we will calculate the *effective* dielectric constant ε_{rp} of the plasma for two important cases in which the magnetic field is assumed to be zero. In case 1 particle collisions are assumed to be unimportant; in case 2 particle collisions are included. Given the effective dielectric constant, application of Eq. 13.32 will yield the relation between k and ω, referred to as the dispersion relation.

Case 1: $B_{0z} = 0$, $\nu_e = 0$, $\nu_i = 0$. We look now at the first case, electromagnetic wave propagation in a plasma in the absence of a static magnetic field and with no collisions. This case will allow us to develop the concept of the electron plasma frequency, the natural frequency of oscillation of the plasma electrons. This frequency provides the boundary between different types of electromagnetic wave behavior in a plasma.

If $B_{0z} = 0$ then the cyclotron frequencies ω_{bi} and ω_{be} are zero, and, in the absence of collisions, the permittivity tensor reduces to a diagonal tensor whose elements are equal. In other words, the permittivity tensor reduces to a scalar *effective* permittivity.

$$\langle \varepsilon \rangle \to \varepsilon_p = \varepsilon_0 \left(1 - \frac{\omega_{pi}^2}{\omega^2} - \frac{\omega_{pe}^2}{\omega^2}\right) \qquad (13.33)$$

where

$$\omega_{pi}^2 = \frac{Z_i^2 |e|^2 n_i}{m_i \varepsilon_0} \quad \text{and} \quad \omega_{pe}^2 = \frac{|e|^2 n_e}{m_e \varepsilon_0} \qquad (13.34)$$

The quantities ω_{pi} and ω_{pe} are called the "ion plasma frequency" and the "electron plasma frequency," respectively. Since m_i is much greater than m_e (ions are much heavier than electrons), the second term on the right of Eq. 13.33 can often be ignored. Equation 13.33 yields the following scalar effective dielectric constant for the plasma:

$$\varepsilon_{rp} = \frac{\varepsilon_p}{\varepsilon_0} \cong 1 - \frac{\omega_{pe}^2}{\omega^2} \qquad (13.35)$$

and the magnitude of the propagation constant, from Eq. 13.32, is

$$k = \frac{\omega}{c}\left(1 - \frac{\omega_{pe}^2}{\omega^2}\right)^{1/2} \qquad (13.36)$$

If the frequency of the electromagnetic wave ω is greater than the electron plasma frequency ω_{pe} then k is a real number. The phase and group velocities of the wave can be determined from Eqs. 13.8 and 13.9. At frequencies $\omega \gg \omega_{pe}$, the propagation constant approaches its free-space value $k = \omega/c$, and the wave behaves as if the plasma were not present. Basically, at very high frequencies the inertia of the charged particles is such that they are unaffected by the wave and the wave is unaffected by them. If the frequency of the electromagnetic wave ω is less than the electron plasma frequency ω_{pe} then k is an imaginary number. In this case, the particles move so rapidly compared with the time variation of the applied electric field that they essentially reflect the field. Since the fields vary spatially as $\exp(i\mathbf{k} \cdot \mathbf{r})$, for $\omega < \omega_{pe}$ the wave will not propagate in the usual sense but will decay without energy loss as it goes forward into the plasma, an evanescent wave. If an electromagnetic wave of angular frequency ω propagating in free space encounters a plasma such that $\omega < \omega_{pe}$ then the wave energy will be reflected.

It is basically in this manner that the ionosphere reflects electromagnetic radiation below 1 MHz or so, such as AM radio waves and sferics, although there are effects due to the geomagnetic field and particle collisions. The electron plasma frequency is the natural frequency of oscillation of a system of plasma electrons and is caused by the self-restoring force of the electric field created when there are regions of net charge (i.e., deviations from equal electron and ion charge densities and hence from overall charge neutrality), as can be derived from Eq. 13.22, a form of Gauss's law in which Newton's first law is contained in the permittivity tensor.

Case 2: $B_{0z} = 0$, $\nu_e \neq 0$, $\nu_i \neq 0$. We consider now the second case of propagation in a plasma with $B_{0z} = 0$ but with collisions. This solution is applicable to the propagation of electromagnetic waves in some parts of the ionosphere and allows us to define formally the concept of a conductivity in terms of plasma parameters.

The effective dielectric permittivity for the plasma from Eqs. 13.5 and 13.15 is

$$\varepsilon_p = \varepsilon_0 \left(1 - \frac{\omega_{pi}^2}{i\omega}\frac{1}{\nu_i - i\omega} - \frac{\omega_{pe}^2}{i\omega}\frac{1}{\nu_e - i\omega}\right) \qquad (13.37)$$

If we reorganize the second and third terms on the right of Eq. 13.37, we can write the dielectric constant of the plasma as

$$\varepsilon_{\text{pr}} = \left(1 + \frac{\omega_{\text{pi}}^2}{\omega} \frac{iv_i - \omega}{v_i^2 + \omega^2} + \frac{\omega_{\text{pe}}^2}{\omega} \frac{iv_e - \omega}{v_e^2 + \omega^2}\right) \quad (13.38)$$

and therefore the propagation constant is

$$k = \frac{\omega}{c}\left[1 - \frac{\omega_{\text{pi}}^2}{v_i^2 + \omega^2} - \frac{\omega_{\text{pe}}^2}{v_e^2 + \omega^2} + \frac{iv_i\omega_{\text{pi}}^2}{\omega(v_i^2 + \omega^2)} + \frac{iv_e\omega_{\text{pe}}^2}{\omega(v_e^2 + \omega^2)}\right]^{1/2} \quad (13.39)$$

When Eq. 13.39 is expanded and with the assumption that $v_e \gg \omega$, the fourth and fifth terms on the right of Eq. 13.39, the imaginary terms, lead to an expression for the "skin depth" of the wave, the distance into the plasma over which the wave is damped to $1/e$ of its original value by energy-dissipating collision processes, similarly to the well-known case where electromagnetic waves penetrate solid conductors (e.g., Inan and Inan 1999, pp. 53–8). The energy of the electromagnetic wave is transferred to the electrons and ions, which dissipate this acquired energy in collisions. For the plasma model we are considering, the low-frequency conductivity of the medium can be found from Eq. 13.5 with $v_e \gg \omega$:

$$\sigma = \frac{|e|^2 n_e}{m_e v_e} \quad (13.40)$$

where the similar conductivity term for the ions is neglected since it is much smaller in magnitude than the term for the electrons. The effective collision frequency for momentum transfer as a function of height is plotted for the ionosphere in Fig. 13.6, and an example of the electron density versus height is given in Fig. 13.5. The effective collision frequency in the lower ionosphere depends primarily on the density of neutral particles, which does not change much with time. As is evident from the example given in Fig. 13.5, the electron density and hence the conductivity vary relatively strongly between day and night. They also vary with latitude and solar activity. Figure 1.3 contains information on the conductivity of the atmosphere from ground level to an altitude of about 120 km. Since collision frequencies in the upper D-region of the ionosphere can be in the range 10^5–10^6 s^{-1}, collisions could play a significant role in the interaction with the ionosphere of radio waves at frequencies near and below 1 MHz.

Equation 13.12, describes "propagation" in a good conductor if $|\mathbf{j}| \cong |\sigma \mathbf{E}| \gg |-i\varepsilon_0\omega\mathbf{E}|$ and free space propagation in a good dielectric if $|\sigma \mathbf{E}| \ll |-i\varepsilon_0\omega\mathbf{E}|$, where $|\sigma \mathbf{E}|$ is the magnitude of the conduction current density and $|-i\varepsilon_0\omega\mathbf{E}|$ is the magnitude of the displacement current density. In view of the above discussion, a common definition of the frequency at which a material can be considered to be a reasonable conductor is that frequency at which the magnitude of the conduction current density equals the magnitude of the displacement current, that is, when

$$\omega\varepsilon_0 = \sigma \quad (13.41)$$

We shall use this definition in discussing the height at which the atmosphere can be considered a reasonable conductor for different frequencies and also in determining the frequencies for which the Earth can be considered a reasonable conductor.

Case 3: $B_{0z} \neq 0, E_z = 0, E_x \neq 0, E_y \neq 0, v_i = 0, v_e = 0$. We consider now an important case of wave propagation in a plasma for which the concept of an effective scalar dielectric constant cannot be used, propagation in the direction of a static magnetic field with no collisions and with a time-varying electric field perpendicular to the direction of the magnetic field. The results will be directly applicable to whistler-mode propagation in the magnetosphere and in the upper parts of the ionosphere.

To gain detailed information about this type of propagation, we must solve Eqs. 13.23 and 13.24 simultaneously. These equations can be written in scalar-component form as follows:

$$k_z E_y + \omega\mu_0 H_x = 0 \quad (13.42)$$

$$k_z E_x - \omega\mu_0 H_y = 0 \quad (13.43)$$

$$-A\omega E_z - GE_y + k_z H_y = 0 \quad (13.44)$$

$$GE_x + A\omega E_y + k_z H_x = 0 \quad (13.45)$$

where

$$A = \varepsilon_0\left(1 + \frac{\omega_{\text{pi}}^2}{\omega_{\text{bi}}^2 - \omega^2} + \frac{\omega_{\text{pe}}^2}{\omega_{\text{be}}^2 - \omega^2}\right) \quad (13.46)$$

and

$$G = i\varepsilon_0\left(\frac{\omega_{\text{bi}}\omega_{\text{pi}}^2}{\omega_{\text{bi}}^2 - \omega^2} + \frac{\omega_{\text{be}}\omega_{\text{pe}}^2}{\omega_{\text{be}}^2 - \omega^2}\right) \quad (13.47)$$

The solution is found by requiring that the following determinant vanish:

$$\begin{Vmatrix} 0 & k_z & \omega\mu_0 & 0 \\ k_z & 0 & 0 & -\mu_0\omega \\ -A\omega & -G & 0 & k_z \\ -G & A\omega & k_z & 0 \end{Vmatrix} \quad (13.48)$$

Equation 13.48 can be expanded to yield

$$k_z^4 - 2k_z^2\omega^2\mu_0 A + \omega^2\mu_0^2(\omega^2 A^2 + G^2) = 0 \quad (13.49)$$

which can in turn be solved for the square of the z component of the propagation constant, that is, the square of

13.2. Theoretical background

propagation constant along the magnetic field:

$$k_z^2 = \omega^2 \mu_0 \varepsilon_0 \left[1 - \frac{\omega_{\text{pi}}^2}{\omega(\omega \pm \omega_{\text{bi}})} - \frac{\omega_{\text{pe}}^2}{\omega(\omega \pm \omega_{\text{be}})} \right] \quad (13.50)$$

The plus and minus signs ± in Eq. 13.50 indicate that the wave can exhibit two different velocities for a given frequency. Thus for plane-wave propagation in the direction of the static magnetic field there are two propagation modes with two distinct velocities. A medium that supports two propagation modes at the same frequency is called anisotropic. Examination of Eq. 13.50 reveals that the propagation constant will exhibit discontinuities, values of zero or infinity, at certain frequencies. Between these discontinuities there will be "pass" bands where k_z is real and propagation can occur and "stop" bands where k_z is imaginary and the wave is exponentially damped. As ω becomes large compared to ω_{bi} and ω_{be}, Eq. 13.50 reduces to Eq. 13.36, the expression for k for the electromagnetic field in a collisionless plasma in the absence of a static magnetic field (case 1).

Consider now the discontinuities in Eq. 13.50. In addition to the "screening" effects ($k_z \to 0$), which can be considered to be due to the plasma's tendency to maintain overall charge neutrality, as discussed for case 1, the plasma–wave interaction is characterized by "resonance" effects ($k_z \to \infty$), which are due to resonant interactions of the charged plasma particles with both the electromagnetic wave and the static magnetic field. In the absence of an electromagnetic wave, the positive ions and the electrons of any velocity will make circular orbits around the static magnetic field lines in opposite directions. The time-varying electric field of the electromagnetic wave will then alter the circular trajectories of the charged particles in a manner dependent upon the relationship between the frequency of the electromagnetic wave and the cyclotron frequencies of the charged particles. We are considering an electromagnetic plane wave with an electric field perpendicular to the magnetic field lines, that is, $E_x \neq 0$, $E_y \neq 0$. The total electric field can be considered as being composed of two waves, one right circularly polarized (polarized clockwise, looking in the direction of propagation) and the other left circularly polarized. Since each of the two circularly polarized waves encounters a different type of plasma-particle behavior, it is to be expected that the characteristics of the two circularly polarized waves will differ. We can, therefore, interpret physically the two modes of propagation (identified by the ± signs in Eq. 13.50) that we have identified mathematically: each mode of propagation corresponds to one of the circularly polarized waves. A resonance in the propagation properties of each circularly polarized wave occurs when the frequency of the wave equals the appropriate cyclotron frequency. (Note that the electron cyclotron frequency is defined as a negative quantity.) Under these circumstances, the wave is "pumping" an oscillating system (one type of orbiting particle) at the resonant frequency of the system. The $\omega = \omega_{\text{bi}}$ resonance is caused by the left circularly polarized wave, since that wave rotates in the direction in which the ions orbit the static magnetic field lines, looking in the direction of the magnetic field, and hence the left circularly polarized wave is represented by the minus signs in Eq. 13.50. Since ω_{be} is negative, the right circularly polarized wave is associated with the plus signs in Eq. 13.50, leading to the electron resonance.

Case 4: $B_{0z} \neq 0$, $E_x = 0$, $E_y = 0$, $E_z \neq 0$, $\nu_e = 0$, $\nu_i = 0$. We examine now the fourth and final special case which, along with case 3, allows us to draw inferences on the general behavior of electromagnetic waves in a collisionless plasma in the presence of a static magnetic field, conditions that exist in the magnetosphere and in the upper ionosphere.

In case 3 we examined the properties of an electromagnetic wave, in the presence of a collisionless plasma and a static magnetic field, as it propagates in the direction of the static magnetic field with the electric field perpendicular to the static magnetic field. For propagation perpendicular to the static magnetic field with the electric field parallel to the static magnetic field, case 4, the electric field forces the charged particles to move parallel to the magnetic field, and hence there is no magnetic force on the particles. The result is an electromagnetic wave that has the same propagation characteristics as a wave in a collisionless plasma in the absence of a magnetic field, case 1.

Between the two extremes, propagation along the static magnetic field with the electric field perpendicular to this magnetic field, as discussed in case 3, and propagation perpendicular to the static magnetic field with the electric field in the direction of this magnetic field, as considered in case 4, there is a wide range of variation in the properties of the electromagnetic waves propagating in the plasma. These properties have been delineated by solving Equation 13.31 (e.g., Radcliffe 1959; Stix 1962; Spitzer 1962; Budden 1988). The reader is referred to these sources for a description of the general case as well as more details regarding the four special cases discussed above.

13.2.4. Reflection and transmission

In order to understand the characteristics of both atmospherics and whistlers it is necessary to examine the theory of the reflection and transmission of an electromagnetic wave propagating from free space inside the Earth–ionosphere cavity into the ionosphere. In doing so, we first define a phase refractive index n, usually just called the refractive index or the index of refraction, for any medium as the ratio of the speed of light in a vacuum, $c = (\mu_0 \varepsilon_0)^{-1/2}$,

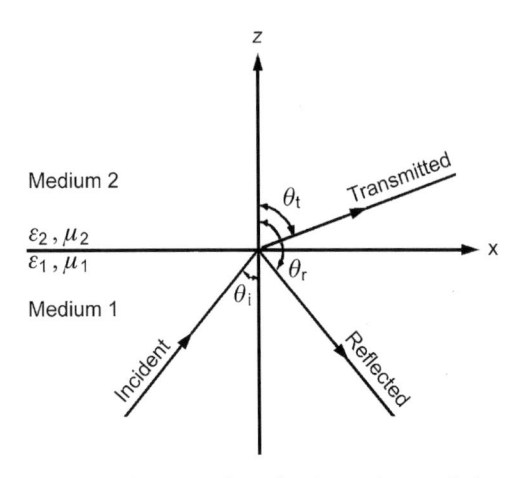

Fig. 13.7. Geometry for reflection and transmission of electromagnetic waves at a sharp planar boundary.

to the phase velocity (Eq. 13.8) of the electromagnetic wave in the medium, that is

$$n = (\mu_0 \varepsilon_0)^{-1/2} \frac{k}{\omega} \tag{13.51}$$

Consider an electromagnetic wave incident at a boundary between two media. Suppose that the wave is a perpendicularly polarized wave (the electric field is perpendicular to the plane of incidence, the xz-plane in Fig. 13.7, and parallel to the boundary $z = 0$) and that the incidence angle is θ_i. For perpendicular polarization, $E_x = 0$, $E_z = 0$, and $H_y = 0$, so that the only component of the electric field is E_y. (For parallel polarization we would have $H_x = 0$, $H_z = 0$, and $E_y = 0$, E_x and E_z being non-zero, that is, the electric field would be in the plane of incidence.) The total fields on either side of the boundary for perpendicular polarization are the sum of the incident and reflected fields:

$$E_y = E_y^i + E_y^r$$
$$= \exp[-ik_0 n_1(x \sin\theta_i + z \cos\theta_i)]$$
$$+ R_\perp \exp[-ik_0 n_1(x \sin\theta_r + z \cos\theta_r)], \quad z < 0 \tag{13.52}$$

$$E_y = E_y^t$$
$$= T_\perp \exp[-ik_0 n_2(x \sin\theta_t + z \cos\theta_t)], \quad z > 0 \tag{13.53}$$

$$H_x = H_x^i + H_x^r = (\varepsilon_1/\mu_1)^{1/2}$$
$$\times \left(-\cos\theta_i E_y^i + \cos\theta_r E_y^r\right), \quad z < 0 \tag{13.54}$$

$$H_x = H_x^t = (\varepsilon_2/\mu_2)^{1/2}\left(-\cos\theta_t E_y^t\right), \quad z > 0 \tag{13.55}$$

where $n_1 = c(\varepsilon_1 \mu_1)^{1/2}$ and $n_2 = c(\varepsilon_2 \mu_2)^{1/2}$; θ_r and θ_t are the angles with respect to the z axis of the directions of propagation of the reflected and transmitted waves; and $R_\perp$ and $T_\perp$ are the reflection and transmission coefficients. The subscript $\perp$ indicates that the polarization is perpendicular to the plane of incidence.

At the boundary $z = 0$, the tangential components of the electric and magnetic fields must be continuous for all values of x, as can be derived from Maxwell's equations (e.g., Inan and Inan 1998, pp. 620–1). This can be true only if

$$n_1 \sin\theta_i = n_1 \sin\theta_r = n_2 \sin\theta_t \tag{13.56}$$

Therefore

$$\theta_r = \pi - \theta_i \tag{13.57}$$

and

$$n_1 \sin\theta_i = n_2 \sin\theta_t \tag{13.58}$$

Equation 13.57 shows that the angle of reflection equals the angle of incidence and Eq. 13.58 is called Snell's law.

Next, matching the fields at the boundary, we find that

$$1 + R_\perp = T_\perp \tag{13.59}$$

$$(\varepsilon_1/\mu_1)^{1/2} \cos\theta_i (1 - R_\perp) = (\varepsilon_2/\mu_2)^{1/2} \cos\theta_t T_\perp \tag{13.60}$$

Solving for $R_\perp$ and $T_\perp$, we obtain

$$R_\perp = \frac{(\varepsilon_1/\mu_1)^{1/2} \cos\theta_i - (\varepsilon_2/\mu_2)^{1/2} \cos\theta_t}{(\varepsilon_1/\mu_1)^{1/2} \cos\theta_i + (\varepsilon_2/\mu_2)^{1/2} \cos\theta_t} \tag{13.61}$$

and

$$T_\perp = \frac{2(\varepsilon_1/\mu_1)^{1/2} \cos\theta_i}{(\varepsilon_1/\mu_1)^{1/2} \cos\theta_i + (\varepsilon_2/\mu_2)^{1/2} \cos\theta_t} \tag{13.62}$$

Between the Earth and the ionosphere, within the ionosphere, and within the magnetosphere, $\mu_1 = \mu_2 = \mu_0$ to a good approximation, so that Eqs. 13.61 and 13.62 can be written as

$$R_\perp = \frac{n_1 \cos\theta_i - n_2 \cos\theta_t}{n_1 \cos\theta_i + n_2 \cos\theta_t} \tag{13.63}$$

$$T_\perp = \frac{2n_1 \cos\theta_i}{n_1 \cos\theta_i + n_2 \cos\theta_t} \tag{13.64}$$

These expressions for the reflection and transmission coefficients are also called the Fresnel formulas.

For parallel polarization, that is, the electric field in the plane of incidence, the reflection and transmission coefficients can be similarly obtained by matching the boundary condition at $z = 0$:

$$R_\parallel = \frac{H_y^r}{H_y^i} = \frac{n_2 \cos\theta_i - n_1 \cos\theta_t}{n_2 \cos\theta_i + n_1 \cos\theta_t} \tag{13.65}$$

$$T_\parallel = \frac{n_1 H_y^t}{n_2 H_y^i} = \frac{2n_1 \cos\theta_i}{n_2 \cos\theta_i + n_1 \cos\theta_t} \tag{13.66}$$

Note that the definitions of the reflection and transmission coefficients for the parallel-polarized wave are different from those for the perpendicularly polarized wave in that the definitions for the former involve the magnetic fields and those for the latter the electric fields. Also, note that there is no coupling at the boundary between waves having the two different polarizations.

Suppose now that for the individual cases of perpendicularly polarized and parallel-polarized waves, we allow that medium 2 be anisotropic, as is the case for a collisionless plasma (such as exists in the ionosphere and the magnetosphere above certain frequencies) in a static magnetic field (e.g., Eq. 13.50). We have seen that, in general, two characteristic modes can propagate in an anisotropic medium. It follows that a linearly polarized wave incident on the boundary may become elliptically polarized after reflection from the boundary, as determined from the boundary conditions on the tangential fields. Hence it is necessary to define four coefficients, $_\perp R_\perp$, $_\perp R_\|$, $_\| R_\|$, $_\| R_\perp$ to indicate the ratios of the various components of the reflected and incident electric fields. The first subscript denotes whether the incident electric field vector is parallel or perpendicular to the plane of incidence, and the second subscript refers in the same way to the reflected electric field vector. In exactly the same manner, we can define four transmission coefficients $_\perp T_\perp$, $_\perp T_\|$, $_\| T_\|$, $_\| T_\perp$.

As an example of how to deal with the problem of reflection and transmission of waves of this nature, consider the system shown in Fig. 13.7, where medium 1 is free space (within the Earth–ionosphere cavity) and medium 2 is a plasma in the presence of a magnetic field in the z direction (the ionosphere). We consider only the effects of electrons and we ignore electron collisions with ions and neutral particles. An electric field linearly polarized in the x-direction is assumed to be incident on the plasma normally from below, that is, in what follows we take $\theta_i = 0$. In the plasma, in general, the refracted wave will consist of two waves, labeled by subscripts a and b, that correspond to the two characteristic modes in the anisotropic medium. Therefore, just above the plane $z = 0$, the field components may be written as

$$E_x = E_{xa} + E_{xb} \tag{13.67}$$

$$E_y = -iE_{xa} + iE_{xb} \tag{13.68}$$

$$H_y = (\varepsilon_0/\mu_0)^{1/2}(n_a E_{xa} + n_b E_{xb}) \tag{13.69}$$

$$H_x = (\varepsilon_0/\mu_0)^{1/2}(in_a E_{xa} - in_b E_{xb}) \tag{13.70}$$

where **H** is related to **E** by Eq. 13.23 and Eq. 13.50 (since the wave is traveling in the same direction as the magnetic field), E_{xa} and E_{xb} are the x components of waves a and b, respectively, and

$$n_a^2 = 1 - \frac{\omega_{pe}^2/\omega^2}{1 - |\omega_{be}|/\omega} \tag{13.71}$$

$$n_b^2 = 1 - \frac{\omega_{pe}^2/\omega^2}{1 + |\omega_{be}|/\omega} \tag{13.72}$$

More details of this derivation, including the complex representation of an elliptically polarized wave, are found, for example, in Yeh and Liu (1972). Just below the plane $z = 0$, the field components are

$$E_x = (1 + {}_\| R_\|)E_x^i \tag{13.73}$$

$$E_y = {}_\| R_\perp E_x^i \tag{13.74}$$

$$H_x = (\varepsilon_0/\mu_0)^{1/2}{}_\| R_\perp E_x^i \tag{13.75}$$

$$H_y = (\varepsilon_0/\mu_0)^{1/2}(1 - {}_\| R_\|)E_x^i \tag{13.76}$$

where E_x^i is the incident wave and $_\| R_\|$, $_\| R_\perp$ are the two reflection coefficients.

Matching the four field components at $z = 0$, we obtain four equations for the four unknowns E_{xa}, E_{xb}, $_\| R_\|$, and $_\| R_\perp$ in terms of E_x^i and n_a, n_b. Solving them, we obtain

$$E_{xa} = \frac{E_x^i}{1 + n_a}, \quad E_{xb} = \frac{E_x^i}{1 + n_b} \tag{13.77}$$

$$\begin{aligned} {}_\| R_\| &= \frac{1}{2}\left[\frac{1 - n_a}{1 + n_a} + \frac{1 - n_b}{1 + n_b}\right], \\ {}_\| R_\perp &= i\left(\frac{1}{1 + n_b} - \frac{1}{1 + n_a}\right) \end{aligned} \tag{13.78}$$

Thus we see the splitting of the incident wave E_x^i into the two characteristic modes E_a and E_b of the plasma at the boundary. Also, the reflected wave consists of two polarizations, $_\| R_\| E_x^i$ in the x direction and $_\| R_\perp E_x^i$ in the y direction. The resultant reflected wave is thus in general elliptically polarized.

More general cases of electromagnetic-wave interaction with model ionospheres, including non-normal wave incidence on the boundary, may be treated in a similar manner, and such derivations have been given, for example, by Budden (1988).

13.3. Atmospherics (sferics)
13.3.1. History and observed characteristics

A number of papers from the 1930s to the early 1940s characterized and attempted to explain the received time-domain electromagnetic signals from individual distant lightning flashes (e.g., Laby et al. 1937, 1940; Schonland et al. 1940; Lutkin 1939; Watson-Watt et al. 1937). Night-time waveforms observed by Schonland et al. (1940) for lightning at distances from 200 km to over 3000 km are given in Figs. 13.8a, b. For night-time observations at the closer ranges, the overall waveform appears to be

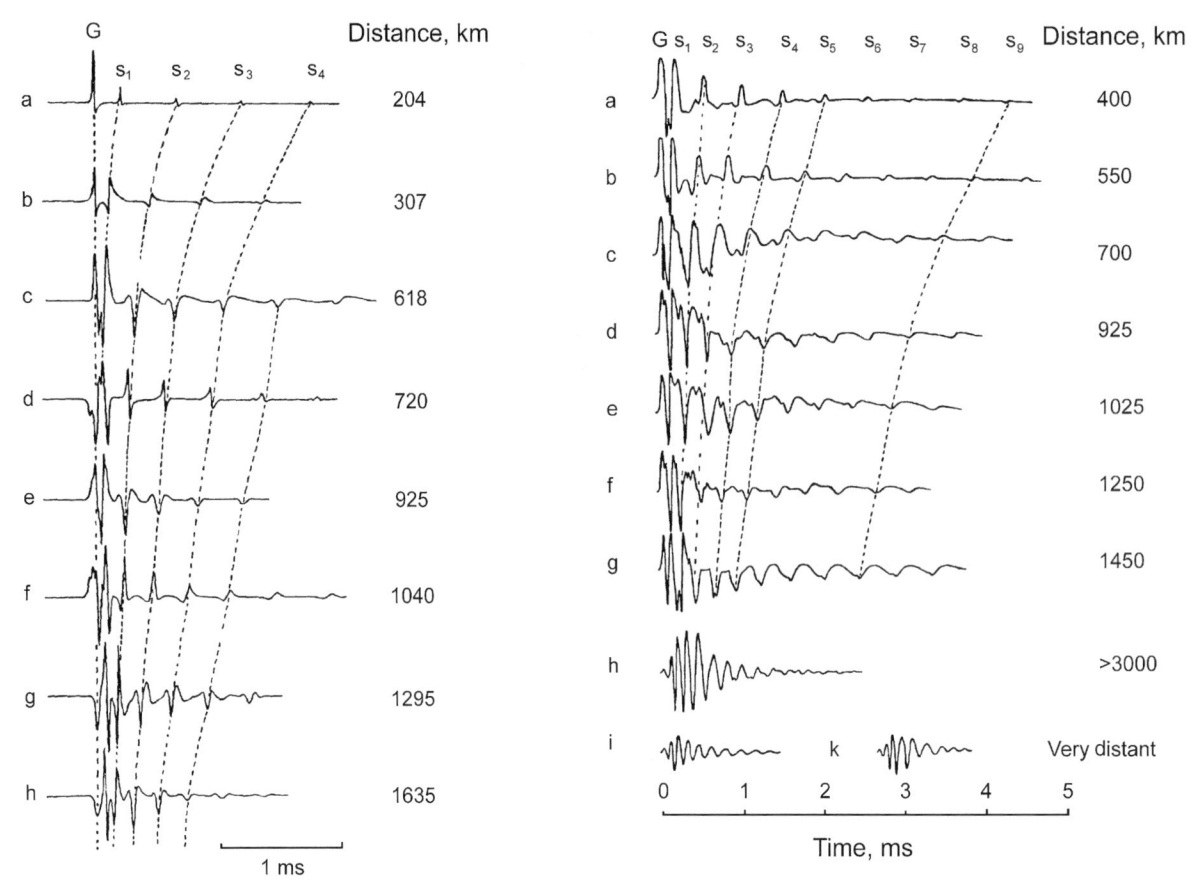

Fig. 13.8. (a) Sferics observed at night in which reflections are separated by quiet intervals. (b) Sferics observed at night in which reflections merge early in the sequence. Adapted from Schonland *et al.* (1940).

composed of several (or a sequence of) individual waveforms, but at greater ranges the individual waveforms appear to overlap. The day-time waveforms are generally observed to consist of damped oscillations or what appear to be overlapping pulses. Examples of day-time waveforms are given in Fig. 13.9. Laby *et al.* (1937) first suggested that the day-time atmospherics that appear as damped oscillations of gradually increasing period were due to multiple ionospheric and ground reflections of a single pulse of short duration. Schonland *et al.* (1940) correctly interpreted the observed features of both day-time and night-time waveforms as due to ionospheric and ground reflections of signals typically produced by one or more non-oscillatory return-stroke current waveforms in a single flash, the variation in the characteristics of the atmospherics being due primarily to the properties of the ionosphere, which are different in the daytime and at night (see Fig. 13.5 for the differences in daytime and night-time ionospheric electron density profiles).

Schonland *et al.* used two-station magnetic direction finding (Section 17.3), at Johannesburg and Durban, South Africa, to estimate independently the lightning location in order to identify the variation in the sferics waveform with distance. Using their measurements of the times of arrival of skywaves of different order relative to that of the groundwave together with some simple theory (see subsection 13.3.2 below), they determined that the mean height of the winter night-time ionospheric reflecting layer was 88 km. Later measurements and discussions of the use of sferics waveforms to determine the distance to the source

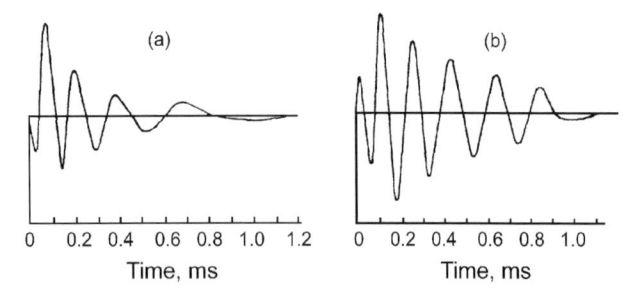

Fig. 13.9. Atmospherics observed during the day in the range 700 to 3000 km. (a) Sources over land to the east of Slough, UK. (b) Sources over the Atlantic Ocean to the west of Slough. The system bandwidth was about 50 kHz. Adapted from Lutkin (1939).

lightning, ionospheric height, and other ionospheric properties are found in, for example, Caton and Pierce (1952), Horner and Clark (1955), Kinzer (1974), and McDonald *et al.* (1979), a theoretical basis for the various features being found in subsection 13.3.2. Caton and Pierce (1952) claimed for their best data a lightning-source-location uncertainty ±6 percent and an ionospheric reflecting-layer-height uncertainty ±2 km, the mean value found being 86 km for lightning about 1000 km away. They state that the reflection height is constant for all skywaves but that occasionally some waveforms show a small height increase with increasing skywave order, corresponding to the more nearly vertical incidence on the ionosphere of the higher-order skywaves. For a given frequency, a steeper angle of incidence on the ionosphere (measured with respect to the normal at the interface) does indeed allow a deeper penetration into the ionosphere before the wave is reflected (e.g., Davies 1990). Caton and Pierce (1952) correctly argued that day-time waveforms and some "regular smooth" night-time waveforms cannot be explained by the simple model of a perfectly reflecting ionosphere, because of interaction of the propagating waveform with an ionosphere whose conductivity varies with height. McDonald *et al.* (1979) studied the night-time sferics produced by lightning about 300 km away (Fig. 13.1). The effective reflecting height for the observed two-hop skywaves was found to be a kilometer or more greater than for the one-hop skywaves, the exact value being dependent on whether the "fast" or the "slow" breakpoints shown in Fig. 13.1 were taken as the beginning of the skywave.

13.3.2. Theory

Two different but equivalent theoretical approaches have been primarily used to describe the propagation of sferics in the Earth–ionosphere cavity. The first approach, the ray or reflection theory, which we used in Section 13.1 and subsection 13.3.1, is the easier of the descriptions to understand physically. Referring to Figs. 13.1a,b, the initial signal arrives via a ground wave, the second signal by reflection from the ionosphere, and succeeding signals by multiple reflections from the ionosphere and the Earth. In the simple model where the ionosphere and the Earth are considered to be well-defined planar reflectors, Eqs. 13.79 and 13.80 below express the distance r from a lightning return stroke and the effective height h of the ionosphere, respectively, in terms of the difference in arrival times of the first-hop skywave and the ground wave t_1 and the difference in arrival times of the second-hop skywave and the ground wave t_2:

$$r = \frac{c}{2}\left(\frac{t_2^2 - 4t_1^2}{4t_1 - t_2}\right) \quad (13.79)$$

$$h = \frac{c}{2}\frac{(t_1 t_2)^{1/2}(t_2 - t_1)^{1/2}}{(4t_1 - t_2)^{1/2}} \quad (13.80)$$

These equations and similar ones for higher-order skywaves for the planar case were used by Kinzer (1974) and earlier researchers to determine the lightning range and the ionospheric height. The planar model of the Earth–ionosphere cavity is reasonable for ranges up to a few hundred kilometers. Beyond that, the spherical geometry of the cavity must be taken into account. Using the spherical geometry shown in Fig. 13.1b, one can derive the following expression for the ionospheric reflecting height (Laby *et al.* 1940; Kessler and Hersperger 1952):

$$h_1 = R_e\left[\cos^2\left(\frac{r}{2R_e}\right) - 1\right] + \left\{R_e^2\left[\cos^2\left(\frac{r}{2R_e}\right) - 1\right]\right.$$
$$\left. + \left(\frac{ct_1 + r}{2}\right)^2\right\}^{1/2} \quad (13.81)$$

where R_e and r represent the mean radius of the Earth and the great-circle distance between the lower portion of the lightning channel and the receiving station, respectively. A similar expression can be derived for the second-hop skywave,

$$h_2 = R_e\left[\cos^2\left(\frac{r}{4R_e}\right) - 1\right] + \left\{R_e^2\left[\cos^2\left(\frac{r}{4R_e}\right) - 1\right]\right.$$
$$\left. + \left(\frac{ct_2 + r}{4}\right)^2\right\}^{1/2} \quad (13.82)$$

As can be seen from the waveforms in Fig. 13.1a, the exact starting points of the skywaves are ambiguous. Higher-frequency components in the lightning electromagnetic signal penetrate further into the ionosphere and the energy loss in the reflection and refraction process is different for different frequency components (e.g., Davies 1990). Either of the first-skywave start times, t_1 or t_1', could be used in Eq. 13.81, and either of the second-hop skywave start times, t_2 or t_2' could be used in Eq. 13.82. Calculation of the reflected waveforms can be done by deriving reflection coefficients for the electromagnetic signals impinging on the ionosphere as a function of frequency and angle of incidence, using the techniques discussed above in subsection 13.2.4, the index of refraction being found from ionospheric models. Because the wavelengths of VLF waves are usually much greater than the thickness of the reflecting ionospheric layer, often the reflection and transmission (refraction) processes can be reasonably approximated by the formulas derived for a sharp boundary. A more sophisticated approach involves treating the ionosphere as a number of sharply bounded slabs of constant characteristics and calculating the properties of the reflected and transmitted waves at each slab boundary (Johler and Harper 1962). This and other approaches are discussed in, for example, the books by Yeh and Liu (1972), Budden (1988), and Davies (1990).

The second common theoretical approach of the propagation of sferics involves viewing the Earth–ionosphere cavity as a waveguide and solving Maxwell's equations for the allowed modes of propagation, in exactly the same manner as is done for metallic waveguides except that the atmospheric waveguide boundaries are more complex than simple uniform conductors; for the latter energy losses occur via ohmic heating effectively within a skin depth (e.g., Inan and Inan 1999), and the geometry is not spherical. If the signal of interest is only a few successive and well-separated skywaves, such as those occurring at night at ranges closer than about 1000 km, the ray or reflection theory discussed above is the preferred theoretical approach. If there are many overlapping skywaves whose frequency content is below about 30 kHz, such as those observed beyond about 1000 km during the day-time, the "waveguide-mode" theory is preferred.

The general theory of waveguide-mode propagation in the Earth–ionosphere cavity is discussed in many books (e.g., Wait 1962; Galejs 1972; Yeh and Liu 1972; Budden 1988; Davies 1990) and is reviewed in Cummer (2000). The wavelength of a 30 kHz signal is 10 km, less than the height of the spherical cavity within which the sferics propagate. For wavelengths shorter than the height of the cavity, a number of TM waveguide modes are possible. The term TM signifies transverse-magnetic, which indicates that there is no component of the magnetic field in the θ direction, the direction of propagation in a spherical coordinate system whose origin is at the center of the Earth, while there is a magnetic field in the spherical ϕ direction and electric fields in the radial and θ directions. The nth TM mode is cut off (will not propagate) when the wavelength $\lambda > 2h/n$, so for the lower frequencies there are fewer modes and below 1 to 3 kHz or so even the first-order mode is cut off. A given waveguide mode can be thought of as composed of two waves propagating at equal angles above and below the θ direction, the direction of overall propagation, and bouncing between the Earth and the ionosphere in such a way that the sum of the two waves satisfies the boundary conditions at the Earth and ionosphere (e.g., Davies 1990). Below the 1 to 3 kHz cutoff for the waveguide modes, quasi-TEM (transverse-electromagnetic) mode propagation allows the existence of slow tails, Q-bursts, and Schumann resonances. These ELF waves have wavelengths considerably greater than the height of the Earth–ionosphere cavity.

Cummer (2000) reviewed both numerical and approximate analytical formulations of waveguide mode theory and additionally employed finite-difference time-domain modeling of the lightning fields in the Earth–ionosphere waveguide, an approach recently made possible by modern computer technology, to investigate the accuracy of previous calculations for different models of the ionosphere boundary.

There are a number of techniques for determining the range to the source of distant sferics from characteristics of the waveforms measured at a single station. Here we address the theoretical basis for these. A method that uses the relative amplitude of the received signal at different frequencies, the so-called spectral amplitude ratio (SAR), was described by Heydt (1982). From theory and multiple-station measurements, the different frequencies are expected to be attenuated differently per unit propagation distance and hence the ratio will depend on the distance over which that attenuation occurs. From a maximum attenuation rate in the range 1 to 3 kHz, the Earth–ionosphere attenuation per unit distance decreases with increasing frequency to about 15 kHz, after which it increases again (Challinor 1967; Hughes and Pappert 1975). In the SAR range-estimation scheme, two narrowband channels of the same bandwidth are tuned to different frequencies, say, 9 and 5 kHz, and the ratio of the signal envelopes is sampled at the instant of the envelope maximum. While the SAR of a single sferic sampled at a single station cannot be used for range estimation since it is dependent on the individual lightning-current frequency spectrum, which is variable, if a large number of SAR values for sferics originating from the same storm center are averaged then the variability of the source frequency spectrum will also be averaged. Thus, with an appropriate propagation model and a "localized" source (less than hundreds of kilometers in size if the source center is at a distance of thousands of kilometers), the mean SAR can provide an indication of the distance to the center of the lightning activity. Propagation models used with the SAR analysis are dependent on the direction to the source and include diurnal variations due to the anisotropic, time-varying properties of the ionosphere.

As noted earlier, the Earth–ionosphere waveguide has a cutoff frequency in the 1 to 3 kHz range and the VLF waves above cutoff propagate in modes different from those of the ELF waves below cutoff. Thus, the different propagation speeds of the VLF and ELF signals lead to an increased time separation with increasing range of the VLF component and the ELF components (the so-called slow tails, see also subsection 13.4.3) of the electromagnetic radiation from an individual stroke; this effect was first studied by Watson-Watt et al. (1937). Examples of sferics waveforms followed by ELF slow tails (or, using a broader definition of sferics than adopted here, sferics waveforms containing slow tails; see Brook 1992) are shown in Figs. 13.10a, b. In Fig. 13.10b the initial portion of the sferics waveform is barely visible owing to limitations of the measuring system; however, to determine the zero time a different measurement system having microsecond-scale

13.3. Atmospherics (sferics)

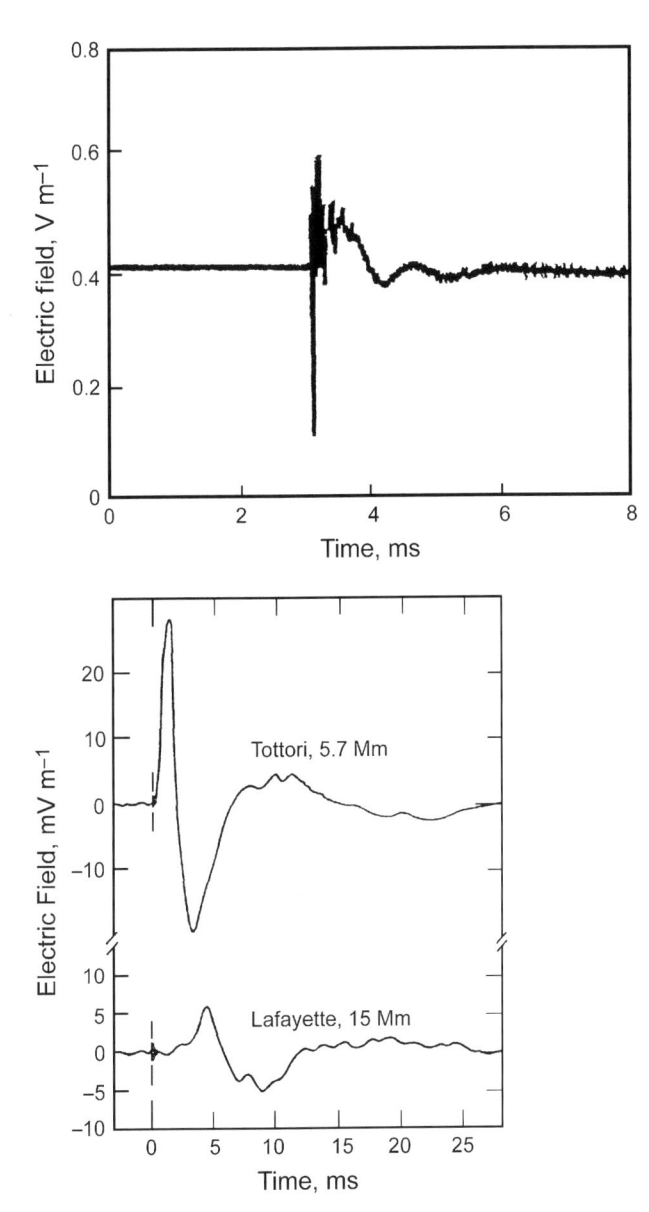

Fig. 13.10. (a) Sferic followed by a slow tail, both produced by the same positive ground discharge located at a distance of 1300 km. Adapted from Brook (1992). (b) Slow-tail waveforms observed in Totteri, Japan and Lafayette, Colorado from the same lightning, which occurred near Java, Indonesia. The preceding sferic is marked and occurs at time zero, as determined primarily from a separate VLF direction finding system. 1 Mm = 10^3 km. Adapted from Taylor and Sao (1970).

resolution was used. In Fig. 13.10a the initial portion is evident. The dependence of the separation time t_s between the initial part of the microsecond-scale sferic and the maximum value of the slow tail for a source at a distance r was derived theoretically by Wait (1962, pp. 313–15):

$$\sqrt{t_s} = 0.3 \left(\frac{r}{2h\sqrt{\omega_r}} + \sqrt{\delta_c} \right) \quad (13.83)$$

where t_s and δ_c are in seconds, h is the effective height of the ionosphere, and $\omega_r = \omega_{pe}^2/\nu_e$, which is proportional to n_e/ν_e (Eq. 13.34), the latter, in turn, being proportional to the effective conductivity of the ionosphere (Eq. 13.40). The parameter δ_c is related, according to Wait (1962), to the width of the source waveform, and is discussed in the following paragraph. Wait's relationship (Eq. 13.83) was used for lightning distance estimation by Sao and Jindoh (1974). Wait showed that the linear relationship between $\sqrt{t_s}$ and the range r in Eq. 13.83 is consistent with the experimental data of Hepburn (1957a) who measured t_s for ranges from about 500 km to over 4000 km during both day-time and night-time. Hepburn and Pierce (1953), using the same data, argued that empirically there is a linear relationship between lightning distance and the time from the initial part of the VLF sferic to the start of the slow tail. They gave the following empirical relations between t, the time from the initial sferic to the start of the slow tail in microseconds, $\tau/4$, the time of the first quarter-cycle of the slow tail in microseconds, and the lightning distance r in kilometers:

$$\begin{aligned} t &= 0.33r - 350, \quad \text{day-time} \\ \tau/4 &= 500 + 0.23r, \quad \text{day-time} \\ t &= 0.13r - 140, \quad \text{night-time} \\ \tau/4 &= 500 + 0.8r, \quad \text{night-time} \end{aligned} \quad (13.84)$$

Apparently, t_s in Eq. 13.83 is roughly equal to $t + \tau/4$ in Eqs. 13.84.

Using the data of Hepburn (1957a), Eq. 13.83, and an assumed h-value of 70 km in the day-time and 90 km at night, Wait (1962) derived a day-time effective ionospheric conductivity of 1×10^{-6} S m^{-1} and a night-time value of 3×10^{-6} S m^{-1}. For a source range of about 4000 km, Hepburn found the time difference t_s between the initial part of the VLF sferic and the ELF slow-tail maximum to be about 1 ms at night and about 2 ms during the day. Slow tails are not always present in distant waveforms, presumably because significant ELF content is not always present in the source. One potential source of such ELF is the short continuing current that immediately follows some return-stroke currents (Shindo and Uman 1989). If the ELF components in the source current occur after a time much greater than a few milliseconds, as is the case for long-continuing current (Section 4.8), the "source" term $\sqrt{\delta_c}$ in Eq. 13.83 will be much larger than the term that varies with range. Wait (1962) found $\delta_c = 5$ ms from analyzing Hepburn's (1957a) data, the same value being found for both day-time and night-time although the measured data for slow-tail separation versus range are different for day-time and night-time.

In addition to predicting the difference in arrival times of the VLF sferics and the following ELF slow tails, as discussed above, waveguide theory indicates that the VLF sferic itself is dispersed, in that the higher-frequency components arrive earlier than the lower-frequency components. The spectral phase function $\varphi(f)$ is not a linear function of frequency f, so that $d\varphi/(2\pi df)$, the so-called group delay, is frequency dependent. The group delay difference (GDD), defined as the difference in the slope of $\varphi(f)$ with respect to f at two different frequencies, is a quantity that can be derived from the waveforms of sferics and then can be related to the distance traveled by the atmospheric (Heydt 1982). As in the case of SAR measurements, it is not possible to estimate the source distance from a single GDD value, but an estimate of the distance to the center of lightning activity can be derived from mean values of the GDD. Rafalsky et al. (1995) suggested an approach to single-station ranging on distant lightning waveforms that belongs to the phase-spectrum class, like the GDD analysis discussed above (e.g., Inkov 1973; Heydt 1982), where the difference in arrival times of the different sferic frequency components in the 5 to 10 kHz range are used with a sferic model to estimate the source distance.

13.3.3. Applications

Since lightning discharges represent the primary source of VLF and ELF electromagnetic background noise in the Earth's atmosphere, a knowledge of the characteristics of this background noise is important in order to be able to design noise-immune communication systems. This need has motivated the studies described in the portion of the sferics literature that involves measurement and calculation of the sferics-caused noise environment at various locations worldwide, as noted at the beginning of subsection 13.3.1 and as discussed below in Section 13.6. Further, in studying sferics, various investigators have been able to devise techniques to find the location of the source lightning, some of which have been discussed in the previous subsections and in Chapter 17, and to determine properties of the ionosphere along the propagation path, also discussed earlier. Finally, changes in global lightning activity, which are in principle derivable from global sferics measurements or global Schumann resonance measurements (Sections 2.6 and 13.4), can potentially provide a measure of global climate change, since increased surface temperatures would result in more convection, producing more thunderstorms and hence more lightning (Williams E.R. 1992, 1994, 1999). Additionally, convection raises water vapor to the upper troposphere, and hence global lightning activity and upper tropospheric water vapor may be linked (Price 2000).

The distance to a lightning source region can be estimated in several different ways from an analysis of the properties of sferics, as discussed in subsection 13.3.2. Conversely, if the proper characteristics of the sferics are measured at several stations, it is possible to derive the ionospheric properties which contribute to the observed characteristics that vary with distance. Calculated and measured propagation characteristics of the Earth–ionosphere waveguide are found, for example, in Volland (1982, 1995), Ferguson and Snyder (1990), Jones (1970), Hughes and Theisen (1970), Hughes and Pappert (1975), and Hughes et al. (1974). Bickel et al. (1970) found that simple exponential models of the ionospheric conductivity (see also subsections 13.3.2 and 13.4.2) could be used to predict the variation in field strength with distance observed from in-flight measurements of continuous-wave (CW) transmissions from a radio transmitter in Hawaii. Morfitt (1973) used exponential ionospheric conductivity profiles to predict multiple-frequency transmission characteristics from Hawaii to San Diego, California, and Pappert and Snyder (1972) used exponential conductivity profiles in predicting VLF signal-fading across the sunrise discontinuity at ionospheric height.

Hughes et. al. (1974) demonstrated that multifrequency propagation data for sferics could be used to deduce an empirical ionospheric exponential-conductivity model over a selected propagation path. Measurements were presented of sferics acquired at three widely separated locations when the propagation path from the source area to the most distant station was under full night-time conditions. The three stations were located near a common great circle of the Earth at Norman, Oklahoma (35.1° N, 97.3° W), La Posta, California (32.7° N, 116.4° W), and Kamuela, Hawaii (20.1° N, 155.7° W). By maintaining time synchronization to better than 10 μs between the stations, the investigators could determine the source location from the difference in arrival times (coupled with the source bearing from Norman) of the waveforms at Norman and La Posta.

Using the numerical techniques of Pappert et al. (1967) for the calculation of the propagation parameters, an appropriate exponential ionospheric conductivity model was chosen that best fitted the data near 10 kHz measured at La Posta. Spectra for the source-current moment (the current times the height of the channel carrying that current) were derived for the measured groundwave at Norman, and the average value was used to predict the average electric field strengths at La Posta and Kamuela for a wide frequency range (2–30 kHz) using the selected conductivity profile. For $\beta = 0.6$ km^{-1} and $h' = 84$ km (see the next paragraph), good agreement between the numerical calculations and the average measured field strength was obtained.

In the discussion above, the ionospheric conductivity parameter $\sigma \varepsilon_0$ (Eqs. 13.34 and 13.40) is of the form (Wait and Spies 1965)

$$\omega_p^2/\nu_e = 2.5 \times 10^5 \exp[\beta(z - h')] \quad (s^{-1}) \quad (13.85)$$

Table 13.1. *Principal characteristics of Schumann resonances. Adapted from Sentman (1995).*

Property	Vertical electric field	Horizontal magnetic field
Resonant frequencies, $f_N < 50$ Hz	7.8, 14, 20, 26, 33, 39, 45 Hz	7.8, 14, 20, 26, 33, 39, 45 Hz
Diurnal variation of f_N	± 0.5 Hz	± 0.5 Hz
Amplitude	~100–200 μV m^{-1} Hz$^{-1/2}$	~0.5–1 pT Hz$^{-1/2}$
Diurnal amplitude variation	± 50–100 μV m^{-1} Hz$^{-1/2}$	± 0.25–0.05 pT Hz$^{-1/2}$
Time of maximum intensity, western hemisphere	2000–2200 UT	2000–2200 UT
Polarization	linear (vertical)	linear (horizontal)
Primary sources of interference to Schumann resonance measurements	power lines, acoustic noise, dust, rain	power lines, acoustic noise, effects of nearby magnetic material (vehicles, trains)

where β^{-1} is a height constant in km, z is the height above ground in km, h' is a reference height in km, ω_p is the angular plasma frequency, and ν_e is the electron–neutral-particle collision frequency defined by

$$\nu_e = 1.816 \times 10^{11} \exp(-0.15z) \quad (\text{s}^{-1}) \qquad (13.86)$$

Hughes and Pappert (1975) made measurements of atmospherics simultaneously in California and on the Island of Hawaii. A third station in the source area in Oklahoma aided in locating each individual lightning event. From the differences in measured field strengths at the two remote stations the transmission loss as a function of frequency was determined for 37 waveform pairs. The average measured transmission loss was then compared to that predicted numerically, using several exponential ionospheric-conductivity profiles described by an inverse scale height β and reference height h' (in the notation of Wait and Spies 1965). Based on a weighted least-squares analysis of the calculated and measured values, the ionospheric conductivity model with $\beta = 0.35$ km^{-1} and $h' = 72$ km was found to predict best the measured day-time east–west transmission loss for the selected frequency band from 7 to 18 kHz.

13.4. Schumann resonances

13.4.1. History and observed characteristics

The idea that the Earth–ionosphere waveguide would support electromagnetic resonances was first proposed on theoretical grounds by Schumann (1952a, b, c). Schumann (1954, 1957) revised his earlier theory to include a more sophisticated treatment of the upper conducting boundary of the spherical cavity. The first measurements of the Schumann resonance phenomena were published by Schumann and König (1954); these were followed by the measurements of Balser and Wagner (1960), who identified the first five Schumann resonant frequencies.

The literature on Schumann resonances and related ELF wave propagation in the Earth–ionosphere cavity is vast. Surveys of Schumann resonance phenomena are found, for example, in the book by Bliokh et al. (1980) and in the review papers by Polk (1982) and by Sentman (1995). The general theory of ELF electromagnetic wave propagation is found, for example, in the books by Budden (1962), Madden and Thompson (1965), Galejs (1972), and Wait (1972).

The principal characteristics of Schumann resonances are given in Table 13.1 and were discussed briefly in Section 13.1. Schumann resonances are ELF standing waves associated with the lowest frequencies of the electromagnetic spectrum of the global lightning activity. The maximum intensity of the Schumann resonances occurs in the afternoon in the Americas, at 2000 to 2200 UT (Table 13.2; note that UT = EST − 5 hours), indicating that the lightning activity in North and South America dominates the global activity. There are also Schumann-resonance intensity peaks at about 0900 UT and 1500 UT, consistent with afternoon thunderstorms (i) over the islands located between the Philippines and Northern Australia and (ii) over sub-Saharan Africa. Frequency–time spectrograms illustrating these effects are found in Fig. 13.11, which is taken from Sentman (1987a) and Sentman and Fraser (1991). Amplitude data providing further illustration are found in Table 13.2 and discussed in subsection 13.3.4. In addition to the diurnal intensity variations, the Schumann resonance intensity varies from day to day in concert with variations in the global thunderstorm activity (e.g., Polk 1969; Clayton and Polk 1974; Heckman et al. 1998). There are differences in the diurnal UT intensity variation in globally separated locations, attributed by Sentman and Fraser

Table 13.2. *Intensity of global thunderstorm activity centers (arbitrary units) as a function of universal time (UT). Adapted from Nickolaenko et al. (1996)*

UT (h)	Africa	America	Asia	UT (h)	Africa	America	Asia
00	0.45	1.75	0.08	12	1.38	0.18	1.14
01	0.51	1.38	0.12	13	2.00	0.17	0.95
02	0.43	1.17	0.15	14	2.40	0.18	0.70
03	0.32	0.91	0.22	15	2.70	0.28	0.60
04	0.24	0.78	0.38	16	2.47	0.47	0.50
05	0.18	0.65	0.60	17	2.03	0.87	0.40
06	0.13	0.52	0.88	18	1.63	1.42	0.30
07	0.14	0.43	1.24	19	1.30	1.90	0.30
08	0.22	0.40	1.46	20	1.02	2.24	0.20
09	0.35	0.35	1.65	21	0.80	2.45	0.15
10	0.55	0.28	1.58	22	0.60	2.30	0.10
11	0.88	0.22	1.37	23	0.51	2.08	0.08

(1991) to differences in the local height of the ionosphere at the different sites.

There is an observed diurnal variation of about ±0.5 Hz about the nominal resonance frequencies given in Table 13.1 (e.g., Balser and Wagner 1962b; Sao 1971). For an ideal Earth–ionosphere cavity (a cavity with perfectly conducting ionosphere and Earth boundaries), Schumann resonances would occur at discrete frequencies determined only by the circumference of the Earth and the speed of light in the cavity. In actuality, the resonance lines are broadened

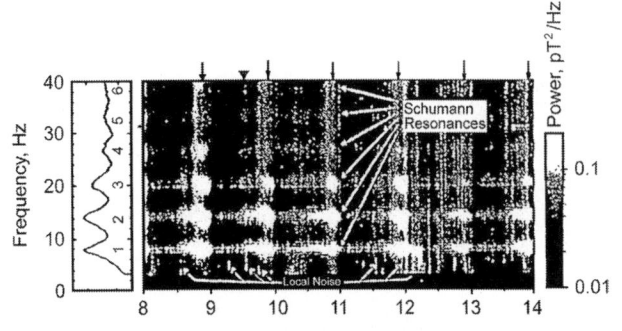

Fig. 13.11. Dynamic frequency–time spectrogram of the horizontal component of Schumann resonances, illustrating the typical diurnal-variation patterns of resonances observed in California during the period 8–14 September 1989. The small vertical arrows indicate the diurnal maxima. At the left is a vertical slice through the spectrogram taken at 1252 UT on 9 September, showing the resonance line structure. Within the spectrogram the power as a function of frequency is coded in a logarithmic gray scale, shown at the right, white corresponding to the most intense power. The resonances appear as the intensifications at 8, 14, 20, 26, ... Hz, indicated by the white arrows. Measurements are occasionally contaminated by various sources of local noise, also indicated in the plot. Adapted from Sentman (1987a) and Sentman and Fraser (1991).

and the observed center frequencies are shifted from the ideal case, the degree of frequency broadening and shift being determined by the conductivity profile of the ionosphere and variations in that profile due to external energy input such as solar X-rays. Additionally, the ionosphere is dissipative to waves at Schumann resonance frequencies, and this dissipation shifts the ideal resonant frequencies, the shift being related to the distance from the observer to the source thunderstorm and hence varying with time at a fixed point on Earth as the afternoon thunderstorm activity moves westward.

The width of a Schumann frequency-resonance signal in the frequency domain is characterized by a Q-factor defined for the Nth mode as

$$Q_N = f_N/\Delta f_N \qquad (13.87)$$

where f_N is the center frequency and Δf_N is the width at half-maximum power. Average Q-factors are in the range 3 to 6, as illustrated in a general way in Fig. 13.11. The Q-factor corresponds approximately inversely to the number of cycles of decaying amplitude of an individual propagating ELF wave at frequency f_N before it is fully damped. For the first Schumann mode at about 8 Hz the damping time is about 0.5 s. Higher modes have comparably shorter decay times, as follows from Eq. 13.87. On average there are of the order of 50 flashes worldwide during the damping time of the first mode, perhaps one-quarter being ground flashes, each ground flash having an average of 3 to 5 strokes; all the flashes occur incoherently in time, and, to a lesser extent, location. Every significant lightning source current occurring during the 0.5 s contributes to the observed first Schumann mode. The effects of individual sources cannot generally be identified in the Schumann resonance records. An exception is the Q-burst (Ogawa *et al.* 1966b), an excitation of the Earth–ionosphere cavity that may exceed the

13.4. Schumann resonances

background resonance amplitude by a factor of 10 or more. There can be 50 to 100 Q-bursts per hour, and they can be detected simultaneously at globally separated stations (Ogawa *et al.* 1967). Examples of Q-bursts from Sentman (1989) are shown in Figure 13.2. Q-bursts have been correlated with the large positive cloud-to-ground flashes (Section 5.3) that produce sprites (Boccippio *et al.* 1995; Huang *et al.* 1999; subsection 14.3.3), and these Q-bursts are apparently similar to the slow tails (two or three cycles of frequency less than about 500 Hz) that follow some sferics (Fig. 13.10a, b, subsections 13.3.2 and 13.4.3), but Q-bursts have components lower in frequency than slow tails. The ELF Q-bursts discussed here should not be confused with the VHF–UHF Q-noise radiated by lightning channels and discussed in subsection 9.3.2.

Schumann resonances are difficult to measure near a frequency of 50 or 60 Hz because of the ubiquitous presence of power-line interference (primary frequency and harmonics), which is generally much larger in amplitude than the Schumann resonances. Typical Schumann-resonance field intensities are given in Table 13.1. Schumann electric fields are measured by metal antennas, generally spheres to minimize corona effects and maximize antenna area, elevated 1 to 10 m above ground level in order to increase the gain (e.g., Ogawa *et al.* 1966b; Sentman 1987a; Polk 1982; Burke and Jones 1996; Heckman *et al.* 1998). The Schumann electric fields are so small that even atmospheric space charge blown by air moving near the ground and nearby trees swaying in the wind and modulating the ambient fair-weather field can provide significant interference, not to mention the effect of any small vertical motion of the antenna in the fair-weather field. Typical instrumental bandwidths are from a few hertz to several tens of hertz with a sensitivity (resolution) within that bandwidth of order 10 μV m^{-1} Hz$^{-1/2}$. Schumann magnetic fields are extremely weak compared with the ambient geomagnetic field, and their picotesla range amplitudes preclude detection by fluxgate or proton-precession magnetometers. In fact they are best detected by wire loops of meter diameter and typically 10 000 to 100 000 turns (Polk 1982), often with high-permeability cores inside the loops to concentrate the field and increase the gain (Heckman *et al.* 1998). Magnetic field measurements can be influenced by the presence of nearby moving objects of high permeability. Wind can directly affect magnetic sensors by causing them to move, thereby generating false signals. Shielding of the loop antenna from electric fields is essential.

13.4.2. Theory

The properties of the ELF waves that propagate within the Earth–ionosphere cavity and produce Schumann resonances are determined by solving Maxwell's equations, for example, in the form given in Eq. 13.30, in both the free-space region and the conducting ionosphere and then applying the appropriate boundary conditions on the fields at the interface. Such derivations are given, for example, by Wait (1960a, b, 1962a, b), Jones (1970a, b), Greifinger and Greifinger (1978, 1979, 1986), Sukhorukov (1993), and Cummer and Inan (2000). The primary complication in doing this is that the conductivity varies with radial distance from the Earth's surface (and, in a less dramatic way, in the two directions, θ and ϕ, perpendicular to the radial vector). Referring to Eq. 13.41, for frequencies in the 3 to 60 Hz range the atmosphere can be considered reasonably conducting at an altitude of 45 to 60 km. There, the electrical conductivity σ typically ranges from 2×10^{-10} to 4×10^{-9} S m^{-1}. At higher altitudes, the conduction current density, $\sigma \mathbf{E}$, exceeds the displacement current density, with a resultant appreciable frequency-dependent energy loss and waveform damping and dispersion, whereas at lower altitudes the magnitude of the displacement current density $\omega \varepsilon_0 \mathbf{E}$ is greater than the conduction current density, allowing essentially free-space wave propagation. Polk (1982) listed the ratios of the conduction and displacement current densities at 45 km and 50 Hz, for day-time or night-time, as 0.05 for a quiet ionosphere and as 5.0 for a disturbed ionosphere. At 75 km and 50 Hz the ratio for a quiet ionosphere is 10^4 in the day-time and 20 at night and for a disturbed ionosphere 4×10^6 in the day-time and 2×10^3 at night. The Earth's surface has conductivities from 10^{-4} S m^{-1} (sand) to 5 S m^{-1} (salt water) and hence $\sigma \gg \omega \varepsilon_0$ at ELF over the entire surface; that is, the Earth can be treated as a perfect conductor at ELF.

For the ideal case in which both the ionosphere and the Earth are considered perfectly conducting spherical boundaries and the region between these boundaries is a perfect dielectric, $\sigma = 0$, the solution of Eq. 13.30 is more or less straightforward. The general solution in the free-space region is subject to the boundary conditions that the tangential component of $\mathbf{E}$ at the two perfectly conducting boundaries is zero, since there can be no field in a perfect conductor and the tangential component of $\mathbf{E}$ must be continuous across any boundary. Thus the electric field is always radial in the spherical cavity and the magnetic field, perpendicular to it, is in the ϕ direction. Since both the electric and magnetic fields are perpendicular to the direction of propagation, the θ direction, the wave is termed a transverse-electromagnetic (TEM) wave. The resonance-frequency values are determined by the Earth's radius (or circumference), since only particular wavelengths can propagate in the fixed cavity, so that they are in phase over the whole globe as they circumnavigate it at the speed of light. These ideal frequencies are

$$f_{N0} = \frac{c}{2\pi R_e}\sqrt{N(N+1)} \qquad (13.88)$$

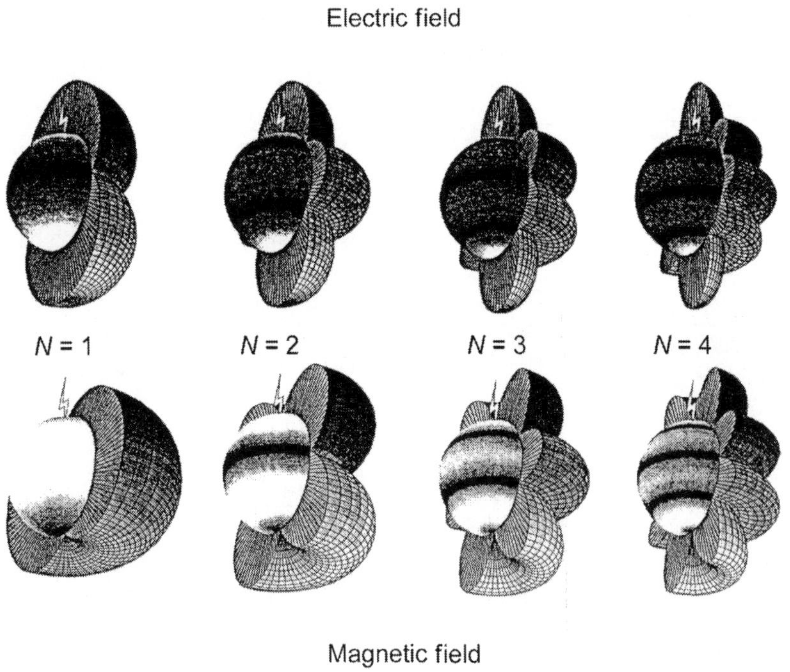

Electric field

Fig. 13.12. Angular structure of the amplitudes of the electric and magnetic components of the lowest four TM$_0$ normal modes of the Earth–ionosphere cavity. The mode structure is azimuthally symmetrical about a hypothetical vertical electric dipole source located at the Pole. Here the unit sphere has been split vertically through the pole into two hemispheres. The mode amplitudes are plotted in different formats for each hemisphere, in one format as the radial distance above the surface of the terrestrial sphere (right-hand hemisphere), and in the other format as a shade of gray (left-hand hemisphere). Adapted from Sentman (1996).

where c is the speed of light, R_e is the Earth's radius ($R_e = 6400$ km, $2\pi R_e = 40\,000$ km) and $N = 1, 2, 3, \ldots$. The first three resonant frequencies in this ideal case are 10.6, 18.3, and 25.9 Hz, a few hertz higher than the observed values (Table 13.1). As noted earlier, the resonance spectral lines suffer no broadening in an ideal cavity because there is no mechanism for energy loss in an ideal system. Examples of the first four Schumann resonant modes for an ideal-ionosphere model are shown in Fig. 13.12, where the single lightning current source is assumed to be at the North Pole.

For the case of an imperfectly conducting ionosphere, there will be an electric field not only in the radial direction but also in the spherical θ direction, that is, tangent to an imaginary sphere concentric with the Earth. Conduction currents associated with the θ component of the electric field, E_θ, cause ohmic heating in the ionosphere and a consequent frequency-dependent energy loss from the propagating wave. The field E_θ is relatively small, however. Thus, when the free-space wavelength λ is large compared with the distance h between the Earth and the bottom of the conducting portion of the atmosphere, a quasi-TEM (quasi-transverse-electromagnetic) mode, sometimes called a zeroth-order TM mode, will propagate. At 10 Hz, $\lambda = 30\,000$ km whereas h is less than 100 km, so the stated condition is easily satisfied. No TE or TM waveguide modes can propagate in the cavity at ELF because the boundary conditions for these waveguide modes cannot be satisfied for wavelengths greater than the height of the ionosphere.

The simplest Schumann resonance model after the ideal model is called the "slab ionosphere" model; in it the conductivity of the Earth is taken as infinite and the conducting atmosphere is modeled as having a sharp lower boundary at a given height and a constant conductivity above, with no effects from the static magnetic field. This relatively simple model allows the modeling of the observed dissipation of ELF waves via ohmic heating, as is observed in practice. For the case of a vertical channel of length ds carrying a frequency-domain current $I(\omega)$ above a perfectly conducting spherical Earth located at the North Pole, $\theta = 0$ in a spherical coordinate system, the radial electric field was derived by Wait (1962a, b) as

$$E_r(\theta, \omega) = \frac{iI(\omega)\,\mathrm{d}s\; m(m+1)}{4\pi\omega\varepsilon_0 h R_e^2} \\ \times \sum_{n=1}^{\infty} P_n(\cos\theta)\frac{2n+1}{n(n+1)-m(m-1)} \quad (13.89)$$

13.4. Schumann resonances

where h is the height of the ionospheric boundary, $P_n(\cos\theta)$ is the nth-order Legendre polynomial, and m is given approximately by

$$m + \frac{1}{2} \approx k_0 R_e S \tag{13.90}$$

with

$$S \approx 1 - \frac{1}{2kh}\left(\frac{1}{n_i} + \frac{1}{n_g}\right) \tag{13.91}$$

where $k_0^2 = \omega^2\mu_0\varepsilon_0$, $k^2 = (\omega^2\mu_0\varepsilon_0 - i\omega\mu_0\sigma_i)$, and n_i, n_g are the refractive indices of the ionosphere and the Earth. The refractive index of the ionosphere is found from

$$n_i \approx \left(\frac{\sigma_i}{i\omega\varepsilon_0}\right)^{1/2} \tag{13.92}$$

and

$$v_p^{-1} = \frac{k}{\omega} \approx (\mu_0\varepsilon_0)^{1/2}\left(1 + \frac{\sigma_i}{i\omega\varepsilon_0}\right)^{1/2} \tag{13.93}$$

Other formulations of the Schumann electric (and magnetic) fields are given in Jones (1967) and Ishaq and Jones (1977), and are discussed in Huang *et al.* (1999). The Schumann resonant frequencies for the model of Wait (1962a, b) are found by dividing Eq. 13.88, the expression for the resonant frequencies for an infinitely conducting ionosphere, by the real part of S from Eq. 13.91. If both the spherical surfaces of the cavity are perfectly conducting then the refractive indices become infinite, $S = 1$, and the solution reduces to that of the ideal case previously discussed. More realistic and more complex models of the ionosphere involve either using a multiple-slab approach, each slab having a different conductivity, or an exponential, multiple-exponential, or other type of conductivity profile that varies smoothly with height and takes account of the effects of the geomagnetic field. For example, Sentman (1996) presented a solution for an ionospheric model involving a two-exponential conductivity profile. In all such models Maxwell's equations are solved in a general way, in the free-space region and in one or more conducting regions, depending on the model; the boundary conditions on the electric and magnetic fields at the edges of the regions mathematically limit the allowed solutions and determine the resonant frequencies, the Q-factors, and the field waveforms.

Some additional theoretical aspects as well as diagnostic uses of Schumann resonance phenomena are given in the following section.

13.4.3. Determination of atmospheric properties, lightning properties, and worldwide thunderstorm activity

Since the Schumann resonance signal represents the summation of the excitations of the Earth–ionosphere cavity by all lightning occurring during a few resonance cycles, which corresponds to a half-second or so for the lowest mode, the signal, if properly analyzed, can potentially serve as a measure of the global lightning activity. As stated earlier, since it is likely that the worldwide lightning flash rate is dependent on the average global temperature, measurements of the magnitude of the Schumann resonance signal may provide a method to estimate global temperature change (Williams 1992, 1994, 1999), as also discussed in Section 2.6. Williams (1992) demonstrated a correlation over a six-year period between the variations in the tropical temperature and the variations in the lowest Schumann-resonance-mode intensity measured in the northeastern United States. Satori and Ziegler (1998) related the increased lightning activity in the Pacific in January 1996 relative to January 1995, determined from Schumann resonance measurements in Central Europe, to a 0.2 °C increase in the surface air temperature and additional meteorological factors.

As noted earlier, Price (2000) presented evidence that there is a link between the upper tropospheric water vapor level and global lightning activity and suggested that Schumann resonance data could be used to infer the former.

Nickolaenko *et al.* (1996, 1999) examined Schumann resonance data from the Tottori Observatory in Japan (Sao *et al.* 1973) from 1967 to 1970 in order to determine the coordinates and intensity of the global thunderstorm centers and the temporal variations of the level of global lightning activity. Theory presented by Polk and Fitchen (1962), Jones (1969), Galejs (1972), Ogawa and Otsuka (1973), and Polk (1982) shows that the amplitude ratios for the different resonances depend weakly on the properties of the ionosphere and strongly on the source distance. Nickolaenko *et al.* (1996) reported on data for one year beginning September 1968. Variations in the effective source–observer distance were estimated using the ratios of the intensities of individual Schumann resonance modes. The three main source locations, the Americas, southeast Asia, and Africa, as noted in subsection 13.4.1, were found to be relatively constant during a given season, moving considerably between seasons, while the global lightning intensity derived from the Schumann field intensities varied substantially during a 24 hour period, maximum lightning activity corresponding to the local afternoon. These effects are illustrated in Table 13.2. Nickolaenko *et al.* (1999) found that Schumann resonance data observed in Japan compared well with the records from Europe except for amplitudes that were larger by a factor 2 in February in Tottori, attributed to nearby winter storms over the Japan Sea (Chapter 8). Satori and Ziegler (1999) found that Schumann resonance data recorded in Nagycenk, Hungary (47.6° N, 16.7° E) indicated a southward shift in the global lightning activity in warm El Niño years and a northward shift in cool La Niña years. Belyaev *et al.* (1999) presented the results of a Schumann-resonance monitoring campaign held at Lehta observatory, Karelia, Russia during July–August 1998. Three electromagnetic field components were recorded simultaneously: the vertical electric and two orthogonal magnetic fields. The purpose of the work was to demonstrate the advantages of the Poynting-vector (PV) technique when studying the space–time dynamics of the worldwide thunderstorm activity from a single observatory. Analysis of the diurnal PV patterns revealed a night-time peak in African thunderstorm activity. This maximum occurred around 0200–0300 UT and reached one-third of normal afternoon level.

Boccippio et al. (1998) located the sources of large transient Schumann resonance signals, Q-bursts, examples of which are shown in Fig. 13.2. They used the predicted range dependence of the ratio of the vertical electric and the horizontal magnetic field in the frequency domain (Ishaq and Jones 1977) to obtain distance estimates and magnetic direction finding for directional information. The resultant locations are within $(1–2) \times 10^3$ km of the locations determined from the optical transient detector (OTD) and the NLDN (Sections 2.5, 17.5, and 17.8). Single-station location using the method of Ishaq and Jones had been previously demonstrated by Kemp (1971) and by Burke and Jones (1995a) using areas of lightning activity as ground truth less well determined than those used by Boccippio et al.

Boccippio et al. (1995) showed that Q-bursts are often related to the large positive discharges (Section 5.3) that produce sprites (subsection 14.3.3). Such discharges also produce a significant ELF slow tail (two or three cycles of frequency less than about 500 Hz) following the initial sferic (subsection 13.3.2, Fig. 13.10a, b). If the slow tails (and even longer-duration ELF-band radiation) are due to continuing currents that follow return strokes (Reising et al. 1996; Sao et al. 1970), then measurements of the ELF along with a source model and a propagation model allow an estimate of the causative continuing current and associated charge-transfer values (e.g., Cummer and Inan 1997, 2000; subsection 14.3.3). Cummer et al. (1998) and Cummer and Füllekrug (2001) have shown that there is simultaneous ELF electromagnetic and optical radiation, both types presumably from the currents in the body of the sprite, making possible estimates of the electrical characteristics of the sprites (subsection 14.3.3), although Huang et al. (1999) questioned this claim except in the case of extraordinarily intense sprites.

Füllekrug and Constable (2000) employed a network of three electromagnetic measurement instruments, located in Germany, California, and Australia, to record simultaneous ELF magnetic field disturbances (4 to 200 Hz) that propagated with little attenuation around the globe within the Earth–ionosphere cavity. The triangulation of individual lightning flashes resulted in a picture of the temporal evolution of intense lightning discharge occurrences on the planetary scale during April 1998. A total of 52 510 events were recorded. According to Füllekrug and Constable (2000), the majority of the triangulated lightning discharges exhibited charge moments with the potential to excite mesospheric sprites and ~5–20 percent could produce air breakdown at sprite altitudes of 50–70 km. The calculated charge moments ranged from 300 to 4200 C km and the location accuracy was estimated at about 1 Mm (10^3 km). One ELF signal, on average, was triangulated every 40 s so that if there were about 100 flashes per second worldwide (Section 2.1), about one in four thousand flashes was detected.

Analysis of Schumann resonance data can yield estimates of the effective conductivity profiles of the ionosphere, as discussed by Galejs (1961a, b), Madden and Thompson (1965), Jones (1967), Ogawa and Murakanni (1973), Tran and Polk (1979a, b), Polk (1982), and Sentman and Fraser (1991). The precise values of the resonance frequencies and the widths of the resonance peaks are a sensitive indicator of ionospheric properties but they also depend on the source-receiver separation distance. For example, Tran and Polk (1979a, b) used source location techniques to find the thunderstorm areas and the method of Wait (1970) for mathematically describing a stratified ionosphere in order to determine iteratively the worldwide average atmospheric conductivity profile.

In practice, both the properties of the local ionosphere and the local flash rate in regions of major lightning activity can be studied in detail in better ways than by using Schumann resonances (e.g., ground-based ionospheric sounders, in situ ionospheric measurements, the US National Lightning Detection Network and similar lightning-locating systems in other countries), except, perhaps, when probing the relatively small electron densities in the lower D-region of the ionosphere. Schumann resonance studies have the advantage of providing a worldwide lightning picture from a relatively simple measurement and may turn out to be very important if global average temperature, upper-tropospheric water vapor, or other global average atmospheric parameters can be related to global lightning activity in a manner suitable for practical application.

13.5. Whistlers
13.5.1. History, observed characteristics, and use to determine magnetospheric properties

Barkhausen (1919) is generally credited with the discovery of whistlers. Reports of whistlers or whistler-like phenomena prior to Barkhausen are reviewed by Helliwell (1965). During World War I, Barkhausen observed whistling tones on German army radio receivers that were used to eavesdrop on Allied army telephone conversations. On the basis of these observations, he suggested

13.5. Whistlers

(Barkhausen 1930) that whistlers were caused by lightning and that their frequency-descending sounds were the result of the propagation of electromagnetic signals within a dispersive medium. Eckersley (1931, 1935) developed the theory of wave propagation in a medium containing charged particles in the presence of a static magnetic field. He derived an equation describing the observed type of whistler dispersion at low frequencies. However, the propagation path length that was required, via the theory, to produce the observed dispersion was very much longer than any potential propagation path known at the time, since the extent and properties of the plasma above the ionospheric F-layer were yet to be discovered. Additional observations of whistlers, including whistler spectrograms, were published by Burton and Boardman (1933), who employed existing submarine cables as antennas. There were few additional significant developments in the understanding of whistlers until the landmark work of Storey (1953). Storey showed conclusively that the whistlers observed at one location originated in lightning discharges in the opposite hemisphere and then propagated in the magnetosphere along the geomagnetic field lines to the hemisphere of the observer. Additionally, he suggested that, in order to explain the observed dispersion, the electron density a few Earth radii away, in the remote part of the propagation path, must be of the order of hundreds of electrons per cubic centimeter, providing the first evidence that plasma was present far beyond the ionospheric F-layer, thought at the time to be the top of the atmosphere. Figures. 13.3a, c show typical propagation paths for a ducted whistler, one that follows the geomagnetic field lines from hemisphere to hemisphere. Also shown, in Fig. 13.3d, is a postulated path for a non-ducted whistler, one that propagates both along and across field lines. Non-ducted whistlers can be measured only *in situ* by rockets and satellites. Following Storey's work, many research stations were established to study whistlers, and the physics of whistler propagation became an important element of space physics (e.g., Helliwell 1965). There is a sizable literature on whistlers, including the book by Helliwell and review papers by Walker (1976), Hayakawa and Tanaka (1978), Park (1982), Al'pert (1990), Hayakawa and Ohta (1992), and Hayakawa (1995).

Lightning-field frequency components from about 100 Hz to about 30 kHz contribute to observed whistlers. Whistlers originate as upward-propagating electromagnetic radiation from the causative lightning discharge. A complete delineating of the whistler phenomena involves understanding (i) the physics of how the pertinent frequency components traverse the ionosphere at both the beginning and end of their journey along the geomagnetic field lines and (ii) the physics of the processes by which electromagnetic waves are guided along geomagnetic field lines from one hemisphere to the other to produce the observed dispersion. We will examine both of these issues in the next section, with reference to Section 13.2.

Not only do whistlers propagate from the source hemisphere to the opposite hemisphere along so-called whistler "ducts", as illustrated in Figs. 13.3a, c; when radiation from, say, a lightning flash in the southern hemisphere arrives at the top of the ionosphere in the northern hemisphere, some of its energy may be reflected back to the southern hemisphere, thus producing a two-hop whistler. This is illustrated by curves 1 and 2 in Fig. 13.3b. Such partial reflections can be repeated many times, so producing whistler echo trains that can be observed in both hemispheres, as shown by the remaining curves in Fig. 13.3b. The time interval from the arrival of the highest-frequency component of the wave to the arrival of the lowest-frequency component for the first hop is of order 1 s, as is illustrated in Fig. 13.4. The subsequent frequency–time curves are increasingly more dispersed because of the longer total path of travel. A cloud-to-ground flash with multiple strokes can produce a whistler with multiple components separated from each other by the interstroke time interval. One lightning discharge can illuminate more than one duct in the magnetosphere, leading to "multi-path" whistlers, which are separated in time owing to the different travel times in the different ducts. Helliwell *et al.* (1956) showed that, in principle, every whistler has a frequency for which the time delay is a minimum, that is, a "nose" in the frequency–time spectrogram, as noted in Section 13.1 and illustrated by Fig. 13.13. This is discussed further in the next section. The nose region is absent in most measured frequency–time curves because the frequency that is potentially the lowest does not survive the trip. The nose frequency can be shown by theory to be roughly proportional to the minimum electron gyrofrequency along the whistler path. Thus each whistler carries information on the geomagnetic field in the path along which it propagates. The nose frequencies of the several components of a multi-path whistler have been used to measure the geomagnetic field in succeedingly distant magnetic field ducts. There is a simple approximate relation between the nose frequency and the geocentric distance of the top of the field duct measured in Earth radii, the so-called L-value, for an assumed dipole magnetic field. Nose time delays can therefore be interpreted in terms of the equatorial electron density as a function of L, an approach which led to the first model of the distribution of magnetospheric electron density as a function of height near the equator (Carpenter 1966; Angerami and Carpenter 1966). Whistler measurements of the electron density in the geomagnetic field ducts were the first measurements to identify an abrupt decrease or "knee" in the equatorial density profile, which is related to "knee" whistlers (Carpenter 1966; Corcuff 1975). This region is now called the plasmapause (subsection 13.2.1). Just below the plasmapause the

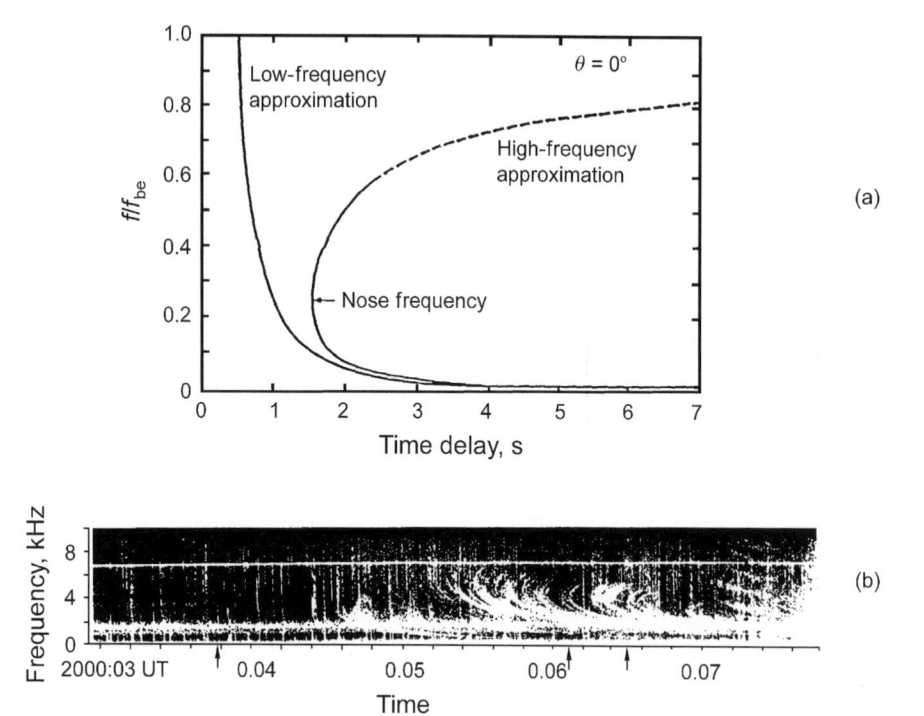

Fig. 13.13. (a) Normalized frequency versus time delay for longitudinal propagation ($\theta = 0°$) over a path of length c/B. Adapted from Helliwell (1965). (b) An example of observed spectrograms of nose whistlers from 24 June 1977. The left-hand arrow indicates the causative sferic for the two whistlers indicated by the right-hand two arrows. Adapted from Sazhin et al. (1990). The figures can also be found in Hayakawa (1995).

electron density is of order 100 cm^{-3}, whereas above it the density is two orders of magnitude less. Most important features of the magnetosphere that were first determined from whistler measurements and accompanying theory have been confirmed and characterized in more detail by in situ measurements using satellites and rocket probes (representative references are given at the end of the next paragraph).

Ducted and non-ducted whistlers interact with energetic electrons (our discussion so far has been about wave propagation through low energy or "cold" electrons) in the radiation belts while propagating through the magnetosphere. Such interactions can result in amplification of the whistlers, triggering of emissions at new frequencies, and precipitation of some of the energetic electrons (ejection of the electrons from their confined paths circling the magnetic field lines). Amplification of the whistler signal by interaction with cyclotron-resonant electrons is illustrated in Fig. 13.3a. Precipitation of energetic electrons can produce enhanced ionization and optical emissions in the lower ionosphere, as well as X-rays detectable down to an altitude of about 30 km. Some of these effects are reviewed in Section 14.5. Burgess and Inan (1993) and Lauben et al. (1999) have examined the role of ducted whistlers in electron precipitation. Armstrong (1987) apparently observed the initiation of a whistler (and hence the causative

lightning) by a preceding whistler, which occurred about 5 seconds earlier. Such two-whistler events do not appear to occur by chance. A possible triggering mechanism in which the precipitated electrons reduce the air resistance was suggested by Armstrong. Hale (1987) also discussed the issue of lightning triggered by whistlers. Whistlers detected in space are generally assumed to be non-ducted unless the measuring probe is inside a duct, which is rare. Non-ducted whistlers, e.g., Fig. 13.3d, apparently follow a variety of paths and exhibit a variety of frequency–time signatures. For example, Carpenter et al. (1964) and Smith R. L. (1964) described "subprotonic whistlers", whose dispersion is less than that for normal whistlers and whose spectrograms involve several successive traces at equal intervals, suggesting that the signal, at frequencies below 4 kHz, travels up and down with repeated reflections between levels of about 100 km and 1000 km. At the upper reflecting layer protons are the dominant ions, as opposed to atomic ions at lower altitudes, hence the name subprotonic for this layer. Calculations of the paths of such whistlers were given by Kimura (1966) and Walker (1968a, b). In situ observations of ducted and non-ducted whistlers from satellites and rockets are reported in, for example, Holzworth et al. (1999), Kelley et al. (1997), Hayakawa (1995), Li et al. (1991), Tixier et al. (1984), Thomson and Dowden (1977a, b), Ondoh (1976), Angerami (1970),

13.5. Whistlers

Shawhan (1970), Walter and Angerami (1969), Smith R. L. and Angerami (1968), and Shawhan and Gurnett (1966).

In addition to pure whistler waves, there is a variety of magnetospheric VLF waves, generated by naturally occurring plasma processes, that enter the Earth–ionosphere waveguide and are observed at ground level. These signals are broadly classified as "chorus", usually composed of a series of discrete discernable tones, and "hiss", a less well-defined wideband noise. Chorus and hiss can both propagate in the whistler mode and can interact with energetic electrons in the same way as whistlers do. Shawhan (1979) reviewed the literature on the satellite observation of chorus and hiss. These signals have been observed and described from ground-based measurements from the time of the earliest whistler studies (e.g., Eckersley 1928; Storey 1953).

Whistler-like signals observed on planets other than Earth are discussed in Chapter 16.

13.5.2. Theory

Many important properties of whistlers can be explained using the theory outlined in Section 13.2 and given in a number of books (e.g., Ratcliffe 1959; Stix 1962; Spitzer 1962). To understand how the electromagnetic waves from lightning propagate into and through the ionosphere, we first examine the application of Snell's law (Eq. 13.58) to the free-space–ionosphere boundary:

$$\sin\theta_i = n_i \sin\theta_t \qquad (13.94)$$

where n_i is the refractive index of the ionosphere and the refractive index of free space is taken as unity. We will show that $n_i \gg 1$ and hence θ_t in the ionosphere must be near zero, that is, the propagation direction in the ionosphere is essentially vertical, along the magnetic field lines, for any incidence angle of the wave in free space impinging on the free-space–ionosphere interface. We now derive n_i for the expected case, a wave in the ionosphere moving along the magnetic field lines. The propagation constant is given by Eq. 13.50 in our discussion of case 3 in Section 13.2. Since $\omega_{pi} \ll \omega_{pe}$ (from Eq. 13.34 this ratio of is m_e/m_i), we can write Eq. 13.50 to good approximation as

$$k_z = \omega(\mu_0\varepsilon_0)^{1/2}\left[1 - \frac{\omega_{pe}^2}{\omega(\omega \pm \omega_{be})}\right]^{1/2} \qquad (13.95)$$

where ω_{be} is defined as a negative quantity (Eq. 13.6). Consider the primary whistler propagation mode, $\omega < \omega_{be} < \omega_{pe}$. For k_z to be real, we must choose the plus sign in the second term on the right-hand side of Eq. 13.95 (recalling that ω_{be} is negative), and hence we are considering a right circularly polarized wave. With this choice of sign, and noting that the second term in Eq. 13.95 will generally be much greater than unity for $\omega < \omega_{be} < \omega_{pe}$, Eq. 13.95 can be written as

$$k_z = \omega(\mu_0\varepsilon_0)^{1/2}\left[-\frac{\omega_{pe}^2}{\omega(\omega - |\omega_{be}|)}\right]^{1/2} \qquad (13.96)$$

so that the refractive index (Eq. 13.51) is

$$n_i = \left[\frac{\omega_{pe}^2}{\omega(|\omega_{be}| - \omega)}\right]^{1/2} \qquad (13.97)$$

Clearly, this refractive index much exceeds that of free space. For example, at the ionospheric F-layer peak, where the electron density is near 10^6 cm^{-3}, n_i is close to 100 for frequencies near 5 kHz.

After the whistler enters the magnetosphere, propagates along a magnetic field line, and reaches the conjugate point on the ionosphere, the reverse interface situation occurs. Snell's law can also be invoked to show that if the wave propagation direction is inside a relatively narrow transmission cone centered on a magnetic field line then it can traverse the ionosphere and hence be detected on Earth. If not, then the wave will be reflected back into the magnetosphere.

Propagation across the free-space–ionosphere interface and through the ionosphere can occur in modes other than that described above. Such coupling from free space to the whistler mode is made possible, for example, if there is a gradient in the ionosphere's refractive index sufficiently steep that the wavelengths under consideration are long compared with the region over which the refractive index changes. This allows an evanescent wave (a wave having amplitude exponentially decreasing with height) to traverse the region of the ionosphere where it cannot "propagate" (because the propagation constant is imaginary) to a region where again it can propagate (because the propagation constant is real). Further, if the effects of electron collisions in the ionosphere are included in the propagation constant, then it is found that coupling to the whistler mode is enhanced although energy is lost in the collision process (Ratcliffe 1959; Budden 1961, 1985; Helliwell 1965). Holzworth et al. (1999) presented evidence from rocket measurements in the magnetosphere that most lightning flashes at high latitudes transmit whistler waves through the ionosphere although most of the whistlers subsequently propagate in an unducted mode.

Once the whistler wave is propagating along a magnetic field line, the time of propagation $t(\omega)$ from source to receiver can be computed from a knowledge of the group velocity of the wave:

$$t(\omega) = \int v_g^{-1}\,dl \qquad (13.98)$$

where, from Eqs. 13.9 and 13.96, v_g^{-1} can be derived as

$$v_g^{-1} = (\mu_0\varepsilon_0)^{1/2} \frac{\omega_{pe}}{2\omega^{1/2}|\omega_{be}|^{1/2}\left(1 - \frac{\omega}{|\omega_{be}|}\right)^{3/2}} \quad (13.99)$$

Thus from Eqs. 13.98 and 13.99, we find that

$$t(\omega) = \frac{(\mu_0\varepsilon_0)^{1/2}}{2}\int \frac{\omega_{pe}}{\omega^{1/2}|\omega_{be}|^{1/2}\left(1 - \frac{\omega}{|\omega_{be}|}\right)^{3/2}} dl \quad (13.100)$$

If ω_{pe} and ω_{be} are specified along a magnetic field line, the integral in Eq. 13.100 can be evaluated numerically to find the whistler travel time for any frequency.

Consider the so-called low-frequency approximation where $\omega \ll \omega_{be}$ along the whole propagation path. Then Eq. 13.100 becomes

$$t(\omega) = \frac{(\mu_0\varepsilon_0)^{1/2}}{2}\int \frac{\omega_{pe}}{\omega^{1/2}|\omega_{be}|^{1/2}} dl \quad (13.101)$$

We can define the dispersion D as

$$D = t(f)f^{1/2} = \frac{(\mu_0\varepsilon_0)^{1/2}}{2\sqrt{2}\pi}\int \frac{\omega_{pe}}{|\omega_{be}|^{1/2}} dl \quad (13.102)$$

which is independent of the wave frequency f. Equation 13.102 is known as the Eckersley dispersion law (Eckersley 1931, 1935) and is valid at frequencies far below the nose frequency. The fact that the dispersion D is independent of frequency is illustrated by Fig. 13.4. Eckersley's law reproduces well the observed functional relationship between wave frequency and time, by which lower whistler frequencies arrive at later times.

For a given path, the integral in Eq. 13.100 has, in general, a frequency f_n for which the propagation time is a minimum, the nose frequency, as noted in Section 13.1 and subsection 13.5.1 above. For a homogeneous plasma the nose frequency $f_n = f_{be}/4$, where $\omega_{be} = 2\pi f_{be}$. In the previous subsection we noted how properties of the magnetosphere can be determined, given some version of the theory outlined above, from the observed nose frequency and from the time delays of all wave frequencies. The ionosphere does not contribute much to the observed time delays; most of the delay is due to the highest part of the path in the magnetosphere. This is so because the integral in Eq. 13.100 is heavily weighted in favor of the high-altitude portion of the propagation path, where ω_{be} is small.

We consider now the conditions for the whistler signal to remain attached to a magnetic field line, the conditions for ducted propagation. Ducting can take place in a variety of ways: (i) for $\omega < \omega_{be}/2$ it will occur when the electron density, linearly related to ω_{pe}^2 (Eq. 13.34), is a maximum along the magnetic field line and decreases perpendicularly to that line; (ii) for $\omega > \omega_{be}/2$ it will occur when the electron density is a minimum along the field line and increases in the outward direction, and (iii) it will occur when the election density decreases only in the outward direction in the presence of curvature of the magnetic field lines (Helliwell 1965; Inan and Bell 1977). The ducting mechanism can be illustrated qualitatively by applying Snell's law (Eq. 13.58). The refractive index in the ionosphere or magnetosphere for propagation along or nearly along the magnetic field lines, after rearranging Eq. 13.95 for a right circularly polarized wave, is found to be

$$n_i = \frac{k_z}{\omega}(\mu_0\varepsilon_0)^{-1/2} = \left(\frac{\omega^2 - \omega|\omega_{be}| - \omega_{pe}^2}{\omega(\omega - |\omega_{be}|)}\right)^{1/2} = \left\{\frac{\left[\omega - \frac{1}{2}|\omega_{be}| + \left(\frac{1}{4}|\omega_{be}|^2 + \omega_{pe}^2\right)^{1/2}\right]\left[\omega - \frac{1}{2}|\omega_{be}| - \left(\frac{1}{4}|\omega_{be}|^2 + \omega_{pe}^2\right)^{1/2}\right]}{\omega(\omega - |\omega_{be}|)}\right\}^{1/2} \quad (13.103)$$

The increase in the refractive index n_i with increasing electron density reverses at $\omega = \omega_{be}/2$, so that the conditions for ducting also change at that frequency. For $\omega < \omega_{be}/2$, a decreasing electron density away from the maximum value found along the magnetic field lines causes, via Snell's law, the direction of propagation of waves propagating into the lower election density, though nearly along the magnetic field direction, to rotate back toward the magnetic field direction. Also, for $\omega < \omega_{be}/2$ there is an additional ducting mechanism requiring a density trough and a wave normal at relatively large angles to the static magnetic field (Gendrin 1960; Helliwell 1965; Park 1982). Note that the expression for the refractive index given in Eq. 13.103 is not valid for propagation at large angles with respect to the magnetic field direction. For $\omega \ll \omega_{be}$ and propagation along the magnetic field, Eq. 13.103 reduces to Eq. 13.97 and hence the index simply varies as the square root of the electron density. From a physical point of view we can consider concentric cylinders for which the index of refraction decreases with increasing radius. Referring to Fig. 13.7, if the wave normal from one cylinder to the next larger cylinder makes an angle θ_i with respect to the radial direction, the transmission angle θ_t will be larger than θ_i since the index of refraction is less in the cylinder with larger radius. Thus a direction of wave propagation that deviates from the direction of the static magnetic field, where the electron density is maximum, will be bent back toward that direction, and the whistler path will be similar to that shown in Fig. 13.3c. For frequencies above $\omega_{be}/2$, the variation in refractive index with electron density is such that a density trough centered on the magnetic field line will allow ducting where wave normals are close to the magnetic field direction, again giving rise to a path such as shown in Fig. 13.3c. For this mode of ducting, the wave frequency must be 0.5 MHz or higher for the wave to be ducted along the whole path so it does not generally occur for whistlers observed on the ground.

It follows from the above discussion that whistlers received on the ground in general require both enhancement ducts and a whistler frequency spectrum whose upper limit is less than half the minimum electron cyclotron frequency along the whole propagation path, a fact confirmed by both ground-based (Carpenter 1968) and satellite-based (Angerami 1970) observations of whistler upper cutoff frequencies. The minimum cyclotron frequency occurs at the highest portion of the path, where the geomagnetic field is a minimum. Duct diameters are apparently of the order of 50 km or so and electron-density-enhancement factors between 10 and 100 percent (Angerami 1970; Hayakawa and Tanaka 1978). Apparently the lifetime of a duct can be from minutes to many hours and may even be periodic (Hansen et al. 1983; Hayakawa et al. 1983). Rogers et al. (1998) reviewed the published theories on whistler duct formation and criticized the theory of Park and Helliwell (1971) involving the plasma interchange of geomagnetic flux tubes in the presence of thunderstorm quasi-electrostatic fields.

The ion cyclotron frequency ω_{bi} is lower than the electron cyclotron frequency by the ratio of the electron mass and the ion mass, a factor of more than three orders of magnitude, as is evident from Eq. 13.6. The features of ducted whistler propagation in the direction of the magnetic field as discussed above are, in general, valid for frequencies below ω_{be} but above ω_{bi}. For frequencies below ω_{bi} (note that there is a different ω_{bi}-value for each type of ion), the full dispersion relation, Eq. 13.50, must be used, thus including the effects of all the ions. The characteristics of whistlers at such low frequencies are influenced by, or even controlled by, the ions. Below the ion cyclotron frequency, an additional mode of whistler propagation exists, the ion cyclotron wave (Gurnett et al. 1965; Gurnett and Brice 1966). Special cases of ion cyclotron whistlers are "proton whistlers" (Smith R.L. et al. 1964) and "helium whistlers" (Barrington et al. 1966). Non-ducted whistlers do not propagate along magnetic field lines, at least for portions of their paths; thus they have modes of propagation more strongly influenced by the ions than those of ducted whistlers, because of the form of the dispersion relation for propagation at significant angles with respect to the magnetic field direction.

13.6. Radio noise

The roughly one hundred cloud and ground flashes per second worldwide (Section 2.1), particularly the return strokes in ground flashes, as well as ionospherically generated chorus and hiss, produce background electromagnetic noise in the Earth–ionosphere cavity that can be observed at any point on Earth, although the noise level varies with location. The fact that lightning is a major source of radio interference was discovered soon after the invention of radio-wave transmission and reception (Popov 1896). As an example of such radio-noise data, monthly average ELF–VLF average spectral amplitudes in the frequency range from 10 Hz to 30 kHz observed at three different locations during the northern hemisphere summer months are given in Fig. 13.14. Two locations are at high latitudes (Arrival Heights, Antarctica and Sondrestromfjord, Norway), and the third is at a relatively low latitude (Kochi, Japan) The amplitudes at Kochi are substantially higher than those for the two high-latitude stations since Kochi is closer to major lightning-source regions in the tropics. The dip in the noise amplitudes in the frequency range 1 to 3 kHz corresponds to the Earth–ionosphere waveguide cutoff as discussed in Section 13.1 and subsection 13.3.2. Below that cutoff, whistlers (Section 13.5), ionospherically–generated hiss and chorus (subsection 13.5.1), and Schumann resonances, Q-bursts, and slow tails are primary contributors to the noise spectrum. Superimposed on the worldwide background-lightning electromagnetic noise, such as is shown in Fig. 13.14, are larger electromagnetic signals from the closer flashes. The background noise from worldwide distant flashes consists primarily of individual

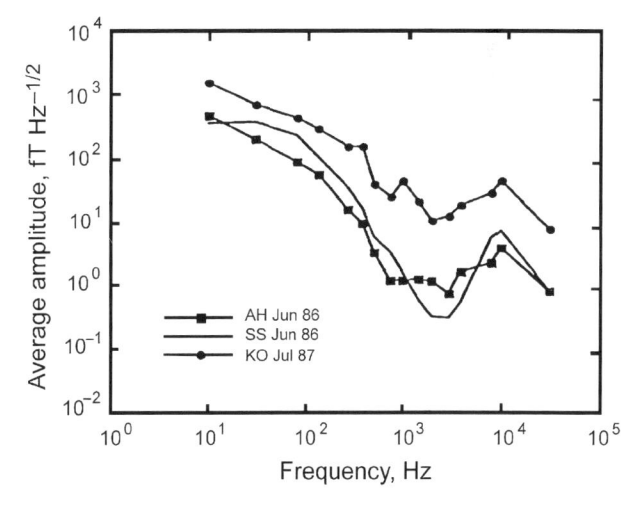

Fig. 13.14. Average ELF–VLF noise amplitudes in femtotesla per square root of hertz at Arrival Heights, Antarctica (AH), Sondrestromfjord, Norway (SS), and Kochi, Japan (KO) for the months of June 1986 (AH and SS) and July 1987 (KO). Overall average amplitudes for 16 narrow-band frequencies are shown. Adapted from Spaulding (1995) and Fraser-Smith et al. (1991).

pulses whose separation time is typically 1 to 100 ms, the longer separation times being associated with pulses of larger amplitude, probably return strokes in ground flashes. The principal source of information on the characteristics of atmospheric radio noise is a series of reports published by the International Radio Consultative Committee (CCIR, an acronym derived from the French name of the committee) (CCIR 1957, 1964, 1966, 1978, 1983, 1988, 1990). These reports draw on many sources of information on radio noise, but the most extensive series of measurements incorporated in the reports were obtained by the US National Bureau of Standards (NBS) from 16 stations distributed over the Earth's surface (e.g., Crichlow et al. 1955; Crichlow 1957). Papers by Watt and Maxwell (1957a, b) extended the NBS work. Later papers by Maxwell and Stone (1963) and Maxwell (1966) on ELF–VLF atmospheric noise are particularly relevant. Watt (1967) incorporated some of the results described in these papers into a comprehensive review of atmospheric radio noise. Reviews of sferics and radio noise are given in Aleksandrov et al. (1972), Pierce (1977), and Remizov (1985). Soderberg (1982) reviewed many different measurements of ELF radio noise. A mini-review by Flock and Smith (1984) presented a selection of radio-noise data in compact form. The most recent comprehensive reviews of the radio noise literature have been by Fraser-Smith (1995) and Spaulding (1995).

A number of statistical measures have been used to describe the properties of atmospheric radio noise. In principle, a knowledge of these measures makes it possible to design long-range communication systems that are essentially immune to the sferics background noise. The most common of these quantities are the average spectral amplitude (Fig. 13.14), the voltage deviation V_d (a measure of the impulsiveness of the noise), the antenna noise factor F_a (used to characterize radio noise, particularly in the CCIR reports), and various time-domain amplitude–probability distributions (APDs). Although they are less often measured and reported, statistical distributions of the time between pulses within the atmospheric noise envelope have considerable practical application. The reason is that often weak radio transmissions can be easily detected between sferic occurrences, even though they may be completely lost when the sferics are occurring. A communication system designed with redundancy based on the statistics of the time between pulses may provide adequate information even during times that would be considered very noisy.

We look now at the definitions of the noise parameters, the voltage deviation, antenna noise factor, and amplitude probability distribution. The voltage deviation V_d is defined as the ratio in decibels of the root-mean-square amplitude to the average (mean) amplitude of the noise envelope (e.g., CCIR 1964). It is a quantitative measure of the impulsiveness, or "spikiness," of the noise. Values of V_d typically range from 2 to 3 dB for moderately impulsive noise to greater than 10 dB for highly impulsive noise (Huntoon and Giordano 1981). Typically, V_d has values near 1 dB for frequencies less than 100 Hz, and its values increase to around 10 dB for frequencies greater than 1 to 2 kHz. Thus the noise is comparatively more impulsive at VLF than it is at ELF.

The antenna noise factor F_a (also referred to as the external noise factor, or the effective antenna noise factor) is often used in the CCIR reports to characterize radio noise. It is defined and discussed in CCIR (1964, 1978), but the most definitive discussion of F_a was provided by Spaulding and Washburn (1985). The basic definition of F_a is given by

$$F_a = 10 \log\left(\frac{p_n}{kT_0 b}\right) = 10 \log\left(\frac{T_a}{T_0}\right) \quad (13.104)$$

where F_a is the antenna noise factor measured in dB, p_n is the external noise power from an equivalent loss-free antenna measured in watts, k is the Boltzmann constant; T_0 is a reference temperature taken to be 288 K, b is the effective receiver noise bandwidth in hertz, and T_a is the effective temperature of the antenna due to external noise in K.

Estimates of the minimum and maximum expected values of F_a were obtained by Spaulding and Hagn (1978) for frequencies in the range 0.1 Hz to 10 kHz. These estimates take into account all seasons and times of day for the entire Earth. There is a more or less steady decline in F_a from around 300 dB at 0.1 Hz to around 160 dB at 10 kHz.

The amplitude probability distribution (APD) is particularly useful for estimating the interference to communication systems caused by atmospheric noise. As a result, there have been a number of important studies, both experimental and theoretical, involving the APD (e.g., Crichlow 1957; Watt and Maxwell 1957a; Crichlow et al. 1960a, b; Spaulding et al. 1962; Galejs 1966; Nakai 1966, 1983, 1986; Watt 1967; Giordano and Haber 1972; Field and Lewinstein 1978; Nakai and Ohba 1984; Remizov 1985). Good examples of measured APDs are found in Watt and Maxwell (1957a) (see also Watt 1967) and in Fraser-Smith and Helliwell (1994).

Among the most recently published data on atmospheric noise, Smith and Jenkins (1998) analyzed VLF radio noise data for the single year 1984 at Halley, Antarctica, where there are no local thunderstorms. They recorded the data on peak, average, and minimum signal levels in frequency bands centered at 0.75, 1.25, 3.2, and 9.6 kHz from 1971 until 1997 with the stated intention of eventually studying any potential long-term changes in global lightning activity. They found that the empirical model of radio noise given in CCIR (1983) did not fit well their observations near 10 kHz, as might have been expected since no measurement stations near Halley contributed to the CCIR model. They presented what they stated to be a simple and better model but which still has deficiencies. Experimentally they found that the peak signal at 3.2 kHz is dominated by sferics, which show a marked diurnal and seasonal variation about Halley local noon and about the solstices. The noise is intense when the ionosphere above Halley is dark, and weak when it is sunlit. The minimum signal level (as opposed to the average or peak signal levels) at 3.2 kHz included few lightning signals. Smith and Jenkins (1998) concluded that attenuation at 3.2 kHz in the Earth–ionosphere waveguide is severe, allowing the reception of only signals from relatively close lightning sources, as would be expected in the waveguide cutoff region. However, the sferics at 9.6 kHz are attenuated much less, and the observed lightning noise is believed to originate from globally distributed distant source regions, particularly near the equator. There is little difference between day and night in the 9.6 kHz signals received, and the diurnal and seasonal variations are not symmetrical about Halley local noon and the solstices but consist of a quasi-sinusoidal diurnal variation in which the phases of the maximum and the minimum vary during the year, being later in July and earlier in January.

Volland *et al.* (1987) recorded VLF atmospherics at 5, 7, and 9 kHz at the German Antarctic station and used the SAR and GDD techniques discussed in subsection 13.3.2 to identify two dominant lightning-producing regions as sources of received sferics, the first to the east of central South America, and the second in the Indian Ocean southeast of South Africa. Grandt (1992) described similar measurements made in Antarctica, in South Africa, and in Germany. Smith and Jenkins (1998) argued that significant thunderstorm regions are present beyond the 4 Mm (4000 km) range limit of Volland *et al.* (1987), and so they included these regions, primarily equatorial, in the atmospheric noise model, noted in the previous paragraph, that they used to describe their measured data. Smith and Jenkins employed the computer program of Ferguson and Snyder (1990) to model Earth–ionosphere waveguide propagation.

13.7. Summary

Lightning-generated electromagnetic waves with primary frequency content in the ELF and VLF bands propagate within the spherical Earth–ionosphere waveguide, often to distances of the order of the Earth's circumference or more. They represent the primary source of atmospheric radio noise. Atmospherics (sferics) are electromagnetic waves, primarily in the VLF range (3–30 kHz) and primarily from lightning return strokes, that propagate in non-TEM waveguide modes in the Earth–ionosphere waveguide.

The upper range of sferic frequencies is hundreds of kilohertz. At frequencies lower than the Earth–ionosphere waveguide cutoff at 1 to 3 kHz, that is, in the ELF band, quasi-TEM waves can propagate. Two classified types of quasi-TEM signal are termed slow tails and Q-bursts. Slow tails follow some sferics (or are the latter portion of a sferic, depending on the sferic definition adopted); it is likely that they are generated by continuing currents flowing in the lightning channel after some return strokes, particularly large positive strokes. Slow tails generally appear as damped oscillations with a primary frequency below 500 Hz. Q-bursts occur at even lower frequencies, those frequencies being characteristic of Schumann resonances, and are the time-domain signals that contribute to Schumann resonances. Schumann resonances are ELF standing waves which occur in the spherical Earth–ionosphere cavity and which can potentially be used for studying global lightning activity. The lowest resonance is at about 8 Hz. Ducted whistlers represent lightning electromagnetic energy, typically in the frequency range from 100 Hz to 10 kHz, that has traversed the ionosphere in the upward direction in one hemisphere, followed the geomagnetic field lines, and traversed the ionosphere in the downward direction in the opposite hemisphere. Lightning-generated electromagnetic signals can also propagate in the magnetosphere as nonducted whistlers.

References and bibliography for Sections 13.1–13.3 and 13.6

Adcock, F., and Clarke, E. 1947. The location of thunderstorms by radio direction finding. *J. Inst. Electr. Eng.* **94B**: 118–25.

Akima, H. 1972. A method of numerical representation for the amplitude–probability distribution of atmospheric radio noise. Office of Telecommunications Research and Engineering Report OT/TRER27, US Government Printing Office, Washington, DC.

Aleksandrov, M.S., Bakleneva, Z.M., Gladshtein, N.D. *et al.* 1972. *Fluctuations of Earth's Electromagnetic Field on VLF*, 195 pp., Moscow: Nauka.

Alfvén, H., and Fälthammer, C.G. 1963. *Cosmical Electrodynamics*, 2nd edition, London and New York: Oxford University Press.

Allis, W.P., Buchsbaum, S.J., and Bers, A. 1963. *Waves in Anisotropic Plasmas*, Cambridge, Massachusetts: MIT Press.

Appleton, E.V., and Barnett, M.A.F. 1925. On some direct evidence for downward atmospheric reflection of electric rays. *Proc. Roy. Soc. A* **109**: 621–41.

Appleton, E.V., and Chapman, S. 1937. On the nature of atmospherics – IV. *Proc. Roy. Soc. A* **158**: 1–22.

Appleton, E.V., Watson-Watt, R.A., and Herd, J.F. 1926. On the nature of atmospherics – II. *Proc Roy. Soc. A* **111**: 615–77.

Baker, S.D., Kelley, M.C., Swenson, C.M., Bonnell, J., Hahn, D.V. 2000. Generation of electrostatic emissions by lightning-induced whistler-mode radiation above thunderstorms. *J. Atmos. Solar-Terr. Phys.* **62**: 1393–404.

Barber, N.F., and Crombie, D.D. 1959. VLF reflections from the ionosphere in the presence of a transverse magnetic field. *J. Atmos. Terr. Phys.* **16**: 37–45.

Barkhausen, H. 1930. Whistling tones from the Earth. *Proc. Inst. Radio Eng. (New York)* **18**: 1155–9.

Barr, R. 1970. The ELF amplitude spectrum of atmospherics with particular references to the attenuation band near 3 kHz. *J. Atmos. Terr. Phys.* **32**: 977–90.

Barr, R. 1974. Multimode propagation in the Earth–ionosphere waveguide. In *ELF–VLF Radio Wave Propagation*, ed. J.A. Holtet, pp. 225–31, Dordrecht, Netherlands: Reidel.

Barr, R. 1977. The effect of sporadic-E on the nocturnal propagation of ELF radio waves. *J. Atmos. Terr. Phys.* **39**: 1379–87.

Barr, R., Jones, D.L., and Rodger, C.J. 2000. ELF and VLF radio waves. *J. Atmos. Solar-Terr. Phys.* **62**: 1689–718.

Beckmann, P. 1962. The amplitude probability distribution of atmospheric radio noise. In *Institute of Radio Engineering and Electronics*, Czechoslovak Academy of Sciences, no. 26.

Beckmann, P. 1964. Amplitude probability distribution of atmospheric radio noise. *Radio Sci.* **68D**: 723.

Bello, P.A., and Esposito, R. 1969. A new method for calculating probabilities of errors due to impulsive noise. *IEEE Trans. Commun. Technol.* **17**: 368.

Bello, P.A., and Esposito, R. 1971. Error probabilities due to impulsive noise in linear and hard-limiting DPSK systems. *IEEE Trans. Commun. Technol.* **19**: 14.

Bickel, J.E., Ferguson, J.A., and Stanley, G.V. 1970. Experimental observations of magnetic field effects on VLF propagation at night. *Radio Sci.* **5**: 19–25.

Boccippio, D.J., Wong, C., Williams, E.R., Boldi, R., Christian, H.J., and Goodman, S.J. 1998. Global validation of single-station Schumann resonance lightning location. *J. Atmos. Solar-Terr. Phys.* **60**: 701–12.

Bowe, P.W.A. 1951. The waveforms of atmospherics and the propagation of very low frequency radio waves. *Phil. Mag.* **42**: 121–38.

Bracewell, R.N., Budden, K.G., Ratcliffe, J.A., Straker, T.W., and Weeks, K. 1951. The ionospheric propagation of low frequency radio waves over distances less than 1000 km. *Proc IEEE*. **98**(3): 221.

Breit, G., and Tuve, M.A. 1926. A test of the existence of the conducting layer. *Phys. Rev.* **28**: 554–75.

Brook, M. 1992. Sferics. In *Encyclopedia of Science and Technology*, pp. 350–52, New York: McGraw Hill.

Budden, K.G. 1951. The propagation of a radio-atmospheric. *Phil. Mag.* **42**: 1–19.

Budden, K.G. 1961. *Radio Waves in the Ionosphere*, London and New York: Cambridge University Press.

Budden, K.G. 1962. *The Waveguide Mode Theory of Wave Propagation*, Englewood Cliffs, New Jersey: Prentice-Hall.

Budden, K.G. 1988. *The Propagation of Radio Waves*, Cambridge, UK: Cambridge University Press.

Burke, C.P., and Jones, D.L. 1992. An experimental investigation of ELF attenuation rates in the Earth–ionosphere duct. *J. Atmos. Terr. Phys.* **54**: 243–50.

Byrne, G.J., Benbrook, J.R., Bering, E.A., Few, A.A., Morris, G.A., Trabucco, W.J., Paschal, E.W. 1993. Ground-based instrumentation for measurements of atmospheric conduction current and electric field at the South Pole. *J. Geophys. Res.* **98**(D2): 2611–8.

Cannon, P.S., Rycroft, M.J. 1982. Schumann resonances frequency variations during sudden ionospheric disturbances. *J. Atmos. Terr. Phys.* **44**: 201–6.

Caton, P.G.F., and Pierce, E.T. 1952. The waveforms of atmospherics. *Phil. Mag.* **43**: 393–409.

CCIR 1957. Revision of atmospheric radio noise data. Report 65, International Radio Consultative Committee, International Telecommunications Union, Geneva.

CCIR 1964. World distribution and characteristics of atmospheric radio noise. Report 322, International Radio Consultative Committee, International Telecommunications Union, Geneva.

CCIR 1966. Operating noise-threshold of a radio receiving system. Report 413, International Telecommunications Union, Geneva.

CCIR 1978. Worldwide minimum external noise levels, 0.1 Hz to 100 GHz. Report 670, in Recommendations and Reports of the CCIR. 1982, 1, International Radio Consultative Committee, International Telecommunication Union, Geneva.

CCIR 1983. Characteristics and applications of atmospheric radio noise data. Report 322-2, International Radio Consultative Committee, Geneva.

CCIR 1988. Characteristics and applications of atmospheric radio noise data. Report 332-3, International Radio Consultative Committee, International Telecommunication Union, Geneva.

CCIR 1990. Man-made noise. Report 258-5, International Radio Consultative Committee, International Telecommunications Union, Geneva.

Challinor, R.A. 1967. The phase velocity and attenuation of audio-frequency electromagnetic waves from simultaneous observations of atmospherics at two spaced stations. *J. Atmos. Terr. Phys.* **29**: 803–10.

Chapman, F.W., and Macario, R.C.V. 1956. Propagation of audio-frequency radio waves to great distances. *Nature* **177**: 930–3.

Chapman, S. 1931. The absorption and dissociative ionizing effect of monochromatic radiation in an atmosphere on a rotating earth, Part I; Part II; Grazing incidence. *Proc. Phys. Soc. London* **43**: 484.

Chrissan, D.A., and Fraser-Smith, A.C. 1996. Seasonal variations of globally measured ELF/VLF radio noise. *Radio Sci.* **31**(5): 1141–52.

Colwell, R.C., and Friend, A.W. 1936. The D region of the ionosphere. *Nature, London* **137**: 782.

Conda, A.M. 1965. The effect of atmospheric noise on the probability of error for an NCFSK system. *IEEE Trans. Commun. Technol.* **13**(3): 280.

Corcuff, Y. 1998. VLF signatures of ionospheric perturbations caused by lightning discharges in an underlying and moving thunderstorm. *Geophys. Res. Lett.* **25**(13): 2385–8.

Crichlow, W.Q. 1957. Noise investigation at VLF by the National Bureau of Standards. *Proc. IRE* **45**: 778.

Crichlow, W.Q., Smith, D.F., Morton, R.N., Corliss, W.R. 1955. Worldwide radio noise level expected in the frequency band 10 kilocycles to 100 megacycles, NBS Circular 557, US Department of Commerce, August.

Crichlow, W.Q., Roubique, C.J., Spaulding, A.D., and Beery, W.M. 1960a. Determination of the amplitude–probability distribution of atmospheric radio noise from statistical moments, *J. Res. Nat. Bur. Stand., Sect. D* **64**(1): 49.

Crichlow, W.Q., Spaulding, A.D., Roubique, C.J., and Disney, R.T. 1960b. Amplitude–probability distributions for atmospheric radio noise. NBS Monograph 23, US Government Printing Office, Washington, DC.

Crombie, D.D. 1966. Further observations of sunrise and sunset fading of very-low-frequency signals. *Radio Sci.* **1**: 47–51.

Croom, D.L. 1964. The frequency spectra and attenuation of atmospherics in the range 1–15 kc/s. *J. Atmos. Terr. Phys.* **26**: 1015–46.

Cummer, S.A. 2000. Modeling electromagnetic propagation in the Earth–ionosphere waveguide. *IEEE Trans. Antennas Propag.* **48**: 1420–9.

Cummer, S.A., and Füllekrug, M. 2001. Unusually intense continuing current in lightning produced delayed mesospheric breakdown. *Geophys. Res. Lett.* **28**: 495–8.

Cummer, S.A., and Inan, U.S. 2000a. Modeling ELF radio atmospheric propagation and extracting lightning currents from ELF observations. *Radio Sci.* **35**: 385–94.

Cummer, S.A., and Inan, U.S. 2000b. Ionospheric E region remote sensing with ELF radio atmospherics. *Radio Sci.* **35**: 1437–44.

David, P., and Voge, J. 1969. *Propagation of Waves* (trans. J. B. Arthur), New York: Pergamon Press.

Davies, K. 1966. *Ionospheric Radio Propagation*, pp. 167–8, New York: Dover Publications.

Davies, K. 1990. *Ionospheric Radio*, London: Peter Peregrinus.

Davis, J.R. 1976. Localized night-time D-region disturbances and ELF propagation. *J. Atmos. Terr. Phys.* **38**: 1309–17.

Field, E.C. Jr, and Lewinstein, 1978. Amplitude–probability distribution model for VLF/ELF atmospheric noise. *IEEE Trans. Commun.* **26**: 83.

Ferguson, J.A., and Snyder, F.P. 1990. Computer programs for assessment of long-wavelength radio communications. (Version 1.0: full FORTRAN code user's guide, 1 April 1990) NOSE Technical Document 1773, National Ocean Systems Center, San Diego, California.

Ferguson, J.A., Morfitt, D.G., and Hansen, P.M. 1985. Statistical model for low-frequency propagation. *Radio Sci.* **20**: 528–34.

Flock, W.L., and Smith, E.K. 1984. Natural radio noise – a minireview. *IEEE Trans. Antennas Propag.* **32**: 762.

Forbes, J.M. 2000. Wave coupling between the lower and upper atmosphere: case study of an ultra-fast Kelvin wave. *J. Atmos. Solar-Terr. Phys.* **62**: 1603–21.

Fraser-Smith, A.C. 1994. The amplitude–probability distributions of ELF/VLF radio noise. Presented at USNC/URSI Meeting, Boulder, Colorado.

Fraser-Smith., A.C. 1995. Low-frequency radio noise. In *Handbook of Atmospheric Electrodynamics*, vol. 1, ed. H. Volland, pp. 297–310, Boca Raton, Florida: CRC Press.

Fraser-Smith, A.C., and Bowen, M.M. 1992. The natural background levels of 50/60 Hz radio noise. *IEEE Trans. Electromagn. Compat.* **34**: 330.

Fraser-Smith, A.C., Helliwell, R.A. 1985. The Stanford University ELF/VLF radiometer project: measurement of the global distribution of ELF/VLF electromagnetic noise. In *Proc. 1985 IEEE Int. Symp. on Electromagnetic Compatibility*, p. 305, IEE Catalog no. 85CH-2116-2.

Fraser-Smith, A.C., and Helliwell, R.A. 1994. Overview of the Stanford University/Office of Naval Research ELF/VLF radio noise survey. In *Proc. 1993 Ionospheric Effects Symp.*, ed., J.M. Goodman, 502, SRI International, Arlington, Virginia.

Fraser-Smith, A.C., Helliwell, R.A., Fortnam, B.R., McGill, P.R., and Teague, C.C. 1988. A new global survey of ELF/VLF radio noise. In *Proc. NATO/AGARD Conf. on Effects of Electromagnetics Noise and Interference on Performance of Military Radio Communications Systems*, vol. 420, pp. 4a1–7, NATO, Brussels.

Fraser-Smith, A.C., McGill, P.R., Bernardi, A., Helliwell, R.A., and Ladd, M.E. 1991. Global measurements of low frequency radio noise. In *Environmental and Space Electromagnetics*, ed. H. Kikuchi, p. 210, Tokyo: Springer-Verlag.

Füllekrug, M., and Constable, S. 2000. Global triangulation of intense lightning discharges. *Geophys. Res. Lett.* **27**: 333–6.

Füllekrug, M., and Fraser-Smith, A.C. 1997. Global lightning and climate variability inferred from ELF magnetic field variations. *Geophys. Res. Lett.* **24**(19): 2411–15.

Füllekrug, M., Fraser-Smith, Bering, E.A., Few, A.A. 1999. On the hourly contribution of global cloud-to-ground lightning activity to the atmospheric electric field in the Antarctic during December 1992. *J. Atmos. Solar-Terr. Phys.* **61**: 745–50.

Füllekrug, M., Constable, S., Heinson, G., Sato, M., Takahashi, Y., Price, C., and Williams, E. 2000. Global lightning acquisition system installed. *Eos, Trans. Am. Geophys. Un.* **81**: 333–43.

Furutsu, I., and Ishida, T. 1960. On the theory of amplitude distribution of impulsive random noise and its application to the atmospheric noise. *J. Radio Res. Lab. Japan* **7**(32): 279.

Galejs, J. 1966. Amplitude distributions of radio noise at ELF and VLF. *J. Geophys. Res.* **71**: 201.

Gajelis, J. 1972. *Terrestrial Propagation of Long Electromagnetic Waves*, New York and Oxford: Pergamon Press.

Giordano, A.A., and Haber, F. 1972. Modeling of atmospheric noise. *Radio Sci.* **7**: 1011.

Ginzburg, V.L. 1964. *Propagation of Electromagnetic Waves in Plasmas*, Oxford: Pergamon Press.

Ginsburg, V.L. 1970a. *The Propagation of Radio Waves*, Cambridge, UK: Cambridge University Press.

Ginzburg, V.L. 1970b. *The Propagation of Electromagnetic Waves in Plasmas*, Oxford, UK: Pergamon Press.

Grandt, C. 1992. Thunderstorm monitoring in South America and Europe by means of very low frequency sferics. *J. Geophys. Res.* **97**: 18 215–26.

Hagn, G.H., and Shepherd, R.A. 1984. Selected radio noise topics. Final Report, SRI International, Arlington, Virginia.

Halton, J.H., and Spaulding, A.D. 1966. Error ranges in differentially coherent phase systems non-Gaussian noise. *IEEE Trans. Commun. Technol.* **14**(5): 594.

Hamelin, J. 1993. Sources of natural noise. In *Electromagnetic Compatibility*, eds. P. Degauque and J. Hamelin, 652 pp., New York: Oxford University Press.

Harth, W. 1982. Theory of low frequency wave propagation. In *Handbook of Atmospherics*, vol. 2, ed. H. Volland, pp. 133–202, Boca Raton, Florida: CRC Press.

Hayakawa, M., Ohta, K., Shimakura, S., and Baba, K. 1995. Recent findings on VLF/ELF sferics. *J. Atmos. Terr. Phys.* **57**: 467–77.

Heaviside, O. 1902. Telegraphy, I. Theory. *Encyclopedia Britannica*, 9th edition, vol. 33: 215.

Helliwell, R.A. 1965. *Whistlers and Related Ionospheric Phenomena*, Stanford, California: Stanford University Press.

Helliwell, R.A. 1997. Whistlers. In *History of Geophysics, Discovery of the Magnetosphere*, vol. 7, pp. 83–94, American Geophysical Union, Washington, DC.

Hepburn, F. 1957a. Atmospheric waveforms with very low frequency components below 1 kc/s known as slow tails. *J. Atmos. Terr. Phys.* **10**: 266–87.

Hepburn, F. 1957b. Waveguide interpretation of atmospheric waveforms. *J. Atmos. Terr. Phys.* **10**: 121–35.

Hepburn, F. 1958. Classification of atmospheric waveforms. *J. Atmos. Terr. Phys.* **12**: 1–7.

Hepburn, F., and Pierce, E.T. 1953. Atmospherics with very low frequency components. *Nature* **171**: 837–8.

Heydt, G. 1982. Instrumentation. In *Handbook of Atmospherics*, vol. 2, ed. H. Volland, pp. 203–56, Boca Raton, Florida: CRC Press.

Hines, C.O., Paghis, I., Hartz, T.R., and Fejer, J.A., eds. 1965. *Physics of the Earth's Upper Atmosphere*, Englewood Cliffs, New Jersey: Prentice-Hall.

Hobara, Y., Iwasaki, N., Hayashida, T., Hayakawa, M., Ohita, K., and Fukunishi, H. 2001. Interrelation between ELF transients and ionospheric disturbances in association with sprites and elves. *Geophys. Res. Lett.* **28**(5): 935–8.

Holden, D.N., Munson, C.P., and Devenport, J.C. 1995. Satellite observations of transionospheric pulse pairs. *Geophys. Res. Lett.* **22**: 889–92.

Hollingworth, J. 1926. The propagation of radio waves. *J. IEEE* **46**: 579–95.

Holzer, R.E., and Deal, D.E. 1956. Low audio frequency electromagnetic signals of natural origin. *Nature* **177**: 536–7.

Horner, F., and Clarke, C. 1955. Some waveforms of atmospherics and their use in the location of thunderstorms. *J. Atmos. Terr. Phys.* **7**: 1–13.

Huang, E., Williams, E., Boldi, R., Heckman, S., Lyons, W., Taylor, M., Nelson, T., and Wong, C. 1999. Criteria for sprites and elves based on Schumann resonance observations. *J. Geophys. Res.* **104**: 16 943–64.

Hughes, H.G. 1971. Differences between pulse trains of ELF atmospherics at widely separated locations. *J. Geophys. Res.* **76**: 2116–25.

Hughes, H.G., and Pappert, R.A. 1975. Propagation prediction model selection using VLF atmospherics. *Geophys. Res. Lett.* **2**: 96–8.

Hughes, H.G., and Theisen, J.F. 1970. Diurnal variations in the apparent attenuations of ELF atmospherics over two different propagation paths. *J. Geophys. Res.* **75**: 2795–801.

Hughes, H.G., Gallenberger, R.J., and Pappert, R.A. 1974. Evaluation of night-time exponential ionospheric models using VLF atmospherics, *Radio Sci.* **9**: 1109–16.

Huntoon, Z.M., and Giordano, A.A. 1981. Rms-to-average deviation ratio for interference and atmospheric noise. In *Proc. 4th Symp. Electromagnetic Compatibility*, ed. T. Dvořák, p. 33.

Ingmann, P., Schaefer, J., Volland, H., Schmolders, M., and Manes, A. 1985. Remote sensing of thunderstorm activity by means of sferics. *Pure Appl. Geophys.* **123**: 155–70.

Inan, U.S., and Inan, A.S. 1998. *Engineering Electromagnetics*, Monlo Park, California: Addison-Wesley.

Inan, U.S., and Inan, A.S. 1999. *Electromagnetic Waves*, Upper Saddle River, New Jersey: Prentice Hall.

Inkov, B.K. 1973. *Phase Methods for the Determination of the Distance to Thunderstorms*, 127 pp., St Petersburg: Gidrometeoizdat.

Jacobson, A.R., Knox, S.O., Franz, R., and Enemark, C.D. 1999. FORTE observations of lightning radio-frequency signatures: capabilities and basic results. *Radio Sci.* **34**: 337–54.

Jean, A.G., Taylor, W.L., Wait, J.R. 1960. VLF phase characteristics deduced from atmospheric wave forms. *J. Geophys. Res.* **65**: 907–12.

Johler, J.R., and Harper, J.D. 1962. Reflection and transmission of radio waves at a continuously stratified plasma with arbitrary magnetic inclination. *J. Res. NBS* **66**: 81.

Jones, D.L. 1970. Propagation of ELF pulses in the earth–ionosphere cavity and application to "slow tail" atmospherics. *Radio Sci.* **5**: 1153–62.

Jones, D.L. 1974. Extremely low frequency (ELF) ionospheric radio propagation studies using natural sources. *IEEE Trans. Commun.* **22**: 477.

Kelso, J.M. 1964. *Radio Ray Propagation in the Ionosphere*, New York: McGraw-Hill.

Kennelly, A.E. 1902. On the elevation of the electrically conducting strata of the earth's atmosphere. *Electr. World and Eng.* **39**: 473.

Kessler, W.J., and Hersperger, 1952. Recent developments in radio location of thunderstorm centers. *Bull. Amer. Meteor. Soc.* **33**: 8–12

Kinzer, G.D. 1974. Cloud-to-ground lightning versus radar reflectivity in Oklahoma thunderstorms. *J. Atmos. Sci.* **31**: 787–99.

Kononov, I.I., Petrenko, I.A., and Snegurov, V.S. 1986. *Radiotechnical Techniques for Locating Thunderstorms*, 222 pp., Leningrad: Gidrometeoizdat.

Laby, T.H., Nicholls, F.G., Nickson, A.F.B. and Webster, H.C. 1937. Reflection of atmospherics at an ionized layer. *Nature* **139**: 837–8.

Laby, T.H., McNeill, J.J., Nicholls, F.G., and Nickson, A.F.B. 1940. Waveform, energy and reflexion by the ionosphere, of atmospherics. *Proc. Roy. Soc. A* **174**: 145–63.

Lanzerotti, L.J., Maclennan, C.G., and Fraser-Smith, A.C. 1990. Background magnetic spectra: $\sim 10^{-5}$ to $\sim 10^5$ Hz. *Geophys. Res. Lett.* **17**: 1593.

Lemmon, J.J. 2001. Wideband model of HF atmospheric radio noise. *Radio Sci.* **36**: 135-91.

Lutkin, F.E. 1939. The nature of atmospherics VI. *Proc. Roy. Soc. A* **171**: 285–313.

Malan, D.J., and Cullens, H. 1937. Progressive lightning III – the fine structure of return lightning strokes. *Proc. Roy. Soc. A* **162**: 175–203.

Malan, D.J., and Schonland, B.F.J. 1947. Progressive lightning VII – directly-correlated photographic and electrical studies of lightning from near thunderstorms. *Proc. Roy. Soc. A* **191**: 485–503.

Massey, R.S., and Holden, 1995. Phenomenology of trans-ionospheric pulse pairs. *Radio Sci.* **30**: 1645–59.

Maxwell, E.L. 1966. Atmospheric noise from 20 Hz to 30 kHz, in *Sub-Surface Communications*, Proc. AGARD Conf., vol. 20, pp. 557.

Maxwell, E.L., and Stone, D.L. 1963. Natural noise fields from 1 cps to 100 kc. *IEEE Trans. Antennas Propag.* **11**: 339.

McDonald, T.B., Uman, M.A., Tiller, J.A., and Beasley, W.H. 1979. Lightning location and lower-ionosphere height determination from two-station magnetic field measurements. *J. Geophys. Res.* **84**: 1727–34.

Morfitt, D.G. 1973. Computer techniques for fitting electron density profiles to oblique-path VLF propagation data, NELC/TR 1854. Available from Code 6700, Naval Electronics Laboratory Center, San Diego, California.

Nakai, T. 1966. The amplitude probability distribution of the atmospheric noise. *Proc. Res. Inst. Atmospherics, Nagoya Univ.* **13**: 23.

Nakai, T. 1983. Modeling of atmospheric radio noise near thunderstorms. *Radio Sci.* **18**: 187.

Nakai, T. 1986. Amplitude probability distributions and impulse amplitude distributions for impulsive noise: atmospheric radio noise from a near thunderstorm and automotive radio noise from a roadway. *Radio Sci.* **21**: 223.

Nakai, T., and Ohba, H. 1984. On the graphical method of drawing APDs for atmospheric radio noise. *IEEE Trans. Electromagn. Compat.* **26**: 71.

NBS 1948. Ionospheric radio propagation. US Department of Commerce, National Bureau of Standards Circular 462.

NBS 1955. World-wide noise levels expected in the frequency band 10 kc/s to 100 Mc/s, US Department of Commerce, National Bureau of Standards Circular 557.

NBS 1959. Quarterly radio Noise Data. Technical Note no. 18-1 through 32, PB 151377, US Department of Commerce, Office of Technical Services, Washington, DC.

Nirenberg, L.M. 1974. Parameter estimation for an adaptive instrumentation of Hall's optimum receiver for digital signals in impulse noise. *IEEE Trans. Commun.* **22**(6): 798.

Nirenberg, L.M. 1975. Low SNR digital communication over certain additive non-Gaussian channels. *IEEE Trans. Commun.* **23**(3): 332.

Norinder, H. 1954. The wave-forms of the electric field in atmospherics recorded simultaneously by two distant stations. *Arkiv Geofysik* **2**(9): 161–95.

Omura, J.K. 1969. Statistical analysis of LF/VLF communications modems, Special Technical Report 1, SRI Project 7045, Stanford Research Institute, Menlo Park, California.

Ovchinnikov, L.M. 1973. Noise immunity of PSK and ASK coherent receivers to quasi-impulsive interference. *Telecommun. Radio Eng.* **28**(10): 64 (English transl).

Pappert, R.A. 1968. A numerical study of VLF mode structure and polarization below an anisotropic ionosphere. *Radio Sci.* **3**: 219–33.

Pappert, R.A., and Snyder, F.P. 1972. Some results of a mode-conversion program for VLF. *Radio Sci.* **7**: 913–23.

Pappert, R.A., Gossard, E.E., and Rothmuller, I.J. 1967. A numerical investigation of classical approximations used in VLF propagation. *Radio Sci.* **2**: 378–400.

Park, C.G. 1982. Whistlers. In *CRC Handbook of Atmospherics*, vol. II, ed. H. Volland, pp. 21–79, Boca Raton, Florida: CRC Press.

Park, C.G., and Carpenter, D.L. 1978. Very low frequency radio waves in the magnetosphere. Upper atmosphere research in Antarctica. In *Antarctic Res. Ser.*, vol. 29, eds. L.J. Lanzerotti and C.G. Park, Washington, DC: American Geophysical Union.

Pierce, E.T. 1977. Atmospherics and radio noise. In *Lightning, vol. 1, Physics of Lightning*, ed. R.H. Golde, Academic Press.

Popov, A.S. 1896. Instrument for detection and registration of electrical fluctuations. *Journal of Russian Physics and Chemistry Society* **XXVIII**, Physics part, I, 1: 1–14.

Price, C. 2000. Evidence for a link between global lightning activity and upper tropospheric water vapour. *Nature* **406**: 290–3.

Rafalsky, V.A., Nickolaenko, A.P., and Shvets, A.V. 1995. Location of lightning discharges from a single station. *J. Geophys. Res.* **100**: 20 829–38.

Ratcliffe, J.A. 1959. *The Magneto-ionic Theory and its Applications to the Ionosphere*, Cambridge, UK: Cambridge University Press.

Ratcliffe, J.A., ed. 1960. *Physics of the Upper Atmosphere*, New York: Academic Press.

Rawer, K., and Suchy, K. 1967. Radio-observations of the ionosphere, in *Handbuch der Physik, Geophysics*, vols. 69/2, 3/2, ed. S. Flügge, pp. 1–546, Berlin and New York: Springer-Verlag.

Remizov, L.T. 1985. *Natural Radio Noise*, 200 pp., Moscow: Nauka.

Rishbeth, H., and Garriott, O.K. 1969. *Introduction to Ionospheric Physics*, New York: Academic Press.

Roussel-Dupré, R.A., Jacobson, A.R., and Triplett, L.A. 2001. Analysis of FORTE data to extract ionospheric parameters. *Radio Sci.* **36**: 1615–30.

Rycroft, M.J., Israelsson, S., and Price, C. 2000. The global atmospheric electric circuit, solar activity and climate change. *J. Atmos. Solar-Terr. Phys.* **62**: 1563–76.

Sao, K., and Jindoh, H. 1974. Real time location of atmospherics by single station techniques and preliminary results. *J. Atmos Terr. Phys.* **36**: 261–6.

Schonland, B.F.J. 1938. Photography of lightning in daytime. *Nature* **141**: 115.

Schonland, B.F.J., Malan, D.J., and Collens, H. 1935. Progressive lightning – II. *Proc. Roy. Soc. A* **152**: 595–624.

Schonland, B.F.J., Hodges, D.B., and Collens, H. 1938. Progressive lightning II. A comparison of photographic and electrical studies of the discharge process. *Proc. Roy. Soc. A* **166**: 56–75.

Schonland, B.F.J., Elder, J.S., van Wyk, J., and Cruickshank, G. 1939. Reflection of atmospherics from the ionosphere. *Nature* **143**: 893–4.

Schonland, B.F.J., Elder, J.S., Hodges, D.G., Phillips, W.E., and van Wyk, J.W. 1940. The wave form of atmospherics at night. *Proc. Roy Soc. A* **176**: 180–202.

Sentman, D.D. 1989. Detection of elliptical polarization and mode splitting in discrete Schumann resonance excitations. *J. Atmos. Terr. Phys.* **52**: 35.

Shindo, T., and M.A. Uman, 1989. Continuing current in negative cloud-to-ground lightning. *J. Geophys. Res.* **94**: 5189–98.

Simpson, G.C. and Scrase, F.J. 1937. The distribution of electricity in thunderclouds. *Proc. Roy. Soc. A* **161**: 309–53.

Smith, A.J., and Jenkins, P.J. 1998. A survey of natural electromagnetic noise in the frequency range $f = 1$–10 kHz at Halley station, Antarctica: 1. Radio atmospherics from lightning. *J. Atmos. Solar-Terr. Phys.* **60**: 263–77.

Smith-Rose, R.L. and Barfield, R.H. 1927. Further measurements on wireless waves received from the upper atmosphere. *Proc. Roy. Soc. A* **116**: 682–93.

Snyder, F.P., and Pappert, R.A. 1969. A parametric study of VLF modes below anisotropic ionospheres. *Radio Sci.* **4**: 213–26.

Soderberg, E.F. 1982. ELF noise – a review. In *Proc. AGARD Conf.*, vol. 305; pp. 15.

Spaulding, A.D. 1964. Determination of error rates for narrow-band communication of binary-coded messages in atmospheric radio noise. *Proc. IEEE* **52**(2): 220.

Spaulding, A.D. 1976. Man-made noise. The problem and recommended steps toward solution. Office of Telecommunications Report OT 76-85, US Department of Commerce, Boulder, Colorado.

Spaulding, A.D. 1995. Atmospheric radio noise and its effects on telecommunication system performance. In *Handbook of Atmospheric Electrodynamics*, vol. 1, ed. H. Volland, Boca Raton, Florida: CRC Press.

Spaulding, A.D., and Disney, R.T. 1974. Man-made radio noise, I. Estimates for business, residential, and rural areas. Office of Telecommunications Report OT 74-38, US Department of Commerce, Boulder, Colorado.

Spaulding, A.D., and Hagn, G.H. 1978. Worldwide minimum environmental radio noise levels (0.1 Hz to 100 GHz). In *Proc. Conf. on Effects of the Ionosphere on Space and Terrestrial Systems*, ed. J.M. Goodman, pp. 177, ONR/NRL, Arlington, Virginia.

Spaulding, A.D., and Middleton, D. 1975. Optimum reception in an impulse interference environment. Office of Telecommunications Report 75-67 (NTIS order no. COM 75-11097/AS).

Spaulding, A.D., and Middleton, D. 1977. Optimum reception in an impulsive interference environment, I. Coherent detection and II. Incoherent reception. *IEEE Trans. Commun.* **25**: 910.

Spaulding, A.D., and Washburn, J.S. 1985. Atmospheric radio noise: worldwide levels and other characteristics. NTIA Report, US Department of Commerce, NTIA/ITS, Boulder, Colorado.

Spaulding, A.D., Roubique, C.J., and Crichlow, W.Q. 1962. Conversion of the amplitude – probability distribution function for atmospheric radio noise from one bandwidth to another. *J. Res. Natl. Bur. Stand., Sect. D* **66**: 713.

Spaulding, A.D., Disney, R.T., and Hubbard, A.G. 1975. Man-made radio noise, II. Bibliography of measurement data, applications, and measurements methods. Office of Telecommunications Report 75-63.

Spitzer, L. Jr 1962. *Physics of Fully Ionized Gases*, 2nd edition, New York: Interscience Publishers (a division of John Wiley and Sons).

Stix, T.H. 1962. *The Theory of Plasma Waves*, New York: McGraw-Hill.

Straker, T.W. 1955. The ionospheric reflection of radio waves of frequency 16 kc/s over short distances. *Proc. IEEE* **102C**: 396.

Suchy, K. 2000. Theory of high-frequency (HF) radio waves in the second half of the 20th century. *J. Atmos. Solar-Terr. Phys.* **62**: 1683–7.

Taylor, W.L. 1960. VLF attenuation for east–west and west–east daytime propagation using atmospherics. *J. Geophys. Res.* **65**: 1933–8.

Taylor, W.L. 1963. Radiation field characteristics of lightning discharges in the band 1 kc/s to 100 kc/s. *J. Res. Natl. Bur. Stand. D* **67**: 539–50.

Taylor, W.L., and Sao, K. 1970. ELF attenuation rates and phase velocities observed from slow-tail components of atmospherics. *Radio Sci.* **5**: 1453–60.

Tomko A.A., and Hepner, T. 2001. Worldwide monitoring of VLF–LF propagation and atmospheric noise. *Radio Science* **36**(2): 363–9.

Ungstrup, E., and Jackerott, I.M. 1963. Observations of chorus below 1500 cycles per second at Godhavn, Greenland, from July 1957 to December 1961. *J. Geophys. Res.* **68**: 2141.

URSI 1962. Special Report no. 7. *The Measurement of Characteristics of Terrestrial Radio Noise*. Amsterdam: Elsevier.

Volland, H. 1982. Low frequency radio noise. In *Handbook of Atmospherics*, vol. 1, ed. H. Volland, pp. 179–250, Boca Raton, Florida: CRC Press.

Volland, H. 1995. Long wave sferics propagation within the atmospheric waveguide. In *Handbook of Atmospheric Electrodynamics*, vol. 2, ed. H. Volland, pp. 65–93, Boca Raton, Florida: CRC Press.

Volland, H., Schmolders, M., Proelss, G.W., and Schaefer, J. 1987. VLF propagation parameters derived from sferics, observations at high southern latitudes. *J. Atmos. Terr. Phys.* **49**: 33–41.

Wait, J.R. 1957. The mode theory of VLF ionospheric propagation for finite ground conductivity. *Proc. IRE* **45**: 760–7.

Wait, J.R. 1958. An extension to the mode theory of VLF ionospheric propagation. *J. Geophys. Res.* **63**: 125–35.

Wait, J.R. 1960. On the theory of the slow-tail portion of atmospheric waveforms. *J. Geophys. Res.* **65**: 1939–46.

Wait, J.R. 1961. Expected influence of a localized change of ionospheric height on VLF propagation. *J. Geophys. Res.* **66**: 3119–23.

Wait, J.R. 1962. *Electromagnetic Waves in Stratified Media*, New York: Pergamon Press.

Wait, J.R., and Spies, K.P. 1965. Influence of finite ground conductivity on the propagation of VLF radio waves. *Radio Sci.* **69D**: 1359–73.

Warber, C.R., and Field, E.C. Jr 1995. A long wave transverse electric–transverse magnetic noise prediction model. *Radio Sci.* **30**: 783–97.

Watkins, N.W., Bharmal, N.A., Clilverd, M.A., and Smith, A.J. 2001. Comparison of VLF sferis intensities at Halley, Antarctica, with tropical lightning and temperature. *Radio Sci.* **36**: 1053–64.

Watson-Watt, R.A. 1927. Supplement. *Q. J. R. Meteor. Soc.*: 53.

Watson-Watt, R.A., and Appleton, E.F. 1923. On the nature of atmospheres – I. *Proc. Roy. Soc. A* **103**: 84–102.

Watson-Watt, R.A., Herd, J.F., and Bainbridge-Bell, L.H. 1933. *The Cathode Ray Oscillograph in Radio Research*. HMSO London.

Watson-Watt, R.A., Bainbridge-Bell, L.H., Wilkins, A.F., and Bowen, E.G., 1936. Return of radio waves from the middle atmosphere. *Nature* **137**: 866.

Watson-Watt, R.A., Herd, J.F., and Lutkin, F.E. 1937. On the nature of atmospheres – V. *Proc. Roy. Soc. A* **162**: 267–91.

Watt, A.D. 1967. Atmospheric radio noise fields. In *VLF Radio Engineering*, vol. 449, New York: Pergamon Press.

Watt, A.D., and Maxwell, E.L. 1957a. Measured statistical characteristics of VLF atmospheric radio noise. *Proc. IRE* **45**: 55–62.

Watt, A.D., and Maxwell, E.L. 1957b. Characteristics of atmospheric noise from 1 to 100 kc. *Proc. IRE* **45**: 787–94.

Williams, E.R. 1992. The Schumann resonance: a global tropical thermometer. *Science* **256**: 1184–7.

Williams, E.R. 1994. Global circuit response to seasonal variations in global surface air temperature. *Mon. Weather Rev.* **122**: 1917–9.

Williams, E.R. 1999. Global circuit response to temperature on distinct time scales: a status report. In *Atmospheric and Ionospheric Electromagnetic Phenomena Associated with Earthquakes*, ed. M. Hayakawa, pp. 939–49, Tokyo: Terra Scientific Publishing Company (TERRAPUB).

Williams, J.L. 1966. Polarisation of atmospherics pulses due to successive reflections at the ionosphere. *J. Atmos. Terr. Phys.* **28**: 199–211.

Wilson, C.T.R. 1920. Investigations on lightning discharges and on the electric field of thunderstorms. *Phil. Trans. A* **221**: 73–115.

Yamashita, M. 1978. Progagation of tweek atmospherics. *J. Atmos. Terr. Phys.* **40**: 151–6.

Yeh and Liu 1972. *Theory of Ionospheric Waves*, New York and London: Academic Press.

Yip, W.-Y., Inan, U.S., Orville, R.E. 1991. On the spatial relationship between lightning discharges and propagation paths of perturbed subionospheric VLF/LF signals. *J. Geophys. Res.* **96**: 249–58.

Zuelsdorf, R.S., Casler, C., Strangeway, R.J., Russell, C.T., and Franz, R. 1998. Ground detection of trans-ionospheric pulse pairs by stations in the National Lightning Detection Network. *Geophys. Res. Lett.* **25**: 481–4.

Zuelsdorf, R.S., Strangeway, R.J., Russell, C.T., Casler, C., Christian, H.J., and Franz, R. 1997. Trans-ionospheric pulse pairs (TIPPs): their geographic distributions and seasonal variations. *Geophys. Res. Lett.* **24**: 3165–8.

References and bibliography for Section 13.4

Abbas, M. 1968. Hydromagnetic wave propagation and excitation of Schumann resonances. *Planet. Space Sci.* **16**: 831–44.

Balser, M., and Wagner, C.A. 1960. Observations of earth–ionosphere cavity resonances. *Nature* **188**: 638–41.

Balser, M., and Wagner, C.A. 1962a. Diurnal power variations of the earth–ionosphere cavity modes and their relationship to worldwide thunderstorm activity. *J. Geophys. Res.* **67**: 619–25.

Balser, M., and Wagner, C.A. 1962b. On frequency variations of the earth–ionosphere cavity modes. *J. Geophys. Res.* **67**: 4081–3.

Balser, M., and Wagner, C.A. 1963. Effect of a high-altitude nuclear detonation on the earth–ionosphere cavity. *J. Geophys. Res.* **68**: 4115–18.

Barr, R., Jones, D.L., and Rodger, C.J. 2000. ELF and VLF radio waves. *J. Atmos. Solar-Terr. Phys.* **62**: 1689–718.

Beamish, D., and Tzanis, A. 1986. High resolution spectral characteristics of the Earth–ionosphere cavity resonances. *J. Atmos. Terr. Phys.* **48**: 187–203.

Behroozi-Toosi, A.B. and Booker, H.G. 1983. Application of a simplified theory of ELF propagation to a simplified worldwide model of the ionosphere. *Space Sci. Rev.* **35**: 91–127.

Bell, T.F., Reising, S.C., and Inan, U.S. 1998. Intense continuing currents following positive cloud-to-ground lightning associated with red sprites. *Geophys. Res. Lett.* **25**: 1285–8.

Belyaev, G.G., Schekotov, A. Yu, Shvets, A.V., and Nickolaenko, A.P. 1999. Schumann resonances observed using Poynting vector spectra. *J. Atmos. Solar-Terr. Phys.* **61**: 751–63.

Bezrodny, V.G., Nickolaenko, A.P., and Sinitsin, V.G. 1977. Radio propagation in natural waveguides. *J. Atmos. Terr. Phys.* **39**: 661–88.

Bliokh, P.V., Nikolayenko, A.P., and Filippov, Yu. F. 1968. Diurnal variations of the natural frequencies of the earth–ionosphere resonator in relation to the eccentricity of the geomagnetic field. *Geomagn. Aeron.* **8**: 198–206.

Bliokh, P.V., Nikolaenko, A.P., and Filippov, Yu. F. 1980. *Schumann Resonances in the Earth–Ionosphere Cavity*, London: Peter Peregrinus.

Boccippio, D.J., Williams, E.R., Heckman, S.J., Lyons, W.A., Baker, I.T., and Boldi, R. 1995. Sprites, Elf transients, and positive ground strokes. *Science* **269**: 1088–91.

Boccippio, D.J., Williams, E.R., Heckman, S.J., Lyons, W.A., Baker, I.T., and Boldi, R. 1996. Sprites, Q-bursts, and positive ground strokes. *Science* **269**: 1088.

Boccippio, D.J., Wong, C., Williams, E.R., Boldi, R., Christian, H.J., and Goodman, S.J. 1998. Global validation of single-station Schumann resonances lightning location. *J. Atmos. Solar-Terr. Phys.* **60**: 701–12.

Budden, K.G. 1962. *The Waveguide Mode Theory of Wave Propagation*, Englewood Cliffs, New Jersey: Prentice-Hall.

Budden, K.G. 1988. *The Propagation of Radio Waves*, Cambridge, UK: Cambridge University Press.

Burke, C.P., and Jones, D.L.I. 1992. An experimental investigation of ELF attenuation rates in the earth–ionosphere duct. *J. Atmos. Terr. Phys.* **54**: 243–50.

Burke, C.P., and Jones, D.L.I. 1994. On the polarity and continuing currents in unusually large lightning flashes deduced from ELF events. *J. Atmos. Terr. Phys.* **58**: 531–40.

Burke, C.P., and Jones, D.L.I. 1995a. Global radio location in the lower ELF frequency band. *J. Geophys. Res.* **100**: 26 263–71.

Burke, C.P., and Jones, D.L.I. 1995b. The geographical distribution of the sources of large ELF atmospherics. In *Proc. Int. Symp. on Electromagnetic Compatibility and Ecology*, St Petersburg, Russia, EMC and EME-95: pp. 251–5.

Burke, C.P., and Jones, D.L.I. 1996. On the polarity and continuing currents in unusually large lightning flashes deduced from ELF events. *J. Atmos. Terr. Phys.* **58**: 531–40.

Burrows, M.L. 1978. *ELF Communications Antennas*, London: Peter Peregrinus.

Cannon, P.S., and Rycroft, M.J. 1982. *J. Atmos. Terr. Phys.* **44**: 201–6.

Chapman, F.W., and Jones, D.L.I. 1964. Earth–ionosphere cavity resonances and the propagation of extremely low frequency radio waves. *Nature* **202**: 654–7.

Clayton, M., and Polk, C. 1974. Diurnal variation and absolute intensity of world-wide lightning activity. September 1970 to May 1971. In *Proc. Conf. Electrical Processes in Atmospheres, Garmisch-Partenkirchen, Germany*.

Cummer, S.A. 2000. Modeling electromagnetic propagation in the Earth–ionosphere waveguide. *IEEE Trans. Antennas Propag.* **48**: 1420–9.

Cummer, S.A., and Inan, U.S. 1997. Measurement of charge transfer in sprite-producing lightning using ELF radio atmospherics. *Geophys. Res. Lett.* **24**: 1731–4.

Cummer, S.A., and Inan, U.S. 2000. Modeling ELF radio atmospheric propagation and extracting lightning currents from ELF observations. *Radio Sci.* **35**: 385–94.

Cummer, S.A., Inan, U.S., Bell, T.F., and Barrington-Leigh, C.P. 1998. ELF radiation produced by electrical currents in sprites. *Geophys. Res. Lett.* **25**: 1281–4.

Cummer, S.A., and Füllekrug, M. 2001. Unusually intense continuing current in lightning produced delayed mesospheric breakdown. *Geophys. Res. Lett.* **28**: 495–8.

Davies, K. 1990. *Ionospheric Radio*. London, UK: Peter Peregrinus.

Egeland, A., and Larsen, T.R. 1968. Fine structure of the earth–ionosphere cavity resonances. *J. Geophys. Res.* **73**: 4986–9.

Etcheto, J., Gendrin, R., Karczewski, J.F. 1966. Simultaneous recording of Schumann resonances at two stations separated by 12 000 km. *Ann. Geophys.* **22**: 646–8.

Fowler, R.A., Kotick, B.J., and Elliot, R.D. 1967. Polarization analysis of natural and artificially induced geomagnetic micropulsations. *J. Geophys. Res.* **72**: 2871–83.

Füllekrug, M., and Constable, S. 2000. Global triangulation of intense lightning discharges. *Geophys. Res. Lett.* **27**: 333–6.

Füllekrug, M., and Reising, S.C. 1998. Excitation of Earth–ionosphere cavity resonances by sprite-associated lightning flashes. *Geophys. Res. Lett.* **25**: 4145–8.

Füllekrug, M., and Sukhorukov, A.I. 1999. The contribution of anistropic conductivity in the ionosphere to lightning flash bearing deviations in the ELF/ULF range. *Geophys. Res. Lett.* **26**: 1109–12.

Galejs, J. 1961a. Terrestrial extremely low frequency noise spectrum in the presence of exponential ionospheric conductivity profiles. *J. Geophys. Res.* **66**: 2787–92.

Galejs, J. 1961b. ELF waves in the presence of exponential ionospheric conductivity profiles. *IRE Trans. Antennas Propag.* **AP-9**: 554–62.

Galejs, J. 1962. A further note on terrestrial extremely low frequency propagation in the presence of isotropic ionospheric with an exponential conductivity height profiles. *J. Geophys. Res.* **67**: 2715–28.

Galejs, J. 1964. Terrestrial extremely-low-frequency propagation. In *Natural Electromagnetic Phenomena Below 30 kc/s*, ed. D.F. Bleil, New York: Plenum.

Galejs, J. 1965a. Schumann resonances. *Radio Sci. J. Res. NBS* **69D**: 1043–55.

Galejs, J. 1965b. On the terrestrial propagation of ELF and VLF waves in the presence of a radial magnetic field. *Radio Sci. J. Res. NBS* **69D**: 705–20.

Galejs, J. 1968. Propagation of ELF and VLF radio waves below an anisotropic ionosphere with a dipping magnetic field. *J. Geophys. Res* **73**: 339–52.

Galejs, J. 1970. Frequency variations of Schumann resonances. *J. Geophys. Res.* **71**: 3237–51.

Galejs, J. 1972. *Terrestrial Propagation of Long Electromagnetic Waves*, New York and Oxford: Pergamon Press.

Ginzburg, V.L. 1970. *The Propagation of Electromagnetic Waves in Plasmas*, Oxford, Pergamon Press.

References and bibliography for Section 13.4

Greifinger, C., and Greifinger, P. 1968. Theory of hydromagnetic propagation in the ionospheric waveguide. *J. Geophys. Res.* **73**: 7473–90.

Greifinger, C., and Greifinger, P. 1973. Wave guide propagation of micropulsation out of the plane of the geomagnetic meridian. *J. Geophys. Res.* **78**: 4611–18.

Greifinger, C., and Greifinger, P. 1978. Approximate method for determining ELF eigenvalues in the earth–ionosphere cavity. *Radio Sci.* **13**: 831–7.

Greifinger, C., and Greifinger, P. 1979. On the ionospheric parameters which govern high-latitude ELF propagation in the earth–ionosphere cavity. *Radio Sci.* **14**: 889–95.

Greifinger, C., and Greifinger, P. 1986. Noniterative procedure for calculating ELF mode constants in the anisotropic earth–ionosphere waveguide. *Radio Sci.* **21**: 981–90.

Heacock, R.R. 1974. Whistler-like pulsation events in the frequency range of 20 to 200 Hz. *Geophys. Res. Lett.* **1**: 77–9.

Heckman, S.J., Williams, E., and Boldi, R. 1998. Total global lightning inferred from Schumann resonance measurements. *J. Geophys. Res.* **103**: 31 775–9.

Hobara, Y., Iwasaki, N., Hayashida, T., Hayakawa, M., Ohita, K., and Fukunishi, H. 2001. Interrelation between ELF transients and ionospheric disturbances in association with sprites and elves. *Geophys. Res. Lett.* **28**(5): 935–8.

Holtham, P.M., and McAskill, B.J. 1988. The spatial coherence of Schumann activity in the polar cap. *J. Atmos. Terr. Phys.* **50**: 83–92.

Holzer, R.E. 1958. World thunderstorm activity and extremely low frequency spherics. In *Recent Advances in Atmospheric Electricity*, ed. L.G. Smith, New York: Pergamon Press.

Holzer, R.E., and Deal, D.E. 1956. Low audio frequency electromagnetic signals of natural origin. *Nature.* **177**: 536–7.

Huang, E., Williams, E., Boldi, R., Heckman, S., Lyons, W., Taylor, M., Nelson, T., and Wong, C. 1999. Criteria for sprites and elves based on Schumann resonance observations. *J. Geophys. Res.* **104**: 16 943–64.

Huzita, A. 1969. Effect of radioactivity fallout upon the electrical conductivity of the lower atmosphere. In *Planetary Electrodynamics*, eds. S. Coroniti and J. Hughes, New York: Gordon & Breach.

Ishaq, M., and Jones, D.L. 1977. Methods of obtaining radiowave propagation parameters for the Earth–ionosphere duct at ELF. *Electron. Lett.* **13**: 254–255.

Johler, J.R., and Berry, L.A. 1962. Propagation of terrestrial radio waves of long wavelength – theory of zonal harmonics with improved summation techniques. *J. Res. Natl. Bur. Stand. Sect. D* **66**: 737–73.

Jones, D.L.I. 1964. The calculation of the Q-factors and frequencies of earth–ionosphere cavity resonances for a two layer ionospheric model. *J. Geophys. Res.* **69**: 4037–46

Jones, D.L.I. 1967. Schumann resonances and ELF propagation for inhomogeneous isotropic ionospheric profiles. *J. Atmos. Terr. Phys.* **29**: 1037–44.

Jones, D.L.I. 1969. The apparent resonance frequencies of the earth–ionosphere cavity when excited by a single dipole source. *J. Geomagn. Geolectr.* **21**: 679–84.

Jones, D.L.I. 1970a. Numerical computations of terrestrial ELF electromagnetic wave fields in the frequency domain. *Radio Sci.* **5**: 803–9.

Jones, D.L.I. 1970b. Propagation of ELF pulses in the earth–ionosphere cavity and application to slow tail sferics. *Radio Sci.* **5**: 1153–62.

Jones, D.L.I. 1970c. Electromagnetic radiation from multiple return strokes of lightning. *J. Atmos. Terr. Phys.* **32**: 1077–93.

Jones, D.L.I. 1974. Extremely low frequency (ELF) ionospheric radio propagation studies using natural sources. *IEEE Trans. Commun.* **22**: 477.

Jones, D.L.I. 1985. Sending signals to submarines. *New Sci.* **37**: XX–XX.

Jones, D.L.I., and Burke, C.P. 1990. Zonal harmonic series expansions of Legendre functions and associated Legendre functions. *J. Phys. A Math. Gem.* **23**: 3159–68.

Jones, D.L.I., and Burke, C.P. 1992. An experimental investigation of ELF attenuation rates in the Earth–ionosphere duct. *J. Atmos. Terr. Phys.* **54**: 243–50.

Jones, D.L.I., and Joyce, G.S. 1989. The computation of ELF radio wave fields in the earth–ionosphere cavity. *J. Atmos. Terr. Phys.* **51**: 233–9.

Jones, D.L.I., and Kemp, D.T. 1970. Experimental and theoretical observations on the transient excitation of Schumann resonances. *J. Atmos. Terr. Phys.* **32**: 1095–108.

Jones, D.L.I., and Kemp, D.T. 1971. The nature and average magnitude of the sources of transient excitation of the Schumann resonances. *J. Atmos. Terr. Phys.* **33**: 557–66.

Kemp, D.T. 1971. The global location of large lightning discharges from single station observations of ELF disturbances in the earth–ionosphere cavity. *J. Atmos. Terr. Phys.* **33**: 919–27.

Kemp, D.T., and Jones, D.L.I. 1971. A new technique for analysis of transient ELF electromagnetic disturbances within the Earth–ionosphere cavity. *J. Atmos. Terr. Phys.* **33**: 567–72 (see also "erratum", *loc. cit.* 1131)

Kessler, W.J., and Hersperger, 1952. Recent developments in radio location of thunderstorm centers. *Bull. Amer. Meteor. Soc.* **33**: 8–12.

Konig, H. 1959. Atmospherics geringster Frequenzen. *Z. Agnew. Phys.* **11**(7): 264.

Large, D.B., and Wait, J.R. 1967. Resonances of the thin-shell model of the earth–ionosphere cavity with a dipolar magnetic field. *Radio Sci.* **2**: 695–702.

Large, D.B., and Wait, J.R. 1968a. Theory of electromagnetic coupling phenomena in the earth–ionosphere cavity. *J Geophys. Res.* **73**: 4335–62.

Large, D.B., and Wait, J.R. 1968b. Influence of a radial magnetic field on the resonances of a spherical plasma cavity. *Radio Sci.* **3**: 663.

Larsen, T.R., and Egeland, A. 1968. Fine structure of the earth–ionosphere cavity resonances. *J. Geophys. Res.* **73**: 4986–9.

Madden, T., and Thompson, W. 1965. Low-frequency electromagnetic oscillations of the Earth–ionosphere cavity. *Rev. Geophys.* **3**: 211–54.

Michalon, N., Nassif, A., Saouri, T., Royer, J.F., and Pontikis, C.A. 1999. Contribution to the climatological study of lightning. *Geophys. Res. Lett.* **26**: 3097–100.

Mitchell, V.B. 1976. Schumann resonance – some properties of discrete events. *J. Atmos. Terr. Phys.* **38**: 77–81.

Nickolaenko, A.P. 1997. Modern aspects of Schumann resonance studies. *J. Atmos. Sol. Terr. Phys.* **59**: 805–16.

Nickoleanko, A.P., and Kudintseva, I.G. 1994. A modified technique to locate the sources of ELF transient events. *J. Atmos. Terr. Phys.* **56**: 1493–8.

Nickolaenko, A.P., and Rabinovich L.M. 1982. Possible global electromagnetic resonances on the planets of the solar system. *Cosmic Res.* **20**: 67–71 (translated from Russian).

Nickolaenko, A.P., and Rabinowicz, L.M. 1995. Study of annual changes of global lightning distribution and frequency variations of the first Schumann resonance mode. *J. Atmos. Terr. Phys.* **57**: 1345–8.

Nickoleanko, A.P., Rafalsky, V.A. Shvets, A.V., and Hayakawa, M. 1994. A time domain direction finding technique for locating wide band atmospherics. *J. Atmos. Electr.* **14**: 97–107.

Nickolaenko, A.P., Hayakawa, M., and Hobara, Y. 1996. Temporal variations of the global lightning activity deduced from the Schumann resonance data. *J. Atmos. Terr. Phys.* **58**: 1699–709.

Nickolaenko, A.P., Satori, G., Zieger, B., Rabinowicz, L.M., and Kudintseva, I.G. 1998. Parameters of global thunderstorm activity deduced from the long-term Schumann resonance records. *J. Atmos. Solar-Terr. Phys.* **60**: 387–99.

Nickolaenko, A.P., Hayakawa, M., and Hobara, Y. 1999. Long-term periodic variations in global lightning activity deduced from the Schumann resonance monitoring. *J. Geophys. Res.* **104**: 27 585–91.

Nickolaenko, A.P., Price, C., and Iudin D.D. 2000. Hurst exponent derived for natural terrestrial radio noise in Schumann resonance band. *Geophys. Res. Lett.* **27**: 3185–8.

Ogawa, T., and Murakami, Y. 1973. Schumann resonance frequencies and the conductivity profiles in the atmosphere. *Contrib. Geophys. Inst. Kyoto Univ.* **13**: 13.

Ogawa, T., and Otsuka, S. 1973. Comparison of observed Schumann resonance frequencies with the single dipole source approximation theories. *Contrib. Geophys. Inst. Kyoto Univ.* **13**: 7–11.

Ogawa, T., and Tanaka, Y. 1970. Q-factors of the Schumann resonances and solar activity. *Contrib. Geophys. Inst. Kyoto Univ.* **10**: 21.

Ogawa, T., Miura, T., Owaki, M., and Tanaka, Y. 1966a. ELF noise bursts and enhanced oscillations associated with the solar-flare of July 7, 1966. *Rep. Ionos. Space Res. Japan* **20**: 528.

Ogawa, T., Tanaka Y., Miura, T., and Yasuhara, M.I. 1966b. Observations of natural ELF and VLF electromagnetic noises by using ball antennas. *J. Geomagn. Geoelectr.* **18**: 443–54.

Ogawa, T., Fraser-Smith, A.C., Gendrin, R., Tanaka, Y., and Yasuhara, M. 1967. Worldwide simultaneity of occurrence of a Q-type ELF burst in the Schumann resonance frequency range. *J Geomag. Geoelectr.* **19**: 377.

Ogawa, T., Tanaka, Y., and Yasuhara, M. 1969a. Schumann resonances and worldwide thunderstorm activity. In *Planetary Electrodynamics vol. 2*, eds. S. Coroniti and J. Hughes, pp. 85–91, New York: Gordon & Breach.

Ogawa, T., Tanaka, Y., and Yasuhara, M. 1969b. Schumann resonances and worldwide thunderstorm activity – diurnal variations of the resonant power of natural noises in the earth–ionosphere cavity, I – Power. *J. Geomagn. Geoelectr.* **21**: 1–30.

Ogawa, T., Kozai, K., and Kawamoto, H. 1979. Schumann resonances observed with a balloon in the stratosphere. *J. Atmos. Terr. Phys.* **41**: 135–42.

Orville, R.E., and Henderson, R. 1986. Global distribution of midnight lightning: September 1977 to August 1978. *Mon. Weather Rev.* **114**: 2640–53.

Pasko, V.P., Inan, U.S., Bell, T.F., and Reising, S.C. 1998. Mechanism of ELF radiation from sprites. *Geophys. Res. Lett.* **25**: 3493–6.

Pierce, E.T. 1963. Excitation of earth–ionosphere resonances by lightning flashes. *J. Geophys. Res.* **68**: 4125–7.

Polk, C. 1969. Relation of ELF noise and Schumann resonances to thunderstorm activity. In *Planetary Electrodynamics*. eds. S. Coroniti and J. Hughes, pp. 55–83, New York: Gordon & Breach.

Polk, C. 1982. Schumann resonances. In *CRC Handbook of Atmospherics*, vol. 1, ed. H. Volland, pp. 111–78, Boca Raton, Florida: CRC Press.

Polk, C. 1983. Natural and man-made noise in the earth–ionosphere cavity at extremely low frequencies (Schumann resonances and man-made "interference"). *Space Sci. Rev.* **35**: 83–9.

Polk, C., and Fitchen, F. 1962. Schumann resonances of the earth–ionosphere cavity: extremely low frequency reception at Kingston. *J. Res. NBS Radio Sci.* **66D**: 313–18.

Price, C. 2000. Evidence for a link between global lightning activity and upper tropospheric water vapour. *Nature* **406**: 290–3.

Raemer, E.T. 1961. On the extra low frequency spectrum of the earth–ionosphere cavity response to electrical storms. *J. Geophys. Res.* **66**: 1580–3.

Raina, B.N., and Raina, R.C. 1988. Diurnal variation of some fair weather electrode effect parameters at Gulmarg. *J. Atmos. Terr. Phys.* **50**: 1–9.

Reeve, N., and Toumi, R. 1999. Lightning activity as an indicator of climate change. *Q.J.R. Meteorol. Soc.* **125**: 893–903.

Reising, S.C., Inan, U.S., Bell, T.F., and Lyons, W.A. 1996. Evidence for continuing current in sprite-producing cloud-to-ground lightning. *Geophys. Res. Lett.* **23**: 3639–42.

Reising, S.C., Inan, U.S., and Bell, T.F. 1999. Sferic energy as a proxy indicator for sprite occurrence. *Geophys. Res. Lett.* **26**: 987–90.

Row, R.V. 1962. On the electromagnetic resonance frequencies of the earth–ionosphere cavity. *IRE Trans. Antennas Propag.* **AP-10**: 766–9.

Rycroft, M.J. 1965. Resonances of the earth–ionosphere cavity observed at Cambridge, England. *Radio Sci. J. Res. NBS* **69**(D): 1071–81.

Rycroft, M.J., Israelsson, S., and Price, C. 2000. The global atmospheric electric circuit, solar activity and climate change. *J. Atmos. Solar-Terr. Phys.* **62**: 1563–76.

Sao, K. 1971. Day to day variation of Schumann resonance frequency and occurrence of Pc1 in view of solar activity. *J. Geomagn. Geoelectr.* **23**: 411.

Sao, K., Yamashita, M., and Tanahashi, S. 1970. Genesis of slow tail atmospherics deduced from frequency analysis and association with VLF components. *J. Atmos. Terr. Phys.* **32**: 1147–51.

Sao, K., Yamashita, M., Tanahashi, S., Jindoh, H., and Ohta, K. 1973. Experimental investigations of Schumann resonance frequencies. *J. Atmos. Terr. Phys.* **35**: 2047–53.

Sapagova, N.A. and Korztorovich, V.M. 1971. The application of group theory to an investigation of the removal of degeneracy in a spherical resonator. *Izv. VUZ Radiofiz.* **14**: 1869.

Satori, G., and Ziegler, B. 1996. Spectral characteristics of Schumann resonances observed in Central Europe. *J. Geophys. Res.* **101**: 29 663–9.

Satori, G., and Zieger, B. 1998. Anomalous behavior of Schumann resonances during the transition between 1995 and 1996. *J. Geophys. Res.* **103**: 14 147–55.

Satori, G., and Ziegler, B. 1999. El Niño related meridional oscillation of global lightning activity. *Geophys. Res. Lett.* **26**: 1365–8.

Satori, G., Szendroi, J., and Vero, J. 1996. Monitoring Schumann resonances – I: methodology. *J. Atmos. Terr. Phys.* **58**: 1475–81.

Schekotov, A. Yu., and Golyavin, A.M. 1978. Watching notch filter for power supply frequency and its harmonics. *Pribory i Tekhika Experimenta* **4**: 175–8. (in Russian).

Schlegel, K., and Füllekrug, M. 1999. Schumann resonance parameter changes during high-energy particle precipitation. *J. Geophys. Res.* **104**: 10 111–18.

Schumann, W.O. 1952a. Über die strahlungslosen Eigenschwingungen einer leitenden Kugel, die von einer Luftschicht und einer Ionosphärenhülle umgeben ist. *Z. Naturforsch.* **7A**: 149–54.

Schumann, W.O. 1952b. Über die Dämpfung der elektromagnetischen Eigenschwingungen des Systems Erde-Luft-Ionosphäre. *Z. Naturforsch.* **7A**: 250.

Schumann, W.O. 1952c. Über die Ausbreitung sehr langer elektrischer Wellen und der Blitzentladung um die Erde. *Z. Agnew. Phys.* **4**(12): 474.

Schumann, W.O. 1954. Über die Oberfelder bei der Austreitung langer, elektrischer Wellen im System Erde-Luft-Ionsphäre und 2 Andwendungen (horizontaler und senkrechter Dipol). *Z. Agnew. Phys.* **6**(1): 35.

Schumann, W.O. 1957. Über elektrische Eigenschwingungen des Hohlraumes Erde-Luft-Ionosphäre, erregt durch Blitzentladungen. *Z. Agnew. Phys.* **9**: 373.

Schumann, W.O. and Koenig, H. 1954. Über die Beobachtung von Atmospherics bei geringsten Frequenzen. *Naturwissenschaften* **41**: 183–4.

Sentman, D.D. 1983. Schumann resonance effects of electrical conductivity perturbations in an exponential atmospheric/ionospheric profile. *J. Atmos. Terr. Phys.* **45**: 55–65.

Sentman, D.D. 1987a. Magnetic elliptical polarization of Schumann resonances, *Radio Sci.* **22**: 595–606.

Sentman, D.D. 1987b. PC monitors lightning worldwide. *Comput. Sci.* **1**: 25.

Sentman, D.D. 1989. Detection of elliptical polarization and mode splitting in discrete Schumann resonance excitations. *J. Atmos Terr. Phys.* **51**: 507–19.

Sentman, D.D. 1990a. Approximate Schumann resonance parameters for a two-scale-height ionosphere. *J. Atmos. Terr. Phys.* **52**: 35–46.

Sentman, D.D. 1990b. Electrical conductivity of Jupiter's shallow interior and the formation of a resonant planetary-ionospheric cavity. *Icarus* **88**: 73.

Sentman, D.D. 1995. Schumann resonances. In *Handbook of Atmospheric Electrodynamics*, vol. 1, ed. H. Volland, pp. 267–95, Boca Raton: CRC Press.

Sentman, D.D. 1996. Schumann resonance spectra in a two-scale-height earth–ionosphere cavity. *J. Geophys. Res.* **101**: 9479–87.

Sentman, D.D. and Ehring, D.A. 1994. Midlatitude detection of ELF whistlers. *J. Geophys. Res.* **99**: 2183–90.

Sentman, D.D., and Fraser, B.J. 1991. Simultaneous observations of Schumann resonances in California and Australia: evidence for intensity modulation by the local height of the D region. *J. Geophys. Res.* **96**: 15 973–84.

Sentman, D.D., and Wescott, E.M. 1993. Observations of upper atmospheric optical flashes recorded from an aircraft. *Geophys. Res. Lett.* **20**: 2857–60.

Sentman, D.D., Wescott, E.M., Osborne, D.L., Hampton, D.L., and Heavner, M.J. 1995. Preliminary results from the Sprites 94 aircraft campaign: 1. Red sprites. *Geophys. Res. Lett.* **22**: 1205–8.

Stefant, R. 1963. Application d'un magnétomètre à l'induction à la détection des fréquences de résonance de la cavité terre-ionosphère. *Ann. Geophys.* **19**: 250–83.

Sukhorukov, A.I. 1991. On the Schumann resonances on Mars. *Planet. Space Sci.* **39**: 1673–6.

Sukhorukov, A.I. 1993. Approximate solution for VLF propagation in an isotropic exponential Earth–ionosphere waveguide. *J. Atmos. Terr Phys.* **55**: 919–30.

Tran, A. 1978. Generalization of ELF mode theory to include unequal surface impedance at the lower ionosphere boundary. *Radio Sci.* **13**(1): 139–45.

Tran, A., and Polk, C. 1976. The earth–ionosphere cavity. *Radio Sci.* **11**: 803–16.

Tran, A., and Polk, C. 1979a. Schumann resonances and electrical conductivity of the atmosphere and lower-ionosphere. I. Effects of conductivity at various altitudes on resonance frequencies and attenuation. *J. Atmos. Terr. Phys.* **41**: 1241–8.

Tran, A., and Polk, C. 1979b. Schumann resonances and electrical conductivity of the atmosphere and lower ionosphere.

II. Evaluation of conductivity profiles from experimental Schumann resonance data. *J. Atmos. Terr. Phys.* **41**: 1249–61.

Tzanis, A., and Beamish, D. 1987a. Audiomagnetotelluric sounding using the Schumann resonances. *J. Geophys.* **61**: 97–109.

Tzanis, A. and Beamish, D. 1987b. Time domain polarization analysis of Schumann resonance waveforms. *J. Atmos. Terr. Phys.* **49**: 217–29.

Vaughan, Jr, O.H., Blakeslee, R., Boeck, W.L., Brook, M., McKune, Jr, J., and Vonnegut, B. 1992. A cloud-to-space lightning as recorded by the space shuttle payload-bay TV cameras. *Mon Weather Rev.* **120**: 1459–61.

Volland, H. 1982. Low frequency radio noise. In *Handbook of Atmospherics*, vol. 1, ed. H. Volland, pp. 179–250, Boca Raton, Florida: CRC Press.

Volland, H. 1995. Long wave sferics propagation within the atmospheric waveguide. In *Handbook of Atmospheric Electrodynamics*, vol. 2, ed. H. Volland, pp. 65–93, Boca Raton, Florida: CRC Press.

Wait, J.R. 1960a. Mode theory and propagation of ELF radio waves. *J. Res. Natl. Bur. Stand., Sect. D* **64**: 387.

Wait, J.R. 1960b. On the propagation of ELF radio waves and the influence of a non-homogeneous ionosphere. *J. Geophys. Res.* **65**: 595–600.

Wait, J.R. 1962a. On the propagation of VLF and ELF radio waves when the ionosphere is not sharply bounded. *J. Res. Natl. Bur. Stand. Sect. D* **66**: 53.

Wait, J.R. 1962b. *Electromagnetic Waves in Stratified Media*, New York, Macmillan.

Wait, J.R. guest ed. 1963. Special issue on electromagnetic waves in the earth. *IEEE Trans. Antennas Propag.* **AP-I I**.

Wait, J.R. 1965a. Earth–ionosphere cavity resonances and the propagation of ELF radio waves. *Radio Sci.* **69D**: 1057–70.

Wait, J.R. 1965b. Cavity resonances for a spherical earth with a concentric anisotropic shell. *J. Atmos. Terr. Phys.* **27**: 81–9.

Wait, J.R. 1970. *Electromagnetic Waves in Stratified Media*, Elmsford, New York: Pergamon Press.

Wait, J.R. 1972. *Electromagnetic Waves in Stratified Media*, 2nd edition, New York, Pergamon Press.

Wait, J.R. 1992. On ELF transmission in the earth–ionosphere waveguide. *J. Atmos. Terr. Phys.* **54**: 109–11.

Williams, E.R. 1992. The Schumann resonance: a global tropical thermometer. *Science.* **256**: 1184–7.

Williams, E.R. 1994. Global circuit response to seasonal variations in global surface air temperature. *Mon. Weather Rev.* **122**: 1917–19.

Williams, E.R. 1999. Global circuit response to temperature on distinct time scales: a status report. In *Atmospheric and Ionospheric Electromagnetic Phenomena Associated with Earthquakes*, ed. M. Hayakawa, pp. 939–49, Tokyo: Terra Scientific Publishing Company (TERRAPUB).

Yeh, K.C., and Liu, C. H. 1972. *Theory of Ionospheric Waves*, New York and London, Academic Press.

References and bibliography for Section 13.5

Aikyo, K., and Ondoh, T. 1971. Propagation of nonducted VLF waves in the vicinity of the plasmapause. *J. Radio Res. Lab. Japan*, **18**: 153.

Akasofu, S.I. 1964. The development of the auroral substorm. *Planet. Space Sci.* **12**: 273–82.

Akasofu, S.-I. 1977. *Physics of Magnetospheric Substorms*, Dordrecht, Holland: Reidel.

Allcock, G. Mck. 1966. Whistler propagation and geomagnetic activity. *J. Inst. Telecomm. Eng.* **12**: 158.

Al'pert, Ya. L. 1990. *Space Plasma*, Cambridge, UK: Cambridge University Press.

Altman, C., and Cory H. 1969. The generalized thin film optical method in electromagnetic wave propagation. *Radio Sci.* **4**: 459–70.

Altman, C., and Cary, H. 1969. The simple thin-film optical method in electromagnetic wave propagation. *Radio Sci.* **4**: 449–57.

Anderson, R.R., Harvey, C.C., Hoppe, M.M., Tsurutani, B.T., Eastman, T.E., and Etcheto, J. 1982. Plasma waves near the magnetopause. *J. Geophys. Res.* **97**: 2087–107.

Angerami, J.J. 1966. A Whistler study of the distribution of thermal electrons in the magnetosphere. *Radio Sci. Lab.*, Stanford Electron Lab., Stanford University, Stanford, California, Technical Report 3412-3417.

Angerami, J.J. 1970. Whistler duct properties deduced from VLF observations made with OGO-3 satellite near the magnetic equator. *J. Geophys. Res.* **75**: 6115–35.

Angerami, J.J., and Thomas, J.O. 1964. Studies of planetary atmosphere. I. The distribution of electrons and ions in the earth's exosphere. *J. Geophys. Res.* **69**: 4537–60.

Angerami, J.J., and Carpenter, D.L. 1966. Whistler studies of the plasmapause in the magnetosphere. II. Equatorial density and total tube electron content near the knee in magnetospheric ionization. *J. Geophys. Res.* **71**: 711–25.

Armstrong, W.C. 1987. Lightning triggered from Earth's magnetosphere as the source of synchronized whistlers. *Nature.* **327**: 405–8.

Aubry, M.P. 1968. Influence des irregularities de densite electronique sur la propagation des ondes TBF dans l'ionosphere. *Ann Geophys.* **24**: 39–48.

Axford, W.I., and Hines, C.O. 1961. A unifying theory of high-latitude geophysical phenomena and geomagnetic storms. *Can. J. Phys.* **39**: 1433–64.

Barkhausen, H., 1919. Zwei mit Hilfe der neuen Verstärker entdeckte Erscheimongen. *Physik A* **20**: 401.

Barkhausen, H. 1930. Whistling tones from the Earth. *Proc. Inst. Radio Eng.* **18**: 1155.

Barrington, R.E. and Belrose, J.S. 1963. Preliminary results from the very-low-frequency receivers aboard Canada's Alouette satellite. *Nature* **198**: 651–6.

Barrington, R.E., Belrose, J.S., and Nelms, G.L. 1965. Ion composition and temperature at 1000 km as deduced from simultaneous observations of a VLF plasma resonance and topside sounder data from the Alouette I satellite. *J. Geophys. Res.* **70**: 1647–64.

Barrington, R.E., Belrose, J.S., and Mather, W.E. 1966. A helium whistler observed in the Canadian Satellite Alouette, II. *Nature* **210**: 80–1.

Bauer, S.J. 1970. Satellite measurements of cold plasma in the magnetosphere. *Progress in Radio Science, 1966–1969*, eds. G.M. Brown, N.D. Clarence, and M.J. Rycroft, vol. 1, p. 159, International Union of Radio Science, Brussels.

Bernard, L.C. 1973. A new nose extension method for whistlers. *J. Atmos. Terr. Phys.* **35**: 871–80.

Bernhardt, P.A. 1979. Theory and analysis of the "super whistler". *J. Geophys. Res.* **84**: 5131–42.

Bernhardt, P.A., and Park, C.G. 1977. Protonospheric–ionospheric modeling of VLF ducts. *J. Geophys. Res.* **82**: 5222–30.

Brice, N.M. 1963. An explanation of triggered very-low-frequency emissions. *J. Geophys. Res.* **68**: 4626–8.

Brice, N.M. 1967. Bulk motion of the magnetosphere. *J. Geophys. Res.* **72**: 5193–211.

Brice, N.M., and Smith, R.L. 1965. Lower hybrid resonance emissions. *J. Geophys. Res.* **70**: 71–80.

Brice, N.M., and Smith, R.L. 1971. Whistlers: diagnostic tools in space plasma. In *Space Physics. vol. 9, Methods of Experimental Physics*, New York: Academic Press.

Buchelet, L.J., and Lefeuvre, R. 1981. One- and two-direction models for VLF electromagnetic waves observed on board GEOS I. *J. Geophys. Res.* **86**: 2377–83.

Budden, K.G. 1961. *Radio Waves in the Ionosphere*, London and New York: Cambridge University Press.

Budden, K.G. 1985. The propagation of radio waves. In *The Theory of Radio Waves of Low Power in the Ionosphere and Magnetosphere*, Cambridge, UK: Cambridge University Press.

Budden, K.G. 1988. *The Propagation of Radio Waves*, Cambridge, UK: Cambridge University Press.

Bullough, K., and Sagredo, J.L. 1973. VLF goniometer observations at Halley Bay, Antarctica – I. The equipment and the measurement of signal bearing. *Planet. Space Sci.* **21**: 899–912.

Burgess, W.C., and Inan, U.S. 1990. Simultaneous disturbance of conjugate ionospheric regions in association with individual lightning flashes. *Geophys. Res. Lett.* **17**: 259–62.

Burgess, W.C., and Inan, U.S. 1993. The role of ducted whistlers in the precipitation loss and equilibrium flux of radiation belt electrons. *J. Geophys. Res.* **98**: 15 643–65.

Burtis, W.J. 1973. Electron concentrations calculated from the lower hybrid resonance noise band observed by OGO-3. *J. Geophys. Res.* **78**: 5515–23.

Burtis, W.J., and Helliwell, R.A. 1976. Magnetospheric chorus: occurrence patterns and normalized frequency. *Planet. Space Sci.* **24**: 1007–24.

Burton, E.T., and Boardman, E.M. 1933. Audio-frequency atmospherics. *Proc. Inst. Radio Eng.* **21**: 1476.

Cain, J.C., Shapiro, I.R., Stolarik, J.D., and Heppner, J.P. 1961. A note on whistlers observed above the ionosphere. *J. Geophys. Res.* **66**: 2677–80.

Carlson, C.R., Helliwell, R.A., and Carpenter, D.L. 1985. Variable frequency VLF signals in the magnetosphere: associated phenomena and plasma diagnostics. *J. Geophys. Res.* **90**: 1507, 1985; correction, *J. Geophys. Res.*, **90**: 6689–92.

Carlson, C.R., Helliwell, R.A., and Inan, U.S. 1990. Space-time evolution of whistler mode wave growth in the magnetosphere. *J. Geophys. Res.* **95**: 73.

Carpenter, D.L. 1962. The magnetosphere during magnetic storms: a whistler analysis. Ph.D. thesis, Radiosci. Lab., Stanford Electron Lab., Stanford University, Stanford, California, Technical Report 12.

Carpenter, D.L. 1963. Whistler evidence of a "knee" in the magnetosphere ionization density profile. *J. Geophys. Res.* **68**: 1675–82.

Carpenter, D.L. 1965. Whistler measurements of the equatorial profile of magnetospheric electron density. *Progress in Radio Science 1960–1963, vol. III, The Ionosphere*, ed. G.M. Brown, p. 76, Amsterdam: Elsevier.

Carpenter, D.L. 1966. Whistler studies of the plasmapause in the magnetosphere – 1. Temporal variations in the position of the knee and some evidence on plasma motions near the knee. *J. Geophys. Res.* **71**: 693–709.

Carpenter, D.L. 1968. Ducted whistler-mode propagation in the magnetosphere: a half-gyrofrequency upper intensity cutoff and some associated wave growth phenomena. *J. Geophys. Res.* **73**: 2919–28.

Carpenter, D.L. 1970. Whistler evidence of the dynamic behavior of the duskside bulge in the plasmasphere. *J. Geophys. Res.* **75**: 3837–47.

Carpenter, D.L. 1983. Some aspects of plasma pause probing by whistlers. *Radio Sci.* **18**: 917–25.

Carpenter, D.L. 1997. Lightning whistlers reveal the plasmapause, an unexpected boundary in space. In *History of Geophysics, vol. 7, Discovery of the Magnetosphere*, pp. 47–59, American Geophysical Union, Washington, DC.

Carpenter, D.L., and Dunckel, N. 1965. A dispersion anomaly in whistlers received on Alouette I. *J. Geophys. Res.* **70**: 3781–6.

Carpenter, D.L., and LaBelle, J.W. 1982. A study of whistlers correlated with bursts of electron precipitation near $L = 2$. *J. Geophys. Res.* **87**: 4427–34.

Carpenter, D.L., and Orville, R.E. 1989. The excitation of active whistler mode signal paths in the magnetosphere by lightning: two case studies. *J. Geophys. Res.* **94**: 8886–94.

Carpenter, D.L., and Park, C.G. 1973. On what ionospheric workers should know about the plasmapause-plasmasphere. *Rev. Geophys. Space Phys.* **11**: 133–54.

Carpenter, D.L., and Seely, N. 1976. Cross-L plasma drifts in the outer plasmaphere: quiet time patterns and some substorm effects. *J. Geophys. Res.* **81**: 2728–36.

Carpenter, D.L., and Stone, K. 1967. Direct detection by a whistler method of the magnetospheric electric field associated with a polar substorm. *Planet. Space Sci.* **15**: 395–7.

Carpenter, D.L., Dunckel, N., and Walkup, J. 1964. A new VLF phenomenon: whistlers trapped below the protonosphere. *J. Geophys. Res.* **69**: 5009–17.

Carpenter, D.L., Stone, K., Siren, J.C., and Crystal, T.L. 1972. Magenetospheric electric fields deduced from drifting whistler paths. *J. Geophys. Res.* **77**: 2819–34.

Carpenter, D.L., Inan, U.S., Trimpi, M.L., Helliwell, R.A., and Katsufrakis, J.P. 1984. Perturbations of subionospheric LF and MF signals due to whistler-induced electron precipitation bursts. *J. Geophys. Res.* **89**: 9857–62.

Carpenter, D.L., Smith, A.J., Gilles, B.L., Chappell, C.R., and Decreau, P.M.E. 1992. A case study of plasma structure in the dusk sector associated with enhanced magnetospheric convection. *J. Geophys. Res.* **97**: 1157–66.

Cerisier, J.C. 1974. Ducted and partly ducted propagation of VLF waves through the magnetosphere. *J. Atmos. Terr. Phys.* **36**: 1443–67.

Chang, H.C., and Inan, U.S. 1985. Test particle modeling of wave-induced energetic electron precipitation. *J. Geophys. Res.* **90**: 6409–18.

Chappell, C.R., Harris, K.K., and Sharp, G.W. 1970. A study of the influence of magnetic activity on the location of the plasmapause as measured by OGO 5. *J. Geophys Res.* **75**: 50–6.

Corcuff, Y. 1961. Variation de la dispersion des sifflements radio-électriques au cours des orages magnétiques. *Ann. Geophys.* **17**: 374–7.

Corcuff, Y. 1975. Probing the plasmapause by whistlers. *Ann. Geophys.* **31**: 53–67.

Corcuff, P., and Corcuff, Y. 1973. Détermination des parametres $f_n - t_n$ caractéristiques des sifflements radio électriques reçus an sul. *Ann. Geophys.* **29**: 273–8.

Cornwall, J.M. 1964. Scattering of energetic trapped electrons by very low frequency waves. *J. Geophys. Res.* **69**: 1251–8.

Cummer, S.A., and Füllekrug, M. 2001. Unusually intense continuing current in lightning produced delayed mesospheric breakdown. *Geophys. Res. Lett.* **28**: 495–8.

Davies, K. 1990. *Ionospheric Radio*. London, Peter Peregrinus.

Delloue, R.L. 1960. La détermination de la direction d'arrivée et de la polarisation des atmospheriques siffleurs, première partie. *J. Phys. Radium* **6**: 514.

Dowden, R.L., and Adams, C.D.D. 1989. Phase and amplitude perturbations or the NWC signal at Dunedin from lightning induced precipitation. *J. Geophys. Res.* **94**: 497–503.

Dowden, R.L., and Allcock, G. McK. 1971. Determination of nose frequency of non-nose whistlers. *J. Atmos. Terr. Phys.* **33**: 1125–9.

Dowden, R.L., Rodgers, C.J., Brundell, J.B., and Cliverd, M.A. 2001. Decay of whistler-induced electron precipitation and cloud – ionosphere electrical discharge Trimpis: observations and analysis. *Radio Sci.* **36**: 151–69.

Draganov, A.B., Inan, U.S., Sonwalker, V.S., and Bell, T.F. 1992. Magnetospherically reflected whistlers as a source of plasmaspheric hiss. *Geophys. Res. Lett.* **19**: 233–6.

Dunckel, N., and Helliwell, R.A. 1969. Whistler-mode emissions on the OGO-1 satellite. *J. Geophys. Res.* **74**: 6371–85.

Dunckel, N., and Helliwell, R.A. 1977. Spacecraft observations man-made whistler-mode signals near the election gyrofrequency. *Radio Sci.* **12**: 821–9.

Dunckel, N., Ficklin, B., Rorden, L.H., and Helliwell, R.A. 1970. Low-frequency noise observed in the distant magnetosphere with OGO-1. *J. Geophys. Res.* **75**: 1854–62.

Dungey, J.W. 1961. Interplanetary magnetic field and the auroral zones. *Phys. Rev. Lett.* **6**: 47.

Dungey, J.W. 1963. Loss of Van Allen electrons due to whistlers. *Planet Space Sci.* **11**: 591–5.

Dungey, J.W. 1967. The theory of the quiet magnetosphere. *Proc. 1966 Symp. on Solar-Terrestrial Physics, Belgrade*, eds. J.W. King and W.S. Newman, p. 91, London and New York, Academic Press.

Eckersley, T.L. 1928. Letter to the editor. *Nature* **122**: 768.

Eckersley, T.L. 1931. 1929–1930 developments in the study of radio wave propagation. *Marconi Rev.* **5**: 1.

Eckersley, T.L. 1935. Musical atmospherics. *Nature* **135**: 104–5.

Edgar, B.C. 1976. The upper- and lower-frequency cutoffs of magnetospherically reflected whistlers. *J. Geophys. Res.* **81**: 205–11.

Eviator, A., Lenchel, A.M., and Singer, S.F. 1964. Distribution of density in an ion-exosphere of a non-rotating planet. *Phys. Fluids* **7**: 1775–9.

Ferencz, C., Bognar, P., Tarcsai, G., Hamar, D., and Smith, A.J. 1996. Whistler-mode propagation: results of model calculations for an inhomogeneous plasma. *J. Atmos. Solar-Terr. Phys.* **58**(5): 625–40.

Galejs, J. 1972. *Terrestrial Propagation of Long Electromagnetic Waves*, Oxford, UK: Pergamon Press.

Gendrin, R. 1960. Guidage des sifflements radio électriques parle champ magnétique terrestre. *Compt. Rend.* **251**: 1085.

Ginzburg, V.L. 1970. *The Propagation of Electromagnetic Waves in Plasma*, Oxford, UK: Pergamon Press.

Gringauz, K.I., Kurth, V.G., Moroz, V.I., and Shklovsky, I.S. 1960. Results of observations of charged particles up to $R = 100\,000$ km with the aid of charged particle traps on Soviet cosmic rockets. *Astron. Zh.*, **716**: (translated as *J. Soviet Astronomy A* **4**: 680–95).

Gurnett, D.A. 1974. The Earth as a radio source: terrestrial kilometric radiation. *J. Geophys. Res.* **79**: 4227–38.

Gurnett, D.A., and Brice, N.M. 1966. Ion temperature in the ionosphere obtained from cyclotron damping of proton whistler. *J. Geophys. Res.* **71**: 3639–52.

Gurnett, D.G., and Scarf, F.L. 1967. Summary report on session on new developments. In *Progress in Radio Science 1963–1966*, p. 1106, International Union of Radio Science, Brussels.

Gurnett, D.A., and Shawhan, D.S. 1966. Determination of hydrogen ion concentration, electron density, and proton gyrofrequency from the dispersion of proton whistlers. *J. Geophys. Res.* **71**: 741–54.

Gurnett, D.A., Shawhan, S.D., Smith, R.L., and Brice, N.M. 1965. Ion cyclotron whistlers. *J. Geophys. Res.* **70**: 1665–88.

Gurnett, D.A., Anderson, R.R., Scarf, F.L., Fredericks, R.W., and Smith, E.J. 1979. Initial results from the ISEE-1 and -2 plasma wave investigations. *Space Sci. Rev.* **23**: 103–22.

Gurnett, D.A., Persoon, A.M., Randall, R.F., Odem, D.L., Remington, S.L., Averkamp, T.F., Debower, M.M., Hospodarsky, G.B., Huff, R.L., Kirchner, D.L., Mitchell, M.A., Pham, B.T., Phillips, J.R., Schintler, W.J., Sheyko P., and Tomash, D.R. 1995. The POLAR plasma wave instrument. *Space Sci. Rev.* **71**: 597–622.

References and bibliography for Section 13.5

Hale, L.C. 1987. Lightning triggering and synchronization. *Nature* **329**: 769.

Hamar, D., Ferencz, C., Lichtenberger, J., Tarcsai, G., Smith, A.J., and Yearby, K.H. 1992. Trace splitting of whistlers: a signature of fine structure or mode splitting in magnetospheric ducts? *Radio Sci.* **27**: 341–6.

Hansen, H.J., Scourfield, M.W. J., and Rash, J.P.S. 1983. Whistler duct lifetimes. *J. Atmos. Terr. Phys.* **45**: 789–94.

Hasegawa, M., Hayakawa, M., and Ohtsu, J. 1978. On the conditions of duct trapping of low latitude whistlers. *Ann. Geophys.* **34**: 317–24.

Hasegawa, M., and Hayakawa, M. 1980. The influence of the equatorial anomaly on the ground reception of whistlers at low latitudes. *Planet Space Sci.* **28**: 17–28.

Hayakawa, M. 1989. Satellite observation of low-latitude VLF radio noises and their association with thunderstorms. *J. Geomagn. Geoelectr.* **41**: 573.

Hayakawa, M. 1991. Observation at Moshiri ($L = 1.6$) of whistler-triggered VLF emissions in the electron slot and inner radiation belt regions. *J. Geomagn. Geoelectr.* **43**: 267–XXX.

Hayakawa, M. 1993. Study of generation mechanisms of magnetospheric VLF/ELF emissions based on the direction findings. In *Proc. Natl. Inst. Planet. Res. Symp. Upper Atmos. Phys.* **6**: 117.

Hayakawa, M. 1995. Association of whistlers with lightning discharges on the earth and on Jupiter. *J. Atmos. Terr. Phys.* **57**: 525–35.

Hayakawa, M. 1995. Whistlers. In *CRC Handbook of Atmospheric Electrodynamics*, vol. II, ed. H. Volland, Boca Raton, CRC Press.

Hayakawa, M., and Ohta, K. 1992. The propagation of low-altitude whistlers: a review. *Planet Space Sci.* **40**: 1339–51.

Hayakawa, M. and Ohtsu, J. 1973. Ducted propagation of low latitude whistlers deduced from the simultaneous observations at multi-stations. *J. Atmos. Terr. Phys.* **35**: 1685–97.

Hayakawa, M., and Sazhin, S.S. 1992. Mid-latitude and plasmaspheric hiss: a review. *Planet Space Sci.* **40**: 1325–38.

Hayakawa, M. and Tanaka, Y. 1978. On the propagation of low-latitude whistlers. *Rev. Geophys. Space Phys.* **16**: 111–23.

Hayakawa, M., Okada, T., and Iwai, A. 1981. Direction findings of a medium-latitude whistlers and their propagation characteristics. *J. Geophys. Res.* **86**: 6939–46.

Hayakawa, M., Tanaka, Y., and Ohtsu, J. 1983. Time scales of formation, lifetime and decay of low-latitude whistler ducts. *Ann Geophys.* **1**: 515–18.

Hayakawa, M., Tanaka, Y., Ohta, K. and Okada, T. 1986a. Absolute intensity of daytime whistlers at low and middle latitudes and its latitudinal variation. *J. Geophys.* **59**: 67–72.

Hayakawa, M., Tanaka, T., Sazhin, S.S., Okada, T., and Kurita, K. 1986b. Characteristics of dawnside mid-latitude VLF emissions associated with substorms as deduced from the two-stationed direction finding measurement. *Planet. Space Sci.* **34**: 225–43.

Hayakawa, M., Ohta, K., and Shimakura, S. 1990. Spaced direction finding of night-time whistlers at low and equatorial latitudes and their propagation mechanism. *J. Geophys. Res.* **95**: 15 091–102.

Hayakawa, M., Ohta, K., and Shimakura, S. 1992. Direction finding techniques for magntospheric VLF waves: recent achievements. *Trends Geophys. Res.* **1**: 157.

Helliwell, R.A. 1949. Ionospheric virtual height measurements at 100 kilocycles. *Proc. IRE* **37**: 887–94.

Helliwell, R.A. 1965. *Whistlers and Related Ionospheric Phenomena*, Stanford: Stanford University Press.

Helliwell, R.A. 1967. A theory of discrete VLF emissions from the magnetosphere. *J. Geophys. Res.* **72**: 4773–90.

Helliwell, R.A. 1970. Intensity of discrete VLF emissions. In *Particles and Fields in the Magnetosphere*, ed. B.M. McCormac, p. 292.

Helliwell, R.A. 1988. VLF wave stimulation experiments in the magnetosphere from Siple Station, Antarctica. *Rev. Geophys.* **26**: 551–78.

Helliwell, R.A. 1997. Whistlers. In *History of Geophysics*, vol. 7, *Discovery of the Magnetosphere*, pp. 83–94, Washington, DC: American Geophysical Union.

Helliwell, R.A., and Carpenter, D.L. 1961. Whistlers – West IGY-IGC synoptic program, final report, NSF grants IGY 6.10/20 and G-8839. Radiosci. Lab., Stanford University.

Helliwell, R.A., and Carpenter, D.L. 1963. Whistlers excited by nuclear explosions. *J. Geophys. Res.* **68**: 4409–20.

Helliwell, R.A., and Gehrels, E. 1958. Observation of magneto-ionic duct propagation using man-made signals of very low frequency. *Proc. IRE* **46**(4): 785–7.

Helliwell, R.A., and Katsufrakis, J.P. 1974. VLF wave injection experiments into the magnetosphere from Siple Station, Antarctica. *J. Geophys. Res.* **79**: 2511–18.

Helliwell, R.A., Mallinkrodt, A.J., and Kruse, F.W. Jr 1951. Fine structure of the lower ionosphere. *J. Geophys. Res.* **56**: 53–62.

Helliwell, R.A., Crary, J.H., Pope, J.H., and Smith, R.L. 1956. The "nose" whistler – a new high-latitude phenomena. *J. Geophys. Res.* **61**: 139–42.

Helliwell, R.A., Katsufrakis, J.P., and Trimpi, M.L. 1973. Whistler-induced amplitude perturbations in VLF propagation. *J. Geophys Res.* **78**: 4679–88.

Helliwell, R.A., Katsufrakis, J.P., Bell, T.F., and Raghurman, R. 1975. VLF line radiation in the earth's magnetosphere and its association with power system radiation. *J. Geophys. Res.* **80**: 4249–58.

Helliwell, R.A., and Katsufrakis, J.P. 1978. Controlled wave–particle interaction experiments. In *Upper Atmosphere Research in Antarctica*, Antarctic Res. Ser., vol. 29, eds. L.J. Lanzerotti and C.G. Park, Washington, DC: American Geophysical Union.

Hines, C.O., 1957. Heavy-ion effects in audio-frequency radio propagation. *J. Atmos. Terr. Phys.* **11**: 36–42.

Ho, D., and Bernard, L.C. 1973. A fast method to determine the nose frequency and minimum group delay of a whistler when the causative sferic is unknown. *J. Atmos. Terr. Phys.* **35**: 881–7.

Hobara, Y., Trakhtengerts, V.Y., Demekhov, A.G., Hayakawa, M. 2000. Formation of electron beams by the interaction of a whistler wave packet with radiation belt electrons. *J. Atmos. Solar-Terr. Phys.* **62**: 541–52.

Holzworth, R.H., Winglee, R.M., Barnum, B.H., Li, Ya Qt., and Kelley, M.C. 1999. Lightning whistler waves in the high-latitude magnetosphere. *J. Geophys. Res.* **104**: 17369–78.

Inan, U.S., and Bell, T. Fl. 1977. The plasmapause as a VLF wave guide. *J. Geophys. Res.* **82**: 2819–27.

Inan, US., and Carpenter, D.L. 1986. On the correlation of whistlers and associated subionospheric VLF/LF perturbations. *J. Geophys. Res.* **91**: 3106–16.

Inan, U.S., and Carpenter, D.L. 1987. Lightning-induced electron precipitation events observed at $L \sim 2.4$ as phase and amplitude perturbations on subionospheric VLF signals. *J. Geophys. Res.* **92**: 3293–303.

Inan, U.S., Bell, T.F., and Helliwell, R.A. 1978. Nonlinear pitch angle scattering of energetic electrons by coherent VLF waves in the magnetosphere. *J. Geophys. Res.* **83**: 3235–53.

Inan, U.S., Carpenter, D.L., Helliwell, R.A., and Katsufrakis, J.P. 1985. Subionospheric VLF/LF phase perturbations produced by lightning-whistler induced particle precipitation. *J. Geophys. Res.* **90**: 7457–69.

Inan, U.S., Shafter, D.C., Yip, W.Y., and Orville, R.E. 1988. Subionospheric VLF signatures of night-time D region perturbations in the vicinity of lightning discharges. *J. Geophys. Res.* **93**: 11455–72.

Inan, U.S., Knifsend, F.A., and Oh, J. 1990. Subionospheric VLF "imaging" of lightning-induced electron precipitation from the magnetosphere. *J. Geophys. Res.* **95**: 17217–31.

Ishikawa, K., Hattori, K., and Hayakawa, M. 1990. A study of ray focusing of whistler-mode waves in the magnetosphere. *Trans. Inst. Electr. Inform. Comm. Eng. Japan* **E73**: 149.

Iwai, A., Okada, T., and Hawakawa, M. 1974. Rocket measurement of wave normal directions of low-latitude sunset whistlers. *J. Geophys. Res.* **79**: 3870–3.

James, H.G. 1972. Refraction of whistler-mode waves by large-scale gradients in the middle-latitude-ionosphere. *Ann. Geophys.* **28**: 301–39.

Johnson, M.P., Inan, U.S., Lauben, D.S. 1999a. Subionospheric VLF signatures of oblique (nonducted) whistler-induced precipitation. *Geophys. Res. Lett.* **26**: 3569–72.

Johnson, M.P., Inan, U.S., Lev-Tov, S.J., Bell, T.F. 1999b. Scattering pattern of lightning-induced ionospheric disturbances associated with early fast VLF events. *Geophys. Res. Lett.* **26**: 2363–6.

Kelley, M.C., Seifring, C.L., Pfaff, R.F., Kintner, P.M., Larsen, M., Green, R., Holzworth, R.H., Hale, L.C., Mitchell, J.D., and LeVine, D. 1985. Electrical measurements in the atmosphere and ionosphere over an active thunderstorm, 1, Campaign overview and initial ionospheric results. *J. Geophys. Res.* **90**: 9815–23.

Kelley, M.C., Ding, J.G., and Holzworth, R.H. 1990. Intense ionospheric electric and magnetic field pulses generated by lightning. *Geophys. Res. Lett.* **17**: 2221–4.

Kelley, N.C., Baker, S.D., Holzworth, R.H., Argo, P., and Cummar, S.A. 1997. LF and MF observations of the lightning electromagnetic pulse at ionospheric altitudes. *Geophys. Res. Lett.* **24**: 1111–14.

Kennel, C.F. and Petschek, H.E. 1966. Limit on stably trapped particle fluxes. *J. Geophys. Res.* **71**: 1–28.

Kimpara, A. and Eguchi, H. 1968. The developments of the study on whistlers and atmospherics in Japan for the last 40 years. *Mem. Chubu Inst. Tech. Japan* **4**: 59–67.

Kimura, I. 1966. Effects of ions on whistler-mode array tracing. *Radio Sci.* **1**: 269–83.

Kimura, I., Smith, R.L., and Brice, N.M. 1965. An interpretation of transverse whistlers. *J. Geophys Res.* **70**: 5961–6.

Kurth, W.S., Craven, J.D., Frank, L.A., and Gurnett, D.A. 1979. Intense electrostatic waves near the upper hybrid frequency. *J. Geophys. Res.* **84**: 4145–64.

LaBelle, J.R., Treumann, R.A., Haerendel, G., Bauer, O.H., Paschmann, G., Baumjohann, W., Luhr, H., Anderson, R.R., Koons, H.C., and Holzworth, R.H. 1987. AMPTE IRM observations of waves associated with flux transfer events in the magnetosphere. *J. Geophys. Res.* **92**: 5827–43.

Lauben, D.S. 1998. Precipitation of radiation belt electrons by obliquely-propagating lightning-generated whistler waves. Ph.D. dissertation, Stanford University, Stanford, California.

Lauben, D.S., Inan, U.S., and Bell, T.F. 1999. Poleward-displaced electron precipitation from lightning-generated oblique whistlers. *Geophys. Res. Lett.* **26**: 2633–6.

Leavitt, M.K. 1975. A frequency-tracking direction finding for whistlers and other VLF signals. Radiosci. Lab. Stanford University, Stanford, California, Technical Report no. 3456-2.

Lefeuvre, F., Parrot, M., and Delannoy, C. 1981. Wave distribution functions estimation of VLF electromagnetic waves observed onboard GEOS-2. *J. Geophys. Res.* **86**: 2359–75.

Lefeuvre, F., Neubert, T., and Parrot, M. 1982. Wave normal directions and wave distribution functions for ground-based transmitter signals observed on GEOS-1. *J. Geophys. Res.* **87**: 6203–17.

Li, Ya Qi, Holzworth, R.H., Hua, Hu, McCarthy, M., Massey, R.D., Kintner, P.M., Rodriguez, J.V., Inan, U.S., and Armstrong, W.C. 1991. Anomalous optical events detected by rocket-borne sensors in the WIPP campaign. *J. Geophys. Res.* **96**: 1315–26.

Likhter, Ya I., and Molchanov, O.A. 1968. Changing in whistler characteristics in the disturbed magnetic field of the magnetosphere. *Geomagn. Aeron.* **8**: 719–23.

Lohrey, B. and Kaiser, A.B. 1979. Whistler induced anomalies in VLF propagation. *J. Geophys. Res.* **84**: 5122–30.

Lyons, L.R., Thorne, R.M., and Kennel, C.F. 1972. Pitch-angle diffusion of radiation belt electrons within the plasmasphere. *J. Geophys. Res.* **77**: 3455–74.

Matsumoto, H., and Kimura, I. 1971. Linear and non linear cyclotron instability and VLF emissions in the magnetosphere. *Planet. Space Sci.* **19**: 567–608.

References and bibliography for Section 13.5

Maynard, N.C., Aggson, T.L., and Heppner, J.P. 1970. Electric field observations of ionospheric whistlers. *Radio Sci.* **5**: 1049–57.

McPherson, D.A., Koons, H.C., Dazey, M.H., Dowden, R.L., Amon, L.E.S., and Thomson, N.R. 1974. Conjugate magnetospheric transmissions at VLF from Alaska to New Zealand. *J. Geophys. Res.* **77**: 1555–7.

Means, J.D. 1972. The use of the three-dimensional covariance matrix in analyzing the properties of plane waves. *J. Geophys. Res.* **77**: 5551–9.

Melrose, D.B. 1986. *Instabilities in Space and Laboratory Plasmas*, New York: Cambridge University Press.

Morgan, M.G. 1980. Some features of pararesonance (PR) whistlers. *J. Geophys. Res.* **85**: 130–8.

Mozer, F.S., and Carpenter, D.L. 1973. Balloon and VLF whistler measurements of electric fields, equatorial electron density, and precipitating particles during a barium cloud release in the magnetosphere. *J. Geophys. Res.* **78**: 5736–44.

Nagano, I., Mambo, M., and Hutatsuishi, G. 1975. Numerical calculation of electromagnetic waves in an anisotropic multilayered medium. *Radio Sci.* **10**: 611–17.

Nagano, I., Wu, X.-Y., Yagatani, S., Miyamura, K., and Matsumoto, H. 1998. Unusual whistler with very large dispersion near the magnetopause: geotail observation and ray-tracing modeling. *J. Geophys. Res.* **102**: 11 827–40.

Nishida, A. 1966. Formation of plasmapause, or magnetospheric plasma knee, by the combined action of magnetospheric convection and plasma escape from the tail. *J. Geophys. Res.* **71**: 5669–79.

Ohta, K., Hayakawa, M., and Tanaka, Y. 1984. Ducted propagation of daytime whistlers at low latitudes deduced from the ground direction finding. *J. Geophys. Res.* **80**: 7557.

Okada, T., Iwai, A., and Hayakawa, M. 1977. The measurement of incident and azimuthal angles and the polarization of whistlers at low latitudes. *Planet. Space Sci.* **25**: 233–41.

Okada, T., Iwai, A., and Hayakawa, M. 1981. A new whistler direction finder. *J. Atmos. Terr. Phys.* **43**: 679–91.

Ondoh, T. 1976. Magnetospheric whistler ducts observed by ISIS satellites. *J. Radio Res. Lab. Tokyo* **23**: 139.

Park, C.G. 1970. Whistler observations of the interchange of ionization between the ionosphere and the protonosphere. *J. Geophys. Res.* **75**: 4249–60.

Park, C.G. 1972. Methods of determining electron concentrations in the magnetosphere from the nose whistlers. Radiosci. Lab., Stanford Electron Lab., Stanford University, Stanford, California, Technical Report no. 3454.

Park, C.G. 1977. VLF wave activity. *J. Geophys. Res.* **82**: 3251–60.

Park, C.G. 1982. Whistlers. In *CRC Handbook of Atmospherics*, vol. II, ed. H. Volland, pp. 21–79, Boca Raton, Florida CRC Press.

Park, C.G., and Carpenter, D.L. 1978. Very low frequency radio waves in the magnetosphere. Upper atmosphere research in Antarctica. In *Antarctic Res. Ser*, vol. 29, eds. L.J. Lanzerotti and C.G. Park, Washington, DC: American Geophysical Union.

Park, C.G., and Chang, D.C.D. 1978. Transmitter simulation of power line radiation effects in the magnetosphere. *Geophys. Res. Lett.* **5**: 861.

Park, C.G., and Helliwell, R.L. 1971. The formation by electric fields of field-aligned irregularities in the magnetosphere. *Radio Sci.* **6**: 299–304.

Park, C.G., and Helliwell, R.A. 1978. Magnetospheric effects of power line radiation. *Science* **200**: 727–30.

Pitteway, M.L.V. 1965. The numerical calculation of wave-fields, reflection coefficients and polarization for long radio waves in the lower ionosphere, I. *Phil. Trans. Roy. Soc. A* **257**: 219–41.

Poeverlin, H. 1948. Strahlwege von Radiowellen in der Ionosphäre. *Sitz Bayerischen Akad. Wiss.* **1**: 175.

Potter, R.K. 1951. Analysis of audio-frequency atmospherics. *Proc. IRE* **39**(9): 1067–9.

Poulsen, W.L., Bell, T.F., Inan, U.S. 1993. The scattering of VLF waves by localized ionospheric disturbances produced by lightning-induced electron precipitation. *J. Geophys. Res.* **98**: 15 553–9.

Poulsen, W.L., Inan, U.S., and Bell, T.F. 1993. A multiple-mode three-dimensional model of VLF propagation in the Earth–ionosphere waveguide in the presence of localized D region disturbances. *J. Geophys. Res.* **98**: 1705–17.

Preece, W.H. 1894. Earth currents. *Nature* **49**: 554.

Price, C. 2000. Evidence for a link between global lightning activity and upper tropospheric water vapour. *Nature* **406**: 290–3.

Price, G.H. 1964. Propagation of electromagnetic waves through a continuously varying stratified anisotropic medium. *Radio Sci. J. Res. NBS* **68D**: 407–18.

Raghuram, R., Smith, R.L., and Bell, T.F. 1974. VLF Antarctic antenna: impedance and efficiency. *IEEE Trans. Antennas Propag.* **AP-22**: 334–8.

Raghuram, R. 1975. A new interpretation of subprotonospheric whistler characteristics. *J. Geophys. Res.* **80**: 4729–31.

Randa, J., Gilliland, D., Gjertson, W., Lauber, W., and McInerney, M. 1995. Catalogue of electromagnetic environment measurements, 30–300 Hz. *IEEE Trans. Electromagn. Compat.* **37**: 16–33.

Rastani, K., Inan, U.S., and Helliwell, R.A. 1985. DE 1 observations of Siple transmitter signals and associated sidebands. *J. Geophys. Res.* **90**: 4128–40.

Ratcliffe, J.A. 1959. *The Magneto-ionic Theory and its Applications to the Ionosphere*, Cambridge, London, and New York: Cambridge University Press.

Ratcliffe, J.A. 1972. *An Introduction to the Ionosphere and Magnetosphere*, Cambridge, UK: Cambridge University Press.

Ristic-Djurovic, J.L., Bell, T.F., and Inan, U.S. 1998. Precipitation of radiation belt electrons by magnetospherically reflected whistlers. *J. Geophys. Res.* **103**: 9249–60.

Rogers, C.J., Thomson, N.R., and Dowden, R.L. 1998. Are whistler ducts created by thunderstorm electrostatic fields? *J. Geophys. Res.* **103**: 2163–9.

Rosenberg, T.J., Helliwell, R.A., and Katsufrakis, J.P. 1971. Electron precipitation associated with discrete very-low-frequency emissions. *J. Geophys. Res.* **76**: 8445–52.

Rycroft, M.J. 1973. Enhanced energetic electron intensities at 100 km altitude and a whistler propagating through the plasmasphere. *Planet. Space Sci.* **21**: 239–51.

Rycroft, M.J., and Mathur, A. 1973. The determination of the minimum group delay of a non-nose whistler. *J. Atmos. Terr. Phys.* **35**: 2177–82.

Rycroft, M.J. 1987. Strange new whistlers. *Nature* **327**: 368–408.

Rycroft, M.J. 1991. Interactions between whistler-mode waves and energetic electrons in the coupled system formed by the magnetosphere, ionosphere and atmosphere. *J. Atmos. Terr. Phys.* **53**: 849–58.

Sagredo, J.L., and Bullough, K. 1973. VLF goniometer observations at Halley Bay, Antarctica, II. Magnetospheric structure deduced from whistler observations. *Planet. Space Sci.* **21**: 913–23.

Sagredo, J.L., Smith, I.D., and Bullough, K. 1973. The determination of whistler nose-frequency and minimum group delay and its implication for the measurement of the east–west electric field and tube content in the magnetosphere. *J. Atmos. Terr. Phys.* **35**: 2035.

Sazhin, S.S., and Hayakawa, M. 1992. Magnetospheric chorus emissions: a review. *Planet. Space Sci.* **40**: 681–97.

Sazhin, S.S., Smith, A.J., and Sazhin, E.M. 1990. Can magnetospheric electron temperature be inferred from whistler dispersion measurements? *Ann. Geophys.* **8**: 273–85.

Sazhin, S.S., Hayakawa, M., and Bullough, K. 1992. Whistler diagnostics of magnetospheric parameters: a review. *Ann. Geophys.* **10**: 293–308.

Sazhin, S.S., Bullough, K., and Hayakawa, M. 1993. Auroral hiss: a review. *Planet. Space Sci.* **41**: 153–66.

Sazhin, S.S., and Hayakawa, M. 1994. Periodic and quasiperiodic emissions. *J. Atmos. Terr. Phys.* **56**: 735–53.

Serbu, G.P., and Meier, E.J.R. 1967. Thermal plasma measurements within the magnetosphere. In *Space Research VII*, eds. R.L. Smith-Rose, S.A. Bowhill, and J.W. King, pp. 527, Amsterdam: North Holland.

Shawhan, S.D. 1970. The use of multiple receivers to measure the wave characteristics of very low frequency noise in space. *Space Sci. Rev.* **10**: 689–736.

Shawhan, S.D. 1979. Magnetospheric plasma waves. In *Solar System Plasma Physics*, eds. C.F. Kennell, L.J. Lanzerotti, and E.N. Parker, Amsterdam: North Holland.

Shawhan, S.D., and Gurnett, D.A. 1966. Fractional concentration of hydrogen ions in the ionosphere from VLF proton whistler measurement. *J. Geophys. Res.* **71**: 46–59.

Shimakura, S., Hayakawa, M., Lefeuvre, F., and Lagoutte, D. 1992. On the estimation of wave energy distribution of magnetospheric VLF waves at the ionospheric base with ground-based multiple electromagnetic field components. *J. Geomagn. Geoelectr.* **44**: 573.

Singh, B. 1976. On the ground observation of whistlers at low latitudes. *J. Geophys. Res.* **81**: 2429–32.

Singh, R.P., Singh, D.K., Singh, A.K., Hamar, D., and Lichtenberger, J. 1999. Application of matched filtering and parameter estimation technique to low latitude whistlers. *J. Atmos. Solar-Terr. Phys.* **61**: 1081–92.

Smith, A.J., and Carpenter, D.L. 1982. Echoing mixed path whistlers near the dawn plasmapause, observed by direction-finding receivers at two Antarctic stations. *J. Atmos. Terr. Phys.* **44**: 973–84.

Smith, A.J., and Cotton, P.D. 1990. The Trimpi effect in Antarctica: observations and models. *J. Atmos. Terr. Phys.* **52**: 341–55.

Smith, A.J., Smith, I.D., and Bullough, K. 1975. Methods of determining whistler nose-frequency and minimum group delay. *J. Atmos. Terr. Phys.* **37**: 1179–92.

Smith, A.J., Smith, I., Deeley, A.M., and Bullough, K. 1979. A semi-automated whistler analyser. *J. Atmos. Terr. Phys.* **41**: 578–600.

Smith, A.J., Cotton, P.D., Robertson, J.S. 1993. Transient (~ 10 s) VLF amplitude and phase perturbations due to lightning-induced electron precipitation into the ionosphere (the "Trimpi effect") In *Proc. AGARD ELF/VLF/LF Radio Propagation and System Aspects.* vol. 529; pp. 8/1–8.

Smith, R.L. 1961a. Propagation characteristics of whistlers trapped in field-aligned columns of enhanced ionization. *J. Geophys. Res.* **66**: 3699–707.

Smith, R.L. 1961b. Properties of the outer ionosphere deduced from nose whistlers. *J. Geophys. Res.* **66**: 3709–16.

Smith, R.L. 1964. An explanation of subprotonospheric whistlers. *J. Geophys Res.* **69**: 5019–21.

Smith, R.L., and Angerami, J.J. 1968. Magnetospheric properties deduced from OGO-1 observations of ducted and non-ducted whistlers. *J. Geophys. Res.* **73**: 1–20.

Smith, R.L., and Carpenter, D.L. 1961. Extension of nose whistler analysis. *J. Geophys. Res.* **66**: 2582–6.

Smith, R.L., and Carpenter, D.L. 1966. Extension of nose whistler analysis. *J. Geophys. Res.* **71**: 3755–66.

Smith, R.L., Helliwell, R.A., and Yabroff, I.W. 1960. A theory of trapping of whistlers in field-aligned columns of enhanced ionization. *J. Geophys. Res.* **65**: 815–23.

Smith, R.L., Brice, N.M., Katsufrakis, J., Gurnett, D.A., Shawhan, S.D., Belrose, J.S., and Barrington, R.E. 1964. An ion gyrofrequency phenomenon observed in satellites. *Nature* **204**: 274–5.

Sonwalkar, V.S., and Carpenter, D.L. 1995. Notes on the diversity of the properties of radio bursts observed on the nightside of Venus. *J. Atmos. Terr. Phys.* **57**: 557–73.

Sonwalkar, V.S., and Inan, U.S. 1989. Lightning as an embryonic source of VLF hiss. *J. Geophys. Res.* **94**: 6986–94.

Spitzer, L. Jr 1962. *Physics of Fully Ionized Gases*, 2nd edition, New York: Interscience Publishers (a division of John Wiley and Sons).

Stix, T.H. 1962. *The Theory of Plasma Waves*, New York, McGraw-Hill.

Storey, L.R.O. 1953. An investigation of whistling atmospherics. *Phil. Trans. Roy. Soc. A* **246**: 113.

Strangeway, R.J. 1995. Plasma wave evidence for lightning on Venus. *J. Atmos. Terr. Phys.* **57**: 537–56.

Strangeways, H.J. 1980. Systematic errors in VLF direction-finding of whistler ducts – I. *J. Atmos. Terr. Phys.* **42**: 995–1008.

References and bibliography for Section 13.5

Strangeways, H.J. 1999. Lightning induced enhancements of D-region ionisation and whistler ducts. *J. Atmos. Solar-Terr. Phys.* **61**: 1067–80.

Strangeways, H.J., and Rycroft, M.J. 1980. Systematic errors in VLF direction-finding of whistler ducts – II. *J. Atmos. Terr. Phys.* **42**: 1009–23.

Stuart, G.F. 1977. Systematic errors in whistler extrapolation. 2. Comparison of methods. *J. Atmos. Terr. Phys.* **39**: 427–31.

Tanaka, Y., Hayakawa, M., and Nishino, M. 1976. Study of auroral VLF hiss observed at Syowa Station, Antarctica. *Mem. Natl. Inst. Polar Res. Ser. A* **13**: 58.

Tarcsai, G. 1975. Routine whistler analysis by means of accurate curve fitting. *J. Atmos. Terr. Phys.* **37**: 1447–57.

Taylor, H.A., Jr, Brinton, H.C., and Smith, C.R. 1965. Positive ion composition in the magnetosphere obtained from OGO-A satellite. *J. Geophys. Res.* **70**: 5769–81.

Thomson, R.J. 1978. The formation and lifetime of whistler ducts. *Planet. Space Sci.* **26**: 423–30.

Thomson, R.J., and Dowden, R.L. 1977a. Simultaneous ground and satellite reception of whistlers. 1. Ducted whistlers. *J. Atmos. Terr. Phys.* **39**: 869–77.

Thomson, R.J., and Dowden, R.L. 1977b. Simultaneous ground and satellite reception of whistlers. 2. PL whistlers. *J. Atmos. Terr. Phys.* **39**: 879–90.

Thomson, R.J., and Dowden, R.L. 1978. Ionospheric whistler propagation. *J. Atmos. Terr. Phys.* **40**: 215–21.

Tixier, M., Charcosset, G., Corcuff, Y., and Okada, T. 1984. Propagation modes of whistlers received aboard satellites over Europe. *Ann. Geophys.* **2**: 211–20.

Trakhtengerts, V.Y., and Rycroft, M.J. 2000. Whistler-electron interactions in the magnetosphere: new results and novel approaches. *J. Atmos. Solar-Terr. Phys.* **62**: 1719–33.

Tsuruda, K. 1973. Penetration and reflection of VLF waves through the ionosphere: full wave calculations with ground effect. *J. Atmos. Terr. Phys.* **35**: 1377–405.

Tsuruda, K., and Hayashi, K. 1975. Direction finding technique for elliptically polarized VLF electromagnetic waves and its application to the low latitude whistlers. *J. Atmos. Terr. Phys.* **37**: 1193–202.

Voss, H.D., *et al.* 1984. Lightning-induced electron precipitation. *Nature* **312**: 740–2.

Wait, J.R. 1970. *Electromagnetic Waves in Stratified Media*, Elmsford, New York: Pergamon Press.

Walker, A.D.M. 1968a. Ray tracing in the ionosphere at VLF – I. *J. Atmos. Terr. Phys.* **30**: 403–9.

Walker, A.D.M. 1968b. Ray tracing in the ionosphere at VLF – II. *J. Atmos. Terr. Phys.* **30**: 411–21.

Walker, A.D.M. 1976. The theory of whistler propagation. *Rev. Geophys. Space Sci.* **14**: 629–38.

Walker, A.D.M. 1978. Formation of whistler ducts. *Planet. Space Sci.* **26**: 375–9.

Walter, F., and Angerami, J.J. 1969. Nonducted mode of VLF propagation between conjugate hemispheres: observations on OGOs -2 and -4 of the "walking-trace" whistler and of Doppler shifts in fixed frequency transmissions. *J. Geophys. Res.* **74**: 6352–70.

Watts, J.M. 1959. Direction finding on whistlers. *J. Geophys. Res.* **64**: 2029–30.

Yeh, K.C., and Liu, C.H. 1972. *Theory of Ionospheric Waves*, New York and London: Academic Press.

Yip, W.-Y., Inan, U.S., and Orville, R.E. 1991. On the spatial relationship between lightning discharges and propagation paths of perturbed subionospheric VLF/LF signals. *J. Geophys. Res.* **96**: 249–58.

14 Lightning effects in the middle and upper atmosphere

> ... it is quite possible that a discharge between the top of the cloud and the ionosphere is a normal accompaniment of a lightning discharge to earth... and many years ago I observed what appeared to be discharges of this kind from a thundercloud below the horizon. There were diffuse, fan-shaped flashes of greenish colour extending up into a clear sky.
>
> C.T.R. Wilson (1956)

14.1. Introduction

For over a century prior to the scientific recordings of the 1990s, there were reports of visually observed lightning-like channels, pillars of light, and diffuse, upward-propagating, optical phenomena above thundercloud tops, sometimes apparently extending between the cloud tops and the ionosphere (e.g., Everett and Everett 1903; Boys 1926; Malan 1937; Ashmore 1950; Wright 1950; Wood 1951; Wilson 1956; Powell 1968; Corliss 1977, 1983; Vonnegut 1980; Vaughan and Vonnegut 1982; Vonnegut et al. 1989; Vaughan and Vonnegut 1989; Fisher 1990; Hammerstrom 1993). On theoretical grounds the existence of discharges in the rarified air between the thundercloud top and the lower ionosphere had been predicted by, for example, Wilson (1925a, b, 1956), Hoffman (1960), and Cole et al. (1966), all of whom viewed runaway electrons as the potential causative mechanism. Runaway electrons are defined as electrons whose initial energy is high enough that they gain more energy from the ambient electric field between collisions with air atoms than they lose in the collision process, and hence they continue to gain energy and increase in speed (Section 14.5). The first recording of luminous phenomena above cloud tops was serendipitously obtained in 1989 by Franz et al. (1990), while testing a new low-light video system. Since 1990 there has been considerable progress in identifying, characterizing, and modeling the variety of low-luminosity optical phenomena that occur in the clear air above thunderstorms. Three general types of transient optical phenomena have been observed: red sprites, blue starters and blue jets, and elves. The location and appearance of sprites, jets, and elves are illustrated in Fig. 14.1. Blue starters and blue jets propagate upward from the top of the cloud, generally at a height of 20 km or less, in the form of blue, cone-shaped structures. Starters propagate upward less than 10 km, jets up to 20 km. Starters and jets are probably variations of the same phenomenon. Apparently, neither is associated with individual cloud-to-ground lightning. Red sprites are most luminous between 40 and 90 km heights and often have faint bluish tendrils extending downward from 50 km or so to altitudes as low as 20 km. Red sprites exhibit a diversity of forms and features and generally occur in association with the larger positive ground flashes (Chapter 5). Elves occur near a height of 90 km, in the lower ionosphere. Elves accompany many lightning flashes and are manifested as circles of light expanding radially across the bottom of the ionosphere from a point above the causative lightning flash. They are likely to be due to the radiation fields of return strokes and perhaps other impulsive processes, such as initial breakdown (subsections 4.3.2 and 9.4.2), that accelerate the electrons in the lower ionosphere.

With or without connection to the optical phenomena discussed above, lightning may produce X-rays, gamma-rays, and runaway electrons, which could be important in coupling energy from the lower atmosphere into the middle atmosphere (defined in Section 15.1) and the ionosphere. Some of the phenomena discussed above change the properties of the lower ionosphere and hence alter the radiowave propagation involving the lower ionosphere, effects well documented at VLF. Further, lightning-produced whistlers (Section 13.5) propagating in magnetic field ducts in the Earth's magnetosphere can cause the precipitation of electrons from those ducts, changing the electron density in the ionosphere and again altering the radiowave propagation conditions. Finally, thundercloud electric fields penetrate the ionosphere and, although these fields are relatively small, they still produce ionospheric heating and potentially can produce a measurable infrared glow at the bottom of the ionosphere. These thundercloud-field-induced changes in the ionosphere may well also affect the propagation of radio waves involving the lower ionospheric boundary.

14.2. Upward lightning channels from cloud tops

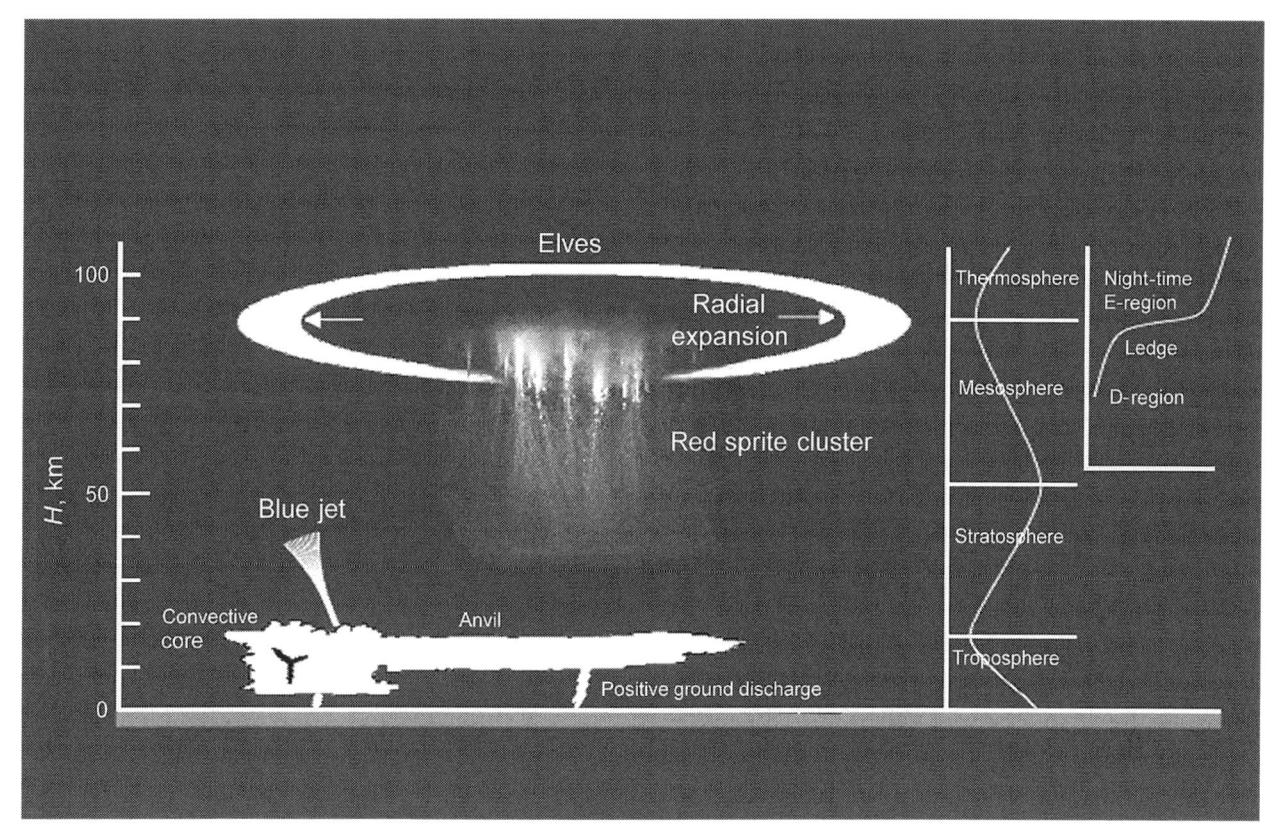

Fig. 14.1. The geometry and environment of blue jets, red sprites, and elves. Plotted horizontally: left-hand inset curve, temperature; right-hand inset curve, electron density, both versus height. Adapted from http://sprite.gi.alaska.edu/html/sprites.htm, with the permission of D.D. Sentman.

In this chapter we first consider ordinary lightning channels extending above the cloud tops (Section 14.2). We then examine the more diffuse blue starters, blue jets, and red sprites that occur in the region above the cloud tops and below the ionosphere (Section 14.3), and the elves that occur in the lower ionosphere (Section 14.4). Next, we consider the thunderstorm as an X-ray, gamma-ray, and runaway-electron producer (Section 14.5). Finally, we examine the interaction of both lightning and thunderstorm electric fields with the ionosphere and the magnetosphere, including the modification of the characteristics of radio-wave propagation and the infrared glow (Section 14.6).

14.2. Upward lightning channels from cloud tops

Upward lightning channels that are similar in appearance to those of the usual cloud-to-air discharges from the sides and bottom of a thundercloud but project upward from the tops of the clouds have been reported, for example, by Everett and Everett (1903), Powell (1968), Wood (1951), Vaughan and Vonnegut (1989), Vonnegut et al. (1989), and Wescott et al. (1995). Viewing conditions are rarely conducive to seeing such discharges from the ground and, if they are seen, it is probably not possible to tell whether such a discharge is truly extending upward from a cloud top or, rather, is horizontally extending from another cloud location toward the observer. Vaughan and Vonnegut (1989) published 15 reports of above-cloud upward lightning flashes sighted from different aircraft at altitudes where viewing was far more favorable than from the ground. Most observations of lightning in the clear air above a storm have been associated with storms having large vertical development. Most reports refer to lightning that is apparently similar to the lightning seen emerging from the sides and bottom of a thundercloud, that is, discharges whose current is confined to a channel of relatively small diameter, probably of the order of centimeters. Some of the reports may be describing more diffuse optical phenomena, but this is often unclear from the narratives presented. Ordinary lightning channels above thunderclouds have been photographed at night from a U-2 aircraft flying at 20 km altitude, as reported by Vonnegut et al. (1989), and two dozen such channels were recorded on videotape from a research aircraft by Wescott et al. (1995). These lightning channels appear to be of short length relative to the more optically diffuse discharge phenomena discussed next.

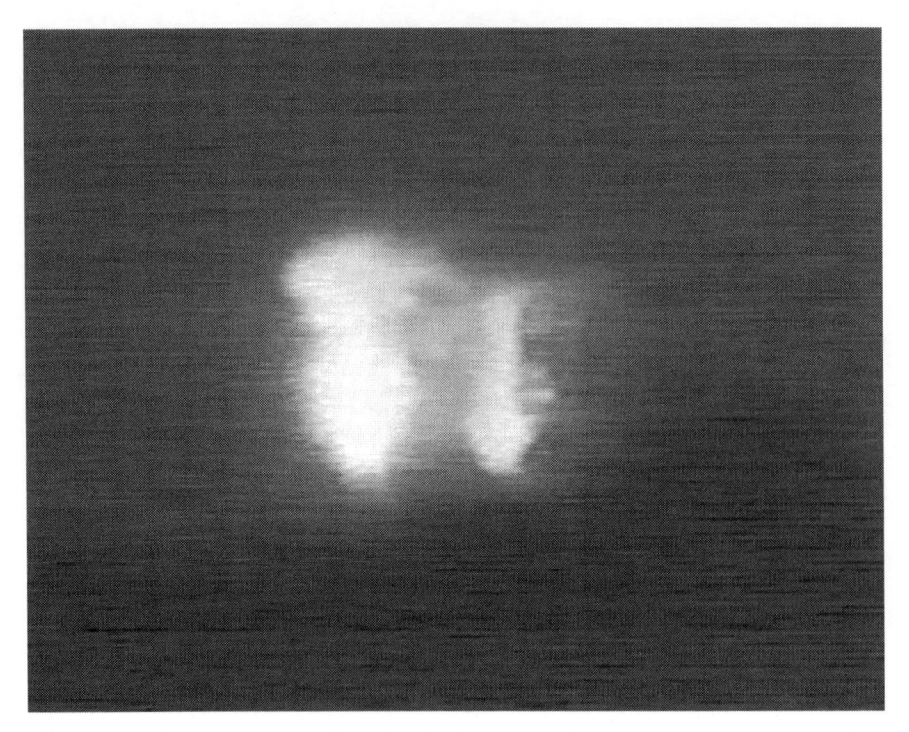

Fig. 14.2. The first photograph of a sprite, from Franz et al. (1990).

14.3. Low-luminosity transient discharges in the mesosphere

In 1989, Franz et al. (1990) obtained a low-light-level black and white video image of two diffuse luminous discharges occurring over a thunderstorm 250 km away. The discovery was made fortuitously in the course of another research project (Winckler et al. 1993). The discharges, shown in Fig. 14.2, were 20 km in vertical extent and were located at a considerable distance above a 14 km cloud top. The diameter of one discharge was about 10 km. This observation of Franz et al. (1990) initiated a new and important area of lightning research. Old eyewitness reports and predictions of luminous discharges above cloud tops were re-examined, and new research was initiated to measure and characterize the observed phenomena, termed "sprites" in 1993 by D. Sentman (Lyons, 1994a). In the process, other faintly luminous transient phenomena with physical characteristics different from sprites, that is, blue starters, blue jets, and elves, were observed and studied. Blue starters (subsection 14.3.1), blue jets (subsection 14.3.2), and red sprites (subsection 14.3.3) are considered in this section. They all take place in the region between the cloud tops and the lower ionosphere. Elves, which occur in the lower ionosphere, are discussed in Section 14.4.

14.3.1. Blue starters

Blue starters were first observed in 1994, and those observations and others made in 1998 have been described and analyzed by Wescott et al. (1996, 1998b, 2001a). Twenty-eight blue starters that occurred above an active Arkansas thunderstorm in 1994 were recorded on videotape from low-light-level monochromatic and color TV cameras by Wescott et al. (1996). A few more blue starters from the same storm were added to the data set by Wescott et al. (1998b). The geometry of the Arkansas blue starters was determined by triangulation from two airplanes (Sentman et al. 1995). Wescott et al. (2001a) gave the spectral properties and the total light output from blue starters observed over two thunderstorm cells in Minnesota in 1998.

Wescott et al. (1996) found the lengths of the blue starters to be less than 10 km, typically a few kilometers. The mean altitude of the origin of the starters was about 18 km with standard deviation 0.9 km. The blue starter tops extended to 18–26 km above ground level. One pixel of the video sensor used by Wescott et al. (1996) represented 230×230 m at the source, defining the spatial resolution of the measurement. Starters propagated upward with a speed that varied widely, from 2.7×10^4 to 1.5×10^5 m s^{-1}. Overall, the blue-starter speeds were similar to the observed range of 15 blue-jet speeds in the same study, 7.7×10^4 to 2.2×10^5 m s^{-1} (see subsection 14.3.2). In most cases the speed of the starters decreased as they propagated upward. The starters were not coincident with cloud-to-ground lightning but occurred in the same general area. The cloud-to-ground flash rate within 50 km of a blue starter decreased significantly for about 3 s following the blue starter.

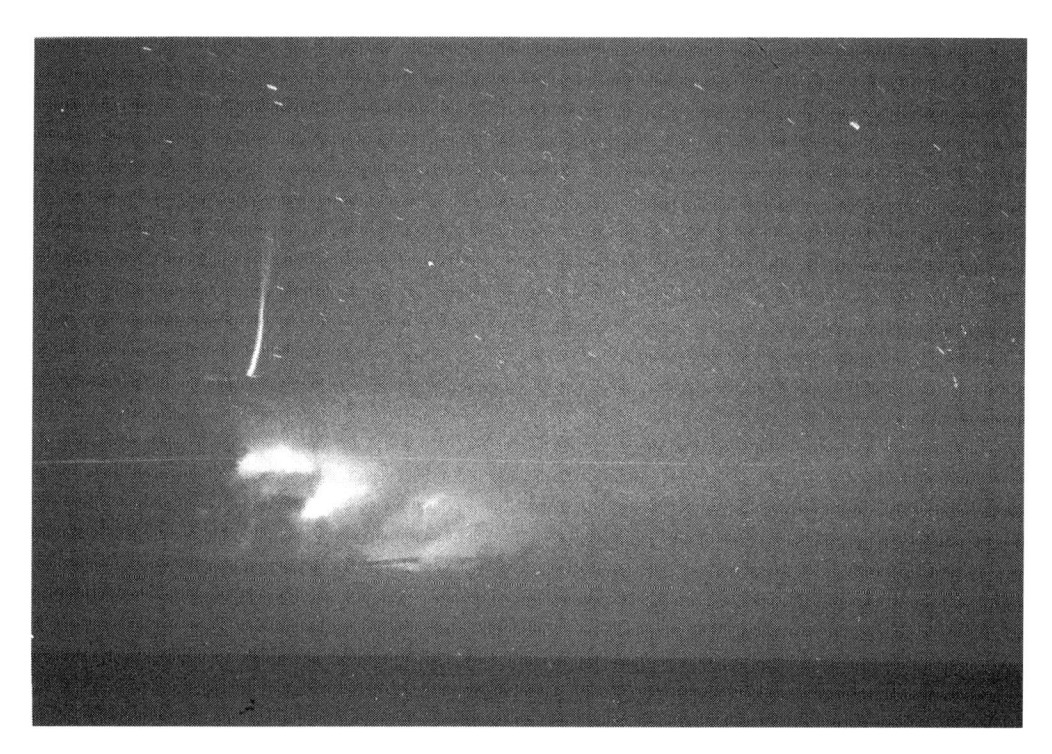

Fig. 14.3. At the time of going to press a colour version of this figure was available for download from http://www.cambridge.org/9780521035415. Photograph of a blue jet over a large thunderstorm in the Indian Ocean north of Réunion Island, taken by Patrice Huet in March 1997. The photograph was a two minute time exposure with 400 speed film at f/4 or f/6. The base of the blue jet above the top of the clouds is at about 19 km altitude, and the jet tip is at 40 km. Wescott et al. (2001a) estimated the light intensity to be over 7 MR. On the original photograph, the stem of the jet has faint upward directed branches. Adapted from Wescott et al. (2001a), who described and analyzed the photograph.

Wescott et al. (1996) viewed blue starters as a phenomenon distinct from the upward lightning channels discussed in Section 14.2, which they also observed (Wescott et al. 1995), but similar to the blue jets discussed in the next subsection, except that they are significantly shorter in length.

According to Wescott et al. (1996), the brightness of the Arkansas blue starters near their points of origin was of order one million rayleigh (1 MR), a result confirmed by Wescott et al. (2001a) for the Minnesota starters. Wescott et al. (2001a) found that only 3 percent of the light output of the starters was from ionized N_2, a result they interpret to indicate that the starters "and by association also blue jets" are "at least partially ionized".

Westcott et al. (1996) observed that most of the Arkansas blue starters originated from one thundercloud anvil, but none originated from a 1 km high turret rising above that anvil. The average blue starter occupied a volume 2.4×10^9 m^3. Wescott et al. (1996) presented evidence that very strong updrafts were present in the areas where blue starters were observed.

Since blue starters and blue jets apparently differ only in their maximum altitudes, any theory to describe the physics of one is probably applicable to the other. Three such theories are presented in the next subsection.

14.3.2. Blue jets

Blue jets were first observed in 1994 in the same overall research program in which blue starters were discovered (Wescott et al. 1995, 1998b, 2001a). A color photograph of a blue jet is found in Fig. 14.3 and a sequence of four images of blue-jet formation simultaneously videotaped from each of two aircraft is given in Fig. 14.4. Figure 14.4 illustrates one of 56 blue jets that were located by triangulation in the 1994 Arkansas storm from two aircraft equipped with low-light-level TV cameras by Wescott et al. (1995). Wescott et al. (1998b) subsequently reclassified five of these as blue starters. The average rate of blue-jet occurrence during the first 22 minutes of the 30 June 1994 Arkansas storm was about 2.8 per minute. The blue jets propagated upward from the tops of thunderclouds at typical speeds of order 10^5 m s^{-1} to terminal altitudes between 33 and 43 km. A typical blue jet propagated upward and outward in a cone-like structure with apex angle between 6.5° and 31.5°, with a mean angle of 14.7°. There was a flare-out at the top of the cone, so that the whole jet resembled a vertical trumpet. Most jets faded away along the whole path simultaneously after about 200 ms of luminosity. All the jets observed were more or less vertical. In some cases upward lightning channels preceded a blue jet by a

Fig. 14.4. A sequence of blue-jet images taken simultaneously from two aircraft. Adapted from Wescott et al. (1995).

few TV frames, the jet originating from the same location as the channel. The maximum jet brightness was estimated to be near 0.5 MR at the base of the jet, decreasing to about 7 kR at the top (Sentman and Wescott 1995).

Wescott et al. (1995) pointed out that the upward speed of a blue jet, about 10^5 m s^{-1}, is similar to that of a stepped leader between the cloud base and ground, but they see no discernable structure or evidence of the stepping or branching that is characteristic of stepped leaders (Section 4.4). Some blue jets exhibited a faint hemispherical "shock front" ahead of the diffuse leading edge.

Rumi (1957) had previously interpreted 28 MHz radar returns as due to upward discharges from the tops of thunderclouds, and Wescott et al. (1995) suggested that the velocity, terminal altitude, and duration of these radar measurements are consistent with the properties of blue jets. Radar returns at 24.4 MHz possibly due to a high-altitude discharge process were discussed by Tsunoda et al. (1998).

Wescott et al. (1998b) performed additional analysis on the blue-jet data of Wescott et al. (1995). Wescott et al. (1998b) reported that the blue jets were not coincident with positive or negative ground flashes but that they occurred in the same general area as negative ground flashes and large hail. The rate of negative ground flashes within 15 km of blue jets increased within a second or so prior to the jets and decreased in the 2 s after, as opposed to situation for blue starters, where there was a flash-rate decrease afterwards but no increase beforehand. For the example that Wescott et al. (1998b) presented, in the 1 s prior to 27 blue jets there were a total of about 10 negative cloud-to-ground lightning events as determined from NLDN records (Section 17.5) whereas in the 2 s following the jets there was a total of two or three such events.

Wescott et al. (2001a) gave low-light-level TV images of a blue jet whose upward-propagating luminosity separates from the stem of the jet and continues to propagate upward. The jet starts at $t = 0$, propagating at 23 km s^{-1}, and brightens at 133 ms. At about 200 ms, the top part of the jet separates and propagates upward at 91 km s^{-1} for an additional 133 ms while the stem luminosity decreases.

Several models have been proposed to explain the physics of blue jets. Any successful theory should explain the jet color, brightness, shape, propagation speed, and duration. Wescott et al. (1995) considered and dismissed a runaway-electron mechanism that could produce a cone-like shape but not the observed propagation speed or overall luminous duration. Sukhorukov and Stubbe (1998) specifically criticized the runaway breakdown theory applied to blue jets by Roussel-Dupré and Gurevich (1996) and Taranenko and Roussel-Dupré (1996). Two research groups have presented models for blue jets based on conventional breakdown theory: Pasko et al. (1996a), and Sukhorukov et al. (1996a) and Sukhorukov and Stubbe (1998). Sukhorukov et al. (1996a) and Sukhorukov and Stubbe (1998) treated the blue jets, and blue starters, as an "attachment-controlled ionization wave" that carries negative charge upward by way of electron avalanches. The electric field change driving the electron avalanches was assumed to result from the rapid disappearance of a large, over 100 C, positive charge at the cloud top caused by an intracloud discharge or a positive cloud-to-ground discharge. Wescott et al. (1998b) found this approach questionable in view of the fact that they had observed no positive cloud-to-ground or significant intracloud flashes preceding blue jets on their video records. However, as noted above, ordinary upward lightning channels of short duration occasionally have preceded the initiation of blue jets by a few hundredths of a second (Wescott et al. 1995), so the association of the jets with some form of lightning discharge cannot be ruled out. Sukhorukov (1996) and Sukhorukov and Stubbe (1998) gave order of magnitude estimates for optical emissions, upward propagation speed, terminal height, and other jet properties based on their model. Pasko et al. (1996a) assumed that a positive charge of 300 to 400 C accumulates in a dish-like volume at 20-km height by normal cloud charging processes. The resulting electrical breakdown was treated as a "streamer-type ionization channel" using the model of Pasko et al. (1995), which may not differ much from the approach of Sukhorukov et al. (1996a) and Sukhorukov and Stubbe (1998) except for the postulated source and the fact that the latter researchers considered a negatively charged upward discharge while Pasko et al. (1996a) considered a positively charged upward discharge. Both models require charges of a magnitude previously thought unreasonably large to be concentrated at the cloud top for some fraction of a second in order to produce the electric fields necessary for breakdown. Sukhorukov and Stubbe (1998) discussed in detail the differences between their theory and that of Pasko et al. (1996a). Wescott et al. (1998b) pointed out that they had not observed the blue-jet spectral emission predicted by Pasko et al. (1996a); whereas only blue wavelengths were predicted in the model optical radiation of Pasko et al. (1996a), significant amounts of red and green optical radiation were observed by Wescott et al. (1998b).

14.3.3. Red sprites

As noted earlier, the first recording of sprites, reproduced in Fig. 14.2, was obtained by Franz et al. (1990) using a low-light-level TV camera. The next sprite recordings were obtained in 1990 and 1991 with Space Shuttle TV cameras, blue jets and elves apparently having been recorded also (Vaughan et al. 1992; Boeck et al. 1992, 1995, 1998). Sprites have been observed in dedicated experiments, both airborne and ground-based, by, for example, Sentman and Wescott (1993), Lyons (1994a, b), Sentman et al. (1995), Rairden and Mende (1995), Winckler (1995, 1998), Winckler et al. (1996), Lyons (1996), Wescott et al. (1998a), Cummer et al. (1998), Stanley et al. (1999), Gerken et al. (2000), Hardman et al. (2000), Hobara et al. (2001), Cummer and Füllerkrug (2001), Barrington-Leigh et al. (2001), and Wescott et al. (2001b). There are as many or more papers which attempt to model sprites as there are papers reporting their measured characteristics. Sprite models are given, for example, by Milikh et al. (1995, 1998a, b), Pasko et al. (1995, 1996b, 1997a, b, 1998b), Fernsler and Rowland (1996), Taranenko and Roussel-Dupré (1996), Lehtinen et al. (1997), Roussel-Dupré et al. (1998), Valdivia et al. (1997, 1998), Cho and Rycroft (1998). Rycroft and Cho (1998), Raizer et al. (1998), Yukhimuk et al. (1998, 1999), Veronis et al. (1999), and Barrington-Leigh et al. (2001).

According to Sentman and Wescott (1996), from 1989, the time of the first sprite image, through 1995 several thousands of images of sprites were recorded with low-light-level video systems and probably the rate of observation has increased since 1995. The early reports referred to the observed transient luminous structures by a variety of names, but they are now universally called red sprites or sprites, "a term that is succinct and whimsically evocative of their fleeting nature and doubly avoids making unwarranted implications about physical processes that have yet to be fully worked out" (Sentman and Wescott 1993).

Observations Sprites are low-luminosity transient discharges that are brightest at altitudes of 40 to 90 km and occur over large thunderstorm systems (e.g., Sentman et al. 1995; Winckler 1995; Lyons 1996). They are primarily associated with positive cloud-to-ground strokes that transfer relatively large quantities of positive charge to ground (e.g., Cummer and Inan 1997; Boccippio et al. 1995), although sprites associated with relatively large negative cloud-to-ground discharges have been observed (Barrington-Leigh et al. 1999). Various terms have been used to describe the diversity of forms of, and features within, sprites as observed primarily on standard-frame-rate (16.7 ms field), low-light-intensity video. A "carrot" sprite is characterized by a relatively bright "head" region of red color, typically at 65 to 75 km altitude, with a red glow or wispy structures ("hair") that extends above the head upward to as high as 95 km, and bluish tendrils below the head often extending downward to as low as 40 km, sometimes apparently reaching cloud anvil tops near 20 km (Sentman et al. 1995), as illustrated in Fig. 14.1. An "angel" sprite also has tendrils, but the head is capped by a diffuse glow, called a "sprite halo" by Barrington-Leigh et al. (2001), who showed that only the brightest of such halos are identifiable on standard-frame-rate, low-light-intensity video. Further, they showed that without high-speed (3000 frames per second) video it is difficult to discriminate between sprite halos, Rayleigh-scattered light from lightning, and elves, all of which occur above the main body of the sprite. "Columniform" sprites are vertically oriented cylinders that do not often show tendrils or hair-like structure (Wescott et al. 1998a). An example is shown in Fig. 14.5. Sprites that assume one of the above forms (or other forms not mentioned here) are almost always part of a "sprite cluster" created by a single positive flash, as illustrated in Figs. 14.1, 14.2, 14.5, and 14.6b. Sentman et al. (1995) described a range of sprite structures,

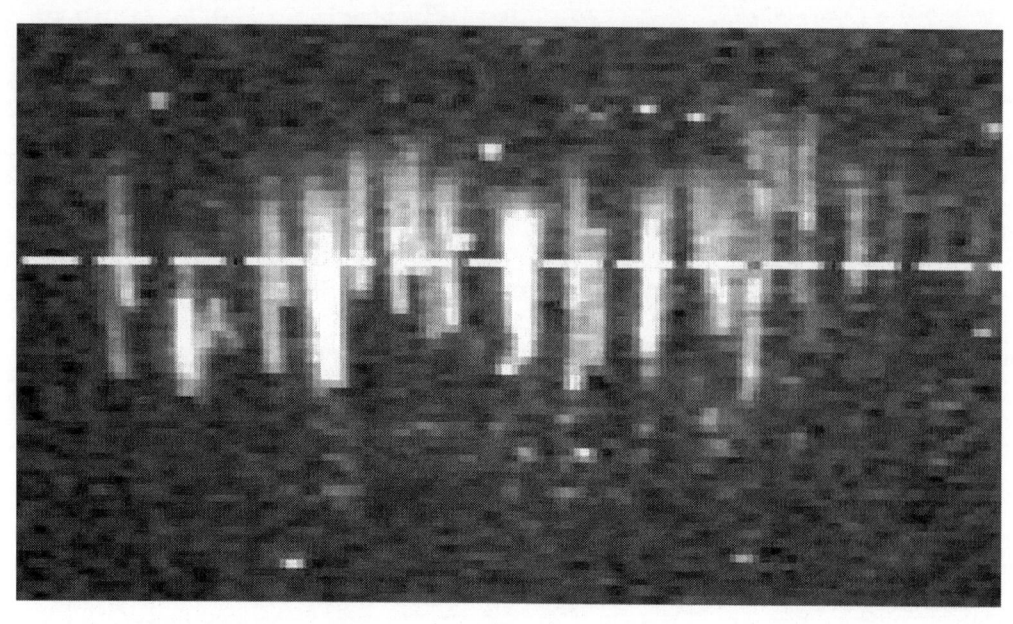

Fig. 14.5. An intensified CCD TV frame showing columniform sprites on 19 June 1996. Adapted from Wescott *et al.* (1998a).

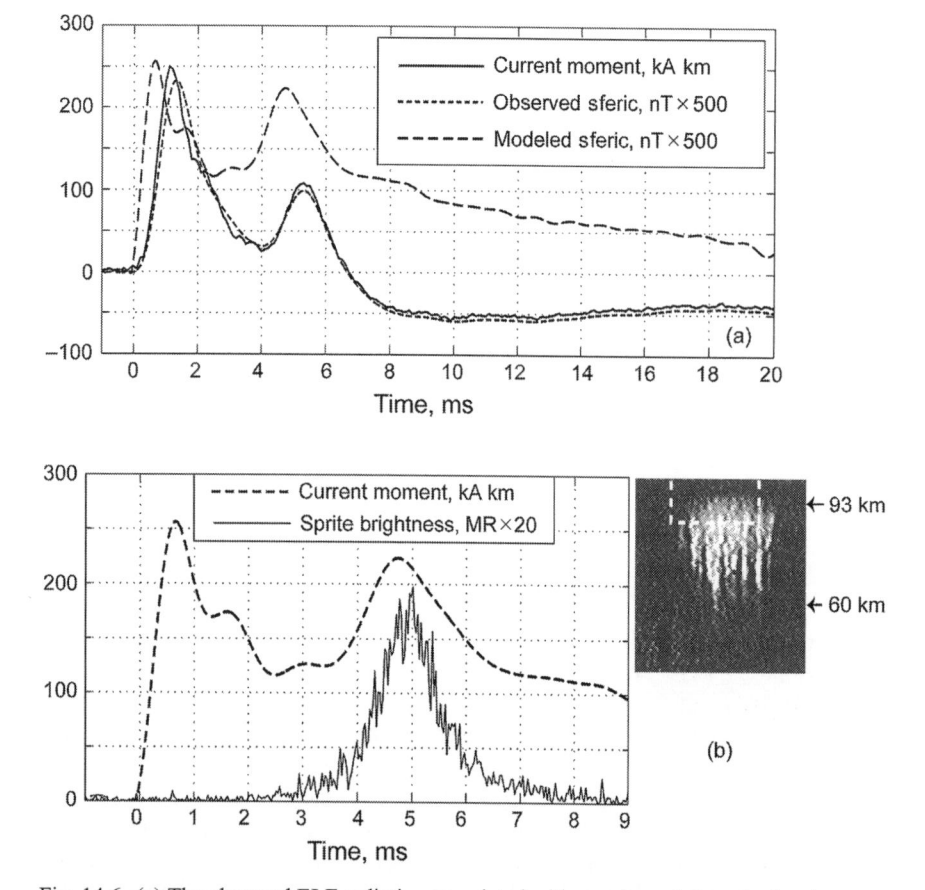

Fig. 14.6. (a) The observed ELF radiation associated with a sprite and the calculated current moment and model sferic; (b) the sprite brightness and calculated current moment. The broken-line box in the photograph shows the field of view of the photometer. Adapted from Cummer *et al.* (1998).

from those as simple as a small single spot to complex structures containing many horizontally and vertically separated elements.

According to Wescott et al. (1998a), columniform sprites are luminous vertical columns about 10 km in height and less than 1 km in diameter, as illustrated in Fig. 14.5. Columniform sprites exhibit a uniform brightness along their length. Some show hair above and tendrils below the main column. The high-speed video records of Stanley et al. (1999) indicated that all types of sprite begin as a vertical columnar form except for a few for which the column appears to grow from a single point. Stanley et al. observed that the initial development of sprites, starting from a columniform shape near 75 km, appears to be dominated by upward- and downward-propagating corona-like streamers with speeds in excess of 10^7 m s^{-1}. Gerken et al. (2000) analyzed telescopic images of sprites and found, for isolated columns, individual luminous elements with transverse dimensions ranging from tens of meters to a few hundred meters at 65 to 80 km altitude.

Clusters of sprites were observed by Sentman et al. (1995) to extend across horizontal distances of over 40 km and to occupy overall volumes in excess of 10^4 km^3. To the naked dark-adapted eye most sprites were reported to be brief, barely detectable, luminous events. The average optical intensity of the sprites observed by Sentman et al., as determined from standard-frame-rate, low-light-level video records, was reported to be comparable to a moderately bright aurora, 25 to 50 kR, with a maximum recorded brightness of 600 kR. It is likely that these values are underestimates of the actual optical intensity since the bright part of a sprite usually occupies only a fraction of the TV field exposure time. Cummer et al. (1998) reported photometer-measured sprite brightness profiles of about 1 ms width and peak brightness values of 10, 1, and 0.5 MR for three events.

Barrington-Leigh et al. (2001) reported that "sprite halos", the diffuse part of the sprite at about 70 to 85 km height, higher than the structured portion, as observed by high-speed image-intensified video and photometers, exhibited brightnesses from tenths of MR to tens of MR. Both Cummer et al. (1998) and Barrington-Leigh et al. (2001) measured brightness using an array of photometers, an instrument called the Fly's Eye, whose spectral passband was 650 to 780 nm and whose time resolution was 30 μs. According to the modeling of Barrington-Leigh et al., the spectral response of the photometer allowed the measurement of only 15 percent of the total optical output of the sprite or the sprite halo. Hence the actual brightness might be nearly an order of magnitude greater than that measured. Barrington-Leigh et al. also indicated that sprites have optical durations that vary greatly, ranging from a few to many tens of milliseconds, while sprite halos have a duration of only about 1 ms and elves have a duration less than 1 ms.

Wescott et al. (2001b) report that sprite halos tend to be centered directly above the causative lightning discharges while the associated sprites are laterally offset by as much as 50 km. They found, for four halos, a center altitude of about 78 km, a thickness of about 4 km, and a diameter of about 66 km.

Winckler et al. (1996), in a continuation of the sprite study of Winckler (1995), reported that all elements of the sprite cluster appear suddenly and have a duration of no more than a few milliseconds. The duration of the sprite light output was roughly 3 ms on their photometer records, whereas the sprite durations on their video records were often several 16.7 ms fields. The difference is probably attributable to the different spectral responses of the two instruments and to the imaging of different portions of the sprite by the different instruments.

Spectral measurements of sprites by Mende et al. (1995) and Hampton et al. (1996) showed that the first positive band of nitrogen produces the observed red light. Measured details of the red spectra and modeling of its atmospheric attenuation were given by Morrill et al. (1998). Measurements of the blue spectrum, notably the N_2 second positive 399.8-nm emission and the N_2^+ first negative 427.8 nm emission, were given by Suszcynsky et al. (1998) and Armstrong et al. (1998), who stated that the blue emission precedes the red and is due to initial energetic ionizing events followed by red emission from secondary processes.

Sprites are associated with luminosity at the cloud top that in turn is almost always associated with the larger positive cloud-to-ground lightning strokes, the exception to date being two sprites observed by Barrington-Leigh et al. (1999), which were produced by negative cloud-to-ground strokes exhibiting unusually large vertical charge-moment changes, up to 1550 C km in 5 ms. Additionally, Barrington-Leigh et al. (2001) reported that lightning discharges of both polarities produce sprite halos with a duration of about 1 ms but that the lower, structured part of the sprite is not often produced by negative discharges. Boccippio et al. (1995) inferred the connection between sprites and large positive-stroke currents by correlating the time of occurrence of 97 sprites with ground flash times from the NLDN (Section 17.5). They studied two mesoscale convective systems in summer 1994 and found that 86 percent and 78 percent, respectively, of the sprites were coincident both with positive ground strokes occurring in the stratiform region of mesoscale convective systems (MCSs) and with Schumann resonance transients in the ELF (Section 13.4). The detection efficiency of the NLDN was probably near 70 percent in 1994 (Section 17.5). The peak currents of these positive strokes, as estimated by the NLDN, were near the upper end of the overall positive peak current distribution. The median current of sprite-producing positive strokes was found to be twice that of non-sprite-producing positive strokes.

Lyons (1996), continuing his earlier sprite study (Lyons 1994a, b), found that the majority of sprites in the US mid-west occur not near the high-reflectivity core of the observed MCSs but above the large stratiform precipitation region and that 34 of the 36 sprites imaged were associated with large positive ground strokes. The NLDN did not respond at the time of the other two sprites, perhaps owing to its less than perfect detection efficiency.

Cummer and Inan (1997) analyzed six ELF (below 2-kHz) electromagnetic signals recorded 1800 km from sprite-producing positive strokes in the US mid-west, to estimate the charge transfer by those strokes. As discussed later (e.g., Cummer et al. 1998; Reising et al. 1999), there is likely to be a component of the overall field waveform that is due to the sprite current rather than to the stroke current, but the studies discussed in this paragraph attribute the total waveform to the current and the resultant charge transfer in the causative positive ground flash. Note that the calculations of current flow and charge transfer from distant field measurements are dependent on a propagation model and so could be subject to errors. Charge transfer from 25 to 325 C were found during the first 5 ms of the positive strokes, charge-transfer values of 145 and 325 C being associated with the two brightest sprites. This work supports the view of Reising et al. (1996) that large charge transfer to ground via continuing currents, as inferred from the large ELF "slow tails" (Sections 13.3 and 13.4) following the initial VLF signal radiated by the return stroke, is responsible for the sprite-producing electric field. In a continuation of the study of Cummer and Inan (1997), Bell et al. (1998) used an improved propagation model and ELF–VLF measurements on 17 waveforms from positive flashes at relatively close range, 600 km, at which distance optical observations of the sprites were also made, to support their view that "intense continuing currents" of about 1 ms duration are responsible for most of the positive charge transfer to ground that precedes the appearance of (and most likely produces) sprites. An additional discussion of ELF propagation and the determination of the lightning currents and charge transfers from measured slow-tail waveforms is found in Cummer and Inan (2000a). Usually a current of millisecond duration is not called a continuing current but rather is considered part of the return-stroke current tail (Sections 4.8 and 5.5), but this is an issue of semantics, not of physics. For one detailed example, Bell et al. (1998) calculated a peak current of 180 kA with a charge transfer of 110 C during a 1 ms duration, a discharge length of 10 km being assumed. For the 17 events, the calculated peak continuing current varied from 6 to 180 kA and the charge transfer from 10 to 112 C. For four events, the currents lasted longer than 15 ms, including the three with the largest charge transfers and the shortest time delays between the NLDN-reported time of the stroke and the time of the sprite's optical appearance. This time delay was longer for smaller events, leading Bell et al. (1998) to suggest that sprites may be produced in two ways: (i) by the rapid removal of about 100 C of positive charge via a ground flash, leading to a short time delay (within a video field of about 16.7 ms), or (ii) by the initial removal of 10 to 50 C of positive charge, attributed to a ground discharge, followed by a slower charge transfer of up to a total of 100 C attributed to an intracloud discharge and leading to a longer time delay, greater than 40 ms, with an observed maximum of about 120 ms.

Since sprites must have some level of internal current flow in order to be rendered luminous, they themselves must radiate electromagnetic energy. Farrell and Desch (1992) calculated the RF electric field spectra expected from sprites assuming that sprite current and optical emission rose to a peak in 10 ms and decayed in about the same time; they obtained maximum electromagnetic fields in the 10 Hz range. However, data presented earlier indicate that the optical risetime is much faster than the 10 ms assumed. Hale (1993) criticized the calculation of Farrell and Desch on the basis that the calculated fields do not couple correctly to the earth–ionosphere waveguide; hence he claimed that the calculated fields were too small. Farrell and Desch (1993) argued against this view.

The ELF signatures of sprite-producing events appear to have a unique feature (Reising et al. 1999) that can be interpreted as due to the sprite current, as discussed further below. An example of such an ELF waveform is found in Fig. 14.6a. It follows that sprites can apparently be detected from their ELF waveforms without optical imaging of the sprite.

Cummer et al. (1998) presented the first experimental evidence that current flowing in a sprite produces ELF radiation, this radiation being comparable to the ELF radiation produced by the causative positive discharge. They showed that peaks in the observed ELF waveforms occurring some milliseconds after the initial VLF sferics signal are coincident with the sprite's peak optical brightness, as illustrated in Fig. 14.6a, b, and thereby infer their causal relationship. The current moments of this optically coincident ELF peak from three separate sprite-clusters were 100 to 200 kA km, yielding peak sprite current estimates of 1 to 4 kA, with charge transfers between 5 and 42 C. The average horizontal dimension for the three sprite clusters (see Fig. 14.6b for one example) was 35 km, yielding a current density j of order 3 μA m^{-2}. For a conductivity σ of 10^{-7} S m^{-1} (10^3 electrons per cm^3) at 80 km, this current density requires an electric field E of 33 V m^{-1} via Ohm's law, $\mathbf{j} = \sigma \mathbf{E}$. This field magnitude is consistent with the removal of a large positive charge at cloud top height. However, at lower altitudes, where sprites are also observed to be luminous, the conductivity is lower (e.g., 10^{-9} S m^{-1} at 60 km) and hence unreasonably high fields (e.g., 3.3 kV m^{-1}

at 60 km) would be required to produce the observed current. Cummer and Füllekrug (2001) gave examples of three events for which the sprite optical emission was essentially coincident with a peak in the ULF (0.1 to 200 Hz) magnetic field signal measured 500–2000 km from the causative positive strokes. They calculated from the measured fields that, following the positive return stroke, a continuing current in the kiloampere range flowed and continued to flow during and after the three sprites were detected, both optically and by their ULF peaks, at times of about 40, 70, and 150 ms after stroke initiation.

Stanley et al. (2000) detected three day-time sprites via their characteristic ELF signatures. During the day-time the ionosphere extends downward into the region of night-time sprite initiation, so that day-time sprites must be initiated at lower altitudes where the air density is higher. Thus day-time initiation presumably requires larger electric fields and hence larger charge transfers by the causative lightning than night-time initiation. For the three daylight sprites, Stanley et al. calculated vertical charge-moment changes of 2800, 1200, and 910 C km, each being a significant fraction of the charge moment changes of 3900 to 6100 C km produced by the causative positive strokes that preceded the sprites by 11 to 13 ms.

Williams (1998) argued against the frequent assumption that the sprite-related positive charge is at the cloud top near 10 km and is associated with a model positive cloud-charge dipole (subsection 3.2.1). He presented evidence that extended MCSs have their positive charge reservoirs in the 4 to 6 km range. If the positive charge is at that height then charge transfers of several hundred coulombs, larger charge values than for higher altitude, are required for consistency between sprite theories and observation.

Suszcynsky et al. (1999) described the possible triggering of a sprite or aspects of the sprite luminosity by a meteor. The sprite developed in the immediate vicinity of the downward-traveling meteor as it reached about 70 km altitude. A wave of luminosity propagated up the ionization trail of the meteor during the late stages of the sprite.

Lyons et al. (1998a) showed that storms occurring in April to June 1998 into which smoke was ingested from forest fires in Mexico exhibited a higher than normal percentage of positive cloud-to-ground lightning, higher than average positive currents, and a higher than normal percentage of flashes that produced sprites (Section 5.2). An MCS over Nebraska on 20 May produced 4250 cloud-to-ground flashes, of which 62 percent were positive (normally about 10 percent), with a 40 kA average peak current (about 10 kA above normal); the data are from NLDN measurements (Section 17.5). The storm produced 380 sprites, three to 10 times more sprites per lightning than previously observed.

Following the work of Cummer et al. (1998), Pasko et al. (1998a) discussed how the sprite, in terms of its ELF signal and current, fits into the global electrical circuit (Section 1.4). Huang et al. (1999) discussed the relation between sprites and Schumann resonance observations (Section 13.4). Fukunishi et al. (1996) and Füllekrug et al. (1998) identified electric field tails with durations of the order of a second associated with the occurrence of sprite-producing lightning strokes. Füllekrug et al. found peaks in the frequency-domain signal at 0.67 Hz and 1.67 Hz and also related the ULF to ionospheric Alfven resonances. Cummer and Inan (2000a, b) discussed the determination of ionospheric electron density from measurement of the lightning ELF signal.

Sukhorukov and Stubbe (1997) argued that if the positive strokes producing sprites were to remove more than 100 C then they would have considerably larger upper ELF (> 300 Hz) magnitudes than observed. Their results are propagation-model dependent.

To study the lightning charges and fields that produce sprites, Marshall et al. (1996) made balloon-borne measurements of field changes (subsection 3.2.3) above MCS stratiform regions when positive lightning occurred. Between 10 and 16 km altitude, field changes of 1 to 4 kV m^{-1} were observed. The quasi-static field was 0.5 to 1.0 kV m^{-1}. Marshall et al. modeled the source of the field changes by uniformly charged cylindrical volumes having charge densities of 1 to 3 nC m^{-3}, a vertical height of 400 m, and diameters from 20 to 200 km. For a 20 km radius the total charge involved is $\pi \times 400 \times 10^6 \times 3 \times 10^{-9}$, or about 10 C, and for 200 km, 1000 C. They claimed that this result supports the suggestion of Boccippio et al. (1998) that sprites are caused by positive cloud-to-ground flashes that discharge a horizontally extensive charge region. Mazur et al. (1998) described "spider" lightning, which may serve that purpose by propagating horizontally near the cloud base for large distances during the decaying stage of a storm, although all sprite models require a relatively large charge transfer if the charge is at cloud top heights and an even larger charge transfer if the charge is located lower in the cloud (Williams 1998), as noted above.

Dowden et al. (1996) reported that rapid-onset, rapid-decay perturbations (a category of "early fast" perturbations) of subionospheric VLF propagation were coincident with one sprite, similar VLF perturbations having about the same occurrence rate as sprites. These perturbations were inferred to be due to scattering from the conducting bodies of sprites. Dowden et al. (2001b) presented additional experimental evidence and supporting theory for the scattering of VLF signals from conducting sprite columns and inferred a sprite electron density of about 10^{10} m^{-3} at 70 km altitude and about 10^{11} m^{-3} at 55 km. In another view, Inan et al. (1996b) discussed how similar scattering is consistent with ionospheric heating due to intense quasi-static fields. Barrington-Leigh *et al.* (2001) suggested that the diffuse upper regions of sprites, the sprite halos, may be responsible for the scattering. Johnson *et al.* (1999b) determined that the lateral extent of the ionospheric disturbance responsible for the rapid-onset, rapid-decay perturbations was 90 ± 30 km. The argument about the origin of these early/fast VLF perturbations is continued in Dowden *et al.* (1998, 2001a, b), Rodger *et al.* (1998a, b, c), and Barrington-Leigh *et al.* (2001), and is discussed further in subsection 14.6.1.

Theory Models of sprite generation involve the production of the required fields in the middle atmosphere by two mechanisms: (i) the electrostatic field produced when positive charge at the cloud top is moved to ground by a positive cloud-to-ground discharge (e.g., Pasko *et al.* 1996b, 1997a; Yukhimuk *et al.* 1998; Fernsler and Roland 1996) and (ii) the radiation field generated by a horizontal cloud discharge (e.g., Milikh *et al.* 1995; Valdivia *et al.* 1997, 1998).

Once the necessary fields are present above the cloud tops, two approaches have been used to account for the observed optical emission: (1) conventional breakdown via Townsend avalanches involving low-energy electrons in the relatively high ambient electric fields (e.g., Pasko *et al.* 1996b, 1997a, 1998b; Raizer *et al.* 1998) and (2) breakdown involving runaway electron avalanches that start with high-energy electrons produced by cosmic rays (e.g., Bell *et al.* 1995; Taranenko and Roussel-Dupré 1996; Lehtinen *et al.* 1996, 1997; Roussel-Dupré *et al.* 1998; Yukhimuk *et al.* 1998). In this latter approach, the secondary low-energy electrons produced by runaway-electron collisions with air molecules will necessarily be involved in some Townsend avalanching, depending on the ambient electric field. All the models require charge transfers and current magnitudes near the upper limit of the measured values. We now discuss them in turn.

(1) The *conventional-breakdown* approach (e.g., Pasko *et al.* 1997a) can be used to model the optical observations of the head of the sprite, at least its upper part, and its spatial structure (Pasko *et al.* 1998b). However, the air density and electric field values below 60 km or so do not allow breakdown at those altitudes without the generation from the sprite head of downward-moving space charge that produces a downward-propagating region of enhanced electric field. Such a space-charge-driven, downward-propagating electric field has been discussed by, for example, Fernsler and Rowland (1996) and modeled by Pasko *et al.* (2000). The observations of Stanley *et al.* (1999) of upward and downward corona-like streamers from a starting point near 75 km would appear to support this model. In the region of the sprite head the conductivity of the atmosphere is sufficiently high that any steady applied electric field will be canceled by the motion of free electrons in a time of order 1 ms; that is, the local relaxation time ε_0/σ (see Fig. 1.3), where ε_0 is the vacuum permittivity and σ is the local conductivity. Therefore any lightning process taking place on an appreciably longer time scale cannot produce a significant electric field in the region of the sprite head.

(2) A breakdown process that starts with *relativistic runaway electrons* (e.g., Lehtinen *et al.* 1996, 1997; Yukhimuk *et al.* 1998) can produce blue light from dissipation of the relativistic electron beam in the air, which results in a trail of secondary, low energy electrons in the atmosphere that facilitate further breakdown in a lower electric field than for conventional breakdown. Hence, blue tendrils can be modeled using this mechanism.

We now consider further the question of the source of the sprite-producing electric field. Milikh *et al.* (1995) proposed that the sprite source is the "electromagnetic pulses" of "horizontal intercloud lightning strokes". In an extension of that mechanism, Valdivia *et al.* (1997, 1998) treated the horizontal discharge as a fractal antenna. Apparently, more than 100 C must be transferred over a 10 km horizontal distance at a 10 km height in 10 ms or so, an average current of about 10 kA, to produce the needed fields in the middle atmosphere. A fractal channel shape is invoked to provide a spatially structured optical pattern simulating the observed sprite structure. Nickolaenko and Hayakawa (2000) argued that at ELF–VLF frequencies such a fractal antenna radiates no differently than a single horizontal wire. Milikh *et al.* (1995) and Valdivia *et al.* (1997, 1998) referred to "electromagnetic pulses", which are commonly associated with radiation fields, but radiation fields from return strokes in ground flashes at a range of 10 km or so are only dominant for tens of microseconds and are followed by more slowly varying predominantly electrostatic fields. Since the currents in the postulated horizontal channels have a duration of about 10 ms, it is apparently the quasi-static electric field that is providing the middle-atmosphere breakdown in the models of Milikh *et al.* and Valdivia *et al.*, so that really the source may not be much different from that in models that rely on an electrostatic field change from the removal of positive charge from the cloud top via a ground discharge. Further, Fernsler and Rowland (1996) discussed radiation field breakdown versus quasi-static electric field breakdown and showed that the former, for either horizontal or vertical lightning discharges, can account only for breakdown between 70 and 95 km altitude and can produce only short-lived optical emission, consistent with the elves to be discussed in Section 14.4 but not consistent with the longer-lived and lower-altitude red sprites.

We now consider further the issue of conventional breakdown versus breakdown initiated by relativistic runaway electrons. Pasko *et al.* (1997a) reviewed the available experimental data on sprites and presented a detailed

14.3. Low-luminosity transient discharges in the mesosphere

update of their quasi-electrostatic model (Pasko *et al.* 1995, 1996b). Pasko *et al.* (1997a) assumed that the causative-discharge duration is about 1 ms. They evaluated the electron density generated by various electron–air-molecule interactions, computed the spectral light output, and examined the altitude profiles of various parameters. The observed optical emission intensities require the removal of 100 to 300 C of positive thundercloud charge from 10 km height in order to simulate the sprite's upper "head" and "hair". When the positive charge at cloud top is rapidly removed by a lightning discharge, the remaining charges of opposite sign in and above the thundercloud produce a field above the thundercloud at each altitude that decays to 1/e of its original value in the local relaxation time, ε_0/σ. The initial field is exactly the field that would be produced by injecting negative charge of equal magnitude at the same height as the positive charge removed by lightning. Injection of the negative charge produces the additional field above the cloud. Pasko *et al.* (1997a) stated that the short-duration electromagnetic pulses from the return strokes do not add significantly to the conventional breakdown due to the quasistatic fields but do produce the observed elves at 80 to 95 km altitude, as discussed in Section 14.4. Pasko *et al.* did not attempt to model the lower sprite features such as the blue tendrils. They did allow that both the conventional breakdown mechanism and the runaway mechanism might operate simultaneously in sprite formation. Pasko *et al.* (1997b), following Rowland *et al.* (1996), invoked vertical gravity waves produced by the thunderstorm systems to induce atmospheric density modulation that in turn produces the vertically striated optical emissions observed in sprites between 50 and 90 km in altitude.

Zabotin and Wright (2001) suggested that the ubiquitous presence of conducting particles of meteoric origin in the regions where sprites form may explain some features of sprite initiation and structure. The conducting micrometeoroids are composed of Fe, Mg, and Si and are 25–100 μm in diameter. They are a source of atmospheric inhomogeneity that could initiate electrical breakdown by electrostatic field amplification and other processes discussed by Zabotin and Wright. Westcott *et al.* (2001b) argued that random micrometeor ionization is necessary to account for the fact that sprites are displaced horizontally a typical distance of about 20 km from the causative lightning discharge, with maximum observed distances of about 50 km.

Yukhimuk *et al.* (1998) reviewed the runaway-electron mechanism of sprite production, which they claim is the only reasonable mechanism for the generation of blue tendrils. They assumed that a relatively large positive charge is removed from the cloud top producing a quasi-static electric field above the cloud. Energetic electrons, in the MeV range, from cosmic ray interactions with the atmosphere are accelerated in this field producing ions and additional electrons as a result of collisions. If the electric field is strong enough (see Section 14.5), the number of high-energy electrons grows exponentially, creating a relativistic electron beam with a mean energy of order 1 MeV and runaway breakdown. The relativistic electron beam in turn produces secondary electrons of low energy, of order 1 eV, whose population also grows exponentially in time and whose mean energy, in the eV range, is determined by the local electric field and the rate of collision of these electrons with air molecules. The secondaries behave in a mode similar to the conventional breakdown discussed earlier. Yukhimuk *et al.* (1998) assumed that a positive cloud charge of 150 to 200 C at about 12 km altitude is removed in a time between 1 and 10 ms. Their simulations suggest that optical emissions from red sprites consist of two components: (i) a short-term emission (0.3 to 2 ms) visible at altitudes from 40 km to 77 km with significant blue radiation near the base of the emission region, this radiation being due to both runaway electrons and secondary electrons at all altitudes but the lower-altitude blue emissions being due primarily to the electron beam dissipation, and (ii) a long-term emission (2 to 10 ms), visible at altitudes above 66 km and primarily red, due to the secondary electrons.

Roussel-Dupré *et al.* (1998) presented a similar runaway-electron model for sprite production that starts with 200 C of positive charge neutralized at 11.5 km altitude. They found blue emission from 40 to 50 km and red from 50 to 77 km. Radio-frequency pulses of 300 μs duration and 20 to 75 V m^{-1} amplitude are predicted by the model to occur 50 km from the sprite at 80 km altitude. They suggested that these pulses, produced as a result of sprite formation, may be responsible for elves and in any case should be sufficient to cause breakdown and heating of the lower ionosphere, all, of course, assuming that the runaway model is correct. Roussel-Dupré *et al.* stated that their sprite simulations yielded results in good agreement with both the amplitude and temporal signatures of the measured gamma- and X-ray fluxes above 30 keV of Fishman *et al.* (1994) (Section 14.5).

The models of runaway-electron effects in the region between cloud tops and the ionosphere published through 1998 (e.g., Lehtinen *et al.* 1997; Taranenko and Roussel-Dupré 1996; Yukhimuk *et al.* 1998; Roussel-Dupré *et al.* 1998) used the results of the runaway avalanche model of Roussel-Dupré *et al.* (1994), which has been found to contain errors (e.g., Symbalisty *et al.* 1998). Apparently more adequate calculations that do not contain these errors have been published by Yukhimuk *et al.* (1999) and Lehtinen *et al.* (1999). Yukhimuk *et al.* (1999) stated that the new calculations of the temporal evolution of sprite optical emissions are in better agreement with ground-based optical observations than were the earlier calculations. In particular, the maximum duration of the detectable optical emissions in the new studies has a value greater than one TV camera field, 16.7 ms, whereas before it was less.

Characteristics of the electron runaway process in air that may be applicable to the sprite models are found in the work of Gurevich et al. (1992, 1994), Roussel-Dupré et al. (1994), Roussel-Dupré and Gurevich (1996), and Taranenko and Roussel-Dupré (1996), although as noted above, there are errors in some of this work. Runaway electrons are discussed further in Section 14.4.

14.4. Elves: low-luminosity transient phenomena in the lower ionosphere

The air density at the bottom of the ionosphere, near 90 km, is such that both (i) transient radiation fields, associated with many return strokes and possibly other impulsive lightning processes, and also (ii) quasistatic thunderstorm fields can accelerate electrons to sufficient energies to cause collisions with molecules that result in some level of optical emission. We consider now the transient events, called elves, that have an observed duration less than 1 ms. Later, in Section 14.6, we discuss the long-lasting infrared glow, which has a duration of the order of that of the thunderstorm, one hour or so.

Boeck et al. (1992) reported video observations from the Space Shuttle of a case of transient brightening in the airglow layer over a tropical oceanic thunderstorm coincident with lightning, the region of enhanced luminosity being 10 to 20 km thick, near 95 km height, and about 500 km in apparent horizontal extent. Such transient luminosity effects were later to be named elves (see below). They are very different from starters, jets, and sprites. They occur at a higher altitude, expand in a circular form, and have a much larger overall horizontal extent. The likely source of elves is the interaction of the lightning return-stroke radiation field, and the fields of other impulsive lightning processes such as initial breakdown (subsections 4.3.2 and 9.4.2), with the electrons of the lower ionosphere. The first time-resolved and spatially resolved documentation of the phenomenon was by Fukunishi et al. (1996), who employed high-speed multichannel photometers and image-intensified CCD cameras. They observed diffuse optical flashes with a duration of less than 1 ms and a horizontal scale of 100 to 300 km occurring at 75 to 105 km altitude just after the onset of cloud-to-ground lightning discharges, but preceding the onset of sprites. Fukunishi et al. (1996) designated these events as "elves" (emissions of light and VLF perturbations due to EMP sources). Fukunishi et al. stated that the luminosity of the elves is 1 to 10 MR, suggesting that the lower ionosphere is significantly heated by the elves.

Inan et al. (1997), using data acquired by an array of horizontally spaced photometers, provided the first measurement of the radial expansion of elves. The narrow individual fields of view of the photometers ($2.2° \times 1.1°$) yielded a spatial resolution of about 20 km at a range of 500 km, providing a time resolution of about 30 μs over a horizontal field of 200 km. A rapid lateral expansion with apparent speed (phase velocity) 3.1 ± 0.8 times the speed of light was observed to be triggered by a positive cloud-to-ground discharge, which according to NLDN records, had a peak current of 150 kA. A sprite was also observed via the video records associated with this event.

Subsequent measurements of elves by Barrington-Leigh and Inan (1999) and Barrington-Leigh et al. (2001) confirmed the observations of Inan et al. (1997) and, further, showed that the optical detectability of elves is related to the field strength in the VLF range, independent of field polarity. Both positive and negative return strokes were observed to produce elves with horizontal spatial extents typically 200 to 700 km.

Most of the observed features of elves are consistent with models in which the optical output is produced as a result of the heating of electrons in the lower ionosphere by the electromagnetic pulses from lightning discharges (Inan et al. 1991, 1993, 1996e; Glukhov and Inan 1996; Taranenko et al. 1993a, b; Rowland et al. 1995, 1996; Sukhorukov et al. 1996b). Optical emissions and significant changes in levels of ionization result from the acceleration of ambient electrons by the electric radiation fields associated with lightning. For example, according to the two-dimensional model of Inan et al. (1996d), the lightning radiation fields produce bright optical emissions at 80 to 95 km altitude, emitted in a cylindrical shell about 30 km thick and expanding to radial distances more than 150 km, with a duration of about 400 μs. The results of Glukhov and Inan (1996) indicate that the peak optical emissions are highly dependent on the radiation field waveform, while the altitude range at which the emissions occur is relatively independent of pulse shape. Results also indicate that even if the radiation field has a fast risetime, the bulk features of the resultant optical emissions (e.g., the altitude range of excitations, the duration, and the optical pulse shapes) are very similar to those calculated by Taranenko et al. (1993a, b) and Inan et al. (1996e) using quasi-stationary models.

Rowland et al. (1995, 1996), Fernsler and Rowland (1996), and Sukhorukov et al. (1996b) presented calculations of the lightning electromagnetic field interaction with the lower ionosphere similar to those discussed above. Rowland et al (1995) use a two-dimensional model to study the lower D-layer breakdown caused by return-stroke electric fields. They found that fields from the larger return strokes in ground flashes cause breakdown of the neutral atmosphere between 80 and 95 km, leading to an order of magnitude increase in electron density; this in turn gives rise to increased reflection and absorption, limiting the pulse strength that propagates higher into the ionosphere. The results are in reasonable agreement with Taranenko et al.

14.5. Runaway electrons, X-rays, and gamma-rays

(1993a, b) except that for the same field magnitudes the level of ionospheric ionization found by Rowland *et al.* (1995) is significantly higher: the electron density at 90 km increases by over an order of magnitude as compared with the value of up to 30 percent found by Taranenko *et al.* The ionization models in the two studies are different. In a continuation of the work of Rowland *et al.* (1995), Rowland *et al.* (1996) studied the dependence of the lower D-layer breakdown on the lightning radiated-field strength, the orientation of the lightning channel, the ambient plasma density, the neutral density, and the ionization model. For horizontal discharges, breakdown occurs for values of E_{100}, the radiation field peak normalized to 100 km, greater than 20 V m^{-1} in the nighttime ionosphere; a vertical discharge, however, requires E_{100} greater than 50 V m^{-1}. As noted in subsection 4.6.3, median values of E_{100} measured at ground level for first strokes are 6 to 8 V m^{-1}, values above 20 V m^{-1} probably being extremely rare. The results are, in general, similar to those of Taranenko *et al.* (1993a, b) and of Glukhov and Inan (1996).

Sukhorukov *et al.* (1996b) used an ionization model different from that of either Rowland *et al.* (1995, 1996) or Taranenko *et al.* (1993a). Sukhorukov *et al.* confirmed the general results of both groups; however, they found ionization values in the night-time ionosphere at 80 to 95 km close to those of Taranenko *et al.* (1993a) but three to 10 times smaller than the values found by Rowland *et al.* (1995). Further, they found that, given the amplitude, shape, and duration of the incident pulse as previously observed in ground-based experiments, there is a large diversity in the space–time characteristics of the ionospheric plasma perturbations.

14.5. Runaway electrons, X-rays, and gamma-rays

Relativistic runaway electrons were considered in the second part of subsections 14.3.3 as a potential source of ionization and optical emission associated with sprites (e.g., Gurevich *et al.* 1992, 1994; Bell *et al.* 1995; Winckler 1995; Taranenko and Roussel-Dupré 1996; Roussel-Dupré and Gurevich 1996; Lehtinen *et al.* 1997; Roussel-Dupré *et al.* 1994, 1998; Yukhimuk *et al.* 1998). If such runaway electrons indeed exist, they will necessarily produce X-rays and gamma-rays in interactions with air particles. As illustrated in Fig. 14.7, for energies less than about 100 MeV these X-rays and gamma-rays have much larger attenuation lengths than the ranges of electrons having equivalent energies, although the X-ray attenuation length does not much exceed 1 km at thunderstorm altitudes (Suszcynsky *et al.* 1996). The terms X-rays and gamma-rays are often used interchangeably, as we do here, and there is some disagreement in the literature as to their precise definitions. Strictly speaking, X-rays comprise the part of the photon spectrum from about 100 eV, the high-energy end of the vacuum ultraviolet, to near 100 keV, gamma-rays occupying the spectrum above 100 keV to 100 MeV or higher; the gamma-ray lower limit is, in some literature, taken as near 1 MeV, the lower limit of the gammas associated with cosmic rays.

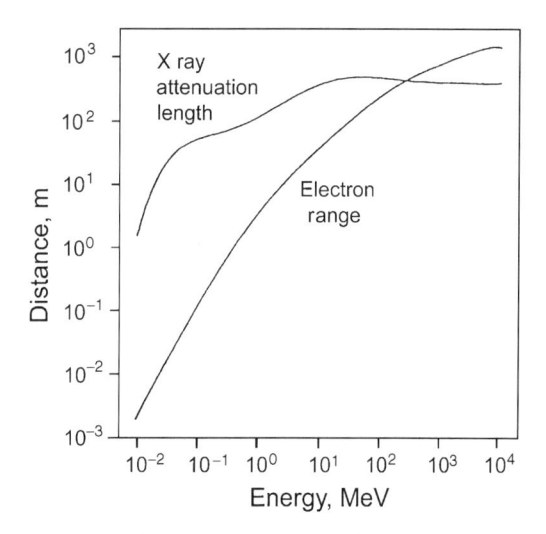

Fig. 14.7. The attenuation length of X-rays and electron range in air at STP as a function of energy. The X-ray attenuation length (electron range) represents the distance at which 1/e of the original flux (original number of electrons) remains. Adapted from Suszcynsky *et al.* (1996) who took data from Berger and Seltzer (1964, 1966) and Hubbell (1969).

Two kinds of runaway electrons can potentially occur: (i) electrons that begin in the eV to 10 eV energy range as a result of local breakdown and find themselves in very high electric fields and relatively low particle densities, this type of runaway being termed thermal runaway, and (ii) electrons that begin in the tens of keV or higher energy range as a result of cosmic ray collisions with air particles, in which case the electric field for runaway can be two orders of magnitude lower than for thermal runaway at the same air density. In either case, in order to "run away", an electron must gain more energy from the electric field between collisions than it loses in a collision, the energy lost in a collision depending on the particle density and type and on the collision cross-section of the particles at the particular electron energy. Without runaway, a stable distribution of electron energies will be developed in a given electric field such that on average the energy gained by the electrons from the field will equal the energy lost in collisions, generally resulting in average energies in the eV to 10 eV range for typical thunderstorm electric fields and air densities.

The so-called breakeven electric field, above which runaway can occur, can be computed, with some assumptions, for a given altitude (and thus air particle density) and starting electron energy, as illustrated in Fig. 14.8. For example, according to McCarthy and Parks (1992) a 10 keV

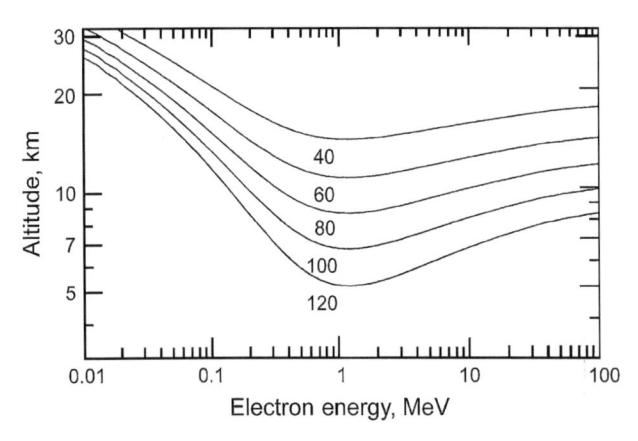

Fig. 14.8. Starting altitudes from which ascending electrons can continuously gain more energy than they lose (so achieving runaway) as a function of initial electron energy for ambient electric field values (top to bottom) 40, 60, 80, 100, and 120 kV.

electron requires a breakeven field of about 100 kV m^{-1} at 25 to 30 km altitude, while a 1 MeV electron requires the same field at 5 km. According to Suszcynsky et al. (1996), for 200 keV electrons the breakeven field at 10 km altitude is 100 kV m^{-1}. Chang and Price (1995) calculated that at 70 km height 500 V m^{-1} is needed for 1 MeV runaway, but Papadoupoulos et al. (1996) claimed that there were significant errors in their calculation. For a given starting electron energy, higher altitudes allow for smaller breakeven fields, owing to the decrease in particle density and the consequent increase in the distance between collisions, resulting in a greater gain in energy from the electric field. Steady fields of 100 kV m^{-1} have been measured in thunderclouds (subsection 3.2.4), while fields an order of magnitude higher are required for avalanche breakdown in clear air at thunderstorm altitudes. Since breakdown obviously does occur in and around thunderclouds, apparently such higher fields are locally present.

Interestingly, for the case of the starter runaway electrons produced by cosmic rays, the highest-energy electrons must be moving toward the earth, in the same direction as the cosmic rays, whereas in generating sprites from the discharge of positive charge near cloud top, electrons must runaway upward, since the electric field points downward after the disappearance of the positive charge.

We now examine the experimental evidence for the existence of runaway electrons in either the transient fields produced by lightning discharges or the steady fields produced by thunderstorm charges. Wilson (1925a, b) hypothesized that the fields in thunderstorms could produce runaways, and subsequently a number of investigators initiated searches for the runaways directly or for the X-rays resulting from them. Suszcynsky et al. (1996) reviewed the ground-based or low-altitude measurements of Schonland (1930), Schonland and Viljoen (1933), Appleton and Bowen (1933), Halliday (1934, 1941), Macky (1934), Clay et al. (1952), Hill (1963), Shaw (1967), Whitmire (1979), and D'Angelo (1987) in the light of their own ground-based measurements of X-rays. They were very skeptical of previous positive results and suggested that their own observed increases in gamma-ray counts of up to 100 percent above background are due to random daughter-ion decay as the daughter ions are precipitated to the ground by rainfall. They could not rule out the existence of X-rays from thunderstorm-field runaway electrons but stated that, if they exist, they are barely above the background level. They found no evidence of X-rays from individual lightning flashes, but only one of the 10 flashes observed was within the 1 km range expected for the X-rays. In fact, such evidence has been reported by Moore et al. (2001), as discussed later in this section.

Suszcynsky et al. (1996) gave alternate explanations for the positive results noted in some of the papers they had reviewed: the modulation of the cosmic ray flux by rain, fluctuations in the power grid due to lightning affecting instrumentation, and electromagnetic coupling to instrumentation from nearby lightning. Brunetti et al. (2000) observed downward gamma-ray bursts, in the MeV energy range, of atmospheric origin on ground 2 km above sea level. Slow increases in radiation below MeV were attributed to rain washout (see above), but the source of minute-duration bursts of gamma rays up to 10 MeV could not be identified, and runaway electrons were suggested as the source.

Suszcynsky et al. (1996) found the available aircraft measurements of X-ray count-rate increases more convincing than the ground-based results. Aircraft measurements have been reported by Parks et al. (1981), McCarthy and Parks (1985), and Eack et al. (1996a, b). McCarthy and Parks (1985) used a six-channel X-ray spectrometer with a measurement every 32 ms. The instrument was flown on an F-106 designed for lightning interception. The results of McCarthy and Parks can be summarized as follows (McCarthy and Parks 1992). Elevated X-ray fluxes, up to three orders of magnitude in energy, from 5 keV to above 110 keV, precede some lightning flashes by several seconds at least and cease immediately and coincidently with the lightning. This result is interpreted to imply, given the aircraft speed, that the elevated X-ray activity occupies a horizontal distance exceeding several hundred meters. The elevated X-ray flux was not observed with all nearby flashes. The scale size and time duration of the enhanced X-ray production indicates that the X-rays are not being produced by the mechanism of Hill (1963) and D'Angelo (1987) in which the large electric field and lowered electron density within the high-temperature return stroke or leader channels represent the region of runaway production. Rather, the observation argues for the production of runaways from radon or cosmic-ray-produced secondaries in the strong thunderstorm field preceding a lightning, the lightning flash when it occurs quenching the field and hence ending the X-ray generation. Calculations by McCarthy and Parks (1992), however, showed that, for the observed X-ray flux, there are 100 times too few electrons from radioactive decay and 10 times too few electrons from cosmic ray secondaries, although the calculations were not sufficiently accurate to rule out this latter source.

Eack et al. (1996a), from balloon-borne measurements within a thunderstorm, reported measurements supporting the observations of McCarthy and Parks (1985) and the analysis of McCarthy and Parks (1992) in which it was concluded that thunderstorm fields, not lightning fields, produce the runaway electrons. When the balloon passed through a field near 100 kV m^{-1}, the breakeven field near 4 km for 1 MeV electrons, the X-ray intensity between 30 and 120 keV was enhanced by a factor 100. The duration of the increased X-ray flux was about one minute and the flux was quenched when lightning occurred. The fluxes were similar in magnitude to those observed by McCarthy and Parks (1985). Gurevich et al. (1997) modeled the role of runaway-electron breakdown in thunderstorm development. They claim that runaway electrons may play a decisive role in triggering lightning and that their model X-ray generation is in qualitative agreement with data of Eack et al. (1996a).

Eack et al. (1996b) flew a balloon with a spectrometer similar to that of Eack et al. (1996a) but at an altitude of 15 km where the pressure was 130 mb, 3 km above the top of the stratiform region of a mesoscale convective system, and observed three X-ray pulses of one second duration that were one to two orders of magnitude above the background X-ray count. No significant electric field was observed at the balloon. No explanation was given for the observations.

A direct association of X-rays and/or electrons having energies exceeding 1 MeV with stepped leaders from three ground flashes observed on a mountain top, two of which struck 34 m and 44 m from the NaI scintillation detector, has been made by Moore et al. (2001). Bursts of MeV radiation started 1 to 2 ms before the first return stroke in each flash and continued until the onset of the return stroke as determined from correlated electric field records. Moore et al. suggested that the strong electric field of the stepped-leader tip combines with a cosmic ray shower to produce "an avalanche of energetic electrons that in turn produced the radiation we detected" via the runaway-electron mechanism.

Fishman et al. (1994) reported gamma-ray bursts of terrestrial origin observed from an Earth-orbiting satellite. The bursts lasted a few milliseconds and were consistent with X-rays generated by 1 MeV electrons at altitudes above 30 km. About 50 such bursts were observed in four years. However, the detector had a 10 min post-trigger dead time, so many bursts could have been missed. Fishman et al. found that a number of the bursts matched the positions of thunderstorms. Inan et al. (1996c) identified one burst observed over central America as occurring within ±1.5 ms of a positive cloud-to-ground flash whose radiated electromagnetic signal at Palmer Station, Antarctica exhibited an ELF slow tail characteristic of positive cloud-to-ground discharges that produce sprites. Another gamma ray burst was found to originate in a region of active thunderstorms. Inan et al. argued that sprites and the gamma-ray bursts are at least sometimes associated with, and most likely involve, the runaway acceleration of energetic electrons. For an origin of the gamma-ray bursts above 30 km (Fishman et al. 1994), rapid charge transfer by lightning would be a likely source of the runaway fields since large steady fields cannot exist there, owing to the relatively short relaxation time of the medium. Lehtinen et al. (1997) calculated that a runaway electron beam that would generate the gamma bursts observed by Fishman et al. could be produced by the removal of more than 240 C of positive charge from 10 km height. Similar gamma-ray yields have been found by Roussel-Dupré and Gurevich (1996) and by Taranenko and Roussel-Dupré (1996), who used a model in which 100 C of positive charge was removed from 15 km height. A model of gamma-ray bursts due to a horizontal lightning discharge with a fractal geometry was proposed by Milikh and Valdivia (1999).

14.6. Interaction of lightning and thundercloud electric fields with the ionosphere and the magnetosphere

14.6.1. Transient effects

There are two general types of transient interaction of the radiated lightning fields with the ionosphere and magnetosphere, direct and indirect. Direct coupling involves the heating of the lower ionosphere by lightning fields, which in turn leads to the production of secondary ionization and optical emission by the same processes that produce the elves discussed in Section 14.4 (e.g., Inan et al. 1991, 1993, 1996e; Taranenko et al. 1993a, b; Glukhov and Inan 1996; Rowland et al. 1995, 1996; Sukhorukov et al. 1996b). The indirect interactions involve the scattering of electrons having energies from tens of keV to over 1 MeV out of their otherwise stable magnetic traps in the Earth's magnetosphere by interactions with ducted whistlers produced by upward-propagating lightning transient fields (Sections 13.1 and 13.5); the precipitating electrons ionize atmospheric molecules at altitudes near and below 90 km (e.g., Voss et al. 1984; Chang and Inan 1985; Goldberg et al. 1986, 1987; Inan et al. 1988a, b, c, 1989; Glukhov et al. 1992; Pasko and Inan 1994; Dowden et al. 1996; Corcuff 1998; Rice and Hughes 1998). Both forms of ionospheric disturbance, direct and indirect, affect the phase and amplitude of VLF signals propagating in the Earth–ionosphere cavity. These disturbances are often called "Trimpi events" after their discoverer (Helliwell et al. 1973), but Trimpi events are now thought to include also the modification of subionospheric VLF transmissions by the conducting bodies of sprites and/or by sprite halos (e.g., Dowden et al. 2001a, b; Barrington-Leigh et al. 2001), phenomena that occur below the ionosphere. From measurements made during winter storms in Japan (Chapter 8), Hobara et al. (2001) found that the modification of subionospheric VLF transmissions associated with sprites was much greater than that associated with elves.

Two classes of Trimpi modulations of the phases and/or amplitudes of the VLF signals in the Earth–ionosphere waveguide have been identified, corresponding to the direct and indirect effects. Trimpi events in the more common class occur with a time delay of a second or more from the causative lightning and are attributed to enhanced D-region ionization created by the precipitation of trapped, energetic electrons through whistler-induced scattering in the magnetosphere (Inan et al. 1988a, b, c). This whistler-induced electron precipitation has been measured directly in satellite and rocket-based experiments (Voss et al. 1984; Goldberg et al. 1986, 1987; Inan et al. 1989). Trimpi events in the second class occur less than 100 ms after the causative lightning (Armstrong 1983) and are attributed to the direct effects discussed above.

The first and second classes of Trimpi events are sometimes described as, respectively, LEP or WEP, (*l*ightning-induced *e*lectron *p*recipitation, e.g., Corcuff 1998, or *w*histler-induced *e*lectron-*p*recipitation, e.g., Dowden et al. 1996) and early/fast, the latter class having both a short (early) time delay from the causative lightning and a fast signal onset duration, less than 100 to 200 ms. In the latter category is the so-called RORD, *r*apid-*o*nset *r*apid-*d*ecay Trimpi; Dowden et al. (1994, 1996, 2001a, b) argued that they are due to scattering from the bodies of conducting sprites, Inan et al. (1996b) attributed them to direct ionospheric heating, and Barrington-Leigh et al. (2001) suggested that they may be the result of scattering from sprite halos, as noted in the second part of subsection 14.3.3. A discussion of the literature regarding the first two proposed mechanisms is given in Barr et al. (2000) as part of a general review of ELF and VLF radiowave phenomena.

Impulsive electric fields have been observed in the ionosphere and at satellite altitudes in the magnetosphere. From rocket-borne ionospheric observations at 200 and 400 km, Kelley et al. (1997) reported on the spectrum of lightning-generated electromagnetic waves at frequencies from 20 kHz to 2 MHz. Apparently, the lightning signal remains pulse-like with 100 μs width even up to 200 km. The higher frequencies radiated by lightning transmit through the ionosphere and have been observed on the Blackbeard payload of the ALEXIS satellite and on the FORTE satellite (both at 800 km altitude). This has led to the identification and characterization of TIPPs, *t*rans*i*onospheric *p*ulse *p*airs. The latter apparently comprise a direct wave and a reflection off the earth, the components of a pulse pair being separated by about 50 μs or so (e.g., Holden et al. 1995; Massey and Holden 1995; Russell et al. 1998; Zuelsdorf et al. 1997, 1998a, b, 2000; Massey et al. 1998, Jacobson et al. 1999, 2000; and Suszcynsky et al. 2000). Kelley et al. (1990) reported the observation of electric and magnetic field pulses at about 300 km. The field magnitudes were tens of mV m^{-1} in the frequency band 10 to 80 kHz. Burke et al. (1992), using the DE2 satellite flying over Hurricane Debbie in 1982 at an altitude of about 300 km, observed a large electric field transient, about 40 mV m^{-1}, which was correlated closely with a burst of highly field-aligned, upward-moving electrons with nearly 1 keV of energy. The event was associated with a lightning discharge since there was energetic electron precipitation from the radiation belts about 2 s later. Burke et al. suggested that the observed electrons are runaways accelerated by an electric field of 1 V m^{-1} and of about 1 ms duration that propagated to the night-time E-region, above 140 km or so. Apparently keV electrons generated in the D-region, under 100 km or so, would be scattered before reaching the F-region where the satellite was flying. A discussion of the effects on the ionosphere of the removal of positive cloud charge by lightning and of the role of sprites in coupling the lower atmosphere to the ionosphere is given in Pasko et al. (1998a).

14.6.2. Steady infrared glow

Picard et al. (1997) predicted that an Earth-orbiting satellite with an appropriate infrared detector would detect a steady glow from the lower ionosphere; this was based on earlier calculations of the sustained heating of lower ionospheric electrons by thundercloud fields throughout the hour or so duration of the storm, as suggested by Inan et al. (1996b). Pasko et al. (1998d) developed a two-dimensional cylindrically symmetric electrostatic (ES) coupling model to examine the effects of the relatively steady electric fields of thunderclouds on the lower ionosphere. In contrast to earlier work (e.g., Tzur and Roble 1985; Velinov and Tonev 1995), their ES-coupling model accounts for the nonlinear dependence of the ionospheric-plasma specific and Pedersen conductivities on the magnitude of electric field due to electron heating. The ES-coupling model predicts the magnitude of the low-intensity electric fields existing at high altitudes above thunderstorms during the time between lightning discharges and allows a detailed quantitative assessment of the geometry and physical characteristics of the modified lower-ionospheric regions. As noted above, Picard et al. (1997) applied the ES-coupling model to the problem of the production of the infrared glow above thunderstorms. In particular, they studied the 4.3 μm CO_2 emission excited by the vibrational pumping of N_2 by heated electrons followed by vibrational energy transfer from N_2 to CO_2. Broadband 4.3 μm enhancements of greater than factor 2 above ambient night-time levels were predicted for tangent heights greater than 77 km for the most perturbed case, with larger enhancements in selected narrower spectral regions.

Measurements of abnormally strong electric fields of unknown origin in the mesosphere are discussed in subsection 1.4.2.

14.7. Summary

Blue starters and blue jets, red sprites, and elves represent a mechanism for energy transfer from lightning and the thunderstorm to the regions of the atmosphere between the thundercloud tops and the lower ionosphere. Sprites have a vertical extent of tens of kilometers and complex spatial structures. They apparently have peak optical intensities that are generally between 0.1 and 10 MR and overall durations from about a millisecond to many tens of milliseconds. Sprites are difficult to see, even with a dark-adapted eye. Blue starters and blue jets propagate upward from the cloud tops at speeds near 10^5 m s^{-1}. The lower, brightest part of blue-starters and blue-jets apparently can have optical intensities above 10 MR, while their tops are considerably less bright. Starters extend less than 10 km above cloud tops, generally only a few kilometers, while blue jets have vertical extents of 20 km or so. Blue-starter and blue-jet durations are up to a few hundred milliseconds. Elves expand outward across the lower ionosphere, in less than a millisecond, to a maximum horizontal extent of 200 to 700 km and are reported to have optical intensities of roughly 1 to 10 MR.

The physical mechanism for the generation of elves would appear to be reasonably well understood. They are caused by the acceleration of electrons by the electromagnetic fields of relatively-high-current impulsive lightning processes, generally return strokes, that spread radially outward across the lower ionosphere from a point above the causative lightning process. Blue starters and blue jets have been modeled as diffuse upward-propagating avalanche-breakdown processes, but the necessary driving charges required by the models are relatively large, and it is not clear why a normal discharge with a centimeter-diameter channel would not always be created in the cloud top environment. The polarity of the starters and jets is unknown. Sprite theory is the most extensive of the theories describing the phenomena discussed in this chapter. Both conventional-breakdown and runaway-electron theories that model sprites require relatively large source charges at high altitudes. In fact, a common thread in the modeling of starters, jets, sprites, and elves is the need to assume source currents and charges that are larger than previously observed maximum values. All such modeling is done with complex computer programs that require as inputs the ambient conductivity and the cross-sections for excitation and ionization of the air between the cloud tops and the lower ionosphere. These quantities are not all well known. Further, the modeling assumes that the salient atmospheric properties do not vary rapidly with position, whereas the diverse sprite morphology may be indicative of a more inhomogeneous atmosphere, perhaps made so by conducting meteoric dust.

References and bibliography for Chapter 14

Adams, C.D.D., and Dowden, R.L. 1990. VLF group delay of LEP echoes from measurement of phase and amplitude perturbations at two frequencies. *J. Geophys. Res.* **95**: 2457–62.

Anderson, H.R., and Few, A.A. 1968. Discussion of paper by Glenn E. Shaw. Background cosmic count increase associated with thunderstorm. *J. Geophys. Res.* **73**: 33–40.

Appleton, E.V., and Bowen, E.G. 1933. Source of atmospherics and penetrating radiation. *Nature* **132**: 965.

Armstrong, W.C. 1983. Recent advances from studies of the Trimpi effect. *Antarctic J.* **18**: 281–3.

Armstrong, W.C. 1987. Lightning triggered from the Earth's magnetosphere as the source of synchronized whistlers. *Nature* **327**: 405–8.

Armstrong, R.A., Shorter, J.A., Taylor, M.J., Suszcynsky, D.M., Lyons, W.A., and Jeong, L.S. 1998. Photometric measurements in the SPRITES '95 & '96 campaigns of nitrogen second positive (399.8 nm) and first negative (427.8 nm) emissions. *J. Atmos. Solar-Terr. Phys.* **60**: 787–800.

Ashmore, S.E. 1950. Unusual lightning. *Weather* **5**: 331–331.

Atlas, D. 1958. Radar lightning echoes and atmospherics in vertical cross section. In *Recent Advances in Atmospheric Electricity*, ed. L.G. Smith, pp. 441–59, Oxford, England: Pergamon Press.

Babich, L.P., Kutsyk, I.M., and Kudryavgsev, A.Yu. 1999. Calculation of x-ray emission of gigantic upward atmospheric discharges governed by relativistic runaway electrons. In *Proc. Int. Conf on Lightning and Static Electricity, Toulouse, France*, pp. 441–4.

Baker, S.D., Kelley, M.C., Swenson, C.M., Bonnell, J., Hahn, D.V. 2000. Generation of electrostatic emissions by lightning-induced whistler-mode radiation above thunderstorms. *J. Atmos. Solar-Terrest. Phys.* **62**: 1393–404.

Barr, R., Jones, D.L., Rodger, C.J. 2000. ELF and VLF radio waves. *J. Atmos. Solar-Terr. Phys.* **62**: 1689–718.

Barrington-Leigh, C.P., and Inan, U.S. 1999. Elves triggered by positive and negative lightning discharges. *Geophys. Res. Lett.* **26**: 683–6.

Barrington-Leigh, C.P., Inan, U.S., Stanley, M., and Cummer, S.A. 1999. Sprites triggered by negative lightning discharges. *Geophys. Res. Lett.* **26**: 3605–8.

Barrington-Leigh, C.P., Inan, U.S., and Stanley, M. 2001. Identification of sprites and elves with intensified video and broadband array photometry. *J. Geophys. Res.* **106**(2): 1741–50.

Berger, M.J., and Seltzer, S.M. 1964. Tables of energy losses and ranges of electrons and positrons. NASA Special Publication SP-3012.

Berger, M.J., and Seltzer, S.M. 1966. Additional stopping power and range tables for protons, mesons, and electrons. NASA Special Publication SP-3036.

Bell, T.F., Pasko, V.P., and Inan, U.S. 1995. Runaway electrons as a source of red sprites in the mesosphere. *Geophys. Res. Lett.* **22**: 2127–30.

Bell, T.F., Reising, S.C., and Inan, U.S. 1998. Intense continuing currents following positive cloud-to-ground lightning associated with red sprites. *Geophys. Res. Lett.* **25**: 1285–8.

Boccippio, D.J., Williams, E.R., Lyons, W.A., Baker, I., and Boldi, R. 1995. Sprites, ELF transients and positive ground strokes. *Science* **269**: 1088–91.

Boccippio, D.J., Wong, C., Williams, E.R., Boldi, R., Christian, H.J., and Goodman, S.J. 1998. Global validation of single-station Schumann resonance lightning location. *J. Atmos. Solar-Terr. Phys.* **60**: 701–12.

Boeck, W.L., Vaughan, O.H. Jr, Blakeslee, R., Vonnegut, B., and Brook, M. 1992. Lightning induced brightening in the airglow layer. *Geophys. Res. Lett.* **19**: 99–102.

Boeck, W.L., Vaughan, O.H. Jr, Blakeslee, R.L., Vonnegut, B., Brook, M., and McKune, J. 1995. Observations of lightning in the stratosphere. *J. Geophys. Res.* **100**: 1465–75.

Boeck, W.L., Vaughan, O.H. Jr, Blakeslee, R., Vonnegut, B., and Brook, M., 1998. The role of the space shuttle videotapes in the discovery of sprites, jets and elves. *J. Atmos. Solar-Terr. Phys.* **60**: 669–77.

Boys, C.V. 1926. Progressive lightning. *Nature* **118**: 749–50.

Brook, M., Stanley, M., Krehbiel, P., Rison, W., Moore, C.B., Barrington-Leigh, C., Suszcynsky, D., Nelson, T., and Lyons, W. 1997. Correlated electric field, video, and photometric evidence of charge transfer within sprites. *Eos, Trans. AGU* **78**: F82–3.

Brunetti, M., Cecchini, S., Galli, M., Giovannini, G., and Pagliarin, A. 2000. Gamma-ray bursts of atmospheric origin in the MeV energy range. *Geophys. Res. Lett.* **27**: 1599–602.

Burgess, W.C., and Inan, U.S. 1990. Simultaneous disturbance of conjugate ionosphere regions in association with individual lightning flashes. *Geophys. Res. Lett.* **17**: 259–62.

Burgess, W.C., and Inan, U.S. 1993. The role of ducted whistlers in the precipitation loss and equilibrium flux of radiation belt electrons. *J. Geophys. Res.* **98**: 15 643–65.

Burke, W.J. 1992. Early Trimpi events from lightning-induced electric fields in the ionosphere. *J. Atmos. Terr. Phys.* **54**: 205–8.

Burke, W.J, Aggson, T.L., Maynard, N.C., Hoegy, W.R., Hoffman, R.A., Candy, R.M., Liebrecht, C., and Rodgers, E. 1992. Effects of a lightning discharge detected by the DE 2 satellite over hurricane Debbie. *J. Geophys. Res.* **97**: 6359–67.

Chang, H.C., and Inan, U.S. 1985. Lightning-induced electron precipitation from the magnetosphere. *J. Geophys. Res.* **90**: 1531–41.

Chang, B., and Price, C. 1995. Can gamma radiation be produced in the electrical environment above thunderstorm? *Geophys. Res. Lett.* **22**: 1117–20.

Chen, J.T., Inan, U.S., and Bell, T.F. 1996. VLF strip holographic imaging of lightning-associated ionospheric disturbances. *Radio Sci.* **31**: 335–48.

Cho, M., and Rycroft, M.J. 1998. Computer simulation of the electric field structure and optical emission from cloud top to the ionosphere. *J. Atmos. Solar-Terr. Phys.* **60**: 871–88.

Clay, J.H., Jongen, H.F., and Aarts, A.J.J. 1952. High energy electrons produced in a thunderstorm. *Physica* **28**: 801–8.

Cole, R.K. Jr, Hill, R.D., and Pierce, E.T. 1966. Ionized columns between thunderstorms and the ionosphere. *J. Geophys. Res.* **71**: 959–64.

Connor, J.W. and Hastie, R.J. 1975. Relativistic limitations on runaway electrons, *Nucl. Fusion* **15**: 415–24.

Cooray, V. 2000. The modeling of positive return strokes in lightning flashes. *J. Atmos. Solar-Terr. Phys.* **62**: 169–87.

Corcuff, Y. 1998. VLF signatures of ionosphere perturbations caused by lightning discharges in an underlying and moving thunderstorm. *Geophys. Res. Lett.* **25**: 2385–8.

Corliss, W.R. 1977. Handbook of unusual natural phenomena. The Sourcebook Project, Glen Arm, Maryland, 542 pp.

Corliss, W.R. 1983. Handbook of unusual natural phenomena. Anchor Books/Doubleday, Garden City, New York, 423 pp.

Cummer, S.A., and Inan, U.S. 1997. Measurement of charge transfer in sprite-producing lightning using ELF radio atmospheric. *Geophys. Res. Lett.* **24**: 1731–4.

Cummer, S.A., and Füllekrug, M. 2001. Unusually intense continuing current in lightning produced delayed mesospheric breakdown. *Geophys. Res. Lett.* **28**: 495–8.

Cummer, S.A., and Inan, U.S. 2000a. Modeling ELF radio atmospheric propagation and extracting lightning currents from ELF observations. *Radio Sci.* **35**: 385–94.

Cummer, S.A., and Inan, U.S. 2000b. Ionospheric E region remote sensing with ELF radio atmospherics. *Radio Sci.* **35**: 1437–44.

Cummer, S.A., and Stanley, M. 1999. Submillisecond resolution lightning currents and sprite development: observations and implications. *Geophys. Res. Lett.* **26**: 3205–8.

Cummer, S.A., Inan, U.S., Bell, T.F., and Barrington-Leigh, C. 1998. ELF radiation produced by electrical currents in sprites. *Geophys. Res. Lett.* **25**: 1281–4.

D'Angelo, N. 1987. On X-rays from thunderclouds. *Ann. Geophys.* **5B**: 119–22.

Daniel, R.R., and Stephens, S.A. 1974. Cosmic-ray-produced electrons and gamma rays in the atmosphere. *Rev. Geophys. Space Phys.* **12**: 233–58.

Dejnakarintra, M., and Park, C.G. 1974. Lightning-induced electric fields in the ionosphere. *J. Geophys. Res.* **79**: 1903–10.

Dowden, R.L., and Adams, C.D.D. 1988. Phase and amplitude perturbations on subionospheric signals explained in terms of echoes from lightning-induced electron precipitation ionization patches. *J. Geophys. Res.* **93**: 11 543–50.

Dowden, R.L., and Adams, C.D.D. 1990. Location of lightning-induced electron precipitation from measurement of VLF phase and amplitude perturbations on spaced antennas and on two frequencies. *J. Geophys. Res.* **95**: 4135–45.

Dowden, R.L., Adams, C.D.D., Brundell, J.B., and Dowden, P.E. 1994. Rapid onset, rapid decay (RORD), phase and amplitude perturbations of VLF subionospheric transmissions. *J. Atmos. Terr. Phys.* **56**: 1513–27.

Dowden, R.L., Brundell, J.B., and Lyons, W.A. 1996. Are VLF rapid onset, rapid decay perturbations and optical sprites produced by scattering off sprite plasma? *J. Geophys. Res.* **101**: 19 175–83.

Dowden, R.L., Hardman, S.F., Rodger, C.L., and Brundell, J.B. 1998. Logarithmic decay and Doppler shift of plasma associated with sprites. *J. Atmos. Solar-Terr. Phys.* **60**: 741–54.

Dowden, R.L., Rodger, C.J., Brundell, J.B., and Cliverd, M.A. 2001a. Decay of whistler-induced electron precipitation and cloud–ionosphere electrical discharge Trimpis: observations and analysis. *Radio Sci.* **36**: 151–69.

Dowden, R.L., Rodger, C.J., and Nunn, D. 2001b. Minimum sprite plasma density as determined by VLF scattering. *IEEE Antennas Propag.* **43**(2): 12–24.

Draganov, A.B., Inan, U.S., and Taranenko, Y.N. 1991. ULF magnetic signatures at the Earth surface due to ground water flow: a possible precursor to earthquakes. *Geophys. Res. Lett.* **18**: 1127–30.

Draganov, A.B., Inan, U.S., Sonwalkar, V.S., and Bell, T.F. 1992. Magnetospherically reflected whistlers as a source of plasmaspheric hiss. *Geophys. Res. Lett.* **19**: 233–6.

Eack, K.B. 1996. Balloon-borne x-ray spectrometer for detection of x-rays produced by thunderstorms. *Rev. Sci. Instr.* **67**: 2005–9.

Eack, K.B., Beasley, W.H., Rust, W.D., Marshall, T.C., and Stolzenburg, M. 1996a. Initial results from simultaneous observations of x-rays and electric-fields in a thunderstorm. *J. Geophys. Res.* **101**: 29 637–40.

Eack, K.B., Beasley, W.H., Rust, W.D., Marshall, T.C., and Stolzenburg, M. 1996b. X-ray pulses observed above a mesoscale convective system. *Geophys. Res. Lett.* **23**: 2915–8.

Eack, K.B., Suszcynsky, D.M., Beasley, W.H., Roussel-Dupré, R., and Symbalisty, E. 2000. Gamma-ray emissions observed in a thunderstorm anvil. *Geophys. Res. Lett.* **27**: 185–8.

Ernstmeyer, J., and Chang T. 1998. Lightning-induced electron heating in the mesosphere. *Geophys. Res. Lett.* **25**: 2389–92.

Everett, J.D., and Everett, W.H. 1903. Rocket lightning. *Nature* **68**: 599.

Farrell, W.M., and Desch, M.D. 1992. Cloud-to-stratosphere lightning discharges: a radio emission model *Geophys. Res. Lett.* **19**: 665–8.

Farrell, W.M., and Desch, M.D. 1993. Reply. *Geophys. Res. Lett.* **20**: 763–4.

Fernsler, R.F., and Rowland, H.L. 1996. Models of lightning-producing sprites and elves. *J. Geophys. Res.* **101**: 29 653–62.

Fisher, J.R. 1990. Upward discharges above thunderstorms. *Weather* **45**: 451–2.

Fishman, G.J., and Inan, U.S. 1988. Observation of an ionospheric disturbance caused by a gamma-ray burst. *Nature* **331**: 418–20.

Fishman, G.J., Bhat, P.N., Mallozzi, R., Horack, J.M., Koshut, T., Kouveliotou, C., Pendleton, G.N., Meegan, C.A., Wilson, R.B., Paciesas, W.S., Goodman, S.J., and Christian, H.J. 1994. Discovery of intense gamma-ray flashes of atmospheric origin. *Science* **264**: 1313–16.

Franz, R.C., Nemzek, R.J., and Winckler, J.R. 1990. Television image of a large upward electrical discharge above a thunderstorm system. *Science* **249**: 48–51.

Fukunishi, H., Takahashi, Y., Kubota, M., Sakanoi, K., Inan, U.S., and Lyons, W.A. 1996. Elves, lightning induced transient luminous events in the lower ionosphere. *Geophys. Res. Lett.* **23**: 2157–60.

Fukunishi, H., Takahashi, Y., and Watanabe, Y. 1997. Ground-based observations of ULF transients excited by strong lightning discharges producing elves and sprites. *Geophys. Res. Lett.* **24**: 2973–6.

Füllekrug, M., and Constable, S. 2000. Global triangulation of intense lightning discharges. *Geophys. Res. Lett.* **27**: 333–6.

Füllekrug, M., and Sukhorukov, A.I. 1999. The contribution of anisotropic conductivity in the ionosphere to lightning flash bearing deviations in the ELF/ULF range. *Geophys. Res. Lett.* **26**: 1109–12.

Füllekrug, M., Fraser-Smith, A.C., and Reising, S.C. 1998. Ultra-slow tails of sprite-associated lightning flashes. *Geophys. Res. Lett.* **25**: 3497–500.

Fullekrug, M., Moudry, D.R., Dawes, G., and Sentman, D.D. 2001. Mesospheric sprite current triangulation. *J. Geophys. Res.* **106**: 20 189–94.

Gerken, E.A., Inan, U.S., and Barrington-Leigh, C.P. 2000. Telescopic imaging of sprites. *Geophys. Res. Lett.* **27**: 2637–40.

Glukhov, V.S., and Inan, U.S. 1996. Particle simulation of the time-dependent interaction with the ionosphere of rapidly varying lightning EMP. *Geophys. Res. Lett.* **23**: 2193–6.

Glukhov, V.S., Pasko, V.P., and Inan, U.S. 1992. Relaxation of transient lower ionospheric disturbances caused by lightning-whistler-induced electron precipitation bursts. *J. Geophys. Res.* **97**: 16 971–9.

Goldberg, R.A., Barcus, J.R., Hale, L.C., and Curtis, S.A. 1986. Direct observation of magnetostatic electron precipitation stimulated by lightning. *J. Atmos. Terr. Phys.* **48**: 293–9.

Goldberg, R.A., Curtis, S.A., and Barcus, J.R. 1987. Detailed spectral structure of magnetospheric electron bursts precipitated by lightning. *J. Geophys. Res.* **92**: 2505–13.

Gomes, C., and Cooray, V. 1998. Long impulse current associated with positive return strokes. *J. Atmos. Solar-Terr. Phys.* **60**: 693–9.

Green, B.D., Fraser, M.E., Rawlins, W.T., Jeong, L., Blumberg, W.A.M., Mende, S.B., Swenson, G.R., Hampton, D.L., Wescott, E.M., and Sentman, D.D. 1996. Molecular excitation in sprites. *Geophys. Res. Lett.* **23**: 2161–4.

Gurevich, A.V. 1961. On the theory of runaway electrons. *Soviet Phys. JETP* **12**: 904–12.

Gurevich, A.V., Milikh, G.M., and Roussel-Dupré, R. 1992. Runaway electron mechanism of air breakdown and preconditioning during a thunderstorm. *Phys. Lett. A* **165**: 463–8.

Gurevich, A.V., Milikh, G.M., and Roussel-Dupré, R. 1994. Nonuniform runaway air breakdown. *Phys. Lett. A* **187**: 197–203.

Gurevich, A.V., Milikh, G.M., and Valdivia, J.A. 1997. Model of X-ray emission and fast preconditioning during a thunderstorm. *Phys. Lett. A* **231**: 402–8.

Gurnett, D.A., and Inan, U.S. 1988. Review of plasma wave observations with the dynamics Explorer 1 spacecraft. *Rev. Geophys.* **26**: 285–316.

Hale, L.C. 1993. Comment on "Cloud-to-stratosphere lightning discharges: a radio emission model" by W.M. Farrell and M.D. Desch. *Geophys. Res. Lett.* **20**: 761–2.

Hale, L.C. 1994. Coupling of ELF/ULF energy from lightning and MeV particles to the middle atmosphere, ionosphere, and global circuit. *J. Geophys. Res.* **99**: 21 089–96.

Hale, L.C., and Baginski, M.E. 1987. Current in the ionosphere following a lightning stroke. *Nature* **329**: 814–16.

Halliday, E.C. 1934. Thunder-storms and the penetrating radiation. *Proc. Cambridge Phil. Soc.* **30**: 206–15.

Halliday, E.C. 1941. The thundercloud as a source of penetrating particles. *Phys. Rev.* **60**: 101–6.

Hammerstrom, J.H. 1993. Mystery lightning. *Aviation Week & Space Tech.* **6** (August 30): p. 6–6.

Hampton, D.L., Heavner, M.J., Wescott, E.M., and Sentman, D.D. 1996. Optical spectral characteristics of sprites. *Geophys. Res. Lett.* **23**: 89–92.

Hansell, S.A., Wells, W.K., and Hunten, D.M. 1995. Optical detection of lightning on Venus. *Icarus* **117**: 345–51.

Hardman, S.F., Dowden, R.L., Brundell, J.B., Bahr, J.L., Kawasaki, Z.i., and Rodger, C.J. 2000. Sprite observations in the northern territory of Australia. *J. Geophys. Res.* **105**: 4689–97.

Helliwell, R.A. 1965. *Whistlers and Related Ionospheric Phenomena*, Stanford University Press: Stanford, California.

Helliwell, R.A., Katsufrakis, J.P., and Trimpi, M.L. 1973. Whistler-induced amplitude perturbation in VLF propagation. *J. Geophys. Res.* **78**: 4679–88.

Hill, R.D. 1963. Investigation of electron runaway in lightning. *J. Geophys. Res.* **68**: 6261–6.

Hobara, Y., Iwasaki, N., Hayashida, T., Hayakawa, M., Ohita, K., and Fukunishi, H. 2001. Interrelation between ELF transients and ionospheric disturbances in association with sprites and elves. *Geophys. Res. Lett.* **28**(5): 935–8.

Hoffman, W.C. 1960. The current-jet hypothesis of whistler generation. *J. Geophys. Res.* **67**: 2047–54.

Holden, D.N., Munson, C.P., and Devenport, J.C. 1995. Satellite observations of transionospheric pulse pairs. *Geophys. Res. Lett.* **22**: 889–92.

Huang, E., Williams, E., Boldi, R., Heckman, S., Lyons, W., Taylor, M., Nelson, T., and Wong, C. 1999. Criteria for sprites and elves based on Schumann resonance observations. *J. Geophys. Res.* **104**: 16 943–64.

Hubbell, J.H. 1969. Photon cross section attenuation coefficients, and energy absorption coefficients from 10 keV to 100 GeV. NSRDS-NABS Special Publication 29.

Inan, U.S. 1990. VLF heating of the lower ionosphere. *Geophys. Res. Lett.* **17**: 729–32.

Inan, U.S., and Bell, T.F. 1991. Pitch angle and energy scattering of energetic particles by oblique whistler waves. *Geophys. Res. Lett.* **18**: 49–52.

Inan, U.S., Burgess, W.C., Wolf, T.G., Shafer, D.C., and Orville, R.E. 1988a. Lightning-associated precipitation of MeV electrons from the inner radiation belt. *Geophys. Res. Lett.* **15**: 172–5.

Inan, U.S., Shafer, D.C., Yip, W.Y., and Orville, R.E. 1988b. Subionospheric VLF signatures of nighttime D-region perturbations in the vicinity of lightning discharges. *J. Geophys. Res.* **93**: 11 455–72.

Inan, U.S., Wolf, T.G., and Carpenter, D.L. 1988c. Geographic distribution of lightning induced electron precipitation observed as VLF/LF perturbation events. *J. Geophys. Res.* **93**: 9841–53.

Inan, U.S., Walt, M., Voss, H., and Imhof, W. 1989. Energy spectra and pitch angle distribution of lightning induced electron precipitation: analysis of an event observed on the S81–1 (SEEP) satellite. *J. Geophys. Res.* **94**: 1379–401.

Inan, U.S., Bell, T.F., and Rodriguez, J.V. 1991. Heating and ionization of the lower ionosphere by lightning. *Geophys. Res. Lett.* **18**: 705–8.

Inan, U.S., Chiu, Y.T., and Davidson, G.T. 1992a. Whistler-mode chorus and morningside aurorae. *Geophys. Res. Lett.* **19**: 653–6.

Inan, U.S., Rodriguez, J.V., Lev-Tov, S., and Oh, J. 1992b. Ionospheric modification with a VLF transmitter. *Geophys. Res. Lett.* **19**: 2071–4.

Inan, U.S., Rodriguez, J.V., and Idone, V.P. 1993. VLF signatures of lightning-induced heating and ionization of the nighttime D-region. *Geophys. Res. Lett.* **20**: 2355–8.

Inan, U.S., Bell, T.F., Pasko, V.P., Sentman, D.D., Wescott, E.M., and Lyons, W.A. 1995. VLF signatures of ionospheric disturbances associated with sprites. *Geophys. Res. Lett.* **22**: 3461–4.

Inan, U.S., Bell, T.F., and Pasko, V.P. 1996a. Reply to comment by R.L. Dowden *et al.* on "VLF signatures of ionospheric disturbances associated with sprites". *Geophys. Res. Lett.* **23**: 3423–4.

Inan, U.S., Pasko, V.P., and Bell, T.F. 1996b. Sustained heating of the ionosphere above thunderclouds as evidenced in 'early-fast' VLF events. *Geophys. Res. Lett.* **23**: 1067–70.

Inan, U.S., Reising, S.C., Fishman, G.J., and Horack, J.M. 1996c. On the association of terrestrial gamma-ray bursts with lightning discharges and sprites. *Geophys. Res. Lett.* **23**: 1017–20.

Inan, U.S., Sampson, W.A., and Taranenko, Y.N. 1996d. Space–time structure of lower ionospheric optical flashes and ionization changes produced by lightning EMP. *Geophys. Res. Lett.* **23**: 133–6.

Inan, U.S. Slingeland, A., Pasko, V.P., and Rodriguez, J. 1996e. VLF signatures of mesospheric/lower ionospheric response to lightning discharges. *J. Geophys. Res.* **101**: 5219–38.

Inan, U.S., Barrington-Leigh, C., Hansen, S., Glukhov, V.S., Bell, T.F., and Rairden, R. 1997. Rapid lateral expansion of optical luminosity in lightning-induced ionospheric flashes referred to as "elves." *Geophys. Res. Lett.* **24**: 583–6.

Jacobson, A.R., Knox, S.O., Franz, R., and Enemark, D.C. 1999. FORTE observations of lightning radio-frequency signatures: capabilities and basic results. *Radio Sci.* **34**: 337–54.

Jacobson, A.R., Cummins, K.L., Carter, M., Klinger, P., Roussel-Dupré, D., and Knox, S.O. 2000. FORTE radio-frequency observations of lightning strokes detected by the National Lightning Detection Network. *J. Geophys. Res.* **105**: 15 653–62.

James, H.G., Inan, U.S., and Rietveld, M.T. 1990. Observations on the DE-1 spacecraft of ELF/VLF waves generated by an ionospheric heater. *J. Geophys. Res.* **95**: 12 187–95.

Jarzembski, M.A. and Srivastava, V. 1995. Low-pressure electrical discharge experiment to simulate high-altitude lightning above thunderclouds. NASA Technical Paper 33578, Marshall Space Flight Center, 29 pp.

Jasna, D., Inan, U.S., and Bell, T.F. 1990. Bell equatorial gyroresonance between electrons and magnetospherically reflected whistlers. *Geophys. Res. Lett.* **17**: 1865–8.

Johnson, M.P., and Inan, U.S. 2000. Sferic clusters associated with early/fast VLF events. *Geophys. Res. Lett.* **27**: 1391–4.

Johnson, M.P., Inan, U.S., and Lauben, D.S. 1999a. Subionospheric VLF signatures of oblique (nonducted) whistler-induced precipitation. *Geophys. Res. Lett.* **26**: 3569–72.

Johnson, M.P., Inan, U.S., Lev-Tov, S.J., and Bell, T.F. 1999b. Scattering pattern of lightning-induced ionospheric disturbances associated with early fast VLF events. *Geophys. Res. Lett.* **26**: 2363–6.

Jones, A.V. 1974. In *Aurora*, pp. 301, Boston. Massachusetts: Reidel.

Kelley, M.C., Farley, D.T., Kudeki, E., and Siefring, C.L. 1984. A model for equatorial explosive spread F. *Geophys. Res. Lett.* **11**: 1168–71.

Kelley, M.C., Siefring, C.L., Pfaff, R.F., Kintner, P.M., Larsen, M., Green, R., Holzworth, R.H., Hale, L.C., Mitchell, J.D., and Le Vine, D. 1985. Electrical measurements in the atmosphere and the ionosphere over an active thunderstorm, 1. Campaign overview and initial ionospheric results. *Geophys. Res. Lett.* **90**: 9815–23.

Kelley, M.C., Ding, J.G., and Holzworth, R.H. 1990. Intense ionospheric electric and magnetic field pulses generated by lightning. *Geophys. Res. Lett.* **17**: 2221–4.

Kelley, M.C., Baker, S.D., Holtzworth, R.H., Argo, P., and Cummer, S.A. 1997. LF and MF observations of the lightning electromagnetic pulse at ionospheric altitudes. *Geophys. Res. Lett.* **24**: 1111–14.

Kutsyk, I.M., Babich, L.P., and Bakhov, K.I. 1999. Selfconsistent computations of optical emissions in the model of gigantic upward atmospheric discharges governed by runaway electrons. In *Proc. Int. Conf. on Lightning and Static Electricity, Toulouse, France*, pp. 457–61.

Lauben, D.S., Inan, U.S., and Bell, T.F. 1998. VLF chorus emissions observed by POLAR during the Jan. 10, 1997 geomagnetic storm. *Geophys. Res. Lett.* **25**: 2995–8.

Lauben, D.S., Inan, U.S., and Bell, T.F. 1999. Poleward-displaced electron precipitation from lightning-generated oblique whistlers. *Geophys. Res. Lett.* **26**: 2633–6.

Lee, M.C., Dalkir, Y.R., and Williams, E.R. 1998. Radar reflectivity of lightning-induced plasmas. *J. Atmos. Solar-Terr. Phys.* **60**: 941–50.

Lee, M.C., Riddolls, R.J., and Moriarty, D.T. 1998. Laboratory study of some lightning-induced effects in the ionospheric plasma. *J. Atmos. Solar-Terr. Phys.* **60**: 965–73.

Lehtinen, N.G., Walt, M., Inan, U.S., Bell, T.F., and Pasko, V.P. 1996. γ-ray emission produced by a relativistic beam of runaway electrons accelerated by quasi-electrostatic thundercloud fields. *Geophys. Res. Lett.* **23**: 2645–8.

Lehtinen, N.G., Bell, T.F., Pasko, V.P., and Inan, U.S. 1997. A two-dimensional model of runaway electron beams driven by quasi-electrostatic thundercloud fields. *Geophys. Res. Lett.* **24**: 2639–42.

Lehtinen, N.G., Bell, T.F., and Inan, U.S. 1999. Monte Carlo simulation of runaway MeV electron breakdown with application to red sprites and terrestrial gamma ray flashes. *J. Geophys. Res.* **105**: 24 699–717.

Lev-Tov, S.J., Inan, U.S., and Bell, T.F. 1995. Altitude profiles of localized D-region density disturbances produced in lightning-induced electron precipitation events. *J. Geophys. Res.* **100**: 21 375–84.

Li, Y.Q., Holzworth, R.H., Hu, H., McCarthy, M., Massey, R.D., Kintner, P.M., Rodrigues, J.D., Inan, U.S., and Armstrong, W.C. 1991. Anomalous optical events detected by rocket-borne sensor in the WIPP campaign. *J. Geophys. Res.* **96**: 1315–26.

Liao, C.P., Freidberg, J.P., and Lee, M.C. 1989. Explosive spread F caused by lightning-induced electromagnetic effects. *J. Atmos. Terr. Phys.* **51**: 751–8.

Lohrey, B. and Kaiser, A.B. 1979. Whistler-induced anomalies in VLF propagation. *J. Geophys. Res.* **84**: 5121–30.

Lyons, W.A. 1994a. Characteristics of luminous structures in the stratosphere above thunderstorms as imaged by low-light video. *Geophys. Res. Lett.* **21**: 875–8.

Lyons, W.A. 1994b. Low-light video observations of frequent luminous structures in the stratosphere above thunderstorms. *Mon. Wea. Rev.* **122**: 1940–6.

Lyons, W.A. 1996. Sprite observations above the U.S. high plains in relation to their parent thunderstorm system, *J. Geophys. Res.* **101**: 29 641–52.

Lyons, W.A., and Armstrong, R.A. 1997. NO_x production within and above thunderstorms: the contribution of lightning and sprites. Preprint, *3rd Conf. on Atmospheric Chemistry, Long Beach*, pp. 3–12, American Meteorological Society.

Lyons, W.A., Eastman, J.L., Pielke, R.A., Biazar, A., and McNider, R. 1994. A preliminary climatology of lightning-generated NO_x and numerical simulations of its redistribution by deep convection. Preprint, *Conf. on Atmospheric*

Chemistry, pp. 193–8, American Meteorological Society, Nashville.

Lyons, W.A., Nelson, T.E., Williams, E.R., Cramer, J.A., and Turner, T.R. 1998a. Enhanced positive cloud-to-ground lightning in thunderstorms ingesting smoke from fires. *Science* **282**: 77–80.

Lyons, W.A., Uliasz, M., and Nelson, T.E. 1998b. Large peak current cloud-to-ground lightning flashes during the summer months in the contiguous United States. *Mon. Wea. Rev.* **126**: 2217–23.

Lyons, W.A., Armstrong, R.A., Gering, E.A., and Williams, E.R. 2000. The hundred year hunt for sprites. *Eos, Trans. Am. Geophys. Union* **81**: 373–7.

Ma, Z.F., Croskey, C.L., and Hale, L.C. 1998. The electrodynamic responses of the atmosphere and ionosphere to the lightning discharge. *J. Atmos. Solar-Terr. Phys.* **60**: 845–62.

Macky, W.A. 1934. An attempt to detect radiation in thunderclouds. *Proc. Cambridge Phil. Soc.* **30**: 70–3.

Malan, D. 1937. Sur les décharges orageuses dans la haute atmosphère. *C.R. Acad. Sci. Paris.* **205**: 812.

Marshall, T.C., Stolzenburg, M., and Rust, W.D. 1996. Electric field measurements above mesoscale convective systems. *J. Geophys. Res.* **101**: 6979–96.

Marshall, L.H., Hale, L.C., Croskey, C.L., and Lyons, W.A. 1998. Electromagnetics of sprite- and elve-associated sferics. *J. Atmos. Solar-Terr. Phys.* **60**: 771–86.

Marshall, T.C., Stolzenburg, M., Rust, W.D., Williams, E.R., and Boldi, R. 2001. Positive charge in the stratiform cloud of a mesoscale convective system. *J. Geophys. Res.* **106**: 1157–63.

Massey, R.S., and Holden, D.N. 1995. Phenomenology of transionospheric pulse pairs. *Radio Sci.* **30**: 1645–59.

Massey, R.S., Knox, S.O., Franz, R.C., Holden, D.N., and Rhodes, C.T. 1998. Measurements of transionospheric radio propagation parameters using the FORTE satellite. *Radio Sci.* **33**: 1739–53.

Maynard, N.C., Aggson, T.L., and Heppner, J.P. 1970. Electric field observations of ionospheric whistlers. *Radio Sci.* **5**: 1049–58.

Mazur, V., Shao, X., and Krehbiel, P.R. 1998. "Spider" lightning in intracloud and positive cloud-to-ground flashes. *J. Geophys. Res.* **103**: 19 811–22.

McCarthy, M.P., and Parks, G.K. 1985. Further observations of X rays inside thunderstorms. *Geophys. Res. Lett.* **12**: 393–6.

McCarthy, M.P., and Parks, G.K. 1992. On the modulation of X ray fluxes in thunderstorms. *J. Geophys. Res.* **97**: 5857–64.

Mende, S.B., Rairden, R.L., Swenson, G.R., and Lyons, W.A. 1995. Sprite spectra: N_2 first positive band identification. *Geophys. Lett.* **22**: 2633–6.

Milikh, G.M., Papadopoulos, K., and Chang, C.L. 1995. On the physics of high altitude lightning. *Geophys. Res. Lett.* **22**: 85–8.

Milikh, G.M., Usikov, D.A., and Valdivia, J.A. 1998a. Model of infrared emission from sprites. *J. Atmos. Solar-Terr. Phys.* **60**: 895–906.

Milikh, G.M., Valdivia, J.A., and Papadopoulos, K. 1998b. Spectrum of red sprites. *J. Atmos. Solar-Terr. Phys.* **60**: 907–16.

Milikh, G.M., and Valdivia, J.A. 1999. Model of gamma ray flashes due to fractal lightning. *Geophys. Res. Lett.* **26**: 525–8.

Mironychev, P.V., and Babich, L.P. 1999. Model of electric field discontinuities occurred above thunderclouds. In *Proc. Int. Conf on Lightning and Static Electricity, Toulouse, France*, pp. 67–9.

Mitchell, J.D. 1985. Electrical measurements in the atmosphere and the ionosphere over an active thunderstorm, 2, Direct current electric fields and conductivity. *J. Geophys. Res.* **90**: 9824–30.

Moore, C.B., Eack, K.B., Aulich, G.D., and Rison, W. 2001. Energetic radiation associated with lightning stepped-leaders. *Geophys. Res. Lett.* **28**: 2141–4.

Morrill, J.S., Bucsela, E.J., Pasko, V.P., Berg, S.L., Heavner, M.J., Moudry, D.R., Benesch, W.M., Wescott, E.M., and Sentman, D.D. 1998. Time resolved N_2 triplet state vibrational populations and emissions associated with red sprites. *J. Atmos. Solar-Terr. Phys.* **60**: 811–29.

Neubert, T., Allin, T.H., Stenbaek-Nielsen, H., and Blanc, E. 2001. Sprites over Europe. *Geophys. Res. Lett.* **28**: 3585–8.

Nickolaenko, A.P., and Hayakawa, M. 2000. Comment on "Model of red sprite due to intracloud fractal lightning discharges" by J.A. Valdivia, G.M Milikh, and K. Papadopoulos. *Radio Science* **35**: 921.

Nunn, D. 1997. On the numerical modelling of the VLF Trimpi effect. *J. Atmos. and Solar-Terr. Phys.* **59**: 537–60.

Nunn, D., and Rodger, C.J. 1999. Modeling the relaxation of red sprite plasma. *Geophys. Res. Lett.* **26**: 3293–6.

Papadopoulos, K., Milikh, G., and Valdivia, J. 1996. Comment on "Can gamma radiation be produced in the electrical environment above thunderstorms". *Geophys. Res. Lett.* **23**: 2283–4.

Park, C.G., and Dejnakarintra, M. 1973. Penetration of thundercloud electric fields into the ionosphere and magnetosphere, 1. Middle and subauroral latitudes. *J. Geophys. Res.* **78**: 6623–33.

Parks, G.K., Mauk, B.H., Spiger, R., and Chin, J. 1981. X ray enhancements detected during thunderstorm and lightning activities. *Geophys. Res. Lett.* **8**: 1176–9.

Pasko, V.P., and Inan, U.S. 1994. Recovery signatures of lightning-associated VLF perturbations as a measure of the lower ionosphere. *J. Geophys. Res.* **99**: 17 523–7.

Pasko, V.P., Inan, U.S., Taranenko, Y.N., and Bell, T.F. 1995. Heating, ionization and upward discharges in the mesosphere due to intense quasi-electrostatic thundercloud fields. *Geophys. Res. Lett.* **22**: 365–8.

Pasko, V.P., Inan, U.S., and Bell, T.F. 1996a. Blue jets produced by quasi-electrostatic pre-discharge thundercloud fields. *Geophys. Res. Lett.* **23**: 301–4.

Pasko, V.P., Inan, U.S., and Bell, T.F. 1996b. Sprites as luminous columns of ionization produced by quasi-electrostatic thundercloud fields. *Geophys. Res. Lett.* **23**: 649–52.

Pasko, V.P., Inan, U.S., Bell, T.F., and Taranenko, Y.N. 1997a. Sprites produced by quasi-electrostatic heating and ionization in the lower ionosphere. *J. Geophys. Res.* **102**: 4529–61.

Pasko, V.P., Inan, U.S., and Bell, T.F. 1997b. Sprites as evidence of vertical gravity wave structures above mesoscale thunderstorms. *Geophys. Res. Lett.* **24**: 1735–8.

Pasko, V.P., Inan, U.S., and Bell, T.F. 1998a. Ionospheric effects due to electrostatic thundercloud fields. *J. Atmos. Solar-Terr. Phys.* **60**: 863–70.

Pasko, V.P., Inan, U.S., and Bell, T.F. 1998b. Spatial structure of sprites. *Geophys. Res. Lett.* **25**: 2123–6.

Pasko, V.P., Inan, U.S., Bell, T.F., and Reising, S. 1998c. Mechanism of ELF radiation from sprites. *Geophys. Res. Lett.* **25**: 3493–6.

Pasko, V.P., Inan, U.S., and Bell, T.F. 1998d. Ionospheric effects due to electrostatic thundercloud fields. *J. Atmos. Solar-Terr. Phys.* **60**: 863–70.

Pasko, V.P., Inan, U.S., and Bell, T.F. 1999. Mesospheric electric field transients due to tropospheric lightning discharges. *Geophys. Res. Lett.* **26**: 1247–50.

Pasko, V.P., Inan, U.S., and Bell, T.F. 2000. Fractal structure of sprites. *Geophys. Res. Lett.* **27**: 497–500.

Pasko, V.P., Inan, U.S., and Bell, T.F. 2001. Mesosphere–troposphere coupling due to sprites. *Geophys. Res. Lett.* **28**: 3821–4.

Petrov, N.I., and Petrova, G.N. 1999. Physical mechanism for the development of lightning discharges between a thundercloud and the ionosphere. *Tech. Phys.* **44**: 472–5.

Phelps, C.T. 1974. Positive streamer system intensification and its possible role in lightning initiation. *J. Atmos. Terr. Phys.* **36**: 103–11.

Picard, R.H., Inan, U.S., Pasko, V.P., Winick, J.R., and Wintersteiner, P.O. 1997. Infrared glow above thunderstorm. *Geophys. Res. Lett.* **24**: 2635–8.

Poulsen, W.L., Bell, T.F., and Inan, U.S. 1993a. The scattering of VLF waves by localized ionospheric disturbances produced by lightning-induced electron precipitation. *J. Geophys. Res.* **98**: 15 553–9.

Poulsen, W.L., Inan, U.S., and Bell, T.F. 1993b. A multiple-mode three-dimensional model of VLF propagation in the Earth–ionosphere waveguide in the presence of localized D region disturbances. *J. Geophys. Res.* **98**: 1705–17.

Powell, G. 1968. Lightning. *Marine Observer* **38**: 173–3.

Rairden, R.L., and Mende, S.B. 1995. Time resolved sprite imagery. *Geophys. Res. Lett.* **22**: 3465–8.

Raizer, Y.P., Milikh, G.M., Shneider, M.N., and Novakovski, S.V. 1998. Long streamers in the atmosphere above thundercloud. *J. Phys. D, Appl. Phys.* **31**: 3255–64.

Rakov, V.A., Crawford, D.E., Rambo, K.J., Schnetzer, G.H., and Uman, M.A. 2001. M-component mode of charge transfer to ground in lightning discharges. *J. Geophys. Res.* **106**: 22 817–31.

Reising, S.C., Inan, U.S., Bell, T.F., and Lyons, W.A. 1996. Evidence for continuing current in sprite-producing cloud-to-ground lightning. *Geophys. Res. Lett.* **23**: 3639–42.

Reising, S.C., Inan, U.S., and Bell, T.F. 1999. ELF sferic energy as a proxy indicator for sprite occurrence. *Geophys. Res. Lett.* **26**: 987–90.

Rice, W.K.M., and Hughes, A.R.W. 1998. Whistlers, Trimpis and evidence that electron precipitation may trigger atmospheric discharges. *J. Atmos. Solar-Terr. Phys.* **60**: 1149–58.

Rodger, C.J. 1999. Red sprites, upward lightning and VLF perturbations. *Rev. Geophys.* **37**: 317–36.

Rodger, C.J., Wait, J.R., and Dowden, R.L. 1997. Electromagnetic scattering from a group of thin conducting cylinders. *Radio Sci.* **32**(3): 907–12.

Rodger, C.J., Wait, J.R., and Dowden, R.L. 1998a. VLF scattering from red sprites–theory. *J. Atmos. Solar-Terr. Phys.* **60**: 755–64.

Rodger, C.J., Wait, J.R., and Dowden, R.L. 1998b. Scattering of VLF from an experimentally described sprite. *J. Atmos. Solar-Terr. Phys.* **60**: 765–70.

Rodger, C.J., Wait, J.R., Dowden, R.L., and Thomson, N.R. 1998c. Radiating conducting columns inside the Earth–ionosphere waveguide: application to red sprites. *J. Atmos. Solar-Terr. Phys.* **60**: 1193–204.

Rodger, C.J., Cho, M., Clilverd, M.A., and Rycroft, M.J. 2001. Lower ionospheric modification by lightning-EMP: Simulation of the night ionosphere over the United States. *Geophys. Res. Lett.* **28**: 1999–202.

Rodriguez, J.V., and Inan, U.S. 1994. Electron density changes in the nighttime D region due to heating by very-low frequency transmitters. *Geophys. Res. Lett.* **21**: 93–6.

Rodriguez, J.V., Inan, U.S., and Bell, T.F. 1992. D region disturbances caused by electromagnetic pulses from lightning. *Geophys. Res. Lett.* **19**: 2067–70.

Rodriguez, J.V., Inan, U.S., and Bell, T.F. 1994. Heating of the nighttime D region by very low frequency transmitters. *J. Geophys. Res.* **99**: 23 329–38.

Rosenberg, T.J., Siren, J.C., Matthews, D.L., Marthinsen, K., Holtet, J.A., Egeland, A., Carpenter, D.L., and Helliwell, R.A. 1981. Conjugacy of electron microbursts and VLF chorus. *J. Geophys. Res.* **86**: 5819–32.

Roussel-Dupré, R.A., and Gurevich, A.V. 1996. On runaway breakdown and upward-propagating discharges. *J. Geophys. Res.* **101**: 2297–311.

Roussel-Dupré, R.A., Gurevich, A.V., Turnell, T., and Milikh, M. 1994. Kinetic theory of runaway air breakdown. *Phys. Rev. E* **49**: 2257–71.

Roussel-Dupré, R.A., Symbalisky, E., Taranenko, Y., and Yukhimuk, V. 1998. Simulations of high-altitude discharges initiated by runaway breakdown. *J. Atmos. Solar-Terr. Phys.* **60**: 917–40.

Rowland, H.L. 1998. Theories and simulations of elves, sprites and blue jets. *J. Atmos. Solar-Terr. Phys.* **60**: 831–44.

Rowland, H.L., Fernsler, R.F., Huba, J.D., and Bernhardt, P.A. 1995. Lightning driven EMP in the upper atmosphere. *Geophys. Res. Lett.* **22**: 361–4.

Rowland, H.L., Fernsler, R.F., and Bernhardt, P.A. 1996. Breakdown of the neutral atmosphere in the D region due to lightning-driven electromagnetic pulses. *J. Geophys. Res.* **101**: 7935–45.

Rumi, G.C. 1957. VHF radar echoes associated with atmospheric phenomena. *J. Geophys. Res.* **62**(4): 547–64.

Russell, C.T., Zuelsdorf, R.S., Strangeway, R.J., and Franz, R. 1998. Identification of the cloud pulse responsible for a trans-ionospheric pulse pair. *Geophys. Res. Lett.* **25**: 2645–8.

Rycroft, M.J., and Cho, M. 1998. Modelling electric and magnetic fields due to thunderclouds and lightning from cloud tops to the ionosphere. *J. Atmos. Solar-Terr. Phys.* **60**: 889–94.

Rycroft, M.J., Israelsson, S., and Price, C. 2000. The global atmospheric electric circuit, solar activity and climate change. *J. Atmos. Solar-Terr. Phys.* **62**: 1563–76.

Schonland, B.F.J. 1930. Thunderstorms and the penetrating radiation. *Proc. Roy. Soc. A* **130**: 37–63.

Schonland, B.F.J. and Viljoen, J.P.T. 1933. On a penetrating radiation from thunderclouds. *Proc. Roy. Soc. A* **140**: 314–33.

Showen, R.L., and Slingeland, A. 1998. Measuring lightning-induced ionospheric effects with incoherent scatter radar or with cross-modulation. *J. Atmos. Solar-Terr. Phys.* **60**: 951–56.

Sentman, D.D. 1998. Effects of thunderstorm activity on the upper atmosphere and ionosphere. *Special Issue, J. Atmos. Solar-Terr. Phys.* **60**: 667–8.

Sentman, D.D., and Wescott, E.M. 1993. Observations of upper atmosphere optical flashes recorded from an aircraft. *Geophys. Res. Lett.* **20**: 2857–60.

Sentman, D.D., and Wescott, E.M. 1995. Red sprites and blue jets: thunderstorm-excited optical emissions in the stratosphere, mesosphere, and ionosphere. *Phys. Plasmas* **2**: 2514–22.

Sentman, D.D., and Wescott, E.M. 1996. Red sprites and blue jets: high altitude optical emissions linked to lightning. *Eos, Trans. AGU* **77**: 1–4.

Sentman, D.D., Wescott, E.M., Osborne, D.L., Hampton, D.L., and Heavner, M.J. 1995. Preliminary results from the Sprites94 aircraft campaign, I, Red sprites. *Geophys. Res. Lett.* **22**: 1205–8.

Shaw, G.E. 1967. Background cosmic ray count increases associated with thunderstorms. *J. Geophys. Res.* **72**: 4623–6.

Showen, R.L., and Slingeland, A. 1998. Measuring lightning-induced ionospheric effects with incoherent scatter radar or with cross-modulation. *J. Atmos. Solar-Terr. Phys.* **60**: 951–6.

Smith, A.J., and Cotton, P.D. 1990. The Trimpi effect in Antarctica: observations and models. *J. Atmos. Terr. Phys.* **52**: 341–55.

Smith, A.J., Cotton, P.D., Robertson, J.S. 1993. Transient (∼10 s) VLF amplitude and phase perturbations due to lightning-induced electron precipitation into the ionosphere (the 'Trimpi effect'). In *Proc. AGARD Conf. on ELF/VLF/LF Radio Propagation and System Aspects*, vol. 529, pp. 8/1–8.

Smith, D.A. and Holden, D.N. 1996. Ground-based observations of sub-ionospheric pulse pairs. *Radio Sci.* **31**: 553–71.

Smith, D.A., Shao, X.M., Holden, D.N., Rhodes, C.T., Brook, M., Krehbiel, P.R., Stanley, M., Rison, W., and Thomas, R.J. 1999. A distinct class of isolated intracloud lightning discharges and their associated radio emissions. *J. Geophys. Res.* **104**: 4189–212.

Sonwalkar, V.S., and Inan, U.S. 1988. Wave normal direction and spectral properties of whistler mode hiss observed on the DE-1 satellite. *J. Geophys. Res.* **93**: 7493–517.

Stanley, M., Krehbiel, P., Brook, M., Moore, C., Rison, W., and Abrahams, B. 1999. High speed video of initial sprite development. *Geophys. Res. Lett.* **26**: 3201–4.

Stanley, M., Brook, M., Krehbiel, P., and Cummer, S.A. 2000. Detection of daytime sprites via a unique sprite ELF signature. *Geophys. Res. Lett.* **27**: 871–4.

Stergis, C.G., Rein, G.C., and Kangas, T. 1957. Electric field measurements in the stratosphere. *J. Atmos. Terr. Phys.* **11**: 77–82.

Strangeways, H.J. 1982. The effect of multi-duct structure on whistler-mode wave propagation. *J. Atmos. Terr. Phys.* **44**: 901–12.

Strangeways, H.J. 1996. Lightning, trimpis and sprites. In *Review of Radio Science*, ed. W.R. Stone, pp. 741–80, Oxford, New York: Oxford University Press.

Strangeways, H.J. 1999. Lightning induced enhancements of D-region ionisation and whistler ducts. *J. Atmos. and Solar-Terr. Phys.* **61**: 1067–80.

Sukhorukov, A.I. 1996. Lightning transient fields in the atmosphere-low ionosphere. *J. Atmos. Terr. Phys.* **58**: 1711–20.

Sukhorukov, A.I., and Stubbe, P. 1997. On ELF pulses from remote lightning triggering sprites. *Geophys. Res. Lett.* **24**: 1639–42.

Sukhorukov, A.I., and Stubbe, P. 1998. Problems of blue jet theories. *J. Atmos. Solar-Terr. Phys.* **60**: 725–32.

Sukhorukov, A.I., Mishin, E.V., Stubbe, P., and Rycroft, M.J. 1996a. On blue jet dynamics. *Geophys. Res. Lett.* **23**: 1625–8.

Sukhorukov, A.I., Rubenchik, E.A., and Stubbe, P. 1996b. Simulation of strong lightning pulse penetration into the lower ionosphere. *Geophys. Res. Lett.* **23**: 2911–14.

Suszcynsky, D.M., Roussel-Dupré, R., and Shaw, G. 1996. Ground-based search for X rays generated by thunderstorms and lightning. *J. Geophys. Res.* **101**: 23 505–16.

Suszcynsky, D.M., Roussel-Dupré, R., Lyons, W.A., and Armstrong, R.A. 1998. Blue-light imagery and photometry of sprites. *J. Atmos. Solar-Terr. Phys.* **60**: 801–10.

Suszcynsky, D.M., Strabley, R., Roussel-Dupré, R., Symbalisty, E.M.D., Armstrong, R.A., Lyons, W.A., and Taylor, M. 1999. Video and photometric observations of a sprite in coincidence with a meteor-triggered jet event. *J. Geophys. Res.* **104**: 31 361–7.

Suszcynsky, D.M., Kirkland, M.W., Jacobson, A.R., Franz, R.C., Knox, S.O, Guillen, J.L.L., and Green, J.L. 2000. FORTE observations of simultaneous VHF and optical emissions from lightning: basic phenomenology. *J. Geophys. Res.* **105**: 2191–201.

Symbalisty, E.M.D., Roussel-Dupré, R.A., and Yukhimuk, V.A. 1998. Finite volume solution of relativistic Boltzmann equation for electron avalanche rates. *IEEE Trans. Plasma Sci.* **26**: 1575–82.

Taranenko, Y.N., and Roussel-Dupré, R.A. 1996. High altitude discharges and gamma-ray flashes: a manifestation of runaway air breakdown. *Geophys. Res. Lett.* **23**: 571–4.

Taranenko, Y.N., Inan, U.S., and Bell, T.F. 1992. Optical signatures of lightning-induced heating of the D region. *Geophys. Res. Lett.* **19**: 1815–18.

Taranenko, Y.N., Inan, U.S., and Bell, T.F. 1993a. Interaction with the lower ionosphere of electromagnetic pulses from lightning: heating, attachment, and ionization. *Geophys. Res. Lett.* **20**: 1539–42.

Taranenko, Y.N., Inan, U.S., and Bell, T.F. 1993b. Interaction with the lower ionosphere of electromagnetic pulses from lightning: excitation of optical emissions. *Geophys. Res. Lett.* **20**: 2675–8.

Tomko A.A., Ferraro, A.J., Lee, H.S., and Mitra, A.P. 1980. A theoretical model of D-region ion chemistry modifications during high power radio wave heating. *J. Atmos. Terr. Phys.* **42**: 275–85.

Toynbee, H., and Mackenzie, T. 1886. Meteorological phenomena. *Nature* **33**: 245–245.

Tsunoda, R.T., Livingston, R.C., Buonocore, J.J., Lyons, W.A., Nelson, T.E., and Kelley, M.C. 1998. Evidence of a high-altitude discharge process responsible for radar echoes at 24.4 MHz. *J. Atmos. Solar-Terr. Phys.* **60**: 957–64.

Tzur, I., and Roble, R.G. 1985. The interaction of a dipolar thunderstorm with its global electrical environment. *J. Geophys. Res.* **90**: 5989–99.

Valdivia, J.A., Milikh, G.M., and Papadopoulos, K. 1997. Red sprites: lightning as a fractal antenna. *Geophys. Res. Lett.* **24**: 3169–72.

Valdivia, J.A., Milikh, G.M., and Papadopoulos, K. 1998. Model of red sprites due to intracloud fractal lightning discharges. *Radio Sci.* **33**: 1655–68.

Vaughan, O.H., and Vonnegut, B. 1982. Lightning to the ionosphere. *Weatherwise* **35**: 70–1.

Vaughan, O.H., and Vonnegut, B. 1989. Recent observations of lightning discharges from the top of a thundercloud into the clear air above. *J. Geophys. Res.* **94**: 13 179–82.

Vaughan, O.H. Jr, Blakeslee, R., Boeck, W.L., Vonnegut, B., Brook, M., and McKune, J. Jr 1992. A cloud-to-space lightning as recorded by the Space Shuttle payload bay TV cameras. *Mon. Wea. Rev.* **120**: 1459–61.

Velinov, P.I., and Tonev, P.T. 1995. Modeling the penetration of thundercloud electric fields into the ionosphere. *J. Atmos. Terr. Phys.* **57**: 687–94.

Veronis, G., Pasko, V.P., and Inan, U.S. 1999. Characteristics of mesospheric optical emissions produced by lightning discharges. *J. Geophys. Res.* **104**: 12 645–56.

Vonnegut, B. 1980. Cloud to stratosphere lightning. *Weather* **35**: 59–60

Vonnegut, B., Vaughan, O.H., and Brook, M. 1989. Nocturnal photographs taken from a U-2 airplane looking down on tops of clouds illuminated by lightning. *Bull. Am. Meteor. Soc.* **70**: 1263–1271.

Voss, H.D., Imhof, W.L., Walt, M., Mobilia, J., Gaines, E.E., Reagan, J.B., Inan, U.S., Helliwell, R.A., Carpenter, D.L., Katsufrakis, J.P., and Chang, H.C. 1984. Lightning-induced electron precipitation. *Nature* **312**: 740–2.

Voss, H.D., Walt, M., Imhof, W.L., Mobilia, J., and Inan, U.S. 1998. Satellite observations of lightning-induced electron precipitation. *J. Geophys. Res.* **103**: 11 725–44.

Wait, J.R. 1991. EM scattering from a vertical column of ionization in the earth–ionosphere waveguide. *IEEE Trans. Antennas Propag.* **39**: 1051–4.

Wait, J.R. 1995. VLF scattering from a column of ionisation in the earth–ionosphere waveguide. *J. Atmos. Terr. Phys.* **57**(8): 955–9.

Wait, J.R. 2000. Scattering from an ionized column in the Earth–ionosphere space. *IEEE Trans. Antennas Propag.* **48**(9): 1496–7.

Wang, C., Yan, M., Liu, X., Zhang, Y., Dong, W., and Zhang, C. 1999. Bidirectional propagation of lightning leader. *Chinese Sci. Bull.* **44**(2): 163–6.

Wescott, E.M., Sentman, D.D., Osborne, D.L., Hampton, D.L., and Heavner, M.J. 1995. Preliminary results from the Sprites94 aircraft campaign 2, Blue jets. *Geophys. Res. Lett.* **22**: 1209–12.

Wescott, E.M., Sentman, D.D., Heavner, M.J., Osborne, D.L., and Vaughan, O.H. 1996. Blue starters: brief upward discharges from an intense Arkansas thunderstorm. *Geophys. Res. Lett.* **23**: 2153–6.

Wescott, E.M., Sentman, D.D., Heavner, M.J., Hampton, D.L., Lyons, W.A., and Nelson, T. 1998a. Observations of 'columniform' sprites. *J. Atmos. Solar-Terr. Phys.* **60**: 733–40.

Wescott, E.M., Sentman, D.C., Heavner, M.J., Hampton, D.L., and Vaughan, O.H. 1998b. Blue jets: their relationship to lightning and very large hailfall, and physical mechanisms for their production. *J. Atmos. Solar-Terr. Phys.* **60**: 713–24.

Wescott, E.M., Sentman, D.D., Stenbaek-Nielsen, H.C., Huet, P., Heavner, M.J., and Moudry, D.R. 2001a. New evidence for the brightness and ionization of blue starters and blue jets. *J. Geophys. Res.* **106**: 10 467–77.

Wescott, E.M., Stenbaek-Nielsen, H.C., Sentman, D.D., Heavner, M.J., Moudry, D.R., and São Sabbas, F.T. 2001b. Triangulation of sprites, associated halos and their possible relation to causative lightning and micrometeors. *J. Geophys. Res.* **106**(A6): 10 467–77.

Whitmire, D.P. 1979. Search for high-energy radiation near lightning strokes. *Lett. Nuovo Cimento* **26**: 497–501.

Williams, E.R. 1998. The positive charge reservoir for sprite-producing lightning. *J. Atmos. Solar-Terr. Phys.* **60**: 689–92.

Williams, E.R. 2001. Sprites, elves, and glow discharge tubes. *Physics Today*, November, 41–7.

Wilson, C.T.R. 1925a. The electric field of a thunderstorm and some of its effects. *Proc. Roy. Soc. D* **37**: 32–7.

Wilson, C.T.R. 1925b. The acceleration of beta-particles in strong electric fields such as those of thunderclouds. *Proc. Cambridge Phil. Soc.* **22**: 534–8.

Wilson, C.T.R. 1956. A theory of thundercloud electricity. *Proc. Roy. Meteor. Soc.* **236**: 297–317.

Winckler, J.R. 1995. Further observations of cloud–ionosphere electrical discharges above thunderstorms. *J. Geophys. Res.* **100**: 14 335–45.

Winckler, J.R. 1997. The cloud ionosphere discharge: a newly observed thunderstorm phenomena. *Proc. Natl. Acad. Sci. USA* **94**: 10512–19.

Winckler, J.R. 1998. Optical and VLF radio observations of sprites over a frontal storm viewed from O'Brien Observatory of the University of Minnesota. *J. Atmos. Solar-Terr. Phys.* **60**: 679–88.

Winckler, J.R. Franz, R.C., and Nemzek, R.J. 1993. Fast low level light pulses from the night sky observed with the SKYFLASH program. *J. Geophys. Res.* **98**: 8775–83.

Winckler, J.R., Lyons, W.A., Nelson, T.E., and Nemzek, R.J. 1996. New high-resolution ground based studies of sprites. *J. Geophys. Res.* **101**: 6997–7004.

Wolf, T.G., and Inan, U.S. 1990. Path-dependent properties of subionospheric VLF amplitude and phase perturbations associated with lightning. *J. Geophys. Res.* **95**: 20 997–1005.

Wood, C.A. 1951. Unusual lightning. *Weather* **6**: 64.

Wright, J.B. 1950. A thunderstorm in the tropics. *Weather* **5**: 230.

Yukhimuk, V., Roussel-Dupré, R.A., Symbalisty, E.M.D., and Taranenko, Y. 1998. Optical characteristics of red sprites produced by runaway air breakdown. *J. Geophys. Res.* **103**: 11 473–82.

Yukhimuk, V., Roussel-Dupré, R.A., and Symbalisty, E.M.D. 1999. On the temporal evolution of red sprites: runaway theory versus data. *Geophys. Res. Lett.* **26**: 679–82.

Zabotin, N.A., and Wright, J.W. 2001. Role of meteoric dust in sprite formation. *Geophys. Res. Lett.* **28**: 2593–6.

Zuelsdorf, R.S., Strangeway, R.J., Russell, C.T., Casler, C., Christian, H.J., and Franz, R.C. 1997. Trans-ionospheric pulse pairs (TIPP): their geographic distribution and seasonal variations. *Geophys. Res. Lett.* **24**: 3165–8.

Zuelsdorf, R.S., Casler, C., Strangeway, R.J., Russell, C.T., and Franz, R. 1998a. Ground detection of trans-ionospheric pulse pairs by stations in the National Lightning Detection Network. Geophys. Res. Lett. **25**: 481–4.

Zuelsdorf, R.S., Strangeway, R.J., Russell, C.T., Franz, R. 1998b. Trans-ionospheric pulse pairs (TIPPs): their occurrence rates and diurnal variation. *Geophys. Res. Lett.* **25**: 3709–12.

Zuelsdorf, R.S., Franz, R.C., Strangeway, R.J., and Russell, C.T. 2000. Determining the source of strong LF/VLF TIPP events: implications for association with NPBPs and NNBPs. *J. Geophys. Res.* **105**(D16): 20 725–36.

15 Lightning effects on the chemistry of the atmosphere

> ... it will be necessary to learn much more about the physics of lightning and its global occurrence before it will be possible to assess accurately its importance as a source of nitrogen oxides.
>
> B. Vonnegut (1982)

15.1. Introduction

The Earth's atmosphere below about 80–90 km, the height above which significant ionization occurs (subsection 13.2.1 and Fig. 13.5), is composed of about 21 percent molecular oxygen, O_2, about 78 percent molecular nitrogen, N_2, and small percentages of carbon dioxide, CO_2, molecular hydrogen, H_2, argon, Ar, and other noble gases, and water vapor, H_2O. Both the pressure and density of the atmosphere decrease approximately exponentially with altitude, with a decay height of about 10 km for all constituents of the air except water vapor. The water vapor content decreases more rapidly with height. The overall pressure at an altitude of 15 km is about one-tenth that at the Earth's surface, at 30 km about one-hundredth, and at 50 km near one-thousandth.

Different regions of the atmosphere have been given different names, based on the characteristics of their temperature profiles, as illustrated in Fig. 14.1. Adjacent to the Earth's surface is the *troposphere*, in which the temperature decreases with height. The troposphere is often called the lower atmosphere. In the mid-latitudes the troposphere extends from the Earth's surface, where the temperature is typically near 300 K, to the *tropopause*, near 15 km, where the temperature in the summer is just above 200 K, although the jet stream can produce a discontinuity in the tropopause leading to a double tropopause. In the tropics the tropopause is a few kilometers higher and 10 to 20 K colder than in temperate regions. The tropopause is also discussed in Section 3.1.

The *stratosphere* extends from the tropopause to about 50 km, near the top of the ozone layer. Some thunderclouds can penetrate the lower stratosphere, as noted in Section 3.1. The stratospheric temperature increases with height, the temperature at 50 km being about 270 K. The atmospheric temperature decreases again vertically through the *mesosphere*, which extends from about 50 km to the bottom of the ionosphere at 80 to 90 km, where the temperature is about 180 K. Together the mesosphere and the stratosphere are often called the middle atmosphere. The ionosphere is sometimes called the upper atmosphere or thermosphere.

Atmospheric electrical discharges, including the corona from thundercloud water drops and ice particles, the various types and phases of lightning discharge (Chapters 4, 5, and 9), and the sprites, blue starters, blue jets, and elves occurring in the region between the cloud tops and the ionosphere (Chapter 14), produce new trace molecules from the ambient constituents of the atmosphere. Nitric oxide, NO, is the most important electric-discharge-produced molecule, primarily because it facilitates chemical reactions in the troposphere and stratosphere that determine the concentrations of ozone, O_3, and of the hydroxyl radical OH. Nitric oxide is also produced by various biogenic and anthropogenic processes. Crutzen (1970) first pointed out the role of the nitrogen oxides in the chemistry of the troposphere and stratosphere. In parts of the lower and upper troposphere, the concentrations of NO and nitrogen dioxide, NO_2, whether from car exhausts, smokestack emissions, biomass burning, lightning, or other sources, are sufficient to induce an increase in the O_3 concentration, whereas in the highest parts of the troposphere photochemical processes destroy ozone (Crutzen 1970). In most of the stratosphere the NO produced by natural processes, primarily the reaction of N_2O with O, acts to decrease the O_3 concentration via the dominant reaction

$$NO + O_3 \rightarrow NO_2 + O_2$$

In the lower stratosphere, lightning and aircraft exhausts contribute to the ambient NO, which can either increase or decrease O_3 levels depending on local conditions. Ozone in high enough concentrations in the lower troposphere can be toxic to plants and animals and is a common component of smog. Ozone in the stratosphere is important to life on Earth because it shields the Earth from the Sun's harmful ultraviolet radiation. Although 90 percent of the ozone resides in the stratosphere, ozone near the tropopause absorbs infrared radiation at wavelengths that stratospheric ozone does not, making tropospheric ozone a greenhouse gas.

Lightning channels in air generate primarily NO as a trace gas with considerably less NO_2 (see Sections 15.2 and 15.3), but NO, once present, is always accompanied by the NO_2 produced from atmospheric oxidation of NO. In the sunlit troposphere, NO_2 is photolyzed (broken into NO and O by light photons); subsequent chemical reactions and the photolyzation of NO_2 establish a balance between NO and NO_2 on a time scale of minutes. The term NO_x is often used in the literature to refer to the mixture of NO and NO_2. Nitric oxide and NO_2 are also members of the fixed-nitrogen family of molecules that includes ammonia, NH_3. Nitrogen is said to be fixed when it is a part of a *less* stable molecule than N_2 and hence is in a form (i) that terrestrial and marine plants can more readily use in photosynthesis and (ii) that can react more readily with other atmospheric gases than N_2. Besides NO_x, trace molecules produced by or generated as a result of atmospheric electrical discharges include CO, N_2O, and the hydroxyl radical OH. Finally, the designation "odd nitrogen" or "reactive nitrogen" is sometimes given to all or a subset of atmospheric molecules containing one nitrogen atom, and the term NO_y is sometimes used to refer to the group including N, NO, NO_2, NO_3, N_2O_5, HNO_2, HO_2NO_2, HNO_3, organo-nitrates, and particulate nitrate, all potentially produced by atmospheric electrical discharges or in interactions of discharge-produced trace gases with ambient air constituents. There are some inconsistencies in these definitions in the literature.

A review of the early research on the production of NO_x by lightning is given by Hutchinson (1954). In the 19th century, von Liebig (1827) proposed that the dissolved nitrate NO_3^- found in rainwater arises from lightning's conversion of molecular nitrogen to NO and the subsequent oxidation of NO in the atmosphere to form nitric acid, via the reactions

$$NO + O_3 \rightarrow NO_2 + O_2$$

and

$$NO_2 + OH \xrightarrow{M} HNO_3$$

Following von Liebig's proposal, a number of investigators working in the late 19th and early 20th centuries attempted to infer the lightning NO_x production rate by determining the degree of correlation between local lightning flash rates and the concentration of nitrate in rainwater. The conclusion of these studies was generally that the lightning was a relatively small source of NO_x compared with biogenic and anthropogenic processes. It is now recognized that these early studies were flawed in that a time much longer than the typical thunderstorm duration is required to convert atmospheric NO to the soluble nitrate incorporated in rain, and hence a strong correlation between local lightning and nitrate in rainwater would not be expected even if lightning were a major producer of atmospheric NO (Tuck 1976; Chameides et al. 1977). In the 1970s, some 150 years after von Liebig's proposal, it was established that NO produced by atmospheric electric discharges can indeed play an important role in the chemistry of the Earth's atmosphere, particularly, as noted earlier, in regulating the concentrations of O_3 and OH (e.g., Crutzen 1970, 1973; Chameides and Walker 1973).

While it is clear that lightning and other electrical discharges in the atmosphere produce a variety of trace gases, the rate of production is a matter of controversy. Various published estimates of the global annual NO production by atmospheric electric discharges range from 1 to over 200 Tg(N) yr^{-1}, as we will discuss later, where 1 Tg(N) = 10^{12} grams or 10^6 metric tons of nitrogen. In estimates that involve a calculated or a laboratory-spark-measured value of NO production per discharge, or per unit energy, and subsequent extrapolation of this value to the NO production by all atmospheric electric discharges, considerable uncertainty is caused by a lack of knowledge of the type, number, and energy of atmospheric discharges and by questions of the applicability of the calculations and laboratory measurements to actual atmospheric discharges. In the approaches to estimating the global NO production that begin with the sampling of trace constituents in the atmosphere in the vicinity of lightning flashes and/or of thunderstorms, the NO production per flash or per thunderstorm is difficult to quantify accurately because assumptions must be made about air transport from the source to the measuring instrument and about the character of the source. Again, extrapolation to the global NO production by all flashes or by all thunderstorms necessarily involves considerable uncertainty.

Table 15.1 gives three estimates for the global NO_x budget of the lower atmosphere, which serve to put the NO_x production by atmospheric electrical discharges in the lower atmosphere in context. Since the concentration of NO_x in the lower atmosphere is more or less constant, the total production (source) rate there must be roughly equal to the total loss (sink) rate. Thus, if the global removal rate of NO_x is fairly accurately known and likewise the biomass burning and fossil fuel NO_x production rates then limits can be placed on the global production by less well-understood sources, atmospheric electric discharges and microbial activity in soils (e.g., Logan 1983). The only known significant loss of atmospheric reactive nitrogen is deposition to the Earth's surface. Nitrate is the most stable thermodynamically of the NO_y species and nearly all NO_x is ultimately oxidized to nitrate. Nitrate, much more soluble than most forms of reactive nitrogen, can be lost from the atmosphere by incorporation into precipitation, a process called rainout, or by direct contact with the Earth's surface in dry

15.1. Introduction

Table 15.1. *Assessments of the global NO_x budget of the lower atmosphere in $Tg(N)\ yr^{-1}$. Adapted from Lawrence et al. (1995)*

	Ehhalt and Drummond (1982)	Logan (1983)	Penner *et al.* (1991)
Sources			
Fossil fuel combustion	13.5 (8.2–18.5)	21 (14–28)	22.4
Biomass burning	11.2 (5.6–16.4)	12 (4–24)	5.8
Soil emissions	5.5 (1–10)	8 (4–16)	10.0
Lightning	5 (2–8)	8 (2–20)	3.0
Other sources	4 (1.7–6.2)	6 (5–11.5)	1.0
Total sources	39 (19–59)	55 (26–100)	42.2
Sinks			
Wet deposition	24 (15–33)	27 (12–42)	
Dry deposition	Very small, but uncertain (0–7)	17 (12–22)	
Total sinks	24 (15–40)	44 (24–64)	

deposition. Over the oceans, nitric acid vapor attaches to sea salt particles, which then return to the ocean as particulate nitrate. Over the continents, the dry deposition of nitric acid vapor is a major sink. Direct measurements of nitrate in precipitation and nitrate aerosol, and estimates of dry deposition based on HNO_3 concentrations, provide estimates of the sink strength (Galloway *et al.* 1982; Albritton *et al.* 1984; Logan 1983; Stedman and Shetter 1983; Jacob 2000; Table 15.1). These studies have established the global NO_x loss as 20 to 60 $Tg(N)\ yr^{-1}$. Because anthropogenic sources are known with reasonable confidence to produce 20 to 50 $Tg(N)\ yr^{-1}$ (Table 15.1), it can be argued that a reasonable upper limit to the NO_x production by lightning is 20 to 30 $Tg(N)\ yr^{-1}$. Higher levels inferred from field measurements are found in Table 15.2 and discussed in Section 15.4, and arguments for higher levels of production are given in Section 15.2.

As a matter of reference, the total Earth-based anthropogenic nitrogen fixation that is *not emitted* into the atmosphere and hence is not listed in Table 15.1 is estimated at 60 $Tg(N)\ yr^{-1}$, primarily from the industrial production of solid and liquid fertilizers (Delwiche 1970a, b). Depending on the actual value of the discharge-generated NO as against that produced by man at the Earth's surface by, for example, industrial smoke stack emissions or biomass burning, lightning and other atmospheric electrical discharges could be locally important sources of NO_x in the lower atmosphere. For example, Biazar and McNider (1995) used the US National Lightning Detection Network (Section 17.5) data for summers 1989 through 1992 together with assumed NO_x production rates to argue that lightning-produced NO_x in the southeastern United States in summer can be comparable to anthropogenic NO_x. Lee *et al.* (1997) estimated that lightning contributes 10 to 20 percent of the global NO_x, which is consistent with the data in Table 15.1. Singh *et al.* (1996) argued that lightning is the dominant NO_x source in the upper troposphere while the upward transport of anthropogenic NO_x produced at the Earth's surface provides only 20 percent of the NO_x found in the upper troposphere. MacGorman and Rust (1998) gave the following three arguments for lightning's being a major source of NO_x in the upper troposphere.

(i) Anthropogenic sources and transport from the stratosphere provide too little NO_x to account for observed concentrations in the upper troposphere, so there must be another natural source (e.g., Ko *et al.* 1986; Murphy *et al.* 1993; Smyth *et al.* 1996; Singh *et al.* 1996; Levy *et al.* 1996; Egorova *et al.* 2000).

(ii) NO_x concentrations in the upper tropical troposphere are substantial, but highly variable, suggesting a transient source (e.g., Davis *et al.* 1987; Murphy *et al.* 1993; Hauglustaine *et al.* 2001).

(iii) Airborne sensors have found substantially enhanced NO_x concentrations in and near thunderstorm anvils, even when the storms had moderate flash rates (e.g., Chameides *et al.* 1987; Davis *et al.* 1987; Ridley *et al.* 1987, 1996; Luke *et al.* 1992; Murphy *et al.* 1993).

Additionally, Pickering *et al.* (1998) calculated from an air transport model with assumed locations for the lightning sources that the maximum value of the ratio of NO_x to nitrogen is found a few kilometers below the tropopause and is about 15 percent by mass. Similar results obtained from modeling by Stockwell *et al.* (1999) and others are discussed in Section 15.8.

Atmospheric trace gases are potentially produced by the primary current-carrying channels of initial breakdown processes, leaders, return strokes, M-components,

Table 15.2. *Some estimates of the global production of NO by atmospheric electrical discharges. See Section 15.5 for definitions of all parameters in the table. Adapted from Lawrence et al. (1995)*

p(NO), 10^{16} NO J^{-1}	E_f, 10^8 J flash^{-1}	P(NO), 10^{25} NO flash^{-1}	F, 10^2 flashes s^{-1}	G(NO), Tg(N) yr^{-1}	References
FEA theoretical estimates					
1–8	—	1.1	5	4.0	Tuck (1976)
3–7	20	6–14	4	18–41	Chameides et al. (1977)
8–17	20	16–34	4	47–100	Chameides (1979a)
6	1.5	0.8	5	3	Dawson (1980)
24	0.5	1.2^a	1	0.9^a	Hill et al. (1980)
9 ± 2	4(1.6–10)	3.6 ± 0.8	1	2.6 ± 0.6	Borucki and Chameides (1984)
32	0.5	1.6	1	1.2	Bhetanabhotla et al. (1985)
FEA laboratory-based estimates					
					Chameides et al. (1977)
6 ± 1	20	12 ± 2	4	35 ± 6	Low energy sparks
8 ± 4	20	16 ± 4	4	47 ± 23	High energy sparks
5 ± 2	1	0.5	5	1.8 ± 0.7	Levine et al. (1981)
4.2	20	8.4	4	25	Peyrous and Lapeyre (1982)
30–40	3.8	3.1	0.3–1	2.5–8.3	Wang et al. (1998)
FEA field-observation-based estimates					
—	—	10	5	37	Noxon (1976, 1978)
—	—	40(10–100)	1	30	Drapcho et al. (1983)
—	—	300	1	230	Franzblau and Popp (1989)
Thunderstorm extrapolation-based estimates					
—	—	—	—	7	Chameides et al. (1987)
—	—	—	—	2–5^c	Ridley et al. (1996)
—	—	—	—	0.7^b	Smyth et al. (1996)
—	—	—	—	0.3–22^d	Huntrieser et al. (1998)
Nuclear explosion extrapolation-based estimate					
2.4	6.25	1.5	5	5.6	Tuck (1976)
Review-based estimates					
—	50	—	1	72 ± 96 (theory) 19 ± 10 (laboratory) 152 ± 60 (field)	Liaw et al. (1990)
—	4	2.3(1–7)	1(0.7–1.5)	2(1–8)	Lawrence et al. (1995)
10	67	67	0.2–03	13(5–25)	Price et al. (1997a, b)

^{a} 6 × 10^{25} molecules per flash and 4.4 Tg(N) yr^{-1} are stated in the original Hill et al. (1980) paper; the values here include a correction factor of 5.2 for the miscalculation noted by Borucki and Chameides (1984), as indicated by Lawrence et al. (1995).
^{b} South Atlantic basin, not global, injected into upper troposphere; probably half from lightning, half from biomass burning.
^{c} NO$_x$ produced or transported above 8 km altitude; includes also a lightning-based estimate (see Section 15.4); most NO$_x$ was NO.
^{d} Value for NO$_x$ of which NO is 50 to 70 percent.

continuing currents, K-processes, and J-processes. Besides this, it is likely that they are produced in the corona streamers associated with some lightning processes, in the corona discharges from water drops and ice particles in high-thunderstorm electric fields, and in discharges above the cloud tops (in the middle atmosphere) such as sprites, blue starters, blue jets, and elves, but the importance of sources other than the primary current-carrying channels is uncertain. Trace-gas production in corona discharges has been studied, for example, by Hill et al. (1984, 1988), Bhetanabhotla et al. (1985), Sisterson and Liaw (1990), Coppens et al. (1998), and Cooray et al. (1998), with varying results. Cooray et al. claimed from laboratory experiments that the NO$_x$ production by corona streamers from metallic electrodes is of the same order of magnitude per unit energy, around 10^{16} molecules per joule, as that from return strokes per joule of return stroke energy (Table 15.2); according to Hill et al. (1988), corona discharges produce about 10^{16} molecules of NO per joule, 10^{17} molecules of N_2O per joule, and 4×10^{17} molecules of O_3 per joule. Airborne measurements in thunderstorm anvils generally show no enhancement of ozone (Section 15.6), although Martin *et al.* (2000) presented evidence for large-scale ozone enhancements presumably associated with the emissions of NO_x from lightning in the tropical troposphere. Coppens *et al.* (1998) argued, from a calculation involving leader and streamer processes coupled to a 35-species chemical kinetic model, that NO_x production from lightning corona is negligible compared with that produced by the return stroke. The fraction of the available energy of a typical flash dissipated in corona streamers relative to the fraction dissipated in the primary lightning current-carrying channels is not known. Cooray (1997), from consideration of the electrostatic energy change when cloud charge is effectively moved by lightning, calculates that about one-third of the combined leader–return-stroke energy in ground flashes is dissipated in the leader and further that, for a given amount of neutralized charge, a cloud flash dissipates more energy than a ground flash. Trace-gas production by sprites (Section 14.3) was discussed by Lyons *et al.* (1994) and Lyons and Armstrong (1997). While many assumptions were made in their analysis that may influence the results, Lyons and Armstrong (1997) calculated that, above a sprite-producing thunderstorm, NO production by the sprites is two orders of magnitude greater than by ambient processes at 30 to 40 km height, while globally the reverse is true because of the relatively small volume that the sprites occupy.

Table 15.2, adopted from Lawrence *et al.* (1995) with some revisions and with information from later papers added, summarizes the methods and results of various estimates of the global NO production rate due to atmospheric electrical discharges. Many papers referenced give a range of results for a range of assumptions so that the numbers in Table 15.2 are necessarily chosen as representative, often a qualitative judgment on our part. In most of the papers referenced in Table 15.2, as noted above, the lightning return stroke is assumed to be the NO source. The so-called flash extrapolation approach (FEA), discussed in Section 15.5, is used to estimate the global NO production starting with one of three inputs: (i) a computation of the number of NO molecules per unit energy for a single return-stroke channel, (ii) a laboratory measurement of the number of NO molecules per input energy for a laboratory spark, or (iii) a ground-based observation of the NO in the vicinity of a natural lightning flash. In (i) and (ii) the extrapolation involves multiplying the number of NO molecules per unit lightning energy by estimates of the energy of a lightning event and by the number of worldwide lightning events per year. In (iii) the number of NO molecules per lightning flash is multiplied by the global number of lightning events per year. As indicated in Table 15.2, there have also been airborne measurements made in the vicinity of thunderstorms from which the NO production rate has been estimated (Section 15.6), and there has been a calculation of the NO production rate based on data from nuclear explosions (Section 15.7). Finally, there are review papers listed in Table 15.2 that attempt to organize the available published data and to draw conclusions about the global annual NO production. As noted in Section 15.1, while it is primarily NO that is created in atmospheric electrical discharges some of that NO is rapidly converted to NO_2, and measurements made away from the source are generally of NO_x. Thus global NO production and global NO_x production are often referred to interchangeably, resulting in some confusion in the literature.

Much of the remainder of this chapter concerns the data listed in Table 15.2. In Sections 15.2–15.7 we examine the methods used to estimate the global NO production rate by atmospheric electrical discharges. In Section 15.8 we touch upon the transport of trace gases produced by atmospheric discharges, and in Section 15.9 we examine the production of trace gases by lightning in the primitive Earth atmosphere and in the atmospheres of other planets. The latter topic is also discussed in Chapter 16.

15.2. Mechanism of NO production by return-stroke channels

As discussed in Section 11.3, some of the early theories describing return-stroke channel development modeled the return stroke as a line source of negligible radius that instantaneously released the return-stroke energy, producing a cylindrical shock wave. These early attempts at modeling the return stroke and its associated shock wave were followed by more realistic numerical simulations that used either a time-varying energy or a time-varying return-stroke current as an input parameter, beginning with an assumed initial radius of the order of a millimeter (subsections 12.2.2 and 12.2.6, first part; Figs. 12.1–12.4). In these more realistic simulations the expanding cylindrical shock wave is driven by ohmic heating from current flow in the channel. The shock wave propagates outward and decays to near-atmospheric pressure in tens of microseconds, leaving behind a hot channel of centimeter radius in pressure equilibrium with the ambient atmosphere. The peak channel temperature is near 30 000 K and is reached within a few microseconds of current initiation. Uman and Voshall (1968) and Picone *et al.* (1981) showed theoretically that when the lightning current ceases, in hundreds of microseconds or in milliseconds, the residual hot channel cools from the 10 000 K or so characteristic of an ambient air arc to a temperature of 3000 K or so in a time of milliseconds to tens of milliseconds.

If the cooling of the hot air of the lightning channel were to take place slowly enough that all chemical reactions could occur a sufficient number of times, at a given temperature level, to establish an equilibrium composition, the final constituents of the cold air would be the same as the constituents prior to the discharge. For example, if NO were to remain in equilibrium, it would reach a maximum concentration of about 10 percent when the air temperature was about 4000 K and would then decrease to negligibly small concentrations as the air slowly cooled to ambient temperatures. Figure 15.1 illustrates the equilibrium concentrations of the major air constituents as a function of temperature. The equilibrium concentrations shown in Fig. 15.1, involving the reactions given below, are not achieved at the lower temperatures because the time for NO to establish an equilibrium concentration increases rapidly with decreasing temperature: whereas only a few microseconds are required for it to reach equilibrium at 4000 K, milliseconds are required at 2500 K, and about 1000 years at 1000 K (Chameides 1986). Thus, as the lightning-heated air cools, a temperature is reached, the "freeze-out" temperature, below which the reactions that produce and destroy NO become too slow to keep NO in the equilibrium shown in Fig. 15.1, and the NO density remains at the value characteristic of equilibrium at that temperature. This freezing-out of molecules during the cooling of hot gases is often referred to as the Zel'dovich mechanism (Zel'dovich and Raizer, 1967). According to Borucki and Chameides (1984), the important temperature-dependent chemical reactions leading to the NO equilibrium are

$$O_2 \rightleftarrows O + O$$

followed by the production of NO via the reaction chain

$$O + N_2 \rightarrow NO + N$$

and

$$O_2 + N \rightarrow NO + O$$

In competition with these NO-producing reactions are those that destroy NO:

$$NO + N \rightarrow N_2 + O$$
$$NO + O \rightarrow N + O_2$$
$$NO \rightleftarrows N + O$$
$$NO + NO \rightarrow N_2O + O$$

There are two primary models for trace-gas production by return strokes.

(i) The first model is the shock-wave model commonly attributed to Chameides et al. (1977) and Chameides (1979a, b), although a similar theory was applied to the pre-biological Earth atmosphere by Bar-Nun and Tauber (1972), Bar-Nun et al. (1970), and Bar-Nun and Shaviv (1975), as discussed in Section 15.9, and an early version of the shock-wave model was developed from nuclear explosion theory by Tuck (1976), as discussed in Section 15.7. In the shock-wave model, the return-stroke input energy, generally taken to be near 10^5 J m^{-1}, is assumed to be dissipated by the expanding shock wave, resulting in a volume of heated air that subsequently cools, freezing out trace molecules.

(ii) The second model is the hot-channel decay model of Hill et al. (1980, 1984) and Hill and Rinker (1981), in which the trace gases are created in the late stages of the discharge as the hot discharge channel cools and in which the energy input to the return stroke is generally taken to be near 10^4 J m^{-1}, as determined from the gas dynamic modeling discussed in subsection 12.2.2. This energy input is

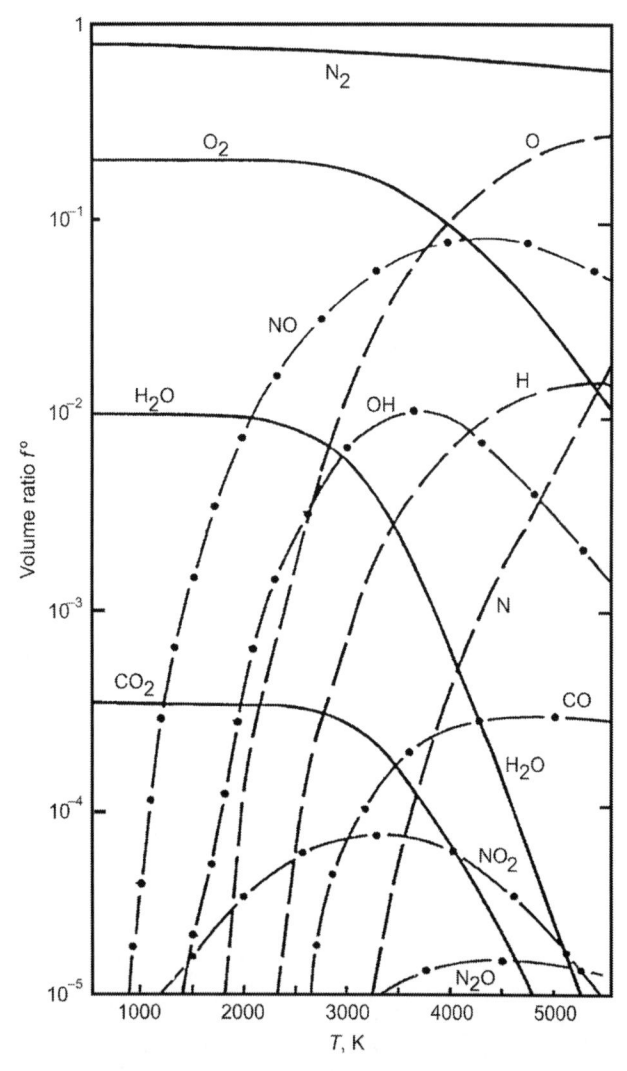

Fig. 15.1. Temperature dependence of the volume ratio in equilibrium, f^o, for several species in heated tropospheric air. Adapted from Chameides (1986).

15.2. Mechanism of NO production by return-stroke channels

used to estimate the channel diameter and, with an assumed channel length, the volume of air involved.

The shock-wave theory has until recently been the theory more widely quoted in the NO-production literature and hence probably the one better accepted by atmospheric chemists, but the hot-channel approach makes better sense, as first argued by Hill et al. (1980), since, according to the more recent gas dynamic modeling discussed above and in subsection 12.2.2, the regions where the expanding shock wave creates high temperatures are only at the edge of the expanding hot channel. The high temperature of the expanding hot channel is maintained by the time-varying current flowing in it. Further, Stark et al. (1996) reported that in laboratory discharges the shock wave is not effective in producing trace chemicals, and hence they concluded that NO must be formed in the cooling channel; and Wang et al. (1998) argued from their gas dynamic modeling that the channel core is the major trace-gas source for the reason given by Hill et al. (1980), although Wang et al. considered only the early dynamics of the channel and, in their theoretical work, they did not study the later-time mixing of the ambient air with the hot, low-density core. Additionally, Stark et al. (1996) disputed the assertion of Goldenbaum and Dickerson (1993) that NO is frozen out by a rapid drop in density in the channel core rather than by a decaying temperature. According to Stark et al., Goldenbaum and Dickerson erroneously assumed in their calculations that the initial channel temperature is 6000 K, whereas 30 000 K is more reasonable, decaying to near 10 000 K and remaining near that value so long as tens to hundreds of amperes are still flowing in the lightning channel.

Chameides et al. (1977) and Chameides (1979a, b) used the early strong-shock theory of Lin (1954) in which, because of the instantaneous energy input, the question of the hot channel's presence behind the expanding shock wave does not arise. Interestingly, Chameides (1979b) predicted measurable N_2O using the shock theory, whereas Hill and Rinker (1981) found negligible N_2O produced by the decaying channel model, motivating them and Hill et al. (1988) to demonstrate experimentally that appreciable N_2O is produced by corona processes as opposed to shock waves.

Hill et al. (1980) and Hill and Rinker (1981) calculated the trace-gas production by assuming that the return-stroke channel decays to a temperature of 3000 K and then mixes with the ambient air at 273 K. A channel radius of 16 cm was calculated from thermodynamic arguments assuming an input energy of 10^4 J m^{-1} and a 3000 K channel temperature. Borucki and Chameides (1984) claimed that this radius is about a factor 2 too large because of a calculation error. For mixing times of the 3000 K air with the ambient air between 1 ms and 20 s, the fractional production of NO was found to be between 1 and 0.5 of the initial amount of NO at 3000 K. Dawson (1980) argued that the shock-wave theory may overestimate the global NO production by up to a factor 100 because of various uncertainties, which he discussed; he advanced a channel-decay model with 500 "stroke equivalents" per second globally, each 10 km long and having 10 cm radius, with a freeze-out temperature of 3000 K at atmospheric pressure and a one percent weight-to-weight conversion of molecular nitrogen and oxygen to NO. (See Section 15.5 for a discussion of strokes and flashes in the context of global NO_x determinations)

Borucki and Chameides (1984) adopted a version of the hot-channel-decay model using a 2.5 ms decay for the cooling channel, as indicated by the laboratory experiments and theory of Picone et al. (1981), who provided qualitative confirmation of the model of Hill et al. (1980) while examining the cooling and mixing mechanisms in more detail. Bhetanabhotla et al. (1985) discussed the question of mixing times and extended the calculations of Hill et al. (1980) to moist air, assuming a channel radius of 10 cm at 3000 K and a mixing time of 10 ms.

If the decaying-channel theory is more realistic than the shock-wave theory, as appears to be the case, then one should take account of the many decaying hot channels in the various processes comprising cloud and ground discharges rather than just those in return strokes; it may be that the total NO production by lightning, or at least the total length of the channels producing NO, has been underestimated in the present literature, which considers primarily return-stroke channels. Laroche et al. (1999) reported that the mean total length of all the channels in over 20 000 cloud and ground flashes, accounting for tortuosity and including branches, as inferred from interferometric measurements (Section 17.6), is 45 km, considerably longer than the 3 to 20 km length usually assumed for the return stroke (Section 15.5). Using this 45 km length, typical flash rates, and two NO_x production rates per unit input energy, they computed a global production of 60 and 10^3 Tg(N) yr^{-1}. Most of the lightning channels detected by the interferometer were within the clouds.

In support of the view that lightning channels other than return strokes produce significant quantities of NO_x, Dye et al. (2000) showed from airborne measurements that the NO_x from a storm on 10 July 1996 that produced almost exclusively cloud discharges was comparable to other observations in storm anvils where both cloud and ground discharges were occurring, implying that cloud flashes are as significant in generating NO as ground flashes. Further, De Caria et al. (2000) modeled airborne NO_x measurements for one thunderstorm on 12 July 1996, including the convective transport of NO_x in the model. They found that intracloud (or the in-cloud portion of cloud-to-ground) lightning was the dominant source of NO_x for that thunderstorm and further that less than 20 percent of the NO_x in the anvil region of the storm was transported from the planetary boundary layer (the bottom kilometer or so of the atmosphere).

The various theoretical estimates of the number of molecules of NO per joule of input energy given in the first column of Table 15.2 range from 1×10^{16} to over 30×10^{16}. There is over an order-of-magnitude variation in the values derived from either the shock-wave theory or the hot-channel-cooling theory. The hot-channel-cooling theory provides slightly higher values, but there is enough overlap in the NO production estimates obtained by different investigators using the two theories to suggest that either technique can be considered to yield about the same results. Basically, both theories are similar, in that in each an assumed lightning input energy heats an assumed or calculated volume of air and the NO production is then determined from the NO concentration at an assumed freeze-out temperature.

15.3. Laboratory determination of NO yield per unit energy

The most recent and best controlled laboratory-spark study of trace-gas production was by Wang et al. (1998), who employed sparks with lightning-like peak currents of up to 30 kA, the typical amplitude of a first return stroke. The sparks were 4 cm long and were produced at atmospheric pressure by breaking down an air gap with a 250 kV pulse and then discharging through that breakdown path a $10^3 \mu F$ capacitor bank charged to 5 to 10 kV. The resultant current risetime was about 30 μs and the decay time about 400 μs. Dissipated energy in the spark, normalized to a 1 m spark length, was 3 to 8 kJ m^{-1}, similar to the values predicted by the gas dynamic models (subsections 12.2.2 and 12.2.6).

Wang et al. (1998) found that the sparks produced a ratio NO/NO$_x$ exceeding 0.9 and that little O$_3$ was produced. They showed that the NO production per joule of spark input energy was a strong function of the input energy which was more or less linearly related to the peak current. The NO production varied from about 15×10^{16} molecules per joule at 10 kA to about 40×10^{16} molecules per joule at 30 kA, as illustrated in the first column of Fig. 15.2. These values are significantly larger than the laboratory-spark measurements made previously, which we now describe.

Chameides et al. (1977) investigated (i) discharges from a 59 pF capacitor charged to 35 kV yielding a discharge energy of the order of 10^{-2} J and (ii) one discharge in a 1-m-long air gap with an energy of about 1 kJ. No current was reported from either experiment. Levine et al. (1981) studied 1 to 10 cm sparks with input energies of about 10 kJ from a 10^4 V source, yielding a spark energy per unit length of 10^5 to 10^6 J m^{-1}, at the upper end of, or higher than, estimates for natural lightning. Interestingly, both Chameides et al. and Levine et al., for the wide range of input energies and spark sizes studied, found that all their sparks produced about 5×10^{16} molecules of NO per joule, as indicated in the first column of Table 15.2. Peyrous and Lapeyre (1982) reported 4.2×10^{16} molecules per joule from spark discharges, but they did not give the characteristics of the discharges studied except for the spark-current values, those being less than 1 A.

As noted in the previous section, Stark et al. (1996) used laboratory-spark experiments and modeling to argue that NO$_x$ is not produced in the shock front and hence must be produced in the cooling channel.

15.4. Ground-based field determination of NO yield per lightning flash

An advantage of the field observation of NO due to lightning over theoretical calculations or laboratory-spark measurements is that field observations represent a more direct approach which does not require an estimate of the lightning energy. However, a major source of uncertainty for all field measurements is that interpretation of the observations generally involves assumptions regarding the mode of dispersion of the trace gases that are measured some distance away from the lightning. Additionally, in the cases where there is more than one lightning flash, an estimate of the number of discharges and perhaps information on the type of lightning flashes that occurred during the measurement period are required.

The first direct observations of enhanced NO concentrations in the vicinity of lightning flashes were made by Noxon (1976, 1978), who used a solar absorption spectrometer to measure NO$_2$ concentrations below the cloud base of lightning-producing and non-lightning-producing storms. Noxon estimated that of order 10^{26} NO$_2$ molecules are produced per lightning flash (the words "flash" and "stroke" appear to be used interchangeably in his work). The observed increase in NO$_2$, about two orders of magnitude above ambient and as high as 100 ppbv (parts per billion by volume, as opposed to by mass), along with estimates of the flash rate in the vicinity of the storm from visual observations and assumptions regarding the volume involved were used to estimate the NO$_2$ yield per flash. The NO$_2$ abundances had decreased by about an order of magnitude one hour after each of two overhead storms had ended. The extrapolated global NO production rate from this approach, using NO$_2$ as a proxy for the NO produced at the source, is 37 Tg(N) yr^{-1}, given in column five of Table 15.2. Noxon (1978) examined storms with and without lightning but with the same relatively high level of corona discharge current at ground, as measured at the Langmuir Laboratory in New Mexico, and found that the ground-based corona did not contribute significantly to the observed ambient NO$_2$ concentrations.

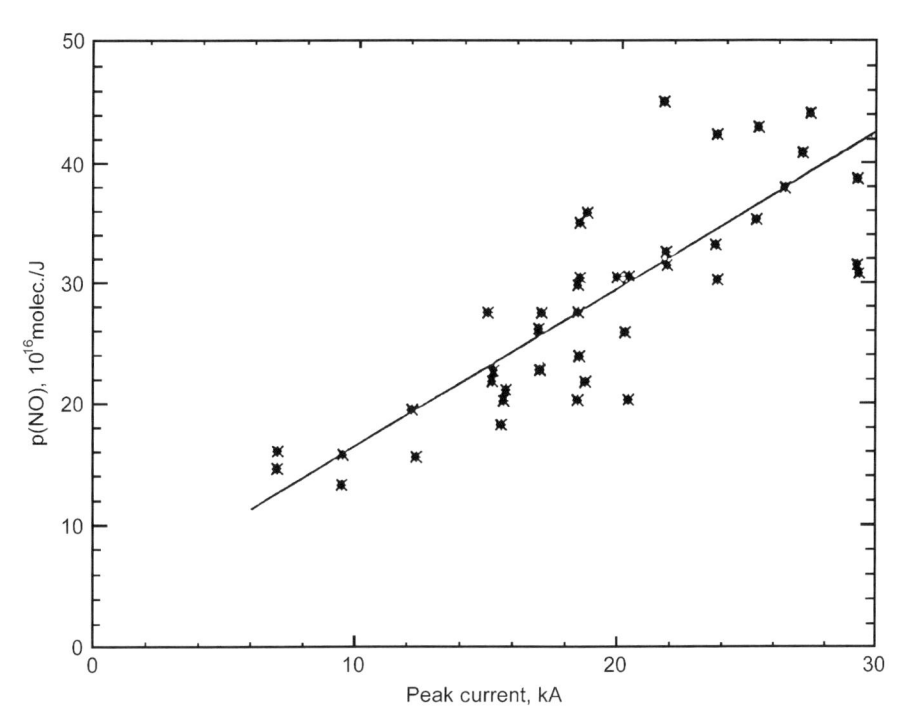

Fig. 15.2. NO production by laboratory sparks per joule of dissipated energy, $p(\text{NO})$, as a function of spark peak current. Adapted from Wang *et al.* (1998).

Drapcho *et al.* (1983) used a chemiluminescent analyzer to measure the NO_x concentrations following one cloud-to-ground flash within 1 km. The production of NO_x from this one flash was then determined, assuming that the flash occurred at 0.5 km and that, when the prevailing winds carried the expanding cylinder of NO_x across the measuring site, the NO_x from the flash was uniformly dispersed in a cylinder of 700 m radius and 3 km height. Experimental data were presented to support these assumptions. The NO_x within the cylinder was equivalent to a uniform value of 15 ppbv, leading to a calculated NO_x production rate of 4×10^{26} molecules per flash, or 30 Tg(N) annually. In this paper by Drapcho *et al.* the terms NO production and NO_x production appear to be used interchangeably. Since mostly NO is produced by the lightning, while NO_x is observed following chemical reactions involving the original NO, we list the 30 Tg(N) yr^{-1} as NO in the fifth column of Table 15.2.

Franzblau and Popp (1989) made two independent measurements of NO_x in the proximity of lightning using a chemiluminescent NO_x analyzer and an absorption spectrometer tuned to the NO_2 absorption. Data were taken at an elevation of 3300 m on a mountain near Socorro, New Mexico. Absorption spectrometer data were obtained from three separate lightning events, two from distant storms and one from direct observation of the plume from a single flash a few hundred meters away, yielding 4×10^{26} to $10\times$ 10^{26} molecules of NO_2 per flash after normalizing to the air density for flashes at sea level. Measurements from the NO_x analyzer during two different thunderstorms indicated that the ratio of NO to NO_x can remain high ($0.7-0.8 \pm 0.1$) for several minutes after a flash.

Combining the data from the NO_2 absorption spectrometer with the NO_x analyzer, Franzblau and Popp (1989) estimated an NO_x production rate of approximately 3×10^{27} molecules per flash. For 100 flashes per second, their assumed global lightning-flash frequency, this molecular production translates to a global production rate of 230 Tg(N) yr^{-1}, among the highest estimates of global NO_x production, although it should be noted that Franzblau and Popp list 100 Tg(N) yr^{-1} "as an order of magnitude estimate" in their paper. Since they indicated that all the observed NO_x was likely to have been originally NO, column five of Table 15.2 lists 230 Tg(N) yr^{-1} as the NO production.

Winterrath *et al.* (1999) used ground-based differential optical absorption spectroscopy and modeling to study the concentrations of O_3 and NO_2 at cloud level in clear-sky, cloudy-sky, and thunderstorm conditions. They found increases of 3 ppbv NO_2 and 38 ppbv of O_3 for thunderstorm conditions. They argued that the NO_2 enhancements are mostly due to lightning but that the enhanced ozone appears to be from the intrusion of stratospheric air and possibly production by non-lightning discharge processes.

15.5. Estimation of global NO production using the flash extrapolation approach (FEA)

As indicated in Table 15.2 and discussed in Section 15.1, many calculations of the global NO production rate have used the flash extrapolation approach (FEA), beginning with one of the three methods outlined in the previous three sections for determining $p(NO)$, the NO production per joule of lightning energy (Sections 15.2 and 15.3), or $P(NO)$, the NO production per flash (Section 15.4). In the flash extrapolation approach, $G(NO)$, the global rate of NO production by lightning, is determined from

$$G(NO) = P(NO) F \tag{15.1}$$

where F is the global lightning flash rate. In the theoretical and laboratory approaches,

$$P(NO) = p(NO)E_f \tag{15.2}$$

where E_f is the flash energy. The global flash rate F is considered in Chapter 2. A reasonable value for F that includes both cloud and cloud-to-ground lightning is $100~s^{-1}$, each ground flash having typically three to five strokes. Nevertheless, the values of F used in previous studies of NO production by lightning vary from 100 to $500~s^{-1}$ (see Table 15.2). Much of the variation can be attributed to confusion about whether cloud or ground flashes or both are described by published flash rates, confusion over the relative importance of cloud-to-ground versus cloud lightning in NO_x production, and confusion between the terms "stroke" and "flash". In discussing this confusion, Dawson (1980) reasonably settled on "500 stroke equivalents per second". In formulating Table 15.2, we provide corrections to the published results for F where there is obvious confusion between strokes and flashes in the original paper.

The energy input to lightning is considered in Chapters 1, 4, and 12. For the total first-leader–return-stroke process, 5×10^8 J would appear to be a reasonable input-energy value. For the return stroke itself, input-energy estimates per unit length range from 10^3 J m^{-1} to 10^5 J m^{-1}, which for a 5 km channel yields 5×10^6 J to 5×10^8 J, the lower value resulting from gas dynamic modeling and the higher value from electrostatic considerations and a comparison of lightning and laboratory sparks. Discussion of the energy issue relative to NO production was given by Dawson (1980). As is evident in Table 15.2, the flash energies, E_f, used in NO production studies range from 0.5×10^8 to 20×10^8 J; they were arrived at generally by finding an energy per return stroke from the product of an assumed energy per unit length and an assumed channel length, usually taken to be in the range 3 to 20 km, and then multiplying that energy per return stroke by an assumed average number of return strokes in a ground flash.

15.6. Estimation of NO production from airborne measurements

The advantage of directly determining the trace-gas production of an entire thunderstorm and then extrapolating that value to the worldwide thunderstorm activity is that no information is required about the number of lightning flashes or the flash characteristics. Furthermore, a potentially more important feature of the direct measurement of trace gases from thunderstorms is that it does not assume *a priori* that hot lightning channels are the only source of those trace gases in thunderstorms and so necessarily includes contributions from other discharge processes that commonly occur in electrified clouds, such as the corona associated with various lightning processes and the corona from water drops and ice particles, although there is no consensus regarding the role of corona in NO production.

Chameides *et al.* (1987) published the first thunderstorm-based estimate of NO production. The amount of NO generated was determined from combining airborne measurements of the elevated NO concentrations in the anvil regions of two active cumulonimbi (Davis *et al.* 1987) with estimates of the typical advective flow from the tops of thunderclouds. This result for a thunderstorm was then multiplied by the assumed global number of thunderstorms per year to obtain a $G(NO)$ of 7 Tg(N) yr^{-1}, as listed in Table 15.2, a value not inconsistent with estimates from other techniques. The peak NO concentrations observed were in the range of 1 ppbv. NO levels outside anvils averaged 20 pptv (parts per trillion by volume) while inside the anvils the average was 440 pptv.

Aircraft measurements of NO, NO_x, NO_y, and O_3 leading to estimates for global NO_x production were made by Ridley *et al.* (1996) in storms over west central New Mexico in summer 1989. There was no evidence of O_3 production by lightning, the same result as for most other studies (Dickerson *et al.* 1987; Luke *et al.* 1992; Ridley *et al.* 1994; Poulida *et al.* 1996; Huntrieser *et al.* 1998). NO enhancements of the order of 2 ppbv were observed. The ratio of NO to NO_2 and of NO to NO_y in anvil regions of the clouds almost always exceeded 0.6. Based on measurements during several anvil penetrations of two different thunderstorms, Ridley *et al.* (1996) estimated the annual global NO_x above 8 km, assuming that it is produced either by lightning or by all potential NO_x sources in the thunderstorm. In the first case they assumed 100 flashes per second worldwide and in the second case 44 000 thunderstorms per day globally. The annual global NO_x produced at or transported to altitudes above 8 km was estimated to be 2 to 3 Tg(N) yr^{-1} from the lightning approach (estimating the number of flashes in the storms and extrapolating to the global number of flashes) and 4 to 5 Tg(N) yr^{-1} using the thunderstorm approach. We list these estimates as quantities of NO in Table 15.2, since it is likely that most NO_x

15.6. Estimation of NO production from airborne measurements

was NO at the source. Additional information on NO, NO_x, NO_y, and O_3 profiles to altitudes of 12 km over New Mexico is found in Ridley et al. (1994) and similar information on NO and O_3 over the Pacific Ocean in Ridley et al. (1987).

The production of NO_x by dissipating thunderclouds has been investigated by Pickering et al. (1996). Using CO as a tracer, they attempted to separate lightning-produced NO_x and convectively transported NO_x for a thunderstorm whose major activity occurred 8 to 9 hours earlier. First the average NO_x/CO ratio without lightning for different altitudes in air was estimated (e.g., at 9.5 km $NO_x/CO =$ 1.93). Then the measured CO in air with lightning was multiplied by the NO_x/CO ratio without lightning for the corresponding altitude to estimate the NO_x without lightning. The lightning NO_x was then estimated as the difference between the measured NO_x in air with lightning and the estimated NO_x without lightning deduced from the CO measurement. Pickering et al. (1996) concluded that about 40 percent of the measured NO_x resulted from lightning. During one flight at 9.5 km, the NO_x contribution from lightning varied between 0.2 and 0.9 ppbv.

Smyth et al. (1996) observed "plumes" of NO with a spatial extension between 100 and 1000 km in the South Atlantic basin. Large measured enhancements, of the order of 1 ppbv NO, were attributed to lightning. Trajectory calculations indicated that deep convective storms occurred near the plume origin several days before the measurements. An annual NO production of 0.7 Tg(N) yr^{-1} was estimated for the South Atlantic Basin, of which half was considered to be from lightning, biomass burning being assumed to supply the remainder.

Huntrieser et al. (1998) made airborne NO_x measurements in the anvils of active thunderstorms over southern Germany and Switzerland (47°–49° N) in summer 1996. About 20 anvil penetrations were made, and measurements of NO, NO_2, CO, CO_2, and O_3 were obtained. In the thunderstorm anvils, the NO_x peak values were 4 ppbv. Some of these enhancements were attributed to the transport of polluted air from the planetary boundary layer, as determined by using CO_2 as a tracer for the planetary boundary layer air. NO_x produced by lightning was obtained by subtracting the fraction of NO_x transported from the planetary boundary layer from the total NO_x measured in the anvil. The NO_x/CO_2 correlation in larger cumuli without lightning was used as a reference for the transport of planetary boundary layer air in the anvils. In smaller thunderstorms, the contributions to the observed NO_x from lightning and from planetary-boundary-layer transport to the anvil were about equal. In medium and large thunderstorms the contribution from lightning was 60 to 75 percent. For these larger thunderstorms it was estimated that $\sim 1.0 \pm 0.5$ ppbv NO_x resulted from lightning. The observations were used to quantify the NO_x production per thunderstorm and to give a rough estimate of the annual production of NO_x. In using different approaches to the calculation of global NO_x production in the upper troposphere, Huntrieser et al. (1998) found values between 0.3 and 22 Tg(N) yr^{-1}, with a best estimate for the global NO_x production rate in the upper troposphere of 4 Tg(N) yr^{-1}. For most of their eight anvil measurements with aircraft, Huntrieser et al. (1998) found that NO comprised 50 to 70 percent of the NO_x concentrations observed. The values given for the NO production rate in Table 15.2 from Huntrieser et al. (1998) are, strictly speaking, for NO_x, but we have assumed that they represent a reasonable estimate of the NO concentrations.

In airborne studies in Colorado, Stith et al. (1999) observed narrow regions, about 100 m to 1 km across, containing up to 19 ppbv of NO, mostly located in or downwind of electrically active storm areas where NO produced by lightning would be expected. They estimated the lightning NO production to be in the range 2×10^{20} to 1×10^{22} molecules of NO per meter of lightning-channel length. Stith et al. stated that these estimates would be "increased somewhat" if the contribution from NO_2 were available. Interestingly, Wang et al. (1998), in the laboratory-spark measurements discussed in Section 15.3, measured NO production rates at atmospheric pressure for peak currents from 10 to 30 kA and found rates between 5×10^{20} and 3×10^{21} molecules per meter of discharge length.

For the same Colorado airborne measurements considered in the previous paragraph, Dye et al. (2000) found that the most likely contribution of lightning to the total NO_x observed in thunderstorm anvils was 60 to 90 percent, with a minimum of 45 percent. They also discussed one storm on 10 July 1996 that had more than 95 percent cloud discharges, that is, almost no lightning to ground. That storm produced NO_x levels comparable to those in storms with a large percentage of ground flashes, as previously noted in Section 15.2.

Crawford et al. (2000) presented data from NASA DC-8 flights over the North Atlantic in Fall 1997 and identified episodes of high NO_x in the upper troposphere, which they linked to specific periods of lightning activity over the US through back-trajectory calculations and data from the US National Lightning Detection Network (Section 17.5). Additional information on this experiment is found in Liu et al. (1999), Thompson et al. (1999), Allen et al. (2000), and Hannan et al. (2000).

Zhang et al. (2000) compared the global NO_2 distribution measured from a satellite in Earth orbit, the Upper Atmosphere Research Satellite, during all of 1993 with global lightning data obtained for five years by the Optical Transient Detector (Section 17.8 and subsection 2.5.3). Enhanced levels of NO_2 in the upper troposphere and lower stratosphere were consistently found in areas of elevated lightning activity.

Martin *et al.* (2000) used 14 years of Nimbus 7 TOMS satellite data on tropical tropospheric ozone enhancement, that enhancement being controlled by the supply of NO_x from all sources, to estimate that 20 percent of the enhancement was due to lightning and 54 percent to biomass burning in Africa and South America.

Hauglustaine *et al.* (2001) used measurements of tropospheric ozone obtained from ozone sondes launched from ships traversing the western Pacific and eastern Indian oceans from 1987 to 1990, along with modeling, to conclude that the large-scale tropospheric ozone plume extending from Africa to the western Pacific across the Indian ocean between mid-November and mid-December was related to a period of high lightning activity and related NO production in the southern hemisphere tropical belt.

15.7. Estimation of NO production from extrapolation of nuclear explosion data

Early estimates of lightning-produced NO were made in analogy to estimates of NO production in low-altitude nuclear explosions. Different approaches were employed in estimating the percentage of N_2 and O_2 converted to nitric oxide by these explosions. Zel'dovich and Raizer (1967) assumed a one percent conversion within a remnant hot sphere defined by the 2000-K-temperature surface. Foley and Ruderman (1973) and Johnston *et al.* (1973) calculated the NO production by mixing the hot gas within a remnant 6000 K sphere with the surrounding air. Goldsmith *et al.* (1973) studied the shock-wave production of NO by using a chemical-kinetic scheme integrated through the pressure–temperature history of the shock wave. Tuck (1976) stated that these and other studies yielded a range of estimates with an average value equivalent to about a one percent production of NO within a 2000 K spherical surface. Given the above information, Tuck (1976) obtained an early estimate of the lightning NO production as follows. Previously, Taylor (1950) had calculated that a nuclear bomb is a blast producer that is half as efficient as a chemical explosive. The energy densities involved in a lightning discharge during the initial development of the shock wave are more similar to a chemical explosion than a nuclear one. The efficiency for a lightning discharge to produce a shock wave, and hence NO, was therefore considered to be twice that of a nuclear bomb, which was estimated to produce 2×10^{30} molecules of NO for an input energy of 10^{14} J, or 2×10^{16} molecules per joule. Tuck (1976) assumed the energy input to a typical lightning stroke to be 10^5 J m^{-1} and a typical stroke length to be 6.25 km, yielding a production of NO per stroke of 2.5×10^{25} molecules. A global frequency of discharges of all types equal to 500 per second was assumed, resulting in an annual production of 4×10^{35} molecules. However, since the nuclear bomb calculations were performed for air near the Earth's surface, Tuck (1976) made a correction for the lower average ambient air density of the lightning by multiplying by 0.6, yielding an annual production of 2.4×10^{35} molecules of NO. The calculation of lightning-produced NO outlined in this section is both overly detailed and rather crude, but it is presented here for completeness and for historical reasons. Interestingly, the global NO production rate extrapolated from the nuclear explosion data, 5.6 Tg(N) yr^{-1}, is not dissimilar from estimates made using other approaches.

15.8. Transport of lightning-produced trace gases

Research involving the global transport of lightning-produced trace atmospheric gases has been published by, for example, Penner *et al.* (1991), Kotamarthi *et al.* (1994), Pickering *et al.* (1998), Lamarque *et al.* (1996), Levy *et al.* (1996), Singh *et al.* (1996), Flatoy and Hov (1997), Kawakami *et al.* (1997), Stockwell *et al.* (1999), Egorova *et al.* (1999, 2000), Smyshlyaev *et al.* (1999), and Hauglustaine *et al.* (2001). Some results from these studies are discussed below.

Most anthropogenic NO_x sources are located on the Earth's surface, and the NO_x released from the Earth's surface to the lower troposphere has a relatively short lifetime. For example, in the planetary boundary layer in summer, that is, below about 1 km, the lifetime of NO_x is only a few hours. Logan (1983) estimated that roughly 30 percent of the NO_x produced by the combustion of fossil fuels is converted to other gases near its point of injection and that about 40 percent is transported away. Anthropogenic sources probably supply most of the NO_x to the lower troposphere, particularly in industrialized countries, although locally lightning may produce comparable emissions (Biazar and McNider 1995). From measurements and transport models, for example, Singh *et al.* (1996) estimated that 65 percent of the tropospheric NO_x is from anthropogenic sources.

The lifetime of NO_x increases with altitude; it is as much as a week in the upper troposphere and about two months in the lower stratosphere. It follows that an NO_x source in the upper troposphere is more effective in increasing the NO_x concentration than the same source in the lower troposphere. Although thunderstorms do lift NO_x from the lower troposphere to the upper troposphere (e.g., Dickerson *et al.* 1987), they do so only locally. Singh *et al.* (1996) estimated that anthropogenic NO_x lifted by convection represents only 20 percent of the NO_x in the upper troposphere, and measurements by several other investigators (e.g., Murphy *et al.* 1993; Smyth *et al.* 1996) have suggested a similar level of NO_x in the upper troposphere from anthropogenic sources. Thus, the elevated location of lightning makes it an important source of NO_x in the upper troposphere. Lightning flashes originate above the 0 °C isotherm (Section 3.2), and lightning channels often extend to the cloud tops and above (Section 14.2). It is likely that there are many more intracloud lightning channels than cloud-to-ground channels, and cloud lightning flashes are more common than ground flashes (Section 2.7). Furthermore, tall vigorous storms are likely to produce much more lightning, primarily cloud flashes, and corona than the small isolated storms in which most NO_x measurements have been made.

The effect of lightning-produced NO_x on the lower stratosphere is apparently a function of latitude, being more pronounced at lower latitudes owing to a stronger upwelling of air between the troposphere and stratosphere there. Ko et al. (1986) found, on the basis of a two-dimensional stratospheric model, that at the equator the concentration of NO_y at 22 km height is an order of magnitude larger than would be expected without the upward transport of nitrogen oxides produced in the troposphere by lightning, which is consistent with available satellite data. Ko et al. (1986) stated that if only five percent of the nitrogen oxides produced in the troposphere entered the stratosphere, that amount would be equivalent to one-third of the global production there from the dominant photochemical reaction involving N_2O and atomic oxygen. At 50° N, however, NO_y concentrations in the lower stratosphere were calculated to be only a factor 2 or so above those expected without lightning. Murphy et al. (1993) measured NO_y and O_3 from the upper troposphere to over 20 km at various latitudes and concluded that there is a source of NO_y in the upper troposphere, the most probable source in the tropics being lightning-produced NO_x. Lamarque et al. (1996), using a three-dimensional global transport model, found that the NO_x concentration in the southern hemisphere is mostly due to lightning while in the northern hemisphere it is due to a combination of sources. For example, in the upper troposphere of the mid-latitudes of the northern hemisphere, they found that 25 to 30 percent of the NO_x is due to aircraft, 15 to 20 percent due to surface sources of fossil fuel combustion such as power plants, and 15 to 20 percent to lightning. Levy et al. (1996), using a global circulation model, calculated that lightning is a major source of NO_x and NO_y in the mid and upper troposphere for a latitude belt extending from 30° N to 30° S, that it is an important contributor to summertime free tropospheric NO_x and NO_y levels over the mid-latitudes, and that it is a major contributor to NO_x and NO_y levels over the remote oceans, where these levels are relatively low. Pickering et al. (1998) employed a transport model for NO_x, with assumed locations for the causative lightning, to compute vertical profiles of the ratio by mass of lightning-produced NO_x to N. For all three regimes studied (mid-latitude continental, tropical continental, and tropical maritime) a maximum was found in the lightning-produced NO_x to N mass-ratio profile of about 15 percent in the upper troposphere, usually within 2 to 4 km of the tropopause.

The mass ratio was a few percent between about 2 and 7 km altitude and greater below 2 km. Stockwell et al. (1999) use a three-dimensional chemical transport model to assess the impact of lightning in producing NO_x and other chemical species in the troposphere. The model with lightning present matches the measured NO_x and O_3 profiles over the Atlantic and Pacific oceans much better than the same model without the lightning included. The model predicts a significant increase in the NO_x concentrations in the upper troposphere, where the NO_x lifetime is long, and a smaller increase in the lower troposphere, where the surface NO_x sources dominate. These changes cause a significant increase in the O_3 there. The model indicates that lightning emissions cause local increases of over 50 pptv in NO_x, 200 pptv in HNO_3 and 20 ppbv (>40 percent) in O_3. In addition, a smaller increase of O_3 in the lower troposphere occurs owing to an increase in the downward transport of O_3. Egorova et al. (2000) used a two-dimensional global model and measured ozone values during the period 1979 to 1994 to conclude that the introduction of NO-producing lightning sources into the model leads to much better agreement between theory and measurement, particularly in the tropical latitudes, than for the model without lightning. They gave a lightning NO production of 20 Tg(N) yr^{-1} as a best estimate.

15.9. Production of trace gases in the primitive Earth atmosphere and in the atmospheres of other planets

Atmospheric electric discharges in the Earth's pre-biological atmosphere may have produced molecules such as hydrogen cyanide (HCN) that contributed to the development of life (Sanchez et al. 1967; see also Section 1.1). Miller and Urey (1959) discussed the role of solar ultraviolet light and atmospheric electric discharges, as well as of presumably less important sources such as cosmic rays and volcano emissions, in producing organic material from an atmosphere of primarily hydrogen, methane, ammonia, and water vapor. They described their previous laboratory-spark experiments in assumed pre-biological atmospheres. These atmospheres are "reducing", that is, they have relatively little oxygen, a requirement for the production of HCN and organic molecules. Other laboratory experiments aimed at producing HCN and organic molecules postulated to be present in the primitive Earth atmosphere include those of Bar-Nun and Shaviv (1975) and Bar-Nun et al. (1970). Bar-Nun et al. used a single-pulse shock tube, simulating the pressure wave from lightning, to test the effects of shock-wave heating on model prebiotic atmospheres. A variety of amino acids were generated in the experiment. From the production rate of the amino acids and an assumed lightning flash rate and input energy, a global rate for the synthesis of organic molecules was estimated. Bar-Nun and

Shaviv (1975) extended the shock experiment of Bar-Nun et al. (1970) to more gases and gas mixtures. They pointed out, as others have done, that two major questions to be answered are: (i) what would be the duration and evolution of the primitive reducing atmosphere, which could itself change over a time scale of a billion years from the effects of the lightning as well as from other effects? and (ii) what would be the rate of accumulation in the oceans of those molecules essential to producing the organic material of life? Yung and McElroy (1979) modeled NO production by lightning in the prebiotic atmosphere by assuming that the equilibrium concentration of NO is frozen at 2000 K when the "shock-heated air mass cools down". Chameides and Walker (1981) modeled the production of trace gases in the primitive Earth atmosphere to estimate quantitatively the generation rates of HCN and NO by lightning. Calculations were performed with their shock-wave model for various mixtures of CH_4, CO_2, CO, N_2, H_2, and H_2O. Atmospheres with more C than O, reducing atmospheres, are shown to have large HCN generation rates, in excess of 10^{17} molecules per joule, and small NO yields. When O is more abundant than C, the situation reverses. The extrapolated global HCN production in a primitive atmosphere rich in hydrocarbons was about 10^{35} molecules per year. Such a production, assuming no sink for HCN, would have led to an oceanic HCN level conducive to the formation of amino acids (Sanchez et al. 1967) but would have depleted the carbon and nitrogen in the atmosphere in a billion years, thus limiting HCN production. As the assumed abundance of CO_2 in the primitive atmosphere increased, that is, as there was more O than C available, HCN levels in the ocean would become inadequate for the formation of amino acids. If the carbon in the primitive atmosphere was principally in CO, the situation is intermediate: about 10^{32} molecules per year, producing significant oceanic HCN in millions of years but not affecting the composition of the atmosphere over billions of years.

While the presence of nitric oxide in the prebiotic Earth's atmosphere is generally thought to have been a result of lightning, as noted above, the production by lightning of nitrogen compounds sufficient to support emerging life forms may well, at some stage, have become inadequate, leading to the onset of the biological production of nutrient forms of nitrogen. Navarro-Gonzalez et al. (2001a) reviewed the literature and the conflicting views of the time of initiation of the biological production of nutrient forms of nitrogen, whether soon after the origin of life when only simple microbial ecosystems were present or much later when the biological demand for such nutrients was great due to the development of higher plant life. Based on experiments with laser-generated laboratory plasmas, Navarro-Gonzalez et al. suggested a scenario by which the biological production of, for example, nitrate and ammonia was triggered by an ecological crisis that occurred about two billion years ago owing to a reduction in the level of lightning-produced nitric oxide. By that time, according to their model, an early atmosphere predominantly of carbon dioxide had been replaced by an atmosphere predominantly of molecular nitrogen, a conversion that occurred over a period of more than a billion years, resulting in the nitric oxide reduction.

As will be discussed in Chapter 16, lightning-like discharges occur in the atmosphere of Jupiter but probably do not occur on Venus or Mars, although they once might have. Further, there may be electrical discharges of some sort on Saturn, Neptune, and Uranus. Chameides et al. (1979) used the shock theory of Lin (1954), as outlined in Section 15.2, to determine that CO, O_2, NO, and O would be produced by lightning in the atmospheres of Venus and Mars if it were present. A similar analysis for Venus was provided by Bar-Nun (1980). Laboratory-spark experiments by Levine et al. (1979, 1982) in a simulated Venus atmosphere of 96 percent CO_2 and 4 percent N_2 yielded a CO production of about 4×10^{17} molecules per joule and NO production of about 4×10^{15} molecules per joule. In relation to Jupiter, Sagan et al. (1967) suggested that the coloration of the Jovian clouds could not be explained in a thermodynamic equilibrium atmosphere and hence must be due to lightning. Ponnamperuma (1966) and Woeller and Ponnamperuma (1969) verified experimentally the synthesis of brightly colored compounds as well as several organic molecules of biological significance via laboratory electrical discharges in Jovian-like atmospheres. Bar-Nun (1975, 1979), Bar-Nun et al. (1984), and Podolak and Bar-Nun (1988) used a shock-wave model to account for observed nonequilibrium amounts of acetylene in the Jovian atmosphere and to compute the amount of amino acids and other chemicals potentially produced in the Jovian water clouds. However, as discussed in subsection 16.4.1, Lewis (1980a, b), considering shock-wave chemistry in a solar-composition gas, found that the synthesis by lightning of organic compounds on Jupiter is negligible compared to their synthesis by photochemical and thermochemical processes.

15.10. Summary

Estimates of the annual NO or NO_x production by electrical discharges in the Earth's atmosphere cover a range of two orders of magnitude, from about 1 to over 200 Tg(N) yr^{-1}. Perhaps this should not be viewed as surprising, given the assumptions and extrapolations necessary to obtain estimates of NO production. The upper end of the range of the production estimates is clearly suspect if the atmospheric chemists have properly identified both the rates of all loss mechanisms for atmospheric NO_x, thought to be 20 to 60 Tg(N) yr^{-1}, and the rates of generation of NO_x

by biomass burning and fossil fuel combustion, thought to be 20 to 50 Tg(N) yr^{-1}. Indeed, from the balance of NO$_x$ sources and sinks, even if lightning is the total remaining source, which it is not, lightning-produced NO$_x$ should be less than 40 Tg(N) yr^{-1} or so. However, if (i) the typical estimates of NO production per unit length of the lightning channel found from theoretical and laboratory-based estimates in Table 15.2 are correct, (ii) the total channel lengths that produce NO$_x$ have been underestimated by a factor 5 to 10 owing to lack of information on the extent of in-cloud channels, as appears to be the case, and (iii) all channels are equally efficient in producing NO, then the actual NO$_x$ production would appear to be considerably larger than 40 Tg(N) yr^{-1}.

Clearly, there are a lot of uncertainties, and the question of how much NO is generated by atmospheric electric discharges and how it is generated has not been satisfactorily answered. In any event, discharge-produced trace gases, especially NO, are clearly of importance to the ozone balance of the upper troposphere and the lower stratosphere and may be locally important in the lower troposphere. Additionally, atmospheric electric discharges may have played an important role in the generation of the organic compounds that made possible life on Earth.

References and bibliography for Chapter 15

Albritton, D.L., Liu, S.C., and Kley, D. 1984. Global nitrate deposition from lightning. In *Proc. Conf. on Environmental Impact of Natural Emissions*, Research Triangle Park, Air Pollut. Cont. Assoc., pp. 100–22.

Allen, D.J., Pickering, K.E., Stenchikov, G., Thompson, A.M., and Kondo, Y. 2000. A three-dimensional total odd nitrogen (NO$_x$) simulation during SONEX using a stretched-grid chemical transport model. *J. Geophys. Res.* **105**: 3851–76.

Anderson, B.E., Gregory, G.L., Collins, J.E. Jr, Sanchse, G.W., Conway, T.J., and Whiting, G.P. 1996. Airborne observations of spatial and temporal variability of tropospheric carbon dioxide. *J. Geophys. Res.* **101**: 1985–97.

Bar-Nun, A. 1975. Thunderstorms on Jupiter. *Icarus* **34**: 86–94.

Bar-Nun, A. 1979. Acetylene formation on Jupiter: photolysis of thunderstorms. *Icarus* **38**: 180–91.

Bar-Nun, A. 1980. Production of nitrogen and carbon species by thunderstorms on Venus. *Icarus* **42**: 338–42.

Bar-Nun, A., and Shaviv, A. 1975. Dynamics of the chemical evolution of Earth's primitive atmosphere. *Icarus* **24**: 197–210.

Bar-Nun, A., and Tauber, M. 1972. "Thunder": shock waves in pre-biological organic synthesis. *Space Life Sci.* **3**: 254–9.

Bar-Nun, A., Bar-Nun, N., Bauer, S.H., and Sagan, C. 1970. Shock synthesis of amino acids in simulated primitive environments. *Science* **168**: 468–73.

Bar-Nun, A., Noy, N., and Podolk, M. 1984. An upper limit to the abundance of lightning-produced amino acid in the Jovian water clouds. *Icarus* **59**: 162–8.

Barth, M.C., Stuart, A.L., and Skamarock, W.C. 2001. Numerical simulations of the July 10, 1996, stratospheric–tropospheric experiment: radiation, aerosols, and ozone (STERAO) deep convection experiment storm: redistribution of soluble tracers. *J. Geophys. Res.* **106**: 12 381–2400.

Berntsen, T.K., and Isasen, I.S.A. 1999. Effects of lightning and convection on changes in tropospheric ozone due to NO$_x$ emissions from aircraft. *Tellus* **51B**: 766–88.

Bhetanabhotla, M.N., Crowell, B.A., Coucouvinos, A., Hill, R.D., and Rinker, R.G. 1985. Simulation of the trace species production by lightning and corona discharge in moist air. *Atmos. Environ.* **19**: 1391–7.

Biazar, A.P., and McNider, R.T. 1995. Regional estimates of lightning production of nitrogen oxides. *J. Geophys. Res.* **100**: 22 861–74.

Bond, D.W., Zhang, R., Tie, X., Brasseur, G., Huffines, G., Orville, R.E., and Boccippio, D.J. 2001. NO$_x$ production by lightning over the continental United States. *J. Geophys. Res.* **106**: 27 701–10.

Borucki, W.J., and Chameides, W.L. 1984. Lightning: estimates of the rates of energy dissipation and nitrogen fixation. *Rev. Geophys. Space Phys.* **22**: 363–72.

Borucki, W.J., Mckay, C.P., and R.C. Whitten, 1984. Possible production by lightning of aerosols and trace gases in Titan's atmosphere. *Icarus* **60**: 260–73.

Bradshaw, J.D., Davis, D., Grodzinsky, G., Smyth, S., Newell, R., Sandholm, S., and Liu, S. 2000. Observed distributions of nitrogen oxides in the remote free troposphere from the NASA Global Tropospheric Experiment programs. *Rev. Geophys.* **38**: 61–116.

Brasseur, G.P., Muller, J.F., and Granier, C. 1996. Atmospheric impact of NO$_x$ emissions by subsonic aircraft: a three-dimensional model study. *J. Geophys. Res.* **101**: 1423–8.

Brunner, D., Staehelin, J., and Jeker, D. 1998. Large-scale nitrogen oxide plumes in the tropopause region and implications for ozone. *Science* **282**: 1305–9.

Brunner, D., Staehelin, J., Jeker, D., Wernli, H., and Schumann, U. 2001. Nitrogen oxides and ozone in the tropopause region of the Northern Hemisphere: measurements from commercial aircraft in 1995/1996 and 1997. *J. Geophys. Res.* **106**: 27 673–99.

Burns, R.C., and Hardy, R.W. 1975. *Nitrogen Fixation in Bacteria and Higher Plants*, 189 pp., Springer-Verlag: Berlin.

Chameides, W.L. 1979a. Effect of variable energy input on nitrogen fixation in instantaneous linear discharges. *Nature* **277**: 123–5.

Chameides, W.L. 1979b. The implication of CO production in electrical discharges. *Geophys. Res. Lett.* **6**: 287–90.

Chameides, WL. 1986. The role of lightning in the chemistry of the atmosphere. In *The Earth's Electrical Environment*, pp. 70–7, Washington, DC: National Academy Press.

Chameides, W.L., and Walker, J.C.G. 1973. A photochemical theory of tropospheric ozone. *J. Geophys. Res.* **78**: 8751–60.

Chameides, W.L., and Walker, J.C.G. 1981. Rates of fixation by lightning of carbon and nitrogen in possible primitive atmospheres. *Origins Life* **11**: 291–302.

Chameides, W.L., Stedman, D.H., Dickerson, R.R., Rusch, D.W., and Cicerone, R.J. 1977. NO_x production in lightning. *J. Atmos. Sci.* **34**: 143–9.

Chameides, W.L., Walker, J.C.G., and Nagy, A.F. 1979. Possible chemical impact of planetary lightning in the atmospheres of Venus and Mars. *Nature* **280**: 820–2.

Chameides, W.L., Davis, D.D., Bradshaw, J., Rodgers, M., Sandholm, S., and Bai, D.B. 1987. An estimate of the NO_x production rate in electrified clouds based on NO observations from the GTE/cite 1 fall 1983 field operation. *J. Geophys. Res.* **92**: 2153–6.

Chatfield, R.B., and Crutzen, P.J. 1984. Sulfur dioxide in remote oceanic air: cloud transport and reactive precursors. *J. Geophys. Res.* **89**: 7111–32.

Cho, M., and Rycroft, M.J. 1997. The decomposition of CFCs in the troposphere by lightning. *J. Atmos. Solar-Terr. Phys.* **59**: 1373–9.

Chyba, C., and Sagan, C. 1991. Electrical energy sources for organic synthesis on the early Earth. *Origin Life Evol. Biosph.* **21**: 3–17.

Cloud, P.E. 1968. Atmospheric and hydrospheric evolution on the primitive Earth. *Science* **160**: 729–36.

Cooray, V. 1997. Energy dissipation in lightning. *J. Geophys. Res.* **102**: 21 401–10.

Cooray, V., Peres, H., Gallarado, L., Oyola, P., and Scuka, V. 1998. Production of NO_x and O_3 by streamer discharges and its application in the global production of these gases by lightning flashes. In *Proc. Int. Conf. on Lightning Protection, Birmingham, UK*.

Coppens, F., Benton, R., Bondiou-Clergerie, A., Gallimberti, I. 1998. Theoretical estimates of NO_x production in lightning corona. *J. Geophys. Res.* **103**: 10 769–85.

Crawford, J., Davis, D., Olson, J., Chen, G., Liu, S., Fuelberg, H., Hannan, J., Kondo, Y., Anderson, B., Gregory, G., Sachse, G., Talbot, R., Viggiano, A., Heikes, B., Snow, J., Singh, H., and Blake, D. 2000. Evolution and chemical consequences of lightning-produced NO_x observed in the North Atlantic upper troposphere. *J. Geophys. Res.* **105**(D15): 19 795–809.

Crutzen, P.J. 1970. The influence of nitrogen oxides on the atmospheric ozone content. *Q.J.R. Meteor. Soc.* **96**: 320–7.

Crutzen, P.J. 1973. A discussion of the chemistry of some minor constituents in the stratosphere and troposphere. *Pure Appl. Geophys.* **106**: 1385–99.

Crutzen, P.J. 1979. The role of NO and NO_2 in the chemistry of the troposphere and stratosphere. *Ann. Rev. Earth Planet. Sci.* **7**: 443–72.

Davis, D.D., Bradshaw, J.D., Rodgers, M.O., Sandholm, S.T., and KeSheng, S. 1987. Free tropospheric and boundary layer measurements of NO over the central and eastern North Pacific Ocean. *J. Geophys. Res.* **92**: 2049–70.

Dawson, G.A. 1980. Nitrogen fixation by lightning. *J. Atmos. Sci.* **37**: 174–8.

De Caria, A.J., Pickering, K.E., Stenchikov, G.L., Scala, J.R., Stith, J.L., Dye, J.E., Ridley, B.A., and Laroche, P. 2000. A cloud-scale model study of lightning generated NO_x in an individual thunderstorm during STERAO-A. *J. Geophys. Res.* **105**(D9): 11 601–16.

Defer, E., Blanchet, P., Théry, Laroche, P., Dye, J.E., Venticinque, M., and Cummins, K.L. 2001. Lightning activity for the July 10, 1996, storm during the stratosphere–troposphere experiment: radiation, aerosol, and ozone-A (STERAO-A) experiment. *J. Geophys. Res.* **106**: 10 151–72.

Delwiche, C.C. 1970a. The nitrogen cycle. *Sci. Am.* **223**(3): 137–46.

Delwiche, C.C. 1970b. The nitrogen cycle. In *The Biosphere*, ed. W.H. Freeman, a *Scientific American* book.

Denning, A.S., Randall, D.A., Collatz, G.J., and Sellers, P.J. 1996. Simulations of terrestrial carbon metabolism and atmospheric CO_2 in a general circulation model. *Tellus* **48B**: 543–67.

Dickerson, R.R. 1984. Measurements of reactive nitrogen compounds in the free troposphere. *Atmos. Environ.* **18**: 2585–93.

Dickerson, R.R., Huffman, G.J., Luke, W.T. *et al.* 1987. Thunderstorms: an important mechanism in the transport of air pollutants. *Science* **235**: 460–5.

Dickerson, R.R., Doddridge, B.G., Rhoads, K.P., Kelley, P. 1995. Large-scale pollution of the atmosphere over the remote north Atlantic Ocean: evidence from Bermuda. *J. Geophys. Res.* **100**: 8945–52.

Doddridge, B.G., Dickerson, R.R., Wardell, R.G., Civerolo, K.L., and Nunnermacker, L.J. 1992. Trace gas concentrations and meteorology in rural Virginia 2. Reactive nitrogen compounds. *J. Geophys. Res.* **97**: 20 631–46.

Drapcho, D.L., Sisterson, D., and Kumar, R. 1983. Nitrogen fixation by lightning activity in a thunderstorm. *Atmos. Environ.* **17**: 729–34.

Drummond, J.W., Enhalt, D.H., and Volz, A. Measurements of nitric oxide between 0–12 km altitude and 67° N to 6° S latitude obtained during STRATOZ III. *J. Geophys. Res.* **93**: 15 831–49.

Dye, J.E., Ridley, B.A., Skamarock, W., Barth, M., Venticinque, M., Defer, E., Blanchet, P., Thery, C., Laroche, P., Baumann, K., Hubler, G., Parrish, D.D., Ryerson, T., Trainer, M., Frost, G., Holloway, J.S., Matejka, T., Bartels, D., Fehsenfeld, F.C., Tuck, A., Rutledge, S.A., Lang, T., Stith, J., and Zerr, R. 2000. An overview of the stratospheric-tropospheric experiment: radiation, aerosols, and ozone (STERAO) deep convection experiment with results for the July 10, 1996 storm. *J. Geophys. Res.* **105**: 10 023–45.

Egorova, T., Zubov, V., Jagovkina, S., and Rozanov, E. 1999. Lightning production of NO_x and ozone. *Phys. Chem. Earth C* **24**: 473–9.

Egorova, T.A., Rozanov, E.V., Zubov, V.A., and Yagovkina, S.V. 2000. Influence of global lightning source of NO_x on atmospheric ozone and odd nitrogen. *Izvestiya, Atmos. Ocean. Phys.* **36**(6): 743–54.

Ehhalt, D.H., and Drummond, 1982. The tropospheric cycle of NO_x. In *Chemistry of the Unpolluted and Polluted*

Troposphere, pp. 219–51, D. Reidel Publishing: Norwell, Massachusetts.

Ehhalt, D.H., Rohrer, H.F., and Walnner, A. 1992. Sources and distribution of NO_x in the upper troposphere at northern mid-latitudes. *J. Geophys. Res.* **97**: 3725–38.

Emmons, R.R., Dickerson, R.R., and 22 others 1997. Climatologies of NO_x and NO_y: a comparison of data and models. *Atmos. Environ.* **31**: 1837–50.

Flatoy, F., and Hov, O. 1997. NO_x from lightning and the calculated chemical composition of the free troposphere. *J. Geophys. Res.* **102**: 21 373–81.

Foley, H.M., and Ruderman, M.A. 1973. Stratospheric NO production from past nuclear explosions. *J. Geophys. Res.* **78**: 4441–50.

Franzblau, E. 1991. Electrical discharges involving the formation of NO, NO_2, NO_3, and O_3. *J. Geophys. Res.* **96**: 22 337–45.

Franzblau, E., and Popp, C.J. 1989. Nigrogen oxides produced from lightning. *J. Geophys. Res.* **94**: 11 089–104.

Frommhold, L. 1964. Uber verzogerte elektronen in elektronenlawinen, insbesondere in sauerstoff und luft, durch bildung und zerfall negativer ionen (O^-). *Fortschr. Phys.* **12**: 597–643.

Galbally, I.E., and Roy, C.R. 1978. Loss of fixed nitrogen from soils by nitric oxide exhalation. *Nature* **275**: 734–5.

Gallardo, L., and Cooray, V. 1996. Could cloud to cloud discharges be as effective as cloud to ground discharges in producing NO_x? *Tellus* **48B**: 641–51.

Galloway, J.N., Likens, G.E., Keene, W.C. *et al.* 1982. The composition of precipitation in remote areas of the world. *J. Geophys. Res.* **87**: 8771–86.

Gambell, A.W., and Fisher, D.W. 1964. Occurrence of sulfate and nitrate in rainfall. *J. Geophys. Res.* **69**: 4203–10.

Gardner, R.M. *et al.* 1997. The ANCAT/EC global inventory of NO_x emissions from aircraft. *Atmos. Environ.* **31**: 1751–66.

Gidel, L.T. 1983. Cumulus cloud transport of transient tracers. *J. Geophys. Res.* **88**: 6587–99.

Goldenbaum, G.C., and Dickerson, R.R. 1993. Nitric oxide production by lightning discharges. *J. Geophys. Res.* **98**: 18 333–8.

Goldsmith, P., Tuck, A.F., Foot, J.S., Simmons, E.L., and Newson, R.L. 1973. Nitrogen oxides, nuclear weapon testing, Concorde and stratospheric ozone. *Nature* **244**: 545–51.

Green, A.E.S., Swada, T., Edgar, B.C., and Uman, M.A. 1973. Production of CO by charged particle deposition mechanisms. *J. Geophys. Res.* **78**: 5284–91.

Greenhut, G.K. 1995. Transport of ozone between boundary layer and cloud layer by cumulus clouds. *J. Geophys. Res.* **91**: 8613–22.

Griffing, G.W. 1977. Ozone and oxides of nitrogen production during thunderstorms. *J. Geophys. Res.* **82**: 943–50.

Hannan, J.R. *et al.* 2000. Atmospheric chemical transport based on high-resolution model-derived winds: a case study. *J. Geophys. Res.* **105**: 3807–20.

Hauf, T., Schulte, P., Alheit, R., and Schlager, H. 1995. *J. Geophys. Res.* **100**: 22 957–70.

Hauglustaine, D., Emmons, L., Newchurch, M., Brasseur, G., Takao, T., Matsubara, K., Johnson, J., Ridley, B., Stith, J., and Dye, J. 2001. On the role of lightning NO_x in the formation of tropospheric ozone plumes: a global model perspective. *J. Atmos. Chem.* **38**: 277–94.

Hill, R.D. 1971. Channels heating in return stroke lightning. *Geophys. Res.* **76**: 637–45.

Hill, R.D. 1979. On the production of nitric oxide by lightning. *Geophys. Res. Lett.* **6**: 945–7.

Hill, R.D., and Rinker, R.G. 1981. Production of nitrate ions and other trace species by lightning. *J. Geophys. Res.* **86**: 3203–9.

Hill, R.D., Rinker, R.G., and Wilson, D. 1980. Atmospheric nitrogen fixation by lightning. *J. Atmos. Sci.* **37**: 179–92.

Hill, R.D., Rinker, R.G., and Coucouvinos, A. 1984. Nitrous oxide production by lightning. *J. Geophys. Res.* **89**: 1411–21.

Hill, R.D., Rahmin, I., and Rinker, R.G. 1988. Experimental study of the production of NO, N_2O, and O_3, in a simulated atmospheric corona. *Ind. Eng. Chem. Res.* **27**: 1264–9.

Holler, H., Finke, U., Huntrieser, H., Hagen, M., and Feigl, C. 1999. Lightning-produced NO_x (LINOX): experimental design and case study results. *J. Geophys. Res.* **104**: 13 911–22.

Huntrieser, H., Schlager, H., Feigl, C., and Holler, H. 1998. Transport production of NO_x in electrified thunderstorms: survey of previous studies and new observations at mid-latitudes. *J. Geophys. Res.* **103**: 28 247–64.

Hutchinson, G.E. 1954. The biogeochemistry of the terrestrial atmosphere. In *The Earth as a Planet*, ed. G.P. Kuiper, University of Chicago Press, Chicago.

Jacob, D.J. 2000. Heterogeneous chemistry and tropospheric ozone. *Atmospheric Environment* **34**: 2131–59.

Jacob, D.J. *et al.* 1996. Origin of ozone and NO_x in the tropical troposphere: a photochemical analysis of aircraft observations over the South Atlantic basin. *J. Geophys. Res.* **101**: 24 235–50.

Jadhav, D.B., Londhe, A.L., and Bose, S. 1996. Observations of NO_x and O_3 during thunderstorm activity using visible spectroscopy. *Adv. Atmos. Sci.* **13**: 359–74.

Jebens, D.S., Lakkaraju, H.S., McKay, C.P., and Borucki, W.J. 1992. Time resolved simulation of lightning by LIP. *Geophys. Res. Lett.* **19**: 273–6.

Johnston, H.S., Whitten, G.Z., and Birks, J. 1973. Effect of nuclear explosions on stratospheric nitric oxide and ozone. *J. Geophys. Res.* **78**: 6107–35.

Jourdain, L., and Hauglustaine, D.A. 2001. The global distribution of lightning NO_x simulated on-line in a general circulation model. *Phys. Chem. Earth.* **26**: 585–91.

Kasibhatla, P.S., Levy, P.S., II, and Moxim, W.J. 1993. Global NO_x, HNO_3, PAN, and NO_y distributions from fossil fuel combustion emissions: a model study. *J. Geophys. Res.* **98**: 7165–80.

Kasting, J.F. 1990. Bolide impacts and the oxidation state of carbon in the Earth's early atmosphere. *Origins Life Evol. Biosph.* **20**: 199–231.

Kasting, J.F., and Walker, J.C.G. 1981. Limits on oxygen concentrations in the prebiological atmosphere and the rate of abiotaic fixation of nitrogen. *J. Geophys. Res.* **86**: 1147–58.

Kawakami, S., Kondo, Y., Koike, M., Nakajima, H., Gregory, G.L., Sachse, G.W., Newell, R.E., Browell, E.V., Blake, D.R., Rodriguez, J.M., and Merrill, J.T. 1997. Impact of lightning and convection on reactive nitrogen in the tropical free troposphere. *J. Geophys. Res.* **102**: 28 367–84.

Kelley, P., Dickerson, R.R., Luke, W.T., and Kok, G.L. 1995. Rate of NO_2 photolysis from the surface to 7.6 km altitude in clear-sky and clouds. *Geophys. Res. Lett.* **22**: 2621–4.

Ko, M.K.W., McElroy, M.B., Weisenstein, D.K., and Sze, N.D. 1986. Lightning: a possible source of stratospheric odd nitrogen. *J. Geophys. Res.* **91**: 5395–404.

Kondo, Y., Kawakami, S., Koike, M., Fahey, D.W., Nakajima, H., Zhao, Y., Toriyama, N., Kanada, M., Sachse, G.W., and Gregory, G.L. 1997a. Performance of an aircraft instrument for the measurement of NO_y. *J. Geophys. Res.* **102**: 28 663–71.

Kondo, Y., Koike, M., Kawakami, S., Singh, H.B., Nakajima, H., Gregory, G.L., Blake, D.R., Sachse, G.W., Merrill, J.T., and Newell, R.E. 1997b. Profiles and partitioning of reactive nitrogen over the Pacific Ocean in winter and early spring. *J. Geophys. Res.* **102**: 28 405–24.

Kotaki, M., Kuriki, I., Kotoh, C., and Sugiuchi, H., 1981. Global distribution of thunderstorm activity observed with ISS-B. In *J. Radio Res. Lab. Tokyo, Japan*, **28** (125/126) 49–71.

Kotamarthi, V.R., Ko, M.K.W., Weisenstein, D.K., Rodriguez, J.M., and Sze, N.D. 1994. Effects of lightning on the concentration of odd nitrogen species in the lower stratosphere: an update. *J. Geophys. Res.* **99**: 8167–73.

Kumar, P.P., Manohar, G.K., and Kandalgaonkar, S.S. 1995. Global distribution of nitric oxide produced by lightning and its seasonal variation. *J. Geophys. Res.* **100**: 11 203–8.

Kumar, R., Singh, V., and Rai, J. 1994. Effect of the reaction $N + NO \rightarrow N_2^* (v = 5) + O$ on the production of NO by lightning. *J. Atmos. Sci.* **51**: 323–5.

Lamarque, J.-F., Brasseur, G.P., Hess, P.G., and Muller, J.F. 1996. Three-dimensional study of the relative contributions of the different nitrogen sources in the troposphere. *J. Geophys. Res.* **101**: 22 955–68.

Lange, L., Hoor, P., Helas, G., Fischer, H., Brunner, D., Scheeren, B., Williams, J., Wong, S., Wohlfrom, K.-H., Arnold, F., Ström, J., Krejci, R., Lelieveld, J. and Andreae, M.O. 2001. Detection of lightning-produced NO in the midlatitude upper troposphere during STREAM 1998. *J. Geophys. Res.* **106**: 27 777–85.

Laroche, P., Defer, E., Blanchet, P., Thery, C. 1999. Evaluation of NO_x produced by storms based on 3D VHF lightning mapping. In *Proc. 11th Int. Conf. on Atmospheric Electricity*, June 7–11, 1999, NASA/CP-1999-209261, Gunthersville, Alabama.

Lawrence, M.G., Chameides, W.L., Kasibhatla, P.S., Levy, H. II, and Moxim, W. 1995. Lightning and atmospheric chemistry: the rate of atmospheric NO production. In *Handbook of Atmospheric Electrodynamics*, vol. 1, ed. H. Volland, pp. 189–202, CRC Press: Boca Raton, Florida.

Lee, D.S., Kohler, I., Grobler, E., Rohrer, F., Sausen, R., Gallardoklenner, L., Oliver, J.G.J., Dentener, F.J., and Bouwman, A.F. 1997. Estimations of global NO_x emissions and their uncertainties. *Atmos. Environ.* **31**: 1735–49.

Levine, J.S., Hughes, R.E., Chameides, W.L., and Howell, W.E. 1979. N_2O and CO production by electric discharge: atmospheric implications. *Geophys. Res. Lett.* **6**: 557–9.

Levine, J.S., Rogowski, R.S., Gregory, G.L., Howell, W.E., and Fishman, J. 1981. Simultaneous measurements of NO_x, NO, and O_3 production in a laboratory discharge: atmospheric implication, *Geophys. Res. Lett.* **8**: 357–360.

Levine, J.S., Gregory, G.L., Harvey, G.A., Howell, W.E., Borucki, W.J., and Orville, R.E. 1982. Production of nitric oxide by lightning on Venus. *Geophys. Res. Lett.* **9**: 893–6.

Levine, J.S., Augustsson, T.R., Anderson, I.C., and Hoell, J.M. Jr 1984. Tropospheric sources of NO_x: Lightning and biology. *Atmos. Environ.* **18**: 1797–804.

Levy, H. II, Mahlman, J.D., and Moxim, W.J. 1980. Stratospheric NO_y: a major source of reactive nitrogen in the unpolluted troposphere. *Geophys. Res. Lett.* **7**: 441–4.

Levy, H. II, Moxim, W.J., and Kasibhatla, S. 1996. A global three-dimensional time-dependent lightning source of tropospheric NO_x. *J. Geophys. Res.* **101**: 22 911–22.

Lewis, J.S. 1980a. Lightning on Jupiter: rate, energetics, and effects. *Science* **210**: 1351–2.

Lewis, J.S. 1980b. Lightning synthesis of organic compounds on Jupiter. *Icarus* **48**: 85–95.

Liaw, Y.P., Sisterson, D.L., and Miller, N.L. 1990. Comparison of field, laboratory, and theoretical estimates of global nitrogen fixation by lightning. *J. Geophys. Res.* **95**: 22 489–94.

Lin, S.C. 1954. Cylindrical shock waves produced by an instantaneous energy release. *J. Appl. Phys.* **25**: 54–7.

Lipschultz, F., Zafirious, C., Wofsy, S.C., McElroy, M.B., Valois, F.W., and Watson, W. 1981. Production of NO and N_2O by soil nitrifying bacteria: a source of atmospheric nitrogen oxides. *Nature* **294**: 641–3.

Liu, S.C. 1977. Possible effects on tropospheric O_3 and OH due to NO emissions. *J. Geophys. Res. Lett.* **4**: 325–8.

Liu, S.C., Yu, H., Ridley, B., Yang, Y., Davis, D.D., Kondo, Y., Koike, M., Anderson, B.E., Voy, S.A., Sachse, G.W., Gregory, G.L., Fuelburg, H., Thompson, A., and Singh, H. 1999. Sources of reactive nitrogen in the upper troposphere during SONEX. *Geophys. Res. Lett.* **26**: 2441–4.

Logan, J.A. 1983. Nitrogen oxides in the troposphere: global and regional budgets. *J. Geophys. Res.* **88**: 10 785–807.

Logan, J.A., Prather, M.J., Wofsy, S.C., and McElroy, M.B. 1981. Tropospheric chemistry. *J. Geophys. Res.* **86**: 7210–54.

Luke, W.T., and Dickerson, R.R. 1987. The flux of reactive nitrogen compounds from eastern North America to the western Atlantic Ocean. *Global Biogeochem. Cycles* **1**: 329–43.

Luke, W.T., Dickerson, R.R., Ryan, W.F., Pickering, K.E., and Nunnermacker, L.J. 1992. Tropospheric chemistry over the lower Great Plains of the United States, 2, Trace gas profiles and distributions. *J. Geophys. Res.* **97**: 20 647–70.

Lyons, W.A., and Armstrong, R.A. 1997. NO_x production within and above thunderstorms: the contribution of lightning and sprites. *Preprint, Proc. 3rd Conf. on Atmospheric Chemistry*, pp. 3–12, Long Beach, California: American Meteorological Society.

Lyons, W.A., Eastman, J.L., Pielke, R.A., Biazar, A., and McNider, R. 1994. A preliminary climatology of lightning-generated NO_x and numerical simulations of its redistribution by deep convection. Preprint, In *Proc. Conf. on Atmospheric Chemistry*, pp. 193–8, Nashville: American Meteorological Society.

MacGorman, D.R., and Rust, W.D. 1998. *The Electrical Nature of Storms*, 422 pp., Oxford, UK: Oxford University Press.

Mackerras, D., and Darveniza, M. 1994. Latitudinal variation of lightning occurrence characteristics. *J. Geophys. Res.* **99**: 10 813–21.

Madronich, S. 1987. Photodissociation in the atmosphere, 1. Actinic flux and the effects of ground reflections and clouds. *J. Geophys. Res.* **92**: 9740–52.

Mancinelli, R.L., and McKay, C.P. 1988. The evolution of nitrogen cycling. *Origins Life Evol. Biosph.* **18**: 311–25.

Martin, R.V., Jacob, D.J., Logan, J.A., Ziemke, J.M., and Washington, R. 2000. Detection of a lightning influence on tropical tropospheric ozone. *Geophys. Res. Lett.* **27**: 1639–42.

Matsueda, H., and Inoue, Y. 1996. Measurements of atmospheric CO_2 and CH_4 using a commercial airliner from 1993 to 1994. *Atmos. Environ.* **30**: 1647–55.

McConnell, J.C. 1973. Atmospheric ammonia. *J. Geophys. Res.* **78**: 7812–21.

McFarland, M.C., Kley, D., Drummond, J.W., Schmeltekopf, H.L., and Winkler, R.H. 1979. Nitric oxide measurements in the equatorial pacific region. *Geophys. Res. Lett.* **6**: 605–8.

McLinden, C.A., Olsen, S.C., Prather, M.J., and Liley, J.B. 2001. Understanding trends in stratospheric NO_y and NO_2. *J. Geophys. Res.* **106**: 27 787–93.

Meijer, E.W., van Velthoven, P.F.J., Brunner, D.W., Huntrieser, H., and Kelder, H. 2001. Improvement and evaluation of the parameterisation of nitrogen oxide production by lightning. *Phys. Chem. Earth.* **26**: 577–83.

Miller, S.L., and Urey, H.C. 1959. Organic compounds synthesis on the primitive Earth. *Science* **130**: 245–51.

Murphy, D.M., Fahey, D.W., Proffitt, M.H., Liu, S.C., Chan, K.R., Eubank, C.S., Kawa, S.R., and Kelley, K.K. 1993. Reactive nitrogen and its correlation with ozone in the lower stratosphere and upper troposphere. *J. Geophys. Res.* **98**: 8751–73.

Nakazawa, T., Sugawara, S., Inoue, G., Machida, T., and Mukai, H. 1976. Aircraft measurements of the concentrations of CO_2, CH_4, N_2O, and CO and the carbon and oxygen isotopic ratios of CO_2 in the troposphere over Russia. *J. Geophys. Res.* **102**: 3843–59.

Navarro-González, R., McKay, C.P., and Mvondo, D.N. 2001a. A possible nitrogen crisis for Archaean life due to reduced nitrogen fixation by lightning. *Nature* **412**: 61–4.

Navarro-González, R., Villagrán-Muniz, M., Sobral, H., Molina, L.T., and Molina, M.J. 2001b. The physical mechanism of nitric oxide formation in simulated lightning. *Geophys. Res. Lett.* **28**: 3867–70.

Nesbitt, S.W., Zhang, R., and Orville, R.E. 2000. Seasonal and global NO_x production by lightning estimated from the optical transient detector (OTD). *Tellus* **52B**: 1206–15.

Noxon, J.F. 1976. Atmospheric nitrogen fixation by lightning. *Geophys. Res. Lett.* **3**: 463–5.

Noxon, J.F. 1978. Tropospheric NO_2. *J. Geophys. Res.* **83**: 3051–7.

Paxton, A.H., Gardner, R.L., and Baker, L. 1986. Lightning return stroke. A numerical calculation of the optical radiation. *Phys. Fluids* **29**: 2736–41.

Pearman, G.I., and Beardsmore, D.J., 1984. Atmospheric carbon dioxide measurements in the Australian region: ten years of aircraft data. *Tellus* **36B**: 1–24.

Penner, J.E., Atherton, C.S., Dignon, J., Ghan, S.J., Walton, J.J., and Hameed, S. 1991. Tropospheric nitrogen: a three-dimensional study of sources, distributions, and deposition. *J. Geophys. Res.* **96**: 959–90.

Peyrous, R., and Lapeyre, R.-M. 1982. Gaseous products created by electrical discharges in the atmosphere and condensation nuclei resulting from gaseous phase reactions. *Atmos Environ.* **16**: 959–68.

Pickering, K.E., Thompson, A.M. Scala, J.R., Tao, W.-K., Dickerson, R.R., and Simpson, J. 1992. Free tropospheric ozone production following entrainment of urban plumes into deep convection. *J. Geophys. Res.* **97**: 17 985–8000.

Pickering, K.E., *et al.* 1996. Convective transport of biomass burning emissions over Brazil during TRACE-A. *J. Geophys. Res.* **101**: 23 993–4012.

Pickering, K.E., Wang, Y., Tao, W.-K., Price, C., and Müller, J.-F. 1998. Vertical distribution of lightning NO_x for use in regional and global chemical transport models. *J. Geophys. Res.* **103**: 31 203–16.

Picone, J.M., Boris, J.P., Grieg, J.R., Rayleigh, M., and Fernsler, R.F. 1981. Convective cooling of lightning channels. *J. Atmos. Sci.* **38**: 2056–62.

Plooster, M.N. 1971. Numerical simulation of spark discharges in air. *Phys. Fluids* **14**: 2111–23.

Podolak, M., and Bar-Nun, A. 1988. Moist convection and the abundances of lightning-produced CO, C_2H_2, and HCN on Jupiter. *Icarus* **75**: 566–70.

Ponnamperuma, C. 1966. Some recent work on prebiological synthesis of organic compounds. *Icarus* **5**: 450–4.

Portman, R.W., Brown, S.S., Gierczak, T., Talukdar, R.K., Burkholder, J.B., and Ravishankara, A.R. 1999. Role of nitrogen oxides in the statosphere: a reevaluation based on laboratory studies. *Geophys. Res. Lett.* **26**: 2387–90.

Poulida, O., Dickerson, R.R., and Heymsfield, A. 1996. Stratosphere-troposphere exchange in a mid-latitude mesoscale convective complex, 1. Observations. *J. Geophys. Res.* **101**: 6823–936.

Prentice, S.A., and Mackerras, D. 1977. The ratio of cloud to cloud-ground lightning flashes in thunderstorms. *J. Appl. Meteorol.* **16**: 545–50.

Price, C. 2000. Evidence for a link between global lightning activity and upper tropospheric water vapour. *Nature* **406**: 290–3.

Price, C., Penner, J., and Prather, M. 1997a. NO_x from lightning 1. Global distribution based on lightning physics. *J. Geophys. Res.* **102**: 5929–41.

Price, C., Penner, J., and Prather, M. 1997b. NO_x from lightning 2. Constraints from the global atmospheric electric circuit. *J. Geophys. Res.* **102**: 5943–51.

Raven, J.A., and Yin, Z.H. 1998. The past, present and future of nitrogenous compounds in the atmosphere, and their interactions with plants. *New Phytol.* **139**: 205–19.

Reiter, R. 1970. On the causal relation between nitrogen–oxygen compounds in the troposphere and atmospheric electricity. *Tellus* **22**: 122–35.

Ridley, B.A., Carroll, M.A., and Gregory, G.L. 1987. Measurements of nitric oxide in the boundary layer and free troposphere over the Pacific Ocean. *J. Geophys. Res.* **92**: 2025–47.

Ridley, B.A., Walega, J.G., Dye, J.E., and Grahek, F.E. 1994. Distributions of NO, NO_x, NO_y, and O_3 to 12 km altitude during the summer monsoon season over New Mexico. *J. Geophys. Res.* **99**: 25 519–34.

Ridley, B.A., Dye, J.E., Walega, J.G., Zheng, J., Grahek, F.E., and Rison, W. 1996. On the production of active nitrogen by thunderstorms over New Mexico. *J. Geophys. Res.* **101**: 20 985–1005.

Rodriguez, V.R., Kotamarthi, J.M., Sze, N.D., Kondo, Y., Pueschel, R., Ferry, G., Bradshaw, J., Sandholm, S., Gregory, G., Davis, D., and Liu, S. 1997. Evidence of heterogeneous chemistry on sulfate aerosols in stratospherically influenced air masses sampled during PEM-West B. *J. Geophys. Res.* **102**: 28 425–36.

Rye, R., and Holland, H.D. 1998. Paleosols and evolution of atmospheric oxygen: a critical review. *Am. J. Sci.* **298**: 621–72.

Rye, R., Kuo, P.H., and Holland, H.D. 1995. Atmospheric carbon dioxide concentrations before 2.2 billion years ago. *Nature* **378**: 603–5.

Sagan, C.E., Lippincott, E.R., Dayhoff, M.O., and Eck, R.V. 1967. Organic molecules and the coloration of Jupiter. *Nature* **213**: 273–4.

Sakurai, A. 1953. On the propagation and structure of the blast wave (1). *J. Phys. Soc. Japan* **8**: 662–9.

Sanchez, R.A., Ferris, J.P., and Orgel, L.E. 1967. Studies in prebiotic synthesis. *J. Mol. Biol.* **30**: 223–53.

Schlager, H., Konopka, P., Schulte, P., Schumann, U., Ziereis, H., Arnold, F., Klemm, M., Hagen, D.E., Whitefield, P.D., and Ovarlez, J. 1997. In situ observations of air traffic emission signatures in the North Atlantic flight corridor. *J. Geophys. Res.* **102**: 10 739–50.

Schulte, P., Schlager, H., Ziereis, H., Schumann, U., Baughcum, S.L., and Deidewig, F. 1997. NO_x emission indices of subsonic long-range jet aircraft at cruise altitude: in situ measurements and prediction. *J. Geophys. Res.* **102**: 21 431–42.

Singh, H.B., Herlth, D., Kolyer, R., Salas, L., Bradshaw, J.D., Sandholm, S.T., Davis, D.D., Crawford, J., Kondo, Y., Koike, M., Talbot, R., Gregory, G.L., Sachse, G.W., Browell, E., Blake, D.R., Rowland, F.S., Newell, R., Merril, J., Heikes, B., Liu, S.C., Crutzen, P.J., and Kanakidou, M. 1996. Reactive nitrogen and ozone over the western Pacific: distribution, partitioning and sources. *J. Geophys. Res.* **101**: 1793–808.

Singh, H.B., Thompson, A.M., and Schlager, H. 1999. SONEX airborne mission and coordinated Polinat-2 activity: overview and accomplishments. *Geophys. Res. Lett.* **26**: 3053–6.

Sinha, A., and Toumi, R. 1997. Tropospheric ozone, lightning and climate change. *J. Geophys. Res.* **102**: 10 667–72.

Siskind, D.E., Nedoluha, G.E., Randall, C.E., Fromm, M., Russell, J.M. III 2000. Correction to "An assessment of southern hemisphere stratospheric enhancements due to transport from the upper atmosphere". *Geophys. Res. Lett.* **27**: 975–5.

Sisterson, D.L., and Liaw, Y.P. 1990. An evaluation of lightning and corona discharge on thunderstorm air and precipitation chemistry. *J. Atmos. Chem.* **10**: 83–90.

Smyshlyaev, S.P., Geller, M.A., and Yudin, V.A. 1999. Sensitivity of model assessments of high-speed civil transport effects on stratospheric ozone resulting from uncertainties in the NO_x production from lightning. *J. Geophys. Res.* **104**: 26 401–17.

Smyth, S.B., Sandholm, S.T., Bradshaw, J.D., Talbot, R.W., Blake, D.R., Blake, N.J., Rowland, F.S., Singh, H.B., Gregory, G.L., Anderson, B.E., Sachse, G.W., Collins, J.E., and Bachmeier, A.S. 1996. Factors influencing the upper free troposphere distribution of reactive nitrogen over the South Atlantic during the TRACE A experiment. *J. Geophys. Res.* **101**: 24 165–86.

Sobral, H., Villagrán-Muniz, M., Navarro-González, R., and Raga, A.C. 2000. Temporal evolution of the shock wave and hot core air in laser induced plasma. *Appl. Phys. Lett.* **77**: 3158–60.

Solomon, S., Portman, R.W., Sanders, R.W., Daniel, J.S., Madsen, W., Bartram, B., and Dutton, E.G. 1999. On the role of nitrogen dioxide in the absorption of solar radiation. *J. Geophys. Res.* **104**: 12 047–58.

Stark, M.S., Harrison, J.T.H., and Anastasi, C. 1996. Formation of nitrogen oxides by electrical discharges and implications for atmospheric lightning. *J. Geophys. Res.* **101**: 6963–9.

Stedman, D.H., and Shetter, R.E. 1983. The global budget of atmospheric nitrogen species. In *Trace Atmospheric Constituents: Properties Transformations, and Fates*, ed. S.E. Schwartz, pp. 411–54, John Wiley and Sons.

Stenchikov, G., Dickerson, R., Pickering, K., Ellis, W. Jr, Doddridge, B., Kondragunta, S., Poulida, O., Scala, J., and Tao, W.-K. 1996. Stratosphere–troposphere exchange in a mid-latitude mesoscale convective complex, 2. Numerical simulations. *J. Geophys. Res.* **101**: 6837–51.

Stith, J., Dye, J., Ridley, B., Laroche, P., Defer, E., Baumann, K., Hübler, G., Zerr, R., and Venticinque, M. 1999. NO signatures from lightning flashes. *J. Geophys. Res.* **104**: 16 081–9.

Stockwell, D.Z., Giannakopoulos, C., Plantevin, P.-H., Carver, G.D., Chipperfield, M.P., Law, K.S., Pyle, J.A., Shallcross, D.E., and Wang, K.-Y. 1999. Modelling NO_x from lightning and its impact on global chemical fields. *Atmos. Environ.* **33**: 4477–93.

Strom, J., Fischer, H., Lelieveld, J., and Schroder, F. 1999. In situ measurements of microphysical properties and trace gases in two cumulonimbus anvils over western Europe. *J. Geophys. Res.* **104**: 12 221–6.

Taylor, G.I. 1950. The formation of a blast wave by a very intense explosion. *Proc. Roy. Soc.* **A201**: 159–86.

Thompson, A.M. 1999. Perspectives on NO, NO_y and fine aerosol sources and variability during sonex. *Geophys. Res. Lett.* **26**: 3073–6.

Thompson, A.M., Pickering, K.E., Dickerson, R.R., Ellis, W.G. Jr, Jacob, J., Scala, J.R., Tao, W.K., McNamara, D.P., and Simpson, J. 1994. Convective transport over the central US and its role in the regional CO and ozone budgets. *J. Geophys. Res.* **99**: 18 703–33.

Thompson, A.M., Tao, W.-K., Pickering, K.E., Scala, J.R., and Simpson, J. 1997. Tropical deep convection and ozone formation. *Bull. Am. Meteor. Soc.* **78**: 1043–54.

Thompson, A.M., Sparling, L.C., Kondo, Y., Anderson, B.E., Gregory, G.L., and Sachse, G.W. 1999. Perspectives on NO, NO_y and fine aerosol sources and variability during SONEX. *Geophys. Res. Lett.* **26**: 3073–6.

Tie, X.X., Zhang, R.Y., Brasseur, G., Emmons, L., and Lei, W.F. 2001. Effects of lightning on reactive nitrogen and nitrogen reservoir species in the troposphere. *J. Geophys. Res.* **106**(D3): 3167–78.

Torres, A.L., and Buchan, H. 1988. Tropospheric nitric oxide measurements over the Amazon Basin. *J. Geophys. Res.* **93**: 1396–406.

Troutman, W.S. 1969. Numerical calculation of the pressure pulse from a lightning stroke. *J. Geophys. Res.* **74**: 4595–6.

Tuck, A.F. 1976. Production of nitrogen oxides by lightning discharges. *Q.J.R. Meteor. Soc.* **102**: 749–55.

Uman, M.A., and Voshall, R.E. 1968. Time interval between lightning strokes and the initiation of dart leaders. *J. Geophys. Res.* **73**: 497–506.

Viemeister, P.E. 1960. Lightning and the origin of nitrates found in precipitation. *J. Meteor.* **17**: 681–3.

Vila-Guerau de Arellano, J., Duynkerke, P.G., and van Weele, M. 1994. Tethered-balloon measurements of actinic flux in a cloud-capped marine boundary layer. *J. Geophys. Res.* **99**: 3699–705.

Von Liebig, J. 1827. Une note sur la nitrification. *Ann. Chem. Phys.* **35**: 329–33.

Vonnegut, B. 1982. The physics of thunderclouds. In *Handbook of Atmospherics*, ed. H. Volland, vol. I, pp. 1–22, Boca Raton, Florida: CRC Press.

Vonnegut, B., Vaughan, O.H. Jr, and Brook, M. 1989. Nocturnal photographs taken from a U-2 airplane looking down on tops of clouds illuminated by lightning. *Bull. Am. Meteor. Soc.* **70**: 1263–71.

Wang, Y., DeSilva, W., Goldenbaum, G.C., and Dickerson, R.R. 1998. Nitric oxide production by simulated lightning: dependence on current, energy, and pressure. *J. Geophys. Res.* **103**: 19 149–59.

Wetselaar, R., and Hutton, J.T. 1963. The ionic composition of rainwater at Katherine, N.T., and its part in the cycling of plant nutrients. *Aust. J. Agr. Res.* **14**: 319–29.

Winterrath, T., Kurosu, T.P., Richter, A., and Burrows, J.P. 1999. Enhanced O_3 and NO_2 in thunderstorm clouds: convection or production? *Geophys. Res. Lett.* **26**: 1291–4.

Woeller, F., and Ponnamperuma, C. 1969. Organic synthesis in a simulated Jovian atmosphere. *Icarus* **10**: 386–92.

Yu, F., and Turaco, R.P. 2001. On the contribution of lightning to ultrafine aerosol formation. *Geophys. Res. Lett.* **28**: 155–8.

Yung, Y.L., and McElroy, M.B. 1979. Fixation of nitrogen in prebiotic atmosphere. *Science* **203**: 1002–4.

Yung, Y.L., Allen, M., and Pinto, J.P. 1984. Photochemistry of the atmosphere of Titan. *Astrophys. J. Suppl. Ser.* **203**: 465–506.

Zahnle, K.J. 1986. Photochemistry of methane and the formation of hydrocyanic acid (HCN) in the Earth's early atmosphere. *J. Geophys. Res.* **91**: 2819–34.

Zhang, R., Sanger, N.T., Orville, R.E., Tie, X., Randel, W., and Williams, E.R. 2000. Enhanced NO_x by lightning in the upper troposphere and lower stratosphere inferred from the UARS global NO_2 measurements. *Geophys. Res. Lett.* **27**: 685–8.

Zel'dovich, Y.B., and Raizer, Y.P. 1966. *Physics of Shock Waves and High-Temperature Hydrodynamic Phenomena*, 445 pp., New York: Academic Press.

Zel'dovich, Y.B., and Raizer, Y.P. 1967. *Physics of Shock Waves and High-Temperature Hydrodynamic Phenomena*, pp. 566–71, San Diego, California: Academic Press.

Zuo, Y., and Deng, Y. 1999. Evidence for the production of hydrogen peroxide in rainwater by lightning during thunderstorms. *Geochimica Cosmochimica Acta* **63**: 3451–5.

16 Extraterrestrial lightning

> In terrestrial lightning the cloud-to-ground discharges are the most intensively studied, and this is a type of lightning that is not expected to occur in other planetary atmospheres.
>
> K. Rinnert (1995)

16.1. Introduction

Most lightning discharges on Earth are produced by convective, precipitating clouds that contain water in both liquid and solid phases (Section 3.2). Apparently, convective clouds whose tops are below the freezing level, so that they do not contain ice, can also sometimes produce lightning (e.g., Foster 1950; Moore et al. 1960; Pietrowski 1960; Michnowski 1963; Rossby 1966). On Earth, lightning-like electrical sparks hundreds of meters long can be produced via charge generation and separation in volcano emissions, and sparks of meter length can occur in sandstorms (subsection 20.7.1). Further, electrical sparks having a length of the order of one meter or less occur in a variety of non-meteorological turbulent particulate media, such as in the material in grain elevators and in the mixture of water and oil present during the water-jet cleaning of oil tanker holds (Pierce 1974). Finally, transient luminous phenomena that may be associated with electrical discharges have been observed during earthquakes (subsection 20.7.2).

Based on the observed conditions for the production of lightning on Earth, the two requirements for the generation of lightning or lightning-like electrical sparks in the extraterrestrial environment are: (i) the interaction of particles of different types or of the same type but with different properties (for example, different temperatures), resulting in local charging, so that charges of opposite sign are acquired by the different classes of particles, and (ii) significant spatial separation of the oppositely charged particles by convection, gravitational forces, or both.

On Earth, apparently both the charge in thunderclouds and the motion of that charge due to lightning can induce electrical discharges in the rarified atmosphere between the cloud tops, the latter usually being at a height of 10 to 20 km (see, for example, Fig. 3.7), and the lower part of the ionosphere, near 100 km (see Fig. 14.1). These recently discovered discharges, discussed in Chapter 14, assume a number of luminous forms termed blue starters, blue jets, red sprites, and elves. Potentially, the atmospheres of other planets may support such discharges or even more exotic electrical discharge phenomena of which we have had no experience. Levin et al. (1983) even speculated that lightning is possible on comets as they approach the Sun.

Electrical discharges in the solar nebula, the cloud of gas and dust from which the solar system was formed some 4.5 billion years ago, have been invoked (e.g., Hecht 1994; Horanyi et al. 1995) to explain the so-called chondrules found in the primitive meteorites formed in the cooling solar nebula prior to the formation of planetary bodies. These chondrules are millimeter-sized beads of glassy silicate that are thought to be the product of the transient heating of nebular material (dust particles) to about 2000 K followed by rapid cooling on time scales of minutes to hours (Hewins 1988; Gibbard et al. 1997). The processes that could have produced large-scale charge separation in the solar nebula are reviewed in Morfill et al. (1993). Gibbard et al. (1997) examined in detail the possibility of nebular lightning and concluded that such lightning was unlikely to have occurred under the conditions then existing. In particular, they stated that the high electrical conductivity of the nebular environment and the relatively low density of solid particles combine to yield a situation in which the large-scale electric fields as well as the electric charges segregated on the particles are short-circuited by the highly mobile electrons and ions. Various models of the generation of lightning in the solar nebula have been presented by Pilipp et al. (1992, 1998), Love et al. (1995), Gibbard et al. (1997), and Desch S.J. and Cuzzi (2000). In this chapter we consider only lightning on presently existing planets of the solar system other than Earth.

A diagram of the major objects that comprise our solar system, the Sun and the nine planets, is found in Fig. 16.1. The eight planets of the solar system nearest to the Sun (excluding Pluto, which is considered separately) can be divided into two groups, one of which comprises the terrestrial planets and the other the jovian planets.

16.1. Introduction

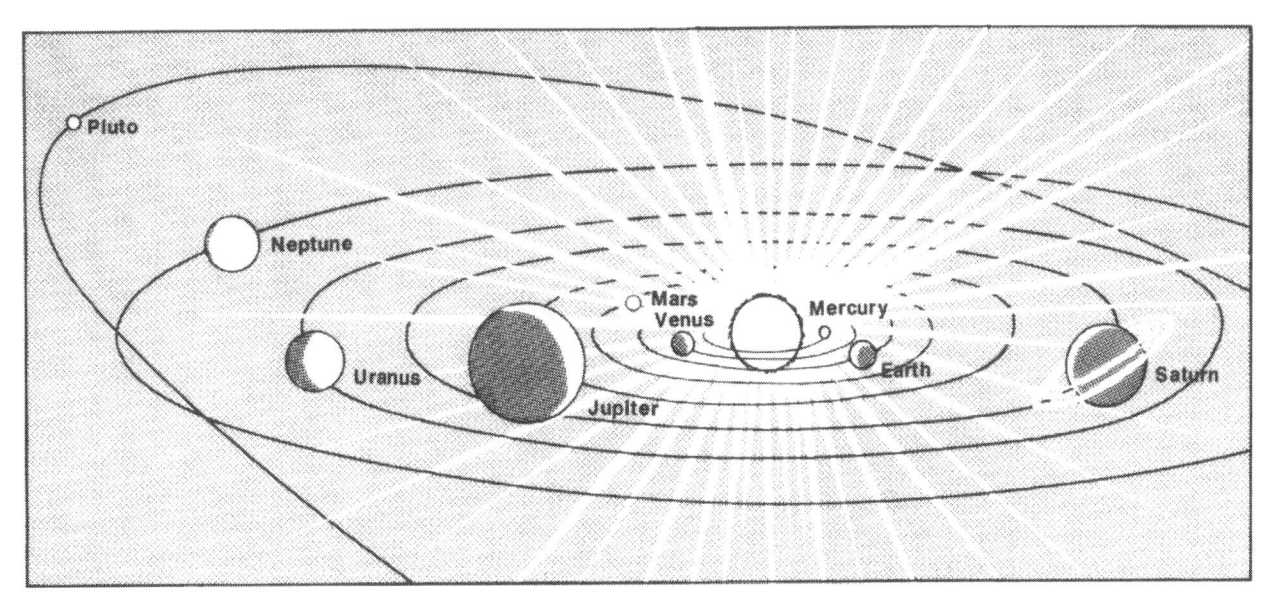

Fig. 16.1. The solar system.

(i) The terrestrial planets, going outwards from the Sun, are Mercury, Venus, Earth, and Mars. Since these planets are primarily composed of rocky materials, they have well-defined solid surfaces, although considerably different atmospheric conditions. Mercury and Mars have relatively thin atmospheres and no present volcanic activity and hence are not expected to host lightning. The severe dust storms that occasionally occur on Mars could, however, produce electrical discharges (e.g., Eden and Vonnegut 1973; Briggs et al. 1977). Venus has an atmosphere potentially capable of supporting lightning activity.

(ii) The jovian (also called giant) planets, going outwards from the Sun, are Jupiter, Saturn, Uranus, and Neptune. Jupiter and Saturn are composed, like the Sun, primarily of hydrogen and helium in various phases, while Uranus and Neptune consist mostly of rock and various "ices" (water, methane, and ammonia, which are presumed to have been in the solid phase at the time of the solar system's formation but are likely to be in the liquid phase now (Hubbard 1984)). The outer gaseous envelopes of Uranus and Neptune are composed primarily of hydrogen (about 83 and 80 percent, respectively), helium, and methane. These four planets are relatively large (radii four to 11 times the Earth's radius). Note that the radii quoted for the giant planets are for the levels corresponding to a pressure of 1 atmosphere on Earth. The giant planets are shrouded with layers of clouds. Calculations by Weidenshilling and Lewis (1973) indicated the existence of three main cloud layers. The outermost cloud layers on Jupiter and Saturn are ammonia (NH_3) ice and on Neptune and Uranus methane ice (CH_4). On all the giant planets the lowest cloud layers are water (H_2O) in solid and liquid states and ammonia water, and the mid-level layers are ammonia or ammonia compounds. The lowest clouds are probably the best candidates for producing lightning, like thunderclouds on Earth. If lightning existed on a jovian planet, it is likely that such lightning would be of the cloud discharge type (Chapter 9) since it is thought that there is no solid or liquid surface beneath the lowest cloud layer to which a discharge could be attracted to form a cloud-to-surface flash.

Completing the list of the known planets is Pluto, which orbits the Sun in a trajectory that is sometimes beyond and sometimes within that of Neptune, and about which relatively little is known.

Two planetary satellites (moons) are known to have atmospheres in which lightning could potentially be generated. Io, one of the 16 known satellites of Jupiter and the closest to Jupiter of the four so-called Galilean satellites, has a relatively thin atmosphere composed primarily of sulfur dioxide. It exhibits high volcanic activity, although there is no evidence of the turbulence necessary for electrical charging within the volcanic plumes. Titan, the largest of at least 18 satellites orbiting Saturn, is covered with thick clouds composed primarily of nitrogen and methane. The Huyghens probe (named after Christian Huyghens, the seventeenth-century Dutch physicist and astronomer who discovered Titan) is scheduled to enter Titan's atmosphere in 2004 and will provide *in situ* measurements. This probe is being carried by the Cassini spacecraft launched on October 15, 1997. Present evidence does not indicate that there is or is likely to be lightning on either Io or Titan (Rinnert 1995). The recent discovery of an ionosphere on another of Jupiter's satellites, Europa, suggests that it also has an atmosphere.

Reviews of the available evidence of possible lightning activity on planets other than Earth have been published by Williams *et al.* (1983), Levin *et al.* (1983), Uman (1987, 2001), Russell (1993), and Rinnert (1982, 1985, 1995). Rinnert (1995) discussed the general characteristics of every planet as well as of Jupiter's moon Io and Saturn's moon Titan and the likelihood of lightning on each. Optical and radio-frequency (RF) signals, including those propagating in the so-called whistler mode (Sections 13.1 and 13.5), recorded by a variety of planetary probes and orbiters have been interpreted to indicate that there is lightning or some form of electrical discharge on five planets: Venus, Jupiter, Saturn, Uranus, and Neptune. In this chapter, we will confine ourselves to a discussion of the evidence, often controversial, for lightning on these five planets; it will be presented, starting with Section 16.3, in the order of the planets given above. In the authors' opinion, at this time the argument for Jovian lightning is convincing, while the arguments for lightning on the other four planets are less than convincing.

16.2. Detection techniques

Lightning on planets other than Earth has never been positively detected from Earth, although Hansell *et al.* (1995), using a telescope on Mount Bigalow in Arizona, reported the possible detection of lightning on Venus. According to Rinnert (1985), there is little possibility of such detection because (i) the optical output from lightning is too low relative to the reflected-sunlight background and (ii) the planets with substantial atmospheres necessarily have ionospheres that would contain, or reflect back toward the planet, lightning electromagnetic signals. As discussed in Sections 13.1–13.3, on Earth electromagnetic waves with frequencies below the plasma frequency of the ionosphere, about 10 MHz for an electron density of 10^{12} m^{-3}, are reflected from it. Examples are AM radio signals and atmospherics (Section 13.3). In the presence of a planetary magnetic field, however, there are electromagnetic wave propagation modes that are confined within the ionosphere and magnetosphere even when the wave frequency is below the plasma frequency, most notably the so-called whistler mode, discussed in detail in Sections 13.1 and 13.5. The whistler propagation mode is possible for electromagnetic waves with frequencies below both the plasma frequency and the electron gyrofrequency (also called the cyclotron frequency), about 1 MHz in the equatorial plane of the ionosphere and magnetosphere of the Earth where the magnetic flux density is about 30 μT (Section 13.2). In the whistler mode, waves tend to propagate along the magnetic field lines, or, more properly, to be confined to magnetic ducts containing electron density gradients perpendicular to the field lines, the wave propagation being highly dispersive so that the higher frequencies propagate significantly faster than the lower ones. There are also non-ducted whistlers, which can propagate both along and across the magnetic field lines. Whistlers recorded both at ground and in the magnetosphere are identified as frequency-dispersed electromagnetic signals, generally in the audio-frequency range (20 Hz to 20 kHz), for which the frequency usually decreases with time over a second or so. The whistler dispersion, which depends on the electron density and magnetic field along the propagation path, can be determined from the curve of observed whistler frequency versus time. Equation 13.102, known as the Eckersley dispersion law, describes the dispersion and is derived in subsection 13.5.2.

Although the term "whistler" in the literature on extraterrestrial lightning is usually used to imply a lightning source, this is not necessarily the case. On Earth, the sources of whistlers are also radiation from VLF transmitters and radiation associated with the kilohertz-range harmonics in power systems. Further, many types of VLF wave propagating in the magnetospheric plasma along the magnetic field lines, that is, in the whistler mode, are generated by processes naturally occurring in the plasma, as opposed to being caused by subionospheric lightning. Examples of such waves, which can have frequency–time characteristics distinctly different from those of the lightning-produced whistlers recorded on Earth, are hiss and chorus (subsection 13.5.1). In general, any electromagnetic wave, regardless of its origin, whose frequency is below both the plasma frequency and the electron gyrofrequency is said to propagate in the whistler mode. Ksanfomality *et al.* (1983) claimed that whistler-mode signals originating in the magnetosphere and detected above the Earth's ionosphere by plasma-wave instruments differ from lightning-produced whistlers. It is reasonable to expect that extraterrestrial lightning would produce whistlers that obey the Eckersley dispersion law (Eq. 13.102), which has been verified experimentally for terrestrial lightning-generated whistlers.

Most whistlers observed with satellites in the Earth's magnetosphere are apparently propagating in the non-ducted mode (subsection 13.5.1); that is, their paths deviate significantly from the Earth's magnetic field lines. Holzworth *et al.* (1999) reported that essentially every lightning event at 1 to 2000 km from an instrumented ionospheric rocket produced upward-going whistler waves. Most of these waves propagated in an non-ducted mode into the outer magnetosphere. Non-ducted whistlers show a wide variety of frequency–time signatures, depending on the location of the receiver and on the medium. Thus, for non-ducted whistlers, the absence of the characteristic dispersion described by Eq. 13.102 would not necessarily allow one to rule out lightning as a possible source of the whistler.

The availability of various types of *in situ* measurements that could be (and have been) interpreted as evidence for lightning on the planets Venus, Jupiter, Saturn,

16.3. Venus

Table 16.1. *Availability of measurements potentially indicative of lightning on planets other than Earth*

	Planet				
Type of data	Venus	Jupiter	Saturn	Uranus	Neptune
Optical	yes	yes	no	no	no
Whistlers	yes	yes	no	no	yes
Other RF	yes	yes	yes	yes	yes

Uranus, and Neptune is indicated in Table 16.1. These measurements will be examined in detail in the following sections. In Table 16.1, the category "Other RF" covers all recorded radio-frequency signals that could not be identified as whistlers. On Venus, only narrowband RF signals have been recorded, some of which have been interpreted as whistlers even though their dispersion generally could not be determined. On Jupiter, Saturn, Uranus, and Neptune, the instrumental bandwidth employed was sufficient to identify whistlers by their characteristic dispersion. No whistlers were detected on Saturn or Uranus. Some of the data to be presented were obtained using detectors specifically designed or intentionally used to search for lightning, while other data were obtained by instruments designed for other projects but unintentionally capable of detecting signals subsequently interpreted as being due to lightning.

16.3. Venus
16.3.1. General information

Venus, with an equatorial radius of 6052 km, is about the size of Earth and has a similar surface topography but no moons. Starting with the first *in situ* observations in 1962, Venus has been the most extensively studied of all the planets, with six spacecraft flybys (Mariners 2, 5, and 10, Galileo, and Cassini), two probes deployed during flybys (Vega 1 and Vega 2), and 15 missions specifically targeting Venus, including the Venera (Venus in Russian) series 4 through 16, Pioneer Venus, and Magellan. Many of these missions were outfitted to look specifically for lightning. Reviews of all aspects of the planet Venus are found in the books *Venus*, edited by Hunten *et al.* (1983), *Venus International Reference Atmosphere (VIRA)*, edited by Kliore *et al.* (1985), and *Venus II*, edited by Bougher *et al.* (1997). The most recent review papers on possible lightning activity on Venus have been by Russell (1991a), Rinnert (1995), Strangeway (1995), and Grebowsky *et al.* (1997).

The Venusian atmosphere consists primarily of carbon dioxide (about 97 percent) and nitrogen (about 3 percent). The surface pressure is about 90 bar (1 bar = 0.987 atm = 10^5 pascal), and the surface temperature is about 740 K (the temperature at which lead melts is 600 K). There is no surface water. Surface winds are light, 0.3 to 1.0 m s^{-1}. While there is considerable evidence in the surface topography of previous volcanic activity, the atmosphere below about 30 km is clear and shows no ejected volcanic material. There is a dense cloud cover completely covering Venus between about 45 and 70 km above the surface, at a pressure near 1 bar, with haze layers above and below this cloud system. The temperature profile through the cloud deck is such that there should be little vertical convection. Horizontal winds at the top of the cloud system can exceed 100 m s^{-1}. Venusian clouds probably consist of sulfuric acid (H_2SO_4). A sketch of the Venusian atmosphere showing the cloud structure and the temperature, pressure, and wind speed profiles is found in Fig. 16.2.

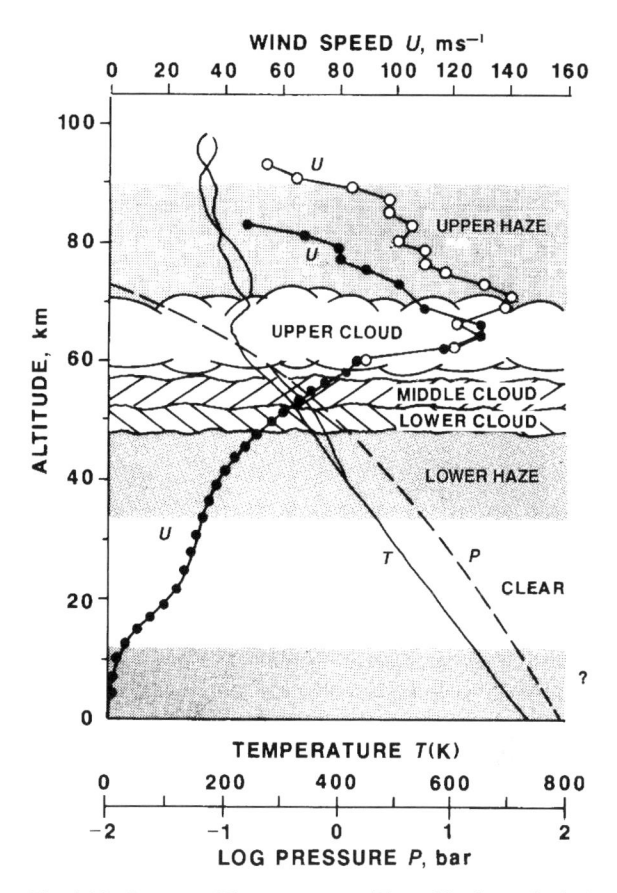

Fig. 16.2. Pressure (P), temperature (T), and horizontal wind speed (U) height profiles for the Venusian atmosphere, inferred from Venera and Pioneer Venus probe measurements, together with a diagram of the global cloud cover between about 45 and 70 km. The Venusian atmosphere contains CO_2, N_2, and also H_2O, O_2, and SO_2. Adapted from Rinnert (1995).

It would appear that Venus is not a good candidate for lightning activity since it has little convection and apparently no present volcanic activity, as will be discussed later. Nevertheless, Venus is the first planet other than Earth where optical and radio-frequency signals have been recorded that, for lack of other credible sources, have been attributed to lightning. A number of the radio-frequency measurements were made by landers on descent below the Venusian ionosphere and even on the surface.

A brief outline of the available information regarding the search for lightning on Venus, which will be considered in more detail in subsections 16.3.2–16.3.4, is as follows. Radio-frequency pulses were detected by the Venera 11 and Venera 12 landers in December 1978 and by the Venera 13 and Venera 14 landers in March 1982. Impulsive electromagnetic signals were observed in the Venusian ionosphere by the Pioneer Venus orbiter plasma-wave instrument at frequencies 100 Hz, 730 Hz, 5.4 kHz, and 30 kHz in 1979. Some of the 100 Hz signals could be attributed to whistlers whose sources were below the ionosphere, while the origin of other 100 Hz signals and the signals recorded in both the 100 Hz channel and in the higher-frequency channels remains unclear. It is possible that the signals that were inconsistent with the hypothesis of the whistler mode of propagation from below the ionosphere were produced by plasma instability mechanisms local to the orbiter. The plasma-wave instrument of the Galileo spacecraft, passing near Venus in 1990 on its way to Jupiter, detected nine radio-frequency events in the range 100 kHz to 5.6 MHz apparently escaping Venus through its ionosphere. Optical pulses detected in 1975 by the spectrometer of the Venera 9 orbiter were interpreted as being due to lightning. However, the Pioneer Venus orbiter star sensor searched but failed to find lightning-generated optical signals on the night side of Venus. Further, the Vega 1 and Vega 2 balloons that probed the Venusian cloud system in 1985 had optical sensors to detect lightning during both day and night but did not find any. Finally, no radio-frequency events were detected in the range 125 kHz to 16 MHz by the Cassini spacecraft passing near Venus in 1998 and 1999 on its way to Saturn.

16.3.2. Optical measurements

Venera 9 spectrometer Krasnopolsky (1980, 1983a, b) reported optical pulses recorded by the scanning spectrometer aboard the Venera 9 orbiter that were interpreted as due to lightning activity occurring over an estimated area of 5×10^4 km^2. The lightning occurrence rate was estimated to be 2×10^{-3} km^{-2} s^{-1} (120 flashes per minute over 1000 km^2), an order of magnitude or so higher than for a localized storm on Earth. Williams *et al.* (1983) pointed out that it was unclear how such a high burst rate could be due to lightning, since the Venusian meteorology is not conducive to the production of active storms. Further, subsequent attempts to detect optical signals on Venus, discussed below, have failed to provide support for the Venera 9 optical data.

Pioneer Venus star sensor The Pioneer Venus Orbiter navigation star sensor was used by Borucki *et al.* (1981, 1991) to search for lightning, although it was not designed for such a search. The Venusian airglow, called the ashen light (Russell and Phillips 1990), often saturated the instrument when the whole detector viewed the dark side of Venus. Therefore, in the lightning search mode the star sensor was programmed to take data when the planet's limb covered only a small part of the detector or when the dark limb was just beyond the detector so that only light from the dark side scattered by the glass of the lens reached the detector element. Comparisons of the signals received when the sensor was pointed in this manner at the night side of Venus with those received when the sensor was pointed into deep space indicated that all signals received were from energetic particle impacts and that, if there were lightning signals, these could not be separated from the energetic-particle impact noise. This noise level set an apparent upper limit of about 3 km^{-2} yr^{-1} or 1×10^{-7} km^{-2} s^{-1} to the lightning that might have been present (Borucki *et al.* 1991). Of course, the actual upper limit could be much less or zero (no lightning). Overall, the star sensor experiment did not answer the question whether there is any lightning activity on Venus. In addressing the apparent conflict between the RF data, to be discussed next, and the star sensor data, Borucki *et al.* (1991) suggested that there may be no lightning on the night side of Venus, that is, lightning may be generated only in daylight.

Williams *et al.* (1982) and Williams and Thomason (1983) showed that attenuation by clouds cannot account for the failure of the star sensor to detect lightning on Venus. They used a Monte Carlo technique to model the effect of the Venusian clouds on a hypothetical optical lightning signal. Williams *et al.* (1982) found that the fraction of photons of visible light (wavelengths of 0.40–0.70 μm) that escapes into space from the Venusian clouds ranges from 0.1 to 0.4 and is about 0.05 for red photons (0.68–0.70 μm) produced near the surface. Williams and Thomason (1983), using an improved model, found about twice as many photons of visible light escaping from clouds and four times as many for red light originating near the surface. About three percent of blue photons (0.40–0.42 μm) emitted by sources at the surface were predicted to escape (Williams and Thomason 1983). Since the star sensor used by Borucki *et al.* (1981) was sensitive to red light (the peak response was at 0.7 μm), lightning luminosity within the clouds or even near the surface should, in principle, have been detectable unless this luminosity was much weaker than that of terrestrial lightning (Williams *et al.* 1982; Williams and Tomason 1983) or unless the lightning luminosity signals were buried within the energetic-particle impact noise.

Vega 1 and Vega 2 balloons In 1985, the Vega 1 and Vega 2 spacecraft each released a balloon into the Venusian

16.3. Venus

atmosphere at midnight near the equator (Sagdeev et al. 1986a, b). The balloons floated westward into the sunlit hemisphere at an altitude of about 54 km, within the middle layer of the global cloud deck shown in Fig. 16.2. Data were received for a total of about 46 hours. The balloons carried a variety of instruments including some to measure the ambient illumination and to detect possible transient optical events including lightning. No reliable indication of lightning flashes was obtained, and no obvious breaks in the cloud deck were observed. The vertical winds were measured as being mainly of order 1 m s^{-1}, but both balloons encountered downdrafts of up to 3.5 m s^{-1}.

16.3.3. RF signals in and above the ionosphere

Although Venus has a conducting liquid interior, similar to Earth, its rotation is not rapid enough to create, as does Earth, its own magnetic field. The magnetic field existing around Venus is generated through the diffusion and transport of the interplanetary magnetic field, which is the Sun's magnetic field carried throughout the solar system by the expanding solar plasma called the solar wind. In other words, Venus has an induced as opposed to an intrinsic magnetosphere (Russell 1991c), the latter being found on Mercury, Earth, Jupiter, Saturn, Uranus, and Neptune, but not on Mars or Venus. The Venusian magnetic field is highly variable in both magnitude and direction, especially on the night side. The Venusian dayside ionosphere is characterized by an electron density profile that has a peak of about 7×10^5 cm^{-3} at an altitude of about 140 km. The night side appears to be patchy and may have layers at about 150 km altitude with peak electron densities up to 2×10^4 cm^{-3} (Rinnert 1995).

Pioneer Venus orbiter The overall mission continued from 1978 to 1992, and the characteristics of the orbit were changed significantly during this period. In 1978–80, the periapsis altitude (the point in the orbit nearest to the planet) was near 150 to 180 km, so that the orbiter penetrated into the ionosphere (Ksanfomality et al. 1983). Later (after the first three seasons), the periapsis altitude was allowed to rise as high as 2900 km, and in 1992 it was held between 150 and 250 km. The plasma-wave instrument of the orbiter, called the orbiter electric field detector (OEFD), detected signals, some of which have been attributed to whistlers presumably propagating along the magnetic field lines and generated below the ionosphere (Taylor W.L. et al. 1979; Scarf et al. 1980a). The instrument had four channels centered at frequencies of 100 Hz, 730 Hz, 5.4 kHz, and 30 kHz. The data from early orbits show no impulsive radio signals, apparently because during this period the orbiter's periapsis was within the day-side ionosphere, in part due to noise associated with sunlight on the orbiter and its solar panels. When the periapsis changed (in a month or so) to the night side, strong impulses, first reported by Taylor W.L. et al. (1979), were detected at low altitudes. The signals apparently lasted less than 500 ms and were recorded in all four frequency channels, although the electron gyrofrequency and plasma frequency were estimated to be such that the propagation of 5.4 and 30 kHz waves could not be supported (Russell 1993).

The data obtained with the four-channel plasmawave instrument on Venus have been examined by many researchers and remain the subject of debate. It appears that due to the limited spectral information an unambiguous interpretation of the data is impossible. The overall data set can be divided into two categories: (i) the solitary 100-Hz signals, that is, those not accompanied by signals in the higher-frequency channels and (ii) the multifrequency signals, that is, those occurring simultaneously in two or more of the four frequency channels. The two categories differ in the hypothetical source (or lack of such), although there exist conflicting views and ambiguities within each category. We will consider next each of the two data categories.

Solitary 100 Hz signals These were first considered by Scarf et al. (1980a), who analyzed data for about 100 orbits and showed that the signals occurred when the local magnetic field was relatively strong (so that the electron gyrofrequency was well in excess of 100 Hz) and steady, the field vector pointing down to the surface of the planet. The signals were attributed to whistlers produced by "atmospheric lightning". From a larger sample, based on 1185 orbits, Scarf and Russell (1983) identified 340 similar events, more than 65 percent of which appeared to originate from sources clustering near the Venusian regions called Beta, Phoebe, and Atla Regios, which are thought to be of volcanic origin. The clustering of source locations was interpreted as an indication that the sources were located at or near the surface, implying a relation to volcanic activity. However, Borucki (1982) pointed out that the Pioneer Venus probes found no evidence of dust or ash in the atmosphere, as would appear to be necessary if the whistler sources were due to volcanic activity, and Levin et al. (1983) argued that the determination of whistler-source location was extremely inaccurate. Taylor et al. (1987) also argued against a volcanic origin and showed that the spatial distribution found by Scarf and Russell (1983) could reflect the instrumental coverage of the surface of the planet. In a later review paper, Russell (1991a) noted that the data are more consistent with a source within the Venus cloud deck than with volcanoes.

Sonwalkar et al. (1991) examined 11 events, each appearing in one or more channels as a group of pulses lasting from 30 s to 4 min, to test the hypothesis that each of these events was the result of whistler-mode propagation from a subionospheric source. Seven out of the 11 events were the 100-Hz-only signals at low altitudes, and the remaining four were the wideband signals discussed later. Various tests, other than the previously used ratio of observed frequency f to gyrofrequency f_g (Scarf *et al.* 1980a; Scarf and Russell 1983) were proposed. The tests made use of the theory of wave propagation in the magnetoplasma (Section 13.2 and subsection 13.5.2) and the available data on electron density, ambient magnetic field, orbiter position and motion, orientation of the spacecraft spin vector, and orientation of the electric field antenna in the spin plane. The primary test, applied to all 11 cases, was to determine whether the angle of the wave vector with respect to the ambient magnetic field vector was within the allowed cone of angles for the whistler mode of propagation (the so-called resonance cone, whose angle θ_r is given by $\cos \theta_r = f/f_g$). Applying the resonance cone test, Sonwalkar *et al.* (1991) found that six out of seven 100-Hz-only signals were consistent with whistler-mode propagation from below the ionosphere. If the signal passed the primary test then it was possible sometimes to compare its dispersion and polarization with those predicted by theory. Note that Strangeway (1991b) found that the solitary 100 Hz signals that satisfied the resonance cone test were polarized perpendicular to the ambient magnetic field, as expected for whistler-mode waves in the Venusian night-side ionosphere, while the solitary 100 Hz signals detected outside the resonance cone tended to be polarized parallel to the ambient magnetic field, similarly to the higher-frequency signals.

Sonwalkar and Carpenter (1995) extended the study of Sonwalkar *et al.* (1991) to include an additional 14 cases. In the overall set of 25 events, 15 events were solitary 100 Hz signals recorded at altitudes ranging from near 150 to 1300 km. Out of those 15 events, 12 were consistent with whistler-mode propagation from subionospheric sources, according to the resonance-cone test. In one case, the signal, recorded at relatively high altitude, passed the resonance-cone test but failed the dispersion test. The three 100 Hz events that failed the test were observed at relatively low altitudes, about 150–400 km. Thus it appears that the majority of the solitary 100 Hz signals might be from subionospheric sources, possibly from lightning. Some of the 100 Hz signals have characteristics inconsistent with whistler-mode propagation, and their generation mechanism is presently unknown.

Taylor H.A. *et al.* (1985, 1986, 1987), Taylor H.A. and Cloutier (1987, 1992), Maeda and Grebowsky (1989), and Huba (1992) have questioned the interpretation of the observed 100 Hz signals as lightning-produced whistlers and have attributed them to local plasma effects. Taylor *et al.* and Taylor and Cloutier found that many of the signals coincided with ion density depletions (troughs), the occurrence of superthermal ions, and magnetic field gradients. They also noted that the signals at 100 Hz were observed primarily when the spacecraft velocity was perpendicular to the ambient magnetic field and argued that the signals might be ion acoustic waves propagating parallel to the ambient magnetic field but shifted to higher frequencies due to the Doppler effect. Scarf and Russell (1988), Strangeway (1991a, b, 1995), and Russell and Strangeway (1992) argued against such *in situ* plasma instabilities being the source of the 100 Hz signals, among other arguments pointing out that the observed signals have an electric field polarized perpendicular to the ambient magnetic field, which is unlikely to be the case for ion acoustic waves.

Other wave instabilities that have been invoked to explain the 100 Hz signals are cyclotron-resonant whistler-mode instabilities (Maeda and Grebowsky 1989) and short-wavelength lower hybrid waves (Huba 1992). Both these instabilities could potentially produce detectable 100 Hz signals. However, Strangeway (1995) argued that (i) cyclotron resonant instability was not a reasonable explanation because the thermal pressure in the Venus ionosphere is comparable to the magnetic pressure, thus suppressing this *in situ* instability, and (ii) the lower hybrid waves require very steep density gradients to overcome the damping due to collisions and have wave properties that are less consistent with the data than hypothetical lightning whistler waves. Cole and Hoegy (1996) found from modeling that the electromagnetic signals observed by the orbiter electric field detector (OEFD) could heat the Venusian ionosphere to unrealistically high electron temperatures if these signals were produced by lightning. Therefore, they argued that the observed signals were more likely to be due to local effects. However, Strangeway (1996) used a more sophisticated model to argue that any excessive heating would be partially offset by electron heat conduction, not adequately accounted for by Cole and Hoegy (1996). Strangeway (1995) concluded that a lightning source remains the most probable explanation of the 100 Hz signals detected in the night-side ionosphere of Venus and that the data are most consistent with whistler-mode waves propagating vertically upward from the cloud deck below the ionosphere. Nevertheless, Strangeway conceded that the interpretation of the 100 Hz signals as due to lightning remains controversial.

Multifrequency signals Such signals have been cited as possibly due to lightning (Singh and Russell 1986; Russell 1991a), although they are clearly not whistler-mode waves. Unlike the solitary 100 Hz signals, the multifrequency signals are detected in regions of mainly horizontal ambient magnetic field (Ho *et al.* 1992). The multifrequency signals are often observed at altitudes greater than 1000 km and frequently occur in regions of locally reduced electron density (Sonwalkar and Carpenter 1995). Sonwalkar and Carpenter (1995) examined in detail 10 multifrequency signals, nine of which were recorded at altitudes ranging from

16.3. Venus

143 to 219 km and one at 1125–1709 km. In all cases the signal appeared in the 100 Hz, 730 Hz, and 5.4 kHz channels, and in four cases also in the 30 kHz channel, while in all 10 cases the gyrofrequency was less than 5.4 kHz. Using the resonance-cone test, Sonwalkar and Carpenter (1995) found all nine events recorded at relatively low altitudes to be inconsistent (for each of the observed frequencies) with whistler-mode propagation from subionospheric sources. They suggested that these events are nonpropagating modes that could have been generated near the orbiter by a mechanism as yet unknown but possibly related to local plasma instabilities. Strangeway (1995) seemed to agree with this interpretation but only for the higher-altitude signals (recorded beyond 1000 km or so). He noted the possibility that the multifrequency signals at low altitudes are due to direct coupling between lightning and the ionosphere, the mechanism of such coupling being presently not understood. One higher-altitude multifrequency signal analyzed in detail by Sonwalkar and Carpenter (1995) passed the resonance-cone test but failed the dispersion test. We will next discuss the morphology of the low-altitude bursts as a function of frequency.

Higher-frequency signals versus 100 Hz signals Singh and Russell (1986) suggested that waves with frequency greater than the electron gyrofrequency may well be produced by lightning in the Venusian atmosphere and leak through the inhomogeneous and patchy night-time ionosphere. The occurrence rates of bursts at 0.73, 5.4, and 30 kHz showed a well-defined maximum at around 21 local time (Russell *et al.* 1988a, c, 1989a; Russell and Scarf 1990; Ho *et al.* 1991) while the 100 Hz bursts were distributed more uniformly throughout the night side. An evaluation by Ho *et al.* (1991) resulted in an average burst rate of 0.14 s^{-1} at 5.4 kHz during the observed rate maximum, at around 21 local time. For an assumed search area of $31\,400 \text{ km}^{-2}$ (a radius of 100 km), this burst rate would yield a hypothetical lightning flash density of $4.5 \times 10^{-6} \text{ km}^{-2} \text{ s}^{-1}$ or $150 \text{ km}^{-2} \text{ yr}^{-1}$ if the activity were continuous throughout the year. The burst occurrence rate versus altitude below 300 km at frequencies higher than the gyrofrequency and at 100 Hz outside the resonance cone differs significantly from that at 100 Hz inside the resonance cone (Ho *et al.* 1992). While there is little variation in occurrence rate with altitude for 100 Hz waves inside the resonance cone, as one would expect for whistler-mode propagation, the rate falls off rapidly, with roughly the same scale length of about 20 km, for both the higher-frequency waves and also the 100 Hz waves outside the resonance cone, implying a common source for these two classes of signals.

Galileo and Cassini flybys During the Galileo spacecraft gravity assisted flyby of Venus on 10 February 1990, the Galileo plasma-wave instrument was used to search for impulsive radio signals from Venus in the frequency range extending well above the local plasma frequency, from about 100 kHz to 5.65 MHz, that is, signals capable of escaping the Venusian ionosphere (Gurnett *et al.* 1991). A total of nine events were detected in the solar wind at a distance of several Venusian radii. Since six out of the nine events were observed at frequencies of 400 kHz and higher, thought to be too high for locally generated plasma waves, and in view of the lack of other credible sources, these events were interpreted by Gurnett *et al.* (1991) as probably being due to lightning in the Venusian atmosphere.

The Cassini spacecraft, on its way to Saturn, made two gravity-assisted flybys of Venus, the first on 26 April 1998 and the second on 24 June 1999. During the flybys, the radio and plasma-wave instrument searched for impulsive radio signals in the frequency range from 125 kHz to 16 MHz. No such signals were detected from Venus, while during the Cassini flyby of the Earth on 18 August 1999 the instrument detected signals at a rate of 70 pulses per second (Gurnett *et al.* 2001).

16.3.4. RF signals below the ionosphere

The Venera 11, 12, 13, and 14 landers carried instruments called Groza (Russian for thunderstorm) to detect and analyze RF noise in the atmosphere of Venus (Ksanfomality 1980; Ksanfomality *et al.* 1983). Each instrument consisted of a loop antenna to respond to the magnetic component of the RF signals, a receiver with narrowband channels centered at 10, 18, 36, and 80 kHz, and a wideband channel from 8 to 90 kHz. The Venera 11 and Venera 12 instruments detected impulsive low-frequency signals with varying intensity during the one-hour descents from 60 km to the surface on 21 and 25 December 1979, respectively. The Venera 11 detector transmitted data for 76 min after landing; Venera 12 transmitted for 110 min. Both probes followed almost the same trajectory during descent. Venera 11 recorded higher pulse rates and higher pulse amplitudes than Venera 12. Venera 11 observed a maximum in the RF intensity when it was between 30 and 15 km, some variations in RF below 15 km, and a gradual decay to zero near the surface. Between 13 and 9 km the noise bursts were grouped in clusters or modulated with a period of roughly 50 s. During the Venera 12 descent the overall RF activity was much less, some activity occurring when the probe was between 50 and 30 km, with a maximum below 9 km, and activity decreased toward the surface. Venera 12, however, detected a burst of about 150 pulses in an eight-second measuring interval 30 min after landing. These low-frequency RF pulse bursts were attributed to lightning for lack of another credible source. Ksanfomality *et al.* (1983) argued against the idea that the observed RF signals were from electrostatic discharging of the probes following triboelectric charging of the spacecraft by the ambient atmosphere. Their arguments were (i) the different height dependences of the RF signal intensities measured by the two different probes descending along similar routes in essentially the same region of the planet, (ii) the periodic nature of the signal on Venera 11, and (iii) the observation of one burst 30 min after landing by the Venera 12 instrument when triboelectric charging and subsequent discharging, if any, should have ceased.

In order to check whether the RF signals observed by the Venera 11 and Venera 12 instruments during their descents could have been due to electrostatic discharges at the probes, the Venera 13 and Venera 14 probes carried additional devices to monitor electrostatic discharges. Venera 13 and Venera 14 entered the Venusian atmosphere on 1 and 5 March 1982, respectively, and again measured noise profiles, finding an overall activity similar to that found by Venera 12. No discharge currents from the probes were detected. The average impulse rate was about 30 s^{-1} but reached a maximum as high as 55 s^{-1}.

Ksanfomality et al. (1983) treated the Venera 11 signals as due to distant lightning and interpreted the intensity modulation observed by Venera 11 as due to spacecraft antenna rotation. With some assumptions they estimated signal-source distances of 700 to 1000 km and, for another period, 1250 to 1500 km for storm regions about 120 to 150 km in extent. With these estimates Ksanfomality et al. found a flash density of 1.5×10^{-3} km^{-2} s^{-1}, a value very similar to the flash density during a typical storm on Earth (Section 2.2).

In summary, the presently existing evidence of lightning activity on Venus is ambiguous. At least some of the RF signals observed in and above the ionosphere are likely to be due to local effects near the spacecraft, although others do not contradict the hypothesis that their source is Venusian lightning. Further, atmospheric conditions on Venus do not appear conducive to lightning activity. It appears that proponents of the existence of Venusian lightning have abandoned the once popular association of this hypothetical lightning with active volcanoes in favor of a cloud origin.

16.4. Jupiter
16.4.1. General information

Jupiter is the largest planet in the solar system, with an equatorial radius of 71 492 km, roughly 11 times that of Earth. It has 16 known satellites. Six spacecraft have observed Jupiter as they flew by: Pioneer 10 and Pioneer 11 in December 1973 and December 1974, respectively, Voyager 1 and Voyager 2 in March and May 1979, respectively, Ulysses in 1992, and Cassini in 2000–2001. The Galileo spacecraft released a probe into the Jovian atmosphere on 7 December 1995 and began its orbital mission around Jupiter at about the same time (the end of the mission is planned for 2003). The book edited by Gehrels (1976) gives a thorough review of all aspects of Jupiter known after Pioneer 10 and Pioneer 11 but before Voyager 1, Voyager 2, and Galileo. Jupiter is thought to have a solid core whose radius is about 10 000 km, or 1.5 times that of Earth, surrounded by a 40 000-km-deep ocean of liquid metallic hydrogen, which is in turn surrounded by a 21 000-km-thick layer of molecular hydrogen. This latter layer gradually changes in the outward direction from liquid to gas as the pressure falls into the range of tens of atmospheres. No gas–liquid boundaries are expected since such transitions are only sharp under relatively low pressures (Morrison and Owen 1988). The above description, generally found in the literature, ignores helium, the second principal constituent of Jupiter. Figure 16.3 illustrates a variety of Jovian cloud and atmospheric properties. Note that clouds are formed in a relatively thin layer of atmosphere, of less than 100 km thickness.

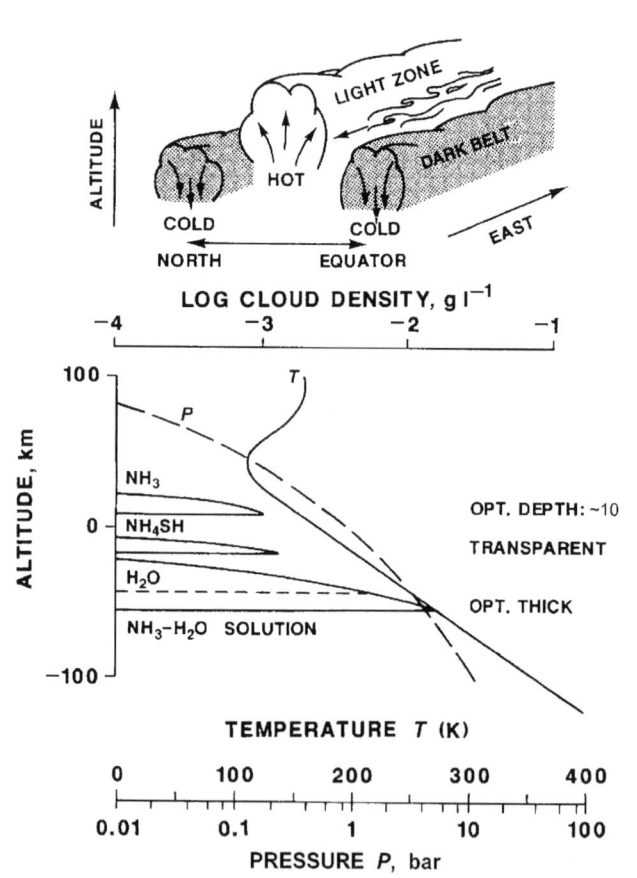

Fig. 16.3. The global convective motions (top panel) and height profiles for pressure (P) and temperature (T) (bottom panel) of the Jovian atmosphere, together with the predictions of the cloud model of Weidenschilling and Lewis (1973). Altitude is measured with respect to the 1 bar pressure level. The broken horizontal line indicates the 6 bar level, at which the H_2O cloud base becomes a solution of NH_3 in liquid water. Adapted from Rinnert (1995).

16.4. Jupiter

Stone (1976), Ingersoll (1976), Smith *et al.* (1979a), West *et al.* (1986), Young *et al.* (1996), and Beebe (1997) have provided reviews of what is known about the Jovian atmosphere. This atmosphere is composed of about 90 percent hydrogen and about 10 percent helium. The atmosphere of Jupiter has a banded structure consisting of about 10 horizontally alternating white zones and dark belts. It is believed that the white zones are upward-moving portions of the atmosphere and the dark belts are downward moving. Eastward and westward winds alternate with latitude. The absolute wind velocities range from about 20 to 150 m s^{-1}. High-resolution imaging of white plumes found near the equator shows that they contain small (100 km in diameter) puffy elements resembling cumuli on Earth. Jupiter has an internal energy source that provides an outward heat flux 1.67 times larger than the heat flux that planet receives from the Sun. This excess heat is thought to be energy left over from the formation of the planet. The interior temperature (about 25 000 K at the center) is far too low to permit nuclear reactions. There is little variation in the temperature of the atmosphere in going from the equator to either pole.

Information on the atmosphere below the upper clouds comes largely from atmospheric models, some using spacecraft or Earth-based optical measurements, since the upper clouds block the view of the lower ones when the observer is outside the cloud system. The Galileo probe provided some *in situ* information that will be discussed later. The solar-composition chemical equilibrium model for the Jovian atmosphere (Weidenschilling and Lewis 1973) is based on an assumed atmospheric composition like the Sun's and an assumed thermal equilibrium of the molecules in the atmosphere. This model predicts uppermost clouds of NH_3 (ammonia) ice near 0.5 bar and 150 K, an intermediate cloud layer of NH_4SH (ammonium hydrosulfide) ice near 3 bar and 200 K, and a bottom cloud of H_2O ice near 5 bar and 270 K, as shown in Fig. 16.3. If the NH_3, H_2O, and H_2S abundances are allowed to increase by a factor 5 then the H_2O cloud base becomes a liquid water and NH_3 solution at the 6 bar level, also indicated in Fig. 16.3. The Weidenschilling and Lewis (1973) model gives an upper limit to the H_2O cloud liquid water content of 10 g m^{-3}, a value substantially higher than the liquid water content of thunderclouds on Earth (typically 3 g m^{-3}). This high water content suggests the presence of precipitation, which in turn could be an element in cloud charge generation and separation mechanisms. Analysis of Earth-based spectral measurements obtained in 1976 and published by Woodman *et al.* (1979) led Sato and Hansen (1979) to conclude (i) that the upper cloud of NH_3 ice crystals has an optical depth of about 10 (the optical depth is defined as the line integral of the opacity over the radiation path; e.g., Rybicki and Lightman 1979) (ii) that the NH_4SH cloud beneath is transparent or nonexistent, (iii) that beneath the "ammonia" cloud region at 3 to 5 bars is an optically thick cloud region, which may be H_2O, and (iv) that an aerosol haze is present above the top, ammonia, cloud, region. The H_2O clouds are thought to be about 30 km thick.

From an analysis of Voyager infrared spectra, Bjoraker *et al.* (1986) argued that water at the 2 to 6 bar level is depleted by a factor of 50 (by two orders of magnitude according to Bjoraker 1985 and West *et al.* 1986) relative to the solar abundance. Such a depletion of water prohibits cloud formation at 5 bar level, where a massive water cloud is predicted by the solar-composition chemical equilibrium model of Weidenschilling and Lewis (1973) discussed above, as shown in Fig. 16.3. The distribution of clouds in Jupiter's atmosphere discussed by Bjoraker (1985), Bjoraker *et al.* (1986), and West *et al.* (1986) postulates the presence of thin NH_4SH–H_2O clouds in the lowest cloud layer, which is at the 2 bar level, and no clouds below that level. Such a cloud distribution is sometimes referred to as the water-depleted model (e.g., Borucki and Williams 1986). The water (and oxygen) depletion inferred from the Voyager infrared spectra can be explained using the Stoker (1986) model of localized updrafts to describe the Jovian equatorial plumes. Starting at the 5 bar level, up-welling moist gas becomes saturated and forms precipitation, and the release of latent heat drives the gas further upward. At roughly the 2 bar level, overturning of the plume occurs, and the dry gas descends. If such localized plumes occupy only two percent of the area, these dry regions would explain the infrared measurements without changing the global water abundance (Lunine and Hunten 1987). Such a system, with strong updrafts and heavy precipitation, could potentially be a lightning generator. The plumes are seen near the equator, but they could exist also at higher latitudes. However, another analysis of the Voyager infrared data (Carlson *et al.* 1992) showed that the abundance of water in the Jovian atmosphere must be at least 1.5 times the solar abundance. Thus, the question of the composition of the Jovian atmosphere is not yet settled.

The Galileo probe that penetrated the Jovian atmosphere to a depth corresponding to an atmospheric pressure of about 24 bars did not detect the distinct water clouds expected near the 5 bar level. It did find evidence of two upper cloud layers at heights corresponding to the expected positions of the ammonia and ammonium hydrosulfide clouds (Ragent *et al.* 1996; Sromovsky *et al.* 1996). It is worth noting that the Galileo probe entered a high-pressure region that was termed an "infrared hot spot".

It has long been speculated that the Jovian atmosphere produces lightning. Sagan *et al.* (1967) argued that the coloration of the Jovian clouds could not be explained in a thermodynamic equilibrium atmosphere. The high temperatures that occur in the lightning discharge allow for chemical interactions that could provide complex colored substances in significant amounts. Ponnamperuma (1966) and Woeller and Ponnamperuma (1969) verified experimentally the synthesis of brightly colored compounds as well as several organic molecules of biological significance by laboratory electrical discharges. The detection of enhanced nonequilibrium amounts of acetylene in the Jovian atmosphere led Bar-Nun (1975) to suggest that intense lightning activity could account for this observation. However, Lewis (1980a, b), considering shock-wave chemistry in a solar-composition gas, found that the synthesis by lightning of organic compounds on Jupiter was negligible compared to their production by photochemical and thermochemical processes. The first solid evidence of Jovian lightning was obtained by Voyager 1 in 1978, in the form of an optical image containing 20 luminous spots (Cook et al. 1979). Similar luminous spots were subsequently detected by Voyager 2 and by the Galileo orbiter. Additional evidence is supplied from RF measurements made with the Galileo probe and from whistler measurements obtained by Voyager 1.

16.4.2. Optical signals

Voyager 1 and Voyager 2 The two spacecraft followed different trajectories, a fact usually invoked to explain the differences in the optical (and whistler) data acquired by their presumably identical instrumentation sets. Voyager 1 passed inside the orbit of Io, approaching within 350 000 km of the center of Jupiter, producing a closer view, a shorter observing period, and a smaller observing area than Voyager 2, which only came within 725 000 km of the center of Jupiter. Since Voyager 1 flew closer to Jupiter, it could, in principal, detect weaker lightning flashes than Voyager 2, but Voyager 2 was able to survey a larger portion of the night side of the planet. Similarly, the closer passage of Voyager 1 apparently allowed the detection of whistlers (discussed in subsection 16.4.3) attributed to lightning (Gurnett et al. 1979; Scarf et al. 1979; Kurth et al. 1985) while the greater distance of Voyager 2 apparently prevented such detection.

Cook et al. (1979) reported the detection of 20 bright spots in the 380 to 580 nm wavelength range on the night side of Jupiter between latitudes 30° and 50° N, most at 45° N, during a time exposure of 192 s. Because of the time exposure, each spot could be the integrated light from multiple lightning flashes spread radially by optical scattering in the clouds surrounding the lightning. Assuming that each spot was an individual lightning flash, Borucki et al. (1982a) calculated an optical energy per flash of $(2.5 \pm 1.9) \times 10^9$ J and a total dissipated energy per flash of $(1.7 \pm 1.3) \times 10^{12}$ J for the 16 brightest Voyager 1 spots. In doing so, they used the optical efficiency factor in the 500 to 1000 nm range derived by Krider et al. (1968) for terrestrial lightning but adapted, assuming that the energy per unit wavelength interval is approximately constant over the 380 to 1000 nm spectral region, to the 380 to 580 nm bandwidth of the Voyager imaging system. These energy values are about 10^3 times larger than those attributed by Borucki et al. (1982a) to terrestrial lightning. Only very rare Earth events, called "superbolts" (Turman 1977), have optical energies comparable to those observed on Jupiter. A later study by Borucki and McKay (1987) lowered the optical efficiency factor by a factor 3.8, so that the estimates of the total energy by Borucki et al. (1982a) should be raised by this amount. A lower limit to the global flash density on Jupiter, based on the 20 bright spots detected by Voyager 1, the exposure time of 192 s, and the viewed area of 10^9 km^2, is about 3×10^{-3} km^{-2} yr^{-1}, about 10^3 times smaller than for Earth. The detected optical signals may have been only those that could transit the less opaque cloud regions, so that the bright spots may be more a measure of cloud opacity than flash density. Magalhaes and Borucki (1991) and Borucki and Magalhaes (1992) presented a map of the optical events detected by Voyager 1 from two 192-s-exposure-time images) and Voyager 2 (from three 96-s-exposure-time images) along with the area covered by each image. This map is reproduced in Fig. 16.4. Voyager 2 detected a number of events similar to the number detected by Voyager 1, but over a wider area. As seen in Fig. 16.4, most of the bright spots detected by both Voyager 1 and Voyager 2 appear aligned at about 50° N. Borucki and Magalhaes (1992) found no unusual cloud features associated with this pattern. A few bright spots were seen at about 60° N (by both Voyager 1 and Voyager 2) and at about 14° N (by Voyager 2). The low-latitude spots could be associated with cloud disturbances in that region. No bright spots were observed in the southern hemisphere, although coverage there was less than in the

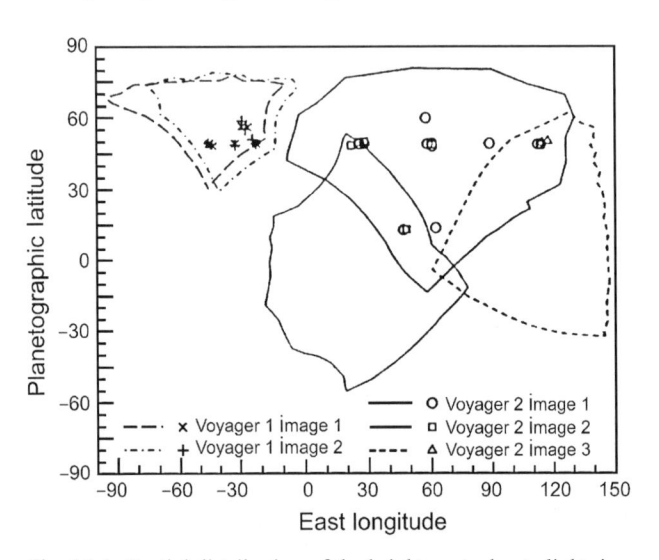

Fig. 16.4. Spatial distribution of the bright spots due to lightning observed on Jupiter by Voyager 1 and Voyager 2. Adapted from Magalhaes and Borucki (1991) and Borucki and Magalhaes (1992).

16.4. Jupiter

northern hemisphere and extended only to 30° S. Note that whistler sources were inferred to exist in both hemispheres (Tokar *et al.* 1982). Magalhaes and Borucki (1991) inferred a total energy of 2×10^{10} to 4×10^{11} J per bright spot from the Voyager 2 images.

Borucki and Williams (1986) used a radiative transfer model and two models of the distribution of clouds in Jupiter's atmosphere, the solar-composition chemical equilibrium model of Weidenschilling and Lewis (1973), illustrated in Fig. 16.3, and the water-depleted model of Bjoraker (1985), Bjoraker *et al.* (1986), and West *et al.* (1986), considered above in subsection 16.4.1. They inferred that the observed bright spots have their origin at the 5 bar level, in massive water clouds predicted by the solar-composition chemical equilibrium model. As noted in subsection 16.4.1, the water-depleted model predicts a thin NH_4SH-H_2O cloud layer at the 2 bar level and no clouds below that level.

Galileo probe The Galileo probe had optical detectors to measure both the ambient background light and the transient optical signals from lightning during its nearly one-hour descent through the Jovian atmosphere from about 0.4 to 22 bar (Lanzerotti *et al.* 1992; Rinnert and Lanzerotti 1998). No transient optical signals associated with lightning were detected (Lanzerotti *et al.* 1996). However, as noted in subsection 16.4.1, the probe trajectory was through an "infrared hot spot" in the atmosphere that apparently could not support lightning. RF data from the probe, to be discussed later, indicate potential lightning at distances greater than 10 000 km (Rinnert and Lanzerotti 1998), consistent with the probe's observation of no clouds other than high thin ones along the descent trajectory (Ragent *et al.* 1996; Sromovsky *et al.* 1996).

Galileo orbiter On 9 November 1996, when the Galileo orbiter was at 2.3 million kilometers from the center of Jupiter, its solid state imaging (SSI) system recorded pictures of eight "knots of light". The largest irregular light spot was more than 500 km across, comparable to the size of the lightning events seen by Voyager 1 in 1977 and much larger than the bright spots detected by Voyager 2. All spots

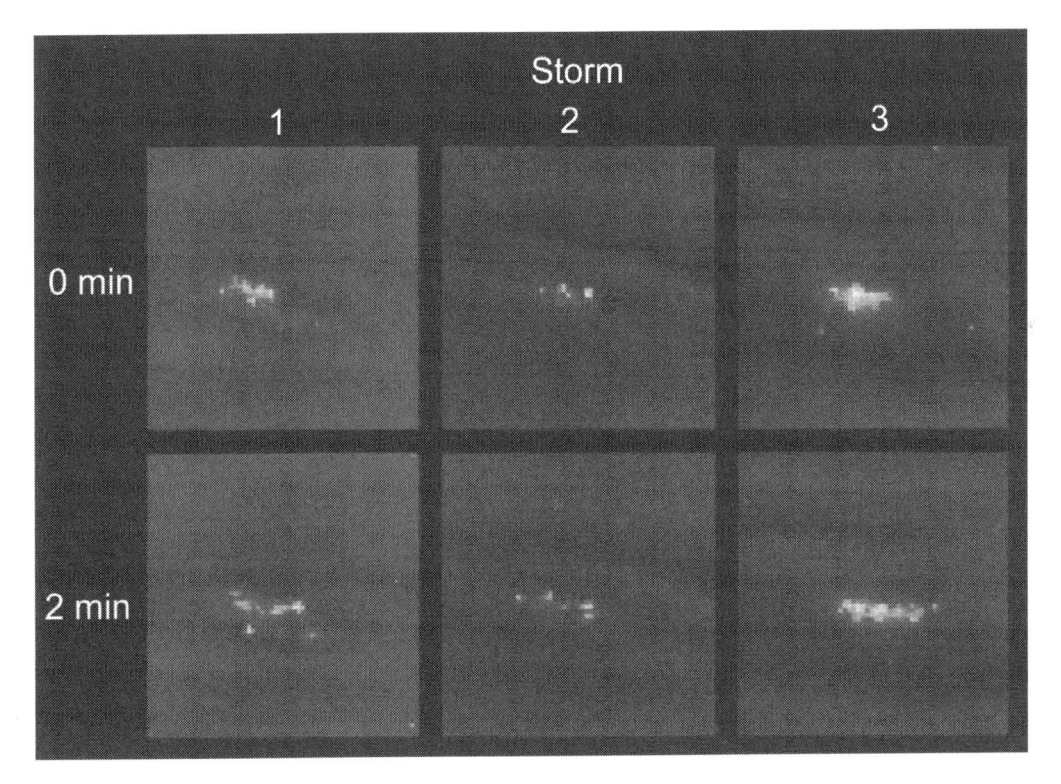

Fig. 16.5. At the time of going to press a colour version of this figure was available for download from http://www.cambridge.org/9780521035415. Lightning storms in three different locations (panels 1, 2, and 3) on Jupiter's night side. Each panel shows multiple lightning strikes, coming from different parts of the same storm. The individual strikes are unresolved in these images, which have a resolution of 133 kilometers per picture element. The bottom row shows the same three storms as the top row but these images were taken two minutes later. The images were taken with the clear filter with an exposure time of 90 s. North is at the top of the picture. The panels are 8000 km on a side. The images were taken on 6 October 1997 by the solid state imaging camera system onboard NASA's Galileo spacecraft. The distance from the planet to the spacecraft was 6.62 million kilometers. This image is posted on the World Wide Web, on the Galileo mission home page at http://galileo.jpl.nasa.gov.

were at latitudes between 43° and 46° N. In October and November of 1997 the Galileo SSI detected lightning from 26 storms on the night side of Jupiter (Little et al. 1999). More than half the surface area of the planet was surveyed. The spatial resolution ranged from 23 to 134 km per pixel, while the storm dimensions were up to about 1500 km. Most storms were imaged more than once, and they typically exhibited many flashes per minute. The storms occurred only in areas of cyclonic shear and near the centers of westward jets. Lightning was most likely to occur near about 50° latitude in both hemispheres and was more common in the northern hemisphere, consistent with Voyager's observations. The greatest optical energy observed in a single flash was 1.6×10^{10} J, which is several times greater than for terrestrial "superbolts" (Turman 1977). Little et al. (1999) inferred that the observed lightning discharges occurred within or below the Jovian water clouds. Examples of the Galileo SSI images of lightning on Jupiter are shown in Fig. 16.5.

16.4.3. Whistlers

Jupiter has a strong magnetic field generated by currents in its liquid metallic hydrogen layer. The magnetic field strength at Jupiter's equator is over ten times greater than that at the Earth's equator. The Jovian magnetosphere extends to 50–100 Jovian radii (6–12 radii for Earth); the Sun could easily fit inside it. Four Galilean satellites orbit deep inside the Jovian magnetosphere. The innermost one of these, Io, has a volcanically produced atmosphere that is constantly being bombarded by the intense radiation belts of Jupiter. This bombardment knocks atoms out of the atmosphere of Io into the magnetosphere of Jupiter, where they become ionized. This process produces a plasma torus (a donut-shaped volume) of hot ions circulating Jupiter near Io's orbit.

Both the Voyager 1 and Voyager 2 spacecraft contained a plasma-wave instrument to record VLF electric fields (Scarf and Gurnett 1977; Scarf et al. 1979; Gurnett et al. 1979). Only Voyager 1, as noted in subsection 16.4.2, detected signals that could be interpreted as lightning-produced whistlers, first reported by Scarf et al. (1979). A detailed summary was given by Kurth et al. (1985) and later analyses were given by Hobara et al. (1995, 1997). Examples of the whistler data are found in Fig. 16.6 along with similar data from Earth. Whistlers were identified as such in the wideband channel (50 Hz to 12 kHz) of the plasma-wave instrument by their unique frequency–time signature, unlike the case of Venus where the identification of whistler-mode signals was ambiguous (subsection 16.3.3). Whistlers in the Jovian magnetosphere were detected in three regions (A, B, and C) along the Voyager 1 trajectory during closest approach (about 5.5 to 6 Jovian radii), in or close to the Io plasma torus. The local gyrofrequency was about 60, 80, and 60 kHz for regions A, B, and C, respectively. Out of a total of 167 detected whistler signals, 90 had a frequency content broad enough, 2 to 7 kHz, to estimate the whistler dispersion, defined in Eq. 13.102. The average dispersion in region B, which is characterized by the largest number of observed whistlers, was smallest, at about 60 to 70 s Hz$^{1/2}$. Whistlers in regions A and C had larger average dispersion values, about 300 and 500 s Hz$^{1/2}$, respectively. Using ray tracing, Menietti and Gurnett (1980) and Hobara et al. (1997) showed that these whistlers originate from Jupiter's high latitudes. Tokar et al. (1982) concluded, from the dispersion and the geometry, that the sources for the A-group whistlers were in the southern hemisphere (60° S) and for the B- and C- groups in the northern hemisphere (55° to 75° N).

On average, one whistler was recorded every 8 s, but the rate was variable. In one 48 s frame as many as 32 whistlers, with a repetition period of 1.5 s, were identified. Scarf et al. (1981) derived an upper bound to the lightning flash density of 40 km^{-2} yr^{-1} assuming (i) that propagation was strictly along the magnetic field lines from Jupiter to Voyager 1, (ii) that the area below the ionosphere over which a lightning can become coupled to a magnetic field line was 10^6 km^2, and (iii) that one lightning in 10 launched a detectable whistler. This value of the lightning-flash density can be compared with the lower limit of 3×10^{-3} km^{-2} yr^{-1} given in subsection 16.4.2 for the optical data. Russell (1993) argued that the value of Scarf et al. (1981) is probably an overestimate. Levin et al. (1983) noted that estimates of the flash rates from the whistler data are uncertain to at least an order of magnitude. From the same whistler data, but using a different set of assumptions, Lewis (1980a) deduced a peak flash density of 4×10^{-2} km^{-2} yr^{-1}, three orders of magnitude lower than the value of Scarf et al. A value about an order of magnitude greater than that of Scarf et al. was deduced from Jovian atmosphere chemistry considerations by Bar-Nun (1979). Estimates from the Galileo probe data (subsection 16.4.4, second part) tend to support the value of Lewis (1980a).

Hobara et al. (1997) used the measured whistler frequency spectrum and the calculated frequency-dependent attenuation of the whistler signal along its propagation path to deduce the frequency spectrum of the source. The strongest whistler, from region A, was inferred to be produced by a source with a spectrum that fell off rapidly with frequency and probably had a peak below 1.5 kHz. However, less reliable estimates for the weaker signals, from region B, suggest sources with local spectral peaks at frequencies higher than 1.5 kHz. These results are not inconsistent with the Galileo probe RF data (subsection 16.4.4). Hobara et al. (1997) speculated that such a source's inferred frequency-spectrum peak below 1.5 kHz could be similar to that of terrestrial sprite-like phenomena (subsection 14.3.3); sprite

16.4. Jupiter

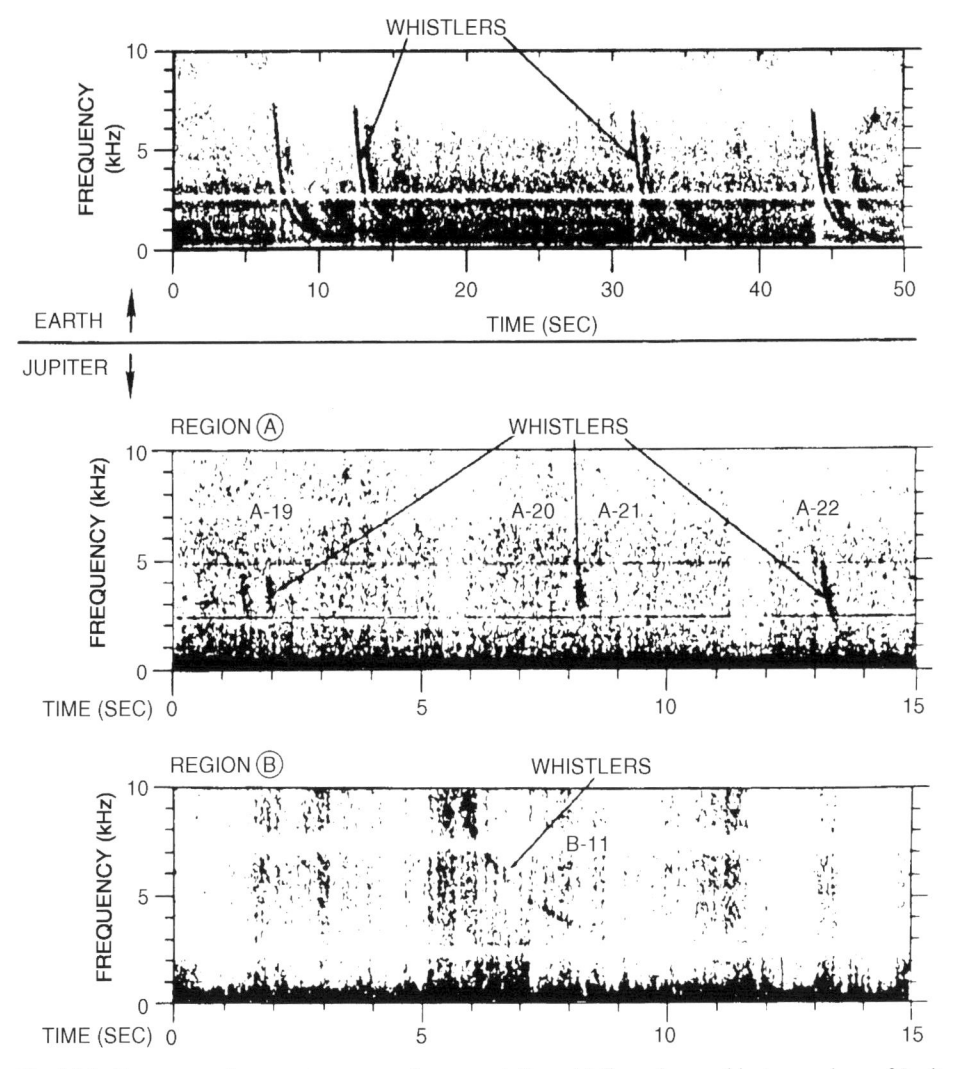

Fig. 16.6. Frequency–time spectrograms of representative whistlers observed in two regions of Jupiter's magnetosphere (middle and bottom panels) and similar data from Earth (top panel). Adapted from Gurnett et al. (1979).

current variations are on a millisecond time scale. The inferred peak values of sprite currents were estimated to be of the order of a few kiloamperes (Cummer et al. 1998).

16.4.4. Other RF signals

Voyager 1 and Voyager 2 The Voyager planetary radio astronomy (PRA) instrument did not detect any impulsive RF signals outside the Jovian ionosphere at frequencies above 1 MHz, whereas such signals were detected at Saturn (subsection 16.5.2) and at Uranus (subsection 16.6.2). One explanation for the absence of this high-frequency radiation would be that the Jovian lightning did not produce RF signals above 1 MHz. The Galileo probe data are consistent with this explanation in that the probe recorded a maximum in the RF spectrum in the kilohertz range or below, as described below. Also, a frequency spectrum with a peak below 1.5 MHz was inferred for the strongest whistler recorded by Voyager 1 by Hobara et al. (1997). It appears that Jovian lightning is a relatively low-frequency phenomenon, suggesting that the higher-frequency signals detected at Saturn and Uranus (subsections 16.5.2 and 16.6.2, respectively) are both produced by distinctly different sources, not necessarily by lightning.

Galileo probe Lanzerotti et al. (1996) and Rinnert et al. (1998) presented the characteristics of the RF signals observed in the Jovian atmosphere beneath the ionosphere. Narrowband signals were measured at 3, 15, and 90 kHz, directional data from the spinning probe being obtained for the 15 kHz channel. Additionally, distributions of statistical characteristics of wideband (100 Hz to 100 kHz)

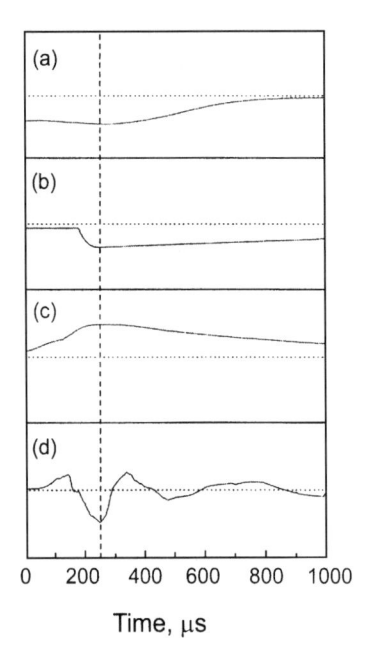

Fig. 16.7. Examples of wideband waveforms recorded in the Jovian atmosphere by the Galileo Probe lightning instrument (LRD). In each panel, the dotted horizontal line indicates the zero amplitude level. The instrument trigger point is indicated by the vertical broken line at 250 μs. The maximum signal amplitudes from the zero amplitude lines are (a) 32.2, (b) 412.5, (c) 37.5, and (d) 32.8 nT (1 T = 1 Wb m^{-2}). Adapted from Lanzerotti et al. (1996).

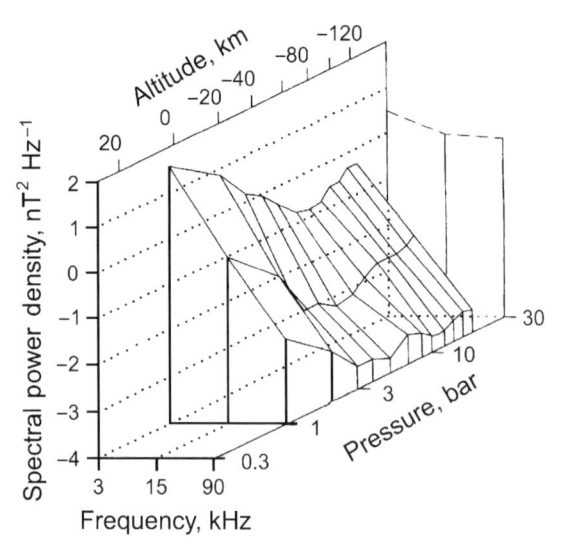

Fig. 16.8. Plot of the spectral power density (log values) measured in each of the three narrowband frequency channels (3, 15, and 90 kHz) of the Galileo Probe lightning instrument (LRD) as a function of atmospheric pressure during probe descent into the Jovian atmosphere. The corresponding atmospheric depth (altitude) is shown on the top axis, zero altitude being placed at the 1 bar pressure level. The power levels at 30 bars (broken line) show a typical LRD response to a close thunderstorm (about 20 km distant) on Earth. Adapted from Lanzerotti et al. (1996).

signals and 11 amplitude-selected wideband waveforms were recorded. A detailed description of the measuring system was given by Lanzerotti et al. (1992). None of the wideband waveforms was accompanied by a transient optical signal (subsection 16.4.2, second part). The distributions of the statistical characteristics of wideband signals were contaminated by the probe noise (Rinnert et al. 1998).

Most of the 11 wideband waveforms obtained were dominated by frequency components near 500 Hz. Four of these are shown in Fig. 16.7. The waveform in Fig. 16.7d, measured at a pressure of about 16 bar, was analyzed by Rinnert et al. (1998) for the properties of its source. The spectral power density of narrowband signals varied with atmospheric depth, as shown in Fig. 16.8. As noted above, the data in the 15 kHz channel were direction dependent. Figure 16.8 shows the 15 kHz data lumped together for all directions. Most energy was concentrated in the lowest frequency range, centered at 3 kHz.

The RF data differ significantly from similar data for Earth lightning: the frequency content of the Jovian RF signals is dominated by lower frequencies and the signal magnitudes are larger. As indicated in the caption of Fig. 16.7, three of the four representative wideband magnetic field waveforms (of a total of 11) have amplitudes near 30 nWb m^{-2}. If the source were at 1000 km, probably the closest reasonable distance considering the absence of a simultaneous optical signal, this magnetic field would correspond to a current of about 10^5 A; if the source were at 10 000 km, the corresponding current would be about 10^6 A. Typical Earth lightning has a peak current of order 10^4 A (Section 4.6). One detected Jovian magnetic field waveform had an amplitude of about 400 nWb m^{-2}. Note, however, that generally only the largest waveform in each data-taking interval (200 to 300 s) was recorded and that about 144 000 wideband pulses were analyzed in the first data-taking interval alone. Rinnert et al. (1998), assuming that the source of the waveform in Fig. 16.7d was at 15 000 km, found a dipole moment of about 10^4 C km, a current of 6×10^6 A, and a charge transfer of 1500 C, all one to two orders of magnitude larger than for Earth lightning. Farrell et al. (1999) proposed a model of the lightning discharge at Jupiter based on the Galileo probe data.

After correcting the observed RF signal rates for probe noise and using the direct line-of-sight distance of about 15 000 km for the effective operating range of the instrument, Lanzerotti et al. (1996) estimated the number of Jovian RF signal sources per unit area per unit time to be about one-tenth that on Earth (6 km^{-2} yr^{-1}, Section 2.1). Rinnert et al. (1998) later estimated this quantity to be one-hundredth of the terrestrial value.

In summary, there is little doubt that lightning exists on Jupiter. Jovian lightning is probably a relatively

16.5. Saturn

16.5.1. General information

Saturn is the second largest planet in the solar system, with an equatorial radius of 60 268 km, more than nine times that of Earth. It has 18 named satellites and exhibits a system of rings that is visible from Earth with a small telescope. Jupiter, Uranus, and Neptune also have rings but they are very faint. Pioneer 11 passed Saturn in September 1979, and Voyager 1 and Voyager 2 flew by in November 1980 and August 1981, respectively. The book edited by Gehrels and Matthews (1984) presents an excellent review of various aspects of Saturn. The internal structure of Saturn is thought to be similar to that of Jupiter. Saturn has a magnetic field whose strength at the equator is of the same order of magnitude as that at the Earth's surface. Saturn, in contrast to all the other planets, has a magnetic field dipole moment that is not tilted with respect to the rotation axis of the planet, although most theories for the generation of planetary magnetic fields require such a tilt. Saturn's magnetosphere extends to 16 to 22 of the planet's radii. The heat flow from the interior of Saturn is two to three times greater than that received from the Sun (Morrison and Owen 1988; Kaufmann 1987) and would be expected to produce atmospheric motions similar to those on Jupiter. The energy radiated by Saturn cannot be explained by the mechanism thought to be acting in the case of Jupiter (subsection 16.4.1) alone. The temperature of Saturn's interior is thought to be low enough to allow the precipitation of helium deep in the planet's interior, a hypothesis supported by the observed deficiency of helium in Saturn's atmosphere (Morrison and Owen 1988; Kaufmann 1987). It is further speculated that as the helium droplets descend through the molecular hydrogen, molecules of the two gases rub against each other, and the resulting friction produces heat that eventually escapes Saturn. The amount of missing helium in the atmosphere is consistent with a rate of helium precipitation that would produce the extra energy that Saturn radiates. This hypothetical mechanism apparently does not operate yet on Jupiter because of its higher interior temperature.

The Saturnian cloud structure as derived from modeling has three main layers, similar to those on Jupiter (Weldenschilling and Lewis 1973). The lower gravity of Saturn leads to a more extended atmosphere than on Jupiter and hence a thicker cloud system: over 200 km versus less than 100 km on Jupiter (Fig. 16.3). The Saturnian cloud system exhibits a band structure, although substantially less distinct than on Jupiter, with wind directions alternating with latitude in the bands. The wind speed reaches a maximum, of order 500 m s^{-1}, in the equatorial region, considerably larger than on Jupiter, and decreases with increasing latitude. The Saturnian atmosphere is apparently less turbulent than the Jovian atmosphere (Smith et al. 1981, 1982). One of the images taken by the Hubble Space Telescope in 1994 shows a white arrowhead-shaped feature near Saturn's equator, similar to a terrestrial thunderhead. The east–west extent of this "storm" was equal to the diameter of the Earth (about 12 700 km).

Although Saturn would appear to be as good a candidate for a lightning-producing planet as Jupiter, the Voyager optical-imaging and plasma-wave instruments, which provided strong evidence for lightning on Jupiter in the form of luminous spots and whistlers, respectively, did not find similar evidence for lightning on Saturn. However, the detection of impulsive optical signals on the night side might have been impossible due to the sunlight scattered from the rings (Smith et al. 1981). The evidence for possible lightning or lightning-like discharges on Saturn comes from the detection by the Voyager planetary radio-astronomy instruments of bursts of radio noise in the frequency range from 20 kHz to 40 MHz, as discussed next.

16.5.2. RF signals

During the Voyager 1 and Voyager 2 flybys of Saturn in November 1980 and August 1981, respectively, the planetary radio-astronomy (PRA) instrument of each spacecraft recorded strong, discrete, wideband bursts of radio emission that were termed Saturn electrostatic discharges (SED). These bursts were analyzed by Warwick et al. (1981, 1982), Evans et al. (1981, 1982, 1983), Kaiser et al. (1983a, b, 1984), Zarka and Pedersen (1983), and Zarka (1985). Zarka and Pedersen (1983) identified 18 000 and 5000 SEDs recorded by Voyager 1 and Voyager 2, respectively. The signals spanned the frequency range from 20 kHz to 40 MHz whereas the entire PRA bandwidth was from 1.2 kHz to 40 MHz. Studies of SEDs have been mainly limited to the higher-frequency range from 1.2 to 40 MHz, to avoid confusion with another type of recorded radiation from Saturn (Zarka and Pedersen 1983). No similar signals were observed on Jupiter. Perhaps the most striking feature of these bursts, besides their broad bandwidth, was their periodic occurrence with a repetition interval of about 10 hours. The duration of one episode of SEDs was about seven hours, during which time the number of detected events rose to a peak value and fell back to the instrument noise level, and the time interval between episodes was about three hours. The SED occurrence depended on the distance of the spacecraft from Saturn (up to over 100 Saturnian radii for Voyager 1 and up to 50 or so Saturian radii for Voyager 2) and was a maximum at closest approach. The SEDs were observed to have similar characteristics during both Voyager missions but were less intense and less frequent during the Voyager 2 mission, nine months later than the Voyager 1 mission. The distribution of the time between these RF bursts during the hour or so of maximum occurrence closely fits a Poisson distribution with a mean period of 5 s. The bursts lasted from less than 30 to about 450 ms with a mean value of about 55 ms. The number of observed bursts decreased exponentially with increasing burst duration, with an e-folding time (the time during which a quantity decreases to 1/e of its original value) of about 40 ms (Zarka and Pedersen 1983). Assuming an isotropic source with 100 MHz bandwidth, Evans et al. (1983) determined an instantaneous radiated power of 10^9 to 10^{10} W for a single burst. For an assumed 40 MHz bandwidth, Zarka and Pedersen (1983) found an instantaneous power of 2×10^9 W for an average burst in the Voyager 1 data and an order of magnitude less in the Voyager 2 data. They also estimated an instantaneous power in excess of 10^{10} W during strong SED events.

It is reasonable to associate the periodic nature of the SEDs with a localized source that would appear only when the rotation of Saturn placed the source near the spacecraft. However, the repetition period of the SEDs is appreciably shorter than the Saturnian rotational period of 10 hr 39.4 min and, in fact, was about 10 min different for the data from Voyager 1 and Voyager 2 (Zarka and Pedersen 1983). Two regions with a rotational period near that of the observed SED repetition period have been identified, one within the B-ring (one of the three rings that can be seen from Earth) at 1.8 Saturnian radii, where the revolution period equals 10 hr 10 min, and the other in the atmosphere at equatorial latitudes, where winds with high speed at the cloud level produce the necessary super-rotation (Smith et al. 1981). Accordingly, two possible locations of the source of the SEDs, within the atmosphere and in the B-ring, have been considered and are discussed next.

Kaiser et al. (1983a, b), Burns et al. (1983), and Zarka (1985) argued that the SEDs originate from an extended long-lived storm system in the equatorial atmosphere of Saturn. To fit the time duration and periodicity of the SED episodes, an active storm region of relatively large longitudinal extent (about 60°) but of narrow latitudinal extent (4° or less) had to be assumed. Further, such a region would have had to remain active for at least nine months to account for the fact that both Voyager 1 and Voyager 2 observed generally similar SEDs. The interburst time interval of about 5 s (see above) and an assumed source area of $60° \times 3°$ yield a mean event rate of 3×10^{-2} km^{-2} yr^{-1}.

One serious problem with the interpretation of the source of the SEDs as atmospheric in origin is that of how the lower end of the SED frequency spectrum can propagate through an ionosphere that should reflect downward all radio waves with frequencies lower than the plasma frequency. Observations indicate that the night ionosphere should not transmit signals below about 200 kHz. Burns et al. (1983) suggested that, because of the shadow of the rings, there exist deep depressions in the ionospheric electron density in the equatorial region where the electrical storms are inferred to occur. If such depressions were to lower the electron density by a factor of about 100, the plasma frequency would be lowered by a factor of about 10, from 200 to 20 kHz, which would be consistent with the lower limit of the frequency range in which PRA signals were observed.

Because of the expected shielding by the ionosphere of some observed frequencies from subionospheric sources and the periodicity of the observed radio noise, Evans et al. (1981, 1982) and Warwick et al. (1981, 1982) initially proposed that the SEDs must originate in a localized source in the B-ring. Weinheimer and Few (1982) discussed the conditions for electrical discharges in the ice of which the B-ring is thought to be composed. Zarka and Pedersen (1983) considered the requirements for a source in the B-ring to be able to produce the observed SED characteristics. They concluded that, similarly to an in-atmosphere source, the SED radiation must emanate from a region greater than 60° in longitude but very much smaller than that in height and radial extent. They also argued that a nonisotropic source is necessary, whether in the B-ring or in the atmosphere, to account for the shape of the SED emission versus time curve.

In view of (i) the lack of optical and whistler observations on Saturn while such observations were made on Jupiter, (ii) the absence in the Jupiter data of RF signals similar to the SEDs on Saturn, and (iii) the absence of SED-like signals from Earth lightning, SEDs alone do not appear a convincing argument for Saturnian lightning. If the SEDs do indeed originate from lightning, this lightning is certainly different from terrestrial or Jovian lightning.

16.6. Uranus
16.6.1. General information

Uranus has an equatorial radius of 25 559 km, about four times that of Earth, and 17 known satellites. Voyager 2 encountered Uranus on 24 January 1986 at a closest approach of about 80 000 km. A comprehensive review of Uranus is given in the book edited by Bergstralh et al. (1991). It is thought that Uranus and Neptune both have rocky cores surrounded by a liquid mantle of water, methane and ammonia, the outer gaseous layer being composed primarily of hydrogen (about 83 percent on Uranus and 80 percent on Neptune), helium, and methane. Overall, hydrogen on Uranus and Neptune is thought to amount to only 15 percent. At a level where the pressure is equal to one atmosphere on Earth, the temperature on Uranus (and Neptune) is about −200 °C, about the temperature of liquid nitrogen. This is considerably colder than on Jupiter

($-98\ °C$) or Saturn ($-133\ °C$) at the same pressure level. In contrast to the other giant planets, Uranus appears to lack a significant internal heat source. Voyager 2 imaging allowed the identification of low-contrast zonal bands, with narrow spacing, from 45° S to 20° S latitude (Smith et al. 1986). A light convective plume and other faint features could be tracked, indicating a super-rotating atmosphere with zonal wind speeds up to 200 m s^{-1}. The cloud model of Weidenschilling and Lewis (1973) for Uranus predicts cloud layers composed (from top to bottom) of methane, ammonia, ammonium hydrosulfide, water, and ammonia water. Both Uranus and Neptune have an excess of heavy elements relative to hydrogen and helium, and therefore 10 times the solar abundances of NH_3, H_2O, CH_4, and H_2S are assumed in the model. The later model of Atreya and Romani (1984) is similar. Voyager 2 measurements verified the methane cloud tops and inferred photochemical haze above the 1 bar level (Rinnert 1995). There were hints of convection but strong motions were not observed. Note, however, that the massive water clouds that could be important for lightning generation are calculated to be 200 to 400 km below the upper methane clouds and the haze.

The evidence for lightning on Uranus is similar to that on Saturn: the observation of broadband radio emissions by the PRA instrument. Very long optical exposures taken of the dark side did not provide evidence of lightning (Rinnert 1995).

16.6.2. RF signals

Within a distance of 600 000 km from Uranus the Voyager 2 PRA instrument recorded 140 impulsive bursts of broadband radio noise spanning frequencies from less than 0.9 to about 40 MHz, these radio emissions being termed Uranian electrostatic discharges (UEDs) by analogy to the SEDs (Zarka and Pedersen 1986). More than half the events were detected near the closest approach to Uranus, about 80 000 km above the cloud tops. The UEDs appear to differ somewhat from the SEDs, although this may be due in part to the much smaller number of UEDs and therefore the less reliable statistics, 140 events compared to about 23 000 for the two Saturn encounters. The UEDs were strongest at closest approach, but no periodicity similar to that found on Saturn was evident. The duration distribution is well described by an exponential law with an e-folding time of about 100 ms (versus 40 ms on Saturn) and average duration 120 ms (versus 55 ms on Saturn). Most of the duration values were from 100 to 300 ms, with a few larger values, up to 500 ms. The wideband spectrum decreases approximately as f^{-2}, while on Saturn it is essentially flat up to 40 MHz. The total instantaneous power, integrated over the whole UED spectrum, is about 10^8 W, less than 0.1 of that estimated for the SEDs. Before encounter, the spacecraft was above the day-side hemisphere of the planet, where the UED low-frequency cutoff at about 7 MHz implies a maximum electron density in the day-side ionosphere of about 6×10^5 cm^{-3}. Similarly, the low-frequency cutoff at 900 kHz or less during closest approach implies a maximum electron density of 10^4 cm^{-3} or less in the night-side ionosphere. There is little doubt that UEDs are natural RF emissions from Uranus. However, their interpretation as due to Uranian lightning, similar to the attribution of SEDs to Saturnian lightning (subsection 16.5.2), is questionable.

16.7. Neptune
16.7.1. General information

Neptune, the fourth largest planet in the solar system, has an equatorial radius of about 24 766 km and eight known satellites. It was visited by Voyager 2 in 1989. The interior structure and chemical composition of Neptune is similar to that of Uranus (subsection 16.6.1). Although Neptune is farther away from the Sun, it has about the same brightness-temperature as Uranus. The ratio of the power emitted from Neptune to its absorbed power from the Sun is 2.61 (Pearl and Conrath 1991). It is thought that neither the Jovian mechanism for the release of energy trapped inside the planet during its formation nor the Saturnian mechanism of the precipitation of helium can operate on Neptune. Therefore, the Neptunian internal heat source remains a mystery. The appreciable energy flow from the interior drives convective motions in the Neptunian atmosphere that are organized in zonal bands. These bands are not as obvious as those on Jupiter or Saturn, but Voyager 2 imaging clearly revealed clouds at preferred latitudes (Smith et al. 1989). Voyager 2 observed long-lived prominent features such as the Great Dark Spot and a smaller dark spot in the southern hemisphere, and also a varying feature called the Scooter. However, Hubble Space Telescope observations in 1994 showed that the Great Dark Spot had disappeared and a few months later showed a new dark spot in the northern hemisphere, indicating that Neptune's atmosphere changes rapidly. Smaller cloud systems evolve significantly in less than a few hours. The zonal winds are eastward at low latitudes with speeds exceeding 600 m s^{-1} and change to westerlies at high latitudes, with speeds up to 300 m s^{-1} (Limaye and Sromovsky 1991). Thus Neptune appears to be a windy planet, with wind speeds up to 2000 km hr^{-1}, the highest in the solar system, and with significant vertical convection. Weidenschilling and Lewis (1973) modeled the Neptunian cloud system, assuming thermal equilibrium and 10 times the solar abundances of H_2O, NH_3, H_2S, CH_4, and argon. Cloud layers composed, from top to bottom, of methane, ammonia, ammonium hydrosulfide, water and ammonia-water were predicted, the overall depth of the cloud system being about 400 km. Other investigators (e.g., Atreya 1992; Baines and Smith 1990) have assumed some hydrocarbon ice-haze layers above the CH_4 clouds. CH_4 ice is probably not a good substance for efficient electrification. However, if the lower ammonia-water clouds exist, conditions for the production of lightning may be fulfilled. The fraction of the convective energy of the Earth's atmosphere that is dissipated by lightning has been estimated to be 4×10^{-7} (Borucki and Chameides 1984). Borucki et al. (1982a) found 4×10^{-5} for Jupiter. Jovian lightning is probably generated well below the dense ammonia cloud top, and therefore the internal heat flux may be a measure of the energy available for lightning. Assuming that Neptune is similar to Jupiter and noting that the Neptunian heat flux is one-nineteenth that of Jupiter, Borucki (1989) estimated the lightning rate on Neptune to be about 2×10^{-4} km^{-2} yr^{-1}. The evidence for lightning on Neptune comprises the 16 weak whistlers detected by the plasma-wave instrument of Voyager 2 (Gurnett et al. 1990) and four marginal RF signals at frequencies above 15 MHz observed by the planetary radio-astronomy instrument (Kaiser et al. 1991).

16.7.2. Optical measurements

Borucki and Pham (1992) evaluated 98 of the best images from the Neptune night side, covering 94 percent of the planet's surface between approximately 60° N and 70° S. No optical signals were detected. The system's sensitivity requires a minimum optical energy of 9×10^9 J (9 GJ) in the 420 to 900 nm band, however, so terrestrial lightning, whose expected total input energy is 1 to 10 GJ (Section 4.2), would not have been detected. If Neptunian lightning had the same characteristics in terms of energy and frequency as those postulated for Jovian lightning, then one might expect to detect such lightning.

16.7.3. Whistlers

Gurnett et al. (1990), from the plasma-wave instrument wideband channel covering the frequency range from 50 Hz to 12 kHz, reported the detection of 16 weak whistlers in the frequency range from 6.1 to 12 kHz when Voyager 2 was within 1.3 to 2.0 Neptunian radii. The upper frequency limit of the whistlers, 12 kHz, is in direct conflict with the Voyager 2 plasma instrument measurements, which yielded electron densities too low for such a whistler mode to propagate. The recorded Neptunian whistlers are much weaker than the whistlers observed in Jupiter's magnetosphere, after accounting for the different radial distance involved. The occurrence rate was 1.6 min^{-1}. The signals were identified as whistlers from frequency–time spectrograms showing a signal decreasing monotonically in frequency with increasing time. The frequency–time slope of the signal also decreased with increasing time, a feature characteristic of lightning-produced whistlers in the Earth's magnetosphere, in good agreement with the Eckersley dispersion law (Eq. 13.102). This latter result provides convincing evidence that the recorded signals are produced via dispersive whistler-mode propagation from an impulsive source, possibly lightning in the Neptunian atmosphere. A problem with this interpretation is the very large dispersion, typically about 26 000 s Hz$^{1/2}$ and in one case about 45 000 s Hz$^{1/2}$, which requires a very long propagation path, from some tens to some thousands of Neptunian radii. Such a large dispersion also implies substantial plasma densities somewhere along the propagation path, and/or a large number of bounces (hops) from one hemisphere to the other. Menietti et al. (1991) modeled the ray paths and showed that the observed large dispersion can be explained by propagation in one particular direction along a non-ducted path. It is nevertheless surprising that the dispersion is so similar for the different events recorded by Voyager 2.

16.7.4. Other RF signals

Kaiser et al. (1991) found four marginal signals at frequencies above 15 MHz in the planetary radio-astronomy (PRA) instrument data covering the frequency range from 1.2 kHz to 40.5 MHz. These were detected within 10 Neptunian radii. The criterion for an event was an impulsive signal, more than 1 dB above the background, detected in three or more consecutive frequency channels. The power level was very low compared to those observed at Saturn and Uranus, but somewhat greater than that of terrestrial lightning. The low level of signals or lack of detected signals imposes some constraints on the characteristics of possible Neptunian lightning. The lightning may be inherently weak and/or the risetime of the discharge current unusually long, with very little electromagnetic power emitted at higher frequencies. The plasma frequency in the Neptunian ionosphere is about 1 MHz, but ionospheric absorption, together with the frequency-dependent sensitivity of the PRA instrument, prevented the detection of signals at frequencies below about 15 MHz (Kaiser et al. 1991). The source of these signals could be lightning, but, whatever the source, it appears to be different from the sources of SEDs on Saturn or UEDs on Uranus.

Because of the detection of whistlers, the argument for Neptunian lightning may be a little stronger than for lightning on either Saturn or Uranus. However, the available RF data (for whistlers and other phenomena) are sparse and in view of the lack of optical signals do not constitute convincing evidence.

16.8. Concluding remarks

The evidence for lightning or lightning-like electrical discharges on Jupiter is convincing. It is difficult to imagine another source for the luminous spots recorded by Voyager 1, Voyager 2, and the Galileo orbiter. The available RF evidence for lightning, including whistlers, if any, on Venus, Saturn, Uranus, and Neptune is less convincing, particularly in view of the absence of optical signals from Saturn, Uranus, and Neptune and the elusiveness of such signals from Venus, where the questionable initial indication has not been confirmed by subsequent searches. A few estimates for lightning flash densities and lightning parameters on planets other than Earth have been given in this chapter. Many more are found in the literature. While those estimates might be useful in designing instruments for future space missions, we would not attach much significance to them due to the many, largely arbitrary, assumptions necessarily made in their derivation.

Russell (1993) stated that when a phenomenon is being studied the first phase is to make sure that the phenomenon exists. Once its existence is proven, its significance is tested by, for example, measuring the rate of occurrence. Further phases involve obtaining first a qualitative understanding, then a quantitative understanding, and finally, a predictive understanding via modeling. In the extraterrestrial lightning studies to date, the first phase is completed only for Jupiter; the second, significance-testing, phase is not completed for any planet; whereas the last phase, modeling, is being worked on vigorously, most of the results being highly speculative, particularly when even the existence of lightning is still uncertain.

It is worth noting that most extraterrestrial lightning studies tend to assume, explicitly or implicitly, a similarity to lightning on Earth, the emphasis being placed on finding various scaling factors. However, the mechanisms behind the various electromagnetic signals observed on planets other than Earth may be qualitatively different from the lightning-generating mechanisms on our planet. In other words, the observed signals may be produced by sources entirely different from what we would call a lightning discharge. It appears that lightning is often invoked when an electromagnetic signal is detected on a planet other than Earth and no other explanation of the origin of that signal is available. Clearly, more measurements are needed. Such measurements will be obtained by Cassini for Saturn.

References and bibliography for Chapter 16

Alexeff, I., and Rader, M. 1992. Observation of closed loops in high-voltage discharges: a possible precursor of magnetic flux trapping. *IEEE Trans. Plasma Sci.* **20**: 669–70.

Anderson, R., Bjornsson, S., Blanchard, D.C., Gathman, S., Hughes, J., Jonasson, S., Moore, C. B., Survilas, H.J., and Vonnegut, B. 1965. Electricity in volcanic clouds. *Science* **148**: 1179–89.

Atreya, S.K. 1992. Hydrocarbons and eddy mixing in Neptune's atmosphere. *Adv. Space Res.* **12**: 11.

Atreya, S.K., and Romani, P.N. 1984. Photochemistry and clouds of Jupiter, Saturn and Uranus. In *Planetary Meteorology*, ed. G.E. Hunt, Cambridge University Press, UK.

Austin, L.W. 1926. Preliminary notes on proposed changes in the constants of the Austin–Cohen transmission formula. *Proc. IRE* **14**: 377–80.

Baines, K.H., and Smith, W.M. 1990. The atmospheric structure and dynamical properties of Neptune derived from ground-based and IUE spectrophotometry. *Icarus* **85**: 65–108.

Bar-Nun, A. 1975. Thunderstorms on Jupiter. *Icarus* **34**: 86–94.

Bar-Nun, A. 1979. Acetylene formation on Jupiter: photolysis or thunderstorms. *Icarus* **38**: 180–91.

Bar-Nun, A. 1980. Production of nitrogen and carbon species by thunderstorms on Venus. *Icarus* **42**: 338–42.

Bar-Nun, A., and Podolak, M. 1985. The contribution by thunderstorms to the abundance of CO, C_2H_2, and HCN on Jupiter. *Icarus* **64**: 112–24.

Bar-Nun, A., Noy, N., and Podolak, M. 1984. An upper limit to the abundance of lightning-produced amino acid in the Jovian water clouds. *Icarus* **59**: 162–8.

Beebe, R. 1997. *Jupiter, the Giant Planet*, Washington, London: Smithsonian Institution Press.

Bergstralh, J., Miner, E., and Matthews, M.S., eds. 1991. *Uranus*, Tucson, Arizona: University of Arizona Press.

Bjoraker, G.L. 1985. The gas composition and vertical cloud structure of Jupiter's troposphere derived from five-micron spectroscopic observations. Ph.D. dissertation, University of Arizona, Tucson.

Bjoraker, G.L., Larson, H.P., and Kunde, V.G. 1986. The abundance and distribution of water vapor in Jupiter's atmosphere. *Astrophys. J.* **311**: 1058–72.

Boeck, W.L., Vaughan, O.H., Blakeslee, R., Vonnegut, B., and Brook, M. 1992. Lightning induced brightening in the airglow layer. *Geophys. Res. Lett.* **19**: 99–102.

Borucki, W.J. 1982. Comparison of Venusian lightning observations. *Icarus* **52**: 354–64.

Borucki, W.J. 1985. Estimate of the probability of a lightning strike to the Galileo probe. *J. Spacecraft and Rockets* **22**: 220–1.

Borucki, W.J. 1987. Lightning on Venus – an alternative view. *Planetary Report* **VII**(4), July/August, p. 6.

Borucki, W.J. 1989. Predictions of lightning activity at Neptune. *Geophys. Res. Lett.* **16**(8): 937–9.

Borucki, W.J., and Chameides, W.L. 1984. Lightning: estimates of the rates of energy dissipation and nitrogen fixation. *Rev. Geophys. Space Phys.* **22**: 363–72.

Borucki, W.J., and Magalhaes, J.A. 1992. Analysis of Voyager 2 images of Jovian lightning. *Icarus* **96**: 1–14.

Borucki, W.J., and McKay, C.P. 1987. Optical efficiencies of lightning in planetary atmospheres. *Nature* **328**: 509–10.

Borucki, W.J., and Pham, P.C. 1992. Optical search for lightning on Neptune. *Icarus* **99**: 384–9.

Borucki, W.J., and Williams, M.A. 1986. Lightning in the Jovian water cloud. *J. Geophys. Res.* **91**: 9893–903.

Borucki, W.J., Dyer, J.W., Thomas, G.Z, Jordan, J.C., and Cornstock, D.A. 1981. Optical search for lightning on Venus. *Geophys. Res. Lett.* **8**: 233–6.

Borucki, W.J., Bar-Nun, A., Scarf, F.L., Cook, A.F., and Hunt, G.E. 1982a. Lightning activity on Jupiter. *Icarus* **52**: 492–502.

Borucki, W.J., Levin, Z., Whitten, R.C., Keesee, R.G., Capone, L.A., Toon, O.B., and Dubach, J. 1982b. Predicted electrical conductivity between 0 and 80 km in the Venusian atmosphere. *Icarus* **51**: 302–21.

Borucki, W.J., Orville, R.E., Levine, J.S., Harvey, G.A., and Howell, W.E. 1983. Laboratory simulation of Venusian lightning. *Geophys. Res. Lett.* **10**: 961–4.

Borucki, W.J., McKay, C.P., and Whitten, R.C. 1984. Possible production by lightning of aerosols and trace gases in Titan's atmosphere. *Icarus* **60**: 260–73.

Borucki, W.J., McKenzie, R.L., McKay, C.P., Duong, M.D., and Boac, D.S. 1985. Spectra of simulated lightning on Venus, Jupiter, and Titan. *Icarus* **64**: 221–32.

Borucki, W.J., Levin, Z., Whitten, R.C., Keese, R.G., Capone, L.A., et al. 1987. Predictions of the electrical conductivity and charging of aerosols in Titan's atmosphere. *Icarus* **72**: 604–22.

Borucki, W.J., Giver, L.P., McKay, C.P., Scattergood, T., and Parris, J.E. 1988. Lightning production of hydrocarbons and HCN on Titan: laboratory measurements. *Icarus* **76**: 125–34.

Borucki, W.J., Dyer, J.W., and Phillips, J.R. 1991. Pioneer Venus orbiter search for Venusian lightning. *J. Geophys. Res.* **96**: 11 033–43.

Borucki, W.J., McKay, C.P., Jebbens, D., Lakkaraju, H.S., and Vanajakshi, C.T. 1996. Spectral irradiance measurements of simulated lightning in planetary atmospheres. *Icarus* **123**: 336–44.

Bougher, S.W., Hunten, D.M., Phillips, R.J., and Bougher, H.U. 1997. Venus II, Tucson, Arizona: University of Arizona Press.

Brace, L.H., and Kliore, A.J. 1991. The structure of the Venus ionosphere. *Space Sci. Rev.* **55**: 81–163.

Brace, L.H., Theis, R.F., Mayr, H.G., Curtis, S.A., and Luhmann, J.G. 1982. Holes in the nightside ionosphere of Venus. *J. Geophys. Res.* **87**: 199–211.

Briggs, G., Klaasen, K., Thorpe, T., Wellman, J., and Baum, J. 1977. Martian dynamical phenomena during June–November 1976: Viking orbiter imaging results. *J. Geophys. Res.* **82**: 4121–49.

Brook, M., Moore, C.B., and Sigurgeirsson, T. 1974. Lightning in volcanic clouds. *J. Geophys. Res.* **79**: 472–5 (Correction, *J. Geophys. Res.* **79**: 3102, 1974).

Brook, M., Stanley, M., Krehbiel, P., Rison, W., Moore, C.B., Barrington-Leigh, C., Suszcynsky, D., Nelson, T., and Lyons, W. 1997. Correlated electric field, video, and photometric evidence of charge transfer within sprites. *Eos, Trans. AGU* **78**: F82–3.

Bruce, C.E.R. 1958. Evolution of extragalactic nebulae and the origin of metagalactic radio noise. Technical Report Z/T117, The British Electrical and Allied Industries Research Association, Leatherhead, Surrey, 7 pp.

Burke, W.J., Aggson, T.L., Maynard, N.C., Hoegy, W.R., Hoffman, R.A., Candy, R.M., Liebrecht, C., and Rodgers E. 1992. Effects of a lightning discharge detected by the DE 2 satellite over Hurricane Debbie. *J. Geophys. Res.* **97**: 6359–67.

Burns, J.A., Showalter, M.R., Cuzzi, J.N., and Durisen, R.H. 1983. Saturn electrostatic discharges: could lightning be the cause? *Icarus* **54**: 280–95.

Cameron, A.G.W. 1966. The accumulation of chondritic material. *Earth Planet. Sci. Lett.* **1**: 93–6.

Campins, H., and Krider, E.P. 1989. Surface discharges on natural dielectrics in the solar system. *Science* **245**: 622–4.

Carlson, B., Lacis, A.A., and Rossow, W.B. 1992. The abundance and distribution of water vapor in the Jovian troposphere as inferred from Voyager Iris data. *Astrophys. J.* **388**: 648–68.

Chameides, W.L., Walker, J.C.G., and Nagy, A.F. 1979. Possible chemical impact of planetary lightning in the atmospheres of Venus and Mars. *Nature* **280**: 820–2.

Cobb, W.E. 1980. Electric fields and lightning in the Mt. St. Helens volcanic cloud. *Eos, Trans. AGU* **61**: 978.

Cole, K.D., and Hoegy, W.R. 1996. Joule heating by ac electric fields in the ionosphere of Venus. *J. Geophys. Res.* **101**: 2269–78.

Colin, L. 1983. Basic facts about Venus. In *Venus*, eds. D. M. Hunter, L. Colin, T.M Donahue, and V.I. Moroz, pp. 10–26, Tucson: University of Arizona Press.

Colin, L., and Hunten, D.M. 1977. Pioneer Venus experiment descriptions. *Space Sci. Res.* **20**: 451–525.

Cook, A.F., Duxbury, T.C., and Hunt, G.E. 1979. First results on Jovian lightning. *Nature* **280**: 794.

Croft, T.A., and Price, G.N. 1983. Evidence for a low-altitude origin of lightning on Venus. *Icarus* **53**: 548–51.

Cummer, S.A., Inan, U.S., Bell, T.F., and Barrington-Leigh, C. 1998. ELF radiation produced by electrical currents in sprites. *Geophys. Res. Lett.* **25**: 1281–4.

Desch, M.D., and Kaiser, M.L. 1990. Upper limit set for level of lightning activity on Titan. *Nature* **343**: 442–4.

Desch, S.J., and Cuzzi, J.N. 2000. The generation of lightning in the solar nebula. *Icarus* **143**: 87–105.

Doyle, L.R., and Borucki, W.J. 1989. Jupiter lightning locations. In *Time–variable Phenomena in the Jovian System*, eds. M. J. S. Belton et al., p. 384, NASA, Washington, DC.

Eckersley, T.L. 1935. Musical atmospherics. *Nature* **135**: 104–5.

Eden, H.F., and Vonnegut, B. 1973. Electrical breakdown caused by dust motion in low pressure atmospheres: considerations for Mars. *Science* **180**: 962–3.

Enomoto, Y., and Zheng, Z. 1998. Possible evidences of earthquake lightning accompanying the 1995 Kobe earthquake inferred from the Nojima fault gouge. *Geophys. Res. Lett.* **25**: 2721–4.

Eshleman, V.R., et al. 1979. Radio science with Voyager 1 at Jupiter: preliminary profiles of the atmosphere and ionosphere. *Science* **204**: 976–8.

Essene, E.J., and Fisher, D.C. 1986. Lightning strike fusion: extreme reduction and metal–silicate liquid immiscibility. *Science* **234**: 189–93.

Evans, D.R., Warwick, J.W., Pearce, J.B., Carr, T.D., and Schauble, J.J. 1981. Impulsive radio discharges near Saturn. *Nature* **292**: 716–18.

References and bibliography for Chapter 16

Evans, D.R., Romig, J.H., Hord, C.W., Simmons, K.E., Warwick, J.W., and Lane, A.L. 1982. The source of Saturn electrostatic discharges. *Nature* **299**: 236–7.

Evans, D.R., Romig, J.H., and Warwick, J.W. 1983. Saturn electrostatic discharges, properties and theoretical considerations. *Icarus* **54**: 267–79.

Farrell, W.M., Kaiser, M.L., and Desch, 1999. A model of the lightning discharge at Jupiter. *Geophys. Res. Lett.* **26**: 2601–4.

Finkelstein, D., and Powell, J. 1970. Earthquake lightning. *Nature* **228**: 759–60.

Foster, H. 1950. An unusual observation of lightning. *Bull. Am. Meteor. Soc.* **31**: 40.

Fujii, N., and Miyamoto, M. 1983. Constraints on the heating and cooling process of chondrule formation. In *Chondrules and Their Origins*, ed. E.A. King, pp. 53–60, Houston: Lunar and Planetary Institute.

Gehrels, T. (ed.) 1976. *Jupiter*, Tucson: University of Arizona Press.

Gehrels, T., and Matthews, M.S. (eds.) 1984. *Saturn*, Tucson: University of Arizona Press.

Gibbard, S., Levy, E.H., and Lunine, J.I. 1995. Generation of lightning in Jupiter's water cloud. *Nature* **378**: 592–5.

Gibbard, S.G., Levy, E.H., and Morfill, G.E. 1997. On the possibility of lightning in the protosolar nebula. *Icarus* **130**: 517–33.

Gibbard, S.G., Levy, E.H., Lunine, J.I., and de Pater, I. 1999. Lightning on Neptune. *Icarus* **139**: 227–34.

Gierasch, P.J., Ingersoll, A.P., Banfield, D., Ewald, S.P., Helfenstein, P., Simon-Miller, A., Vasavada, A., Breneman, H.H., Senske, D.A., and the Galileo Imaging Team 2000. Observation of moist convection in Jupiter's atmosphere. *Nature* **403**: 628–30.

Grard, R. 1998. Electrostatic charging processes of balloon and gondola surfaces in the Earth atmosphere. *J. Geophys. Res.* **103**: 23 315–20.

Grard, R., Svedhem, H., Brown, V., Falkner, P., and Hamelin, M. 1995. An experimental investigation of atmospheric electricity and lightning activity to be performed during the descent of the Huygens Probe on Titan. *J. Atmos. Terr. Phys.* **57**: 575–85.

Grebowsky, J.M., Curtis, S.A., and Brace, L.H. 1991. Small-scale plasma irregularities in the nightside Venus ionosphere. *J. Geophys. Res.* **96**: 21 347–59.

Grebowsky, J.M., Strangeway, R.J., and Hunten, D.M. 1997. Evidence for Venus lightning. In *Venus II, Geology, Geophysics, Atmosphere, and Solar Wind Environment*, eds. S. W. Bougher, D. M. Hunten, and R.J. Phillips, pp. 125–57, Tucson, Arizona: University of Arizona Press.

Gurnett, D.A., Shaw, R.R., Anderson, R.R., Kurth, W.S., and Scarf, F.L. 1979. Whistlers observed by Voyager 1: detection of lightning on Jupiter. *Geophys. Res. Lett.* **6**: 511–14.

Gurnett, D.A., Kurth, W.S., and Scarf, F.L. 1981. Plasma waves near Saturn: initial results from Voyager 1. *Science* **212**: 235–9.

Gurnett, D.A., Grun, E., Gallagher, D., Kurth, W.S., and Scarf, F.L. 1983. Micron-size particles detected near Saturn by the Voyager plasma wave instrument. *Icarus* **53**: 236–56.

Gurnett, D.A., Kurth, W.S., Cairns, I.H., and Granroth, L.J. 1990. Whistlers in Neptune's magnetosphere: evidence of atmospheric lightning. *J. Geophys. Res.* **95**: 20 967–76.

Gurnett, D.A., Kurth, W.S., Roux, A., Gendrin, R., Kennel, C.F., and Bolton, S.J. 1991. Lightning and plasma wave observations from the Galileo flyby Venus. *Science* **253**: 1522–5.

Gurnett, D.A., Zarka, P., Manning, R., Kurth, W.S., Hospodarsky, G.B., Averkamp, T.F., Kaiser, M.L., and Farrell, W.M. 2001. Non-detection at Venus of high-frequency radio signals characteristic of terrestrial lightning. *Nature* **409**: 313–15.

Hale, L.C., and Baginski, M.E. 1987. Current to the ionosphere following lightning stroke. *Nature* **329**: 814–16.

Handel, P.H., and James, P.B. 1983. Polarization catastrophe model of static electrification and strokes in the B-ring of Saturn. *Geophys. Res. Lett.* **10**: 1–4.

Hansell, S.A., Wells, W.K., and Hunten, D.M. 1995. Optical detection of lightning on Venus. *Icarus* **117**: 345–51.

Hayakawa, M. 1995a. Whistlers. In *Handbook of Atmospheric Electrodynamics*, vol. II, ed. H. Volland, pp. 155–93, Boca Raton, Florida: CRC Press.

Hayakawa, M. 1995b. Association of whistlers with lightning discharges on the Earth and on Jupiter. *J. Atmos. Terr. Phys.* **57**: 525–35.

Hecht, J. 1994. Cosmic lightning bolts heralded birth of planets. *New Scientist* **17**: 16.

Helliwell, R.A. 1965. *Whistlers and Related Ionospheric Phenomena*, Stanford, California: Stanford University Press.

Hewins, R.H. 1988. Experimental studies of chondrules. In *Meteorites and the Early Solar System*, pp. 660–79, Tucson, Arizona: University of Arizona Press.

Ho, C.-M., Strangeway, R.J., and Russell, C.T. 1991. Occurrence characteristics of VLF bursts in the nightside ionosphere of Venus. *J. Geophys. Res.* **96**: 21 361–9.

Ho, C.-M., Strangeway, R.J., and Russell, C.T. 1992. Control of VLF burst activity in the nightside ionosphere of Venus by the magnetic field orientation. *J. Geophys. Res.* **97**: 11 673–80.

Ho, C.-M., Strangeway, R.J., and Russell, C.T. 1993. Evidence of Langmuir oscillations and a low density cavity in the Venus magnetotail. *Geophys. Res. Lett.* **20**: 1775–8.

Hobara, Y., and Hayakawa, M. 1997. Ducted propagation of lightning-generated whistlers in the Jovian magnetosphere. *J. Atmos. Electr.* **17**: 33–45.

Hobara, Y., Molchanov, O.A., Hayakawa, M., and Ohta, K. 1995. Propagation characteristics of whistler waves in the Jovian ionosphere and magnetosphere. *J. Geophys. Res.* **100**: 23 523–31.

Hobara, Y., Kanemaru, S., Hayakawa, M., and Gurnett, D.A. 1997. On estimating the amplitude of Jovian whistlers observed by Voyager 1 and implications concerning lightning. *J. Geophys. Res.* **102**: 7115–25.

Hoblitt, R.P. 1994. An experiment to detect and locate lightning associated with eruptions of Redoubt Volcano. *J. Volcanology Geothermal Res.* **62**: 499–517.

Holzworth, R.H., Winglee, R.M., Barnum, B.H., Li, Y.Q., and Kelly, M.C. 1999. Lightning whistler waves in the high-latitude magnetosphere. *J. Geophys. Res.* **104**: 17369–78.

Horanyi, M., Morfill, G., Goertz, C.K., and Levy, E.H. 1995. Chondrule formation in lightning discharges. *Icarus* **144**: 174–85.

Horanyi, M., and Robertson, S. 1996. Chondrule formation in lightning discharges: status of theory and experiments. In *Chondrules and the Protoplanetary Disk*, eds. R.H. Hewins, R.H. Jones, and E.R.D. Scott, pp. 303–10, Cambridge: Cambridge University Press.

Huba, J.D. 1992. Theory of small scale density and electric field fluctuations in the nightside Venus ionosphere. *J. Geophys. Res.* **97**: 43–50.

Huba, J.D., and Grebowsky, J.M. 1993. Small-scale density irregularities in the nightside Venus ionosphere: comparison of theory and observations. *J. Geophys. Res.* **98**: 3079–86.

Huba, J.D., and Rowland, H.L. 1993. Propagation of electromagnetic waves parallel to the magnetic field in the nightside Venus ionosphere. *J. Geophys. Res.* **98**: 5291–300.

Hubbard, W.B. 1984. Interior structure of Uranus. In *Uranus and Neptune*, ed. J.T. Bergstralh, pp. 291–325, NASA Conf. Publ. 2330.

Hubbard, W.B., and MacFarlane. 1980. Theoretical predictions of deuterium abundances in the jovian planets. *Icarus* **44**: 676–82.

Hunten, D.M., Colin, L., Donahue, T.M., and Moroz, V.I. 1983. *Venus*, Tucson, Arizona: University of Arizona Press.

Ingersoll, A.P. 1976. The atmosphere of Jupiter. *Space Sci. Rev.* **18**: 603–39.

Ingersoll, A.P., Gierasch, P.J., Banfield, D., Vasavada, A.R., and the Galileo Imaging Team. 2000. Moist convection as an energy source for the large-scale motions in Jupiter's atmosphere. *Nature* **403**: 630–2.

Kaiser, M.L., Connerney, J.E.P., and Desch, M.D. 1983a. The source of Saturn electrostatic discharges: atmospheric storms. NASA Technical Memorandum 849666.

Kaiser, M.L., Connerney, J.E.P., and Desch, M.D. 1983b. Atmospheric storm explanation of Saturnian electrostatic discharges. *Nature* **303**: 50–3.

Kaiser, M.L., Desch, M.D., Kurth, W.S., Lecacheux, A., Genova, F., and Pedersen, B.M. 1984. Saturn as a radio source. In *Saturn*, ed. T. Gehrels, Tucson, Arizona: University of Arizona Press.

Kaiser, M.L., Zarka, P., Desch, M.D., and Farrell, W.M. 1991. Restrictions on the characteristics of Neptunian lightning. *J. Geophys. Res.* **96**: 19043–7.

Kamra, A.K. 1972a. Visual observation of electric sparks on gypsum dunes. *Nature* **240**: 143–4.

Kamra, A.K. 1972b. Measurements of the electrical properties of dust storms. *J. Geophys. Res.* **77**: 5856–69.

Katahira, O. 1992. Observation of volcanic lightning of Mt. Fugen, Unzen. *Res. Lett. Atmos. Electr.* **12**: 225–34.

Kaufmann, W.J. 1987. *Discovering the Universe*, 381 pp., New York: W.H. Freeman and Co.

Kelley, M.C., Ding, J.G., and Holzworth, R.H. 1990. Intense ionospheric electric and magnetic field pulses generated by lightning. *Geophys. Res. Lett.* **17**: 122221–4.

Kelley, M.C., Sierfring, C.L., Praff, R.F., Kintner, P.M., Larsen, M., Green, R., Holzworth, R.H., Hale, L.C., Mitchell, J.D., and Le Vine, D. 1985. Electrical measurements in the atmosphere and the ionosphere over an active thunderstorm. 1. Campaign overview and initial ionospheric results. *J. Geophys. Res.* **90**: 9815–23.

Kerr, R.A. 1991. Lightning found on Venus at last? *Science* **253**: 1492.

Kliore *et al.* 1985. Venus International Reference Atmosphere (VIRA).

Knollenberg, R., and Hunten, D.M. 1979. Clouds of Venus: a preliminary assessment of microstructure. *Science* **205**: 70–4.

Knollenberg, B.G., and Hunten, D.M. 1980. The microphysics of the clouds of Venus: results of the Pioneer Venus particle size spectrometer experiment. *J. Geophys. Res.* **85**: 8039–58.

Knollenberg, R., Travis, L., Tomasko, M., Smith, P., Ragent, B., Esposito, L., McCleese, D., Martewchick, J., and Beer, R. 1980. The clouds of Venus: a synthesis report. *J. Geophys. Res.* **85**: 8059–81.

Krasnopolsky, V.A. 1980. On lightning in the Venus atmosphere according to the Venera 9 and 10 data. Report of the Space Res. Inst. Acad. Sci. of USSR, Moscow.

Krasnopolsky, V.A. 1983a. Venus spectroscopy in the 3000–8000 Å region by Veneras 9 and 10. In *Venus*, eds. D.M. Hunten, L. Colin, T.M. Donahue, and V.I. Moroz, pp. 459–83, Tucson, Arizona: University of Arizona Press.

Krasnopolsky, V.A. 1983b. Lightnings and nitric oxide on Venus. *Planetary Space Sci.* **31**: 1363–9.

Krider, E.P., Dawson, G.A., and Uman, M.A. 1968. The peak power and energy dissipation in a single-stroke lightning flash. *J. Geophys. Res.* **73**: 3335–9.

Ksanfomality, L.V. 1979. Lightning in the cloud layer of Venus. *Kosm. Issled.* **17**: 747–62.

Ksanfomality, L.V. 1980. Discovery of frequent lightning discharges in clouds on Venus. *Nature* **284**: 244–6.

Ksanfomality, L.V. 1983. Electrical activity in the atmosphere of Venus 1. Measurements on descending probes. *Kosm. Issled.* **21**: 279–96.

Ksanfomality, L.V., Vasil'chikov, N.M., Ganpantserova, O.F., Petrova, E.V., Souvorov, A.O., Fillipov, G.F., Vablonskaya, O.V., and Yabrova, L.V. 1979. Electrical discharges in the atmosphere of Venus. *Pisma Astron. Zh.* **5**: 229–36.

Ksanfomality, L.V., Scarf, F.L., and Taylor, W.L. 1983. The electrical activity of the atmosphere of Venus. In *Venus*, eds. D.M. Hunten, L. Colin, T.M. Donahue, and V.I., Moroz, pp. 565–603, Tucson, Arizona: University of Arizona Press.

Kurth, W.S. 1993. The low-frequency interplanetary radiation. *Adv. Space Res.* **13**(6): 209–15.

Kurth, W.S., Gurnett, D.A., and Scarf, F.L. 1983a. A search for Saturn electrostatic discharge in the Voyager plasma data. *Icarus* **53**: 255–61.

Kurth, W.S., Scarf, F.L., Gurnett, D.A., and Barbosa, D.D. 1983b. A survey of electrostatic waves in Saturn's magnetosphere. *J. Geophys. Res.* **88**: 8959–70.

Kurth, W.S., Strayer, B.D., Gurnett, D.S., and Scarf, F.L. 1985. A summary of whistlers observed by Voyager 1 at Jupiter. *Icarus* **61**: 497–507.

Lanzerotti, L.J., Rinnert, K., Krider, E.P., Uman, M. A., Dehmel, G., Gliem, F.O., and Axford, W. I. 1983. Planetary lightning and lightning measurements on the Galileo Probe to Jupiter's atmosphere. In *Proceedings in Atmospheric Electricity*, eds. L.N. Ruhnke, and J. Latham, pp. 408–13, Hampton, Virginia: A. Deepak.

Lanzerotti, L.J., Thomson, D.J., Maclennan, C.G., Rinnert, K., Krider, E.P., and Uman, M.A. 1989. Power spectra at radio-frequency of lightning return stroke waveforms. *J. Geophys. Res.* **94**: 13 221–7.

Lanzerotti, L.J., Rinnert, K., Dehmel, G., Gliem, F.O., Krider, E.P., Uman, M.A., Umlauft, G., and Bach, J. 1992. The lightning and radio emission detector (LRD) instrument. *Space Sci. Rev.* **60**: 91–109.

Lanzerotti, L.J., Rinnert, K., Dehmel, G., Gliem, F.O., Krider, E.P., Uman, M.A., and Bach, J. 1996. Radio-frequency signals in Jupiter's atmosphere. *Science* **272**: 858–60.

Levin, Z., and Tzur, I. 1986. Models of the development of the electrical structure of clouds. In *The Earth's Electrical Environment*, eds. E.P. Krider and R.G. Roble, pp. 131–45, Washington, DC: National Academy Press.

Levin, Z., Borucki, W.J., and Toon, O.B. 1983. Lightning generation in planetary atmospheres. *Icarus* **56**: 80–115.

Levine, J.S. 1969. The ashen light: an auroral phenomenon on Venus. *Planet. Space Sci.* **17**: 1081–7.

Levine, J.S., Hughes, R.E., Chameides, W.L., and Howell, W.E. 1979. N_2O and CO production by electric discharge, atmospheric implications. *Geophys. Res. Lett.* **6**: 557–9.

Levine, J.S., Gregory, G.L., Harvey, G.A., Howell, W.E., Borucki, W.J., and Orville, R.E. 1982. Production of nitric oxide by lightning on Venus. *Geophys. Res. Lett.* **9**: 893–6.

Levy, E.H. 1988. Energetics of chondrule formation. In *Meteorites and the Early Solar System*, eds. J.F. Kerringe and M.S. Matthews, pp. 697–711, Tucson: University of Arizona Press.

Lewis, J.S. 1980a. Lightning on Jupiter: rate, energetics, and effects. *Science* **210**: 1351–2.

Lewis, J.S. 1980b. Lightning synthesis of organic compounds on Jupiter. *Icarus* **48**: 85–95.

Limaye, S.S., and Sromovsky, L.A. 1991. Winds on Neptune: Voyager observations of cloud motions. *J. Geophys. Res.* **96**: 18 941–60.

Lindal, G.G., Wood, G.E., Hotz, H.B., Sweetnam, D.N., Eshleman, V.R., and Tyler, G.L. 1983. The atmosphere of Titan: an analysis of the Voyager 1 radio occultation measurements. *Icarus* **53**: 348–68.

Lindley, B.C. 1970. Dr. Bruce and astrophysics. *Nature* **226**: 985–6.

Little, B., Anger, C.D., Ingersoll, A.P., Vasavada, A.R., Senske, D.A., Breneman, H.H., Borucki, W.J., and the Galileo SSI team 1999. Galileo images of lightning on Jupiter. *Icarus* **142**: 306–23.

Love, S.G., Keil, K., and Scott, E.R.D. 1995. Electrical discharge heating of chondrules in the solar nebula. *Icarus* **115**: 97–108.

Luhmann, J.G., and Nagy, A.P. 1986. Is there lightning on Venus? *Nature* **319**: 266.

Lunine, J.I., and Hunten, D.M. 1987. Moist convection and the abundance of water in the troposphere of Jupiter. *Icarus* **69**: 566–70.

MacFarlane, J.J., and Hubbard, W.B. 1982. Internal structure of Uranus. In *Uranus and the Outer Planets*, ed. G. Hunt, pp. 111–24, London: Cambridge University Press.

Maeda, K., and Grebowsky, J.M. 1989. VLF emission bursts in the terrestrial and Venusian nightside troughs. *Nature* **34**: 219–21.

Magalhaes, J.A., and Borucki, W.J. 1991. Spatial distribution of visible lightning on Jupiter. *Nature* **349**: 311–13.

Marten, A., Rouan, D., Baluteau, J.P., Gautier, D., Conrath, B.J., Hanel, R.A., Kunde, V., Samuelson, R., Chedin, A., and Scott, N. 1981. Study of the ammonia ice cloud layer in the equatorial region of Jupiter from the infrared interferometric experiment on Voyager. *Icarus* **46**: 233–48.

Meinel, A.B., and Hoxie, D.T. 1962. On the spectrum of lightning in the atmosphere of Venus. *Commun. Lunar Planet. Lab.* **1**: 35–8.

Menietti, J.D., and Gurnett, D.A. 1980. Whistler propagation in the Jovian magnetosphere. *Geophys. Res. Lett.* **7**: 49–52.

Menietti, J.D., Tsintikidis, D., Gurnett, D.A., and Curran, D.B. 1991. Modeling of whistler ray paths in the magnetosphere of Neptune. *J. Geophys. Res.* **96**: 19 117–22.

Michnowski, S. 1963. On the observation of lightning in warm clouds. *Indian J. Meteorol. Geophys.* **14**: 320–2.

Moore, C.B., Vonnegut, B., Stein, B.A., and Survilas, H.J. 1960. Observations of electrification and lightning in warm clouds. *J. Geophys. Res.* **65**: 1907–10.

Morfill, G., Spuit, H., and Levy, E.H. 1993. Physical processes and conditions associated with the formation of protoplanetary disks. In *Protostars and Planets III*, pp. 939–88, Tucson, Arizona: University of Arizona Press.

Morrison D., and Owen, T. 1988. *The Planetary System*, 519 pp., Reading, Massachusetts: Addison-Wesley.

Nagai, K., Ohta, K., Hobara, Y., and Hayakawa, M. 1993. Transmission characteristics of VLF/ELF radio waves through the Jovian ionosphere. *Geophys. Res. Lett.* **20**: 2435–8.

Napier, W.M. 1971. The ashen light on Venus. *Planet. Space Sci.* **19**: 1049–51.

Orton, G.S., Appleby, J.F., and Martonchik, J.V. 1982. The effect of ammonia ice on the outgoing thermal radiance from the atmosphere of Jupiter. *Icarus* **52**: 94–116.

Park, C.G. 1982. Whistlers. In *Handbook of Atmospherics*, vol. II, ed. H. Volland, pp. 21–77, Boca Raton, Florida: CRC Press.

Parrot, M. 1995. Electromagnetic noise due to earthquakes. In *Handbook of Atmospheric Electrodynamics*, vol. II, ed. H. Volland, pp. 95–116, Boca Raton, Florida: CRC Press.

Pearl, J.C., and Conrath, B.J. 1991. The albedo, effective temperature, and energy balance of Neptune, as determined from Voyager data. *J. Geophys. Res.* **96**: 18 921.

Pierce, E.T. 1957. Lightning. *Science Progr.* **43**: 62–75.

Pierce, E.T. 1974. Atmospheric electricity – some themes. *Bull. Am. Meteor. Soc.* **55**: 1186–94.

Pietrowski, E.L. 1960. An observation of lightning in warm clouds. *J. Meteor.* **17**: 562–3.

Pilipp, W., Hartquist, T.W., and Morfill, G.E. 1992. Large electric fields in acoustic waves and the stimulation of lightning discharges. *Astrophys. J.* **387**: 364–71.

Pilipp, W., Hartquist, T.W., Morfill, G.E., and Levy, E.H. 1998. Chrondule formation by lightning in the Protosolar Nebula? *Astron. Astrophys.* **331**: 121–46.

Podolak, M., and Bar-Nun, A. 1988. Moist convection and the abundances of lightning-produced CO, C_2H_2, and HCN on Jupiter. *Icarus* **75**: 566–70.

Ponnamperuma, C. 1966. Some recent work on prebiological synthesis of organic compounds. *Icarus* **5**: 450–4.

Pounder, C. 1980. Volcanic lightning. *Weather* **35**: 357–60.

Ragent, B., Colburn, D.S., Avrin, P., and Rages, K. 1996. Results of the Galileo Probe nephelometer experiment. *Science* **272**: 854–6.

Rinnert, K. 1982. Lightning within planetary atmospheres. In *Handbook of Atmospherics*, vol. II, ed. H. Volland, pp. 99–132, Boca Raton, Florida: CRC Press.

Rinnert, K. 1985. Lightning on other planets. *J. Geophys. Res.* **90**: 6225–37.

Rinnert, K. 1995. Lightning within planetary atmospheres. In *Handbook of Atmospheric Electrodynamics*, vol. I, ed. H. Volland, pp. 203–33, Boca Raton, Florida: CRC Press.

Rinnert, K., and Lanzerotti, L.J. 1998. Radio wave propagation below the Jovian ionosphere. *J. Geophys. Res. – Planets* **103**: 22 993–9.

Rinnert, K., Lanzerotti, L.J., Krider, E.P., Uman, M.A., Dehmel, G., Gliem, F.O., and Axford, W.I. 1979. Electromagnetic noise and radio wave propagation below 100 kHz in the Jovian atmosphere, 1, The equatorial region. *J. Geophys. Res.* **84**: 5181–8.

Rinnert, K., Lanzerotti, L.J., Dehmel, G., Gliem, F.O., Krider, E.P., and Uman, M.A. 1985. Measurements of the RF characteristics of Earth lightning with the Galileo probe lightning experiment. *J. Geophys. Res.* **90**: 6239–44.

Rinnert, K., Lauderdale, R., Lanzerotti, L.J., Krider, E.P., and Uman, M.A. 1989. Characteristics of magnetic field pulses in Earth lightning measured by the Galileo probe instrument. *J. Geophys. Res.* **94**: 13 229–35.

Rinnert, K., Lanzerotti, L.J., Uman, M.A., Dehmel, G., Gliem, F.O., Krider, E.P., and Bach, J. 1998. Measurements of radio-frequency signals from lightning in Jupiter's atmosphere. *J. Geophys. Res. – Planets* **103**: 22 979–92.

Ross, M. 1981. The ice layer in Uranus and Neptune–diamonds in the sky? *Nature* **292**: 435–6.

Rossby, A.A. 1966. Sferics from lightning in warm clouds. *J. Geophys. Res.* **71**: 3807–9.

Rulenko, O.P. 1985. Electrification of volcanic clouds. *Vulkanologiya i Seismologiya* **2**: 71–83.

Rulenko, O.P., Tokarev, P.I., and Firstov, P.P. 1976. Electricity of volcanoes. *Bull. Vulkanol. Stan.* **52**: 11–7.

Russell, C.T. 1991a. Venus lightning. *Space Sci. Rev.* **55**: 317–56.

Russell, C.T. 1991b. Reply to Taylor and Cloutier. *Geophys. Res. Lett.* **18**: 755–8.

Russell, C.T. 1991c. Planetary magnetospheres. *Sci. Progress* **75**: 93–105.

Russell, C.T. 1993. Planetary lightning. *Ann. Rev. Earth Planet. Sci.* **21**: 43–87.

Russell, C.T., and Phillips, J.L. 1990. The ashen light. *Adv. Space Res.* **10**(5): 137–41.

Russell, C.T., and Scarf, F.L. 1990. Evidence for lightning on Venus. *Adv. Space Res.* **10**(5): 137–41.

Russell, C.T., and Singh, R.N. 1989. A re-examination of impulsive VLF signals in the night ionosphere of Venus. *Geophys. Res. Lett.* **16**: 1481–4.

Russell, C.T., and Strangeway, R.J. 1992. Venus lightning: an update. *Adv. Space Res.* **12**: 43.

Russell, C.T., von Dornum, M., and Scarf, F.L. 1988a. The altitude distribution of impulsive signals in the night ionosphere of Venus. *J. Geophys. Res.* **93**: 591–21.

Russell, C.T., von Dornum, M., and Scarf, F.L. 1988b. Planetocentric clustering of low altitude impulsive electric signals in the night ionosphere of Venus. *Nature* **331**: 591–4.

Russell, C.T., von Dornum, M., and Scarf, F.L. 1988c. VLF bursts in the night ionosphere of Venus. *Planet. Space Sci.* **36**: 1211–18.

Russell, C.T., von Dornum, M., and Scarf, F.L. 1989a. Source locations for impulsive electric signals seen in the night ionosphere of Venus. *Icarus* **80**: 390–415.

Russell, C.T., von Dornum, M., Scarf, F.L., and Strangeway, R.J. 1989b. VLF bursts in the night ionosphere of Venus: estimates of the Poynting flux. *Geophys. Res. Lett.* **16**: 579–82.

Russell, C.T., von Dornum, M., and Scarf, F.L. 1990. Impulsive signals in the night ionosphere of Venus: comparison of results obtained below the local electron gyro-frequency with those above. *Adv. Space Res.* **10**(5): 37–40.

Rybicki, G.B., and Lightman, A.P. 1979. *Radiative Processes in Astrophysics*, 400 pp., New York: Wiley.

Sagan, C.E., Lippincott, E.R., Dayhoff, M.O., and Eck, R.V. 1967. Organic molecules and the coloration of Jupiter. *Nature* **213**: 273.

Sagdeev, R.Z., Linkin, V.M., Blamont, J.E., and Preston, R.A. 1986a. The VEGA Venus balloon experiment. *Science* **231**: 1407–8.

Sagdeev, R.Z., Linkin, V.M., Kerzhanovich, V.V., Lipatov, A.N., Shurupov, A.A., Blamont, J.E., Crisp, D., Ingersoll, A.P., Elson, L.S., Preston, R.A., Hildebrand, C.E., Ragent, B., Seiff, A., Young, R.E., Petit, G., Boloh, L., Alexandrov, Yu. N., Armand, N.A., Bakitko, R.V., and Selivanov, A.S. 1986b. Overview of VEGA Venus balloon in situ meteorological measurements. *Science* **231**: 1411–14.

Sato, M., and Hansen, J.E. 1979. Jupiter's atmospheric composition and cloud structure deduced from absorption bands in reflected sunlight. *J. Atmos. Sci.* **36**: 1133–67.

Scarf, F.L. 1986. Comment on "Venus nightside ionospheric troughs: implications for evidence of lightning and volcanism" by Taylor, H.A. Jr, Grebowsky, J.M., and Cloutier, P.A. *J. Geophys. Res.* **91**: 4594–8.

Scarf, F.L., and Gurnett, D.A. 1977. A plasma wave investigation for the Voyager mission. *Space Sci. Rev.* **21**: 289–308.

Scarf, F.L., and Russell, C.T. 1983. Lightning measurements from the Pioneer Venus orbiter. *Geophys. Res. Lett.* **10**: 1192–5.

Scarf, F.L., and Russell, C.T. 1988. Evidence of lightning and volcanic activity on Venus: pro and con. *Science* **240**: 222–4.

Scarf, F.L., Gurnett, D.A., and Kurth, W.S. 1979. Jupiter plasma wave observations: an initial Voyager 1 overview. *Science* **204**: 991–5.

Scarf, F.L., Taylor, W.L., Russell, C.T., and Brace, L.H. 1980a. Lightning on Venus: orbiter detection of whistler signals. *J. Geophys. Res.* **85**: 8158–66.

Scarf, F.L., Taylor, W.L., and Virobik, P.F. 1980b. The Pioneer Venus orbiter plasma wave investigation. *IEEE Trans. Geosci. Remote Sens.* **GE-18**: 36–8.

Scarf, F.L., Gurnett, D.A., Kurth, W.S., Anderson, R.R., and Shaw, R.R. 1981. An upper bound to the lightning flash rate in Jupiter's atmosphere. *Science* **213**: 684–5.

Scarf, F.L., Gurnett, D.A., Kurth, W.S., and Poynter, R.L. 1982. Voyager 2 plasma wave observations at Saturn. *Science* **215**: 587–94.

Scarf, F.L., Gurnett, D.A., Kurth, W.S., and Poynter, R.L. 1983. Voyager plasma wave measurements at Saturn. *J. Geophys. Res.* **88**: 8971–84.

Scarf, F.L., Jordan, K. F, and Russell, C.T. 1987. Distribution of whistler mode bursts at Venus. *J. Geophys. Res.* **92**: 12 407–11.

Singh, R.N. 1991. Venus lightning. *Science* **251**: 1298–9.

Singh, R.N. and Russell, C.T. 1986. Further evidence for lightning on Venus. *Geophys. Res. Lett.* **13**: 1051–4.

Singh, R.N. and Russell, C.T. 1987. Reply to Taylor and Cloutier. *Geophys. Res. Lett.* **14**: 571–2.

Smith, B.A. *et al.* 1979a. The Jupiter system through the eyes of Voyager 1. *Science* **204**: 951–72.

Smith, B.A. *et al.* 1979b. The Galilean satellites and Jupiter: Voyager 2 imaging science results. *Science* **206**: 927–50.

Smith, B.A. *et al.* 1981. Encounter with Saturn: Voyager 1 imaging science results. *Science* **212**: 163–91.

Smith, B.A. *et al.* 1982. A new look at the Saturn system: the Voyager 2 images. *Science* **215**: 504–37.

Smith, B.A. *et al.* 1986. Voyager 2 in the Uranian system: imaging science results. *Science* **233**: 43–64.

Smith, B.A. *et al.* 1989. Voyager 2 at Neptune: imaging science results. *Science* **246**: 1422–49.

Sonwalkar, V.S., and Carpenter, D.L. 1995. Notes on the diversity of the properties of radio bursts observed on the nightside of Venus. *J. Atmos. Terr. Phys.* **57**: 557–73.

Sonwalkar, V.S., Carpenter, D.L., and Strangeway, R.J. 1991. Testing radio bursts observed on the nightside of Venus for evidence of whistler-mode propagation from lightning. *J. Geophys. Res.* **96**: 17 763–78.

Sromovsky, L.A., Best, F.A., Collard, A.D., Fry, P.M., Revercomb, H.E., Freedman, R.S., Orton, G.S., Hayden, J.L., Tomasko, M.G., and Lemmon, M.T. 1996. Solar and thermal radiation in Jupiter's atmosphere: initial results of the Galileo Probe net flux radiometer. *Science* **272**: 851–4.

Stoker, C.R. 1986. Moist convection: a mechanism for producing the vertical structure of the Jovian equatorial plumes. *Icarus* **67**: 106–25.

Stone, P.H. 1973. The dynamics of the atmosphere of the major planets. *Space Sci. Rev.* **14**: 444–59.

Stone, P.H. 1976. The meteorology of the Jovian atmosphere. In *Jupiter*, ed. T. Gehrels, pp. 586–618, Tucson, Arizona: University of Arizona Press.

Strangeway, R.J. 1990. Radioemission source disputed. *Nature* **345**: 213–14.

Strangeway, R.J. 1991a. Plasma waves at Venus. *Space Sci. Rev.* **55**: 275–316.

Strangeway, R.J. 1991b. Polarization of the impulsive signals observed in the nightside ionosphere of Venus. *J. Geophys. Res.* **96**: 22 741–52.

Strangeway, R.J. 1992. An assessment of lightning or in situ instabilities as a source for whistler mode waves in the night ionosphere of Venus. *J. Geophys. Res.* **97**: 12 203–15.

Strangeway, R.J. 1995. Plasma wave evidence for lightning on Venus. *J. Atmos. Terr. Phys.* **57**: 537–56.

Strangeway, R.J. 1996. Collisional Joule dissipation in the ionosphere of Venus: the importance of electron heat conduction. *J. Geophys. Res.* **101**: 2279–95.

Strangeway, R.J., Russell, C.T., and Ho, C.M. 1993. Observation of intense wave bursts at very low altitudes within the Venus nightside ionosphere. *Geophys. Res. Lett.* **20**: 2771–4.

Su, Y. 1995. Testing of the engineering model of the Galileo Lightning and Radio Emission Detector. M.S. thesis, Univ. Florida, Gainesville, 146 pp.

Taylor, H.A. Jr 1987. Auroras at Venus? Taming the Venus Dragon. *Planetary Report* **VII**(4), July/August, pp. 4–6.

Taylor, H.A. Jr, and Cloutier, P.A. 1987. Comment on "Further evidence for lightning at Venus." *Geophys. Res. Lett.* **14**: 568–70.

Taylor, H.A. Jr, and Cloutier, P.A. 1988. Telemetry interference incorrectly interpreted as evidence for lightning and present-day volcanism at Venus. *Geophys. Res. Lett.* **15**: 729–32.

Taylor, H.A. Jr, and Cloutier, P.A. 1991. Comment on "A re-examination of impulsive VLF signals in the night ionosphere of Venus". *Geophys. Res. Lett.* **18**: 753–4.

Taylor, H.A. Jr, and Cloutier P.A. 1992. Non-evidence of lightning and associated volcanism at Venus. *Space Sci. Rev.* **61**: 387–91.

Taylor, H.A. Jr, and Cloutier, P.A. 1994. Optical searches for Venusian lightning: implications for nightside field and plasma relationships. *Earth, Moon, and Planets* **64**: 201–5.

Taylor, H.A. Jr, Grebowsky, J.M., and Cloutier, P.A. 1985. Venus nightside ionospheric thoughts: implications for evidence of lightning and volcanism. *J. Geophys. Res.* **90**: 7415–26.

Taylor, H.A. Jr, Grebowsky, J.M., and Cloutier, P.A. 1986. Reply. *J. Geophys. Res.* **91**: 4599–605.

Taylor, H.A. Jr, Cloutier, P.A., and Zheng, Z. 1987. Venus "lightning" signals reinterpreted as in situ plasma noise. *J. Geophys. Res.* **92**: 9907–19.

Taylor, W.L., Scarf, F.L., Russell, C.T., and Brace, L.H. 1979. Evidence for lightning on Venus. *Nature* **279**: 614–16.

Tokar, R.L., Gurnett, D.A., and Bagenal, F. 1982. The proton concentration in the vicinity of the Io plasma torus. *J. Geophys. Res.* **87**: 10 395–400.

Toon, O.B., Ragent, B., Colburn, D., Blamont, J., and Cot, C. 1984. Large solid particles in the clouds of Venus: do they exist? *Icarus* **57**: 143–60.

Turman, B.N. 1977. Detection of lightning superbolts. *J. Geophys. Res.* **82**: 2566–8.

Turman, B.N. 1978. Analysis of lightning data from the DMSP satellite. *J. Geophys. Res.* **83**: 5019–24.

Uman, M.A. 1987. *The Lightning Discharge*, 377 pp., San Diego: Academic Press.

Uman, M.A. 2001. *The Lightning Discharge*, 377 pp., Mineola, New York: Dover.

Vonnegut, B. 1995. Jovian lightning after comet impacts? *Science* **268**: 1829.

Wang, K., Thorne, R.M., and Horne, R.B. 1995. The propagation characteristics and Landau damping of Jovian whistlers in the Io torus. *J. Geophys. Res.* **100**: 21 709–16.

Warwick, J.W., Pearce, J.B., Evans, D.R., Carr, T.D., Schaubli, J.J., Alexander, J.K., Kaiser, M.L., Desch, M.D., Pederson, M., Lecacheux, A., Daigne, G., Boischot, A., and Barrow, C.H. 1981. Planetary radio astronomy observations from Voyager 2 near Saturn. *Science* **212**: 239–43.

Warwick, J.W., Evans, D.R., Ronig, J.H., Alexander, J.K., Desch, M.D., Kaiser, M.L., Aubier, M., Leblanc, Y., Lecacheux, A., and Pedersen, B.M. 1982. Planetary radio astronomy observations from Voyager 2 near Saturn. *Science* **215**: 582–7.

Weidenschilling, S.J., and Lewis, J.S. 1973. Atmospheric and cloud structure of the Jovian planets. *Icarus* **20**: 465–76.

Weinheimer, A.J., and Few, A.A. Jr 1982. The spokes in Saturn's rings: a critical evaluation of possible electrical processes. *Geophys. Res. Lett.* **9**: 1139–42.

West, R.A. and Tomasko, M.G. 1980. Spatially resolved methane band photometry of Jupiter. III. Cloud vertical structures for several axisymmetric bands and the great red spot. *Icarus* **41**: 278–92.

West, R.A., Strobel, D.F., and Tomasko, M.G. 1986. Clouds, aerosols, and photochemistry in the Jovian atmosphere. *Icarus* **65**: 161–217.

Whipple, F.L. 1966. Chondrules: suggestion concerning their origin. *Science* **153**: 54–6.

Williams, M.A. 1983. On the energy of possible Saturnian lightning. *Icarus* **56**: 611–12.

Williams, M.A., and Thomason, L.W. 1983. Optical signature of Venus lightning as seen from space. *Icarus* **55**: 185–6.

Williams, M.A., Thomason, L.W., and Hunten, D.M. 1982. The transmission to space of the light produced by lightning in the clouds of Venus. *Icarus* **52**: 166–70.

Williams, M.A., Krider, E.P., and Hunten, D.M. 1983. Planetary lightning: Earth, Jupiter, and Venus. *Rev. Geophys. Space Phys.* **21**: 892–902.

Woeller, F., and Ponnamperuma, C. 1969. Organic synthesis in a simulated Jovian atmosphere. *Icarus* **10**: 386–92.

Woodman, J.H., Cochran, W.D., and Starsky, D.B. 1979. Spatially resolved reflectivities of Jupiter during the 1976 opposition. *Icarus* **37**: 73–8.

Yair, Y., Levin, Z., and Tzivion, S. 1995a. Microphysical processes and dynamics of a Jovian thundercloud. *Icarus* **114**: 278–99.

Yair, Y., Levin, Z., and Tzivion, S. 1995b. Lightning generation in a Jovian thundercloud: results from an axisymmetric numerical cloud model. *Icarus* **115**: 421–34.

Yair, Y., Levin, Z., and Tzivion, S. 1998. Model interpretation of Jovian lightning activity and the Galileo Probe results. *J. Geophys. Res.* **103**: 14 157–66.

Young, R.E., Smith, M.A., and Sobeck, C.K. 1996. Galileo probe: in situ observations of Jupiter's atmosphere. *Science* **272**: 837–8.

Zarka, P. 1985. On detection of radio bursts associated with Jovian and Saturnian lightning. *Astron. Astrophys.* **146**: 115–18.

Zarka, P., and Pedersen, B.M. 1983. Statistical study of Saturn electrostatic discharges. *J. Geophys. Res.* **88**: 9007–18.

Zarka, P., and Pedersen, B.M. 1986. Radio detection of Uranian lightning by Voyager 2. *Nature* **323**: 605–8.

17 Lightning locating systems

> Lightning detection and mapping systems can be used to minimize the harmful effects of lightning by providing early warnings of such hazards. These systems can also be used to determine how much lightning actually occurs within a given region and the statistics of discharge parameters that are important in research and in the design of lightning protection systems.
>
> E.P. Krider (1994)

17.1. Introduction

There are many individual physical processes in cloud and ground flashes. Each of these processes has associated with it characteristic electric and magnetic fields. Lightning is known to emit significant electromagnetic energy in the frequency range from below 1 Hz to near 300 MHz, with a peak in the frequency spectrum near 5 to 10 kHz for lightning at distances beyond 50 km or so. As noted in Section 1.3, at frequencies higher than that of the spectrum peak, the spectral amplitude varies roughly in inverse proportion to the frequency up to 10 MHz or so and in inverse proportion to the square root of frequency from about 10 MHz to 10 GHz. Further, electromagnetic radiation from lightning is detectable at even higher frequencies, for example, in the microwave, 300 MHz to 300 GHz and, obviously, in visible light, roughly 10^{14} to 10^{15} Hz. In addition to electromagnetic radiation, lightning produces the acoustic radiation discussed in Chapter 11. In general, any observable signal from a lightning source can be used to detect and locate the lightning process that produced it. Acoustic locating techniques, including acoustic-signal time of arrival and acoustic ray tracing, are described in Section 11.5 and hence are not discussed further here.

For the three most common electromagnetic radio-frequency locating techniques, magnetic direction finding (MDF), time of arrival (TOA), and interferometry, the type of locating information obtained depends on the frequency f (or equivalently on the wavelength λ) of the radiation detected ($f\lambda = c$, where c is the speed of light). For detected signals whose wavelengths are very short compared with the length of a radiating lightning channel, that is, the very-high-frequency (VHF) range where $f = 30$ to 300 MHz and $\lambda = 10$ to 1 m, the whole lightning channel can, in principle, be imaged in three dimensions. For wavelengths that are a significant fraction of the lightning channel length, that is, the very-low-frequency (VLF) range where $f = 3$ to 30 kHz and $\lambda = 10$ to 1 km and the low-frequency (LF) range where $f = 30$ to 300 kHz and $\lambda = 1$ km to 100 m, generally only one location can be usefully obtained. This location is usually interpreted as some approximation to the ground strike point in the case of a cloud-to-ground discharge. The best electromagnetic-channel imaging methods at VHF and the best ground-strike-point locating techniques at VLF and LF have accuracies (actually location errors or uncertainties) of the order of a hundred meters. On the other end of the accuracy scale, long-range VLF systems that operate in a narrow frequency band somewhere between 5 and 10 kHz and detect lightning at distances up to thousands of kilometers have uncertainties in locating individual lightning flashes of tens to hundreds of kilometers. These latter systems are often called thunderstorm locators. They are discussed in subsection 13.3.3.

For those electromagnetic locating techniques involving the measurement of field-change amplitudes at multiple stations, the bandwidth of the measurement is not directly related to the locating accuracy: it is only necessary to have a measurement system that can reproduce the field changes of the process of interest faithfully. Hence, for example, from measuring the electrostatic field change in the frequency range from a fraction of a hertz to a few hertz at multiple stations, one can locate an "average" position for the charge source of a complete cloud-to-ground flash (Fig. 3.6); and with a system bandwidth from a few hertz to a few kilohertz, so as to be able to resolve electrostatic field changes on a millisecond time scale, one can locate the charge sources for individual strokes in the flash as well as for the continuing current (Figs. 3.5 and 4.9). Lightning location using the return-stroke electric or magnetic radiation field peaks, as in the case of location using the electrostatic field change, only requires that the system faithfully reproduces those peaks.

Accurate lightning locating systems, whether they image the whole lightning channel or locate only the ground strike points or the cloud-charge centers, necessarily employ multiple sensors. Single-station ground-based sensors, such as the lightning flash counters discussed in subsection 2.5.1, detect the occurrence of lightning but cannot be used to locate it on an individual flash basis, nor are they designed to do so, because of the wide range of amplitudes and waveshapes associated with individual events. Nevertheless, with single-station sensors one can assign groups of flashes to rough distance ranges if data are accumulated and "averaged" for some period of time. Single-station lightning (or thunderstorm) locating methods and systems used mostly for research are described, for example, by Pierce (1956), Ruhnke (1962, 1971), Heydt and Volland (1964), Kemp (1971), Inkov (1973), Harth and Pelz (1973), Sao and Jindoh (1974), Ryan and Spitzer (1977), Harth et al. (1978), Kononov et al. (1986), Petrenko and Kononov (1992), Grandt (1992), Rafalsky et al. (1995), Panyukov (1996), and Mikhailov and Dubovoy (1997). Some of these are discussed in Section 13.3. There are many relatively simple commercially available single-station devices that purport to locate lightning. Most operate like AM radios: the amplitude of the radio static is used to gauge the distance to the individual lightning flashes, a technique characterized by inherently large errors. Some commercial single-station devices employ, in addition to field-amplitude detectors, optical detectors and/or magnetic direction finders. Single-station optical sensors on Earth-orbiting satellites detect the light scattered by the volume of cloud that produces the lightning and hence cannot locate to an accuracy better than about 10 km, about the diameter of a small cloud. Additionally, satellite-based sensors cannot distinguish between cloud and ground discharges.

In this chapter we will primarily discuss how individual sensors measuring various properties of the lightning electromagnetic radiation have been combined into systems to provide practical lightning locating. Also discussed will be the detection and location of lightning from satellites and by ground-based radars.

17.2. Electric and magnetic field amplitude techniques

If a mathematical formula for the electric or magnetic field amplitude of a particular lightning process can be constructed containing the distance from a sensor to the location of the process as an unknown parameter then an appropriate number of remote measurements by the sensors, at least equal to the number of unknowns in the equation, will allow a determination of those unknowns, including the distance. With this approach, the measured electrostatic field changes for whole lightning flashes have been used successfully to locate both the overall charge sources for cloud and for ground flashes and also the separate charge sources for individual strokes and the continuing current in ground flashes. Further, the initial, predominantly radiation field peaks associated with the return stroke in ground flashes have been used, not particularly successfully, to locate the ground-stroke termination point.

17.2.1. Electrostatic field change

It is usually assumed that a predominantly electrostatic field component is recorded when the measuring system's frequency bandwidth is in the range from 1 Hz to 1 kHz (Pierce 1977a, b). The use of multiple-station electrostatic field measurements to determine the location of the lightning-caused changes in the cloud charge distribution and the magnitude of those changes has been discussed by Jacobson and Krider (1976), Krehbiel et al. (1979), Koshak and Krider (1989, 1994), and Murphy et al. (1996). Jacobson and Krider (1976) first presented a least-square optimization method for fitting the parameters of assumed models of lightning-neutralized cloud charges to the electric field changes measured at multiple stations. That method has been generally adopted for the analysis of multiple-station electrostatic field changes measured using either field mill networks such as at the Kennedy Space Center, Florida (e.g., Jacobson and Krider 1976) or networks of flat-plate electric field antennas, which have a higher upper-frequency response (e.g., Krehbiel et al. 1979). The frequency response of a typical field mill is inadequate to resolve the field changes produced by individual lightning processes within a flash, only the total field change produced by the flash being measured in this way. An example of the application of the technique to a cloud-to-ground discharge follows. A chi-squared (χ^2) function for the cloud-to-ground discharge is defined as

$$\chi^2 = \sum_{i=1}^{N} \frac{(\Delta E_{mi} - \Delta E_{ci})^2}{\sigma_i^2} \quad (17.1)$$

where ΔE_{mi} is the measured field change at the ith ground station, ΔE_{ci} is the model-predicted (calculated) field change at the ith station, σ_i^2 is the variance of the measurement at the ith station due to experimental error, and N is the number of stations. The factor $1/\sigma_i^2$ can be viewed as a weighting factor for the data from the ith station in the sum on the right-hand side of Eq. 17.1. Thus, that factor is a measure of the quality of the data from the ith station. We will start with the simplest case, in which the model-predicted field change ΔE_{ci} at each ground-based station ($z_i = 0$) is assumed to be due to the neutralization of a single, spherically symmetric charge region whose unknown charge is Q and whose center is at an unknown location (x, y, z) above an assumed flat and perfectly conducting ground. Such a

17.2. Electric and magnetic field amplitude techniques

model is called the monopole or point-charge model, and the corresponding equation for ΔE_{ci} is

$$\Delta E_{ci} = \frac{2Qz}{4\pi\varepsilon_0[(x-x_i)^2+(y-y_i)^2+z^2]^{3/2}} \quad (17.2)$$

which is a generalized form of Eq. 3.2. This equation can be applied to cloud-to-ground strokes, continuing currents (or portions of continuing currents), and ground flashes as a whole. There are four unknowns, x, y, z, and Q, in Eq. 17.2, and hence measurements at four or more stations are required. Additional measurements allow the evaluation of errors in estimating the unknowns. Equation 17.2 is substituted in Eq. 17.1 and the four unknowns x, y, z, and Q are iteratively adjusted until χ^2 is a minimum using, for example, the Marquardt algorithm described by Bevington (1969). The values of x, y, z, and Q corresponding to the χ^2 minimum are considered as the best fit to the measurements and the value of the χ^2 minimum is considered as a measure of the adequacy of the fit.

Jacobson and Krider (1976), Koshak and Krider (1989), and Murphy et al. (1996) multiplied the sum on the right-hand side of Eq. 17.1 by $1/\nu$, where ν is the number of degrees of freedom found as the number of measurements minus the number of unknowns in the model, the result being a normalized χ^2. Jacobson and Krider (1976) considered a solution valid when values of the normalized χ^2 at the χ^2 minimum were equal to or less than 10.0. Equation 17.2 can be easily extended (using the principle of superposition) to model cloud flashes or in-cloud processes of ground flashes by postulating that the calculated field change is due to the destruction of two charges of equal magnitude but of opposite polarity, a dipole model (subsection 3.2.1, second part). The number of unknowns in this case is seven, including three coordinates for each of the two charges and their magnitude. Accordingly, the minimum number of stations is seven. When the spacing between the two charges is small compared to the distance from the charges to each of the measurement stations, the "point dipole" approximation (e.g., Krehbiel et al. 1979) can be used, in which

$$\Delta E_{ci} = \frac{1}{4\pi\varepsilon_0}\left[\frac{2\Delta p_z}{R_i^3} - \frac{6z}{R_i^5}\mathbf{R}_i \cdot \Delta \mathbf{p}\right] \quad (17.3)$$

where $\Delta \mathbf{p} = Q\Delta \mathbf{l} = \Delta p_x \mathbf{a}_x + \Delta p_y \mathbf{a}_y + \Delta p_z \mathbf{a}_z$ is the vector dipole moment change, $\Delta \mathbf{l}$ being the vector distance from the negative to the positive charge and $\mathbf{R}_i = (x - x_i)\mathbf{a}_x + (y - y_i)\mathbf{a}_y + z\mathbf{a}_z$ is the slant-range vector from the observer to the source (point dipole) location. There are six unknowns, x, y, z, Δp_x, Δp_y, Δp_z, and hence at least six measurements are required. Clearly, the point-dipole model provides less information than the dipole model, and the solutions are usually presented graphically as arrows whose direction indicates the direction of the effective positive-charge transport and whose length indicates the magnitude of the point-dipole moment change $|\Delta \mathbf{p}|$. The middle point on the arrow, usually marked by a dot in the graphical representation, indicates a single position for the neutralized positive and negative charges, the actual positions being unresolved in this approximation. Examples of graphical representations of the point-dipole solutions are found in Fig. 3.6.

Krehbiel et al. (1979) compared the χ^2 minimization method exemplified by Eq. 17.1 to an analytical inversion technique in which Eq. 17.2 or Eq. 17.3 is solved directly. They found good agreement between the results provided by the two approaches, but prefer the χ^2 minimization method. Koshak and Krider (1994) developed yet another technique, which involves the use of an iterative search method (Landweber 1951) to find that charge distribution on a large grid of charges which provides the best fit to the measured field changes. Koshak and Krider (1994) found from modeling that this technique tended to arrive at a reasonable solution as long as the flash was directly over the network of electric field sensors, with typical errors in the location of the centroid of charge of 1 to 2 km and in the charge magnitude of 20 to 35 percent. Koshak et al. (1999) introduced an algorithm in which the change in the cloud charge is represented by a truncated multipole expansion, so that the unknown source distribution is described by a combination of both monopole and dipole terms. Krehbiel et al. (1979) indicated that to determine the location of a charge centroid with an uncertainty not exceeding 100 m it is necessary to measure electric field changes to within one percent at the different stations. They estimated that a typical error in their field-change measurements was approximately two percent.

Lightning-charge locations and magnitudes derived from multiple-station measurements of the electrostatic field changes for whole flashes have been published by Jacobson and Krider (1976), Lhermitte and Williams (1985), Krider (1989), Nisbet et al. (1990a, b), and Koshak and Krider (1989, 1994), all of whom used measurements made with the Kennedy Space Center network of 20 to 30 field mills, and for individual lightning processes by Krehbiel et al. (1979), from measurements made with a network of flat-plate electric field antennas. Some of these results are described in subsection 3.2.2.

17.2.2. Electric and magnetic radiation field peaks

When observed beyond a range of a few kilometers, the return stroke exhibits an initial, predominantly radiation-field peak in both its electric and magnetic field records (Fig. 4.38). The theory given in subsection 4.6.4 (Eq. 4.13) predicts that the radiation field amplitude over a flat, perfectly conducting Earth surface decreases inversely with the radial distance from the lightning channel. Thus, in principle, assuming (i) that the measured fields are essentially radiation fields and (ii) that the initial field peak is radiated from near the ground strike point (x, y), one can form and minimize a χ^2-function from multiple-station measurements similar to that in Eq. 17.1, the model-predicted electric field E_{ci} being of the form

$$E_{ci} = \frac{E_0}{[(x - x_i)^2 + (y - y_i)^2]^{1/2}} \tag{17.4}$$

where E_0 is the electric field at unit distance from the lightning channel (the source function). The source function is a variable to be determined along with the source location via the minimization procedure. A similar equation can be constructed for the magnetic radiation field. A minimum of three stations is needed.

Unfortunately, the Earth is not a perfect conductor and since the terrain is, in general, neither uniform nor flat and its character can vary with the direction to the lightning, the field peaks will decrease faster than the assumed inverse-distance relation in generally unknown ways. Thus, the use of field-peak data alone does not in practice produce locations as accurate as, for example, magnetic direction finding. However, the field-peak data have been used in conjunction with magnetic direction finding, and in combined magnetic direction finding and time-of-arrival systems, taking into account signal attenuation due to propagation effects, to improve the system's locating accuracy (Section 17.5).

17.3. Magnetic field direction finding

Two vertical and orthogonal loops with planes oriented NS and EW, each measuring the magnetic field from a given vertical radiator, can be used to obtain the direction to the source. This is the case because the output voltage of a given loop, by Faraday's law, is proportional to the cosine of the angle between the magnetic field vector and the normal vector to the plane of the loop. For a vertical radiator the magnetic field lines are horizontal circles coaxial with the source. Hence, for example, a loop whose plane is oriented NS (perpendicular to the EW direction) receives a maximum signal if the source is north or south of the antenna, while an orthogonal EW loop at the same position receives no signal. The signal in the NS loop varies as the cosine of the angle between north and the source as viewed from the antenna, while the signal in the EW loop varies as the sine of the same angle. It follows that the ratio of the two signals from the loops is proportional to the tangent of the angle between north and the source as viewed from the antenna (the azimuth angle to the source).

Crossed-loop magnetic direction finders (DFs) used for lightning detection can be divided into two general types: narrowband (tuned) DFs and gated wideband DFs. In both cases the direction-finding technique involves an implicit assumption that the radiated electric field is oriented vertically, so that the associated magnetic field is oriented horizontally and is perpendicular to the propagation path. Narrowband DFs have been used to detect distant lightning since the 1920s (Horner 1954, 1957). They generally operate in a narrow frequency band with a center frequency in the range 5 to 10 kHz, where attenuation in the Earth–ionosphere waveguide is relatively low (Section 13.3) and where the lightning signal energy is relatively high (Section 4.2). Before the development of weather radars in the 1940s, lightning locating systems were the primary means of identifying and mapping thunderstorms at medium and long ranges. In the 1920s, Watson-Watt and Herd (1926) developed a DF using a pair of orthogonal loop antennas tuned to a frequency near 10 kHz. The azimuth angle to the discharge was obtained by displaying the north–south and east–west antenna outputs simultaneously on an xy oscilloscope, so that the resulting line (in the case of a properly polarized signal) on the oscilloscope screen had an orientation that indicated the direction to the discharge. This line is sometimes called a direction vector (e.g., Krider et al. 1980). Two such direction finders at known positions are sufficient to determine the location of a discharge from the intersection of simultaneous direction vectors. Similar systems were used before and during World War II in many countries. For example, during World War II the British Meteorological Office operated a narrowband DF network (a 250 Hz bandwidth centered at 9 kHz), containing seven sensors located in the United Kingdom and the Mediterranean region, in support of the Royal Air Force (WMO 1955). That network tracked thunderstorms from the United Kingdom to as far south as North Africa. According to Lee (1986b), the British Meteorological Office's system operated for forty years. Narrowband DFs and other related methods for locating lightning have been reviewed by Keen (1938), Adcock and Clarke (1947), Norinder (1953), Kashprovsky (1966), Pierce (1977b), Rakov (1990), and Cummins et al. (1998a, b).

A major disadvantage of narrowband DFs is that for lightning at ranges less than about 200 km, these DFs have inherent azimuthal errors, called polarization errors, of order 10° (Nishino et al. 1973; Kidder 1973). These errors are caused by the detection of magnetic field components from (i) non-vertical channel sections, whose magnetic field lines form circles in a plane perpendicular to the non-vertical channel section, and (ii) skywaves produced by reflection from the ionosphere (Sections 13.1 and 13.3), whose magnetic fields are similarly improperly oriented for direction finding of the ground strike point (Yamashita and Sao 1974a, b; Uman et al. 1980). Additionally, unwanted magnetic field components may occur due to other effects such as non-horizontal conducting topography near the DF and reradiation by nearby buried conductors (e.g., Horner 1953, 1954). Errors due to the latter effect, usually called site errors, have been reported to be as large as 30° for narrowband DFs (Horner 1954). However, Taylor and Jean (1959) and Taylor (1963) reported using crossed loops with a relatively wide bandwidth (1 kHz to 100 kHz) to locate successfully lightning in the range 100 to 500 km by displaying on an xy oscilloscope only the first 100 μs or so of the received signal corresponding to the return-stroke groundwave and hence omitting from the received signal the improperly polarized skywaves, which arrive later. Taylor (1963) gave ±2° as the azimuthal error but presented no data to support this claim. Taylor and Jean (1959) stated that only discharges producing straight lines (as opposed to elliptical patterns indicative of significant polarization errors) on the xy oscilloscope were accepted for analysis.

To overcome the problem of large polarization errors at short ranges inherent in the operation of narrowband DFs, gated wideband DFs were developed in the early 1970s. Direction finding is accomplished by sampling (gating on) the NS and EW components of the initial peak of the return-stroke magnetic field, this peak being radiated from the bottom hundred meters or so of the channel in the first microseconds of the return stroke. Since the bottom of the channel tends to be straight and vertical, the magnetic field is horizontally polarized. Additionally, a gated DF does not record ionospheric reflections since those reflections arrive long after the initial peak magnetic field is sampled. The operating bandwidth of the gated wideband DF is from a few kilohertz to about 500 kHz.

The gated wideband DF was described by Krider *et al.* (1976) along with a determination of azimuthal errors made by comparing direction-finder vectors with video recordings of the causative return strokes. Krider *et al.* (1976) determined the distributions of the difference between the azimuths determined via magnetic field and video direction finding techniques for two groups of strokes, mostly between 10 and 100 km. For one group (325 strokes over flat terrain), the mean value of the difference was zero and the standard deviation was 1.8°. For the other group (164 strokes over or behind mountains), the mean value was 1° and the standard deviation was 2°. To demonstrate the advantages of gating on the early part of the magnetic field waveform, Herrman *et al.* (1976) showed that, for lightning between 3 and 12 km, DF azimuthal error (both mean value and standard deviation) progressively increased as sampling of the NS and EW magnetic fields took place at increasingly greater times up to 155 μs after the beginning of the waveform. Thus, the gated wideband DF, in addition to eliminating polarization errors from ionospheric reflections, clearly minimizes the polarization errors associated with non-vertical channel sections that are characteristic of narrowband DFs. Further, Krider *et al.*

(1980) designed a gated wideband DF that responds primarily to return strokes in ground flashes. The magnetic field waveforms of return strokes are electronically separated from the waveforms of in-cloud processes and various non-lightning sources by measurement of the return stroke waveform's rise and fall times, peak structure, magnitude of opposite-polarity overshoot after the zero crossing, and field variation prior to the primary signal, and by comparison of those characteristics with the known characteristics of first and subsequent return strokes. Details of this waveform-discrimination procedure are found in Krider *et al* (1980).

The gated wideband DF system was originally designed to detect only negative ground flashes, but in the late 1980s it was modified to accept both negative and positive ground flashes. Since it is not known *a priori* whether a stroke to ground lowers positive or negative charge, there is an 180° ambiguity in stroke azimuth from the measurement of only the orthogonal magnetic fields. This ambiguity is resolved in all wideband DF systems by measurement of the associated electric field, whose polarity indicates the sign of the charge transferred to ground.

The random errors in gated wideband DFs are thought to be due primarily to superimposed noise preferentially appearing on one antenna output and to imperfect instrumental processing and digitizing of the two signals. Interestingly, although an upper frequency response of many megahertz is needed to assure accurate reproduction of the incoming radiation field peak, particularly if the propagation is over salt water, practical DFs need an upper frequency response of only a few hundred kilohertz in order to obtain an azimuthal error of about 1°. This is the case because the ratio of the peak signals in the two loops is insensitive to the identical distortion produced by the identical associated electronic circuits of the two loops. Thus the gated wideband DF can operate at frequencies below the AM radio band and below the frequencies of some aircraft navigational transmitters, either of which could otherwise cause unwanted directional noise.

Gated wideband DFs, like narrowband DFs, are susceptible to site errors. Site errors are a systematic function of direction but generally are time invariant. These errors are caused by the presence of unwanted magnetic fields due to non-flat terrain and to nearby conducting objects, such as underground and overhead power lines and structures, which are excited to radiate by the incoming lightning fields. In order to eliminate site errors completely, the area surrounding a DF must be flat and uniform, without significant conducting objects, including buried ones, nearby. These requirements are usually difficult to satisfy, so it is often easier to measure the DF site errors and to compensate for any that are found than to find a location exhibiting tolerably small site errors. The causes of site errors in gated wideband DFs and methods of correction are considered,

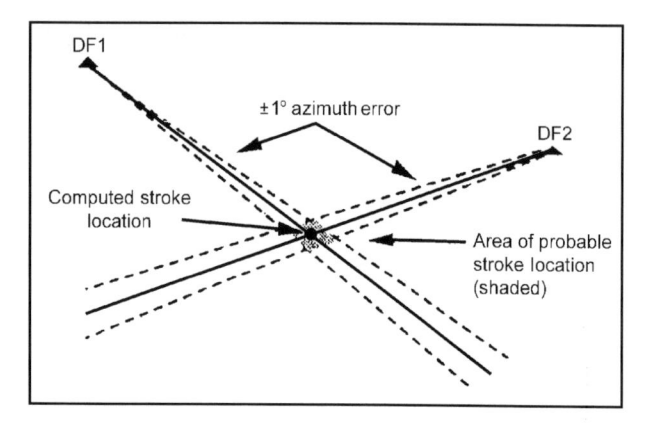

Fig. 17.1. Determination of lightning stroke location when only two direction finders (DFs) detect the stroke. The solid lines represent the measured azimuths to the stroke; the broken lines represent the $\pm 1°$ angular random error in the azimuth measurements. The solid circle indicates the computed stroke location; the shaded region indicates the uncertainty in location of the stroke. Adapted from Holle and Lopez (1993).

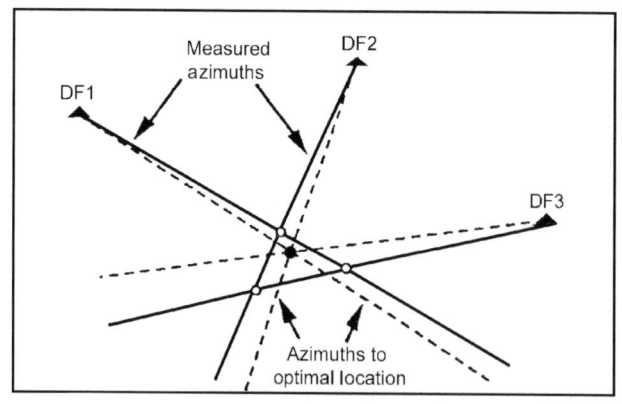

Fig. 17.2. Determination of lightning stroke location when three DFs detect the stroke. The solid lines represent the measured azimuths to the stroke. The open circles indicate the three possible locations defined by the three different intersections of the azimuth vectors. The optimal stroke location (solid circle) is determined by minimizing the χ^2 function. The broken lines show the azimuth vectors to the computed optimal location. Adapted from Holle and Lopez (1993).

for example, in Mach et al. (1986), Hiscox et al. (1984), Schütte et al. (1987b), Orville Jr (1987), Passi and Lopez (1989), Lopez and Passi (1991), and Petersen et al. (1996). Generally, in order to correct site errors, the data from a network of at least three DFs are recorded and analyzed for self-consistency, for an appreciable fraction of a storm season, to determine a site correction curve as a function of angle for each DF, the site-error corrections thereafter being made in the software. Once corrections are made, the residual errors have been reported (using independent optical data) to be usually less than two to three degrees (e.g., Mach et al. 1986).

As illustrated in Fig. 17.1 the intersection of two direction (azimuth) vectors, lines from the DF to the apparent source, provides a stroke location, but a location containing error because each azimuth vector has some random angular error and may have some systematic site error. If a three-DF system is employed, each pair of DFs provides a location, so there are three locations, the distance between the locations providing some measure of the system error, as illustrated in Fig. 17.2. For three or more DF responses to a return stroke, the optimal estimate of the location is best found using a χ^2 minimization technique.

Most of the existing DFs and DF networks are a commercial outgrowth of the original work of Krider et al. (1976). In the mid-1970s, E.P. Krider, M.A. Uman, and A.E. Pifer formed the company Lightning Location and Protection, Inc. (LLP) to develop commercial DFs as well as "position analyzers", computers that process the DF data to determine location, probable error, and various lightning stroke parameters. The first commercial gated wideband DF networks required operators at each station to record the DF vectors that were later combined to determine locations. Krider et al. (1980) described the commercial gated wideband DFs used in 1976 in Alaska for forest fire detection, where operators of individual DFs were in phone contact with each other to determine lightning location in virtually real time. Boulanger and Maier (1977) used two of the first commercial LLP direction finders to measure the location and frequency of cloud-to-ground lightning flashes in south Florida. The independently recorded direction vectors were correlated at the end of the summer season. Similarly, McDonald et al. (1979) used the peaks of the NS and EW magnetic field waveforms determined from research oscilloscope traces in 1976 at the Kennedy Space Center and in Gainesville, Florida, 200 km apart, to locate lightning over the Atlantic Ocean and over the Gulf of Mexico. The locations were found to be in good agreement with those of radar echoes.

Among the first journal papers describing LLP DF networks employing position analyzers, that is, producing the stroke locations automatically, were Krider et al. (1980), McGraw (1982), Rust et al. (1982), Orville et al. (1983), Maier et al. (1984), and Peckham et al. (1984). The data of Peckham et al. (1984) from the two-DF network operated in Tampa in 1979 are discussed in subsection 2.2.1 (see Table 2.1 and Fig. 2.1). These early networks used the ratio of the initial peak fields recorded at different stations, that is, the radiation field-amplitude method (subsection 17.2.2) to improve the locating accuracy, particularly when the flash was near a DF baseline, that is, the line joining two responding DFs, where direction finding

17.3. Magnetic field direction finding

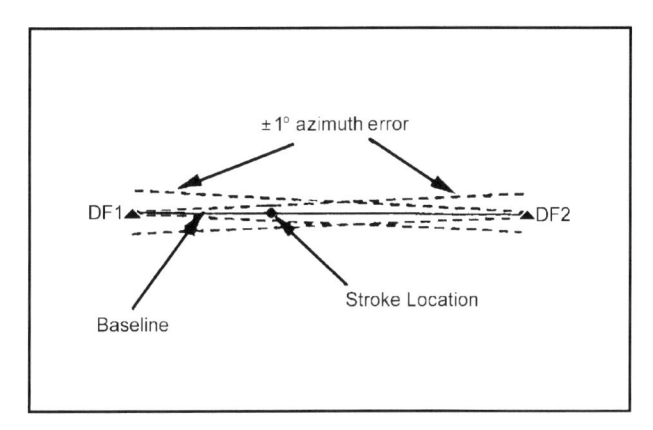

Fig. 17.3. Determination of lightning stroke location when only two DFs detect a stroke close to the baseline. Because of the large errors in using the intersection of azimuth vectors to determine the location in this case, the stroke is assumed to be on the baseline, and the ratio of the signal strengths is used to determine its position.

alone is characterized by relatively large errors, as illustrated in Fig. 17.3.

While LLP DF networks in the mid-1980s were typically composed of three stations communicating to the central position analyzer by dedicated telephone lines, larger networks were being developed. Starting in 1976 and continuing until 1996 when it was superseded by the US National Lightning Detection Network (NLDN), the US Bureau of Land Management (BLM) developed an LLP DF system covering 11 states in the western United States and Alaska for the purpose of advanced forest-fire prediction (Krider et al. 1980). Starting in 1979, researchers at the US National Severe Storm Laboratory (NSSL) developed a network covering Oklahoma and parts of adjacent states (Mach et al. 1986). In 1982–6, researchers at the State University of New York at Albany (SUNYA) established the US East Coast network (e.g., Orville et al. 1983; Orville et al. 1986; Orville and Songster 1987; Orville 1987, 1990b). By 1989, ground flashes in the contiguous United States were being monitored by an integrated system composed of the three large networks of DFs discussed above (Orville 1991a). This system and its further development are described in Section 17.5.

The important performance characteristic of DF networks, or any lightning-locating networks for that matter, are the detection efficiency (the percentage of the flashes or strokes occurring that are detected), the false alarm rate, and the locating accuracy. Such performance characteristics of DF networks have been studied by Mach et al. (1986), MacGorman and Rust (1988), Brook et al. (1989), MacGorman and Taylor 1989), Hojo et al. (1988, 1989), Lopez et al. (1992), and Diendorfer et al. (1998b). Most LLP DF networks with baselines of a few hundred kilometers have a ground flash detection efficiency of about 70 percent within a nominal range of approximately 400 km (e.g., Orville et al. 1987), although the detection efficiency may vary significantly with the position of the storm relative to the network. Small lightning events whose fields fall below the system thresholds are undetected as are strokes that do not meet the waveform-discrimination criteria listed earlier. False alarms are relatively rare since it is unlikely that two or more widely separated DFs will respond to the same non-lightning source and the discrimination criteria further decrease this number and filter out most but not all signals from cloud flashes. The location error for a typical LLP DF network with sensors separation of order 100 km is 2 to 4 km, although Maier (1991) estimated 8 to 10 km and a 60 percent detection efficiency for the NLDN in the vicinity of KSC. The KSC six-DF network with distances between sensors of order 10 km is characterized by a typical location error of 0.5 km and a flash detection efficiency of 90 percent (Maier and Wilson 1996). Detection efficiency apparently depends on polarity and season (Hojo et al. 1989), and on storm type (Peckham et al. 1984).

DF networks with two stations must necessarily compute stroke locations by the intersection technique illustrated in Figs. 17.1, 17.2, and 17.3, as must larger networks when only two DFs respond to a given event, but in the larger networks most locations are obtained by minimizing a χ^2 function:

$$\chi^2 = \sum_{i=1}^{N} \left(\frac{\theta_{mi} - \theta_i}{\sigma_{\theta i}} \right)^2 + \sum_{i=1}^{N} \left(\frac{E_{mi} - E_i}{\sigma_{Ei}} \right)^2 \quad (17.5)$$

where θ_i and E_i are the unknown azimuth and electric field peak values, θ_{mi} and E_{mi} are the azimuth and electric field peak measured at the ith station, and the σ's are the measurement error estimates. Note that beyond 5 km or so, where both electric and magnetic field initial peaks are dominated by their radiation components, the electric field peak can be found by multiplying the measured magnetic flux density peak by the speed of light. The values of unknowns found via minimization of the χ^2 function provide the most probable location and, additionally, allow estimation of the errors in this most probable location. The error estimates are usually represented by confidence ellipses within which there is a high, for example, 99 percent, probability that the stroke was actually located. Different measures of location errors have also been used (e.g., Murphy et al. 1996). The determination of confidence ellipses, assuming that the errors in the measured parameters obey a Gaussian distribution, was discussed by Stansfield (1947). This approach implies that the major systematic errors, such as site errors, are eliminated from the measurements, leaving the random errors, network geometry, and number of responding DFs to determine the size of the confidence ellipses.

LLP DF networks manufactured before about 1990 determined the location of only the first stroke in a ground flash, the polarity and peak current of that stroke, and the number of strokes per flash (called the "multiplicity"). The current peak was determined using the measured magnetic field peak and the linear field-peak–current-peak relationship predicted by the transmission line (TL) model (subsection 12.2.5). The proportionality coefficient between the current and field was initially found either (i) by using the results of calibration of the LLP magnetic field antenna in a Helmholz coil and the TL model or (ii) by equating the product of this coefficient and the median value of the peak field measured by the LLP system to the median peak current, 30 kA, reported by Berger *et al.* (1975) from direct measurements on towers (subsection 4.6.1). Later, for the NLDN discussed in Section 17.5, an equation based on the linear regression equations relating simultaneously measured peak currents and peak fields for triggered lightning was used (Orville 1991b; Idone *et al.* 1993). Since the early 1990s, the location, polarity, and peak current of all detected strokes in a flash as well as the time intervals between strokes, the confidence ellipses, and the χ^2 values have been the outputs of LLP systems.

When the US National Lightning Detection Network (Section 17.5) was formed in 1989, its sensors were LLP DFs only. Presently, the network also employs the time-of-arrival technique discussed below in Section 17.4. Surveys of literature describing various applications of data from the DF networks were published by Rakov (1993) and Holle and Lopez (1993).

17.4. Time-of-arrival technique

A single time-of-arrival sensor provides the time at which some portion of the lightning electromagnetic field signal arrives at the sensing antenna. Time-of-arrival systems for locating lightning can be divided into three general types: (i) very-short-baseline (tens to hundreds of meters), (ii) short-baseline (tens of kilometers), and (iii) long-baseline (hundreds to thousands of kilometers). Very-short- and short-baseline systems generally operate at VHF, that is, at frequencies from 30 to about 300 MHz, while long-baseline systems generally operate at VLF and LF, 3 to 300 kHz. It is generally thought that VHF radiation is associated with air breakdown processes, while VLF signals are due to current flow in already existing lightning channels. Short-baseline systems provide images of lightning channels and are used to study the spatial and temporal development of discharges. Long-baseline systems are usually employed to identify the ground strike point or the "average" location of the flash.

A general analysis of locating algorithms and the associated errors for time-of-arrival systems is given in Koshak and Solakiewicz (1996). Additional discussions of this subject are found in Hager and Wang (1995) and in many of the papers referenced in the following sections.

17.4.1. Very-short-baseline (tens to hundreds of meters) systems

A very-short-baseline system is composed of two or more VHF time-of-arrival (TOA) receivers whose spacing is such that the time difference between the arrival of an individual VHF pulse from lightning at those receivers is short compared with the time between pulses, which is some microseconds to hundreds of microseconds. The locus of all source points capable of producing a given time difference between two receivers is, in general, a hyperboloid, but if the receivers are very closely spaced then the hyperboloid degenerates, in the limit, into a plane on which the source is found. Two time differences from three very closely spaced receivers yield two planes whose intersection gives the direction to the source, that is, its azimuth and elevation. To find the source location, as opposed to the direction to the source, two or more sets of three closely spaced receivers, the sets being separated by tens of kilometers or more, must be used. Each set of receivers is basically a TOA direction finder, and the intersection of two or more direction vectors yields the location.

Oetzel and Pierce (1969) were the first to suggest that a very-short-baseline TOA technique could be used for line-of-sight locating of lightning VHF sources. They considered a three-antenna array for determining the direction to the source, the separation between antennas being 30 to 300 m, and used a frequency range from 30 to 100 MHz. Basically, with a very-short-baseline system, waveform identification is no problem since the same pulse arrives at each of the closely spaced receivers in a time that is short compared with the time between pulses and thus sequences of pulses arrive at each receiver in the same order. However, as noted above, a short-baseline system with three receivers produces azimuth and elevation only, that is, the direction to the source, not its location. Oetzel and Pierce mentioned the potential of a network of such "direction finders" for finding source locations, although this would involve providing accurate timing at two stations separated by tens of kilometers, which at the time would have been difficult and expensive to accomplish. Note that such time synchronization is now readily available using clocks synchronized to global positioning system (GPS) signals from Earth-orbiting satellites.

Cianos *et al.* (1972) successfully tested the TOA direction finding technique suggested by Oetzel and Pierce (1969) using one pair of antennas initially separated by 300 m and later by 122 m. The system operated at a central frequency of 30 MHz with a 10 MHz bandwidth. Murty and MacClement (1973) tested a similar system with antenna separation 115 m and center frequency 85 MHz, the bandwidth being 6 MHz. A modification of this system, in which a third antenna was added (antenna separations were of order 100 m) and the central frequency was changed to 69 MHz, was tested by MacClement and Murty (1978). Taylor (1978) employed a two-station system operating at frequencies from 20 to 80 MHz, the separation between the stations being about 18 km. Each station had a pair of horizontally spaced antennas to measure the azimuth angle and a pair of vertically spaced antennas to measure the elevation angle. The baseline of each pair was 13.74 m. Taylor (1978) gave examples of data recorded simultaneously at the two stations, but noted the difficulties of identifying the same events at those stations, owing to the lack of adequate techniques for synchronization (within 10 μs or so) between stations in the 1970s. Ray et al. (1987) used an upgraded version of Taylor's system with two sets of receivers separated by 42 km, time being synchronized between the two sites to within 1 ms during data acquisition. During analysis, the time was reportedly synchronized to within about 20 μs by comparing data from the two sites. Coincident events from the two sites (typically 20 to 30 percent of the signals detected at one site) provided three-dimensional locations. This VHF system was also described by Rust and MacGorman (1988).

17.4.2. Short-baseline (tens of kilometers) systems

Two types of short-baseline VHF TOA system have been extensively used since the early 1970s: (i) the 253 and 355 MHz systems developed in South Africa for research (Proctor 1971, 1976, 1981a, 1983, 1991; Proctor et al. 1988) and (ii) the lightning detection and ranging (LDAR) system developed at the Kennedy Space Center and operated at a central frequency between 56 and 75 MHz (Lennon and Poehler 1982; Maier et al. 1995). Further, Thomson et al. (1994) have designed and used at the Kennedy Space Center a short-baseline wideband TOA research system operating in the frequency range between 800 Hz and 4 MHz. Finally, a portable version of the LDAR has been developed by researchers at the New Mexico Institute of Mining and Technology (NMIMT) (Rison et al. 1999; Krehbiel et al. 2000; Thomas et al. 2000, 2001).

Proctor (1971, 1976, 1981a, 1983, 1991) and Proctor et al. (1988) employed the short-baseline time-of-arrival technique using five ground stations with baselines ranging from 10 to 40 km. The resultant lightning channel reconstructions have provided much valuable information on the development of lightning channels, particularly in the cloud (subsection 9.3.2). The system was operated at 253 MHz (Proctor 1981a) or 355 MHz (Proctor 1983; Proctor et al. 1988) with a 5 MHz bandwidth (the central station had a bandwidth of 10 MHz). The spatial resolution of the system used in these studies was of order 100 m. Errors in determining x and y coordinates were about 25 m, and errors in determining heights, z, were of order 100 m but approached 1 km when the x and y coordinates were greater than several kilometers or the height of the source was small. Recorded VHF-envelope pulse widths ranged from about 0.2 μs, the system limit, to about 2 μs. On average, a VHF location was obtained every 70 μs. Four stations were needed to produce an unambiguous source location. Basically, the location can be viewed as the intersection of three hyperboloids, each being the locus of all possible source points for a constant measured time difference between two stations. The fifth station was employed for redundancy, to confirm the adequacy of the four-station location. Locations were found via manual examination of the recorded data. Wideband electric field records were additionally obtained and proved to be very useful in interpreting the VHF source location data.

A similar short baseline system at the Kennedy Space Center, Florida was described by Lennon and Poehler (1982) and Maier et al. (1995). The system uses a network of seven receivers (six outlying stations and one central station) presently operating at 66 MHz with a 6 MHz bandwidth. The receivers are located in an area about 20 km in diameter and provide reasonable lightning surveillance out to a range of 100 km or more. The system operates in near real time and has a spatial resolution of about 100 m in its interior. At a pre-established signal level, the central station triggers the recording of about 82 μs of data at all seven stations. The time of occurrence of the largest peak within this time window (different time windows were used at different times) is determined for each station and corrected for the known time delay due to the microwave transmission link. The differences between the times of the largest peaks within the time window at different stations are used in an algorithm that automatically computes the three-dimensional source location. According to Starr et al. (1998), LDAR discards about 60 percent of detected events because it cannot find a self-consistent solution. Boccippio et al. (2001b), using total lightning observed from space by the LIS (subsection 2.5.3 and Section 17.8), estimated that LDAR flash detection efficiency remained above 90 percent out to 90 to 100 km range, and was below 25 percent at 20 km range. While the LDAR system was primarily designed for operational use, a number of research papers presenting results from this system have been published (e.g., Rustan et al. 1980; Uman et al. 1978; Lhermitte and Krehbiel 1979; Krehbiel 1981; Lhermitte and Williams 1985; Nisbet et al. 1990a, b; Mazur et al. 1997).

The NMIMT portable version of LDAR referred to earlier (Rison et al. 1999) takes advantage of the recent availability of (i) inexpensive, high-accuracy GPS technology to achieve the required timing accuracy and (ii) low-power serial communication systems to transmit the data to a central station for sensor data management and for real-time processing. Some observations of storms in central Oklahoma and in New Mexico using this TOA lightning-locating system were presented by Krehbiel et al. (2000) and Thomas et al. (2000, 2001).

The research TOA system deployed at the Kennedy Space Center (Thomson et al. 1994) detected the broadband time derivative of the electric field (dE/dt) at a central (master) station (800 Hz to 2 MHz) and at each of four peripheral (slave) stations (800 Hz to 4 MHz) located about 10 km from the central station. The signals from the peripheral stations were sent to the central station for digitizing and recording. The absolute time (trigger time) was provided with 1 μs resolution, and the relative time between stations was adjusted to within 400 ns from television synch signals. The relative times were further adjusted as part of the analysis to within better than 50 ns. Once the system was triggered, signals from each of the five stations were digitized by a processor at the central station at a sampling rate of 20 MHz in 204.8 μs segments. The system could record up to 25 data segments per flash, with 40 μs of dead time between consecutive segments (a total of about 6 ms per flash). Of course, the time intervals (gaps) between segments could be much longer, depending on the time intervals between the pulses that exceed the system's trigger threshold level.

The time of arrival of the signal to be located by the system of Thomson et al. (1994) was determined for three different features of the dE/dt waveform: (i) the rising-portion half peak, (ii) the peak, and (iii) the falling-portion half peak. The time of arrival used in the location algorithm was the mean of these three times. As noted earlier, measurements of signals from any four of the stations are sufficient to determine the source location and the time of occurrence, so signals from five stations provided an overdetermined set of equations. Two different optimization techniques were used to determine locations, both indicating comparable errors. The location error was reported to be less than 100 m for sources within the network. Some results from the system of Thomson et al. (1994) were reported by Davis (1999); they are discussed in subsections 4.7.5, 4.7.6, and 9.5.2.

17.4.3. Long-baseline (hundreds to thousands of kilometers) systems

The first long-baseline TOA system operated at VLF and LF. Lewis et al. (1960) used a pair of receiving stations in Massachusetts with a bandwidth of 4 to 45 kHz and separated by over 100 km (the overall network was composed of four stations) to compare differences in the times of arrival of the signals at each station and hence to determine directions to the causative lightning discharge in western Europe. The two-station system was basically a direction finder similar to the very-short-baseline systems described in subsection 17.4.1 but operating at lower frequencies and with a longer baseline. The resultant "directions" compared favorably with the locations reported by the British Meteorological Office's narrowband DF network (Section 17.3). For the long-range TOA direction finding employed by Lewis et al. (1960), spherical geometry was used to account for propagation over the Earth's surface in finding the locus of points for a constant measured arrival-time difference between receivers.

Lee (1986a, b, 1989a, 1990) described the replacement of the British Meteorological Office's narrowband magnetic direction finding network by a long-baseline VLF time-of-arrival system. The system has seven stations with separations ranging from 250 to 3300 km and operates in the 2 to 18 kHz frequency range. The stations are in the United Kingdom, Gibraltar, and Cyprus (Lee 1986b). The system's coverage area extends from 30° N to 70° N and from 40° W to 40° E, the stated flash-location error is 2 to 20 km, and the detection efficiency is limited by the system's capability to locate flashes at a rate not exceeding about 400 per hour (Lee 1986b). Thus the system is useful in practice only for the detection of storm areas. No attempt is made to distinguish between cloud and ground flashes.

A commercial long-baseline TOA system called the lightning positioning and tracking system (LPATS) was developed in the 1980s by Atlantic Scientific Corporation, which later became Atmospheric Research Systems, Inc. (ARSI). The LPATS, operating at LF and VLF, uses electric field whip antennas at four or more stations 200 to 400 km apart to determine locations via the measured differences between signal arrival times at the stations. In the frequency band used, the return-stroke waveforms are generally the largest and hence the most easily identified. Early versions of LPATS were synchronized by measuring the Loran-C or other Earth-based standard timing signal at each of the individual stations, while later versions used the global positioning system (GPS) to synchronize station clocks. Unlike the LLP gated wideband DF system, LPATS, at least in its early versions, did not employ any method of discrimination against cloud discharges and non-lightning sources. Further, it neither distinguished between first and subsequent strokes nor combined strokes into flashes. The relatively small magnitude of the signals from cloud discharges served as some discriminator against them, although Ishii et al. (1991) found that about 20 percent of the reported negative ground strokes within 200 km were actually cloud discharges, a percentage decreasing to less than five percent for negative ground strokes beyond 400 km. Ishii et al. also reported that a greater percentage of misidentified cloud-flash pulses were contained in LPATS-reported positive ground strokes than in negative ground strokes. Later versions of LPATS have interpreted signals as due to either a ground stroke or a cloud discharge depending on whether

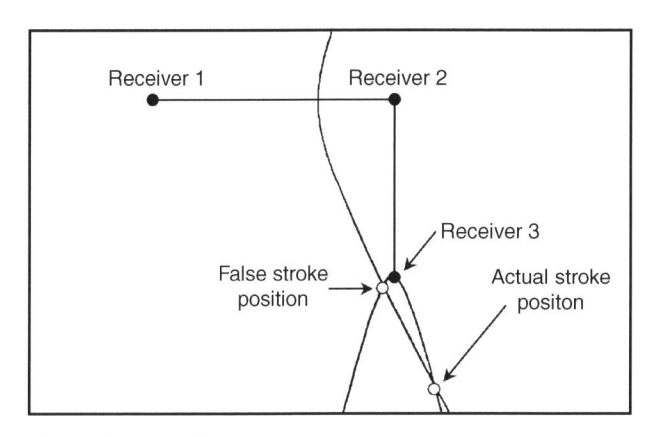

Fig. 17.4. Determination of lightning stroke location by three TOA receivers when the solution is not unique. Shown are two hyperbolas, defined by the TOA differences, that intersect at two points (open circles); one point corresponds to the actual stroke position and the other is a false solution. Adapted from Holle and Lopez (1993).

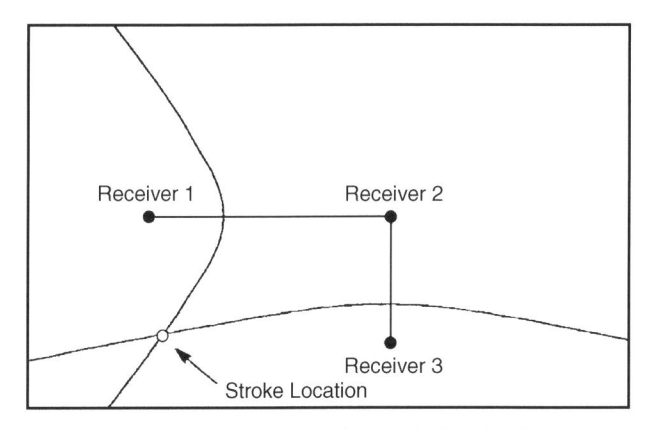

Fig. 17.5. Determination of lightning stroke location by three TOA receivers when the solution is unique. Shown are two hyperbolas, defined by the TOA differences, that intersect to define the unique location of the stroke (open circle). Adapted from Holle and Lopez (1993).

the waveform width did or did not exceed 10 μs or so. Pulses wider than 10 μs have been attributed to ground strokes.

In principle, responses from four stations (three time differences) are needed to produce a unique location since the hyperbolae on the earth's surface from only two time differences can, in general, intersect at two different points, as illustrated in Fig. 17.4. For cloud-to-ground lightning near or within the network, there is often only one solution, as illustrated in Fig. 17.5, so in this case the three-station solution suffices. Early versions of LPATS required a few four-station solutions to identify the storm region and then used the three-station solutions in that region.

LPATS networks, including a US national network established in the late 1980s, are described by Lyons *et al.* (1989), Rakov (1990), Ishii *et al.* (1991), and Casper and Bent (1992). From this literature it appears that the detection efficiency for ground flashes in early versions of LPATS was limited by the data communication bandwidth to an average of about 50 percent. Errors in locations are caused by anything that changes the arrival time of an ideal signal, such as (i) the identification of different parts of the received waveform at different stations, owing to signal distortion in propagation or other effects, (ii) path elongation due to mountains, and (iii) inadequate time synchronization between stations. The problems in early versions of LPATS were mostly ameliorated in later versions with the adoption of GPS timing. In the latest version, location errors can be expected to be less than 1 km. The results from LPATS systems are described by Montandon *et al.* (1992), Pinto *et al.* (1992b, 1996, 1999a, b), Redelinghuys *et al.* (1996), Lyons and Keen (1994), Roohr and Vonder Haar (1994), and Watson *et al.* (1995).

In the early 1990s, ARSI and LLP, along with Geomet Data Services, Inc. (GDS), the latter being a commercial company distributing LLP lightning data, were merged to form Global Atmospherics, Inc. (GAI). At about that time, LLP had incorporated the measurement of arrival time in its gated wideband DFs via a GPS clock, the resulting device being termed an IMPACT (improved accuracy from combined technology) sensor. As discussed in Section 17.5, the commercial US National Lightning Detection Network (NLDN) was originally based only on DF technology, but it was superseded in 1995 by a commercial nationwide network incorporating both DF and TOA technologies, still called the NLDN.

17.5. The US National Lightning Detection Network

The origin of the present NLDN can be traced to 1987 when data from the BLM, NSSL, and SUNYA regional networks of LLP gated wideband magnetic direction finders described in Section 17.3 were combined on an experimental basis by researchers at SUNYA to provide national coverage. The NLDN DF network began real-time operation in 1989 (Orville 1991a). At about the same time an independent commercial nationwide network of ARSI time-of-arrival sensors was also being installed (Lyons *et al.* 1989; Casper and Bent 1992). The ARSI network was composed of 11 regional clusters having six TOA sensors each, the sensors typically being separated by 200 to 400 km. By 1991 there was sufficient interest in the nationwide lightning data for Geomet Data Services, Inc. (GDS) to be formed, to disseminate the lightning data from the nationwide network of DFs. As noted earlier, in the mid-1990s LLP, GDS, and ARSI combined to form Global Atmospherics, Inc. (GAI). There followed a significant renovation of the NLDN to optimize the use of the two major locating technologies, direction finding and time-of-arrival, thus

Fig. 17.6. Location of the sensors in the US National Lightning Detection Network in 2001. The triangles denote LLP IMPACT sensors, which combine gated wideband magnetic direction finding with the time-of-arrival technique. The circles denote ARSI TOA sensors. Courtesy of Global Atmospherics, Inc.

providing increased detection efficiency and almost an order of magnitude improvement in locating accuracy. Data are presently available from the NLDN within 40 s of occurrence, and available archivally back to 1989 in a version corrected for identified deficiencies in the real-time data. A summary of the cloud-to-ground lightning activity reported by the NLDN for the Blizzard of 93 is given in Fig. 2.3. The complete status of the NLDN as of 1998 is described in Cummins et al. (1998a, b). Since 1998, the NLDN and the similar Canadian Lightning Detection Network (CLDN) have been combined to form the North American Lightning Detection Network (NALDN), providing continental-scale coverage (Cummins 2000a).

After the 1994 upgrade (which was completed in 1995), the NLDN contained 106 sensors, 47 LLP IMPACT sensors and 59 ARSI LPATS sensors, versus over 130 DFs before the upgrade. All sensors use GPS clocks. Data from each sensor are sent via satellite to a central station where locations and other parameters are calculated and forwarded by satellite links and phone lines to users of the data. Both LLP and ARSI sensors were modified from their previous versions as part of the 1994 network upgrade. The IMPACT gated wideband DFs with timing had their gains increased, trigger thresholds reduced, and waveform acceptance criteria changed (apparently narrower waveforms were accepted than previously) to allow detection of lower peak currents and more distant lightning. The LPATS time-of-arrival sensors had their gains reduced and had waveform acceptance criteria implemented similar to those used in the IMPACT sensors, in order to reduce the undesirable cloud-discharge triggers and to make the sensitivity of the LPATS sensors to ground flashes similar to that of the IMPACT sensors. With these changes, the calibration factors were adjusted so that the derived peak current distributions remained essentially the same (Cummins et al. 1998a). At the time of writing, the total number of NLDN sensors is 106, and the typical distance between sensors is about 300 km. The locations of the NLDN sensors and their type in 2001 (44 IMPACT sensors and 62 LPATS sensors) are shown in Fig. 17.6.

Generally, data from the NLDN sensors are used to produce an optimum lightning location using the χ^2 minimization technique described by Hiscox et al. (1984) and Cummins et al. (1993). The χ^2 to be minimized is given by

$$\chi^2 = \sum_{i=1}^{N1} \left(\frac{\theta_{mi} - \theta_i}{\sigma_{\theta i}} \right)^2 + \sum_{i=1}^{N2} \left(\frac{t_{mi} - t_0}{\sigma_{ti}} \right)^2 + \sum_{i=1}^{N2} \left(\frac{E_{mi} - E_i}{\sigma_{Ei}} \right)^2 \quad (17.6)$$

where t_0 is the unknown time of the event, t_{mi} is the measured time of arrival at the ith station minus the time required for the signal to propagate from the source to the station, σ_{ti} is the expected error in the time measurement, and the other parameters are the same as in Eq. 17.5. The summation in the first term of Eq. 17.6 is over the IMPACT sensors (total number $N1$) reporting azimuth angle θ_{mi}, while the summation in the second and third terms is over both IMPACT and LPATS sensors (total number $N2$) reporting time t_{mi} and electric field peak E_{mi}. For the IMPACT sensors, E_{mi} is computed from the measured magnetic field peak assuming that the field peaks are essentially

17.5. The US National Lightning Detection Network

determined by their radiation components. It appears that only the first two terms on the right-hand side of Eq. 17.6 are routinely used for lightning locating in the NLDN (e.g., Cummins et al. 1998a). The timing error is assumed to be about 1.5 μs and the azimuthal error is assumed to be about 0.9° according to Cummins et al. (1998b). The expected error in field measurement σ_{Ei} is assumed to be 10 percent of E_{mi} (K. Cummins, personal communication, 2000b). From analysis of the NLDN output, it has been determined that strokes having a peak current below 5 kA are usually not detected because their signals fall below the trigger threshold of multiple sensors, whereas 100 kA strokes are typically detected by 20 or more sensors. A common first return-stroke current of 25 kA is detected by six to eight sensors. Because of the effective threshold near 5 kA, a significant fraction of the subsequent strokes in a multiple-stroke flash is missed.

Prior to 1992, average location errors in the NLDN were theoretically and experimentally determined to be 8 to 16 km, often being due to inadequately corrected site errors (Cummins et al. 1998a, b). In early 1992, with such corrections done properly, average location errors were experimentally shown to be 2 to 4 km in the vicinity of the Kennedy Space Center (Cummins et al. 1992) and predicted to be the same in the rest of the country. Presently, average location error is defined as the semi-major axis of the 50 percent confidence ellipse surrounding the optimum location obtained in minimizing the χ^2 function. By definition, provided that there are no significant systematic errors and that random errors obey a normal distribution, there is a 50 percent chance that the located stroke is actually within the 50 percent confidence ellipse and a 50 percent chance that it is outside this ellipse. Since 1995, and with this new definition of location error, the median location errors are 500 m to 1 km by both calculation and measurement, as discussed in the next paragraph. Ninety-nine percent confidence ellipses are about twice the size of the 50 percent ellipses and give the area within which there is a 99 percent chance for the stroke to have occurred.

Locating accuracy has been determined from comparing NLDN locations with (i) those of rocket-triggered lightning (Section 7.2) at Camp Blanding (Cummins et al. 1998a), (ii) locations from the NASA Kennedy Space Center network, which has been shown to have a 0.6 km average location accuracy (Maier and Wilson 1996), and (iii) locations provided by multiple video cameras near Albany, New York, both before and after the 1995 upgrade (Idone et al. 1998 a, b). The results of these comparisons are consistent with the model-based prediction that for most of the continental United States the median error is 500 m (Cummins et al. 1998a). Only those events for which the semi-major axis of the error ellipse is less than 50 km and the χ^2 value is less than 15 are accepted by the system.

The NLDN detection efficiency for flashes with first-stroke peak currents greater than 5 kA was about 70 percent prior to 1992 and 65 to 80 percent between 1992 and 1995. After the 1994 upgrade the expected detection efficiency is 80 to 90 percent for strokes with peak currents higher than 5 kA (Cummins et al. 1998a, b; Idone et al. 1998a). A 5 kA stroke is assumed to correspond to an initial electric field peak of 1.5 V m^{-1} normalized to 100 km (Cummins et al. 1998b). According to the formula of Rakov et al. (1992), Eq. 4.9, this field peak yields a current peak of 4 kA. Note that the detection efficiency falls off rapidly outside the perimeter of the network.

In addition to locating individual strokes, the NLDN groups the strokes into flashes and determines polarity and a peak current estimate for each stroke, using the measured electric and magnetic radiation field peaks, respectively. Before the 1994 upgrade, the number of strokes in a flash (the multiplicity) was defined as the maximum number of strokes seen by any responding DF within 2.5° of the first stroke and within 1 s after the first stroke. In the upgraded NLDN, strokes are assigned to a given flash if they occur within 10 km of the first stroke (apparently based on the observations of Thottappillil et al. 1992) and within a time interval from the previous stroke of 500 ms, the maximum flash duration still being 1 s. Additionally, a stroke is included in the flash if it is located within 10 to 50 km of the first stroke and the location error ellipses (50 percent) of these two strokes overlap. The maximum allowed multiplicity is 15, the 16th stroke being treated as the first stroke in a new flash. Note that the percentage of flashes that have more than 15 strokes has been observed to be 2.6 percent in Florida and 4.8 percent in New Mexico (Fig. 4.5). Since many small subsequent strokes fall below the sensor threshold, the multiplicity reported by the NLDN is likely to be an underestimate (e.g., Rakov et al. 1994).

Note that the magnitudes of subsequent strokes reported by the NLDN are not much different from their first-stroke counterparts, while independent measurements typically give a ratio in the range 1 : 2 to 1 : 3 (subsection 4.6.1). This apparent discrepancy can be explained partially by the poor detection of relatively small subsequent strokes. Another possible reason is rejection of the actual first stroke by the waveform discrimination algorithm and acceptance of the second stroke as a first stroke.

In order to translate the peak radiation (electric or magnetic) field into a peak current, a formula based on the linear regression equation relating the NLDN-measured field peak to the directly measured current peak for triggered-lightning strokes (Orville 1991b; Idone et al. 1993) is employed. This formula may be viewed as equivalent to Eq. 12.4, derived for the transmission-line (TL) model of the lightning return stroke. The field peak used in the formula is the range-normalized signal strength (RNSS) in so-called LLP units averaged over all sensors that reported the event within 625 km. The range-normalization procedure is based on the assumption of a power-law distance dependence with an exponent equal to -1.13, which is presently taken to be the same for all sensors, although it should vary from sensor to sensor and may depend on the direction to the source from a given station (Diendorfer et al. 1998a; Mair et al. 1998). An absolute value of the exponent greater than unity is used to model signal attenuation due to propagation over lossy ground. Propagation effects as applied to the operation of lightning locating systems have been discussed by Schütte et al. (1988), Herodotou (1990), Herodotou et al. (1993), Idone et al. (1993), and Cummins et al. (1998a). The conversion from range-normalized peak field to peak current prior to 1996 was made using the formula (a linear regression equation) proposed by Idone et al. (1993): $I = 5.20 + 0.148 RNSS$. Since 1996, a modification of this formula in which the intercept is forced to be zero (in order to satisfy the condition that zero current corresponds to zero field) (Cummins et al. 1998a) has been used: $I = 0.185 RNSS$. Note that forcing the intercept (5.2 kA in the original equation) to be zero changes the slope from the original 0.148 to 0.185 kA/LLP units, this change introducing an error with respect to the original, best-fit, equation. According to Cummins et al. (1998a), the NLDN has a median error in the current estimation of 20 to 30 percent, the larger errors being associated with the smaller currents. This current error estimate is based on a comparison of NLDN-reported currents with the measured currents of negative rocket-triggered lightning strokes. No "ground truth" data for the NLDN current estimates exist for (i) first strokes in natural lightning, (ii) natural positive strokes (Orville 1999), or (iii) triggered-lightning peak currents exceeding 60 kA, triggered-lightning strokes probably being representative of natural subsequent strokes. It follows that the NLDN current estimates should be viewed with caution for strokes other than negative subsequent strokes.

The effects of the 1994 NLDN upgrade on network-reported lightning characteristics have been discussed by Wacker and Orville (1996, 1999a, b) and Huffines and Orville (1999). The upgraded NLDN has recorded a new class of small positive events not detected before the upgrade and identifies these as positive ground strokes with peak currents generally less than 10 kA (Cummins et al. 1998a). The number of events identified by the NLDN as positive flashes after the upgrade (in 1995–6) increased by a factor 2–4 (Huffines and Orville 1999). Because of the increased sensitivity of the upgraded system, it is likely that most of these small positive events are actually cloud discharges that were not recorded by the NLDN before the 1994 upgrade (Cummins et al. 1998a). The effects of the NLDN instrumentation and the network configuration on the NLDN-reported data were discussed by Orville and Huffines (1999).

There are lightning-locating networks similar to the NLDN in nearly 40 countries (K. Cummins, personal communication, 2000b), including Canada (e.g., Janischewskyj and Chisholm 1992; Cummins 2000a), Sweden (e.g., Schütte et al. 1988), Austria (e.g., Diendorfer et al. 1998b), France (e.g., Le Boulch and Plantier 1990), Japan (e.g., Hojo et al. 1989; Honma et al. 1998; Shindo and Yokoyama 1998), and Brazil (e.g., Pinto et al. 1999a, b). The Canadian Lightning Detection Network (CLDN) and the NLDN have been combined since 1998, as noted earlier, and the resultant North American Lightning Detection Network (NALDN) employs 187 sensors and monitors 20 million square kilometers (Cummins 2000a). A bibliography on real-time lightning detection networks worldwide, including network instrumentation, applications of lightning data, and network performance, is maintained on the Web (www.nssl.noaa.gov/~holle/) by R.L. Holle.

17.6. Interferometry

In addition to radiating isolated pulses, lightning also produces noise-like bursts of electromagnetic radiation lasting for tens to hundreds of microseconds (e.g., Richard et al. 1986; Rhodes et al. 1994). These bursts are hard to locate using TOA techniques, owing to the difficulty in identifying individual pulses. In the case of interferometry, no identification of individual pulses is needed, since the interferometer measures the phase difference between narrow-band signals corresponding to these noise-like bursts received by two or more closely spaced sensors. The simplest lightning interferometer consists of two antennas some meters apart, each antenna being connected via a narrowband filter to a receiver. The antennas, filters, and receivers are identical. The outputs of the two receivers are sent to a phase detector that produces a voltage proportional to the difference in phase between the two quasi-sinusoidal signals. Like the time difference in very-short-baseline TOA systems, the phase difference defines a plane on which the source is located, that is, one direction angle to the VHF source. To find the azimuth and elevation of a source, three receiving antennas with two orthogonal baselines are needed at least. To locate the source in three dimensions, two or more synchronized interferometers, each effectively acting as a direction finder and separated by a distance of the order of 10 km, are needed.

The first interferometer for studying lightning was designed by Warwick et al. (1979). Hayenga and Warwick (1981) described a version of that instrument which operated at 34.3 MHz with a 3.4 MHz bandwidth. The interferometer had two pairs of sensors, each pair separated by about 15 m, that is, by two wavelengths (2λ) of the 34.3 MHz signal, the two baselines being perpendicular to each other.

17.6. Interferometry

There was no second site so only azimuth and elevation were determined. The phase difference data were recorded on magnetic tape. Angular errors were estimated as being a few degrees. The azimuth and elevation values were obtained as 2.5 μs averages over the phase-difference data. Since the time-domain pulses are "destroyed" by the interferometer circuitry, it is generally necessary to use time-domain electric field measuring equipment synchronized with the interferometer in order to identify the physical processes that are being located.

Advanced versions of the interferometric system described in the previous paragraph that operate at 274 MHz with a 6 MHz bandwidth were used by the New Mexico Institute of Mining and Technology (NMIMT) research group (Rhodes et al. 1994; Shao et al. 1995; and Shao and Krehbiel 1996). In each version the interferometer included five antennas arranged to form two sets of receivers along each of two orthogonal horizontal directions (a total of four interferometric pairs). Rhodes et al. (1994) and Shao et al. (1995) used baselines of 4λ and $\lambda/2$ along each of the orthogonal directions and Shao and Krehbiel (1996) used 4.5λ and λ, where λ is the wavelength corresponding to the center frequency at which the system operates, $c/(2.74 \times 10^8)$ or about 1 m. Next, we explain the reason for the use of two baselines. In principle, if a given baseline is longer than $\lambda/2$ then the phase difference cannot be determined uniquely because the output direction angle of the interferometer is a trigonometric function of the phase difference and hence has a cycle of 2π radians. These cycles in the phase difference are often referred to as "fringes", in analogy to the visible fringe in optical interferometers. For a given baseline d, there are $2d/\lambda$ fringes. Shorter baselines, while decreasing the fringe ambiguities, increase the phase measurement error and therefore the angular error. Thus, narrowband interferometers are often designed to include two different baselines, a longer one equal to several times the operating wavelength and a shorter one equal to half or one wavelength; the long baseline provides accurate but ambiguous measurements and the short baseline provides coarse but unambiguous, or less ambiguous, measurements used to resolve the long-baseline ambiguities.

The interferometers used by Rhodes et al. (1994), Shao et al. (1995), and Shao and Krehbiel (1996) averaged over a running window of 1 μs and sampled the output each 1 μs. They provided two-dimensional locations (azimuth and elevation) with 1 μs time resolution and an angular resolution of a few degrees. Angular uncertainties due to random phase measurement errors were about 1° at high elevation angles. The elevation error rapidly increase with decreasing elevation angles. Rhodes et al. and Shao et al. used a computer display to remove, when possible, fringe ambiguities manually from the recorded data, while Shao and Krehbiel's choice of baselines reportedly allowed all the fringe ambiguities to be removed automatically. Electric field changes and the amplitude of the RF radiation signal were obtained from separate measuring systems. The interferometer locates the centroid of the lightning radiation as a function of time, which is meaningful only when the source is localized or, in the presence of many spatially separate and simultaneously active sources, when one source is clearly dominant. At least in some cases, neither of these two conditions is satisfied, for example, during the development of a stepped leader (Rhodes et al. 1994). Mazur et al. (1998) reported that the NMIMT interferometer failed to resolve simultaneously developing branches and to detect "fast negative leaders" in the cloud occurring during a "spider" discharge characterized by extensive horizontal channels near the cloud base. Inferences on lightning processes derived by the NMIMT group from interferometric measurements are discussed in subsections 4.3.1, 4.9.5, and 9.3.1 (see also Fig. 4.16).

Shao et al. (1996) described an interferometer, operating at frequencies from 40 to 350 MHz, which has only a single pair of antennas. Such a broad-band system determines phase differences at different frequencies and therefore is equivalent to a narrowband interferometer having multiple baselines (ratios of baseline to wavelength). Lower frequencies correspond to shorter baselines and higher frequencies to longer baselines. Like narrowband systems, the broad-band interferometer operates on the assumption that the radiation which is being processed for the angle of its arrival originates from a single localized source. A broadband interferometer operating at frequencies from 25 to 250 MHz has been described by Ushio et al. (1997), Mardiana et al. (1998), Mardiana and Kawasaki (2000a), and Kawasaki et al. (2000).

Richard and Auffray (1985) and Richard et al. (1986) described the operating principles of a 300 MHz lightning interferometer with a 10 MHz bandwidth, first used for research in 1981 in the Ivory Coast, Africa and then in New Mexico in 1982. A single-site measurement is obtained from two interferometric antenna pairs within a three-antenna system: a 10-wavelength baseline for an accurate but ambiguous location of sources and a short-baseline system for ambiguity removal, its baseline being one wavelength in 1981 and a half-wavelength in 1982. At the center frequency of 300 MHz, the baselines are 1 m or 50 cm long for the small system and 10 m long for the large system. The time resolution is 1.6 μs. Richard et al. (1986, 1988), and Richard (1990, 1992) described a commercial lightning interferometer called SAFIR (surveillance et alerte foudre par interférométrie radioélectrique) derived from the French research interferometer described above. A typical SAFIR system has three interferometric stations separated by 10 to 100 km. Each station is basically a VHF direction finder. To map lightning sources in three dimensions the separation between stations is near 10 km so as to provide sufficient vertical resolution. If the stations are separated by 100 km or so then the vertical resolution deteriorates and two-dimensional plan-view maps are produced.

The SAFIR system operates at a center frequency between 110 and 118 MHz with a 1 MHz bandwidth. The center frequency is adjustable to avoid interference from local noise sources. For three-dimensional locations, each station has one array of eight horizontally separated antennas for the azimuth and another independent array of 16 horizontally and vertically separated antennas for the elevation. For systems that map the plan view of lightning, a station requires only a single horizontal array to determine the azimuth from that station. Within the perimeter of a 100-km-baseline network, typical location errors are reported to be about 2 km. The central processor that determines the intersection of the vectors from the individual interferometric stations is apparently capable of determining locations at a rate up to 4×10^4 s^{-1}. Kawasaki et al. (1994) evaluated the performance of the SAFIR system installed in Japan, comparing its output with corresponding meteorological-radar-echo patterns.

Mazur et al. (1997) compared the output of a three-dimensional SAFIR system with that of the LDAR (subsection 17.4.2), that is, the outputs of a VHF interferometric system and a VHF TOA system, and found significant differences between the locations and propagation speeds reported by the two systems. The LDAR detected sources more or less continuously through a flash while the SAFIR detected sources sporadically. LDAR sources extended higher than SAFIR sources, but SAFIR provided more detailed source sequences during dart leaders and return strokes. Lightning development recorded by SAFIR had speeds of 10^7 to 10^8 m s^{-1} while LDAR-imaged channels extended at 10^4 to 10^5 m s^{-1}. Richard et al. (1986) had previously noted that the stepped-leader and dart-leader sources detected with their interferometric system moved faster than those typically observed with optical techniques. Apparently, SAFIR and LDAR observe different aspects of the complex lightning discharge, perhaps neither giving a complete picture. As noted earlier, TOA systems are best suited for locating sources of impulsive radiation, while interferometric systems deal best with longer-duration noise-like radiation bursts.

17.7. Ground-based optical direction finding

Lightning locating using lightning optical signals has many similarities to methods employing radio-frequency signals. Often, however, the use of optical systems is severely limited because of channel obscuration by topography, precipitation, clouds, trees, and buildings. In addition to standard cameras using photographic film, two classes of optical sensors have been used to detect and to determine the direction to lightning, television cameras and photoelectric detectors. Two or more optical detectors can be used to find the position of the lightning by triangulation. Kidder (1973) described a system employing several photographic cameras to give bearings and the subsequent triangulation to locate the discharges. Idone et al. (1998a, b) used a network of video cameras in the Albany, New York area to locate the ground-strike points of flashes to Earth in order to obtain ground-truth data for testing the NLDN. Standard television cameras and associated videotape recorders can be used not only to locate lightning but also to measure its properties on a 16.7 ms time scale, the standard TV-field rate. A number of investigators (e.g., Winn et al. 1973; Brantley et al. 1975; Clifton and Hill 1980; Thottappillil et al. 1992; Rakov et al. 1994) have used television camera records to determine a variety of lightning properties such as the number of strokes per flash, the number of separate channels to ground per flash, interstroke intervals, and flash duration (Section 4.2). The television technique has the advantage over photography of not missing portions of the event due to shutter action and of having relatively high optical sensitivity (e.g., Clifton and Hill 1980), but it is limited in time resolution to about 1 ms in even the latest high-speed TV cameras.

17.8. Detection from satellites

With the advent of Earth-orbiting satellites, it has become possible, in principle, to chart systematically worldwide lightning activity from Earth orbit by detecting the light or radio-frequency signals emitted in the upward direction by both cloud and cloud-to-ground discharges. A number of satellites have contained optical sensors capable of detecting lightning activity, some with this as their primary objective and others being aimed at the Earth for different reasons. The satellites used to date have recorded only a small fraction of the discharges occurring because these satellites were in relatively low orbit and hence spent a relatively short time over any given storm. Further, published data obtained prior to the 1990s involved lightning that occurred only at dusk, dawn, and midnight, and trigger thresholds were such as to cause the smaller events to be missed. Nevertheless, it has been possible to estimate total flash densities and to determine ratios of activity in different geographical locations and in different seasons (see subsection 2.5.3). It has not been possible to distinguish between the optical signals of cloud and cloud-to-ground discharges. Orville and Spencer (1979) and Orville (1981) used photographic data from two satellites in the Defense Meteorological Satellite Program (DMSP) orbiting at 830 km altitude to study lightning activity between 60° S and 60° N with about 100 km resolution at local dusk and midnight. Turman and Edgar (1982), also using the two

17.8. Detection from satellites

DMSP satellites, found that their optical detectors recorded about two percent of the lightning within their fields of view. Turman (1978) used an array of 12 photodiodes, each having a field of view of 700 km by 700 km, flown on a DMSP satellite to observe 10 000 flashes from 24 storm complexes during 15 orbits in September 1974 and March 1975. Some results from the three experiments discussed above are found in subsection 2.5.3.

During the 1990s NASA researchers developed a lightning mapper designed for geostationary orbit. Characteristics of this sensor were given by Davis M.H. et al. (1983) and Christian et al. (1989a), as outlined next. The sensor is a CCD (charge-coupled device) optical array with electronics capable of detecting transient luminosity from lightning, even during the day. It is designed to detect lightning from geostationary altitudes with a spatial resolution of 10 km and with a temporal resolution of 1 ms. This sensor has been packaged with lenses to provide coverage of much of North America, including all of the contiguous United States and the nearby ocean areas, Central America, South America, and the intertropical convergence zone. The system is designed to detect 90 percent of the flashes that occur. Such a system would enable mapping of both cloud and ground lightning indistinguishably and continuously on continental scales. The first geostationary lightning mapper covering much of North and South America is expected to be launched in the 2002–5 time frame.

Optical detectors derived from the design of the geostationary mapper have been flown on two satellites in low Earth orbits, where they observed swaths of the Earth as they circled it. The optical transient detector (OTD), also discussed in subsection 2.5.3, was launched on the Microlab-1 (recently renamed OV-1) satellite in 1995 into an Earth orbit of 735 km altitude with an inclination of 70° with respect to the equator, a near polar orbit. The OTD operated for five years and stopped sending data in April 2000. It had an 100° field of view and hence observed 1300×1300 km^2, about 1/300 of the Earth's surface at any instant, orbiting the Earth in 100 minutes with a nominal spatial resolution of about 10 km and a nominal time resolution of 2 ms (Christian et al. 1992, 1996).

Boccippio et al. (2000b) estimated that the OTD had, on average, about 20–40 km spatial error and less than 100 ms temporal error. They also reported, from a comparison of OTD and NLDN outputs, that the OTD detection efficiency for ground flashes was about 46 to 69 percent and that it was likely to be slightly higher for cloud flashes. Because of its orbit, the OTD never observed a given location for more than a few minutes per day. Therefore, data averaging over 55 or 110 days was required to remove the diurnal lightning-cycle bias (Boccippio et al. 2000b). Data from the OTD are found on the Web site http://thunder.msfc.nasa.gov/otd.html. Flash density maps for July–August 1995 and January–February 1996 are found in Christian and Latham (1998). A global lightning flash density map based on data from two satellite detectors, OTD (five years) and LIS (three years), the latter being discussed next, is shown in Fig. 2.12. The second orbital lightning mapper, also discussed in subsection 2.5.3, is the lightning imaging sensor (LIS), which was launched aboard the Tropical Rainfall Measuring Mission (TRMM) observatory in 1997. It observes a 600×600 km^2 region with a resolution of the size of the storm that is potentially less than 10 km. A given point on the Earth is observed for almost 90 seconds as the TRMM satellite moves at 7 km s^{-1}. The TRMM observatory orbit has an inclination of 35° so that LIS can observe lightning between latitudes 35° S and 35° N. Its estimated flash detection efficiency is near 90 percent. Thomas et al. (2000) compared lightning detected by LIS and by a VHF TOA lightning imaging system. Out of 101 flashes that extended above 7 km altitude, 99 were detected by LIS (eight of them only in the raw data). Out of 27 flashes that were confined below 7 km, only nine were detected (five in the raw data); the NLDN located 16 out of the 27 flashes. Data from the LIS are found on the website http://thunder.msfc.nasa.gov/lis.html (see also Fig. 2.12).

Kotaki et al. (1981a, b) and Kotaki and Katoh (1983) reported on lightning detected by the Japanese Ionospheric Sounding Satellite ISS-b, which sensed HF radiation at 2.5, 5, 10, and 25 MHz. They presented worldwide lightning maps derived from the HF data for two years. Some results from these experiments are presented in subsection 2.5.3.

Lightning RF emissions in the VHF band (30 to 300 MHz) have been observed by the Blackbeard receiver aboard the Alexis satellite (Holden et al. 1995; Massey and Holden 1995; Massey et al. 1998; Zuelsdorf et al. 1997, 1998a, b, 2000). Blackbeard observed almost exclusively high-energy, narrow-pulse VHF emissions that occurred in pairs and came to be known as "transionospheric pulse pairs" or TIPPs (see also subsection 14.6.1). TIPPs are apparently generated by "compact intracloud discharges", which produce narrow bipolar pulses and strong HF emissions observed on ground (subsection 9.4.2, second part). The first pulse of the pair is thought to be from an in-cloud process and the second from a ground reflection. In part to explore further the origin of TIPPs, the FORTE satellite, containing both RF and optical sensors, was launched in 1997 (Jacobson et al. 1999, 2000; Suszcynsky et al. 2000). The correlation between lightning optical radiation and the corresponding VHF radiation was discussed by Suszcynsky (2000), who inferred that the detected light, even for return strokes, is primarily produced in the cloud. Details on TIPPs observed by FORTE are found in Jacobson et al. (1999). Of primary interest to this chapter, however, is the correlation by Jacobson et al. (2000) of VHF emissions observed at FORTE with lightning detected by the ground-based NLDN. For this study, raw NLDN sensor data were reprocessed without the normal discrimination criteria so as to be able to locate both cloud and ground flashes. When all contributing sensors reported a pulse width, the time from signal peak to the following zero-crossing, of less than 10 µs, a cloud discharge was said to have been identified. Jacobson et al. (2000) found that there was a statistically significant correlation between a subset of FORTE events and NLDN-identified cloud discharges. Considerably fewer cloud-to-ground discharges were correlated with satellite-observed VHF emissions, and these were more likely to be positive discharges than negative ones. TIPPs associated with NLDN events are essentially all from cloud discharges. NLDN–FORTE associations that rise above the accidental correlation level are within 30 µs or so of each other, after account is taken of propagation-time differences. Tierney et al. (2001) proposed a method of locating groups of FORTE-observed VHF signals from lightning that emanate from isolated storm regions. Using this method, they assigned approximate locations to the sources of 6131 VHF signals that could not be located using timing coincidences with NLDN-located events.

Bondiou-Clergerie et al. (1999) reported on the "ORAGES" project for space-borne detection of lightning flashes using VHF–UHF interferometry, which has an expected launch date in 2004.

17.9. Radar

Radar can be used to locate lightning, determine the physical characteristics of discharge channels, and relate lightning to storm evolution. Observations of transient radar returns from lightning have been reported by Ligda (1950, 1956), Browne (1951), Marshall (1953), Miles (1953), Pawsey (1957), Hewitt (1957), Atlas (1958), Cerni (1976), Holmes et al. (1980), Szymanski and Rust (1979), Szymanski et al. (1980), Proctor (1981b), Zrnić et al. (1982), Stepanenko and Galperin (1983), Mazur and Rust (1983), Mazur et al. (1984a, b, 1985, 1986a, b), and Mazur (1986). Dawson (1972) applied the scattering theory originally developed for meteor trails to cloud-to-ground return-stroke channels. Additional theory was presented by Mazur and Walker (1982) and Mazur and Doviak (1983). The use of microwave radar to observe thunderstorms was reviewed by Atlas (1964), Battan (1973), and Browning (1977).

Williams et al. (1990), in discussing various radar studies of lightning, noted that a radar suitably configured to map lightning in three dimensions would have an advantage over passive techniques such as the measurement of the time of arrival of radiated signals (Section 17.4) or interferometry (Section 17.6), because the current-carrying channels reflect incident signals even if their electromagnetic radiation is undetectable by a passive system. Further, there is evidence that the radiation from certain lightning processes such as the positive leader is relatively low, making passive detection difficult (subsection 5.3.2). What is required for optimal radar mapping in three dimensions is a radar system capable of both high spatial resolution and short time resolution. Such a system does not exist at present although it is thought to be technically feasible. Radars can be used to map lightning, because lightning channels are highly reflective at radio-frequencies for a significant fraction of their duration. Reflections can often be received for hundreds of milliseconds during which time the temperature probably exceeds 5000° K (Williams et al. 1989a). Lightning channels can be considered to be similar to conducting wires; they are easy targets to detect in the clouds if the precipitation echo does not overpower them. As a lightning channel cools, the electron density decreases, as does the channel's reflecting ability. The reflectivity power often decays at 0.2 dB ms^{-1} (Holmes et al. 1980). The longer the wavelength, the less reflective the precipitation and the easier the detection of lightning in precipitation. From a practical point of view, the wavelength needs to be at least 10 cm to avoid the masking of lightning echoes by precipitation. At 10 cm, lightning will usually be detectable in precipitation, at least outside the most intense precipitation cores. The use of wavelengths of up to a meter or two will further reduce the relative precipitation return. In principle, polarization-diversity radar techniques could also be used to suppress the precipitation echo and hence enhance lightning detection at a given wavelength.

The most thorough work to date on the location of lightning and the measurement of its properties via radar is that by Holmes et al. (1980) in New Mexico and Proctor (1981b) in South Africa. Holmes et al. used a 10.9 cm radar with a 1 ms pulse-repetition period to detect echoes from 156 lightning flashes. The flashes were located at altitudes ranging from 5 to 14 km above sea level with extents along the fixed radar beam of less than 300 m to over 2 km. Most echoes rose to peak intensity in less than the 1 ms resolution of the radar and had a duration between 10 and 600 ms.

Proctor (1981b) observed radar echoes from lightning at 5.5, 50, and 111 cm. From an analysis of his measurements and those of Pawsey (1957), he reached the same conclusion as Holmes et al. (1980): that the echoes are due to many reflectors distributed throughout a volume of cloud. Proctor (1981b) gave values for measured effective radar cross-sections and compared those results with the theory of Dawson (1972). Proctor (1981b) reported that lightning echoes at 50 and 111 cm, unlike echoes from precipitation, did not fluctuate greatly in amplitude from one pulse to the next except for sudden rises in amplitude when strokes occurred followed by a smooth echo decay in tens of milliseconds, as previously reported by Hewitt (1957). It is interesting to note that Holmes et al. (1980) found that the echoes at 10.9 cm exhibited short-term fluctuations in amplitude similar to those from precipitation.

Szymanski et al. (1980), as part of the study of Holmes et al. (1980), described one lightning echo that was quickly followed by the development of precipitation in the same area of the cloud, and they referenced previous literature on this so-called rain-gush phenomenon.

Although not directly a lightning observation, Krehbiel et al. (1996) used a dual-channel circular-polarization radar to observe the buildup of the electric field inside the thunderclouds and the sudden collapse of the field at the time of lightning. These observations are based on the detection of particles (probably small ice crystals) which are aligned by the electrostatic field of the thundercloud.

17.10. Summary

There exists a variety of lightning-locating techniques based on the detection of lightning electromagnetic signals, accurate locating being possible only by using multiple-station systems. When a single location per cloud-to-ground lightning stroke, typically the ground strike point, is required, magnetic field direction finding, the time-of-arrival technique, or a combination of the two can be employed. Location accuracies of the order of 1 km and detection efficiencies approaching 90 percent are possible. The NALDN, operating on a continental scale, is an example of a network combining both magnetic direction finding and time-of-arrival techniques. When electromagnetic imaging of the developing channels of any type of lightning flash is required, the VHF–UHF time-of-arrival technique or VHF–UHF interferometry can be used. The interpretation of observations obtained using the VHF–UHF techniques is not always straightforward, and the two different techniques produce somewhat different results when imaging the same lightning event. The use of satellite-based optical sensors is probably the most promising approach to the global monitoring of lightning activity, although presently existing sensors have limited spatial resolution and cannot distinguish between cloud and cloud-to-ground flashes.

References and bibliography for Chapter 17

Adcock, F., and Clarke, E. 1947. The location of thunderstorms by radio direction finding. *J. Inst. Electr. Eng.* **94B**: 118–25.

Aleksandrov, M.S. and Orlov, A.V. 2000. A comparative analysis of range-difference and direction-finding methods for lightning location. *J. Commun. Technol. Electron.* **46**(3): 279–87.

Aleksandrov, M.S., Bakleneva, Z.M., Gladshtein, N.D. et al. 1972. *Fluctuations of Earth's Electromagnetic Field on VLF*, 195 pp., Moscow: Nauka.

Atlas, D. 1958. Radar lightning echoes and atmospherics in vertical cross section. In *Recent Advances in Atmospheric Electricity*, ed. L.G. Smith, pp. 441–59, Oxford, UK: Pergamon.

Atlas, D. 1964. Advances in radar meteorology. In *Advances in Geophysics*, vol. 10, ed. H. Landsberg, pp. 317–481, New York: Academic.

Baru, N.V., Kononov, I.I., and Solomonik, M.E. 1976. *Radio Direction Finders as Ranging Devices for Close Thunderstorms*, 143 pp., Leningrad, Russia: Gidrometeoizdat.

Battan, L.J. 1973. *Radar Observation of the Atmosphere*, 324 pp., Chicago: University of Chicago Press.

Bent, R.B. 1969. Investigations of global thunderstorm activity from the Ariel III satellite. In *Planetary Electrodynamics*, eds. S.C. Coroniti and J. Huges, pp. 111–27, Newark, New Jersey: Gordon and Breach.

Bent, R.B., and Lyons, W.A. 1984. Theoretical evaluations and initial operational experiences of LPATS (Lightning Positioning and Tracking System) to monitor lightning strikes using a time-of-arrival (TOA) technique. In *Proc. 7th Int. Conf. on Atmospheric Electricity, Albany, New York*, pp. 317–24.

Berger, K., Anderson, R.B., and Kroninger, H. 1975. Parameters of lightning flashes. *Electra* **80**: 23–37.

Bernstein, R., Samm, R., Cummins, K., Pyle, R., and Tuel, J. 1996. Lightning detection network averts damage and speeds restoration. *IEEE Computer Appl. Power* **9**: 12–17.

Bevington, P.R. 1969. *Data Reduction and Error Analysis for the Physical Sciences*, New York, NY: McGraw-Hill.

Biswas, K.R. and Hobbs, P.V. 1990. Lightning over the Gulf Stream. *Geophys. Res. Lett.* **17**: 941–3.

Boccippio, D.J., Williams, E.R., Heckman, S.J., Lyons, W.A., Baker, I.T., and Boldi, R. 1995. Sprites, extreme-low-frequency transients, and positive ground strokes. *Science* **269**: 1088–91.

Boccippio, D.J., Wong, C., Williams, E.R., Boldi, R., Christian, H.J., and Goodman, S.J. 1998. Global validation of single-station Schumann resonance lightning location. *J. Atmos. Solar-Terr. Phys.* **60**: 701–12.

Boccippio, D.J., Goodman, S.J., and Heckman, S. 2000a. Regional differences in tropical lightning distributions. *J. Appl. Meteor.* **39**: 2231–48.

Boccippio, D.J., Koshak, W., Blakeslee, R., Driscoll, K., Mach, D., Buechler, D., Boeck, W., Christian, H.J., and Goodman, S.J. 2000b. The Optical Transient Detector (OTD): instrument characteristics and cross-sensor validation. *J. Atmos. Ocean. Technol.* **17**: 441–58.

Boccippio, D.J., Heckman, S., and Goodman, S.J. 2001a. A diagnostic analysis of the Kennedy Space Center LDAR network 1. Data characteristics. *J. Geophys. Res.* **106**(D5): 4769–86.

Boccippio, D.J., Heckman, S., and Goodman, S.J. 2001b. A diagnostic analysis of the Kennedy Space Center

LDAR network 2. Cross-sensor studies. *J. Geophys. Res.* **106**(D5): 4787–96.

Bochkovskii, B.B., Rogozhin, I.B., Rosanov, N.I., and Timashova, L.V. 1997. Recording of lightning strike locations and determination of lightning current amplitudes. *Elektrichestvo* **8**: 24–9 (English translation in *Electrical Technology*, no. 3, pp. 57–67, 1997).

Bondiou, A., Taudièrre, I., Richard, P., and Helloco, F. 1990. Analyse spatio-temporelle du rayonnement VHF–UHF associé à l'éclair. *Rev. Phys. Appl.* **25**: 147–57.

Bondiou-Clergerie, A. Blanchet, P., Théry, C., Delannoy, A., Lojou, J.Y., Soulage, A., Richard, P., Roux, F., and Chauzy, S. 1999. "ORAGES": a project for space-borne detection of lightning flashes using interferometry in the VHF–UHF band. In *Proc. 11th Int. Conf. on Atmospheric Electricity, Guntersville, Alabama*, pp. 184–7.

Boulanger, A.G., and Maier, M.W. 1977. On the frequency of cloud-to-ground lightning from tropical cumulonimbus clouds. In *Proc. 11th Technical Conf. on Hurricanes and Tropical Meteorology, Miami Beach, Florida*, pp. 450–4, Am. Meteor. Soc., Boston.

Branick, M.L., and Doswell, C.A. III 1992. An observation of the relationship between supercell structure and lightning ground-strike polarity. *Wea. Forecast.* **7**: 143–9.

Brantley, R.D., Tiller, J.A., and Uman, M.A. 1975. Lightning properties in Florida thunderstorms from video tape records. *J. Geophys. Res.* **80**: 3402–6.

Brock, M. 1991. The observation and implications of a storm dominated by positive lightning. Master's thesis, State University of New York at Albany, Albany, New York.

Brook, M., Henderson, R.W., and Pyle, R.B. 1989. Positive lightning strokes to ground. *J. Geophys. Res.* **94**: 13 295–303.

Browne, I.C. 1951. A radar echo from lightning. *Nature* **167**: 438.

Browning, K.A. 1977. The structure and mechanisms of hailstorms. In *Hail: A Review of Hail Science and Hail Suppression*, eds. G.B. Foote and C.A. Knight, pp. 1–43, Boston, Massachusetts: American Meteorological Society.

Buechler, D.E., Driscoll, K.T., Goodman, S.J., and Christian, H.J. 2000. Lightning activity within a tornadic thunderstorm observed by the Optical Transient Detecter (OTD). *Geophys. Res. Lett.* **27**: 2253–6.

Carpenter, D.L., and Orville, R.E. 1989. The excitation of active whistler mode signal paths in the magnetosphere by lightning: two case studies. *J. Geophys. Res.* **94**: 8886–94.

Casper, P.W., and Bent, R.B. 1992. Results from the LPATS USA national lightning detection and tracking system for the 1991 lightning season. In *Proc. 21st Int. Conf. on Lightning Protection, Berlin, Germany*, pp. 339–42.

Cerni, T.A. 1976. Experimental investigation of the radar cross section of cloud-to-ground lightning. *J. Appl. Meteor.* **15**: 795–8.

Changnon, S.A. 1989. Relations of thunderstorms and cloud-to-ground lightning frequency. *J. Clim.* **2**: 897–921.

Changnon, S.A. 1992. Temporal and spatial relations between hail and lightning. *J. Appl. Meteor.* **31**: 587–604.

Changnon, S.A. 1993. Relationships between thunderstorms and cloud-to-ground lightning in the United States. *J. Appl. Meteor.* **32**: 88–105.

Changnon, S.A., Changnon, D., and Pyle, R.B. 1988. Thunder events and cloud-to-ground lightning frequencies. *J. Geophys Res.* **93**: 9495–502.

Chisholm, W.A., and Janischewskyj, W. 1988. Evaluation of detection efficiency and location accuracy for Canadian lightning location systems, 10 pp., Canadian Electrical Association.

Christensen, U., and Israelsson, S. 1987. Relationships between radar echo characteristics and lightning parameters for a thunderstorm in Sweden. *Weather* **42**: 165–76.

Christian, H.J., and Goodman, S. 1992. Global observations of lightning from space. In *Proc. 9th Int. Conf. on Atmospheric Electricity, St Petersburg, Russia*, pp. 316–21.

Christian, H.J., and Latham, J. 1998. Satellite measurements of global lightning. *Q.J.R. Meteor. Soc.* **124**: 1771–3.

Christian, H.J., Frost, R.L., Gillaspy, P.H., Goodman, S.J., Vaughan, O.H. Jr, Brook, M., Vonnegut, B., and Orville, R.E. 1983. Observations of optical lightning emissions from above thunderstorms using U-2 aircraft. *Bull. Amer. Meteor. Soc.* **64**: 120–3.

Christian, H.J., Blakeslee, R.J., and Goodman, S.J. 1989a. The detection of lightning from geostationary orbit. *J. Geophys. Res.* **94**: 13 329–37.

Christian, H.J., Mazur, V., Fisher, B.D., Ruhnke, L.H., Crouch, K., and Perala, R.P. 1989b. The Atlas/Centaur lightning strike incident. *J. Geophys. Res.* **94**: 13 169–77.

Christian, H.J., Blakesee, R.J., and Goodman, S.J. 1992. Lightning imaging sensor for the Earth observing system. NASA Technical Memorandum 4350.

Christian, H.J., Driscoll, K.T., Goodman, S.J., Blakeslee, R.J., Mach, D.A., and Buechler, D.E. 1996. The Optical Transient Detector (OTD). In *Proc. 10th Int. Conf. on Atmospheric Electricity, Osaka, Japan*, pp. 368–71.

Cianos, N., Oetzel, G.N., and Pierce, E.T. 1972. A technique for accurately locating lightning at close ranges. *J. Appl. Meteor.* **11**: 1120–7.

Cianos, N., Oetzel, G.N., and Pierce, E.T. 1973. Reply. *J. Appl. Meteor.* **12**: 1421–3.

Clegg, R.J. 1971. A photoelectric detector of lightning. *J. Atmos. Terr. Phys.* **33**: 1431–9.

Clifton, K.S., and Hill, C.K. 1980. Low-light-level television measurement of lightning. *Bull. Amer. Meteor. Soc.* **61**: 987–92.

Cook, B., and Casper, P. 1992. USA national lightning data service. In *Proc. 21st Int. Conf. on Lightning Portection, Berlin, Germany*, pp. 351–6.

Cooray, V. 1986. Errors in direction finding due to nonvertical lightning channels: effect of the finite ground conductivity. *Radio Sci.* **21**: 857–62.

Cooray, V., and Orville, R.E. 1989. Loran-C timing errors caused by propagation over finitely conducting ground. *Radio Sci.* **24**: 179–82.

Crozier, C.L., Herscovitch, H.N., and Scott, J.W. 1988. Some observations and characteristics of lightning ground discharges in southern Ontario. *Atmos. Ocean* **26**(3): 399–436.

Cummins, K.L. 2000a. Continental-scale detection of cloud-to-ground lightning. *Trans. IEE Japan* **120-B**: 2–5.

Cummins, K.L. 2000b. Personal communication.

Cummins, K.L., Hiscox, W.L., Pifer, A.E., and Maier, M.W. 1992. Performance analysis of the US National Lightning Detection Network. In *Proc. 9th Int. Conf. on Atmospheric Electricity, St Petersburg, Russia*, pp. 914–19.

Cummins, K.L., Burnett, R.O., Hiscox, W.L., and Pifer, A.E. 1993. Line reliability and fault analysis using the National Lightning Detection Network. Preprint, *Proc. Conf. on Precise Measurements in Power Systems, Arlington, Virginia*, II-4.1–15.

Cummins, K.L., Murphy, M.J. Bardo, E.A., Hiscox, W.L., Pyle, R.B., and Pifer, A.E. 1998a. A combined TOA/MDF technology upgrade of the US National Lightning Detection Network. *J. Geophys. Res.* **103**: 9035–44.

Cummins, K.L., Krider, E.P., and Malone, M.D. 1998b. The US National Lightning Detection Network™ and applications of cloud-to-ground lightning data by electric power utilities. *IEEE Trans. Electromagn. Compat.* **40**(II): 465–80.

Curran, E.B., and Rust, W.D. 1992. Positive ground flashes produced by low-precipitation thunderstorms in Oklahoma on 26 April 1984. *Mon. Wea. Rev.* **120**: 544–53.

Dalkir, Y.R., and Lee, M.C. 1993. Radar studies of lightning-induced plasmas with potential applications to radio communications and space surveillance. *Radio Sci.* **28**: 1039–47.

Davis, M.H., Brook, M. Christian, H., Heikes, B.G., Orville, R.E., Park, C.G., Roble, R.G. and Vonnegut, B. 1983. Some scientific objectives of a satellite-borne lightning mapper. *Bull. Amer. Meteor. Soc.* **64**: 114–19.

Davis, S.M. 1999. Properties of lightning discharges from multiple-station wideband electric field measurements. Ph.D. dissertation, University of Florida, Gainesville, 228 pp.

Dawson, G.A. 1972. Radar as a diagnostic tool for lightning. *J. Geophys. Res.* **77**: 4518–28.

de la Rosa, F., and Velazquez, R. 1989. Review of ground flash density measuring devices regarding power system applications. *IEEE Trans. Pow. Del.* **4**: 921–37.

Diendorfer, G., and Schulz, W. 1995. Möglichkeiten und Grenzen der Bestimmung von Blitzstromparametern mit Hilfe von Blitzortungssystemen. *Elektrotechnik und Informationstechnik* **112**(6): 279–83.

Diendorfer, G., and Schulz, W. 1998. Lightning incidence to elevated objects on mountains. In *Proc. 24th Int. Conf. on Lightning Protection, Birmingham, UK*, pp. 173–5.

Diendorfer, G., Schulz, W., and Fuchs, F. 1998a. Comparison of correlated data from the Austrian lightning locating system and measured lightning currents at the Peissenberg tower. In *Proc. 24th Int. Conf. on Lightning Protection, Birmingham, UK*, pp. 168–72.

Diendorfer, G., Schulz, W., and Rakov, V.A. 1998b. Lightning characteristics based on data from the Austrian lightning locating system. *IEEE Trans. Electromagn. Compat.* **40**(II): 452–64.

Dodge, P.P., and Burpee, R.W. 1993. Characteristics of rainbands, radar echoes, and lightning near the North Carolina coast during GALE. *Mon. Wea. Rev.* **121**: 1936–55.

Dotzek, N., Höller, H., Théry, C., and Fehr, T. 2001. Lightning evolution related to radar-derived microphysics in the 21 July 1998 EULINOX supercell storm. *Atmos. Res.* **56**: 335–54.

Dubovoy, E.I., Mikhailov, M.S., Ogonkov, A.L., and Pryazhinsky, V.I. 1995. Measurement and numerical modeling of radio sounding reflection from a lightning channel. *J. Geophys. Res.* **100**: 1497–502.

Edgar, B.C. 1978. Global lightning distribution at dawn and dusk for August–December, 1977 as observed by the DMSP lightning detector. Report SSL-78 (3639-02)-1, Aerospace Corporation Space Science Laboratory, Los Angeles, California.

Elsom, D.M., Meaden, G.T., Reynolds, D.J., Rowe, M.W., and Webb, J.D.C. 2001. Advances in tornado and storm research in the United Kingdom and Europe: the role of the Tornado and storm research organisation. *Atmos. Res.* **56**: 19–29.

Engholm, C.D. 1988. Positive lightning and bipolar lightning patterns: observational characteristics. M.Sc. thesis, Massachusetts Institute of Technology, Cambridge, Massachusetts, 214 pp.

Engholm, C.D., Williams, E.R., and Dole, R.M. 1990. Meteorological and electrical conditions associated with positive cloud-to-ground lightning. *Mon. Wea. Rev.* **118**: 470–87.

Frisius, J., Heydt, G., and Harth, W. 1970. Observations of parameters characterizing the VLF atmospherics activity as functions of the azimuth. *J. Atmos. Terr. Phys.* **32**: 1403–22.

Füllekrug, M., and Constable, S. 2000. Global triangulation of intense lightning discharges. *Geophys. Res. Lett.* **27**: 333–6.

Füllekrug, M., and Sukhorukov, A.I. 1999. The contribution of anisotropic conductivity in the ionosphere to lightning flash bearing deviations in the ELF/ULF range. *Geophys. Res. Lett.* **26**: 1109–12.

Füllekrug, M., Constable, S., Heinson, G., Sato, M., Takahashi, Y., Price, C., and Williams, E. 2000. Global lightning acquisition system installed. *Eos, Trans. AGU* **81**: 333, 343.

Galperin, S.M., and Stepanenko, W.D. 1968. Effectiveness of thunderstorm detection by radar and thunderstorm bearing indicator. In *Proc. 13th Radar Meteorology Conf.*, pp. 552–6, American Meteorological Society, Boston, Massachusetts.

Gething, P.J.D. 1978. *Radio Direction Finding and the Resolution of Multicomponent Wave-Fields*. 329 pp., London: Peter Peregrinus.

Goodman, S.J., and MacGorman, D.R. 1986. Cloud-to-ground lightning activity in mesoscale convective complexes. *Mon. Wea. Rev.* **114**: 2320–8.

Goodman, S.J., Buechler, D.E., and Meyer, P.J. 1988a. Convective tendency images derived from a combination of lightning and satellite data. *Wea. Forecast.* **3**: 173–88.

Goodman, S.J., Buechler, D.E., Wright, P.D., and Rust, W.D. 1988b. Lightning and precipitation history of a microburst-producing storm. *Geophys. Res. Lett.* **15**: 1185–8.

Goodman, S.J., Christian, H.J., and Rust, W.D. 1988c. A comparison of the optical pulse characteristics of intracloud and cloud-to-ground lightning as observed above clouds. *J. Appl. Meteor.* **27**: 1369–81.

Graham, B.L., Holle, R.L., and Lopez, R.E. 1997. Lightning detection and data use in the United States. *Fire Management Notes* **57**: 4–9.

Grandt, C. 1992. Thunderstorm monitoring in South Africa and Europe by means of very low frequency sferics. *J. Geophys. Res.* **97**: 18 215–26.

Griffiths, R.F., and Vonnegut, B. 1975. Tape recorder photocell instrument for detection and recording lightning strokes. *Weather* **30**: 254–7.

Guillo, P.Y. 1985. Study of lightning parameters using advanced analytical and statistical tools, M. Appl. Sci. thesis, University of Toronto, Ontario, Canada.

Hager, W.W., and Wang, D. 1995. An analysis of errors in the location, current, and velocity of lightning. *J. Geophys. Res.* **100**: 25 721–9.

Hamelin, J. 1993. Sources of natural noise. In *Electromagnetic Compatibility*, eds. P. Degauque and J. Hamelin, 652 pp., New York: Oxford.

Harth, W. 1982. Theory of low frequency wave propagation. In *Handbook of Atmospherics*, vol. II, ed. H. Volland, pp. 133–202, Boca Raton, Florida: CRC Press.

Harth, W., Hoffmann, C.A., Falcoz, H., and Heydt, G. 1978. Atmospherics measurements in San Miguel, Argentina. *J. Geophys. Res.* **83**: 6231–7.

Harth, W., and Pelz, J. 1973. Eastern thunderstorms located by VLF atmospherics parameters. *Radio Sci.* **8**: 117–22.

Hayakawa, M., Ohta, K., Shimakura, S., and Baba, K. 1995. Recent findings on VLF/ELF sferics. *J. Atmos. Terr. Phys.* **57**: 467–77.

Hayenga, C.O. 1979. Positions and movement of VHF lightning sources determined with microsecond resolution by interferometry. Ph.D. thesis, University of Colorado, Boulder.

Hayenga, C.O. 1984. Characteristics of lightning VHF radiation near the time of return strokes. *J. Geophys. Res.* **89**: 1403–10.

Hayenga, C.O., and Warwick, J.W. 1981. Two-dimensional interferometric positions of VHF lightning sources. *J. Geophys. Res.* **86**: 7451–62.

Herodotou, N. 1990. Study of peak currents due to lightning in Ontario using an LLP system. M. Appl. Sc. thesis, University of Toronto, Ontario, Canada, 215 pp.

Herodotou, N., Chisholm, W.A., and Janischewskyj, W. 1993. Distribution of lightning peak stroke currents in Ontario using an LLP system. *IEEE Trans. Pow. Del.* **8**: 1331–9.

Herrman, B.D., Uman, M.A., Brantley, R.D., and Krider, E.P. 1976. Tests of the principle of operation of a wideband magnetic direction finder for lightning return strokes. *J. Appl. Meteor.* **15**: 402–5.

Hewitt, F.J. 1953. The study of lightning streamers with 50 cm radar. *Proc. Phys. Soc. London B* **66**: 895–7.

Hewitt, F.J. 1957. Radar echoes from interstroke processes in lightning. *Proc. Phys. Soc. London B* **70**: 961–79.

Heydt, G. 1974. Observation of thunderstorm activity at close range by means of VLF radio meteorology. *Kleinheubacher Berichte* **17**: 433–41.

Heydt, G. 1982. Instrumentation. In *Handbook of Atmospherics*, vol. II, ed. H. Volland, pp. 203–56, Boca Raton, Florida: CRC Press.

Heydt, G., and Volland, H. 1964. A new method for locating thunderstorms and counting their lightning discharges from a single observing station. *J. Atmos. Terr. Phys.* **26**: 780–2.

Hiscox, W.L., Krider, E.P., Pifer, A.E., and Uman, M.A. 1984. A systematic method for identifying and correcting 'site errors' in a network of magnetic direction finders. Preprint, *Proc. Int. Aerospace Ground Conf. on Lightning and Static Electricity, Orlando, Florida*, pp. 7-1–5, National Interagency Coordination Group.

Hoblitt, R.P. 1994. An experiment to detect and locate lightning associated with eruptions of Redoubt volcano. *J. Volcan. Geoth. Res.* **62**: 499–517.

Hohl, R., and Schiesser, H.-H., 2001. Cloud-to-ground lightning activity in relation to the radar-derived hail kinetic energy in Switzerland. *Atmos. Res.* **56**: 375–96.

Hojo, J., Ishii, M., Kawamura, T., Suzuki, F., Komuro, H., and Shiogama, M. 1988. Characteristics and evaluation of lightning field waveforms. *Electr. Eng. Japan*, **108**: 55–65 (Translated from *Denki Gakkai Ronbunshi*, **108B**: 165–72, 1988).

Hojo, J., Ishii, M., Kawamura, T., Suzuki, F., Komuro, H., and Shiogama, M. 1989. Seasonal variation of cloud-to-ground lightning flash characteristics in the coastal area of the Sea of Japan. *J. Geophys. Res.* **94**: 13 207–12.

Holden, D.N., Munson, C.P., and Devenport, J.C. 1995. Satellite observations of transionospheric pulse pairs. *Geophys. Res. Lett.* **22**: 889–92.

Holle, R.L., and Lopez, R.E. 1993. Overview of real-time lightning detection systems and their meteorological uses. NOAA Technical Memorandum ERL NSSL-102, 68 pp.

Holle, R.L., Watson, A.I., Lopez, R.E., MacGorman, D.R., Ortiz, R., and Otto, W.D. 1994. The life cycle of lightning and severe weather in a 3–4 June 1985 PRE-STORM mesoscale convective system. *Mon. Wea. Rev.* **122**: 1798–808.

Holmes, C.R., Szymanski, E.W., Szymanski, S.J., and Moore, C.B. 1980. Radar and acoustic study of lightning. *J. Geophys. Res.* **85**: 7517–32.

Hondl, K.D., and Eilts, M.D. 1994. Doppler radar signatures of developing thunderstorms and their potential to indicate the onset of cloud-to-ground lightning. *Mon. Wea. Rev.* **122**: 1818–36.

Honma, N., Suzuki, F., Miyake, Y., Ishii, M., and Hidayat, S. 1998. Propagation effect on field waveforms in relation to time-of-arrival technique in lightning location. *J. Geophys. Res.* **103**: 14 141–5.

Horner, F. 1953. Radio direction finding: influence of buried conductors on bearings. *Wireless Engineer* **30**: 187–91.

Horner, F. 1954. The accuracy of the location sources of atmospherics by radio direction finding. *Proc. IEEE* **101**: 383–90.

Horner, F. 1957. Very-low-frequency propagation and direction finding. *Proc. IEEE* **101B**: 73–80.

Huffines, G.R., and Orville, R.E. 1999. Lightning ground flash density and thunderstorm duration in the continental United States: 1989–96. *J. Appl. Meteor.* **38**: 1013–19.

Hunter, S.M., Schuur, T.J., Marshall, T.C., and Rust, W.D. 1992. Electric and kinematic structure of the Oklahoma mesoscale convective system of 7 June 1989. *Mon. Wea. Rev.* **120**: 2226–39.

Idone, V.P., Orville, R.E., and Henderson, R.W. 1984. Ground truth: a positive cloud-to-ground lightning flash, *J. Clim. Appl. Meteor.* **23**: 1148–51.

Idone, V.P., and Orville, R.E. 1990. Notes and correspondence – delimiting "thunderstorm watch" periods by real-time lightning location for a power utility company. *Wea. Forecast.* **5**: 139–47.

Idone, V.P., Saljoughy, A.B., Henderson, R.W., Moore, P.K., and Pyle, R.B. 1993. A reexamination of the peak current calibration of the National Lightning Detection Network. *J. Geophys. Res.* **98**: 18 323–32.

Idone, V.P., Davis, D.A., Moore, P.K., Wang, Y., Henderson, R.W., Ries, M., and Jamason, P.F. 1998a. Performance evaluation of the US National Lightning Detection Network in eastern New York, 1. Detection efficiency. *J. Geophys. Res.* **103**: 9045–55.

Idone, V.P., Davis, D.A., Moore, P.K., Wang, Y., Henderson, R.W., Ries, M., and Jamason, P.F. 1998b. Performance evaluation of the US National Lightning Detection Network in eastern New York, 2. Location accuracy. *J. Geophys. Res.* **103**: 9057–69.

Inan, U.S., Burgess, W.C., Wolf, T.G., Shater, D.C., and Orville, R.E. 1988. Lightning-associated precipitation of MeV electrons from the inner radiation belt. *Geophys. Res. Lett.* **15**: 172–5.

Inkov, B.K. 1973. *Phase Methods for the Determination of the Distance to Thunderstorms*, 127 pp., Leningrad, Russia: Gidrometeoizdat.

Ishii, M., and Hojo, J.-I. 1989. Statistics on fine structure of cloud-to-ground lightning field waveforms. *J. Geophys. Res.* **94**: 13 267–74.

Ishii, M., Yamanoto, T., Sawada, J., Hojo, J.-I., Zaima, E.-I., and Fukiyama, N. 1991. On the distribution of lightning current amplitude observed by LPATS. In *Proc. 1991 Annual Conf. Power Energy Society*, pp. 325–6, Fukuoka, Japan: IEE Japan.

Israelsson, S., Schutte, T., Pisler, E., and Lundquist, S.1987. Increased occurrence of lightning flashes in Sweden during 1986. *J. Geophys. Res.* **92**: 10 996–8.

Ito, Y., and Goto, M. 1957. *Radio Direction Finder*, Tokyo, Japan: Corona-Sha.

Iwai, A., Kashiwagi, M., Nishino, M., and Sato, M. 1979. Triangulation direction finding network for fixing the sources of atmospherics. *Proc. Res. Inst. Atmospherics Nagoya Univ., Japan* **26**: 1–16.

Iwai, A., Kashiwagi, M., Nishino, M., Katoh, Y., and Kengpol, A. 1982. On the accuracy of direction finding methods for atmospheric sources in South-East Asia. *Proc. Res. Inst. Atmospherics Nagoya Univ., Japan* **29**: 35–46.

Jacobson, E.A., and Krider, E.P. 1976. Electrostatic field changes produced by Florida lightning. *J. Atmos. Sci.* **33**: 113–17.

Jacobson, A.R., Knox, S.O., Franz, R., and Enemark, D.C. 1999. FORTE observations of lightning radio-frequency signatures: capabilities and basic results. *Radio Sci.* **34**: 337–54.

Jacobson, A.R., Cummins, K.L., Carter, M., Klingner, P., Roussel-Dupré, D., and Knox, S.O. 2000. FORTE radio-frequency observations of lightning strokes detected by the National Lightning Detection Network. *J. Geophys. Res.* **105**: 15 653–62.

Janhunen, P. 1992. Propagation of the electromagnetic signal from lightning over a non-planar inhomogeneous Earth. *J. Atmos. Terr. Phys.* **54**: 251–64.

Janischewskyj, W., and Chisholm, W.A. 1992. Lightning ground flash density measurements in Canada, March 1, 1984 to December 31, 1991. Canadian Electrical Association Report 179 T382, Montreal, Quebec, 106 pp.

Johnson, R.L., Janota, D.E., and Hay, J.E. 1982. An operational comparison of lightning warning systems. *J. Appl. Meteor.* **21**: 703–7.

Kane, R.J. 1991. Correlating lightning to severe local storms in the northeastern United States. *Wea. Forecast.* **6**: 3–12.

Kashiwagi, M., Iwai, A., and Nishino, M. 1981. Fixing of the sources of atmospherics using the measurement of the arrival time difference of atmospherics between Toyokawa and Bangkok. *Res. Lett. Atmos. Electr.* **1**: 119–24.

Kashprovsky, V.E. 1966. *Locating Thunderstorms with Radiotechnical Methods*, 248 pp., Moscow: Nauka.

Katahire, O. 1992. Observation of volcanic lightning of Mt Fugen, Uzen. *Res. Lett. Atmos. Electr.* **12**: 225–34.

Kawamura, T., Ishii, M., and Miyake, Y. 1988. Site errors of magnetic direction finder for lightning flashes. In *Proc. Int. Aerospace and Ground Conf. on Lightning and Static Electricity, Oklahoma City, Oklahoma*, pp. 487–93.

Kawasaki, Z.-I., Yamamoto, K., Matsuura, K., Richard, P., Matsui, T., Sonoi, Y., and Shimokura, N. 1994. SAFIR operation and evaluation of its performance. *Geophys. Res. Lett.* **21**: 1133–6.

Kawasaki, Z., Mardiana, R., and Ushio, T. 2000. Broadband and narrowband RF interferometers for lightning observations. *Geophys. Res. Lett.* **27**: 3189–92.

Keen, R. 1938. *Wireless Direction Finding*, 3rd edition, 803 pp., London: Iliffe and Sons.

Keighton, S.J., Bluestein, H.B., and MacGorman, D.R. 1991. The evolution of a severe mesoscale convective system: cloud-to-ground lightning location and storm structure. *Mon. Wea. Rev.* **119**: 1533–56.

Kelley, M.C., Ding, J.G., and Holzworth, R.H. 1990. Intense ionospheric electric and magnetic field pulses generated by lightning. *Geophys. Res. Lett.* **17**: 2221–4.

Kemp, D.T. 1971. The global location of large lightning discharges from single station observations of ELF disturbances in the Earth–ionosphere cavity. *J. Atmos. Terr. Phys.* **33**: 919–27.

Kidder, R.E. 1973. The location of lightning flashes at ranges less than 100 km. *J. Atmos. Terr. Phys.* **35**: 283–90.

Kidder, R.E. 1975. Location of lightning flashes to ground with a single camera. *Weather* **30**: 72–7.

King, T.S. and Balling, R.C. Jr 1994. Diurnal variations in Arizona monsoon lightning data. *Mon. Wea. Rev.* **122**: 1659–64.

Kingwell, J., Shimizu, J., Narita, K., Kawabata, H., and Shimizu, I. 1991. Weather factors affecting rocket operations: a review and case history. *Bull. Amer. Meteor. Soc.* **72**: 778–93.

Kirkland, M.W., Suszcynsky, D.M., Guillen, J.L.L., and Green, J.L. 2001. Optical observations of terrestrial lightning by the FORTE satellite photodiode detector. *J. Geophys. Res.* **106**: 33 499–509.

Kohl, D.A. 1969. A 500 kHz sferics range detector. *J. Appl. Meteor.* **8**: 610–17.

Kononov, I.I., Petrenko, I.A., and Snegurov, V.S. 1986. *Radiotechnical Methods for Locating Thunderstorms*, 222 pp., Leningrad, Russia: Gidrometeoizdat.

Kononov, I.I., and Petrenko, I.A. 1996. Experience of lightning location systems elaboration in Russia. In *Proc. 23rd Int. Conf. on Lightning Protection, Florence, Italy*, pp. 236–40.

Koral, M.A., and Nickolaenko, A.P. 1993. A technique to derive the distance from near discharges. *J. Atmos. Electr.* **13**: 1–7.

Koshak, W.J., and Krider, E.P. 1989. Analysis of lightning field changes during active Florida thunderstorms. *J. Geophys. Res.* **94**: 1165–86.

Koshak, W.J., and Krider, E.P. 1994. A linear method for analyzing lightning field changes. *J. Atmos. Sci.* **51**: 473–88.

Koshak, W.J., and Solakiewicz, R.J. 1996. On the retrieval of lightning radio sources from time-of-arrival data. *J. Geophys. Res.* **101**: 26 631–9.

Koshak, W.J., and Solakiewicz, R.J. 1999. Electro-optic lightning detector. *Appl. Optics* **38**: 4623–34.

Koshak, W.J., and Solakiewicz, R.J. 2001. TOA lightning location retrieval on spherical and oblate spheroidal Earth geometries. *J. Atmos. Ocean. Technol.* **18**: 187–99.

Koshak, W.J., Solakiewicz, R.I., Phanord, D.D., and Blakeslee, R.J. 1994. Diffusion model for lightning radiative transfer. *J. Geophys. Res.* **99**: 14 361–71.

Koshak, W.J., Krider, E.P., and Murphy, M.J. 1999. A multipole expansion method for analyzing lightning field changes. *J. Geophys. Res.* **104**: 9617–33.

Koshak, W.J., Stewart, M.F., Christian, H.J., Bergstrom, J.W., Hall, J.M., and Solakiewicz, R.J. 2000a. Laboratory calibration of the Optical Transient Detector and the Lightning Imaging Sensor. *J. Atmos. Ocean. Technol.* **17**: 905–15.

Koshak, W.J., Blakeslee, R.J., and Bailey, J.C. 2000b. Data retrieval algorithms for validating the Optical Transient Detector and the Lightning Imaging Sensor. *J. Atmos. Ocean. Technol.* **17**: 279–97.

Kotaki, M., and Katoh, C. 1983. The global distribution of thunderstorm activity observed by the ionospheric sounding satellite (ISS-B). *J. Atmos. Terr. Phys.* **45**: 833–47.

Kotaki, M., Kuriki, I., Katoh, C., and Sugiuchi, H. 1981a. Global distribution of thunderstorm activity observed with ISS-b. *J. Radio Res. Lab. Tokyo* **28**: 49–71.

Kotaki, M., Sugiuchi, H., and Katoh, C. 1981b. World distribution of thunderstorm activity obtained from Ionosphere Sounding Satellite-b observations June 1978 to May 1980. Radio Research Laboratories, Ministry of Posts and Telecommunications, Japan.

Kozak, L.E. 1987. Network tracks lightning for New Jersey. *Bell. Teleph. Eng. Man.* **91**: 58–61.

Krehbiel, P.R. 1981. An analysis of the electric field change produced by lightning. Ph.D. thesis, University of Manchester Institute of Science and Technology, Manchester, UK (available as Report T-11, Geophys. Res. Ctr., New Mexico Inst. Min. and Tech., Socorro, 87801, 1981).

Krehbiel, P.R., Brook, M., and McCrory, R.A. 1979. An analysis of the charge structure of lightning discharges to the ground. *J. Geophys. Res.* **84**: 2432–56.

Krehbiel, P., Chen, T., McCrary, S., Rison, W., Gray, G., and Brook, M. 1996. The use of dual channel circular-polarization radar observations for remotely sensing storm electrification. *Meteor. Atmos. Phys.* **59**: 65–82.

Krehbiel, P.R., Thomas, R.J., Rison, W., Hamlin, T., Harlin, J., and Davis M. 2000. GPS-based mapping system reveals lightning inside storms. *Eos, Trans. AGU* **81**(3): 21–5.

Krider, E.P. 1989. Electric field changes and cloud electrical structure. *J. Geophys. Res.* **94**: 13 145–9.

Krider, E.P. 1994. Physics of lightning today. *Extraits de la Revue générale de l'Electricité* no. 6, 7 pp.

Krider, E.P., and Noggle, R.C. 1975. Broadband antenna systems for lightning magnetic fields. *J. Appl. Meteor.* **14**: 252–6.

Krider, E.P., Noggle, R.C., and Uman, M.A. 1976. A gated wideband magnetic direction finder for lightning return strokes. *J. Appl. Meteor.* **15**: 301–6.

Krider, E.P., Noggle, R.C., Pifer, A.E., Vance, D.L. 1980. Lightning direction-finding systems for forest fire detection. *Bull. Amer. Meteor. Soc.* **61**: 980–6.

Krider, E.P., Leteinturier, C., and Willett, J.C. 1996. Submicrosecond fields radiated during the onset of first return strokes in cloud-to-ground lightning. *J. Geophys. Res.* **101**: 1589–97.

Kumar, P.P., Manohar, G.K., and Kandalgaonkar, S.S. 1995. Global distribution of nitric oxide produced by lightning and its seasonal variation. *J. Geophys. Res.* **100**: 11 203–8.

Labaune, G., Richard, P., and Bondiou, A. 1987. Electromagnetic properties of lightning channels formation and propagation. *Electromagnetics* **7**: 361–93.

Landweber, L. 1951. An iteration formula for Fredholm integral equations of the first kind. *Am. J. Math.* **73**: 615–24.

Latham, D. 1991. Lightning flashes from a prescribed fire-induced cloud. *J. Geophys. Res.* **96**: 17 151–7.

Le Boulch, M., and Plantier, T. 1990. The Meteorage thunderstorm monitoring system: A tool for new EMC protection strategies. In *Proc. 20th Int. Conf. on Lightning Protection, Interlaken, Switzerland*, paper 6.13P.

Lee, A.C.L. 1986a. An experimental study of the remote location of lightning flashes using a VLF arrival time difference technique. *Q.J.R. Meteor. Soc.* **112**: 203–29.

Lee, A.C.L. 1986b. An operational system for the remote location of lightning flashes using a VLF arrival time difference technique. *J. Atmos. Ocean. Tech.* **3**: 630–42.

Lee, A.C.L. 1989a. The limiting accuracy of long wavelength lightning flash location. *J. Atmos. Ocean. Tech.* **6**: 43–9.

Lee, A.C.L. 1989b. Ground truth confirmation and theoretical limits of an experimental VLF arrival time difference lightning flash locating system. *Q.J.R. Meteor. Soc.* **115**: 1146–66.

Lee, A.C.L. 1990. Bias elimination and scatter in lightning location by the VLF arrival time difference technique. *J. Atmos. Ocean. Tech.* **7**: 719–33.

Lennon, C.L. 1975. LDAR – a new lightning detection and ranging system. *Eos, Trans. AGU* **56**: 991.

Lennon, C., and Maier, L. 1991. Lightning mapping system. In *Proc. 1991 Int. Aerospace and Ground Conf. on Lightning and Static Electricity, Cocoa Beach, Florida*, vol. II, pp. 89/1–10, NASA Conf. Publ. 3106.

Lennon, C.L., and Poehler, H.A. 1982. Lightning detection and ranging. *Astronautics and Aeronautics* **20**: 29–31.

Le Vine, D.M., Willett, J.C., and Bailey, J.C. 1989. Comparison of fast electric field changes from subsequent return strokes of natural and triggered lightning. *J. Geophys. Res.* **94**: 13 259–65.

Lewis, E.A., Harvey, R.B., and Rasmussen, J.E. 1960. Hyperbolic direction finding with sferics of transatlantic origin. *J. Geophys. Res.* **65**: 1879–905.

Lhermitte, R., and Krehbiel, P.R. 1979. Doppler radar and radio observations of thunderstorms. *IEEE Trans. Geosci. Electron.* **GE-17**: 162–71.

Lhermitte, R., and Williams, E. 1985. Thunderstorm electrification: a case study. *J. Geophys. Res.* **90**: 6071–8.

Lin, Y.T., Uman, M.A., Tiller, J.A., Brantley, R.D., Beasley, W.H., Krider, E.P., and Weidman, C.D. 1979. Characterization of lightning return stroke electric and magnetic fields from simultaneous two-station measurements. *J. Geophys. Res.* **84**: 6307–14.

Ligda, M.G.H. 1950. Lightning detection by radar. *Bull. Amer. Meteor. Soc.* **31**: 279–83.

Ligda, M.G.H. 1956. The radar observation of lightning. *J. Atmos. Terr. Phys.* **9**: 329–46.

Light, T.E., Suszcynsky, D.M., and Jacobson, A.R. 2001a. Coincident radio-frequency and optical emissions from lightning observed with the FORTE satellite. *J. Geophys. Res.* **106**: 28 223–31.

Light, T.E., Suszcynsky, D.M., Kirkland, M.W., and Jacobson, A.R. 2001b. Simulations of lightning optical waveforms as seen through clouds by satellites. *J. Geophys. Res.* **106**: 17 103–14.

Lopez, R.E., and Holle, R.L. 1986. Diurnal and spatial variability of lightning activity in northeastern Colorado and central Florida during the summer. *Mon. Wea. Rev.* **114**: 1288–312.

Lopez, R.E., and Passi, R.M. 1991. Simulations in site error estimation for direction finders. *J. Geophys. Res.* **96**: 15 287–96.

Lopez, R.E., Maier, M.W., and Holle, R.L. 1991. Comparison of the signal strength of positive and negative cloud-to-ground lightning flashes in northeastern Colorado. *J. Geophys. Res.* **96**: 22 307–18.

Lopez, R.E., Holle, R.L., Ortiz, R., and Watson, A.I. 1992. Detection efficiency losses of networks of direction finders due to flash signal attenuation with range. In *Proc. Int. Aerospace and Ground Conf. on Lightning and Static Electricity, Atlantic City, New Jersey*, paper 75, 18 pp.

Lyons, W.A., and Keen, C.S. 1994. Observations of lightning in convective supercells within tropical storms and hurricanes. *Mon. Wea. Rev.* **122**: 1897–916.

Lyons, W.A., Moon, D.A., Schuh, J.A., Pettit, N.J., and Eastman, J.R. 1989. The design and operation of a national lightning detection network using time-of-arrival technology. In *Proc. 1989 Int. Conf. on Lightning and Static Electricity, Bath, England*, pp. 2B.2.1–8.

Lyons, W.A., Uliasz, M., and Nelson, T.E. 1998. Large peak current cloud-to-ground lightning flashes during the summer months in the contiguous United States. *Mon. Wea. Rev.* **126**: 2217–33.

MacClement, W.D., and Murty, R.C. 1978. VHF direction finder studies of lightning. *J. Appl. Meteor.* **17**: 786–95.

MacGorman, D.R., and Burgess, D.W. 1994. Positive cloud-to-ground lightning in tornadic storms and hailstorms. *Mon. Wea. Rev.* **122**: 1671–97.

MacGorman, D.R., and Nielsen, K.E. 1991. Cloud-to-ground lightning in a tornadic storm on 8 May 1986. *Mon. Wea. Rev.* **119**: 1557–74

MacGorman, D.R., and Rust, W.D. 1988. An evaluation of the LLP and LPATS lightning ground strike mapping systems. In *Proc. 8th Int. Conf. on Atmospheric Electricity, Uppsala, Sweden*, pp. 668–73.

MacGorman, D.R., and Taylor, W.L. 1989. Positive cloud-to-ground lightning detection by a direction-finder network. *J. Geophys. Res.* **94**: 13 313–18.

MacGorman, D.R., Burgess, D.W., Mazur, V., Rust, W.D., Taylor, W.L., and Johnson, B.C. 1989. Lightning rates relative to tornadic storm evolution on 22 May 1981. *J. Atmos. Sci.* **46**: 221–50.

Mach, D.M. 1984. Evaluation of an LLP ground strike locating system. M.Sc. thesis, University of Oklahoma, Norman, Oklahoma, 55 pp.

Mach, D.M., MacGorman, D.R., Rust, W.D., and Arnold, R.T. 1986. Site errors and detection efficiency in a magnetic direction-finder network for locating lightning strikes to ground. *J. Atmos. Ocean. Tech.* **3**: 67–74.

Maier, M.W. 1991. Preliminary evaluation of National Lightning Detection Network performance at Cape Canaveral during August 1990, CSR Report CDRL 137A2, CSR-322-0007, 24 pp. (Comput. Sci. Raytheon Instrumentation Syst. Eval. Test.)

Maier, M.W., and Jafferis, W. 1985. Locating rocket-triggered lightning using the LLP lightning location system at the NASA Kennedy Space Center. In *Proc. 10th Int. Conf. on Lightning and Static Electricity, Paris, France*, pp. 337–45.

Maier, M.W., and Krider, E.P. 1982. A comparative study of the cloud to ground lightning characteristics in Florida and Oklahoma thunderstorms. Preprint, *Proc. 12th Conf. on Severe Local Storms, San Antonio, Texas*, pp. 334–7, American Meteorological Society, Boston, Massachusetts.

Maier, M.W., and Wilson, M.B. 1996. Accuracy of the NLDN real-time data service at Cape Canaveral, Florida. Preprints, *Proc. Int. Lightning Detection Conf.*, 11 pp., GAI, 2705 East Medina Road, Tucson, Arizona 85706-7155.

Maier, L.M., Krider, E.P., and Maier, M.W. 1984. Average diurnal variation of summer lightning over the Florida peninsula. *Mon. Wea. Rev.* **112**: 1134–40.

Maier, L., Lennon, C., Britt, T., and Schaefer, S. 1995. LDAR system performance and analysis. In *Proc. Int. Conf. Cloud Physics, Dallas, Texas*, paper 8.9, American Meteorological Society, Boston, Massachusetts.

Mair, M., Hadrian, W., Diendorfer, G., and Schulz, W. 1998. Effect of signal attenuation on the peak current estimates from lightning location systems. In *Proc. 24th Int. Conf. on Lightning Protection, Birmingham, UK*, pp. 193–8.

Malan, D.J., and Schonland, B.F.J. 1950. An electrostatic fluxmeter of short response-time for use in studies of transient field-changes. *Proc. Phys. Soc. London B* **63**: 402–8.

Mardiana, R., and Kawasaki, Z. 2000a. Broadband radio interferometer utilizing a sequential triggering technique for locating fast-moving electromagnetic sources emitted from lightning. *IEEE Trans. Instrum. Meas.* **49**: 376–81.

Mardiana, R., and Kawasaki, Z. 2000b. Dependency of VHF broad band lightning source mapping on Fourier spectra. *Geophys. Res. Lett.* **27**: 2917–20.

Mardiana, R., Ohta, Y., Murakami, M., Ushio, T., Kawasaki, Z., and Matsuura, K. 1998. A broadband radio interferometer for observing lightning discharge processes. *J. Atmos. Electr.* **18**: 111–17

Mardiana, R., Morimoto, T., and Kawasaki, Z.-I. 2001. Imaging lightning progression using VHF broadband radio infererometry. *IEICE Trans. Electron.* **E84-C**: 1892–99.

Mardiana, R., Kawasaki, Z.-I., and Morimoto, T. 2002. Three-dimensional lightning observations of cloud-to-ground flashes using broadband interferometers. *J. Atmos. Solar-Terr. Phys.* **64**: 91–103.

Marshall, J.S. 1953. Frontal precipitation and lightning observed by Radar. *Can. J. Phys.* **31**: 194–203.

Marshall, T.C., and Rust, W.D. 1993. Two types of vertical electrical structures in stratiform precipitation regions of mesoscale convective systems. *Bull. Amer. Meteor. Soc.* **74**: 2159–70.

Massey, R.S., and Holden, D.N. 1995. Phenomenology of trans-ionospheric pulse pairs. *Radio Sci.* **30**: 1645–59.

Massey, R.S., Holden, D.N., and Shao, X.-M., 1998. Phenomenology of trans-ionospheric pulse pairs: further observations. *Radio Sci.* **33**: 1755–61.

Mazur, V. 1982. Associated lightning discharges. *Geophys. Res. Lett.* **9**: 1227–30.

Mazur, V. 1986. Rapidly occurring short duration discharges in thunderstorms, as indicators of a lightning-triggering mechanism. *Geophys. Res. Lett.* **13**: 355–8.

Mazur, V. 1989. Triggered lightning strikes to aircraft and natural intracloud discharges. *J. Geophys. Res.* **94**: 3311–25.

Mazur, V., and Doviak, R. 1983. Radar cross section of a lightning element modeled as a plasma cylinder. *Radio Sci.* **18**: 381–90.

Mazur, V. and Rust, W.D. 1983. Lightning propagation and flash density in squall lines as determined with radar. *J. Geophys. Res.* **88**: 1495–502.

Mazur, V., and Walker, G.B. 1982. The effect of polarization on radar detection of lightning. *Geophys. Res. Lett.* **9**: 1231–4.

Mazur, V., Fisher, B.D., and Gerlach, J.C. 1984a. Lightning strikes to an airplane in a thunderstorm. *J. Aircraft* **21**: 607–11.

Mazur, V., Gerlach, J.C., and Rust, W.D. 1984b. Lightning flash density versus altitude and storm structure from observations with UHF- and S-band radars. *Geophys. Res. Lett.* **11**: 61–4.

Mazur, V., Zrnić, D.S., and Rust, W.D. 1985. Lightning channel properties determined with vertically pointing Doppler radar. *J. Geophys. Res.* **90**: 6165–74.

Mazur, V., Fisher, B.D., and Gerlach, J.C. 1986a. Lightning strikes to a NASA airplane penetrating thunderstorms at low altitudes. *J. Aircraft* **23**: 499–505.

Mazur, V., Rust, W.D, and Gerlach, J.C. 1986b. Evolution of lightning flash density and reflectivity structure in a multicell thunderstorm. *J. Geophys. Res.* **91**: 8690–700.

Mazur, V., Zrnić, D.S., and Rust, W.D. 1987. Transient changes in Doppler spectra of precipitation associated with lightning. *J. Geophys. Res.* **92**: 6699–704.

Mazur, V., Krehbiel, P.R., and Shao, X-M. 1995. Correlated high-speed video and radio interferometric observations of a cloud-to-ground lightning flash. *J. Geophys. Res.* **100**: 25 731–53.

Mazur, V., Williams, E., Boldi, R., Maier, L., and Proctor, D.E. 1997. Initial comparison of lightning mapping with operational time-of-arrival and interferometric systems. *J. Geophys. Res.* **102**: 11 071–85.

Mazur, V., Shao, X.-M., and Krehbiel, P.R. 1998. "Spider" lightning in intracloud and positive cloud-to-ground flashes. *J. Geophys. Res.* **103**: 19 811–22.

McDonald, T.B., Uman, M.A., Tiller, J.A., and Beasley, W.H. 1979. Lightning location and lower ionospheric height determination from two station magnetic field measurements. *J. Geophys. Res.* **84**: 1727–34.

McGraw, M.G. 1982. 'On-line' lightning maps lead crews to 'trouble'. *Electr. World* **196**: 111–14.

McNulty, R.P., Schaefer, J.T., Sunkel, W.E., and Townsend, T.A. 1990. On the need for augmentation in automated surface observations. *Natl. Wea. Digest.* **15**: 9–16.

Michnowski, S., Israelsson, S., Parfiniewicz, J., Enaytollah, M.A., and Pisler, E. 1987. A case of thunderstorm system development inferred from lightning distribution. *Publs. Inst. Geophys. Pol. Acad. Sc. D-26* **198**: 3–54.

Mikhailov, M.S., and Dubovoy, E.I. 1997. Possibility of estimating the coordinates of a lightning discharge by single-point measurement of its electromagnetic radiation. *Elektrichestvo* **12**: 8–15 (English translation in *Electrical Technology*, no. 4, pp. 95–109, 1997).

Miles, V.H. 1953. Radar echoes associated with lightning. *J. Atmos. Terr. Phys.* **3**: 258–63.

Molinari, J., Moore, P.K., Idone, V.P., Henderson, R.W., and Saljoughy, A.B. 1994. Cloud-to-ground lightning in Hurricane Andrew. *J. Geophys. Res.* **99**: 16 665–76.

Molinie, J., and Pontikis, C.A. 1995. A climatological study of tropical thunderstorm clouds and lightning frequencies on the French Guyana coast. *Geophys. Res. Lett.* **22**: 1085–8.

Molinie, G., Soula, S., and Chauzy, S. 1999. Cloud-to-ground lightning activity and radar observations of storms in the Pyrenees range area. *Q.J.R. Meteor. Soc.* **125**: 3103–22.

Montandon, E. 1995. Messung und Ortung von Blitzeinschlaegen und ihren Auswirkungen am Fernmeldeturm "St Chrischona" bei Basel der Schweizerischen Telecom PTT. *Elektrotechnik und Informationstechnik* **112**: 283–9.

Montandon, E., Ahnebrink, T., and Bent, R.B. 1992. Analysis of lightning strike density and recorded waveforms by the Swiss lightning position and tracking system. In *Proc. 21st Int. Conf. on Lightning Protection, Berlin, Germany*, pp. 313–18.

Moore, C.B., and Vonnegut, B. 1991. Comments on "A radar study of the plasma and geometry of lightning". *J. Atmos. Sci.* **48**: 369–70.

Moore, P.K., and Orville, R.E. 1990. Lightning characteristics in lake-effect thunderstorms. *Mon. Wea. Rev.* **118**: 1767–82.

Moreau, J.P. and Rustan, P.L. 1987. A study of lightning initiation based on VHF radiation. *Electromagnetics* **7**: 333–52.

Murphy, M.J., Krider, E.P., and Maier, M.W. 1996. Lightning charge analyses in small convection and precipitation electrification (CAPE) experiment storms. *J. Geophys. Res.* **101**: 29 615–26.

Murphy, M.J., Cummins, K.L., and Maier, L.M. 2000. The analysis and interpretation of three-dimensional lightning flash information. In *Proc. 80th American Meteorological Society Meeting, Long Beach, California*, paper 4.8.

Murty, R.C., and MacClement, W.D. 1973. VHF direction finder for lightning location. *J. Appl. Meteor.* **12**: 1401–5.

Nesbitt, S.W., Zhang, R., and Orville, R.E. 2000. Seasonal and global NO_x production by lightning estimated from the optical transient detector (OTD). *Tellus Chem. Phys. Meteor.* **52B**: 1206–15.

Newman, L.E. 1988. The relationship between radar reflectivity and cloud-to-ground lightning frequency in Minnesota. Master's thesis, College of Liberal Arts, University of Minnesota, 75 pp.

Nielsen, K.E. 1988. Lightning ground rates relative to mesocyclone evolution on 8 May 1986. Master's thesis, University of Oklahoma, Norman, Oklahoma, 91 pp.

Nielsen, K.E., Maddox, R.A., and Vasiloff, S.V. 1994. The evolution of cloud-to-ground lightning within a portion of the 10–11 June 1985 squall line. *Mon. Wea. Rev.* **122**: 1809–17.

Nisbet, J.S., Barnard, T.A., Forbes, G.S., Krider, E.P., Lhermitte, R., and Lennon, C.L. 1990a. A case study of the Thunderstorm Research International Project storm of July 11, 1978 – 1. Analysis of the data base. *J. Geophys. Res.* **95**: 5417–33.

Nisbet, J.S., Kasha, J.R., and Forbes, G.S. 1990b. A case study of the Thunderstorm Research International Project storm of July 11, 1978 – 2. Interrelations among the observable parameters controlling electrification. *J. Geophys. Res.* **95**: 5435–45.

Nishino, M., Iwai, A., and Kashiwagi, M. 1973. Location of the sources of atmospherics in and around Japan. In *Proc. Res. Inst. Atmospherics, Nagoya Univ., Japan* **20**: 9–21.

Nishizawa, Y., Iwai, A., and Satoh, M. 1980. VHF direction finding for lightnings at close ranges. In *Proc. Res. Inst. Atmospherics, Nagoya Univ., Japan* **27**: 11–24.

Norinder, H. 1953. Long-distance location of thunderstorms. In *Thunderstorm Electricity*, ed. H. R. Byers, pp. 276–327, Chicago: University of Chicago Press.

Oetzel, G.N., and Pierce, E.T. 1969. VHF technique for locating lightning. *Radio Sci.* **4**: 199–201.

Orville, R.E. 1981. Global distribution of midnight lightning – September to November 1977. *Mon. Wea. Rev.* **109**: 391–5.

Orville, R.E. 1986. Lightning phenomenology. In *The Earth's Electrical Environment*, eds. E.P. Krider and R.G. Roble, pp. 23–9, Washington, DC: National Academy Press.

Orville, R.E. 1987. Meteorological applications of lightning data (US National Report to International Union of Geodesy and Geophysics 1983–6). *Rev. Geophys.* **25**: 411–14.

Orville, R.E., Jr 1987. An analytical solution to obtain the optimum source location using multiple direction finders on a spherical surface. *J. Geophys. Res.* **92**: 10 877–86.

Orville, R.E. 1990a. Peak-current variations of lightning return strokes as a function of latitude. *Nature* **343**: 149–51.

Orville, R.E. 1990b. Winter lightning along the east coast. *Geophys. Res. Lett.* **17**: 713–15.

Orville, R.E. 1991a. Annual summary – lightning ground flash density in the contiguous United States – 1989. *Mon. Wea. Rev.* **119**: 573–7.

Orville, R.E. 1991b. Calibration of a magnetic direction finding network using measured triggered lightning return stroke peak currents. *J. Geophys. Res.* **96**: 17 135–42.

Orville, R.E. 1993. Cloud-to-ground lightning in the Blizzard of '93. *Geophys. Res. Lett.* **13**: 1367–70.

Orville, R.E. 1994. Cloud-to-ground lightning flash characteristics in the contiguous United States: 1989-1991. *J. Geophys. Res.* **99**: 10 833–41.

Orville, R.E. 1995. Lightning detection from ground and space. In *Handbook of Atmospheric Electrodynamics*, vol. 1, ed. H. Volland, pp. 137–49, Boca Raton, Florida: CRC Press.

Orville, R.E. 1999. Comments on "Large peak current cloud-to-ground lightning flashes during the summer months in the contiguous United States". *Mon. Wea. Rev.* **127**: 1937–8.

Orville, R.E., and Huffines, G.R. 1999. Lightning ground flash measurements over the contiguous United States: 1995–1997. *Mon. Wea. Rev.* **127**: 2693–703.

Orville, R.E., and Songster, H. 1987. The east coast lightning detection network. *IEEE Trans. Pow. Del.* **PWDR-2**: 899–904.

Orville, R.E., and Spenser, D.W. 1979. Global lightning flash frequency. *Mon. Wea. Rev.* **107**: 934–43.

Orville, R.E., Henderson, R.W., and Bosart, L.F. 1983. An east coast lightning detection network. *Bull. Amer. Meteor. Soc.* **64**: 1029–37.

Orville, R.E., Pyle, R.B., and Henderson, R.W. 1986. The east coast lightning detection network. *IEEE Trans. Pow. Syst.* **PWRS-1**: 243–6.

Orville, R.E., Weisman, R.A., Pyle, R.B., Henderson, R.W., and Orville, R.E. Jr 1987. Cloud-to-ground lightning flash characteristics from June 1984 through May 1985. *J. Geophys. Res.* **92**: 5640–4.

Orville, R.E., Henderson, R.W., and Bosart, L.F. 1988. Bipole patterns revealed by lightning locations in mesoscale storm systems. *Geophys. Res. Lett.* **15**: 129–32.

Panyukov, A.V. 1996. Estimation of the location of an arbitrary oriented dipole under single point direction finding. *J. Geophys. Res.* **101**: 14 977–82.

Panyukov, A.V., and Strauss, V.A. 1996. A method to determine parameters of a linear functional equations set and its application to the lightning location problem. In *Parameter Identification and Inverse Problems in Hydrology, Geology, and Ecology*, ed. J. Gottlieb and P. DuChateau, pp. 199–209, New York: Kluwer Academic.

Passi, R.M., and Lopez, R.E. 1989. A parametric estimation of systematic errors in networks of magnetic direction finders. *J. Geophys. Res.* **94**: 13 319–28.

Pawsey, J.L. 1957. Radar observation of lightning on 1.5 meters. *J. Atmos. Terr. Phys.* **11**: 289–90.

Peckham, D.W., Uman, M.A., and Wilcox, C.E. Jr 1984. Lightning phenomenology in the Tampa Bay Area. *J. Geophys. Res.* **89**: 11 789–805.

Pensky, M., and Cannon, J.R. 1999. Statistical estimation of locations of lightning events. *J. Geophys. Res.* **104**: 9635–41.

Perez, A.H., Wicker, L.J., and Orville, R.E. 1997. Characteristics of cloud-to-ground lightning associated with violent tornadoes. *Wea. Forecast.* **12**: 428–37.

Petersen, W.A., and Rutledge, S.A. 1992. Some characteristics of cloud-to-ground lightning in tropical northern Australia. *J. Geophys. Res.* **97**: 11 553–60.

Petersen, W.A., S.A. Rutledge, and R.E. Orville, 1996. Cloud-to-ground lightning observations from TOGA COARE: selected results and lightning location algorithms. *Mon. Wea. Rev.* **124**: 602–20.

Petrenko, I.A., and Kononov, I.I. 1992. Methods of lightning location. In *Proc. 9th Int. Conf. on Atmospheric Electricity, St. Petersburg, Russia*, vol. 1, pp. 292–5.

Pierce, E.T. 1956. Some techniques for locating thunderstorms from a single observing station. In *Vistas in Astronomy*, ed. A. Beer, vol. 2, pp. 850–5. London and New York: Pergamon Press.

Pierce, E.T. 1977a. Atmospherics and radio noise. In *Lightning, vol. 1, Physics of Lightning*, ed. R.H. Golde, pp. 351–84, New York: Academic Press.

Pierce, E.T. 1977b. Lightning warning and avoidance. In *Lightning, vol. 2, Lightning Protection*, ed. R.H. Golde, pp. 497–519, New York: Academic Press.

Pierce, E.T. 1982. Sferics and other electrical techniques for storm investigations in thunderstorms. In *Instruments and Techniques for Thunderstorm Observation and Analysis*, vol. 3, ed. E. Kessler, pp. 135–47, US Department of Commerce, Boulder, Colorado.

Pifer, A.E., Hiscox, W.L., Cummins, K.L., and Neumann, W.T. 1991. Range estimation techniques in single-station thunderstorm warning sensors based upon gated, wideband, magnetic direction finder technology. In *Proc. 1991 Int. Aerospace and Ground Conf. on Lightning and Static Electricity, Cocoa Beach, Florida*, vol. I, pp. 21/1–10, NASA Conf. Publ. 3106.

Pinto, I.R.C.A., Pinto, O. Jr, Gin, R.B.B., and Mendes, O. Jr 1992a. A coordinated study of a storm system over the South American continent, 1. Weather information and quasi-DC stratospheric electric field data. *J. Geophys. Res.* **97**: 18 195–204.

Pinto, I.R.C.A., Pinto, O. Jr, Gin, R.B.B., Diniz, J.H., de Araujo, R.L., and Carvalho, A.M. 1992b. A coordinated study of a storm system over the South American continent, 2. Lightning-related data. *J. Geophys. Res.* **97**: 18 205–13.

Pinto, O. Jr, Gin, R.B.B., Pinto, I.R.C.A., and Mendes, O. Jr 1996. Cloud-to-ground lightning flash characteristics in Southeastern Brazil for the 1992–1993 summer season. *J. Geophys. Res.* **101**: 29 627–35.

Pinto, O. Jr, Pinto, I.R.C.A., Gomes, M.A.S.S., Vitorello, I., Padilha, A.L., Diniz, J.H., Carvalho, A.M., and Cazetta, A. 1999a. Cloud-to-ground lightning in southeastern Brazil in 1993: 1. Geographical distribution. *J. Geophys. Res.* **104**: 31 369–79.

Pinto, I.R.C.A., Pinto O. Jr, Rocha, R.M.L., Diniz, J.H., Carvalho, A.M., and Cazetta, A. 1999b. Cloud-to-ground lightning in southeastern Brazil in 1993: 2. Time variations and flash characteristics. *J. Geophys. Res.* **104**: 31 381–7.

Poehler, H.A., and Lennon, C.L. 1979. Lightning detection and ranging system, LDAR, system description and performance objectives. NASA Technical Memorandum 741005.

Popov, M., and He, S. 2000. Identification of a transient electric dipole over a conducting half space using a simulated annealing algorithm. *J. Geophys. Res.* **105**: 20 821–31.

Proctor, D.E. 1971. A hyperbolic system for obtaining VHF radio pictures of lightning. *J. Geophys. Res.* **76**: 1478–89.

Proctor, D.E. 1973. Comments on 'A technique for accurately locating lightning at close range'. *J. Appl. Meteor.* **12**: 1419–23.

Proctor, D.E. 1974a. Sources of cloud-flash sferics. CSIR Special Report no. TEL 118, Pretoria, South Africa.

Proctor, D.E. 1974b. VHF radio pictures of lightning. CSIR Special Report no. TEL 120, Pretoria, South Africa.

Proctor, D.E. 1976. A radio study of lightning. Ph.D. thesis, University of Witwatersrand, Johannesburg, South Africa.

Proctor, D.E. 1981a. VHF radio pictures of cloud flashes. *J. Geophys. Res.* **86**: 4041–71.

Proctor, D.E. 1981b. Radar observations of lightning. *J. Geophys. Res.* **86**: 12 109–14.

Proctor, D.E. 1983. Lightning and precipitation in a small multicellular thunderstorm. *J. Geophys. Res.* **88**: 5421–40.

Proctor, D.E. 1984. Correction to "Lightning and precipitation in a small multicellular thunderstorm". *J. Geophys. Res.* **89**: 11 826.

Proctor, D.E. 1991. Regions where lightning flashes began. *J. Geophys. Res.* **96**: 5099–112.

Proctor, D.E., Uytenbogaardt, R., and Meredith, B.M. 1988. VHF radio pictures of lightning flashes to ground. *J. Geophys. Res.* **93**: 12 683–727.

Rafalsky, V.A., Nickolaenko, A.P., Shvets, A.V., and Hayakawa, M. 1995. Location of lightning discharges from a single station. *J. Geophys. Res.* **100**: 20 829–38.

Rakov, V.A. 1990. Modern passive lightning locating systems. *Meteor. Gidrol.* **11**: 118–23.

Rakov, V.A. 1993. Data acquired with the LLP lightning locating systems. *Meteor. Gidrol.* **7**: 105–14.

Rakov, V.A., and Uman, M.A. 1990. Some properties of negative cloud-to-ground lightning flashes versus stroke order. *J. Geophys. Res.* **95**: 5447–53.

Rakov, V.A., Uman, M.A., Thottappillil, R., and Shindo, T. 1991. Statistical characteristics of negative ground flashes as derived from electric field and TV records. *Proc. USSR Academy of Sciences (Izvestiya AN SSSR, ser. Energetika i Transport)* **37**: 61–71.

Rakov, V.A., Thottappillil, R., and Uman, M.A. 1992. On the empirical formula of Willett *et al.* relating lightning return stroke peak current and peak electric field. *J. Geophys. Res.* **97**: 11 527–33.

Rakov, V.A., Uman, M.A., and Thottappillil, R. 1994. Review of lightning properties from electric field and TV observations. *J. Geophys. Res.* **99**: 10 745–50.

Ray, P.S., MacGorman, D.R., Rust, W.D., Taylor, W.L., and Rasmussen, L.W. 1987. Lightning location relative to storm structure in a supercell storm and a multicell storm. *J. Geophys. Res.* **92**: 5713–24.

Reap, R.M. 1986. Evaluation of cloud-to-ground lightning data from the western United States for the 1983–84 summer seasons. *J. Clim. Appl. Meteor.* **25**: 785–99.

Reap, R.M. 1991. Climatological characteristics and objective prediction of thunderstorms over Alaska. *Wea. Forecast.* **6**: 309–19.

Reap, R.M. 1994. Analysis and prediction of lightning strike distribution associated with synoptic map types over Florida. *Mon. Wea. Rev.* **122**: 1698–1715.

Reap, R.M., and MacGorman, D.R. 1989. Cloud-to-ground lightning – climatological characteristics and relationships to model fields, radar observations, and severe local storms. *Mon. Wea. Rev.* **117**: 518–35.

Reap, R.M., and Orville, R.E. 1990. The relationships between network lightning locations and surface hourly observations of thunderstorms. *Mon. Wea. Rev.* **118**: 94–108.

Redelinghuys, M.G., van der Merwe, W.C., Jandrell, I.R., and Redelinghuys, D. 1996. The effects of flash sort parameters on lightning design statistics from a lightning position and tracking system. In *Proc. 23rd Int. Conf. Lightning Protection, Florence, Italy*, pp. 161–5.

Rhodes, C.T., Shao, X.M., Krehbiel, P.R., Thomas, R.J., and Hayenga, C.O. 1994. Observations of lightning phenomena using radio interferometry. *J. Geophys. Res.* **99**: 13 059–82.

Richard, P. 1990. SAFIR system: an application of real-time VHF lightning localization to thunderstorm monitoring. Preprint, *Proc. 16th Conf. on Severe Local Storms and Conf. on Atmospheric Electricity, Kananaskis Provincial Park, Alberta, Canada*, pp. J21–6, Am. Meteor. Soc., Boston.

Richard, P. 1992. Severe thunderstorm nowcasting. In *Proc. Int. Conf. on Lightning and Static Electricity, Atlantic City, New Jersey*, FAA Report DOT/FAA/CT-92/20, 77-1–77-9.

Richard, P., and Auffray, G. 1985. VHF–UHF interferometric measurements, applications to lightning discharge mapping. *Radio Sci.* **20**: 171–92.

Richard, P., Delannoy, A., Labaune, G., and Laroche, P. 1986. Results of spatial and temporal characterization of the VHF–UHF radiation of lightning. *J. Geophys. Res.* **91**: 1248–60.

Richard, P., Soulage, A., Laroche, P., and Appel, J. 1988. The SAFIR lightning monitoring and warning system, application to aerospace activities. In *Proc. Int. Aerospace and Ground Conf. on Lightning and Static Electricity, Oklahoma City, Oklahoma*, pp. 383–90, National Interagency Coordination Group.

Rison, W., Thomas, R.J., Krehbiel, P.R., Hamlin, T. and Harlin, J. 1999. A GPS-based three-dimensional lightning mapping system: initial observations in central New Mexico. *Geophys. Res. Lett.* **26**: 3573–6.

Roohr, P.B., and Vonder Haar, T.H. 1994. A comparative analysis of the temporal variability of lightning observations and GOES imagery. *J. Appl. Meteor.* **33**: 1271–90.

Rorig, M.L., and Ferguson, S.A. 1999. Characteristics of lightning and wildland fire ignition in the Pacific Northwest. *J. Appl. Meteor.* **38**: 1565–75.

Rottger, J., Liu, C.H., Pan, C.J., and Su, S.Y. 1995. Characteristics of lightning echoes observed with VHF ST radar. *Radio Sci.* **30**: 1085–97.

Roussel-Dupré, R.A., Jacobson, A.R., and Triplett, L.A. 2001. Analysis of FORTE data to extract ionospheric parameters. *Radio Sci.* **36**: 1615–30.

Ruhnke, L.H. 1962. Distance to lightning strokes as determined from electrostatic field strength measurements. *J. Appl. Meteor.* **1**: 544–7.

Ruhnke, L.H. 1971. Determining distance to lightning strokes from a single station. NOAA Technical Report ERL 195-APCL 16, US Department of Commerce.

Rust, W.D., and Doviak, R.J. 1982. Radar research on thunderstorms and lightning. *Nature* **297**: 461–8.

Rust, W.D., and MacGorman, D.R. 1988. Techniques for measuring electrical parameters of thunderstorms. In *Thunderstorms, vol. 3, Instruments and Techniques for Thunderstorm Observation and Analysis*, ed. E. Kessler, pp. 91–118, Norman, Oklahoma: University of Oklahoma Press.

Rust, W.D., MacGorman, D.R., and Arnold, R.T. 1981. Positive cloud to ground lightning flashes in severe storms. *Geophys. Res. Lett.* **8**: 791–4.

Rust, W.D., Taylor, W.L., and MacGorman, D. 1982. Preliminary study of lightning location relative to storm structure. *AIAA J.* **20**: 404–9.

Rustan, P.L. Jr 1979. Properties of lightning derived from time series analysis of VHF radiation data. Ph.D. thesis, University of Florida, Gainesville.

Rustan, P.L., Uman, M.A., Childers, D.G., Beasley, W.H., and Lennon, C.L. 1980. Lightning source locations from VHF radiation data for a flash at Kennedy Space Center. *J. Geophys. Res.* **85**: 4893–903.

Rutledge, S.A., Lu, C., and MacGorman, D.R. 1990. Positive cloud-to-ground lightning in mesoscale convective systems. *J. Atmos. Sci.* **47**: 2085–100.

Rutledge, S.A., and MacGorman, D.R. 1988. Cloud-to-ground lightning activity in the 10–11 June 1985 mesoscale convective system observed during Oklahoma-Kansas PRE-STORM Project. *Mon. Wea. Rev.* **116**: 1393–408.

Rutledge, S.A., and Petersen, S.A. 1994. Vertical radar reflectivity structure and cloud-to-ground lightning in the stratiform region of MCSs: further evidence for in situ charging in the stratiform region. *Mon. Wea. Rev.* **122**: 1760–76.

Ryan, P.A., and Spitzer, N. 1977. Stormscope. US Patent no. 4023 408, 12 pp.

Sadiku, M.N.O. 1994. *Elements of Electromagnetics*, 821 pp., Orlando, Florida: Sounders College.

Samsury, C.E., and Orville, R.E. 1994. Cloud-to-ground lightning in tropical cyclones: a study of hurricanes Hugo (1989) and Jerry (1989). *Mon. Wea. Rev.* **122**: 1887–96.

Sao, K., and Jindoh, H. 1974. Real time location of atmospherics by single station techniques and preliminary results. *J. Atmos. Terr. Phys.* **36**: 261–6.

Schütte, T.E. 1987. Optimum performance of lightning localization systems. Ph.D. thesis, Uppsala University, Sweden.

Schütte, T., Pisler, E., Filpovic, D., and Israelsson, S. 1987a. Acceptance of lightning detectors and localization systems under different damping conditions. *J. Atmos. Ocean. Tech.* **4**: 401–10.

Schütte, T., Pisler, E., and Israelsson, S. 1987b. A new method for the measurement of the site error of lightning localization systems – description and first results. *J. Atmos. Ocean. Tech.* **4**: 305–11.

Schütte, T., Cooray, V., and Israelsson, S. 1988. Recalculation of lightning localization system acceptance using a refined damping model. *J. Atmos. Ocean. Tech.* **5**: 375–80.

Schuur, T.J., Smull, B.F., Rust, W.D., and Marshall, T.C. 1991. Electrical and kinematic structure of the stratiform precipitation region trailing an Oklahoma squall line. *J. Atmos. Sci.* **48**: 825–42.

Seimon, A. 1993. Anomalous cloud-to-ground lightning in an F5-tornado-producing supercell thunderstorm on 28 August 1990. *Bull. Amer. Meteor. Soc.* **74**: 189–203.

Shao, X.M. 1993. The development and structure of lightning discharges observed by VHF radio interferometer. Ph.D. dissertation, New Mexico Inst. of Min. and Technol., Socorro.

Shao, X.-M., and Jacobson, A.R. 2001. Polarization observations of broadband VHF signals by the FORTE satellite. *Radio Sci.* **36**: 1573–89.

Shao, X.M., and Krehbiel, P.R. 1996. The spatial and temporal development of intracloud lightning. *J. Geophys. Res.* **101**: 26 641–68.

Shao, X.M., Krehbiel, P.R., Thomas, R.J., and Rison, W. 1995. Radio interferometric observations of cloud-to-ground lightning phenomena in Florida, *J. Geophys. Res.* **100**: 2749–83.

Shao, X.M., Holden, D.N., and Rhodes, C.T. 1996. Broad band radio interferometry for lightning observations. *Geophys. Res. Lett.* **23**: 1917–20.

Shao, X.M., Rhodes, C.T., and Holden, D.N. 1999. RF radiation observations of positive cloud-to-ground flashes. *J. Geophys. Res.* **104**: 9601–8.

Shchukin, G.G., Stepanenko, V.D., Yegorov, A.D., Galperin, S.M., and Karavayev, D.M. 1999. Radiophysical studies of atmosphere and underlying surface. In *Contemporary Investigation at Main Geophysical Observatory*, vol. 1, eds. M.E. Berlyand and V.P. Meleshko, pp. 172–90, St. Petersburg, Russia: Gidrometeoizdat.

Sheridan, S.C., Griffiths, J.F., and Orville, R.E. 1997. Warm season cloud-to-ground lightning–precipitation relationships in the South-Central United States. *Wea. Forecast.* **12**: 449–58.

Shindo, T., and Yokoyama, S. 1998. Lightning occurrence data observed with lightning location systems in Japan: 1992–1995. *IEEE Trans. Pow. Del.* **13**: 1468–74.

Smith, S.B. 1993. Comments on "Lightning ground flash density in the contiguous United States – 1989". *Mon. Wea. Rev.* **121**: 1572–5.

Soriano, L.R., de Pablo, F., and Diez, E.G. 2001. Cloud-to-ground lightning activity in the Iberian Peninsula: 1992–1994. *J. Geophys. Res.* **106**: 11 891–901.

Sparrow, J.G., and Ney, E.P. 1968. Discrete light sources observed by satellite OSO-B. *Science* **161**: 459–60.

Sparrow, J.G., and Ney, F.E. 1971. Lightning observations by satellite. *Nature* **232**: 540–1.

Stansfield, R.G. 1947. Statistical theory of D.F. fixing. *J. Instr. Electr. Eng.* **2A**: 762–70.

Starr, S., Sharp, D., Merceret, F., Madura, J., and Murphy, M. 1998. LDAR, a three-dimensional lightning warning system: its development and use by the government, and transition to public availability. In *Proc. 35th Space Congress, Horizons Unlimited, Cocoa Beach, Florida*, pp. 299–305, Canaveral Council of Technical Societies.

Stepanenko, V.D., and Galperin, S.M. 1983. *Radiotechnical Methods for Studying Thunderstorms*, 204 pp., Leningrad: Gidrometeoizdat.

Stern, A.D., Brady, R.H. III, Moore, P.D., and Carter, G.M. 1994. Identification of aviation weather hazards based on the integration of radar and lightning data. *Bull. Amer. Meteor. Soc.* **75**: 2269–80.

Stolzenburg, M. 1990. Characteristics of the bipolar pattern of lightning locations observed in 1988 thunderstorms. *Bull. Amer. Meteor. Soc.* **71**: 1331–8.

Stolzenburg, M. 1994. Observations of high ground flash densities of positive lightning in summertime thunderstorm. *Mon. Wea. Rev.* **122**: 1740–50.

Stolzenburg, M., Marshall, T.C., Rust, W.D., and Smull, B.F. 1994. Horizontal distribution of electrical and meteorological conditions across the stratiform region of a mesoscale convective system *Mon. Wea. Rev.* **122**: 1777–97.

Suszcynsky, D.M., Kirkland, M.W., Jacobson, A.R., Franz, R.C., Knox, S.O., Guillen, J.L.L., and Green, J.L. 2000. FORTE observations of simultaneous VHF and optical emissions from lightning: basic phenomenology. *J. Geophys. Res.* **105**: 2191–201.

Suszcynsky, D.M., Light, T.E., Davis, S., Green, J.L., Guillen, J.L.L., and Myre, W. 2001. Coordinated observations of optical lightning from space using the FORTE photodiode detector and CCD imager. *J. Geophys. Res.* **106**: 17 897–906.

Szymanski, E.W., and Rust, W.D. 1979. Preliminary observations of lightning radar echoes and simultaneous electric field changes. *Geophys. Res. Lett.* **6**: 527–30.

Szymanski, E.W., Szymanski, S.J., Holmes, C.R., and Moore, C.B. 1980. An observation of a precipitation echo intensification associated with lightning. *J. Geophys. Res.* **85**: 1951–3.

Taylor, W.L. 1963. Radiation field characteristics of lightning discharges in the band 1kc/s to 100 kc/s. *J. Res. National Bureau of Standards* **67D**: 539–50.

Taylor, W.L. 1969. Determining lightning stroke height from ionospheric components of atmospheric waveforms. *J. Atmos. Terr. Phys.* **32**: 983–90.

Taylor, W.L. 1973. An electromagnetic technique for tornado detection. *Weatherwise* **26**: 70–1.

Taylor, W.L. 1978. A VHF technique for space–time mapping of lightning discharge processes. *J. Geophys. Res.* **83**: 3575–83.

Taylor, W.L., and A.G. Jean 1959. Very low frequency radiation spectra of lightning discharges. *J. Res. National Bureau of Standards* **63D**: 199–204.

Taylor, W.L. Brandes, E.A., Rust, W.D., and MacGorman, D.R. 1984. Lightning activity and severe storm structure. *Geophys. Res. Lett.* **11**: 545–8.

Théry, C. 2001. Evaluation of LPATS data using VHF interferometric observations of lightning flashes during the EULINOX experiment. *Atmos. Res.* **56**: 397–409.

Thomas, R.J., Krehbiel, P.R., Rison, W., Hamlin, T., Boccippio, D.J., Goodman, S.J., and Christian, H.J. 2000. Comparison of ground-based 3-dimensional lightning mapping observations with satellite-based LIS observations in Oklahoma. *Geophys. Res. Lett.* **27**: 1703–6.

Thomas, R.J., Krehbiel, P.R., Rison, W., Hamlin, T., Harlin, J., and Shown, D. 2001. Observations of VHF source powers radiated by lightning. *Geophys. Res. Lett.* **28**: 143–6.

Thomson, E.M., Stone, J.W., Uman, M.A., and Beasley, W.H. 1981. Location of lightning sources at VHF using cross correlation techniques. *Trans. AGU* **62**: 880.

Thomson, E.M., Medelius, P.J., and Davis, S. 1994. System for locating the sources of wideband dE/dt from lightning. *J. Geophys. Res.* **99**: 22 793–802.

Thottappillil, R, Rakov, V.A., Uman, M.A., Beasley, W.H., Master, M.J., and Shelukhin, D.V. 1992. Lightning subsequent stroke electric field peak greater than the first stroke peak and multiple ground terminations. *J. Geophys. Res.* **97**: 7503–9.

Thunderstorm Direction and Range Finder "Ochag-2П". 1988, Leningrad: Gidrometeoizdat.

Tierney, H.E., Jacobson, A.R., Beasley, W.H., and Argo, P.E. 2001. Determination of source thunderstorms for VHF emissions observed by the FORTE satellite. *Radio Sci.* **36**: 79–96.

Tsuruda, K., and Ikeda, M. 1979. Comparison of three different types of VLF direction-finding techniques. *J. Geophys. Res.* **84**: 5325–32.

Turman, B.N. 1977. Detection of lightning superbolts. *J. Geophys. Res.* **82**: 2566–8.

Turman, B.N. 1978. Analysis of lightning data from the DMSP satellite. *J. Geophys. Res.* **83**: 5019–24.

Turman, B.N. 1979. Lightning detection from space. *Amer. Scientist* **67**: 321–9.

Turman, B.N., and Edgar, B.C. 1982. Global lightning distributions at dawn and dusk. *J. Geophys. Res.* **87**: 1191–206.

Turman, B.N., and Tettelbach, R.J. 1980. Synoptic-scale satellite observations in conjunction with tornadoes. *Mon. Wea. Rev.* **108**: 1878–82.

Turman, B.N., Cummings, T.B., and Deabenderfer, P.R. 1976. Measurements of optical power radiated by Florida lightning. Headquarters U.S. Air Force, Airforce Technical Applications Center, Report no. 76–6.

Tyahla, L.J., and Lopez R.E. 1994. Effect of surface conductivity on the peak magnetic field radiated by first return stroke in cloud-to-ground lightning, *J. Geophys. Res.* **99**: 10 517–25.

Uman, M.A. 1978. Criticism of "Comments on 'Detection of lightning superbolts'" by B.N. Turman. *J. Geophys. Res.* **83**: 5523.

Uman, M.A. 1986. Application of advances in lightning research to lightning protection. In *The Earth's Electrical*

Environment, eds. E.P. Krider and R.G. Roble, pp. 61–9, Washington, DC: National Academy Press.

Uman, M.A., Beasley, W.H., Tiller, J.A., Lin, Y.T., Krider, E.P., Weidmann, C.D., Krehbiel, P.R., Brook, M., Few, A.A. Jr, Bohannon, J.L., Lennon, C.L., Poehler, H.A., Jafferis, W., Gulick, J.R., and Nicholson, J.R. 1978. An unusual lightning flash at Kennedy Space Center. *Science* **201**: 9–16.

Uman, M.A., Lin, Y.T., and Krider, E.P. 1980. Errors in magnetic direction finding due to non-vertical lightning channels. *Radio Sci.* **15**: 35–9.

Uman, M.A., Krider, E.P., Rustan, P.L., Kuhlman, B.P., Moreau, J.P., Thomson, E.M., Stone, J.W., and Beasley, W.H. 1983. Airborne and ground based lightning electric and magnetic fields and VHF source location for three nearby lightning flashes. In *Proc. Int. Aerospace and Ground Conf. on Lightning and Static Electricity, Fort Worth, Texas*, pp. 51-1-8.

Ushio, T., Kawasaki, Z.-I., Ohta, Y., and Matsuura, K. 1997. Broadband interferometric measurement of rocket triggered lightning in Japan. *Geophys. Res. Lett.* **24**: 2769–72.

Vera, N.L. 1985. Study of Ontario lightning intensity. M. Eng. thesis, University of Toronto, Ontario, Canada.

Volland, H. 1982. Low frequency radio noise. In *Handbook of Atmospherics*, vol. I, ed. H. Volland, pp. 179–250, Boca Raton, Florida: CRC Press.

Volland, H., Schafer, J., Ingmann, P., Harth, W., Heydt, G., Eriksson, A.J., and Manes, A. 1983. Registration of thunderstorm centers by automatic atmospherics stations. *J. Geophys. Res.* **88**: 1503–18.

Vorpahl, J.A., Sparrow, J.G., and Ney, E.P. 1970. Satellite observations of lightning. *Science* **169**: 860–2.

Wacker, R.S., and Orville, R.E. 1996. Peak current estimates in the NLDN? *Eos, Trans. AGU* **77**(47): Fall Meeting Suppl., p. F89.

Wacker, R.S., and Orville, R.E. 1999a. Changes in measured lightning flash count and return stroke peak current after the 1994 U.S. National Lightning Detection Network upgrade; 2. Theory. *J. Geophys. Res.* **104**: 2159–62.

Wacker, R.S., and Orville, R.E. 1999b. Changes in measured lightning flash count and return stroke peak current after the 1994 U.S. National Lightning Detection Network upgrade; 1: Observations. *J. Geophys. Res.* **104**: 2151–7.

Wangsness, D.L. 1973. A new method of position estimation using bearing measurements. *IEEE Trans. Aerosp. Electron. Syst.* **AES-9**: 959–60.

Warwick, J.W., Hayenga, C.O., and Brosnahan, J.W. 1979. Interferometric position of lightning sources at 34 MHz. *J. Geophys. Res.* **84**: 2457–68.

Watson, A.I., Lopez, R.E., Holle, R.L., and Daugherty, J.R. 1987. The relationship of lightning to surface convergence at Kennedy Space Center: a preliminary study. *Wea. Forecast.* **2**: 140–57.

Watson, A.I., Holle, R.L., Lopez, R.E., Ortiz, R., and Nicholson, J.R. 1991. Surface wind convergence as a short-term predictor of cloud-to-ground lightning at Kennedy Space Center. *Wea. Forecast.* **6**: 49–64.

Watson, A.I., Holle, R.L., and Lopez, R.E. 1994a. Cloud-to-ground lightning and upper-air patterns during bursts and breaks in the southwest monsoon. *Mon. Wea. Rev.* **122**: 1726–39.

Watson, A.I., Lopez, R.E., and Holle, R.L. 1994b. Diurnal cloud-to-ground lightning patterns in Arizona during the southwest monsoon. *Mon. Wea. Rev.* **122**: 1716–25.

Watson, A.I., Holle, R.L., and Lopez, R.E. 1995. Lightning from two national detection networks related to vertically integrated liquid and echo-top information from WSR-88D radar. *Wea. Forecast.* **10**: 592–605.

Watson-Watt, R.A., and Herd, J.F. 1926. An instantaneous direct-reading radio goniometer. *J. Inst. Electr. Eng.* **64**: 611–22.

White, J., and Driggans, R. 1990. TVA's experience with the SUNYA lightning detection network. *IEEE Trans. Pow. Del.* **5**: 2054–62.

Whitney, B. F., and Asgeirsson, H. 1991. Lightning location and storm severity display system. *IEEE Trans. Pow. Del.* **6**: 1715–20.

Williams, E.R., Geotis, S.G., and Bhattacharya, A.B. 1989a. A radar study of the plasma and geometry of lightning. *J. Atmos. Sci.* **46**: 1173–85.

Williams, E.R., Weber, M.E., and Orville, R.E. 1989b. The relationship between lightning type and convective state of thunderclouds. *J. Geophys. Res.* **94**: 13 213–20.

Williams, E.R., Mazur, V., and Geotis, S.G. 1990. Lightning investigation. In *Radar in Meteorology*, ed. D. Atlas, pp. 143–50, Boston, Massachusetts: Amer. Meteor. Soc.

Winckler, J.R. 1993. Fast low-level light pulses from the night sky observed with the skyflash program. *J. Geophys. Res.* **98**: 8775–83.

Winckler, J.R. 1995. Further observations of cloud–ionosphere electrical discharges above thunderstorms. *J. Geophys. Res.* **100**: 14 335–45.

Winn, W.P., Aldridge, T.V., and Moore, C.B. 1973. Video-tape recordings of lightning flashes. *J. Geophys. Res.* **78**: 4515–19.

(WMO 1955) World Meteorological Organization 1955. Technical Note 12, Atmospheric techniques, Secretariat of the World Meteorological Organization, Geneva, Switzerland.

Wolfe, W.L. 1983. Aircraft-borne lightning sensor. *Optical Eng.* **22**: 456–9.

Yamashita, M., and Sao, K. 1974a. Some considerations of the polarization error in direction-finding of atmospherics II. Effect of the inclined electric dipole. *J. Atmos. Terr. Phys.* **36**: 1633–41.

Yamashita, M., and Sao, K. 1974b. Some considerations of the polarization error in direction-finding of atmospherics I. Effect of the Earth's magnetic field. *J. Atmos. Terr. Phys.* **36**: 1623–32.

Yamashita, H., Iwai, A., Satoh, M., and Katoh, T. 1983. Technical note: VHF direction finder for locating lightnings at close ranges. In *Proc. Res. Inst. Atmospherics, Nagoya Univ., Japan.* **30**: 15–24.

Zipser, E.J. 1994. Deep cumulonimbus cloud systems in the tropics with and without lightning. *Mon. Wea. Rev.* **122**: 1837–51.

References and bibliography for Chapter 17

Zrnić, D.S., Rust, W.D., and Taylor, W.L. 1982. Doppler radar echoes of lightning and precipitation at vertical incidence. *J. Geophys. Res.* **87**: 7179–91.

Zuelsdorf, R.S., Strangeway, R.J., Russel, C.T., Casler C., Christian, H.J., and Franz, R. 1997. Trans-ionospheric pulse pairs (TIPPs): their geographic distribution and seasonal variations. *Geophys. Res. Lett.* **24**: 3165–8.

Zuelsdorf, R.S., Casler, C., Strangeway, R.J., and Russel, C.T. 1998a. Ground detection of trans-ionospheric pulse pairs by stations in the National Lightning Detection Network. *Geophys. Res. Lett.* **25**: 481–4.

Zuelsdorf, R.S., Strangeway, R.J., Russell, C.T., and Franz, R. 1998b. Trans-ionospheric pulse pairs (TIPPs): their occurrence rates and diurnal variation. *Geophys. Res. Lett.* **25**: 3709–12.

Zuelsdorf, R.S., Franz, R.C., Strangeway, R.J., and Russell, C.T. 2000. Determining the source of strong LF/VLF TIPP events: implications for association with NPBPs and NNBPs. *J. Geophys. Res.* **105**(D16): 20 725–36.

18 Deleterious effects of lightning and protective techniques

How to secure Houses Etc. from Lightning.
It has pleased God in his goodness to mankind, at length to discover to them the means of securing their habitations and other buildings from mischief by thunder and lightning. The method is this: Provide a small iron rod (it may be made of the rod-iron used by the nailers) but of such a length, that one end being three or four feet in the moist ground, the other may be six or eight feet above the highest part of the building. To the upper end of the rod fasten about a foot of brass wire, the size of a common knitting-needle, sharpened to a fine point; the rod may be secured to the house by a few small staples. If the house or barn be long, there may be a rod and point at each end, and a middling wire along the ridge from one to the other. A house thus furnished will not be damaged by lightning, it being attracted by the points, and passing thro the metal into the ground without hurting any thing. Vessels also, having a sharp pointed rod fix'd on the top of their masts, with a wire from the foot of the rod reaching down, round one of the shrouds, to the water, will not be hurt by lightning.

B. Franklin (1753)

18.1. Introduction

Lightning does damage to a wide range of objects and systems, including miniature electronic integrated circuits, trees, overhead and underground electric power and communication systems, buildings, boats, and aircraft and launch vehicles in flight. While the general principles of lightning protection apply to all objects and systems, there is often a specific focus in the techniques used to protect different classes of objects or systems. In this chapter we will examine the basic mechanisms of lightning damage (Section 18.2), discuss the various aspects of lightning protection (Section 18.3), examine lightning's interaction with a variety of objects and systems (Section 18.4), and consider lightning test standards (Section 18.5). The interaction of lightning with aircraft and launch vehicles was discussed in Chapter 10, and lightning hazards to humans and animals constitute the subject of Chapter 19; so those topics, while they would be appropriate for this chapter, are not considered in any detail here. Some historical aspects of lightning protection are additionally considered in Section 1.1, a discussion of lightning incidence to structures in Section 2.9, and a detailed characterization of the lightning attachment process is found in Section 4.5.

18.2. Basic mechanisms of lightning damage

The type and amount of lightning damage that an object suffers depends on both the characteristics of the lightning discharge and the properties of the object. The physical characteristics of lightning of most interest are various properties of the current waveform and of the RF electromagnetic fields. Damage can also occur from electromagnetic radiation in other frequency bands and from the acoustic shock wave (Sections 11.3 and 19.3).

Four distinct properties of the lightning current waveform can be considered important in producing damage: (i) the peak current, (ii) the maximum rate of change of current, (iii) the integral of the current over time (i.e., the charge transferred), and (iv) the integral of the current squared over time, the "action integral", all of these properties having been discussed in Chapters 4, 5, and 6. We now briefly examine each of these properties and the type of damage to which they are thought to be related.

(i) *Peak current* For objects or systems that present an essentially resistive impedance such as, under certain conditions, a ground rod driven into the earth, a long power line, or a tree, the voltage V on the object or system with respect to remote ground will be proportional to the current

I via Ohm's law, $V = RI$, where R is the effective resistance at the strike point. For example, a 30 kA peak current injected into a power line phase conductor with a 500 Ω characteristic impedance (the effective resistance is 250 Ω, since 500 Ω is "seen" in each direction) produces a line voltage of 7.5 MV with respect to the earth. Such a large voltage can lead to an electric discharge from the struck phase conductor to adjacent phase or neutral conductors or to ground, across insulating materials or through the air.

(ii) *Maximum rate of change of current* For objects that present an essentially inductive impedance such as, under some circumstances, wires in an electronic system, the peak voltage will be proportional to the maximum rate of change of the lightning current ($V = L\,dI/dt$, where L is the inductance of the length of wire and V is the voltage difference between the two ends of the wire). For example, if a "ground" wire connecting two electronic systems (for example, in a communications tower and in an adjacent electronics building) has an inductance per unit length of 10^{-6} H m^{-1} and if 10 percent of the direct lightning current flows in the wire, producing $dI/dt = 10^{10}$ A s^{-1}, then 10 kV will be produced across each meter of the wire. It is easy to understand how even a very small fraction of the lightning current circulating in grounding and bonding wires can cause damage to solid-state electronic circuits that have communication, power and other inputs "grounded" at different locations.

(iii) *Integral of the current over time* The severity of heating or burn-through of metal sheets such as airplane wing surfaces and metal roofs is, to a first approximation, proportional to the lightning charge transferred, which is in turn proportional to the energy delivered to the surface (e.g., Bellaschi 1941; McEachron and Hagenguth 1942; Hagenguth 1949; Kofoid 1970; Testé *et al.* 2000). This is the case because the input power to the conductor surface is the product of the current and the more or less current-independent voltage drop at the arc–metal interface, this voltage drop being typically 5 to 10 V. The voltage drop at the interface can be thought of as a contact potential between the two materials or the difference in the work functions of the two materials, although the situation is considerably more complex than the case of two contiguous solid conductors in that the metal surface is partially melted and the "air" arc contains metal vapor. Generally, large charge transfers are due to long-duration (tens to hundreds of milliseconds) lightning currents, such as long continuing currents (Section 4.8), whose magnitude is in the tens to hundreds of amperes range, rather than return strokes having larger currents but relatively short duration and hence producing relatively small charge transfers. Additionally, even those impulse currents which do have relatively large charge transfers cause only relatively minor surface damage on metal sheets, apparently because the current duration is too short to allow penetration of heat into the metal.

(iv) *Action integral* The heating and melting of resistive materials, which may or may not be relatively good conductors, and the explosion of poorly conducting materials are, to a first approximation, related to the value of the action integral, that is, the time integral of the Joule heating power $I^2(t)R$ for the case that $R = 1\,\Omega$. Thus the action integral is a measure of the ability of the lightning current to generate heat in a strike object characterized by a resistance R. About five percent of negative first strokes in ground flashes have action integrals exceeding 5.5×10^5 A^2 s (Table 4.4); about five percent of positive strokes have action integrals exceeding 10^7 A^2 s (Table 5.1). In the case of most materials that are poorly conducting, this heat vaporizes the internal material and the resultant gas pressure causes an explosive fracture. In addition to heating effects, the action integral is also a measure of some mechanical effects such as the crushing of hollow metal tubes carrying lightning current. The effect is a function of both the instantaneous force, which is proportional to the square of the current, and the duration of application of the force. In this case, the applied force must also exceed some threshold value.

Electromagnetic fields from lightning that impinge on any conducting objects induce currents and resultant voltages in those objects. Two properties of the electromagnetic fields are sufficient to describe most of the important damaging effects, commonly the destruction of electronic components: (i) the peak values of the electric and magnetic fields and (ii) the maximum rates of change of the fields. For certain types of unintended antennas, such as elevated conductors that are capacitively coupled to ground, the peak induced voltage on the conductors with respect to ground is proportional to the peak electric field. For other unintended antennas, such as a loop of wire in an electronic circuit, some underground communication cables, and elevated conductors resistively coupled to ground, the peak voltage is proportional to the maximum rate of change of the electric or the magnetic field. The degree of coupling of fields through holes or apertures in the metal skins of aircraft and spacecraft is generally proportional to the rate of change of the electric and magnetic fields.

18.3. Protection
18.3.1. Types of protection

There are generally two aspects of lightning protection design: (i) diversion and shielding, primarily intended for structural protection but also serving to reduce the lightning electric and magnetic fields within the structure, and (ii) the limiting of currents and voltages on electronic, power, and communication systems via surge protection.

A: 50 ft (15 m) maximum spacing
B: 20 ft (6 m) or 25 ft (7.6 m) maximum spacing

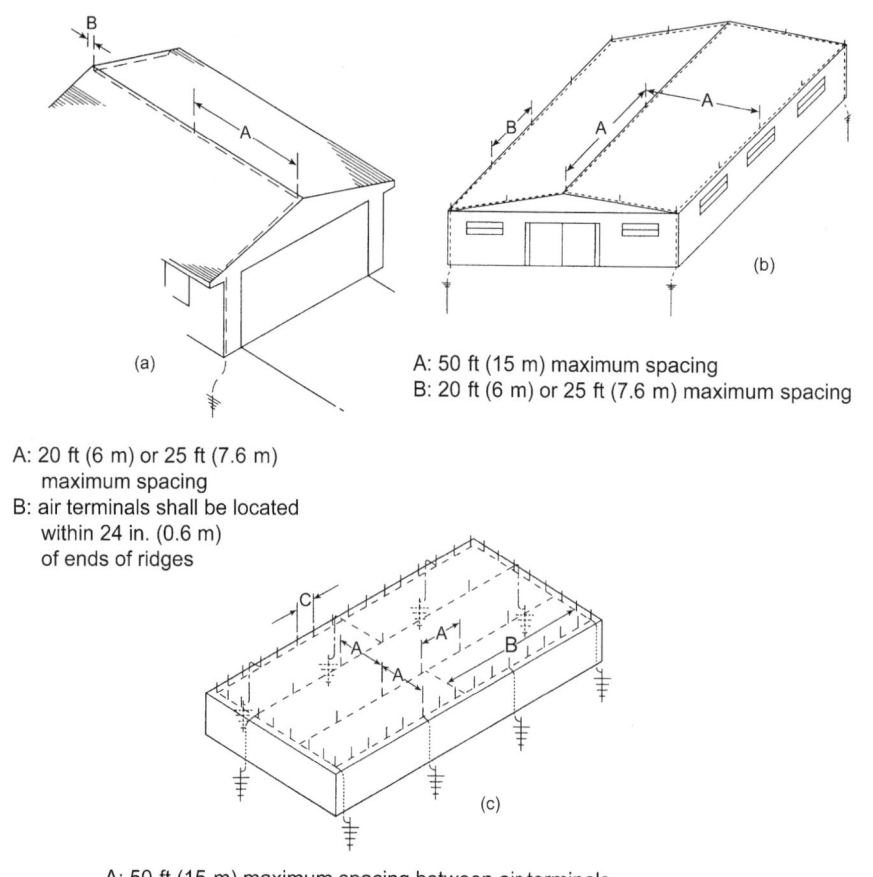

A: 20 ft (6 m) or 25 ft (7.6 m)
   maximum spacing
B: air terminals shall be located
   within 24 in. (0.6 m)
   of ends of ridges

A: 50 ft (15 m) maximum spacing between air terminals
B: 150 ft (45 m) maximum length of cross run conductor permitted
   without a connection from the cross run conductor to the main
   perimeter or down conductor
C: 20 ft (6 m) maximum spacings between air terminals along edge

Fig. 18.1. Lightning protection of ordinary structures via diversion as recommended by NFPA-780.

We now summarize each of these aspects of protection design.

(i) On a building or other structure to be protected, the diversion of lightning currents to ground is accomplished by (a) either one or more connected vertical lightning rods (also referred to as air terminals, Franklin rods, or lightning conductors) or a system of connected horizontal wires intended for the same purpose, that of intercepting the descending lightning leader (Sections 2.9 and 4.5), (b) down conductors to carry the lightning current to the ground terminal connections, and (c) ground terminals. This general protection scheme for structures was originally proposed by Benjamin Franklin (see also Section 1.1) and is specified, for example, in the US lightning protection standard, National Fire Protection Association Standard NFPA-780, and in many other national and international lightning protection standards. Note that the term "Franklin rod" is sometimes used to mean only a sharp-pointed rod, consistent with Franklin's view of the proper rod geometry (see the quote at the beginning of this chapter). If the current is brought to ground by a number of down conductors symmetrically placed around the structure (an approximation to an all-metal building), as opposed to a single down conductor or an unsymmetrical arrangement of down conductors, such a diversion system will also decrease the potentially harmful effects to electronic equipment inside the structure from induced voltages primarily produced by the time-varying magnetic flux density whose source is the time-varying current in the down conductors. Examples of structural lightning protection systems using lightning rods are shown in Fig. 18.1. It is worth emphasizing that the lightning rods, or alternatively a mesh of horizontal wires covering the top of the structure, is solely intended to intercept the imminent lightning leader, after which lightning current is directed to ground. The intercepting conductors and the down conductors of the diversion system are discussed further in subsections 18.3.2 and 18.3.3. The ground terminals of the diversion system,

18.3. Protection

a critical and integral part of the structural protection system, and aspects of grounding in general are considered in subsection 18.3.4. All elements of the structural lightning protection system must be well connected electrically and all significant nearby conductors, including the ground wires on incoming utilities, must be bonded to the overall system for reasons discussed in subsection 18.3.3.

(ii) The protection of electronic, power, or communication equipment within a structure must include the control of currents and voltages resulting from direct strikes to the structure containing the equipment as well as from lightning-induced current and voltage surges propagating into the structure on electric-power, communication, or other, wires and metal pipes entering the structure from outside. Four general types of current- and voltage-limiting techniques are commonly used. (a) *Voltage crowbar* devices limit the deleterious voltages to values small compared with the operating voltage and short-circuit the associated current to ground. The gas-tube arresters used by telephone companies are a good example of crowbar devices. Silicon controlled rectifiers (SCRs) and triacs are other examples. Crowbar devices do not operate instantaneously and, in general, do not operate as rapidly as the voltage clamps discussed next. (b) *Voltage clamps* are solid-state devices, such as metal oxide varistors (MOVs), Zener diodes, and p–n junction transistors, that both reflect and absorb energy while clamping the applied voltage across their terminals to a more or less safe value, generally 30 to 50 percent above the system operating voltage. Voltage clamps are, in general, nonlinear devices that can handle less energy than crowbar devices before failing. Both voltage clamps and voltage crowbar devices are often labeled "surge protective devices" or SPDs. (c) *Circuit filters* are linear circuits that both reflect and absorb the frequencies contained in the damaging lightning transient pulses while passing the operating waveforms, which may be communication signals of 50 Hz or 60 Hz power. A series inductor whose impedance is higher to the frequencies of the unwanted transient than the electronics following it forms the simplest circuit filter. Frequently, crowbar devices, clamps, and filters are used together in a coordinated way. More details are found in subsection 18.3.5. (d) *Isolating devices* such as optical isolators and isolation transformers are capable of suppressing relatively large transients. Isolators are connected in series with the equipment to be protected and represent a large series impedance to the unwanted transient signals. The SPDs discussed above are connected at the input terminals of the electronics or directly on circuit boards, in parallel with the electronics to be protected. A review of all devices and circuits used in the protection of electronics is given in the book by Standler (1989a).

While the lightning protection design discussed above is generally effective, essentially perfect protection

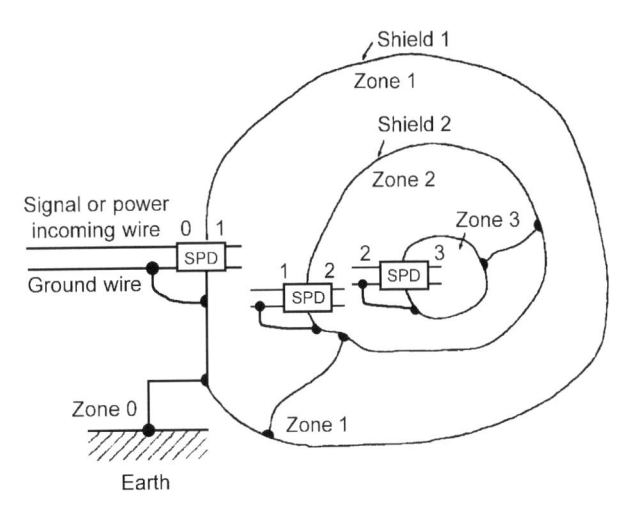

Fig. 18.2. The general principles of topological shielding. Adapted from Vance (1980).

from lightning is provided by an all-metal structure with a wall thickness of many skin depths, no apertures in the walls, and no wall penetrations by conductors. Such a container is often called an electrodynamic Faraday cage or just a Faraday cage or Faraday shield. The concept of ideal protection afforded by a Faraday cage is of limited practical value. However, a systematic approach to lightning protection is available that makes use of the principle that time-varying currents and electromagnetic fields cannot reach the interior of a Faraday cage, while additionally employing the voltage-limiting devices (SPDs) discussed above to control the magnitude of signals entering the cage via wires and pipes and through apertures. This "topological shielding" with transient protection allows an optimal lightning protection system to be designed for most structures and their contents (e.g., Vance 1980). The technique consists of identifying shields that can be nested within each other and "grounding" the outside of each shield to the inside of the one enclosing it. Wires entering a given shield pass through a transient protection device whose ground is connected to the outside of that shield. Therefore, at each successive inner shield, deleterious voltage, energy, and power levels are reduced. In Fig. 18.2 we illustrate the concept of a topological shielding system. Further discussion of topological shielding is found in subsection 18.3.6.

While there are many individual examples illustrating the value of lightning protection systems, statistical data on their efficacy are limited and there are apparently no recent statistical studies. Kellogg (1912) provided overall statistics on structure losses from the Iowa Farmer's Mutual Fire Insurance Associations from 1908 to 1911, with detailed data provided for 1908 and 1909. It was estimated that, over the time period studied, about 25 percent of significant rural buildings in Iowa had lightning protection.

In 1908 and 1909, buildings in Iowa without lightning rod systems suffered $81 077 in damage whereas buildings with such systems suffered $1078 in damage. For 1908 to 1911, the ratio of the cost of damage in unprotected versus protected buildings was about 60, and Kellogg (1912) stated that in most cases of damage to buildings with rods "the cause was found to be defective or incomplete rodding". Taking account of the fact that only about 25 percent of the structures had lightning protection, Kellogg argued that the ratio of the costs of damage to unprotected and protected structures was about 15.

Covert (1930) gave detailed data on lightning fires in Iowa from 1919 to 1924. Almost 80 percent of the losses for unprotected structures were to barns, grain elevators, and other farm structures. He stated that about half the structures in rural Iowa had lightning protection during the time considered. Of about $2 million in lightning-caused fire losses over the six years, 93 percent was to structures without lightning protection, the ratio of the costs of damage to unprotected and protected buildings being about 14.

McEachron and Patrick (1940) reported on a study of lightning protection systems that began in the 1920s in Ontario, Canada.

A 10-year survey in the Province of Ontario, in Canada, disclosed that during the period covered, 10 079 lightning fires took place in structures not equipped with lightning rods, while only 60 such fires occurred in buildings with lightning rod systems of protection. Of these sixty fires, it was found that many were started in structures equipped with improper lightning rods, or rods in bad condition because of poor maintenance. It is safe to say today that a lightning rod system practically eliminates the chance of damage from a stroke, although it will not prevent the stroke itself...

A study in Poland by Szpor (1959) (reported in English by Müller-Hillebrand 1962a) showed that there were about six fires per 10 000 houses from lightning for unprotected houses in Poland. Between 1956 and 1960, there were 97 percent fewer fires for houses with a lightning protection system than for houses without protection.

Holle *et al.* (1996) extrapolated insurance claims for lightning losses in Colorado, Utah, and Wyoming from 1987 to 1993 to the whole United States and found that there was about $330 million in annual insured losses due to lightning in the United States. Since 20 to 30 million cloud-to-ground flashes occur annually (subsection 2.5.2), the financial cost per ground flash is $10 to $15. In fact, according to Holle *et al.* (1996) there is one claim for every 57 ground flashes. The average claim is about $1000 with significant damage to structures resulting in claims up to the $100 000 range. Additional information on lightning damage in the US is found in Curran *et al.* (2000).

18.3.2. Zones of protection

In this section we examine the first element of a structural lightning protection system, the vertical rods or horizontal wires designed to intercept the lightning leader so that it cannot terminate on critical parts of the structure being protected. A discussion of the other aspects of structural lightning protection systems is found in subsection 18.3.3, and a discussion of grounding is found in subsection 18.3.4.

The determination of the volume or zone in which lightning cannot strike has been a subject of discussion since the time of Benjamin Franklin, that is, for over two centuries. Golde (1977b) showed a sketch of a building near London that in 1777 sustained lightning damage to the edge of its roof, which was at a horizontal distance of about 38 feet (11.6 m) from a vertical lightning rod whose top was 24 feet (7.3 m) above the roof edge. The ratio of the maximum horizontal distance from the rod within which the lightning was thought to not be able to strike (in this case, near 38 feet) to the rod height (24 feet), was termed the protective ratio, which for this case was equal to $38/24 \cong 1.6$ but in actuality could be smaller if other lightning leaders were able to attach to the roof closer than 38 feet from the rod. The lightning protection system on this structure had been designed by a committee, of which Benjamin Franklin was a member (Cavendish *et al.* 1773). According to Golde (1977b), this was the first recorded case of the limitation of the protection afforded by such a lightning protection system, although the damage was minor considering that the building housed explosives. About 50 years later, the French Academy of Sciences, under the chairmanship of Gay-Lussac (Gay-Lussac and Pouillet 1823), concluded that a vertical rod would protect a circular area around its base whose radius is twice the rod height, a protective ratio of 2. Schwaiger (1938) found from a literature survey that the proposed protective ratios from the time of Franklin to the date of his study varied between 0.125 and 9.0. Some of these proposed ratios and associated zones of protection are depicted in Fig. 18.3, as originally published

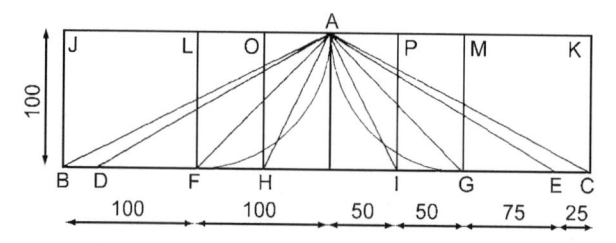

Fig. 18.3. Zones of protection of a vertical lightning rod. JBCK, cylinder, Gay Lussac (1823); BAC, cone, DeFonville (1874); DAE, cone, Paris Commission (1875); LFGM, cylinder, Chapman (1875); FAG, cone, Adams (1881); OHIP, cylinder, hypothesis, and FAG, special cone, Preece (1880); HAI, cone, Melsens. Adapted from Lodge (1892) and Golde (1977b).

18.3. Protection

by Lodge (1892) and reproduced and discussed by Golde (1977b). Note that the zones of protection were viewed as either cones or cylinders with the exception of a special cone having a concave lateral surface (Preece 1880). Anderson R. (1879), Lodge (1892), and Walter (1937), from evidence available to them, reached the conclusion that no protective zone associated with a vertical rod could be specified with absolute confidence. They were correct, as we shall discuss in subsection 18.3.3.

There have been many laboratory studies of the zone of protection carried out using small-scale models and laboratory voltage sources (e.g., Peek 1926; DeBlois 1933; Rühling 1972), but there is a question regarding the validity of any laboratory test designed to simulate the physics of the lightning stepped leader and the attachment process, since typical leader step lengths, tens of meters, exceed the size of the laboratory experiment (e. g., Golde 1941). Additionally, the type of voltage waveform used to simulate lightning in the laboratory test and the impedance of the voltage generator will, in general, influence the laboratory results.

The three approaches most commonly used to describe the region or zone where lightning cannot strike are (a) the cone-of-protection method, (b) the rolling-sphere method, and (c) various physics-based and "electrogeometric" models. The physics-based and electrogeometric models provide the theoretical foundation for simpler, more practical approaches to structural lightning protection such as the rolling-sphere method and related methods listed at the end of this subsection. The physics-based and electrogeometric models have at their cores the concept of a "striking distance", which we will discuss in the next paragraph. Before we do that, let us review the cone-of-protection method. Historically, the protected volume provided by a lightning rod or other grounded vertical conductor has most often been considered to be an imaginary cone, a cone-of-protection, whose apex is at the top of the conductor. This concept is adequate for the design of the protection for small structures; that is, even though there may be failures for a given cone angle, for reasons to be discussed later, protection design using the cone of protection is not much different from that using other techniques. For a protective ratio of unity, the angle between the rod and the lateral surface of the cone at its apex is 45°, and the protected zone is generally referred to as a 45° cone. A protective ratio 2 corresponds approximately to a 60° cone. In specifying a cone angle for the cone of protection prior to the latter part of the twentieth century, no account was taken of the physics of the lightning leader or the attachment process since that information was not available. Nevertheless, as we shall see in subsection 18.3.3, recent attempts have been made to relate the cone angle to the lightning physics.

We now discuss the concept of a "striking distance". From a physical point of view, the descending stepped leader, when it is tens to hundreds of meters above ground, will generate a relatively large electric field via the leader charge and the opposite polarity charge induced on the Earth and on objects connected to the Earth, leading to an electrical breakdown between the leader tip and Earth or objects connected to it. The physics of this lightning attachment, which determines the "striking distance" (defined below), involves an upward connecting leader or leaders from the ground or from grounded objects. In order to calculate the electric field produced by the stepped leader charge, it is first necessary to specify the leader charge distribution along the channel. Wagner and McCann (1942) assumed a uniform charge distribution, as inferred by Schonland *et al.* (1938) from remote measurements of electric field changes. Bruce and Golde (1941) argued that the charge distribution must be such as to produce a fairly uniform voltage drop along the channel and thus that the charge density must decrease exponentially from the leader tip towards the cloud charge source. Modern models of lightning leaders are considered in Sections 12.3 and 12.4. Golde (1945, 1967) used an exponential charge distribution to calculate the electric field beneath the leader as a function of the height above ground of the leader tip for different values of the charge deposited on the leader channel. The "striking distance" was defined as the height of the leader tip above ground at which a critical breakdown electric field is reached across the final gap. An alternative definition of the striking distance would be the height of the leader tip for which an upward connecting leader is initiated (see also Rakov and Lutz 1990). Golde (1945, 1973) adopted a critical breakdown field of 500 kV m^{-1} (the average field in the gap) for negative stepped leaders and 300 kV m^{-1} for positive leaders. Figure 18.4, curve 1, shows the variation in striking distance with the return-stroke peak current for a negative leader where the peak current is assumed by Golde (1945, 1973) to be proportional to the charge on the leader, a peak current of 20 kA corresponding to a leader charge of 1 C. Similar analyses were presented by Schwab (1965) and by Braunstein (1970). Wagner and Hileman (1961) assumed that the leader consisted of a thin conducting core surrounded by a negative space charge (corona sheath) of several tens of meters in diameter; with this assumption and an assumed breakdown electric field strength between 500 and 600 kV m^{-1}, Wagner (1963) derived curve 2 in Fig. 18.4. A few years later Wagner (1967), on the basis of another approach, modified his view of the striking distance from that given in curve 2 to about 40 m for a current of 10 kA and about 100 m for 50 kA. Davis (1962) calculated the potential of the leader tip by assuming a uniform charge distribution along the leader channel, but he further assumed values of breakdown voltage as a function of gap spacing. Only three

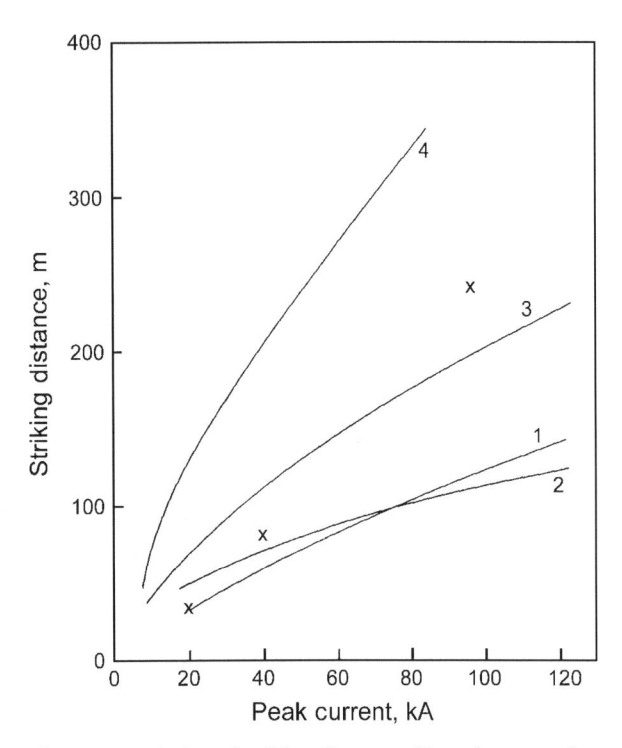

Fig. 18.4. Variation of striking distance with peak current for negative lightning: curve 1, Golde (1945); curve 2, Wagner (1963); curve 3, Love E.R. (1973); curve 4: Rühling (1972); x : Davis (1962). Adapted from Golde (1977b).

striking distances were calculated, based on the assumption of a current of 100 kA being associated with a charge of 5 C. They are indicated in Fig. 18.4 by the crosses.

Horváth (1969, 1971) presented a method of calculating the striking distance different from those just considered. Noting that the point-discharge (corona) current through a grounded conductor increases rapidly with increasing electric field, he proposed that, when this current reaches a value of between 10 and 50 mA, the (corona) discharge undergoes a glow-to-arc transition and an upward connecting leader is initiated. The striking distance was determined by equating the electric field under the leader with the field required to produce a glow-to-arc transition. Horváth's (1969, 1971) striking distances are within the range of those determined by earlier investigators.

One major factor affecting the results in Fig. 18.4, and other studies of the striking distance in which a particular average breakdown field in the gap is assumed to initiate the attachment process, is that the value of the actual average field is not known. Such breakdown field values are necessarily extrapolated from laboratory measurements of rod–rod and rod–plane gaps and vary with the waveshape of the applied voltage as well as with other factors such as the generator circuitry. For negative first strokes, Wagner (1963) adopted a value between 500 and 600 kV m^{-1}, derived from laboratory measurements with the standard (1/50 µs) impulse waveform, and Golde (1963) adopted a similar value, 500 kV m^{-1}. The breakdown values used subsequently appear to have decreased with time. Armstrong and Whitehead (1968), while not specifically stating any particular value, apparently used a value of about 350 kV m^{-1} (Love E.R. 1973). Rühling (1973) suggests a value as low as 100 kV m^{-1}, and the resulting striking distance is plotted as a function of peak current in Fig. 18.4, curve 4. Chowdhuri (1996) reproduced plots of the critical breakdown voltage for a variety of voltage waveshapes and gap configurations. The lowest values of average breakdown field, near 100 kV m^{-1}, are for long, up to 25 m, rod–plane gaps with applied voltage risetimes in the 200 µs to 1000 µs range.

Eriksson (1987a, b) avoided the problem of choosing an average breakdown field in computing the striking distance to a structure by requiring the enhanced electric field at the structure extremity due to the leader to exceed the breakdown field in a uniform gap at standard temperature and pressure, 3×10^6 V m^{-1}, over a critical "corona" radius, assumed to be the determining factor in launching an upward connecting leader. The leader model used was one whose charge density decreased linearly with height. The approach of Eriksson (1987a, b) takes better account of the physics of the attachment process but necessitates a knowledge of the actual electric field enhancement at the structure, the critical corona radius, and the leader charge distribution, parameters that are not, in general, known adequately. More about this approach is given later in this section. Some discussion of the method of Eriksson (1987a, b) is given by Chowdhuri and Kotapalli (1989).

It is clearly more practical to express the striking distance as a function of the return-stroke peak current, as done in Fig. 18.4, than as a function of the leader charge distribution, since charge distributions cannot be measured directly while peak current distributions are among the lightning parameters for which experimental data are available (Section 4.6). Golde (1945) was apparently the first to suggest an analytical expression for this relationship of the general form

$$r = AI^b \qquad (18.1)$$

where r is the striking distance in meters, I is the peak current in kiloamperes, and A and b are constants. Most of the proposed expressions of the type given in Eq. 18.1 were developed as part of electrogeometric models intended for application in electric power-line-protection studies. Armstrong and Whitehead (1968) determined the constants A and b by applying the theoretical approach of Wagner (1963). From such an approach, Wagner (1967) revised his striking distances from those in curve 2 of Fig. 18.4, as noted earlier. Additional discussion of Eq. 18.1 is given in

18.3. Protection

Rühling (1972), Eriksson (1974), Chowdhuri and Katapalli (1989), and Bazelyan and Raizer (2000a).

In the simplest version of the electrogeometric model applied to power-line-protection studies, the striking distances to the phase wires and to the shield or ground wires of a power line and the striking distance to the Earth are all assumed to be equal, so that for an assumed peak current the lightning is predicted to terminate on whichever object the tip of the leader first approaches within the striking distance for that peak current as determined from Eq. 18.1. The assumption of equal striking distances ignores the different electric field enhancements at the various potential strike points. The striking distances to the shield wire, phase conductor, and Earth are not, in general, taken as equal in the more sophisticated electrogeometric models (e.g., Currie *et al.* 1971; Chowdhuri and Kotapilli 1989; Mousa and Srivastava 1990; Rizk 1994a, b). Table 18.1 gives the constants A and b of the striking distance formula, Eq. 18.1, from different studies, for strikes to different elements of a power line. Combined electrogeometric models and physics-based theory describing this attachment process is given in Eriksson (1987a, b), Dellera and Garbagnati (1990), and Rizk (1989a, b, 1990, 1994a, b). More information on the physics of the processes by which lightning attaches to ground or to objects on the ground is found in Section 4.5. Eriksson (1987a, b) used observations of the frequency of lightning strikes to various structures and a somewhat more sophisticated electrogeometric model than his predecessors (e.g. Wagner and Hileman 1961; Armstrong and Whitehead 1968), as noted earlier in this section, to describe the lightning interaction with vertical structures located on ground. He considered the speeds of both the downward-moving leader and the upward connecting leader in his analysis. Initiation of the upward leader is assumed to take place when the electric field at the structure top, calculated including the enhancement effect of the structure, exceeds the breakdown field of air over a critical "corona" radius. As a part of this work, he found that the incidence of strikes to structures of 10 to 30 m height had been underestimated by more than 50 percent and that the incidence of strikes to taller structures could have been overestimated by more than 30 percent by the previously used techniques involving Eq. 18.1.

We discuss now the rolling-sphere method, which is the simplest practical application of the electrogeometric approach. The rolling sphere method has been primarily applied to the lightning protection of buildings. In this method, the tip of the leader is considered to be located at the center of an imaginary sphere whose radius is the striking distance. If such an imaginary sphere of a given radius, corresponding to a given peak current, is rolled along the Earth and over objects on the Earth, every point touched by the sphere is a possible point of strike, whereas points not touched are not, as illustrated in Fig. 18.5. Thus, the sphere should touch lightning rods and other lightning-intercepting conductors but should not touch any part of the structure to be protected. Since the radius of the rolling sphere depends on the peak current, within this concept there will always be a small value of peak current below which the system protection will fail because a sphere of small enough radius will roll between the lightning-intercepting conductors. However, there may be a minimum value for the peak lightning current below which the charge on the leader is too low for the leader to propagate downward all the way to the Earth. According to IEC 61024-1 (see subsections 18.3.3 and 18.5), 99 percent of ground flashes can be accounted for by using a sphere radius equal to 20 m, and 84 percent

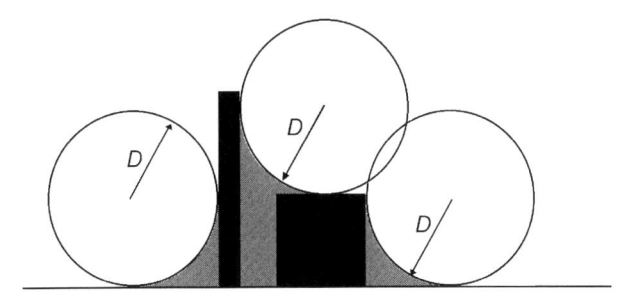

Fig. 18.5. Illustration of the rolling-sphere technique. The shaded area is that area into which, it is postulated, lightning cannot enter. Adapted from Szczerbinski (2000).

Table 18.1. *Constants A and b of the striking distance equation, $r = AI^b$. Here the striking distance to the phase conductor, r_c, and that to the shield wire, r_s, are taken as equal*

	Striking distance to			
	Earth		phase conductor and shield wire	
Source	A	b	A	b
Young et al. (1963)	27.0	0.32	(1)	0.32
Armstrong and Whitehead (1968)	6.0	0.80	6.7	0.80
Brown and Whitehead (1969)	6.4	0.75	7.1	0.75
Love E.R. (1973)	10.0	0.65	10.0	0.65
IEEE Committee Report (1985)	(2)	0.65	8.0	0.65
Eriksson (1987b)	$r_c, r_s = 0.67 H^{0.6} I^{0.74}$			

(1) $A = 27.0$ for $H < 18$ m, where H is the line height; $A = 27(444)/(462 - H)$ for $H > 18$ m.
(2) $A = 8.0(H/22)$ and must be greater than 4.8 but less than 7.2.

by using a radius equal to 60 m. For simple vertical rods or for elevated horizontal wires, the volume determined by rolling a sphere of a given radius can be used to define a cone angle for the cone of protection (subsection 18.3.3). The rolling sphere technique for use in modern lightning protection is most often attributed to Lee (1978). Related practical protection techniques have been proposed that are more complex than the rolling sphere, in that they take account of the fact that different system elements have different striking distances because of the different electric field enhancement at those elements. These are the geometric zone of capture (e.g., Eriksson 1987a, b), the collection volume method (e.g., D'Alessandro and Gumley 2001), and the collection surface method (e.g., Hartono and Robiah 2000).

18.3.3. Protection systems: application to structures

A lightning protection system that is attached directly to the structure to be protected is shown in Fig. 18.1. Similarly effective protection systems can be isolated from the structure, such as the tall conducting masts that often surround oil storage tanks, with or without wires between mast tops. In general, isolated protection systems are used for structures which demand the highest level of safety, such as fuel or explosive storage facilities, and structures for which a close lightning channel or close lightning-induced sparking cannot be tolerated.

Ideally, any air terminal should be designed to intercept lightning without significant damage to itself. Conducting spires, conducting roof edges, metal roofs, and various vertical metal components above roof level can be used as air terminals if potential melting or burn-through of these metallic objects is not a problem for other reasons. The US standard, National Fire Protection Association (NFPA) 780, and many other national standards and codes recommend installing multiple lightning rods, but similar protective effectiveness is apparently obtained by covering the insulating roof of a building with a mesh of metallic conductors. Many national codes (e.g., British Standards Institute 1992) specify metal meshes. The usual size of the mesh recommended is between 5 and 20 m, as discussed later. Clearly, the ultimate air terminal is a solid metal roof, provided it is of adequate thickness to avoid lightning penetration at the point of strike if such penetration would be a problem because of (i) fire below roof level that potentially could be caused by dripping hot metal or (ii) water damage to the interior via rain passage through the hole or holes produced in the roof.

Steel supports and well-bonded rebar in the reinforced concrete of walls and roofs can be used both for part of the air-terminal system and for the down conductors. Air terminals should be well bonded to the metal supports and rebar at roof level.

Tall structures may be subjected to lightning strikes from the side, as has often been observed. Buildings exceeding 20 to 30 m in height should therefore be provided with additional air terminals on their side walls. Metallic facades, window-frames, railings, and exposed down conductors can be used for this purpose.

A specification of materials for air terminals and down conductors is given in Table 18.2. According to IEC-61024-1 (Section 18.5), the minimum cross section of wire to avoid its melting by the ohmic heating associated with large action integrals is 16 mm^2 for copper, 25 mm^2 for aluminum, 50 mm^2 for steel, and 100 mm^2 for stainless steel. NFPA 780 specifies a cross-sectional area of lightning conductor cable for structures under 23 m tall equal to 29 mm^2 for copper and 50 mm^2 for aluminum. For taller structures, NFPA 780 specifies 58 mm^2 for copper and 97 mm^2 for aluminum. According to Golde (1968), a no. 4 AWG (American Wire Gage) copper wire (cross-sectional area about 21 mm^2) will have a temperature rise of about 100 °C for an action integral of 5×10^6 A^2s and will melt at an action integral of about 2×10^7 A^2s. A no. 8 AWG copper wire (about 8.4 mm^2) will melt at an action integral of about 5×10^6 A^2s. As noted in Section 18.2, Berger *et al.* (1975) reported that the upper five percent of negative first strokes exhibit an action integral above 5.5×10^5 A^2 s, and the upper five percent of positive strokes exhibit an action integral above 1.5×10^7 A^2s. The maximum values of flash action integrals might be expected to be near 10^6 A^2s for negative flashes and near 5×10^7 A^2s for positive flashes.

There appears to be no generally accepted view of the proper tip geometry for an air terminal. Franklin recommended a sharp-tipped rod (see the quote at beginning of this chapter). The question of whether sharp-tipped rods are preferable to blunt rods has been repeatedly raised since Franklin. Moore *et al.* (2000a, b) performed field experiments where pairs of sharp and blunt rods of equal height separated by 5 to 20 m were erected on a mountain top in New Mexico. They report that over a five-year period no sharp rods were struck, but that twelve blunt rods with diameters from 12.7 to 25.4 mm were struck. Apparently, five to eight pairs of rods were used each year. Moore *et al.* (2000a, b) concluded that the strike probability for lightning rods is increased when their tips are made moderately blunt (with a tip height to tip radius-of-curvature ratio of about 680:1) as opposed to sharper rods or very blunt ones.

For the last decade or so, the placement of air terminals on structures has generally been determined using the rolling sphere approach, as specified, for example, in the latest versions of NFPA 780. Rolling sphere radii specified by IEC 61024-1 for its four levels of protection, taking into account both positive and negative lightning, are given in Table 18.3. According to the theory, for a given rolling sphere radius R, all flashes with first-stroke peak current

18.3. Protection

Table 18.2. *Material, configuration, and minimum cross-sectional area of air termination conductors, air termination rods, and down conductors. Adapted from IEC 61024-1*

Material	Configuration	Minimum cross-sectional area, mm^2	Comments
Copper	solid tape	50	2 mm min. thickness
	solid round[g]	50	8 mm diameter
	stranded	50	1.7 mm min. diameter of each strand
	solid round[c,d]	200	16 mm diameter
Tin plated	solid tape	50	2 mm min. thickness
Copper[a]	solid round[g]	50	8 mm diameter
	stranded	50	1.7 mm min. diameter of each strand
Aluminum	solid tape	70	3 mm min. thickness
	solid round	50	8 mm diameter
	stranded	50	1.7 mm min. diameter of each strand
Aluminum alloy	solid tape	50	2.5 mm min. thickness
	solid round	50	8 mm diameter
	stranded	50	1.7 mm min. diameter of each strand
	solid round[c]	200	16 mm diameter
Hot dip	solid tape	50	2.5 mm min. thickness
Galvanized steel[b]	solid round	50	8 mm diameter
	stranded	50	1.7 mm min. diameter of each strand
	solid round[c,d]	200	16 mm diameter
Stainless steel[e]	solid tape[f]	60	2 mm min. thickness
	solid round[f]	50	8 mm diameter
	stranded	70	1.7 mm min. diameter of each strand
	solid round[c,d]	200	16 mm diameter

[a] Hot dipped or electroplated minimum thickness coating of 2 microns.
[b] The coating should be smooth, continuous and free from flux stains with a minimum thickness coating of 50 microns.
[c] Applicable for air termination rods only.
[d] Applicable for Earth lead-in rods only.
[e] Chromium $\geq 16\%$, nickel $\geq 8\%$, carbon 0.1% max.
[f] For stainless steel embedded in concrete, and/or in direct contact with flammable material the minimum sizes should be increased to 75 mm^2 (10 mm diameter) for solid round and 75 mm^2 (3 mm minimum thickness) for solid tape.
[g] 50 mm^2 (8 mm diameter) may be reduced to 28 mm^2 (6 mm diameter) in certain applications where mechanical strength is not an essential requirement. Consideration should, in this case, be given to reducing the spacing of the fasteners.

Table 18.3. *Protection levels defined in IEC 61024-1 and corresponding rolling-sphere radius, minimum peak current, and interception probability*

Protection level	Rolling sphere radius R, m	Minimum peak current I, kA	Interception probability, percent
I	20	2.9	99
II	30	5.4	97
III	45	10.1	91
IV	60	15.7	84

values higher than the corresponding minimum peak current value I will be intercepted by the air terminals. Note from Table 18.3 that if there are first stroke-peak currents smaller than about 3 kA, estimated in Table 18.3 to be one percent of the population of first-stroke currents, use of a 20-m-radius rolling sphere for protection design will not be sufficient to assure complete protection. As noted in subsection 18.3.2, protection system design using a 60 m sphere is predicted to protect against only 84 percent of flashes. Table 18.4 gives a suggested mesh size for the same four protection levels, while Fig 18.6 gives equivalent angles for the cone of protection as a function of structure height for the four protection levels.

Table 18.4. *Relation between rolling sphere radius and protection system mesh size for different protection levels. Adapted from IEC 61024-1*

Protection level	Rolling sphere radius, m	mesh size, m × m
I	20	5 × 5
II	30	10 × 10
III	45	15 × 15
IV	60	20 × 20

The air terminals (the first element of the structural protection system) should be connected by the shortest possible route via the maximum possible number of down conductors (the second element of the system) to the grounding system (the third element of the system), so as both to minimize inductive voltage drops in the down-conductor system that can potentially lead to voltage differences between objects or equipment connected to the down-conductor system at different heights and also to minimize the magnetic field in the building due to the lightning current in the down conductors. Loops in down conductors are best avoided since the lightning might arc over them, potentially through flammable material, due to the loop inductance. Hence, down conductors should, in general, be passed through large overhanging projections on buildings instead of being bent around them. Even a small structure should have a minimum of two down conductors, generally placed at opposite sides of the structure (e.g., Kithil and Rakov 2000). Large metallic bodies within about 5 m of the down conductors should be bonded to the down conductors in order to reduce the possibility of side flashes. A justification of this particular distance within which such bonding should be carried out follows. Consider a point on a vertical down conductor at height h above ground. The bottom of the down conductor is grounded in the earth through a resistance R_{gr} (see subsection 18.3.4). The down conductor has an inductance per unit length L. If current $I(t)$ flows along the down conductor, the voltage at height h on the down conductor with respect to distant ground is

$$V(t) = R_{gr}I(t) + Lh\frac{dI(t)}{dt} \quad (18.2)$$

A potential difference $V(t)$ will exist between any separately grounded piece of metal (or a human being standing on the ground) and the point at height h. If $V(t)$ exceeds the breakdown voltage between the two conductors, there will be an electrical discharge between them. As mentioned above, the discharge is usually called a side flash. We can roughly estimate the distance over which a side flash can occur by assuming an average breakdown field of $500 \, kV \, m^{-1}$ for a voltage risetime in the microsecond or less range (Chowdhuri 1996), a conservative value of the down conductor inductance of $2 \, \mu H \, m^{-1}$, a grounding resistance of 25 Ω, a maximum current of 100 kA, a maximum current derivative of $10^{11} \, A \, s^{-1}$, and $h = 10$ m. The resistive component of $V(t)$ and its inductive component will peak at different times since the current derivative peak precedes the current peak. The resistive and inductive voltage waveforms will nevertheless overlap. The resistive peak voltage is 2.5 MV and the inductive peak voltage is 2.0 MV. Thus a side flash of roughly 5 m length will occur for the parameters assumed. The distance will be less if (i) the ground resistance is lower, perhaps naturally made so via ground surface arcing (subsections 18.3.4, 7.2.5, 19.2), a close relative of the through-the-air side flash, (ii) if there are a number of parallel down conductors reducing the inductance, (iii) if the height h is less than 10 m, and (iv) if the current has the typical value of 30 kA rather than the high value of 100 kA purposefully chosen for the calculation.

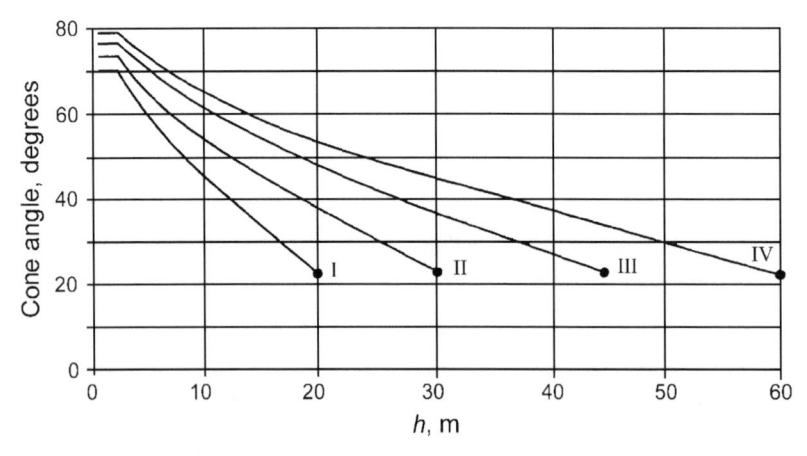

Fig. 18.6. Angles of the equivalent cone of protection for structures of various heights with each of four levels of protection, I–IV, as specified in Table 18.3. Adapted from IEC 61024-1.

However the current derivative peak value can be a factor of 2 to 4 higher than assumed, but it is of relatively short duration and hence a larger average breakdown field may be required than assumed.

In relation to lightning-protection-system grounding, a preferred way to conduct lightning current into the Earth while minimizing the potential differences between different parts of the overall system is to bury a closed loop of wire around the structure. Such a ring grounding electrode should be placed at a distance of about 1 to 2 m from the building and be buried at a depth of at least 0.5 m. A ring grounding electrode can serve as a suitable location to bond metallic service installations, although it is preferable that all service grounds be bonded at one point. A ring grounding electrode has the further advantage of tending to equalize the potential at all points within the ring (assuming voltage differences associated with the ring electrode inductance are not significant and that lightning current is not directly injected into the ground within the ring) via the principle of electrostatics that states that if the potential is equal at all points on a closed surface then the potential has the same value within the surface in the absence of charges within the surface. Thus, achieving a low value of grounding resistance is not as critical when a ring grounding electrode is employed since the likelihood of side flashes is minimized when potential differences are minimized. The optimum grounding system for an ordinary structure is a metal mesh buried beneath the structure to which the down conductors and utility grounds are connected, because of both the mesh's relatively low grounding resistance and the minimization of potential differences between different points on the mesh due to the relatively low inductance of the multiple paths between those points; however, such grounding meshes are not often practical. A recent review of the lightning protection of ordinary structures, including golf shelters, is found in Kithil and Rakov (2000) and in Rakov (2000).

18.3.4. Grounding

The primary purpose of a lightning grounding system is to provide a means to direct the lightning current from the down conductors into the earth with a minimum rise in the potential of the above-ground part of the protection system. For example, if a peak current of 100 kA is injected into a grounding electrode whose impedance is 25 Ω then the component of the potential of the above-ground protection system due to the grounding resistance (Eq. 18.2) will rise to 2.5 MV, which will lead to side flashes from the above-ground system to any isolated (or grounded at a distance) metal bodies or wires within a few meters of the conductors of the protection system (see also subsection 18.3.3). For this reason, as noted in subsection 18.3.3, nearby metallic objects must be bonded to the lightning-protection-system conductors. Since about half the peak lightning currents for first strokes are larger than about 30 kA and since it is difficult to obtain grounding resistances below about 10 Ω, voltages on the down-conductor system in the hundreds of kilovolts range are common, owing to the component of the potential associated with the grounding resistance. To reduce the potential differences within a structure caused by the grounding resistance of the commonly used driven ground rod of 2 to 3 m length, grounding by a buried ring electrode encircling the structure or by a buried metal mesh beneath the structure, as discussed in the previous subsection, or the use of topological shielding (subsection 18.3.6), is preferable to grounding via one or a few ground rods.

The calculation of the grounding resistance of a given piece of buried metal in the dc or low-frequency cases is more or less straightforward. Basically, Laplace's equation is solved in the conducting earth for the potential distribution around the grounding electrode, given the boundary conditions at the conducting electrode. The grounding resistance is then found as the potential difference between the grounding electrode and a point far away (at infinity) divided by the assumed current. We can illustrate a simple approach to this calculation by considering a conducting hemisphere of radius a whose flat surface is flush with the surface of the Earth. The Earth conductivity is σ and the resistivity ρ, where $\sigma = 1/\rho$. If a current I is injected into the electrode then the current density in the Earth, assuming that current flows uniformly outward from the hemispherical surface, is

$$\mathbf{j} = \frac{I}{2\pi r^2}\mathbf{a}_r \quad (18.3)$$

where $r \geq a$ is the radial distance from the center of the circular flat surface of the hemisphere, and $2\pi r^2$ represents the underground surface area of an imaginary hemisphere concentric with the hemispherical electrode surface whose area is $2\pi a^2$. Combining Ohm's law in point form

$$\mathbf{j} = \sigma \mathbf{E} \quad (18.4)$$

where $\mathbf{E}$ is the electric field intensity, with Eq. 18.3 we obtain

$$\mathbf{E} = \frac{I}{2\pi r^2 \sigma}\mathbf{a}_r, \quad r \geq a \quad (18.5)$$

The potential difference between two points in the earth at two different radii, $r = a$ and $r = b$, is

$$V_{ab} = -\int_a^b \mathbf{E} \cdot d\mathbf{r} = \frac{I}{2\pi r\sigma}\left(\frac{1}{a} - \frac{1}{b}\right) \quad (18.6)$$

The grounding resistance with respect to a distant point ($b \gg a$) is given by

$$R_{gr} = \frac{V_{ab}}{I} \cong \frac{1}{2\pi\sigma a} = \frac{\rho}{2\pi a} \quad (18.7)$$

Thus the grounding resistance is inversely proportional to both the conductivity and the radius of the hemispherical grounding electrode. The potential difference between any two points on the Earth's surface at different radii a and b, the "step voltage", is given by Eq. 18.6. Step voltages cause shocks to humans and animals and can kill four-legged animals standing near objects, for example trees, struck by lightning (Sections 19.2 and 19.3).

The most common grounding electrode is the vertical ground rod whose resistance, according to Dwight (1936), is

$$R_{gr} = \frac{1}{2\pi\sigma l}\left(\ln\frac{8l}{d} - 1\right) \qquad (18.8)$$

where l is the length of the rod and d is the rod diameter. A close approximation to this expression can be derived by assuming that the current flows uniformly from the ground rod through concentric imaginary surfaces, the shape of each of which is that of a cylinder of length l and radius r, concentric with the rod and having a hemispherical cap of radius r on its bottom end (Liew and Darveniza 1974). The equipotential surfaces for such current flow are illustrated in Fig. 18.7. Hence the magnitude of the current density is

$$j = \frac{I}{2\pi rl + 2\pi r^2} \qquad (18.9)$$

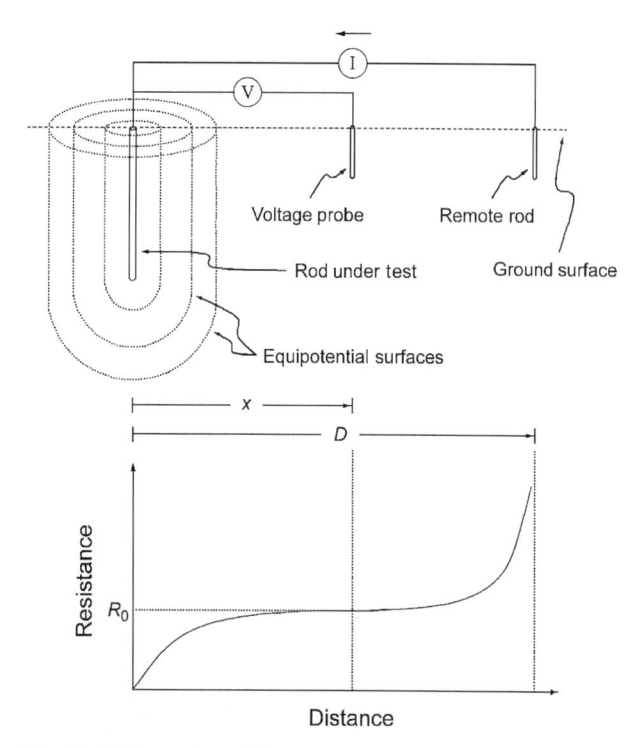

Fig. 18.7. Illustration of the measurement of the low-frequency, low-current grounding resistance of a ground rod using the fall-in-potential method.

from which **E** can be found using Eq. 18.4 and the potential difference between the rod to infinity from Eq. 18.5. The result of this calculation is

$$R_{gr} = \frac{1}{2\pi\sigma l}\ln\left(\frac{2l}{d} + 1\right) \qquad (18.10)$$

To a good approximation, the low-frequency grounding resistance of any vertical ground rod is inversely proportional to both the length of the rod and the ground conductivity and is relatively independent of the diameter of the rod.

The low-frequency grounding resistance of buried horizontal conductors such as a long straight wire or a ring electrode encircling a structure (discussed earlier in this section and in subsection 18.3.3) is also inversely proportional to σl (proportional to ρ/l), where l is the length of the wire; it differs from Eqs. 18.8 or 18.10 by the factor within parentheses in those equations, the difference in the logarithms of the different factors in the parentheses being of order unity. Since a buried horizontal conductor of some tens of meters length is often easier to install than a vertical rod of similar length (usually installed as a sequence of connected vertical rods), buried horizontal electrodes are often preferable from a practical point of view. A vertical ground rod, however, may extend downward into an area of higher soil conductivity than exists near the surface, perhaps below the water table, thus providing a lower grounding resistance compared with a horizontal conductor of similar length buried near the surface. Expressions for the grounding resistance of a variety of grounding electrodes of different shapes are given by Saraoja (1977).

In our discussion of grounding in this section, we have thus far assumed that the grounding electrode impedance is resistive. This is a good approximation for frequencies under 1 kHz or so. For frequencies above 100 kHz, in general it is necessary to take account of the inductances and capacitances of the grounding system, as we shall discuss later in this section.

In the simple calculation of grounding resistance presented above, it is assumed that Ohm's law in point form applies and that the Earth conductivity is uniform. If the electric field at the grounding electrode surface given, for example, by Eq. 18.5, exceeds a value typically in the range 100 to 500 kV m^{-1} (Petropoulos 1948; Liew and Darveniza 1974; Mousa 1994), electrical breakdown will occur from the electrode both across the Earth's surface and into the surrounding soil, lowering the grounding resistance by effectively enlarging the size of the electrode. It follows from Eq. 18.5 that for a hemispherical grounding conductor and assumed uniform conditions there will be an ionized region in the earth within a radius r_0 surrounding the grounding conductor, where

$$r_0 = \sqrt{\frac{I}{2\pi\sigma E_b}} \qquad (18.11)$$

and E_b is the soil breakdown electric field. It is likely that thermal instabilities will cause the current in any uniformly ionized region to collapse into one or more localized arcs. The behavior of electrodes when earth breakdown occurs has been studied, for example, by Petropoulos (1948), Liew and Darveniza (1974), Chisholm and Janischewskyj (1989), and Geri et al. (1992). A review of a number of such studies is given in Mousa (1994). Perhaps the best work on this subject was by Liew and Darveniza (1974). They showed experimentally, in the laboratory, and from model calculations that soil breakdown around a ground rod significantly reduces the dc ground resistance in a few microseconds, a time comparable to the risetime of the applied current in the experiments.

Soils of higher resistivities were shown to have greater decreases in the ratio of the breakdown-reduced resistance to the initial resistance. For example, for current above 50 kA, a 300 Ω dc ground rod resistance in a soil of resistivity 10^3 Ω m was reduced by about a factor 5, whereas a 30 Ω dc ground rod resistance in a soil of resistivity 100 Ω m was reduced by only about a factor 2. The calculations of Liew and Darveniza (1974), which model the experimental data well, assume an ionization time constant of 2 μs and a de-ionization time constant (relating to the time in which the resistance recovers its dc value after the electric field has fallen below the breakdown value) of 4.5 μs. Model breakdown fields between 100 to 300 V m^{-1} for different soils provided the best agreement with the experimental data.

Ground surface arcing in triggered lightning experiments was discussed by Fisher et al. (1994) and by Rakov et al. (1998) (subsections 7.2.5 and 19.2).

We now examine the issue of the transient response of grounding electrodes in situations where the grounding system cannot be assumed to be purely resistive. When transient current is injected into an extended earth electrode, the electrode initially exhibits a surge impedance, which, for a long buried horizontal wire (a counterpoise), is 150 to 200 Ω (Bewley 1963). The transient signal propagates along the conductor, generally with a velocity about one-third the speed of light due to the relatively high permittivity of the earth and the ohmic losses, and reflects back and forth between the ends of the electrode while being damped by radial current flow into the earth (Bewley 1963). The variation in the potential at the source with time is the result of a combination of the rising voltage at the source due to the rising current of the source and the reflected voltage waves from the end of the electrode, the first one arriving in about 1 μs for an electrode of 50 m length. If the dc value of the grounding resistance of a long buried wire is less than its surge impedance, as is generally the case, the effect of the multiple reflections is to reduce the input impedance at the source to its dc value after a few reflections.

As noted above, all low-frequency grounding resistances are directly proportional to the value of the local Earth resistivity. According to Saraoja (1977), the soil resistivity depends largely on the water content of the soil and the resistivity of the water, the relation being given by Hummel's empirical formula:

$$\rho = \left(\frac{1.5}{p} - 0.5\right)\rho_v \qquad (18.12)$$

where ρ is the soil resistivity in Ω m, ρ_v the resistivity of the water in the soil in Ω m, and p the relative volume of the water in the soil. If $p = 0.1$ then $\rho = 14.5\rho_v$, which, according to Saraoja (1977), is easily verified experimentally. From Eq. 18.12, if $p = 0$ then ρ would be infinite, so the equation fails in this limit although the resistivity of dry soil is indeed very high. Since the soil resistivity varies according to the content and characteristics of water in the soil, a knowledge of the resistivity of natural waters also gives some insight into the question of soil resistivities. Table 18.5 gives the resistivity for different kinds of water and soil.

Completely dry concrete has a very high resistivity, but when concrete is embedded in earth, moisture penetrates into it and its resistivity becomes about the same as that of the surrounding soil (Saraoja 1977). For this reason, well-bonded reinforcing bars in concrete foundations should be used as Earth electrodes where possible.

Apparently, the resistivity of the Earth can be reduced by adding chemicals to the soil around the grounding

Table 18.5. *Water and soil resistivities. Adapted from Saraoja (1977)*

Types of water or soil	Resistivity, Ω m
Water in oceans	0.1 to 0.5
Sea water at the coasts of Finland	1 to 5
Ground water, well, and spring water	10 to 150
Lake and river water	100 to 400
Rain water	800 to 1300
Commercial distilled water	1000 to 4000
Chemically clean water	250 000
Clay	25 to 70
Sandy clay	40 to 300
Peat, marsh soil, and cultivated soil	50 to 250
Sand	1000 to 3000
Moraine	1000 to 10 000
Ose (calcereous remains)	3 000 to 30 000

electrode in order to make the soil more conducting. These chemicals, available commercially, are (i) poured onto the soil, (ii) combined with soil dug from the hole in which the electrode is to reside and the combination placed in the hole, or (iii) released by special ground rods, which are hollow and are filled with materials that diffuse through holes in the walls of the hollow rods (subsection 18.3.7, second part). The efficacy of these approaches is not documented in the reviewed literature, although the approaches are widely advertised.

Sunde (1968) and many other authors have discussed the techniques for measuring low-frequency grounding resistance and Earth resistivity. The fall-in-potential method for measuring grounding resistance is illustrated in Fig. 18.7.

18.3.5. Surge protective devices

Surge protective devices (SPDs), also called arresters, surge arresters, or lightning arresters, are intended to limit transient overvoltages, usually to protect electrical or electronic equipment and to divert surge currents away from that equipment. Equipment containing solid-state electronics as well as motors and transformers, which have many closely spaced turns of wire, are most susceptible to lightning surges. SPDs must remain basically open circuits or high impedances to the normal signals on the circuit being protected, whether these are 5 V digital logic signals on a communication line, 240 V ac on a 60 Hz household power circuit, or higher-voltage ac on distribution and transmission lines. As noted in subsection 18.3.1, transient overvoltages can reduce the voltages across the SPD to near zero if the SPD is a crowbar device such as an air spark gap, gas tube, thyristor (silicon-controlled rectifier), or triac. Crowbar devices are usually not appropriate on dc power circuits, where the short-circuit arc could destroy the SPD or where the short circuit could damage the dc power source. Crowbar devices are used in some ac power circuits where the design is such that the resulting 50 Hz or 60 Hz arc extinguishes following a current zero, restoring the device to its previous high-impedance state. Voltage-clamping devices such as metal oxide varistors (MOVs) and junction diodes attempt to hold the voltage at a near constant value while conducting and diverting transient currents of any magnitude. In general, diodes are used in electronic circuits with protection levels up to 10 V or so, whereas MOVs are available from protection levels of a few volts to the megavolt protection levels needed for power transmission lines. MOVs are commercially available in a range from small single disks of finger-nail size for use on 5 V circuits to disks 10 cm in diameter and some centimeters thick, stackable to several meters length, for use on power transmission lines. Circuits containing both clamping and crowbar devices combine the positive features of the two types of device and will be discussed later in this section. Filters, with or without nonlinear elements, can absorb and reflect unwanted high-frequency transients while passing the desired signals.

SPDs are rated by the maximum continuous operating voltage (rated voltage) they can withstand, by the total transient power and energy they can withstand, by the peak current they can withstand, and by the peak voltage that will appear across the arrester terminals when a current of a certain magnitude and waveshape is passed through the arrester. Typical test current waveforms are the 8/20 μs (8 μs risetime, 20 μs duration to half-peak value) and the 10/350 μs.

The voltage–current characteristics of a spark gap arrester and of an MOV are shown in Figs. 18.8a, b, respectively. The use of these two devices in combination is illustrated in Fig. 18.9. The decoupling inductor shown in this figure allows the spark gap to operate before the MOV reaches its voltage, current, energy, or power limit. Clearly the waveshape of the transient is critical in determining the value of the inductance required, since the voltage across the inductor, $L\,dI/dt$, is the difference between the voltage across the crowbar device and the voltage across the MOV. At relatively high initial rates of change of current, the spark gap will always operate first, reflecting the bulk of the surge energy. At relatively low rates of rise of the current (perhaps not the transient case we are anticipating) the MOV will reach its voltage limit first, the spark gap will not fire, and perhaps the MOV will fail. Clearly, in designing the simplest two-stage system it is critical to know the lowest value of current derivative associated with a transient waveform of sufficient magnitude to do damage.

Several stages (more than the two shown in Fig. 18.9) of SPDs are often used to obtain optimum protection. The first stage can be a crowbar device or an MOV that can handle the largest signal expected. The various stages must be decoupled by series impedances, generally inductances, although sometimes series inductance and resistance combinations are used. The voltage level at the final stage is that necessary to protect the equipment. Such networks can be designed theoretically, but it is always best to test them in the laboratory to make sure they can withstand the range of transient signals expected. Transient signals used for testing are specified in various standards, as discussed in Section 18.5. The voltage and current waveforms for communication networks are different from those for power networks. Unfortunately, lightning does not always produce waveforms similar to those specified in the standards. In fact, it may seldom do so. Some features of non-standard waves are discussed in Section 18.5.

If a number of SPDs are to be used for power-system protection in separated locations within one structure, it is best that the voltage clamping levels of the SPDs be about

18.3. Protection

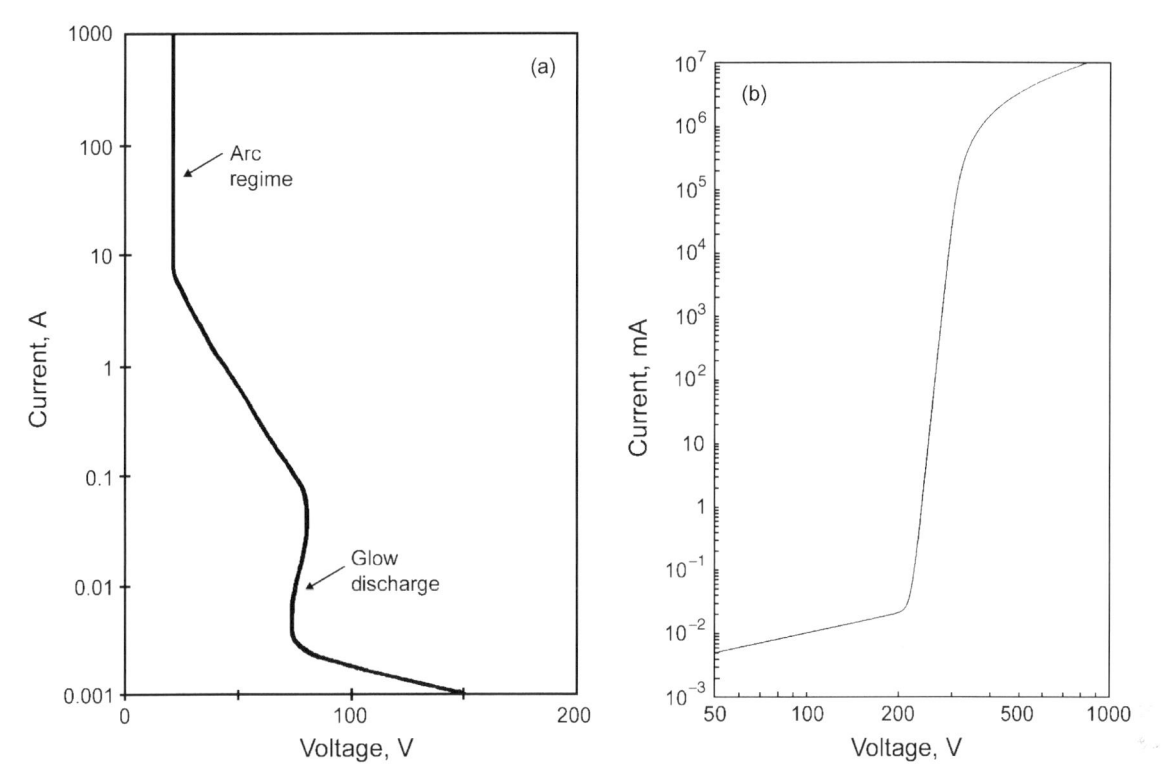

Fig. 18.8. (a) The idealized voltage–current characteristic of a gas tube (spark gap) arrester with a dc firing voltage of 150 V. The glow-to-arc transition can be seen. The arrester is designed to operate in the arc regime with an arrester voltage near 20 V. (b) The idealized voltage–current characteristic of a metal oxide varistor on a log-log scale. At currents less than about 100 μA, the varistor behaves like a simple resistor; at relatively large currents, over 100 A, it is also resistive. In the clamping voltage range of 200 to 300 V, the varistor is highly nonlinear and can carry currents between about 100 μA and 100 A. Adapted from Standler (1989a).

the same, or even better, that a lower clamping voltage be used for the primary, heavy-duty MOV and somewhat higher clamping voltages for the secondary MOVs. On a 120 V 60 Hz circuit, the primary SPD should typically be rated at 180 V, with the secondary SPDs at that level or higher. Otherwise, the SPD with the lowest rated voltage

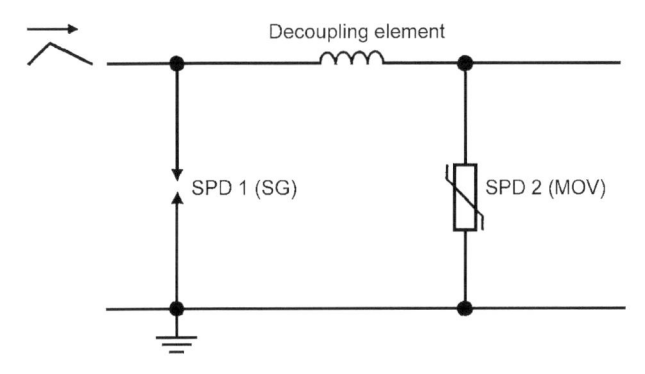

Fig. 18.9. A circuit illustrating the coordinated use of a spark gap, SG, arrester (SPD 1) and an MOV (SPD 2) to limit the output voltage to a value characteristic of the MOV without destroying it.

(perhaps at or within an electronics device such as a hi-fi amplifier) may operate first. This will be the case if the decoupling impedances provided by, for example, the inductance of the sections of power cable in the structure do not decouple the SPDs. A low-voltage-rated SPD could well be destroyed as it attempts to "protect" all the higher-voltage rated SPDs in the system. The primary SPD should have the highest energy and power rating and should be on the outside of the structure, at the service entrance; smaller energy, power, and current ratings are appropriate inside at, for example, wall outlets supplying power for consumer electronics.

The most common type of SPD now used in power or in communication circuits is the MOV. In power-line applications, MOVs have replaced the previously standard silicon carbide (SiC) arrester containing an air gap (the so-called gapped arrester) on distribution and transmission power lines. Such MOV arresters operate at voltage levels from near 10 kV to 1 MV. MOVs are manufactured by sintering zinc oxide (ZnO) powder together with small amounts of other oxides to obtain a solid disk with the highly nonlinear voltage–current characteristic shown in

Fig. 18.8b. Thus, metal oxide varistors are also called zinc oxide varistors or zinc oxide arresters and also, sometimes, metal oxide arresters.

For increased current-carrying capability, the diameter of the disk is increased. For increased voltage-handling capability, many disks are stacked in series. The performance, including failure, of MOVs on power lines subjected to lightning has been studied by Barker et al. (1993), Nakada et al. (1997), Darveniza et al. (1997), Fernandez et al. (1999) and Mata et al. (2000).

In the presence of a time-varying input current, the typical V–I characteristic of an MOV exhibits some degree of hysteresis. On a fast-rising current front the voltage attains a higher value for a given current magnitude than on the slower current decay to zero after peak. The differences in the voltages on the upper and lower parts of the hysteresis curve can be 20 percent or more for a current risetime of 1 μs or less, whereas for the 8/20 μs current waveform used in most arrester tests the hysteresis effect is minimal (Fig. 18.8b).

Circuit models have been developed to simulate this observed hysteresis and the overall nonlinear behavior (IEEE W.G. 3.4.11 1992; Pinceti and Giannettoni 1999). The IEEE W.G. 3.4.11 (1992) circuit model contains two simple nonlinear elements for which $I = BV^q$, where B and q are constants, a capacitor, two inductors and two resistors. The model of Pinceti and Giannettoni (1999) is simpler, with two such nonlinear elements, two inductors, and a resistor, but it provides similar results. The IEEE model is illustrated in Fig. 18.10. To model a given MOV, one needs test data for that MOV in order to calculate the values of the model circuit elements, including B and q. The resulting circuit models are suitable for use in large-scale circuit programs such as EMTP or SPICE.

Bartkowiak et al. (1999) examined the failure modes and energy absorption capability of ZnO varistors. They also reviewed the literature on the subject. They assumed current pulses of various waveshapes and computed the temperature profile and thermal stress as a function of time within the varistor disks by solving the heat transfer equation coupled to a temperature-dependent varistor E–j (electric-field–current-density) characteristic. Their model conforms to the available published failure data and hence can potentially be used to estimate the thermomechanical behavior, and the energy handling capability, of various types of varistor disks without resorting to performing destructive experiments. Bartkowiak et al. identified four different failure modes for MOVs: puncture, thermal runaway, cracking under tension, and cracking under compression. Cracking and puncture are caused by localization of the current, which causes local heating leading to nonuniform thermal expansion and thermal stresses. Puncture is most likely in varistor disks with low thickness-to-diameter ratios, when the current density has intermediate values. Cracking dominates at higher current densities and for disks with high thickness-to-diameter ratios. Puncture and cracking do not occur when the current is small because the time evolution of the nonuniform heating is slow enough for the temperature distribution to flatten. They are also unlikely at the very large currents corresponding to the upturn region of the E–j characteristic, since in this case the current becomes uniformly distributed. For low and for very high current densities the most likely failure mode is thermal runaway.

18.3.6. Topological shielding

Descriptions of the principles of topological shielding are found in Vance (1980), Baker et al. (1992), and Baum (1992). The use of this approach, involving shielding with appropriately located transient protection, is intended to reduce to non-destructive values any lightning-caused voltage differences in the interior of structures. If the walls of a building, say Shield 1 in Fig. 18.2, are perfectly conducting and contain no apertures then there can be no penetration of lightning current or of lightning electric or magnetic fields through the walls and into the volume enclosed by Shield 1. All external fields impinging on the shield will be totally reflected by the walls and no current or charge injected on the outside surface will penetrate to the interior (the skin depth for a perfect conductor is zero). A perfectly conducting shield completely isolates the enclosed space from external electromagnetic influences. In the absence of penetration of the shield by wires or pipes (one penetration is shown in Fig. 18.2), the potential of all the portions of the conducting building and all the space inside it will be equal, regardless of whether that potential is zero or has a high value. If there are no potential differences within the building, the potential of the building with respect to the Earth may assume any value without deleterious consequences to the interior. Hence, the topological shielding technique minimizes the importance of achieving a low grounding resistance.

It is not possible to make a structure a perfect shield since metal building materials have a finite conductivity and since metallic power and communication cables generally

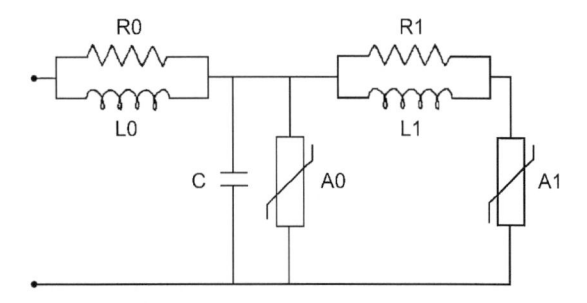

Fig. 18.10. Circuit models for an MOV arrester according to IEEE WG 3.4.11 (1992).

18.3. Protection

must enter the shield, although the use of fiber optic cables for communication systems can ameliorate this problem. Additionally, most buildings contain doors and windows as well as openings for plumbing and for heating and air conditioning. As an example of the imperfection of a seemingly perfect conducting enclosure without openings, Johnson et al. (1995) computed, in the time domain, the fields diffusing to the interior of a simple hollow cylinder of constant conductivity, permittivity, and permeability and the resultant internal voltages when lightning current was injected at the top, flat, surface of the cylinder, flowed down the side walls, and exited the bottom flat surface. Warne et al. (1995) have provided a similar calculation for a hollow ferromagnetic shield. As a result of the various imperfections of any realistic shield, a single shield is usually not adequate to provide sufficient reduction in potentially deleterious internal signals, even with the use of transient protection on the incoming wires.

To achieve a sufficient reduction in internal voltage differences due to either direct or nearby lightning via topological shielding, generally more than one shield is required. As shown in Fig. 18.2, shields are nested within each other with the outside of an inner shield "grounded" to the inside of the next enclosing shield. The voltage differences present become smaller, progressing from the outer shield to the inner ones. The shielded zones may be irregular in shape or they may be interconnected. For example, from a topological shielding point of view two shielded structures interconnected with a shielded cable, where the shield of the cable is bonded to the structures around the apertures through which the signal cable enters the structures, form one continuous shielded zone. Similarly, equipment cabinets, together with their shielded interconnecting cables or ducts, can form a continuous zone, perhaps the most interior one. Two levels of shielding can easily be designed, and often easily retrofitted, in existing facilities. The first level could be a conducting liner of appropriate thickness (a few skin depths) installed in a ground-based facility, an already present steel-reinforced concrete building, a structure with an enclosing metal facade, the existing metal skin of an aircraft or rocket, or the existing steel hull of a ship. The second level of shielding would then be the electrical and electronic equipment cabinets and their associated interconnecting cable shields or conduits.

Referring to Fig. 18.2, the first-level shield separates the external environment of zone 0 from the room environment of zone 1. The first-level shield accomplishes this separation by diverting or absorbing as much as possible of the unwanted signals that would otherwise enter the system, and it should provide at least enough reduction to prevent insulation breakdown or other damage to the components in zone 1, perhaps limiting peak voltages to, say, 500 to 1000 V. However, there is little benefit in reducing externally generated interference to levels much smaller than those produced by internal sources. In many facilities, peak voltages of a few hundred volts are generated in zone 1 by switches, relays, solenoids, rectifiers, and other similar devices, hence reduction of the externally generated interference to levels much below these internal noise levels is not warranted. In a two-level shielding system, the second-level shield, perhaps the equipment cabinets, separates the room environment, zone 1, from the sensitive small-signal circuits located in zone 2. Since these circuits may generally be upset or may develop an unacceptable error rate at peak signal levels of volts or tens of volts, the second-level shield must be capable of reducing the voltages induced on zone-2 conductors to less than a few volts.

Inherent in the approach of topological shielding is the fact that the current flowing in a conductor that is attached to a shield also flows on or near the surface to which the conductor is attached, a manifestation of the skin effect. Thus, currents flowing on conductors outside a shield can be diverted to the outside surface of that shield. A grounding conductor should never pass through a shield and be connected to the "inside" of the next shield, where the conductor current can flow on the "inside" surface and interact with internal components. Similarly, surge arresters and filters must divert unwanted currents to the outside surface of a shield, thereby preventing these currents from entering the next protected region. Thus, the "ground" lead of the SPD or the filter must be attached to the outside of the shield at the aperture where the protected cable enters the shield. As noted above, the outside of each successively smaller shield is to be connected (grounded) to the inside of the enclosing shield, a design feature that cannot be overemphasized.

18.3.7. Non-conventional protection techniques: lightning elimination and early streamer emission systems

Overview Most non-conventional structural-lightning-protection systems that are available commercially fall into two general classes. These classes are often referred to as (i) lightning elimination (or, more recently, charge transfer) systems and (ii) early streamer emission systems. Hardware for each type of system is sold worldwide by a number of manufacturers under a variety of trade names. Presently, there is no scientific evidence that either of these non-conventional techniques provides better protection than the conventional approaches previously discussed.

(i) In lightning elimination systems, one or more arrays of sharp points, often similar to barbed wire, are installed on structures and connected to grounded down conductors. The qualitative theory of operation of such systems, according to their proponents, is that the charge released via corona discharge at the sharp points in the presence of the electric field of an overhead thundercloud either (a) will deflect a downward-moving leader from the vicinity of the structure by reducing the electric field near the array or (b) will serve to discharge the cloud, thereby eliminating any possibility of lightning. Evidence that this is not the case is presented below.

(ii) Early streamer emission systems employ air terminals that, according to their proponents, launch an upward connecting leader earlier than would a standard lightning rod of the same height. Based on the assumption of a higher than typically observed upward leader speed, early streamer emission terminals are claimed to produce significantly longer upward leaders and hence to provide significantly larger zones of protection than, say, protection systems designed using the rolling sphere method (subsections 18.3.2 and 18.3.3). There are convincing arguments that this is not the case, as we shall discuss in the third part of this subsection. There is a variety of types of early streamer emission systems. All, however, involve ionizing the air around the tip of the air terminal.

Lightning elimination systems. Mousa (1998) provided a review of what is known about lightning elimination devices that are claimed to employ corona discharge, also known as point discharge, at a large number of sharp-points for lightning protection. According to Müller-Hillebrand (1962a), the idea of using multiple-point discharge to neutralize cloud charge was first suggested in 1754 by the Czech scientist Prokop Divisch, who constructed a "machina meteorologica" with over 200 sharp points. A few years later, Lichtenberg (1775) suggested that a catenary barbed wire suspended above a structure might prevent lightning from striking. According to Hughes (1977), a patent for such a system was issued in 1930 to J.M. Cage of Los Angeles. The patent described the use of point-bearing wires suspended from a steel tower to protect petroleum storage tanks from lightning. A commercial version of this system, a so-called dissipation array, has been available since 1971 although the product name and the name of the company that marketed the original system have changed from time to time (Carpenter 1977; Carpenter and Auer 1995). Most elimination devices were originally designed for use on tall communication towers, but recently they have been applied to a wide range of systems and facilities including airports and power lines. A paper by Mousa (1998) shows illustrations of six so-called dissipators produced by five different manufacturers. One of these, the umbrella dissipator, was described by Bent and Llewellyn (1977) as about 300 m of barbed wire wrapped spirally around the frame of a 6 m diameter umbrella. The barbed wire has 2 cm barbs, four barbs separated by 90° being placed every 7 cm along the wire. The umbrella dissipator described by Bent and Llewellyn was mounted on a 30.5 m tower in Merritt Island, Florida. Mousa (1998) also described a ball dissipator, a barbed power-line shield wire, a conical barbed wire array, a cylindrical dissipator, a panel dissipator (fakir's bed of nails), and a doughnut dissipator.

Carpenter and Auer (1995) gave their view of how the dissipation system marketed by Carpenter's company works, stating that it had been used for 20 years as a strike preventor. The system described had an ionizer with many hundreds of points, a "ground charge collector", and a grounding subsystem. The ground charge collector was claimed to neutralize the positive charge on the ground that would otherwise be there in the presence of negative cloud charge overhead. In addition to the claimed "ground neutralization", it was stated that "millions of ionized air molecules" from the ionizer are drawn away from the site toward the thundercloud by the high electrostatic field, and, in the process, a protective 'space charge' or ion cloud is formed between the site and the storm. According to Carpenter and Auer, many proponents consider the space charge the primary protective mode, saying its function is much like a "Faraday shield providing a second mode of protection".

In this description, by the major manufacturer of dissipation arrays, quantitative arguments are certainly absent. In a comment attached to the paper of Carpenter and Auer, Zipse (see also Zipse 1994) pointed out that trees and blades of grass generate corona discharge, often exceeding that of dissipation arrays, without apparently inhibiting lightning. This same point was previously made by Golde (1977b) and by Zeleny (1934). Zeleny observed that "During a storm in Switzerland the top of a whole forest was seen to take on a vivid glow, repeatedly, which increased in brilliance until a lightning bolt struck".

Mousa (1998) discussed the extensive grounding procedures used by the manufacturers and installers of lightning elimination devices. The leading manufacturer typically uses a buried encircling ground ring with 1-m-long ground rods located at 10 m intervals along the ring. In poor-conductivity soils, the same dissipation array company uses chemical rods, which it manufactures. These are hollow copper tubes filled with a chemical that leaches into the soil in order to increase the soil conductivity surrounding the grounding system (subsection 18.3.4). This same manufacturer highly recommends the installation of surge protective devices on sensitive electronics, a service performed by that manufacturer, at the same time that the dissipation array system is installed. As pointed out by Mousa (1998) and many others, elimination systems are, in principle, capable of providing conventional lightning protection; that is, they can intercept and survive a lightning strike without damage to the protected structure if there is sufficient coverage of the structure by arrays (air terminals) and potentially without significant damage to the electronics within the structure because of the presence of surge protective devices and good grounding. Carpenter (1977) listed many customers

18.3. Protection

who had reported a cessation of lightning-caused damage after installation of the system he manufactures.

We summarize now the records of observed lightning strikes to dissipation arrays. Kuwabara *et al.* (1998) reported on a study of dissipation array systems installed in summer 1994 on the tops of two communication towers on the roof of a communication building in Japan. Current waveforms were measured using current probes located under the arrays and on the antenna tower legs. Measurements of lightning current waveforms during strikes to the towers were made prior to the installation of dissipation arrays, from winter 1991 to winter 1994, and were subsequently made after the installation, from winter 1995 to winter 1996. Additionally, six direct strikes to the towers after the arrays had been installed were photographed between December 1997 and January 1998. Kuwabara *et al.* (1998) stated that the dissipation "system was not installed [as] per the manufacturer's recommendations as a result of the building construction conditions in Japan". Twenty-six lightning current waveforms were recorded in the three years before installation of the dissipation arrays and 16 in the year or so after installation. The distribution of peak currents was essentially the same before and after installation. Estimated peak currents varied from 1 kA to 100 kA. Kuwabara *et al.* (1998) stated that after installing the dissipation array, improving the grounding, and improving the equipment protection in summer 1994 "malfunctions of the telecommunications system caused by lightning direct strike have not occurred", whereas they were common before. Apparently, the presence of the dissipation arrays neither stopped the lightning strikes nor changed the character of the lightning stroke current, and equipment damage was eliminated by means other than elimination of lightning strikes.

In 1988 and 1989 the Federal Aviation Administration undertook studies at three Florida airports of the performance of dissipation arrays in relation to conventional lightning rods (FAA 1990). The umbrella dissipator installed on the central tower of the Tampa International Airport was struck by lightning on 27 August 1989, as evidenced by video and current records (FAA 1990). Carpenter and Auer (1995) disputed the findings of FAA (1990), and Mousa (1998) reviewed the attempts of the dissipation array manufacturer to suppress FAA (1990). Additional lightning strikes to dissipation arrays are described in Durrett (1977), Bent and Llewellyn (1977), and Rourke (1994). The former two references detail strikes to towers protected by dissipation arrays at the Kennedy Space Center, Florida and at the Eglin Air Force Base, Florida, respectively. Rourke (1994) described a nuclear power plant that was struck by lightning a total of three times in 1988 and 1989 before having dissipators installed in the expectation of preventing further strikes. After installation of the dissipation arrays, the nuclear power plant was struck a total of three times in 1991 and 1992. Rourke (1994) noted that "there has been no evidence that lightning dissipation arrays can protect a structure by dissipating electric charge prior to the creation of the lightning".

Golde (1977b, c) suggested that lightning elimination arrays installed on tall structures, typically towers, will inhibit upward lightning flashes (Chapter 6) by modifying the needle-like shape of the structures to a shape exhibiting a less pronounced field-enhancing effect. In this view, dissipation arrays will inadvertently eliminate some of these upward flashes, which represent the majority of flashes to very tall towers (Table 2.3). The upward flashes contain initial continuous current and are often followed by subsequent strokes similar to those in normal downward lightning flashes (Chapter 6), thus providing the potential for damage to electronics. The reason suggested above for preventing some upward-initiated flashes clearly does not involve charge dissipation. The view of Golde (1977b, c) was expanded upon by Mousa (1998), who argued that the elimination of upward flashes is particularly effective for towers of 300 m height or more and that charge dissipators will have no effect on the frequency of strikes to smaller structures such as power substations and transmission line towers.

While it is likely that corona charge generated by dissipation arrays near ground level will have no effect on the development or lightning-producing capability of thunderclouds, there is evidence that the injection of corona-producing chaff (typically aluminum-covered fiberglass needles thinner than a human hair) into developing thunderstorms can reduce the lightning flash rate. The effects of chaff seeding on thunderstorms were explored during two field studies in Colorado in 1972 and 1973 (Holitza and Kasemir 1974; Kasemir *et al.* 1976). Rust and Krehbiel (1977) discussed the experimental detection by ground-based microwave radiometry of corona from chaff in thunderstorms. The studies of Holitza and Kasemir (1974) and Kasemir *et al.* (1976) appeared to indicate that seeding ongoing thunderstorms with chaff can reduce the number of observed cloud-to-ground flashes to one-third or less of those observed in a control group of non-seeded thunderstorms. Maddox *et al.* (1997) reported that an inadvertent release of chaff from a military base in Arizona apparently resulted in the modification of 30 cells out of a total of 88 in an overall storm system, so that the 30 each produced only three, or fewer, cloud-to-ground flashes. Radar indicated that it was likely that the chaff cloud had been ingested by the 30 cells that subsequently produced little lightning. On weather radar, the cells that produced little lightning were indistinguishable from those producing lightning at a normal rate.

Early streamer emission systems. Early streamer emission (ESE) systems employ specially designed air terminals that are claimed to create enhanced ionization of the surrounding air, either by employing radioactive sources, by a special arrangement of passive electronics and electrodes that facilitate the electrical breakdown of small spark gaps in a high electric field, or by the application of a voltage source periodically, or at an appropriate time, to the terminal. The first early streamer emission devices were radioactive rods, although when these were initially marketed the term "early streamer emission" had not been invented. According to Baatz (1972), in 1914 the Hungarian physicist L. Szilard first raised the question of whether the attractive effect of a lightning rod could be increased by the addition of a radioactive source. Various tests in the field and the laboratory have shown that there is little or no difference under thunderstorm conditions between the action of a radioactive rod and of a conventional rod of the same height in the same location. For example, Müller-Hillebrand (1962b) measured the emission current of radioactive (radium 226) rods and conventional rods of the same dimensions under fair weather and under thunderstorm conditions. In fair weather the radioactive rods emitted a current of 10^{-8} A while the current from the conventional rods was unmeasurably small. However under a moderate thunderstorm field of 1 to 2 kV m^{-1} (subsection 3.2.2), the currents were about equal from the radioactive and conventional rods, roughly 10^{-6} A, and for larger fields, the current from the conventional rod exceeded that of the radioactive rod. In the laboratory, Baatz (1972) measured the current emitted by identical radioactive (americium 241) and conventional rods. While the radioactive rods emitted more current at very low electric fields, the rod with and the rod without a radioactive source emitted about the same current, for currents above 10^{-5} A. Laboratory tests purporting to show the superiority of radioactive rods over conventional rods were published by Heary et al. (1989), but, in discussions accompanying the paper, five researchers (Carraca, Grant, Liew, Menemenlis, and Mousa) used the paper's stated results to claim otherwise. Mackerras et al. (1987) gave examples of the failure of radioactive protective systems in Singapore when, at the time of their research, over 100 radioactive protection systems were in operation. Golde (1977b) cited the case of the failure of a radioactive lightning rod to prevent the papal crest on the Vatican's Bernini Colonnade in Rome from being knocked off by lightning on 6 March 1976, the crest being located about 150 m from a 22-m-high radioactive rod which was supposed to protect it. Radioactive rods are, as noted above, a class of early streamer emitters and, as such, are included in the following general discussion.

Claims for the superiority of ESE devices over Franklin rods are based on questionable theory, inconclusive laboratory experiments that are questionably extrapolated to natural lightning, and two inconclusive field experiments using triggered lightning. For these reasons, the ESE approach has been rejected by most scientists in the field of lightning physics and protection (e.g., Mackerras et al. 1997; Chalmers et al. 1999). Surveys of the ESE literature have been provided by Van Brunt et al. (1995) and Bryan et al. (1999); they were commissioned by the US National Fire Protection Association as part of an independent investigation to determine whether there should be a US national standard for early streamer emission systems, comparable with NFPA 780 (subsection 18.3.1) for conventional rod systems. The conclusion was that doing so was not appropriate. Nevertheless, there is presently both a French standard (1995) and a Spanish standard (1996) for the qualification of early streamer emission systems.

ESE proponents argue that ESE devices emit a positive upward-moving connecting leader (intended to meet the downward-moving negative stepped leader that initiates the usual cloud-to-ground lightning flash) at an earlier time, by a time interval Δt, than do conventional rods. They claim that this earlier launching time is achieved in a smaller electric field than that necessary for the conventional rod to emit a leader. However, they present no evidence that the ESE upward connecting leader can propagate in the required manner in this smaller field. Despite this lack of evidence, they translate the claimed time advantage Δt into a length advantage ΔL for the ESE connecting leader via the equation $\Delta L = v \Delta t$, where v is the speed of the upward connecting leader. ESE proponents assume that the speed of the upward connecting leader is of the order of 10^6 m s^{-1} (e.g., French standard 1995). The only existing measurements of upward positive-leader speeds for natural lightning found in the peer-reviewed literature are those presented in Berger and Vogelsanger (1966, 1969), McEachron (1939), and Yokoyama et al. (1990). Berger and Vogelsanger (1966, 1969) gave speeds for seven upward positive leaders of between 4×10^4 and about 10^6 m s^{-1}, the lengths of individual leader steps ranging from 4 to 40 m (Fig. 5.8b and Table 6.2). Further, for four of the seven leaders Berger and Vogelsanger (1966) gave speeds from 4 to 7.5×10^4 m s^{-1} and step lengths from 4 to 8 m at altitudes ranging from 40 to 110 m from the top, where a connection between a downward leader and an upward connecting leader would be expected. McEachron (1939) reported that the upward positive leaders initiated from the Empire State Building propagated at speeds ranging from 5.2×10^4 to 6.4×10^5 m s^{-1}, the step lengths ranging from 6.2 to 23 m (Table 6.2). The measurements by Yokoyama et al. (1990) showed, for three cases, upward positive-leader speeds from 0.8 to 2.7×10^5 m s^{-1} (subsection 4.5.1, second part). Yokoyama et al. (1990) reported that the lengths of the upward connecting leaders whose speeds they measured were from some tens of meters to over 100 m at the time that a connection was made with the downward moving stepped leader. They showed figures in which the

18.3. Protection

stepping of both the upward and downward leaders is apparent. Their measurements are apparently the only ones of the speeds of upward connecting leaders that actually connect to downward stepped leaders. It should be noted that positive upward connecting leaders in laboratory spark experiments typically have speeds of 10^4 m s^{-1}, an order of magnitude less than in the field measurements and two orders of magnitude less than the field value of 10^6 m s^{-1} assumed by ESE proponents.

It is necessary for the proponents of ESE devices to assume a value for Δt of about 100 μs and a value for v of 10^6 m s^{-1} in order to be able to claim a significant length advantage ΔL of 100 m for the upward connecting leader from an ESE rod over that for a conventional rod. If, however, the typical value for the upward connecting leader speed of 10^5 m s^{-1} that is accepted by the general scientific community is used, even allowing a 100 μs time advantage and even allowing that the leader could propagate in the lower field in which initiation is claimed to occur, the length advantage would be only $\Delta L = 10$ m, which is not significant. The value $v = 10^6$ m s^{-1} is sometimes referenced to an inapplicable measurement of Yokoyama et al. (1990), attributed by them to the optical step-formation speed, of the order of 10^6 m s^{-1}, while ignoring their measurements of upward connecting leader speed cited above. These step-formation speeds do not apply to the determination of the length advantage of ESE air terminals. The extension speed during the step-formation process in downward negative leaders has been estimated by other researchers to be between 10^7 and 10^8 m s^{-1} (subsection 4.4.6).

Two triggered-lightning tests of an ESE system known as a Prevectron are sometimes cited as support for ESE-terminal efficacy. The Prevectron has several spark gaps at its air terminal tip, which are intended to be activated in a sufficiently high electric field. The two triggered-lightning experiments involving the Prevectron have been described by Eybert-Berard et al. (1998). The first test, at Camp Blanding, Florida (Chapter 7), showed a current pulse of about 0.8 A peak amplitude and 2 μs duration from an ESE rod 85 μs prior to a triggered-lightning return stroke to ground at an unstated distance. No appreciable current followed the initial pulse in the ESE record. The ESE rod was not struck, and there was no evidence that the observed current pulse was associated with the initiation of an upward leader. Thus, this triggered-lightning experiment proves little except that the ESE rod emitted a small current pulse 85 μs prior to the return stroke in a flash that terminated nearby. The second triggered-lightning experiment, conducted in France in 1996 and described in the same paper, involved lightning that was triggered in close proximity to an ESE rod, a conventional rod being located further away. The ESE rod was the termination point of a leader–return-stroke sequence. The critical experiment, in which the positions of the ESE and conventional rods would have been interchanged to see whether only the rod closer to the rocket launcher was hit or whether a more distant ESE rod could compete with a conventional rod placed closer to the launcher, was not performed. Both triggered-lightning experiments described above involved dart-leader–return-stroke sequences in classical triggered lightning (a grounded triggering wire) and did not simulate the attachment process in virgin air that occurs in natural-lightning first strokes (subsection 4.5.1, first part).

Thus, in fact, no support for the proposed ESE system operation is given by any triggered- or natural-lightning study. On the contrary, natural-lightning studies have shown ESE systems not to work in the way that their proponents claim. Moore et al. (2000a, b) reported no advantage of ESE rods over ordinary lightning rods. In fact, they reported that in seven years of observations on a mountain top in New Mexico neither ESEs nor sharp conventional rods were struck, while 12 blunt conventional rods with tip diameters ranging from 12.7 mm to 25.4 mm were (subsection 18.3.3). Studies in Indonesia by Hartono and Robiah (1995, 1999) showed lightning damage to buildings protected by ESE systems indicating that failures of such systems are common. They gave many case studies with photographs. Their studies of buildings with conventional protection showed similar lightning damage on or near roof tops. Hartono and Robiah concluded that no advantage is obtained by using an ESE system.

Strong arguments can be made that no laboratory-spark test can be extrapolated to describe the case of natural lightning. For example, the step length of a lightning stepped leader is of the order of tens of meters, a distance considerably greater than the length of laboratory-spark gaps, of the order of a meter, that are specified to test and certify ESE systems (see, for example, French standard 1995, which requires a gap no smaller than 2 m with the air terminal being between 0.25 and 0.5 times the gap size). It is not likely that one can simulate in a 2 m laboratory gap the natural-lightning attachment process (Section 4.5). As another example, in natural lightning the downward negative leader from the cloud has a length of many kilometers while the positive upward connecting leader from ground or from elevated objects is generally some tens of meters long. However, for laboratory sparks between two identical electrodes, the positive leaders are always much longer than the negative leaders.

Besides the fact that the scaling of laboratory phenomena to natural lightning is questionable, the laboratory studies of ESE systems often do not even support the claims made by ESE proponents. Tests conducted by Berger G. (1992) showed that in the laboratory both ESE and conventional-rod upward connecting leaders propagate at the same speed, 10^4 m s^{-1}, but exhibit different modes of development. According to Chalmers et al. (1999), in the test described by Berger the ESE rod begins emitting current 50 μs prior to the conventional rod, but both upward leaders eventually travel about the same length, the corona ahead of the conventional-rod leader extending further than that ahead of the ESE rod leader. Berger showed that there is essentially no difference in the average triggering time of an ESE air-spark rod and a conventional rod. In the data presented by him, an ESE surface trigger rod appears to have a slight average time advantage over a conventional rod in leader inception, but the spread in leader inception times for the ESE rod is from 680 to 840 μs while that for the Franklin rod is from 720 to 860 μs, so that for any given event it is impossible to say which rod will initiate a leader earlier. As mentioned above, it is only by averaging all the initiating times that the ESE shows a slight advantage. It follows that for one isolated lightning event it is impossible to determine from the laboratory data which rod will be struck, although one could determine a probability which would be nearly equal for the two rods, assuming that the laboratory tests could be extrapolated to lightning. As noted above, there are strong arguments against the validity of such an extrapolation.

Abdel-Salam and Al-Abdul-Latif (1997) presented some calculations on the behavior of an active lightning rod to which a high voltage is applied at an appropriate time. Allen et al. (1998a) presented laboratory data in which such an active rod is simultaneously subjected to a uniform field and a superimposed impulse field, simulating the cloud field and the negative leader field, respectively. To provide the "early" streamer emission, a positive impulse voltage with a 1 μs risetime and 50 μs fall time to half-value was applied to the rod, this voltage being variable in amplitude up to 40 kV, with adjustable delay from the start of the simulated negative leader. Such a system, with proper adjustment of variables, was able to produce an initiation-time advantage over a conventional rod, as might be expected. It is certainly not obvious how this system would be operated in the field. Allen et al. (1998a) also tested a commercial ESE device with internal spark gaps (as have all such devices that do not use the radioactive principle) but also containing a sealed "power unit". This ESE device in general showed little difference in its breakdown characteristics in the laboratory gap from a conventional rod.

Mackerras et al. (1997) and Chalmers et al. (1999) both critically reviewed the claimed principles of operation of ESE devices. Both raised the important question whether an upward connecting leader, if launched earlier than for a conventional rod (assuming this can be achieved) and hence in a lower electric field, is able to propagate in the required manner in this lower field and so compete with a conventional rod whose leader is launched in a larger field. According to Mackerras et al., once the upward connecting leader propagates into the space remote from the air terminal, its further progress depends upon the supply of energy from the electric field in the space near the tip of the discharge and upon the dielectric properties of the air undergoing breakdown, neither of these factors being influenced by the air terminal. Using this and geometrical arguments, they concluded that "it is not possible to gain a significant improvement in lightning interception performance by causing the early emission of a streamer from an air terminal". Additionally, Chalmers et al. pointed out that any length advantage ΔL that could potentially be associated with early streamer emission is not dependent just on the upward leader velocity but also on the ratio of the speeds of the downward and upward leaders.

Mackerras et al. (1997) proposed a field experiment in which an ESE air terminal would be placed in the center of an elevated ring of wire whose plane was parallel to the Earth and whose radius was less than the protective radius claimed by the ESE manufacturer. Current recorders would be placed on the wire and on the ESE system. If the wire were struck then the ESE system would have failed to exhibit the advertised protective radius. Presumably if many such test systems were placed in a region of high lightning activity, significant information would be obtained in the course of a year or so.

18.4. Lightning interaction with specific objects and systems

In this section we consider the interaction of lightning with boats, trees, solid-state electronic systems, communication systems, and overhead and underground power systems. As noted earlier, lightning hazards to humans and animals will be considered in Chapter 19 and the interaction of lightning with aircraft in Chapter 10.

18.4.1. Boats

Ten years or so after Franklin proposed a method for the lightning protection of houses and vessels (see the quote at the beginning of the chapter), as discussed in subsections 18.3.2 and 18.3.3, those principles were applied to the protection of wooden ships. The history of the lightning protection of the ships of the British Royal Navy is particularly interesting and has been reviewed by Bernstein and Reynolds (1978), from which the following discussion is taken. In 1762, William Watson, one of Britain's early electrical scientists, wrote to George Anson, First Lord of the Admiralty, recommending the installation of a lightning protection system on British Royal Navy vessels. Watson (1761) had previously suggested protecting sailing ships by connecting a brass wire conductor about "the thickness of a large goose quill" to the masts, and leading it from there, by the most convenient path, into the water. Anson was apparently receptive to the suggestion, and the British Royal

Navy adopted lightning protection, but it used an inferior system due to Winn (1770) that consisted of a series of copper rods connected every few feet (every 60 cm or so) by connecting links. This chain of rods was attached to a rope hung from a metal spike at the top of the mast and loosely dangled into the sea. Neither the spike nor the conductor was kept permanently in place; that is, they were installed only when storms were imminent. Winn's protection system had major drawbacks. Often the chain was not in place when lightning struck, and the system interfered with seamen working on the rigging. Nevertheless, that system was the dominant method of marine lightning protection in the period 1770 to 1840 and it was the system that became the standard in the British Royal Navy. A more adequate system, which contained most of the basic elements subsequently recommended in modern codes and which ultimately replaced Winn's chain conductors, was devised by William Snow Harris in 1820 (Harris 1834, 1843). Fixed conducting plates were routed along the mast down through the hull to a copper sheathing on the bottom of the ship. Harris spent almost 25 years trying to persuade the British Admiralty to adopt his system. It took a successful trial installation on 11 ships, an extensive campaign by Harris to publicize the extent of the previous lightning damage to the British Royal Navy, the favorable recommendations of two study committees, and administrative changes in the Admiralty before the British Royal Navy finally adopted the Harris system in 1842.

The adoption of Harris' system by the British Navy did not, at first, lead to the wider use of permanently installed conductors. Chain conductors remained the standard in the United States, while many wooden merchant vessels, both British and foreign, continued to carry no protection at all. Not long after Harris succeeded in getting adequate lightning protection installed on British Royal Navy ships, the wooden sailing ships of the Navy and that of the navies of other countries were replaced by iron steam ships. Iron ships are self-protected in that the metal comprising the ship provides the air terminal, the down conductors, and the means of grounding.

The US National Fire Protection Association's lightning protection standard NFPA 780 devotes a section to boats. Some of the specifications from this section of the US standard follow. The placement of air terminals, as with the placement for other structures considered in NFPA 780, is determined (or justified) using the rolling-sphere approach, with the sphere radius chosen to be 30 m. Copper is the best conductor for the protection system because of its resistance to corrosion in a salty atmosphere. Aluminum masts, which may be used as down conductors, are an exception to the recommendation of copper. As is the case with other structural protection, junctions between aluminum and copper should be made with special stainless steel connectors in order to reduce potential corrosion at those junctions. Copper down conductors connecting the air terminals to the grounding system should be of diameter not less than no. 4 AWG (about 21 mm^2 cross-sectional area), according to recent versions of NFPA 780, unless there are a number of down leads, no. 6 AWG (about 13 mm^2 cross-sectional area) being acceptable for two separate down conductors in parallel. All metal on the ship should be interconnected to the lightning protection system. A metal hull is the ideal grounding system. For boats with nonmetallic hulls, conducting grounding plates or strips should be installed on the underside of the hull to provide a path for the lightning current to flow into the water. Plates should be of minimum size 1 ft^2 × 3/16 in. thick (0.093 m^2 × 4.8 mm). Through-hull connectors should be metallic and have a cross-sectional area equivalent to a no. 4 AWG copper conductor. As we shall discuss later, such grounding plates may provide a satisfactory ground in salt water, but they are probably not sufficient to inhibit side flashes in fresh water. It follows from the above that, except for the grounding system, the protection of boats is essentially the same as the protection of ground-based structures.

Thomson (1991) examined lightning damage to sailboats and discussed the relation of that damage to several protection codes published by boating agencies and to the 1980 version of NFPA 780. The codes he discussed specify a no. 8 AWG (about 8.4 mm^2 cross-section area) copper wire for the case where a single down lead is to be used. Thomson (1991) showed from calculations referenced to Golde (1968) that the temperature of a no. 8 AWG copper wire will rise to near melting for a relatively large action integral, 5×10^6 A^{2}s (see subsection 18.3.3), whereas the no. 4 AWG recommended by the more recent versions of NFPA 780, possibly in response to the criticism of Thomson (1991), will rise in temperature by only about 100°C for the same action integral. Thomson calculated that in fresh water a lightning current of 30 kA flowing out of the typically recommended 1 ft^2 (0.093 m^2) ground plate gives rise to a potential difference of up to about 45 MV between the sailboat rigging and the water, potentially leading to destructive side-flashes. For an assumed water resistivity of 10^3 Ω-m, that for fresh water, the grounding resistance is 1.5 kΩ. Arcing from the ground plate into the water may lower the ground resistance, as discussed later, but hazardous voltages may still appear between the bonded conductors above water level and the water. The situation is clearly better in salt water, where the conductivity is three orders of magnitude greater and hence, above water, voltages will be in the tens of kilovolts range for the example given above, insufficient to cause a spark of more than a few centimeters.

Thomson (1991) analyzed 71 reports of sailboat damage by lightning. The reports met the following criteria: (i) the boat hull was constructed of fiberglass, (ii) the

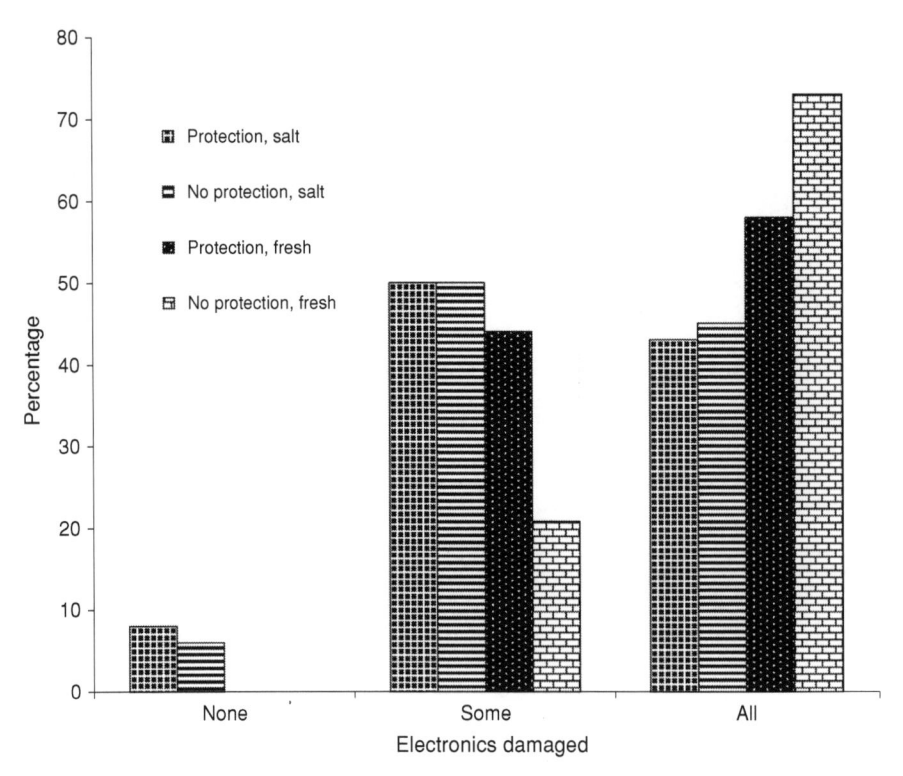

Fig. 18.11. Frequency distribution for proportion of boats with electronics systems that had none, some, or all of these systems damaged as a result of a direct lightning strike. Adapted from Thomson (1991).

mast was aluminum, and (iii) there was clear evidence of lightning attachment to the top of the mast, usually in the form of damage to a masthead antenna. The data were divided into four categories depending on whether the boats were in salt or fresh water and whether they had lightning protection systems in place when struck. A boat was considered to have a protection system if a connection existed between the base of the mast or shrouds to either a metallic keel or a ground plate below the hull.

Thomson found that annually about three percent of all moored sailboats in southwest Florida suffer lightning-induced damage to marine electronics. Bar graphs showing the frequency of occurrence of the three degrees of electronics damage are given in Fig. 18.11. The frequency of occurrence is given as the percentage of all boats in each category falling in the particular damage class. The number of boats in each category are as follows: 26 with protection in salt water, 16 with no protection in salt water, 14 with protection in fresh water, and 11 with no protection in fresh water. Apparently, the present state of lightning protection is particularly ineffective for marine electronics. Boats in fresh water sustained slightly more electronics damage than boats in salt water: 16 out of 25 (64 percent) versus 18 out of 42 (43 percent), respectively, with damage to all systems.

Figure 18.12 gives bar graphs showing the frequency of occurrence of the four classes of hull damage. Damage to the boat hull was classified on a 0 to 4 severity index scale according to the following criteria: 0, no discernible burns or fractures; 1, small non-leaking cracks or burns; 2, small holes (typically described as "pinholes" of a millimeter or less in diameter) that did not pose a threat of serious leaks; 3, large (several millimeters in diameter) holes above the waterline; and 4, large holes (several millimeters in diameter) below the waterline. Indices 2 to 4 represented electrical breakdown through the hull, that is, a serious failure of the lightning protection system. Boats with hull damage in category 4 were in sinking condition. The number of boats in each category are as follows: 28 with protection in salt water, 16 with no protection in salt water, 15 with protection in fresh water, and 12 with no protection in fresh water. A major difference between boat strikes in fresh and salt water was apparent in the severity of hull damage, as is evident from Fig. 18.12.

As is evident from Figs. 18.11 and 18.12, protection systems are of value and, consistent with the availability of a better ground in salt water, there is a much lower occurrence of serious hull damage from side flashes in salt water than in fresh water. In fact, since the voltages developed for fresh water are much larger than those required for breakdown, it is surprising that 60 percent of protected boats in fresh water experienced no through-hull electrical breakdown. In this regard, the mitigating factor may be the

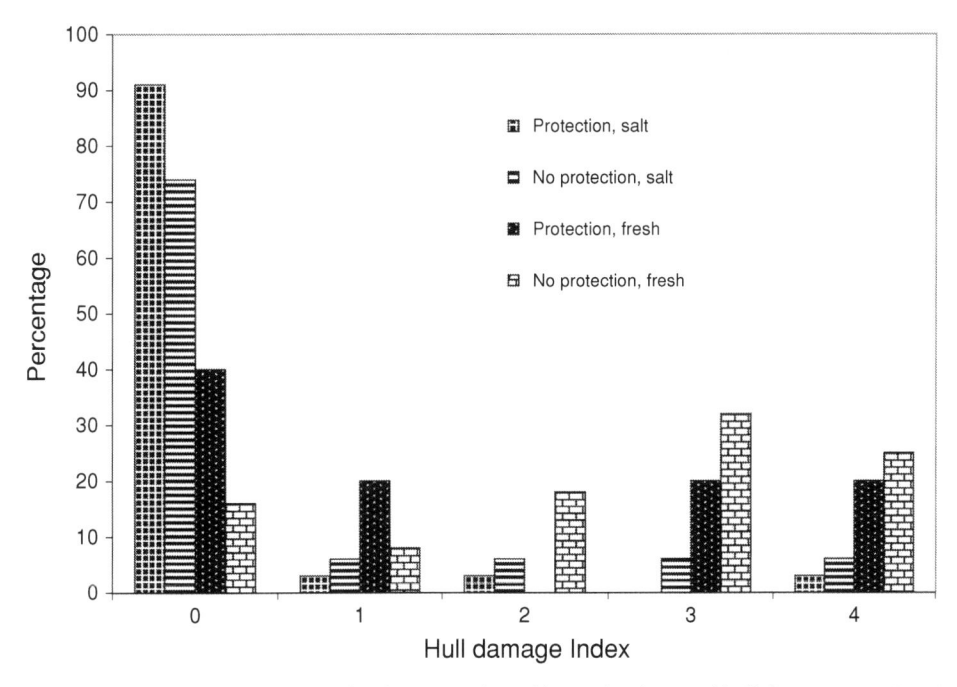

Fig. 18.12. Frequency distribution for the proportion of boats that incurred hull damage, on a 0 to 4 scale, as a result of a direct lightning strike. The index values are defined in the text. Adapted from Thomson (1991).

dynamic ground resistance (subsection 18.3.4) that arises when breakdown occurs in the water as the lightning current flows out of the ground plate, this resistance being smaller than the dc ground resistance. Applying Eq. 18.11 (subsection 18.3.4) to a hemispherical grounding conductor in fresh water, assuming $I = 30$ kA, $\sigma = 10^{-3}$ S m^{-1}, and $E_b = 1 \times 10^6$ V m^{-1} we find that there will be breakdown arcs in the water within a radius r_0 equal to about 2 m. Petropoulos (1948) found experimentally that a 5-cm-radius sphere in "town water" did not produce arcing when subjected to an impulse voltage of 50 kV, so that E_b in this case was in excess of 10^6 V m^{-1}. For a single ground plate, the effective area caused by the enlarging of the conducting area of the plate via arcing is about 30 m^2 and is independent of the actual area of the plate, according to Thomson (1991). The effective resistance is 72 Ω, and the maximum voltage is 2.2 MV, which is a factor about 20 smaller than that developed for a circular plate of 1 ft^2 (0.093 m^2) area. It is, however, still large enough to make side flashes of meter length in air. The assumed value of 1×10^6 V m^{-1} for the breakdown electric field E_b in water in the example above is apparently a lower limit for a spherical electrode. A larger breakdown electric field for the same input current will result in a smaller r_0 via Eq. 18.11 and a higher maximum voltage than that calculated above.

The shape of the ground plate is also important. Petropoulos (1948) found that a 5-cm-radius spherical electrode equipped with seven 4-cm-long points started producing small sparks at an impulse crest voltage of 28 kV, whereas the 5 cm sphere without points did not produce sparks even at 50 kV. Thus, a grounding conductor with sharp corners or points may well initiate discharges at a lower voltage, with the result, according to Petropoulos, that there is a decrease in grounding resistance at a lower level of current than that for a grounding conductor with a smooth surface.

18.4.2. Trees

Lightning can strike a tree and leave it apparently unharmed (e.g., Orville 1968; Uman 1991), it can cause considerable structural damage to the tree without noticeable burning, or it can set the tree on fire. The detailed effects of lightning on trees were documented in studies by Taylor A.R. (1964, 1965, 1969a, b) and by Schmitz and Taylor (1969). They found that most trees that were struck were not killed. The majority recovered from whatever lightning damage they may have sustained, though many were weakened and ultimately succumbed to attacks by insects and disease. Visible damage to tree trunks ranged from superficial bark flaking, to strip-like furrowing along the trunk, to almost total destruction. Typical damage to evergreens and rough-barked deciduous trees is shown in Fig. 18.13. Taylor A.R. (1964) examined 1000 lightning-damaged Douglas firs in western Montana. Most had shallow continuous scars a few inches wide along their trunks. About 20 percent had two or more scars, 10 percent had severed tops, and about one percent had been reduced to slabs and slivers. Most of the scars were spiral, a few were

Fig. 18.13. At the time of going to press a colour version of this figure was available for download from http://www.cambridge.org/9780521035415. Typical lightning scars on a loblolly pine tree (right) and on a live oak tree (left) in north central Florida. The trees are two of six on the second author's 10 acre (4×10^4 m^2) property that have suffered lightning damage in the past six years. The pine tree died and the oak tree lived. Overall half the trees showing lightning scars died.

straight. The average scar extended along 80 percent of the tree height, but none extended to the very tops of the trees. Scars either reached to ground level or close to ground level. Often along the center line of the lightning scar was a crack which penetrated into the tree, and, when wood was removed from a tree by lightning, it was usually ejected as two parallel slabs, separated along this crack. Sometimes, in place of the crack, the lightning had left a narrow strip of shredded inner-bark fiber fixed in a smooth shallow groove about 1/16 in (0.16 cm) wide. These damage characteristics are typical not only for Douglas firs but for all conifers (cone-bearing trees, mostly evergreens) throughout the United States. The descriptions also appear valid for most rough-barked species of deciduous trees such as oaks. The relatively smooth barks of other deciduous trees, for example birches, present quite different damage characteristics. For these trees the bark is not removed in narrow, uniform strips but rather is torn off in large, somewhat irregular patches or sheets as shown in Fig. 18.14.

Sometimes a single lightning discharge can kill a group of trees. In a typical group kill, obvious lightning damage is visible on only one or two trees, often near the center of the dying group. As many as 160 trees in a group have been reported as killed in this way, but in most cases the groups are probably smaller. It is unclear whether lightning does unseen damage to the roots of trees surrounding the struck tree or whether the aerial parts are affected by the discharge (Minko 1966). After reviewing the worldwide literature on the group killing of trees by lightning, Taylor A.R. (1969a) noted that, while very few cases have been reported in the United States, many instances have been reported in Europe, Australia, Malaysia, and other parts of the world. A possibly related phenomenon involving lightning and tree groups occurs frequently in parts of the United States. Entomologists report that several species of bark beetle attack single trees damaged by lightning and then proceed to attack other trees surrounding the damaged one (e.g., Schmitz and Taylor 1969; Komarek 1964; Coulson et al. 1983, 1986; Schowalter et al. 1981; Lovelady et al. 1991). Often the eventual result is a group kill similar to those not involving bark beetles. Possibly, lightning is playing a hidden role in this type of group kill in that it may have done unseen damage to the trees surrounding an obviously struck tree, thereby reducing their natural resistance to attack by the bark beetles.

We now discuss the question of forest fires. To set fire to a piece of wood, the flame must contact the wood for a sufficient period of time, since the initiation of burning

18.4. Lightning interaction with specific objects and systems

Fig. 18.14. Lightning removed bark from this paper birch in large, irregular patches. US Forest Service photo.

depends on the amount of time during which a sufficiently high temperature is applied. In general, the higher the temperature, the shorter the time. The lightning continuing current (Section 4.8) provides a high enough temperature for a sufficient time to ignite woody fuels. Fuquay *et al.* (1967, 1972) showed conclusively that continuing currents in ground flashes are likely to cause forest fires. Since positive flashes generally if not always have a continuing-current component with associated large charge transfer (Section 5.3), positive flashes are prime candidates for setting forest fires (Fuquay *et al.* 1972). Trees may not necessarily be set on fire directly by lightning. Rather, the lightning current may ignite the leaves, moss, and other flammable material on the forest floor surrounding a tree, leading to eventual ignition of the tree.

Studies of forest fuel ignition using laboratory discharges to simulate the lightning current have been performed by Latham and Schlieter (1989) and Darveniza and Zhou (1994). There are so many variables in the fuel–discharge interaction that it is difficult to extrapolate the published laboratory results to the case of lightning.

While lightning without continuing current may or may not set fires, it can nevertheless be very destructive. The bottom end of the lightning channel is often viewed as a current source that forces the lightning current into the struck object. This current generates heat in the object through which it flows, the amount of heat depending on the object's ohmic resistance (Section 18.2). If the object has a relatively high resistance, as a tree does, there can be a great deal of heating although not necessarily enough to cause burning. The rapid lightning current rise inside a high-resistance material causes rapid heating and consequent vaporization of some of the internal material. As a result, a very high pressure is quickly generated within the material, and this pressure blows the material apart. For many of the tree scars examined by Taylor A.R. (1964), the lightning apparently followed a path through the cambium (a thin layer of living cells between the inner bark and the wood) or through the moist inner bark tissue. These zones were apparently chosen as the lightning's paths because they offered lower resistance than the outer bark or the wood. The pressure generated in the cambial region expelled a bark strip, creating the vertical scar shown in Fig. 18.13. Such pressure also could cause a split in a tree, but it is not clear why a strip of inner-bark fiber is pasted along the centerline of the scar, as is often observed. In some of the trees examined by Taylor, wood as well as bark was blown out from the tree. In these cases the lightning current apparently traveled deeper within the trunk. Taylor found that older trees (over 200 years) were more likely to suffer wood-loss scars and suggested that perhaps the wood in old trees offers less electrical resistance to current flow than does the cambial zone.

Until recently, it was generally felt that lightning-caused forest fires should be quickly suppressed, particularly those fires occurring near populated areas. Nevertheless, lightning has been a significant part of the fire ecology since there have been forests and lightning, long before man was a significant feature of the landscape (e.g., Love 1970). Before suppression became a general policy, frequent fires in the western United States kept the forest floor clean; the fires themselves were small and did not damage the trees. Efforts to prevent and contain forest fires in the American West have enabled the brush to grow more thickly and now most fires are big ones. The record fires of summer 2000 renewed the view that regular prescribed burns and mechanical clearing of forest underbrush are essential to inhibit large fires. Prescribed burns serve the same purpose as lightning-caused fires and are, in general, easier to control, although the disastrous prescribed burn which got badly out of control near Los Alamos, New Mexico in summer 2000 illustrates that the imitation of regular lightning burns is not a trivial pursuit. Interestingly, we may be indebted to ancient forest fires for California's giant sequoias. The seedlings of these trees can germinate in ashes but are suppressed under the thick layer of needles that might cover an unburned forest floor.

Rising smoke from forest fires can produce clouds that can in turn produce lightning (Latham 1991). Smoke from fires, when ingested by existing clouds, can change the characteristics of the lightning produced by these clouds (Section 5.2).

In order to protect a tree from the effects of lightning, the tree must be outfitted with a lightning protective system similar to that used for a structure. This protective system consists of metal wire or a metal strip attached to the tree from a well-established ground and run up the tree trunk to the top of the tree and out along the major branches. Allowance must be made for wind sway and tree growth in affixing the wire or down-conductor strips. The proper placement of the air terminals (or the tops of the down conductors) can be determined using the rolling sphere approach (subsections 18.3.2 and 18.3.3). According to NFPA 780, grounding is optimally done as follows. Extend three or more radial conductors in trenches 1 ft (0.3 m) deep, spaced at equal intervals about the tree base to a distance extending to the branch line but not less than 25 ft (7.6 m). Either have ground rods at the end of the radials or have the ends of the radials bonded to a conductor that encircles the tree (a ring electrode) at depth of not less than 1 ft (0.3 m). Bond the grounding system to any underground metallic water pipe within 25 ft (7.6 m) of the branch line.

18.4.3. Distribution and transmission power lines
General information Transmission lines, power lines that operate at voltage levels from near 50 kV to near 1 MV and transport electric power cross-country, necessarily have relatively high insulation levels and usually have overhead ground wire protection so that they suffer less damage and fewer outages from lightning than do the lower-voltage power distribution lines, typically operating near 10 kV, that deliver power within municipalities. Because of the differences in the two types of power delivery system, their lightning protection is usually considered separately. The lightning effects on both transmission and distribution lines include (i) flashovers from phase conductors, when they are struck, to metallic towers, ground or shield wires, neutral wires, or other phase conductors, and (ii) flashover from metallic tower or shield wires, when they are struck, to the phase conductors, so-called back-flashover. Both types of flashover can potentially be followed by power-frequency current flow through the lightning flashover path. Such "power-follow" current is generally interrupted by the action of a circuit breaker, resulting in a momentary power interruption. Back-flashovers occur because metallic towers and shield wires are connected to Earth by the impedance of the down conductor or tower to which the shield wires are attached and also by the grounding resistance, a deleterious voltage on the shield wires being produced as described in subsections 18.3.3 and 18.3.5 in the discussion of side flashes. Lightning-caused voltages, in addition to leading to flashovers, can cause damage to line insulators and to equipment on the line such as transformers and lightning arresters. Lightning arresters are used in an attempt to limit the lightning voltages to levels that do not produce

18.4. Lightning interaction with specific objects and systems

equipment failure or flashovers, but they are not always capable of achieving this goal; some arrester-failure mechanisms are discussed in subsection 18.3.5.

Distribution lines Flashovers and line-hardware failures on distribution lines may result from direct lightning strikes to the line or from induced voltages due to nearby strikes. Direct strikes are the most difficult to protect against because the overvoltages can be many megavolts: as noted in Section 18.2, a typical return-stroke peak current of 30 kA times the parallel combination of the typical line surge impedances "seen" in each direction of 500 Ω yields 7.5 MV, whereas the distribution-line insulation usually fails between 100 and 500 kV. Induced voltages from nearby strikes are thought to be generally less than 300 kV, so they are much easier to protect against. Measurements of and methods for calculating such induced voltages are discussed later in this section. Direct stroke protection can be achieved by one or more of the following methods: (i) use of the highest economically reasonable insulation level, (ii) use of overhead shield wires with good grounding, and (iii) use of grounded arresters between phase conductors and neutral, spaced as close as economically reasonable, typically every third pole. While nearby trees may protect a distribution line from direct strikes, the induced voltages from these close strikes may still represent a problem. Improvement options for induced effects are the same as for direct-strike protection. The question of the proper arrester spacing to protect against either direct or nearby strikes is a matter of controversy. An arrester of high enough quality at every pole should provide reasonable protection against flashovers, although there will be arrester failures if the lightning imparts, for example, a much higher action integral to an individual arrester than specified by the standards to which the arrester was manufactured and tested (Section 18.5).

As parameters of distribution line performance, the basic insulation level (BIL) and the related critical flashover voltage (CFO) of a distribution line are defined as the crest value of the impulse wave that, under specified conditions, causes flashover for a specified percent of applications. The CFO of a structure is determined by various system elements such as ceramic insulators, wood cross-arms, and the placement of guy wires and other conductors. It is usually difficult to estimate the CFO of a structure. The CFO may be lower than its normal value when the distribution structure is wet. For wood distribution poles and crossarms with associated insulators, the CFO can be determined roughly by adding 250 kV per meter of wood to the insulator CFO. The CFO is often degraded at "weak-link" poles, where equipment on the poles reduces the clearances. System elements that may degrade the CFO include the improper placement of guy wires, fuse cutouts, conducting supports, and other conductors in the system.

As discussed in subsection 2.9.2, the number of strikes to a distribution line can be calculated roughly by using the concept of equivalent attractive area. If we assume that all flashes that would have hit ground within two line heights on either side of the line center will strike the line, we can estimate the number of strikes per 100 km per year, N, as

$$N = 0.4 N_g h \qquad (18.13)$$

where N_g is the ground flash density in $km^{-2} \, yr^{-1}$ and h is the line height in meters. Eriksson (1987a) proposed a more complex formula, for which N in open terrain (no significant trees or buildings nearby) is given by

$$N = N_g \left(\frac{28 h^{0.6} + b}{10} \right) \qquad (18.14)$$

where b is the width of the distribution-line structure. For a distribution-line height of 10 m with $b = 0$, Eq. 18.14 yields $N = 11$ flashes to the line per 100 km per year for $N_g = 1 \, km^{-2} \, yr^{-1}$. Thus for $N_g = 10 \, km^{-2} \, yr^{-1}$, as in Florida, about one flash per kilometer per year is predicted. For the same line, Eq. 18.13 predicts a line strike rate about 0.4 times that of Eq. 18.14.

Nearby buildings and, as noted earlier, trees may play a major role in the lightning performance of distribution lines by intercepting many lightning flashes that otherwise would have directly hit a line.

Calculated values of the flashover probability for a direct strike to a distribution line as a function of arrester placement, arrester grounding resistance, and line CFO are given in Fig. 18.15, taken from IEEE Standard 1410-1997 (Section 18.5). These calculations involve many idealizations and assumptions and hence may not be accurate.

To estimate voltages induced on distribution lines by nearby strikes, three basic coupling models have been used: those of Rusck (1958, 1977), Chowdhuri and Gross (1967), and Agrawal *et al.* (1980). All of these models are based on transmission-line theory. The model of Agrawal *et al.* and its equivalent formulations (Taylor C.D. *et al.* 1965; Rachidi 1993; Nucci and Rachidi 1995) can be considered to be accurate within the limits of the transmission-line theory (Nucci and Rachidi 1995), but in both the Rusck and the Chowdhuri-Gross models, some source terms have been omitted (Nucci *et al.* 1995a, b; Cooray 1994; Cooray and Scuka 1998).

However, for the case of the electromagnetic field radiated from a vertical lightning channel, the Rusck model is equivalent to the model of Agrawal *et al.* (Cooray 1994). Besides his detailed model, Rusck also proposed a simplified analytical formula for the peak induced voltage on an infinitely long overhead line above a perfectly conducting ground, as a function of the return-stroke peak current. A discussion of the validity of the transmission-line theory

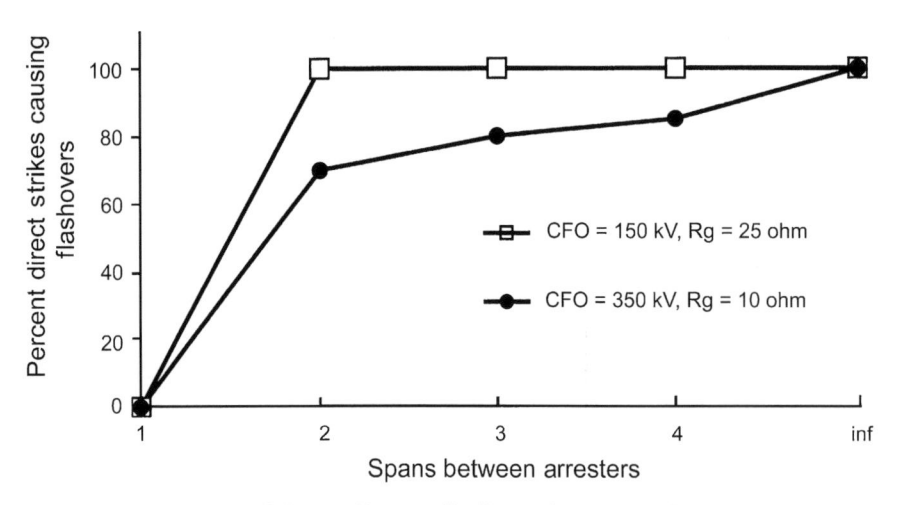

Fig. 18.15. The percent of direct strikes to a distribution line causing flashover of the line. The span length was 75 m. Adapted from IEEE Standard 1410-1997.

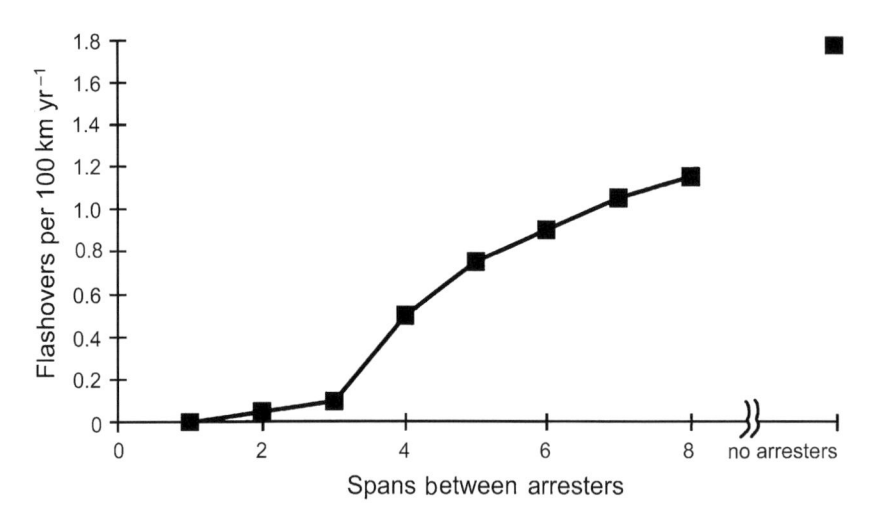

Fig. 18.16. Distribution-line flashovers per 100 km per year for a flash density N_g of 1 km^{-2} yr^{-1}, a line height h of 10 m, and a CFO of 150 kV as a function of arrester spacing. The span length was 75 m. Adapted from IEEE Standard 1410-1997.

versus the more sophisticated scattering theory is found in Tesche (1992).

Coupling models have been tested by means of natural lightning (Eriksson 1987a; Yokoyama et al. 1983; Master et al. 1984; De la Rosa et al. 1988) and triggered lightning (Rubinstein et al. 1989; Georgiadis et al. 1992; Barker et al. 1996). Laboratory tests have also been performed in a more controlled environment using NEMP (nuclear electromagnetic pulse) simulators (e.g., Guerrieri et al. 1995) and also using reduced scale models (e.g., Ishii et al. 1994; Piantini and Janiszewski 1992; Nucci et al. 1998). Nucci (1995) presented a survey of these tests.

Calculations of the voltages induced on overhead lines under various conditions have been given by Master and Uman (1984), Diendorfer (1990), Chowdhuri (1989a, b, 1990), Georgiadis et al. (1992), Nucci et al. (1993a, b), Ishii et al. (1994), Rachidi et al. (1997a, b), and Michishita et al. (1996, 1997), to name a few. Calculated values of the flashovers of a distribution line per 100 km per year due to induced voltages as a function of arrester spacing is found in Fig. 18.16, taken from IEEE standard 1410-1997 (Section 18.5). As noted above, the calculations involve many assumptions and idealizations. A small-scale model experiment on induced voltages was performed by Ishii et al. (1999).

Barker et al. (1996) measured voltages on a 10-m-high test distribution line when lightning was triggered 145 m from the line (subsection 7.2.7, first part). Stroke current, line voltage, and nearby electric and magnetic fields versus time were recorded with submicrosecond time resolution. For 63 strokes having peak currents between a few kA and 44 kA, peak induced voltages at the center of the line relative to a grounded wire about 2 m below ranged between 8 and 100 kV, the peak values being linearly correlated with correlation coefficient 0.75. The waveshapes of the induced voltage were unipolar with a median width at half-peak value of about 4 μs, whereas the typical width of the current was nearly an order of magnitude larger. According to Barker et al. (1996), the induced voltage waveform resembles the derivative of the vertical electric field, the theory of Agrawal et al. (1980) and other equivalent approaches accounting well for the observations. The peak induced voltages were roughly 60 percent larger than predicted by the simplified formula of Rusck (1958, 1977) discussed above.

Transmission lines Since insulation levels on transmission lines are relatively high, induced voltages from nearby lightning will seldom result in flashover. The primary protection against direct strikes to the phase conductors of transmission lines is generally an overhead shield (ground) wire or wires, although arresters are also used. Such overhead shield wires are intended to intercept lightning leaders before they attach to the phase conductors. However, lightning strikes to an overhead shield wire may result in back-flashover if the resultant voltage on the shield wire is high enough, as determined by transmission-line tower impedances, including ground connections, and traveling wave theory. The "shielding angle" is defined as the angle between an imaginary vertical line extending downward from the shield wire and an imaginary line connecting the shield and phase conductor. Shielding angles of about 30 degrees have been used successfully on the great majority of transmission lines for which tower heights are in the range of 24 m (Wagner et al. 1941, 1942). The much higher than expected rate of shielding failures (when lightning directly strikes the phase conductors) of the first EHV (345 kV) line in the United States, which had a tower height of 46 m and employed a shielding angle of 33 degrees (Price et al. 1956), prompted a re-examination of the shielding issue. Observed outage rates were in the range of four to six per 100 km per year. The 345 kV line was designed for an outage rate of less than 0.3 per 100 km per year in accordance with procedures found in AIEE (1950). The development of an electrogeometric model (subsection 18.3.2) based on the concept of a current-dependent striking distance (Eq. 18.1) was instrumental in the understanding of both shielding failures and back-flashovers on the EHV line and in the development of improved protection (Whitehead 1977).

Flashovers due to shielding failure can be reduced to relatively rare events, by providing properly located shielding conductors. Note that even imperfectly located shield wires intercept most of the flashes to the line and even ideally located (via electrogeometric theory) shield wires fail to intercept some of the flashes to the line. In the latter case, however, insulation flashover may not occur because the current to the phase conductor must be, according to the electrogeometric model, small in order for the striking distance to be small enough that the shield wire does not produce a successful upward connecting leader; and it is likely that such small phase-conductor currents will not produce large enough voltages on the phase conductors to lead to flashover at the high insulation levels of transmission lines. The primary problem, therefore, involves identifying an appropriate position for the shield wires so that they intercept strokes having prospective currents above some minimum amplitude. Gilman and Whitehead (1973) and Whitehead (1977) discussed the various difficulties and uncertainties in solving this problem.

Back-flashovers too can be reduced to relatively rare events by coordinating the impulse insulation level of the line with the impulse voltage developed across the line insulators in response to the lightning current in the ground wire and transmission tower. The framework for the study of the back-flashover mode is that of distributed-circuit theory with time-domain waveforms considered on a microsecond or smaller time scale. Whitehead (1977) described in detail an analytical model to calculate shield-wire voltages and transmission-line failure rates due to direct lightning strikes to the shield wire.

Various versions of the electrogeometric model have been applied to power lines with the work of Whitehead and his collaborators as their basis, some discussion being found in subsection 18.3.2. Sargent and Darveniza (1967, 1970), Currie et al. (1971), Anderson (1981), and Liew and Darveniza (1971, 1982a, b) applied a Monte Carlo method to the electrogeometric model to determine the frequency of strikes to transmission lines and the resulting shielding failure and back flashover rates. In these studies the leader location and approach angle and the return-stroke peak current were chosen randomly from assumed distributions of these parameters, and the actual transmission-line geometry and electrical characteristics were modeled. After a sufficient number of random events, the line performance could be predicted.

18.4.4. Underground cables

When lightning strikes the Earth, the strike point can be considered to be a small volume of ground of characteristic dimension about 10 cm or less. In subsection 18.3.4 we showed that lightning will cause the breakdown field in the ground E_b to be exceeded to a radius

$$r_0 = \sqrt{\frac{I}{2\pi\sigma E_b}}$$

assuming a hemispherical electrode. From Eq. 18.11 with a breakdown field $E_b = 300$ kV m^{-1}, a soil conductivity

Fig. 18.17. At the time of going to press a colour version of this figure was available for download from http://www.cambridge.org/9780521035415. Evidence of surface arcing on a golf course green in Arizona. Courtesy of E.P. Krider.

10^{-3} S m^{-1}, and a peak current 30 kA, the hemispherical breakdown radius is about 4 m. For non-uniform arc formation, longer arcs than r_0 will form. Figure 18.17 shows the results of surface arcing on a golf course green (see also Fig. 7.24). Ground arcs and furrows in the soil many tens of meters long that are due to lightning are common. If there is buried metal in the vicinity of a strike to ground, such as an underground cable, the non-uniformity of the underground electric fields will serve to attract the arc to the conductor over a longer path than would otherwise occur.

Underground lightning arcs vaporize and ionize poorly conducting soil, which is not unlike what happens in a lightning channel in air. When sandy soil cools after much an event, a fulgurite, a hollow glassy tube, is formed, leaving a record of the lightning path. A discussion of different aspects of fulgurite formation is found in Rakov (1999). An example of an excavated 5 m fulgurite is shown in Fig. 18.18 (see Fig. 7.31 for two other photographs of fulgurites) There is considerable evidence from excavated fulgurites that single channels in sandy soil may carry the bulk of the lightning current as they are attracted to underground conductors.

Figure 7.31 shows excavated fulgurites produced by triggered lightning, the lower end of the fulgurite in Fig. 7.31b being attached to an underground 15 kV power cable buried at about 1 m depth. In the tests that produced the fulgurites shown in Fig. 7.31a, b, lightning strikes were triggered to the earth above underground distribution cables to investigate how such strikes impact the underground system. The description of these experiments to follow is taken from Barker and Short (1996a, b, c); additional information can be found in subsection 7.2.7. The cables tested included an unjacketed direct-buried cable (coaxial, stranded neutral in contact with the earth), a jacketed direct-buried cable (insulation covering the neutral), and a jacketed cable within PVC conduit. During the experiment, 30 lightning flashes were triggered to the test cables with stroke currents ranging from a few kiloamperes up to 44 kA. All flashes contained initial continuous current and some contained long-continuing current. The available evidence indicates that during most tests breakdown occurred from the Earth's surface to the cable, the lightning "arc" being formed in the soil and directly attaching to the cable. All cable types tested showed significant damage at the lightning attachment point. The damage to jacketed cables ranged from minor punctures of the cable jacket up to extensive puncturing of the jacket and melting of some neutral strands. For unjacketed cable, some discharges melted nearly all neutral strands. Some damage to the cable insulation shield was also observed. In the case of the PVC-conduit cable installation, the wall of the conduit was blown open (see Fig. 7.31a) and the lightning channel attached to the cable inside. On an actual operating power cable, a delayed failure occurring days, weeks, or even years after the lightning event might be the result of such damage. A punctured jacket could allow water to leak into the cable, which could

18.4. Lightning interaction with specific objects and systems

Fig. 18.18. At the time of going to press a colour version of this figure was available for download from http://www.cambridge.org/9780521035415. A Florida fulgurite of about 5 m length excavated by the University of Florida lightning research group.

then cause corrosion of the neutral and shorten the cable's life. Damage to the insulation shield could cause a dielectric stress point which might also lead to premature cable failure. Damage to the cable neutral is important additionally if the neutral unintendedly carries a large fraction of the power frequency return current, since the cable could become hot at the point where the neutral was partially severed, degrading the cable insulation. Fifteen to 25 percent of the total lightning return-stroke current reached padmount transformers located 70 m on either side of the strike point. The current is likely to be highest near the strike point and probably "bleeds" off along the way to the padmount transformer by punctures through the jacket or by direct contact with the soil in the case of unjacketed cable. Measurements show that most of the lightning current entered the cable neutral at the strike point.

The voltages occurring across the cable insulation, between phase and neutral, in the triggered-lightning experiment were relatively modest. The largest was about 17 kV and apparently was due to the axial electric field caused by the flow of lightning current through the resistance of the neutral. The surge voltage across the phase-neutral cable insulation was well below the system BIL rating. Surge voltages up to nearly 4 kV were measured at the secondary terminals of one padmount transformer. These voltage levels could pose a threat to residential appliances. None of the voltages measured was large enough to threaten the primary of the transformer winding although lightning attachment to a buried secondary power conductor might well cause failure of a transformer.

According to Barker and Short (1996c), in the triggered-lightning tests lightning was attracted to the cable for strikes as far as 10 m on either side of the cable. Since first-stroke currents in natural flashes are generally larger than triggered-lightning stroke currents, it is not unreasonable to expect that any natural flash within 10 m horizontal distance (or perhaps more) of a buried cable would have a high likelihood of attaching to the cable under the soil conditions which existed in the experiment, a conductivity of 2.5×10^{-4} S m^{-1} (Table 7.6). A cable located in an area with a flash density of about 10 km^{-2} yr^{-1} and similar soil conditions might experience about 0.2 lightning attachments per kilometer of cable per year. This incidence can be compared with strikes to overhead distribution lines. The incidence of strikes to overhead lines is dependent on the flash density, the line height, and its shielding factor (due to trees, buildings, etc.). A line shielding factor approaching unity represents a line imbedded in very tall trees and would not be expected to experience a direct hit. A shielding factor of zero represents an exposed line erected in open terrain with no trees and would receive the maximum number of direct strikes. No matter what object the lightning strikes, such as a tree, some of the lightning current can still potentially contact an underground cable. For a 10-m-high overhead distribution line in an open area assumed to attract lightning from 20 m on either side (the simplest model) in a region having $N_g = 10$ km^{-2} yr^{-1}, there will be 0.4 lightning strikes per kilometer per year (Eq. 18.13), twice the value for a buried cable. If the overhead line has a typical shielding factor of 0.5, the strike rates to overhead and underground lines are similar. While there is no accurate way to determine the distance over which lightning is attracted to either overhead lines or underground cables, the estimates we have presented in this section indicate that they are not much different.

18.4.5. Telecommunications systems

General information Telecommunication systems of interest here are primarily those which involve the transmission of data, such as audio and video, on metallic cables. A review of various aspects of the lightning protection of telecommunication systems is given in Boyce (1977). Fiber optic data transmission, above or under the ground, is immune to many of the lightning problems suffered by metal-wired systems, although fiber optic amplifiers and other signal processors can be subject to electronic damage as can accompanying metallic wires paralleling the fiber optic cables for support or strength. We consider two aspects of lightning interaction with telecommunication systems: (i) injury or death to individuals using telephones or similar devices operated near the human head (see also Section 19.3), (ii) damage and interference on (a) overhead lines and (b) underground cables, including damage to the electronics at the terminations of the lines and cables. In subsections 18.3.4 and 18.4.4, we considered the attractive distance of any type of buried cable for a lightning strike to the Earth's surface. Those results are applicable to buried telecommunication cables. In subsection 18.4.3 we discussed the voltages and currents on overhead conductors due to both nearby and lightning direct strikes. These results are also applicable to overhead telecommunication lines, although few such studies have been aimed specifically at telecommunication lines, one exception being that of Zeddam and Degauque (1987). Many telecommunication lines are shielded by overhead distribution power lines, so that some aspects of the discussion of the shielding of overhead power lines by grounded wires found in subsection 18.4.3 may be applicable to telecommunications lines shielded by power lines.

Electric and acoustic shock from telephones Telephone signal wires are required by law in most countries to be protected from overvoltages due to both lightning effects and power line contact. This is generally accomplished by SPDs (subsection 18.3.5), which are most often gas tubes or carbon-block air gap arresters that reduce the voltage to near zero after it exceeds 500 to 1000 V. These SPDs are generally placed between the signal wires and a grounding system (typically a ground rod) at the point that the wires enter a structure. Boyce (1977) discussed the types of SPDs specifically used on telecommunication circuits. Unfortunately, significant voltages can still appear on telephone handset wiring if the grounding resistance of the SPD is not sufficiently low, the voltage depending on the product of the grounding resistance and the current produced by lightning that flows into the grounding resistance. Voltages above the 5 to 10 kV range will exceed the insulation level of most telephones. Voltage differences between the telephone ground and the power and other grounds, including plumbing, should be minimized by bonding, ideally at a common entrance point for all utilities. If the bonding connectors are too long, lightning currents flowing in them can create inductive voltages (Section 18.2) that could cause a flashover between, say, a telephone headset and a power-grounded refrigerator, perhaps through the telephone user's head. The general principles of protection given in Section 18.3 are applicable. The use of a cordless or cellular phone allows one to avoid the hazard associated with hard-wired phone. In addition to electrical injury caused by voltage differences between the wires in the headset and an external conductor, perhaps the human body, which is a fairly good conductor (Section 19.2), relatively large voltage differences between the signal wires can produce an acoustic shock from the acoustic transducer of the telephone. Telephone headsets contain SPDs that are intended to limit the voltages between signal wires and hence to limit any acoustic shock, but this approach is not always successful. Further information

18.4. Lightning interaction with specific objects and systems

on the lightning hazard to telephone users is found in Chapter 19.

Overhead lines Barker et al. (1996a) showed from triggered-lightning studies that return-stroke peak currents near 20 kA at a distance of 145 m from an overhead phase wire at 10 m height produce voltages between the phase wire and a grounded wire about 2 m below that are near 50 kV, the observed range of 63 peak voltage measurements being between 8 and 100 kV for currents between a few and 44 kA (subsection 18.4.3). The causative electric and magnetic fields decrease roughly inversely with distance beyond 145 m. Thus lightning at a distance of a kilometer or so from an overhead telephone line would be expected to induce more than 1 kV between the line and ground. Boyce (1977) stated that strokes to ground between 25 and 3000 m generally induce more than 1 kV on overhead communication lines with respect to ground and that strokes closer than 25 m are likely to terminate on the line. The maximum induced voltage will occur at a point nearest the lightning, and surge current and voltage will propagate outward in both directions with reflections from discontinuities and the line ends, resulting in overall surge durations up to several milliseconds. According to Boyce, the current induced in a single telecommunication line by a large close lightning current may be hundreds of amperes, with a pulse width of a few microseconds, decaying to tens of amperes with a pulse width of 10 microseconds or so after propagating over one kilometer of line. If flashover of the communication line insulators does not occur, SPDs on the line (if SPDs are there) will therefore conduct currect in the range of hundreds of amperes in the first tens of meter of line to tens of amperes in the line one kilometer away. Insulation levels on telecommunication lines are, from a practical point of view, generally in the 50 kV range although design values are up to a factor 2 higher (Boyce 1977). The previous discussion is primarily for single-wire telecommunication conductors. The surge impedance of a single overhead wire is about 500 Ω. According to Boyce, for a 32-wire bundle the overall surge impedance is about 150 Ω, and hence the total current will be about 3.5 times what it would have been for the same lightning-induced voltage and a single wire, but this total current will be divided between the 32 wires. Thus, for multiconductor overhead routes the currents discharged through SPDs will be about 10 percent of the value for a single wire. Boyce (1977) states that typical observed current waveshapes on long lines are 10/1000 μs while on short lines 2/100 μs is more representative, multiple reflections occurring on long lines.

Overhead telecommunication lines will be struck directly five or so times per 100 km of line per year for a flash density of 1 km^{-2} yr^{-1}, if we estimate the strike rate as in subsection 18.4.3. Wooden poles can be protected with lightning rods and grounded down-conductors. Often extensions of the down-conductor wire above the pole top are used in place of the rods. Telecommunication lines suspended below power lines are well protected from direct strikes and are also partially shielded from the induced effect of nearby strikes. In some cases, an overhead shield wire similar to that used in power-line protection may be needed. SPDs are essential to protect equipment on the line and at the line terminations.

Underground lines Underground telecommunication cables may be directly buried or may be run in ducts or pipes or conduits, either metallic or more commonly plastic (PVC). The depth at which a telecommunication cable is buried has little effect on its susceptibility to lightning since within a meter or so of the surface, which is the practical depth at which cables can be buried, no immunity is obtained from direct strikes and the electric and magnetic fields coupled to underground cables from nearby strikes do not vary much with depth because the skin depth at frequencies of interest in the earth is of the order of a meter. The use of PVC pipes provides no protection against lightning, but it does provide some protection against rodents and misplaced shovels. As in the case of underground power cables (subsection 18.4.4), the parameters of importance to lightning effects on underground communication cables are the incidence of lightning, the earth resistivity, the degree of shielding afforded by buildings, and the number and placement of other cables and metal pipes that can share the lightning current. Underground cable protection is provided by the shielding of metallic sheaths, ducts, and pipes, the use of SPDs, the use of buried "shield" wires over the signal wires, and a high dielectric strength of the signal conductor insulation. In general, underground cables in open areas can be expected to be directly struck at a not much different rate, perhaps a factor 2 less, than an overhead distribution line in the same area, as noted in subsection 18.4.4. Boyce (1977) discussed 14 different types of cable, each with different types and levels of insulation and shielding.

The ability of a cable to withstand lightning damage is optimized by reducing the "transfer impedance" of the metallic sheath, duct, pipe, or conduit containing the signal wires and by increasing the dielectric strength of the insulation. The transfer impedance is defined as the ratio, in the frequency domain, of the voltage per unit length induced on an internal wire to the lightning current or lightning-induced current flowing on the sheath, duct, pipe, or conduit. Tesche et al. (1998) examined, both theoretically and via laboratory experiments, the electrical response of signal wires within four different types of buried conduit when lightning-like currents flowed in the conduits. They discussed both the case of a metallic conduit in contact with the earth and that of a conduit covered with a non-conducting dielectric jacket. As noted above, the key element in the calculations is the transfer impedance, since it determines, along with the conduit current, the voltage appearing on the signal wires.

Kinsler and Hmurcik (1999) discussed the case of lightning flashover from the grounds of power lines to buried communication cables located beneath the power lines. In this case, a long-duration 50 Hz or 60 Hz power-follow current may flow through the lightning-initiated arc, potentially damaging the underground cable.

Sato and Kuramoto (1996) made measurements of the division of lightning current striking a mountain-top radio relay station. They found that 90 percent of the current flowed down the antenna-tower legs and waveguides, but only about 60 to 75 percent of this current flowed to ground at the antenna-tower legs and other relay-station structural components; the remainder flowed away from the relay station on the incoming telecommunication cables and power lines, presenting a hazard to the telecommunication system.

Radio and television broadcast towers, police and fire 911 facilities containing broadcast towers, microwave relay stations, and other tall towers associated with communication services are, because of their height, particularly exposed to the lightning hazard. Adequate protection for electronics and for individuals working in such facilities can only be obtained using the principles of topological shielding and transient protection discussed in subsection 18.3.6. Grounding and bonding is seldom sufficient to limit deleterious voltages between the separated but bonded grounds for the antenna tower and for the supporting structure containing communication electronics because of the inductive voltage drop in the bonding wires, which are necessarily exposed to a significant fraction of the total lightning current and typically extend over a distance of tens of meters (Section 18.2). Thus the antenna-tower voltage wave entering the electronic-housing structure on coaxial cables and waveguides may be very different from the voltages on consoles (the metal tables at which equipment operators sit), the consoles being typically bonded to ring electrodes some distance away from the tower. Transient protection is essential in this situation in order to eliminate electrical shock to equipment operators, as well as to ameliorate damage to the electronics. Seldom, however, is transient protection provided on enough of the voluminous amount of wiring in such facilities to solve the problem of voltage differences appearing between contiguous metal when the antenna tower is struck by lightning. Thus, individuals using hard-wired communication headsets and typing at metallic consoles in fire and police 911 centers are particularly susceptible to serious shock.

18.5. Lightning test standards

Various test methods and standards have been developed to enable system designers to evaluate the effectiveness of proposed protective measures or to verify the adequacy of final protection designs. These range from the simulation of lightning currents, voltages, and electric and magnetic fields to the generation of current and voltage transients at the terminals of electronic equipment typical of those induced by lightning. Often, the specifications for testing are divided into "direct" and "indirect" effects, as noted in Section 10.2 for the case of aircraft. Direct effects include damage to metal and insulator surfaces, ignition of flammable vapors or detonation of explosive materials due to a direct lightning attachment; indirect effects include the currents and voltages induced on internal circuits by lightning which has struck on or near the exterior of a structure.

There are separate standards for the many different types of lightning arresters, transformers, and circuit breakers used in 50 and 60 Hz power systems; and there are separate standards for line-powered and mounted telecommunication equipment and for the gas tubes and carbon block arresters found on the telephone service entrance at residential structures. An IEEE standards listing is available from IEEE, Service Center, PO Box 1331, Piscataway, NJ 08855-1331. Some pertinent IEEE lightning standards are found at http://grouper.ieee.org/groups/spd/. The European-based International Electrotechnical Commission (IEC) promulgates worldwide standards. Pertinent international lightning standards are listed in the reference list under IEC and some are found at http://www.iec.ch/cgi-bin/procgi.pl/www/iecwww.p?wwwlang=E&wwwprog=dirdet.p&committee=TC&number=81; http://www.iec.ch/cgi-bin/procgi.pl/www/iecwww.p?wwwlang=E&wwwprog=dirdet.p&commmittee=TC&number=37; and http://www.iec.ch/cgi-bin/procgi.pl/www/iecwww.p?wwwlang=E&wwwprog=dirdet.p&committee=SC&number=37A.

For many types of equipment that operate on power systems, an open-circuit test voltage waveform of 1.2/50 µs (front time/half-peak value) and a short-circuit test current waveform of 8/20 µs with appropriate amplitudes are used. Both direct and induced lightning voltages and currents can have waveshapes with, for example, much faster risetimes and much longer time durations, and can have multiply repeated waveforms so that the loss to lightning of equipment in power systems is not particularly surprising. The electrical industry's voltage waveform standard for dielectric tests and the aerospace industry's standard for fuel ignition tests is the 1.2/50 µs waveform. Gas tube telephone protectors are tested to the following three standards: a 10/1000 µs current waveform for currents from 50 to 500 A, an 8/20 µs current waveform 5 to 20 kA, and linear voltage ramps of 100, 500, 5000, 10 000 V µs^{-1} up to sparkover (IEEE STD 465.1).

One of the better known standards for the simulation of the effect of lightning is IEEE C62.41.2-2000, IEEE

18.5. Lightning test standards

Recommended Practice on Characterization of Surges in Low-voltage (1000 V or less) AC Power Circuits. Two related documents are IEEE Std C62.41.1-2000, IEEE Guide on the Surge Environment in Low-voltage (1000 V or less) AC Power Circuits, which includes a data base, and IEEE Std C62.45-2000, IEEE Recommended Practice on Surge Testing for Equipment Connected to Low-voltage AC Power Circuits. IEEE C62.41.2-2000 is the result of 20 years of evolution from the initial document published in 1980, IEEE STD 587-1980, IEEE Guide for Surge Voltages in Low-voltage AC Power Circuits, which was also published as IEEE/ANSI Std C62.41, with the same title. This guide was updated in 1991 as IEEE C62.41-1991, Recommended Practice on Surge Voltages in Low-voltage AC Power Circuits, reflecting new data on the surge environment and experience in the use (and misuse) of the original guide.

The 1980 version, based on data available up to 1979, advanced two novel ideas:

(i) The reduction of a complex data base to two representative surges: a new "ring wave" featuring a decaying 100 kHz oscillation, and the combination of the "classical" 1.2/50 μs voltage waveform and 8/20 μs current waveform into a combination wave to be delivered by a surge generator having a well-defined open-circuit voltage and short-circuit current.

(ii) The idea that "location categories" could be defined within an installation, where surge voltages impinging upon the service entrance of an installation or generated within an installation would propagate, unabated, in the branch circuits while the associated currents, impeded by (mostly) the inductance of the conductors, would be reduced from the service entrance to the end of long-branch circuits.

The 1991 version, based on additional data as well as experience in the use of the 1980 guide, maintained the concepts of the location categories and recommendation of representative surge waveforms. The two test waves, ring wave and combination wave, were designated as "standard surge-testing waveforms" and three new waveforms "additional surge-testing waveforms" were added to the guide. Meanwhile, a companion document was developed, IEEE Std C62.45, IEEE Guide on Surge Testing for Equipment Connected to Low-voltage AC Power Circuits, outlining procedures for safe and error-free application of the waveforms defined by C62.41. The perceived need to justify the expansion of the two-only waveforms into the five waveforms listed in the next paragraph led to the growth in the document volume, from the 25-page IEEE Std 587-1980 to the 111-page IEEE Std C62.41-1991. Additional information collected in order to update the 1991 version would have increased further the volume of the document. Instead, a new approach was taken: to create a "trilogy" separating the information into three distinct documents. These three are listed in the first two sentences of the previous paragraph.

As noted above, IEEE C62.41.2-2000 specifies five types of voltage waveform. The two "standard" waveforms are (i) the ring wave, a voltage wave with a 0.5 μs rise to the first crest followed by a 100 kHz oscillation that decays in amplitude with time and (ii) the combination wave, which provides an open-circuit voltage of 1.2/50 μs and a short-circuit current of 8/20 μs. The peak values for voltages and currents depend on the perceived exposure and are discussed in C62.41.1-2000. The three additional waveforms are (i) the electrical fast transient waveform, which consists of bursts of individual voltage pulses of 5 ns risetime and 50 ns duration, with a repetition rate of 2.5 to 5 kHz during a 15 ms burst duration, the overall burst being repeated every 300 ms; (ii) the unidirectional 10/1000 μs long wave, this waveform describing both open-circuited voltage and short-circuited current; and (iii) the 10/350 μs waveform, proposed as a current test for SPDs at structure utility service entrances. As noted earlier, international lightning standards equivalent to the IEEE standards, and in fact standards from which some of the IEEE standards are derived, are published by the IEC and are listed in the references at the end of this chapter.

While the insulation capability of most power-related equipment, be it low voltage, distribution level, or transmission level, is tested using the standard 1.2/50 μs voltage wave, lightning-caused transient voltages seldom have this waveshape. IEEE Task Force 15.09 (1994) reviewed most research on electrical breakdown by non-standard voltage waveforms. Chowdhuri et al. (1994a, b) also reviewed the literature on non-standard voltage wave breakdown and provided their own tests with eight different voltage waveshapes in small gaps. In general, for any electrode shape, the breakdown voltage increases as the risetime and duration to half-peak value of a 1.2/50 μs wave are decreased. There is a minimum breakdown voltage for voltage waveshapes near 10/1000 μs, similar waveshapes being characteristic of switching surges in power systems; breakdown voltages in general decrease as the voltage time to half-peak value increases. However, various investigators have obtained different results for the same voltage waveshape and polarity, and clearly the characteristics of the voltage generator, electrodes, and gap size affect the results obtained. The effects of electrode shape and of ambient humidity on the breakdown voltage have also been extensively studied, as noted in IEEE Task Force 15.09 (1994).

Perhaps because of the obvious potential for catastrophic effects, the requirements and standards for aircraft lightning protection, discussed separately in Section 10.5, have kept better pace both with advancements in air-vehicle system design and with improvements in our understanding of lightning characteristics than have requirements for most ground-based systems.

18.6. Summary

The lightning attachment process is a complex and relatively poorly understood event. The designs of most structural lightning-protection systems are predicated on simple models of this complex process. While the general principles are the same for all lightning protection, the particulars and emphasis are different in the protection of such specific objects and systems as structures, boats, trees, and overhead and underground power and communication lines. Nevertheless, experience indicates that such efforts are by and large effective in preventing lightning damage. Topological shielding with surge suppression provides the optimal approach to both structural and electronics protection, but is often costly. There is no scientific evidence that non-conventional protection methods such as lightning elimination systems or early streamer emitter methods are more effective than conventional lightning protection.

References and bibliography for Chapter 18

Abdel-Salam, M., and Al-Abdul-Latil, U.S. 1997. Simulation of energized Franklin rods for lightning protection. *IEEE Trans. Ind. Appl.* **33**: 651–9.

Agrawal, A.K., Price, H.J., and Gurbaxani, S.J. 1980. Transient response of a field. *IEEE Trans. Electromagn. Compat.* **22**: 119–29.

AIEE 1950. A method of estimating the lightning performance of transmission lines. *Trans. Am. Inst. Electr. Eng.* **69** (II): 1187–96, American Institute of Electrical Engineers Committee Report.

Allen, N.L., Cornick, K.J., Faircloth, D.C., and Kouzis, C.M. 1998a. Tests of the "early streamer emission" principle for protection against lightning. *IEE Proc. Sci. Meas. Techol.* **145**: 200–6.

Allen, N.L., Huang, C.F., Cornick, K.J., and Greaves, D.A. 1998b. Sparkover in the rod–plane gap under combined direct and impulse voltages. *IEE Proc. Sci. Meas. Technol.* **145**: 207–14.

Almeida, M.E., and Correeia, de B. 1994. Tower modelling for lightning surge analysis using electromagnetic transient program. *IEE Proc. Gener. Transm. Distrib.* **141**: 637–9.

American Boat and Yacht Council 1985. Recommended practices and standards covering lightning protection. Project E-4, ABYC E-4 85.

American Institute of Electrical Engineers Committee Report 1950. A method of estimating the lightning performance of transmission lines. *Trans. Am. Inst. Electr. Eng.* **69** (II): 1187–96.

Anderson, J.G. 1981. Monte-Carlo computer calculation of transmission line lightning performance. *IEEE Trans. Pow. Appar. Syst.* **80**: 414–20.

Anderson, J.G., and Short, T.A. 1993. Algorithms for calculation of lightning induced voltages on distribution lines. *IEEE Trans. Pow. Del.* **8**: 1217–25.

Anderson, R. 1879. *Lightning Conductors*, London: E. and F. N. Spon.

Anderson, R. 1885. *Lightning Conductors*, pp. 349–50, London: E. and F. N. Spon.

Anderson, R.B. 1970. The lightning protection of high structures. *S. Afr. Electr. Rev.* **61**: 4–14.

Anderson, R.B. 1985. Lightning performance criteria for electric power systems. *IEE Proc.* **132C**: 298–306.

Anderson, R.B. 1987. Discussion on lightning performance criteria for electric power systems. *IEEE Proc. Gen. Trans. Distrib.* **134C**: 328–30.

Anderson, R.B. and Eriksson, A.J. 1980. Lightning parameters for engineering application. *Electra* **69**: 65–102.

Angeli, M., and Cardelli, E. 1997. An approach to the analysis of the electromagnetic interferences radiated by metallic grids struck by lightning. *IEEE Trans. Mag.* **33**: 1804–7.

Antonini, G., Christina, S. and Orlandi, A. 1998. PEEC modeling of lightning protection systems and coupling to coaxial cables. *IEEE Trans. Electromagn. Compat.* **40**: 481–91.

Araujo, A.E.A., Paulino, J.O.S., Silva, J.P., and Dommel, H.W. 2001. Calculation of lightning-induced voltages with RUSCK'S method in EMTP. Part I: Comparison with measurements and Agrawal's coupling model. *Electr. Pow. Syst. Res.* **60**: 49–54.

Armstrong, H.R., and Whitehead, E.R. 1968. Field and analytical studies of transmission line shielding – III. *IEEE Trans. Pow. Appar. Syst.* **87**: 270–81.

Aubrecht, L., Koller, J., and Stanek, Z. 2001. Onset voltages of atmospheric corona discharges on coniferous tree. *J. Atmos. Solar-Terr. Phys.* **63**: 1901–6.

Austin, K.A. 1987. Discussion on lightning performance criteria for electric power systems. *IEEE Proc. Gen. Trans. Distrib.* **134**: 328–30.

Baatz, H. 1972. Radioaktive Isotope verbesern nich den Blitzschutz. *Elektrotech. Z.* **93**: 101–4.

Baatz, H. 1977. Protection of structures. In *Lightning, vol. 2, Lightning Protection*, ed. R.H. Golde, pp. 599–632, New York: Academic.

Baba, Y., and Ishii, M. 2000. Numerical electromagnetic field analysis on lightning surge response of tower with shield wire. *IEEE Trans. Pow. Del.* **15**: 1010–15.

Baginski, M.E., Riggs, L.S., and German, F.J. 1988. Electrical breakdown of soil about earthed conductors resulting from late time EMP effects. *IEEE Trans. Electromag. Compat.* **30**: 380–5.

Baker, G., Castillo, J.P., and Vance, E.V. 1992. Potential for a unified topological approach to electromagnetic effects protection. *IEEE Trans. Electromagn. Compat.* **34**: 267–74.

Baker, M.B., Christian, H.J., and Latham, J. 1995. A computational study of the relationships linking lightning frequency and other thundercloud parameters. *Q.J.R. Meteor. Soc.* **121**: 1525–48.

Baker, P.P., Short, T.A., Eybert-Berard, A., and Belandis, J.B. 1996. Induced voltage measurements on an experimental distribution line during nearby rocket triggered lightning flashes. *IEEE Trans. Pow. Del.* **11**: 980–95.

Ball, L.M. 1974. The laser lightning rod system: thunderstorm domestication. *Appl. Opt.* **13**: 2292–6.

Bandinelli, M., Bessi, F., Chiti, S., Infantino, M., and Pomponi, R. 1996. Numerical modeling for LEMP effect evaluation inside a telecommunication exchange. *IEEE Trans. Electromagn. Compat.* **38**: 265–73.

Baranov, M.I., and Bondina, N.N. 1992. Electromagnetic and thermal transients in cylindrical conductors subjected to lightning current impacts. *Electrichestvo* **10**: 9–15 (in Russian).

Barker, P.P., and Mancao, R.T. 1992. Lightning research advances with digital surge recordings. *IEEE Comptr. Appl. Pow.* 11–16.

Barker, P.P., and Short, T.A. 1996a. Lightning effects studied: the underground cable program. In *Transmission and Distribution World*, pp. 24–33.

Barker, P.P., and Short, T.A. 1996b. Lightning measurements lead to an improved understanding of lightning problems on utility power systems. In *Proc. 11 CESPSI, Kuala Lumpur, Malaysia*, vol. 2, pp. 74–83.

Barker, P., and Short, T.A. 1996c. Findings of recent experiments involving natural and triggered lightning. Panel Session Paper presented at Transmission and Distribution Conference, Los Angeles, California, September 16–20, 1996.

Barker, P.P., Mancao, R.T., Kvaltine, D.J., and Parrish, D.E. 1993. Characteristics of lightning surges measured at metal oxide distribution arresters. *IEEE Trans. Pow. Del.* **8**: 301–10.

Barker, P.P., Short, T.A., Eybert-Berard, A.R., and Berlandis, J.P. 1996. Induced voltage measurements on an experimental distribution line during nearby rocket-triggered lightning flashes. *IEEE Trans. Pow. Del.* **11**: 980–95.

Bartkowiak, M., and Mahan, G.D. 1995. Nonlinear currents in Voronoi networks. *Phys. Rev. B* **51**: 10 825–32.

Bartkowiak, M., Comber, M.G., and Mahan, G.D. 1999. Failure modes and energy absorption capability of ZnO varistors. *IEEE Trans. Pow. Del.* **14**: 152–62.

Bartkowiak, M., Comber, M.G., and Mahan, G.D. 2001. Influence of nonuniformity of ZnO varistors on their energy absorption capability. *IEEE Trans. Pow. Del.* **16**: 591–8.

Bartlett, C.J. 1963. *Great Britain and Sea Power 1815–1853*, pp. 13–21, Oxford: Clarendon.

Baum, C.E. 1992. From the electromagnetic pulse to high-power electromagnetics. *Proc. IEEE* **80**: 789–817.

Bazelyan, E.M. 1974. Selecting points for lightning strokes (in Russian). *Elektrichestvo* **10**: 15–19.

Bazelyan, E.M., and Raizer, Yu. P. 2000a. *Lightning Physics and Lightning Protection*, 325 pp., Bristol: IOP Publishing.

Bazelyan, E.M., and Raizer, Yu. P. 2000b. Lightning attraction mechanism and the problem of lightning initiation by lasers. *UFN* **170**(7): 753–69.

Beck, E., Mcnutt, H.R. Jr, Shankle, D.F., and Tirk, C.J. 1969. Electric fields in the vicinity of lightning strokes. *IEEE Trans. Pow. Appar. Syst.* **88**: 904–10.

Bell, E., and Price, A.L. 1931. Lightning investigation on the 220-kV system of the Pennsylvania Power and Light Company (1930). *AIEE Trans.* **50**: 1101–10.

Bellaschi, P.L. 1941. Lightning strokes in field and laboratory III. *AIEE Trans.* **60**: 1248–56.

Bellaschi, P.L., Armington, R.E., and Snowden, A.E. 1942. Impulse and 60-cycle characteristics of driven grounds Pt. II. *AIEE Trans.* **61**: 349–63.

Benn, I.M., and Shanahan, S.T. 1991. Of lightning rods, charged conductors, curvature, and things. *Am. J. Phys.* **59**: 658–60.

Bennison, E., et al. 1973. Lightning surges in open wire, coaxial and paired cables. *IEEE Trans. Comm.* **COM-21**(10): 1136.

Bent, R.B., and Llewellyn, S.K. 1977. An investigation of the lightning elimination and strike reduction properties of dissipation arrays. In *Review of Lightning Protection Technology for Tall Structures*, pp. 149–241, ed. J. Hughes, Publication no. AD-A075 449, Office of Naval Research, Arlington, Virginia.

Berger, G. 1992. The early emission lightning rod conductor. In *Proc. 15th ICOLSE, Atlantic City, United States*, paper 38, 9 pp.

Berger, G. 1995. Inception electric field of the lightning upward leader initiated from a Franklin rod in laboratory. In *Proc. 11th Int. Conf. on Gas Discharges and their Applications, Chuo University, Tokyo, Japan*, Sept. 11–15, 1995.

Berger, G. 1996. Leader inception field from a vertical rod conductor – efficiency of electrical triggering techniques. In *Proc. IEEE-EI Conf.*, Montreal, Canada.

Berger, G., Senouci, B., Goldman A., and Goldman, M. 1988. A physical approach to lightning protection. In *Proc. 9th Int. Conf. on Gas Discharges and Their Applications, Venice, Italy*, pp. 435–8.

Berger, K. 1967. Novel observations on lightning discharges: results of research on Mount San Salvatore. *J. Franklin Inst.* **283**: 478–525.

Berger, K. 1971a. Blitzforschung und Personen-Blitzschutz. *ETZ (A)* **92**: 508–11.

Berger, K. 1971b. Zum Problem des Personenblitzschutzes. *Bull. Schweiz Elektrotech. Ver.* **62**: 397–99.

Berger, K. 1977. Protection of underground blasting operations. In *Lightning, vol. 2, Lightning Protection*, ed. R.H. Golde, pp. 633–58, New York: Academic Press.

Berger, K. and Vogelsanger, E. 1966. Photographische Blitzuntersuchungen der Jahre 1955–1965 auf dem Monte San Salvatore. *Bull. Schweiz Elektrotech. Ver.* **57**: 599–620.

Berger, K., and Vogelsanger, E. 1969. New results of lightning observations. In *Planetary Electrodynamics*, eds. S.C. Coroniti and J. Hughes, pp. 489–510, New York: Gordon and Breach.

Berger, K., Anderson, R.B., and Kroninger, H. 1975. Parameters of lightning flashes, *Electra* **80**: 23–37.

Berjo, G. 1970. Lightning rods go radioactive. *Electl Wld* **173**: 36–7.

Bernstein, R., Samm, R., Cummins, K., Pyle, R., and Tuel, J. 1996. Lightning detection network averts damage and speeds restoration. *IEEE Comptr Appl. Pow.* **9**: 12–17.

Bernstein, T, and Reynolds, T.S. 1978. Protecting the Royal Navy from lightning – William Snow Harris and his struggle with the British Admiralty for fixed lightning conductors. *IEEE Trans. Education* **21**: 7–14.

Bewley, L.V. 1963. *Traveling Waves on Transmission Systems*. New York: Dover.

Bishop, D. 1990. Lightning devices undergo tests at Florida airports. In *Mobile Radio Technology*, pp. 16–26.

Block, R.R. 1988. Dissipation arrays: do they work? In *Mobile Radio Technology*, pp. 9–14.

Bogges, S., and Hideyasu, A. 2001. A statistical approach to prediction of ZnO arrester element characteristics. *IEEE Trans. Pow. Del.* **16**: 604–10.

Bondarenko, E.J. 1975. Effects of lightning on subscribers services. *Telecom. Aust.*, Perth, New South Wales, Research Laboratory Report 6990.

Bondarenko, E.J. 1976. Effects of lightning on subscribers services – Sydney, NSW. *Telecom. Aust.*, Research Laboratory Report 7078.

Bondarenko, E.J. 1980. Effects of lightning on subscribers services, Minyip Victoria, Canberra Act and Brisbane Qld. *Applied Sciences Brisbane Research Laboratory*, Paper 0018.

Bondiou, A., and Gallimberti, I. 1994. Theoretical modelling of the development of the positive spark in long gaps. *J. Phys. D. Appl. Phys.* **27**: 1252–66.

Boyce, C.F. 1977. Protection of telecommunication systems. In *Lightning, vol. 2, Lightning Protection*, ed. R. H. Golde, pp. 793–829, New York: Academic Press.

Braunstein, A. 1970. Lightning strokes to power transmission lines and the shielding effect of groundwires. *IEEE Trans. Pow. Appar. Syst.* **89**: 1900–10.

British Standards Institute 1992. Protection of structures against lightning. BS 6651.

Brown, G.W. 1978. Lightning performance – I, Shielding failures simplified. *IEEE Trans. Pow. Appar. Syst.* **97**: 33–8.

Brown, G.W. and Whitehead, E.R. 1969. Field and analytical studies of transmission line shielding II. *IEEE Trans. Pow. Appar. Syst.* **88**: 617–26.

Bruce, C.E.R., and Golde, R.H. 1941. The lightning discharge. *J. Inst. Electr. Eng.* **88**(II): 487–520.

Brunk, I.W. 1958a. Ben Franklin was lucky. *Weatherwise* **11**: 92–3.

Brunk, I.W. 1958b. Lightning death during kite flight (or, Radiosondes are safer). *Weatherwise* **11**: 204–5.

Bryan, J.L., Biermann, R.G., and Erikson, G.A. 1999. Report of the third-party independent evaluation panel on the early streamer emission lightning protection technology. Submitted to the National Fire Protection Association Standards Council on 1 September 1999 in response to a legal agreement of settlement and release between the National Fire Protection Association, Heary Bros. Lightning Protection Company, Inc., and Lightning Prevention of America, Inc.

Buccella, C., Cristina, S., and Orlandi, A. 1992. Frequency analysis of the induced effects due to the lightning stroke radiated electromagnetic field. *IEEE Trans. Electromagn. Compat.* **34**: 338–44.

Burkhardt, K. 1987. Protect your boat from lightning. *Sail* **18**: 49–52.

Cabrera, M.V.M., Lundquist, S., and Cooray, V. 1993. On the physical properties of discharges in sand under lightning impulses. *J. Electrostatics* **30**: 17–28.

Cannel, H. 1979. Struck by lightning – the effects upon the men and the ships of HM Navy. *J. Roy. Nav. Med. Serv.* **65**: 165–70.

Carpenter, R.B. 1976. Lightning prevention – practical and proven. *Measurements Data* **12**: 90–6.

Carpenter, R.B. 1977. 170 system years of guaranteed lightning prevention. In *Review of Lightning Protection Technology for Tall Structures*, ed. J. Hughes, pp. 1–23, Publication no. AD-A075 449, Office of Naval Research, Arlington, Virginia.

Carpenter, R.B. and Auer, R.L. 1995. Lightning and surge protection of substations. *IEEE Trans. Ind. Appl.* **31**: 162–74.

Cassie, A.M. 1969. The effect of a radio-active source on the path of a lightning stroke. *Electrical Research Association*, Report no. 5262, Leatherhead, Surrey.

Cavendish, H., Watson, W., Franklin, B., and Robertson, J. 1773. Report of the committee appointed by the Royal Society to consider a method of securing the powder magazine at Purfleet. *Phil. Trans. Roy. Soc.* **63**: 42–7.

Celozzi, S., and Feliziani, M. 1995. Time-domain solution of field-excited multiconductor transmission line equations. *IEEE Trans. Electromagn. Compat.* **37**: 421–32.

Chalmers, I.D., Evans, J.C., and Siew, W.H. 1999. Emission lightning protection. *IEE Proc. Sci. Meas. Technol.* **146**: 57–63.

Changnon, S.A. 2001. Damaging thunderstorm activity in the United States. *Bull. Amer. Meteorol. Soc.* **82**: 597–608.

Chisholm, W.A., and Janischewskyi, W. 1989. Lightning surge response of ground electrodes. *IEEE Trans. Pow. Del.* **4**: 1329–37.

Chowdhuri, P. 1989a. Parametric effects on the induced voltages on overhead lines by lightning strokes to nearby ground. *IEEE Trans. Pow. Del.* **4**: 1185–94.

Chowdhuri, P. 1989b. Analysis of lightning-induced voltages on overhead lines. *IEEE Trans. Pow. Del.* **4**: 479–92.

Chowdhuri, P. 1989c. Estimation of flashover rates of overhead power distribution lines by lightning strokes to nearby ground. *IEEE Trans. Pow. Del.* **4**: 1982–9.

Chowdhuri, P. 1990. Lightning-induced voltages on multiconductor overhead lines. *IEEE Trans. Pow. Del.* **5**: 658–66.

Chowdhuri, P. 1994. Shielding of substations against direct lightning strokes by shield wires. *IEEE Trans. Pow. Del.* **9**: 314–22.

Chowdhuri, P. 1996. *Electromagnetic Transients in Power Systems*, New York: John Wiley and Sons.

Chowdhuri P., and Gross, E.T.B. 1967. Voltage surges induced on overhead lines by lightning strokes. *Proc. IEE* **114**: 1899–907.

Chowdhuri, P., and Kotapalli, A.K. 1989. Significant parameters in estimating the striking distance of lightning strokes to overhead lines. *IEEE Trans. Pow. Del.* **4**: 1970–81.

Chowdhuri, P., and Mehairjan, S. 1997. Estimation of lightning incidence to overhead lines. *Proc. Gen. Transm. Distrib.* **144**: 129–31.

Chowdhuri, P., Baker, A.C., Carrara, G., Chisholm, W.A., Feser, K., Grzybowski, S., Lux, A., and Newman, F.R. 1994b. Bibliography of research on nonstandard lightning voltage waves. *IEEE Trans. Pow. Del.* **9**: 1982–90.

Chowdhuri, P., Baker, A.C., Carrara, G., Chisholm, W.A., Feser, K., Grzybowski, S., Lux, A., and Newman, F.R. 1994a. Review of research on nonstandard lightning voltage waves. *IEEE Trans. Pow. Del.* **9**: 1972–81.

Chowdhuri, P., Mishra, A.K., Martin, P.M., and McConnell, B.W. 1994c. The effects of nonstandard lightning voltage waveshapes on the impulse strength of short air gaps. *IEEE Trans. Pow. Del.* **9**: 1991–9.

Choy, L.A., and Darveniza, M. 1971. A sensitivity analysis of lightning performance calculations for transmission lines. *IEEE Trans. Pow. Appar. Syst.* **90**: 1443–51.

CIGRE Task Force 33.01.03 1997. Lightning exposure of structures and interception efficiency of air terminals. Report 118.

Cinieri, E., and Muzi, F. 1996. Lightning induced overvoltages, improvement in quality of service in MV distribution lines by addition of shield wires. *IEEE Trans. Pow. Del.* **11**: 361–72.

Clayton, R.E., Grant, I.S., Hedman, D.E. and Wilson, D.D. 1983. Surge arrester protection and very fast surges. *IEEE Trans. Pow. Appar. Syst.* **102**.

Cohen, I.B. 1941. Benjamin Franklin's experiments (letter to D'Alibard of 29 June 1755). Cambridge, Massachusetts: Harvard University Press.

Colominas, I., Gómez-Calviño, J., Navarrina, F., and Casteleiro, M. 2001. Computer analysis of earthing systems in horizontally of vertically layered soils. *Electr. Pow. Syst. Res.* **59**: 149–56.

Cooray, V. 1994. Calculating lightning-induced overvoltages in power lines: a comparison of two coupling models. *IEEE Trans. Electromagn. Compat.* **36**: 179–82.

Cooray, V., and Scuka, V. 1998. Lightning-induced overvoltages in power lines: validity of various approximations made in overvoltage calculations. *IEEE Trans. Electromagn. Compat.* **40**: 355–63.

Cortina, R., and Porrino, A. 1992. Calculation of impulse current distributions and magnetic fields in lightning protection structures – a computer program and its laboratory validation. *IEEE Trans. Mag.* **28**: 1134–7.

Coulson, R.N., Hennier, P.B., Flamm, R.O., Rykiel, E.J., Hu, K.C., and Payme, T.L. 1983. The role of lightning in the epidemiology of the southern pine beetle. *Z. Agnew. Entomol.* **96**: 182–93.

Coulson, R.N., Flamm, R.O., Pulley, P.E., Payne, T.L., Rykiel, E.J. and Wagner, T.L. 1986. Response of the southern pine park beetle guild to host disturbance. *Environ. Entomol.* **15**: 859–68.

Covert, R.N. 1930. Protection of buildings and farm property from lightning. US Department of Agriculture Farmers Bulletin no. 1512, issued November 1926, revised August 1930, Washington, DC.

Cristina, S. and Orlandi, A. 1992. Calculation of the induced effects due to a lightning stroke. *IEE Proc. B* **139**: 374–80.

Curran, E.B., Holle, R.L. and López, R.E. 1997. Lightning fatalities, injuries and damage reports in the United States from 1959–1994. NOAA Technical Memorandum NWS SR-193, 64 pp. (Available from National Weather Service Southern Region, 819 Taylor St., Fort Worth, Texas 76102.)

Curran, E.B., Holle, R.L., and Lopez, R.E. 2000. Lightning casualties and damages in the United States from 1959 to 1994. *J. Climate* **13**: 3448–64.

Currie, J.R., Choy, L.Ah., and Darveniza, M. 1971. Monte Carlo determination of the frequency of lightning strokes and shielding failures. *IEEE Trans. Pow. Appar. Syst.* **90**: 2305–12.

D'Alessandro, F., and Gumley, J.R. 2001. A "collection volume method" for the placement of air terminals for the protection of structures against lightning. *J. Electrost.* **50**: 279–302.

Darveniza, M. 1980. Electrical properties of wood and line design. University of Queensland Press.

Darveniza, M., and Mercer, D.R. 1993. Laboratory studies of the effects of multipulse lightning currents on distribution surge arresters. *IEEE Trans. Pow. Del.* **8**: 1035–44.

Darveniza, M., and Uman, M.A. 1984. Research into lightning protection of distribution systems II – results from Florida field work 1978 and 1979. *IEEE Trans. Pow. Appar. Syst.* **103**: 673–82.

Darveniza, M., and Zhou, Y. 1994. Lightning-initiated fires: energy absorbed by fibrous materials from impulse current arcs. *J. Geophys. Res.* **99**: 10 663–70.

Darveniza, M., Popolansky, F., and Whitehead, E.R. 1975. Lightning protection of UHV transmission lines. *Electra* **41**: 39–69.

Darveniza, M., Tumma, L.R., Richter, B., and Roby, D.A. 1997. Multipulse lightning currents and metal-oxide arresters. *IEEE Trans. Pow. Del.* **12**: 1168–75.

Davis, R. 1962. Frequency of lightning flashover on overhead lines. In *Gas Discharges and the Electricity Supply Industry*, pp. 125–38, London: Butterworths.

DeBlois, I. A. 1933. Some investigations on lightning protection of buildings. *AIEE Trans. I* **33**: 519–35.

De la Rosa, F., Valdivia, R., Perez, H., and Loza, J. 1988. Discussion about the inducing effects of lightning in an experimental power distribution line in Mexico. *IEEE Trans. Pow. Del.* **3**: 1080–9.

Dellera, L., and Garbagnati, E. 1990. Lightning stroke simulation by means of the leader progression model: 1. Description of the model and evaluation of exposure of free standing structures. *IEEE Trans. Pow. Del.* **5**: 2009–22.

Denno, K. 1982. Dynamic modeling for the process of inducing and induced voltage surges due to lightning. *J. Electrostatics* **13**: 55–69.

Dibner, B. 1977. Benjamin Franklin. In *Lightning, vol. 1, Physics of Lightning*, pp. 23–49, New York: Academic Press.

Diendorfer, G. 1990. Induced voltage on an overhead line due to nearby lightning. *IEEE Trans. Electromagn. Compat.* **32**: 292–9.

Drabkin, M.M., and Carpenter, R.B. 1988. Lightning protection devices: how do they compare? In *Mobile Radio Technology*, pp. 24–32.

Dudley, D.G., and Casey, K.F. 1989. Pulse propagation on a horizontal wire above ground: far-zone radiated fields. *Radio Sci.* **24**: 224–34.

Dulzon, A.A. 1996. Lightning as a source of forest fires. *Combustion, Explosion, and Shock Waves* **32**: 587–94.

Dunscombe-Haniball, O. 1900. Accidents and injuries caused by lightning. *JMA S. Afr.* **1**: 1153.

Durham, M.O., and Durham, R.A. 1995. Lightning, grounding and protection for control systems. *IEEE Trans. Ind. Appl.* **31**: 45–54.

Durrett, W.R. 1977. Dissipation arrays at Kennedy Space Center. In *Review of Lightning Protection Technology for Tall Structures*, ed. J. Hughes, pp. 24–52, Publication no. AD-A075 449, Office of Naval Research, Arlington, Virginia.

Dvorak, S.L., and Dudley, D.G. 1992. Pulse propagation on a semi-infinite horizontal wire above ground: near-zone fields. *Radio Sci.* **24**: 899–910.

Dwight, H.B. 1936. Calculation of resistances to ground. *Trans. Am. Inst. Electr. Eng.* **55**: 1319–28.

Eda, K. 1984. Destruction mechanism of ZnO varistors due to high currents. *J. Appl. Phys.* **56**: 2948–55.

Elmendorf, F., King, L., and Ingram, M. 2001. Correlating voltage sags with line faults and lightning. *IEEE Computer Applications in Power*, April, pp. 22–4.

Erdin, I., Dounavis, A., Achar, R., and Nakhla, M.S. 2001. A SPICE model for incident field coupling to lossy multiconductor transmission lines. *IEEE Trans. Electromagn. Compat.* **43**: 485–94.

Eriksson, A.J. 1974. The striking distance of a lightning flash. NEERI Report ELEK 46, Pretoria, South Africa.

Eriksson, A.J. 1976. Lightning overvoltages on high voltage transmission lines: investigation of wave shape characteristics. *Electra* **47**: 87–110.

Eriksson, A.J. 1978. Lightning and tall structures. *Trans. SAIEE* **69**(8): 238–252.

Eriksson, A.J. 1987a. The incidence of lightning strikes to power lines. *IEEE Trans. Pow. Del.* **2**: 859–70.

Eriksson., A.J. 1987b. An improved electrogeometric model for transmission line shielding analysis. *IEEE Trans. Pow. Del.* **2**: 871–86.

Eskow, D. 1983. Striking back at lightning. In *Popular Mechanics*, pp. 55–7 and 116.

Evans, G. 1987. Discussion on lightning performance criteria for electric power systems. *IEEE Proc. Gen. Transm. Distrib. C* **134**: 328–30.

Eybert-Berard, A., Lefort, A., and Thirion, B. 1998. On-site tests. In *Proc. 24th International Conference on Lightning Protection, Birmingham, UK*, pp. 425–35.

FAA (1990). 1989 lightning protection multipoint discharge systems tests: Orlando, Sarasota and Tampa, Florida, Federal Aviation Administration, FAATC T16 Power Systems Program, ACN-210 Final Report, 48 pp. (Includes a review of findings by M.A. Uman and E.P. Krider.)

Farag, A.S., Cheng, T.C., and Penn, D. 1998. Ground terminations of lightning protection systems. *IEEE Trans. Dielectrics and Electrical Insulation* **5**: 869–77.

Fernandez, M.I., Rambo, K.J., Rakov, V.A., and Uman, M.A. 1999. Performance of MOV arresters during very close, direct lightning strikes to a power distribution system. *IEEE Trans. Pow. Del.* **14**: 411–18.

Fisher, R.J., Schnetzer, G.H., and Morris, M.E. 1994. Measured fields and earth potentials at 10 and 20 meters from the base of triggered-lightning channels. In *Proc. 22nd Int. Conf. on Lightning Protection, Budapest, Hungary*, paper R 1c-10, 6 pp.

Flores, D., Jordà, X., Hidalgo, S., Fernández, J., Rebollo, J. Millán, J., Sierra, I., and Mazarredo, I. 1999. An optimized bidirectional lightning surge protection semiconductor device. *IEEE Trans. Electromagn. Compat.* **41**: 30–8.

Fofana, I., and Béroual, A. 1998. Induced effects on an overhead line due to nearby positive lightning downward leader. *Electr. Pow. Syst. Res.* **48**: 105–19.

Fofana, I., Ben Rhouma, A., Béroual, A., and Auriol, Ph. 1998. Modelling a positive lightning downward leader to study its effects on engineering systems. *IEEE Proc. Gen., Transm. and Distrib.* **145**: 392–403.

Fontaine, J.M., Umbert, A., Djebari, B., and Hamelin, J. 1982. Ground effects in the response of a single-wire transmission line illuminated by an E.M.P. *Electromagnetics* **2**: 43–54.

Franklin, B. 1774. *Experiments and Observations on Electricity Made at Philadelphia*. London: E. Cave, ed.

Franklin, B. 1753. Poor Richard's almanack for 1753. In *The Papers of Benjamin Franklin*, eds. L.W. Labaree, W.B. Wilcox, C.A. Lopez, B.B. Oberg et al., vol. 1, dated 1959, to vol. 31, dated 1995, New Haven, Connecticut: Yale University Press.

Franklin, B. 1767. Letter XXIV. In *Benjamin Franklin's Experiments*, I.B. Cohen, 1941, pp. 388–92, Cambridge, Massachusetts: Harvard University Press.

Franklin, B. 1961a. *The Autobiography and Other Writings*, Signet Paperback, p. 167, New York: The New American Library.

Franklin, B. 1961b. *The Autobiography and Other Writings*, Signet Paperback, pp. 234–5, New York: The New American Library. (Priestley's account of Franklin's kite flight, taken from Priestley, *J. History of the Present State of Electricity*, Dodsley, Johnson, Davenport, and Cadell, London, 1967.)

French standard 1995. Protection of structures and of open areas against lightning using ESE air terminals. French standard NFC 17 102.

Fruhauf, G. 1974. Erkennung und Beurteilung von Blitzwirkungen. *Bull. Schweiz. Elektrotech. Ver.* **65**: 1903–8.

Fuquay, D.M., Baughman, R.G., Taylor, A.R., and Hawe, R.G. 1967. Characteristics of seven lightning discharges that caused forest fires. *J. Geophys. Res.* **72**: 6371–3.

Fuquay, D.M., Taylor, A.R., Hawe, R.G., and Schmidt, C.W. Jr 1972. Lightning discharges that have caused forest fires. *J. Geophys. Res.* **77**: 2156–8.

Galvan, A., Cooray, V., and Thottappillil, R. 2001. A technique for the evaluation of lightning-induced voltages in complex low-voltage power installation networks. *IEEE Trans. Electromagn. Compat.* **43**(3): 402–9.

Gänger, B. 1971. Elektrische Festigkeit von Luftisolierstrecken bei hohen Schaltspannungen. *Bull. Schweiz Elektrotech. Ver.* **62**: 227–36.

Gardner, R.L., Baker, L., Gilbert, J.L., Baum, C.E., and Andersh, D.J. 1985. Comparison of published HEMP and natural lightning on the surface of an aircraft. In *Proc. 10th Int. Aerospace Ground Conf. on Lightning and Static Electricity, Paris, France*, pp. 1–8.

Gary, C., Cimador, A., and Fieux, R. 1975. La foudre: étude du phénomène, applications à la protection des lignes de transport. *Rev. Gen. Electr.* **84**: 25–62.

Gay-Lussac, F., and Pouillet, C. 1823. Introduction sur les paratonnères. Adopteé par l'*Académie des Sciences*, Paris, France.

Georgiadis, N., Rubinstein, M., Uman, M.A., Medelius, P.J., and Thomson, E.M. 1992. Lightning-induced voltages at both ends of a 448-m power-distribution line. *IEEE Trans. Electromagn. Compat.* **34**: 451–60.

Geri, A. 1999. Behaviour of grounding systems excited by high impulse currents: the model and its validation. *IEEE Trans. Pow. Del.* **14**: 1008–17.

Geri, A., and Veca, G.M. 1988. Magnetic field generated by lightning protection system. *J. Appl. Phys.* **63**: 3191–3.

Geri, A., Veca, G.M., Garbagnati, E., and Sartorio, G. 1992. Non-linear behavior of ground electrodes under lightning surge currents: computer modeling and comparison with experimental results. *IEEE Trans. Magnetics* **28**: 1442–5.

Gillespie, P.J. 1965. Ionizing radiation: a potential lightning hazard? *Nature* **208**: 577–8.

Gilman, D.W. and Whitehead, E.R. 1973. The mechanism of lightning flashover on high-voltage and extra-high-voltage transmission lines. *Electra* **27**: 69–89.

Golde, R.H. 1941. The validity of lightning tests with scale models. *J. Inst. Electr. Eng.* **88**(II): 67–8.

Golde, R.H. 1945. On the frequency of occurrence and the distribution of lightning flashes to transmission lines. *AIEE Trans.* **64**(III): 902–10.

Golde, R.H. 1946. Lightning currents and potentials on overhead transmission lines. *J. Inst. Electr. Eng.* **93**(2): 559–69.

Golde, R.H. 1947. Occurrence of upward streamers in lightning discharges. *Nature* **160**: 395.

Golde, R.H. 1963. The attractive effect of a lightning conductor. *J. Inst. Electr. Eng.* **9**: 212–3.

Golde, R.H. 1967. The lightning conductor. *J. Franklin Inst.* **283**: 451–77.

Golde, R.H. 1968. Protection of structures against lightning. *Proc. IEE Contr. Sci.* **115**: 1523–9.

Golde, R.H. 1969. A plain man's guide to lightning. *Electronics Power* **15**: 84–6.

Golde, R.H. 1971. The British code of practice on the protection of structures against lightning. *Elek. Tech. Zeit.* **92**: 516–20.

Golde, R.H. 1973. *Lightning Protection*. 254 pp., London: Edward Arnold.

Golde, R.H. (ed.) 1977a. *Lightning: vol. 1, Physics of Lightning, vol. 2, Lightning Protection*, New York: Academic Press.

Golde, R.H., 1977b. The lightning conductor. In *Lightning, vol. 2: Lightning Protection*, ed. R.H. Golde, pp. 545–76 New York: Academic Press.

Golde, R.H. 1977c. The lightning protection of tall structures. In *Review of Lightning Protection Technology for Tall Structures*, ed. J. Hughes, pp. 242–9, Publ. no. AD-A075 449, Office of Naval Research, Arlington, Virginia.

Golde, R.H., and Lee, W.R. 1976. Death by lightning. *Proc. Inst. Electr. Eng.* **123**: 1163–80.

Grcev, L.D. 1996. Computer analysis of transient voltages in large grounding systems. *IEEE Trans. Pow. Del.* **11**: 815–23.

Grcev, L.D., and Dawalibi, F. 1990. An electromagnetic model for transients in grounding system. *IEEE Trans. Pow. Del.* **5**: 1773–81.

Grcev, L.D., and Heimbach, M. 1997. Frequency dependent and transient characteristics of substation grounding system. *IEEE Trans. Pow. Del.* **12**: 172–8.

Greetsai, V.N., Kozlovsky, A.H., Kuvshinnikov, V.M., Loborev, V.M., Parfenov, Y.V., Tarasov, O.A., and Zdoukhov, L.N. 1998. Response of long lines to nuclear high-altitude electromagnetic pulse (HEMP). *IEEE Trans. Electromagn. Compat.* **40**: 348–54.

Griscomb, S.B. 1960. The Kine–Klydonograph – a transient waveform recorder. *AIEE Trans. Pow. Appar. Syst.* **79**: 603–10.

Griscomb, S.B., Caswell, R.W., Graham, R.E., Mcnutt, H.R., Schlomann, R.H., and Thornton, J.K. 1965. Five-year field investigation of lightning effects on transmission lines. *IEEE Trans. Pow. Appar. Syst.* **84**: 257–80.

Gross, I.W., and Cox, J.H. 1931. Lightning investigation on the Appalachian Electric Power Company's transmission system. *AIEE Trans.* **50**: 1118–31.

Guerrieri, S., Rachidi, F., Ianoz, M., Zweiacker, P., Nucci, C.A. 1995. A time-domain approach to evaluate induced voltages on tree-shaped electrical networks by external electromagnetic fields. In *Proc. 11th EMC Symp. on Electromagnetic Compatibility*, pp. 433–8, Zurich, March 1995.

Gumley, J.R., and Berger, G. 1998. A review of lightning attachment process and requirements to achieve improved modeling. In *Proc. 24th Int Conf. on Lightning Protection, Birmingham, UK*, pp. 442–6.

Gumley, J.R., Invernizzi, C.G., and Khaled, M. 1976. Lightning protection – a proven system. *Telecommunications* **10**: 115–20.

Gumley, J.R., D'Alessandro, F., and Austin, M.A. 1998. Experimental arrangements to study lightning attachment characteristics in Northern Australia. In *Proc. 24th Int. Conf. on Lightning Protection, Birmingham, UK*, pp. 1–4.

Gupta, B.G., and Thapar, B. 1980. Impulse impedance of grounding grids. *IEEE Trans. Pow. Appar. a Syst.* **99**: 2357–62.

Gupta, T.K. 1990. Application of zinc oxide varistors. *J. Am. Ceram. Soc.* **73**: 1817–40.

Hagenguth, J.H. 1949. Lightning stroke damage to aircraft. *AIEE Trans.* **68**(II): 1036–46.

Hama, M., Ishizaka, H., Shimizu, J., Nishiki, S. and Kaku, I. 2001. Development of advanced built-in surge arresters for distribution system with new zinc-oxide elements. *IEEE Trans. Pow. Del.* **16**: 576–81.

Hara, T., and Yamamoto, O. 1996. Modelling of a transmission tower for lightning-surge analysis. *IEE Proc. Gen. Transm. Distrib.* **143**: 283–9.

Harris, W.S. 1834. On the protection of ships from lightning. *Nautical Magazine* **3**: 151–6, 225–33, 353–8, 402–7, 477–84, 739–44, 781–7. (These separate articles were collected by Harris under the title *A series of papers on the defence of ships and buildings from lightning*, dated 1835, and available.)

Harris, W.S. 1838a. State of the question relating to the protection of the British Navy from lightning. Plymouth: T.J. Bond University of Illinois, Urbana Library.

Harris, W.S. 1838b. Illustrations of cases of damage by lightning in the British Navy. *Nautical Magazine* (enlarged series) **2**: 590–5, 747–8.

Harris, W.S. 1843. *On the Nature of Thunderstorms*, pp. 140–56, London: John W. Parker.

Hartono, Z.A., and Robiah, I. 1995. A method of identifying a lightning strike location on a structure. In *Proc. Int. Conf. on Electromagnetic Compatibility, Kuala Lumpur*, paper 4.5, pp. 112–17.

Hartono, Z.A., and Robiah, I. 1999. A long term study of the performance of early streamer emission air terminals in a high isokeraunic region. Submitted to NFPA Codes and Standard Administration, 1 Batterymarch Park, Quincy, Massachusetts. Available from hartono@pe.jaring.my.

Hartono, Z.A., and Robiah, I. 2000. The collection surface concept as a reliable method for predicting lightning strike location. In *Proc. 25th Int. Conf. on Lightning Protection, Rhodes, Greece, 18–22 September 2000*, pp. 328–33.

Hasse, P., and Wiesinger, J. 1992. *Handbuch für Blitzschutz und Erdung*, 4th edition, München: Pflaum Verlag and Berlin Offenbach: VDE-Verlag.

Hasse, P. and Wiesinger, J. 1993. EMV Blitz-Schutzzonen-Konzept. München: Pflaum-Verlag; Berlin/Offenbach: VDE-Verlag.

Heary, K.P., Chaberski, A.Z. Gumley, S., Gumley, J.R., Richens, F., and Moran, J.H. 1989. An experimental study of ionizing air terminal performance. *IEEE Trans. Pow. Del.* **4**: 1175–84.

Heidler, F. 1996. Coordination of surge protection devices using the current data from lightning field measurements. *ETEP* **6**: 441–4.

Heimbach, M., and Grcev, L.D. 1997. Grounding system analysis in transients programs applying electromagnetic field approach. *IEEE Trans. Pow. Del.* **12**: 186–93.

Henley, W., Lane, T., Nairne E., and Planta, J. 1778. The report of the committee appointed by the Royal Society, for examining the effect of lightning, 15 May 1777, on the parapet-wall of the house of the Board of Ordnance, at Purfleet, Essex. *Phil. Trans. Roy. Soc.* **68**: 236–8.

Hermosillo, V.F. 1995. Calculation of fault rates of overhead power distribution lines due to lightning-induced voltages including the effect of ground conductivity. *IEEE Trans. Electromagn. Compat.* **37**: 392–9.

Heymann, D. 1998. Search for C_{60} fullerene in char produced on a Norway spruce by lightning. *Fullerene Science and Technology* **6**(6): 1079–86.

Hileman, J.R., and Weck, K.H. 1990. Protection performance of metal oxide surge arresters. *Electra* **133**: 132–43.

Hill, P.G. 1989. Lightning at sea. In *Saltwater Sportsman (Florida edition)*, pp. 19–22.

Hodges, J.U.D., and Pickard, L.S. 1971. Lightning in the ecology of the southern pine beetle, *Dendroctonus frontalis* (Coleoptera: Scolytidae). *Can. Entomol.* **103**: 44–51.

Hohenberger, G., Tomandl, G., Ebert, R., and Taube, T. 1991. Inhomogeneous conductivity in varistor ceramics: methods of investigation. *J. Am. Ceram. Soc.* **74**: 2067–72.

Hoidalen, H.K. 1997. Lightning induced voltages in low-voltage systems with emphasis on lossy ground effects. In *Proc. IPST '97 Int. Conf. on Power Systems and Transients, Seattle*, pp. 336–41.

Hoidalen, H.K., Sletbak, J., and Henriksen, T. 1997. Ground effects on induced voltages from nearby lightning. *IEEE Trans. Electromagn. Compat.* **39**: 269–78.

Holitza, F.J. and Kasemir, H.W. 1974. Accelerated decay of thunderstorm electric fields by chaff seeding. *J. Geophys. Res.* **79**: 425–9.

Holle, R.L., Lopez, R.E., Arnold, L.J., and Endries, J. 1996. Insured lightning-caused property damage in three western states. *J. Appl. Meteorol.* **35**: 1344–51.

Hoole, P.R.P., and Hoole, S.R.H. 1993. Simulation of lightning attachment to open ground, tall towers and aircraft. *IEEE Trans. Pow. Del.* **8**: 732–40.

Horváth, T. 1969. Eine theoretische Betrachtung des Entstehens der Frangentladung. In *Proc. Conf. on Int. Lightning Protection, Budapest*.

Horváth, T. 1971. Gleichwertige Fläche und relative Einschlagungsgefahr als charakteristische Ausdrücke des Schutzeffektes von Blitzableitern. In *Proc. Int. Conf. on Lightning Protection, Munich*.

Horváth, T. 1989. Computation of the lightning stroke probability and the effectiveness of the air termination. *J. Electrostat.* **23**: 305–22.

Horváth, T. 1991. *Computation of Lightning Protection*, 204 pp., Taunton, Somerset, England: Research Studies Press (also New York: John Wiley and Sons).

Howard, K.W., and Holle, R.L. 1995. Peligro de rayos! US Dept. Commerce, Environmental Res. Labs., National Severe

Storms Lab., 1 p. (Available from NSSL, 1313 Halley Circle, Norman, Oklahoma 73069.)
Howard, K.W., Holle, R.L. and López, R.E. 1998. Lightning danger! US Dept Commerce, Environmental Res. Labs., National Severe Storms Lab, 1 p.
Hughes, J. 1977. Introduction. Pp. i–iv of Review of lightning protection technology for tall structures, Publ. no. AD-A075 449, Office of Naval Research, Arlington, Virginia.
Ianoz, M., Nucci, C.A., and Tesche, F.M. 1988. Transmission line theory for field-to-transmission line coupling calculations. *Electromagnetics* **8**: 171–211.
Idone, V.P., and Orville, R.E. 1990. Notes and correspondence – delimiting "thunderstorm watch" periods by real-time lightning location for a power utility company. *Wea. Forecast.* **5**: 139–47.
IEC 1990. Protection of structures against lightning, Part 1: general principles *IEC* 61024–1.
IEC 1993. Protection of structures against lightning, Part 1: general principles, section 1: guide A: selection of protection levels for lightning protection systems. IEC 61024-1-1.
IEC 1994. High voltage test technology, Part 2: measuring systems. IEC 60060-2.
IEC 1995. Protection against lightning electromagnetic impulse, Part 1: general principles. IEC 61312-1.
IEC 1995. Assessment of the risk of damage due to lightning. IEC/TR2 61662.
IEC 1996. Surge arresters, Part 5: selection and application recommendations, section 1: general. IEC 60099-5.
IEC 1998. Protection of structures against lightning, Part 1–2: general principles, guide B: design installation, maintenance and inspection of lightning protection systems. IEC 61024-1-2.
IEC 1998. Protection against lightning electromagnetic impulse, Part 4: protection of equipment in existing structures. IEC/TS 61312-4.
IEC 1998. Surge protection protective devices connected to low-voltage power distribution systems, Part 1: performance requirements and testing methods. IEC 61643-1.
IEC 1999. Protection against lightning electromagnetic impulse (LEMP), Part 2: shielding of structures, bonding inside structures and earthing. IEC/TS 61312-2.
IEC 1999. Lightning protection – telecommunication lines, Part 1: fibre optic installations. IEC 61663-1.
IEC 2000. Protection against lightning electromagnetic impulse, Part 3: requirements of surge protective devices (SPDs). IEC/TC 61312-3.
IEC 2000. Low voltage surge protection, Part 2-1: surge protective devices connected to telecommunications and signalling networks – performance requirements and testing methods. IEC 61643-21.
IEEE Committee Report 1985. A simplified method for estimating lightning performance of transmission lines. *IEEE Trans. Pow. Appar. Syst.* **104**: 919–32.
IEEE W.G. 3.4.11 of Surge Protection Devices Committee 1992. Model of metal oxide surge arrestors. *IEEE Trans. Pow. Del.* **7**: 301–9.
IEEE Task Force 15.09 1994. Review of research on nonstandard lightning voltage waves. *IEEE Trans. Pow. Del.* **9**: 1972–81
Inoue, A., and Kanao, S.-I. 1996. Observation and analysis of multiple-phase grounding faults caused by ligntning. *IEEE Trans. Pow. Del.* **11**: 353–60.
Ishii, M. 1992. Lightning-induced voltages on overhead wire. *Res. Lett. Atmos. Electr.* **12**: 77–81.
Ishii, M. and Shindo, T. 1995. Recent topics on research of lightning and lightning protection in Japan. *Elektrotechnik und Informationstechnik* **112** (JG., H.6): 269–72.
Ishii, M., Ohsaki, E., Kawamura, T., Murotani K., Kouno, T., Higuchi, T. 1991. Multistory transmission tower model for lightning surge analysis. *IEEE Trans. Pow. Del.* **6**: 1327–35.
Ishii, M., Michishita, K., Hongo, Y., and Oguma, S. 1994. Lightning-induced voltage on an overhead wire dependent on ground conductivity. *IEEE Trans. Pow. Del.* **9**: 109–18.
Ishii, M., Michishita, K., and Hongo, Y. 1999. Experimental study of lightning-induced voltage on an overhead wire over lossy ground. *IEEE Trans. Electromagn. Compat.* **41**: 39–45.
Jankov, V. 1997. Estimation of the maximal voltage induced on an overhead line due to the nearby lightning. *IEEE Trans. Pow Del.* **12**: 315–24.
Jernegan, M.W. 1928. Benjamin Franklin's "electrical kite" and lightning rod. *New England Quarterly* **1**: 180–96.
Johnson, W.A., Warne, L.K., Chen, K.C., and Gurrola, E.M. 1995. Linear diffusion and internal voltages in conducting enclosures subjected to a direct lightning strike. *Electromagnetics* **15**: 189–207.
Kan, M., Nishiwaki, S., Sato, T., Kojima, S., and Yanabu, S. 1983. Surge discharge capability and thermal stability of a metal oxide surge arrester. *IEEE Trans. Pow. Appar. Syst.* **102**: 282–9.
Kasemir, H.W., Holitza, F.J., Cobb, W.E., and Rust, W.D. 1976. Lightning suppression by chaff seeding at the base of thunderstorm. *J. Geophys. Res.* **81**: 1965–70.
Kawamura, T. 1992. Lightning protection and standards. *Res. Lett. Atmos. Electr.* **12**: 97–100.
Kellogg, E.W. 1912. The use of metal conductors to protect buildings from lightning. University of Missouri Bulletin no. 7, Engineering Experiment Station, vol. 3, no. 1, Columbia, Missouri.
Kilgore, B.M., and Briggs, G.S. 1972. Restoring fire to high elevation forests in California. *J. Forestry* **70**: 266–70.
Kim, I., and Funabashi, O. 1996. Study of ZnO arrester model for steep front wave. *IEEE Trans. Pow. Del.* **7**: 301–9.
Kinsler, M., and Hmurcik, L.V. 1999. A damage mechanism: lightning-initiated fault-current arcs to communication cables buried beneath overhead electric power lines. *IEEE Trans. Ind. Appl.* **35**: 163–8.
Kirkby, P., Erven, C.C., and Nigol, O. 1988. Long-term stability and energy discharge capacity of metal oxide value elements. *IEEE Trans. Pow. Del.* **3**: 1656–65.

Kitagawa, N. 2000. The actual mechanisms of so-called step voltage injuries. In *Proc. 25th Int. Conf. on Lightning Protection, Rhodes, Greece*, pp. 781–5.

Kithil, R., and Rakov,V. 2000. Small shelters and safety from lightning. *Golf Course Management* **68**: 104–12.

Kofoid, M.J. 1970. Lightning discharge heating of aircraft skins. *J. Aircraft* **1**: 21–6.

Komarek, E.V. Sr 1964. The natural history of lightning. In *Proc. 3rd Annual Conf. on Tall Timbers Fire Ecol.* pp. 139–86.

Komarek, E.V. 1968. Lightning and lightning fires as ecological forces. In *Proc. Annual Tall Timbers Fire Ecology Conf.*, vol. 8, pp. 169–97. (Available from Tall Timbers Research Station, Tallahassee, Florida.)

Komarek, E.V. 1973. Introduction to lightning ecology. In *Proc. Annual Tall Timbers Fire Ecology Conf.*, vol. 13, pp. 421–7. (Volume 13 contains many additional papers concerning lightning as both a predator and fire-starter and is available from Tall Timbers Research Station, Tallahassee, Florida.)

Kostenko, M.V. 1995. Electrodynamic characteristics of lightning and their influence on disturbances of high-voltage lines. *J. Geophys. Res.* **100**: 2739–47.

Krider, E.P. 1977. On lightning damage to a golf course green. *Weatherwise* **30**: 111.

Krider, E.P. 1996. Lightning rods in the 18th century. In *Proc. 23rd Int. Conf. on Lightning Protection, Florence, Italy*, 9 pp.

Krider, E.P., and Ladd, C.G. 1975. Upward streamers in lightning discharges to mountainous terrain. *Weather* **30**: 77–81.

Krider, E.P. and Uman, M.A. 1995. Cloud-to-ground lightning: mechanisms of damage and methods of protection. *Seminars in Neurology* **15**: 227–32.

Kruse, V.J., Tesche, F.M., Liu, T.K., and Barnes, P.R. 1990. Flashover vulnerability of transmission and distribution lines to high-altitude electromagnetic pulse (HEMP). *IEEE Trans. Pow. Del.* **5**: 1164–69.

Kuwabara, N., Tominaga, T., Kanazawa, M., and Kuramoto, S. 1998. Probability occurrence of estimated lightning surge current at lightning rod before and after installing dissipation array system (DAS). In *Proc. IEEE Electromagn. Compat. Int. Symp.*, pp. 1072–7, Denver, Colorado.

Larmor, J., and Larmor, J.S.B. 1914. On protection from lightning and the range of protection afforded by lightning rods. *Proc. Roy. Soc.* **90**: 312–7.

Lat, M.V. 1983. Thermal properties of metal oxide surge arresters. *IEEE Trans. Pow. Appar. Syst.* **102**: 2194–202.

Latham, D. 1991. Lightning flashes from a prescribed fire-induced cloud. *J. Geophys. Res.* **96**: 17 151–17.

Latham, D.J., and Schlieter, J.A. 1989. Ignition probability of wildland fuels based on simulated lightning discharges. USDA Forest Service Research Paper INT-411.

Lattarulo, F. 1996. Comments on "Calculation of fault rates of overhead power distribution lines due to lightning-induced voltages including the effect of ground conductivity. *IEEE Trans. Electromagn. Compat.* **38**: 110.

Lee, H., and Mousa, A.M. 1996. GPS travelling wave fault locator systems: investigation into the anamalous measurements related to lightning strikes. *IEEE Trans. Pow. Del.* **11**: 1214–23.

Lee, K.S.H. (ed.) 1986. *Emp Interaction: Principles, Techniques, and Reference Data*. Washington, DC: Hemisphere.

Lee, R.H. 1978. Protection zone for buildings against lightning strokes using transmission line protection practice. *IEEE Trans. Ind. Appl.* **14**: 465–70.

Lejop, D. 1993. Lightning causes interference to telephone and computer networks. *Telecommunication J.* **60**: 119–25.

LeRoy J.B. 1790. Mémoire sur un voyage... pour y établir des paratonnerres... *Académie des Sciences, Paris, Mémoires*, 472–48

Les Renardières Group 1972. *Electra* **23**: 53–157

Les Renardières Group 1974. *Electra* **35**: 49–56

Les Renardières Group 1977. Positive discharges in long air gaps at Les Renardières, 1975 results and conclusions. *Electra* **53**: 31–153.

Les Renardières Group 1981. Negative discharges in long air gaps at Les Renardières, 1978 results. *Electra* **74**: 67–216.

Les Renardières Group 1986. *IEE Proc. A* **133**.

Lichtenberg, L.C. 1775. Verhaltungsregeln bey nahen Donnerwettern. Gotha.

Liew, A.C., and Darveniza, M. 1971. A sensitivity analysis of lightning performance for transmission lines. *IEEE Trans. Pow. Appar. Syst.* **90**: 1443–51.

Liew, A.C., and Darveniza, M. 1974. Dynamic model of impulse characteristics of concentrated earths. *Proc. Inst. Electr. Eng.* **121**: 123–5.

Liew, A.C., and Darveniza, M. 1982a. Calculation of the lightning performance of unshielded transmission lines. *IEEE Trans. Pow. Appar. Syst.* **101**: 1471–7.

Liew, A.C., and Darveniza, M. 1982b. Lightning performance of unshielded transmission lines. *IEEE Trans. Pow. Appar. Syst.* **101**: 1478–82.

Liu, Y., Zitnik, M., and Thottappillil, R. 2001. An improved transmission-line model of grounding system. *IEEE Trans. Electromagn. Compat.* **43**: 348–55.

Lodge, Oliver J. 1892. *Lightning Conductors and Lightning Guards*, London: Whittaker and Co.

Lopez, R.E., Holle, R.L., Watson, A.I., and Skindlov, J. 1997. Spatial and temporal distributions of lightning over Arizona from a power utility perspective. *J. Appl. Meteor.* **36**: 825–31.

Love, E.R. 1973. Improvements on lightning stroke modelling and applications to the design of EHV and UHV transmission lines. M.Sc. thesis, University of Colorado.

Love, R.M. 1970. The rangelands of the western United States. *Sci. Am.* **222**: 88–96.

Lovelady, C.N., Pulley, P.E., Coulson, R.N., and Flamm, R.O. 1991. Relation of lightning to herbivory by the southern pine bark beetle guild (Coleoptera: Scolytidae). *Environ. Entomol.* **20**: 1279–84.

Lowrey, J. 1991. Protecting substations. *Rural Electrification*, p. 5.

Loyka, S.L. 1999. A simple formula for the ground resistance calculation. *IEEE Trans. Electromagn. Compat.* **41**: 152–4.

Lundholm, R. 1957. Induced overvoltage surges on transmission lines. *Chalmers. Tek. Hoegsk. Handl.* **188**: 1–117.

Lyons, W.A., Nelson, T.E., Williams, E.R., Cramer, J.A., and Turner, T.R. 1998. Enhanced positive cloud-to-ground lightning in thunderstorms ingesting smoke from fires. *Science* **282**: 77–80.

Mackerras, D., Darveniza, M., and Liew, A.C. 1987. Standard and non-standard lightning protection methods. *J. Electr. Eng. Aust.* **7**: 133–40.

Mackerras, D., Darveniza, M., and Liew, A.C. 1997. Review of claimed enhanced lightning protection of buildings by early streamer emission air terminals. *IEE Proc. Sci. Meas. Tech.* **144**: 1–10.

Maddox, R.A, Howard, K.W., and Dempsey, C.L. 1997. Intense convective storms with little or no lightning over central Arizona: a case of inadvertent weather modification? *J. Appl. Meteorol.* **36**: 302–14.

Makalsky, L.M., Orlov, A.V., Temnikov, A.G. 1996. Possible mechanism of lightning strokes to extra-high-voltage power transmission lines. *J. Electrost.* **37**: 249–60.

Malan, D.J. 1969. Lightning and its effects on high structures. *Trans. S. Afr. Inst. Electr. Engrs* **60**(II): 241–2.

Mamis, M.S., and Köksal, M. 2001. Lightning surge analysis using nonuniform, single-phase line model. *IEE Proc. Gen. Trans. Distrib.* **148**: 85–90.

Mansoor, A., and Martzloff, F. 1998. The effect of neutral earthing practices on lightning current dispersion in a low-voltage installation. *IEEE Trans. Pow. Del.* **13**: 783–92.

Mansoor, A., Martzloff, F.D., and Phipps, K.O. 1998. Gapped arresters revisited: a solution to cascade coordination. *IEEE Trans. Pow. Del.* **13**: 1174–84.

Maruvada, P.S., Nguyen, D.H., and Hamadany-Zadeh, H. 1989. Studies on modeling corona attenuation of dynamic overvoltages. *IEEE Trans. Pow. Del.* **4**: 1441–9.

Master, M.J., and Uman, M.A. 1984. Lightning induced voltage on power lines: theory. *IEEE Trans. Pow. Appar. Syst.* **103**: 2502–18.

Master, M.J., Uman, M.A., Beasley, W.H., and Darveniza, M. 1984. Lightning induced voltages on power lines: experiment. *IEEE Trans. Pow. Appar. Syst.* **103**: 2519–29.

Master, M.J., Uman, M.A., Beasley, W.H., and Darveniza, M. 1986. Voltages induced on an overhead line by the lightning stepped leader. *IEEE Trans. Electromagn. Compat.* **28**: 158–61.

Mata, C.T. 2000. Interation of lightning with power distribution lines. Ph.D. dissertation, 388 pp., University of Florida, Gainesville.

Mata, C.T., Fernandez, M.I., Rakov, V.A., and Uman, M.A. 2000. EMTP modeling of a triggered-lightning strike to the phase conductor of an overhead distribution line. *IEEE Trans. Pow. Del.* **15**(4): 1175–81.

Matsumoto, Y., Sakuma, O., Shinjo, K., Saiki, M., Wakai, T., Sakai, T., Nagasaka, H., Motoyama, H., Ishii, M. 1996. Measurement of lightning surges on test transmission line equipped with arrester's struck by natural and triggered lightning. *IEEE Trans. Pow. Del.* **11**: 996–1002.

Maxwell, J.C. 1877. On the protection of buildings from lightning. British Association Report, London.

Mazur, V., Ruhnke, L.H., Bondiou-Clergerie, A., and Lalande, P. 2000. Computer simulation of a downward negative stepped leader and its interaction with a ground structure. *J. Geophys. Res.* **105**: 22 361–9.

McEachron, K.B. 1939. Lightning to the Empire State Building. *J. Franklin Inst.* **227**: 147–217.

McEachron, K.B. 1941. Lightning to the Empire State Building. *AIEE Trans.* **60**: 1379–80.

McEachron, K.B., and Hagenguth, J.H. 1942. Effects of lightning on thin metal skins. *AIEE Trans.* **61**: 559–64.

McEachron, K.B., and Morris, W.A. 1936. The lightning stroke: mechanism of discharge. *Gen. Electr. Rev.* **39**: 487–96.

McEachron, K.B., and Patrick, K.G. 1940. *Playing with Lightning*, 231 pp. New York: Random House.

McDermott, T.E., Short, T.A., and Anderson, J.G. 1994. Lightning protection of distribution lines. *IEEE Trans. Pow. Del.* **9**: 138–52.

Meliopulos, A.P., and Moharam, M.G. 1983. Transient analysis of grounding systems. *IEEE Trans. Pow. Appar. Syst.* **102**: 389–99.

Michishita, K., Ishii, M., and Hongo, Y. 1996. Induced voltage on an overhead wire associated with inclined return-stroke channel-model experiment on finitely conductive ground. *IEEE Trans. Electromagn. Compat.* **38**: 508–13.

Michishita, K., Ishii, M., and Hongo, Y. 1997. Lightning-induced voltage on an overhead wire influenced by a branch line. *IEEE Trans. Pow. Del.* **12**: 296–301.

Minko, G. 1966. Lightning in radiata pine stands in northeastern Victoria. *Australian Forester* **30**: 257–67.

Miyake, K., Suzuki, T., and Shinjou, K. 1992. Characteristics of winter lightning current on Japan sea coast. *IEEE Trans. Pow. Del.* **7**: 1450–6.

Mizukoshi, A., Ozawa, J., Shirakawa, S., and Nakano, K. 1983. Influence of uniformity on energy absorption capabilities of zinc oxide elements as applied in arresters. *IEEE Trans. Pow. Appar. Syst.* **102**: 1384–90.

Moini, R., Kordi, B., and Abedi, M. 1998. Evaluation of LEMP effects on complex wire structures located above a perfectly conducting ground using electric field integral equation in time domain. *IEEE Trans. Electromagn. Compat.* **40**: 154–62.

Montandon, E., and Rubinstein, M. 1998. Some observations on the protection of buildings against the induced effects of lightning. *IEEE Trans. Electromagn. Compat.* **40**: 505–12.

Moore, C.B. 1983. Improved configurations of lightning rods and air terminals. *J. Franklin Inst.* **315**: 61–85.

Moore, C.B., Brook, M., and Krider, E.P. 1981. A study of lightning protection systems. Office of Naval Research, Arlington, Virginia, AD-A158258.

Moore, C.B., Aulich, G.D., and Rison, W. 2000a. Measurement of lightning rod response to nearby strikes. *Geophys. Res. Lett.* **27**: 1487–90.

Moore, C.B., Rison, W., Mathis, J., and Aulich, G. 2000b. Lightning rod improvement studies. *J. Appl. Meteor.* **39**: 593–609.

Moore, C.B., Eack, K.B., Aulich, G.D., and Rison, W. 2001. Energetic radiation associated with lightning stepped-leaders. *Geophys. Res. Lett.* **28**: 2141–4.

Morgan, Z., Headley, R., Alexander, E.A., and Sawyer, C.G. 1954. A trial fibrillation and epidural hematoma associated with lightning stroke. *N. Engl. J. Med.* **259**(20): 956–9.

Morris, M.E., Fisher, R.J., Schnetzer, G.H., Merewether, K.O., and Jorgenson, R.E. 1994. Rocket-triggered lightning studies for the protection of critical assets. *IEEE Trans. Ind. Appl.* **30**: 791–804.

Motoyama, H., Shinjo, K., Matsumoto, Y., and Itamoto, N. 1998. Observation and analysis of multiphase back flashover on the Okushishiku test transmission line caused by winter lightning. *IEEE Trans. Pow. Del.* **13**: 1391–8.

Mousa, A.M. 1994. The soil ionization gradient associated with discharge of high currents into concentrated electrodes. *IEEE Trans. Pow. Del.* **9**: 1669–77.

Mousa, A.M. 1998. The applicability of lightning elimination devices to substations and power lines. *IEEE Trans. Pow. Del.* **13**: 1120–7.

Mousa, A.M., and Srivastava, K.D. 1988. A revised electrogeometric model for the termination of lightning strokes on ground objects. In *Proc. Int. Aerospace and Ground Conf. on Lightning and Static Electricity, Oklahoma City, Oklahoma*, pp. 342–52.

Mousa, A.M. and Strivastava, K.D. 1989. The implications of the electrogeometric model regarding effect of height of structure on the median amplitude of collected strokes. *IEEE Trans. Pow. Del.* **4**: 1450–60.

Mousa, A.M., and Srivastava, K.D. 1990. Modelling of power lines in lightning incidence calculations. *IEEE Trans. Pow. Del.* **5**: 303–10.

Mousa, A.M., and Wehling, R.J. 1993. A survey of industry practices regarding shielding of substations against direct lightning strokes. *IEEE Trans. Pow. Del.* **8**: 38–47.

Müller-Hillebrand, D. 1962a. The protection of houses by lightning conductors – an historical review. *J. Franklin Inst.* **273**: 35–44.

Müller-Hillebrand, D. 1962b. Beeinflussung der Blitzbahn durch radio-aktive Strahlen und durch Raumladungen. *Elektrotech. Z. Aus.* **83**: 152–7.

Murooka, Y. 1992. A survey of lightning interaction with aircraft in Japan. *Res. Lett. Atmos. Electr.* **12**: 101–6.

Murray, N.D., Orville, R.E., and Huffines, G.R. 2000. Effect of pollution from Central American fires on cloud-to-ground lightning in May 1998. *J. Geophys. Res.* **27**: 2249–52.

Nakada, K., Yokota, T., Yokoyama, S., Asakawa, A., Nakamura, M., Taniguchi, H., and Hashimoto, A. 1997. Energy absorption of surge arresters on power distribution lines due to direct lightning strokes – effects of an overhead ground wire and installation position of surge arresters. *IEEE Trans. Pow. Del.* **12**: 1779–85.

Nakada, K., Yokoyama, S., Yokota, T., Asakawa, A., and Kawabata, T. 1998. Analytical study on prevention methods for distribution arrester outages caused by winter lightning. *IEEE Trans. Pow. Del.* **13**: 1399–404.

Nakahori, K., Egawa, T., and Mitani, H. 1982. Characteristics of winter lightning currents in Hokuriku district. *IEEE Trans. Pow. Appar. Syst.* **101**: 4407–12.

Nakamura, K., Horii, K., Kito, Y., Wada, A., Ikeda, G., Sumi, S., Yoda, M., Aiba, S., Sakurano, H., and Wakamatsu, K. 1991. Artificially triggered lightning experiments to an EHV transmission line. *IEEE Trans. Pow. Del.* **6**: 1311–18.

Nanevicz, J.E., Vance, E.F., Radsky, W., Uman, M.A., Soper, G.K., and Pierre, J.M. 1988. EMP susceptibility insights from aircraft exposure to lightning. *IEEE Trans. Electromagn. Compat.* **30**: 463–72.

Narita, K., Goto, Y., Komuro, H., and Sawada, S. 1989. Bipolar lightning in winter at Maki, Japan. *J. Geophys. Res.* **94**: 13 191–5.

NFPA 780 (National Fire Protection Association) 1997. Standard for the Installation of Lightning Protection Systems. Available from NFPA, 1 Batterymarch Park, PO Box 9101, Quincy, Massachusetts 02269–9101.

Nickson, E. 1778. Letter to the Right Honourable Lord Amherst, Lieutenant-General of his Majesty's Ordnance. *Phil. Trans. Roy. Soc.* **68**: 234–5.

Nordgard, J.D., and Chen, C.L. 1979. Lightning-induced transients on buried shielded transmission lines. *IEEE Trans. Electromagn. Compat.* **21**: 171–81.

Novak, T., and Fisher, T.J. 2001. Lightning propagation through the earth and its potential for methane ignitions in abandoned areas of underground coal mines. *IEEE Trans. Ind. Appl.* **37**: 1555–62.

Nucci, C.A. 1995. Lightning-induced voltages on overhead power lines, Part II: coupling models for the evaluation of the induced voltages. *Electra* **162**: 121–45.

Nucci, C.A., and Rachidi, F. 1995. On the contribution of the electromagnetic field components in field-to-transmission line interaction. *IEEE Trans. Electromagn. Compat.* **37**: 505–8.

Nucci, C.A., Mazzetti, C., Rachidi, F., and Ianoz, M. 1988. Analyse du champ electromagnétique d'une décharge de foudre dans les domaines temporel et fréquentiel. *Ann. Telecommun.* **43**: 11–12, 625–37.

Nucci, C.A., Rachidi, F., Ianoz, M.V., and Mazzetti, C. 1993a. Lightning-induced voltages on overhead lines. *IEEE Trans. Electromagn. Compat.* **35**: 75–86.

Nucci, C.A., Rachidi, F., Ianoz, M., and Mazzetti, C. 1993b. Corrections to "Lightning-induced voltages on overhead lines." *IEEE Trans. Electromagn. Compat.* **35**: 488.

Nucci, C.A., Ianoz, M., Rachidi, R., Rubinstein, M., Tesche, F.M., Uman, M.A., and Mazzetti, C. 1995a. Modelling of lightning-induced voltages on overhead lines: recent developments. *Elektrotechnik und Informationstechnik* **112**(6): 290–6.

Nucci, C.A., Rachidi, F., Ianoz, M., and Mazzetti, C. 1995b. Comparison of two coupling models for lightning-induced overvoltage calculations. *IEEE Trans. Pow. Del.* **10**: 330–8.

Nucci, C.A. Borghetti, A., Piantini, A., and Janiszewski, J.M. 1998. Lightning-induced voltages on distribution overhead lines: comparison between experimental results from a reduced-scale model and most recent approaches. In *Proc. 24th Int. Conf. on Lightning, Birmingham, UK*, pp. 314–20.

Nucci, C.A., Guerrieri, S., de Barros, M.T.C., and Rachidi, F. 2000. Influence of corona on the voltages induced by nearby lightning on overhead distribution lines. *IEEE Trans, Pow. Del.* **15**: 1264–73.

Oakeshott, D.F. 1987. Discussion on lightning performance criteria for electric power systems. *IEE Proc. C, Gen. Transm. Distrib.* **134**(5): 328–30.

Omid, M., Kami, Y., and Hayakawa, M. 1997. Field coupling to nonuniform and uniform transmission lines. *IEEE Trans. Electromagn. Compat.* **39**: 201–11.

Orlandi, A. and Schietroma, F. 1996. Attenuation by a lightning protection system of induced voltages due to direct strikes to a building. *IEEE Trans. Electromagn. Compat.* **38**: 43–50.

Orlandi, A., Mazzetti, C., Flisowski, Z., and Yarmarkin, M. 1998. Systematic approach for the analysis of the electromagnetic environment inside a building during lightning strike. *IEEE Trans. Electromagn. Compat.* **40**: 521–35.

Orville, R.E. 1968. Photograph of a close lightning flash. *Science* **162**: 666–7.

Pan, E., and Liew, A.C. 2000. Analysis of a novel method of current sharing in a resistive lightning protection terminal. *IEEE Trans. Pow. Del.* **15**: 948–52.

Papalexopoulos, A.D., and Meliopulos, A.P. 1987. Frequency dependent characteristics of grounding systems. *IEEE Trans. Pow. Del.* **2**: 1073–81.

Parker, J.W. 1848. *A History of the Royal Society...*, vol. 2, pp. 392–4, London.

Parrish, D.E., and Kvaltine, D.J. 1989. Lightning faults on distribution lines. *IEEE Trans. Pow. Del.* **4**: 2179–86.

Parrish, D.E., et al. 1990. Working group report: calculating the lightning performance of distribution lines. *IEEE Trans. Pow. Del.* **5**: 1408–17.

Paul, C.R. 1994. A SPICE model for multiconductor transmission lines excited by an incident electromagnetic field. *IEEE Trans. Electromagn. Compat.* **36**: 342–54.

Paul, C.R. 1995. Literal solutions for the time-domain response of a two-conductor transmission line excited by an incident electromagnetic field. *IEEE Trans. Electromagn. Compat.* **37**: 241–51.

Paulino, J.O.S., de Araújo, A.E.A., and de Miranda, G.C. 1998. Lightning induced voltage calculation in lossy transmission lines. *IEEE Trans. Magnetics* **34**: 2807–10.

Peek, F.W. 1926. Lightning: a study of lightning rods and cages, with special reference to the protection of oil tanks. *AIEE Trans.* **45**: 1131–46.

Peters, O.S. 1915. Protection of life and property against lightning. Technologic Papers of the Bureau of Standards, no. 56, Washington Government Printing Office, 27 pp.

Petropoulos, G.M. 1948. The high voltage characteristics of earth resistances. *JIEEE* **95**: 59–70.

Petrov, N.I., Avanskii, V.R., and Bombenkova, N.V. 1994. Measurements of the electric field in the streamer zone and in the sheath of the channel of a leader discharge. *Tech. Phys.* **39**: 546–51.

Petrov, N.I., and Waters, R.T. 1995. Determination of the striking distance of lightning to earthed structures. *Proc. Roy. Soc. London A* **450**: 589–601.

Pettersson, P. 1989. A method for study of transmission line shield wire potential for long front lightning current. *IEEE Trans. Pow. Del.* **4**: 508–14.

Pettersson, P. 1991. A unified probabilistic theory of the incidence of direct and indirect lightning strikes. *IEEE Trans. Pow. Del.* **6**: 1301–10.

Philipp, H.R., and Levinson, L.M. 1989. Watts loss and conductivity processes in Zno varistors. In *Ceramic Transactions, vol. 3: Advances in Varistor Technology*, ed. L.M. Levinson, pp. 155–68, Westerville, Ohio: The American Ceramic Society.

Piantini, A., and Janiszewski, J.M. 1992. An experimental study of lightning induced voltages by means of a scale model. In *Proc. 21st Int. Conf. on Lightning Protection, Berlin*, pp. 195–9.

Piantini, A., and Janiszewski, J.M. 1998. Induced voltages on distribution lines due to lightning discharges on nearby metallic structures. *IEEE Trans. Magnetics* **34**: 2799–802.

Pinceti, P., and Giannettoni, M. 1999. A simplified model for zinc oxide surge arresters. *IEEE Trans. Pow. Del.* **14**: 393–7.

Plumer, J.A. 1992. Aircraft lightning protection design and certification standards. *Res. Lett. Atmos. Electr.* **12**: 83–96.

Plummer, C.W., Goedde, G.L., Pettit, E.L. Jr, Godbee, J.S., and Hennessey, M.G. 1995. Reduction in distribution transformer failure rates and nuisance outages using improved lightning protection concepts. *IEEE Trans. Pow. Del.* **10**: 768–77.

Podporkin, G.V., and Sivaev, A.D. 1997. Lightning impulse corona characteristics of conductors and bundles. *IEEE Trans. Pow. Del.* **12**: 1842–7.

Podporkin, G.V., and Sivaev, A.D. 1998. Lightning protection of distribution lines by long flashover arresters (LFA). *IEEE Trans. Pow. Del.* **13**: 814–23.

Preece, W.H., 1880. On the space protected by a lightning conductor. *Phil. Mag.* **9**: 427–30.

Price, C., and Rind, D. 1994. The impact of a $2 \times CO_2$ climate on lightning-caused fires. *J. Climate* **7**: 1484–94.

Price, W.S., Bartlett, S.C., and Zobel, E.S. 1956. Lightning and corona performance of 330 kV lines of the American Gas and Electric and the Ohio Valley Electric Corporation Systems. *AIEE Trans.* **75**(III): 270–81.

Puri, J.L., Abi-Samra, N.C., Dionise, T.J., and Smith, D.R. 1988. Lightning induced failures in distribution transformers. *IEEE Trans. Pow. Del.* **3**: 1784–801.

Rachidi, F. 1993. Formulation of the field-to-transmission line coupling equations in terms of magnetic excitation field. *IEEE Trans. Electromagn. Compat.* **35**: 404–7.

Rachidi, F., Nucci, C.A., Ianoz, M., and Mazzetti, C. 1996. Influence of a lossy ground on lightning-induced voltages on overhead lines. *IEEE Trans. Electromagn. Compat.* **38**: 250–64.

Rachidi, F., Rubinstein, M., Guerrieri, S., and Nucci, C.A. 1997a. Voltages induced on overhead lines by dart leaders and subsequent return strokes in natural and rocket-triggered lightning. *IEEE Trans. Electromagn. Compat.* **39**: 160–6.

Rachidi, F., Nucci, C.A., Ianoz, M., and Mazzetti, C. 1997b. Response of multiconductor power lines to nearby lightning return stroke electromagnetic fields. *IEEE Trans. Pow. Del.* **12**: 1404–11.

Rakov, V.A. 1999. Lightning makes glass. In *1999 Journal of the Glass Art Society*, pp. 45–50.

Rakov, V.A. 2000. Lightning protection of structures and personal safety. In *Proc. 2000 Int. Conf. on Lightning Detection, Tucson, Arizona*, 7–8 November, 10 pp. (Available from Global Atmospherics, Inc., 2705 E. Medina Road, Tucson, Arizona 85706-7155.)

Rakov, V.A., and Lutz, A.O. 1990. A new technique for estimating equivalent attractive radius for downward lightning flashes. In *Proc. 20th Int. Conf. on Lightning Protection, Interlaken, Switzerland*, paper 2.2, 5 pp.

Rakov, V.A., Uman, M.A., Rambo, K.J., Fernandez, M.I., Fisher, R.J., Schnetzer, G.H., Thottappillil, R., Eybert-Berard, A., Berlandis, J.P., Lalande, P., Bonamy, A., Laroche, P., and Bondiou-Clergerie, A. 1998. New insights into lightning processes gained from triggered-lightning experiments in Florida and Alabama. *J. Geophys. Res.* **103**: 14 117–30.

Ramamoorty, M., Babu Narayanan, M.M., Parameswaran, S., *et al.* 1989. Transient performance of grounding grids. *IEEE Trans. Pow. Del.* **4**: 2053–9.

Randa, J., Gilliland, D., Gjertson, W., Lauber, W., and McInerney, M. 1995. Catalogue of electromagnetic environment measurements, 30–300 Hz. *IEEE Trans. Electromagn. Compat.* **37**: 16–33.

Ratnamahilan, P., Hoole, P., Ratnajeevan, S., and Hoole, H. 1993. Simulation of lightning attachment to open ground, tall towers and aircraft. *IEEE Trans. Pow. Del.* **8**: 732–40.

Ringler, K.G., Kirkby, P., Erven, C.C., Lat, M.V., and Malkiewicz, T.A. 1997. The energy absorption capability and time-to-failure of varistors used in station-class metal-oxide surge arresters. *IEEE Trans. Pow. Del.* **12**: 203–12.

Risler, W.T. 1977. Lightning elimination associates (LEA) array on top of 150 meter tower. In *Review of Lightning Protection Technology for Tall Structures*, ed. J. Hughes, pp. 53–63, Publication no. AD-A075 449, Office of Naval Research, Arlington, Virginia.

Rizk, F.A.M. 1989a. A model for switching impulse leader inception and breakdown of long air-gaps. *IEEE Trans Pow. Del.* **4**: 596–606.

Rizk, F.A.M. 1989b. Switching impulse strength of air insulation: leader inception criterion. *IEEE Trans. Pow. Del.* **4**: 2187–95.

Rizk, F.A.M. 1990. Modeling of transmission line exposure to direct strokes. *IEEE Trans. Pow. Del.* **5**: 1983–90.

Rizk, F.A.M. 1994a. Modeling of lightning incidence to tall structures Part I: theory. *IEEE Trans. Pow. Del.* **9**: 162–71

Rizk, F.A.M. 1994b. Modeling of lightning incidence to tall structures Part II: application. *IEEE Trans. Pow. Del.* **9**: 172–93.

Rorig, M.L., and Ferguson, S.A. 1999. Characteristics of lightning and wildland fire ignition in the Pacific Northwest. *J. Appl. Meteor.* **38**: 1565–75.

Rourke, C. 1994. A review of lightning-related operating events at nuclear power plants. *IEEE Trans. Energy Conversion* **9**: 636–41.

Rubinstein, M., Tzeng, A., Uman, M.A., Medelius, P.J., and Thomson, E.M. 1989. An experimental test of a theory of lightning induced voltages on an overhead wire. *IEEE Trans. Electromagn. Compat.* **31**: 376–83.

Rubinstein, M., Uman, M.A., Medelius, P.J., and Thomson, E.M. 1994. Measurements of the voltage induced on an overhead power line 20 m from triggered lightning. *IEEE Trans. Electromagn. Compat.* **36**: 134–40.

Rühling, F. 1972. Modelluntersuchungen über den Schutzraum und ihre Redeutung für Gebäudeblitzableiter. *Bull. Schweiz. Elektrotech. Ver.* **63**: 522–8.

Rühling, F. 1973. Der Schutzraum von Blitzfangstangen und Derseilen. Ph.D. thesis, Technical University of Munich.

Rusck, S. 1958. Induced lightning overvoltages on power transmission lines with special reference to the overvoltage protection of low voltage networks. *Trans. Roy. Inst. Tech. (K. Tek. Högsk. Handl.), Stockholm*, **120**.

Rusck, S. 1977. Protection of distribution systems. In *Lightning, vol. 2, Lightning Protection*, ed. R.H. Golde, pp. 747–72, New York: Academic.

Rust, W.D. and Krehbiel, P.R. 1977. Microwave radiometric detection of corona from chaff within thunderstorms. *J. Geophys Res.* **82**: 3945–50.

Rustan, P.L. 1987. Description of an aircraft lightning and simulated nuclear electromagnetic pulse (HEMP) threat based on experimental data. *IEEE Trans. Electromagn. Compat.* **29**: 49–63.

Sadovic, S., Joulie, R., Tartier, S., and Brocard, E. 1997. Use of line surge arresters for the improvement of the lightning performance of 63 kV and 90 kV shielded and unshielded transmission lines. *IEEE Trans. Pow. Del.* **12**: 1232–40.

Sakshaug, E.C., Burke, J.J., and Kresge, J.S. 1989. Metal oxide arresters on distribution systems: fundamental considerations. *IEEE Trans. Pow. Del.* **4**: 2076–89.

Saraoja, E.K. 1977. Lightning earths. In *Lightning, vol. 2, Lightning Protection*, ed. R.H. Golde, pp. 577–98, New York: Academic Press.

Sargent, M.A. 1972. The frequency distribution of current magnitudes of lightning strokes to tall structures. *IEEE Trans. Pow. Appar. Syst.* **91**: 2224–9.

Sargent, M.A., and Daveniza, M. 1967. The calculation of the double-circuit outage rate of transmission lines. *IEEE Trans. Pow. Appar. Syst.* **86**: 665–78.

Sargent, M.A., and Daveniza, M. 1969. Tower surge impedance. *IEEE Trans. Pow. Appar. Syst.* **88**: 680–7.

Sargent, M.A., and Daveniza, M. 1970. Lightning performance of double circuit lines. *IEEE Trans. Pow. Appar. Syst.* **89**: 913–25

Sarto, M.S. 2001. Innovative absorbing-boundary conditions for the efficient FDTD analysis of lightning-interaction problems. *IEEE Trans. Electromagn. Compat.* **43**(3): 368–81.

Sartori, C.A., Cardosa, J.R., and Orlandi, A. 1998. Transient induced voltage computation in a high building struck by lightning. *IEEE Trans. Magnetics* **34**: 2815–18.

Sato, M., and Kuramoto, S. 1996. Observed direct lightning current distribution at a mountain-top radio relay station. *IEICE Trans. Commun.* **E79-B**: 522–7.

Schmitz, R.F., and Taylor, A.R. 1969. An instance of lightning damage and infestation of Ponderosa Pines by the pine engraver beetle in Montana. USDA Forest Service Research Note INT-88, 8 pp.

Schonland, B.F.J. 1956. The lightning discharge. In *Encyclopaedia of Physics*, vol. 22, pp. 576–628, Berlin: Springer Verlag.

Schonland, B.F.J., Hodges D.B., and Collens, H. 1938. Progressive lightning V. *Proc. Roy. Soc. A* **161**: 654–74.

Schowalter, T.D., Coulson, R.N., and Crossley, D.A. Jr 1981. The role of southern pine beetle and fire in maintenance of structure and function of southeastern coniferous forests. *Environ. Entomol.* **10**: 821–5.

Schwab, F. 1965. Berechung der Schutzwirkung von Blitzableitern und Türmen. *Bull. Schweiz. Elektrotech. Ver.* **56**: 678–83.

Schwaiger, A. 1938. *Der Schutzbereich von Blitzableitern*, Munich: R. Oldenbourg.

Scuka, V. 1987. On the interception of the lightning discharge. *Electromagnetics* **7**: 353–60.

Seckel, A., and Edwards, J. 1986. The revolt against the lightning rod. *Free Inquiry*, summer, 54–5.

Shindo, T., and Aihara, Y. 1993. A shielding theory for upward lightning. *IEEE Trans. Pow. Del.* **8**: 318–24.

Short, T.A. 1992. Lightning protection: analyzing distribution designs. *IEEE Computer Appl. Pow.* **5**: (2): 51–5.

Silva, J.P., Araujo, A.E.A., and Paulino, J.O.S. 2002. Calculation of lightning-induced voltages with Rusck's method in EMTP – Part II: effects of lightning parameter variations. *Electr. Pow. Syst. Res.* **61**: 133–7.

Smeloff, N.N., and Price, A.L. 1930. Lightning investigation on 220-kv system of the Pennsylvania Power and Light Company (1928 and 1929). *J. AIEE* **49**: 771–5.

Snyder, R.E. 1973. New protection system may eliminate lightning damage. *Wld Oil* **176**: 59–61.

Spanish standard 1996. Protection of structure and of open areas against lightning using early streamer emission air terminal. UNE 21186, June 1996.

Sporn, P., and Lloyd, W.L. Jr 1930. Lightning investigation on 132-kV system of the Ohio Power Company. *J. AIEE* **49**: 259–62.

Sporn, P., and Lloyd, W.L. Jr 1931. Lightning investigations on the transmission system of the American Gas and Electric Company. *AIEE Trans.* **49**: 1111–17.

Standards Association of Australia 1983. Lightning protection. AS 1768–1983.

Standler, R.B. 1989a. *Protection of Electronic Circuits from Overvoltages*. New York: John Wiley and Sons.

Standler, R.B. 1989b. Transients on the mains in a residential environment. *IEEE Trans. Electromagn. Compat.* **31**: 170–6.

Sternberg, B.K., and Levitskaya, T.M. 2001. Electrical parameters of soils in the frequency range from 1 kHz to 1 GHz, using lumped-circuit methods. *Radio Sci.* **36**: 709–19.

Stockum, F.R. 1994. Simulation of the nonlinear thermal behavior of metal oxide surge arresters using a hybrid finite difference and empirical model. *IEEE Trans. Pow. Del.* **9**: 306–13.

Sunde, E.D. 1968. *Earth Conduction Effects in Transmission Systems*, 373 pp. New York: Dover.

Suzuki, S., Koyama, K., Hayami, T., and Murooka, Y. 1992. On the lightning protection of ground crews. *Res. Lett. Atmos. Electr.* **12**: 123–6.

Suzuki, T., Miyake, K., and Kishizima, I. 1981. Study on experimental simulation of lightning strokes. *IEEE Trans. Pow. Appar. Syst.* **4**: 1703–11.

Szczerbinski, M. 2000. A discussion of 'Faraday cage' lightning protection and application to real building structures. *J. Electrostatics* **48**: 145–54.

Szpor, S. 1959. Paratonnerres ruraux de type légér. *Revue Générale de l'Électricité* **68**: 263–70.

Szpor, S. 1972. Negative protective angles, illusions and reality. *Archiwum Elektrotechniki* **21**: 17–18.

Tanabe, K. 2001. Novel method for analyzing dynamic behavior of grounding systems based on the finite-difference time-domain method. *IEEE Pow. Eng. Rev.*, September, 55–7.

Taylor, A.R. 1964. Lightning damage to trees in Montana. *Weatherwise* **17**: 62–5.

Taylor, A.R. 1965. Diameter of lightning as indicated by tree scars. *J. Geophys. Res.* **70**: 5693–5.

Taylor, A.R. 1969a. Lightning effects on the forest complex. In *Proc. 9th Annual Tall Timbers Fire Ecol. Conf.*, pp. 127–50.

Taylor, A.R. 1969b. Tree-bole ignition in superimposed lightning scars. USDA Forest Service Research Note INT-90, 4 pp.

Taylor, C.D., Satterwhite, R.S., and Harrison, C.W. 1965. The response of a terminated two-wire transmission line excited by a non-uniform electromagnetic field. *IEEE Trans. Antennas Propag* **AP-13**: 987–9.

Ten Dius, H.J. 1998. Lightning strikes: danger overhead. *British J. Sports Medicine* **32**: 276–8.

Tesche, F.M. 1992. Comparison of the transmission line and scattering models for computing the HEMP response of overhead cables. *IEEE Trans. Electromagn. Compat.* **34**: 93–9.

Tesche, F.M., Kälin, A.W., Brändli, B., Reusser, B., Ianoz, M., Tabara, D., and Zweiacker, P. 1998. Estimates of lightning-induced voltage stresses within buried shielded conduits. *IEEE Trans. Electromagn. Compat.* **40**: 492–504.

Testé, Ph., Leblanc, T., Uhlig, F., and Chabrerie, J.-P. 2000. 3D modeling of the heating of a metal sheet by a moving arc: application to aircraft lightning protection. *Eur. Phys. J.* **AP11**: 197–204.

Thione, L. 1980. The dielectric strength of large air insulation. In *Surges in High Voltage Networks*, ed. K. Ragaller, pp. 165–206, New York: Plenum.

Thomas, D.W.P., Chistopoulos, C., and Pereira, E.T. 1994. Calculation of radiated electromagnetic fields from cables using time-domain simulation. *IEEE Trans. Electromagn. Compat.* **36**: 201–5.

Thomson, E. 1991. A critical assessment of the U.S. Code for lightning protection of boats. *IEEE Trans. Electromagn. Compat.* **33**: 132–8.

Thum, P.C. Liew, A.C., and Wong, C.M. 1982. Computer simulation of the initial stages of the lightning protection mechanism. *IEEE Trans. Pow. Appar. Syst.* **101**: 4370–7.

Tkatchenko, S., Rachidi, F., and Ianoz, M. 1995. Electromagnetic field coupling to a line of finite length: theory and fast iterative solutions in frequency and time domains. *IEEE Trans. Electromagn. Compat.* **37**: 509–18.

Tkatchenko, S., Rachidi, F., and Ianoz, M. 2001. High-frequency electromagnetic field coupling to long terminated lines. *IEEE Trans. Electromagn. Compat.* **43**(2): 117–29.

Tobias, J.M. 1996. Testing of ground conductors with artificially generated lightning current. *IEEE Trans. Ind. Appl.* **32**: 594–8.

Toepler, M. 1944. Bisher unbekannte Formen des Elmsfeuers. *Die Naturwissenschaften* **31**: 365–8.

Towne, H.M. 1956. Lightning, its behavior and what to do about it. Published by United Lightning Protection Association, Inc., Box 9, Onondaga, New York.

Udo, T. 1993. Estimation of lightning current wave front duration by the lightning performance of Japanese EHV transmission line *IEEE Trans. Pow. Del.* **8**: 660–71.

Udo, T. 1998. Contents of large current flashes among all the lightnings measured on transmission lines. *IEEE Trans. Pow. Del.* **13**: 1432–6.

Uman, M.A. 1986. Application of advances in lightning research to lightning protection. In *The Earth's Electrical Environment*, pp. 61–9, Washington, DC: National Academy Press.

Uman, M.A. 1988. Natural and artificially initiated lightning and lightning test standards. *Proc. IEEE* **76**: 1548–65.

Uman, M.A. 1989. Discussion on "Review of ground flash density measuring devices regarding power system applications" by F. De la Rosa and R. Velazquez. *IEEE Trans. Pow. Del.* **4**: 926.

Uman, M.A. 1991. The best lightning picture I've ever seen. *Weatherwise* **44**: 8–9.

Uman, M.A., Master, M.J., and Krider, E.P. 1982. A comparison of lightning electromagnetic fields with the nuclear electromagnetic pulse in the frequency range 104 to 107 Hz. *IEEE Trans. Electromagn. Compat.* **24**: 410–16.

US Coast Guard 1979. Lightning: cone of protection. MICHU-SG-79-114, Michigan Sea Grant.

US Lightning Protection Code 1980. Publication 78, National Fire Protection Association, Inc. 470 Atlantic Avenue, Boston, Massachusetts 02210.

Van Brunt, R.J., Nelson, T.L., and Firebaugh, S.L. 1995. Early streamer emission air terminals lightning protection systems. National Institute of Standards and Technology (Gaithersburg, Maryland), Report for National Fire Protection Research Foundation, Batterymarch Park, Quincy, Massachusetts, 31 January 1995.

van der Laan, P.C.T., and van Deursen, A.P.J. 1998. Reliable protection of electronics against lightning: some practical applications. *IEEE Trans. Electromagn. Compat.* **40**: 513–20.

Vance, E.F. 1980. Electromagnetic interference control. *IEEE Trans. Electromagn. Compat.* **22**: 319–28.

Vance, E.F., and Uman, M.A. 1988. Differences between lightning and nuclear electromagnetic pulse interactions. *IEEE Trans. Electromagn. Compat.* **30**: 54–62.

Vonnegut, B., Latham, D.L., Moore, C.B., and Hunyady, S.J. 1995. An explanation for anomalous lightning from forest fire clouds. *J. Geophys. Res.* **100**: 5037–50.

Wagner, C.F. 1963. Relation between stroke current and velocity of the return stroke. *AIEE Trans. Pow. Appar. Syst.* **82**: 609–17.

Wagner, C.F. 1967. Lightning and transmission lines. *J. Franklin Inst.* **283**: 558–94.

Wagner, C.F., and Hileman, A.R. 1961. The lightning stroke, II. *AIEE Trans.* **80**: 622–42.

Wagner, C.F. and McCann, G.D. 1942. Induced voltages on transmission lines. *AIEE Trans.* **61**: 916–30.

Wagner, C.F., McCann G.D., and MacLane, G.L. 1941. Shielding of transmission lines. *AIEE Trans.* **60**: 313–28.

Wagner, C.F., McCann G.D., and Lear, C.M. 1942. Shielding of substations. *AIEE Trans.* **61**: 96–100.

Wagner, C.F., Cross, I.W., and Lloyd, B.L. 1954. High voltage impulse tests on transmission lines. *AIEE Trans. Pt. III* **73**: 196–209.

Wait, J.R. 1999. Upward traveling current wave excitation of overhead cable. *IEEE Trans. Electromagn. Compat.* **41**: 75–7.

Walsh, K.M., Hanley, M.J., Granger, S.J., Bean, D., and Bazluki, J. 1997. A survey of lightning policy in selected division I colleges. *J. Athletic Training* **32**: 206–10.

Walter, B. 1937. Von wo ab steuert der Blitz auf seine Einschlagstelle los? *Z. Tech. Phys.* **18**: 105–9.

Warne, L.K., Johnson, W.A., and Chen, K.C. 1995. Nonlinear diffusion and internal voltages in conducting ferromagnetic enclosures subjected to lightning currents. *IEEE Trans. Electromagn. Compat.* **37**: 145–54.

Watson, W.P. 1761. Some suggestions concerning the preventing the mischiefs which happen to ships and their masts by lightning. *Phil. Trans. Roy. Soc.* **52**: 629–35.

Wei-Gang, H., and Semlyen, A. 1987. Computation of electromagnetic transients on three-phase transmission lines with corona and frequency-dependent parameters. *IEEE Trans. Pow. Del.* **2**(3): 887–98.

Weld, C.R. 1822. Electrical conductors for ships. *Phil. Mag., Series 1*, **60**: 256.

Weller, S. 1988. Can you really protect your boat from lightning? *Practiced Sailor* **14**: 8–14.

Whitehead, E.R. 1972. Mechanism of lightning flashover research project, Final report of the Edison Electric Institute, Publication no. 72-900, Edison Electric Institute, New York.

Whitehead, E.R. 1974. CIGRE survey of the lightning performance of extra-high voltage transmission lines. *Electra* **33**: 63–89.

Whitehead, E.R. 1977. Protection of transmission lines. In *Lightning, vol. 2, Lightning Protection*, ed. R.H. Golde, pp. 697–746, New York: Academic.

Whitehead, J.T., Chisholm, W.A., Anderson, J.G., Clayton, R., Elahi, H., Eriksson, A.J., Grzybowski, S., Hileman, A.R., Janischewskyj, W., Longo, V.J., Moser, C.H., Mousa, A.M., Orville, R.E., Parrish, D.E., Rizk, F.A.M., and Renowden, J.R. 1993. IEEE Working group report estimating lightning performance of transmission lines II – updates to analytical models (ed. W.A. Chisholm). *IEEE Trans. Pow. Del.* **8**: 1254–67.

Whitney, B.F., and Asgeirsson, H. 1991. Lightning location and storm severity display system. *IEEE Trans. Pow. Del.* **6**: 1715–20.

Wiesinger, J. 1972. *Blitzforschung und Blitzschutz*, Munchen: R. Oldenbourg Verlag.

Winn, J.L. 1770. A letter to Dr. Benjamin Franklin, F.R.S. giving an account of the appearance of lightning on a conductor fixed from the summit of the mainmast of a ship, down to the water. *Phil. Trans.* **60**: 188–91.

Xiaoqing, Z. 1998. Simulation of lightning transients in a class of multiconductor systems. *Electric Machines and Power Systems* **26**: 545–56.

Xiong, W., and Dawalibi, F.P. 1994. Transient performance of substation grounding systems subjected to lightning and similar surge currents. *IEEE Trans. Pow. Del.* **9**: 1412–20.

Yamada, T., Mochizuki, A., Sawada, J., Zaima, E., Kawamura, T., Ametani, A., Ishii, M., and Kato, S. 1995. Experimental evaluation of a UHV tower model for lightning surge analysis. *IEEE Trans. Pow. Del.* **10**: 393–402.

Yoda, M., Miyachi, I., Kawashima, T., and Katsuragi, Y. 1992. Lightning current protection of equipments in winter. *Res. Lett. Atmos. Electr.* **12**: 117–21.

Yokoyama, S. 1992. Lightning induced voltages on power distribution lines – observation results at Fukui steam power station. *Res. Lett. Atmos. Electr.* **12**: 71–6.

Yokoyama, S., Miyake, K., Mitani, H., and Takanishi, A. 1983. Simultaneous measurement of lightning induced voltages with associated stroke currents. *IEEE Trans. Pow. Appar. Syst.* **102**: 2420–9.

Yokoyama, S., Miyake, K., and Suzuki, T. 1990. Winter lightning on Japan sea coast – development of measuring systems on progressing feature of lightning discharge. *IEEE Trans. Pow. Del.* **5**: 1418–25.

Young, F.S., Clayton, J.M., and Hileman, A.R. 1963. Shielding of transmission lines. *IEEE Trans. Pow. Appar. Syst.* **82** (special supplement): 132–54.

Zeddam, A., and Degauque, P. 1987. Current and voltage induced on telecommunication cables by a lightning stroke. *Electromagnetics* **7**: 541–64.

Zeddam, A., Degauque, P., and Leray, R. 1988. Etude des perturbations induites par une décharge orageuse sur un cable de télécommunication. *Ann. Telecommun.* **43**: 11–12, 638–48.

Zeleny, J. 1934. Do lightning rods prevent lightning? *Science* **79**: 269–71.

Zhu, H., and Raghuveer, M.R. 2001. Influence of representation model and voltage harmonics on metal oxide surge arrester diagnostics. *IEEE Trans. Pow. Del.* **16**: 599–603.

Zipse, D. 1994. Lightning protection systems: advantages and disadvantages. *IEEE Trans Ind. Appl.* **30**: 1351–61.

Zipse, D.W. 2001. Lightning protection methods: an update and a discredited system vindicated. *IEEE Trans. Ind. Appl.* **37**(2): 407–14.

19 Lightning hazards to humans and animals

Lightning struck a supporting steel pole of a tent in which 26 schoolgirls, two adult supervisors, and seven dogs were sleeping. There were injuries to 23 girls, four of which were fatal, and four of the dogs died.

R.B. Anderson (2001)

19.1. Statistics

The annual average death toll, according to the US National Oceanographic and Atmospheric Administration (NOAA) publication *Storm Data*, for each of the four major storm-related hazards, floods, lightning, tornadoes, and hurricanes, is given in Fig. 19.1. Interestingly, warnings and similar forecast products issued by the US National Weather Service generally cover only floods, tornadoes, and hurricanes, three of the four most important causes of storm-related fatalities (Holle *et al.* 1999). The fourth cause, lightning, is, however, the second most efficient weather-related killer in the United States.

According to *Storm Data*, the average number of deaths per year from lightning in the United States from 1966 through 1995 was 85 (Fig. 19.1). Distributions of lightning deaths and injuries by the place of occurrence are shown in Fig. 19.2. Over 30 percent of all lightning deaths involve people who work outdoors, mostly farmers and ranchers, and over 25 percent involve outdoor recreationists (Uman 1986, Table 3.1). The *Storm Data* statistics are primarily derived from newspaper clippings describing weather-related injury and death. Shearman and Ojala (1999) compared the statistics in *Storm Data Summary* (an adjunct to *Storm Data*) with those from the Michigan Department of Public Health Records and found that the *Storm Data Summary* reported 17 percent fewer lightning mortality cases in Michigan, during 1978 to 1994, but slightly more lightning injuries. Lopez *et al.* (1993) inferred an under-reporting of lightning deaths for the United States by 25 to 30 percent from comparing Colorado medical data with *Storm Data* statistics. In view of the apparent under-reporting of lightning deaths by *Storm Data*, probably about 100 individuals, on average, are killed annually by lightning in the United States.

According to *Storm Data*, about 300 individuals are injured by lightning each year in the United States. If there is an under-reporting by about 40 percent of the lightning injuries compiled by Storm Data, as estimated by Lopez *et al.* (1993) from Colorado medical data, then more than 500 people, on average, are injured each year. Note, however, that Shearman and Ojala (1999), as indicated above, found that lightning injuries in Michigan were slightly over-reported by *Storm Data*. Cherington *et al.* (1999) found from Colorado medical records that the lightning injury to death ratio is about 10. If this is generally true, then there are about 1000 individuals injured annually in the United States.

Curran *et al.* (2000) extracted information from *Storm Data* on lightning-related fatality, injury, and damage in the United States for a period of 36 years. They presented many tables and figures of statistical data by state and by year. From 1959 through 1994 there were 3239 deaths, 9818 injuries, and 19 814 property-damage reports in *Storm Data*. Florida led the nation in both the number of deaths, 345, and the number of injuries, 1178. The largest number of damage reports came from Pennsylvania: 1441. Nationwide, there were 0.42 lightning deaths per million people per year. Wyoming and New Mexico had the highest death/injury rates per million state residents in the United States, 1.47/5.74 and 1.88/13.84, respectively.

All lightning reports analyzed by Curran *et al.* (2000) indicated that there was a maximum occurrence in July. Two-thirds of the casualties occurred between 12 and 16 local standard time (LST). The number of casualties showed a steady increase toward a maximum at 16 LST, followed by a somewhat faster decrease. Curran *et al.* reported that for incidents involving deaths only, 91 percent of the cases had a single fatality, while 68 percent of the injury cases had a single injury. Males were five times more likely to be killed or injured than females. Curran *et al.*, making use of NLDN data (Section 17.5), estimated that there was one fatality for every 86 000 cloud-to-ground flashes.

Lopez and Holle (1998) presented data on the annual number of lightning-caused deaths in the United States for a period of nearly a century. These data are reproduced in Fig. 19.3. Prior to about 1920 not all states consistently

19.1. Statistics

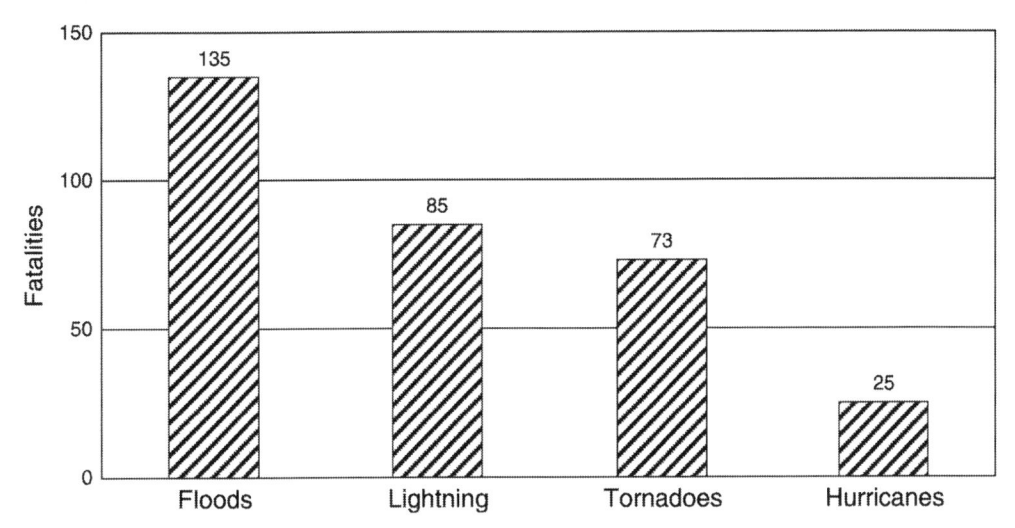

Fig. 19.1. Average annual number of storm-related deaths in the United States, 1966–95. Adapted from *Storm Data*, US National Climatic Data Center, NOAA, Ashville, North Carolina.

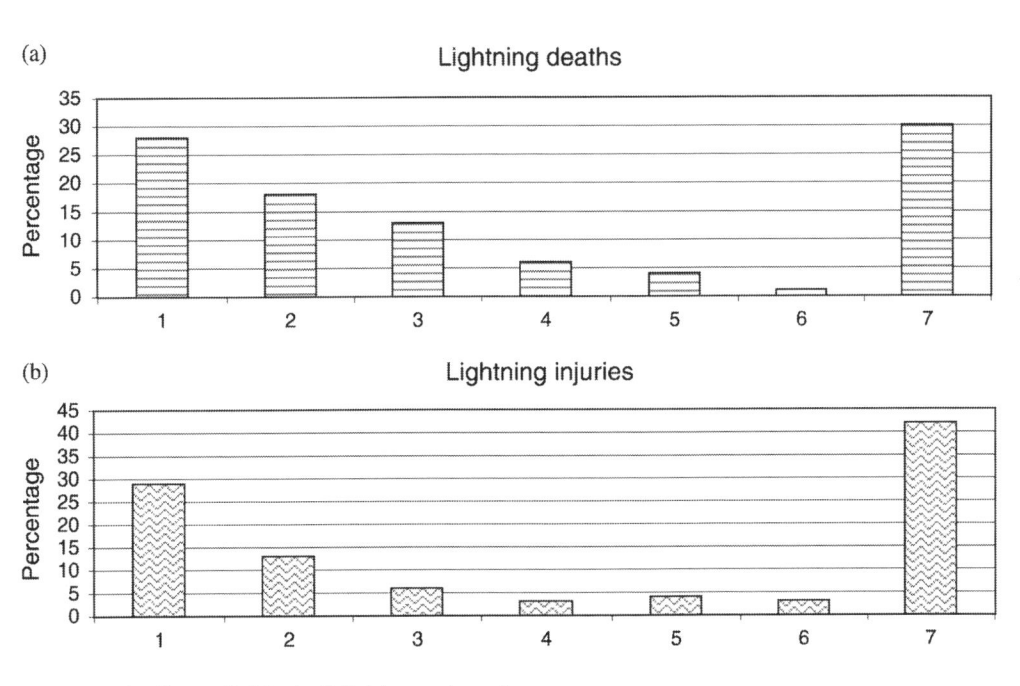

(a) Lightning deaths

(b) Lightning injuries

1. Open fields, ball fields, parks, etc.
2. Under trees.
3. Boating, fishing, water-related, etc.
4. Golf courses.
5. Farming, construction, near heavy machinery, etc.
6. On telephone, radios, electronics, etc.
7. Various other/unknown.

Fig. 19.2. The places of occurrence (1–7) of (a) lightning deaths 1959–96 and (b) lightning injuries 1960–96 in the United States. Adapted from *Storm Data*, US National Climatic Data Center, NOAA, Ashville, North Carolina.

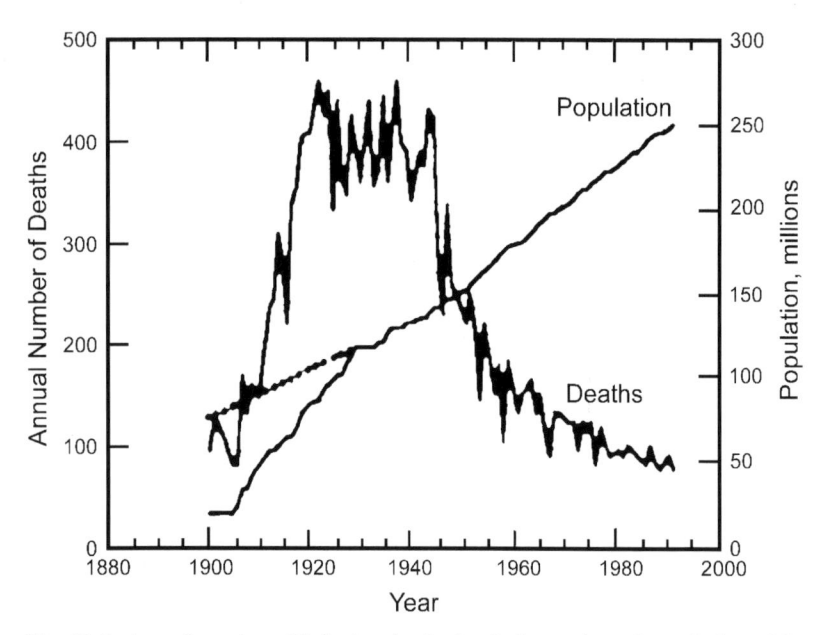

Fig. 19.3. Annual number of lightning deaths (peaked curve), total population of the reporting states (left branch of solid line), and total population of the contiguous USA (dotted line and right branch of solid line). Adapted from Holle and Lopez (1998).

compiled lightning death records, so more deaths occurred than are shown. There has been a steady decline in the annual death rate since about 1930, even though the population has steadily increased, as is also illustrated in Fig. 19.3. According to Lopez and Holle (1998), the decrease in death rate with time parallels a decrease in the percentage of people living in rural areas, that is, the fraction of the population regularly performing farming and ranching duties outdoors. In addition, the introduction of metallic plumbing and wiring into residential and other buildings has helped to provide unintended protection against direct lightning strikes. Finally, improved communication and transportation allow medical aid to arrive faster, and there have been advances in resuscitation techniques and greater public awareness of those techniques, particularly cardiopulmonary resuscitation (CPR).

Lightning death statistics are available for a number of countries besides the United States. For example, data on the number of individuals killed by lightning in the Netherlands between 1910 and 1995 were published by Ten Duis (1998). An annual number of deaths above 20 occurred in the 1920s, but since about 1970 there have been only one to five deaths per year. Elsom (2001) compiled the annual death toll in England and Wales from 1852 to 1999. The average annual number of deaths was 19 between 1852 and 1899, 13 for the period 1900 to 1949, and 5 between 1950 and 1999, 80 to 90 percent of the fatalities during these three periods being male. The average annual number of deaths in the United Kingdom (England, Wales, Scotland, and Northern Ireland) between 1993 and 1999 was 0.05 per million people, eight times less than in the United States (Curran et al. 2000). In all countries for which data are available, there has been an apparent decrease in the annual number of lightning deaths during the course of the twentieth century.

From the available statistics on lightning deaths in various countries and the worldwide distribution of lightning, the authors of this book estimate that the annual death toll for lightning worldwide is a few thousand individuals. The annual worldwide number of lightning injuries is probably five to 10 times the annual number of deaths.

The issue whether there have been injuries or deaths associated with (or due to) ball lightning is considered in Chapter 20.

19.2. Electrical aspects

Berger (1971a, b) was the first to provide a reasonable scenario of what happens, from an electrical point of view, when a human being is struck by lightning. This scenario is reviewed by Golde (1973), from whom the following discussion and the values of the parameters presented are taken; a more detailed description is found in Andrews et al. (1992). Assume that lightning current is flowing between an individual's head and his feet (about five times as many males are killed and injured by lightning as females in the United States and in the United Kingdom, as noted in Section 19.1) and that the current is increasing with time. Since a 1 to 2 m air gap experiences electrical breakdown at an average electric field of about $500\ kV\ m^{-1}$ (see also subsection 18.3.2), although surface breakdown may occur at a lower field value, a flashover across the surface of a 1.8-m-tall individual's body can be expected

19.2. Electrical aspects

when the potential between the individual's head and ground reaches about 900 kV (500 kV m^{-1} × 1.8 m). If the individual's body resistance is assumed to be 700 Ω, the actual value apparently varying between 300 Ω and 5000 Ω depending on skin wetness (Andrews *et al.* 1992), a flashover will occur when the lightning current through the body is about 1.3 kA (900 kV/700 Ω), that is, early in the return-stroke current's rise to a peak value of typically tens of kiloamperes.

Once surface flashover has taken place, the electric field along the arc formed on the surface of the body is about 2 kV m^{-1}, this value being more or less independent of the arc current, as determined from laboratory arc experiments (King 1962). Thus, the surface flashover reduces the potential difference across the body to 3.6 kV (2 kV m^{-1} × 1.8 m), and the current flowing through the person's body is therefore reduced from about 1.3 kA to about 5 A (3.6 kV/700 Ω), a value that will be maintained as long as there is arc current flowing along the surface of the body. The bulk of the lightning current therefore flows outside the human body by virtue of the surface flashover. Thus, for each stroke in a flash, the interior of the human body will be subjected to a short current pulse of near 1 kA with a duration of some microseconds or less, followed by a relatively constant current of 5 A or so maintained for milliseconds or longer in the case of continuing current flow (Section 4.8).

The external flashover mechanism of Berger (1971a, b) is consistent with the typical observations of burn marks on the clothing and the body surface of lightning-strike victims and the melting of their metal jewelry, as further discussed in Section 19.3. The current calculated to flow in the body is certainly sufficient, if directed into the heart or particular parts of the nervous system, to cause either cardiac or respiratory arrest or both (Section 19.3). Apparently, the relatively short duration of the lightning current makes such an outcome less likely than with a typically longer accidental exposure to 50 or 60 Hz alternating current of equivalent amplitude. Before the return-stroke current, the body will first be subjected to the current of an upward-propagating connecting leader that is part of the attachment process, and some individuals may experience only unconnected upward discharges occurring in response to nearby lightning flashes (Section 4.5). The currents associated with such upward discharges could be of the order of a hundred amperes and could last for tens of microseconds.

In addition to direct strikes and unconnected upward discharges, individuals can experience a significant fraction of the total lightning current by way of side flashes and surface arcing. Side flashes and surface arcs are also considered in subsections 7.2.5, 18.3.3, 18.3.4, and 18.4.4.

Many cases have been reported in which groups of individuals who were well separated from each other were simultaneously shocked and injured or killed (e.g., Fahmy *et al.* 1999; Kitagawa 2000). One or two of these individuals may, in fact, have been subjected to a direct lightning strike (for example, branches of the same flash in the case of two individuals) or to a side flash or surface arc from a directly struck individual or object, but the rest were likely to have been victims of step voltages occurring between their feet due to a uniform current flow in the Earth (subsection 18.3.4). Kitagawa (2000) reported on a lightning accident in which 11 mountain hikers were apparently killed by a single surface arc. For the case of uniform current flow, the magnitude of the step voltage can be determined from Eq. 18.6. Assuming a lightning current of 20 kA, a relatively low soil resistivity of 100 Ω m^{-1}, a distance between an individual's feet in the direction of current flow of 0.5 m, and a distance from the lightning strike point to the individual of 10 m, the step voltage, from Eq. 18.6, is 1.5 kV. If the resistance of a current path through both legs is assumed, for example, to be about 1000 Ω (the value will vary with the type of shoe and the degree of shoe and skin wetness), the current up one leg and down the other will reach about 1.5 A. If the resistance of the current path is higher than 1000 Ω, the current through the legs will be less than 1.5 A.

Kitagawa (2000) divided all injuries attributed to step voltage into two categories. In the first category are the injuries caused by uniform current flow in the ground, the type of step voltage described in subsection 18.3.4. Life-threatening currents generally do not reach the heart or critical parts of the nervous system, but an appreciable electric shock may be experienced. Although the lightning current density in the ground decreases inversely with the square of the distance from the source, the lightning currents flowing in the Earth in the vicinity of the strike point to ground can be high enough, particularly for relatively high-resistivity soils (soils with resistivity of order 10^3 Ω m^{-1} or more), to produce widespread electric shocks to nearby groups of people. The second type of injury that has been attributed to step voltage, according to Kitagawa (2000), involves surface arc discharges (subsection 7.2.5, Fig. 7.24; subsection 18.4.4, Fig. 18.7). Figure 7.24 shows surface arcing in a triggered-lightning experiment, and Fig. 18.17 shows the results of surface arcing on a golf course green due to a natural lightning strike to a metal pole. These arcs are similar to side flashes but the arc current flows on the surface of the Earth. Thus injury or death from these surface arcs should more properly be viewed as due to a form of side flash. Encounters with surface arcs can lead to burns, shocks resulting in temporary paralysis, and even death as in the case of the accident involving mountain hikers described by Kitagawa (2000).

19.3. Medical aspects

Reviews of the medical literature on lightning death and injury are given in the book by Andrews et al. (1992) and in review papers, for example, by Cooper (1980, 1995a) and Cooper and Andrews (1995). A complete bibliography on the medical aspects of lightning injury and death is found at http://home.pacific.net.au/~candy.hub/rsch.htm. Cooper and Andrews (1995) listed four primary mechanisms of lightning-related death and injury: direct strike, side flash, step voltage, and blunt trauma. To this list could be added surface arc discharges, which as noted in the previous section can be viewed as a form of side flash, and "touch voltage", the voltage encountered when touching a metallic object such as a wire fence that has been raised in voltage by lightning. The electrical mechanisms of injury or death from a direct strike and of injury from a step voltage have been discussed in the previous section and in subsection 18.3.4. A discussion of side flashes and their suppression by proper lightning protection is found in subsections 18.3.3 and 18.3.4. There are many medical reports of side flashes occurring indoors from metallic objects such as plumbing to individuals (e.g., Edwards 1925; Spencer 1934; Morgan et al. 1954). There is a considerable literature on side flashes from wired telephones (not portable or cellular phones), as we shall discuss later in this section. Side flashes to individuals also occur outdoors, from fences or other long conductive objects that are hit by lightning some distance away from the individuals. Furthermore, as noted in Section 19.2, side flashes can occur from person to person when several people are standing close together (e.g., Lynch and Shorthouse 1949; Arden et al. 1956; Buechner et al. 1961).

Injuries apparently also occur from the blunt trauma related to the so-called "sledgehammer effect" (Spencer 1934). It is not uncommon for a lightning strike victim to be thrown many meters, apparently by opisthotonic contractions caused by the lightning strike; furthermore, individuals are injured indirectly by lightning, for example, by falling or being thrown off a horse, being hit by flying tree bark or falling masonry, or by smoke inhalation from a lightning-caused fire (Bergstrom et al. 1974; Panse 1975; Auerbach 1980). Lightning victims can thereby incur a variety of fractures, including those of the skull, ribs, extremities, and spine (Clark et al. 1872; Pritchard 1934; Barthlome et al. 1975; Strasser et al. 1977; Baker 1978). Individuals may also be injured by the high pressure of the acoustic wave produced by nearby lightning (Chapter 11).

Some of the most common consequences of lightning injury are listed in Table 19.1. According to Cooper and Andrews (1995), for reasons of medical prognosis lightning-strike victims can generally be placed in one of three categories, which they label (i) minor, (ii) moderate, and (iii) severe. We consider these three categories now.

Table 19.1. *Signs and symptoms of lightning injury. Adapted from Cooper (1980)*

Immediate
   ventricular standstill
   chest pain, muscle aches
   neurologic signs
      seizures
      deafness
      confusion
      blindness
   contusions from shock wave
   tympanic membrane rupture
Late
   cutaneous burns
   cataracts
   myoglobinuria and hemoglobinuria (rare)

(i) In the *minor* category are patients who awake and report a feeling of having been hit on the head or having been in an explosion. They may or may not remember the lightning or thunder. They are often confused and amnesic, with temporary deafness, blindness, or unconsciousness (Cannel 1979). They seldom demonstrate any cutaneous burns or paralysis, but may complain of paresthesia, muscular pain, and confusion lasting from hours to days. Vital signs are usually stable, although patients occasionally demonstrate transient mild hypertension (Taussig 1968; Hanson and McIwraith 1973). Recovery is usually gradual and may or may not be complete.

(ii) Patients in the *moderate* category may be disoriented, combative, or comatose. They frequently exhibit motor paralysis, with mottling of the skin and diminished or absent pulses, particularly of the lower extremities. Occasionally, these individuals have suffered temporary cardiopulmonary standstill, although it is seldom documented (Spencer 1934). Spontaneous recovery of a pulse is attributed to the heart's inherent automaticity. However, the respiratory arrest that often occurs with lightning injury may be prolonged and lead to a secondary cardiac arrest from hypoxia. Seizures may also occur. First- and second-degree burns may not be immediately prominent but may evolve over the first several hours. Third-degree burns occur rarely. Patients in this category often have at least one tympanic membrane ruptured as a result of the lightning-produced pressure wave. Tympanic membrane rupture or hemotympanum may also indicate the presence of a basilar skull fracture. Patients may exhibit long-lasting sleep disorders, irritability, difficulty with fine psychomotor functions, chronic pain, and general weakness. A few cases of atrophic spinal paralysis have been reported (Buechner and Rothbaum 1961).

19.3. Medical aspects

(iii) In the *severe* category, patients exhibit cardiac arrest with either ventricular standstill or fibrillation. Cardiac resuscitation may not be successful if the patient has suffered a long period of cardiac and central nervous system ischemia. Direct brain damage may occur from the lightning-strike current or the high pressure of its acoustic wave. Tympanic membrane rupture with hemotympanum and cerebrospinal fluid otorrhea are common. Patients with other signs of blunt injury are likely to be the victim of a direct strike, although sometimes no burns are noted. The prognosis is usually poor in this group, but this is usually due as much to a delay in cardiopulmonary resuscitation as to the severity of the direct lightning damage.

It is thought that the only immediate cause of death in a lightning victim is cardiopulmonary arrest. In fact, according to Cooper (1980), a victim is highly unlikely to die unless a cardiopulmonary arrest is suffered as an immediate effect of the strike. Until recently, nearly 75 percent of those who suffered cardiopulmonary arrest from lightning injuries died, many because cardiopulmonary resuscitation was not attempted. The second major cause of lightning injury and death is central nervous system damage. When current traverses the brain, there can be coagulation of the brain substance, formation of epidural and subdural hematomas, paralysis of the respiratory center, and intraventricular hemorrhage (Morgan et al. 1954; Strasser et al. 1977). Autopsy findings include extravasation of blood in the meninges and cerebral tissues, petechiae, dural tears, extravasations of blood beneath the scalp, and skull fractures (Clark and Brigham 1872; Knaggs 1894; Dunscombe-Hannibal 1900; Critchley 1934; Crawford and Hoopes 1941; Skan 1949; Buechner and Rothbaum 1961).

Much of the general public believes that a direct-lightning-strike victim will be seriously burned internally and externally (Cannel 1979). This is not the case, as discussed in Section 19.2. The relatively short duration of the overall lightning current, typically a fraction of a second or less, saves all but a few victims from more than minor burns. While extensive third- and fourth-degree burns may occur, they are very rare (Clark and Brigham 1872; Elwell 1934; Skan 1949). Overall, the burns observed in lightning accidents may be divided into five categories: linear burns, punctate burns, full-thickness burns, feathering or flowers, and thermal burns from clothing or heated metal.

Explosive effects typically occur at the feet, where boots or socks may be torn or blown off. Presumably this is due to the high pressures associated with the vaporization of the moisture confined within the shoes.

Many types of eye damage have been reported (e.g., Cannel 1979). Direct thermal or electrical damage, intense light, contusion from the high-pressure acoustic wave, and combinations of these factors contribute to injuries which may occur in as many as 55 percent of those struck by lightning (Castren and Kytila 1963).

Temporary deafness is not uncommon (West 1955; Gabrielli 1965; Kleinot et al. 1966; Barthlome et al. 1975). It has been postulated that the intense sound associated with close lightning may be responsible for sensorineural loss (Bergstrom et al. 1974). At least 50 percent of lightning victims incur the rupture of one or both tympanic membranes, which may be the result of the lightning pressure wave or, perhaps, the consequence of a basilar skull fracture (Cooper 1980). Direct nerve damage from the lightning may cause facial palsies (Bergstrom et al. 1974). A detailed examination of lightning's effect on the auditory system of four individuals has been presented by Jones et al. (1991).

Only a few percent of lightning deaths and injuries occur indoors (see Fig. 19.2). These are associated with voltages from lightning strikes on or near outdoor conductors such as telephone wiring, metallic plumbing, or power system wiring, which enter a structure and result in side flashing to the individual. There is considerable literature on telephone-related injury and death, one of the most common indoor lightning events. Andrews et al. (1989) estimated that in Australia about 60 people report telephone injuries annually in a population of 16 million. They reported no telephone-related deaths, although one to three such telephone-related lightning deaths occur annually in the United States. Andrews et al. (1989) tabulated victims' observations and subsequent medical histories for telephone-injury reports received for the period 1980–5. About 35 percent of the sample reported hearing a loud noise and feeling an electric shock. Twenty percent reported sparks emanating from the telephone, and 40 percent reported being physically thrown. A thorough examination of one case of telephone-related injury, where the patient sustained paralysis, sensory symptoms, ostological disturbance, and pathognomonic dendriform cutaneous marks (fern-like patterns on the skin also known as Lichtenberg figures), is presented in Johnstone et al. (1986). According to Elsom (2001), in the United Kingdom from 1993 to 1999 there were an average of about 25 lightning incidents inside buildings per year, one-quarter of the indoor incidents involving the telephone. More information on the interaction of lightning with telephones and telecommunication systems is found in subsection 18.4.5.

As discussed in Section 19.2, humans are not often killed by step voltages, apparently because the lightning current does not easily find its way to life-critical areas of the human body. Interestingly, four-legged animals, particularly those who happened to be seeking shelter from the rain under struck trees, can have fatal current forced through their hearts and through critical parts of their nervous systems by the step voltages that appear between their front and hind legs. There are many examples of groups of animals

Fig. 19.4. A photograph of a group lightning kill of 23 Holstein heifers near a fence in Island Park, Idaho. Courtesy of Ruth and Kent Bateman.

(e.g., cows, horses, sheep, elk) being killed simultaneously, often without visible lightning effects on their bodies, typically when they have taken shelter under a tree or have been near a fence. A photograph of one such a group kill is found in Fig. 19.4. Animals can also be killed by direct strikes, side flashes, or surface arcs.

19.4. Personal safety

In Holle et al. (1999), a group of lightning safety experts has summarized their consensus views of personal safety from lightning as follows.

No place is absolutely safe from the lightning threat; however, some places are safer than others.

- Large enclosed structures (substantially constructed buildings) tend to be much safer than smaller or open structures. The risk for lightning injury depends on whether the structure incorporates lightning protection, construction materials used, and the size of the structure.
- In general, fully enclosed metal vehicles such as cars, trucks, buses, vans, fully enclosed farm vehicles, etc., with the windows rolled up provide good shelter from lightning. Avoid contact with metal or conducting surfaces outside or inside the vehicle.
- *Avoid* being in or near high places and open fields, isolated trees, unprotected gazebos, rain or picnic shelters, baseball dugouts, communications towers, flagpoles, light poles, bleachers (metal or wood), metal fences, convertibles, golf carts, and water (ocean, lakes, swimming pools, rivers, etc.) When inside a building *avoid* use of the telephone, taking a shower, washing your hands, doing dishes, or any contact with conductive surfaces with exposure to the outside such as metal door or window frames, electrical wiring, telephone wiring, cable TV wiring, plumbing, etc.

Where groups of people are involved, an action plan for getting to a lightning-safe place must be made in advance by the responsible individuals.

Many lightning strike victims can survive their encounter with lightning, especially with timely medical treatment, including cardiopulmonary resuscitation (CPR) (Andrews et al. 1992; Cooper 1995a, b; Cooper and Andrews 1995). Individuals struck by lightning do not carry a charge so it is safe to touch them, and indeed it is imperative to do so in order to administer CPR.

The proper lightning protection of structures within which individuals may seek safety is discussed in Section 18.3.

19.5. Summary

Lightning is the second most efficient storm-related killer, floods being the first. Lightning can injure or kill by way of direct strikes, side flashes, surface arcs, step voltages, and touch voltages. Although the lightning death rate in industrialized countries has decreased during the course of the twentieth century by a factor 5 or more, there are still about 100 individuals killed annually in the United States and perhaps 1000 injured. Worldwide, we estimate that annually there are a few thousand lightning deaths and tens of thousands of lightning injuries. There is considerable medical literature on the effects of lightning on humans and on the treatment of lightning strike victims. Contrary to what is thought by the general public, more individuals involved in lightning accidents survive than die, and direct-lightning-strike victims are rarely burned badly. Following relatively simple personal safety rules can considerably reduce the chances of being killed or injured by lightning.

References and bibliography for Chapter 19

Abt, J.L. 1985. The pupillary responses after being struck by lightning. *JAMA* **254**: 23 312.

Akahane, T. *et al.* 1983. Lightning injury: report of two cases. *Burns* **10**: 45.

Alcock, R. *et al.* 1949. Abdominal injury from lightning. *Lancet* **1**: 823–5.

Alexandropoulos, J. 1969. Electric cataract after being struck by lightning. *Opth. Chron. Athens* **61**: 166–71.

Amy, B.W., Mcmanus, W.F., Goodwin, C.W., and Pruitt, B.A. 1985. Lightning injury with survival in five patients. *J. Am. Med. Assoc.* **253**(2): 243–5.

Anderson, R.B. 2001. Does a fifth mechanism exist to explain lightning injuries? *IEEE Eng. Med. Biol.* January/February: 105–13.

Andrews, C.J. 1988. Myocardial infarction after electrocution. *Med. J. Aust.* **149**: 343.

Andrews, C.J. 1992. Telephone-related lightning injury. *Med. J. Austr.* **157**: 823–6.

Andrews, C.J., and Darveniza, M. 1989. Telephone-mediated lightning injury: an Australian survey. *J. Trauma* **29**: 665–71.

Andrews, C.J., Darveniza, M., and Mackerras, D. 1989. Lightning – a review of clinical aspects, pathophysiology, and treatment. *Adv. Trauma* **4**: 241–88.

Andrews, C.J., Cooper, M.A., Darveniza, M., and Mackerras, D. 1992. *Lightning Injuries: Electrical, Medical and Legal Aspects*, 195 pp., Boca Raton, Florida: CRC Press.

Appleberg, D.B., Masters, F.W., and Robinson, D.W. 1974. Pathophysiology and treatment of lightning injuries. *J. Trauma* **14**: 453–60.

Arden, G.P., Harrison, S.H., Lister, J., and Maudsley, R. 1956. Lightning accident at Ascot. *Br. Med. J.* **1**: 1450–3.

Ashby, H.T. 1919. Loss of hair due to lightning. *Br. Med. J.* **1**: 307.

Auerbach, P. 1980. Lightning strike. *Top Emerg. Med.* **2**: 129–35.

Baker, R. 1978. Paraplegia as a result of lightning injury. *Br. Med. J.* **4**: 1464–5.

Barthlome, C.W., Jacoby, W.D., and Ramchand, S.C. 1975. Cutaneous manifestations of lightning injury. *Arch. Dermatol.* **111**: 1466–8.

Bellucci, R.J. 1983. Traumatic injuries of the middle ear. *Otolar. Clin. N. Am.* **16**(3): 633–50.

Bennett, B.L. 1997. A model lightning safety policy for athletics. *J. Athletic Training* **32**: 251–3.

Bennett, B.L., Holle, R.L., and López, R.E. 1997. Lightning safety. In *1998–99 NCAA Sports Medicine Handbook*, 11th edition, ed. M.V. Earle, pp. 12–14, National Collegiate Athletic Association.

Bennison, E. *et al.* 1973. Lightning surges in open wire, coaxial and paired cables. *IEEE Trans. Comm.* **COM-21**(10): 1136.

Berger, K. 1971a. Blitzforschung und Personen-Blitzschutz. *ETZ(A)*: **92**: 508–11.

Berger, K. 1971b. Zum Problem des Personenblitzschutzes. *Bull. SEV* **62**: 397–99.

Bergstrom, L. 1975. Some pathologies of sensory and neural hearing loss. *Can. J. Otolaryngol.* **4**(2): Supplement 2.

Bergstrom, L., Neblett, L., Sando, I., Hemenway, W., and Harrison, G.D. 1974. The lightning damaged car. *Arch. Otolaryngol.* **100**: 117–21.

Bernstein, T. 1973. Effects of electricity and lightning on man and animals. *J. For. Sco.* **18**(1).

Bertelson, S. 1973. Traumatic facial palsy (quoted in A. Richards). *Proc. Roy. Soc. Med.* **66**: 28–9.

Black, R.C., Clark, G.M., and Patrick, J.F. 1981. Current distribution measurements within the human cochlea. *IEEE Trans. BME* **28**(10): 721–5.

Blake-Pritchard, E.A. 1934. Changes in the central nervous system due to electrocution. *Lancet* **1**: 1163–7.

Bondarenko, E.J. 1975. Effects of lightning on subscribers services. Telecom Australia, Perth, New South Wales, Res. Lab. Report 6990.

Bondarenko, E.J. 1976. Effects of lightning on subscribers services – Sydney, NSW. Telecom Australia, Res. Lab. Report 7078.

Bondarenko, E.J. 1980. Effects of lightning on customer services, Minyip Victoria, Canberra Act and Brisbane Qld. App. Sci. Brisbane Res. Lab., Paper 0018.

Bridges, J.E. (ed.) 1985. *Electrical Shock Safety Criteria*. New York: Pergamon.

Buechner, H.A., and Rothbaum, J.C. 1961. Lightning stroke injury – a report of multiple casualties from a single lightning bolt. *Milit. Med.* **126**: 775–62.

Burda, C.D. 1966. Electrocardiographic changes in lightning stroke. *Am. Heart J.* **72**(4): 521–4.

Butler, E.D., and Grant, T.D. 1977. Electrical injuries, with special reference to the upper extremities. *Am. J. Surg.* **134**: 95–101.

Callagan, J. 1999. Medical aspects of lightning injuries. *J. Meteorol.* **24**(242): 280–4.

Campo, R.V. 1984. Lightning induced macular hole, *Am. J. Oph.* **97**(6): 792–4.

Cannel, H. 1979. Struck by lightning – the effects upon the men and the ships of HM Navy. *J. R. Nav. Med. Serv.* **65**: 165–70.

Castren, J.A., and Kytila, J. 1963. Eye symptoms caused by lightning. *ACTA Opthalmol.* **41**: 139–43.

Chan, Y.F., and Sivasamboo, R. 1972. Lightning accidents in pregnancy. *J. Obstet. Gynecol. Br. Comm.* **79**: 761–2.

Changnon, S.A. Damaging thunderstorm activity in the United States. *Bull. Am.* **82**(4): 597–608.

Chao, T.C. *et al.* 1981. A study of lightning deaths in Singapore. *Singapore Med. J.* **22**(3): 150–7.

Charcot, J.M. 1889. Des accidents neureux provoqués par la fondre. *Bull. Med. (Paris)* **3**: 1323–6.

Cheng, T.O. 1988. Lightning injury causing prolongation of the QT interval. *Postgrad. Med. J.* **64**: 335.

Cherington, M. 2001. Lightning injuries in sports. Situations to avoid. *Sports Medicine* **31**: 301–8.

Cherington, M. and Vervalin, C. 1990. Lightning injuries – who is at greatest risk? *Physician Sportsmed.* **18**: 58–61.

Cherington, M., Watchel, H., and Yarnell, P.R. 1998. Could lightning injury be magnetically induced? *Lancet* **351**: 1788.

Cherington, M., Yarnell, P., and Lammereste, D. 1992. Lightning strikes: nature of neurological damage in patients evaluated in hospital emergency departments, *Ann. Emer. Med.* **21**(5): 575/133–578/136.

Cherington, M., Krider, E.P., Yarnell, P.R., and Breed, D.W. 1997. A bolt from the blue. Lightning strike to the head. *Neurology* **48**: 683–6.

Cherington, M., Breed, D.W., Yarnell, P.R., and Smith, W.E. 1998. Lightning injuries during snowy conditions. *British J. Sports Med.* **32**: 333–5.

Cherington, M., Walker, J., Boyson, M., Glancy, R., Hedegaard, H., and Clark, S. 1999. Closing the gap on the actual numbers of lightning casualties and deaths. In *Proc. 11th Conf. on Applied Climatology, Dallas, Texas*, pp. 379–80, American Meteorological Society.

Chia, B.L. 1981. Electrocardiographic abnormalities and congestive cardiac failure due to lightning stroke. *Cardiology* **68**: 49–53.

Christensen, J.A., Sherman, R.T., Balis, G.A., *et al.* 1980. Delayed neurological injury secondary to high-voltage current, with recovery. *J. Trauma* **20**: 166–8.

Clark, R.O., and Brigham, J.K. 1872. Death from lightning. *Lancet* **1**: 77.

Clore, E.R., and House, M.A. 1987. Prevention and treatment of lightning injuries. *Nurs. Pract.* **12**(12): 37–45.

Coates, L., Blong, R., and Siciliano, F. 1993. Lightning fatalities in Australia, 1824–1991. *Nat. Hazards* **8**: 217–33.

Coleman, T. 1974. Death by lightning. *Gen. Pract.* **37**: 81.

Coleman, T. 1969. Death by lightning. *PA Med.* **72**: 56–8.

Connole, J.V. 1935. Lightning and electric cataract. *PA Med. J.* **38**: 939.

Cooper, M.A. 1980. Lightning injuries, prognostic signs for death. *Ann, Emerg. Med.* **9**: 134–8.

Cooper, M.A. 1983a. Lightning strikes. *Ann Emerg. Med.* **12**: 113.

Cooper, M.A. 1983b. Lightning injuries. *Em. Med. Clin. N. Am.* **1**(3): 639.

Cooper, M.A. 1983c. Lightning injuries. In *Management of Wilderness and Environmental Emergencies*, eds. P.S. Auerbach and E.C. Geehr, Chap. 18, pp. 500–21, New York: McMillan.

Cooper, M.A. 1984. Electrical and lightning injuries. *Em. Med. Clin. N. Am.* **2**(3): 489.

Cooper, M.A. 1989. Lightning injuries. In *Management of Wilderness and Environmental Emergencies*, eds. P.S. Auerbach and E.C. Geehr, pp. 173–93, St. Louis: C.V. Mosby.

Cooper, M.A. 1992. Treatment of lightning injury. In *Lightning Injuries: Electrical, Medical, and Legal Aspects*, eds. Andrews, C.J., Cooper, M.A., Darveniza, M., and Mackerras, D., pp. 115–40, Boca Raton, Florida: CRC Press.

Cooper, M.A. 1995a. Emergent care of lightning and electrical injuries. *Sem. Neurol.* **15**: 268–78.

Cooper M.A. 1995b. Myths, miracles, and mirages. *Sem. Neurol.* **15**: 358–61.

Cooper, M.A. 2002. A fifth mechanism of lightning injury. *Academic Emergency Medicine* **9**: 172–4.

Cooper, M.A., and Andrews, C.J. 1995. Lightning injuries. In *Wilderness Medicine*, 3rd edition, ed. P. Auerbach, pp. 261–89, Mosby: St. Louis (see also 4th edition, pp. 73–110).

Craig, S.R. 1986. When lightning strikes. Pathophysiology and treatment of lightning injuries. *Postgrad Med* **79**(4): 109–12, 121–4.

Crawford, A.S., and Hoopes, B.F. 1941. The surgical aspects of lightning stroke. *Surgery* **9**: 80.

Critchley, M. 1932. The effects of lightning with especial reference to the nervous system. *Bristol Med. Chir. J.* **49**: 285.

Critchley, M. 1934. Neurological effects of lightning and of electricity. *Lancet* **1**: 68–72.

Critchley, M. 1935. Injuries from electricity and lightning. *Br. Med. J.* **2**: 1217.

Cudennec, Y.F., De Rotalier, P., and Aubert, C. 1986. Fulguration d'orielle. *Ann. Othlaryngol. (Paris)* **103**: 343–9.

Curran, E.B., and Holle, R.L. 1997. Lightning fatalities, injuries and damage reports in the United States, 1959–1994, NOAA Technical Memorandum NWS SR-193, NOAA scientific services division, Fort Worth, Texas.

Curran, E.B., Holle, R.L., and Lopez, R.E. 2000. Lightning casualties and damages in the United States from 1959 to 1994. *J. Climate* **13**: 3448–3464.

Currens, J.H. 1945. Arterial spasm and transient paralysis from lightning striking an aeroplane. *Aerospace Med. Fmly Aviation Med.* **16**: 275.

Cwinn, A.A. *et al.* 1985. Lightning injuries. *J. Emerg. Med.* **2**(5): 379–88.

Dalziel, C.F. 1953. A study of the hazards of impulse currents. *Trans. IEE* **72**: 1032.

Dalziel, C.F. 1968. Re-evaluation of lethal electric currents. *IEEE Trans. IGA* **4**(5): 467.

Dalziel, C.F. 1972. Electric shock hazard. *IEEE Spectrum*, February, p. 41.

Dalziel, C.F., *et al.* 1969. Lethal electric currents. *IEEE Spectrum*, February, p. 44.

Darling, Pedlow, and Frier 1905. Lightning stroke. *Br. Med. J.* 1522–3.

Davidson, G.S., and Deck, J.H. 1988. Delayed myelopathy following lightning stroke: a demyelinating process. *ACTA Neuropathol.* **77**: 104–8.

Day, P. 1980. List of reported customers injuries caused by lightning. Telecom Australia Data, 26/5/80 (quoted in AHMED).

Dilip, K.G-R. 1979. Fetal death at term due to lightning. *Am. J. Obstet. Gynecol.* **134**: 103–5.

Dill, A.V. 1942. Notes on a case of death by lightning. *Br. Med. J.* 10 October, p. 426.

Dinakaran, S., Desai, S.P., and Elsom, D.M. 1998. Telephone-mediated lightning injury causing cataract. *Injury* **29**: 645–6.

Divakov, G.M. 1966. ECG changes in persons struck by lightning (title trans. from Russian). pp. 95–6. Moscow: Klin. Meditsima.

Dollinger, S.P., O'Donnell, J.P., and Stanley, A.A. 1984. Lightning strike disaster: effects on children's fears and worries. *J. Cons. Clin. Psych.* **52**: 1028–38.

Dowling, J., Byrne, K., Barry, O.C.D., Long, J.P., and Lennon, F. 1984. Lightning injury. *Irish Med. J.* **77**(8): 249–50.

Du Pasquier, G., and Freeman, J. 1986. Le foudroiement. *Ann. Fr. Anesth. Reanim.* **4**: 601–4.

Du Toit, J.S. 1927. A case of bilateral cataract caused by lightning. *J. Med. Soc. S. Afr.* **1**: 503.

Duclos, P.J., and Sanderson, P.J. 1990. An epidemiological description of lightning-related deaths in the United States. *Int. J. Epidemiol.* **19**: 673–9.

Duclos, P.J., Sanderson, L.M., and Klontz, K.C. 1990. Lightning-related mortality in Florida. *Public Health Rep.* **105**: 276–82.

Dudley, S.C., *et al.* 1997. The human discharge chain. *Am. J. Phys.* **65**: 553.

Dunscombe-Hannibal, O. 1900. Accidents and injuries caused by lightning. *Br. Med. J.*, May 12, pp. 153–5.

Dupasquier, G., and Freemen, J. 1986. Cardiac injury due to lightning. *Ann. Fr. Anesth.* **5**: 601–4.

Dutt, J. 1979. Overvoltages in subscribers lines. *Electrical Comm.* **54**(2): 115.

Eaton, R.D.P. 1983. Lightning injury. *Can. Med. Assoc. J.* **128**: 893–4.

Eber, B., Himmel, G., Schubert, B., *et al.* 1989. Myokardiale schaedigung nach blitzschlag. *Z. Kardiol.* **78**: 402–4.

Edwards, W. 1925. Injury due to lightning striking a wireless aerial. *Br. Med. J.*, August 25, p. 294.

Ehsan, M., Waxman, J., and Finley, J.M. 1981. Delayed gangrene after lightning strike. *AFP* **24**(5): 117–9.

Ekoe, J.M., Cunningham, M., Jaques, O., *et al.* 1985. Disseminated intravascular coagulation and acute myocardial necrosis caused by lightning. *Int. Care. Med.* **11**: 160–2.

Elsom, D.M. 1989. Learn to live with lightning. *New Scientist* **122**: 54–8.

Elsom, D.M. 1993. Deaths caused by lightning in England and Wales, 1852–1990. *Weather* **48**: 83–90.

Elsom, D.M. 1994. Injuries and deaths caused by lightning in the United Kingdom: a new national database. *J. Meteor.* **19**: 322–7.

Elsom, D.M. 1996. Surviving being struck by lightning: a preliminary assessment of the risk of lightning injuries and death in the British Isles. *J. Meteor.* **21**(209): 197–206.

Elsom, D.M. 2001. Deaths and injuries caused by lightning in the United Kingdom: analyses of two databases. *Atmos. Res.* **56**: 325–34.

Elsom, D.M., Meaden, G.T., Reynolds, D.J., Rowe, M.W., and Webb, J.D.C. 2001. Advances in tornado and storm research in the United Kingdom and Europe: the role of the tornado and storm research organisation. *Atmos. Res.* **56**: 19–29.

Elwell, E.G. 1934. Non fatal lightning burns. *Br. Med. J.*, October 27, p. 771.

Eng, L.I.L., and Sinnadurai, C. 1965. Syndrome of erythremia de guglielmo after lightning injury with autoimmune antibodies and terminating in acute monocytic leukemia. *Blood* **25**: 845–8.

Factor, S.M., and Cho, S. 1985. Smooth muscle contraction bands in the media of coronary arteries: a postmortem marker of antemortem coronary spasm. *JACC* **6**(6): 1329–37.

Fahmy, F.S., Brisden, M.D., Smith, J., and Frame, J.D. 1999. Lightning: the multisystem group injuries. *J. Trauma: Injury, Infection, and Critical Care* **46**: 937–40.

Farrell, D.F., and Starr, A. 1968. Delayed neurological sequelae of electrical injuries. *Neurology* **18**: 601–6.

Ferrett, R.L., and Ojala, C.F. 1992. The lightning hazard in Michigan. *Michigan Academician* **24**: 427–41.

Flannery, D.B., and Wiles, M. 1982. Follow-up of a survivor of intrauterine lightning exposure. *Am. J. Obstet. Gynecol.* **142**(2): 238–239.

Flisowski, Z., and Mazzetti, C. 1985. An approximate method of assessment of the electric shock hazard by lightning strike. In *Proc. 18th Int. Conf. on Lightning, Munich*.

Frauenfelder, F., *et al.* 1972. Electric cataract. *Arch. Ophthalmol.* **87**: 179.

Frauenfelder, F., and Hanna, C. 1972. Electric cataracts: I. Sequential changes, unusual and prognostic finding. *Arch Opthalmol.* **87**: 179–83.

Frayne, J., and Gilligan, B.S. 1988. Neurological sequelae of lightning. *Clin. Exp. Neurol.* **24**: 195–200.

Fulde, G.W.O., and Marsden, S.J. 1990. Lightning strikes. *Med. J. Aust.* **153**: 496–8.

Funiak, S., Hendrich, F., Drimal, J., *et al.* 1984. Srdcovy infarkt pri poraneni bleskom. *Vnitri Lekarstvi* **30**: 1118–21.

Gabrielli, L. 1965. Unusual clinical picture of intermittent deafness in a subject struck by lightning. *Otorhinolaryngology* **31**: 79–90.

Gallagher, T.J. 1988. Lightning on the links. *Weatherwise* **41**: 212–13.

Gathier, J.C. 1960. Neurological changes in a soldier struck by lightning. *Psych. Neurol. Neurochirurg* **63**: 125–38.

Gem, W. 1913. A case of lightning stroke, followed by recovery. *Lancet* **2**: 88.

Girard, P.F., *et al.* 1954. *Lyon Med.* **191**: 209.

Golde, R.H. 1969. A plain man's guide to lightning. *Electronics Power* **15**: 84–6.

Golde, 1973. *Lightning Protection*, London: Edward Arnold.

Golde, R.H., and Lee, W.R. 1976. Death by lightning. *Proc. IEEE* **123**: 1163–79.

Gourbière, E., Lambrozo, J., Virenque, C., Menthonnex, P., and Cabane, J. 1997. Lightning injured people in France: the first French national inquiry with regard to the strike of people – objectives, methods, first results. In *Proc. Conf. on Lightning and Mountains 97, Chamonix Mont Blanc, France*, Socièté des Electriciens et des Electroniciens and Fédèration des Clubs Alpins.

Guha-Ray, D.K. 1979. Fetal death at term due to lightning. *Am. J. Obstet. Gynecol.* **134**: 103–5.

Gupta, G.B., Gupta, S.R., Somani, P.N., and Argrawal, B.V. 1988. Atrial fibrillation with inferior wall myocardial ischaemia following lightning. *J. Assoc. Phys. India* **36**: 354–5.

Gusakov, A.D., and Zaiko, N.G. 1979. Lesion of the human middle ear by a stroke of lightning. *Zh. Ushn. Nos. Gorl. Bolezn.* **5**: 67–8.

Hammer, Z.F., Klett, U.I., and Wegener, R. 1997. Myocardial injury due to lightning. *Int. J. Legal Med.* **110**: 326–8.

Hann, J. 1957. Symptomatic psychosis of the schizophrenic type after injury by lightning. *Nervenarzt* **28**(3): 127.

Hanna, C., and Frauenfelder, F.T. 1972. Electric cataracts: II. Ultrastructural lens changes. *Arch. Ophthalmol.* **87**: 184–91.

Hanson, G.C., and McIwraith, G.R. 1973. Lightning injury: two case histories and a review of management. *Br. Med. J.* **4**: 271–4.

Haraldsson, P.O., and Bergstedt, M. 1983. Unconscious and persistent tinnitus caused by lightning injury to the ear during telephoning. *Lakartidningen* **80**: 2024.

Hasse, P., and Wiesinger, J. 1982. *Handbuch Für Blitzchutz and Erdung*, Munich, Germany: Richard Pflaum.

Harrison, S. 1955. Lightning burns. *Br. J. Plast. Surg.* **8**: 154–7.

Harwood, S.J., Catrow, P.G., and Cole, G.W. 1978. Creative phosphokinase isoenzyme fractions in the serum of a patient struck by lightning. *Arch. Intern. Med.* **138**: 645–6.

Heffernan, D. 1877. Autopsy in a case of death by lightning. *Lancet* **2**: 266.

Hegner, C.F. 1917. Lightning – some of its effects. *Ann. Surg.* **65**: 401–9.

Hocking, B., and Andrews, C.J. 1989. Lightning injury and fractals. *Med. J. Aust.* **150**: 409–10.

Holle, R.L., and Lopez, R.E. 1998. Lightning – impacts and safety. *Bull. World Meteor. Soc.* **47**: 148–55.

Holle, R.L., López, R.E., Ortiz, R., Paxton, C.H., Decker, D.M., and Smith, D.L. 1993. The local meteorological environment of lightning casualties in central Florida. In *Proc. 17th Conf. on Severe Local Storms and Atmospheric Electricity, St. Louis, Minnesota*, pp. 779–84, Am. Meteor. Soc.

Holle, R.L., López, R.E., Howard, K.W., Vavrek, J., and Allsopp, J. 1995. Safety in the presence of lightning. *Sem. Neurol.* **15**: 375–80.

Holle, R.L., Lopez, R.E., Arnold, L.J., and Endries, J. 1996. Insured lightning-caused property damage in three western states. *J. Appl. Meteorol.* **35**: 1344–51.

Holle, R.L., Lopez, R.E., Howard, K.W., Cummins, K.L., Malone, M.D., and Krider, E.P. 1997. An isolated winter cloud-to-ground lightning flash causing damage and injury in Connecticut. *Bull. Am. Meteor. Soc.* **78**: 437–41.

Holle, R.L., López, R.E., and Zimmerman, C. 1999. Updated recommendations for lightning safety. *Bull. Am. Meteor. Soc.* **80**: 2035–41.

Holle, R.L., López, R.E., and Navarro, B.C. 2001. US lightning deaths, injuries, and damages in the 1890s compared to 1990s. NOAA Technical Memorandum OAR NSSL-106, National Severe Storms Laboratory, Norman, Oklahoma.

Hornstein, R.A. 1962. Canadian lightning deaths and damage. Meteorological Branch, Dept Transport Canada, CIR-3719, TEC-423, 5 pp. (Available from Meteorological Service of Canada, 4905 Dufferin St, Downsview, ON O3H 5T4, Canada.)

Howard, J.R. 1966. Effects of lightning and simulated lightning on tissues of animals. Ph.D. thesis, Iowa State University.

Hunt, J.L., McManus, W.F., and Haney, W.P. 1974. Vasular lesions in acute electric injuries. *J. Trauma* **14**: 461–73.

Imboden, L.E., and Newton, C.B. 1952. Myocardial infarction following electric. *US Armed Forces Med. J.* **3**: 497–502.

Ishikawa, T. 1982. Prevention against lightning accidents in Japan. *Nihon. Univ. J. Med.* **24**: 1–14.

Ishikawa, T., Miyazawa, T., Ohashi, M., *et al.* 1981. Experimental studies on the effect of artificial respiration after lightning accidents. *Res. Exp. Med. (Berlin)* **179**: 59–68.

Ishikawa, T., Ohashi, M., Kitagawa, N., *et al.* 1985. Experimental study on the lethal threshold value of multiple successive voltage impulses to rabbits simulating multi stroke lightning flash. *Int. J. Biometeorol* **29**(2): 157–68.

Jackson, S.H.D., and Parry, D.J. 1980. Lightning and the heart. *Br. Heart J.* **43**: 454–7.

Jellinek, S. 1942. Death by lightning (Letter). 1942. *Br. Med. J.*: 714.

Jemmi, A. 1965. Cochleovestibular lesions caused by electrical energy. *L'otorhinolar. Ital. Bologna* **34**: 231–65.

Jex-Blake, A.J. 1913. Goulstonian lectures: death by electric currents and by lightning. *Br. Med. J.*, 1, 8, 15, 22 March: 425–30, 492–8, 548–52, 601–3.

Johnstone, B.R., Harding, D.L., and Hocking, B. 1986. Telephone-related lightning injury. *Med. J. Aust.* **144**: 706–9.

Jones, D.T., Ogren, F.P., Roh, L.H., and Moore, G.F. 1991. Lightning and its effects on the auditory system. *Larynogoscope* **101**: 830–4.

Junkins, E.P., *et al.* 1999. Lightning strike to the head of a helmeted motorcyclist. *Am. J. Emerg. Med.* **17**: 213–14.

King, L.A. 1962. The voltage gradient of the free burning arc in air or nitrogen. In *Proc. 5th Int. Conf. on Ionization Phenomena in Gases, Munich*, pp. 871–7, North Holland.

Kitagawa, N. 2000. The actual mechanisms of so-called step voltage injuries. In *Proc. 25th Int. Conf. on Lightning Protection, Rhodes, Greece*, pp. 781–5.

Kitagawa, N. *et al.* 1971. Experimental studies on artificial lightning stroke with dummies and rabbits. *Kagaku* **40**: 486–94.

Kitagawa, N., Kinoshita, K., and Ishikawa, T. 1973. Discharge experiments using dummies and rabbits simulating lightning strokes on human bodies. *Int. J. Biometeorol.* **17**: 239–41.

Kitagawa, N., Turumi, T., Isikawa, T., *et al.* 1985. The nature of lightning discharges on human bodies and the basis for safety and protection. In *Proc. 18th Conf. on Lightning, Munich.*

Kithil, R., and Rakov, V.A. 2001. Small shelters and safety from lightning. *Golf Course Management* **68**: 104–12.

Kleiner, J.P., and Wilkin, J.H. 1978. Cardiac effects of lightning stroke. *J. Am. Med. Assoc.* **240**: 2757–9.

Kleinot, S., Klachko, D.M., and Kelley, K.J. 1966. The cardiac effects of lightning injury. *S. Afr. Med. J.* **40**: 1141–3.

Knaggs, R.H.E. 1894. Unusual injuries caused by lightning strike. *Lancet* **1**: 1216.

Kobayashi, T. 1974. Aural trauma caused by electric shock by lightning. *Otolaryngol. Tokyo* **40**: 529.

Kotogal, S., Rawlings, C.A., Chen, S., *et al.* 1982. Neurologic, physiatric and cardiovascular complications in children struck by lightning. *Ped.* **70**: 190–2.

Krasheminnikov, V.F. 1965. Lightning injuries of the ear. *Zh. Ushn. Nos. Gorl. Bolezn.* **25**: 75.

Krasnov, A.K. 1967. A contribution to case histories on lightning injuries. *For. Med. Ex.* **10**: 45.

Kravitz, R., Wasserman, M.J., Valaitis, J., Anzinger, R.E., and Naidu, S.H. 1977. Lightning injury – management of a case with ten-day survival. *A. J. Dis. Child* **131**: 413–5.

Kristensen, S., and Tveteras, K. 1985. Lightning induced acoustic rupture of the tympanic membrane. *J. Laryng. Otol.* **99**: 711–3.

Krob, M.J., and Cram, A.E. 1983. Lightning injuries: a multisystem trauma. *J. Iowa. Med. Soc.* **73**: 221–5.

Langley, R.L., Dunn, K.A., and Esinhart, J.D. 1991. Lightning fatalities in North Carolina 1972–1988. *N. Carolina Med. J.* **52**: 281–4.

Langworthy, O.R. 1936. Neurological abormalities produced by electricity. *J. Nerve Ment. Dis.* **84**: 13–26.

Lea, J.A. 1920. Paresis of accommodation following injury by lightning. *Br. J. Ophthalmol.* **4**: 417.

Lee, W.R. 1977. Lightning injuries and death. In *Lightning, vol. 2, Lightning Protection*, ed. R.H. Golde, pp. 521–44, New York: Academic.

Leveille-Nizeroll, E.M., Lintzer, J.P., and Berezin, A. 1978. Auricular consequences of lightning. *Ann. Otolarynol. Chir. Cervicofac.* **95**: 695–702.

Lopez, R.E., and Holle, R.L. 1995. Demographics of lightning casualties. *Sem. Neurology* **15**: 286–95.

Lopez, R.E., and Holle, R.L. 1996. Fluctuations of lightning casualties in the United States: 1959–1990. *J. Climate* **9**: 608–15.

Lopez, R.E., and Holle, R.L. 1998. Changes in the number of lightning deaths in the United States during the twentieth century. *J. Climate* **11**: 2070–7.

Lopez, R.E., Holle, R.L., Heitkamp, T.A., Boyson, M., Cherington, M., and Langford, K. 1993. The underreporting of lightning injuries and deaths in Colorado. *Bull. Am. Meteorol. Soc.* **74**: 2171–8.

Lopez, R.E., Holle, R.L., and Heitkamp, T.A. 1995. Lightning casualties and property damage in Colorado from 1950 to 1991 based on storm data. *Wea. Forecast.* **10**: 114–26.

Lynch, M.J.G., and Shorthouse, P.H. 1949. Injuries and death from lightning. *Lancet* **1**: 473–8.

Mackerras, D. 1983. Observations of lightning-caused surges in a telephone line for the period 1979–1983. University of Queensland, Dept Electrical Engineering, Report EE83/12.

Maggied, S.M. 1973. Lightning's incredible attack on American football in 1970. *Ohio State Med. J.* **69**: 603–6.

Mann, H., Kozic, Z., and Boulos, M. 1983. Account of lightning injury. *Am. J. Neuroradiol.* **4**: 976–7.

Massello, W. 1988. Lightning deaths. *Med. Leg. Bull.* **37**.

Matthews, M.S., and Fahey, A.L. 1997. Plastic surgical considerations in lightning injuries. *Ann. Plast. Surg.* **39**: 561–5.

McCarthy, L.J., and Parker, C. 1981. Positive antiglobulin tests in a boy struck by lightning (Letter). *New Engl. J. Med.* **305**(5): 283.

McCrady-Kahn, V.L., and Kahn, A.M. 1981. Lightning burns. *West. J. Med.* **134**: 215–19.

Milward, J.M. 1975. Prolonged gastric dilatation as a complication of lightning injury. *Burns* **1**: 175–8.

Moore, M.C. 1956. Ocular injury from lightning current and lightning flash. *Trans. Oph. Soc. Aust.* **16**: 87–92.

Moran, K.T., Thupari, J.N., and Munster, A.M. 1986. Electric- and lightning-induced cardiac arrest reversed by prompt cardiopulmonary resuscitation (Letter). *JAMA* **255**(16): 2157.

Moran, K.T., Thupari, J.N., and Munster, A.M. 1986. Lightning injury, physics, pathophysiology and clinical features. *Ir. Med. J.* **79**(5): 120–2.

Morgan, Z., Headley, R., Alexander, E.A., and Sawyer, C.G. 1954. A trial fibrillation and epidural hematoma associated with lightning stroke. *New Engl. J. Med.* **259**(20): 956–9.

Morgan, Z.V., Headley, R.N., Alexander, E.A., and Sawyer, C.G. 1958. Atrial fibrillation and epidural hematoma associated with lightning stroke. *New Engl. J. Med.* **259**(20): 946–9.

Morikawa, S., Steichen, F. 1960. Successful resuscitation after 'death' from lightning. *Anesthesiol.* **21**: 222–4.

Moser, M.Y. 1986. When lightning strikes. *Am. J. Nurs.* **86**: 802–3.

Moulson, A.M. 1984. Blast injury to the lungs due to lightning. *Br. Med. J.* **289**: 1270–1.

Mournier-Kuhn, P. 1963. Clinical study of lesions of the auditory apparatus caused by electricity. *Rev. Otoneuroophthal.* **35**: 165.

Msonge, B., and Evans, R.L. 1976. Lightning stroke on Ukerewe Island. *East Afr. Med. J.* **53**(6): 350–1.

Moto, T. 1972. Marks of lightning on the skin: clinical observations and its experimental simulation. *Keio Igaku* **49**: 51–9.

Myers, G.J., Colgan, M.T., and Van Dyke, D.H. 1977. Lightning-strike disaster among children. *J. Am. Med. Assoc.* **238**: 1045–6.

Nagai, Y., Ishikawa, T., Ohashi, M., *et al.* 1982. Study of lethal effects of multiple stroke flash. *Res. Lett. Atmos. Electr.* **2**: 87–90.

Nesmith, M.A. 1971. A case of lightning stroke. *J. Fla. Med. Assoc.* **58**: 36.

Ness, C.E., and Pilgrim-Larsen, J. 1984. Personkader ved lynnedslag, *Tiddskr. Nor. Laegeforen* **104**: 1031–3.

Nippon Telegraph and Telephone Public Corporation 1979. Lightning surges on subscribers lines. *CCITT Study Group V* **77**, December.

NOAA 1970. Lightning. NOAA Public Affairs 70005, 5 pp.

NOAA 1985. Thunderstorms and lightning. NOAA Public Affairs NOAA/PA 830001, 6 pp.

NOAA 1994. Thunderstorms and lightning ... the underrated killers! NOAA Public Affairs NOAA/PA 92053 and American Red Cross ARC 5001, 12 pp. (Available from local National Weather Service offices and Red Cross chapters.)

Noel, L.P., Clarke, W.N., and Addison, D. 1980. Ocular complication of lightning. *J. Ped. Ophthal. Stralismis.* **17**: 245–6.

Norman, M.E., and Younge, B.R. 1999. Association of high-dose intravenous methylprednisolone with reversal of blindness from lightning in two patients. *Ophthalmology* **106**: 743–5.

Ohashi, M., Kitagawa, N., and Ishikawa, I. 1986. Lightning injury caused by discharges accompanying flashovers: a clinical and experimental study of death and survival. *Burns* **12**: 496–501.

Ost, L.G. 1978. Behavioural treatment of thunder and lightning phobias. *Behav. Res. Ther.* **16**: 197–207.

Pakiam, J.E., Chao, T.C., and Chia, J. 1981. Lightning fatalities in Singapore. *Meteor. Mag.* **110**: 175–87.

Palmer, A.B.D. 1987. Lightning injury causing prolongation of the Q-T internal. *Post. Med. J.* **63**: 891–4.

Panse, F. 1975. Electrical trauma. In *Handbook of Clinical Neurology 23*, eds. P.J. Vinken, G.W. Bruyn, pp. 683–729, Amsterdam: Elsevier.

Paterson, J.H., and Turner, J.W. 1944. Lightning and the central nervous system. *J. Roy. Arm. Med. Corps* **82**: 73–5.

Peters, O.S. 1915. Protection of life and property against lightning. Technologic Papers of the Bureau of Standards no. 56, Washington Government Printing Office, 27 pp.

Peters, W.J. 1983. Lightning injury. *Can. Med. Assoc. J.* **128**: 148–50.

Pierce, M.R., Henderson, R.A., and Mitchell, J.M. 1986. Cardiopulmonary arrest secondary to lightning injury in a pregnant woman. *Ann. Emerg. Med.* **15**(5): 597–9.

Plueckhahn, V. 1986. Injury and death caused by lightning (Editorial). *Med. J. Aust.* **144**: 673.

Poulsen, P., and Knudstrup, P. 1986. Lightning caused inner ear damage and intracranial haematoma. *J. Laryngol. Otolaryngol.* **100**: 1067–70.

Prentice, S.A. 1972. Lightning fatalities in Australia. *EE Trans. IE Aust.* **8**(2): 55–63.

Primeau, M., Engelstatter, G.H., and Bares, K.K. 1995. Behavioral consequences of lightning and electrical injury. *Sem. Neurology* **15**: 279–85.

Pritchard, E.A.B. 1934. Changes in the central nervous system due to electrocution. *Lancet* **1**: 1163–7.

Rakov, V.A. 2000. Lightning protection of structures and personal safety. In *Proc. Int. Lightning Detection Conf.*, 10 pp., GAI, 2705 East Medina Road, Tucson, Arizona 85706-7155.

Ravitch, M.M., Lane, R., Safar, P., Steichen, F.M., and Knowles, P. 1961. Lightning stroke – report of a case with recovery after cardiac massage and prolonged artifical respiration. *New Engl. J. Med.* **264**: 36–8.

Raymond, L.F. 1967. Specific treatment of uveitis, lightning induced: an autoimmune disease. *Ann Allergy* **27**: 242–44.

Read, J.M. 1966. Man struck by lightning reveals marked ECG changes hours later. *Med. Trib.*, 28 March: 3.

Rees, W.D. 1965. Pregnant woman struck by lightning. *Br. Med. J.* **1**: 103–4.

Rich, J. 1986. Lightning injury. *Nursing* **16**(6): 25.

Richards, A. 1973. Traumatic facial palsy. *Proc. Roy. Soc. Med.* **66**: 28–9.

Robinson, M., and Seward, P.N. 1986. Electrical and lightning injuries in children. *Paediatr. Emerg. Care* **2**: 186–90.

Royer, J., and Gainet, F. 1972. Consequences oculaires de la fulguration: à propos d'un cas. *Bull. Soc. Opth. Fr.* **72**: 491–7.

Rush, M., and Glasser, J. 1981. Lightning risk. *Rev. Environ. Health* **3**: 301–13.

Saddler, M.C., and Thomas, J.E. 1990. Temporary bulbar palsy following lightning stroke. *Cent. Afr. J. Med.* **36**: 161–2.

Sances, A., Klebust, J., Szablya, J., *et al.* 1983. Current pathways in high voltage injuries. *IEEE Trans. Biomed. Eng.* **30**: 118–23.

Schallock, G. 1952. *Zol. Allg. Path. Anat.* **88**: 245.

Schmidt, W., Gruetzner, A., and Schoen, H.R. 1957. Beobachtungen bei Blitzschlagverletzungen unter Beruecksichtigung von EKG and EEG. *Deutsches Arch. Klin. Med.* **204**: 307–24.

Schneider, R. 1956. Blitzchlag als teleponunfall. *Med. Klinik* **51**: 2208–10.

Shan, G.P., and Atkinson, L. 1981. Hearing loss secondary to lightning strike. *J. South Carolina Med. Assoc.* **77**: 233–5.

Shapiro, M.B. 1984. Lightning cataracts. *Univ. Wisc. Med. J.* **83**: 23–4.

Sharma, M., and Smith, A. 1958. Paraplegia as a result of lightning injury. *Br. Med. J.* **2**: 1464–5.

Shaw, D., and York-Moore, M.E. 1957. Neuropsychiatric sequelae of lightning stroke. *Br. Med. J.* **1**: 1152–5.

Shearman, K.M., and Ojala, C.F. 1999. Some causes for lightning data inaccuracies: the case of Michigan. *Bull. Am. Meteor. Soc.* **80**: 1883–91.

Sheppard, L.B. 1945. Report of an eye injured by lightning. *Am. J. Ophth.* **28**: 195–8.

Silverman, N.M. 1961. Unilateral deafness as sequel to non-fatal lightning stroke. *J. Indiana Med. A.* **29**: 530.

Simpson, G.C. 1929. Lightning. *Nature* **123**: 801–12.

Singleton, G.T., Keim, R.J., Whitaker, D.L., *et al.* 1984. Cordless telephones: a threat to hearing. *Ann. Otolrhinolaryngol.* **93**: 565–8.

Sinha, A.K. 1985. Lightning induced myocardial injury – A case report with management. *Angiol.* **36**(5): 327–31.

Skan, D.A. 1949. Death from lightning stroke, with multiple injuries. *Br. Med. J.* **1**: 666.

Slingerland, I. 1914. Lightning stroke. *Med. Soc. NY* **14**: 466.

References and bibliography for Chapter 19

Smith, J. 1983. Lightning injuries. *J. Em. Nurs.* **9**(5): 248–50.
Smith, S.B. 1993. Comments on "Lightning ground flash density in the contiguous United States – 1989" *Mon. Wea. Rev.* **121**: 1572–5.
Spencer, H.A. 1934. Lightning stroke and its treatment. In *Lightning*, p. 91, London: Bailliere Tindall and Cox.
Spirov, A. 1968. Damage of the ear by thunderbolt. *Vojnosanit. Pregl.* **25**: 648–51.
Stanley, L.D., and Suss, R.A. 1985. Intracerebral hematoma secondary to lightning stroke: case report and review of literature. *Neurosurg.* **16**(5): 686–8.
Stanley-Turner, H.M. 1942. Death by lightning (Letter). *Br. Med. J.*, 24 October: 503.
Strasser, E.J., Davis, R.M., and Menchley, M.J. 1977. Lightning injuries. *J. Trauma* **17**(4): 315–19.
Struzewski, B., and Struzewski, 1985. Lightning discharge not causing death as a significant hazard for human hearing organ. In *Proc. 18th Int. Conf. on Lightning, Munich.*
Subramanian, N., Somasundaram, B., and Periasamy, J.K. 1985. Cardiac injury due to lightning – report of a survivor. *Indian Heart J.* **37**(1): 72–3.
Suri, M.L., and Vijayan, G.P. 1978. Neurological sequelae of lightning. *J. Assoc. Phys. Ind.* **26**: 209–12.
Taussig, H.B. 1968. Death from lightning – and the possibility of living again. *Ann. Int. Med.* **68**(6): 1345–53.
Taussig, H.B. 1969. "Death" from lightning and the possibility of living again. *Am. Sci.* **57**(3): 306–16.
Ten Duis, H.J. 1998. Lightning strikes: danger overhead. *Br. J. Sport Med.* **32**: 276–8.
Ten Duis, H.J., Klasen, H.J., and Reenalda, P.E. 1985. Keraunoparalysis, a 'specific' lightning injury. *Burns Incl. Therm. Inj.* **12**(1): 54–7.
Ten Duis, H.J., Klasen, H.J., Jijsten, M.W.N., *et al.* 1987. Superficial lightning injuries: their "fractal" shape and origin. *Burns* **13**: 141–6.
Terranova, S. 1957. Case of myocardial infarction caused by atmospheric electrocution. *Riforme Medica (Napoli)* **71**(4): 93–6.
Tribble, G.C., Persing, J.A., Morgan, R.F., *et al.* 1985. Lightning injuries. *Compr. Ther.* **11**: 32–40.
Tveteras, K., and Kristenson, S. 1984. *Lynnedslagulykker tidsskr nor laegeforen* **104**: 1413–14.
Tveteras, K., Kristenson, S., and Clausen, I.H. 1984. Fatal lynnedslagsulykker. *Ugeskr. Laeger* **146**: 1858–9.
Uman, M.A. 1986. *All About Lightning*, New York: Dover.
Veimeister, P.E. 1972. *The Lightning Book*, Cambridge, Massachusetts: MIT Press.
Vigansky, H.N. 1985. General summary of lightning. *NOAA Storm Data* **27**: 37–44.
Villard, C., Boissonnot, M., Riss, J., *et al.* 1985. Complication oculaire par fulguration discussion à propos de deux cas simultanes. *Bull. Soc. Opht. Fr.* **85**: 1027–8, 1031–4.
Wakasugi, C., and Masui, M. 1986. Secondary brain hemorrhages associated with lightning stroke: report of a case. *Japan J. Legal Med.* **40**(1): 42–6.
Walsh, K.M., Hanley, M.J., Graner, S.J., Beam, D., and Bazluki, J. 1997. A survey of lightning policy in selected division 1 colleges. *J. Athletic Train.* **32**: 206–10.
Webb, J., Srinivasan, J., Fahmy, F., and Frame, J.D. 1996. Unusual skin injury from lightning. *Lancet* **347**: 321.
Weeks, A.W., Alexander, L. 1939. Distribution of electric current in the animal body: an experimental investigation of 60 cycle alternating current. *J. Indust. Hyg.* **21**: 517–25.
Wehe, W. 1918. Report of a case of burns from lightning. *Int. J. Surg.*, June, 191–4.
Weinstein, L. 1979. Lightning: a rare cause of intrauterine death with maternal survival. *S. Med. J.* **72**: 632–3.
Weiss, K.S. 1980. Otological lightning bolts. *Am. J. Otolaryngol.* **1**: 334–7.
West, G. 1955. Lightning as a cause of hearing loss. *Maryland State Med. J.* **4**: 35–7.
Wright, J.W., and Silk, K.J. 1974. Acoustic and vestibular defects in lightning survivors. *Laryngoscope* **84**: 1378–87.
Yost, J.W., and Holmes, F.F. 1974. Myoglobinuria following lightning stroke. *J. Am. Med. Assoc.* **228**(9): 1147–8.
Youngs, R., Deck, J., Kwok, P., *et al.* 1988. Severe sensorineural hearing loss caused by lightning. *Arch. Otolaryngol Head Neck Surg.* **114**: 1189–97.
Zeana, C.D. 1984. Acute transient myocardial ischaemia after lightning injury. *Int. J. Cardiol.* **5**: 207–9.
Zegel, F.H. 1967. Lightning deaths in the United States: a seven year survey from 1959 to 1965. *Weatherwise* **20**: 168–73, 179.

20 Ball lightning, bead lightning, and other unusual discharges

> ... a foot away from the iron rod, [he] looked at the electrical indicator again; just then a palish blue ball of fire, as big as a fist, came out of the rod without any contact whatsoever. It went right to the forehead of the professor, who in that instant fell back without uttering a sound ...
>
> B. Dibner (1977) quoting the witness of a fatality attributed to ball lightning, the death of Georg Richmann in 1753 while repeating the lightning rod experiment originally suggested by Benjamin Franklin in 1750 and first performed by Thomas-François D'Alibard in 1752.

20.1. Introduction

Ball lightning (in French, boules de feu, éclair en boule, foudre globulaire; in German, Kugelblitz; in Russian, sharovaya molniya) is a phenomenon for which there exist numerous witness reports but little, if any, scientific documentation such as photographs, videotapes, or other scientific recordings. A reproduction of a woodcut depicting ball lightning "crossing a kitchen and a barn" is shown in Fig 20.1. Barry (1980b) and Stenhoff (1999) in their books presented a variety of photographs purported to be of ball lightning. Stenhoff (1999) stated that 65 such photographs have been published and attributes most of these to causes other than actual ball lightning, including street or car lights photographed by a shaking camera, fireworks, hoaxes, and film blemishes. Jennison et al. (1990) discussed and showed frames from a possible video recording of ball lightning, although the individual who took the video did not see the ball lightning through the viewfinder when he was recording a nearby storm. Despite the lack of scientific documentation, the properties of ball lightning are relatively well known from statistical analyses of observers' reports spanning a period of three centuries. Table 20.1 lists nine compilations of eyewitness observations of ball lightning, involving almost 5000 reports. Examples of witness reports are given in Section 20.2. Apparently, from a few percent to about 10 percent of individuals who live in regions with appreciable thunderstorm activity have seen ball lightning, so ball lightning is probably not as rare as often assumed (Rayle 1966; also, an observation of this book's authors, drawn from reports of ball lightning by attendees of their public lectures on lightning).

There may be more than one type of ball lightning and more than one mechanism by which ball lightning is generated, but the most commonly reported observation is of an orange- to grapefruit-size sphere (the range for the vast majority of reports is from the size of a golf ball to that of a basketball), which is usually red, orange, or yellow in color and is about as bright as a 60 watt light bulb. Ball lightning most often has a duration of a few seconds, during which time it generally moves more or less horizontally (it does not rise as would hot air) after which it decays either slowly and silently or abruptly and explosively. Detailed statistics on size, color, duration, and other properties are found in Section 20.3. The luminosity of ball lightning is roughly constant until it extinguishes. It is most often seen spatially close to and just after a cloud-to-ground flash. It appears to be able to pass through glass windows and through metal window screens. There is a significant number of credible reports of ball lightning occurring within metal (aluminum) aircraft. Ball lightning is sometimes reported to have an odor and sometimes to leave burn marks. Human beings are seldom, if ever, injured or killed by ball lightning; most reports of injury and death are from the 18th and 19th centuries and can probably be attributed to ordinary lightning or to meteors. Since ball lightning and meteors are both referred to in the literature as fireballs, it is not surprising that some reports of the effects of fireballs that actually refer to meteors have been misinterpreted as being due to ball lightning. Interestingly, until the late eighteenth century, meteors, like ball lightning, were thought by scientists to originate in the atmosphere. Additionally, the term "thunderbolt" has been used interchangeably for both ordinary lightning and ball lightning, so that some reports of thunderbolts that were actually lightning have been misinterpreted as ball lightning. The report of Richmann's death from ball lightning quoted at the beginning of this chapter

20.1. Introduction

Table 20.1. *Ball lightning reports. Adapted from Stenhoff (1999)*

Survey	Number of reports	Notes
Brand (1923)	215	Selected from a collection of 600 ball lightning reports in the university libraries of Berlin, Marburg, and Gottingen
Humphreys (1936)	280	
McNally (1961, 1966)	498	Based on a survey of 15 923 Union Carbide Nuclear Company personnel in Oak Ridge, Tennessee
Barry (1967a)	400	
Rayle (1966)	112	Based on a survey of personnel at NASA Lewis Research Center, Cleveland, Ohio
Charman (1979)	68	
Egely (1989a)	380	Analysis of Hungarian ball lightning reports
Bychkov, Smirnov, and Stridjev (1993)	2500	Russian–Austrian data bank
Hubert (1996)	253	

Fig. 20.1. Ball lightning in a nineteenth century woodcut. The original title, translated from the French, reads "Ball lightning crossing a kitchen and a barn". Perhaps the ball lightning came down a chimney used as an exhaust for the cooking fires. Courtesy, Burndy Library.

(see also Section 1.1) may have been due to ordinary lightning, particularly in view of the fact that the physical effects on Richmann's body were similar to those observed in normal lightning strokes (Section 19.3); in addition to the wound on his forehead, the shoe on his left foot was torn open and there was a mark on that foot (Lee 1977). Such effects have not been attributed to ball lightning in other cases. Ball lightning or ball-lightning-like phenomena are occasionally reported to be generated from high-power electrical equipment, as discussed in subsection 20.2.9.

There have been many theories devised to explain ball lightning. None is completely satisfactory, a number violate accepted laws of physics (Finkelstein and Rubinstein 1964), most do not simulate the observed characteristics of ball lightning, and some are clearly unrealistic (e.g., electron–positron annihilation, miniature black holes, cosmic rays focused by cloud electric fields, quantum mechanical plasmas). Ball lightning models can be divided into two classes, those that are internally powered and those that are externally powered (Uman 1969, 1984), as we shall discuss in Section 20.4.

Ball lightning has apparently not been generated in the laboratory; at least such generation has not been documented in the literature in a manner that inspires confidence and with sufficient detail that it can be reproduced by other researchers, although many luminous phenomena created in the laboratory are claimed by their creators to be ball lightning or to have some relation to ball lightning. The laboratory simulation of ball lightning is discussed in Section 20.5.

The roughly 2400 references to papers on ball lightning given in the book by Stenhoff (1999), a thorough up-to-date examination of all aspects of ball lightning, represent a significant fraction of the total literature on lightning. (The book you are reading contains reference and bibliography lists comprising over 6000 publications on various aspects of lightning and its effects.) Other books on ball lightning containing similar material have been written by Barry (1980b) and by Singer (1971).

Bead lightning (in French, éclair en chapelet; in German, Perlschnurblitz; in Russian, chetochnaya molniya) is often considered to be in the same category as ball lightning, both being unusual and unexplained luminous phenomena related to lightning. While there is no consensus view of the physical mechanism for bead lightning, it is clearly related to processes occurring when the luminous cloud-to-ground lightning channel extinguishes, resulting in a beaded or string-of-pearls appearance (subsection 4.7.8). The bead-lightning literature has been reviewed by Uman (1969, 1984) and by Barry (1980b). Bead lightning is discussed further in Section 20.6.

While ball and bead lightning are apparently related to natural lightning discharges, charge separation in various non-thunderstorm conditions is responsible for the lightning-like discharges produced in the material emitted from volcanoes, the upward-initiated discharges occurring in nuclear explosions, the meter-length sparks reported in sand storms, and apparently the luminous phenomena associated with earthquakes. These unusual lightning or lightning-like discharges are considered in Section 20.7.

20.2. Witness reports of ball lightning

In this section we present a number of eyewitness reports of ball lightning: observations of ball lightning outdoors, in houses, and in aircraft, and ball lightning, or a phenomenon similar to storm-related ball lightning, generated by electrical machinery. If no reference is given in conjunction with the observation then it was received by the authors directly from the observer.

20.2.1. Outdoors in Australia

Near Murray's dairy, a property on the Queanbeyan road on the then outskirts of Canberra, and not far from what is now the industrial area of Fyshwick, I was riding along the right-hand side of the road, just off the paving to prevent a possible slip and fall on the wet surface, and about 20 to 25 yards [or metres] ahead, on the left-hand side and also off the bitumen, a farm employee was leading a Shorthorn bull. There was very little, if any rain at that time. I can clearly recall that there was one of those periods of "quietness" that sometimes precedes a downpour. Just as I drew level with the bull there was one very loud bang or explosion and immediately down the white traffic line in the center of the road appeared the fireball. It seemed to be about 6 or 8 inches [15–20 cm] off the ground, was about the size of a basketball, like very golden butter in color, and had the appearance of being "spun" or "fuzzy", like silk threads or wool, as distinct from a "molten" liquid look. It did not sparkle – just a ball of fuzz. It came straight down say three of the white marker lines. These have about equal distance of unmarked road between them, so I should say it traveled about 18 to 20 ft [6 m]. Then it simply disappeared. It did not break apart. There was no further noise like an explosion. It was there one moment and not there the next.

The whole thing would have been over in probably two or three seconds, before the horse had time to be startled. The young men leading the bull cried out – in pure Australian – "What the bloody hell was that?" As there was another downpour shortly, we did not stop to discuss it.

20.2.2. Outdoors in Germany

I was on an evening hunt on the Schöntal preserve after a very hot day toward the end of June 1874. At about 9 pm, on my way back home, I was suddenly taken unawares by a very violent thunderstorm, which forced me to seek shelter in an old wooden cabin directly situated on a narrow macadamized road. While the storm was raging with unbelievable force, I noticed bluish-colored balls, which were rolling about on the road outside and burst into a crackling spray of sparks partly in front and partly behind the place where I was standing. The size of the balls corresponded to that of an average-size skittle ball. The bursting of these balls, which took place several times in my direct vicinity (so that I could very clearly see what was happening), did not cause any loud detonation, but instead such a dazzling light that I was momentarily blinded. The speed at which the balls traveled was not very great, no greater than that of a firmly cued skittle ball. These balls moved in the same direction, all the time following the course of the road, in varying time intervals (at times in rapid succession); in a period of half an hour I counted 25–30 of these balls. The thunderstorm descended upon a basin-shaped forest complex, where it raged at

20.2. Witness reports of ball lightning

full force for two hours. The ball lightnings appeared at the very beginning of the thunderstorm, while the rain was pouring down and ordinary lightning streaks accompanied by rolling thunder were coming down from all sides and struck several times in the vicinity of the cabin. Since I dared not go out onto the road, not because of the ball lightning but because of the ordinary lightning discharges which were striking the high trees along the road, I stayed inside the old hut until the storm died down, which was at about 11 pm. I was happy, when I could finally leave the clammy hut without having suffered any injury.

This report is taken from Brand (1923) and is not a typical observation.

20.2.3. Indoors in Virginia

My experience with ball lightning happened about 12 years ago when I was 16 years old, one summer evening on my mother's farm in Barhamsville, Virginia, a small town about 30 miles outside of Williamsburg. It was typical that summer for my mom and me to get up in the middle of the night and watch the spectacular lightning storms that passed over the farm. During one of these storms, my mom, myself, and a friend of mine who was spending the night, got up to go and sit out on the screened-in porch to watch the lightning. My friend and I had just gotten up from bed and were about to go out on the porch. I had blown my nose and was reaching out to toss the tissue into the trash basket, when a round blue light, about the size of a grapefruit, came through the metal screen of the window, about 4 ft [1.3 m] away, and touched my hand. This felt like the shock of sticking your finger into a Christmas tree light socket for a period of a second. I had no idea what it was, and I asked my friend if she saw it and she said she had. I do not remember any particularly loud or dramatic thunder or lightning before this event, however it was frequently doing that at that time. The window was an average height from the floor (about 3 ft [1 m]) and the ball lowered its altitude slightly to touch my hand from where it came in from the window. After it touched me, it traveled at a constant speed and in a straight line down the hall toward the other end of the house, at the same altitude at which it touched my hand. I did not notice any strange smell afterwards but there was a lot of wind coming into the room; although it moved quickly, I was able to follow its course through the house. I saw it for about 4 seconds as it covered a distance of about 15 meters. I am not sure how I even came to believe what had happened was ball lightning, but in telling the story to my friends in Gainesville, one suggested that I contact Dr Uman and share my experience with him.

This observation was confirmed in writing by the friend referred to in the report.

20.2.4. Indoors in Nebraska

I was standing in the kitchen of my home in Omaha, Nebraska while a terrible thunderstorm was in progress. A sharp cracking noise caused me to look toward a window screen to my left. Then I saw a round, iridescent (mostly blue) object, baseball size, coming toward me. It curved over my head and went through the isinglass [mica] door of the kitchen range, striking the back of the oven and spattering into brilliant streamers. There was no sound and no effect on me except a tingle as it passed over my hair. Later examination showed a tiny hole with scorched edges in the screen and isinglass, and scorch-like marks on the back of the oven.

20.2.5. Indoors in Virginia

I was laying in bed after a thunderstorm, happy that the storm was over. Suddenly I could see a bluish-colored ball, nearly perfectly round, about the size of a softball or slightly larger, floating outside my bedroom window, perhaps 20 feet above the ground. I could see it through the plastic mini-blinds covering the window, which were closed but not tightly closed (as we usually keep them such that I can frequently see the moon from my bed). After a few seconds, the ball lightning floated into my bedroom through the window. It came through the glass window and the plastic mini-blinds as though the window and blinds were not even there. There was no sound of it hitting the window and no change in direction or the shape of the object as it came through the glass and plastic blinds. It continued floating, quite slowly (slower than soap bubbles blown with a bubble wand) in my direction. After a few seconds, it was about two feet from my head. It 'poofed' – not a loud explosion but a definite poof, and then I smelled something, as though something had burned. I'm sure about the burning smell. Had I not just seen that thing explode, and been sure that it was the cause of the smell, I would have gotten up and checked the furnace and other things for a source of a possible fire. The ball of lightning had totally disappeared with that poof. The explosion, however, was considerably softer in its sound than the popping of a balloon. It never occurred to me at the time to check the window for damage, because the ball lightning came through the window as though nothing at all was in its path and without making any sound whatsoever at the point. Checking recently, I saw no damage. The window was a storm window, with one pane of glass, some air space, plus another pane of glass on the outside. The frame of the window is a combination of wood and aluminum. There are no power lines near our house (for at least 1/2 mile) as all power, phone, and cable lines are underground.

20.2.6. Indoors in Washington state

The following account was recorded by a professional engineer who, in October 1993, was sitting at a desk in his home.

1. My desk is situated on the upper floor of a 2-story wood frame house near Seattle, Washington. 2. I was facing the window which was 4 ft [1.3 m] from my seated location. 3. The drapes were open, but the window was shut. 4. My Macintosh computer was in front of me but was not switched on, although it was connected to an outlet. 5. It was early evening, and it was raining quite heavily. 6. I heard a faint thunderclap. 7. About 2–4 minutes later, I was astonished to observe a bright white ball of light, 1 ft away from my face, and about 6 inches [15 cm] above my Mac. It seemed to form out of nowhere, and I estimate

that the duration of its existence was 3 seconds. It was 9–12 inches [23–30 cm] in diameter. 8. The ball "burst" and disappeared with a moderately loud, electrical sounding "click" and there was an instantaneous, loud thunderclap that sounded as if it was right above my house. 9. I jumped up startled and I ran outside to see if my house was damaged. It was not. 10. No trace was left by the phenomenon, no smoke and no odor, and my computer was undamaged when I switched it on.

20.2.7. In a KC-97 aircraft

I was at the controls of a KC-97 USAF tanker aircraft, heavily loaded with JP-4 fuel for offload to B-47 bombers. En route to the refueling rendezvous (Elko, Nevada vicinity) we were in the clouds at 18 000 ft. There was light precipitation, temperature was above freezing and there was no turbulence. I recall that St Elmo's fire was dancing around the edges of the aircraft front windows. (This is a not too uncommon occurrence but may have some significance to you.) The crew was experienced in all phases of all-weather operation and not concerned or apprehensive about any portion of the mission to be accomplished. As I was concentrating on the panel (no outside visual references were visible) a ball of yellow-white color approximately 18 inches [0.5 m] in diameter emerged through the windshield center panels and passed at a rate about that of a fast run between my left seat and the co-pilot's right seat, down the cabin passageway past the navigator and engineer. I had been struck by lightning two times through the years in previous flights and recall waiting for the explosion of the ball of light! I was unable to turn around and watch the progress of the ball as it proceeded to the rear of the aircraft, as I was expecting the explosion with a full load of JP-4 fuel aboard and concentrated on flying the aircraft. After approximately 3 seconds of amazingly quiet reaction by the four crew members in the flight compartment, the boom operator sitting in the rear of the aircraft called on the interphone in an excited voice describing a ball of fire that came rolling through the aft cargo compartment abeam the wings, then danced out over the right wing and rolled off into the night and clouds! No noise accompanied the arrival or departure of the phenomenon.

20.2.8. In a commercial aircraft

This communication records the observation of ball lightning in unusual circumstances. I was seated near the front of the passenger cabin of an all-metal airliner (Eastern Airlines Flight EA 539) on a late night flight from New York to Washington. The aircraft encountered an electrical storm during which it was enveloped in a sudden bright and loud electrical discharge (0005 h EST 29 March 1963). Some seconds after this a glowing sphere a little more than 20 cm [8 inches] in diameter emerged from the pilot's cabin and passed down the aisle of the aircraft approximately 50 cm [20 inches] from me, maintaining the same height and course for the whole distance over which it could be observed.

...the relative velocity of the ball to that of the containing aircraft was 1.5 ± 0.5 meters [or yards] per second... the object did not seem to radiate heat... the optical output could be assessed as 5 to 10 watts and its color was blue-white... the course was straight down the whole central aisle of the aircraft.

This report is found in Jennison (1969).

20.2.9. Electrical-apparatus-related ball lightning observations

A phenomenon very similar to storm-related ball lightning has been reported to occur in submarines due to the discharge of a high current (about 150 kA dc from a 260 V source) across a circuit breaker (Silberg 1962, 1965). In addition, the authors have received a number of reports of ball-lightning-like phenomena being initiated in electrical equipment, both with and without lightning present. One such incident was described by a retired professor and engineering department chairman at a major state university, another from an individual in the military, as reported below.

20.2.10. From a power circuit breaker switchboard operator

In 1914 I was working for the Interurban Co. and happened to be in the main power house during a very severe thunderstorm. We had recently installed a 600 volt dc rotary converter which supplied power to both the city and Interurban lines. I was working on the main switchboard and watching a bit carefully as there were many breakers opening. I looked up just in time to see a ball of fire about 15 to 18 inches in diameter come off the converter commutator and float at a good speed down the length of the switchboard, maybe 3 ft [1 m] above my head. Some 100 ft [30 m] from the converter it struck the ceiling of an office room where a cat was asleep. The ball splashed in all directions like water and the cat sprang high in the air. The cat squalled loudly but seemed to be uninjured. In both cases [i.e., rooms] there was the strong odor of ozone, but due to the constant lightning I would not attribute it to the 'ball lightning'.

20.2.11. From a radio transmitter operator

Back in the year 1950, midsummer, and in a place called Degendorf nach Brannenburg, Bavaria, I was stationed there as a US army soldier. I was a radio operator with the 3rd Battalion 6th AC Regiment. This particular day, it was about 2 p.m. mid-afternoon, and the sky was clear and sunny. I was on radio duty sitting in an enclosed housing on the back of an army $2\frac{1}{2}$ ton truck. The power supply was a hookup line from a nearby building (220 V). Normal mobile power was from an towed generator trailer. The interior comprised one wall counter unit, with two receivers and a counter top with mounted telegrapher's key, a long bench or box seat running lengthwise with the truck bed, storage units behind the seating, and at the far end interior a radio transmitter approx. 30 inches × 25 inches [75 cm × 63 cm] and 36 inches [1 m] high. As the truck was in "fixed location" status, we were using a long wire antenna strung between two 2-story building roofs (approx. 200 ft length). The lead-in wire ran from its [the antenna's] center, down and into the truck, and fastened to the antenna post connector on the back of

20.2. Witness reports of ball lightning

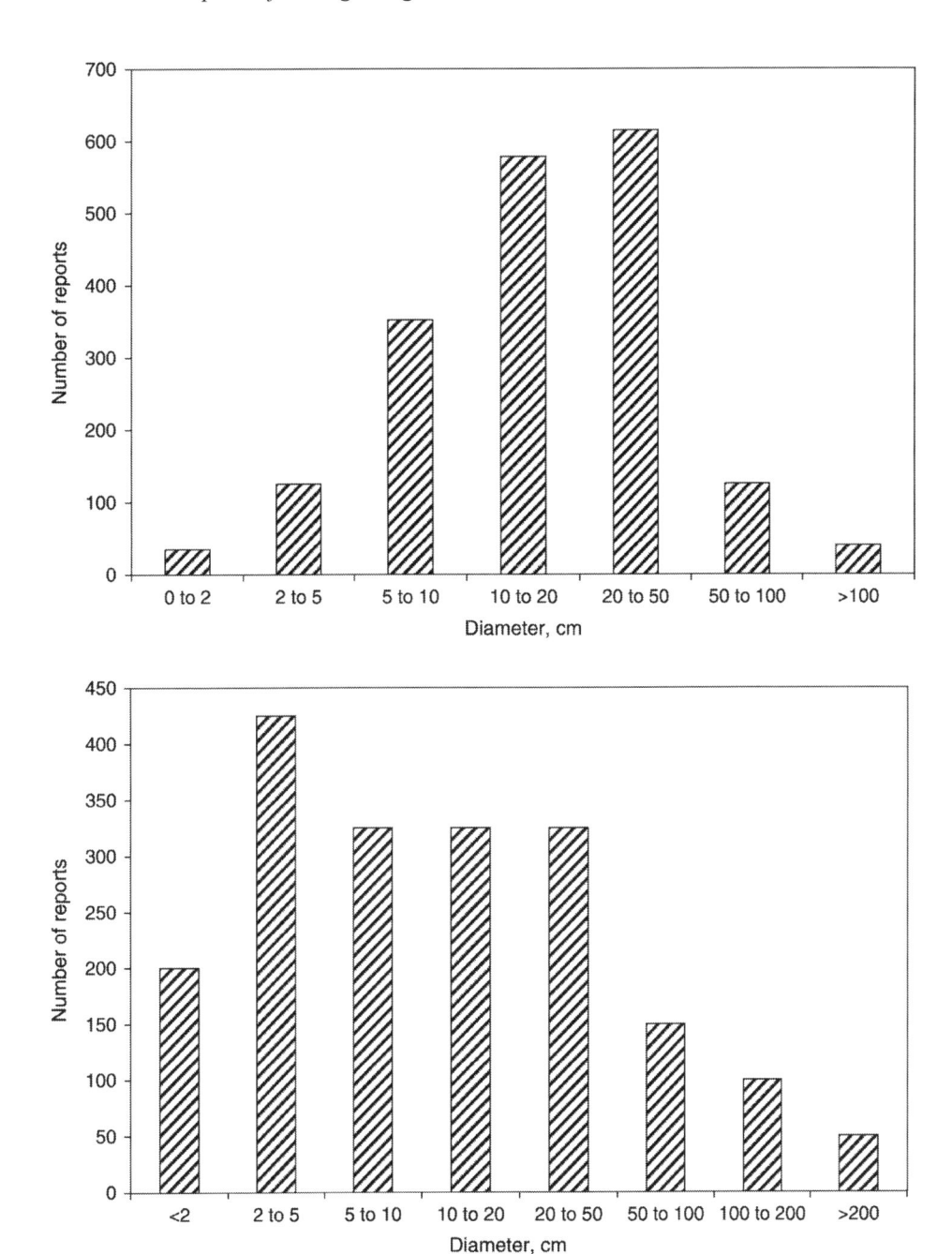

Fig. 20.2. (a) Frequency distribution of ball lightning diameter from 1869 reports. Adapted from Bychkov *et al.* (1993). (b) Frequency distribution of ball lightning diameter from 446 reports. Adapted from McNally (1966).

the transmitter. The antenna wire was skinned back of its insulation the first $1\frac{1}{2}$ inches and threaded thru an aperture of the post and fastened with a set screw. I was sitting inside with a fellow operator, rear door wide open and playing cards while monitoring radio traffic. Locking the telegraph key down to the "send" position, I turned the transmitter on, leaned over it and [taking] a common wooden lead pencil, I put it close to the antenna coupling until I "drew forth" an arc whereof I then leaned in, with my cigarette to mouth, and lit it from the arcing. As I leaned back to sitting again, up from behind the transmitter floated that "shimmering fireball". Shimmering, pulsing, blue of its fire, it then floated right at us (18 inches [0.5 m] in diameter) (of a truth, to now I can't say for sure whether I had yet shut the transmitter off or not). We two young men that we were, this experience was beyond us, and we fought ourselves to bail out of that truck before that object could touch us. Hitting the ground, we turned to look back, and there was nothing to be seen of it anymore. In retrospect, I would have to say that the fireball had to originate at the same antenna tie-in as where I had lit the cigarette from. It floated up from that very side where the antenna post lead was connected. And then bee-lined straight for the open doorway that we were in front of.

20.3. Ball lightning statistics

The authors of the surveys listed in Table 20.1 have compiled a variety of statistical information from the reports collected. Some of those statistics follow. In two different surveys, nearly 85 percent of observers according to McNally (1966) and 73 percent of observers according to Rayle (1966) reported that ball lightning was seen following a lightning flash, and in each survey nearly 90 percent said that the flash was near the ground, presumably implying a cloud-to-ground discharge. Ball lightning is described as spherical in about 90 percent of reports (Rayle 1966). Other shapes described include ellipsoids, rings, rods, and irregular forms (Rayle 1966; Amirov et al. 1995). Statistical surveys of ball lightning reports have given fairly consistent data concerning reported diameters, which are usually between 2 and 50 cm (see Figs. 20.2a, b). Rayle (1966) found that estimates of ball lightning diameter followed a log-normal distribution, with a median value of 36 cm. A log-normal distribution was also found by Dijkhuis (1992a, b). McNally's (1966) distribution of ball lightning durations is given in Fig. 20.3. Typical durations are 1 to 4 s. Rayle (1966) noted that durations refer to the time the ball was in sight, not the lifetime of the ball because the majority of observers did not see the formation and/or the decay of the ball.

Most reports describe ball lightning as "bright enough to be clearly visible in daylight". Rayle (1966) found that this was so in 60 percent of reports. He found that 76 percent of reports stated that the ball was uniformly illuminated across its surface. In only 11 percent was there an indication of limb-darkening. Dijkhuis (1992a, b) reported a log-normal distribution for ball lightning luminosity. Amirov and Bychkov (1997) found that the reported diameter, lifetime, and radiation power of ball lightning depended on whether it was seen indoors or outdoors. The frequency distribution of ball lightning color from four surveys is given in Fig. 20.4. The average distribution is labeled "sum".

Rayle (1966) indicated that the size, luminosity, and appearance of ball lightning are fairly constant with time. While most often ball lightning moves more or less horizontally, various other kinds of motion of ball lightning have been described, including (i) from clouds to near the earth, or from near the earth to clouds, (ii) a spiral, zigzag, or random path above the earth, (iii) motionless above the earth, and (iv) movement between clouds (Barry 1966). Ball lightning does not usually rise as would heated air in the absence of other forces.

20.4. Ball lightning theories

A successful ball lightning theory should at minimum account for the first four of the following eight characteristics or sets of characteristics:

(i) ball lightning's association with thunderstorms or with cloud-to-ground lightning;
(ii) its reported shape, diameter, and duration, and the fact that its size, luminosity, and appearance generally do not change much throughout its lifetime;
(iii) its occurrence in both open air and in enclosed spaces such as buildings or aircraft;

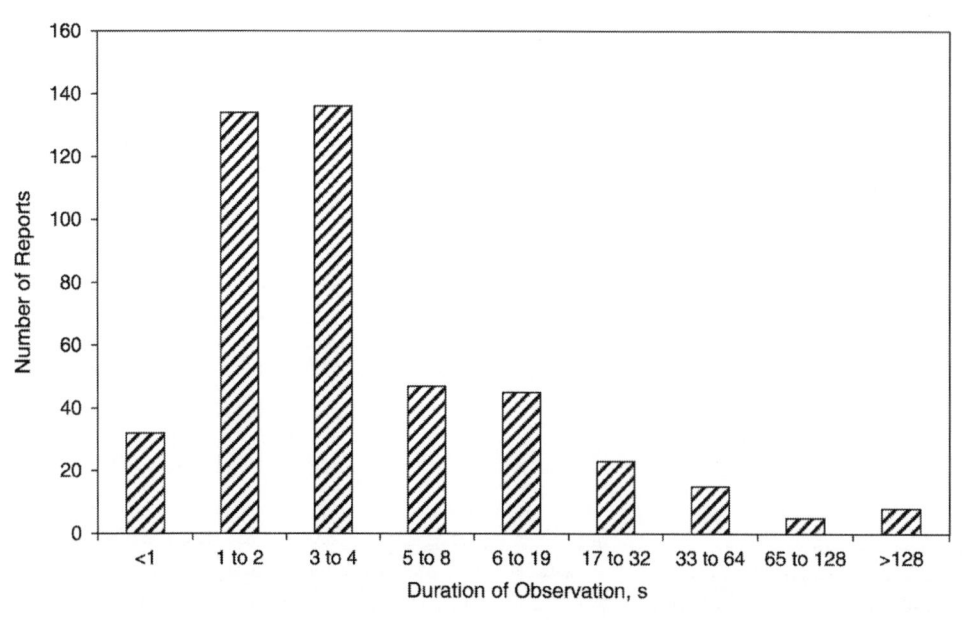

Fig. 20.3. Frequency distribution of ball lightning duration of observation from 445 reports. Adapted from McNally (1996).

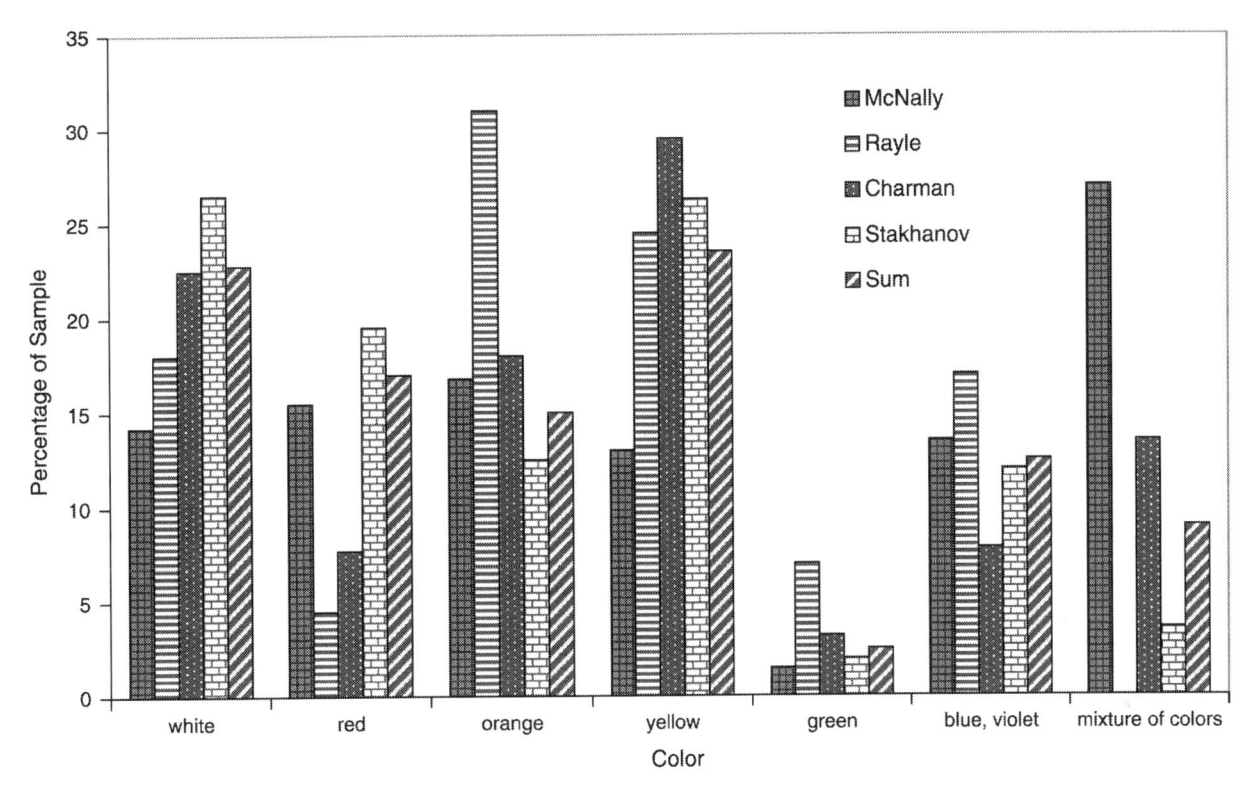

Fig. 20.4. Frequency distribution of reported ball lightning color from four surveys. Adapted from Smirnov (1987c).

(iv) the fact that ball lightning motion is inconsistent with the convective behavior of a hot gas;
(v) the fact that it may decay either silently or explosively;
(vi) the fact that ball lightning does not often cause damage;
(vii) the fact that it appears to pass through small holes, through metal screens, and through glass windows;
(viii) the fact that it is occasionally reported to produce acrid odors and/or to leave burn marks, is occasionally described as producing hissing, buzzing, or fluttering sounds, and is sometimes observed to rotate, roll, or bounce off the ground.

Theories of ball lightning are summarized in Table 20.2, where the various models are divided into two classes, internally powered and externally powered. The most reasonable candidates for the true mechanism or mechanisms of ball lightning, in the authors' opinion, are in the category postulating internal energy sources and involve the processes listed in the first three items for such sources, involving air with material in it, that material being produced when the preceding lightning strikes a tree, metal, or the ground. The most recent of the internally powered theories is that of Abrahamson and Dinniss (2000), who have postulated that silicon nanoparticles evaporated from sand by the lightning form the ball and by their oxidation illuminate the ball. Detailed, if not always sufficiently critical, discussions of most of the theories referenced in Table 20.2 are found in Stenhoff (1999) and in Barry (1980b). Since no published theory is adequate, or at least is adequately developed, and since a proper discussion of the many proposed theories would fill another book, the reader is referred for further information to Stenhoff (1999), Barry (1980b), and the references therein.

20.5. Laboratory simulation of ball lightning

There have apparently been no convincing and independently repeated laboratory simulations of ball lightning. It is certainly possible to produce glowing regions of air in high-power microwave cavities, in low-pressure dc gas discharges, in transient electrical discharges across ablative materials, and in combustion or explosion experiments. None of these discharges exhibits the observed salient characteristics (see items (i) to (iv) in Section 20.4) of ball lightning, however. Barry (1968a) described a "combustion" experiment in which a 0.5 cm spark ignited a mixture of 98.6 to 98.2 percent air and 1.4 to 1.8 percent propane to produce a yellow-green ball of a few centimeters diameter and 2 s lifetime. He gives references to earlier similar experiments. Stenhoff (1999) stated that the company Convectron succeeded in the 1980s in reproducing discharges very similar to the ball lightning reported to occur in high-current

Table 20.2. *Some ball-lightning-model categories with representative references*

Internal energy source	
1. Heated spheres of air or air mixed with trace materials at atmospheric pressure (Lowke et al. 1969)	6. Air vortex that provides containment for luminous gases (Coleman 1993; Dauvillier 1965; Voitsekhovskii, B.V. and Voitsekhovskii, B.B.1974.)
2. Dust, droplets, aerosols, aggregations (Frenkel 1940; Schwegler 1951; Aleksandrov et al. 1982; Mukharev 1986; Bychkov 1993, 1994; Bychkov et al. 1996b; Smirnov 1987a, b, c, 1993a; Abrahamson and Dinniss 2000)	7. Electromagnetic fields contained within a thin spherical shell of plasma (Dawson and Jones 1969; Jennison 1973, 1987, 1990; Edean 1976, 1992a, 1993; Zheng 1990)
3. Chemical reactions or combustion (Smirnov 1975, 1976, 1977; Benedicts 1951; Nauer 1953; Barry 1967a, b, 1968a, b, c; Blok et al. 1981; Lowke et al. 1969)	8. Nuclear reactions, matter–antimatter annihilation (Dauvillier 1957; Altschuler et al. 1970)
4. Very-high-density plasma, which exhibits quantum-mechanical properties characteristic of the solid state (Neugebauer 1937; Dijkhvis 1980, 1981, 1982)	9. Miniature black holes (Rabinowitz 1999)
	10. Charge separation (Puhringer 1967; de Tessan 1859a, b; Hill 1960; Stakhanov 1973, 1974, 1975; Mesenyashin 1991; Sanduloviciu and Lozheanu 2000)
5. Closed-loop current contained by its own magnetic field (Wooding 1963; Shafranov 1957a, b; Alexeff and Rader 1992; Ranada and Trueba 1996; Ranada et al. 1998)	11. Maser theory: population inversion of rotational energy levels of water vapor (Handel and Leitner 1994)

External energy source	
1. Focused atmospheric high-frequency (hundreds of megahertz) electromagnetic fields (Kapitza 1955; Cerrillo 1943; Handel and Leitner 1994).	3. Focused cosmic rays (Rice-Evans 1982; Arabadji 1956; Chalmers 1976)
	4. Antimatter meteors (Ashby and Whitehead 1971; Crawford 1972; Cecchini et al. 1974)
2. Steady and locally focused current flow from cloud to ground (Finkelstein and Rubinstein 1964; Powell and Finkelstein 1969, 1970; Uman and Helstrom 1966)	5. Electric fields at ground level produced by long-lasting lightning-deposited charge on Earth (Lowke 1996)

submarine battery bank discharges (Silberg 1962, 1965). Stenhoff stated that the currents necessary to produce the balls were over 150 kA, that the balls created had 10 cm diameter and 1 s lifetime, and that they were recorded on film. No reference was given. Golka (1994) stated that he had created similar balls by 1.2 kA 60 Hz discharges between copper and aluminum electrodes under water. Supporting data were not presented. Andrianov and Sinitsyn (1977a, b) simulated the gas discharge effects occurring when lightning strikes and melts sand. They used a low-pressure chamber whose walls were erodable and in which the electrical discharge ruptured a diaphram producing luminous vortex rings 4 cm in diameter with lifetimes of order 0.1 s. Similar experiments have been reported by Bychkov et al. (1997). Avramenko et al. (1990, 1992) reported making discharges in cylindrical chambers containing erodable walls and claimed that some of the discharges had properties reminiscent of ball lightning. Powell and Finkelstein (1969, 1970) described discharges in air produced by a 75 MHz radio-frequency generator and attempted to relate the observed air luminosity to ball lightning. Ohtsuki and Ofuruton (1991) produced air fireballs in a metal cavity using microwave radiation, and Ofuruton et al. (2001) generated fireballs in air without a resonant cavity by illuminating a high-voltage discharge in a 3 mm gap with 2.45 GHz microwaves. Sanduloviciu and Lozneanu (2000) described experiments in which a dc source is used to initiate a spark and a 90 MHz radio-frequency generator is then used to produce a fireball from the spark. Since many of the laboratory simulations involve radio-frequency fields, it is worth noting that radio-frequency fields of the magnitude used in the laboratory have never been reported to occur in the near vicinity of a lightning flash, although it is not evident that any serious effort has been made to obtain such measurements.

20.6. Bead lightning

Bead lightning is a well-documented optical phenomenon in which the lightning channel to ground breaks up, or appears to break up, into luminous fragments generally reported to be some tens of meters in length. These beads appear to persist for a longer time than does the usual cloud-to-ground discharge channel. Bead lightning occurs relatively infrequently, or at least is reported relatively infrequently. Still-camera photographs purported to be of bead lightning were published by Scheminzky and Wolf (1948) and by Matthias and Buchsbaum (1962). In neither case was the phenomenon in question seen by the observer. The photograph of Scheminzky and Wolf shows 73 beads each of about 10 m length centered along what appears to be a black channel. The photograph of Matthias and Buchsbaum shows three beads, each of a different length, at the top of the photograph with no visible channel beneath to ground. A high-speed motion picture, taken by US Navy personnel, of lightning initiated from a plume of water rising from a depth charge (Brook et al. 1961; Young 1962; Uman 1986, 1987) shows, in the final frames of each of three strokes, a channel that has broken up into light and dark sections, probably an example of bead lightning. The sections are a few meters in length. One frame of the motion picture is reproduced in Fig. 20.5. Similar observations were reported by Fisher et al. (1993) for rocket-triggered lightning (see subsection 4.7.8). Various theories of bead lightning have been set forth and some of these are now described.

(i) Bead lightning arises because the observer views portions of the lightning channel end-on. That is, if the lightning channel is coming toward or going away from an observer, the observer will see a greater length of the channel within a given viewing angle than would be the case were the channel perpendicular to his or her viewing direction. The greater length of channel thus appears to the eye as a normal channel of greater-than-normal brightness. To give the beaded effect, the channel must periodically be slanted toward or away from the observer.

(ii) Bead lightning is due to the periodic masking of a normal cloud-to-ground channel by clouds or rain.

(iii) Bead lightning is due to a magnetic pinch-effect instability by which the current-carrying channel is distorted into a "string of sausages", the strong light emission coming from the pinched regions of high current density and high particle density (Uman 1962). It is possible for the magnetic-pinch effect to occur if the lightning current is high enough while the current-carrying radius is small. For example, a current of 80 kA flowing in a channel of 1 cm radius produces a radially inward magnetic pressure of 10 atm at the surface (see also subsection 12.2.2). If this exceeds the kinetic pressure of the channel, the channel will pinch. The magnetic pressure varies as the square of the current. Whether conditions conducive to a significant magnetic pinch effect ever occur in lightning is not known. To obtain a series of beads by the pinch-effect mechanism, the channel must somehow be conditioned periodically as a

Fig. 20.5. Bead lightning as it appears on one frame of a motion picture showing a three-stroke flash initiated from the top of a plume of water rising from the explosion of a depth charge. Courtesy, US Naval Laboratory Ordnance.

function of height in the final stages of a cloud-to-ground flash.

(iv) Bead lightning is a series of spherical arcs formed out of a defunct lightning channel by a resurgence of channel current (Uman and Helstrom 1966). Calculations describing one of these spherical arcs, as part of a theory of ball lightning, are presented by Uman and Helstrom (1966) (see Table 20.2, external energy source, second item). Each spherical arc is formed by the constriction of a diffuse cloud-to-ground current that produces little light into a region of relatively high temperature (near 5000 °K) and relatively high conductivity (10^{-2} S m^{-1}), a sort of thermal pinch effect operating in space. The constricted current and the energy input associated with it serve to maintain the arc.

(v) Bead lightning is due to the long lifetimes of luminous sections of the lightning channel of exceptionally large radius (Uman and Voshall 1968). As noted in subsection 4.7.8, channels of large radius take longer to cool than channels of small radius. To a first approximation the cooling time is proportional to the square of the radius. If the lightning-channel radius were somehow periodically modulated as a function of height then, as the channel luminosity decayed, the channel would take on the appearance of a string of beads. Perhaps such modulation occurs accidentally when the channel consists of a large number of kinks or bends.

In order to account for the observed long persistence of the beads, those theories which require current flow should invoke the existence of long-continuing currents following a final stroke in a cloud-to-ground flash. Of the theories outlines above, (v) is probably the most reasonable. It is not impossible that the long persistence of bead lightning reported by observers is due in part to the characteristics of the human eye, in addition to the lightning-channel physics.

20.7. Other types of unusual lightning and lightning-like discharges

20.7.1. Volcano lightning

Lightning-like discharges not associated with thunderstorms have been observed to originate in the ejected material (primarily ash) above some active volcanoes (e.g., Anderson et al. 1965; Brook et al. 1974a, b; Pounder 1980; Katahira 1992; McNutt and Davis 2000), as illustrated in Fig. 20.6. These discharges are typically hundreds of meters in length and occur both inside the ash cloud and between that cloud and ground.

McNutt and Davis (2000) listed in their Table 1 about 60 volcanoes that had been reported to have produced lightning upon eruption, along with the dates of those events and corresponding references. Some of the more detailed observations are discussed below.

Anderson et al. (1965) studied the electrical phenomena associated with the formation of the volcanic island Surtsey in the Westmann Islands off the south coast of Iceland in 1964. The electric field intensity measured near the volcano rapidly intensified during vigorous eruptions of black clouds. Lightning was observed only during these eruptions. The observed overhead charge was positive, and all discharges reduced that positive charge and returned the ambient field to near the fair-weather value (subsection 1.4.2), indicating that most of the available volcano cloud charge was neutralized. High electric fields were often regenerated in 10 to 40 s. Lightning occurred both in the clouds and between the cloud and the island or the ocean. Most discharges to ground struck the island near the source of the eruptions. Anderson et al. (1965) estimated that 0.1 to 0.5 C of positive charge was neutralized per event in many of the discharges.

Another volcanic eruption in the Westmann Islands occurred on the island of Heimaey in 1973, and the resulting lightning was studied by Brook et al. (1974a, b). They reported that the positive charge carried aloft by the clouds was formed when molten lava came in contact with seawater, the resultant electric field values on the ocean surface sometimes exceeding 7 kV m^{-1}. Lightning flashes were associated with the more violent steam eruptions. The flashes lowered positive charge to earth in single strokes. As inferred from measured electric field changes and from time-lapse photographs, the charge appeared to be distributed in a vertical column extending upward from the sea to a height not exceeding about 100 m. As noted in Section 5.2 and subsection 18.4.2, smoke from forest fires can form clouds that produce positive lightning to Earth in situations where no clouds were previously present.

While only positive discharges were observed in the two volcanic eruptions in the Atlantic Ocean near Iceland in 1964 and 1973, Hoblitt (1994) and McNutt and Davis (2000) have reported that both positive and negative discharges were detected at Redoubt volcano and Mount Spurr volcano, both located in Alaska, from measurements made with the BLM magnetic direction finder network (Sections 17.3 and 17.5). Such measurements are clearly more subject to misidentification of stroke polarity than the close electric field measurements of Anderson et al. (1965) and Brook et al. (1974a, b). That is, magnetic direction finding systems designed to determine the polarity of natural ground strokes could potentially have difficulty making such a determination properly if the volcano-lightning field waveshapes were significantly different from those of natural-lightning strokes. At both the Redoubt and Spurr volcanoes, the magnetic direction finder network indicated that the lightning charge lowered to ground was negative for the initial lightning flashes and later changed

20.7. Other types of unusual lightning and lightning-like discharges

Fig. 20.6. Lightning produced by the volcano cloud over Surtsey, near Iceland, in December 1963. Volcanic eruptions were observed on 14 November 1963 off the southern coast of Iceland in water 130 m deep. Within 10 days an island nearly 1 km long and about 100 m above sea level was formed. The island was named Surtsey by the Icelandic government. Additional discussion is found in Anderson et al. (1965). The photograph originally appeared on the cover of the issue of Science (28 May 1965) containing that article. Copyright 1965 by the AAAS. Courtesy of Sigurgeir Jónasson, Icelandair.

to positive. Hoblitt (1994) stated that both positive and negative polarity discharges also were detected during the 1980 eruption of Mount St Helens in Washington state and that most of the discharges at Mount St. Helens appeared to terminate at the top of the volcano. Information derived from the measurement of remote electromagnetic fields from volcano lightning was also given by Katahira (1992) for the Mount Fugen volcano and by Ondoh (1992) for the Izu-Oshima volcano, both located in Japan.

McNutt and Davis (2000) reported that lightning occurred during all three of Mount Spurr's eruptions in 1992, and that in each case there was a time delay of 21 to 26 minutes from the beginning of the eruption to the first recorded lightning discharge. This delay is two to five times longer than that observed by Hoblitt (1994) for the Redoubt volcano. Interestingly, the tephra compositions of Spurr and Redoubt are different, and McNutt and Davis suggested that the higher silica content at Redoubt provided more efficient electrical charging and hence shorter time delays. McNutt and Davis found no relation between the size of the eruptions at Spurr and the number of lightning discharges produced, but Hoblitt (1994) found that at Redoubt these quantities were positively correlated.

The magnetic direction finder network in Iceland recorded some of the lightning discharges produced during the eruptions of Grímsvötn and Hekla in 1998 and 2000, respectively (Arason et al. 2000). Thirty-three negative and 18 positive flashes were detected at Grímsvötn. Interestingly, the geometric-mean peak currents of either polarity for volcano lightning were considerably (a factor 2–3) lower than those for ordinary lightning recorded by the same network.

Proponents of the existence of lightning on Venus initially associated this hypothetical lightning with active volcanoes (Section 16.3). Later, this hypothesis was abandoned in favor of a cloud origin. As discussed in Section 16.3, the argument for Venusian lightning is presently less than convincing.

In another type of charging related to turbulent particulate material, shorter discharges than volcano lightning, about 1 m in length, have been observed in New Mexico gypsum sandstorms (Kamra 1972a, b).

20.7.2. Earthquake lightning

Transient luminous phenomena that may be due to electrical discharges have long been reported during earthquakes and more recently attributed to the electric fields generated by seismic strain (e.g., Finkelstein and Powell 1970; Ikeya and Takaki 1996; Enomoto and Zheng 1998). What is called earthquake lightning is a very poorly understood phenomenon. Earthquake lights, steady atmospheric glows associated with earthquakes, are a more commonly reported phenomena, however. Some observations of steady or transient luminosity associated with earthquakes in the twentieth century are likely to be due to faults (flashovers) in the local electric power systems rather than from other causes associated more directly with the earthquake.

20.7.3. Nuclear lightning

By virtue of the detonation of thermonuclear devices (H-bombs) at ground level, negative charge is deposited in the atmosphere, which results in kilometer-length discharges such as those shown in Fig. 20.7 (e.g., Uman *et al.* 1972; Hill 1973; Grover 1981; Gardner *et al.* 1984; Williams *et al.* 1988). The upward-going leaders associated with this so-called nuclear lightning appear to be similar to the upward unconnected discharges that are initiated from grounded objects in response to nearby or distant natural or rocket-triggered lightning (subsection 4.5.2).

Uman *et al.* (1972) studied the lightning flashes associated with a thermonuclear detonation at Eniwetok Atoll in the Pacific in 1952, seen in Fig. 20.7. The five discharges were upward propagating and were apparently initiated from instrumentation stations projecting above sea level. The likely mechanism for the necessary charge and electric-field generation, Compton electrons produced by gamma-rays from the detonation, is described in Uman *et al.* (1972). The surface electric-field intensity calculated, 30 kV m^{-1}, is indeed near the value needed to initiate lightning from surface projections. The charge estimated to be available, however, appeared to be insufficient to account for the observed lightning. Detailed theoretical studies of the nuclear lightning were subsequently carried out by Grover (1981), Longmire *et al.* (1982) and Gardner *et al.* (1984). Colvin *et al.* (1987) measured the lightning-channel luminous

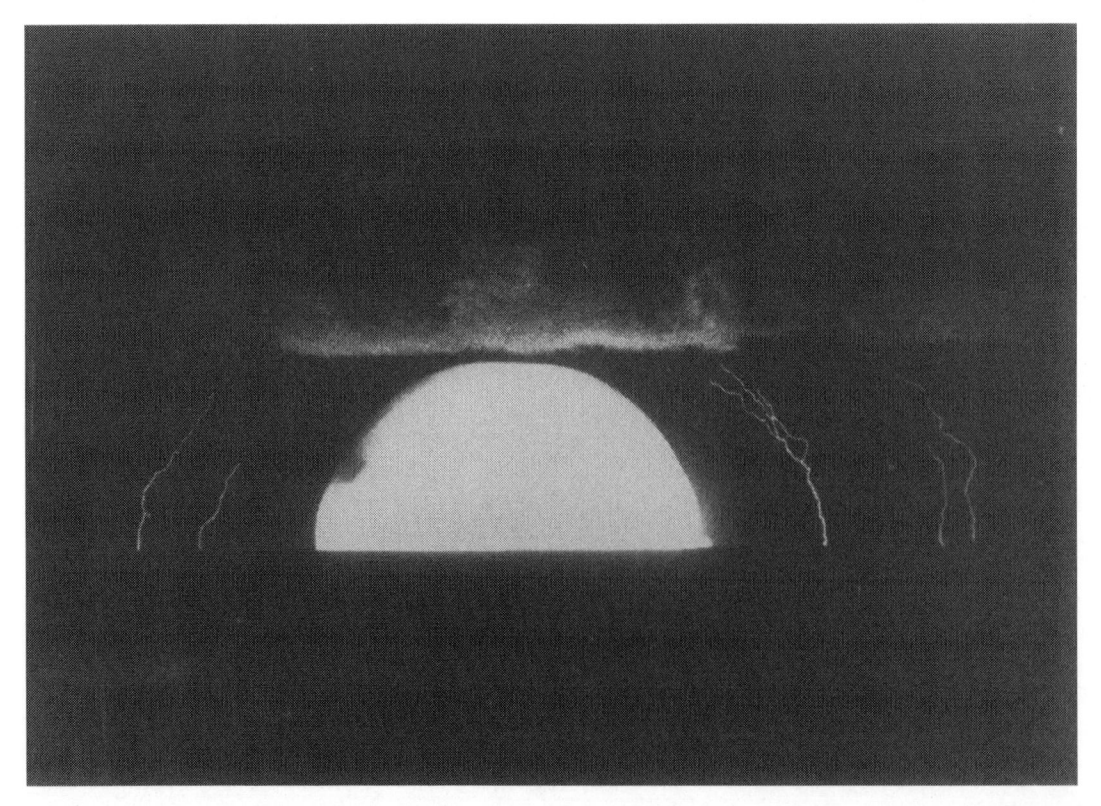

Fig. 20.7. Five lightning flashes produced by an experimental thermonuclear device exploded on 31 October 1952, at Eniwetok in the Pacific. The photograph is frame number 72 (detonation occurred in frame 1) of a 2000 frames per second movie taken from about 30 km away. The five flashes were initiated immediately after detonation, probably from instrumentation stations projecting above sea level. The tops of the lightning channels bend toward the fireball. Scattered trade-wind cumulus clouds are visible with bases at about 600 m, roughly the radius of the fireball in frame 72. The clouds were not lightning producers. The photograph was originally published in Uman *et al.* (1972). The lightning channels in the photograph were enhanced for illustrative purposes. Courtesy of US Atomic Energy Commission.

intensity from photographic records of the nuclear lightning and compared this intensity with the measured intensity of laboratory arcs. On the basis of these comparisons they estimated that the likely peak currents in the nuclear lightning were 250 kA ± 50 kA, consistent with the theoretical estimates of Gardner *et al.* (1984) but considerably larger than the initial estimates of Uman *et al.* (1972) and Hill (1973).

20.8. Concluding comments

Ball lightning is a visually well-documented phenomenon. Given the wide range of observed characteristics, conditions under which it occurs, and locations where it occurs, there may well be more than one type of ball lightning and more than one mechanism that creates it. Supporting this point of view is the fact that similar phenomena are generated accidentally from high-power electrical apparatus in the absence of nearby thunderstorms. As of this writing, there is no widely accepted theory of ball lightning and no confirmed laboratory simulation. Bead lightning is a little less mysterious than ball lightning since it clearly involves an already existing lightning channel, but no mechanism for its formation is generally accepted. Finally, charge separation in a variety of non-thunderstorm situations, from sand storms to volcano eruptions to nuclear explosions, can result in the production of lightning-like electrical discharges.

References and bibliography for Chapter 20

Abrahamson, J., and Dinnis, J. 2000. Ball lightning caused by oxidation of nanoparticle networks from normal lightning strikes on soil. *Nature* **403**: 519–521.

Aleksandrov, V.Ya., Golubev, E.M., and Podmoshenskii, I.V. 1982. Aerosol nature of ball lightning. *Zh. Tekh. Fiz.* **52**: 1987 (in Russian).

Alexeff, I., and Rader, M. 1992. Observation of closed loops in high-voltage discharges: a possible precursor of magnetic flux trapping. *IEEE Trans. Plasma Sci.* **20**: 669–70.

Alexeff, I., and Rader, M. 1995. Possible precursors of ball lightning – observation of closed loops in high-voltage discharges. *Fusion Tech.* **27**: 271–3.

Altschuler, M.D., House, L.L., and Hildner, E. 1970. Is ball lightning a nuclear phenomen? *Nature* **228**: 545–7.

Amirov, S.K., Bychkov, V.L., and Strizhev, A.Yu. 1995. Principles of creating and processing data banks: meteorological applications – illustrated with regard to ball-lightning data processing. *J. Meteorol.* **20**: 85.

Amirov, A.Kh., Bobkov, S.E., and Bychkov, V.L. 1998. On the dependence lifetime-diameter of ball lightnings. *Physica Scripta* **58**: 56–60.

Amirov, S.K., and Bychkov, V.L. 1997. Influence of thunderstorm atmospheric conditions on the properties of ball lightning. *Zh. Tekh. Fiz.* **67**: 117 (in Russian).

Anderson, F.J., and Freier, G.D. 1972. A report on ball lightning. *J. Geophys. Res.* **77**: 3928–30.

Anderson, R., Bjornsson, S., Blanchard, D.C., Gathman, S., Hughes, J., Jonasson, S., Moore, C.B., Survilas, H.J., and Vonnegut, B. 1965. Electricity in volcanic clouds. *Science* **148**: 1179–89.

Andrianov, A.M., and Sinitsyn, V.I. 1977a. Erosion-discharge model for ball lightning. *Sov. Phys. Tech. Phys.* **22** (11): 1342.

Andrianov, A.M., and Sinitsyn, V.I. 1977b. Ispol'zovanie erozionnogo raziada dlia modelirovaniia odnogo iz vozmuzhnykh vidor sharovoi molnii. *Zh. Tekh. Fiz.* **47**: 2318.

Arabadji, W.I. 1956. K teorii yavlenii atmosfernogo elektrischestva. *Minskogo Gosudarstvennogo Pedagogicheskogo Instituta* **5**: 77.

Arabadzhi, V.I. 1976. On the problem of ball lightning. *J. Geophys. Res.* **81**: 6455.

Arason, P., Sigurdsson, E., Kristmundsson, G.M., Johannsdottir, H., and Juliusson, G. 2000. Volcanogenic lightning during a sub-glacial eruption in Iceland. In *Proc 25th Int. Conf. on Lightning Protection, Rhodes, Greece*, pp. 100–2.

Argyle, E. 1971. Ball lightning as an optical illusion. *Nature* **230**: 179–80.

Ashby, D.E.T.F., and Whitehead, C. 1971. Is ball lightning caused by antimatter meteorites? *Nature* **230**: 180–2.

Avramenko, R.F., Bakhtin, B.I., Nikolaeva, V.I., Poskacheeva, L.P., and Shirokov, N.N. 1990. A study of the plasma formations produced in an erosion discharge (English transl.). *Sov. Phys. Tech. Phys.* **35**: 1396–1400.

Avramenko, R.F., Gridin, A.Yu., Klimov, A.I., and Nikoleava, V.I. 1992. Experimental study of energetic compact plasma formations (English transl.). *High Temperature* **30**: 870.

Bailey, B.H. 1977. Ball lightning. *Weatherwise* **30**: 99–105.

Barry, J.D. 1966. Ball lightning – a natural phenomenon in atmospheric physics. M.S. thesis, Los Angeles: California State College.

Barry, J.D. 1967a. Ball lightning. *J. Atmos. Terr. Phys.* **29**: 1095–101.

Barry, J.D. 1967b. A model for ball lightning. *Wiss. Zeit. Elektro.* **8**: 202.

Barry, J.D. 1968a. Laboratory ball lightning. *J. Atmos Terr. Phys* **30**: 313–17.

Barry, J.D. 1968b. Fireball, ball lightning and St. Elmo's fire. *Weather* **23**: 180.

Barry, J.D. 1968c. Ball lightning in the laboratory. *Wiss. Zeit. Elektro* **12**: 7.

Barry, J.D. 1974. Bibliography of ball lightning. *J. Atmos. Terr. Phys.* **36**: 1577.

Barry, J.D. 1979. Frequency of ball lightning reports. *J. Geophys. Res.* **84**: 308–12.

Barry, J.D. 1980a. On the energy density and forms of ball lightning. *J. Geophys. Res.* **85**: 4111–14.

Barry, J.D. 1980b. *Ball Lightning and Bead Lightning*, New York: Plenum Press.

Benedicks, C. 1951. Theory of lightning balls and its application to the atmospheric phenomenon called 'flying saucers'. *Ark. Geofys.* **2**: 1.

Berger, K. 1973. Kugelblitz und Blitzforschung. *Naturwissenschaften* **60**: 485–92.

Blair, A.J.F. 1973. Magnetic fields, ball lightning and campanology. *Nature* **243**: 512–13.

Blok, V.R., Kogan, A.Y., Kogan, B.Y., Krochik, G.M., and Lebedev, O.L. 1981. Stability of the ball lightning. *Zh. Tekh. Fiz.* **51**: 2190.

Boichenko, A.M. 1996. On the nature of bead lightning. *Plasma Phys. Rep.* **22**: 1012–16.

Brand, W. 1923. *Der Kugelblitz*, Hamburg, Germany: Grand.

Brook, M., Armstrong G., Winder, R.P.H., Vonnegut, B., Moore, C.B. 1961. Artificial initiation of lightning discharges. *J. Geophys. Res.* **66**: 3967–9.

Brook, M., Moore, C.B., and Sigurgeirsson, T. 1974a. Lightning in volcanic clouds. *J. Geophys. Res.* **79**: 472–5.

Brook, M., Moore, C.B., and Sigurgeirsson, T. 1974b. Correction to "Lightning in volcanic clouds". *J. Geophys. Res.* **79**: 3102.

Brovetto, P., Maxia, V., and Bussetti, G. 1976. On the nature of ball lightning. *J. Atmos. Terr. Phys.* **38**: 921–34.

Bychkov, V.L. 1993. Ball lightning as a polymer formation. *J. Meteorol.* **18**: 149.

Bychkov, V.L. 1994. Polymer ball lightning model. *Physica Scripta* **50**: 591–9.

Bychkov, V.L., Smirnov, B.M., and Stridjev, A. Ju. 1993. Analysis of the Russian–Austrian ball lightning data banks. *J. Meteorol.* **18**: 113.

Bychkov, V.L., Bychkov, A.V., and Standnik, S.A. 1996a. Polymer fire balls in discharge plasma. *Physica Scripta* **53**: 749–59.

Bychkov, V.L., Bychkov, A.V., Vasiliev, M.N., and Klimov, A.I. 1996b. Evaluation of possible ball lightning energy by analysing an event involving damage to a metal pot. *J. Meteorol.* **21**: 77.

Bychkov, V.L., Emelin, S.E., Klimov, A.I., and Semenov, V.S. 1997. Approaches to ball lightning modelling with erosive discharge. In *Proc. 5th Int. Symp. on Ball Lightning (ISBL97)*, eds. Y.H. Ohtsuki and H. Ofuruton, p. 188.

Cecchini, S., Diocco, G., and Mandolesi, N. 1974. Positron annihilation in EAS and ball lightning. *Nature* **250**: 637–8.

Cerrillo, M. 1943. Sobre las posibles interpretaciones electromagneticas del fenomeno de las centellas (The possible electromagnetic interpretations of the phenomenon of lightning). Comision Impulsora Coordinadora Invest. Client., Mexico, *Ann.* **1**: 151–78.

Chalmers, J.A. 1976. *Atmospheric Electricity*, London: Pergamon Press.

Charman, W.N. 1971a. Perceptual effects and the reliability of ball lightning reports. *J. Atmos. Terr. Phys.* **33**: 1973–6.

Charman, W.N. 1971b. After-images and ball lightning. *Nature* **230**: 576.

Charman, W.N. 1972. The enigma of ball lightning. *New Scientist* **56**: 632–5.

Charman, W.N. 1976. Ball lightning photographed? *New Scientist* **60**: 444–7.

Charman, W.N. 1979. Ball lightning. *Phys. Rep.* **54**: 261–306.

Chown, M. 1993. Fire and water: a recipe for ball lightning. *New Scientist* **20**: 18.

Coleman, P.F. 1993. An explanation of ball lightning? *Weather* **48**: 30.

Colvin, J.D., Mitchell, C.K., Greig, J.R., Murphy, D.P., Pechacek, R.E., and Raleigh, M. 1987. An empirical study of the nuclear explosion-induced lightning seen on ivy-mike. *J. Geophys. Res.* **92**: 5695–712.

Covington, A.E. 1970. Ball lightning. *Nature* **226**: 252–3.

Crawford, J.F. 1972. Antimatter and ball lightning. *Nature* **239**: 395.

Dauvillier, A. 1957. Foudre globulaire et reactions thermonucléaires. *C. R. Hebd. Séances Acad. Sci.* **245**: 2155–6.

Dauvillier, A. 1965. Sur la nature de la foudre globulaire. *C. R. Hebd. Séances Acad. Sci.* **260**: 1707–8.

Davidov, B. 1958. Rare photograph of ball lightning. *Priroda* **47**: 96–7 (in Russian).

Davies, D.W., and Standler, R.B. 1972. Ball lightning. *Nature* **240**: 144.

Davies, P.C.W. 1971. Ball lightning or spots before the eyes. *Nature* **230**: 567–77.

Dawson, G.A., and Jones, R.C. 1969. Ball lightning as a radiation bubble. *Pure Appl. Geophys.* **75**: 247–62.

de Tessan, M. 1859a. Sur la foudre en boule. *C.R. Hebd. Séances Acad. Sci.* **49**: 189–98.

de Tessan, M. 1859b. Sur la foudre en boule (Summary). *Fortschr. Phys.* **15**: 62.

Dewan, E.M. 1964. Eyewitness accounts of Kugelblitz. Microwave Physics Laboratory, Air Force Cambridge Research Laboratories, CRD-125 (March 1964).

Dibner B. 1977. Benjamin Franklin. In *Lightning, vol. 1, Physics of Lightning*, ed. R.H. Golde, pp. 23–49, New York: Academic Press.

Dijkhuis, G.C. 1980. A model for ball lightning. *Nature* **285**: 150–1.

Dijkhuis, G.C. 1981. Reply to E.A. Witalis. *Nature* **290**: 166.

Dijkhuis, G.C. 1982. Threshold current for fireball generation. *J. Appl. Phys.* **53**: 3516–19.

Dijkhuis, G.C. 1992a. Statistics and structure of ball lightning. In *Proc. 4th TORRO Conf. on Ball Lightning*, ed. M. Stenhoff, p. 61, Tornado and Storm Research Organization, Oxford, UK: Oxford Brooks University.

Dijkhuis, G.C. 1992b. Statistics and structure of ball lightning. In *Proc. 3rd Int. Symp. on Ball Lightning*.

Dmitriyev, M.T. 1969. Stability mechanism for ball. *Sov. Phys. Tech. Phys.* **14**: 284–9 (English translation).

Edean, V.G. 1976. Ball lightning as electromagnetic energy. *Nature* **263**: 753–5.

Edean, V.G. 1978. Ball lightning. In *Proc. 5th Int. Conf. on Gas Discharges*, p. 116.

Edean, V.G. 1992a. Electromagnetic-field energy containment. In *IEEE Proc. A (Sci. Meas. and Tech.)* **139**: 137–.

Edean, V.G. 1992b. Electromagnetic field energy models – some recent developments. In *Proc. 4th TORRO Conf. on Ball Lightning*, ed. M. Stenhoff p. 75. Tornado and Storm Research Organisation, Oxford, United Kingdom, Oxford Brookes University.

Edean, V.G. 1993. Spinning electric dipole model of ball lightning. *IEEE Proc.* **140A**: 474.

Edean, V.G. 1977a. Development of the radiation bubble model of ball lightning. *J. Meteor.* **22**: 98.

Edean, V.G. 1977b. The virial theorem with an extension for statically electrified gaseous spheres. In *Proc. 5th Int. Symp. on Ball Lightning (ISBL97)*, eds. Y.H. Ohtsuki and H. Ofuruton, p. 96.

Egely, G. 1989a. Hungarian ball lightning observations in 1987. In *Science of Ball Lightning (Fireball)*, pp. 19–30, Singapore: World Scientific.

Egely, G. 1989b. Physical problems and physical properties of ball lightning. In *Science of Ball Lightning (Fireball)*, pp. 81–7, Singapore: World Scientific.

Enomoto, Y., and Zheng, Z. 1998. Possible evidences of earthquake lightning accompanying the 1995 Kobe earthquake inferred from the Nojima fault gouge. *Geophys. Res. Lett.* **25**: 2721–4.

Enomoto, Y., Asuke, F., Zheng, Z., and Ishigaki, H. 2001. Hardened foliated fault gouge from the Nojima Fault zone at Hirabayashi: evidence for earthquake lightning accompanying the 1995 Kobe earthquake. *The Island Arc* **10**: 447–56.

Eriksson, A.J. 1977. Video tape recording of a possible ball lightning event. *Nature* **268**: 35–6.

Felsher, M. 1970. Ball lightning. *Nature* **227**: 982.

Finkelstein, D., and Powell, J. 1970. Earthquake lightning. *Nature* **228**: 759–60.

Finkelstein, D., and Rubinstein, J. 1964. Ball lightning. *Phys. Rev. A* **135**: 390–6.

Finkelstein, D., Hill, R.D., and Powell, J.P. 1973. The piezoelectric theory of earthquake lightning. *J. Geophys. Res.* **78**: 992–3.

Fishcher, E. 1981. Ball lightning – a combustion phenomenon. *Naturwissenschafften* **68**: 568.

Fisher, R.J., Schnetzer, G.H., Thottappillil, R., Rakov, V.A., Uman, M.A. and Goldberg, J.D. 1993. Parameters of triggered-lightning flashes in Florida and Alabama. *J. Geophys. Res.* **98**: 22 887–902.

Frenkel, Y.I. 1940. O prirode sharovoi molnii. *Zh. Eksp. Teor. Fiz.* **10**: 1424.

Gardner, R.L., Frese, M.H., Gilbert, J.L., and Longmire, C.L. 1984. A physical model of nuclear lightning. *Phys. Fluids* **27**: 2694–8.

Garrett, A.J.M. 2001. Comment on "A model of ball lightning as a magnetic knot with linked streamers" by A.F. Rañada et al. *J. Geophys. Res.* **106** (D3): 2977.

Gibbs-Smith, C.H. 1971. On fireballs. *Nature* **232**: 187.

Golka, R.K. Jr 1994. Laboratory-produced ball lightning. *J. Geophys. Res.* **99**: 10 679–81.

Grover, M.K. 1981. Some analytical models for quasi-static source region EMP: Application to nuclear lightning. *IEEE Trans. Nuclear Sci.* **28**: 990–4.

Handel, P.H., and Leitner, J.F. 1994. Development of the maser-caviton ball lightning theory. *J. Geophys. Res.* **99**(D5): 10 689–91.

Hill, C.R. 1970. Radiation from ball lightning. *Nature* **228**: 1007.

Hill, E.L. 1960. Ball lightning as physical phenomenon. *J. Geophys. Res.* **65**: 1947–52.

Hill, E.L. 1970. Ball lightning. *Am. Scientist* **58**: 479.

Hill, R.D. 1973. Lightning induced by nuclear bursts. *J. Geophys. Res.* **78**: 6355–8.

Hensen, A., and Van der Hage, J.C.H. 1994. Development of the Maser-caviton ball lightning theory. *J. Geophys. Res.* **99**: 10 689–91.

Hermant, A., and Hubert, P. 1988. Ball lightning photographed? *Atmos. Res.* **22**: 275–80.

Hoblitt, R.P. 1994. An experiment to detect and locate lightning associated with eruptions of Redoubt volcano. *J. Volcanol. Geothermal Res.* **62**: 499–517.

Hubert, P. 1996. Nouvelle enquête sur la foudre en boule – analyse et discussion des résultats. Rapport PH/SC/96001, Commisariat a l'Energie Atomique, Service d'Electronique Physique, Centre d'Etudes Nucléaires de Saclay, France.

Humphreys, W.J. 1936. Ball lightning. *Proc. Am. Phil. Soc.* **76**: 613–26.

Ikeya, M. and Takaki, S. 1996. Electromagnetic fault for earthquake lightning. *Japan J. Appl. Phys.* **35**: L355–7.

Janhunen, P. 1991. Magnetically dominated plasma models of ball lightning. *Ann. Geophys.* **9**: 377–80.

Jennings, R.C. 1962. Path of a thunderbolt. *New Scientist* **13**: 156.

Jennison, R.C. 1969. Ball lightning. *Nature* **224**: 895.

Jennison, R.C. 1971. Ball lightning and after-images. *Nature* **230**: 576.

Jennison, R.C. 1973. Can ball lightning exist in a vacuum? *Nature* **245**: 95–6.

Jennison, R.C. 1987. The non-particulate nature of matter and the universe. In *Problems in Quantum Physics: Gdansk '87, Recent and Future Experiments and Interpretations*, eds. L. Kostro, A. Posiewnik, J. Pykacz, and M. Zukowski, p. 163, Singapore: World Scientific.

Jennison, R.C. 1990. Relativistic phase-locked cavity model of ball lightning. In *Physical Interpretations of Relativity Theory: Proceedings*, vol. II, p. 359, London: British Society for the Philosophy of Science.

Jennison, R.C., Lobeck, R., and Cahill, R.J. 1990. A video recording of ball lightning. *Weather* **45**: 151–2.

Jensen, J.C. 1933. Ball lightning. *Physics* (now *J. Appl. Phys.*) **4**: 372–4.

Kamra, A.K. 1972a. Visual observation of electric sparks on gypsum dunes. *Nature* **240**: 143–4.

Kamra, A.K. 1972b. Measurements of the electrical properties of dust storms. *J. Geophys. Res.* **77**: 5856–69.

Kapitsa, P.L. 1969. Ball lightning. *Physics* (now *J. Appl. Phys.*) **4**: 372–4.

Kapitza, P.L. 1955. The nature of ball lightning. *Dokl. Akad. Nauk SSSR* **101**: 245–8 (in Russian).

Katahira, O. 1992. Observation of volcanic lightning of Mt Fugen, Unzen. *Res. Lett. Atmos. Electr.* **12**: 225–34.

Keul, A.G. 1981. Ball lightning reports. *Naturwissenschaften* **68**: 134–6.

Kozlov, B.N. 1975. Principles of the relaxation theory of ball lightning. *Sov. Phys. Dokl.* **20**: 261–2 (English transl.).

Kozlov, B.N. 1978. Maximum energy liberated by ball lightning. *Sov. Phys. Dokl.* **23**: 41–2 (English transl.).

Kuhn, E. 1951. Ein Kugelblitz auf einer Moment – Aufnahme?. *Naturwissenschaften* **38**: 518–19.

Lee, W.R. 1977. Lightning injuries and death. In *Lightning, vol. 2, Lightning Protection*, ed. R.H. Golde, pp. 521–44, New York: Academic Press.

Longmire, C.L., Gardner, R.L., Gilbert, J.L., and Frese, M.H. 1982. A physical model of nuclear lightning. In *Lightning Phenomenology Notes*, no. 4.

Lowke, J.J. 1996. A theory of ball lightning as an electric discharge. *J. Phys. D: Appl. Phys.* **29**: 1237–44.

Lowke, J.J., Uman, M.A., and Lieberman, R.W. 1969. Toward a theory of ball lightning. *J. Geophys. Res.* **74**: 6887–98.

Manykin, E.A., Ozhovan, M.I., and Polyenktov, P.P. 1983. On the nature of ball lightning. In *Sov. Phys. Tech. Phys. (1982 Am. Inst. Phys.)*, pp. 905–6.

Matthias, B.T., and Buchsbaum, S.J. 1962. Pinched lightning. *Nature* **194**: 327.

McGinley, J., Ziegler, C., and Ziegler, D. 1982. Observation of steady glow and multicolored flashes associated with a thunderstorm. *Bull. Am. Meteor. Soc.* **63**: 189–90.

McNally, J.R. 1961. Ball lightning – a survey. *Bull Amer. Phys. Soc.* **6**: 202.

McNally, J.R. Jr 1966. *Preliminary Report on Ball Lightning*, Oak Ridge Natl Lab, ORNL-3938, UC-34-Phys.

McNutt, S.R., and Davis, C.M. 2000. Lightning associated with the 1992 eruptions of Crater Peak, Mount Spurr Volcano, Alaska. *J. Volcanol. Geothermal Res.* **102**: 45–65.

Mesenyashin, A.I. 1991. Electrostatic and bubble nature of ball lightning. *Appl. Phys. Lett.* **58**: 2713–15.

Mills, A.A. 1971. Ball lightning and thermoluminescence. *Nature, Phys. Sci.* **233**: 131–2.

Morris, W.A. 1936. *A Thunderstorm Mystery*, letters to the editor of *Daily Mail*, London (5 November 1936).

Mukharev, L.A. 1986. The nature of ball lightning. *Sov. J. Commun. Technol. Electron.* **30**: 77.

Muldrew, D.B. 1990. The physical nature of ball lightning. *Geophys. Res. Lett.* **17**: 2277–80.

Müller-Hildebrand, D. 1963. Zur Frage des Kugelblitzes. *Elektrie* **17**: 211–14.

Navarro-Gonzalez, R., Molina, M.J., and Molina, L.T. 1998. Nitrogen fixation by volcanic lightning in the early Earth. *Geophys. Res. Lett.* **25**: 3132–6.

Neugebauer, T. 1937. Zu dem Problem des Kugelblitzes. *Z. Physik* **106**: 474–84.

Nguyen, M.D. 1987. Lightning observation at Tam Dao. *Publs. Inst. Geophys. Pol. Acad. Sc. D-26* **198**: 75–83.

Ofuruton, H., Kondo, N., Kamogawa, M., Aoki, M., and Ohtsuki, Y.-H. 2001. Experimental conditions for ball lightning creation by using air gap discharge embedded in a microwave field. *J. Geophys. Res.* **106**: 12 367–9.

Ohtsuki, Y.H., and Ofuruton, H. 1991. Plasma fireballs formed by microwave interference in air. *Nature* **350**: 139–41.

Ondoh, T. 1992. Observation of LF atmospherics associated with lightning discharges in volcano eruption smoke. *Res. Lett. Atmos. Electr.* **12**: 235–51.

Pierce, E.T. 1971. Triggered lightning and some unsuspected lightning hazards. Stanford Research Institute, Menlo Park, California, 20 pp.

Pierce, E.T. 1974. Atmospheric electricity – some themes. *Bull. Amer. Meteor. Soc.* **55**: 1186–94.

Pounder, C. 1980. Volcanic lightning. *Weather* **35**: 357–60.

Powell, J.R., and Finkelstein, D. 1969. Structure of ball lightning. In *Advances in Geophys, vol. 13*, eds. H.E. Landsberg and J. van Mieghem, pp. 141–89, New York: Academic Press.

Powell, J.R., and Finkelstein, D. 1970. Ball lightning. *Am. Scientist* **58**: 262–79.

Protasevic, E. 1988. On possibilities of modelling ball lightning. *Scripta Fac. Sci. Nat. Univ. Purk. Brun. (Physica)* **18**: 335–40.

Puhringer, A. 1967. Boebachtung eines Kugelblitzes im Salzkammergut und ein Erklarungsversuch. *Wetter und Leben* **19**: 57.

Rabinowitz, M. 1999. Little black holes: dark matter and ball lightning. *Astrophys. Space Sci.* **262**: 391–410.

Rañada, A.F., and Trueba, J.L. 1996. Ball lightning an electromagnetic knot? *Nature* **383**: 32.

Rañada, A.F., Soler, M., and Trueba, J.L. 1998. A model of ball lightning as a magnetic knot with linked streamers. *J. Geophys. Res.* **103**: 23 309–13.

Rayle, W.D. 1966. Ball lightning characteristics. NASA Technical Note D-3188.

Rice-Evans, P.C. 1982. Ball lightning in the laboratory. *Nature* **299**: 774.

Rodewald, M. 1954. Kugelblitzbeobachtungen. *Z. Meteorol.* **9**: 27–9.

Rulenko, O.P. 1985. Electrification of volcanic clouds. *Vulkanologiya i Seismologiya* **2**: 71–83

Rulenko, O.P., Tokarev, P.I., and Firstov, P.P. 1976. Electricity of volcanoes. *Bull. Vulkanol. Stan.* **52**: 11–17.

Sanduloviciu, M., and Lozneanu, E. 2000. Ball lightning as a self-organization phenomenon. *J. Geophys. Res.* **105**: 4719–27.

Scheminzky, F., and Wolf, F. 1948. Photographie eines Perlschnurblitzes. *Sitzber. Akad. Wiss. Wien Abt. IIA* **156**: 1–8.

Schmidt, D.S., Schmidt, R.A., and Dent, J.D. 1998. Electrostatic force on saltating sand. *J. Geophys. Res.* **103**: 8997–9001.

Schwegler, H. 1951. Uber der Kugelblitz. *Naturwiss. Rund.* **4**: 169.

Shafranov, V.D. 1957a. On magnetohydrodynamical equilibrium configurations. *Sov. Phys. JETP* **6**: 545–54. (English transl.).

Shafranov, V.D. 1957b. On magnetohydrodynamical equilibrium configurations. *Zh. Eksp. Teor. Fiz. SSSR* **33**: 710 (in Russian).

Silberg, P.A. 1962. Ball lightning and plasmoids. *J. Geophys. Res.* **67**: 4941–2.

Silberg, P.A. 1965. A review of ball lightning. In *Problems of Atmospheric and Space Electricity*, ed. S.C. Coroniti, pp. 436–54, New York: American Elsevier.

Singer, S. 1971. *The Nature of Ball Lightning*, New York: Plenum Press.

Singer, S. 1980. Ball lightning – the persistent enigma. *Naturwissenschaften* **67**: 332–7.

Singer, S. 1991. Great balls of fire. *Nature* **350**: 108–9.

Sinitsyn, V.I. 1977a. Properties of a low pressure spherical discharge. *Sov. Phys. Tech. Phys.* **22**: 576.

Sinitsyn, V.I. 1977b. O nekotorykh svoistvakh sfericheskogo razriada pri ponizhennom davlenii. *Zh. Tekh. Fiz.* **47**: 966.

Smirnov, B.M. 1975. Analysis of the nature of ball lightning. *Sov. Phys. Usp.* **18**: 636–40 (English transl.).

Smirnov, B.M. 1976. Formation of ball lightning. *Sov. Phys. Doklady* **21**: 89–90. (English transl.).

Smirnov, B.M. 1977. Ball-lightning model. *Sov. Phys. Tech. Phys.* **22**: 488–92 (English transl.).

Smirnov, B.M. 1987a. Ball lightning – what is it? In *Priroda*, no. 2, p. 15.

Smirnov, B.M. 1987b. Electrical phenomena in ball lightning. *Doklady Akad. Nauk. SSSR* **292**: 1363 (in Russian).

Smirnov, B.M. 1987c. The properties and the nature of ball lightning. *Phys. Reports* (Review Section of *Phys. Lett.*) **152**: 177–226.

Smirnov, B.M. 1989a. Electrical and radiative properties of ball lightning. In *Science of Ball Lightning (Fireball)*, ed. Y.H. Ohtsuki, pp. 192–219, Singapore: World Scientific.

Smirnov, B.M. 1989b. The candle flame as a model of ball lightning. In *Science of Ball Lightning (Fireball)*, ed. Y.H. Ohtsuki, pp. 220–8, Singapore: World Scientific.

Smirnov, B.M. 1989c. Electrical and radiative processes in ball lightning. *Nuovo Cimento* **12C**: 575–95.

Smirnov, B.M. 1990a. Physics of ball lightning (in English). *Sov. Phys. Usp.* **33**: 261–8.

Smirnov, B.M. 1990b. The properties of fractal clusters. *Phys. Rep.* **188**: 1–78.

Smirnov, B.M. 1991. A tangle of fractal fibers as a new state of matter. *Usp. Fiz. Nauk.* **161**: 141.

Smirnov, B.M. 1992. Observational properties of ball lightning (English transl.). *Sov. Phys. Usp.* **35**: 650–70.

Smirnov, B.M. 1993a. Observational parameters of ball lightning. *Physica Scripta* **48**: 638–40.

Smirnov, B.M. 1993b. Observational parameters of ball lightning. In *Progress in Ball Lightning Research: Proc. VIZOTUM*, ed. A.G. Keul, p. 85, The Vizotum Project, Salzburg, Austria.

Smirnov, B.M. 1993c. Physics of ball lightning. *Phys. Reports* (Review Section of *Phys. Lett.*) **224**: 151–236.

Smirnov, B.M. 1993d. Gas dynamics of a fractal ball. *Zh. Tekh. Fiz.* **63**: 190.

Smirnov, B.M. 1993e. Radiation of some fractal structures. *Int. J. Theoret. Phys.* **32**: 1453–64.

Smirnov, B.M. 1993f. Physics of ball lightning. *Phys. Reports* **224**: 151–236.

Smirnov, B.M. 1993g. Radiative processes involving fractal structures (English translation). *Physics – Usp.* **36**: 592.

Smirnov, B.M. 1996. Comparison of histograms of data-banks. *Physica Scripta* **54**: 125–8.

Smirnov, B.M., and Strizhev, A.Ju. 1994. Analysis of observational ball lightning by correlation methods. *Physica Scripta* **50**: 606–8.

Stakhanov, I.P. 1973. Concerning the nature of ball lightning. *Zh. Etf. Fiz.* **18**: 193 (in Russian).

Stakhanov, I.P. 1974. Concerning the nature of ball lightning (English transl.). *JETP Lett.* **18**: 114–16.

Stakhanov, I.P. 1975. Stability of ball lightning. *Sov. Phys. Tech. Phys.* **19**: 861–3 (English transl.).

Stakhanov, I.P. 1976. Cluster plasma and radiation from ball lightning. *Sov. Phys. Tech. Phys.* **21**: 44–8 (English transl.).

Stenhoff, M. 1976. Ball lightning. *Nature* **260**: 596–7.

Stenhoff, M. 1999. *Ball Lightning: Unsolved Problem in Atmospheric Physics*, 349 pp., New York: Kluwer Academic and Plenum Publishers.

Takaki, S., and Ikeya, M. 1998. A dark discharge model of earthquake lightning. *Jpn J. Appl. Phys.* **37**: 5016–20.

Taylor, H.A. Jr, and Cloutier, P.A. 1992. Non-evidence of lightning and associated volcanism at Venus. *Space Sci. Rev.* **61**: 387–91.

Tompkins, D.R., and Gooding, R. 1975. Photographic observations of ball lightning. *Bull. Am. Phys. Soc.* **20**: 659.

Tompkins, D.R., and Rodney, R.F. 1980. Possible photographic observations of ball lightning. *Nuovo Cimento* **3C**: 200–5.

Turner, D.J. 2002. The fragmented science of ball lightning (with comment). *Phil. Trans. Roy. Soc. A* **360**: 107–52.

Uman, M.A. 1962. Bead lightning and the pinch effect. *J. Atmos. Terr. Phys.* **24**: 43–5.

Uman, M.A. 1968. Some comments on ball lightning. *J. Atmos. Terr. Phys.* **30**: 1245–6.

Uman, M.A. 1969. *Lightning*, 264 pp., New York: McGraw-Hill.

Uman, M.A. 1984. *Lightning*, 298 pp., New York: Dover.

Uman, M.A. 1986. *All About Lightning*, New York: Dover.

Uman, M.A. 1987. *The Lightning Discharge*, 377 pp., San Diego: Academic Press.

Uman, M.A. 2001. *The Lightning Discharge*, 377 pp., Mineola, New York: Dover.

Uman, M.A., and Helstrom, C.W. 1966. A theory of ball lightning. *J. Geophys. Res.* **71**: 1975–84.

Uman, M.A., and Voshall, R.E. 1968. Time interval between lightning strokes and the initiation of dart leaders. *J. Geophys. Res.* **73**: 497–506.

Uman, M.A., Seacord, D.F., Price, G.H., and Pierce, E.T. 1972. Lightning induced by thermonuclear detonations. *J. Geophys. Res.* **77**: 1591–6.

Voitsekhovskii, B.V. 1975. Nature of ball lightning. *Sov. Phys. Dokl.* **19**: 580–1. (English transl.).

Voitsekhovskii, B.V., and Voitsekhovskii, B.B. 1974. Priroda sharrovoy molnii. *Akad. Nauk. SSSR Dokl. Mat. Fiz.* **218**: 77 (in Russian).

Wagner, G.A. 1971. Optical and acoustic detection of ball lightning. *Nature* **232**: 187.

Williams, E.R., Cooke, C.M., and Wright, K.A. 1988. The role of electric space charge in nuclear lightning. *J. Geophys. Res.* **93**: 1679–88.

Wittman, A. 1971. In support of a physical explanation of ball lightning. *Nature* **232**: 625.

Wolf, F. 1956. Interessante Aufnahme eines Kugelblitzes. *Naturwissenschaften* **43**: 415–17.

Wooding, E.R. 1963. Ball lightning. *Nature* **199**: 272–3.
Wooding, E.R. 1972. Laser analogue to ball lightning. *Nature* **239**: 394–5.
Wooding, E.R. 1976. Ball lightning to Smetwick. *Nature*: **262**: 379–80.
Young, G.A. 1962. *A Lightning Strike of an Underwater Explosion Plume*, US Naval Ordinance Lab., NOLTR 61–43 (March 1962).

Zabotin, N.A., and Wright, J.W. 2001. Role of meteoric dust in sprite formation. *Geophys. Res.* **28**(13): 2593–6.
Zaitsev, A.V. New theory of ball lightning. *Sov. Phys. Tech. Phys.* **17**: 173–5 (English transl.).
Zheng, X.-H. 1990. Quantitative analysis for ball lightning. *Phys. Lett. A* **148**: 463–9.
Zimmerman, P.D. 1970. Energy content of Covington's lightning ball. *Nature* **228**: 853.

Appendix: Books on lightning and related subjects

Aleksandrov, M.S., Bakleneva, Z.M., Gladshtein, N.D. et al. 1972. *Fluctuations of Earth's Electromagnetic Field on VLF*, 195 pp., Moscow: Nauka.

Alfvén, H. and Fälthammer, C.G. 1963. *Cosmical Electrodynamics*, 2nd edition, London and New York: Oxford University Press.

Allis, W.P., Buchsbaum, S.J., and Bers, A. 1963. *Waves in Anisotropic Plasmas*. Cambridge, Massachusetts: MIT Press.

Al'pert, Ya. L. 1990. *Space Plasma*. Cambridge, UK: Cambridge University Press.

Andrews, C.J., Cooper, M.A., Darveniza, M., and Mackerras, D. 1992. *Lightning Injuries: Electrical Medical and Legal Aspects*, 195 pp., Boca Raton, Florida: CRC Press.

Austin, B. 2001. *Schonland, Scientist and Soldier*, 639 pp., Bristol: IoP.

Barry, J.D. 1980. *Ball Lightning and Bead Lightning*, New York: Plenum.

Baru, N.V., Kononov, I.I., and Solomonik, M.E. 1976. *Radio Direction Finders for Ranging Close Thunderstorms*, 143 pp., Leningrad: Gidrometeoizdat.

Battan, L.J. 1959. *Radar Meteorology*, Chicago: University of Chicago Press.

Battan, L.J. 1961. *The Nature of Violent Storms*, 158 pp., Garden City, New York: Anchor Books.

Battan, L.J. 1964. *The Thunderstorm*, 128 pp., New York: Signet.

Battan, L.J. 1973. *Radar Observation of the Atmosphere*, 324 pp., Chicago: University of Chicago Press.

Bazelyan, E.M., Gorin, B.N., and Levitov, V.I. 1978. *Physical and Engineering Foundations of Lightning Protection*, 223 pp., Leningrad: Gidrometeoizdat.

Bazelyan, E.M., and Raizer, Yu. P. 1998. *Spark Discharge*, 294 pp., Boca Raton, Florida: CRC Press.

Bazelyan, E.M., and Raizer, Yu. P. 2000a. *Lightning Physics and Lightning Protection*, 325 pp., Bristol: IOP.

Bazelyan, E.M., and Razhansky, I.M. 1988. *Spark Discharge in Air*, 164 pp., Novosibirsk: Nauka.

Bell, T.H. 1962. *Thunderstorms*, London: Dobson.

Bewley, L.V. 1951. *Travelling Waves on Transmission Systems*, 2nd edition, New York: Dover.

Bliokh, P.V., Nikolaenko, A.P., and Filippov, Yu.F. 1980a. *Schumann Resonances in the Earth–ionosphere Cavity*, London: Peter Peregrinus.

Bluestein, H.B. 1999. *Tornado Alley*, 180 pp., New York: Oxford.

Brand, W. 1923. *Der Kugelblitz*, Hamburg, Germany: Grand.

Budden, K.G. 1961. *Radio Waves in the Ionosphere*, London and New York: Cambridge University Press.

Budden, K.G. 1962. *The Waveguide Mode Theory of Wave Propagation*. Englewood Cliffs, New Jersey: Prentice-Hall.

Budden, K.G. 1988. *The Propagation of Radio Waves*. Cambridge, UK: Cambridge University Press.

Burgsdorf, V.V., and Yakobs, A.I. 1987. *Grounding in Electrical Installations*, 400 pp., Moscow: Energoatomizdat.

Byers, H.R. (ed.) 1953. *Thunderstorm Electricity*, Chicago: University of Chicago Press.

Byers, H.R., and Braham, R.R. 1949. *The Thunderstorm*, Washington, DC: US Weather Bureau.

Cade, C.M., and Davis, D. 1969. *The Taming of Thunderbolts*, New York: Abelard-Schuman.

Chalmers, J.A. 1967. *Atmospheric Electricity*, 2nd edition, New York: Pergamon Press.

Chew, J, 1987. *Storms Above the Desert – Atmospheric Research in New Mexico 1935–1985*, 153 pp., Albuquerque: University of New Mexico Press.

Chowdhuri, P. 1996. *Electromagnetic Transients in Power Systems*, 400 pp., New York: Wiley.

Cobine, J.D. 1958. *Gaseous Conductors. Theory and Engineering Applications*, 606 p., New York: Dover.

Cooray, G.V. 2002. *The Lightning Flash*, 400 pp. (approx), London: The Institute of Electrical Engineers (in press).

Coroniti, S.C. (ed.) 1965. *Problems of Atmospheric and Space Electricity*, 616 pp., New York: American Elsevier.

Coroniti, S.C., and J. Hughes (eds.) 1969. *Planetary Electrodynamics*, I and II, 503 and 587 pp., New York: Gordon and Breach.

Cotton, W.R., and Anthes, R.A. 1989. *Storm and Cloud Dynamics*, 883 pp., London: Academic Press.

David, P., and Voge, J. 1969. *Propagation of Waves* (translated by J. B. Arthur), New York: Pergamon Press.

Davies, K. 1966. *Ionospheric Radio Propagation*, New York: Dover.

Davies, K. 1990. *Ionospheric Radio*. London, UK: Peter Peregrinus.

Degauque, P., and Hamelin, J. 1993. *Electromagnetic Compatibility*, 652 pp., New York: Oxford.

Dolezalek, H., and R. Reiter (eds.) 1977. *Electrical Processes in Atmospheres*, Darmstadt, Germany: Dr. Dietrich Steinkopff.

Doswell, C.A. (ed.) 2001. *Severe Convective Storms*. 561 pp., Boston, Massachusetts: American Meteorological Society.

Dragan, G. 1992. *Lightning Overvoltages in Electric Power Systems*, 408 pp., Buharest: Academy of Romania.

Dulzon, A.A., and Kalyatsky, I.I. 1970. *Lightning Protection of Substations*, 219 pp., Tomsk: Tomsk State University Press.

Filippov, A.H. 1974. *Thunderstorms in Eastern Siberia*, 75 pp., Leningrad: Gidrometeoizdat.

Fisher, F.A., Plumer, J.A., and Perala, R.A. 1999. *Lightning Protection of Aircraft*. Pittsfield, Massachusetts: Lightning Technologies Inc.

Fleagle, R.G., and Businger, J.A. 1963. *An Introduction to Atmospheric Physics*, New York: Academic Press.

Fleming, J.R., and Goodman, R.E. (eds.) 1994. *International Bibliography of Meteorology*, 701 pp., Upland, Pennsylvania: Diane Publishing Co.

Forrest, J.S., Howard, P.R. and Littler, D.J. (eds.) 1962. *Gas Discharges and the Electricity Supply Industry*, London: Butterworths.

Franklin, B. 1774. *Experiments and Observations of Electricity, Made at Philadelphia in America*. 5th edition, 530 pp., London: F. Newberry.

Gabrielson, B.C. 1988. *The Aerospece Engineer's Handbook of Lightning Protection*, 218 pp., Gainesville, Virginia: Interference Control Technologies.

Gajelis, J. 1972. *Terrestrial Propagation of Long Electromagnetic Waves*, New York and Oxford: Pergamon Press.

Gardner, R.L. (ed.) 1990. *Lightning Electromagnetics*, 540 pp., New York: Hemisphere.

Gary, C. 1994. *La Foudre, des Mythologies Antiques à la Récherche Moderne*, 208 pp., Paris: Masson.

Gething, P.J.D. 1978. *Radio Direction Finding and the Resolution of Multicomponent Wave-Fields*. 329 pp., London: Peter Pergrinus.

Gindullin, F.A., Goldstein, V.G., Dulzon, A.A., Khalilov, F.H. 1989. *Overvoltages in 6-35 kV Networks*, 190 pp., Moscow: Energoatomizdat.

Ginzburg, V.L. 1970. *The Propagation of Radio Waves*. Cambridge, UK: Cambridge University Press.

Ginzburg, V.L. 1970. *The Propagation of Electromagnetic Waves in Plasmas*, Oxford, UK: Pergamon Press.

Golde, R.H. 1973. *Lightning Protection*, 254 pp., London: Edward Arnold.

Golde, R.H. (ed.) 1977. *Lightning: vol. 1, Physics of Lightning, vol. 2, Lightning Protection*, New York: Academic Press.

Gunn, R. 1953. *Thunderstorm Electricity*, Chicago: University of Chicago Press.

Hart, W.C., and Malone, E.W. 1979. *Lightning and Lightning Protection*, Gainesville, Virginia: Don White Consultants.

Hasse, P. 2000. *Overvoltage Protection of Low Voltage Systems*, 2nd edition, 358 pp., London: The Institution of Electrical Engineers.

Hasse, P., and Weisinger, J. 1982. *Handbuch für Blitzschutz und Erdung*, München, Germany: Richard Pflaum Verlag.

Haydon, S.C. (ed.) 1964. *Discharge and Plasma Physics*, 509 pp., Armidale, Australia: University of New England.

Helliwell, R.A. 1965. *Whistlers and Related Ionospheric Phenomena*, Stanford, California: Stanford University Press.

Hileman, A.R. 1999. *Insulation Coordination for Power Systems*, 767 pp., New York: Marcel Dekker.

Hines, C.O., Paghis, I., Hartz, T.R., and Fejer, J.A. (eds.) 1965. *Physics of the Earth's Upper Atmosphere*, Englewood Cliffs, New Jersey: Prentice-Hall.

Horvath, T. 1991. *Computation of Lightning Protection*, 204 pp., Taunton, Somerset, England: Research Studies Press (also New York: John Wiley & Sons).

Houghton, J.T. 1986. *The Physics of Atmospheres*, 2nd edition, 271 pp., Cambridge: Cambridge University Press.

Houze, R.A., Jr 1993. *Cloud Dynamics* 573 pp., San Diego, California: Academic Press.

Huck, M.V. 1995. *Lightning and Boats*, 65 pp., Brookfield, Wisconsin: Seaworthy Publications.

Imyanitov, I.M. 1957. *Instrumentation and Methods for Studying the Electricity of the Atmosphere*, 484 pp., Moscow: Gostekhizdat.

Imyanitov, I.M. 1970. *Electrification of Aircraft in Clouds and in Precipitation*, 211 pp., Leningrad: Gidrometeoizdat.

Imyanitov, I.M. 1974. *Physics of Thunderstorms*, 243 pp., Leningrad: Gidrometeoizdat.

Imyanitov, I.M., and Chubarina, E.V. 1965. *Electricity of Free Atmosphere*, 210 pp., Leningrad: Gidrometeoizdat.

Imyanitov, I.M., Chubarina, E.V., and Shvarts, Ya. M. 1971. *Electricity in Clouds*, 93 pp., Leningrad: Gidrometeoizdat.

Imyanitov, I.M., Evteev, B.F., and Kamaldina, I.I. 1981. *Physical and Meteorological Conditions Leading to Atmospheric Electrical Discharges Through Aircraft*, Leningrad: Gidrometeoizdat.

Inkov, B.K. 1973. *Phase Methods for the Determination of the Distance to Thunderstorms*, 127 pp., St Petersburg: Gidrometeoizdat.

Israel, H. 1970. *Atmospheric Electricity, vol. I, Fundamentals, Conductivity, Ions*, published for the National Science Foundation by the Israel Program for Scientific Translation, Jerusalem, 317 pp.

Israel, H. 1973. *Atmospheric Electricity, vol. II, Fields, Charges, Currents*, published for the National Science Foundation by the Israel Program for Scientific Translation, Jerusalem, 796 pp.

Kachurin, L.G. 1990. *Physical Foundations of Influences on Atmospheric Processes*, 463 pp., Leningrad: Gidrometeoizdat.

Kashprovsky, V.E. 1966. *Locating Thunderstorms with Radiotechnical Methods*, 248 pp., Moscow: Nauka.

Kelso, J.M. 1964. *Radio Ray Propagation in the Ionosphere*, New York: McGraw-Hill.

Kessler, E. (ed.) 1983. *Thunderstorms: A Social, Scientific, and Technological Documentary, vol. 1, The Thunderstorm in Human Affairs*, 2nd edition, Norman, Oklahoma: University of Oklahoma Press.

Kessler, E. (ed.) 1986. *Thunderstorms: A Social, Scientific, and Technological Documentary, vol. 2, Morphology and Dynamics*, 2nd edition, Norman, Oklahoma: University of Oklahoma Press.

Kessler, E. (ed.) 1988. *Thunderstorms: A Social, Scientific, and Technological Documentary, vol. 3, Instruments and Techniques for Thunderstorm Observation and Analysis*, 2nd edition, Norman, Oklahoma: University of Oklahoma Press.

Appendix: Books on lightning and related subjects

Kinney, G.F. 1962. *Explosive Shocks in Air.* 198 pp., New York: Macmillan.
Kodali, V.P. 1996. *Engineering Electromagnetic Compatibility,* 369 pp., New York: IEEE Press.
Kononov, I.I., Petrenko, I.A., and Snegurov, V.S. 1986. *Radiotechnical Methods for Locating Thunderstorms,* 222 pp., Leningrad: Gidrometeoizdat.
Kostenko, M.V. 1949. *Lightning Overvoltages and Lightning Protection of High Voltage Installations,* 333 pp., Moscow: Gosenergoizdat.
Kravchenko, V.I. 1991. *Lightning Protection of Radioelectronic Devices,* 264 pp., Moscow: Radio i Svyaz.
Krider, E.P., and Roble, R.G. (eds.) 1986. *The Earth's Electrical Environment, Studies in Geophysics,* 263 pp., Washington, DC: National Academy Press.
Lewis, W.W. 1965. *The Protection of Transmission Systems Against Lightning,* 422 pp., New York: Dover.
Lightning and the Protection of Electric Systems, 1939, Collection of Papers, 141 pp., General Electric Company.
Llewellyn-Jones, F. 1966. *Ionization and Breakdown in Gases,* 176 pp., London: Associated Book Publishers.
Loeb, L.B., and Meek, J.M. 1941. *The Mechanism of the Electric Spark,* London: Oxford University Press.
Ludlam, F.H. 1980. *Clouds and Storms,* 405 pp., University Park: Pennsylvania State University Press.
MacGorman, D.R., and Rust, W.D. 1998. *The Electrical Nature of Thunderstorms,* 422 pp., New York: Oxford University Press.
Malan, D.J. 1963. *Physics of Lightning,* 176 pp., London: English Universities Press.
Magono, C. 1980. *Thunderstorms,* 261 pp., New York: Elsevier.
Marshall, J.L. 1973. *Lightning Protection,* 190 pp., New York: John Wiley and Sons.
Mason, B.J. 1957. *The Physics of Clouds,* Oxford: Clarendon Press.
Mason, B.J. 1962. *Clouds, Rain and Rainmaking,* 145 pp., Cambridge: Cambridge University Press.
Mason, B.J. 1971. *The Physics of Clouds,* 2nd edition, 671 pp., London: Oxford University Press.
McEachron, K.B., and Patrick, K.G. 1940. *Playing With Lightning,* 231 pp., New York: Random House.
Meek, J.M., and Craggs, J.D. 1953. *Electrical Breakdown of Gases,* 507 pp., Oxford: Clarendon Press.
Meliopoulos, A.P.S., 1988. *Power Sysyem Grounding and Transients,* 450 pp., New York: Marcel Dekker.
Melrose, D.B. 1986. *Instabilities in Space and Laboratory Plasmas,* New York: Cambridge University Press.
Muchnik, V.M. 1974. *Physics of Thunderstorms,* 351 pp., Leningrad: Gidrometeoizdat.
Norinder, H. *Studies of Lightning Discharges,* 30 pp., Moscow-Leningrad: Gosenergoidzat.
Orville, R.E. (ed.) 1984. *Preprints from Seventh International Conference on Atmospheric Electricity, Albany, NY, 3–8 June 1984,* Boston, Massachusetts: American Meteorological Society.
Papoular, R. 1965. *Electrical Phenomena in Gases,* New York: American Elsevier.
Paul, C.R. 1992. *Introduction to Electromagnetic Compatibility,* 765 pp., New York: John Wiley and Sons.
Pierce, E.T., and Wormell, T.W. 1953. *Thunderstorm Electricity,* Chicago: University of Chicago Press.
Pruppacher, H.R., and Klett, J.D. 1997. *Microphysics of Clouds and Precipitation,* 2nd edition, 954 pp., Kluwer.
Raether, H. 1964. *Electron Avalanches and Breakdown in Gases,* London: Butterworth.
Raizer, Yu.P. 1997. *Gas Discharge Physics,* 449 pp., New York: Springer.
Ratcliffe, J.A., ed. 1960. *Physics of the Upper Atmosphere,* New York: Academic Press.
Ratcliffe, J.A. 1972. *An Introduction to the Ionosphere and Magnetosphere,* Cambridge, UK: Cambridge University Press.
Razevig, D.V. 1959. *Lightning Overvoltages on Electric Power Lines,* 216 pp., Moscow: Gosenergoizdat.
Reiter, R. 1992. *Phenomena in Atmospheric and Environmental Electricity,* 541 pp., New York: Elsevier.
Remizov, L.T. 1985. *Natural Radio Noise,* 200 pp., Moscow: Nauka.
Rishbeth, H. and Garriott, O.K. 1969. *Introduction to Ionospheric Physics,* New York: Academic Press.
Ruhnke, L.H., and J. Latham J. (eds.) 1983. *Proceedings in Atmospheric Electricity.* Hampton, Virginia: A. Deepak.
Ryabkova, E. Ya. 1978. *Grounding in High Voltage Installations,* 224 pp., Moscow: Energiya.
Salanave, L.E. 1980. *Lightning and Its Spectrum,* 136 pp., Tucson: University of Arizona Press.
Schonland, B.F.J. 1953. *Atmospheric Electricity,* 2nd edition, 95 pp., London: Metheun and Co.
Schonland, B.F.J. 1964. *The Flight of Thunderbolts,* 2nd edition, 182 pp., Oxford: Clarendon Press.
Singer, S. 1971. *The Nature of Ball Lightning,* 169 pp., New York: Plenum.
Smirnov, B.M. 1993. *Physics of Ball Lightning,* Elsevier.
Smith, A.A. 1977. *Coupling of External Electromagnetic Fields to Transmission Lines,* 132 pp., New York: John Wiley and Sons.
Smith, L.G. (ed.) 1959. *Recent Advances in Atmospheric Electricity,* New York: Pergamon Press.
Spitzer, L. Jr 1956. *Physics of Fully Ionized Gases,* 105 pp., New York: Interscience Publishers.
Spitzer, L. Jr 1962. *Physics of Fully Ionized Gases,* 2nd edition, New York: Interscience Publishers (a division of John Wiley and Sons).
Stachanov, I.P. 1985. *On the Physical Nature of the Ball Lightning,* 208 pp., Moscow: Energoizdat.
Stekolnikov, I.S. 1940. *Lightning,* 327 pp., Moscow, Leningrad: AN SSSR Press.
Stekolnikov, I.S. 1943. *Physics of Lightning and Lightning Protection,* 226 pp., Moscow, Leningrad: AN SSSR Press.
Stekolnikov, I.S. (ed.) 1951. *Lightning Protection of Industrial Structures and Buildings,* 202 pp., Moscow: AN SSSR Press.
Stekolnikov, I.S. 1960. *The Nature of the Long Spark,* 272 pp. Moscow: AN SSSR Press.

Stenhoff, M. 1999. *Ball Lightning: An Unsolved Problem in Atmospheric Physics*, 349 pp., New York: Kluwer and Plenum.

Stepanenko, V.D., and Galperin, S.M. 1983. *Radar Techniques for Studying Thunderstorms*, 204 pp., Leningrad: Gidrometeoizdat.

Stix, T.H. 1962. *The Theory of Plasma Waves*, New York: McGraw-Hill.

Sunde, E.D. 1967. *Earth Conduction Effects in Transmission Systems*, 373 pp., New York: Dover.

Thunderstorm Direction and Range Finder "Ochag - 2П". 1988, Leningrad: Gidrometeoizdat.

Tverskoy, P.N. 1949. *Atmospheric Electricity*, 252 pp., Leningrad: Gidrometeoizdat.

Uman, M.A. 1964. *Introduction to Plasma Physics*, 226 pp., New York: McGraw-Hill.

Uman, M.A. 1969. *Lightning*, 264 pp., New York: McGraw-Hill.

Uman, M.A. 1971. *Understanding Lightning*, 166 pp., Pittsburgh: Bek Technical Publications.

Uman, M.A. 1984. *Lightning*, 298 pp., New York: Dover.

Uman, M.A. 1986. *All About Lightning*, 167 pp., New York: Dover.

Uman, M.A. 1987. *The Lightning Discharge*, 377 pp., San Diego: Academic Press.

Uman, M.A. 2001. *The Lightning Discharge*, 377 pp., Mineola, New York: Dover.

Vance, E.F. 1978. *Coupling to Shielded Cables*, 183 pp., New York: John Wiley and Sons.

Viemeister, P.E. 1961. *The Lightning Book*, 316 pp., New York: Doubleday and Co.

Volland, H. (ed.) 1982. *Handbook on Atmospherics, vol. 1 and 2*, Boca Raton, Florida: CRC Press.

Volland, H. 1984. *Atmospheric Electrodynamics*, 205 pp., New York: Springer-Verlag.

Volland, H. 1995 (ed.). *Handbook of Atmospheric Electrodynamics, vol. I and II*, Boca Raton, Florida: CRC Press.

Wahlin, L. 1986. *Atmospheric Electrostatics*, 120 pp., New York: John Wiley.

Wait, J.R. 1962. *Electromagnetic Waves in Stratified Media*, New York: Pergamon Press.

Wait, J.R. 1972. *Electromagnetic Waves in Stratified Media*, 2nd edition New York: Pergamon Press.

Wait, J.R. 1982. *Geo-electromagnetism*, 268 pp., New York: Academic Press.

Waters, W.E. 1983. *Electrical Induction From Distant Current Surges*, 168 pp., Englewood Cliffs, New Jersey: Prentice-Hall.

Wiesinger, J. 1972. *Blitzforschung und Blitzschutz*, Munchen: R. Oldenburg Verlag.

Index

acoustic energy 376–377, 381, 397
acoustic imaging 316, 387–389
acoustic radiation, *see also* thunder, 177, 260, 374–393
acoustic waves 386–387
action integral 12, 146, 215, 249, 277, 280–282, 363, 588, 589
adiabatic cooling 67
advection 308
    lightning location associated with 33
    of charge 33, 92, 93
    of hydometeors 33
    of NO 516–519
aerosols 67, 297, 298
air discharge, *see* cloud-to-air lightning
air–earth current 10
aircraft
    accidents involving lightning 364–368
    enhancement factor 347
    lightning strikes to 348–350, 353–362
    measurement of
        conductivity 7
        electric field 73, 80, 84, 91, 353–362
    protection from lightning 362–364
    research programs 350–353
    static discharge 347–348
altitude triggering, *see* rocket-triggered lightning
altocumulus 91
altostratus 91
amplitude probability distribution 460
angle
    of incidence 441–443, 457
    of reflection 441–443, 457
    of refraction (transmission) 441–443, 457
antenna noise factor 460
antenna theory model 398, 406
anvil 81, 84
Apollo 12 364, 368
arc, voltage–current characteristic 603
arrester 591, 602–604
arrival-time-difference mapping system 562–565
artificially initiated lightning, *see* rocket-triggered lightning
ashen light 532
Atlas Centaur 364, 369
atmosphere
    electrical structure of 6–9, 436, 481
    role of lightning in chemistry of 507–511
atmospheric boundary layer 518

atmospheric chemicals, due to lightning 507–519, 537–538
atmospheric electrical generator 10–12
atmospheric electricity sign convention 8–9
atmospheric ions, *see* ions
atmospheric waveguide
    cutoff frequency 433, 436, 446
    mode parameters 446
    mode theory 446–448
    ray theory 445–448
atmospherics, *see* sferics
attempted leader 127, 185
attraction of lightning 49–50, 137–138
    by aircraft 346–362, 364–369
    by lightning rod 592–599
    by tall object 50, 617
attachment process 49–50, 137–143, 592–599
attractive area 50–52, 592–599, 617
attractive distance 51
attractive radius 51
average noise amplitude 460
azimuthal error 558–561, 567

B (breakdown) field change 116–118
ball lightning
    death from 656, 657
    diameter 661
    duration 662
    laboratory simulations 663–664
    observations 656–662
    statistics 662, 657
    theories 662–664
balloon-borne measurement
    conductivity 7–8
    electric field 73, 79–82, 84, 92, 489
    individual particle charges 84, 87–88
    potential (voltage) 11, 82, 92, 111
    total charge density 84, 87–88, 92
    X-rays 495
balloons, lightning to 3, 76, 232, 241, 260
basalt 3
basic insulation level (BIL) 616–619
bead lightning 658, 665–666
BG model, *see* Bruce–Golde model
bidirectional leader 90, 228, 269–272, 275, 322, 346, 353–354, 416–417
    in step formation process 137
bidirectional return stroke 138, 140, 414
BIL, *see* basic insulation level
BIL (breakdown-intermediate-leader) field waveform 116–118

bipolar lightning flash 232–234, 247–249, 311–313, 316
bipolar pattern 31–34, 218
blue jet 480, 481, 483–485
blue starter 480, 481–483
boats 610–613
Boys camera, *see also* streak photographs, 3
branch component 111, 148
branching, lightning channel 109, 148, 161–162, 172, 225–226, 247, 252, 385
breakdown electric field, *see also* critical electric field 82, 592–594
breakdown processes, *see also* initial breakdown 6, 121, 173
breakdown voltage 227, 624–626
breakeven electric field 84, 493, 494
break-through phase 138, 228, 414
broadband interferometer 569
Bruce–Golde model 129, 401–403, 407–409
Buddha 1
buildings, lightning protection for 596–599
buoyancy 67

cable, underground 293–295, 619–624
Camp Blanding lightning triggering facility (ICLRT) 266, 293–296
capacitance per unit length of lightning channel, *see* channel capacitance per unit length
capacitance of aircraft 359
capacitor, as a model of the global circuit 9–10
Carnegie curve, *see* diurnal variation of the global circuit
Cassini 535, 547
cell, thunderstorm, *see also* supercell, 24, 68
CGR1 39
CGR3 39, 43, 45
channel
    acoustic reconstruction, *see* acoustic imaging
    capacitance per unit length 398–399, 418
    conductivity 164, 397
        dart leader 164
    diameter, stepped leader 134
    electron density 163, 404
    inductance per unit length 398–399
    pressure 111, 163, 396, 404
    radius 7, 134, 163–164, 395
    resistance per unit length 164, 395, 398–400
    temperature 7, 111, 163–164, 396, 404
        dart leader 164
        stepped leader 134

679

channel core 134, 399, 413
"chaotic" leader 164, 171
charge
   conservation of 128, 416
   deposited on
      dart leader, *see* dart leader
      stepped leader, *see* stepped leader
   distribution of
      along leader channel 128–131, 169, 291, 416–417
      in thundercloud 79
   effect on measurements 78
   electric field due to 68–72
   gross distribution in cloud 68–69, 310–311
   height of 75–82
   image charge 68–71
   in anvil 81, 84
   in the global circuit 9–11
   of leader step 134, 171
   on cloud particles 88
   on precipitation 68, 84, 87–88
   per unit length, *see* line charge density
   transferred by
      continuing current, *see* continuing current
      first return stroke, *see* return stroke
      flash 7, 113, 214–215, 221–222, 247–250, 274, 314, 323, 325, 338
      subsequent return stroke, *see* return stroke
   using Gauss's law to find 80, 92
charge density
   line, *see* line charge density
   volume 84, 92
charge moment, *see* dipole moment
charging mechanisms, *see* cloud electrification mechanisms
chi squared 556–558, 561–562, 566–567
chorus 457, 530
CIGRE, 10 kHz 35, 38–39
CIGRE, 500 Hz 38–39, 43
"classical" triggering, *see* rocket-triggered lightning
clear-air lightning 73
cloud base 67
cloud boundary charging 76, 78, 88, 91
cloud electrification mechanisms 84–88, 92, 528
   convection 85
   graupel–ice 86–87, 92
cloud lightning flash
   charge transfer 323, 325, 338
   comparison with ground flash 340–341
   current in 330–331, 338
   definition 321
   development 326–329
   dipole moment 75, 77, 557
   duration 323, 325
   electric field change 72–75, 323–325
   fraction, *see also* cloud-to-ground-flash ratio, 44, 321
   height of origin 323–324, 327
   initiation 322, 326, 327, 340
   K-processes 322, 338
   microsecond-scale pulses 8, 331–340

   optical observations 323, 324, 326
   propagation velocity 322, 326–331, 340
   recoil streamer 322, 338
   structure 322
   VHF imaging 326–329
cloud particles 68, 88
clouds
   altostratus 91
   charging mechanisms in, *see* cloud electrification mechanisms
   cumulonimbus 67–91
   cumulus 67
   dissipating stage 81, 217, 310
   electric field in 346, 354
   electrical development of 73–75, 310–311
   mature stage 310
   nimbostratus 91
   stratiform 91–93
   stratocumulus 91
   stratus 91
   warm 528
   winter, in Japan, 77, 217–218, 308–311
cloud-to-air lightning 4, 321, 338
cloud-to-cloud lightning flash, *see* cloud lightning flash
cloud-to-ground-flash ratio 44–46
cloud-to-ground lightning flash, *see* ground lightning flash
CN tower 242, 255–256, 259–260, 411–412
collision frequency 436–437, 439–440
communication line 620–624
compact intracloud discharge, *see also* narrow bipolar pulses, 334
condensation 67
conductivity, electrical
   atmospheric 6–9, 541
   caused by cosmic rays 6–7
   caused by radioactive decay 6–7
   due to small ions 6
   ground 7, 287, 541, 601
   ionospheric 7–9, 447, 473
   of channel between strokes 5–6, 164, 172
   of cloudy air versus clear air 91
   of dart leader channel, *see* dart leader
   of return-stroke channel, *see* return stroke
conductor 440
connecting leader, *see* upward connecting leader
conservation of charge 128, 416
conservation of energy 394, 395, 415
conservation of mass 394, 395
conservation of momentum 394, 395, 415
continental versus oceanic storms 31, 34–35, 48–49
continuing current 4–7, 112, 173–176, 488
   charge transfer 7
   duration 7
   electric field change 114, 174
   fires set by 613–616
   frequency of occurrence 112, 174
   long 112, 173–174
   magnitude 7

   mode of charge transfer 4–6
   positive flashes 222
   types 175–176
continuity equation, *see* current continuity equation
convection 67–68
convection current density 10–11
convection mechanism 85
convective region of storm systems 80–81, 83, 93
corona discharge 5, 92, 357
   at ground 10, 78, 84, 85, 88
   at leader channel, *see* corona sheath
   at leader tip, *see also* streamer zone, 135–136
   at lightning rods 596, 605–610
   initial impulsive 136–137, 227–228
   on hydrometeors 82–83, 121
corona sheath 134, 176, 399, 413
cosmic rays 6–7, 78
Coulomb's law 128
critical electric field, *see also* breakdown electric field; breakeven electric field
   for corona streamers from hydrometeors 82–83, 121
   for lightning initiation 82, 89–90
   for sustained propagation of streamers 83, 121
critical flashover voltage (CFO) 617, 618
cumulative statistical distribution 145–146, 215, 249, 251, 253, 254, 278–285
cumulonimbus, *see* thundercloud
cumulus 67
current, electric
   above thunderstorms 10
   continuing, *see* continuing current
   corona 10, 357, 510, 511
   of dart leader, *see* dart leader
   fair weather 10
   in lightning strikes to aircraft 353–362
   in sprites 489
   initial continuous, *see* initial continuous current
   Maxwell 10–11
   M-component, *see* M-components
   precipitation 10
   return stroke, *see* return stroke
   stepped leader, *see* stepped leader
cutoff of channel current to ground 172
cyclotron frequency, *see also* gyrofrequency, 434, 437
cyclotron resonance (gyroresonance) 441

damage from lightning, *see* lightning protection
dart leader 4–7, 164–169, 171–173, 415–417
   charge deposited by 7, 169
   conductivity 164
   conductivity in channel prior to 164
   current 7, 168
   duration 7, 168, 225
   electric field change 127, 131, 169, 288–292
   electric potential 7
   luminosity 149, 168–169, 172, 173, 271
   mechanism 171–173

Index

radius 164
RF (narrowband) radiation 171
speed of propagation 7, 165–168
temperature 7, 164
dart length 173
dart-stepped leader 7, 8, 165, 169–171, 271
death from lightning 642–649, 656, 657
dE/dt, rate of rise of field pulse, 134, 154, 157–158, 170, 171, 228–230, 256, 280, 281, 289, 339, 404, 413–414
dI/dt, rate of rise of current, 7, 146, 147–148, 215, 221, 249, 251, 254, 256, 278, 279, 280–281, 284, 285, 286
Defense Meteorological Satellite Program (DMSP) 41–42, 570–571
Diendorfer–Uman (DU) model 402–404, 407–412
diffusion 605
dipole 72, 78, 117
dipole cloud charge distribution 68, 73, 77
dipole moment 68
dirac delta function 416
direction finding (DF) systems, *see* magnetic direction finding (MDF) systems
discharge time constant, *see also* Diendorfer–Uman (DU) model, 402
displacement current density 10–11
distributed-circuit models 394, 398–400, 406, 415–416, 418
diurnal variation
  of the global circuit 10
  of the ionosphere 8
  of lightning occurrence 42, 571
Doppler radar 30, 88
double-exponential current pulse waveform 161
downdrafts, *see also* microbursts, 85
drift velocity, electron 415
DU model, *see* Diendorfer–Uman model
duration
  of cloud lightning flash 323, 325
  of continuing current 7
  of dart leader 7, 168, 225
  of ground lightning flash 7, 112–113, 215, 221, 247–249, 274
  of interstroke interval 7, 47, 49, 112, 115, 282
  of M-components 179
  of return stroke 7, 146, 215, 221, 249
  of stepped leader 7, 123, 124, 125
  of storm 68
dust storm 667

Earth–ionosphere cavity 432, 445–454
Earth–ionosphere waveguide 432, 445–454
earthquake light (seismo-electromagnetic effect), 667–668
Earth's magnetic field (geomagnetic field) 7, 434
East Coast network (SUNYA network) 46, 49, 561, 565
Eckersley's law 435–436, 458, 530
effective height 50, 144, 241

effective range of lightning flash counter 38–39
electric breakdown, *see* breakdown processes
electric conductivity, *see* conductivity, electric
electric current, *see* current, electric
electric dipole, *see* dipole, dipole cloud charge distribution
electric field
  above storms 73
  across dart leader front 172
  across M-component front 172
  across return-stroke front 172
  aloft 81–84
  at the ground 70–74, 310
  boundary conditions for 71
  breakdown, *see* breakdown electric field
  fair-weather 6, 8–9
  in non-cumulonimbus 91–92
  in thundercloud 81–84
  ionospheric 9–10, 495, 496
  lightning charge calculated from 75–77, 556–557
  measurements of 3, 73, 79–80
  over water 78
  polarity reversal 71, 130–131, 134, 183
  reversal distance 71
  sign convention 8–9
  superposition principle 68
  under thundercloud 74, 78
electric field antenna 114
electric field change, *see also* electrostatic field change, 72
electric field meter 24, 75, 556–557
electric field mill, *see* electric field meter
electric field thresholds, *see* critical electric field
electric potential (voltage)
  of electrosphere 9–10
  of leader tip 7, 111
  within cloud 11, 82, 92, 111
electric power line
  distribution 292–295, 616–619
  transmission 296, 314–315, 616, 619
  underground 293–295, 619–624
electric space charge 78, 85
electrification mechanisms, 84–88, 92, 528
electrification processes 84–88, 92, 528
electrogeometric method (EGM) 51, 593–596
electromagnetic coupling 589
electromagnetic field 159–161
  electric induction component 160
  electric radiation component 160–161, 403–404, 557–558
  electrostatic component, *see also* electrostatic field change, 160–161
  magnetic induction component 129–130
  magnetic radiation component 160–161, 404, 557–558
electromagnetic models 394, 398, 406, 415
electromagnetic signals 159–161
electron, energetic 493–495
electron
  avalanche 415

density 163, 404, 436
drift velocity 415
gyrofrequency 437, 440–441, 455–459, 530
plasma frequency 439, 530
precipitation 9, 456, 480, 491, 493–495
runaway, *see* runaway electrons
thermal velocity 415
electronic equipment, lightning protection for 589, 591, 602–605
electrosphere 8–10
electrostatic approximation 126, 128–129, 161
electrostatic field 69–71, 310
electrostatic field change 71–72, 128–131, 556–557
ELF (extremely low frequencies) 43, 432–434
elves 480, 481, 492–493
Empire State Building 3, 241, 242, 243, 245, 248
energetic electron 493–495
energy 7, 12, 114–116, 376–377, 397, 405–406, 538–540, 546
energy source, lightning as 12
engineering models 394, 400–404, 406–412
enhanced lightning activity over cities 48
equalization layer of the atmosphere (electrosphere) 8–10
equations
  gas-dynamic 394, 395
  Maxwell's 394, 398, 415
  telegrapher's 398–399
equipotential surface 599, 604, 605
expansion of lightning channel 378–380, 395–397
external noise figure (F_a) 460
extraterrestrial lightning 528–547

F- (final) processes 112
fair-weather electric field 6, 8–9
Faraday cage 3, 590, 591, 604, 605
fast transition 154–157, 223, 229
field mill, *see* electric field meter
field-aligned duct 457–459
final jump, *see* break-through phase
fine structure of field waveform
  negative ground flash 154–158
  positive ground flash 229–230
fine-weather electric field, *see* fair-weather electric field
fires caused by lightning 613–616
flash
  cloud, *see* cloud lightning flash
  ground, *see* negative ground flash; positive ground flash
  rocket-triggered, *see* rocket-triggered lightning
  upward, *see* upward lightning
flash rate 24–26, 31
forest fires 489, 613–616
FORTE 496, 571–572
Franklin, Benjamin 3, 588, 590, 592, 656
freeze-out temperature, chemical 512, 513

freezing level 31, 67
frequency spectrum
    of acoustic radiation 374–377, 381
    of electromagnetic radiation 6, 134, 158–159, 555
fulgurites 288, 295–296, 620, 621
full wave theory, *see* mode theory

Galileo 535, 539–543
gamma rays 492–495, 480, 481
gas-dynamic models 394–397, 404–406
Gaussian (normal) probability distribution 145, 567
Gauss's law 80, 92
geomagnetic field (Earth's magnetic field) 7, 434
global electric circuit 6–12
global lightning flash frequency 24, 41–43
global nitrogen production 508–518
global temperature 43, 448, 453
    related to worldwide lightning activity, *see also* Schumann resonances, 453–454
glow discharge 603
glow-to-arc transition 603
graupel 31, 86–88, 310
gravity 86, 528
ground lightning flash *see* negative ground flash; positive ground flash
ground termination 37–38, 114–115, 288–289, 599–602
grounding 282–288, 599–602
grounding impedance 283–284, 599–602
grounding resistance 286–287, 599–602
ground surface arcing 288–289, 601, 620
ground wave 433
group delay time 457–458
group velocity 437, 457–458
gyrofrequency 437, 440–441, 455–459, 530

H_α line 163, 404
hail 28–29, 31, 219
half-peak width
    current 179–180, 273, 280, 285, 286
    dE/dt 154, 157–158, 230
    field waveform 132, 134, 135, 289, 291, 332–333, 337
hazard, lightning
    to aircraft 346–369
    to animals 647–648
    to communication systems 620–624
    to distribution lines 616–619
    to humans 642–649, 656, 657
    to structures 596–599
    to transmission lines 616, 619
    to underground cables 619–622
HCN, atmospheric production of 519, 520
Heaviside layer 435
Heaviside function 400–401, 416
Heidler function 161
height
    of charge in thundercloud 75–82

of charge reversal for graupel–ice charging 86
of electrosphere 8
of ionospheric regions 432–436, 443–445, 481
of lightning charge centers 75–79
of lightning initiation 323–324, 327
of tropopause 68, 308, 507
high-frequency (HF) radiation 334–337, 340, *see also* narrowband radiation
production by lightning processes 6
hiss 457, 530
history 1–4
horizontal electric field 159, 162
hurricanes 26–28
hydrometeors 68, 82–83, 85, 88, 121, 310
hydrometeor charge 84, 87–88, 310

ice crystal alignment 573
ice crystals 86–87, 573
IC flash, *see* cloud lightning flash
ICLRT, *see* Camp Blanding lightning triggering facility
ICV, *see* initial current variation
image charge 68–71
images, method of 68–70
impact velocity 87
incidence
    downward flashes 50–52
    to areas 24–49
    to structures 49–52
    upward flashes 52
inductance per unit length 398–399
induction field 129–130, 160
infrared glow 480, 481, 496, 497
infrasonic thunder 374, 385–386
initial breakdown
    in cloud flashes 8, 89–90, 322, 329–337, 340
    in ground flashes 8, 82–84, 116–122, 230, 340
initial continuous current
    in natural upward lightning 243, 247–250
    in rocket-triggered lightning 267, 270, 272–274
initial current variation 274
initial peak of field waveform 152–156, 175, 228–229
initial stage of rocket-triggered lightning 267–270
injury from lightning 642–649
interferometer 568–570, 572
interstroke interval 7, 47, 49, 112, 115, 282
intracloud lightning flash, *see* cloud lightning flash
Io 540
ionization 6–7, 396
ionospheric electron number density, scale height for 436, 481
ions 6–7
isokeraunic map 35–37

Japan, winter thunderstorms in 308–316, 349
J-changes, *see* J-processes

Joule heating, *see also* ohmic heating, 396, 588, 589, 596, 597, 611
J-processes 111–112, 182–184
junction processes, *see* J-processes
Jupiter 529, 536–543
    atmospheric properties 536–538
    lightning 538–543
    magnetosphere 540
    optical signals 538–540
    RF signals 540–542
    whistlers 540–541

K-changes, *see* K-processes
Kennedy Space Center (KSC) 24–25, 30, 75, 232, 253–254, 265–266, 269, 271, 272, 278–280, 285–288, 293, 323, 335, 337, 388, 556–557, 560–561, 563–564, 567
K-processes
    definition 184–185
    in cloud discharges 322, 338
    in ground discharges 111–112, 184–188
    mechanism 186–188
    time interval between 185

laboratory simulation of thunderstorm electrification processes 86–87
Langmuir Laboratory 266
laser 296–299
    infrared 297–298
    ultraviolet 298–299
latitude dependence
    of cloud-to-ground-flash ratio 44–46, 48
    of peak current 48
LDAR (lightning detection and ranging system) 30, 563
leader
    attempted 127, 185
    "chaotic", *see* "chaotic" leader
    comparison of cloud and ground flashes 340
    comparison of positive and negative 224–226
    dart, *see* dart leader
    dart-stepped, *see* dart-stepped leader
    in cloud flashes 322, 327, 330–331, 340
    negative, mechanism 136–137
    positive, mechanism 226–228
    stepped, *see* stepped leader
    streamer zone, *see* streamer zone
    upward, *see* upward leader
    upward connecting, *see* upward connecting leader
leader–return-stroke mode of charge transfer 4–6
lightning
    ball, *see* ball lightning
    bead, *see* bead lightning
    bipolar, *see* bipolar lightning flash
    cloud, *see* cloud lightning flash
    cloud-to-cloud, *see* cloud lightning flash
    cloud-to-ground, *see* ground lightning flash
    extraterrestrial 528–547
    global frequency 24, 41–43
    in clear air 73

Index

in earthquakes, *see* seismo-electromagnetic effect
in sandstorms 664
intercloud, *see* cloud lightning flash
intracloud, *see* cloud lightning flash
negative, *see* negative ground flash
nuclear 668–669
on Jupiter 538–543
planetary 528–547
positive, *see* positive ground flash
rocket-triggered, *see* rocket-triggered lightning
in solar nebula 528
spider 326, 489, 569
triggered, *see* rocket-triggered lightning
upward, *see* upward lightning
volcanic 564–565, 666–667
winter 311–316
lightning channel, *see* channel; tortuosity; branching, lightning channel; ground termination
lightning charge centers and dipoles 76–78, 117
lightning currents 7, 143–148, 161, 176, 177–181, 215, 221, 249–253, 272–278, 283, 312–313
lightning detection systems
  interferometer 568–570, 572
  magnetic direction finding (MDF) 558–562
  time of arrival (TOA) 562–565
lightning discharge, *see* lightning
lightning flash, *see* lightning
lightning flash counter (LFC) 38–39
lightning imaging sensor (LIS) 35, 42–43, 571
lightning incidence, *see* incidence
lightning initiation
  breakdown field for, *see* breakdown electric field
  breakeven field for, *see* breakeven electric field
  height of 323–324, 327
  of cloud flash, *see* initial breakdown in cloud flashes
  of ground flash, *see* initial breakdown in ground flashes
  on aircraft 346–347, 353–359
lightning locating techniques
  acoustic, *see* acoustic imaging
  charge center analysis 556–557
  interferometric 568–570, 572
  long-baseline, time-of-arrival 564–565
  magnetic direction finding 558–562
  radar 572–573
  satellite 570–572
  short-baseline, time-of-arrival 563–564
  thunder ranging 387–389
  thunder ray tracing 387–389
  very short baseline, time-of-arrival 562–563
  VHF–UHF channel mapping 563–564, 568–570
lightning luminosity 149–150, 177–178, 271–272
lightning mapping techniques, *see* lightning locating techniques

lightning models, *see* models
lightning processes
  attachment process 49–50, 137–143, 592–599
  bidirectional leader 90, 228, 269–272, 275, 332, 346, 353–354, 416–417
  break-through phase 138, 228, 414
  "chaotic" leader 164, 171
  continuing current, *see* continuing current
  dart leader, *see* dart leader
  dart-stepped leader 7, 8, 165, 169–171, 271
  initial breakdown, *see* initial breakdown
  initial continuous current, *see* initial continuous current
  J- (junction) processes 111–112, 182–184
  K-processes, *see* K-processes
  M-components, *see* M-components
  negative leader, *see* stepped leader; subsequent leader
  positive leader, *see* positive leader
  preliminary breakdown, *see* initial breakdown
  recoil streamer 187, 322, 329, 338, 357
  return stroke, *see* return stroke
  stepped leader, *see* stepped leader
  subsequent leader, *see* dart leader; dart-stepped leader
lightning protection
  air terminals 589–591, 596–599
  airborne vehicles 362–364
  arrester 591, 602–604
  boats 610–613
  common structures 596–599
  communication 620–624
  cone-of-protection method 593
  counterpoise 599–602
  crowbar circuit 591, 602–604
  dissipation array 605–607
  down conductor 589–591, 596–599
  early streamer emission (ESE) 605, 607–610
  earthing system 599–602
  electrogeometric method 51, 593
  electromagnetic shielding 604, 605
  Faraday cage 3, 590, 591, 604, 605
  ground rod 599–602
  grounding system 599–602
  melting energy 596, 597, 611
  metal oxide varistor (MOV) 591, 602–604
  peak current 588
  personal protection 648
  physics-based methods 593
  potential equalization 599
  power lines 616–619
  protection angle 592–596, 598
  protection zone 592–596, 598
  rolling-sphere method 593–598
  safety distance 598–599
  shielding 589–591, 594–596
  surge protective device (SPD) 591, 602–604
  telecommunications systems 620–624
  topological shielding 591
  trees 613–616

underground cables 619–622
underground communication lines 623–624
lightning stroke
  first 7, 122–159
  subsequent 7, 164–173
line charge density 84, 123–126, 129, 161, 292, 327, 330–331, 416
line-of-sight locating 562
LLP (Lightning Location and Protection Inc.) 560–562, 565–566, 568
local thermodynamic equilibrium 163, 395
log normal distribution 145
long-continuing current 7, 173–175
long spark 136–137, 226–228
Lorentz force 395–396
low-frequency (LF) radiation 555
lower atmosphere 507
lower positive charge center 88
lightning positioning and tracking system (LPATS) 564–566
luminous phenomena above storms
  elves 480, 481, 482, 492–493
  jets 480, 481, 482, 483–485
  sprites 480, 481, 482, 485–490
  starters 480, 481, 482, 483

magnetic direction finding (MDF) systems, *see also* NLDN, 558–562
magnetic field 129–130, 153, 160–161
magnetosphere 454–459
magnetospheric plasma 454–459
magnetostatic approximation 129–130
magnetostatic field 129–130, 160
Mars 529
Maxwell current density 10–11
Maxwell's equations 394, 398, 415
M-components
  charge transferred 7, 179
  current 177–181, 183
  current peak 7, 179
  current risetime 7, 179
  duration 179
  electric field change 180, 181, 183
  luminosity 177–178
  mechanism 181–182
  mode of charge transfer 4–6
  time interval between 179
  VHF–UHF radiation 181
medium frequency (MF) 340
mesoscale convective complexes or systems (MCCs or MCSs) 27–28, 31–34, 45, 218, 231
  definition 28
  lightning bipolar pattern, *see* bipolar pattern
  positive ground flashes in 218
  red sprites above 485–490
  spider lightning in, *see* spider lightning
mesosphere 481, 507
meteors 489, 491
microbursts 30
microphysical processes causing electrification, *see* cloud electrification mechanisms
microwave beam 299

middle atmosphere 507
mixed-phase region 31, 67
mode theory (full wave theory), 446–448
models
   of bidirectional leader 416
   of dart leader 415–417
   of evolution of clouds, lightning parameterization 88–91
   of global electric circuit 11
   of luminous phenomena in middle and upper atmosphere 484, 485, 490–493
   of M-components 419–420
   of return stroke 394–415
   of stepped leader 417–419
   of thunder 379–381
modified Diendorfer–Uman (MDU) model 403
modified transmission line model
   with exponential current decay with height (MTLE) 401–404, 407–412
   with linear current decay with height (MTLL) 401–403, 406–410
monsoon storms 34, 47, 48
Monte San Salvatore 139, 144, 214, 232, 241–242, 244–248
MULS model 395
multiparameter radar 31
multiple ground terminations 37–38, 114–115
multiplicity, see number of strokes per flash
mythology 1

NALDN, see North American Lightning Detection Network
narrow bipolar pulses, see also compact intracloud discharges, 8, 334–337
narrowband radiation 120, 158–159, 171, 185–186, 187–188, 226, 334, 341
National Lightning Detection Network (NLDN) 487, 565–568
negative ground flash 5, 7, 8, 108–191, 244–245
negative leader versus positive leader 224–226
Neptune 529
   atmosphere 545–546
   radio emission 546
nimbostratus 91
nitric oxide (NO)
   concentration 514–519
   equilibrium 512
   freeze-out temperature 512
   production 507–519
nitrogen fixation, see NO production
nitrogen oxides (NO_x)
   definition 508
   production 508–519
NLDN, see National Lightning Detection Network
NO production 507–519
NO_x 507–519
NO_y 508, 516
noise power (P_n) 460
non-cumulonimbus 91–93
non-inductive charging mechanisms 85, 86–87

non-thunderstorm clouds, see non-cumulonimbus
North American Lightning Detection Network (NALDN) 568
nose whistler 345, 355, 358
NSSL network 561
nuclear lightning 668
number of strokes per flash 7, 47, 48, 49, 113, 222, 562, 567
numerical cloud models 88–91
N-wave 379–381

O_3, see ozone
oceanic storms 31, 34–35, 49
ohmic heating, see also Joule heating, 378
OH radical 87, 507, 508
ohmic losses 396, 588, 589, 596, 597, 611
Ohm's law 440, 488
optical locating techniques 570
optical properties of lightning 163–164
optical scattering 150
optical thickness 163
Optical Transient Detector (OTD) 30, 35, 42–43, 571
Ostankino tower 242, 252, 255
ozone 507, 508, 516–518

particle charge, see charge on cloud particles; charge on precipitation; hydrometeor charge
peak current of lightning return strokes 7, 143–148, 215, 221, 249, 251–254
Peissenberg tower 242, 250–252, 256
phase velocity 437
photoionization 418
physics sign convention 8–9
Pioneer Venus orbiter 531, 532, 533–535
planetary atmospheres
   Jupiter 536–538
   Mars 529
   Mercury 529
   Neptune 545–546
   Saturn 543
   Uranus 544–545
   Venus 531
planetary lightning 528–547
planetary radio emission
   Jupiter 540–542
   Neptune 546
   Saturn, electrostatic discharge (SED) 543–544
   Uranus, electrostatic discharge (UED) 545
   Venus 533–536
planets
   giant (jovian) 529
   terrestrial 529
plasma wave, evanescent 439, 457
plasma wave
   absorption 439–440
   cutoff frequency 441
   dispersion 437–441
   frequency 439
   group velocity 437

mode
   chorus 457
   electromagnetic 437–443
   electrostatic 439
   hiss 457
   ion cyclotron 459
   whistler 432, 434, 435, 454–459
   phase velocity 437
   polarization 442–443
   propagation 437–441
      ducted 435, 455–459
      field-aligned ducts in magnetosphere 435, 455–459
      full wave theory 446–448
      geomagnetic guiding 435, 455–459
      group delay time 458
      non-ducted in magnetosphere 435, 455, 456, 459
   reflection 441–443
   resonance 439
   transmission 441–443
plasma frequency
   electron 439
   ion 439
plasmasphere 435–437, 455
Pluto 529
point dipole 77, 117, 557
point discharge, see corona discharge
Poisson distribution 52
polarimetric radar 31
polarization
   circular 440–443
   elliptic 443
   of hydrometeors 82
   parallel 441–443
   perpendicular 441–443
polarization errors 558, 559
positive dipole 68, 73
positive flash percentage 214, 216, 311
positive ground flash 5, 214–234, 245–247, 311–316
   charge transfer 214–215, 221–222
   definition 214
   electric field change 223, 226
   fine structure of field waveforms 229–230
   frequency of occurrence 26, 214, 216, 311
   in dissipating storms 217
   in hurricanes 28, 218
   in mesoscale convective systems 31–34, 218
   in severe storms 29, 218, 219
   in winter thunderstorms 216–218, 312–313
   initial breakdown in 230
   peak current 215, 221, 230–231
   related to forest fires 218, 220
   related to volcano eruptions 666–667
   return-stroke velocity 231–232
   seasonal dependence 216–218
positive leader versus negative leader 224–226
positive lightning, see positive ground flash
positive triggered lightning 267, 272

potential
  of electrosphere 9–10
  of leader tip 7, 111
  within cloud 11, 82, 92, 111
  potential difference between cloud and ground 111
power 12, 116, 152–153, 163–164
power lines or systems
  lightning protection for 616–622
Poynting vector 162
precipitation
  amount per lightning 34
  charge carried by 68, 84, 87–88
precursor current pulses 268, 274–276
preliminary breakdown, *see* initial breakdown
propagation effects 162–163
propagation speed
  of acoustic signals 374
  of "chaotic" leader 171
  of dart leader 6, 7, 165–167
  of dart-stepped leader 7, 170–171
  of electromagnetic signals (speed of light) 374, 555
  of in-cloud fast streamer 181, 186
  of return stroke 6, 7, 150–152, 231–232, 413–414
  of stepped leader 7, 122–123, 327, 331, 340
  of upward connecting leader 139–140
  of upward positive leader 224, 245, 270
protection, *see* lightning protection

Q-bursts 432, 433, 450, 451, 459
Q-noise 186, 328–329
Q-factor of Schumann resonance lines 450, 453
Q-streamers 186, 327–329
quiet period 159

radar 31, 572–573
radiation
  at HF, *see also* narrowband radiation, 344–337, 340
  at LF 555
  at UHF and VHF 6, 340, 555, 563–564, 568–570
  at VLF 555, 564–565
  optical 570
radiation field, lightning locating 557–558
radio noise 459–461
radioactive emanation 6–7
radius or diameter
  of dart leader 164
  of return stroke 7, 163–164
  of stepped leader 134
rainbands 26–28, 218
rainfall amount per lightning 34
rain gush 573
rate of rise (steepness)
  of current 7, 134, 146–147, 161, 215, 221, 249, 251, 254, 256, 278, 279, 284, 285, 286
  of field pulse 134, 154, 157–158, 170, 171, 228–230, 256, 280, 281, 289, 339, 404, 413–414

ratio of leader to return-stroke electric field change 130–131, 132, 133, 134, 169, 408
recoil streamers 187, 322, 329, 338, 357
red sprites 480–482, 485–492
reflection 439, 441–443
refractive index 441–443
regular pulse bursts
  in cloud flashes 8, 338–340
  in ground flashes 8, 188–190
relaxation time 9, 490, 491
resistance, *see also* grounding resistance, 599–602
resistance per unit length 164, 395, 398–400
retardation effects 128, 160–161
return stroke
  charge transferred by 7, 146, 215, 221, 249, 254, 280, 281
  conductivity 164
  current 7, 143–148, 215, 221, 249, 251–254
  duration 7, 146, 215, 221, 249
  electric and magnetic fields 8, 152–163, 229–230, 277, 278
  luminosity 148–150, 223
  propagation speed 6, 7, 150–152, 231–232, 413–414
  radius 7, 163–164
return-stroke models
  AT (antenna theory) 394, 398, 406, 415
  BG (Bruce–Golde) 129, 401–403, 407–409
  distributed-circuit (RLC) 394, 398–400, 406, 415–416, 418
  DU (Diendorfer–Uman) 402–404, 407–412
  electromagnetic 394, 398, 406, 415
  engineering 394, 400–404, 406–412
  gas-dynamic 394–397, 404–406
  MDU (modified Diendorfer–Uman), *see* modified Diendorfer–Uman (MDU) model 403
  MTLE (modified transmission line with exponential current decay with height) 401–404, 407–412
  MTLL (modified transmission line with linear current decay with height) 401–403, 406–410
  TCS (traveling current source) 129, 401–404, 407–412
  transmission line 161, 400–403, 407–413, 567
return-stroke speed versus peak current 151–152
reversal distance 71
riming 86–87
risetime (front duration)
  of electric and magnetic field 132, 154, 229
  of luminosity 134, 149–150, 172
  of M-component current 177, 179–180
  of return-stroke current 147, 215, 249, 252, 280, 283, 284, 286
RLC models, *see* distributed-circuit models

rocket-triggered lightning
  altitude triggering 222, 269–270, 271–272, 346
  classical triggering 267–269, 270–271
rockets
  electric field measurements 80
  lightning triggering 266–296
rolling-sphere method 592–598
runaway electrons 84, 121, 480–481, 484, 490–495

SAFIR 30, 569–570
sandstorms, lightning in 667
satellites 41–43, 570–572
Saturn 529
  atmosphere 543
  radio emission 543–544
scale height for ionospheric electron number density 436, 481
scattering, optical 150
Schumann resonances 432–434, 449–454
screening layers 69, 76, 78, 80, 85, 88, 91
screening layer charge 84
seasonal variations
  of lightning characteristics 46–47
  of percentage of positive ground flashes 216–218
SED (Saturn electrostatic discharges) 543–544
seismo-electromagnetic effect (earthquake light) 667–668
severe storms 25, 28–30, 84, 218
severe weather phenomena
  hail 29, 219
  microbursts 30
  tornadoes 29–30, 219
  strong winds 30
sferics 3, 432–433, 443–449
  definition 432
  measurements 443–445
shock wave, *see* thunder
sign convention for electric field 8–9
site errors 559, 560
skin depth 440
sky wave 433, 443–448, 558–559
slow front 139, 144, 154–155, 223, 229
slow tail 432, 434, 446–448, 459, 488
small ions 6
Snell's law 442, 457–458
snowflakes 310
soil conductivity 7, 287, 541, 601
solar cycle 43
space charge 78, 85
space stem 136–137
spark, long laboratory 136–137, 226–228
SPD, *see* surge protective device
spectrogram 335–336
spectroscopy, lightning 3, 134, 163, 168
speed of propagation, *see* propagation speed
spider lightning 326, 489, 569
split in lightning channel 140–141
sprites, red 480–482, 485–492
St Elmo's fire 347

standards, test
  for aircraft, 362–364
  for communication systems 624, 625
  for power systems 624, 625
Stark effect 163, 404
stepped leader 7, 8, 122–137, 331, 340, 417–419
  charge deposited by 7, 123–125
  diameter 134
  duration 7, 123, 124, 125
  speed of propagation 7, 122–123, 327, 331, 340
storm area 68
stratocumulus 91
stratosphere 68, 481
stratus 91
streak photographs, *see also* lightning luminosity, 122, 139, 149, 177, 223, 225, 245, 246, 271, 272
streamer, definition 5–6
streamer zone 135–137, 227–228, 247, 272
striking distance 592–596
strokes
  number per flash 7, 47, 48, 49, 113, 222, 562, 567
  time interval between 7, 47, 49, 112, 115
SUNYA network, *see also* East Coast network, 46, 49
supercell 29, 68, 80–81, 83, 326
superposition principle 68
surge protective device (SPD) 591, 602–604

TCS model, *see* traveling current source model
telegrapher's equations 398–399
TEM radio wave 398
temperature
  during interstroke interval 163–164
  in charge regions of thundercloud 75–78, 79, 80–82, 83
  of atmosphere versus height 481, 507
  of dart leader channel 7, 164
  of return-stroke channel 7, 163–164
terminology 4–6
test standards
  for aircraft 362–364
  for communication systems 624, 625
  for power systems 624, 625
thermal conduction 396
thermal gradient, vertical (lapse rate) 67, 386
thermodynamic equilibrium, local 163, 395
thermosphere 481
Thor 1
thunder
  attenuation of, in air 387
  clap 375, 385
  close sounds 375
  cloud flashes 376–377, 381
  distance heard 374, 386–387
  duration 374
  energy 376–377, 381, 397
  frequency spectrum 374–377, 381
  generation of
    electrostatic mechanism 385–386
    heating mechanism 378–385
  infrasonic 374, 385–386
  leader 383, 386
  models of 379–381
  N-wave 379–381
  peal 375
  power spectrum of 381
  rocket-triggered lightning 377, 383–385
  roll 375
  rumble 375
  shock wave 378, 379
    relaxation radius 379
  source location of, *see* acoustic imaging
  time interval between thunder and lightning 374
  upward lightning 260
  volcano lightning 666, 667
thunder events 47
Thunderbird 1
thundercloud 67–91
  charge location 75–82, 83
  dipolar structure 68, 73, 77
  electrical charges of p, N, and P regions 79
  electrification 84–88, 92, 528
  models 88–91
  potential difference from ground 111
  properties related to flash rates 30–31
  tripolar structure 68–70, 79, 80, 81
thunderday, *see* thunderstorm days
thunderhour, *see* thunderstorm hours
thunderstorm days 35–36, 37, 43–44
thunderstorm hours 36–37, 43
Thunderstorm Research International Program (TRIP) 297
thunderstorms
  air-mass 49, 68
  as generators in global electric circuit 10
  charge distributions 75–82, 83
  charging mechanisms 84–88, 92, 528
  convective 68
  current above 10
  evolution 73–75, 310–311
  flash rates 24–26, 31
  frontal 49, 68
  severe 25, 28–30, 84, 218
  supercell 29, 68, 80–81, 83, 326
  tornadic 29–30, 219
  winter 77, 217–218, 308–311
time constant, *see* relaxation time
time-of-arrival mapping systems 562–568
TIPPs 334, 496, 571–572
TL model, *see* transmission line model
topological shielding 591, 604, 605
tornadic storms 25, 28–30, 84, 218
tornadoes 29–30, 219
tortuosity
  description of 161
  electromagnetic radiation affected by 161–162
  thunder affected by 378, 379, 385

transient flame 299
transmission line 616–619
transmission line equations (telegrapher's equations) 398–399
transmission line (TL) model 161, 400–403, 407–413, 567
traveling current source (TCS) model 129, 401–404, 407–412
triggered lightning, *see* rocket-triggered lightning
Trimpi event 496
troposphere 67, 481, 507

UED (Uranian electrostatic discharges) 545
unconnected upward discharge 141–143
updrafts 68, 34
updraft velocity 34, 68
upper atmosphere 507
upward connecting leader 110–111, 137–143, 220–222, 275–276
upward leader 224–226, 244–247, 270–272
upward lightning 50, 52, 214, 232–233, 241–260, 311–312
Uranus 529
  atmosphere 544–545
  radio emission 545

velocity, *see* propagation speed
velocity of electron drift 415
Venera 531, 532, 535–536
Venus 529
  atmospheric properties 531
  ionosphere 533
  optical measurements 532–533
  radio emission 533–536
vertical wind shear 217, 311, 386–387
VHF (very high frequency) 340, 555
VHF–UHF mapping systems 563–564, 568–570
VHF radiation produced by lightning processes 6
VLF (very low frequency) mapping systems 555, 564–565
volcanic lightning 564–565, 666–667
voltage deviation 460
Voyager 536, 538–539, 540–541, 543–546

warm cloud 528
water droplets 86–87
water jet 299
waveform
  BIL 116
  bipolar lightning current 248, 312, 313, 316
  cloud flash electric field 323, 324, 325
  continuing current 176, 179, 183, 281
  dE/dt 157, 256, 336
  downward negative flash current 111
  downward negative flash electric field 114, 117, 174
  ICC pulses 273

initial breakdown pulse 119, 120, 334, 335, 336
initial continuous current 273
initial current variation (ICV) 273
K-change 184
leader field 118, 126, 127, 130, 131, 135, 157, 169, 180, 184, 276, 290, 291
leader luminosity pulse 150
M-component current 179, 180, 181, 183
M-component electric field 180, 181, 183, 184
M-component luminosity 178
M-component magnetic field 181
narrow bipolar pulse 335, 336
positive lightning current 215, 221, 246, 312, 313, 314
positive lightning field 223, 226
precursor current pulses 275, 276
regular pulse burst 187, 188, 189, 190, 339
return-stroke current 144, 147, 215, 221, 252, 255, 256, 281–285
return-stroke field 119, 126, 127, 135, 153, 157, 169, 180, 184, 223, 226, 256, 277, 290, 291
return-stroke luminosity 150
triggered-lightning flash current 273
upward connecting leader current 276
upward flash current 243, 245, 246, 248, 250
winter lightning current 215, 312, 313, 314, 315, 316
wave-tilt formula 162
wet bulb temperature 43
whistlers 432, 434–435, 454–459, 530
coupling from atmosphere to magnetosphere 457
dispersion parameter 435–436, 455–458
duct 435, 455–459
frequency–time characteristics 458, 435–436
induced electron precipitation 456, 480, 491, 493–495
nose 435, 455, 458
observation
ground-based 434–435, 454–456
satellite-based 456–457, 459
occurrence rate 435
partial reflection 435
path 357–359
propagation
multi-path 455
non-ducted 435, 455–456, 459
triggered emission 456
X-ray 456, 480, 491, 493–495
two-hop 435, 455
wind shear 217, 311, 386–387
winter lightning 311–316
winter thunderstorms 77, 217–218, 308–311, 349

X-rays 121, 480, 491, 493–495

zero-crossing time of field waveforms 154, 229, 407–408
Zeus 1